D1591976

Neurobiology of Disease

Neurobiology of Disease

Edited by

Sid Gilman, MD, FRCP

AMSTERDAM • BOSTON • HEIDELBERG • LONDON • NEW YORK • OXFORD
PARIS • SAN DIEGO • SAN FRANCISCO • SINGAPORE • SYDNEY • TOKYO

Academic Press is an imprint of Elsevier

Elsevier Academic Press
30 Corporate Drive, Suite 400, Burlington, MA 01803, USA
525 B Street, Suite 1900, San Diego, California 92101-4495, USA
84 Theobald's Road, London WC1X 8RR, UK

This book is printed on acid-free paper. ∞

Copyright © 2007, Elsevier Inc. All rights reserved.

No part of this publication may be reproduced or transmitted in any form or by any means, electronic or mechanical, including photocopy, recording, or any information storage and retrieval system, without permission in writing from the publisher.

Permissions may be sought directly from Elsevier's Science & Technology Rights Department in Oxford, UK: phone: (+44) 1865 843830, fax: (+44) 1865 853333, E-mail: permissions@elsevier.co.uk. You may also complete your request on-line via the Elsevier homepage (http://elsevier.com), by selecting "Customer Support" and then "Obtaining Permissions."

Library of Congress Cataloging-in-Publication Data
Application submitted.

British Library Cataloguing in Publication Data
A catalogue record for this book is available from the British Library

ISBN-13: 978-0-12-088592-3
ISBN-10: 0-12-088592-1

For all information on all Elsevier Academic Press publications
visit our Web site at www.books.elsevier.com

Printed in China
06 07 08 09 10   9 8 7 6 5 4 3 2 1

Working together to grow
libraries in developing countries

www.elsevier.com | www.bookaid.org | www.sabre.org

ELSEVIER   BOOK AID International   Sabre Foundation

Contents

Contributors ix
Editors xvi
Preface xvii

PART A. CENTRAL NERVOUS SYSTEM

Section I. Metabolic Diseases
Hugo W. Moser, MD

1. Lysosomal Disorders of the Nervous System 1
   Steven U. Walkley, DVM, PhD
2. Neurobiology of Peroxisomal Disorders 19
   James M. Powers, MD
3. Creatine Deficiency Syndromes 33
   Peter W. Schütz, MD, and Sylvia Stöckler, MD
4. Leukodystrophies 43
   Shoji Tsuji

Section II. Neurodegenerative Disorders
Gregor K. Wenning, MD, PhD

5. Parkinson's Disease 51
   Néstor Gálvez-Jiménez, MD, MSc, MHSA, FACP
6. Alzheimer's Disease 69
   Kurt A. Jellinger, MD
7. Multiple System Atrophy 83
   Carlo Colosimo, MD; Felix Geser, MD, PhD; Eduardo E. Benarroch, MD, DSci; and Gregor K. Wenning, MD, PhD
8. Olivopontocerebellar Atrophy (OPCA) 95
   José Berciano
9. Neurobiology of Progressive Supranuclear Palsy 105
   Pratap Chand, DM, FRCP, and Irene Litvan, MD

Section III. Genetic Diseases
Christopher Gomez, MD, PhD

10. Protein Aggregation Disorders 111
    Tiago Fleming Outeiro, PhD; Pamela J. McLean, PhD; and Bradley T. Hyman, MD, PhD
11. RNA-Based Disorders of Muscle and Brain 125
    Paul J. Hagerman, MD, PhD
12. Ion Channel Disorders 135
    Alan L. Goldin, MD, PhD
13. Spinocerebellar Ataxia Type 1 149
    Harry T. Orr, PhD
14. Mitochondrial Genetic Diseases 157
    Alice Wong, PhD, and Gino Cortopassi, PhD

Section IV. Neuroimmunological Disorders
Anthony T. Reder, MD

15. Paraneoplastic Neurological Disorders 163
    Steven Vernino, MD, PhD, and Josep Dalmau, MD, PhD
16. Systemic Lupus Erythematosus: Descriptive Past and Mechanistic Future 171
    Czeslawa Kowal, PhD; Cynthia Aranow, MD; Meggan Mackay, MD; Betty Diamond, MD; and Bruce T. Volpe, MD
17. Progressive Multifocal Leukoencephalopathy 185
    Benjamin Brooks, MD
18. Immunopathogenesis of Multiple Sclerosis 197
    Bernhard Hemmer, MD; Nicole Töpfner; and Hans-Peter Hartung, MD
19. Immune-Mediated Neuropathies 205
    Michael Schroeter, MD; Bernd C. Kieseier, MD; Hubertus Köller, MD; and Hans-Peter Hartung, MD
20. Hashimoto Encephalopathy 217
    Ji Y. Chong, MD

Section V. Cerebrovascular Diseases
Lewis B. Morgenstern, MD

21. Vascular Cognitive Impairment — 223
Velandai Srikanth, PhD; Michael M. Saling, PhD; and Amanda G. Thrift, PhD
22. Cardioembolism — 235
Karen Furie, MD, MPH
23. Clinical and Neurobiological Aspects of Stroke Recovery — 241
Bruce T. Volpe, MD, and Rajiv R. Ratan, MD, PhD
24. Nonatherosclerotic Cerebral Vasculopathies — 255
Rima M. Dafer, MD, MPH, and Jose Biller, MD
25. Subarachnoid Hemorrhage — 265
David Palestrant, MD, and E. Sander Connolly, Jr., MD
26. Cerebral Ischemia: Molecular Mechanisms and Protective Therapies — 271
Kenneth R. Wagner, PhD
27. Intracerebral Hemorrhage and Intraventricular Hemorrhage–Induced Brain Injury — 281
Richard F. Keep, PhD; Guohua Xi, MD; and Julian T. Hoff, MD

Section VI. Paroxysmal Disorders
Timothy A. Pedley, MD

28. Idiopathic Generalized Epilepsy — 289
Jose-Luis Perez-Velasquez, PhD; Edward H. Bertram, III, MD; and O. Carter Snead, III, MD
29. Paroxysmal Dyskinesia — 297
Laurent Vercueil, MD, PhD; Anne de Saint Martin, MD; and Edouard Hirsch, MD
30. Myoclonus — 305
Pietro Mazzoni, MD, PhD, and Samay Jain, MD
31. Channelopathies of the Nervous System — 319
James C. Cleland, MBChB, and Robert C. Griggs, MD
32. Migraine as a Cerebral Ionopathy with Abnormal Central Sensory Processing — 333
Michel D. Ferrari, MD, PhD, and Peter J. Goadsby, MD, PhD, DSc
33. Temporal Lobe Epilepsy — 349
Helen E. Scharfman, PhD, and Timothy A. Pedley, MD

Section VII. Neoplastic Diseases
John J. Laterra, MD, PhD

34. Central Nervous System Metastases — 371
Robert J. Weil, MD, FCAS
35. Meningioma — 381
Katrin Lamszus, MD; Christian Hagel, MD; and Manfred Westphal, MD
36. Primary Central Nervous System Lymphoma — 395
Maciej Mrugala, MD, PhD, MPH; Anne Newcomer; and Tracy Batchelor, MD, MPH
37. Neurofibromatosis 1 — 413
Linda Piersall, MS, and David H. Gutmann, MD, PhD
38. Medulloblastoma and Primitive Neuroectodermal Tumors — 425
Said Elshihabi, MD, and James T. Rutka, MD, PhD
39. Glioma — 433
Anders I. Persson, PhD; QiWen Fan, PhD; Joanna J. Phillips, PhD; and William A. Weiss, PhD

Section VIII. Infectious Diseases
Richard T. Johnson, MD, FRCP

40. Bacterial and Fungal Infections of the Nervous System — 445
Thomas P. Bleck, MD, FCCM
41. Parasitic Infections — 453
Ana-Claire Meyer, MD, and Gretchen L. Birbeck, MD, MPH
42. Prion Diseases — 473
Adriano Aguzzi, MD, PhD, DVM, hc, FRCP, FRCPath; Frank L. Heppner, MD; Ingo M. Westner, MD; and Markus Glatzel, MD
43. Central Nervous System Viral Infections: Clinical Aspects and Pathogenic Mechanisms — 485
Christopher Power, MD, FRCP (C), and Farshid Noorbakhsh, MD, PhD

Section IX. Motor Neuron Diseases
Eva L. Feldman, MD, PhD

44. Spinal Muscular Atrophy — 501
Charlotte J. Sumner, MD, and Kenneth H. Fischbeck, MD
45. Amyotrophic Lateral Sclerosis–Like Syndromes Associated with Malignancy — 513
Zachary Simmons, MD
46. Amyotrophic Lateral Sclerosis: Idiopathic and Inherited — 521
Nicholas J. Maragakis, MD, and Jeffrey D. Rothstein, MD, PhD
47. Hereditary Spastic Paraplegia and Primary Lateral Sclerosis — 537
Philip A. Wilkinson, BSc, MBBS, MRCP, and Michael Swash, MD, FRCP, FRCPath

Contents

48. Poliomyelitis — 545
    Eric J. Sorenson, MD
49. Spinobulbar Muscular Atrophy (Kennedy's Disease) — 553
    Jeffrey D. Zajac, MBBS, PhD, FRACP, and Karen J. Greenland, BSc, (Hons) PhD

Section X. Malformations and Developmental Disorders
Michael V. Johnston, MD

50. Neurobiology of Genetic Mental Retardation — 563
    Walter E. Kaufmann, MD; John C. Carter; Irena Bukelis; and David N. Lieberman, MD, PhD
51. Cerebral Palsy — 575
    Alexander H. Hoon, Jr., MD, MPH
52. Autism — 581
    Martha R. Herbert, MD, PhD
53. Neurobiology of Dyslexia — 593
    Sheryl L. Rimrodt, MD, and Laurie E. Cutting, PhD
54. Neonatal Brain Injuries — 599
    Steven P. Miller, MD, MAS, and Donna M. Ferriero, MD, MS
55. Spina Bifida — 611
    Stephen L. Kinsman, MD
56. Circuits to Synapses: The Pathophysiology of Tourette Syndrome — 619
    Harvey S. Singer, MD, and Kendra Harris, MSc
57. Attention-Deficit Hyperactivity Disorder — 631
    Paul H. Lipkin, MD, and Stewart Mostofsky, MD
58. Congenital Hydrocephalus — 641
    Stephen L. Kinsman, MD

Section XI. Neurologic Manifestations of Medical Diseases
John J. Caronna, MD

59. Neurological Manifestations of Hematological Disease — 649
    Babette B. Weksler, MD
60. Neurological Manifestations of Renal Disease — 659
    Ajay K. Singh, MB, MRCP, and Radhika Bhatia, MD
61. Diabetes and Endocrine Disorders — 669
    Rodica Pop-Busui, MD, PhD; Zachary London, MD; and Aaron Kellogg, BS
62. Mechanisms and Consequences of Central Nervous System Hypoxia — 681
    Meredith L. Turetz, MD, and Ronald G. Crystal, MD
63. Neurological Manifestations of Gastrointestinal and Hepatic Diseases — 689
    Brain P. Bosworth, MD, and Brian R. Landzberg, MD
64. Neurosarcoidosis — 703
    John J. Caronna, MD

Section XII. Sleep Disorders
Emmanuel Mignot, MD, PhD

65. REM Sleep Behavior Disorder — 709
    Carlos H. Schenck, MD, and Mark W. Mahowald, MD
66. Neurobiology of Narcolepsy and Hypersomnia — 715
    Emmanuel Mignot, MD, PhD, and Jamie M. Zeitzer, PhD
67. Restless Legs Syndrome and Periodic Limb Movements in Sleep — 723
    Richard P. Allen, PhD
68. Neurobiology of Insomnia — 735
    Michael L. Perlis, PhD; Wilfred R. Pigeon, PhD; and Sean Patrick Andrews Drummond, PhD

Section XIII. Substance Abuse and Basic Toxicology
John C. M. Brust, MD

69. Organic Chemicals — 745
    John L. O'Donoghue, VMD, PhD, DABT
70. Metals — 759
    Luigi Manzo, MD
71. Neurobiology of Drug Addiction — 771
    Timothy P. Condon, PhD, and Curtis W. Balmer, PhD

Section XIV. Imaging the Nervous System
John C. Mazziotta, MD, PhD

72. Assessment of Neurobiological Diseases with Magnetic Resonance Spectroscopy — 781
    Jeffry R. Alger, PhD
73. Magnetic Resonance Imaging — 793
    John A. Detre, MD
74. Neurovascular Computed Tomography Angiography — 801
    Stuart R. Pomerantz, MD, and Michael H. Lev, MD
75. PET Imaging in Parkinson's Disease and Other Neurodegenerative Disorders — 821
    Vijay Dhawan, PhD, and David Eidelberg, MD

76. Single-Photon Emission Computed Tomography 829
Joseph C. Masdeu, MD, PhD
77. Functional Magnetic Resonance Imaging 839
Gereon R. Fink, MD, PhD

PART B. PERIPHERAL NERVOUS SYSTEM

Section XV. Peripheral Neuropathies
James W. Russell, MD

78. Impaired Glucose Regulation and Neuropathy 849
James W. Russell, MD, MS, FRCP; A. G. Smith, MD; and J. R. Singleton, MD
79. Acquired Inflammatory Demyelinating and Axonal Neuropathies 859
James W. Teener, MD, and James W. Albers, MD, PhD
80. Toxic and Drug-Induced Neuropathies 871
Guido Cavaletti, MD
81. Inherited Peripheral Neuropathies 885
Angelo Schenone, MD, and Lucilla Nobbio, PhD
82. Neurological Manifestations of Vasculitis 901
Safwan S. Jaradeh, MD
83. Neuropathies Associated with Infections 913
Paul Twydell, DO, and David N. Herrmann, MBBCh

Section XVI. Myopathies and Neuromuscular Junction Disorders
Charles A. Thornton, MD

84. Muscular Dystrophies 925
Leland E. Lim, MD, PhD; Charles A. Thornton, MD; and Thomas A. Rando, MD, PhD
85. Myasthenia Gravis and Myasthenic Syndromes 935
Angela Vincent, MA (MB, BS, MSc Lond.), FRCPath, FMedSci
86. Metabolic Myopathies 947
Michio Hirano, MD
87. Immunobiology of Autoimmune Inflammatory Myopathies 957
Marinos C. Dalakas, MD

Section XVII. Autonomic Disorders
Phillip A. Low, MD

88. Central Autonomic Network 969
Peter Novak, MD, PhD
89. Autonomic Neuropathies 979
Steven Vernino, MD, PhD, and Phillip A. Low, MD
90. Thermoregulation and Its Disorders 987
Robert D. Fealey, MD
91. Control of Blood Pressure—Normal and Abnormal 997
Michael J. Joyner, MD; William G. Schrage, PhD; and John H. Eisenach, MD

Section XVIII. Pain
Kenneth L. Casey, MD

92. Neoplasm-Induced Pain 1007
Sebastiano Mercadante, MD
93. Pain Associated with the Autonomic Nervous System 1021
Wilfrid Jänig
94. Postherpetic Neuralgia 1031
Misha-Miroslav Backonja, MD
95. Central Post-Stroke Pain 1039
David Bowsher, MD, ScD, PhD, FRCPEd, FRCPath

Index 1047

Contributors

Adriano Aguzzi, MD, PhD, DVM, hc, FRCP, FRCPath
Department of Pathology, University Hospital Zurich
Zurich, Switzerland

James W. Albers, MD, PhD
Department of Neurology, University of Michigan Medical School
Ann Arbor, Michigan

Jeffry R. Alger, PhD
Department of Neurology, Ahmanson-Lovelace Brain Mapping Center, UCLA Brain Research Institute
Jonsson Comprehensive Cancer Center,
David Geffen School of Medicine,
University of California–Los Angeles
Los Angeles, California

Richard P. Allen, PhD
Department of Neurology and Sleep Medicine,
Johns Hopkins University
Baltimore, Maryland

Cynthia Aranow, MD
Department of Medicine,
Columbia University Medical Center
New York, New York

Misha-Miroslav Backonja, MD
Department of Neurology, University of Wisconsin
Madison, Wisconsin

Curtis W. Balmer, PhD
JBS International, Inc., National Institute on Drug Abuse, National Institutes of Health
Bethesda, Maryland

Tracy Batchelor, MD, MPH
Department of Neurology, Harvard Medical School
Stephen E. and Catherine Pappas
Center for Neuro-Oncology,
Massachusetts General Hospital
Boston, Massachusetts

Eduardo E. Benarroch, MD, DSci
Department of Neurology, Mayo Clinic
Rochester, Minnesota

José Berciano
Service of Neurology, University Hospital
Marqués de Valdecilla
Santander, Spain

Edward H. Bertram, III, MD
Department of Neurology, University of Virginia
Charlottesville, Virginia

Radhika Bhatia, MD
Renal Research, Brigham and Women's Hospital
Boston, Massachusetts

Jose Biller, MD
Department of Neurology, Loyola University of Chicago, Stritch School of Medicine
Chicago, Illinois

Gretchen L. Birbeck, MD, MPH
International Neuropsychiatric Epidemiology Program (INPEP), Michigan State University
East Lansing, Michigan

Thomas P. Bleck, MD, FCCM
Department of Neurology, Evanston Northwestern Healthcare
Evanston, Illinois

Brian P. Bosworth, MD
Department of Gastroenterology, Division of Gastroenterology and Hepatology, Weill Medical College of Cornell University
New York, New York

David Bowsher, MD, ScD, PhD, FRCPEd, FRCPath
Pain Research Institute, University Department of Neurological Science, University Hospital Aintree
Liverpool, United Kingdom

Benjamin Brooks, MD
Department of Neurology, University of Wisconsin
Madison, Wisconsin

Irena Bukelis
Center for Genetic Disorders of Cognition and Behavior, Kennedy Krieger Institute
Baltimore, Maryland

John J. Caronna, MD
Department of Neurology and Neuroscience, Weill Medical College of Cornell University
New York, New York

John C. Carter
Center for Genetic Disorders of Cognition and Behavior, Kennedy Krieger Institute
Baltimore, Maryland

Guido Cavaletti, MD
Department of Neurosciences and Biomedical Technologies, University of Milan–Bicocca
Monza, Italy

Pratap Chand, DM, FRCP
Department of Neurology, University of Louisville School of Medicine
Louisville, Kentucky

Ji Y. Chong, MD
Department of Neurology, Columbia University
New York, New York

James C. Cleland, MBChB
Department of Neurology, Highland Hospital, University of Rochester School of Medicine and Dentistry
Rochester, New York

Carlo Colosimo, MD
Department of Neurological Sciences, University of Rome "La Sapienza"
Rome, Italy

Timothy P. Condon, PhD
National Institute on Drug Abuse, National Institutes of Health
Bethesda, Maryland

E. Sander Connolly, Jr., MD
Department of Neurological Surgery, Columbia Unversity
New York, New York

Gino Cortopassi, PhD
Department of Molecular Biosciences, University of California–Davis School of Veterinary Medicine
Davis, California

Ronald G. Crystal, MD
Department of Genetic Medicine, Division of Pulmonary and Critical Care Medicine, Weill Medical College of Cornell University
New York, New York

Laurie E. Cutting, PhD
Department of Neurology, Johns Hopkins University School of Medicinenl
Department of Education,
Johns Hopkins University
Developmental Cognitive Neurology,
Kennedy Krieger Institute
Baltimore, Maryland

Rima M. Dafer, MD, MPH
Department of Neurology, Loyola University of Chicago, Stritch School of Medicine
Chicago, Illinois

Marinos C. Dalakas, MD
National Institutes of Health, National Institute of Neurological Disorders and Stroke, Neuromuscular Diseases Section
Bethesda, Maryland

Josep Dalmau, MD, PhD
Division of Neuro-Oncology, University of Pennsylvania
Philadelphia, Pennsylvania

Anne de Saint Martin, MD
Department of Pediatric Neurology, Strasbourg University Hospital
Strasbourg, France

John A. Detre, MD
Department of Neurology, Department of Radiology, University of Pennsylvania
Philadelphia, Pennsylvania

Vijay Dhawan, PhD
Center for Neurosciences, Feinstein Institute for Medical Research, North Shore–Long Island Jewish Health System Department of Neurology and Medicine, New York University School of Medicine
Manhasset, New York

Betty Diamond, MD
Department of Medicine, Columbia University Medical Center
New York, New York

Sean Patrick Andrews Drummond, PhD
Department of Psychiatry, San Diego Veterans Affairs Medical Center
San Diego, California

David Eidelberg, MD
Center for Neurosciences, Feinstein Institute for Medical Research, North Shore–Long Island Jewish Health System Department of Neurology and Medicine, New York University School of Medicine
Manhasset, New York

John H. Eisenach, MD
Department of Anesthesiology, Mayo Clinic College of Medicine
Rochester, Minnesota

Said Elshihabi, MD
Division of Neurosurgery, University of Toronto
Toronto, Ontario, Canada

QiWen Fan, PhD
Departments of Neurology, Pediatrics, and Neurological Surgery; Brain Tumor Research Center and Comprehensive Cancer Center; University of California–San Francisco
San Francisco, California

Contributors

Robert D. Fealey, MD
Department of Neurology, Mayo Clinic
Rochester, Minnesota

Michel D. Ferrari, MD, PhD
Department of Neurology, Leiden University
Medical Center
Leiden, Netherlands

Donna M. Ferriero, MD, MS
Department of Neurology, Department of Pediatrics,
University of California–San Francisco
San Francisco, California

Gereon R. Fink, MD, PhD
Department of Neurology–Cognitive Neurology,
Universitätsklinikum Aachen, Rheinisch-Westfälische
Technische Hochschule Aachen
Aachen, Germany

Kenneth H. Fischbeck, MD
Neurogenetics Branch, National Institute of
Neurological Disorders and Stroke, National Institutes
of Health
Bethesda, Maryland

Karen Furie, MD, MPH
Stroke Service, Massachusetts General Hospital
Boston, Massachusetts

Néstor Gálvez-Jiménez, MD, MSc, MHSA, FACP
Department of Neurology, Florida Atlantic
University–University of Miami Miller School
of Medicine
Weston, Florida

Felix Geser, MD, PhD
Department of Neurology, Innsbruck Medical University
Innsbruck, Austria

Markus Glatzel, MD
Department of Pathology, University Hospital Zurich
Zurich, Switzerland

Peter J. Goadsby, MD, PhD, DSc
Institute of Neurology, The National Hospital for
Neurology and Neurosurgery
London, United Kingdom

Alan L. Goldin, MD, PhD
Department of Microbiology and Molecular Genetics,
Department of Anatomy and Neurobiology,
University of California–Irvine
Irvine, California

Karen J. Greenland, BSc, (Hons) PhD
Center for Hormone Research, Murdoch Children's
Research Institute, University of Melbourne, Royal
Children's Hospital Victoria, Australia
Pediatric Endocrinology Section, University of
Children's Hospital
Tuebingen, Germany

Robert C. Griggs, MD
Department of Neurology,
University of Rochester
School of Medicine and Dentistry
Rochester, New York

David H. Gutmann, MD, PhD
Department of Neurology, Washington University
Neurofibromatosis Center, Washington University
Siteman Cancer Center
St. Louis, Missouri

Christian Hagel, MD
Institute of Neuropathology, University Medical Center
Hamburg–Eppendorf
Hamburg, Germany

Paul J. Hagerman, MD, PhD
Department of Biochemistry and Molecular
Medicine, University of California–Davis
School of Medicine
Davis, California

Kendra Harris, MSc
Department of Pediatric Neurology, Johns Hopkins
University School of Medicine
Baltimore, Maryland

Hans-Peter Hartung, MD
Department of Neurology, Heinrich Heine University
Düsseldorf, Germany

Bernhard Hemmer, MD
Department of Neurology, Heinrich Heine University
Düsseldorf, Germany

Frank L. Heppner, MD
Department of Pathology, University Hospital Zurich
Zurich, Switzerland

Martha R. Herbert, MD, PhD
Department of Pediatric Neurology, Center for
Morphometric Analysis, Massachusetts General
Hospital, Harvard Medical School
Charlestown, Massachusetts

David N. Herrmann, MBBCh
Department of Neurology, University of Rochester
School of Medicine and Dentistry
Rochester, New York

Michio Hirano, MD
Department of Neurology, Columbia University
Medical Center
New York, New York

Edouard Hirsch, MD
Department of Neurology, Strasbourg
University Hospital
Strasbourg, France

Julian T. Hoff, MD
Department of Neurosurgery, University of Michigan
Ann Arbor, Michigan

Alexander H. Hoon, Jr., MD, MPH
Department of Pediatrics, Johns Hopkins University
School of Medicine
Phelps Center for Cerebral Palsy and
Neurodevelopmental Medicine,

Kennedy Krieger Institute
Baltimore, Maryland

Bradley T. Hyman, MD, PhD
Department of Neurology, Massachusetts General Hospital, Harvard Medical School
Charlestown, Massachusetts

Samay Jain, MD
Division of Movement Disorders, Department of Neurology, Columbia University
New York, New York

Wilfrid Jänig
Physiologisches Institut,
Christian-Albrechts-Universität
Kiel, Germany

Safwan S. Jaradeh, MD
Department of Neurology, Medical College of Wisconsin
Milwaukee, Wisconsin

Kurt A. Jellinger, MD
Institute of Clinical Neurobiology
Vienna, Austria

Michael J. Joyner, MD
Department of Anesthesiology, Mayo Clinic College of Medicine
Rochester, Minnesota

Walter E. Kaufmann, MD
Center for Genetic Disorders of Cognition and Behavior, Kennedy Krieger Institute
Departments of Pathology, Neurology, Pediatrics, Psychiatry and Behavioral Sciences, and Radiology and Radiological Science, Johns Hopkins University School of Medicine
Baltimore, Maryland

Richard F. Keep, PhD
Department of Neurosurgery, Department of Integrative and Molecular Physiology,
University of Michigan
Ann Arbor, Michigan

Aaron Kellogg, BS
Molecular Basis of Disease, Medical College of Ohio
Toledo, Ohio

Bernd C. Kieseier, MD
Department of Neurology, Heinrich Heine University
Düsseldorf, Germany

Stephen L. Kinsman, MD
Departments of Pediatrics and Neurology, University of Maryland School of Medicine
Baltimore, Maryland

Hubertus Köller, MD
Department of Neurology, Heinrich Heine University
Düsseldorf, Germany

Czeslawa Kowal, PhD
Department of Medicine,
Columbia University Medical Center
New York, New York

Katrin Lamszus, MD
Department of Neurosurgery, University Medical Center Hamburg–Eppendorf
Hamburg, Germany

Brian R. Landzberg, MD
Division of Gastroenterology and Hepatology, Weill Medical College of Cornell University
New York, New York

Michael H. Lev, MD
Department of Radiology, Massachusetts General Hospital, Harvard Medical School
Boston, Massachusetts

David N. Lieberman, MD, PhD
Center for Genetic Disorders of Cognition and Behavior, Kennedy Krieger Institute
Division of Pediatric Neurology, Department of Neurology, Johns Hopkins University School of Medicine
Baltimore, Maryland

Leland E. Lim, MD, PhD
Department of Neurology and Neurological Sciences, Stanford University School of Medicine
Stanford, California

Paul H. Lipkin, MD
Department of Pediatrics, Johns Hopkins University School of Medicine
Center for Development and Learning, Kennedy Krieger Institute
Baltimore, Maryland

Irene Litvan, MD
Department of Neurology,
University of Louisville School of Medicine
Louisville, Kentucky

Zachary London, MD
Department of Neurology,
University of Michigan
Ann Arbor, Michigan

Phillip A. Low, MD
Department of Neurology, Mayo Clinic
Rochester, Minnesota

Meggan Mackay, MD
Department of Medicine, Columbia University Medical Center
New York, New York

Mark W. Mahowald, MD
Department of Psychiatry, University of Minnesota, Hennepin County Medical Center
Minneapolis, Minnesota

Luigi Manzo, MD
Department of Internal Medicine,
University of Pavia,
Maugeri Foundation Medical Center
Pavia, Italy

Contributors

Nicholas J. Maragakis, MD
Department of Neurology, Johns Hopkins University
Baltimore, Maryland

Joseph C. Masdeu, MD, PhD
Department of Neurological Sciences, University of Navarra Medical School
Pamplona, Spain

Pietro Mazzoni, MD, PhD
Division of Movement Disorders, Department of Neurology, Columbia University
New York, New York

Pamela J. McLean, PhD
Department of Neurology,
Massachusetts General Hospital,
Harvard Medical School
Charlestown, Massachusetts

Sebastiano Mercadante, MD
Anesthesia and Intensive Care Unit, Pain Relief and Palliative Care Unit, La Maddalena Cancer Center
Palermo, Italy

Ana-Claire Meyer, MD
Partners Neurology Program, Massachusetts General Hospital
Boston, Massachusetts

Emmanuel Mignot, MD, PhD
Howard Hughes Medical Institute and Stanford University Center for Narcolepsy
Palo Alto, California

Steven P. Miller, MD, MAS
Department of Pediatrics, University of British Columbia Division of Neurology, British Columbia's Children's Hospital Vancouver, British Columbia, Canada Department of Neurology and Pediatrics, University of California–San Francisco
San Francisco, California

Stewart Mostofsky, MD
Department of Neurology and Psychiatry, Johns Hopkins University School of Medicine Department of Developmental Cognitive Neurology, Center for Autism and Related Disorders, Kennedy Krieger Institute Baltimore, Maryland

Maciej Mrugala, MD, PhD, MPH
Stephen E. and Catherine Pappas Center for Neuro-Oncology, Massachusetts General Hospital
Boston, Massachusetts

Anne Newcomer
Stephen E. and Catherine Pappas Center for Neuro-Oncology, Massachusetts General Hospital
Boston, Massachusetts

Lucilla Nobbio, PhD
Department of Neuroscience, Ophthalmology and Genetics, University of Genova
Genova, Italy

Farshid Noorbakhsh, MD, PhD
Department of Immunology,
Tehran University of Medical Sciences
Tehran, Iran

Peter Novak, MD, PhD
Department of Neurology, Boston University School of Medicine
Boston, Massachusetts

John L. O'Donoghue, VMD, PhD, DABT
Department of Environmental Medicine, Environmental Health Sciences Center, University of Rochester School of Medicine and Dentistry
Rochester, New York

Harry T. Orr, PhD
Department of Laboratory Medicine and Pathology, University of Minnesota
Minneapolis, Minnesota

Tiago Fleming Outiero, PhD
Department of Neurology, Massachusetts General Hospital, Harvard Medical School
Charlestown, Massachusetts

David Palestrant, MD
Department of Neurological Surgery,
Columbia University
New York, New York

Timothy A. Pedley, MD
Neurological Institute of New York, Columbia University Medical Center
New York, New York

Jose-Luis Perez-Velasquez, PhD
Division of Neurology and the Brain and Behavior Program, The Hospital for Sick Children Department of Pediatrics, University of Toronto
Toronto, Ontario, Canada

Michael L. Perlis, PhD
Department of Psychiatry, University of Rochester Sleep and Neurophysiology Laboratory
Rochester, New York

Anders I. Persson, PhD
Departments of Neurology, Pediatrics, and Neurological Surgery; Brain Tumor Research Center and Comprehensive Cancer Center; University of California–San Francisco
San Francisco, California

Joanna J. Phillips, PhD
Department of Pathology (Neuropathology), University of California–San Francisco
San Francisco, California

Linda Piersall, MS
Division of Genetics and Genomic Medicine, Department of Pediatrics, Washington University School of Medicine
St. Louis, Missouri

Wilfred R. Pigeon, PhD
Department of Psychiatry, University of Rochester Sleep and Neurophysiology Laboratory
Rochester, New York

Stuart R. Pomerantz, MD
Department of Radiology, Massachusetts General Hospital, Harvard Medical School
Boston, Massachusetts

Rodica Pop-Busui, MD, PhD
Division of Metabolism, Endocrinology and Diabetes, Department of Internal Medicine,
University of Michigan
Ann Arbor, Michigan

Christopher Power, MD, FRCP (C)
Laboratory for Neurological Infection and Immunity, Department of Medicine,
University of Alberta
Edmonton, Alberta, Canada

James M. Powers, MD
Department of Pathology, University of Rochester School of Medicine and Dentistry
Rochester, New York

Thomas A. Rando, MD, PhD
Department of Neurology and Neurological Sciences, Stanford University School of Medicine
Stanford, California

Rajiv R. Ratan, MD, PhD
Department of Neurology and Neuroscience, The Burke Medical Research Institute, Weill Medical College of Cornell University
White Plains, New York

Sheryl L. Rimrodt, MD
Department of Pediatrics, Johns Hopkins University School of Medicine Developmental Cognitive Neurology, Kennedy Krieger Institute
Baltimore, Maryland

Jeffrey D. Rothstein, MD, PhD
Department of Neurology, Johns Hopkins University
Baltimore, Maryland

James W. Russell, MD, MS, FRCP
Department of Neurology, University of Michigan
Ann Arbor, Michigan

James T. Rutka, MD, PhD
Division of Neurosurgery,
University of Toronto
Toronto, Ontario, Canada

Michael M. Saling, PhD
Department of Psychology, University of Melbourne
Parkville, Australia

Helen E. Scharfman, PhD
Departments of Pharmacology and Neurology, Columbia University and Center for Neural Recovery and Rehabilitation Research, Helen Hayes Hospital, New York State Department of Health
West Haverstraw, New York

Carlos H. Schenck, MD
Department of Psychiatry, University of Minnesota, Hennepin County Medical Center
Minneapolis, Minnesota

Angelo Schenone, MD
Department of Neuroscience, Ophthalmology and Genetics, University of Genova
Genova, Italy

William G. Schrage, PhD
Department of Anesthesiology, Mayo Clinic College of Medicine
Rochester, Minnesota

Michael Schroeter, MD
Department of Neurology, Heinrich Heine University
Düsseldorf, Germany

Peter W. Schütz, MD
Department of Pathology and Laboratory Medicine, Department of Pediatrics, University of British Columbia Division of Biochemical Diseases, British Columbia's Children's Hospital
Vancouver, British Columbia, Canada

Zachary Simmons, MD
Department of Neurology, Penn State Hershey Medical Center
Hershey, Pennsylvania

Harvey S. Singer, MD
Department of Pediatric Neurology,
Johns Hopkins University
School of Medicine
Baltimore, Maryland

Ajay K. Singh, MB, MRCP
Renal Division, Brigham and Women's Hospital
Boston, Massachusetts

J. R. Singleton, MD
Department of Neurology, University of Utah
Salt Lake City, Utah

A. G. Smith, MD
Department of Neurology, University of Utah
Salt Lake City, Utah

O. Carter Snead, III, MD
Division of Neurology and the Brain and Behavior Program, The Hospital for Sick Children
Department of Pediatrics, University of Toronto
Toronto, Ontario, Canada

Eric J. Sorenson, MD
Department of Neurology, Mayo Clinic
Rochester, Minnesota

Velandai Srikanth, PhD
Department of Medicine, Southern Clinical School, Monash University, Monash Medical Center Clayton, Melbourne, Australia Menzies Research Institute, University of Tasmania
Hobart, Australia

Sylvia Stöckler, MD
Department of Pediatrics,
University of British Columbia
Division of Biochemical Diseases,
British Columbia's Children's Hospital Vancouver, British Columbia, Canada

Contributors

Charlotte J. Sumner, MD
Neurogenetics Branch, National Institute of Neurological Disorders and Stroke, National Institutes of Health
Bethesda, Maryland

Michael Swash, MD, FRCP, FRCPath
Institute of Neuroscience, Queen Mary School of Medicine, University of London
Department of Neurology, Royal London Hospital
London, United Kingdom

James W. Teener, MD
Department of Neurology, University of Michigan Medical School
Ann Arbor, Michigan

Charles A. Thornton, MD
Department of Neurology, University of Rochester Medical Center
Rochester, New York

Amanda G. Thrift, PhD
Division of Epidemiology, National Stroke Research Institute, Heidelberg Repatriation Hospital, Austin Health Heidelberg Heights,
Victoria, Australia

Nicole Töpfner
Department of Neurology, Heinrich Heine University
Düsseldorf, Germany

Shoji Tsuji
Department of Neurology, The University of Tokyo
Tokyo, Japan

Meredith L. Turetz, MD
Division of Pulmonary and Critical Care Medicine, Weill Medical College of Cornell University
New York, New York

Paul Twydell, DO
Department of Neurology, University of Rochester School of Medicine and Dentistry
Rochester, New York

Laurent Vercueil, MD, PhD
Department of Neurology, Grenoble University Hospital Grenoble, France

Steven Vernino, MD, PhD
Department of Neurology, University of Texas Southwestern Medical Center
Dallas, Texas

Angela Vincent, MA (MB, BS, MSc Lond.), FRCPath, FMedSci
Department of Clinical Neurology and Weatherall Institute of Molecular Medicine, University of Oxford, John Radcliffe Hospital
Oxford, United Kingdom

Bruce T. Volpe, MD
Department of Neurology and Neuroscience, The Burke Medical Research Institute, Weill Medical College of Cornell University
White Plains, New York

Kenneth R. Wagner, PhD
Department of Neurology, University of Cincinnati College of Medicine Medical Research Service, Department of Veterans Affairs Medical Center
Cincinnati, Ohio

Steven U. Walkley, DVM, PhD
Department of Neuroscience, Rose F. Kennedy Center for Research in Mental Retardation and Human Development, Albert Einstein College of Medicine
Bronx, New York

Robert J. Weil, MD, FCAS
Brain Tumor Institute, Cleveland Clinic
Cleveland, Ohio

William A. Weiss, PhD
Departments of Neurology, Pediatrics, and Neurological Surgery; Brain Tumor Research Center and Comprehensive Cancer Center; University of California–San Francisco
San Francisco, California

Babette B. Weksler, MD
Department of Medicine, Division of Hematology and Medical Oncology, Weill Medical College of Cornell University
New York, New York

Gregor K. Wenning, MD, PhD
Department of Neurology, Innsbruck Medical University
Innsbruck, Austria

Ingo M. Westner, MD
Department of Pathology, University Hospital Zurich
Zurich, Switzerland

Manfred Westphal, MD
Department of Neurosurgery, University Medical Center Hamburg–Eppendorf
Hamburg, Germany

Philip A. Wilkinson, BSc, MBBS, MRCP
Department of Neurology, Royal London Hospital
London, United Kingdom

Alice Wong, PhD
Department of Molecular Biosciences, University of California–Davis School of Veterinary Medicine
Davis, California

Guohua Xi, MD
Department of Neurosurgery, University of Michigan
Ann Arbor, Michigan

Jeffrey D. Zajac, MBBS, PhD, FRACP
Department of Medicine, Austin Health and Northern Health, University of Melbourne
Heidelberg, Victoria, Australia

Jamie M. Zeitzer, PhD
Department of Psychiatry and Behavioral Sciences, Stanford University
Palo Alto, California

Editors

Editor in Chief

Sid Gilman, MD, FRCP
Department of Neurology, University of Michigan
Ann Arbor, Michigan

Section Editors

John C. M. Brust, MD (Substance Abuse and Basic Toxicology)
Department of Neurology, Harlem Hospital Center
and Columbia University College of Physicians & Surgeons
New York, New York

John J. Caronna, MD (Neurologic Manifestations of Medical Diseases)
Department of Neurology and Neuroscience,
Weill Medical College of Cornell University
New York, New York

Kenneth L. Casey, MD (Pain)
Departments of Neurology and Molecular and Integrative
Physiology, University of Michigan and VA Medical Center
Ann Arbor, Michigan

Eva L. Feldman, MD, PhD (Motor Neuron Diseases)
Department of Neurology, University of Michigan
Ann Arbor, Michigan

Christopher Gomez, MD, PhD (Genetic Diseases)
Department of Neurology, The University of Chicago
Chicago, Illinois

Richard T. Johnson, MD, FRCP (Infectious Diseases)
Department of Neurology, Microbiology, and Neuroscience;
The Johns Hopkins University School of Medicine and the
Bloomberg School of Public Health
Baltimore Maryland

Michael V. Johnston, MD (Malformations and Developmental Disorders)
Kennedy Krieger Institute and Departments of
Neurology, Pediatrics, and Physical Medicine and
Rehabilitation; Johns Hopkins University School of
Medicine; The Kennedy Krieger Institute
Baltimore, Maryland

John J. Laterra, MD, PhD (Neoplastic Diseases)
The Johns Hopkins School of Medicine
and The Kennedy Krieger Institute
Baltimore, Maryland

Phillip A. Low, MD (Autonomic Disorders)
Department of Neurology, Mayo Clinic
Rochester, Minnesota

John C. Mazziotta, MD, PhD (Imaging the Nervous System)
Brain Mapping Center, Department of Neurology,
David Geffen School of Medicine, University of
California–Los Angeles School of Medicine
Los Angeles, California

Emmanuel Mignot, MD, PhD (Sleep Disorders)
Howard Hughes Medical Institute and Stanford
University Center For Narcolepsy
Palo Alto, California

Lewis B. Morgenstern, MD (Cerebrovascular Diseases)
Departments of Neurology, Epidemiology,
Emergency Medicine, and Neurosurgery;
University of Michigan Health System
Ann Arbor, Michigan

Hugo W. Moser, MD (Metabolic Diseases)
Department of Neurology and Pediatrics,
Johns Hopkins University, Department of Neurogenetics,
Kennedy Krieger Institute
Baltimore, Maryland

Timothy A. Pedley, MD (Paroxysmal Disorders)
Neurological Institute of New York,
Columbia University Medical Center
New York, New York

Anthony T. Reder, MD (Neuroimmunological Disorders)
Department of Neurology, The University of Chicago
Chicago, Illinois

James W. Russell, MD (Peripheral Neuropathies)
Department of Neurology, University of
Michigan Health System
Ann Arbor, Michigan

Charles A. Thornton, MD (Myopathies and Neuromuscular Junction Disorders)
Department of Neurology,
University of Rochester Medical Center
Rochester, New York

Gregor K. Wenning, MD, PhD (Neurodegenerative Disorders)
Department of Neurology, Innsbruck Medical University
Innsbruck, Austria

Preface

This book is focused on the basic mechanisms underlying neurological disorders, presenting clinical information where needed to place into perspective the discussion of mechanisms. The book is aimed at nonclinician neuroscientists, who may possess the conceptual and technical expertise required for productive investigations in translational research but lack frequent contact with scientifically oriented clinicians who can inform them about current concepts and recent developments. My own experience with many basic scientists interested in neurological disorders is that in their home institutions, they can learn from clinicians only the clinical aspects of neurological diseases and not the fundamental underpinnings or areas of opportunity relevant to specific diseases. This book developed from a conception of Dr. Jasna Markovac, currently Senior Vice President, Global Academic and Customer Relations, Elsevier, who perceived the need for a book providing basic neuroscientists with a comprehensive view of the mechanisms underlying specific neurological diseases. As Dr. Markovac originally envisioned, this book compiles in a single volume the major areas of neurological disease in which there are active, ongoing research and in which the basic scientist's interest may be high. The book is not intended to be a textbook of neurology or even a book for the clinician, but rather a book focused on areas of interest to basic scientists. To this end it is highly selective, dealing with areas of either current activity or great promise for the immediate future. The book is not intended to be all-inclusive. Despite this orientation on basic mechanisms, the contributors and I hope that many clinicians and clinicians in training will find the book both interesting and useful.

In assembling the team to write this material, we selected thought leaders in the specific areas included, asking them to choose 4 to 10 topics that they consider key to the understanding of basic concepts within their areas of expertise and asking them to select authors for each topic. We requested that they recruit mature, experienced authors, preferably clinician scientists, capable of presenting information regarding the mechanisms underlying neurological disorders with only sufficient clinical material for the reader to comprehend the full picture of each disorder. As each author completed a chapter, the responsible section editor reviewed the material for content, accuracy, and clarity, and I have done this as well. There is naturally a good deal of variability among chapters, as certain topics are far advanced with respect to our comprehension of underlying mechanisms, whereas others are less so.

The book is divided somewhat arbitrarily into two parts, covering topics relevant to the central nervous system (Part A) and the peripheral nervous system (Part B). Within these divisions the topics include many of the major categories of disease process with individual chapters describing the mechanisms underlying specific disorders. We required that the authors limit their chapters in size and number of references, using principally recent references and reviews rather than original articles. Hence, we have been able to present a large number of individual topics within a single volume.

I am greatly indebted to Dr. Jasna Markovac for suggesting this project and encouraging me along the way as it unfolded; to Phil Carpenter of Elsevier for his diligent work in monitoring progress bringing the chapters into review; to the section editors who selected the topics, recruited the authors of the chapters, and critically reviewed the chapters; and most of all to the individual authors who wrote the chapters.

Sid Gilman, MD, FRCP

1

Lysosomal Disorders of the Nervous System

Steven U. Walkley, DVM, PhD

Keywords: *autophagosome, axon, axonal spheroid, cholesterol, dendrite, endosome, ganglioside, glycosphingolipid, lysosome, raft, sphingomyelin, synapse*

I. Introduction
II. Classifying Lysosomal Storage Diseases by Their Clinicopathological Features
III. Classifying Lysosomal Diseases by Their Defective Proteins
IV. The Greater Lysosomal System as a Central Coordinator of Neuron Metabolism and Function
V. Dysfunction of the Lysosomal System in Neurons Leads to Complex Pathogenic Cascades
VI. Understanding Lysosomal Disease Pathogenesis Provides Key Insight for Development of Therapy
References

I. Introduction

A variety of degenerative diseases, many affecting children and consisting of brain, visceral, and sometimes skeletal involvement, began to be identified clinically and pathologically in the later part of the nineteenth century. One of the most carefully delineated of these disorders was initially referred to as *amaurotic family idiocy,* to denote its tendency to run in families and to cause blindness and mental retardation. Later, this condition came to be known as *Tay-Sachs disease* and is today viewed as the prototypic example of a lysosomal disorder of the brain. The earliest pathological studies of Tay-Sachs disease were carried out by Bernard Sachs in New York, and patients exhibiting similar neurodegenerative changes and organomegaly were described by Albert Niemann and Ludwig Pick in Berlin, Phillipe Gaucher in Paris, and Fredrick Batten and others in London. What became readily apparent was that even though several types of clinical diseases could be identified, they all exhibited similar distortion of cellular cytoplasm in neurons, hepatocytes, and other cells. This material was often described as lipid-like and the disorders as *dyslipidoses* and later simply as *storage diseases* to emphasize the idea that something appeared to be accumulating in cells. Numerous theories about what was causing this accumulation emerged, including ideas that too much of the material was being produced by cells and the prescient view that affected cells might be missing some "ferment" normally responsible for metabolizing this material. In spite of the near simultaneous development of the concept of inborn errors of metabolism by Archibald Garrod in London in 1909, a clear understanding of the nature of storage diseases had to await the discovery of the lysosome by Christian de Duve in the 1950s, followed within a decade by H. G. Hers' insightful hypothesis that storage diseases represented conditions in which particular lysosomal enzymes were genetically absent.

Today, we recognize that the molecular causes of lysosomal storage disorders are far more complex than the simple absence of an individual lysosomal enzyme and include defects in a wide variety of soluble and transmembrane proteins, some of which lack any known enzyme activity [1–3]. Analysis of the intimate association of lysosomes with vesicular traffic belonging to the endosomal, autophagosomal, and secretory systems has likewise expanded our understanding of the substantial role played by this system in the normal functioning of cells and in the full extent of abnormalities that can be brought into play when lysosomal proteins are defective. Growth of knowledge in the neurosciences has also added considerable insight into possible pathogenic cascades set in motion when the lysosomal system is disturbed in specialized cells such as neurons. Here, intimate association of the endosomal–lysosomal (E/L) and autophagosomal/lysosomal (A/L) systems with vital vesicular trafficking events ranging from synaptic vesicle turnover in axons to recycling of neurotransmitter receptors at synapses in dendrites promises to revolutionize our appreciation of the many ways by which lysosomal diseases can have an effect on neuron function. The goals of this chapter are to describe storage disorders as they are currently recognized, to define key pathogenic features affecting neurons in these diseases, and to identify links between these abnormalities and known functions of the greater lysosomal system.

II. Classifying Lysosomal Storage Diseases by Their Clinicopathological Features

As described briefly in the preceding section, the accumulation of substances in cells in lysosomal diseases has been viewed, along with the clinical features exhibited by affected patients, to be of paramount importance for classifying and understanding these diseases. The generality of lipid-like materials gave way to much more specific characterization of storage material during the first half of the twentieth century. Given the absence of any established cause for this accumulation (until 1965), a system of nomenclature developed that identified many storage diseases by the individuals who first described them (hence, Tay-Sachs, Niemann-Pick, and Gaucher diseases), with these groupings later being organized according to the types and subtypes of materials documented as "stored" in cells, and still later, by differences in defective proteins. Thus, as shown in Table 1, we can recognize major categories of storage diseases as, for example, the sphingolipidoses, the mucopolysaccharidoses (MPSs), and the glycoproteinoses, to denote primary storage of lipids and glycolipids, of mucopolysaccharides (now referred to as *glycosaminoglycans*), and of glycoproteins, respectively. These major groupings can then be further divided into more than 50 individual types of disorders based on gene/protein defects, subtle differences in primary storage and/or organs affected, and variations in clinical onset/progression. Thus, for the sphingolipidoses, the GM_1 gangliosidoses denote one major category of disease characterized by the storage of ganglioside G_{M1}. Although the same lysosomal enzyme (β-galactosidase) is defective in all individuals with GM_1 gangliosidosis, multiple clinical subtypes (variants) are recognized ranging from a rapidly advancing infantile disease to more chronic forms in which onset occurs in juveniles or even in adults. This pattern of clinical progression is typical of most types of storage diseases, that is, the later the clinical onset, the longer the disease course. Clinical variants for the GM_1 gangliosidoses, again like other storage diseases, may in part be due to modifier genes expressed in different individuals. Of likely greater importance, however, is the large number of mutations found in the gene for β-galactosidase and the fact that most patients with this (and other lysosomal diseases) are actually compound heterozygotes for their gene defect. As a result of carrying two different mutant alleles, a variable amount of the defective protein is expressed that may retain some degree of residual function. Depending on the amount of this function, storage and related metabolic events can be temporarily delayed and provide months or years of protection. Eventually, however, the tendency toward intracellular storage and its consequences overwhelms this residual function and clinical disease is initiated.

For lysosomal diseases in which ganglioside G_{M2} rather than G_{M1} is the primary storage material, three major subcategories are delineated with different genes being involved for each (see Table 1). Most cases of GM_2 gangliosidosis are caused by defective activity of the lysosomal hydrolase known as β-hexosaminidase. This enzyme is composed of two subunits, α and β, which are coded for by different genes with defects in each leading to Tay-Sachs and Sandhoff diseases, respectively. A third type of GM_2 gangliosidosis, known as *AB variant*, is caused by mutations in a gene coding for a small (nonenzymatic) soluble protein of

Table 1 Lysosomal Storage Diseases

Disease Categories and Types[a]	Protein Defect	Location	Function/Molecular Defect[b]
Sphingolipidoses			
GM$_1$ Gangliosidosis			
Infantile, late infantile/juvenile, late onset	β-Galactosidase	LE/LY	Lysosomal hydrolase[1]
GM$_2$ Gangliosidoses			
Tay-Sachs disease (infantile, juvenile, late-onset)	β-Hexosaminidase A	LE/LY	Lysosomal hydrolase[1]
Sandhoff disease (infantile, juvenile, late-onset)	β-Hexosaminidase A+B	LE/LY	Lysosomal hydrolase[1]
G$_{M2}$ Activator deficiency	G$_{M2}$ Activator	LE/LY	Nonenzymatic glycoproteins critical for G$_{M2}$ degradation[5]
Gaucher disease			
Acute and subacute neuropathic (types II and III)	β-Glucosidase	LE/LY	Lysosomal hydrolase[1]
Fabry disease	α-Galactosidase A		
Krabbe disease (globoid cell leukodystrophy)	β-Galactosylceramidase	LE/LY	Lysosomal hydrolase[1]
Infantile, late infantile, late onset			
Metachromatic leukodystrophy (MLD)	Arylsulfatase A	LE/LY	Lysosomal hydrolase[1]
Late infantile, juvenile, late onset			
Sphingolipid activator protein (SAP) deficiencies	SAP precursor, SAPs A, B, C, D	LE/LY	Nonenzymatic glycoproteins critical for degradation[5]
Niemann-Pick disease			
Types A and B	Acid sphingomyelinase	LE/LY	Lysosomal hydrolase[1]
Type C1	NPC1 protein	LE/LY	Cholesterol transport protein?[6]
Type C2	NPC2 protein	LE/LY	Cholesterol-binding protein[6]
Farber disease	Acid ceramidase	LE/LY	Lysosomal hydrolase[1]
Mucopolysaccharidoses			
Hurler (MPS IH)	α-L-Iduronidase	LE/LY	Lysosomal hydrolase[1]
Scheie (MPS IS)	α-L-Iduronidase	LE/LY	Lysosomal hydrolase[1]
Hurler/Scheie (MPS H/S)	α-L-Iduronidase	LE/LY	Lysosomal hydrolase[1]
Hunter (MPS II)	Iduronate sulfatase	LE/LY	Lysosomal sulfatase[1]
Sanfilippo A (IIIA)	Heparan N-sulfatase	LE/LY	Lysosomal sulfatase[1]
Sanfilippo B (IIIB)	α-N-Acetyl-glucosaminidase	LE/LY	Lysosomal hydrolase[1]
Sanfilippo C (IIIC)	Acetyl-CoA:α-glucosaminide acetyltransferase	LE/LY	Lysosomal transferase[1]
Sanfilippo D (IIID)	N-Acetylglucosamine 6-sulfatase	LE/LY	Lysosomal sulfatase[1]
Morquio A (MPS IVA)	Galactose 6-sulfatase	LE/LY	Lysosomal sulfatase[1]
Morquio B (MPS IVB)	β-Galactosidase	LE/LY	Lysosomal hydrolase[1]
Maroteaux–Lamy (MPS VI)	Arylsulfatase B	LE/LY	Lysosomal hydrolase[1]
Sly (MPS VII)	β-Glucuronidase	LE/LY	Lysosomal hydrolase[1]
MPS IX	Hyaluronidase	LE/LY	Lysosomal hydrolase[1]
Mucolipidoses (MLs)			
ML II (I-cell disease)	N-Acetylglucosamine-phosphoryltransferase (αβ-subunit)	Golgi	Lysosomal enzyme targeting[2]
ML IIIA (pseudo-Hurler polydystrophy)	N-Acetylglucosamine-phosphoryltransferase (αβ-subunit)	Golgi	Lysosomal enzyme targeting[2]
ML IIIC	N-Acetylglucosamine-phosphoryltransferase (γ-subunit)	Golgi	Lysosomal enzyme targeting[2]
ML IV	Mucolipin-1	LE/LY	TRP cation channel?[7]
Glycoproteinoses			
α-Mannosidosis	α-Mannosidase	LE/LY	Lysosomal hydrolase[1]
β-Mannosidosis	β-Mannosidase	LE/LY	Lysosomal hydrolase[1]
Fucosidosis	α-L-Fucosidase	LE/LY	Lysosomal hydrolase[1]
Aspartylglucosaminuria	Aspartylglucosaminidase	LE/LY	Lysosomal hydrolase[1]
Galactosialidosis	PPCA	LE/LY	Protective protein[4]
Sialidosis (mucolipidosis I)	Neuraminidase	LE/LY	Lysosomal hydrolase[1]
Schindler–Kanzaki	N-Acetylgalactosamindase	LE/LY	Lysosomal hydrolase[1]
Glycogenoses			
Pompe disease	Acidic α-glucosidase	LE/LY	Lysosomal hydrolase[1]
Danon disease	LAMP2	LE/LY	Lysosomal membrane integrity, autophagic events?[7]

(Continued)

Table 1 (Continued)

Disease Categories and Types[a]	Protein Defect	Location	Function/Molecular Defect[b]
Neuronal Ceroid Lipofuscinoses (NCLs)			
Infantile NCL (CLN1 disease)	CLN1 (Palmitolyl protein thioesterase)	LE/LY	Lysosomal thioesterase[(1)]
Classic late infantile NCL (CLN2 disease)	CLN2 (Tripeptidyl peptidase)	LE/LY	Lysosomal protease[(1)]
Late infantile NCL (Finnish variant) (CLN5 disease)	CLN5	LE/LY	Unknown?[(7)]
Late infantile NCL (Atypical variant) (CLN 6 disease)	CLN6	ER	Unknown
Juvenile NCL (Batten) (CLN3 disease)	CLN3	LE/LY	Unknown?[(7)]
Early juvenile variant (CLN7 disease)	CLN7	?	Unknown
Adult-onset form (Kufs) (CLN4)	CLN4	?	Unknown
Northern epilepsy (CLN8)	CLN8	?	Unknown
Others			
Salla disease	Sialin	LE/LY	Sialic acid transporter[(6)]
Infantile free sialic acid storage (ISSD)	Sialin	LE/LY	Sialic acid transporter[(6)]
Cystinosis	Cystinosin	LE/LY	Cysteine transporter[(6)]
Wolman disease	Acid lipase	LE/LY	Lysosomal hydrolase[(1)]
Cholesterol ester storage disease	Acid lipase	LE/LY	Lysosomal hydrolase[(1)]
Multiple sulfatase deficiency (MSD)	Formylglycine-generating enzyme	ER	Sulfatase catalytic activation[(3)]

[a] Detailed descriptions of individual diseases can be found in individual chapters provided in Reference 3.
[b] Major categories of proteins according to function (see also Fig. 1 and text).
[(1)] Enzymes with acidic pH optima directly involved in lysosomal degradation.
[(2)] Golgi-associated enzyme critical for correct trafficking of enzymes to the lysosome.
[(3)] Lysosomal enzyme maturation/processing/catalytic activation mechanism.
[(4)] Lysosomal protein critical for lysosomal enzyme protection.
[(5)] Soluble nonenzyme glycoprotein critical for enzyme degradation of substrate.
[(6)] Transmembrane protein involved with substrate transfer.
[(7)] Transmembrane protein involved with vesicle fusion, autophagy.

the lysosome (the G_{M2} activator) that is essential for the enzyme-G_{M2} substrate interaction. Interestingly, elucidation of the AB variant form of G_{M2} gangliosidosis followed that of Tay-Sachs and Sandhoff diseases by many years and was the key disease that led to the serendipitous discovery of ectopic dendritogenesis, a unique feature of many neuronal storage diseases, as described later.

Individuals with G_{M2} gangliosidosis, just as those with G_{M1} gangliosidosis, exhibit major brain disease but minimal visceral involvement simply because of the abundance of gangliosides in cells of the brain versus other organs. This theme—that the extent of disease impact on cells or organs occurs according to the prominence of the primary storage substrate to that type of cell or tissue—like that of compound heterozygosity and clinical variation, is one that repeats itself across all lysosomal storage diseases. In addition to the storage material present in greatest amounts, smaller amounts of additional compounds may also be present. In many cases, as in the accumulation of oligosaccharides in Sandhoff disease, this is due to the missing lysosomal enzyme having activity toward the same molecular linkages in these substrates. In other cases, however, as with the secondary accumulation of gangliosides (G_{M2}, G_{M3}) and cholesterol in widespread types of storage diseases, the reason for the elevation is less clear but likely is not due directly to the primary enzyme defect.

In addition to G_{M1} and G_{M2} gangliosidoses, numerous other sphingolipid storage diseases are also documented (see Table 1). Gaucher disease occurs secondary to the absence of the lysosomal enzyme β-glucosidase (glucocerebrosidase) and exhibits storage of glucosylceramide, a degradative product of gangliosides and globoside. Although type 1 is primarily a liver disease, types 2 (acute form) and 3 (subacute form) exhibit central nervous system (CNS) storage and neurological deterioration. However, as with other storage diseases, considerable heterogeneity exists in these forms across ethnic and demographic groups. Unlike primary ganglioside storage diseases, the major cell storage in the neuropathic forms of Gaucher disease appears not to involve neurons but instead macrophages, so-called Gaucher cells. Neuron death can be widespread, however, and the lack of major intraneuronal storage has suggested the presence of a neurotoxic agent, possibly the deacylated analogue of glucosylceramide (glucosylsphingosine). As described in a later section, however, there is more recent evidence that defective calcium homeostasis secondary to altered distribution of glucosylceramide in the endoplasmic reticulum (ER) and other cell membranes is responsible for neuron death [4].

Fabry disease is the only known X-linked storage disorder and is caused by deficiency of α-galactosidase A resulting in widespread tissue accumulation of crystalline glycosphingolipids (GSLs) exhibiting birefringence. Lysosomes of endothelial and muscle cells are predominately affected, even in the brain, and neuronal storage is not reported. Patients exhibit onset in childhood or adolescence with bouts of severe pain in the extremities and with skin lesions. CNS damage may occur secondary to transient ischemic attacks and strokes.

Krabbe disease or globoid cell leukodystrophy (galactosylceramide lipidosis) and metachromatic leukodystrophy (MLD) are two types of glycosphingolipidoses that exhibit severe white-matter disease in the CNS. Gray matter, in contrast, is relatively spared in both disorders. Krabbe disease results from deficiency of galactosylceramidase, which normally degrades galactosylceramide to ceramide and galactose. The buildup of a toxic compound, galactosylsphingosine, has been suggested to kill oligodendroglia, which normally myelinate axons. Additionally, there is a massive cellular reaction and the presence of so-called *globoid cells* is characteristic. As with lysosomal diseases with predominately gray-matter involvement, there is again variation in the age at onset, with infantile, late infantile, and adult onset variants described. The disease MLD is caused by a defect in degradation of sulfated glycolipids that occur in white matter as a result of deficiency of the arylsulfatase A lysosomal enzyme. Sulfatide accumulation in cells leads to the presence of characteristic metachromatic granules in a variety of cell types, particularly in the CNS and peripheral nervous system (PNS). Arylsulfatase A is also functionally compromised along with other sulfatases in a disorder known as multiple sulfatase deficiency (MSD), as described later in this chapter. A condition resembling MLD can also be caused by absence of saposin B, the sulfatide activator protein, one of a family of proteins essential for enabling specific lysosomal hydrolases to degrade sphingolipid substrates that have short hydrophilic head groups, as described in the next section.

Niemann-Pick disorders represent a separate category of sphingolipidoses (see Table 1). These disorders were classified as types A, B, C, and D in the 1960s based on clinical and pathological criteria. Types A and B were subsequently recognized as acid sphingomyelinase deficiencies with primary storage of sphingomyelin occurring predominately in brain and liver, respectively. Gangliosides and cholesterol were also found to accumulate secondarily in the brain of patients with type A. Types C and D, in spite of their clinical similarity to types A and B, were not found to exhibit a primary defect in sphingomyelinase. Subsequently, most type C and D cases were shown to be caused by a mutation in NPC1, a gene coding for a transmembrane protein believed critical for the retroendocytic transport of cholesterol. Earlier work had established a defect in cholesterol esterification in type C patient fibroblasts in culture and this had been exploited as a successful diagnostic tool. Since the discovery of NPC1, a small percentage of cases showed normal NPC1 expression and this led to the discovery of another protein, NPC2, also known as *HE1*. Like NPC1, NPC2 protein has also been implicated in cholesterol processing in cells, although the basis of cholesterol accumulation remains unknown. In spite of their distinctly different molecular defects, types A and C diseases both exhibit the same severe CNS abnormalities and storage of cholesterol, as well as several GSLs, most notably gangliosides G_{M2} and G_{M3}.

Another disorder of sphingolipid catabolism is Farber lipogranulomatosis or Farber disease. In these individuals, there is a defect in lysosomal acid ceramidase and primary accumulation of ceramide in lysosomes. Although exceedingly rare, multiple clinical variants have been described, with age at onset ranging from severe infantile to intermediate and mild forms. In some cases, CNS involvement is substantial and involves seizures and progressive motor dysfunction secondary to neuronal storage and death of neurons.

The MPSs, in contrast to some of the aforementioned disease groups, are subdivided solely on the basis of the lysosomal enzyme that is defective, with a variety of the same glycosaminoglycans (e.g., heparan sulfate, keratan sulfate, and dermatan sulfate) stored in each form based on specific characteristics of the missing enzyme (see Table 1). Glycosaminoglycans are split from the parent proteoglycan molecules and degraded within the lysosome. Here, not only is brain involvement significant in many of these diseases, but bone and cartilage are similarly involved because of the importance of proteoglycans in these tissues. In addition to storage of particular glycosaminoglycans, most MPS diseases are also known to undergo significant accumulation of certain gangliosides (most notably G_{M2} and G_{M3}) and cholesterol.

Other categories of storage diseases reflect significant diversity in the types of proteins that are defective and in the nature of stored compounds. The so-called mucolipidoses (MLs), for example, were grouped together historically (and perhaps prematurely), largely because they were found to exhibit clinical similarities intermediate between the MPSs and the sphingolipid storage diseases. The infantile form of sialidosis, caused by a defect in α-neuraminidase (sialidase), was originally called *ML I* but has now been removed from this category and classified with other glycoproteinoses, as described later in this chapter. ML disease types II and III, also known as *I-cell disease* and *pseudo-Hurler polydystrophy*, respectively, likewise affect glycoprotein processing, but in this case, the ML classification has withstood change. ML II and ML III diseases are actually allelic variants of the same disorder and as discussed later are due to a unique mechanism that alters the targeting of multiple lysosomal enzymes to their appropriate location in lysosomes, not to primary defects in the enzymes themselves. The other disease placed in the ML category (ML IV) is also unique in its disease features and in the nature of the protein defect that again is not a lysosomal enzyme. It, too, is discussed in the next section.

Another major group of lysosomal diseases alluded to are the so-called *glycoproteinoses*. This category also represents a diverse grouping of disorders that are linked simply on the basis of defects affecting lysosomal processing of the carbohydrate chains of glycoproteins, with subsequent tissue storage and urinary secretion of noncatabolized sugar chains and small glycopeptides [5]. In addition to sialidosis and I-cell disease, this category includes defects in several lysosomal enzymes, including α-mannosidase (α-mannosidosis), β-mannosidase (β-mannosidosis), α-fucosidase (fucosidosis), α-N-acetylgalactosaminidase (Schindler disease), and aspartylglucosaminidase (aspartylglucosaminuria). All of these disorders involve CNS storage of glycoproteins and in some cases secondary storage of glycolipids.

The category known as *glycogen storage diseases* (see Table 1) also contains two diversely different conditions, both of which are known to involve to some degree the accumulation of glycogen within the lysosome. Pompe disease, or glycogen storage disease type II, is a true lysosomal disorder (indeed, the first to be recognized as such) caused by deficiency of acid α-glucosidase. Storage affects multiple cell types including those of muscle, liver, and brain. In addition to Pompe disease, glycogen storage has also been reported to occur in another disorder known as Danon disease, which has been found to be caused by a genetic defect in lysosomal-associated membrane protein 2 (LAMP2). Here, mechanisms of storage may involve abnormalities in vesicle fusion, particularly as concerns autophagic vacuoles with lysosomes.

Perhaps the most enigmatic of all the groups of lysosomal disease are the so-called *neuronal ceroid lipofuscinoses* (NCLs) (see Table 1), originally named on the basis of an ill-defined ceroid-lipofuscin type of storage and the belief that only neurons were involved in storage. Although neither finding has stood the test of time, the name has persisted. By the 1990s, the NCL disorders had essentially been left behind by advances delineating defective enzymes and other proteins causing the other major types of storage diseases. Only in the past decade have defective proteins been identified and individual disease variants recognized according to specific gene designations (CLN2 as the classic late infantile form, CLN3 as the juvenile form or Batten disease, etc.). Although two of these proteins are clearly lysosomal enzymes (palmitoyl protein thioesterase for CLN1 disease and tripeptidyl peptidase for CLN2 disease), others are not and remain ill-defined and poorly understood. Furthermore, the relationship between the defective proteins of the NCL disorders and the putative primary storage material believed present in most of these conditions, a lipoprotein known as the *c-subunit* of mitochondrial adenosine triphosphate (ATP) synthase, remains unexplained.

Finally, as seen in Table 1, some lysosomal diseases do not fit neatly into any major category and are listed as "other." Included here are two conditions caused by defects in the lysosomal hydrolase known as *acid lipase* (Wolman disease and cholesterol ester storage diseases). Although caused by defects in the same protein, the two conditions exhibit clear-cut clinical differences, due likely, as explained earlier, to differences in particular allelic mutations present in cells. The other diseases in this category are not due to defects in enzymes of the lysosome at all, but to a processing enzyme of the ER (MSD) or to transmembrane proteins involved with substrate transport out of the lysosome, not substrate degradation (Salla disease, infantile free sialic acid storage disease, and cystinosis). These and other "oddball" storage diseases mentioned earlier, when analyzed and grouped according to the identified functions of the defective proteins, provide considerable insight into how far-reaching the E/L system is in terms of controlling the overall metabolism of cells.

III. Classifying Lysosomal Diseases by Their Defective Proteins

With Hers' hypothesis (in 1965) potentially explaining all storage diseases as being secondary to the absence of individual degradative enzymes of the lysosome, the challenge became to identify which enzymes were deficient in each recognized type of storage disease. In the ensuing 2 decades, defective lysosomal enzymes were discovered for many lysosomal diseases, and one by one, diseases were defined and categorized as given in Table 1. However, as already alluded to in the preceding discussion, some storage diseases proved not to be so straightforward in that the primary defective proteins were found either not to be enzymes at all or to be enzymes located outside the E/L system (Fig. 1). In some cases, it was found that diseases classified together (e.g., Niemann–Pick types A and C) had completely different types of molecular defects, and other diseases believed distinct (ML II and ML III) were found to actually involve the same gene and protein. Indeed, the MLs are a good example of an early exception to Hers' hypothesis (at least in one sense of the hypothesis) in that they reveal a second mechanism by which a specific protein defect could cause widespread lysosomal storage. ML II disease was first recognized as a condition resembling Hurler's disease (MPS I) but lacking the presence of mucopolysaccharides in the urine. The original name, I-cell disease, came from the presence of inclusions in fibroblasts and other cells ("inclusion cells"), and these were noted to be similar to ones seen in yet another Hurler-like condition, pseudo-Hurler polydystrophy, now known as ML III. Instead of showing an absence of function of a single lysosomal enzyme, studies of these conditions revealed that

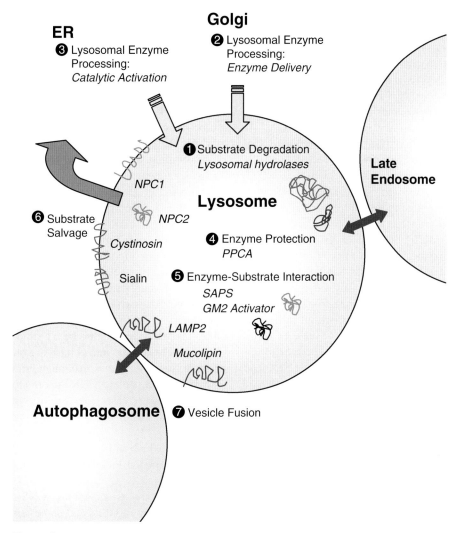

Figure 1 Types of proteins implicated in lysosomal storage diseases. Lysosomal storage diseases are caused by defects in a wide range of types of proteins. Illustrated here are seven categories ranging from defects involving lysosomal enzymes themselves ① and in enzymes that are part of the normal trafficking ②, catalytic activation ③, or protection of known lysosomal enzymes ④. Other proteins include those that assist lysosomal enzymes in substrate processing ⑤ or in removal of substrate from lysosomes ⑥. Finally, newly discovered proteins involved as cation channels and other transmembrane constituents may influence vesicle fusion and ultimately substrate processing ⑦. See text and Table 1 for additional details.

numerous lysosomal enzymes were deficient in activity in cultured fibroblasts. At the same time, however, their activity in culture media was elevated, as was their activity in sera from patients with these diseases. In a series of remarkable experiments, it was shown that I-cell disease fibroblasts were capable of taking up enzymes secreted by normal cells but that normal cells could not take up I-cell–secreted enzymes. This important finding led to the discovery that lysosomal enzymes possessed a recognition marker, mannose 6-phosphate, which was responsible for their targeting to lysosomes and for their uptake by cells (in association with an appropriate mannose 6-phosphate receptor) (see Fig. 1 [②]). The enzyme responsible for attaching this residue was found to be N-acetylglucosamine-1-phosphotransferase, located in the Golgi complex, and it was functionally absent in both ML II and ML III diseases. Most forms of ML II and III (type A for the latter) are allelic and caused by a defect in a single gene coding for the α- and β-subunits of the phosphotransferase enzyme. The subtype of ML III known as *ML IIIC* has been shown to be caused by a separate gene coding for the subunit (γ) of phosphotransferase involved in lysosomal enzyme recognition.

Individuals with the other designated form of mucolipidosis, type IV, also exhibit clinical features common to both the MPSs and the sphingolipidoses, but the molecular mechanism responsible for storage is mechanistically unrelated to others in the ML group. The defective protein in this case is a transmembrane protein (mucolipin-1), which exhibits homology to members of the transient receptor potential (TRP) family of cation channels. The precise role of the mucolipin-1 protein in cells is not understood, but possibilities include control of vesicle fusion during the trafficking of late endosomes and lysosomes (see Fig. 1), as is further described later.

The finding of reduced activity of multiple lysosomal enzymes is characteristic not only of ML II and III diseases but also of another condition known as MSD, although the mechanism responsible is completely different. Patients with MSD resemble those with late infantile MLD but display defective function not only in arylsulfatase A (characteristic of MLD) but also of the 11 additional types of sulfatase enzymes of lysosomes. The reason for defective function of all 12 enzymes has been shown to result from a failure to generate the key catalytic residue unique to all sulfatases (formylglycine) from a cysteine precursor. This process is normally carried out in the ER by the formylglycine-generating enzyme. Absence of sulfatase activity in turn leads to storage in cells consisting of a wide variety of sulfated glycosaminoglycans, GSLs, and glycoproteins (see Fig. 1 [③]).

Another mechanism leading to deficiency of more than one lysosomal enzyme, in this case a combined deficiency of β-galactosidase and neuraminidase, involves a defect in a lysosomal protein known as *protective protein/cathepsin A* (PPCA) (see Fig. 1 [④]). PPCA associates with these two enzymes during synthesis and aids in their correct routing to the lysosome, as well as in their protection from intralysosomal proteolysis. In addition to the protective function, PPCA also has catalytic activity against a number of neuropeptides. Patients lacking PPCA exhibit wide-ranging features involving defects in brain and viscera, as well as skeletal dysplasia. Storage of sialyloligosaccharides is found in many tissues, hence, the common name of this disorder *galactosialidosis*.

Yet a fourth mechanism leading to lysosomal storage that does not involve defects in individual lysosomal enzymes is caused by mutations in genes coding for nonenzyme lysosomal proteins known as *sphingolipid activator proteins* (SAPs), which are critical for enzyme–substrate interactions (see Fig. 1 [⑤]). The substrates in this case are sphingolipids with short hydrophilic head groups. The two recognized groups of SAPs are the saposins and the G_{M2} activator protein. Saposins are coded for by one gene, with the prosaposin precursor molecule being transported to the cell surface and subsequently endocytosed and processed to individual SAPs (A, B, C, and D). There is also evidence that prosaposin is released from cells and may function as a neurotrophic factor. SAP-A has been found to stimulate the degradation of galactosylceramide by glucosylceramidase and galactosylceramidase. SAP-B shows less specificity and stimulates hydrolysis of multiple glycolipids by different enzymes, with defects leading to disorders resembling juvenile MLD. SAP-C is essential for degradation of glucosylceramide and its absence leads to a condition resembling type 3 Gaucher disease. SAP-D has been reported to be essential for degradation of ceramide by acid ceramidase. The other type of SAP is the G_{M2} activator protein found to be deficient in a form of GM_2 gangliosidosis known as the *AB variant* (to indicate that the A and B isoforms of β-hexosaminidase are present in normal amounts). The G_{M2} activator is believed to act as a "liftase" that allows for the degradation of ganglioside G_{M2} by this enzyme.

Another mechanism that has emerged as a cause of lysosomal storage is illustrated by proteins that appear to be essential for normal trafficking of specific substrates out of lysosomes (see Fig. 1 [⑥]). Sialin, for example, has been shown critical for removal of free sialic acid, a degradative product of gangliosides and some glycoproteins, from the lysosome. The two clinical variants that occur as a result of mutations in this gene are Salla disease and infantile free sialic acid storage disease. Cystinosin is another carrier protein of lysosomes and in this case moves the amino acid, cystine, from lysosomes. Although principally a disorder of renal function, severe CNS abnormalities have also been reported in some cases. Yet a third type of protein that has been viewed within the context of substrate transfer from the E/L system is the transmembrane protein known as NPC1, defects in which are responsible for the vast majority of cases of Niemann-Pick type C (NPC) disease, as described earlier.

NPC disease is a striking CNS disorder exhibiting massive storage of GSLs and cholesterol. Faulty esterification of cholesterol in fibroblasts from affected individuals suggested that this is a primary disorder of cholesterol trafficking, a view reinforced by the finding that the NPC1 protein contains a sterol-sensing domain. The NPC1 protein has also been shown to be homologous to certain prokaryotic "permeases" and thus may share a common function of moving substrates across membranes [6]. The identity of the primary substrate moved by NPC1 remains uncertain, however, as studies failed to find evidence of cholesterol transfer by NPC1 and other studies have indicated that the overall level of cholesterol sequestration in NPC-affected neurons is controlled by the degree of ganglioside accumulation.

A small number of cases of NPC disease are caused by defects in a second protein, NPC2 (HE1), which has been shown to be a cholesterol-binding protein delivered to lysosomes by the same mannose 6-phophate recognition moiety described for lysosomal enzymes. The NPC2 protein has been compared to the G_{M2} activator in that both are soluble proteins of the lysosome that interact with lysosomal substrates (cholesterol and ganglioside G_{M2}, respectively). Since the disease states caused by absence of either NPC1 or

NPC2 appear essentially identical, it has been suggested that the two proteins interact in a common mechanism linked to cholesterol-GSL homeostasis. This may occur through the regulation of retroendocytic transport of one or both types of compounds from the E/L system to the Golgi/trans-Golgi network (TGN) or other internal membrane sites, as is discussed in a later section.

Another mechanism that when deranged may lead to lysosomal storage involves vesicle fusion within the greater E/L system (see Fig. 1 [⑦]). This mechanism was mentioned in relation to mucolipin-1 in ML IV disease and may conceivably also be related to NPC1 function. Another protein implicated here is LAMP2, also mentioned earlier as being defective in the glycogen storage disease known as Danon disease. This disorder is characterized by intracellular accumulation of glycogen (in spite of normal α-glucosidase activity), along with large numbers of autophagic vacuoles, suggestive of fusion defects between autophagosomes and vesicles of the E/L system.

Advances in the identification of molecular mechanisms underlying individual types of storage diseases have shown that defects in a wide variety of proteins are capable of disrupting normal function of the lysosomal system and of causing storage of materials within cells. Many of these newly recognized proteins, particularly those affecting vesicle fusion and/or substrate movement within cells, remain poorly understood and, indeed, may be viewed quite differently as their specific functions are fully elucidated. The roles played by some of these proteins (e.g., CLN6 and CLN3 for late infantile variant and juvenile forms of neuronal ceroid lipofuscinosis, respectively) remain essentially unknown and may involve new or novel mechanisms not yet considered as part of lysosomal function.

Lysosomal diseases thus represent a widely diverse family of disorders caused by defects in many types of proteins but united by the common thread of E/L and A/L system dysfunction and subsequent storage. Within this family of conditions, different mutations (or combinations of mutations in compound heterozygotes) affecting the *same* gene can in some cases lead to dramatically different disease states. Examples of this are seen with Niemann-Pick disease types A versus B, and in the Hurler versus Scheie forms of MPS type I. Conversely, defects in dramatically *different* types of proteins are known to cause near-identical disease states at the clinical and pathological level, as is seen for NPC1 and NPC2 proteins and for the G_{M2} activator protein and β-hexosaminidase. As discussed later, remarkably similar disease cascades also may occur in lysosomal diseases of completely separate categories, undoubtedly reflecting recruitment of the same or similar downstream metabolic processes by the defective proteins. Notable examples of such pathological features are ectopic dendritogenesis (a feature unique to lysosomal disorders) and neuroaxonal dystrophy. These diverse changes in neuron structure and connectivity, which occur across a wide spectrum of lysosomal diseases, dramatically illustrate the importance of proteins linked to E/L and A/L function, as well as the central influence that the greater lysosomal system maintains over cells of the nervous system.

IV. The Greater Lysosomal System as a Central Coordinator of Neuron Metabolism and Function

The tendency to refer to the lysosome as the "garbage can" of the cell unfortunately remains alive and well in the writings of some modern cell biologists. Although understandable from a historical point of view, such an analogy suggests an essentially isolated and passive function for these organelles, a view that is no longer tenable based on what modern studies of lysosomes are telling us. Rather than viewing lysosomes in this limited manner, this chapter posits a much more vital role for this organelle and its protein machinery—namely that the lysosome and its closely allied endosomal system (which also contains lysosomal enzymes and related proteins), and in coordinated action with autophagosomal and secretory vesicular systems, serves as a central metabolic coordinator, a vital recycling center that in one way or another is involved with literally every aspect of the life of the cell from signal transduction following receptor-mediated endocytosis to gene expression. The complexity of the pathogenic cascades observed in lysosomal storage diseases is indicative of this plethora of functions, and nowhere is this more evident than in neurons.

The lysosomal system of neurons represents one part of an intricate and extensive vesicular networking system (Fig. 2). Lysosomes are historically defined as cellular organelles with an acidic pH (pH 5.0–5.5) rich in hydrolases with acidic pH optima. They are in direct continuity with late endosomes, which are also acidified (pH 5–6) and not only contain hydrolytic enzymes but also possess receptors for the cation-independent mannose 6-phosphate moiety used to deliver lysosomal enzymes to these vesicles. Thus, lysosomes and late endosomes share the feature of being sites of degradation of a variety of complex molecules. These molecules may be delivered to late endosomes and then to lysosomes through interaction with early endocytic compartments, often referred to as *sorting* and *recycling endosomes*, which likewise exhibit acidified environments (pH 6.0–6.5). It is also likely that early endosomes will contain some hydrolytic enzymes derived, for example, from extracellular sources secondary to secretion. Receptor-mediated endocytosis, as well as a variety of other mechanisms (phagocytosis, pinocytosis, etc.), result in internalization of an array of complex molecules. Included here are not only cell surface

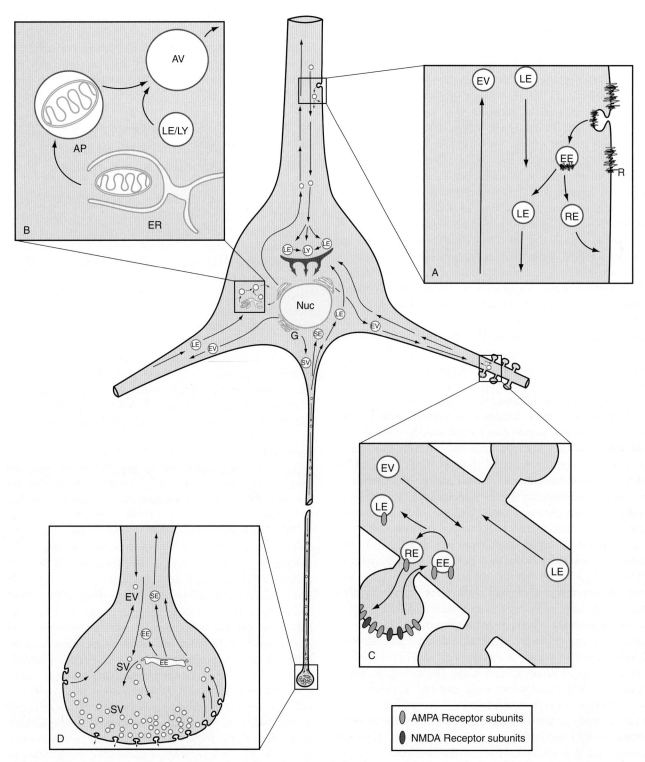

Figure 2 The greater lysosomal system as a central coordinator for cell metabolism. Lysosomes in close continuity with endosomes, autophagosomes, and related vesicles comprise a system that is essential not only for the degradation and recycling of complex molecules but also for control of signaling cascades involving events in axons, dendrites, and neuronal perikarya. Further, byproducts of lysosomal system processing may participate in the control of biosynthetic and related homeostatic mechanisms essential for cell function and survival. Illustrated here are examples of such functions ranging from metabolic salvage (red arrows, center), to signal transduction events (**A**), to autophagy (**B**), to neurotransmitter receptor recycling (**C**), and to vesicular trafficking within axons (**D**). See text for details. AP, autophagosome; APL, autophagolysosome; EE, early endosome; ER, endoplasmic reticulum; EV, exocytic vesicles; G, Golgi; LE, late endosome; LY, lysosome; Nuc, nucleus; R, raft microdomain; RE, recycling endosome; SE, signaling endosome (retrosome); SV, synaptic vesicle.

receptors and their ligands (largely glycoproteins), but also GSLs (including gangliosides) and proteoglycans.

Substantial evidence now supports a key role for the E/L system in intracellular signal transduction involving a wide range of receptors and their ligands [7,8]. Cell surface receptors are endocytosed following ligand binding and subsequently transit through the early sorting endosomal compartments. This process may eventually lead to dissociation of receptors and their ligands with recycling of receptors to the plasmalemma, or to degradation of both receptors and ligands. Such selective trafficking events would be anticipated to have significant influence over signaling. For example, as well documented for epidermal growth factor and its receptor, as well as other receptor–ligand complexes, the process of internalization does not necessarily terminate the signaling event (as originally believed) but can be a mechanism allowing for sustained signaling. Furthermore, as shown for tumor necrosis factor and its receptor, the resulting signaling responses (along with their ultimate biological consequences) may differ according to the particular subcellular domain (e.g., plasmalemma vs endosome) where the receptor–ligand complex is found. In concert, these and related findings argue that the E/L system plays a central, indeed crucial, role in many aspects of intracellular signaling.

The E/L system in neurons is likewise implicated in signaling events at synapses involving neurotransmitter receptors, for example, in the prominent excitatory glutamatergic synapses present on dendritic spines of cortical pyramidal neurons [9]. The postsynaptic density of these receptors is known to contain both α-amino-3-hydroxy-5-methyl-4-isoxazole propionic acid (AMPA) and N-methyl-D-aspartate (NMDA) receptors. AMPA receptors mediate the characteristic postsynaptic depolarization responsible for action potentials in neurons, whereas NMDA receptors are linked to synaptic plasticity. Although NMDA receptors are relatively fixed within the postsynaptic density, there is evidence that specific subunits of AMPA receptors cycle in and out of the synapse in a manner that can be either dependent (i.e., regulated by) or independent of NMDA activation and resulting influx of Ca^{2+}. A large number of proteins have been identified that influence the cycling of AMPA receptors and thus are believed critical to controlling synaptic strength and, in turn, synaptic plasticity. Removal of AMPA receptor subunits from synapses occurs by endocytosis (see Fig. 2). Subsequent rephosphorylation of the subunits while in association with recycling endosomes leads to reinsertion at the synapse. Alternatively, AMPA receptor subunits removed from the synapse may be trafficked from early endosomes to late endosomes for degradative processing. In this way, the E/L system plays a crucial role in influencing events at glutamatergic synapses and in the all-important mechanisms believed associated with learning and memory (long-term potentiation and long-term depression) and thus with synaptic and dendritic plasticity.

In addition to direct endocytic events summarized earlier in this chapter, another significant source of materials flowing into the lysosomal system is linked to autophagy. The autophagic process is believed to start with an inner membrane leaflet, for example, in the ER, encircling a portion of cytoplasm so that it forms a double membrane-bound vacuole known as the *autophagosome* or *autophagocytic vacuole* (see Fig. 2). Maturation of this structure is believed to include fusion with vesicles of the lysosomal system, which leads to formation of autophagolysosomes, which are acidified and contain lysosomal enzymes. Autophagy is increasingly identified as a significant event in many neurodegenerative diseases, with increases in autophagic vacuoles shown to occur in Alzheimer's and Huntington's diseases [10] and in some lysosomal diseases, including Danon disease and the neuronal ceroid lipofuscinoses [2].

Although the preceding discussion has principally focused on events in neuronal perikarya and dendrites that directly involve the greater lysosomal system, there is also evidence of a substantial functional role for this system in axons (see Fig. 2). Axons are characterized by complex anterograde and retrograde transport of large numbers of molecules and organelles. Fast anterograde transport, mediated by the molecular motor kinesin, is known to facilitate the translocation of mitochondria, Golgi-derived vesicles, synaptic vesicles, and related organelles, neurotransmitters and their synthetic enzymes, neuropeptides, ion channels, and so forth. Retrograde transport, mediated by dyneins, involves movement of a variety of organelles, including early and late endosomal vesicles. The latter are often referred to as *multivesicular bodies* or as *signaling endosomes* or "*retrosomes*." Such terms aptly reflect the function attributed to these organelles, that is, they contain ligand-bound growth factor receptors derived from postsynaptic sites, which are being transported back to the cell body [11]. It has been widely assumed that it would be here, rather than in axons per se, where these retrogradely transported vesicles could recruit appropriate second-messenger cascades for signal transduction purposes.

The axon terminal also exhibits the presence of pools of synaptic vesicles as required for neurotransmitter secretion at synapses. Axon potential propagation into the synaptic terminal leads to synaptic vesicle fusion with the presynaptic membrane and to release of neurotransmitter in a Ca^{2+}-triggered exocytic process. Endocytosis in nearby membrane of the axon terminal leads to development of additional local pools of synaptic vesicles for subsequent cycles of neurotransmitter release [12]. Newly derived endosomes may also fuse with nearby early endosomal compartments with the possibility of initiating retrograde transport back to the cell body, as described already [11].

A more controversial issue for axons involves the possible presence of acid hydrolase–containing late endosomes and lysosomes. Historically, it has been believed that these organelles were restricted to neuronal perikarya. There is evidence, however, of increasing acidification of many (though not all) retrogradely transported endocytic organelles from axon terminal to cell body, with proximal areas of axons exhibiting significant numbers of endosomes with an internal pH (5.0–6.2) consistent with identification as late endosomes and lysosomes [13]. These vesicles are believed acidified through contact with anterogradely transported vesicles derived from the Golgi/TGN. The extent to which lysosomal enzymes actually reach retrogradely transported endosomes is less clear. There are, however, two possible mechanisms for delivery of these enzymes, one being anterograde axonal transport from the perikaryon and the other through endocytosis of locally secreted enzymes at the nerve terminal.

Whether endocytosis occurs in dendrites, axons, or parts of the perikaryon, transit of materials from early endosomes into late endosomes and to lysosomes ultimately results in their degradation and in the recycling of breakdown products for subsequent biosynthetic purposes, a process aptly referred to as "metabolic salvage" [14]. For some of these byproducts of lysosomal catabolism (e.g., free sialic acid and cystine), specific transport proteins have been identified (sialin and cystinosin, respectively) whose importance is made apparent by diseases in which they are absent, as discussed earlier. Yet, there is ample evidence that complex molecules entering late endosomes and lysosomes do not have to be degraded to their simplest components prior to exit from the lysosome.

The pathway from synthesis to salvage and reutilization of complex molecules is well illustrated by gangliosides and related GSLs. Gangliosides are synthesized in the Golgi/TGN in sequential fashion by a series of well-documented sialyl transferases, after which they are believed delivered to the cell surface by vesicular transport in exocytic vesicles. Here, they insert into the outer leaflet of the membrane where they are believed to reside in close association with cholesterol in membrane microdomains known as *rafts*. Rafts are also thought to contain numerous other types of molecules including GPI-anchored proteins, and as an overall assembly to function as a signaling platform linking available ligands in the extracellular environment to signal transduction elements within the cell. One study has shown that drugs that block or slow the synthesis of GSLs can also influence the transport of GPI-anchored proteins to the cell surface, suggesting the presence of possible co-transport mechanisms for gangliosides and these proteins. In spite of the abundance of GSL expression in brain and the demonstrated presence of gangliosides in neurons, the functions of these compounds are not well understood. Although there is evidence that certain gangliosides may act to modulate the dimerization and thus the activation of some types of receptors (e.g., those for growth factors), the exact nature of these relationships remains poorly defined, particularly in terms of *in vivo* studies.

Eventually, gangliosides at the plasmalemma enter the E/L system by way of endocytosis where they eventually come into contact with an array of hydrolytic enzymes and activator proteins that facilitate their catabolism. The mature mammalian brain exhibits significant levels of several complex gangliosides (e.g., G_{M1}, G_{D1a}, G_{D1b}, G_{T1b}, G_{D2}) but very low or essentially undetectable levels of others (e.g., G_{M2}, G_{M3}). The disialogangliosides and trisialogangliosides are transformed in lysosomes to corresponding monosialogangliosides (G_{M1}, G_{M2}) by sialidase. Subsequently, G_{M1} is degraded to G_{M2} (by β-galactosidase) and G_{M2} to G_{M3} (by β-hexosaminidase). Interestingly, a membrane-associated sialidase has also been reported at synapses, which potentially allows for similar conversion of gangliosides at this site. In some species, sialic acid may be removed from G_{M1} and G_{M2} residing in lysosomes by another sialidase and thus generate the asialo-derivatives of these compounds. Importantly, however, numerous studies on the fate of "simple" gangliosides (G_{M2} and G_{M3}, as well as G_{M1}) and of neutral GSLs (lactosylceramide, glucosylceramide) generated from complex gangliosides by lysosomal degradation suggest that they may be recycled (or "salvaged") after endocytosis prior to full degradation to their simplest components (i.e., monosaccharides, fatty acids, etc.). Thus, gangliosides like G_{M2} and G_{M3} may exit the E/L system and be trafficked to the Golgi/TGN where they would be sialosylated and delivered again to the plasmalemma. Importantly, these findings suggest that there are two potential routes available to gangliosides as they transit through the E/L system: one involving direct recycling to the Golgi/TGN and the other leading to further degradation. Factors controlling such trafficking into one path or another, as well as the consequences for cells of recycling partially degraded vs fully degraded substrates, remain largely unexplored. As described later, lysosomal diseases with documented defects in substrate catabolism vs substrate trafficking may provide unique opportunities to understand the importance of these mechanisms.

V. Dysfunction of the Lysosomal System in Neurons Leads to Complex Pathogenic Cascades

Most lysosomal diseases, but certainly not all, affect the nervous system and do so by direct impact on neurons or in some cases by damaging oligodendroglia and therefore myelin formation or maintenance. The latter conditions, known as *leukodystrophies*, are largely discussed elsewhere in this book and are therefore not covered in detail here.

Rather, this chapter focuses on lysosomal diseases characterized by primary neuronal involvement.

Neurons undergoing lysosomal storage demonstrate the presence of abnormal storage bodies principally in perikarya and to a degree in larger dendrites, consistent overall with the known distribution of lysosomes. At the ultrastructural level, storage bodies vary greatly and their morphologies are believed to reflect the nature of the materials accumulating in the disease. For primary gangliosidoses, storage bodies appear highly membranous, hence, the term *membranous cytoplasmic bodies* (MCBs), reflecting primary accumulation of ganglioside G_{M1} or G_{M2} (Fig. 3). For the MPSs, some inclusions (but not all) exhibit stacks of membranes often referred to as "zebra bodies," with these being the sites of secondary ganglioside sequestration, whereas other inclusions contain floccular or amorphous material, presumably consistent with primary storage of glycosaminoglycans. The neuronal ceroid lipofuscinoses variably exhibit aptly named "fingerprint bodies" or "Finnish snowballs" to depict the largely proteinaceous nature of the storage material here. Still other diseases with soluble storage compounds, like α-mannosidosis, exhibit mostly open, clear inclusions indicative of substrate washout during processing. In neurons in these diseases where secondary ganglioside storage occurs, membrane leaflets similar to those seen in MPS disorders are commonly found. Regardless of the types of inclusions, their distribution in neurons is largely perikaryal, with linear arrays of individual storage bodies in some cases trailing out into mainstem dendrites, as often seen with the apical dendrites of cortical pyramidal neurons. Characteristic storage bodies, interestingly, are not routinely observed in axons of affected neurons.

A provocative notion is that storage bodies in lysosomal diseases largely represent collection points for nondegraded membrane rafts ("raft log jams") within cells since many lysosomal diseases are in fact disorders in the degradation and recycling of specific raft components (e.g., GSLs and cholesterol) [15]. If correct, this would suggest that a key aspect of cell dysfunction could be altered signal transduction secondary to raft accumulation in the E/L system. Confocal analysis of storage compounds in individual neurons, however, has revealed a more complex arrangement of constituent storage components than might be anticipated as part of a general stasis of raft processing. For example, the two key gangliosides (G_{M2} and G_{M3}) that accumulate in many types of storage diseases, presumably derived from G_{M1} or other complex gangliosides present in rafts, are in fact largely sequestered independently of one another [16]. This unexpected finding indicates the existence of sequential processing steps in different endosomal compartments or other mechanisms to account for their independent locations. Similarly, unesterified cholesterol, a common co-constituent of so many lysosomal diseases with ganglioside storage, likewise is not systematically co-sequestered in all sites of ganglioside elevation. Although surprising at first, this finding of heterogeneity of storage in vesicles of the E/L system in neurons is not unlike that reported for the maturational process of individual phagosomes in macrophages, where heterogeneity appears the rule, not the exception [17]. In addition to the sequestration of rafts and raft components in the E/L system in storage diseases, a second disturbance of raft function conceivably may arise secondary to the failed recycling of materials to the Golgi/TGN, as discussed later, with resulting changes in the synthesis, assembly, or transport of new rafts and/or raft constituents to the plasmalemma.

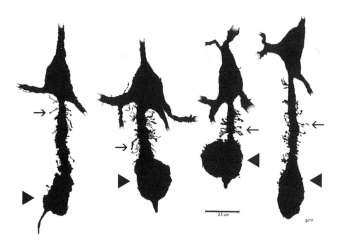

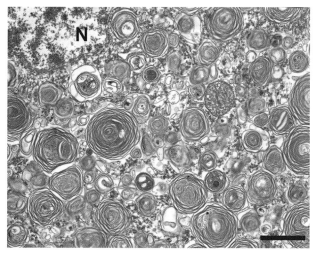

Figure 3 Ectopic dendritogenesis occurs on select types of neurons affected by lysosomal disease. (Top) Camera lucida drawings of Golgi-impregnated cortical pyramidal neurons in an animal model of GM_1 gangliosidosis illustrating the presence of enlarged axon hillocks (meganeurites; arrowheads) covered with dendritic spines and neuritic sprouts (arrows). Normal basilar and apical dendrites have been truncated for purposes of illustration. (Reprinted, with permission, from Purpura, D. P. and Baker, H. G. [1978]. Meganeurites and other aberrant processes of neurons in feline GM1-gangliosidosis: a Golgi study. *Brain Res.* **143**, 13–26.) (Bottom) Electron micrograph showing typical storage bodies (membranous cytoplasmic bodies) as found in both meganeurites and perikarya at this stage of disease. N, nucleus. Calibration bar equals 1 μm. See text for additional details.

The specialized sorting mechanisms mentioned earlier may be responsible for generating the largely separate sites of sequestration for gangliosides G_{M2} and G_{M3}, as occurs in many storage diseases [16]. Yet, why do these two particular gangliosides, normally expressed at undetectable levels in mature mammalian brain, accumulate in the first place, and why is this storage commonly accompanied by sequestration of unesterified cholesterol in these same cells? The lysosomal disorder that may offer the greatest insight into this question is NPC disease, a condition in which not only cholesterol but also gangliosides G_{M2} and G_{M3} and the neutral GSLs, lactosylceramide, and glucosylceramide are documented to accumulate. The salvage pathway believed defective in this disease, as described earlier, has been widely assumed to chiefly involve unesterified cholesterol. That is, lack of NPC1 (or NPC2) protein is believed to lead to a failure of retroendocytic trafficking of unesterified cholesterol from the E/L system to the Golgi/TGN and related internal cell membranes. Yet several studies have shown that the degree of cholesterol sequestration in affected neurons is largely dependent on the types of gangliosides expressed by those cells, suggestive of some type of co-dependency in the sequestration process. Importantly, for gangliosides G_{M2} and G_{M3}, there is no clear explanation for their storage since all known catabolic elements essential for their degradation are present in affected neurons and believed functional. As mentioned in the previous section, after complex gangliosides enter the E/L system and are degraded to simpler gangliosides (G_{M2}, G_{M3}; also G_{M1}), these catabolic products appear to face two alternative paths: one involving metabolic salvage, the other degradation. What is suggested by NPC disease is that the salvage pathway for simple gangliosides may be dependent on an NPC1-mediated mechanism, just as proposed for cholesterol.

Although the function of a lysosomal catabolic pathway for GSLs appears obvious, the role for direct shunting of partially degraded GSLs to the Golgi/TGN is less clear. Discoveries regarding the consequences of recycling defects for unesterified cholesterol in NPC disease, however, may provide clues to the importance of this process. NPC fibroblasts, which have long been believed to have a defect in retroendocytic trafficking of unesterified cholesterol, have been reported deficient in normal byproducts of cholesterol metabolism known as *oxysterols*. Oxysterols, in turn, are known to be key regulators of cholesterol availability in cells as they act as ligands for liver-X-receptors (LXRs), which themselves are transcription factors controlling genes regulating cholesterol homeostasis. Evidence suggests that failure to provide feedback through this system may be a cause of the elevated cholesterol in NPC-affected cells [18]. This finding suggests that retroendocytic movement of materials from the E/L system provides critical feedback for regulation of specific metabolic pathways, including synthesis, by acting through changes in gene expression.

The above scenario thus may explain the cholesterol dysregulatory events in NPC disease, but what about ganglioside storage? If ganglioside homeostatic mechanisms are also in some way regulated by metabolic salvage, as suggested for oxysterols and LXRs for cholesterol, the absence of this signal may likewise lead to changes in ganglioside synthesis in the Golgi/TGN and exacerbate the overabundance of gangliosides not unlike that found for cholesterol. Overproduction of complex gangliosides conceivably could lead to ganglioside storage simply by overwhelming the catabolic capacity of neurons. This speculative scenario aptly illustrates the complexity of recycling/homeostatic mechanisms associated with function of the E/L system.

In addition to deficient production of oxysterols, there is also evidence that NPC neurons lack adequate production of neurosteroids, including allopregnanolone, which are normally synthesized from cholesterol [19]. This finding, viewed in concert with the preceding discussion, strongly suggests that a key aspect of the pathogenesis of NPC disease, and possibly other lysosomal disorders, is not just one of overabundance (i.e., storage) of nondegraded substrate(s) but also a state of deficiency of particular downstream metabolic products normally derived from the stored compounds. If correct, this insight may not only provide a basis for understanding the complex pathogenic cascades occurring in these diseases but also provide a new basis for therapy, namely replenishment of missing compounds, as described in the next section.

Another area of speculation in terms of pathogenic mechanisms affecting neurons in storage diseases involves ganglioside distribution in diseased cells and the possibility that the E/L system defect leads to an abnormal distribution of specific GSLs in internal membranes of cells [4]. This could occur secondary to "leakage" or to misrouting of sequestered gangliosides, or indeed, as a result of abnormal synthesis, as described earlier. Evidence suggests that the presence of GSLs in these internal membranes can have profound effects on calcium homeostasis. For example, it has been reported that glucosylceramide elevation in neurons affected by Gaucher disease causes increases in Ca^{2+} mobilization via ryanodine receptors located in the ER and that as a result these neurons show greater sensitivity to cell death induced by drugs (e.g., glutamate) that mobilize Ca^{2+}. Altered Ca^{2+} homeostasis has also been reported in brain cells in Sandhoff disease and in Niemann-Pick type A disease. Here, the defect appears to involve the rate of Ca^{2+} uptake rather than its release by way of ryanodine receptors, with the result again being one of elevated cytosolic Ca^{2+} levels, which in turn render these cells more sensitive to stress-induced changes [4]. These studies raise important questions about the role of GSLs in normal cell physiology and the consequences of their sequestration in storage diseases.

Although changes in the GSL constituents of intracellular membranes may be one consequence of the catabolic defects inherent in lysosomal diseases, the more prominent disease effect is the conspicuous sequestration of large amounts of nonrecycled materials within vesicular compartments of the E/L system. Neuronal perikarya may eventually swell to accommodate this need for increased volume, with the nuclei of neurons being pushed into nearby apical dendrites (e.g., in the case of cortical pyramidal neurons). Chronic increases in perikaryal volume also lead to a preferential expansion of the axon hillocks of many cells, although the reason for the specificity of this local change is unknown. Enlarged axon hillocks, referred to as *meganeurites*, may in some cases exhibit the presence of neuritic sprouts and/or dendritic spines, complete with well-formed asymmetrical synapses (see Fig. 3). Other neurons may lack meganeurites but nonetheless sprout new dendritic neurites from the axon hillock region. The growth of these new dendrites, a phenomenon known as "ectopic dendritogenesis," has been found not only in diseases like GM_2 gangliosidosis (where it was initially discovered) but also in other storage diseases ranging from the MPSs to some types of glycoproteinoses [20]. The phenomenon, however, is unique to lysosomal storage disorders and undoubtedly is further evidence for a critical role of the E/L system in regulating dendritic plasticity, as described earlier.

Interestingly, not just any disturbance in the integrity of the E/L system is adequate to initiate new dendrite growth, as there are some types of lysosomal diseases (e.g., the neuronal ceroid lipofuscinoses) that clearly lack this phenomenon. Early studies also clearly established that ectopic dendritogenesis in storage diseases was not panneuronal. Rather, it is an event restricted to specific subsets of neurons, which include pyramidal cells of the cerebral cortex and multipolar cells of the amygdala and claustrum. Other types of neurons, including Purkinje cells and motor neurons, do not exhibit this phenomenon in any type of storage disease. The incidence of ectopic dendritogenesis in some storage diseases (e.g., the primary gangliosidoses) is such that nearly every pyramidal neuron in the cerebral cortex in late-stage disease will show some level of altered dendritic sprouting. A wide range of studies have implicated the presence of one ganglioside, G_{M2}, as the critical marker for this event across all types of storage diseases exhibiting this phenomenon, with the single possible exception of GM_1 gangliosidosis (where elevated G_{M1} appears sufficient to generate the change). Thus, it appears that there are three critical components linked to the reinitiation of dendritic sprouting in storage diseases: involvement of select types of neurons (excitatory, multipolar, dendritic spine-covered possessing neurons) of the cerebral cortex and subcortex that receive glutamatergic inputs, altered expression of gangliosides (particularly G_{M2}) in the affected neurons, and disordered function of the E/L system (including that caused by defects in lysosomal hydrolases, transmembrane proteins, and activator proteins). This triad of involvement suggests that glutamatergic receptor subunit recycling through the E/L system, already described earlier as a critical component of synaptic plasticity, may also in some manner be associated with GSL trafficking in normal neurons.

Just as certain types of neurons show the presence of renewed dendrite growth following lysosomal storage, other neurons exhibit unusual axonal alterations [20]. In this case, the abnormality is observed as localized swellings along the axon and is referred to as *neuroax-*

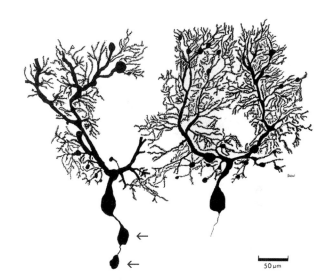

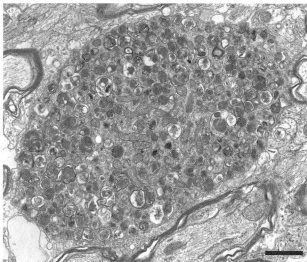

Figure 4 Neuroaxonal dystrophy occurs on select types of neurons affected by lysosomal disease. (Top) Camera lucida drawings of Golgi-impregnated cerebellar Purkinje cells in an animal model of the glycoproteinosis known as α-mannosidosis. Illustrated are conspicuous enlargements in the proximal axon known as *axonal spheroids* or *neuroaxonal dystrophy*. (Reprinted, with permission from Walkley, S. U., Blakemore, W. F., and Purpura, D. P. [1981]. *Acta Neuropathol. [Berl.]* **53**, 75–79.) (Bottom) Electron micrograph of an axonal spheroid in the granule cell layer of the cerebellum in α-mannosidosis showing ultrastructural morphology typical of all lysosomal disorders, namely accumulation of dense bodies, tubulovesicular profiles, mitochondria, and related organelles. Calibration bar equals 1 μm. See text for additional details.

onal dystrophy or *axonal spheroid formation*. Ultrastructural studies have revealed that these enlargements consist of collections of tubulovesicular profiles, mitochondria, dense and multivesicular-type bodies, and only occasional neurofilaments, hence, the descriptive term, *granular spheroids* (Fig. 4). They do not, however, contain the characteristic lysosomal inclusions found in neuronal perikarya. Interestingly, spheroids in a wide variety of storage diseases, unlike the characteristic storage granules of perikarya, routinely resemble one another, consistent with a generic cause for their formation.

Spheroids have been studied extensively in animal models of lysosomal disorders where they are found to be abundant in primary ganglioside storage diseases, in Niemann-Pick disorders (types A and C) and in the glycoproteinosis α-mannosidosis. Their incidence and distribution correlate closely with the onset and type of motor system defects exhibited by affected animals, suggesting that they are major players in generating neuronal dysfunction [20]. Spheroids are most readily visualized in neurons using immunocytochemistry to label components trafficked in axons, including enzymes (glutamic acid decarboxylase), as well as calcium-binding proteins (parvalbumin, calbindin). Using such cell-selective markers, spheroids were found to occur most commonly in neurons that use γ-aminobutyric acid (GABA) as a neurotransmitter, including Purkinje cells, select neuronal populations in basal ganglia, and nonpyramidal (intrinsic) cells of the cerebral cortex, hippocampus, and numerous other brain regions. The similarity of these spheroids of storage diseases with those found distal to nerve crush experiments in normal animals has suggested that a defect in retrograde transport is the most likely explanation for their occurrence. Several studies have also suggested that spheroids are initially found in distal portions of axons and only later develop in more proximal areas. The prominence of spheroids in axons of Purkinje cells and the finding that these cells are prone to die in many storage diseases suggests that the two events may be related. For example, spheroids may cause this early demise secondary to a block in retrograde transport of a critical substrate (e.g., a growth factor) essential for cell survival.

A key question involving spheroids is why a disturbance in function of the E/L system, presumably largely confined to the perikaryal region, can lead to such dramatic cellular pathology in distant axons. As reviewed earlier, controversy remains as to the extent of lysosomal components within the axon and most studies favor the retrograde transport of materials in endosomal compartments for subsequent fusion with the lysosomal system after complete retrograde transport. Whether the E/L defect within the perikaryon itself deprives the axon of critical components for successful retrograde transport is a largely unexplored issue. From the previous discussion, it is clear that the dysfunction and death of neurons in storage diseases can potentially be caused by numerous events ranging from altered calcium homeostasis to defects in dendritic and axonal integrity, all attributable in one way or another to errors in substrate processing within the E/L system. Defects in membrane fusion events and in autophagy likewise represent mechanisms of disease linked to neurodegeneration in several types of lysosomal disorders.

A recent study [21] has shown that cathepsin D deficiency in mice resembles NCL disorders in terms of storage of the c-subunit of mitochondrial ATP synthase and also exhibits abnormalities specifically in autophagocytosis. Because autophagosomes and endosomes may engage lysosomes through essentially parallel but different "streams" (see Fig. 2), it is possible that this difference explains why the NCL disorders lack many of the features seen in other lysosomal diseases [20]. The dominance of autophagosomal rather than endosomal defects may likewise provide the common mechanistic thread linking the pathologically similar but genetically diverse NCL disorders.

Two other mechanisms that are emerging as critical players in the development of neurological disease in storage disorders are apoptosis and inflammation. Lysosomal diseases affecting brain vary widely in terms of the amount of neuron death that accompanies the storage process. In some, like the infantile form of neuronal ceroid lipofuscinosis (CLN1 deficiency), massive loss of neurons occurs very early in the cerebral cortex, leading to conspicuous cortical atrophy, but similar changes are less prominent in other brain regions. In NPC disease, loss of neurons in the cerebral cortex is less prominent in early disease, but selective death of Purkinje cells of the cerebellum is well documented, particularly in the mouse model of this disorder. Several studies have indicated that loss of cells in storage diseases is accompanied by both molecular and cellular attributes of apoptosis, but it does not appear that this is the sole mechanism accounting for neuron death. Inflammation has also been documented in brain tissues of many types of lysosomal diseases, particularly those involving ganglioside storage [4]. For example, gene expression profiling and immunohistochemical studies of diseases such as Tay-Sachs, Sandhoff, and NPC disease have revealed prominent involvement of activated microglia, macrophages, and astrocytes. Clinical benefit also has been observed in the murine model of Sandhoff disease treated with nonsteroidal anti-inflammatory agents, providing additional evidence for the importance of inflammation in neuron death.

VI. Understanding Lysosomal Disease Pathogenesis Provides Key Insight for Development of Therapy

Lysosomal diseases affecting the brain present formidable difficulties in terms of treatment. Early studies on I-cell disease led to the discovery of lysosomal enzyme secretion and delivery mechanisms and to the concept of "cross-correction," whereby diseased cells could be normalized if presented with the missing enzyme (e.g., by co-culturing diseased and normal cells). Enzyme replacement therapy (ERT) and bone marrow transplantation (BMT), as treatments, both evolved from this simple concept and have been extensively tested for storage diseases caused by defects in lysosomal enzymes. ERT, today, is being applied in the clinic to a variety of lysosomal diseases, but generally these are only lysosomal diseases without significant neuronal involvement (Gaucher type 1, Fabry, MPS I-S, MPS VI, etc.) since the blood–brain barrier would be anticipated to block entry of therapeutic proteins to the brain parenchyma. BMT for lysosomal diseases with brain involvement, although controversial at the outset, was found to be surprisingly successful in rare circumstances, as for the glycoproteinosis known as α-mannosidosis [reviewed in 22]. Here, it appears that normal BMT-derived monocytes invade the brain, differentiate as microglia, and secrete lysosomal α-mannosidase in significant amounts, followed by substantial correction of storage in neurons. Most lysosomal diseases, however, appear less readily correctable by this procedure, because of either limited secretion of the pivotal enzyme or its lack of stability in the extracellular space. These inadequacies have driven other studies looking for ways to breech the blood–brain barrier (osmotic treatments, fusion proteins, direct brain injections) to get enzyme into brain, but with limited success. Gene therapy approaches are also being widely tested in animal models of lysosomal diseases, with some successes at least at the experimental level. However, a key issue (in addition to safety) is again whether the missing protein for the particular disease is secreted. If it is, widespread correction after successful gene delivery is more likely, whereas if it is not, the need for therapeutic transduction of literally every cell in the brain would present a formidable problem.

Indeed, these difficulties, coupled with the realization that many lysosomal diseases are caused by defects in nonsecreted membrane proteins, have driven development of alternative methods of treatment. One method, enzyme enhancement therapy (EET), is an attempt to rescue misfolded or mis-trafficked proteins prior to their elimination by cells through the use of chemical or pharmacological chaperones. Another method, substrate reduction therapy (SRT), is the use of drugs that have the ability to limit the buildup of offending substrates in brain and in other organs [23]. An initial approach here has been centered on the use of inhibitors of GSL synthesis, for example, the imino sugar known as N-butyldeoxynorjirimycin (NB-DNJ), which has the ability to slow the rate of GSL storage in a variety of disorders, including Gaucher, Tay-Sachs, and NPC diseases. The latter was considered a candidate for this approach because there was no evidence of any catabolic defect causing storage of gangliosides, suggesting that the accumulation was secondary to abnormalities in GSL synthesis and/or trafficking in cells, as described earlier.

Another closely related approach, also under consideration for NPC disease, is based on the discovery, mentioned earlier, that a driving force behind the cholesterol dysregulatory events in this disease may lie in the lack of production of oxysterols as byproducts of endocytosed cholesterol. Failure of feedback of oxysterols through binding to LXR may be responsible for cholesterol dyshomeostasis in affected cells. If correct, drugs that could effectively substitute for the missing oxysterols may help normalize cells [18]. Similarly, replacement of the deficient neurosteroid, allopregnanolone, in the NPC disease mouse model led to increased longevity and reduced cellular storage of both cholesterol and gangliosides [19]. Although the mechanism to account for this improvement remains unknown, the clear suggestion is again that storage diseases, instead of just being conditions where too much of a particular substrate is present, may also be conditions in which cells are being deprived of downstream products of lysosomal system processing. Hence, a form of *byproduct replacement therapy* (BRT) is suggested that is focused on replenishment of these missing compounds or on providing suitable substitutes. Thus, for NPC disease BRT could consist of administration of allopregnanolone (e.g., through the use of synthetic analogues of allopregnanolone) or of drugs substituting for missing oxysterols, or combinations of such drugs. The use of BRT would be which may prove that this type of therapy would thus be beneficial in helping cells normalize homeostatic mechanisms that ultimately control sterol and lipid balance through gene transcription. Such findings once again clearly reveal the overall importance of the greater lysosomal system, not simply as a mechanism to degrade complex molecules but as an essential recycling center whose function ultimately has an impact on all aspects of cell metabolism.

Acknowledgments

The author would like to thank the many investigators, most of whom could not be cited here because of space constraints, whose work has contributed so enormously to our understanding of lysosomal diseases. Gratitude is also expressed to the National Institutes of Health and the Ara Parseghian Medical Research Foundation for support of research in the author's laboratory.

References

1. Platt, F. M., and Walkley, S. U. (2004). Lysosomal defects and storage. *In*: "Lysosomal Disorders of the Brain" (F. M. Platt and S. U. Walkley, eds.), Chapter 2, pp. 32–49. Oxford University Press, Oxford, UK.

2. Eskelinen, E. L., Tanaka, Y., and Saftig, P. (2003). At the acidic edge: emerging functions for lysosomal membrane proteins. *Trends Cell Biol.* **13**, 137–145.
3. Scriver, C. R., Beaudet, A. L., Sly, W. S., and Valle, D. (2001). "The Metabolic and Molecular Bases of Inherited Disease," 8th edition. McGraw-Hill, New York.
4. Rass-Rothschild, A., Pankova-Kholmyansky, I., Kacher, Y., and Futerman, A. H. (2004). Glycosphingolipidoses: beyond the enzyme defect. *Glycoconjugate. J.* **21**, 295–304.
5. Winchester, B. (2005). Lysosomal metabolism of glycoproteins. *Glycobiology* **15**, 1R–115.
6. Ioannou, Y. A. (2001). Multidrug permeases and subcellular cholesterol transport. *Nat. Rev. Mol. Cell Biol.* **2**, 657–668.
7. Miaczynska, M., Pelkmans, L., and Zerial, M. (2004). Not just a sink: endosomes in control of signal transduction. *Curr. Opin. Cell Biol.* **16**, 400–406.
8. Pol, S., Di Fiore, P. P. (2006). Endocytosis conducts the cell signaling orchestra. *Cell* **124**, 897–900.
9. Barry, M. F., and Ziff, E. B. (2002). Receptor trafficking and the plasticity of excitatory synapses. *Curr. Opin. Neurobiol.* **12**, 279–286.
10. Shintani, T., and Klionsky, D. J. (2004). Autophagy in health and disease: a double-edged sword. *Science* **306**, 990–995.
11. Parton, R. G., and Dotti, C. G. (1993). Cell biology of neuronal endocytosis. *J. Neurosci. Res.* **36**, 1–9.
12. Südof, T. (2004). The synaptic vesicle cycle. *Ann. Rev. Neurosci.* **27**, 509–547.
13. Overly, C. C., and Hollenbeck, P. J. (1996). Dynamic organization of endocytic pathways in axons of cultured sympathetic neurons. *J. Neurosci.* **16**, 6056–6064.
14. Tettamanti, G., Bassi, R., Viani, P., and Riboni, L. (2003). Salvage pathways in glycosphingolipid metabolism. *Biochemie* **85**, 432–437.
15. Simons, K., and Gruenberg, J. (2000). Jamming the endosomal system: lipid rafts and lysosomal storage diseases. *Trends Cell Biol.* **10**, 459–462.
16. Walkley, S. U. (2004). Secondary increases in ganglioside expression in lysosomal disorders. *Semin. Cell Dev. Biol.* 15:433–444, 2004.
17. Griffiths, G. (2004). On phagosome individuality and membrane signaling networks. *Trends Cell Biol.* **14**, 343–351.
18. Frolov, A., Zielinski, S. E., Crowley, J. R., Dudley-Rucker, N., Schaffer, J. E., and Ory, D. S. (2003). NPC1 and NPC2 regulate cellular cholesterol homeostasis through generation of low density lipoprotein cholesterol-derived oxysterols. *J. Biol. Chem.* **278**, 25517–25525.
19. Griffin, L. D., Gong, W., Verot, L., and Mellon, S. H. (2004). Niemann–Pick type C disease involves disrupted neurosteroidogenesis and responds to allopregnanolone. *Nat. Med.* **10**, 704–711.
20. Walkley, S. U. (2004). Pathogenic cascades and brain dysfunction. *In*: "Lysosomal Disorders of the Brain" (F. M. Platt and S. U. Walkley, eds.), Chapter 12, pp. 290–324. Oxford University Press, Oxford, UK.
21. Koike, M., Shibata, M., Waguri, S., Yoshimura, K., Tanida, I., Kominami, E., Gotow, T., Peters, C., von Figura, K., Mizushima, N., Saftig, P., and Uchiyama, Y. (2005). Participation of autophagy in storage of lysosomes in neurons from mouse models of neuronal ceroid-lipofuscinoses (Batten disease). *Am. J. Pathol.* **167**, 1713–1728.
22. Dobrenis, K. (2004). Cell-mediated delivery systems. *In:* "Lysosomal Disorders of the Brain" (F. M. Platt and S. U. Walkley, eds.), Chapter 14, pp. 339–380. Oxford University Press.
23. Platt, F. M., and Butters, T. D. (2004). Inhibition of substrate synthesis: a pharmacological approach for glycosphingolipid storage disease therapy. *In:* "Lysosomal Disorders of the Brain" (F. M. Platt and S. U. Walkley, eds.), Chapter 15, pp. 381–408. Oxford University Press.

2

Neurobiology of Peroxisomal Disorders

James M. Powers, MD

Keywords: *adreno-leukodystrophy, adrenomyeloneuropathy, fatty acids, genes, knockout mice, lipid metabolism, mouse models, peroxins, peroxisome, peroxisome biogenesis disorders, PEX genes, plasmalogens, Refsum's disease, very long chain fatty acids, Zellweger spectrum, Zellweger syndrome*

I. History and Nomenclature
II. The Peroxisomal Disorders
III. Etiology
IV. Molecular Pathogenesis and Pathophysiology
V. Human Pathology and Pathogenesis
VI. Mouse Models and Pathogenesis
VII. Clinicopathological Correlations
References

I. History and Nomenclature

A. The Peroxisome

Peroxisomes, discovered with the electron microscope in 1954 by Rhodin and called "*microbodies*," are cytoplasmic organelles found in all mammalian cells with the exception of mature erythrocytes. In humans its basic ultrastructure consists of a single unit membrane that encloses a finely granular matrix of mild to moderate electron density (Fig. 1). Peroxisomes are normally smaller than mitochondria, at least in hepatocytes; they can be definitively identified by membrane-bound or particulate staining either with diaminobenzidine (DAB) because of the presence of catalase in their matrix or with antibodies to one of the peroxisomal membrane or matrix proteins. However, catalase is one of the most inefficiently imported matrix enzymes; hence, the failure to detect particulate catalase does not exclude the

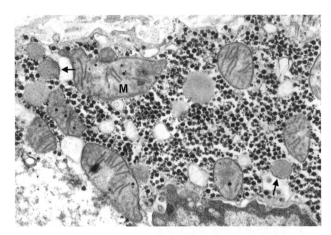

Figure 1 Ultrastructural appearance of peroxisomes (arrows) and mitochondria (M) in hepatocytes. Biopsy. Uranyl acetate–lead citrate.

energy-dependent import of other matrix moieties. Before 1973 the peroxisome was considered a somewhat vestigial organelle in humans even though it was known to be capable of generating hydrogen peroxide through the activity of a few oxidases, which then could be neutralized by catalase. To underscore this metabolic function, DeDuve renamed them "peroxisomes" in the mid-1960s. A seminal observation by Goldfischer et al. was reported in 1973: The hepatocytes of patients with the lethal Zellweger cerebrohepatorenal syndrome (ZS) (also discovered in the mid-1960s) lacked peroxisomes; this report demonstrated the previously unsuspected importance of peroxisomes for normal cell function and host viability. It is noteworthy that they also reported mitochondrial abnormalities, originally considered the primary defect, in these same patients. A direct numerical association of mitochondria with peroxisomes had been observed by DeDuve, and as we will see later in this chapter, this peroxisome–mitochondria connection has assumed a prominent contemporary pathogenetic focus. We now know that peroxisomes have an alkaline pH, in striking contrast to the acidic lysosome, and contain more than 50 matrix enzymes that play important catabolic and anabolic roles, particularly in lipid metabolism: the β- and α-oxidation of fatty acids (straight and branched chain), and the biosynthesis of ether phospholipids (plasmalogens), polyunsaturated fatty acids (e.g., docosahexanoic acid [DHA]), bile acids, cholesterol, and dolichols.

The number, size, shape, and major functions of peroxisomes vary between different cell types, with the microperoxisome being the major form seen in neural cells. Most of the available functional information has been derived from fibroblasts and hepatocytes in mammals and yeast. These data do not necessarily translate to the microperoxisomes in the nervous system, and some hepatic data reveal evidence of peroxisomal mosaicism even in adjacent cells. In the developing mammalian central nervous system (CNS), microperoxisomes have been identified in neuroblasts, immature neurons, and oligodendrocytes, whereas in the mature mammalian CNS, they seem to be largely restricted to oligodendrocytes. Catalase-positive neurons are observed in the basal ganglia, thalamus, and cerebellum of 27- to 28-week human fetuses and by 35 weeks in the frontal cortex. The deep white matter exhibits catalase-positive glia at approximately 31–32 weeks of gestation, and they later appear to shift from a deeper to a more superficial location in the white matter.

Both the matrix and the membrane proteins are encoded by nuclear genes and synthesized on free polyribosomes before being imported into the peroxisome. The normal importation of matrix proteins results in normal peroxisomal morphology. Matrix proteins are imported through the activities of two distinct peroxisomal targeting signals: PTS1 and PTS2. The majority of matrix proteins utilize the PTS1 pathway, whereas the PTS2 pathway seems to be restricted to the importation of phytanoyl-coenzyme A (CoA) hydroxylase, peroxisomal thiolase and alkyl dihydroxyacetone phosphate (DHAP) synthase. The targeting of the four integral proteins known to constitute the relatively impermeable peroxisomal membrane (ABCD1/ALDP, ABCD2/ALDR, ABCD3/PMP70, ABCD4/PMPR) occurs by a different, currently unknown, mechanism. The designation "ABC" refers to their inclusion in the superfamily of *a*denosine triphosphate (*AT*P) *b*inding *c*assette proteins, which play important transport roles; all the known ABC proteins in peroxisomes are half-transporters that must homodimerize or heterodimerize to become fully functional. At least 25 nuclear genes are involved in the formation or biogenesis of the peroxisome; they are referred to as *PEX genes* and their gene products are designated "peroxins" (PEX#p). All of the peroxins, except for 3, 11, 16, and 19, play a role in matrix import; peroxins 3, 16, and 19 are needed for the import of integral membrane proteins. PEX5p is the cytoplasmic receptor for PTS-1 sequences, whereas PEX7p is the same for PTS-2 import. It is beyond the scope of this chapter to discuss the intricacies of peroxisome biogenesis, biochemistry, or matrix enzyme importation. Several excellent reviews can provide the relevant details, including a web site maintained by Dr. Stephen Gould's laboratory (www.peroxisome.org) [1,2].

II. The Peroxisomal Disorders

Peroxisomal disorders are divided into two major groups: those in which peroxisome biogenesis is defective and the peroxisomes are morphologically abnormal or absent, which has an adverse impact on numerous peroxisomal functions (Group 1); and those in which a single protein, almost invariably a matrix enzyme, is deficient and their peroxisomes usually are morphologically intact (Group 2). Group 1, the peroxisomal biogenesis disorders (PBDs), formerly called *general* or *generalized peroxisomal diseases,* includes ZS, neonatal adrenoleukodystrophy (NALD), infantile Refsum's disease (IRD) or phytanic acid storage disease, and rhizomelic chondrodysplasia punctata (RCDP), type 1, classic. Hyperpipecolic acidemia was formally included in this group, but the apparent illegitimacy of a specific nosological position for this common biochemical abnormality in the Group 1 disorders has resulted in its exclusion, at least presently. ZS, NALD, and IRD are now included under the rubric of the Zellweger spectrum, with the most severe phenotype being ZS and least severe being IRD. The Zellweger spectrum is most faithful to all the inclusionary criteria of the PBDs; RCDP mainly affects plasmalogen synthesis.

The most common disease in Group 2 is X-linked adreno-leukodystrophy (XALD, ALD) and its adult variant adrenomyeloneuropathy (AMN). The "hyphen" in *adrenoleukodystrophy* is both historically (i.e., true to Michael Blaw's original appellation) and grammatically correct but has become a casualty of editorial indiscretions and is currently omitted by most authors. The "X-linked" modifier sets it apart from the autosomal-recessive "neonatal adrenoleukodystrophy" mentioned earlier, but I find this modifier to be superfluous. It is also of historical interest that our group was unwittingly exposing the most common peroxisomal disease, ALD/AMN, at the same time and in the same place, the Pathology Department in Albert Einstein College of Medicine, where Goldfischer's group was identifying ZS as a peroxisomal disease; yet, it took another decade for us to appreciate this ironic twist. Group 2 also contains diseases that have a single matrix enzyme deficiency but have dysmorphic features that simulate the PBDs. The commonest of these pseudo-PBDs is D-bifunctional protein (D-BP) deficiency; the single reported case of thiolase deficiency (pseudo-ZS) is now known to be D-BP deficiency. Other pseudo-PBDs include acyl-CoA oxidase deficiency (previously pseudo-NALD), RCDP type 2 (DHAP acyltransferase [DHAPAT] deficiency), RCDP type 3 (alkyl DHAP synthase deficiency), and most recently mulibrey (muscle-liver-brain-eye) nanism. Adult Refsum's disease (ARD), peroxisomal 2-methylacyl-CoA racemase deficiency, glutaric aciduria type 3, acatalasemia, and hyperoxaluria type 1 (alanine glyoxylate aminotransferase deficiency) complete the current list of Group 2 peroxisomal disorders. ALD/AMN is an exception to the Group 2 rule in that its defective gene product is an integral membrane protein, not a matrix enzyme, and D-BP and acyl-CoA oxidase deficiency are exceptional because they display defects in peroxisomal structure. ALD/AMN, D-BP deficiency, and acyl-CoA oxidase deficiency were believed to reflect primary abnormalities in the catabolic β-oxidation system, whereas the RCDPs mainly affect the anabolic plasmalogen system. As mentioned later in this chapter and discussed in a separate chapter on ALD, the assignment of ALD/AMN to this catabolic category appears to have been somewhat unjustified. With the exception of glutaric aciduria type 3, hyperoxaluria type 1, and acatalasemia, all peroxisomal disorders involve the nervous system. The focus of this chapter is on the pathophysiological basis of neurological abnormalities in the most common peroxisomal disorders, but one should know that they are systemic diseases in which the adrenal, ears, eyes, kidneys, liver, and skeleton also may display clinical and morphological abnormalities; many of the latter are expressed in affected fetuses. The mildest clinical phenotypes can have signs or symptoms limited to a sensorineural hearing deficit. It is again beyond the scope of this chapter to discuss the systemic manifestations; the reader is referred to the references cited previously [3–5].

III. Etiology

The peroxisomal disorders are heredofamilial diseases and, with the exception of ALD/AMN, are transmitted in an autosomal-recessive pattern; ALD/AMN is X linked. The Zellweger spectrum is caused by a mutation in 1 of at least 13 PEX genes, particularly *PEX1*, which interferes with the normal assembly of peroxisomes and usually with matrix protein import. RCDP, type 1, classic is due to a mutation in *PEX7*, which blocks PTS2-mediated import. The causes of the Group 2 disorders are more heterogeneous. Some are obvious in that the mutated gene is the same as the name of the disease, such as in acyl-CoA oxidase and D-BP deficiencies. ARD can be caused by a mutation in either the gene that encodes phytanoyl-CoA hydroxylase or in *PEX7*. The genetic defect in ALD/AMN resides in *ABCD1*, located at Xq28, which codes for an integral peroxisomal membrane protein, ABCD1 (formerly ALDP) [5].

IV. Molecular Pathogenesis and Pathophysiology

First and foremost, these are diverse neurometabolic diseases caused by a malfunction of peroxisomes. Consequently, they can display variable plasma and tissue elevations of very long ($\geq C22$) chain fatty acids (VLCFAs), a marker of virtually all peroxisomal disorders, dihydroxycholestanoic acid (DHCA) and trihydroxycholestanoic acid (THCA) (bile acid precursors), pipecolic acid, and, dependent on a sufficient dietary intake, phytanic and pristanic acids; as well as decreases in DHA, bile acids, and plasmalogens. How their molecular pathogeneses relate to their biochemical pathologies has become clearer, but how both translate to their pathophysiologies and pathological abnormalities is still a matter of much discussion [5].

Our understanding of the pathogenesis of the Group 1 disorders, at least at the molecular level, has grown by quantum leaps during the past decade, thanks in large part to molecular studies in yeast and to complementation analysis of affected patients. Complementation analysis involves the fusion of genetically incompetent fibroblasts from two patients into a multinucleated heterokaryon. If one can correct or "complement" the defect in the other, they must have different genotypes. The PBDs (Group 1) have been separated into 12 major complementation groups (CGs), all of whose genetic defects have been identified: CG1 (*PEX1*), CG2 (*PEX5*), CG3 (*PEX12*), CG4 (*PEX6*), CG7 (*PEX10*), CG8 (*PEX26*), CG9 (*PEX16*), CG10 (*PEX2*), CG11 (*PEX7*), CG12 (*PEX3*), CG13 (*PEX13*), and CG14 (*PEX19*). The identification of a mutated *PEX14* in a ZS patient has increased the genotypic total of PBD groups to 13. Most CGs, except CG11 (RCDP), display all three of the Zellweger spectrum phenotypes (ZS, NALD, IRD). CGs 9, 12, and 14 lack identifiable peroxisomal structures, whereas CGs 1, 2, 3, 4, 7, 8, 10, 11, and 13 exhibit defects in matrix protein import but can form peroxisomal membranes that are usually defective (e.g., "ghosts"). The commonest group, CG1, due to a defect in *PEX1* that consists of 24 exons and is located at 7q21-22, includes approximately one-half to two-thirds of all PBD patients, whereas patients with RCDP constitute as much as 20–25% in some series. PEX1p is an AAA adenosine triphosphatase (ATPase) whose members are participants in many cellular processes, such as unfolding, refolding, or disassembling protein complexes. PEX1p and PEX6p, another AAA ATPase, stabilize PEX5p, the PTS1 receptor. PEX1p and PEX6p also interact with PEX7p, the PTS2 receptor. PEX5p and PEX7p are necessary for PTS2-mediated import. Thus, the loss of PEX1p will impair both PTS1 and PTS2 import, whereas the loss of PEX7p will only interfere with PTS2 import. Consequently, in CG1, elevations in VLCFA, DHCA, THCA, L-pipecolic acid, and to a variable degree due to the dietary source, phytanic/pristanic acids occur, as well as deficiencies in plasmalogens and DHA. The more severe the clinical phenotype (i.e., ZS), the more abnormal is the biochemical profile. In RCDP, these fatty acid elevations are not present and the deficiencies are restricted to plasmalogens, alkyl DHAP synthase, phytanoyl-CoA hydroxylase, peroxisomal thiolase, and a partial secondary deficiency of DHAPAT.

In one sense, genotype–phenotype correlations in the PBDs have been poor, except for *PEX7*, because defects in the same gene can produce different clinical phenotypes and defects in different genes can produce the same clinical phenotype. However, if one considers the severity of the phenotype, good genotype–phenotype correlations do exist. In CG1 either one of two mutated *PEX1* alleles (G843D and 2097insT) is found in approximately 80% of patients. The G843D missense mutation in *PEX1* causes only a partial loss of PEX1p, which results in some matrix enzyme import and the milder phenotypes (NALD and IRD), whereas the frameshift insertion 2097insT in exon 13 of *PEX1* causes a complete loss of PEX1p, no matrix import, and the severe ZS phenotype. A comparable situation exists for *PEX 5, 6, 7, 10*, and *12*. In RCDP a common mutant allele, *PEX7* L292ter, is found in approximately one-half of the patients, which results in reduced amounts of PEX7 with no residual activity. On the other hand, some missense mutations in *PEX7* (e.g., A218V) code for mutant PTS2 receptors that have some residual activity and milder RCDP phenotypes [2,4,5].

The pathogenetic scenario of the major Group 2 disorder, ALD/AMN, has a much less complicated biochemical pathology but lacks the molecular lucidity of the Group 1 disorders. The only significant and consistent biochemical abnormality is the accumulation of saturated straight VLCFA. These are found at highest concentrations not only in cholesterol esters of brain macrophages and adrenocortical cells, but also within gangliosides, the proteolipid protein (PLP), and phospholipids, particularly phosphatidylcholine. *ABCD1* at Xq28 consists of approximately 25 kilobases (kb) of genomic DNA, which contains 10 exons that transcribe a messenger RNA (mRNA) of 4 kb and translates a small protein of 745–750 amino acids. As mentioned earlier, it is an integral membrane protein with the properties of an ABC half-transporter, yet its precise physiological function remains unknown. The transcript is present in all tissues but is highest in the adrenal cortex, intermediate in the brain, and almost undetectable in the liver. ABCD1 is primarily expressed in microglia, astrocytes, and endothelial cells in brain and adrenocortical cells; oligodendrocytes appear to have little, except in the corpus callosum and internal capsule, the sites that typically exhibit the earliest signs of myelin breakdown. Many mutations, usually private, have been reported: Approximately 54% are missense, 25% frameshift, 10% nonsense, and 7% large deletions. All exons are vulnerable, but exon 5 appears to be a hot spot. All mutations, except approximately one-third of the missense, do not express ABCD1 in affected patients' fibroblasts. There is no correlation between the site or type of mutation in *ABCD1* and the clinical phenotype, because all clinical phenotypes (e.g., the inflammatory demyelinative ALD, the spinal axonopathic AMN, or pure Addison's disease) can exist in the same family. Thus, both autosomal modifier genes and environmental factors have been implicated in the pathogenesis of ALD/AMN. Brain trauma has been contributory in a few patients with ALD, at least to the onset of inflammatory demyelination, and hormonal (corticosteroids and androgens) changes have been suspected of playing an important role in the onset of ALD and AMN. In spite of extensive investigations for over a decade, "a" or "the" modifier gene has not been identified [1,6].

On the other hand, a few genotype–phenotype correlations have been found in the second most common Group 2 disorder, D-BP deficiency, but the biochemical pathology is more involved. D-BP plays a role in the peroxisomal β-oxidation of VLCFAs, DHCA, THCA, and pristanic acid. Hence, all of these metabolites accumulate in most patients with D-BP [5].

V. Human Pathology and Pathogenesis

Before the molecular age of gene discovery and yeast manipulations, attempts to unravel the pathogenetic secrets of peroxisomal disorders had centered on neuropathological and biochemical–neuropathological correlative studies in affected patients; since then the emphasis has been on modern imaging techniques and mouse models. Our current goals are to integrate the molecular pathogenetic insights gained in the PBDs with their neuropathologies and pathophysiologies and, conversely, to gain further insights into the pathogenesis of ALD/AMN through neuropathological studies of mice and men because of a relative molecular stagnation in that area. The neuropathological lesions of human peroxisomal disorders consist of three major types: (a) abnormalities in neuronal migration or positioning, (b) abnormalities in the formation or maintenance of myelin, and (c) postdevelopmental neuronal degenerations.

Group 1 and Group 2 disorders have both shared and distinctive neuropathological features. Their shared neuropathological features include abnormalities of myelin, usually of CNS and particularly cerebral, and postdevelopmental neuronal or axonal degenerations. ARD primarily exhibits a hypertrophic (onion bulb) demyelinating peripheral neuropathy. The lesions in CNS white matter vary greatly between these diseases but can be divided into three basic types: (a) a nonspecific reduction in myelin volume or myelin staining with or without a mild reactive astrocytosis (hypomyelination); (b) noninflammatory dysmyelination; and (c) inflammatory demyelination. For the sake of this discussion, dysmyelination (Greek: *dys* for *ill*) refers to a genetic disorder of myelin in which either the myelin is biochemically abnormal or the myelin-forming cells have a molecular abnormality that affects either the formation or the maintenance of myelin. The neuropathological distinctions between delayed and arrested types of hypomyelination and between hypomyelination and hypomyelinative dysmyelination are imprecise and subjective. Demyelination (Latin: *de* for *after*) refers to the destruction, usually inflammatory and immune mediated, of myelin that has already formed and is usually both morphologically and biochemically normal. In ZS, IRD and RCDP (Group 1), and acyl-CoA oxidase and D-BP deficiencies (Group 2), the fundamental myelin problem appears to be mainly a lack or delay of CNS myelin formation, that is, hypomyelination, which is seen as myelin pallor microscopically (Fig. 2) and reduced volume of white matter grossly. A cerebellar preference seems to

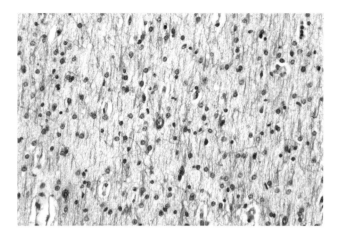

Figure 2 A paucity of myelin (blue linear staining) in cerebral white matter of Zellweger syndrome (ZS). Luxol fast blue periodic acid–Schiff (PAS) staining.

exist for D-BP deficiency. Other microscopic features may include a slight reactive astrocytosis but no macrophage or classic inflammatory (e.g., lymphocytes) response. In NALD, however, true inflammatory demyelination also occurs, as in ALD (see later discussion), but on a much smaller scale (Fig. 3). Patients with AMN (even "pure" without clinical or radiological involvement of CNS white matter), either acyl-CoA oxidase or D-BP deficiencies, and "normal" white matter in ALD display dysmyelinative foci. Magnetic resonance spectroscopy (MRS) confirms abnormalities in "normal" white matter of ALD in the form of increases in choline (demyelination) and decreases in N-acetylaspartate (axonal and/or oligodendrocytic loss). Neuropathologically, these foci consist of myelin pallor, variable axonal and oligodendrocytic loss, periodic acid–Schiff (PAS)–positive macrophages, a few reactive astrocytes and few or no detectable lymphocytes (Fig. 4). On

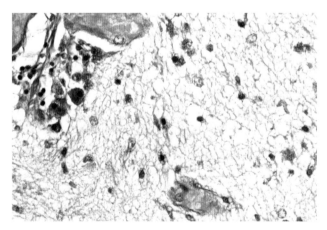

Figure 3 Loss of myelin and oligodendrocytes with replacement by fibrillary astrogliosis in neonatal adrenoleukodystrophy (NALD). A perivascular cluster of periodic acid–Schiff (PAS)–positive macrophages in top left portion of the field. Luxol fast blue PAS.

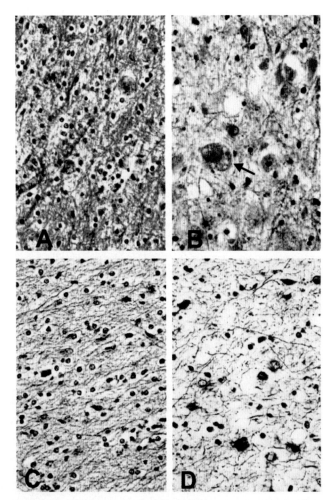

Figure 4 Variable degrees of dysmyelination in cerebral white matter of adrenomyeloneuropathy (AMN). The same can be seen in "normal" white matter of adreno-leukodystrophy (ALD). (**A**) Mild myelin pallor and normal number of oligodendroglial nuclei. (**B**) Marked loss of myelin and oligodendroglia with a periodic acid–Schiff (PAS)–positive macrophage (arrow). (**C**) Axonal sparing in a serial section of (A). (**D**) Greater axonal loss, but still relative sparing, in a serial section of (B). A and B: Luxol fast blue PAS. C and D: Bodian. (Reprinted, with permission, from Powers, J. M., DeCiero, D. P., Ito, M., Moser, A. B., and Moser, H. W. [2000]. Adrenomyeloneuropathy: a neuropathologic review featuring its noninflammatory myelopathy. *J. Neuropathol. Exp. Neurol.* **59**, 89–102.)

consider briefly the possible cause of hypomyelination and dysmyelination in the PBDs and the pseudo-PBDs, and why it may become inflammatory. As we have just discussed, biochemical abnormalities of fatty acids are prominent markers of peroxisomal disorders and, hence, would be expected to play dominant roles in their pathophysiologies and pathogeneses. Therefore, it has great appeal to implicate the fatty acid(s) that is (are) both elevated in and common to all these hypomyelinative/dysmyelinative conditions. In AMN, ALD, and acyl-CoA oxidase deficiency, the biochemical abnormality is restricted to elevations of saturated VLCFA. In D-BP deficiency, both saturated VLCFA and cholestanoic acid (bile acid precursors) are elevated. In ZS, NALD, and IRD, these, pipecolic acid, and perhaps phytanic/pristanic acids are elevated. In ZS, the excessive VLCFAs are both saturated and monounsaturated, and polyunsaturated fatty acids are elevated; in NALD and IRD, the VLCFAs also are saturated and monounsaturated. The common biochemical denominator in all these white-matter lesions is saturated VLCFA and, hence, a logical candidate for the offending dysmyelinative agent. The causes of the cerebral hypomyelination in ZS, however, are probably multifactorial and also due to (a) severe neuronal migration abnormalities with a consequent reduction in the normal complement of axons, (b) marked deficiencies in plasmalogens, (c) a gliopathy due to the accumulation of abnormal unidentified lipids (Fig. 5), (d) abnormal gangliosides lacking the proper signaling characteristics, and (e) superimposed hypoxia/ischemia/acidosis. Why then do certain PBDs and patients with pseudo-PBD with dysmyelinative lesions also develop inflammatory demyelination? In the

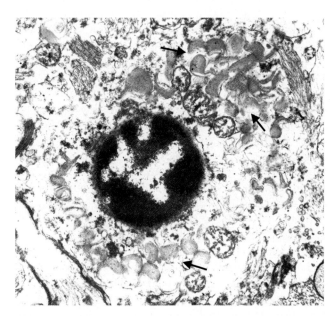

Figure 5 Ultrastructural view of a probable oligodendrocyte in white matter of Zellweger syndrome (ZS) containing a variety of abnormal lipid cytosomes (arrows). Uranyl acetate–lead citrate.

the other hand, some prolonged or clinically progressive cases of acyl-CoA oxidase or D-BP deficiencies have had imaging findings more consistent with demyelination, including inflammation due to the finding of contrast enhancement. I have personally examined slides from two patients with D-BP deficiency of the same complementation group: The 6-month-old showed only hypomyelination/dysmyelination, and the 14-month-old also showed inflammatory demyelination [3].

Before proceeding to a discussion of the pathomechanisms of the dysmyelination in AMN or ALD and the overwhelmingly inflammatory demyelination of ALD, we might case of AMN that becomes ALD, or ALD, we believe that a modifier gene or brain trauma may play a role. Could this also explain why the leukoencephalopathy of ZS has never been reported to be inflammatory while that of its phenotypic first-cousin NALD is, or why some acyl-CoA and D-BP deficiencies are? Perhaps. But I have come to believe that it is more likely that the presence of an inflammatory element in these infantile settings is due to the older age of the inflamed group that has developed a more competent immune system [1,3].

Based on our human studies, we have postulated that the dysmyelinative foci described earlier and illustrated in Fig. 4 constitute the initial myelin lesion of AMN and ALD and might be due to the incorporation of saturated VLCFA into myelin, which can lead to its spontaneous breakdown. Free saturated VLCFAs are extremely insoluble, particularly at normal body temperature; they adversely affect the viscosity of erythrocyte and adrenocortical cell membranes, disrupt model membranes, and are toxic to a number of cell types. In tissue, their presence, when in excess, is detected as linear clearings or striations with the light microscope and clear, fine acicular, or mildly curved clefts (thinner and more pliant appearing than those of cholesterol) and lamellae ultrastructurally, because they are removed during processing by nonpolar lipid solvents such as xylene. In some settings, they appear to originate as slitlike dilatations of rough endoplasmic reticulum. The toxicity of free fatty acids varies directly with their length and degree of saturation. Most of the emphasis in ALD has been on C26:0 and C24:0, but these patients accumulate a considerable amount of saturated VLCFA with chain lengths more than C26. In tissue culture, at least, even C16:0 is cytotoxic. The sources for the VLCFA are both endogenous and exogenous but are discussed fully in Chapter 1. Whereas a mild to moderate excess of VLCFA is present in all brain lipids, the greatest excess occurs in ganglioside, PLP, cholesterol ester and phosphatidylcholine fractions, the latter even in "normal" white matter. The cholesterol esters are not found in "normal" white matter, but in macrophages of actively demyelinating areas, which indicates that they are secondary players in the dysmyelination. VLCFA in any of the other three myelin components would be reasonable candidates to destabilize the myelin sheaths, once a certain threshold is reached. I have always found PLP to be the most appealing candidate, both for the dysmyelination and particularly for the transition to inflammatory demyelination (see later discussion). Some AMN dysmyelinative lesions remain indolent, whereas others elicit a mild inflammatory response (Fig. 6) and others (about 20%) progress to the classic confluent inflammatory demyelination of ALD (AMN-ALD). If an autosomal modifier gene is responsible for the phenotypic divergence of the AMN and ALD phenotypes, then what factor converts AMN to AMN-ALD? We have some anecdotal evidence that brain trauma may be one such factor. Irrespective of the cause or

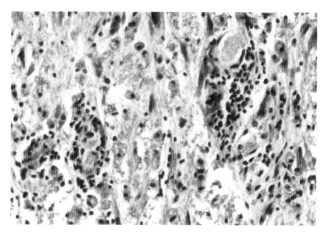

Figure 6 Mild and limited lymphocytic inflammatory response in adrenomyeloneuropathy (AMN) dysmyelinative lesion. Hematoxylin-eosin. (Reprinted, with permission, from Powers, J. M., DeCiero, D. P., Ito, M., Moser, A. B., and Moser, H. W. [2000]. Adrenomyeloneuropathy: a neuropathologic review featuring its noninflammatory myelopathy. *J. Neuropathol. Exp. Neurol.* **59**, 89–102.)

trigger for inflammatory demyelination in the peroxisomal disorders, only that of ALD or AMN-ALD has been studied in enough neuropathological detail to justify some pathogenetic speculations. We assume, however, that the mechanism of inflammatory demyelination in other peroxisomal settings would have comparable participants and probably the same inciting agent [1,3,6].

The inflammatory demyelination of ALD or AMN-ALD is distinctive. Neither its presence nor age at onset appears to correlate with the plasma or fibroblast levels of VLCFA or with adrenal or testicular insufficiency. Confluent and bilaterally symmetrical loss of myelin typical of most leukodystrophies involves the cerebral and cerebellar white matter. The cerebral lesions usually begin in the parietooccipital regions (Fig. 7) and progress more asymmetrically toward

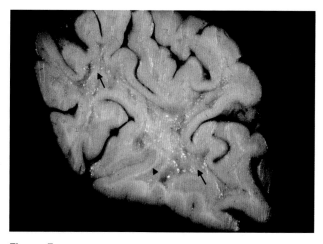

Figure 7 Confluent demyelination and replacement astrogliosis (arrows) with some arcuate fiber sparing (arrowhead) in parietooccipital white matter of adreno-leukodystrophy (ALD).

the frontal and temporal lobes. A topographical progression has not been recognized in the cerebellum, and the cerebellar lesions generally are less severe and lag behind the cerebral lesions. Arcuate fibers are generally spared, except in chronic cases. The loss of myelin exceeds that of axons, but there is considerable axonal loss. The earliest change at the advancing edge appears to be enlargement of the extracellular space with a few astrocytes and macrophages at the ultrastructural level; scant vacuoles and mild myelin swelling, reactive astrocytosis, and microglial/macrophage infiltration can be appreciated at the light microscopic level. Perivascular lymphocytes, and much less frequently plasma cells, are generally found behind this advancing edge where numerous lipophages (macrophages with myelin debris) are identified (Fig. 8). This localization, as well as the presence of a lymphocytic perivasculitis within the corticospinal tract degeneration secondary to cerebral axonal loss, led to a pathogenetic proposal: First myelin breaks down (dysmyelination), then the dominant second phase of inflammatory demyelination ensues [7]. The predominant cells that participate in the demyelination of ALD are reactive astrocytes, microglia/macrophages, and, to a lesser extent and probably subsequently, T lymphocytes. The T cells are predominantly CD8 cytotoxic lymphocytes (CTLs) with the α/β receptor; most CD8 CTLs are CD44 positive. They appear to kill oligodendrocytes by lysis, rather than apoptosis, via the granule exocytosis pathway. We have confirmed that oxidative damage also occurs in the lesion; more importantly, we have provided evidence for a TH-1 response and the presence of peroxynitrite and 4-hydroxynonenal (HNE), the latter being a toxic byproduct of lipid peroxidation. Both peroxynitrite and HNE also can rapidly kill cells via a nonapoptotic lytic mechanism. Lipid peroxidation is a self-propagating process and, therefore, could be one explanation for the confluent nature of the demyelinative lesion [Powers et al., in press]. The demonstration of CD1b and CD1c molecules on astrocytes and macrophages suggests that lipid antigen presentation is a significant pathogenetic event in ALD and makes the CD1 gene a plausible candidate for a modifier gene. Both ALD lipids (cholesterol esters, ganglioside, and phospholipids) and PLP contain the abnormal VLCFA and, hence, would be suitable molecules for CD1 presentation. Perivascular cells and macrophages express major histocompatibility complex (MHC) class 1 and class 2 molecules, as well as a number of chemokines and cytokines, particularly tumor necrosis factor-α (TNF-α). Reactive hypertrophic astrocytes in the same area can be even more immunoreactive for TNF-α. Although frank vasogenic edema usually is not seen microscopically, small blood vessels demonstrate upregulation of intercellular adhesion molecules (I-CAMs), and increased permeability of the blood–brain barrier can be inferred from contrast enhancement in the imaging studies. Within the demyelinative lesion, these immunoreactivities gradually diminish and interstitial lipophages move to a perivascular location. Oligodendrocytes are reduced around the advancing edge. At the end stage of the lesion, there is almost total loss of myelin and oligodendrocytes with a significant loss of axons. A few interstitial or perivascular lipophages may persist with isomorphic or anisomorphic fibrillary astrogliosis [1,8].

Ultrastructurally, lamellae and lipid profiles, highly similar to those in adrenocortical and Leydig cells (Fig. 9), have been identified rarely within cells consistent with oligodendrocytes; lamellar inclusions are commonly observed within CNS macrophages containing myelin debris. Spicular or trilaminar inclusions, more typical of PBDs, also may be found within CNS macrophages. The lamellae have been postulated to be bilayers of free VLCFA, whereas the lipid profiles are the abnormal cholesterol esters. The spicular inclusions, found within angulate lysosomes, probably represent biochemically modified abnormal cholesterol esters that are taken up by lysosomes [1,7].

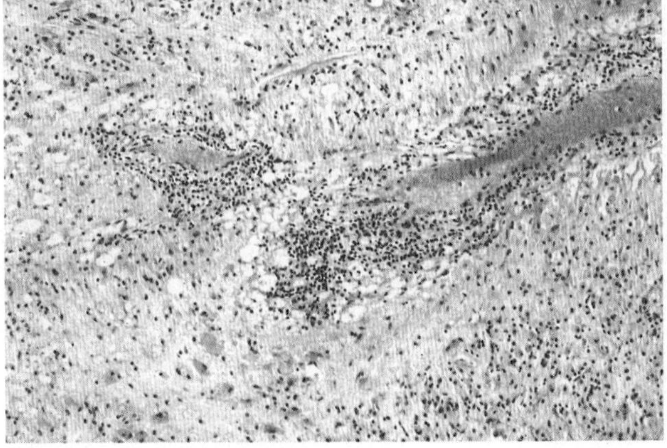

Figure 8 Robust lymphocytic and macrophage infiltrations in cerebral demyelinative lesion of adreno-leukodystrophy (ALD). Hematoxylin-eosin.

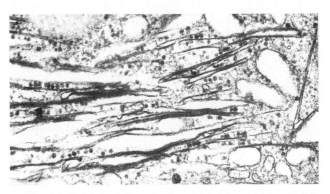

Figure 9 Ultrastructural appearance of lamellae and clear lipid profiles or clefts in adrenocortical cells of adreno-leukodystrophy (ALD). Similar to identical structures are seen in Schwann and Leydig cells, some macrophages and rare oligodendrocytes of ALD and adrenomyeloneuropathy (AMN). Biopsy. Uranyl acetate–lead citrate.

To summarize, based on our human neuropathological studies, we have postulated that the inflammatory demyelination of both ALD and AMN-ALD involves a two-stage pathogenetic scheme: (1) dysmyelination as described previously and (2) inflammatory demyelination. The inflammatory demyelination appears to involve an initial innate immune response to the insoluble lipids that may simulate a bacterial pathogen, in which macrophages and astrocytes produce cytokines, particularly TNF-α; this promotes a compromise of the blood–brain barrier and lymphocytic influx. Following that an adaptive immune response occurs in which several pathogenic elements seem to participate: an MHC-dependent TH-1 response, MHC-unrestricted CD1 lipid presentation, CD8 CTLs (probably unconventional), and oxidative damage by peroxynitrite and HNE, with resultant oligodendroglial lysis and loss of myelin. We again believe that free saturated VLCFAs play a pivotal role and that their incorporation into PLP may be the most defining alteration. It is noteworthy that, despite biochemical and ultrastructural evidence for the involvement of brain, peripheral nerve, adrenal cortex, and testis, the only inflammatory site in ALD or AMN-ALD is the brain. Hence, a CNS-specific antigen, such as PLP, has particular appeal [8; Powers et al., in press]. The genetic deficiency of ABCD1 is glaringly absent from the pathogenetic scheme. Perhaps, when we know the precise physiological role of ABCD1, we will learn how its absence causes the pathogenic elevations in saturated VLCFA.

The next group of neuropathologic lesions shared by disorders of both Group 1 and Group 2 are postdevelopmental, noninflammatory neuronal degenerations and lipid storage, of which the myeloneuropathy of AMN is the most common. The myeloneuropathy of AMN, and perhaps that of D-BP deficiency, displays the pattern of a central-peripheral distal (dying-back) axonopathy. The severity of the myeloneuropathy does not correlate with the duration or severity of endocrine dysfunction. The spinal lesions display equivalent losses of axons and myelin sheaths, most commonly in the gracile and corticospinal tracts; the most severe losses are usually observed in the cervical gracile tracts and the lumbar lateral corticospinal tracts, that is, in the most distal parts of the axons furthest from the nutritive parent cell body (a "dying-back" pattern) (Fig. 10). The spinal lesions are bilateral, usually symmetrical, and lack reactive astrocytes or lymphocytes, but they do exhibit a prominent microglial infiltration. Peripheral nerve lesions are milder, more variable, and more nonspecific than the myelopathy, except for the diagnostic lamellar and lipid inclusions in Schwann cells that are highly similar to identical to those in adrenocortical cells. Additionally, the largest myelinated fibers in peripheral nerve are the most severely affected. A comparable situation has been reported in a 4-year-old with D-BP deficiency, both in sural and trigeminal nerves. Surprisingly, at

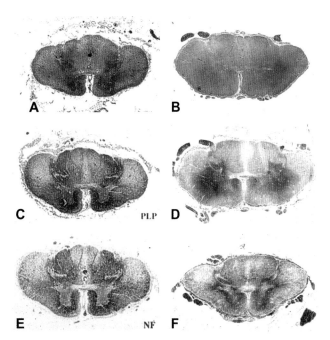

Figure 10 Adrenomyeloneuropathy (AMN) spinal cord. **(A, C, E)** Cross sections of lower cervical cord showing decreased myelin staining in posterior and lateral columns. The immunoreactivity for phosphorylated neurofilaments (NFs) **(E)** is less than or equal to the immunoreactivity for myelin proteolipid protein (PLP) in fasciculus gracilis; the same is true for dorsal spinocerebellar and lateral corticospinal tracts in the lateral columns. **(B, D, F)** Cross sections of cervical cord (B, D) and lower thoracic cord (F) showing greater reduction of axons in cervical fasciculus gracilis (D) than lower thoracic (F). Lateral corticospinal tracts are about the same. Luxol fast blue PAS (A, B), modified Bielschowsky (D, F), anti-PLP immunostain (C), antiphosphorylated neurofilament immunostain (E). (Reprinted, with permission, from Powers, J. M., DeCiero, D. P., Ito, M., Moser, A. B., and Moser, H. W. (2000). Adrenomyeloneuropathy: a neuropathologic review featuring its noninflammatory myelopathy. *J. Neuropathol. Exp. Neurol.* **59**, 89–102.)

autopsy in lumbar dorsal root ganglia (DRG), there is no apparent neuronal loss, necrosis or apoptosis, or obvious atrophy; nodules of Nageotte are inconspicuous. Morphometric studies, however, reveal neuronal atrophy with a decrease in the number of large neurons and a corresponding increase in neurons less than 2000 μm^2, especially in the 1500–1999 μm^2 range. No consistent immunohistochemical differences in the size or number of specific cell types are observed. Many mitochondria in the AMN neurons contain lipidic inclusions at the ultrastructural level, raising the possibility that, in addition to the well-known peroxisomal defect, impaired mitochondrial function may contribute to the myelopathy through a failure of ATP-dependent axoplasmic transport in AMN spinal tracts with consequent "dying-back" axonal degeneration [9–11].

The second postdevelopmental neuronal degeneration is cerebellar atrophy, which has been reported in a few protracted RCDP cases, ages 4, 9 and 11 years and one 4-year-old with D-BP deficiency and seizures. It is important to note that the patients with RCDP did not have a significant seizure or hypoxic history. These patients have demonstrated diffuse cortical atrophy due to severe losses of Purkinje and granule cells with focal losses of basket cells. The cerebellar atrophy in our two patients with RCDP was uneven in that the distal and superficial folia were more severely affected, and the vermis was more affected in one patient. Surviving Purkinje cells show a variety of progressive nuclear and cytoplasmic degenerative changes characterized predominantly by contraction and distortion of their nuclei. Extensive study of this lesion predominantly with immunohistochemical stains reveals that both Purkinje and internal granule cells are lost, at least in part, by an apoptotic mechanism. Surviving Purkinje cells demonstrate prominent decreases in parvalbumin and other proteins involved in calcium homeostasis and energy production. Cerebellar atrophy also was noted in IRD, but the authors interpret it as a hypoplasia [1].

The third postdevelopmental neuronal degeneration is a progressive loss of hearing that has been classified as sensorineural in type. It can be seen in ZS, NALD, IRD, ARD, RCDP, acyl-CoA oxidase deficiency, and D-BP deficiency. Neuropathological data to explain the nature of this deafness have been meager. One case of NALD showed severe atrophy of the sensory epithelium and the tectorial membrane, whereas a single case of IRD showed again severe atrophy of the sensory epithelium and the stria vascularis. A 4-year-old patient with D-BP deficiency had a reduction of nerve fiber density in the cochlear nerve [11]. No morphological abnormalities have been reported in the relevant central neurons and pathways. The fourth lesion is retinal pigmentary degeneration, which has been reported in ZS, NALD, IRD, and ARD. A few cases of ZS and NALD have been studied thoroughly and have revealed ganglion cell loss, gliosis of the nerve fiber layer and optic nerve, optic atrophy, extensive loss of outer and inner segments of photoreceptor cells, and thinning of the nuclear layer. D-BP deficiency also may demonstrate optic atrophy. At the ultrastructural level, spicular inclusions were present in retinal macrophages and ganglion cells contained some electron-opaque membranous cytoplasmic bodies. A single case of IRD showed comparable light microscopic changes. Thus, in the sensorineural deafness and pigmentary retinopathy, the lesion appears to reside in specialized sensory neurons [3].

The fifth lesion is restricted to the neurons of the dorsal nuclei of Clarke and the lateral cuneate nuclei, second-order sensory neurons, which demonstrate striated perikarya because of the accumulation or storage of lamellar and lipid profiles of abnormal cholesterol esters containing VLCFA. This is accompanied by lamellae and lipid clefts within neighboring axonal spheroids virtually identical to those seen in infantile neuroaxonal dystrophy. Thus far, this CNS neuronal lesion has been seen only in ZS. The sixth lesion is multifocal; clusters of striated and swollen PAS-positive cells containing abnormal lipid cytosomes are scattered within cerebral and cerebellar gray matter and white matter of patients with ZS. These cells appear to be both macrophages and neuroglia [1,3] (Fig. 11).

Based on our autopsy studies, we have postulated that the fundamental lesion in AMN is axonal degeneration, again perhaps related to the saturated VLCFA, particularly in gangliosides, that might interfere with axonal membrane–trophic factor interactions. Electrophysiological data and MRS data confirm the fundamental axonopathic nature of AMN [12]. We also have proposed that the cerebellar degeneration in RCDP involves apoptotic cell death of Purkinje and granule cells because of abnormalities in calcium homeostasis that might be precipitated by the excess phytanic acid in the cerebrospinal fluid of these patients, perhaps in concert with a deficiency in tissue plasmalogens [1,3]. Kahlert et al. [13] have provided evidence for this proposal by demonstrating that phytanic acid affects calcium homeostasis, depolarizes mitochondria, generates reactive oxygen species, and kills rat hippocampal astrocytes.

The last group of neuropathological abnormalities are those of neuronal migration or positioning, which are almost restricted to the PBDs of Group 1 disorders. Abnormalities in neuronal migration are most prominent in ZS, which is characterized by a unique combination of parasylvian or centrosylvian pachygyria/polymicrogyria. Pachygyria affects medial gyri, whereas polymicrogyria is disposed more laterally and extends into the lateral frontal lobe and parieto-temporooccipital region (Fig. 12). Other foci of cortical dysgenesis, such as frontal polar pachygyria, have been observed. On coronal sectioning of the cerebrum, the abnormal gyri are seen as a thickened cortex with either excessive superficial plications or large subcortical heterotopias, the latter being more typical of pachygyric foci. Polymicrogyric cortex typically demonstrates fusion

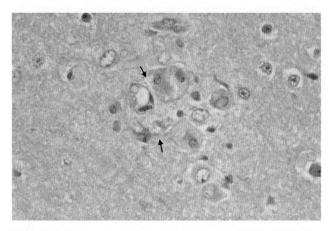

Figure 11 Swollen, lipid-filled cells (arrows), probably macrophages, in putamen of Zellweger syndrome (ZS). Hematoxylin-eosin.

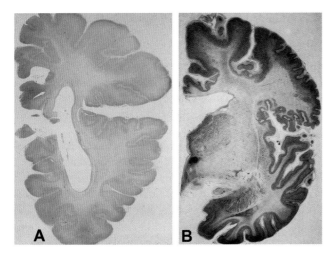

Figure 12 Whole-mount coronal views of Zellweger syndrome (ZS) cerebra from two patients showing the characteristic lateral centrosylvian or parasylvian micropolygyria (right side of field) and medial pachygria with prominent subcortical heterotopias (left top of field). (**A**) Nissl. (**B**) Black-and-white photograph of Nissl. A. (Reprinted, with permission, from Powers, J. M., and Moser, H. W. (1998). Peroxisomal disorders: genotype, phenotype, major neuropathologic lesions, and pathogenesis. *Brain Pathol.* **8**, 101–120.) B. [14]

of the molecular layers with medium to large pyramidal cells destined for the deep cortex admixed with decreased numbers of layer 2 and 3 neurons in the outer cortex. Many layer 2 and 3 neurons are located in the deep cortex and within heterotopias of subcortical white matter. The pachygyric cortical plate demonstrates similar, but more severe, architectonic changes, including the subcortical heterotopias. Migration of all neuronal classes appears to be affected, but particularly those destined for the outer layers [14]. Less severe cerebral migratory abnormalities, usually in the form of polymicrogyria, are seen in NALD and D-BP deficiency as diffuse, focal, or multifocal lesions that may be associated with subcortical heterotopias. Cerebral neuronal migration problems have not been identified in IRD, ARD, acyl-CoA oxidase deficiency, ALD, or AMN. RCDP has lacked evidence of cerebral neuronal migration lesions, except for one case that was reported before the era of biochemical or genetic confirmation. In this patient with putative RCDP, there was a large focus of pachygyria of the posterior frontal lobe in the rolandic region and a focus of microgyria of the frontal pole. Minor neuronal migration problems, apparently asymptomatic, are found in the form of heterotopic Purkinje cells or clusters of disordered granule-Purkinje cells (heterotaxias) in cerebellar white matter in ZS, NALD, and D-BP deficiency. Classic heterotaxias (of cerebellar cortex) have been seen in ZS, especially in the nodulus, and malpositioned Purkinje cells in D-BP deficiency. Two cases of IRD have displayed Purkinje cells abnormally distributed in the molecular layer [1,3].

Defects in neuronal positioning or terminal migration are common and usually involve the principal nuclei of the inferior medullary olives and less frequently the dentate nuclei and claustra. Dysplastic inferior medullary olives, not heterotopic, as in some other types of pachygyria, have been reported in ZS, RCDP, and D-BP deficiency. They exhibit a simplification of, or discontinuities in, the normal convolutional pattern, sometimes with an apparent reduction in neuronal number and their alignment in a single row along the periphery of the olive. A milder peripheral palisading of olivary neurons without convolutional abnormalities has been reported in IRD. Dysplastic dentate nuclei, usually seen as a simplification of the normal serpiginous pattern, has occurred in ZS and D-BP deficiency. Neuronal loss has been reported in the dentate and olivary nuclei of two patients with NALD and one with D-BP deficiency, but not dysplasias [1,3,11,14].

Fetuses at risk to develop ZS disclose neocortical migratory defects in the form of micropolygyric ripples and subtle subcortical heterotopias. Thin abnormal cortical plates with more obvious subcortical heterotopias occur later in gestation (22 weeks postmenstrual estimated gestational age) (Fig. 13). Astrocytes, neuroblasts, immature neurons, radial glia, and PAS-positive macrophages contain abnormal pleomorphic cytosomes; these include membranous cytoplasmic bodies that are electron opaque (Fig. 14). The latter seem to be typical of ZS and may represent gangliosides containing saturated VLCFA. Some neurites also exhibit lamellae and lipid profiles (Fig. 15). Dysplastic alterations of the

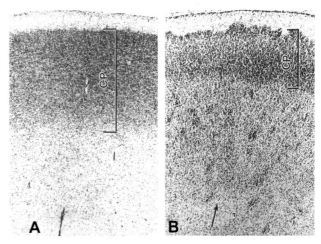

Figure 13 Fetal cerebrum from control (adreno-leukodystrophy [ALD] at risk) and Zellweger syndrome (ZS) (at risk). (**A**) Control cortical plate (CP) as compared with thin irregular CP and heterotopic zones (arrow) of fetal ZS. (**B**) Twenty-two weeks estimated gestational age. Hematoxylin-eosin. (Reprinted, with permission, from Powers, J. M., Tummons, R. C., Caviness, V. S., Jr., Moser, A. B., and Moser, H. W. [1989]. Structural and chemical alterations in the cerebral maldevelopment of fetal cerebro-hepato-renal (Zellweger) syndrome. *J. Neuropathol. Exp. Neurol.* **48**, 270–289.)

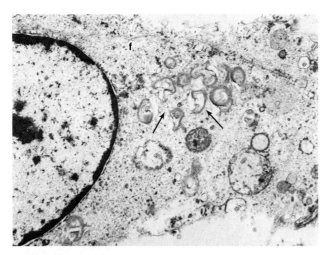

Figure 14 Radial glial cell from centrosylvian area contains pale lamellated cytoplasmic inclusions (arrows) and cytofilaments (f) in a fetal Zellweger syndrome of 22 weeks estimated gestational age. Uranyl acetate–lead citrate. (Reprinted, with permission, from Powers, J. M., Tummons, R. C., Caviness, V. S., Jr., Moser, A. B., and Moser, H. W. [1989]. Structural and chemical alterations in the cerebral maldevelopment of fetal cerebro-hepato-renal (Zellweger) syndrome. *J. Neuropathol. Exp. Neurol.* **48**, 270–289.)

that again tissue elevations in saturated VLCFA are a key pathogenic element through their incorporation into migrating cell membranes and acylated adhesion molecules. The stereotypic centrosylvian localization and the combination of both pachygria and polymicrogyria also suggest that regional tissue constraints, such as the density of radial glia and interweave of axons, act in concert to impede neuronal migration. This anatomical pathogenetic component has particular relevance when one considers the validity of the ZS mouse models mentioned in the next section [1,3]. Both PBDs and D-BP deficiency display abnormalities in neuronal migration and both demonstrate increases in saturated VLCFA. However, ALD/AMN and acyl-CoA oxidase also have elevated saturated VLCFA and yet do not demonstrate the migratory lesions. Consequently, if one is to maintain this pathogenetic perspective, then one might have to implicate a threshold effect in which the highest elevations in VLCFA should have the most severe and consistent neuropathological lesion. ZS does have both the highest elevations and the most severe pachygyria-micropolygyria; hence, this postulate gains some support. Alternatively, one might also implicate the bile acid precursors that are common to the PBDs and some patients with D-BP deficiency.

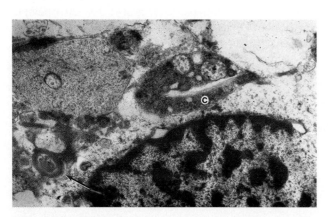

Figure 15 Centrosylvian heterotopic zone containing migrating neuroblasts with lamellated inclusions (arrow), radial glial processes and probable neuritic process (C) with clear cleft, the latter comparable to some of those in Fig. 9. Fetal Zellweger syndrome (ZS), 22 weeks estimated gestational age. Uranyl acetate–lead citrate. (Reprinted, with permission, from [1989]. *J. Neuropathol. Exp. Neurol.* **48**, 270–289.)

inferior olivary and dentate nuclei are present. These fetal lesions confirm the prediction that the insult (presumably metabolic) causing the neocortical migration defect is operating throughout the entire neocortical neuronal migratory period [14]. Immunohistochemical deficiencies of the cell adhesion molecule L1 and doublecortin have been reported in fetal Zellweger brains [1].

Based on the human postmortem studies of ZS, particularly of the fetuses at risk, we have proposed

VI. Mouse Models and Pathogenesis

Pathogenetic insights and hypotheses can arise from human morphological and biochemical studies, but true cause-and-effect relationships cannot be established through this method of scientific investigation. The most that one can hope to achieve is sound associative data with variable levels of confidence, even when reinforced with relevant correlative data. Through these methods, we have offered a unifying pathogenetic hypothesis for the peroxisomal disorders. Abnormal fatty acids, particularly saturated VLCFA and phytanic acid, accumulate because of their respective genetic defects and are incorporated into cell membranes resulting in a perturbation of their microenvironments and the dysfunction, atrophy, and death of vulnerable cells. Cell vulnerability in the PBDs may be determined, in part, by low tissue levels of plasmalogens and DHA [1,3]. However, under completely controllable laboratory conditions, one can establish the cause-and-effect relationships necessary for scientifically valid pathogenetic formulations. The advent of transgenic and knockout (KO) mouse models affords us this opportunity, but only when the mouse model faithfully recapitulates the human disease. Many KO mouse models of peroxisomal disorders have been produced, but mice are not men and none of these models, in my opinion, faithfully recapitulates their respective human diseases. Nonetheless, they do provide some interesting, and sometimes unanticipated, pathogenetic insights.

Three laboratories have generated KO mice lacking *ABCD1*. None of these have displayed the cerebral inflammatory demyelination typical of human ALD. Rather, spinal axonal degeneration, more reminiscent of AMN, occurs, particularly if one also knocks out *ABCD2* (ALDR) [15]. We found mitochondrial abnormalities in an *ABCD1* KO and we found that the rate of peroxisomal VLCFA β-oxidation is directly related to the rate of mitochondrial long-chain fatty acid oxidation. Thus, although this KO has not enhanced our understanding of the pathogenesis of inflammatory demyelination, it has provided another link to a mitochondrial abnormality in ALD that was first demonstrated in the adrenal cortex in 1974. It also has suggested that ABDC1 may not participate directly in the degradation of VLCFA, and that an alternative possibility, increased microsomal elongation, is a better explanation for the elevated VLCFA. (See Chapter 4, for details.) The percentage of C26:0 in total fatty acids of the KO *whole brain* was much higher than that of the human ALD "normal" white matter (fivefold vs twofold elevation) [16]. Thus, the mouse data cast a shadow over our hypothesis that a threshold level of saturated VLCFA leads to dysmyelination. However, both the status of the longer chain lengths (more than C26) and the concentrations of saturated VLCFA in the different subclasses of myelin lipids or PLP were not included. Finally, in contrast to the widespread oxidative damage in the human brain mentioned earlier, the *ABCD1* KO brain displayed none [Powers et al., in press].

In respect to the Group 1 disorders, many KOs have been produced, particularly ZS KOs, but space limitations compel me to confine this recognition to one RCDP KO and two ZS KOs. The RCDP KO did show an abnormality in cerebrocortical neuronal migration and, surprisingly, temporary elevations in saturated VLCFA (not seen in patients with RCDP). The authors speculated that the migration defect might have been caused by the elevated saturated VLCFA and low tissue plasmalogens [17]. This speculation parallels that suggested for the cerebellar atrophy in chronic RCDP in which phytanic acid elevations and low tissue plasmalogens were implicated [3].

Two mouse models of ZS, the *PEX5* KO and *PEX2* KO, exhibit abnormal cerebrocortical laminations and subcortical heterotopic neurons, features that they share with ZS. However, these lesions are subtle in comparison to ZS and do not have a stereotypic topography. Further studies on the *PEX5* KO reveal mitochondrial abnormalities with evidence of oxidative stress, that elevated levels of saturated VLCFA alone do not disturb neocortical migration (in mice), and that the plasmalogen (but not DHA) deficiency may be a necessary pathogenic cofactor. These studies also suggest that, in addition to local biochemical abnormalities, neocortical migration also may be affected by systemic metabolic abnormalities, such as bile acid precursors [18–20]. These data and their implications resonate with some of the predictions from human neuropathological studies related to the leukoencephalopathy in ZS and cerebellar degeneration in chronic RCDP.

VII. Clinicopathological Correlations

In addition to providing pathophysiological insights and generating pathogenetic hypotheses, classic neuropathological studies explain to some degree the neurological symptomatology of the affected patients. For example, it is reasonable to ascribe the abnormalities in neuronal migration for the neonatal hypotonia, seizures, failure to thrive, psychomotor retardation, and early deaths typically seen in the PBDs. The myelin abnormalities cause the spasticity, ataxia, and aphasias seen primarily in the Group 2 disorders, especially ALD. The postdevelopmental neuronal degenerations help to explain the sensorineural hearing loss and pigmentary retinopathy of the PBDs and the pseudo-PBDs, as well as, most importantly, the spasticity, weakness, and sensory ataxia of AMN. Hemizygous males present with difficulty in walking due to a combination of sensory ataxia and spastic paraparesis in the third decade; heterozygous females often display a milder myelopathy later in life.

References

1. Powers, J. M., and De Vivo, D. C. (2002). Peroxisomal and mitochondrial disorders. *In:* "Greenfield's Neuropathology" (P. L. Lantos and D. L. Graham, eds.), Vol. I, pp. 737–797. Arnold-Oxford University Press, New York.
2. Gould, S. J., Raymond, G. V., and Valle, D. (2001). The peroxisome biogenesis disorders. *In:* "The Metabolic & Molecular Bases of Inherited Disease" (C. R. Scriver, A. L. Beaudet, W. S. Sly, and D. Valle, eds.), 8th edition, pp. 3181–3218. The McGraw-Hill Companies, Inc., New York.
3. Powers, J. M., and Moser, H. W. (1998). Peroxisomal disorders: genotype, phenotype, major neuropathologic lesions, and pathogenesis. *Brain Pathol.* **8**, 101–120.
4. Wanders, R. J. A., Barth, P. G., and Heymans, H. S. A. (2001). Single peroxisomal enzyme deficiencies. *In:* "The Metabolic & Molecular Bases of Inherited Disease" (C. R. Scriver, A. L. Beaudet, W. S. Sly, and D. Valle, eds.), 8th edition, pp. 3219–3257. The McGraw-Hill Companies, Inc., New York.
5. Wanders, R. J. A. (2004). Metabolic and molecular basis of peroxisomal disorders: a review. *Am. J. Med. Genet.* **126A**, 355–375.
6. Moser, H. W., Smith, D. K., Watkins, P. A., Powers, J. M., and Moser, A. B. (2001). X-linked adrenoleukodystrophy. *In:* "The Metabolic & Molecular Bases of Inherited Disease" (C. R. Scriver, A. L. Beaudet, W. S. Sly, and D. Valle, eds.), 8th edition, pp. 3257–3301. The McGraw-Hill Companies, Inc., New York.
7. Schaumburg, H. H., Powers, J. M., Raine, C. S., Suzuki, K., and Richardson, E. P. Jr. (1975). Adrenoleukodystrophy. A clinical and pathological study of 17 cases. *Arch. Neurol.* **33**, 577–591.
8. Ito, M., Blumberg, B. M., Mock D. J., Goodman, A. D., Moser, A. B., Moser, H. W., Smith, K. D., and Powers, J. M. (2001). Potential environmental and host participants in the early white matter lesion of adreno-leukodystrophy: morphologic evidence for CD8 cytotoxic

T cells, cytolysis of oligodendrocytes and CD1-mediated lipid antigen presentation. *J. Neuropathol. Exp. Neurol.* **60**, 1004–1090.

9. Powers, J. M., DeCiero, D. P., Ito, M., Moser, A. B., and Moser, H. W. (2000). Adrenomyeloneuropathy: a neuropathologic review featuring its noninflammatory myelopathy. *J. Neuropathol. Exp. Neurol.* **59**, 89–102.

10. Powers, J. M., DeCiero, D., Cox, C., Richfield, E. K., Ito, M., Moser, A. B., Moser, and H. W. (2001). The dorsal root ganglia in adrenomyeloneuropathy: neuronal atrophy and abnormal mitochondria. *J. Neuropathol. Exp. Neurol.* **60**, 493–501.

11. Schröder, J. M., Hackel, V., Wanders, R. J. A., Gohlich-Ratmann, G., and Voit, T. (2004). Optico-cochleo-dentate degeneration associated with severe peripheral neuropathy and caused by peroxisomal D-bifunctional protein deficiency. *Acta Neuropathol.* **108**, 154–167.

12. Dubey, P., Fatemi, A., Barker, P. B., Degaonkar, M., Troeger, M., Zackowski, K., Bastian, A., Smith, S. A., Pomper, M. G., Moser, H. W., and Raymond, G. V. (2005). Spectroscopic evidence of cerebral axonopathy in patients with "pure" adrenomyeloneuropathy. *Neurology* **64**, 304–310.

13. Kahlert, S., Schönfeld, P., and Reiser, G. (2005). The Refsum disease marker phytanic acid, a branched chain fatty acid, affects Ca^{2+} homeostasis and mitochondria, and reduces cell viability in rat hippocampal astrocytes. *Neurobiol. Dis.* **18**, 110–118.

14. Evrard, P., Caviness V. S., and Prats-Vinas, J. (1978). The mechanism of arrest of neuronal migration in the Zellweger malformation: an hypothesis based upon cytoarchitectonic analysis. *Acta Neuropathol.* **41**, 109–117.

15. Pujol, A., Ferrer, I., Camps, C., Metzger, E., Hindelang, C., Callizot, N., Ruiz, M., Pampols, T., Giros, M., and Mandel, J. L. (2004). Functional overlap between ABCD1 (ALD) and ABCD2 (ALDR) transporters: a therapeutic target for X-adrenoleukodystrophy. *Hum. Mol. Genet.* **13**, 2997–3006.

16. Heinzer, A. K., McGuinness, M. C., Lu, J-F., Stine, O. C., Wei, H., Van der Vlies, M., Dong, G. X., Powers, J., Watkins, P. A., and Smith, K. D. (2003). Mouse models and genetic modifiers in X-linked adrenoleukodystrophy. *In:* "Peroxisomal Disorders and Regulation of Genes" (F. Roels, M. Baes, and S. de Bie, eds.). *Advances in Experimental Biology* **544**, pp. 75–93. Kluwer Plenum Press, New York.

17. Brites, P., Motley, A. M., Gressens, P., Mooyer, P. A., Ploegaert, I., Everts, V., Evrard, P., Carmeliet, P., Dewerchin, M., Schoonjans, L., Duran, M., Waterham, H. R., Wanders, R. J., and Baes, M. (2003). Impaired neuronal migration and endochondral ossification of Pex7 knockout mice: a model for rhizomelic chondrodysplasia punctata. *Hum. Mol. Genet.* **12**, 2255–2267.

18. Baumgart, E., Vanhorebeek, I., Grabenbauer, M., Borgers, M., Declercq, P. E., Fahimi, H. D., and Baes, M. (2001). Mitochondrial alterations caused by defective peroxisomal biogenesis in a mouse model for Zellweger syndrome (PEX5 knockout mouse). *Am. J. Pathol.* **159**, 1477–1494.

19. Baes, M., Gressens, P., Huyghe, S., De, N. K., Qi, C., Jia, Y., Mannaerts, G. P., Evrard, P., Van, V. P., Declercq, P. E., and Reddy, J. K. (2002). The neuronal migration defect in mice with Zellweger syndrome (Pex5 knockout) is not caused by the inactivity of peroxisomal β-oxidation. *J. Neuropathol. Exp. Neurol.* **61**, 368–374.

20. Faust, P. L., Banka, D., Siriratsivawong, R., Ng, V. G., and Wikander, T. M. (2005). Peroxisome biogenesis disorders: the role of peroxisomes and metabolic dysfunction in developing brain. *J. Inherit. Metab. Dis.* **28**, 369–383, 2005.

3

Creatine Deficiency Syndromes

Peter W. Schütz, MD
Sylvia Stöckler, MD

Keywords: *AGAT deficiency, GAMT deficiency, SLC6A8 deficiency, guanidinoacetate, epilepsy, extrapyramidal symptoms, mental retardation*

I. History and Nomenclature
II. Etiology and Biochemical Pathogenesis
III. Pathophysiology
IV. Natural History
V. Prospects
   References

I. History and Nomenclature

Creatine deficiency syndromes (CDSs) are a novel group of inborn errors of creatine metabolism including the biosynthetic disorders arginine:glycine amidinotransferase deficiency (AGAT; MIM 602360) and guanidinoacetate methyltransferase deficiency (GAMT; MIM 601240), and a defect of the creatine transporter, SLC6A8 deficiency (SLC6A8; MIM 300036). In all these disorders, the common clinical hallmark is mental retardation, expressive speech delay, and epilepsy; the common biochemical hallmark is cerebral creatine deficiency as detected by proton magnetic resonance spectroscopy (H-MRS).

The first patient with GAMT deficiency was described in 1994. This boy was considered normal until 4 months of age when he was noted to have developmental arrest, hypotonia, hyperkinetic (hemiballistic) extrapyramidal movements, and head nodding. On oral creatine substitution, cerebral creatine deficiency was almost completely reversed and his clinical symptoms improved significantly. Biochemical and molecular genetic analysis finally revealed that the underlying condition was a genetically inactivated GAMT [1]. To date, 27 patients (among them 10 adults 18 years or older) are known with this disorder [2]. AGAT deficiency is the second disorder affecting creatine synthesis and has been diagnosed so far in four patients with delayed psychomental development and occasional seizures. SLC6A8 deficiency was first reported in 2001. The index patient had mental retardation and speech delay. So far more than 100 patients from more than 50 families have been diagnosed. The prevalence of SLC6A8 deficiency is relatively high and may be responsible for approximately 2% of males with X-linked mental retardation. For a review of CDS, see reference 3.

II. Etiology and Biochemical Pathogenesis

Etiologically, CDSs are genetic disorders, with autosomal-recessive inheritance in GAMT and AGAT deficiency and X-chromosomal inheritance in SLC6A8 deficiency. Pathogenetically, CDSs are determined by biochemical changes including both creatine deficiency and changes in the concentrations of substrates occurring elsewhere in the creatine biosynthetic pathway. For a better understanding of the biochemical pathogenesis, main features of (1) creatine physiology and metabolism, (2) the genetic basis of CDSs, and (3) biochemical changes observed in CDSs are described in this section.

A. Creatine Metabolism

Creatine is synthesized in a two-step process: (1) transfer of the amidino group from arginine to glycine, yielding guanidinoacetic acid and ornithine (this reaction is catalyzed by L-AGAT); (2) transfer of a methyl group from S-adenosylmethionine to guanidinoacetate, yielding creatine and S-adenosylhomocysteine (this reaction is catalyzed by S-adenosyl-L-methionine:N-GAMT) (Fig. 1).

AGAT activity is the main regulatory site of creatine synthesis. It is repressed by high creatine concentrations at a pretranslational level. Additional allosteric inhibition is effected by high ornithine concentrations. No *in vivo* regulatory mechanism is known for GAMT activity. Investigations in chicken embryos have rather shown that suprophysio-

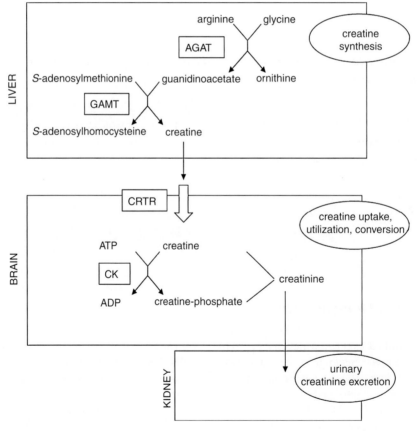

Figure 1 Metabolic pathways for creatine and creatine phosphate, schematically, with organ-specific disjunction of synthesis and usage. AGAT, arginine:glycine amidinotransferase; GAMT, guanidinoacetate methyltransferase; CRTR, creatine transporter (SLC6A8); CK, creatine kinase. (Reproduced, with permission, from Stockler-Ipsiroglu, S., and Salomons, G. S. [2005]. Creatine deficiency syndromes. *In:* "Inborn Metabolic Diseases" [J. Fernandes, et al., eds.]. Springer, Heidelberg, New York [in press].)

logical doses of guanidinoacetate lead to an uncontrolled production of creatine. *In vitro*, GAMT is inhibited allosterically by high *S*-adenosylhomocysteine concentrations.

Creatine synthesis primarily occurs in the kidney and pancreas, which have high AGAT activity, and in liver, which has high GAMT activity. From these organs of synthesis, and from nutritional sources, creatine is transported via the bloodstream to the organs of usage (mainly muscle and brain), where it is taken up by a sodium- and chloride-dependent creatine transporter (SLC6A8).

Intracellular creatine is reversibly converted into the high-energy compound creatine phosphate by the action of creatine kinase (CK). Two dimer forming cytosolic isoforms, brain type B-CK and muscle type M-CK, and two octamer forming mitochondrial isoforms, ubiquitous uMit-CK and sarcomeric sMit-CK, exist. Cytosolic CK transphosphorylates glycolytically generates adenosine triphosphate (ATP) to creatine phosphate, or it uses pooled creatine phosphate of either cytosolic or mitochondrial origin to regenerate ATP in the vicinity of energy expending adenosine triphosphatases (ATPases). Mitochondrial CK forms contact sites in the intermembrane space and catalyzes the phosphate transfer from mitochondrial ATP to creatine in a concerted manner with the ATP/adenosine diphosphate (ADP) translocator (ANT) of the inner mitochondrial membrane. This arrangement facilitates intracellular delivery of high-energy phosphates and buffers energy resources during peak demands.

Creatine and creatine phosphate are nonenzymatically converted into creatinine, with a constant daily turnover of 1.5% of body creatine. Creatinine is mainly excreted in urine and its daily excretion is directly proportional to total body creatine, and in particular to muscle mass (i.e., 20–25 mg/kg/24 h in children and adults, and lower in infants younger than 2 years).

For reviews of creatine metabolism and the creatine kinase system, see references 4 and 5.

B. Genetic Basis of Creatine Deficiency Syndromes

The GAMT and AGAT gene have been mapped to chromosomes 19p13.3 and 15q15.3, respectively. The SLC6A8 gene has been mapped to Xq28. As SLC6A8 deficiency is an X-linked disorder, males are mainly affected, but females may become symptomatic as well. The mother of an index patient is usually a carrier for the mutation. However, there is evidence that *de novo* mutations are responsible for approximately 5% of the affected families [3].

In GAMT deficiency, 14 mutations occurring in 27 patients have been published, including nonsense and missense mutations, splice errors, insertions, deletions, and frameshifts. There is no evidence for a hotspot region, although certain mutations appear to occur more frequently than others (c.327G>A and c.59G>A). Comparison of the homozygous mutations with the type and severity of clinical presentation does not show any genotype–phenotype correlation [2]. In the four AGAT patients identified, only two mutations have been found: a nonsense mutation and a splice error [3; unpublished data]. In the SLC6A8 gene, numerous mutations have been identified, with two of them occurring more frequently (c.319_321delCTT and c.1221_1223delTTC) [3]. Analysis of genotype–phenotype correlation has not been done.

C. Biochemical Changes in Creatine Deficiency Syndrome

All CDSs result in an almost complete lack of creatine in the brain, which is shown *in vivo* by brain H-MRS (Fig. 2).

(A) 22 Months

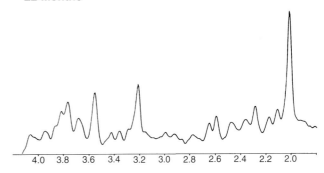

(B) 48 Months

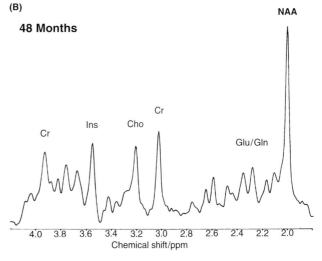

Figure 2 *In vivo* proton magnetic resonance spectroscopy (^{1}H-MRS) of the brain of a patient with cerebral creatine deficiency due to guanidinoacetate methyltransferase (GAMT) deficiency (Patient 1 in reference 2). (**A**) Complete lack of creatine resonance. (**B**) Normalization of creatine spectrum after 6 months of treatment with oral creatine monohydrate. Cr, creatine and creatine phosphate; Ins, myoinositol; Cho, choline; Glu, glutamate; Gln, glutamine; NAA, *N*-acetylaspartate. (Reproduced, with permission, from Mercimek-Mahmutoglu, S., and Stockler-Ipsiroglu, S. [2005]. Clinical and biochemical characteristics of creatine deficiency syndromes. *In:* "Clinical and Molecular Aspects of Defects in Creatine & Polyol Metabolism" [C. Jakobs and S. Stockler-Ipsiroglu, eds.]. SPS Publications, Heilbronn, Germany.)

Interestingly, creatine deficiency in muscle, as compared with the brain, is much less severe. Although skeletal muscular creatine levels have not been investigated extensively in patients with CDS and insights are based on case studies only, muscle creatine levels seem to vary from normal to moderately reduced in all three disorders. No data on creatine levels in heart muscle of affected patients are available. Clinically, there is no evidence of cardiac dysfunction in the patients who underwent cardiological evaluation.

Guanidinoacetate is the main intermediary product in creatine synthesis. In GAMT deficiency, guanidinoacetate accumulates in urine and plasma 2- to 30-fold, and in cerebrospinal fluid (CSF) about 200-fold [2]. The high guanidinoacetate concentrations in GAMT deficiency are due to two mechanisms: First, defective GAMT activity causes an accumulation of guanidinoacetate per se. Second, creatine deficiency and subsequent absence of AGAT repression causes an increased rate of guanidinoacetate synthesis. Creatine's regulatory feedback mechanism on guanidinoacetate synthesis is confirmed in GAMT patients who show a significant decrease (but not normalization) of guanidinoacetate concentrations as a response to oral creatine substitution [2,6]. In AGAT deficiency, guanidinoacetate is low as a result of defective synthesis.

In SLC6A8 deficiency, the main site of metabolic disturbance is the intracellular creatine pool, whereas plasma guanidinoacetate and creatine levels are in the normal range. However, the ratio of creatine to creatinine concentrations in urine is increased, seemingly because of elevated creatine excretion due to impaired tubular reabsorption and a reduced creatinine excretion due to reduced intracellular creatine pools [3].

The biochemical profile of the individual CDS determines its diagnostic algorithms. In practical terms, determination of urinary guanidinoacetate concentrations allows to differentiate between AGAT (low concentration) and GAMT deficiency (high concentration). The determination of the urinary creatine/creatinine ratio has been found to be a useful diagnostic marker of SLC6A8 deficiency. For review of biochemical changes in CDS, see reference 3.

III. Pathophysiology

The pathophysiology of CDS is determined (1) by the disjunction of creatine synthesis and usage, (2) by cellular creatine depletion and the consequential derangement of cellular functions normally supported by creatine, and (3) by changes in cellular concentrations of intermediates of creatine's biosynthetic pathway (mainly guanidinoacetate) and biochemical changes secondary to this. These aspects are discussed in this section in an attempt to explain the clinical symptoms observed in patients with CDS.

A. Disjunction of Creatine Synthesis and Usage

The previous concept of organ-specific disjunction of creatine synthesis (occurring in liver, pancreas, and kidney) and of creatine usage (occurring in brain and muscle) [4] has been revised with regard to the brain by the demonstration of endogenous cerebral creatine synthesis and by *in situ* hybridization and immunohistochemical studies in rat brain, which clearly show that both AGAT and GAMT are expressed in brain tissue. The cellular distribution has still not been resolved completely, with one study reporting ubiquitous neuronal and glial messenger RNA (mRNA) expression of both enzymes [7] and another reporting a distinct pattern of predominantly oligodendroglial and moderate astroglial but no neuronal GAMT protein expression [8]. On the latter basis, the view has been advanced that creatine synthesis and usage in brain may be disjoint on a cellular basis with synthesis in glial cells and neuronal uptake. SLC6A8 is found in capillary endothelial cells, neurons, and oligodendrocytes [7]. The balance between endogenous creatine synthesis and transport across the capillary endothelium *in vivo* is unknown.

These results are important for an understanding of the behavior of cerebral creatine concentrations on replacement therapy in the biosynthetic CDSs and in SLC6A8 deficiency. Oral substitution of creatine in patients with AGAT and GAMT deficiency results in a 60–80% restoration of the cerebral creatine pool. Increasing the dosage or changing the time intervals of creatine administration does not result in any further increase of brain creatine levels [6]. The incomplete restoration of the cerebral creatine pool despite abundant exogenous creatine suggests different intracerebral compartments of creatine metabolism including sites predominated by creatine synthesis (effected by intracerebral GAMT and AGAT expression, e.g., in glial cells) and others with predominantly active creatine transport (affected by SLC6A8 expression) but without synthetic activity (e.g., in neuronal cells). The incomplete correction of cerebral creatine deficiency in GAMT and AGAT deficiency on high-dosage oral creatine supplementation might be one explanation why clinical manifestations are not completely reversible despite treatment. In SLC6A8 deficiency, cerebral creatine levels are very low or undetectable on *in vivo* H-MRS measurement. This suggests that the contribution of intracerebral creatine synthesis to the total pool is rather small. As expected, oral creatine supplementation has no effect on cerebral creatine levels in SLC6A8 deficiency.

B. Intracellular Creatine Depletion

An important role for creatine during embryogenesis is suggested by the appearance of AGAT and GAMT in various organs of rats as early as embryonic day 12.5 [9]. Experiments on cell cultures of developing neurons have revealed that the arginine kinase system, an arthropod analogue of the CK system, is selectively expressed in growth cones during various migrational phases [10], and that the regular growth of cholinergic axons of primary rat brain cells during exposure to ammonia depends on a normalization of the ammonia-induced intracellular depletion of creatine [11]. These results indicate that the creatine/phosphocreatine system is necessary for axonogenesis and that creatine deficiency during this phase of brain development could cause impoverished axonal networks and reduced synaptic density. This might contribute to the pathogenesis in SLC6A8 deficiency where creatine uptake into the brain is defective. Whether this speculation also applies to GAMT and AGAT deficiency remains to be seen; as in these disorders, the transplacental creatine supply will compensate for the defective synthesis in the fetus and might provide at least part of the cerebral creatine pool *in utero*. The detailed analysis of axonal and dendritic morphology and stereological measurements of synaptic density in autopsy material or transgenic animal models of AGAT or SLC6A8 deficiency will be necessary to study this hypothesis. Microstructural anomalies of dendrites and neuronal networks have been suggested as a morphological correlate of mental retardation in Down's syndrome, fragile X syndrome, or infantile autism, and it may also partly underlie mental retardation characteristic of SLC6A8 deficiency.

Creatine has a protective effect on neuronal survival. This has been shown in neuronal cultures during exposure to high glutamate concentrations and in animal models of various neurodegenerative diseases. The mechanisms by which intracellular creatine reduces neuronal death during excitotoxic stress have been studied, including effects on apoptotic triggers and signaling [12], but no definitive results are available. Regardless of the cellular mediators, if the presence of creatine confers advantages for neuronal survival, its absence may tip the balance to neuronal death. This could result in heightened apoptotic rates during the regressive organizational phase of brain development or in an exaggerated apoptotic response to excitotoxic or oxidative stress. The latter may contribute to the brain atrophy and deteriorating behavioral symptoms seen in some SLC6A8 patients.

Creatine deficiency may also cause a disruption of cellular energy homeostasis. The absence of creatine and phosphocreatine from the cytosol is expected to reduce the buffering capacity for peak ATP demands and the transport capacity for high-energy compounds. The pathological consequences of this are difficult to predict. Because creatine phosphate is partly responsible for a smooth energy transport from mitochondria to ATPases elsewhere in the cell, its lack may resemble mitochondrial dysfunction. Moreover, experimental evidence from cardiomyocyte preparations indicates a role for mitochondrial CK in the regulation of respiratory activity [13]. Mitochondrial CK catalyzes the transphosphorylation of ATP to creatine phosphate in the intermembrane space and, by this, regenerates ADP, which is in turn transported into the mitochondrial matrix by ANT, and partly regulates the availability of ADP for ATP synthase. According to this model, insufficient intracellular creatine hinders the mitochondrial regeneration of ADP, reduces its availability for ATP synthase, and thus limits the respiratory rate. Pathological signal intensities in the globus pallidus reversible with creatine supplementation (Fig. 4) have been reported in single patients with GAMT deficiency. These changes are most likely caused by a disturbed cerebral energy homeostasis. Comparable changes are found in patients with mitochondrial encephalopathies (e.g., cytochrome *c* oxidase deficiency) and in other states of disrupted maintenance of cerebral energy state such as hypoxic-ischemic events in newborns and carbon monoxide intoxication. Observations of patients with GAMT and SLC6A8 deficiencies presenting clinically with signs of myopathy and biochemically with signs of mitochondrial dysfunction (lactic academia and decreased ATP production in muscle biopsy samples) suggest that oxidative phosphorylation may indeed be deregulated in these conditions.

C. Interstitial and Intracellular Accumulation of Substances Other Than Creatine

Guanidinoacetate is a neurotoxic substance that accumulates exclusively in GAMT deficiency. The epileptogenic effects of guanidinoacetate are held to be the main cause for the intractable seizures that occur in about one-third of GAMT patients [2] but have not been observed in patients with AGAT or SLC6A8 deficiency [3]. A clear correlation between seizure activity and guanidinoacetate levels in plasma was shown in one patient who despite creatine replacement therapy continued to have intractable seizures. It was only after her guanidinoacetate levels had been brought down by additional dietary restriction of arginine (which is the rate-limiting substrate for guanidinoacetate synthesis) that her seizures improved significantly (Patient 2 in reference 2).

There is also evidence that guanidinoacetate plays a pathogenetic role in the development of extrapyramidal movement disorder, which occurs in about 50% of GAMT patients and manifests as athetosis, chorea, and/or ataxia [2]. Experiments with murine brain slices indicate that guanidinoacetate partially activates γ-aminobutyric acid A ($GABA_A$) receptors at concentrations in the pathogenetically significant micromolar range. In the presence of GABA, guanidinoacetate may thus have either agonistic or antagonistic effects on $GABA_A$ receptors, depending on the relative concentrations. Chronically increased guanidinoacetate could also cause down-regulation of $GABA_A$ receptors and thus a desensitization of GABAergic neurons. Because basal ganglia rely on an interconnected GABAergic system for inhibitory purposes, in particular projections from the globus pallidus to the substantia nigra pars reticulata, and because they are involved in movement regulation, it has been proposed that this effect could be responsible for the extrapyramidal symptoms in GAMT deficiency [14].

Guanidinoacetate within cells, like creatine, is partly phosphorylated, as has been demonstrated by ^{31}P-MRS studies, and serves as an alternative source of high-energy phosphates in skeletal muscle of GAMT-deficient mice [15], providing a certain degree of compensation for energy buffering and transport. In humans, guanidinoacetate phosphate has been detected magnetic resonance spectroscopically in brain [16] (Fig. 3) and skeletal muscle of GAMT-deficient patients [Patient 10 in reference 2; 17]. Whether this mechanism attenuates the clinical effects of phosphocreatine depletion is an open question.

Changes in guanidino compounds other than guanidinoacetate secondary to a deranged creatine metabolism may also play a role in the development of neurological manifestations in GAMT deficiency. Analysis of guanidino compounds in one GAMT patient has shown threefold increased guanidinosuccinate and homoarginine levels in CSF, and the excretion of guanidinosuccinate, β-guanidinopropionate, and γ-guanidinobutyrate in urine was elevated [6]. The observed changes in CSF might only partly reflect the true intraparenchymal quantitative changes, as the increase of guanidinosuccinate, β-guanidinopropionate, and γ-guanidinobutyrate was found to be much higher in brain homogenates of homozygous

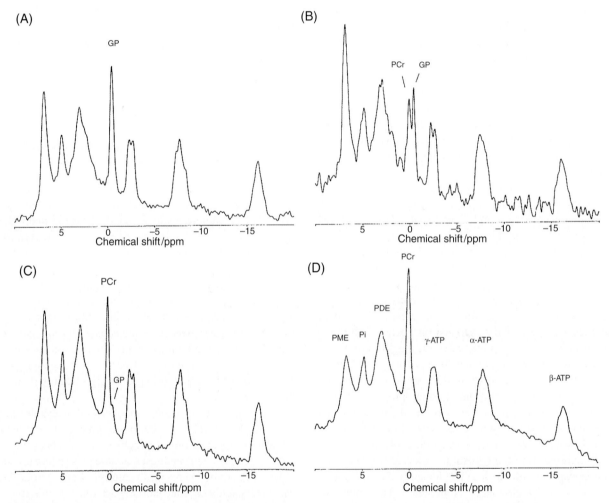

Figure 3 *In vivo* phosphorus magnetic resonance spectroscopy of the brain of a 2-year-old male patient with guanidinoacetate methyltransferase (GAMT) deficiency (nonlocalized spectra, 2.0 Tesla, repetition time 12 seconds): (**A**) 4 weeks of oral L-arginine (data are from the first patient with GAMT deficiency [Patient 1 in reference 2]). At the time, L-arginine was given to facilitate the biochemical diagnosis of a suspected inborn error of creatine metabolism [16]. (**B**) 6 weeks of oral creatine monohydrate and L-arginine substitution followed by (**C**) additional 6 weeks of oral creatine monohydrate (400 mg/kg) substitution alone resulted in a significant increase of PCr and reduction of GP; (**D**) normal brain. ATP, adenosine triphosphate; PDE, phosphodiester; P$_i$, inorganic phosphate; PME, phosphomonoester). (Reproduced, with permission, from Stockler, S. (1995). Guanidinoazetat Methyltransferase mangel. Habilitationsschrift, Georg-August Universitaet, Goettingen, Germany.)

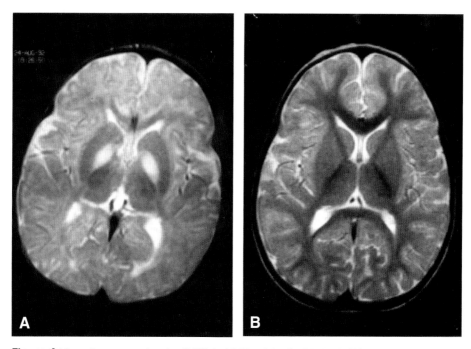

Figure 4 Magnetic resonance imaging (MRI) (axial T2 weighted) of the brain (**A**) of a 12-month-old male patient with GAMT deficiency (Patient 1 in reference 2) and bilaterally high-signal intensities of the globus pallidus; (**B**) resolution of the pathological signal intensities in the same patient at the age of 35 months after 12 months of treatment with oral creatine supplementation. (Reproduced, with permission, from Figura, K., Hanefeld, F., Isbrandt, D., and Stockler-Ipsiroglu, S. [2001]. Guanidinoacetate methyltransferase deficiency. *In:* "The Metabolic and Molecular Bases of Inherited Disease" [C. R. Scriver, et al., eds.], 8th edition, pp. 1897–1908. McGraw-Hill, New York).

GAMT knockout mice [18]. Guanidinosuccinate has been investigated as uremic neurotoxin and has been shown to activate N-methyl-D-aspartate (NMDA) receptors and to induce seizures. β-Guanidinopropionate is a competitive inhibitor of the creatine transporter SLC6A8, and γ-guanidinobutyrate has been ascribed convulsive properties. The contribution of these substances to the pathogenesis of neurological symptoms in GAMT deficiency is unclear, and synergistic effects may contribute to the total neurotoxic load [19]. In addition, the slow increase of cerebral creatine levels in response to oral creatine supplementation in GAMT patients [20] may be partly caused by β-guanidinopropionate or guanidinoacetate, both of which inhibit creatine transport.

Beside guanidinoacetate, AGAT produces ornithine in an equimolar ratio. Extracerebral ornithine is easily metabolized by urea cycle enzymes in the liver, and no changes in plasma ornithine concentrations are seen in GAMT patients. In the brain, the metabolic fate of ornithine is less certain and the effect of increased intracerebral production in GAMT deficiency is unknown. Ornithine decarboxylase catalyzes polyamine synthesis from ornithine in the brain. One may speculate that disturbances of ornithine metabolism could have repercussions for the cellular concentrations of these compounds. Polyamines (putrescine, spermine, spermidine) have effects on DNA structure and apoptosis, cell proliferation, NMDA channels, and signaling pathways. The role of ornithine and related metabolites in the pathogenesis of CDS (mainly GAMT deficiency) has not been investigated.

A high percentage of total body methyl-group transfer is used for creatine synthesis [4]. Therefore, one might expect that in GAMT deficiency, S-adenosylmethionine, which acts as a methyl donor for the GAMT reaction, accumulates. Investigations in one GAMT patient have shown unremarkable plasma concentrations of S-adenosylmethionine and its degradation product adenosine [20], suggesting systemic alternative pathways for accumulating S-adenosylmethionine. Investigation of these compounds in CSF of GAMT patients and in brain homogenates in the GAMT knockout mice will address whether changes in the methylation/remethylation pathway might contribute to the pathophysiology of GAMT deficiency.

So far only few patients have been identified with CDS, and availability of data on biochemical and structural detail in the brain is extremely limited. Genetically engineered animal models have the potential to fill the gaps of understanding the pathophysiology of these disorders. The only animal model available for CDS is the GAMT knockout mouse. This mouse model shares all the biochemical features of human GAMT deficiency, but because it does not show major clinical/neurological effects, its usefulness for the research on pathophysiology of CDS is limited [18].

IV. Natural History

The knowledge of the natural history of CDS stems from the clinical presentations of only a few patients identified with the condition. As most of the known patients are children or young adults who have been followed for a few years only, no definite conclusions can be drawn in terms of disease progression and life expectancy. Furthermore, so far only patients with the most severe clinical phenotypes may have been diagnosed and milder phenotypes might still be underrecognized. Nevertheless, some preliminary data on the disease course are available.

An overview of the known 27 cases with GAMT deficiency shows a broad clinical spectrum from mild to severe mental retardation, occasional to drug-resistant seizures, and in the most severe cases, additional extrapyramidal movement disorder and pathological signal intensities in the basal ganglia on brain magnetic resonance imaging [2]. The clinical onset of the disease has been reported between the ages of 4 months and 3 years. The course of the untreated disease is variable from relative stability after the development of symptoms in infancy to a progression of aggressive and self-injurious behavior or of the extrapyramidal movement disorder. In some patients with severe seizure disorder, high-amplitude theta delta background activity and multifocal spike activity have been observed, but other patterns like hypsarrhythmia have been described as well. In some but not all patients with extrapyramidal movement disorder, pathological signal intensities high in T2 in the globus pallidus have been observed (see Fig. 4). Apart from a nonspecific delay in myelination, no other findings have been reported. Interestingly, in one adult who suffered from pharmacologically intractable epilepsy and severe mental retardation, the MRI scan of the brain was normal, without brain atrophy or any other specific changes [17].

In the three patients with AGAT deficiency described in the literature, presenting features include delayed psychomotor and language development, autistic-like behavioral disturbances, and occasional seizures. Ages at diagnosis were between 2 and 6 years [3]. No gross structural abnormalities have been described in MRI scans of the brain.

The clinical phenotype of SLC6A8 deficiency varies from mild to severe mental retardation associated with speech delay and epileptic seizures. A neuropsychological profile analyzed in four affected Dutch boys from two unrelated families included hyperactive impulsive attention deficit and a semantic-pragmatic language disorder with oral dyspraxia. Dysmorphic features, microcephaly, and brain atrophy have been described as accompanying structural characteristics. Brain atrophy may become more marked during the course of the disease, as may the behavioral symptoms. The diagnosis has been made in both children and adults (age range 2–66 years) [3]. Approximately 50% of heterozygous females have learning and behavioral problems. Skewed X-inactivation may cause pronounced clinical manifestations in these cases, similar to the male phenotype, whereas others remain completely asymptomatic.

V. Prospects

The prominent CNS involvement in all CDS patients has prompted speculation on additional roles of creatine in the brain, for example, as neuromodulator. The incongruent expression patterns of AGAT and GAMT during fetal development in rats may also suggest a specific function of guanidinoacetate or ornithine, both products of AGAT, during development and possibly in the adult brain. The role of intracerebral ornithine synthesis has hardly been explored in this context but may be relevant as substrate generation for ornithine decarboxylase and polyamine synthesis in the brain.

The phenotypic variability of GAMT- and SLC6A8-deficient patients is considerable and does not seem to correlate with the genotype. Different individual susceptibilities to creatine deficiency should be explained by a more comprehensive understanding of the disease. The lack of neurological symptoms in the GAMT knockout mouse model makes the exploration of this issue more difficult. Transgenic animals as models for AGAT and SLC6A8 deficiency may become available and provide an opportunity to investigate the effects of creatine depletion in the absence of guanidino compound toxicity. Importantly, these models may help to devise new and improved therapeutic strategies. In the case of the transporter deficiency, creatine replacement is notoriously ineffective and new therapeutic approaches are urgently needed.

References

1. Stockler, S., Isbrandt, D., Hanefeld, F., Schmidt, B., and von Figura, K. (1996). Guanidinoacetate methyltransferase deficiency: the first inborn error of creatine metabolism in man. *Am. J. Hum. Genet.* **58**, 914–922.
2. Mercimek-Mahmutoglu, S., Stockler-Ipsiroglu, S., Adami, A., Appleton, R., Caldeira Araujo, H., Duran, M., Ensenauer, R., E., F.-A., Garcia, P., Grolik, C., Item, C. B., Leuzzi, V., Marquardt, I., Muhl, A., Saelke-Kellermann, R. A., Schulze, A., Salomons, G. S., Surtees, R. A., van der Knaap, M. S., Vasconcelos, R., Vilarinho, L., Verhoeven, N. M., Wilichowski, E., and Jakobs, C. (2005). Guanidinoacetate methyltransferase deficiency: features and outcome of the first 27 patients. *Neurology* (in press).
3. Stockler-Ipsiroglu, S., and Salomons, G. S. (2005). Creatine deficiency syndromes. *In:* "Inborn Metabolic Diseases" (J. Fernandes, et al., eds.). Springer, Heidelberg, New York (in press).
4. Wyss, M., and Kaddurah-Daouk, R. (2000). Creatine and creatinine metabolism. *Physiol. Rev.* **80**, 1107–1213.
5. Wallimann, T., and Hemmer, W. (1994). Creatine kinase in non-muscle tissues and cells. *Mol. Cell Biochem.* **133-134**, 193–220.

6. Stockler, S., Marescau, B., De Deyn, P. P., Trijbels, J. M., and Hanefeld, F. (1997). Guanidino compounds in guanidinoacetate methyltransferase deficiency, a new inborn error of creatine synthesis. *Metabolism* **46**, 1189–1193.
7. Braissant, O., Henry, H., Loup, M., Eilers, B., and Bachmann, C. (2001). Endogenous synthesis and transport of creatine in the rat brain: an in situ hybridization study. *Brain Res. Mol. Brain Res.* **86**, 193–201.
8. Tachikawa, M., Fukaya, M., Terasaki, T., Ohtsuki, S., and Watanabe, M. (2004). Distinct cellular expressions of creatine synthetic enzyme GAMT and creatine kinases uCK-Mi and CK-B suggest a novel neuron-glial relationship for brain energy homeostasis. *Eur. J. Neurosci.* **20**, 144–160.
9. Braissant, O., Henry, H., Villard, A. M., Speer, O., Wallimann, T., and Bachmann, C. (2005). Creatine synthesis and transport during rat embryogenesis: spatiotemporal expression of AGAT, GAMT and CT1. *BMC Dev. Biol.* **5**, 9.
10. Wang, Y. E., Esbensen, P., and Bentley, D. (1998). Arginine kinase expression and localization in growth cone migration. *J. Neurosci.* **18**, 987–998.
11. Braissant, O., Henry, H., Villard, A. M., Zurich, M. G., Loup, M., Eilers, B., Parlascino, G., Matter, E., Boulat, O., Honegger, P., and Bachmann, C. (2002). Ammonium-induced impairment of axonal growth is prevented through glial creatine. *J. Neurosci.* **22**, 9810–9820.
12. Juravleva, E., Barbakadze, T., Mikeladze, D., and Kekelidze, T. (2005). Creatine enhances survival of glutamate-treated neuronal/glial cells, modulates Ras/NF-kappaB signaling, and increases the generation of reactive oxygen species. *J. Neurosci. Res.* **79**, 224–230.
13. Kay, L., Nicolay, K., Wieringa, B., Saks, V., and Wallimann, T. (2000). Direct evidence for the control of mitochondrial respiration by mitochondrial creatine kinase in oxidative muscle cells *in situ*. *J. Biol. Chem.* **275**, 6937–6944.
14. Neu, A., Neuhoff, H., Trube, G., Fehr, S., Ullrich, K., Roeper, J., and Isbrandt, D. (2002). Activation of GABA(A) receptors by guanidinoacetate: a novel pathophysiological mechanism. *Neurobiol. Dis.* **11**, 298–307.
15. Kan, H. E., Renema, W. K., Isbrandt, D., and Heerschap, A. (2004). Phosphorylated guanidinoacetate partly compensates for the lack of phosphocreatine in skeletal muscle of mice lacking guanidinoacetate methyltransferase. *J. Physiol.* **560**, 219–229.
16. Frahm, J., Requart, M., Helms, G., Haenicke, W., Stoeckler, S., Holzbach, U., and Hanefeld, F. (1994). Creatine deficiency in the brain: A new treatable inborn error of metabolism identified by proton and phosphorus MR spectroscopy *in vivo*. *Proc. Soc. Magn. Res. 2nd Ann. Meet.* **340**.
17. Schulze, A., Bachert, P., Schlemmer, H., Harting, I., Polster, T., Salomons, G. S., Verhoeven, N. M., Jakobs, C., Fowler, B., Hoffmann, G. F., and Mayatepek, E. (2003). Lack of creatine in muscle and brain in an adult with GAMT deficiency. *Ann. Neurol.* **53**, 248–251.
18. Torremans, A., Marescau, B., Possemiers, I., Van Dam, D., D'Hooge, R., Isbrandt, D., and De Deyn, P. P. (2005). Biochemical and behavioural phenotyping of a mouse model for GAMT deficiency. *J. Neurol. Sci.* **231**, 49–55.
19. De Deyn, P. P., D'Hooge, R., Van Bogaert, P. P., and Marescau, B. (2001). Endogenous guanidino compounds as uremic neurotoxins. *Kidney Int. Suppl.* **78**, S77–S83.
20. Stockler, S., Hanefeld, F., and Frahm, J. (1996). Creatine replacement therapy in guanidinoacetate methyltransferase deficiency, a novel inborn error of metabolism. *Lancet* **348**, 789–790.
21. Stockler, S. (1995). Stockler, Guanidinoazetat Methyltransferasemangel. Habilitation-s schrift, Georg-August-Universitaet, Goettingen.

4

Leukodystrophies

Shoji Tsuji

Keywords: *demyelination, hypomyelination, leukodystrophy, leukoencephalopathy, myelin, white matter*

I. Introduction
II. Lipid Metabolism Disorders
III. Myelin Protein Disorders
IV. Organic Acid Disorders
V. Other Leukodystrophies
References

I. Introduction

Leukodystrophies are inherited disorders characterized by progressive breakdown (demyelination) or hypomyelination (dysmyelination) of the white matter of the central nervous system (CNS), which is caused by various mechanisms involving lipid metabolism, myelin proteins, organic acid metabolisms, or other mechanisms. On magnetic resonance imaging (MRI) findings, abnormalities in the white matters are usually detected as the hyperintense signal areas on T2-weighted images. The pathological classification can be divided into (1) demyelinating (confluently destructing) or (2) hypomyelinating (abnormalities in myelin formation). The biochemical basis of leukodystrophies can be classified as (1) lipid metabolism disorders, (2) myelin protein disorders, (3) organic acids disorders, and (4) other causes. Leukodystrophies include X-linked adrenoleukodystrophy, metachromatic leukodystrophy (MLD), globoid cell leukodystrophy (Krabbe's disease), Alexander's disease, Canavan's disease, Pelizaeus-Merzbacher disease (PMD), and other diseases. Advancement in molecular genetics has revealed causative genes for these diseases and further brought better understanding of the pathophysiological processes caused by mutations of the causative genes.

II. Lipid Metabolism Disorders

A. Adrenoleukodystrophy

Adrenoleukodystrophy (ALD) is an X-linked disorder characterized by progressive demyelination of the CNS occasionally accompanied by adrenal insufficiency. ALD is the most common form among the leukodystrophies with the incidence being estimated to be 1 in 20,000–200,000 males [1,2]. Accumulation of esterified cholesterols with very long chain saturated fatty acids (VLCSFAs) with the carbon chain length exceeding 24 in the adrenal cortex and the cerebral white matter with demyelination was first identified as the specific biochemical change. Subsequently, increased content of the VLCSFAs was identified in red blood cells (RBCs), plasma, and leukocytes, establishing the biochemical diagnosis for ALD [3–5].

The gene maps to Xq28, and the causative gene was identified by positional cloning [6], which is referred to as ABCD1. The gene product, ALDP, codes for a member of the adenosine triphosphate-binding cassette (ABC) transporter superfamily. ALDP is localized in the peroxisomal membrane [7], suggesting that ALDP is involved in transport across peroxisomal membranes. The exact functions of the ALDP, however, remain to be elucidated. Decreased activity of very long chain fatty acid acylcoenzyme A (acyl-CoA) synthase has been suggested as the cause for increased (VLCSFAs) [8], but it still remains unclear how defects in ALDP lead to decreased very long chain fatty acid acyl-CoA synthase activities. Furthermore, the mechanisms of demyelination and adrenal insufficiency still remain to be elucidated.

The clinical presentations of ALD are widely variable. Childhood cerebral form is the most common of ALD with the age at onset usually from 5 to 15 years [2,9]. Other clinical forms of ALD include adrenomyeloneuropathy (AMN), adult cerebral form, Addison only, and symptomatic heterozygote.

Initial clinical presentations of childhood cerebral form of ALD are characterized by decline in their school performance, change in their personal character, visual deterioration, and hearing impairment. Progression is rather rapid, and they usually enter a vegetative state within 1–2 years. MRI findings reveal hyperintense signal areas on T2-weighted images, usually in the parietooccipital regions, although in some cases demyelination starts in the frontal lobe.

AMN is characterized by onset in adolescence through adulthood and a noninflammatory distal axonopathy involving mainly the spinal cord long tracts and to a lesser extent peripheral nerves [10,11]. Slowly progressive pyramidal weakness in the lower extremities is the major symptom of AMN, and sensory disturbance, not infrequently associated with sensory levels suggesting myelopathy, is occasionally present. Neurophysiological studies reveal mildly reduced nerve conduction velocities.

The clinical presentations of the adult cerebral form are similar to those of the childhood cerebral form. It should be noted that approximately half of patients with AMN eventually develop cerebral involvement. Once cerebral involvement gets started, the progression is rather rapid [2]. Adult patients with predominant involvement of the cerebellum and brainstem have also been described. Although the frequency is rare, the clinical presentation can be Addison's disease without any neurological abnormalities [12,13].

The female carriers usually do not present with any neurological abnormalities but may develop neurological abnormalities such as spastic paraparesis later in their adulthood [14–16]. Skewed inactivation of the X chromosome has been suggested for symptomatic heterozygotes [17].

The diagnosis of ALD is usually made based on an increase in VLCSFAs in plasma. It should be noted that diagnosis of the carrier state based on the plasma fatty acid analysis is not perfect because 15% of obligate carriers show plasma levels of VLCSFAs overlapping with those of healthy individuals. Mutational analyses of the ABCD1 gene have revealed numerous mutations including missense, nonsense, and deletion/insertion mutations [18,19]. Quite interestingly, there are no clear genotype–phenotype correlations regarding the mutations in the ABCD1 gene, suggesting that other genetic or environmental factors are involved as the modifying factors.

MRI has been proven to be highly sensitive for evaluating CNS involvement (Fig. 1). The scores for quantitative estimation of CNS involvement have been developed, which is quite useful for longitudinal follow-up [20]. Occipital and parietal white matters are preferentially involved in ALD, but inflammatory demyelinating processes can start from frontal white matter in some cases [21,22].

As described earlier, once cerebral involvement gets started, the progression of ALD is rather rapid. For the treatment of ALD, hematopoietic stem cell transplantation (HSCT) has been tried for the treatment of the childhood cerebral form of ALD. Reports on the compiled data on cases with HSCT strongly support its clinical benefit. Clinical progress seems to be halted in 1.0–1.5 years after the transplantation if HSCT is conducted at the early stage

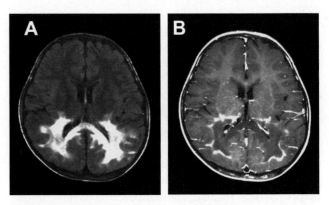

Figure 1 Magnetic resonance imaging (MRI) findings of a patient with childhood adrenoleukodystrophy (ALD). The fluid-attenuated inversion recovery (FLAIR) image shows confluent high-intensity signal areas in the parietooccipital white matter (**A**). Gadolinium (Gd)-DTPA-enhancement is observed at the margins surrounding the high intensity areas (**B**).

with performance IQ more than 80 [23,24]. The efficacy of HSCT for the adult cerebral form awaits compilation of cases with HSCT. Lorenzo's oil has long been tried for the treatment of ALD. Lorenzo's oil contains unsaturated fatty acids (four parts oleic acid and one part erucic acid). Administration of Lorenzo's oil has been shown to decrease plasma VLCSFA levels presumably by competitive inhibition of fatty acid elongation system [25]. A report on the follow-up of 89 asymptomatic patients with ALD treated with Lorenzo's oil suggests some association between the development of MRI abnormalities and the plasma hexacosanoic acid levels [26,27].

B. Globoid Cell Leukodystrophy

Globoid cell leukodystrophy is an autosomal-recessive disorder caused by a deficiency of galactocerebrosidase (galactosylceramide β-galactosidase), which is responsible for the degradation of galactosylceramide, a major constituent of myelin of the CNS and peripheral nervous system (PNS), and psychosine (galactosylsphingosine). It should be noted that galactocerebrosidase is distinct in terms of its substrate specificity from GM_1 β-galactosidase, which is deficient in GM_1 gangliosidosis.

The gene locus maps to 14q21-q31 and the structure of the causative gene coding for galactocerebrosidase has been identified [28]. Although the exact prevalence rate is unknown, globoid cell leukodystrophy is a rare disease.

The major substrate for galactocerebrosidase is galactosylceramide, but this enzyme also has a catalytic activity toward galactosylsphingosine (psychosine). Because psychosine has a potent cytotoxicity, inability to degrade psychosine is assumed to be involved in extensive demyelination or oligodendroglial cell loss. In accordance with this, neuropathological findings are characterized by extensive demyelination associated with large multinucleated macrophages containing galactosylceramide (globoid cells).

In the infantile form, the age at onset is 3–6 months, and the clinical presentations are characterized by irritability, crying, increased muscle tone, and arrest of psychomotor development, followed by rapid psychomotor deterioration, increasing hypertonia, opisthotonus, hyperreflexia, and optic atrophy. In the late infantile variant, the age at onset is 6 months to 3 years. Patients gradually develop ataxia, weakness, spasticity, and later dysarthria [29]. The juvenile form is characterized by later age at onset. The adult form is rare and is characterized by mild dementia and slowly progressive spasticity [30].

MRI findings reveal diffuse hyperintense signals on T2-weighted images. Cerebrospinal protein level is mildly elevated. Nerve conduction studies show uniformly decreased conduction velocities.

Diagnosis of globoid cell leukodystrophy is suggested by the aforementioned clinical features, the MRI findings, and nerve conduction study results. The biochemical diagnosis is made by measuring galactocerebrosidase activities in leukocytes or cultured skin fibroblasts.

Curative treatment is not available, and the treatment is basically symptomatic.

C. Metachromatic Leukodystrophy

MLD is a rare autosomal-recessive disorder with a deficiency in arylsulfatase A that removes sulfate moiety from sulfatide (cerebroside sulfate), a major glycosphingolipid of the CNS and PNS, leading to accumulation of sulfatide [31,32]. Accumulation of sulfatide is observed in oligodendrocytes and macrophages throughout the white matter and in Schwann's cells in the peripheral nerves. Neuropathological findings are characterized by loss of myelin accompanied with numerous periodic acid–Schiff (PAS)–positive macrophages. The macrophages contain numerous granules that stain metachromatically brown with acid cresyl violet. The gene coding for arylsulfatase A maps to 22q13.31-qter.

In the late infantile form, the most common, patients present with psychomotor delay, tonic seizures, visual disturbance with onset at 15–18 months, followed by progressive deterioration. In the juvenile form, patients present with psychomotor deterioration, behavioral abnormalities, gait disturbance, and spastic paraparesis. Adult patients may present with dementia and character changes. Increased tendon reflexes with pathological changes and foot deformity such as pes cavus are frequently observed [33]. Because of the psychiatric symptoms, patients are occasionally misdiagnosed as being schizophrenic.

MRI findings reveal diffuse hyperintense signals on T2-weighted images. Cerebrospinal protein level is mildly elevated. Nerve conduction studies show uniformly decreased conduction velocities that are characteristic findings for MLD and globoid cell leukodystrophy.

Diagnosis is established based on marked decrease in the cerebroside arylsulfatase A activities in leukocytes or cultured skin fibroblasts.

Curative treatment is not available, and the treatment is basically symptomatic.

D. Cerebrotendinous Xanthomatosis

Cerebrotendinous xanthomatosis (CTX) is a rare autosomal-recessive disorder characterized by lipid storage due to a sterol 27-hydroxylase deficiency. The deficiency of sterol 27-hydroxylase enzyme leads to marked increase in cholestanol [34]. Clinical features of CTX are characterized by xanthoma of Achilles tendons, cataracts, and neurological symptoms including dementia, pyramidal weakness, cerebellar ataxia, and psychiatric symptoms [35]. MRI findings typically include a bilateral and almost symmetrical increase of the signal intensity on the T2-weighted images in the cerebellar and periventricular cerebral white matter, the basal ganglia, the dentate nuclei, and the brainstem accompanied with cerebellar and cerebral atrophy [36].

Chenodeoxycholic acid (CDCA) has been demonstrated to suppress abnormal bile acid synthesis [37]. 3-Hydroxy-3-methylglutaryl (HMG) CoA reductase inhibitor and low density lipoprotein (LDL) apheresis have also been employed for the treatment of CTX [38–40].

E. Sjögren-Larsson Syndrome

Sjögren-Larsson syndrome is a rare autosomal-recessive disorder caused by a deficiency of fatty aldehyde dehydrogenase (FALDH) and characterized by congenital ichthyosis, mental retardation, and spastic diplegia or tetraplegia. The primary enzymatic defect in Sjögren-Larsson syndrome is deficient activity in the FALDH [41,42]. The MRI findings are consistent with a leukoencephalopathy.

III. Myelin Protein Disorders

A. Pelizaeus-Merzbacher Disease

PMD is an X-linked recessive hypomyelinating leukoencephalopathy characterized by mutations of the gene coding for proteolipid protein (PLP), a major constituent of myelin protein of the white matter of the CNS [43]. Pathological findings are characterized by lack of myelin in large parts of white matter with the preservation of islands of intact myelin [44].

The classic form of PMD is characterized by psychomotor delay, nystagmus, feeding difficulties, hypotonia, choreoathetosis, ataxia, and spastic tetraparesis with the age at onset in late infantile to juvenile age. The co-natal form of Seitelberger is characterized by severe hypotonia, psychomotor delay, sucking difficulty, and rapid deterioration followed by death in infancy or childhood. The transitional form is intermediate between the classic and the co-natal forms [45].

Molecular genetic analyses of the PLP gene have revealed that mutations in the PLP gene can also lead to clinical presentations of familial spastic paraplegia (SPG2) [46,47]. Interestingly patients with classic PMD and spastic paraplegia have also been identified in the same family [48]. Patients with neonatal spinal muscular atrophy presenting with profound neonatal hypotonia and hyporeflexia and electrophysiological findings consistent with lower motor neuron disease [49], or those with demyelinating peripheral neuropathy [50] have also been reported

to be caused by mutations in the PLP gene. Progressive personality change and a gait disorder in the mid-20s were described in a heterozygous female [51].

MRI findings include diffuse high-intensity areas in the white matter of the hemispheric areas or patchy involvement of hemispheric white matters [52].

The diagnosis is made by mutational analysis of the PLP gene. The majority of classic-onset cases have a duplication of the proteolipid gene [53].

IV. Organic Acid Disorders

A. Canavan's Disease

Canavan's disease is a rare autosomal-recessive disorder associated with macrocephaly, severe mental retardation, and spongiform degeneration of brain. Canavan's disease is caused by a deficiency in aspartoacylase that converts N-acetylaspartic acid (NAA) to aspartic acid and acetic acid [54]. The deficiency of aspartoacylase leads to accumulation of its substrate, NAA, which can be measured by magnetic resonance spectroscopy (MRS). Urinary excretion of NAA is increased to 200 times.

The age at onset is infantile to juvenile, and the clinical presentations are characterized by severe psychomotor delay, hypotonia, seizures, optic atrophy, and macrocephaly [55]. Pathological findings are characterized by spongiform degeneration of the white matter. MRI findings show a diffuse leukoencephalopathy that extends to subcortical white matter [56].

The diagnosis is made based on increased urinary excretion of NAA. Measurement of NAA in the brain using MRS is also an ancillary finding [57].

V. Other Leukodystrophies

A. Alexander's Disease

Alexander's disease is a rare disorder characterized by development of macrocephaly in infancy accompanied by progressive spasticity, psychomotor delay, and seizures. Based on the age at onset, three forms of Alexander's disease, including infantile, juvenile, and adult forms, have been recognized. Patients with onset in infancy present with seizures, megalencephaly, developmental delay, and spasticity. Patients with juvenile onset present with more variable symptoms, including dysphagia and a slowly progressive gait disorder manifested by ataxia and progressive spasticity. Cognitive decline occurs only late in the disease [56,58].

Computed tomographic (CT) and MRI findings are characterized by diffuse white-matter lesions that are predominant in the frontal lobe white matter associated with enlargement of ventricles. Neuropathological findings are characterized by severe degeneration of white matter, astrogliosis, and Rosenthal fibers, as well as abnormal inclusions within astrocytes. Molecular genetic studies have revealed *de novo* heterozygous mutations in the gene coding for glial fibrillary acidic protein (GFAP) [59].

B. Vanishing White-Matter Disease

Leukoencephalopathy with vanishing white matter (VWM), also called *childhood ataxia with CNS hypomyelination*, is a rare autosomal-recessive disorder. The name VWM was proposed because of the characteristic MRI findings, which include increasing areas of diffuse abnormal white matter eventually leading to signal intensities close to those of cerebrospinal fluid (CSF) on all pulse sequences. Neuropathological findings show extensive cystic degeneration of the cerebral white matter with reactive change and a preserved cortex [60,61]. The onset of disease is usually in childhood, and the clinical presentations of this disease are characterized by chronic progressive cerebellar ataxia and spasticity, often accompanied by episodes of rapid deterioration of clinical symptoms following a minor head trauma or an infection. Although most patients with VWM exhibit neurological deterioration in their childhood, patients with adult onset have been described [62]. The patient reported by Ohtake et al. [62] had been well until the age of 40 years, when she became progressively disorganized, forgetful, delusional, and emotionally unstable. By age 52 years, she had developed spastic gait, hyperreflexia, and dementia with defective planning and confabulation. Thus, the clinical spectrum of VWM is wider than previously reported.

Interestingly, VWM has been found to be caused by mutations in any of the five genes encoding subunits of the translation initiation factor eIF2B, including EIF2B1, EIF2B2, EIF2B3, EIF2B4, or EIF2B5 [63–65]. The mechanisms of diffuse involvement of cerebral white matter and focal rarefaction and cystic degeneration of the white matter, however, remain to be elucidated.

C. Cerebral Autosomal-Dominant Arteriopathy Subcortical Infarcts and Leukoencephalopathy

Cerebral autosomal-dominant arteriopathy subcortical infarcts and leukoencephalopathy (CADASIL) is an autosomal-dominant leukoencephalopathy. The age at onset is in late adulthood (mean age at onset is 46 years) and the clinical presentations include multi-infarct dementia and MRI findings with diffuse white-matter abnormalities [66,67]. Recurrent ischemic episodes (transient ischemic attack [TIA] or stroke) were the most frequent presentation. Dementia, gait disturbance, urinary incontinence, and pseudobulbar palsy are the major symptoms. Interestingly, 38% of the patients had a history of migraine, which was classified as migraine with aura in 87% of the cases. The MRI findings show hyperintense areas on T2-weighted images in periventricular and deep white matters, as well as in basal ganglia and brainstem [68].

The gene locus for the causative gene for CADASIL was mapped to 19p13.1 and the causative gene was identified to be Notch3 [69]. Notch3 is implicated as a receptor important in embryonic development. Although the pathophysiological mechanisms caused by mutations in Notch3 are not fully understood, interestingly the mutations that have been identified in patients with CADASIL always involve cysteine residues with either a gain or a loss of cysteine residues, raising the possibility that aberrant dimerization of Notch3, due to abnormal disulfide bridging with another Notch3 molecule or with another protein, may be involved in the pathogenesis of this disorder [70].

Neuropathological studies revealed a nonatherosclerotic small-vessel angiopathy with PAS-positive granular changes in the media and white-matter gliosis, with unremarkable cortex. Involvement of the vascular system in CADASIL is systemic, as demonstrated by the presence of granular osmiophilic material (GOM) on skin biopsy. Although the presence of GOM on skin biopsy is diagnostic for CADASIL, detection GOM can be negative, indicating low sensitivity [71].

References

1. Bezman, L., Moser, A. B., Raymond, G. V., Rinaldo, P., Watkins, P. A., Smith, K. D., Kass, N. E., and Moser, H. W. (2001). Adrenoleukodystrophy: incidence, new mutation rate, and results of extended family screening. *Ann. Neurol.* **49**(4), 512–517.
2. Suzuki, Y., Takemoto, Y., Shimozawa, N., Imanaka, T., Kato, S., Furuya, H., Kaga, M., Kato, K., Hashimoto, N., Onodera, O., and Tsuji, S. (2005). Natural history of X-linked adrenoleukodystrophy in Japan. *Brain Dev.* **27**(5), 353–357.
3. Tsuji, S., Suzuki, M., Ariga, T., Sekine, M., Kuriyama, M., and Miyatake, T. (1981). Abnormality of long-chain fatty acids in erythrocyte membrane sphingomyelin from patients with adrenoleukodystrophy. *J. Neurochem.* **36**(3), 1046–1049.
4. Moser, H. W., Moser, A. B., Frayer, K. K., Chen, W., Schulman, J. D., O'Neill, B. P., and Kishimoto, Y. (1981). Adrenoleukodystrophy: increased plasma content of saturated very long chain fatty acids. *Neurology* **31**(10), 1241–1249.
5. Molzer, B., Bernheimer, H., and Toifl, K. (1981). Fatty acid patterns in brain, fibroblast, leukocyte and body fluid lipids in adrenoleukodystrophy. *Acta. Neuropathol. Suppl. (Berl.)* **7**, 211–214.
6. Mosser, J., Douar, A. M., Sarde, C. O., Kioschis, P., Feil, R., Moser, H., Poustka, A. M., Mandel, J. L., and Aubourg, P. (1993). Putative X-linked adrenoleukodystrophy gene shares unexpected homology with ABC transporters [see Comments]. *Nature* **361**(6414), 726–730.
7. Mosser, J., Lutz, Y., Stoeckel, M. E., Sarde, C. O., Kretz, C., Douar, A. M., Lopez, J., Aubourg, P., and Mandel, J. L. (1994).

The gene responsible for adrenoleukodystrophy encodes a peroxisomal membrane protein. *Hum. Mol. Genet.* **3**(2), 265–271.
8. Hashmi, M., Stanley, W., and Singh, I. (1986). Lignoceroyl-CoASH ligase: enzyme defect in fatty acid beta-oxidation system in X-linked childhood adrenoleukodystrophy. *FEBS Lett.* **196**(2), 247–250.
9. Takemoto, Y., Suzuki, Y., Tamakoshi, A., Onodera, O., Tsuji, S., Hashimoto, T., Shimozawa, N., Orii, T., and Kondo, N. (2002). Epidemiology of X-linked adrenoleukodystrophy in Japan. *J. Hum. Genet.* **47**(11), 590–593.
10. Powers, J. M., DeCiero, D. P., Ito, M., Moser, A. B., and Moser, H. W. (2000). Adrenomyeloneuropathy: a neuropathologic review featuring its noninflammatory myelopathy. *J. Neuropathol. Exp. Neurol.* **59**(2), 89–102.
11. Powers, J. M., DeCiero, D. P., Cox, C., Richfield, E. K., Ito, M., Moser, A. B., and Moser, H. W. (2001). The dorsal root ganglia in adrenomyeloneuropathy: neuronal atrophy and abnormal mitochondria. *J. Neuropathol. Exp. Neurol.* **60**(5), 493–501.
12. Ohno, T., Tsuchida, H., Fukuhara, N., Yuasa, T., Harayama, H., Tsuji, S., and Miyatake, T. (1984). Adrenoleukodystrophy: a clinical variant presenting as olivopontocerebellar atrophy. *J. Neurol.* **231**(4), 167–169.
13. Waragai, M., Takaya, Y., Hayashi, M., Shibata, N., and Kobayashi, M. (1996). MRI of adrenoleukodystrophy involving predominantly the cerebellum and brain stem. *Neuroradiology* **38**(8), 788–791.
14. O'Neill, B. P., Moser, H. W., Saxena, K. M., and Marmion, L. C. (1984). Adrenoleukodystrophy: clinical and biochemical manifestations in carriers. *Neurology* **34**(6), 798–801.
15. Schlote, W., Molzer, B., Peiffer, J., Poremba, M., Schumm, F., Harzer, K., Schnabel, R., and Bernheimer, H. (1987). Adrenoleukodystrophy in an adult female. A clinical, morphological, and neurochemical study. *J. Neurol.* **235**(1), 1–9.
16. Holmberg, B. H., Hägg, E., and Hagenfeldt, L. (1991). Adrenomyeloneuropathy—report on a family. *J. Intern. Med.* **230**(6), 535–538.
17. Maier, E. M., Kammerer, S., Muntau, A. C., Wichers, M., Braun, A., and Roscher, A. A. Symptoms in carriers of adrenoleukodystrophy relate to skewed X inactivation. *Ann. Neurol.* **52**(5), 683–688.
18. Takano, H., Koike, R., Onodera, O., Sasaki, R., and Tsuji, S. (1999). Mutational analysis and genotype-phenotype correlation of 29 unrelated Japanese patients with X-linked adrenoleukodystrophy. *Arch. Neurol.* **56**(3), 295–300.
19. Kemp, S., Pujol, A., Waterham, H. R., van Geel, B. M., Boehm, C. D., Raymond, G. V., Cutting, G. R., Wanders, R. J., and Moser, H. W. (2001). ABCD1 mutations and the X-linked adrenoleukodystrophy mutation database: role in diagnosis and clinical correlations. *Hum. Mutat.* **18**(6), 499–515.
20. Loes, D. J., Hite, S., Moser, H., Stillman, A. E., Shapiro, E., Lockman, L., Latchaw, R. E., and Krivit, W. (1994). Adrenoleukodystrophy: a scoring method for brain MR observations. *AJNR Am. J. Neuroradiol.* **15**(9), 1761–1766.
21. Castellote, A., Vera, J., Vazquez, E., Roig, M., Belmonte, J. A., and Rovira, A. (1995). MR in adrenoleukodystrophy: atypical presentation as bilateral frontal demyelination. *AJNR Am. J. Neuroradiol.* **16(4 Suppl)**:814–815.
22. Larner, A. J. (2003). Adult-onset dementia with prominent frontal lobe dysfunction in X-linked adrenoleukodystrophy with R152C mutation in ABCD1 gene. *J. Neurol.* **250**(10), 1253–1254.
23. Peters, C., Charnas, L. R., Tan, Y., Ziegler, R. S., Shapiro, E. G., DeFor, T., Grewal, S. S., Orchard, P. J., Abel, S. L., Goldman, A. I., Ramsay, N. K., Dusenbery, K. E., Loes, D. J., Lockman, L. A., Kato, S., Aubourg, P. R., Moser, H. W., and Krivit, W. (2004). Cerebral X-linked adrenoleukodystrophy: the international hematopoietic cell transplantation experience from 1982 to 1999. *Blood* **104**(3), 881–888.
24. Shapiro, E., Krivit, W., Lockman, L., Jambaqué, I., Peters, C., Cowan, M., Harris, R., Blanche, S., Bordigoni, P., Loes, D., Ziegler, R., Crittenden, M., Ris, D., Berg, B., Cox, C., Moser, H., Fischer, A., and Aubourg, P. (2000). Long-term effect of bone-marrow transplantation for childhood-onset cerebral X-linked adrenoleukodystrophy. *Lancet* **356**(9231), 713–718.
25. Koike, R., Tsuji, S., Ohno, T., Suzuki, Y., Orii, T., and Miyatake, T. (1991). Physiological significance of fatty acid elongation system in adrenoleukodystrophy. *J. Neurol. Sci.* **103**(2), 188–194.
26. Moser, H. W., Raymond, G. V., Lu, S. E., Muenz, L. R., Moser, A. B., Xu, J., Jones, R. O., Loes, D. J., Melhem, E. R., Dubey, P., Bezman, L., Brereton, N. H., and Odone, A. (2005). Follow-up of 89 asymptomatic patients with adrenoleukodystrophy treated with Lorenzo's oil. *Arch. Neurol.* **62**(7), 1073–1080.
27. Siva, N. (2005). Positive effects with Lorenzo's oil. *Lancet Neurol.* **4**(9), 529.
28. Chen, Y. Q., Rafi, M. A., de Gala, G., and Wenger, D. A. (1993). Cloning and expression of cDNA encoding human galactocerebrosidase, the enzyme deficient in globoid cell leukodystrophy. *Hum. Mol. Genet.* **2**(11), 1841–1845.
29. Kolodny, E. H., Raghavan, S., and Krivit, W. (1991). Late-onset Krabbe disease (globoid cell leukodystrophy): clinical and biochemical features of 15 cases. *Dev. Neurosci.* **13**(4-5), 232–239.
30. Furuya, H., Kukita, Y., Nagano, S., Sakai, Y., Yamashita, Y., Fukuyama, H., Inatomi, Y., Saito, Y., Koike, R., Tsuji, S., Fukumaki, Y., Hayashi, K., and Kobayashi, T. (1997). Adult onset globoid cell leukodystrophy (Krabbe disease): analysis of galactosylceramidase cDNA from four Japanese patients. *Hum. Genet.* **100**(3-4), 450–456.
31. Austin, J., McAfee, D., Armstrong, D., O'Rourke, M., Shearer, L., and Bachhawat, B. (1964). Low sulfatase activities in metachromatic leukodystrophy (MLD). A controlled study of enzymes in 9 living and 4 autopsied patients with MLD. *Trans. Am. Neurol. Assoc.* **89**, 147–150.
32. Stein, C., Gieselmann, V., Kreysing, J., Schmidt, B., Pohlmann, R., Waheed, A., Meyer, H. E., O'Brien, J. S., and von Figura, K. (1989). Cloning and expression of human arylsulfatase A. *J. Biol. Chem.* **264**(2), 1252–1259.
33. Kondo, R., Wakamatsu, N., Yoshino, H., Fukuhara, N., Miyatake, T., and Tsuji, S. (1991). Identification of a mutation in the arylsulfatase A gene of a patient with adult-type metachromatic leukodystrophy. *Am. J. Hum. Genet.* **48**(5), 971–978.
34. Oftebro, H., Björkhem, I., Skrede, S., Schreiner, A., and Pederson, J. I. (1980). Cerebrotendinous xanthomatosis: a defect in mitochondrial 26-hydroxylation required for normal biosynthesis of cholic acid. *J. Clin. Invest.* **65**(6), 1418–1430.
35. Berginer, V. M., Salen, G., and Shefer, S. (1989). Cerebrotendinous xanthomatosis. *Neurol. Clin.* **7**(1), 55–74.
36. Vanrietvelde, F., Lemmerling, M., Mespreuve, M., Crevits, L., De Reuck, J., and Kunnen, M. (2000). MRI of the brain in cerebrotendinous xanthomatosis (van Bogaert-Scherer Epstein disease). *Eur. Radiol.* **10**(4), 576–578.
37. Batta, A. K., Salen, G., Shefer, S., Tint, G. S., and Batta, M. (1987). Increased plasma bile alcohol glucuronides in patients with cerebrotendinous xanthomatosis: effect of chenodeoxycholic acid. *J. Lipid Res.* **28**(8), 1006–1012.
38. Berginer, V. M., and Salen, G. (1994). LDL-apheresis cannot be recommended for treatment of cerebrotendinous xanthomatosis. *J. Neurol. Sci.* **121**(2), 229–232.
39. Mimura, Y., Kuriyama, M., Tokimura, Y., Fujiyama, J., Osame, M., Takesako, K., and Tanaka, N. (1993). Treatment of cerebrotendinous xanthomatosis with low-density lipoprotein (LDL)-apheresis. *J. Neurol. Sci.* **114**(2), 227–230.
40. Ito, S., Kuwabara, S., Sakakibara, R., Oki, T., Arai, H., Oda, S., and Hattori, T. (2003). Combined treatment with LDL-apheresis, chenodeoxycholic acid and HMG-CoA reductase inhibitor for cerebrotendinous xanthomatosis. *J. Neurol. Sci.* **216**(1), 179–182.
41. Rizzo, W. B., Carney, G., and Lin, Z. (1999). The molecular basis of Sjögren–Larsson syndrome: mutation analysis of the fatty aldehyde dehydrogenase gene. *Am. J. Hum. Genet.* **65**(6), 1547–1560.

42. De Laurenzi, V., Rogers, G. R., Hamrock, D. J., Marekov, L. N., Steinert, P. M., Compton, J. G., Markova, N., and Rizzo, W. B. (1996). Sjögren–Larsson syndrome is caused by mutations in the fatty aldehyde dehydrogenase gene. *Nat. Genet.* **12**(1), 52–57.
43. Willard, H. F., and Riordan, J. R. (1985). Assignment of the gene for myelin proteolipid protein to the X chromosome: implications for X-linked myelin disorders. *Science* **230**(4728), 940–942.
44. Sasaki, A., Miyanaga, K., Ototsuji, M., Iwaki, A., Iwaki, T., Takahashi, S., and Nakazato, Y. (2000). Two autopsy cases with Pelizaeus-Merzbacher disease phenotype of adult onset, without mutation of proteolipid protein gene. *Acta Neuropathol. (Berl.)* **99**(1), 7–13.
45. Renier, W. O., Gabreëls, F. J., Hustinx, T. W., Jaspar, H. H., Geelen, J. A., Van Haelst, U. J., Lommen, E. J., and Ter Haar, B. G. (1981). Connatal Pelizaeus-Merzbacher disease with congenital stridor in two maternal cousins. *Acta Neuropathol. (Berl.)* **54**(1), 11–17.
46. Saugier-Veber, P., Munnich, A., Bonneau, D., Rozet, J. M., Le Merrer, M., Gil, R., and Boespflug-Tanguy, O. (1994). X-linked spastic paraplegia and Pelizaeus-Merzbacher disease are allelic disorders at the proteolipid protein locus. *Nat. Genet.* **6**(3), 257–262.
47. Kobayashi, H., Hoffman, E. P., and Marks, H. G. (1994). The rumpshaker mutation in spastic paraplegia. *Nat. Genet.* **7**(3), 351–352.
48. Naidu, S., Dlouhy, S. R., Geraghty, M. T., and Hodes, M. E. (1997). A male child with the rumpshaker mutation, X-linked spastic paraplegia/Pelizaeus-Merzbacher disease and lysinuria. *J. Inherit. Metab. Dis.* **20**(6), 811–816.
49. Kaye, E. M., Doll, R. F., Natowicz, M. R., and Smith, F. I. (1994). Pelizaeus-Merzbacher disease presenting as spinal muscular atrophy: clinical and molecular studies. *Ann. Neurol.* **36**(6), 916–919.
50. Garbern, J. Y., Cambi, F., Lewis, R., Shy, M., Sima, A., Kraft, G., Vallat, J. M., Bosch, E. P., Hodes, M. E., Dlouhy, S., Raskind, W., Bird, T., Macklin, W., and Kamholz, J. (1999). Peripheral neuropathy caused by proteolipid protein gene mutations. *Ann. NY Acad. Sci.* **883**, 351–365.
51. Nance, M. A., Boyadjiev, S., Pratt, V. M., Taylor, S., Hodes, M. E., and Dlouhy, S. R. (1996). Adult-onset neurodegenerative disorder due to proteolipid protein gene mutation in the mother of a man with Pelizaeus-Merzbacher disease. *Neurology* **47**(5), 1333–1335.
52. Nezu, A., Kimura, S., Takeshita, S., Osaka, H., Kimura, K., and Inoue, K. (1998). An MRI and MRS study of Pelizaeus–Merzbacher disease. *Pediatr. Neurol.* **18**(4), 334–337.
53. Sistermans, E. A., de Coo, R. F., De Wijs, I. J., and Van Oost, B. A. (1998). Duplication of the proteolipid protein gene is the major cause of Pelizaeus-Merzbacher disease. *Neurology* **50**(6), 1749–1754.
54. Kaul, R., Gao, G. P., Balamurugan, K., and Matalon, R. (1993). Cloning of the human aspartoacylase cDNA and a common missense mutation in Canavan disease. *Nat. Genet.* **5**(2), 118–123.
55. Matalon, R., Kaul, R., and Michals, K. (1993). Canavan disease: biochemical and molecular studies. *J. Inherit. Metab. Dis.* **16**(4), 744–752.
56. Kaye, E. M. (2001). Update on genetic disorders affecting white matter. *Pediatr. Neurol.* **24**(1), 11–24.
57. Wittsack, H. J., Kugel, H., Roth, B., and Heindel, W. (1996). Quantitative measurements with localized 1H MR spectroscopy in children with Canavan's disease. *J. Magn. Reson. Imaging* **6**(6), 889–893.
58. Gorospe, J. R., Naidu, S., Johnson, A. B., Puri, V., Raymond, G. V., Jenkins, S. D., Pedersen, R. C., Lewis, D., Knowles, P., Fernandez, R., De Vivo, D., van der Knaap, M. S., Messing, A., Brenner, M., and Hoffman, E. P. (2002). Molecular findings in symptomatic and pre-symptomatic Alexander disease patients. *Neurology* **58**(10), 1494–1500.
59. Brenner, M., Johnson, A. B., Boespflug-Tanguy, O., Rodriguez, D., Goldman, J. E., and Messing, A. (2001). Mutations in GFAP, encoding glial fibrillary acidic protein, are associated with Alexander disease. *Nat. Genet.* **27**(1), 117–120.
60. van der Knaap, M. S., Barth, P. G., Gabreëls, F. J., Franzoni, E., Begeer, J. H., Stroink, H., Rotteveel, J. J., and Valk, J. (1997). A new leukoencephalopathy with vanishing white matter. *Neurology* **48**(4), 845–855.
61. Schiffmann, R., and van der Knaap, M. S. (2004). The latest on leukodystrophies. *Curr. Opin. Neurol.* **17**(2), 187–192.
62. Ohtake, H., Shimohata, T., Terajima, K., Kimura, T., Jo, R., Kaseda, R., Iizuka, O., Takano, M., Akaiwa, Y., Goto, H., Kobayashi, H., Sugai, T., Muratake, T., Hosoki, T., Shioiri, T., Okamoto, K., Onodera, O., Tanaka, K., Someya, T., Nakada, T., and Tsuji, S. (2004). Adult-onset leukoencephalopathy with vanishing white matter with a missense mutation in EIF2B5. *Neurology* **62**(9), 1601–1603.
63. Leegwater, P. A., Vermeulen, G., Könst, A. A., Naidu, S., Mulders, J., Visser, A., Kersbergen, P., Mobach, D., Fonds, D., van Berkel, C. G., Lemmers, R. J., Frants, R. R., Oudejans, C. B., Schutgens, R. B., Pronk, J. C., and van der Knaap, M. S. (2001). Subunits of the translation initiation factor eIF2B are mutant in leukoencephalopathy with vanishing white matter. *Nat. Genet.* **29**(4), 383–388.
64. van der Knaap, M. S., Leegwater, P. A., Könst, A. A., Visser, A., Naidu, S., Oudejans, C. B., Schutgens, R. B., and Pronk, J. C. (2002). Mutations in each of the five subunits of translation initiation factor eIF2B can cause leukoencephalopathy with vanishing white matter. *Ann. Neurol.* **51**(2), 264–270.
65. Leegwater, P. A., Pronk, J. C., and van der Knaap, M. S. (2003). Leukoencephalopathy with vanishing white matter: from magnetic resonance imaging pattern to five genes. *J. Child Neurol.* **18**(9), 639–645.
66. Sourander, P., and Wålinder, J. (1977). Hereditary multi-infarct dementia. Morphological and clinical studies of a new disease. *Acta Neuropathol. (Berl.)* **39**(3), 247–254.
67. Dichgans, M., Mayer, M., Uttner, I., Brüning, R., Müller-Höcker, J., Rungger, G., Ebke, M., Klockgether, T., and Gasser, T. (1998). The phenotypic spectrum of CADASIL: clinical findings in 102 cases. *Ann. Neurol.* **44**(5), 731–739.
68. Chabriat, H., Levy, C., Taillia, H., Iba-Zizen, M. T., Vahedi, K., Joutel, A., Tournier-Lasserve, E., and Bousser, M. G. (1998). Patterns of MRI lesions in CADASIL. *Neurology* **51**(2), 452–457.
69. Joutel, A., Corpechot, C., Ducros, A., Vahedi, K., Chabriat, H., Mouton, P., Alamowitch, S., Domenga, V., Cécillion, M., Marechal, E., Maciazek, J., Vayssiere, C., Cruaud, C., Cabanis, E. A., Ruchoux, M. M., Weissenbach, J., Bach, J. F., Bousser, M. G., and Tournier-Lasserve, E. (1996). Notch3 mutations in CADASIL, a hereditary adult-onset condition causing stroke and dementia. *Nature* **383**(6602), 707–710.
70. Joutel, A., Vahedi, K., Corpechot, C., Troesch, A., Chabriat, H., Vayssière, C., Cruaud, C., Maciazek, J., Weissenbach, J., Bousser, M. G., Bach, J. F., and Tournier-Lasserve, E. (1997). Strong clustering and stereotyped nature of Notch3 mutations in CADASIL patients. *Lancet* **350**(9090), 1511–1515.
71. Markus, H. S., Martin, R. J., Simpson, M. A., Dong, Y. B., Ali, N., Crosby, A. H., and Powell, J. F. (2002). Diagnostic strategies in CADASIL. *Neurology* **59**(8), 1134–1138.

5

Parkinson's Disease

Néstor Gálvez-Jiménez, MD, MSc, MHSA, FACP

Keywords: α-synuclein, basal ganglia, dementia, tremor, Lewy body, parkin, parkinsonism, rigidity, tau protein

I. Brief History
II. Epidemiology
III. General Pathology (and Basic Structural Detail)
IV. Pathogenesis
V. Pathophysiology
VI. Pathophysiology of Symptoms
VII. Pharmacology, Biochemistry, Molecular Mechanism
VIII. Natural History of Parkinson's Disease and Its Realationship to Pharmacology
References

I. Brief History

In 1817 the British physician James Parkinson, in a monograph called "An Essay on the Shaking Palsy," described the disease that now bears his name [1]. In this landmark work, he practically described the symptoms and signs of this condition. He noted the progressive nature of the disease and the imperceptible onset of the disease. "...So slight and nearly imperceptible are the first inroads of this malady, and so extremely slow is its progress, that it rarely happens, that the patient can form any recollection of the precise period of its commencement." He described the onset of symptoms as "...a slight sense of weakness, with a proneness to trembling in some particular part; sometimes in the head, but most commonly in one of the hands and arms. These symptoms gradually increase in the part first affected; and at an uncertain period, but seldom in less than twelve months or more, the morbid influence is felt in some other part." He further explains difficulties with gait and postural instability as "...a propensity to bend the trunk forwards, and to pass from a walking to a running pace." He differentiated Parkinson's disease (PD) from other movements such as chorea, athetosis, and alcoholic and senile tremor. He hinted at neuroprotection stating that "...there appears to be sufficient reason for hoping that some remedial process may ere long be discovered, by which, at least, the progress of the disease may be stopped." Bloodletting, vesicatories, ointments resulting in blisters, and bowel cleansing were recommended. The French physician Charcot was the first to use the eponym Parkinson's disease and further clarified its clinical features [2]. During Gower's time (circa 1897), arsenic, Indian hemp, and opium were the treatments of choice. In 1913, Frits Heinrich Lewy described the body that bears his name [3], and by 1919, Tretiakoff localized the disease to the substantia nigra (SN). S. A. Kinnier Wilson described hepatolenticular degeneration in 1912 and correlated alterations of the basal ganglia with rigidity, slowness, and tremor [4]. He postulated the existence of an "extrapyramidal system" different from the "pyramidal" motor system and associated the "extrapyramidal hypertonia" of hepatolenticular degeneration to "paralysis agitans" and the basal ganglia. He further differentiated rigidity from dystonia and had the notion that some cases of torsion dystonia were due to lenticular degeneration in the context of liver cirrhosis. Subsequently, Derek Denny-Brown published a series of monographs on the role of basal ganglia, subthalamic nuclei, and cerebral cortex on motor control, paving the way for the current approach in studying basal ganglia physiology [5–7]. Some of the ideas we hold today on the genesis of movement owe their beginnings to Denny-Brown's initial studies. In the early 1960s, Ehringer and Hornykiewicz [8] demonstrated that dopamine was decreased in the striatum of patients with PD, and Birkmayer and Hornykiewicz [9] first demonstrated improvement of akinesia in patients with PD after the use of intravenous L-dopa. Cotzias et al. [10] further confirmed these observations, when they demonstrated the successful use of oral L-dopa for the treatment of PD. It was not until 1985, when an animal model for the study of parkinsonism became available, that a new era opened for PD research [11].

A. Surgical History

During the twentieth century, many surgical procedures were developed for the treatment of PD and other movement disorders (Fig. 1A). Cortical excision, capsulotomies, caudotomies, ansotomies, pedunculotomies, pyramidotomies, and ramicotomies were performed with variable clinical benefit and high morbidity. In 1952, Cooper accidentally ligated the anterior choroidal artery in a 39-year-old postencephalitic patient with PD who had severe rigidity, tremor, and retrocollis, resulting in improvements in tremor and rigidity [12] (Fig. 1B). Since then, the role of stereotactic procedures for the treatment of parkinsonism has evolved as new neuroimaging procedures and improved intraoperative techniques allow surgeons better anatomical localization for the treatment of the cardinal features of PD. With time, ventrolateral thalamotomy became the procedure of choice for the treatment of tremor, despite evidence for the role of posteroventral pallidotomy by Leksell [13]. He showed that lesions at the point of origin of the ansa lenticularis resulted in a 95% improvement in tremor, rigidity, and bradykinesia for up to 5 years after treatment. During the latter part of the 1960s, L-dopa became established as the treatment of choice for PD and the number of lesioning procedures such as pallidotomies diminished. Advances in intraoperative recording devices and neurophysiological and imaging techniques provided the grounds for the resurgence of functional neurosurgery, especially deep brain stimulation (DBS) [14]. The twenty-first century began with hopes for the potential benefits of embryonic dopamine neurons as treatment for severe PD. The authors confirmed survival of the embryonic transplanted cells, but the clinical benefits were harder to demonstrate, and the potential long-term complications of such procedures were worrisome [15].

II. Epidemiology

PD has an annual prevalence of 48–69 per 1000 persons per year, with an average age at onset of 60 years. In the United States, there are 500,000 to 1 million cases of PD. The annual incidence is between 25 and 55% per year affecting up to 0.3% of the general population. Prevalence increases with age, thus those persons older than 65 years have a PD prevalence of 14.9%, increasing to 29.5% and 52.4% by ages 75 and 85 years, respectively [16]. Only 5% will have symptoms before age 40 years. In a Chinese study, Zhang et al. [17] reported a PD prevalence of 1.7% in those older than 65 years. Considering that China has the world's largest population, the numbers are staggering. There are approximately 1.7 million persons with this neurodegenerative condition in China. The burden to Chinese society is tantamount to the PD burden known to the United States, but in a country with a dissimilar medical system and lack of access to Western-style medical care, this burden is greater than in the United States, where the absolute number of PD cases is smaller. In this study, the number of persons unaware of having the disease was 48%, with a higher incidence in those living in rural areas, suggesting that a large segment of the world population has no access to medical care. In a European prevalence study, the overall prevalence for those older that 65 years was 1.8, increasing by 0.6 for those age 65–69 years and to 2.6 for those 85–89 years [18]. Depending on the criteria used for the diagnosis of PD, the actual incidence of this condition may vary across different age-groups, but Bower et al. [19] demonstrated that using the most accepted diagnostic clinical criteria for PD, the incidence indeed increases with age. Von Campenhausen et al. [20] studied the prevalence and incidence for PD in Europe and found a PD prevalence of 108–257/100,000 and an incidence of 11–19/100,000 per year. When only older age-groups (those older than 60 years) were included, rates of prevalence and incidence were

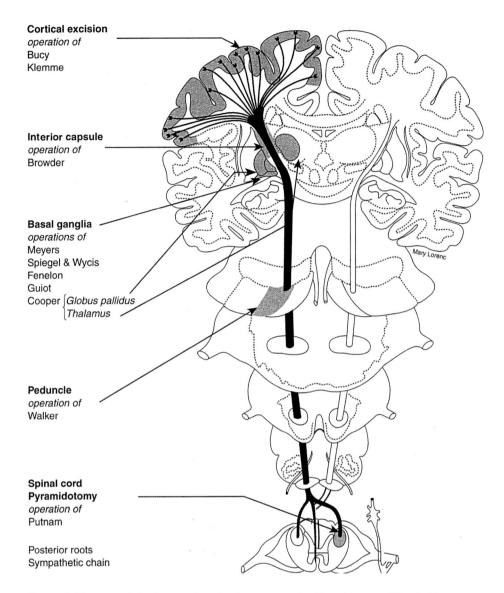

Figure 1 (A) Surgical lesioning procedures for the treatment of parkinsonism now of historical interest.

much higher (1280–1500/100,000 and 346/100,000, respectively). The authors noted that the large variations in rates might have resulted from environmental and genetic factors and diagnostic criteria and differences in methodology, survey design, case-finding strategies, and age distributions.

PD is mostly a disease of middle age and the elderly, so substantial direct and indirect costs may be associated with the illness, including lack of productivity, uncompensated informal care-giving by family members, as well as direct medical care, medications, and supplies. Increased mortality has been a concern for many patients with PD. It has been argued that the introduction of L-dopa decreased the higher than expected mortality rate observed before the use of this medication. Marras et al. [21] addressed survival in patients with PD in a 13-year follow-up study using the Deprenyl and Tocopherol Antioxidative Therapy of Parkinson's disease (DATATOP) cohort. The authors found the mortality rate of patients with PD to be the same as that of the general population. In the same study, the authors found increasing age, male gender, more pack-years smoking, lower Mini-Mental State Examination (MMSE) scores, extensor–plantar responses, and the presence of cardiac, pulmonary, and urological co-morbidities strongly associated with a higher than expected mortality in those patients with PD who have some of these characteristics. Patients with PD who were full-time employees had a better survival. Duration of disease and tremor were not associated with increased mortality. PD-specific variables associated with increased mortality were increased symmetry of parkinsonism, presence and severity of gait dysfunction, rate of progression of symptoms before baseline evaluation, higher Hoehn and Yahr stage, and increased Unified

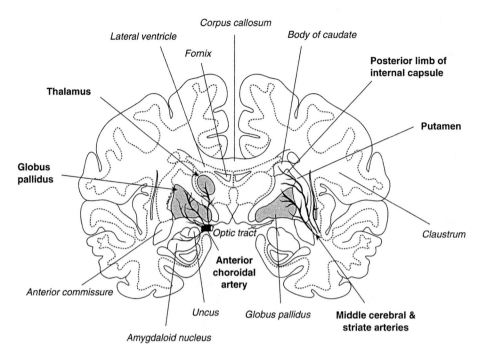

Figure 1 (**B**) In October 1952, Cooper accidentally ligated the anterior choroidal artery (AchA) in a 39-year-old man with severe tremor, rigidity, and retrocollis due to postencephalitic parkinsonism. Cooper planned a left pedunculotomy for the control of tremor by producing a hemiparesis. This is Cooper's own diagram demonstrating a pallidal and, to a lesser degree, a thalamic lesion after the accidental ligation of the AchA. (Modified from 12.)

Parkinson's Disease Rating Scale (UPDRS) total, motor, and activities of daily living (ADL) scores. In addition, those patients whom the investigator thought that PD was the most likely diagnosis did better than those in which the diagnosis was multiple-system atrophy, progressive supranuclear palsy, atypical parkinsonism of uncertain type, and diffuse Lewy body disease (Fig. 2).

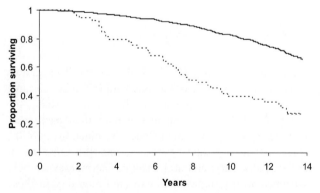

Figure 2 Kaplan-Meier curves of time to death stratified by retention of a high probability of a diagnosis of Parkinson disease (PD). Retained a >60% likelihood of PD: no (dashed line); yes (solid line). Log rank test, $p = 0.0004$. (Reproduced, with permission, from Marras, C., McDermott, M. P., Rochon, P. A., et al. (2005). Survival in Parkinson's Disease. Thirteen-year follow-up of the DATATOP cohort. *Neurology* **64**, 87–93.)

Since 1988, PD with dementia (PDD) has been recognized as perhaps the second most common cause of dementia after Alzheimer's disease (AD), adding to an already escalating late-life health expense with growing burden to family and society. In some clinicopathological series, PDD has been found to be present in between 7 and 11% [22,23]. In a case series from rural Japan, PDD was found in 0.1% of the population, but AD and vascular dementia were found in 2.1 and 1%, respectively.

III. General Pathology (and Basic Structural Detail)[1]

In PD, the gross pathology of the brain is normal. On sectioning of the brainstem, there may be decreased pigmentation evident in SN and locus ceruleus [24] (Fig. 3). In the elderly, age-related cortical and subcortical brain atrophy may be present, especially in those patients with PD with overlapping Alzheimer's pathology. Microscopically, there is marked loss of pigmented neurons in the SN pars compacta (SNc) and locus ceruleus, with phagocytic cells, and a preponderance of gliosis and astrocytosis in advanced cases. A pathological hallmark is the Lewy body, an intracytoplasmic eosinophilic inclusion body 8–30 nm in

---

[1] Relevant structural details are discussed in the sections that follow.

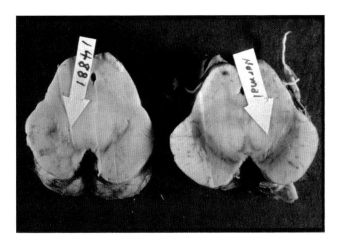

Figure 3 Cross-section of the midbrain demonstrating decreased pigmentation of the substantia nigra in a patient with Parkinson's disease on the left when compared with a normal control (right). (Photograph courtesy Dr. Abraham Lieberman, Professor of Neurology, University of Miami.)

diameter with a core composed of tightly pack intermediate neurofilaments and a surrounding halo chiefly composed of ubiquitin and radially but loosely arranged 7–20 nm intermediated neurofilaments. In addition to ubiquitin and synuclein, these "classic" or "brainstem" Lewy bodies are composed of tau protein, gelsolin, microtubule-associated protein, tubulin, β-amyloid precursor protein, calmodulin-dependent protein kinase II, and hydrolases and proteases associated with ubiquitin-mediated proteolysis [25]. In addition, these intracytoplasmic (classic or brainstem) bodies may be present in the dorsal nucleus of the vagus, intermediolateral cell column of the spinal cord, olfactory tubercle, Meissner plexus, and the so-called cortical Lewy body in cerebral cortex layers V and VI. These poorly defined, polymorphic, hematoxylin-eosin (H&E) staining "cortical Lewy bodies" tend to cluster in the temporal and insular cortices. The halo that characterizes brainstem Lewy bodies is not present in the cortical Lewy body variant.

α-Synuclein and neurofilament members of the cytoskeletal proteins are the most common components of the Lewy body. These bodies accumulate after phosphorylation and fragmentation by the ubiquitin–proteolytic system. The abnormally phosphorylated α-synuclein interacts with membrane lipids, resulting in anomalous membrane fluidity, intracellular and axonal transport alterations, and abnormal cell metabolism, leading to synaptic dysfunction and eventually cell death.

IV. Pathogenesis

The causes that trigger cell death in PD are unknown. *Neuromelanin accumulation, mitochondrial dysfunction, oxidative stress, exposure to iron and other metals, mutations in the α-synuclein gene, trauma,* and *dysfunction of the ubiquitin–proteasome system* have all been implicated in the pathogenesis of PD.

A. Neuromelanin

Neuromelanin is a dark polymer pigment present in greatest quantities in the SN. Why neuromelanin accumulates in the cells is not known, but current hypothesis links the production of melanin to an autoxidation pathway [26]. Disruption of this pathway may lead to accumulation of neuromelanin, leading to increased susceptibility to *oxidative stress*, further compounded by the interaction with *iron* and other metals. Melanocytes, the peripheral cell counterpart of the SN cells, store melanin and have the ability to transfer these granules to other cells, avoiding the continuous exposure to melanin byproducts, which are thought to trigger neuronal cell death. The nigropallidal system is known to have high concentrations of iron; consequently, the interaction between melanin, iron and other metals may lead to *free radical formation* and *mitochondrial dysfunction*.

B. α-Synuclein, Parkin, and the Cytoskeletal Microtubule-Associated Tau Protein

Wild-type α-synuclein, a presynaptic ubiquitous protein, may play a role in oxidative stress and neuroprotection. On the other hand, mutated α-synuclein has been linked to the genesis of PD. In animals overexpressing mutated A30P α-synuclein, abnormal function of the mitochondrial-associated metabolic proteins has been demonstrated by Fai Poon et al. [27], suggesting that mitochondrial respiratory chain dysfunction may play a role in the genesis of PD. Two other α-synuclein mutations, A53T and E46K, have been reported. Lo Bianco et al. [28] showed α-synuclein aggregation and selective dopaminergic neuronal cell loss in a rat model when SN cells were injected with α-synuclein using a lentivirus as vector. Elkon et al. [29] confirmed specific protein-to-protein interaction between α-synuclein and the mitochondrial complex IV enzyme, cytochrome *c* oxidase. Excessive phosphorylation of tau protein prevents cross-linking of microtubules affecting axonal transport, which in turn affects α-synuclein transfer from the cytoplasm to the synaptic axonal bouton. In addition, α-synuclein when interacting with tau protein polymerizes into amyloid fibrils, forming intraneuronal filamentous inclusions, as seen in other neurodegenerative conditions such as AD [30]. Similar findings have been reported in PD [31]. SN extracts containing neurofilaments staining positively for α-synuclein have also been demonstrated by Crowthers et al. [32] and Kirik et al. [33]. In purified Lewy bodies, using rabbit polyclonal antibody to α-synuclein and tau epitopes, Ishizawa et al. [34] were able to demonstrate the presence of both proteins in locus ceruleus, nucleus basalis of Meynert, and dorsal motor nucleus of the vagus. This co-aggregation with tau and α-synuclein further supports the role of both proteins in the pathogenesis of PD.

Mutations in the α-synuclein genes have been shown to be responsible for about 5% of familial cases of PD. Chromosomes 1, 2, 4, 6, and 12 have been recognized as containing 11 loci responsible for the small proportion of familial cases of PD. The protein products of such genes have been identified in some cases (Table 1) and the clinical features in others (Table 2) [35–37].

Age at onset of disease was linked to a region on chromosome 1p overlapping with the PARK10 locus. In the same study, the authors found the human immunodeficiency enhancer-binding protein 3 gene (HIVEP3) to be associated with an increased risk for susceptibility to PD [38]. In relatives of patients with early onset parkinsonism due to PARK2 mutation (Chr6) and PARK6 (Chr1) mutation have been reported to have nigrostriatal dysfunction in positron emission tomography (PET) scanning [39].

C. Neurofilaments

In addition to the microtubule-associated tau protein mutations, other cytoskeletal structures may play a role in neurodegeneration and PD. A single base-pair substitution (G1747A) mutation of the neurofilament M gene has been reported in patients with early onset PD [40], but not with neurofilament M mutation G336S variant or with mutations in neurofilament L [41,42]. Others have found similar

Table 1 Chromosomal Loci and Gene Products in Familial Parkinson's Disease[a]

Locus	Chromosome	Protein Product	Function	Inheritance
PARK1	4q21	α-Synuclein		Dominant
PARK2	6q25.2-q27	Parkin	Ubiquitin ligase	Recessive
PARK3	2p13	Unknown		Dominant
PARK4	4p15	α-Synuclein		Dominant
PARK5	4p14	UCH-L1	Ubiquitin ligase, ubiquitin hydrolase	Dominant
PARK6	1p35	PINK1	Mitochondrial kinase	Recessive
PARK7	1p36	DJ-1	Chaperone, mitochondrial kinase	Recessive
PARK8	12p11.2-q13.1	LRRK2		Dominant
PARK9	1p36	Unknown		Recessive
PARK10	1p32	Unknown		
PARK11	2q22-23	Nurr-1		Dominant

Note: UCH-L1, ubiquitin C-terminal hydrolase-L1; PINK1, PTEN–induced putative kinase-1; LRRK2, leucine-rich repeat kinase-2.

[a] Modified, with permission from Vila, M., and Przedborski, S. (2004). Genetic clues to the pathogenesis of Parkinson's disease. *Nat. Med.* **10(Suppl)**, S58–S62.

[b] New mutation recently described.

Table 2 Clinical Features in Familial Parkinson's Disease[a]

Locus	Clinical Features	Lewy Bodies
PARK1	Early onset, predominantly akinetic-rigid, dementia	Yes
PARK2	Juvenile onset; slow progression; focal dystonia and drug-induced dyskinesias	Absent
PARK3	Dementia and parkinsonism; rapid progression, late onset	Yes
PARK4	Dementia, dysautonomia, postural tremor; early onset with rapid progression	Yes
PARK5	None	Unknown
PARK6	Early onset, slow progression	Unknown
PARK7	Early onset, slow progression; neuropsychiatric side effects	Unknown
PARK8	Late-onset parkinsonism	Absent
PARK9	Juvenile onset; dementia, supranuclear gaze paresis, spasticity and behavioral side effects	Unknown
PARK10	Late onset	Unknown
PARK11	Late onset parkinsonism	Yes

[a] Modified, with permission from Vila, M., and Przedborski, S. (2004). Genetic clues to the pathogenesis of Parkinson's disease. *Nat. Med.* **10(Suppl)**, S58–S62; and from Huang, Y., Cheung, L., Rowe, D., and Halliday, G. (2004). Genetic contributions to Parkinson's disease. *Brain Res. Brain Res. Rev.* **46**, 44–70.

confounding results [43]. Therefore, the extent to which neurofilament abnormalities contribute to the pathogenesis of PD remains to be determined [44].

D. Ubiquitin–Proteasome Pathway

In most neurodegenerative diseases such as PD, neurons have intracellular misfolded aggregated proteins, containing increasing amounts of *ubiquitin*, which is the intracellular signal for proteolysis by the 26S proteasome. Chaperones, also known as *heat shock proteins* (HSPs), attach to proteins to assist in the proper folding and conformation of their molecules. When cells are exposed to stress or heat, proteins denature, resulting in increasing concentrations of HSPs as a compensatory mechanism. The HSP families consist of HSP60, HSP70, and HSP90. The compensatory increase in HSP helps in reshaping and folding of such denatured proteins. Proteasomes have the opposite function of ribosomes helping degrade proteins that are no longer functional, damaged, or misfolded. Ubiquitin first tags proteins destined for degradation; the dysfunction of this ubiquitin–proteasome system is believed by some to be the cause of the intracellular protein accumulation responsible for the neurodegeneration that characterizes PD [45].

Excitotoxicity results from the excessive activation of glutamate receptors. In glutamate-induced neuronal cell death, there is a massive swelling and influx of electrolytes such as Na^+, Cl^-, Ca^{2+}, enzyme activation such

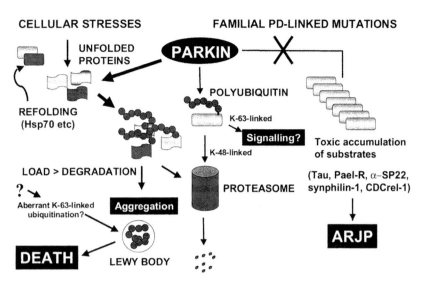

Figure 4 See text for discussion. (Modified, with permission, from Lim, K. L., Dawson, V. L., and Dawson, T. M. [2003]. The cast of molecular characters in Parkinson's disease: felons, conspirators, and suspects. *Ann. NY Acad. Sci.* **991**, 80–92.)

as phospholipase A2, calpain, nitric oxide (NO) synthase, mitochondrial failure, and cytoskeletal and membrane disruption. This slow pathway has been implicated in PD and other neurodegenerative diseases such as AD and motor neuron disease [46,47] (Fig. 4).

V. Pathophysiology

The basal ganglia consist of the caudate and putamen (striatum), as well as the globus pallidus or pallidum. Other structures functionally and/or morphologically related to the basal ganglia are the ventral striatum or accumbens in the region of the rostral substantia innominata, subthalamic nucleus, SN, motor thalamus, cerebellum, pedunculopontine nuclei, and cerebral cortex. The basal ganglia receive glutamatergic excitatory inputs from all areas of the cerebral cortex in a somatotopic fashion, and most of the intrinsic GABAergic inhibitory output from the basal ganglia is via the globus pallidus/SN to motor thalamus and cerebral cortex. The striatum is composed predominantly of medium spiny neurons that utilize γ-amino butyric acid (GABA) as their principal neurotransmitter and that are the main efferents of the striatum (Fig. 5A). The remainder of cells are interneurons with varied morphology such as the large aspiny cholinergic interneurons, medium aspiny somatostatin producing neurons, and small aspiny GABAergic interneurons. Most of the input to the medium spiny neurons occurs in an axodendritic and axosomatic fashion, resulting in modulatory influences mediated by the complex interaction of dopamine, acetylcholine, substance P, somatostatin, GABA, and enkephalin (Fig. 5B). Although it has been suggested that there are segregated parallel loops within and between different structures of the basal ganglia, clinical observations in patients who had undergone surgery for movement disorders have demonstrated overlap between such loops. The medium spiny neurons have large dendritic trees, resulting in similar large convergence and divergent "fields" from widespread neuronal regions. The striatal neurons are distributed in modules based on their immunostaining characteristics into an acetylcholinesterase-rich and acetylcholinesterase-poor matrix and striosome compartments, respectively, with differential distributions among them where the smaller striosomes are embedded in the larger matrix compartment. The striosomes receive cortical connections from the limbic cortex and prefrontal cortex, sending efferents to the SNc, whereas the matrix medium spiny neuron inputs arise from other cortical regions, sending axonal connections to the globus pallidus and SNc. Matrix efferents to the lateral globus pallidus contain enkephalin and GABA, those to the internal segment of the globus pallidus contain GABA and substance P, and the cholinergic neurons are present in both compartments. This is in keeping with the current model of segregated basal ganglia direct and indirect striatopallidal motor loops where the direct pathway to the internal segment of the globus pallidus uses GABA, dynorphin, and substance P, and such medium spiny neurons express the D_1/D_5 dopamine family receptors, and the indirect pathway to the external segment of the globus pallidus uses GABA and enkephalin as neurotransmitters expressing $D_2/D_3/D_4$ dopamine receptors. Cortically mediated impulses are driven by the interaction between the direct striatopallidothalamocortical loop and the indirect striatosubthalamic-pallidothalamocortical loop (Fig. 6A). In parkinsonism, the decrease in dopaminergic innervation results in underactivity of the direct pathway and

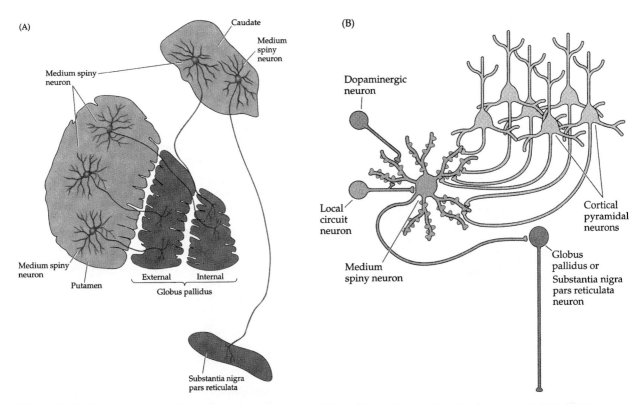

Figure 5 (A) The primary output of the medium spiny cells is to the globus pallidus and to the substantia nigra pars reticularis. (B) Convergent inputs onto a medium spiny neuron from cortical neurons, dopaminergic cells of the substantia nigra, and local circuit neurons. (Modified, with permission, from Purves, D., Agustin, G. J., Fitzpatrick, D., Katz, L. C., LaMantina, A. S., McNamara, J. O., and Williams, S. M., eds. [2001]. Modulation of movement by the basal ganglia. *In* "Neuroscience," 2nd edition, pp. 391–407. Sunderland, Massachusetts.)

simultaneous overactivity of the indirect pathway driving the subthalamic nucleus (STN). The inhibitory GABAergic output from the external segment of the globus pallidus (GPe) to the sensorimotor portion of the STN and soon after to globus pallidus interna/SN pars reticularis complex (GPi/SNr) results in a net inhibitory drive to thalamus and motor cortex. The resultant inhibition of cortically mediated impulses leads to bradykinesia and rigidity (Fig. 6B).

During development, the STN originates from the dorsocaudal part of the lateral hypothalamic cell column alongside both segments of the pallidum. It is located on the inner surface of the peduncular portion of the internal capsule, has reciprocal connections with the pallidum, SN, and receives excitatory afferents from the motor cortex. Efferent projections from the STN exit to both segments of the pallidum and the SNr. STN neurons are tonically active, discharging at an average rate of 20 Hz. Because of a widespread topographical distribution, cortically mediated impulses from the supplementary motor and sensorimotor cortex influence STN activity, resulting in varied differential neuronal activity according to movement, particularly before the initiation of movement. The discharge pattern increases 90% 50 ms before the onset of motor activity, helping control, regulate, or stop ongoing and/or direction of movement. The influence exerted by the STN on the limbic system and processing or transmission of emotional information has been underscored by various reports [48]. Depression, anxiety, emotional lability, mirthful laughter, mania, and aggression have all been reported after DBS [49–59].

A strategically placed lesion in the STN or pallidum results in improvements in bradykinesia and rigidity. The more ventrally located the lesion, the superior bradykinesia response. This probably relates to the destruction of fibers constituting the ansa lenticularis, which represent the major pallidofugal pathway projecting to the ventral anterior and ventrolateral thalamic nuclei. The expected worsening of dyskinesias such as drug-induced chorea or dystonia and other related abnormal involuntary movement as predicted by this model are not observed after surgery [49].

Parkinsonian tremor has a different pathophysiological basis. Synchronization of oscillatory discharges within the basal ganglia–thalamocortical loop may be the cause of tremor. Modulation of these oscillatory discharges using dopaminergic pharmacological manipulation or disruption of these oscillations with destructive or neuromodulatory surgical techniques such as DBS may result in tremor control. These procedures have shown the role that the

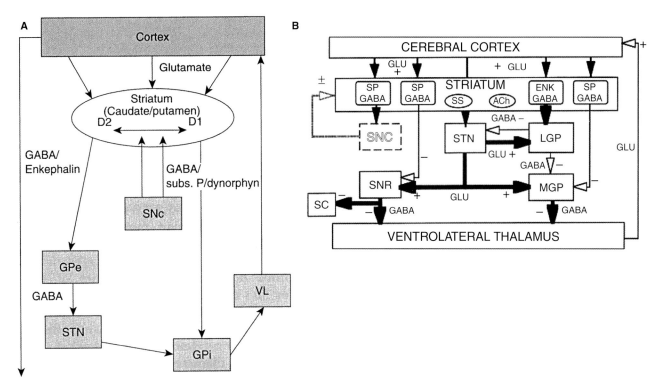

Figure 6 (**A**) Normal Motor Circuit. See text for discussion. (Modified, with permission, from Galvez-Jimenez, N. [2005]. Advances in the surgical treatment of Parkinson's disease: emphasis on pallidotomy and deep brain stimulation. *In:* "Scientific Basis for the Treatment of Parkinson's Disease," 2nd edition [N. Galvez-Jimenez, ed.], pp. 151–168. Taylor & Francis, London). (**B**) Motor circuit alteration in Parkinson's disease. (Young, A. B., and Penney, J. B. [2002]. Biochemical and functional organization of the basal ganglia. *In:* "Parkinson's Disease and Movement Disorders" [J. Jankovic and E. Tolosa, eds.], 4th edition, pp. 1–10. Lippincott Williams & Wilkins, Philadelphia.)

cerebellum and thalamus may play on tremor origin as strategically placed lesions in the ventral intermediate thalamic relay nuclei (VIM), a cerebellar relay center, results in amelioration of tremor. Indeed, PET studies have demonstrated activation of the cerebellar nuclei in patients with some forms of tremor. Transcranial magnetic stimulation (TMS) has shown the additional role the motor cortex may play in tremor genesis, as rest tremor may retune after TMS. What is clear is the existence of a central oscillator driving via a long-loop reflex arc, the tremor seen in PD. In advanced PD, akinesia and gait disturbances may be due to involvement of the pedunculopontine nucleus (PPN) Ch5; other brainstem, spinal, and cortical systems may be responsible for the development of autonomic dysfunction, gait difficulties, and cognitive decline.

The PPN (Ch5) is located in the pontomesencephalic region of the brainstem, embedded in the reticular formation, and fibers of the superior cerebellar peduncle. It has an important anatomical position connecting the basal ganglia and limbic systems with motor execution structures in the pons and spinal cord [60]. It is predominantly a cholinergic nucleus, 60% of PPN cholinergic neurons send efferent fibers to the thalamus, comprising 90% of the tegmentothalamic cholinergic projections co-localizing with GABA, glutamatergic, and aminergic neurons [61]. It receives glutamatergic and GABAergic inputs from and has additional reciprocal connections with mesencephalic dopaminergic neurons, suggesting a key role in fronto-striatal information processing [62]. In addition, the PPN receives dopaminergic input from the SNc and the ventral tegmental area. The main afferents come from the GPi/SNr using pallidotegmental projections, notwithstanding [63–65]. These fibers descend along the ventrolateral border of the red nucleus sweeping dorsolaterally, terminating in the PPN pars disipatus. Furthermore, the PPN receives afferents from the ipsilateral prefrontal and motor cortices, STN, and thalamus. According to Olszewski and Baxter [66], the PPN is composed of approximately 20,000 cholinergic neurons divided into two main regions: PPN pars compacta (PPNc) and pars disipatus (PPNd) based on size and density of its neurons. Moreover, there is a differential distribution of neurons in the PPN according to neurotransmitter expression and firing pattern. Bursting and nonbursting neurons have been identified based on their neurophysiological characteristics. The burst-firing neurons located in the PPNd are glutamatergic, receiving predominantly GABAergic input from the GPi/SNr, and may be involved in initiation of movement, whereas the nonbursting predominantly cholinergic neurons located in the PPNc may be important in the maintenance of gait. Ascending PPN efferents to the thalamus and basal ganglia go along the ansa lenticularis and lenticular fasciculus, whereas descending efferents go to the cerebellum, reticular formation, locus ceruleus, raphe nuclei, SNc, and spinal cord. In patients with PD, the degeneration of dopaminergic neurons may lead to alterations in the PPN–STN and PPN–GPi/SNr loops, resulting in gait disturbances in patients with PD. This is in keeping with the clinical observation of drug-resistant gait disturbances in patients with advanced parkinsonism despite aggressive treatment with dopaminergic therapy.

VI. Pathophysiology of Symptoms

The archetypal basal ganglia disease is PD, characterized by bradykinesia, rigidity, a 4- to 6-Hz rest tremor, and, in advanced disease, gait and postural instability. Most patients have a masklike facial expression, with poverty of movement giving a deliberate character, a stooped posture, and decreased arm swing on the initial most affected side. The tremor usually begins in one hand and may spread to other areas such as the lips, tongue, jaw, or lower extremities. On occasion, the rest tremor may begin as a footrest tremor. In the early stages of the disease, the rest tremor may disappear with movement. Some patients complain that they cannot keep up with their friends when walking, feel clumsy, and slow with the use of their hands, find it difficult to dance or keep a beat, had their signature questioned by banks, or had others complain they cannot hear the patient during conversations. The tremor may have a postural component in more advanced cases. During the examination, some patients may find it difficult to relax, especially the elderly. When relaxed, passive movement of a limb demonstrates increased resistance and at times cogwheel rigidity, representing the presence of tremor superimposed on rigidity. In patients suspected of having PD, the presence of rigidity can be elicited using Froment's maneuver, consisting of having the patient move a limb when the contralateral limb is examined or having the person open and close his hands while the leg is examined. While standing, postural instability can be elicited by pushing or pulling the patient at the shoulders, observing for reactive postural responses and rescue reactions. When the patients are asked to get out of a low chair, righting reflexes may be assessed [67].

With degeneration of the nigrostriatal pathway, as well as the lack of dopaminergic stimulation to the cyclic adenosine monophosphate (cAMP) D_1-dependent direct GABAergic–dynorphin pathway and the D_2 indirect GABAergic–enkephalin pathway, there is an overactivity of the GABAergic STN, inhibiting the ventral thalamus and motor cortex, resulting in slowness of movement and increased tone. Although this model does not clearly explain all features of parkinsonism such as tremor, gait, and postural instability, it provides the framework in which new hypotheses and therapeutic modalities may be tested.

The cholinergic nigrospinal PPN pathways are probably responsible for the abnormal gait and righting reflexes in PD and are just beginning to be understood (see preceding discussion). Clinical observations in patients during the pontomedullary phase of central herniation have demonstrated the existence of a locomotor center in humans. In one such report, 0.2–0.5 Hz automatic stepping was noted in two patients after massive hemispheric infarctions [68]. The automatic stepping was believed to have been generated at the most primitive spinal level after release from higher brain modulation [69]. A mesencephalic locomotor center has also been reported after superior collicular transection. Thus, the PPN predominantly cholinergic modulatory projections to the spinal cord, perhaps through reticulospinal pathways, may play a role in the abnormal gait and tone observed in PD.

1. Rest Tremor

A strategically placed lesion in the ventromedian tegmentum of experimental animals results in rest tremor. These lesions involved the nigrostriatal, cerebellofugal (rubrothalamic), and rubroolivary tracts. Only when these combined nigrostriatal and cerebellar lesions occur do tremors develop. PD rest tremor has a frequency of 3–5 Hz, with alternating activation of agonist and antagonist muscles. At times the firing pattern is such that desynchronization occurs giving rise to activation of agonist and antagonist muscles resulting in co-contraction and a more irregular fast frequency tremor. Classic PD rest tremor may be associated with a 6-Hz postural tremor in some advanced cases. Although some details of the anatomical substrate and neurotransmitter dysfunction are known, much of the generation of tremor is not well understood. The selective destruction of the nigrostriatal pathway, as seen in patients stricken with MPTP parkinsonism, results only in bradykinesia and rigidity, suggesting a different mechanism for tremor.

2. Rigidity

Tone may appear increased when mechanical and local factors such as degenerative joint disease, muscle atrophy, and fibrosis exist, especially in the elderly. In addition, some patients are unable to relax. Notwithstanding, the increased tone observed in PD is due to enhanced long-latency stretch (polysynaptic) reflexes, influenced by the abnormal downstream modulation from the basal ganglia and sensorimotor cerebral cortex. The prominence of the long-latency stretch reflexes in PD may represent increased excitability of the transcortical sensorimotor pathways. The short-latency stretch (or monosynaptic) reflex is abnormal when there is involvement of the pyramidal tract resulting in spasticity, which is not a feature of PD. Spasticity is velocity dependent, whereas rigidity is stretch or length dependent, as noted during passive movement of a joint. Other possible causes of increased tone in patients with PD include diminished spinal reciprocal inhibition, increased activity of the nucleus gigantocellularis, reduced excitability of the nucleus reticularis pontis caudalis, abnormal cutaneous reflexes, and abnormal muscle spindle activity, especially the gamma neurons.

3. Bradykinesia

In parkinsonism, the overactive STN inhibits the motor thalamus and cortex, resulting in slowness of movements or bradykinesia. According to Sohn and Hallett [70] and Hallett et al. [71], abnormally slow movement in PD may be separated into the time it takes to perform an ongoing movement and measured as "prolongation of movement time" and akinesia or failure of initiation of movement and measured as "reaction time" [72]. Many paradigms have been devised to study and measure these two phenomena. Reaction time paradigms, the time it takes to respond and perform a movement after stimulus, correlate with akinesia or onset of movement. Prolongation of movement time, or the time it takes from initiation of movement to complete the movement, correlates with bradykinesia. At the bedside, prolongation of movement time manifests as bradykinesia. Alterations in simultaneous and sequential movements and failure to energize muscles up to the level necessary to complete a movement have all been determined to play a role in the genesis of bradykinesia.

In patients with PD, there is inhibition of the dorsolateral prefrontal cortex, supplementary motor area, and primary motor area. Ceballos-Baumann et al. [73] have conclusively demonstrated correction of these alterations after surgery using fluorodeoxyglucose-PET (Fig. 7). In addition, using transcranial magnetic cortical stimulation, correction of excessive movement-related cortical neuronal discharges was demonstrated in patients with PD after surgery [74]. Ren et al. [75] reported lengthening of the cortical stimulation silent period after surgery, suggesting a decrease in the activation of the cortical motor inhibitory circuits, resulting in symptom improvement.

VII. Pharmacology, Biochemistry, Molecular Mechanism

Dysfunction of the monoaminergic neurons responsible for the cholinergic, dopaminergic, and serotonergic systems may result in disorders such as PD and other parkinsonian syndromes, depression, and dementia. These neurons are characterized by long branched axons co-localizing in the brainstem (pons and midbrain) and basal forebrain (SN, locus ceruleus, and nucleus basalis of Meynert). The

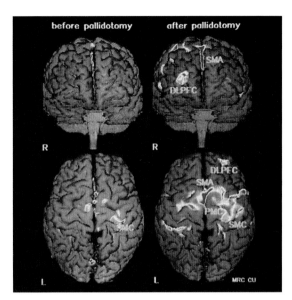

Figure 7 Before pallidotomy, a patient with Parkinson's disease had a fluorodeoxyglucose (FDG)–positron emission tomography (PET) scan demonstrating decreased activation in the dorsolateral prefrontal (DLPF) cortex, supplementary motor area (SMA), primary motor cortex, (PMC) and sensorimotor cortex (SMC), as expected based on the current model of basal ganglia physiology (see text for discussion). After surgery (right) there is a marked increase in the uptake of FDG-PET, suggesting "correction" of these alterations after surgery. (Modified, with permission, from Ceballos-Baumann, A. O., Obeso, J. A., Vitek, J. L., et al. [1994]. Restoration of thalamocortical activity after posteroventral pallidotomy in Parkinson's disease. *Lancet* **344**, 814.)

SN provides 70% of the dopaminergic innervation to the brain. The serotonergic system localizes in the dorsal raphe nuclei, the noradrenergic system in the locus ceruleus, and the cholinergic system in the basal forebrain region–nucleus basalis of Meynert. The locus ceruleus contains about 30,000 neurons and the dorsal raphe nuclei approximately 250,000 [76]. Therefore, these systems are at a disadvantage, as any degenerative process will lead to a rapid loss of neurons despite compensatory mechanism, such as neuronal hypertrophy, when compared with other brain regions such as the cerebral cortex, which contains a large number of neurons. The dopaminergic and cholinergic systems are more restricted in their distribution, leading to more specific symptoms such as parkinsonism or cognitive decline. These systems exert their function via metabotropic receptors.

A. Cholinergic System

The cholinergic neurons, predominantly localized in the substantia innominata, are divided into eight neuronal groups, Ch1–Ch8, and are chiefly located in the basal cholinergic nuclei such as the nucleus basalis of Meynert, nucleus of the diagonal band, and septal area. Ch4 localizes in the nucleus basalis of Meynert innervating the amygdala, and the pedunculopontine (Ch5) and dorsolateral tegmental nuclei (Ch6) innervate the cerebral cortex through the thalamus. Acetylcholine is synthesized from choline and acetylcoenzyme A, a reaction catalyzed by choline acetyltransferase (ChAT). Although the striatum is highly dense in dopamine and acetylcholine, it receives widespread cholinergic input from Ch4 to Ch6, most of the striatal acetylcholine localizes in the type II large aspiny interneurons, and the SN and pallidum receive their cholinergic innervation from Ch5 and Ch6. The amygdala and entorhinal cortex have dense cholinergic innervation. Ch1–Ch4 extends from the reticular activating system to the limbic system and thalamus influencing level of alertness, sleep–wake cycle, memory, and cognition. The main effect of the cholinergic innervation is the reduction in K^+ conductance increasing the susceptibility of these neurons to excitatory as well as inhibitory influences via GABAergic interneurons. With degeneration of Ch5 and Ch6, there is widespread cholinergic denervation of the cerebral cortex. In patients with PDD, there is a 70% reduction in cholinergic innervation, predominantly from degeneration of the nucleus basalis of Meynert. In this regard, the decreased cholinergic innervation seen in patients with PD may be similar to that seen in patients with AD. Approximately 30% of patients with AD may develop extrapyramidal symptoms, namely rigidity and bradykinesia. Using N-methyl-4-[^{11}C] piperidyl acetate PET, Shinotoh et al. [77] demonstrated a 17% reduction of cerebral cortex ChAT activity in patients with PD and a 38% reduction of thalamic ChAT activity in patients with progressive supranuclear palsy (PSP). The authors concluded that there is a significant differential reduction of cholinergic innervation to the cerebral cortex in patients with PD and PSP via the thalamus. They further concluded that PET might differentiate between these two conditions. Using radioenzymatic assay for ChAT activity in autopsied brains, Masliah et al. [78] demonstrated similar cortical cholinergic denervation in patients with parkinsonism–dementia complex in Guam. In these patients, the distribution of cholinergic denervation was present predominantly in the midfrontal, inferior parietal, and temporal cortices.

B. Dopaminergic System

The dopaminergic neurons of the SNc and ventral tegmental area project to the striatum, amygdala, hippocampus, and prefrontal cortex via long highly branched axons with collateral connections to the SN and pallidum. These axons are rich in presynaptic vesicles containing levodopa but lacking dopamine β-hydroxylase (DOH), the enzyme responsible for the conversion of dopamine to noradrenaline. Hydroxylation of tyrosine into

Tyrosine→L-Dopa →Dopamine →Noradrenaline →Adrenaline
    ↑   ↑   ↑   ↑
Tyrosine OH* DOPA Dopamine Phenilethanolamine
            Decarboxilase β-OH N-transferase

*OH = hydroxilase

Figure 8 Dopamine and catecholamine pathways.

levodopa by tyrosine hydroxylase is the rate-limiting step in the production of dopamine (Fig. 8). Levodopa is converted into dopamine by the aromatic amino acid decarboxylase (AAD), an enzyme present in central and peripheral tissues, responsible for some of the side effects observed when therapeutic doses of dopamine are given for the treatment of the parkinsonian syndrome.

Before the introduction of the AAD inhibitor α-methyldopa hydrazine (carbidopa), high doses of levodopa were needed to penetrate the blood–brain barrier and obtain striatal therapeutic concentrations. Dopamine is stored in presynaptic vesicles via an energy-dependent membrane transporter. Once an action potential activates a synaptobrevin/calmodulin/calcium complex, the presynaptic vesicle anchors to the inner cell membrane and dopamine is released. In most noradrenergic neurons, dopamine is converted into noradrenaline by hydroxylation of the lateral chain by the enzyme dopamine β-hydroxylase. The absence of the enzyme dopamine β-hydroxylase in the presynaptic vesicle allows for the storage of dopamine for further release.

Dopamine effects are regulated via an energy Na-dependent dopamine reuptake transport system and degradation by monoamine oxidase (MAO) and catechol-O-methyltransferase (COMT) enzymes (Fig. 9). Cocaine is a presynaptic dopamine transporter blocker resulting in increasing concentrations of synaptic dopamine, especially in the accumbens and prefrontal cortex. Amphetamines both block the presynaptic dopamine transporter and increase the release of dopamine at the presynaptic level. Studies in knockout mice for the dopamine transporter have shown an increase in the firing frequency of the dopaminergic cells, decreased sensibility of the presynaptic receptors with a corresponding increase in sensibility at the postsynaptic receptor and an increase in sensibility to cocaine and amphetamines.

A highly complex interaction exists between the dopaminergic and cholinergic system, maintaining a "balance" between both systems. The striatum is rich in dopamine and acetylcholine innervation. The highly branched medium spiny neurons are the predominant cell type in the striatum having long axons, multiple dendrodendritic and axosomatic synapses, and large synaptic boutons with high concentrations of presynaptic vesicles rich in GABA, acetylcholine, substance P, and dynorphin.

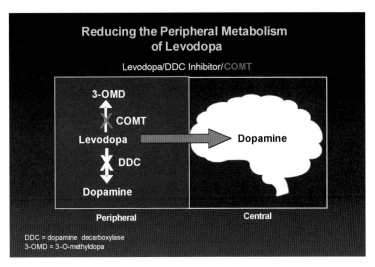

Figure 9 L-Dopa is metabolized into dopamine by dopa-decarboxylase (DCC) and 3-ortho-methyldopa (3-OMD). The peripheral blockade of DDC by carbidopa results in fewer less peripheral side effects and the use of lower doses of the drug. Blocking peripheral catechol-O-methyltransferase (COMT) allows for a more "physiological" plasma level of L-dopa and a longer half-life.

C. Serotonergic System

The raphe nucleus is responsible for most of the serotonergic cortical and striatal innervation. Tryptophan, the precursor of serotonin, is first converted into 5-hydroxytryptophan, by the rate-limiting step catalyzed by tryptophan hydroxylase, and promptly transformed into serotonin by AAD. In addition, serotonin release depends on the neuronal firing pattern of the raphe nuclei and the concentration and interaction between Ca^{2+} and calmodulin. A decrease in neuronal firing pattern results in a corresponding reduction of serotonin release. With *in situ* hybridization, it has been demonstrated a differential distribution of these enzymes where tryptophan-hydroxylase is present only in serotonergic neurons and AAD present in both serotonergic and catecholaminergic cells. Most serotonergic receptors are coupled to G proteins and are excitatory. Receptors $5\text{-}HT_{1B}$, $5\text{-}HT_{1D}$, $5\text{-}HT_{1E}$, $5\text{-}HT_{2A}$, $5\text{-}HT_{2C}$, $5\text{-}HT_4$, and $5\text{-}HT_6$ are present in the basal ganglia, and $5\text{-}HT_{2A}$ and $5\text{-}HT_{2C}$ are present in greatest concentration in the striatum. In PD, there is a 50% reduction of raphe neurons, with a corresponding decrease of serotonergic innervation of the cortex and striatum.

D. Histaminergic System

Histamine concentration in the putamen, SNc, and globus pallidus increases in patients with PD when compared with that seen with other parkinsonian syndromes such as multiple system atrophy [79]. The significance of these findings is unknown, but histamine may be implicated in the motor and behavioral alterations in PD. Histamine binds to four types of G-protein–coupled transmembrane receptors (H_1–H_4) with possible modulatory influences at the dopamine D_2 family of receptors. Most of the histaminergic afferents originate in the tuberomammillary hypothalamic nuclei, composed of 64,000 neurons, projecting widely to the central nervous system. Of all histamine receptors, H_3 has caught the attention of most researchers because of its unique function as a presynaptic autoreceptor with high expression in the striatum, cerebellum, limbic system, and thalamus, hence the interest as a new therapeutic target in PD [80]. Studies have shown high levels of H_3 autoreceptors in medium-size interneurons, co-localizing with the medium spiny cholinergic striatal neurons resulting in modulation of the direct and indirect nigrostriatal pathways, and dopamine, acetylcholine, and glutamate release [81]. Using *in situ* hybridization, Anichtchik et al. [82] have shown an increase of H_3 receptor expression in the SN, putamen, globus pallidus, and cortex of patients with PD. The authors hypothesized that the increase expression of H_3 receptors results as a consequence of the modulatory effects of dopamine on the expression of messenger RNA (mRNA) in the striatum and SN. The authors further concluded that the modulation of the histaminergic system results in alterations of other neurotransmitter such as GABA, serotonin, acetylcholine, and dopamine. Histamine has been implicated in the regulation of hibernation, circadian rhythm, locomotion, movement, memory, and cognition. Also, Goodchild et al. [83] have shown the presence of H_3 receptors in the striatonigral projection neurons to the direct and indirect pathways, further supporting the influence histamine may have over the activity of these motor circuits.

During aging, there is a decrease in histamine receptor mRNA levels, particularly H_1, H_2, and H_3, especially in the cortex, hypothalamus, hippocampus, and medulla in 3-month-old rats. In addition, H_2 levels are decreased in the cerebellum and pons [84]. We can hypothesize that the activation of the histaminergic system with its resultant inhibition of dopamine release may be partly responsible for the motor and behavioral alterations seen in PD. This will await additional study results.

VIII. Natural History of Parkinson's Disease and Its Relationship to Pharmacology

Treatment options are not the topic of this chapter, but emerging treatment modalities for symptom control in PD have changed the natural history of the disease, especially the development of motor complications. Before the use of L-dopa, most patients died from medical complications such as aspiration pneumonia, deep vein thrombosis, skin infections, decubitus ulcers, poor nutrition, and other causes. PD is a progressive neurological illness resulting in severe morbidity and disability in some patients. Once L-dopa became the mainstay of treatment, the degree of disability and dead decreased, but sadly many patients developed motor complications such as wearing off and dyskinesias. Unfortunately, there is no appreciable change on the rate of progression of the disease even when treatment is instituted. Symptoms are controlled, at times, giving a false sense of reduced progression. Rarely, especially the young patient with PD may express an inclination to discontinue treatment because of a sense of improvement and no need for further therapy. Patients should be educated to avoid the occasional confusion about long-term expectations.

Although many patients exposed to L-dopa may develop motor complications after 5 years, it was not until recently that the natural history of these abnormal involuntary movements became clear [85]. In a 5-year study comparing ropinirole with L-dopa, the onset of dyskinesias was observed as early as 12 months in approximately 10% of patients in the L-dopa group [86]. In a similar study, pramipexole was compared with L-dopa as initial treatment for PD. A total of 301 patients were randomized to either 0.5 mg of pramipexole three times daily with L-dopa–placebo or 25/100 mg three times daily of carbidopa/L-dopa with pramipexole–placebo. The authors observed similar findings as described for ropinirole, namely a decreased reduction of motor complications such as wearing off and dyskinesias in those patients treated with pramipexole when compared with those treated with carbidopa/L-dopa. Side effects such as freezing, somnolence, and leg edema where more common in the pramipexole group. The motor scores as noted in the motor subsection of the UPDRS were better in the L-dopa group than in the pramipexole group. The authors concluded that L-dopa and pramipexole both appear to be reasonable options as initial dopaminergic therapy for PD. As these studies have shown, the motor benefit obtained from L-dopa is greater than that obtained when a dopamine agonist is used. Some have questioned [87] the benefit for the early use of dopamine agonist versus the fear for the development of dyskinesias. They argue that the motor scores, as shown in these studies, are better than those obtained with dopa agonist. Patients should be made aware of such differences.

A double-blind, placebo-controlled study looked at the effect of L-dopa on progression of PD symptoms [88]. A total of 361 patients with early PD were assigned to either a low versus a high L-dopa daily dose for 10 months, followed by a washout period of 2 weeks. The primary outcome was a change on the motor subsection of the UPDRS between baseline and 42 weeks. In addition, 142 patients had an iodine-123–labeled 2-β-carboxymethoxy-3-β-(4-iodophenyl) tropane ([^{123}]β-CIT) dopamine transporter density study as a surrogate marker for neurodegeneration induced by the use of L-dopa. The authors found conflicting results. When patients were exposed to L-dopa, there was a higher decline of β-CIT uptake in those patients treated with L-dopa than placebo. The expectation was that if L-dopa hastens neurodegeneration, the worsening of the uptake of β-CIT would have been correlated with a similar worsening of PD symptoms when patients were treated long term with L-dopa. The reverse was found. Patients in the L-dopa group have improved symptoms even after a washout period of 2 weeks. Those patients receiving the highest L-dopa dose had more dyskinesias, hypertonia, infection, headache, and nausea than those receiving placebo. The authors concluded that L-dopa may slow the progression of the disease or has a prolonged effect on symptoms. In contrast, the neuroimaging data suggest that L-dopa either accelerates the loss of nigrostriatal dopamine nerve terminals or that its pharmacological effects modify the dopamine transporter. Therefore, the authors were unable to determine the potential long-term effects of L-dopa on PD. Ongoing studies will help clarify these inconsistencies. A report from Hilker et al. [89] demonstrated continuing progression of symptoms despite clinically effective subthalamic DBS for the treatment of PD.

As these studies have shown, treating PD has changed the natural history of the disease. Once a condition characterized by progressive relentless disability, current treatments allow for better symptom control, but with the emergence of new motor complications not seen until L-dopa was introduced. This has resulted in new treatment algorithms in hopes of decreasing such motor complications. Unfortunately, the goal of neuroprotection has not yet been achieved.

Acknowledgments

The author thanks Vislava T. Tylman MLS, AHIP, chief librarian of the Goldblatt Medical Library, Cleveland Clinic–Florida, for reviewing the accuracy and obtaining those hard-to-find references, and to my dear secretary Ms. Lissette Rojas, BA, for careful review of the grammatical correctness of the manuscript.

References

1. Parkinson, J. (1817). "An Essay on the Shaking Palsy." Sherwood, Neely, & Jones, London.
2. Goetz, C. G. (1987). "Charcot, the Clinician: The Tuesday Lessons." Raven Press, New York.
3. Sweeney, P. J., Lloyd, M. F., and Daroff, R. B. (1997). What's in a name? Dr. Lewey and the Lewy body. *Neurology* **49**, 629–630.
4. Kinkier Wilson, S. A. (1912). Progressive lenticular degeneration: a familial nervous disease associated with cirrhosis of the liver. *Brain* **34**(4), 295–309.
5. Denny-Brown, D. (1966). "The Cerebral Control of Movement." Liverpool University Press, Liverpool.
6. Denny-Brown, D. (1962). "The Basal Ganglia and Their Relationship to Disorders of Movement." Oxford University Press, London.
7. Denny-Brown, D. (1946). "Disease of the Basal Ganglia and Subthalamic Nuclei." Oxford University Press, New York.
8. Ehringer, H., and Hornykiewicz, O. (1960). Veteilung von noradrenalin und dopamine (3-hydroxytyramin) im Gehirn des Menschen und ihr Verhalten bei Erkrankungen des extrapyramidalen system. *Klin. Wochenschr.* **38**, 1236–1239.
9. Birkmayer, W., and Hornykiewicz, O. (1961). Der L-3,4-Dioxyphenylalanin (L-DOPA)–Effekt bei der Parkinson-Akinese. *Wien. Klin. Wochenschr.* **73**, 787–788.
10. Cotzias, G. C. (1971). Levodopa in the treatment of parkinsonism. *JAMA* **218**, 1903–1908.
11. Burns, R. S., LeWitt, P. A., Ebert, M. H., Pakkenberg, H., and Kopin, I. J. (1985). The clinical syndrome of striatal dopamine deficiency. Parkinsonism induced by 1-methyl-4-phenyl-1,2,3,6-tetrahydropydine (MPTP). *N. Engl. J. Med.* **312**, 1418–1421.
12. Cooper, I. S. (1961). "Parkinsonism. Its Medical and Surgical Therapy." Charles C. Thomas, Springfield, IL.
13. Svennilson, E., Torvik, A., Lowe, R., and Leksell, L. (1960). Treatment of parkinsonism by stereotactic thermolesions in the pallidal region. *Acta Psychiatr. Scand.* **35**, 358–377.
14. Lang, A. E., and Lozano, A. M. (1988). Medical progress: Parkinson's disease. *N. Engl. J. Med.* **339**, 1044–1053.
15. Freed, C. R., Greene, P. E., Breeze, R. E., Tsai, W. Y., DuMouchel, W., Kao, R., Dillon, S., Winfield, H., Culver, S., Trojanowski, J. Q., Eidelberg, D., and Fahn, S. (2001). Transplantation of embryonic dopamine neurons for severe Parkinson's disease. *N. Engl. J. Med.* **344**, 710–719.
16. Galvez-Jimenez, N., and Lang, A. E. (2004). The perioperative management of Parkinson's disease revisited. *Neurol. Clin. North Am.* **22**, 367–377.
17. Zhang, Z. X., Roman, G. C., Hong, Z., Wu, C. B., Qu, Q. M., Huang, J. B., Zhou, B., Geng, Z. P., Wu, J. X., Wen, H. B., Zhao, H., and Zahner, G. E. (2005). Parkinson's disease in China: prevalence in Beijing, Xian, and Shanghai. *Lancet* **365**(9459), 595–596.
18. Poewe, W. H., and Wenning, G. K. (1996). The natural history of Parkinson's disease. *Neurology* **47**, S146–S152.
19. Bower, J. H., Maraganore, D. M., McDonnell, S. K., and Rocca, W. A. (2000). Influence of strict, intermediate, and broad diagnostic criteria on the age- and sex-specific incidence of Parkinson's disease. *Mov. Disord.* **15**, 819–825.
20. von Campenhausen, S., Bornschein, B., Wick, R., Botzel, K., Sampalo, C., Poewe, W., Oertel, W., Siebert, U., Berger, K., and Dodel, R. (2005). Prevalence and incidence of Parkinson's disease in Europe. *Eur. Neuropsychopharmacol.* **15**, 473–490.
21. Marras, C., McDermott, M. P., Rochon, P. A., Tanner, C. M., Naglie, G., Rudolph, A., Lang, A. E., and Parkinson Study Group. (2005). Survival in Parkinson's disease. Thirteen-year follow-up of the DATATOP cohort. *Neurology* **64**, 87–93.
22. Yamada, T., Hattori, H., Miura, A., Tanabe, M., and Yamori, Y. (2001). Prevalence of Alzheimer's disease, vascular dementia and dementia with Lewy bodies in a Japanese population. *Psychiatry Clin. Neurosci.* **55**, 21–25.
23. Shergill, S., Mullan, E., D'ath, P., and Katona, C. (1994). What is the clinical prevalence of Lewy body dementia. *Int. Geriatr. Psychiatry* **8**, 571–576.
24. Prayson, R. A. (2001). "Neuropathology Review." Humana Press, Totowa, NJ.
25. Graham, D. I., and Lantos, P. L., eds. (2002). "Greenfield's Neuropathology," 7th edition. Arnold, London.
26. Fedorow, H., Tribl, F., Halliday, G., Gerlach, M., Riederer, P., and Double, K. L. (2005). Neuromelanin in human dopamine neurons: comparison with peripheral melanins and relevance to Parkinson's disease. *Progr. Neurobiol.* **75**, 109–124.
27. Poon, H. F., Frasier, M., Shreve, N., Calabrese, V., Wolozin, B., and Butterfield, D. A. (2005). Mitochondrial associated metabolic proteins are selectively oxidized in A30P α-synuclein transgenic mice—a model of familial Parkinson's disease. *Neurobiol. Dis.* **18**, 492–498.
28. Lo Bianco, C., Ridet, J. L., Schneider, B. L., Deglon, N., and Aebischer, P. (2002). α-Synucleinopathy and selective dopaminergic neuron loss in a rat lentiviral-based model of Parkinson's disease. *Proc. Natl. Acad. Sci. USA* **99**, 10813–10818.
29. Elkon, H., Don, J., Melamed, E., Ziv, I., Shirvan, A., and Offen, D. (2002). Mutant- and wild-type alpha-synuclein interact with mitochondrial cytochrome c oxidase. *J. Mol. Neurosci.* **18**, 229–238.
30. Giasson, B. I., Forman, M. S., Higuchi, M., Golbe, L. I., Graves, C. L., Kotzbauer, P. T., Trojanowski, J. Q., and Lee, V. M. (2003). Initiation and synergistic fibrillization of tau and alpha-synuclein. *Science* **300**, 636–640.
31. Mamah, C. E., Lesnick, T. G., Lincoln, S. J., Strain, K. J., de Andrade, M., Bower, J. H., Ahiskog, J. E., and Rocca, W. A., Farrer, M. J., and Maraganore, D. M. (2005). Interaction of α-synuclein and tau genotypes in Parkinson's disease. *Ann. Neurol.* **57**, 439–443.
32. Crowther, R. A., Daniel, S. E., and Goedert, M. (2000). Characterization of isolated α-synuclein filaments from substantia nigra of Parkinson's disease brain. *Neurosci. Lett.* **292**, 128–130.
33. Kirik, D., Rosenblad, C., Burger, C., Lundberg, C., Johansen, T. E., Muzyczka, N., Mandel, R. J., and Bjorklund, A. (2002). Parkinson-like neurodegeneration induced by targeted overexpression of α-synuclein in the nigrostriatal system. *J. Neurosci.* **22**, 2780–2791.
34. Ishizawa, T., Mattila, P., Davies, P., Wang, D., and Dickson, D. W. (2003). Colocalization of tau and alpha-synuclein epitopes in Lewy bodies. *J. Neuropathol. Exp. Neurol.* **62**, 389–397.
35. Vila, M., and Przedborski, S. (2004). Genetic clues to the pathogenesis of Parkinson's disease. *Nat. Med.* **10(Suppl)**, S58–S62.
36. Huang, Y., Cheung, L., Rowe, D., and Halliday, G. (2004). Genetic contributions to Parkinson's disease. *Brain Res. Brain Res. Rev.* **46**, 44–70.
37. Cookson, M. R. (2005). The biochemistry of Parkinson's disease. *Annu. Rev. Biochem.* **74**, 29–52.
38. Oliveira, S. A., Li, Y. J., Noureddine, M. A., Zuchner, S., Qin, X., Pericak-Vance, M. A., and Vance, J. M. (2005). Identification of risk and age-at-onset genes on chromosome 1p in Parkinson's disease. *Am. J. Hum. Genet.* **77**, 252–264.

39. Khan, N. L., Scherfler, C., Graham, E., Bhatia, K. P., Quinn, N., Lees, A. J., Brooks, D. J., Wood, N. W., and Piccini, P. (2005). Dopaminergic dysfunction in unrelated, asymptomatic carriers of a single parkin mutation. *Neurology* **64**, 134–135.
40. Han, F., Bulman, D. E., Panisset, M., and Grimes, D. A. (2005). Neurofilament M gene in a French-Canadian population with Parkinson's disease. *Can. J. Neurol. Sci.* **32**, 68–70.
41. Perez-Olle, R., Lopez-Toledano, M. A., and Liem, R. (2004). The G336S variant in the human neurofilament-M gene does not affect its assembly or distribution: importance of the functional analysis of neurofilament variants. *J. Neuropathol. Exp. Neurol.* **63**, 759–774.
42. Rahner, N., Holzmann, C., Kruger, R., Schols, L., Berger, K., and Riess, O. (2002). Neurofilament L gene is not a genetic factor of sporadic and familial Parkinson's disease. *Brain Res.* **951**, 82–86.
43. Kruger, R., Fischer, C., Schulte, T., Strauss, K. M., Muller, T., Woitalla, D., Berg, D., Hungs, M., Gobbele, R., Berger, K., Epplen, J. T., Riess, O., and Schols, L. (2003). Mutation analysis of the neurofilament M gene in Parkinson's disease. *Neurosci. Lett.* **351**, 125–129.
44. Julien, J.-P., and Mushynski, W. E. (1998). Neurofilaments in health and disease. *Prog. Nucleic Acid Res. Mol. Biol.* **61**, 1–23.
45. Ross, C. A., and Pickart, C. M. (2004). The ubiquitin-proteasome pathway in Parkinson's disease and other neurodegenerative diseases. *Trends Cell Biol.* **14**, 703–711.
46. Benarroch, E. E. (2005). Neuron-astrocyte interactions: partnership for normal function and disease in the central nervous system. *Mayo Clin. Proc.* **80**(10), 1326–1338.
47. Burke, R. E. (2003). Postnatal developmental programmed cell death in dopamine neurons. *Ann. NY Acad. Sci.* **991**, 69–79.
48. Kuhn, A. A., Hariz, M. I., Silberstein, P., Tisch, S., Kupsch, A., Schneider, G. H., Limousin-Dowsey, P., Yarrow, K., and Brown, P. (2005). Activation of the subthalamic region during emotional processing in Parkinson's disease. *Neurology* **65**, 707–713.
49. Benabid, A. L., Pollak, P., Gross, C., Hoffmann, D., Benazzouz, A., Gao, D. M., Laurent, A., Gentil, M., and Perret, J. (1994). Acute and long-term effects of subthalamic nucleus stimulation in Parkinson's disease. *Stereotact. Funct. Neurosurg.* **62**, 76–84.
50. Bejjani, B. P., Damier, P., Arnulf, I., Thivard, L., Bonnet, A. M., Dormont, D., Cornu, P., Pidoux, B., Samson, Y., and Agid, Y. (1999). Transient acute depression induced by high frequency deep brain stimulation. *N. Engl. J. Med.* **340**, 1476–1480.
51. Houeto, J. L., Mesnage, V., Mallet, L., Pillon, B., Gargiulo, M., du Moncel, S. T., Bonnet, A. M., Pidoux, B., Dormont, D., Cornu, P., and Agid, Y. (2002). Behavioral disorders, Parkinson's disease, and subthalamic stimulation. *J. Neurol. Neurosurg. Psychiatry* **72**, 701–707.
52. Funkiewiez, A., Ardouin, C., Caputo, E., Krack, P., Fraix, V., Klinger, H., Chabardes, S., Foote, K., Benabid, A. L., and Pollak, P. (2004). Long term effects of bilateral subthalamic nucleus stimulation on cognitive function, mood, and behavior in Parkinson's disease. *J. Neurol. Neurosurg. Psychiatry* **75**, 834–839.
53. Okun, M. S., Raju, D. V., Walter, B. L., Juncos, J. L., DeLong, M. R., Heilman, K., McDonald, W. M., and Vitek, J. L. (2004). Pseudobulbar crying induced by stimulation in the region of the subthalamic nucleus. *J. Neurol. Neurosurg. Psychiatry* **75**, 921–923.
54. Krack, P., Kumar, R., Ardouin, C., Dowsey, P. L., McVicker, J. M., Benabid, A. L., and Pollak, P. (2001). Mirthful laughter induced by subthalamic nucleus stimulation. *Mov. Disord.* **16**, 867–875.
55. Romito, L. M., Raja, M., Daniele, A., Contarino, M. F., Bentivoglio, A. R., Barbier, A., Scerrati, M., and Albanese, A. (2002). Transient mania with hypersexuality after surgery for high frequency stimulation of the subthalamic nucleus in Parkinson's disease. *Mov. Disord.* **17**, 1371–1374.
56. Bejjani, B. P., Houeto, J. L., Hariz, M., Yeinik, J., Mesnage, V., Bonnet, A. M., Pidoux, B., Dormont, D., Cornu, P., and Agid, Y. (2002). Aggressive behavior induced by intraoperative stimulation in the triangle of Sano. *Neurology* **59**, 1425–1427.
57. Berney, A., Vingerhoets, F., Perrin, A., Guex, P., Villemure, J. G., Burkhard, P. R., Benkelfat, C., and Ghika, J. (2002). Effect on mood of subthalamic DBS for Parkinson's disease: a consecutive series of 24 patient. *Neurology* **59**, 1427–1479.
58. Funkiewiez, A., Ardouin, C., Krack, P., Fraix, V., Van Blercom, N., Xie, J., Moro, E., Benabid, A. L., and Pollak, P. (2003). Acute psychotropic effects of bilateral subthalamic nucleus stimulation and levodopa in Parkinson's disease. *Mov. Disord.* **18**, 524–530.
59. Galvez-Jimenez, N. (2005). Advances in the surgical treatment of Parkinson's disease: emphasis on pallidotomy and deep brain stimulation. *In* "Scientific Basis for the Treatment of Parkinson's Disease," 2nd edition (N. Galvez-Jimenez, ed.), pp. 151–168. Taylor & Francis, London.
60. Lee, M. S., Rinne, J. O., and Marsden, C. D. (2000). The pedunculopontine nucleus: its role in the genesis of movement disorders. *Yonsei Med. J.* **41**, 167–184.
61. Bevan, M. D., and Bolam, J. P. (1995). Cholinergic, GABAergic, and glutamate-enriched inputs from the mesopontine tegmentum to the subthalamic nucleus in the rat. *J. Neurosci.* **15**, 7105–7120.
62. Steiniger, B., and Kretschmer, B. D. (2003). Glutamate and GABA modulate dopamine in the pedunculopontine tegmental nucleus. *Exp. Brain Res.* **149**, 422–430.
63. Pahapill, P. A., and Lozano, A. M. (2000). The pedunculopontine nucleus and Parkinson's disease. *Brain* **123**, 1767–1783.
64. Scarnati, E., Proia, A., Di Loreto, S., and Pacitti, C. (1987). The reciprocal electrophysiological influence between the nucleus tegmenti pedunculopontinus and the substantia nigra in normal and decorticated rats. *Brain Res.* **423**, 116–124.
65. Garcia-Rill, E. (1991). The pedunculopontine nucleus. *Progr. Neurobiol.* **36**, 363–389.
66. Olszewski, J., and Baxter, D. (1954). "Cytoarchitecture of the Human Brainstem." Lippincott, Philadelphia.
67. Nutt, J. G., Marsden, C. D., and Thompson, P. D. (1993). Human walking and higher-level gait disorders, particularly in the elderly. *Neurology* **43**, 268–279.
68. Hanna, J. P., and Frank, J. I. (1995). Automatic stepping in the pontomedullary stage of central herniation. *Neurology* **45**, 985–986.
69. Miller, S., and Scott, P. D. (1980). Spinal generation of movement in a single limb: functional implications of a model based on the cat. *Prog. Clin. Neurophysiol.* **8**, 263–281.
70. Sohn, Y. H., and Hallett, M. (2005). Basal ganglia, motor control and parkinsonism. *In* "The Scientific Basis for the Treatment of Parkinson's Disease" (N. Galvez-Jimenez, ed.), 2nd edition, pp. 33–51. Taylor & Francis, London.
71. Hallett, M., Daube, J. R., and Mauguuiere, F., eds. (2003). Movement disorders. *In* "Handbook of Clinical Neurophysiology," Vol. 1. Elsevier, Amsterdam.
72. Rothwell, J. (1995). "Control of Human Voluntary Movement," 2nd edition. Chapman & Hall, London.
73. Ceballos-Baumann, A. O., Obeso, J. A., Vitek, J. L., Delong, M. R., Bakay, R., Linazasoro, G., and Brooks, D. J. (1994). Restoration of thalamocortical activity after posteroventral pallidotomy in Parkinson's disease. *Lancet* **344**, 814.
74. Strafella, A., Ashby, P., Lozano, A., and Lang, A. E. (1997). Pallidotomy increases cortical inhibition in Parkinson's disease. *Can. J. Neurol. Sci.* **24**, 133–136.
75. Ren, Y., Zhao, J., and Feng, J. (2003). Parkin binds to α/β tubulin and increases their ubiquitination and degradation. *J. Neurosci.* **23**, 3316–3324.
76. Paxinos, G., and Mai, J. K., eds. (2004). "The Human Nervous System," 2nd edition. NeuroScience Associates, Knoxville, TN.
77. Shinotoh, H., Namba, H., Yamaguchi, M., Fukushi, K., Nagatsuka, S., Iyo, M., Asahina, M., Hattori, T., Tanada, S., and Irie, T. (1999). Positron emission tomographic measurement of acetylcholinesterase activity reveals differential loss of ascending cholinergic systems in

Parkinson's disease and progressive supranuclear palsy. *Ann. Neurol.* **46**, 62–69.

78. Masliah, E., Alford, M., Galasko, D., Salmon, D., Hansen, L. A., Good, P. F., Perl, D. P., and Thal, L., (2001). Cholinergic deficits in the brains of patients with parkinsonism-dementia complex of Guam. *Neuroreport* **12**, 3901–3903.

79. Rinne, J. O., Anichtchik, O. V., Eriksson, K. S., Kaslin, J., Tuomisto, L., Kalimo, H., Roytta, M., and Panula, P. (2002). Increased brain histamine levels in Parkinson's disease but not in multiple system atrophy. *J. Neurochem.* **81**, 954–960.

80. Chazot, P. L., and Hann, V. (2001). Overview: H3 histamine receptor isoforms: new therapeutic targets in the CNS? *Curr. Opin. Investig. Drugs* **2**, 1428–1431.

81. Anichtchik, O. V., Rinne, J. O., Kalimo, H., and Panula, P. (2000). An altered histaminergic innervation of the substantia nigra in Parkinson's disease. *Exp. Neurol.* **163**, 20–30.

82. Anichtchik, O. V., Peitsaro, N., Rinne, J. O., Kalimo, H., and Panula, P. (2001). Distribution and modulation of histamine H3 receptors in basal ganglia and frontal cortex of healthy controls and patients with Parkinson's disease. *Neurobiol. Dis.* **8**, 707–716.

83. Goodchild, R. E., Court, J. A., Hobson, I., Piggott, M. A., Perry, R. H., Ince, P., Jaros, E., and Perry, E. K. (1999). Distribution of histamine H3-receptor binding in the normal human basal ganglia: comparison with Huntington's disease and Parkinson's disease cases. *Eur. J. Neurosci.* **11**, 440–456.

84. Terao, A., Steininger, T. L., Morairty, S. R., and Kilduff, T. S. (2004). Age-related changes in histamine receptor mRNA levels in the mouse brain. *Neurosci. Lett.* **355**, 81–84.

85. Block, G., Liss, C., Reines, S., Irr, J., and Nibbelink, D. (1997). Comparison of immediate-release and controlled release carbidopa/levodopa in Parkinson's disease. A multicenter 5-year study. The CR First Study Group. *Eur. Neurol.* **37**, 23–27.

86. Rascol, O., Brooks, D. J., Korczyn, A. D., DeDeyn, P. P., Clarke, C. E., and Lang, A. E. (2000). A five-year study of the incidence of dyskinesias in patients with early Parkinson's disease who were treated with ropinirole or levodopa. *N. Engl. J. Med.* **342**, 1484–1491.

87. Weiner, W. J., and Factor, S. A. (2000). Ropinirole as compared with levodopa in Parkinson's disease. *N. Engl. J. Med.* **343**, 884–885.

88. The Parkinson Study Group. (2004). Levodopa and the progression of Parkinson's disease. The Parkinson Study Group. *N. Engl. J. Med.* **351**, 2498–2508.

89. Hilker, R., Portman, A. T., Voges, J., Staal, M. J., Burghaus, L., van Laar, T., Koulousakis, A., Maguire, R. P., Pruim, J., de Jong, B. M., Herholz, K., Sturm, V., Heiss, W. D., and Leenders, K. L. (2005). Disease progression continues in patients with advanced Parkinson's disease and effective subthalamic nucleus stimulation. *J. Neurol. Neurosurg. Psychiatry* **76**, 1217–1221.

90. Lim, K. L., Dawson, V. L., and Dawson, T. M. (2003). The cast of molecular characters in Parkinson's disease. felons, conspirators, and suspects. *Ann. NY Acad. Sci.* **991**, 80–92.

91. Purves, D., Agustin, G. J., Fitzpatrick, D., Katz, L. C., LaMantina, A. S., McNamara, J. O., and Williams, S. M., eds. (2001). Modulation of movement by the basal ganglia. *In* "Neuroscience," 2nd edition, pp. 391–407. Sunderland, Massachusetts.

92. Young, A. B., and Penney, J. B. (2002). Biochemical and functional organization of the basal ganglia. *In* "Parkinson's Disease and Movement Disorders" (J. Jankovic and E. Tolosa, eds.), 4th edition, pp. 1–10. Lippincott Williams & Wilkins, Philadelphia.

6

Alzheimer's Disease

Kurt A. Jellinger, MD

Keywords: *amyloid deposits, cholinesterase inhibitors, clinicopathologic relations, pathogenesis, (pre)senile, tau pathology*

I. History and Nomenclature
II. Epidemiology
III. Etiology and Risk Factors
IV. Pathogenesis
V. Neuropathology of AD (Relevant Structural Details)
VI. Pathophysiology
VII. Biochemistry and Molecular Mechanisms
VIII. Explanation of Symptoms in Relation to Pathophysiology
IX. Natural History
X. Management of AD
References

I. History and Nomenclature

The clinical entity known as *Alzheimer's disease* (AD) clearly existed long before 1907, when Alois Alzheimer described the clinical course and changes in the brain of a 55-year-old woman dying after a 4-year history of progressive dementia. In his report of Auguste D., he demonstrated neurofibrillary tangles (NFTs) using the newly developed Bielschowsky silver impregnation method and observed cortical "miliary foci" or senile plaques (SPs) [1], described 15 years later by Blocq and Marinesco. Sections of the brain of Auguste D. have been recovered in Munich, and micrographs from these historic slides confirmed the presence of typical histopathological changes of AD in the neocortex. In the second case, Johann F., described by Alzheimer in 1911 [3], only plaques and no tangles were found, corresponding to the "plaque only" type of AD, and DNA extracted from this material revealed an ApoE ε3/3 genotype (see reference 2). Even after Alzheimer's original description of the pathological nature of the disease, it was decades until the relationships between pathology and clinical features were understood. Dementia of the Alzheimer type or AD summarizes those types of late-life dementia that are not related to Lewy body (LB) disease, frontotemporal and vascular dementias. AD, now recognized as the most common cause of dementia in the elderly, accounting in the Western world between 50 and 80% of dementia patients, has been the focus of intensive research during only the past 3 decades.

II. Epidemiology

The prevalence and incidence of AD increase with advancing age in an exponential fashion until the ninth decade. The prevalence increases from approximately 1% in 65- to 69-year-olds, doubling every subsequent 5 years, to 20–50% in those older than 85 years. The incidence exceeds 1/100 persons per year from age 70 to 80 years and reaches 2/100 persons per year until age 80 years, now affecting 20–30 million people worldwide. The lifetime risk of AD in nondemented 65-year-old men and women are 6.3 and 12%, respectively. The greater risk of women to develop AD is due to greater longevity and estrogen deficits [4].

III. Etiology and Risk Factors

The etiology of AD is unknown, but a number of risk and protective factors are known. Important risk factors are aging and familial disposition/history. Familial forms of AD occur in 7–10%; rare dominant early onset AD (EOAD) is caused by mutations of various genes: amyloid precursor protein (APP) at chromosome 21, presenilin (PS) 1 and 2 (chromosome 14 and 1), enhancing the production of β-amyloid (Aβ) from APP, whereas the gene encoding apolipoprotein E (ApoE) linked to chromosome 19 is a risk factor for late-onset AD (LOAD). Genetic epidemiology has confirmed a 10- to 12-fold increased risk for AD in homozygote carriers of ApoEε4. There are other AD linkage regions on chromosomes 10, 12.9, and other chromosomes [5]. Further risk factors are atherosclerosis, low educational level, head injury, cardiovascular diseases, elevated serum homocysteine and cholesterol levels, and female gender (estrogen deficit). Putative risk factors are nicotine abuse, low socioeconomic status, and episodic depression. Putative protective factors include life-long activity, ApoE ε2, antioxidant substances (vitamins E and C), estrogen replacement, low caloric diet, nonsteroidal anti-inflammatory agents (NSAIDs), and statins.

IV. Pathogenesis

To understand the pathogenesis of AD, it is necessary to determine the sequence of events that occurs over the life of a patient. Amyloid-β(Aβ) peptide is considered seminal to the pathogenesis of AD. Derived from the proteolysis of a large transmembrane glycoprotein precursor, APP, Aβ is released as soluble peptides of 39–46 residues, with Aβ-40 and Aβ-42 as the most abundantly produced isoforms. Its sequence is hydrophobic, explaining the strong tendency of this small protein to self-aggregate and form clusters of fibrils. Aβ-42 can form fibrils more rapidly and at lower concentrations than Aβ-40 cleavage of APP by β- and γ-secretases. The balance between biogenesis versus catabolism of Aβ may be an important factor in the pathogenesis of AD (Fig. 1). The balance is disturbed between overproduction of Aβ through enhancement of β- and γ-secretase activity or an altered conformation of substrate found in some inherited mutations of APP, and failure to adequately degrade or clear Aβ from its intracellular and extracellular compartment in the brain [6]. In LOAD, there are compelling reasons that a failure of degradation or clearance of Aβ from the brain underlies its accumulation (Fig. 2). Clearance of aggregated Aβ is a complex process, for which microglia, macrophages, and bulk flow across the blood–brain barrier are important.

Fine diffuse plaques, initial stages of Aβ deposition with accumulation of nonfibrillary Aβ-42, are followed by deposition of Aβ-40, and oligomerization of both, later evolving into primitive and neuritic plaques (NPs) containing fibrillary Aβ, tau-positive dystrophic neurites, activated glia, and subsequent development of NFTs (see Fig. 2). Whereas soluble Aβ oligomers (protofibrils) and spherical aggregates of Aβ-42 (amyloid spheroids), rather than Aβ fibrils, are neurotoxic and may be the principal cause of synaptotoxicity, nonfibrillar accumulation of tau protein in premature NFTs is followed by fibrillary tau in the cell soma and processes, binding to and destabilizing microtubules, which are important transport structures in the neuron. Misregulation of APP and tau by abnormalities in APP, presenilins, ApoE ε4, and so on, can lead to impairment in fast axonal transport, leading to axonal depletion of critical components, and neurodegeneration (see reference 7). Although it is widely believed that an increase in the production of Aβ is central to the pathogenesis of AD, little is known about the relationship between APP-Aβ and tau changes in sporadic AD, where the pathogenic trigger has not been identified. Both lesions induce tissue damage; their interrelationship is under discussion, but a link between the development of amyloid and NFT pathologies may be activation of caspases by Aβ, which in turn cleaves tau and may initiate or accelerate the development of tangle pathology [8].

The molecular causes of AD are unresolved, but a variety of pathogenic factors are implicated in the complex cascade of events leading to neuronal degeneration: (a) reduced cerebral energy and glucose metabolism; (b) oxidative stress with accumulation of cytotoxic free superoxide radicals and advanced glycation end products (AGEs), lipid

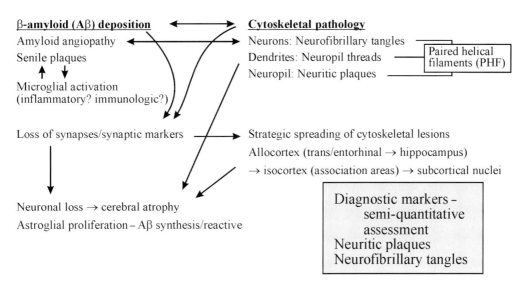

Figure 1 Histopathology of Alzheimer's disease—diagnostic markers.

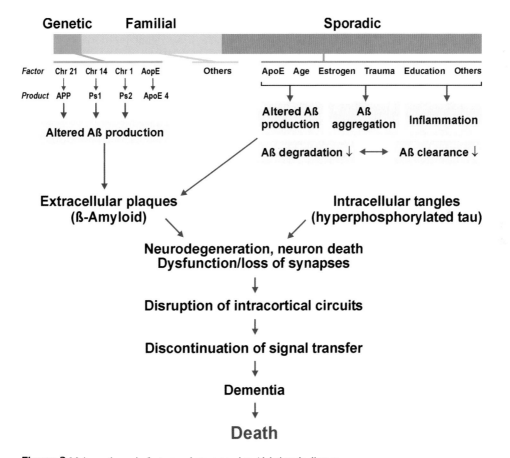

Figure 2 Major pathogenic factors and steps causing Alzheimer's disease.

peroxidation products, and abnormally oxidated cytoskeletal proteins; (c) trace elements such as iron may serve as catalyzers for free radical formation, leading to oxidative modification of tau and Aβ, increased oxidative damage associated with ApoE ε4; (d) reduced defense enzyme activity; (e) antioxidant activity, such as upregulation of the enzyme heme oxygenase-2 in tangle-bearing neurons; (f) impaired calcium homeostasis; and (g) dysfunction of mitochondria leading to a cellular bioenergetic crisis that may induce cell death. In addition, these factors contribute to both abnormal phosphorylation of cytoskeletal tau protein and proteolytic processing of APP by PSs with increased production of Aβ-42, the main component of SPs, which via excess glutamine production and other excitatory toxins may lead to neurodegeneration [9].

The central role of oxidative damage and other stress factors in the pathogenesis of AD and its close relationship to increased AGE production is supported by findings in transgenic (tg) animal models of AD [10]. Region-specific differences in the severity of oxidative stress–mediated injury, an early event in the brain [11], are likely to contribute to the variable intensity of neurodegeneration in different brain areas. Phospholipid breakdown of neuronal membranes and Aβ deposition, both exaggerated by ApoEε4, activate microglia-secreting cytokines, acute-phase reactants, nitric oxide, and other substances that mediate inflammatory reactions. The pathogenic cascade of AD, finally resulting in neuronal cell loss exceeding 50% in certain regions, is suggested to be initiated 30–40 years before onset of clinical symptoms, and manifest AD is considered a late stage of these processes. However, little is known about the mode and molecular course of neuronal degeneration and its relationship to plaques and tangles. Studies demonstrating DNA fragmentation and an upregulation of proapoptotic and cell death regulating proteins have raised the question of whether apoptosis, a specific gene-directed form of programed cell death, may be a general mechanism in AD. In spite of a severe increase in DNA fragmentation in the AD brain, mainly in glial cells, only few hippocampal neurons display the typical morphology of apoptosis and express the proapoptosis-executing enzyme-activated caspase-3, suggesting a proapoptotic environment causing increased vulnerability of specific neurons in predominantly involved areas, for example, hippocampus. While in EOAD and rodent models, neuronal damage has been related to APP and/or PS genes, in human AD, it is only partly induced by either Aβ deposits or tangles, causing a 5.6- and 3-fold higher risk of neurodegeneration. However, progressive neuronal injury is associated with plaque formation, and expression in hippocampal AD neurons of some antiapoptotic proteins as a possible response to oxidative stress or an attempt to repair damaged DNA counteracting the proapoptotic environment may prevent some of these neurons from irreversible death (see reference 12).

V. Neuropathology of AD (Relevant Structural Details)

The brain may look relatively unremarkable but often shows considerable atrophy with loss of brain weight and volume. Atrophy may be particularly severe in EOAD, less in early stages of LOAD, with overlaps to age-matched controls, whereas in late stages, brain weight is decreased by 14 (frontal lobes) to 41% (temporal lobes), with prominent gyral atrophy, especially in the temporal and frontal lobes, and ventricular dilatation [13–15]. The histological features of AD include extracellular deposition of Aβ in plaques and cerebral vasculature (cerebral amyloid angiopathy/CAA), and accumulation of hyper-phosphorylated tau protein forming paired helical filaments (PHFs) in neurons (NFTs), dendrites (NTs), and around plaques (NPs) (Fig. 3). Although most of these changes are nonspecific, the diagnosis of AD depends on their semiquantitative assessment (Fig. 4). Other lesions are loss of synapses (in late stages of AD ranging from 10 to 60% [16], most severe in the frontal region and greater in presenile than in senile-onset AD, and loss of neurons [20–40% in neocortex, 25–65% in hippocampus], causing cerebral volume loss/atrophy) and progressive disruption/disconnection of neuronal circuits as major substrates of dementia, microglial activation (inflammatory cascades), astroglial proliferation, granulovacuolar degeneration, and LBs.

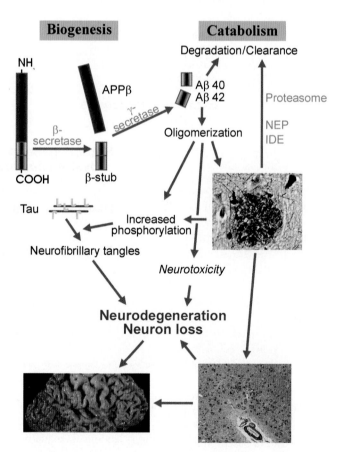

Figure 3 Molecular pathogenesis of Alzheimer's disease. NEP: Neprolysin; IDE: Insulin degrading enzyme.

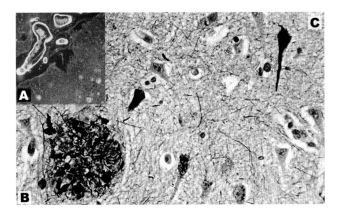

Figure 4 Major histopathological changes in Alzheimer's disease. (**A**) Amyloid deposits in the neuropil (plaques) and vasculature (CAA). (**B**) Neuritic plaque with dystrophic changes. (**C**) Neurofibrillary tangles and neuropil threads.

The cytoskeletal lesions associated with conformational changes of tau protein show a distinct/predictable spreading from the transentorhinal cortex in the mediobasal temporal lobe via hippocampus to neocortical association areas and, later, to subcortical nuclei [17]. These changes have been divided into six stages, including transentorhinal (stages I and II), hippocampal (III and IV), and neocortical (V and VI) (Fig. 5A), and have been confirmed by neurochemical detection of tau proteins [18]. The extension of tau pathology is different from the spread of Aβ deposition that is related to diffusion of soluble Aβ in the extracellular space, leading to neurodegeneration. Amyloidosis usually begins in the neocortex and later progresses to allocortical regions expanding anterogradely into regions that receive neuronal projections from already affected areas [19] (Fig. 5B). In contrast to the precise pathway of tau pathology, it is more diffuse and less predictably distributed. Data suggest nonoverlapping but synergistic action of both pathologies in sporadic AD [18]. The sequence that is believed to be comparable with a cascade of events beginning with the deposition of Aβ and ending with NFT pathology (amyloid cascade hypothesis) awaits further confirmation. The spatial relationship in early AD stages indicating that SPs lay in the terminal fields of NFT-bearing neurons suggests that NFTs either antecede plaques or, less likely, are independently formed.

Criteria for the neuropathological diagnosis of AD are based on (semi)quantitative assessment of plaques and tangles, considered histological hallmarks of the disease (see references 13, 15, and 20):

1. National Institute for Aging (NIA) criteria emphasizing neocortical SPs per unit corrected for age; they include both diffuse and neuritic types.
2. Criteria based on semiquantitative assessment of plaques and NFTs in neocortex and hippocampus, which, in part, disregard neocortical AD lesions.
3. The Consortium to Establish a Registry of AD (CERAD) criteria, using semiquantitative NP counts with adjustment for age and a clinical history of dementia.
4. Topographical staging of neurofibrillary changes distinguishing six different stages [17].
5. The Washington University quantitative criteria for diagnosis of AD, modifications of the 1985 NIA criteria, using a modified Bielschowsky silver stain that better visualizes the whole range of SPs, including the diffuse variety and a counting protocol that evaluates the total number of plaques in 10 contiguous microscopic fields in each brain region, assessing a total average plaque distribution across the extent of cortex, thus precluding a diagnosis of AD based on only one to three selected fields. This protocol differs from both the NIA and the CERAD strategies [21].
6. The guidelines of the NIA and the Ronald and Nancy Reagan Institute of the Alzheimer's Association (RI) combine the CERAD and the Braak scores. Assessment by NIA-RI guidelines leads to a probability statement for a *low* (CERAD 0–1; Braak I–II), *intermediate* (CERAD B, Braak III–IV), and *high* likelihood (CERAD C, Braak V–VI) that dementia is due to AD. These categories apply only to individuals with dementia, but the underlying guiding principle is that any degree of Alzheimer changes is abnormal.
7. These algorithms that only consider the classic "plaque-and-tangle" phenotype of AD do not recognize other subtypes; for example, the "plaque-predominant" type with abundant amyloid plaques, no or very little neuritic AD pathology restricted to the hippocampus, and abnormal phosphorylated tau in neocortical pyramidal neurons but lacking overt tangle formation accounts for 3.4–8.0% of demented subjects older than 85 years [22]. The "tangle-predominant" type with NFT pathology in the limbic areas, absence of neuritic plaques, and no or very little amyloidosis accounts for 5–7% of oldest olds with female preponderance [15], and the LB variant of AD (LBV-AD) displaying cortical and subcortical LBs with severe AD pathology. Multivariate analysis of consecutive autopsy cohorts of aged individuals revealed significant correlations between cognitive function assessed by the Mini-Mental State Examination (MMSE) and Blesed test, and both the CERAD criteria and Braak staging, respectively, much weaker for both Tierney and Khachaturian criteria [15].
8. The evaluation of the NIA-RI criteria demonstrated fairly good correlations with clinical dementia and good agreement with pathological methods, their easy and rapid use in AD and nondemented subjects, but much less reliability for other dementing disorders. Although the sensitivity and specificity of the earlier algorithm is suggested to be more than 90%, only about 40% of the brains with the clinical diagnosis of AD show "pure" AD

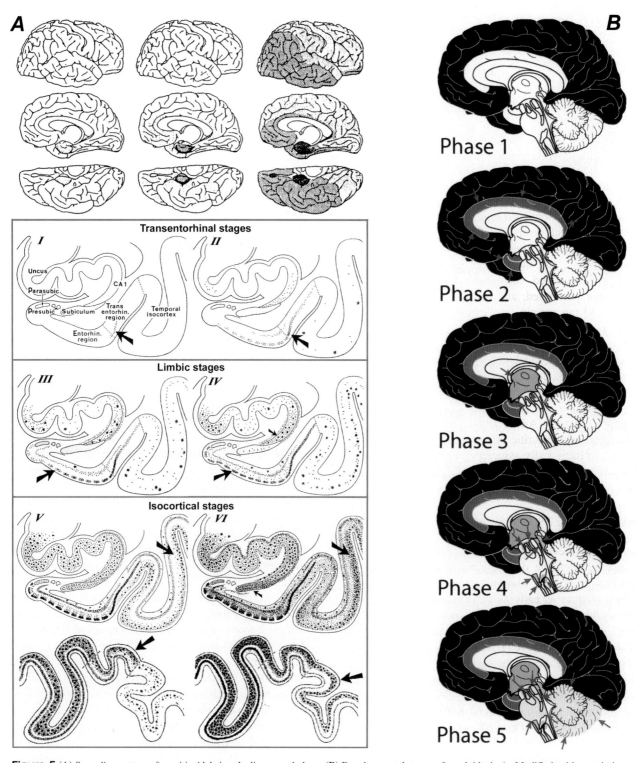

Figure 5 (**A**) Spreading pattern of neuritic Alzheimer's disease pathology. (**B**) Developmental stages of amyloidosis. (a, Modified, with permission, from Braak, H., and Braak, E. [1991]. Neuropathological staging of Alzheimer-related changes. *Acta Neuropathol. [Berl.]* **82**, 239–259. b, Modified, with permission, from Thal, D. R., Rub, U., Orantes, M., et al. [2002]. Phases of A beta-deposition in the human brain and its relevance for the development of AD. *Neurology* **58**, 1791–1800.)

pathology, whereas between 20 and 80% have other coexisting or superimposed pathologies, most frequently cerebrovascular lesions (20–60%), LB pathology (12–70%), and others, thus reducing considerably the predictive value of current neuropathological criteria for AD and contributing to the clinical heterogeneity of the disease (see reference 15).

VI. Pathophysiology

The application of molecular genetics and biochemical and morphological techniques to AD has produced a large and complex body of data on the pathophysiology of the disease, showing a complex cascade of events that precedes clinical symptoms. The quality and pattern of brain lesions, their spreading along well-defined anatomical pathways, and the consecutive biochemical changes have been described, correlations with the clinical course have been performed, and insights into the cell biology of Aβ and tau protein are beginning to offer hope for a better understanding of the underlying processes. However, many fundamental questions concerning the pathology of AD, its relations to clinical phenotypes, and the lesion threshold for cognitive impairment remain unanswered. Aβ is the principal constituent of both meningeal and capillary (parenchymal) CAA, as well as of SPs, the latter both containing mainly Aβ-42 and showing statistically high correlation with neuritic AD pathology [23]. Four genes that are unequivocally associated with the development of AD have been identified, APP, PS-1, and PS-2, which are causative for FEOAD, whereas ApoE ϵ4 is a major risk factor for LOAD, causing increased frequency and earlier onset of the disease. The localization of the APP gene to human chromosome 21 offered an explanation for the invariant development of AD pathology (often accompanied by cognitive decline) in patients with trisomy 21 (Down's syndrome) who show immature Aβ deposits or diffuse plaques as early as the second decade of life and later develop NPs and NFTs indistinguishable from those found in AD. Both Aβ in plaques and in tau as the principal component of NFTs show a possible link to these protein abnormalities in the genetics of FEOAD. The identification of PS as the active site of γ-secretase provides a linchpin for the amyloid hypothesis of AD; all of the mutations known to cause autosomal familial AD occur either in the substrate of APP or in its proteases (PSs) of the reaction that produces Aβ. The long-standing debate about the priority of NFTs versus plaques in AD pathogenesis has been largely resolved by the discovery of mutations in tau protein (chromosome p17). The phenotypes associated with tau mutations, including frontotemporal dementia, progressive supranuclear palsy, and other tauopathies, but not AD, indicate that a primary alteration of tau can cause severe neurodegeneration and profound dementia, in the absence of Aβ deposition. In AD, the accumulation of tangles cannot by itself explain the development of Aβ deposits. Studies in double tg mice suggest that the presence of Aβ-cleaving APP mutations augments the formation of tau deposits rather than the reverse, and the introduction of Aβ into tau-tg mice led to enhanced tau pathology with no change in Aβ deposition, but most double tg animal models show abnormal synaptic morphology before Aβ pathology. In triple tg mice harboring PS-1, APP, and tau (F301 L) antigens, progressively developing SPs and NFTs, synaptic dysfunction manifests in an age-related manner before plaque and tangle pathology. These and other AD animal models showing hippocampal neuronal loss with intraneuronal Aβ-42 accumulation confirm its critical role for synaptic plasticity and neuronal loss, implying that the major causative synaptotoxic insult in AD occurs during the early steps of amyloid processing and fibrillary tau aggregation [24]. Such changes, first subtle and then increasingly robust, are accompanied by an array of cellular and biochemical alterations, discussed previously, ultimately leading to neuronal death.

Although the relationship between plaques, NFTs, and neuronal/synaptic loss remains to be elucidated, there is a cascade of reactions, and most clinicopathological studies have shown that both lesions, if present in sufficient amounts, particularly in the neocortex, are the best markers for dementia, and that all clinically defined AD cases have large amounts of amyloid and widespread tau pathology. However, widely distributed diffuse and neuritic plaques in allocortex are frequently present in both nondemented elderly subjects and those with mild cognitive impairment (MCI). Preclinical AD or "pathological aging" often has more neuritic pathology mainly in the hippocampus than "normal" aging with few Aβ deposits and no to moderate NFTs in hippocampus and frontal cortex[25]. They also differ in both PHF tau and amyloid biochemistry.

Cognitively intact elderly individuals and those with MCI show abundant diffuse amyloid plaques; they may or may not have NPs and NFTs in the entorhinal and perirhinal cortex and hippocampus, whereas NPs and NFTs in the neocortex are often absent in nondemented elderly controls. Although Braak stages I and II have been described as "clinically silent," some subjects in these stages had already an identifiable cognitive impairment. Only noncognitively impaired subjects were classified as either Braak stage 0 or II-III, whereas those with MCI were found in all Braak stages, but others found no or only little pathological differences between persons with MCI and cognitively normals, and between MCI and AD. Several studies indicate that the densities of total neocortical (neuritic and diffuse) or neuritic plaques, although correlating moderately and not as robustly as neocortical NFTs with dementia severity constitute the best marker to differentiate AD, even in the earliest symptomatic stages, from nondemented controls, whereas neither neocortical nor hippocampal NFTs discriminate nearly as well between these two groups [21]. Although the mesolimbic cortex and hippocampus are thought to be central in the integration of emotional responses, behavior, and intellect, NPs and NPTs in these areas do not appear to be a reliable predictor of dementia, as 20–100% of mentally intact elderly persons show neuritic lesions in these regions and cannot be clearly distinguished from demented persons. However, all demented patients had dystrophic neurites both in the neuropil or surrounding plaques, and the number of NFTs, and much less than that of NTs, usually correlated with the clinical severity of dementia, while even widespread Aβ deposition may be necessary, but by itself it is often insufficient for the development of dementia. Only up to 20% of aged cognitively intact subjects have no or only few AD changes, whereas all the others show numerous AD lesions, including 9% with isocortical Braak stages V or VI, indicating that some individuals tolerate the insult of AD pathology, probably due to neurocognitive reserve of compensation. In other cohorts, 7–50% of individuals with no or only MCI met one or more of the morphological AD criteria or had tau pathology in the perforant path target zone and loss of entorhinal neurons, whereas only 10–20% were free of AD-related lesions [26]. About 20% of cases pathologically defined by the CERAD criteria as AD do not have clinical dementia, whereas more than half of the octogenarians without dementia meet CERAD criteria for pathologically confirmed AD (see reference 15).

VII. Biochemistry and Molecular Mechanisms

Neurochemistry findings in AD concern (1) levels and progression patterns of Aβ and tau pathologies in the brain; (2) chemical changes in plaques and NFTs; and (3) involvement of neurotransmitter systems, related to clinical signs and symptoms.

1. Biochemical studies of tau protein confirmed the morphological pathways of neurofibrillary degeneration, distinguishing 12 successive stages, with onset in the entorhinal region, progressing to hippocampus and, later, to neocortical association areas with comparative preservation of the occipital visual centers or involving the whole brain [18]. Amyloidosis usually begins in the neocortex and later progresses to allocortical regions, thus differing considerably from the extension of tau pathology [19]. More biochemical studies in brains from elderly people at Braak stages I to III showed a rise of tau levels and possibly Aβ-42 in the entorhinal cortex at age 75 years, with increasing levels of insoluble tau from state I to II, and small increase of Aβ-42 with increasing SPs, whereas Aβ-40 increased continuously with advancing Braak stage. There was no significant correlation between the levels of tau and Aβ in the entorhinal cortex, and substantial accumulation of tau occurred even in the absence of significant Aβ accumulation [27]. An explanation for the independent accumulation of tau and Aβ in the entorhinal cortex is that the amyloid cascade hypothesis [28] is valid in the neocortex but not in the limbic area. This cannot explain accelerated alteration of both regions of the brain affected by familial EOAD (FEOAD), probably because of additional effects of mutant APP and PS 1/2.

2. SPs contain many proteins and peptides, for example, antichymotrypsin, ubiquitin, Aβ, ApoE, endoplasmic reticulum–associated Aβ-binding protein (ERAB), nitric oxide synthase, cholinergic markers, multiple synaptic–associated proteins, such as synaptophysin, and β-synuclein, as well as several neuroproteins (somatostatin, substance P, neuropeptide Y, etc.). This points to the senile plaque as a "graveyard" of degenerating debris containing nerve terminals in various stages of demise, in a milieu of inflammatory reaction, indicated by reactive astrocytes and increasing numbers of microglia NFTS, in addition to hyperphosphorylated tau, ubiquitin, ApoE, and iron appear not to be involved in transmitter changes.

3. Neurotransmitter changes in AD involve many neuronal systems with variable intensity [29,30]. Neuronal disorganization and disruption of connections appear important for dysfunction of neurotransmitter systems. The decline of the cholinergic system is strong and has been observed long ago as the basic principle of the "cholinergic hypothesis" [31] and the current acetyl-cholinesterase (AChe) inhibitor treatment of AD. It should be born in mind, however, that AD is a multisystemic disease involving many neuronal and transmitter networks, the cholinergic depletion being only one among a very large number of pathological alterations in AD (Fig. 6) to be considered in therapeutic options.

A. Cholinergic System

Neurochemical and behavioral studies support the possibility that aging reveals the vulnerability of an abnormally regulated cortical cholinergic input system. Its decline and that in cognitive functions is further accelerated as a result of interactions with APP processing. Extracellular Aβ aggregation is suggested to affect cholinergic terminations before progression onto other neurotransmitter systems.

Noncognitively impaired subjects and early stages of AD (average Braak stage <2) in plaque-containing cases showed a significant decrease of ChAT activity (71–80%) compared with plaque-free cases and in inferior temporal

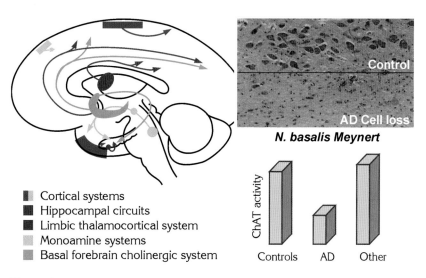

Figure 6 Involvement of major neurotransmitter systems in Alzheimer's disease.

cortex, it had an inverse correlation with Aβ concentration, suggesting a preclinical onset of the cholinergic deficit in AD. NFTs and AT8 immunoreactive neurons in the nucleus basalis of Meynert (NBM) were seen even in cognitively intact subjects, but the percentage of tau-positive NBM neurons was greater in MCI cases and showed a significant correlation with premortem memory scores. Increased medial temporal NFT correlates strongly and better than hippocampal ChAT activity with impairment of episodic memory. These results and reduced cerebral AChe activity in MCI indicate that the cholinergic deficit is established at an early histopathological stage of AD before the onset of clinical symptoms.

In fully developed AD, the cholinergic system shows decreased activity of the acetylcholine (ACh)-synthesizing enzyme choline acetyltransferase (ChaT), with an overall decline of 35–75%, which is most severe in the temporal cortex (52–75%), followed by the visual system in the occipital cortex (35–66%), the somatosensory system in the parietal cortex (44–68%), the limbic loop (hippocampus, amygdala—around 56%), and the NBM (62%). The enzyme AChe shows reduction by 60–67% in the cortex and 64–79% in the basal forebrain. Whereas the muscarinic M_1 receptors are either intact or show slight increase, the M_2 receptors show a decline of 27–38% in various cortical areas, with increases of M_1 receptors in the limbic loop and decrease of M_2 receptor density by around 50%. Cholinergic changes are relatively small in basal ganglia.

B. Serotonergic System

Lesions are heterogeneous, with decreases by 6–77% in various cortical areas, 6–59% in the limbic loop, 25–43% in the striatal loop, and similar reduction of 5-HT and 5-HT$_2$ receptor densities.

C. Noradrenergic System

Noradrenergic content is either unchanged or shows decrease by 4–44% in cortical regions, and noradrenaline (NA) content in the limbic loop from 38 to 114% of controls, without involvement of the adrenergic receptors, and increase of 3-methoxy-4-hydroxyphenylglycol (MHPG), and of the enzyme phenylethanolamine-N-methyltransferase (PNMT) activity pointing to an increase in NA turnover in AD.

D. Dopaminergic System

The amygdala complex, the main target of the mesolimbic dopamine (DA) projection, shows a strong increase of DA content and a decrease of both metabolites 3,4-dihydroxyphenylalanine (DOPAC) and homovanillic acid (HVA), pointing to increased activity in AD. Despite neuron loss in substantia nigra, frequently seen in AD, DA content and turnover are intact, whereas DA receptor densities in this striatonigral system are lowered, and tyrosine hydroxylase (TH), the key enzyme of catecholaminergic synthesis, is strongly reduced in caudate and putamen, but less severe than in Parkinson's disease. DA 1–3 receptor densities are decreased. The cortex shows a subtle decrease in DA content.

E. Glutamatergic System

Both excitatory neurotransmitters glutamate and aspartate are decreased by 2–52% in AD cortex, with severe reduction of transporter (excitatory amino acid [EAA]) transport, and aspartate/glutamate uptake systems show reduction by 21–30%, with similar decline in limbic and striatal systems. While the transmitter content is slightly reduced, uptake systems are broken down.

F. GABAergic System

γ-Aminobutyric acid (GABA) concentration is decreased by 12–50% in various cortical areas, with strongest decline in the limbic system and variable decreases of the unspecified and more severe, of the benzodiazepine GABA receptors in all areas. Within the striatal loop, GABA concentrations are increased, because of increased GABA synthesis by glutamate decarboxylase (Glu-DC) and a slight decrease of GABA uptake.

G. Neuropeptides

Cortical levels of somatostatin (SS) and their receptors are considerably reduced in AD, as are corticotrophin-releasing factor (CRF), with increased CRF receptor numbers in deficient areas, mainly in the frontal and temporal but not in the parietal cortex, whereas vasopressin intestinal polypeptide (VIP) and cholecystokinin (CCK) remain unchanged. Vasopressin has been found to be reduced in the cortex and increased in the hypothalamus, whereas substance P (SP) was either reduced or unchanged. Galanin, a powerful inhibitor of ACh, is increased in cerebral cortex and nucleus basalis, whereas neuropeptide Y (NPY) is unchanged in the AD cortex or even significantly increased in the basal ganglia.

VIII. Explanation of Symptoms in Relation to Pathophysiology

AD is a heterogeneous disorder with variable clinical and pathological phenotypes. Several stages in the development of AD can be distinguished that usually show an inverse relation between neuropsychological status and progression of AD pathology (Fig. 7).

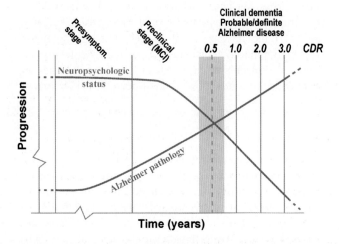

Figure 7 Theoretical development curves of Alzheimer's disease.

MCI, comparable to clinical dementia rating (CDR) scale score of 0–0.5, is a term given to memory impairment and relative sparing of other cognitive domains that is intermediate between cognitively normals and early (subclinical) AD without causing impairment of everyday functions or signs of dementia in most nonmemory tests that, however, show a mild decline of previous capacities [32]. Only some patients have language, visuospatial, or executive deficits. MCI is a high-risk predementia state that in 50–70% converts into overt dementia, with a conversion rate of 12–15% per year. The morphological basis of MCI is heterogeneous and includes abundant diffuse plaques with and without NPs or NFTs limited to the limbic system and usually, but not necessarily, without neocortical neuritic AD pathology, and patients with cerebrovascular lesions, in whom memory and executive performance impairment precedes the likelihood to develop dementia. The earliest stages of AD show increasing Aβ depositions in the brain and subtle alterations of hippocampal synaptic function before frank neurological degeneration and tau pathology with early NFTs in layer II of the (trans-) entorhinal cortex, which represents the portal for neocortical information that enters the limbic loop. Initially, there are no changes in neuronal numbers in the entorhinal and superior temporal cortices. Parahippocampal tau pathology is less severe in cognitively normal aged subjects compared with MCI and fully developed AD, and the NFT density is significantly correlated with memory deficits but no other cognitive abilities. The initial symptoms of MCI may develop when a threshold level of neuronal and synaptic loss, concurrent with NFT formation and gliosis, is reached in the entorhinal cortex. Subsequent disease progression and conversion to clinical AD are associated with progressive neuronal and synaptic loss, as well as NFT formation on the background of elevated Aβ that reaches a "ceiling" early in the disease, whereas NFT formation continues throughout its course. This indicates that substantial involvement of the entorhinal and hippocampal regions is often the key step in the development of dementia, irrespective of the age of the patient. Tau pathology in the central mediotemporal temporal lobe is suggested to develop before dysfunction of episodic memory, caused by the disruption of the GABAergic "perforant path" projecting from the entorhinal pyramidal neurons to the molecular layer of the dentate nucleus and, further, by the mossy fiber system to hippocampal subareas CA 3 and 4. This leads to a disconnection of the hippocampus from the association and limbic cortices. Although strong correlations are found between the density of NFTs in the hippocampus and the neocortex and both the degree of cholinergic deficits and dementia, many elderly subjects with MCI or early/very mild AD do not display limbic and cortical ChAT deficits but show a cholinergic upregulation with elevation of hippocampal and frontal cortex

ChAT, reflecting a compensatory response to the progressive denervation of the hippocampus by loss of entorhinal input [33]. This suggests that earliest cognitive deficits (e.g., episodic memory loss in MCI) are not involved by cholinergic deficits but, more likely, relate to disturbed entorhinal–hippocampal connectivity. Dementia conversion rates increase as cognitive impairment increases and the likelihood of subsequent conversion of MCI to dementia is usually related to the severity of local AD pathology.

Mild AD: The early stages of AD (CDR stage 1, duration 1–3 years) are characterized by clearcut deficits of episodic memory and in at least one of the other cognitive domains, which include new learning and encoding of information and delayed recall, while the ability to retrieve information from long-term memory is intact. An important diagnostic feature is the deficit of new learning. The patient forgets names, telephone numbers, and important dates and may show visuospatial disorientation. Functional loss may take the form of difficulties in financial affairs or an inability to do tasks in one's job or around the home. Changes in personality and behavioral capacities may occur, ranging from increased apathy and social withdrawal to disinhibition and irritability, although they may still a rather intact "façade" of personality. Depression also is common, whereas delusions and hallucinations are rare. On mental state examination, patients show between 20 and 26–30 points on the MMSE score, and memory performance may be the most abnormal portion of cognitive impairment. The patient may have largely intact conversational comprehension and spontaneous speech, but family members may report word-finding difficulties. Abstract reasoning deficits may be detectable with more difficult tasks that require mental agility. Many patients have some constructional difficulties, and widespread deficits in visuospatial processing are often observed. The motor neurological examination results are usually normal, whereas some subtle extrapyramidal signs may be seen. The morphological substrates of mild AD include a wide variety of Aβ deposition in the limbic and neocortical systems, without definite correlations between plaque density and the extent of cognitive impairment, and neuritic tau pathology ranging from entorhinal and hippocampal to early neocortical Braak stages (Fig. 8).

Moderate AD: In later stages of AD (CDR stage 2), there is a progression of disorders of memory, thinking, orientation, with impairment of personality and behavioral impairment, clearly impaired orientation, assessment, disturbances of language function, with deficits of word fluency and name finding that are obvious in conversation. Information on recent events is often almost instantly forgotten. Patients are dependent on others for higher level daily living activities such as finances, shopping, or transportation may occasionally need to be reminded to bathe and to dress appropriately and may need supervised living situations. They may fail to recognize persons who are not part of their daily routine. Some assistance becomes needed for patients and caregivers. Patients in this stage should no longer operate motor vehicles or devices such as lawn mowers, power saws, or even stoves. Neuropsychiatric disturbances may become prominent:

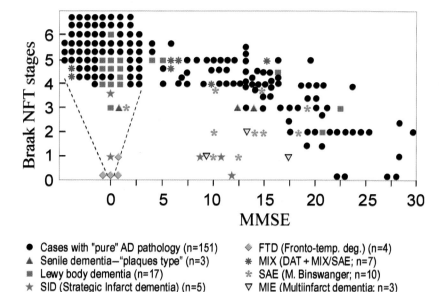

Figure 8 Relationship between Mini-Mental State Examination and Braak neuritic Alzheimer's disease stages in 207 consecutive autopsies of aged individuals (mean age at death 81.4 ± 8.6 years).

Delusions and hallucinations are common. Patients often misrecognize their own home, and irritability and paranoia are common symptoms. Disrupted sleep may also occur. On examination with the MMSE, patients will score between 10 and 19. Even in this stage of illness, no motor symptoms may be present, except for signs of mild rigidity or bradykinesia. The morphological lesions of this stage of AD correspond to Braak stage IV up to V, with frequent neuritic pathology in the neocortex (see Fig. 8).

Severe AD (CDR stage 3): In the terminal stages of illness, patients have negligible memory for events, conversation, and even close family members. There are substantial word-finding difficulties and impoverishment of spontaneous speech. Patients may be virtually mute or their speech incomprehensible. They need 24-hour supervision and extensive assistance with bathing, dressing, eating, and toileting. They usually become aggressive and may experience depression, anxiousness, and fear. Occasional seizures occur, and patients may exhibit marked rigidity, bradykinesia, gait and balance difficulties, masked face, and bladder and bowel incontinence. They typically score lower than 10 on the MMSE.

Morphologically, most patients show fully developed AD pathology corresponding to CERAD "definite" AD with Braak stages V and VI (see Fig. 8), and according to the NIA-RI criteria, dementia has a "high likelihood" to be due to AD. Frequent concurrent pathologies, in particular cerebrovascular lesions, may exacerbate the clinical manifestations of AD. Small vascular lesions in fully developed stages of AD, however, have no essential impact on global cognitive decline, which mainly depends on the severity of AD pathology, but in early stages of AD, they may influence and promote the development of dementia. Brains with vascular lesions may contain lower densities of plaques and tangles, and similar severity of dementia may be related to fewer AD lesions in brain with than without additional vascular lesions.

The major clinicopathological correlates of AD can be summarized as follows:

1. Early, often preclinical, involvement of the allocortex causes deafferentation of the hippocampus due to involvement of the (trans)entorhinal region by alterations of synaptic function and neuronal degeneration due to cytoskeletal PHF changes and Aβ deposition clinically presenting as early cognition, olfactory, and amnestic deficits without overt dementia.

2. Spreading of AD pathology to the isocortex with involvement of isocortical association areas causing the following syndromes (see reference 15):

   a. *Temporal isocortex* induces degeneration and disruption of corticocortical connections to prefrontal and parietal areas and causes visuospatial, speech, and auditive disorders.

   b. Frontal *cortex*: high NFT and Aβ load in the frontal variant of AD with isolated executive impairment.

   c. Orbitofrontal *and cingulate pathology* often causes agitation.

   d. Major *NFT load in areas 7, 24, and CA 1* induces spatial-temporal disorientation.

   e. Occipital *cortex* induces rare visual variant of AD (Balint's syndrome), referred to as *progressive posterior cortical dysfunction* involving selective visuospatial deficits [34].

   f. Progressive *degeneration of intracortical, corticocortical connections* causes global corticocortical discontinuation, with loss of higher cortical functions and dementia.

3. Late spreading of AD pathology to subcortical areas causes degeneration of corticosubcortical neuronal systems and induces primary sensory deafferentation with affective, extrapyramidal, and vegetative signs or symptoms.

IX. Natural History

The natural history of AD shows considerable individual variability, but there are some approximate values for the different phases of the illness. The clinical onset of LOAD is between 50 and 80 years of age, whereas most cases of FEOAD show earlier onset. The average length from onset of symptoms until diagnosis is 2–3 years. The average time from diagnosis to nursing home placement in progressed dementia stages is roughly 3–6 years, and patients with AD spend about 3 years in nursing homes before death. Thus, the total duration of AD is roughly 7–12 years [4]. Very elderly patients occasionally show a slower progression with plateau phases (transient stabilization), whereas early onset cases usually have more rapid progression. Across patients with initial MMSE scores ranging from 10 to 26, the average rate of change per years is 2–3 points. Slower rates of decline occur among milder and more severe cases, whereas faster rates of decline occur in those in the midportion of the scales, but a decline over a 6- to 12-month period does not predict the rate of decline over subsequent time intervals. Parkinsonian signs, hallucinations, and delusions are associated with a more rapid decline. Mortality in AD averages about 10% per year. Causes of death include pneumonia, urinary tract infections, pulmonary embolism, and other common causes of mortality in the elderly, such as cardiovascular disease. Progress in pharmacotherapy, medical management, and caregiving has increased the survival time of patients with AD.

X. Management of AD

Although the cause of AD is still unknown, management and treatment of the disease have improved considerably. Multidimensional pharmacological and psychosocial management includes the following options [see references 35 and 36]:

1. Pharmacotherapy of cognitive symptoms using cholesterin inhibitors (donepezil, rivastigmine, galantamine) may cause a transient stabilization of the disease with delay of its progression for a 12- to 18-month period and have modest but clinically relevant effects on cognition and daily function, delaying nursing home placement. In late stages of AD, memantine, an N-methyl-D-aspartate (NMDA) antagonist, has shown similar effects and a combination therapy has been proposed [37]. The effects of other antidementive/nosotropic substances (e.g., piracetam, ginkgo biloba, and vasoactive drugs) are less well documented. The efficacy of anti-inflammatory drugs, antioxidants (vitamin E/α-tocopherol), estrogen replacement therapy, and statins is under discussion because of contradictory trial results. The treatment of MCI is under discussion [39]. Human studies with secretase inhibitors, β-sheet breakers, tau phosphorylation blockers, and other disease-modifying substances are underway, as are different types of anti-Aβ immunotherapy, which because of severe (lethal) side effects, had to be reorganized [4,38]. Based on the results of modern molecular pathobiology, patients will be offered potentially disease-slowing or modifying therapies.

2. Medical treatment of noncognitive (psychiatric, behavioral) symptoms includes modern atypical neuroleptics, as well as antidepressive and anticonvulsive drugs to be carefully applied.

3. Integrated psychosocial management and caregiving are important factors to increase the quality of life for patients with AD and should be combined with other treatments.

4. Among protective factors, physical and mental activity and prevention and treatment of risk factors, such as hypertension, diabetes, hypercholesterinemia, and diet, are important.

References

1. Alzheimer, A. (1907). Über eine eigenartige Erkrankung der Hirnrinde. *Allgem. Z. Psychiatr.* **64**, 146–148.
2. Bick, K. L. (1999). The early history of Alzheimer's disease. *In* "Alzheimer Disease." (R. D. Terry, R. Katzman, K. L. Bick, and S. S. Sisodia, eds.), pp. 1–9. Lippincott, Williams and Wilkins, Philadelphia.
3. Alzheimer, A. (1911). Ueber eigenartige Krankheitsfaelle des spaeteren Alters. *Z. Ges. Neurol. Psychiatrie* **4**, 356–385.
4. Knopman, D. S. (2003). Alzheimer type dementia. *In* "Neurodegeneration: The Molecular Pathology of Dementia and Movement Disorders" (D. W. Dickson, ed.), pp. 24–39. ISN Neuropath Press, Basel.
5. Hardy, J. (2004). Genetics of Alzheimer's disease and related disorders. *In* "Current Clinical Neurology. Alzheimer's Disease: A Physician's Guide to Practical Management." (R. W. Richter, and B. Zoeller-Richter, eds.), pp. 3–20. Humana Press, Totowa, NJ.
6. Masters, C. L., and Beyreuther, K. (2003). Molecular pathogenesis of Alzheimer's disease. *In* "Neurodegeneration: The Molecular Pathology of Dementia and Movement Disorders" (D. W. Dickson, ed.), pp. 69–73. ISN Neuropath Press Basel.
7. Roy, S., Zhang, B., Lee, V. M., and Trojanowski, J. Q. (2005). Axonal transport defects: a common theme in neurodegenerative diseases. *Acta Neuropathol. (Berl)* **109**, 5–13.
8. Cotman, C. W., Poon, W. W., Rissman, R. A., and Blurton-Jones, M. (2005). The role of caspase cleavage of tau in Alzheimer disease neuropathology. *J. Neuropathol. Exp. Neurol.* **64**, 104–12.
9. Mattson, M. P., and Magnus, T. (2006). Ageing and neuronal vulnerability. *Nat. Rev. Neurosci.* **7**, 278–94.
10. McGowan, E., Pickford, F., and Dickson, D. W. (2003). Alzheimer animal models: models of A-beta deposition in transgenic mice. *In* "Neurodegeneration: The Molecular Pathology of Dementia and Movement Disorders" (D. W. Dickson, ed.), pp. 74–79. ISN Neuropath Press, Basel.
11. Markesbery, W. R., Kryscio, R. J., Lovell, M. A., and Morrow, J. D. (2005). Lipid peroxidation is an early event in the brain in amnestic mild cognitive impairment. *Ann. Neurol.* **58**, 730–5.
12. Jellinger, K. A. (2003). Apoptosis vs nonapoptotic mechanisms in neurodegeneration. *In* "Neuroinflammation. 2nd Edition" (P. L. Wood, ed.), pp. 29–88. Humana Press, Totowa, NJ.
13. Duyckaerts, C., and Dickson, D. W. (2003). Neuropathology of Alzheimer's disease. *In* "Neurodegeneration: The Molecular Pathology of Dementia and Movement Disorders" (D. W. Dickson, ed.), pp. 47–65. ISN Neuropath Press, Basel.
14. Terry, R. D., Katzman, R., Bick, K. L., and Sisodia, S. S., eds. (1999). "Alzheimer Disease." Lippincott, Williams and Wilkins, Philadelphia.
15. Jellinger, K. A. (2006). A view on early diagnosis of dementias from neuropathology. *In* "Early Diagnosis of Dementias" (K. Herholz, C. Morris, and D. Perani, eds.), pp. 311–428. Marcel Dekker, New York.
16. Scheff, S. W., and Price, D. A. (2003). Synaptic pathology in Alzheimer's disease: a review of ultrastructural studies. *Neurobiol. Aging* **24**, 1029–46.
17. Braak, H., and Braak, E. (1991). Neuropathological stageing of Alzheimer-related changes. *Acta Neuropathol. (Berl)* **82**, 239–59.
18. Delacourte, A., Sergeant, N., Champain, D., Wattez, A., Maurage, C. A., Lebert, F., Pasquier, F., and David, J. P. (2002). Nonoverlapping but synergetic tau and APP pathologies in sporadic Alzheimer's disease. *Neurology* **59**, 398–407.
19. Thal, D. R., Rub, U., Orantes, M., and Braak, H. (2002). Phases of A beta-deposition in the human brain and its relevance for the development of AD. *Neurology* **58**, 1791–800.
20. Mirra, S. S., and Hyman, B. T. (2002). Aging and dementia. *In* "Greenfield's Neuropathology, 7th Ed." (D. I. Graham, and P. L. Lantos, eds.), pp. 195–271. E. Arnold, London.
21. McKeel, D. W., Price, J. L., Miller, J. P., Grant, E. A., Xiong, C., Berg, L., and Morris, J. C. (2004). Neuropathologic criteria for diagnosing Alzheimer disease in persons with pure dementia of Alzheimer type. *J. Neuropathol. Exp. Neurol.* **63**, 1028–1037.
22. Tiraboschi, P., Sabbagh, M. N., Hansen, L. A., Salmon, D. P., Merdes, A., Gamst, A., Masliah, E., Alford, M., Thal, L. J., and Corey–Bloom, J. (2004). Alzheimer disease without neocortical neurofibrillary tangles: "a second look." *Neurology* **62**, 1141–7.
23. Attems, J., Lintner, F., and Jellinger, K. A. (2004). Amyloid beta peptide 1-42 highly correlates with capillary cerebral amyloid angiopathy and Alzheimer disease pathology. *Acta Neuropathol. (Berl)* **107**, 283–91.

24. Wirths, O., Multhaup, G., and Bayer, T. A. (2004). A modified beta-amyloid hypothesis: intraneuronal accumulation of the beta-amyloid peptide—-the first step of a fatal cascade. *J. Neurochem.* **91**, 513–20.
25. Markesbery, W. R., Schmitt, F. A., Kryscio, R. J., Davis, D. G., Smith, C. D., and Wekstein, D. R. (2006). Neuropathologic substrate of mild cognitive impairment. *Arch Neurol* **63**, 38–46.
26. Riley, K. P., Snowdon, D. A., and Markesbery, W. R. (2002) Alzheimer's neurofibrillary pathology and the spectrum of cognitive function: findings from the Nun Study. *Ann. Neurol.* **51**, 567–77.
27. Katsuno, T., Morishima–Kawashima, M., Saito, Y., Yamanouchi, H., Ishiura, S., Murayama, S., and Ihara, Y. (2005). Independent accumulations of tau and amyloid beta-protein in the human entorhinal cortex. *Neurology* **64**, 687–92.
28. Hardy, J. A., and Higgins, G. A. (1992). Alzheimer's disease: the amyloid cascade hypothesis. *Science* **256**, 184–5.
29. Gsell, W., Jungkunz, G., and Riederer, P. (2003). Functional neurochemistry of Alzheimer's disease. *Curr. Pharm. Design* **9**, 265–293.
30. Lyness, S. A., Zarow, C., and Chui, H. C. (2003). Neuron loss in key cholinergic and aminergic nuclei in Alzheimer disease: a meta–analysis. *Neurobiol. Aging* **24**, 1–23.
31. Francis, P. T., Palmer, A. M., Snape, M., and Wilcock, G. K. (1999). The cholinergic hypothesis of Alzheimer's disease: a review of progress. *J. Neurol. Neurosurg. Psychiatry* **66**, 137–47.
32. Petersen, R. C., ed. (2003). "Mild Cognitive Impairment: Aging to Alzheimer's Disease." Oxford University Press, Oxford, UK.
33. DeKosky, S. T., Ikonomovic, M. D., Styren, S. D., Beckett, L., Wisniewski, S., Bennett, D. A., Cochran, E. J., Kordower, J. H., and Mufson, E. J. (2002). Upregulation of choline acetyltransferase activity in hippocampus and frontal cortex of elderly subjects with mild cognitive impairment. *Ann. Neurol.* **51**, 145–55.
34. Renner, J. A., Burns, J. M., Hou, C. E., McKeel, D. W., Jr., Storandt, M., and Morris, J. C. (2004). Progressive posterior cortical dysfunction: a clinicopathologic series. *Neurology* **63**, 1175–80.
35. Richter, R. W., and Zoeller-Richter, B., eds. (2005). "Alzheimer's Disease: A Physician's Guide to Practical Management." Humana Press, Totowa, NJ.
36. Lleo, A., Greenberg, S. M., and Growdon, J. H. (2006). Current pharmacotherapy for Alzheimer's disease. *Annu. Rev. Med.* **57**, 513–33.
37. Dantoine, T., Auriacombe, S., Sarazin, M., Becker, H., Pere, J. J., and Bourdeix, I. (2006). Rivastigmine monotherapy and combination therapy with memantine in patients with moderately severe Alzheimer's disease who failed to benefit from previous cholinesterase inhibitor treatment. *Int. J. Clin. Pract.* **60**, 110–8.
38. Citron, M. (2004). Strategies for disease modification in Alzheimer's disease. *Nat. Rev. Neurosci.* **5**, 677–85.
39. Jelic, V., Kivipelto, M., and Winblad, B. (2006). Clinical trials in mild cognitive impairment: lessons for the future. *J. Neurol. Neurosurg. Psychiatry* **77**, 429–38.

7

Multiple System Atrophy

Carlo Colosimo, MD
Felix Geser, MD, PhD
Eduardo E. Benarroch, MD, DSci
Gregor K. Wenning, MD, PhD

Keywords: *ataxia, autonomic failure, dopamine, parkinsonism*

I. Brief History and Nomenclature
II. Epidemiology
III. Pathogenesis
IV. Structural Details
V. Biochemical and Neuropharmacological Findings
VI. Molecular Biology
VII. Animal Models
VIII. Clinical Picture
IX. Clinical Diagnosis and Clinical Diagnostic Criteria
X. Natural History of the Disease
XI. Conclusion
    References

I. Brief History and Nomenclature

The term *multiple system atrophy* (MSA) was introduced by Graham and Oppenheimer in 1969 to denote a neurodegenerative disease characterized clinically by various combinations of autonomic, parkinsonian, cerebellar, or pyramidal symptoms and signs, and pathologically by cell loss and gliosis in the basal ganglia and olivopontocerebellar system. Previously, cases of MSA were reported under the rubrics of olivopontocerebellar atrophy (OPCA), idiopathic orthostatic hypotension or progressive autonomic failure, Shy-Drager syndrome (SDS), and striatonigral degeneration (SND). Although Dejerine and Thomas were the first to introduce the term *OPCA* in 1900, reporting two sporadic cases of late-onset ataxia, the case of Stauffenberg in 1918, who was diagnosed as having OPCA, was the first to associate cerebellar, parkinsonian, and autonomic features with identified pathological lesions not only of olives, pons, and cerebellum, but also of the basal ganglia. The term *SND* was introduced in 1960 by van der Eecken, Adams, and van Bogaert who noted pronounced shrinkage and brownish discoloration of the putamen and pallidum, as well as depigmentation of the substantia nigra in three patients with progressive and severe parkinsonism associated with cerebellar, pyramidal, and autonomic features. In 1960 Shy and Drager reported two cases with marked autonomic failure, slurred speech, ataxia, parkinsonism, pyramidal signs, and distal muscle wasting. Postmortem examination of case 2 demonstrated pathological lesions consistent with MSA, including involvement of the intermediolateral cell column. Subsequently, the term *SDS* was erroneously widened to include cases of Parkinson's disease (PD) and autonomic failure. Its further use has, therefore, been discouraged [1,2]. In 1989 Papp et al. reported glial cytoplasmic inclusions (GCIs) in the brains of patients with MSA [3] regardless of clinical presentation. GCIs were not present in a large series of patients with other neurodegenerative disorders. The abundant presence of GCIs in all clinical subtypes of MSA introduced grounds for considering SDS, SND, and sporadic OPCA as one disease entity characterized by neuronal multisystem degeneration based on unique oligodendroglial inclusion pathology. Subsequently, neuronal and axonal inclusions were also identified in MSA brains; however, they were less numerous compared with GCIs [4]. In the late 1990s, α-synuclein immunostaining was recognized as the most sensitive marker of inclusion pathology in MSA, being superior to the previously used ubiquitin immunostaining [5]. Because of these advances in molecular pathogenesis, MSA has been firmly established as an α-synucleinopathy along with PD and dementia with Lewy bodies (LBs). In parallel, the clinical recognition greatly improved after the introduction of diagnostic criteria. In 1989 Quinn [6] was the first to propose a list of diagnostic criteria for MSA, but in the late 1990s, a number of pitfalls associated with the Quinn criteria were identified. As a result, consensus diagnostic criteria were developed in 1998 [7]. These criteria have since been widely established in the research community and in movement disorders clinics. More recently, the European MSA-Study Group has developed the Unified MSA Rating Scale (UMSARS) for use in natural history studies and therapeutic trials [8]. Furthermore, transgenic MSA mouse models have become available to study the pathogenesis and identify promising targets for interventions.

II. Epidemiology

Determining the incidence and prevalence of MSA is difficult, as only a few epidemiological studies have been reported. Estimates of the prevalence of MSA (per 100,000 in the population) in different studies ranged from 1.9 to 4.9 [9]. The annual incidence of MSA was estimated to be about 0.6 cases per 100,000 persons or 3.0/100,000 people older than 50 years [9]. These figures are similar to those of other well-known neurodegenerative disorders such as Huntington's disease or motor neuron disease. Analytical epidemiology of MSA is even poorer; a case–control study in North America showed an increased risk of MSA associated with occupational exposure to organic solvents, plastic monomers and additives, pesticides, and metals [10]. A further case–control study in Italy showed an increased risk of MSA associated with farming [11]. Smoking habits seem to be less common in MSA cases (as in PD cases) than in healthy controls. The fact that the inverse association with smoking found previously in PD is shared by MSA, but not by progressive supranuclear palsy (PSP), lends epidemiological support to the notion that different smoking habits are associated with separate groups of neurodegenerative diseases [12].

III. Pathogenesis

The α-synuclein pathway appears to be the key pathway to selective loss of glia and neurons in MSA. The differential distribution of α-synuclein deposits and associated neuronal pathology (SND, OPCA, and spinal cord) suggests variability of pathogenetic mechanisms underlying the multifaceted disease process of MSA.

MSA, as reflected in its current definition, is regarded as a sporadic disease and no confirmed familial cases of MSA have been described yet; notwithstanding, it is conceivable that genetic factors may play a role in the etiology of the disease. However, initial screening studies for candidate genes reveal no risk factors [13]. Other studies have further looked for polymorphisms or mutations in candidate genes that may predispose an individual to developing MSA. The apolipoprotein ϵ4 allele is not overrepresented in MSA when compared with controls, and there have been conflicting reports of the association of a cytochrome P-450 (CYP) 2D6 polymorphism with MSA [1].

The pathogenetic role of α-synuclein is still unclear. Inactivation of the α-synuclein gene by homologous recombination does not lead to a severe neurological phenotype. So, loss of function of the α-synuclein protein is unlikely to account for its role in neurodegeneration. Mice that lack α-synuclein were found to show increased release of striatal dopamine, indicating that this protein could function as an activity-dependent negative regulator of neurotransmission in the striatum [1]. Although there is strong evidence that α-synuclein participates in the pathogenesis of some types of familial PD, no mutations have been found in the entire coding region of the α-synuclein gene in MSA [14].

The polymorphisms in the α-synuclein gene, which have been identified in PD, may also increase the risk of developing MSA by promoting α-synuclein protein aggregation. Polymorphisms in codons 1–39 of the α-synuclein gene, a domain related to interaction with synphilin-1, or indeed polymorphisms in the synphilin-1 gene itself, or in the genes of other protein-interacting partners of α-synuclein, may also need to be considered in the pathogenesis of MSA [1]. The number of α-synuclein protein–interacting partners has expanded to include 14-3-3 protein chaperones, protein kinase C, extracellular-regulated kinase, and BAD, a molecule that regulates cell death. Nevertheless, association study results with genetic polymorphisms for α-synuclein have been negative in MSA.

Gilman et al. [15] have reported an MSA-like phenotype including GCIs in one spinocerebellar ataxia type 1 (SCA1) family. Other SCA mutations (except for SCA2) have not been reported to present with MSA-like features [1]. Conversely, the majority of patients with MSA-C (i.e., MSA with cerebellar ataxia as the predominant motor disorder) do not appear to have expanded SCA1 and SCA3 alleles [16]. Indeed, MSA-C appears to be a common form of sporadic cerebellar ataxia of late onset because 29% of patients with sporadic adult-onset ataxia may suffer from MSA [1]. This finding corresponds well with data of a study of patients with sporadic OPCA who were followed for 10 years [17]. Within this period, 17 of 51 patients developed autonomic failure or parkinsonism, indicating a diagnosis of MSA.

To summarize, the up-to-date knowledge about the role of α-synuclein and its aggregation in neurodegenerative disorders remains unclear. It is yet not known how the expression and aggregation of α-synuclein in glial cells affect their biology, as well as the glia–neuron interactions, which might be a critical step in the pathogenesis of α-synucleinopathies. Whether environmental factors influence α-synuclein aggregation and the survival of glial and neuronal cells remains unknown as well.

IV. Structural Details

A. Gross Neuropathology of MSA

External examination of the brain is often normal in MSA. However, when there is a significant involvement of the olivopontocerebellar system, the appearances are characteristic. The cerebellum is small, with the hemispheres far from covering the occipital poles. The white matter of the cerebellum appears gray and the folia atrophic. The basis pontis and middle cerebellar peduncles are reduced. In the medulla, the protuberance of the inferior olives may be reduced. Occasionally, macroscopic abnormality is confined to the brainstem pigmented nuclei, and in these instances, it is impossible to make a distinction from PD on naked-eye appearance alone [18]. In SND, the putamen is shrunken with gray-green discoloration. When pathology is severe, there may be a cribriform appearance. Atrophy and discoloration of the caudate nucleus and pallidum are less common. The substantia nigra invariably shows decreased pigmentation and the locus ceruleus may also appear pale [2].

B. Microscopic Neuropathology of MSA

1. Cellular Inclusions

Glial inclusion formation is a prominent feature of MSA pathology [3,4]. Although inclusions have been described in five cellular sites (i.e., in oligodendroglial and neuronal cytoplasm and nuclei, as well as in axons), GCIs are most ubiquitous and appear to represent the subcellular hallmark lesion of MSA. Their distribution selectively involves the basal ganglia, the supplementary and primary motor cortices, the reticular formation, the basis pontis, the middle cerebellar peduncles, and the cerebellar white matter [9]. Staining with antibodies against ubiquitin, α- and β-tubulin, and tau indicates an origin from cytoskeletal proteins. MSA has been recognized as an α-synucleinopathy with prominent GCI pathology [5]. Accompanying cytoplasmic inclusions in oligodendrocytes, there also may be rodlike nuclear inclusions [4].

Nuclear and cytoplasmic inclusions also occur in neurons and are ubiquitinated but distinguishable from those in oligodendrocytes by a lack of immunoreactivity for cytoskeletal proteins. Despite the different immunohistochemical reactions of the various inclusions described, electron microscopy invariably shows irregular filaments composed of 20- to 30-μm tubules [2].

2. Striatonigral Degeneration

The striatonigral system is the main site of pathology, but less severe degeneration can be widespread and usually includes the olivopontocerebellar system. In early stages, the putaminal lesion shows a distinct topographical distribution with a predilection for the caudal and dorsolateral regions. GCIs predominate, whereas nerve cell loss and gliosis can be difficult to identify without the aid of glial fibrillary acidic protein (GFAP) immunostaining. The majority of reports indicate that the small neurons, which normally outnumber the large by about 170 to 1, are preferentially affected [4]. Later on during the course of disease, the entire putamen is usually affected, with the result that bundles of striatopallidal fibers are narrowed and poorly stained for myelin. When atrophy is severe, the neuropil becomes rarefied, with few remaining nerve cells lying among hypertrophic astrocytes. Brown pigment granules accumulate in astrocytes, macrophages, and around blood vessel walls; they have been identified as containing lipofuscin, neuromelanin, and iron [18].

In the caudate nucleus, there is nerve cell loss and gliosis but rarely to the extent found in the putamen; the dorsomedial region is usually most affected, with gliosis extending through the striatal bridges. The globus pallidus may appear uninvolved or show a reduction of myelinated fibers with gliosis and variable neuronal depletion. Pallidal atrophy is generally considered secondary to loss of putaminal efferents; however, in occasional cases where there is definite neuronal degeneration, this may represent a transsynaptic effect or primary involvement.

Degeneration of pigmented nerve cells occurs in the substantia nigra pars compacta (SNc), whereas nonpigmented cells of the substantia nigra pars reticulata (SNr) are reported as normal. The caudal-lateral region is most severely affected, with neuronal depletion, glial hyperplasia, and granules of neuromelanin lying free in the neuropil and within macrophages. The appearances are usually of a more active type of degeneration when compared with that of PD; the neuropil is often vacuolated, and occasionally there are microglial nodules and evidence of neuronophagia, in addition to a diffuse increase of microglia. The topographical patterns of neurodegeneration involving the motor neostriatum, efferent pathways, and nigral neurons reflect their anatomical relationship and suggest a common denominator or "linked" degeneration [18].

3. Olivopontocerebellar Atrophy

The brunt of pathology is in the olivopontocerebellar system, whereas the involvement of the striatum and substantia nigra is less severe. The basis pontis is atrophic, with loss of pontine neurons and transverse pontocerebellar fibers. In sections stained for myelin, the intact descending corticospinal tracts stand out against the degenerate transverse fibers and the middle cerebellar peduncles. Many authors report a disproportionate depletion of fibers from the middle cerebellar peduncles compared with the loss of pontine neurons, an observation that led to the suggestion of a "dying back" process [19].

In MSA, cerebellar atrophy is often stated to be greatest in the neocerebellum, whereas the paleocerebellum is involved in primary cerebellar cortical degenerations. However, in several examples of MSA, pathology is most severe in the vermis [20], or, alternatively, the vermis and hemispheres may be equally affected. There is loss of Purkinje nerve cells and accompanying astrocytosis resulting in isomorphic gliosis in the molecular layer. Both the folial and the central hemispheric white matter are reduced, whereas that around the dentate nucleus and within the hilus is well preserved. Because of the loss of Purkinje cell axon terminals, increased gliosis occurs in the dentate nucleus, but there is usually no neuronal depletion at this site [18]. In the medulla, there is loss of neurons in the inferior and accessory olivary nuclei with increased gliosis. A lack of topographical relationship between neuronal cell loss in inferior olives and cerebellar cortex suggests that these may be primary unrelated degenerations [2].

4. Autonomic Failure

Autonomic failure in MSA is caused by dysfunction of (1) central and preganglionic efferent autonomic activity, (2) neuronal networks in the brainstem controlling cardiovascular and respiratory function, and (3) the neuroendocrine component of the autonomic regulation via the hypothalamic–pituitary axis.

Degeneration of sympathetic preganglionic neurons in the intermediolateral column of the thoracolumbar spinal cord is considered contributory to orthostatic hypotension [18]. It is noteworthy that there is not always a strong correlation between nerve cell depletion or gliosis and the clinical degree of autonomic failure. It is estimated that more than 50% of cells within the intermediolateral column need to decay before symptoms become evident [1]. A supraspinal contribution to the autonomic failure of MSA is now well established [18]. Cell loss is reported in the dorsal motor nucleus of the vagus, as well as catecholaminergic neurons of the ventrolateral medulla [21] and serotonergic neurons of the medullary raphe [22] (Fig. 1). Both the ventrolateral medulla and the medullary raphe send direct projections to the intermediolateral cell column. The ventrolateral medulla exerts a tonic excitatory control on sympathetic neurons that maintain blood pressure and mediate the baroreflex and other cardiovascular reflexes. The medullary raphe controls sympathetic neurons involved in thermoregulation.

Papp and Lantos [23] have shown marked involvement of brainstem pontomedullary reticular formation with GCIs, providing a supraspinal histological counterpart for impaired visceral function. Neuronal loss has also been described for the Edinger-Westphal nucleus and posterior hypothalamus [4], including the tuberomammillary nucleus.

Disordered bladder, rectal, and sexual function in SND and OPCA have been associated with cell loss in parasympathetic preganglionic nuclei of the spinal cord. These neurons are localized rostrally in Onuf's nucleus between sacral segments S2 and S3 and more caudally in the inferior intermediolateral nucleus, chiefly in the S3 to S4 segments [1]. Loss of neurons in the pontine micturition area producing corticotrophin-releasing factor may contribute to neurogenic bladder dysfunction in MSA [24].

In the peripheral component of the autonomic nervous system, atrophy of the glossopharyngeal and vagus nerves is described [2]. No pathology has been reported in the visceral enteric plexuses or in the innervation of glands, blood vessels, or smooth muscles. Sympathetic ganglia have not often been examined in pathological studies of autonomic failure and have seldom been described quantitatively. In a few cases, there were either no obvious or mild abnormalities in sympathetic ganglia. Any morphological changes reported in sympathetic ganglionic neurons in MSA

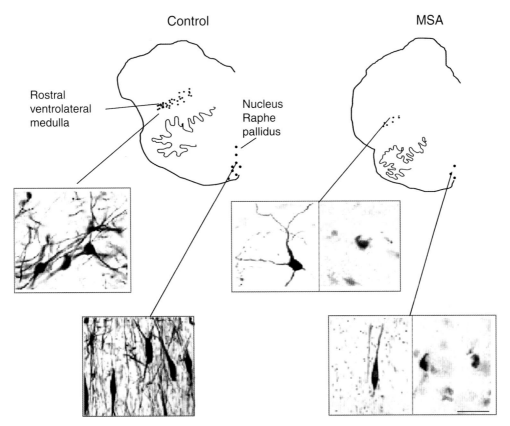

Figure 1 Picture showing cell loss of catecholaminergic neurons of ventrolateral medulla and serotonergic neurons of the medullary raphe in a patient affected by multiple system atrophy (MSA) (right) as compared with a control (left).

have tended to be nonspecific, falling within the normal age-related range of appearances [2].

5. Additional Sites of Pathology

A variety of other neuronal populations are noted to show cell depletion and gliosis with considerable differences in vulnerability from case to case. Various degree abnormalities in the cerebral hemisphere, including Betz cell loss, were detected in pathologically proven MSA cases [1].

In three autopsy cases of MSA, cerebellar cortical lesions were more conspicuous in the vermis than in the hemisphere [2]. These neuropathological findings differ from the established theory that cerebellar lesions of MSA are more pronounced in the hemisphere than in the vermis. The degree of cerebellar cortical lesions in these cases increased in relation to the duration of the disease.

Furthermore, anterior horn cells may show some depletion but rarely to the same extent as that occurring in motor neuron disease [1]. Laryngeal stridor is a common feature of MSA and may occur as a presenting sign or, more often, in later stages of the disease. Depletion of large myelinated nerve fibers in the recurrent laryngeal nerve, which innervates intrinsic laryngeal muscles, has been demonstrated in patients with MSA with vocal cord palsy. However, loss of motor neurons in the nucleus ambiguus has been reported in some studies [25], but not in others [26].

V. Biochemical and Neuropharmacological Findings

A. Biochemical Findings

Neurochemical studies have shown alterations consistent with sites of major pathology. Calcineurin, a marker for medium-sized spiny neurons, is decreased in striosomes of the putamen and in the efferent pathway of the globus pallidus and substantia nigra. Others also reported that regardless of clinical presentation, there is reduced immunoreactivity for additional markers of the striatal efferent system, including met-enkephalin, substance P (SP), and calbindin. In the SNc, tyrosine hydroxylase (TH)–containing dopaminergic neurons are depleted. Similar neurons in the C1 and A2 regions of the medulla also showed reduced TH activity, which has been associated with orthostatic hypotension [1].

Biochemical analyses have found only minor differences in reduced striatal and nigral dopamine content in MSA when compared with PD [1]. However, unlike PD, mitochondrial respiratory chain function in the substantia nigra is normal in MSA. An increase in total iron content appears to reflect sites of primary damage and occurs in both PD and MSA substantia nigra, as well as in MSA striatum. Decreased noradrenaline levels are reported in septal nuclei, nucleus accumbens, hypothalamus, and locus ceruleus, whereas a consistent deficit of choline acetyltransferase is found in red nucleus, dentate, pontine, and inferior olivary nuclei, with variable involvement of the striatum and additional areas. Cerebellar and, in particular, Purkinje cell damage has been indicated by reduced levels of glutamate dehydrogenase, amino acid–binding sites, and cerebrospinal fluid (CSF) calbindin-D [2].

It is shown that the content of neurofilament in CSF is significantly higher both in PSP and MSA compared with PD patients, reflecting the degree of ongoing neuronal degeneration affecting mainly the axonal compartment [1].

There are several studies measuring CSF content of biogenic amine metabolites/derivatives, thiamine, neuropeptide Y, SP, or corticotropin-releasing hormone in patients with MSA [1,2]. In particular, the levels of the dopamine metabolite homovanillic acid (HVA), the serotonin metabolite 5-hydroxyindoleacetic acid (5-HIAA) and precursor tryptophan, and the noradrenaline metabolite 3-methoxy-4-hydroxyphenylethylene glycol (MHPG) and thiamine were measured in the CSF of patients with OPCA (among others), as compared with sex- and age-matched control subjects. CSF HVA, MHPG, and thiamine values were markedly lower than those in control subjects, whereas CSF 5-HIAA values showed only a trend toward lower levels than control subjects.

B. Neuropharmacological Findings

The combination of nigral and striatal degeneration is the core pathology underlying parkinsonism in MSA (MSA-P). The degenerative process affects nigrostriatal dopaminergic transmission at both presynaptic and postsynaptic sites. Pathologically, the loss of dopaminergic neurons in MSA-P is comparable to that found in PD. Only a few patients with MSA exhibit a presynaptic pattern with minimal putaminal changes. There is a close anatomical relationship between nigral and striatal degeneration in MSA-P. Degeneration of pigmented dopaminergic neurons begins and predominantly involves the ventrolateral tier of the SNc, which in turn projects to the dorsolateral posterior putamen, the latter of which is the predominant site of striatal degeneration in MSA. Several postmortem immunohistochemical, autoradiographical, and *in vivo* neuroimaging studies suggest that both striatal outflow pathways are affected: enkephalin-containing striatal neurons projecting to the external globus pallidus that carry dopamine D_2 receptors (indirect pathway) and SP-containing cells projecting to the internal globus pallidus and SNr that carry D_1 receptors (direct pathway) [1].

In accordance with the topographical projection of the putamen onto pallidal segments, the posterolateral portions of the external and internal globus pallidus and the ventrolateral portion of the substantia nigra are deafferented from striatal projections [2].

Progressive loss of striatal dopamine receptors and striatal output systems might explain levodopa unresponsiveness in most patients with MSA-P. Those patients with a good initial response to levodopa would, thus, have less striatal damage than those with absent or poor initial response. However, evidence suggests that the response to levodopa does not always depend solely on the degree of striatal cell loss [1]. *In vivo* positron emission tomographic studies have also failed to clearly correlate therapeutic response with striatal D_2 receptor status. Additional loss of D_1 and opiate receptors could also be an important factor underlying dopaminergic unresponsiveness in MSA-P, as well as other changes downstream of striatum itself [2].

VI. Molecular Biology

The discovery of GCIs in MSA brains firmly established glial pathology as a biological hallmark of this disorder, akin to the Lewy bodies (LBs) of PD. GCIs are argyrophilic and half-moon, oval, or conical in shape [3,4]. They consist of 20- to 30-nm diameter filaments and contain ubiquitin and tau, which are the classic cytoskeletal antigens.

Furthermore, α-synuclein, a presynaptic protein that is affected by point mutations in some families with autosomal-dominant PD and that is present in LBs, has also been observed in both neuronal and GCIs [5] in brains of patients with MSA. The accumulation of α-synuclein into filamentous inclusions appears to play a mechanistic role in the pathogenesis of several α-synucleinopathies including PD, dementia with Lewy bodies, Down's syndrome, familial Alzheimer's disease (AD), sporadic AD, MSA, and other synucleinopathies.

The α-synuclein accumulation in these inclusions appears to precede their ubiquitination, because α-synuclein antibodies detect a greater number of inclusions than ubiquitin antibodies. Importantly, α-synuclein, but not ubiquitin, antibodies also reveal numerous degenerating neurites in the white matter of MSA cases [1]. This suggests that an unrecognized degree of pathology may be present in the axons of MSA cases. However, whether neuronal/axonal α-synuclein pathology precedes glial α-synuclein pathology has not been examined.

VII. Animal Models

The growing number of animal models for MSA reflects the search for a preclinical testbed for elucidating the pathogenesis and developing novel effective therapeutic interventions. The available MSA animal models (toxin model, α-synuclein transgenic model, and the combined toxin–α-synuclein transgenic model) have been reviewed in detail by Stefanova et al. [27].

VIII. Clinical Picture

A. Presenting Features

Patients with MSA may present with parkinsonism that usually responds poorly to levodopa. This has been identified as the most important early clinical discriminator of MSA and PD, although a subgroup of patients with MSA may show a good or, rarely, excellent, but usually short-lived, response to levodopa [1]. Progressive ataxia may also be the presenting feature of MSA, and in Japan — as compared with Western countries — this presentation (i.e., MSA-C) appears to be more than the parkinsonian variant (i.e., MSA-P) [28]. Autonomic failure with symptomatic orthostatic hypotension and/or urogenital disturbance may accompany the motor disorder in up to 50% of patients at disease onset.

Unusual presentations in pathologically proven cases have included strokelike episodes evolving into parkinsonism, rapid eye movement (REM) sleep behavior disorder, pseudo-transient ischemic attacks in the anterior or posterior circulation, as well as limb-shaking attacks [1].

B. Features of Established Disease

1. Parkinsonism

Bradykinesia, rigidity, postural tremor, dysequilibrium, and gait unsteadiness characterize parkinsonism associated with MSA. Jerky postural tremor and, less commonly, tremor at rest may be superimposed. Frequently, patients exhibit orofacial dystonia associated with a characteristic quivering high-pitched dysarthria. Postural stability is compromised early on; however, recurrent falls at disease onset are unusual in contrast to PSP. According to the published literature, the vast majority (84–100%) of patients developed parkinsonism. Pure MSA-P (parkinsonism without cerebellar signs) represented the single most common motor subtype, ranging from 40% through 48% to 66%. In contrast, pure MSA-C (i.e., cerebellar ataxia without parkinsonism) was just present in up to 16% of patients [1]. It has been initially suggested that a symmetrical akinetic-rigid picture without tremor might distinguish MSA-P from PD. However, motor disturbance was asymmetrical in 74% of patients and unilateral at onset in 47% of cases. Also, some sort of tremor was present in 64–80% of cases, and tremor present at rest was observed in 29–40% of cases. Even so, a classic pill-rolling resting tremor was reported in only 7–9% of subjects [1]. Therefore, the differential diagnosis of MSA-P and PD may be quite difficult in the early stages because of a number of overlapping features such as rest tremor or asymmetrical akinesia and rigidity. Furthermore, levodopa-induced improvement of parkinsonism may be seen in 30% of patients with MSA-P; however, the benefit is transient in most of them. Levodopa-induced dyskinesia affecting orofacial and neck muscles occurs in 50% of patients with MSA-P, sometimes in the absence of motor benefit [9]. In most instances, a fully developed clinical picture of MSA-P evolves within 5 years of disease onset, allowing a clinical diagnosis during follow-up [1].

2. Dysautonomia

Dysautonomia is characteristic of both MSA subtypes, primarily comprising urogenital and orthostatic dysfunction.

a. Urogenital Dysfunction Urinary incontinence (71%) or retention (27%), often early in the course or as presenting symptoms, are frequent [1,29]. Disorders of micturition in MSA generally occur more commonly, earlier, and to a more severe degree than in PD. Urinary retention can be caused or exacerbated by benign prostatic hypertrophy or by perineal laxity secondary to difficult childbirth or uterine descent in women. In men, the urological symptoms of pollakiuria, urgency, nocturia, and incontinence together with hesitancy and incomplete emptying or chronic retention may simulate those of prostatic outflow obstruction. In a series of patients with probable MSA, 43% of men had undergone futile prostatic or bladder neck surgery before the correct diagnosis was made, although more than half had neurological symptoms or signs at the time of the procedure [29]. Stress incontinence occurred in 57% of the women, and of these, half had undergone surgery. The results of surgery were also poor [29]. Fecal incontinence was much rarer (2–12%), despite frequent severe denervation of the external anal sphincter, suggesting that the mechanisms of urinary and fecal continence are distinct in MSA [1].

Early impotence is virtually universal in men with MSA. In a series of 62 patients with MSA, impotence occurred in 96% of the men and was the first symptom alone in 37% [29]. In addition, patients with MSA may note severe constipation and hypohidrosis or anhidrosis.

b. Orthostatic Dysfunction Recurrent syncopal attacks are commonly regarded as a typical feature of MSA. However, severe orthostatic hypotension with recurrent (more than three) syncopes was only reported in 15% of subjects, whereas postural faintness was present, but only to a mild or moderate degree, in up to 53% of cases [1]. The analysis of a detailed questionnaire and autonomic function tests in a series of 121 patients with clinically diagnosed MSA showed that urinary symptoms (96%) were more common than orthostatic symptoms (43%) [1]. Orthostatic hypotension is frequently associated with impaired or absent reflex tachycardia on standing. Dopaminergic drugs may provoke or worsen orthostatic hypotension. Recumbent arterial hypertension, mainly due to loss of baroreflexes, may be seen in a few patients with severe cardiovascular autonomic failure [9].

3. Cerebellar Disorder

The cerebellar disorder comprises gait ataxia, limb kinetic ataxia, scanning dysarthria, and cerebellar oculomotor disturbances. Cerebellar signs, most commonly manifesting as a wide-based ataxic gait, developed in 34–59% of patients [1]. However, a subgroup of patients presented with narrow-based unsteady gait due to more marked impairment of postural reflexes. Spontaneous and/or gaze-evoked nystagmus, often subtle, was detected in 23–25% of patients [1]. Cerebellar scanning dysarthria may also occur. The finding of a mixed dysarthria with combinations of hypokinetic, ataxic, and spastic components is consistent with both the overall clinical and the neuropathological changes in MSA [1].

Patients with MSA-C usually develop additional noncerebellar symptoms and signs but, before doing so, may be indistinguishable from other patients with idiopathic late-onset cerebellar ataxia, many of whom have a disease restricted clinically to cerebellar signs and pathologically to degeneration of the cerebellum and olives [9].

4. Pyramidal Signs

Although pyramidal signs may be elicited in up to 61% of patients with MSA [1], obvious spastic paraparetic gait or significant pyramidal weakness should cast doubt on the clinical diagnosis of MSA.

C. Other Clinical Features

Besides the poor response to levodopa and the additional presence of pyramidal or cerebellar signs or autonomic failure as major diagnostic clues, certain other features may either raise suspicion of MSA or at least suggest that one might not be dealing with PD [1,9]. These early warning signs of MSA, so-called "red flags," are described in the following paragraphs.

Camptocormia (bent spine) has been associated with PD for a long time. It is characterized by severe forward flexion of the thoracolumbar spine, which increases while walking and disappears in the recumbent position. In contrast with other skeletal disorders of the spine such as kyphosis, the deformity in camptocormia is not fixed and is corrected by passive extension or lying in the supine position.

Subacute *Pisa syndrome*, a form of severe axial dystonia, has also been reported in MSA [9]. However, Pisa syndrome is a nonspecific feature of parkinsonism and it may also emerge in patients with other neurodegenerative conditions.

Tremulous myoclonic jerks, of small amplitude and often stretch-sensitive, usually affecting the fingers, occur in a number of patients with MSA but are otherwise rare in nondemented patients with PD [1].

A more common and characteristic feature suggesting MSA is the development of a disproportionate *antecollis* [1], hampering feeding, communication, and vision. The pathophysiological basis remains uncertain (myopathy, focal dystonia?). Botulinum toxin injections into sternocleidomastoid muscles are usually unrewarding and worsen dysphagia.

Other manifestations of focal dystonia are less common but may include *facial and hand dystonia, torticollis,* and *dystonic toe movements.* Levodopa exposure may worsen dystonic movements in the absence of antiparkinsonian benefit.

Nighttime respiratory stridor, commonly attributed to vocal cord paralysis, but perhaps reflecting dystonia of the vocal cords, is also a helpful diagnostic pointer. Nocturnal stridor has been considered a poor prognostic feature. Analysis of survival curves of 30 patients with follow-up information showed a significantly shorter survival from the sleep evaluation, but not from disease onset, for patients with stridor compared with those without [1]. Inspiratory stridor was documented in 9–34% of patients and occurred at any point in the disease process [1]. In fact, several cases have presented acutely with laryngeal palsy requiring tracheostomy or nasotracheal intubation. Tracheostomy in later disease stages may be more controversial.

Speech impairment develops in virtually all patients with MSA and is probably largely related to laryngeal dysfunction. It tends to be dominated by hypophonic monotony or a scanning quality according to clinical subtype. As well as the low volume monotony of parkinsonism, a quivering, irregular, severely hypophonic, or slurring *dysarthria* is often so characteristic that the diagnosis can be suggested by listening to the patient on the telephone. Dysarthria tends to develop earlier, be more severe, and is associated with more marked dysphagia in MSA compared with PD. Patients with MSA are prone to sleep-related breathing disorders, often resulting in nighttime oxygen desaturation [2]. They also often show disrupted sleep with REM phase alterations and may present with isolated *REM sleep behavior disorder* [1]. Sleep disorders are more common in patients with MSA than in those with PD after the same disease duration, reflecting the more diffuse underlying pathological process in MSA [21].

Abnormal sudomotor function, sympathetic skin response, impaired heat tolerance, and skin temperature regulation have been described in MSA [2]. Furthermore, patients with MSA often have cold, dusky, violaceous hands, with poor circulatory return after blanching by pressure, suggesting a defect in neurovascular control of distal extremities.

IX. Clinical Diagnosis and Clinical Diagnostic Criteria

The clinical diagnosis of MSA is fraught with difficulty and there are no pathognomonic features to discriminate the common (80% of cases) parkinsonian variant (MSA-P) from PD. In a clinicopathological study conducted by Litvan et al. [30], primary neurologists (who followed the patients clinically) identified only 25% of patients with MSA at the first visit (42 months after disease onset). Even at last neurological follow-up (74 months after disease onset), half of the patients were misdiagnosed and the correct diagnosis in the other half was established on average 4 years after disease onset. Mean rater sensitivity for movement disorder specialists was higher but still suboptimal at the first (56%) and last (69%) visit. Consistent with this observation, most of the patients with MSA identified in epidemiological surveys were only diagnosed during the study, suggesting that MSA is poorly recognized in clinical practice and is commonly mistaken for PD or other atypical parkinsonian disorders because of a number of overlapping features [1].

Clinical diagnostic criteria for MSA were proposed by Quinn in 1989 [6]. These criteria were slightly modified in 1994 to include an investigation (sphincter electromyogram [EMG]); furthermore, the category "possible MSA-OPCA" was defined for the first time [31]. According to this schema, patients are classified as either "SND-" or "OPCA-" type MSA depending on the predominance of parkinsonism or cerebellar ataxia. These criteria define two levels of clinical diagnostic certainty (possible, probable) and reserve a definite diagnosis to neuropathological confirmation. So far, sensitivity and specificity of the Quinn criteria have never been prospectively determined. Retrospective validation studies demonstrated have shown good specificity, but poor sensitivity of these criteria [1]. In April 1998, an International Consensus Conference was convened to develop optimized criteria for a clinical diagnosis of MSA [7]. The consensus criteria have since been widely established in the research community and in movement disorders clinics. They also define three diagnostic categories of increasing certainty: possible, probable, and definite. The diagnoses of possible and probable MSA are based on the presence of specific clinical features. In addition, several well-defined exclusion criteria must be considered [7]. A definite diagnosis requires a typical neuropathological lesion pattern with α-synuclein–positive GCIs [5]. The diagnostic levels are similar to those of Quinn criteria, but there are a number of major differences between the two sets of criteria [2]. These differences are the lack of investigations in Gilman criteria and the avoidance of potential confusion of possible MSA-P and probable MSA-C as occurs in Quinn criteria. Although such formal diagnostic criteria are important for certain types of clinical research, they add little to the problem of detecting early cases and improved screening instruments are certainly needed.

X. Natural History of the Disease

MSA usually manifests in middle age, affects both sexes, and progresses relentlessly with a mean survival significantly shorter than PD [2].

A. Onset

MSA is a disease that commonly causes clinical symptoms beginning in the sixth decade, although occasionally symptoms commence as early as the fourth decade. In a series of 100 cases of MSA, the median age at onset was 53 years and the range was 33–76 years [1]. In a meta-analysis of 433 cases, mean age at onset was 54.2 years (range 31–78 years) [1].

B. Progression

MSA is a chronically progressive disease characterized by the gradual onset of neurological symptoms and accumulation of disability reflecting involvement of the systems initially unaffected. Thus, patients who present initially with extrapyramidal features commonly progress to develop autonomic disturbances, cerebellar disorders, or both. Conversely, patients who begin with symptoms of cerebellar dysfunction often progress to develop extrapyramidal disorders, autonomic disorders, or both. Patients whose symptoms initially are autonomic may later develop other neurological disorders.

The progression to different Hoehn and Yahr (HY) stages was evaluated in 81 patients with pathologically confirmed parkinsonism. Patients with PD showed significantly longer latencies to each HY stage than patients with atypical parkinsonian disorders (APDs). In fact, development of an HY-III within 1 year of motor onset accurately predicted an APD. However, the progression to each HY stage was unhelpful in distinguishing the APD from each other. Once patients with PD and APD became wheelchair bound, both had equally short survival times [32].

In another study [28], median intervals from onset to aid-requiring walking, confinement to a wheelchair, a bedridden state, and death were 3, 5, 8, and 9 years, respectively.

XI. Conclusion

During the past 15 years, there have been major advances in our understanding of the cellular and molecular pathology of MSA. At the same time, the first multicenter intervention trials have been launched in Europe, and others are planned in North America. Although therapeutic options are limited, there is a real hope for a radical change in our approach to this devastating illness.

References

1. Geser, F., Colosimo, C., and Wenning, G. K. (2005). Multiple system atrophy. In "Neurodegenerative Diseases" (M. Flint Beal, A. E. Lang, and A. Ludolph, eds.), pp. 623–662. University Press, Cambridge.
2. Colosimo, C., Geser, F., and Wenning, G. K. (2005). Clinical spectrum and pathological features of multiple system atrophy. In "Animal Models of Movement Disorders" (M. LeDoux, ed.), pp. 541–570. Elsevier, Amsterdam.
3. Papp, M. I., Kahn, J. E., and Lantos, P. L. (1989). Glial cytoplasmic inclusions in the CNS of patients with multiple system atrophy (striatonigral degeneration, olivopontocerebellar atrophy and Shy–Drager syndrome). *J. Neurol. Sci.* **94**, 79–100.
4. Papp, M. I., and Lantos, P. L. (1992). Accumulation of tubular structures in oligodendroglial and neuronal cells as the basic alteration in multiple system atrophy. *J. Neurol. Sci.* **107**, 172–182.
5. Spillantini, M. G., Crowther, R. A., Jakes, R., Cairns, N. J., Lantos, P. L., and Goedert, M. (1998). Filamentous alpha-synuclein inclusions link multiple system atrophy with Parkinson's disease and dementia with Lewy bodies. *Neurosci. Lett.* **251**, 205–208.
6. Quinn, N. (1989). Multiple system atrophy—the nature of the beast. *J. Neurol. Neurosurg. Psychiatry* **52(Suppl)**, 78–89.
7. Gilman, S., Low, P. A., Quinn, N., Albanese, A., Ben-Shlomo, Y., Fowler, C. J., Kaufmann, H., Klockgether, T., Lang, A. E., Lantos, P. L., Litvan, I., Mathias, C. J., Oliver, E., Robertson, D., Schatz, I., and Wenning, G. K. (1998). Consensus statement on the diagnosis of multiple system atrophy. *J. Auton. Nerv. Syst.* **74**, 189–192.
8. Wenning, G. K., Tison, F., Seppi, K., Sampaio, C., Diem, A., Yekhlef, F., Ghorayeb, I., Ory, F., Galitzky, M., Scaravilli, T., Bozi, M., Colosimo, C., Gilman, S., Shults, C. W., Quinn, N. P., Rascol, O., and Poewe, W. (2004). Multiple System Atrophy Study Group. Development and validation of the Unified Multiple System Atrophy Rating Scale (UMSARS). *Mov. Disord.* **19**, 1391–1402.
9. Wenning, G. K., Colosimo, C., Geser, F., and Poewe, W. (2004). Multiple system atrophy. *Lancet Neurol.* **3**, 93–103.
10. Nee, L. E., Gomez, M. R., Dambrosia, J., Bale, S., Eldridge, R., and Polinsky, R. J. (1991). Environmental-occupational risk factors and familial associations in multiple system atrophy: a preliminary investigation. *Clin. Auton. Res.* **1**(1), 9–13.
11. Vanacore, N., Bonifati, V., Fabbrini, G., Colosimo, C., De Michele, G., Marconi, R., Stocchi, F., Nicholl, D., Bonuccelli, U., De Mari, M., Vieregge, P., Meco, G., and the ESGAP Consortium. (2005). Case–control study of multiple system atrophy. *Mov. Disord.* **20**, 158–163.
12. Vanacore, N., Bonifati, V., Fabbrini, G., Colosimo, C., Marconi R, Nicholl D., Bonuccelli U., Stocchi F., Lamberti P., Volpe G., De Michele G., Iavarone I., Bennett P., Vieregge P., and Meco G (2000). Smoking habits in multiple system atrophy and progressive supranuclear palsy. *Neurology* **54**, 114–119.
13. Nicholl, D. J., Bennett, P., Hiller, L., Bonifati, V., Vanacore, N., Fabbrini, G., Marconi, R., Colosimo, C., Lamberti, P., Stocchi, F., Bonuccelli, U., Vieregge, P., Ramsden, D. B., Meco, G., and Williams, A. C. (1999). A study of five candidate genes in Parkinson's disease and related neurodegenerative disorders. European Study Group on Atypical Parkinsonism. *Neurology* **53**, 1415–1421.
14. Ozawa, T., Takano, H., Onodera, O., Kobayashi, H., Ikeuchi, T., Koide, R., Okuizumi, K., Shimohata, T., Wakabayashi, K., Takahashi, H., and Tsuji, S. (1999). No mutation in the entire coding region of the alpha-synuclein gene in pathologically confirmed cases of multiple system atrophy. *Neurosci. Lett.* **270**, 110–112.
15. Gilman, S., Sima, A. A., Junck, L., Kluin, K. J., Koeppe, R. A., Lohman, M. E., and Little, R. (1996). Spinocerebellar ataxia type 1 with multiple system degeneration and glial cytoplasmic inclusions. *Ann. Neurol.* **39**, 241–255.
16. Bandmann, O., Sweeney, M. G., Daniel, S. E., Wenning, G. K., Quinn, N., Marsden, C. D., and Wood, N. W. (1997). Multiple-system atrophy is genetically distinct from identified inherited causes of spinocerebellar degeneration. *Neurology* **49**, 1598–1604.
17. Gilman, S., Little, R., Johanns, J., Heumann, M., Kluin, K. J., Junck, L., Koeppe, R. A., and An, H. (2000). Evolution of sporadic olivopontocerebellar atrophy into multiple system atrophy. *Neurology* **55**, 527–532.
18. Daniel, S. E. (1999). The neuropathology and neurochemistry of multiple system atrophy. In "Autonomic Failure: A Textbook of Clinical Disorders of the Autonomic Nervous System" (C. J. Mathias and R. Bannister, eds.), pp. 321–328. Oxford University Press, Oxford.
19. Oppenheimer, D. (1984). Diseases of the basal ganglia, cerebellum and motor neurons. In "Greenfield's Neuropathology" (J. H. Adams, J. A. N. Corsellis, and L. W. Duchen, eds.), pp. 699–747. Wiley, New York.
20. Takei, Y., and Mirra, S. (1973). Striatonigral degeneration: a form of multiple system atrophy with clinical parkinsonism. In "Progress in Neuropathology" (H. M. Zimmermann, ed.), Vol. 2, pp. 217–251. Grune & Stratton, New York.
21. Benarroch, E. E., Smithson, I. L., Low, P. A., and Parisi, J. E. (1998). Depletion of catecholaminergic neurons of the rostral ventrolateral medulla in multiple systems atrophy with autonomic failure. *Ann. Neurol.* **43**, 156–163.
22. Benarroch, E. E., Schmeichel, A. M., Low, P. A., and Parisi, J. E. (2004). Involvement of medullary serotonergic groups in multiple system atrophy. *Ann. Neurol.* **55**, 418–422.
23. Papp, M. I., and Lantos, P. L. (1994). The distribution of oligodendroglial inclusions in multiple system atrophy and its relevance to clinical symptomatology. *Brain* **117**, 35–243.
24. Benarroch, E. E., and Schmeichel, A. M. (2001). Depletion of corticotrophin-releasing factor neurons in the pontine micturition area in multiple system atrophy. *Ann. Neurol.* **50**, 640–645.
25. Isozaki, E., Matsubara, S., Hayashida, T., Oda, M., and Hirai, S. (2000). Morphometric study of nucleus ambiguus in multiple system atrophy presenting with vocal cord abductor paralysis. *Clin. Neuropathol.* **19**, 213–220.
26. Benarroch, E. E., Schmeichel, A. M., and Parisi, J. E. (2003). Preservation of branchimotor neurons of the nucleus ambiguus in multiple system atrophy. *Neurology* **60**, 115–117.
27. Stefanova, N., Tison, F., Reindl, M., Poewe, W., and Wenning, G. K. (2005). Animal models of multiple system atrophy. *Trends Neurosci.* **28**, 501–506.

28. Watanabe, H., Saito, Y., Terao, S., Ando, T., Kachi, T., Mukai, E., Aiba, I., Abe, Y., Tamakoshi, A., Doyu, M., Hirayama, M., and Sobue, G. (2002). Progression and prognosis in multiple system atrophy: an analysis of 230 Japanese patients. *Brain* **125**, 1070–1083.
29. Beck, R. O., Betts, C. D., and Fowler, C. J. (1994). Genitourinary dysfunction in multiple system atrophy: clinical features and treatment in 62 cases. *J. Urol.* **151**, 1336–1341.
30. Litvan, I., Goetz, C. G., Jankovic, J., Wenning, G. K., Booth, V., Bartko, J. J., McKee, A., Jellinger, K., Lai, E. C., Brandel, J. P., Verny, M., Chaudhuri, K. R., Pearce, R. K., and Agid, Y. (1997). What is the accuracy of the clinical diagnosis of multiple system atrophy? A clinicopathologic study. *Arch. Neurol.* **54**, 937–944.
31. Quinn, N. (1994). Multiple system atrophy. *In* "Movement Disorders" (C. D. Marsden and S. Fahn, eds.), 3rd edition, pp. 262–281. Butterworth-Heinemann, London.
32. Müller, J., Wenning, G. K., Jellinger, K., McKee, A., Poewe, W., and Litvan, I. (2000). Progression of Hoehn and Yahr stages in parkinsonian disorders: a clinicopathologic study. *Neurology* **55**, 888–891.

8

Olivopontocerebellar Atrophy (OPCA)

José Berciano

Keywords: *cerebellum, olivopontocerebellar atrophy, polyglutamine diseases, spinocerebellar ataxia*

I. Brief History and Nomenclature
II. Etiology and Pathogenesis
III. Diagnosis
IV. Treatment
   References

I. Brief History and Nomenclature

In 1900, the term *olivopontocerebellar atrophy* (OPCA) was introduced by Dejerine and Thomas to designate the pathological framework in a sporadic case with idiopathic late-onset progressive cerebellar ataxia [reviewed in reference 1]. Nine years before, however, Menzel had reported a family with a complex clinical picture characterized by progressive ataxia, spasmodic dysphonia, rigidity in the lower limbs, dysphagia, and dystonic posture of the neck [reviewed in reference 1]. Onset of symptoms was at about 30 years of age. There were four affected members over two generations. Autopsy revealed olivopontocerebellar lesions, together with degeneration of posterior and Clarke's columns, pyramidal and spinocerebellar tracts, and substantia nigra. Menzel found "very flattened and reduced subthalamic nuclei," but unfortunately he did not give any microscopic description of these structures; demonstration of luysian atrophy would have been of great interest in view of the dystonic postures of the patient. Be that as it may, this family is a good example of autosomal-dominant cerebellar ataxia (ADCA) type I in Harding's classification *(vide infra)*.

Dejerine and Thomas [see reference 1] described a sporadic case with progressive ataxic gait, dysarthria, impassive face, hypertonia, hyperreflexia, and urinary incontinence beginning at the age of 53 years. Autopsy 2 years later showed an advanced degeneration of the basis pontis, inferior olives, middle cerebellar peduncles, and to a lesser degree inferior cerebellar peduncles. There was severe atrophy of Purkinje cells, more marked in the cerebellar hemispheres than in the vermis. Neither the basal ganglia nor the substantia nigra is mentioned. According to the authors, OPCA is a nonfamilial disease that should be included among primary cerebellar degenerative disorders. Berciano [1] revised the pathological material of this case ("Vais D.V." Dejerine Laboratory, Paris), available preparations stained with the Weigert-Pal or carmin methods being as follows: seven transverse sections of the spinal cord, six transverse sections of the brainstem and cerebellum through medulla, pons and *isthmus rhombencephali*, and one horizontal section of the basal ganglia through anterior commissure. Although confirming the reported olivopontocerebellar lesions (Fig. 1) and the absence of apparent lesions of the putamen, it was not possible to establish whether the substantia nigra had degenerated. This finding would have been of great interest because the patient had had an incipient parkinsonism. Dejerine and Thomas's clinicopathological study probably represents the first description of multiple system atrophy (MSA) (see Chapter 7, which is devoted to this disorder).

The early reports of Dejerine and Thomas and, later, Loew's thesis [1], developed under the tutelage of Dejerine himself, considered OPCA to be atypical when there was a hereditary factor (as is the case for the aforementioned family reported by Menzel), lesions extending beyond the olivopontocerebellar framework, or a clinical presentation not limited to cerebellar symptoms. However, the concept of atypical OPCA fell into disuse with the recognition of familial OPCA (FOPCA) and of the many lesions that frequently accompany olivopontine degeneration.

In 1954, Greenfield [reviewed in reference 1] divided the pathological framework of ataxias into six main categories: (a) familial cerebellar, Menzel type (FOPCA); (b) familial cerebellar, type Holmes (cortical cerebellar atrophy [CCA]); (c) sporadic cerebellar, Dejerine–Thomas type (sporadic OPCA [SOPCA]); (d) sporadic cerebellar, Marie–Foix–Alajouanine type (sporadic CCA); (e) spinal forms (Friedreich's ataxia); and (f) dentatorubral atrophy (Ramsay Hunt syndrome). In this way, OPCA was divided into two forms: sporadic (Dejerine–Thomas type) and familial (Menzel type).

Using genetic, clinical, and pathological data, Konigsmark and Weiner classified OPCA into five categories (type I, dominant; type II, recessive; type III, with retinal degeneration; type IV, Schut and Haymaker type; and type V, with dementia, ophthalmoplegia, and extrapyramidal signs) [reviewed in reference 1]. They added a further two categories for sporadic observations and for those that do not fit the previous five, although in their opinion such cases probably belong to type II. Berciano [1] indicated that OPCA is a complex clinicopathological syndrome, which makes it difficult to sustain any classification based on clinical and pathological criteria. Thus, for example, the creation of "special types" or "variants" ignores the fact that mental deterioration or atrophy of the anterior gray horn is seen in half the cases of FOPCA. Furthermore, he indicated several omissions in the study by Konigsmark and Weiner, making the borderlines of their "types" somewhat hazy.

Pathological classification of the ataxias has several drawbacks [1,2]. It is not particularly helpful to clinicians who, not unnaturally, prefer to make some sort of working diagnosis before the autopsy results are available. Pathological classification ignores the fact that genetic heterogeneity affects not only the clinical picture but also the pathological framework, that is, this classification is impossible within reported families in which autopsy findings were not consistent.

We have seen that for almost a century clinicopathological studies in hereditary ataxias contributed to a compartmentalized but also confused nosology of these syndromes. To find a new classification was a pressing need. This task was achieved by Harding, culminating in a series of exceptional contributions to the field of hereditary ataxias and related disorders [2]. In 1983, she proposed starting from genetic and clinical features, which are, certainly, the tools used by neurologists in clinical practice. In this way, she proposed the clinicogenetic classification appearing in Table 1, which was soon universally accepted. In the past decade, this clinicogenetic classification has been drastically modified with the molecular genetic advances; in fact,

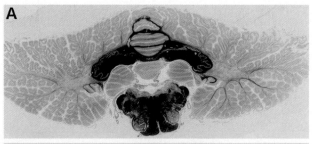
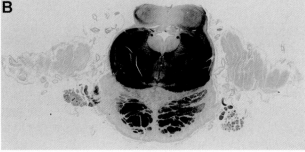

Figure 1 Olivopontocerebellar lesions in the case reported by Dejerine and Thomas in 1900 (*Nouv. Iconog. Salpêt.* **13**, 330–370) and reviewed by Berciano (*Doctoral thesis*, University of the Basque Country, Spain, 1978; see also reference 1). Both transverse sections are stained with the Weigert-Pal method. (**A**) This section through the medulla and cerebellum shows demyelination of the cerebellar white matter and olivocerebellar fibers. (**B**) This section through the upper half of the pons shows demyelination of the middle cerebellar peduncles.

Table 1 Harding's Clinicogenetic Classification of the Hereditary Ataxias and Paraplegias[a]

I. Congenital disorders of unknown etiology
II. Ataxic disorders with known metabolic of other cause
III. Ataxic disorders of unknown etiology
    A. Early onset cerebellar ataxia (EOCA) (onset usually before 20 years)
        i. Friedreich's ataxia
        ii. Early onset cerebellar ataxia with retained tendon reflexes
        iii. With hypogonadism ± deafness and/or dementia
        iv. With myoclonus (Ramsay Hunt syndrome, Baltic myoclonus)
        v. With pigmentary retinal degeneration ± mental retardation and/or deafness
        vi. With optic atrophy ± mental retardation
        vii. With cataracts and mental retardation (Marinesco–Sjögren syndrome)
        viii. With childhood-onset deafness and mental retardation
        ix. With congenital deafness
        x. With extrapyramidal features
        xi. X-linked recessive spinocerebellar ataxia
    B. Late-onset cerebellar ataxia (onset usually after 20 years)
        i. Autosomal-dominant cerebellar ataxia (ADCA) with optic atrophy/ophthalmoplegia/dementia/extrapyramidal features/amyotrophy (probably includes Azorean ataxia) (ADCA type I)
        ii. ADCA with pigmentary retinal degeneration ± ophthalmoplegia and/or extrapyramidal features (ADCA type II)
        iii. "Pure" ADCA of later onset (over 50 years) (ADCA type III)
        iv. Periodic autosomal-dominant ataxia
        v. "Idiopathic" late-onset cerebellar ataxia (ILOCA) (with either "pure" or cerebellar-plus syndrome)
IV. Hereditary spastic paraplegia (HSP)
    A. "Pure" HSP
    B. Complicated forms of HSP

[a] Adapted, with permission, from Harding, A. E. (1983). Classification of the hereditary ataxias and paraplegias. *Lancet* **i**, 1151–1155.

hereditary ataxias and paraplegias now comprise about 80 *loci*, 25 of them belonging to ADCA and being designated SCA1 through SCA25 (from *spinocerebellar ataxia* and numbers indicating the order of locus discovery) [3]. In 12 of these syndromes, the underlying mutations are known. Seven SCA subtypes (SCA1, SCA2, SCA3, SCA6, SCA7, and dentatorubral-pallidoluysian atrophy [DRPLA]) are caused by CAG trinucleotide expansions (polyglutamine; polyQ) in the respective genes. OPCA might be the pathological hallmark in every subgroup of Harding's classification except for Friedreich's ataxia, ADCA III, idiopathic late-onset cerebellar ataxia (ILOCA) with persistent "pure" cerebellar semeiology, and "pure" hereditary spastic paraplegia. According to neuroimaging or pathological findings, OPCA is the usual pathological background of SCA1, SCA2, SCA7, SCA12, SCA13, and DRPLA; furthermore, mild OPCA may also occur in SCA3, although spinopontine atrophy is the most common pathological framework.

To summarize, OPCA is a pathological label applicable to an increasing number of neurodegenerative syndromes included under the umbrella of congenital ataxias, early onset cerebellar ataxia (EOCA), ADCA I and III, periodic ataxia, ILOCA with cerebellar-plus syndrome, complicated hereditary spastic paraplegia, and MSA.

II. Etiology and Pathogenesis

OPCA is a type of primary neuronal degeneration of the spinocerebellar systems, namely, a form of sporadic or hereditary neurodegenerative disorder of unknown etiology [1]. The molecular advances in the field of hereditary ataxias have demonstrated that degeneration of olivopontine and cerebellar systems may occur in a vast number of genetic defects, including SCA caused by polyQ expansions (*vide supra*). Major insights have been attained into the molecular pathology of the trinucleotide repeat neurodegenerative diseases over the past decade [3,4]. For OPCA associated with SCA caused by CAG repeat expansion in the coding region of a gene, the functions of the affected proteins (ataxin-1 for SCA1, ataxin-2 for SCA2, ataxin-3 for SCA3, ataxin-7 for SCA7, and atrophin for DRPLA) are still unknown. Expanded polyQ repeats can form insoluble aggregates that are the hallmark of all polyQ diseases. Neuronal intranuclear inclusions (NIIs) characteristically occur. It is generally held that polyQ disorders are due to a toxic gain of function of mutant expanded proteins, although haploinsufficiency remains a tenable possibility according to which polyQ domains would alter proteasomal degradation in a malignant manner. The most intriguing possibility is transcriptional dysregulation. PolyQ sequences are relatively hydrophobic and can interact with each other in "polar zipper" configurations. Many transcriptionally active proteins have polyQ domains, and the expanded sequences resulting from expanded CAG repeats are suggested to sequester transcriptionally active proteins and alter cellular functions. This does not necessarily imply that the NIIs are themselves pathogenic agents of neurodegeneration; in fact, NIIs may be epiphenomenal or even protective [3,4]. The proposed pathogenic cascade of neurodegeneration probably includes protein misfolding, interference with DNA transcription and RNA processing, activation of apoptosis, and dysfunction of cytoplasmic elements [4]. The molecular and cellular bases for the differential vulnerability of olivopontine and cerebellar neuronal populations in the referred SCA polyQ expansion disorders are poorly understood. It has been speculated that expanded polyQ-containing proteins differ in their affinity for different transcriptionally active proteins and that the intersection of a given protein with an expanded polyQ domain and the repertoire of transcriptionally active proteins expressed by specific populations of neurons may determine regional effects [4]. Animal models of SCA1, SCA2, SCA3, or SCA7 have not reliably reproduced OPCA pathology. This is not the case with MSA in which combined mitochondrial inhibition and overexpression of oligodendroglial α-synuclein in a transgenic mouse generates a novel model of MSA, including OPCA and striatonigral degeneration; furthermore, in MSA degeneration of the olivopontocerebellar system significantly correlates with the frequency of glial (oligodendroglial) cytoplasmic inclusions (GCIs) (see Chapter 7).

A. Anatomical Underpinning

For any pathological and clinical reference, I will start from my comprehensive literature OPCA review comprising

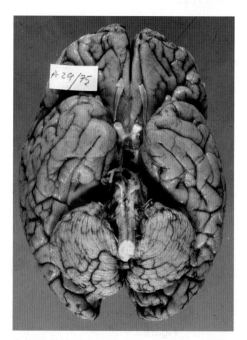

Figure 2 Brain macroscopic appearance in a patient belonging to an autosomal-dominant familial olivopontocerebellar atrophy (FOPCA) pedigree (taken from Berciano, *Doctoral thesis*, University of the Basque Country, Spain, 1978). Afterwards, molecular analysis in a patient of this pedigree showed CAG repeat expansion in the SCA2 gene. Note marked wasting of the cerebellum and brainstem.

117 cases reported by 1982 [1]. Macroscopic appearance is illustrated in Figure 2. The fundamental lesions (Fig. 3) are those localized in the cerebellum and in structures derived from Essick bands: the pontine, inferior olivary, arcuate, and pontobulbare nuclei. Systematic neuron loss in these nuclei and cerebellar cortex leads to demyelination, fibrillary gliosis of the cerebellar white matter, and formation of olivocerebellar, external arcuate, and pontocerebellar fibers. There may be evidence of dying-back axonopathy, especially in the middle cerebellar peduncles and white matter of the cerebellar hemispheres. Cerebellar cortical atrophy fundamentally involves Purkinje cells and predominates in the neocerebellum; silver stains reveal empty baskets where Purkinje cells have disappeared (see Fig. 3). In OPCA associated with MSA, the presence of GCIs, immunoreactive for both ubiquitin and α-synuclein, is constant (see Chapter 7). Similar inclusions have rarely been reported in OPCA associated with SCA1 or SCA2, but conversely to MSA, they lack immunoreactivity for α-synuclein [5].

Revision of the most detailed histological descriptions led me to conclude that associated lesions are constant. The main associated lesions are shown in Figure 4; differences between FOPCA and SOPCA are significant for lesions of the locus ceruleus, dentate nucleus, and spinal localizations except those of spinal tracts.

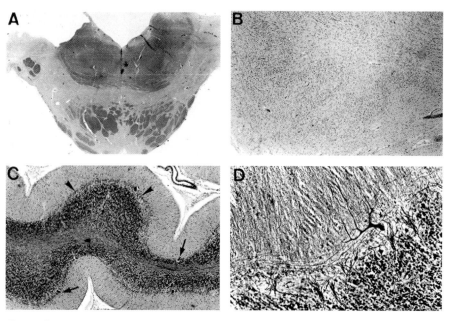

Figure 3 Microscopic olivopontocerebellar lesions in the case illustrated in Fig. 2. (**A**) Transverse section through the middle pons showing complete demyelination of transverse pontine fibers (Spielmeyer). (**B**) Disappearance of neurons and severe gliosis in the inferior olive (Nissl). (**C**) Marked loss of Purkinje cells in the cerebellum (arrows indicate a few cells remaining) with prominence of the Bergmann glia (arrowheads) (phosphotungstic hematoxylin). (**D**) Empty baskets (Naoumenko–Feigin).

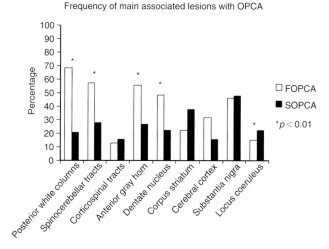

Figure 4 Histogram showing frequencies of the main associated lesions in olivopontocerebellar atrophy. (Adapted, with permission, from Table 5 in Berciano, J. [1982]. Olivopontocerebellar atrophy. A review of 117 cases. *J. Neurol. Sci.* **53**, 253–272.)

B. Epidemiological, Clinical, and Pathophysiological Underpinning

OPCA is not explicitly considered in general epidemiological surveys on spinocerebellar syndromes. We found in Cantabria (Spain) that the prevalence ratios of ADCA and ILOCA were 1.2 and 2.2 cases per 100,000, respectively [6]. Some 60% of patients included in these groups had a cerebellar-plus syndrome and their computed tomographic (CT) or magnetic resonance imaging (MRI) scans revealed cerebellar and brainstem atrophy, allowing a presumptive diagnosis of OPCA. According to these estimations, the prevalence ratio of OPCA is about 2 per 100,000.

The age at onset of disease ranges from 2 months to 53 years in FOPCA and from a congenital onset to 66 years in SOPCA [1]. Such great variability is explained by the fact that OPCA is the pathological background of many different neurodegenerative syndromes *(vide supra)*. By definition, a congenital or infantile onset occurs in OPCA associated with congenital OPCA; onset in the first two decades is characteristic of OPCA associated with EOCA and occasionally with ADCA, and adult onset is a hallmark of OPCA associated with ADCA, ILOCA, and MSA.

The clinical presentation of OPCA usually begins with cerebellar ataxia, especially involving gait [1]. When dementia or extrapyramidal rigidity is among the first manifestations, they remain dominant symptoms throughout the disease. In OPCA associated with SCA7, visual defect is usually an early symptom, which may precede gait ataxia [3]. Spasticity or psychomotor retardation is the most common presenting semeiology of OPCA associated with congenital ataxia.

The established clinical picture of OPCA is essentially a progressive cerebellar-plus syndrome. I will briefly revise the clinical picture, paying special attention to its pathophysiology. It is worth noting that complex pathological background of OPCA (vide supra) accounts for clinical variability, that is, for the appearance of noncerebellar semeiology. The frequencies of FOPCA and SOPCA symptoms and signs are summarized in Figure 5; the differences between them are significant for higher percentages of involuntary movements, dementia, ophthalmoplegia, and reduced vision in FOPCA, as well as sphincter disturbances in SOPCA. It is also notable that the presence of additional noncerebellar semeiology is much lower than that observed in a series of wheelchair-bound patients with ADCA I, where the usual presumptive pathological framework is OPCA (vide supra); between 70% and 95% of such patients had ophthalmoplegia, sphincter disturbances, dysphagia, and spinal or pyramidal signs [7], indicating that disease duration is an essential factor in the appearance and worsening of semeiology.

Progressive cerebellar disturbance, including gait and limb ataxia and dysarthria, is OPCA's most outstanding clinical feature (see Fig. 5). Extensive cerebellar wasting involving both the paleocerebellum and the neocerebellum accounts for this global cerebellar semeiology (see Figs. 1–3). PET scans revealed marked hypometabolism in the cerebellar hemispheres, cerebellar vermis, and brainstem of patients with OPCA compared with control subjects, and this abnormality correlated with the severity of cerebellar ataxia and dysarthria [8].

Parkinsonian symptoms, with rigidity and akinesia predominant, occur in 39% of FOPCA and 55% of SOPCA cases (see Fig. 5). The usual pathological substrate of OPCA parkinsonism is atrophy of the substantia nigra without Lewy bodies. Striatal lesions are usually absent or, if present, are mildly to moderately intense. A few patients with SOPCA with early extrapyramidal manifestations display severe striatal lesions; leaving aside the question of the presence of GCIs immunoreactive for α-synuclein (vide supra), it is impossible to differentiate these cases from those with parkinsonian MSA. Autoradiographic labeling with [^{3}H]-spiperone in a patient with SOPCA with parkinsonism showed normal densities of D_2 dopamine striatal receptors [9]. This finding is concordant with a PET study showing normal [^{11}C]-diprenorphine uptake despite reduced striatal accumulation of [^{18}F]-fluorodopa, which suggests that in some patients with OPCA, nigrostriatal projections degenerate in the absence of intrinsic loss of striatal neurons [9]. In OPCA associated with SCA2, Estrada et al. [11] found subclinical marked loss (ranging between 91% and 33% of normal; mean reduction, 73%) of nigral neurons. In our ADCA series comprising 73 examined patients coming from 30 pedigrees, bradykinesia was observed in just 12% of SCA2 cases [12]. These observations indicate the difficulty in detecting parkinsonian signs in severely ataxic patients.

Abnormal movements considered as a whole are significantly more common in FOPCA (see Fig. 5). These may include myoclonus, spasmodic torticollis, choreoballistic jerks, blepharospasm, and choreiform or athetotic dyskinesias. They are not always obvious manifestations, usually appearing late in the clinical course. A few cases merit comment. Bonduelle et al. reported an autosomal-dominant FOPCA pedigree clinically diagnosed as having dyssynergia cerebellaris myoclonica (Ramsay Hunt syndrome), because their patients exhibited a severe myoclonic postural syndrome; autopsy study, however, revealed OPCA lesions with normal or minimal involvement of the cerebellar dentatus nuclei [see reference 1]. In Roshenagen's eighth patient, suffering from FOPCA, the triad of choreiform movements, dementia, and ataxia led to a diagnosis of Huntington's chorea; nevertheless, no lesions of the striatum or cerebral cortex were found [see reference 1]. The pathophysiology of dyskinesias in these two pedigrees remains unestablished. Spasmodic torticollis in Menzel's observation [see reference 1] might be related to associated pallidoluysian degeneration to OPCA (vide supra). Blepharospasm in OPCA has been correlated with rostral brainstem lesions disrupting central dopaminergic and cholinergic pathways, resulting in disinhibition of brainstem reflexes or denervation supersensitivity of the facial nuclear complex. In patients with FOPCA, there may be twitching of the cheeks and perioral muscles induced by facial movements [13]. At rest, electromyographic recordings of the orbicularis oris and risorius muscles revealed myokymic discharges in the absence of visible movements. With voluntary contraction, electromyography showed synchronous discharges in these muscles ipsilaterally associated with visible twitching. There were enhanced long-latency facial reflex responses to stimuli applied to the facial or trigeminal nerves. The duration of electromyographic bursts was 10–75 ms, with a frequency of 8–25 Hz, consistent with

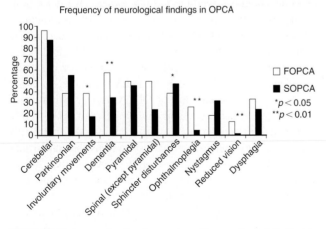

Figure 5 Histogram showing frequencies of the neurological findings in OPCA. (Adapted, with permission, from Table 4 in Berciano, J. [1982]. Olivopontocerebellar atrophy. A review of 117 cases. *J. Neurol. Sci.* **53**, 253–272.)

a myoclonic disorder. Because it was induced by activation of the facial muscles, this movement disorder represents a form of action facial myoclonus, probably due to a global brainstem functional derangement associated with pontine degeneration [13]. In our series of OPCA associated with SCA, facial action myoclonus occurred in 44% of SCA2, 12% of SCA3, and 20% of SCA7 cases, differences between SCA2 and SCA3 or SCA7 being significant ($p < .01$) [12].

Nonsevere cognitive decline occurs in up to 57% of cases (see Fig. 5). Descriptions of the mental state in OPCA are usually sparse, but most patients undergo personality and affective changes, as well as memory loss without aphasia, apraxia, or agnosia, all of which suggest that dementia is subcortical. In a series of 57 patients with FOPCA, detailed neuropsychological testing disclosed previously unrecognized deficits in verbal and nonverbal intelligence, memory, and frontal system [14]. The clinicopathological correlation is poor [1], as indicated by frequencies of dementia and atrophy of cerebral cortex (see Figs. 4 and 5). Nevertheless, subcortical structures implicated in subcortical dementia, namely striatum, thalamus, and brainstem nuclei, sometimes are involved in OPCA. Substantial neuronal reduction in the magnocellular population of the basal nucleus of Meynert has been reported in an FOPCA case with mental deterioration [15]. Moreover, Kish et al. [14] demonstrated in OPCA a marked cortical cholinergic deficiency, evidenced by reduced activity of the cholinergic marker enzymes choline acetyltransferase and acetylcholinesterase, and reduced cholinergic innervation of the caudate and dorsomedial nucleus accumbens. Prospective neuropsychological studies in patients with SCA1, SCA2, and SCA3 have shown that there may be mild deficits of verbal memory, probably resulting from disruption of a cerebrocerebellar circuitry at the pontine level.

In most reports, pyramidal signs, observed in 56 of 117 cases reviewed (see Fig. 5), are considered merely semeiological findings in the form of enhanced stretch reflexes, extensor-plantar responses, or both; rarely was spastic paraplegia or pseudobulbar dysarthria observed [1]. On serial examination, tendon reflexes can become hypoactive or, on the contrary, enhanced after initial normality. Degeneration of the pyramidal tract has only been reported in 17 cases (see Fig. 4), in 8 of which there were no or doubtful pyramidal signs. This gross clinicopathological dissociation had been pointed out earlier [1]. Pyramidal signs may be explained by a disease-specific pattern of neuronal loss in the spinal ventral horn involving small neurons of the dorsomedial zone [16].

Nonpyramidal spinal signs, such as amyotrophy, fasciculations, hypopallesthesia in lower limbs, lightnings pains, areflexia, kyphoscoliosis, and pes cavus, are significantly less common in SOPCA than in FOPCA (see Fig. 5). Amyotrophy and sensory loss often predominate distally in a pattern of peripheral neuropathy. Nerve conduction studies show normal or slightly slowed motor and sensory conduction velocities with reduction or disappearance of distal sensory potentials. Some patients display electromyographic evidence of collateral reinnervation, giant motor units, loss of motor units, or spontaneous activity at rest. Loss of large myelinated fibers and secondary demyelination to axonal loss has been found on sural nerve biopsies. These features, together with high proportion white column demyelination and anterior gray horn atrophy (see Fig. 4), suggest that peripheral neuropathy in OPCA is due to degeneration of posterior root ganglion and anterior horn cells in the spinal cord, that is, a type of spinal sensory and motor neuronopathy [1]. Spinal nonpyramidal signs are significantly more common in OPCA associated with SCA2 than in SCA3 or SCA7, although areflexia restricted to lower limbs is characteristic of SCA3 [12]. Amyotrophy and fasciculations occur in 20% of OPCA associated with SCA2 [12], but again a gross clinicopathological dissociation is evident, as autopsy studies have demonstrated a systematic reduction of lumbar motoneurons ranging between 33 and 83% (mean, 57%) and of thoracic motoneurons between 27 and 64% (mean, 48%) [11].

Sphincter disturbances are present in 39% of FOPCA cases and 48% of SOPCA cases (see Fig. 5), with similar frequencies being recorded in any OPCA associated with SCA [12]. They usually appear late in the clinical course. Bladder dysfunction is initially manifested as urinary urgency and afterward combined with incontinence. Double incontinence rarely develops, but urinary retention is uncommon. Erectile dysfunction was not recorded in our patients with OPCA SCA [12], a departure feature from MSA-OPCA, where this is an early cardinal manifestation (see Chapter 7). The pathophysiology of OPCA-related sphincter disturbances has not been completely established. According to Bakker, they should result from an interruption of cerebellar control over the sympathetic nervous system [see reference 1]. Urodynamic mechanisms of bladder dysfunction are probably similar to those reported in MSA, which have been studied in more detail. Kirby et al. [17] identified three fundamental lower urinary tract abnormalities in MSA: (a) involuntary detrusor contractions in response to bladder filling, which may result from a loss of inhibitor influences from the corpus striatum and substantia nigra; (b) loss of the ability to initiate voluntary micturition reflex, perhaps reflecting the degeneration of motor nuclei situated in the rostral pons and the lateral parts of the medulla, including corticotrophin-releasing factor neurons in the putative pontine micturition center [18]; and (c) profound urethral dysfunction, which appears to result in part from a loss of proximal urethral sphincter tone. In addition, the function of the striated component of the urethral sphincter is impaired, probably as a result of degeneration of the Onuf nucleus, which innervates the external muscles of the anus and urethra. However, among the 51 patients with OPCA with urinary incontinence reviewed by Berciano [1], the following features were observed: dementia in 28, posterior column ataxia or degeneration in 27, and pyramidal signs or degeneration of the tract in 7. Only four patients were free of clinical signs or anatomical lesions, which theoretically could cause sphincter disorders. Be that as it may, these data indicate that various factors are involved in the pathophysiology of urinary incontinence in OPCA.

Nuclear or supranuclear ophthalmoplegia is mainly associated with FOPCA (see Fig. 5). The entire spectrum of cerebellar disorders of ocular motility can be seen in OPCA, the most characteristic being slowing of saccades or the combined loss of pursuit and vestibular function [1]. Slowed saccades are observed in 56% of SCA2 and 60% of SCA7 cases but in just 29% of SCA3 cases [12]. It has been suggested that degeneration of the paramedian reticular formation might be the anatomic cause of saccadic anomalies in heredoataxic patients [1]. This hypothesis has been confirmed with the descriptions of damage of several precerebellar nuclei, including pontis centralis caudalis, reticulotegmental and raphe interpositus, implicated in the generation or regulation of saccadic movements [19].

With the exception of peripheral tapetoretinal degeneration, OPCA is associated with all varieties of visual defect described as occurring in the course of heredoataxia [1]. Visual defects are characteristic of FOPCA (see Fig. 5). ADCA II combines pigmentary macular degeneration and progressive cerebellar ataxia. Families with ADCA II usually have CAG expansions in the SCA7 gene [3].

Dysphagia is a relatively common manifestation of OPCA (see Fig. 5) and OPCA associated with SCA. Swallowing disorders are characteristic of intermediate and advanced disease. Dysphagia has been related to dysfunction of the superior esophageal sphincter [1]. Like the striated muscles of the urethral and the posterior cricoarytenoid muscle, the cricopharyngeus muscle is tonically and rhythmically active as a result of spontaneous discharge of reticular interneurons adjacent to the nucleus ambiguus. Conceivably the tonic firing of these motor neurons (Onuf's nucleus and nucleus ambiguus), which differs from other skeletal muscles, accounts for their conjoint tendency to degenerate in OPCA.

Sleep disorders, including nocturnal stridor, characteristic of OPCA associated with MSA (see Chapter 7), have exceptionally been reported in FOPCA. Restless legs syndrome and impaired sleep appear to be relatively common manifestations in SCA3 [20]. In our patients with FOPCA associated with SCA1, SCA2, and SCA3, neither nocturnal stridor nor prominent parasomnias occurred [12].

III. Diagnosis

Diagnosis of OPCA relies on the clinical picture, essentially characterized by a progressive cerebellar-plus syndrome *(vide supra)*. The nature, sporadic or familial, and age at onset of disease allow planning of genetic molecular studies to detect dynamic or point mutations associated with dominant ataxias and EOCA. (It is out of the scope of this chapter to address this question; see references 1 and 2 for review.) Electrophysiological studies, including nerve conduction studies, multimodal evoked potentials, eye movement recordings, and central motor conduction time investigation determination using magnetic stimulation, are useful to assess the participation of the corresponding neural system in OPCA. Neuroimaging techniques are the gold standard proof for delineating brainstem and cerebellar atrophy (Fig. 6), which is the hallmark of OPCA.

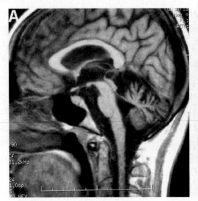
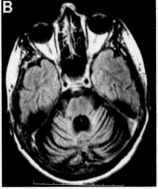
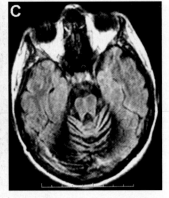

Figure 6 T1-weighted magnetic resonance imaging pictures from a patient with SCA2 with cerebellar-plus syndrome. **(A)** Sagittal section showing extensive wasting of the brainstem and cerebellar vermis. **(B)** Axial section through the mid-pons illustrating marked atrophy of the pons and cerebellar hemispheres with enlargement of the fourth ventricle. **(C)** This higher axial section shows atrophy of the pons and cerebellar vermis.

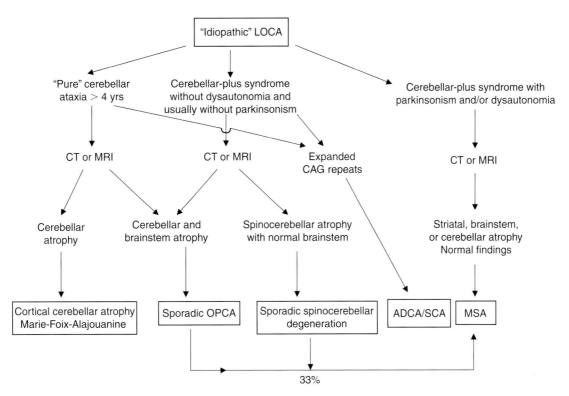

Figure 7 Algorithm outlining the author's suggested approach to the diagnosis of a patient with idiopathic late-onset cerebellar ataxia (ILOCA). The period of 4 years for "pure" cerebellar syndrome is based on the fact that additional noncerebellar symptoms usually come later. Patients with sporadic olivopontocerebellar atrophy may also have presynaptic parkinsonism, though rarely (see text).

A. Natural History

Because OPCA may be the pathological hallmark of many neurodegenerative ataxic disorders *(vide supra)*, the natural history varies accordingly with such nosological heterogeneity. In our series of dominant ataxias [12], the study population for survival analyses included 45 deceased patients belonging to the following subtypes; 24 of SCA2, 11 of SCA3, 2 of SCA7, and 8 with unknown mutations. Mean disease duration until death was 17.2 ± 9.1 years in SCA2 (range, 2–43), 18.4 ± 7.3 years in SCA3 (range, 11–35), and 22.1 ± 13.5 years in patients with unknown mutations (range, 4–40); the two deceased patients with SCA7 mutation died at 6 and at 93 years. Mean ages at death were similar in the different groups: 54.4 ± 18.6 years in SCA2 (range, 23–88), 59.3 ± 11.8 years in SCA3 (range, 40–80), and 58.1 ± 20.1 years in patients with unknown mutations (range, 20–88). No significant differences were observed in survival probability between the different groups ($p = .16$; log-rank test) (see Figure 2 in reference 12).

The clinical picture of ILOCA may be pure cerebellar ataxia or cerebellar-plus syndrome, with onset usually after the age of 20 years [6,22]. Neuroimaging studies usually show isolated cerebellar atrophy in the first case, suggesting that the pathological framework is CCA. Contrariwise in cases with additional noncerebellar semeiology, neuroimaging generally reveals a combination of brainstem and cerebellar atrophy suggestive of OPCA. Difficulties of nosological delimitation appear when confronting MSA, defined following *ad hoc* criteria (see Chapter 7), and sporadic OPCA as a variant of ILOCA. As there is no definitive clearcut limit between ILOCA and MSA, our diagnostic approach to the patient with an ILOCA syndrome is summarized in Figure 7; it is now clear that at least one-third of ILOCA-SOPCA cases evolve into MSA [23].

IV. Treatment

Symptomatic treatment is only possible for a selection of clinical manifestations. Levodopa may provide some benefit for extrapyramidal rigidity or bradykinesia. As in Parkinson's disease, treatment may be initiated with Sinemet 25/100, 0.5–1.0 tablet twice a day, and increased every few days to efficacy or toxicity. Good levodopa responsiveness could be linked to relative putaminal preservation *(vide supra)*. If there is no response, dopamine agonists in maximally tolerable doses may also be used.

Urinary disturbances due to detrusor hyperreflexia are treated with a peripherally acting anticholinergic agent such as oxybutynin (5–10 mg at bedtime) or propantheline. In any case, treatment of either urinary or sexual disturbances should be carried out in close collaboration with an expert in neurourology.

Although numerous reports have discussed the treatment of ataxia, most authors agree that results obtained with a variety of therapies generally are disappointing; in fact, none of the trials has produced results convincing enough to justify routine use of any drug in OPCA.

Psychosocial support is most important. Physical, occupational, and speech therapies may be essential to reduce the patient's disability and to maintain independent functioning longer. Gait training and assistive devices to prevent falling will avoid further debilitation of the patient.

Acknowledgment

Supported by "Centro de Investigación de Enfermedades Neurológicas (CIEN), Nodo HUMV/UC (CO3/CO6), Instituto de Salud Carlos III" (Madrid, Spain).

References

1. Berciano, J. (1982). Olivopontocerebellar atrophy. A review of 117 cases. *J. Neurol. Sci.* **53,** 253–272.
2. Harding, A. E. (1983). Classification of the hereditary ataxias and paraplegias. *Lancet* **i,** 1151–1155.
3. Schöls, L., Bauer, P., Schmidt, T., Schulte, T., and Riess, O. (2004). Autosomal dominant cerebellar ataxias: clinical features, genetics, and pathogenesis. *Lancet Neurol.* **3,** 291–304.
4. Albin, R. L. (2003). Dominant ataxias and Friedreich ataxia: an update. *Curr. Opin. Neurol.* **16,** 507–514.
5. Berciano, J., and Ferrer, I. (2005). Glial cell cytoplasmic inclusions in SCA2 do not express α-synuclein. *J. Neurol.* **252,** 742-744
6. Polo, J. M., Calleja, J., Combarros, O., and Berciano, J. (1991). Hereditary ataxias and paraplegias in Cantabria, Spain. An epidemiological and clinical study. *Brain* **114,** 855–866.
7. Dürr, A., Chneiweiss, H., Kathi, C., Stevanin, G., Cancel, G., Feingold, J., Agid, Y., and Brice, A. (1993). Autosomal dominant cerebellar ataxia type I is unrelated to genetic heterogeneity. *Brain* **116,** 1497–1508.
8. Roshenthal, G., Gilman, S., Koeppe, R. A., Kluin, K. J., Martel, D. S., Junck, L., and Gebarski, S. S. (1988). Motor dysfunction in olivopontocerebellar atrophy is related to cerebral metabolic rate with positron emission tomography. *Ann. Neurol.* **24,** 414–419.
9. Pascual, J., Pazos, A., Del Olmo, E., Figols, J., Leno, C., and Berciano, J. (1991). Presynaptic parkinsonism in olivopontocerebellar atrophy: clinical, pathological and neurochemical evidence. *Ann. Neurol.* **30,** 425–428.
10. Rinne, J. O., Burn, D. J., Mathias, C. J., Quinn, N. P., Marsden, C. D., and Brooks, D. J. (1995). Positron emission tomography studies on the dopaminergic system and striatal opioid binding in the olivopontocerebellar atrophy variant of multiple system atrophy. *Ann. Neurol.* **37,** 568–573.
11. Estrada, R., Galarraga, J., Orozco, G., Nodarse, A., and Auburger, G. (1999). Spinocerebellar ataxia 2 (SCA2): morphometric analyses in 11 autopsies. *Acta Neuropathol. (Berl.)* **97,** 303–311.
12. Infante, J., Combarros, O., Volpini, V., Corral, J., Llorca J, and Berciano J. (2005). Autosomal dominant cerebellar ataxias in Spain. Molecular and clinical correlations, prevalence estimation and survival analysis. *Acta Neurol. Scand.* **111,** 391–399.
13. Valls-Solé, J., Lou, J., and Hallet, M. (1994). Brainstem reflexes in patients with olivopontocerebellar atrophy. *Muscle Nerve* **17,** 1439–1448.
14. Kish, S. J., Robitaille, Y., El-Awar, M., Deck, J. H., Simmonds, J., Schut, J., Chang, L. J., DiStefano, L., and Freeman, M. (1989). Non-Alzheimer type pattern of brain choline acetyltransferase reduction in dominantly inherited olivopontocerebellar atrophy. *Ann. Neurol.* **26,** 362–367.
15. Tagliavini, F., and Pilleri, G. (1985). Neuronal loss in the basal nucleus of Meynert in a patient with olivopontocerebellar atrophy. *Acta Neuropathol. (Berl.)* **66,** 127–133.
16. Terao, S., Sobue, G., Hashizume Y., Mitsuma, T., and Takahashi, A. (1994). Disease-specific patterns of neuronal loss in the spinal ventral horn in amyotrophic lateral sclerosis, multiple system atrophy and X-linked recessive bulbospinal neuronopathy, with special reference to the loss of small neurons in the intermediate zone. *J. Neurol.* **241,** 196–203.
17. Kirby, R., Fowler, C., Gosling, J., and Bannister, R. (1986). Urethral dysfunction in progressive autonomic failure with multiple system atrophy. *J. Neurol. Neurosurg. Psychiatry* **49,** 554–562.
18. Benarroch, E. E., and Schmeichel, A. M. (2001). Depletion of corticotrophin-releasing factor neurons in the pontine micturition area in multiple system atrophy. *Ann. Neurol.* **50,** 640–645.
19. Rüb, U., Bürk, K., Schöls, L., Brunt, F. G., de Voss, R. A., Diez, G. O., Gierga, K., Ghebremedhin, E., Schultz, C., Del Turco, D., Mittelbronn, M., Auburger, G., Deller, T., and Braak, H. (2004). Damage to the reticulotegmental nucleus of the pons in spinocerebellar ataxia type 1, 2, and 3. *Neurology* **63,** 1258–1263.
20. Schöls, L., Haan, J., Riess, O., Amoiridis, G., and Przuntek, K. H. (1998). Sleep disturbance in spinocerebellar ataxias. Is the SCA3 mutation a cause of restless legs syndrome? *Neurology* **51,** 1603–1607.
21. Van de Warrenburg, B. P., Sinke, R. J., and Kremer, B. (2005). Recent advances in hereditary spinocerebellar ataxia. *JNEP* **64,** 171–180.
22. Harding, A. E. (1981). "Idiopathic" late onset cerebellar ataxia. A clinical and genetic study of 36 cases. *J. Neurol. Sci.* **51,** 259–271.
23. Gilman, S., Little., R., Johanns, J., Heumann M., Kluin, K. J., Junck, K. L., Koeppe, R. A., and An, H. (2000). Evolution of sporadic olivopontocerebellar atrophy into multiple system atrophy. *Neurology* **55,** 527–532.

9

Neurobiology of Progressive Supranuclear Palsy

Pratap Chand, DM, FRCP
Irene Litvan, MD

Keywords: *atypical parkinsonism, mitochondrial I inhibitors, neurodegeneration, progressive supranuclear palsy, tau*

I. History and Nomenclature
II. Etiology
III. Pathogenesis
IV. Relevant Structural Detail
V. Pharmacology, Biochemistry, Molecular Mechanisms
VI. Explanation of Symptoms in Relation to Pathophysiology
VII. Natural History
   References

I. History and Nomenclature

Progressive supranuclear palsy (PSP) was first described as a distinct clinicopathological entity in 1964 by neurologist Dr. J. Clifford Richardson (1909–1986) when he realized that a series of patients evaluated in Toronto since 1955 had an unusual combination of symptoms that seemed to correspond to a disease of which he was unaware [1]. When John C. Steele (1934) joined the neurology residency in 1961, he started to investigate the clinical features and progression of these patients during the following 2 years. Pathologist Dr. Jerry Olszewski (1913–1964) described in detail the pathological findings of seven cases that came to autopsy [1]. The disease was later called Steele–Richardson–Olszewski (SRO) syndrome after these investigators, but difficulties in using such a lengthy eponym led to it being called *PSP*. British investigators have coined the term *Richardson's syndrome* for the most typical presentation of this disease [2], and *PSP-parkinsonism* for atypical presentations.

II. Etiology

The cause of PSP remains unknown, but it is hypothesized that genetic and environmental factors contribute to its development.

A. Genetic Aspects

The neuronal proteins that bind microtubules and provide cytoskeletal scaffolding have a strong influence on the morphology and physiology of neurons. The microtubule-associated protein tau (MAPT) promotes assembly and stabilization of microtubules under normal physiological conditions (3). However, in PSP, tau protein undergoes a modification through phosphorylation, resulting in the generation of aberrant aggregates that are toxic to neurons.

The gene for the MAPT is located on chromosome 17, where it occupies more than 100 kb and contains at least 16 exons (3). The chromosomal region containing MAPT is composed by two haplotypes, H1 and H2, which are defined by linkage disequilibrium (LD) among several polymorphisms over the entire MAPT gene (3). PSP is associated with the H1 tau haplotype. The H1/H1 genotype is present in approximately 90% of patients with this disorder and approximately 60% of healthy Caucasians. It is unclear whether inheritance of the H1/H1 tau genotype represents a predisposition to develop PSP requiring other environmental or genetic factors, whether the H2/H2 tau genotype is protective, or whether a relatively rare mutation with low penetrance could contribute to the abnormal tau aggregation in PSP. The H1 haplotype dosage does not influence age at onset, severity, or survival of patients with PSP [4]. Using single nucleotide polymorphisms, Pittman et al. [5] mapped the LD regions flanking MAPT and established the maximum extent of the haplotype block on chromosome 17q21.31 as a region covering 2 million base pairs. The entire fully extended haplotype is associated with PSP and could implicate several other genes in addition to MAPT as candidate pathogenic loci [5]. Pastor et al. [6] also identified single nucleotide polymorphisms associated with higher risk for the disease in the homozygous state in a critical region of the tau gene that delimit a region of more than 1 Mb. Subhaplotype analyses in their sample showed that the H1E' is the most common subhaplotype in patients with PSP, and that H1E'A, present in 16% of patients with PSP but not in controls, suggests that a pathogenic allele exists in a subgroup of PSP [6]. Both studies suggest that the H2 haplotype may be protective.

Although most cases of PSP appear to be sporadic, rare genetically determined forms may exist. Garcia De Yebenes et al. [7] studied a five-generation family in which PSP was transmitted as an autosomal-dominant trait and described six other families with multiple affected individuals. Some cases of PSP may be an autosomal-recessive condition that maps to a polymorphism in the tau gene and Rojo et al. [8] reported 12 pedigrees with familial PSP. Stanford et al. [9] identified a novel silent mutation (S305S) in the *tau* gene in a subject with pathological features of PSP, neurofibrillary tangles concentrating within the subcortical regions of the basal ganglia. Two affected family members presented with symptoms of dementia and later developed abnormality of vertical gaze and extrapyramidal signs. The third presented with dystonia of the left arm and dysarthria, and later developed a supranuclear gaze palsy and falls. The mutation was located in exon 10 of the *tau* gene and formed part of a stem–loop structure at the 5' splice donor site. Although the mutation did not give rise to an amino acid change in the tau protein, functional exon-trapping experiments showed that it resulted in a significant 4.8-fold increase in the splicing of exon 10, resulting in the presence of tau containing four microtubule-binding repeats. This provides direct molecular evidence for a functional mutation that causes PSP pathology and demonstrates that mutations in the *tau* gene are pleiotropic [9].

B. Environmental Factors

Caparros-Lefebvre et al. [10] identified an increase prevalence of patients with PSP and atypical parkinsonism in the French West Indies in Guadeloupe. They also found that patients with this disorder had an increased consumption of tropical fruits (paw paw) and herbal tea (boldo), which contain tetrahydroisoquinolones (TIQs) and acetogenins, which are neurotoxic in animal models [10–11]. Neuropathological examination of three of these patients with PSP, homozygous for the H1 tau haplotype, showed an accumulation of 4-repeat tau protein predominating in the midbrain [10]. Cellular studies have shown that TIQs and acetogenins exert a direct toxicity to dopaminergic neurons through inhibition of complex 1 enzymes—a mitochondrial mechanism similar to MPTP or rotenone exposure [12]. It is unlikely that patients with PSP in the United States or Europe are past consumers of these tropical fruits or herbs, but TIQs and acetogenins have been found in foods common in the Western diet, specifically in cheese, milk, eggs, cocoa, and bananas [1].

A number of other environmental factors—traumatic brain injury, hypertension, and organic solvents—were reported as associated with PSP occurrence but remain unproven as etiological factors [1]. It is likely that sporadic PSP is partially due to genetic susceptibility and to other stressors such as environmental toxins and oxidative stress.

III. Pathogenesis

The organization of the neuronal cytoskeletal elements is critical for axonal formation and neuronal migration. Tau protein is abundant in normal neurons and has several roles: It maintains neuronal morphology through microtubule binding, is responsible for axonal transport of vesicles, interacts with the plasma membrane, and is a crucial link in communication of the neuronal cytoskeleton with the external environment [3]. In PSP the normal soluble tau protein collects in insoluble protease-resistant helical filaments composed of aggregated 4-repeat forms that accumulate in neuronal and glial inclusions in the brain (4R-tauopathy). This accumulation of misfolded hyperphosphorylated pathological tau as neurofibrillary and glial fibrillary tangles is the major substrate of PSP and is thought to be central to the neuronal degeneration that occurs. The exact triggers for the conversion from normal to the aggregate form are not completely understood [13]. The mechanisms of neuronal degeneration and cell death are postulated to be due to the mechanisms discussed in the following subsections.

A. Oxidative Injury

Oxidative stress results from an imbalance in the prooxidant/antioxidant systems in intact cells and leads to oxidative damage to lipids, proteins, carbohydrates, and nucleic acids, resulting in cell death. Lipid peroxidation is an indicator of oxidative stress in cells. It leads to formation of unstable lipid peroxides derived from polyunsaturated fatty acids, such as malondialdehyde (MDA), which can be used as an indicator of lipid peroxidation [14]. Several lines of evidence suggest that lipid peroxidation and mitochondrial abnormalities may play a role in the neurodegeneration in PSP. Albers et al. [15] found increased levels of MDA in the subthalamic nucleus and superior frontal cortex in patients with PSP and significant decreases in the α-ketoglutarate dehydrogenase/glutamate dehydrogenase ratio. These findings suggest that lipid peroxidation may explain region-specific neurodegeneration in PSP. They also observed significant decreases in complex I+III activity in cell lines expressing mitochondrial genes from patients with PSP, as well as significant increases in markers of lipid oxidative damage in response to mitochondrial toxins as compared with control cybrid cell lines, which strongly suggests mitochondrial dysfunction and oxidative damage in PSP [16]. Oxidative injury may be the mechanism of action of the previously discussed environmental toxins in PSP.

B. Inflammation

There is strong evidence that in PSP, neurodegeneration is accompanied by specific inflammatory mechanisms including augmented complement activation in the brain (with higher CSF levels of C4) and activation of the microglia [17]. Levels of matrix metalloproteinases (MMPs) and tissue inhibitors of metalloproteinases (TIMPs) are significantly increased in the substantia nigra of PSP cases as compared with controls [18]. MMP was also increased in the frontal cortex, consistent with the possibility that alterations in MMPs/TIMPs may contribute to disease pathogenesis. As in other neurodegenerative diseases, it remains speculative whether the inflammation is a secondary event to a primary neuronal degeneration.

It is likely that irrespective of the primary cause of PSP, oxidative stress, mitochondrial dysfunction, lipid peroxidation, and inflammation are common mechanisms by which neuronal degeneration and death occur and that they contribute to disease progression [1,19].

IV. Relevant Structural Detail

Pathologically, gross examination of the brain in PSP shows midbrain atrophy. There are neuronal loss and neurofibrillary tangles in the basal ganglia, diencephalon, and brainstem. The substantia nigra, subthalamic nucleus, and pontine base are typically involved, as well as the ventral anterior and lateral thalamic nuclei. The cerebellar dentate nucleus may show degeneration. Cortical pathology is minimal except for motor areas [20]. PSP is characterized by abundant neurofibrillary tangles and/or neuropil threads, particularly in the striatum, especially globus pallidus interna, subthalamic nucleus, substantia nigra, oculomotor complex, reticular formation, periaqueductal gray, superior colliculi, basis pontis, dentate nucleus, and prefrontal cortex. There is uniform presence of tau-positive cortical lesions. These were found in highest concentration in the precentral and angular gyrus, primarily affecting the deep cortical layers and involved both small and large neurons. Neuronal loss and gliosis are variable [20]. Tau-positive glial inclusions, tufted astrocytes, are a consistent pathological finding. Coiled bodies, which are tau deposits in oligodendrocytes found in the white matter, are also seen in a widespread distribution. Early pathology is evident primarily in the midbrain, perhaps explaining the relatively early vertical eye movement abnormalities. The pontine nucleus raphe interpositus, pedunculopontine, and deep pontine nuclei are also affected. Statistical analysis has suggested that both cortical and subcortical neurofibrillary tangles are linked to the pedunculopontine nucleus, which may play a prominent role in spreading the lesions.

The distribution and ultrastructure of neurofibrillary tangles in PSP are distinct from those found in Alzheimer's disease. PSP is associated with more subcortical involvement, with 15- to 20-nm wide single tubules, compared with the cortically based paired helicoidal filaments of Alzheimer's disease.

V. Pharmacology, Biochemistry, Molecular Mechanisms

Neurochemical studies indicate that the degenerative process in PSP involves the nigrostriatal dopaminergic neurons that innervate the striatum, as well as cholinergic and GABAergic efferent neurons in the striatum and other basal ganglia and brainstem nuclei, thereby explaining the lack of or transient levodopa response [21]. Although in Parkinson's disease D_1 and D_2 receptors remain preserved as compared with controls, in PSP, IBZM single-photon emission computed tomographic (SPECT) studies show that there is a reduction of nigral D_1 receptors and striatal D_2 receptors [22].

Cholinergic dysfunction is related to loss of interneurons in the striatum, compounded by reduced inputs into the circuits from other cholinergic nuclei, such as the pedunculopontine and nucleus basalis of Meynert. Normal cholinergic transmission requires the presence of intact cholinergic neurons capable of releasing sufficient acetylcholine, as well as functional muscarinic and nicotinic receptors. Although there is evidence from autopsy and *in vivo* studies of loss of cholinergic neurons in PSP, the receptor status is unknown. This may be critical to understanding the basis for the poor therapeutic response to cholinomimetics. Symptomatic treatment using cholinergic drugs may, thus, be improved by more specific targeting of cholinergic receptors or nuclei [23]. A study using immunohistochemistry for tryptophan hydroxylase, phosphorylated tau, and α-synuclein in the nucleus centralis superior, nucleus raphe obscurus, and pallidus in PSP showed a significant increase in the percentage of neurons synthesizing serotonin in the nucleus centralis superior in PSP when compared with controls [24].

Degeneration of multiple neurotransmitter systems leads to a more diffuse disorder than in Parkinson's disease. The cerebral cortex is also affected in PSP, and positron emission tomography (PET) scan studies show decreased metabolism of cerebral glucose in the frontal lobes and reduced cortical benzodiazepine receptors that correlate with dementia [25].

At a molecular level, the tau deposits in PSP are different from those observed in Alzheimer's disease (similar ratio of 3-R and 4-R tau isoforms) and Pick's disease (mostly 3-R tau), both in morphology and tau isoform content. On the other hand, the tau in PSP resembles that in corticobasal degeneration (CBD) [26]. Williams et al. [2] showed that the clinical variability in PSP relates to a different molecular tau isoform type. These authors describe two distinct clinical phenotypes with different tau isoform deposition in 103 pathologically confirmed cases of PSP.

The first type, observed in 54% of cases and labeled "Richardson's syndrome," as it corresponds to the cases originally described by Steele, Richardson, and Olszewski, had at onset postural instability and falls, supranuclear vertical palsy, and cognitive dysfunction and a predominant 4-repeat tau isoform composition of insoluble tangle tau in the basal pons (4-R:3-R of 2.84). The second type, observed in 33% of the cases labeled "PSP-parkinsonism," had an asymmetrical onset of tremor, moderate response to levodopa, and tau isoform ratio of 4-R:3-R of 1.63. Hence, there is more clinical and molecular phenotype variation between the Richardson's syndrome and the PSP-parkinsonism than between PSP and CBD, raising the question of whether these atypical presentations should be considered part of the spectrum of PSP or a different nosological disorder altogether. If there are more commonalities between Richardson's syndrome and CBD than between these two newly described PSP phenotypes, it is time to reassess whether PSP and CBD should be considered different nosological entities or two phenotypes of the same disorder.

VI. Explanation of Symptoms in Relation to Pathophysiology

A. Postural Instability and Falls

Postural instability and falls are the most common symptom presentation in PSP. In the National Institute of Neurological Disorders and Stroke (NINDS) study, 96% of 24 patients with PSP had postural instability and 83% had a history of falls at the first visit, which generally occurred within 3.0–3.5 years after symptom onset. Cholinergic deficits are thought to underlie the postural instability and cognitive impairment of PSP, but trials of cholinergic agonists and cholinesterase inhibitors have failed to show improvement in motor function, quality of life, or cognitive impairment. The five frontal cortical–subcortical loops, linking functionally related areas of the brain, are damaged in PSP, leading to specific clinical deficits [23].

B. Ocular Motor Abnormalities

Supranuclear gaze palsy is the hallmark of PSP. Slowing of vertical saccades precedes the development of vertical gaze palsy. Vertical gaze palsy for either upward or downward gaze is rarely present at symptom onset (8%), usually takes 3–4 years to develop, and precedes the development of horizontal gaze palsy. The dissociation between an impaired voluntary and pursuit gaze and preservation of oculocephalic reflexes is the evidence of the supranuclear origin of the oculomotor disorder in PSP. The vertical gaze palsy of PSP has been attributed to the midbrain pathology affecting the tectum and the superior colliculi. Mechanisms involve loss of cholinergic neurons and glycinergic omnipause neurons in the nucleus raphe interpositus that are so important for ocular motility and are reduced by 50% in patients with PSP with supranuclear gaze palsy [27]. However, in a recent study, comparisons between cases of PSP with and without gaze palsy at autopsy revealed a 40% greater decrease in the number of substantia nigra neurons in cases with gaze palsy compared with those without. As the substantia nigra projects to the superior colliculus, degeneration of this basal ganglia structure may disrupt eye movements in PSP [28].

Reduced blink rate and eyelid apraxia—a difficulty in opening closed eyelids accompanied by compensatory eyebrow elevation and frontalis overactivity and blepharospasm—are also features observed in PSP.

C. Behavioral and Cognitive Frontal Features

Frontal lobe symptoms usually manifest early in PSP. Apathy may be the initial symptom and is present in more than 80% of the patients. Neuropsychological studies show an early executive dysfunction that includes difficulties with planning, problem solving, concept formation, and social cognition. Impaired abstract thought, decreased verbal fluency, motor perseveration, apathy, and disinhibition are all features observed at an early stage in the illness.

D. Extrapyramidal Signs

A predominant akinetic rigid syndrome with bilateral bradykinesia, axial rigidity, poor response to levodopa, and the presence of axial dystonia such as retrocollis, blepharospasm, and oromandibular dystonia are observed. Limb dystonia may be observed, but usually at later stages.

E. Speech, Swallowing, and Other Neurological Features

Patients with PSP may present with early speech and swallowing problems, and classically a hypokinetic-spastic dysarthria is observed. Speech perseveration with repetition of words and phrases and anomia are also present. The hypokinetic dysarthria of PSP may result from neuronal loss and gliosis-degenerative changes in the substantia nigra pars compacta and pars reticulata and not from changes in the striatum or globus pallidus [29].

Swallowing disturbances and sialorrhea also occur early in PSP. Patients also often misjudge the amount of food they can swallow and take oversized mouthfuls or stuff their mouths. Pyramidal signs are seen in a third of patients and are often a late feature. Early or late insomnia and difficulties in maintaining sleep are reported, but in contrast to what occurs in the synucleinopathies, REM behavioral disorder is unusual.

VII. Natural History

In general, symptoms and signs apparent early in the course of PSP progress steadily. Most patients eventually require a wheelchair and a feeding tube and speech may become unintelligible, palilalic, or mute. Goetz et al. [30] reported that these three milestones (wheelchair bound, unintelligible speech, and need for a feeding tube) occur rapidly in PSP and can be monitored with standardized rating scales such as the Unified Parkinson's Disease Rating Scale. In their study, 88% of their sample (50 patients) met at least one of the three milestones, but nasogastric tube was never the first [30]. The median time from symptom onset to the first key motor impairment was 48 months. Gait disturbances occurred at a median symptom duration of 57 months and unintelligible speech at 71 months. As a composite endpoint, speech and gait abnormalities accounted for 98% of this key first motor impairment. These indices could be used as outcome measures in clinical trials to assess how interventions alter anticipated disease progression [30]. Litvan et al. [31] studied the progression of PSP in patients selected from the research and clinical files of seven medical centers involving tertiary centers of Austria, England, France, and the United States. The patients' mean age at onset of PSP was 63 (range 45–73) years and median survival time was 5.6 (range 2.0–16.6) years. Onset of falls during the first year, early dysphagia, and incontinence predicted a shorter survival time. Age at onset, gender, and early onset of dementia, vertical supranuclear palsy, or axial rigidity had no effect on prognosis of survival. Pneumonia was the most common immediate cause of death [31]. In another clinical cohort study, Nath et al. [32] evaluated clinical predictors of survival in PSP after a mean of 6.4 years. Older age at onset and classification as probable PSP were associated with poorer survival. Onset of falls, speech, or diplopia within 1 year and swallowing problems within 2 years were associated with a worse prognosis and predicted reduced survival [32].

References

1. Litvan, I. (2005). Progressive supranuclear palsy. *In* "Current Clinical Neurology: Atypical Parkinsonian Disorders" (I. Litvan, ed.), Vol. **18**, pp. 288. Humana Press, Inc., Totowa, NJ.
2. Williams, D. R., de Silva, R., Paviour, D. C., Pittman, A., Watt, H. C., Kilford, L., Holton, J. L., Revesz, T., and Lees, A. J. (2005). Characteristics of two distinct clinical phenotypes in pathologically proven progressive supranuclear palsy: Richardson's syndrome and PSP parkinsonism. *Brain* **128**, 1247–1258.

3. Avila, J., Lucas, J. J., and Perez, M., Hernandez, F. (2004). Role of tau protein in both physiological and pathological conditions. *Physiol. Rev.* **84**, 261–284.
4. Litvan, I., Baker, M., and Hutton, M. (2001). Tau genotype: no effect on onset, symptom severity, or survival in progressive supranuclear palsy. *Neurology* **57**, 138–140.
5. Pittman, A. M., Myers, A. J., Duckworth, J., Bryden, L., Hanson, M., Abou-Sleiman, P., Wood, N. W., Hardy, J., Lees, A., and de Silva, R. (2004). The structure of the tau haplotype in controls and in progressive supranuclear palsy. *Hum. Mol. Genet.* **13**, 1267–1274.
6. Pastor, P., Ezquerra, M., Perez, J. C., Chakraverty, S., Norton, J., Racette, B. A., McKeel, D., Perlmutter, J. S., Tolosa, E., and Goate, A. M. (2004). Novel haplotypes in 17q21 are associated with progressive supranuclear palsy. *Ann. Neurol.* **56**, 249–258.
7. De Yebenes, J. G., Sarasa, J. L., Daniel, S. E., and Lees, A. J. (1995). Familial progressive supranuclear palsy. Description of a pedigree and review of the literature. *Brain* **118**, 1094–1103.
8. Rojo, A., Pernaute, R. S., Fontan, A., Ruiz P. G., Honnorat, J., Lynch, T., Chin, S., Gonzalo, I., Rabano, A., Martinez, A., Daniel, S., Pramstaller, P., Morris, H., Wood, N., Lees, A., Tabernero, C., Nyggard, T., Jackson, A. C., Hanson, A., and de Yebenes, J. G. (1999). Clinical genetics of familial progressive supranuclear palsy. *Brain* **122**, 1233–1245.
9. Stanford, P. M., Halliday, G. M., Brooks, W. S., Kwok, J. B., Storey, C. E., Creasey, H., Morris, J. G., Fulham, M. J., and Schofield, P. R. (2000). Progressive supranuclear palsy pathology caused by a novel silent mutation in exon 10 of the tau gene: expansion of the disease phenotype caused by tau gene mutations. *Brain* **123**, 880–893.
10. Caparros-Lefebvre, D., Sergeant, N., Lees, A., Camuzat, A., Daniel, S., Lannuzel, A., Brice, A., Tolosa, E., Delacourte, A., and Duyckaerts, C. (2002). Guadelopean parkinsonism: a cluster of progressive supranuclear palsy-like tauopathy. *Brain* **125**(Pt 4), 801–811.
11. Champy, P., Hoglinger, G. U., and Feger, J. (2004). Annonacin, a lipophilic inhibitor of mitochondrial complex I, induces nigral and striatal neurodegeneration in rats: possible relevance for atypical parkinsonism in Guadeloupe. *J. Neurochem.* **88**(1), 63–69.
12. Lannuzel, A., Michel, P., and Abaul, M. J. (2000). Neurotoxic effects of alkaloids from *Annona muricata* (sour-sop) on dopaminergic neurons: potential role in etiology of atypical parkinsonism in the French West Indies. *Mov. Disorders* **13**(Suppl 3), 28.
13. Spillantini, M. G., and Goedert, M. (1998). Tau protein pathology in neurodegenerative diseases. *Trends Neurosci.* **21**, 428–433.
14. Droge, W. (2003). Oxidative stress and aging. *Adv. Exp. Med. Biol.* **543**, 191–200.
15. Albers, D. S., Augood, S. J., Park, L. C., Browne, S. E., Martin, D. M., Adamson, J., Hutton M., Standaert, D. G., Vonsattel, J. P., Gibson, G. E., and Beal, M. F. (2000). Frontal lobe dysfunction in progressive supranuclear palsy: evidence for oxidative stress and mitochondrial impairment. *J. Neurochem.* **74**, 878–881.
16. Chirichigno, J. W., Manfredi, B., Beal, M. F., and Albers, D. S. (2002). Stress induced mitochondrial depolarization and oxidative damage in PSP cybrids. *Brain Res.* **951**, 31–35.
17. Yamada, T., Moroo, I., Koguchi, Y., Asahina, M., and Hirayama, K. (1994). Increased concentration of C4d complement protein in the cerebrospinal fluids in progressive supranuclear palsy. *Acta Neurol. Scand.* **89**, 42–46.
18. Lorenzl, S., Albers, D. S., and Chirichigno, J. W. (2004). Elevated levels of matrix metalloproteinases-9 -1 and of tissue inhibitors of MMPs, TIMP-1 and TIMP-2 in postmortem brain tissue of progressive supranuclear palsy. *J. Neurol. Sci.* **15**, 39–45.
19. Odetti, P., Garibaldi, S., Norese, R., Angelini, G., Marinelli, L., Valentini, S., Menini, S., Traverso, N., Zaccheo, D., Siedlak, S., Perry G., Smith M A., and Tabaton, M. (2000). Lipoperoxidation is selectively involved in progressive supranuclear palsy. *J. Neuropathol. Exp. Neurol.* **59**, 393–397.
20. Litvan, I., Hauw, J. J., Bartko, J. J., Lantos, P. L., Daniel, S. E., Horoupian, D. S., McKee, A., Dickson, D., Bancher, C., Tabaton, M., Jellinger, K., and Anderson, D. W. (1996). Validity and reliability of the preliminary NINDS neuropathological criteria for progressive supranuclear palsy and related disorders. *J. Neuropathol. Exp. Neurol.* **55**(1), 97–105.
21. Ruberg, M., Javoy-Agid, F., Hirsch, E., Scatton, B., Lheureux, R., Hauw, J. J., Duyckaerts, C., Gray, F., Morel-Maroger, A., and Rascol, A. (1985). Dopaminergic and cholinergic lesions in progressive supranuclear palsy. *Ann. Neurol.* **18**, 523–529.
22. Plotkin, M., Amthauer, H., Klaffke, S., Kuhn, A., Ludemann, L., Arnold, G., Wernecke, K. D., Kupsch, A., Felix, R., and Venz, S. (2005). Combined (123)I-FP-CIT and (123)I-IBZM SPECT for the diagnosis of parkinsonian syndromes: study on 72 patients. *J. Neural. Transm.* **112**, 677–692.
23. Warren, N. M., Piggott, M. A., Perry, E. K., and Burn, D. J. (2005). Cholinergic systems in progressive supranuclear palsy. *Brain* **128**(2), 239–249.
24. Kovacs, G. G., Kloppel, S., Fischer, I., Dorner, S., Lindeck-Pozza, E., Birner, P., Botefur, I. C., Pilz, P., Volk, B., and Budka, H. (2003). Nucleus-specific alteration of raphe neurons in human neurodegenerative disorders. *Neuroreport* **20**(14), 73–76.
25. Foster, N. L., Minoshima, S., and Johanns, J. (2000). PET measures of benzodiazepine receptors in progressive supranuclear palsy. *Neurology* **54**(9), 1768–1773.
26. Morris, H. R., Gibb, G., Katzenschlager, R., and Wood, N. W. (2002). Pathological, clinical and genetic heterogeneity in progressive supranuclear palsy. *Brain* **125**, 969–975.
27. Revesz, T., Sangha, H., and Daniel, S. E. (1996). The nucleus raphe interpositus in the Steele-Richardson-Olszewski syndrome (progressive supranuclear palsy). *Brain* **119**, 1137–1143.
28. Halliday, G. M., Hardman, C. D., Cordato, N. J., Hely, M. A., and Morris, J. G. (2000). A role for the substantia nigra pars reticulata in the gaze palsy of progressive supranuclear palsy. *Brain* **123**, 724–732.
29. Kluin, K., Gilman, S., and Foster, N. (2001). Neuropathological correlates of dysarthria in progressive supranuclear palsy. *Arch. Neurol.* **58**, 265–269.
30. Goetz, C. G., Leurgans, S., Lang, A. E., and Litvan, I. (2003). Progression of gait, speech and swallowing deficits in progressive supranuclear palsy. *Neurology* **60**(6), 919–922.
31. Litvan, I., Mangone, C. A., McKee, A., Verny, M., Parsa, A., Jellinger, K., D'Olhaberriague, L., Chaudhuri, K. R., and Pearce, R. K. (1996). Natural history of progressive supranuclear palsy (Steele–Richardson–Olszewski syndrome) and clinical predictors of survival: a clinicopathological study. *J. Neurol. Neurosurg Psychiatry* **60**, 615–620.
32. Nath, U., Ben-Schlomo, Y., and Thomson, R. G. (2003). Clinical features and natural history of progressive supranuclear palsy: a cohort study. *Neurology* **60**, 910–916.

10

Protein Aggregation Disorders

Tiago Fleming Outeiro, PhD
Pamela J. McLean, PhD
Bradley T. Hyman, MD, PhD

Keywords: *aggregation, amyloid, conformation, neurodegenerative disease, protein misfolding*

I. Protein Folding and Misfolding
II. Protein Aggregation Disorders
III. "Natively Unfolded" Proteins and Other Structural Determinants of Protein Aggregation
IV. Cellular Quality Control Systems for Protein Folding as Targets for Therapeutic Intervention in Neurodegenerative Diseases
V. Conclusions
   References

I. Protein Folding and Misfolding

Structurally, a nascent polypeptide is an unstable entity that is present in environments that are not always favorable for adopting the proper fold and can often be destabilizing [1,2]. Protein folding is a complex process in which a variety of factors play specific roles. The classic work of Christian Anfinsen in the 1950s on the enzyme ribonuclease (RNA) revealed the relationship between the primary amino acid sequence of a protein and its conformation. Anfinsen showed that the information encoded in the primary sequence of RNA was sufficient for correct refolding after denaturation (i.e., the formation of the native protein fold from the unfolded state is a spontaneous process determined by the global free energy minimum) [3]. It was later found that this was not the case for every protein, although the primary structure is an important factor in the folding process.

More than 30 years ago, Levinthal pointed out that proteins cannot fold by randomly checking all possible conformations of their unfolded states because that process would take longer than the age of the universe. This is known as the *Levinthal paradox*. Traditionally, the native state of a protein was seen as the most thermodynamically stable conformation of the polypeptide chain under physiological conditions.

A newer view of folding *in vitro*, however, says that the pathway from an unfolded polypeptide chain to the folded native state is a stochastic process that occurs on a rather flat energy landscape [4,5]. As with any stochastic process, there is a finite possibility for polypeptides to misfold and adopt nonnative states in which they might be at least transiently stable (kinetic traps). Proteins could be trapped in nonnative states in various ways, for example, as partially folded intermediates or as a result of interactions with other molecules or proteins with the same or different sequences.

Protein folding *in vivo* is, naturally, a quite different story. In a living cell, folding conditions have been optimized by millions of years of natural selection and evolution. This explains the ability of proteins to fold under conditions that could appear to be counterproductive for efficient folding because of the high temperature, the extremely crowded milieu reaching extreme concentrations of about 100–400 mg/ml, and the large number of "nonnative" proteins present [6,7]. Given the circumstances, it is quite an accomplishment for cells to avoid the accumulation of aggregated proteins.

II. Protein Aggregation Disorders

Protein aggregation disorders are a group of diseases that arise because of or are associated with the misfolding and aggregation of one or more proteins. They are believed to result from the failure of proteins to reach their active state or from the accumulation of abnormally folded proteins. Proteins, as the main effectors in the cell, play underpinning roles in all biological processes. It is, therefore, not surprising that the number of diseases identified as protein aggregation disorders is expanding, in parallel to the discovery and functional characterization of new proteins.

Despite the strong connection between protein misfolding and aggregation and disease, the manner by which it results in disease is not understood. In some cases, it seems that the deposition of protein aggregates may physically disrupt the functioning of specific organs. In other cases, it seems that the lack of functional protein, because of its recruitment into the aggregates, results in the failure of crucial cellular processes [8]. However, for neurodegenerative diseases, such as Alzheimer's disease (AD), Parkinson's disease (PD), or the prion diseases, it appears that the symptoms arise from the destruction of cells by a "gain of toxic function" that results from the aggregation process (oligomers, protofibrils, amyloid fibrils, or other intermediates) or by a combination of this gain of toxic function and a loss of normal function of the protein [9,10].

A. Alzheimer's Disease

AD is a progressive neurodegenerative disorder characterized by global cognitive decline and the accumulation of protein aggregates in the brain (Aβ plaques and neurofibrillary tangles [NFTs]) (Fig. 1).

AD was first described by Alois Alzheimer, a German neuropathologist and psychiatrist, in 1906. Alzheimer reported the disease in a 51-year-old woman with presenile dementia who displayed diffuse cortical atrophy, neuronal cell loss, plaques, and tangles. AD, along with many other neurodegenerative disorders, is an age-related disorder, that is, it is found with increasing prevalence as a function of age. The larger mean lifespan of the population has led to a large increase in the number of AD cases, approaching 20–30 million people worldwide. Aging is the greatest risk factor for AD, but family history also plays a role. In 1984, Glenner and Wong [11] reported the amino acid sequence of amyloid β (Aβ), which gradually accumulates and aggregates in the brains of patients with AD. The discovery of Aβ (later found to be a proteolytic fragment of the amyloid

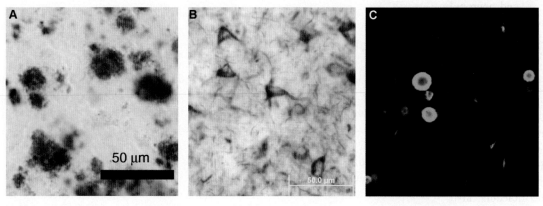

Figure 1 Pathological lesions in Alzheimer's disease (AD) and dementia with Lewy bodies (DLB). (**A**) Aβ plaques in AD. (**B**) Neurofibrillary tangles in AD. (**C**) Lewy bodies in DLB.

precursor protein [APP]) preceded the first genetic discoveries. AD is complex and heterogeneous: It can present as a rare (<5%) highly penetrant early onset familial disease (EOFAD) or as a late-onset disease (LOAD) without an obvious genetic component. Notwithstanding, the growing understanding of AD genetics has been central to the knowledge of the pathogenic mechanisms leading to the disease. In 1986, the APP gene was the first AD-causing gene to be identified (on chromosome 21) (reviewed in reference 12). Today, about 20 missense dominantly inherited mutations in APP are known to lead to an early and aggressive form of the disease.

The next two EOFAD genes to be discovered were presenilin 1 (PSEN1) on chromosome 14 and presenilin 2 (PSEN2) on chromosome 1. The most frequently mutated (>140 missense mutations) gene, PSEN1, accounts for the majority of AD cases with onset before age 50 years. Roughly, 10 missense mutations have been described in the PSEN2 gene, which is a rare cause of AD.

Although these AD-causing mutations occur in three different genes, the mechanism of the disease is identical; it involves the altered production of Aβ, leading to either an increase of Aβ or an excess of the Aβ42 species, which is associated with Aβ plaque deposition, neuronal cell death, and, ultimately, dementia.

The E4 allele of the apolipoprotein E (apoE) gene, on chromosome 19, has been identified as a major risk factor for conventional LOAD. Genetic studies have consistently confirmed an enhanced risk for developing AD in patients who are either heterozygous or homozygous for E4, with a two to threefold increased risk for heterozygotes and a six to tenfold increased risk for homozygotes. Much of the risk for AD that is familial, with an age at onset from 60 to 70 years, reflects apoE4-mediated risk [13].

1. Pathology

The "amyloid hypothesis" of AD sustains that the accumulation of Aβ in the brain is the primary culprit of AD-related pathogenesis, including NFT formation, synapse loss, and neuronal cell death [14]. Evidence from neuropathological, genetic, molecular, and animal modeling studies suggests that the gradual accumulation of Aβ42 (Aβ ending at position 42) in the limbic and associated cortices leads to its aggregation into oligomers, protofibrils, and amyloid fibrils. These various Aβ species, in particular those that precede the formation of "mature" amyloid fibrils, appear to be able to induce synaptic and dendritic dysfunction, activating microglia and astrocytes, which are signs of local inflammation. An early feature in AD is the selective loss in layer II of the entorhinal cortex, which then leads to subsequent deafferentation and loss of the synaptic input to the molecular layer of the dentate gyrus [15].

Additionally, NFT and neuronal cell loss are evident in the CA1 area of the hippocampus and in deeper layers of entorhinal cortex in AD, affecting these memory-related neuronal systems [16].

NFTs are made of a highly phosphorylated, misfolded version of the microtubule-associated protein called *tau* (see Fig. 1). Antibodies directed against phosphorylated tau protein dramatically stain several dysfunctional regions in the AD brain, highlighting the possibility that AD pathology is partially due to abnormalities in the normal biology of tau, due to misfolding and aggregation, which might also contribute to neuronal cell death [17].

As previously stated, the precise mechanism by which Aβ and tau exert the putative neurotoxic effects on neurons remains unclear. Several studies of Aβ in cell-based and animal models suggest that the assembly of the peptide into oligomers is a central step toward toxicity [18,19]. The toxic oligomeric species is also unknown, but some data suggest that it can disrupt synaptic plasticity [20].

2. But How Is Aβ Generated?

APP is an integral membrane protein with a large ectodomain, a single transmembrane-spanning domain, and a short cytosolic region. APP, which exists as several isoforms of 695, 751, and 770 amino acids, is cleaved by selective proteases—α-, β- and γ-secretases—into various fragments.

Proteases with proposed α-secretase function belong to the ADAM (a desintegrin and metalloprotease) family of proteins. Cleavage by α-secretase results in the production of a soluble APP and should prevent the generation of Aβ peptide.

The formation of amyloidogenic Aβ requires the activity of β- and γ-secretases (Fig. 2).

The main β-secretase activity belongs to a novel transmembrane aspartic protease called β-*site APP-cleaving*

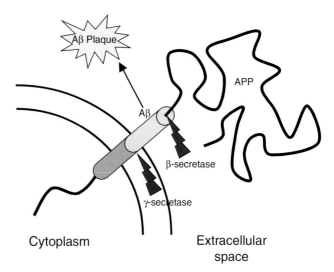

Figure 2 Proteolytic processing of amyloid precursor protein (APP). γ- and β-secretase cleave APP generating Aβ, which is then released in the extracellular space where it forms plaques.

enzyme 1 (BACE1). BACE activity appears to be elevated in AD [21] and in aging [22]. BACE knockout mice do not form Aβ and appear to be healthy, suggesting BACE1 might constitute an attractive drug target for AD. However, endogenous BACE1 substrates are being identified, advising for a certain caution.

γ-Secretase is still somewhat elusive, but the consensus is that PSENs are involved in the amyloidogenic processing of APP. Other essential members of the γ-secretase complex include aph-1, pen-2, and nicastrin. PSENs are clearly genetically linked to AD, as previously mentioned, causing EOAD. They are members of a family of integral membrane proteases that may use a catalytic mechanism similar to classic aspartic proteases, such as pepsin, rennin, and cathepsin D [23]. PSENs are able to cleave substrate polypeptides within a transmembrane region and appear to be the catalytic subunit of multiprotein complexes that possess γ-secretase activity. Cleavage of APP by γ-secretase results in the formation of Aβ fragments with 40, 42, and 43 amino acids, of which Aβ40 is the most abundant form, whereas Aβ42 is more toxic and less soluble and tends to form deposits.

Mutations in the PSEN1 and PSEN2 dramatically affect the ratio of Aβ40/Aβ42. However, γ-secretase as a drug target has some limitations because PSEN1 knockout mice had developmental abnormalities that caused their failure to survive.

3. Aβ Clearance and Degradation

The levels of Aβ in the brain are much higher than in the periphery because of both increased generation of the peptide in EOAD and LOAD and decreased clearance and degradation. Increasing evidence suggests that the low-density lipoprotein receptor–related protein (LRP) mediates the efflux of Aβ to the periphery, and LRP antagonists specifically reduce the efflux from the brain by up to 90%. Aβ forms a complex with the LRP ligands a2M or apoE. The complexes then bind to LRP, are internalized to late endosomes, and either are delivered to lysosomes for degradation or undergo transcytosis across the blood–brain barrier into the plasma. Soluble Aβ can also be exported by directly binding LRP. LRP also interacts directly with APP and has an impact on Aβ synthesis [24].

Several proteases such as neprilysin (NEP), insulin-degrading enzyme (IDE), endothelin-converting enzyme (ECE-1 and ECE-2), and plasmin have been found to degrade Aβ [25]. Downregulation of these enzymes could, therefore, predispose to accumulation of Aβ in the brain and consequent development of AD. Thus, a possible therapeutic approach for intervention in AD might be to upregulate these proteinases, either pharmacologically or through gene therapy.

Degenerating neurons in AD exhibit increased oxidative damage, impaired energy metabolism, and perturbed cellular calcium homeostasis. Aβ seems to be an important instigator of these abnormalities.

Oxidative stress (unbalanced production of highly reactive oxygen species) and impaired cellular energy metabolism are also observed in several other age-related disorders. In AD brains, cells exhibit abnormally high amounts of oxidatively damaged proteins, lipids, and DNA, especially in the environment of plaques and NFTs. Oxidative stress may, therefore, play a role in amyloid-mediated cell damage. Aβ and redox-active species such as Fe^{2+} and Cu^+ are two possible sources of oxidative stress in AD. Aggregation of Aβ leads to the generation of H_2O_2 in a process that is greatly enhanced by Fe^{2+} and Cu^+. Lipid peroxidation caused by Aβ impairs the function of several proteins and membrane transporters that negatively affect cellular ion homeostasis and energy metabolism, rendering neurons vulnerable to excitotoxicity and apoptosis.

Imaging of brains of patients with AD has demonstrated deficits in glucose use, an abnormality thought to occur before the onset of clinical symptoms.

4. Alzheimer's Disease Symptoms and Pathophysiology

In AD, brain regions involved in learning and memory processes, including the temporal and frontal lobes, are reduced in size as a result of the degeneration and neuronal death. Brain regions bearing Aβ plaques exhibit a reduced number of synapses, and neurites associated with the plaques are often dystrophic [26]. Learning and memory, the neurological functions most affected in AD, require exchange of information between neurons. In the hippocampus, the convergence of signals received, for example, from the visual or auditory pathways, leads to changes in synaptic strength that facilitate memory. It remains uncertain why neuronal circuits critical for learning and memory are particularly sensitive in AD. However, drugs such as acetylcholinesterase inhibitors, which enhance activation of synapses in those circuits, can improve cognition during the early stages of the disease.

The death of populations of neurons in AD brains appears to take place over a long period (several years), hence, the progressive nature the disease.

5. Role of Protein Misfolding and Aggregation in Tau and Aβ Accumulation in Alzheimer's Disease

Evidence from a variety of diseases suggests that specific alterations in the primary sequence of cellular proteins, posttranslational modifications, or defective interactions with other proteins can impose conformational constraints that alter the normal function/biology of these proteins and facilitate aggregation. Plaques are composed primarily of a proteolytic fragment of APP, which misfolds and adopts conformations that are associated with the disease. Likewise,

NFTs are intracellular aggregates of misfolded filamentous tau polymers that occur in the regions of the brain that are compromised during the course of AD. Moreover, the discovery of tau mutations, in genetic forms of frontotemporal dementia (FTD), that cause it to form filaments more readily, strongly supports the idea that aggregation into filaments results in a toxic gain of function, much like what had been hypothesized for Aβ accumulation in AD. Even if Aβ and tau filaments are not, per se, the culprits, it is likely that the toxic species is generated during fibrillization.

B. Synucleinopathies

α-Synuclein, a member of the synuclein family of proteins, was initially identified on the electric lobe of *Torpedo californica* for reacting to antiserum against purified cholinergic vesicles [27]. The name "synuclein" was chosen because the protein was initially found in the *syn*apse and in the *nucl*ear envelope [27]. α-Synuclein was not found in the nucleus in several subsequent studies, so it appears to be a presynaptic protein. In 1994, two synucleins of 140 and 134 amino acids were purified from human brain and were named α- and β-synuclein, respectively [28]. γ-Synuclein, a protein of 127 amino acids, was later discovered as a third member of the synuclein family and was expressed in adult brain and in ovarian tumors.

The synucleins became connected to several neurodegenerative disorders after the initial report of a non-Aβ component of AD amyloid (NAC). NAC was a peptide of a minimum of 35 amino acids generated by cleavage of a protein of 140 amino acids, initially called *NACP,* for NAC precursor protein and now known as α-*synuclein.* The detection of the NAC fragment of α-synuclein in amyloid plaques stimulated interest among researchers in a putative role for this protein in AD, although in retrospect NAC may have been isolated from Lewy bodies (LBs) rather than amyloid plaques.

In 1997, the α-synuclein gene, *PARK1,* became associated with PD, when a point mutation (G209A) was found in an Italian kindred afflicted by autosomal-dominant PD. The mutation causes a threonine for alanine substitution at position 53 (A53T) in α-synuclein. This discovery was then followed by a report identifying α-synuclein in LBs (proteinaceous fibrillar cytoplasmic inclusions) from brain tissue of sporadic PD cases.

Shortly thereafter, another familial form of PD was linked to a mutation (G88C) in α-synuclein causing a proline for alanine substitution at position 30 (A30P) [29]. A third mutation consisting of a lysine for glutamate substitution at position 46 (E46K) was discovered to be associated with familial PD (FPD). The dominant nature of the inherited mutants is thought to reflect a gain rather than a loss of function in the α-synuclein proteins. Haploinsufficiency of the WT allele in heterozygotes also seems to play a role in the severity of the PD phenotype [30].

After the initial discovery of α-synuclein in LBs in PD, the protein was detected in cellular inclusions in several other neurodegenerative diseases, including dementia with LBs (DLB), multiple system atrophy (MSA), amyotrophic lateral sclerosis (ALS), and Hallervorden–Spatz syndrome, now called *neurodegeneration with brain iron accumulation type 1* (NBIA). The neurodegenerative diseases that share α-synuclein pathology as a primary feature are collectively known as *synucleinopathies.*

In this chapter, we focus primarily on PD because it is the most common synucleinopathy.

PD is one of the most common progressive neurodegenerative disorders, affecting about 2% of people older than 65 years and 4–5% of people older than 85 years (between 1 and 1.5 million Americans). It was James Parkinson, an English physician, who first described the disease in 1817 (in the days of the Industrial Revolution).

Were there any patients who suffered from PD before the nineteenth century? PD, as well as other neurodegenerative diseases, is mostly a disease of the elderly, so it is not surprising that only when life expectancy started increasing it became more common. The Industrial Revolution represents an important landmark in terms of improving the quality of life. Nonetheless, it is interesting to note that some believe that Leonardo da Vinci, who lived in the fifteenth and sixteenth centuries, had already described the abnormalities that appear in patients with PD [31].

PD is characterized by loss of dopaminergic neurons in the *substantia nigra pars compacta* and is accompanied by muscle rigidity, bradykinesia, resting tremor, and postural instability. Although the underlying causes for neuronal cell loss are unknown, the symptoms of PD can be effectively treated by replacement of dopamine (DA) (via L-DOPA) or by DA agonists [32]. The neuropathological hallmark of idiopathic PD is the presence of concentric hyaline cytoplasmic inclusions called LBs, first described by Friederich H. Lewy in 1912. LBs are spherical inclusions of 5–25 μm in diameter, seen as a dense eosinophilic core with a pale surrounding halo in the cytoplasm of affected neurons (see Fig. 1). LBs were first seen in the substantia innominata and the dorsal vagal nucleus in PD [33]. All LBs were shown to contain the protein α-synuclein [34,35].

It is now known that LBs can be seen in pigmented neurons of the substantia nigra in almost every case of PD [36]. However, LBs have also been observed in the brains of asymptomatic individuals [37]. It is not clear whether so-called incidental LB disease is actually preclinical PD or is a case in which rapid fibril formation precludes accumulation of other toxic intermediates, such as protofibrils.

LBs were identified in many regions of the brain, not only in familial and sporadic PD but also in DLB, a neurodegenerative disorder that is closely related to PD. In DLB, however, LBs are abundant in cortical brain areas, such as the entorhinal and cingulate cortices, underlying clinical symptoms of dementia, which coincide with parkinsonian motor dysfunction in patients with DLB.

1. Genetics of Parkinson's Disease

A genetic component in PD was thought to be unlikely until recently, because of the long preclinical phase that makes a family history difficult to discern. A high percentage of the nigral neurons (50–60%) can be lost with no obvious clinical consequence. Moreover, initial studies of twins showed equally low rates of concordance in monozygotic and dizygotic twins [38]. Only when improved positron emission tomography (PET) imaging methods became available to monitor the number of dopaminergic neurons in the substantia nigra did it became clear that the concordance among monozygotic twins was significant, even in presymptomatic PD.

FPD with specific genetic defects may account for fewer than 10% of all cases of PD [39]; however, the identification of these rare genetic defects and the functions of those genes has provided tremendous insight into the pathogenesis of PD. This knowledge is essential for developing new avenues for therapeutic intervention.

Several genes have now been linked to PD, in both dominant and recessive cases, and a number of other genetic linkages have been identified that may cause PD [40].

2. α-Synuclein Structure and Interacting Partners

The structure of α-synuclein is still unknown, but under physiological conditions it seems to be mostly unstructured (random coil) [41,42]. All synuclein proteins consist of a highly conserved amino-terminal domain that appears to adopt an α-helical conformation under certain conditions [43]. α-Synuclein has a middle hydrophobic domain (residues 61–95), also referred to as *NAC* (non Aβ component), as well as a C-terminal region rich in prolines and the acidic residues glutamate and aspartate. The α-helical N-terminal domain has an 11-amino acid repeat with a highly conserved motif (KTKEGV). This domain is thought to adopt amphipathic α-helical structure similar to that of apolipoproteins of the class A_2. The N-terminal region also harbors one of two characteristic signatures for fatty acid–binding proteins (FABPs) and is important for lipid binding.

α-Synuclein seems to bind exclusively to acidic phospholipids, especially phosphatidic acid, and to vesicles with small diameters. This is likely to target the protein to specific populations of membranes or vesicles [44]. The N-terminal domain is also likely to be responsible for the protein's ability to disrupt lipid bilayers [45]. The conservation of the α-helical domain among the synuclein proteins suggests that all three synucleins may be involved in similar biological processes that involve lipid binding.

The central region of α-synuclein (NAC) comprises the highly amyloidogenic part of the protein. This amyloidogenic part is responsible for properties that distinguish α-synuclein from the other members of the synuclein family: (a) the ability to undergo a conformational change from random coil to β-sheet structure [46,47], (b) the ability to form single cylindrical β-sheets [48], and (3) the ability to form Aβ-like protofibrils and fibrils [46,49–54].

The acidic C-terminal region of α-synuclein does not seem to associate with vesicles, remaining free and unfolded [55], despite its similarity to cytosolic FABP.

Although α-synuclein is intrinsically unstructured in solution, it is unlikely that this is the case in the context of a cell. α-Synuclein binds a number of ligands and proteins, which likely alter its conformation. α-Synuclein also shares some sequence similarity with the signaling chaperone protein 14-3-3 [56]. Like the 14-3-3 proteins, α-synuclein binds to extracellular-regulated kinase (ERK), dephospho-BAD (a Bcl-2 homolog), and protein kinase C (PKC). It also binds to the 14-3-3 proteins themselves. These interactions suggest that α-synuclein may be involved in the regulation of cell viability in ways that are not fully understood. A yeast two-hybrid screen identified synphilin-1 as an α-synuclein–interacting protein [57]. Synphilin-1 is a novel protein with little similarity to other known proteins, so the significance of this interaction is not totally clear. Synphilin-1 may act as an adaptor protein and may be involved in vesicle transport or cytoskeletal function. Its presence in LBs and its connection to parkin suggest it may play a role in PD pathology as well [57,58].

Casein kinases 1 and 2 (CK-1 and CK-2), but not protein kinase A (PKA) or C (PKC), phosphorylate α-synuclein on serine 129, and to a lesser extent on serine 87 [59]. Phosphorylation of α-synuclein seems to be tightly regulated and it may influence its binding to lipid membranes or to phospholipase D [59,60]. Notably, phosphorylation of α-synuclein affects its aggregation and toxicity in a fly model of synucleinopathies [61].

3. Putative Functions of α-Synuclein

α-Synuclein is widely expressed in the brain, in both neuronal and nonneuronal cell types, including dopaminergic neurons, cortical neurons, noradrenergic neurons, endothelial cells, and platelets [62–64]. The structural features and interacting partners of α-synuclein, together with evidence from other experimental approaches, allow speculation about the normal function of α-synuclein. The putative functions of α-synuclein include the binding of fatty acids, physiological regulation of certain enzymes, transporters and neurotransmitter vesicles, as well as roles in neuronal survival.

The involvement of α-synuclein in synaptic plasticity is suggested by its presence in presynaptic regions, where it interacts with brain vesicles and phospholipid membranes, affecting DA storage [65]. Specifically, the involvement of α-synuclein in synaptic plasticity may be mediated by the inhibition of phospholipase D (PLD) [60,66,67] because PLD was shown to be implicated in cell growth and differentiation [68].

Interestingly, the absence of α-synuclein in mice causes only mild behavioral changes. Mice lacking α-synuclein show only subtle changes in DA-dependent behavior and do not seem to be impaired in spatial learning [63,69]. It is possible that there is some degree of redundancy between α-synuclein and other members of the synuclein family. Double and triple knockouts should yield useful information about the function of α-synuclein and the other members of this family of proteins.

α-Synuclein seems to negatively regulate DA release through yet undefined mechanisms. Tyrosine hydroxylase (TH), the rate-limiting enzyme in DA biosynthesis, seems to be inhibited by α-synuclein [70]. α-Synuclein increases the ratio of the nonphosphorylated form of TH over the more active phosphorylated form. α-Synuclein may also modulate the releasable DA pool through its interactions with vesicle membranes, which may also be related to PLD inhibition. Another possibility is that membrane fusion and lysis are inhibited by the membrane-stabilization effects that class A α-helices, such as those formed by α-synuclein, have on membranes [44]. Studies in HEK293 cells have also indicated that α-synuclein facilitates an increase in the number of plasma membrane DA transporters (DATs) [71].

The finding of a triplication of the α-synuclein locus in a family with FPD suggests the levels of expression of α-synuclein are critical for any toxic effects that might arise [72]. This idea was confirmed independently using yeast as a model organism [73]. Nevertheless, measured levels of α-synuclein mRNA or protein are not elevated in PD or DLB.

4. α-Synuclein Misfolding, Aggregation, and Toxicity

The "amyloid hypothesis" (developed originally for AD) states that the aggregation of proteins into an ordered fibrillar structure is causally related to aberrant protein interactions that culminate in neuronal dysfunction and ultimately neurodegeneration [74]. However, the actual nature of the toxic species is unknown. Evidence from PD suggests the hallmark inclusions (i.e., LBs) may actually be a protective mechanism neurons develop to preclude the accumulation of the pathogenic intermediates, which have been proposed to be α-synuclein oligomers [51].

Aggregation of α-synuclein and the putative gain of toxic function may contribute to haploinsufficiency by entrapping the wild-type protein, thus reducing the amount of available functional α-synuclein. Studies in yeast suggest that once α-synuclein starts to aggregate, it may recruit α-synuclein away from other sites in the cell, harmonizing the gain and loss of function hypotheses [73]. One hypothesis is that molecular crowding could first accelerate the formation of α-synuclein spherical protofibrils, a process that involves β-sheet formation [75] and then chainlike and annular protofibrils [51]. *In vitro* studies with purified α-synuclein demonstrated that protofibril and fibril formation require different critical concentrations, suggesting there are ranges under which one may form preferentially over the other. If annular protofibrils do indeed exist *in vivo*, they could exert toxicity because of their binding and permeabilization of vesicles. These porelike structures formed, *in vitro*, by α-synuclein have been visualized by electron microscopy [76,77] and by atomic force microscopy [51], and they are structurally similar to a subset of Aβ protofibrils and other unregulated pore-forming toxins [75–78].

Findings in cellular systems are consistent with the toxic protofibril hypothesis. The accumulation of prefibrillar aggregates in the membrane fraction, before the appearance of α-synuclein inclusions, was associated with Golgi fragmentation and a reduction in cell viability [79]. Lysosomes may also be disrupted by α-synuclein protofibrils in a similar fashion. If lysosomal degradation of α-synuclein is critical in α-synuclein turnover, this could start a vicious toxic cycle [80].

In PD flies, the disconnection between pathogenesis and α-synuclein inclusion formation adds support for roles of α-synuclein oligomers rather than fibrillar inclusions in toxicity and pathogenesis [81].

Notwithstanding the findings that favor roles for α-synuclein oligomers in cellular toxicity, it is likely that fibrillar and/or amorphous α-synuclein aggregates also contribute to physical damage in neurons [82].

Several groups described differences in the aggregation kinetics of three forms of α-synuclein (WT, A53T, and A30P) *in vitro* [46,49,50,83–85]. Structural differences between these α-synuclein alleles have also been reported [86]. All three forms of α-synuclein have the ability to form fibers with the typical properties of other amyloids. However, the A30P mutant seems to form fibrils slower than WT α-synuclein, whereas the A53T fibrillizes faster than WT [50]. These findings are suggestive of a toxic role for small α-synuclein oligomers (not the fibers themselves) in disease. The A30P mutant, unlike A53T and WT α-synuclein, seems to be defective in binding to vesicles and membranes [87,88]. This suggests that the conformation of the A30P mutant differs from that of WT and A53T α-synuclein [89]. These differences also point at disease mechanisms that might differ depending on the mutant form of α-synuclein.

5. Selective Neuronal Degeneration in PD: α-Synuclein and Dopamine

The α-synuclein protein with 140 amino acids is predominantly expressed in the brain, but there is also a 112 amino acid splice variant that is predominantly expressed in the heart, skeletal muscle, and pancreas [90].

Why are dopaminergic neurons selectively affected in PD? This is still an open question, but studies suggest that DA metabolism may explain, at least in part, the selective neurodegeneration. Nigral dopaminergic neurons are particularly exposed to oxidative stress because the metabolism of DA generates several potentially toxic molecules, such as DA-quinone, superoxide radicals, and hydrogen peroxide [91]. DA-quinone molecules have been shown to form covalent adducts with α-synuclein, inhibiting fibrillization and leading to the accumulation of protofibrils [92]. If α-synuclein protofibrils are toxic, as explained in the previous section, increased levels of cytoplasmic DA could lead to the accumulation of the toxic forms of α-synuclein. Likewise, α-synuclein protofibrils may themselves increase the levels of cytoplasmic dopamine by binding to and permeabilizing synaptic vesicles through the formation of pores.

Although this hypothesis explains the selective degeneration of dopaminergic neurons, it still lacks ultimate proof because the DA adducts with α-synuclein have yet to be found *in vivo*. Moreover, α-synuclein toxicity has been observed in cells other than just dopaminergic neurons, suggesting events upstream of the putative α-synuclein–DA interaction also contribute to toxicity.

C. Prion Diseases

The prion diseases, also known as *transmissible spongiform encephalopathies* (TSEs), are a group of rare but inevitably fatal neurodegenerative diseases that affect both humans and a wide variety of animals. Prion diseases include *Creutzfeldt-Jakob* disease (CJD), *Gerstmann–Sträussler–Scheinker* (GSS) syndrome, *fatal familial insomnia* (FFI), and *kuru* in humans; *chronic wasting disease* (CWD) in deer and elk; *scrapie* in sheep; and *bovine spongiform encephalopathy* (BSE), or "mad cow" disease, in cattle.

Scrapie, the prototypic prion disease of sheep and goats, has been known since the nineteenth century. Scrapie is a subacute progressive ataxia that shepherds regarded as an inherited degenerative disease of the brain and spinal cord. In the twentieth century, Gajdusek showed that kuru, which affected aborigines of Papua New Guinea, was a TSE.

The causative agent of TSEs displays unusual biological properties, especially in that it cannot be destroyed by nucleases or ultraviolet radiation but is sensitive to agents that inactivate proteins. These properties led Griffith to propose, in 1967, that the infectious agent could be entirely made of protein–protein-based infectious agent, abbreviated later to *prion*. The conundrum of how infectivity could be concealed in a protein was solved when it was found to be an endogenous host protein, termed *prion protein* (PrP^C),

encoded by the *PRNP* gene on chromosome 20. In 1982, the disease-associated protease-resistant prion protein was isolated and termed PrP^{Sc}, leading Prusiner to formulate the "protein-only hypothesis" [93].

Prion diseases differ from other progressive encephalopathies such as AD and PD in that they are transmissible by inoculation or ingestion of prion-contaminated material. The most common human TSE is CJD, which can be sporadic (sCJD), familial (fCJD), iatrogenic (iCJD), and variant (vCJD) [94]. The incidence of sCJD is low—about one in a million people—and its etiology is unclear. Familial forms of the disease are autosomal dominant in nature, co-segregating with mutations in the *PRNP* gene. Iatrogenic cases are associated with neurosurgical intervention, tissue transplantation, or administration of hormones derived from individuals with undetected TSEs. A new form of CJD was identified in the United Kingdom in 1996 and was called vCJD. This variant form of the disease is thought to represent transmission of BSE to humans. The incidence of vCJD rose between 1996 and 2001 in the United Kingdom, evoking fears of a large upcoming epidemic. Fortunately, since 2001 the numbers appear to be stabilizing.

1. Clinical Disease

About 80% of sCJD cases are diagnosed between the fifth and seventh decade of life. Initially, about one-third of patients express vague feelings of fatigue, disordered sleep, or decreased appetite. Another third have neurological symptoms, such as memory loss, confusion, or uncharacteristic behavior. The last third have focal signs such as ataxia, aphasia, visual loss, hemiparesis, or amyotrophy. The typical inexorable progression of the disease leads to death in only 5–12 months of onset. Patients with vCJD are usually much younger than those with sCJD, also presenting psychiatric symptoms and a longer duration of the illness (~14 months).

2. Prion Protein

Human PrP is a protein of 253 amino acids encoded by exon 3 of the *PRNP* gene. The first 22 amino acids encode a signal peptide that is cleaved off during translation. The region between residues 51 and 91 contains a nonapeptide followed by four identical octarepeats, thought to function as copper-binding sites. The cellular prion protein is expressed in neurons and other cells of the central nervous system (CNS), but also in the lymphoreticular system and the skeletal or heart muscle. Nevertheless, PrP^C levels are highest in the CNS [94].

The structure of PrP^C was initially determined by nuclear magnetic resonance (NMR) imaging and was found to consist of a highly structured C-terminal region, containing three α-helices and two short antiparallel β-strands, as well as an unstructured N-terminus of 120 amino acids.

The protein is posttranslationally modified by N-linked glycosylation at residues 181 and 197 and the addition of a C-terminal glycosyl phosphatidylinositol anchor (GPI) at residue 230. PrP^C is attached to the cell surface in specialized domains called *lipid rafts* and is thought to cycle between the cell surface and endosomes. PrP^{Sc}, a misfolded conformer of PrP^C, has only been amenable to low-resolution methods, because of its insolubility. What we know is that PrP^{Sc} consists mainly of β-sheets and that PrP^{Sc} aggregates are extremely ordered in structure [95].

The function of PrP is still unknown. PrP knockout mice show only subtle phenotypes, besides their resistance to prion infectivity. Several functions have been proposed for PrP, including its involvement in signal transduction, protease activity, or as a form of superoxide dismutase.

3. Pathology

Diagnosis of prion disease ultimately requires histological examination of the brain and immunostaining for PrP. Crucial features are spongiform change accompanied by extreme neuronal loss and gliosis. Amyloid plaques can also be found in about 10% of sporadic cases of the disease. The tonsillar tissue may also be stained for PrP^{Sc} at autopsy.

GSS syndrome is characterized by a slowly progressive cerebellar ataxia, accompanied by cognitive decline. A unique neuropathological feature of GSS syndrome is the presence of multicentric PrP plaques. The mutations P102L and G131V are the most commonly found in this prion disease.

FFI is characterized by a profound disruption of the normal sleep–wake cycle, insomnia, and sympathetic overactivity.

The accumulation of PrP^{Sc} in the brain is the hallmark of prion diseases, but, as in other neurodegenerative diseases, whether it is directly responsible for the devastating pathology is still unclear. It seems unlikely that PrP^{Sc} accumulation fully accounts for prion pathology. The gain of toxic function hypothesis is plausible, as depletion of PrP^C does not seem to cause any pathology [96].

Our lack of knowledge about the biochemical pathways leading to brain damage makes it difficult to devise ways for therapeutic intervention.

4. Yeast Prions

The budding yeast *Saccharomyces cerevisiae* is one of the most extensively characterized eukaryotes and is, therefore, often used as a model organism in which to investigate different biological aspects of living cells, including those associated with the molecular basis of the protein aggregation disorders [97]. The transmissible character of mammalian prions proved instrumental for explaining two non-Mendelian yeast traits: the yeast prions [*PSI*+] and [*URE3*]. Reed Wickner was the first to propose a prion-like behavior for two yeast proteins that determine [*PSI*+] and [*URE3*], Sup35p and Ure2p, respectively. The study of these and other yeast prions, which are not associated with any yeast "disease" but are rather part of the normal biology of yeast cells, has yielded tremendous insight into the biology of mammalian prions. Most notably, prion conformations have been proposed to store information required for long-term memory formation, working in synapses to maintain their functional state [98].

III. "Natively Unfolded" Proteins and Other Structural Determinants of Protein Aggregation

Despite the lack of similarity at the level of the primary sequences between proteins associated with aggregation and misfolding, some structural motifs and properties emerge as potential clues as to how toxicity might arise. For example, both Aβ and α-synuclein have amphipathic regions that bind lipids and perturb membranes. α-Synuclein has the amphipathic domain near the N-terminus and a highly acidic C-terminal domain, suggesting it may fold onto itself, rather than existing as a natively unfolded protein, as appears to be the case under physiological conditions in the test tube [43] (Fig. 3). The tau protein, another candidate member of the "natively unfolded" family of proteins, was found to undergo a large conformation change before the aggregation/polymerization process. A very useful antibody

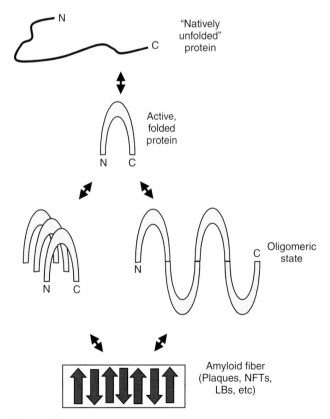

Figure 3 Folding, intramolecular, and intermolecular interactions of "natively unfolded" proteins.

(Alz50) was found to recognize a version of tau in which its N-terminus comes into contact with the microtubule-binding repeats, suggesting tau, like α-synuclein, also folds onto itself [99,100]. Also, both proteins seem to multimerize in the presence of polyunsaturated fatty acids, emphasizing the notion that these proteins adopt specific conformations according to their cellular environment and that some of those conformations may be more prone to enable aggregation/fibrillization, and hence, toxicity, somewhere along the pathway.

IV. Cellular Quality Control Systems for Protein Folding as Targets for Therapeutic Intervention in Neurodegenerative Diseases

Protein aggregates characteristic of AD, PD, prion diseases, and other neurodegenerative diseases share common morphological and biochemical features. In addition, they co-localize with several of the same proteins, including ubiquitin, proteasome and lysosome subunits, and molecular chaperones [101]. This sequestration of the cell's quality control machinery with inclusions might lead to a loss of function, rendering the cell less likely to refold/degrade other misfolded/aggregated proteins and causing a series of events that may ultimately result in cell death.

An alternate possibility is that with aging, environmental insults, mutations, or other unidentified triggers, the activity of the quality control systems becomes compromised, causing aggregation prone proteins to misfold and be left "unattended." This would lead to their accumulation and the consequent pleiotropic effects, including cell death.

Thus, identifying and targeting the misfolded state of each of the proteins involved in these disorders holds promise to constitute a useful therapeutic strategy (Fig. 4). It will, therefore, be important to develop cell models in which the conformation of these proteins can be readily assessed and manipulated, and yeast cells are already proving to be valuable "living test tubes" for these types of studies because of their great ease of manipulation.

V. Conclusions

Many studies have highlighted the overlap between neurodegenerative disorders associated with protein misfolding and aggregation, mainly because of the coexistence of clinical and/or pathological features of more than one disorder in the same individual case. Another factor contributing to overlap between neurodegenerative disorders is the disease heterogeneity, which "blurs" the definition of individual diseases.

Several of the proteins that aggregate in the different diseases appear to be "natively unfolded," a characteristic that may bear connections with their abilities to adopt altered conformations and form aggregates. This process likely generates the toxic species that cause neuronal cell dysfunction and, ultimately, death. The common theme (i.e., protein misfolding and aggregation) strongly suggests that understanding the molecular events underpinning one of these disorders might prove instrumental for devising therapeutics with potential to be used across a spectrum of neurodegenerative disorders.

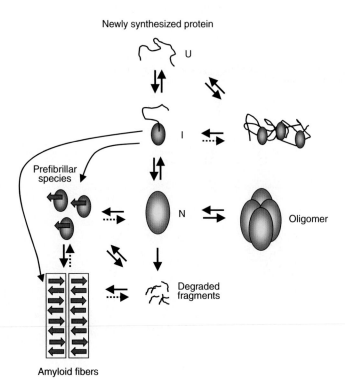

Figure 4 Protein Misfolding as a Therapeutic Target. Once proteins are synthesized, they face a wide variety of challenges in the crowded environment of the cell. Throughout their lives, proteins need to change their conformations, allowing unwanted species to accumulate, which may lead to cytotoxicity. Therapeutic intervention should be possible at different levels along these pathways. I, intermediate; N, native; U, unfolded.

Acknowledgments

Supported by grant P50NS38372. P. J. M. is supported by Harvard Medical School Scholars in Medicine Fellowship Award (Claflin Distinguished Scholars Award) and the American Parkinson Disease Association Grant. T. F. O. acknowledges the MBRC for a Tosteson Award Fellowship. We also thank the Massachusetts Alzheimer's Disease Research Center Brain Bank for access to human neuropathological tissues.

References

1. Barral, J. M., Broadley, S. A., Schaffar, G., and Hartl, F. U. (2004). Roles of molecular chaperones in protein misfolding diseases. *Semin. Cell Dev. Biol.* **15**, 17–29.
2. Dobson, C. M. (2004). Principles of protein folding, misfolding and aggregation. *Semin. Cell Dev. Biol.* **15**, 3–16.
3. Anfinsen, C. B. (1973). Principles that govern the folding of protein chains. *Science* **181**, 223–230.
4. Dobson, C. M. (2003). Protein folding and misfolding. *Nature* **426**, 884–890.
5. Daggett, V., and Fersht, A. (2003). The present view of the mechanism of protein folding. *Nat. Rev. Mol. Cell Biol.* **4**, 497–502.
6. Ellis, R. J. (2001). Macromolecular crowding: an important but neglected aspect of the intracellular environment. *Curr. Opin. Struct. Biol.* **11**, 114–119.
7. Minton, A. P. (2001). The influence of macromolecular crowding and macromolecular confinement on biochemical reactions in physiological media. *J. Biol. Chem.* **276**, 10577–10580.
8. Thomas, P. J., Qu, B. H., and Pedersen, P. L. (1995). Defective protein folding as a basis of human disease. *Trends Biochem. Sci.* **20**, 456–459.
9. Caughey, B., and Lansbury, P. T. (2003). Protofibrils, pores, fibrils, and neurodegeneration: separating the responsible protein aggregates from the innocent bystanders. *Annu. Rev. Neurosci.* **26**, 267–298.
10. Ross, C. A., and Poirier, M. A. (2004). Protein aggregation and neurodegenerative disease. *Nat. Med.* **10(Suppl)**, S10–S17.
11. Glenner, G. G., and Wong, C. W. (1984). Alzheimer's disease and Down's syndrome: sharing of a unique cerebrovascular amyloid fibril protein. *Biochem. Biophys. Res. Commun.* **122**, 1131–1135.
12. Price, D. L., Tanzi, R. E., Borchelt, D. R., and Sisodia, S. S. (1998). Alzheimer's disease: genetic studies and transgenic models. *Annu. Rev. Genet.* **32**, 461–493.
13. Strittmatter, W. J., Weisgraber, K. H., Huang, D. Y., Dong, L. M., Salvesen, G. S., Pericak-Vance, M., Schmechel, D., Saunders, A. M., Goldgaber, D., and Roses, A. D. (1993). Binding of human apolipoprotein E to synthetic amyloid beta peptide: isoform-specific effects and implications for late-onset Alzheimer disease. *Proc. Natl. Acad. Sci. USA* **90**, 8098–8102.
14. Selkoe, D. J. (2001). Alzheimer's disease: genes, proteins, and therapy. *Physiol. Rev.* **81**, 741–766.
15. Gomez-Isla, T., Price, J. L., McKeel, D. W., Jr., Morris, J.C., Growdon, J. H., and Hyman, B. T. (1996). Profound loss of layer II entorhinal cortex neurons occurs in very mild Alzheimer's disease. *J. Neurosci.* **16**, 4491–4500.
16. Hyman, B. T., Van Hoesen, G. W., and Damasio, A. R. (1990). Memory-related neural systems in Alzheimer's disease: an anatomic study. *Neurology* **40**, 1721–1730.
17. Wolozin, B., and Davies, P. (1987). Alzheimer-related neuronal protein A68: specificity and distribution. *Ann. Neurol.* **22**, 521–526.
18. Podlisny, M. B., Ostaszewski, B. L., Squazzo, S. L., Koo, E. H., Rydell, R. E., Teplow, D. B., and Selkoe, D. J. (1995). Aggregation of secreted amyloid beta-protein into sodium dodecyl sulfate-stable oligomers in cell culture. *J. Biol. Chem.* **270**, 9564–9570.
19. Selkoe, D. J. (2002). Deciphering the genesis and fate of amyloid beta-protein yields novel therapies for Alzheimer disease. *J. Clin. Invest.* **110**, 1375–1381.
20. Klyubin, I., Walsh, D. M., Lemere, C. A., Cullen, W. K., Shankar, G. M., Betts, V., Spooner, E. T., Jiang, L., Anwyl, R., Selkoe, D. J., and Rowan, M. J. (2005). Amyloid beta protein immunotherapy neutralizes Abeta oligomers that disrupt synaptic plasticity *in vivo*. *Nat. Med.* **11**, 556–561.
21. Fukumoto, H., Rosene, D. L., Moss, M. B., Raju, S., Hyman, B. T., and Irizarry, M. C. (2004). Beta-secretase activity increases with aging in human, monkey, and mouse brain. *Am. J. Pathol.* **164**, 719–725.
22. Fukumoto, H., Cheung, B. S., Hyman, B. T., and Irizarry, M. C. (2002). Beta-secretase protein and activity are increased in the neocortex in Alzheimer disease. *Arch. Neurol.* **59**, 1381–1389.
23. Martoglio, B., and Golde, T. E. (2003). Intramembrane-cleaving aspartic proteases and disease: presenilins, signal peptide peptidase and their homologs. *Hum. Mol. Genet.* **12(Spec No 2)**, R201–R206.
24. Kounnas, M. Z., Moir, R. D., Rebeck, G. W., Bush, A. I., Argraves, W. S., Tanzi, R. E., Hyman, B. T., and Strickland, D. K. (1995). LDL receptor-related protein, a multifunctional ApoE receptor, binds secreted beta-amyloid precursor protein and mediates its degradation. *Cell* **82**, 331–340.
25. Selkoe, D. J. (2001). Clearing the brain's amyloid cobwebs. *Neuron* **32**, 177–180.
26. Spires, T. L., and Hyman, B. T. (2004). Neuronal structure is altered by amyloid plaques. *Rev. Neurosci.* **15**, 267–278.
27. Maroteaux, L., Campanelli, J. T., and Scheller, R. H. (1988). Synuclein: a neuron-specific protein localized to the nucleus and presynaptic nerve terminal. *J. Neurosci.* **8**, 2804–2815.
28. Jakes, R., Spillantini, M. G., and Goedert, M. (1994). Identification of two distinct synucleins from human brain. *FEBS Lett.* **345**, 27–32.
29. Kruger, R., Kuhn, W., Muller, T., Kruger, R., Kuhn, W., Muller, T., Woitalla, D., Graeber, M., Kosel, S., Przuntek, H., Epplen, J. T., Schols, L., and Riess, O. (1998). Ala30Pro mutation in the gene encoding alpha-synuclein in Parkinson's disease. *Nat. Genet.* **18**, 106–108.
30. Kobayashi, H., Kruger, R., Markopoulou, K., Wszolek, Z., Chase, B., Taka, H., Mineki, R., Murayama, K., Riess, O., Mizuno, Y., and Hattori, N. (2003). Haploinsufficiency at the alpha-synuclein gene underlies phenotypic severity in familial Parkinson's disease. *Brain* **126**, 32–42.
31. Calne, D. B., Dubini, A., and Stern, G. (1989). Did Leonardo describe Parkinson's disease? *N. Engl. J. Med.* **320**, 594.
32. Parkinson Study Group. (2000). Pramipexole vs levodopa as initial treatment for Parkinson disease: a randomized controlled trial. Parkinson Study Group. *JAMA* **284**, 1931–1938.
33. E. Forster, E., and Lewy, F. H. (1912). Paralysis agitans. In "Pathologische Anantomie. Handbuch der Neurologie" (M. Lewandowsky, ed.). pp. 920–933. Springer Verlag, Berlin.
34. Spillantini, M. G., Schmidt, M. L., Lee, V. M., Trojanowski, J. Q., Jakes, R., and Goedert, M. (1997). Alpha-synuclein in Lewy bodies. *Nature* **388**, 839–840.
35. Irizarry, M. C., Growdon, W., Gomez-Isla, T., Newell, K., George, J. M., Clayton, D. F., and Hyman, B. T. (1998). Nigral and cortical Lewy bodies and dystrophic nigral neurites in Parkinson's disease and cortical Lewy body disease contain alpha-synuclein immunoreactivity. *J. Neuropathol. Exp. Neurol.* **57**, 334–337.
36. Jellinger, K. (1987). Overview of morphological changes in Parkinson's disease. *Adv. Neurol.* **45**, 1–18.
37. Nussbaum, R. L., and Polymeropoulos, M. H. (1997). Genetics of Parkinson's disease. *Hum. Mol. Genet.* **6**, 1687–1691.
38. Tanner, C. M., Ottman, R., Goldman, S. M., Ellenberg, J., Chan, P., Mayeux, R., and Langston, J. W. (1999). Parkinson disease in twins: an etiologic study. *JAMA* **281**, 341–346.
39. Gasser, T. (2001). Genetics of Parkinson's disease. *J. Neurol.* **248**, 833–840.
40. Cookson, M. R. (2004). The biochemistry of Parkinson's disease. *Annu. Rev. Biochem.* **74**, 29–52.
41. Kim, J. (1997). Evidence that the precursor protein of non-A beta component of Alzheimer's disease amyloid (NACP) has an extended structure primarily composed of random-coil. *Mol. Cells* **7**, 78–83.
42. Weinreb, P. H., Zhen, W., Poon, A. W., Conway, K. A., and Lansbury, P. T., Jr (1996). NACP, a protein implicated in Alzheimer's disease and learning, is natively unfolded. *Biochemistry* **35**, 13709–13715.
43. George, J. M., Jin, H., Woods, W. S., and Clayton, D. F. (1995). Characterization of a novel protein regulated during the critical period for song learning in the zebra finch. *Neuron* **15**, 361–372.

44. Davidson, W. S., Jonas, A., Clayton, D. F., and George, J. M. (1998). Stabilization of alpha-synuclein secondary structure upon binding to synthetic membranes. *J. Biol. Chem.* **273**, 9443–9449.
45. Jo, E., McLaurin, J., Yip, C. M., St. George-Hyslop, P., and Fraser, P. E. (2000). Alpha-synuclein membrane interactions and lipid specificity. *J. Biol. Chem.* **275**, 34328–34334.
46. Serpell, L. C., Berriman, J., Jakes, R., Goedert, M., and Crowther, R. A. (2000). Fiber diffraction of synthetic alpha-synuclein filaments shows amyloid-like cross-beta conformation. *Proc. Natl. Acad. Sci. USA* **97**, 4897–4902.
47. el-Agnaf, O. M., and Irvine, G. B. (2002). Aggregation and neurotoxicity of alpha-synuclein and related peptides. *Biochem. Soc. Trans.* **30**, 559–565.
48. Perutz, M. F., Pope, B. J., Owen, D., Wanker, E. E., and Scherzinger, E. (2002). Aggregation of proteins with expanded glutamine and alanine repeats of the glutamine-rich and asparagine-rich domains of Sup35 and of the amyloid beta-peptide of amyloid plaques. *Proc. Natl. Acad. Sci. USA* **99**, 5596–5600.
49. Conway, K. A., Harper, J. D., and Lansbury, P. T. (1998). Accelerated *in vitro* fibril formation by a mutant alpha-synuclein linked to early-onset Parkinson disease. *Nat. Med.* **4**, 1318–1320.
50. Conway, K. A., Lee, S. J., Rochet, J. C., Ding, T. T., Williamson, R. E., and Lansbury, P. T., Jr. (2000). Acceleration of oligomerization, not fibrillization, is a shared property of both alpha-synuclein mutations linked to early-onset Parkinson's disease: implications for pathogenesis and therapy. *Proc. Natl. Acad. Sci. USA* **97**, 571–576.
51. Ding, T. T., Lee, S. J., Rochet, J. C., and Lansbury, P. T., Jr. (2002). Annular alpha-synuclein protofibrils are produced when spherical protofibrils are incubated in solution or bound to brain-derived membranes. *Biochemistry* **41**, 10209–10217.
52. Harper, J. D., Wong, S. S., Lieber, C. M., and Lansbury, P. T. (1997). Observation of metastable Abeta amyloid protofibrils by atomic force microscopy. *Chem. Biol.* **4**, 119–125.
53. Harper, J. D., Lieber, C. M., and Lansbury, P. T., Jr. (1997). Atomic force microscopic imaging of seeded fibril formation and fibril branching by the Alzheimer's disease amyloid-beta protein. *Chem. Biol.* **4**, 951–959.
54. Harper, J. D., Wong, S. S., Lieber, C. M., and Lansbury, P. T., Jr. (1999). Assembly of A beta amyloid protofibrils: an in vitro model for a possible early event in Alzheimer's disease. *Biochemistry* **38**, 8972–8980.
55. Eliezer, D., Kutluay, E., Bussell, R., Jr., and Browne, G. (2001). Conformational properties of alpha-synuclein in its free and lipid-associated states. *J. Mol. Biol.* **307**, 1061–1073.
56. Yaffe, M. B. (2002). How do 14-3-3 proteins work?—Gatekeeper phosphorylation and the molecular anvil hypothesis. *FEBS Lett.* **513**, 53-57.
57. Engelender, S., Kaminsky, Z., Guo, X., Sharp, A. H., Amaravi, R. K., Kleiderlein, J. J., Margolis, R. L., Troncoso, J. C., Lanahan, A. A., Worley, P. F., Dawson, V. L., Dawson, T. M., and Ross, C. A. (1999). Synphilin-1 associates with alpha-synuclein and promotes the formation of cytosolic inclusions. *Nat. Genet.* **22**, 110–114.
58. Wakabayashi, K., Engelender, S., Yoshimoto, M., Tsuji, S., Ross, C. A., and Takahashi, H. (2000). Synphilin-1 is present in Lewy bodies in Parkinson's disease. *Ann. Neurol.* **47**, 521–523.
59. Okochi, M., Walter, J., Koyama, A., Nakajo, S., Baba, M., Iwatsubo, T., Meijer, L., Kahle, P. J., and Haass, C. (2000). Constitutive phosphorylation of the Parkinson's disease associated alpha-synuclein. *J. Biol. Chem.* **275**, 390–397.
60. Jenco, J. M., Rawlingson, A., Daniels, B., and Morris, A. J. (1998). Regulation of phospholipase D2: selective inhibition of mammalian phospholipase D isoenzymes by alpha- and beta-synucleins. *Biochemistry* **37**, 4901–4909.
61. Chen, L., and Feany, M. B. (2005). Alpha-synuclein phosphorylation controls neurotoxicity and inclusion formation in a *Drosophila* model of Parkinson disease. *Nat. Neurosci.* **8**, 657–663.
62. Hashimoto, M., Yoshimoto, M., Sisk, A., Hsu, L. J., Sundsmo, M., Kittel, A., Saitoh, T., Miller, A., and Masliah, E. (1997). NACP, a synaptic protein involved in Alzheimer's disease, is differentially regulated during megakaryocyte differentiation. *Biochem. Biophys. Res. Commun.* **237**, 611–616.
63. Abeliovich, A., Schmitz, Y., Farinas, I., Choi-Lundberg, D., Ho, W. H., Castillo, P. E., Shinsky, N., Verdugo, J. M., Armanini, M., Ryan, A., Hynes, M., Phillips, H., Sulzer, D., and Rosenthal, A. (2000). Mice lacking alpha-synuclein display functional deficits in the nigrostriatal dopamine system. *Neuron* **25**, 239–252.
64. Tamo, W., Imaizumi, T., Tanji, K., Yoshida, H., Mori, F., Yoshimoto, M., Takahashi, H., Fukuda, I., Wakabayashi, K., and Satoh, K. (2002). Expression of alpha-synuclein, the precursor of non-amyloid beta component of Alzheimer's disease amyloid, in human cerebral blood vessels. *Neurosci. Lett.* **326**, 5–8.
65. Lotharius, J., and Brundin, P. (2002). Pathogenesis of Parkinson's disease: dopamine, vesicles and alpha-synuclein. *Nat. Rev. Neurosci.* **3**, 932–942.
66. Ahn, B. H., Rhim, H., Kim, S. Y., Sung, Y. M., Lee, M. Y., Choi, J. Y., Wolozin, B., Chang, J. S., Lee, Y. H., Kwon, T. K., Chung, K. C., Yoon, S. H., Hahn, S.J., Kim, M. S., Jo, Y. H., and Min do, S. (2002). Alpha-synuclein interacts with phospholipase D isozymes and inhibits pervanadate-induced phospholipase D activation in human embryonic kidney-293 cells. *J. Biol. Chem.* **277**, 12334–12342.
67. Payton, J. E., Perrin, R. J., Woods, W. S., and George, J. M. (2004). Structural determinants of PLD2 inhibition by alpha-synuclein. *J. Mol. Biol.* **337**, 1001–1009.
68. Klein, J., Chalifa, V., Liscovitch, M., and Loffelholz, K. (1995). Role of phospholipase D activation in nervous system physiology and pathophysiology. *J. Neurochem.* **65**, 1445–1455.
69. Chen, P. E., Specht, C. G., Morris, R. G., and Schoepfer, R. (2002). Spatial learning is unimpaired in mice containing a deletion of the alpha-synuclein locus. *Eur. J. Neurosci.* **16**, 154–158.
70. Perez, R. G., Waymire, J. C., Lin, E., Liu, J. J., Guo, F., and Zigmond, M. J., (2002). A role for alpha-synuclein in the regulation of dopamine biosynthesis. *J. Neurosci.* **22**, 3090–3099.
71. Lee, M., Hyun, D., Halliwell, B., and Jenner, P. (2001). Effect of the overexpression of wild-type or mutant alpha-synuclein on cell susceptibility to insult. *J. Neurochem.* **76**, 998–1009.
72. Singleton, A. B., Farrer, M., Johnson, J., Singleton, A., Hague, S., Kachergus, J., Hulihan, M., Peuralinna, T., Dutra, A., Nussbaum, R., Lincoln, S., Crawley, A., Hanson, M., Maraganore, D., Adler, C., Cookson, M. R., Muenter, M., Baptista, M., Miller, D., Blancato, J., Hardy, J., and Gwinn-Hardy, K. (2003). Alpha-synuclein locus triplication causes Parkinson's disease. *Science* **302**, 841.
73. Outeiro, T. F., and Lindquist, S. (2003). Yeast cells provide insight into alpha-synuclein biology and pathobiology. *Science* **302**, 1772–1775.
74. Hardy, J., and Selkoe, D. J. (2002). The amyloid hypothesis of Alzheimer's disease: progress and problems on the road to therapeutics. *Science* **297**, 353–356.
75. Volles, M. J., and Lansbury, P. T., Jr. (2002). Vesicle permeabilization by protofibrillar alpha-synuclein is sensitive to Parkinson's disease-linked mutations and occurs by a pore-like mechanism. *Biochemistry* **41**, 4595–4602.
76. Lashuel, H. A., Petre, B. M., Wall, J., Simon, M., Nowak, R. J., Walz, T., and Lansbury, P. T., Jr. (2002). Alpha-synuclein, especially the Parkinson's disease-associated mutants, forms pore-like annular and tubular protofibrils. *J. Mol. Biol.* **322**, 1089–1102.
77. Lashuel, H. A., Hartley, D., Petre, B. M., Walz, T., and Lansbury, P. T., Jr. (2002). Neurodegenerative disease: amyloid pores from pathogenic mutations. *Nature* **418**, 291.
78. Volles, M. J., Lee, S. J., Rochet, J. C., Shtilerman, M. D., Ding, T. T., Kessler, J. C., and Lansbury, P. T., Jr. (2001). Vesicle permeabilization by protofibrillar alpha-synuclein: implications for the pathogenesis and treatment of Parkinson's disease. *Biochemistry* **40**, 7812–7819.

79. Gosavi, N., Lee, H. J., Lee, J. S., Patel, S., and Lee, S. J. (2002). Golgi fragmentation occurs in the cells with prefibrillar alpha-synuclein aggregates and precedes the formation of fibrillar inclusion. *J. Biol. Chem.* **277**, 48984–48992.
80. Stefanis, L., Larsen, K. E., Rideout, H. J., Sulzer, D., and Greene, L. A. (2001). Expression of A53T mutant but not wild-type alpha-synuclein in PC12 cells induces alterations of the ubiquitin-dependent degradation system, loss of dopamine release, and autophagic cell death. *J. Neurosci.* **21**, 9549–9560.
81. Auluck, P. K., Chan, H. Y., Trojanowski, J. Q., Lee, V. M., and Bonini, N. M. (2002). Chaperone suppression of alpha-synuclein toxicity in a Drosophila model for Parkinson's disease. *Science* **295**, 865–868.
82. Neumann, M., Kahle, P. J., Giasson, B. I., Ozmen, L., Borroni, E., Spooren, W., Muller, V., Odoy, S., Fujiwara, H., Hasegawa, M., Iwatsubo, T., Trojanowski, J. Q., Kretzschmar, H. A., and Haass, C. (2002). Misfolded proteinase K-resistant hyperphosphorylated alpha-synuclein in aged transgenic mice with locomotor deterioration and in human alpha-synucleinopathies. *J. Clin. Invest.* **110**, 1429–1439.
83. Li, J., Uversky, V. N., and Fink, A. L. (2002). Conformational behavior of human alpha-synuclein is modulated by familial Parkinson's disease point mutations A30P and A53T. *Neurotoxicology* **23**, 553–567.
84. Narhi, L., Wood, S. J., Steavenson, S., Jiang, Y., Wu, G. M., Anafi, D., Kaufman, S. A., Martin, F., Sitney, K., Denis, P., Louis, J. C., Wypych, J., Biere, A. L., and Citron, M. (1999). Both familial Parkinson's disease mutations accelerate alpha-synuclein aggregation. *J. Biol. Chem.* **274**, 9843–9846.
85. Ostrerova, N., Petrucelli, L., Farrer, M., Mehta, N., Choi, P., Hardy, J., and Wolozin, B., (1999). Alpha-synuclein shares physical and functional homology with 14-3-3 proteins. *J. Neurosci.* **19**, 5782–5791.
86. Bussell, R., Jr., and Eliezer, D. (2003). A structural and functional role for 11-mer repeats in alpha-synuclein and other exchangeable lipid binding proteins. *J. Mol. Biol.* **329**, 763–778.
87. Jo, E., Fuller, N., Rand, R. P., St. George-Hyslop, P., and Fraser, P. E. (2002). Defective membrane interactions of familial Parkinson's disease mutant A30P alpha-synuclein. *J. Mol. Biol.* **315**, 799–807.
88. Jensen, P. H., Nielsen, M. S., Jakes, R., Dotti, C. G., and Goedert, M. (1998). Binding of alpha-synuclein to brain vesicles is abolished by familial Parkinson's disease mutation. *J. Biol. Chem.* **273**, 26292–26294.
89. McLean, P. J., Kawamata, H., and Hyman, B. T. (2001). Alpha-synuclein-enhanced green fluorescent protein fusion proteins form proteasome sensitive inclusions in primary neurons. *Neuroscience* **104**, 901–912.
90. Ueda, K., Saitoh, T., and Mori, H. (1994). Tissue-dependent alternative splicing of mRNA for NACP, the precursor of non-A beta component of Alzheimer's disease amyloid. *Biochem. Biophys. Res. Commun.* **205**, 1366–1372.
91. Graham, D. G. (1978). Oxidative pathways for catecholamines in the genesis of neuromelanin and cytotoxic quinones. *Mol. Pharmacol.* **14**, 633–643.
92. Conway, K. A., Rochet, J. C., Bieganski, R. M., and Lansbury, P. T, Jr. (2001). Kinetic stabilization of the alpha-synuclein protofibril by a dopamine-alpha-synuclein adduct. *Science* **294**, 1346–1349.
93. Prusiner, S. B. (1982). Novel proteinaceous infectious particles cause scrapie. *Science* **216**, 136–144.
94. Glatzel, M., Stoeck, K., Seeger, H., Luhrs, T., and Aguzzi, A. (2005). Human prion diseases: molecular and clinical aspects. *Arch. Neurol.* **62**, 545–552.
95. Collinge, J. (2005). Molecular neurology of prion disease. *J. Neurol. Neurosurg. Psychiatry* **76**, 906–919.
96. Unterberger, U., Voigtlander, T., and Budka, H. (2005). Pathogenesis of prion diseases. *Acta Neuropathol. (Berl.)* **109**, 32–48.
97. Outeiro, T. F., and Muchowski, P. J. (2004). Molecular genetics approaches in yeast to study amyloid diseases. *J. Mol. Neurosci.* **23**, 49–60.
98. Shorter, J., and Lindquist, S. (2005). Prions as adaptive conduits of memory and inheritance. *Nat. Rev. Genet.* **6**, 435–450.
99. Binder, L. I., Guillozet-Bongaarts, A. L., Garcia-Sierra, F., and Berry, R. W. (2005). Tau, tangles, and Alzheimer's disease. *Biochim. Biophys. Acta* **1739**, 216–223
100. Hyman, B. T., Augustinack, J. C., and Ingelsson, M. (2005). Transcriptional and conformational changes of the tau molecule in Alzheimer's disease. *Biochim. Biophys. Acta* **1739**, 150–157.
101. Muchowski, P. J., and Wacker, J. L. (2005). Modulation of neurodegeneration by molecular chaperones. *Nat. Rev. Neurosci.* **6**, 11–22.

11

RNA-Based Disorders of Muscle and Brain

Paul J. Hagerman, MD, PhD

Keywords: *ataxia, fragile X syndrome, FXTAS, myotonic dystrophy, neuropathy, RNA, SCA, trinucleotide repeat*

I. Overview
II. Myotonic Dystrophy Type 1 and Type 2
III. Fragile X–Associated Tremor/Ataxia Syndrome
IV. Prospects for RNA Pathogenic Mechanisms in Other Neuromuscular Disorders
References

I. Overview

The disorders described in this chapter reflect an emerging concept in the pathogenesis of genetic disease, namely, that an expanded repeat element within a noncoding portion of an RNA transcript can be directly responsible for a disease phenotype. In this instance, the RNA itself displays a "toxic," or pathological, gain of function that is unrelated to the protein product of the gene from which the RNA was transcribed. Such an RNA-based mechanism is remotely analogous to the toxic gain of function exhibited for a number of the CAG-repeat (polyglutamine; polyQ) disorders, such as Huntington's disease and several of the spinocerebellar ataxias, which are addressed elsewhere in this book. The closely related myotonic dystrophies, DM1 (OMIM #160900) and DM2 (OMIM #602688), both oligonucleotide repeat disorders, collectively represent the paradigm for this class of disorders, and because DM2 helped to define the pathogenic mechanism for DM, these two disorders are presented together (for reviews, see references 1–4).

A third disorder, fragile X–associated tremor/ataxia syndrome (FXTAS; OMIM +309550), also appears to be caused by an RNA-based pathogenesis due to an expanded (noncoding) trinucleotide repeat in the fragile X mental retardation 1 *(FMR1)* gene (for review, see reference 5). Interestingly, larger expansions of the same gene give rise to fragile X syndrome (OMIM +309550) due to an entirely separate mechanism (gene silencing). DM and FXTAS are discussed mainly from the standpoint of their pathogenic mechanisms and how the expanded-repeat RNAs can give rise to the clinical phenotypes.

Finally, two additional repeat-expansion spinocerebellar ataxias, SCA8 (OMIM #608768) and SCA10 (OMIM +603516), are considered briefly, because evidence for an RNA pathogenesis for these disorders is unresolved.

Throughout this chapter, we frequently point to the Online Mendelian Inheritance in Man (OMIM), in which detailed information and extensive background citations are available.

II. Myotonic Dystrophy Type 1 and Type 2

A. History and Nomenclature

Myotonic dystrophy (dystrophia myotonica, DM) is an autosomal-dominant disorder and the most common form of muscular dystrophy in adults. DM is of historical significance as the first recognized example of genetic anticipation, the more frequent occurrence and/or earlier onset of disease in succeeding generations within a pedigree. Although myotonia and muscle weakness are cardinal features of DM, it is clearly a multisystem disorder, with involvement of the eye (characteristic iridescent cataracts), endocrine function (hypogonadism, insulin resistance), heart (cardiac conduction defects), immune function (hypogammaglobulinemia), and central nervous system (CNS) (cognitive and behavioral abnormalities, white-matter changes, cortical atrophy). The significance of these findings to this discussion is that they are highly diverse and cannot be readily ascribed to the altered function of a single gene (protein) product. This multisystem character of DM, coupled with the phenomenon of anticipation, points to a novel type of pathogenic mechanism that is now known to be linked to oligonucleotide repeat expansions in noncoding portions of the host genes.

With the identification in 1992 of the DM protein kinase gene (*DMPK*, OMIM 605377) and the discovery of a second DM gene, the zinc-finger protein 9 gene (*ZNF9*; OMIM 116955), the designation of DM was subdivided into DM1 (*DMPK*) and DM2 (*ZNF9*). The discovery of the second gene on a separate chromosome from the *DMPK* gene, which nevertheless gives rise to many of the same clinical features, has strengthened considerably the argument for an RNA-based mechanism for DM pathogenesis.

Historically, different designations have been given to DM based on both the age at onset and the presenting clinical features. These designations include congenital (neonatal) myotonic dystrophy, a juvenile-onset form, and adult-onset DM. The neonatal- and juvenile-onset forms appear to be almost exclusively DM1 and are both associated with cognitive deficits that are not characteristic of adult-onset DM2. Other clinical features differ between DM1 and DM2, which are addressed briefly in the discussion of symptoms in relation to pathophysiology. However, in light of the pathogenic mechanism, involving oligonucleotide repeat expansions within the *DMPK* and *ZNF9* genes, the highly variable age at onset and degree of severity can be thought of as manifestations of differing degrees of repeat expansion.

B. Etiology of DM

In 1992, the genetic basis for DM1 was determined to be an abnormal expansion of a $(CTG)_n$ element located in the 3' untranslated region (UTR) of the *DMPK* gene (19q13.2-q13.3) (OMIM *605377). The $(CTG)_n$ tract can vary from fewer than 50 (as few as four) CTG repeats in unaffected individuals to more than several thousand repeats in the most severe forms of congenital DM. For repeat expansions above ~400–500 CTG repeats, there is little additional correlation of repeat size with disease severity. The size of the CTG repeat tract is subject to substantial intergenerational expansion, as well as significant somatic instability. This tendency to expand on transmission is the basis of the genetic anticipation in DM. Nearly 98% of DM cases are caused by CTG repeat expansions in the *DMPK* gene and are, therefore, classified as DM1.

Numerous population studies have been performed to define both the general prevalence and the population stratification of expanded repeat alleles. Two basic conclusions emerge: First, the near absence of expanded disease-forming alleles in native sub-Saharan African populations suggests that most cases of DM1 in European and North American populations are due to founder alleles that appeared after migration of the host populations out of Africa. Second, among the populations harboring the expanded alleles, prevalence estimates range from a high of ~1/500 individuals in Quebec, to more typical numbers of ~1/10,000 to 1/20,000 in European populations.

In 2001, the genetic basis for an additional ~2% cases of adult-onset DM, not linked to the *DMPK* gene, was identified as being due to CCTG (tetranucleotide) repeat expansions in intron 1 of the *ZNF9* gene (3q21) [6]. Although less epidemiological work has been conducted with this much less common form of DM, it is noteworthy that the intronic $(CCTG)_n$ expansions of the DM2 alleles can attain much larger sizes (>10 kb) than for the CTG repeat expansions of DM1, even though such expansions do not result in a severe congenital form of DM.

C. Molecular Pathogenesis

Following the discovery of the *DMPK* mutation, a number of hypotheses were advanced to explain the pathogenesis of DM, including *DMPK* haploinsufficiency models and contiguous gene models that would involve either altered regulation of neighboring genes (e.g., the downstream *SIX5* gene) or repeat-expansion–coupled gene silencing. These alternative models are discussed elsewhere [4; OMIM #160900]. None of those models, however, can explain the fundamental multisystem character of DM, nor can they account for the phenotypic similarities of DM1 and DM2.

A fundamental clue to the role played by the expanded repeat came from the observation of Philips et al. [7] that the heterogeneous nuclear ribonucleoprotein (CUG binding protein; CUGBP) was upregulated in DM1 muscle [8], and that this upregulation led to abnormal splicing of cardiac troponin T (cTNT). The importance of this observation is that it provides a causal link between the CUG expansion (in RNA) and disturbed cellular function, namely, abnormal cTNT. Thus, the basic pathogenic mechanism can be considered to comprise three parts (Fig. 1): (1) the initial expansion of the CUG repeat (as noncoding RNA); (2) one or more events whereby the effects of the expanded repeat are "transduced" through the altered expression and/or function of one or more key proteins (e.g., CUGBP); (3) the downstream actions of the key proteins—in the case studied by Philips et al. [7], the abnormal splicing of cTNT.

It has been demonstrated that human homologs of *Drosophila muscleblind* (MBNL1, MBLL, and MBXL; also designated MBNL1-3; MBNL) are recruited to the expanded CUG repeats in DM1, and that these proteins (but not CUGBP) co-localize with the *DMPK* mRNA in nuclear foci of patients with DM [9]. Thus, MBNL and CUGBP represent two RNA binding proteins that appear to mediate (transduce) the effect of the expanded repeat. Interestingly, CUG repeat expansion leads, by unknown mechanisms, to increased levels of CUGBP, and to reduced levels of MBNL, presumably due to sequestration of the latter protein by the expanded CUG repeat. The strength of the RNA gain-of-function model for DM lies in the fact that both CUGBP and MBNL act in *trans* to modulate the splicing of the transcripts of at least six different genes. Moreover, both increased CUGBP and decreased MBNL result in a shift to splice isoforms that are normally present in early development. The abnormal distribution of splice isoforms for at least six different genes, resulting from the expanded CUG repeat, provides a unifying feature of the RNA-pathogenesis model.

The argument for an RNA-based mechanism for DM pathogenesis is based on the following lines of evidence:

1. Two unlinked genes of entirely unrelated function, *DMPK* on chromosome 19 and *ZNF9* on chromosome 3, give rise to remarkably similar multisystem phenotypes, DM1 and DM2, respectively. This observation provides a powerful argument against contiguous gene effects, because the regions flanking the two genes are quite different.

2. No coding mutation that recapitulates the clinical features of DM has been described for either *DMPK* or *ZNF9*. This observation indicates that abnormal forms (or haploinsufficiency) of either gene product are not responsible for DM. Interestingly, although a number of coding mutations do exist within the *FMR1* gene, such mutations result in the neurodevelopmental disorder (fragile X syndrome), not FXTAS.

3. Both genes possess noncoding repeats, $(CTG)_n$ in the 3'UTR of the *DMPK* gene and $(CCTG)_n$ in the first intron of the *ZNF9* gene, which are expanded in the disease state. These observations argue against any participation of altered *DMPK* or *ZNF9* protein products in the development of DM.

4. Expression of a ~250 CUG-repeat element (i.e., as RNA) in a transgenic mouse model [10] reproduced a core feature (myotonia) of DM. In a related experiment,

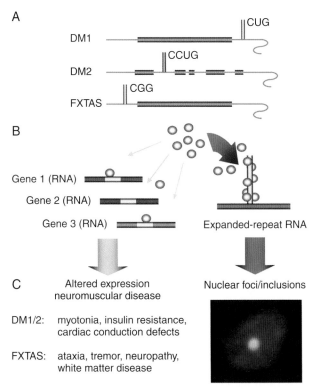

Figure 1 Outline of the basic features of the RNA gain-of-function model as envisioned for myotonic dystrophy types 1 and 2 (DM1/2) and fragile X-associated tremor/ataxia syndrome (FXTAS). (**A**) Expanded repeats, represented as vertical bars, are located in noncoding portions of the RNA transcripts for DM1 (CUG repeat, 3'UTR of *DMPK* gene), DM2 (CCUG repeat, intron 1 of *ZNF9* gene), FXTAS (CGG repeat, 5'UTR of *FMR1* gene). (**B**) Excess recruitment of one or more RNA-binding proteins to the expanded repeat element causes depletion (loss of function) of those proteins from binding sites in other RNAs. (**C**) Loss of function of the sequestered proteins results in altered processing/function of other RNAs, leading to specific features of the disease phenotype; excess binding of proteins to the expanded repeat leads to accumulation of intranuclear foci (inclusions) that contain both the expanded-repeat RNA and one or more of the associated proteins.

expression of an expanded ~200 CUG-repeat 3'UTR in cultured myoblasts inhibited myoblast differentiation [2]. This set of observations provides a strong argument for the gain of function of the expanded repeats *at the RNA level*.

5. In accord with a model in which the expanded C(C)UG repeat sequesters one or more specific proteins, the *DMPK* and *ZNF9* mRNAs are each sequestered in nuclear foci that co-localize with MBNL, as noted earlier. This set of observations provides a physical link between the primary causal abnormality (the expanded RNA repeat) and one or more proteins (e.g., the MBNL proteins) that are believed to transduce the effect of the RNA repeat.

6. Both CUGBP and MBNL are known to influence the distribution of splice isoforms of at least six (and perhaps more) genes that are themselves coupled to various components of the clinical phenotype. This last aspect of DM is discussed in the following section.

D. Pathophysiology

In the RNA pathogenesis model for DM1 and DM2, all of the diverse features of those disorders are ascribable to altered levels of a small number of proteins (e.g., MBNL and CUGBP) that are, in turn, responsible for the altered expression of a number of additional genes. The products of at least six genes are known to be altered as a consequence of the altered expression of CUGBP and MBNL. These genes are discussed below in turn; however, it is important at the outset to consider some important differences between DM1 and DM2 in terms of the underlying pathophysiology. First, although the size of the repeat expansion in DM2 can often vastly exceed the CUG repeat sizes that cause the severe congenital form of DM1, no comparable level of severity of clinical phenotype, including the defects in muscle development, has been observed in DM2. Second, the cognitive impairment that generally accompanies the early onset forms of DM1 is not present in DM2. Although there are additional differences between DM1 and DM2, most of these are of a more quantitative nature.

If DM1 and DM2 are caused by the same pathogenic mechanism, even involving the same transducing proteins, how are these differences in phenotype to be understood? Although all of the data are not available on this issue, there are at least three explanations for the differences between DM1 and DM2. The first and most obvious possibility is that the transducing proteins (e.g., MBNL) do not bind in the same manner to the CUG and CCUG repeats. Thus, even though the CCUG repeat expansions are larger, they exert a less profound effect on the transducing proteins. The second possibility is that there are differences in levels of expression of *DMPK* and *ZNF9* in various tissues, so that differing levels of the expanded-repeat RNAs are present in target tissues (e.g., muscle, heart) for DM1 and DM2. This second possibility is an important consideration when thinking about the differences between DM1/2 and FXTAS, which does not have an associated myotonia. In both DM1/2 and FXTAS, there is evidence for the presence of MBNL in the intranuclear foci/inclusions [11]; however, there is very little, if any, expression of the *FMR1* gene in muscle; thus, one would not expect to see altered function of the muscle chloride channel (ClC-1) in FXTAS. This principle of phenotypic variation due to differing cell-type expression is likely to be responsible for many of the differences among the CAG-repeat disorders.

A third possible explanation for the phenotypic differences between DM1 and DM2 is that the timing of splicing/removal of the CCUG-containing intronic sequence in the *ZNF9* transcript is itself regulated differently in different cell types, resulting in quite different levels of CCUG-containing intronic RNA despite similar final levels of *ZNF9* mRNA. This intriguing possibility needs to be explored further, particularly in light of the fact that the pathogenesis of both disorders is related to altered splice regulation.

The differences between DM1 and DM2 notwithstanding, their similarities are remarkable and represent a fundamental strength of the RNA pathogenesis model. Generally, the dysregulation of CUGBP and MBNL appears to alter the splicing of target transcripts so the products of splicing are biased toward RNAs that are normally present at earlier developmental stages. From the perspective of the altered muscle electrophysiology, altered splicing of the transcripts for the principal chloride channel of muscle (ClC-1; OMIM *118425) leads to functional loss of this channel. It is remarkable that the noncoding expansions of the two very different genes (*DMPK* and *ZNF9*) lead to this highly specific form of muscle pathology. Allelic expansion also leads to abnormal splicing of myotubularin-related 1 (*MTMR1*; OMIM *300171) protein, which could be responsible for the failure of proper muscle differentiation, a feature of the congenital form of DM. As noted previously, altered levels of CUGBP and MBNL are also responsible for abnormal splice regulation of cTNT (retention of a fetal exon 5), which is likely to contribute to the cardiac conduction defects observed in both DM1 and DM2. It is also likely that the altered distribution of splice isoforms of the microtubule-associated protein tau (OMIM +157140) may be involved with the cognitive/behavioral features present in congenital and juvenile forms of DM1. Finally, altered splicing of the insulin receptor (preferential exclusion of exon 11) toward an insulin-insensitive form of the insulin receptor is consistent with the insulin resistance associated with DM1/2.

It is likely that a better understanding of the interactions of MBNL with the CUG and CCUG repeats and a more complete description of the ranges and levels of expression of *DMPK* and *ZNF9* will provide an explanation of the differences in pathophysiology and clinical phenotype between DM1 and DM2.

E. Explanation of Symptoms in Relation to Pathophysiology

Many of the clinical features of DM1 and DM2 can be understood in terms of the abnormalities of the splice isoforms, as noted earlier. In particular, the core feature of myotonia (ClC-1 channel), the associated features of insulin resistance (insulin receptor), the functional cardiac abnormalities (cTNT), the muscle developmental pathology (MTMR1), and at least some of the CNS effects (tau) can be interpreted in terms of the unifying RNA pathogenesis model. However, several additional clinical features remain unexplained, including the distinctive iridescent cataracts, the hypogonadism, and the hypogammaglobulinemia, features associated with both DM1 and DM2. Thus, it seems likely that as additional genes are identified that are dysregulated in DM, the explanations for these additional clinical features will be revealed.

F. Natural History

As noted earlier, the natural histories of DM1 and DM2 are highly variable, with some but not all aspects of the clinical presentations being interpretable in terms of the degree of repeat expansion. More severe and earlier onset (congenital) DM correlates with larger expansions, which are maternally inherited. Clinical presentation with most severe neonatal cases of DM1 can be associated with substantial musculoskeletal abnormalities that are thought to be due to an underlying defect in muscle differentiation. Interestingly, myotonia is not a feature of this early onset phenotype, although it generally appears by 10 years of age. Cataracts also develop later. CNS involvement includes significant cognitive delays, as well as cerebral atrophy. Juvenile-onset cases (DM1) also can display cognitive delays. Adult-onset cases generally present with myotonia, muscle pain, or cataracts; they are often ascertained in a family through a proband with congenital DM. Although all of these features can be qualitatively understood in terms of the size and expression levels of the repeat expansions, further understanding of the age at onset and some of the CNS, ocular, and endocrinological correlates of DM will require refinement in the understanding of the RNA and its interacting proteins.

III. Fragile X–Associated Tremor/Ataxia Syndrome

A. History and Nomenclature

Fragile X-associated tremor/ataxia syndrome (FXTAS) was first identified in a developmental pediatrics clinic when it became evident that the carrier grandfathers of children being seen for the neurodevelopmental disorder fragile X syndrome were themselves often experiencing neurological problems, principally involving progressive intention tremor and disturbances of gait. Although fragile X syndrome had been known for more than 25 years [12], and the gene responsible for fragile X syndrome known since 1991, the neurological disorder now known as FXTAS was not recognized until the late 1990s and was not reported until 2001 [13]. Part of the reason for the delay in identification of FXTAS was that the genetic model for fragile X syndrome posited that carriers of "premutation" (CGG) trinucleotide expansions were clinically unaffected. Moreover, adult neurologists were generally unaware of the genetics of fragile X syndrome, which would often involve the grandchildren (or grandnieces/nephews) of the patients being seen in neurology clinic for tremor, balance problems, or parkinsonism.

Abnormal expression of the fragile X mental retardation 1 (*FMR1*) gene (OMIM +309550) leads to both the neurodegenerative FXTAS and the leading heritable form of cognitive impairment, fragile X syndrome. Despite being caused by the same gene, the two syndromes differ in genetic mechanism, affect different groups of individuals, and are manifest in entirely different age ranges. Both FXTAS and fragile X syndrome result from trinucleotide (CGG) repeat expansions that lie within the 5′UTR (noncoding) of the *FMR1* gene, that is, they do not affect the sequence of the protein product of the gene. However, the ranges of CGG repeat expansions differ for the two disorders. Most individuals in the general population harbor *FMR1* alleles with approximately 5–44 CGG repeats. Larger expansions are designated "gray zone" (45–54 repeats), "premutation" (55–200 repeats), and "full mutation" (>200 to several thousand repeats). Most full mutation alleles are hypermethylated in the promoter and adjacent CGG repeat regions, which leads to transcriptional silencing and the consequent absence of *FMR1* protein (FMRP). The lack of FMRP is responsible for the neurodevelopmental disorder fragile X syndrome.

B. Epidemiology of FXTAS

FXTAS is a movement disorder that affects some carriers of premutation alleles of the *FMR1* gene. The disorder generally presents with progressive intention tremor and/or gait ataxia and parkinsonism. Onset is generally after 50 years and is much more common in male carriers, although female carriers can develop FXTAS. Associated (more variable) clinical features include autonomic dysfunction (incontinence of bladder or bowel, impotence); cognitive deficits that involve loss of memory and executive function, often progressing to dementia; and peripheral neuropathy, with decreased sensation in the distal lower extremities.

Studies of the penetrance of tremor and ataxia among adult premutation carriers in known fragile X families suggest that at least one-third of male carriers older than 50 years have both tremor and ataxia, and that at least one-half of male carriers older than 70 years have these features [14]. Based on these observations and the male premutation carrier frequency in the general population of ~1/800 (95% confidence interval [CI]: 1/530 to 1/1800), the general population prevalence of FXTAS among males older than 50 years is estimated to be approximately 1/3000. However, unpublished studies suggest that the general prevalence is lower, perhaps by as much as twofold, because of ascertainment bias within known fragile X families; that is, larger premutation alleles have a greater likelihood of expanding to full mutation alleles (and a fragile X syndrome proband) in two generations. In this regard, of the more than one dozen screening studies for premutation alleles among adult neurology populations (e.g., ataxia, essential tremor, Parkinson's disease, multiple system atrophy), most premutation alleles (detected in ~2–5% of ataxia patients) have been in the 80–100 CGG repeat range, whereas most of the premutation alleles in the general population are below 80 CGG repeats. This last set of observations is consistent with the notion that there is a bias toward larger CGG repeats among cases of FXTAS, thus supporting a somewhat lower estimate for the prevalence within the general population for males older than 50 years.

The penetrance of FXTAS among female premutation carriers is not known; it is clearly lower than the penetrance in males, although some females are definitely affected by FXTAS [15]. This difference may be due in part to the effects of random X inactivation, so many females will have a substantial fraction of their normal *FMR1* allele as the active allele. Clearly, more epidemiological work is needed in this area. Finally, although fragile X syndrome has been identified in nearly all of the major populations throughout the world, similar studies have not yet been performed for FXTAS.

C. Neuropathology

The hallmark neuropathological feature of FXTAS is the presence of eosinophilic intranuclear inclusions that are present in neurons and astrocytes in broad distribution throughout the brain and brainstem [11,16]. The inclusion counts are highest in the hippocampus, occupying 40–50% of hippocampal neurons and astrocytes in some cases, whereas counts are lower in the cerebral cortex (~5–10%) and are not detected in cerebellar Purkinje cells. The inclusions are spherical (2–5 μ), generally solitary, and are periodic acid–Schiff (PAS), tau, silver, polyglutamine, and synuclein negative. Thus, the inclusions of FXTAS are fundamentally distinct from the cytoplasmic Lewy bodies, from the glial cytoplasmic inclusions of multiple system atrophy, and from the intranuclear inclusions of the CAG-repeat (polyglutamine) disorders. Additional immunochemical staining of the inclusions is positive for the stress-related proteins, αB-crystallin and heat shock protein 70 (Hsp70), for ubiquitin, and for at least one of the proteasomal subunits (11S).

Other features of the neuropathology of FXTAS include general cerebral and cerebellar atrophy, spongiosis of the deep cerebellar white matter extending into the middle cerebellar peduncles, severe cerebellar Purkinje cell dropout, and Bergmann gliosis. FXTAS is neuroanatomically distinct from fragile X syndrome in at least two respects: No intranuclear inclusions have been detected in the brains of older adult males with fragile X syndrome, and in stark contrast to the loss in cerebral volume associated with FXTAS, the overall brain volume tends to be increased in the full mutation.

Findings on magnetic resonance imaging (MRI) are consistent with the neuropathology of FXTAS [17] and include moderate to severe global brain atrophy, as well as white-matter changes in the subcortical regions of the cerebrum. The most distinct finding, found in roughly 60% of carriers with clinical involvement, is a symmetrical increase in T2 signal intensity in the middle cerebellar peduncles (MCP sign), which reflects the underlying spongiform changes in that region. The presence of this MRI finding is used as a criterion for establishing a diagnosis of "definite" FXTAS [15].

D. Molecular Pathogenesis

As is the case with the myotonic dystrophies, there are several lines of reasoning that strongly favor an RNA "toxic" gain-of-function model for FXTAS (see Fig. 1). First, the disorder appears to be confined to carriers of premutation alleles of the *FMR1* gene; it has not been reported among older adults with full mutation alleles, which are generally transcriptionally silent. The apparent absence of FXTAS among adult patients with fragile X syndrome, who possess full mutation alleles, mitigates against pathogenic mechanisms that involve either the absence of FMRP or large CGG repeat expansions that operate at the DNA level; thus, the gene must be transcriptionally active in patients with FXTAS. Second, expression of premutation alleles of the *FMR1* gene is dysregulated in at least three respects: (1) for premutation alleles, *FMR1* mRNA levels are elevated by as much as eightfold over the levels found for normal alleles; (2) the mRNA itself is altered due the presence of the expanded CGG repeat; (3) the start site for transcription is itself altered by the presence of the expanded repeat, such that the 5'end of the message is extended by about 50 nucleotides. Third, both mouse and *Drosophila melanogaster* models that harbor the CGG repeat expansions in the premutation range (~90–100 CGG repeats) manifest features of the neuropathology of FXTAS [15]. In the case of the fly model, neuropathic features are present even when the expanded CGG repeat is transcribed upstream of a heterologous promoter. Therefore, the expanded repeat, as RNA, is capable of inducing several of the features of the human disease. Fourth, adding further direct support for an RNA-based pathogenesis for FXTAS is the presence of *FMR1* mRNA within the inclusions [18]. This last observation provides another parallel with the foci of myotonic dystrophy, which also contain the expanded (CUG or CCUG) repeat RNAs.

It should be noted that there are distinct differences in the nature of the inclusions of FXTAS and those found in the mouse and fly, suggesting that certain features of the pathogenesis of FXTAS are different from the mechanisms operating in the animal models. In the mouse, no inclusions are detected in astrocytes, whereas in humans, the percentages of inclusions in astrocytes are often greater than in neurons. If the pathogenesis of FXTAS were due in part to altered astrocytic function, the absence of overt astrocyte pathology in the mouse may explain its much milder neuropathology and corresponding phenotype. In the *Drosophila* model, the distribution and physical characteristics of the inclusions are quite distinct from those in cases of FXTAS. For example, there are substantial numbers of cytoplasmic inclusions in the fly model, although no cytoplasmic inclusions are detected in FXTAS cases.

Microgram quantities of purified inclusions have been isolated using automated particle sorting of immunofluorescence-tagged inclusions, following the fractionation of intact inclusion-bearing nuclei from postmortem human brain tissue of FXTAS cases. Subsequent mass spectrometric analysis of the protein complement of the inclusions has revealed the presence of more than 30 proteins [11]. Several of these proteins are of potential interest to the pathogenesis of FXTAS, including two RNA binding proteins, heterogeneous nuclear ribonuclear protein A2 (hnRNP A2) and MBNL1; and lamin A/C (OMIM *150330), as well as several of the neurofilament proteins. Although the functional importance of these proteins is not known, it is interesting that MBNL1 is associated with the pathogenesis of DM, and lamin A/C mutations are associated with an autosomal-recessive form of Charcot–Marie–Tooth disease (see following discussion).

E. Pathophysiology

Although there are important parallels between the basic molecular mechanisms of FXTAS and myotonic dystrophy, much less is known about FXTAS regarding either the proteins that are likely to be transducing the effect of the expanded CGG repeat or the consequent pathophysiology. Interestingly, although MBNL1 appears to be present in the inclusions of FXTAS, there is no evident myotonia associated with FXTAS. However, even if MBNL1 were dysregulated in FXTAS, the absence of myotonia would not be surprising because, based on the DM model, significant *FMR1* expression would be required in muscle, which is not the case. However, the possibility of MBNL1 dysregulation in other tissues, such as brain, remains an open question.

What is clearly lacking is an understanding of the relationship between the expanded CGG repeat and two key neuropathological findings in FXTAS, namely, intranuclear inclusions and white-matter disease. Although further analysis of the inclusions is likely to reveal some of the key protein players in the pathophysiology of FXTAS, a greater understanding at the systems level is also required. With respect to the latter, it is possible that the degenerative (spongiform) changes in the middle cerebellar peduncles reflect damage to the neurons within the pontine nuclei or afferents to those neurons. Because the middle cerebellar peduncles constitute the major afferent pathway to the cerebellum, the dramatic Purkinje cell dropout in FXTAS could be a consequence of reduced input from the pontine neurons. Of course, the loss of Purkinje neurons could be a more direct consequence of RNA toxicity, perhaps reflecting the absence of a (hypothetical) protective effect of inclusion formation in those cells.

F. Explanation of Symptoms in Relation to Pathophysiology

At the clinical level, FXTAS is fundamentally a movement disorder with a prominent cerebellar component. The core features of cerebellar tremor and gait ataxia are entirely consistent with the loss of cerebellar Purkinje cells, the presence of dystrophic axons, and the gross spongiform changes in the cerebellar white matter. Indeed, the cerebellum is one of the most severely affected regions of the brain in patients with FXTAS, with the most consistent and severe neuropathology. Because an important role of the cerebellum is to coordinate movement, with Purkinje cells constituting the only output neurons of the cerebellar cortex, both the gait ataxia and the tremor are interpretable in terms of the evident cerebellar pathology.

Another clinical component of FXTAS, the development of cognitive deficits with loss of memory and executive function, may also be interpretable, at least in part, as a consequence of the cerebellar pathology. Apart from its function in motor coordination, the cerebellum contributes to the coordination of "frontal" executive functions such as planning and reasoning. Thus, the cerebellar pathology in patients with FXTAS may contribute to their prominent executive deficits and their motor dysfunction. The memory loss and emotional lability experienced by many patients with FXTAS may be related to the substantial numbers of intranuclear inclusions in the hippocampus. However, it is important to bear in mind that the relationship between inclusion load and cellular/system dysfunction is not known. To underscore this point, there is relatively little evident neural cell loss in the hippocampus, where the numbers of inclusions are greatest; however, cerebellar Purkinje cell loss is often profound, although only rarely have inclusions been detected in those cells.

Finally, the presence of lamin A/C within the inclusions raises the interesting possibility that lamin dysfunction could be contributing to either (both) the white-matter disease or (and) the peripheral neuropathy. Mutations in *LMNA* (1q21), the gene encoding the lamin A/C splice isoforms, give rise to an autosomal-recessive form of Charcot–Marie–Tooth disease (designated CMT2B1) (OMIM #605588) that involves sensory impairment of the distal lower extremities. Some patients may also have difficulty swallowing and weakness of the muscles of respiration—all features associated with some cases of FXTAS. Thus, it is a formal possibility that at least some of the features of FXTAS may reflect a disturbance of lamin A/C function.

G. Natural History

Based on studies of roughly 100 patients to date, predominantly from families identified through fragile X syndrome probands, the principal neurological complaints among older adult (premutation) carriers are gait ataxia and intention tremor, each present in approximately 70–90% of cases. Patients also reported more variable neurological symptoms, including parkinsonism, numbness/pain in the lower extremities, and autonomic dysfunction (urinary and/or bowel incontinence, impotence). The gait ataxia generally begins as difficulty with tandem walking, with progressive difficulty with walking and balance problems to the point where use of a cane, walker, or wheelchair is required. The tremor is a progressive, intention type, but a resting component is sometimes observed. Tremor generally involves both upper extremities and usually begins (or is first noticed) in the dominant hand. Deficits in cognitive function are associated with behavioral problems generally characterized by disinhibition and impairments in working memory with relative sparing of language.

The natural history of FXTAS is characterized by substantial variability in both the age at onset of neurological symptoms and the rate of progression of the disorder. Although most individuals experience onset of symptoms after 50 years of age, a few have reported earlier onset. The average age at onset is around 60 years; however, onset can occur as late as the 70s or 80s, and some premutation carriers in their 90s exhibit no symptoms of FXTAS. For individuals who do have clinical involvement, some remain stable with mild symptoms for many years, whereas others experience a much more rapid downhill course (5–6 years). The life expectancy of patients affected with FXTAS appears to be shortened somewhat, with death often due to secondary difficulties with respiration/pneumonia and/or cardiovascular disease, although the influence of FXTAS on life expectancy has not yet been studied in a systematic fashion.

There is, as yet, no clear understanding of the variable penetrance, the broad variability in rate of progression, or the much smaller numbers of females affected with FXTAS. It is likely that these features reflect background gene effects that are operating in conjunction with the primary *FMR1*-based genetic mechanism.

IV. Prospects for RNA Pathogenic Mechanisms in Other Neuromuscular Disorders

There is now substantial evidence from both human tissues/cell cultures and animal models to support an RNA-based pathogenesis for both DM and FXTAS. The closely related RNA gain-of-function models for DM and FXTAS are all the more plausible in light of the fact that the pronounced oligonucleotide repeat expansions are present in noncoding portions of the respective transcripts; that is, the possibility of a protein-based pathogenesis could be more readily excluded than would be the case if the repeat expansions were also coding. However, the converse situation (coding repeat expansion) raises the following question: Do neuromuscular diseases exist in which coding (protein) expansions (e.g., polyQ) are not responsible for disease pathogenesis? This is an important issue, particularly in view of the broad spectrum of clinical phenotypes even among the polyQ (CAG-repeat) disorders. Thus, it remains a formal possibility that for one or more of those disorders, the expanded-repeat RNA could also be contributing. However, for most if not all of the polyQ disorders, relatively small expansions are required for clinical involvement—much smaller expansions than for either DM1 or DM2, and somewhat smaller expansions than for FXTAS.

So, the question remains open as to whether there are additional neuromuscular disorders whose pathogenesis is driven at the RNA level. Interestingly, such a possibility has been raised for both SCA8 (OMIM #603680; for a review, see reference 19) and SCA10 (OMIM #603516; for a review, see reference 20). In SCA8, the noncoding CTG repeat expansion (CUG repeat as RNA) is located in the 3′UTR of the Kelch-like 1 (*KLHL1*) gene. Although the RNA element has been postulated to be involved in disease pathogenesis, the SCA8 transcripts do include a small open reading frame (ORF) that encompasses the repeat region. Interestingly, although CUG repeat RNA has not been detected in the small intranuclear inclusions found in SCA8 autopsy tissue, the inclusions are reactive to antibodies (1C2) that recognize polyQ tracts [21]. Thus, SCA8 may actually reflect a protein-level pathogenesis. It will be interesting to see how this mechanism is resolved. In SCA10, the expanded pentanucleotide (ATTCT) repeat is located in intron 9 of the SCA10 gene. Even though RNA foci have been observed in cell culture transfection experiments [20], no clear link has been established between the expanded intronic repeat and disease pathogenesis.

References

1. Mankodi, A., and Thornton, C. A. (2002). Myotonic syndromes. *Curr. Opin. Neurol.* **15**, 545–552.
2. Amack, J. D., and Mahadevan, M. S. (2004). Myogenic defects in myotonic dystrophy. *Dev. Biol.* **265**, 294–301.
3. Meola, G., and Moxley, R. T., 3rd. (2004). Myotonic dystrophy type 2 and related myotonic disorders. *J. Neurol.* **251**, 1173–1182.
4. Day, J. W., and Ranum, L. P. (2004). RNA pathogenesis of the myotonic dystrophies. *Neuromusc. Disord.* **15**, 5–16.
5. Hagerman, R. J., Leavitt, B. R., Farzin, F., Jacquemont, S., Greco, C. M., Brunberg, J. A., Tassone, F., Hessl, D., Harris, S. W., Zhang, L., Jardini, T., Gane, L. W., Ferranti, J., Ruiz, L., Leehey, M. A., Grigsby, J., and Hagerman, P. J. (2004). Fragile-X-associated tremor/ataxia syndrome (FXTAS) in females with the *FMR1* premutation. *Am. J. Hum. Genet.* **74**, 1051–1056.
6. Liquori, C. L., Ricker, K., Moseley, M. L., Jacobsen, J. F., Kress, W., Naylor, S. L., Day, J. W., and Ranum, L. P. (2001). Myotonic dystrophy type 2 caused by a CCTG expansion in intron 1 of ZNF9. *Science* **293**, 864–867.
7. Philips, A. V., Timchenko, L. T., and Cooper, T. A. (1998). Disruption of splicing regulated by a CUG-binding protein in myotonic dystrophy. *Science* **280**, 737–741.
8. Timchenko, L. T., Miller, J. W., Timchenko, N. A., DeVore, D. R., Datar, K. V., Lin, L., Roberts, R., Caskey, C. T., and Swanson, M. S. (1996). Identification of a (CUG)n triplet repeat RNA-binding protein and its expression in myotonic dystrophy. *Nucleic Acids Res.* **24**, 4407–4414.
9. Jiang, H., Mankodi, A., Swanson, M. S., Moxley, R. T., and Thornton, C. A. (2004). Myotonic dystrophy type 1 is associated with nuclear foci of mutant RNA, sequestration of muscle blind proteins and deregulated alternative splicing in neurons. *Hum. Mol. Genet.* **13**, 3079–3088.
10. Mankodi, A., Logigian, E., Callahan, L., McClain, C., White, R., Henderson, D., Krym, M., and Thornton, C. A. (2000). Myotonic dystrophy in transgenic mice expressing an expanded CUG repeat. *Science* **289**, 1769–1773.
11. Iwahashi, C.K., Yasui, D. H., An, H. J., Greco, C. M., Tassone, F., Nannen, K., Babineau, B., Lebrilla, C. B., Hagerman, R. J., and Hagerman, P. J. (2006). Protein composition of the intranuclear inclusions of FXTAS. *Brain* **129**, 256–271.
12. Hagerman, R. J., and Hagerman, P. J., eds. (2002). "Fragile X Syndrome: Diagnosis, Treatment, and Research," 3rd edition. The Johns Hopkins University Press, Baltimore.
13. Hagerman, R. J., Leehey, M., Heinrichs, W., Hagerman, R. J., Leehey, M., Heinrichs, W., Tassone, F., Wilson, R., Hills, J., Grigsby, J., Gage, B., and Hagerman, P. J. (2001). Intention tremor, parkinsonism, and generalized brain atrophy in male carriers of fragile X. *Neurology* **57**, 127–130.
14. Jacquemont, S., Hagerman, R. J., Leehey, M. A., Hall, D. A., Levine, R. A., Brunberg, J. A., Zhang, L., Jardini, T., Gane, L. W., Harris, S. W., Herman, K., Grigsby, J., Greco, C. M., Berry-Kravis, E., Tassone, F., and Hagerman, P. J. (2004). Penetrance of the fragile X-associated tremor/ataxia syndrome in a premutation carrier population. *JAMA* **291**, 460–469.
15. Hagerman, P. J., and Hagerman, R. J. (2004). The fragile-X premutation: a maturing perspective. *Am. J. Hum. Genet.* **74**, 805–816.
16. Greco, C. M., Hagerman, R. J., Tassone, F., Chudley, A. E., Del Bigio, M. R., Jacquemont, S., Leehey, M., and Hagerman, P. J. (2002). Neuronal intranuclear inclusions in a new cerebellar tremor/ataxia syndrome among fragile X carriers. *Brain* **125**, 1760–1771.
17. Brunberg, J. A., Jacquemont, S., Hagerman, R. J., Berry-Kravis, E. M., Grigsby, J., Leehey, M. A., Tassone, F., Brown, W. T., Greco, C. M., and Hagerman, P. J. (2002). Fragile X premutation carriers: characteristic MR imaging findings of adult male patients with progressive cerebellar and cognitive dysfunction. *AJNR Am. J. Neuroradiol.* **23**, 1757–1766.
18. Tassone, F., Iwahashi, C., and Hagerman, P. J. (2004). *FMR1* RNA within the intranuclear inclusions of fragile X-associated tremor/ataxia syndrome (FXTAS). *RNA Biol.* **1**, 103–105.
19. Ranum, L. P., and Day, J. W. (2004). Pathogenic RNA repeats: an expanding role in genetic disease. *Trends Genet.* **20**, 506–512.
20. Lin, X., and Ashizawa, T. (2005). Recent progress in spinocerebellar ataxia type-10 (SCA10). *Cerebellum* **4**, 37–42.
21. Ikeda, Y., Moseley, M. L., Dalton, J. C., et al. (2004). Purkinje cell degeneration and 1C2 positive neuronal intranuclear inclusions in SCA8 patients. Paper presented at the American Society of Human Genetics, 54th Annual Meeting. Toronto, CA, October 26–29, 2004.

12

Ion Channel Disorders

Alan L. Goldin, MD, PhD

Keywords: *acetylcholine receptors, ataxia, epilepsy, GABA receptors, neuromuscular disorders, sodium channels*

I. Introduction
II. Epilepsy
III. Ataxia
IV. Neuromuscular Disorders
   References

I. Introduction

Electrical excitability and the propagation of impulses in the nervous system are determined by the coordinated function of many different ion channels. These channels are opened and closed by either voltage or ligands, and different subsets allow the passage of different populations of cations or anions. Given the critical importance of ion channels to neuronal excitability, it is not surprising that abnormalities of their function lead to a variety of neurological diseases. Because ion channels are so diverse, it is also not surprising that the clinical manifestations of channel dysfunction are variable. On the other hand, similar syndromes are often caused by abnormal function of different ion channels, so it is impossible to predict the clinical effect based on the channel abnormality. A common feature of neurological disorders caused by ion channel malfunction is that the disorders are paroxysmal, usually with long periods of normal activity punctuated by brief episodes of abnormal activity. The neurological disorders discussed in this chapter result from mutations in the genes encoding either voltage-gated or ligand-gated ion channels, so they have been called *channelopathies*. In all cases, the underlying abnormality is present at birth, although the clinical manifestations may not develop until later in life.

Ion channels can be divided into two major categories: those gated by voltage and those gated by ligands [1]. Voltage-gated ion channels include those selective for cations such as sodium, potassium, and calcium. These channels share many functional and structural characteristics, including the fact that they are opened by membrane depolarization. There are also voltage-gated channels that are selective for anions such as chloride. The voltage-gated ion channels are responsible for the shape and propagation of action potentials, as well as modulation of membrane potential and excitability (Fig. 1). Ligand-gated

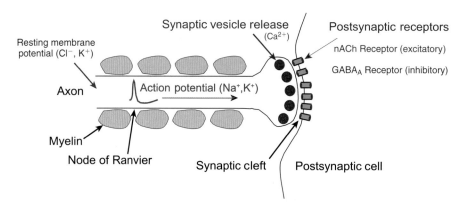

Figure 1 Diagram of a neuron showing the role of different ion channels and receptors in the conduction of electrical impulses. The resting membrane potential is determined primarily by the activity of voltage-gated chloride and potassium channels. The shape and frequency of action potentials depend on the activity of voltage-gated sodium and potassium channels. When the action potentials reach the axon terminus, depolarization of the membrane causes voltage-gated calcium channels to open, which initiates fusion of neurotransmitter-containing vesicles with the axonal membrane. The neurotransmitters are released into the synaptic cleft, through which they diffuse and activate receptors on the postsynaptic cell membrane. Activation of nicotinic acetylcholine (ACh) receptors depolarizes the postsynaptic cell and is excitatory, whereas activation of $GABA_A$ receptors hyperpolarizes the postsynaptic cell and is inhibitory.

ion channels include the acetylcholine (ACh) receptor and the γ-aminobutyric acid (GABA) receptor. These receptor channels are present in postsynaptic membranes, and they are responsible for the transmission of excitatory (ACh) or inhibitory (GABA) impulses across the synapse (see Fig. 1). Mutations in the genes encoding members of all of these channel families have been identified as causing human neurological diseases, including epilepsy, ataxia, and neuromuscular disorders.

II. Epilepsy

A. History and Nomenclature

Epilepsy is a condition characterized by recurrent unprovoked seizures, which are paroxysmal events of synchronous electrical activity in the central nervous system (CNS). The condition is relatively common, with an incidence of approximately 3% when it is defined as the occurrence of two or more seizures [2]. Seizures are classified based on the clinical characteristics and the electroencephalographic (EEG) findings, as described by the International League Against Epilepsy. The classification is not based on etiology, although some channelopathies cause specific types of seizures. There are three main categories. The first category consists of partial seizures, which occur within a discrete region of the brain. Partial seizures can be simple, complex (consciousness is impaired), or simple with secondary generation (seizures begin in a discrete region and then spread throughout the cortex). The second category consists of generalized seizures, which initiate in both cerebral hemispheres. These include absence or petit mal (brief loss of consciousness), tonic-clonic or grand mal (muscle contraction followed by alternations of muscle relaxation and contraction, tonic (muscle contraction), atonic (sudden loss of postural control), and myoclonic (brief periods of muscle contraction). The third category consists of unclassified seizures, which includes many types that occur in neonates and infants. Epilepsy syndromes are defined based on the types of seizures involved and other characteristics such as neurological abnormalities, age at onset, and familial occurrence.

B. Etiology

Although epilepsy is a condition defined by the single characteristic of recurrent seizures, it is actually a collection of many different syndromes that are caused by a variety of factors [3]. The causes vary by age, with approximately one-third due to acquired pathologies and the rest being idiopathic, meaning that the cause has not been identified. Most of the idiopathic epilepsies are presumed to be due to genetic mutations, but specific mutations have been identified as the basis for only a small number of syndromes. Within this small number, mutations have been identified in more than 70 different genes, and that number is increasing rapidly. The genes can be divided into three functional categories: those that cause primary defects of membrane and synaptic signaling, those that alter neuronal plasticity and metabolism, and those that affect network development. At the present time, channelopathies account for about one third of the single gene mutations.

Table 1 Ion Channelopathies Associated with Epilepsy

Type of Channel	Gene	Channel	Subunit	Syndrome
Sodium channel	SCN1A	$Na_v1.1$	α	Generalized epilepsy with febrile seizures plus
Sodium channel	SCN1A	$Na_v1.1$	α	Intractable childhood epilepsy with generalized tonic-clonic seizures
Sodium channel	SCN1A	$Na_v1.1$	α	Severe myoclonic epilepsy of infancy
Sodium channel	SCN1B		$β_1$	Generalized epilepsy with febrile seizures plus
Sodium channel	SCN2A	$Na_v1.2$	α	Benign familial neonatal-infantile seizures
Sodium channel	SCN2A	$Na_v1.2$	α	Generalized epilepsy with febrile seizures plus
$GABA_A$ receptor	GABRA1		$α_1$	Autosomal dominant juvenile myoclonic epilepsy
$GABA_A$ receptor	GABRG2		γ	Childhood absence epilepsy and febrile seizures
$GABA_A$ receptor	GABRG2		γ	Generalized epilepsy with febrile seizures plus
Potassium channel	KCNQ2	$K_v7.2$		Benign familial neonatal seizures
Potassium channel	KCNQ3	$K_v7.3$		Benign familial neonatal seizures
Calcium channel	CACNA1A	$Ca_v2.1$	$α_{1A}$	Generalized epilepsy with episodic ataxia
Calcium channel	CACNB4		$β_4$	Juvenile myoclonic epilepsy
Chloride channel	CLCN2			Childhood absence epilepsy
Chloride channel	CLCN2			Epilepsy with grand mal seizures on awakening
Chloride channel	CLCN2			Idiopathic generalized epilepsy
Chloride channel	CLCN2			Juvenile absence epilepsy
Chloride channel	CLCN2			Juvenile myoclonic epilepsy
Nicotinic ACh receptor	CHRNA4		$α_4$	Autosomal-dominant nocturnal frontal lobe epilepsy
Nicotinic ACh receptor	CHRNB2		$β_2$	Autosomal-dominant nocturnal frontal lobe epilepsy

Mutations in six different types of channels or receptors have been identified as causing epilepsy [4–6] (Table 1). These include voltage-gated sodium, potassium, calcium, and chloride channels and receptors for ACh and GABA. Mutations in a single type of channel can cause more than one type of syndrome, and the same syndrome can be caused by mutations in more than one type of channel (see Table 1). For example, mutations in three different genes encoding subunits of the voltage-gated sodium channel and in one gene encoding a subunit of the $GABA_A$ receptor cause generalized epilepsy with febrile seizures plus (GEFS+). Mutations in two of the same sodium channel genes cause other epilepsy syndromes, including intractable childhood epilepsy with generalized tonic-clonic seizures (ICEGTC), severe myoclonic epilepsy of infancy (SMEI), and benign familial neonatal-infantile seizures (BFNIS). Similarly, other mutations in the same $GABA_A$ receptor subunit gene cause childhood absence epilepsy and febrile seizures, and mutations in a different $GABA_A$ receptor subunit gene cause autosomal-dominant juvenile myoclonic epilepsy. Mutations in either voltage-gated calcium or chloride channels cause other epilepsy syndromes, including juvenile myoclonic epilepsy, juvenile absence epilepsy, childhood absence epilepsy, and idiopathic generalized epilepsy.

All of the epilepsy syndromes that have been identified thus far as resulting from ion channel mutations are inherited in an autosomal-dominant fashion, meaning that mutation of just one copy of the gene is sufficient to cause the syndrome. Most of the mutations change a single amino acid so that the channel is still active with altered functional properties. However, some of the mutations insert a nonsense codon and terminate the protein early in the reading frame, resulting in complete loss of protein. For example, missense mutations in the SCN1A sodium channel gene are the most common cause of GEFS+, whereas nonsense mutations in the same gene are the most common cause of SMEI. The severity of the epileptic syndrome resulting from mutations in SCN1A depends on the type of mutation, with the most severe phenotype being caused by complete loss of sodium channel protein, which occurs in SMEI. There are multiple mechanisms by which mutations in a single ion channel gene can cause epilepsy. In addition, it is likely that some ion channel mutations increase susceptibility to seizures rather than directly cause epilepsy. In a similar fashion, ion channel mutations may be involved in some epileptic syndromes that have complex inheritance.

C. Pathogenesis and Pathophysiology

The hallmark of a seizure is synchronization of neuronal firing. However, the mechanisms that generate that synchronization are not known and most likely vary depending on the seizure type and underlying pathology. For example, absence seizures probably involve an alteration in communication between the thalamus and the cerebral cortex, resulting in rhythmic cortical activation typical of normal non–rapid eye movement (non-REM) sleep [2]. The chloride channel and $GABA_A$ receptor mutations that cause absence epilepsy presumably alter that communication to facilitate aberrant rhythmic activation of the cortex during wakefulness. The abnormal circuitry leading to other types of generalized seizures (tonic/clonic, myoclonic, astatic) have not been defined but undoubtedly differ from each other; yet, all of these seizure types are observed in family members with GEFS+. Therefore, the sodium channel or GABA$_A$ receptor mutations causing GEFS+ probably lower the threshold for seizure generation in multiple but independently triggered circuits [3].

D. Structural Detail and Pharmacology

Voltage-gated sodium, calcium, and potassium channels share many similarities and are members of a single superfamily of voltage-gated cation channels [1,7]. The primary pore-forming α subunit of each channel consists of four homologous domains termed I–IV, with each domain containing six transmembrane segments called S1–S6 and a hairpin-like loop between S5 and S6 that forms part of the channel pore (Fig. 2). Potassium channels consist of tetramers of single-domain α subunits (Fig. 3), whereas sodium and calcium channels contain four homologous domains within a single α subunit (Figs. 2 and 4). The primary α subunit is associated with different accessory subunits for each channel type.

Voltage-gated sodium channels in the CNS consist of a pore-forming α subunit associated with two of four accessory subunits termed β_1, β_2, β_3, and β_4 [7] (see Fig. 2). The β_2 or β_4 subunit is covalently linked to the α subunit by a disulfide bond, and the β_1 or β_3 subunit is noncovalently attached. The β subunits are expressed in a complementary fashion, so that α subunits are associated with either β_1 or β_3, and β_2 or β_4. There are nine genes encoding different α subunits, with four expressed at high levels in the CNS (*SCN1A, SCN2A, SCN3A, SCN6A*), and there is one gene encoding each of the β subunits (*SCN1B, SCN2B, SCN3B,* and *SCN4B*). Mutations in two of the α subunit genes (*SNC1A* and *SCN2A*) and in one of the β subunit genes (*SNC1B*) have been shown to cause epilepsy syndromes (see Table 1).

Voltage-gated calcium channels contain a pore-forming subunit termed α_1 that is associated with an intracellular β subunit, a disulfide-linked $\alpha_2\delta$ subunit, and a γ subunit in some tissues [7] (see Fig. 4). The primary functional properties of the channel are determined by the α_1 subunit, with the accessory subunits serving to modulate the channel. There are 10 genes encoding different α_1 subunits, divided into three families termed Ca$_v$1, Ca$_v$2, and Ca$_v$3. Most of the calcium channel subtypes are expressed in some populations of neurons. Mutations in the *CACNA1A* gene encoding the Ca$_v$2.1 channel and in the *CACNB4* gene encoding the β_4 subunit have been shown to cause epilepsy syndromes (see Table 1).

Voltage-gated potassium channels are the most diverse group of channels, with 12 different gene families termed K$_v$1–K$_v$12, each of which contains multiple subtypes [7]. Many of the potassium channels are also associated with accessory subunits that have multiple names. Mutations in two genes (*KCNQ2* and *KCNQ3*) encoding K$_v$7.2 and K$_v$7.3, respectively, cause benign familial neonatal seizures (see Table 1).

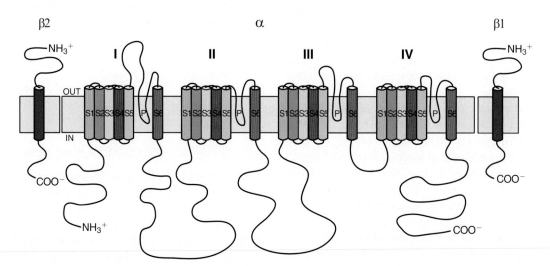

Figure 2 Schematic diagram of the voltage-gated sodium channel. The α subunit of the channel consists of four homologous domains labeled I–IV, with six transmembrane spanning segments termed S1–S6 in each domain. The P region between S5 and S6 in each domain forms part of the channel pore. Sodium channels in the central nervous system (CNS) also contain two associated β subunits, shown in this figure as β_1 and β_2. The β_1 subunit is noncovalently attached to the α subunit and the β_2 subunit is covalently attached via a disulfide linkage. All of the β subunits have a similar structure that consists of a small, carboxy-terminal cytoplasmic region, a transmembrane spanning region, and a larger external amino-terminal region that contains immunoglobulin-like domains. The α subunit includes the channel pore and gating machinery, and the β subunits modulate the properties of the channel complex. (Diagram modified, with permission, from Goldin, A. L. [2003]. Mechanisms of sodium channel inactivation. *Curr. Opin. Neurobiol.* **13**, 284–290.)

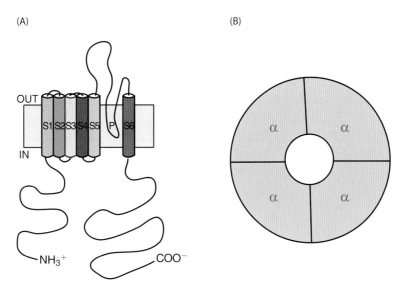

Figure 3 (**A**) Diagram of a voltage-gated potassium channel monomer. A monomer is comparable to one domain of the voltage-gated sodium channel and contains six transmembrane spanning segments termed S1–S6 with a region between S5 and S6 that forms part of the pore (P). Potassium channels associate with one of a variety of accessory cytoplasmic subunits, which are not shown in this diagram. (**B**) Four potassium channel monomers combine to form a functional channel. A detailed crystal structure of the rat $K_v1.2$ voltage-gated potassium channel in association with the rat β_2 subunit has been determined [19].

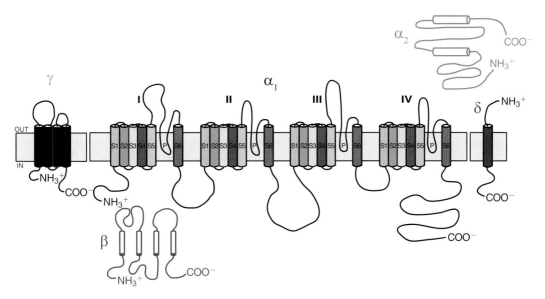

Figure 4 Diagram of the voltage-gated calcium channel. The α_1 subunit of the channel is comparable to the α subunit of the voltage-gated sodium channel and consists of four homologous domains labeled I–IV, each containing six transmembrane spanning segments termed S1–S6 and a P region. The α_1 subunit includes the channel pore and gating machinery. Most calcium channels also contain an intracellular β subunit (red), an extracellular α_2 subunit (blue) that is attached to a membrane-spanning δ subunit (black) by a disulfide linkage, and a membrane-spanning γ or related subunit (green). The accessory subunits modulate the properties of the channel complex. (Diagram modified, with permission, from Catterall, W. A., Chandy, K. G., and Gutman, G. A. [2002]. "The IUPHAR Compendium of Voltage-Gated Ion Channels." IUPHAR Media, Leeds, UK.)

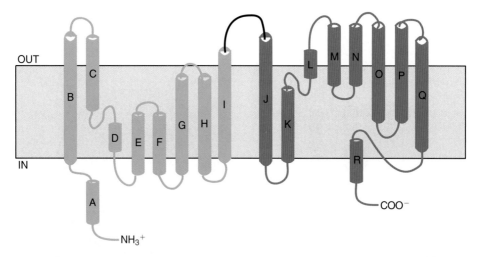

Figure 5 Diagram of one subunit of the voltage-gated CLC chloride channel. The subunit consists of 18 α-helical segments labeled A–R with the amino-terminal half (green) being homologous to the carboxy-terminal half (blue), but in the opposite orientation in the membrane. The two halves surround a common center that forms the selectivity filter. The functional channel consists of two identical subunits and contains two independent pores through which chloride ions permeate. (Diagram modified, with permission, from Dutzler, R., Campbell, E. B., Cadene, M., Chait, B. T., and MacKinnon, R. [2002]. X-ray structure of a ClC chloride channel at 3.0 Å reveals the molecular basis of anion selectivity. *Nature* **415**, 287–294, which also shows the detailed crystal structure of a prokaryotic CLC chloride channel.)

Voltage-gated chloride channels are members of the CLC gene family [8]. These channels consist of two identical subunits, each containing of 18 α helices and an ion permeation pathway (Fig. 5), so that two chloride ions can pass through the channel independently. There are nine genes encoding CLC channels, and mutations in one of these genes (*CLCN2*) cause a variety of epilepsy syndromes (see Table 1).

The nicotinic ACh and $GABA_A$ receptors are members of a large family of ligand-gated channels that also includes receptors for serotonin and glycine. They are composed of pentamers of different subunits, with each subunit having a comparable structure consisting of four transmembrane segments and a large extracellular amino-terminus that contains the ligand-binding site (Fig. 6). The nicotinic ACh receptor complexes are composed of two different subunits (α and β) and can consist either entirely of α subunits or of heteromers containing two α and three β subunits [9]. The receptor forms a channel that is permeable to cations, including sodium, calcium, and potassium, so that it serves to depolarize the membrane and to initiate the firing of action potentials. There are nine genes encoding α subunits (α_2–α_{10}) and three genes encoding β subunits (β_2–β_4). The α_1 and β_1 subunits are only present in muscle nicotinic ACh receptors. Mutations in the *CHRNA4* gene encoding the α_4 subunit and in the *CHRNB2* gene encoding the β_2 subunit cause autosomal-dominant nocturnal frontal lobe epilepsy (Table 1). The mutations are located close to the second or third transmembrane domains that form the pore of the channel.

The $GABA_A$ receptor pentamers contain combinations of different subunits including α, β, γ, δ, ε, π, and θ, with most receptors consisting of α, β, and γ or α, β, and δ subunits [10]. The receptor forms a channel that is selective for anions, mainly chloride, so that it functions to hyperpolarize the membrane and inhibit excitability. At least 19 different subunits are encoded by different genes. Mutations in the *GABRG2* gene encoding the γ subunit cause GEFS+ and childhood absence epilepsy and febrile seizures, whereas mutations in the *GABRA1* gene encoding the α_1 subunit cause autosomal-dominant juvenile myoclonic epilepsy.

Given that ion channels play such a fundamental role in the regulation of neuronal excitability, it is not surprising that most pharmaceuticals used to treat epilepsy affect ion channels [4]. These drugs can act to block either the initiation or the spread of seizure activity, so that the mechanism of action is usually nonspecific. Most of the drugs act in a use-dependent manner, meaning that they more effectively block channels that are rapidly opening, which occurs during a seizure. The commonly used drugs affect the same ion channels and receptors that are aberrant in epilepsy syndromes. Drugs that are used to treat epilepsy include those that inhibit voltage-gated sodium channels (phenytoin, carbamazepine, valproate, lamotrigine), those that inhibit voltage-gated calcium channels (ethosuximide), and those that increase $GABA_A$ receptor function (increasing inhibition) either directly (phenobarbital, clonazepam) or by increasing the availability of GABA (gabapentin).

Because different types of seizures result from alterations in different neuronal pathways, the same drugs are

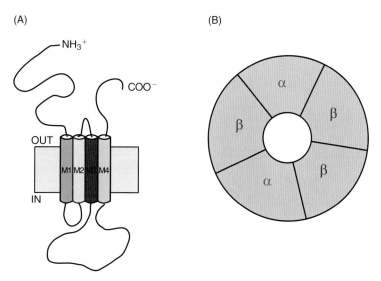

Figure 6 (**A**) Diagram of one subunit of the nicotinic acetylcholine receptor. A subunit consists of four α-helical segments termed M1–M4, with the amino- and carboxy-terminal regions on the outside of the membrane. The acetylcholine binding site is contained within the amino-terminal region. (**B**) Five subunits combine to form a functional receptor, which can be either monomeric (all α subunits) or heteromeric (two α and three β subunits). The subunits surround a central pore through which cations permeate. The predicted structure of the $GABA_A$ receptor is comparable, with the GABA binding site in the amino-terminal region. The functional $GABA_A$ receptor also consists of a pentamer, with the most common composition being two α, two β, and one γ subunit, although channels can alternatively contain a variety of other subunits. The $GABA_A$ subunits surround a central pore through which chloride ions permeate. (Diagram modified, with permission, from Gotti, C., and Clementi, F. [2004]. Neuronal nicotinic receptors: from structure to pathology. *Prog. Neurobiol.* **74**, 363–396.)

not effective for all syndromes. For example, ethosuximide suppresses absence seizures, but it is not effective for other types of seizures, which may be because ethosuximide blocks T-type calcium channels that are involved in bursting during non-REM sleep. The specificity of other drugs is less clear. In addition, it is not known whether drugs that block a specific channel are more effective in treating epilepsy syndromes caused by mutations in one of the genes encoding that channel. Because neuronal excitability is carefully tuned, drugs that modulate ion channel activity can also induce or exacerbate seizures. For example, benzodiazepines such as clonazepam activate a $GABA_A$ receptor subtype to increase neuronal inhibition. Given this mechanism of action, it is not surprising that some benzodiazepines suppress absence seizures. Somewhat paradoxically, other benzodiazepines increase the bursting mode characteristic of absence seizures, thus exacerbating the disease [2].

E. Explanation of Symptoms in Relation to Pathophysiology

The exact mechanism by which channelopathies cause epilepsy is unknown, although there is a significant amount of information regarding how the mutations alter the function of the channels. Surprisingly, the functional effects vary among the different mutations. The sodium channel mutations that cause GEFS+ have been studied in a number of heterologous systems including *Xenopus* oocytes and transfected human embryonic kidney cells [5,11]. In those studies, some of the mutations were found to increase channel activity, whereas others were found to decrease activity. The effects leading to increased activity include a negative shift in the voltage-dependence of activation (making it easier to open the channel), increased persistent current (leading to sustained depolarization), and faster recovery from inactivation (making it possible to sustain higher frequency bursts of action potentials). On the other hand, some of the mutations shift the voltage dependence of activation in the positive direction (making it more difficult to open the channels), enhance slow inactivation (resulting in fewer functional channels in the membrane), or demonstrate no functional activity in the heterologous expression system. All of these alterations would decrease neuronal excitability. These results indicate that GEFS+ can be caused both by mutations that increase sodium channel activity and by those that decrease activity. The physiologically important effect is not what the mutations do to the channel, but how they affect neuronal firing. This property has been examined using computer modeling, and all of the mutations examined in this way were predicted to increase repetitive firing of action potentials [12]. Studies to confirm these predictions are being performed by constructing mice expressing the human mutations. These animal models should provide more information about how the alterations in channel activity lead to the generation of seizure activity.

Mutations in genes encoding the $GABA_A$ receptor also cause GEFS+, but by a completely different mechanism. The $GABA_A$ receptor forms a chloride channel that functions to inhibit membrane excitability. The mutations decrease channel activity, which decreases inhibition resulting in neuronal hyperexcitability [5].

Mutations in genes encoding the $Ca_v2.1$ calcium channel and the β_4 subunit cause two different epilepsy syndromes. These mutations reduce calcium currents, so that they do not directly increase excitability by lowering the threshold for action potential initiation. Because calcium channels are critical mediators of neurotransmitter release, one hypothesis for the mechanism of seizure generation is that the mutations alter synaptic strength to increase synchronization. The effects of the mutations most likely depend on the specific synapse because different synapses use different types of calcium channels for neurotransmitter release [3]. Therefore, the effects of decreasing just $Ca_v2.1$ channel function could specifically decrease inhibitory synapses, leading to greater synchronization and seizure development.

The mechanism by which mutations in the *CLCN2* gene encoding the voltage-gated chloride channel cause epilepsy is unknown. The chloride channel plays an important role in maintaining the normal membrane potential, so disruption of that channel would lead to membrane depolarization and hyperexcitability. The effects of some mutations are consistent with this mechanism of action in that they decrease activity of the channel. However, other mutations appear to alter activation of the channel in a manner that is not consistent with decreased activity, so there may be multiple mechanisms by which chloride channel mutations cause epilepsy [5].

The potassium channel gene mutations that cause benign familial neonatal seizures are located in genes encoding two inward rectifier channels ($K_v7.2$ and $K_v7.3$). These channels co-assemble to form the M current, which helps to determine the subthreshold excitability of neurons and to limit sustained membrane depolarization. All of the mutant proteins act as dominant-negative subunits that reduce currents through the channels. The decreased potassium conductance leads to prolonged membrane depolarization and the development of seizures [5].

The mutations in the genes encoding the α_4 and β_2 subunits of the ACh receptors are all located near the pore of the channel, although they have different effects on channel function. Some mutations inhibit function of the receptor either by decreasing currents through the channel or by enhancing desensitization, which closes the channel. However, other mutations appear to increase receptor function, so the mechanism by which ACh receptor mutations cause epilepsy is unknown. There are at least two hypotheses. One possibility is that the mutations increase the sensitivity of the receptor to ACh, which enhances neuronal excitability [13]. An alternative explanation is that the mutations interfere with calcium modulation of the receptor, thus preventing a negative feedback mechanism by which glutamate receptors deplete local extracellular calcium to reduce receptor potentiation [4,5].

F. Natural History

The clinical course of epilepsy has been described based on the syndrome rather than the etiology, as it has only recently been possible to determine the specific mutations causing some types of epilepsy. Therefore, the course does not necessarily correlate with the type of mutation. Epilepsy syndromes resulting from mutations in the voltage-gated sodium channel demonstrate a wide range of severity. GEFS+ is a complex syndrome that encompasses a variety of phenotypes within a single family, so the diagnosis can only be made by examining the entire family rather than a single individual. The phenotype ranges from febrile seizures that persist beyond 6 years of age to generalized epilepsies such as absences, myoclonic seizures, atonic seizures, and myoclonic-astatic epilepsy. There has not been any distinction made in the clinical course for GEFS+ resulting from sodium channel versus $GABA_A$ receptor mutations. SMEI is a more severe syndrome in which the seizures are usually uncontrolled and are associated with intellectual disability. It is characterized by febrile hemiclonic seizures or generalized status epilepticus starting at approximately 6 months, with other seizure types including partial, absence, atonic, and myoclonic seizures occurring after 1 year. Development is delayed and patients often experience motor impairment including spasticity and ataxia [6].

Potassium channel mutations have thus far been shown to cause a single epilepsy syndrome, benign familial neonatal seizures (also known as benign familial neonatal convulsions). The symptoms begin within the first 2–3 days, manifested as either clonic or apneic (cessation of breathing) seizures. The initial seizures subside within a few weeks, but approximately 11% of individuals demonstrate seizures later in life [6].

Chloride channel mutations cause childhood absence epilepsy, epilepsy with grand mal seizures on awakening, juvenile absence epilepsy, idiopathic generalized epilepsy, and juvenile myoclonic epilepsy. Calcium channel mutations also cause juvenile myoclonic epilepsy, as well as generalized epilepsy with episodic ataxia. Juvenile myoclonic epilepsy is a generalized seizure disorder appearing early in adolescence, with bilateral myoclonic jerks and preservation of consciousness. Patients can also experience generalized tonic-clonic seizures or absence seizures. The seizures respond well to anticonvulsants.

Mutations in the nicotinic ACh receptor genes cause autosomal-dominant nocturnal frontal lobe epilepsy. In this syndrome, clusters of motor seizures arise in sleep, usually starting between 8 and 12 years of age. The clusters occur within a few hours and are preceded by nonspecific auras. Most individuals respond well to treatment [5,6].

III. Ataxia

A. History and Nomenclature

Ataxia is a disorder characterized by lack of coordination, disturbed gait, unclear speech, and tremor with movement. There is a wide variety of hereditary ataxias that were originally defined pathologically and clinically, but they can be divided into five main groups based on pathogenetic mechanisms. The five general causes are mitochondrial, metabolic, defective DNA repair, abnormal protein folding, and channelopathies [14]. The channelopathies that have been identified thus far as causing ataxia include disorders of potassium and calcium channels, resulting in episodic ataxia types 1 (EA-1) and 2 (EA-2), and spinocerebellar ataxia 6 (SCA6).

B. Etiology

All of the channelopathies that cause ataxia are inherited in an autosomal-dominant fashion, although the types of mutations that cause the three diseases are quite different. EA-1 is caused by mutations in the *KCNA1* gene that encodes the $K_v1.1$ potassium channel (Table 2). This channel is expressed in many regions of the CNS, including the cerebellum. Most of the mutations are missense changes resulting in channels that either traffic aberrantly or demonstrate abnormal functional properties [14]. EA-2 is caused by mutations in the *CACNA1A* gene encoding the $Ca_v2.1$ calcium channel, which is expressed at high levels in the cerebellar cortex, granule, and Purkinje cells. In this disorder, most of the mutations are nonsense changes resulting in a truncated protein that is probably nonfunctional [14]. SCA6 is caused by mutations in the same calcium channel gene, but in this case, the mutations are expansions of the triplet sequence cytosine-adenine-guanine (CAG). Triplet expansions are the genetic mechanism for many of the SCAs, but for SCA6, a relatively short expansion consisting of as few as 20 repeats can cause the disease (normal individuals have from 4 to 16 triplet repeats). Longer expansions generally lead to a younger age at onset. The triplet expansions result in a polyglutamine sequence, which causes the protein to fold incorrectly and thus become degraded [15]. Thus, both EA-2 and SCA6 result from complete loss of functional protein from one allele of the *CACNA1A* gene. Not surprisingly, there is some overlap between these two syndromes, and the definitive diagnosis is based on identification of the causative mutation.

C. Pathogenesis and Pathophysiology

The ataxias caused by channelopathies result from primary cerebellar dysfunction. Because Purkinje cells are the sole output neuron from the cerebellum, the ultimate effect of these disorders is to alter the output firing of those cells. Purkinje cells have a rhythmic firing pattern and they continually fire action potentials that are inhibitory to the deep cerebellar nuclei. Decreased inhibition from the cerebellar Purkinje cells results in increased and aberrant firing from the cells in the deep cerebellar nuclei, which is the direct cause of the primary symptoms of ataxia [16]. The ion channel mutations that cause ataxia alter this pathway either by altering channels expressed in the Purkinje cells (potassium channels in EA-1) or by altering channels expressed in cells that modulate Purkinje cell firing (calcium channels in EA-1 and SCA6). Additional symptoms in these disorders result from abnormalities in peripheral neurons, which most likely are due to aberrant ion channel function in those cells.

D. Structural Detail and Pharmacology

The ataxia syndromes are caused by mutations in genes encoding the primary pore-forming subunits of two ion channel genes that were described in Section II, earlier in this chapter. Mutations in the *KCNA1* gene encoding the α subunit of the $K_v1.1$ potassium channel cause EA-1, and mutations in the *CACNA1A* gene encoding the α subunit

Table 2 Ion Channelopathies Associated with Ataxia

Type of Channel	Gene	Channel	Subunit	Syndrome
Calcium channel	*CACNA1A*	$Ca_v2.1$	$α_{1A}$	Episodic ataxia type 2
Calcium channel	*CACNA1A*	$Ca_v2.1$	$α_{1A}$	Spinocerebellar ataxia 6
Potassium channel	*KCNA1*	$K_v1.1$	α	Episodic ataxia type 1

of the $Ca_v2.1$ calcium channel cause EA-2 and SCA6 (see Table 2).

There is no general treatment for the hereditary ataxias, but patients with the syndromes resulting from channelopathies often respond to pharmacological intervention that affects ion channel function [15]. The potassium channel blocker 4-aminopyridine has been used to effectively treat some individuals with EA-2. Potassium channel blockers increase the electrical excitability of neuronal cells. This therapy may seem counterintuitive, because the *KCNA1* mutations that cause EA-1 also reduce potassium channel activity. However, the critical factor is the location of the potassium channels that are being blocked. The therapeutic effect of 4-aminopyridine results from reduced potassium channel activity in cerebellar Purkinje cells, which counteracts the decreased excitability resulting from the loss of $Ca_v2.1$ calcium channels in those cells. In contrast, the pathogenic effect of the *KCNA1* mutations that cause EA-1 results from loss of potassium channel activity in basket cells and interneurons that inhibit Purkinje cell firing. A second drug that has been used to effectively treat individuals with EA-2 is acetazolamide, which is a carbonic anhydrase inhibitor. In this case, the mechanism of action is unknown, as the drug does not directly affect the functional properties of either mutant or wild-type $Ca_v2.1$ channels.

E. Explanation of Symptoms in Relation to Pathophysiology

EA-1 results from missense mutations in the *KCNA1* gene encoding the $K_v1.1$ potassium channel. Each of the families that has been studied demonstrates a different single site mutation, but all of the mutations are dominant, so all individuals are heterozygous for mutant and wild-type channels. The effect of the mutations is to decrease potassium channel currents, which should increase neuronal excitability. However, the mechanisms by which the mutations decrease potassium currents vary considerably. Some mutations alter channel assembly or trafficking, thus resulting in fewer functional channels in the membrane, whereas others alter the gating kinetics of the channel, resulting in decreased channel activity. The $K_v1.1$ potassium channel is localized with GABA in the axons of basket cells and interneurons, and one mechanism by which decreased potassium current might lead to ataxia is by increasing GABA release at the synaptic terminals, thus leading to greater inhibition of cerebellar Purkinje cells [14].

The mutations in *CACNA1A* that cause EA-2 are also single nucleotide changes. Most of the mutations are nonsense changes that result in premature termination of translation, but there are also some missense mutations that alter the properties of the channel. These changes include reduced current density, faster inactivation, and more positive activation, all of which decrease calcium currents. Thus, the common denominator of EA-2 mutations appears to be decreased calcium currents through the $Ca_v2.1$ channel, which is the most highly expressed calcium channel in cerebellar Purkinje cells. Because calcium channels mediate an excitatory voltage-gated current, decreased $Ca_v2.1$ channel activity decreases Purkinje cell firing, which in turn decreases inhibition of the deep cerebellar nuclei. Consistent with this mechanism of action, the missense mutations that decrease $Ca_v2.1$ channel activity generally have a milder phenotype than the mutations that result in complete loss of channel protein [14].

SCA6 results from expanded CAG triplet repeats in the same *CACNA1A* gene encoding the $Ca_v2.1$ channel. The polyglutamine stretches cause the protein to form insoluble aggregates, resulting in decreased numbers of functional $Ca_v2.1$ calcium channels. Therefore, the mechanism of action of the mutations causing SCA6 is comparable to those causing EA-2 [14].

F. Natural History

EA-1 demonstrates onset in late childhood or early adolescence, with episodes of cerebellar incoordination that usually last seconds to minutes. The episodes are often triggered by stress, emotion, or sudden movements, and the symptoms usually subside by the second decade. EA-2 also starts in childhood or early adolescence, but the symptoms are more severe than in EA-1. The episodes last for hours to days and consist of disabling attacks of ataxia, vertigo, diplopia, nystagmus, and nausea. The attacks are triggered by emotional and physical stress, caffeine, alcohol, or phenytoin [14].

SCA6 has a late onset that occurs after 50 years of age in most patients. The symptoms are mainly cerebellar and include bradykinesia, dystonia, and a spastic gait. These symptoms are often associated with oculomotor disturbances including nystagmus and diplopia. Some patients experience mild peripheral neuropathy. The course is generally milder than that of the other SCAs, most likely because brainstem functions are spared. Therefore, patients generally have normal life expectancy, although the disability can still be severe [15].

IV. Neuromuscular Disorders

A. History and Nomenclature

Ion channel mutations that alter the electrical excitability of skeletal muscle result in two related disorders: periodic paralyses and nondystrophic myotonias. The distinguishing feature of periodic paralysis is episodic attacks of flaccid weakness that are often accompanied by myotonia (delayed muscle relaxation after contraction), whereas nondystrophic myotonia is characterized by transient muscle weakness, severe myotonia, and muscle hypertrophy without paralysis. These disorders have been classified into different syndromes based on clinical features, initiating factors, and serum potassium levels [17,18].

There are three types of periodic paralysis [17]. The first type is hyperkalemic periodic paralysis (HyperPP), in which serum potassium levels are either elevated (>5 mEq/liter) or normal and attacks can be precipitated by potassium. The second type is hypokalemic periodic paralysis (HypoPP), in which serum potassium levels are low (<3 mEq/liter) and the symptoms can be alleviated by potassium administration. The final type is Andersen's syndrome (AS), in which tissues other than skeletal muscle are also involved. The symptoms of AS include periodic paralysis of skeletal muscle that is not affected by serum potassium levels, ventricular arrhythmia, and dysmorphic developmental abnormalities.

There are four types of nondystrophic myotonias [18]. Paramyotonia congenita (PMC) is characterized by symptoms that are similar to those of HyperPP, including weakness, but there is no paralysis and the predominant finding is myotonic stiffness that gets worse with repeated muscle contraction. Potassium aggravated myotonia (PAM) also involves myotonic stiffness, but in this syndrome the condition is aggravated by administration of potassium. The final two forms of nondystrophic myotonia are subtypes of myotonia congenita (MC), an autosomal-dominant form called *Thomsen's disease* and an autosomal-recessive form called *Becker's disease*. These disorders have similar clinical features, with myotonia that is worse after rest, improves with activity, and is precipitated by voluntary contraction or emotional stimulation. Becker's disease is generally more severe and involves greater myotonia in the legs compared with the face and arms, but the major distinguishing feature is the mode of inheritance.

B. Etiology

Although the periodic paralyses and nondystrophic myotonias have been divided into different syndromes based on clinical features, there is considerable overlap among the syndromes. In addition, mutations in one ion channel cause multiple syndromes, and the same syndrome is caused by mutations in different ion channels.

The periodic paralyses are caused by mutations in three different ion channel genes [17] (Table 3). Mutations in the *SCN4A* gene encoding the $Na_v1.4$ sodium channel cause HyperPP and HypoPP. The $Na_v1.4$ channel is expressed primarily in skeletal muscle and it is the only sodium channel that is highly expressed in that tissue, which explains the localization of the symptoms. All of the changes are missense mutations that alter the functional properties of the channel. HyperPP and HypoPP can also be caused by mutations in the *KCNE3* gene that encodes the MiRP2 subunit of the Kv7.1 potassium channel. Mutations in a different potassium channel gene, *KCNJ2* encoding the $K_v2.1$ channel, cause AS. Mutations in the *CACNA1S* gene encoding the $Ca_v1.1$ calcium channel cause HypoPP but not HyperPP. All of these disorders are inherited in an autosomal-dominant fashion.

The nondystrophic myotonias are caused by mutations in two genes [18] (see Table 3). Mutations in the *SCN4A* sodium channel gene cause both PMC and PAM. These mutations are similar to those that cause HyperPP and HypoPP, that is, they are all missense changes that alter the functional properties of the sodium channel. In addition, they are all inherited in an autosomal-dominant fashion. The two forms of MC are caused by mutations in the *CLCN1* gene encoding a voltage-gated chloride channel. These alterations are also missense mutations that change the functional properties of the channel. There is no way to predict whether a particular mutation will cause the autosomal-dominant (Thomsen's) or autosomal-recessive (Becker's) form of this disease.

C. Pathogenesis and Pathophysiology

Periodic paralysis and myotonia result from the same physiological changes in membrane excitability of skeletal muscle [17]. During normal muscle contraction, repetitive depolarization leads to potassium accumulation in the transverse tubules because of potassium efflux in the recovery phase of each action potential (the inside of the transverse tubules are comparable to the extracellular environment). Sustained membrane depolarization increases the amount of potassium that is trapped in the transverse tubules and brings the membrane potential closer to the threshold for action potential initiation. The increased potassium accumulation results in depolarization of the transverse tubule membrane, which causes a repetitive and self-sustaining train of action potentials. The repetitive depolarizations cause the muscle fibers to repeatedly contract without relaxation, which is myotonia. The phenomenon is specific to muscle because it is the only cell type that has a transverse tubule architecture.

If the membrane becomes sufficiently depolarized, the potential reaches a point at which normal sodium channels become "slow inactivated." Sodium channels demonstrate at least two forms of inactivation [11,17]. *Fast inactivation* occurs within milliseconds and is the mechanism responsible for closing the channels during each action potential. *Slow inactivation* occurs over a period of seconds and determines the availability of functional channels. Thus, increased slow inactivation decreases the number of functional sodium channels, which makes the muscle membrane less excitable. If a sufficient percentage of sodium channels

Table 3 Ion Channelopathies Associated with Neuromuscular Disorders

Type of Channel	Gene	Channel	Subunit	Syndrome
Calcium channel	CACNA1S	$Ca_v1.1$	α_{1S}	Hypokalemic periodic paralysis
Chloride channel	CLCN1			Myotonia congenita (Becker's autosomal recessive)
Chloride channel	CLCN1			Myotonia congenita (Thomsen's autosomal dominant)
Potassium channel	KCNE3	$K_v7.1$	MiRP2	Hyperkalemic periodic paralysis
Potassium channel	KCNE3	$K_v7.1$	MiRP2	Hypokalemic periodic paralysis
Potassium channel	KCNJ2	$K_{ir}2.1$		Andersen's syndrome
Sodium channel	SCN4A	$Na_v1.4$	α	Hyperkalemic periodic paralysis
Sodium channel	SCN4A	$Na_v1.4$	α	Hypokalemic periodic paralysis
Sodium channel	SCN4A	$Na_v1.4$	α	Paramyotonia congenita
Sodium channel	SCN4A	$Na_v1.4$	α	Potassium aggravated myotonia

become slow inactivated, then the muscle membrane cannot fire action potentials, resulting in paralysis.

D. Structural Detail and Pharmacology

The neuromuscular disorders are caused by mutations in genes encoding subunits of four ion channel genes that were described in Section II, earlier in this chapter. Mutations in three of the channels occur in the primary pore-forming α subunits [17,18]. Mutations in the SCN4A gene encoding the α subunit of the voltage-gated sodium channel cause HyperPP, HypoPP, PMC, and PAM. These mutations are located throughout the channel, with no correlation between the location of the mutation and the type of disorder. Mutations in the CACNA1S gene encoding the α_{1S} subunit of the $Ca_v1.1$ calcium channel cause HypoPP. Mutations in the CLCN1 gene encoding the chloride channel cause both Becker's and Thomsen's MC. As with the sodium channel mutations, there is no correlation between the location of the mutation and the specific syndrome. Mutations in the KCNE3 gene encoding the MiRP2 accessory subunit of the $K_v7.1$ potassium channel cause both HyperPP and HypoPP. Again, it is not possible to predict the type of disorder from the location or type of mutation.

An effective treatment for some patients with HyperPP, HypoPP, and PMC is to use carbonic anhydrase inhibitors such as acetazolamide [18]. The mechanism of action is unknown, although the drug has two effects that may be responsible for the therapeutic efficacy in this case. First, inhibition of carbonic anhydrase acidifies the intracellular environment. Second, acetazolamide opens some potassium channels in a mechanism unrelated to the inhibition of carbonic anhydrase. Both of these effects may result in normalization of serum potassium levels.

E. Explanation of Symptoms in Relation to Pathophysiology

Although periodic paralysis is caused by mutations in a variety of ion channel genes, the ultimate pathogenesis of the disorder is similar in all cases. The final common denominator that causes paralysis is a depolarization of the membrane potential. The more positive membrane potential causes some fraction of the normal sodium channels to become slow inactivated. When the number of functional channels decreases past a critical point, the membrane can no longer propagate action potentials from the motor endplate, resulting in flaccid paralysis [17].

The calcium channel mutations that cause HypoPP occur in the gene encoding the L-type calcium channel, which is localized in the transverse tubule membrane. This channel functions as a voltage-gated calcium channel and as a voltage sensor for calcium release from the sarcoplasmic reticulum. All of the mutations that cause HypoPP decrease calcium currents, so defects in channel function rather than calcium release function are most likely responsible for the periodic paralysis. The mechanism by which decreased calcium currents cause periodic paralysis is unknown, but it may be linked to inward rectifier potassium channels, whose currents are impaired in muscle fibers from patients with the calcium channel mutations. Decreased potassium currents will depolarize the resting membrane potential and could lead to hypokalemia [17].

The potassium channel mutations that cause AS occur in the gene encoding $K_v2.1$, which encodes an inwardly rectifying potassium channel. This channel is involved in setting the resting membrane potential and in reuptake of potassium from the transverse tubules. The mutations decrease function of the channel, sometimes in a dominant-negative fashion, which results in depolarization of the resting membrane potential leading to periodic paralysis. This channel is also involved in repolarization of the cardiac action potential, which accounts for the cardiac symptoms in AS [17].

The potassium channel mutations that cause HyperPP occur in the gene encoding the MiRP2 subunit that associates with $K_v7.1$, a delayed rectifier voltage-gated potassium channel. These mutations also decrease potassium current, causing membrane depolarization and periodic paralysis. The MiRP2 subunit is not necessary for cardiac action potential repolarization, explaining the lack of cardiac symptoms in HyperPP compared with AS.

The sodium channel mutations that cause HyperPP, PMC, and PAM lead to sustained membrane depolarization by a variety of mechanisms. The most common alteration is an impairment of fast inactivation. Many of the HyperPP mutations make the inactivation process incomplete, resulting in a persistent current that depolarizes the membrane. Other mutations cause a persistent current by shifting the voltage dependence of inactivation in the positive direction, thus increasing the overlap of activation and inactivation and the window current. In contrast, many of the PMC mutations slow the kinetics of fast inactivation, although the channels eventually inactivate completely. The slower inactivation broadens the action potential, which increases the amount of potassium that enters the transverse tubules. In this case, it is the higher potassium concentration in the transverse tubules that depolarizes the membrane potential. Other effects of sodium channel mutations include a shift in the voltage dependence of activation and an impairment of slow inactivation. The type of mutation determines whether the primary effect is myotonia or paralysis. The sodium channel mutations that cause HypoPP cause paralysis by enhancing sodium channel inactivation, resulting in a decreased number of available sodium channels at the normal resting membrane potential. The decreased number of functional channels results in an inability to propagate action potentials [17].

The chloride channel mutations causing MC occur in the *CLC1* gene encoding a voltage-gated chloride current. Normally, a high chloride conductance attenuates the depolarization of the transverse tubule membrane that results from potassium accumulation. This chloride conductance is decreased in MC, leading to depolarization of the transverse tubules and myotonia [18].

F. Natural History

HyperPP is usually first apparent in infancy or early childhood, being manifested as episodic weakness. The weakness can be consistently initiated by potassium administration, but it is often triggered by other factors such as exercise, stress, and muscle cooling. The individual attacks begin with muscle irritability and myotonia and then progress to weakness and complete loss of excitability (paralysis). The episodes last from minutes to hours and can occur two to three times daily. The frequency of attacks usually decreases with age, although permanent weakness can occur [17].

HypoPP generally begins later than HyperPP, around adolescence, but the more severe cases may begin in late childhood. The episodes are initiated by rest after exercise, by stress, or by ingestion of carbohydrates or ethanol. Individual episodes do not involve myotonia but begin with weakness. The duration of each episode varies but is generally longer than that for HyperPP (hours to days), and the frequency is usually less but also quite variable (daily to yearly). The disorder progresses with age, often resulting in permanent muscle weakness [17].

AS similarly begins in early childhood, but it is unique among the periodic paralyses because tissues other than skeletal muscle are also involved. The triad of symptoms includes periodic paralysis, ventricular arrhythmias ranging from long QT to frank tachycardia, and dysmorphic developmental abnormalities that generally involve the facial structures. The episodes are similar to those of HypoPP and consist of weakness without myotonia. The duration and frequency of episodes are also similar to those of HypoPP, with each episode lasting hours to days and occurring daily to yearly. Permanent muscle weakness can often be observed, and the cardiac abnormalities can be life threatening [17].

PMC is in many ways similar to HyperPP. The symptoms begin at about the same time (early childhood) and the duration and frequency are also comparable. However, in PMC the most prominent symptom is myotonia, although weakness can still occur. Episodes can be initiated by cold, which also tends to result in weakness following the myotonia. The myotonia gets worse with repeated muscle contractions, which is in contrast to the other nondystrophic myotonias (PAM and MC), in which the myotonia improves with muscle use ("warmup" phenomenon) [18].

The symptoms of PAM and the two forms of MC are similar and include myotonia that improves with use. The symptoms begin in infancy or early childhood and decrease in the third or fourth decade. Becker's disease has the most severe myotonia that is also associated with weakness and prominent muscle hypertrophy. Weakness is rarely seen in PAM, and muscle hypertrophy is variable in both PAM and Thomsen's disease [18].

References

1. Hille, B. (2001). "Ion Channels of Excitable Membranes." Sinauer Associates, Inc., Sunderland, MA.
2. Chang, B. S., and Lowenstein, D. H. (2003). Epilepsy. *N. Engl. J. Med.* **349**, 1257–1266.
3. Noebels, J. L. (2003). The biology of epilepsy genes. *Annu. Rev. Neurosci.* **26**, 599–625.
4. Gourfinkel-An, I., Baulac, S., Nabbout, R., Ruberg, M., Baulac, M., Brice, A., and LeGuern, E. (2004). Monogenic idiopathic epilepsies. *Lancet Neurol.* **3**, 209–218.
5. George, A. L., Jr. (2004). Inherited channelopathies associated with epilepsy. *Epilepsy Curr.* **4**, 65–70.
6. Scheffer, I. E., and Berkovic, S. F. (2003). The genetics of human epilepsy. *Trends Pharmacol. Sci.* **24**, 428–433.
7. Catterall, W. A., Chandy, K. G., and Gutman, G. A. (2002). "The IUPHAR Compendium of Voltage-Gated Ion Channels." IUPHAR Media, Leeds, UK.
8. Chen, T.-Y. (2005). Structure and function of CLC channels. *Annu. Rev. Physiol.* **67**, 809–839.

9. Gotti, C., and Clementi, F. (2004). Neuronal nicotinic receptors: from structure to pathology. *Prog. Neurobiol.* **74**, 363–396.
10. Jones-Davis, D. M., and MacDonald, B. T. (2003). $GABA_A$ receptor function and pharmacology in epilepsy and status epilepticus. *Curr. Opin. Pharmacol.* **3**, 12–18.
11. Goldin, A. L. (2003). Mechanisms of sodium channel inactivation. *Curr. Opin. Neurobiol.* **13**, 284–290.
12. Spampanato, J., Aradi, I., Soltesz, I., and Goldin, A. L. (2004). Increased neuronal firing in computer simulations of sodium channel mutations that cause generalized epilepsy with febrile seizures plus. *J. Neurophysiol.* **91**, 2040–2050.
13. Steinlein, O. K. (2004). Genetics mechanisms that underlie epilepsy. *Nat. Rev. Neurosci.* **5**, 400–408.
14. De Michele, G., Coppola, G., Cocozza, S., and Filla, A. A. (2004). A pathogenetic classification of hereditary ataxias: is the time ripe? *J. Neurol.* **251**, 913–922.
15. Schöls, L., Bauer, P., Schmidt, T., Schulte, T., and Riess, O. (2004). Autosomal dominant cerebellar ataxias: clinical features, genetics, and pathogenesis. *Lancet* **3**, 291–304.
16. Orr, H. T. (2004). Into the depths of ataxia. *J. Clin. Invest.* **113**, 505–507.
17. Cannon, S. C. (2002). An expanding view for the molecular basis of familial periodic paralysis. *Neuromusc. Disord.* **12**, 533–543.
18. Davies, N. P., and Hanna, M. G. (2003). The skeletal muscle channelopathies: distinct entities and overlapping syndromes. *Curr. Opin. Neurol.* **16**, 559–568.
19. Long, S. B., Campbell, E. B., and MacKinnon, R. (2005). Crystal structure of a mammalian voltage-dependent *Shaker* family K^+ channel. *Science* **309**, 897–903.
20. Dutzler, R., Campbell, E. B., Cadene, M., Chait, B. T., and MacKinnon, R. (2002). X-ray structure of a ClC chloride channel at 3.0 Å reveals the molecular basis of anion selectivity. *Nature* **415**, 287–294.

13

Spinocerebellar Ataxia Type 1

Harry T. Orr, PhD

Keywords: *ataxia, polyglutamines, Purkinje cells, transgenic mice*

I. SCA1: The Disease
II. SCA1 Pathogenesis: Regional Involvement
III. SCA1 Pathogenesis: Molecular Mechanisms
IV. SCA1: Recovery from Disease
V. Linking Pathology to Pathophysiology
VI. Concluding Comments
 References

I. SCA1: The Disease

Spinocerebellar ataxia type 1 (SCA1) is a member of a large group of dominantly inherited SCAs that have a population frequency of about 1 per 100,000 and are characterized by neural degeneration in the cerebellum, spinal tracts, and brainstem. The genetics of SCA1 extend to the last half of the nineteenth century with Friedreich's description of a recessive form of hereditary ataxia that today bears his name. In 1893, Pierre Marie noticed a form of ataxia distinct from that described by Friedreich. The designation "SCA1" for one form of dominant ataxia reflects the fact that the genetic locus for this disease was the first to be localized to a specific chromosomal region, the short arm of chromosome 6, by virtue of its linkage to the human leukocyte antigen (HLA) complex [1]. Today the ataxias recognized by Marie are known as the *autosomal-dominant* or *spinocerebellar ataxias*.

The clinical features seen in patients with SCA1 usually include ataxia, dysarthria, and bulbar dysfunction. Additional features of the disease often vary depending on disease stage. Early in the course of SCA1, patients usually have a mild loss of limb and gait coordination, slurred speech, and poor handwriting skills. Some have hyperreflexia, hypermetric saccades, and nystagmus. As disease progresses, the ataxia worsens; dysmetria, dysdiadochokinesis, and hypotonia develop; and patients with SCA1 often experience vibration and proprioceptive loss. At the advanced stage, the ataxia becomes severe and brainstem dysfunction results in facial weakness and swallowing and breathing problems. Patients typically die from the loss of the ability to cough effectively, food aspiration, and respiratory failure. In SCA1, death usually occurs between 10 and 15 years after the onset of symptoms.

A study of 11 patients with SCA1 from a single kindred emphasized the olivopontocerebellar degeneration in most cases [2]. The most commonly seen and most severe alterations were loss of Purkinje cells in the cerebellar cortex and loss of neurons in the inferior olivary nuclei, the cerebellar dentate nuclei, and the red nuclei. The cerebellar granular layer was spared or had slight loss of neurons. Involvement of the basal pontine nuclei ranged from mild to severe. The posterior columns and spinocerebellar tracts of the spinal cord typically were moderately to severely affected. The number of motor neurons of the anterior horns was decreased in all cases in which the cord was examined. Nuclei of the 3rd, 10th, and 12th cranial nerves also had variable involvement, with the hypoglossal nuclei most commonly and severely affected. The striatonigral system was minimally affected or unaffected when examined by classic anatomical methods, but biochemical studies on the same population of patients revealed evidence of decreased dopaminergic terminals in the striatum without changes in the cellularity of the nigra. The cerebral cortex and hippocampus had slight loss of cellularity in some of the cases.

In general, SCA1 appears to have a defined neuropathological pattern that separates it from other dominantly inherited ataxias and from sporadic entities such as multiple system atrophy. Because of the range of neuropathological findings in SCA1 and the other forms of hereditary ataxia, individual cases of SCA1 may still have neuropathological features that overlap with other diseases.

A. The SCA1 Gene

Using a positional cloning strategy, the *SCA1* gene was identified and isolated in 1993 [3]. The *SCA1* gene consists of nine exons spanning over 450 kb of genomic DNA, with the largest intron extending for at least 150 kb. The 2448-bp coding region is located within exons 8 and 9. Thus, the 935-bp 5' untranslated region (UTR) is contained within the first seven exons and a portion of the eighth exon. The 3'UTR is contained within the ninth exon and remarkably includes more than 7000 bp, yielding an *SCA1* transcript of 10,660 bp.

The disease-causing mutation is expansion of an unstable CAG trinucleotide repeat within the coding region of the gene [3]. Disease is caused by the expansion of this repeat producing an expanded tract of glutamine amino acids within the *SCA1* gene product, ataxin-1. There is an inverse relationship between the lengths of the CAG repeat tract on affected chromosomes and the age at onset and severity of disease. The longer the mutant CAG repeat, the younger the age at onset and the more severe the disease. The number of CAG repeats in *SCA1* is highly variable in the general population, ranging from 6 to 44. Interestingly, normal alleles of *SCA1* having more than 20 repeats typically have one to three CAT triplet interruptions within the CAG tract [4]. In contrast, affected alleles of *SCA1* have expansions between 39 and 82 of pure CAG repeat tract without a CAT interruption. The CAT interruptions are likely to be an important factor in the intergenerational stability of the longer normal alleles. Perhaps an early event in the conversion of a normal stable allele of *SCA1* to an unstable allele is the loss of the CAT interruptions.

II. SCA1 Pathogenesis: Regional Involvement

Research into the cellular changes associated with SCA1 has been limited primarily to the Purkinje cell, a principal target of the disease process. A Golgi impregnation study revealed reduced dendritic arborization in Purkinje cells with decreased numbers of spiny branchlets [5]. A pathological hallmark of SCA1 is the occurrence of frequent proximal axonal dilatations (torpedoes), a feature not found in the *SCA1* mice (see later discussion). An electron microscopic study of cerebellar cortical biopsies was performed on two patients with SCA1 [6]. A variety of mild ultrastructural abnormalities were identified, although the pathogenic significance of these changes was uncertain. Some of the more severe alterations could be attributed to artifact, although there was evidence of axonal degenerative changes within afferent fibers in the granular layer as well as somatic and dendritic degeneration in some of the Purkinje cells.

It is evident from autopsy studies of patients with SCA1 that morphological alterations antedate cell death in at least some of the Purkinje cells of these patients. The results of these investigations have been difficult to interpret. Because of the complexity of the disease process and the involvement of multiple neuronal populations, it is difficult to distinguish between trans-synaptic effects and events in the cell that are the direct consequence of expression of a mutant *SCA1* gene. In addition, autopsy studies are typically performed late in the course of the disease, when cell degeneration is more extensive. Directing the expression of the *SCA1* mutation to a single cell type known to be involved in the disease, the Purkinje cell, has more readily allowed an evaluation of the relative effects of structural alterations and cellular loss on the development of ataxia in the *SCA1* transgenic mice.

III. SCA1 Pathogenesis: Molecular Mechanisms

The molecular basis of SCA1 pathogenesis has been largely pursued using material from experimental models. In the case of SCA1, mice carrying an allele of ataxin-1 with 82 glutamines whose expression was directed specifically to Purkinje cells were developed shortly after the cloning of the *SCA1* gene [7]. Transgenic mice expressing high levels of a wild-type allele of *SCA1* with 30 CAG repeats failed to develop any signs of Purkinje cell pathology or ataxia. Only those animals expressing the expanded allele of SCA1 developed Purkinje cell pathology and an associated ataxia. Disease progression in the transgenic mice expressing a mutant allele of SCA1 has been examined in considerable detail.

Like the human disease, the *SCA1* transgenic mice presented with a progressive neurological disorder [7,8]. Results of behavioral tests demonstrated that at 5 weeks of age, cerebellar impairment appeared to be limited to a decreased ability to improve motor performance as assessed by the accelerating rotating rod apparatus. At this age, the *SCA1* mutant animals perform as well as the wild-type littermate controls on the first day of trials, and impairment was noted in the mutant mice only on successive days of trials. The latter observation indicates that a training or learning phase is required for the deficiency in the *SCA1* mutant mice to manifest itself on the rotating rod. The absence of gait abnormalities, as well as normal motor activity, balance, and coordination, supports the conclusion that the impairment on the rotating rod was due to a decreased ability of the mutant mice to learn the task as opposed to impairment in motor activity, fine motor control, or coordination.

With increasing age, as the cerebellar impairments of *SCA1* mutant mice worsened to reach a stage of severe ataxia, deficiencies in motor activity and gait became apparent. By 1 year of age when there was substantial loss of cerebellar function, and the mutant mice were never able to match the performance of wild-type animals on the rotating rod even on the first day of trials and did not demonstrate any ability to improve their performance with training. These results suggest that in the *SCA1* mutant mice cerebellar dysfunction can be divided into two phases. In the first phase, dysfunction is limited to an impairment of motor learning. At a later stage, impairment advances to a point at which motor activity and coordination become abnormal and severe ataxia ensues.

In the *SCA1* mice, there were no histological alterations during the first 3 weeks of postnatal life. The morphological alterations noted in the *SCA1* mice, after they were ataxic by home-cage behavior at 12 weeks, developed after the cerebellar cortex had matured and, therefore, were not a result of maldevelopment induced by expression of the mutant transgene. At the time of decreased motor learning on the rotating rod apparatus, a loss of proximal dendritic branches and a decrease in the number of dendritic spines was observed in Purkinje cells of *SCA1* mice. At the time that ataxia, as assessed by home-cage behavior, was established in the *SCA1* mice (12–15 weeks), there was little evidence of loss of Purkinje cells, but there were significant alterations in dendritic and perikaryal morphology, as well as evidence of perikaryal heterotopia that was not seen in younger animals. These changes subsequently become more widespread and severe as the animals age.

In the *SCA1* mutant mice, onset of ataxia occurred at a time when loss of Purkinje cells was barely detectable. This observation, coupled with the numerous morphological abnormalities found in Purkinje cells from *SCA1* transgenic animals, indicates that expression of the mutant form of ataxin-1, in a cell that is vulnerable in the human disease, can lead to cellular dysfunction sufficient to induce ataxia without causing death of the affected neuronal population. Therefore, it is likely that in the *SCA1* transgenic mice the disease is not caused by cell loss. Rather, loss of Purkinje cells seen at later stages of disease is most likely the result of the dysfunction induced at an earlier stage.

An interesting feature of expanded polyglutamine proteins observed in both transgenic mice and patient material is the presence of neuronal intranuclear inclusions or aggregates that contain the polyglutamine protein. The two most striking characteristics of these inclusions are their specific occurrence in cells affected by the disease, despite widespread expression of the expanded proteins, and their generally consistent nuclear location, despite widely varying subcellular localizations of the nonaggregating forms of the protein. Besides containing the polyglutamine protein, these intranuclear aggregates are also ubiquitinated and contain elements of the proteasome apparatus [9]. Thus, it has been proposed that the formation of the aggregates reflects an impairment of cellular proteolytic degradation by the polyglutamine disease protein.

To examine directly the role of nuclear ataxin-1 aggregates in SCA1 pathogenesis, two additional series of transgenic mice were established. The development of these SCA1 mice was based on work performed using transfected COS cells in tissue culture. In contrast to intact ataxin-1, a deletion mutant of ataxin-1 that lacks the self-association region was transfected into COS cells and did not form nuclear aggregates [10]. The same result was observed in transgenic Purkinje cells, in which the nuclei stained in a diffuse pattern, without evidence of inclusions [10]. Interestingly, mice expressing the ataxin-1 deletion mutant with an expanded number of glutamines developed an ataxic phenotype despite the absence of microscopic aggregates. These mice developed both the histopathological changes and the associated motor disability characteristics of the original SCA1 mouse lines [7,8]. These results demonstrate that aggregation of mutant ataxin-1 is not a primary component of pathogenesis (i.e., the aggregates do not cause the disease). Because the mice have not yet been followed to advanced age, it remains quite possible that the aggregates do have a secondary effect, which augments the primary pathogenesis of the disease.

Another fundamental point regarding the molecular basis of SCA1 pathogenesis that stems from work with transfected COS cells is the importance of the subcellular localization of ataxin-1 in pathogenesis. As in Purkinje cells, ataxin-1 localizes to the nucleus of COS cells. A series of COS cell transfections demonstrated that ataxin-1 contains a functional arginine-lysine nuclear localization sequence (NLS) centered at lysine 772 near the carboxy-terminus [10]. The mutation K772T completely eliminated the activity of this NLS in COS cells. Transgenic mice expressing ataxin-1[82Q]^{K772T} also accumulated the protein primarily in the cytoplasm. In contrast to the previous *SCA1* transgenic mice, which expressed nuclear ataxin-1, the ataxin-1[82Q]^{K772T} mice did not develop the characteristic histopathological changes and associated phenotype of the disease, even at advanced ages. Thus, nuclear localization of intact ataxin-1[82Q] protein is essential for SCA1 pathogenesis, at least in the transgenic mouse model.

The essential molecular processes governing the initiation and progression of polyglutamine diseases in general, and SCA1 in particular, remain uncertain. Yet, it is possible, taking into consideration the available data, to generate a hypothetical model. In a vulnerable neuron, after translation a substantial portion of ataxin-1 is transported into the nucleus. However, in the cytoplasm, it exerts no pathogenic effects and forms no aggregates even when present in the cytoplasm at elevated concentrations for extended periods, as in the ataxin-1[82Q]^{K772T} mice. On recognition of the intact NLS by the transport system and docking with the nuclear membrane pore complex, cytoplasmic ataxin-1 is translocated into the nucleus. The process of recognition and translocation is very likely to be regulated by cell-specific mechanisms, because ataxin-1 has been observed to distribute quite differently in various cell types [11].

Once mutant ataxin-1 is in the nucleus, there is a change associated with the gain in function that initiates SCA1 pathogenesis, perhaps based on the interaction of ataxin-1 with other nuclear proteins. This gain in function may affect any one of various nuclear processes, such as transcription or RNA processing, or it may more generally affect nuclear architecture [12,13]. Regardless of the pathway affected by mutant ataxin-1 in the nucleus, it is becoming increasingly clear that change associated with gain-in-function state of mutant ataxin-1 is related to an altered conformation of the protein. Ataxin-1 within the aggregates likely has a conformation or binding configuration different from either the cytoplasmic state or the soluble nuclear state. The presence of ubiquitin, proteasome components, and the chaperone protein HDJ-2/HSDJ within aggregates indicates that the aggregates arise from ataxin-1 misfolding and a decreased turnover by the proteasome system. It is possible that the aggregates mediate a pathogenic effect that is distinct from that of the soluble pathogenic form of ataxin-1. If so, then this effect is not necessary for the initiation of pathogenesis but may have a role in disease progression [14].

Two lines of experimental evidence illustrate that the ubiquitin-proteasome system and clearance of mutant ataxin-1 have an impact on pathogenesis. First, Cummings et al. [15] crossed ataxin-1[82Q] transgenic mice with mice lacking expression of the *Ube3a* gene. *Ube3a* encodes the E6-associated protein (E6-AP), a member of the E3 class of ubiquitin ligases. Two important observations were made in the mice expressing the full-length ataxin-1[82Q] protein in the absence of *Ube3a* expression. The presence of nuclear inclusions of ataxin-1[82Q] was reduced significantly in terms of both their frequency and their size. Yet, Purkinje cell pathology was considerably worse compared with that seen in the ataxin-1[82Q] mice. These studies show *in vivo* the importance of the ubiquitin-proteasome pathway for the formation of the nuclear inclusions and indicate that inclusion formation is an active cellular process. Furthermore, they showed that pathology is not dependent on the formation of nuclear inclusions.

Secondly, taking a cue from the fact that phosphorylation is a prominent means by which protein degradation and subcellular localization are regulated, purified ataxin-1 was subjected to mass spectroscopy analysis to identify potential sites of phosphorylation. Toward this end, the serine at residue 776 of ataxin-1 was found to be phosphorylated [16]. To investigate the role of this phosphorylation event in mutant ataxin-1 subcellular localization and neuronal dysfunction, the serine residue was mutated to alanine, thus preventing its phosphorylation. Mutation of this residue to an alanine in transgenic mice (A776) resulted in a significant dampening of the pathogenic effects of mutant ataxin-1. By both behavioral analyses, and by assessing Purkinje cell morphology, mutation of serine 776 to alanine rendered mutant ataxin-1 in effect innocuous. Mutation of this residue did not affect the subcellular localization of mutant ataxin-1, indicating that phosphorylation of this residue is not required for nuclear transport.

In tissue culture cells, mutation of serine 776 to alanine drastically reduced that ability of mutant ataxin-1 to aggregate. In cells expressing ataxin-1[82Q]-A776, less than 1 in 1000 cells contained nuclear aggregates. In contrast, in cells expressing ataxin-1[82Q]-S776, at least 60% of the cells contained aggregates. As expected, this reduction in aggregate formation coincided with an increase in extractable soluble mutant ataxin-1. In ataxin-1[82Q]-A776 transgenic mice, the ability of the mutant protein to aggregate was similarly diminished. The reduction of ataxin-1[82Q]-A776 aggregates suggests that the cell is able to more efficiently handle the mutant protein. Consistent with this is the increased solubility of the ataxin-1[82Q]-A776 protein.

Polyglutamine tract expansion and nuclear localization of mutant ataxin-1 within Purkinje cells are not sufficient to induce disease SCA1 pathogenesis. A single amino acid within mutant ataxin-1, serine 776, plays a critical role in pathogenesis.

Toward elucidating the potential mechanism by which S776 mediates ataxin-1–induced pathogenesis, a screen was conducted to identify proteins that interact with ataxin-1 in an S776-specific manner. In this screen, an immunoprecipitation assay from COS cell lysates followed by mass spectrometry was used to identify proteins that interact with ataxin-1 [82Q]-S776 but not with ataxin-1 [82Q]-A776. A yeast two-hybrid screen was used to confirm the interaction. Through this approach, 14-3-3 was identified as an ataxin-1-S776–specific interactor. 14-3-3 is a multifunctional regulatory protein family with many isoforms. Several of these isoforms, including β, ε, and ζ, interact with ataxin-1 in a Neuro2A cell line. 14-3-3 interacts with both mutant and wild-type ataxin-1-S776; however, the strength of this interaction appears to be mediated by the length of the polyglutamine tract. In COS cells, the relative level of 14-3-3 bound to ataxin-1 was found to increase with increasing lengths of the polyQ tract, independent of the phosphorylation state of ataxin-1 [17].

As described earlier, S776 is an important mediator of both ataxin-1 solubility and the formation of nuclear inclusions. Because the interaction between 14-3-3 and ataxin-1 is dependent on the presence of S776, it is possible that 14-3-3 may play an active role in determining ataxin-1 solubility and nuclear inclusion formation. To address this question, the localization of 14-3-3 in COS cells was examined.

The subcellular localization of 14-3-3 in transfected COS cells was altered by the presence of ataxin-1 [82Q]-S776 but not ataxin-1 [82Q]-A776. When expressed alone, 14-3-3 distributed to the cytoplasm and nucleoplasm of the COS cells. Co-transfection of both ataxin-1 [82Q]-S776 and 14-3-3 resulted in the redistribution of the 14-3-3 protein to the nuclear inclusions formed by ataxin-1 [82Q]-S776. This redistribution of 14-3-3 was not seen when 14-3-3 was co-transfected with ataxin-1 [82Q]-A776 (a protein that rarely forms nuclear inclusions when transfected alone).

One explanation for the observed aggravated nuclear inclusion formation in the presence of 14-3-3 is that 14-3-3 is acting to stabilize the ataxin-1 protein. To test this hypothesis, the steady state level of ataxin-1 in HeLa cells was determined in the presence or absence of 14-3-3. When ataxin-1 and 14-3-3 were co-transfected into HeLa cells, the steady state level of ataxin-1 increased in the presence of 14-3-3. This increase in protein levels was dependent on polyglutamine tract length, supporting the data suggesting that the interaction of 14-3-3 with ataxin-1 is dependent on polyglutamine tract length. 14-3-3 did not increase the levels of ataxin-1 [82Q]-A776, suggesting that the ability of 14-3-3 to stabilize ataxin-1 is the result of its direct interaction with S776 [17].

The results from these experiments suggest that 14-3-3, through its ability to bind to and stabilize ataxin-1, might be an important player in SCA1 pathogenesis. To look at the *in vivo* role of 14-3-3 in SCA1 neurodegeneration, a *Drosophila* model of SCA1 was used. Double transgenic flies were generated that expressed both 14-3-3 and ataxin-1 [82Q]-S776 in a retinal-specific manner. These flies presented with a thin disorganized retinal layer. The phenotype was much more severe than that observed in transgenic flies expressing ataxin-1 [82Q]-S776 alone. These data support a role for 14-3-3 in SCA1 pathogenesis, perhaps by binding to and stabilizing the mutant ataxin-1 protein [17].

IV. SCA1: Recovery from Disease

Demonstrating in the SCA1 mice that there is a long period of Purkinje dysfunction preceding neuronal loss raised the issue of whether, if given the chance, Purkinje cells might be able to recover from the deleterious effects of mutant ataxin-1. If so, is there a stage beyond which SCA1 pathology becomes irreversible? To address these points, an SCA1 conditional mouse model was developed [18]. The approach used the regulatory region from the *Pcp2 (L7)* gene to direct expression of the tetracycline-responsive activator, tTA, to cerebellar Purkinje cells, which then drove the expression of an *SCA1[82Q]* transgene. Oral administration of dox, an inhibitor of tTA, was used to terminate expression of the *SCA1[82Q]* transgene. On cessation of ataxin-1 expression, Purkinje cells were able to recover from polyglutamine-induced disease. Importantly, the extent of recovery decreased with age and progression of disease. Clearance of diffuse nuclear mutant ataxin-1 from Purkinje cells was rapid after *SCA1* expression was halted. In addition, the ability of Purkinje cells to clear mutant ataxin-1 did not seem to vary with disease progression from 8 to 30 weeks, a span of time when there was a dramatic decrease in neurological function and worsening of pathology. Mutant ataxin-1 localized to the nuclear inclusions was also readily cleared from Purkinje cells, having an apparent half-life of 2 days at 16 weeks of age. This result supports the concept that polyglutamine nuclear inclusions are dynamic.

In this study [18], the capacity of Purkinje neurons to recover from the adverse effects of ataxin-1[82Q] was examined at three stages of disease: at an early stage after 6 weeks of *SCA1[82Q]* expression, at a mid-stage after 12 weeks of *SCA1[82Q]* expression, and at a late stage after 32 weeks of *SCA1[82Q]* expression. The extent to which Purkinje cells recovered decreased as the length of their exposure to ataxin-1[82Q] increased and pathology became progressively worse. When expression of ataxin-1[82Q] was stopped early in disease, after 6 weeks of *SCA1[82Q]* expression, disease progression was quickly halted and neurological status improved so that the mice recovered to an essentially normal neurological phenotype and Purkinje cell morphology. One abnormal morphological feature that failed to show signs of improvement was the presence of heterotopic Purkinje cells in the molecular layer. The failure of the mice to correct this feature is consistent with the suggestion that the heterotopic Purkinje cells are a compensatory response to the dendritic pruning induced by mutant ataxin-1.

In contrast to the full recovery from an early stage of disease, at later stages, a more protracted Purkinje cell recovery was reported. Conditional *SCA1[82Q]* mice in which gene expression was ceased at 12 weeks of age, at mid-stage, after ataxia was evident by both Rotarod and home-cage assessments, demonstrated a partial recovery within the 12-week gene-off period. Mice in which the gene remained "on" until a much later stage of disease, 32 weeks when the animals had severe ataxia by Rotarod and cage behavior in addition to very extensive Purkinje cell pathology, showed the least amount of recovery during the 12 weeks of gene off. Thus, the ability of the conditional *SCA1* mice to recover decreased with age and disease progression. The extent to which the decreased ability of the older mice to recover was due to effects of disease progression versus effects of age remains to be determined. Nevertheless, it is important to note that within the confines of this experimental design, the disease process never became completely irreversible.

Overall, the results from the conditional *SCA1[82Q]* mice support the concept that most if not all of the pathological changes seen in polyglutamine disease, excluding cell loss, are reversible. These results also show that the earlier an intervention is administered the more complete will be the restoration of normal function. Yet, even at a very late stage of disease, the pathological effects of mutant ataxin-1 were partially reversible. This demonstrates that Purkinje cells have a sustained ability to recover from long-term effects of mutant ataxin-1. A therapeutic intervention targeted at reducing the expression of mutant ataxin-1 showed that *SCA1* RNAi was effective in suppressing Purkinje cell pathology and neurological signs when *SCA1* mice were injected at 7 weeks of age [19].

V. Linking Pathology to Pathophysiology

By following recovery in the conditional *SCA1[82Q]* mice, a sequence of events leading to repair and restoration of neurological function can be assembled. In particular, following recovery after stopping *SCA1[82Q]* expression at a mid-stage of disease (12 weeks) is especially informative. Within days of stopping *SCA1[82Q]* expression, mutant ataxin-1 was cleared by Purkinje neurons. After 4 weeks of the gene off, mice improved their neurological function in that by home-cage behavior ataxia was no longer evident. Yet, their ability to perform the more challenging accelerating Rotarod task showed no signs of improvement. During the initial 4-week gene-off period, Purkinje cell dendritic arborization and spine density improved considerably. However, compared with wild-type mice, many of the Purkinje dendritic spines in the 4-week gene-off mice were not positive for postsynaptic mGluR1. With an additional 4 weeks of the *SCA1[82Q]* gene being off, improvement on the Rotarod was seen. Improvement on the Rotarod at 8 weeks of gene off correlated with a consistent appearance of the mGluR1 receptor at the Purkinje neuron/parallel fiber synapse in the molecular layer.

The importance of mGluR1 for proper motor function is consistent with previous data from mice with a targeted deletion of mGluR1. Mice lacking mGluR1 were unable to perform on the Rotarod, whereas introduction of mGlur1α expression into Purkinje cells of mGluR1 —/— mice restored normal Rotarod performance [20]. The conditional *SCA1[82Q]* mice revealed that, in addition to being key for synapse formation and plasticity during development, dendritic mGluR1 also has a role in Purkinje cell plasticity associated with repair from neurodegeneration in the adult.

VI. Concluding Comments

Current data strongly support the notion that SCA1 pathogenesis is induced by some toxic function gained by the mutant polyglutamine protein as a result of the expanded glutamine tract. Polyglutamine expansion seems to cause mutant ataxin-1 to adopt an altered folding state, leading to its ubiquitination, aggregation, and resistance to proteasomal degradation. Over the course of the disease, the aggregated protein may in turn contribute to disease progression by altering the turnover of other critical proteins that may encumber the ubiquitin-proteasome system. Yet, the protein context in which the disease-causing polyglutamine tract is located plays an important role in defining the disease and its process. Most dramatic of these is the effect that altering the subcellular localization has on pathogenesis in the SCA1 transgenic mice. A single amino acid substitution that alters the ability of ataxin-1 to be transported into the nucleus prevented the development of disease. Thus, the disruption of nuclear structures and/or essential nuclear functions in a neuron appears to be key to SCA1 pathogenesis. Another important role of protein context in pathogenesis is likely to be in determining the selective neuronal vulnerability seen in SCA1.

Although considerable progress has been made in understanding the disease process in SCA1, several important questions remain to be answered. These include the following: What is the molecular basis of the cell-specific determinants that underlie the selectivity of neuronal degeneration? To what extent is the toxic function gained by mutant ataxin-1 linked to the natural function of this protein? Can the gains made in altering the expression of ataxin-1 in animal models of SCA1 be extended and adapted for human therapeutic trials?

Acknowledgments

My apologies to the many colleagues whose original work could not be cited because of space limitations. The research from the author's laboratory described in this chapter was supported by grants NS22920 and NS45667 from the NINDS/NIH.

References

1. Yakura, H., Wakisaka, A., Fujimoto, S., and Itakura, K. (1974). Hereditary ataxia and HLA genotypes. *N. Engl. J. Med.* **291**, 154–155.
2. Robitaille Y., Schut, L., and Kish, S. J. (1995). Structural and immunocytochemical features of olivopontocerebellar atrophy caused by the spinocerebellar ataxia type 1 (SCA-1) mutation define a unique phenotype. *Acta Neuropathol.* **90**, 572–581.
3. Orr, H. T., Chung, M.-y., Banfi, S., Kwiatkowski, T. J., Jr., Servadio, A., Beaudet, A. L., McCall, A. E., Duvick, L. A., Ranum, L. P. W., and Zoghbi, H. Y. (1993). Expansion of an unstable trinucleotide CAG repeat in spinocerebellar ataxia type 1. *Nat. Genet.* **4**, 221–226.
4. Chung, M.-y., Ranum, L. P. W., Duvick, L., Servadio, A., Zoghbi, H. Y., and Orr, H. T. (1993). Analysis of the CAG repeat expansion in spinocerebellar ataxia type I: evidence for a possible mechanism predisposing to instability. *Nat. Genet.* **5**, 254–258.
5. Ferrer, I., Genis, D., Davalos, A., Bernado, L., Sant, F., and Serrano, T. (1994). The Purkinje cell in olivopontocerebellar atrophy. A Golgi and immunocytochemical study. *Neuropathol. Appl. Neurobiol.* **20**, 38–46.
6. Landis, D. M. D., Rosenberg, R. N., Landis, S. C., Schut, L. J., and Nyhan, W. L. (1974). Olivopontocerebellar degeneration. Clinical and ultrastructural abnormalities. *Arch. Neurol.* **31**, 295–307.
7. Burright, E. N., Clark, H. B., Servadio, A., Matilla, A., Feddersen, R. M., Yunis, W. S., Duvick, L. A., Zoghbi, H. Y., and Orr, H. T. (1995). SCA1 transgenic mice: a model for neurodegeneration caused by an expanded CAG trinucleotide repeat. *Cell* **82**, 937–948.
8. Clark, H. B., Burright, E. N., Yunis, W. S., Larson, S., Wilcox, C., Hartman, B., Matilla, A., Zoghbi, H. Y., and Orr, H. T. (1997). Purkinje cell expression of a mutant allele of SCA1 in transgenic mice leads to disparate effects on motor behaviors, followed by a progressive cerebellar dysfunction and histological alterations. *J. Neurosci.* **17**, 7385–7395.
9. Cummings, C. J., Mancini, M. A., Antalffy, B., DeFranco, D. B., Orr, H. T., and Zoghbi, H. Y. (1998). Chaperone suppression of aggregation and altered subcellular proteasome localization imply protein misfolding in SCA1. *Nat. Genet.* **19**, 148–154.
10. Klement, I. A., Skinner, P. J., Kaytor, M. D., Yi, H., Hersch, S. M., Clark, H. B., Zoghbi, H. Y., and Orr, H. T. (1998). Ataxin-1 nuclear localization and aggregation: role in polyglutamine-induced disease in SCA1 transgenic mice. *Cell* **95**, 41–53.
11. Servadio, A., Koshy, B., Armstrong, D., Antalffy, B., Orr, H. T., and Zoghbi, H. Y. (1995). Expression analysis of the ataxin-1 protein in tissues from normal and spinocerebellar ataxia type 1 individuals. *Nat. Genet.* **10**, 94–98.
12. Irwin, S., Vandelft, M., Howell, J. L., Graczyk, J., Orr, H. T., and Truant, R. (2005). Ataxin-1 RNA association *in vivo* and nucleocytoplasmic shuttling. *J. Cell Sci.* **118**, 233–242.
13. Skinner, P. J., Koshy, B. T., Cummings, C. J., Klement, I. A., Helin, K., Servadio, A., Zoghbi, H. Y., and Orr, H. T. (1997). Ataxin-1 with an expanded glutamine tract alters nuclear matrix-associated structures. *Nature* **389**, 971–974.
14. Skinner, P. J., Vierrra-Green, C. A., Emamian, E., Zoghbi, H. Y., and Orr, H. T. (2002). Amino acids in a region of ataxin-1 outside of the polyglutamine tract influence the course of disease in *SCA1* transgenic mice. *NeuroMol. Med.* **1**, 33–42.
15. Cummings, C. J., Reinstein, E., Sun, Y., Antalffy, B., Jiang, Y.-h., Ciechanover, A., Orr, H. T., Arthur, L., Beaudet, A. L., and Zoghbi, H. Y. (1999). Mutation of the E6-AP ubiquitin ligase reduces nuclear inclusion frequency while accelerating polyglutamine-induced pathology in SCA1 transgenic mice. *Neuron* **24**, 879–892.
16. Emamian, E. S., Kaytor, M. D., Duvick, L. A., Zu, T., Susan, K., Tousey, S. K., Zoghbi, H. Y., Clark, H. B., and Orr, H. T. (2003). Serine 776 of ataxin-1 is critical for polyglutamine-induced disease in *SCA1* transgenic mice. *Neuron* **38**, 975–987.
17. Chen, H.-K., Fernandez-Funez, P., Acevedo, S. F., Lam, Y. C., Kaytor, M. D., Fernandez, M. H., Aitken, A., Skoulakis, E. M. C., Orr, H. T., Botas, J., and Zoghbi, H. Y. (2003). Interaction of Akt-phosphorylated ataxin-1 with 14-3-3 mediates neurodegeneration in spinocerebellar ataxia type 1. *Cell* **133**, 457–468.
18. Zu, T., Duvick, L. A., Kaytor, M. D., Berlinger, M., Zoghbi, H. Y., Clark, H. B. and Orr, H. T. (2004). Recovery from polyglutamine-induced neurodegeneration in conditional *SCA1* transgenic mice. *J. Neurosci.* **24**, 8853–8861.
19. Xia, H., Mao, Q., Eliason, S. I., Harper, S. Q., Martins, I. H., Orr, H. T., Paulson, H. L., Yang, L., Kotin, R. M., and Davidson, B. I. (2004). RNAi suppresses polyglutamine-induced neurodegeneration in a model of spinocerebellar ataxia. *Nat. Med.* **10**, 816–820.
20. Ichise, T., Kano, M., Hashimoto, K., Yanagihara, D., Nakao, K., Shigemoto, R., Katsuki, M., and Aiba, A. (2000). mGluR1 in cerebellar Purkinje cells essential for long-term depression, synapse elimination, and motor coordination. *Science* **288**, 1832–1835.

14

Mitochondrial Genetic Diseases

Alice Wong, PhD
Gino Cortopassi, PhD

Keywords: *Dominant Optic Atrophy; Friedreich's ataxia; Kearns-Sayre syndrome/chronic progressive external ophthalmoplegia, Leber's hereditary optic neuropathy; Mitochondrial Encephalomyopathy, Lactic Acidosis, and Strokelike episodes; Myoclonic Epilepsy and Ragged Red Fibers; mitochondria; Neuropathy, Ataxia, and Retinitis pigmentosa*

I. Mutations in Mitochondrial Protein-Encoding Genes
II. Mutations in Mitochondrial Protein Synthesis Genes
III. Defective Mitochondrial Function Caused by Nuclear DNA Mutations
References

The concept of mitochondrial disorders was first introduced in 1962 [1]. Biochemical assays during the 1970s provided evidence for defects in mitochondrial metabolism, including substrate transport, substrate utilization, Krebs cycle, electron transport chain, and oxidation–phosphorylation coupling. It was not until 1988, however, that the first mitochondrial DNA (mtDNA) point mutations and deletions were demonstrated to cause disease [2,3], defining the first mitochondrial genetic diseases. Mitochondrial genetic diseases are defined as those caused by mutations in mtDNA, or in nuclear-encoded genes targeted to the mitochondria.

Between 1988 and 2004, 118 mitochondrial mutations associated with disease were recognized, with new mutations literally discovered on a monthly basis. Nuclear DNA mutations affecting mitochondrial function were also discovered [4–6] in the mid-1990s.

In this chapter, we present a brief overview of the major diseases that result from mtDNA mutations, divided into mutations that affect mitochondrial protein-encoding genes and those that affect mitochondrial protein synthesis.

I. Mutations in Mitochondrial Protein-Encoding Genes

A. Leber's Hereditary Optic Neuropathy

Leber's hereditary optic neuropathy (LHON) results from three primary point mutations occurring at nt-11778, nt-3460, or nt-14484 in the mitochondrial genome. These mutations occur in subunits ND4, ND1, and ND6 of complex I in the respiratory chain, respectively. The signs of LHON include subacute bilateral blindness, with onset in early adulthood. Blindness is the result of degeneration of the optic nerve, which is thought to originate at the retinal ganglion cell layer. Men are affected more than women, in an approximate 3:1 ratio. Patients with LHON are usually homoplasmic for the mutation, that is, the mutation occurs in 100% of mtDNA. There are usually no other clinical signs in patients with LHON. Strikingly, there are incontrovertible demonstrations of recovery from this disease, most often for the "weaker" 14484 mutation [7].

Multiple hypotheses for an LHON pathogenic mechanism have been advanced, but none is universally accepted. The earliest hypothesis proposed was a bioenergetic one, and studies using patient cells or cybrids (cells that contain the same nuclear genome but mtDNA from a patient's sample) demonstrated that LHON mutations result in decreased oxygen consumption in whole mitochondria supplied with complex I substrates. In addition, data demonstrate that LHON mutations cause a decrease in complex I–driven adenosine triphosphate (ATP) synthesis [8]. A potential problem with a simple bioenergetic hypothesis, however, is that isolated complex I activity, assayed in submitochondrial particles, varies considerably among LHON mutations [7] and is not decreased in some studies of LHON.

Several other hypotheses have been proposed for LHON pathophysiology, including the involvement of apoptosis and reactive oxygen species (ROS). LHON cells undergo cell death when grown in galactose media [9] and are also more susceptible to Fas-induced apoptosis [10]. These cells are resistant to both rotenone and other quinols, which are thought to bind at the "quinone-binding pocket," suggesting that all three primary LHON mutations most likely interfere with the interaction of complex I with the ubiquinone substrate coenzyme Q (CoQ) [11]. Increased superoxide production at complex I has been demonstrated by inhibition of complex I with rotenone [12] and LHON mutations in complex I in a neural environment generate increased superoxide [13]; thus, increased ROS might be an important neurodegenerative mechanism in LHON [7].

B. Neuropathy, Ataxia, and Retinitis Pigmentosa

T-to-G or T-to-C mutations at nt-8993 of the ATPase6 gene cause neuropathy, ataxia, and retinitis pigmentosa (NARP). The clinical signs of NARP include retinitis pigmentosa, dementia, seizures, ataxia, proximal muscle weakness, and sensory neuropathy [5]. The T-to-G mutation is more clinically severe [4]. NARP occurs when greater than 90% of the mutant mtDNA is present. Both ATP synthesis and the assembly of ATP synthase (F1F0-ATPase) are reduced in NARP; however, ATP hydrolysis is not affected. It has been proposed that NARP mutations cause a defect in proton flow through the F0 portion [11,14]. Fibroblasts and cybrids with the NARP mutation have increased ROS production that may lead to impaired oxidative phosphorylation and ATP synthesis. Further, partial rescue of ATP synthesis and respiration occurs with the addition of antioxidants [15].

II. Mutations in Mitochondrial Protein Synthesis Genes

A. Kearns-Sayre Syndrome/Chronic Progressive External Ophthalmoplegia/Progressive External Ophthalmoplegia

Deletions of the mtDNA cause Kearns–Sayre syndrome (KSS) and chronic progressive external ophthalmoplegia (CPEO), which consist of muscle weakness, lactic acidosis, pigmentary degeneration of the retina, and cardiac conduction defects. Ataxia, deafness, limb-girdle myopathy, diabetes mellitus, and dementia can also occur, though more rarely. KSS and PEO are probably the same disorder but differ in the degree of severity. Individuals with KSS are heteroplasmic, as complete homoplasmy for a mtDNA deletion would be lethal. The severity of the disease depends on the distribution and amount of wild-type and mutant mtDNA [16]. In the most common deletion of 4977 bp, seven protein-encoding genes (four from complex I) and five tRNA genes are lost. The diagnosis of KSS is usually supported by the presence of ragged red fibers in muscle biopsy sections stained with Gomori trichrome, with positive succinate dehydrogenase staining, and/or negative cytochrome oxidase staining.

Several biochemical defects have been characterized in muscle tissue from patients with KSS or cybrids heteroplasmic for the mutation. In muscle tissue, complex I deficiencies have been described, as well as decreased cytochrome c oxidase activity. Cybrids with greater than 65% 4977-bp mtDNA deletion have increased oxidative stress and reduced mitochondrial membrane potential. In addition, decreased ATP synthesis and reduced ATP/adenosine diphosphate (ADP) ratio was demonstrated in cybrids with more than 55% deletion [17]. Because tRNA and rRNAs are nearly always deleted, defects in mitochondrial protein synthesis are observed in KSS/CPEO.

B. Myoclonic Epilepsy and Ragged Red Fibers

Myoclonic epilepsy and ragged red fibers (MERRF) is caused in 80% of cases by an A-to-G mutation in the tRNA^{Lys} gene (*MTTK*) at position nt-8344 of the mtDNA. Two other mutations, 8356T-to-C and 8363G-to-A, have been found in the same gene but occur infrequently. MERRF is characterized by myoclonus, generalized epilepsy, ataxia, and ragged red fibers in muscle biopsy. The disease starts after the second or third decade of life and clinical presentations are extremely variable, even among family members. The most common clinical symptoms include myopathy, neuropathy, hearing loss, dementia, short stature, and optic atrophy. Other clinical presentations include cardiomyopathy, pigmentary retinopathy, pyramidal signs, ophthalmoparesis, multiple lipomas, and diabetes mellitus. Patients are heteroplasmic for the mutation, possessing both wild-type and mutant alleles and a threshold amount is needed for the disease to develop clinically. Proportions of mutant mtDNA can vary in tissues from the same patient with MERRF and among family members with MERRF [18]. Muscle biopsy material from patients shows reduced respiratory chain activities [19]. Cybrids that contain more than 10% wild-type mtDNA have normal respiratory rates, consistent with results from a study demonstrating that clinical symptoms appeared if mutant mtDNA was greater than 85% [17]. In MERRF, the result of a tRNALysine mutation, there is a good correlation between lysine content and translation impairment, suggesting a straightforward disease of mitochondrial protein synthesis [20]. The mechanism appears to include both an amino-acylation deficiency explanation [20], which has been disputed (see reference 21), and a wobble modification defect that disturbs the codon–anticodon interaction [22], resulting in decreased mitochondrial translation. In MERRF, cybrid studies show decreased respiratory chain activities, presumably as a result of the reduced protein synthesis [23]. ROS overproduction may be associated with MERRF, as increased antioxidant enzyme activities have been reported [24,25].

C. Mitochondrial Encephalomyopathy, Lactic Acidosis, and Strokelike Episodes

An (A-to-G) nt-3243 in the tRNA$^{Leu(UUR)}$ gene causes mitochondrial encephalomyopathy, lactic acidosis, and stroke-like episodes (MELAS), although other mutations also cause MELAS. MELAS occurs when more than 85% of the mutant mtDNA is present [17]. MELAS presents in young adults after normal development with clinical signs that include recurrent vomiting, migraine-like headache, and strokelike episodes causing cortical blindness, hemiparesis, or hemianopia.

Cybrids with the 3243 mtDNA mutation show severe respiratory deficiency, along with reduced synthesis of mitochondrial translation products. Other biochemical defects include decreased mitochondrial membrane potential and reduced amino acylation [17,21]. However, unlike in MERRF, there is no strict correlation between leucine content and translation impairment [26]. Inhibition of mitochondrial translation may be caused by dimerization of the mutant tRNA$^{Leu(UUR)}$ [27] or by reduced association of mRNA with mitochondrial ribosomes [26]. Overproduction of ROS may be associated with this disease as well, as increased ROS production and increased antioxidant enzyme activities have been demonstrated [24,25].

III. Defective Mitochondrial Function Caused by Nuclear DNA Mutations

Nuclear gene defects have been organized according to their function: genes encoding structural components of the mitochondrial respiratory chain (NDUF, SDHA); genes encoding assembly proteins for the mitochondrial respiratory chain (BCS1L, USRF1, SCO1, SCO2, COX10, COX15, ATP12); genes involved in mitochondrial proteins (TIMM8A, HSP60); genes affecting the stability or maintenance of mtDNA (ANT1, TP, POLG, Twinkle, TK2, dGKz); and genes indirectly involved in mitochondrial function (TAZ). This discussion focuses on two nuclear-encoded genes: OPA1, which causes a similar endpoint pathology to LHON, and frataxin, a mitochondrial protein whose precise function remains unknown.

A. Dominant Optic Atrophy

Mutations in OPA1, a mitochondrial dynamin-related guanosine triphosphatase (GTPase), have been demonstrated in patients with dominant optic atrophy (DOA), one of the most frequently occurring optic neuropathies [28]. This results in variable visual loss, resulting from degeneration of retinal ganglion cells (RGCs) and loss of myelin and nerve tissue in the optic nerve. The pathological process is similar to that found in LHON. The gene encodes a protein that localizes to the mitochondria. Studies in yeast suggest that OPA1 may be involved in maintenance of mtDNA and the mitochondrial network. In human cells, the protein localizes in the mitochondrial inner membrane space, anchored to the cristae inner membrane. Knockdown of OPA1 by siRNA results in mitochondrial network fragmentation and inner membrane perturbation in HeLa cells, suggesting a role for OPA1 in mitochondrial fusion. Although OPA1 is ubiquitously expressed, RGCs appear to be the only tissue affected in patients with DOA. Patients with DOA have reduced oxidative phosphorylation ability [29], similar to results from LHON cybrids, which have reduced ATP synthesis given complex I substrates [8].

B. Friedreich's Ataxia

Friedreich's ataxia (FRDA) results from an abnormal GAA triplet repeat expansion in the first intron of the frataxin gene, resulting in decreased frataxin expression. FRDA is the most common recessive inherited ataxia, with a frequency of 1 in 50,000. The disease is an autosomal-recessive degenerative disorder characterized by progressive gait and limb ataxia, distal sensory loss, loss of limb deep tendon reflexes, extensor–plantar responses, and hypertrophic cardiomyopathy. Neuropathologically, there is degeneration of large sensory neurons in the dorsal root ganglia. Normal GAA expansion ranges are between 8 and 22 repeats, and symptomatic patients have 100–1700 repeats. There is a correlation between the size of the expansion and the severity and age at onset of clinical symptoms, with longer expansions associated with younger age at onset and greater severity of clinical expression. Point mutations instead of repeats have been reported in some patients [30].

The FRDA gene encodes for the protein frataxin, which localizes to the mitochondria and has the highest levels of expression in the heart and spinal cord. The function of frataxin is still unknown, but studies have suggested several possibilities that could result in defective mitochondrial function, including reduced ATP synthesis. The first studies to suggest an ROS hypothesis came from knockout of the yeast frataxin homolog, which resulted in increased ROS production and mitochondrial iron. A reduction in aconitase (an ROS-sensitive enzyme) activity has been demonstrated, and postmortem cardiac muscle samples from patients with FRDA revealed reduced mitochondrial complexes I, II, and III activities. In support of an ROS hypothesis, overexpression of human frataxin causes increased activation of glutathione peroxidase and reduced thiols, suggesting that frataxin may assist in the detoxification of ROS [30].

Frataxin may act as a chaperone for Fe(II) and a storage compartment for iron. Frataxin may have a role in iron export, Fe-S cluster assembly, and heme biosynthesis. Complete frataxin knockdown is lethal at the embryonic stage, but selective inactivation causes neurological symptoms and cardiomyopathy. Decreased mitochondrial Fe-S cluster containing enzyme deficiencies have been reported, as well as accumulation of mitochondrial iron. A 230-GAA expansion in the first intron of the mouse frataxin gene results in a mild phenotype [30].

References

1. Luft, R., Ikkos, D., Palmieri, G., Ernster, L., and Afzelius, B. (1962). A case of severe hypermetabolism of nonthyroid origin with a defect in the maintenance of mitochondrial respiratory control: a correlated clinical, biochemical, and morphological study. *J. Clin. Invest.* **41**, 1776–1804.
2. Wallace, D. C., Singh, G., Lott, M. T., Hodge, J. A., Schurr, T. G., Lezza, A. M., Elsas, L. J. II, and Nikoskelainen, E. K. (1988). Mitochondrial DNA mutation associated with Leber's hereditary optic neuropathy. *Science* **242**, 1427–1430.
3. Holt, I. J., Harding, A. E., and Morgan-Hughes, J. A. (1988). Deletions of muscle mitochondrial DNA in patients with mitochondrial myopathies. *Nature* **331**, 717–719.
4. DiMauro, S. (2004). Mitochondrial diseases. *Biochim. Biophys. Acta* **1658**, 80–88.
5. Jacobs, H. T., and Turnbull, D. M. (2005). Nuclear genes and mitochondrial translation: a new class of genetic disease. *Trends Genet.* **21**, 312–314.
6. Zeviani, M., and Carelli, V. (2003). Mitochondrial disorders. *Curr. Opin. Neurol.* **16**, 585–594.
7. Carelli, V., Rugolo, M., Sgarbi, G., Ghelli, A., Zanna, C., Baracca, A., Lenaz, G., Napoli, E., Martinuzzi, A., and Solaini, G. (2004). Bioenergetics shapes cellular death pathways in Leber's hereditary optic neuropathy: a model of mitochondrial neurodegeneration. *Biochim. Biophys. Acta.* **1658**, 172–179.
8. Baracca, A., Solaini, G., Sgarbi, G., Lenaz, G., Baruzzi, A., Schapira, A. H., Martinuzzi, A., and Carelli, V. (2005). Severe impairment of complex I-driven adenosine triphosphate synthesis in Leber hereditary optic neuropathy cybrids. *Arch. Neurol.* **62**, 730–736.
9. Ghelli, A., Zanna, C., Porcelli, A. M., Schapira, A. H., Martinuzzi, A., Carelli, V., and Rugolo, M. (2003). Leber's hereditary optic neuropathy (LHON) pathogenic mutations induce mitochondrial-dependent apoptotic death in transmitochondrial cells incubated with galactose medium. *J. Biol. Chem.* **278**, 4145–4150.
10. Danielson, S. R., Wong, A., Carelli, V., Martinuzzi, A., Schapira, A. H., and Cortopassi, G. A. (2002). Cells bearing mutations causing Leber's hereditary optic neuropathy are sensitized to Fas-induced apoptosis. *J. Biol. Chem.* **277**, 5810–5815.
11. Lenaz, G., Baracca, A., Carelli, V., D'Aurelio, M., Sgarbi, G., and Solaini, G. (2004). Bioenergetics of mitochondrial diseases associated with mtDNA mutations. *Biochim. Biophys. Acta.* **1658**, 89–94.
12. Genova, M. L., Ventura, B., and Giuliano, G. (2001). The site of production of superoxide radical in mitochondrial Complex I is not a bound ubisemiquinone but presumably iron-sulfur cluster N2. *FEBS Lett.* **505**, 364–368.
13. Wong, A., Cavelier, L., Collins-Schramm, H. E., Seldin, M. F., McGrogan, M., Savontaus, M. L., and Cortopassi, G. A. (2002). Differentiation-specific effects of LHON mutations introduced into neuronal NT2 cells. *Hum. Mol. Genet.* **11**, 431–438.
14. Schon, E. A., Santra, S., and Pallotti, F. (2001). Pathogenesis of primary defects in mitochondrial ATP synthesis. *Semin. Cell Dev. Biol.* **12**, 441–448.
15. Mattiazzi, M., Vijayvergiya, C., Gajewski, C. D., DeVivo, D. C., Lenaz, G., Wiedmann, M., and Manfredi, G. (2004). The mtDNA T8993G (NARP) mutation results in an impairment of oxidative phosphorylation that can be improved by antioxidants. *Hum. Mol. Genet.* **13**, 869–879.
16. Moraes, C. T., Sciacco, M., Ricci, E., Tengan, C. H., Hao, H., Bonilla, E., Schon, E. A., and DiMauro, S. (1995). Phenotype-genotype correlations in skeletal muscle of patients with mtDNA deletions. *Muscle Nerve* **3**, S150–S153.
17. McKenzie, M., Liolitsa, D., and Hanna, M. G. (2004). Mitochondrial disease: mutations and mechanisms. *Neurochem. Res.* **29**, 589–600.
18. Shoffner, J. M., and Wallace, D. C. (1992). Mitochondrial genetics: principles and practice. *Am. J. Hum. Genet.* **51**, 1179–1186.
19. Wallace, D. C., Zheng, X. X., Lott, M. T., Shoffner, J. M., Hodge, J. A., Kelley, R. I., Epstein, C. M., and Hopkins, L. C. (1988). Familial

mitochondrial encephalomyopathy (MERRF): genetic, pathophysiological, and biochemical characterization of a mitochondrial DNA disease. *Cell* **55**, 601–610.
20. Enriquez, J. A., Chomyn, A., and Attardi, G. (1995). MtDNA mutation in MERRF syndrome causes defective aminoacylation of tRNA(Lys) and premature translation termination. *Nat. Genet.* **10**, 47–55.
21. Borner, G. V., Zeviani, M., Tiranti, V., Carrara, F., Hoffmann, S., Gerbitz, K. D., Lochmuller, H., Pongratz, D., Klopstock, T., Melberg, A., Holme, E., and Paabo, S. (2000). Decreased aminoacylation of mutant tRNAs in MELAS but not in MERRF patients. *Hum. Mol. Genet.* **9**, 467–475.
22. Yasukawa, T., Suzuki, T., and Ishii, N. (2001). Wobble modification defect in tRNA disturbs codon-anticodon interaction in a mitochondrial disease. *EMBO J.* **20**, 4794–4802.
23. Chomyn, A. (1998). The myoclonic epilepsy and ragged-red fiber mutation provides new insights into human mitochondrial function and genetics. *Am. J. Hum. Genet.* **62**, 745–751.
24. Vives-Bauza, C., Gonzalo, R., and Manfredi, G. (2006). Enhanced ROS production and antioxidant defenses in cybrids harboring mutations in mtDNA. *Neurosci. Lett.* **391**(3), 136–141.
25. Wei, Y. H., Lu, C. Y., and Wei, C. Y. (2001). Oxidative stress in human aging and mitochondrial disease-consequences of defective mitochondrial respiration and impaired antioxidant enzyme system. *Chin. J. Physiol.* **44**, 1–11.
26. Chomyn, A., Enriquez, J. A., Micol, V., Fernandez-Silva, P., and Attardi, G. (2000). The mitochondrial myopathy, encephalopathy, lactic acidosis, and stroke-like episode syndrome-associated human mitochondrial tRNALeu(UUR) mutation causes aminoacylation deficiency and concomitant reduced association of mRNA with ribosomes. *J. Biol. Chem.* **275**, 19198–19209.
27. Wittenhagen, L. M., and Kelley, S. O. (2002). Dimerization of a pathogenic human mitochondrial tRNA. *Nat. Struct. Biol.* **9**, 586–590.
28. Kamei, S., Chen-Kuo-Chang, M., Cazevieille, C., Lenaers, G., Olichon, A., Belenguer, P., Roussignol, G., Renard, N., Eybalin, M., Michelin, A., Delettre, C., Brabet, P., and Hamel, CP. (2005). Expression of the Opa1 mitochondrial protein in retinal ganglion cells: its downregulation causes aggregation of the mitochondrial network. *Invest. Ophthalmol. Vis. Sci.* **46**, 4288–4294.
29. Lodi, R., Tonon, C., Valentino, M. L., Iotti, S., Clementi, V., Malucelli, E., Barboni, P., Longanesi, L., Schimpf, S., Wissinger, B., Baruzzi, A., Barbiroli, B., and Carelli, V. (2004). Deficit of *in vivo* mitochondrial ATP production in OPA1-related dominant optic atrophy. *Ann. Neurol.* **56**, 719–723.
30. Calabrese, V., Lodi, R., Tonon, C., D'Agata, V., Sapienza, M., Scapagnini, G., Mangiameli, A., Pennisi, G., Stella, A. M., and Butterfield, D. A. (2005). Oxidative stress, mitochondrial dysfunction and cellular stress response in Friedreich's ataxia. *J. Neurol. Sci.* **233**, 145–162.

15

Paraneoplastic Neurological Disorders

Steven Vernino, MD, PhD
Josep Dalmau, MD, PhD

Keywords: *antibody, autoimmunity, cancer, cerebellar ataxia, cytotoxic T cells, ion channels, limbic encephalitis, lymphocytes, myasthenia, thymoma*

I. Introduction
II. History and Nomenclature
III. Etiology and Pathogenesis
IV. Pathophysiology
V. Symptoms and Natural History
VI. Summary
   References

I. Introduction

Paraneoplastic syndromes are remote complications of cancer that cannot be attributed to direct effects of the neoplasm or its metastases. Broadly defined, paraneoplastic complications might include common problems that occur in patients with cancer such as fatigue, opportunistic infections, and side effects of chemotherapy and radiation. Other common paraneoplastic syndromes are related to inappropriate secretion of hormone-like compounds by the tumor, leading to hypercalcemia, hyponatremia, hyperadrenalism, or cancer-associated coagulopathy. The diagnosis of a paraneoplastic neurological disorder (PND), however, is generally reserved for a disorder of the central or peripheral nervous system where other causes (especially toxic, infectious, and metabolic causes) have been excluded. Ideally, the diagnosis should be reserved for patients with a histologically proven cancer. However, certain clinical scenarios, imaging findings, and laboratory study results make the diagnosis of PND highly probable even when the presence of cancer cannot be proven.

It is generally accepted that PNDs have an autoimmune etiology, but the details of the pathogenesis are not fully known. Further, the pathophysiology appears to differ from one disorder to another. An immunological pathogenesis readily explains many of the clinical and laboratory features, including the subacute time course, characteristic autoantibodies, tissue histopathology, and the highly specific targeting of individual neuronal populations in some cases. Overall, PNDs are quite rare (estimated at 0.1% of patients with cancer) [1]. However, studies of these rare disorders have advanced our understanding of natural tumor immunity and of autoimmune neurological disorders in general.

II. History and Nomenclature

PNDs were first clinically recognized more than 50 years ago with descriptions of sensory neuronopathy, limbic encephalitis, and cerebellar degeneration in patients with cancer. It is now recognized that any part of the nervous system can be affected (Table 1), and often multiple levels of the nervous system are involved simultaneously. For any clinical presentation, however, similar or identical clinical syndromes may occur in the absence of cancer due to idiopathic neurological autoimmunity or other causes. Some presentations of PND may be indistinguishable from common neurological disorders (e.g., peripheral neuropathy, myasthenia gravis, motor neuron disease, and myelitis). In these cases, the association with cancer may be underrecognized because of lack of clinical suspicion.

Because PND can affect any part of the nervous system, the symptoms and signs of the disease can be quite varied. A few clinical features are typical. Most PNDs have a subacute and progressive course with the onset of neurological symptoms over weeks or months. The age at onset has a wide range, with a median at about 65 years. In North America, similar to most other autoimmune disorders, there is a strong female predominance (about 2:1) even when cases of sex-specific tumors (breast, ovary, and testes) are not considered [2]. In most cases, the neurological presentation precedes the diagnosis of cancer. In patients with a history of cancer, the neurological illness may herald cancer recurrence. The tumors, when found, tend to be histologically malignant but limited in stage [3]. Spread to regional lymph nodes is common, but widespread metastasis is unusual at the time of diagnosis.

Certain tumor types are more likely to be associated with PND. About 30% of patients with thymoma have myasthenia gravis or some other form of neurological autoimmunity [4]. Small cell carcinoma, most commonly arising in the lung (small cell lung carcinoma [SCLC]), is associated with one or more PND in up to 3% of cases [1]. PND also occurs regularly in association with gynecological malignancies (breast, ovary, fallopian tube, or peritoneal primary), Hodgkin's and non-Hodgkin's lymphoma, testicular cancer, and neuroblastoma. These various malignancies have some features that probably contribute to the association with paraneoplastic disorders. First, these cancer cells have the capacity to aberrantly express proteins that are usually restricted to the nervous system. Second, some of these tumors (especially ovary, breast, and SCLC) frequently spread to regional lymph nodes, which may promote early and more vigorous recognition by the immune system. In testicular tumors, the antigen presentation may occur in testis in cases of "carcinoma *in situ*" or intratubular germ cell tumor [5]. In the case of thymoma and lymphoma, autoimmunity may reflect a primary disturbance in immune regulation because the malignant cells derive from components of the immune system.

III. Etiology and Pathogenesis

PND appears to represent an aberrant manifestation of an effective antitumor immune response. Patients with PND have a favorable cancer survival compared with those with tumors of similar stage and grade without PND [3,6]. Several clinicopathological studies have shown that tumors that are heavily infiltrated with lymphocytes have a better prognosis [7]. There are, therefore, two aspects of the pathogenesis of PND: (1) the generation of a vigorous immune

Table 1 Paraneoplastic Neurological Syndromes

Brain	
	Cerebellar degeneration
	Limbic encephalitis
	Brainstem encephalitis
	Opsoclonus-myoclonus
	Chorea
Eye and cranial nerve	
	Optic neuritis
	Retinal degeneration
	Subacute hearing loss
Spinal cord	
	Myelopathy
	Stiff-person and "stiff-limb" syndromes
	Motor neuronopathy
Nerve	
	Sensory neuronopathy (pure sensory neuropathy)
	Sensorimotor peripheral neuropathy
	Polyradiculoneuropathy
	Mononeuritis multiplex
	Autonomic neuropathy, gastrointestinal dysmotility
Neuromuscular Junction/Muscle	
	Lambert-Eaton myasthenic syndrome
	Myasthenia gravis
	Dermatomyositis
	Neuromyotonias

response to cancer and (2) a break in immune tolerance that allows this response to target the nervous system.

Although antibodies may contribute, effective tumor immunity probably relies predominantly on a cytotoxic (CD8+) T-lymphocyte (CTL) response against the cancer cells [8,9]. CTLs express a T-cell receptor, which recognizes antigenic peptides bound to surface major histocompatibility complex (MHC) class I molecules. The target peptides derive from the intracellular degradation of cellular proteins. Nearly all cells of the body express MHC class I molecules, but the immune system is normally "tolerized" (not reactive) to the repertoire of proteins expressed by normal cells. Cells that begin to express novel or unrecognized proteins (because of viral infection or malignant transformation) become potential targets for CTLs. Antigen-presenting cells (APCs), including dendritic cells, are responsible for processing and presenting peptides to lymphocytes. The precise mechanism by which cancer antigens are presented to the immune system is an area of active investigation but probably requires the participation of APCs. Cancer antigens may be liberated into the extracellular compartment by necrosis of cancer cells or presented to lymphocytes directly by cancer cells that have traveled to regional lymph nodes. Most likely, the relevant antigenic proteins liberated by apoptotic and necrotic cancer cells are taken up by dendritic cells, which then migrate to regional lymph nodes to interact with CD4+ and CD8+ T cells. These important processes are described in greater detail elsewhere [8,9].

The activation of the CTL response also requires signals from helper (CD4+) cells, which have been activated through recognition of similar peptide antigens. Without these costimulatory signals from helper T cells, presentation of antigens to CD8+ T cells will result in tolerance. In addition to participating in the CTL response, the activation of helper T cells also provides a signal to antigen-specific B cells to differentiate into antibody-producing plasma cells. Through these mechanisms, a cancer cell is identified because of its abnormal expression of cellular proteins. In response, the immune system generates antibodies against these proteins, as well as activated CTLs, which can recognize peptide antigens bound to MHC class I on the surface of the tumor cell [8]. If foreign peptide antigens (viral or neoplastic) bound to MHC class I are recognized by the activated CTLs, the cell is killed either by enzymatic lysis (via release of perforin and granzyme) or by induction of apoptosis (via Fas/Fas-ligand signaling).

Under ideal circumstances, the antitumor immune response should specifically target the cancer cells and leave nonmalignant cells intact. However, neurons do not normally express MHC class I, so proteins that are exclusively restricted to neurons may be immunologically privileged (i.e., not normally exposed to the immune system for the development of tolerance). Some tumor cells inappropriately express onconeural proteins (cancer proteins that are ordinarily restricted to the nervous system) [8,10]. A prime example is SCLC, a tumor of neuroendocrine lineage that can express neuronal nuclear, cytoplasmic, or membrane proteins including neurotransmitter receptors and voltage-gated ion channels [10,11]. The immune response against such tumors may target these proteins as novel antigens. Whether this occurs will depend on many factors including the genetic determinants that influence the repertoire of the T-cell receptor and the binding properties of the human leukocyte antigen molecules. Many patients with PND have a personal or family history of other autoimmune conditions, which may reflect a genetic predisposition to autoimmunity. Although the generation of immunity against onconeural antigens is part of the antitumor immune response, it does not always lead to PND. A few studies have shown that paraneoplastic autoantibodies may be found in 20–40% of patients with SCLC or ovarian cancer without neurological symptoms and that the presence of the antibodies is associated with a favorable oncological status [12,13]. Similarly, antibodies against various neuronal and muscle ion channels can be found in patients with thymoma even in the absence of neurological disease [4]. Despite laboratory evidence of antitumor immunity, as well as the aforementioned studies, most clinical series question whether the T cells are effective against the tumor, and, if so, whether the antitumor effect is sustained enough to be clinically efficient [14].

IV. Pathophysiology

Once an immune response to cancer involving neuronal antigens has been established, neurological disease can arise in one of two ways related to the two arms of the immune response. PND can be divided into two groups based on the type of associated autoantibodies. One group of antibodies that may be associated with PND are antibodies against cell surface antigens, including neuronal ion channels (Table 2). Ion channel antibodies are usually very sensitive and quite specific for a particular neurological disorder, but not specific for paraneoplastic disease. Antibodies against neuronal P/Q-type voltage-gated calcium channels are found in nearly all patients with the Lambert-Eaton myasthenic syndrome, but only about 60% of adult patients with Lambert-Eaton syndrome have cancer [15]. Antibodies against muscle acetylcholine receptors are found in more than 80% of patients with myasthenia gravis, but only 15% have thymoma. Hence, the antibody is a marker of the disease but not a marker of cancer. Other PNDs associated with ion channel antibodies include neuromyotonia, autoimmune autonomic neuropathy, limbic encephalitis, and Morvan's syndrome.

Autoantibodies have direct access to ion channels because these are integral membrane proteins with an extracellular domain. The antibodies bind and interfere with

Table 2 Antibodies against Ion Channels (Not Specific for Paraneoplastic Disease)

Antibody	Antigen(s)	Tumor	Associated Syndromes
Ion channel antibodies			
Muscle AChR	Muscle acetylcholine receptor	Thymoma (15%)	Myasthenia gravis
Ganglionic AChR	Neuronal acetylcholine receptor	SCLC (15%)	Autonomic neuropathy
P/Q-type VGCC[a]	Neuronal Ca^{2+} channel	SCLC (60%)	Lambert-Eaton syndrome
VGKC	Neuronal K^+ channels	Thymoma or SCLC	Neuromyotonia; limbic encephalitis
MGluR1	Metabotropic glutamate receptor	Hodgkin's lymphoma	Paraneoplastic cerebellar degeneration

[a]In patients with cerebellar degeneration, with or without associated Lambert-Eaton myasthenic syndrome, detection of P/Q voltage-gated calcium channels should prompt the search of a small cell lung cancer.

proper function or cause internalization and degradation of these channels. In many cases, the antibodies have been proven to be directly pathogenic; injection of purified antibodies into experimental animals reproduces key features of the disease. Removal of the antibodies using plasmapheresis improves the clinical symptoms. The antibody-mediated PNDs predominantly affect the peripheral nervous system. This may reflect the limited ability for serum antibodies to cross the blood-brain barrier in large quantities.

The second and larger group consists of antibodies directed against intracellular antigens in the nucleus or cytoplasm of neurons (Table 3). In some cases, the protein antigen has been definitively identified and characterized. These include proteins involved in nuclear RNA and DNA binding and proteins involved in synaptic function. In other cases, the identity of the protein antigen remains unknown, and the antibody is defined descriptively based on the pattern of immunohistochemical staining of brain sections and Western blots. In most cases, these antibodies recognize antigens both in neurons and in tumor cells, but their role in the pathogenesis of disease is unclear. It seems unlikely that the antibodies are directly pathogenic because their specific antigens are intracellular and not readily accessible in living neurons. Passive transfer of the autoantibodies into animals or immunization of animals against these antigens fails to reproduce neurological symptoms. In vitro studies showing direct cytotoxic effects of paraneoplastic antibodies are quite limited. Furthermore, the antibodies often bind to a wide variety of neurons even though the associated clinical syndromes are usually restricted to a specific region or cell type of the nervous system. The anti-Hu antibody (also known as *ANNA-1*), for example, binds to the nuclei of all neurons but is most commonly associated with limbic encephalitis or neuropathy [2,10,14].

Instead, the antibodies are important surrogate markers of the specific immune response to cancer where CTLs

Table 3 Neuronal Autoantibodies Associated with Paraneoplastic Disorders

Antibody	Antigen(s)	Tumor	Associated Syndromes
Well-characterized paraneoplastic antibodies[a]			
Anti-Hu ANNA-1	*HuD, HuC, Hel-N1*	SCLC	Encephalomyelitis, sensory neuronopathy, autonomic and sensorimotor neuropathies, PCD
CRMP-5 anti-CV2	*CRMP-5* (66 kDa)	SCLC or thymoma	Encephalomyelitis, chorea, neuropathy, optic neuropathy
Anti-Yo PCA-1	*CDR34* and *CDR62*	Ovarian or breast cancer	PCD
Anti-Ma2[b]	*Ma2*	Lung or testicular	Limbic and brainstem encephalitis
Amphiphysin	*Amphiphysin*	Lung or breast cancer	Encephalomyelitis, neuropathy, "stiff-person syndrome"
Anti-Ri ANNA-2	*Nova*	Lung or breast cancer	Ataxia, opsoclonus-myoclonus, neuropathy, brainstem encephalitis
Recoverin	*Recoverin*	SCLC	Cancer-associated retinopathy
Partially characterized paraneoplastic antibodies			
Tr		Hodgkin's disease	PCD
Zic4		SCLC	PCD
PCA2		SCLC	Several
ANNA3		SCLC	Several

Note: PCD, paraneoplastic cerebellar degeneration; SCLC, small cell lung carcinoma.

[a]These antibodies are listed in approximate order of decreasing clinical frequency. When alternative nomenclature exists, both names are given.
[b]The presence of Ma1 antibody usually associates with predominant brainstem, cerebellar involvement, and tumors other than testicular neoplasms. The prognosis in those cases is poorer than that of patients with Ma2 antibodies and testicular neoplasms.

are thought to be the principal effectors [8,9]. Presumably, these CTLs recognize autoantigen peptides in the context of MHC class I on both tumor cells and neurons [16]. CTLs are exquisitely sensitive to minimal antigen expression and can identify and kill target cells even before antigen can be biochemically detected [17]. Activated lymphocytes are capable of crossing the blood–brain barrier, but neurons might still be protected from CTL damage by their lack of constitutive expression of MHC class I molecules. However, evidence has shown that neurons can express these molecules in an activity-dependent manner or in response to cytokine signals. As neuronal damage occurs, liberation of neuronal antigens and inflammatory cytokines may help perpetuate autoimmunity locally in the nervous system. Similarly, muscle cells in dermatomyositis show a dramatic increase in MHC class I expression compared with normal muscle.

Consistent with a cell-mediated mechanism, the pathological findings in PND are often nonspecific and do not reflect intense inflammation. Cerebrospinal fluid analysis may be normal or may show a mild elevation of protein and a mild lymphocytic pleocytosis. In the central nervous system, the histopathological changes can resemble those seen in viral encephalitis, namely perivascular lymphocytic cuffing, small collections of tissue lymphocytes (CD8+), and microglial activation (forming microglial nodules) (Fig. 1). These findings are found in the affected neuroanatomical areas (i.e., cerebellum in paraneoplastic ataxia and mesial temporal lobe in limbic encephalitis). More dramatic, but still nonspecific, changes are seen late in the course of the disease when neuronal loss and gliosis become predominant. In paraneoplastic cerebellar degeneration, the cerebellum may show a complete loss of Purkinje cells, with little or no change in the adjacent molecular and granule cell layers [18]. In the peripheral nervous system, inflammatory infiltrates of lymphocytes are often more readily defined. These can be seen in dorsal root ganglia (in cases of paraneoplastic sensory neuronopathy) or in myenteric plexus (in cases of paraneoplastic enteric ganglionopathy).

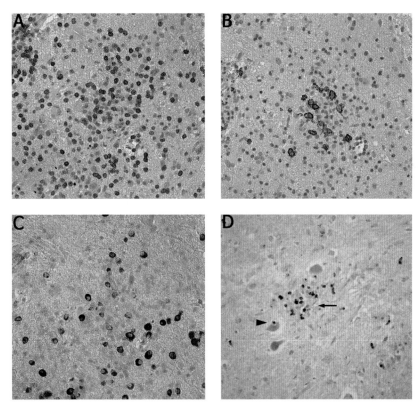

Figure 1 Neuropathological Findings Panels A, B, and C correspond to a temporal lobe biopsy of a patient with anti-Ma2–associated encephalitis and non–small cell lung cancer. (**A**) Extensive infiltrates of CD3+ T cells. (**B**) Consecutive section of the same area demonstrating infiltrates of CD20+ cells (B lymphocytes). (**C**) Additional infiltrates of CD79a+ cells (plasma cells). (**D**) A section of brainstem from the autopsy of a patient with anti-Hu–associated encephalomyelitis; the dark-brown cells correspond to T cells expressing TIA-1, an RNA-binding protein specific for activated cytotoxic T cells. Arrow indicates a neuronophagic node composed of TIA-1–positive cells; arrowhead indicates a TIA-1–positive cell indenting a neuron. Magnification: A, B, C, ×400; D, ×200. All panels have been mildly counterstained with hematoxylin.

V. Symptoms and Natural History

The symptoms of PND vary and follow from the area of the nervous system affected. The loss of Purkinje cells in paraneoplastic cerebellar degeneration results in severe cerebellar ataxia. The loss of dorsal root ganglia cells in paraneoplastic sensory neuronopathy results in profound sensory loss, sensory ataxia, and pseudoathetosis. In the cell-mediated forms of PND, neuronal loss is irreversible, and the neurological deficits are largely permanent. With early and aggressive treatment of the tumor combined with immunotherapy, one hopes to prevent further neuronal loss. There have been very few prospective trials, and no clearly effective treatments have emerged. Treatments aimed at reducing antibody levels (plasma exchange) appear to be ineffective for most paraneoplastic disorders. Suppression of cell-mediated immunity with cytotoxic agents (such as cyclophosphamide) or lymphocyte-specific drugs (such as tacrolimus and mycophenolate mofetil) may be more appropriate treatment options. Modest improvement of function may be seen in some cases—perhaps because of plasticity of remaining neurons. Unfortunately, many cases of PND continue to progress even after diagnosis and successful treatment of the underlying cancer. About half of patients ultimately die from complications of the neurological disability, whereas the other half die from their malignancy.

The situation is clearly different for the handful of antibody-mediated PNDs. Myasthenia gravis associated with thymoma tends to be more severe and refractory to treatment than nonthymomatous myasthenia gravis. When the thymoma is removed, the disease becomes easier to manage. Similarly, patients with paraneoplastic Lambert-Eaton syndrome tend to improve after cancer chemotherapy. Also, these antibody-mediated disorders are more likely to respond to traditional immunosuppressants or to immunomodulatory treatments like plasma exchange or intravenous immunoglobulin.

The tumors in patients with PND are often limited in extent and difficult to find. In up to 10% of PND cases, no tumor is ever found despite frequent and comprehensive follow-up examinations. When found, the tumors are often infiltrated with plasma cells and T lymphocytes. This reflects the vigorous antitumor immune response and may explain the relatively better cancer prognosis in patients with PND. The failure to find the tumor in some cases may represent spontaneous tumor regression mediated by immune mechanisms.

VI. Summary

Paraneoplastic neurological syndromes are rare, often devastating, autoimmune disorders affecting any part of the nervous system. PNDs highlight the delicate balance between an effective immune response against cancer and destructive neurological autoimmunity.

References

1. Darnell, R. B., and Posner, J. B. (2003). Paraneoplastic syndromes involving the nervous system. *N. Engl. J. Med.* **349**(16), 1543–1554.
2. Lucchinetti, C. F., Kimmel, D. W., and Lennon, V. A. (1998). Paraneoplastic and oncologic profiles of patients seropositive for type 1 antineuronal nuclear autoantibodies. *Neurology* **50**(3), 652–657.
3. Graus, F., Dalmou, J., Rene, R., Tora, M., Malats, N., Verschuuren, J. J., Cardenal, F., Vinolas, N., Garcia del Muro, J., Vadell, C., Mason, W. P., Rosell, R., Posner, J. B., and Real, F. X. (1997). Anti-Hu antibodies in patients with small-cell lung cancer: association with complete response to therapy and improved survival. *J. Clin. Oncol.* **15**(8), 2866–2872.
4. Vernino, S., and Lennon, V. A. (2004). Autoantibody profiles and neurological correlations of thymoma. *Clin. Cancer Res.* **10**, 7270–7275.
5. Bataller, L., Wade, D. F., Graus, F., Stacey, H. D., Rosenfeld, M. R., and Dalmau, J. (2004). Antibodies to Zic4 in paraneoplastic neurologic disorders and small-cell lung cancer. *Neurology* **62**(5), 778–782.
6. Dalmau, J., Graus, F., Rosenblum, M. K., and Posner, J. B. (1992). Anti-Hu associated paraneoplastic encephalomyelitis/sensory neuronopathy. A clinical study of 71 patients. *Medicine* **71**, 59–72.
7. Zhang, L., Conejo-Garcia, J. R., Katsaros, D., Gimotty, P. A., Massobrio, M., Regnani, G., Makrigiannakis, A., Gray, H., Schlienger, K., Liebman, M. N., Rubin, S. C., and Coukos, G. (2003). Intratumoral T cells, recurrence, and survival in epithelial ovarian cancer. *N. Engl. J. Med.* **348**(3), 203–213.
8. Roberts, W. K., and Darnell, R. B. (2004). Neuroimmunology of the paraneoplastic neurological degenerations. *Curr. Opin. Immunol.* **16**(5), 616–622.
9. Albert, M. L., and Darnell, R. B. (2004). Paraneoplastic neurological degenerations: keys to tumour immunity. *Nat. Rev. Cancer* **4**(1), 36–44.
10. Dalmau, J., Furneaux, H. M., Cordon-Cardo, C., and Posner J. B. (1992). The expression of the Hu (paraneoplastic encephalomyelitis/sensory neuronopathy) antigen in human normal and tumor tissues. *Am. J. Pathol.* **141**(4), 881–886.
11. Oguro-Okano, M., Griesmann, G. E., Wieben, E. D., Slaymaker, S. J., Snutch, T. P., and Lennon, V. A. (1992). Molecular diversity of neuronal-type calcium channels identified in small cell lung carcinoma. *Mayo Clin. Proc.* **67**(12), 1150–1159.
12. Drlicek, M., Bianchi, G., Bogliun, G., Casati, B., Grisold, W., Kolig, C., Liszka-Setinek, U., Marzorati, L., Wondrusch, E., and Cavaletti, G. (1997). Antibodies of the anti-Yo and anti-Ri type in the absence of paraneoplastic neurological syndromes: a long-term survey of ovarian cancer patients. *J. Neurol.* **244**(2), 85–89.
13. Galanis, E., Frystak, S., Rowland, K. M., Jr., Sloan, J. A., and Lennon, V. A. (1999). Neuronal autoantibody titers in the course of small-cell lung carcinoma and platinum-associated neuropathy. *Cancer Immunol. Immunother.* **48** (2-3), 85–90.
14. Bataller, L., and Dalmau, J. O. (2004). Paraneoplastic disorders of the central nervous system: update on diagnostic criteria and treatment. *Semin. Neurol.* **24**(4), 461–471.

15. Lennon, V. A., Kryzer, T. J., Griesmann, G. E., O'Suilleabhain, P. E., Windebank, A. J., Woppmann, A., Miljanich, G. P., and Lambert, E. H. (1995). Calcium-channel antibodies in the Lambert-Eaton syndrome and other paraneoplastic syndromes. *N. Engl. J. Med.* **332**(22), 1467–1474.
16. Dalmau, J., Graus, F., Cheung, N. K., Rosenblum, M. K., Ho, A., Canete, A., Delattre, J. Y., Thompson, S. J., and Posner, J. B. (1995). Major histocompatibility proteins, anti-Hu antibodies, and paraneoplastic encephalomyelitis in neuroblastoma and small cell lung cancer. *Cancer* **75**(1), 99–109.
17. Skias, D. D., Kim, D. K., Reder, A. T., Antel, J. P., Lancki, D. W., and Fitch, F. W. (1987). Susceptibility of astrocytes to class I MHC antigen-specific cytotoxicity. *J. Immunol.* **138**(10), 3254–3258.
18. Greenlee, J. E., and Brashear, H. R. (1983). Antibodies to cerebellar Purkinje cells in patients with paraneoplastic cerebellar degeneration and ovarian carcinoma. *Ann. Neurol.* **14**, 609–613.

16

Systemic Lupus Erythematosus: Descriptive Past and Mechanistic Future

Czeslawa Kowal, PhD
Cynthia Aranow, MD
Meggan Mackay, MD
Betty Diamond, MD
Bruce T. Volpe, MD

Keywords: *blood-brain barrier, brain-reactive antibodies, neuropsychiatric lupus, systemic lupus*

I. Introduction
II. Epidemiology, Etiology, and Pathogenesis
III. Neuropsychiatric SLE
IV. Neuroimaging in SLE
V. Pathogenesis of NPSLE
VI. Anti-Phospholipid Antibodies and Anti-Phospholipid Syndrome
VII. Summary
References

I. Introduction

Systemic lupus erythematosus (SLE) is an autoimmune disease of diverse and complex clinical manifestations. Patients with this disease were first identified by the presence of characteristic cutaneous manifestations; in fact, the name refers to the classic malar rash that was thought to make affected individuals resemble a wolf (*lupus* means wolf in Latin). By the mid-nineteenth century, physicians recognized that patients with the lupus rash also experienced noncutaneous manifestations of disease, such as arthritis, lymphadenopathy, subcutaneous nodules, fever, and weight loss [1]. It is clear, a century and a half later, that neuropsychiatric features are also manifestations of this systemic disease. Two physicians, Kaposi and Osler, are credited with first identifying central nervous system (CNS) features of lupus. Kaposi reported the occurrence of stupor and coma in some patients during the final stages of lupus and Osler described delirium and recurrent episodes of aphasia and hemiplegia in others. Osler additionally noted that the pathological changes in affected brain tissue were similar to those observed in the skin [2]. This chapter discusses neuropsychiatric SLE (NPSLE) emphasizing a novel pathophysiological paradigm after a brief review of disease epidemiology and current understanding of the diagnosis and pathogenetic mechanisms of NPSLE.

II. Epidemiology, Etiology, and Pathogenesis

The incidence and prevalence of SLE are significantly influenced by gender, ethnicity, and geography. The reported incidence is approximately 1.8–5.5 cases per 100,000 population, and prevalence figures range from 14.6 to 130 per 100,000 individuals. Differences in reported figures result from different methods of case ascertainment and from actual differences across ethnic and racial groups [3–5], as SLE is two to three times more common in non-Caucasians [6–8]. Although race and ethnicity have been identified as predictors of disease severity and mortality, these data are confounded by factors such as socioeconomic status and income, which have also been implicated as independent predictors of disease severity and mortality [9–11]. The disease primarily affects young women with a striking female-to-male ratio of 9:1 [3,12]. There are data to suggest that the incidence of SLE has increased during the last 4 decades, although the reasons for this increase are not known [4,13].

The American College of Rheumatology (ACR) initially established classification criteria for SLE in 1971 to facilitate research and epidemiological studies of SLE [14]. These criteria were modified in 1982 [15] and again in 1997 [16]. A patient may be classified with SLE if a minimum of four (of 11) criteria have been fulfilled (Table 1). Validation of the 1982 revised criteria by several groups has yielded a sensitivity of 78–97% and specificity of 89–99% [17–21]. Individuals fulfilling two or three criteria are defined as having "incomplete lupus" and approximately 60% of these patients will progress to SLE over a period of 5–6 years [22]. Although the components of the classification criteria are useful for identifying major clinical features of the disease (mucocutaneous, serosal, renal,

Table 1 The ACR Classification Criteria for Lupus (Revised in 1982)

1. Malar rash
2. Discoid rash
3. Photosensitivity
4. Oral ulcers
5. Arthritis
6. Serositis
7. Renal disorder
8. Neurological disorder
9. Hematological disorder
10. Immunological disorder
11. Anti-nuclear antibody

articular, and some neurological), many of the commonly seen and infrequent manifestations of the disease are not captured by the criteria. These include the frequently seen constitutional symptoms (fever, fatigue, weight loss, and lymphadenopathy), alopecia, and more unusual findings such as peripheral neuropathies, aplastic anemia, and interstitial nephritis.

Over the past several decades, survival of patients with lupus has improved, presumably because of more aggressive immunosuppressive intervention and better adjunctive therapies. However, with improved prognosis, an appreciation of the significance of other comorbid conditions such as accelerated atherosclerosis, thrombosis, and neuropsychiatric disorders has emerged [23–27].

Although clinical manifestations of SLE are diverse, a common finding in patients is the presence of autoantibodies. Two of the eleven classification criteria are positive serologic tests: a positive test for anti-nuclear antibodies (ANAs) and an immunological disorder defined as the presence of an autoantibody to Smith antigen (Sm; a ribonucleoprotein complex consisting of multiple antigenic proteins that constitute part of the spliceosome), double-stranded DNA (dsDNA), or phospholipid. Most of the autoantibodies found in lupus have specificity against nuclear antigens. These are commonly directed against dsDNA, histone, and ribonucleoproteins. The ANA test is extremely sensitive but not specific and, therefore, is useful as a screening test; however, it is important to recognize that positive ANA test results are found in other autoimmune and non-autoimmune diseases, as well as in normal individuals. In contrast, anti-Sm and anti-dsDNA antibodies are highly specific for SLE and are only rarely seen in other diseases. It has been demonstrated that autoantibodies may be present in the peripheral blood for several years before clinical diagnosis [28]. Anti-dsDNA antibodies are believed to be pathogenic because they have been implicated in tissue damage in glomerulonephritis and have been eluted from lupus kidneys [29,30]. Anti-Ro and anti-La antibodies have been associated with congenital fetal heart block, neonatal lupus, and subacute cutaneous lupus. Both Ro and La are ribonucleoproteins.

La is involved in the transcription termination of RNA polymerase III and in translation initiation of some RNA transcripts. The Ro complex consists of a 52- and 60-kDa protein and small hY RNAs, but its function remains uncertain. Because the Ro and La antigens have been identified in fetal myocardial conducting tissue and on the surface of apoptotic keratinocytes, anti-Ro and anti-La antibodies are felt to be actively involved in disease pathogenesis [31–34].

Autoantibodies in SLE may mediate organ dysfunction by several mechanisms [35]. Antibodies binding to cells mark them for elimination by the reticuloendothelial system; for example, the autoantibodies bound to red blood cells, platelets, and leukocytes are thought to be responsible for the immune-mediated hematological abnormalities of SLE. Alternatively, autoantibodies may cause organ dysfunction indirectly by immune complex deposition in tissue followed by complement fixation and subsequent activation of an inflammatory cascade. Autoantibodies also bind directly to tissue antigens and mediate tissue damage through either complement fixation or alteration of cellular function. Anti-DNA antibodies in the kidney, for example, both bind directly to cross-reactive glomerular antigens and deposit as immune complexes. Interestingly, both anti-DNA and anti-Sm autoantibodies are associated with renal disease, but titers of anti-DNA antibodies fluctuate with disease activity, whereas titers of anti-Sm antibodies do not.

Critical factors in the pathogenesis of SLE include a genetic predisposition, environmental exposures, and hormonal factors. The interaction of these elements may, in some genetically predisposed individuals, result in autoimmunity with loss of self-tolerance and production of autoantibodies. Autoantibody production may precede the development of disease expression by years or may occur, persist, and never progress to clinical disease [28]. The development of pathogenic autoimmunity requires an interplay of environmental and, perhaps, hormonal factors and genetics. Genetic and nongenetic factors not only contribute to the initiation of immune hyperreactivity and an autoreactive antibody response, but also determine disease progression and target organ vulnerability.

Antibodies are an effector mechanism of the immune system designed to protect the organism from pathogens. To defend against an extremely broad world of microbial pathogens, an individual's antibody repertoire must exhibit unparalleled diversity and this diversity must be achieved while maintaining tolerance to self-antigens. A breakdown in tolerance leads to autoimmunity. Autoreactive T and B cells are normally deleted in the thymus and bone marrow, respectively, by a process known as *negative selection,* which is an essential mechanism for the maintenance of self-tolerance [36]. Self-recognition can result from permissive or leaky negative selection of self-reactive B or T cells as they mature in the bone marrow and thymus. Both mouse models and a study of patients with lupus suggest altered selection of the B-cell repertoire with an increased frequency of autoreactive B cells maturing through "developmental checkpoints" [37]. Once B cells mature and exit the bone marrow to encounter antigen, somatic hypermutation of immunoglobulin genes occurs. Thus, an essential process in B-cell activation and memory cell formation leads to improved affinity for microbial antigens but can also generate B-cell autoreactivity. In a non-autoimmune individual, autoreactive B cells will not mature to immunocompetence and will not enter the memory B-cell compartment.

The basis for the abnormal B-cell selection in SLE and the ability of autoreactive B cells to achieve immunocompetence is not known. Multiple abnormalities of receptors, receptor signaling, cellular subsets, and cytokines have been reported, as well as enhanced expression of costimulatory molecules that mediate T-cell–B-cell interactions [35,38,39]. Once autoantibody production is initiated, engagement of toll-like receptors (TLRs) 9 and 7 in B cells by DNA- and RNA-containing immune complexes, respectively, leads to B-cell proliferation and antibody secretion [40,41].

There is a growing interest in dendritic cells (DCs) in SLE [42]. Activated DCs are efficient antigen-presenting cells (APCs) that augment an immune response, whereas nonactivated DCs promote tolerance. TLRs in these cells, as well as in B cells, are believed to be activated by immune complexes containing DNA or RNA. Once activated, the DCs migrate into inflamed tissue where they help regulate the immune response with their efficient antigen presentation and increased production of the proinflammatory cytokine interferon-α (IFN-α). The role of IFN-α in the pathogenesis of SLE has become a major focus of research. Elevated serum levels of IFN-α were noted in SLE 23 years ago [43] and, in some studies, have correlated with increased disease activity [44]. Increased transcription of IFN-α–inducible genes in peripheral blood mononuclear cells has been detected by microarray analysis, and up-regulation of these IFN-inducible genes (giving rise to the "alpha interferon signature") has been correlated with disease severity [45], activity [46], and a subset of phenotypically similar patients with SLE [47]. However, often an "alpha interferon signature" is detected in the peripheral blood mononuclear cells of patients with SLE without measurable serum levels of IFN-α [38,45,46,48–51]. The lack of correlation between the "alpha interferon signature" and elevations in serum IFN-α levels may reflect the presence of masking antibodies, the migration of IFN-α–producing DCs into tissue, possibly the activation of IFN-inducible genes by factors that operate downstream of IFN, or the technical consideration that the enzyme-linked immunosorbent assay (ELISA) for IFN-α is less sensitive than the bioassay for IFN-inducible genes.

Finally, down-regulation of an immune response is as important as negative selection in maintaining self-tolerance. There are data suggesting that lupus may be associated with a deficiency in regulatory T-cell number. There are also genetic analyses suggesting that polymorphisms of inhibitory receptors may be susceptibility factors for the development of SLE.

Many clinical clues suggest a genetic predisposition for SLE, including familial aggregation of disease, a higher prevalence in certain ethnic groups, and twin studies that demonstrate a greater concordance in monozygotic twins than in dizygotic twins. It is increasingly clear that SLE is a multigenic disease and that a number of different constellations of genes can contribute to the disease phenotype [52,53]. Murine studies with chemical mutagenesis suggest that more than 130 genes may contribute to a lupus-like phenotype [54]. Both human leukocyte antigen (HLA) and non-HLA genes have been examined as potential candidates conferring susceptibility to the development of SLE. There are HLA class I and class II associations with SLE, highlighting the potential significance of antigen presentation via the major histocompatability complex (MHC) to the pathogenesis of SLE. Additionally, specific class II molecules are associated with the production of particular autoantibodies [55]. Studies of non-MHC genes have produced several interesting candidates that may associate with an increased risk of SLE. These include but are certainly not limited to polymorphisms of the genes encoding the activating Fcγ receptors RIIa and IIIa [56], the inhibitory receptors Fcγ IIb, CTLA4 and PD1, the phosphatase PTPN22 and the phosphatase recruiter CD22, the anti-apoptotic molecule Bcl-2, interleukin-10 (IL-10), and the estrogen receptor alpha gene [57,58]. The importance of an interaction between susceptibility genes in SLE is exemplified by a study of IL-10 and Bcl-2 polymorphisms. When particular alleles of IL-10 and Bcl-2 are present in combination, they confer a 40-fold increased risk of disease. Additional genetic susceptibilities are present in small numbers of patients with lupus-like diseases. Individuals with deficiencies of genes controlling pathways for removal of apoptotic debris and individuals with deficiencies of complement components, particularly components of the early phases of complement activation, are predisposed to development of a lupus-like syndrome.

Although SLE is believed to have a strong genetic component, other factors are clearly required, as even monozygotic twins are only 30% concordant for disease expression [59]. The only definitively identified environmental trigger for disease onset is exposure to ultraviolet (UV) light. UV light causes cells to undergo apoptosis, and impaired clearance of apoptotic cells and cellular debris has been suggested as a potential explanation for the source of autoantigens in SLE. Thus, UV light has been postulated to activate autoimmunity by providing a source of autoantigen. Subsequently, ingestion of immune complexes containing DNA or RNA by DCs and autoreactive B cells leads to activation of TLR-9 (for DNA) or TLR-3 and TLR-7 (for RNA). Downstream signaling of TLRs induces secretion of inflammatory cytokines and, in the case of B cells, autoantibodies, thereby generating a cascade of autoreactivity.

An alternative hypothesis proposes that autoimmunity is based on observations that bacterial and viral infections may trigger flares of disease activity. Autoantibodies may arise by molecular mimicry during the course of an antibody response to a microbial challenge. Anti-DNA antibodies, for example, have been shown to cross-react with bacterial polysaccharide and with Epstein-Barr viral antigens. Other environmental triggers might include diet or chemicals; however, there are no clear associations between SLE and these agents. Dysregulation in either the activation of the immune response, its subsequent down-regulation, or both is a necessary precondition for the development of SLE following exposure to provocative environmental antigens, as most individuals do not sustain autoantibody production following infection.

Another factor that may contribute to triggering autoimmunity in a genetically susceptible host is a change in hormonal milieu, particularly in estrogenic hormones. The influence of estrogens on autoimmunity is suggested by observations that SLE has a strong female predominance, peak disease incidence occurs during the childbearing years, and disease activity usually dampens after menopause. Additionally, there are many case reports of pregnancy-related flares in known patients with lupus, as well as new diagnoses in the peripartum period. Studies in murine models of SLE have shown that estrogen impairs self-tolerance, allowing maturation of autoreactive B cells to immunocompetence. Some epidemiological data in humans support a link of SLE with prior exogenous estrogenic exposure [60], although administration of exogenous estrogen to patients with SLE does not clearly result in exacerbation of disease activity [61].

The disease, once initiated, is characterized by remissions and flares. Triggers for disease flares are not defined, except again for exposure to UV light, which can precipitate disease flares in approximately 40% of patients. It is also possible that changes in hormonal milieu activate lupus flares; some patients with SLE appear to be exquisitely sensitive to estrogen exposure, although the reasons for this are not known.

III. Neuropsychiatric SLE

There is a remarkable diversity of neuropsychiatric symptoms in SLE. They range from focal neurological symptoms such as seizure, stroke, and peripheral neuropathies to the more diffuse CNS symptoms such as depression, confusion, disorientation, and frank psychosis. Symptoms can be isolated, they may coexist with other organ system involvement, or they may occur sequentially. Because of the vast array of clinical presentations and a lack of understanding regarding pathogenesis, terminology for NPSLE has been confusing. The term "cerebritis" has been used to convey a variety of symptoms, and in an effort to standardize the nomenclature of the neuropsychiatric manifestations of SLE for further NPSLE study, precise consensus definitions for 19 neuropsychiatric syndromes were developed by an *ad hoc* committee convened by the American College of Rheumatology in 1999 [62] (Table 2). The 19 syndromes include the two classic CNS manifestations—seizures and psychosis—which are part of the classification criteria, and inflammatory syndromes known to be related to SLE such as aseptic meningitis and mononeuropathy; common and relatively nonspecific symptoms such as cognitive dysfunction, headache, and mood disorder are also included. A population-based study in Finland suggests that as many as 91% of patients with lupus fulfilled at least one of the NPSLE criteria [63]. However, at least one NPSLE criterion was present in 56% of the healthy matched controls. The most frequent neuropsychiatric manifestation in SLE was cognitive dysfunction, occurring in 81% of patients. Similarly, in the San Antonio Lupus Study of Neuropsychiatric Disease, 80% of 128 patients with lupus had one or more of the NPSLE syndromes. The most common reported NPSLE syndromes in that study were headache, cognitive dysfunction, and mood disorders [64].

Table 2 Neuropsychiatric Syndromes in Systemic Lupus Erythematosus

Central Nervous System

1. Aseptic meningitis
2. Cerebrovascular disease
3. Demyelinating syndrome
4. Headache (including migraine and benign intracranial hypertension)
5. Movement disorder (chorea)
6. Myelopathy
7. Seizure disorders
8. Acute confusional state
9. Anxiety disorder
10. Cognitive dysfunction
11. Mood disorder
12. Psychosis

Peripheral Nervous System

13. Acute inflammatory demyelinating polyradiculoneuropathy (Guillain-Barré syndrome)
14. Autonomic disorder
15. Mononeuropathy, single/multiplex
16. Myasthenia gravis
17. Neuropathy, cranial
18. Plexopathy
19. Polyneuropathy

IV. Neuroimaging in SLE

Because of their noninvasive nature, neuroimaging technologies are potentially important aids in the diagnosis of neuropsychiatric disease. Many modalities have been used to detect structural changes in the brain that may correlate with the various clinical manifestations of NPSLE; however, most neuroimaging techniques lack sensitivity and specificity for detecting changes associated with the diffuse presentations of NPSLE. For example, computed tomography (CT) is a fairly sensitive measure for brain atrophy, calcifications, hemorrhage, or mass lesions in 29–59% of patients [65,66] but is not useful for detecting change in patients suffering from cognitive impairment, psychosis, or mood disorders.

Magnetic resonance imaging (MRI) is considered a more sensitive measurement for detecting structural abnormalities. However, it, too, is best for abnormalities that associate primarily with focal neurological symptoms such as strokes and seizures, especially prevalent in the anti-phospholipid syndrome (see later discussion). The most common findings on MRI in patients with lupus are "UBOs," small punctate white-matter lesions that are believed to reflect a water signal from dilated Virchow-Robin spaces occurring from local atrophy and neuronal injury [67]. Atrophy, periventricular white-matter changes, and gross infarction are also seen. Unfortunately, many of these changes are not specific to NPSLE, are present in patients with lupus without neuropsychiatric syndromes [68], and are usually not helpful in differentiating between previous injury and active CNS disease [69,70]. When MRI is combined with cerebrospinal fluid (CSF) signal suppression in fluid-attenuating inversion recovery (FLAIR) sequence, there is increased sensitivity for detecting small lesions. However, the sensitivity for detecting abnormalities in patients with diffuse disease such as cognitive dysfunction, confusional states, depression, and anxiety is low [71–79].

Magnetic resonance spectroscopy (MRS) is a tool that uses the same technology as MRI to measure the biochemical composition of tissue and, thereby, provides a powerful metabolic adjunct to conventional MRI for assessment of function. Most MRS studies have used proton MRS (^{1}H-MRS) because the hydrogen proton exists in high concentration in living tissue. This technology is particularly useful in neuropsychiatric manifestations of lupus, as it provides information about brain injury not detected by modalities that are sensitive only to structural lesions. Several neurochemical signals measured by ^{1}H-MRS are analyzed in assessing brain integrity: *N*-acetylaspartate (NAA), creatine (Cr), choline-related metabolites, myoinositol, glutamate, glutamine, lactate, and lipid macromolecules. NAA is present in mature neurons and is not present in mature glial cells. It is well accepted as a marker for neuronal viability and integrity. Increased choline (Cho) metabolites are usually related to stroke or inflammation. Lactate indicates anaerobic metabolism and is detected in ischemic tissue, usually during and after stroke. Finally, an elevated concentration of lipids/macromolecules usually indicates the loss of cell integrity. The most commonly observed change indicating neuronal death or damage in NPSLE is a reduced NAA signal, usually presented as the ratio of NAA:Cr. Reduced NAA peaks have been observed in areas of brain correlating with the small focal lesions observed on MRI but have also been reported in brain tissue seemingly unaffected on MRI of patients with lupus [80,81]. One longitudinal study of a small group of patients with SLE reported areas of decreased neurometabolite ratios on ^{1}H-MRS that correlated with hypoperfusion on single-photon emission CT (SPECT) scan but were initially normal on MRI. Four to six years later, those same areas had evolved from normal to abnormal on repeated MRI scans and correlated clinically with increased headache and cognitive decline, suggesting that ^{1}H-MRS may have increased sensitivity for detecting early neuronal death [82]. Unfortunately, the cause of neuronal death remains elusive on the ^{1}H-MRS scan; a reduced NAA:Cr ratio is not specific to NPSLE and may be seen in other conditions such as chronic neurodegenerative diseases.

Clinically, cognitive defects have been associated with reduced NAA:Cho ratios. Thrombotic events and the presence of anti-phospholipid antibodies have been closely associated with an increased Cho:Cr level in SLE. Elevated lactate peaks might be expected to be present because multiple chronic microinfarctions (secondary to a thrombotic state often associated with SLE) leading to a multiinfarct-like dementia has been one of the leading theories of cognitive dysfunction in SLE. However, MRS studies have not detected elevated lactate peaks, even in patients hospitalized with acute diffuse neuropsychiatric manifestations [78]. A study of patients with NPSLE suggests that MRS may be useful for identifying patients with brain tissue at risk of imminent neuronal death. Patients with differing degrees of severity of neuropsychiatric symptoms were assessed using ^{1}H-MRS [83]. In patients with minor symptoms, a significant increase in the total choline peak and a reversible myoinositol peak were observed. In contrast, patients with major NPSLE symptoms displayed significantly and permanently reduced NAA levels, near-normal choline, and increased myoinositol peaks. Both groups were significantly different from normal healthy controls. These observations were confirmed in a prospective study of 50 patients with SLE and 9 normal controls imaged with MRI/MRS at baseline and again after 12 months. Those patients with active CNS disease at baseline who improved clinically demonstrated improvement in the NAA:Cr ratios. In contrast, those with active CNS disease at baseline who continued to deteriorate clinically, or those who were asymptomatic at baseline and subsequently developed clinical CNS disease, showed reduced NAA:Cr ratios at the follow-up study [84]. Collectively, these data suggest that there is a metabolic window of opportunity for therapeutic intervention before irreversible neuronal death.

Other neuroimaging modalities have also been used to assess patients with NPSLE. Positron emission tomography (PET) and SPECT scans measure cerebral glucose uptake and blood flow, respectively. They have been correlated to a limited extent with diffuse CNS disease. Their utility in detecting active ongoing disease may be less than originally anticipated because chronic damage with atrophy and neuronal cell loss will also lead to changes in metabolism and blood flow [71,76,85]. In summary, the results of each of the available testing modalities (CT, PET, SPECT, MRI, MRS) have been found to be abnormal in some patients with NPSLE; however, none has demonstrated consistent correlates with cognitive dysfunction, a major source of disability in patients with lupus. As MRS reflects changes in tissue metabolites, it appears to be the most promising tool for detecting active CNS disease in NPSLE. However, these modalities cannot provide information on the cause of neuronal death or on the pathophysiology of cognitive dysfunction.

V. Pathogenesis of NPSLE

Although an understanding of many aspects of disease pathogenesis increases, the etiology of most manifestations of NPSLE remains unknown. Vasculitis can occur within the CNS and may contribute to the pathophysiology of some neuropsychiatric syndromes, characteristically involving mononuclear inflammatory infiltrates and fibrous thickening of the vessel walls. However, although common in several organ systems in SLE, vasculitis is distinctly rare in the CNS in patients with NPSLE, occurring in 3–8% of cases [27,86–88]. More commonly seen is a noninflammatory small vessel vasculopathy, which is present particularly in association with anti-phospholipid antibodies and thrombosis with multiple foci of infarction. Cardiac emboli and thrombotic thrombocytopenic purpura may also cause stroke [89]. Microinfarcts are frequently seen in imaging studies and on postmortem analysis of brain tissue and may in some patients contribute to the cognitive dysfunction in SLE [87,88,90].

Proinflammatory cytokines have been observed in CSF of patients with lupus experiencing NPSLE manifestations. Elevated concentrations of IL-1, IL-6, IL-10, transforming growth factor-β (TGF-β), and IFN-γ have been reported in CSF of patients with NPSLE [91–94]. Increased levels of these cytokines could potentially result from either peripheral overproduction concomitant with a breach in the blood-brain barrier or local production of cytokines by activated cells within the brain. Cytokine-producing cells within the brain tissue might be resident brain cells or they might originate from the peripheral blood.

Some of the medications given to patients with lupus may contribute to neuropsychiatric syndromes. Cytotoxic drugs are routinely used in severe disease. When given at higher doses to patients with malignancies, these drugs have been implicated in cognitive impairment. Corticosteroids are also given to patients with lupus to control the inflammatory process. Studies in rodents have suggested that hippocampal neurons, in particular, may be susceptible to corticosteroid-induced toxicity [95], although clinical studies have not found significant correlations between corticosteroids and cognitive dysfunction or abnormalities in brain volume in SLE [96–101].

SLE is characterized by the production of autoantibodies. Many of these autoantibodies are capable of binding to tissue and inducing damage. The brain, however, is an immunologically privileged site and protected from antibody-mediated attack by the blood-brain barrier. Therefore, for antibodies to mediate brain damage, not only must brain-reactive antibodies exist, but these antibodies must be able to access the cerebral tissue. Antibody access to the brain might result from breach of the blood-brain barrier or by local secretion of the antibodies by B cells that have previously penetrated into brain tissue.

The presence of antibodies in patients with lupus that are capable of binding to brain tissue (anti-neuronal antibodies) was demonstrated several decades ago. These antibodies bound neuronal tissue, as well as antigens on glial cells. There have been several studies examining the association of anti-neuronal antibodies with clinical disease. In one cross-sectional study [102], patients with anti-neuronal antibodies were more likely to have significant cognitive impairment than patients without these antibodies. When followed longitudinally, changes in cognitive function corresponded with changes in serum titers of anti-neuronal antibodies [103]. Anti-neuronal antibodies have additionally been detected within the CSF of patients with lupus. However, other investigations have reported conflicting results regarding the presence of serum anti-neuronal antibodies in patients with SLE with impaired cognition [104–107]. Studies of clinical associations with diffuse cognitive dysfunction and autoantibodies in the CSF have also yielded inconsistent results [108–112]. Although the presence of antibodies with potential neurotoxic capability is recognized, the exact mechanisms leading to tissue injury are generally unknown.

VI. Anti-Phospholipid Antibodies and Anti-Phospholipid Syndrome

The anti-phospholipid syndrome (APLS) is characterized by recurrent thrombotic events including arterial or venous thrombosis and miscarriages, presumably secondary to placental thrombosis and thrombocytopenia. It may occur as a primary syndrome or as part of SLE. Autoantibodies to phospholipid are required for the diagnosis and these include anti-cardiolipin antibodies (immunoglobulin [Ig] G, IgM, and IgA), anti-phosphatidylcholine, and anti-phosphatidylserine antibodies. All of the anti-phospholipid antibodies are nonspecific and can be found in a variety of infections and other autoimmune diseases. β_2-Glycoprotein-1 (β2GP1) is a circulating protein with anticoagulant properties that is felt to be the antigenic target for the pathogenic anti-phospholipid antibodies, and anti-β2GP1 antibodies are a more specific but less sensitive marker for the APLS. The "lupus anticoagulant" is an unfortunate misnomer for a functional *in vitro* assay where anti-phospholipid antibodies prolong the clotting time by competing with coagulation factors for binding to phospholipid. Although these antibodies occur in approximately 30–40% of patients with lupus, only a few patients experience thrombotic events.

The major neurological manifestation of the APLS is cerebrovascular disease with ischemic stroke due to thrombosis or embolism in small vessels in the brain. Other neuropsychiatric symptoms in which anti-phospholipid antibodies have been implicated include seizures, dementia, chorea, transverse myelopathy, migraines, and cognitive dysfunction. Multiple microinfarcts resulting from thrombosis of small vessels is an attractive pathophysiological explanation for cognitive dysfunction, but investigations attempting to correlate cognitive impairment and anti-phospholipid antibodies have yielded inconsistent results [101,113–116].

Neurological and behavioral deficits have been observed in BALB/c mice immunized with a human anti-cardiolipin antibody. Hyperactivity in an open field, impairment in placing, postural reflexes, and, in some, locomotion tests were described in these mice. Electron microscopy of cerebral tissue revealed thrombotic occlusion of capillaries in combination with mild meningeal inflammation [117]. The molecular mechanisms leading to this behavioral pathology remain unknown.

A. Anti-Ribosomal P Protein Antibodies

Anti-ribosomal P antibodies have been studied for 20 years. These antibodies recognize an immunodominant epitope of 22 amino acids present at the carboxyl-terminal of three conserved proteins: P_0 (38 kDa), P_1 (19 kDa), and P_2 (17 kDa) of human ribosomal protein, 60S [118]. The early clinical studies of anti-ribosomal P antibodies showed a strong correlation between elevated titers of these antibodies and psychosis [109,119]; however, later studies demonstrated cross-reactivity with dsDNA [120,121] and found associations between the anti-ribosomal P antibodies and renal and liver disease [122].

Two studies of an anti-ribosomal P antibody reactive with a synthetic peptide showed significant correlation with NPSLE [110,123]. The second study demonstrated a stronger correlation of anti-ribosomal P protein antibodies in CSF with neuropsychiatric symptoms than serum anti-ribosomal P protein antibodies. Anti-ribosomal P antibody was present in the CSF of all patients with concurrent focal CNS disease plus a diffuse neuropsychological syndrome.

The mechanism by which anti-ribosomal P antibodies result in psychosis is not clear. Anti-ribosomal P protein antibodies are reported to bind to the surface of neuronal cells, suggesting that they may directly alter neuronal function [110], perhaps through binding to cross-reactive antigen. Another mechanism of anti-ribosomal P antibody–mediated neuronal pathophysiology may be secondary to stimulation of nonneuronal cells. Anti-ribosomal P antibodies stimulate production of tumor necrosis factor-α (TNF-α) and IL-6, proinflammatory cytokines, by activated human peripheral blood monocytes [124]. It will be interesting to determine whether these antibodies are also capable of activating microglial cells.

B. Anti-NR2 Antibodies

Anti-dsDNA antibodies are specific for SLE. They have been found in a number of different tissues in patients with SLE and are unequivocally implicated in tissue damage in kidneys. A subset of anti-DNA antibodies have been found to cross-react with a pentapeptide consensus sequence D/EWE/DYS/G (DWEYS) [125]. Competition assays have demonstrated that the peptide binds at or near the DNA binding site. This pentapeptide consensus sequence is present in the NR2A and NR2B subunits of glutamatergic *N*-methyl-D-aspartate (NMDA) receptor, and anti-DNA that antibodies cross-react with peptide also bind the NMDA receptor. Approximately 50% of patients with lupus display serum titers of anti-peptide antibody. Thus, the fine specificity of this antibody appears to be common in SLE and, therefore, potentially may be of broad clinical significance.

The NMDA receptors play a crucial role in excitatory synaptic transmission. The NMDA receptor is a heterotetrameric structure consisting of two NR1 subunits and two NR2A-D subunits. NR1 subunits are expressed on neurons uniformly throughout the brain, whereas the expression of NR2 subunits is developmentally and spatially regulated. NR2A and NR2B are expressed in high density in the hippocampus and the amygdala. Activation of the NMDA receptor by its natural ligand, glutamate, and co-ligand, glycine, results in calcium (Ca^+) flux in the postsynaptic cell. However, overstimulation of NMDA receptors leads to excitatory death of neural cells. Long-term potentiation (LTP) and long-term depression (LDP) are cellular processes of learning and memory achieved through activation of NMDA receptors. They have been best studied in the hippocampus but also occur in other brain regions.

Cross-reactive anti-DNA and anti-NR2 antibodies have been demonstrated to cause *in vitro* and *in vivo* neuronal death [125–127]. The mechanism of neuronal loss is independent of complement- or cell-mediated cytotoxicity and is mediated by antibody-mediated activation of intracellular signaling pathways. A mouse model has been generated in which peptide immunization induces high titers of anti-DNA, anti-NMDA receptor lupus-like antibodies. These mice have normal brains and normal behavior until there is a breach in the blood-brain barrier. Administration of lipopolysaccharide (LPS) causes a breach in the blood-brain barrier and leads to immunoglobulin deposition primarily in the CA1 domain of the hippocampus and to neuronal stress and apoptosis. Neither cellular infiltration nor complement activation is required for antibody-mediated cell death. Mice with anti-NMDA receptor antibody–mediated death of hippocampal neurons display a marked defect in tasks requiring an intact hippocampus for memory and learning. Thus, lupus antibodies present in circulation can result in cognitive impairment if access to neuronal tissue is provided.

Epinephrine is released under stressful conditions and can also result in a penetrable blood–brain barrier. Epinephrine causes regional increases in cerebral blood flow; there is a 30% increase in the amygdala but only a 15% increase in the hippocampus. Mice harboring anti-NMDA receptor antibodies display death of amygdala neurons after exposure to epinephrine. The mice showed a diminished response to a fear-conditioning paradigm, which is dependent on intact function of the lateral and central nuclei of the amygdala.

Thus, murine studies show that a subset of lupus anti-DNA antibodies that cross-react with NMDA receptors can cause cognitive and behavioral symptoms following a breach in the blood-brain barrier. The agent used to open the blood-brain barrier determines the regional vulnerability.

Other spontaneous murine models of lupus develop behavioral abnormalities with disease progression. NZB/NZWxF1 mice demonstrate increased anxiety and learning disturbances concordant with disease progression. However, these mice additionally develop multiple other abnormalities such as reduced exploration, impaired motor function, and reduced sensitivity to pain stimuli. The presence of both antibody and cellular infiltrates in the brain additionally makes any analysis of casual relationships difficult [128]. The MRL/lpr mouse displays impaired exploration and "emotional reactivity," both perhaps signs of depression. Histological examination of the CA1 region of the hippocampus reveals atrophy of pyramidal neurons. Autoantibody-mediated neuronal toxicity is a possible pathophysiological mechanism explaining the observed neuronal atrophy, as the blood-brain barrier is known to become increasingly permeable in aging MRL/lpr mice and there are increased levels of immunoglobulin and albumin in the CSF. Furthermore, the CSF from older animals is toxic to neurons *in vitro* [129].

C. Therapy

Management of the diverse clinical manifestations of NPSLE remains one of the most difficult challenges in the treatment of SLE. Because of the paucity of randomized controlled clinical trials, treatment decisions are based on case reports, anecdotal experience, retrospective series, and expert opinion. Before attributing signs and symptoms to "NPSLE," it is imperative that a broad differential diagnosis be generated, including consideration of infection, drug toxicity, metabolic derangements, and hypertension as potential causes for the abnormal neuropsychiatric function. Analysis of CSF and serum in conjunction with brain imaging are all recommended as aids in the diagnosis of NPSLE and exclusion of other etiologies. The CSF of a patient with active NPSLE may be normal or show evidence of increased protein along with a lymphocytosis. Complement levels in the CSF may also vary and are generally not useful in the diagnosis. If the patient has evidence of active lupus in other organ systems (e.g., rash, arthritis, serositis, and nephritis) and/or the serum serologies are abnormal with low complement and elevated titers of anti-DNA antibodies, the diagnosis of NPSLE may be a little easier. However, patients often have neuropsychiatric symptoms as their only manifestation of disease activity.

Once the diagnosis of NPSLE has been established with reasonable certainty, treatment decisions are based on the nature and severity of symptoms, with careful attention to underlying potential pathological mechanisms. Treatment strategies include the use of supportive medication for seizures, mood disorders, and psychosis along with immunosuppressive therapies for severe diffuse disease, and anticoagulant therapies for thrombosis and focal disease. Milder CNS symptoms such as anxiety, mild forms of depression, and headache are commonly managed with standard regimens of anxiolytics, antidepressants, and analgesia. Steroids are rarely, if ever, used in the treatment of mild, diffuse CNS symptoms.

Severe CNS disease, whether focal or diffuse, requires more aggressive therapy. Severe focal CNS disease (strokes, cerebral venous thrombosis, chorea, transverse myelitis, aseptic meningitis) is often associated with a secondary anti-phospholipid syndrome and presence of pro-thrombotic autoantibodies. Treatment consists of long-term anticoagulation with fractionated heparin or warfarin, along with aspirin or platelet inhibitors. The target international normalized ratio (INR) for anticoagulation with warfarin has been greatly disputed; however, data from a randomized controlled trial suggest that an INR of 2–3 may be safe for venous thromboses [130]. The recommended INR for arterial thromboses remains 3–4. Generally, immunosuppression with steroids and other agents is not used for focal CNS disease unless the underlying mechanism is thought to be vasculitis. Focal gadolinium enhancement on MRI or increased protein and lymphocytes in the CSF may help determine this, but biopsy proof is understandably not possible and patients may be treated with both anticoagulation and immunosuppression when the distinction is not clear.

Severe diffuse CNS disease with psychosis, delirium, and even catatonia is treated with high doses of steroids (intravenous methylprednisolone [500–1000 mg daily for 3 days] or oral prednisone [1–2 mg/kg]) along with a secondary immunosuppressive agent for additional treatment and steroid-sparing effects. Cyclophosphamide remains the recommended treatment for severe diffuse disease despite the fact that there have been no definitive prospective studies. Azathioprine has been used as an alternative, but perhaps less effective, treatment. The efficacy of mycophenolic mofetil has been established for lupus nephritis, but reports of its use in NPSLE are few preliminary studies without follow-up randomized controlled trials. Plasmapheresis, intravenous immunoglobulin, and intrathecal methotrexate have been used in very ill patients refractory to other forms of treatment but have not been assessed in controlled studies. Rituximab is a monoclonal antibody directed against the B-cell marker CD20 that results in B-cell deletion and is approved for use in the treatment of B-cell lymphomas. Although there are ongoing randomized clinical trials of rituximab for SLE, there are only few anecdotal reports of its use in NPSLE [131,132].

Seizure disorders associated with SLE are treated with standard anticonvulsant therapy. Depending on whether they are felt to be the result of a thrombotic or inflammatory mechanism, anticoagulation or immunosuppressive therapy may be added.

There are no recommended specific treatments for the cognitive disorders associated with NPSLE other than close attention to comorbid illnesses such as hypertension and medication side effects that may exacerbate symptoms. If the cognitive impairment is associated with multiple microinfarcts on imaging and anti-phospholipid antibodies, anticoagulation may be recommended. As the evidence for antibody-mediated neuronal toxicity grows, newer therapies, such as the NMDA receptor antagonist memantine, may be successfully employed to halt progression of this insidious and debilitating manifestation of NPSLE. It is vitally important to recognize that many of the NPSLE symptoms can overlap and distinctions between inflammatory and thrombotic mechanisms may not be possible. Treatment of NPSLE is, therefore, tailored to the individual patient following careful exclusion of other etiologies and it is likely that multiple treatment strategies will be employed simultaneously.

VII. Summary

Cognitive impairment and neuropsychiatric manifestations of SLE contribute to the morbidity of patients with SLE. The etiologies are not known, but associations of circulating autoantibodies with neuropsychiatric manifestations abound. The clinical correlation of anti-phospholipid antibodies in thrombotic events is well documented, and the correlation of anti-ribosomal P antibodies with psychosis is of considerable interest. Previous studies do not directly implicate autoantibodies as neurotoxic agents. Evidence for the direct mechanistic involvement of the immune system in NPSLE comes from clinical practice where systemic immunosuppressive therapy improves the neuropsychiatric symptoms, as well as from mouse models of lupus where the pretreatment of young animals with immunosuppression during the onset of disease prevents brain damage.

Newer studies provide a crucial link between peripheral autoimmunity and brain manifestations of lupus. Brain-specific autoantibodies might arise through molecular mimicry with a foreign antigen or with other self-antigens or as a result of direct interaction with brain antigens. Anti-NMDA receptor antibodies cross-react with dsDNA, but this cross-reactivity is unlikely to be a characteristic of all brain-reactive antibodies in SLE. Other brain antigens like microtubule-associated protein-2 (MAP-2) [133] or fractalkine have begun to emerge as potential targets in NPSLE studies [134]. Brain-reactive antibodies can gain access to brain antigens through the abrogation of the blood-brain barrier, providing an explanation for the clinical observation that NPSLE may progress at times of lupus quiescence. The subsequent behavioral pathology may be dependent on the mechanisms involved in opening the blood-brain barrier and the region of endothelial leakiness. Alternatively, intrathecal production of autoantibodies may occur but is more likely to be in the context of active disease and inflammatory syndromes.

These studies raise the question of antibody-mediated neuronal dysfunction in multiple conditions characterized by changes in cognition or behavior, suggesting that antibody-mediated brain dysfunction in the absence of inflammation may not be uncommon. Finally, studies that demonstrate a mechanism for neuronal dysfunction in SLE provide the hope of new and novel therapeutic interventions. With the identification of pathogenic antibodies, proper blocking agents may be developed to protect exposed neurons.

References

1. Kaposi, M. (1872). Neue eitrage zur Kenntis des lupus erythematosus. *Arch. Derm. Syph.* **4**, 36–78.
2. Osler, W. (1904). On the visceral manifestations of the erythema group of skin diseases. *Am. J. Med. Sci.* **127**, 1–23.
3. McCarty, D. J., Manzi, S., Medsger, T. A., Jr., Ramsey-Goldman, R., LaPorte, R. E., and Kwoh, C. K. (1995). Incidence of systemic lupus erythematosus. Race and gender differences. *Arthritis Rheum.* **38**(9), 1260–1270.
4. Michet, C. J., Jr., McKenna, C. H., Elveback, L. R., Kaslow, R. A., and Kurland, L. T. (1985). Epidemiology of systemic lupus erythematosus and other connective tissue diseases in Rochester, Minnesota, 1950 through 1979. *Mayo Clin. Proc.* **60**(2), 105–113.
5. Uramoto, K. M., Michet, C. J., Jr., Thumboo, J., Sunku, J., O'Fallon, W. M., and Gabriel, S. E. (1999). Trends in the incidence and mortality of systemic lupus erythematosus, 1950–1992. *Arthritis Rheum.* **42**(1), 46–50.
6. Alarcon, G. S., Friedman, A. W., Straaton, K. V., Moulds, J. M., Lisse, J., Bastian, H. M., McGwin, G., Jr., Bartolucci, A. A., Roseman, J. M., and Reveille, J. D. (1999). Systemic lupus erythematosus in three ethnic groups: III. A comparison of characteristics early in the natural history of the LUMINA cohort. LUpus in MInority populations: Nature vs. Nurture. *Lupus* **8**(3), 97–209.
7. Alarcon, G. S., Roseman, J., Bartolucci, A. A., Friedman, A. W., Moulds, J. M., Goel, N., Straaton, K. V., and Reveille, J. D. (1998). Systemic lupus erythematosus in three ethnic groups: II. Features predictive of disease activity early in its course. LUMINA Study Group. Lupus in minority populations, nature versus nurture. *Arthritis Rheum.* **41**(7), 1173–1180.
8. Rivest, C., Lew, R. A., Welsing, P. M., Sangha, O., Wright, E. A., Roberts, W. N., Liang, M. H., and Karlson, E. W. (2000). Association between clinical factors, socioeconomic status, and organ damage in recent onset systemic lupus erythematosus. *J. Rheumatol.* **27**(3), 680–684.
9. Alarcon, G. S., McGwin, G., Jr., Bartolucci, A. A., Roseman, J., Lisse, J., Fessler, B. J., Bastian, H. M., Friedman, A. W., and Reveille, J. D. (2001). Systemic lupus erythematosus in three ethnic groups. IX. Differences in damage accrual. *Arthritis Rheum.* **44**(12), 2797–2806.
10. Sutcliffe, N., Clarke, A. E., Gordon, C., Farewell, V., and Isenberg, D. A. (1999). The association of socio-economic status, race, psychosocial factors and outcome in patients with systemic lupus erythematosus. *Rheumatology (Oxford)* **38**(11), 1130–1137.
11. Ward, M. M., Pyun, E., and Studenski, S. (1995). Long-term survival in systemic lupus erythematosus. Patient characteristics associated with poorer outcomes. *Arthritis Rheum.* **38**(2), 274–283.
12. Azizah, M. R., Ainol, S. S., Kong, N. C., Normaznah, Y., and Rahim, M. N. (2001). Gender differences in the clinical and serological features of systemic lupus erythematosus in Malaysian patients. *Med. J. Malaysia* **56**(3), 302–307.
13. Siegel, M., and Lee, S. L. (1973). The epidemiology of systemic lupus erythematosus. *Semin. Arthritis Rheum.* **3**(1), 1–54.
14. Cohen, A. S. (1971). SLE clinical criteria. *Bull. Rheum. Dis.* **21**, 643.
15. Eng, M. (1982). SLE criteria. *Arthritis Rheum.* **25**, 1271.
16. Hochberg, M. C. (1997). Updating the American College of Rheumatology revised criteria for the classification of systemic lupus erythematosus. *Arthritis Rheum.* **40**(9), 1725.
17. Davatchi, F., Chams, C., and Akbarian, M. (1985). Evaluation of the 1982 American Rheumatism Association revised criteria for the classification of systemic lupus erythematosus. *Arthritis Rheum.* **28**(6), 715.
18. Gilboe, I. M., and Husby, G. (1999). Application of the 1982 revised criteria for the classification of systemic lupus erythematosus on a cohort of 346 Norwegian patients with connective tissue disease. *Scand. J. Rheumatol.* **28**(2), 81–87.
19. Levin, R. E., Weinstein, A., Peterson, M., Testa, M. A., and Rothfield, N. F. (1984). A comparison of the sensitivity of the 1971 and 1982 American Rheumatism Association criteria for the classification of systemic lupus erythematosus. *Arthritis Rheum.* **27**(5), 530–538.

20. Passas, C. M., Wong, R. L., Peterson, M., Testa, M. A., and Rothfield, N. F. (1985). A comparison of the specificity of the 1971 and 1982 American Rheumatism Association criteria for the classification of systemic lupus erythematosus. *Arthritis Rheum.* **28**(6), 620–623.
21. Yokohari, R., and Tsunematsu, T. (1985). Application, to Japanese patients, of the 1982 American Rheumatism Association revised criteria for the classification of systemic lupus erythematosus. *Arthritis Rheum.* **28**(6), 693–698.
22. Stahl Hallengren, C., Nived, O., and Sturfelt, G. (2004). Outcome of incomplete systemic lupus erythematosus after 10 years. *Lupus* **13**(2), 85–88.
23. Abu-Shakra, M., Urowitz, M. B., Gladman, D. D., and Gough, J. (1995). Mortality studies in systemic lupus erythematosus. Results from a single center. I. Causes of death. *J. Rheumatol.* **22**(7), 1259–1264.
24. Cervera, R., Khamashta, M. A., Font, J., Sebastiani, G. D., Gil, A., Lavilla, P., Aydintug, A. O., Jedryka-Goral, A., de Ramon, E., Fernandez-Nebro, A., Galeazzi, M., Haga, H. J., Mathieu, A., Houssiau, F., Ruiz-Irastorza, G., Ingelmo, M., and Hughes, G. R. (1999). Morbidity and mortality in systemic lupus erythematosus during a 5-year period. A multicenter prospective study of 1,000 patients. European Working Party on Systemic Lupus Erythematosus. *Medicine (Baltimore)* **78**(3), 167–175.
25. Duffy, K. N., Duffy, C. M., and Gladman, D. D. (1991). Infection and disease activity in systemic lupus erythematosus: a review of hospitalized patients. *J. Rheumatol.* **18**(8), 1180–1184.
26. Rosner, S., Ginzler, E. M., Diamond, H. S., Weiner, M., Schlesinger, M., Fries, J. F., Wasner, C., Medsger, T. A., Jr., Ziegler, G., Klippel, J. H., Hadler, N. M., Albert, D. A., Hess, E. V., Spencer-Green, G., Grayzel, A., Worth, D., Hahn, B. H., and Barnett, E. V. (1982). A multicenter study of outcome in systemic lupus erythematosus. II. Causes of death. *Arthritis Rheum.* **25**(6), 612–617.
27. Scolding, N. J., and Joseph, F. G. (2002). The neuropathology and pathogenesis of systemic lupus erythematosus. *Neuropathol. Appl. Neurobiol.* **28**(3), 173–189.
28. Arbuckle, M. R., McClain, M. T., Rubertone, M. V., Scofield, R. H., Dennis, G. J., James, J. A., and Harley, J. B. (2003). Development of autoantibodies before the clinical onset of systemic lupus erythematosus. *N. Engl. J. Med.* **349**(16), 1526–1533.
29. Sasaki, T., Hatakeyama, A., Shibata, S., Osaki, H., Suzuki, M., Horie, K., Kitagawa, Y., and Yoshinaga, K. (1991). Heterogeneity of immune complex-derived anti-DNA antibodies associated with lupus nephritis. *Kidney Int.* **39**(4), 746–753.
30. Winfield, J. B., Faiferman, I., and Koffler, D. (1977). Avidity of anti-DNA antibodies in serum and IgG glomerular eluates from patients with systemic lupus erythematosus. Association of high avidity anti-native DNA antibody with glomerulonephritis. *J. Clin. Invest.* **59**(1), 90–96.
31. Alexander, E., Buyon, J. P., Provost, T. T., and Guarnieri, T. (1992). Anti-Ro/SS-A antibodies in the pathophysiology of congenital heart block in neonatal lupus syndrome, an experimental model. *In vitro* electrophysiologic and immunocytochemical studies. *Arthritis Rheum.* **35**(2), 176–189.
32. Horsfall, A. C., Venables, P. J., Taylor, P. V., and Maini, R. N. (1991). Ro and La antigens and maternal anti-La idiotype on the surface of myocardial fibres in congenital heart block. *J. Autoimmun.* **4**(1), 165–176.
33. Lopez-Longo, F. J., Monteagudo, I., Gonzalez, C. M., Grau, R., and Carreno, L. (1997). Systemic lupus erythematosus: clinical expression and anti-Ro/SS–a response in patients with and without lesions of subacute cutaneous lupus erythematosus. *Lupus* **6**(1), 32–39.
34. Niimi, Y., Ioannides, D., Buyon, J., and Bystryn, J. C. (1995). Heterogeneity in the expression of Ro and La antigens in human skin. *Arthritis Rheum.* **38**(9), 1271–1276.
35. Davidson, A., and Diamond, B. (2001). Autoimmune diseases. *N. Engl. J. Med.* **345**(5), 340–350.
36. Goodnow, C. C., Sprent, J., Fazekas de St Groth, B., and Vinuesa, C. G. (2005). Cellular and genetic mechanisms of self tolerance and autoimmunity. *Nature* **435**(7042), 590–597.
37. Jacobi, A. M., and Diamond, B. (2005). Balancing diversity and tolerance: lessons from patients with systemic lupus erythematosus. *J. Exp. Med.* **202**(3), 341–344.
38. Crow, M. K., Kirou, K. A., and Wohlgemuth, J. (2003). Microarray analysis of interferon-regulated genes in SLE. *Autoimmunity* **36**(8), 481–490.
39. Datta, S. K., Zhang, L., and Xu, L. (2005). T-helper cell intrinsic defects in lupus that break peripheral tolerance to nuclear autoantigens. *J. Mol. Med.* **83**(4), 267–278.
40. Lau, C. M., Broughton, C., Tabor, A. S., Akira, S., Flavell, R. A., Mamula, M. J., Christensen, S. R., Shlomchik, M. J., Viglianti, G. A., Rifkin, I. R., and Marshak-Rothstein, A. (2005). RNA-associated autoantigens activate B cells by combined B cell antigen receptor/Toll-like receptor 7 engagement. *J. Exp. Med.* **202**(9), 1171–1177.
41. Pawar, R. D., Patole, P. S., Zecher, D., Segerer, S., Kretzler, M., Schlondorff, D., and Anders, H. J. (2006). Toll-like receptor-7 modulates immune complex glomerulonephritis. *J. Am. Soc. Nephrol.* **17**(1), 141–149.
42. Hardin, J. A. (2005). Dendritic cells: potential triggers of autoimmunity and targets for therapy. *Ann. Rheum. Dis.* **64**(**Suppl 4**), iv86–90.
43. Ytterberg, S. R., and Schnitzer, T. J. (1982). Serum interferon levels in patients with systemic lupus erythematosus. *Arthritis Rheum.* **25**(4), 401–406.
44. Baechler, E. C., Gregersen, P. K., and Behrens, T. W. (2004). The emerging role of interferon in human systemic lupus erythematosus. *Curr. Opin. Immunol.* **16**(6), 801–807.
45. Baechler, E. C., Batliwalla, F. M., Karypis, G., Gaffney, P. M., Ortmann, W. A., Espe, K. J., Shark, K. B., Grande, W. J., Hughes, K. M., Kapur, V., Gregersen, P. K., and Behrens, T. W. (2003). Interferon-inducible gene expression signature in peripheral blood cells of patients with severe lupus. *Proc. Natl. Acad. Sci. USA* **100**(5), 2610–2615.
46. Bennett, L., Palucka, A. K., Arce, E., Cantrell, V., Borvak, J., Banchereau, J, and Pascual, V. (2003). Interferon and granulopoiesis signatures in systemic lupus erythematosus blood. *J. Exp. Med.* **197**(6), 711–723.
47. Kirou, K. A., Lee, C., George, S., Louca, K., Peterson, M. G., and Crow, M. K. (2005). Activation of the interferon-alpha pathway identifies a subgroup of systemic lupus erythematosus patients with distinct serologic features and active disease. *Arthritis Rheum.* **52**(5), 1491–1503.
48. Bave, U., Alm, G. V., and Ronnblom, L. (2000). The combination of apoptotic U937 cells and lupus IgG is a potent IFN-alpha inducer. *J. Immunol.* **165**(6), 3519–3526.
49. Kirou, K. A., Lee, C., George, S., Louca, K., Papagiannis, I. G., Peterson, M. G., Ly, N., Woodward, R. N., Fry, K. E., Lau, A. Y., Prentice, J. G., Wohlgemuth, J. G., and Crow, M. K. (2004). Coordinate overexpression of interferon-alpha-induced genes in systemic lupus erythematosus. *Arthritis Rheum.* **50**(12), 3958–3967.
50. Lovgren, T., Eloranta, M. L., Bave, U., Alm, G. V., and Ronnblom, L. (2004). Induction of interferon-alpha production in plasmacytoid dendritic cells by immune complexes containing nucleic acid released by necrotic or late apoptotic cells and lupus IgG. *Arthritis Rheum.* **50**(6), 1861–1872.
51. Vallin, H., Blomberg, S., Alm, G. V., Cederblad, B., and Ronnblom, L. (1999). Patients with systemic lupus erythematosus (SLE) have a circulating inducer of interferon-alpha (IFN-alpha) production acting on leucocytes resembling immature dendritic cells. *Clin. Exp. Immunol.* **115**(1), 196–202.

52. Block, S. R., Winfield, J. B., Lockshin, M. D., D'Angelo, W. A., and Christian, C. L. (1975). Studies of twins with SLE: a review of the literature and a presentation of 12 new cases. *Am. J. Med.* **59**, 533–552.
53. Deapan, D., Escalante, A., and Weinrib, L. (1992). A revised estimate of twin concordance in SLE. *Arthritis Rheum.* **35**, 311.
54. Vinuesa, C. G., and Goodnow, C. C. (2004). Illuminating autoimmune regulators through controlled variation of the mouse genome sequence. *Immunity* **20**(6), 669–679.
55. Scofield, R. H., Farris, A. D., Horsfall, A. C., and Harley, J. B. (1999). Fine specificity of the autoimmune response to the Ro/SSA and La/SSB ribonucleoproteins. *Arthritis Rheum.* **42**(2), 199–209.
56. Reveille, J. D. (2005). Genetic studies in the rheumatic diseases: present status and implications for the future. *J. Rheumatol. Suppl.* **72**, 10–13.
57. Kassi, E., Vlachoyiannopoulos, P. G., Kominakis, A., Kiaris, H., Moutsopoulos, H. M., and Moutsatsou, P. (2005). Estrogen receptor alpha gene polymorphism and systemic lupus erythematosus: a possible risk? *Lupus* **14**(5), 391–398.
58. Lee, Y. J., Shin, K. S., Kang, S. W., Lee, C. K., Yoo, B., Cha, H. S., Koh, E. M., Yoon, S. J., and Lee, J. (2004). Association of the oestrogen receptor alpha gene polymorphisms with disease onset in systemic lupus erythematosus. *Ann. Rheum. Dis.* **63**(10), 1244–1249.
59. Grennan, D. M., Parfitt, A., Manolios, N., Huang, Q., Hyland, V., Dunckley, H., Doran, T., Gatenby, P., and Badcock, C. (1997). Family and twin studies in systemic lupus erythematosus. *Dis. Markers* **13**(2), 93–98.
60. Sanchez-Guerrero, J., Liang, M. H., Karlson, E. W., Hunter, D. J., and Colditz, G. A. (1995). Postmenopausal estrogen therapy and the risk for developing systemic lupus erythematosus. *Ann. Intern. Med.* **122**(6), 430–433.
61. Buyon, J. P. (2005). Dispelling the preconceived notion that lupus pregnancies result in poor outcomes. *J. Rheumatol.* **32**(9), 1641–1642.
62. The American College of Rheumatology nomenclature and case definitions for neuropsychiatric lupus syndromes. *Arthritis Rheum.* (1999) **42**(4), 599–608.
63. Ainiala, H., Loukkola, J., Peltola, J., Korpela, M., and Hietaharju, A. (2001). The prevalence of neuropsychiatric syndromes in systemic lupus erythematosus. *Neurology* **57**(3), 496–500.
64. Brey, R. L., Holliday, S. L., Saklad, A. R., Navarrete, M. G., Hermosillo-Romo, D., Stallworth, C. L., Valdez, C. R., Escalante, A., del Rincon, I., Gronseth, G., Rhine, C. B., Padilla, P., and McGlasson, D. (2002). Neuropsychiatric syndromes in lupus: prevalence using standardized definitions. *Neurology* **58**(8), 1214–1220.
65. Gaylis, N. B., Altman, R. D., Ostrov, S., and Quencer, R. (1982). The selective value of computed tomography of the brain in cerebritis due to systemic lupus erythematosus. *J. Rheumatol.* **9**(6), 850–854.
66. Shapeero, L. G. (1992). Imaging in systemic lupus erythematosus. In: "Systemic Lupus Erythematosus" (R. G. Lahita, ed.), pp. 447–526. Churchill Livingstone, New York.
67. Stimmler, M. M., Coletti, P. M., and Quismorio, F. P., Jr. (1993). Magnetic resonance imaging of the brain in neuropsychiatric systemic lupus erythematosus. *Semin. Arthritis Rheum.* **22**(5), 335–349.
68. Kozora, E., West, S. G., Kotzin, B. L., Julian, L., Porter, S., and Bigler, E. (1998). Magnetic resonance imaging abnormalities and cognitive deficits in systemic lupus erythematosus patients without overt central nervous system disease. *Arthritis Rheum.* **41**(5), 41–47.
69. McCune, W. J., MacGuire, A., Aisen, A., and Gebarski, S. (1988). Identification of brain lesions in neuropsychiatric systemic lupus erythematosus by magnetic resonance scanning. *Arthritis Rheum.* **31**(2), 159–166.
70. Sibbitt, W. L., Jr., Sibbitt, R. R., Griffey, R. H., Eckel, C., and Bankhurst, A. D. (1989). Magnetic resonance and computed tomographic imaging in the evaluation of acute neuropsychiatric disease in systemic lupus erythematosus. *Ann. Rheum. Dis.* **48**(12), 1014–1022.
71. Huang, J. L., Yeh, K. W., You, D. L., and Hsieh, K. H. (1997). Serial single photon emission computed tomography imaging in patients with cerebral lupus during acute exacerbation and after treatment. *Pediatr. Neurol.* **17**(1), 44–48.
72. Jarek, M. J., West, S. G., Baker, M. R., and Rak, K. M. (1994). Magnetic resonance imaging in systemic lupus erythematosus patients without a history of neuropsychiatric lupus erythematosus. *Arthritis Rheum.* **37**(11), 1609–1613.
73. Kao, C. H., Lan, J. L., ChangLai, S. P., Liao, K. K., Yen, R. F., and Chieng, P. U. (1999). The role of FDG-PET, HMPAO-SPET and MRI in the detection of brain involvement in patients with systemic lupus erythematosus. *Eur. J. Nucl. Med.* **26**(2), 129–134.
74. Sabbadini, M. G.; Manfredi, A. A.; Bozzolo, E. Ferrario, L., Rugarli, C., Scorza, R., Origgi, L., Vanoli, M., Gambini, O., Vanzulli, L., Croce, D., Campana, A., Messa, C., Fazio, F., Tincani, A., Anzola, G., Cattaneo, R., Padovani, A., Gasparotti, R., Gerli, R., Quartesan, R., Piccirilli, M., Farsi, A., Emmi, E., Domeneghetti, M., Piccini, C., Massacesi, L., Pupi, A., De Cristoforis, M., Danieli, M., Candela, M., Fraticelli, P., Bartolini, M., Salvolini, U., Danieli, G., and Passaleva, A. (1999). Central nervous system involvement in systemic lupus erythematosus patients without overt neuropsychiatric manifestations. *Lupus* **8**(1), 11–19.
75. Sabet, A., Sibbitt, W. L., Jr., Stidley, C. A., Danska, J., and Brooks, W. M. (1998). Neurometabolite markers of cerebral injury in the antiphospholipid antibody syndrome of systemic lupus erythematosus. *Stroke* **29**(11), 2254–2260.
76. Sailer, M., Burchert, W., Ehrenheim, C., Smid, H. G., Haas, J., Wildhagen, K., Wurster, U., and Deicher, H. (1997). Positron emission tomography and magnetic resonance imaging for cerebral involvement in patients with systemic lupus erythematosus. *J. Neurol.* **244**(3), 186–193.
77. Sanna, G., Piga, M., Terryberry, J. W., Peltz, M. T., Giagheddu, S., Satta, L., Ahmed, A., Cauli, A., Montaldo, C., Passiu, G., Peter, J. B., Shoenfeld, Y., and Mathieu, A. (2000). Central nervous system involvement in systemic lupus erythematosus: cerebral imaging and serological profile in patients with and without overt neuropsychiatric manifestations. *Lupus* **9**(8), 573–583.
78. Sibbitt, W. L., Jr., Haseler, L. J., Griffey, R. R., Friedman, S. D., and Brooks, W. M. (1997). Neurometabolism of active neuropsychiatric lupus determined with proton MR spectroscopy. *AJNR Am. J. Neuroradiol.* **18**(7), 1271–1277.
79. Teh, L. S., Hay, E. M., Amos, N., Black, D., Huddy, A., Creed, F., Bernstein, R. M., Holt, P. J., and Williams, B. D. (1993). Anti-P antibodies are associated with psychiatric and focal cerebral disorders in patients with systemic lupus erythematosus. *Br. J. Rheumatol.* **32**(4), 287–290.
80. Brooks, W. M., Jung, R. E., Ford, C. C., Greinel, E. J., and Sibbitt, W. L., Jr. (1999). Relationship between neurometabolite derangement and neurocognitive dysfunction in systemic lupus erythematosus. *J. Rheumatol.* **26**(1), 81–85.
81. Sibbitt, W. L., Jr., Haseler, L. J., Griffey, R. H., Hart, B. L., Sibbitt, R. R., and Matwiyoff, N. A. (1994). Analysis of cerebral structural changes in systemic lupus erythematosus by proton MR spectroscopy. *AJNR Am. J. Neuroradiol.* **15**(5), 923–928.
82. Castellino, G., Govoni, M., Padovan, M., Colamussi, P., Borrelli, M., and Trotta, F. (2005). Proton magnetic resonance spectroscopy may predict future brain lesions in SLE patients: a functional multi-imaging approach and follow up. *Ann. Rheum. Dis.* **64**(7), 1022–1027.
83. Axford, J. S., Howe, F. A., Heron, C., and Griffiths, J. R. (2001). Sensitivity of quantitative (1)H magnetic resonance spectroscopy of the brain in detecting early neuronal damage in systemic lupus erythematosus. *Ann. Rheum. Dis.* **60**(2), 106–111.
84. Appenzeller, S., Li, L. M., Costallat, L. T., and Cendes, F. (2005). Evidence of reversible axonal dysfunction in systemic lupus erythematosus: a proton MRS study. *Brain* **128**, 2933–2940.

85. Carbotte, R. M., Denburg, S. D., Denburg, J. A., Nahmias, C., and Garnett, E. S. (1992). Fluctuating cognitive abnormalities and cerebral glucose metabolism in neuropsychiatric systemic lupus erythematosus. *J. Neurol. Neurosurg. Psychiatry* **55**(11), 1054–1059.
86. Ellis, S. G., and Verity, M. A. (1979). Central nervous system involvement in systemic lupus erythematosus: a review of neuropathologic findings in 57 cases, 1955–1977. *Semin. Arthritis Rheum.* **8**(3), 212–221.
87. Hanly, J. G., Walsh, N. M., and Sangalang, V. (1992). Brain pathology in systemic lupus erythematosus. *J. Rheumatol.* **19**(5), 732–741.
88. Johnson, R. T., and Richardson, E. P. (1968). The neurological manifestations of systemic lupus erythematosus. *Medicine (Baltimore)* **47**(4), 337–369.
89. Devinsky, O., Petito, C. K., and Alonso, D. R. (1988). Clinical and neuropathological findings in systemic lupus erythematosus: the role of vasculitis, heart emboli, and thrombotic thrombocytopenic purpura. *Ann. Neurol.* **23**(4), 380–384.
90. Sibbitt, W. L., Jr., Sibbitt, R. R., and Brooks, W. M. (1999). Neuroimaging in neuropsychiatric systemic lupus erythematosus. *Arthritis Rheum.* **42**(10), 2026–2038.
91. Alcocer-Varela, J., Aleman-Hoey, D., and Alarcon-Segovia, D. (1992). Interleukin-1 and interleukin-6 activities are increased in the cerebrospinal fluid of patients with CNS lupus erythematosus and correlate with local late T-cell activation markers. *Lupus* **1**(2), 111–117.
92. Hirohata, S., Tanimoto, K., and Ito, K. (1993). Elevation of cerebrospinal fluid interleukin-6 activity in patients with vasculitides and central nervous system involvement. *Clin. Immunol. Immunopathol.* **66**(3), 225–229.
93. Svenungsson, E., Andersson, M., Brundin, L., van Vollenhoven, R., Khademi, M., Tarkowski, A., Greitz, D., Dahlstrom, M., Lundberg, I., Klareskog, L., and Olsson, T. (2001). Increased levels of proinflammatory cytokines and nitric oxide metabolites in neuropsychiatric lupus erythematosus. *Ann. Rheum. Dis.* **60**(4), 372–379.
94. Trysberg, E., Hoglund, K., Svenungsson, E., Blennow, K., and Tarkowski, A. (2004). Decreased levels of soluble amyloid beta-protein precursor and beta-amyloid protein in cerebrospinal fluid of patients with systemic lupus erythematosus. *Arthritis Res. Ther.* **6**(2), R129–R136.
95. Sapolsky, R. M. (2000). Glucocorticoids and hippocampal atrophy in neuropsychiatric disorders. *Arch. Gen. Psychiatry* **57**(10), 925–935.
96. Ginsburg, K. S., Wright, E. A., Larson, M. G., Fossel, A. H., Albert, M., Schur, P. H., and Liang, M. H. (1992). A controlled study of the prevalence of cognitive dysfunction in randomly selected patients with systemic lupus erythematosus. *Arthritis Rheum.* **35**(7), 776–782.
97. Hanly, J. G., Fisk, J. D., Sherwood, G., Jones, E., Jones, J. V., and Eastwood, B. (1992). Cognitive impairment in patients with systemic lupus erythematosus. *J. Rheumatol.* **19**(4), 562–567.
98. Hay, E. M., Huddy, A., Black, D., Mbaya, P., Tomenson, B., Bernstein, R. M., Lennox Holt, P. J., and Creed, F. (1994). A prospective study of psychiatric disorder and cognitive function in systemic lupus erythematosus. *Ann. Rheum. Dis.* **53**(5), 298–303.
99. Kozora, E., Thompson, L. L., West, S. G., and Kotzin, B. L. (1996). Analysis of cognitive and psychological deficits in systemic lupus erythematosus patients without overt central nervous system disease. *Arthritis Rheum.* **39**(12), 2035–2045.
100. Papero, P. H., Bluestein, H. G., White, P., and Lipnick, R. N. (1990). Neuropsychologic deficits and antineuronal antibodies in pediatric systemic lupus erythematosus. *Clin. Exp. Rheumatol.* **8**(4), 417–424.
101. Waterloo, K., Omdal, R., Sjoholm, H., Koldingsnes, W., Jacobsen, E. A., Sundsfjord, J. A., Husby, G., and Mellgren, S. I. (2001). Neuropsychological dysfunction in systemic lupus erythematosus is not associated with changes in cerebral blood flow. *J. Neurol.* **248**(7), 595–602.
102. Long, A. A., Denburg, S. D., Carbotte, R. M., Singal, D. P., and Denburg, J. A. (1990). Serum lymphocytotoxic antibodies and neurocognitive function in systemic lupus erythematosus. *Ann. Rheum. Dis.* **49**(4), 249–253.
103. Hanly, J. G., Behmann, S., Denburg, S. D., Carbotte, R. M., and Denburg, J. A. (1989). The association between sequential changes in serum antineuronal antibodies and neuropsychiatric systemic lupus erythematosus. *Postgrad. Med. J.* **65**(767), 622–627.
104. Bresnihan, B., Oliver, M., Grigor, R., and Hughes, G. R. (1977). Brain reactivity of lymphocytotoxic antibodies in systemic lupus erythematosus with and without cerebral involvement. *Clin. Exp. Immunol.* **30**(3), 333–337.
105. Denburg, S. D., Carbotte, R. M., Long, A. A., and Denburg, J. A. (1988). Neuropsychological correlates of serum lymphocytotoxic antibodies in systemic lupus erythematosus. *Brain Behav. Immun.* **2**(3), 222–234.
106. How, A., Dent, P. B., Liao, S. K., and Denburg, J. A. (1985). Antineuronal antibodies in neuropsychiatric systemic lupus erythematosus. *Arthritis Rheum.* **28**(7), 789–795.
107. Wilson, H. A., Winfield, J. B., Lahita, R. G., and Koffler, D. (1979). Association of IgG anti-brain antibodies with central nervous system dysfunction in systemic lupus erythematosus. *Arthritis Rheum.* **22**(5), 458–462.
108. Bluestein, H. G., Williams, G. W., and Steinberg, A. D. (1981). Cerebrospinal fluid antibodies to neuronal cells: association with neuropsychiatric manifestations of systemic lupus erythematosus. *Am. J. Med.* **70**(2), 240–246.
109. Bonfa, E., Golombek, S. J., Kaufman, L. D., Skelly, S., Weissbach, H., Brot, N., and Elkon, K. B. (1987). Association between lupus psychosis and anti-ribosomal P protein antibodies. *N. Engl. J. Med.* **317**(5), 265–271.
110. Isshi, K., and Hirohata, S. (1998). Differential roles of the anti-ribosomal P antibody and antineuronal antibody in the pathogenesis of central nervous system involvement in systemic lupus erythematosus. *Arthritis Rheum.* **41**(10), 1819–1827.
111. Lai, N. S., and Lan, J. L. (2000). Evaluation of cerebrospinal anticardiolipin antibodies in lupus patients with neuropsychiatric manifestations. *Lupus* **9**(5), 353–357.
112. Mevorach, D., Raz, E., and Steiner, I. (1994). Evidence for intrathecal synthesis of autoantibodies in systemic lupus erythematosus with neurological involvement. *Lupus* **3**(2), 117–121.
113. Hanly, J. G., Hong, C., Smith, S., and Fisk, J. D. (1999). A prospective analysis of cognitive function and anticardiolipin antibodies in systemic lupus erythematosus. *Arthritis Rheum.* **42**(4), 728–734.
114. Menon, S., Jameson-Shortall, E., Newman, S. P., Hall-Craggs, M. R., Chinn, R., and Isenberg, D. A. (1999). A longitudinal study of anticardiolipin antibody levels and cognitive functioning in systemic lupus erythematosus. *Arthritis Rheum.* **42**(4), 735–741.
115. Sastre-Garriga, J., and Montalban, X. (2003). APS and the brain. *Lupus* **12**(12), 877–882.
116. Waterloo, K., Omdal, R., Husby, G., and Mellgren, S. I. (2002). Neuropsychological function in systemic lupus erythematosus: a five-year longitudinal study. *Rheumatology (Oxford)* **41**(4), 411–415.
117. Ziporen, L., Polak-Charcon, S., Korczyn, D. A., Goldberg, I., Afek, A., Kopolovic, J., Chapman, J., and Shoenfeld, Y. (2004). Neurological dysfunction associated with antiphospholipid syndrome: histopathological brain findings of thrombotic changes in a mouse model. *Clin. Dev. Immunol.* **11**(1), 67–75.
118. Elkon, K., Skelly, S., Parnassa, A., Moller, W., Danho, W., Weissbach, H., and Brot, N. (1986). Identification and chemical synthesis of a ribosomal protein antigenic determinant in systemic lupus erythematosus. *Proc. Natl. Acad. Sci. USA* **83**(19), 7419–7423.
119. Schneebaum, A. B., Singleton, J. D., West, S. G., Blodgett, J. K., Allen, L. G., Cheronis, J. C., and Kotzin, B. L. (1991). Association of psychiatric manifestations with antibodies to ribosomal P proteins in systemic lupus erythematosus. *Am. J. Med.* **90**(1), 54–62.

120. Caponi, L., Chimenti, D., Pratesi, F., and Migliorini, P. (2002). Anti-ribosomal antibodies from lupus patients bind DNA. *Clin. Exp. Immunol.* **130**(3), 541–547.
121. Takeda, I., Rayno, K., Movafagh, F. B., Wolfson-Reichlin, M., and Reichlin, M. (2001). Dual binding capabilities of anti-double-stranded DNA antibodies and anti-ribosomal phosphoprotein (P) antibodies. *Lupus* **10**(12), 857–865.
122. Hulsey, M., Goldstein, R., Scully, L., Surbeck, W., and Reichlin,M. (1995). Anti-ribosomal P antibodies in systemic lupus erythematosus: a case-control study correlating hepatic and renal disease. *Clin. Immunol. Immunopathol.* **74**(3), 252–256.
123. Yoshio, T., Hirata, D., Onda, K., Nara, H., and Minota, S. 2005). Antiribosomal P protein antibodies in cerebrospinal fluid are associated with neuropsychiatric systemic lupus erythematosus. *J. Rheumatol.* **32**(1), 34–39.
124. Nagai, T., Arinuma, Y., Yanagida, T., Yamamoto, K., and Hirohata, S. (2005). Anti-ribosomal P protein antibody in human systemic lupus erythematosus up-regulates the expression of proinflammatory cytokines by human peripheral blood monocytes. *Arthritis Rheum.* **52**(3), 847–855.
125. DeGiorgio, L. A., Konstantinov, K. N., Lee, S. C., Hardin, J. A., Volpe, B. T., and Diamond, B. (2001). A subset of lupus anti-DNA antibodies cross-reacts with the NR2 glutamate receptor in systemic lupus erythematosus. *Nat. Med.* **7**(11), 1189–1193.
126. Huerta, P. T., Kowal, C., DeGiorgio, L. A., Volpe, B. T., and Diamond, B. (2006). Immunity and behavior: antibodies alter emotion. *Proc. Natl. Acad. Sci. USA* **103**(3), 678–683.
127. Kowal, C., DeGiorgio, L. A., Nakaoka, T., Hetherington, H., Huerta, P. T., Diamond, B., and Volpe, B. T. (2004). Cognition and immunity; antibody impairs memory. *Immunity* **21**(2), 179–188.
128. Brey, R. L., Sakic, B., Szechtman, H., and Denburg, J. A. (1997). Animal models for nervous system disease in systemic lupus erythematosus. *Ann. NY Acad. Sci.* **823**, 97–106.
129. Sidor, M. M., Sakic, B., Malinowski, P. M., Ballok, D. A., Oleschuk, C. J., and Macri, J. (2005). Elevated immunoglobulin levels in the cerebrospinal fluid from lupus-prone mice. *J. Neuroimmunol.* **165**(1-2), 104–113.
130. Finazzi, G., Marchioli, R., Brancaccio, V., Schinco, P., Wisloff, F., Musial, J., Baudo, F., Berrettini, M., Testa, S., D'Angelo, A., Tognoni, G., and Barbui, T. (2005). A randomized clinical trial of high-intensity warfarin vs. conventional antithrombotic therapy for the prevention of recurrent thrombosis in patients with the antiphospholipid syndrome (WAPS). *J. Thromb. Haemost.* **3**(5), 848–853.
131. Marks, S. D., Patey, S., Brogan, P. A., Hasson, N., Pilkington, C., Woo, P., and Tullus, K. (2005). B lymphocyte depletion therapy in children with refractory systemic lupus erythematosus. *Arthritis Rheum.* **52**(10), 3168–3174.
132. Tokunaga, M., Fujii, K., Saito, K., Nakayamada, S., Tsujimura, S., Nawata, M., and Tanaka, Y. (2005). Down-regulation of CD40 and CD80 on B cells in patients with life-threatening systemic lupus erythematosus after successful treatment with rituximab. *Rheumatology (Oxford)* **44**(2), 176–182.
133. Williams, R. C., Jr., Sugiura, K., and Tan, E. M. (2004). Antibodies to microtubule-associated protein 2 in patients with neuropsychiatric systemic lupus erythematosus. *Arthritis Rheum.* **50**(4), 1239–1247.
134. Yajima, N., Kasama, T., Isozaki, T., Odai, T., Matsunawa, M., Negishi, M., Ide, H., Kameoka, Y., Hirohata, S., and Adachi, M. (2005). Elevated levels of soluble fractalkine in active systemic lupus erythematosus: potential involvement in neuropsychiatric manifestations. *Arthritis Rheum.* **52**(6), 1670–1675.

17

Progressive Multifocal Leukoencephalopathy

Benjamin Brooks, MD

Keywords: *AIDS, demyelinating disease, HIV, JC virus, polyomavirus, progressive multifocal leukoencephalopathy*

I. Brief History and Nomenclature
II. Etiology
III. Pathogenesis
IV. Lymphocyte Control on Latency
V. Pathophysiology
VI. Molecular Mechanisms
VII. Explanation of Symptoms in Relation to Pathophysiology
VIII. Natural History
    References

I. Brief History and Nomenclature

Progressive multifocal leukoencephalopathy (PML) is a human, virus-induced, subacute, fatal, demyelinating, neurodegenerative disease. It was first described, nearly 50 years ago by K. E. Astrom, E. Mancall, and E. P. Richardson, in patients with chronic lymphocytic leukemia and lymphoma who were participating in the nascent clinical attempts to employ chemotherapy in oncology. In the ensuing 2 decades, the viral etiology of this disease was established as a neuropathogenic strain of human polyomavirus isolated in human fetal glial cells from a patient with Hodgkin's disease JC with virological techniques by B. L. Padgett and D. L. Walker working with G. M. ZuRhein, who had been the first to note viral particles in PML lesions after employing novel controversial electron microscopic techniques.

Initially recognized as an endemic complication of chemotherapy in cancer patients and organ transplant recipients, PML reached epidemic proportions when an increasing proportion of the population was becoming immunosuppressed with the onset of the acquired immune deficiency syndrome (AIDS), which is caused by human immunodeficiency virus type 1 (HIV-1). There is no curative treatment for PML, but clinical features improve with restitution of immune competence in AIDS-associated and chemotherapy-associated cases. The dependence on the integrity of the immune system in limiting activation of human polyomavirus was demonstrated with the development of PML in interferon (IFN)-β1a–treated multiple sclerosis (MS) patients who also received natalizumab, a humanized recombinant murine myeloma cell–produced immunoglobulin G4κ(IgG4κ) monoclonal antibody that binds to the α_4-subunit of $\alpha_4\beta_1$- and $\alpha_4\beta_7$-integrins, which is expressed on the surface of all leukocytes except neutrophils and inhibits the α_4-mediated adhesion of leukocytes to their counter-receptor(s) (Fig. 1).

II. Etiology

Polyomaviruses are small (45-nm) non-enveloped icosahedral virions that contain a small genome of approximately 5000 bp of circular double-stranded DNA (Fig. 2). Polyomaviruses were isolated from rodents in 1953, monkeys in 1960, and humans in 1971. These viruses infect hosts from humans to birds, with some viruses of this family inducing tumors naturally or in experimental animals. It is uncertain whether gliomas may be similarly induced in humans.

During primary infection, five RNA transcripts are produced with both strands of DNA used for transcription. Five proteins produced include T antigen and t protein as early proteins. The latter replicating structural proteins are VP1 (39 kDa) (viral antibody and induction), VP2 (35 kDa), and VP3 (25 kDa). The DNA of JC virus (JCV), BK virus (BKV), and SV40 exhibit homology. The early replicating nuclear T antigens of JCV, BKV, and SV40 cross-react serologically and functionally in chimeras studied under laboratory conditions. The late replicating regulatory agnoprotein is not incorporated into virions. Agnoprotein interacts with large T antigen and a cellular transcription factor, YB-1, in a reciprocal pattern. Initiation of polyomavirus DNA replication requires the participation of the viral early protein T antigen, cellular replication factors, and DNA polymerases.

JCV exhibits a narrow host range, due to species specificity of the DNA polymerase and a tissue specificity that restricts its replication to glial cells, coupled with glial cell–specific transcription of the viral early promoter. Infection of cells takes place via α2, 6-linked sialic acid residues that are neuraminidase sensitive and may be blocked by 5-HT$_{2A}$ serotonin receptor antagonists. DNA binding proteins with binding sites in the Non-Coding Control Region (NCCR) (In-1, Sp1, NF-1, CRE, TAT, p53, AP-1) appear to determine regulation of JCV DNA replication and transcription following infection. Other cellular transcription factors (c-Jun, nuclear factor-κB [NF-κB], Tst-1, YB-1,

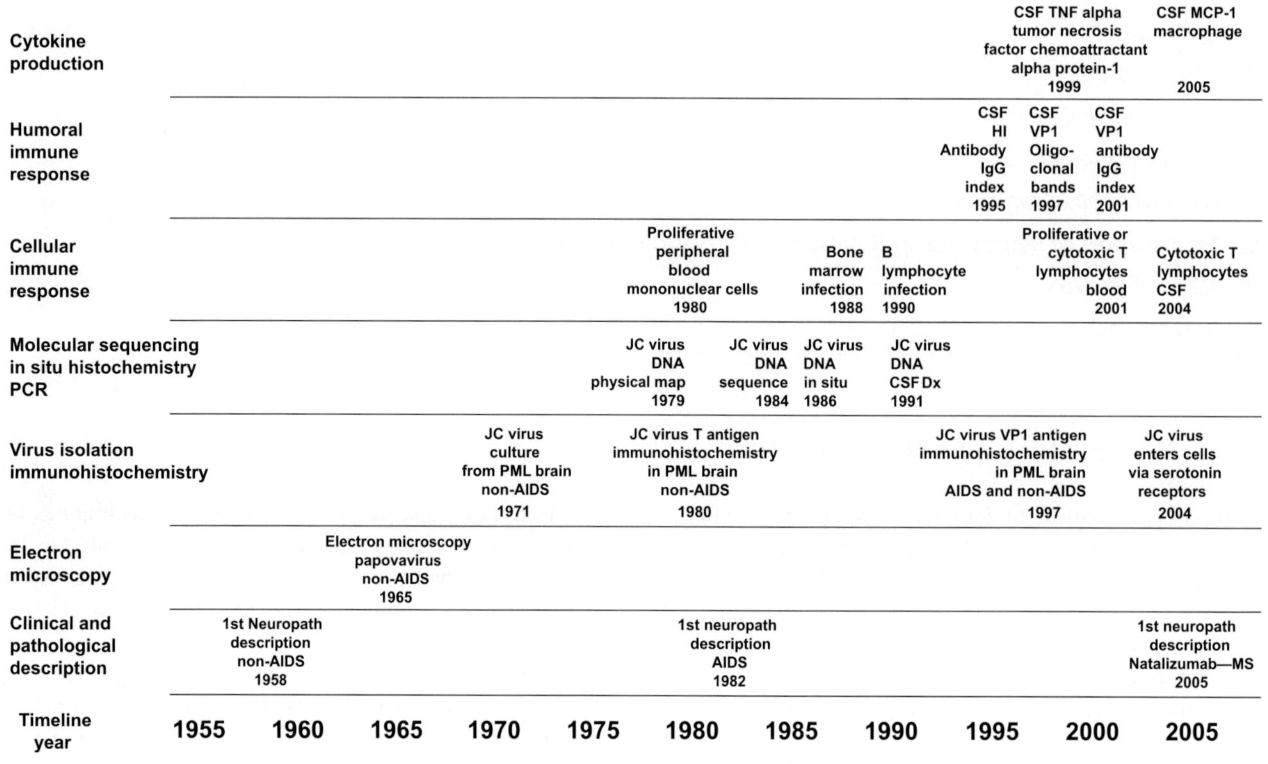

Cytokine production								CSF TNF alpha tumor necrosis factor chemoattractant alpha protein-1 1999		CSF MCP-1 macrophage 2005
Humoral immune response								CSF HI Antibody IgG index 1995	CSF VP1 Oligo-clonal bands 1997	CSF VP1 antibody IgG index 2001
Cellular immune response						Proliferative peripheral blood mononuclear cells 1980	Bone marrow infection 1988	B lymphocyte infection 1990	Proliferative or cytotoxic T lymphocytes blood 2001	Cytotoxic T lymphocytes CSF 2004
Molecular sequencing in situ histochemistry PCR						JC virus DNA physical map 1979	JC virus DNA sequence 1984	JC virus DNA in situ 1986	JC virus DNA CSF Dx 1991	
Virus isolation immunohistochemistry				JC virus culture from PML brain non-AIDS 1971		JC virus T antigen immunohistochemistry in PML brain non-AIDS 1980			JC virus VP1 antigen immunohistochemistry in PML brain AIDS and non-AIDS 1997	JC virus enters cells via serotonin receptors 2004
Electron microscopy			Electron microscopy papovavirus non-AIDS 1965							
Clinical and pathological description	1st Neuropath description non-AIDS 1958						1st neuropath description AIDS 1982			1st neuropath description Natalizumab—MS 2005
Timeline year	1955	1960	1965	1970	1975	1980	1985	1990	1995	2000	2005

Figure 1 Progressive multifocal leukoencephalopathy timeline pre-AIDS and post-AIDS.

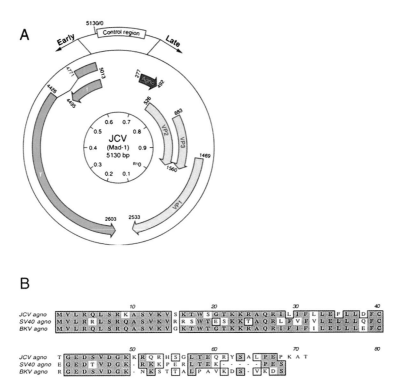

Figure 2 JC virus genome. (Reproduced, with permission, from Safak, M., Barrucco, R., Darbinyan, A., Okada, Y., Nagashima, K., and Khalili, K. [2001]. Interaction of JC virus agnoprotein with T antigen modulates transcription and replication of the viral genome in glial cells. *J. Virol.* **75**, 1476–1486.)

GBP-I, Pur-alpha) are also indirectly involved. The NCCR is separate from the structural genes in sequenced brain and cerebrospinal fluid (CSF) isolates. The NCCR can be further segregated into six boxes containing binding sites for cellular transcription factors—A (Tst-1), B (Sp1), C (NF-1, CRE, upTAR), D (Sp-1), E, and F (NF-1, p53, AP-1), which are conserved in all sequenced isolates. Sp1-binding site in box B activates JCV early promoter and increases JCV replication, whereas JCV-produced T antigen mediates transactivation of JCV early genes through interactions with SP1 binding site in box D. Isolates from nonbrain regions show less tendency to rearrangements in the NCCR. The NCCR of the JCV isolates from HIV-1–positive patients with PML show particular increases of the JCV upTAR sequence that is important for the HIV-1 Tat protein stimulation of JCV later promoter genes. HIV-1 Tat protein enhances Pur-alpha binding, which increases JCV transcription. Different rearrangements are present in isolates from non-HIV-1–coinfected PML patients. The isolates from brain or CSF of HIV-1 and –negative patients without PML do not show such rearrangements.

Primary infections with the human polyomaviruses BKV and JCV occur during childhood and are largely asymptomatic, with both BKV and JCV persisting as latent infections in the kidneys and B lymphocytes throughout early adulthood.

Infection increases throughout childhood with BKV strains more representative than JCV (Fig. 3). Increased *reactivation* from latent persistent infection of both JCV and BKV occurs with aging when measured as increasing *viruria* in antibody-positive individuals (47% > 40 years; 80% > 79 years). *JCV viruria* is significantly higher in chemotherapy, including treated patients (67%, corticosteroids as well as cytotoxic drugs) than that in the noncytotoxic chemotherapy-treated patients (28%). *Reactivated infection* can be confirmed by measuring nested PCRs for BKV and JCV NCCR and VP1 DNA sequences in peripheral blood mononuclear cells (PBMCs). BKV NCCR DNA sequences were found in nearly 25% of control blood donors, but less than 10% had BKV VP1 DNA sequences indicative of active infection. In contrast, JCV NCCR and VP1 DNA sequences were present in less than 1% of control blood donors. Control subjects show decreasing positivity for virus DNA sequences with increasing age. This is consistent with the idea that there are other sites for latent infection, and that PBMCs are a secondary site indicating recent infection or reactivation. JCV early T antigen DNA sequences were detected in PBMCs from less than 20% of HIV-1–positive immunocompetent patients, from less than 25% of HIV-1–positive immunocompromised patients, and from more than 50% of patients with AIDS and PML.

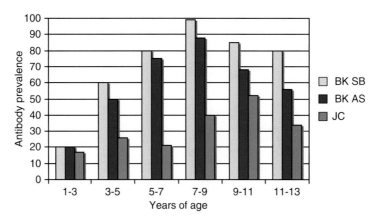

Figure 3 Development of JC virus and BK virus antibody with age. (Redrawn, with permission, from Stolt, A., Sasnauskas, K., Koskela, P., Lehtinen, M., and Dillner, J. [2003]. Seroepidemiology of the human polyomaviruses. *J. Gen. Virol.* **84**, 1499–1504.)

Table 1 JC Virus Genotypes

1	2	3	4	6	7	8
Europe		West Africa	Europe	West Africa	Asia	Western Pacific
1A Europe 1B Europe Northern Japan Northeast Siberia, Northern Canada	2A/2C Japanese Native Americans 2B Eurasians 2D Indians 2E Australians Western Pacific	Ethiopia Tanzania South Africa Biaka Pygmies Bantus	Europe	Ghana Biaka Pygmies Bantus	7A Southern China Southeast Asia 7B Northern China Mongolia Japan 7C Northern and Southern China	Papua New Guinea Pacific Islands

The promoter/enhancer region of JCV, unlike the remainder of the viral genome structural genes, differs much more among the various clinical isolates. JCV has been classified into several types on the basis of the differences within the viral NCCR regulatory region, as well as changes in other virus genes. Genotyping of urine specimens looking at type-specific mutations from individuals allows definition of the population history for different ethnic groups. JCV is a valuable marker to trace human dispersal because of its strong genetic stability, lack of pathogen power (causing disease only in 5–7% of immunosuppressed people), and absence of significant genetic recombination due to its circular DNA. The mechanism of transmission is mainly within the family. The main mode of transmission of JCV is from parents to children through long-term cohabitation. Populations that are in the same geographical area but do not intermingle have different JCV infections.

Seven genotypes differing in DNA sequence not only in the NCCR but also other portions of the genome by 1–3% characterize several population groups (African, European, and Asian, as well as Oceania). It is possible to follow populations from Africa, Europe, and Asia into North and South America by the JCV genotypes they carried. The first population to settle in the Americas, the Native Americans, brought with them type 2A from Northeast Asia. European settlers arriving after Columbus carried primarily type 1 and type 4. Africans brought by the slave trade carried type 3 and type 6 (Table 1).

JCV and BKV can agglutinate type O human and guinea pig erythrocytes (red blood cells [RBCs]) and are able to transform certain cells *in vitro* or induce tumors in newborn Syrian hamsters (JCV: medulloblastoma, glioblastoma, meningioma; BKV, ependymoma, osteosarcoma, pancreatic islet tumor) and adult owl monkeys, but tumor promotion is not established in humans.

III. Pathogenesis

JCV and BKV are ubiquitous throughout the world. The natural route of infection and transmission has not been definitively established. Seroepidemiology studies using hemagglutination inhibition or JCV-like VLP protein–based enzyme-linked immunoassay (EIA) for human polyomavirus antibodies have shown early childhood infection followed by stable antibodies over time with only a low incidence of new infections in immunocompetent subjects. Primary JCV infection in immunocompetent patients is not well described. JCV viral DNA is present in tonsillar B lymphocytes and bone marrow cells of various types. Primary BKV infection may be characterized by asymptomatic mild urinary tract infection or upper respiratory tract infection. Serological evidence for reactivation of JCV and/or BKV by polymerase chain reaction (PCR) or virus isolation occurs in 5–10% of women during pregnancy. It is unclear whether human intrauterine infection with JCV or BKV occurs. Congenital infections also may occur, as children with various congenital disorders were found to have BKV-specific immunoglobulin M (IgM). There is no evidence for the existence of animal reservoirs. The exact route of primary infection is still unknown. By analogy with murine polyoma virus and SV40, infection may occur by aerosol inhalation or oral ingestion of virus with excretion occurring via the urinary tract. JCV and BKV have also been measured in sewer water sources (Table 2).

JCV DNA sequences are present in tonsillar lymphocytes, PBMCs, kidney, brain, heart, spleen, bone marrow, lung, liver, and colon. JCV and BKV DNA sequences are found in nearly 2 and 4% of CSF samples from clinically suspected undiagnosed meningitis or encephalitis in both immunocompromised and immunocompetent patients. Taken together, these observations suggest that primary infection may precede latent infection in lymphocytes, urogenital tract, and brain—with reactivation to cause disease in immunocompromised hosts.

The oncogenicity of JCV is attributed to the viral early gene product T antigen. T antigen has the ability to associate with and functionally inactivate well-studied tumor suppressor proteins including p53 and pRb.

PML is uncommon in children and young persons, and it more often develops in people in the fifth and sixth decades of life, consistent with the higher reactivation of JCV with age. It is not clear whether there is primary infection of the brain following reactivation of JCV in nonbrain reservoirs or reactivation of JCV from latent infection of the brain by immunosuppression caused by disease or chemotherapy. The distribution of brain involvement in PML is consistent with hematogenous dissemination of the virus, either right after initial primary infection or after reactivation of latent infection in sites after original infection of the central nervous system (CNS). Regional localization of PML to the brainstem and cerebellum or to the brainstem and spinal cord compared with bilateral periventricular centrum semiovale or subcortical lesions is consistent with this mechanism.

JCV is conveyed in peripheral blood as latent or reactivated virus in or on immune cells. Nested PCR assays identify JCV T antigen–related DNA and VP1 structural protein mRNAs, which confirm viral replication may occur in blood. VP1 mRNAs are rarely identified in blood of non-PML patients but are present in nearly 4 in 10 JCV T antigen DNA–positive urine samples and in nearly all CSF samples from patients with both AIDS and PML.

PML infections are associated with elevated CSF antibody index and oligoclonal bands specific for JCV VP1 capsid antigen. Antibody levels are highest in PML patients. However some non-PML patients, including those who were given the diagnosis of MS, also have JCV anti-VP1 antibody, although at lower levels (Table 3).

The incidence of PML increased after the development of the AIDS epidemic caused by HIV-1. The equilibrium in humans between low-level pathogenicity caused by JCV in a population was not disturbed, causing an endemic disease largely associated with chemotherapy and disease-induced immunosuppression to become a near-epidemic disease with a 20-fold increase in the incidence of PML (Fig. 4). With the increasing development of transplantation programs, the endemic rate is also increasing slightly.

Table 2 Urban Sewer Water Virus Content–Particle Equivalents/ml

	JC virus	BK virus	Adenovirus
Barcelona, Spain		100	100–1000
Nancy, France		10	100–1000
Umea, Sweden	100	1000	100–1000
Pretoria, South Africa	100	1000	100–1000

IV. Lymphocyte Control on Latency

The nature of the immune response that contains the replication of this virus is unknown. JCV-specific cellular immune responses are present in patients with PML and control subjects. JCV antigen-stimulated PBMCs of HIV-1–infected patients who were survivors of PML, and one HIV-uninfected patient recently diagnosed with PML lysed autologous B-lymphoblastoid cell lines expressing either the

Table 3 Cerebrospinal Fluid JC Virus Immune Response in PML and Non-PML Patients [cited in Weber T]

	JCV HI Antibody (CSF Index)	JCV HI Antibody (CSF OCB)	JCV Anti-VP1 Antibody (CSF Index)	JCV Anti-VP1 Antibody (CSF OCB)	
London PML	12/18 (67%)	7/7 (100%)	14/18 (78%)	10/18 (55%)	Knowles 2003 Sindic 1997
Non-PML	0/71 (0%)	—	0/31 (0%)	2/31 (6%)	MS, Lyme
Hamburg PML	—	—	47/62 (76%)	—	Weber 1997
Non-PML	—	—	5/155 (3%)	—	

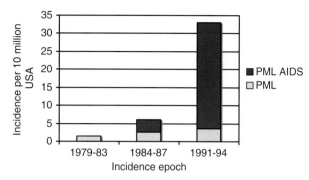

Figure 4 Progressive multifocal leukoencephalopathy incidence in the pre-AIDS and post-AIDS epoch. (Redrawn, with permission, Holman, R. C., Torok, T. J., Belay, E. D., Janssen, R. S., and Schonberger, L. B. [1998]. Progressive multifocal leukoencephalopathy in the United States, 1979–1994: increased mortality associated with HIV infection. *Neuroepidemiology* **17,** 303–309; and Holman, R. C., Janssen, R. S., Buehler, J. W., Zelasky, M. T., and Hooper, W. C. [1991]. Epidemiology of progressive multifocal leukoencephalopathy in the United States: analysis of national mortality and AIDS surveillance data. *Neurology* **41**(11), 1733–1736.)

JCV T-regulatory protein or the VP1 major capsid protein. This cytotoxic lysis is mediated by CD8+ T lymphocytes and is major histocompatibility complex (MHC) class I–restricted, consistent with cytotoxic T lymphocyte (CTL) restriction of JCV virus infection in addition to antibody responses that control JCV infection. JCV-specific CTLs which might control JCV replication in healthy subjects are measured by ^{51}Cr release and tetramer staining assays, and are present in over nearly 75% of immunocompetent control subjects. These JCV-specific CD8+ CTLs recognize one or two epitopes of JCV VP1 protein, the HLA-A*0201–restricted VP1p36 and VPp1100 epitopes. The observed frequency of JCV VP1 epitope–specific CTLs varied from less than 1/100,000 to slightly more than 1/2500 PBMCs.

JCV-specific CTLs are rarely detected in PBMCs of HIV-infected PML patients who have progressive neurological disease and eventual fatal outcome. These CTLs may also prove useful as a favorable prognostic marker in the clinical management of these patients.

Genomes of JCV strains isolated from different PML patients frequently demonstrate specific sequence alterations involving the transcriptional control elements. These variations reflect an adaptation process, by which JCV changes from a kidney-specific to a brain-specific virus. Persistently infected patients during prolonged immunosuppression have shown new types of JCV in the urine that exhibit deletions and duplications within their promoter-enhancer region and that have changes similar to JCV isolates found in PML brain. JCV DNA is present in most kidneys from non-PML subjects. The sequence of isolates from one kidney is similar to isolates from the paired kidney, consistent with hematogenous infection. JCV DNA amplified from the kidneys of non-PML patients is "archetype," whereas JCV DNA from the kidney of adult PML patients can bear resemblance to JCV DNA isolated in the PML brain. However, JCV DNA is present in many "normal" brains and is not archetype. The presence of JCV DNA in PBMCs raises questions concerning whether the changes in the NCCR regions leading to the adapted form of JCV arise during the primary infection or whether lymphocyte infection causes the rearrangement process.

V. Pathophysiology

PML is a unique demyelinating disease with distinctive pathology consisting of multiple foci of demyelination of varying size, from pinpoint lesions to areas of several centimeters. The lesions may occur anywhere but are usually in the cerebral hemispheres, less often in the cerebellum and brainstem, and rarely in the spinal cord. The neuropathological examination of freshly sectioned brain shows focal areas of excavation in the white matter that may be as small as 1 mm and range to many centimeters in diameter. At the microscopic level, there is the triad of multifocal demyelination, dense, basophilic twofold to threefold swollen oligodendroglial nuclei, together with ballooned astrocytes demonstrating irregularly lobulated hyperchromatic nuclei. Structural viral clusters are seen at the electron microscopic level as groups of 28- to 45-nanometer papovavirus particles in dense crystalline arrays. Smaller groups of virus particles may also be noted as well, primarily in oligodendroglial nuclei, with additional smaller groups of virus particles in macrophages that do not usually have the crystalline structures. The bizarre astrocytes do not have the large crystalline arrays seen in infected oligodendroglial cells (Fig. 5).

The significantly higher incidence of PML in AIDS is due to molecular interactions between HIV-1 and JCV, via the HIV-1–encoded trans-regulatory Tat protein, which are responsible for the activation of the JCV enhancer-promoter in the NCCR part of the JCV genome. An indirect mechanism through activation of cytokines, such as transforming growth factor (TGF)-β1 and Smad-3 and Smad-4, may also be responsible for the enhancement of JCV gene expression. Evidence for involvement of TGF-β1 signaling pathway in activation of JCV is shown neuropathologically (Table 4).

The role of the immune constitution in the prognosis of PML was brought into focus in PML cases without AIDS, where there was general impairment of cell-mediated immune response, along with severe and selective impairment of cell-mediated immune response to JCV virus antigen. However, although immune response to viral antigen is decreased in AIDS, there is also altered cytokine production by PBMCs in response to JCV viral antigen (Table 5).

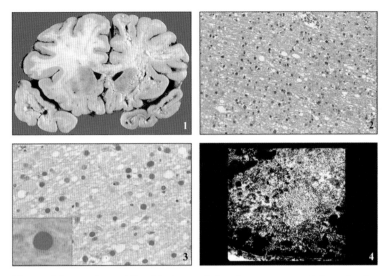

Figure 5 Progressive multifocal leukoencephalopathy gross neuropathology showing cavitating lesions with demyelination (1), oligodendrocytes with dark deep-stained nuclei and homogenous nuclear inclusions enlarged in the nuclei of oligodendrocytes (2). Also shown are antibody stains against polyomavirus demonstrated strong nuclear staining of the oligodendrocytes (3) and numerous particles, which were spherical and approximately 35–45 nm, within the nuclei of oligodendrocytes (4). (Reproduced, with permission, from Wang, H. Y. [2004]. Pathologic quiz case: a 54-year-old deceased man with diffuse subcortical lesions of the central nervous system. Progressive multifocal leukoencephalopathy. *Arch. Pathol. Lab. Med.* **128**(4), e70–e72.)

Table 4 Activation of Cytokines in PML Brain in PML Associated with HIV-1 Infection[a]

Neuropathology	JCV VP1	JCV Agno	HIV p24	HIV Tat	TGF -beta-R	TGF -beta	Smad-3	Smad-4
Astrocytes	++n	–	–	–	–	–	–	–
HIV-PML	+cy	+cy	+cy	+cy	++cy	++cy	+cy	+cy
Astrocytes	++n	–	–	–	–	–	–	–
PML	+cy	+cy	–	–	++cy	++cy	+cy	+cy
Astrocytes	–	–	–	–	–	–	–	–
Normal	–	–	–	–	+cy	+cy	+cy	+cy
Oligodendrocytes	+++n	–	–	++n	–	–	++n	++n
HIV-PML	+cy	++cy	–	–	++cy	+++cy	+cy	+cy
Oligodendrocytes	+++n	–	–	–	–	–	+n	+n
PML	–	++cy	–	–	+cy	++cy	+cy	+cy
Oligodendrocytes	–	–	–	–	–	–	–	–
Normal	–	–	–	–	+cy	+cy	+cy	+cy

[a] From Enam, S., Sweet, T. M., Amini, S., Khalili, K., and Del Valle, L. (2004). Evidence for involvement of transforming growth factor beta1 signaling pathway in activation of JC virus in human immunodeficiency virus 1-associated progressive multifocal leukoencephalopathy. *Arch. Pathol. Lab. Med.* **128**, 282–291.

Table 5 Decreased PBMC Proliferation and Altered Cytokine Production in PML Associated with HIV-1 Infection

Viral or Bacterial Antigen or Lectin Activation	IL-10 Production in PML	IL-10 Production in HIV	IL-10 Production in HIV-PML	Reference
VP1-VLP recombinant capsid	Normal	Increased	Increased	*Ann Neurol* 2001; 49:636–642
Tetanus toxoid	Normal	Normal	Increased	*Ann Neurol* 2001; 49:636–642
Phytohemagglutinin	Normal	Decreased	Decreased	*Ann Neurol* 2001; 49:636–642

VI. Molecular Mechanisms

JCV DNA sequences are detected at higher levels in the CSF samples from the progressive PML patients, but at lower levels in the CSF from those with nonprogressive PML. The frequency of JCV DNA detection in the PBMCs and urine did not differ among the three groups studied (progressive PML in AIDS patients, nonprogressive PML in AIDS on highly active antiretroviral therapy [HAART], PML patients without AIDS). The isolated JCV strains have hypervariable NCCR-regulatory regions. In the urine samples of healthy and nonimmunocompromised individuals, the conserved archetype NCCR form is found, whereas in the brains of PML patients, forms with tandem repeats and deletions in the NCCR are found. The NCCR, seen in the Mad-1 JCV isolate, the first sequenced strain of JCV, contains two 98-bp tandem repeats each containing a TATA box. The archetype has additional 23-bp and 66-bp inserts or fragments thereof and only one TATA box. The NCCR regions from isolates cloned from different anatomic compartments of PML patients and controls subjects differ. JCV isolates from PML patients with poor clinical outcome have higher proportions of NCCR clones with both tandem repeats in plasma (>50%) and brain or CSF (>80%). In those PML patients who became longer survivors of PML, archetype sequences predominated in plasma (>75%) and brain or CSF (>90%). In patients with advanced AIDS without PML, less than 10% of JCV NCCR clones obtained in the plasma contained tandem repeats. Therefore, rearrangement of JCV NCCR is associated with the presence of tandem repeats in plasma and CNS poor clinical outcome in patients with PML.

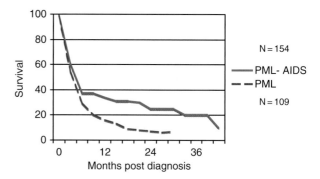

Figure 6 Survival in progressive multifocal leukoencephalopathy patients from the pre-AIDS era and the AIDS era. (Redrawn, with permission, from Brooks, B. R., and Walker, D. L. [1984]. Progressive multifocal leukoencephalopathy. *Neurol. Clin.* **2**, 299–313; and Berger, J. L., Pall, L., Lanska, D., and Whiteman, M. [1998]. *J. Neurovirol.* **4**, 59–68.)

VII. Explanation of Symptoms in Relation to Pathophysiology

PML was initially described in the context of immunosuppression caused by chemotherapy. With the development of HIV-1–induced immunosuppression, the prevalence of PML increased and further insights into the interaction between cellular transcription factors and the replication of JCV within CNS and non-CNS cells were identified. The occurrence of longer survival in some PML patients associated with effectively treated HIV-1–induced AIDS immunosuppression indicates that immune constitution is an important determinant of the occurrence of this disease (Fig. 6).

However, there are increasing reports of individuals without clearcut immunosuppression developing "primary PML." In a proportion of these cases, there may be a history of immunosuppression or even prolonged treatment with interferon-α. An example of this difficulty was reported with the development of PML in two MS patients treated with interferon and natalizumab. The interpretation that this is due to reactivation of JCV in the context of excessive immunosuppression may be complicated by concerns that a small proportion of patients diagnosed with MS may actually have primary PML occurring in immunocompetent hosts. The role of other infections such as Epstein-Barr virus (EBV) that may be present in patients with more severe MS and that may play a role in activating JCV has not been fully explored.

The acute infection of a young patient may result in clinical or asymptomatic disease. After establishment of latent infection, reactivation later in life by alteration in immune status due to malignancy, infection, or chemotherapy may lead to specific stereotyped disease expression (Fig. 7). PML may manifest itself as a cerebral-predominant disease, a

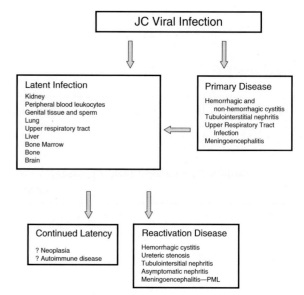

Figure 7 Proposed steps in natural history of JC virus infection and the development of progressive multifocal leukoencephalopathy.

brainstem- and spinal cord–predominant disease, and even a cerebellar-predominant disease. The possible role of JCV in cerebral malignant disease is still being actively researched.

VIII. Natural History

PML occurring before the onset of AIDS and PML occurring in AIDS patients is similar in the lack of optic nerve involvement, together with the presence of visual field defects at presentation. Weakness and speech disturbances, especially dysarthria, were common presentations of PML in both groups of patients. However, compared with PML without AIDS, patients with both PML and AIDS are younger, as were the PML cases associated with natalizumab plus IFN-β1a–treated MS patients. Safety analysis of MS patients treated with natalizumab indicates that JCV is uncommon in patients treated with monoclonal antibody alone and only occurred in the context of concurrent IFN treatment or previous immunosuppression. Significant differences in the neuroimaging presentations of PML and MS should allow for early recognition of PML if patients are carefully monitored (Fig. 8).

The selectivity afforded by monoclonal antibodies in attacking the pathogenesis of MS at well-defined and limited points will help advance their use in the treatment of such diseases. The potential drawbacks, illuminated by the occurrence of PML in MS patients also receiving IFN-β1a

Features Visualized on Magnetic Resonance Imaging to Be Considered in the Differential Diagnosis of Multiple Sclerosis and Progressive Multifocal Leukoencephalopathy.*

Feature	Multiple Sclerosis	Progressive Multifocal Leukoencephalopathy
Location of new lesions	Mostly focal; may affect entire brain and spinal cord, in white and possibly gray matter; posterior cranial fossa lesions are rarely seen	Diffuse lesions, mainly subcortical and rarely periventricular, located almost exclusively in white matter, although occasional extension to gray matter has been seen; posterior fossa frequently involved (cerebellum)
Borders	Sharp edges; mostly round or finger-like in shape (especially periventricular lesions), confluent with other lesions; U-fibers may be involved	Ill-defined edges; infiltrating; irregular in shape; confined to white matter, sparing gray matter; pushing against the cerebral cortex; U-fibers destroyed
Mode of extension	Initially focal, lesions enlarge within days or weeks and later decrease in size within months	Lesions are diffuse and asymmetric, extending homogeneously; no confluence with other lesions; confined to white-matter tracks, sparing the cortex; continuous progression
Mass effect	Acute lesions show some mass effect	No mass effect even in large lesions (but lesion slightly abuts cerebral cortex)
On T_2-weighted sequence	Acute lesions: hyperintense center, isointense ring, discrete hyperintensity outside the ring structure Subacute and chronic lesions: hyperintense, with no ring structure	Diffuse hyperintensity, slightly increased intensity of newly involved areas compared with old areas, little irregular signal intensity of lesions
On T_1-weighted sequence	Acute lesions: densely hypointense (large lesions) or isointense (small lesions); increasing signal intensity over time in 80 percent; decreasing signal intensity (axonal loss) in about 20 percent	Slightly hypointense at onset, with signal intensity decreasing over time and along the affected area; no reversion of signal intensity
On FLAIR sequence	Hyperintense, sharply delineated	Hyperintensity more obvious, true extension of abnormality more clearly visible than in T_2-weighted images
With enhancement	Acute lesions: dense homogeneous enhancement, sharp edges Subacute lesions: ring enhancement Chronic lesions: no enhancement	Usually no enhancement even in large lesions; in patients with HIV, some peripheral enhancement is possible, especially under therapy
Atrophy	Focal atrophy possible, due to focal white-matter degeneration; no progression	No focal atrophy

* FLAIR denotes fluid-attenuated inversion recovery, and HIV human immunodeficiency virus.

Figure 8 Neuroimaging features that distinguish multiple sclerosis from progressive multifocal leukoencephalopathy. (Reproduced, with permission, from Tarek, A. Y., Major, E. O., Ryschkewitsch, C., Fahle, G., Fischer, S., Hou, J., Curfman, B., Miszkiel, K., Mueller-Lenke, N., Sanchez, E., Barkhof, F., Radue, E.-W., Jäger, H. R., and Clifford, D. B. [2006]. Evaluation of patients treated with natalizumab for progressive multifocal leukoencephalopathy. *N. Engl. J. Med.* **354**, 924–933.)

together with natalizumab, could be replicated in other clinical settings requiring long-term immunosuppressive therapy of autoimmune disease with agents that could lead to reactivation of latent viruses. A more complete understanding of the interaction of JCV with humans over their lifetime will be needed to evaluate the role of selective immunosuppression versus other therapies in the safe treatment of autoimmune disease patients.

References

Ahsan, N. (ed.) (2006). Polyomaviruses and human diseases. In: "Advances in Experimental Medicine and Biology" (N. Ahsan, ed.), pp. 1–387. Springer, New York.

Astrom, K., Mancall, E., and Richardson, E. P. (1958). Progressive multifocal leukoencephalopathy: a hitherto unrecognized complication of chronic lymphatic leukaemia and Hodgkin's disease. Brain 81, 93–111.

Khalili, K., and Stoner, G. L. (eds.) (2001). "Human Polyomaviruses: Molecular and Clinical Perspectives." John Wiley & Sons, New York.

Nath, A., and Berger, J. R. (eds.) (2003). "Clinical Neurovirology." Marcel Dekker, New York.

Beck, R. C., Kohn, D. J., Tuohy, M. J., Prayson, R. A., Yen-Lieberman, B., and Procop, G. W. (2004). Detection of polyoma virus in brain tissue of patients with progressive multifocal leukoencephalopathy by real-time PCR and pyrosequencing. Diagn. Mol. Pathol. 13, 15–21.

Berger, J. R. (2003). JCV-specific CD4 T cell response: another piece of the puzzle in explaining some aspects of AIDS associated PML. AIDS 17, 1557–1559.

Berger, J. R. (2003). Progressive multifocal leukoencephalopathy in acquired immunodeficiency syndrome: explaining the high incidence and disproportionate frequency of the illness relative to other immunosuppressive conditions. J. Neurovirol. 9(suppl. 1), 38–41.

Berger, J. R., Levy, R. M., Flomenhoft, D., and Dobbs, M. (1998). Predictive factors for prolonged survival in acquired immunodeficiency syndrome-associated progressive multifocal leukoencephalopathy. Ann. Neurol. 44, 341–349.

Berger, J. R., Pall, L., Lanska, D., and Whiteman, M. (1998). Progressive multifocal leukoencephalopathy in patients with HIV infection. J. Neurovirol. 4, 59–68.

Bofill-Mas, S., Clemente-Casares, P., Major, E. O., Curfman, B., and Girones, R. (2003). Analysis of the excreted JC virus strains and their potential oral transmission. J. Neurovirol. 9, 498–507.

Bofill-Mas, S., and Girones, R. (2003). Role of the environment in the transmission of JC virus. J. Neurovirol. 9(suppl. 1), 54–58.

Brooks, B. R., and Walker, D. L. (1984). Progressive multifocal leukoencephalopathy. Neurol. Clin. 2, 299–313.

Darbinyan, A., Siddiqui, K. M., Slonina, D., Darbinian, N., Amini, S., White, M. K., and Khalili, K. (2004). Role of JC virus agnoprotein in DNA repair. J. Virol. 78, 8593–8600.

Davies, N. W., Brown, L. J., Gonde, J., Irish, D., Robinson, R. O., Swan, A. V., Banatvala, J., Howard, R. S., Sharief, M. K., and Muir, P. (2005). Factors influencing PCR detection of viruses in cerebrospinal fluid of patients with suspected CNS infections. J. Neurol. Neurosurg. Psychiatry 76, 82–87.

Delbue, S., Sotgiu, G., Fumagalli, D., Valli, M., Borghi, E., Mancuso, R., Marchioni, E., Maserati, R., and Ferrante, P. (2005). A case of a progressive multifocal leukoencephalopathy patient with four different JC virus transcriptional control region rearrangements in cerebrospinal fluid, blood, serum, and urine. J. Neurovirol. 11, 51–57.

Du Pasquier, R. A., Kuroda, M. J., Zheng, Y., Jean-Jacques, J., Letvin, N. L., and Koralnik, I. J. (2004). A prospective study demonstrates an association between JC virus-specific cytotoxic T lymphocytes and the early control of progressive multifocal leukoencephalopathy. Brain 127, 1970–1978.

Enam, S., Sweet, T. M., Amini, S., Khalili, K., and Del Valle, L. (2004). Evidence for involvement of transforming growth factor beta1 signaling pathway in activation of JC virus in human immunodeficiency virus 1-associated progressive multifocal leukoencephalopathy. Arch. Pathol. Lab. Med. 128, 282–291.

Fedele, C. G., Ciardi, M. R., Delia, S., Contreras, G., Perez, J. L., De Ona, M., Vidal, E., and Tenorio, A. (2003). Identical rearranged forms of JC polyomavirus transcriptional control region in plasma and cerebrospinal fluid of acquired immunodeficiency syndrome patients with progressive multifocal leukoencephalopathy. J. Neurovirol. 9, 551–558.

Gallia, G. L., Houff, S. A., Major, E. O., and Khalili, K. (1997). Review: JC virus infection of lymphocytes—revisited. J. Infect. Dis. 176, 1603–1609.

Gasnault, J., Kahraman, M., de Goer de Herve, M. G., Durali, D., Delfraissy, J. F., and Taoufik, Y. (2003). Critical role of JC virus-specific CD4 T-cell responses in preventing progressive multifocal leukoencephalopathy. AIDS 17, 1443–1449.

Goldmann, C., Petry, H., Frye, S., Ast, O., Ebitsch, S., Jentsch, K. D., Kaup, F. J., Weber, F., Trebst, C., Nisslein, T., Hunsmann, G., Weber, T., and Luke, W. (1999). Molecular cloning and expression of major structural protein VP1 of the human polyomavirus JC virus: formation of virus-like particles useful for immunological and therapeutic studies. J. Virol. 73, 4465–4469.

Holman, R. C., Janssen, R. S., Buehler, J. W., Zelasky, M. T., and Hooper, W. C. (1991). Epidemiology of progressive multifocal leukoencephalopathy in the United States: analysis of national mortality and AIDS surveillance data. Neurology 41, 1845–1846.

Holman, R. C., Torok, T. J., Belay, E. D., Janssen, R. S., and Schonberger, L. B. (1998). Progressive multifocal leukoencephalopathy in the United States, 1979–1994: increased mortality associated with HIV infection. Neuroepidemiology 17, 303–309.

Holmes, E. C. (2004). The phylogeography of human viruses. Mol. Ecol. 13, 745–756.

Isella, V., Marzorati, V., Curtò, N., Cappellini, A., and Appollonioa, I. (2005). Primary progressive multifocal leukoencephalopathy: report of a case. Funct. Neurol. 20, 139–142.

Katz-Brull, R., Lenkinski, R. E., Du Pasquier, R. A., and Koralnik, I. J. (2004). Elevation of myoinositol is associated with disease containment in progressive multifocal leukoencephalopathy. Neurology 63, 897–900.

Kim, J., Woolridge, S., Biffi, R., Borghi, E., Lassak, A., Ferrante, P., Amini, S., Khalili, K., and Safak, M. (2003). Members of the AP-1 family, c-Jun and c-Fos, functionally interact with JC virus early regulatory protein large T antigen. J. Virol. 77, 5241–5252.

Kleinschmidt-DeMasters, B. K., and Tyler, K. L. (2005). Progressive multifocal leukoencephalopathy complicating treatment with natalizumab and interferon beta-1a for multiple sclerosis. N. Engl. J. Med. 353, 369–374.

Knowles, W. A., and Sasnauskas, K. (2003). Comparison of cell culture-grown JC virus [primary human fetal glial cells and the JCI cell line] and recombinant JCV VP1 as antigen for the detection of anti-JCV antibody by haemagglutination inhibition. J. Virol. Methods 109, 47–54.

Komagome, R., Sawa, H., Suzuki, T., Suzuki, Y., Tanaka, S., Atwood, W. J., and Nagashima, K. (2002). Oligosaccharides as receptors for JC virus. Journal of Virology 76, 12992–13000.

Koralnik, I. J., Schellingerhout, D., and Frosch, M. P. (2004). Case records of the Massachusetts General Hospital. Weekly clinicopathological exercises. Case 14-2004. A 66-year-old man with progressive neurologic deficits. N. Engl. J. Med. 350, 1882–1893.

Koralnik, I. J., Wuthrich, C., Dang, X., Rottnek, M., Gurtman, A., Simpson, D., and Morgello, S. (2005). JC virus granule cell neuronopathy: A novel clinical syndrome distinct from progressive multifocal leukoencephalopathy. Ann. Neurol. 57, 576–580.

Koralnik, I. J. (2002). Overview of the cellular immunity against JC virus in progressive multifocal leukoencephalopathy. J. Neurovirol. 8(suppl. 2), 59–65.

Li, T. C., Takeda, N., Kato, K., Nilsson, J., Xing, L., Haag, L., Cheng, R. H., and Miyamura, T. (2003). Characterization of self-assembled virus-like particles of human polyomavirus BK generated by recombinant baculoviruses. *Virology* **311**, 115–124.

Major, E. O., Amemiya, K., Tornatore, C. S., Houff, S. A., and Berger, J. R. (1992). Pathogenesis and molecular biology of progressive multifocal leukoencephalopathy, the JC virus-induced demyelinating disease of the human brain. *Clin. Microbiol. Rev.* **5**, 49–73.

Marzocchetti, A., Cingolani, A., Giambenedetto, S. D., Ammassari, A., Giancola, M. L., Cauda, R., Antinori, A., and Luca, A. D. (2005). Macrophage chemoattractant protein-1 levels in cerebrospinal fluid correlate with containment of JC virus and prognosis of acquired immunodeficiency syndrome–associated progressive multifocal leukoencephalopathy. *J. Neurovirol.* **11**, 219–224.

Munoz-Marmol, A. M., Mola, G., Fernandez-Vasalo, A., Vela, E., Mate, J. L., and Ariza, A. (2004). JC virus early protein detection by immunohistochemistry in progressive multifocal leukoencephalopathy: a comparative study with in situ hybridization and polymerase chain reaction. *J. Neuropathol. Exp. Neurol.* **63**, 1124–1130.

Pavesi, A. (2005). Utility of JC polyomavirus in tracing the pattern of human migrations dating to prehistoric times. *J. Gen. Virol.* **86**, 1315–1326.

Pietropaolo, V., Videtta, M., Fioriti, D., Mischitelli, M., Arancio, A., Orsi, N., and Degener, A. M. (2003). Rearrangement patterns of JC virus noncoding control region from different biological samples. *J. Neurovirol.* **9**, 603–611.

Polo, C., Perez, J. L., Mielnichuck, A., Fedele, C. G., Niubo, J., and Tenorio, A. (2004). Prevalence and patterns of polyomavirus urinary excretion in immunocompetent adults and children. *Clin. Microbiol. Infect.* **10**, 640–644.

Sabath, B. F., and Major, E. O. (2002). Traffic of JC virus from sites of initial infection to the brain: the path to progressive multifocal leukoencephalopathy. *J. Infect. Dis.* **186(suppl. 2)**, S180–S186.

Safak, M., Barrucco, R., Darbinyan, A., Okada, Y., Nagashima, K., and Khalili, K. (2001). Interaction of JC virus agno protein with T antigen modulates transcription and replication of the viral genome in glial cells. *J. Virol.* **75**, 1476–1486.

Safak, M., and Khalili, K. (2003). An overview: human polyomavirus JC virus and its associated disorders. *J. Neurovirol.* **9(suppl. 1)**, 3–9.

Santagata, S., and Kinney, H. C. (2005). Mechanism of JCV entry into oligodendrocytes. *Science* **309**, 381–382.

Seth, P., Diaz, F., and Major, E. O. (2003). Advances in the biology of JC virus and induction of progressive multifocal leukoencephalopathy. *J. Neurovirol.* **9**, 236–246.

Sindic, C. J., Trebst, C., Van Antwerpen, M. P., Frye, S., Enzensberger, W., Hunsmann, G., Luke, W., and Weber, T. (1997). Detection of CSF-specific oligoclonal antibodies to recombinant JC virus VP1 in patients with progressive multifocal leukoencephalopathy. *J. Neuroimmunol.* **76**(1–2), 100–104.

Stolt, A., Sasnauskas, K., Koskela, P., Lehtinen, M., and Dillner, J. (2003). Seroepidemiology of the human polyomaviruses. *J. Gen. Virol.* **84**, 1499–1504.

Sweet, T. M., Valle, L. D., and Khalili, K. (2002). Molecular biology and immunoregulation of human neurotropic JC virus in CNS. *J. Cell. Physiol.* **191**, 249–256.

Tarek, A. Y., Major, E. O., Ryschkewitsch, C., Fahle, G., Fischer, S., Hou, J., Curfman, B., Miszkiel, K., Mueller-Lenke, N., Sanchez, E., Barkhof, F., Radue, E.-W., Jäger, H. R., and Clifford, D. B. (2006). Evaluation of patients treated with natalizumab for progressive multifocal leukoencephalopathy. *N. Engl. J. Med.* **354**, 924–933.

Vendrely, A., Bienvenu, B., Gasnault, J., Thiebault, J. B., Salmon, D., and Gray, F. (2005). Fulminant inflammatory leukoencephalopathy associated with HAART-induced immune restoration in AIDS-related progressive multifocal leukoencephalopathy. *Acta Neuropathol.* **109**, 449–455.

von Einsiedel, R. W., Samorei, I. W., Pawlita, M., Zwissler, B., Deubel, M., and Vinters, H. V. (2004). New JC virus infection patterns by in situ polymerase chain reaction in brains of acquired immunodeficiency syndrome patients with progressive multifocal leukoencephalopathy. *J. Neurovirol.* **10**, 1–11.

Wang, H. Y. (2004). Pathologic quiz case: a 54-year-old deceased man with diffuse subcortical lesions of the central nervous system. Progressive multifocal leukoencephalopathy. *Arch. Pathol. Lab. Med.* **128**, e70–e72.

Wang, M., Tzeng, T. Y., Fung, C. Y., Ou, W. C., Tsai, R. T., Lin, C. K., Tsay, G. J., and Chang, D. (1999). Human anti-JC virus serum reacts with native but not denatured JC virus major capsid protein VP1. *J. Virol. Methods* **78**, 171–176.

Weber, F., Goldmann, C., Kramer, M., Kaup, F. J., Pickhardt, M., Young, P., Petry, H., Weber, T., and Luke, W. (2001). Cellular and humoral immune response in progressive multifocal leukoencephalopathy. *Ann. Neurol.* **49**, 636–642.

Weber, T., Trebst, C., Frye, S., Cinque, P., Vago, L., Sindic, C. J., Schulz-Schaeffer, W. J., Kretzschmar, H. A., Enzensberger, W., Hunsmann, G., and Luke, W. (1997). Analysis of the systemic and intrathecal humoral immune response in progressive multifocal leukoencephalopathy. *J. Infect. Dis.* **176**(1), 250–254.

Wyen, C., Hoffmann, C., Schmeisser, N., Wohrmann, A., Qurishi, N., Rockstroh, J., Esser, S., Rieke, A., Ross, B., Lorenzen, T., Schmitz, K., Stenzel, W., Salzberger, B., and Fatkenheuer, G. (2004). Progressive multifocal leukoencephalopathy in patients on highly active antiretroviral therapy: survival and risk factors of death. *J. Acquired Immune Defic. Syndrome JAIDS* **37**, 1263–1268.

18

Immunopathogenesis of Multiple Sclerosis

Bernhard Hemmer, MD
Nicole Töpfner
Hans-Peter Hartung, MD

Keywords: *B cells, EAE, experimental encephalomyelitis, immune system, immunotherapy, multiple sclerosis, pathogenesis, T cells*

I. Introduction
II. Immune Surveillance of the CNS
III. Animal Models of CNS Inflammation
IV. Immunology of the Multiple Sclerosis Lesion
   References

I. Introduction

Multiple sclerosis (MS) is a chronic disease of the brain and spinal cord that affects women more often than men. It starts during early adulthood and, despite advances in treatment, remains a leading cause of disability [1]. MS is not generally associated with other diseases, although certain autoimmune disorders (e.g., diabetes) are slightly more common in relatives of patients with MS. The etiology of MS is unknown, but genes and environmental factors influence the risk of developing the disease. The impact of genes in MS has been demonstrated by epidemiological and family studies. Prevalence rates vary depending on the genetic background. Although MS is rare among Africans and Asians (1–10 per 100,000), it is prevalent in Caucasians (30–300 per 100,000) [2]. Moreover, concordance rates of 30% in monozygotic twins and 3% in siblings of patients with MS strongly support the role of genes in this disease. Genetic studies indicate a highly polygenetic mode of inheritance. Despite major efforts, the human leukocyte antigen (HLA) region has remained the only established disease gene locus [3]. In the HLA locus, the HLA-DR1501/DQ0601 alleles code for proteins that are important for T-cell recognition. They are associated with a two to four times higher risk of developing MS in Caucasians. No other MS-associated genes have been unequivocally identified.

The role of environmental factors is supported by migration studies, a few putative MS epidemics, and the association between clinical relapses and viral infections [4]. Although a number of environmental factors have been implicated in the pathogenesis of MS, infectious agents are still the most likely candidates. However, the search for an MS-associated microbial organism has been unsuccessful.

II. Immune Surveillance of the CNS

Based on a reduced rejection of transplants, the central nervous system (CNS) was long considered an immunoprivileged organ, but now this view has been revised [5]. Any disruption of CNS tissue activates resident microglial cells, which generate and maintain an inflammatory milieu in the damaged area. These cells up-regulate major histocompatibility complexes (MHCs; or HLA in humans) and costimulatory molecules. Microglia also secrete inflammatory mediators that attract leukocytes to the damaged CNS area. In parallel, antigens from the lesion gain access to the periphery. Microbial antigens introduced into the CNS during infection are rapidly detected in cervical or paraspinal lymph nodes (LNs). How these antigens get into the lymphoid tissue is still poorly understood. They may be passively drained or actively carried to the LNs by phagocytosing cells.

Dendritic cells (DCs) are essential to capture antigens and induce an acquired immune response in the LNs. Because DCs are present in the CNS, they might also be involved in antigen capture in this compartment. After processing the proteins, DCs present peptides bound to MHC molecules to T cells, which enter the LN. CD8+ T cells recognize individual T cell receptor (TCR) peptides bound to MHC class I molecules, whereas CD4+ T cells recognize peptides bound to MHC class II molecules. A T cell becomes activated by high-affinity binding of its TCRs to the specific MHC-bound peptides in concert with ligation of its costimulatory receptors by molecules expressed on the antigen-presenting cell (APC). These events result in massive expansion of the antigen-specific T cells in the LN, up-regulation of adhesion molecules, and acquisition of effector functions. After release from the LNs, antigen-specific T cells can cross the blood-brain barrier (BBB) and infiltrate the inflamed CNS tissue. For antigen presentation in the CNS and local reactivation of T cells, DCs also seem to play the key role [6,7]. Although MHC class II is constitutively expressed only on DCs and mature B cells at high levels ("professional APCs"), it is inducible on cells of the monocyte/macrophage lineage ("nonprofessional APCs"), MHC class I molecules are ubiquitously expressed on all intrinsic CNS cells, including neurons and oligodendrocytes.

Corresponding to the expression pattern of MHC molecules in the brain, CD4+ cells are predominantly found in perivascular cuffs and the meninges, whereas CD8+ T cells also seem to invade the inflamed parenchyma. In the CNS, the physical contact of T cells with their target peptide–MHC complex induces their effector functions and retains the cells in the lesion. CD4+ T cells release cytokines that attract macrophages. Activated macrophages secrete proinflammatory cytokines and toxic molecules (e.g., nitric oxide) and are considered the main effector cell population in CNS inflammation. In contrast to CD4+ T cells, CD8+ T cells may also directly attack CNS cells that express peptide–MHC class I molecules such as oligodendrocytes and neurons.

The cellular immune response is supported by the humoral immune response, which is also primarily initiated in the LNs. In contrast to T cells, B cells recognize by their B cell receptor (BCR) conformational or linear epitopes of proteins that are displayed by DCs. B cells that bind with their BCR such proteins with high affinity undergo activation and clonal expansion if they are supported by the presence of antigen-specific T cells that provide a proinflammatory milieu and appropriate costimulatory signals. Activated matured B cells leave the LNs and, attracted by chemokines, can cross the BBB and infiltrate the perivascular space and meninges. On rechallenge of the specific antigen, B cells release soluble immunoglobulins corresponding to the specificity of their BCR. These immunoglobulins can bind membrane-associated or soluble antigens. Membrane-associated protein–antibody complexes lead to death of the expressing cell by complement activation. Alternatively, membrane-bound or soluble antigens are inactivated by phagocytosis through binding of the antibody-Ag complex to the Fc receptor of macrophages ("opsonization").

III. Animal Models of CNS Inflammation

A number of animal models have been developed to study the role of the immune system in CNS diseases. These models are induced through autoimmune, infectious, and neurodegenerative processes. The most widely used animal model is experimental autoimmune encephalomyelitis (EAE) (Fig. 1A). Disease is elicited by immunization of susceptible animals with myelin antigens and adjuvant. This results in a T cell–mediated inflammatory disease of the spinal cord and brain with variable degrees of demyelination and axonal damage. Myelin-specific CD4+ T helper 1 (Th1) cells are central in the pathogenesis of EAE [8]. These T cells become activated in the LNs. After acquisition of effector functions and up-regulation of adhesion molecules, these lymphocytes cross the BBB. By recruiting macrophages, they precipitate inflammatory

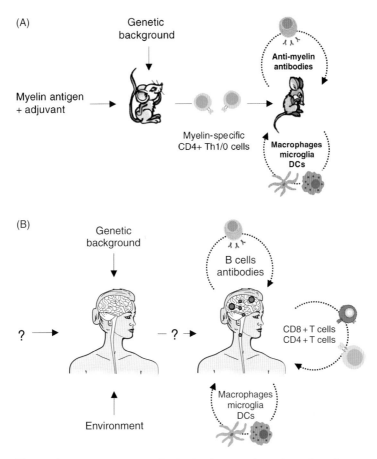

Figure 1 Pathogenetic concept for the development of experimental autoimmune encephalomyelitis (EAE) (**A**) and multiple sclerosis (MS) (**B**). The figure displays the similarities and differences in pathogenic pathways between EAE and MS. EAE is caused by myelin-specific CD4+T cells, which are supported by the innate immune system (dendritic cells [DCs], macrophages, microglia) and in some instances by B cells and antibodies. The cause of MS is unknown, but B cells, T cells, and the innate immune system seem to contribute to demyelination and neurodegeneration in MS.

damage, especially in the myelinated areas of the spinal cord and to a lesser extent in the brain.

B cells and antibodies apparently do not exert a major role in EAE because disease can be induced in B cell–deficient animals. However, autoantibodies targeting oligodendrocyte membrane proteins, such as myelin oligodendrocyte protein, enhance demyelination in some models [9]. CD8+ T cells also seem to play no disease exacerbating role in most EAE models, because β_2-microglobulin knockout mice, which cannot express MHC class I, show enhanced disease severity. Only in highly artificial models can myelin-specific CD8+ T cells induce CNS inflammation and demyelination.

Because EAE does not occur spontaneously but requires immunization with autoantigens, several hypotheses have been advanced to explain the induction of autoimmunity in MS. Cross-reactivity between foreign (viral, bacterial) proteins and autoantigens (e.g., myelin proteins), termed *molecular mimicry*, has been one appealing hypothesis of how autoreactive T cell responses may emerge after infection. The release of self-antigens from the brain might then break tolerance and thereby promote additional autoreactive T cells that target multiple myelin epitopes. This process is termed *epitope spreading* and seems to be crucial in chronic EAE models [10].

In a number of animal models, infection with neurotropic viruses induces inflammation and demyelination in the CNS [11]. In all these models, the immune system has an important role in controlling viral infection and mediating inflammatory damage to the tissue. Depending on the virus, the immune response might be beneficial or harmful. An antiviral immune response may clear the virus from the CNS and thus prevent death of the animal. By contrast, viruses with low neurotoxicity (e.g., Borna virus) might induce little pathology in the brain per se, but virus-specific T cells that target virally infected cells could cause severe damage to the CNS. Suppressing this immune response can prevent CNS damage in some models of viral demyelination. This is also true for models in which chronic viral infection induces secondary autoimmune T cell responses. This process is similar to the epitope spreading observed in EAE. Chronic CNS infection in the Theiler's murine encephalomyelitis virus model primes autoreactive T cells that significantly contribute to CNS damage [10]. In the infectious disease models of the CNS, CD8+ and CD4+ T cells are important for early virus control and significantly contribute to tissue destruction [11]. Although B cells do not seem to be crucial during the acute disease phase, they are important for viral control during subacute or chronic CNS infection. The contribution of the humoral immune response to tissue damage in this model is largely unknown.

Inflammation is seen not only during infection and autoimmunity but also following any other tissue damage in the CNS. Many studies have addressed the role of the immune system in models of primary neurodegeneration. These studies have consistently demonstrated activation of microglial cells and infiltration of immune cells from the periphery into the lesioned area. However, the impact of the immune response on neurodegeneration has remained controversial, because it largely depends on the model, the genetic background of the animals, and the timing and quality of the immune response. A detrimental effect of the immune response has been well established for some models, although in others, it may be beneficial. The beneficial effect may be due to the capability of some immune cells to produce anti-inflammatory cytokines and even neurotrophic factors (e.g., brain-derived neurotrophic factor). Whether an autoimmune response that secretes neurotrophic factors prevents damage in primary neurodegenerative in vivo remains controversial.

IV. Immunology of the Multiple Sclerosis Lesion

One of the major difficulties in assessing the evolution of MS lesions is the limited access to biopsies from acute lesions and the impossibility to obtain serial biopsies from patients. Few studies have, therefore, addressed lesion evolution in MS [12]. The changes that seem to occur during the first days of lesion development are activation of microglia and macrophages. Microglia and macrophages up-regulate HLA class II molecules. Complement receptor C3d–immunoglobulin complexes are found on the surface of these cells. Only a few cell infiltrates are seen at that time. The BBB seems to be intact, and demyelination and astrogliosis are largely absent. The early changes that occur may also include extensive oligodendrocyte apoptosis, although whether this represents a common phenomenon or whether it is restricted to a few atypical patients is not clear [13].

Subacute lesions, which are 1–4 months old, represent the most active MS lesions. They are characterized by the presence of cell infiltrates and the release of a large array of different cytokines and chemokines. Demyelination, BBB leakage, reactive astrocytes, and proliferating oligodendrocytes are observed during this phase. With time, inflammatory infiltrates decrease in the center of the lesion and later at the lesion border. Axonal damage and demyelination are seen in all phases of disease. Although axonal damage may occur independent of inflammation, it is most pronounced in the lesions, correlating with the extent of cellular infiltrates [14].

Proliferation of oligodendrocytes and remyelination are detectable in many lesions but usually result only in partially myelinated axons. The extent of inflammation, demyelination, neurodegeneration, and remyelination varies in individual patients, raising the question of whether the variable clinical phenotype of disease may be reflected by differences in the quality of the lesion pathology. This prompted studies to stratify patients with MS in subgroups according to histopathology of acute demyelinating lesions. Four subgroups were described on the basis of relative quantity and quality of inflammation, antibody deposition, and oligodendrocyte dystrophy [15]. Interestingly, all acute lesions in one patient seem to belong to one subtype. This may suggest differences in the underlying pathomechanisms. However, it remains to be determined whether these subgroups correspond to different phases of disease or represent distinct pathogenetic entities, and how the distinct subtypes relate to clinical disease parameters. Clearly, development of appropriate biomarkers is needed.

A. T Cells in Multiple Sclerosis

CD4+ and CD8+ T cells are present in MS lesions. CD4+ T cells are found predominantly in the perivascular cuff, whereas CD8+ T cells are more prevalent in the center and border zone of the lesion [16]. Early studies suggested that part of the lesion-infiltrating T cells in MS originate from the same precursor T cell, reflecting what has been termed *clonotypic accumulation*. This observation was confirmed by studies on MS lesions and cerebrospinal fluid (CSF) of patients with MS. Interestingly, T cell clonotypes were most dominant in CD8+ T cells and were rarely observed in CD4+ T cells [17]. A similar T cell accumulation is noted in infectious CNS diseases, with a dominance of CD8+ T cells in the lesion center and CD4+ T cells in the perivascular cuff. Persistence of clonotypes in the local compartment was confirmed by longitudinal studies in MS. Given the low abundance of the T cell clonotypes in the blood, a specific antigen-driven enrichment of the clonotypes in the affected organ compartment seems likely.

Myelin-specific T cells are central to the autoimmune pathogenesis of EAE. Based on this experimental paradigm, MS has been considered an autoimmune disease mediated by autoreactive T cells. Myelin proteins are processed in MS lesions, and peptides thereof presented on HLA molecules to infiltrating T cells [18]. Accordingly, a number of studies have addressed the role of myelin-specific T cells in MS [19]. Although myelin-specific CD4+ T cells (e.g., specific for MBP and MOG) can be retrieved from the blood or CSF of patients with MS, the frequency and the phenotype of these cells do not significantly differ between patients and controls. Nevertheless, the pathogenic potency of a myelin basic protein in the context of HLA-DR15–specific CD4+ T cell derived from patients with MS was formally demonstrated in a transgenic mouse model. These mice expressing the HLA-DR15 molecule and the human autoreactive TCR on the majority of T cells developed EAE after immunization with the autoantigen [20]. Little is known about myelin-protein–specific CD8+ T cells. Although these cells are present in the T cell repertoire of patients with MS, no study has systematically addressed differences between patients and controls. Although many findings support the central role of T cells in the pathogenesis of MS, their target antigens still remain elusive. Myelin antigens are promising candidates, but their role in the pathogenesis of MS, in contrast to EAE, remains vague.

B. Humoral Immunity in Multiple Sclerosis

The identification of elevated immunoglobulin G (IgG) antibody levels in the CSF of patients with MS was the first immunological laboratory marker that proved useful in the diagnosis of MS. Further studies revealed that 95% of all patients with MS show oligoclonal IgG bands (OCBs) in the CSF. OCBs indicate an activation of a limited number of B cells in the CNS with local release of antibodies. They are seen in the CSF of not only patients with MS but also in patients with a number of inflammatory CNS diseases, in particular viral or atypical bacterial infections. Interestingly, in infectious diseases, the locally produced antibodies are specific for the causative pathogen.

Beside the role of antibodies in the diagnostic workup in MS, B cells were neglected in MS research for decades because of their dispensable role in EAE [21]. Only recently have B cells received more attention. Similar to T cells, B cells in MS lesions or CSF contain dominant clonotypes, which are less prevalent in the peripheral blood compartment. The BCRs of these clonotypes contain replacement mutations, suggesting an antigen-driven selection process. Serial studies on CSF disclosed that, in a given patient, the same B cell clonotypes are present in the CNS throughout the course of the disease. This implies that the same clonotypic B cells are periodically recruited, or that they persist in the CNS compartment.

This concept is in line with other CSF findings. The intrathecal immunoglobulin (predominantly IgG1) production in MS persists over long periods, suggesting a continuous release of these antibodies by local B cells. Indeed plasma blasts and plasma cells are present in the CSF throughout the course of disease [22]. These terminally differentiated B cells are very efficient producers of antibodies and seem to correspond to the inflammatory activity in the CNS [23]. Interestingly, these cells are usually only found in acute infectious CNS diseases. New findings point out that this local B cell activation is supported by the presence of B cell cytokines (e.g., BAFF) and chemokines (e.g., lymphotoxin-alpha, CXC ligand [CXCL] 12, and CXCL13) [22,24].

Although these findings strongly support a role for B cells and antibodies in MS, the target antigens of the humoral immune response and their contribution to disease remain uncertain. Autoantigens (e.g., neuronal or myelin proteins or protein lipids) and microbial antigens (e.g., derived from possibly MS-associated infectious agents) are possible candidates. In neuromyelitis optica, a rare variant of MS, a diagnostic serum antibody that targets aquaporin has been described [25]. However, this finding awaits confirmation by independent groups. In addition, it remains to be determined whether the antibody has any pathogenic properties or only functions as a diagnostic marker.

C. Immune-Mediated Neurodegeneration

Despite early descriptions of axonal damage in MS lesions, MS had been considered a primarily demyelinating disease of the CNS for decades. A paradigm shift has occurred during recent years, based on pathological and MRI studies that highlighted the additional importance of axonal damage and axonal loss as key determinants of permanent disability in MS. Even during early stages of disease axonal damage is quite prominent [26]. Although axonal damage is not always associated with inflammation, it is most prominent in acute inflammatory lesions.

Several concepts have been advanced to explain the relation between inflammation and neurodegeneration. Axonal damage and loss may be the result of a direct impact of the immune mediators to the neuron or via the loss of protective or nutritive support from glial cells. Immediate mechanisms include CD8+ T cells that directly attack neurons, CD4+ T cells that initiate release of inflammatory mediators, and toxic molecules by recruited macrophages or antibodies that bind to neuronal surface antigens, activating the complement cascade or antibody-mediated phagocytosis. Immune-mediated damage of oligodendrocytes and astrocytes may, via multiple pathways, lead to a loss of protective myelin, mitochondrial dysfunction, and release of glutamate or nitric oxide. Although neurodegeneration has been considered the key for the progression of disease, the underlying molecular mechanisms in MS remain controversial.

D. Lessons from Immunotherapy

Immunopharmacotherapies have been widely used in MS, because they are the only approved treatment strategies that have an effect on the course of disease [27]. Global immunosuppression was the first immune therapy that was applied to patients with relapsing–remitting MS (RR-MS). Small beneficial effects were demonstrated for immunosuppressants such as azathioprine and cyclosporine. Later the immunosuppressant mitoxantrone proved its efficacy and is now widely used for worsening forms of MS. However, the most commonly used drugs in MS are the immunomodulatory agents interferon-β (IFN-β) and glatiramer acetate (GA). IFN-β has multiple immunomodulatory effects (e.g., anti-migratory and shift of cytokine profiles of T cells) and exhibits antiviral properties [28]. GA is a synthetic polypeptide composed of the most prevalent amino acids in myelin basic protein. It is supposed to modulate autoreactive T cells and interfere with monocyte activation [28]. Although the specific mode of action of these two immunomodulatory drugs remains uncertain, they reduce relapse rates, inflammatory activity, and lesion load as measured by magnetic resonance imaging (MRI) [27]. IFN-β also seems to slow disease progression, although the effects are, at best, marginal.

With the introduction of humanized monoclonal antibodies and small specific molecules (e.g., receptor agonists or antagonists), ablation of distinct immune populations or selective blockade/activation of immune molecules has become possible. Antibodies that target cell-specific molecules on the surface allow depletion of T cells, B cells, or other immune-cell subsets via antibody binding and complement-mediated cell lysis. Depletion of B cells or antibodies seems to be beneficial in a subgroup of patients with high humoral activity. Depletion of T cells using an anti-CD52 antibody (Campath-1H) markedly diminished relapse rates, as well as inflammatory MRI activity, and is now being investigated in a large phase II trial.

The first therapies that interfere with cell migration have entered the clinic [29]. Blockade of adhesion molecules prevents leukocyte binding to the vessel wall and eventually prevents their passage across the BBB. On the basis of positive results in the EAE model, the humanized monoclonal antibody natalizumab was developed to block α_4-integrin. Natalizumab was efficient in phase II and two phase III trials, producing a profound reduction in MRI activity, relapse rates, and disease progression [30]. Marketing of the drug was temporarily suspended after three patients developed progressive multifocal leukoencephalopathy, an opportunistic infection of the CNS induced by JC virus.

Although some of these therapeutic strategies have been developed in the EAE model (e.g., GA and natalizumab), others were brought to the clinic without supportive results from experiments in the EAE model (e.g., IFN-β). In contrast, a number of therapies have proven efficacy in EAE but failed in MS [31]. Depletion of CD4+ T cells cures EAE but has no effect on MS. Antigen-based therapies that specifically target T cell responses against myelin proteins were quite efficient in EAE. Those strategies included the tolerization of autoreactive T cells by oral administration of myelin antigens or by the administration of an altered peptide ligand (APL) based on myelin basic protein. Both phase III clinical trials produced negative results, and the APL trial had to be stopped because of side effects. A small study even indicated worsening of some patients treated with the APL. TCR-specific therapies (e.g., TCR-peptide vaccination) were also successful in EAE but had no significant impact on the disease course of MS.

Even more discordant were the results obtained with anti-tumor necrosis factor-α (TNF-α) therapies in EAE and MS [31]. TNF-blocking antibodies and TNF-receptor antagonists showed strong ameliorative effects in the EAE model, but they increased disease activity in MS.

Overall, the predictive value of EAE experiments for the efficacy of pharmacotherapies in MS is limited, raising substantial questions about whether both diseases share the same pathomechanism.

E. Immunopathogenic Concept

MS is a chronic disabling disease and its cause remains uncertain, despite considerable progress in research during the past decade (see Fig. 1B). The immune system seems to play a central role in the disease pathogenesis.

The involvement of CD8+ T cells and B cells strongly supports the primary inflammatory nature of MS in most patients. Although T cells are recruited to the CNS tissue following acute or chronic neurodegeneration, this does not result in recruitment of activated B cells and persistent release of antibodies in the CNS. CD8+ T cell infiltrates that persist over time are also uncommon in primary neurodegenerative diseases in humans (e.g., Alzheimer's disease) or their animal models.

It is not clear whether the invasion of the CNS by T and B cells is the primary event or whether it is secondary to an infection or to the death of oligodendrocytes and activation of the microglia/macrophage system with the release of self-antigens or foreign antigens (Fig. 2). However, it is likely that the highly focused and persisting T and B cell infiltration in MS is driven by a small number of antigens that are present in the CNS. Although we still do not know these antigens, it is likely that they are expressed in neurons or glial cells. The involvement of B cells, which predominately release IgG1 antibodies, would indicate that their targets are proteins either released or displayed on the surface of CNS cells. Myelin or neuronal antigens, and antigens from infectious agents that have been epidemiologically associated with MS, are possible but still unconfirmed targets of the humoral immune response in the CNS of patients with MS.

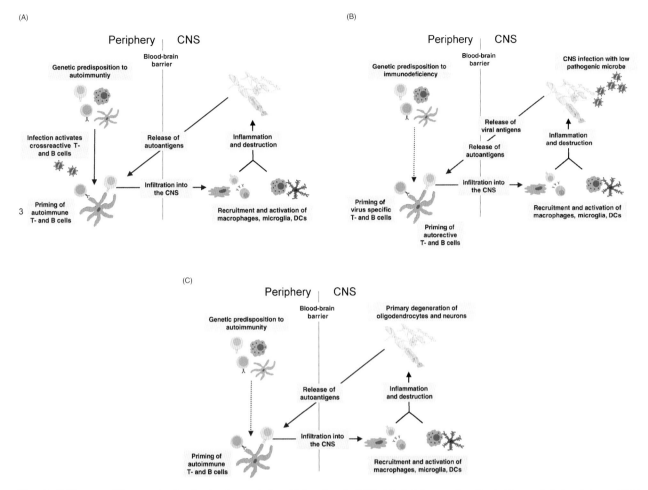

Figure 2 Three possible scenarios for the development of multiple sclerosis (MS). (**A**) *Autoimmune hypothesis*. Disease is induced on a genetic predisposition by autoreactive T cells that are activated by cross-reactive antigens from infectious agents. These cells cross the blood–brain barrier and, in concert with the innate immune response, mediate demyelination and neurodegeneration. The release of additional central nervous system (CNS) antigens facilitates the autoimmune response. (**B**) *Infectious disease hypothesis*. MS is induced by a yet unknown microbe that persists in the CNS. Antigens from the microbe get to the periphery and induce a specific immune response. These cells cross the blood–brain barrier and not only damage infected cells but also cause collateral damage to neighboring cells. The persistence of the microbe leads to chronic CNS inflammation. The process may be enhanced by the generation of autoimmune responses induced by the continuous release of self-antigens (epitope spreading). (**C**) *Neurodegeneration hypothesis*. Disease is the result of a primary neurodegenerative process that emerges on the background of a genetic predisposition. The release of autoantigens induces secondary autoimmune responses that contribute to the disease process.

Although we still do not know the target antigens of the acquired immune response, most studies suggest a detrimental effect of CNS inflammation in MS. This is supported not only by a number of findings from experimental animal models, but also by the relationship between inflammation and neurodegeneration in active MS lesions. However, it is important to keep in mind that such a detrimental effect of the immune response is not proof of the autoimmune nature of MS, but may also be observed in chronic CNS infection with microbes that have an inherently low pathogenic potential. The chronic infection not only may induce an antiviral immune response that directly damages infected cells but also could induce collateral damage to neighboring CNS cells. In addition, via epitope spreading, chronic infection may lead to the occurrence of autoimmune T cells that contribute to the destructive inflammatory process in MS.

The evidence reviewed in this chapter also points out important differences between MS in humans and the currently available animal models, EAE in particular. CD4+ T cells are key to the development of EAE, but their depletion does not have an impact on the course of MS. CD8+ T cells play an important role in MS, although these cells seem to be more protective than encephalitogenic in most EAE models. Likewise, B cells play a dominant role in MS but are not considered to contribute to tissue damage in most EAE models. Furthermore, the role of some inflammatory molecules, in particular TNF-α, is discrepant in MS and EAE. Although blocking TNF-α ameliorates EAE, it exacerbates MS. These findings suggest that immunological pathways and target antigens may significantly differ between the human disease and the experimental animal model. Although animal models are crucial to decrypt mechanisms of CNS inflammation and neurodegeneration, the value of the currently available models to determine the cause of MS and develop new therapies is limited.

Acknowledgments

B. H. and N. T. were supported by grants from the Deutsche Forschungs gemeinschaft, the Gemeinnützige Hertie-Stiftung, and the German MS Society.

References

1. Noseworthy, J. H., Lucchinetti, C., Rodriguez, M., and Weinshenker, B. G. (2000). Multiple sclerosis. *N. Engl. J. Med.* **343**, 938–952.
2. Rosati, G. (2001). The prevalence of multiple sclerosis in the world: an update. *Neurol. Sci.* **22**, 117–139.
3. Oksenberg, J. R., Baranzini, S. E., Barcellos, L. F., and Hauser, S. L. (2001). Multiple sclerosis: genomic rewards. *J. Neuroimmunol.* **113**, 171–184.
4. Kurtzke, J. F. (2000). Epidemiology of multiple sclerosis. Does this really point toward an etiology? Lectio Doctoralis. *Neurol. Sci.* **21**, 383–403.
5. Cserr, H. F., and Knopf, P. M. (1992). Cervical lymphatics, the blood-brain barrier and the immunoreactivity of the brain: a new view. *Immunol. Today* **13**, 507–512.
6. Heppner, F. L., Greter, M., Marino, D., Falsig, J., Raivich, G., Hovelmeyer, N., Waisman, A., Rulicke, T., Prinz, M., Priller, J., Becher, B., and Aguzzi, A. (2005). Experimental autoimmune encephalomyelitis repressed by microglial paralysis. *Nat. Med.* **11**, 146–152.
7. Greter, M., Heppner, F. L., Lemos, M. P., Odermatt, B. M., Goebels, N., Laufer, T., Noelle, R. J., and Becher, B. (2005). Dendritic cells permit immune invasion of the CNS in an animal model of multiple sclerosis. *Nat. Med.* **11**, 328–334.
8. Zamvil, S. S., and Steinman, L. (1990). The T lymphocyte in experimental allergic encephalomyelitis. *Annu. Rev. Immunol.* **8**, 579–621.
9. Linington, C., Bradl, M., Lassmann, H., Brunner, C., and Vass, K. (1988). Augmentation of demyelination in rat acute allergic encephalomyelitis by circulating mouse monoclonal antibodies directed against a myelin/oligodendrocyte glycoprotein. *Am. J. Pathol.* **130**, 443–454.
10. Vanderlugt, C. L., and Miller, S. D. (2002). Epitope spreading in immune-mediated diseases: implications for immunotherapy. *Nat. Rev. Immunol.* **2**, 85–95.
11. Stohlman, S. A., and Hinton, D. R. (2001). Viral induced demyelination. *Brain Pathol.* **11**, 92–106.
12. Gay, F. W., Drye, T. J., Dick, G. W., and Esiri, M. M. (1997). The application of multifactorial cluster analysis in the staging of plaques in early multiple sclerosis. Identification and characterization of the primary demyelinating lesion. *Brain* **120**, 1461–1483.
13. Barnett, M. H., and Prineas, J. W. (2004). Relapsing and remitting multiple sclerosis: pathology of the newly forming lesion. *Ann. Neurol.* **55**, 458–468.
14. Kuhlmann, T., Lingfeld, G., Bitsch, A., Schuchardt, J., and Bruck, W. (2002). Acute axonal damage in multiple sclerosis is most extensive in early disease stages and decreases over time. *Brain* **125**, 2202–2212.
15. Lucchinetti, C., Bruck, W., Parisi, J., Scheithauer, B., Rodriguez, M., and Lassmann, H. (2000). Heterogeneity of multiple sclerosis lesions: implications for the pathogenesis of demyelination. *Ann. Neurol.* **47**, 707–717.
16. Hemmer, B., Archelos, J. J., and Hartung, H. P. (2002). New concepts in the immunopathogenesis of multiple sclerosis. *Nat. Rev. Neurosci.* **3**, 291–301.
17. Babbe, H., Roers, A., Waisman, A., Lassmann, H., Goebels, N., Hohlfeld, R., Friese, M., Schroder, R., Deckert, M., Schmidt, S., Ravid, R., and Rajewsky, K. (2000). Clonal expansions of CD8(+) T cells dominate the T cell infiltrate in active multiple sclerosis lesions as shown by micromanipulation and single cell polymerase chain reaction. *J. Exp. Med.* **192**, 393–404.
18. Krogsgaard, M., Wucherpfennig, K. W., Cannella, B., Hansen, B. E., Svejgaard, A., Pyrdol, J., Ditzel, H., Raine, C., Engberg, J., and Fugger, L. (2000). Visualization of myelin basic protein (MBP) T cell epitopes in multiple sclerosis lesions using a monoclonal antibody specific for the human histocompatibility leukocyte antigen (HLA)-DR2-MBP 85-99 complex. *J. Exp. Med.* **191**, 1395–1412.
19. Sospedra, M., and Martin, R. (2005). Immunology of multiple sclerosis. *Annu. Rev. Immunol.* **23**, 683–747.
20. Madsen, L. S., Andersson, E. C., Jansson, L., Krogsgaard, M., Andersen, C. B., Engberg, J., Strominger, J. L., Svejgaard, A., Hjorth, J. P., Holmdahl, R., Wucherpfennig, K. W., and Fugger, L. (1999). A humanized model for multiple sclerosis using HLA-DR2 and a human T-cell receptor. *Nat. Genet.* **23**, 343–347.
21. Cross, A. H., Trotter, J. L., and Lyons, J. (2001). B cells and antibodies in CNS demyelinating disease. *J. Neuroimmunol.* **112**, 1–14.
22. Corcione, A., Casazza, S., Ferretti, E., Giunti, D., Zappia, E., Pistorio, A., Gambini, C., Mancardi, G. L., Uccelli, A., and Pistoria, V. (2004). Recapitulation of B cell differentiation in the central nervous system of patients with multiple sclerosis. *Proc. Natl. Acad. Sci. USA* **101**, 11064–11069.
23. Cepok, S., Rosche, B., Grummel, V., Vogel, F., Zhou, D., Sayn, J., Sommer, N., Hartung, H. P., and Hemmer, B. (2005). Short-lived plasma blasts are the main B cell effector subset during the course of multiple sclerosis. *Brain* **128**, 1667–1676.
24. Krumbholz, M., Theil, D., Derfuss, T., Rosenwald, A., Schrader, F., Monoranu, C. M., Kalled, S. L., Hess, D. M., Serafini, B., Aloisi, F., Wekerle, H., Hohlfeld, R., and Meinl, E. (2005). BAFF is produced by astrocytes and up-regulated in multiple sclerosis lesions and primary central nervous system lymphoma. *J. Exp. Med.* **201**, 195–200.
25. Lennon. V. A., Kryzer, T. J., Pittock, S. J., Verkman, A. S., and Hinson, S. R. (2005). LgG marker of optic-spinal multiple sclerosis binds to the aquaporin-4 water channel. *J. Exp. Med.* **202**, 473–477.
26. Trapp, B. D., Peterson, J., Ransohoff, R. M., Rudick, R., Mork, S., and Bo, L. (1998). Axonal transection in the lesions of multiple sclerosis. *N. Engl. J. Med.* **338**, 278–285.
27. Neuhaus, O., Archelos, J. J., and Hartung, H. P. (2003). Immunomodulation in multiple sclerosis: from immunosuppression to neuroprotection. *Trends Pharmacol. Sci.* **24**, 131–138.
28. Yong, V. W. (2002). Differential mechanisms of action of interferon-beta and glatiramer acetate in MS. *Neurology* **59**, 802–808.
29. Engelhardt, B., and Ransohoff, R. M. (2005). The ins and outs of T-lymphocyte trafficking to the CNS: anatomical sites and molecular mechanisms. *Trends Immunol.* **26**, 485–495.
30. Miller, D. H., Khan, O. A., Sheremata, W. A., Blumhardt, L. D., Rice, G. P., Libonati, M. A., Willmer-Hulme, A. J., Dalton, C. M., Miszkiel, K. A., O'Connor, P. W., International Natalizumab Multiple Sclerosis Trial Group. (2003). A controlled trial of natalizumab for relapsing multiple sclerosis. *N. Engl. J. Med.* **348**, 15–23.
31. Wiendl, H., and Hohlfeld, R. (2002). Therapeutic approaches in multiple sclerosis: lessons from failed and interrupted treatment trials. *BioDrugs* **16**, 183–200.

19

Immune-Mediated Neuropathies

Michael Schroeter, MD
Bernd C. Kieseier, MD
Hubertus Köller, MD
Hans-Peter Hartung, MD

Keywords: *chronic inflammatory demyelinating, Guillain-Barré syndrome, neuritis, paraproteinemias, physiopathology, polyradiculoneuritis, vasculitis*

I. Introduction
II. Acute Immune-Mediated Neuropathies
III. Chronic Immune-Mediated Neuropathies
IV. Systemic Autoimmune Disease Affecting the Peripheral Nerve
V. Immune-Mediated Exacerbation of Nonimmune Peripheral Nerve Disease
References

I. Introduction

Some 30 years ago, Dyck coined the term *chronic inflammatory demyelinating polyneuropathy* (CIDP) to describe a steroid-responsive polyneuropathy. Immune-mediated neuropathies are treatable conditions but are still likely to be underdiagnosed. Establishing animal models of acute and chronic polyneuritis such as experimental autoimmune neuritis (EAN) has fostered the understanding of human disease pathophysiology and has helped to dissect different forms of immune-mediated neuropathies [1].

The nomenclature and classification of immune-mediated neuropathies, though still under debate, is converging (Table 1). Starting from animal models, we will discuss pathophysiological and clinical aspects of acute polyneuritis (Guillain-Barré syndrome [GBS] and variants), as opposed to chronic polyneuritis, mainly CIDP. By contrast, in disimmune polyneuropathy, endoneural deposition of immunoglobulins leads to malfunctioning, and a local inflammatory reaction may be secondary or even absent. The latter holds true in multifocal motor neuropathy

Table 1 Classification of Immune-Mediated Neuropathies

Acute polyneuritis (Guillain-Barré syndrome [GBS])
  Acute inflammatory demyelinating polyradiculoneuropathy (AIDP)
  Acute motor axonal neuropathy (AMAN)
  Acute motor and sensory axonal neuropathy (AMSAN)
  Miller Fisher syndrome (MFS)
  Acute pandysautonomia
Chronic polyneuritis
  Chronic inflammatory demyelinating polyneuropathy (CIDP)
  CIDP with MGUS of the IgA, IgG type
  Multifocal motor neuropathy (MMN) with or without conduction blocks
  Multifocal acquired demyelinating sensory and motor neuropathy (MADSAM, Lewis Sumner syndrome)
  Polyneuropathy associated with IgM-paraprotein (IgM-PN)
  Polyneuropathy, organomegaly, endocrinopathy, M protein, skin changes (POEMS syndrome)
  Mixed and polyclonal cryoglobulinemia
  Disimmune neuropathy in systemic rheumatic disease and vasculitis
Chronic polyneuritis in other polyneuropathies
  Polyneuritis with underlying hereditary neuropathy
  Polyneuritis with underlying diabetic neuropathies (DM-CIDP)

(MMN) and polyneuropathy associated with immunoglobulin M (IgM) paraprotein. Peripheral nerve injury may be secondary to systemic inflammation, namely vasculitis. Last, but not least, the immune-mediated disease process may be superimposed on a preexisting nerve disorder, as exemplified by hereditary and diabetic neuropathies.

The immune system is characterized by a well-balanced network of immunological factors. Heuristically, the immune system has to differentiate "self" from "nonself" to maintain self-tolerance and defeat attack from "nonself," for example, viruses and bacteria. Dysbalance may be two sided: an immunosuppressed state may facilitate infections. By contrast, when self-tolerance breaks down, heightened adaptive immune responses governed by T and B cells damage specific organs, as in the case of immune-mediated neuropathies [2]. Principles of the pathophysiology of immune-mediated neuritis may be translated to and verified in other human autoimmune diseases. For instance, in GBS and the animal model of EAN, the hypothesis of molecular mimicry has been studied extensively and has substantially contributed to our understanding of the initial phase of autoimmune disease [3].

II. Acute Immune-Mediated Neuropathies

A. Experimental Autoimmune Neuritis

EAN resembles human GBS in many clinical, electrophysiological, and immunological aspects and, as such, has been widely exploited as a model for disease mechanisms in inflammatory demyelination of the peripheral nervous system (PNS) [4,5]. EAN is an acute inflammatory demyelinating polyradiculoneuropathy that can be induced in rats, mice, rabbits, and monkeys by active immunization with whole peripheral nerve homogenate, myelin, and myelin proteins P0 and P2 or peptides thereof. Other autoantigens have been identified, such as myelin basic protein (MBP), peripheral myelin protein-22 (PMP22), and myelin-associated glycoprotein (MAG). EAN can also be produced by adoptive transfer of P2, P2 peptide–specific, P0, and P0 peptide–specific CD4+ T cell lines. It appears conceivable that a variety of other antigens become targets of an immune attack.

Besides the classic models of this experimental disease, EAN can be elicited by immunization with nonneural antigens. Injection of CD4+ T lymphocytes reactive to the nonneural antigen ovalbumin-induced mononuclear inflammatory infiltrates the tibial nerve of Lewis rats that had received a previous intraneural injection of ovalbumin. Clinically and morphologically, these animals resembled those with P2-induced EAN. The observation that activated T cells of nonneural specificity prompt the breakdown of the blood-nerve barrier and produce clinical symptoms similar to human GBS underlines the pathogenetic relevance of the concept of molecular mimicry, as is discussed later.

The pathological hallmark of EAN is the infiltration of the PNS by lymphocytes and macrophages, which results in multifocal demyelination of axons predominantly around venules. As shown by electron microscopy, macrophages actively strip myelin lamellae from axons, induce vesicular disruption of the myelin sheath, and phagocytose both intact and damaged myelin. Transfer of increasing numbers of activated autoreactive T cells produces significant axonal damage admixed with the demyelinative changes in recipient animals as EAN develops.

B. Guillain-Barré Syndrome

GBS, in its classical phenotype of acute inflammatory demyelinating polyneuropathy (AIDP), was first described by Guillain, Barré, and Strohl in 1916. GBS is characterized by areflexia and a rapidly ascending flaccid paralysis. Without treatment, the nadir is reached within 4 weeks. Subsequently, recovery occurs over intervals unpredictable in length. Recovery may be complete, but a proportion of patients (some 15%) will suffer from persistently disabling deficits. The disease is even associated with a mortality of 2–5%. Its clinical course is monophasic. Relapses or a chronic-progressive course falsify the diagnosis of GBS, but repeat episodes have been reported in single cases. Results of electrophysiological and cerebrospinal fluid (CSF) analyses support the clinical diagnosis of GBS. Prolonged distal latencies in nerve conduction mirror involvement of the most distal part of the nerves, whereas absent or delayed F waves point to demyelination of proximal nerve segments.

As a correlate of the clinical loss of the ankle jerk, the loss of the H response reveals the involvement of highly myelinated Ia fibers from muscle spindles early in the disease process. Elevated CSF protein levels in the presence of normal cell counts ("dissociation albuminocytologique"), which occurs in half the cases during the first week of the disease, point to the disruption of the blood-nerve barrier at the most proximal parts of nerve roots.

The pathogenesis and pathophysiology of GBS are prototypic for immune-mediated neuropathies. It is caused by an inflammatory attack on the myelin sheath (acute inflammatory demyelinating polyneuritis [AIDP]) or axons (acute motor axonal polyneuritis, [AMAN]).

The concept of molecular mimicry in animal models contributes to our understanding of how self-tolerance is broken and autoimmunity evolves and provides a framework for understanding GBS in humans. Certain bacterial species, above all *Campylobacter jejuni* and *Haemophilus influenzae*, share epitopes with nerve fibers and myelin sheaths. Congruent with an adaptive B cell response and subsequent antibody production, GBS often develops 10–14 days after enteric infection with *C. jejuni*. Experimentally, molecular similarities of *C. jejuni* carbohydrates and human ganglioside GM_1 and related glycoconjugates exist. Immunization of experimental animals with *C. jejuni* lipo-oligosaccharides causes a GBS-like disease with antibody deposition *in situ* and anti-GM_1 antibodies that disrupt neuromuscular activity. These observations provide compelling evidence for the relevance of molecular mimicry of carbohydrates from the surface of *C. jejuni* for the pathogenesis of GBS [6]. However, only a small percentage of patients with *C. jejuni* enteritis develop GBS. According to the Swedish national laboratory register, only 9 or 30,000 patients with *C. jejuni* enteritis developed GBS, indicating a 100-fold risk of GBS in this cohort versus a background risk of 0.3 per 100,000 in a two-month period [7]. Further subanalysis did not strengthen the relationship between specific *C. jejuni* serotypes and GBS. Thus, *C. jejuni* may be one of many factors involved in a cascade of extrinsic and intrinsic events that ultimately cause GBS.

Searching for an intrinsic disposition to develop GBS, attempts have been made to establish immunogenetic factors for this disease. Unlike many immune-mediated diseases, however, predisposing human leukocyte antigen (HLA) class I and II antigens have not been unequivocally established. However, because humoral immune responses to carbohydrates are T cell independent and, therefore, not restricted by class I and II antigens, it is possible that B cells on antibody-mediated recognition of antigens is important in GBS.

For the sake of clarity, we separate the role of humoral and cellular immunity in the pathogenesis of GBS.

There is compelling evidence that humoral immunity is directly involved in GBS pathogenesis. There are clear therapeutic effects of plasmapheresis and administration of intravenous immunoglobulins (IVIGs). In sera of patients with GBS, there are circulating glycolipid antibodies, and some glycolipid antibodies cross-react with carbohydrates of *C. jejuni*. A broad range of antibodies to one or more of the gangliosides/glycoconjugates, GM_1, $GM_{1(NeuGc)}$, GM_{1b}, GalNAc-GM_{1b}, GD_{1a}, $GD_{1\alpha}$, GalNAcGD$_{1a}$, GD_{1b}, 9-0-acetyl-GD_{1b}, GD_3, GT_{1a}, GT_{1b}, GQ_{1b}, $GQ_{1\beta}$, GalC, LM1, and SGPG, have been repetitively demonstrated in a high percentage of patients with GBS (Fig. 1). Autoantibodies exhibit extensive cross-reactivity among gangliosides. Thus, these antibodies are capable of initiating a broad autoimmune response against gangliosides and related structures. There is a spectrum of clinical GBS variants (Table 2), and the type of anti-ganglioside antibody is weakly associated with the

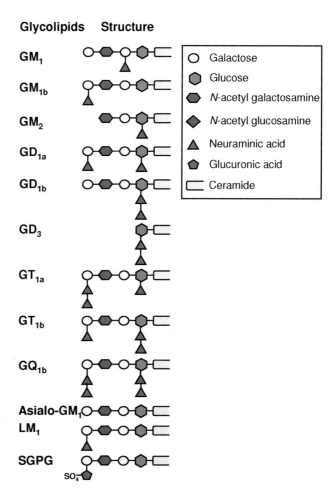

Figure 1 Principal Structure of Gangliosides and Related Glycoconjugates, Potential Autoantigens in Guillain-Barré Syndrome (GBS) and Chronic Inflammatory Demyelinating Polyneuropathy (CIDP). Structural homologies cause extensive cross-reactivity of anti-ganglioside antibodies involved in GBS and CIDP pathophysiology. (Adapted, with permission, from Kieseier, B. C., Kiefer, R., Gold, R., Hemmer, B., Willison, H. J., and Hartung, H. P. [2004]. Advances in understanding and treatment of immune-mediated disorders of the peripheral nervous system. *Muscle Nerve* **30**, 131–156.)

Table 2 Variants of the Guillain-Barré Syndrome

Type	Symptoms	Electrophysiological findings	Anti-ganglioside antibody
AIDP	Ascending flaccid paresis, areflexia	Demyelinating motor neuropathy, and loss of F waves and H response	Any
AMAN	Flaccid tetraparesis, atrophy, slow recovery	Axonal motor neuropathy, and extensive denervation of muscles	GM_1, GM_{1a}, GM_{1b}, GD_{1a}-GalNac
AMSAN	Flaccid tetraparesis, sensory loss, poor prognosis	Axonal motor neuropathy, denervation, and loss of sensory action potentials	GM_1
Miller Fisher	Ophthalmoplegia, ataxia, areflexia	Reduction of sensory action potentials (or normal)	GQ_{1b}
Acute pandysautonomia	Loss of sympathetic and parasympathetic functions	Loss of respiratory arrhythmia and sympathetic skin response	Not known
Pure ataxic	(Sensory) ataxia	Normal or reduced sensory action potentials	GQ_{1b}
Pharyngeal, cervico-bulbar	Dysfunction of cranial nerves, swallowing, respiration	Normal or reduced sensory action potentials	GT_{1a}, GT_{1b}

Note: AIDP, acute demyelinating polyneuritis; AM(S)AN, acute motor (and sensory) axonal neuropathy.

clinical phenotype [8]. This especially holds true for the Miller Fisher variant of GBS, which is almost invariably associated with GQ_{1b} antibodies. Congruently, the oculomotor nerves of patients with the Miller Fisher variant of GBS accumulate GQ_{1b} antibody.

In the sequence of events in GBS pathogenesis, autoantibodies first access peripheral nerve sites where a functional blood-nerve barrier is not existent (i.e., most proximal and distal parts) or pass through the barrier along with invading T cells. Autoantibodies may also be locally produced by invading or resident B cells. Antibodies bind to nerve and myelin structures. This leads to binding of complement and complement activation via the classical pathway and results in the formation of the lytic membrane attack complex, C5b-9. Antibodies attaching to Fc-γ receptors may mediate demyelination by antibody-dependent cellular toxicity of invading macrophages.

Deposition of ganglioside antibodies may directly disrupt nerve conductance at the nodes of Ranvier or at the neuromuscular junction *in vitro* (Figs. 2 and 3). Anti-GD_{1a}

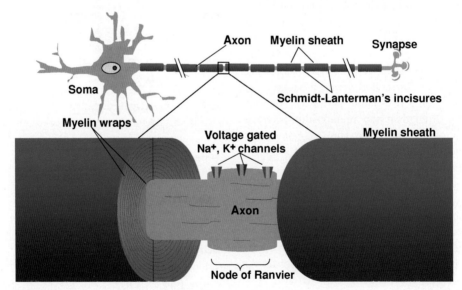

Figure 2 Principal Anatomical Structures at the offspring of the axon from the nerve cell soma, the axonal hillock, a nerve potential is generated that is conducted via the axons to the synaptic endings. Schwann cells normally generate the myelin sheath by wrapping multiple myelin lamellae around the axon. The myelin sheath is interrupted at regular distances, forming the Schmidt–Lanterman incisures. Here, the axon diameter is increased (at the node of Ranvier), and the axonal membrane is densely packed with voltage-gated Na^+ and K^+ channels to enable action potential generation. In the diseased nerve, ganglioside antibodies may block these channels and thereby block the saltatory conduction of action potentials.

antibodies cause severe ultrastructural damage at the presynaptic membrane and the nodes of Ranvier, and motor nerve terminal function is blocked [9].

The electrophysiological effects of antibodies to gangliosides do not necessarily parallel the immune attack. Antibodies are sufficient to induce conduction blockade or to disrupt neuromuscular transmission, but the immune attack needs a critical interplay of humoral and cellular immune responses.

Regarding cellular immune responses in GBS, research has focused on the detection of a specific T cell repertoire in GBS. Peptides are recognized by the traditional T cell receptor (TCR). However, glycolipids are recognized by the nonclassical class I antigen-presenting molecule CD1d expressed by a certain subset of T cells, the natural killer (NK) T cell, which expresses both TCRs and surface molecules typical of NK cells [10]. CD1-restricted T cells/NK T cells recognize sulfatide and glycolipids *in vitro*. The structural basis of glycolipid loading of the CD1-binding groove has been characterized in detail [11]. NK T cells can drive the immune response toward the Th1 and the Th2 direction (see Fig. 3). CD1 molecules have been observed in nerves from patients with GBS, suggesting that the nonclassical pathway could participate in an autoimmune response in GBS.

Although still controversial, there appears to be increased use of the V15 T cell chain in GBS. A gut-associated Vγ8/δ1 TCR has been identified in a sural biopsy from a patient with GBS, underlining the importance of a gut-associated process (i.e., a preceding *C. jejuni* enteritis). CD8+ T lymphocytes were found in autopsy series of fatal acute GBS, giving rise to the notion that an acute cytotoxic T cell response may occur in concert with the established humoral responses.

Major histocompatibility complex class I and II expression and macrophage accumulation are hallmarks in the histopathology of acutely inflamed nerve specimens. Conventionally, macrophages are thought to act somewhat downstream from the initial immune response cascades and play an important role in removing debris of fragmented myelin. However, experimental data suggest a central role in EAN pathophysiology, because depletion of macrophages will completely prevent disease.

Cellular responses initiate an immune attack and are expected to be tightly regulated by a network of cytokine, chemokines, and other signaling factors (see Fig. 3). In patients with GBS, there is up-regulation of the proinflammatory cytokines tumor necrosis factor-α (TNF-α) and interferon-γ (IFN-γ), and conversely, down-regulation of counteracting Th2 cytokines such as interleukin (IL)-4, IL-10, and transforming growth factor-β (TGF-β). Concerning the chemokine system, the CXCR3 receptor is preponderant. CCR1 and CCR5 were detected on macrophages, and CCR-2, CCR3, and CCR4 were on T cells. Intracellularly, activation of transcriptional factors augments the rate of specific gene transcription. Among them, nuclear factor-κB (NF-κB) is up-regulated in macrophages that invade the diseased sural nerve. Cellular responses may also be critically involved in the down-regulation of the autoimmune disease. Patients with GBS have an early shift toward the Th2 T cell phenotype in the peripheral blood. It is intriguing to speculate that this Th2 response, particularly up-regulation of TGF-β, may contain the immunoinflammatory response and establish the self-limiting and monophasic course of the disease.

III. Chronic Immune-Mediated Neuropathies

Originally, Dyck et al. described CIDP as a steroid-responsive polyneuropathy. This condition typically affects both upper and lower extremities fairly symmetrically, with involvement of both distal and proximal muscles. Its dynamics differentiate it from GBS; CIDP is defined arbitrarily by clinical worsening for a period of more than 2 months. The clinical presentation may mimic the nonimmune chronic polyneuropathies such as hereditary, metabolic, diabetic, or toxic neuropathies that clearly outnumber immune-mediated neuropathies in industrialized countries. In immune-mediated chronic neuropathies, however, immunosuppressive therapy offers a therapeutic option—with efficacy rates up to 70%.

A. Models of Chronic Immune-Mediated Neuropathy

There is no animal model available that mirrors the histopathology, electrophysiology, and the clinical course of CIDP. Only two models have been described that replicate some features of CIDP. The autoimmune diabetes-prone nonobese diabetic (NOD) mouse strain, deficient in B7-2 costimulation, develops a spontaneous autoimmune polyneuropathy. It exhibits clinical, electrophysiological, and morphological similarities to human CIDP and underscores the role of endoneural macrophages as local antigen-presenting cells in the pathogenesis of inflammatory demyelination of the PNS [12]. It remains elusive why spontaneous CIDP develops in the transgenic animal model. The multiple immunological modification in the NOD model system carries limitations for studying the physiological immune response in the context of CIDP. In dark agouti (DA) rats, however, a biphasic form of EAN can be established by a single immunization with bovine peripheral nerve myelin. This is a model of relapsing forms of inflammatory demyelinating polyneuropathies, as seen in CIDP. Notably, the chronicity in this model is limited to two relapses and hence may represent an early and

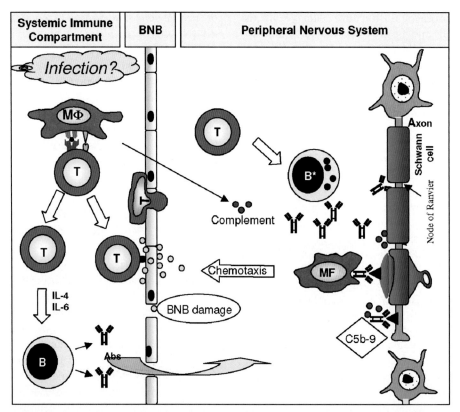

Figure 3 Pathophysiological Events in Acute Polyneuritis Schematic illustration of pathophysiological events orchestrating the humoral immune response in acutely inflamed peripheral nervous tissue. Triggered by a preceding infection, T cell–driven B cells differentiate into plasma cells producing antibodies (Abs) locally or in the systemic compartment. Ab deposition at the node of Ranvier may block nerve conduction or incite antibody-dependent cytotoxicity (right, upper part). Macrophages (Mφ) are recruited that produce an array of chemoattractants and factors damaging the blood–nerve barrier (BNB). Complement activation via the classical pathway leads to formation of the terminal membrane attack complex C5b-9 (right, lower part), which effects myelin or axonal damage.

a limited time frame in the course of chronic immune-mediated polyradiculoneuropathies [13].

B. CIDP

CIDP is clinically heterogeneous [14]. The disease is characterized by a subacute onset of weakness and sensory symptoms. Later on, weakness may be progressive in a stepwise fashion, and subacute worsening occurs, as well as spontaneous remissions. Alternatively, the disease may take an insidiously chronic progressive course. Symptoms affect all four limbs in a distal and symmetrical fashion, and this typical pattern responds reasonably well to anti-inflammatory and immunosuppressive treatment [15]. Symptoms in hereditary neuropathies usually have a steep distal-proximal gradient. However, in CIDP, proximal and distal muscles are commonly affected. Sensory symptoms typically predominate, but motor deficits may equal or even dominate the clinical picture in some cases. Electrophysiological examination reveals a demyelinating or axonal-demyelinating polyneuropathy. Prolonged distal motor latencies suggest that the most distal aspects of the nerves are affected. Segmental slowing of nerve conduction and the quite typical feature of conduction blocks correlate to segmental demyelination [16]. Proximal conduction blocks mirror involvement of proximal nerve segments, plexus, and nerve roots. Elevated protein content of the CSF without raised cell numbers ("dissociation albuminocytologique") is caused by the disturbance of the blood-nerve barrier at the most proximal parts of peripheral nerves. Magnetic resonance studies can visualize the blood-nerve barrier leakage at the nerve roots *in vivo*.

Sural nerve biopsy can be diagnostic of CIDP. Histopathology shows endoneural edema and ongoing or recurrent demyelination, paralleled by remyelination and Schwann cell proliferation with onion bulb formation. There is a preferentially endoneural accumulation of T cells and macrophages, axonal degeneration, and overall axonal loss.

The pathogenesis of CIDP is unknown. The concept of molecular mimicry may be a key event in CIDP pathogenesis, although the temporal correlation of a chronic disease to an acute antecedent illness is clearly more difficult to establish [17]. Antecedent infections appear much less common in CIDP compared with GBS and may occur in less than one-third of patients.

Ganglioside antigens have been implicated in the pathophysiology of CIDP. Again, they are detectable with lesser frequency than in GBS (some estimates go up to 15%). These antibodies may act in a pharmacological fashion to directly disturb nerve conduction by suppressing Na^+ and K^+ currents. Once deposited on the myelin sheath, they may also cause structural damage [9]. In sera of patients with CIDP, there are antibodies to peripheral myelin proteins such as PMP22, P0, and connexin 32 [18] (Fig. 4). Ganglioside antibodies have also been detected in nonimmune neuropathies. In diabetic neuropathies, they correlate with substantial motor neuron damage. In these nonimmune neuropathies, neural damage may expose antigens to immune surveillance and a subsequent autoimmune response may be superimposed onto the nonimmune neuropathy (see following discussion).

Cellular immune responses are aberrant in CIDP. Myelin-specific T cell responses have been documented in few patients with CIDP [19]. Biopsies have shown that T cells, NK cells, and macrophages accumulate within the endoneural compartment and along the basal membranes. Macrophages phagocytose degraded myelin fragments, and axons stripped of their myelin sheath may succumb to degeneration. Axonal loss is secondary to demyelination, which is subsequent to an inflammatory response. Immunosuppressant therapy may stop autoimmune responses but will not reverse axonal loss. However, axonal degeneration, not inflammation, correlates with long-term prognosis and disability of the patient and is blamed for incomplete remission of symptoms.

C. Multifocal Motor Neuropathy and Variants

Pestronk et al. [20] have described characteristics of the disease and coined the term *MMN* as late as 1988. Typical features clearly distinguish MMN from CIDP. MMN presents with chronically progressive motor deficits, mostly without significant involvement of sensation. The Lewis Sumner syndrome describes a rare clinical syndrome with symptoms of MMN and additional sensory involvement.

In most cases, involvement of the upper limbs is more severe. Paresis, muscle fasciculations, and wasting are

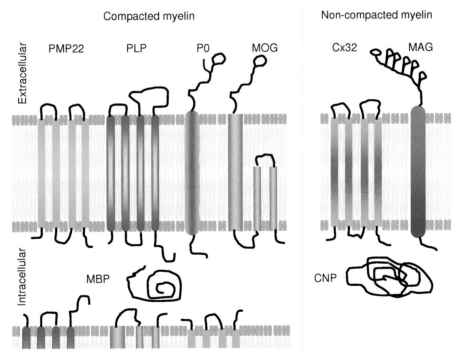

Figure 4 Myelin Proteins Schematic drawing of the principal myelin proteins in compacted and noncompacted myelin. Note that myelin basic protein (MBP) is situated intracellularly and lacks a transmembrane domain. Myelin-associated glycoprotein (MAG) is present in noncompacted myelin and the innermost layers of the myelin sheath. CNP, myelin cyclic nucleotide phosphodiesterase; Cx32, connexin 32; MOG, myelin oligodendrocyte protein; P0, protein zero; PLP, proteolipid protein; PMP22, peripheral myelin protein 22. (Modified, with permission, from Tzakos, A. G., Troganis, A., Theodorou, V., Tselios, T., Svarnas, C., Matsoukas, J., Apostolopoulos, V., and Gerothanassis, I. P. [2005]. Structure and function of the myelin proteins: current status and perspectives in relation to multiple sclerosis. *Curr. Med. Chem.* **12**, 1569–1587.)

asymmetrical and predominantly distal and follow the distribution of individual nerves.

Conduction blocks are the hallmark of the electrophysiological workup of MMN. Conduction blocks are situated at sites not responsible for typical nerve entrapment syndromes, may be multiple, and predominantly affect proximal nerve segments. Conduction blocks usually persist for months and may affect different nerves and nerve segments in the long term. The investigator will be guided by manifestation of paresis to detect conduction blocks in the respective nerves, but conduction blocks may be additionally present in clinically unaffected nerves. The presence of conduction blocks proves MMN to be a demyelinating nerve disorder, but the longer the course of disease the more axonal damage of motor nerves is present.

MMN is frequently associated with anti-ganglioside antibodies. In some 50%, anti-GM_1 antibodies are identified. Anti-GD_{1a} antibodies have been described, too, and it is likely that further subspecies of anti-ganglioside antibodies will be discovered. GM_1 gangliosides are located within the myelin sheath close to the nodes of Ranvier and at the unmyelinated nerve terminals. *In vitro* data suggest that the antibody alters Na^+ and K^+ conductance (see Fig. 3). It is intriguing to speculate about a tight relationship between the antibody attachment to gangliosides, the disturbance of nerve conductance, and the electrophysiological finding of nerve conduction block. Many patients with MMN complain about a distinct fatigability in motor tasks, which suggests that ganglioside deposition causes dysfunction at the neuromuscular junction.

Pathological and histopathological data in MMN are scarce beyond the detection of ganglioside antibodies. In multifocal disease, biopsies are taken from sensory nerves, which may show minimal or no changes.

In one autopsy case, the perivascular demyelination did not essentially differ from CIDP findings [21]. The positive response to immunoglobulins or cyclophosphamide provides indirect evidence that MMN is an immune-mediated neuropathy.

D. Paraproteinemic Neuropathies and Neuropathies with Anti-MAG Antibodies

In 1938, Scheinker first described the presence of polyneuropathy in a patient with osteosclerotic myeloma. In the past decades, the association of paraproteinemia and polyneuropathy has been increasingly recognized. About 10% of patients with paraproteinemia develop polyneuropathy, and this risk increases with rising age. Disimmune neuropathies of this group are associated with monoclonal gammopathy of unknown significance (MGUS), low-grade non-Hodgkin's lymphoma with paraproteinemia, Waldenström's disease, osteosclerotic myeloma, and Castleman's disease. MGUS is the most common entity. In MGUS, low levels of M protein, low or absent Bence Jones proteinuria, less than 5% bone marrow plasma cells, the lack of bone lesions, anemia, or hypercalcemia exclude the presence of low-grade lymphoma.

Among the patients with paraproteinemias, patients with IgG or IgA monoclonal gammopathy have clinical and electrophysiological features indistinguishable from "idiopathic" CIDP. These patients respond well to therapy with immunoglobulins, steroids, or plasmapheresis.

Polyneuropathies with IgM-paraproteinemia (IgM-PN) have several features that set them apart from CIDP. Sensory symptoms clearly predominate and may develop months to years before motor deficits occur. Sensory ataxia, deafferentation symptoms of the upper extremity, and action tremor point to the involvement of large sensory fibers and of sensory dorsal root ganglia (see following discussion). A clinical syndrome of sensory or sensorimotor deficits without proximal weakness has been termed *distal acquired demyelinating symmetric (DADS) neuropathy* and is highly suggestive of IgM-PN. The course of the disease is normally chronic progressive, although dynamics may change, and in some patients, disease progression may be favorably slow over decades. A consistently relapsing-remitting course raises doubts about the diagnosis.

In a high percentage of patients with IgM-PN, antibodies against the MAG can be detected and may precede manifestation of Waldenström's disease by years. It is still subject to discussion whether IgM-PN or MAG antibodies are more specific in differentiating IgM-PN from CIDP.

Electrophysiological assessment reveals a predominantly demyelinating sensory neuropathy with early loss of sensory action potentials. Additionally, some axonal degeneration with reduction of amplitudes will be present in a variable degree. Motor responses of the lower limbs are usually absent, and the so-called terminal latency index is reduced in the arms, reflective of significant damage to the most distal nerve segments. The reduced terminal latency index characterizes IgM-PN as a distal polyneuropathy with "dying-back" pathology. By contrast, conduction blocks pointing at demyelination in proximal nerve segments are less frequent than in CIDP.

Sural nerve biopsy findings have contributed much to our understanding of IgM-PN pathophysiology [22]. In 60–90% of patients with IgM-PN, the paraprotein reacts with an epitope recognized with the monoclonal mouse antibody HNK-1. This epitope is expressed by NK cells. It is also shared by several myelin glycoproteins, neural cell adhesion molecule (NCAM), and basal laminae [23].

Biopsies reveal the deposition of paraprotein in the outer myelin lamellae, with consecutive widening. MAG is not expressed in these outer myelin lamellae but is located predominantly in the periaxonal myelin. However, the reactive epitopes of MAG are shared by several other gangliosides and sulfatides, including P0, PMP-22, and SGPG that are located on the outer myelin lamellae. Therefore, some MAG antibodies will substantially cross-react with other gangliosides, and in turn, anti-ganglioside and anti-sulfatide antibodies may be responsible for MAG-negative IgM-PN. Another characteristic feature of biopsy findings is the massive precipitation of paraprotein at the nodes of Ranvier and the paranodal myelin lamellae. IgM deposits are also found at the basal laminae and the noncompacted myelin. Widening of the outer myelin lamellae and massive deposition of paraprotein at the nodes of Ranvier strongly suggest a causative role for paraprotein in the demyelinating pathology of IgM-PN. There is a highly significant negative correlation of MAG expression in myelin with serum anti-MAG antibody titers. This supports the idea that MAG is the predominant target of an autoimmune process, and that the autoimmune attack leads to down-regulation of MAG expression. Myelin loss may be induced without recruitment of macrophages, because most biopsies do not feature inflammatory cells. Therapy and prognosis differ markedly from CIDP. In IgM-PN the prognosis depends on the underlying hematological disease, so the differentiation between MGUS and Waldenström's disease is of superior importance. Overall, the natural course of IgM-PN associated with MGUS is milder than in CIDP, but response to therapy is often incomplete or lacking. A new advent in therapy of IgM-PN is a B cell–depleting monoclonal antibody against the CD20 antigen, rituximab. It suppresses the production of paraprotein for months, and this is correlated with clinical improvement, as reported in a certain proportion of patients. Still, some patients fail to respond, and a controlled clinical trial is required to judge efficacy of this therapeutic approach [24].

E. Disimmune Sensory Ganglionopathies

In sensory ganglionopathies, dorsal root ganglion cells degenerate [25,26]. For this reason, they should be separated from polyneuropathies, as they have more diffuse, segmental, or distal sensorimotor patterns of damage. All sensory ganglionopathies are frequently associated with disimmune or neoplastic disorders and toxic effects. The latter two are not the subject of this review; an overview of these variants is given in reference 25.

Sensory ataxia is the hallmark of the clinical syndrome. Position sense is impaired, most likely due to degeneration of the sensory axons from the muscle spindles. Sensory symptoms such as paresthesias or numbness reflect damage in both proximal and distal nerves, sometimes in an asymmetrical or patchy fashion. Loss of afferents causes pseudoathetoid movements, especially of the arms and hands. Electrophysiology reveals unexcitable sensory nerve action potentials, in line with the clinical findings. The underlying pathology is unknown. Dorsal root ganglia are only rarely biopsied, and autopsy material is scarce. Sural biopsies from these patients reveal axonal loss but no obvious inflammatory or autoimmune process. Because of the bipolar architecture of dorsal root ganglia cells, wallerian degeneration of ascending fibers will occur in parallel and on top of the distal degeneration. Sjögren's syndrome is the most common autoimmune disease associated with sensory ganglionopathies, but any other connective tissue disorder may occasionally give rise to a sensory ganglionopathy. Sensory ganglionopathy almost invariably predates the manifestation of Sjögren's syndrome or other connective tissue disorders. Sjögren's syndrome is characterized by the presence of anti-nuclear antibodies with typical anti-SS-A and anti-SS-B subdifferentiation. The detection of autoantibodies and therapeutic responses to high-dose IVIGs and plasma exchange, in some patients, supports a disimmune pathophysiology.

IV. Systemic Autoimmune Disease Affecting the Peripheral Nerve

A. Vasculitic Neuropathies

Involvement of the panarteritis nodosa (PAN) is prototypic for vasculitic neuropathies. PAN depicts a syndrome of multisystem vasculitic changes with skin (livedo, cutaneous necrosis, and modules), pulmonary (asthma), and renal manifestations. Clinical involvement of peripheral nerves occurs in as many as 75% of all patients. PNS problems are less frequent in Churg-Strauss vasculitis, rheumatoid arthritis, Sjögren's syndrome, Wegener's granulomatosis, scleroderma, and systemic lupus erythematosus.

Vasculitic neuropathies affect subacutely sensorimotor nerves in a multiplex fashion with stepwise progression. Asymmetry is typical at the beginning, but as the disease evolves, asymmetry may be lost when an increasing number of nerves are affected and deficits are confluent. Pain may be more disabling than other sensorimotor symptoms and may occur or exacerbate suddenly. The peroneal nerve is the predilection site for vasculitic neuropathies, but both upper and lower extremities, and even cranial nerves, may be affected. Pain involves hands and feet but extends beyond the area of numbness toward the proximal thighs and arms. Myalgic pain and pain located at the big joints may point to additional muscular involvement or rheumatological manifestation of the disease.

The hallmark of vasculitic neuropathies is the large-scale production of immune complexes. Intriguingly, the highest titers of circulating immune complexes occur in PAN, the disease most frequently associated with vasculitic neuropathy. The amount of circulating immune complex is a powerful laboratory marker of disease activity. On biopsy, the pathological process primarily affects the *vasa nervorum*. Immune complexes adhere to the vessel wall, endothelial cells express MHC class I and II antigen, and nonspecific tissue damage is mediated by complement activation, matrix metalloproteinases, proteases, free radicals, and other effector molecules. Transmural inflammation narrows vessel lumen, generates fibrinoid necrosis, and causes destruction of the endothelium. This leads to microthrombosis and occlusion of the vessel. Endothelium and internal elastic lamellae are disrupted and microbleeding frequently occurs. This vasculitic neuropathy is essentially ischemic, and ischemia is likely to cause pain. The occurrence of sudden pain attacks supports the ischemic pathogenesis of pain. The individual occurrence of ischemia in some nerves explains the patchy or multiplex pattern, with a tendency for confluent deficits when disease progresses. Inflammatory cells secrete toxic mediators and cytokines that further damage neighboring nerve fascicles. Although no direct immune attack of axons and myelin sheaths occurs, activated matrix metalloproteinases and the severely damaged blood-nerve barrier will endanger neural integrity by aggravating ischemic damage. The normal vascular architecture of nerves gives rise to a well-collateralized network of endoneural vessels, and epineural and perineural vessels enter the nerve at several segments. Watershed zones between bigger supplying arteries are prone to ischemic damage in vasculitis. Collateralization of epineural vessels is best at the proximal and distal nerve segments, but the middle parts of the nerves are at higher risk for ischemia. Consequently, pain frequently locates to the middle parts of the nerve—at the thigh or the forearm, respectively.

B. Cryoglobulinemic Neuropathy

Cryoglobulinemia is characterized by antibodies that reversibly precipitate in the vasculature at low temperatures. Cryoglobulins may be rarely monoclonal (type I) in Waldenström's disease but far more often are of mixed polyclonal and monoclonal type (II) or of polyclonal antibodies (type III). As much as 80% of patients with mixed cryoglobulinemia suffer from hepatitis C virus (HCV) infection. Peripheral neuropathies occur in up to 70% of patients with cryoglobulinemia. Immune complexes occlude small and medium arteries, and systemic complement is depleted from immune complex formation. Microscopically, there are intraluminal thrombi composed of precipitated immune complexes, as well as capillary deposits of IgM cryoglobulins. Polyneuropathy is most frequently multiplex but can be of the distal symmetrical type. Pathophysiologically, occlusion of *vasa nervorum* causes the peripheral nerve involvement. This explains the multiplex pattern, similar to vasculitic neuropathy.

V. Immune-Mediated Exacerbation of Nonimmune Peripheral Nerve Disease

A. Hereditary Neuropathies

Thanks to progress in molecular diagnostics, an increasing number of hereditary neuropathies can be confidently diagnosed. First recognized in CMT1B, hereditary neuropathies can be accompanied by substantial lymphocytic infiltration in sural nerve biopsies. In a group of patients with subacute exacerbation of long-standing and genetically proven hereditary neuropathies [28], coexistence of inflammatory and hereditary neuropathies was more common than expected by chance. In sural nerve biopsies, lymphocytic infiltration in the endoneural and epineural space was paralleled by severe loss of myelinated fibers, congruent to the diagnosis of hereditary neuropathy.

In animal models of hereditary neuropathy, inflammatory response with macrophage and lymphocyte infiltration also appears to foster axonal degeneration. Hence, a chronic inflammatory process seemingly accompanies myelin depletion and axonal degeneration in hereditary neuropathies. It is tempting to speculate that the degenerative process may unravel epitopes on myelin and axons that normally escape from immune surveillance, engendering an autoimmune response that will exacerbate nerve damage. Antibodies against myelin proteins (e.g., PMP22) have been reported in patients with hereditary neuropathies. However, patients with or without antibodies did not differ in the course of their disease. In hereditary neuropathy, the formation of myelin, especially the compaction of myelin lamellae, may be disturbed. Intriguingly, this means widening of myelin lamellae, as seen in IgM-PN and in several hereditary neuropathies as well. Poorly compacted myelin may expose normally sequestered or hidden antigens to systemic immune attack (see Fig. 4).

At least a part of these patients with exacerbated hereditary neuropathies benefited clinically from immunotherapy with high-dose IVIGs or corticosteroids [29].

B. Chronic Polyneuritis in Diabetic Neuropathies (DM-CIDP)

Similar to the situation in hereditary neuropathies, an acquired nonimmune neuropathy may be complicated by an inflammatory response. Again, the coincidence of diabetic and CIDP is far more common than would be expected by chance. The presentation of axonal and myelin antigens that normally escape immune surveillance and immune tolerance may initiate an autoimmune response.

Sural biopsy in diabetic neuropathy may prove the existence of endoneural lymphocytic and monocytic infiltrates, contrasting with the expected microangiopathy and degenerative axonal loss. In a pilot study of patients with histological evidence of inflammation, the combination of high-dose IVIG and antidiabetic therapy was superior to antidiabetic therapy alone. Hence, coexistent inflammation may contribute significantly to neural damage, even in diabetic neuropathy [29].

Patients with immune-mediated neuropathy superimposed on nonimmune neuropathy have added a fascinating aspect to our understanding of peripheral nerve disease. Further elucidation of pathophysiological events in immune-mediated neuropathies is needed to identify new therapeutic strategies [15], enhance therapeutic efficacy, and minimize side effects of long-term therapy in these oftentimes crippling chronic diseases.

References

1. Gold, R., Kieseier, B. C., and Hartung, H. P. (2006). Immunology properties of the peripheral nervous system. *In:* "Clinical Neuroimmunology" (J. Antel, G. Birnbaum, H. P. Hartung, and A. Vincent, eds.), pp. 77–86. Oxford University Press, New York.
2. Christen, U., and von Herrath, M. G. (2005). Infections and autoimmunity—good or bad? *J. Immunol.* **174**, 7481–7486.
3. Yuki, N. (2005). Carbohydrate mimicry: a new paradigm of autoimmune diseases. *Curr. Opin. Immunol.* **17**, 577–582.
4. Gold, R., Hartung, H. P., and Toyka, K. V. (2000). Animal models for autoimmune demyelinating disorders of the nervous system. *Mol. Med. Today* **6**, 88–91.
5. Gold, R., Archelos, J. J., and Hartung, H. P. (1999). Mechanisms of immune regulation in the peripheral nervous system. *Brain Pathol.* **9**, 343–360.
6. Yuki, N., Susuki, K., Koga, M., Nishimoto, Y., Odaka, M., Hirata, K., Taguchi, K., Miyatake, T., Furukawa, K., Kobata, T., and Yamada, M. (2004). Carbohydrate mimicry between human ganglioside G_{M1} and *Campylobacter jejuni* lipooligosaccharide causes Guillain-Barré syndrome. *Proc. Natl. Acad. Sci. USA* **101**, 11404–11409.
7. McCarthy, N., and Giesecke, J. (2001). Incidence of Guillain-Barré syndrome following infection with *Campylobacter jejuni*. *Am. J. Epidemiol.* **153**, 610–614.
8. Usuki, S., Thompson, S. A., Rivner, M. H., Taguchi, K., Shibata, K., Ariga, T., and Yu, R. K. (2006). Molecular mimicry: sensitization of Lewis rats with *Campylobacter jejuni* lipopolysaccharides induces formation of antibody toward GD3 ganglioside. *J. Neurosci. Res.* **83**, 274–284.
9. Goodfellow, J. A., Bowes, T., Sheikh, K., Odaka, M., Halstead, S. K., Humphreys, P. D., Wagner, E. R., Yuki, N., Furukawa, K., Furukawa, K., Plomp, J. J., and Willison, H. J. (2005). Overexpression of GD1a ganglioside sensitizes motor nerve terminals to anti-G_{D1a} antibody-mediated injury in a model of acute motor axonal neuropathy. *J. Neurosci.* **25**, 1620–1628.
10. Godfrey, D. I., and Kronenberg, M. (2004). Going both ways: immune regulation via CD1d-dependent NKT cells. *J. Clin. Invest.* **114**, 1379–1388.
11. Zajonc, D. M., Maricic, I., Wu, D., Halder, R., Roy, K., Wong, C. H., Kumar, V., and Wilson, I. A. (2005). Structural basis for CD1d presentation of a sulfatide derived from myelin and its implications for autoimmunity. *J. Exp. Med.* **202**, 1517–1526.
12. Salomon, B., Rhee, L., Bour-Jordan, H., Hsin, H., Montag, A., Soliven, B., Arcella, J., Girvin, A. M., Padilla, J., Miller, S. D., and Bluestone, J. A. (2001). Development of spontaneous autoimmune peripheral polyneuropathy in B7-2-deficient NOD mice. *J. Exp. Med.* **194**, 677–684.
13. Jung, S., Gaupp, S., Korn, T., Kollner, G., Hartung, H. P., and Toyka, K. V. (2004). Biphasic form of experimental autoimmune neuritis in dark Agouti rats and its oral therapy by antigen-specific tolerization. *J. Neurosci. Res.* **75**, 524–535.
14. Lewis, R. A. (2005). Chronic inflammatory demyelinating polyneuropathy and other immune-mediated demyelinating neuropathies. *Semin. Neurol.* **25**, 217–228.
15. Koller, H., Kieseier, B. C., Jander, S., and Hartung, H. P. (2005). Chronic inflammatory demyelinating polyneuropathy. *N. Engl. J. Med.* **352**, 1343–1356.
16. European Federation of Neurological Societies/Peripheral Nerve Society. (2005). Guideline on management of chronic inflammatory demyelinating polyradiculoneuropathy. Report of a joint task force of the European Federation of Neurological Societies and the Peripheral Nerve Society. *J. Peripher. Nerv. Syst.* **10**, 220–228.
17. Kieseier, B. C., Kiefer, R., Gold, R., Hemmer, B., Willison, H. J., and Hartung, H. P. (2004). Advances in understanding and treatment of immune-mediated disorders of the peripheral nervous system. *Muscle Nerve* **30**, 131–156.
18. Allen, D., Giannopoulos, K., Gray, I., Gregson, N., Makowska, A., Pritchard, J., and Hughes, R. A. (2005). Antibodies to peripheral nerve myelin proteins in chronic inflammatory demyelinating polyradiculoneuropathy. *J. Peripher. Nerv. Syst.* **10**, 174–180.
19. Csurhes, P. A., Sullivan, A. A., Green, K., Pender, M. P., and McCombe, P. A. (2005). T cell reactivity to P0, P2, PMP-22, and myelin basic protein in patients with Guillain-Barré syndrome and chronic inflammatory demyelinating polyradiculoneuropathy. *J. Neurol. Neurosurg. Psychiatry* **76**, 1431–1439.
20. Pestronk, A., Cornblath, D. R., Ilyas, A. A., Baba, H., Quarles, R. H., Griffin, J. W., Alderson, K., and Adams, R. N. (1988). A treatable multifocal motor neuropathy with antibodies to GM1 ganglioside. *Ann. Neurol.* **24**, 73–78.
21. Oh, S. J. LaGanke, C., Powers, R., Wolfe, G. I., Quinton, R. A., and Burns, D. K. (2005). Multifocal motor sensory demyelinating neuropathy: inflammatory demyelinating polyradiculoneuropathy. *Neurology* **65**, 1639–1642.
22. Vital, A. (2001). Paraproteinemic neuropathies. *Brain Pathol.* **11**, 399–407.
23. Stefansson, K., Reder, A. T., and Antel, J. P. (1986). An epitope shared by central nervous system myelin and peripheral blood macrophages. *J. Neuroimmunol.* **12**, 49–55.
24. Nobile-Orazio, E. (2004). IgM paraproteinaemic neuropathies. *Curr. Opin. Neurol.* **17**, 599–605.
25. Kuntzer, T., Antoine, J. C., and Steck, A. J. (2004). Clinical features and pathophysiological basis of sensory neuronopathies (ganglionopathies). *Muscle Nerve* **30**, 255–268.
26. Sinnreich, M., Klein, C. J., Daube, J. R., Engelstad, J., Spinner, R. J., and Dyck, P. J. (2004). Chronic immune sensory polyradiculopathy: a possibly treatable sensory ataxia. *Neurology* **63**, 1662–1669.
27. Younger, D. S. (2004). Vasculitis of the nervous system. *Curr. Opin. Neurol.* **17**, 317–336.
28. Ginsberg, L., Malik, O., Kenton, A. R., Sharp, D., Muddle, J. R., Davis, M. B., Winer, J. B., Orrell, R. W., and King, R. H. (2004). Coexistent hereditary and inflammatory neuropathy. *Brain* **127**, 193–202.
29. Sharma, K. R., Cross, J., Farronay, O., Ayyar, D. R., Shebert, R. T., and Bradley, W. G. (2002). Demyelinating neuropathy in diabetes mellitus. *Arch. Neurol.* **59**, 758–765.
30. Tzakos, A. G., Troganis, A., Theodorou, V., Tselios, T., Svarnas, C., Matsoukas, J., Apostolopoulos, V., and Gerothanassis, I. P. (2005). Structure and function of the myelin proteins: current status and perspectives in relation to multiple sclerosis. *Curr. Med. Chem.* **12**, 1569–1587.

20

Hashimoto Encephalopathy

Ji Y. Chong, MD

Keywords: *autoimmune thyroiditis, Hashimoto encephalopathy, steroid responsive encephalopathy*

I. History and Nomenclature
II. Clinical Features
III. Etiology
IV. Pathogenesis
V. Clinical Course
    References

I. History and Nomenclature

In 1966, Lord Brain described a patient with relapsing-remitting encephalopathy, psychosis, and stroke-like events in association with Hashimoto thyroiditis. This patient continued to have recurrent events despite treatment with anticoagulants and corticosteroids and eventually became asymptomatic on only thyroxine [1]. Over the years, more case reports and case series have been published that have better defined the syndrome that has been termed *Hashimoto encephalopathy* (HE). Although a syndrome of reversible encephalopathy with thyroid antibodies that may respond to corticosteroid treatment has evolved, the proper nomenclature and overall classification of this disorder is under great debate.

This debate over terminology stems from the paucity of knowledge of the disease pathophysiology and the lack of a well-defined clinical syndrome. A wide range of neuropsychiatric abnormalities have been reported in association with anti-thyroid antibodies: myelopathy, hallucinations, subacute cognitive decline, and coma [2]. The uniform feature that has been used to identify patients with HE is thyroid antibodies. However, these antibodies may have a high prevalence in normal subjects. In one study, the prevalence of high serum anti-thyroid antibody concentrations (>1:200 on hemagglutination assay) was 11% in euthyroid young adults, 23% in euthyroid women older than 60 years, and 67% in older women with hypothyroid [3]. It is clear that defining a clinically heterogenous syndrome by a relatively common antibody may be erroneous, and the term *Hashimoto encephalopathy*, misleading. For this reason, some have proposed changing the name *Hashimoto encephalopathy* to *steroid-responsive encephalopathy associated with autoimmune thyroiditis* (SREAT) [4]. This would replace the name that implies a causal relationship between Hashimoto thyroiditis and encephalopathy with a name that suggests association. This term may also be misleading, however, with the emphasis on steroid responsiveness, which may not be a necessary feature.

Some authors have proposed creating a new category of disease, nonvasculitic autoimmune inflammatory meningoencephalitis (NAIM) [5], as a larger group of reversible autoimmune encephalopathies. NAIM would include disorders such as systemic lupus erythematosus and Sjögren-associated encephalopathy. HE, or SREAT, would be classified as a subset of NAIM. All these disorders share a presumed autoimmune mechanism of brain disease without vasculitis. Although these disorders may have another associated autoimmune disease, the mechanism of brain dysfunction and its relationship with the primary autoimmune disease is no clearer than in HE.

It is likely that there is a broad range of autoimmune encephalopathies. One group described a new syndrome of autoimmune limbic encephalitis with voltage-gated potassium channel antibodies [6]. Well-defined pathophysiological markers may help clarify the diagnoses and proper terminology. Until then, the term *Hashimoto encephalopathy* is accepted for patients with unexplained encephalopathy and thyroid antibodies.

II. Clinical Features

HE appears to be a clinically heterogeneous syndrome. Some have described a relapsing and remitting course with recurrent strokelike symptoms. Patients may also have a diffuse progressive pattern with gradual cognitive decline [7]. Other clinical features include seizure, tremor, and myoclonus (Table 1) and patient presentation may be very similar to Creutzfeldt–Jakob disease [8].

Table 1 Clinical Features of Patients with Hashimoto Encephalopathy[a]

Female	85%
Mean age (years)	45
Range	9–78
Focal deficit	27%
Seizures	64%
Myoclonus	40%
Psychosis	40%
Relapsing–remitting	55%
High cerebrospinal fluid protein content	76%
Abnormal imaging	46%
Abnormal electroencephalogram	97%

[a]*Source:* Chong, J. Y. (2005). Hashimoto encephalopathy. *In:* "Merrit's Neurology" (L. P. Rowland, ed.), 11th edition, pp. 1120–1121. Lippincott Williams & Wilkins, Philadelphia.

III. Etiology

The etiology of this disorder is not known. Because of the association with Hashimoto thyroiditis, some investigation has centered around the relationship between thyroid hormone dysfunction and encephalopathy. There are two thyroid hormones: thyroxine (T_4) and 3,5,3-triiodothyronine (T_3). Thyroid hormones are highly protein bound and it is the 1% unbound T_4 and T_3 that are biologically active. These hormones bind to nuclear receptors; however, T_3 binds preferentially over T_4, and therefore, T_4 typically has little direct physiological effect. It must be deiodinated to T_3 to be active [9]. Thyroid hormone nuclear receptors are found in brain, heart, liver, and kidney, as well as the pituitary gland and hypothalamus, so deficiency of thyroid hormone can cause multisystem dysfunction that includes hypothermia, bradycardia, neuropathy, hyponatremia, and skin changes [10].

Severe hypothyroidism may cause neurological symptoms such as decreased level of alertness, cognitive dysfunction, and even psychosis ("myxedema madness") and coma. It has also been associated with diffuse slowing on electroencephalogram (EEG) and high cerebrospinal fluid (CSF) protein concentrations [11]. These nonspecific changes may be due to systemic metabolic disarray. However, there have also been radiographic studies suggesting direct metabolic effects of hypothyroidism on the brain. These studies show global decreased cerebral blood flow and brain glucose metabolism in patients with hypothyroidism [12]. Another postulated mechanism of thyroid hormone effects causing neurological symptoms is related to thyrotropin-releasing hormone (TRH). Some have suggested TRH causes symptoms of HE. TRH administration to a patient with epilepsy induced a seizure [13]. In one patient who had relapses of symptoms associated with her menstrual cycle, symptoms could be induced with TRH. Her symptoms were treated using thyroxine [14]. TRH levels are not commonly measured in patients with encephalopathy and it is possible that they are abnormal in patients with HE. TRH acts on the pituitary gland. Any relationship between this hormone and HE is speculative.

Most patients had subclinical or overt hypothyroidism, further suggesting thyroid dysregulation may contribute to symptoms [2]. Among 83 patients reported with HE, 55% had subclinical or overt hypothyroidism; 9% improved in response to thyroxine; and 30% improved in response to thyroxine and corticosteroids. However, some patients with hypothyroidism did not respond to thyroxine therapy. Furthermore, the neurological findings were similar in the patients with euthyroidism, with or without thyroxine therapy, and those with subclinical or overt hypothyroidism.

Subclinical hypothyroidism has not been shown to have the effects of overt hypothyroidism on cerebral function, EEG tracings, and CSF protein content [15]. Subclinical hypothyroidism was associated with increased prevalence of neuropsychiatric symptoms in one review, but all symptoms were mild and none had encephalopathy [16]. Additionally, six patients with encephalopathy in this review had overt or subclinical hyperthyroidism that has not been associated with encephalopathy. Considering the variability of thyroid function and that most patients did not have hypothyroidism to the degree that could explain the neurological symptoms, thyroid dysfunction alone cannot account for the syndrome of HE.

An autoimmune etiology is more likely rather than any direct thyroid effects in this disorder. Chronic autoimmune thyroiditis has two forms: Hashimoto thyroiditis (goitrous) and atrophic thyroiditis. Both are associated with thyroid antibodies. Hashimoto thyroiditis (Hashimoto's disease) was first described as struma lymphomatosa (lymphomatous goiter) in 1912. It is characterized by the presence of high serum concentrations of anti-thyroid antibodies, and pathologically by lymphocytic infiltration, lymphoid germinal centers, and fibrosis in the thyroid gland [2]. Patients may be hypothyroid, subclinically hypothyroid, or even transiently hyperthyroid. Middle-aged women are more likely to be affected by Hashimoto thyroiditis.

The hypothyroidism is thought to be caused primarily by T cell cytotoxicity in which CD4 cells react with thyroid antigen, but antibodies that block the action of thyroid-stimulating hormone (TSH) may play a role in some patients. The antibodies formed in Hashimoto thyroiditis are against thyroglobulin, thyroid peroxidase, and TSH receptor. Some thyroglobulin is present in serum, but otherwise, it and thyroid peroxidase are found only in thyroid tissue. Thyroglobulin and thyroid peroxidase are both large glycoproteins that are necessary for synthesis of thyroid hormone. Anti-thyroid peroxidase antibodies (formerly called *anti-thyroid microsomal antibodies*) and anti-thyroglobulin antibodies are most commonly measured, but these have not been found to have any biological activity. Both are usually present in Hashimoto thyroiditis, but the presence of either alone is sufficient to make a diagnosis of Hashimoto thyroiditis [2].

Hashimoto thyroiditis is clearly an autoimmune process, but thyroid antibodies may not be causally associated with the brain symptoms. There is no evidence that any anti-thyroid antibody reacts with brain tissue or affects nerve function. The search for thyroid antibodies in CSF has not yielded a consistent relationship with brain disease. One series of patients found anti-thyroid antibodies in the CSF of patients with HE but none in control patients with other neurological disorders [17]. However, the role of the intrathecal antibodies was not clear and clinical improvement in this small series was independent of CSF antibody levels [17].

In another study, all patients admitted with unexplained acute or subacute encephalopathy or myelopathy or those with a history of thyroid disorder were screened for thyroid antibodies. Of 143 patients screened for antibodies, 9 were given a diagnosis of HE. However, only six of these nine had Hashimoto thyroiditis based on the presence of antibodies and typical features on thyroid biopsy. These authors suggest Hashimoto thyroiditis is not necessary for HE because not all patients had thyroiditis [18].

It is possible the high serum anti-thyroid antibody concentrations are a response to an environmental exposure, such as an infectious agent that injured the brain and that is structurally similar to thyroid peroxidase or thyroglobulin. Molecular mimicry has been postulated as a possible explanation for autoimmune disease in general. The primary hypothesis of this mechanism is that an epitope of an infectious agent stimulates an immune response against a similar self-epitope. The external infectious antigen may not be detectable when the autoimmune disease develops. An example is a murine encephalomyelitis virus that causes encephalitis by autoimmune reaction after the initial infection [19].

The autoimmune etiology is supported by the higher prevalence of HE in women, the reported response to corticosteroids, and its apparent association with other autoimmune disorders. Autoimmune disorders have been known to cluster [2]. Many patients who have autoimmune diseases have high serum concentrations of one or more antibodies directed against tissues not affected by the particular autoimmune disease. For example, patients with myasthenia gravis may have high serum anti-thyroid antibody concentrations. In a review, 8 of 85 patients with HE (though not all were tested) had another autoimmune antibody [2]. Therefore, Hashimoto thyroiditis may be a marker of an autoimmune state in which the brain autoimmune process has not yet been elucidated.

In fact, several investigators have described patients with encephalopathy clinically similar to HE but with high serum concentrations of antibodies to other self-antigens. One patient with myasthenia gravis had two episodes of unexplained encephalopathy [20]. Four patients were described with serologic evidence of other autoimmune diseases, including high serum concentrations of rheumatoid factor, anti-nuclear antibodies, anti-RNA antibodies (SSA, SSB), or cardiolipin antibodies. Brain biopsies revealed only perivascular lymphocytic infiltration [5]. Others have suggested that this syndrome may actually be an undiagnosed encephalopathy with voltage-gated potassium channel antibodies [21].

The search for antibodies that are possibly pathogenic in the brains of patients with HE has revealed an anti-neuronal autoantibody in the serum of a patient with clinical HE, but no reactivity with serum of a patient with Hashimoto thyroiditis without encephalopathy [22]. Another study detected anti-α-enolase antibodies in patients with HE. This autoantigen was found in the thyroid gland and brain. It was not found in healthy patients or patients with other known neurological disorders [23]. Future testing of these antibodies in patients with unexplained encephalopathy may help determine whether they are indeed pathogenic in autoimmune encephalopathy.

IV. Pathogenesis

There have been few case reports describing pathological findings in HE and these have not been revealing of pathogenesis. Autopsies have shown lymphocytic infiltration of brainstem veins, as well as lymphocytic infiltration around venules and arterioles [2]. Biopsy data suggest similar findings [2]. Although these may support an autoimmune mechanism, none has shown true vasculitis. Others have found spongiform change with gliosis, perivascular mononuclear cells, and parenchymal microglia, and in some cases, there were no histological abnormalities [2].

The lymphocytic infiltration of small blood vessels has been postulated to cause hypoperfusion or cerebral edema. Autoimmune deposits may contribute to cerebral vessel abnormalities, which may then lead to hypoperfusion. One case report showed single-photon emission computed tomography (SPECT) abnormalities with evidence of hypoperfusion in HE [24]. Another case report described a patient with HE with diffuse homogenous hypoperfusion on SPECT. The authors speculated cerebral microvasculature involvement with vasculitis led to the global hypoperfusion [25].

An alternative hypothesis is that an unknown antibody directly affects cortical neurons and causes global cerebral dysfunction. EEG findings in HE have shown generalized slow-wave abnormalities, which also suggest a global process [26]. Autoantibodies that have been suggested as pathogenic in HE include an anti-neuronal antibody and antibody against α-enolase, as described earlier.

V. Clinical Course

Overall, there appears to be a low mortality associated with HE. Because there are no large epidemiological studies, this is based on case reports and series. In one review of the literature, 3 of 85 patients with the diagnosis of HE died. One had no neurological symptoms at the time of death and the two other patients died while being treated with corticosteroids [2].

Patients also appear to improve with corticosteroid treatment. Of the 85 patients reviewed, 44 were treated with corticosteroids (some were already taking thyroxine), and 43 (98%) improved. Definitions of improvement differed between published reports. Twenty-three patients were treated with corticosteroids and thyroxine, of whom 21 (91%) improved. Although steroid responsiveness has been suggested as a feature of this disorder, some patients improved with thyroxine alone: of 12 patients, 8 (67%) improved on thyroxine alone. The clinical characteristics of the patients who improved in response to these different regimens were similar. The clinical features of the nine treated patients who did not improve were also similar and treatment regimens in those who failed to improve varied: 1 patient was treated with corticosteroids, 2 with corticosteroids and thyroxine, 4 with thyroxine, and 1 with intravenous immunoglobulins. One patient received no treatment [2].

Most patients with HE improved in association with, but not necessarily due to, corticosteroid treatment. The patients who responded had no distinguishing clinical characteristics, but too few were treated in other ways for meaningful comparison. The natural history of untreated HE is not well described.

Although there is a growing body of literature regarding HE, much is unknown with respect to etiology and pathophysiology. A better clinical definition of this syndrome will help guide investigation into these areas.

References

1. Brain, L., Jellinek, E. H., and Ball, K. (1966). Hashimoto's disease and encephalopathy. *Lancet* **2**, 512–514.
2. Chong, J. Y., Rowland, L. P., and Utiger, R. D. (2003). Hashimoto encephalopathy: syndrome or myth? *Arch. Neurol.* **60**, 164–171.
3. Sawin, C., Bigos, S. T., Land, S., and Bacharach, P. (1985). Relationship between elevated serum thyrotropin level and thyroid antibodies in elderly patients. *Am. J. Med.* **79**, 591–595.
4. Castillo, P. R., Boeve, B. F., Caselli, R. J., et al. (2002). Steroid responsive encephalopathy associated with thyroid autoimmunity: clinical and laboratory findings. *Neurology* **58(Suppl 3)**, A248.
5. Caselli, R. J., Boeve, B. F., Scheithauer, B. W., O'Duffy, J. D., and Hunder, G. G. (1999). Nonvasculitic autoimmune inflammatory meningoencephalitis (NAIM): a reversible form of encephalopathy. *Neurology* **53**, 1579–1581.
6. Thieben, M. J., Lennon, V. A., Boeve, B. F., Aksamit, A. J., Keegan, M., and Vernino, S. (2004). Potentially reversible autoimmune limbic encephalitis with neuronal potassium channel antibody. *Neurology* **62**, 1177–1182.
7. Kothbauer-Margreiter, I., Sturzenegger, M., Komor, J., Baumgartner, R., and Hess, C. W. (1996). Encephalopathy associated with Hashimoto thyroiditis: diagnosis and treatment. *J Neurol* **243**, 585–593.
8. Seipelt, M., Zerr, I., Nau, R., Mollenhauer, B., Kropp, S., Steinhoff, B. J., Wilhelm-Gossling, C., Bamberg, C., Janzen, R. W., Berlit, P., Manz, F., Felgenhauer, K., and Poser, S. (1999). Hashimoto's encephalitis as a differential diagnosis of Creutzfeldt-Jakob disease. *J. Neurol. Neurosurg. Psychiatry* **66**, 172–176.
9. Taurog, A. (2000). Hormone synthesis: thyroid hormone metabolism. *In:* "The Thyroid: Fundamental and Clinical Text" (L. E. Braverman and R. D. Utiger, eds.), 8th edition, p. 61. Lippincott Williams & Wilkins, Philadelphia.
10. Larsen, P. R., Silva, J. E., and Kaplan, M. M. (1981). Relationships between circulating and intracellular thyroid hormones. Physiological and clinical implications. *Endocr. Rev.* **2**, 87.

11. Whybrow, P. C. (2000). Behavioral and psychiatric aspects of hypothyroidism. *In:* "The Thyroid: A Fundamental and Clinical Text" (L. E. Braverman and R. D. Utiger, eds.), 8th edition, pp. 837–842. Lippincott Williams & Wilkins, Philadelphia.
12. Constant, E. L., De Volder, A. G., Ivanoiu, A., Bol, A., Labar, D., Seghers, A., Cosnard, G., Melin, J., and Daumerie, C. (2001). Cerebral blood flow and glucose metabolism in hypothyroidism: A positron emission tomography study. *J. Clin. Endocrinol. Metab.* **86**, 3864–3870.
13. Maeda, K., and Tanimoto, K. (1981). Epileptic seizures induced by thyrotropin releasing hormone. *Lancet* **1**, 1058–1059.
14. Ishii, K., Hayashi, A., Tamaoka, A., Usuki, S., Mizusawa, H., and Shoji, S. (1995). Case report: thyrotropin-releasing hormone-induced myoclonus and tremor in a patient with Hashimoto's encephalopathy. *Am. J. Med. Sci.* **310**, 202–205.
15. Ross, D. S. (2000). Subclinical hypothyroidism. *In:* "The Thyroid: A Fundamental and Clinical Text" (L. E. Braverman and R. D. Utiger, eds.), 8th edition, pp. 1001–1006. Lippincott Williams & Wilkins, Philadelphia.
16. Ayala, A. R., Danese, M. D., and Ladenson, P. W. (2000). *Endocrinol. Metab. Clin. North Am.* **29**, 399–415.
17. Ferracci, F., Moretto, G., Candeago, R. M., Cimini, N., Conte, F., Gentile, M., Papa, N., and Carnevale, A. (2003). Antithyroid antibodies in the CSF: Their role in the pathogenesis of Hashimoto's encephalopathy. *Neurology* **60**, 712–714.
18. Ferracci, F., Bertiato, G., and Moretto, G. (2004). Hashimoto's encephalopathy: epidemiologic data and pathogenetic considerations. *J. Neurol. Sci.* **217**, 165–168.
19. Davies, J. M. (2000). Introduction: Epitope mimicry as a component cause of autoimmune disease. *Cell Mol. Life Sci.* **57**, 523–526.
20. Maher, D., and Fritz, V. (1996). Autoimmune encephalopathy after treatment of thymoma-associated myasthenia gravis. *Br. J. Clin. Pract.* **50**, 406–407.
21. Schott, J. M., Warren, J. D., and Rossor, M. N. (2003). The uncertain nosology of Hashimoto encephalopathy. *Arch. Neurol.* **60**, 1812.
22. Oide, T., Tokuda, T., Yazaki, M., Watarai, M., Mitsuhashi, S., Kaneko, K., Hashimoto, T., Ohara, S., and Ikeda, S. (2004). Anti-neuronal autoantibody in Hashimoto's encephalopathy: neuropathological, immunohistochemical, and biochemical analysis of two patients. *J. Neurol. Sci.* **217**, 7–12.
23. Ochi, H., Horiuchi, I., Araki, N., Toda, T., Araki, T., Sato, K., Murai, H., Osoegawa, M., Yamada, T., Okamura, K., Ogino, T., Mizumoto, K., Yamashita, H., Saya, H., Kira, J. (2002). Proteomic analysis of human brain identifies alpha enolase as a novel autoantigen in Hashimoto's encephalopathy. *FEBS Lett.* **528**, 197–202.
24. Kalita, J., Misra, U. K., Rathore, C., Pradhan, P. K., Das, B. K., and Gandhi, S. (2003). Hashimoto's encephalopathy: clinical, SPECT, and neurophysiological data. *Q. J. Med.* **96**, 453–457.
25. Forchetti, C., Katsamakis, G., and Garron, D. (1997). Autoimmune thyroiditis and a rapidly progressive dementia: global hypoperfusion on SPECT scanning suggests a possible mechanism. *Neurology* **49**, 623–626.
26. Schauble, B., Castillo, P. R., Boeve, B. F., and Westmoreland, B. F. (2003). EEG findings in steroid-responsive encephalopathy associated with autoimmune thyroiditis. *Clinical Neurophysiology* **114**, 32–37.

21

Vascular Cognitive Impairment

Velandai Srikanth, PhD
Michael. M. Saling, PhD
Amanda G. Thrift, PhD

Keywords: *cerebrovascular disease, dementia, stroke, vascular cognitive impairment*

I. Introduction
II. Nomenclature, Epidemiology, and Natural History
III. Etiology of Vascular Cognitive Impairment
IV. Causes of Hereditary Vascular Cognitive Impairment
V. Causes of Sporadic Vascular Cognitive Impairment
VI. Cardiovascular Causes of Sporadic Vascular Cognitive Impairment Independent of Structural Brain Abnormality
VII. Pathophysiology of Vascular Cognitive Impairment
VIII. Clinical Features
IX. Summary
  References

In this chapter, we provide a concise update on the neurobiology of vascular cognitive impairment (VCI), a heterogenous entity. The scope of this topic is extremely large, and we have attempted to describe as best as possible the current state of play in the understanding of this group of conditions. We begin by introducing the concept of VCI, with an emphasis on nomenclature, epidemiology, and natural history, and then progress to etiology, pathophysiology, and clinical features. We recognize that this is a field of many uncertainties and have dealt with the sections emphasizing the nature of these uncertainties and potential ways forward to understanding VCI.

I. Introduction

Cerebrovascular disease, cognitive impairment, and dementia are highly prevalent conditions affecting older people [1]. When combined, they are responsible for a large portion of the burden of disability in aging populations. Cerebrovascular disease and cognitive impairment often coexist, leading to much speculation about potential common mechanisms that may be involved in causing these disorders. Given their prevalence and effect on individuals and society at large, research into such potential common mechanisms is critical if appropriate therapeutic and preventive strategies are to be planned. Such research efforts have begun to unravel the complex interrelationship between these disorders. "Vascular cognitive impairment" refers to the wide spectrum of cognitive disorders that may be ascribed to vascular causes [2]. The concept of VCI arose in an attempt to resolve many uncertainties surrounding the entity of vascular dementia (VaD). However, knowledge regarding the cerebrovascular mechanisms involved in causing cognitive impairment and the interplay between such mechanisms is still in its infancy. This chapter deals with current state of knowledge of VCI, with particular emphasis on potential underlying mechanisms that may deserve future attention from a research perspective.

II. Nomenclature, Epidemiology, and Natural History

VaD, though an extremely difficult entity to define, was the predominant nosological entity linking cerebrovascular disease and cognitive dysfunction. Diagnostic criteria for VaD, such as the *National Institute of Neurological Diseases and Stroke—Association Internationale pour la Recherche et l'Enseignement en Neurosciences criteria* (NINDS-AIREN) [2] (Table 1), were developed without strong empirical evidence linking the proposed vascular cause (most commonly brain infarction) and the occurrence of dementia. These criteria depended solely on subjective clinical judgment to link the presence of brain infarction on imaging to the presence of dementia, with dementia being modeled on a pattern of deficits seen commonly in Alzheimer's disease (AD), namely memory plus other domain impairment. Apart from this, the criteria also required the clinician to make "gut-judgments" regarding the relative importance of physical signs (such as gait slowing and incontinence) toward making a diagnosis of VaD. Other VaD criteria (such as from the *Diagnostic and Statistical Manual of Mental Disorders*, fourth edition [DSM-IV] [4]) also specify the need for the presence of a "decline from a previously higher level of functioning" for diagnosis. This would only allow the detection of cases of dementia severe enough to cause decline in everyday functional abilities. Finally, there are pathological data showing that most people dying with dementia have a combination of neurodegenerative and vascular changes in their brains, thus making the antemortem diagnosis of a "pure" VaD extremely difficult [5]. In fact, such findings have led to speculation over whether cerebrovascular disease predisposes to the clinical expression of AD or even whether AD is caused by such vascular disease. Given such a flawed modeling of VaD criteria, some investigators have suggested a move away from the traditional concept of VaD [2] to that of "VCI."

Bowler and Hachinski [2] proposed the concept of VCI, referring to all grades of cognitive impairment (from very mild to severe) that can be ascribed to a broad range of vascular causes. Furthermore, in contrast to existing criteria, they suggested that severity of physical disability or functional decline not be emphasized as a requirement for case finding. They also suggested that memory not be stipulated as an essential criterion in diagnosis, enabling the inclusion of cases with different patterns of cognitive loss. Using these modifications, the investigators of the population-based Canadian Study of Health and Aging (CHSA) found a prevalence rate of VCI of 5.0% in people 65 years or older compared with a prevalence rate of 5.1% cases of AD [6]. Among the VCI cases, 2.6% were classified as "vascular cognitive impairment–not dementia" (vascular CIND), 1.5% as VaD, and 0.9% as "mixed" AD with a vascular component. After 5 years of follow-up, cases of VCI were found to have poorer outcomes in terms of institutionalization (relative risk [RR] 3.1, 95% confidence interval [CI], 2.1–4.6) or death (RR 1.8, 95% CI, 0.9–1.2) compared with cognitively normal individuals. This indicates that cases of VCI may represent a group distinct from nondemented cases in terms of clinical prognosis. In a follow-up of the 149 cases of "vascular CIND," 58 (46%) had developed dementia in 5 years, indicating that these cases may represent a predementia group [7]. Baseline impairment in memory and category fluency tests predicted progression to dementia in this cohort, similar to the prediction-to-progression profile seen in AD [8]. This still raises the question of whether cases of VCI identified in this study are distinct from other predementia entities such as mild cognitive impairment (MCI). Moreover, the process of ascribing a cause-and-effect relation between "vascular" lesion and "cognitive impairment" was mainly by consensus between physicians involved in the CHSA, thereby still not providing empirical evidence for a causal link between the two. Finally, in the absence of pathological data, it becomes extremely difficult to separate out true VCI from cognitive impairment due to a mixture of

Table 1 The National Institute of Neurological Diseases and Stroke–Association Internationale pour la Recherche et l'Enseignement en Neurosciences (NINDS-AIREN) Criteria for Vascular Dementia[a]

1. Dementia:
   Impairment in memory and at least two other domains
   - Orientation
   - Attention
   - Language
   - Visuospatial functions
   - Executive functions, motor control, and praxis
2. Cerebrovascular disease
   Focal signs on neurological examination (hemiparesis, lower facial weakness, Babinski's sign, sensory deficit, hemianopia, and dysarthria)
   Evidence of relevant cerebrovascular disease by brain imaging (computed tomography [CT])
   - Large-vessel infarcts
   - Single strategically placed infarct
   - Multiple basal ganglia and white-matter lacunes
   - Extensive periventricular white-matter lesions
   - Combinations thereof
3. A relationship between the above disorders manifested or inferred by the presence of at least one of the following:
   Onset of dementia within 3 months of a recognized stroke
   Abrupt deterioration in cognitive functions
   Fluctuating stepwise progression of cognitive deficits
4. Clinical features consistent with a diagnosis of probable vascular dementia:
   Early presence of a gait disturbance
   History of unsteadiness or frequent unprovoked falls
   Early urinary incontinence
   Pseudobulbar palsy
   Personality and mood changes
5. Features that make a diagnosis of vascular dementia uncertain:
   Early onset of memory deficit and progressive worsening of memory and other cognitive functions in the absence of focal neurological signs and cerebrovascular lesions on CT or magnetic resonance imaging (MRI) scans

[a]Derived from Roman, G. C., Tatemichi, T. K., Erkinjuntti, T., Cummings, J. L., Masdeu, J. C., Garcia, J. H., Amaducci, L., Orgogozo, J. M., Brun, A., and Hofman, A., (1993). Vascular dementia: diagnostic criteria for research studies. Report of the NINDS-AIREN International Workshop. *Neurology* **43**(2), 250–260.

causes. Prospective cohort studies and interventional studies of cognitive outcome in well-defined samples deemed to be at a greater cerebrovascular risk combined with postmortem studies would provide the ideal way to identify the risk of incident cognitive impairment attributable to vascular causes, establish the natural history of VCI, and unravel the complex relationship between cerebrovascular disease and dementia.

III. Etiology of Vascular Cognitive Impairment

There are a number of cerebrovascular disorders that may potentially lead to cognitive dysfunction [1] (Table 2). The contribution of some of these conditions (such as stroke) to cognitive dysfunction has been relatively well studied, whereas it is difficult to accurately quantify the causal contribution of others because of a paucity of good prospective or pathological data. Moreover, given the heterogeneity in the types of cerebrovascular disease, it is not possible to discuss a unifying mechanism for VCI. Broadly, VCI may be termed "hereditary" (with a strong primary genetic basis) or "sporadic." In the following section, we deal with the different forms of cerebrovascular disease that may contribute to the development of cognitive impairment.

IV. Causes of Hereditary Vascular Cognitive Impairment

A. Cerebral Autosomal-Dominant Arteriopathy with Subcortical Infarcts and Leukoencephalopathy

Cerebrovascular disorders with a strong genetic basis and associated cognitive impairment include cerebral autosomal-dominant arteriopathy with subcortical infarcts and leukoencephalopathy (CADASIL) and the hereditary forms of cerebral amyloid angiopathy (CAA). CADASIL is a monogenic disorder in young adults caused by a mutation in the Notch3 gene on chromosome 19q12 [9] that codes

Table 2 Observational Studies of the Risk of Dementia after Stroke (Includes Only Studies Using a Comparison Group)

Study (Reference No.)	Sample Size (Stroke/Nonstroke)	Study Type	Exposure	Outcome	Results and Comments
Tatemichi et al., 1992* [45]	251/249	Hospital based	Ischemic stroke (all severity)	Modified DSM-III-R dementia at 3 months after stroke	OR for prevalent dementia 9.4 (95% CI, 4.2–21.1)
Tatemichi et al., 1994 [46]	185/241	Hospital based	As above	Modified DSM-III-R dementia; annual diagnosis; 52 months total follow-up	RR for incident dementia 5.5 (95% CI, 2.5–11.1)
Desmond et al., 2002 [47]	334/241	Hospital based	As above	Modified DSM-III-R dementia; annual diagnosis; 21 months median follow-up	RR for incident dementia 3.83 (95% CI, 2.1–6.8); age, intercurrent illness, baseline MMSE, recurrent stroke independently associated with incident dementia; confounding by education and healthy volunteer effect possible
Ivan et al., 2004 [23]	212/1060	Population-based nested case–cohort study	Stroke (all severity)	DSM-IV dementia; 10-year risk estimation	HR for incident dementia 2.4; 95% CI, 1.6–3.7 after adjustment for demographics, stroke features, and stroke risk factors; stroke recurrence strongly associated with incident dementia
Schneider et al., 2003 [24]	58/106	Autopsy study of Catholic nuns	All ischemic stroke	CERAD criteria	OR for single infarctions 1.69, 95% CI, 0.70–4.09; OR for multiple infarctions 2.12, 95% CI, 1.08–6.61; the effect was seen only for infarctions clinically evident during life
Srikanth et al., 2004 [16]	99/99	Population-based matched cohort study	Mild to moderate first-ever stroke; nonaphasic cases	DSM-IV criteria (blinded rating of progression over 1 year)	No association detected between stroke and dementia; RR for prevalent dementia 1.1, 95% CI, 0.5–2.2

Note: CI, confidence interval; DSM, *Diagnostic and Statistical Manual of Mental Disorders*; HR, hazard ratio; MMSE, Mini-Mental State Examination; OR, odds ratio; RR, relative risk.

for a transmembrane receptor predominantly expressed in vascular smooth muscle cells. The pathological hallmark of CADASIL is a nonamyloid, nonatherosclerotic microangiopathy affecting the leptomeningeal and perforating arteries of the brain between 100 and 400 mm in diameter. On electron microscopy, osmiophilic immunoglobulin-like deposits are seen in close proximity to vascular smooth muscle cells. Magnetic resonance imaging (MRI) shows multiple subcortical infarcts and diffuse white-matter abnormalities, with some predilection for the anterior temporal and external capsular white matter [10]. Clinical manifestations of CADASIL include transient ischemic attacks (TIAs), strokes, migraine, and epilepsy with consequent progressive, global cognitive and behavioral decline.

B. Hereditary Cerebral Amyloid Angiopathy

Hereditary cerebral amyloid angiopathy (HCAA) is a group of familial dementing disorders in which there is deposition of amyloid in the walls of leptomeningeal and cortical blood vessels [11]. There are several autosomal dominantly inherited variants of hereditary CAA, and these are characterized by deposition of different types of amyloid. These include β- in hereditary AD due to mutations in genes encoding amyloid precursor protein (APP) and presenilin 1 and 2 (PSEN1, PSEN2), the mutant cystatin C (ACys) in hereditary cerebral hemorrhage with amyloidosis of Icelandic type, variant transthyretins (ATTR) in meningovascular amyloidoses, mutated gelsolin

(AGel) in familial amyloidosis of Finnish type, disease-associated prion protein (PrP[Sc]) in a variant of the Gerstmann–Sträussler–Scheinker syndrome, and ABri and ADan in HCAAs observed in familial British dementia and familial Danish dementia, respectively. Large amounts of amyloid deposition have been associated with severity of cognitive decline in such cases of HCAA [12]. Such amyloid deposits may lead to disruption of the vessel wall, microaneurysm formation, and fibrinoid necrosis potentially leading to lobar intracerebral hemorrhage. Apart from these hemorrhagic lesions, it has been also hypothesized that such deposits may lead to chronic damage to deeper brain structures leading to alterations in cerebral white matter and consequently contribute to cognitive decline [13]. However, there is uncertainty in the literature regarding the relative contributions of vascular amyloid, focal brain lesions, and white-matter alterations to the risk of cognitive decline in such families [14].

V. Causes of Sporadic Vascular Cognitive Impairment

The cerebrovascular pathologies most commonly implicated in the development of cognitive impairment in older people are stroke and white-matter lesions (WMLs).

A. Stroke

Stroke is the prototype sporadic cerebrovascular disease commonly linked with cognitive impairment. By definition, stroke is associated with an acute-onset focal neurological deficit due to vascular causes. Among cerebrovascular lesions, pathogenetic mechanisms are best understood for stroke and most commonly involve large artery atherosclerosis, embolism, and small vessel (lacunar) disease. There is little doubt that clinical stroke episodes carry a high risk of early and delayed cognitive impairment. A higher risk of any cognitive impairment (RR 1.5, 95% CI, 1.1–2.1) was found 3 and 12 months after the event in a population-based study of first-ever nonaphasic stroke patients [15,16], and this risk would be greater if aphasic patients were included. In older stroke patients not suffering from dementia, the most common early impairments were found in the areas of attention, processing speed, working memory, and executive functions [17]. In the same sample, however, the predictive value of such impairments in determining whether a future dementia may develop over 15 months was found to be poor [18].

The location of stroke lesion may play a role in determining the severity and type of cognitive impairment irrespective of the size of the lesion. Complex cognitive syndromes have been well described for stroke lesions occurring in certain strategic locations in the brain such as thalamus, caudate nucleus, left capsular genu, angular gyrus, inferior mesial temporal lobe, and basal forebrain. Infarcts affecting these locations are thought to lead to global cognitive changes by disrupting important connections to other cortical regions [19]. Such strategic strokes are, however, relatively uncommon. The more commonly occurring major hemispherical stroke syndromes are clearly associated with a broad range of cognitive deficits. In addition, the less severe lacunar strokes have also been associated with the presence of subtle cognitive impairments, particularly in tasks demanding greater effort (e.g., dual-tasking) with tract disconnection as a possible underlying mechanism [20]. However, there is a need for the systematic study of the effect of stroke location on patterns and severity of cognitive impairment using available sophisticated imaging techniques.

Although stroke has been recognized to cause cognitive impairment in general terms, there remains much uncertainty regarding its contribution to the risk of a dementia syndrome. Building on earlier pathological studies by Tomlinson and Roth [21], as well as concepts of multi-infarct dementia developed by Hachinski [22], Tatemichi [19] proposed the possibility of a "chronic global brain failure" or dementia that may be set in motion by the occurrence of a stroke event, otherwise referred to as a "post-stroke dementia." There have been a number of studies with conflicting reports on the magnitude of the risk for such a dementia after stroke. The results of studies in which a stroke-free comparison group was used are summarized in Table 2. If one considers the only two population-based studies on this matter, the magnitude of the association varies between no increase [16] and an approximate doubling of the risk [23], a variation that may be explained largely by the differences in stroke severity between the samples, the exclusion of aphasic individuals in the former study, and the inclusion of recurrent strokes in the latter study. In particular, the risk of dementia after stroke appears to be consistently greater for multiple strokes [24], possibly reflecting an additive or multiplicative effect of brain infarction superimposed on preexisting neurodegenerative change [25]. Consistent with this observation, treatment with antihypertensive medications resulted in a reduced risk of incident dementia only among those with additional strokes on follow-up in an interventional study of stroke patients [26]. However, longitudinal data describing trajectories of cognitive performance in stroke patients compared with a stroke-free reference group are lacking.

1. Silent Brain Infarcts

Silent or asymptomatic brain infarcts were found on brain scans in approximately 20% of the general population older than 60 years in the Rotterdam Study, with the risk increasing with age by 8% per year [27]. The incidence of silent brain infarcts strongly increased with age and was five times higher than that of symptomatic stroke, with the majority being lacunar lesions. A prevalent silent brain infarct strongly predicted a new silent infarct (age- and sex-adjusted odds ratio, 2.9; 95% CI, 1.7–5.0). Age, blood pressure, diabetes mellitus, cholesterol and homocysteine levels, carotid intima-media thickness, carotid plaques, and smoking were associated with incident silent brain infarcts [27], suggesting a vascular risk profile similar to that of symptomatic stroke. However, the presence of silent brain infarcts was associated with a nearly fourfold increase in the risk of future symptomatic stroke over a 4-year period independent of other stroke risk factors, suggesting that there may be unknown mechanisms underlying the pathogenesis of silent infarcts.

Although termed "silent," these infarcts are not necessarily without adverse effect. Cognitive impairments have been found more frequently among those with silent infarcts in large population-based series including the Cardiovascular Health Study [28] and the Rotterdam Scan Study [29]. In the Rotterdam Scan Study, the presence of silent brain infarcts at baseline more than doubled the risk of dementia (hazard ratio, 2.26, 95% CI 1.09–4.70) over a 4-year period [29]. The presence of silent brain infarcts on baseline MRI was associated with poorer performance on neuropsychological tests and a steeper decline in global cognitive function. When participants with silent brain infarcts at baseline were stratified into those with and those without additional infarcts at follow-up, cognitive decline was restricted to those with additional silent infarcts occurring over the follow-up period. This is similar to previous observations regarding dementia after stroke [24], possibly reflecting cumulative load of structural damage to the brain.

B. White-Matter Lesions

WMLs are usually detected as hyperintense (bright) signals on T2-weighted MRI scans (Fig. 1) in most people older than 65 years [30] and are much more common in people with chronic untreated hypertension. Their pathological features are varied and may include myelin rarefaction, reactive gliosis, axonal loss, infarction, venular collagenosis, and accompanying arteriosclerotic small vessel changes. The commonly held view is that WMLs represent insidious structural damage to the brain occurring as a result of small vessel disease of the brain, although there may be other, as yet unexplained, reasons for their development. WMLs may theoretically lead to disconnection of axonal connections between subcortical structures and the cortex, potentially leading to disorders of mobility and cognition. However, the magnitude and nature of their association with such disorders is by no means well established. If they are proven to be of sufficient clinical

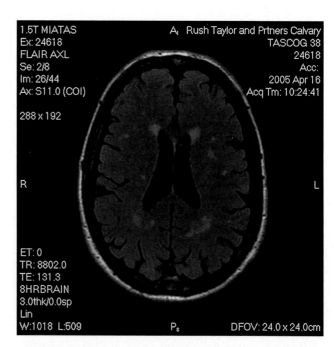

Figure 1 Axial flair magnetic resonance imaging scan showing periventricular and subcortical white-matter lesions

importance, they may be a worthwhile therapeutic target in the prevention or treatment of dementia, particularly given the increasing potential to successfully treat or modify vascular risk factors.

Many assume that WMLs play an important role in causing cognitive decline and dementia in older people. Contrary to this assumption, however, available data on this issue are inconclusive and conflicting. Results of many previous studies show only weak or nonsignificant correlations between WML estimates and cognition. Among population-based studies, the strongest evidence for a link between WMLs and cognitive decline has been shown in the Rotterdam Study, where the risk of incident dementia was elevated (risk ratio 1.42, 95% CI, 1.04–1.94 per standard deviation increase in WML grade) [31], the risk being partly explained by other structural brain changes such as stroke and brain atrophy. The weak or modest associations between WMLs and cognitive scores in most previous studies may reflect either the relative lack of importance of WMLs, the use of brief and insensitive measures of cognitive function, or the use of insensitive and inaccurate visual rating measurements of MRI scans. In the only study using volumetric MRI and a comprehensive cognitive battery in healthy older people by DeCarli et al. [32], relatively strong associations were found (correlation coefficients ranging from 0.20 to 0.72) between WML volume and cognitive scores. There are also few studies of the location-specific effects of WMLs on cognition. Lesions located adjacent to the brain ventricles (periventricular WMLs), but not subcortical lesions, were found to be associated with reduced speed of processing in the Rotterdam Study [33]. In contrast, only weak or inconsistent associations were detected in the Sydney Older Person's Study for both locations [34]. The inconsistency between these few studies may be partly because of differences in cohort demographics or variable study methods. In addition, many of these studies were restricted to samples of people with lower grades of severity of WMLs. It is possible that severe WMLs have more of an impact on cognitive loss, but more studies are needed to examine this question. The lack of high-resolution volumetric MRI may also have significantly limited the ability of these studies to detect stronger and more consistent regional associations. Moreover, the division of WMLs into periventricular and subcortical locations may also be arbitrary. Using volumetric MRI, DeCarli et al. [35] failed to demonstrate specific subpopulations of WMLs in a volunteer sample, calling into question the basis for categorizing WMLs in this fashion. Diffusion tensor imaging (DTI) has shown promise in studying the link between WMLs and cognition [36], with DTI measures correlating more strongly with executive dysfunction than conventional MRI measures. In summary, more detailed studies of WMLs using state-of-the-art volumetric imaging and newer techniques such as DTI may shed more light on the effect of WMLs on cognition.

VI. Cardiovascular Causes of Sporadic Vascular Cognitive Impairment Independent of Structural Brain Abnormality

Although there is reasonable evidence to suspect that vascular brain lesions may be causally related to cognitive dysfunction, there is a body of evidence suggesting that there may be other cardiovascular causes for such impairment independent of structural brain abnormality. Such causes are hypothesized to work by means of reduction of cerebral blood flow to the brain in the absence of infarction or WMLs, although good evidence for this hypothesis is lacking. In the large population-based Cardiovascular Health Study, asymptomatic high-grade left (but not the right) internal *carotid artery stenosis* was found to be associated with an increased risk of prevalent cognitive impairment (odds ratio [OR] 6.7, 95% CI, 2.4–18.1) and longitudinal decline in Mini-Mental State Examination (MMSE) scores (OR 2.6, 95% CI, 1.1–6.3) over 5 years [37]. This association persisted even after adjustment for contralateral stenosis, demographic characteristics, and vascular risk factors, suggesting that the underlying mechanism for the decline may be only partly due to known vascular mechanisms. In this study, the increased risk for prevalent impairment was found even among those without MRI signs of brain infarction, but not so in the longitudinal study. These results were also based on a small number of cases of high-grade stenosis. Therefore, some doubt still remains whether carotid artery disease can lead to cognitive decline in the absence of structural brain disease. Clarification of this issue would need powerful neuroimaging techniques incorporating measures of cerebral blood flow, arterial morphology, and neurodegeneration.

Coronary artery bypass grafting (CABG) has also been described as a risk factor for cognitive impairment [38]. However, the etiology and mechanisms of this phenomenon remain unresolved, and they are probably multifactorial. The relatively old age of the patients who undergo CABG and their widespread atherosclerotic disease are possible reasons for vascular sequelae, particularly those leading to neurological dysfunction. However, there may be a subgroup of people who suffer cognitive sequelae independent of structural lesions in the brain. Such people, by virtue of their age, may be more prone to hypoperfusion injury of critical areas of the brain such as the hippocampus or the basal ganglia. Further research is required to better identify this subgroup and to resolve uncertainties regarding underlying mechanisms.

VII. Pathophysiology of Vascular Cognitive Impairment

There are multiple putative pathophysiological mechanisms underlying VCI, and this is obviously a result of the heterogeneity in the types of cerebrovascular diseases that may be associated with it. Some mechanisms are well understood, particularly those leading to large cerebral vessel occlusion, whereas others remain unclear. Current interest centers on pathophysiological pathways that may contribute to the development of cerebral small vessel disease and the interplay of such pathways with those leading to neurodegenerative diseases like AD.

A. Large Vessel Disease and Cardioembolism

Large cerebral vessel occlusion may occur as a result of atherosclerosis and subsequent artery-to-artery embolism or to embolism from cardiogenic sources. In either case, the end-effect is that of single or multiple cortical infarction. Large vessel atherosclerosis begins as fatty streaks at a young age, and over many years, this develops into an atherosclerotic plaque. The progression from fatty streak to unstable plaque is a dynamic process and may initially begin as a response to endothelial injury that may be inflammatory in origin. This process of evolution into a plaque involves inflammatory cell migration, lipid deposition, cytokine activation, and smooth muscle cell proliferation. Thereafter, a cascade of events may occur to cause instability of the plaque, plaque rupture, platelet aggregation, and thrombus formation, as well as subsequent distal embolization leading to brain infarction. The development of atheroma may be influenced by genetic factors (presumed polygenic inheritance), environmental factors (e.g., hypertension, diabetes, physical activity, and diet) or an interaction between the two. Cardioembolism may result from a number of causes, usually secondary to chronic atrial fibrillation, valvular heart disease, or cardiomyopathy. The development of cardiac thrombi involves well-known coagulation pathways activated by hemostasis.

B. Small Vessel Disease

Cerebral small vessel disease has assumed a position of emerging importance in the field of VCI. Lacunar infarcts and, to a certain extent, WMLs represent the structural brain abnormalities linked to small vessel disease. Well-known risk factors for small vessel disease include increasing age, hypertension, diabetes, and other known risk factors for atherosclerosis such as smoking. Final pathways underlying small vessel occlusion as a result of these risk factors include arteriolosclerosis, atherosclerosis, and lipohyalinosis. Arteriolosclerosis is a uniform circumferential hyaline thickening of the arteriolar wall frequently seen in autopsy in older vasculopathic patients and is not a proven cause of lacunar infarction. Atherosclerosis involves the larger perforating arteries ($>300\,\mu m$) of the circle of Willis and is strongly linked with lacunar infarction. Lipohyalinosis is a complex entity involving vessels that are smaller than 300 μm with collections of intimal xanthomata and fibrinoid necrosis, and are implicated in causing lacunar infarctions. Newer risk factors for small vessel disease described in the Rotterdam Study also include serum homocysteine levels, serum C-reactive protein, and plasma amyloid β [39]. These results suggest that endothelial cytotoxicity, inflammation, and disruption of the blood-brain barrier may have important roles to play in causing lacunar lesions or WMLs. Such results remain to be replicated and examined further in other studies. The influence of genetic polymorphisms underlying vascular risk such as apolipoprotein E (ApoE), angiotensin-converting enzyme (ACE), and methylene tetrahydrofolate reductase (MTHFR) also need to be considered in future studies for a fuller understanding of the pathophysiology of small vessel disease.

C. Common Pathways between Vascular Disease and Neurodegeneration

Some researchers have proposed the hypothesis that classic neurodegenerative cognitive disorders such as AD may have a primary vascular cause [40]. This is based on several observations: (1) risk factor association between AD and VaD; (2) improvement of cerebral perfusion obtained from pharmacotherapy used to reduce the symptoms or progression of AD; (3) detection of regional cerebral hypoperfusion with the use of neuroimaging techniques to preclinically identify AD candidates; (4) presence of regional brain microvascular abnormalities in AD brains; (5) common overlap of clinical AD and VaD cognitive symptoms; and (6) presence of cerebral hypoperfusion preceding hypometabolism, cognitive decline, and neurodegeneration in AD. However, the opposing view to AD pathogenesis, namely the amyloid hypothesis, has a strong basis, and at this stage, more research needs to be performed to show more definitively the cause–effect relationship or interactions between vascular disease and AD.

VIII. Clinical Features

Is there a distinct neuropsychological profile in VCI? The data in this regard are predominantly from studies of VaD [41]. Despite a large number of studies typically comparing VaD with AD, it is still not clear whether VaD produces a unique pattern of neurocognitive impairment. Comparative research of this type is fraught with a number of methodological difficulties. These include the difficulty in defining VaD, the lack of a standard definition across studies, patient selection, and circularity bias. For example, if patients for such studies were selected based on already existing flawed criteria for VaD, then the differences in neurocognitive profiles between the groups (VaD and AD) would tend to be biased. This may be due to a skewed adherence to an Alzheimer-type cognitive model in VaD diagnosis that requires the primacy of memory impairment. On the other hand, if VaD subjects were identified using executive impairment as the primary criterion, then it will be self-fulfilling that executive impairment will be the distinguishing feature between VaD and AD. A similar conceptual argument may be made for VCI, in which even fewer data exist in this regard.

Notwithstanding such methodological difficulties, there appears to be wide consensus that VaD causes impairments in executive and attentional functions [41]. Such a view is also being adopted in the more general entity of VCI. Studies of stroke patients with healthy controls indicate that nondemented stroke patients have a greater frequency of attentional and executive impairments than controls [17]. The term "executive function" has, however, come to be overapplied to a rather broad range of cognitive phenomena. In essence, however, the concept of executive function implies a high-level assignment of priority to competing response tendencies, allowing for the inhibition or suppression of irrelevant or inappropriate actions. Executive function defined in this way ultimately underlies all forms of behavior and cognition, and impairments lead to the well-described executive dysfunctions of inability to shift from one conceptual set to another, reduced capacity to deal with information from multiple sources ("multitasking"), poor planning and monitoring of behavior, inability to suppress strongly competing but task-irrelevant responses (e.g., Stroop effect), reduced word generation, tangentiality, and social inappropriateness. Executive systems are also capacity limited, with the result that adequate selective attention is crucial to their operation. Executive function is mediated by an extensive network, the cortical component of which lies in the dorsolateral prefrontal, lateral orbitofrontal, and anterior cingulate cortices [42].

"Executive" refers to the highest or supervisory level of control and does not encompass the full range of neurocognitive phenomena that may be seen in VCI. A more useful conceptual distinction is the *fundamental* versus *instrumental* dichotomy in cognitive ability [43]. Instrumental functions relate to the storage and processing of content and include language and other symbolic functions, spatial cognition, consolidated long-term memory, praxis, and gnosis. They are mediated by the neocortex, with a large contribution from temporal, parietal, and occipital regions. Fundamental functions, on the other hand, provide basic and organizational resources necessary for the normal integrated cognitive processing. These include arousal, attention and concentration, sequencing, retrieval processes, and at the highest level, executive function. Fundamental functions are hierarchically organized and distributed over fronto-subcortical circuitry, with the result that even small subcortical vascular lesions may exert widespread effects irrespective of location [44].

The literature is consistent with the view that subcortical ischemic vascular disease has a relatively greater and early effect on fundamental functions, by compromising the subcortical components (striatum, globus pallidus, substantia nigra, and thalamus) of a number of fronto-subcortical circuits [42]. These effects include reduced speed of information processing, impaired attention, sequencing, retrieval, and executive function (including working memory), with relative preservation of memory encoding and other instrumental functions. Fundamental functions, in line with their domain nonspecific nature, are capable of exerting a *secondary* effect on domain-specific systems (such as the instrumental functions), resulting in a degree of neuropsychological overlap between VCI and AD and subnormative performance on a wide variety of cognitive tests in VCI when compared with normal populations.

IX. Summary

In summary, VCI is a heterogeneous entity that can be attributed to a wide range of vascular disorders. These disorders may be hereditary, such as CADASIL and HCAA, or sporadic. The hereditary disorders are characterized by nonamyloid and amyloid deposits within cerebral vessel walls. Sporadic VCIs are mainly ascribed to stroke, WMLs, and "silent" infarcts, usually in nonstrategic locations, with most risk factors for such conditions (such as hypertension) being relatively well characterized. Progression of VCI with these conditions may be due to recurrences of these events rather than the presence of such lesions per se, possibly reflecting a cumulative load of structural damage to the brain. Other sporadic causes independent of structural brain abnormality are thought to be due to reduction of cerebral blood flow to the brain, such as in asymptomatic high-grade left internal carotid artery stenosis and carotid artery bypass grafting. The cognitive profile of VCI is yet to be clarified but may involve disruption of fundamental functions such as attention, processing speed, and executive ability. A better understanding of the incidence, natural history, and cognitive profile of VCI and the complex relationship between cerebrovascular disease and cognitive impairment could be unraveled using a combination of prospective cohort and intervention study designs. This could be further supplemented by the use of sophisticated imaging techniques, including state-of-the-art volumetric imaging and DTI, to determine the contribution of lesion type and lesion location to the pattern and severity of cognitive impairment, and postmortem studies to help separate true VCI from cognitive impairment due to a mixture of causes.

References

1. O'Brien, J. T., Erkinjuntti, T., Reisberg, B., Roman, G., Sawada, T., Pantoni, L., Bowler, J. V., Ballard, C., DeCarli, C., Gorelick, P. B., Rockwood, K., Burns, A., Gauthier, S., and DeKosky, S. T. (2003). Vascular cognitive impairment. *Lancet Neurol.* **2**(2), 89–98.
2. Bowler, J. V., and Hachinski, V. (1995). Vascular cognitive impairment: a new approach to vascular dementia. *Baillieres Clin. Neurol.* **4**(2), 357–376.
3. Roman, G. C., Tatemichi, T. K., Erkinjuntti, T., Cummings, J. L., Masdeu, J. C., Garcia, J. H., Amaducci, L., Orgogozo, J. M., Brun, A., and Hofman, A., (1993). Vascular dementia: diagnostic criteria for research studies. Report of the NINDS-AIREN International Workshop. *Neurology* **43**(2), 250–260.
4. American Psychiatric Association. (1994). *Diagnostic and Statistical Manual of Mental Disorders,* 4th edition. American Psychiatric Association, Washington, DC.
5. CFAS M. (2001). Pathological correlates of late-onset dementia in a multicentre, community-based population in England and Wales. Neuropathology Group of the Medical Research Council Cognitive Function and Ageing Study (MRC CFAS). *Lancet* **357**(9251), 169–175.
6. Rockwood, K., Wentzel, C., Hachinski, V., Hogan, D. B., MacKnight, C., and McDowell, I. (2000). Prevalence and outcomes of vascular cognitive impairment. Vascular Cognitive Impairment Investigators of the Canadian Study of Health and Aging. *Neurology* **54**(2), 447–451.
7. Wentzel, C., Rockwood, K., MacKnight, C., Hachinski, V., Hogan, D. B., Feldman, H., Ostbye, T., Wolfson, C., Gauthier, S., Verreault, R., and McDowell, I. (2001). Progression of impairment in patients with

vascular cognitive impairment without dementia. *Neurology* **57**(4), 714–716.

8. Ingles, J. L., Wentzel, C., Fisk, J. D., and Rockwood, K. (2002). Neuropsychological predictors of incident dementia in patients with vascular cognitive impairment, without dementia. *Stroke* **33**(8), 1999–2002.

9. Tournier-Lasserve, E., Joutel, A., Melki, J., Weissenbach, J., Lathrop, G. M., Chabriat, H., Mas, J. L., Cabanis, E. A., Baudrimont, M., Maciazek, J., Bach, M. A., and Bousser, M. G. (1993). Cerebral autosomal dominant arteriopathy with subcortical infarcts and leukoencephalopathy maps to chromosome 19q12. *Nat. Genet.* **3**(3), 256–259.

10. O'Sullivan, M., Jarosz, J. M., Martin, R. J., Deasy, N., Powell, J. F., and Markus, H. S. (2001). MRI hyperintensities of the temporal lobe and external capsule in patients with CADASIL. *Neurology* **56**(5), 628–634.

11. Revesz, T., Ghiso, J., Lashley, T., Plant, G., Rostagno, A., Frangione, B., and Holton, J. L. (2003). Cerebral amyloid angiopathies: a pathologic, biochemical, and genetic view. *J. Neuropathol. Exp. Neurol.* **62**(9), 885–898.

12. Castellani, R. J., Smith, M. A., Perry, G., and Friedland, R. P. (2004). Cerebral amyloid angiopathy: major contributor or decorative response to Alzheimer's disease pathogenesis. *Neurobiol. Aging* **25**(5), 599–604.

13. Haan, J., Algra, P. R., and Roos, R. A. (1990). Hereditary cerebral hemorrhage with amyloidosis-Dutch type. Clinical and computed tomographic analysis of 24 cases. *Arch. Neurol.* **47**(6), 649–653.

14. Bornebroek, M., Van Buchem, M. A., Haan, J., Brand, R., Lanser, J. B., de Bruine, F. T., and Roos, R. A. (1996). Hereditary cerebral hemorrhage with amyloidosis-Dutch type: better correlation of cognitive deterioration with advancing age than with number of focal lesions or white matter hyperintensities. *Alzheimer Dis. Assoc. Disord.* **10**(4), 224–231.

15. Srikanth, V. K., Thrift, A. G., Saling, M. M., Anderson, J. F., Dewey, H. M., Macdonell, R. A., and Donnan, G. A. (2003). Increased risk of cognitive impairment 3 months after mild to moderate first-ever stroke. A community-based prospective study of nonaphasic English-speaking survivors. *Stroke* **34**, 1136–1143.

16. Srikanth, V. K., Anderson, J. F., Donnan, G. A., Saling, M. M., Didus, E., Alpitsis, R., Dewey, H. M., Macdonell, R. A., and Thrift, A. G. (2004). Progressive dementia after first-ever stroke: a community-based follow-up study. *Neurology* **63**(5), 785–792.

17. Ballard, C., Stephens, S., Kenny, R., Kalaria, R., Tovee, M., and O'Brien, J. (2003). Profile of neuropsychological deficits in older stroke survivors without dementia. *Dement. Geriatr. Cogn. Disord.* **16**(1), 52–56.

18. Ballard, C., Rowan, E., Stephens, S., Kalaria, R., and Kenny, R. A. (2003). Prospective follow-up study between 3 and 15 months after stroke: improvements and decline in cognitive function among dementia-free stroke survivors >75 years of age. *Stroke* **34**(10), 2440–2444.

19. Tatemichi, T. K. (1990). How acute brain failure becomes chronic: a view of the mechanisms of dementia related to stroke. *Neurology* **40**(11), 1652–1659.

20. Van Zandvoort, M. J., De Haan, E. H., and Kappelle, L. J. (2001). Chronic cognitive disturbances after a single supratentorial lacunar infarct. *Neuropsychiatry Neuropsychol. Behav. Neurol.* **14**(2), 98–102.

21. Tomlinson, B. E., Blessed, G., and Roth, M. (1970). Observations on the brains of demented old people. *J. Neurol. Sci.* **11**(3), 205–242.

22. Hachinski, V. C., Lassen, N. A., and Marshall, J. (1974). Multi-infarct dementia. A cause of mental deterioration in the elderly. *Lancet* **2**(7874), 207–210.

23. Ivan, C. S., Seshadri, S., Beiser, A., Au, R., Kase, C. S., Kelly-Hayes, M., and Wolf, P. A. (2004). Dementia after stroke: the Framingham Study. *Stroke* **35**(6), 1264–1268.

24. Schneider, J. A., Wilson, R. S., Cochran, E. J., Bienias, J. L., Arnold, S. E., Evans, D. A., and Bennett, D. A. (2003). Relation of cerebral infarctions to dementia and cognitive function in older persons. *Neurology* **60**(7), 1082–1088.

25. Snowdon, D. A., Greiner, L. H., Mortimer, J. A., Riley, K. P., Greiner, P. A., Markesbery, W. R. (1997). Brain infarction and the clinical expression of Alzheimer disease. The Nun Study. *JAMA* **277**(10), 813–817.

26. Tzourio, C., Anderson, C., Chapman, N., Woodward, M., Neal, B., MacMahon, S., and Chalmers, J. (2003). Effects of blood pressure lowering with perindopril and indapamide therapy on dementia and cognitive decline in patients with cerebrovascular disease. *Arch. Intern. Med.* **163**(9), 1069–1075.

27. Vermeer, S. E., Den Heijer, T., Koudstaal, P. J., Oudkerk, M., Hofman, A., and Breteler, M. M. (2003). Incidence and risk factors of silent brain infarcts in the population-based Rotterdam Scan Study. *Stroke* **34**(2), 392–396.

28. Price, T. R., Manolio, T. A., Kronmal, R. A., Kittner, S. J., Yue, N. C., Robbins, J., Anton-Culver, H., and O'Leary, D. H. (1997). Silent brain infarction on magnetic resonance imaging and neurological abnormalities in community-dwelling older adults. The Cardiovascular Health Study. CHS Collaborative Research Group. *Stroke* **28**(6), 1158–1164.

29. Vermeer, S. E., Prins, N. D., den Heijer, T., Hofman, A., Koudstaal, P. J., and Breteler, M. M. (2003). Silent brain infarcts and the risk of dementia and cognitive decline. *N. Engl. J. Med.* **348**(13), 1215–1222.

30. Breteler, M. M., van Swieten, J. C., Bots, M. L., Grobbee, D. E., Claus, J. J., van den Hout, J. H., van Harskamp, F., Tanghe, H. L., de Jong, P. T., and van Gijn, J. (1994). Cerebral white matter lesions, vascular risk factors, and cognitive function in a population-based study: the Rotterdam Study. *Neurology* **44**(7), 1246–1252.

31. Prins, N. D., van Dijk, E. J., den Heijer, T., Vermeer, S. E., Koudstaal, P. J., Oudkerk, M., Hofman, A., and Breteler, M. M. (2004). Cerebral white matter lesions and the risk of dementia. *Arch. Neurol.* **61**(10), 1531–1534.

32. DeCarli, C., Murphy, D. G., Tranh, M., Grady, C. L., Haxby, J. V., Gillette, J. A., Salerno, J. A., Gonzales-Aviles, A., Horwitz, B., and Rapoport, S. I. (1995). The effect of white matter hyperintensity volume on brain structure, cognitive performance, and cerebral metabolism of glucose in 51 healthy adults. *Neurology* **45**(11), 2077–2084.

33. de Groot, J. C., de Leeuw, F. E., Oudkerk, M., van Gijn, J., Hofman, A., Jolles, J., and Breteler, M. M. Cerebral white matter lesions and cognitive function: the Rotterdam Scan Study. *Ann. Neurol.* **47**(2), 145–151.

34. Piguet, O., Ridley, L., Grayson, D. A., Bennett, H. P., Creasey, H., Lye, T. C., and Broe, G. A. (2003). Are MRI white matter lesions clinically significant in the "old-old?" Evidence from the Sydney Older Persons Study. *Dement. Geriatr. Cogn. Disord.* **15**(3), 143–150.

35. DeCarli, C., Fletcher, E., Ramey, V., Harvey, D., and Jagust, W. J. (2005). Anatomical mapping of white matter hyperintensities (WMH): exploring the relationships between periventricular WMH, deep WMH, and total WMH burden. *Stroke* **36**(1), 50–55.

36. O'Sullivan, M., Morris, R. G., Huckstep, B., Jones, D. K., Williams, S. C., and Markus, H. S. (2004). Diffusion tensor MRI correlates with executive dysfunction in patients with ischaemic leukoaraiosis. *J. Neurol. Neurosurg. Psychiatry* **75**(3), 441–447.

37. Johnston, S. C., O'Meara, E. S., Manolio, T. A., Lefkowitz, D., O'Leary, D. H., Goldstein, S., Carlson, M. C., Fried, L. P., and Longstreth, W. T., Jr. (2004). Cognitive impairment and decline are associated with carotid artery disease in patients without clinically evident cerebrovascular disease. *Ann. Intern. Med.* **140**(4), 237–247.

38. Royter, V., Bornstein, N. M., and Russell, D. (2005). Coronary artery bypass grafting (CABG) and cognitive decline: a review. *J. Neurol. Sci.* **229-230,** 65–67.

39. van Dijk, E. J. (2004). Causes of cerebral small vessel disease. Erasmus University, Rotterdam.

40. de la Torre, J. C. (2002). Alzheimer disease as a vascular disorder: nosological evidence. *Stroke* **33**(4), 1152–1162.
41. Desmond, D. W. (2004). The neuropsychology of vascular cognitive impairment: is there a specific cognitive deficit? *J. Neurol. Sci.* **226**(1-2), 3–7.
42. Cummings, J. L. (1993). Frontal-subcortical circuits and human behavior. *Arch. Neurol.* **50**(8), 873–880.
43. Albert, M. L. (1978). Subcortical dementia. *In:* "Alzheimer's Disease, Senile Dementia and Related Disorders" (R. Katzman, R. D. Terry, and K. L. Bick, eds.), pp. 173–179. Raven Press, New York.
44. Tullberg, M., Fletcher, E., DeCarli, C., Mungas, D., Reed, B. R., Harvey, D. J., Weiner, M. W., Chui, H. C., and Jagust, W. J. (2004). White matter lesions impair frontal lobe function regardless of their location. *Neurology* **63**(2), 246–253.
45. Tatemichi, T. K., Desmond, D. W., Mayeux, R., Paik, M., Stern, Y., Sano, M., Remien, R. H., Williams, J. B., Mohr, J. P., and Hauser, W. A. (1992). Dementia after stroke: baseline frequency, risks, and clinical features in a hospitalized cohort. *Neurology* **42**(6), 1185–1193.
46. Tatemichi, T. K., Paik, M., Bagiella, E., Desmond, D. W., Stern, Y., Sano, M., Hauser, W. A., and Mayeux, R. (1994). Risk of dementia after stroke in a hospitalized cohort: results of a longitudinal study. *Neurology* **44**(10), 1885–1891.
47. Desmond, D. W., Moroney, J. T., Sano, M., and Stern, Y. (2002). Incidence of dementia after ischemic stroke: results of a longitudinal study. *Stroke* **33**(9), 2254–2262.

22

Cardioembolism

Karen Furie, MD, MPH

Keywords: *atrial fibrillation, cardiac, cardioembolic, cerebrovascular, embolic, stroke*

I. Nomenclature
II. Etiology
III. Pathophysiology
IV. Molecular Mechanisms
   References

I. Nomenclature

Ischemic stroke is a heterogeneous disorder that can be classified according to mechanism: large artery atherothrombotic, cardioembolic, small vessel lacunar, or "other" less common etiologies such as dissection, vasculitis, and so on. Up to 40% of stroke cases have no identifiable cause established by conventional diagnostic tests, and these are classified as having an "undetermined" or "cryptogenic" mechanism. Unproven embolism is often suspected in these cases.

Cardioembolism accounts for approximately 30% of all stroke (60% if the cryptogenic cases are included) and 25–30% of strokes in the young (younger than 45 years) [1]. A significant proportion of these strokes (15–25%) are due to atrial fibrillation (AF).

An ischemic stroke is considered cardioembolic if the clinical features and neuroimaging findings described later in this chapter support this diagnosis and a cardiac source of embolism is identified. Table 1 lists potential sources of cardioembolism. Other stroke subtypes should be excluded before assigning a cardioembolic etiology by ensuring that there is not a 50% or more stenosis or occlusion of a large artery supplying the ischemic territory, a clinical and radiographic syndrome consistent with a small vessel (lacunar) stroke, an established diagnosis of vasculitis or other unusual cause of stroke, or an atheroma more than 4 mm of the aortic arch.

II. Etiology

Stroke is predominantly a disease of the elderly, a population with high rates of cardiac and systemic atherosclerotic disease. It is important to recognize that there may be competing stroke mechanisms. For example, up to

Table 1 Sources of Cardioembolism

I. High-risk sources
   A. Atrial fibrillation
   B. Left ventricular aneurysm or thrombus
   C. Left atrial thrombus
   D. Recent transmural anterior myocardial infarction
   E. Rheumatic valvular disease
   F. Mechanical prosthetic valve
   G. Nonbacterial thrombotic endocarditis
   H. Bacterial endocarditis
   I. Primary intracardiac tumors (myxoma, papillary fibroelastoma)

II. Moderate risk sources
   A. Mitral annular calcification
   B. Mitral valve prolapse
   C. Cardiomyopathy
   D. Segmental wall motion abnormality
   E. Patent foramen ovale with atrial septal aneurysm
   F. Atrial flutter
   G. Sick sinus syndrome
   H. Valve strands
   I. Left atrial spontaneous echo contrast

Table 2 Stroke Mechanism and Outcomes[a]

	Recurrent Stroke/Death		
	7 day	90 day	1 year
Cardioembolism	2.4%/15.2%	8.6%/37.9%	13.7%/53.0%
Large artery	8.5%/4.1%	21.4%/8.1%	24.4%/10.8%
Lacunar	1.4%/0%	1.4%/2.8%	7.1%/6.9%
Undetermined	1.9%/7.3%	4.8%/17.7%	13.2%/25.6%

[a] Used, with permission, from Petty, G. W., Brown, R. D., Jr., Whisnant, J. P., Sicks, J. D., O'Fallon, W. M., and Wiebers, D. O. (2000). Ischemic stroke subtypes: a population-based study of functional outcome, survival, and recurrence. *Stroke* **31**, 1062–1068.

18% of patients with lacunar infarctions have identifiable cardioembolic sources. In these situations, it is incumbent on the clinician to reconcile the most likely cause. Factors such as radiographic evidence of prior large embolic-appearing or small vessel infarctions may provide useful evidence in support of a particular mechanism.

Repetitive, stereotyped transient ischemic attacks (TIAs), associated most commonly with low flow due to large vessel atherosclerotic disease, are unusual in embolic stroke. The classic presentation is a sudden onset of maximal symptoms. Although TIA and transient monocular blindness can occur as a result of cardioembolism, less than one-third of patients experience transient ischemic symptoms before stroke [2]. Sudden loss of consciousness has also been described but is an insensitive and nonspecific characteristic. "Non-sudden" cerebral embolism with a stuttering course has been described and could be due to intermittent flow obstruction by the embolus, as it causes a "ball-valve" effect in the vessel or a result of small embolic debris migrating and undergoing downstream autolysis. Seizures related to acute stroke are more common in patients with cardiac embolism.

The size of the embolic material determines, in part, which vessels become occluded. Small emboli can cause retinal ischemic or lacunar symptoms. Posterior cerebral artery territory infarcts, in particular, are often due to cardiac embolism. This predilection is not completely consistent across the various cardiac structural abnormalities that predispose to stroke and may be due to patterns of blood flow associated with specific cardiac pathology.

Neuroimaging data can support a diagnosis of cardioembolism. Multifocal infarctions that involve more than one vascular territory favor a proximal source of embolism. Recurrent ischemic events in a single vascular territory in the absence of a proximal large artery stenosis may also be due to cardioembolism. Embolic-appearing infarcts on neuroimaging may be clinically "silent."

Compared with other stroke mechanisms, the prognosis after cardioembolic stroke is poor (Table 2). The 7-day recurrence risk is 6.5%, in-hospital mortality 27.3%, and 5-year mortality as high as 80% [1]. These statistics highlight the importance of aggressive cardiovascular secondary prevention in this high-risk population.

III. Pathophysiology

A. Atrial Fibrillation

AF is the most common cause of cardioembolic stroke [3,4]. Thrombi in AF arise from the left atrium and atrial appendage. Event-loop recording and Holter monitoring are more sensitive than a standard 12-lead electrocardiogram (ECG) for detecting AF in stroke patients.

The rate of AF-related stroke increases with age. Younger patients with AF free of cardiac disease, diabetes, or hypertension have an extremely low rate of stroke, 1.3% per 15 years. At the age of 65 years, there is a 3–5% annual risk of stroke, which increases to 10% per year or greater by the age of 80 years. The age-related risk is independent of other major risk factors (diabetes, hypertension, previous stroke, and congestive heart failure).

Two clinical risk stratification schemes have been developed, the CHADS2 index and the Framingham AF risk score [5,6]. The CHADS2 classifies patients according to congestive heart failure, hypertension, age of 75 years or older, diabetes mellitus, or prior cerebral ischemic episode. A point increase correlated to a 1.5-fold increase in stroke rate per 100 patient-years, ranging from 1.9 for a score of 0 to 18.2 for a score of 6. The Framingham score is based on age, sex, hypertension, diabetes, and prior cerebral ischemic events in new-onset AF.

Five randomized primary and secondary prevention trials have demonstrated the efficacy and safety of warfarin in preventing AF-related stroke [4]. Pooled data from these trials demonstrate a 68% reduction in ischemic stroke (95% confidence interval [CI], 50–79%) and an intracerebral hemorrhage rate of less than 1% per year. The data for aspirin suggest that it has a lesser effect, with a 36% risk reduction (95% CI, 4–57%). Aspirin appears to be equal or superior to warfarin for preventing non-cardioembolic stroke in patients with AF. In addition, there is a small but definite risk of hemorrhage, particularly intracerebral hemorrhage, which is more common in the elderly population (older than 75 years). Independent risk factors associated with an increased risk of stroke include age older than 65 years, history of stroke or TIA, hypertension, and diabetes mellitus. Impaired left ventricular function was identified as an additional risk factor in the Stroke Prevention in Atrial Fibrillation (SPAF) population. Despite these findings, only 30–60% of patients with AF receive appropriate anticoagulation therapy. Risk factors for warfarin-associated intracranial hemorrhage include older age, prior cerebrovascular disease, prosthetic valve, and increased intensity of anticoagulation.

Echocardiographic findings can provide data for risk stratification. Moderate to severe impairment of left ventricular function is associated with a 2.5-fold greater risk of stroke. A left atrial anteroposterior diameter of more than $2.5\,cm/m^2$ was also found to be a predictor of cerebral embolism. In the SPAF study, 26% of patients had no clinical risk factor and had a normal echocardiogram. These patients had a low (<1% per year) risk of stroke. Approximately one-third of patients considered low risk by clinical criteria will be reclassified as high risk based on echocardiographic findings. Transesophageal echocardiography (TEE) has greater sensitivity for detecting abnormalities in left atrium, such as spontaneous echo contrast, left atrial thrombus, reduced left atrial appendage velocity, and complex aortic arch atheroma, which are markers of increased stroke risk.

Many patients have a rhythm that varies between atrial flutter and AF. Atrial flutter is associated with a 40% higher risk of stroke. Given that the concordance of the AF and atrial flutter is high, anticoagulation should be considered in patients with atrial flutter and coexisting cardiac pathology predisposing to left atrial thrombus.

Patients with sick sinus syndrome (SSS) often exhibit runs of intermittent atrial flutter and AF in addition to other cardiac arrhythmias. SSS is associated with a higher risk of stroke. There has not been a trial of antithrombotic agents for stroke prevention in SSS alone and the main intervention is usually pacemaker placement.

B. Cardiomyopathy

The annual stroke rates of 1.3–3.5% derived from congestive heart failure clinical trials is likely an underestimate of the true risk of stroke [7]. The risk of stroke is inversely related to the ejection fraction (EF), with as much as a 58% increase in thromboembolic events for every 10% decrease in EF. In the SAVE study, patients with an EF of 29–35% had a stroke rate of 0.8% per year; the rate in patients with an EF of 28% or less was 2.5% per year. There does not appear to be a correlation between the functional classification (New York Heart Association [NYHA] class) and stroke risk. Although cardiomyopathy can be due to ischemic and nonischemic etiologies, the stroke risk appears similar in the two groups. Cardiomyopathic patients often have AF as a complicating factor that further increases the risk of stroke.

Warfarin has not consistently been shown to significantly reduce the risk of stroke compared with placebo by 40–55% in patients in ischemic and nonischemic cardiomyopathy. A study comparing the efficacy of warfarin with aspirin for the prevention of stroke in patients with low EFs is currently underway [7].

C. Acute Myocardial Infarction

The short-term (2- to 4-week) stroke risk after acute myocardial infarction (AMI) is 2.5% [8]. Stroke is usually an early (within 14 days) complication of AMI and is more common in anterior wall (4–12%) than inferior wall infarction (1%) [8–10]. Approximately 40% of patients with an anterior wall myocardial infarction develop left ventricular thrombus, usually in the first 2 weeks. Patients with low EF after AMI have a cumulative 5-year stroke risk of 8.1%.

The WARIS II trial compared warfarin or aspirin or both in 3630 patients younger than 75 years with AMI [11]. The primary outcome, a 2-year composite of death, nonfatal reinfarction, or thromboembolic cerebral stroke, occurred in 241 (20%) of 1206 patients receiving aspirin, 203 (1216) of 1216 receiving warfarin (16.7%), and 181 (15%) of 1208 receiving warfarin and aspirin. Compared with aspirin alone, the reduction in nonfatal thromboembolic stroke was the same in those receiving warfarin or warfarin plus aspirin. However, a significantly higher rate of major nonfatal bleeding occurred in patients receiving warfarin (0.62% per treatment-year) than in patients treated with aspirin (0.17%).

Although there are no data from rigorous randomized controlled clinical trials, acute anticoagulation maintained for at least 3–6 months is the standard of care when left ventricular thrombus is detected. Intravenous tissue plasminogen activator (t-PA) has been used safely in patients with acute ischemic stroke and left ventricular thrombus.

D. Patent Foramen Ovale and Atrial Septal Aneurysm

Patent foramen ovale (PFO), a persistence of an embryonic defect in the interatrial septum, is present in up to 27%

of the general population. Atrial septal aneurysms (ASAs), defined as more than 10-mm excursions of the interatrial septum, are less common, affecting approximately 2% of the population. The size of a PFO can be quantified based on the degree of shunt. The specific definitions vary by study, but in general, fewer than 10 bubbles with 3–5 cardiac cycles is usually classified as "small," 10–30 bubbles "moderate," and more than 30 bubbles "large." In addition, the actual size of the PFO—the maximum distance between the septum primum and secundum—can be measured using biplane TEE. The presence of an ASA or a large right-to-left shunt has been reported to increase the risk of stroke in patients with PFO [12].

There is considerable controversy over whether a PFO-associated stroke should be classified as "cardioembolic." The mechanism of stroke most commonly ascribed to this structural defect is "paradoxical embolism," a venous thromboembolic event that results in stroke due to passage of the embolus across the PFO. More speculative mechanisms include thrombus formation within the PFO or ASA, arrhythmias, or migrainous stroke (Fig. 1). Because PFOs are so prevalent, it may be more appropriate to consider these strokes "cryptogenic with PFO." It is important to look for other potential causes of stroke before implicating the PFO.

There is evidence of an association between PFO and cryptogenic stroke, particularly in the young [13,14]. In 581 patients younger than 55 years with cryptogenic stroke, the prevalence of PFO was 46%. In contrast, in the Patent Foramen Ovale in Cryptogenic Stroke Study (PICSS) patients between the ages of 30 and 85 years with non-cardioembolic stroke, there was a PFO prevalence of 34% overall, 39% in patients with cryptogenic stroke, and 29% in those with a defined mechanism [14]. Estimates for the rates of annual stroke recurrence in cryptogenic stroke patients with PFO range from 1.5 to 1.8% [12–14].

Therapeutic options include antithrombotic medications and percutaneous catheter-based or surgical closure, or a combination. Optimal medical therapy has not been established. The only randomized trial to compare warfarin and aspirin in patients with PFO was the PICSS [14]. In PICSS, 33.8% of 630 patients found to have a PFO on TEE were randomized to either aspirin (325 mg) or warfarin (target international normalized ratio [INR] range, 1.4–2.8) and followed for 2 years. There was no significant difference in rates of recurrent stroke or death in patients with PFO versus no PFO. Event rates among the cryptogenic stroke patients with PFO treated with aspirin (17.9%) and warfarin (9.5%) were not statistically significant. Percutaneous closure of PFO has been available since 1989 and has largely replaced surgical approaches. Although complications are rare, the devices have been reported to embolize, fracture, cause tissue erosion, and stimulate inflammation and thrombosis. The existing PFO closure studies vary in patient selection criteria, medical treatments, choice of device, and outcome definitions. Therefore, it is impossible to draw definitive conclusions from existing data.

E. Left Atrial Myxoma

Myxomas are the most common of the primary cardiac tumors. They are usually benign, solitary, pedunculated, and situated in the left atrium. They are more common in women. An embolic event occurs in roughly 35–50% of patients with atrial myxoma. The friability of the tumor lends itself to embolism of tumor fragments [15]. The diagnosis should be suspected in cases where patients, particularly the young, present with murmur, heart failure, or systemic emboli. Myxomas can be detected on transthoracic echocardiogram or TEE. The treatment is surgical resection.

F. Valvular Disease

1. Mitral Annulus Calcification

Mitral annulus calcification (MAC) is associated with older age, ischemic heart disease, and cardiac arrhythmias,

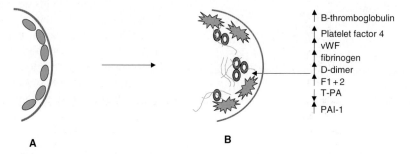

Figure 1 Intracardiac Thrombus Formation (A) Normal endocardium. (B) Endothelial injury caused by chamber dilatation and aberrant blood flow causes von Willebrand's factor (vWF) release, initiating activation of surface platelet receptors and the coagulation cascade resulting in thrombus formation.

as well as a higher risk of ischemic stroke [16]. Patients with MAC are, therefore, at risk of stroke by mechanisms other than valvular cardioembolism. It is unclear whether there is a direct causal relationship between MAC and stroke or merely an association based on its being a marker of systemic atherosclerosis.

2. Prosthetic Valves

The rates of embolism in patients with mechanical prosthetic valves on anticoagulation and non-anticoagulated patients with bioprosthetic valves are approximately 3–4% annually for mitral valves and 1.3–3.2% for aortic valves [17]. After bioprosthetic valve insertion, at least 3 months of anticoagulation is recommended. Long-term anticoagulation is warranted if there is evidence of AF, left atrial thrombus, or previous embolic events. Combination aspirin and warfarin therapy is superior to warfarin alone in reducing vascular mortality and systemic embolism (1.9 vs 8.5% per year respectively) in patients with mechanical valves and high-risk bioprosthetic valves, without a significant difference in major bleeding complications [18].

G. Mitral Stenosis

Left atrial thrombus is found in 15–17% of autopsy cases of mitral stenosis. Mitral stenosis is often due to rheumatic heart disease and can be associated with concomitant AF. The annual rate of stroke in patients with mitral stenosis is approximately 2%. Recurrent embolism is common.

H. Infective Endocarditis

Stroke occurs in 15–20% of cases of infective endocarditis, usually within the first 48 hours. Appropriate antibiotic therapy dramatically reduces the risk of stroke. Late embolism occurs in less than 5% of cases [19]. An elevated erythrocyte sedimentation rate in the setting of cerebral ischemic symptoms and fever or a new murmur should trigger a diagnostic evaluation including blood culture; a transthoracic echocardiogram; and if a high level of suspicion remains, a TEE. The most common organisms causing native valve endocarditis are streptococci, staphylococci, and enterococci, although other species of bacteria, fungi, spirochetes, and rickettsiae can infect valves. Echocardiography has not been shown to be useful in predicting risk of embolism. The risk of subarachnoid hemorrhage from mycotic aneurysms represents a contraindication to the use of anticoagulation in infectious endocarditis.

1. Nonbacterial Thrombotic Endocarditis

Nonbacterial thrombotic endocarditis (NBTE), also known as *marantic endocarditis,* is associated with malignancy and autoimmune disorders [20]. Libman–Sacks (L-S) endocarditis is an atypical form of NBTE associated with systemic lupus erythematosus. NBTE can be diagnosed using echocardiography (transthoracic and/or transesophageal). Although unproven, anticoagulation may be effective in preventing recurrent embolism, especially if there is a coexisting hypercoagulable state.

2. Papillary Fibroelastoma

Papillary fibroelastomas are rare, small (<20 mm) primary cardiac neoplasms that are attached to the aortic or, less commonly, the mitral valve. Although there are case reports of embolic strokes in patients with these tumors, whether the embolic material is thrombotic or tumor particles is unclear.

IV. Molecular Mechanisms

Virchow postulated that abnormalities in the vessel wall, blood flow, and coagulation system interacted to result in thrombus formation. This dynamic interaction occurs in the heart, as well as the blood vessels, and can predispose to cardioembolism. Flow conditions in the heart can be either high shear stress (e.g., mitral stenosis) or low shear stress (left atrial appendage stasis). Studies examining coagulation parameters across high stroke-risk conditions find patterns of hemostatic activation and fibrinolytic activity similar to those seen with severe atherosclerotic disease. Because of the high prevalence of systemic atherosclerosis in a stroke or cardiac population, the interpretation of peripheral marker levels is complicated.

The composition of the thrombus and the cerebrovascular response (endogenous fibrinolytic potential) may vary based on the pathophysiology and affect the probability of autolysis and recanalization. Interestingly, t-PA-induced middle cerebral artery recanalization has been shown to be more rapid and complete with MCA occlusion due to cardioembolism than to large artery or undetermined mechanisms.

In specific cardiac conditions, such as PFO, where a source of venous thromboembolism is suspected, hypercoagulability studies can contribute to the diagnostic evaluation. Venous thrombosis in the deep veins of the lower extremity or pelvis is documented in approximately 5% of stroke patients with PFO. The standard hypercoagulable panel to detect hemostatic abnormalities associated with venous thromboembolism includes activated protein C resistance, protein S, protein C, antithrombin III, antiphospholipid antibodies, and homocysteine. More extensive testing can include factor V Leiden and prothrombin gene mutations. A minority of stroke patients with PFO will have demonstrable prothrombotic abnormalities [13]. The presence of a hypercoagulable state has been associated with a higher risk of recurrence.

AF, a model for embolic stroke, has been linked to a systemic hypercoagulable state. Hemostatic regulation in the endocardium likely contributes to the development of left atrial thrombus. As in other vascular beds, the regulation of this signaling is under genetic control, although a variety of external stimuli modulate the expression. In patients with AF, the level of F1+2, a marker of thrombin generation, increases with age in parallel to the rate of stroke.

Other studies have confirmed that prothrombotic factor levels are increased in AF (factor VIII, fibrinogen, thrombin–antithrombin complex), as are those involved in fibrinolysis (t-PA, D-dimer). There does seem to be endothelial damage in AF, as well, with higher levels of von Willebrand factor (vWF). Platelet activation has been shown to play a role in AF. Elevations in levels of platelet factor-4, β-thromboglobulin, and p-selectin have been demonstrated in patients with AF. Markers of endothelial injury correlate with clinical stroke risk calculator assessments (CHADS2 and Framingham risk score). This increased hemostatic activity is not unique to AF and suggests a common mechanism of stroke among other cardiac conditions associated with risk of embolism (e.g., left ventricular aneurysm, mitral stenosis, and heart failure).

The dramatic stroke preventive efficacy of warfarin correlates well with a reduction in the level of activity of the hemostatic system. In addition, aspirin had little effect on F1+2, corresponding to its relative lack of efficacy in stroke prevention in patients with AF. Therapy with conventional strength warfarin (INR 2–3) has also been shown to effectively reduce fibrinogen and fibrin D-dimer levels in patients with AF, whereas aspirin (300 mg) alone or in combination with fixed low-dose warfarin (2 mg/day) resulted in higher levels of plasminogen activator inhibitor-1.

References

1. Petty, G. W., Brown, R. D., Jr., Whisnant, J. P., Sicks, J. D., O'Fallon, W. M., and Wiebers, D. O. (2000). Ischemic stroke subtypes: a population-based study of functional outcome, survival, and recurrence. *Stroke* **31**, 1062–1068.
2. Bogousslavsky, J., Cachin, C., Regli, F., Despland, P. A., Van Melle, G., and Kappenberger, L. (1991). Cardiac sources of embolism and cerebral infarction–clinical consequences and vascular concomitants: the Lausanne Stroke Registry. *Neurology* **41**, 855–859.
3. Hart, R. G., Halperin, J. L., Pearce, L. A., Anderson, D. C., Kronmal, R. A., McBride, R., Nasco, E., Sherman, D. G., Talbert, R. L., and Marler, J. R. (2003). Lessons from the stroke prevention in atrial fibrillation trials. *Ann. Intern. Med.* **138**, 831–838.
4. Atrial Fibrillation Investigators. (1994). Risk factors for stroke and efficacy of antithrombotic therapy in atrial fibrillation. Analysis of pooled data from five randomized controlled trials. *Arch. Intern. Med.* **154**, 1449–1457.
5. Gage, B. F., Waterman, A. D., Shannon, W., Boechler, M., Rich, M. W., and Radford, M. J. (2001). Validation of clinical classification schemes for predicting stroke: results from the National Registry of Atrial Fibrillation. *JAMA* **285**, 2864–2870.
6. Wang, T. J., Massaro, J. M., Levy, D., Vasan, R. S., Wolf, P. A., D'Agostino, R. B., Larson, M. G., Kannel, W. B., and Benjamin, E. J. (2003). A risk score for predicting stroke or death in individuals with new-onset atrial fibrillation in the community: the Framingham Heart Study. *JAMA* **290**, 1049–1056.
7. Pullicino, P. M., Halperin, J. L., and Thompson, J. L. (2000). Stroke in patients with heart failure and reduced left ventricular ejection fraction. *Neurology* **54**, 288–294.
8. Komrad, M. S., Coffey, C. E., Coffey, K. S., McKinnis, R., Massey, E. W., and Califf, R. M. (1984). Myocardial infarction and stroke. *Neurology* **34**, 1403–1409.
9. Cerebral Embolism Task Force. (1986). Cardiogenic brain embolism. *Arch. Neurol.* **43**, 71–84.
10. Cerebral Embolism Task Force. (1989). Cardiogenic brain embolism. The second report of the cerebral embolism task force. *Arch. Neurol.* **46**, 727–743.
11. Hurlen, M., Abdelnoor, M., Smith, P., Erikssen, J., and Arnesen, H. (2002). Warfarin, aspirin, or both after myocardial infarction. *N. Engl. J. Med.* **347**, 969–974.
12. Mas, J. L., Arquizan, C., Lamy, C., Zuber, M., Cabanes, L., Derumeaux, G., and Coste, J. (2001). Recurrent cerebrovascular events associated with patent foramen ovale, atrial septal aneurysm, or both. *N. Engl. J. Med.* **345**, 1740–1746.
13. Bogousslavsky, J., Garazi, S., Jeanrenaud, X., Aebischer, N., and Van Melle, G. (1996). Stroke recurrence in patients with patent foramen ovale: the Lausanne Study. Lausanne stroke with paradoxical embolism study group. *Neurology* **46**, 1301–1305.
14. Homma, S., Sacco, R. L., Di Tullio, M. R., Sciacca, R. R., and Mohr, J. P. (2002). Effect of medical treatment in stroke patients with patent foramen ovale: patent foramen ovale in cryptogenic stroke study. *Circulation* **105**, 2625–2631.
15. Sandok, B. A., von Estorff, I., and Giuliani, E. R. (1980). CNS embolism due to atrial myxoma: clinical features and diagnosis. *Arch. Neurol.* **37**, 485–488.
16. Nair, C. K., Thomson, W., Ryschon, K., Cook, C., Hee, T. T., and Sketch, M. H. (1989). Long-term follow-up of patients with echocardiographically detected mitral anular calcium and comparison with age- and sex-matched control subjects. *Am. J. Cardiol.* **63**, 465–470.
17. Kuntze, C. E., Blackstone, E. H., and Ebels, T. (1998). Thromboembolism and mechanical heart valves: a randomized study revisited. *Ann. Thorac. Surg.* **66**, 101–107.
18. Salem, D. N., Stein, P. D., Al-Ahmad, A., Bussey, H. I., Horstkotte, D., Miller, N., and Pauker, S. G. (2004). Antithrombotic therapy in valvular heart disease—native and prosthetic: The seventh ACCP conference on antithrombotic and thrombolytic therapy. *Chest* **126**, 457S–482S.
19. Hart, R. G., Foster, J. W., Luther, M. F., and Kanter, M. C. (1990). Stroke in infective endocarditis. *Stroke* **21**, 695–700.
20. Rogers, L. R., Cho, E. S., Kempin, S., and Posner, J. B. (1987). Cerebral infarction from non-bacterial thrombotic endocarditis. Clinical and pathological study including the effects of anticoagulation. *Am. J. Med.* **83**, 746–756.

23

Clinical and Neurobiological Aspects of Stroke Recovery

Bruce T. Volpe, MD
Rajiv R. Ratan, MD, PhD

Keywords: *motor learning, phosphodiesterase inhibition, plasticity, robotics, stroke recovery, task-specific training*

I. Introduction
II. Brief History: The "General Course of Recovery"
III. Emergence after Brain Injury of Progressive and Mutable Reflexive Motor Power
IV. Unmasking: Phenomena of Altering Ineffective Synaptic Potential
V. Bench to Bedside: Experimental Precedents Fuel Clinical Treatment
VI. Evidence for Functional Reorganization in Stroke Recovery: Neuroimaging Tools
VII. Motor Systems Cortical Physiology: Influences from the Bench
VIII. From Mutable Motor Maps to Neuroimaging to Magnetism
IX. Motor Learning as a Guide to Pharmacological Interventions
X. Conclusions
References

I. Introduction

Seminal observations by Twitchell and Brodal in patients following a stroke affecting the motor cortex or related corticospinal pathway led to the notion that central neural activation in motor pathways is engaged in a spatially and temporally stereotyped manner for those who recover motor deficits. The notion of a "motor recovery program" has been reinforced by animal experiments in which adaptive plasticity of cortical activity underlying new motor behavior has been monitored using direct electrophysiological techniques, as well as studies involving novel clinical neuroimaging and unique applications of bioengineering applied to humans. The emerging conclusions from these preclinical and clinical studies are that motor recovery after stroke involves pathways identical to those engaged in motor learning. Motor recovery should be viewed as a learning opportunity modifiable by the same behavior, stimulation, or drugs that enhance motor learning. It follows then that the history of dim or severely limited expectations for motor recovery following stroke might give way to more aggressive novel approaches based on years of data accumulated on motor learning. For example, there has been much written about the early poststroke period of most rapid improvement and the prediction of outcome based on lesion size, location, and initial impairment. These remain a useful platform for further study of recovery by understanding how negative prognostic factors for recovery adversely affect motor learning and how this can be overcome by pharmacological, biological, cell-based, or robotic approaches known to augment established motor learning pathways.

II. Brief History: The "General Course of Recovery"

Although entire volumes have been written on the history of the brain and its normal functions, and so by extension the history of recovery from brain injury, it is prudent to note how long ago this evidence began to gather, and to take from the more recent histories a few observations that have garnered experimental support and clinical relevance. This chapter focuses on the data that demonstrate that the structural underpinnings in neurons and groups of neurons for movement are flexible and mutable; that the recovery of any formerly well learned or overlearned activity requires enormous and persistent effort, in part because the focal motor impairments are accompanied by a diffuse impairments; and that motor recovery elicits aspects of motor learning. It is with a great sense of anticipation that the environment may alter each of these aspects of the motor system and motor behavior, especially when the organism must interact with the environment in an activity-dependent manner.

Physicians have been observing patients recovering from brain injury since, at least, 3000 BC. Egyptian glyphs were discovered on ancient papyrus in which an unanticipated craniotomy revealed an organ "fluttering" and "throbbing" like "corrugation in metallic slag" (reviewed in reference 1. It would seem certain that these injuries were descriptions of soldiers sustaining battle trauma, and whether the damage was primarily cortical or more diffuse does not detract from the earliest correlations of structure and function. These records may also provide the earliest views of "neurological recoverology," for there were bold and clear predictions or verdicts called favorable, uncertain, or unfavorable.

The unfavorable verdicts, also described as an "ailment not to be treated," were reserved for those with paralyzed limbs, especially those who "walk fumbling the ground," or in the words of the surgeon, "he shuffles..." or those with upper extremity spasticity, described as "not released in the head of the shoulder fork." Those ancients with strokelike symptoms would today be the focus of intense scrutiny and treatment options. Now the clinical question in stroke recovery focuses on the optimal timing and the most efficient duration of poststroke therapy, and the fundamental question focuses on a mechanism for the observed change.

III. Emergence after Brain Injury of Progressive and Mutable Reflexive Motor Power

The deterioration of motor function and the natural history of recovery of the nervous system after a stroke were first and most clearly recorded by Twitchell [2]. In a series of elegant observations, he documented the emergence or nonemergence of movement in the arms of patients who were paralyzed after stroke. He followed more than 100 patients, some 19 for longer periods and in greater detail. There were 17 in that group who could not move their arm at all soon after stroke; however, 5 progressed over 4–10 weeks to complete recovery through stages marked by stretch reflexes that were facilitated by posture and traction on the paralyzed limb. Those who failed to develop these facilitated stretch reflexes sustained no recovery. Among the gems of observation, perhaps the most apt for modern day is the remark that the "problem" of the increasing appearance of facilitated stretch reflexes is not so much "to abolish" them, but "to harness [their] diffuse hyperactivity" [2]. Studies suggest the focus on increased tone in the form of spasticity as a deleterious intermediate in stroke recovery may be overemphasized, and current management practices for spasticity, which might be necessary for the infrequent, severe, and local problem, often result in improvement of spasticity but with unwanted associated weakness [3]. However, Twitchell's early observations on motor recovery, as progressive organization and facilitation of stretch reflexes, emphasize the point that the reorganization and repair of the motor system reflect a basic feature of the system—namely that it undergoes great spontaneous fluctuations. Putative mechanisms that underlie these fluctuations come from basic experiments that show that the sensory and motor representation in cortical neurons appears to be continuously modulated depending on sensory input from the periphery, among other things (Fig. 1).

COMPLETE RECOVERY – 5, (20%)

- 3/5 started with hemiplegia, 2/5 started with hemiparesis;
- Reflexes, including finger-jerks, elicited within hours to days;
- Increased tone, upper and lower extremity, within hours to days, never severe;
- Comprehensive stretch-reflexes facilitated within 2 weeks;
- Tone appears proximal before distal, and
- Tone appears in upper extremity before lower extremity within 2 weeks;
- Discrete muscle control after, and finger grasp before digit control, 1-2.5 mo.

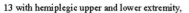
13 with hemiplegic upper and lower extremity,
4 with hemiplegic upper extremity,
8 with hemiparesis of the upper and lower extremity

INCOMPLETE RECOVERY – 7, (28%)

- Usually prolonged flaccidity;
- Spasticity within days, moderate to severe;
- Delayed comprehensive stretch reflexes, days to weeks;
- Failure to develop individual discrete proximal control, 1-2 mo.

NO RECOVERY – 6, (24%)

- Often prolonged flaccidity, days to weeks;
- Delayed appearance spasticity, then moderate to severe;
- Delayed reflexes, days to weeks, 1-2 mo.

Figure 1 In this diagram, patients who had "complete" recovery demonstrated early appearance of intact muscle proprioception, exemplified by increased tone. Twitchell used facilitation techniques to trigger the reflexes in the upper extremity that, when present, appeared to signal optimal recovery. Tone is the force with which muscle resists lengthening; spasticity is the phasic resistance to passive force that increases as a function of the increase in speed of the stretch. However, he also recorded one patient who demonstrated the progressive appearance of facilitated reflexes and who did not completely recover, and another who had incomplete recovery despite a failure to develop comprehensive reflexes.

A. Recovered Motor Function May Rely on Specific and Diffuse Structure Function Correlations and Will Require Great New Effort

Brodal [4] provides another important historical and observational account of stroke recovery. This right-handed professor of anatomy had a stroke, presumably in the right internal capsule that caused a left-sided hemiparesis without vision or sensory loss. The dramatic description of attempting to move a paralyzed limb focuses on the relation between the "greater the degree of paresis... the greater was the mental effort needed to make..." the paralyzed limb segment move [4]. Of note is his description that the "sensory information produced by passive movement" of the paralyzed limb "helped to direct the force of innervation to the proper channels" and presumably ease his subsequent attempt to move. His scholarly pursuits required writing, and this unexpectedly was impaired. Indeed, there was no modern neuroimaging to discount the possibility of additional lesion, but this writing difficulty, at least, argues for the insufficiency of intact neurons in the left hemisphere to subserve particularly complex but overlearned motor tasks carried out by the presumably normal right arm and hand. There was diminution of his verbal expressive ability that was different from the early poststroke dysarthria and probably more related to the easy fatigue and lack of mental endurance. Finally, his habit of tying a bowtie for 40 years was perturbed by inadequate speed and coordination and summed up in his words that "his fingers did not know the next move." The recovery, which proceeded over 10 months, recorded improved gross and fine motor abilities. Recovery of the powers of concentration and short-term memory was less apparent, and in particular, there was decreased endurance and easy fatigue. It might be argued that the poststroke depression played some role and whether the right hemisphere location of the stroke influenced the depression [5]. Alternatively, it may be that fatigue results from the inordinate amount of work required after the incomplete and inefficient reassembly of motor control. Movements that are generally termed *voluntary* so often fall into a category of overlearned and automatic performances [6]. A review of the literature on motor recovery after stroke may provide insight into these possibilities.

B. Recovery Depends on the Mutability of Sensory-Motor Maps and on Neuronal Plasticity: Background from the Bench

After a brain injury, the recovery processes depend on electrophysiological reorganization of sensory motor

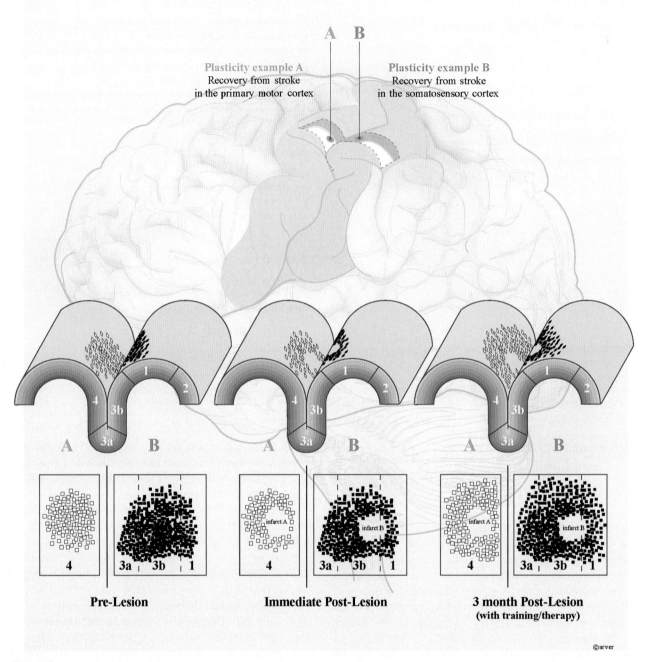

Figure 2 This view of the sensory and motor cortex after local cortical injury depicts post-lesional reorganization as represented by the increased dimensions of local field electrophysiological activity. Experiments have demonstrated that whether the injury or loss is peripheral or cortical, the respective cortical regions reorganize so that new cortical areas demonstrate electrophysiological activity formerly dedicated to other somatic distributions or motor neurons. The clinical point is that some studies have demonstrated that post-injury training had an additional facilitatory effect on the area of increased activation, and, most relevant, the real-world motor behavior was significantly improved.

processing or the biological flexibility of the neuron and the substrates of altered synapses, neurites, or even entire cells. Nineteenth-century neuroscientists suggested injury caused depression of functionally connected but potentially anatomically distant circuits—so-called diaschisis—and that recovery depended on the resolution of this depression [7]. Recovery from this depression may require the unmasking of functionally inactive connections. Another statement of this mechanism of recovery is that regions of brain that had been dedicated to contributing to one aspect of motor function could participate in a new function, also known as vicariation [8]. This form of reorganization can take place in nearby regions of the same cortical area or in cortical, subcortical, or cerebellar areas farther afield. For example, this form of recovery might depend on motor control emanating from the ipsilateral motor cortex or the cerebellum. The precise mechanisms for this plasticity are unclear, but several lines of investigation provide testable hypotheses. One possibility is that diaschisis represents diminished blood flow secondary to neural deactivation in areas that are functionally connected to the primary infarct. Studies suggest that neural deactivation results in vasoconstriction in remote areas rather than a reduction in oxygen metabolism, and that recovery might involve mechanisms that overcome local vasoconstriction [9]. Another possibility is that recovery from diaschisis involves general mechanisms of neural plasticity and regeneration. Such mechanisms might involve, alone or collectively, changes in dendritic sprouting, leading to enhanced synaptic strength and enhanced neural progenitor proliferation, leading to increased neural numbers, neutralization of molecules nonpermissive for growth of axons, or a number of other targets. One line of basic experiments with the hippocampus led to exciting possibilities of redirection and regrowth of dendrites under the influence of endogenously produced growth factor molecules. The precedent of persistent neurogenesis in the olfactory bulb led to rediscovery of neurogenesis in the adult human brain [10,11]. Injury-induced neurogenesis may be an important mechanism for enhancing functional outcome after stroke [12]. And finally, work in the spinal cord and cerebellum has also been revealing for so-called collateral sprouting and new synaptic potential in mature brain [13,14]. All of these biological correlates of recovery or plasticity may be important mechanisms for enhancing functional outcome after stroke.

IV. Unmasking: Phenomena of Altering Ineffective Synaptic Potential

Stimulation and injury to the peripheral nervous system coupled with electrophysiological recording demonstrated that some of the previously uncommitted fibers could effectively deliver a signal to a deafferented spinal cord or thalamus [15]. Deactivation of inhibition tangentially spreading from the stimulation site immediately enhanced the normal input. Furthermore, the hierarchically organized sensory systems could reorganize, and these reorganization responses are not just the result of alternate subcortical relay paths that might occur immediately. Experiments show that there were responses of the deafferented cortical areas to new stimuli, even after degeneration had occurred over longer periods [16]. Work with similar techniques in the motor system reflected plasticity in the sensorimotor cortex that now is commensurate with the recovery of motor performance in experimental animals recovering from ischemic brain injury [17]. This experimental approach requires intracortical microstimulation mapping of cortical regions activated during specific movement behaviors. In animals that were trained following injury, results demonstrate an advantage both in functional recovery and in expansion of the cortical maps thought to underlie the motor activity into new areas [18]. Further studies have manipulated the size of the cortical injury and so influenced the expansion, or the vicariation, of the cortical area mediating the movement behavior [19]. Improved behavior is persistently associated with the expanded cortical maps (Fig. 2).

V. Bench to Bedside: Experimental Precedents Fuel Clinical Treatment

Because the currency of the coordinated activity of neurons is behavior, it is no surprise that there is much information about the effect of environment and, more particularly, the training or nurturing aspects of environment on the development of motor behavior. Although leaving aside the extensive story of the effect of environmental enrichment on neural development [20], the implications from these early studies have generalized to include models of recovery. Further, the experimental designs that employ so-called enriched environments have expanded to include physical exercise and task-specific training [21]. A noble experimental psychological tradition demonstrated improvements in perceptual judgments that were a function of training [22], or controlled practice. These experimental developments proceeded in concert with a conceptual framework that includes neuron performance and synaptic activity [23]. Neuronal changes that occur during recovery from injury can be altered in positive respects when stimulated by enriched environments and task-specific training [24]. It is this background in part that has led trained therapists first to move passively the paralyzed limbs and then actively guide the performance of patients with stroke.

Stroke is a generic term referring to the sudden alteration in brain function, made apparent by an abrupt change in behavior caused by a diminishment or complete stoppage of blood flow to one or more brain regions. Decreased blood flow results from vessel or blood-clotting pathology or both and may also lead to vessel rupture. The downstream cellular and molecular events after a critical blood flow diminishment play out over minutes, hours, and days, but recovery processes play out over months and perhaps years. In some sense, however, stroke, like spinal cord injury, for example, does not represent a homogeneous category of injury. Disability or impairment scales that capture overall motor outcome characterize recovery from stroke, and the actual recovery depends on the severity [25] and details of the location (see reference 26, for example) and type of injury and concurrent medical condition [27]. Treatment for stroke and stroke recovery also proceeds in stages that require multimodal treatment programs. Stabilization of the crisis presented by a stroke syndrome gives way to correcting reversible pathology and decreasing the risk of a second event. Therapy for recovery more often proceeds with general efforts to encourage the patient to move and to exercise in a wide array of training protocols and now with an equally broad array of devices and programs. Given the potential for functional reorganization, new devices and training protocols will likely include new drugs and biomolecules and eventually generate more specific programs targeted to specific outcome goals.

A. Do Strategies for Poststroke Care Based on Enriched Experience and Task-Specific Training Matter?

Despite the success of preventive strategies that control blood pressure and treat atrial fibrillation and that encourage the cessation of smoking, attention to weight control, and a modest exercise regimen, studies have revised upward by 20% the incidence of stroke to more than 700,000 new cases per year. In the United States, the ranks of the more than 5 million survivors of stroke are likely to swell considerably, with increasing life expectancy and the coincidence of the "baby boom" generation growing older, coupled with improved medical treatment of the complications caused by acute stroke. For example, new figures demonstrate that less than 10% die from stroke in the first week to month, and the mortality in the first year is about 25%, rising in 5 years to about 50%. But after 5 years, more than half of the survivors are disabled, so approximately 175,000 Americans develop long-term disabilities each year and the cost for all health care–related to stroke now exceeds $50 billion. Stroke is the leading cause of permanent impairment and disability, so treatment of recovery matters.

Although there have been some promising developments in the treatment of acute stroke, most survivors are left with significant and permanent residual physical, cognitive, and psychological impairments, and the combination of these impairments results in disability. Randomized controlled trials in which patients after stroke were cared for in a medical unit or a specialized stroke unit have demonstrated that treatment in a specialized stroke unit conferred lower mortality rates, shorter hospitalization, and greater likelihood of discharge home (see, for example, reference 28). The positive influences persisted over a 10-year follow-up period [29]. These results have begun to set a standard for the rehabilitation of patients after stroke.

In rehabilitation hospitals, current interdisciplinary treatment programs for patients with stroke focus on both impairment and disability reduction; however, the rush to discharge the patient from the rehabilitation hospital more quickly has prompted a shift toward encouraging functional improvement by learning compensatory techniques. Although disability reduction with real-world outcome differences remains the crucial goal of any rehabilitation experience [30], data suggest that hasty compensation for a disability engenders a pattern of disuse in the impaired limbs that extinguishes temporarily that aspect of recovery and mutes the potential for future impairment change, as well as disability reduction [31,32] Constraint-induced programs in which the patient's attention is focused on the impaired limb through a variety of techniques that bind or render the unimpaired limb useless have generally, but not uniformly, demonstrated a positive effect on impairment and disability (see reference 33 for an example and reference 34 for a review).

The renewed interest in impairment reduction has provoked a number of innovative training protocols that have demonstrated that additional task-specific sensorimotor training has improved outcome. For example, the addition to a program of poststroke rehabilitation of 30 minutes of a pushing exercise for the proximal paretic upper limb over 30 sessions facilitated motor recovery of the paretic limb [35], and in a similar approach with a simple device [36], or in a program targeted to the hand [37]. Focused training and specific measurement of outcome based on the trained performance—for example, in designs to train the upper or lower limb—led to improved outcome for the treated limbs and not especially for the "untreated" limbs [38]. Along those lines, there have been a number of studies testing whether body weight supported gait training on a treadmill can abet recovery of walking in patients with stroke [39]. Results of pilot and controlled trials of gait training with a number of devices, with always improving configuration, have been generally positive [40], although not in all centers [41]. So the number of recent protocols that insist on task-specific training targeted to the impairment have found success by refocusing attention on the impaired limbs or increased practice intensity, or both.

Whether the environmental enrichment and the intensive training programs can further enhance motor recovery using "smart" interactive robots [42], position-sensitive industrial-designed robots [43], or devices intended to train functional reach [44] remain undetermined. Robots increase the intensity of the training experience, and the compliant "smart" interactive robots have the advantage of delivering reproducible forces [42]. These studies and others demonstrate that motor recovery can be trained in patients with chronic stroke. Patients, on average 3.5 years after stroke with a partially paralyzed upper limb, demonstrated significantly improved motor function after 18 1-hour training periods [45] (Fig. 3). Other novel approaches to motor recovery

Figure 3 Interactive robotic devices are new tools for therapists that employ the latest in visual displays and haptic control. If a patient cannot move, the robot will move the limb passively. As the patient begins to move, the robot "gets out of the way." After some attempts to reach the target, the robot will guide and correct the movement trajectory along an optimal path and finish the movement to a target if necessary. Task-specific training can be intensive, so in a 50-minute session, the patient will make more than 1000 movements. In patients with acute and chronic stroke, successful pilot experiments have improved movement of the shoulder and elbow. Training programs are now testing the effectiveness of robotic devices that move and guide the arm against gravity, and that move and guide the wrist and hand. Investigators are beginning to exploit the power of the robot to record a "dose" of training and to provide a temporal record of position and speed information over the duration of the training period.

after stroke include transcranial magnetic stimulation [46], direct motor cortex stimulation [47], or drugs that have general arousal effects or specific effects on glutamate or γ-aminobutyric acid (GABA) transmission [48,49] all remain active current areas of investigation.

VI. Evidence for Functional Reorganization in Stroke Recovery: Neuroimaging Tools

There are currently few promising experimental methods to study collateral sprouting in the brain of humans recovering from stroke; however, neuroimaging tools with standard magnetic resonance sequences and protocols are readily available to study potential functional reorganization. Detailed understanding of the anatomy of the motor cortex and the descending and integrative paths from work in nonhuman primates and in the clinic have focused on the primary motor cortex and the premotor (dorsal and ventral), supplemental motor (rostral and caudal), rostral cingulate (anterior and posterior) (for example, see reference 50). Clearly the anatomical relations with the thalamus, basal ganglia, cerebellum, and the insula and parietal cortices are also crucial for different classes of motor behavior under variable conditions [51]. Given the more frequent location of strokes in the subcortical area, it may be that the elucidation of the cortical representations in the descending fiber tracts has additional currency for understanding stroke recovery [52]. For example, the well-known path of the primary motor cortex projection descends in the middle third of the posterior limb of the internal capsule, but it is also important to note that the axons of the premotor cortex pass through the capsular genu, and those of the supplemental motor cortex pass through the anterior limb. These anatomic substrates may contribute to a new central organization that underlies functional outcome.

For the purpose of understanding the structure–function relationship of motor behavior and of motor recovery, several dozen studies using positron emission tomography (PET) or functional magnetic resonance imaging (fMRI) with blood oxygenation level–dependent sequences (BOLD) in the past 10 years have produced some important new clinical information (reviewed in reference 53). The history of applying imaging technology to capture brain surrogate markers that correlate with behavior is long and the interested reader should know that although the clinical application story began more than 20 years preceding the latest review [54], critical experiments that demonstrate the quantitative relationship between the fMRI signal and the underlying neural activity occurred recently in animals [55] and in humans as well [56].

VII. Motor Systems Cortical Physiology: Influences from the Bench

To appreciate the complexity of the motor system, it is worth dispelling some simple metaphors that blur important details. The motor cortex is unlikely to be a static control repository for muscle activity, the "static" neural keyboard on which movements are "played out." Early analysis of single neuron activity from the motor cortex stressed the relationship between neural activity and simple motor properties, first force [57], then later variables called *preferred direction*, but indicating target direction, position, and distance, movement goal, and even apparent movement [58]. The control of voluntary movements appears to result from the collective activity of distributed networks of activated neurons, although some evidence describes conditions under which the motor cortex can function to plan and produce the complete motor behavior [59]. These networks are overlapping and capable of modification under even normal environmental conditions [60]. Although the rules of somatotopy generated from the work on animals cited previously holds, it holds only for major body divisions. Not surprisingly, if the motor maps are mutable, the primary motor cortex is likely involved in motor learning and, given the wide connectivity, is also likely involved in a variety of cognitive tasks [61]. This analysis leaves aside the question of whether modern computational models are adequate to capture the general rules for motor control, developed and practiced over long periods, that describe the relationship between outcome and the various parameters of amplitude and timing, and so the reader is referred to some proposals [62,63].

Experimental findings in the motor skill acquisition of normal subjects may inform the understanding of motor recovery after stroke for some patients. These findings suggest that some patients who have recovered motor performance have functional imaging characteristics on activation tasks that have similarities to those of normal subjects who have learned a motor skill. These findings provoke the hypothesis that motor recovery depends in part on the acquisition of new motor skills. Motor skill is the process by which a motor act or sequence of motor acts becomes effortless to perform. Acquisition of motor skill depends on practice and interaction with the environment. Training normal subjects to perform finger thumb opposition in a repeated or novel sequence for a few minutes each day led to different patterns of fMRI activation. Early in the training, there was widespread cortical activation that became more localized for the repeated sequence compared with the novel sequence, and the performance accuracy improved for the repeated sequence. Improvement in accuracy was associated with increased fMRI signal in the primary motor cortex and less signal in widespread cortical areas [61,64]. These studies were controlled for path length, velocity, and accuracy, as well as all movement parameters that alter the fMRI activation. These and related studies show that improved performance and increased motor cortex fMRI signal occurred in stages, followed by consolidation so that the profit from learning was persistent many weeks after the training was discontinued. There was also significant retention of the skill even 1 year after the training had stopped. Recent work has led investigators to explore the interaction of the several anatomic systems that subserve the different aspects of motor performance, learning and consolidation. Skilled motor performance is acquired in several stages: "fast" learning, an initial, within-session improvement phase, followed by a period of consolidation of hours' or days' duration, and then "slow" learning, consisting of delayed incremental gains in performance emerging long after practice ceased [65]. To continue the argument for the hypothesis that motor recovery depends on motor learning, it is necessary to review the functional neuroimaging of patients with stroke.

A. Functional Neuroimaging in Patients with Stroke

Some two dozen cross-sectional and longitudinal studies of patients with stroke scanned over time or during the performance of motor tasks demonstrate significant alteration in the PET or fMRI patterns of activation [53]. The studies have been longitudinal and cross-sectional; they have focused on patients who have recovered, those who have not recovered, and those in the process of recovering. The timing of the studies with respect to the onset of the stroke also ranges from patients studied days after stroke to those who are years after stroke. The patients have been as young as 21 years and as old as 86 years. Patients have had capsular lesions, other subcortical lesions, cortical lesions, or all three. There have been variable techniques and different attention to mirror movements. In activation studies, patients have opposed thumb to fingers, or tapping of thumb and finger; there have been hand-tapping tasks and elbow flexion tasks. Some studies have controls; many studies do not have controls. Although these studies have generated detailed information that has refined the techniques of the general approach to neuroimaging, the range of variables has pointed to a few important conclusions.

In recovered or recovering patients, the results demonstrate that there is overactivation of the PET or fMRI signal. This overactivation is often contralateral; however, several studies demonstrate ipsilateral overactivation that occurs in primary and secondary motor cortical areas. In the initial PET studies in completely recovered patients, investigators demonstrated that at rest there was decreased signal in the contralateral motor areas and during activation there was increased activation in premotor, cingulate, and parietal regions [66,67]. Leaving aside the ipsilateral activation as several of the patients had mirror movements, patients with recovered motor function demonstrated increased PET signal when performing a simple movement. Although the recovery was listed as "complete," the issues raised by the recovering anatomist, described earlier, make understanding the details of complete recovery important. Were the recovered movements performed with individual muscle group control and deftness? What was the patient's subjective sense of ease of movement? And importantly what were the fluctuations in these activation maps over longer periods? As in the motor skill learning in normal subjects, widespread activation could have been followed by more specific activation as performance improved and the task was mastered.

Other work with fMRI demonstrates similar results. Investigators used a design that recorded mirror movements and actively discouraged them in cross-sectional BOLD-fMRI studies of patients who had recovered finger movement. These studies show increased signal in the contralateral and ipsilateral hemisphere during a finger-tapping task (for example, see reference 68). Because mirror movements appear frequently in recovering patients, these investigators introduced a quantitative approach that normalizes the activity in the contralateral and ipsilateral region of interest [68], although they also developed other approaches to identifying the imaging signal from mirror movements [69]. Another controlled fMRI study of patients recovering from subcortical strokes demonstrated that stroke patients also had increased activation. Group analysis based on change in laterality index from the acute to the chronic period showed significant increased ipsilateral activation for the recovered motor ability in the stroke patients compared with controls, but nearly all the patients had mirror movements during the activation task [70]. Longitudinal controlled studies in which patients were scanned on multiple occasions in different states of recovery strongly suggest that increased fMRI activation correlated with poor recovery. These data propose that a return to more normal brain activation patterns, showing less widespread activation, may be the reliable surrogate marker for improved recovery. In one study, patients at least 3 months after stroke participated in a visually paced handgrip task. The degree of recovery as measured in the real world was negatively correlated with the overactivation of the fMRI signal obtained during a motor task [71,72]. Those patients with greater fMRI activation demonstrated the least recovery. When this approach was applied to patients who were 10–14 days after stroke, there was again a negative correlation between overactivation and outcome [73]. Patients after stroke who can participate in a handgrip task so that both flexion and extension occur with some attention to speed and pace may have actually acquired a skill. The trend of brain activation toward more normal regional patterns in patients recovering motor function has similarity to the shrinking of widespread activation apparent during normal learning of a motor skill. Recalling the anecdotal report of the recovering anatomist, it might be important to note the quality and ease of movement execution. The recovering anatomist's recollection that the extraordinary energy and effort required to complete ordinary and formerly overlearned motor tasks early after stroke, eased, later, as the movement became more accurate and automatic, even effortless suggests that skill learning and motor learning share features for those recovering from stroke. The hypothesis would test whether this subjective awareness and objective motor improvement signaled decreased fMRI or PET activation, as in the case of motor skill learning.

B. Functional Neuroimaging after an Intervention in Patients Recovering from Stroke

Neuroimaging studies of patients with stroke scanned before and after a training program support this idea. Patients recovering from stroke were randomized to intensive training using a finger-tracking protocol, undoubtedly more difficult and probably closer to skill learning than a handgrip task. Aged-matched controls also participated in this protocol. Patients, treated and control, and age-matched well subjects were scanned before and after training with BOLD-fMRI. Before training, patients showed increased ipsilateral activation compared with the well elderly. For the group of stroke patients exposed to the training protocol, fMRI activation switched to the contralateral hemisphere and demonstrated a pattern comparable to the well elderly, whose pattern of activation did not change significantly after the training. The untreated control patients eventually participated in the active treatment protocol and demonstrated activation results comparable to those of the initially treated stroke patients. Of great import, transfer of this training effect correlated with improved functional skills [74]. Along these lines, a PET study randomized patients with stroke to an intensive training program or standard therapy. Passive range of motion of the paretic upper extremity elicited PET activation before and after training. The group treated with the intensive program showed increased PET activation compared with the pretraining activation [75]. This activation was contralateral and ipsilateral and was comparable to that of control subjects.

VIII. From Mutable Motor Maps to Neuroimaging to Magnetism

Noninvasive stimulation of the cortex can occur with the application of a magnetic field generated by passing a current through a coil placed near the skull [46]. Brief high-current electrical pulse or high-frequency repetitive pulses produce a graded magnetic field and subsequent cortical stimulation (transcranial magnetic stimulation [TMS]). As the coil can be moved over the skull, it is theoretically possible to aim the magnetic field and, in a specific region of cortex, inhibit with slow rates of repetitive stimulation or excite with high rates of repetitive stimulation. Measurement or interruption of central motor conduction time had been used to study, for example, the contribution of the occipital cortex to Braille reading in blind subjects [76]. Using a similar approach but in the sensory motor system, investigators demonstrated that application of the magnetic field to an area in the premotor cortex disrupted the motor performance whether it was applied to the ipsilesional or contralesional hemisphere. These disruptions were dependent on stroke severity and time from stroke. Ipsilesional TMS to the premotor region disrupted performance in mildly impaired patients, whereas contralesional premotor TMS disrupted movements in the severely impaired patients early after stroke and in the chronic phase (for example, see reference 77). TMS had no effect on controls. Other studies

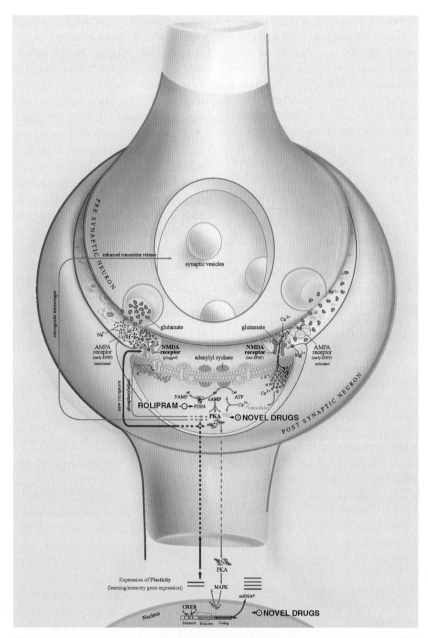

Figure 4 Pharmacological Targets for Enhancing Motor Learning at a Central Nervous System Synapse The presynaptic neuron is triggered to relapse neurotransmitter (glutamate) via depolarization-induced changes in calcium and fusion of synaptic vesicles. Released glutamate activates AMPA receptors and N-methyl-D-aspartate (NMDA) receptors. NMDA receptors mediate influx of calcium into the postsynaptic neuron via their associated ion channels. Increased cytosolic calcium activates adenylate cyclase (postsynaptic membrane), leading to an increase in levels of the second-messenger, cyclic adenosine monophosphate (cAMP). Elevated levels of postsynaptic cAMP lead to increased activation of protein kinase A (PKA), a kinase that phosphorylates proteins primarily on serine and threonines. PKA can directly or indirectly lead to the phosphorylation of the nuclear transcription factor, cAMP response element binding protein (CREB) at serine 133. This phosphorylation event leads to the recruitment of transcriptional coactivators such as CREB-binding protein (CBP) and increased transcription of genes containing a binding site for CREB (cAMP response element) in their promoters. A number of these genes are involved in plasticity and learning such as brain-derived neurotrophic factor (BDNF). Levels of cAMP can be enhanced in the brain by inhibiting the enzyme involved in its degradation, phosphodiesterase 4. This can be achieved with the antidepressant rolipram. Other targets for enhancing motor learning include direct activators of PKA (none identified to date) or agents that directly enhance CREB-mediated transcription (histone deacetylase inhibitors, CBP activators).

have used TMS to measure cortical change after task-specific intervention, specifically the constraint-induced movement therapy. TMS maps of the cortex, which were activated with hand movement, were significantly larger after therapy during a time when hand function had improved significantly [31]. The functional reversible disruption of motor behavior using targeted TMS supports the idea generated by the neuroimaging data and prompted by a variety of animal experiments that alternate motor cortical regions contribute to functional motor recovery.

That the frequency of repetitive TMS can alter the inhibitory activity of the cortex raises another experimental possibility [78]. Early animal experiments demonstrated that washing activated cortical neurons that underlie activity restricted either to the forepaw or to a vibrissa (whisker) with bicuculline (a GABA antagonist) blocked local inhibitory circuits. GABA inhibition led to expanded adjacent cortical regions when preexisting lateral excitatory connections were unmasked, and the cortical test site stimulation elicited a wider activation, so both forepaw and vibrissae moved [79]. These data together with the increase of TMS maps following transient deafferentation or using anesthesia to the proximal limb reduced inhibition at the cortical level and led to increased excitability of the motor system and potentially increased plasticity [80]. More study will be required to test whether chronic deafferentation of the proximal affected limb during training of the distal affected limb will abet recovery.

IX. Motor Learning as a Guide to Pharmacological Interventions

The use of the concept of motor learning as a guide to motor recovery in stroke has implications for novel pharmacological interventions in the subacute or chronic period after a stroke. The molecular details of this type of learning, also known as *implicit* or *nondeclarative memory*, have emerged from the work of a number of groups, especially that of Kandel [81]. These elegant studies have identified a phylogenetically conserved "core signaling pathway" involved in mediating synaptic changes associated with long-term memories such as acquisition of a motor skill. The "core-signaling pathway" involves the second-messenger cyclic adenosine monophosphate (cAMP) and some of its downstream targets including protein kinase A and the transcriptional activator cAMP response element–binding protein (CREB) (see Fig. 1). Levels of cAMP could be modulated in the motor cortex following a stroke via systemic administration of low-molecular-weight substances (drugs) that increase the activity of enzymes involved in cAMP generation (adenylate cyclase) or decreased activation of enzymes involved in cAMP breakdown (phosphodiesterase). The latter are more attractive therapeutic targets for stroke, as cAMP hydrolyzing phosphodiesterases constitute a family of enzymes with distinct tissue distributions [82] (Fig. 4). It is, therefore, theoretically possible to identify phosphodiesterase inhibitors that would increase cAMP, activation of protein kinase A, and CREB activation in the motor cortex but not in other areas of the brain or body. Prior studies have shown that phosphodiesterase inhibitors that inhibit isoform 4 (PDE_4) can decrease the amount of time required to form a memory and enhance regeneration in the central nervous system [82]. An example of a PDE_4 inhibitor that may be well suited for testing in stroke recovery includes the antidepressant rolipram. Clinical use of this agent has been limited by the side effect of nausea; however, dosing schedules and routes of administration are being identified to limit this problem. Rolipram (or some other novel PDE4 inhibitor), in conjunction with standard or novel motor training strategies described in this chapter, may a viable approach to alter the amplitude or the kinetics of motor recovery following stroke for the benefit of patients. Other biological targets in the core signaling pathway involved in memory (e.g., cAMP, protein kinase A, and CREB) are under active investigation in academic and industry laboratories and these studies should yield a spectrum of drugs appropriate for testing in stroke recovery.

X. Conclusions

Data from animal experiments and functional neuroimaging in patients strongly demonstrate that cortical maps in animals and humans are mutable. Multiple cortical and several other systems like the basal ganglia and the cerebellum contribute to motor behavior and provide a wide substrate for reorganization. Mechanisms to increase synaptic efficiency depend on experience. The size and shape of the synapses can be further modified with drugs and biomolecules. Current clinical treatment of stroke recovery is driven by economic decisions and has not kept pace with scientific evidence. Devices or protocols to focus on task-specific and more intensive training protocols significantly alter impairment. Whether impairment alters functional outcome and causes reduced disability is a testable hypothesis, and early analysis suggests that impairment reduction on the order of a 5% change on a standard validated and reliable scale translates into a minimal clinically meaningful change—namely, a real-world significant improvement. Many of the studies with advanced training protocols have met these minimal criteria. Whether drugs, biomolecules, and stem cell transplants will also aid stroke recovery are experiments for the future. It is clear that clinical evidence is available that more can be done to treat stroke recovery.

References

1. Plum, F., and Volpe, B. T. (1987). Neuroscience and higher brain function: from myth to public responsibility. In: "Handbook of Physiology" (V. B. Mountcastle, ed.). American Society of Physiology, Bethesda, MD.
2. Twitchell, T. E. (1951). The restoration of motor function following hemiplegia in man. Brain **74**(4), 443–480.
3. Sommerfeld, D. K., Eek, E. U., Svensson, A. K., Holmqvist, L. W., and von Arbin, M. H. (2004). Spasticity after stroke: its occurrence and association with motor impairments and activity limitations. Stroke **35**(1), 134–139.
4. Brodal, A. (1973). Self-observations and neuro-anatomical considerations after a stroke. Brain **96**(4), 675–694.
5. Narushima, K., Kosier, J. T., and Robinson, R. G. (2003). A reappraisal of poststroke depression, intra- and inter-hemispheric lesion location using meta-analysis. J. Neuropsychiatry Clin. Neurosci. **15**(4), 422–430.
6. Granit, R. (1970). "The Basis of Motor Control." Academic Press, New York.
7. Von Monakow, C. (1914). Die localisation im groshirn und der abbau der funktion durch kortikale herde. In: "Brain and Behaviour" (P. K. Harris, ed.), pp. 27–36. Penguin Books, London.
8. Munk, H. (1881). "Uber die Funktionen der Grosshirnrinde. Gesammelte Mittei-lungen aus den Jahren 1877–1880." Hirshwald, ed., Berlin.
9. Ito, H., Kanno, I., Shimosegawa, E., Tamura, H., Okane, K., and Hatazawa, J. (2002). Hemodynamic changes during neural deactivation in human brain: a positron emission tomography study of crossed cerebellar diaschisis. Ann. Nucl. Med. **16**(4), 249–254.
10. Altman, J., and Das, G. D. (1965). Autoradiographic and histological evidence of postnatal hippocampal neurogenesis in rats. J. Comp. Neurol. **124**(3), 319–335.
11. Eriksson, P. S., Perfilieva, E., Bjork-Eriksson, T., Alborn, A. M., Nordborg, C., Peterson, D. A., and Gage, F. H. (1998). Neurogenesis in the adult human hippocampus. Nat. Med. **4**(11), 1313–1317.
12. Lichtenwalner, R. J., and Parent, J. M. (2004). Adult neurogenesis and the ischemic forebrain. J. Cereb. Blood Flow Metab. 2005.
13. Cesa, R., and Strata, P. (2005). Axonal and synaptic remodeling in the mature cerebellar cortex. Prog. Brain Res. **148**, 45–56.
14. Raineteau, O., and Schwab, M. E. (2001). Plasticity of motor systems after incomplete spinal cord injury. Nat. Rev. Neurosci. **2**(4), 263–273.
15. Wall, P. D. (1977). The presence of ineffective synapses and the circumstances which unmask them. Philos. Trans. R. Soc. Lond. B Biol. Sci. **278**(961), 361–372.
16. Kaas, J. H. (1991). Plasticity of sensory and motor maps in adult mammals. Annu. Rev. Neurosci. **14**, 137–167.
17. Buonomano, D. V., and Merzenich, M. M. (1998). Cortical plasticity: from synapses to maps. Annu. Rev. Neurosci. **21**, 149–186.
18. Nudo, R. J., Wise, B. M., SiFuentes, F., and Milliken, G. W. (1996). Neural substrates for the effects of rehabilitative training on motor recovery after ischemic infarct. Science **272**(5269), 1791–1794.
19. Frost, S. B., Barbay, S., Friel, K. M., Plautz, E. J., and Nudo, R. J. (2003). Reorganization of remote cortical regions after ischemic brain injury: a potential substrate for stroke recovery. J. Neurophysiol. **89**(6), 3205–3214.
20. Diamond, M. C., Krech, D., and Rosenzweig, M. R. (1964). The effects of an enriched environment on the histology of the rat cerebral cortex. J. Comp. Neurol. **123**, 111–120.
21. Will, B., Galani, R., Kelche, C., and Rosenzweig, M. R. (2004). Recovery from brain injury in animals: relative efficacy of environmental enrichment, physical exercise or formal training (1990–2002). Prog. Neurobiol. **72**(3), 167–182.
22. Gibson, E. J. (1953). Improvement in perceptual judgments as a function of controlled practice or training. Psychol. Bull. **50**(6), 401–431.
23. Hebb, D. O. (1949). "The Organization of Behavior." Wiley, New York.
24. van Praag, H., Kempermann, G., and Gage, F. H. (2000). Neural consequences of environmental enrichment. Nat. Rev. Neurosci. **1**(3), 191–198.
25. Mohr, J. P., Foulkes, M. A., Polis, A. T., Hier, D. B., Kase, C. S., Price, T. R., Tatemichi, T. K., and Wolf, P. A. (1993). Infarct topography and hemiparesis profiles with cerebral convexity infarction: the Stroke Data Bank. J. Neurol. Neurosurg. Psychiatry **56**(4), 344–351.
26. Miyai, I., T. Suzuki, J. Kang, K. Kubota, and B. T. (1999). Volpe, Middle cerebral artery stroke that includes the premotor cortex reduces mobility outcome. Stroke **30**(7), 1380–1383.
27. Dromerick, A., and Reding, M. (1994). Medical and neurological complications during inpatient stroke rehabilitation. Stroke **25**(2), 358–361.
28. Jorgensen, H. S., Kammersgaard, L. P., Houth, J., Nakayama, H., Raaschou, H. O., Larsen, K., Hubbe, P., and Olsen, T. S. (2000). Who benefits from treatment and rehabilitation in a stroke unit? A community-based study. Stroke **31**(2), 434–439.
29. Indredavik, B., Bakke, F., Slordahl, S. A., Rokseth, R., and Haheim, L. L. (1999). Stroke unit treatment. 10-year follow-up. Stroke **30**(8), 1524–1527.
30. Wade, D. (1999). Rehabilitation therapy after stroke. Lancet **354**(9174), 176–177.
31. Liepert, J., Bauder, H., Wolfgang, H. R., Miltner, W. H., Taub, E., and Weiller, C. (2000). Treatment-induced cortical reorganization after stroke in humans. Stroke **31**(6), 1210–1216.
32. Taub, E., Miller, N. E., Novack, T. A., Cook, E. W., 3rd, Fleming, W. C., Nepomuceno, C. S., Connell, J. S., and Crago, J. E. (1993). Technique to improve chronic motor deficit after stroke. Arch. Phys. Med. Rehabil. **74**(4), 347–354.
33. Wittenberg, G. F., Chen, R., Ishii, K., Bushara, K. O., Eckloff, S., Croarkin, E., Taub, E., Gerber, L. H., Hallett, M., and Cohen, L. G. (2003). Constraint-induced therapy in stroke: magnetic-stimulation motor maps and cerebral activation. Neurorehabil. Neural. Repair **17**(1), 48–57.
34. van Der Lee, J. H. (2001). Constraint-induced therapy for stroke: more of the same or something completely different? Curr. Opin. Neurol. **14**(6), 741–744.
35. Feys, H. M., De Weerdt, W. J., Selz, B. E., Cox Steck, G. A., Spichiger, R., Vereeck, L. E., Putman, K. D., and Van Hoydonck, G. A. (1998). Effect of a therapeutic intervention for the hemiplegic upper limb in the acute phase after stroke: a single-blind, randomized, controlled multicenter trial. Stroke **29**(4), 785–792.
36. Whiteall, J., McCombe-Waller, S., Silver, K. H., and Macko, R. J. (2000). Repetitive bilateral arm training with rhythmic auditory cueing improves motor function in chronic hemiparetic stroke patients. Stroke **31**, 2390–2395.
37. Butefisch, C., Hummelsheim, H., Denzler, P., and Mauritz, K. H. (1995). Repetitive training of isolated movements improves the outcome of motor rehabilitation of the centrally paretic hand. J. Neurol. Sci. **130**(1), 59–68.
38. Kwakkel, G., Wagenaar, R. C., Twisk, J. W., Lankhorst, G. J., and Koetsier, J. C. (1999). Intensity of leg and arm training after primary middle-cerebral-artery stroke: a randomised trial. Lancet **354**(9174), 191–196.
39. Visintin, M., Barbeau, H., Korner-Bitensky, N., and Mayo, N. E. (1998). A new approach to retrain gait in stroke patients through body weight support and treadmill stimulation. Stroke **29**(6), 1122–1128.
40. Hesse, S., and Werner, C. (2003). Partial body weight supported treadmill training for gait recovery following stroke. Adv. Neurol. **92**, 423–428.
41. Kosak, M. C., and Reding, M. J. (2000). Comparison of partial body weight-supported treadmill gait training versus aggressive bracing assisted walking post stroke. Neurorehabil. Neural. Repair **14**(1), 13–19.

42. Hogan, N., Krebs, H. I., Fasoli, S., Rohrer, B., Stein, J., and Volpe, B. T. (2003). Technology for recovery after stroke. In: "Recovery After Stroke" (B. M. Bogousslavsky, I, and B. Dobkin, ed.), Chapter 30. Cambridge University Press, New York.
43. Lum, P. S., Burgar, C. G., and Shor, P. C. (2003). Evidence for strength imbalances as a significant contributor to abnormal synergies in hemiparetic subjects. *Muscle Nerve* **27**(2), 211–221.
44. Reinkensmeyer, D. J., Kahn, L. E., Averbuch, M., McKenna-Cole, A., Schmit, B. D., and Rymer, W. Z. (2000). Understanding and treating arm movement impairment after chronic brain injury: progress with the ARM guide. *J. Rehabil. Res. Dev.* **37**, 653–662.
45. Ferraro, M., Palazzolo, J. J., Krol, J., Krebs, H. I., Hogan, N., and Volpe, B. T. (2003). Robot-aided sensorimotor arm training improves outcome in patients with chronic stroke. *Neurology* **61**(11), 1604–1607.
46. Hallett, M. (2000). Transcranial magnetic stimulation and the human brain. *Nature* **406**(6792), 147–150.
47. Brown, J. A., Lutsep, H., Cramer, S. C., and Weinand, M. (2003). Motor cortex stimulation for enhancement of recovery after stroke: case report. *Neurol. Res.* **25**(8), 815–818.
48. Ziemann, U., Chen, R., Cohen, L. G., and Hallett, M. (1998). Dextromethorphan decreases the excitability of the human motor cortex. *Neurology* **51**(5), 1320–1324.
49. Butefisch, C. M., Davis, B. C., Wise, S. P., Sawaki, L., Kopylev, L., Classen, J., and Cohen, L. G. (2000). Mechanisms of use-dependent plasticity in the human motor cortex. *Proc. Natl. Acad. Sci. USA* **97**(7), 3661–3665.
50. He, S. Q., Dum, R. P., and Strick, P. L. (1995). Topographic organization of corticospinal projections from the frontal lobe: motor areas on the medial surface of the hemisphere. *J. Neurosci.* **15**(5 Pt 1), 3284–3306.
51. Middleton, F. A., and Strick, P. L. (2000). Basal ganglia and cerebellar loops: motor and cognitive circuits. *Brain. Res. Brain Res. Rev.* **31**(2-3), 236–250.
52. Fries, W., Danek, A., Scheidtmann, K., and Hamburger, C. (1993). Motor recovery following capsular stroke. Role of descending pathways from multiple motor areas. *Brain* **116**(Pt 2), 369–382.
53. Calautti, C., and Baron, J. C. (2003). Functional neuroimaging studies of motor recovery after stroke in adults: a review. *Stroke* **34**(6), 1553–1566.
54. Roland, P. E., Larsen, B., Lassen, N. A., and Skinhoj, E. (1980). Supplementary motor area and other cortical areas in organization of voluntary movements in man. *J. Neurophysiol.* **43**(1), 118–136.
55. Logothetis, N. K., Pauls, J., Augath, M., Trinath, T., and Oeltermann, A. (2001). Neurophysiological investigation of the basis of the fMRI signal. *Nature* **412**(6843), 150–157.
56. Mukamel, R., Gelbard, H., Arieli, A., Hasson, U., Fried, I., and Malach, R. (2005). Coupling between neuronal firing, field potentials, and FMRI in human auditory cortex. *Science* **309**(5736), 951–954.
57. Evarts, E. V. (1968). Relation of pyramidal tract activity to force exerted during voluntary movement. *J. Neurophysiol.* **31**(1), 14–27.
58. Merchant, H., Battaglia-Mayer, A., and Georgopoulos, A. P. (2004). Neural responses in motor cortex and area 7a to real and apparent motion. *Exp. Brain Res.* **154**(3), 291–307.
59. Lu, X., and Ashe, J. (2005). Anticipatory activity in primary motor cortex codes memorized movement sequences. *Neuron* **45**(6), 967–973.
60. Sanes, J. N., Donoghue, J. P., Thangaraj, V., Edelman, R. R., and Warach, S. (1995). Shared neural substrates controlling hand movements in human motor cortex. *Science* **268**(5218), 1775–1777.
61. Sanes, J. N., and Donoghue, J. P. (2000). Plasticity and primary motor cortex. *Annu. Rev. Neurosci.* **23**, 393–415.
62. Poggio, T., and Girosi, F. (1998). A sparse representation for function approximation. *Neural. Comput.* **10**(6), 1445–1454.
63. Wolpert, D. M., and Ghahramani, Z. (2000). Computational principles of movement neuroscience. *Nat. Neurosci.* **3**(**Suppl**), 1212–1217.
64. Karni, A., Meyer, G., Rey-Hipolito, C., Jezzard, P., Adams, M. M., Turner, R., and Ungerleider, L. G. (1998). The acquisition of skilled motor performance: fast and slow experience-driven changes in primary motor cortex. *Proc. Natl. Acad. Sci. USA* **95**(3), 861–868.
65. Korman, M., Raz, N., Flash, T., and Karni, A. (2003). Multiple shifts in the representation of a motor sequence during the acquisition of skilled performance. *Proc. Natl. Acad. Sci. USA* **100**(21), 12492–12497.
66. Weiller, C., Ramsay, S. C., Wise, R. J., Friston, K. J., and Frackowiak, R. S. (1993). Individual patterns of functional reorganization in the human cerebral cortex after capsular infarction. *Ann. Neurol.* **33**(2), 181–189.
67. Weiller, C., Chollet, F., Friston, K. J., Wise, R. J., and Frackowiak, R. S. (1992). Functional reorganization of the brain in recovery from striatocapsular infarction in man. *Ann. Neurol.* **31**(5), 463–472.
68. Cramer, S. C., Nelles, G., Benson, R. R., Kaplan, J. D., Parker, R. A., Kwong, K. K., Kennedy, D. N., Finklestein, S. P., and Rosen, B. R. (1997). A functional MRI study of subjects recovered from hemiparetic stroke. *Stroke* **28**(12), 2518–2527.
69. Cramer, S. C., Finklestein, S. P., Schaechter, J. D., Bush, G., and Rosen, B. R. (1999). Activation of distinct motor cortex regions during ipsilateral and contralateral finger movements. *J. Neurophysiol.* **81**(1), 383–387.
70. Marshall, R. S., Perera, G. M., Lazar, R. M., Krakauer, J. W., Constantine, R. C., and DeLaPaz, R. L. (2000). Evolution of cortical activation during recovery from corticospinal tract infarction. *Stroke* **31**(3), 656–661.
71. Ward, N. S., Brown, M. M., Thompson, A. J., and Frackowiak, R. S. (2003). Neural correlates of outcome after stroke: a cross-sectional fMRI study. *Brain* **126**(**Pt 6**), 1430–1448.
72. Ward, N. S., Brown, M. M., Thompson, A. J., and Frackowiak, R. S. (2003). Neural correlates of motor recovery after stroke: a longitudinal fMRI study. *Brain* **126**(**Pt 11**), 2476–2496.
73. Ward, N. S., Brown, M. M., Thompson, A. J., and Frackowiak, R. S. (2004). The influence of time after stroke on brain activations during a motor task. *Ann. Neurol.* **55**(6), 829–834.
74. Carey, J. R., Kimberley, T. J., Lewis, S. M., Auerbach, E. J., Dorsey, L., Rundquist, P., and Ugurbil, K. (2002). Analysis of fMRI and finger tracking training in subjects with chronic stroke. *Brain* **125**(**Pt 4**), 773–788.
75. Nelles, G., Jentzen, W., Jueptner, M., Muller, S., and Diener, H. C. (2001). Arm training induced brain plasticity in stroke studied with serial positron emission tomography. *Neuroimage* **13**(**6 Pt 1**), 1146–1154.
76. Sadato, N., Pascual-Leone, A., Grafman, J., Ibanez, V., Deiber, M. P., Dold, G., and Hallett, M. (1996). Activation of the primary visual cortex by Braille reading in blind subjects. *Nature* **380**(6574), 526–528.
77. Fridman, E. A., Hanakawa, T., Chung, M., Hummel, F., Leiguarda, R. C., and Cohen, L. G. (2004). Reorganization of the human ipsilesional premotor cortex after stroke. *Brain* **127**(**Pt 4**), 747–758.
78. Chen, R., Cohen, L. G., and Hallett, M. (2002). Nervous system reorganization following injury. *Neuroscience* **111**(4), 761–773.
79. Jacobs, K. M., and Donoghue, J. P. (1991). Reshaping the cortical motor map by unmasking latent intracortical connections. *Science* **251**(4996), 944–947.
80. Ziemann, U. (2004). TMS induced plasticity in human cortex. *Rev. Neurosci.* **15**(4), 253–266.
81. Kandel, E. R. (2001). The molecular biology of memory storage: a dialog between genes and synapses. *Biosci. Rep.* **21**(5), 565–611.
82. Tully, T., Bourtchouladze, R., Scott, R., and Tallman, J. (2003). Targeting the CREB pathway for memory enhancers. *Nat. Rev. Drug Discov.* **2**(4), 267–277.

24

Nonatherosclerotic Cerebral Vasculopathies

Rima M. Dafer, MD, MPH
Jose Biller, MD

Keywords: *arteriopathy, dissection, stroke, transient ischemic attack, vasculitis*

I. Introduction
II. Noninflammatory Arteriopathies
III. Inflammatory Arteriopathies
IV. Conclusion
References

I. Introduction

Nonatherothrombotic vasculopathies are uncommon but well-recognized causes of stroke mainly in children and young adults. These conditions are more commonly recognized with the advent of neuroimaging and genetic testing. Underlying etiologies can be divided into inflammatory and noninflammatory conditions, the former are commonly encountered in the setting of infections, autoimmune disorders, and connective tissue diseases.

II. Noninflammatory Arteriopathies

Fibromuscular dysplasia (FMD) is a nonatherosclerotic segmental noninflammatory angiopathy of unknown etiology. The disease usually affects the medium and small-sized vessels of every arterial bed, predominantly the renal and extracranial segment of the internal carotid arteries (ICAs). Intracranial extension of the disease is usually rare. Cerebrovascular involvement in FMD occurs in 20–30% of patients, with the majority remaining asymptomatic. Patients may present with headaches, dizziness, pulsatile tinnitus, Horner's syndrome, vertigo, cranial-nerve palsies, transient ischemic attacks (TIAs), or focal neurological symptoms [1,2]. Cerebral infarctions occur due to stenosis or dissection of the major arteries or artery-to-artery embolism from a critically stenosed major cerebral artery (Fig. 1). Subarachnoid hemorrhage (SAH) is not an uncommon presentation due to rupture of an associated intracranial aneurysm. Renal FMD is not uncommon, especially among hypertensive patients. Though generally benign, the natural history of cerebrovascular FMD is

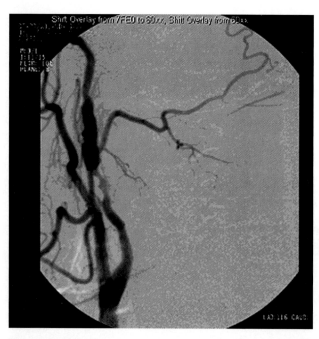

Figure 1 Carotid dissection in FMD. Notice string of beads in the mid-segment of the ICA.

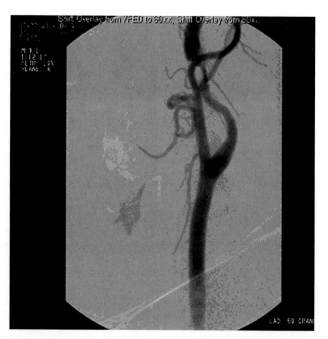

Figure 2 Dissection of the left ICA, 2–3 cm distal to its origin.

unknown. Angiographic studies reveal multifocal sausage-like mural dilatation (string of beads) predominantly in the middle portion of the ICA, 2–3 cm distal to its origin (type I), or irregular patterns of stenosis and aneurysms (types II and III). Differential diagnoses include atherosclerosis, vasospasm, and giant cell arteritides [3]. Symptomatic patients may be treated with antithrombotic therapy or anticoagulation. Percutaneous intervention for symptomatic stenosis with balloon angioplasty and stenting is emerging as a promising treatment options in FMD of the ICA.

Cervicocephalic arterial dissection is a well-recognized cause of ischemic stroke in young and middle-aged individuals. Dissection usually involves the extracranial segments of the carotid and vertebral arteries (Fig. 2). Intracranial dissections are rare and more commonly seen in the second decade of life. Dissection usually arises from a subintimal tear of an extracranial vessel, with development of intramural hematoma within the layers of the tunica media. Subintimal extension of the hematoma subsequently leads to luminal narrowing, vessel stenosis, or occlusion. A false lumen or pseudoaneurysmal formation occurs when the dissection extends between the medial and adventitial layers [4]. The pathogenesis of dissections is unknown. Although spontaneous dissection has often been reported, patients may report a remote history of minor trauma. Mechanical factors that may contribute to dissection include abrupt neck rotation, chiropractic manipulation, head or neck trauma, or cervical spine hyperextension. Patients with underlying arteriopathy, including collagen-vascular disorders such as Marfan's syndrome or Ehlers-Danlos, fibrocystic dysplasia, FMD, homocystinuria, α_1-antitrypsin deficiency, cystic medial necrosis, infections such as syphilis, Takayasu's disease, and migraine are more predisposed to dissection. Bilateral involvement is seen in 5–10% of patients and usually raises the possibility of underlying arteriopathy such as FMD. Patients usually present with unilateral headache, neck pain, Horner's syndrome, pulsatile tinnitus, amaurosis fugax, TIA, or focal ischemic neurological deficit. Magnetic resonance angiography (MRA) or conventional catheter cerebral angiogram may reveal evidence of pseudoaneurysm, intimal flap, vessel stenosis, or occlusion [1,4]. The majority of cervicocephalic dissections heal spontaneously. There is indirect evidence that short-term anticoagulation therapy for 3–6 months is beneficial to prevent embolization from a stenosed segment, but the validity of such treatment has not been proven. Anticoagulation is contraindicated in patients with intracranial dissections. Surgical or endovascular treatment, including vessel ligation or extracranial-to-intracranial bypass, should be reserved for patients with recurrent symptoms despite adequate anticoagulation.

Moyamoya is a progressive obstructive and occlusive intracranial cerebral arteriopathy affecting the distal segments of the ICAs at the bifurcation, resulting in the formation of a fine network of collateral neovasculature. The condition is usually idiopathic, although it has been reported in patients with sickle cell anemia, FMD, and with radiation-induced and postinfectious vasculitides. Prominent features include intimal thickening of the cerebral arterial trunks and abundant angiogenesis for collateral blood supplies. Women are more commonly affected than men. The etiology of the disease remains unknown. Genetic predisposition has been found in Japanese families, with the gene located on chromosome 17q25. Studies have shown increased amounts of basic fibroblast growth factor in the tissues of patients with the disease [2,5]. Diagnosis is usually established by the presence of bilateral stenosis or occlusion of the distal segments of the intracranial ICAs, and the classic appearance of abnormal hazy tuft of collateral blood supplies or "puff of smoke" (Fig. 3). Medical therapies for moyamoya are unknown. Patients may benefit from antiplatelet therapy. Recurrent progressive focal cerebral ischemic symptoms may require surgical intervention, which is usually aimed at cerebral revascularization. This includes (1) superficial temporal artery to middle cerebral artery (MCA) bypass, (2) encephalomyosynangiosis, (3) omental pedicle transposition, and (4) encephaloduroarteriosynangiosis [2,5].

Migraine is a common disorder affecting more than 28 million Americans in the United States. Migraine is an uncommon independent risk factor for ischemic stroke. Stroke in migraineurs occurs secondary to other coexisting conditions such as patent foramen ovale, dissection, or conditions resembling migraine, such as cerebral autosomal-dominant arteriopathy with subcortical infarcts and leukoencephalopathy (CADASIL), mitochondrial encephalopathy with lactic acidosis and strokelike episodes (MELAS), or cerebral vasculitis. Pure migrainous infarctions occur during a typical attack of migraine with aura, when the aura symptoms last beyond 60 minutes, with evidence of ischemic lesion on neuroimaging, and in the absence of other causes of stroke [6]. Cigarette smoking and oral contraceptives carry a higher risk. Stroke has a predilection to the cerebellum and occipital lobes, and white-matter lesions have been commonly reported.

CADASIL is an important risk for stroke in young patients. It is an inherited condition usually affecting the media of small cerebral arteries, characterized by recurrent subcortical ischemic episodes, migraine attacks, dementia, and diffuse white-matter abnormalities on MRA. CADASIL mutations affect the epidermal growth factor-like repeats located in the extracellular domain of the Notch3 receptor on chromosome 19. Pathologic examination shows multiple small, deep cerebral infarcts; leukoencephalopathy; and nonatherosclerotic nonamyloid angiopathy involving the media of small cerebral arteries. To date, there is no clear treatment for the condition. Antiplatelets may be beneficial in reducing the symptoms of recurrent strokes [7].

Hereditary endotheliopathy with retinopathy nephropathy and stroke, hereditary vascular retinopathy, and *cerebroretinal vasculopathy* are autosomal-dominant occlusive microangiopathies characterized by progressive visual loss, migraine headache, focal neurological deficits, depression, seizures, and dementia. Renal insufficiency and Raynaud's phenomenon are prominent. Linkage analysis mapped these conditions to a common locus on chromosome 3p21 [7]. Magnetic resonance imaging (MRI) scans usually show diffuse, multiple, deep, white-matter infarcts, although contrast-enhancing brain mass lesions may be seen. Electron microscopy shows multilayered basement membrane of capillary and arteriolar endothelial cells in the brain and other tissues, including the kidney and skin.

Metabolic disorders such as Fabry's, homocystinuria, α_1-antitrypsin deficiency, and *Pseudoxanthoma elasticum* are rare causes of stroke in children and young adults [8]. *Fabry's disease* is an X-linked lysosomal disorder characterized by deposition of neutral glycosphingolipids in the vascular endothelium of multiple organs, including skin, eye, kidney, heart, brain, and peripheral nervous system. Patients may present with progressive visual loss, angiokeratomata, numbness and tingling in the extremities, unexplained cardiac manifestations, and recurrent focal neurological deficits. Cerebral infarctions result from deposition of sphingolipids in the distal small arteries or arterioles, leading to occlusive arteriopathies. Diagnosis is made by biochemical testing or muscle biopsy. *Homocystinuria* is an autosomal-recessive disorder of methionine metabolism, leading to an abnormal accumulation of homocysteine and its metabolites in blood and urine, disruption of methionine metabolism, deficiency in the cystathionine B-synthase enzyme, defective methylcobalamin synthesis, or abnormality in methylenetetrahydrofolate reductase. Accelerated vascular disease results from direct endothelial cell damage or smooth muscle cell proliferation. In patients with *P. elasticum,* deposition of calcium in the coronary, retinal, and cerebral arteries leads to progressive blindness, cardiac rhythm abnormalities, and occlusive cerebrovascular disease. Low plasma levels of α_1-*antitrypsin*

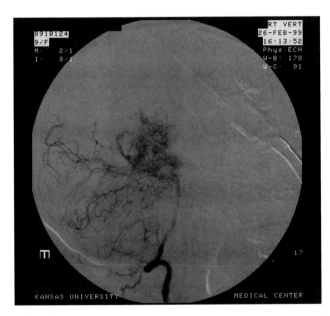

Figure 3 Left vertebral injection shows dilated medial striate arteries and collateral vessels (puff of smoke) in a patient with moyamoya. (Courtesy Dr. Donald Eckard.)

deficiency are a risk factor for carotid artery dissection and, thus should be suspected in young patients with unexplained cerebral infarction and arterial dissection [8].

Sneddon syndrome is a rare noninflammatory progressive disorder affecting the small and medium-caliber blood vessels, characterized by livedo reticularis, headache, dizziness, vertigo, elevated blood pressure, arterial or venous thrombosis, and elevated antiphospholipid antibodies. Central nervous system (CNS) manifestations include TIA, multiple small predominantly cortical infarcts, and fibrotic occlusion of cortical and leptomeningeal vessels. Other rare causes of noninflammatory nonatherothrombotic vasculopathies include carotid stenosis from direct trauma, radiation injury, and exposure to certain sympathomimetics and vasospastic drugs [9,10].

Other causes of hemorrhagic and ischemic stroke with residual vasculopathy and focal vasospasm include *radiation-induced carotid stenosis,* which is seen in 30–50% of patients following head and neck radiation, excessive exposure to *phenylpropanolamine* or other vasopressor amines in cough suppressants, CNS stimulants such as amphetamine and phenylpropanolamine (PCP), and *ergot alkaloids* commonly used to treat acute migraines headaches [1].

III. Inflammatory Arteriopathies

CNS vasculitides represent a heterogeneous group of inflammatory diseases that primarily affect the small leptomeningeal and parenchymal blood vessels of the brain, resulting in vessel inflammation and destruction. Primary idiopathic CNS vasculitis is a rare, but not uncommon, distinct entity that primarily manifests as CNS injury. Secondary CNS vasculitides are encountered in the setting of systemic infections or secondary to exposure to toxins or ionizing radiation [7–11] (Table 1). More commonly, they are neurological manifestations of disseminated multisystem autoimmune diseases such as systemic lupus erythematosus, polyarteritis nodosa, Sjögren's disease, rheumatoid arthritis, scleroderma, Wegener's granulomatosis, allergic vasculitides, or giant cell arteritis (Table 2). These may be classified by the size and type of vessel involvement or by underlying etiology.

CNS vasculitis should be suspected in children and young adults with otherwise absent cardioembolic, atherosclerotic, or thrombophilic causes of ischemic stroke or unexplained intracranial hemorrhage. The clinical features of vasculitis syndromes often overlap. Patients can present with nonspecific constitutional symptoms such as headache, fatigue, and malaise. Neuropsychiatric manifestations include seizures, cognitive impairment, meningitis, cranial neuropathies, peripheral neuropathies, intracranial hemorrhages, or recurrent cerebral infarctions. Systemic manifestations due to multiorgan involvements usually raise the suspicion of secondary vasculitides [9,10,12,13].

Table 1 Infections Vasculitides

Viral
  Herpes zoster virus
  Human immunodeficiency virus
  Coxsackie-9

Bacterial
  Cat-scratch disease
  Purulent meningitis

Spirochetal
  Rocky Mountain spotted fever
  Syphilis
  Lyme disease

Fungal
  Aspergillosis
  Mucormycosis
  Coccidiosis

Parasitic
  Amebiasis

The pathogenesis of CNS vasculitides remains unclear. The response to immunosuppressive agents suggests cell-mediated immune mechanisms, although immune complex deposition has also been implicated in the disease process. Evaluation in suspected inflammatory vasculitis can be approached in a stepwise fashion. Ancillary biochemical evaluation, with complete blood cell count, serum chemistry, erythrocyte sedimentation rate (ESR), Venereal Diseases Research Laboratory (VDRL) analyses, C-reactive protein, and urinalysis are important in determining organ involvement and excluding the presence of other diseases [9,12]. Specific laboratory testing is directed toward the diagnosis of systemic vasculitides. This includes antinuclear antibodies (ANAs) and, when positive, anti-double-stranded DNA, rheumatoid factor (RA), serum complement levels (C3, C4, and CH50), anti-neutrophil cytoplasmic antibodies (pANCA and cANCA), cryoglobulins, anti-SSA and anti-SSB antibodies, human immunodeficiency virus (HIV), and hepatitis serology. Chest radiography and pulmonary function tests may reveal asymptomatic pulmonary involvement. Urinalysis may show proteinuria, and direct immunofluorescence microscopy is helpful in selected patients. Nerve conduction studies may reveal evidence of underlying peripheral neuropathy. Electroencephalographic findings are nonspecific, with focal or diffuse slowing or rarely epileptiform activity [12,14]. Despite its limited specificity, catheter cerebral angiography remains a vital diagnostic modality for CNS vasculitis, predominantly for medium and large vessel vasculitides. In the setting of a suspicious clinical picture, a skin or sural nerve biopsy is usually diagnostic in systemic vasculitis. The diagnosis of primary or isolated CNS angiitis remains limited by the low specificity of cortico-leptomeningeal biopsy. Nevertheless,

Table 2 Noninfectious Inflammatory Vasculitides[a]

Necrotizing Vasculitides	Giant Cell Arteritides
Wegener granulomatosus	Temporal arteritis
Polyarteritis nodosa	Takayasu's disease
Churg-Strauss allergic angiitis	**Hypersensitivity Vasculitides**
Lymphomatoid granulomatosis	Henoch–Schönlein purpura
Vasculitis Associated with Collagen-Vascular Diseases	Drug-induced vasculitides
Systemic lupus erythematosus	Essential mixed cryoglobulinemia
Scleroderma	Chemical vasculitides
Sjögren's disease	**Miscellaneous**
Rheumatoid arthritis	Vasculitis associated with neoplasia
Vasculitis Associated with Other Systemic Illnesses	Vasculitis associated with radiation
Inflammatory bowel diseases	Cogan's syndrome
Behçet's disease	Dermatomyositis polymyositis
Degos—Köhlmeier disease	Buerger's disease
Relapsing polychondritis	X-linked lymphoproliferative syndrome
Sarcoidosis	Kawasaki syndrome
Susac syndrome	
	Primary CNS angiitis

[a] Adapted, with permission, from Biller, J., and Grau, R. G.

a histological diagnosis should be attempted before initiation of aggressive treatment with immunosuppressant therapies. A leptomeningeal wedge biopsy of the nondominant temporal lobe or of an actively enhancing lesion proves of higher yield [9,10,12,14,15].

A. Infectious Inflammatory Vasculopathies

CNS vasculitis is a rare complication of viral infections, including varicella zoster virus (VZV), HIV, enteroviruses, mumps, and coxsackie-9 virus. Many bacterial and spirochetal infections including syphilis, cat-scratch *Bartonellosis, Mycoplasma pneumoniae* infections, lyme disease, and Rocky Mountain spotted fever are rare causes of CNS vasculitis. Cerebral vasculitis is an uncommon, serious complication of VZV infections [9,10]. VZV is a neurotropic herpesvirus that can spread to the CNS via the hematogenous route, through direct extension, and invasion of the cerebral vessels. Varicella vasculopathy is more common among children who contract chickenpox infection and in immunocompromised adult patients. The anterior circulation, predominantly the MCA, is the most commonly involved. The diagnosis is made by angiographic findings of multiple focal arterial stenosis, and the cerebrospinal fluid (CSF) findings of oligoclonal varicella immunoglobulin G (IgG) antibodies. Early diagnosis and rapid initiation of antiviral therapy is associated with favorable outcome. The mechanism of stroke in HIV-seropositive patients is multifactorial, including coagulopathies, hyperviscosity, hypertension, and mycotic (infective) aneurysms. Cerebral vasculitis occurs secondary to underlying opportunistic infections, including syphilis, tuberculosis, aspergillosis, cryptococcosis, VZV, and toxoplasmosis. It can also be the result of irradiation for CNS lymphoma. Many bacterial and spirochetal infections such as syphilis, lyme disease, and Rocky Mountain spotted fever caused by the intracellular pathogen *Rickettsiae rickettsii* are rare causes of CNS vasculitis. Syphilitic infection is associated with aortitis, with predisposition to arterial dissection [7–11,14]. Invasive aspergillosis and mucormycosis are relatively common fungal infections in allogeneic stem cell transplant recipients and in immunocompromised patients. Although the lungs are most commonly affected, CNS involvement is not uncommon and is seen in 1.2–3.0% of patients. Patients present with ischemic strokes, with hemorrhagic transformation due to invasion of the leptomeningeal vessels by the fungus (Fig. 4). Protozoal infestations with toxoplasmosis, cysticercosis, and amebiasis have been associated with infective CNS vasculitis, meningitis, endocarditis, and SAH [9].

B. Noninfectious Inflammatory Vasculopathies

1. Necrotizing Autoimmune Vasculitides

Wegener's granulomatosis (WG) is a small vessel inflammatory necrotizing granulomatosis, primarily involving the upper and lower respiratory tracts and kidneys. It is distinguished from other vasculitides by the pattern of organ involvement and by the histological features of granulomatosis and necrotizing inflammation. Patients usually present with nasal septal perforation, glomerulonephritis, and pulmonary symptoms. Cranial and peripheral neuropathies are the most common neurological manifestations of the disease, occurring in about

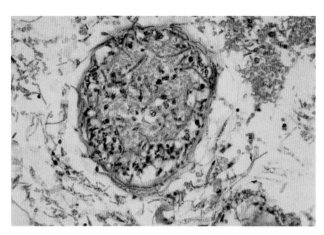

Figure 4 *Aspergillus vasculitis.* Histological evaluation of a leptomeningeal vessel shows inflammatory damage and infiltration of the vessel wall by hyphae.

35% of cases. Stroke is an uncommon complication of WG resulting from focal and segmental infiltration and necrotizing changes of small to medium-sized vessels, leading to cerebral infarction, SAH, and intraparenchymal hemorrhage. The diagnosis is usually made by the classic presentations of multiorgan involvement and the presence of C-antineutrophil cytoplasmic antibody (ANCA) levels. If untreated, the disease carries a poor prognosis [9–11].

Polyarteritis nodosa (PAN) is a systemic necrotizing vasculitis of small and medium-sized arteries. PAN affects multiple areas of the body, including the renal, musculoskeletal, nervous, gastrointestinal (GI), integument, cardiac, and genitourinary systems. Typical presentation includes fever, malaise, fatigue, skin rash, and arthralgias. Neurological complications include peripheral neuropathy, mononeuritis multiplex, encephalopathy, myelopathy, seizure, and transient monocular blindness. Cerebral vasculitis is rare, occurring late in the course of the disease, with intraparenchymal hemorrhage or SAH [9,11,14].

Churg-Strauss syndrome (eosinophilic granulomatosis) is a rare multiple-organ disorder predominantly affecting the lungs. It is characterized by necrotizing vasculitis of small and medium-sized blood vessels, inflammatory nodular granulomas, and abnormal clustering of eosinophilic cells [11]. The disorder is more common among women. Patients usually present with asthmatic attacks, fever, and weight loss. Cranial nerve involvement and optic neuropathy may occur, whereas stroke is rare.

Lymphomatoid granulomatosis is a rare systemic angiocentric and angiodestructive T cell–rich B cell–lymphoproliferative disorder, which is caused by neoplastic transformation of B lymphocytes by Epstein-Barr virus (EBV) infection. The disease primarily affects the lungs, with variable involvement of skin, renal, GI, and, rarely, the lymphatic system. CNS involvement is seen in one third of patients, and usually carries a poor prognosis [9]. Patients usually present with systemic constitutional symptoms such as fever, weight loss, malaise, cough, dyspnea, and hemoptysis. Erythematous macular rash typically involves the gluteal regions and extremities. Central and peripheral nervous system involvement has been reported, due to infiltration of the leptomeninges with B-cell. Isolated neurological manifestations have also been reported. Laboratory findings are non-specific. Diagnosis is confirmed by histological findings of vascular polymorphic lymphocytic infiltrate, with destruction and occlusion of blood vessels, followed by central necrosis. Prognosis is poor, with a 5 year survival rate of 40%.

Other rare causes of CNS vasculitides include microscopic polyangiitis and overlap syndrome.

2. Autoimmune Associated with Collagen Vascular Disorders

Systemic lupus erythematosus (SLE) is a complex autoimmune disease characterized by the production of ANAs and deposition of immune complexes in multiple organs, causing inflammatory injury to tissues and end-organ damage. The underlying etiology of the disease remains unclear. Neuropsychiatric manifestations have been reported in 20–25% of patients [9,10,13]. Virtually all levels of the nervous system involvement have been reported, including aseptic meningitis, optic neuropathy, strokes, seizures, memory impairment, myelopathy, peripheral nerves, and mononeuropathy. Stroke occurs secondary to various mechanisms, including premature atherosclerosis, hypercoagulability, presence of circulating antiphospholipid antibodies, cardiac embolization, Libman-Sacks endocarditis, and arterial hypertension. CNS vasculitis is less common, primarily due to deposition of circulating immune complex with residual vasculitis and small vessel thrombosis.

Scleroderma or systemic sclerosis is a systemic autoimmune disorder of unknown etiology characterized by progressive severe aberrations of extracellular matrix deposition, cutaneous and multiorgan visceral fibrosis, alterations in microvasculature, and cellular and humoral immunological abnormalities. The disease is associated with high morbidity and mortality, secondary to extensive organ fibrosis and microvascular alteration. Neurological complications of systemic sclerosis are relatively rare [9,14,16]. A variety of neurological manifestations have been reported, most commonly seizures. Other neurological symptoms and signs such as cranial nerve abnormalities, peripheral neuropathy, autonomic peripheral neuropathy, and psychological disorders are not uncommon. TIA, transient global amnesia, and focal neurological defects result from other organ involvement. CNS vasculitis in scleroderma remains a rare occurrence.

Sjögren's syndrome is a chronic multisystem autoimmune disorder characterized by dryness of the eyes, mouth, and other mucous membranes. The disease is characterized by lymphocytic infiltration of exocrine glands with acinar atrophy and fibrosis [9,10,14,16]. Other clinical manifestations include arthralgias, Raynaud's syndrome, and pulmonary and renal involvement. The most common neurological symptom in Sjögren's syndrome is peripheral neuropathy. Presence of white-matter changes due to small vessel vasculitis is increasingly encountered.

Rheumatoid arthritis is a chronic inflammatory condition characterized by articular and extraarticular manifestations. Neurological manifestations are rare and may include meningitis, subcortical infarctions, and diffuse white-matter changes.

3. Autoimmune Leukocytoclastic Vasculitis or Hypersensitivity Vasculitis

Henoch-Schönlein purpura is a nonthrombocytopenic small vessel systemic vasculitis, predominantly seen in children, with a male-to-female ratio of 2:1. The disease represents a variety of leukocytoclastic hypersensitivity conditions, characterized by immunoglobulin A, C3, and immune complex deposition in arterioles, capillaries, and venules. Patients typically present with a palpable purpuric rash over the lower extremities, arthritis, abdominal pain, and glomerulonephropathy [8,13,17]. The disease may occur in response to systemic infections with group A streptococci, *Mycoplasma*, EBV, HZV, parvovirus, typhoid, measles, cholera, yellow fever, and *Campylobacter enteritis*; after exposure to allergens in food; or insect bites. Neurological involvement is rare. Cerebral infarction and intracranial hemorrhage may occur due to arteritis.

Mixed cryoglobulinemia is another form of hypersensitivity vasculitis, characterized by immune complex deposition in small vessels. Clinical manifestations associated with type I cryoglobulinemia include acrocyanosis, retinal hemorrhage, severe Raynaud's phenomenon, pseudotumor cerebri, peripheral neuropathy, and arterial thrombosis. Joint involvement, nephropathy, and vascular purpuric lesions are more common with cryoglobulinemia types II and III. CNS involvement is rare, and cerebral infarctions have been reported [9,10].

Various chemical and illicit drugs have also been associated with hypersensitivity CNS vasculitis and intracranial hemorrhages, including silica, Talwin, methamphetamine, cocaine, and PCP [1,9].

4. Giant Cell Arteritides

Giant cell arteritis is a chronic systemic vasculitis of the medium and large vessels. The disease predominantly affects women in their sixties. Clinical presentation includes fever, nonspecific headache, weight loss, and malaise. The usual alarming findings are anterior ischemic optic neuropathy and focal neurological deficit. Tenderness over the temporal artery may be elicited on examination. The diagnosis is usually made in the setting of typical clinical presentation, elevated ESR, and histological findings on temporal artery biopsy including intimal thickening, scar tissue, infiltration by multinucleated giant cell, and smooth muscle disruption in the layers of the media [9,10,12,18].

Takayasu's arteritis is a rare inflammatory disease of the large arteries, predominantly the aorta and its branches. The disease is predominantly seen in Asian young women, between the ages of 10 and 30 years. Symptoms include fever, weight loss, fatigue, absent or weak peripheral pulses, dizziness, and focal neurological deficit. Blood workup reveals anemia and elevated ESR. Angiography usually reveals irregular dilatation of the aortic arch and its major branches, with focal narrowing at the origin of the major arteries [11,16,19].

5. Vasculitis Associated with Systemic Illnesses

Ulcerative and *Crohn's disease* are chronic inflammatory diseases of the GI tract of unknown etiology. CNS manifestations of inflammatory bowel diseases include thromboembolic stroke, cerebral venous thrombosis, and cerebral vasculitis. Hypercoagulable state may occur, resulting in deep venous thromboses, pulmonary embolism, and arterial thromboses [16].

Sarcoidosis is a multisystem inflammatory disease process of unknown etiology, characterized by the presence of granulomas. The disease usually involves the lungs and musculoskeletal system. Neurological involvement occurs in 5–10% of cases, predominantly mononeuropathy, peripheral neuropathy, meningitis, seizures, and headaches. Cerebral infarctions and TIAs may rarely be the presenting manifestations of the disease. Serum angiotensin-converting enzyme concentration may be elevated in blood and CSF. The pathogenesis of sarcoid neuropathology is not completely clear. Diagnosis of neurosarcoidosis is usually difficult; tissue biopsy finding of noncaseous granulomas associated with vasculitis is diagnostic [9,10,12].

Susac syndrome or retinocochleocerebral vasculopathy is an extremely rare neurological disorder of young women consisting of the triad of vision loss, hearing loss, and focal neurological deficits. It is an autoimmune disorder characterized by small vessel arteriopathy, predominantly involving the brain, retina, and cochlea. Small vessel ischemic strokes are rare [7].

Malignant atrophic papulosis or Degos-Köhlmeier disease is a rare systemic occlusive vasculopathy of the medium and small-size vessels of unknown etiology. About 35% of patients present with papular rash with peripheral telangiectatic rim, which gradually heal leaving a scar with pathognomonic atrophic centers. Symptoms include abdominal pain, diarrhea, weight loss, and visceral perforation. Intracranial hemorrhage occurs secondary to intracranial vessel rupture [8].

Behçet's disease is a multisystem inflammatory disease of unknown etiology, commonly involving the mucocutaneous system, ocular system, joint, and blood vessels. Other systems may also be involved. The pathogenesis of the disease remains obscure. Lymphocytic vasculitis of the small vessels has been observed, although vessels of all sizes may be affected. It is characterized by recurrent oral or genital ulcers, synovitis, uveitis, and vasculitic rash in the absence of collagen-vascular disorders. A broad spectrum of neurological involvement has been reported in 5–30% of cases, including severe nonspecific headache, meningoencephalitis, and pseudotumor cerebri. Cerebral vein thrombosis, myelopathy, pseudobulbar palsy, cognitive impairment, and rarely stroke secondary to cerebral vasculitis have been reported [14,16].

Relapsing polychondritis is a rare condition characterized by recurrent inflammation of cartilaginous structures and specific sense organs. CNS involvement is rare; meningoencephalitis and cerebral infarctions due to arterial vasculitis have been reported [9,10].

6. Primary Isolated CNS Angiitis

Primary isolated CNS angiitis is a leptomeningeal and cortical vasculitis involving the small and medium leptomeningeal and cortical arteries. Pathological findings include classic granulomatous angiitis with Langhans' giant cells, lymphocytic, and necrotizing vasculitis. The etiology of the disease remains unknown. A variety of pathogens have been documented in association with CNS arteritis, including VZV, HIV, and cytomegalovirus [12,15,17,20]. The most common presenting features include nonspecific headaches and mental status changes. Other neurological symptoms include seizures, TIA, visual changes, chronic meningitis, and recurrent unexplained strokes. The lack of constitutional symptoms and systemic involvement differentiates this condition from secondary vasculitides. Laboratory tests and CSF analysis are nondiagnostic. Neuroimaging studies may reveal multiple bilateral cortical and subcortical infarctions with or without contrast enhancement. Angiographic yield is low, with a specificity of less than 25%, but when present, the findings of vessel stenosis or occlusion in the setting of suspected clinical picture is vital for the diagnosis. Despite the limited sensitivity for brain biopsy, histological confirmation is necessary before initiation of therapy, with a sampling of the leptomeninges and underlying cortex of an enhancing lesion, or an open wedge biopsy of the tip of the nondominant temporal lobe. If untreated, the disease carries a poor prognosis [12,15].

C. Miscellaneous

Buerger's disease or thromboangiitis obliterans is an uncommon cell-mediated autoimmune peripheral vascular disorder of the small and medium-sized arteries and veins. The disease, which predominantly affects smokers, is characterized by inflammation of the extremities, intermittent claudication, and Raynaud's disease. Stroke in Buerger's disease is rare and may result from prolonged vessel vasoconstriction [1].

Kawasaki's syndrome is a multisystem vasculitis that almost exclusively affects young children [8]. Symptoms and signs include benign febrile exanthema, prolonged fever, and cardiac manifestations, including coronary aneurysms, myocardial infarction, and sudden cardiac death. Intracranial hemorrhages from ruptured cerebral aneurysms and cerebral infarctions from small vessel vasculitis are rare neurological manifestations.

Cogan's syndrome is a rare syndrome of nonluetic interstitial keratitis and vestibulo-auditory characterized by abrupt onset of vertigo, tinnitus, and rapid development of bilateral deafness. Neurological symptoms, including headache, seizure, cognitive impairment, and subcortical infarction, occur in 50% of cases. The pathogenesis of the disease is unknown [9,14]. Pathological findings include vasculitis and fibrosis of medium and large-sized vessels.

IV. Conclusion

Identification, diagnosis, and management of CNS vasculitis remain a challenging task for the treating physician. Proper diagnosis and appropriate therapy reduce the morbidity and mortality of these progressive diseases and improve outcome. Ancillary laboratory testing is mandatory to identify underlying organ involvement. Specific laboratory testing is helpful in differentiating among various secondary and systemic vasculitides. Despite its low specificity, conventional catheter angiography is vital in differentiating between the vasculitides based on the caliber of the vessels involved. Tissue diagnosis remains the gold standard. Management is targeted at suppressing the immune system. Aggressive treatment with corticosteroids and immunosuppressive therapies such as cyclophosphamide or azathioprine should be initiated early in the course of the disease to prevent disease progression and end-organ damage.

References

1. Biller, J., Mathews, D. K., and Love, B. B. (1999). Non-atherosclerotic vasculopathies. *In:* "Stroke in Children and Young Adults" (J. Biller, and D. K. Mathew, eds.), pp. 57–81. Butterworth–Heinemann, Woburn.
2. Biller, J. (1995). Strokes in the young. *In:* "Cerebrovascular Disorders" (J. F. Toole, ed.), 5th edition, pp. 283–316. Lippincott Williams & Wilkins, Philadelphia.

3. Slovut, D. P., and Olin, J. W. (2004). Fibromuscular dysplasia. *N. Engl. J. Med.* **350**, 1862–1871.
4. Beletsky, V., and Norris, J. W. (2001). Spontaneous dissection of the carotid and vertebral arteries. *N. Engl. J. Med.* **345**, 467.
5. Scott, R. M. (2000). Moyamoya syndrome: a surgically treatable cause of stroke in the pediatric patient. *Clin. Neurosurg.* **47**, 378–384.
6. Etminan, M., Takkouche, B., Isorna, F. C., and Samii, A. (2005). Risk of ischaemic stroke in people with migraine: systematic review and meta-analysis of observational studies. *BMJ* **330**, 63.
7. Bousser, M. G., and Biousse, V. (2004). Small vessel vasculopathies affecting the central nervous system. *J. Neuroophthalmol.* **24**, 56–61.
8. Benseler, S., and Schneider, R. (2004). Central nervous system vasculitis in children. *Curr. Opin. Rheumatol.* **16**, 43–50.
9. Carolei, A., and Sacco, S. (2003). Central nervous system vasculitis. *Neurol. Sci.* **24(Suppl 1)**, S8–S10.
10. West, S. G. (2003). Central nervous system vasculitis. *Curr. Rheumatol. Rep.* **5**, 116–127.
11. Frankel, S. K., Sullivan, E. J., and Brown, K. K. (2002). Vasculitis: Wegener granulomatosis, Churg–Strauss syndrome, microscopic polyangiitis, polyarteritis nodosa, and Takayasu arteritis. *Crit. Care Clin.* **18**, 855–879.
12. Calabrese, L. H. (2002). Diagnostic strategies in vasculitis affecting the central nervous system. *Cleve. Clin. J. Med.* **69(Suppl 2)**, SII105–SII108.
13. Lie, J. T. (1997). Classification and histopathologic spectrum of central nervous system vasculitis. *Neurol. Clin.* **15**, 805–819.
14. Borhani, Haghighi, A., Pourmand, R., and Nikseresht, A. R. (2005). Neuro Behcet disease. A review. *Neurologist* **11**, 80–89.
15. Calabrese, L. (2001). Primary angiitis of the central nervous system: the penumbra of vasculitis. *J. Rheumatol.* **28**, 465–466.
16. Moore, P. M., and Cupps, T. R. (1983). Neurological complications of vasculitis. *Ann. Neurol.* **14**, 155–167.
17. Calabrese, L. H., Duna, G. F., and Lie, J. T. (1997). Vasculitis in the central nervous system. *Arthritis Rheum.* **40**, 1189–1201.
18. Nordborg, E., and Nordborg, C. (2003). Giant cell arteritis: epidemiological clues to its pathogenesis and an update on its treatment. *Rheumatology (Oxford)* **42**, 413–421.
19. Kohrman, M. H., and Huttenlocher, P. R. (1986). Takayasu arteritis: a treatable cause of stroke in infancy. *Pediatr. Neurol.* **12**, 154–158.
20. Ceccarelli, A., De Blasi, R., Pavone, I., Lamberti, P., Carella, A., Livrea, P., and Simone, I. L. (2005). Primary angiitis of the central nervous system: a misinterpreted clinical onset of CNS vasculitis. *Eur. Neurol.* **53**, 40–42.

25

Subarachnoid Hemorrhage

David Palestrant, MD
E. Sander Connolly, Jr., MD

Keywords: *cerebral aneurysm, pathophysiology, subarachnoid hemorrhage*

I. Introduction
II. Etiology
III. Epidemiology and Genetics
IV. Aneurysm Pathogenesis
V. Pathophysiology of Aneurysm Rupture
VI. Pathophysiology of Delayed Cerebral Ischemia
VII. Cerebral Salt Wasting Syndrome
VIII. Conclusion
    References

I. Introduction

A unique pathophysiological cascade is triggered with the sudden appearance of blood in the subarachnoid space after cerebral aneurysm rupture, often leading to death or significant morbidity for the victim. The elements of this cascade are pathophysiologically particular to subarachnoid hemorrhage (SAH), occur commonly, and require careful clinical observation to treat as they emerge. These events include cerebral vasospasm, delayed cerebral ischemia (DCI), hydrocephalus, and the cerebral salt wasting syndrome. Confounding outcomes further, intraparenchymal hematomas and intraventricular hemorrhage (IVH) may also develop. The pathophysiological basis for these events are discussed in this chapter.

II. Etiology

SAH is due to rupture of a saccular aneurysm at the base of the brain in 80% of cases. Usually berry shaped or multilobulated, the aneurysms occur at arterial bifurcations around the circle of Willis. The majority (85–95%) form in the anterior portion of the circle of Willis along the internal carotid artery (ICA) or one of its branches; 40% occur at the posterior communicating artery (Pcom) bifurcation, 30% at the anterior communicating artery (Acom) bifurcation, and 20% at the middle cerebral artery (MCA) bifurcation [1]. Most of the remainder are found in the posterior circulation. Multiple intracranial aneurysms are a common finding in patients [2].

Other types of aneurysms include fusiform, mycotic, and dissecting aneurysms. Saccular aneurysms are distinguished from fusiform aneurysms and dilated atherosclerotic arteries by the presence of a neck and by not being in line with the artery. Mycotic aneurysms, with focal areas of vessel necrosis, are seen in bacterial and fungal infections. Dissecting aneurysms, with separation of blood vessel layers, are caused by trauma [1]. Causes other than aneurysm rupture for SAH include arterial dissection without aneurysm formation, arteriovenous malformation (AVM) rupture, dural AVM rupture, perimesencephalic SAH, head trauma, anticoagulation, and sickle cell disease [3].

III. Epidemiology and Genetics

A combination of genetic, congenital, and acquired factors is responsible for aneurysm formation. Race and gender appear to influence predisposition; women are 1.6 times more likely to suffer an SAH, blacks have 2.1 times the rates of whites, and SAH incidence is higher in Finland and Japan [3–5]. For women, risks are highest in the fifth decade, suggesting a hormonal influence [6]. Heavy alcohol use, hypertension, smoking, anticoagulation, and oral contraceptive use are all associated [3]. Congenital blood vessel abnormalities, which disrupt blood flow hemodynamics, such as a hypoplastic A1 segments and AVMs, also appear to predispose [7]. Heritable connective tissue disorders, including polycystic kidney disease, Ehlers-Danlos syndrome type IV, Marfan's syndrome, neurofibromatosis type I, and α_1-antitrypsin deficiency, carry a strong association [8,9].

First-degree relatives of patients with SAH have seven times the risk of developing an SAH and about 7–20% of patients with aneurysmal SAH have a first- or second-degree relative with a history of SAH or an unruptured aneurysm [5,10]. Although some small pedigrees of autosomal-dominant and autosomal-recessive modes of inheritance have been found, SAH appears to follow complex inheritance patterns and investigation has been complicated by likely polygenetic mechanisms and environmental factors [5,10]. Linkage studies have identified four possible genetic loci, three of which code for extracellular matrix proteins. The most promising of these genes code for elastin and collagen type 1A2 [5].

IV. Aneurysm Pathogenesis

Aneurysms histologically demonstrate abnormalities in multiple blood vessel layers. The endothelial layer is sparse, the internal elastic lamina is absent or fragmented, the muscularis layer of the media ends at the aneurysm neck, and atherosclerotic changes are found especially in the intimal pads at the aneurysm entrance [11]. Ruptured aneurysms demonstrate further degradation. The continuous endothelial lining, which is seen in unruptured aneurysms, is lost. The internal elastic lamina, responsible for most of the mechanical strength of the blood vessel wall, is absent. Collagen is decreased, disorganized, and poorly cross-linked when compared with both normal blood vessels, and unruptured aneurysms, the number of smooth muscle cells, and actin content is also decreased [11].

What causes intracranial aneurysm formation and rupture remains ill defined. Gaps in the tunica media and intimal "pads" have been assumed to be central. Gaps in the tunica media are located at arterial bifurcations in the circle of Willis where aneurysms form [12]. Prior theories postulated that it was through these defects that the aneurysm herniated and, therefore, were essential to aneurysm formation. Recent findings have questioned the importance of the media gaps. Evidence suggest that the earliest defects in aneurysm development are found distal to these gaps and only involve the gap region later in development [12]. Furthermore, tunica media gaps are strongly reinforced with collagen and are not pathological, but are normal intersections of two muscle layers being found equally in both persons with and persons without aneurysms [12].

Given their location, intimal pads have been thought to play an essential role in pathogenesis [12]. Consisting of a fragmented elastic layer often with atherosclerotic changes, they are found both distal and proximal to arterial bifurcations [7]. Hemodynamic forces acting on the intimal pads may be a necessary precipitant for aneurysm formation. Evidence for this comes from the findings that aneurysms develop over time, seldom appear before adulthood, and may appear during follow-up angiographic monitoring. Also, aneurysms are common in AVMs, where regional blood flow is increased, and in experimental models, where hemodynamic manipulation has been shown to induce aneurysm formation [13]. Theoretically, pressure exerted on the intimal pads by hemodynamic stressors is transmitted to the underlying gaps in the tunica media or areas of altered cell wall morphology with subsequent vessel outpouching and aneurysm formation [1,13]. Another possibility is that the intimal pads are not related to aneurysm formation but are compensatory for atrophy in the media found distally [7,12].

Abnormalities in the extracellular matrix are emerging as key factors in aneurysm pathogenesis. The extracellular matrix creates the stabilizing framework of the arterial wall. Secreted by fibroblasts, it is found in all layers of the artery and consists of the macromolecules collagen, elastin, glycoproteins, and proteoglycans [14]. Hereditable connective tissue disorders, which are strongly associated with aneurysm formation, often have abnormalities in genes coding for proteins involved in extracellular matrix formation and maintenance. The mutation for polycystic kidney disease is in the PDK1 or PDK2 genes, which code for the proteins polycystin-1 and polycystin-2, respectively [9,15]. The proteins are involved in extracellular matrix maintenance through protein–protein and protein–carbohydrate interactions [9,15]. Ehlers-Danlos type IV, a disease with a high rate of blood vessel rupture, dissection, and aneurysm formation, carries a mutation for the gene COL3A1 encoding type III procollagen, a structural component of distensible tissues like blood vessels [9,16]. Likewise, a mutation in the gene FBN1 encoding for fibrillin-1, a glycoprotein that is a component of microfibrils, an important constituent of the extracellular matrix, is found in Marfan's syndrome [9].

In patients without known connective tissue diseases, extracellular matrix abnormalities are also found. Structural proteins such as collagen type 3 and 4 and elastin are decreased in intracranial and extracranial arteries, ruptured aneurysms, and skin biopsies of patients with aneurysm. These deficiencies are explained by abnormalities in multiple pathways involved in degradation, formation, and regulation [5]. Collagen type III metabolism is reported to be abnormal in up to 50% of patients with intracranial aneurysm who do not have Ehlers-Danlos type IV, although mutations in COL3A1 gene do not appear to be responsible [9].

Whatever the factors are that predispose to aneurysm formation, aneurysm appears to form over a relatively short period and then either stabilize or go on to rupture. What causes an aneurysm to rupture is also unknown. Size of the aneurysm is certainly important, as is growth rate, location, shape, wall properties, and local flow dynamics [7,17].

V. Pathophysiology of Aneurysm Rupture

The rupture of the aneurysm with its high-pressure extravasation of blood leads to blood and blood clot formation, predominantly in the subarachnoid space, but can also lead to intracerebral hemorrhage (ICH), intraventricular hemorrhage (IVH), and occasionally subdural hemorrhage.

A. Intracerebral Hemorrhage

ICH is most commonly seen with rupture of the ACA and MCA aneurysm, as these lesions often lie in direct contact with the brain with little intervening subarachnoid space. The blood clot is found in the basal frontal lobe with the former and in the medial temporal lobe, lateral frontal lobe, or external capsule with the latter. Pcom aneurysm rupture may lead to temporal lobe hematomas. Prior small hemorrhages with subsequent development of adhesions, large aneurysm expanding into the brain parenchyma, and aneurysmal domes lying between cerebral surfaces are all potential mechanisms for ICH seen in SAH [7]. Large hematomas contribute to brain shift, herniation, and intracranial pressure (ICP) elevation.

B. IVH

IVH is associated with a worse outcome, with an increased incidence of sudden death at rupture and with the development of DCI and hydrocephalus [18–21]. Acom aneurysms are commonly the cause, with rupture through the lamina terminalis or extension of the frontal ICH. Aneurysms at other locations may directly rupture into the third ventricle (e.g., basilar apex aneurysm) or reflux blood into the ventricles (Pcom or choroidal aneurysm through the choroidal fissure into the temporal horn; posterior inferior cerebellar artery [PICA] aneurysm through the foramen of Luschka into the fourth ventricle) [7].

C. Hydrocephalus

Hydrocephalus is present in 15–30% of patients with SAH [22–24]. It is associated with a poorer grade, elevated ICP, and a worse outcome [22–24]. The cause is multifactorial and remains somewhat controversial. It appears to be largely obstructive with blockage of the foramina and arachnoid granulations creating pressure differentials between the CSF compartments, leading to the development of hydrocephalus [24]. Patient characteristics such as age, sex, and alcohol use have all been associated [25]. The presence of blood in the ventricle is the most consistent associated factor [19,22,23] Other associated factors include ICH, causing shift with ventricular obstruction and the quantity of blood in the subarachnoid space [19,22]. Procollagen propeptides were increased in the CSF after SAH, which may explain later leptomeningeal fibrosis and blockage of the arachnoid granulations [26].

VI. Pathophysiology of Delayed Cerebral Ischemia

A major cause of morbidity and mortality in patients with SAH is the development of DCI. Evidence of DCI is common, with 80% of patients with SAH demonstrating infarctions on magnetic resonance imaging (MRI) scans [27]. Symptomatic cerebral vasospasm (CVS) occurs in

33% of patients in the first 2 weeks after SAH [28]. The cause of DCI and CVS remains obscure, and although DCI is clearly related to CVS, it is not sufficiently explained by the presence of CVS alone, with other poorly understood factors likely contributing to the development of both.

A. Cerebral Vasospasm Pathophysiology

Although exact mechanisms leading to vasospasm are not completely understood, the presence of formed components of blood in the subarachnoid space is a necessary prerequisite. Clinical evidence has demonstrated a correlation between subarachnoid clot thickness and location on computed tomographic (CT) scan and the development of CVS in the adjacent arteries [29,30]. Despite the presence of blood in the subarachnoid space, CVS is seldom seen in SAH from causes other than aneurysmal rupture, such as AVM rupture, ICH with subarachnoid extension, or perimesencephalic hemorrhage [3]. In addition, in many patients with aneurysmal SAH, the spasm does not correlate with the distribution or the side of the subarachnoid blood, and other factors like the presence of ICH and IVH also have a strong association with its development [21,31]. Therefore, the presence of blood and its breakdown products, though clearly central to the formation of CVS, is insufficient to explain the entire picture.

Oxyhemoglobin, and especially the mixture of oxyhemoglobin and methemoglobin, are powerful precipitants of CVS [32], but how these cause CVS is less clear. Powerful free radicals are generated with the breakdown of oxyhemoglobin. The radicals have wide effect, peroxidizing membrane phospholipids and generating vasoactive compounds through lipid peroxidation such as the arachidonic acid metabolite 8-iso-PGF2α, a powerful vasoconstrictor [32,33]. The radicals also damage cerebral endothelium and smooth muscle cells, increase intracellular Ca, cause cell depolarization, and change the balance between prostaglandin I_2 and prostaglandin E_2 toward vasoconstriction [28,34]. Hemoglobin byproducts also scavenge nitric oxide (NO), a powerful vasodilator [32]. The cascade triggered by the presence of blood components disrupts multiple layers of the blood vessel including the endothelium and smooth muscle and precipitates inflammatory factors, all of which contribute to the development of CVS.

Equilibrium in normal cerebral arterial blood vessel tone is a balance between vasoconstrictor and dilator systems. Modulation of smooth muscle relaxation and dilatation is mediated by endothelial-derived relaxing factors (EDRFs) including NO and prostaglandin derivatives. Vasoconstriction is mediated by endothelial-derived contracting factors (EDCFs) such as peptide endothelin (ET) [34]. Disruption of any of these pathways is thought to result in the spasm seen after SAH.

Endothelial-mediated mechanisms appear to be central to the disruption seen in vasospasm. Relaxation of the blood vessel is impaired due to decreased levels of EDRFs, the most potent of these being NO. Normally, NO is released from endothelial cells, diffuses across to the adjacent smooth muscle cell, and activates guanylate cyclase, which through a cyclic guanosine monophosphate (cGMP)–mediated cascade eventually sequesters Ca in the sarcoplasmic reticulum, relaxing the smooth muscle [34]. Within minutes after SAH, NO levels drop precipitously secondary to the scavenging by oxyhemoglobin metabolites and destruction of NO, producing neurons by oxyhemoglobin [33,34]. Furthermore, hemoglobin directly inactivates guanylate cyclase, hindering the cGMP-mediated cascade [34]. The net result is an attenuation of the vasodilatory effect of NO.

On the opposite side of cerebral vessel tone regulation, endothelial damage increases EDCFs. The prime EDCF-mediated vasoconstriction candidate is the ET isoform ET-1, which is produced by endothelial cells, glial cells, leukocytes, and mononuclear cells [35] and may have both vasoconstrictive and direct ischemic effects. The molecule is found in CSF and plasma of patients with SAH and has a marked effect on cerebral arteries, causing a dose-dependent and long-lasting vasoconstriction [36]. Disputing the vasoconstrictive effect of ET-1, other studies have indicated that ET-1 may be more a marker of ischemia rather than a cause of vasospasm [37].

Disruption of the smooth muscle function with the induction of sustained contraction plays an important role in vasospasm. Potassium and Ca ion channels may be important. Potassium channels under normal conditions are usually involved in membrane hyperpolarization and smooth muscle relaxation. After SAH, vascular smooth muscle cells appear to depolarize possibly due to potassium channel dysfunction [34]. There also appears to be an influx of Ca through voltage-independent Ca channels, after intracellular Ca is depleted by oxyhemoglobin, leading to smooth muscle contraction [34]. Other factors, such as the loss of thin filament-associated proteins, responsible for inhibiting contractile proteins, may also exist [34]. Furthermore, contraction of smooth muscle may be stimulated by signaling molecules such as endothelin, oxyhemoglobin, epidermal growth factor (EGF), and platelet-derived growth factor (PDGF), which act via a G protein–coupled receptor and a receptor tyrosine kinase [32]. G protein receptors downstream activate Rho proteins and protein kinase C (PKC), both of which directly act on smooth muscle components contributing to contraction [32]. PKC may, in addition, feedback up-regulating ETB and 5-HT1B receptors, increasing the sensitivity of the smooth muscle cells to ET-1 and 5-HT—both triggers of vasoconstriction [32]. Receptor tyrosine kinase and PKC activate mitogen-activated protein kinase (MAPK) pathways downstream, which likewise triggers muscle contraction [32].

Inflammatory mechanisms have also been invoked. Leukocyte adhesion to the blood vessel endothelium and the factors contributing to it has been an active area of experimental investigation in SAH given it is an important first step in the initiation of inflammation [35]. Levels of adhesion molecules and immune-mediating cytokines are shown to be altered in SAH. The adhesion molecules E-selectin, soluble CAM-1, and soluble VCAM-1 are all elevated in the CSF of patients with SAH, and their decay over time suggests a correlation with vasospasm [35]. Moreover, experimental models show that blocking ICAM-1 with monoclonal antibodies reduces cerebral vasospasm [35]. Likewise, increased levels of the cytokines interleukin (IL)-1β, IL-6, and tumor necrosis factor-α (TNF-α) correlate closely in time and extent with elevated blood flow velocities seen on transcranial Doppler ultrasonography [35]. Some cytokines like IL-6 and platelet-activating factor are elevated in blood drawn from the jugular vein starting on day 4 and remain elevated until day 14, corresponding with the time course of vasospasm [35]. As mentioned, ET-1 is synthesized by activated mononuclear cells in the CSF of patients with SAH, indicating a possible link between inflammation, vasospasm, and cerebral ischemia in SAH [33].

B. Delayed Cerebral Ischemia Pathophysiology

The direct relationship between vasospasm and delayed ischemia, with narrowed cerebral blood vessels resulting in a decreased regional CPP with subsequent infarction, may not be as linear as originally thought. Physicians treating SAH know that directly treating vasospasm with endovascular techniques or "triple H" therapy often leads to symptomatic improvement. However, in patients who do develop vasospasm, many do not develop secondary ischemia, and in patients with only slight spasm, ischemia may develop [38]. Therefore factors other than vasospasm alone likely contribute to the DCI seen in patients with SAH.

There is mounting evidence that the combination of both vasospasm and disrupted cerebral autoregulation is necessary for ischemia to occur [32]. Cerebral blood flow (CBF) remains constant over a wide range of perfusion pressures due to autoregulation. In SAH, autoregulation may be disrupted and, therefore, perfusion dependent. The combination of narrowed blood vessels and disrupted autoregulation hinders perfusion, leading to ischemia [32]. This would explain why some patients with severely narrowed vessels do not develop ischemia (autoregulation intact), whereas others with only mild spasm do (disrupted autoregulation).

Factors other than vasospasm contributing to delayed ischemia are less well characterized. Glutamate, an excitatory amino acid, is a neurotoxin involved in ischemic damage. Human studies using intracerebral microdialysis demonstrate an increase in extracellular glutamate concentration in SAH [34]. Similar findings have been noted in an animal SAH model. An inflammatory mechanism as well as free radical production may also be important [34]. In experimental models, application of ET-1 or products of hemolysis with inhibition of NO in the subarachnoid space over the cortex induces cortical-spreading ischemic depressions [39].

VII. Cerebral Salt Wasting Syndrome

Hyponatremia is the most common electrolyte disorder in SAH [39]. Historically it was thought to be the syndrome of inappropriate antidiuretic hormone (SIADH). Later work demonstrated that the hyponatremia seen in SAH was a volume-depleted state in contrast to the volume-expanded state seen in SIADH, and the hyponatremia in SAH was associated with a significant natriuresis and a negative sodium balance. Based on this, the cerebral salt wasting (CSW) syndrome was defined [29,40]. The mechanism by which SAH provokes the natriuresis is not completely known. Hyponatremia is significantly associated with dilatation of the third ventricle on initial head CT. More recent work has focused on the role of brain natriuretic factors, particularly atrial natriuretic factor (ANP) and brain natriuretic factor (BNP), as both are able to precipitate a natriuretic response [39]. Both factors increase in the plasma in the days following SAH, and ANP reduces ADH levels and inhibits renin–aldosterone release [39]. On the other hand, in a study of 10 patients with SAH, BNP was more closely related to Na secretion than ANP. The source of ANP is the ventricle and BNP is made in the heart. How these factors are secreted is not known [40]. The loss of sympathetic innervation to the proximal nephron, where the majority of Na is reabsorbed, may be another potential mechanism. Autonomic and hormonal factors may be related, as ANP and BNP are capable of decreasing autonomic outflow through the brainstem.

VIII. Conclusion

The pathobiology of SAH, in particular that caused by cerebral aneurysm rupture, is complex and incompletely understood. Nevertheless, great strides have been made in understanding the processes responsible for the formation and rupture of brain aneurysms, as well as the key processes responsible for DCI and cerebral vasospasm. Although outcomes remain less than optimal, these insights have led to marked improvement in the care of patients and have set the stage for even greater improvements in the years ahead.

References

1. Zhang, B., Fugleholm, K., Day, L. B., Ye, S., Weller, R. O., and Day, I. N. (2003). Molecular pathogenesis of subarachnoid haemorrhage. *Int. J. Biochem. Cell Biol.* 2003. **35**(9), 1341–1360.
2. Kaminogo, M., Yonekura, M., and Shibata, S. (2003). Incidence and outcome of multiple intracranial aneurysms in a defined population. *Stroke* **34**(1), 16–21.
3. van Gijn, J., and Rinkel, G. J. (2001). Subarachnoid haemorrhage: diagnosis, causes and management. *Brain* **124(Pt 2)**, 249–278.
4. Broderick, J. P., Brott, T., Tomsick, T., Huster, G., and Miller, R. (1992). The risk of subarachnoid and intracerebral hemorrhages in blacks as compared with whites. *N. Engl. J. Med.* **326**(11), 733–736.
5. Ruigrok, Y. M., Rinkel, G. J., and Wijmenga, C. (2005). Genetics of intracranial aneurysms. *Lancet Neurol.* **4**(3), 179–189.
6. Kongable, G. L., Lanzino, G., Germanson, T. P., Truskowski, L. L., Alves, W. M., Torner, J. C., and Kassell, N. F. (1996). Gender-related differences in aneurysmal subarachnoid hemorrhage. *J. Neurosurg.* **84**(1), 43–48.
7. Burnett, M. G., Danish, S. F., McKann II, G. M., and Le Roux, P. D. (2004). Pathology and pathophysiology of aneurysmal subarachnoid hemorrhage. *In:* "Management of Cerebral Aneurysms" (P. D. Le Roux, H. R. Winn, and D. W. Newell, eds.), pp. 127–137. Elsevier, Inc., Philadelphia.
8. Schievink, W. I. (1998). Genetics and aneurysm formation. *Neurosurg. Clin. North Am.* **9**(3), 485–495.
9. Schievink, W. I. (2004). Genetic and acquired reasons for aneurysm formation. *In:* "Management of Cerebral Aneurysms" (P. D. Le Roux, H. R. Winn, and D. W., Newell, eds.). Elsevier, Inc., Philadelphia.
10. Schievink, W. I. (2004). "Management of Cerebral Aneurysms." (P. D. Le Roux, ed.), pp. 169–182. Elsevier, Inc., Philadelphia.
11. Chyatte, D. (2004). Pathology of intracranial aneurysms. *In:* "Management of Cerebral Aneurysms" (P. D. Le Roux, H. R. Winn, and D. W. Newell, eds.). Elsevier, Inc., Philadelphia.
12. Krex, D., Schackert, H. K., and Schackert, G. (2001). Genesis of cerebral aneurysms–an update. *Acta Neurochir. (Wien.)* **143**(5), 429–449.
13. Kalimo, H., Kaste, M., and Haltia, M. (2002). Vascular disease. *In:* "Greenfield's Neuropathology" (P. L. Lantos and D. I. Graham, eds.), 7th edition, pp. 281–356, Hodder Arnold Publishers, London, United Kingdom.
14. Gibbons, G. H., and Dzau, V. J. (1994). The emerging concept of vascular remodeling. *N. Engl. J. Med.* **330**(20), 1431–1438.
15. Harris, P. C. (1999). Autosomal dominant polycystic kidney disease: clues to pathogenesis. *Hum. Mol. Genet.* **8**(10), 1861–1866.
16. Pepin, M., Schwarze, U., Superti-Furga, A., and Byers, P. H. (2000). Clinical and genetic features of Ehlers–Danlos syndrome type IV, the vascular type. *N. Engl. J. Med.* **342**(10), 673–680.
17. (1998). Unruptured intracranial aneurysms—risk of rupture and risks of surgical intervention. International Study of Unruptured Intracranial Aneurysms Investigators. *N. Engl. J. Med.* **339**(24), 1725–1733.
18. Schievink, W. I., Wijdicks, E. F., Parisi, J. E., Piepgras, D. G., and Whisnant, J. P. (1995). Sudden death from aneurysmal subarachnoid hemorrhage. *Neurology* **45**(5), 871–874.
19. Jartti, P., Karttunen, A., Jartti, A., Ukkola, V., Sajanti, J., and Pyhtinen, J. (2004). Factors related to acute hydrocephalus after subarachnoid hemorrhage. *Acta Radiol.* **45**(3), 333–339.
20. Gruber, A., Reinprecht, A., Bavinzski, G., Czech, T., and Richling, B. (1999). Chronic shunt-dependent hydrocephalus after early surgical and early endovascular treatment of ruptured intracranial aneurysms. *Neurosurgery* **44**(3), 503–512.
21. Claassen, J., Bernardini, G. L, Kreiter, K., Bates, J., Du, Y. E., Copeland, D., Connolly, E. S., Mayer, S. A. (2001). Effect of cisternal and ventricular blood on risk of delayed cerebral ischemia after subarachnoid hemorrhage: the Fisher scale revisited. *Stroke* **32**(9), 2012–2020.
22. Demirgil, B. T., Tugcu, B., Postalci, L., Guclu, G., Dalgic, A., and Oral, Z. (2003). Factors leading to hydrocephalus after aneurysmal subarachnoid hemorrhage. *Minim. Invasive Neurosurg.* **46**(6), 344–348.
23. Dorai, Z., Hynan, L. S., Kopitnik, T. A., and Samson, D. (2003). Factors related to hydrocephalus after aneurysmal subarachnoid hemorrhage. *Neurosurgery* **52**(4), 763–771.
24. Suarez-Rivera, O. (1998). Acute hydrocephalus after subarachnoid hemorrhage. *Surg. Neurol.* **49**(5), 563–565.
25. Sheehan, J. P., Polin, R. S., Sheehan, J. M., Baskaya, M. K., and Kassell, N. F. (1999). Factors associated with hydrocephalus after aneurysmal subarachnoid hemorrhage. *Neurosurgery* **45**(5), 1120–1128.
26. Sajanti, J., Heikkinen, E., and Majamaa, K. (2000). Transient increase in procollagen propeptides in the CSF after subarachnoid hemorrhage. *Neurology* **55**(3), 359–363.
27. Kivisaari, R. P., Salonen, O., Servo, A., Autti, T., Hernesniemi, J., and Ohman, J. (2001). MR imaging after aneurysmal subarachnoid hemorrhage and surgery: a long-term follow-up study. *AJNR Am. J. Neuroradiol.* **22**(6), 1143–1148.
28. Harrod, C. G., Bendok, B. R., and Batjer, H. H. (2005). Prediction of cerebral vasospasm in patients presenting with aneurysmal subarachnoid hemorrhage: a review. *Neurosurgery* **56**(4), 633–654.
29. Fisher, C. M., Kistler, J. P., and Davis, J. M. (1980). Relation of cerebral vasospasm to subarachnoid hemorrhage visualized by computerized tomographic scanning. *Neurosurgery* **6**(1), 1–9.
30. Kistler, J. P., Crowell, R. M., Davis, K. R., Heros, R., Ojemann, R. G., Zervas, T., and Fisher, C. M. (1983). The relation of cerebral vasospasm to the extent and location of subarachnoid blood visualized by CT scan: a prospective study. *Neurology* **33**(4), 424–436.
31. Brouwers, P. J., Wijdicks, E. F., and Van Gijn, J. (1992). Infarction after aneurysm rupture does not depend on distribution or clearance rate of blood. *Stroke* **23**(3), 374–379.
32. Hansen-Schwartz, J. (2004). Cerebral vasospasm: a consideration of the various mechanisms involved in the pathophysiology. *Neurocrit. Care* **1**(2), 235–246.
33. Pluta, R. M. (2005). Delayed cerebral vasospasm and nitric oxide: review, new hypothesis, and proposed treatment. *Pharmacol. Ther.* **105**(1), 23–56.
34. Grasso, G. (2004). An overview of new pharmacological treatments for cerebrovascular dysfunction after experimental subarachnoid hemorrhage. *Brain Res. Brain Res. Rev.* **44**(1), 49–63.
35. Dumont, A. S., Dumont, R. J., Chow, M. M., Lin, C. L., Calisaneller, T., Ley, K. F., Kassell, N. F., and Lee, K. S. (2003). Cerebral vasospasm after subarachnoid hemorrhage: putative role of inflammation. *Neurosurgery* **53**(1), 123–135.
36. Zimmermann, M., and Seifert, V. (1998). Endothelin and subarachnoid hemorrhage: an overview. *Neurosurgery* **43**(4), 863–876.
37. Pluta, R. M., Boock, R. J., Afshar, J. K., Clouse, K., Bacic, M., Ehrenreich, H., and Oldfield, E. H. (1997). Source and cause of endothelin-1 release into cerebrospinal fluid after subarachnoid hemorrhage. *J. Neurosurg.* **87**(2), 287–293.
38. Clyde, B. L., Resnick, D. K., Yonas, H., Smith, H. A., and Kaufmann, A. M. (1996). The relationship of blood velocity as measured by transcranial Doppler ultrasonography to cerebral blood flow as determined by stable xenon computed tomographic studies after aneurysmal subarachnoid hemorrhage. *Neurosurgery* **38**(5), 896–905.
39. Wijdicks, E. F. (2003). Acid-base disorders and sodium handling. *In:* "The Clinical Practice of Critical Care Neurology" (E. F. Wijdicks, ed.), 2nd edition, pp. 501–516. Oxford University Press, Oxford.
40. Palmer, B. F. (2000). Hyponatraemia in a neurosurgical patient: syndrome of inappropriate antidiuretic hormone secretion versus cerebral salt wasting. *Nephrol. Dial. Transplant.* **15**(2), 262–268.

26

Cerebral Ischemia: Molecular Mechanisms and Protective Therapies

Kenneth R. Wagner, PhD

Keywords: *drug treatments, gene expression, intracellular signaling, neurovascular unit, penumbra*

I. Introduction
II. Definitions: Ischemic Core and Penumbra
III. Cellular and Biochemical Responses to Focal Cerebral Ischemia
IV. The Neurovascular Unit
V. Blood-Brain Barrier Damage in Cerebral Ischemia
VI. Postischemic Necrosis and Apoptosis
VII. Molecular Events Underlying Ischemic Cell Death
VIII. Gene Expression Following Cerebral Ischemia
IX. The Inflammatory Response to Cerebral Ischemia
X. Hyperglycemia and Hemorrhagic Infarct Conversion
XI. Acute Treatments for Cerebral Ischemia
XII. Global Cerebral Ischemia
References

I. Introduction

Brain injury following focal or global cerebral ischemia is a major cause of mortality and morbidity worldwide. Great strides have been made in delineating the various cellular and molecular events that occur following cerebral ischemia and lead to edema, selective neuronal necrosis, and tissue infarction. However, despite decades of extensive investigations by devoted researchers in both academia and the pharmaceutical industry, no neuroprotective drugs are available to treat acute stroke or cardiac arrest patients. Only thrombolysis with Food and Drug Administration (FDA)–approved tissue plasminogen activator (t-PA) administered within 3 hours of symptom onset has benefited acute stroke patients [1,2]. To develop more effective therapeutic targets, additional understanding of the complex processes underlying ischemia-induced brain tissue injury is required.

In this chapter, we present a broad overview of the pathophysiological, cellular, biochemical, and molecular events that occur in the brain in response to cerebral ischemia and reperfusion. Most of our understanding of those processes that are initiated following cerebral ischemia and that lead to tissue infarction have been derived from animal models. Although we mainly focus on focal rather than global cerebral ischemia, similarities exist in the injury processes. We have included a brief discussion of global ischemia and a recent reference at the end of the chapter. We are limited in the number of references that can be cited, so all citations are from recent review papers that contain the additional details and references to the original work. The reader who seeks further information is directed to these cited reviews.

II. Definitions: Ischemic Core and Penumbra

Focal cerebral ischemia or stroke is initiated by an embolus or thrombus that occludes a major cerebral artery. The degree of ischemia—that is, the reduction in local cerebral blood flow (CBF) and specifically its severity and duration—can precisely predict the infarct size (for more details and references, see references 3–5). Within the ischemic core, CBF may only be 5–20% of control, whereas in brain tissue regions surrounding the core, an intermediate reduction in blood flow of approximately 20–40% defines the ischemic penumbra [3]. Although the consequence of severe blood flow reduction to the core is rapid tissue damage, the penumbral area undergoes brain cell death that evolves more slowly. This area, though electrically silent, has sufficient oxygen delivery to be metabolically active and to sustain membrane potentials. Though functionally impaired, the nerve cells within the penumbra remain alive and are generally believed to be salvageable if CBF is restored within a reasonable time. However, if reperfusion fails, the penumbral tissue is incorporated into the developing infarct. Thus, the ultimate fate of penumbral neurons depends on two essential variables: CBF and ischemia duration. If CBF reductions in the core are below 25%, the risk for infraction is more than 95% in the absence of early reperfusion. In contrast, in brain regions where the CBF is more than 50% of normal, the risk falls to 5% [5]. The importance of the penumbra and the reason for its focus in neuroprotection studies is that the penumbral tissue can add up to 50% to the final infarct volume [5].

III. Cellular and Biochemical Responses to Focal Cerebral Ischemia

At the cellular and biochemical levels, the reduction in oxygen delivery to the tissue is rapidly sensed and initiates a series of well-defined biochemical events [3] (Fig. 1). Detailed reviews of these events have been published [4,5]. Within seconds following marked ischemia, the loss of oxygen and glucose delivery, coupled with the brain's very high metabolic rate and relatively low energy and glycogen stores, causes a rapid depletion of high-energy phosphates, accumulation of lactate, and a reduction of intracellular pH. The loss of high-energy phosphate compounds leads to an inability of cells to maintain their ionic gradients. Membrane depolarization then occurs with the resulting release of neurotransmitters including the excitotoxic amino acid glutamate. With energy depletion, glutamate uptake is also impaired. Glutamate binding to its postsynaptic receptors promotes excessive entry of extracellular calcium into cells, as well as intracellular calcium release. The consequences of unregulated elevated calcium concentrations include activation of degrading enzymes such as phospholipases, arachidonic acid release, free radical formation, and activation of poly-ADP ribose polymerase (PARP), proteases, and endonucleases (see Fig. 1). As outlined by Lo et al. [6], signaling through glutamate receptors is also coupled to multiple intracellular kinases that conduct the cellular stress response to alterations in gene expression. Indeed, a host of subcellular signaling pathways are activated by ischemia and are described in more detail later (Fig. 2).

Careful studies of the pathophysiological events in the penumbra have been conducted by several research groups (for details and references, see references 3–5). Recordings of direct-current (DC) potentials from the cortical penumbra demonstrate peri-infarct depolarizations that resemble normal spreading depression but actually originate from critically hypoperfused tissue. These episodic cellular depolarizations are associated with neuronal potassium efflux and sodium and calcium influx. Tissue deoxygenation, along with episodic adenosine triphosphate (ATP) depletion, occurs during these depolarizations. Metabolic energy is required to restore these ionic gradients. Unless reperfusion occurs, high-energy phosphates eventually become depleted, the penumbra becomes irreversibly depolarized, and the tissue is irreversibly damaged.

Early following ischemia, despite reduced CBF, the penumbral local glucose metabolic rate is not decreased, leading to a markedly elevated metabolism/blood flow ratio [3]. However, after recirculation, glucose metabolism is

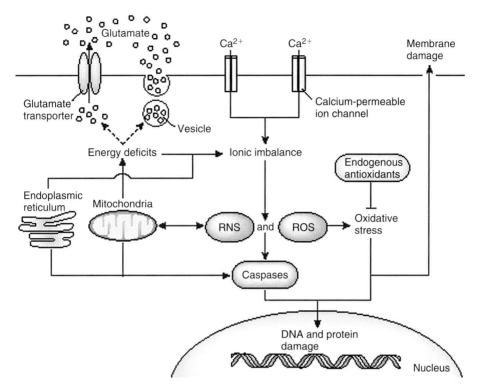

Figure 1 Figure depicts the major pathways implicated in ischemic cell death that interact and overlap. Cerebral ischemia leads to reductions in oxygen delivery and energy substrate concentrations, resulting in ionic imbalances, excitotoxic glutamate efflux and intracellular calcium accumulation, oxidative and nitrosative stresses, and apoptotic-like mechanisms. Additionally, energy deficits induce mitochondrial dysfunction and the formation of reactive oxygen species (ROS) and reactive nitrogen species (RNS). Downstream pathways include free radical damage to membrane lipids, cellular proteins, and DNA. Calcium-activated proteases and caspase cascades dismantle a wide range of homeostatic, reparative, and cytoskeletal proteins. (Reproduced, with permission, from Lo, E. H., Dalkara, T., and Moskowitz, M. A. [2003]. Mechanisms, challenges and opportunities in stroke. *Nat. Rev. Neurosci.* **4**, 399–415.)

markedly reduced in the previously ischemic hemisphere, thereby normalizing the metabolism/flow ratio. Ginsberg et al. [3] have nicely used multimodal image analysis and histopathology to show that infarction probability is determined by the degree of CBF reduction during ischemia. Furthermore, these findings from animal stroke models conform very well with positron emission tomography (PET) results in acute stroke patients [3].

IV. The Neurovascular Unit

The concept of the neurovascular unit (i.e., the endothelium, astrocyte, and neuron) is a recently instituted framework that focuses on the cell–cell and cell–matrix signaling that participates in the overall tissue response to cerebral ischemia and its treatments [4,6]. This more integrative approach to cell death processes in brain following ischemia considers the interaction between these cellular elements. These cellular elements interact during reperfusion when oxidative stress, vascular activation, leukocyte infiltration, and dysregulated extracellular proteolysis are potential triggers of hemorrhagic transformation [4,5,7]. In addition, these pathophysiological/pathochemical events contribute to the development of inflammation and apoptosis. In their paper, Lo et al. [6] interestingly describe the three major pathways that underlie brain cell death as "exciting (glutamate and excitotoxicity), radical (oxidative stress and free radicals), and suicidal (apoptotic-like pathways)." They conclude that because these pathways are inextricably linked, future studies should employ a "systems biology" approach to the neurovascular unit when examining the relationships between cell death and survival signaling.

V. Blood-Brain Barrier Damage in Cerebral Ischemia

The integrity of the blood-brain barrier (BBB) depends on the interaction between cerebral vascular endothelial cells, astrocytic end-feet, and the extracellular matrix.

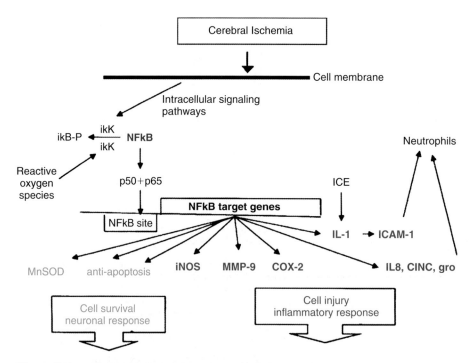

Figure 2 Figure depicts secondary molecular events following the induction of cerebral ischemia. Transduction of the pathophysiological events depicted in Fig. 1 by intracellular signaling pathways lead to activation of the transcription factor (nuclear factor-κB [NF-κB]), which plays a central role in postischemic regulation of both cell injury and cell survival responses. COX-2, cyclooxygenase-2; iNOS, inducible nitric oxide synthase; IL-8, interleukin-8; CINC, cytokine-induced neutrophil chemoattractant; Gro, chemokine CXCL1; ikK, i-κB kinase; ikB-P is the phosphorylated NF-κB inhibitory subunit; p50+p65 are NF-κB subunits, which act on target genes. (Used, with permission, from Sharp, F. R., Lu, A., Tang, Y., and Millhorn, D. E. [2000]. Multiple molecular penumbras after focal cerebral ischemia. *J. Cereb. Blood Flow Metab.* **20**(7), 1011–1032.)

Enhanced expression and activation of proteases, especially the matrix metalloproteases (MMPs), play an important role in ischemic injury [4,5,7]. MMP2 and MMP9 are induced after cerebral ischemia and their expression correlates with increases in BBB permeability (see Fig. 2). Damage to the BBB enables leukocyte transmigration, vasogenic edema formation, hemorrhagic infarct conversion, and neuronal injury. Genetic manipulations in MMP knockout animal models, as well as pharmacological targeting with MMP inhibitors, reduce BBB damage and infarct volumes [4,7].

VI. Postischemic Necrosis and Apoptosis

Neuronal cell death can occur through necrotic or apoptotic mechanisms, depending on the nature and severity of the insult [5,6,8]. It should be noted that neurons and oligodendrocytes appear to be more vulnerable to ischemic injury than astroglial or endothelial cells [4]. Furthermore, specific neuronal populations seem to be especially susceptible in global ischemic insults (e.g., CA1 hippocampal pyramidal neurons and the cerebellar Purkinje cells). These vulnerable neurons may have genotypes and phenotypes involving complex factors including intracellular signal transduction pathways that make them susceptible to ischemia [6].

In the ischemic core, excitotoxicity can rapidly lead to necrotic tissue injury. Excitation-induced ionic imbalances result in marked intracellular edema. Failure of intracellular volume regulation ultimately can cause osmotic lysis [5]. A programmed pathway leading to ischemic neuronal necrosis in the nematode, c-elegans, remains to be studied in the mammalian central nervous system (CNS) (cited in reference 6).

On the other hand, in the penumbra, both mild excitotoxic and inflammatory mechanisms underlie delayed cell death that appears to occur through apoptotic mechanisms [4,5,8,9]. These apoptotic cell death mechanisms include both caspase-dependent and caspase-independent processes. "Executioner" molecules (e.g., caspases) are activated and degrade cytoplasmic and nuclear processes, thereby promoting apoptotic cell death. The role of caspases in postischemic cell death is supported by findings that administration of caspase inhibitors or gene deletions reduces ischemic injury and neurological deficits. However, as described by Lo et al. [4], approaches to reduce caspase activities are not universally successful. This may be due in part to a caspase-independent apoptotic pathway that involves apoptosis-inducing factor (AIF), a key signaling molecule in this cascade [4].

Apoptosis has been described as "programmed cell death" with triggers that include excessive generation of free radicals, activation of death receptors, damage to DNA, and activations of lysosomal proteases [4,10]. Apoptosis is an active process that requires high-energy phosphates, gene expression, and caspase activity and occurs normally during CNS development, in neurodegenerative diseases, and especially in the ischemic penumbra [4,5,10]. Neuropathologically, apoptosis is seen as nuclear shrinkage, chromatin clumping, and cytoplasmic blebbing in dying cells. Biochemically, active endonuclease-mediated DNA cleavage leads to discrete bands on gels called *apoptotic ladders*. At the cellular level, double- and single-strand DNA breaks can be labeled with staining reagents [4]. It should be noted that some controversy exists that is related to the prevalence of apoptosis after cerebral ischemia, as well as to the details of some of the specific intracellular events (see citations in references 5, 10).

VII. Molecular Events Underlying Ischemic Cell Death

Considerable evidence implicates oxidative stress in secondary pathochemical events following cerebral ischemia [10,11] (see Figs. 1 and 2). Various molecules including proteins, lipids, and DNA are important targets for reactive oxygen species (ROS). Indeed, in the brain, levels of endogenous antioxidant enzymes (including superoxide dismutase [SOD], catalase, and glutathione) and antioxidant vitamins (e.g., α-tocopherol and ascorbic acid) appear to be insufficient to handle excess ROS generation, especially early during reperfusion [4]. It is noteworthy that more recent work indicates that direct attack by ROS on susceptible molecules is not their only mode of operation. Genetic manipulation in transgenic overexpressing mice and in knockout mice of antioxidant enzymes and molecules in intracellular signaling pathways have provided support and understanding for the role of oxygen radicals in cell death/survival signaling pathways and in ischemic brain injury [10,11].

Cellular death signaling pathways following cerebral ischemia have been described as intrinsic (mitochondria-dependent) and extrinsic (receptor-mediated) pathways of apoptosis [4,5,10]. In the mitochondrial-dependent pathway, ROS facilitate formation of the mitochondrial transition pore (MTP) that dissipates the proton motive force required for oxidative phosphorylation and ATP generation. The result of MTP formation is a marked increase in membrane permeability and the release of the mitochondrial apoptosis-related proteins (i.e., cytochrome *c* and AIF) [4,10]. The Bcl-2 protein family controls the release of these proteins.

Up-regulation or overexpression of Bcl-2 or Bcl-xL can suppress cell death through various mechanisms, including MTP stabilization, suppression of cytochrome *c* or AIF release, and the silencing of proapoptotic Bcl-2 family members [4,10].

The extrinsic death receptor pathway of apoptosis is initiated by the Fas receptor and the tumor necrosis factor (TNF) receptor [4,10]. In the Fas receptor pathway, the extracellular Fas ligand (FasL) binds to its receptor and an adaptor molecule, Fas-associated death domain (FADD) protein, which then activates procaspase-8. The cleaved product, caspase-8, then activates caspase-3, and this effector caspase cleaves PARP and activates caspase-activated DNAase (CAD), resulting in DNA damage and apoptotic cell death [10].

Support for this death receptor signaling pathway is based on findings from hybrid mice that are deficient in both Fas and tumor necrosis factor (TNF) expression and are strongly resistant to ischemic injury [4,10]. Significant protection from ischemia is also observed with treatment of wild-type mice with neutralizing antibodies against both FasL and TNF [4]. Additional details of these apoptotic pathways are described elsewhere [4,5,10].

VIII. Gene Expression Following Cerebral Ischemia

Cerebral ischemia is a very powerful stimulator of CNS gene expression [3,12,13]. Interestingly, although protein synthesis is markedly reduced in the postischemic brain, genomic studies demonstrate that the expression of a large number of genes is actively up-regulated. Many of the genes are up-regulated in the penumbra both during ischemia and on reperfusion and reoxygenation include stress response, heat shock, and apoptosis and inflammation related (see Fig. 2).

The up-regulation of gene expression in the brain after focal ischemia includes not only those that may be considered deleterious (i.e., proinflammatory cytokine genes), but also those involved in the survival response (cellular attempts at self-preservation) (see Fig. 2). *In situ* hybridization autoradiographs for gene expression have been correlated with autoradiographic CBF data and histological infarct maps and demonstrate that perfusion gradients determine differential patterns of gene expression in ischemia [3].

In other studies, DNA microarray technology has been used to screen for thousands of expressed genes with the demonstration of down-regulated genes in addition to the expected up-regulated ones (see citations in references 3, 12, 13). Immediate early genes (IEGs) such as c-fos can also be up-regulated in nonischemic regions of the ipsilateral and the contralateral hemisphere. Detailed findings from gene expression studies demonstrate the complexity of postischemic cellular events.

It is also noteworthy that up-regulated gene expression (and new protein synthesis) occurs in the setting of hypoxic or ischemic preconditioning [5,13]. Preconditioning is an endogenous process that follows an insult that is below threshold to induce damage and is neuroprotective against an otherwise damaging insult [5,13]. A better understanding of the neuroprotective pathways induced by preconditioning and continued examination of the details of ischemia-induced gene expression are expected to yield novel mechanisms and targets that could be exploited for future interventions.

IX. The Inflammatory Response to Cerebral Ischemia

The events of cerebral ischemia and reperfusion elicit a robust inflammatory response [5,9]. The elements of this response include activation of microglia and leukocyte infiltration into the brain from the systemic circulation. This process is dynamic with activation of the nuclear transcription factor-κB (NF-κB) (see Fig. 2) in microglia, leading to cytokine gene expression, which in turn propagates proinflammatory signaling [9]. Interleukin (IL)-1α and IL-1β up-regulation appears to play a central role in the injury response and is the current focus of potential new therapies [14] (see Fig. 2). IL-1β–converting enzyme (ICE; caspase-1) is involved in the activation of cytokines and links inflammation with apoptosis [5] (see Fig. 2).

During postischemic recirculation, cerebral vascular endothelial cells express intercellular adhesion molecule-1 (ICAM-1), which enables peripheral circulating leukocytes to migrate into the brain parenchyma (see Fig. 2). These cells appear to amplify inflammatory signaling cascades. It should be noted, however, that literature evidence regarding the inflammatory response following cerebral ischemia suggests it has apparent benefits and deleterious effects. This area of research is actively trying to define the various relationships including temporal and topographical, as well as the specific mRNA and protein levels of the mediators [9,14].

In addition to cytokines, recent work is uncovering the pathophysiological role of chemokines and their receptors in ischemic brain injury [5,9] (see Fig. 2). These structurally related small cytokines, which were originally identified as factors regulating leukocyte migration in inflammatory and immune responses, are also produced by the brain. Monocyte chemoattractant protein-1 (MCP-1) and macrophage inflammatory protein-1α (MIP-1α), members of the CC chemokines, are induced after focal cerebral ischemia. The ability of chemokine receptor antagonists to reduce infarct volumes underscores their important role in brain tissue injury after stroke [5,9].

X. Hyperglycemia and Hemorrhagic Infarct Conversion

Extensive studies in experimental animals including anoxia/ischemia and middle cerebral artery occlusion studies in cats from our laboratory, as well as findings from various other laboratories and cerebral ischemia models, have demonstrated that hyperglycemia in the setting of cerebral ischemia is detrimental (for review, see reference 15). Clinical studies have supported the finding in animal studies of the strong correlation between hyperglycemia and hemorrhagic infarct conversion. Ongoing clinical trials are testing the effectiveness of actively reducing elevated serum glucose concentrations to normoglycemic levels with insulin treatment (www.clinicaltrials.gov; The Internet Stroke Center at www.strokecenter.org).

XI. Acute Treatments for Cerebral Ischemia

Despite considerable progress in defining the cellular and molecular events that are initiated by cerebral ischemia and reperfusion, and despite the outstanding efficacy of many previously tested therapies in the preclinical testing, no neuroprotective drugs have been successful in clinical trials for stroke patients. Gladstone et al. [16] have carefully reviewed the reasons for the failures. An important concept that they raise is the "disconnect" between the traditional assessments of neuroprotective drugs in animal models versus their testing in the clinical setting. For example, in preclinical studies, they state that efficacy has relied on "infarct volume (instead of functional outcome) measures, short-term instead of long-term endpoints, transient instead of permanent ischemia models, short (instead of extended) time windows for drug administration, and protection of cerebral gray matter (instead of both gray and white matter)" [16]. Similarly, clinical trials have had their shortcomings with "long time windows prior to drug administration, insufficient statistical power, insensitive outcome measures, inclusion of protocol violators, failure to target specific stroke subtypes, and failure to target the ischemic penumbra" [16]. In the second part of their review, these authors consider new paradigms for stroke treatment including trials that address multiagent therapy and neuroplasticity and repair.

A. Tissue Plasminogen Activator

Of the more than 700,000 new strokes that occur in the United States each year, 80% are ischemic. Without treatment, about one-third of patients will remain moderately or severely disabled at 3 months. The FDA-approved therapy for clot thrombolysis in ischemic stroke, recombinant t-PA (rt-PA), when administered intravenously within the first 3 hours, significantly improves patient outcome (for review, see reference 1). The efficacy of rt-PA administration has even been shown to have an important temporal relationship to outcome within the FDA-approved 3-hour treatment window. Stratifying treated patients to 0–90 and 91–180 minutes after stroke onset showed odds ratios of 2.8 and 1.5 for improved patient outcome, respectively [2]. The efficacy of rt-PA for acute stroke treatment has supported the concept combinations of a neuroprotective therapy with rt-PA that enables recanalization and reperfusion of the ischemic territory are an important direction for future acute stroke therapy. If the protective agent could be administered early, even pre-hospital after stroke, it could reduce the risk of thrombolysis-induced intracerebral hemorrhage and may even extend the treatment window beyond 3 hours (for review, see reference 17).

B. Antioxidants/Nitrone Spin Trap Agents

As described earlier, considerable evidence implicates oxidative stress in the secondary pathochemical events that occur during reperfusion following cerebral ischemia. In a previous report, we described that the nitrone spin trap agents have been effective neuroprotectives in various stroke models [17]. These agents reduced mortality, infarct size, edema, and neurological deficits in global and focal ischemia models and t-PA–induced hemorrhage as well. The recently developed nitrone free radical trapping agent, disodium 2,4-disulfophenyl-N-tert-butylnitrone (NXY-059, CEROVIVE), has also been effective in preclinical trials. Renovis, Inc. (www.renovis.com), announced in May 2005 its first analysis of data from the Phase III SAINT I (Stroke-Acute Ischemic-NXY-059-Treatment) trial conducted by its licensee AstraZeneca. This study, which has enrolled about 1700 patients, demonstrated a statistically significant improvement versus placebo ($p = .038$) in the primary outcome measure of disability, the Modified Rankin Scale (MRS). However, no significant reduction was present in neurological deficit as measured by the National Institutes of Health Stroke Scale (NIHSS). Half of the trial's 1700 patients were randomized to receive CEROVIVE (NXY-059) and the incidence and profile of adverse events were similar to placebo. The final clinical significance of these findings is being assessed in the ongoing SAINT II trial.

A newer, novel, second-generation and potentially more effective nitrone, stilbazulenyl nitrone (STAZN), has been developed and is being studied in cerebral ischemia by Dr. Myron Ginsberg and colleagues at the University of Miami (see citations in reference 17). STAZN is about 300 times more potent *in vitro* in inhibiting free radical–mediated peroxidation than first-generation nitrones. In a rat focal ischemia model of MCA suture occlusion, Ginsberg and colleagues found STAZN treatment beginning at a clinically significant time of 2 hours of reperfusion reduced mean cortical infarct volumes by 64–97% and total infarct volumes by 42–72% at 72 hours. In more than one-half of STAZN-treated animals, cortical infarction was virtually abolished.

C. Targeting the Neurovascular Unit

As described by Lo et al. [6], future approaches to stroke treatment should also consider the neurovascular unit in which neuroprotective therapy includes the glial and endothelial cells in the integrative brain response of signaling between cells and between cells and the intracellular matrix. Thus, an ideal therapeutic target would be one that has efficacy in multiple brain cell types. In this regard, the stress-activated protein kinase p38 and the arachidonic acid cascade enzyme 12/15-lipoxygenase, as discussed by Lo et al. [6], could be important targets.

The statins are a class of lipid-lowering drugs that act as 3-hydroxy-3-methylglutaryl (HMG)-coenzyme A (CoA) reductase inhibitors and are promising for stroke prophylaxis, although their role in acute stroke treatment is unclear. As outlined by Lo et al. [4], statins have pleiotropic effects including up-regulating endothelial nitric oxide synthase, suppression of prothrombotic activity, and anti-inflammatory properties. A meta-analysis of several statin clinical trials determined that statins decreased stroke risk and reduced mortality among treated stroke patients (see citations in reference 4). Other potential newer therapies and hypothermia are described in the following subsections.

1. Albumin

A strategy that is proving extremely promising for ischemic stroke is human albumin therapy [3]. In animal focal ischemia models, albumin infusions reduce infarct volume by 60% and markedly decrease the extent of brain swelling, with a therapeutic window extending to 4 hours. A Phase III trial of albumin therapy in human stroke patients has been awarded by the NINDS in June 2005.

2. Erythropoietin

Erythropoietin (EPO) has shown promise to be both safe and possibly efficacious in a limited clinical trial [6]. An advantage of EPO is that it targets multiple pathways of cross-talk signaling in cell death. Furthermore, EPO appears to enhance neurogenesis and angiogenesis during stroke recovery. An EPO derivative with neuroprotective but without erythropoietic properties has been developed.

3. Activated Protein C

As reviewed by Lo et al. [6], co-administering rt-PA with APC may be a new approach to treating acute ischemic stroke. Data suggest that APC may enhance thrombolysis, reduce glutamate and oxidative stress-induced apoptosis, and down-regulate endothelial cell adhesion molecules, thereby reducing the inflammatory response and providing BBB protection. Because of its pleiotropic effects, APC may be a suitable candidate that addresses the goal of protection aimed at the neurovascular unit.

4. Hypothermia

Hypothermia is a neuroprotective approach that antagonizes multiple injury mechanisms in all cells. In a review of hypothermia in stroke treatment and the description of a new approach and experimental studies with local hypothermia, we have cited many current references in both clinical and animal model studies [18]. The efficacy of hypothermia in reducing ischemia-induced brain injury is very well described. However, the challenges for hypothermia-induced neuroprotection in the clinical setting include the best route of delivery (surface cooling or endovascular), the difficulties in patient management, and the potential secondary adverse events [18]. In this previous report, we also reviewed findings from combined studies of drug therapies with hypothermia including work by Aronowski and Grotta and colleagues on the robust effects of caffeinol and other agents, including free radical scavengers, magnesium, and growth factors [18]. Overall, the preclinical findings of hypothermia alone and hypothermia plus adjunctive drugs have been quite remarkable.

5. Targeting Neuroprotective Rather than Cell Death Pathways

As described by Chan et al. [10], an alternative and novel therapeutic approach for neuroprotection in stroke would be to target survival signaling pathways rather than cell death pathways. One potential strategy proposed by these workers is to target a neuronal survival signaling pathway involving the phosphatidylinositol 3-kinase (PI3-K)/Akt (protein kinase B) pathway. PI3-K/Akt and downstream phosphorylated Bad and proline-rich Akt substrate survival signaling cascades are up-regulated in surviving neurons in the ischemic brain that overexpresses copper-zinc SOD activity.

6. Stem Cell Therapies/Neurogenesis for Stroke Treatment

Experimental evidence demonstrating that stem cells can generate neurons for transplantation and studies in animal models showing that neurogenesis is stimulated following cerebral ischemia raise hopes for new and exciting treatments for stroke patients [19]. Twelve stroke patients have received implants of neurons generated from the human teratocarcinoma cell line NTera-2 (NT-2) into infarct areas and some have shown improvements that correlate with increased metabolic activity at the graft site. In one patient, populations of these grafted cells were demonstrated to express a neuronal marker 2 years after surgery (see citations in reference 19). Lindvall and Kokaia [19] present "a road map to the clinic" in their review and describe the various scientific accomplishments that are necessary for stem cell–based approaches to be applied to stroke patients.

XII. Global Cerebral Ischemia

Although this review has mainly focused on focal cerebral ischemia, certain similarities exist in the pathophysiological, cellular, and molecular events following both focal and global cerebral ischemic insults [20]. Global cerebral ischemia can occur in patients who suffer cardiac arrest, anesthesia accidents, airway obstruction, drug intoxication, or hemorrhagic shock. Treatment strategies aimed at excitotoxicity, free radical generation, neuroinflammation, and apoptosis have met with limited success. Hypothermia, which targets multiple molecular pathways, significantly protects the brain in animal models, as well as when instituted in out-of-hospital cardiac arrest patients (see citations in references 18, 20). As described in their review, Zhang et al. have found that hyperbaric oxygen (HBO), which increases oxygen delivery to the ischemic brain, also enables pan-brain protection therapy, preserves the brain functional activity, and attenuates "secondary brain injury" in animals including after global ischemia [20]. Additional ongoing work with HBO both experimentally and clinically is expected to answer various outstanding questions with HBO therapy over the next few years.

References

1. del Zoppo, G. J. (2004). Thrombolysis: from the experimental findings to the clinical practice. *Cerebrovasc. Dis.* **17(suppl 1)**, 144–152.
2. Hacke, W., Donnan, G., Fieschi, C., Kaste, M., von Kummer, R., Broderick, J. P., Brott, T., Frankel, M., Grotta, J. C., Haley, E. C. Jr, Kwiatkowski, T., Levine, S. R., Lewandowski, C., Lu, M., Lyden, P., Marler, J. R., Patel, S., Tilley, B. C., Albers, G., Bluhmki, E., Wilhelm, M., Hamilton, S.; ATLANTIS Trials Investigators; ECASS Trials Investigators; and NINDS rt-PA Study Group Investigators. (2004). Association of outcome with early stroke treatment: pooled analysis of ATLANTIS, ECASS, and NINDS rt-PA stroke trials. *Lancet* **363**, 768–774.
3. Ginsberg, M. D. (2003). Adventures in the pathophysiology of brain ischemia: penumbra, gene expression, neuroprotection: the 2002 Thomas Willis Lecture. *Stroke* **34**, 214–223.
4. Lo, E. H., Dalkara, T., and Moskowitz, M. A. (2003). Mechanisms, challenges and opportunities in stroke. *Nat. Rev. Neurosci.* **4**, 399–415.
5. Mergenthaler, P., Dirnagl, U., and Meisel, A. (2004). Pathophysiology of stroke: lessons from animal models. *Metab. Brain Dis.* **19**, 151–167.
6. Lo, E. H., Moskowitz, M. A., and Jacobs, T. P. (2005). Exciting, radical, suicidal: how brain cells die after stroke. *Stroke* **36**, 189–192.
7. Rosenberg, G. A. (2002). Matrix metalloproteinases in neuroinflammation. *Glia* **39**, 279–291.

8. Zhang, F., Yin, W., and Chen, J. (2004). Apoptosis in cerebral ischemia: executional and regulatory signaling mechanisms. *Neurol. Res.* **26**, 835–845.
9. Zheng, Z., and Yenari, M. A. (2004). Post-ischemic inflammation: molecular mechanisms and therapeutic implications. *Neurol. Res.* **26**, 884–892.
10. Sugawara, T., Fujimura, M., Noshita, N., Kim, G. W., Saito, A., Hayashi, T., Narasimhan, P., Maier, C. M., and Chan, P. H. (2004). Neuronal death/survival signaling pathways in cerebral ischemia. *NeuroRx* **1**, 17–25.
11. Chan, P. H. (2001). Reactive oxygen radicals in signaling and damage in the ischemic brain. *J. Cereb. Blood Flow Metab.* **21**, 2–14.
12. Weinstein, P. R., Hong, S., and Sharp, F. R. (2004). Molecular identification of the ischemic penumbra. *Stroke* **35**, 2666–2670.
13. Sharp, F. R., Ran, R., Lu, A., Tang, Y., Strauss, K. I., Glass, T., Ardizzone, T., and Bernaudin, M. (2004). Hypoxic preconditioning protects against ischemic brain injury. *NeuroRx* **1**, 26–35.
14. Gibson, R. M., Rothwell, N. J., and Le Feuvre, R. A. (2004). CNS injury: the role of the cytokine IL-1. *Vet. J.* **168**, 230–237.
15. Kagansky, N., Levy, S., and Knobler, H. (2001). The role of hyperglycemia in acute stroke, *Arch. Neurol.* **58**, 1209–1212.
16. Gladstone, D. J., Black, S. E., and Hakim, A. M. (2002). Toward wisdom from failure: lessons from neuroprotective stroke trials and new therapeutic directions. *Stroke* **33**, 2123–2136.
17. Wagner, K. R., and Jauch, E. C. (2004). Extending the window for acute stroke treatment: thrombolytics plus CNS protective therapies. *Exp. Neurol.* **188**, 195–199.
18. Wagner, K. R., and Zuccarello, M. (2005). Local brain hypothermia for neuroprotection in stroke treatment and aneurysm repair. *Neurol. Res.* **27**, 238–245.
19. Lindvall, O., and Kokaia, Z. (2004). Recovery and rehabilitation in stroke: stem cells. *Stroke* **35**, 2691–2694.
20. Zhang, J. H., Lo, T., Mychaskiw, G., and Colohan, A. (2005). Mechanisms of hyperbaric oxygen and neuroprotection in stroke. *Pathophysiology* **12**, 63–77.

27

Intracerebral Hemorrhage and Intraventricular Hemorrhage–Induced Brain Injury

Richard F. Keep, PhD
Guohua Xi, MD
Julian T. Hoff, MD

Keywords: *brain edema, complement, hydrocephalus, inflammation, intracerebral hemorrhage, intraventricular hemorrhage, iron, stroke, thrombin*

I. Introduction
II. Etiology
III. Pathophysiology
IV. Mechanisms of Intracerebral Hemorrhage–Induced Brain Injury
V. Mechanisms of Intraventricular Hemorrhage–Induced Brain Injury
VI. Therapeutic Interventions for Intracerebral Hemorrhage–Induced Brain Injury
VII. Therapeutic Interventions for Intraventricular Hemorrhage–Induced Brain Injury
VIII. Summary
References

I. Introduction

Hemorrhagic strokes account for about 15% of all stroke deaths in Western countries and a higher percentage in Asian countries. The effect of such hemorrhagic strokes is disproportionate, with a greater mortality rate in hemorrhagic compared with ischemic strokes. Even in those patients who survive a hemorrhagic stroke, many will have long-term disabilities [1,2].

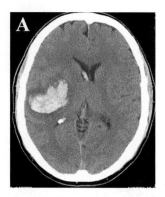 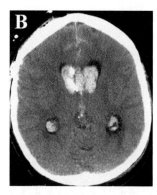

Figure 1 (**A**) Computed tomographic (CT) scan, noncontrast, axial view of an intracerebral hemorrhage. Note the small intraventricular hemorrhage in this example. (**B**) CT scan, noncontrast, axial view of an intraventricular hemorrhage.

This chapter focuses on two forms of hemorrhagic stroke: intracerebral hemorrhage (ICH) (Fig. 1A) and intraventricular hemorrhage (IVH) (Fig. 1B). Subarachnoid hemorrhage is dealt with in a separate chapter. There are approximately 40,000 cases of ICH in the United States annually [3]. Although there are much fewer cases of pure IVH in adults, approximately 40% of ICH cases have intraventricular extension [3]. In addition, IVH is a major problem in premature infants, in whom it is a leading cause of cerebral palsy, cognitive abnormalities, and other forms of neurological impairment [4].

II. Etiology

Hypertension is the major underlying cause of spontaneous ICH. In addition, amyloid angiopathy, aneurysms, arteriovenous malformations, cavernous angiomas, and arteriovenous fistulae are major causes of ICH [2]. Apolipoprotein E alleles e2 and e4 have been found to be associated with the occurrence of lobar hemorrhages. People with defects in the coagulation cascade or who are taking anticoagulants are also at risk of ICH [2].

It should also be noted that apart from primary ICH, parenchymal hemorrhages are a major component of traumatic brain injury. Similarly, hemorrhagic conversion, with the formation of a hematoma, can occur after cerebral ischemia and is associated with a poor prognosis. Thus, an understanding of the mechanisms underlying hemorrhage induced–brain injury is also important for those conditions.

The location of the ICH is important in terms of outcome, with hindbrain hemorrhages being particularly devastating. About 30–50% of ICHs are in the basal ganglia, 10–20% in the thalamus, 25% are lobar, 5–10% pontine, and about 10% cerebellar [5].

IVH is a common occurrence in premature infants (15–45% of premature babies weighing < 1.5kg). The fetal germinal matrix, adjacent to the lateral ventricles, is supplied by large thin-walled blood vessels that are prone to rupture in response to hypoxia or changes in blood pressure [4,6]. In the adult, parenchymal cerebral hematomas close to the ventricular system often extend into the ventricles. It has been estimated that about 40% of ICH cases have an intraventricular component and such extension is a predictor of poor outcome [3,7]. Other causes of adult IVH include insertion or removal of ventricular catheters and intraventricular vascular malformations, aneurysms, or tumors [7].

III. Pathophysiology

A. Intracerebral Hemorrhage

In massive intracerebral hematomas (volume > 150 ml), cerebral perfusion pressure usually falls to zero and the resultant cessation in cerebral blood flow results in death. With smaller hematomas, the patient often survives the ictus, the bleeding normally stops, and a clot forms [2]. However, in about 20–40% of patients, there may be continued bleeding over the first day [8].

As a result of the ICH, there is deformation of the brain, perihematomal necrosis (within 6 hours), edema formation, and blood-brain barrier (BBB) disruption. There is an inflammatory reaction in the surrounding brain shortly after ictus, which peaks several days later. These early events are followed by a period of hematoma absorption (complete in ~2–3 weeks to months in humans) and finally formation of a scar or cavity [8]. There is evidence that an ICH can result in brain atrophy with a dilated ipsilateral cerebral ventricle. In humans, patients who survive an ICH often suffer permanent disability.

A number of animal models have been developed to examine mechanisms of brain injury following ICH. Thus, direct injection of blood into the brain parenchyma (usually the caudate in rodents models or cortical white matter in larger animals) is useful to study mechanisms of ICH-induced neurotoxicity and mass effect. Results from such models have generally recapitulated the findings in humans, although the overall time course may be shorter, reflecting differences in hematoma size. Such animal studies have also found evidence that perihematomal death involves both apoptosis and necrosis.

Apart from direct parenchymal injection of blood, the most common animal model of ICH involves intracerebral injection of collagenase to disrupt the vasculature. This has the advantage of vessel disruption as the basis of the hematoma. There are concerns, however, that the form of vascular disruption does not match that of most human ICHs and that collagenase may have effects other than simply disrupting blood vessels. There is a great need for animal models of spontaneous ICH, but current models (e.g., the spontaneously hypertensive stroke-prone rat [SHRSP], on a high-salt diet) do not mimic the human condition.

B. Intraventricular Hemorrhage

In many preterm infants, mild IVH is often benign. However, about 15% of preterm infants with IVH develop posthemorrhagic ventricular dilatation (PHVD) and most of these will either die or become shunt dependent. PHVD has a high risk of developmental disabilities. Approximately two-thirds have motor dysfunction and one-third have cognitive or other neurological abnormalities [4,6]. A major cause of the disabilities is damage to the periventricular white matter. There is also damage to the ependymal lining of the ventricles, gliosis, and enhanced extracellular matrix deposition. Ventricular dilatation may result from tissue destruction, due to periventricular hemorrhagic infarction or periventricular leukomalacia, or obstruction to cerebrospinal fluid (CSF) flow pathways [4,6].

The course of the illness following an adult IVH varies greatly with location and the severity of the initial bleeding. Thus, large acute hematomas are associated with coma and early death. Patients may survive smaller and slower IVHs. IVH may result in increased intracranial pressure, impacting cerebral blood flow and blockage of CSF outflow pathways, resulting in hydrocephalus. As with neonatal hemorrhage, however, adult IVH may also cause white-matter and ependymal damage, fibrosis, inflammation, and gliosis [7]. Although computed tomographic (CT) scans suggest that intraventricular hematomas resolve in 2–3 weeks, post-mortem samples suggest that hematomas can persist for weeks after ictus, a reflection of the low fibrinolytic activity of CSF [7].

The most common animal model of IVH involves direct infusion of blood into the lateral ventricle. This has been done in a number of species (mice, rats, pigs, dogs) in either neonates or adults. In addition, there have been models of spontaneous IVH in preterm infants. This involves inducing a variety of insults such as hypotension and hypercarbia in rabbit kits, beagle pups, or fetal sheep. These models have had various degrees of success in modeling the human condition [4].

IV. Mechanisms of Intracerebral Hemorrhage–Induced Brain Injury

A. Physical Disruption/Mass Effect

The normal functioning of the brain is crucially dependent on spatial relationships, either at the microscopic level, in terms of synapses linking individual cells, or at the gross level, in terms of the fiber tracts linking areas of the brain. The entry of blood into the brain parenchyma can disrupt these physical relationships causing cerebral dysfunction.

The physical mass of blood in brain tissue increases intracranial pressure because the brain is enclosed by the skull. The increased intracranial mass will, in part, be compensated for by an expulsion of CSF from the intracranial compartment, but in large hematomas, the increase in intracranial mass may result in compression of the brain, brain herniation, and death.

Although many of the physical effects of a hematoma occur at ictus, a number of studies have shown that hematoma enlargement continues over the first day in 20–40% of patients, with a higher incidence in the first several hours after ictus [8]. Prevention of secondary hematoma enlargement is the basis of a current trial of activated factor VII (NovoSeven) for ICH [9].

The physical effects of the hematoma may be compounded by perihematomal edema. In humans, perihematomal edema develops within 3 hours of symptom onset and peaks between 10 and 20 days after the ictus. The volume of edema can exceed the hematoma volume and it contributes to poor outcome [10].

B. Ischemia

The role of ischemia in ICH-induced injury remains controversial. Animal studies, with direct injection of blood into the brain, have generally shown transient and modest declines in perihematomal blood flow that would not usually cause ischemic brain damage [8]. In humans, a perihematomal zone of hypoperfusion has been reported in many studies [8,10]. However, Zazulia et al. [11] found that this zone also has a low oxygen extraction fraction, suggesting that ischemic blood flow does not occur in that region.

A caveat to conclusions about the role of ischemia in ICH-induced brain injury is that the presence of ischemic blood flow may also depend on the nature of the initial vessel rupture. Thus, rupture of a blood vessel supplying an area of brain with limited collateral supply may reduce blood flow to that region irrespective of the secondary effects of the hematoma.

C. Clot-Induced Toxicity

Apart from the physical effects of an ICH, there is considerable evidence for clot-derived neurotoxicity in ICH. Thus, in animal studies, inert masses with the same volume as intracerebral hematomas do not induce as much brain injury [10]. A number of pathways can induce brain injury and might be amenable to therapeutic intervention, including iron, thrombin, inflammation, and complement (Fig. 2).

D. Thrombin

Although thrombin is an essential part of the coagulation cascade that stops the expansion of a cerebral

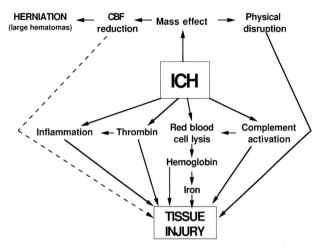

Figure 2 Schematic of potential injury mechanisms induced by an intracerebral hemorrhage. Hemorrhages can induce cell damage through mass effect, particularly in large hematomas that can cause increased intracranial pressure and herniation. The role of a transient reduction in cerebral blood flow in smaller hematomas is less certain (dashed line). Hemorrhages may also cause tissue damage via "neurotoxic" mechanisms, as outlined in the bottom half of the figure. Not all the potential interactions between neurotoxic mechanisms are shown (e.g., thrombin may potentiate iron-induced injury and the complement and inflammatory systems overlap).

hematoma, there is also considerable evidence that thrombin participates in ICH-induced brain injury [10,12]. Intraparenchymal infusions of thrombin can cause inflammation, brain edema, BBB disruption, cell death, and seizures. At high concentrations, thrombin can also kill neurons and astrocytes *in vitro*. It induces apoptosis in cultured neurons and astrocytes, potentiates N-methyl-D-aspartate (NMDA) receptor function, activates rodent microglia, and potentiates iron-induced neurotoxicity. In animals, inhibition of thrombin reduces ICH-induced brain injury [13]. There is also less ICH-induced brain edema in patients who are anticoagulated [10].

In contrast to the effects of high concentrations of thrombin, low concentrations may activate protective pathways [12]. This is discussed further in section I, Protective Pathways.

E. Erythrocyte Lysis, Hemoglobin, and Iron

During clot resolution, there is lysis of red blood cells (RBCs) within the hematoma. In rat, injection of lysed RBCs into the brain (in contrast to intact RBCs) causes massive brain injury. This injury appears to be due to hemoglobin and its breakdown products, including iron [12]. Intracerebral injection of hemoglobin, heme, bilirubin, or iron induces brain injury in the rat. In addition, inhibiting heme oxygenase-1, an enzyme that converts heme to biliverdin (which is then converted to bilirubin), carbon monoxide, and iron reduces ICH-induced brain injury [12,14].

Iron, in particular, may be important in ICH-induced injury [12,14]. Brain non-heme iron increases threefold after ICH in rats. Via the Fenton reaction, iron can result in the generation of the toxic hydroxyl radical. Iron can cause lipid peroxidation, oxidative DNA damage, and protein oxidation. Iron has been implicated in inducing brain injury in a number of diseases such as Alzheimer's and Parkinson's. However, the iron content of a parenchymal clot suggests that the potential for iron overload and iron-induced injury is much greater after an ICH. Deferoxamine, an iron chelator, attenuates ICH-induced brain injury [15]. These results indicate that hemoglobin and its degradation products are toxic and probably important causes of brain damage after ICH.

Although most blood iron is bound in hemoglobin within RBCs, some is present in plasma primarily bound to transferrin (holo-transferrin). Although an intracerebral infusion of holo-transferrin alone does not cause brain injury, it does when infused in combination with thrombin [16]. This may reflect thrombin-induced up-regulation of the transferrin receptor in the brain causing enhanced iron uptake into cells and increased toxicity.

F. Glutamate/γ-Aminobutyric Acid

Glutamate plays a major role in ischemic brain injury through so-called *excitotoxicity*. Its role in ICH-induced injury, however, is still uncertain. There are high concentrations of glutamate in RBCs and plasma, and there is an interaction between thrombin and glutamate receptors [14]. However, the effects of glutamate receptor antagonists in animal models of ICH have not been compelling [10]. For example, Lyden et al. [17] found no protection with the glutamate antagonist MK801 in a rat ICH model. They did, however, find some protection with the γ-aminobutyric acid (GABA) receptor agonist muscimol.

G. Inflammation

An inflammatory response occurs in the surrounding brain shortly after ICH and peaks several days later. Neutrophil infiltration develops within 2 days in rats and activated microglial cells persist for a month. These changes are associated with an up-regulation of proinflammatory mediators (e.g., tumor necrosis factor [TNF]-α and interleukin [IL]-1β). This inflammatory reaction is important in terms of clot resolution, but as in cerebral ischemia, evidence from animal studies also indicates that inflammation participates in secondary brain injury after ICH [8].

H. Complement

Elements of the complement system are up-regulated in the brain following ICH [10]. They may be derived from the blood through a disrupted BBB or generated *de novo* in the brain. The complement system may be important in clot resolution through activating erythrocyte lysis by formation of the membrane attack complex. However, that lysis may result in the release of potentially harmful compounds (hemoglobin and hemoglobin breakdown products including iron). In addition, membrane attack complex insertion may also occur in neurons, glia, and endothelial cells (a bystander effect) where it may cause neuronal death and BBB disruption. Complement inhibition in animal ICH models has resulted in reduced brain injury, although some elements of the complement system may be beneficial [10].

The effect of the complement system on ICH is likely controlled by endogenous inhibitors. Thus, clusterin, a membrane attack complex inhibitor, is also markedly up-regulated after ICH. It should be noted that there are also interactions between the complement system and inflammation. In particular, C5a is a potent chemoattractant for polymorphonuclear leukocytes [10].

I. Protective Pathways

This section has focused on mechanisms that induce brain injury following ICH and that may, therefore, be therapeutic targets to limit injury. It should be noted, however, that there are ICH-induced changes in the brain that may also serve to limit injury [12,14]. For example, there is an up-regulation in a variety of iron-handling proteins including the iron-binding protein ferritin that may limit iron toxicity.

Endogenous protective pathways may have important therapeutic implications: (1) Differences between patients in the up-regulation of protective pathways may alter the natural progression of ICH-induced injury and explain why some patients with apparently similar hematomas have very different outcomes; (2) therapeutic interventions may inadvertently alter the induction of protective pathways; and (3) endogenous protective pathways may alter therapeutic time windows. For example, although release of iron from an ICH may peak several days after ictus, the degree of iron-induced injury will depend on the levels of endogenous iron chelators (e.g., ferritin), which also increase after ICH. Thus, the timing for delivery of an exogenous iron chelator will depend on the imbalance between iron load and endogenous chelators rather than iron load alone.

An ICH may trigger endogenous defense mechanisms by multiple pathways. However, thrombin may play a pivotal role. In contrast to high concentrations, low concentrations of thrombin are neuroprotective [12]. Prior intracerebral infusion of thrombin at a low concentration reduces ICH-induced brain injury, as well as a number of other forms of injury (an effect termed *thrombin preconditioning*). Intracerebral injection of thrombin up-regulates a variety of heat-shock proteins, iron-handling proteins, and thrombin inhibitors that may participate in protection.

V. Mechanisms of Intraventricular Hemorrhage–Induced Brain Injury

A. Physical Disruption/Mass Effect

As with ICH, the presence of the IVH mass will transiently increase intracranial pressure with the potential for causing ischemia in periventricular tissues. In addition, the hematoma and the microthrombi that are generated by the IVH may block CSF drainage pathways and contribute to post-PHVD. The physical distortion of periventricular white-matter fiber tracts may also cause damage independent of ischemia [4,6].

B. Fibrosis

IVH is associated with meningeal fibrosis and there is increased deposition of extracellular matrix around periventricular blood vessels. These changes may impede the flow of CSF and absorption across the arachnoid villi or other drainage sites. Transforming growth factor (TGF)-β, the levels of which are increased with IVH, may participate in these extracellular matrix changes [4]. Transgenic animals overexpressing TGF-β1 in brain develop hydrocephalus, seizures, and die prematurely.

C. Free Radicals

As with ICH, the presence of hemoglobin and iron, its breakdown product, is potentially deleterious in IVH. There is evidence for the release of non–heme-bound iron into CSF, hemosiderin deposition in the ventricular wall, and free radical–mediated damage following neonatal IVH. In particular, oligodendrocyte progenitor cells appear to be highly susceptible to oxidative damage [4].

D. Inflammation

IVH increases the level of a number of cytokines in CSF in infants with PHVD including TNF-α, IL-1β, IL-6, IL-8, and interferon (IFN)-γ. The up-regulation of proinflammatory cytokines suggests that inflammation may be a therapeutic target, although it should be realized that infiltrating leukocytes may be involved in hematoma removal [4].

VI. Therapeutic Interventions for Intracerebral Hemorrhage–Induced Brain Injury

The mass and neurotoxic effects of an ICH suggest that surgical removal of the hematoma should improve clinical outcome. However, there have been several clinical trials that have failed to show such a benefit from surgery, including the recent Surgical Trial in Intracerebral Haemorrhage (STICH) trial [18]. This failure may be due to complications arising from the surgical procedure or it may reflect the time course of the ICH-induced injury necessitating early removal of the hematoma. Ultra-early surgical removal is, however, associated with the risk of rebleeding [19].

Apart from surgical removal, hematomas may also be removed by injection of a fibrinolytic (such as tissue plasminogen activator [t-PA]) and aspiration. Although this method has the benefit of less surgical trauma, it does not fully remove the hematoma. There are also concerns that t-PA may have harmful effects on brain tissue. A clinical trial is underway.

There is a current trial of activated factor VII to prevent post-ictus hematoma enlargement, the initial results of which are promising [9]. There have also been trials of dexamethasone for ICH, the results of which were negative [8].

VII. Therapeutic Interventions for Intraventricular Hemorrhage–Induced Brain Injury

One approach to IVH has been early, repeated tapping of lumbar or ventricular CSF. In preterm infants with PHVD, however, a large multicenter trial found no benefit from lumbar punctures, judged by the need for shunting and long-term neurological disabilities. In addition, lumbar punctures are associated with an increased rate of CNS infections [6].

In both adults and infants, there have been clinical trials of fibrinolytics to assess their potential for removing intraventricular blood and, thus, diminishing the mass and neurotoxic components and improving CSF drainage [4,6,20]. Preclinical and adult human IVH trials have shown the feasibility of removing hematoma with thrombolytics [4,20] and clinical trials are continuing. For preterm infants with PHVD, there have been a number of small clinical trials with thrombolytics, which have yet to show a clear benefit [4,6]. Rebleeding and infection are clinical risks [6].

One approach to ameliorate PHVD is to reduce CSF production. This formed the basis for treatment with acetazolamide and furosemide in preterm infants with PHVD. However, a clinical trial found that this treatment did not improve the need for shunt replacements and worsened neurological outcome [6].

VIII. Summary

There are no proven therapies for ICH and IVH. There has been rapid expansion in understanding the mechanisms underlying hemorrhage-induced brain injury. It is to be hoped that such advances will lead to new therapies for these devastating forms of stroke.

Acknowledgments

This work was supported by grants NS-17760, NS-34709, NS-39866, and NS-47245 from the National Institutes of Health.

References

1. Kase, C. S., and Caplan, L. R. (1994). "Intracerebral Hemorrhage." Butterworth-Heinemann, Boston.
2. Xi, G., and Hoff, J. T. (2005). The pathophysiology of hemorrhagic lesions. *In:* "Imaging of the Nervous System: Diagnosis & Therapeutic Application" (R. E. Latchaw, J. Kucharczyk, and M. E. Moseley, eds.), pp. 519–534. Elsevier Mosby, Philadelphia.
3. Broderick, J. P., Brott, T., Tomsick, T., Miller, R., and Huster, G. (1993). Intracerebral hemorrhage more than twice as common as subarachnoid hemorrhage. *J. Neurosurg.* **78**, 188–191.
4. Cherian, S., Whitelaw, A., Thoresen, M., and Love, S. (2004). The pathogenesis of neonatal post-hemorrhagic hydrocephalus. *Brain Pathol.* **14**, 305–311.
5. Voelker, J. L., Schochet, S. S., and Kaufman, H. H. (1999). Pathology of primary cerebral hemorrhage. *In:* "Neurosurgery: The Scientific Basis for Clinical Practice" (A. Crockard, R. Hayward, and J. T. Hoff, eds.), pp. 695–703. Blackwell, Oxford.
6. Tortorolo, G., Luciano, R., Papacci, P., and Tonelli, T. (1999). Intraventricular hemorrhage: past, present and future, focusing on classification, pathogenesis and prevention. *Child's Nerv. Syst.* **15**, 652–661.
7. Engelhard, H. H., Andrews, C. O., Slavin, K. V. and Charbel, F. T. (2003). Current management of intraventricular hemorrhage. *Surg. Neurol.* **60**, 15–21.
8. Xi, G., Fewel, M. E., Hua, Y., Thompson, B. G., Jr., Hoff, J. T., and Keep, R. F. (2004). Intracerebral hemorrhage: pathophysiology and therapy. *Neurocrit. Care* **1**, 5–18.
9. Mayer, S. A., Brun, N. C., Begtrup, K., Broderick, J., Davis, S., Diringer, M. N., Skolnick, B. E., Steiner, T., and Recombinant Activated Factor VII Intracerebral Hemorrhage Trial Investigators. (2005). Recombinant activated factor VII for acute intracerebral hemorrhage. *N. Engl. J. Med.* **352**, 777–785.
10. Xi, G., Keep, R. F., and Hoff, J. T. (2002). Pathophysiology of brain edema formation. *Neurosurg. Clin. North Am.* **13**, 371–383.
11. Zazulia, A. R., Diringer, M. N., Videen, T. O., Adams, R. E., Yundt, K., Aiyagari, V., Grubb, R. L. Jr., and Powers, W. J. (2001). Hypoperfusion without ischemia surrounding acute intracerebral hemorrhage. *J. Cereb. Blood Flow Metab.* **21**, 804–810.
12. Xi, G., Reiser, G., and Keep, R. F. (2003). The role of thrombin and thrombin receptors in ischemic, hemorrhagic and traumatic brain injury: deleterious or protective? *J. Neurochem.* **84**(1), 3–9.
13. Kitaoka, T., Hua, Y., Xi, G., Hoff, J. T., and Keep, R. F. (2002). Delayed argatroban treatment reduces edema in a rat model of intracerebral hemorrhage. *Stroke* **33**, 3012–3018.

14. Wagner, K. R., Sharp, F. R., Ardizzone, T. D., Lu A., and Clark, J. F. (2003). Heme and iron metabolism: role in cerebral hemorrhage. *J. Cereb. Blood Flow Metab.* **23**, 629–652.
15. Nakamura, T., Keep, R. F., Hua, Y., Schallert, T., Hoff, J. T., and Xi, G. (2004). Deferoxamine attenuates brain edema and neurological deficits after experimental intracerebral hemorrhage. *J. Neurosurg.* **100**, 672–678.
16. Nakamura, T., Xi, G., Park, J. W., Hua, Y., Hoff, J. T., and Keep, R. F. (2005). Holo-transferrin and thrombin can interact to cause brain damage. *Stroke* **36**, 348–352.
17. Lyden, P. D., Jackson-Friedman, C., and Lonzo-Doktor, L. (1997). Medical therapy for intracerebral hematoma with the gamma-aminobutyric acid-A agonist muscimol. *Stroke* **28**, 387–391.
18. Mendelow, A. D., Gregson, B. A., Fernandes, H. M., Murray, G. D., Teasdale, G. M., Hope, D. T., Karimi, A., Shaw, M. D., and Barer, D. H.; STICH investigators. (2005). Early surgery versus initial conservative treatment in patients with spontaneous supratentorial intracerebral haematomas in the International Surgical Trial in Intracerebral Haemorrhage (STICH): a randomised trial. *Lancet* **365**, 387–397.
19. Morgenstern, L. B., Demchuk, A. M., Kim, D. H., Frankowski, R. F., and Grotta, J. C. (2001). Rebleeding leads to poor outcome in ultra-early craniotomy for intracerebral hemorrhage. *Neurology* **56**, 1294–1299.
20. Naff, N. J., Hanley, D. F., Keyl, P. M., Tuhrim, S., Kraut, M., Bederson, J., Bullock, R., Mayer, S. A., and Schmutzhard, E. (2004). Intraventricular thrombolysis speeds clot resolution: results of a pilot, prospective, randomized, double-blind, controlled trial. *Neurosurgery* **54**, 577–583.

28

Idiopathic Generalized Epilepsy

Jose-Luis Perez-Velasquez, PhD
Edward H. Bertram, III, MD
O. Carter Snead, III, MD

Keywords: *absence, epilepsy, generalized, idiopathic, mechanisms, models*

I. Clinical Overview
II. Animal Models
III. Genetic Models of Absence Seizures
IV. Pharmacological Models of Absence Seizures
V. The Nature of Thalamocortical Synchronized Activity
VI. Genetics
VII. Concluding Remarks
    References

I. Clinical Overview

The collection of clinical syndromes that fall under the heading "idiopathic generalized epilepsies" (IGEs) is clinically and physiologically quite diverse. From benign familial neonatal convulsions to the childhood absence syndromes to juvenile myoclonic epilepsy (JME) and a number of convulsive epilepsy syndromes, there is a clear and distinct combination of age at onset, clinical symptoms, and electroencephalographic (EEG) patterns during seizures (a reflection of the underlying pathophysiology) that distinguishes them from one another. They are placed in this classification, however, because of an apparent generalized (probably more correctly *bilateral* or *bilateral multifocal*) onset of seizures. However, this neat picture is muddied a bit by the occasional occurrence of more than one distinct seizure type in the same individual, such as absence in childhood followed by a later convulsive disorder or JME (or all three) in a single individual [1]. The IGE syndromes have a clear but complex genetic component. Although there are some families with a well-defined Mendelian pattern of inheritance, in other families there is an overall increased incidence of the idiopathic epilepsies, but the genetic basis for the inheritance of the disorder is less clear, in part because the affected family members can display different phenotypes [2]. Many of the patients' seizures resolve over time, and many of the rest have their seizures brought under complete or reasonable control through currently available therapies. For these reasons, many of the syndromes have been designated *benign*, which has resulted in a lowered research interest in them, either for understanding the pathophysiology or for developing new treatments, because many view the disorders as problems solved. Unfortunately, many of the patients do not have a benign outcome and many carry the tendency for epilepsy for their entire lives, often with a different clinical manifestation (such as a child with absence epilepsy who develops convulsions as an adult well after the absence seizures have resolved).

Because this collection of seizure disorders continues to be a problem, many questions remain to be answered about them. What is the underlying cause for each syndrome? For every seizure phenotype, how many different genetic etiologies exist? Is the circuitry (and therefore the potential therapeutic targets) the same for all of the seizure types? For all seizures with a similar phenotype, is the pathophysiology of seizure initiation the same? Is a mutation associated with an IGE phenotype the primary cause of the disorder, or does it result in an alteration of the cascade of developmental events that eventually leads to a different expression of normal genes and a subsequent heightened susceptibility to seizures? In this chapter, we review what is known about the pathophysiology of these disorders as it relates to the aforementioned questions. Unfortunately, for many of the IGE syndromes such as JME, there are no well-validated models, so our understanding of the basis for that particular disorder is limited to what we can derive from clinical observations. We outline the key features of several of the syndromes and then examine what is known about the neurobiology.

A. Brief Overview of the Major IGE Syndromes

The IGEs are divided, as a first step, into convulsive and nonconvulsive. The latter generally comprise the absence or petit mal disorders and the former, myoclonic and convulsive. Among the absence syndromes there are two main groups: childhood and juvenile, depending partly on age at onset.

Those with the *childhood absence epilepsy* (CAE) syndrome tend to have many more but much briefer seizures over the day. The seizures are also more likely to stop with a lesser risk for the development of other types of seizures later. Those with the *juvenile absence epilepsy* (JAE) syndrome tend to have fewer but longer absence attacks and are more likely to develop other types of seizures later such as myoclonic and convulsive. Although both groups have generalized spike–wave patterns on EEG, the basis for the different age at onset and natural history is largely unknown, in part because the significant overlap in the age at onset and clinical phenotype makes clear clinical distinctions, at times, difficult.

Benign familial neonatal convulsions have a clear autosomal-dominant inheritance, usually appear within the first week of life, and always remit without increased risk for later epilepsy. It is associated with several mutations in potassium channels.

JME is defined clinically by the presence of brief myoclonic jerks that can occur at varying intervals (days, weeks, or months) and with varying severities (barely perceptible brief jerks to falling to the floor). This disorder tends to begin in adolescence to early adulthood and may only come to attention when a full convulsion occurs. JME may be preceded by one of the absence syndromes and may evolve to include occasional convulsions. JME is considered a lifelong condition.

There are a variety of sporadic *convulsive syndromes*, one of which has seizures that occur at or shortly after awakening. The generalized convulsions tend to begin in adolescence or early adulthood and rarely remit.

II. Animal Models

Of all the IGE syndromes, there are an overwhelming number of models for spike–wave (absence) seizures, but few good ones for the others, especially if we want to replicate the physiology, behavior, and pharmacology of each syndrome. For this reason, this discussion focuses on models of spike–wave or absence epilepsy. The generation of a variety of animal models of synchronized thalamocortical activity, as occurs during absence seizures, has provided an enormous amount of fundamental information regarding potential mechanisms of bilaterally synchronous epileptiform discharges in generalized absence epilepsy. In this brief overview of some absence seizure animal models, a few queries are more specifically addressed. Of particular interest is the relation between the anatomy of the thalamocortical circuitry and the biophysical characteristics of neurons in these circuits, because absence seizures are a clear expression of a specific brain rhythm (a paroxysm in these cases) that is determined by the cellular and synaptic properties of the neurons within the involved circuit, as well as the anatomic location of the involved circuitry. In the following paragraphs, we describe some genetic and pharmacological models that fulfill the pharmacological, electrophysiological, and behavioral criteria for valid absence seizure models. As will become clear from this discussion, there are many ways to create these models reflecting the complex network of interactions that underlie this group of disorders. A more complete description of the different approaches can be found in Pitkänen et al. [3].

III. Genetic Models of Absence Seizures

The genetic absence epilepsy rat from Strasbourg (GAERS) has been a widely used model, developed by selective breeding within the Wistar rat colony. These animals develop spontaneous 7- to 11-Hz spike and wave discharges (SWDs) that are recorded in thalamus and cortex, associated with staring, behavioral immobility, facial myoclonus, and rhythmic twitching of the vibrissae. No abnormalities are observed during interictal EEG recordings. The seizures are responsive to anti-absence drugs such as ethosuximide, valproate, and benzodiazepines. Thus, the behavioral, pharmacological, and electrophysiological manifestations correlate well with those of human absence seizures, except for the frequency of the SWD, that is, 3–4 Hz in humans. This difference may stem from distinct brain anatomy and physiology of thalamocortical circuitry in the rodent, although slower SWDs, more analogous to the 3 Hz seen in humans, have been reported in rats. This indicates that similar waveforms can be generated by different thalamocortical circuitries, and that the differences in specific waveforms may have to do more with the particular underlying pathophysiology in each model. This could be a potential subject of debate [4], and some computer simulation studies have revealed that γ-aminobutyric acid B ($GABA_B$) transmission may be relevant in the determination of oscillation frequencies.

The circuitry responsible for SWD in the GAERS appears to be the same at the macrolevel as that in other rodent models of absence seizures and includes the thalamus and cortex [4,5], the nigral inhibitory system [6], the brainstem cholinergic system, and possibly, noradrenergic systems in brain [3]. However, notably absent in involvement of SWDs in rodent models of absence seizures is hippocampal circuitry [3]. There are a few exceptions to this rule, but those rodents that manifest hippocampal circuitry involvement in bilaterally synchronous SWDs show pharmacological resemblance to absence but behavioral similarities to partial complex seizures.

A rodent model of absence similar to the GAERS is the WAG/Rij (Wistar albino Glaxo/Rijwijk) strain of rats where SWDs at similar frequencies (7–9 Hz) with an identical pharmacological and behavioral profile have been observed [3]. The GAERS and WAG/Rij models have been widely used to address the mechanisms of epileptogenesis in absence seizures. Of note for these two models, in contrast to the human absence syndromes, which begin in childhood or adolescence and frequently remit by early adulthood, the seizures in these rats usually first appear in adulthood and persist for the remainder of the animal's life.

Other genetic models that have been useful in the investigation of fundamental mechanisms of epileptogenesis in absence seizures include a number of single gene mutations in mice, such as the tottering mouse, the stargazer, and the lethargica. Of some note in these animals, in comparison with humans, there are frequent behavioral or neurological abnormalities, and they are associated with mutations in voltage-gated calcium channels, many channels that have not been associated with a human epilepsy syndrome [7]. The implications of these observations are discussed in Section VI, later in this chapter.

Although most animal seizure models show ictal activity of only one type, it would be illuminating to see the evolution of different seizures in the same animal, something that does happen in people with IGE. A rodent model that seems promising along these lines is the mouse line deficient in succinate semialdehyde dehydrogenase (SSADH). These mice show a progression from absence seizures to tonic–clonic convulsive seizures. Although the pathophysiology has not been completely elucidated, a possible mechanism related to inhibitory $GABA_A$-mediated transmission has been advanced [8]. As well, a subpopulation of WAG/Rij rats with both absence and audiogenic seizures has been identified [9].

IV. Pharmacological Models of Absence Seizures

One of the most useful models of generalized absence seizures relies on the administration of penicillin to cats, that is, the feline generalized penicillin epilepsy model. It has been useful not only for providing some information about one potential pathophysiology underlying the spike–wave syndrome, but it has also given us significant insight into the circuitry of the discharges. Neither cortex nor thalamus alone is able to sustain the SWD in this model, thus illustrating the importance of a recurrent thalamocortical loop for the maintenance of SWD. In rodents, there are a number of pharmacological models of absence seizures, most of which are also acute. The most widely used of these are the pentylenetetrazole model and the γ-hydroxybutyric acid (GHB) model [3]. The GHB model has been widely characterized. The specific molecular mechanism by which GHB causes SWDs remains obscure but probably relates to a $GABA_B$ receptor agonist effect of GHB [10].

A chronic model of SWD has also been described and relies on the administration of the cholesterol synthesis inhibitor AY9944 to rats at early postnatal ages [3]. The blockade of cholesterol synthesis early in life results in chronic SWDs that not only propagate through thalamus and cortex, but, unlike SWD in other models of absence, can also be recorded from limbic structures such as hippocampus and amygdala in a manner similar to the seizures observed in *atypical* absence seizures in humans. Similarly, AY9944-treated rats present behavioral and cognitive manifestations that differ from those seen with typical absence seizures, but that are comparable to those symptoms associated with *atypical* absence seizures in children [11]. Hence, the AY 9944-treated rat is termed a model of *atypical absence* seizures. The mechanism by which blockade of cholesterol synthesis in developing brain leads to the aforementioned constellation of electroclinical symptoms remains a mystery.

Another "pharmacological" model of absence seizures deserves mention, as it has yielded impressive information on the mechanisms of absence seizures and the transition from spindle activity to SWD. Cats under ketamine/xylazine anesthesia develop spontaneous SWD, a natural consequence of the highly synchronized thalamocortical activity that this anesthetic mixture causes. Observations derived from this model are extensively reviewed in Steriade [12], and we remit the interested reader to this publication.

V. The Nature of Thalamocortical Synchronized Activity

The aforementioned models have been used to investigate mechanisms of synchronized activity in thalamocortical circuitry and have shed light onto the generation, maintenance, and termination of absence seizures. As described, these animal models display rhythmic synchronized SWDs, which are associated with behavioral immobility and are attenuated or abolished by antiabsence anticonvulsant drugs such as ethosuximide or valproic acid. Whether we decide to call these "absence seizures" is a matter of definition, which may impose limitations on broader perspectives. The point is that the study of these synchronized activities as natural phenomena is important in and of itself because such investigation affords us a window on the fundamental mechanisms of normal brain function. Whether these models represent a close approximation to human absence seizures is another issue.

A. The Basic Circuitry

Although a number of possible primary seizure circuits have been proposed over the decades [13], there is now a general consensus, as noted earlier, on the neuronal components that underlie the rhythmic SWD of an SWD: corticothalamic neurons, thalamocortical neurons, and the neurons of the reticular nucleus of the thalamus (Fig. 1). In this scenario, the primary rhythm takes place between the projection neurons of the thalamus and cortex, but the GABAergic neurons of the reticular nucleus play a key role in modulating and maintaining the rhythmic discharge. The circuit, in a simplified explanation, works like this: The thalamocortical neurons send excitatory axons to the thalamic relay nucleus and the reticular neurons. The stimulation of the thalamic relay (thalamocortical) neurons

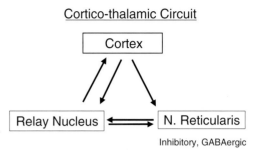

Figure 1 Simplified absence seizure circuit. Volleys from cortical neurons simultaneously stimulate thalamic relay neurons and nucleus reticularis neurons. Synchronized inhibitory discharges from the nucleus reticularis hyperpolarize a large population of the thalamic relay neurons, which activates the calcium T currents, which becomes a key factor in the rhythmic discharges of the absence seizure. Recurrent excitatory bursts from the cortex and thalamic relay neurons continue to stimulate the reticular nucleus neurons, which reciprocate with inhibitory volleys back to the thalamic relay neurons. The terminals from the cortical and thalamic relay neurons are glutamatergic excitatory and the terminals from the nucleus reticularis neurons GABAergic inhibitory.

may or may not result in a return stimulation of the cortical neurons, depending on the strength and synchronization of the stimulation. The simultaneous stimulation of the reticular neurons by the cortical neurons, however, results in an inhibitory GABAergic input to the thalamic relay nucleus. This inhibitory input becomes key to the rhythmic discharge because, when of sufficient magnitude and sufficiently synchronized over a large enough population of neurons, this hyperpolarizing stimulation, as will be explained later, allows for the activation of the calcium T currents, which are a key component for the rhythmicity of the SWD.

One of the interesting aspects of these spike–wave seizures is the apparent propensity for them to occur in particular regions of the brain. Although described as generalized seizures, it has been recognized that the distribution of the seizure activity is not uniform and usually has a propensity, in people, for the more anterior regions of the head. Meeren et al. [13] studied this issue in the WAG/Rij rat model and concluded that the seizure initiation is limited to a restricted region of the cortex that then recruits other nearby areas of the cortex and related thalamic nuclei. It is not clear from this observation whether this pattern of onset is model specific or represents a general liability of these regions to initiate SWD. In either case, interesting but unanswered questions arise about the pathophysiology of absence epilepsy. How restricted is the discharge in this and other models? What is it about the intrinsic properties of the neurons, the receptors, and the nature of the connections in the circuits that restricts the distribution of the discharge? The AY9944 rat model, which has a more widely distributed discharge that includes the hippocampus, may help answer some of these questions at least in part.

B. What Initiates Synchronized Thalamocortical Activity?

There has been some debate over whether the thalamus or the cortex is the culprit in triggering SWDs. There is evidence for both, but the evidence tends to favor the cortex as the initiator of seizures [13]. For example, in the GAERS, bursts of action potentials in cortical neurons tend to lead the thalamic cell bursting firing pattern and the concomitant EEG paroxysm. We propose that it really does not matter where the seizure is initiated, because the SWD will be maintained by the thalamocortical loop. This is clearly demonstrated by the fact that a single intracerebral electrical stimulation, either to the thalamus or to the cortex, can generate SWDs in one of the aforementioned pharmacological models, indicating the propensity of this circuitry to seize. This observation complements older experimental data that demonstrated that cortical stimulation is most effective at triggering spindle oscillation. Extending this scenario to a patient with absence epilepsy, we can conclude that in the hyperexcitable brain, any increased neuronal activity in the thalamus or cortex will be amplified, resulting in a SWDs through the thalamus and cortex. Recognizing that the seizures represent a circuit phenomenon is also important for choosing therapeutic targets. Altering circuit behavior at any point along the way expands the number of potential targets for therapy.

It is serendipitous that so many of the animal models of absence seizures are in rodents, because the rodent brain has a propensity to express rhythmic discharges. In general, whatever the mechanism that promotes hyperpolarization of thalamic cells (relay and reticular cells) enhances the probability of developing a synchronized activity within the thalamocortical loop. This is due to a special characteristic of thalamic cells; namely, they possess low-threshold calcium channels that activate by hyperpolarization of the membrane, resulting in a calcium spike of considerable amplitude as the cell repolarizes (Fig. 2). Thus, in thalamic relay cells, we find an interesting combination of intrinsic cellular properties and extrinsic factors, that is, synaptic inputs and circuitry, which contribute to the almost rhythmic features of these cells. In particular, the depolarization conferred by low-threshold calcium spikes and the hyperpolarization-activated cation conductance Ih, which is a non-inactivating sodium/potassium current (both intrinsic cellular properties), are important factors contributing to rhythmic activity of thalamic cells. Hence, for thalamic neurons, more hyperpolarization does not necessarily mean less activity. These factors, along with synaptic inputs (network properties) from cortical and thalamic reticular cells, can create a variety of thalamocortical rhythms.

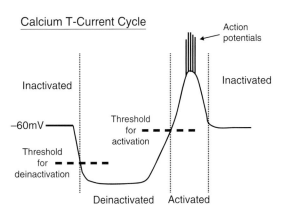

Figure 2 Calcium T current. This voltage-gated current is one of the key components in the generation and maintenance of the spike-wave discharge. At a normal resting state ($-60\,\text{mV}$), this channel is inactivated and cannot be activated until it is de-inactivated by hyperpolarization to below the de-inactivation threshold. While the neuron is hyperpolarized but de-inactivated, the channel remains inactive until the cell returns towards its normal resting state. During that process, the membrane crosses the activation threshold potential, at which point the channel activates, or opens, which further depolarizes the cell, often causing action potentials. The channel then is inactivated and remains so until the neuron is hyperpolarized below the de-inactivation threshold.

Thalamic reticular neurons are particularly important in establishing and maintaining these rhythms because they are inhibitory and, therefore, cause deep hyperpolarization of relay cells because of the reticular burst firing in synchronous neurons. This hyperpolarization ("inhibition") of relay cells promotes further calcium spikes and the possibility of synchronized activity in the thalamocortical loop. In this regard, the involvement of inhibitory $GABA_B$ neurotransmission in absence seizures has been demonstrated [14], as shown in a number of animal models including the GAERS, lethargica, and the GHB and PTZ models, because slow $GABA_B$ inhibitory potentials are long-lasting and, therefore, increase the probability of low-threshold calcium spike activation. However, it is the fast $GABA_A$ conductance that is most clearly recorded in relay neurons during spindles and SWDs. The strength of inhibition on relay thalamic cells is manifest in the "inhibitory dominance" that is observed on relay neurons on cortical stimulation, which is also the basis of the thalamic "augmenting responses" that occur after stimulation of thalamus at frequencies between 7 and 14 Hz. This is one reason that a strong synaptic input from the neocortex to the thalamus can elicit synchronized activity by these inhibitory mechanisms and gives rise to the hypothesis of a "cortical focus" of SWD. One should bear in mind, however, that other strong inputs (e.g., cholinergic pathways from the brainstem to thalamus) also can start the synchronized loop.

The progression from "normal" thalamocortical rhythms (slow oscillation and spindles) to paroxysmal SWDs is quite interesting and has been extensively studied in the feline penicillin model [15] and in anesthetized cats [4,12]. The data from these investigations suggest that the paroxysms of SWDs are but an extension of normal physiological brain activity. Indeed, cellular studies have determined that epilepsy-related thalamic cellular activities are similar to those found in normal activities. In general, most evidence suggests that what transforms spindles into SWDs is a "diffuse state of cortical hyperexcitability." But then, why is the SWD long lasting? And, more interestingly, why are the SWDs constrained within thalamocortical circuitry during typical absence seizures but recruit hippocampal circuitry in atypical absence seizures?

C. What Maintains Thalamocortical Synchronized Activity?

After a set of cortical neurons sends a relatively strong synchronized input to thalamic cells, the corticothalamic feedback mechanism, determined by the anatomical connections, engages the loop into synchronized activity. Some thalamic cells project in a more diffuse fashion than others, a finding that led to the concept of the "thalamic matrix" [16]. In general, the anatomy and synaptic mechanisms of these circuitries, combined with intrinsic cellular properties, such as the tendency of some thalamic nuclei, most importantly the reticular nucleus, to exhibit both tonic and oscillatory rhythms, promote the sustained synchronization of the thalamocortical loops.

The investigation of the relation of individual neuronal firing to the SWD can provide important clues as to the waveform recorded as field potentials. In this regard, the relation of neuronal spike firing to the field potential recordings (the SWD) has been done in both GAERS and WAG/Rij rats. In those experiments, unit recordings (individual action potentials) showed that thalamic cells tend to fire at the time of the spike component recorded in the field potential recording. An area that has not been investigated relates to the presumed synchronized action potential firing or bursts of spikes in neighboring thalamic neurons during the SWD. Are most of those recorded cells participating, either with a spike or a burst, in the SWD waveform? These issues can be partially addressed with multielectrode recordings, notwithstanding the complexity of these experiments.

Although the SWDs in most seizure models are restricted to the thalamocortical circuitry, rats treated with AY9944, as described earlier, show 7- to 9-Hz synchronized discharges also in limbic areas, such as hippocampus and amygdala. Why do the seizures in AY-treated animals spread to these brain areas? Conversely, what protects these limbic areas from becoming entrained in the paroxysm in the typical absence seizure? Based on the observation that there is, in fact, synchronization but no paroxysms in specific frequency bands between the two hippocampi in GHB-treated rats that exhibit typical SWDs (Perez Velazquez, Garcia Dominguez, and Snead, unpublished observations), we propose that the treatment with AY9944 early during development may cause specific damage to the hippocampal formation. More specifically, the inhibitory interneurons located in the dentate gyrus area could be selectively involved in this model, so the hippocampus is unprotected from the high synchronous input from cortical and thalamic areas and, thus, is entrained into the paroxysm. We are investigating these issues, as they may reveal features of neuronal networks in developing brains that make them more susceptible to the epileptogenic process.

D. What Terminates Thalamocortical Synchronized Activity?

The answer to this question will most likely turn out to be multifactorial. Anything that promotes depolarization of thalamic cells, ranging from depletion of synapses to excessive depolarization of neurons, contributes to the ending of the paroxysm. In this regard, a specific action of the cation conductance, Ih, in thalamic neurons has been advanced [17], because this mixed Na/K current slowly but continuously depolarizes neurons, an action that can inactivate the low-threshold calcium spike. As a result, fewer and reduced calcium spikes will result in fewer bursts of action potentials in thalamic neurons, reduced input to neocortical areas, and ultimately, termination of the paroxysm.

VI. Genetics

That this group of epilepsies has a genetic component has been recognized for decades. Within families there is an incidence of epilepsy much higher than would be expected at random, and the concordance for IGE among monozygotic twins is placed variably between 70 and 85%, whereas for dizygotic twins, it is closer to 20–25% [18]. For this reason, there has been an active search among this group of patients for "the epilepsy gene," but there has been a number of confounding issues that have limited success. First, although the incidence of epilepsy in family lines with members who have some form of IGE is higher than expected in the general population, the seizure phenotype can be quite variable across the affected family members, at times creating difficulties in identifying affected and unaffected members. Second, although there are some families in which a classic mendelian pattern of inheritance is seen, in many other families, the pattern is less well defined and usually gets classified as complex or polygenic [2]. The revolution in genetics in the past decades has helped considerably in improving our understanding of the genetics behind these disorders, but the many new discoveries have also emphasized the difficulties in understanding the genetic basis of IGE. In this section, we review some of the genes that have been associated with IGE in people, as well as the exploding number of findings from mouse genetics. We also discuss the complexities in linking genetic mutations and variations to specific seizure disorders.

For human IGE, there have been a limited number of mutations associated with the disorder (Table 1), but it is interesting, in part for our understanding of the pathophysiology of the disorder and the circuitry of the seizures, that the mutations have involved a number of ion channels, ligand and voltage gated [7]. Sodium, potassium, and chloride channels have all been associated with different forms of IGE, but currently for most patients with IGE, no well-defined genetic abnormality has been discovered. With the growing recognition that there is a variable number of polymorphisms for each gene that exist within the population, many of which have functional significance, it is quite possible, even likely, that for many patients the seizures arise from the chance association of a sufficient number of these polymorphisms to the point that the balance of the system is tipped toward seizures. In this scenario, determining the likely multiple combinations of normal polymorphisms that may lead to the many forms of IGE will be a labor of extraordinary complexity.

As discussed in more detail in Section II, earlier in this chapter, some of the earliest rodent models of spike–wave seizures came with the chance observations of spike–wave bursts on EEGs associated with behavioral arrest seen in several of the inbred rat strains. The two most widely used strains, the WAG/Rij and the GAERS, have yet to have the genetic basis for their seizures elucidated. Of some note, however, as these rats are studied, there is increasing evidence that the strains, other than spike–wave seizures on EEG, are phenotypically and likely genotypically different.

Table 1 Genes Associated with Idiopathic Generalized Epilepsy Syndromes

Syndrome	Gene
Benign familial neonatal convulsions	KCNQ2
	KCNQ3
Childhood absence with febrile seizures	GABRG2
Juvenile myoclonic epilepsy	GABRA1
	EFHC1
Autosomal-dominant idiopathic generalized epilepsy	CLCN2
Generalized epilepsy with febrile seizures plus	SCN1A
	SCN1B
	SCN2A
	GABRG2

See reference 2 for complete references for these mutations.

Transgenic mice, on the other hand, have provided a significant opportunity to examine the complexities of epilepsy genetics, by allowing us to discover, many times by chance, the potential of particular mutations or deletions to cause spike–wave epilepsy. Investigators have also replicated some of the human mutations that are associated with IGE. A number of interesting observations have arisen. Studying mice that have an induced mutation that does not necessarily have a known relationship with human epilepsy has revealed that a number of genetic changes can cause absence-like seizures, many of which are not directly related to ion channels. This suggests that, in some cases, the change in system excitability may be secondary to more widespread effects of the mutation either in channel subtype expression or on development. These mutations tell us that there are many genetic paths to the SWDs [7].

The replication in mice of human mutations associated with specific generalized epilepsy syndromes has led to other instructive observations, perhaps the most important of which is that not all human mutations associated with IGE cause an epileptic phenotype in mice. More importantly, it has been observed that the presence or absence of epilepsy in the transgenic mouse depends on its overall genetic background. One mouse strain may show no overt effects of the mutations, whereas another strain may have significant seizures that are worse than what occurs in people with the same mutation. We have had direct experience with this phenomenon. A mouse strain with a mutation in the voltage-gated sodium channel SCN1A that has been associated with generalized epilepsy/febrile seizures syndrome was observed to have intermittent behaviors that suggested seizures. Mice with this mutation underwent continuous EEG monitoring. After weeks of monitoring during the age when the animals were known to show the behavior, we found no abnormalities, either in the EEG or in the mice themselves. These mice were a different strain from the mice in which the altered behaviors were originally seen. When the mutation was rederived in the original strain, the previously observed behaviors returned, and there were clear seizures on the EEG [19]. This observation was extremely important in providing new insights into the relation between a mutation and the expression of the epilepsy phenotype: that not only must the mutation be present, but also the mutation must be in the right environment. In the pursuit of the genetic bases for human epilepsy, it is an observation that must be considered.

So, how does one then make sense of the many transgenic mice that have epileptogenic mutations that have never been associated with seizures in people or the human "epilepsy" mutations that do nothing when recreated in mice or only when placed in particular strains? Does that mean we should largely disregard the results of mouse genetic studies in regard to their implications for human work? Definitely not, but the lessons are clear. First, the current information shows that there are multiple and often unexpected ways to destabilize the system and cause epilepsy, as evidenced by the myriad mutations that can cause absence-like seizures. Mutations and knockouts that clearly have nothing to do with ion channels can cause SWDs. For the moment, little is known about the physiology in these animals, but it is likely that the "non-channelopathy" mutations influence the expression of normal channels and receptors or the structure of the neurons in a way that results in the seizures. These findings also tell us that the search for "epileptogenic" genes must go beyond a focus on ion channels. The second message is that there are other factors that influence the development of epilepsy besides the simple presence of a mutation. Thus, the failure of a candidate mutation to cause seizures does not mean that in other strains (or human families) it will not. The influence of the genetic background (or other unrecognized environmental factors) may cloud the potential effect of a mutation or polymorphism. Finally, the rat strains that have an unrecognized genetic component to their seizures strongly suggest, as mentioned earlier, that the concurrent occurrence of a sufficient number of normal polymorphisms may result in absence seizures. There is already good evidence from human studies of the multiplicity of genetic contributions, and it is likely that ultimately we will find that much of the global information we have gained from the genetics of rodent epilepsy will apply to people as well.

VII. Concluding Remarks

From the early "reverberating thalamocortical circuits" to the current conceptualization of thalamocortical rhythmicity, much information has been obtained that relates cellular, anatomical, and synaptic characteristics to specific oscillations of neuronal activity in the brain. The picture has emerged that absence seizures, and in general, the wide display of thalamocortical synchronized activity, depend on the interaction among those characteristics. If one, or a combination of them, does not show the proper "balance" (and this word should be interpreted carefully, because most of the time it is not really clear what the "balance" should be), the recurrent loops of the thalamocortical circuitry fosters the development of paroxysmal activity. Even though the direct tie to human epilepsy is not always clear, the explosion of genetic information has also provided many new insights into how the system can be tilted in the direction of epilepsy. It is certain, however, that use of *in vivo* animal models has contributed enormously to our understanding of the mechanisms of absence epilepsy, as, ultimately, they point to a dysfunction in the thalamocortical system.

References

1. Janz, D. (1997). The idiopathic generalized epilepsies of adolescence with childhood and juvenile age of onset. *Epilepsia.* **38**, 4–11.
2. Ottman, R. (2005). Analysis of genetically complex epilepsies. *Epilepsia* **46(suppl. 10)**, 7–14.
3. Pitkänen, A., Schwartzkroin, P. A., and Moshé, S. L. (eds.) (2006). "Models of Seizures and Epilepsy." Elsevier, Oxford.
4. McCormick, D. A., and Contreras, D. (2001). On the cellular and network bases of epileptic seizures. *Annu. Rev. Physiol.* **63**, 815–846.
5. Blumenthal, H. (2005). Cellular and network mechanisms of spike-wave seizures. *Epilepsia* **46(suppl. 9)**, 21–33.
6. Depaulis, A., Vergnes, M., and Marescaux, C. (1994). Endogenous control of epilepsy: the nigral inhibitory system. *Prog. Neurobiol.* **42**, 33–52.
7. Noebels, J. L. (2003). The biology of epilepsy genes. *Annu. Rev. Neurosci.* **26**, 599–625.
8. Wu, Y., Buzzi, A., Frantseva, M. V., Perez-Velazquez, J. L., Cortez, M., Liu, C., Shen, L., Gibson, K. M., and Snead, O. C. III. (2006). Status epilepticus in mice deficient for succinate semialdehyde dehydrogenase: $GABA_A$ receptor-mediated mechanisms. *Ann. Neurol.* **59**, 42–52.
9. Midzyanovskaya, I. S., Kuznetsova, G. D., Vinogradova, L. V., Shatskova, A. B., Coenen, A. M. L., and van Luijtelaar, G. (2004). Mixed forms of epilepsy in a population of WAG/Rij rats. *Epilepsy Behav.* **5**, 655–661.
10. Wong, C. G. T., Gibson, K. M., and Snead, O. C. (2004). From the street to the brain: neurobiology of the recreational drug γ-hydroxybutyric acid. *Trends Pharmacol. Sci.* **25**, 29–35.
11. Chan, K. F., Jia, Z., Murphy, P. A., Burnham, M. W., Cortez, M. A., and Snead, O. C. III. (2004). Learning and memory impairment in rats with chronic atypical absence seizures. *Exp. Neurol.* **190**, 328–336.
12. Steriade, M. (2001). Impact of network activities on neuronal properties in corticothalamic systems. *J. Neurophysiol.* **86**, 1–39.
13. Meeren, H., van Luijtelaar, G., Lopes da Silva, F., and Coenen, A. (2005). Evolving concepts on the pathophysiology of absence seizures. The cortical focus theory. *Arch. Neurol.* **62**, 371–376.
14. Crunelli, V., and Laresche, N. (2002). Childhood absence epilepsy: genes, channels, neurons, and networks. *Nat. Rev. Neurosci.* **3**, 371–382.
15. Gloor, P., and Fariello, R. G. (1988). Generalized epilepsy: some of its cellular mechanisms differ from those of focal epilepsy. *Trends Neurosci.* **11**, 63–68.
16. Jones, E. G. (2001). The thalamic matrix and thalamocortical synchrony. *Trends Neurosci.* **24**, 595–601.
17. Bal, T., and McCormick, D. A. (1996). What stops synchronized thalamocortical oscillations? *Neuron* **17**, 297–308.
18. Berkovic, S. F., Howell, R. A., Hay, D. A., and Hopper, J. L. (1998). Epilepsies in twins: genetics of the major epilepsy syndromes. *Ann. Neurol.* **43**, 435–445.
19. Kearney, J. A., Plummer, N. W., Smith, M. R., Kapur, J., Cummins, T. R., Waxman, S. G., Goldin, A. L., and Meisler, M. H. (2001). A gain-of-function mutation in the sodium channel gene Scn2a results in seizures and behavioral abnormalities. *Neuroscience* **102**, 307–317.

29

Paroxysmal Dyskinesia

Laurent Vercueil, MD, PhD
Anne de Saint Martin, MD
Edouard Hirsch, MD

Keywords: *channelopathies, dystonia, epilepsy, exercise, kinesigenic, paroxysmal dyskinesia*

I. Brief History and Nomenclature
II. Paroxysmal Kinesigenic Dyskinesia (Formerly Paroxysmal Kinesigenic Choreoathetosis)
III. Paroxysmal Nonkinesigenic Dyskinesia (Formerly Familial Paroxysmal Dystonia or Paroxysmal Dystonic Choreoathetosis)
IV. Paroxysmal Exercise–Induced Dystonia
V. Paroxysmal Hypnogenic Dyskinesia
VI. Conclusion
    References

I. Brief History and Nomenclature

Paroxysmal dyskinesias comprise a heterogenous group of disorders that share the recurrence of sudden abnormal involuntary movements out of a normal background. The abnormal movements may be choreic, ballistic, or dystonic but are more frequently a combination of these, thus giving strong support to the use of the term "dyskinesia" to account for this complexity. The episodes last for a few seconds to several hours and are purely motor, without alteration in consciousness. Recovery from attacks is usually rapid and complete. The neurological examination is typically normal between attacks, at least in idiopathic cases, and diagnosis relies on descriptions given by observers because attacks are usually not witnessed by a physician.

The presence or absence of precipitating factors has been considered crucial in delineating different forms of the disorder. Whereas stress, alcohol, fatigue, and coffee seem to act nonspecifically as promoting factors for all subtypes of paroxysmal dyskinesia, attacks in some patients can be selectively induced by sudden voluntary movements, such as standing up rapidly or making a ballistic limb movement. The term "kinesigenic" refers to this reflex component. Other patients report the occurrence of paroxysmal dyskinesia after sustained exercise, which might be of relatively low intensity (e.g., 5–10 minutes walking). In 1977, Lance [1] proposed to distinguish between three main categories: paroxysmal kinesigenic dyskinesia (PKD), paroxysmal nonkinesigenic dyskinesia (PNKD), and paroxysmal exercise-induced dyskinesia (PED). (Table 1 summarizes the main clinical characteristics of each type.) Twenty years later, Dermikiran and Jankovic [2] extended this classification, adding paroxysmal hypnogenic dystonia as a fourth category.

Whatever the subtype, paroxysmal dyskinesia usually starts during childhood (between ages 5 and 15 years) and tends to improve with age; adults rarely suffer from attacks. An association with different forms of epilepsy has been reported, such as benign afebrile infantile convulsions or absence epilepsy. The age-dependent pattern of such disorders, with seizures generally preceding the development of paroxysmal dyskinesia, is strongly suggestive of an inherited functional deficit, whose expression varies depending on brain maturation. This point will be developed later in the chapter.

From an etiological point of view, paroxysmal dyskinesias can be classified as primary (without any evident structural or degenerative cause) and secondary. Familial cases of paroxysmal dyskinesias and most sporadic PKD cases are categorized as primary paroxysmal dyskinesias, and genetic causes are increasingly identified (Table 2). Most secondary paroxysmal dyskinesias are variants of PNKD, but PKD can also be observed. These are usually related to structural abnormalities of the brain (stroke, multiple sclerosis) or metabolic/toxic disorders (Table 3). In one case series, secondary cases accounted for 22% of all patients with paroxysmal dyskinesia [3]. No heredodegenerative disease has been described as a cause of a "pure" paroxysmal dyskinesia phenotype (i.e., without neurological abnormality between the attacks), although paroxysmal dyskinesia has been reported in association with otherwise typical Huntington's disease and juvenile Parkinson's disease.

There are scant data about the frequency of paroxysmal dyskinesia in the general population, but the overall rate is believed to be low. In a study examining the relative frequency of different paroxysmal nonepileptic events in children and adolescents who underwent video electroencephalographic monitoring for suspected seizures over a 6-year period, only 5 (0.5%) of 883 patients had paroxysmal movement disorders [4]. In another study, only 1 of 285 children with nonepileptic paroxysmal events was diagnosed with paroxysmal dystonic choreoathetosis [5]. Such reports are suggestive of a low prevalence rate. Among 12,063 patients evaluated over 19 years in an U.S. movement disorders unit, 92 patients (0.76%) were given the diagnosis of paroxysmal dyskinesia [3].

This chapter deals separately with each subtype of paroxysmal dyskinesia, because etiologies and presumed pathophysiology appear to be substantially different.

II. Paroxysmal Kinesigenic Dyskinesia (Formerly Paroxysmal Kinesigenic Choreoathetosis)

When sudden movement provokes attacks, the term "paroxysmal kinesigenic dyskinesia" (PKD) is used. Typically, startle or sudden movement after a period of rest is easily identified as the precipitating factor and can be reproduced at will. There is often a refractory period after an attack, during which sudden movement may not succeed in eliciting another episode. Usually only one limb or side of the body is involved with rapid patterned dystonic movements in attacks that are brief, frequent (up to 100 daily), and exquisitely responsive to small doses of carbamazepine. A clinical reappraisal proposed new diagnostic criteria for PKD: identified kinesigenic trigger, short duration (<1 minute), no loss of consciousness or pain, normal neurological examination findings and exclusion of other

Table 1 Clinical Features of Different Forms

	PKD	PNKD	PED	PHD
Age at onset	Childhood	Childhood	Childhood	
Triggering factors	Sudden movement, startle	None	Sustained exercise	Nocturnal
Precipitating factors	Stress, coffee, alcohol	Stress, coffee, alcohol, tea, anxiety		
Duration of attacks	Seconds to 1 min	Minutes to several hours	Minutes (5–30)	Short, seconds to minutes
Frequency of attacks	Daily, up to 100	Rare, monthly	Rare	Frequent
Semiology	Choreoathetosis	Sensitive aura preceding dystonia	Dystonia	Choreoathetosis
Treatment	Carbamazepine +++	Clonazepam +/−	Clonazepam +/−	Carbamazepine

Table 2 Genetics of Paroxysmal Dyskinesia

Syndrome	Type of PxD	Epilepsy	Mutation/locus	References
ICCA	PKD, PED	Infantile partial epilepsy	Chromosome 16p12-q12	Szepetowski et al., 1997
EKD2	PKD	No	Chromosome 16p12-11.2	Valente et al., 2000
	PKD	No	Chromosome 16p11.2-q12.1	Tomita et al., 1999
EKD1	PKD	No	Chromosome 16p11.2-q11.2	Bennett et al., 2000
RE-PED-WC	PED	Rolandic epilepsy	Chromosome 16p12-11.2	Guerrini et al., 1999
	PKD	Infantile convulsion	Chromosome 16 D16S3131-D16S3396	Swodoba et al., 2000
Autosomal dominant PNKD	PNKD	No	Chromosome 2q	Fink et al., 1996; Fouad et al., 1996
Autosomal dominant CSE	PNKD, spasticity	No	Chromosome 1p	Auburger et al., 1996
	PKD, migraine, hemiplegic migraine	Febrile seizures (GEFS+)		Singh et al., 1998
X-linked PD	PKD, PHD severe global retardation	No epilepsy	Mutated MCT8 gene	Brockmann et al., 2005
GEPD	PNKD	Generalized epilepsy	Mutated *KCNMA1* gene encoding BK channel α subunit	Du et al., 2005

Table 3 Secondary Paroxysmal Dyskinesia: Etiological Factors

Paroxysmal kinesigenic dyskinesia:
Demyelinating disease (multiple sclerosis)
Idiopathic hypoparathyroidism
Nonketotic hyperglycemia
Stroke (putaminal, thalamic infarcts)
Head trauma
Cerebral palsy (delayed onset)
Progressive supranuclear palsy
Peripheral trauma
Monocarboxylase transporter 8 (*MCT8*) gene mutation

Paroxysmal nonkinesigenic dyskinesia:
Multiple sclerosis
Hypoglycemia
Thyrotoxicosis
Familial idiopathic hypoparathyroidism
Head trauma
Transient ischemic attacks
Nonketotic hyperglycemia
De Vivo syndrome
Kernicterus
Cytomegalovirus encephalitis
Migraine

Paroxysmal exertion-induced dyskinesia:
Head trauma

---

organic diseases, control of attacks with phenytoin or carbamazepine, and age at onset between 1 and 20 years if there is no family history of PKD [6].

A. Etiology

Primary PKD: PKD is usually inherited in an autosomal-dominant fashion, although cases appear to be sporadic. In families with autosomal-dominant PKD, a locus called *EKD1* has been mapped to the pericentromeric region of chromosome 16, 16p11.2-q12.1 [7], overlapping regions linked to other paroxysmal neurological disorders, such as autosomal-dominant infantile convulsions and choreoathetosis (ICCA) [8,9] and rolandic epilepsy, writer's cramp, and paroxysmal exercise-induced dyskinesias [10]. Another region on chromosome 16 (EKD2), outside the pericentromeric region, has been linked to PKD and epilepsy in one family [11]. To date, no gene has been identified.

Secondary PKD: There are various etiologies for secondary PKD, but two broad categories of conditions predominate: acquired structural brain lesions and metabolic disorders (see Table 3). The most common cause of secondary PKD is demyelinating disease. In these cases, attacks are painful and consistently triggered by hyperventilation. In patients with multiple sclerosis, the distinction between PKD and tonic spasms (painful, isometric, sustained muscular contractions) is often difficult to determine, and an overlap between the two clinical conditions is likely.

B. Pathogenesis and Pathophysiology

As a group, paroxysmal dyskinesias have been considered to represent transient dysfunction within the basal ganglia, mainly based on the observation that attacks replicate typical movement disorders and do not impair consciousness. However, conclusive evidence supporting this hypothesis is lacking. Given the exquisite response to antiepileptic drugs, the short duration of attacks, and triggering factors, PKD has been viewed by many as representing striatal seizures. In one patient with PKD, monitoring with depth electrodes showed ictal discharges in the supplementary motor area and the caudate nucleus during a typical attack [12]. In several aspects, however (mental retardation, tonic posturing during attacks, and noise-induced dyskinesia), this case differed substantially from typical PKD. On the other hand, medial frontal lobe epilepsy, which is responsible for brief and frequent hypermotor seizures sparing consciousness, which are often difficult to ascertain with only scalp electroencephalograms (EEG), has been suggested as a cause of symptoms in at least some patients [13].

Alternatively, several paroxysmal neurological disorders that are not of epileptic origin (episodic ataxia type 1 and 2, familial hemiplegic migraine) have been associated with different ion channel gene mutations, and these disorders exhibit the same pattern of age-dependent phenotype expression. In this context, the co-occurrence in some families of infantile epilepsy (benign infantile convulsion) and PKD suggests that a common genetically determined pathophysiological abnormality of ion channel function is variably expressed during brain mutation in both the cerebral cortex and the basal ganglia.

The role of sudden movement in triggering paroxysmal dyskinesia could be relevant for understanding the pathophysiology of PKD. Sudden movement causes rapid changes in primary motor and premotor cortex excitability, as demonstrated by transcranial magnetic studies. A similar increase in focal excitability has also been provoked simply by imagining the movement without actually performing it. In some patients, imagination of movement has triggered attacks, thus suggesting a role for the premotor cortex. A discrete pattern of abnormalities in cortical and spinal inhibitory circuits has been demonstrated in 11 patients with primary PKD [14]: Normal silent period and reduced short-interval cortical inhibition, suggesting impaired cortical γ-aminobutyric acid A (GABA$_A$) inhibition with sparing of GABA$_B$-mediated inhibition. This pattern is similar to what has been reported in untreated patients with epilepsy but different from what has been observed in patients with primary dystonia.

C. Relevant Structural Detail

Little is known about structural brain abnormalities underlying PKD. In 1967, Kertesz published an autopsy study of a young boy from his case series of PKD, who died after self-hanging. In this case, there were only slight changes in the substantia nigra. Brain imaging studies of patients with secondary PKD have shown structural lesions in the putamen, right frontotemporal region, globus pallidus, dorsal medulla oblongata, cervical spinal cord, and thalamus. In a patient without abnormalities on magnetic resonance imaging (MRI), single-photon emission computed tomography (SPECT) scans demonstrated a prominent increase in cerebral blood flow in the left posterolateral part of the thalamus during a PKD attack [15]. In a study of two young sisters with primary short-lasting paroxysmal dyskinesias, SISCOM (substraction of ictal SPECT co-registered to MRI) demonstrated a *decrease* in perfusion in bilateral premotor cortical areas during attacks [16]. These findings differ from what is usually described on ictal SPECT studies during frontal lobe seizures, namely unilateral frontal *hyperperfusion* associated with ipsilateral basal ganglia and thalamic hyperperfusion. However, the lack of functional imaging data makes impossible a definite conclusion regarding levels of brain activity and perfusion associated with PKD. In a patient with PKD caused by idiopathic primary hypoparathyroidism, calcium treatment improved ventral striatum hypometabolism, as documented by fluorodeoxyglucose–positron emission tomography (FDG-PET) studies [17].

D. Pharmacology, Biochemistry, and Molecular Mechanisms

The exquisite response of PKD to low doses of carbamazepine has been repeatedly emphasized. This drug acts by inhibiting the excitability of cellular membranes through use-dependent inactivation of the sodium channel, another indication of possible abnormal ion channel function. Despite the lack of rigorous drug trials in this rare clinical condition, several small case series have reported on the clinical efficacy of various antiepileptic drugs. This contrasts with what is observed in other forms of paroxysmal dyskinesia (PNKD, PED), in which drug treatment is generally disappointing.

III. Paroxysmal Nonkinesigenic Dyskinesia (Formerly Familial Paroxysmal Dystonia or Paroxysmal Dystonic Choreoathetosis)

Mount and Reback introduced the term "paroxysmal dystonic choreoathetosis" to describe choreodystonic attacks observed in a young adult. Alcohol, coffee, and fatigue—but not sudden movement—brought on these episodes, which could last several hours. Other members of the family were affected in a pattern indicating an autosomal-dominant pattern of inheritance. Attacks in PNKD tend to be dystonic and of long duration but with a low frequency of recurrence. This contrasts with what is usually observed in PKD.

A. Etiology

Primary PNKD: In the mid-1990s, the familial "Mount and Reback type" of PNKD was linked to chromosome 2q31-36 in different pedigrees. More recently, missense mutations in the myofibrillogenesis regulator 1 (*MR-1*) gene have been reported to co-segregate with PNKD in 50 individuals from 8 families [18]. The mutations (Ala to Val) cause changes in the N-terminal region of two MR-1 isoforms. In two unrelated families, the same single-nucleotide mutations predicted substitution of valine for alanine in residue 7 in one family and residue 9 in the other [19].

A different gene mutation has been found in a large family with coexistent generalized epilepsy (mainly absence seizure) and PNKD. The mutation occurs in a subunit of the BK channel on chromosome 10q22 (GEPD) [20]. Another form of PNKD, associated with interictal spasticity (called *CSE*), has been mapped to a 2cM region on chromosome 1p, but the gene is unknown [21].

Secondary PNKD: Various etiological factors can be responsible for secondary PNKD (see Table 3). In one case series, secondary PNKD was related to stroke, central trauma, kernicterus, cytomegalovirus (CMV) encephalitis, multiple sclerosis, and migraine [3].

B. Pathogenesis and Pathophysiology

Genetic findings in PNKD suggest a new conception of the molecular basis of the condition. The two mutations found in the *MR-1* gene in four unrelated families with PNKD probably disrupt the amino terminal helix, which indicates that this region of the gene is critical for proper gene function under stressful conditions. The MR-1L isoform is specifically expressed in brain and localized to the cell membrane, whereas the MR-1S isoform is ubiquitously expressed and shows diffuse cytoplasmic and nuclear localization. The *MR-1* gene is homologous to the hydroxyacylglutathione hydrolase *(HAGH)* gene, which functions in a pathway to detoxify methylglyoxal, a compound present in coffee and alcoholic beverages and produced as a by-product of oxidative stress. A mechanism whereby alcohol, coffee, and stress act as precipitants of PNKD attacks mediated by abnormal stress response pathways remains to be elucidated.

The discovery of a mutation in the BK channel, a large conductance calcium-sensitive potassium channel, in GEPD has promoted *in vitro* studies that suggest another pathophysiological model. The primary effect of this mutation is to increase the Ca^{2+} sensitivity of the BK channel (threefold to fivefold), which leads to greater macroscopic potassium conductance. Thus, the D434G mutation leads to a gain in function of the α subunit: more rapid repolarization of action potential by D434G mutant channels. Enhancing this repolarization enables faster recruitment (removal of inactivation) of sodium channels, allowing neurons to fire at higher frequency. Alternatively, enhancing some inhibitory currents could promote a bursting mode of discharge within a neuronal circuit. Ethanol can directly activate the BK channel *in vivo* in *C. elegans*. Thus, enhancement of BK channels should lead to increased excitability by inducing rapid repolarization of action potentials, resulting in generalized epilepsy and paroxysmal dyskinesia by allowing neurons to fire at a faster rate.

C. Relevant Structural Detail

A brain imaging study of a large British family with chromosome 2q-linked PNKD showed no lesions using MRI, MR spectroscopy (MRS), and striatal D_2 receptor binding measured by ^{11}C-raclopride [22]. Two autopsy cases have been published showing only a slight asymmetry in the substantia nigra and abnormal melanin pigment noted in macrophages of the locus ceruleus [23]. Relatively few studies have explored changes in brain perfusion during PNKD attacks. In a patient studied by ^{99m}Tc-EDC SPECT, there was increased blood flow in the right caudate and thalamus but reduced perfusion in the left inferior dorsal frontal area [24]. Insufficient data make any conclusion regarding this issue impossible.

D. Pharmacology and Biochemistry

In contrast to PKD, PNKD usually is not responsive to anticonvulsants, with the exception of benzodiazepine and GABA-potentiating drugs, which could be of some help. On the other hand, transient dopaminergic overactivity has also been suggested to play a role in PNKD. However, a normal (^{11}C)dihydrotetrabenazine PET study has been carried out in a family with PNKD, suggesting that nigrostriatal dopaminergic nerve terminals are spared in this disorder [25], as already mentioned in a British family [22].

The tottering mutant mouse, resulting from a mutation within the gene encoding the α_1 subunit of P/Q-type calcium channels, manifests an episodic movement disorder similar to paroxysmal dyskinesia, in addition to absence epilepsy. In this model, attacks of dyskinesia impair motor function approximately once or twice per day and last from 20 to 40 minutes. Although attacks are often preceded by increased activity, the main triggering factor appears to be a stress such as short-term restraint [26]. Caffeine administration and ethanol provoked attacks, and carbamazepine appears therapeutically ineffective, because, conversely, this drug significantly induced attacks [26]. In this model, nimodipine, an L-type calcium channel blocker, and MK801, a noncompetitive N-methyl-D-aspartate (NMDA) receptor antagonist, blocking calcium entry in the cell, prevented attacks provoked by caffeine, but only nimodipine was effective against attacks provoked by ethanol [26].

IV. Paroxysmal Exercise–Induced Dystonia

Paroxysmal exercise–induced dystonia (PED) is distinct from PKD because attacks develop after several minutes, or hours, of continuous exercise (such as modest walking or running). Attacks resolve within a few minutes of stopping the activity. Dystonic features predominate, with frequent fixed postures and sometimes pain that is misinterpreted as muscular cramps. The legs are mainly involved, but generalized attacks can also occur. It has been suggested that PED may be an intermediate form between PKD and PNKD or belong to the spectrum of PNKD, given the relatively long duration of the episodes and the lack of response to anticonvulsant drugs. More than other paroxysmal dyskinesias, PED must be recognized and distinguished from action-induced dystonia occurring in juvenile parkinsonism or dopamine-responsive dystonia. Since Lance first described the syndrome in 1977, it has been clear that the PED group of patients with paroxysmal dyskinesia is the smallest group, with only a few families and single cases reported in the literature.

A. Etiology

Primary PED: Guerrini et al. [10] described a family with rolandic epilepsy, writer's cramp, and PED with linkage to chromosome 16p12-11.2, an area included in the ICCA critical region. Because some of the patients reported in ICCA families have exhibited exercise-induced paroxysmal dyskinesia, a continuum could exist between the two entities. It also remains open whether PED is an intermediate form between PKD and PNKD, or a *forme fruste* of PNKD.

Secondary PED: Few cases of PED are reported as secondary. Head trauma has been mentioned in one case, but none was observed among a case series of secondary paroxysmal dyskinesia [3].

B. Pathophysiology

Experimental studies in animals have shown that exercise increases serum calcium, which is transported to the brain, where it stimulates dopamine synthesis through a calcium/CaM-dependent system. Such data favor a role for dopamine transmission in the pathophysiology of PED. Exercise-induced foot dystonia is a classic sign in Parkinson's disease. However, there are no direct data that support a dopaminergic hypothesis of PED. On the other hand, one case study showed that a patient with PED had normal motor cortex excitability as measured by transcranial magnetic stimulation, whereas there was a facilitation of cortically elicited responses in forearm muscles during an attack [27].

C. Relevant Structural Detail

There are no autopsy studies of PED. A SPECT study showed lack of cortical hyperperfusion during exercise-induced attacks, suggesting that an epileptic cause was unlikely [28]. In this particular case, increased cerebellar and decreased frontal perfusion were found during an attack. Further functional imaging studies are needed to confirm this finding.

D. Pharmacology and Biochemistry

The pharmacology of PED is even less clear than for the other paroxysmal dyskinesias. The role of serum calcium rise induced by exercise could play a role in the occurrence of attacks, because some secondary paroxysmal dyskinesias occur in patients with hypoparathyroidism. On the other hand, primary motor cortex excitability changes have been shown to occur secondary to prolonged exercise. The coincidence of membrane excitability changes and the varying ionic concentration could be the key factor for paroxysmal dyskinesia to occur.

V. Paroxysmal Hypnogenic Dyskinesia

The first description of paroxysmal hypnogenic dyskinesia was published in the early 1980s. This report was of five patients who had paroxysmal dyskinesia that occurred during sleep almost every night. Patients exhibited large-amplitude dystonic movements, brief in duration, and no loss of consciousness, which could be repeated throughout the night. Because there were no detectable accompanying EEG abnormalities, the authors concluded that this entity was a nocturnal form of a paroxysmal movement disorder, or alternatively, a new form of parasomnia. However, it has become increasingly clear since the early 1990s that in a large proportion of these cases, especially those with a strong familial component, the nocturnal attacks are related to mesial frontal lobe seizures. The identification of autosomal-dominant nocturnal frontal lobe epilepsy (ADNFLE) in families exhibiting typical PHD, and the further identification of the causal mutation, adds further doubt on the validity of the syndrome. It is now generally accepted that attacks of PHD are manifestations of a seizure disorder, usually frontal lobe epilepsy.

VI. Conclusion

Despite advances, important details of the pathophysiology of paroxysmal dyskinesia remain unknown. Progress in gene identification and understanding the link between a mutated gene product and probable transient ion channel dysfunction are becoming key targets for investigation. However, it is already possible to formulate various hypotheses that can be tested and that promote a general perspective of the disorder.

At the brain level, paroxysmal dyskinesias may be viewed as prominent features of basal ganglia disorders or frontal lobe seizures arising from deep mesial foci. In both cases, attacks are brief, consciousness is not obviously impaired, and EEG abnormalities are not evident on scalp recordings. Carbamazepine offers effective treatment, at least when attacks are short in duration. There is no evidence that favors one hypothesis exclusively over the other. Indeed, there is growing evidence, mainly from familial forms of Parkinson's disease that are associated with early onset generalized epilepsy (either benign infantile convulsions or absence epilepsy), which suggests a continuum between the disorders, based mainly on different stages of brain maturation.

At the molecular level, a mutation found in families with both epilepsy and paroxysmal dyskinesia suggests that paroxysmal dyskinesia belongs to the group of episodic nervous system disorders that now include periodic paralysis, cardiac arrhythmias, episodic ataxia, hemiplegic migraine, and some rare idiopathic forms of epilepsy. All of these are characterized by relatively brief paroxysmal attacks of neurological dysfunction, normal behavior between episodes, and, often, precipitating factors (stress, fatigue, alcohol, altered serum potassium). The list of gene mutations that result in episodic neurological phenotypes is growing rapidly. All of these disorders result either from direct ion channel gene mutations or from genes encoding proteins that modulate ion channel function.

References

1. Lance, J. W. (1977). Familial paroxysmal dystonic choreoathetosis and its differentiation from related syndromes. *Ann. Neurol.* **2**(4), 285–293.
2. Dermikiran, M., and Jankovic, J. (1995). Paroxysmal dyskinesias: clinical features and classification. *Ann. Neurol.* **38**(4), 571–579.
3. Blakeley, J., and Jankovic, J. (2002). Secondary paroxysmal dyskinesias. *Mov. Disord.* **17**, 726–734.
4. Kotagal, P., Costa, M., Wyllie, E., and Wolgamuth, B. (2002). Paroxysmal nonepileptic events in children and adolescents. *Pediatrics* **110**(4), e46.
5. Bye, A., Kok, D., Ferenschild, F. T. J., and Vles, J. (2000). Paroxysmal nonepileptic events in children: a retrospective study of a period of 10 years. *J. Paediatr. Child Health* **36**, 244–248.
6. Bruno, M. K., Hallett, M., Gwinn-Hardy, K., Sorensen, B., Considine, E., Tucker, S., Lynch, D. R., Mathews, K. D., Swoboda, K. J., Harris, J., Soong, B. W., Ashizawa, T., Jankovic, J., Renner, D., Fu, Y. H., and Ptacek, L.J. (2004). Clinical evaluation of idiopathic paroxysmal kinesigenic dyskinesia. New diagnostic criteria. *Neurology* **63**, 2280–2287.
7. Tomita, H., Nagamitsu, S., Wakui, K. Fukushima, Y., Yamada, K., Sadamatsu, M. Masui, A., Konishi, T., Matsuishi, T., Aihara, M., Shimizu, K., Hashimoto, K., Mineta, M., Matsushima, M., Tsujita, T., Saito, M., Tanaka, H., Tsuji, S., Takagi, T., Nakamura, Y., Nanko, S., Kato, N., Nakane, Y., and Niikawa, N. (1999). Paroxysmal kinesigenic choreoathetosis locus maps to chromosome 16p11.2-q12.1. *Am. J. Hum. Genet.* **65**(6), 1688–1697.
8. Szepetowski, P., Rochette, J., Berquin, P., Piussan, C., Lathrop, G. M., and Monaco, A. P. (1997). Familial infantile convulsions and paroxysmal choreoathetosis: a new neurological syndrome linked to the pericentromeric region of human chromosome 16. *Am. J. Hum. Genet.* **61**(4), 889–898.
9. Swodoba, K. J., Soong, B., McKenna, C., Brunt, E. R., Litt, M., Bale, J. F., Jr., Ashizawa, T., Bennett, L. B., Bowcock, A. M., Roach, E. S., Gerson, D., Matsuura, T., Heydemann, P. T., Nespeca, M. P., Jankovic, J., Leppert, M., and Ptacek, L. J. (2000). Paroxysmal kinesigenic dyskinesia and infantile convulsions: clinical and linkage studies. *Neurology* **55**(2), 224–230.
10. Guerrini, R., Bonanni, P., Nardocci, N., Parmeggiani, L., Piccirilli, M., De Fusco, M., Aridon, P., Ballabio, A., Carrozzo, R., and Casari, G. (1999). Autosomal recessive rolandic epilepsy with paroxysmal exercise-induced dystonia and writer's cramp: delineation of the syndrome and gene mapping to chromosome 16p12-11.2. *Ann. Neurol.* **45**(3), 344–352.
11. Valente, E. M., Spacey, S. D., Wali, G. M., Bhatia, K. P., Dixon, P. H., Wood, N. W., and Davis, M. B. (2000). A second paroxysmal kinesigenic choreoathetosis locus (EKD2) mapping on 16q13-q22.1 indicates a family of genes which give rise to paroxysmal disorders on human chromosome 16. *Brain* **123**, 2040–2045.
12. Lombroso, C. T. (1995). Paroxysmal choreoathetosis: an epileptic or non-epileptic disorder? *Ital. J. Neurol. Sci.* **16**(5), 271–277.
13. Fish, D. R., and Marsden, C. D. (1994). Epilepsy masquerading as a movement disorder. *In:* "Movement Disorders" (C. D. Marsden and S. Fahnn, eds.), 3rd edition. pp. 346–359. Butterworth-Heinemann, Oxford.
14. Mir, P., Huang, Y.-Z., Gilio, F., Edwards, M. J., Berardelli, A., Rothwell, J. C., and Bhatia, K. P. (2005). Abnormal cortical and spinal inhibition in paroxysmal kinesigenic dyskinesia. *Brain* **128**, 291–299.

15. Shirane, S., Sasaki, M., Kogure, D., Matsuda, H., and Hashimoto, T. (2001). Increased ictal perfusion of the thalamus in paroxysmal kinesigenic dyskinesia. *J. Neurol. Neurosurg. Psychiatry* **71**, 408–410.
16. Thiriaux, A., de Saint Martin, A., Vercueil, L., Battaglia, F., Armspach, J. P., Hirsch, E., Marescaux, C., and Namer, I. J. (2002). Co-occurrence of infantile epileptic seizures and childhood paroxysmal choreoathetosis in one family: clinical, EEG, and SPECT characterization of episodic events. *Mov. Disord.* **17**, 98–104.
17. Volonte, M. A., Perani, D., Lanzi, R., Poggi, A., Anchisi, D., Balini, A., Comi, G., and Fazio, F. (2001). Regression of ventral striatum hypometabolism after calcium/calcitriol therapy in paroxysmal kinesigenic choreoathetosis due to idiopathic primary hypoparathyroidism. *J. Neurol. Neurosurg. Psychiatry* **71**, 691–695.
18. Lee, H. Y., Xu, Y., Huang, Y., Ahn, A. H., Auburger, G. W., Pandolfo, M., Kwiecinski, H., Grimes, D. A., Lang, A. E., Nielsen, J. E., Averyanov, Y., Servidei, S., Friedman, A., Van Bogaert, P., Abramowicz, M. J., Bruno, M. K., Sorensen, B. F., Tang, L., Fu, Y. H., and Ptacek, L. J. (2004). The gene for paroxysmal non-kinesigenic dyskinesia encodes an enzyme in a stress response pathway. *Hum. Mol. Genet.* **13**(24), 3161–3170.
19. Chen, D. H., Matsushita, M., Rainier, S., Meaney, B., Tisch, L., Feleke, A., Wolff, J., Lipe, H., Fink, J., Bird, T. D., and Raskind, W. H. (2005). Presence of alanine-to-valine substitutions in myofibrillogenesis regulator 1 in paroxysmal nonkinesigenic dyskinesia: confirmation in 2 kindreds. *Arch. Neurol.* **62**(4), 597–600.
20. Du, W., Bautista, J. F., Yang, H., Diez-Sampedro, A., You, S. A., Wang, L., Kotagal, P., Luders, H. O., Shi, J., Cui, J., Richerson, G. B., and Wang, Q. K. (2005). Calcium-sensitive potassium channelopathy in human epilepsy and paroxysmal movement disorder. *Nat. Genet.* **37**, 733–738.
21. Auburger, G., Ratzlaff, T., Lunkes, A., Nelles, H. W., Leube, B., Binkofski, F., Kugel, H., Heindel, W., Seitz, R., Benecke, R., Witte, O. W., and Voit, T. (1996). Gene for autosomal dominant paroxysmal choreoathetosis/spasticity (CSE) maps to the vicinity of a potassium channel gene cluster on chromosome 1p, probably within 2 cM between D1S443 and D1S197. *Genomics* **31**, 90–94.
22. Jarman, P. R., Bhatia, K. P., Davie, C., Heales, S. J. R., Turjanski, N., Taylor-Robinson, S. D., Marsden, C. D., and Wood, N. W. (2000). Paroxysmal dystonic choreoathetosis: clinical features and investigation in a large family. *Mov. Disord.* **15**(4), 648–657.
23. Goodenough, D., Fariello, R., Annis, B., and Chun, R. (1978). Familial and acquired paroxysmal dyskinesias. *Arch. Neurol.* **35**, 827–831.
24. Del Carmen Garcia, M., Intruvini, S., Vazquez, S., Beserra, F., and Rabinowicz, A. L. (2000). Ictal SPECT in paroxysmal non-kinesigenic dyskinesia. Case Report and review of the literature. *Parkinsonism Rel. Disord.* **6**, 119–121.
25. Bohen, N. I., Albin, R. L., Frey, K. A., and Fink, J. K. (1999). (+)-alpha-(^{11}C)dihydrotetrabenazine PET imaging in familial paroxysmal dystonic choreoathetosis. *Neurology* **52**, 1067–1069.
26. Fureman, B. E., Jinnah, H. A., and Hess, E. J. (2002). Triggers of paroxysmal dyskinesia in the calcium channel mouse mutant tottering. *Pharmacol. Biochem. Behav.* **73**, 631–637.
27. Meyer, B.-U., Irlbacher, K., and Meierkord, H. (2001). Analysis of stimuli triggering attacks of paroxysmal dystonia induced by exertion. *J. Neurol. Neurosurg. Psychiatry* **70**, 247–251.
28. Kluge, A., Kettner, B., Zschenderlein, R., Sandrock, D., Munz, D. L., Hesse, S., and Meierkord, H. (1998). Changes in perfusion pattern using ECD-SPECT indicate frontal lobe and cerebellar involvement in exercise-induced paroxysmal dystonia. *Mov. Disord.* **13**, 125–134.
29. Bennett, L. B., Roach, E. S., and Bowcock, A. M. (2000). Locus for paroxysmal kinesigenic dyskinesia maps to human chromosome 16. *Neurology* **54**, 125–130.
30. Fink, J. K., Rainer, S., Wilkowski, J., Jones, S. M., Kume, A., Hedera, P., Albin, R., Mathay, J., Girbach, L., Varvil, T., Otterud, B., and Leppert, M. (1996). Dystonic choreoathetosis: tight linkage to chromosome 2q. *Am. J. Hum. Genet.* **59**, 140–145.
31. Fouad, G. T., Servidei, S., Durcan, S., Bertini, E., and Ptacek, L. J. (1996). A gene for familial paroxysmal dyskinesia (FPD1) maps to chromosome 2q. *Am. J. Hum. Genet.* **59**, 135–139.
32. Singh, R., Macdonell, R. A., Scheffer, I. E., Crossland, K. M., and Berkovic, S. F. (1999). Epilepsy and paroxysmal movement disorders in families: evidence for shared mechanisms. *Epileptic Disord.* **1**, 93–99.
33. Brockmann, K. Dumitrescu, A. M., Best, T. T., Hanefeld, F., and Refetoff, S. (2005). X-linked paroxysmal dyskinesia and severe global retardation caused by defective MCT8 gene. *J. Neurol.* **252**, 663–666.

30

Myoclonus

Pietro Mazzoni, MD, PhD
Samay Jain, MD

Keywords: *asterixis, electrophysiology, EMG, myoclonus, startle*

I. History and Nomenclature
II. Pathophysiology
III. Pharmacology and Molecular Genetics
IV. Relationship between Pathophysiology and Symptoms
V. Natural History
   References

I. History and Nomenclature

A. History

The study of myoclonus has been the stimulus for a series of elegant investigations of motor system electrophysiology that led to significant new insights into the relationship between cortical activity and movement [1]. The original description by E. A. Carmichael in 1947 of a patient with stimulus-sensitive limb jerks was followed by the investigations of G. D. Dawson and P. A. Merton, and later by C. D. Marsden's group, which led to the identification of transcortical reflexes. Under Marsden's leadership, the MRC Unit at the National Hospital in Queen Square, London, established a vast body of knowledge on the neurophysiology of myoclonus. Subsequently, this research effort has involved many laboratories around the world and now includes investigations into the pharmacological, biochemical, and genetic bases of this condition. Following the approach pioneered by Marsden, the study of myoclonus remains tightly linked to our understanding of normal motor control processes in the central nervous system (CNS).

B. Definition

Myoclonus is a sudden, rapid, shock-like, involuntary movement that can appear in a variety of diseases [2]. (*Note*: Citations in this chapter are limited to review articles and selected studies. References to individual experiments mentioned in the text are contained in the relevant review articles.) The most common form of myoclonus (*positive myoclonus*) consists of sudden brief muscle activations that lead to fast, shock-like muscle contractions. When the activated muscles span a joint, contractions result in jerks of limb segments. In other cases, movements may be limited to more focal body regions, as seen in the contractions of the cheek and the lower lid in hemifacial spasm. A variety of muscles may be involved in myoclonus, either independently or concurrently. Observed patterns include synchronous activation of agonists and antagonists, spread of activation from one muscle group to adjacent ones, and frequent independent contractions of multiple small muscles of the distal limb *(polyminimyoclonus)*. Myoclonus usually occurs at irregular time intervals, but it can appear rhythmically in certain conditions, such as familial cortical tremor.

A distinct form of myoclonus is *negative myoclonus*, which consists of a transient pause in activation of contracting muscles, leading to brief loss of muscle tension. In patients being asked to hold their arms extended forward at shoulder level, with wrist and fingers fully extended, negative myoclonus of the finger extensors produces brief forward flexion of the fingers *(asterixis)*. In a patient who is standing, negative myoclonus of proximal leg muscles is manifest as transient loss of posture (posture lapse). Positive and negative myoclonus can coexist in the same patient.

C. Classification

Myoclonus is a neurological sign. Although there are syndromes of idiopathic myoclonus in which this sign is the only clinical manifestation of the disease, myoclonus also appears as a clinical finding in many other medical conditions. Therefore, multiple classification schemes exist. One scheme is based on the *involved muscles*, dividing myoclonus into focal/segmental, multifocal, and generalized. Another classification is based on its *site of origin*: cortical, brainstem, spinal. A *pathophysiological* classification divides myoclonus according to its electrophysiological nature. Such a scheme identifies specific types of myoclonus, including (but not limited to) cortical positive myoclonus (CPM), negative myoclonus (defined earlier), reticular reflex myoclonus (originating in the brainstem), hyperekplexia (excessive startle), propriospinal myoclonus (repetitive, nonrhythmic myoclonus of axial muscles from nerve impulses transmitted along the slow propriospinal pathway), and peripheral nerve myoclonus.

Another classification scheme reflects the *diseases* that can cause myoclonus. A group of conditions have myoclonus as their only or principal symptom, including essential myoclonus and familial cortical tremor. The clinical triad of myoclonus, ataxia, and epilepsy, previously described as the Ramsay Hunt syndrome, includes specific diseases that have since been identified, including mitochondrial encephalomyopathy with ragged red fibers (MERRF), Lafora's disease (LD), and Kufs' disease. It also includes progressive myoclonic epilepsies of unknown cause and progressive myoclonus ataxias. Myoclonus can appear as part of systemic metabolic derangement, as in uremic renal failure and hepatic encephalopathy; neurodegenerative diseases, such as corticobasal degeneration; and infectious conditions, such as Creutzfeldt-Jakob disease. Focal lesions, as well as medications, can cause myoclonus.

In this chapter, we review the current understanding of the neurobiological abnormalities underlying myoclonus. The emphasis is on experimental and clinical evidence, for specific pathogenetic mechanisms. Clinical aspects of the myoclonic syndromes are mentioned only briefly.

II. Pathophysiology

Insight into the mechanisms and pathogenesis underlying myoclonus derives from three major areas of investigation: neurophysiological (electrophysiological) studies, pharmacological experiments, and molecular genetic analysis.

A. Neurophysiology

Neurophysiologically, myoclonus can be classified based on type of discharge (positive and negative) and location of origin (cortex, brainstem, spinal cord).

1. Positive Myoclonus

The neurophysiological hallmark of myoclonus is a sudden burst of electromyographic (EMG) activation, which is as brief as 30–60 ms in myoclonus of cortical origin but can be much longer in other forms (Fig. 1).

a. Cortical Positive Myoclonus CPM is caused by transient electrical discharges in a focal cortical area [3,4]. The EMG discharge of CPM is preceded by a small scalp discharge seen on electroencephalographic (EEG) recordings. This EEG signal is often difficult to identify among electrographic noise and EEG signals unrelated to myoclonus. The *back-averaging* technique, introduced by Shibasaki et al. [5], allows identification of EEG discharges related to myoclonic jerks. After recording the EEG during the occurrence of several myoclonic jerks, the stored traces are synchronized to the EMG discharge and averaged. In the averaging process, EEG activity that is not related to the muscle jerk is attenuated, whereas activity that is time-locked to the jerk is enhanced. This method reveals a clear EEG signal, the pre-myoclonus spike, occurring shortly before the EMG discharge (Fig. 2). Its latency is consistent with nerve conduction time from the cortex to the involved muscles, and it arises from cortex adjacent to the central sulcus in the contralateral hemisphere. Various approaches have been employed to investigate the possibility that CPM is due to hyperexcitability of either the primary motor cortex or the primary somatosensory cortex.

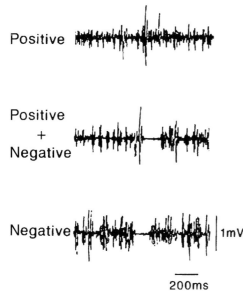

Figure 1 Electromyogram (EMG) correlates of cortical myoclonus, positive, negative, or combination of the two, recorded during sustained muscle contraction from a patient with progressive myoclonus epilepsy. (Reprinted, with permission, from Shibasaki, H., and Hallett, M. [2005]. Electrophysiological studies of myoclonus. *Muscle Nerve* **31**[2], 157–174.)

Dipole source localization allows spatial analysis of scalp recordings of signals obtained via EEG or magnetoencephalography. Studies using this approach suggest that most pre-myoclonus spikes originate from primary motor cortex, with fewer coming from primary somatosensory cortex.

Motor cortical hyperexcitability is also supported by paired transcranial magnetic stimulation (TMS) studies. A single magnetic pulse applied on the scalp over the hand area of the primary motor cortex normally elicits a motor evoked potential (MEP) in hand muscles when the pulse's intensity is sufficiently high (suprathreshold). The magnitude of the resulting MEP can be modulated by applying a preceding subthreshold pulse. The first pulse acts as a conditioning pulse for the second one (test pulse), which is the one that produces the MEP. In healthy subjects, when the interval between these pulses is less than 6 ms, the MEP resulting from the second pulse is reduced (Fig. 3). This reduction reflects the activation of intracortical inhibitory mechanisms. In patients with cortical myoclonus, the MEP produced by the second pulse is reduced by a smaller amount than in control subjects, suggesting impaired intracortical inhibition. The inhibitory mechanisms unmasked by paired-pulse TMS studies are mediated by the inhibitory neurotransmitter γ-aminobutyric acid (GABA). In cortical positive myoclonus, abnormal GABA transmission has thus been implicated as a possible mechanism for this condition (also see later discussion).

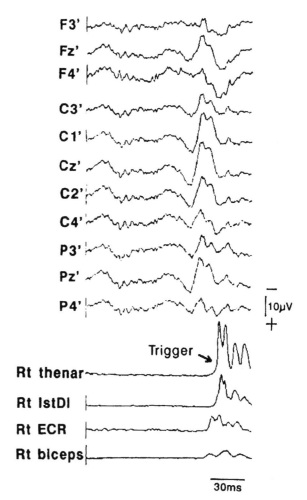

Figure 2 Records of jerk-locked back-averaging obtained from a patient with progressive myoclonic epilepsy. Surface electromyograms (EMGs) were recorded from four different muscles of the right upper extremity, and the onset of the EMG discharge from the right thenar muscle was used as a trigger pulse to back-average multichannel electroencephalogram (EEG) recording. EEG recorded in reference to ipsilateral earlobe electrode. A positive–negative biphasic EEG spike is seen maximally near the midline vertex, slightly shifted to the left (C1–Cz), and widespread over the scalp. Note that the myoclonic EMG discharge, which was also averaged with respect to the same fiducial point, spreads rapidly from the proximal muscles to the distal ones. ECR, extensor carpi radialis muscle; 1st DI, first dorsal interosseous muscle; Rt, right. (Reprinted, with permission, from Shibasaki, H., and Hallett, M. [2005]. Electrophysiological studies of myoclonus. *Muscle Nerve* **31**[2], 157–174.)

Other studies point to hyperexcitability of the somatosensory cortex as a possible cause of CPM. Somatosensory evoked potentials (SSEPs) are EEG discharges elicited by electrical stimulation of peripheral nerves and recorded over the postcentral gyrus. They normally do not lead to movement. Many patients with cortical myoclonus have markedly enlarged SSEPs *(giant SSEPs)* that are followed by reflex muscle contractions. The latter occur with a delay

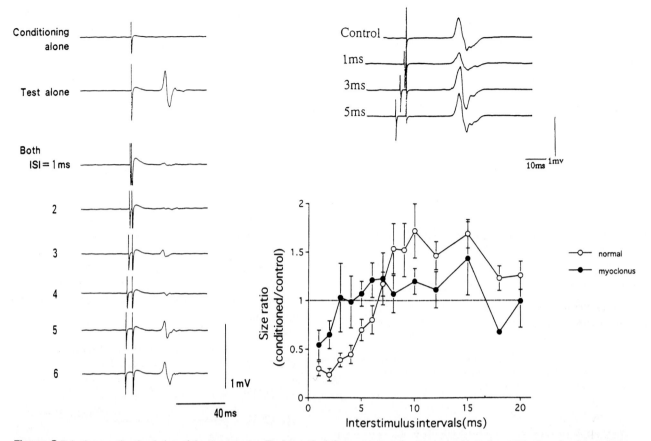

Figure 3 Paired magnetic stimulation of the motor cortex. Traces on the left are responses in a normal subject. A conditioning stimulus evoked no response (subthreshold) (top trace). A test stimulus given alone elicited a response (~0.1–0.3 mV) (second trace). The conditioning stimulus reduced the size of response to the stimulus when it preceded the test stimulus by 1–6 ms. This suppression is considered an event at the motor cortex and should reflect GABAergic-inhibitory circuits in the motor cortex. Responses from a cortical myoclonic patient are shown in the upper traces on the right. In contrast to normal subjects, the suppression was not evoked at interstimulus intervals (ISI) of 4 and 5 ms, whereas it was evoked at an ISI of 1 ms. Mean (±SE) time courses of intracortical inhibition are shown in the lower figure on the right. Inhibition at short intervals (1–5 ms) was seen in normal subjects but reduced in patients with cortical myoclonus. (Reprinted, with permission, from Ugawa, Y., Hanajima, R., Okabe, S., Yuasa, K. [2002]. Neurophysiology of cortical positive myoclonus. *Adv. Neurol.* **89**, 89–97.)

long enough to suggest that the sensory signal travels to the cortex and triggers a discharge from the motor cortex, causing the reflex movement (long loop reflex). Such a loop may be the basis for the phenomenon of *reflex myoclonus*: somatic stimulation, such as a light tap or pinprick of the hand, elicits myoclonus, either in the same body region or in a different one. This feature is characteristic of CPM.

Overall, it remains unclear whether CPM is due to hyperexcitability of motor cortex, somatosensory cortex, or both. An additional consequence of abnormal intracortical inhibition in the motor cortex of one hemisphere may be a reduction of inhibition in the contralateral motor cortex. In healthy subjects, the hand motor area has a facilitatory connection to the contralateral homotopic area with surround inhibition. In CPM, abnormalities of intracortical inhibition might release the inter-hemispheric facilitatory connection, which is masked by surround inhibition in normal subjects. This mechanism, though not proven, could explain how myoclonus can appear bilaterally even with unilateral cortical lesions.

b. *Brainstem Myoclonus* Two types of myoclonus originating in the brainstem [6] are reticular reflex myoclonus and exaggerated startle (hyperekplexia). A syndrome of rhythmic movements of muscles around the palate, also termed "palatal myoclonus," has been reclassified as a type of tremor and is not discussed here.

i. Reticular Reflex Myoclonus Reticular reflex myoclonus (RRM) consists of brief generalized EMG bursts, lasting 10–30 ms, which are triggered by sensory stimulation (touch, muscle stretch) and action. It was first described in the chronic posthypoxic state, as part of the Lance-Adams syndrome [7]. (Note, however, that this syndrome can also produce cortical myoclonus.) RRM can also appear in the setting of renal insufficiency. EEG correlates are not always present but include generalized spike–wave discharges that are not time-locked to muscle activation. If cranial muscles are involved, the sternocleidomastoid is activated first and muscles innervated by other cranial nerves are activated in sequence, from caudal to rostral.

Some insight into the mechanism of RRM comes from an animal model of brainstem myoclonus, for example, urea-induced myoclonus in the rat. In this model, the electrical activity of cells in the nucleus reticularis gigantocellularis (in the medulla) undergoes spontaneous fluctuations, known as "paroxysmal depolarization shifts," which may be due to reduced binding of the inhibitory neurotransmitter glycine. Glycine receptors in the brainstem and spinal cord have been implicated in the generation of myoclonus and startle response (as in hereditary hyperekplexia, discussed later in this chapter). Other evidence pointing to a brainstem source is that thyrotropin-releasing hormone, a hormone that stimulates medullary reticular neurons, enhances posthypoxic myoclonus in patients with this syndrome.

ii. Startle Syndromes (Hyperekplexia) Hyperekplexia is a pathological exaggeration of the physiological startle response [8]. It consists of an exaggerated response to unexpected stimuli, especially sounds. Compared with normal startle, the response is more intense and longer lasting; it can be triggered more easily; and it usually does not habituate. Unlike normal startle, the response in hyperekplexia almost always involves the lower limbs in a sitting subject.

Hyperekplexia can be inherited, almost always in an autosomal-dominant fashion. Genetic forms of hyperekplexia are due to mutations in the glycine receptor gene and are discussed later in this chapter (see Section II.A.3, later in this chapter). Sporadic cases may be symptomatic of brainstem pathology, such as infarct, hemorrhage, or encephalitis.

The site of origin of the startle response in hyperekplexia has not been definitively established. The presence, in a minority of patients, of giant cortical evoked potentials and evidence of cortical neuronal loss on magnetic resonance spectroscopy (MRS) support a cortical origin. However, pathology in symptomatic cases is often confined to the brainstem. Moreover, familial cases are due to mutations in the α_1 subunit of the glycine receptor, which is concentrated in the brainstem and spinal cord. Finally, the latency of EMG responses in facial muscles to taps to the head or face is often less than 20 ms, compatible only with relay within the brainstem. Thus, more evidence supports brainstem rather than cortical origin.

The normal startle response is thought to arise in the caudal reticular formation, within the nucleus reticularis pontis caudalis, and results in EMG activity that starts in the sternocleidomastoid muscle and sequentially appears in trunk and limb muscles. Hyperekplexia has a similar EMG recruitment pattern, consistent with a similar site of origin.

Electrophysiological evidence implicates abnormalities of glycine neurotransmission in hyperekplexia. The first period of spinal reciprocal inhibition, which is mediated by glycinergic Ia inhibitory interneurons, is deficient in patients with familial hyperekplexia. Additional evidence comes from molecular genetic studies (see later discussion). Abnormal glycine transmission could explain the presence of other clinical features of the startle syndromes, such as stimulus-induced tonic spasm and slowing of horizontal eye movements, as glycine receptors are found throughout the brainstem and spinal cord.

c. Spinal Myoclonus Myoclonus can originate in the spinal cord [9]. In animal models, inoculation of isolated spinal cord segments with virus or penicillin can lead to rhythmic myoclonus of the hind limbs, which persists after thoracic transection and deafferentation. The generator of myoclonus in these models is intrinsic to the cord and can discharge in the absence of peripheral or suprasegmental input. Over time, the jerks can spread to other limbs, showing that myoclonus can be conducted to other segments via a spinospinal pathway. Cases of myoclonus in patients with complete spinal transection have been reported.

Spinal myoclonus may be due either to increased excitability of facilitatory mechanisms or to reduced activity of inhibitory mechanisms at the level of interneurons or motoneurons. Clinical studies suggest that motoneuron involvement is rare, because weakness and denervation potentials are usually absent. Thus, dysfunction of spinal interneuronal circuits is thought to underlie its pathogenesis. Such dysfunction could result from direct effects of a local disease process or through alteration of descending or peripheral inputs.

Causes of spinal myoclonus include arteriovenous malformations, tumors, cysts, spondylosis, trauma, multiple sclerosis, amyotrophic lateral sclerosis, and viral infection. The resulting myoclonus can be of two types: spinal segmental and propriospinal.

Spinal segmental myoclonus: The distribution of muscle jerking is limited to muscles innervated by one to two contiguous spinal segments. It can be rhythmic or irregular and range from 1 to 4 Hz. The jerks are usually synchronous, although asynchronous activation has been observed. This type of myoclonus is often stimulus sensitive. The latency of reflex jerks can vary from 44 to 100 ms. Such latencies include conduction times for slow spinal and fast supraspinal pathways. Studies using paired pulse stimulation suggest hyperactivity of dorsal horn interneurons.

Propriospinal myoclonus consists of jerks of spinal origin with a wide distribution. The jerks start in muscles innervated by one spinal segment and spread to muscles supplied by progressively more rostral and more caudal segments. Multichannel surface EMG recordings show myoclonic EMG bursts starting in one group of muscles and appearing, over time, in adjacent spinal myotomes. The bursts appear with increasing delay in muscles that are increasingly farther, in both the rostral and the caudal direction, from the site of the

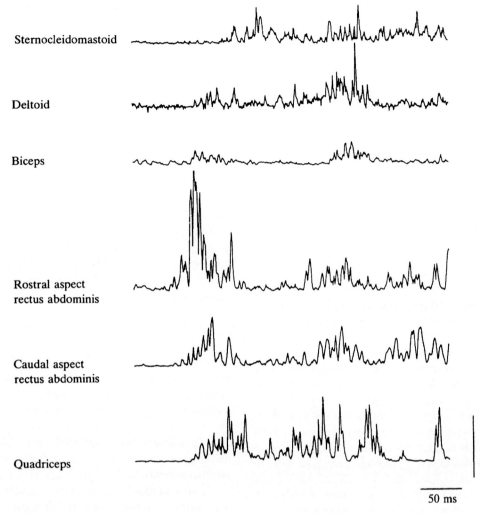

Figure 4 Electromyogram (EMG) record of a single spontaneous jerk in a patient with propriospinal myoclonus. Muscles are sternocleidomastoid, deltoid, biceps, rostral aspect of rectus abdominis, caudal aspect of rectus abdominis, and quadriceps. Contraction of sternocleidomastoid, rectus abdominis, and quadriceps occurred in all jerks, but activity in deltoid and biceps (evident here) was variable. Vertical calibration line = 200 μV and 100 μV for bottom and top three channels, respectively. (Reprinted, with permission from Brown, P., Thompson, P. D., Rothwell, J. C., Day, B. L., Marsden, C. D. [1991]. Axial myoclonus of propriospinal origin. *Brain* **114**[Pt 1A], 197–214.)

initial burst (Fig. 4). This pattern suggests that myoclonic activity starts in one spinal segment and then spreads rostrally and caudally along the cord via spinospinal pathways. EMG bursts occur irregularly, with frequencies up to 2 Hz, and can last between 40 ms and 4 seconds. When jerks occur in response to somatic stimulation, latencies of 100 ms or more are observed, suggesting conduction through long propriospinal pathways.

2. Negative Myoclonus

Negative myoclonus consists of a brief pause in the activation of a muscle that is contracting in a sustained fashion [10]. This pause leads to brief loss of muscle tone, usually manifesting as transient loss of a maintained posture. The original description was of the fingers showing intermittent "flapping" (asterixis) when patients with hepatic encephalopathy were asked to hold their arms outstretched with the hands and fingers pointing upward.

The physiological correlate of negative myoclonus is a brief lapse in EMG discharge in a contracting muscle. Etiologically, at least three groups of negative myoclonus have emerged: epileptic negative myoclonus (ENM), asterixis with toxic/metabolic states, and unilateral asterixis due to focal CNS lesions. Like its positive counterpart, negative myoclonus can arise from various brain structures. Cortical involvement, especially the sensorimotor cortex, underlies the pathogenesis of ENM. Subcortical pathology is thought to underlie asterixis, but the cortex may be involved in some cases. Unilateral asterixis has been reported most commonly with thalamic lesions, although the sites of pathology are variable.

a. Epileptic Negative Myoclonus ENM, also known as *cortical negative myoclonus*, is characterized by an abrupt increase in EMG discharge preceding the EMG silent period (see Fig. 1). The initial EMG burst and following silent period are related to an EEG spike and slow wave, EEG slow wave having comparable duration to the EMG silent period. H reflexes can be elicited during the silent period, suggesting that the inhibition does not take place in the spinal cord.

ENM may be absent at rest and appear only when the patient attempts an action, such as reaching or walking. The reach is then repeatedly interrupted by the transient loss of muscle tone, leading to a characteristic irregular trajectory of the hand. Involvement of axial and leg muscles in ENM leads to falls, or "drop attacks."

ENM can arise from several locations in the contralateral sensorimotor cortex. These generators have been identified through analysis of the EEG spike associated with the EMG silent period of ENM, as well as through single-photon emission computed tomography (SPECT) studies of blood flow during episodes of ENM. Identified locations include the primary motor, primary sensory, and premotor cortex. Stimulation of the scalp and of the cortical surface over the motor and premotor cortex can produce interruption of movement and EMG silent periods. The role of these stimulation sites in cortical myoclonus, however, has not been well established.

A possible mechanism for ENM is abnormal enhancement of inhibitory activity in the primary motor cortex. TMS studies in patients with cortical myoclonus show increased threshold for eliciting EMG activation, abnormally long ensuing silent periods, and prolonged recovery of the SSEP with paired pulse stimulation. It is thought that excessive input into the motor cortex as the result of epileptic activity or enhanced excitability of the sensory cortex may activate an inhibitory motor system that is already hyperactive, leading to suppression of cortical motor output.

Involvement of the somatosensory cortex is likely in cortical reflex negative myoclonus (i.e., ENM evoked by somatosensory stimulation). The EMG pause is preceded by a giant SSEP, and a significant positive correlation is seen between the amplitude of the giant SSEP and the duration of the EMG silent period. Figure 5 summarizes possible mechanisms of ENM generation.

b. Asterixis in Metabolic Encephalopathy The term "asterixis" is generally used to refer to the intermittent "flapping" tremor of the fingers, arms, or lower extremities when they are held outstretched in a sustained posture. It typically accompanies hepatic encephalopathy, although it can also be seen with uremic encephalopathy and can follow hypoxic brain injury. EEG correlates are usually not found. These

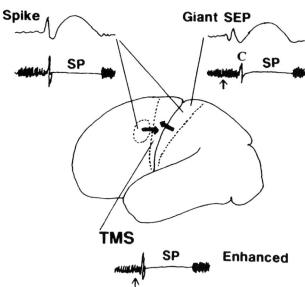

Figure 5 Schematic diagram showing possible mechanisms underlying the generation of epileptic-negative myoclonus. The inhibitory function of the primary motor cortex is enhanced, as demonstrated by transcranial magnetic stimulation. Inputs into the primary motor cortex from the spontaneous epileptic activity arising either in the premotor area or in the somatosensory cortex, or as a result of sensory stimulation via enhanced excitability of the sensory cortex, might activate the inhibitory system of the motor cortex, producing negative myoclonus of cortical origin. (Reprinted, with permission, from Shibasaki, H. [2002]. Physiology of negative myoclonus. *Adv. Neurol.* **89**, 103–113.)

findings suggest origin in the brainstem. However, evidence for cortical origin has also been reported in some cases [11].

c. Unilateral Asterixis with Focal CNS lesions Unilateral asterixis can be caused by focal brain lesions, the most common by far being infarct and hemorrhage. The most common lesion location associated with focal asterixis is the thalamus, and the ventrolateral and lateral posterior nuclei are candidate sites.

III. Pharmacology and Molecular Genetics

A. Pharmacology

1. Serotonin

The role of serotonin in myoclonus (reviewed in reference 12) was first demonstrated by Lhermitte, who treated patients with posthypoxic myoclonus with 5-hydroxytryptophan (5-HTP), based on the observation that these patients have low levels of cerebrospinal fluid (CSF) 5-hydroxyindole acetic acid (5-HIAA), a metabolite of serotonin. On the other hand, other forms of myoclonus, including drug-induced myoclonus, may be due to excessive serotonin levels.

Serotonin is synthesized from tryptophan and 5-HTP by neurons in the raphe nuclei, which project to regions throughout the CNS, including the spinal cord. Many types of serotonin receptors exist and are unequally distributed [13]. The synaptic action of serotonin is terminated by presynaptic reuptake and is degraded by monoamine oxidase to 5-HIAA.

a. Animal Models

i. 5-HTP–Induced Myoclonus in Young Guinea Pigs Symmetrical generalized rhythmic myoclonus develops in young guinea pigs over a 60- to 90-minute interval after subcutaneous injections of 5-HTP. Severity of myoclonus is dose dependent, and severity and duration correlate with brain serotonin levels. The myoclonus is augmented by serotonin agonists, selective serotonin reuptake inhibitors (SSRIs), and tryptophan with monoamine oxidase inhibition. Myoclonus is blocked by serotonergic receptor antagonists in a dose-dependent fashion. It is modulated by dopamine receptor blockers, which diminish the myoclonus acutely but may augment it with longer treatment duration.

This model supports high serotonin levels as the cause of human myoclonus caused by SSRIs, serotonin agonists, and dopaminergic drugs such as has been reported with levodopa, bromocriptine, and amantadine. High levels of serotonin may also play a role in the pathogenesis of infantile spasms [14]. Finally, excessive production of catecholamines or indole-aminergic products may explain myoclonus associated with neuroblastoma [14].

ii. Posthypoxic Audiogenic Myoclonus in the Rat Cardiac arrest in rats produces marked enhancement of myoclonus in response to sudden sounds, a naturally occurring phenomenon. Impaired serotonin release appears to be a key step in the pathogenesis of this condition. Treatment with 5-HTP improves myoclonus, and severity scores correlate inversely with striatal serotonin and cortical 5-HIAA levels. Two serotonergic antagonists (mesulergine and methiothepin), however, block myoclonus in a dose-dependent manner. These drugs primarily act on the 5-HT1 and 5-HT2 receptor subsystems. Serotonergic antagonists that block other serotonin receptors do not inhibit myoclonus in the posthypoxic rat model. Myoclonus in this model is reduced by 5-HT1c agonists in the acute phase and by 5-HT1a agonists in the chronic phase. The anatomic substrate of posthypoxic myoclonus is thus likely to be intimately related to the distribution of type 1 and type 2 serotonin receptors.

An unexpected feature of this model is that myoclonus is aborted by agents that augment serotonergic activity and by agents that antagonize it. A full explanation of this finding remains elusive, but it may involve nonlinear relationships between serotonin levels and depolarization of selected neuronal populations.

iii. GEPR-3: A Rat Model of Myoclonic Epilepsy The genetically epilepsy-prone rat type 3 (GEPR-3) has a genetic form of myoclonic epilepsy in which single myoclonic jerks can be triggered by auditory stimulation. This rat also experiences myoclonic seizures consisting of episodes of rhythmic myoclonus. Serotonin levels are severely reduced in most brain structures, and drugs that increase serotonin levels block myoclonus in this model. Although the anatomic substrate of audiogenic myoclonic jerks has not been identified, rhythmic myoclonus in this model seems specifically due to reduction of serotonin in the inferior olive. This results in unconstrained rhythmic firing of neurons in this structure [15], which may lead to periodic spiking producing rhythmic myoclonus.

2. GABA

Decreased levels of **GABA** have been measured in patients with myoclonus [16]. GABA is a major inhibitory neurotransmitter. It is ubiquitous in the CNS and can act via three receptor subtypes: A, B, and C. The $GABA_A$ receptor subtype has been implicated in the pathogenesis of myoclonus. Myoclonus results from injections of $GABA_A$ antagonists in a variety of brain structures in animals, including contralateral cortex, caudate, putamen, and the thalamic nucleus reticularis [16]. Similarly, myoclonus in humans can result from lesions to various CNS structures, both focal and diffuse.

The fact that myoclonus can result from disruption of GABA transmission at several points along the neuraxis suggests that dysfunction of neural circuits, rather than structural abnormalities in discrete regions, is the cause of the abnormal movements. The corticospinal tract is likely to be one such circuit, given the reliable appearance of myoclonus following disruption of GABAergic transmission in the motor cortex. Another circuit may involve the GABAergic projections from striatum to globus pallidus, which in turn projects to the subthalamic nucleus as part of the basal ganglia's indirect pathway. Loss of GABAergic inhibition from striatum to globus pallidus in this circuit could release involuntary movements such as myoclonus. Involvement of this circuit may also underlie myoclonus in Huntington's disease, a condition characterized by degeneration of striatal GABAergic neurons.

Given that myoclonus is sometimes thought of as an "epileptic fragment" associated with deficits in GABAergic function, anticonvulsants with GABAergic properties have been tested in patients with myoclonus. Among the most effective agents have been clonazepam and valproic acid. Clonazepam is a benzodiazepine that enhances postsynaptic GABA inhibition by increasing the frequency of opening of the $GABA_A$ receptor complex. Valproic acid enhances GABA activity by increasing the activity of the enzyme glutamic acid decarboxylase, which synthesizes GABA, and by inhibiting enzymes that catabolize GABA.

3. Glycine

Glycine is the major inhibitory neurotransmitter in the spinal cord and brainstem. Acute poisoning with strychnine, a glycine receptor antagonist, results in generalized hypertonia. Disruption of glycinergic transmission by lower doses leads to exaggerated responsiveness to sensory stimuli, similar to the abnormal startle responses seen in hyperekplexia. Studies of the molecular genetics of hyperekplexia (reviewed in reference 17) have led to significant insights into the role of glycine in the pathogenesis of the abnormal startle response.

The locus for hyperekplexia has been mapped to 5q32, a region known to possess several neurotransmitter receptor genes. One such gene codes for the ligand-binding α_1 subunit of the glycine receptor. Six different mutations within the α_1 subunit have been identified. The residue is located at the extracellular mouth of a membrane-spanning domain.

Four animal models with mutations in the glycine receptor's gene support the importance of this subunit in the pathogenesis of hyperekplexia. Three of these are lines of mutant mice with a recessively inherited startle syndrome. In the "spasmodic" mouse, a missense mutation in the α_1 subunit leads to reduced sensitivity of the glycine receptor to activation by glycine, in spite of normal binding. In the "oscillator" mouse, a frame-shift mutation results in a truncated α_1 subunit unable to assemble into functional receptors. "Spastic" mice have an intronic insertion that results in aberrant splicing of β-subunit messenger RNAs (mRNAs). This prevents efficient receptor assembly. Bovine myoclonus is an autosomal-recessive disease with spontaneous and stimulus-sensitive myoclonic jerks. It is caused by single base-pair substitution leading to a truncated α_1-glycine receptor subunit that lacks ligand-binding and membrane-spanning domains. The result is absence of glycine receptors from the cell membrane.

Thus, mutations in glycine receptor genes result in disease phenotypes through one of two mechanisms. Missense mutations in the α_1 subunit cause a loss of glycine receptor sensitivity either by reduction of agonist affinity or by defective signal transduction. This underlies all dominantly and some recessively inherited human hyperekplexia mutations and the spasmodic mouse model. Alternatively, mutations result in a reduction in the expression of the glycine receptor as in spastic mice, oscillator mice, bovine myoclonus, and a human null allele. Some mutations in this category are lethal. Figure 6 illustrates the effect of mutations in glycine receptor genes causing excessive startle or myoclonus.

B. Genetics

The study of genetic diseases that cause myoclonus promises to help elucidate its pathogenesis. However, a unifying picture has yet to emerge. Here we summarize the current understanding of selected genetic conditions causing myoclonus. Idiopathic myoclonic epilepsies are not included.

1. Myoclonus-Dystonia

Inherited myoclonus dystonia (MD, DYT11) refers to a group of syndromes associated with mutations on chromosome 7q21 (reviewed in reference 18). These syndromes include a portion of cases previously referred to as hereditary essential myoclonus and alcohol-responsive myoclonic dystonia. The term *dystonia* refers to sustained contractions of muscles resulting in a twisting or posturing of a limb or other body region. MD appears in the first or second decade, has a benign course, and is inherited as autosomal dominant with partial penetrance. Myoclonus is the predominant feature in most cases of MD and predominantly affects the upper body, sparing the legs. In addition to normal EEG scans, SSEPs are also normal, distinguishing MD from familial cortical myoclonic tremor.

Several mutations responsible for MD have been identified, and all affect the ε-sarcoglycan (SCGE) gene. The resulting phenotype is not clinically distinguishable based on different mutations. SCGE is the fifth member of the sarcoglycan family, a group of proteins that are part of the dystrophin–glycoprotein complex that links the cytoskeleton to the extracellular matrix. It is not clear how abnormalities of SCGE lead to myoclonus or dystonia.

2. Progressive Myoclonic Epilepsy

Progressive myoclonic epilepsy (PME) refers to a group of rare inherited disorders that manifest with epilepsy, myoclonus, and progressive neurological deterioration such as ataxia and dementia. The five main forms of PME are Unverricht-Lundborg disease, LD, myoclonic epilepsy with RRFs (MERRF), neuronal ceroid lipofuscinosis, and type I sialidosis. The first four are discussed here.

a. Unverricht-Lundborg Disease Major features of *Unverricht-Lundborg disease* (EPM1) include autosomal-recessive inheritance, higher incidence in Finland, onset at age 6–15 years, stimulus-sensitive myoclonus, and generalized tonic-clonic seizures [19]. There may be progressive ataxia, action tremor, dysarthria, and dementia. The EEG findings are always abnormal with a slow background and symmetrical, generalized, and high-voltage spike–wave and polyspike–wave patterns. There is marked photosensitivity. EPM1 is one of the most common PME.

EPM1 is due to a mutation in the gene for cystatin B (CSTB) on chromosome 21q22.3, which results in greatly reduced levels of this cysteine protease inhibitor. The most common mutation is an unstable expansion of a dodecamer (12 base-pair) repeat unit located upstream of the promoter region of the CSTB gene. The repeat contains at least 30 copies, instead of the normal two to three. Unlike diseases

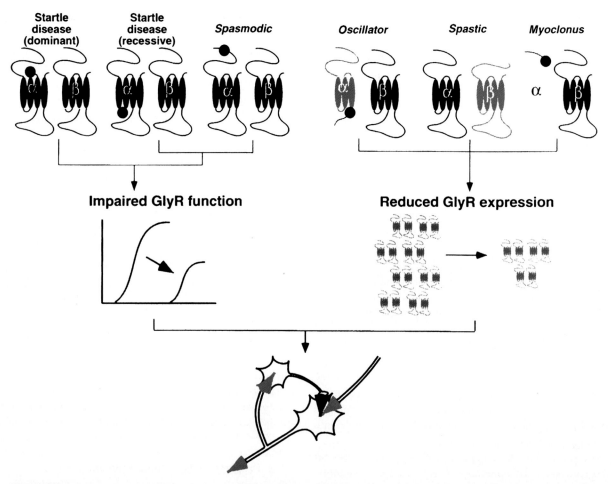

Figure 6 The molecular mechanisms of the various startle disorders converge physiologically. Both dominant and recessive human startle disease mutations, as well as the murine mutant *spasmodic*, all lead to impaired receptor function, and the murine mutants *oscillator* and *spastic* and the bovine mutant *myoclonus* all lead to loss of glycine receptor expression. Phenotypically, all mutations, whether reducing glycine receptor sensitivity or reducing glycine receptor expression, diminish glycinergic inhibition and result in hypertonia, myoclonus, and an excessive startle response. (Reprinted, with permission, from Schofield, P. R. [2002]. The role of glycine and glycine receptors in myoclonus and startle syndromes. *Adv. Neurol.* **89**, 263–274.)

with trinucleotide repeat expansion, there is no correlation between the repeat size and age at onset or disease severity. Homozygosity is common.

The repeat expansion mutation results in reduced transcriptional activity, which is consistent with the reduced CSTB mRNA seen in patients. CSTB belongs to a family of proteins that inhibit papain-family cysteine proteases. Papain inhibition is reduced in lymphoblastoid cells of EPM1 patients. Increased protease activity may thus contribute to the pathogenesis of EPM1.

Mice deficient for CSTB show progressive ataxia and seizures, which are associated with apoptotic death of cerebellar granule cells. CSTB may have a neuroprotective role for specific cells, possibly by inhibiting activators of caspases, which are involved in the initiation of apoptosis.

b. Lafora's Disease Also known as *progressive myoclonus epilepsy with polyglucosan bodies*, LD is characterized by myoclonus, generalized seizures, ictal visual hallucinations, and dementia [20]. Myoclonus progresses to the point that patients are unable to walk.

Neurophysiology: Myoclonus associated with LD can be either positive or negative and is primarily of cortical origin. The EEG is characterized by generalized irregular spike–wave discharges with occipital predominance. Myoclonus can be induced by light flashes, which trigger electrical discharges that travel from the occiput to the ipsilateral precentral gyrus and from there to the contralateral sensorimotor cortex. Visual evoked potentials and SSEPs are characterized by high-voltage and prolonged central latencies.

Pathology: Lafora bodies accumulate in neurons, near the nucleus. They are composed of fibrils of polyglucosan, a glucose polymer without the regular branching pattern of glycogen. The absence of a normal branching pattern prevents polyglucosan from being suspended in the cytoplasm. Instead, Lafora bodies remain attached to the endoplasmic reticulum. As the disease progresses, there is neuronal loss in the cerebral hemispheres, cerebellum, basal ganglia, thalamus, hippocampus, retina, and spinal cord.

Genetics: LD is caused by mutations (>20 reported) in the coding region of the EPM2A gene, located on chromosome 6q24, which codes for *laforin*, a 331 amino acid protein. Laforin is a tyrosine phosphatase whose phosphorylation target is unknown. Its N-terminus contains a carbohydrate-binding module, potentially allowing it to interact with starch and glycogen.

Pathogenesis: A hypothesis for the pathogenesis of LD is as follows. An early event may be abnormal phosphoregulation by laforin of unknown components of the glycogen branching mechanism, leading to the formation of glucose polymers with abnormal branching patterns. These polymers (the polyglucosan bodies) are insoluble and accumulate near the cell's endoplasmic reticulum. Because the latter structure is localized near the nucleus and dendrites, accumulation of polyglucosan bodies in the soma and dendrites may result in the abnormal excitability of affected neurons that leads to myoclonus.

c. Myoclonic Epilepsy with Ragged Red Fibers MERRF was the first human disease in which maternal inheritance was clearly recognized, indicating a defect in mitochondrial DNA (mtDNA) (reviewed in reference 21). It was also the first disorder that linked an epilepsy syndrome to a molecular defect. MERRF is one of the most common mitochondrial encephalomyopathies. Its major manifestations are myoclonus, generalized epilepsy, ataxia, and ragged red fibers in muscles. Other common findings are hearing loss, peripheral neuropathy, dementia, short stature, exercise intolerance, and optic atrophy. Laboratory findings include increased blood lactate and pyruvate at rest, which increase excessively during exercise; increased CSF protein; myopathic pattern on EMG (sometimes with accompanying peripheral neuropathy); generalized spike–wave discharges with background slowing and focal epileptiform discharges on EEG; and brain atrophy with basal ganglia calcifications on brain magnetic resonance imaging (MRI) scans. Biopsied muscle fibers, when stained with the modified Gomori trichrome stain, show ragged red fibers in more than 90% of typical patients.

Genetics: Three mutations are associated with MERRF, all affecting the tRNA-lys gene of the mtDNA. The most common mutation, present in about 90% of patients, is an A-to-G transition at the nt8344 locus. This mutation is heteroplasmic and has to be present in abundance to cause symptoms.

Neuropathology: Histopathological changes in MERRF consist of neuronal loss and gliosis in several CNS regions, including the cerebellum (especially the dentate nucleus), brainstem (especially inferior olive, red nucleus, substantia nigra), and spinal cord. Demyelination affects the superior cerebellar peduncle and the posterior spinocerebellar tracts.

Pathogenesis: Mitochondrial encephalomyopathies, such as MERRF, result from disruption of the electron-transport chain that is part of oxygen-based energy production in cells of all tissues. These reactions are intimately linked to the mitochondrial membrane, and the genes for several key enzymes reside in mtDNA. Immunohistochemical studies imply that selective decrease in the expression of the mtDNA-encoded subunit of cytochrome c oxidase (COX-2) expression may underlie some of the pathological changes. Biochemical studies of respiratory chain enzymes in muscle extracts usually show decreased activities of respiratory chain complexes that contain mtDNA-encoded subunits, especially COX deficiency. Disruption of the mitochondrial electron-transport apparatus may make cells vulnerable to damage from oxidative stress. It is not known why cells in certain tissues are preferentially affected in individual mitochondrial diseases. In the case of MERRF, myoclonus may be due in part to the involvement of the inferior olivary nucleus and the cerebellar dentate nucleus, which have both been implicated in reticular reflex and segmental myoclonus.

d. Neuronal Ceroid Lipofuscinosis The neuronal ceroid lipofuscinoses (NCL) are a group of neurodegenerative disorders characterized by the accumulation of lipopigment in neurons and other cell types (reviewed in reference 22). Clinical manifestations include psychomotor retardation, visual failure, and epilepsy. Five types include progressive myoclonus epilepsy: classic late infantile (CLN2; Jansky-Bielschowsky disease), juvenile (CLN3; Spielmeyer-Vogt, Sjögren, or Batten's disease), adult (CLN4; Kufs' or Parry's disease), infantile Finnish variant (CLN5), and late infantile variant (CLN6). All have autosomal recessive inheritance except for the adult form, which is autosomal dominant.

Key phenotypic and genetic features of selected NCLs are listed in Table 1. Juvenile NCL (CLN3) is the most common childhood neurodegenerative disease, with incidence 1/25,000 births and increased prevalence in northern Europe. The EEG in CLN2 may include a giant visual evoked potential with flash stimulation. CLN6 is a variant of late infantile NCL. CLN8 (northern epilepsy) does not cause myoclonus epilepsy.

3. Dentatorubral Pallidoluysian Atrophy

Dentatorubral pallidoluysian atrophy (DRPLA), like Huntington's disease and several spinocerebellar ataxias, is a member of a group of genetic diseases in which the mutation is an expansion of a sequence of three nucleotide base pairs, cytosine-adenosine-guanine (CAG trinucleotide repeat) [23]. Because CAG codes for the amino acid glutamine, these conditions are also known as *polyglutamine expansion diseases*. DRPLA manifests as various combinations of cerebellar ataxia, choreoathetosis, myoclonus,

Table 1 Features of Neuronal Ceroid Lipofuscinoses Causing Myoclonus and Epilepsy

		Initial Symptoms	Late Clinical Features	Specific Features	Gene Locus, No. of Mutations	Gene Product and Postulated Role	Assay
Infantile NCL	CLN1 Haltia–Santavuori disease	Blindness in infancy	Dystonia, choreoathetosis	Rapid progression	1p32 (>30 mutations)	PPT1; lysosomal enzyme	PPT1 activity in leukocytes or fibroblasts
Late infantile NCL	CLN2 Classic late infantile	Epilepsy at 2–3 yr	Ataxia, dementia	Giant VEP with flash stimulation	11p15 (>30 mutations)	Tripeptidyl peptidase (TPP1); lysosomal enzyme	TPP1 activity in leukocytes or fibroblasts
	CLN5 Finnish variant	Clumsiness, hypotonia at 5 yr	Ataxia, dystonia, extrapyramidal signs	Found almost exclusively in Finland	13q21-q32 (3 mutations)	Unknown	
Juvenile NCL	CLN3 Batten's disease	Visual failure at 5–10 yr	Parkinsonism, dysarthria, hallucinations	Vacuolated lymphocytes	Chromosome 16 (14 mutations)	Unknown	

Note: NCL, neuronal ceroid lipofuscinosis; PPT1, palmitoyl protein thioesterase; TPP1, tripeptidyl peptidase; VEP, visual evoked potential.

epilepsy, dementia, and psychiatric symptoms. The clinical presentation can resemble progressive myoclonic epilepsy, especially if onset is before age 20 years. The most common syndrome after age 20 is ataxia, choreoathetosis, and dementia. Brain MRI scans show characteristic atrophy of the pontine tegmentum and cerebellum. DRPLA is autosomal dominant with high penetrance, and prevalence is higher in the Japanese population.

Genetics: The DRPLA gene is on chromosome 12 (12p13.31) and codes for a protein of unknown function, which appears to localize to the nucleus. DRPLA is characterized by prominent *anticipation*: the offspring inheriting the mutation begin to experience symptoms at a younger age than their parent. Anticipation is more marked when the mutation is inherited from the father rather than the mother. As in several other diseases caused with a CAG repeat expansion, the length of the expansion (the CAG repeat "load") relates to the phenotype. The longer the expansion, the earlier the age at onset of symptoms. In addition, age at the time of imaging and size of CAG repeat seem to independently influence the degree of cerebellar and pontine atrophy.

Pathogenesis: It is likely that mutant DRPLA proteins with expanded polyglutamine repeats are toxic to neurons. Autopsy studies show accumulation of mutant DRPLA protein in neuronal nuclei. Mutant DRPLA may interact with nuclear protein, interfering with transcription and leading to neuronal death. Myoclonus likely results from degeneration of specific brain structures, but the widespread nature of this degeneration precludes confident identification of the pathogenesis of myoclonus in this condition.

IV. Relationship between Pathophysiology and Symptoms

The neurophysiological nature of myoclonus was described earlier in this chapter. At the CNS level, myoclonus consists of a relatively uniform phenomenon: a sudden, inappropriate discharge of action potentials in motor neurons (or, for negative myoclonus, an inappropriate pause in such discharges). Once this inappropriate discharge reaches the involved muscles, the outward manifestations of myoclonus are dictated by a variety of factors. Chief among these are the number and size of the muscles involved, the context in which myoclonus occurs, and its persistence. Physiological myoclonus of the diaphragm (hiccups), for example, is usually transient and causes only minor annoyance. Hypnic jerks (generalized myoclonus during sleep onset), although generalized to the entire body, are not dangerous because the person is usually already in a recumbent position. On the other hand, action myoclonus can be incapacitating even if it is confined to one hand, as its appearance with action can preclude functional use of the hand. Action-induced negative myoclonus of the lower body can be particularly disabling; the legs give way when the person stands or walks, leading to falls and often resulting in wheelchair confinement.

V. Natural History

The course of myoclonus is entirely tied to the underlying condition that causes myoclonus. Although certain conditions, such as renal failure and hepatic encephalopathy, can often be reversed, there is no cure for many of the progressive neurodegenerative diseases that cause myoclonus. Symptomatic drug treatment is effective in many patients, but even in these cases the effectiveness is often only partial or limited in time. In nonprogressive conditions with isolated myoclonus, such as essential myoclonus, symptomatic drug treatment can provide satisfactory control of symptoms for decades.

References

1. Obeso, J. A., Bhatia, K., and Rothwell, J. C. (2002). The contribution of C. David Marsden to the study and treatment of myoclonus. *Adv. Neurol.* **89**, 1–12.
2. Fahn, S. (2002). Overview, history, and classification of myoclonus. *Adv. Neurol.* **89**, 13–17.
3. Shibasaki, H., and Hallett, M. (2005). Electrophysiological studies of myoclonus. *Muscle Nerve* **31**(2), 157–174.
4. Ugawa, Y., Hanajima, R., Okabe, S., and Yuasa, K. (2002). Neurophysiology of cortical positive myoclonus. *Adv. Neurol.* **89**, 89–97.
5. Shibasaki, H., Yamashita, Y., Tobimatsu, S., and Neshige, R. (1986). Electroencephalographic correlates of myoclonus. *Adv. Neurol.* **43**, 357–372.
6. Hallett, M. (2002). Neurophysiology of brainstem myoclonus. *Adv. Neurol.* **89**, 99–102.
7. Lance, J. W., and Adams, R. D. (1963). The syndrome of intention or action myoclonus as a sequel to hypoxic encephalopathy. *Brain* **86**, 111–136.
8. Brown, P. (2002). Neurophysiology of the startle syndrome and hyperekplexia. *Adv. Neurol.* **89**, 153–159.
9. Rothwell, J. C. (2002). Pathophysiology of spinal myoclonus. *Adv. Neurol.* **89**, 137–144.
10. Shibasaki, H. (2002). Physiology of negative myoclonus. *Adv. Neurol.* **89**, 103–113.
11. Artieda, J., Muruzabal, J., Larumbe, R., Garcia de Casasola, C., and Obeso J. A. (1992). Cortical mechanisms mediating asterixis. *Mov. Disord.* **7**(3), 209–216.
12. Goetz, C. G., Carvey, P. M., Pappert, E. J., Vu, T. Q., and Leurgans S.E. (2002). The role of the serotonin system in animal models of myoclonus. *Adv. Neurol.* **89**, 244–248.
13. Pauwels, P. J. (2000). Diverse signalling by 5-hydroxytryptamine (5-HT) receptors. *Biochem. Pharmacol.* **60**(12), 1743–1750.
14. Klawans, H. L., Jr., Goetz, C., and Weiner, W. J. (1973). 5-Hydroxytryptophan-induced myoclonus in guinea pigs and the possible role of serotonin in infantile myoclonus. *Neurology* **23**(11), 1234–1240.
15. Welsh, J. P., Placantonakis, D. G., Warsetsky, S. I., Marquez, R. G., Bernstein, L., and Aicher, S. A. (2002). The serotonin hypothesis of myoclonus from the perspective of neuronal rhythmicity. *Adv. Neurol.* **89**, 307–329.
16. Matsumoto, R. R. (2002). Involvement of gamma-aminobutyric acid in myoclonus. *Adv. Neurol.* **89**, 249–262.
17. Schofield, P. R. (2002). The role of glycine and glycine receptors in myoclonus and startle syndromes. *Adv. Neurol.* **89**, 263–274.
18. Saunders-Pullman, R., Ozelius, L., and Bressman, S. B. (2002). Inherited myoclonus-dystonia. *Adv. Neurol.* **89**, 185–191.
19. Lehesjoki, A. E. (2002). Clinical features and genetics of Unverricht–Lundborg disease. *Adv. Neurol.* **89**, 193–197.
20. Minassian, B. A. (2002). Progressive myoclonus epilepsy with polyglucosan bodies: Lafora disease. *Adv. Neurol.* **89**, 199–210.
21. DiMauro, S., Hirano, M., Kaufmann, P., Tanji, K., Sano, M., Shungu, D. C., Bonilla, E., and DeVivo, D. C. (2002). Clinical features and genetics of myoclonic epilepsy with ragged red fibers. *Adv. Neurol.* **89**, 217–229.
22. Gardiner, R. M. (2002). Clinical features and molecular genetic basis of the neuronal ceroid lipofuscinoses. *Adv. Neurol.* **89**, 211–215.
23. Tsuji, S. (2002). Dentatorubral-pallidoluysian atrophy: clinical aspects and molecular genetics. *Adv. Neurol.* **89**, 231–239.
24. Brown, P., Thompson, P. D., Rothwell, J. C., Day, B. L., and Marsden, C. D. (1991). Axial myoclonus of propriospinal origin. *Brain* **114**(Pt 1A), 197–214.

31

Channelopathies of the Nervous System

James C. Cleland, MBChB
Robert C. Griggs, MD

Keywords: *arrhythmia, ataxia, channelopathy, epilepsy, paralysis*

I. Introduction
II. Channelopathies: Defects of Ion Channel Function
III. Channelopathies of the Peripheral Nervous System
IV. Channelopathies of the Central Nervous System
V. Acquired Channelopathies of the Nervous System
References

I. Introduction

Few neurological disorders can claim as long a passage of time between initial description and subsequent molecular characterization as the "channelopathies": acquired or inherited disorders of membrane ion channel function. The first channelopathy, *myotonia congenita*, was described more than 100 years ago, but not until the past 30 years have patch-clamping and expression cloning techniques allowed full characterization of ion channel defects. Mutations have now been identified for most channelopathies, and ongoing research suggests that these disorders are relatively common. For example, ion channel dysfunction has been recognized as a cause of disorders as diverse as migraine (calcium channel), epilepsy (nicotinic acetylcholine receptor [nAchR] and potassium and sodium channels), and malignant hyperthermia (ryanodine receptor and calcium channel). This chapter focuses on the inherited disorders of ion channel function. However, acquired ion channel disorders due to autoimmune (e.g., acquired neuromyotonia), paraneoplastic (e.g., Lambert-Eaton myasthenic syndrome), and toxic causes (e.g., ciguatera poisoning) are well described and the pathophysiology often well characterized.

II. Channelopathies: Defects of Ion Channel Function

Ion channels facilitate rapid communication of electrical information in the central, peripheral, and autonomic nervous systems. So fundamental is this function that interference by channel-blocking toxins can have rapidly lethal effects (e.g., tetrodotoxin from the puffer fish can kill humans within 10 minutes of a sufficient exposure in the improperly prepared Japanese sushi delicacy, *fugu*).

Many of the early discoveries of ion channel structure and function were based on electrophysiological and pharmacological studies of individual ion channels, thus forming the foundation of our understanding of ion channel physiology even today. With the development of molecular techniques, channel structure has been further elucidated through electrophysiological studies after cloning and expression in animal systems (e.g., *Xenopus* oocytes). Based on DNA sequence homology, channels initially thought to be electrophysiologically and pharmacologically distinct are now recognized as members of the same superfamily and may have evolved from a common ancestor (possibly a potassium channel).

Voltage-gated calcium, sodium, and potassium channels are all of similar structure, typically consisting of an α subunit co-assembled with an auxiliary subunit that may further modify channel gating or permeability, interact with second messenger systems, and modify channel expression levels. The presence (or absence) and type of auxiliary subunit for each channel type (sodium, potassium, chloride, or calcium) are believed to confer many of the site-specific electrophysiological characteristics of ion channels in their diverse roles throughout the central and peripheral nervous systems. Although disease-causing mutations in genes coding for auxiliary subunits occur, they are much rarer than those arising in the α subunit.

The α subunits of voltage-gated sodium, potassium (K_v type), and calcium channels are all members of the "S4 superfamily" of cation channels that are composed of individual domains each consisting of six transmembrane (TM) spanning segments, designated S1–S6, with a pore-forming loop between S5 and S6. The fourth segment (S4) comprises a specific "S4 motif" of basic amino acid residues every third or fourth amino acid that is believed to represent the voltage sensor. The α subunits of sodium and calcium channels comprise a single polypeptide chain of four domains to make a total of 24TM segments per α subunit (Fig. 1).

For potassium channels, there is greater structural diversity and increasing functional complexity across different subtypes of potassium channel, supporting the contention that it is the channel from which others evolved. The simplest potassium channels are inwardly rectifying K_{ir} subtypes that are not voltage dependent and are composed of 2TM segments. Certain K_{ir} channels are composed of multiple 2TM segment domains (e.g., adenosine triphosphate [ATP]-dependent subtypes, in which four 2TM subunits co-assemble with four auxiliary subunits; such channels facilitate coupling of cellular metabolic activity to cellular excitability and are thought to be involved in

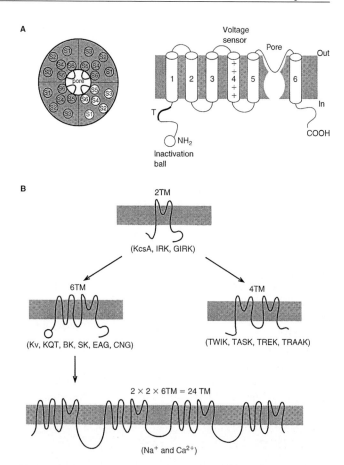

Figure 1 Membrane topology of the voltage-gated ion channels. (**A**) Top view looking down on the plane of the membrane (left) shows the fourfold symmetry of subunits (K^+ subchannels) or homologous repeats (Na^+ or Ca^{2+} channels) about a central pore. Each subunit contains six transmembrane segments, shown opened into a linear membrane-folding diagram (right). The loop between S5 and S6 forms the pore and the voltage sensor is denoted by the positive charges in S4. For K^+ channels, the cytoplasmic amino-terminus acts as an "inactivation ball" that plugs the pore in open channels. (**B**) Evolution of the voltage-gated ion channel superfamily is believed to have resulted from duplications of a primordial 2 transmembrane (2TM) K^+ channel gene. The 2TM group includes bacterial (KcsA) and inward rectifier K^+ (IRK) channels (and G-protein–coupled IRK [GIRK]). A single duplication produced the 4TM group, with two pore regions in each α subunit. These channels (TWIK, TASK, TREK, TRAAK) act as "leakage" K^+-selective channels that are open at all membrane voltages and are thought to play a major role in setting the resting potential. Strong voltage sensitivity first appears in the 6TM group, which includes voltage-gated (Kv, KQT, EAG), Ca^{2+}-activated (BK, SK), and cyclic-nucleotide gated (CNG) channels. CNG channels are nonselective cation channels, whereas all other members of the 6TM group are highly K^+ selective. Two rounds of gene duplication gave rise to the 24TM group, which includes voltage-gated Na^+ and Ca^{2+} channels. Each of the four internal repeats is homologous to a 6TM subunit of Kv channels, and a single α subunit is capable of forming a functional channel. (Reproduced, with permission, from Rose, M. R., and Griggs, R. C. [eds.]. [2001]. "Channelopathies of the Nervous System." Butterworth-Heinemann, Oxford, UK.)

the pathophysiology of the periodic paralyses). Through evolution, the 2TM channel likely duplicated to form the 4TM segment "leak" channel that also lacks voltage dependency. Both K_{ir} and "leak" potassium channels lack the

S4 voltage–sensing motif and, therefore, are not members of the S4 superfamily. The most highly evolved potassium channels are the K_v ($K_v1.1$, 1.2, etc.) and Ca-activated voltage-gated types (e.g., BK, SK) whose α subunits consist of tetramers of polypeptides, each consisting of a single 6TM segment domain (see Fig. 1).

The CLC-1 voltage-gated chloride channel in skeletal muscle is structurally unrelated to other voltage-gated ion channels and comprises a homodimer formed from two identical pore-forming units each containing a fast gate, with a shared voltage sensor and slow gate [1]. Each pore-forming subunit comprises a minimum of 10 transmembrane spanning segments, as compared with the 6 segments present in each domain of voltage-gated sodium and calcium channels (Fig. 2). Although the CLC-1 chloride channel is the only known cause of human skeletal muscle disease, multiple other subtypes exist that are also expressed in muscle and other organs (e.g., kidney and heart); however, the role of chloride channels in these organs is unclear in many cases.

Ion channels modify the passage of ions across membranes in a highly controlled fashion through *gating* that is controlled by changes in membrane voltage ("voltage gated"), binding of a specific molecule ("ligand gated"), or other mechanisms such as pressure-induced changes in membrane conformation. Some channels are freely permeable to certain ions (i.e., nongated). Channelopathies of the central and peripheral nervous systems involve both voltage- and ligand-gated ion channels.

When an ion channel opens, permeant ions flow down their electrochemical gradient towards their *equilibrium potential* (for sodium, the major extracellular cation, this potential is approximately +60 mV, and conversely for potassium, the major intracellular cation, it is approximately −90 mV). The membrane potential at any given point is defined by the balance of inward and outward currents contributed by individual ion conductances, as defined by the *Hodgkin–Katz–Goldman equation*. The *resting membrane potential* is the membrane potential in the steady non-depolarized state; this is between −65 and −75 mV for neurons, as compared with −80 to −90 mV in skeletal muscle.

Action potential generation begins with an initiating capacitative current that depolarizes the membrane to the activation threshold for voltage-gated sodium channels (approximately −55 mV). Opening of these channels rapidly depolarizes the membrane toward the equilibrium potential for sodium, thus forming the upstroke of the action potential. Within a few tenths of a millisecond, sodium channels inactivate via the closure of intracellular loop between the third and fourth domains of the α subunit that "swings" to close the intracellular portion of the channel ("fast inactivation"). Thus, the same stimulus that initiates depolarization of the membrane also leads to its termination. This form

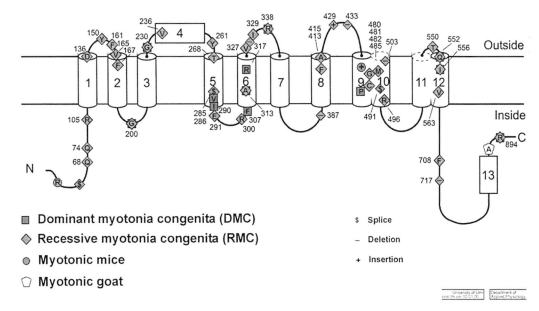

Figure 2 Membrane topology model of the skeletal muscle chloride channel monomer, CLC-1. The functional channel is a homodimer. The different symbols used for the known mutations leading to dominant Thomsen-type myotonia and recessive Becker-type myotonia are explained on the left-hand bottom. Conventional one-letter abbreviations were used for replaced amino acids. (Reproduced, with permission, from Rose, M. R., and Griggs, R. C. [eds.]. [2001]. "Channelopathies of the Nervous System." Butterworth-Heinemann, Oxford, UK.)

of control mechanism limits potentially deleterious excessive and prolonged depolarization. Many of the mutations causing periodic paralysis interfere with channel inactivation through steric effects of amino acid substitutions in the inactivation arm (Figs. 3 and 5). Following inactivation, the channel must be returned to the fully repolarized state for several milliseconds before further activation can occur: This is called the *refractory period*. Although opening and closing of individual voltage-gated channels is an "all or none" phenomenon, the membrane consists of a population of channels, a proportion of which are in the open state at any given instant that is determined by the membrane potential and the voltage-dependent gating characteristics for the channel. Alteration in voltage dependency due to altered channel protein conformation, thus, forms a second pathomechanism for many of the mutations associated with channelopathies.

As well as being responsible for generating the action potential, ion channels also play critical roles in the maintenance of the resting membrane potential within a precise range despite rapid ion fluxes that occur with high-frequency action potential generation. Such control is achieved through ionic currents that are most active at voltages at or near the resting membrane potential, major contributors include *inwardly rectifying* potassium currents, currents from potassium "leakage" channels, and

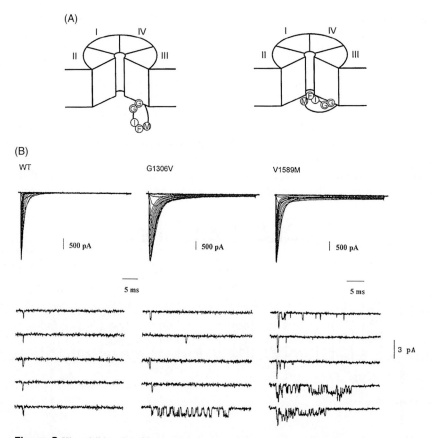

Figure 3 Hinged-lid model of fast inactivation of sodium channels and the effects of mutations at various locations on the current decay. (**A**) Bird's eye view on the channel consisting of four similar repeats (I–IV). The channel is cut and spread open between repeats II and III to allow the view on the intracellular loop between repeats III and IV. The loop acts as the inactivation gate whose "hinge" GG (= a pair of glycines) allows it to "swing" between two positions (i.e., the nonactivated channel state [pore open, left sequence isoleucine, phenylalanine, methionine, right panel]). (**B**) Two examples of faulty mutant sodium channel inactivation. Patch-clamp recordings from normal (WT) Gly-1306-Val and Val-1589-Met channels expressed in HEK-293 cells. Upper panels: Families of sodium currents recorded at various test potentials in the whole-cell mode show slowed decay and failure to return completely to baseline. Slowed inactivation is more pronounced with Gly1306Val, and persistent inward sodium current is larger for Val1589Met. Lower panels: Traces of five single-channel recordings, each obtained by clamping the membrane potential to −20 mV. Mutant channels show reopenings, which are the "macroscopic" current alterations shown in the upper panels. (Modified, with permission, from Rose, M. R., and Griggs, R. C. [eds.]. [2001]. "Channelopathies of the Nervous System." Butterworth-Heinemann, Oxford, UK.)

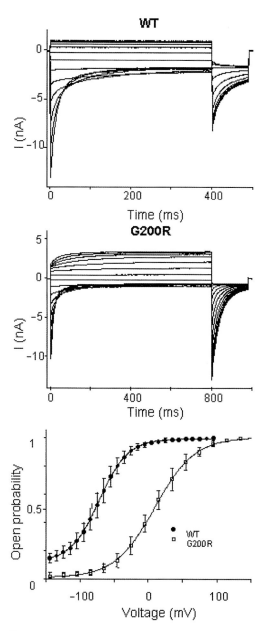

Figure 4 Recordings from human skeletal CLC-1 channels expressed in a mammalian cell line. Compared are currents from normal (WT) and dominant myotonia-causing mutant (Gly200Arg) channels. Upper and middle panels: Macroscopic currents, recorded in the whole-cell mode, were activated from a holding potential of 0 mV by voltage steps to potentials of −145 to +95 mV and deactivated after 400 ms by polarization to −105 mV. Lower panel: Voltage dependence of the relative open probability that is much reduced for the mutant channel in the physiological potential range. All mutations that cause such a voltage shift have dominant effects. (Modified, with permission, from Wagner, S., Deymeer, F., Kurz, L., Benz, S., Schleithoff, L., Lehmann-Horn, F., Serdaroglu, P., Ozdemir, C., and Rüdel, R. [1998]. The dominant chloride channel mutant G200R causing fluctuating myotonia: clinical findings, electrophysiology, and channel pathology. *Muscle Nerve* 21, 1122–1128.)

the chloride current. The physiological basis for such rectification ("anomalous") current is complex and beyond the scope of this review. Fundamentally, rectifying-type channels preferentially conduct ions in one direction across the membrane; for example, K_{ir} channels facilitate *inward rectification* because the inward conductance of potassium at potentials negative to the equilibrium potential is greater than the outward conductance at potentials positive to the equilibrium potential; this is "anomalous" to the typical potassium conductance. The mechanism of inward rectification has been partially elucidated; polyamines such as spermine and cations such as magnesium modify the degree of potassium channel inward rectification by partially blocking the passage of potassium ions through the channel. This rectification function is further modified by alterations in intracellular and extracellular potassium ion concentrations. Whereas inward rectification currents are most important in the setting and the maintaining of the resting membrane potential, conversely, outward rectification currents are most active during action potential generation. The "delayed-rectifier" potassium current mediated by voltage-gated potassium channels (VGKCs) is an outwardly rectifying current that is a major contributor to the depolarization-activated potassium conductance that facilitates repolarization during action potential generation.

III. Channelopathies of the Peripheral Nervous System

A. Non-dystrophic Myotonias

The non-dystrophic myotonias include *myotonia congenita*, both recessive (Becker's) and dominant (Thomsen's), *paramyotonia congenita* (PMC), and *potassium-aggravated myotonia* (PAM). Hyperkalemic periodic paralysis (HyperPP), PAM, and PMC are allelic disorders of the α subunit of the sodium channel Skm-1, encoded by the SCN4A gene on chromosome 17, whereas myotonia congenita is a disorder of the chloride channel CLC-1 encoded by CLCN-1 on chromosome 7q35. Distinguishing these disorders from the other major group of myotonic disorders, the myotonic dystrophies (DM1 and DM2), is usually straightforward on clinical grounds; patients with one of the non-dystrophic myotonias (as the name implies) do not manifest progressive muscle weakness or any of the ocular, endocrine, or cardiac features characteristic of the myotonic dystrophies.

1. Myotonia Congenita: Disorder of Chloride Channels

Both Becker-type and Thomsen-type myotonia congenita are characterized clinically by myotonia that improves with repeated contractions ("warmup phenomenon"), but

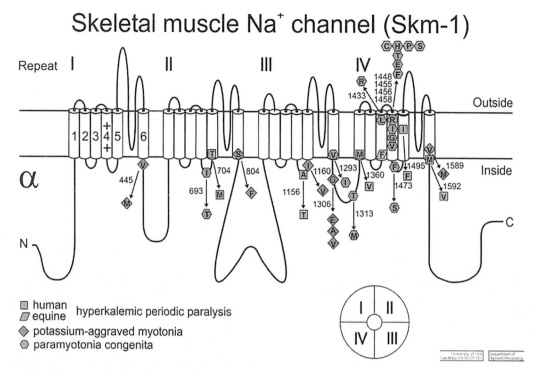

Figure 5 Subunits of the voltage-gated sodium channel of skeletal muscle. The α subunit consists of four highly homologous domains (repeats I–IV) containing six transmembrane segments each (S1–S6). The S5–S6 loops from the ion selective pore and the S4 segments contain positively charged residues conferring voltage dependence to the protein. The repeats are connected by intracellular loops; one of them, the III–IV, contains the supposed inactivation particle of the channel. When inserted in the membrane, the four repeats of the protein fold to generate a central pore, as schematically indicated of the right-hand bottom of the figure. (Reproduced, with permission, from Rose, M. R., and Griggs, R. C. [eds.]. [2001]. "Channelopathies of the Nervous System." Butterworth-Heinemann, Oxford, UK.)

the recessive form is typically more disabling because of severe muscle stiffness and weakness. Lid myotonia (manifested as difficulty opening the eyes after sustained forced eye closure) and percussion myotonia are seen in both forms. Pioneering experiments in myotonic goats identified a reduction in chloride channel conductance in the early 1970s, and in the 1990s, mutations were identified in CLCN-1 on chromosome 7q35, the gene encoding for the CLC-1 skeletal muscle chloride channel (see Fig. 2).

The gating characteristics of CLC-1 result in chloride ion flux that represents 70–85% of the total membrane conductance at the resting membrane potential [2]. Chloride conductance plays a major role in the stabilization of the resting membrane potential despite the rapid fluctuations in both intracellular and sarcoplasmic ion concentrations that occur with sustained muscle activity. This is particularly relevant in the transverse tubules where the accumulation of potassium with sustained muscle contraction tends to cause resting membrane depolarization that might otherwise inactivate sodium channels and result in membrane inexcitability. More than 40 mutations have been described in CLC-1, most of which interfere with its voltage dependency. For example, the G200R mutation shifts the voltage dependence toward marked depolarization, meaning a greater depolarization stimulus is required for channel opening as compared with wild-type channels (Fig. 4). The reduction in chloride conductance results in a depolarized resting membrane potential that produces hyperexcitability manifest as repetitive discharges recorded on the electromyogram (EMG) as myotonia.

Despite our understanding of the pathophysiology underlying membrane hyperexcitability and resultant myotonia, the relationship between genotype and phenotype has proven to be complex. Experimental evidence indicates that chloride conductance must fall to between 25 and 40% of normal before myotonia occurs. Recessive mutations generally are of the premature termination or frameshift type, thereby producing a partially or nonfunctional protein, and both alleles must be mutated to reduce chloride conductance sufficiently to cause myotonia. In the case of dominant mutations such as G200R and others, the mutation does not appear to alter channel expression but instead imposes a "dominant negative" effect on chloride conductance that is likely due to suppression of function of wild-type subunits when co-assembled with mutant subunits as chloride channel homodimers [3].

2. Hyperkalemic Periodic Paralysis, Paramyotonia Congenita, and Potassium-Aggravated Myotonia: Disorders of Sodium Channels

These two allelic disorders of the sodium channel Skm-1 are clinically distinct. PMC is characterized by myotonia that paradoxically worsens with repeated contraction (hence, paramyotonia is used to distinguish it from typical myotonia that improves with repeated contractions) and with cold exposure. Affected individuals become profoundly stiff in the cold (often manifest in the eyelids and hands) and develop cold-induced weakness; both may take hours to reverse after warming. The stiffness caused by myotonia may be successfully treated with agents that block sodium channels (e.g., carbamazepine or mexiletine).

PAM is characterized by myotonia that is worsened by the intake of potassium-rich foods or other agents that cause partial depolarization of the muscle membrane (e.g., depolarizing neuromuscular blocking agents such as suxamethonium). Whereas patients with PAM may develop severe exercise-induced myotonia, unlike PMC, they do not develop weakness. The clinical spectrum of PAM is broad, and various subtypes have been described with varying degrees of myotonia, for example, *myotonia permanens*, in which myotonia may be so severe to interfere with ventilation from involvement of intercostal muscles, and *myotonia fluctuans*, in which patients experience dramatic day-to-day and month-to-month spontaneous fluctuations in myotonia without clear explanation [4]. The term "sodium channel myotonia" is used to describe such patients who have myotonia without attacks of weakness; however, the situation is further complicated by the fact that current evidence suggests the same phenotype may be seen with chloride channel mutations usually associated with myotonia congenita [5].

HyperPP is an autosomal-dominant disorder similar to HypoPP, with episodes of transient weakness lasting minutes to hours. The age at onset is typically earlier than for HypoPP, with attacks during early childhood being the rule. The attacks typically become less frequent with age and may be replaced with fixed weakness, but this appears less frequent than with HypoPP. Attacks may be precipitated by prolonged rest, stress, or fasting and potassium supplementation or potassium-sparing diuretics also provoke attacks in HyperPP. Like those with HypoPP, patients with HyperPP may respond to acetazolamide and other carbonic anhydrase inhibitors.

The first evidence of dysfunction of muscle voltage-gated sodium channels came from pioneering electrophysiological studies by Lehmann-Horn and others who demonstrated persistent depolarization in biopsied muscle fibers from patients with HyperPP. This depolarization is probably due to persistent inward sodium channel current given that the depolarization could be prevented with tetrodotoxin (a sodium channel blocker) [6]. Further electrophysiological studies identified the cause for HyperPP and the sodium channel myotonias PMC and PAM as being due to defective inactivation of the Skm-1 sodium channel in 1987, and the gene encoding it (SCN4A) was cloned in 1990 [7].

For the sodium channel disorders (HyperPP, PMC, PAM), the relationship between genotype and phenotype is complicated because two phenotypes may exist in the same family, even in the same patient. For example, many patients with HyperPP have EMG (and sometimes clinical) myotonia, an association that serves as a useful diagnostic feature because EMG myotonia is typically not seen in HypoPP. Understanding how mutations in the same sodium channel (albeit at different sites) cause, on the one hand, PMC, and, on the other, HyperPP has been difficult, but on the basis of electrophysiological studies, plausible explanations have been suggested: Relatively mild persistent depolarizations (e.g., 5–10 mV) of the muscle membrane cause activation of a proportion of sodium channels that are not inactivated, thereby facilitating repetitive action potentials. By contrast, more substantial depolarizations in the 20- to 30-mV range cause more extensive sodium channel inactivation that result in an electrically inexcitable membrane and paralysis. The pathophysiological explanation for differences in the degree of membrane depolarization responsible for myotonia versus paralysis is explained by the mutation's differing effects on the kinetics and the completeness of sodium channel inactivation gating. For example, mutations resulting in the myotonic phenotype typically interfere with the rapidity of fast inactivation, but it remains complete (see Figs. 3 and 5). Recovery from fast inactivation is normal or, in the case of PMC mutations, more rapid than normal. The combination of slowed fast inactivation prolongs action potential duration, and the increased rapidity of recovery shortens the refractory period. These two factors promote membrane hyperexcitability and facilitate the repetitive discharges observed clinically and electrically as myotonia. By contrast, the most common mutation causing HyperPP produces a defect of slow sodium channel inactivation, whereas fast inactivation remains normal. This causes persistent inward sodium current that depolarizes the membrane, inactivates wild-type sodium channels, and results in paralysis that is reversed when the sodium-potassium adenosine triphosphatase (ATPase) pump restores the resting membrane potential to normal.

3. Hypokalemic Periodic Paralysis: Disorder of Calcium and Sodium Channels

HypoPP is probably the most common inherited periodic paralysis. It is autosomal dominant with complete penetrance in men; however, the penetrance may be decreased in women. The disorder is characterized by recurrent episodes of weakness ("attacks") lasting minutes to hours, with sparing of respiratory and bulbar muscles. During an attack, reflexes are diminished or absent, and muscles are electrically inexcitable. Power returns to normal or near normal between attacks, at least early in the course of the disease, but most patients demonstrate weakness later in life. Common precipitants of attacks include carbohydrate-rich meals, exercise, a high-sodium diet, and prolonged rest.

The causative mutation HypoPP was first identified in the α_1 subunit of the voltage-gated calcium channel $Ca_v1.1$, a dihydropyridine (DHP)-sensitive L-type calcium channel primarily found on the surface of T tubules that is encoded by CACNA1S (also known as *CACNL1A3*) on chromosome 1q. It is intimately associated with excitation–contraction coupling through its physical association with the ryanodine receptor (RyR1), a channel facilitating calcium release from the sarcoplasmic reticulum. CACNA1S acts as a voltage sensor for RyR1, and mutations in CACNA1S have been identified as conveying malignant hyperthermia (MH) susceptibility similar to ryanodine receptor mutations; HypoPP and MH may, therefore, be allelic conditions (Fig. 6). Three mutations have been identified in CACNA1S, all occurring in the S4 motif of either domain 2 or domain 4: R528H (domain 2) and R1239G and R1239H (domain 4) (Fig. 7). Collectively these three mutations account for at least 70–80% of all PP patients screened for mutations. In 1999, a mutation was

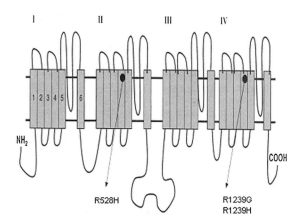

Figure 7 A model of the calcium channel with the three defined HypoPP mutations. This calcium channel is known to interact with the ryanodine receptor at the sarcoplasmic reticulum membrane. (Reproduced, with permission, from Rose, M. R., and Griggs, R. C. [eds.]. [2001]. "Channelopathies of the Nervous System." Butterworth-Heinemann, Oxford, UK.)

identified in the α_1 subunit of the voltage-gated sodium channel SCN4A in one kindred [8], and further mutations in the α subunit of the sodium channel have now been identified in other kindreds [9,10], confirming voltage-gated sodium channel mutations as a less common cause of HypoPP. Thus, HypoPP has now been further subdivided into HypoPP type 1 (calcium channel mutation) and type 2 (sodium channel mutation). Whereas HypoPP1 and HypoPP2 are essentially indistinguishable clinically, anecdotal evidence suggests patients who have myalgias with or before attacks or who worsen with CAI therapy may be more likely to possess sodium channel mutations (i.e., HypoPP2).

Early electrophysiological studies by Rüdel and colleagues in *ex vivo* studies of muscle fibers from patients with HypoPP identified a paradoxical membrane depolarization response to hypokalemia as compared with the normal hyperpolarization response to hypokalemia as predicted by the Nernst equation. This effect is potentiated by insulin and is not attenuated by tetrodotoxin (a sodium channel blocker), indicating that the depolarization is not mediated by inward sodium channel current. The pathophysiological basis for this finding remains a mystery. Despite mutations in DHP-sensitive calcium channels, electrophysiological studies have demonstrated inconsistent abnormalities in calcium currents in mutant channels. For example, the mutation R528H produced no identifiable abnormalities in L-type calcium currents or changes in gating properties of mutant channels expressed in cultured myotubules. However, in human myotubules from carriers of the R528H mutation, a 6-mV shift in the inactivation and activation voltages towards more hyperpolarized potentials was identified. A second study demonstrated reductions in calcium current density in mouse cells expressing mutant R528H

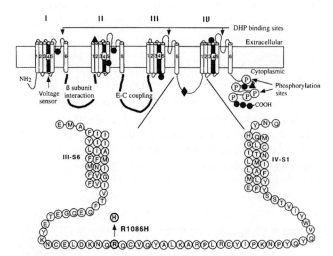

Figure 6 Schematic representation of the α_1 subunit of a skeletal muscle dihydropyridine receptor. Sequences involved in both EC coupling and interaction with the β subunit of the dihydropyridine receptor are shown in bold. S4 transmembrane segments of the different domains that operate as voltage sensors are blackened, and dihydropyridine binding sites and phosphorylation sites present in the C-terminal region are indicated. Positions of the sequence modification are represented: Black dots denote homozygous corrections; black triangles denote heterozygous polymorphisms; and black diamond denotes Arg1086His mutation. (Reproduced, with permission, from Monnier, N., Procaccio, V., Stieglitz, P., and Lunardi, J. [1997]. Malignant hyperthermia susceptibility is associated with a mutation of the α_1 subunit of the human dihydropyridine-sensitive L-type voltage dependent calcium-channel receptor in skeletal muscle. *Am. J. Hum. Genet.* **60**, 1316–1325.)

channels [11]. How these abnormalities in calcium current relate to the paralysis of HypoPP is unclear. Although the identification of mutations in the α subunit of the sodium channel SCN4A as a less common cause of HypoPP has complicated the picture, similar abnormal membrane responses to hypokalemia have been described in patients with HypoPP, suggesting a common pathogenesis. All α-subunit sodium channel mutations except one have been identified in the voltage-sensor (S4) portion of the protein. Jurkat-Rott et al. [9] identified two major pathophysiological effects of sodium channel mutations Arg672His and Arg672Gly: (1) a 10-mV left shift in the steady-state inactivation curve, resulting in fewer sodium channels available for activation at the resting membrane potential of −80 mV, and (2) reduced sodium current density. The combined effect of these two factors leads to reduced muscle fiber hypoexcitability that presumably causes paralysis. A reduced sodium current density may be explained by either reduced channel expression levels or alternatively reduced channel open probability, or both [9]. Similar findings were reported by Kuzmenkin et al. [10] who studied the three mutations Arg669His, Arg672His, and Arg672Gly: Enhanced fast inactivation was identified in all three, and two were associated with enhanced slow inactivation (Arg669His and Arg672Gly) [10]. However, the situation is further complicated by the report of an SCN4A HypoPP mutation (Pro1158Ser) arising in a location distinct from the voltage sensor, in an individual with otherwise typical HypoPP but with myotonia, thus representing the first evidence of myotonia in HypoPP (myotonia is typically considered only to occur in HyperPP) [12]. Attention has focused on the role of the inwardly rectifying ATP-sensitive potassium current as playing a central role in the pathogenesis of HypoPP, given that agents that promote opening of such channels (e.g., cromakalim) have increased contractile force in muscle fibers from HypoPP patients [13]; however, *in vivo* studies have not been performed to confirm this finding.

4. Andersen-Tawil Syndrome: Disorder of Potassium Channels

Andersen-Tawil syndrome (ATS), previously termed long-QT syndrome type 7 (LQT7), comprises less than 10% of all cases of periodic paralysis. In its typical form, the disorder is characterized by a triad of periodic paralysis, cardiac conduction defects (ventricular arrhythmias), and characteristic physical features that include short stature and a shortened index finger, among other features. Some individuals demonstrate only fragments of the syndrome, and patients have been described with the cardiac manifestations but no neuromuscular manifestations. For unclear reasons, death from cardiac arrhythmia is less common than for other causes of long-QT syndrome, despite the presence of sometimes dramatic electrocardiographic repolarization abnormalities and frequent nonsustained ventricular tachycardia. Typically patients are hypokalemic during attacks of weakness, but normokalemia and even hyperkalemia have been reported. Mutations in KCNJ2, the gene encoding the α subunit of the inwardly rectifying potassium channel $K_{ir}2.1$ have been identified in ATS patients [14], and more than 20 mutations have now been described. Approximately 70% of patients with ATS have an identifiable mutation in KCNJ2, such cases are designated ATS1. The remaining 30% have an as yet unidentified mutation and are designated ATS2. ATS1 and ATS2 are phenotypically indistinguishable.

$K_{ir}2.1$ channels have been identified in skeletal and cardiac muscle, the developing fetus, and in brain. Skeletal and cardiac muscle and embryological manifestations of ATS are relatively well characterized; however, central nervous system manifestations have not been identified. In cardiac muscle, the $K_{ir}2.1$ channel is responsible for the major terminal repolarization current I_{K1} that plays a major role in the stabilization of the cardiac resting membrane potential. In general, mutations causing ATS result in a loss of function (i.e., suppression of inwardly rectifying potassium current); however, the mechanism by which attenuation of I_{K1} causes ventricular arrhythmias and periodic paralysis is not entirely clear. On physiological grounds, attenuation of inward rectifying potassium current would be expected to prolong action potential duration, thereby prolonging QT interval and acting as a substrate for ventricular arrhythmias. In skeletal muscle, $K_{ir}2.1$ mutations might result in either (or both) loss of inward rectification current or excessive potassium trapping in the transverse tubules that might be expected to cause membrane depolarization and sodium channel inactivation. This remains to be proven. Potassium supplementation has improved the cardiac rhythm abnormalities and reduced QT prolongation in some patients; however, its role in treating neuromuscular manifestations is unclear.

5. Malignant Hyperthermia: Disorder of Ryanodine Calcium Release Channel

Malignant hyperthermia (MH) is characterized by an abnormal response to inhalational anesthetics (e.g., halothane): hyperthermia, rigidity, and autonomic instability. The disorder has a high mortality if not recognized rapidly. With appropriate treatment including discontinuation of the offending agent, cooling, and the use of muscle relaxants such as dantrolene that block release of calcium from the sarcoplasmic reticulum, the mortality has fallen. MH susceptibility is due in most cases to mutations in the skeletal muscle ryanodine receptor RyR1, encoded on chromosome 19q. In some cases, MH susceptibility has also been linked to loci on chromosomes 1, 3, 5, 7, and 17; however, disease-causing mutations have not been identified for these loci. In addition to RyR1 mutations, a single mutation in CACNA1S has been identified as a second cause of MH susceptibility (CACNA1A encodes the α_1 subunit of the skeletal muscle DHP-sensitive calcium channel). This is perhaps explained by the role of the DHP-sensitive calcium channel in excitation–contraction coupling through its physical association with RyR1 (Fig. 7). The CACNA1S mutation arises in the linker between domains III and IV, a corresponding location to the inactivation gate of the voltage-gated sodium channel (in which mutations cause PMC); however, the exact role of the III–IV linker in CACNA1S is unclear [15]. MH is allelic to central core disease (CCD), a slowly progressive proximal myopathy with characteristic biopsy features in which RyR1 mutations have also been identified. Mutations in RyR1 causing CCD are believed to cause weakness through the following mechanisms: (1) depletion of sarcoplasmic calcium stores through calcium efflux via excessively leaky channels, and/or (2) disruption of excitation–contraction coupling [16]. Finally, MH susceptibility has also been demonstrated in a kindred in which some members exhibited HyperPP, and others HyperPP and MH. Both were linked to SCN4A (the gene encoding the Skm-1 voltage-gated sodium channel), suggesting that like other channelopathies, considerable genotypic heterogeneity may exist in this condition [17].

IV. Channelopathies of the Central Nervous System

A. Familial Hyperekplexia: Disorder of the Ligand-Gated Chloride Channel

Familial hyperekplexia was the first channelopathy of the central nervous system to be identified. It is characterized by generalized stiffness and an exaggerated startle response; the stiffness is provoked by unexpected stimuli so profound as to cause affected individuals to fall suddenly. Most cases are dominantly inherited, but recessive inheritance is also reported. The disorder was linked to chromosome 5q in 1992, and multiple mutations in GLRA1, the gene encoding the α subunit of the glycine receptor, have been reported. The glycine receptor is a ligand-gated chloride channel and a member of the same ion channel superfamily as the γ-aminobutyric acid A (GABA$_A$), serotonin, and nAchR. It is most prominently expressed in inhibitory interneurons and Renshaw cells in the brainstem and spinal cord. The glycine receptor α-subunit structure comprises an extracellular N-terminal domain, four transmembrane segments (M1–M4), and an extracellular C-terminus. Mutations have been identified in one of the four M domains that result in reduced chloride flux through impaired signal transduction. Given that glycine is an inhibitory neurotransmitter, the reduction in chloride flux results in disinhibition of repetitive discharges in spinal motor neurons that is manifest as an exaggerated motor response to startle.

B. Paroxysmal and Progressive Ataxias: Disorders of Potassium and Sodium Channels

Paroxysmal and progressive ataxias include episodic ataxia type 1 (EA-1), a disorder of the α subunit of the potassium channel Kv$_\alpha$1.1 (KCNA1 gene), episodic ataxia type 2 (EA-2), and spinocerebellar ataxia type 6 (SCA-6). The latter two are disorders of the P/Q-type calcium channel Ca$_v$2.1 (CACNA1A gene). SCA-6 is a CAG-repeat disorder characterized by progressive isolated cerebellar ataxia without paroxysmal features.

Episodic ataxias, as suggested by the name, are characterized by paroxysmal and often disabling ataxia that occurs spontaneously and lasts minutes (EA-1) to hours (EA-2). Attacks in EA-2 are more likely to be complicated by vertigo, nausea, vomiting, headache, and sometimes brainstem symptoms [18]. Symptoms of both EA-1 and EA-2 typically begin in childhood; EA-2 often and EA-1 occasionally respond to acetazolamide. Patients with EA-1 are typically normal between attacks, whereas those with EA-2 may exhibit interictal nystagmus and occasionally a progressive ataxia syndrome resembling SCA-6 [19]. Symptoms may be precipitated or exacerbated (often dramatically) by anxiety or exercise for unclear reasons. EA-1 is associated with facial myokymia ("rippling" effect visible on careful inspection of facial muscles) and, on rare occasions, neuromyotonia. Both EA-1 and EA-2 are associated with epilepsy. Overlap exists with familial hemiplegic (FHP) migraine, and FHP and EA-2 are now regarded as allelic disorders of the CACNA1A gene on chromosome 19p; individuals within the same FHM kindred due to CACNA1A mutations have also been recognized as having persistent nystagmus and, on occasion, episodic ataxia. Notwithstanding this overlap, mutations responsible for EA-2 are distinct from those associated with FHM, the former usually caused by missense mutations in conserved portions of the sequence and the latter usually caused by mutations resulting in a truncated peptide. Pathomechanisms for EA-2 and FHM have yet to be fully characterized. For EA-1, expression studies of mutated Kv$_\alpha$1.1 channels in *Xenopus* oocytes have yielded conflicting results. Overall, studied mutations affected Kv$_\alpha$1.1 delayed-rectifier current in several ways: (1) a reduced peak current from a depolarizing stimulus and (2) alterations in channel gating kinetics, causing a shift in channel activation threshold to more positive potentials (Fig. 8). Presumably, suppression of delayed-rectifier potassium current is central to the pathogenesis of EA-1, as supported by the correlation between disease severity and degree of suppression of such current [20]. As with other dominantly inherited channelopathies, mutations have been identified as likely causing a dominant negative effect through co-assembly of mutated α subunits with wild type. Presumably, attenuation of delayed-rectifier potassium current may result in persistent depolarization and neuronal hyperexcitability, but how this causes brief

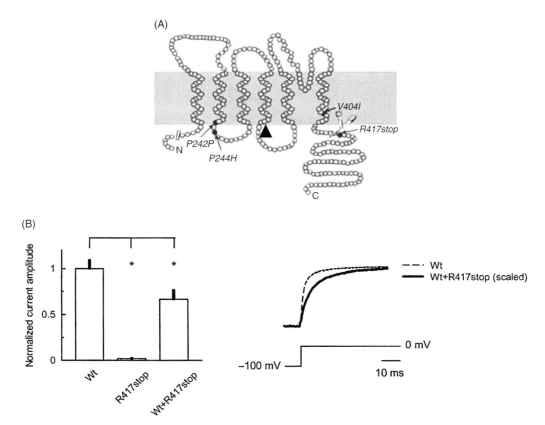

Figure 8 KCNA1 mutations impair potassium channel function. (**A**) Structure of the Kvα1.1 subunit, showing the six transmembrane domains, the voltage-sensing S4 domain (filled triangle), and the pore-lining loop between S5 and S6. Each circle represents a single amino acid, and filled circles indicate positions of known missense mutations. (**B**) Example of K$^+$ channel dysfunction caused by mutation caused by a mutation associated with severe drug-resistant episodic ataxia type 1. The R417 stop mutation truncates the C-terminus and gives rise to a nonfunctional channel. However, the mutant subunit can assemble with the wild-type (Wt) subunit, as shown by a significant ($^*p < .05$) reduction in potassium current obtained by coexpressing the two alleles together (a dominant negative effect; histogram of left) and by a change in the activation rate (current traces on right). In addition, the mutant shifted the activation threshold to more positive voltages and accelerated deactivation (not shown). These effects are predicted to result in a profound reduction in potassium flux and impaired repolarization of neurons. (Reproduced, with permission, from Kullmann, D. [2002]. The neuronal channelopathies. *Brain* **125**, 1177–1195.)

recurrent episodes of ataxia is unclear. Moreover, Kv$_\alpha$1.1 is certainly expressed in Purkinje cells in the cerebellum but is also widely expressed throughout other areas of the brain. This has prompted an intensive and ongoing search for other central nervous system manifestations in both EA-1 and EA-2.

For the mouse model of EA-2, knockout of Ca$_v$2.1 produced no discernible phenotype, whereas mutations in Ca$_v$2.1 and its accessory subunits produced ataxia and absence type seizures. Although the wide expression of Ca$_v$2.1 in human brain and the absence epilepsy phenotype in the mouse have suggested a role for CACNA1A mutations in the pathogenesis of human primary generalized epilepsies, no definite association has been proven. This is further discussed in Section IV.D, later in this chapter.

SCA-6 is the only known polyglutamine repeat disorder of ion channel function. Unlike the paroxysmal ataxias (see Section IV.B, earlier in this chapter), individuals with SCA-6 experience slow progressive rather than episodic symptoms, and unlike most of the other dominantly inherited SCAs, they experience isolated cerebellar ataxia without pyramidal, extrapyramidal, or peripheral nervous system disease. Individuals with SCA-6 have an abnormally elevated (>21) triplet repeats occurring in the open reading frame of CACNA1A, whereas normal individuals have 19 or fewer repeats. How these repeats cause disease is the subject of debate; however, unlike other triplet repeat disorders (e.g., Huntington's disease), it is not clear that the abnormal protein accumulation in the cytoplasm is the major pathomechanism. Nonetheless, midline cerebellar atrophy indicative of widespread Purkinje cell loss is typical in SCA-6; one plausible mechanism for such neuronal loss includes abnormal calcium entry into Purkinje cells leading to excitotoxic cell death or initiation of neuronal apoptosis. This hypothesis is in part supported by the identification of abnormal P/Q calcium channel currents in channels expanded with increasing polyglutamine repeats in baby hamster cells [21] but is unproven in humans.

C. Familial Hemiplegic Migraine: Disorder of Calcium Channels

FHM is a rare form of dominantly inherited migraine accompanied by reversible hemiparesis and other features, including transient coma. The symptoms are typically but not universally responsive to acetazolamide. In approximately 50% of cases, FHM is caused by mutations in the voltage-gated P/Q-type calcium channel CACNA1A on chromosome 19p. In other kindreds, linkage has been demonstrated to chromosome 1q in North American and French families, but the candidate gene is unidentified. In a small proportion, no linkage has been identified. In cases harboring a CACNA1A mutation, there is overlap with the ataxia syndromes EA-2 and SCA-6; some individuals with CACNA1A mutations demonstrate marked persistent nystagmus, and others experience episodic ataxia. Neuroimaging may demonstrate atrophy of the anterior cerebellar vermis. This is further discussed in Section IV.B, earlier in this chapter.

D. Inherited Epilepsy Syndromes: Disorder of Sodium and Potassium Channels and the Nicotinic Acetylcholine Receptor

Mutations in ion channels have been identified as causes of the rare inherited epilepsy syndromes autosomal–dominant nocturnal frontal lobe epilepsy (ADNFLE), benign familial neonatal convulsions (BFNC), and generalized epilepsy with febrile convulsions "plus" (GEFS+). For the majority of more common epilepsy syndromes, a large body of research has failed to definitively prove channelopathies as playing a causative role. However, it seems plausible that ion channel dysfunction contributes in some way based on effects on channels of many of the therapies effective in epilepsy. The complex polygenic inheritance patterns for the majority of the common epilepsy syndromes have made major progress in this area difficult. Based on the presumed presence of susceptibility genes, attempts have been made to identify candidate genes for epilepsy by association studies: the identification of statistically significant differences in the allele frequency in the disease population as compared with a suitable control population. Many association studies suggesting causative genes in epilepsy have not been able to be replicated, for a variety of reasons; however, mutations in calcium, sodium, and potassium channels have been suggested. Only the association between opioid receptor μ-subunit (OPRM) gene polymorphisms and idiopathic generalized epilepsy is supported by reproducible and consistent study evidence; however, plausible associations have been suggested between CACNA1A single nucleotide polymorphisms (SNPs) and idiopathic generalized epilepsy and between CACNA1H missense mutations and childhood absence epilepsy [22].

BFNC is autosomal dominant with high penetrance that is characterized by seizures in the first week of life. Affected individuals do not appear to have a higher likelihood of seizures in later life compared with the general population. Linkage studies identified in some families KCNQ2 on chromosome 20q and later, KCNQ3 on 8q. KCNQ2 and KCNQ3 encode channels that play a major role in determining the excitability of neurons. Unlike the dominant negative effect typical of other channelopathies, the pathomechanism of these dominantly inherited mutations is believed to result in a variable reduction in functional protein expression; only a 25% reduction in M current may be sufficient to produce the BFNC phenotype, based on expression studies in *Xenopus* oocytes [23]. Although the transient nature of this disease is unclear, plausible explanations have been suggested: (1) KCNQ2 and KCNQ3 are only transiently expressed, and other channels soon take over the same function after birth, and/or (2) that wild-type alleles may become up-regulated soon after birth when one of the alleles is nonfunctional. Evidence consistent in part with these hypotheses is suggested from the phenotype of KCNQ2 knockout mice; the homozygous deletion is lethal, but the fact that heterozygotes develop normally but do not have seizures is unexplained. Such a finding perhaps highlights the partial limitations of currently available animal models of human channelopathies.

ADNFLE is characterized by frontal lobe seizures, often partial, that occur during sleep. In the mid-1990s, linkage to chromosome 20q was identified, followed by the discovery of mutations in CHRNA4, the gene encoding the α_4 subunit of the nAchR. Several mutations have now been reported, all in the M2 domain. A second locus on chromosome 1p was also identified, and the discovery of mutations in CHRNB2 (the gene encoding the M2 domain of the nAchR β_2 subunit) confirms this as an additional cause of ADNFLE. The M2 domain is the pore-forming unit of the nAchR; missense mutations in this domain probably alter the permeability and gating characteristic of the channel and cause dramatically reduced calcium flux and an increase in rapidity of desensitization, as well as prolongation of recovery from desensitization.

The third epilepsy syndrome associated with a channelopathy, GEFS+, was described in a family with an autosomal dominantly inherited epilepsy syndrome and is remarkable in its phenotypic heterogeneity. The disorder has been linked to SCN1A (chromosome 2q), the gene encoding the α subunit of the voltage-gated sodium channel $Na_v1.1$, and SCN1B (chromosome 19q), the gene encoding the sodium channel β_1 subunit. Linkage to 2q but without identifiable mutations in SCN1A has also been reported, raising the possibility that mutations in other sodium channels (e.g., SCN2A encoding the $Na_v1.2$ channel) may also be a cause of generalized epilepsy; however, this is unproven. Several mutations have now been reported in the S4 motif of $Na_v1.1$, arising in a similar region to SCN4A mutations in HypoPP. Yet, although sodium channel mutations in HypoPP have been demonstrated in HypoPP to cause inactivation defects, this has not been proven for $Na_v1.1$. With respect to the SCN1B mutations that produce a defective β_1 subunit, such mutations have been shown to slow sodium channel inactivation and presumably cause neuronal hyperexcitability based on the ability of the β_1 subunit to accelerate sodium channel inactivation. Mutations in the γ_2 subunit of a third channel, the $GABA_A$ receptor (a chloride channel), have now been identified in some GEFS+ families, proving that there is considerable genetic heterogeneity and phenotypic heterogeneity in this disorder [24].

V. Acquired Channelopathies of the Nervous System

The acquired channelopathies of the nervous system comprise both central and peripheral nervous system disorders. For some, ion channel defects have been proven, but for others, defects have been proposed and supported with evidence but are not yet proven. The majority are immune mediated and include the disorders autoimmune neuromyotonia (AINMT) (Isaac's syndrome), cramp-fasciculation syndrome (VGKCs for both conditions), and Lambert–Eaton syndrome (voltage-gated P/Q-type calcium channel). Whereas considerable electrophysiological evidence in patients with the central and peripheral nervous system disorders multiple sclerosis and Guillain-Barré syndrome has suggested a pathogenetic role for ion channels, a central role for ion channel dysfunction in these disorders remains to be proven.

AINMT is a disorder of peripheral nerve hyperexcitability, probably at the level of the terminal axon based on Isaac's original report that curare abolished neuromyotonic discharges, but distal or proximal nerve block by local anesthesia did not. Up to 90% of patients with AINMT express autoantibodies to VGKCs, using advanced molecular immunohistochemical assays [25,26]. Typical symptoms of AINMT are muscle stiffness, fasciculations, myokymia, weight loss, and hyperhidrosis. High-frequency (40- to 400-Hz) neuromyotonic discharges, fasciculations, and fibrillations may be seen on the EMG, sometimes accompanied by demonstrable motor axon loss. Occasionally, central nervous system symptoms will accompany otherwise typical AINMT, ranging from sleep disturbance to frank psychosis (Morvan's syndrome).

Although AINMT is distinct, the phenotypic spectrum of immune-mediated potassium channelopathies is steadily widening; other disorders also associated with a voltage-gated potassium antibody in some patients include "cramp-fasciculation syndrome," which is characterized by similar symptoms to AINMT (fasciculations, myalgias), but without neuromyotonia or myokymia, and "rippling muscle disease," which is characterized by slow, rolling, localized muscle movements that are precipitated by stretch or percussion [26]. Rippling muscle disease is also caused by mutations in caveolin-3, a protein important in transverse tubule development in skeletal muscle. All three disorders have been associated with VGKC antibodies in up to 20–55% of cases, and with ganglionic AchR antibodies in 10–20%, and all three have associations with other autoimmune disorders (e.g., myasthenia gravis) and tumors (e.g., small cell lung cancer, thymoma). The pathophysiology of VGKC dysfunction due to circulating VGKC antibodies in these disorders remains to be proven; however, insights have come from two lines of evidence: (1) the addition of 4-aminopyridine (4-AP, a specific VGKC blocker) to *Xenopus* oocytes and cultured dorsal root ganglion cells expressing VGKC results in membrane hyperexcitability and spontaneous high-frequency action potential discharges similar to AINMT through lowering of the excitation threshold, and (2) patch-clamping studies demonstrated that the addition of AINMT antibodies to cultured cells expressing VGKC suppressed potassium ion flux. Moreover, in patients with syndromes associated with VGKC antibodies, symptoms often improve with immune-based treatments such as plasma exchange, supporting the proposed autoimmune pathogenesis in VGKC-associated syndromes. However, why antibodies to both ganglionic AchRs and VGKCs have been identified in some patients and exactly how VGKC antibodies produce neuromyotonia *in vivo* both remain unclear [26].

References

1. Colding-Jørgensen, E. (2005). Phenotypic variability in myotonia congenital. *Muscle Nerve* **32**, 19–34.
2. Bryant, S. H., and Morales-Aguilera, A. (1971). Chloride conductance in normal and myotonic muscle fibres and the action of monocarboxylic aromatic acids. *J. Physiol.* **219**, 367–383.
3. Barchi, R. L. (1998). Phenotype and genotype in the myotonic disorders. *Muscle Nerve* **21**, 1119–1121.
4. Ricker, K., Moxley, R. T., Heine, R., and Lehmann-Horn, F. (1994). Myotonia fluctuans, a third type of muscle sodium channel disease. *Arch. Neurol.* **51**, 1095–1102.
5. Wagner, S., Deymeer, F., Kurz, L., Benz, S., Schleithoff, L., Lehmann-Horn, F., Serdaroglu, P., Ozdemir, C., and Rüdel, R. (1998). The dominant chloride channel mutant G200R causing fluctuating myotonia: clinical findings, electrophysiology, and channel pathology. *Muscle Nerve* **21**, 1122–1128.
6. Lehmann-Horn, F., Rüdel, R., Ricker, K., Lorkovic, H., Dengler, R., and Hopf, H. C. (1983). Two cases of adynamia episodica hereditaria: in vitro investigation of muscle cell membrane and contraction parameters. *Muscle Nerve* **6**, 113–121.
7. Fontaine, B., Khurana, T. S., Hoffman, E. P., Bruns, G. A., Haines, J. L., Trofatter, J. A., Hanson, M. P., Rich, J., McFarlane, H.,

Yasek, D. M., et al. (1990). Hyperkalemic periodic paralysis and the adult muscle sodium channel alpha-subunit gene. *Science* **250**, 1000–1002.

8. Bulman, D. E., Scoggan, K. A., van Oene, M. D., Nicolle, M. W., Hahn, A. F., Tollar, L. L., and Ebers, G. C. (1999). A novel sodium channel mutation in a family with hypokalemic periodic paralysis. *Neurology* **53**, 1932–1936.

9. Jurkat-Rott, K., Mitrovic, N., Hang, C., Kouzmekine, A., Iaizzo, P., Herzog, J., Lerche, H., Nicole, S., Vale-Santos, J., Chauveau, D., Fontaine, B., and Lehmann-Horn. F. (2000). Voltage-sensor sodium channel mutations cause hypokalemic periodic paralysis type 2 by enhanced inactivation and reduced current. *PNAS* **97**, 9549–9554.

10. Kuzmenkin, A., Muncan, V., Jurkat-Rott, K., Hang, C., Lerche, H., Lehmann-Horn, F., and Mitrovic, N. (2002). Enhanced inactivation and pH sensitivity of Na^+ channel mutations causing hypokalemic periodic paralysis type II. *Brain* **125**, 835–843.

11. Jurkat-Rott, K., Uetz, U., Pika-Hartlaub, U., Powell, J., Fontaine, B., Melzer, W., and Lehmann-Horn, F. (1998). Calcium currents and transients of native and heterologously expressed mutant skeletal muscle DHP alpha1 subunits (R528H). *FEBS Lett.* **423**, 198–204.

12. Sugiura Y, Makita N, Li L, Noble PJ, Kimura J, Kumagai Y, Soeda T, and Yamamoto T. (2003). Cold induces shifts of voltage dependence in mutant SCN4A, causing hypokalemic periodic paralysis. *Neurology* **61**, 914–918.

13. Grafe, P., Quasthoff, S., Strupp, M., and Lehmann-Horn, F. (1990). Enhancement of K+ conductance improves in vitro the contraction force of skeletal muscle in hypokalemic periodic paralysis. *Muscle Nerve* **30**, 451–457.

14. Plaster, N. M., Tawil, R., Tristani-Firouzi, M., Canun, S., Bendahhou, S., Tsunoda, A., Donaldson, M. R., Iannaccone, S. T., Brunt, E., Barohn, R., Clark, J., Deymeer, F., George, A. L., Jr., Fish, F. A., Hahn, A., Nitu, A., Ozdemir, C., Serdaroglu, P., Subramony, S. H., Wolfe, G., Fu, Y.-H., and Ptacek, L. J. (2001). Mutations in Kir2.1 cause the developmental and episodic electrical phenotypes of Andersen's syndrome. *Cell* **105**, 511–519.

15. Monnier, N., Procaccio, V., Stieglitz, P., and Lunardi, J. (1997). Malignant hyperthermia susceptibility is associated with a mutation of the α_1 subunit of the human dihydropyridine-sensitive L-type voltage dependent calcium-channel receptor in skeletal muscle. *Am. J. Hum. Genet.* **60**, 1316–1325.

16. Lyfenko, A. D., Goonasekara, S. A., and Dirksen, R. T. (2004). Dynamic alterations in myoplasmic Ca^{2+} in malignant hyperthermia and central core disease. *Biochem. Biophys. Res. Commun.* **322**, 1256–1266.

17. Moleshi, R., Langlois, S., Yam, I., and Friedman, J. M. (1998). Linkage of malignant hyperthermia and hypokalemic periodic paralysis to the adult skeletal muscle sodium channel (SCN4A) gene in a large pedigree. *Am. J. Med. Genet.* **76**, 21–27.

18. Baloh, R. W., Yue, Q., Furman, J. M., and Nelson, S. F. (1997). Familial episodic ataxia: clinical heterogeneity in four families linked to chromosome 19p. *Ann. Neurol.* **41**, 8–16.

19. Jen, J. (1999). Calcium channelopathies in the central nervous system. *Current Opin. Neurobiol.* **9**, 274–280.

20. Eunson, L. H., Rea, R., Zuberi, S. M., Youroukos, S., Panayiotopoulos, C. P., Ligouri, R., Avoni, P., McWilliam, R. C., Stephenson, J. B., Hanna, M. G., Kullmann, D. M., and Spauschus, A. (2000). Clinical, genetic, and expression studies of mutations in the potassium channel gene KCNA1 reveal new phenotypic variability. *Ann. Neurol.* **48**, 647–656.

21. Matsuyama, Z., Wakamori, M., Mori, Y., Kawakami, H., Nakamura, S., and Imoto. K. (1999). Direct alteration of the P/Q-type Ca^{2+} channel property by polyglutamine expansion in spinocerebellar ataxia 6. *J. Neurosci.* **19**, 1–5.

22. Tan, N. C. K., Mulley, J. C., and Berkovic, S. F. (2004). Genetic association studies in epilepsy: "the truth is out there." *Epilepsia* **45**, 1429–1442.

23. Schroeder, B. C., Kubisch, C., Stein, V., and Jentsch, T. J. (1998). Moderate loss of function of cyclic-AMP-modulated KCNQ2/KCNQ3 K^+ channels causes epilepsy. *Nature* **396**, 687–690.

24. Wallace, R. H., Marini, C., Petrou, S., Harkin, L. A., Bowser, D. N., Panchal, R. G., Williams, D. A., Sutherland, G. R., Mulley, J. C., Scheffer, I. E., and Berkovic, S. F. (2001). Mutant GABA(A) receptor γ_2-subunit in childhood absence epilepsy and febrile seizures. *Nature Genet.* **28**, 49–52.

25. Hart, I. K. (2000). Acquired neuromyotonia: a new autoantibody mediated neuronal potassium channelopathy. *Am. J. Med. Sci.* **319**, 209–216.

26. Vernino, S., and Lennon, V. A. (2002). Ion channel and striational antibodies define a continuum of autoimmune neuromuscular excitability. *Muscle Nerve* **26**, 702–707.

27. Rose, M. R., and Griggs, R. C. (eds.). (2001). "Channelopathies of the Nervous System." Butterworth-Heinemann.

28. Kullmann, D. (2002). The neuronal channelopathies. *Brain* **125**, 1177–1195.

32

Migraine as a Cerebral Ionopathy with Abnormal Central Sensory Processing*

Michel D. Ferrari, MD, PhD
Peter J. Goadsby, MD, PhD, DSc

Keywords: *brainstem, calcium channels, channelopathy, serotonin, transgenic mouse models, trigeminal*

I. Migraine: A Common, Disabling, Episodic Disorder
II. The Migraine Attack: Clinical Phases and Pathophysiology
III. The Migraine Trigger Threshold: Repeated Recurrence of Attacks
IV. Conclusions
References

I. Migraine: A Common, Disabling, Episodic Disorder

Migraine has been described since antiquity. Written accounts date from about 3000 BC. Alverez quotes a couplet from a Sumerian poem, which in some sort of heaven or Abode of the Blessed:

The sick-eyed says not "I am sick-eyed"

The sick-headed (says) not "I am sick-headed." [2]

It can be seen even from the earliest writings that migraine presents with more than headache; indeed, headache and nausea have been recognized for many years. Moreover, the term itself, which is derived from the Greek word *hemicrania*, acknowledges the fascinating key feature of lateralized pain. Migraine is a primary headache in which the problem is a disorder itself, not a secondary headache, whereas other pathology such as brain tumor leads to headache. Migraine is in essence a familial *episodic* disorder whose marker is headache, although the key to recognition

---
*Some portions of the pathophysiology have appeared fully referenced [1].

Table 1A. International Headache Society Features of Migraine*

Repeated episodic headache (4–72 hrs) with the following features:

Any two of:	Any one of:
• unilateral	• throbbing
• worsened by movement	• moderate or severe nausea/vomiting
• photophobia and phonophobia	

Table 1B. Triggers Believed to Precipitate Migraine Attacks

- Altered sleep patterns: becoming tired or over-sleep
- Skipping meals
- Over-exertion
- Weather change
- Stress or relaxation from stress
- Hormonal change, such as menstrual periods
- Excess afferent stimulation: bright lights, strong smells
- Chemicals: alcohol or nitrates

*From the Headache Classification Committee of the International Headache Society. (2004). "The International Classification of Headache Disorders, second edition." *Cephalgia* **24(1)**, 1–160.

Table 2 Neuroanatomical Processing of Vascular Head Pain

	Structure	Comments
Target innervation:		
• Cranial vessels	Ophthalmic branch of	
• Dura mater	trigeminal nerve	
1st	Trigeminal ganglion	Middle cranial fossa
2nd	Trigeminal nucleus	Trigeminal nerve caudalis and C_1/C_2 dorsal horns
	(*quintothalamic tract*)	
3rd	Thalamus	Ventrobasal complex
		Medial nerve of posterior group
		Intralaminar complex
Modulatory	Midbrain	Periaqueductal gray matter
	Hypothalamus	?
Final	Cortex	• Insulae
		• Frontal cortex
		• Anterior cingulate cortex
		• Basal ganglia

is in the associated features of the attack (Table 1A) and the biology of the triggering of exacerbations (Table 1B).

The clinical definition is found in the International Headache Society criteria [3]. Migraine is broadly classified into 1) *migraine with aura* (previously called "classic(al) migraine"), in which at least some of the attacks are temporally associated with distinct aura symptoms that suggest focal brain dysfunction such as flashing lights or visual loss (suggesting occipital cortex involvement), or 2) *migraine without aura*, in which there are no associated neurological symptoms of a focal nature [4]. Migraine without aura is more common, representing two-thirds of all patients. In the past, it was called "common migraine," perhaps because it is such a common disease.

Indeed, at least 12% of the general population (2/3 females) suffer from a median of 18 migraine attacks per year, each lasting 1–3 days; about 10% of migraineurs have attacks at least once weekly [5]. Migraine is rated by the World Health Organization (WHO) as one of the most disabling chronic disorders [6] and has been estimated to be the most costly neurological disorder in the European Community at more than €27 billion per year [7]. Migraine patients also have an increased risk (comorbidity) of epilepsy, depression, anxiety disorders, and stroke [5]. Those with high attack frequency have an increased risk for white matter and cerebellar lesions visible on magnetic resonance imaging (MRI) [8].

II. The Migraine Attack: Clinical Phases and Pathophysiology

Migraine attacks may consist of up to four distinct phases, although not every patient will experience all of the following:

1. Up to one-third of patients may, at least sometimes, experience premonitory symptoms for several hours before the aura or headache phase begins. These warning symptoms may include mood changes (e.g., depression or irritation), hyperactivation, fatigue, yawning, neck pain, smell disturbances, craving for particular foods such as sweets, and water retention resulting in swollen ankles or breasts

2. Up to one-third of patients may have aura symptoms in at least some of their attacks. These symptoms usually last as long as 1 hour but sometimes are much longer in duration.

3. The headache phase may include headache and associated symptoms such as nausea, vomiting, and sensitivity to light and sound. This phase may range from 4–72 hours but usually lasts 1 day.

4. The recovery phase may take several hours to sometimes several days [10].

The pathophysiology of the individual phases of the attack, once the attack has started, is reasonably well understood and is discussed in the following section. However, the questions of why and how migraine attacks are triggered is essentially unknown. These topics will be discussed in the chapter on the molecular biology and genetics of migraine.

A. The Premonitory Phase

Little is known about the pathogenesis of the premonitory symptoms. Patients report a distinctive collection of symptoms in the hours before an attack, known as premonitory symptoms [9]. These symptoms are remarkably stereotyped and can be reproducibly triggered by nitroglycerin infusion [11]. In the rodent, *in vivo* D_2 receptor activation can lead to experimentally induced yawning [12–15]. Similarly, apomorphine, a dopamine agonist, elicits yawning in migraineurs at doses that do not affect age-matched control groups [16]. Apomorphine has also been reported to induce headache in 86% of migraine sufferers but not in age-matched control individuals [17]. Dopamine receptor agonist administration was reported to markedly worsen the headache in two patients with prolactinoma-associated headache [18]. Conversely the dopamine receptor antagonist domperidone, when taken during the premonitory phase, prevented the occurrence of migraine in uncontrolled trials [19–21]. It has recently been reported that, similar to dorsal horn neurons [22–24], dopamine receptors are present in the trigeminocervical complex of the rat [25]. These dopamine receptors are inhibitory in function [26]. The only dopaminergic neurons known to innervate the spinal cord come from the hypothalamic nucleus A11 region [27], with electrical stimulation suppressing the firing of spinal wide-dynamic range neurons through D_2 receptors [28]. Taken together, the available (albeit limited) data point to a dopaminergic/hypothalamic involvement in the premonitory phase of migraine.

B. The Migraine Aura

Migraine aura is defined as a spreading, focal, neurological disturbance manifested as visual, sensory, or motor symptoms [3]. It is seen in about one-third of patients, and clearly it is neurally driven. The case for the aura being the human equivalent of the cortical spreading depression (CSD) of Leao has been well made [29]. In humans, visual aura has been described as affecting the visual field, suggesting involvement of the visual cortex, and it starts at the center of the visual field and propagates to the periphery at a speed of 3 mm/min. This is very similar to spreading depression described in rabbits. Blood flow studies in patients have also shown that a focal hyperemia tends to precede the spreading oligemia; again, this is similar to what would be expected with spreading depression. After this passage of oligemia, the cerebrovascular response to hypercapnia in patients is blunted, although autoregulation remains intact. Again, this pattern is repeated with experimental spreading depression. Human observations have rendered the arguments reasonably sound that human aura has its equivalent in animal CSD. Whether CSD may, in fact, trigger the rest of the attack through activation of the trigeminovascular system is controversial (see the following section). Although there is some evidence from animal experiments, so far human evidence for this intriguing hypothesis is still lacking [30]. Therapeutic developments may shed further light on these relationships.

Tonabersat (SB-220453) is a CSD inhibitor that has been used in clinical trials for migraine. Tonabersat inhibits CSD, CSD-induced nitric oxide (NO) release, and cerebral vasodilation. Tonabersat does not constrict isolated human blood vessels, but it does inhibit trigeminally induced craniovascular effects [31]. Remarkably, topiramate, a proven preventive agent in migraine, also inhibits CSD in cats and rats [32]. Tonabersat is inactive in the human NO model of migraine, as is propranolol [33], although valproate showed some activity in that model [34]. Topiramate inhibits trigeminal neurons activated by nociceptive intracranial afferents [35], but not by a mechanism local to the trigeminocervical complex [36], and thus CSD inhibition may be a model system to contribute to the development of preventive medicines.

C. The Headache Phase

1. The Trigeminal Innervation of Pain-Producing Intracranial Structures

Surrounding the large cerebral vessels, pial vessels, large venous sinuses, and dura mater is a plexus of largely unmyelinated fibers that arise from the ophthalmic division of the trigeminal ganglion and, in the posterior fossa, from the upper cervical dorsal roots. Trigeminal fibers that innervate cerebral vessels arise from neurons in the trigeminal ganglion containing substance P (SP) and calcitonin gene–related peptide (CGRP), both of which can be released when the trigeminal ganglion is stimulated either in humans or cats. Stimulation of the cranial vessels, such as the superior sagittal sinus (SSS), is certainly painful in humans [37]. Human dural nerves that innervate the cranial vessels largely consist of small-diameter myelinated and unmyelinated fibers that almost certainly subserve a nociceptive function.

2. Peripheral Connections: Plasma Protein Extravasation

Moskowitz et al. have provided a series of experiments to suggest that the pain of migraine may be a form of sterile neurogenic inflammation [38]. Although this is as yet clinically unproven, the model system has been helpful for understanding some aspects of trigeminovascular physiology [38]. Neurogenic plasma extravasation can be seen during electrical stimulation of the trigeminal ganglion in the rat. Plasma extravasation can be blocked by ergot alkaloids, indomethacin, acetylsalicylic acid, and serotonin – $5HT_{1B/1D}$ agonists (triptans) such as sumatriptan [38]. Structural changes in the dura mater are observed after trigeminal ganglion stimulation [39]. These changes include mast cell degranulation and changes in postcapillary venules such as platelet aggregation [40]. Although it is generally accepted that such changes, particularly the initiation of a sterile inflammatory response, would cause pain, it is not clear whether this is sufficient in and of itself or if it requires other stimulators or promoters. Preclinical studies suggest that CSD may be a sufficient stimulus to activate trigeminal neurons [41], although this finding is controversial [30].

Although plasma extravasation in the retina, which is blocked by sumatriptan, can be seen after trigeminal ganglion stimulation in experimental animals, no changes are seen with retinal angiography during acute attacks of migraine or cluster headache [42]. A limitation of this study was the probable sampling of both retina and choroid elements in rats, given that choroidal vessels have fenestrated capillaries. Clearly, however, blockade of neurogenic plasma protein extravasation is not completely predictive of antimigraine efficacy in humans, as evidenced by the failure in clinical trials of SP; neurokinin-1 antagonists; specific PPE blockers; CP122, 288, and 4991w93; Bosentan, an endothelin antagonist; and ganaxolone, a neurosteroid [43,44].

a. Sensitization and Migraine Although it is as yet unproven that there is a significant sterile inflammatory response in the dura mater during migraine, it is clear that some form of sensitization takes place during migraine because allodynia is common. About two-thirds of patients complain of pain from nonnoxious stimuli, or allodynia [45,46]. A particularly interesting aspect is the demonstration of allodynia in the upper limbs, ipsilateral and contralateral to the pain. This finding is consistent with at least third-order neuronal sensitization (e.g., sensitization of thalamic neurons) and firmly places important parts of the pathophysiology of migraine within the central nervous system. Sensitization in migraine may be peripheral with local release of inflammatory markers, which would certainly activate trigeminal nociceptors. More likely in migraine is a form of central sensitization, which may be classical central sensitization, or a form of disinhibitory sensitization with dysfunction of descending modulatory pathways. Just as dihydroergotamine (DHE) can block trigeminovascular nociceptive transmission [47], probably at least by a local effect in the trigeminocervical complex, DHE can block central sensitization associated with dural stimulation by an inflammatory soup [48].

3. Neuropeptide Studies

Electrical stimulation of the trigeminal ganglion in both humans and cats leads to increases in extracerebral blood flow and local release of both CGRP and SP. In the cat, trigeminal ganglion stimulation also increases cerebral blood flow by a pathway traversing the greater superficial petrosal branch of the facial nerve, again releasing a powerful vasodilator peptide, vasoactive intestinal polypeptide (VIP) [49]. Interestingly, the VIP-ergic innervation of the cerebral vessels is predominantly anterior rather than posterior, which may contribute to this region's vulnerability to spreading depression and explain why the aura is so often commences posteriorly. Stimulation of the more specifically vascular pain–producing SSS increases cerebral blood flow and jugular vein CGRP levels. Human evidence that CGRP is elevated in the headache phase of migraine, cluster headache, and chronic paroxysmal hemicrania supports the view that the trigeminovascular system may be activated in a protective role in these conditions. Moreover, NO donor–triggered migraine, which is in essence typical migraine, also results in increases in CGRP that are blocked by sumatriptan, just as in spontaneous migraine. The recent development of nonpeptide, highly specific CGRP antagonists, and the announcement of proof-of-concept for a CGRP antagonist in acute migraine [50], firmly establishes this as a novel and important emerging treatment principle for acute migraine. At the same time, the lack of any effect of CGRP blockers on plasma protein extravasation explains in some part why that model has proved inadequate at translation into human therapeutic approaches.

4. Central Connections: The Trigeminocervical Complex

Fos immunohistochemistry is a method of looking at activated cells by plotting the expression of Fos protein. After meningeal irritation with blood Fos expression is noted in the trigeminal nucleus caudalis, whereas after stimulation of the SSS Fos-like immunoreactivity is seen in the trigeminal nucleus caudalis and in the dorsal horn at the C_1 and C_2 levels in cats and monkeys. These latter findings are in accord with similar data using 2-deoxyglucose measurements with SSS stimulation. Similarly, stimulation of a branch of C_2, the greater occipital nerve, increases metabolic activity in the same regions (i.e., trigeminal nucleus caudalis and $C_{1/2}$ dorsal horn). In experimental animals, one can record directly from trigeminal neurons with both supratentorial trigeminal input and input from the greater occipital nerve, a branch of the C_2 dorsal root. Stimulation of the greater occipital nerve for 5 minutes results in substantial increases in responses to supratentorial dural stimulation, which can last for more than 1 hour. Conversely, stimulation of the middle meningeal artery dura mater with the

C-fiber irritant mustard oil sensitizes responses to occipital muscle stimulation. Taken together, these data suggest convergence of cervical and ophthalmic inputs at the level of the second-order neuron. Moreover, stimulation of a lateralized structure, the middle meningeal artery, produces Fos expression bilaterally in both cat and monkey brains. This group of neurons from the superficial laminae of trigeminal nucleus caudalis and $C_{1/2}$ dorsal horns should be regarded functionally as the *trigeminocervical* complex [51].

These data demonstrate that trigeminovascular nociceptive information comes by way of the most caudal cells. This concept provides an anatomical explanation for the referral of pain to the back of the head in migraine. Moreover, experimental pharmacological evidence suggests that some abortive anti-migraine drugs, such as ergot derivatives, acetylsalicylic acid, and several triptans, can have actions at these second-order neurons that reduce cell activity and suggest a further possible site for therapeutic intervention in migraine [52]. This action can be dissected out to involve each of the $5-HT_{1B}$, $5-HT_{1D}$, and $5-HT_{1F}$ receptor subtypes [53] and is consistent with the localization of these receptors on peptidergic nociceptors. Interestingly, triptans also influence the CGRP promoter, as well as regulate CGRP secretion from neurons in culture. Furthermore, the demonstration that some part of this action is postsynaptic with either $5-HT_{1B}$ or $5-HT_{1D}$ receptors located *non*-presynaptically offers a prospect of highly anatomically localized treatment options [52].

5. Higher-Order Processing

Following transmission in the caudal brainstem and high cervical spinal cord, information is relayed rostrally.

a. Thalamus Processing of vascular nociceptive signals in the thalamus occurs in the ventroposteromedial (VPM) thalamus, medial nucleus of the posterior complex, and intralaminar thalamus. It has been shown by application of capsaicin to the SSS that trigeminal projections with a high degree of nociceptive input are processed in neurons, particularly in the ventroposteromedial thalamus and in its ventral periphery. These neurons in the VPM can be modulated by activation of $GABA_A$ inhibitory receptors, and perhaps of more direct clinical relevance by propranolol though a β_1-adrenoceptor mechanism. Remarkably, through $5-HT_{1B/1D}$ mechanisms, triptans can also inhibit VPM neurons locally, as demonstrated by microiontophoretic application, suggesting a hitherto-unconsidered locus of action for triptans in acute migraine (Fig. 1). Human

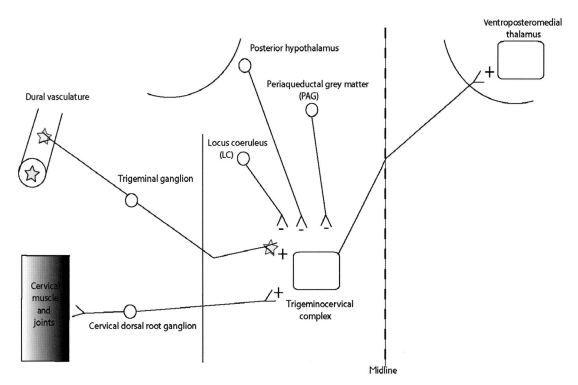

Figure 1 Illustration of some elements of migraine biology. Dural vessels and dura mater, and upper ($C_{1/2}$) cervical muscles and joints all project to neurons on the upper cervical spinal dorsal horn and trigeminal nucleus caudalis that may be labeled the trigeminocervical complex. The input is excitatory (+). This input then projects to trigeminovascular neurons in the ventroposteromedial thalamus (+) after crossing the midline. Descending brain systems from, for example, the locus coeruleus, posterior hypothalamus (PH), and periaqueductal gray matter (PAG) have inhibitory influences (−) on trigeminocervical neurons. Although the sum of stimulating these structures is inhibitory, it is noteworthy that individual neurons from PAG or PH can facilitate nociception.

imaging studies have confirmed activation of thalamus contralateral to pain in acute migraine, cluster headache, and short-lasting unilateral neuralgiform headache with conjunctival injection and tearing (SUNCT) [54].

b. Activation of Modulatory Regions Stimulation of nociceptive afferents by stimulation of the SSS in the cat activates neurons in the ventrolateral periaqueductal gray matter (PAG). PAG activation in turn provides feedback to the trigeminocervical complex with an inhibitory influence. PAG is clearly included in the area of activation seen in PET studies in migraineurs. This typical negative feedback system will be further considered in the following as a possible mechanism for the symptomatic manifestations of migraine [54].

Another potentially modulatory region activated by stimulation of nociceptive trigeminovascular input is the posterior hypothalamic gray matter. This area is crucially involved in several primary headaches, notably cluster headache, short-lasting unilateral neuralgiform headache attacks with conjunctival injection and tearing (SUNCT), paroxysmal hemicrania, and hemicrania continua. Moreover, the clinical features of the premonitory phase and other features of the disorder suggest dopamine neuron involvement. Orexinergic neurons in the posterior hypothalamus can be both pro- and anti-nociceptive, offering a further possible region where dysfunction might involve the perception of head pain.

6. Central Modulation of Trigeminal Pain

a. Brain Imaging in Humans Functional brain imaging with positron emission tomography (PET) has demonstrated activation of the dorsal midbrain, including the PAG and in the dorsal pons, near the locus coeruleus, in studies of migraine without aura. Dorsolateral pontine activation is seen with PET in spontaneous episodic and chronic migraine and with nitroglycerin-triggered attacks (Fig. 2). These areas are active immediately after successful treatment of the headache but are not active interictally. The activation corresponds with the brain region that causes migraine-like headache when stimulated in patients with electrodes implanted for pain control. Similarly, excess iron is present in the PAG of patients with episodic and chronic migraine, and chronic migraine can develop after a bleed into a cavernoma in the region of the PAG or due to a lesion of the pons. What could result from dysfunction of these brain areas?

b. Animal Experimental Studies of Sensory Modulation It has been shown in the experimental animal that stimulation of nucleus locus coeruleus, the main central noradrenergic nucleus, reduces cerebral blood flow in a frequency-dependent manner [55] through an α_2-adrenoceptor–linked mechanism. This reduction is greatest in the occipital cortex. Although a 25% overall reduction in cerebral blood flow is seen, extracerebral vasodilatation occurs in parallel. In addition, the main serotonin-containing nucleus in the brainstem, the midbrain dorsal raphe nucleus, can increase cerebral blood flow when activated. Furthermore, stimulation of PAG will inhibit sagittal sinus–evoked trigeminal neuronal activity in cats, whereas blockade of P/Q-type voltage-gated Ca^{2+} channels in the PAG facilitates trigeminovascular nociceptive processing with the local GABAergic system in the PAG still intact.

c. Electrophysiology of Migraine in Humans Studies of evoked potentials and event-related potentials provide some link between animal studies and human functional imaging [56]. Authors have shown changes in neurophysiological measures of brain activation but there is much discussion as to how to interpret such changes [57]. Perhaps the most reliable theme is that the migrainous brain does not habituate to signals in a normal way. Similarly, contingent negative variation (CNV), an event-related potential, is abnormal in migraineurs compared with controls. Changes in CNV predict attacks, and preventive therapies alter (normalize) such changes.

D. Migraine Symptoms in Relationship to the Pathophysiology

Migraine is a multifactorial, episodic disorder involving sensory sensitivity [5]. Patients complain of pain in the head that is throbbing, but there is no reliable relationship between vessel diameter and the pain or its treatment. Patients also complain of discomfort from normal lights and the unpleasantness of routine sounds. Some mention that otherwise pleasant odors are unpleasant. Normal movement of the head causes pain, and many mention a sense of unsteadiness similar to having just stepped off a boat, having been nowhere near the water.

The anatomical connections of the pain pathways, for example, are clear: The ophthalmic division of the trigeminal nerve subserves sensation within the cranium and

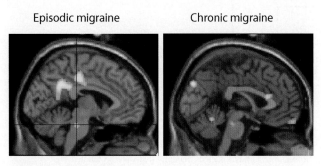

Figure 2 Positron emission tomography (PET) scans from groups of patients with episodic and chronic migraine studied during acute pain. Activation in the dorsal rostral pons is a constant finding in PET studies in migraine and may represent a crucial area that mediates important aspects of the pathophysiology of the disorder.

explains why the top of the head is headache, and the maxillary division is *facial pain*. The convergence of cervical and trigeminal afferents explains why neck stiffness or pain is so common in primary headache. The genetics of channelopathies (see later discussion) is becoming a plausible way to think about the episodic nature of migraine. However, where is the lesion, and what is actually the pathology?

If one considers what patients say, then perhaps they can provide the answer to this question. Migraine aura cannot be the trigger in all patients because it occurs in no more than one-third of migraine patients and can be experienced without subsequent pain. Migraine patients do not receive a single photon of extra light, so for that symptom, as well as phonophobia and osmophobia, the basis of the problem must be abnormal central processing of a normal signal. Perhaps electrophysiological changes in the brain have been mislabeled as *hyperexcitability*, whereas *dishabituation* might be a simpler explanation. If migraine was basically an attentional problem with changes in cortical synchronization [58], *hypersynchronization*, then all its manifestations could be accounted for in a single, over-arching, pathophysiological hypothesis of a disturbance of subcortical sensory modulation systems. Although it seems likely that the trigeminovascular system and its cranial autonomic reflex connections, the trigeminal-autonomic reflex [49], act as a feed-forward system to facilitate the acute attack, the fundamental problem in migraine is in the brain. Unraveling its basis will deliver great benefits to patients and considerable understanding of some fundamental neurobiological processes.

III. The Migraine Trigger Threshold: Repeated Recurrence of Attacks

A. How are Migraine Attacks Triggered?

The *disease* migraine is defined as at least five attacks of migraine without aura or at least two attacks of migraine with aura. One of the reasons for this is that many people may experience one or two sporadic migraine headaches or auras during their lifetime, without repeated recurrence of further attacks. This suggests it is not the migraine attack itself, but the repeated recurrence of attacks that is abnormal. In this respect, migraine is similar to that other classical episodic brain disorder, epilepsy: "One attack doesn't make a patient" [59].

To understand the *disease* migraine, we must understand how migraine attacks are triggered and why patients have recurrent attacks. It is not unlikely that at least part of the clue is in an imbalance between the individual's trigger threshold ("defense") and the "attack" by migraine triggers. This section discusses the growing evidence that migraine patients are likely to have a genetically determined reduced threshold for migraine triggers and that attacks may occur when 1) migraine triggers are particularly strong or frequent; 2) there is a temporarily further reduction of the threshold due to endogenous factors (e.g., menstruation or sleep deprivation), facilitating the triggering of an attack; or 3) there is a temporal coincidence of both triggering and facilitating factors. Understanding the mechanisms involved in the triggering of migraine attacks will help identify novel treatment targets for urgently needed specific prophylactic agents to prevent migraine attacks. An important initial step to achieve these goals is to unravel the genetic basis of the migraine threshold and decipher the common pathways for triggering migraine attacks.

B. Genetic Epidemiology

It is clear from clinical practice and historical reports that many patients have first-degree relatives who also suffer from migraine [60]. Twin studies are the classical method to investigate the relative importance of genetic and environmental factors. A Danish study included 1013 monozygotic and 1667 dizygotic twin pairs of the same gender, obtained from a population-based twin register. The pairwise concordance rate was significantly higher among monozygotic than dizygotic twin pairs ($P < 0.05$) but never reached 100%. These results, combined with population-based epidemiological surveys, strongly suggest that migraine is a multifactorial disorder caused by a combination of multiple genetic and environmental factors.

C. Finding Genes for Complex Disorders: The Migraine Highway

Identifying genes for complex disorders is hampered by the multiplicity and the relatively low penetration of the genes involved. Basically there are two main approaches to tackle this issue. The traditional approach is hunting for genes for rare monogenic subtypes that may serve as models for the common complex disorder. More recently, genome-wide association studies are used to test hundreds of thousands of genetic markers in extended and clinically homogenous populations. Once genes for monogenic subtypes have been cloned, the functional consequences of the disease-related mutations can be analyzed for common disease pathways leading to the attack. Thus far, genes for three monogenic subtypes of migraine have been identified: familial hemiplegic migraine (FHM), sporadic hemiplegic migraine (SHM), and cerebral autosomal dominant arteriopathy with subcortical infarcts and leukoencephalopathy (CADASIL).

1. Familial Hemiplegic Migraine

FHM is a rare, severe, monogenic subtype of migraine with aura and is characterized by at least some degree of hemiparesis during the aura [59]. The hemiparesis may last from minutes to several hours or even days. Quite often patients are initially misdiagnosed with epilepsy. Apart from the hemiparesis, the other headache and aura features of the FHM attack are identical to those of attacks of the common types of migraine. In addition to hemiparetic attacks, the majority of FHM patients also experience attacks of "normal" migraine with or without aura [61,62]. As in normal migraine, attacks of FHM may be triggered by mild head trauma. Thus, from a clinical point of view, although not perfect, there are many similarities, and FHM seems a valid model for the common forms of migraine [59]. Major differences, apart from the hemiparesis, are that FHM also may be associated with cerebellar ataxia (in about 20% of patients) and other neurological symptoms such as epilepsy, mental retardation, brain edema, and (fatal) coma. Thus far, three genes for FHM have been published, and based on unpublished linkage results in several families, there are more to come.

a. The FHM1 CACNA1A Gene The first gene identified for FHM is the CACNA1A gene on chromosome 19p13. It is responsible for approximately 50% of all families with FHM (Fig. 3). The FHM1 gene encodes the ion-conducting, pore-forming alpha1A subunit of $Ca_v2.1$ (P/Q) type, voltage-gated, neuronal calcium channels [63]. The main function of neuronal P/Q type calcium channels is to modulate the release of neurotransmitters, both at peripheral neuromuscular junctions and central synapses, mainly within the cerebellum, brainstem, and cerebral cortex [64]. Many different CACNA1A mutations have been found and associated with a wide range of clinical phenotypes [65]. These include pure forms of FHM, combinations of FHM with various degrees of cerebellar ataxia or fatal coma due to excessive cerebral edema, and disorders not associated with FHM such as episodic ataxia type 2, progressive ataxia, spinocerebellar ataxia type 6, and generalized epilepsy [65].

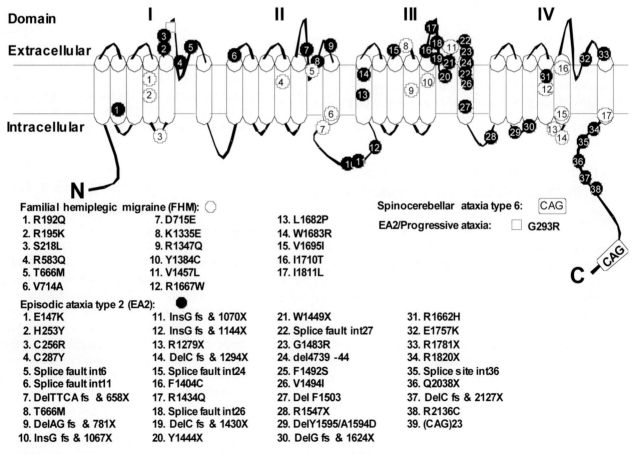

Figure 3 The *CACNA1A* gene with mutations. The $Ca_v2.1$ pore-forming subunit of P/Q-type voltage-gated calcium channels is located in the neuron membrane and contains four repeated domains, each encompassing six transmembrane segments. Positions of mutations identified in this gene are given. (*CACNA1A* ref. seq.: Genbank Ac. nr. X99897.)

b. The FHM2 ATP1A2 Gene The ATP1A2 FHM2 gene on chromosome 1q23 encodes the α2 subunit of a Na^+, K^+ pump ATPase. This catalytic subunit binds Na^+, K^+, and ATP and uses ATP hydrolysis to extrude Na^+ ions. Na^+ pumping provides the steep Na^+ gradient essential for the transport of glutamate and Ca^{2+}. The gene is predominantly expressed in neurons at neonatal age and in glial cells at adult age [66,67]. In adults, an important function of this specific ATPase is to modulate the re-uptake of potassium and glutamate from the synaptic cleft into the glial cell. Mutations in the ATP1A2 gene are responsible for 20% or more of FHM cases (Fig. 4) and have been associated with pure FHM and FHM in combination with cerebellar ataxia, alternating hemiplegia of childhood, benign focal infantile convulsions, and other forms of epilepsy [65]. In an Italian family a variant was found in the ATP1A2 gene that segregated with basilar migraine, a subtype of migraine with aura characterized by aura symptoms attributable to the brainstem and both occipital lobes [68]. The variant was not observed in 200 controls. However, because no functional studies were described, a definite conclusion as to whether this gene variation is also *causally* linked to basilar migraine cannot be established.

c. The FHM3 SCNA1 Gene The SCNA1 gene on chromosome 2q24 encodes the alpha subunit of a neuronal voltage-gated sodium ($Na_v1.1$) channel. The $Na_v1.1$ channel is mainly responsible for the generation and propagation of neuronal action potentials. Different mutations in this gene are known to be associated with epilepsy and febrile seizures. Dichgans et al. [69] found the novel Q1489K mutation in three German FHM families of common ancestry. This missense mutation is located within a domain critical for fast inactivation of the sodium channel and causes a two- to four-fold increased acceleration of the recovery from fast inactivation. This would predict enhanced neuronal excitation and release of neurotransmitter. Another new SCNA1 mutation was found in an American FHM family, confirming the relationship between SCNA1 and FHM3 [70].

2. Sporadic Hemiplegic Migraine

Hemiplegic migraine patients are not always clustered in families. Sporadic patients without affected family members are often seen and may sometimes represent the first "FHM patient" (spontaneous mutation) in a family [71]. Apart from sharing the clinical phenotype with the common forms of migraine as FHM displays, SHM and normal migraine also

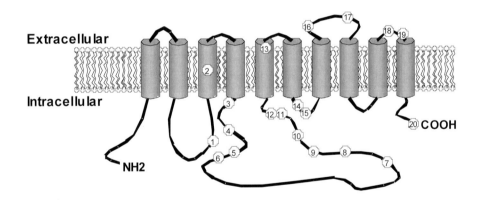

Familial hemiplegic migraine

1. T263M
2. G301R
3. T345A
4. T376M
5. T378N (AHC?)
6. R383H
7. del1804 -1820 & ins TT
8. R689Q
9. D718N
10. M731T
11. R763H
12. L764P
13. P796R
14. M829R
15. R834Q
16. W887R
17. E902K
18. del2897 -2898 CT
19. P979L
20. X1021R

Figure 4 The *ATP1A2* gene with mutations. The alpha2 subunit of sodium potassium pumps is located in the plasma membrane and contains 10 transmembrane segments. Positions of mutations identified in this gene are given. *AHC*, Alternating hemiplegia of childhood. (*ATP1A2* ref. seq.: Genbank Ac. nr. NM_000702.)

show a remarkable genetic epidemiological relationship. SHM patients have a greatly increased risk of also suffering from typical migraine with aura, and their first-degree relatives have a greatly increased risk of both migraine with and without aura [72]. Although in an initial study, CACNA1A mutations were found in only 2 of 27 SHM patients [73], a much larger study did find several mutations in all three FHM genes in a substantial proportion of SHM patients, confirming a genetic relationship between FHM and SHM [74].

3. Migraine as a Cerebral Ionopathy

We now need to consider whether a channelopathy, or perhaps more precisely a dysfunction of ion flux or transportation, is involved in migraine pathogenesis. At this point in time, the evidence is only circumstantial, although it is growing and ranges from clinical arguments to genetic, clinical neurophysiological, and neuropharmacological evidence.

First, migraine shares strikingly similar clinical characteristics with established channelopathies such as FHM and SHM (see previous discussion), as well as, quite remarkably, episodic neuromuscular disorders such as myotonia and periodic paralysis [75]. The clinical similarities include the primarily paroxysmal presentation of symptoms, the similar duration and frequency of the attacks, the triggers for attacks, and the gender-related expression of attack frequency with an onset mostly around puberty and amelioration after age 40.

Secondly, some linkage and association studies do suggest a role of FHM genes in SHM and normal migraine, although other studies could not replicate these findings [65]. Size and homogeneity of the study populations clearly are main factors complicating such genetic studies in complex disorders.

A third line of evidence comes from clinical neurophysiological studies in migraineurs. Ambrosini et al. [68] found single-fiber abnormalities, suggesting an altered release of acetylcholine at the neuromuscular junction, which is mainly controlled by P/Q type $Ca_v2.1$ channels, and Sandor et al. [76] found evidence of subclinical cerebellar dysfunction. Migraine patients were more likely to show a systematic horizontal deviation when subjected to a highly sensitive test for cerebellar coordination. As P/Q type $Ca_v2.1$ channels are highly expressed in cerebellar Purkinje cells, this finding seems to suggest a dysfunction of $Ca_v2.1$ channels in Purkinje cells of migraineurs.

Finally, a number of neuropharmacological animal experiments suggest that application of selective blockers of P/Q type $Ca_v2.1$ channels, within certain areas within the brainstem, can modulate important migraine mechanisms. These include inhibition of the release of CGRP and neurogenic inflammation [77], facilitation of trigeminal firing [78], and modulation of nociceptive transmission in the trigeminocervical complex [79].

4. Functional Consequences of FHM Gene Mutations

Understanding the functional consequences of gene mutations is crucial to the understanding of the disease pathways. For the FHM genes, this has been studied in cellular models, as well as knock-in mouse models carrying human mutations.

a. Functional Studies in Cellular Models for FHM1 Mutations Several FHM mutations have been analyzed with electrophysiological techniques in neuronal and nonneuronal cells [80–86]. Although episodic ataxia type 2 mutations all show a dramatic decrease or even complete loss of current density, FHM mutations cause different effects on channel conductance, kinetics, and/or expression in transfected cells. The most consistent change found with FHM mutations seems to be a hyperpolarizing shift of about 10 mV of the activation voltage. Although this effect *in theory* leads to easier opening of the channels in neurons, the overall change in calcium influx is difficult to predict. It is determined by the delicate interplay of different effects of a particular mutation on the different channel properties and the cellular environment. Some phenomena might contribute to the episodic nature of symptoms because calcium influx may vary, especially during high neuronal activity. Mutant T666M and V714A channels have a low conductance mode that sometimes switches to the wild-type state. For other FHM mutations, such as R583Q and D715E, accumulation of inactivated channels was observed during repetitive stimulation.

b. Functional Studies in Knock-in Mouse Models for FHM1 Mutations Recently, a knock-in mouse model was generated, carrying the human pure FHM1 R192Q mutation [87]. Unlike the natural CACNA1A mutant mouse models [65], transgenic R192Q mice exhibit no overt clinical phenotype or structural abnormalities. This is similar to the human situation. However, multiple gain-of-function effects were found, including increased Ca^{2+} influx in cerebellar neurons, increased release of neurotransmitters both spontaneous and on stimulation, and in the intact animal, a reduced threshold and increased velocity of CSD. It seems that whole-animal studies may be better suited to dissect the effects of mutations and to understand the integrated physiology of the disease. Other studies aiming at functional changes within the brainstem are underway and will shed more light on the important question of whether mechanisms within the trigeminocervical complex are also affected.

c. Functional Studies in Cellular Models for FHM2 Mutations Functional analysis of mutated proteins revealed inhibition of pump activity and decreased affinity for K^+, with decreased catalytic turnover resulting in reduced re-uptake of K^+ and glutamate into glial cells [66,88]. Two groups have generated α2-subunit deficient mice [89,90], but thus far, no FHM2 knock-in mouse models have been generated.

5. A Common Mechanism for Triggering FHM Attacks

Three different genes encoding for three different proteins and mechanisms have now been associated with FHM. How can we fit these apparently diverse mechanisms into one (final) common pathway? There is one obvious candidate, CSD [91]. Mutations in the FHM1 calcium gene cause increased neuronal release of neurotransmitters, more specifically, the neuroexcitatory amino acid glutamate [92] that can induce, maintain, and propagate CSD. Mutations in the FHM2 sodium potassium pump gene cause reduced re-uptake of K^+ and glutamate from the synaptic cleft into the glial cell. Mutations in the FHM3 sodium channel gene result in hyperexcitability and, most likely, increased release of neurotransmitter in the synaptic cleft. The overall result is increased levels of glutamate and K^+ in the synaptic cleft resulting in an increased propensity for CSD. This would easily explain the aura of FHM attacks. More controversial, however, is whether the enhanced tendency for CSD might also be responsible for triggering the headache phase; for example, by activation of the trigeminovascular system (see previous discussion). Lastly, the most important question that remains to be answered is whether the same mechanisms are also involved in the common forms of migraine with and without aura. Alternative explanations might be alterations within the trigeminocervical complex that result in reduced modulation of trigeminal signal transmission (see previous discussion).

6. CADASIL

CADASIL is a severe arteriopathy caused by mutations in the Notch3 gene on chromosome 19p13. CADASIL is clinically characterized by recurrent subcortical infarcts, white matter lesions on MRI, dementia and other neuropsychiatric symptoms, and most relevant here, migraine with aura in 40% of patients [93,94]. Migraine is usually the presenting symptom, occurring approximately 10 years before the other symptoms become apparent. The Notch3 gene is primarily involved in the regulation of arterial differentiation and maturation of vascular smooth muscle cells [95]. Neuronal effects have not been found. Why mutations in this gene would cause migraine is as yet unclear. Because of the apparent primarily vascular effects of the gene, the CADASIL-migraine relationship may be seen as a support for an important vascular contribution to migraine pathogenesis, although there is no direct evidence for this hypothesis. Alternately, cerebral ischemia may serve as a focus for CSD, which would then set off migraine aura and migraine attacks. Notch3 CADASIL mouse models are available, and they do not seem to express a CADASIL-like cerebral phenotype [96,97]. Studying both neuronal and vascular changes in these mice may prove invaluable in further dissecting the triggering mechanisms for migraine attacks.

7. Vascular Retinopathy and Migraine

A rare syndrome, clinically characterized by a combination of cerebroretinal vasculopathy, Raynaud's phenomenon, migraine, pseudotumor cerebri, and various other forms of vascular dysfunction, has been linked to chromosome 3p21 in three families [98,99]. Migraine is clearly part of the syndrome [99]. Identification of the responsible gene for this neurovascular syndrome will evidently be important for a wide range of vascular disorders, including the pathogenesis of migraine.

8. Other Candidate Genes and Loci for Migraine

a. Linkage Studies A number of genome-wide linkage studies have found significant or suggestive linkage for migraine and non-FHM loci, two of which have been replicated in independent samples. Linkage to chromosome 6p12.2-p21.1 in a large family with migraine with and without aura from northern Sweden [101] was confirmed in Australian patients [102]. Linkage to 4q24 in 50 Finnish families with migraine with aura [103] was confirmed in a study in 289 Icelandic patients with migraine without aura [104]: both the Finnish and Icelandic populations are considered genetic isolates. Other migraine loci that have been found but not yet replicated are 11q24 [105], 14q21.1-q22.3 [106], Xq24-28 [107], and 15q11-q13 [108]. The variety in loci reported is probably a reflection of the genetic heterogeneity of migraine.

Nyholt et al. [105,109] used an interesting new approach to classify migraine patients in a large Australian sample of 12,245 twins. Rather than dividing them into migraine with or without aura, they applied latent class analysis (LCA) to identify patient subgroups based on the cluster patterns of their symptoms. Using a quantitative trait analysis in 790 independent subpairs concordant for LCA migraine class, they found significant linkage for chromosome 5q21 and a severe migraine phenotype with pulsating headache. Interestingly, they also found associations between certain loci and specific migraine characteristics such as phonophobia, photophobia, nausea/vomiting, and pulsating quality of the headache. The relevance of this latter finding remains to be proven. We feel that such a linkage might only reflect general sensitivities rather than migraine-specific relationships. That is, patients showing linkage to the photophobia gene locus, for example, might show a tendency for photophobia under a variety of conditions such as influenza or stomach pain. Another critical point is that migraine patients show a variable and changing pattern of symptoms over their lifetimes. For instance, they may have severe nausea and vomiting together with aura as part of their migraine attacks at a young age, may "lose" the aura and vomiting in their 20s and have attacks of migraine without aura, and later have attacks of isolated auras without headache or other associated symptoms. It seems that the presence of the individual clinical characteristics are "time locked" rather than "gene locked" and that patients who are asked for their symptoms may give different answers at different stages in their lives.

b. Association Studies In complex diseases, multiple genes are expected to contribute to the phenotype. Each gene only has a limited contribution. Finding such genes by using the classical linkage approach is therefore difficult. Association studies may be more powerful in the detection of genes that confer increased disease susceptibility. However, there are a number of important pitfalls when conducting such studies. For example, many association studies in migraine have been conducted with insufficient sample sizes, inadequate definition of patients, and inadequate control samples. Detailed overviews of the numerous association studies in migraine have been published [65,110]. The latter review [110] also provides an excellent overview of the many association studies that have been performed on the relationship between migraine and genes for dopamine receptors and genes involved in the metabolism and transportation of serotonin (5-HT). In brief, none of the associations have been convincingly replicated. Here we shall only briefly discuss those associations that have been replicated at least once.

An association between the C677T variant in the methylenetetrahydrofolate reductase gene (MTHFR) and migraine with aura has been found in several clinic-based (and therefore selected) study populations [111–114] but also, and most importantly, in a large sample of the general population [115]. This makes MTHFR the first true migraine-risk gene at the population level. The association was found to be enhanced in the presence of another variant (C1298C) in the same gene [112] and in combination with an ACE variant [116]. If replicated, this would indicate also the first gene-gene interactions to be involved in modulating the risk for migraine.

Other replicated associations in selected clinic-based samples include associations with a progesterone receptor insert in two independent populations [117], the tumor necrosis factor gene in two separate studies [118,119], and with variants in the angiotensin converting enzyme (ACE) [116,120,121]. Remarkably, in one study the ACE-DD variant seemed to have a slight protective effect against migraine in male patients [122].

IV. Conclusions

Migraine is a highly prevalent, multifactorial, episodic disorder of the brain, with high impact on patient and society. Changes within the trigeminocervical complex and trigeminovascular system, mainly leading to abnormal central sensory processing, are crucial pathophysiological mechanisms of the migraine attack. Interference with these neurovascular mechanisms offers new avenues for novel, specific, *acute* treatments of the migraine attack, hopefully not associated with potential cardiovascular complications. Effective and well-tolerated treatments to *prevent* attacks, rather than abort them after they begin, are dearly needed. New insights into the genetics and molecular biology of the migraine trigger threshold suggest that migraine might be a cerebral ionopathy resulting in enhanced propensity for cortical spreading depression as a triggering mechanism for attacks. Similar mechanisms might be involved in changing the modulatory role of the trigeminocervical complex on central trigeminal pain and other sensory signal transmission. Pharmacological interferences aimed at normalizing the disturbed ion homeostasis might offer new avenues for the development of specific migraine prophylactic treatments.

Acknowledgments

The work of Peter J. Goadsby has been supported by the Wellcome Trust.

References

1. Goadsby, P. J. (2006). Pathophysiology of migraine. *In:* "Migraine and Other Headache Disorders" (R. B. Lipton et al., eds.). Marcel Dekker, Taylor & Francis, New York.
2. Lance, J. W., and Goadsby, P. J. (2005). "Mechanism and Management of Headache," 7th edition. Elsevier, New York.
3. Headache Classification Committee of the International Headache Society. (2004). "The International Classification of Headache Disorders, 2nd edition." *Cephalgia* **24(1)**, 1–160.
4. Silberstein, S. D., Lipton, R. B., and Goadsby, P. J. (2002). "Headache in Clinical Practice," 2nd edition. Martin Dunitz, London.
5. Goadsby, P. J., Lipton, R. B., and Ferrari, M. D. (2002). Migraine: current understanding and treatment. *New Engl. J. Med.* **346**, 257–270.
6. Menken, M., Munsat, T. L., and Toole, J. F. (2000). The global burden of disease study—implications for neurology. *Arch. Neurol.* **57**, 418–420.
7. Andlin-Sobocki, P., Jonsson, B., Wittchen, H. U., and Olesen, J. (2005). Cost of disorders of the brain in Europe. *Eur. J. Neurol.* **12(1)**, 1–27.
8. Kruit, M. C., van Buchem, M. A., Hofman, P. A., Bakkers, J. T., Terwindt, G. M., Ferrari, M. D., and Launer, L. J. (2004). Migraine as a risk factor for subclinical brain lesions. *JAMA* **291**, 427–434.
9. Giffin N. J., Ruggiero, L., Lipton, R. B., Silberstein, S. D., Tvedskov, J. F., Olesen, J., Altman, J., Goadsby, P. J., and Macrae, A.

(2003). Premonitory symptoms in migraine: an electronic diary study. *Neurology* **60**, 935–940.
10. Giffin, N. J., Lipton, R. B., Silberstein, S. D., Tvedskov, J. F., Olesen, J., and Goadsby, P. J. (2005). The migraine postdrome: an electronic diary study. *Cephalalgia* **25**, 958.
11. Afridi, S., Kaube, H., and Goadsby, P. J. (2004). Glyceryl trinitrate triggers premonitory symptoms in migraineurs. *Pain* **110**, 675–680.
12. Mogilnicka, E., and Klimek, V. (1977). Drugs affecting dopamine neurons and yawning behavior, *Pharmacol. Biochem. Behav.* **7**, 303–305.
13. Protais, P., Dubuc, I., and Costentin, J. (1983). Pharmacological characteristics of dopamine receptors involved in the dual effect of dopamine agonists on yawning behaviour in rats. *Eur. J. Pharmacol.* **94**, 271–280, 1983.
14. Serra, G., Collu, M., and Gessa, G. L. (1986). Dopamine receptors mediating yawning: are they autoreceptors? *Eur. J. Pharmacol.* **120**, 187–192.
15. Yamada, K., Tanaka, M., Shibata, K., and Furukawa, T. (1986). Involvement of septal and striatal dopamine D-2 receptors in yawning behavior in rats. *Psychopharmacology (Berl)* **90**, 9–13.
16. Blin, O., Azulay, J., Masson, G., Aubrespy, G., and Serratrice, G. (1991). Apomorphine-induced yawning in migraine patients: enhanced responsiveness. *Clin. Neuropharmacol.* **14**, 91–95.
17. del Bene, E., Poggonioni, M., and de Tommasi, F. (1994). Video assessment of yawning induced by sublingual apomorphine in migraine. *Headache* **34**, 536–538.
18. Levy, M. J., Matharu, M.S., and Goadsby, P.J. (2003). Prolactinomas, dopamine agonist and headache: two case reports. *Eur. J. Neurol.* **10**, 169–174.
19. Amery, W. K., and Waelkens, J. (1983). Prevention of the last chance: an alternative pharmacological treatment of migraine. *Headache* **23**, 37–38.
20. Waelkens, J. (1981). Domperidone in the prevention of complete classical migraine. *Brit. Med. J.* **284**, 944.
21. Waelkens, J. (1984). Dopamine blockade with domperidone: bridge between prophylactic and abortive treatment of migraine? A dose-finding study. *Cephalalgia* **4**, 85–90.
22. Levant, B., and McCarson, K. E. (2001). D(3) dopamine receptors in rat spinal cord: implications for sensory and motor function. *Neurosci. Lett.* **303**, 9–12.
23. Levey, A. I., Hersch, S. M., Rye, D. B., Sunahara, R. K., Niznik, H. B., Kitt, C. A., Price, D. L., Maggio, R., Brann, M. R., Ciliax, B. J., et al. (1993). Localization of D1 and D2 dopamine receptors in brain with subtype-specific antibodies. *Proc. Natl. Acad. Sci. USA* **90**, 8861–8865.
24. van Dijken, H., Dijk, J., Voorn, P., and Holstege, J. C. (1996). Localization of dopamine D2 receptor in rat spinal cord identified with immunocytochemistry and in situ hybridization. *Eur. J. Neurosci.* **8**, 621–628.
25. Bergerot, A., and Goadsby, P. J. (2005). Distribution of the dopamine D1 and D2 receptors in the rat trigeminocervical complex. *Cephalalgia* **25**, 872.
26. Bergerot, A., Storer, R. J., and Goadsby, P. J. (2005). Dopamine inhibits trigeminovascular transmission in the rat. *Cephalalgia* **25**, 862.
27. Skagerberg, G., Bjorklund, A., Lindvall, O., and Schmidt, R. H. (1982). Origin and termination of the diencephalo-spinal dopamine system in the rat. *Brain. Res. Bull.* **9**, 237–244.
28. Fleetwood-Walker, S. M., Hope, P. J., and Mitchell, R. (1988). Antinociceptive actions of descending dopaminergic tracts on cat and rat dorsal horn somatosensory neurones. *J. Physiol.* **399**, 335–348.
29. Lauritzen, M. (1994). Pathophysiology of the migraine aura. The spreading depression theory. *Brain* **117**, 199–210.
30. Goadsby, P. J. (2001). Migraine, aura, and cortical spreading depression: why are we still talking about it? *Ann. Neurol.* **49**, 4–6.
31. Goadsby, P. J. (2005). Can we develop neurally acting drugs for the treatment of migraine? *Nature Rev. Drug Discov.* **4**, 741–750.
32. Akerman, S., and Goadsby, P. J. (2005). Topiramate inhibits cortical spreading depression in rat and cat: impact in migraine aura. *Neuro. Report* **16**, 1383–1387.
33. Tvedskov, J. F., Thomsen, L. L., Iversen, H. K., Williams, P., Gibson, A., Jenkins, K., Peck, R., and Olesen, J. (2004). The effect of propranolol on glyceryltrinitrate-induced headache and arterial response. *Cephalalgia* **24**, 1076–1087.
34. Tvedskov, J. F., Thomsen, L. L., Iversen, H. K., Gibson, A., Williams, P., and Olesen, J. (2004). The prophylactic effect of valproate on glyceryltrinitrate induced migraine. *Cephalalgia* **24**, 576–585.
35. Storer, R. J., and Goadsby, P. J. (2004). Topiramate inhibits trigeminovascular neurons in the cat. *Cephalalgia* **24**, 1049–1056.
36. Storer, R. J., and Goadsby, P. J. (2005). Topiramate has a locus of action outside of the trigeminocervical complex. *Cephalalgia* **25**, 934.
37. Wolff, H. G. (1963). "Headache and Other Head Pain," 3rd edition. Oxford University Press, New York.
38. Moskowitz, M. A., and Cutrer, F. M. (1993). Sumatriptan: a receptor-targeted treatment for migraine. *Ann. Rev. Med.* **44**, 145–154.
39. Dimitriadou, V., Buzzi, M. G., Moskowitz, M. A., and Theoharides, T. C. (1991). Trigeminal sensory fiber stimulation induces morphological changes reflecting secretion in rat dura mater mast cells. *Neuroscience* **44**, 97–112.
40. Dimitriadou, V., Buzzi, M. G., Theoharides, T. C., and Moskowitz, M. A. (1992). Ultrastructural evidence for neurogenically mediated changes in blood vessels of the rat dura mater and tongue following antidromic trigeminal stimulation. *Neuroscience* **48**, 187–203.
41. Bolay, H., Reuter, U., Dunn, A. K., Huang, Z., Boas, D. A., and Moskowitz, M. A. (2002). Intrinsic brain activity triggers trigeminal meningeal afferents in a migraine model. *Nature Med.* **8**, 136–142.
42. May, A., Shepheard, S., Wessing, A., Hargreaves, R. J., Goadsby, P. J., and Diener, H. C. (1998). Retinal plasma extravasation can be evoked by trigeminal stimulation in rat but does not occur during migraine attacks. *Brain* **121**, 1231–1237.
43. May, A., and Goadsby, P. J. (2001). Substance P-receptor antagonists in the therapy of migraine. *Expert Op. Investigat. Drug* **10**, 1–6.
44. Peroutka, S. J. (2005). Neurogenic inflammation and migraine: implications for therapeutics. *Molecul. Intervent.* **5**, 306–313.
45. Selby, G., and Lance, J. W. (1960). Observations on 500 cases of migraine and allied vascular headache. *J. Neurol. Neurosurg. Psychiat.* **23**, 23–32.
46. Burstein, R., Yarnitsky, D., Goor-Aryeh, I., Ransil, B. J., and Bajwa, Z. H. (2000). An association between migraine and cutaneous allodynia. *Ann. Neurol.* **47**, 614–624.
47. Hoskin, K. L., Kaube, H., and Goadsby, P. J. (1996). Central activation of the trigeminovascular pathway in the cat is inhibited by dihydroergotamine. A c-Fos and electrophysiology study. *Brain* **119**, 249–256.
48. Pozo-Rosich, P., and Oshinsky, M. (2005.) Effect of dihydroergotamine (DHE) on central sensitisation of neurons in the trigeminal nucleus caudalis. *Neurology* **64(1)**, A151.
49. May, A., and Goadsby, P. J. (1999). The trigeminovascular system in humans: pathophysiological implications for primary headache syndromes of the neural influences on the cerebral circulation. *J. Cerebr. Blood Flow Metab.* **19**, 115–127.
50. Olesen, J., Diener, H.-C., Husstedt, I.-W., Goadsby, P. J., Hall, D., Meier, U., Pollentier, S., and Lesko, L. M. (2004). Calcitonin gene–related peptide (CGRP) receptor antagonist BIBN4096BS is effective in the treatment of migraine attacks. *New Engl. J. Med.* **350**, 1104–1110.
51. Bartsch, T., and Goadsby, P. J. (2005). Anatomy and physiology of pain referral in primary and cervicogenic headache disorders. *Headache Currents* **2**, 42–48.
52. Goadsby, P. J. (2005). New targets in the acute treatment of headache. *Curr. Opin. Neurol.* **18**, 283–288.

53. Goadsby, P. J. (2004). Prejunctional and presynaptic trigeminovascular targets: what preclinical evidence is there? *Headache Currents* **1**, 1–6.
54. Cohen, A. S., and Goadsby, P. J. (2004). Functional neuroimaging of primary headache disorders. *Curr. Neurol. Neurosci. Rep.* **4**, 105–110.
55. Goadsby, P. J., Lambert, G. A., and Lance, J. W. (1982). Differential effects on the internal and external carotid circulation of the monkey evoked by locus coeruleus stimulation. *Brain Res.* **249**, 247–254.
56. Kaube, H., and Giffin, N. J. (2002). The electrophysiology of migraine. *Curr. Opin. Neurol.* **15**, 303–309.
57. Schoenen, J., Ambrosini, A., Sandor, P. S., and Maertens de Noordhout, A. (2003). Evoked potentials and transcranial magnetic stimulation in migraine: published data and viewpoint on their pathophysiologic significance. *Clin. Neurophys.* **114**, 955–972.
58. Niebur, E., Hsiao, S. S., and Johnson, K. O. (2002). Synchrony: a neural mechanism for attentional selection? *Curr. Opin. Neurobiol.* **12**, 190–194.
59. Ferrari, M. D. (1998). Migraine. *Lancet* **351**, 1043–1051.
60. Kors, E. E., Vanmolkot, K. R., Haan, J., Frants, R. R., van den Maagdenberg, A. M., and Ferrari MD. (2004). Recent findings in headache genetics. *Curr. Opin. Neurol.* **17**, 283–288.
61. Ducros, A., Denier, C., Joutel, A., Cecillon, M., Lescoat, C., Vahedi, K., Darcel, F., Vicaut, E., Bousser, M. G., and Tournier-Lasserve, E. (2001). The clinical spectrum of familial hemiplegic migraine associated with mutations in a neuronal calcium channel. *New Engl. J. Med.* **345**, 17–24.
62. Terwindt, G. M., Ophoff, R. A., Haan, J., Vergouwe, M. N., van Eijk, R., Frants, R. R., and Ferrari, M. D. (1998). Variable clinical expression of mutations in the P/Q-type calcium channel gene in familial hemiplegic migraine. *Neurology* **50**, 1105–1110.
63. Ophoff, R. A., Terwindt, G. M., Vergouwe, M. N., van Eijk, R., Oefner, P. J., Hoffman, S. M., Lamerdin, J. E., Mohrenweiser, H. W., Bulman, D. E., Ferrari, M., Haan, J., Lindhout, D., van Ommen, G. J., Hofker, M. H., Ferrari, M. D., and Frants, R. R. (1996). Familial hemiplegic migraine and episodic ataxia type-2 are caused by mutations in the Ca^{2+} channel gene CACNL1A4. *Cell* **87**, 543–552.
64. Catterall, W. A. (1998). Structure and function of neuronal Ca^{2+} channels and their role in neurotransmitter release. *Cell Calcium* **24**, 307–323.
65. Haan, J., Kors, E. E., Vanmolkot, K. R., van den Maagdenberg, A. M., Frants, R. R., and Ferrari, M. D. (2005). Migraine genetics: an update. *Curr. Pain. Headache Rep.* **9**, 213–220.
66. De Fusco, M., Marconi, R., Silvestri, L., Atorino, L., Rampoldi, L., Morgante, L., Ballabio, A., Aridon, P., and Casari, G. (2003). Haploinsufficiency of ATP1A2 encoding the Na^+/K^+ pump alpha2 subunit associated with familial hemiplegic migraine type 2. *Nature Genetics* **33**, 192–196.
67. Vanmolkot, K. R. J., Kors, E. E., Hottenga, J. J., Terwindt, G. M., Haan, J., Hoefnagels, W. A., Black, D. F., Sandkuijl, L. A., Frants, R. R., Ferrari, M. D., and van den Maagdenberg, A. M. (2003). Novel mutations in the Na^+, K^+–ATPase pump gene ATP1A2 associated with familial hemiplegic migraine and benign familial infantile convulsions. *Ann. Neurol.* **54**, 360–366.
68. Ambrosini, A., D'Onofrio, M., Grieco, G. S., Di Mambro, A., Fortini, D., Nicoletti, F., Nappi, G., Sances, G., Schoenen, J., Buzzi, M. G., Santorelli, F. M., and Pierelli, F. (2005). A new mutation on the ATPA2 gene in one Italian family with basilar-type migraine linked to the FHM2 locus. *Neurology* **64(1)**, A132.
69. Dichgans, M., Freilinger, T., Eckstein, G., Babini, E., Lorenz-Depiereux, B., Biskup, S., Ferrari, M. D., Herzog, J., van den Maagdenberg, A. M., Pusch, M., and Strom, T. M. (2005). Mutation in the neuronal voltage-gated sodium channel *SCN1A* causes familial hemiplegic migraine. *Lancet* **366**, 371–377.
70. Vanmolkot, K., et al. Unpublished data.
71. Thomsen, L. L., Ostergaard, E., Olesen, J., and Russell, M. B. (2003). Evidence for a separate type of migraine with aura: sporadic hemiplegic migraine. *Neurology* **60**, 595–601.
72. Thomsen, L. L., Ostergaard, E., Romer, S. F., Andersen, I., Eriksen, M. K., Olesen, J., and Russell, M. B. (2003). Sporadic hemiplegic migraine is an aetiologically heterogeneous disorder. *Cephalalgia* **23**, 921–928.
73. Terwindt, G., Kors, E., Haan, J., Vermeulen, F., van den Maagdenberg, A., Frants, R., and Ferrari, M. (2002). Mutation analysis of the CACNA1A calcium channel subunit gene in 27 patients with sporadic hemiplegic migraine. *Arch. Neurol.* **59**, 1016–1018.
74. deVries et al. Unpublished data.
75. Lehmann-Horn, F., and Ferrari, M. D. (1997). International Headache Congress HIS, Amsterdam.
76. Sandor, P. S., Mascia, A., Seidel, L., de Pasqua, V., and Schoenen, J. (2001). Subclinical cerebellar impairment in the common types of migraine: a three-dimensional analysis of reaching movements. *Ann. Neurol.* **49**, 668–672.
77. Asakura, K., Kanemasa, T., Minagawa, K., Kagawa, K., Yagami, T., Nakajima, M., and Ninomiya, M. (2000). α-Eudesmol, a P/Q-type Ca^{2+} channel blocker, inhibits neurogenic vasodilation and extravasation following electrical stimulation of trigeminal ganglion. *Brain Res.* **873**, 94–101.
78. Knight, Y. E., Bartsch, T., Kaube, H., and Goadsby, P. J. (2002). J. P/Q-type calcium channel blockade in the PAG facilitates trigeminal nociception: a functional genetic link for migraine? *J. Neurosci.* **22(213)**, 1–6.
79. Shields, K. G., Storer, R. J., Akerman, S., and Goadsby, P. J. (2005). Calcium channels modulate nociceptive transmission in the trigeminal nucleus of the cat. *Neuroscience* **135**, 203–212.
80. Jouvenceau, A., Eunson, L. H., Spauschus, A., Ramesh, V., Zuberi, S. M., Kullmann, D. M., and Hanna, M. G. (2001). Human epilepsy associated with dysfunction of the brain P/Q-type calcium channel. *Lancet* **358**, 801–807.
81. Imbrici, P., Jaffe, S. L., Eunson, L. H., Davies, N. P., Herd, C., Robertson, R., Kullmann, D. M., and Hanna, M. G. (2004). Dysfunction of the brain calcium channel CaV2.1 in absence epilepsy and episodic ataxia. *Brain* **127**, 2682–2692.
82. Cao, Y. Q., Piedras-Renteria, E. S., Smith, G. B., Chen, G., Harata, N. C., and Tsien, R. W. (2004). Presynaptic Ca^{2+} channels compete for channel type–preferring slots in altered neurotransmission arising from Ca^{2+} channelopathy. *Neuron* **43**, 387–400.
83. Hans, M., Luvisetto, S., Williams, M. E., Spagnolo, M., Urrutia, A., Tottene, A., Brust, P. F., Johnson, E. C., Harpold, M. M., Stauderman, K. A., and Pietrobon, D. (1999). Functional consequences of mutations in the human alpha (1A) calcium channel subunit linked to familial hemiplegic migraine. *J. Neurosci.* **19**, 1610–1619.
84. Kraus, R. L., Sinnegger, M. J., Glossmann, H., Hering, S., and Striessnig, J. (1998). Familial hemiplegic migraine mutations change alpha (1A) Ca^{2+} channel kinetics. *J. Biol. Chem.* **273**, 5586–5590.
85. Kraus, R. L., Sinnegger, M. J., Koschak, A., Glossmann, H., Stenirri, S., Carrera, P., and Striessnig, J. (2000). Three new familial hemiplegic migraine mutants affect P/Q-type Ca(2+) channel kinetics. *J. Biol. Chem.* **275**, 9239–9243.
86. Tottene, A., Fellin, T., Pagnutti, S., Luvisetto, S., Striessnig, J., Fletcher, C., and Pietrobon, D. (2002). Familial hemiplegic migraine mutations increase Ca^{2+} influx through single human CaV2.1 channels and decrease maximal $Ca_V2.1$ current density in neurons. *Proc. Natl. Acad. Sci. USA* **99**, 13284–13289.
87. van den Maagdenberg, A. M., Pietrobon, D., Pizzorusso, T., Kaja, S., Broos, L. A., Cesetti, T., van de Ven, R. C., Tottene, A., van der Kaa, J., Plomp, J. J., Frants, R. R., and Ferrari, M. D. (2004). A Cacna1a knockin migraine mouse model with increased susceptibility to cortical spreading depression. *Neuron* **41**, 701–710.

88. Segall, L., Mezzetti, A., Scanzano, R., Gargus, J. J., Purisima, E., and Blostein, R. (2005). Alterations in the alpha2 isoform of Na,K–ATPase associated with familial hemiplegic migraine type 2. *Proc. Natl. Acad. Sci USA* **102**, 11106–11111.

89. Ikeda, K., Onimaru, H., Yamada, J., Inoue, K., Ueno, S., Onaka, T., Toyoda, H., Arata, A., Ishikawa, T. O., Taketo, M. M., Fukuda, A., and Kawakami, K. (2004). Malfunction of respiratory–related neuronal activity in Na+, K+–ATPase alpha2 subunit–deficient mice is attributable to abnormal Cl- homeostasis in brainstem neurons. *J. Neurosci.* **24**, 10693–10701.

90. James, P. F., Grupp, I. L., Grupp, G., Woo, A. L., Askew, G. R., Croyle, M. L., Walsh, R. A., and Lingrel, J. B. (1999). Identification of a specific role for the Na,K–ATPase alpha 2 isoform as a regulator of calcium in the heart. *Mol. Cell* **3**, 555–563.

91. Moskowitz, M. A., Bolay, H., Dalkara, T. (2004). Deciphering migraine mechanisms: clues from familial hemiplegic migraine genotypes. *Ann. Neurol.* **55**, 276–280.

92. Pietroben, D., et al. Unpublished data.

93. Tournier–Lasserve, E., Joutel, A., Melki, J., Weissenbach, J., Lathrop, G. M., Chabriat, H., Mas, J. L., Cabanis, E. A., Baudrimont, M., Maciazek, J., Bach, M. A., and Bousser, M-G. (1993). Cerebral autosomal dominant arteriopathy with subcortical infarcts and leukoencephalopathy maps to chromosome 19p12. *Nature Genetics* **3**, 256–259.

94. Joutel, A., Corpechot, C., Ducros, A., Vahedi, K., Chabriat, H., Mouton, P., Alamowitch, S., Domenga, V., Cecillion, M., Marechal, E., Maciazek, J., Vayssiere, C., Cruaud, C., Cabanis, E. A., Ruchoux, M. M., Weissenbach, J., Bach, J. F., Bousser, M. G., and Tournier-Lasserve, E. (1996). Notch3 mutations in CADASIL, a hereditary adult-onset condition causing stroke and dementia. *Nature* **383**, 707–710.

95. Domenga, V., Fardoux, P., Lacombe, P., Monet, M., Maciazek, J., Krebs, L. T., Klonjkowski, B., Berrou, E., Mericskay, M., Li, Z., Tournier-Lasserve, E., Gridley, T., and Joutel, A. (2004). Notch3 is required for arterial identity and maturation of vascular smooth muscle cells. *Genes Dev.* **18**, 2730–2735.

96. Ruchoux, M. M., Domenga, V., Brulin, P., Maciazek, J., Limol, S., Tournier-Lasserve, E., and Joutel A. (2003). Transgenic mice expressing mutant Notch3 develop vascular alterations characteristic of cerebral autosomal dominant arteriopathy with subcortical infarcts and leukoencephalopathy. *Am. J. Pathol.* **162**, 329–342.

97. Lundkvist, J., Zhu, S., Hansson, E. M., Schweinhardt, P., Miao, Q., Paul Beatus, P., Dannaeus, K., Karlström, H., Johansson, C. B., Viitanen, M., Rozell, B., Spenger, C., Mohammed, A., Kalimo, H., and Lendahl, L. (2005). Mice carrying a R142C Notch 3 knock-in mutation do not develop a CADASIL-like phenotype. *Genesis* **41**, 13–22.

98. Terwindt, G. M., Haan, J. Ophoff, R. A., Groenen, S. M., Storimans, C. W., Lanser, J. B., Roos, R. A., Bleeker-Wagemakers, E. M., Frants, R. R., and Ferrari, M. D. (1998). Clinical and genetic analysis of a large Dutch family with autosomal dominant vascular retinopathy, migraine and Raynaud's phenomenon. *Brain* **121**, 303–316.

99. Ophoff, R. A., DeYoung, J., Service, S. K., Joosse, M., Caffo, N. A., Sandkuijl, L. A., Terwindt, G. M., Haan, J., van den Maagdenberg, A. M., Jen, J., Baloh, R. W., Barilla-LaBarca, M. L., Saccone, N. L., Atkinson, J. P., Ferrari, M. D., Freimer, N. B., and Frants, R. R. (2001). Hereditary vascular retinopathy, cerebroretinal vasculopathy, and hereditary endotheliopathy with retinopathy, nephropathy and stroke map to a single locus on chromosome 3p21.1-p21.3. *Am. J. Hum. Genet.* **69**, 447–453.

100. Hottenga, J. J., Vanmolkot, K. R., Kors, E. E., Kheradmand, Kia S., de Jong, P. T., Haan, J., Terwindt, G. M., Frants, R. R., Ferrari, M. D., and van den Maagdenberg, A. M. (2005). The 3p21.1-p21.3 hereditary vascular retinopathy locus increases the risk for Raynaud's phenomenon and migraine. *Cephalalgia* **25**, 1168–1172.

101. Carlsson, A., Forsgren, L., Nylander, P. O., Hellman, U., Forsman-Semb, K., Holmgren, G., Holmberg, D., and Holmberg, M. (2002). Identification of a susceptibility locus for migraine with and without aura on 6p12.2-p21.1. *Neurology* **59**, 1804–1807.

102. Nyholt, D. R., Morley, K. I., Ferreira, M. A., Medland, S. E., Boomsma, D. I., Heath, A. C., Merikangas, K. R., Montgomery, G. W., and Martin, N. G. (2005). Genomewide significant linkage to migrainous headache on chromosome 5q21. *Am. J. Hum. Genet.* **77**, 500–512.

103. Wessman, M., Kallela, M., Kaunisto, M. A., Marttila, P., Sobel, E., Hartiala, J., Oswell, G., Leal, S. M., Papp, J. C., Hamalainen, E., Broas, P., Joslyn, G., Hovatta, I., Hiekkalinna, T., Kaprio, J., Ott, J., Cantor, R. M., Zwart, J. A., Ilmavirta, M., Havanka, H., Farkkila, M., Peltonen, L., and Palotie, A. (2002). A susceptibility locus for migraine with aura, on chromosome 4q24. **70**, 652–662.

104. Bjornsson, A., Gudmundsson, G., Gudfinnsson, E., Bjornsson, A., Gudmundsson, G., Gudfinnsson, E., Hrafnsdottir, M., Benedikz, J., Skuladottir, S., Kristjansson, K., Frigge, M. L., Kong, A., Stefansson, K., and Gulcher, J. R. (2003). Localization of a gene for migraine without aura to chromosome 4q21. *Am. J. Hum. Genet.* **73**, 986–993.

105. Cader, Z. M., Noble-Topham, S., Dyment, D. A., Cherny, S. S., Brown, J. D., Rice, G. P., and Ebers, G. C. (2003). Significant linkage to migraine with aura on chromosome 11q24. *Hum. Mol. Genet.* **12**, 2511–2517.

106. Soragna, D., Vettori, A., Carraio, G., Marchioni, E., Vazza, G., Bellini, S., Tupler, R., Savoldi, F., and Mostacciuolo, M. L. (2003). A locus for migraine without aura maps on chromosome 14q21.2-q22.3. *Am. J. Hum. Genet.* **72**, 161–167.

107. Nyholt, D. R., Dawkins, J. L., Brimage, P. J., Goadsby, P. J., Nicholson, G. A., and Griffiths, L. R. (1998). Evidence for an X-linked genetic component in familial typical migraine. *Hum. Mol. Genet.* **7**, 459–463.

108. Russo, L., Mariotti, P., Sangiorgi, E., Russo, L., Mariotti, P., Sangiorgi, E., Giordano, T., Ricci, I., Lupi, F., Chiera, R., Guzzetta, F., Neri, G., and Gurrieri, F. (2005). A new susceptibility locus for migraine with aura in the 15q11-q13 genomic region containing three GABA-A receptor genes. *Am. J. Hum. Genet.* **76**, 327–333.

109. Lea, R. A., Nyholt, D. R., Curtain, R. P., Ovcaric, M., Sciascia, R., Bellis, C., Macmillan, J., Quinlan, S., Gibson, R. A., McCarthy, L. C., Riley, J. H., Smithies, Y. J., Kinrade, S., and Griffiths, L. R. (2005). A genome-wide scan provides evidence for loci influencing a severe heritable form of common migraine. *Neurogenetics* **6**, 67–72.

110. Montagna, P., Pierangeli, G., Cevoli, S., Mochi, M., and Cortelli, P. (2005). Pharmacogenetics of headache treatment. *Neurologic. Sci.* **26(2)**, S143–S147.

111. Kowa, H., Yasui, K., Takeshima, T., Urakami, K., Sakai, F., and Nakashima, K. (2000). The homozygous C677T mutation in the methylenetetrahydrofolate reductase gene is a genetic risk factor for migraine. *Am. J. Med. Genet.* **96**, 762–764.

112. Kara, I., Sazci, A., Ergul, E., Kaya, G., and Kilic, G. (2003). Association of the C677T and A1298C polymorphisms in the 5,10 methylenetetrahydrofolate reductase gene in patients with migraine risk. *Brain Res. Mol. Brain Res.* **111**, 84–90.

113. Lea, R. A., Ovcaric, M., Sundholm, J., MacMillan, J., and Griffiths, L. R. (2004). The methylenetetrahydrofolate reductase gene variant C677T influences susceptibility to migraine with aura. *BMC Med.* **2**, 3.

114. Oterino, A., Valle, N., Bravo, Y., Munoz, P., Sanchez-Velasco, P., Ruiz-Alegria, C., Castillo, J., Leyva-Cobian, F., Vadillo, A., and Pascual, J. (2004). MTHFR T677 homozygosis influences the presence of aura in migraineurs. *Cephalalgia* **24**, 491–494.

115. Scher, A. I., Terwindt, G. M., Verschuren, W. M., Kruit, M. C., Blom, H. J., Kowa, H., Frants, R. R., van den Maagdenberg, A. M., van Buchem, M., Ferrari, M. D., and Launer, L. J. (2006). Migraine

and MTHFR C677T genotype in a population-based sample. *Ann. Neurol.* **59**, 372–375.

116. Lea, R. A., Ovcaric, M., Sundholm, J., Solyom, L., Macmillan, J., and Griffiths, L. R. (2005). Genetic variants of angiotensin converting enzyme and methylenetetrahydrofolate reductase may act in combination to increase migraine susceptibility. *Brain Res. Mol. Brain Res.* **136**, 112–117.

117. Colson, N. J., Lea, R. A., Quinlan, S., MacMillan, J., and Griffiths, L. R. (2005). Investigation of hormone receptor genes in migraine. *Neurogenetics* **6**, 17–23.

118. Trabace, S., Brioli, G., Lulli, P., Morellini, M., Giacovazzo, M., Cicciarelli, G., and Martelletti, P. (2002). Tumor necrosis factor gene polymorphism in migraine. *Headache* **42**, 341–345.

119. Rainero, I., Grimaldi, L. M., Salani, G., Valfre, W., Rivoiro, C., Savi, L., and Pinessi, L. (2004). Association between the tumor necrosis factor-alpha-308 G/A gene polymorphism and migraine. *Neurology* **62**, 141–143.

120. Paterna, S., Di Pasquale, P., D'Angelo, A., Seidita, G., Tuttolomondo, A., Cardinale, A., Maniscalchi, T., Follone, G., Giubilato, A., Tarantello, M., and Licata, G. (2000). Angiotensin–converting enzyme gene deletion polymorphism determines an increase in frequency of migraine attacks in patients suffering from migraine without aura. *Eur. Neurol.* **43**, 133–136.

121. Kowa, H., Fusayasu, E., Ijiri, T., Ishizaki, K., Yasui, K., Nakaso, K., Kusumi, M., Takeshima, T., and Nakashima, K. (2005). Association of the insertion/deletion polymorphism of the angiotensin I–converting enzyme gene in patients of migraine with aura. *Neurosci. Lett.* **374**, 129–131.

122. Lin, J. J., Wang, P. J., Chen, C. H., Yueh, K. C., Lin, S. Z., and Harn, H. J. (2005). Homozygous deletion genotype of angiotensin converting enzyme confers protection against migraine in man. *Acta. Neurol. Taiwan* **14**, 120–125.

33

Temporal Lobe Epilepsy

Helen E. Scharfman, PhD
Timothy A. Pedley, MD

Keywords: *Ammon's horn sclerosis, anticonvulsant, cortical dysplasia, cortical malformation, endfolium sclerosis, dentate gyrus, epileptogenesis, febrile seizure, gliosis, granule cell, granule cell dispersion, hilus, hippocampus, interneuron, intractable, mossy cell, mossy fiber sprouting, perinatal injury, plasticity, resection, seizure, status epilepticus, temporal lobe, temporal lobectomy, transgenic*

I. Historical Background
II. Pathology of Temporal Lobe Epilepsy
III. Pathophysiological Mechanisms
IV. Summary and Future Directions: Developing a Comprehensive Hypothesis for the Pathology and Pathophysiology of Temporal Lobe Epilepsy
References

I. Historical Background

The word *epilepsy* is derived from the Greek verb *epilavainem*, meaning "to be seized" or "to be taken hold of." This reflected an ancient but enduring idea that epilepsy, like other diseases, occurred as the result of actions by gods or evil spirits, often as punishment. Epilepsy was considered "the sacred disease" because it was the most dramatic example of demonic possession. Greek writings, which date from about 400 BCE, and are traditionally attributed to Hippocrates, clearly convey the concept that epilepsy was a disease arising from the brain and should be treated by diet and drugs, not religious incantations. However, the history of epilepsy for more than two millennia was largely one of beliefs based on religious views, magic, superstitions, and various taboos. During this time, people with epilepsy were almost always avoided and viewed with disgust and horror. The modern era of epilepsy as a neurological disorder arising from brain dysfunction only dates to the end of the 19th century and the contributions of John Hughlings Jackson, Jean-Martin Charcot, and William R. Gowers. Jackson is credited with the first biological definition of epilepsy as "an occasional, an excessive, and a disorderly discharge of nerve tissue." He was also one of the first to propose that the different clinical manifestations of seizures resulted from specific, localized areas of the brain that became corrupted by this abnormal "disorderly discharge."

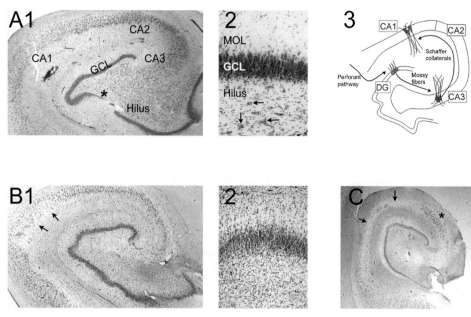

Figure 1 Pathology in hippocampus of patients with intractable temporal lobe epilepsy. **A**, Normal human hippocampus: *1*, Subfields of the hippocampus and dentate gyrus are shown in a Nissl-stained section from an individual without temporal lobe epilepsy. The human dentate gyrus granule cell layer is continuous, but in this section there is an appearance of discontinuity (*) because of the tangential plane of section through one of the many folds of the dentate gyrus. *2*, Higher magnification of a cross-section through the normal dentate gyrus showing the relatively cell-free molecular layer (MOL), densely packed granule cell layer (GCL), and hilus, which contains diverse types of neurons (*arrows*). *3*, Schematic of the trisynaptic glutamatergic circuit of the hippocampus. Layer II neurons of the entorhinal cortex innervate granule cells of the dentate gyrus (the perforant pathway) in the outer two-thirds of the molecular layer. Granule cell axons, the mossy fibers, innervate the proximal dendrites of the CA3 pyramidal cells, as well as other neurons that are not shown, such as the hilar glutamatergic neurons ("mossy" cells) and dentate inhibitory (GABAergic) neurons. CA3 pyramidal cells have axons that innervate CA1 pyramidal cells (Schaffer collaterals), as well as other CA3 neurons. **B**, Characteristic hippocampal pathology in temporal lobe epilepsy. *1*, A Nissl-stained section from the hippocampus of a patient who had a medial temporal lobe resection to control intractable temporal lobe epilepsy. The hallmarks of Ammon's horn sclerosis are shown: 1) cell loss in area CA1 (*arrows*), 2) cell loss in the endfolium (hilus/area CA3), and 3) relative preservation of area CA2 and the dentate gyrus granule cells. Note that the classic definition of Ammon's horn sclerosis is associated with more substantial loss of area CA1 than is shown here. (Classic Ammon's horn sclerosis is illustrated schematically in Fig. 2, *A*.) *2*, A higher magnification of the dentate gyrus in a different tissue specimen shows the typical loss of hilar neurons (compare with A2), which are replaced by large numbers of glial cells (gliosis). **C**, Severe pathology in temporal lobe epilepsy. A Nissl-stained section from a different patient with long-standing intractable temporal lobe epilepsy exemplifies more severe sclerosis, including almost complete loss of area CA1 and the hilus/area CA3, dentate granule cell loss, and some loss in area CA2. Note the much smaller size of the cross-section as compared with A1 and B1, illustrating the shrinkage that is typical of such cases, which is often called "total" Ammon's horn sclerosis.

The idea that a characteristic type of seizure was associated with the temporal lobe evolved from several key developments in science and medicine in the 19th and 20th centuries. The first of these developments was neuroanatomical: distinguishing the unique features of the temporal lobe and its circuitry from other areas of the cerebrum. For example, the temporal lobe is anatomically heterogeneous compared with other parts of the cerebral hemisphere, containing the classic six-layered isocortex (i.e., neocortex), as well as cortical areas with three layers (allocortex, or "simple" cortex, exemplified by the hippocampus), and areas with four to five layers (mesocortex, or "transitional" cortex, such as the entorhinal cortex). The temporal lobe also includes "corticoid" areas that are composed of subcortical nuclei adjacent to allocortex (e.g., amygdala and septal nuclei). This was an essential step for attributing particular signs and symptoms to temporal lobe dysfunction. However, this anatomical characterization had not been widely embraced when Jackson was first developing his pathophysiological concepts of epilepsy. In addition, the evolution of the concept of the limbic system was only introduced much later, first by Papez in 1937 [1] and then elaborated more fully by MacLean [2].

A second development was the recognition that some seizures seemed clinically intermediate between those

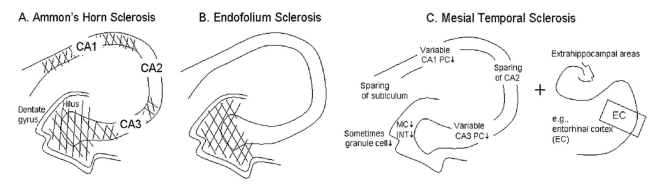

Figure 2 Schematic illustrations of pathology in temporal lobe epilepsy. **A**, Ammon's horn sclerosis is defined as the loss of area CA1 and the endfolium (including the hilus and area CA3), with relative sparing of area CA2 and the granule cells of the dentate gyrus. **B**, Endfolium sclerosis is defined by the selective loss of neurons in the endfolium. **C**, Mesial temporal sclerosis (MTS) is a term that has evolved after recognition that pathology in the hippocampus is often variable and damage often exists in extrahippocampal sites. MTS is typically defined by cell loss in area CA1 and the endfolium, although it may not be complete (see Fig. 1A). There is sparing of CA2 and the granule cell layer, similar to AHS, but some cell loss may be present. In addition, neuronal damage in extrahippocampal areas (see Table 1) is often present. One example is the parahippocampal gyrus, which includes the entorhinal cortex (*EC*) (see text for details).

considered to be *grand mal* and *petit mal* [3,4]. That is, they had motor movements that were more complex and less convulsive than those seen with grand mal seizures (e.g., ambulatory automatisms), as well as hallucinatory phenomena (both psychic and sensory) and disturbances in mood, memory, awareness, and other cognitive functions. VanGieson [5] termed this constellation of features *psychomotor epilepsy*. It was Jackson, however, who linked "dreamy states" [6] and, later, sensory hallucinations with "discharging lesions" in the uncus (so-called "uncinate fits") [7]. Other clinicopathological studies confirmed these associations [8].

A third development was neurophysiological: correlating particular electroencephalogram (EEG) patterns with clinical observations of seizure phenomena. Pioneering studies in Boston by Frederick and Erna Gibbs, often in collaboration with William G. Lennox, demonstrated that the majority of patients with epilepsy manifested by mood changes, psychic symptoms, and automatisms (psychomotor seizures) exhibited sharp wave discharges that could be recorded interictally over the anterior temporal region during sleep [9,10]. Furthermore, during typical attacks, some patients showed a characteristic ictal pattern: initial flattening of activity in the temporal leads of one side followed by rhythmic 4–6 Hz discharges, either unilaterally in the same area or involving both temporal regions [11,12]. The combination of distinctive clinical and EEG features led to the classification of "psychomotor epilepsy" as a third major type of seizure disorder (in addition to grand mal and petit mal). Subsequently, intracranial electrodes demonstrated that the electrical abnormalities arose from the temporal lobe [13–15], and electrical stimulation of the temporal lobe in patients with epilepsy reproduced the symptoms [16]. As evidence continued to accumulate that the unique clinical and EEG features of psychomotor seizures had their origin in the temporal lobe, the terms *temporal lobe epilepsy* [12] or *temporal lobe seizures* supplanted the older terminology.

II. Pathology of Temporal Lobe Epilepsy

There is a strong association between seizures and temporal lobe pathology, especially in the hippocampus. Different lesions are found, including tumors (e.g., gliomas, dysembryoplastic neuroepithelial tumors [DNETs], and gangliogliomas), cavernous malformations, encephalomalacia following trauma or stroke, encephalitis, and developmental malformations such as cortical heterotopias. By far, however, the most commonly encountered specific type of pathology is hippocampal sclerosis. Discussion of this entity is complicated by the use of a number of different terms that, although related, are not pathologically synonymous. These include *mesial temporal sclerosis, Ammon's horn sclerosis*, and *endfolium sclerosis*. This complex terminology has its origins in various descriptions that have been published over the years, beginning with that of Bouchet and Cazauvieilh in 1825 [17]. Differences have arisen because of the nature of the specimens (e.g., autopsy vs surgical), the extent and rigor of the pathological examination, the particular emphasis or bias of the investigators, and the complexity of the pathology itself. In addition, early descriptions were often terse. For example, the hippocampus might be simply described as "shrunken" or having an altered "consistency." Other investigators, however, provided careful and detailed reports that led to an increasing appreciation for the relationship between hippocampal sclerosis and temporal lobe epilepsy.

A. Hippocampal Sclerosis

The first pathological studies were focused less on understanding the mechanisms of epilepsy and more on the abnormal behavior of patients with epilepsy. The research of Bouchet and Cazauvieilh [17] originated from clinical observations about the coexistence of epilepsy and psychiatric disorders. Almost half of their specimens obtained from epilepsy patients with "insanity" revealed neuronal loss in the hippocampus and cerebellum. Sommer [18] made three important observations. First, he noted that hippocampal pathology, based on gross macroscopic observations, was often unilateral. Second, in a case he examined microscopically, he observed that pyramidal cell loss within the hippocampus was not uniform, pointing out that it was maximal in the region now designated CA1 (commonly referred to as "Sommer's sector"). Third, he inferred that the hippocampal pathology was the origin of the sensory hallucinations and illusions that were characteristic features of his patients' seizures. This was the first time that hippocampal sclerosis was linked to clinical symptomatology. In 1899 Bratz observed hippocampal sclerosis in 50% of autopsies performed in patients with epilepsy and described the pathological findings more fully [19]. He pointed out that the pathology was not limited to the hippocampus proper but also involved the amygdala and parahippocampal gyrus. He also commented on the nonuniform neuronal loss and identified CA2 as a "resistant zone." Bratz concluded that the sclerosis played a pathogenic role in the development of seizures. This assertion became the basis for a continuing, often bitter debate as to whether hippocampal sclerosis was the cause or the consequence of seizures. An important, and conceptually quite modern, theory was introduced by Oscar and Cécile Vogt [20,21]. They proposed that particular "physicochemical properties" of neurons were the basis for specific and differing susceptibilities to injury, a concept they termed "pathoclisis." Spielmeyer [22] concluded that pathological changes were the result of epilepsy, not the cause, and he attributed them to vascular spasm. A theory that enjoyed brief popularity was that of Earle, Baldwin, and Penfield [23], who hypothesized that damage to the hippocampus occurred as a result of neuronal hypoxia caused by deformation of the skull and herniation of the medial temporal lobe during birth ("incisural sclerosis"). Little more than a decade later, however, Ounsted, Lindsay, and Norman [24] emphasized that the pattern of brain damage seen with epilepsy was radically different from that resulting from hypoxia and arterial occlusion. Furthermore, Veith [25] performed a detailed examination of several hundred infant brains and found no evidence of temporal lobe herniation. In the modern era, the most important pathological studies have been those of Corsellis and his colleagues, especially Margerison [26] and Bruton [27]. Much of the information in the following sections relies heavily on their work.

A fundamental aspect of hippocampal sclerosis is the relative selective vulnerability of different types of neurons to injury. Thus there is early and often substantial loss of neurons in CA1 and the *endfolium*, a somewhat variously defined area that includes the hilus and the area of CA3 lying between the blades of the dentate gyrus (an area including CA4 by Ramón y Cajal and CA3c by Lorenté de No). In contrast, areas adjacent to these vulnerable neuronal populations appear relatively resistant, notably area CA2 and the granule cell layer of the dentate gyrus. This striking pattern of juxtaposed vulnerable and resistant areas is the essential feature of hippocampal sclerosis (*classical Ammon's horn sclerosis*). There are other characteristic changes as well, most notably astrocytic proliferation and scarring and atrophy of the hippocampal formation, which, when severe, can be perceived by the naked eye. The degree of abnormality varies among cases, ranging from minimal neuronal loss limited to the endfolium (*endfolium sclerosis*) to almost complete loss of neurons throughout the entire hippocampus (Fig. 3). Additionally, in most patients the pathological changes extend, to some degree, beyond the hippocampus

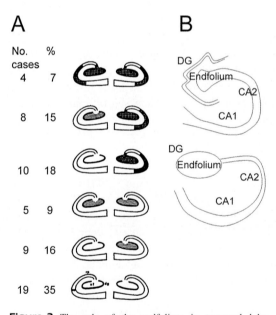

Figure 3 The role of the endfolium in temporal lobe epilepsy. **A**, A diagram from Margerison and Corsellis [26] summarizing their evaluation of autopsy material from patients with temporal lobe epilepsy (listed on the left), illustrates that neuronal loss in the endfolium is common (reflected by the diagrams on the right). This work changed prevailing views that damage to area CA1 was most critical in the etiology of temporal lobe epilepsy, although observations of endfolium damage had been noted as early as Sommer [18] and Bratz [19]. **B**, Schematic of the human hippocampus and dentate gyrus (*top*) for comparison to the simplified diagram of the human hippocampus by Margerison and Corsellis (*bottom*). Margerison and Corsellis represented the endfolium and part of area CA3 as a large oval, with the edge of the oval as the dentate gyrus granule cell layer.

into the amygdala and the mesial temporal cortex (parahippocampal, inferior, middle and fusiform gyri), a pattern termed *mesial temporal sclerosis* [27–31].

The reasons for the variable severity of hippocampal sclerosis among patients with temporal lobe epilepsy is not well understood. A relationship between duration and severity of epilepsy and severity of brain damage has not been established. More than 60% of Falconer's temporal lobectomy cases with hippocampal sclerosis had histories of a prolonged convulsion, usually febrile, in early childhood [32,33]. Bruton's more extensive review of Falconer's surgical material found that 81% of temporal lobectomy patients had some form of early "cerebral insult," of which febrile convulsions in infancy were the most common [27]. This association has been confirmed by many subsequent investigators as well [34–38]. In contrast to surgical series in which an association among febrile seizures, hippocampal sclerosis, and temporal lobe epilepsy has been demonstrated repeatedly, epidemiological studies of epilepsy have failed to show the same correlations [39]. Although febrile seizures are common, occurring in up to 5% of healthy children in the United States and Europe, temporal lobe epilepsy is much less frequent. This suggests that only a small subset of children with febrile seizures has a selective vulnerability to hippocampal damage leading to temporal lobe epilepsy, as revealed by the various surgical series. Magnetic resonance imaging (MRI) scans immediately after febrile seizures have demonstrated increased hippocampal T2-weighted signal and increased hippocampal volume consistent with edema in children with complex febrile seizures [40,41]. The changes were most pronounced in the hippocampus contralateral to the lateralized or focal clinical features. Rarely, such acute changes progress to hippocampal atrophy and gliosis [42]. Systematic longitudinal studies are still lacking, however, so the evolution of these changes, incidence of permanent hippocampal abnormalities, and frequency with which temporal lobe epilepsy develops in those patients with abnormal hippocampi remain unknown.

B. Extrahippocampal Pathology

Major areas showing pathological changes in temporal lobe epilepsy outside the hippocampus are shown in Table 1. Indeed, when the whole brain is available for study, nearly all patients with hippocampal sclerosis will have damage elsewhere as well. Beginning with the studies of Feindel and Penfield [43,44], it became clear that neuronal loss and gliosis were common in the amygdala, and some reports have suggested that damage to the amygdala can occur in isolation, a situation termed *amygdala sclerosis* by Miller and colleagues [45]. Experimental studies in laboratory animals have supported the vulnerability of the amygdala and have identified the nuclei of the amygdala that are most susceptible [46]. In addition to the amygdala, the parahippocampal region and other sites in the temporal lobe are frequently affected as well, as mentioned earlier. The entorhinal cortex is notable because entorhinal neuronal loss is common in temporal lobe epilepsy [29], and damage may exist without evidence of atrophy in the adjacent hippocampus [30,31]. These studies have led to new emphasis on extrahippocampal areas [47]. Interestingly, the concept that extrahippocampal sites are relevant to temporal lobe epilepsy was conceived even by the first pathologists, and has repeatedly emerged since then [48]. Other areas that exemplify extrahippocampal damage include the ventral limbic cortex and perirhinal and piriform cortices [49]. Cerebellar pathology occurs in about 45% of patients with hippocampal sclerosis epilepsy, but the degree of damage is quite variable [26]. Loss of Purkinje cells is the most common finding, but destruction of granule cells can also occur. Humans with temporal lobe epilepsy demonstrate variable degrees of neuronal injury in the cerebral cortex (about 22% of cases) and thalamus (25% of cases) [26]. Animals models also demonstrate striking and consistent change in the thalamus [50] and more variable changes in the cerebral cortex [51].

C. Mossy Fiber Sprouting and Other Changes

In addition to neuronal loss and gliosis, more complex changes occur. One change that has received great attention is axonal and synaptic reorganization of the mossy fiber axons of the granule cells in the dentate gyrus (Fig. 4). First identified in animal models of epilepsy [52], mossy fiber sprouting was subsequently demonstrated in temporal lobe resections in patients with intractable epilepsy [53,54]. Interestingly, mossy fiber sprouting occurs even in animal models that do not demonstrate spontaneous recurrent seizures, such as the kindling model. Furthermore, mossy fiber sprouting occurs after lesions that do not result in persistent seizures. This, combined with the observation that sprouting is not present in all tissue specimens from patients with intractable temporal lobe epilepsy, has required modification of the hypothesis that mossy fiber sprouting is a cause of temporal lobe epilepsy.

Figure 4 provides a simplified diagram of the circuitry of the dentate gyrus and the changes that occur when mossy fiber axons sprout. These new axon collaterals establish new synapses at targets that are quite different from those of normal mossy fiber projections. The most obvious abnormality is the innervation of the inner molecular layer, which contains the proximal dendrites of granule cells as well as processes of other cell types (see Fig. 4). However, new mossy fiber collaterals also project to other areas, including the hilar region where they normally terminate. A number of investigators have suggested that the new connections onto granule cells resulting from sprouted

Table 1 Extrahippocampal Areas and Temporal Lobe Epilepsy

Extrahippocampal Areas	Evidence of Structural Change in Patients with TLE	References
Limbic cortex		
Parahippocampal gyrus		
Entorhinal cortex	+	[30,145]
Perirhinal	+	[30,145]
Temporal pole	+	[145,146]
Insula	+	[147]
Cingulate gyrus	+	[147]
Orbitofrontal cortex	+	[147]
Amygdala	+	[46]
Olfactory cortex		
Piriform cortex	+	[148]
Periamygdaloid cortex	?	
Diencephalon		
Hypothalamus	?	
Thalamus	+	[147,149]
Anterior nucleus		
Mediodorsal nucleus		
Habenula		
Basal ganglia	(Metabolism)*	[150–152]
Ventral striatum		[153]
Ventral pallidum		
Basal forebrain	+	[147]
Septal nuclei	?	
Fimbria/fornix	+	[135]
Mamillary bodies	+	[154]
Brainstem	?	

Extrahippocampal areas, with emphasis on the limbic system, are listed on the left, and each is marked if there is evidence from MRI studies of structural changes in temporal lobe epilepsy. +, Positive evidence; ?, no available evidence to date. References for the positive evidence are listed on the right. In addition, references 26 and 155 provide information about multiple areas based on pathology of surgical specimens.

*No available MRI data report structural changes to the basal ganglia in temporal lobe epilepsy; however, metabolic markers suggest altered basal ganglia metabolism in temporal lobe epilepsy.

axons create a recurrent excitatory circuit that plays an important role in epileptogenesis [52,55–57]. This remains a point of considerable controversy, however, because some sprouted fibers innervate inhibitory neurons. Given the widely divergent axonal terminations of most inhibitory interneurons, it is possible that even a few might be sufficient to inhibit a large number of neurons within the dentate gyrus network and reduce excitability. In addition to sprouting, other changes also occur concurrently both within and around the mossy fibers, such as alterations in proteins and receptors of the mossy fiber pathway, other pathways that sprout or retract, neuronal loss, neurogenesis, and vascular changes. A comprehensive understanding of all these changes, as well as their implications for epilepsy, is still emerging.

Another characteristic finding in the dentate gyrus of patients with temporal lobe epilepsy is dispersion of the granule cell layer [54] (Fig. 5). This refers to a decrease in the density of the granule cell layer, in which the neurons are normally tightly packed. Why granule cell dispersion occurs is currently a matter of debate. One hypothesis is that new neurons born after seizures do not migrate normally, resulting in greater dispersion and a less compact granule cell layer [58]. However, recent data indicate that granule cell dispersion may involve granule cells already in existence, rather than impaired movement of granule cells born after seizures This hypothesis proposes that granule cells move along a radial glial scaffold [59] that is altered by seizures. *RELN*, a gene that influences normal granule cell migration during development in rodents [60], could play a critical role because its expression changes in patients with intractable temporal lobe epilepsy [61]. This idea is supported by the broad granule cell layer of the *reeler* mouse, which has a mutation in *RELN* [62]. These and related studies have led to a new emphasis on seizure-induced changes in gene expression that are relevant to the development of temporal lobe epilepsy (see later discussion).

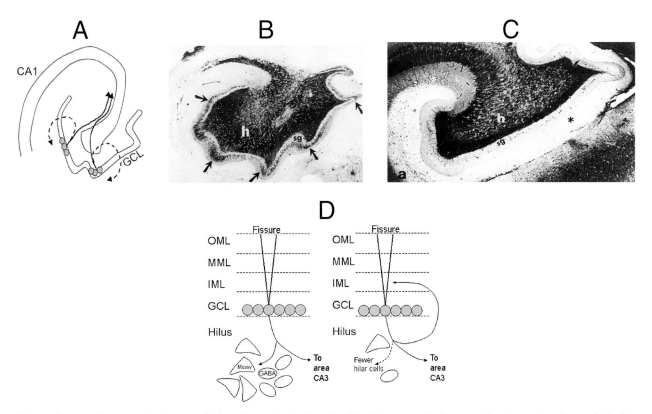

Figure 4 Mossy fiber sprouting in temporal lobe epilepsy. **A**, A schematic of the hippocampus in the same orientation as the micrographs in **B**, a Timm's stained section from an individual with intractable temporal lobe epilepsy, and **C**, a Timm's stained section from a normal primate. Sections were stained with Timm's stain for heavy metals to illustrate the mossy fiber pathway, which stains because of its high concentration of zinc. In the individual with temporal lobe epilepsy, Timm's stain is present not only in the hilus and area CA3, but also in the molecular layer, where mossy fiber axons form new collaterals (*arrows in B*). These new collaterals in the molecular layer define mossy fiber sprouting, although there is evidence from animal models of temporal lobe epilepsy that mossy fibers may also sprout within the hilar region [182]. Note that the scale for B and C is not the same. (From [53]). **D**, A schematic of the dentate gyrus illustrates the alteration in the mossy fiber pathway due to mossy fiber sprouting. *Left*, Normal granule cell axons target mossy cells ("mossy") and various GABAergic neurons ("GABA"), as well as area CA3 pyramidal cells (*not shown*). *Right*, After mossy fiber sprouting, new collaterals of granule cell axons innervate dendrites in the inner molecular layer. The new collaterals innervate granule cell dendrites as well as the dendrites of GABAergic neurons (*not shown*).

D. Implications of Hilar Cell Loss for Development of Temporal Lobe Epilepsy

Considering the variability in extent and type of pathological changes in patients with temporal lobe epilepsy, a vital question is "What is the minimal lesion?" This is important because, as will be discussed later, hippocampal sclerosis is likely to be both a consequence and a cause of temporal lobe epilepsy. Implicit in the work of Margerison and Corsellis [26] is the suggestion that the minimal lesion is endfolium sclerosis. Many studies have now examined the cells and circuitry of the endfolium and considered ways in which cell loss in this region might contribute to development of temporal lobe epilepsy. In both surgical specimens from humans with temporal lobe epilepsy and animal models, there is striking and early loss of hilar neurons [63–66], suggesting that they are among the neurons in the temporal lobe most vulnerable to injury. Possible reasons for the vulnerability of these cells come from animal studies. In rodents, localized discharges of granule cells lead to selective hilar cell damage that is coincident with, or actually precedes, an increase in excitability of granule cells both *in vivo* [63,67] and *in vitro* [68,69]. This supports the view that hilar cell loss results in development of epileptiform discharges. The specific neurons that appear to be most vulnerable are the hilar "mossy" cells and subsets of GABAergic hilar neurons, such as the GABAergic neurons that express the neuropeptide somatostatin [67,70]. Other GABAergic hilar neurons, such as those that express the calcium-binding protein parvalbumin and GABAergic neurons with cell bodies in lamina within the dentate gyrus (but outside the hilus), seem to be far less vulnerable in these particular studies. Hilar mossy cells and somatostatinergic neurons are also extremely vulnerable to hypoxia and ischemia [71,72] and traumatic brain injury [73], which are other causes of temporal lobe epilepsy.

At the same time, there are arguments against the concept that endfolium sclerosis is the most basic pathological element in temporal lobe epilepsy or in appropriate animal

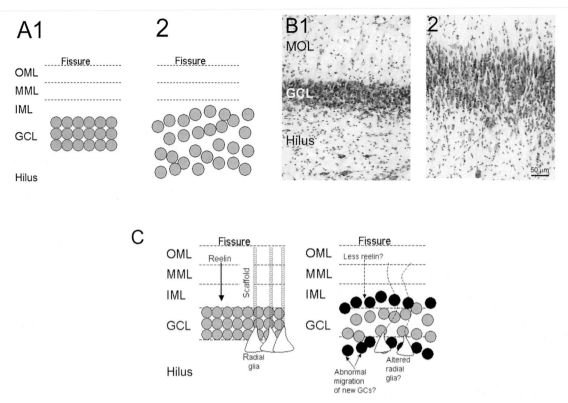

Figure 5 Granule cell dispersion in temporal lobe epilepsy. **A**, Diagrammatic illustration of granule cell dispersion. Granule cells that are normally densely packed (*1*) become dispersed (*2*). OML, Outer molecule layer; MML, middle molecule layer; GCL, granule cell layer. **B**, Nissl-stained section from a normal human dentate gyrus (*1*) and the dentate gyrus of a patient with intractable temporal lobe epilepsy showing granule cell dispersion (*2*). **C**, Hypotheses that have been proposed to explain granule cell dispersion are diagrammed. *Left*, The normal granule cell layer is compact due to many factors, including a radial glial scaffold for migrating cells and substances such as reelin that provide "stop" signals. *Right*, It has been suggested that granule cell dispersion results from a decrease in reelin and the generation of new granule cells after seizures, which migrate inappropriately due to the loss of reelin. In addition, radial glia may be altered by seizures and contribute to granule cell dispersion. Other hypotheses have also been suggested (see text).

models. From animal studies, a major stumbling point is that although hilar neuronal damage clearly alters excitability in the dentate gyrus, there is no evidence that this alone is sufficient to cause recurrent spontaneous seizures. Furthermore, there are animal models of epileptogenesis, such as kindling, that do not involve destruction of large numbers of hilar neurons. Thus, epileptogenesis may not require hilar cell loss as a *sine qua non*. On the other hand, kindling involves some loss of neurons in the hilus, and an association has been made between the degree of hilar cell loss and altered excitability [74]. The extent of hilar cell loss has also been related to other changes associated with increased excitability, such as mossy fiber sprouting [75]. In summary, the hypothesis that hilar neuronal loss, without other changes, is both necessary and sufficient to induce temporal lobe epilepsy, is attractive but by no means definitely proved.

Another issue is the mechanism underlying the selective vulnerability of hilar neurons. Indeed, this was a question that concerned the earliest investigators of hippocampal sclerosis. One of the first ideas was that the blood supply to the region was marginal, thus rendering the hilus particularly vulnerable even to a mild ischemic insult [22]. No difference in blood supply has ever been demonstrated between vulnerable and resistant areas (for example, comparing CA1 to CA2, or hilar neurons to granule cells). However, more recent studies have identified an unusually fine network of blood vessels in the hilus [76], which may contribute to the vulnerability of neurons in this area as Spielmeyer and others had postulated many years ago. One possible mechanism might be that neurotoxins are able to enter the hilar region when the blood-brain barrier is altered by seizures. At the present time, the role of this unique local vasculature is simply unclear.

A second hypothesis to explain selective vulnerability is that specific subsets of neurons are especially susceptible to excitotoxic injury. This might be due to 1) a particularly dense innervation by glutamatergic afferents, 2) weak inhibition from GABAergic neurons, or 3) other factors such as concentrations of intracellular calcium-binding proteins insufficient to protect against calcium-mediated injury during intense or prolonged excitation. This last theory received great attention when it was shown that vulnerable hilar neurons lacked calcium–binding proteins such as calbindin D28K and parvalbumin, whereas resistant GABAergic neurons and granule cells expressed either parvalbumin or calbindin [77]. Addition of calcium chelators transformed vulnerable cells to resistant ones [78]. Freund and colleagues [79] argued that this could not explain vulnerability completely because some vulnerable neurons did express calcium-binding proteins. Furthermore, calbindin itself is not necessarily protective [80]. Nonetheless, the basic idea that selective vulnerability is due to inherent cellular constituents that weaken a neuron's capacity to withstand challenge, be it a period of prolonged excitation, hypoxia, or something else, is still valid. Indeed, each of the vulnerable cell types may be susceptible for different reasons. Thus hilar neurons could be susceptible because of their limited calcium-binding capacity, a particular repertoire of ion channels that reduce their ability to repolarize, or other reasons. CA1 neurons might be selectively vulnerable because of their tight packing or propensity for NMDA receptor activation.

III. Pathophysiological Mechanisms

A. Historical Perspective

Although there is clearly an association between hippocampal sclerosis and temporal lobe epilepsy, a causal relationship has never been clearly established. Gowers [81] reflected the views of many at the time when he wrote "It is more than doubtful whether any importance is to be ascribed to the induration of the cornu Ammonis to which so much weight has been attached...," and he referred to the observed pathological changes as "trifling." The early prevailing view was that epilepsy was due to inherent causes, what might now be considered "genetic" factors.

One of the theories prevailing in the early 20th century was that vascular compromise led to hippocampal sclerosis and seizures, a view argued primarily by Spielmeyer [22] and Scholz [82]. Spielmeyer based his "vascular theory" on the notion that the CA1 region (Sommer's sector) of the hippocampus had a relatively poor vascular supply, and as a result, the hippocampus was more likely to be injured than other areas of the brain. However, this theory could not explain the particular vulnerability of the endfolium and relative sparing of CA2 and the dentate gyrus, all of which have a blood supply that is similar to that of CA1. Scholz [82] proposed that vasospasm led to hypoxia which, in turn, caused damage, but there has been little evidence for this.

Given the inability of the vascular hypothesis to explain key features of the pathological findings in temporal lobe epilepsy, an alternative proposal, pathoclisis, was developed by Vogt and Vogt [21]. *Pathoclisis* referred to inherent properties of neurons that rendered them selectively vulnerable to injury. Details of the cell types involved, the properties that rendered them vulnerable, the consequences of injury, and how their injury conferred a predisposition to seizures were not well defined. Although rarely mentioned in today's discussions, current ideas about the development of hippocampal sclerosis have aspects that resemble pathoclisis, especially the emerging appreciation over the past several decades of the robust and very different plasticity of limbic neurons after injury. We now understand that the brain regions most associated with temporal lobe epilepsy include some of the neurons and neuronal pathways in the brain that are the most prone to damage but also are the most dynamic. The granule cells of the dentate gyrus, for example, have perhaps the greatest capacity of any of the neuronal subtypes to initiate growth, alter their protein content, and change their neurotransmitter and co-transmitters. Furthermore, these changes are most apt to occur after insult or injury. It is likely that the modifications that arise in cells and circuits as a result of injury underlie epileptogenesis.

B. Current Hypotheses about the Pathogenesis of Temporal Lobe Epilepsy

Current discussions about the pathophysiology of temporal lobe epilepsy usually assume three stages of development: 1) an initial precipitating event that injures temporal lobe neurons, 2) a subsequent series of reactive or evolutionary ("plastic") changes in limbic regions, and 3) the emergence of a state of spontaneous seizures (Fig. 6). This three-stage hypothesis will be elaborated on in the next section, followed by a discussion of its weaknesses. It is important to state at the outset, however, that although this is perhaps the most commonly cited hypothesis for temporal lobe epilepsy, it is by no means accepted universally. Other hypotheses view temporal lobe epilepsy as a developmental disorder or as a consequence of a particular genetic susceptibility. These various hypotheses are not necessarily mutually exclusive.

1. The Three-Stage Hypothesis

a. Stage 1: The Initial Event In surgical and autopsy series of patients with temporal lobe epilepsy, it is common to obtain a history of an illness or injury in infancy or early childhood that potentially affects the brain [83]. Some of the more common insults include perinatal events ("birth injuries"), prolonged or complex febrile seizures, head trauma, and meningitis or encephalitis. Because these events are of such a diverse nature, they generally have been referred to simply as *initial precipitating events*. It is assumed that this initial event is critical to the development of temporal lobe epilepsy because 1) its occurrence is

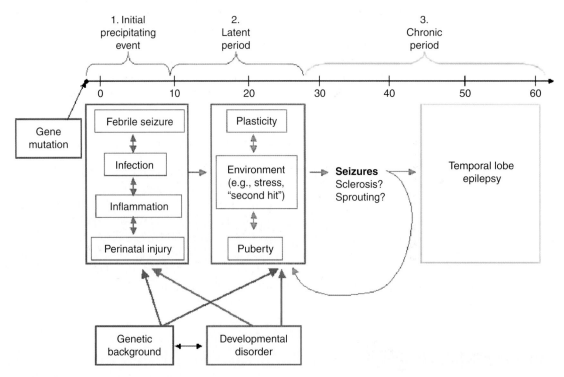

Figure 6 Diagram illustrating the three-stage hypothesis for the development of temporal lobe epilepsy. Modifying factors that influence the three stages are indicated. These include an individual's genetic background and the presence of occult developmental abnormalities.

common and 2) because it would be likely to initiate changes in the brain that might permanently alter excitability.

However, there is no proof that temporal lobe epilepsy actually begins with an identifiable precipitating event. And even if this were true, little is understood about how, and under what circumstances, such events initiate changes that lead to epilepsy. What is the critical window for such precipitating events to occur? Most occur in early childhood, but even so, over a relatively long period. Furthermore, some patients with temporal lobe epilepsy seem to sustain an initial precipitating event as adults. One example is injury sustained in war, which has been reported in older patients when they present with temporal lobe epilepsy later in life. The evidence for this, although generally circumstantial, is nonetheless most compelling when the war injury involves the temporal lobe or areas immediately adjacent to it, and there is nothing to suggest that the brain was abnormal before the injury. Thus, the initial event may not have to occur in the first years of life.

Another reason for the uncertainty about the significance of an initial precipitating event is that among individuals with similar insults, some develop temporal lobe epilepsy and others do not. One of the best examples of this has been mentioned previously: The majority of children with febrile seizures do not have epilepsy later in life. This may, of course, mean that the children with febrile seizures who develop temporal lobe epilepsy have additional predisposing factors, whereas febrile seizures have no lasting consequence in those children without these other factors. Differences in genetic susceptibility, as well as specific gene-environment interactions, may be important. Such variables would be relevant to the question of how similar insults could lead to quite different outcomes. Another consideration that may be important is the brain's reaction to the initial precipitating event, rather than the event itself. This may be germane to the observation that most initial events occur early in life. The explanation for this might be the recognition that, although plasticity is greatest at a young age, the phenomena that underlie it never cease completely. There is evidence, for example, that plasticity can be increased in adult animals, including humans, by environmental factors such as stress [84–86]. McGaugh and colleagues [86] have shown that epinephrine can facilitate hippocampal memory. However, the effects of stress are complicated. Although glucocorticoids increase the likelihood of neuronal damage in the CA1 region of the hippocampus, they preserve granule cells in the dentate gyrus [87,88]. It is likely that stress plays some role in the development of temporal lobe epilepsy because glucocorticoid receptors, which mediate stress responses, are particularly dense in CA1, on granule cells, and on hilar mossy cells [89, see also 90].

b. Stage 2: The Latent Period As suggested by the foregoing discussion, the sequence of pathophysiological changes that lead to temporal lobe epilepsy requires a period of time after the initial precipitating event. Historically, this interval was termed "the silent period," implying that little activity occurred during this time. Penfield, however, suspected that a "ripening" process was occurring that ultimately resulted in seizures (i.e., epileptogenesis) [16]. More data, mainly from animal models of temporal lobe epilepsy, have indicated that this time is anything but "silent" [91], and the term *latent period* is now used. These animal models exhibit an interval between the initial event and development of seizures that is analogous to the latent period in humans, although it is measured in weeks, not years. This briefer latent period most likely reflects the short lifespan of the rodents that are commonly used (about 2 years).

In animal models, the latent period is characterized by many morphological and functional changes, which are summarized in Table 2. They include changes in neuronal structure and function as well as alterations in glia, blood vessels, and the blood-brain barrier. In animal models, epileptiform discharges begin to appear, increasing in frequency and complexity over time. The changes listed in Table 2 are likely to be central to epileptogenesis and the development of temporal lobe epilepsy. The challenge now is to identify, from among these many changes, those that are critical to the development of seizures. This is not an easy task, partly because of the sheer number of changes that have been described, but also because both direct and indirect effects have to be considered.

c. Stage 3: The Chronic Period The chronic period is defined by spontaneous, recurrent seizures. An important point, however, is that onset of spontaneous seizures is not necessarily the end of the pathophysiological process. Although it is often assumed that epileptogenesis has been completed at this point, data from both patients and animals indicate that ongoing changes in the brain continue to occur, affecting seizure frequency and severity.

The issue of whether "seizures beget seizures," as Gowers believed, remains a matter of ongoing debate. However, there is growing evidence that temporal lobe epilepsy is a progressive disorder, at least in some patients. Continued cell loss [92] and synaptic reorganization indicate ongoing changes [93,94], as do MRI studies that show progressive volume loss not only in the hippocampus [95,96] but in the amygdala and entorhinal cortex as well [31]. Cognitive function also declines with increasing duration of temporal lobe epilepsy, consistent with progressive neuropsychological morbidity [97]. There has been growing attention to the role of inflammation in the development of epilepsy. Inflammatory processes involving proinflammatory cytokines, chemokines, cell-adhesion molecules, prostaglandins, complement factors, and other molecules may contribute to seizure-induced changes in the brain [98]. It is possible that they also contribute to the brain's response to the initial precipitating event.

In some patients, continuing changes, at least initially, seem to indicate what might be termed an ongoing "dynamic" state. Findings in both animals and humans vary: chronic seizures may or may not lead to steadily increasing brain damage or worsening seizure severity [93,94]. Seizure frequency and severity waxes and wanes, possibly due to changes in the environment, age-related alterations in hormone levels, and other factors. In humans with epilepsy, development of concurrent medical illness, especially in the elderly, can have dramatic effects on seizures.

Another challenge is that the onset of the chronic period is not necessarily easy to define in the course of the disease. In animal models of epilepsy, the problem can be particularly difficult. This is because animals may have seizures in the days immediately after the stimulus that is used to model the precipitating event, which is typically status epilepticus, induced by convulsant drugs or electrical stimulation. Although not common, such seizures can occur as soon as 24 hours after the status epilepticus. Early seizures are difficult to interpret, but are thought to be unrelated to the chronic stage. They are generally viewed as a transient manifestation of acute brain dysfunction caused by the status epilepticus. Metabolic factors may also play a role, as some animals do not resume normal food and water consumption immediately after status epilepticus. In contrast, later seizures representing the chronic stage reflect a persistent epileptic state.

2. Arguments for and against the Three-Stage Hypothesis

Although the three-stage hypothesis has substantial support from clinical and animal studies, it is not universally accepted as an explanation for temporal lobe epilepsy. One reason is that not all cases of human temporal lobe epilepsy have three well-defined stages. For example, in some patients, no initial precipitating event can be identified. Although the event may have been forgotten or overlooked, this observation has nonetheless raised doubt in the minds of some that the first stage is necessary. Also, in some animal models, recurrent seizures develop in the absence of an initial precipitating event. For example, a transgenic mouse with a mutation in one gene may develop a syndrome similar to temporal lobe epilepsy at some point in its lifetime (see Alternate Hypotheses, later). In fact, the gene mutation might be viewed as the initial precipitating event. There are also patients who develop temporal lobe epilepsy almost immediately after a precipitating event, suggesting that a latent period is also not always necessary.

The long latent period in humans is another issue. Sometimes decades pass before a patient reaches the chronic

Table 2 Changes during the Latent Period That are Associated with Epileptogenesis

Type of Change	Example	Effect	Reference(s)
1. Changes in neuronal circuits			
Axon sprouting and synaptogenesis			
↑ Glutamatergic input	Mossy fiber sprouting to granule cells	+	[53]
↓ Glutamatergic input	Deafferentation of GABAergic cells	+	[67]
↑ Inhibitory input	Sprouting of GABAergic cells to granule cells	−	[156]
↓ Inhibitory input	Loss of vulnerable GABAergic cells	+	[70,79]
Change to modulatory inputs	Cholinergic sprouting	?	[157]
Addition of new neurons	Seizure-induced neurogenesis of granule cells	+, ?	[158]
Altered gap junctions between neurons	↓ Gap junction protein expression (Connexins)	−, ?	[159]
Alteration in laminar organization/cytoarchitecture	Granule cell dispersion	+, ?	[54]
	Cell loss in CA1 leads to ↓ ephaptic interactions	−	[160]
2. Changes in synaptic transmission			
Structural changes	Active zone postsynaptic density changes	?	[161]
	Dendritic spine loss to CA3 pyramidal cells	?	[162]
Presynaptic mechanisms	Transmitter release machinery changes	?	[163]
	Axonal transport due to neurofilament changes	?	[164]
Postsynaptic mechanisms: neurotransmitters/receptors, neuromodulators			
↑↓ synthesis of transmitter/modulator	GABA in granule cells	−	[165]
De novo expression of transmitter/modulator	NPY in granule cell axons	−	[166]
↑↓ synthesis and subunit composition of receptor	GABA receptor subunit changes	+, ?	[167]
Signalling cascades	MAPK phosphorylation after seizures	?	[168]
Glial uptake and metabolism of transmitters			
Changes in glutamate or GABA transporters	GABA uptake decreases	+, −	[169]
3. Changes in voltage-gated ion channels			
Ion channel (Na^+, K^+, Ca^{2+}) density, distribution, subunit composition	CA1 changes in H channel expression, function	+, ?	[170]
Alteration in modulators of ion channels	↓ K^+Cl^- cotransporters that regulate Cl^- E_{rev}	+	[171]
4. Changes in homeostatic mechanisms (neuronal and glial)			
Energy metabolism	Persistent ATP depletion following status epilepticus	+	[172]
Intracellular regulation of cellular energy	Mitochondrial changes	?	[173]
Altered transcription, translation, DNA repair	Jerky gene regulates RNA; mutation proconvulsant	+	[174]
Ionic balance, ionic microenvironment	Defective Na^+/K^+ ATPase	+	[175]
Water balance	Redistribution of Aquaporin 4 in astrocytes	+	[176]
Acid-base balance (pH)	pH shifts that uncouple gap junctions	−	[177]
Oxidation	Increased free radicals	+	[178]
Vascular changes	Increased VEGF	−	[179]
5. Progression of damage			
Slow evolution of neuronal damage in target structures of area where damage first occurred		+	[94]
Slow degeneration of axons from damaged neurons		+, ?	
Reaction to edema and inflammation during initial insult	Cytokine production, microglial invasion	+	[180]
Gliosis		?	[181]
Transformation to "reactive" glia		?	
Influx of microglia	Release of various neuromodulators	?	
Gliosis-induced changes to extracellular matrix	↓ Extracellular diffusion/permeation of neurochemicals	+, ?	

Changes known to occur during the latent period in animal models of epilepsy are shown on the left. Specific examples are provided in the center, and on the right is the predicted pro- (+) or anti- (−) convulsant effect of the specific example. References provide reviews of research that discuss the specific example. ?, Data regarding a pro- vs anti-convulsant effect are unclear or dependent on experimental paradigm.

For the purposes of the table, the *latent period* is defined as the time between the initial insult, typically status epilepticus in a rat or mouse, until the time that spontaneous seizures become recurrent. Status epilepticus usually is induced by chemoconvulsants or electrical stimulation and is typically modified by anticonvulsant administration 1–3 hours after the onset of status to decrease the severity and duration of status. Most investigators suggest the latent period lasts for 1–3 weeks after the induction of status epilepticus, but a spontaneous seizure may occur during this so-called latent period.

stage. In the Vietnam Head Injury study, most patients developed epilepsy within 2 years of the head injury. In some patients, however, seizures did not appear until much later, sometimes 15 or more years after the trauma [99]. How could any disease process require so much time? There is no answer to this question at the present time, but it is possible that circumstances or events may delay epileptogenesis. For example, it has been suggested that the chronic period is attained only after a "second hit," such as another injury, hormonal changes (e.g., puberty), exposure to toxins, infections and inflammation, or other causes. If this second hit does not occur, epileptogenesis may never be completed. Against the requirement of a second hit is the observation that animal models develop epilepsy after a single insult and rarely, if ever, require second hits. Still, laboratory animals often require a greater degree of neuronal damage to develop recurrent seizures than is usually apparent in humans. Thus, one may speculate that unless the initial precipitating event produces a sufficient injury, a second hit is required to complete epileptogenesis. A related issue, one that is the subject of active debate at present, concerns the role of neuronal injury. Is brain damage both necessary and sufficient to produce epilepsy? If so, what is its role? An important point is that all animal models of temporal lobe epilepsy include some degree of neuronal injury, cell loss, and gliosis. Furthermore, most, if not all, of this damage is complete before spontaneously recurrent seizures begin. Although there is no question that severe, prolonged, or recurrent seizures themselves can damage the brain, some preexisting pathological abnormality that leads to changes in excitability seems to be a requirement for temporal lobe epilepsy, and probably other focal lesional epilepsies as well. Some investigators have pointed out that the kindling model of epileptogenesis does not depend on neuronal injury and cell loss, or if it does, the degree is substantially less than is observed in the animal models involving an initial period of status epilepticus, as already described. However, kindling is not, strictly speaking, an animal analog of temporal lobe epilepsy because spontaneous seizures do not typically develop. This might suggest that in some circumstances, neuronal loss can be minimal but still sufficient to result in temporal lobe epilepsy.

3. Alternate Hypotheses

a. Temporal Lobe Epilepsy as a Genetic Disorder The three-stage hypothesis discussed previously is not without flaws, and additional factors have been identified that may play a role in development of temporal lobe epilepsy. One such factor is genetics. There is considerable evidence from both clinical and laboratory studies that genes influence the risk of both seizures and epilepsy [100]. For example, epilepsy cases aggregate in families, and first-degree relatives of persons with idiopathic or cryptogenic epilepsy have a two- to four-fold increased risk of epilepsy [101].

In addition, specific gene mutations have been identified in several rare monogenic forms of human epilepsy [100]. However, it is likely that in sporadic, non-Mendelian forms of epilepsy, genetic influence on risk is most often conferred by complex disease genes, although *de novo* mutations may play a role as well [101]. Today, temporal lobe epilepsy is not usually thought of as a "genetic disorder," but a few, rare temporal lobe epilepsy syndromes are familial. Ottman et al. [102] first described a familial form of lateral temporal lobe epilepsy, which they named Autosomal Dominant Partial Epilepsy with Auditory Features (ADPEAF). This is due to a mutation in the gene *LGI1* on chromosome 10 (10q22-24) [103,104]. Initially identified as a tumor-suppressor gene, recent studies suggest that LGI-1 may play a role in matrix metalloproteinases or possibly the MAP kinase pathway [104]. Although LGI-1 mutations thus far have been specific for ADPEAF, it is not yet known how this gene alters normal limbic excitability. Other authors have described additional cases [105–107]. Most of the patients with ADPEAF have a variety of auditory auras (e.g., buzzing, clicking, or roaring) preceding complex partial and secondarily generalized seizures [102,108], although visual symptoms and aphasia may also occur [105,108]. Berkovic et al. [109] described a familial form of mesial temporal lobe epilepsy, and although some of these patients had a benign course, others developed hippocampal atrophy and required surgery [110,111].

Other genes potentially relevant to temporal lobe epilepsy include those leading to other types of partial seizures as well as those causing febrile seizures. A gene on chromosome 22 (22q11-12) has been associated with familial partial epilepsy with variable foci [112]. Persons with episodic ataxia with myokymia have partial seizures, and in mouse models of this disorder, the seizures appear to be localized to limbic areas. The mutated gene is in *KCNA1*, which encodes a delayed rectifier-type of potassium channel [113].

Several genes have been associated with seizures occurring early in life. These merit consideration because many patients with temporal lobe epilepsy have a history of febrile seizures during childhood. A subset of individuals with benign neonatal febrile seizures (BNFS) develop temporal lobe epilepsy later in life. This syndrome is caused by a mutation in either the gene *KCNQ2* on chromosome 20q13 or *KCNQ* on chromosome 8q24 [114]. These mutations cause a defect in the potassium channel that is regulated by muscarinic receptors, the so-called M-channel, which is constructed from subunits encoded by a family of five KCNQ genes. In the brain, and specifically in the hippocampus, KCNQ2/3 gene products form channels that limit the extent of electrical discharge occurring in response to a sustained input [115]. Thus, KCNQ mutations lead to increased rates of action potential discharge, which presumably causes increased excitability. However, a number of

GABAergic neurons express KCNQ channels, and therefore the increased firing rates could result in more GABA release that would enhance inhibition. More studies are required to define exactly how this mutation leads to BNFS and why some affected individuals develop temporal lobe epilepsy later in life.

Another cause of febrile seizures, generalized epilepsy with febrile seizures plus (GEFS+), has not been linked to temporal lobe epilepsy, but it is a good example of a genetic cause of febrile seizures. In GEFS+ there is a mutation in one of the subunits that form the voltage-dependent sodium channel. Mutations may arise in one of several genes encoding the major subunits (*SCN1A*, the α1 subunit, *SCN2A*, the α2 subunit) or accessory subunits (*SCN1B*, the β1 subunit), leading to delayed inactivation of sodium channels and persistent inward (depolarizing) current that increases action potential discharge. GEFS+ has also been linked to a mutation in an entirely distinct gene, *GABRG2*, on chromosome 5q34, which encodes the γ2 subunit of the $GABA_A$ receptor.

Just as human studies identified important mutations in SCN genes, experiments in rodents demonstrated the importance of persistent sodium current in regulating neuronal depolarizations [116]. Persistent sodium current may be particularly relevant to temporal lobe epilepsy because subicular neurons from tissue resected from patients with intractable temporal lobe epilepsy showed increased persistent sodium current [117]. This has led to an even greater appreciation of the role of persistent sodium current in temporal lobe epilepsy than identification of the SCN genes alone might have suggested. From an investigative perspective, it underscores the point that identified genes not specifically related to temporal lobe epilepsy might nonetheless have relevance for the condition.

An area of animal research that has been an extremely important complement to genetic studies in humans has been the analysis of mice with single gene mutations causing seizures. The availability of these mice offers the opportunity to study mutations that might be informative in patients and families with temporal lobe epilepsy. An impressive number of transgenic mice with seizures now exist (see www.jax.org or www.neuromice.org), although not all are relevant to temporal lobe epilepsy. Many of the mouse mutations are in fact associated with bilateral spike-wave discharges, even when the clinical seizures seem to be focal. Nevertheless, some of the mice suggest novel potential candidate genes to investigate in temporal lobe epilepsy. For example, the p35 mouse has a mutation in p35, which is involved in cell cycle regulation. This mouse develops seizures, and also develops some of the pathological changes observed in human temporal lobe epilepsy, such as granule cell dispersion [118,119].

Another example is the BK channel-β4 knock-out mouse, which lacks the β4 accessory subunits of the BK channel. In dentate gyrus granule cells, the BK channel acts indirectly via SK channels to control dentate granule cell firing frequency [120]. Therefore, the knock-out demonstrates increased firing frequency of dentate granule cells *in vitro* [120]. These animals also exhibit seizures of temporal origin *in vivo* [120], and the authors suggest that this is due to an increase in granule cell firing, which would weaken the normal gate function of the dentate gyrus [121]. Thus, when granule cells discharge more, the inhibitory "gate" is lowered, leading to increased activity in the hippocampus and other limbic targets.

Application of microarray technology has suggested other genes that may be informative. Microarray analysis has been performed on tissue obtained from humans during resective surgery for intractable temporal lobe epilepsy [122,123], and from animals that developed recurrent spontaneous seizures following status epilepticus as an initial precipitating event [122–125]. The animals were examined during the latent period, and again once spontaneous seizures had been established. Collectively, the microarray data indicate that a large number of genes may contribute to the different stages in the development of temporal lobe epilepsy. These genes have been classified into functional groups: genes involved in cell death or survival; genes that regulate gliosis; genes associated with inflammation; and genes that are involved in dynamic changes of neurons and networks [126]. At the very least, the findings emphasize that genetic background is an essential element of the three-stage hypothesis. For example, expression of a specific gene, combination of genes, or a gene mutation may be the critical determinant of whether a child with a febrile seizure or other cerebral insult in early life develops hippocampal sclerosis and temporal lobe epilepsy. It is clear from the categories of genes revealed by microarray analysis that the identified genes contribute to every aspect implicit in the three-stage hypothesis: cell survival/resistance, reaction to injury (gliosis, inflammation), and neuronal plasticity. That one of the major categories includes genes involved in neuronal plasticity lends support to an important aspect of the conceptual framework for the three-stage hypothesis: that temporal lobe epilepsy is, in one sense, a disorder of neuronal plasticity. It also provides a basis for concluding that the apparently divergent views of temporal lobe epilepsy as either a consequence of an initial injury followed by a cascade of subsequent changes adversely affecting excitability (the three-stage hypothesis) or a genetic disorder are matters of perspective and emphasis, not fundamental differences.

b. Temporal Lobe Epilepsy as a Neurodevelopmental Disorder A third perspective on the pathophysiology of temporal lobe epilepsy that is important to consider is the neurodevelopmental perspective, as argued persuasively by Susan Spencer in her critical review of the biological substrates of temporal lobe epilepsy [127]. In the clinical setting, a number of studies have supported an association between developmental abnormalities affecting the hippocampus and the occurrence of temporal lobe epilepsy. Lehericy and colleagues [128] at the Hôpital de la Salpetriere in Paris identified malformations of the temporal lobe in the MRI scans of 7.2% of consecutive patients with temporal lobe epilepsy. These consisted mainly of heterotopias, focal dysgenesis, and structural abnormalities of the hippocampus. Some patients exhibited more than one type of malformation. This same group of investigators subsequently described the malformative changes in the hippocampus more fully in 19 patients, reporting disturbances in neuronal migration, heterotopias, abnormal gyration, and reduced volume [129]. Four of the patients also had hippocampal sclerosis. Other investigators have described developmental abnormalities in patients with hippocampal sclerosis [130–132]. The most common finding has been cortical dysplasia, which could be either microscopic or macroscopic. Fernandez et al. [133] used quantitative visual analysis and hippocampal volumetry of MRI scans from the members of two families. About half of these family members had a history of febrile seizures, and one person from each family subsequently developed temporal lobe epilepsy. Results were compared with matched controls. All of the family members who had a history of febrile convulsions, including the two with temporal lobe epilepsy, had left hippocampi with smaller volumes. Changes in individuals who had febrile seizures but did not develop epilepsy included atrophy of the hippocampus, loss of definition of the internal structure of the hippocampus, or flattening of the body of the hippocampus. Among family members without a history of febrile seizures or epilepsy, six had similar findings in the left hippocampus. Fernandez et al. [133] interpreted their findings as indicating that these families had genetically determined, subtle developmental malformations that made the hippocampus more vulnerable to injury by febrile seizures and increased the risk of developing hippocampal sclerosis and temporal lobe epilepsy.

Sloviter et al. [134] analyzed hippocampi that had been resected en bloc for treatment of refractory temporal lobe epilepsy. They found focal areas of microscopic dysplasia in 16 of 190 (8%) consecutive patients undergoing temporal lobe resection. The most common abnormality was a localized expansion of the CA1 pyramidal cell layer and subiculum with invagination of the adjacent granule cell layer of the dentate gyrus. None of the patients had abnormalities in hippocampal shape, location, or orientation that could be detected by MRI scan, even in retrospect. Sloviter et al. also postulated that such developmental structural alterations might create a predisposing substrate for febrile convulsions, temporal lobe epilepsy, or both. In 2005, Eriksson and colleagues also found that microdysgenesis is common in patients with temporal lobe epilepsy. It is now evident that many patients with temporal lobe epilepsy and hippocampal sclerosis have such "dual pathology" [135,136]; in fact, it may be much more common than has usually been assumed.

In recent years, a number of single gene mutations have been identified as causes of several human syndromes characterized by abnormal cortical development [137–139], and it is now clear, from both human and animal studies, that abnormalities can occur during any of the phases required for normal cortical development. These include 1) the stage when neurons and glia proliferate, 2) the subsequent stage of neuronal migration, and 3) a later stage of cortical elaboration and organization [139,140]. Although the association of developmental cortical malformations and epilepsy is well known, the underlying epileptogenic mechanisms are not fully understood. Nonetheless, in recent years animal models and, increasingly, additional observations in humans, are providing insights into possible ways that anomalies of cortical development promote epileptogenesis [119,141]. The different animal models all have advantages and disadvantages, and no single model is suitable for studying all aspects of human developmental malformations. Current animal models of cortical developmental malformations include genetic mutations (spontaneous, knock-outs/transgenics), cortical irradiation, cortical injury (undercutting, freezing), and exposure to teratogens.

Many of the consequences of cortical malformations are on adjacent cortex and neuronal networks, and both direct and indirect effects are involved. It is beyond the scope of this review to provide a comprehensive discussion of the effects cortical malformations can have on neuronal excitability, but several examples from both animal and human studies are useful in illustrating the range of functional changes that can occur. Specific alterations in inhibitory function have been identified, and they may occur by multiple mechanisms, including reduced GABAergic neuronal number after gamma-irradiation at early stages in development [142], as well as changes in $GABA_A$ receptor subunit composition, which has been reported in neocortical microgyria that are experimentally induced by neonatal freeze lesions [143]. Alterations in glutamatergic transmission have also been identified; again, these may be due to changes in glutamatergic neuronal number and distribution as well as changes in glutamate receptors. For example, in dysplastic human neocortex, the NR2B subunit of the NMDA receptor was increased [144]. The wide range of potential changes (for review, see reference 119) creates a challenge to potential therapeutic intervention.

The foregoing observations are relevant to the three-stage hypothesis because developmental abnormalities, whether within or adjacent to the temporal lobe, may create additional susceptibility for the development of temporal lobe epilepsy in individuals who sustain an initial precipitating event. One argument against this perspective is that some animals with specific defects in genes that control development have malformations and epilepsy but lack an early initial event (e.g., febrile seizure). In such cases, however, the initial precipitating event may be the gene mutation if the period of early gestation is included within the framework. Although developmental abnormalities by themselves rarely produce the typical syndrome of temporal lobe epilepsy, the evidence is accumulating that they may increase susceptibility to other factors more directly causative of hippocampal sclerosis and temporal lobe epilepsy.

IV. Summary and Future Directions: Developing a Comprehensive Hypothesis for the Pathology and Pathophysiology of Temporal Lobe Epilepsy

As evidenced from the foregoing discussion, views of temporal lobe epilepsy have evolved substantially since the beginning of the modern era with the pioneering work of Jackson, Charcot, and Gowers. One possible theory for the development of temporal lobe epilepsy, based on the three-stage hypothesis, is shown schematically in Figure 6. The key components are an initial precipitating event such as febrile seizure, infection, or status epilepticus. The brain's response to this is influenced by other factors, including the individual's genetic background and the presence of associated developmental malformations, which, by themselves, may be trivial and asymptomatic (microdysgenesis). Details of the latent period are currently unknown but involve dynamic functional changes (i.e., plasticity) within the limbic system and may involve a "second hit." Stress and hormonal changes (exemplified by puberty) may also contribute. Opportunities for prevention of epilepsy will depend on delineation of specific epileptogenic mechanisms active in the latent period and identification of possible therapeutic targets. Clinical seizures manifest when intrinsic inhibitory mechanisms are no longer sufficient to constrain the level of excitability below a threshold critical for generation of seizures. At this point—which clinically is typically during the second decade but extends into the 20s and 30s, in some cases appearing even later—a diagnosis of temporal lobe epilepsy is made and symptomatic treatment is started. The frequency and intensity of seizures, whether they secondarily generalize, the brain's response to them, and response to treatment are influenced by a combination of physiological or environmental factors (e.g., sleep, alcohol, medications, stress, hormones) as well as the individual's genetic background.

Of course, not all cases of temporal lobe epilepsy are due to mesial temporal sclerosis. Tumors and vascular malformations can cause temporal lobe epilepsy, although sometimes there is dual pathology with hippocampal sclerosis occurring as well. Even in cases in which the MRI scan is normal and neuronal loss is undetectable by current noninvasive means, the temporal lobe may well be abnormal in its microscopic structure, neuroglia, or expression of proteins that regulate brain excitability.

It may be no coincidence that neuronal plasticity, a hallmark of neurons in the limbic system, is critical to this process. Perhaps the remarkable ability of neurons in the temporal lobe to change dynamically has become a double-edged sword. It seems that limbic system neurons and networks were designed to optimize the capacity for change in response to new experiences and the challenge of different environmental cues. That ability, however, may have come with a risk: the possibility of developing temporal lobe epilepsy.

References

1. Papez, J. W. (1937). A proposed mechanism of emotion. *Arch. Neurol. Psychiat.* **38**, 725–743.
2. MacLean, P. D. (1952). Some psychiatric implications of physiological studies of fronto-temporal portion of limbic system (visceral brain). *Electroencephalogr. Clin. Neurophysiol.* **4**, 407–418.
3. Hammond, W. A. (1871). "Diseases of the Nervous System." D. Appleton and Company, New York.
4. Morel, B. A. (1860). D'une forme de delire suite d'une surexcitation nerveuse se rattachant a une variete non encore decrite d'epilepsie (Epilepsie larvee). *Gaz. Hebdon. Med. Chir.* **7**, 773–841.
5. VanGieson, I. (1924). A case of psychomotor epilepsy. *In:* "Semi-Centennial Volume of the American Neurological Association (1875–1924)." Boyd Printing, Albany, NY.
6. Jackson, J. H. (1988). On a particular variety of epilepsy ("intellectual aura"), one case with symptoms of organic brain disease. *Brain* **11**, 179–207.
7. Jackson, J. H. (1899). Epileptic attacks in a patient who had symptoms pointing to gross organic disease of the right temporo-sphenoidal lobe. *Brain* **22**, 534–549.
8. Kennedy, F. (1911). The symptomatology of temporo-sphenoidal tumors. *Arch. Int. Med.* **8**, 317–351.
9. Gibbs, E. L., and Gibbs, F. A. (1947). Diagnostic and localizing value of electronencephalographic studies in sleep. *Publ. Assoc. Res. Nerv. Ment. Dis.* **26**, 366–376.
10. Gibbs, E. L., Fuster, B., and Gibbs, F. A. (1948). Peculiar low temporal localization of sleep-induced seizure discharges of psychomotor epilepsy. *Arch. Neurol.* **60**, 95–97.
11. Gibbs, F. A., Gibbs, E. L., and Lennox, W. G. (1937). Epilepsy: a paroxysmal cerebral dysrhythmia. *Brain* **60**, 377–388.
12. Jasper, H. H., Pertuiset, B., and Flanigin, H. (1951). EEG and cortical electrograms in patients with temporal lobe seizures. *Arch. Neurol. Psychiat.*, **65**, 272–290.
13. Brazier, M. A. B. (1956). Depth recordings from the amygdaloid region in patients with temporal lobe epilepsy. *EEG Clin. Neurophysiol.* **8**, 532–533.
14. Kellaway, P. (1956). Depth recording in focal epilepsy. *EEG Clin. Neurophysiol.* **8**, 527–728.
15. Ajmone-Marsan, C., and Van Buren, J. M. (1958). Epileptiform activity in cortical and subcortical structures in the temporal lobe of man. *In:* "Temporal Lobe Epilepsy" (M. Baldwin et al., eds.), pp. 78–108. C. C. Thomas, Springfield, IL.
16. Penfield, W., and Jasper, H. (1954). "Epilepsy and the Functional Anatomy of the Human Brain." Little, Brown & Co., Boston.

17. Bouchet, C., and Cazauvieilh, M. (1825). De l'epilepse consideree dans ses raports avec l'alienation mentale. Recherche sur la nature et le siege de ces deux maladies. *Arch. Gen. Med.* **9**, 510–542.
18. Sommer, W. (1880). Erkrankung des Ammonshornes als aetiolgisches Moment der Epilepsie. *Arch. Psychiat. Nervendrankh.* **10**, 631–675.
19. Bratz, E. (1899). Ammonshornebefunde bei epileptischen. *Arch. Psychiat. Nervenkr.* **31**, 820–835.
20. Vogt, O. (1925). Der begriff der pathoklise. *J. Psychol. Neurol. (Leipzig)* **31**, 245–255.
21. Vogt, C., and Vogt, O. (1937). Sitz und wesen der krankheiten im lichte der topistischen hirnforschung und des varienens der tiere. I. Teil: befunde der topistischen hirnforschung als beitrag zur lehre vom krankheitssitz. *J. Psychol. Neurol. (Leipzig)* **47**, 237–457.
22. Spielmeyer, W. (1927). Die pathogenese des epileptischen krampfes. Histopathologischer teil., *Ztschr. F. D. Gest. Neurol. U. Psychiat.* **209**, 501–520.
23. Earle, K. M., Baldwin, M., and Penfield, W. (1953). Incisural sclerosis and temporal lobe seizures produced by hippocampal herniation at birth. *Arch. Neurol. Psychiat.* **69**, 27–42.
24. Ounsted, C., Lindsay, J., and Norman, R. (1966). Biological factors in temporal lobe epilepsy. *In:* "Clinics in Developmental Medicine," vol. 22. William Heinemann Medical, London.
25. Veith, G. (1970). Anatomische studie uber die Ammonshornsklerose im epileptikergehiern. *Dt. Z. Herv. Heilk.* **197**, 293–314.
26. Margerison, J. H., and Corsellis, J. A. N. (1966). Epilepsy and the temporal lobes: a clinical, electroencephalographic, and neuropathological study of the brain in epilepsy, with particular reference to the temporal lobes. *Brain* **89**, 499–530.
27. Bruton, C. J. (1988). The neuropathology of temporal lobe epilepsy. *In:* "Maudsley Monograph No. 31." Oxford University Press, Oxford.
28. Cavanagh, J. B., and Meyer, A. (1956). Aetiological aspects of Ammon's horn sclerosis associated with temporal lobe epilepsy. *Brit. Med. J.* **2**, 1403–1407.
29. Du, F., Whetsell, W. O., Abou-Khalil, B., Blumenkopf, B., Lothman, E. W., and Schwarcz, R. (1993). Preferential neuronal loss in layer III of the entorhinal cortex in patients with temporal lobe epilepsy. *Epilepsy Res.* **16**, 223–244.
30. Bernasconi, N., Bernasconi, A., Caramanas, Z., Antel, S. B., Andermann, F., and Arnold, D. L. (2003). Mesial temporal damage in temporal lobe epilepsy: a volumetric MRI study of the hippocampus, amygdala and parahippocampal region. *Brain* **126**, 462–469.
31. Bernasconi, N., Natsume, J., and Bernasconi, A. (2005). Progression in temporal lobe epilepsy: differential atrophy in mesial temporal lobe structures. *Neurology* **65**, 223–228.
32. Falconer, M. A., Serafetinides, E. A., and Corsellis, J. A. N. (1964). Etiology and pathogenesis of temporal lobe epilepsy. *Arch. Neurol.* **10**, 233–248.
33. Falconer, M. A. (1974). Mesial temporal (Ammon's horn) sclerosis as a common cause of epilepsy. Aetiology, treatment, and prevention. *Lancet* **ii**, 767–770.
34. French, J. A., Williamson, P. D., Thadani, V. M., Darcey, T. M., Mattson, R. H., Spencer, S. S., and Spencer, D. D. (1993). Characteristics of medial temporal lobe epilepsy: I. Results of history and physical examination. *Ann. Neurol.* **34**, 774–780.
35. Gloor, P. (1997). "The Temporal Lobe and Limbic System." Oxford University Press, New York.
36. Sagar, H. J., and Oxbury, J. M. (1987). Hippocampal neuron loss in temporal lobe epilepsy; correlation with early childhood convulsions. *Ann. Neurol.* **22**, 334–340.
37. Kim, J. H., Guimaraes, P. O., Shen, M. Y., Masukawa, L. M., and Spencer, D. D. (1990). Hippocampal neuronal density in temporal lobe epilepsy with and without gliomas. *Acta. Neuropathol. (Berl.)* **80**, 41–45.
38. Chabardes, S., Kahane, P., Monotti, L., Tassi, L., Grand, S., Hoffmann, D., and Benabid, A. L. (2005). The temporopolar cortex plays a pivotal role in temporal lobe seizures. *Brain* **128**, 1818–1831.
39. Hauser, W. A. (1998). Incidence and prevalence. *In:* "Epilepsy: A Comprehensive Textbook" (J. Engel et al., eds)., pp. 47–57. Lippincott-Raven, Philadelphia.
40. Van Landingham, K. E., Heinz, E. R., Cavazos, J. E., and Lewis, D. V. (1998). Magnetic resonance imaging evidence of hippocampal injury after prolonged febrile convulsions. *Ann. Neurol.* **43**, 413–426.
41. Scott, R. C., Gadian, D. G., King, M. D., Chong, W. K., Cox, T. C., Neville, B. G., and Connelly, A. (2002). Magnetic resonance imaging findings within 5 days of status epilepticus in childhood. *Brain* **125**, 1951–1959.
42. Lewis, D. V., and Barboriak, D. P. (2002). Do prolonged febrile seizures produce medial temporal sclerosis? Hypotheses, MRI evidence, and unanswered questions. *Prog. Brain Res.* **135**, 263–278.
43. Feindel, W., Penfield, W., and Jasper, H. (1952). Localization of epileptic discharge in temporal lobe automatism. *Trans. Am. Neurol. Assoc.* **77**, 14–17.
44. Feindel, W., and Penfield, W. (1954). Localization of discharge in temporal lobe automatism. *Arch. Neurol. Psychiat.* **72**, 605–630.
45. Miller, L. A., McLachlan, R. S., Bouwer, M. S., Hudson, L. P., and Munoz, D. G. (1994). Amygdalar sclerosis: preoperative indicators and outcome after temporal lobectomy. *J. Neurol. Neurosurg. Psychiat.* **57**, 1099–1105.
46. Pitkanen, A., Tuunanen, J., Kalvianen, R., Partanen, K., and Salmenpera, T. (1998). Amygdala damage in experimental and human temporal lobe epilepsy. *Epilepsy Res.* **32**, 233–253.
47. Schwarcz, R., Scharfman, H. E., and Bertram, E. H. (2001). Temporal lobe epilepsy: renewed emphasis on extrahippocampal areas. *In:* "ACNP 5th Generation of Progress," (K. L. Davis, et al., eds.), pp. 1843–1855. Lippincott Williams and Wilkins, Philadelphia.
48. Gastaut, H. (1956). Colloque sur les problemes d'anatomie normale et pathologique poses par les dechares epileptiques. *In:* "Editions Acta Medica Belgica," pp. 5–20. Brussels.
49. Niessen, H. G., Angenstein, F., Vielhaber, S., Frisch, C., Kudin, A., Elger, C. E., Heinze, H. J., Scheich, H., and Kunz, W. S. (2005). Volumetric magnetic resonance imaging of functionally relevant structural alterations in chronic epilepsy after pilocarpine-induced status epilepticus in rats. *Epilepsia* **46**, 1021–1026.
50. Bertram, E. H., and Scott, C. (2000). The pathological substrate of limbic epilepsy: neuronal loss in the medial dorsal thalamic nucleus as the consistent change. *Epilepsia* **41**, S3–S8.
51. Menini, C., Meldrum, B. S., Riche, D., Silva-Comte, C., and Stutzmann, J. M. (1980). Sustained limbic seizures induced by intraamygdaloid kainic acid in the baboon: symptomatology and neuropathological consequences. *Ann. Neurol.* **8**, 501–509.
52. Tauck, D. L., and Nadler, J. V. (1985). Evidence of functional mossy fiber sprouting in hippocampal formation of kainic acid-treated rats. *J. Neurosci.* **5**, 1016–1022.
53. Sutula, T., Cascino, G., Cavazos, J., Parada, I., and Ramirez, L. (1989). Mossy fiber synaptic reorganization in the epileptic human temporal lobe. *Ann. Neurol.* **26**, 321–330.
54. Houser, C. R., Miyashiro, J. E., Swartz, B. E., Walsh, G. O., Rich, J. R., and Delgado-Escueta, A. V. (1990). Altered patterns of dynorphin immunoreactivity suggest mossy fiber reorganization in human hippocampal epilepsy. *J. Neurosci.* **10**, 267–282.
55. Sutula, T. P., Golarai, G., and Cavazos, J. (1992). Assessing the functional significance of mossy fiber sprouting. *Epilepsy Res. Suppl.* **7**, 251–259.
56. Dudek, F. E., Obenaus, A., Schweitzer, J. S., and Wuarin, J. P. (1994). Functional significance of hippocampal plasticity in epileptic brain: electrophysiological changes of the dentate granule cells associated with mossy fiber sprouting. *Hippocampus* **4**, 259–265.
57. Nadler, J. V. (2003). The recurrent mossy fiber pathway of the epileptic brain. *Neurochem. Res.* **28**, 1649–1658.

58. Jessberger, S., Romer, B., Babu, H., and Kempermann, G. (2005). Seizures induce proliferation and dispersion of doublecortin-positive hippocampal progenitor cells. *Exp. Neurol.* **196**, 342–351.
59. Frotscher, M., Haas, C. A., and Forster, E. (2003). Reelin controls granule cell migration in the dentate gyrus by acting on the radial glial scaffold. *Cereb. Cortex* **13**, 634–640.
60. Zhao, S., Chai, X., Forster, E., and Frotscher, M. (2004). Reelin is a positional signal for the lamination of dentate granule cells. *Development* **131**, 5117–5125.
61. Haas, C. A., Dudeck, O., Kirsch, M., Huszka, C., Kann, G., Pollak, S., Zentner, J., and Frotscher, M. (2002). Role for reelin in the development of granule cell dispersion in temporal lobe epilepsy. *J. Neurosci.* **22**, 5797–5802.
62. Drakew, A., Deller, T., Heimrich, B., Gebhardt, C., Del Turco, D., Tielsch, A., Forster, E., Herz, J., and Frotscher, M. (2002). Dentate granule cells in reeler mutants and VLDLR and ApoER2 knockout mice. *Exp. Neurol.* **176**, 12–24.
63. Sloviter, R. S. (1983). "Epileptic" brain damage in rats induced by sustained electrical stimulation of the perforant path. I. Acute electrophysiological and light microscopic studies. *Brain Res. Bull.* **10**, 675–697.
64. Meldrum, B. S. (1986). Cell damage in epilepsy and the role of calcium in cytotoxicity. *Adv. Neurol.* **44**, 849–855.
65. de Lanerolle, N. C., Kim, J. H., Robbins, R. J., and Spencer, D. D. (1989). Hippocampal interneuron loss and plasticity in human temporal lobe epilepsy. *Brain Res.* **495**, 387–395.
66. Scharfman, H. E. (1999). The role of nonprincipal cells in dentate gyrus excitability and its relevance to animal models of epilepsy and temporal lobe epilepsy. *In:* "Basic Mechanisms of the Epilepsies: Molecular and Cellular Approaches," (A. V. Delgado-Esqueta, et al., eds.), 3rd edition, pp. 805–820. Lippincott-Raven, New York.
67. Sloviter, R. S. (1991). Permanently altered hippocampal structure, excitability, and inhibition after experimental status epilepticus in the rat: the "dormant basket cell" hypothesis and its possible relevance to temporal lobe epilepsy. *Hippocampus* **1**, 41–66.
68. Scharfman, H. E., and Schwartzkroin, P. A. (1990). Consequences of prolonged afferent stimulation of the rat fascia dentata: epileptiform activity in area CA3 of hippocampus. *Neuroscience* **35**, 505–517.
69. Scharfman, H. E., and Schwartzkroin, P. A. (1990). Responses of cells of the rat fascia dentata to prolonged stimulation of the perforant path: sensitivity of hilar cells and changes in granule cell excitability. *Neuroscience* **35**, 491–504.
70. Houser, C. R., and Esclapez, M. (1996). Vulnerability and plasticity of the GABA system in the pilocarpine model of spontaneous recurrent seizures. *Epilepsy Res.* **26**, 207–218.
71. Crain, B. J., Westerkam, W. D., Harrison, A. H., and Nadler, J. V. (1988). Selective neuronal death after transient forebrain ischemia in the Mongolian gerbil. A silver impregnation study. *Neuroscience* **27**, 387–402.
72. Bering, R., Draguhn, A., Diemer, N. H., and Johansen, F. F. (1997). Ischemia changes the coexpression of somatostatin and neuropeptide Y in hippocampal interneurons. *Exp. Brain Res.* **115**, 423–429.
73. Lowenstein, D. H., Thomas, M. J., Smith, D. H., and McIntosh, T. K. (1992). Selective vulnerability of dentate hilar neurons following traumatic brain injury: a potential mechanistic link between head trauma and disorders of the hippocampus. *J. Neurosci.* **12**, 4846–4853.
74. Cavazos, J. E., Das, I., and Sutula, T. P. (1994). Neuronal loss induced in limbic pathways by kindling: evidence for induction of hippocampal sclerosis by repeated brief seizures. *J. Neurosci.* **14**, 106–121.
75. Masukawa, L. M., O'Connor, W. M., Lynott, J., Burdette, L. J., Uruno, K., McGonigle, P., and O'Connor, M. J. (1995). Longitudinal variation in cell density and mossy fiber reorganization in the dentate gyrus from temporal lobe epileptic patients. *Brain Res.* **678**, 65–75.
76. Palmer, T. D., Willhoite, A. R., and Gage, F. H. (2000). Vascular niche for adult hippocampal neurogenesis. *J. Comp. Neurol.* **425**, 479–494.
77. Sloviter, R. S. (1989). Calcium-binding protein (calbindin-D28k) and parvalbumin immunocytochemistry: localization in the rat hippocampus with specific reference to the selective vulnerability of hippocampal neurons to seizure activity. *J. Comp. Neurol.* **280**, 183–196.
78. Scharfman, H. E., and Schwartzkroin, P. A. (1989). Protection of dentate hilar cells from prolonged stimulation by intracellular calcium chelation. *Science* **246**, 257–260.
79. Freund, T. F., Ylinen, A., Miettinen, R., Pitkanen, A., Lahtinen, H., Baimbridge, K. G., and Riekkinen, P. J. (1992). Pattern of neuronal death in the rat hippocampus after status epilepticus. Relationship to calcium binding protein content and ischemic vulnerability. *Brain Res. Bull.* **28**, 27–38.
80. Airaksinen, L., Virkala, J., Aarnisalo, A., Meyer, M., Ylikoski, J., and Airaksinen, M. S. (2000). Lack of calbindin-D28k does not affect hearing level or survival of hair cells in acoustic trauma. *ORL J. Otorhinolaryngol. Relat. Spec.* **62**, 9–12.
81. Gowers, W. R. (1881). "Epilepsy and other chronic convulsive diseases. Their causes, symptoms and treatment." William Wood and Co., New York.
82. Scholz, W. (1993). Uber die entstehung des hirnbefundes bei der epilepsie. *Z. Ges. Neurol. Psychiat.* **145**, 471.
83. Baram, T., and Shinnar, S. (2002). "Febrile Seizures." Academic Press, New York.
84. Radley, J. J., and Morrison, J. H. (2005). Repeated stress and structural plasticity in the brain. *Aging Res. Rev.* **4**, 271–287.
85. Berger, S., Wolfer, D. P., Selbach, O., Alter, H., Erdmann, G., Reichardt, H. M., Chepkova, A. N., Welzl, H., Haas, H. L., Lipp, H. P., and Schutz, G. (2005). Loss of the limbic mineralocorticoid receptor impairs behavioral plasticity. *Proc. Natl. Acad. Sci. USA* **103**, 195–200.
86. McGaugh, J. L., and Roozendaal, B. (2002). Role of adrenal stress hormones in forming lasting memories in the brain. *Curr. Opin. Neurobiol.* **12**, 205–210.
87. Sapolsky, R. M. (1990). Glucocorticoids, hippocampal damage and the glutamatergic synapse. *Prog. Brain Res.* **86**, 13–23.
88. Sloviter, R. S., Valiquette, G., Abrams, G. M., Ronk, E. C., Sollas, A. L., Paul, L. A., and Neubort, S. (1989). Selective loss of hippocampal granule cells in the mature rat brain after adrenalectomy. *Science* **243**, 535–538.
89. Patel, A., and Bulloch, K. (2003). Type II glucocorticoid receptor immunoreactivity in the mossy cells of the rat and the mouse hippocampus. *Hippocampus* **13**, 59–66.
90. Brunson, K. L., Chen, Y., Avishai-Eliner, S., and Baram, T. Z. (2003). Stress and the developing hippocampus: a double-edged sword? *Mol. Neurobiol.* **27**, 121–136.
91. White A. M., Clark S., Williams, P. A., et al. (2005). Relationship between frequency of interictal spikes and spontaneous recurrent seizures in an animal model of temporal lobe epilepsy. *Epilepsia* **45(S7)**, 51–52.
92. Mathern, G. W., Babb, T. L., Vickrey, B. G., Melendez, M., and Pretorius, J. K. (1995). The clinical-pathogenic mechanisms of hippocampal neuron loss and surgical outcomes in temporal lobe epilepsy. *Brain* **118**, 105–118.
93. Pitkanen, A., and Sutula, T. P. (2002). Is epilepsy a progressive disorder? Prospects for new therapeutic approaches in temporal lobe epilepsy. *Lancet Neurol.* **1**, 173–181.
94. Sutula, T. P. (2004). Mechanisms of epilepsy progression: current theories and perspectives from neuroplasticity in adulthood and development. *Epilepsy Res.* **60**, 161–171.
95. Tasch, E., Cendes, F., Li, L. M., Dubeau, F., Andermann, F., and Arnold, D. L. (1999). Neuroimaging evidence of progressive neuronal

loss and dysfunction in temporal lobe epilepsy. *Ann. Neurol.* **45**, 568–576.
96. Fuerst, D., Shah, J., Shah, A., and Watson, C. (2003). Hippocampal sclerosis is a progressive disorder: a longitudinal volumetric MRI study. *Ann. Neurol.* **53**, 413–416.
97. Ovegbile, T. O., Dow, C., Jones, J., Bell, B., Rutecki, P., Sheth, R., Seidenberg, M., and Hermann, B. P. (2004). The nature and course of neuropsychological morbidity in chronic temporal lobe epilepsy. *Neurology* **62**, 1736–1742.
98. Vezzani, A. (2005). Inflammation and epilepsy. *Epilepsy Curr.* **5**, 1–6.
99. Salazar, A. M., Jabbari, B., Vance, S. C., Grafman, J., Amin, D., and Dillon, J. D. (1985). Epilepsy after penetrating head injury. I. Clinical correlates: a report of the Vietnam Head Injury Study. *Neurology* **35**, 1406–1414.
100. Meisler, M. H., Kearney, J., Ottman, R., and Escayg, A. (2001). Identification of epilepsy genes in human and mouse. *Annu. Rev. Genet.* **35**, 567–588.
101. Ottman, R. (2005). Analysis of genetically complex epilepsies. *Epilepsia* **46**, 7–14.
102. Ottman, R., Risch, N., Hauser, W. A., Pedley, T. A., Lee, J. H., Barker-Cummings, C., Lustenberger, A., Nagle, K. J., Lee, K. S., and Scheuer, M. L. (1995). Localization of a gene for a partial epilepsy to chromosome 10q. *Nat. Genet.* **10**, 56–60.
103. Kalachikov, S., Evgrafov, O., Ross, B., Winawer, M., Barker-Cummings, C., Martinelli Boneschi, F., Choi, C., Morozov, P., Das, K., Teplitskaya, E., Yu, A., Cayanis, E., Penchaszadeh, G., Kottmann, A. H., Pedley, T. A., Hauser, W. A., Ottman, R., and Gilliam, T. C. (2002). Mutations in LGI1 cause autosomal-dominant partial epilepsy with auditory features. *Nat. Genet.* **30**, 335–341.
104. Kunapuli, P., Kasyapa, C. S., Hawthorn, L., and Cowell, J. K. (2004). LGI1, a putative tumor metastasis suppressor gene, controls in vitro invasiveness and expression of matrix metalloproteinases in glioma cells through the ERK1/2 pathway. *J. Biol. Chem.* **279**, 23151–23157.
105. Poza, J. J., Saenz, A., Martinez-Gil, A., Cheron, N., Cobo, A. M., Urtasun, M., Marti-Masso, J. F., Grid, D., Beckmann, J. S., Prud'homme, J. F., and Lopez de Munain, A. (1999). Autosomal dominant lateral temporal epilepsy: clinical and genetic study of a large Basque pedigree linked to chromosome 10q. *Ann. Neurol.* **45**, 182–188.
106. Brodtkorb, E., Gu, W., Nakken, K. O., Fischer, C., and Steinlein, O. K. (2002). Familial temporal lobe epilepsy with aphasic seizures and linkage to chromosome 10q22-q24. *Epilepsia* **43**, 228–235.
107. Cendes, F., Kobayashi, E., and Lopes-Cendes, I. (2005). Familial temporal lobe epilepsy with auditory features. *Epilepsia* **46**, 59–60.
108. Ottman, R., Winawer, M. R., Kalachikov, S., Barker-Cummings, C., Gilliam, T. C., Pedley, T. A., and Hauser, W. A. (2004). LGI1 mutations in autosomal dominant partial epilepsy with auditory features. *Neurology* **62**, 1120–1126.
109. Berkovic, S. F., McIntosh, A., Howell, R. A., Mitchell, A., Sheffield, L. J., and Hopper, J. L. (1996). Familial temporal lobe epilepsy: a common disorder identified in twins. *Ann. Neurol.* **40**, 227–235.
110. Cendes, F., Lopes-Cendes, I., Andermann, E., and Andermann, F. (1998). Familial temporal lobe epilepsy: a clinically heterogeneous syndrome. *Neurology* **50**, 554–557.
111. Kobayashi, E., Sousa, S. C., Lopes-Cendes, I., Sousa, S. C., Guerreiro, M. M., and Cendes, F. (2001). Seizure outcome and hippocampal atrophy in familial mesial temporal lobe epilepsy. *Neurology* **56**, 166–172.
112. Xiong, L., Labuda, M., Li, D. S., Hudson, T. J., Desbiens, R., Patry, G., Verret, S., Langevin, P., Mercho, S., Seni, M. H., Scheffer, I., Dubeau, F., Berkovic, S. F., Andermann, F., Andermann, E., and Pandolfo, M. (1999). Mapping of a gene determining familial partial epilepsy with variable foci to chromosome 22q11-q12. *Am. J. Hum. Genet.* **65**, 1698–1710.
113. Cooper, E. C., and Jan, L. Y. (2003). M-channels: neurological diseases, neuromodulation, and drug development. *Arch. Neurol.* **60**, 496–500.
114. Singh, N. A., Charlier, C., Stauffer, D., DuPont, B. R., Leach, R. J., Melis, R., Ronen, G. M., Bjerre, I., Quattlebaum, T., Murphy, J. V., McHarg, M. L., Gagnon, D., Rosales, T. O., Peiffer, A., Anderson, V. E., and Leppert, M. (1998). A novel potassium channel gene, KCNQ2, is mutated in an inherited epilepsy of newborns. *Nat. Genet.* **18**, 25–29.
115. Rogawski, M. A. (2000). KCNQ2/KCNQ3 K+ channels and the molecular pathogenesis of epilepsy: implications for therapy. *Trends Neurosci.* **23**, 393–398.
116. Crill, W. E. (1996). Persistent sodium current in mammalian central neurons. *Annu. Rev. Physiol.* **58**, 349–362.
117. Vreugdenhil, M., Hoogland, G., vanVeelen, C. W., and Wadman, W. J. (2004). Persistent sodium current in subicular neurons isolated from patients with temporal lobe epilepsy. *Eur. J. Neurosci.* **19**, 2769–2778.
118. Patel, L. S., Wenzel, H. J., and Schwartzkroin, P. A. (2004). Physiological and morphological characterization of dentate granule cells in the p35 knock-out mouse hippocampus: evidence for an epileptic circuit. *J. Neurosci.* **24**, 9005–9014.
119. Schwartzkroin, P. A., Roper, S. N., and Wenzel, H. J. (2004). Cortical dysplasia and epilepsy: animal models. *Adv. Exp. Med. Biol.* **548**, 145–174.
120. Brenner, R., Chen, Q. H., Vilaythong, A., Toney, G. M., Noebels, J. L., and Aldrich, R. W. (2005). BK channel beta4 subunit reduces dentate gyrus excitability and protects against temporal lobe seizures. *Nat. Neurosci.* **8**, 1752–1759.
121. Lothman, E. W., Stringer, J. H., and Bertram, E. H. (1992). The dentate gyrus as a control point for seizures in the hippocampus and beyond. *Epilepsy Res. Suppl.* **7**, 301–313.
122. Becker, A. J., Wiestler, O. D., and Blumcke, I. (2002). Functional genomics in experimental and human temporal lobe epilepsy: powerful new tools to identify molecular disease mechanisms of hippocampal damage. *Prog. Brain Res.* **135**, 161–173.
123. Lukasiuk, K., Kontula, L., and Pitkanen, A. (2003). cDNA profiling of epileptogenesis in the rat brain. *Eur. J. Neurosci.* **17**, 271–279.
124. Tang, Y., Lu, A., Aronow, B. J., Wagner, K. R., and Sharp, F. R. (2002). Genomic responses of the brain to ischemic stroke, intracerebral haemorrhage, kainate seizures, hypoglycemia, and hypoxia. *Eur. J. Neurosci.* **15**, 1937–1952.
125. Elliott, R. C., Miles, M. F., and Lowenstein, D. H. (2003). Overlapping microarray profiles of dentate gyrus gene expression during development and epilepsy-associated neurogenesis and axon outgrowth. *J. Neurosci.* **23**, 2218–2227.
126. Lukasiuk, K., and Pitkanen, A. (2004). Large-scale analysis of gene expression in epilepsy research: is synthesis already possible? *Neurochem. Res.* **29**, 1169–1178.
127. Spencer, S. S. (1998). Substrates of localization-related epilepsies: biologic implications of localizing findings in humans. *Epilepsia* **39**, 114–123.
128. Lehericy, S., Dormont, D., Semah, F., Clemenceau, S., Granat, O., Marsault, C., and Baulac, M. (1995). Developmental abnormalities of the medial temporal lobe in patients with temporal lobe epilepsy. *Am. J. Neuroradiol.* **16**, 617–626.
129. Baulac, M., De Grissac, M., Hasboun, D., Oppenheim, C., Adam, C., Arzimanoglou, A., Semah, F., Lehericy, S., Clemenceau, S., and Berger, B. (1998). Hippocampal developmental changes in patients with partial epilepsy: magnetic resonance imaging and clinical aspects. *Ann. Neurol.* **44**, 223–233.
130. Cendes, F., Cook, M. J., Watson, C., Oppenheim, C., Adam, C., Arzimanoglou, A., Semah, F., Lehericy, S., Clemenceau, S., and Berger, B. (1995). Frequency and characteristics of dual pathology in patients with lesional epilepsy. *Neurology* **45**, 2058–2064.

131. Diehl, B., Najm, I., LaPresto, E., Prayson, R., Ruggieri, P., Mohamed, A., Ying, Z., Lieber, M., Babb, T., Bingaman, W., and Lüders, H. O. (2004). Temporal lobe volumes in patients with hippocampal sclerosis with or without cortical dysplasia. *Neurology* **2004**, 1729–1735.

132. Raymond, A. A., Fish, D. R., Stevens, J. M., Cook, M. J., Sisodiya, S. M., and Shorvon, S. D. (1994). Association of hippocampal sclerosis with cortical dysgenesis in patients with epilepsy. *Neurology* **44**, 1841–1845.

133. Fernandez, G., Effenberger, O., Vinz, B., Steinlein, O., Elger, C. E., Dohring, W., and Heinze, H. J. (2001). Hippocampal malformation as a cause of familial febrile convulsions and subsequent hippocampal sclerosis. *Neurology* **57**, S13–S21.

134. Sloviter, R. S., Kudrimoti, H. S., Laxer, K. D., Barbaro, N. M., Chan, S., Hirsch, L. J., Goodman, R. R., and Pedley, T. A. (2004). "Tectonic" hippocampal malformations in patients with temporal lobe epilepsy. *Epilepsy Res.* **59**, 123–153.

135. Kuzniecky, R., Bilir, E., Gilliam, F., Faught, E., Martin, R., and Hugg, J. (1999). Quantitative MRI in temporal lobe epilepsy: evidence for fornix atrophy. *Neurology* **53**, 496.

136. Ho, S. S., Kuzniecky, R. I., Gilliam, F., Faught, E., and Morawetz, R. (1998). Temporal lobe developmental malformations and epilepsy: dual pathology and bilateral hippocampal abnormalities. *Neurology* **50**, 748–754.

137. Mochida, G. H., and Walsh, C. A. (2004). Genetic basis of developmental malformations of the cerebral cortex. *Arch. Neurol.* **61**, 637–640.

138. Roper, S. N., and Yachnis, A. T. (2002). Cortical dysgenesis and epilepsy. *Neuroscientist* **8**, 356–371.

139. Crino, P. (2004). Malformations of cortical development: molecular pathogenesis and experimental strategies. *Adv. Exp. Med. Biol.* **548**, 175–191.

140. Barkovich, A. J., Kuzniecky, R. I., and Dobyns, W. B. (2001). Radiologic classification of malformations of cortical development. *Curr. Opin. Neurol.* **14**, 145–149.

141. Jacobs, K. M., Kharazia, V. N., and Prince, D. A. (1999). Mechanisms underlying epileptogenesis in cortical malformations. *Epilepsy Res.* **36**, 165–188.

142. Deukmedjiam, A. J., King, M. A., Cuda, C., and Roper, S. N. (2004). The GABAergic system of the developing neocortex has a reduced capacity to recover from in utero injury in experimental cortical dysplasia. *J. Neuropathol. Exp. Neurol.* **63**, 1265–1273.

143. Hablitz, J. J., and DeFazio, R. A. (1999). Altered receptor subunit expression in rat neocortical malformations. *Epilepsia* **41**, S82–S85.

144. Moddel, G., Jacobson, B., Ying, Z., Janigro, D., Bingaman, W., Gonzalez-Martinez, J., Kellinghaus, C., Prayson, R. A., and Najm, I. M. (2005). The NMDA receptor NR2B subunit contributes to epileptogenesis in human cortical dysplasia. *Brain Res.* **1046**, 10–23.

145. Jutila, L., Ylinen, A., Partanen, K., Alafuzoff, I., Mervaala, E., Partanen, J., Vapalahti, M., Vainio, P., and Pitkanen, A. (2001). MR volumetry of the entorhinal, perirhinal, and temporopolar cortices in drug-refractory temporal lobe epilepsy. *Am. J. Neuroradiol.* **22**, 1490–1501.

146. Bernasconi, N., Duchesne, S., Janke, A., Lerch, J., Collins, D. L., and Bernasconi, A. (2004). Whole-brain voxel-based statistical analysis of gray matter and white matter in temporal lobe epilepsy. *Neuroimage* **23**, 717–723.

147. Duzel, E., Schiltz, K., Solbach, T., Peschel, T., Baldeweg, T., Kaufmann, J., Szentkuti, A., and Heinze, H. J. (2006). Hippocampal atrophy in temporal lobe epilepsy is correlated with limbic systems atrophy. *J. Neurol.*, **253**, 294–300.

148. Goncalves Pereira, P. M., Insausti, R., Artacho-Perula, E., Artacho-Perula, E., Salmenpera, T., Kalviainen, R., and Pitkanen, A. (2005). MR volumetric analysis of the piriform cortex and cortical amygdala in drug-refractory temporal lobe epilepsy. *Am. J. Neuroradiol.* **26**, 319–332.

149. Natsume, J., Bernasconi, N., Andermann, F., and Bernasconi, A. (2003). MRI volumetry of the thalamus in temporal, extratemporal, and idiopathic generalized epilepsy. *Neurology* **60**, 1296–1300.

150. Dlugos, D. J., Jaggi, J., O'Connor, W. M., Ding, X. S., Reivich, M., O'Connor, M. J., and Sperling, M. R. (1999). Hippocampal cell density and subcortical metabolism in temporal lobe epilepsy. *Epilepsia* **40**, 408–413.

151. Amorim, B. J., Etchebehere, E. C., Camargo, E. E., Rio, P. A., Bonilha, L., Rorden, C., Li, L. M., and Cendes, F. (2005). Statistical voxel-wise analysis of ictal SPECT reveals pattern of abnormal perfusion in patients with temporal lobe epilepsy. *Arq. Neuropsiquiatr.* **63**, 977–983.

152. Shin, W. C., Hong, S. B., Tae, W. S., Seo, D. W., and Kim, S. E. (2001). Ictal hyperperfusion of cerebellum and basal ganglia in temporal lobe epilepsy: SPECT subtraction with MRI coregistration. *J. Nucl. Med.* **42**, 853–858.

153. Bouilleret, V., Semah, F., Biraben, A., Taussig, D., Chassoux, F., Syrota, A., and Ribeiro, M. J. (2005). Involvement of the basal ganglia in refractory epilepsy: an 18F-fluoro-L-DOPA PET study using 2 methods of analysis. *J. Nucl. Med.* **46**, 540–547.

154. Ng, S. E., Lau, T. N., Hui, F. K., Chua, G. E., Lee, W. L., Chee, M. W., Chee, T. S., and Boey, H. K. (1997). MRI of the fornix and mamillary body in temporal lobe epilepsy. *Neuroradiology* **39**, 551–555.

155. Chan, S., Erickson, J. K., and Yoon, S. S. (1997). Limbic system abnormalities associated with mesial temporal sclerosis: a model of chronic cerebral changes due to seizures. *Radiographics* **17**, 1095–1110.

156. Bausch, S. B. (2005). Axonal sprouting of GABAergic interneurons in temporal lobe epilepsy. *Epilepsy Behav.* **7**, 390–400.

157. Holtzman, D. M., and Lowenstein, D. H. (1995). Selective inhibition of axon outgrowth by antibodies to NGF in a model of temporal lobe epilepsy. *J. Neurosci.* **15**, 7062–7070.

158. Scharfman, H. E. (2004). Functional implications of seizure-induced neurogenesis. *Adv. Exp. Med. Biol.* **548**, 192–212.

159. Nemani, V. M., and Binder, D. K. (2005). Emerging role of gap junctions in epilepsy. *Histol. Histopathol.* **20**, 253–259.

160. Jefferys, J. G. (1995). Nonsynaptic modulation of neuronal activity in the brain: electric currents and extracellular ions. *Physiol. Rev.* **75**, 689–723.

161. Wyneken, U., Smalla, K. H., Marengo, J. J., Soto, D., de la Cerda, A., Tischmeyer, W., Grimm, R., Boeckers, T. M., Wolf, G., Orrego, F., and Gundelfinger, E. D. (2001). Kainate-induced seizures alter protein composition and N-methyl-D-aspartate receptor function of rat forebrain postsynaptic densities. *Neuroscience* **102**, 65–74.

162. Swann, J. W., Al-Noori, S., Jiang, M., and Lee, C. L. (2000). Spine loss and other dendritic abnormalities in epilepsy. *Hippocampus* **10**, 617–625.

163. Hinz, B., Becher, A., Mitter, D., Schulze, K., Heinemann, U., Draguhn, A., and Ahnert-Hilger, G. (2001). Activity-dependent changes of the presynaptic synaptophysin-synaptobrevin complex in adult rat brain. *Eur. J. Cell. Biol.* **80**, 615–619.

164. Haglid, K. G., Wang, S., Qiner, Y., and Hamberger, A. (1994). Excitotoxicity. Experimental correlates to human epilepsy. *Mol. Neurobiol.* **9**, 259–263.

165. Gutierrez, R. (2005). The dual glutamatergic-GABAergic phenotype of hippocampal granule cells. *Trends Neurosci.* **136**, 550–554.

166. Tu, B., Timofeeva, O., Jiao, Y., and Nadler, J. V. (2005). Spontaneous release of neuropeptide Y tonically inhibits recurrent mossy fiber synaptic transmission in epileptic brain. *J. Neurosci.* **25**, 1718–1729.

167. Coulter, D. A. (2001). Epilepsy-associated plasticity in gamma-aminobutyric acid receptor expression, function, and inhibitory synaptic properties. *Int. Rev. Neurobiol.* **45**, 237–252.

168. Ferrer, I. (2002). Cell signaling in the epileptic hippocampus. *Rev. Neurol.* **34**, 544–550.

169. Patrylo, P. R., Spencer, D. D., and Williamson, A. (2001). GABA uptake and heterotransport are impaired in the dentate gyrus of epileptic rats and humans with temporal lobe sclerosis. *J. Neurophysiol.* **85**, 1533–1542.
170. Santoro, B., and Baram, T. Z. (2003). The multiple personalities of h-channels. *Trends Neurosci.* **26**, 550–554.
171. Woo, N. S., Lu, J., England, R., McClellan, R., Dufour, S., Mount, D. B., Deutch, A. Y., Lovinger, D. M., and Delpire, E. (2002). Hyperexcitability and epilepsy associated with disruption of the mouse neuronal-specific K-Cl cotransporter gene. *Hippocampus* **12**, 258–268.
172. Gupta, R. C., Milatovic, D., and Dettbarn, W. D. (2001). Depletion of energy metabolites following acetylcholinesterase inhibitor-induced status epilepticus: protection by antioxidants. *Neurotoxicology* **22**, 271–282.
173. Patel, M. (2004). Mitochondrial dysfunction and oxidative stress: cause and consequence of epileptic seizures. *Free Radic. Biol. Med.* **37**, 1951–1962.
174. Liu, W., Seto, J., Donovan, G., and Toth, M. (2002). Jerky, a protein deficient in a mouse epilepsy model, is associated with translationally inactive mRNA in neurons. *J. Neurosci.* **22**, 176–182.
175. Grisar, T., Guillaume, D., and Delgado-Escueta, A. V. (1992). Contribution of Na+, K(+)-ATPase to focal epilepsy: a brief review. *Epilepsy Res.* **12**, 141–149.
176. de Lanerolle, N. C., and Lee, T. S. (2005). New facets of the neuropathology and molecular profile of human temporal lobe epilepsy. *Epilepsy Behav.* **7**, 190–203.
177. de Curtis, M., Manfridi, A., and Biella, G. (1998). Activity-dependent pH shifts and periodic recurrence of spontaneous interictal spikes in a model of focal epileptogenesis. *J. Neurosci.* **18**, 7543–7551.
178. Costello, D. J., and Delanty, N. (2004). Oxidative injury in epilepsy: potential for antioxidant therapy? *Expert Rev. Neurother.* **4**, 541–553.
179. Croll, S. D., Goodman, J. H., and Scharfman, H. E. (2004). Vascular endothelial growth factor (VEGF) in seizures: a double-edged sword. *Adv. Exp. Med. Biol.* **548**, 57–68.
180. Vezzani, A., Moneta, D., Richichi, C., Perego, C., and De Simoni, M. G. (2004). Functional role of proinflammatory and anti-inflammatory cytokines in seizures. *Adv. Exp. Med. Biol.* **548**, 123–133.
181. Heinemann, U., Gabriel, S., Schuchmann, S., and Eder, C. (1999). Contribution of astrocytes to seizure activity. *Adv. Neurol.* **79**, 583–590.
182. Sutula, T., Zhang, P., Lynch, M., Sayin, U., Golarai, G., and Rod, R. (1998). Synaptic and axonal remodeling of mossy fibers in the hilus and supragranular region of the dentate gyrus in kainate-treated rats. *J. Comp. Neurol.* **390**, 578–594.

34

Central Nervous System Metastases

Robert J. Weil, MD, FCAS

Keywords: *angiogenesis, blood-brain barrier, blood-tumor barrier, brain metastasis, chemokines, extravasation, growth factors, Her-2, integrins, intravasation, matrix metalloproteinases, neurotrophins*

I. Overview
II. Brief History and Nomenclature
III. Natural History of CNS Metastasis
IV. Etiology and Pathogenesis: The Brain Metastatic Process
V. Pathophysiology: An Unique Environment?
VI. Conclusion
   References

I. Overview

Metastatic brain tumors represent a significant health burden and portend a poor prognosis because they are usually found in conjunction with disseminated systemic cancer. Although precise numbers of new cases of metastasis to the central nervous system (CNS) are not known, brain metastases are approximately 10 times more common than primary brain tumors, which suggests an incidence as high as 200,000 cases per year [1]. At autopsy, as many as 25% of all patients with systemic cancer have intracranial metastases [1,2]. Given advances in early diagnosis, locoregional cancer control, and more effective treatment of systemic disease, the incidence of brain metastases may be increasing, which accentuates the need to both understand the process of metastasis to the CNS and develop more effective and safer treatments.

II. Brief History and Nomenclature

The field of cancer metastasis is an old one, and it has drawn speculations and investigations dating back to antiquity [3,4]. The modern age of inquiry into the pathobiology of metastasis began with Stephen Paget's publication in 1889 of his "seed and soil" hypothesis [5]. Paget's keen eye noted the nonrandom pattern of metastasis to both visceral organs and bones at autopsy. In more than 900 autopsies of patients with a variety of tumors, he found a discrepancy between the potential blood supply to certain sites and the frequency with which tumors metastasized to them. Rather than being a function of chance, Paget hypothesized that some tumor cells (which he called the "seed") had a particular affinity for the milieu of certain organs, the "soil." For a metastasis to occur and grow required a match of the seed and the soil [3–5].

Forty years later, the pathologist James Ewing challenged Paget's theory [3,4]. Ewing postulated that metastasis occurred principally as a function of vascular flow and a result of mechanical factors related to the vascular system, which might explain why the liver and lungs were such common sites for tumor deposition. However, over the past few decades, a synthesis of sorts has occurred, which posits that locoregional metastases are mostly the result of anatomical or mechanical factors such as venous or lymphatic drainage, whereas metastases to distant sites from a variety of cancers are more likely to be site specific [3,4,6]. This may be especially true of metastases to the CNS, where most tumors are originally found within vascular border zones and at the gray-white junction [7].

Thus modern perspectives of metastasis suggest that a continuous, shifting balance between the two dominant mechanisms—the tumor cells (the seeds) and the vascular, homeostatic, and end-organ sites (the soil)—is more likely to be a rational explanation for the metastatic process [8–10]. As summarized by Fidler in a recent review [4], the metastatic process may be understood on the basis of three principles. First, both the host cancer and the metastatic tumor consist of a mixture of neoplastic and nonneoplastic peritumoral cells, including epithelial and stromal cells, fibroblasts, endothelial cells, and infiltrating lymphocytes, which creates a biologically heterogeneous "organism." Second, metastasis may be a selective process wherein only select cells, which may differ from those that constitute the mass of the primary tumor, are capable of surviving the process of metastasis (Fig. 1). Third, the

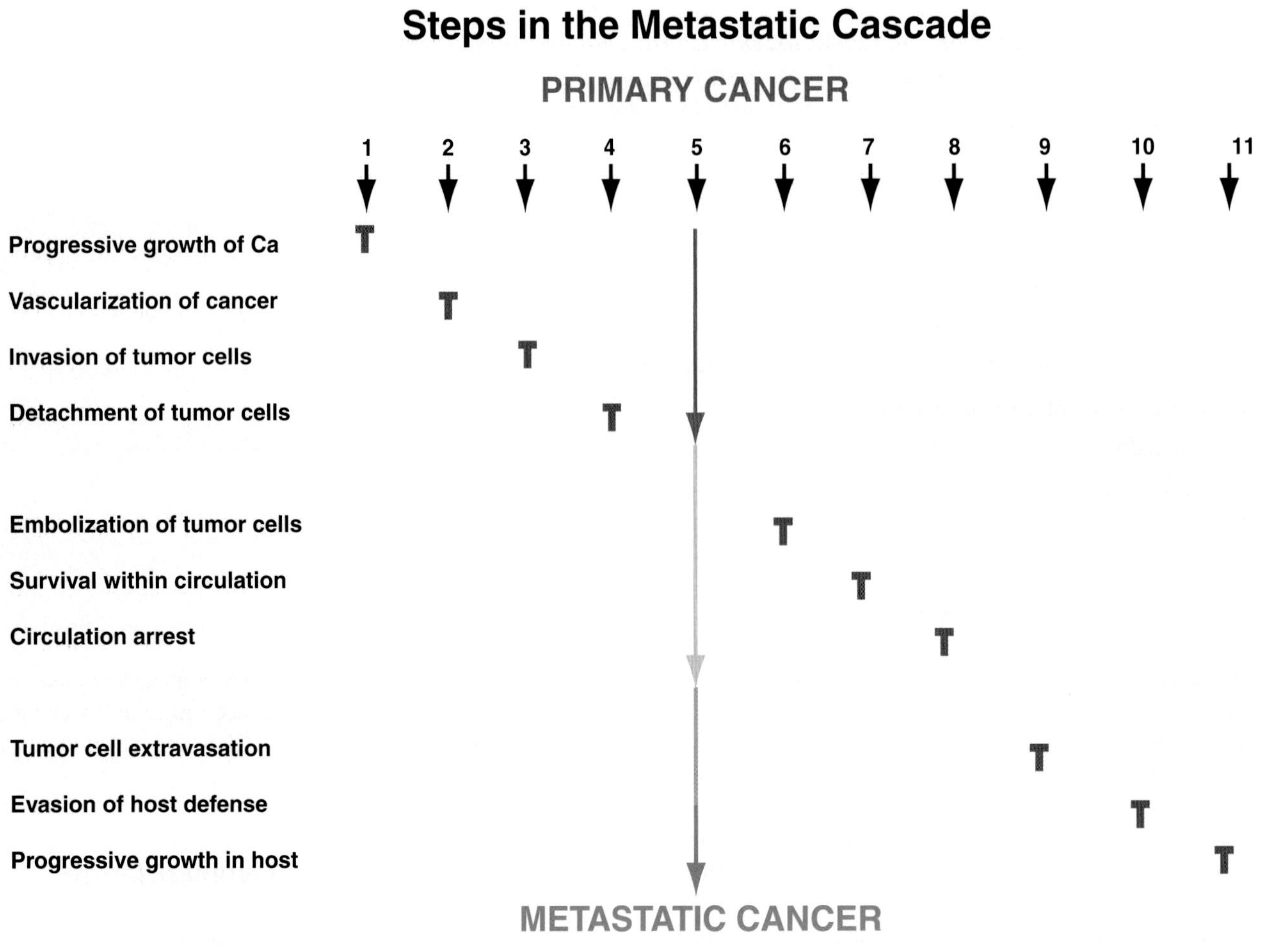

Steps in the Metastatic Cascade

PRIMARY CANCER

Figure 1 Schematic of the multi-stage process of metastasis. *Top*, The primary cancer has numerous clones (1–11) of cells that could escape the primary tumor and create a viable metastasis.

microenvironments of different organs such as the brain are biologically unique, which in turn may be welcoming or hostile to tumor cells from various primary sites. These local factors, which include homeostatic factors that induce tumor cell growth and survival, angiogenesis, invasion, and dissemination, may be exceedingly unique and may have little in common with the factors at another site. Such differences need to be explored to prevent or hinder development of metastases from specific primary tumors as well as to prevent their subsequent deposition and growth in individual metastatic haven(s) [2–6,8–10].

III. Natural History of CNS Metastasis

Of the nearly 1.3 million people diagnosed with cancer in the U.S. each year, roughly 100,000–170,000 will develop brain metastases, for an annual incidence of approximately 4.1 to 11.1 per 100,000 population [1,2]. Large autopsy studies suggest that between 20% and 40% of all patients with metastatic cancer will have brain metastases [1,2]. Given their overall greater frequency, lung and breast cancer are by far the most common tumors to present with brain metastases, although renal cell carcinoma, thyroid cancers, and malignant melanomas have a higher individual incidence: as much as 50% of patients with metastatic melanoma, for example [1,2]. The median latency between the initial diagnosis of breast or lung cancer and the onset of brain metastasis is approximately 1 to 3 years, although many years' latency is possible [1,2]. In most cases, cancer patients develop brain metastases after metastases have appeared systemically in the lung, liver, and/or bone. Nonetheless a brain metastasis can be the first clinical manifestation of an extra-CNS malignancy in as many as 10% of individuals. For the purposes of this chapter, the terms *central nervous system (CNS)* and *brain* are used interchangeably.

The metastasis of cancer to the CNS, either the brain parenchyma or the leptomeninges, is generally a late feature of metastatic disease [2,4,9]. Metastases to the brain parenchyma are thought to be hematogenous in origin. In a retrospective survey of breast cancer patients with brain metastases, for example, 78% had multiple intracerebral metastases, 14% had a solitary intracerebral metastasis, and the remaining 8% had leptomeningeal metastases [9,11]. Breast cancer is the most common solid tumor to exhibit leptomeningeal colonization [9,11]. Within the three membranous coverings, or meninges, that surround the brain, leptomeningeal metastases arise on the innermost covering (pia) and the middle membrane (arachnoid) or in the cerebrospinal fluid (CSF)–filled space between the arachnoid and the pia (subarachnoid space). Spreading to the leptomeninges may occur via multiple routes, including hematogenous, direct extension, transport through the venous plexus, and extension along nerves or perineural lymphatics. Once the tumor cells reach the leptomeninges, they are thought to spread via the CSF.

IV. Etiology and Pathogenesis: The Brain Metastatic Process

The nature of site-specific metastasis has been pondered since even before Paget first theorized that metastasis is a function of the interplay between both the "seed" (the tumor cells) and the "soil" (the host) [2–5]. However, most transgenic and xenograft systems model only a fraction of these sites simultaneously. Certain general steps appear to be necessary for metastasis and have been described in a variety of recent reviews [2–4,8,9]. This highly selective, nonrandom process consists of a series of linked, sequential events; it appears that if one step or even a few steps only are carried out, metastasis will not succeed [4,8,9,12]. Various molecular and genetic changes define the multistep tumor dissemination process, which has been described as the "metastatic cascade." These include invasion of the primary tumor border and intravasation of the circulatory system, survival and persistence/quiescence in the circulation, extravasation at a distant site, formation of a micrometastasis, and then progressive colonization and growth to form a metastasis (see Fig. 1) [4,8,9,12].

Hematogenous metastasis to the CNS is an inherently inefficient process [4,8,9,12]. Successful metastasis depends on the interaction of tumor cells with host defenses and the brain microenvironment. Nonetheless, several characteristics of primary tumors, such as non–small cell lung cancer (NSCLC), breast cancer, and melanoma, among others, are predisposed to hematogenous dissemination. An immature neovascular system, high interstitial pressure, and the close proximity of cancer cells to blood vessels favor tumor cell intravasation. Once in the circulation, the intravascular tumor cells are subject to a variety of nonspecific mechanical forces, such as hemodynamic turbulence, which might cause mechanical disruption of the cells either before or during extravasation. They are also subject to immune surveillance, recognition, and destruction.

A. Peritumoral Brain Events

The borders between a brain metastasis and the surrounding brain parenchyma are usually distinct. However, unlike gliomas, which diffusely infiltrate the brain parenchyma, metastatic tumors are usually demarcated from the surrounding brain tissues by a pseudocapsule that appears to enclose the growing tumor. Nonetheless the tumor cells react with the surrounding brain parenchyma and blood vessels, resulting in a series of pathological and biochemical changes. At this point, a variety of local factors, including secretion of paracrine and endocrine growth factors; proteolytic factors such as matrix metalloproteinases; and angiogenic factors and their receptors, which enhance neovascularization, are balanced to promote or inhibit growth at the final site [13–15].

B. Local Brain Invasion, Chemokines, and Cytokines

Local brain invasion is a multifaceted process of interconnected and potentially conflicting mechanisms that include cell motility, adhesion, and enzymatic re-modeling of extracellular matrix (ECM) components [4,6,9,16–19]. Brain invasion requires paracrine interactions between brain stromal and endothelial cells and the invading metastatic tumor cells. Proteolytic degradation of the ECM appears to promote tumor invasion by clearing a pathway for the invading neoplastic cells; proteolysis is concentrated in the region of the cell membrane on the advancing edge of the invading tumor cell. In addition, the proteolytic activity likely releases several factors from the ECM that modulate cell proliferation, adhesion, migration/invasion, and angiogenesis.

A variety of molecules and pathways have been shown to contribute to local tumor invasion [4,6,9,16–19]. For example, the E-cadherin–catenin complex is a prime mediator of cell-cell adhesion and is crucial for intercellular adhesiveness and the maintenance of normal and malignant tissue architecture. Reduced expression of this complex has been associated with tumor invasion, metastasis, and an unfavorable prognosis. It was shown that E-cadherin expression correlates with a high MIB-1 index for metastatic adenocarcinomas with consistent E-cadherin expression, regardless of the degree of differentiation or the extent of spread of the disease. Integrins are major adhesion and signaling receptor proteins that mediate cell migration and invasion. They also trigger a variety of signal transduction pathways and regulate cytoskeletal organization, specific gene expression, growth control, and apoptosis (programmed cell death). For example, blockade of integrin $\alpha 3\beta 1$ in an animal model using a human NSCLC cell line has been shown to decrease brain metastasis [6,9,16].

Several other cytokines, chemokines, and growth factors have been implicated in brain metastases [4,6,9,13,14]. Chemokines have been reported to contribute to breast cancer metastasis and may contribute to organ specificity. CXCL12 (stromal cell-derived factor 1a, SDF-1a), a ligand for the CXCR4 chemokine receptor, has been reported to be expressed in the brain. CXCL12 has been demonstrated to facilitate the propensity of breast carcinoma cells (the cell line MDA-MB-231) to invade through human brain microvascular endothelial cells. When compared with brain-homing tumor cells, bone-seeking breast cancer cells produce greater levels of parathyroid hormone-related protein (PTHrP) and plasminogen activator inhibitor 1 (PAI-1), exhibit greater insulin-like growth factor (IGF)-1R phosphorylation upon ligand stimulation, and are resistant to tumor growth factor (TGF)-b inhibition of soft agar colonization. These data may reflect a general insensitivity of brain metastases to endocrine signals, given their poor penetration of the blood-brain barrier (BBB). Alternatively, brain metastases may show exquisite reactivity to distinct signals. Potential candidate-enhancing signals include a variety of locally produced factors, including nerve growth factor, transferrin, gangliosides, and other enzymes [4,6,9,13,14].

Neurotrophins (NTs) may play an important role because they have been shown to stimulate brain invasion [17]. In brain metastatic melanoma cells, for example, NTs can promote invasion by enhancing the production of ECM degradative enzymes such as heparinase, an enzyme capable of locally destroying both the ECM and the basement membrane of the BBB [17]. Heparinase cleaves the heparin sulfate (HS) chains of ECM proteoglycans, and it is a unique metastatic determinant because it is the dominant mammalian HS degradative enzyme. Furthermore, tumor cell expression of the p75 neurotrophin receptor, a receptor broadly shared by several members of the NT family, favors the development of brain metastasis [9,17].

Since the initial report that urokinase-type plasminogen activator (uPA) is produced and released from cancer cells, strong clinical and experimental evidence has accumulated that the tumor-associated serine protease plasmin, its activator uPA, the receptor uPA-R (CD87), and the inhibitors plasminogen activator inhibitor type 1 and 2 (PAI-1/2) have been linked to cancer invasion and metastasis.[18] uPA converts the zymogen plasminogen to plasmin, a trypsin-like enzyme with broad substrate specificities. uPA binds to the surface of the cell membrane and causes localized cell surface proteolytic activity, which is required for the destruction of the ECM, a vital step in tumor cell invasion. The proteolytic activity of uPA is modulated by its cell surface receptor and by PAI-1. Plasmin degrades multiple components of the ECM such as fibrin, fibronectin, proteoglycans, and laminin, thus facilitating tumor cell migration through the host tissues. Plasmin also activates several other proteolytic enzymes, one important group of which comprises the metalloproteases [13,18].

Matrix metalloproteinases (MMPs) comprise a family of at least 20 enzymes that have variable ability to degrade the ECM, with resultant enzymatic breakdown of connective tissue barrier proteins including collagens, laminins, fibronectin, vitronectin, and HS proteoglycans [13]. Most MMPs are made by stromal cells (e.g., inflammatory cells, endothelial cells, and fibroblasts), although their expression is regulated at least in part by tumor cells via cytokine production. ECM degradation enhances tumor cell motility. For a variety of metastatic tumors, MMP activity correlates with invasiveness, metastasis, and poor prognosis. MMP-2 and MMP-9 up-regulation have been shown to correlate with a poorer prognosis in some patients, whereas another MMP, tissue inhibitor of metalloproteinases 1 (TIMP-1), exerts a suppressive effect [6,13,16].

Neoangiogenesis and Metastasis

Angiogenesis is necessary for continued tumor growth [4,9,12,15]. Because the limit of oxygen diffusion capacity through tissue is approximately 200 microns, failure of vascular growth will restrict the tumor mass to this size [4]. The onset of angiogenesis within small clusters of tumor cells, known as the *angiogenic switch*, is influenced by a complex interplay of angiogenic molecules, such as vascular endothelial growth factor (VEGF) and members of the angiopoietin family, as well as several antiangiogenic molecules, such as angiostatin [4,6,9]. Furthermore, multiple MMPs and TIMP-1 influence the angiogenic process, both altering the release of proangiogenic peptides from the ECM and degrading the ECM, which promotes ingrowth of new blood vessels. One potent example of the angiogenic switch comes from recent experimental work that showed that MMP-9/gelatinase B is up-regulated in angiogenic tumor-cell islets, which induced tumor cells to release VEGF from an extracellular reservoir. Blocking MMP-9 inhibits tumor growth [4,6,13].

Experimental metastases to the brain have been found to be more angiogenic than metastases to systemic sites. The potential role of angiogenesis in cancer metastasis to the brain has been studied, particularly the role of VEGF, a prominent angiogenic factor [6,9,15]. For example, when the ZR75-1 human breast cell line was injected either into the mammary fat pad (mfp) or intracranially into nude mice bearing estrogen pellets, the resulting cranial tumors had a higher vascular density than the identical tumors located in the mfp [9,15]. In addition, administration of exogenous VEGF promotes penetration of metastatic breast carcinoma cells in migration studies and also appears to be able to modulate the permeability of endothelial cells *in vitro*. Administration of PTK787, a VEGF receptor tyrosine kinase inhibitor can decrease brain metastasis of tumor cells injected into the carotid artery by more than 60%, with the development of fewer microvessels, a decrease in the number of proliferating cell nuclear antigen (PCNA)–staining tumor cells, and greater numbers of apoptotic tumor cells in the experimental brain metastases [4,6,9,15]. These data not only functionally link VEGF to brain metastasis but also demonstrate the potential utility of model systems for preclinical validation studies.

V. Pathophysiology: An Unique Environment?

The BBB is hypothesized to create and/or interact with the unique brain microenvironment and influence metastatic colonization [7,9,19]. The BBB consists of capillary endothelial cells that lack fenestrations, are interconnected by continuous tight junctions, and generate high transendothelial electrical resistance.[19] Pericytes, basement membrane, and astrocytic endfeet line the abluminal surface of microvessel endothelial cells. Low permeability to ions and small molecules and virtual impermeability to macromolecules and peptides (except through specific receptor-mediated transcytosis) is a feature of the BBB. A lack of pinocytosis, which facilitates the transport of molecules via cellular transcytosis, contributes to selectivity. Both adenosine triphosphate (ATP)-binding cassette C1 (ABCC1) and ABCB1 (P-glycoprotein) are present on the luminal membrane of the cerebral endothelium and prevent most drugs from entering the brain parenchyma. The BBB works in concert with the blood-CSF barrier to protect the neural environment [7,9,16,19].

Furthermore, the brain's microenvironment also differs from that of the lung, liver, or bone, which are other common loci for metastasis [9,19]. It is bathed in an interstitial fluid that is high in chloride, an environment that may not be conducive to potential metastatic clones from tumors of epithelial origin. It may be that the unique milieu of the brain attracts cells of neuroepithelial origin, such as small cell carcinoma of the lung or melanoma, which are two to three times more likely to metastasize to the CNS than breast cancer or NSCLC. It has been suggested that the common involvement of the CNS in metastatic melanoma may be the result of a "homing" influence because melanocytes and neuronal subpopulations share a common embryological origin [4,6,9,16,19]. Alternatively, the CNS site may be more conducive to the survival (i.e., colonization and expansion) of extravasated cancer cells of neuroepithelial origin. To produce brain metastases, tumor cells must reach the brain vasculature by attaching to the endothelial cells of brain microvessels at the BBB, extravasate into the brain parenchyma, induce angiogenesis, and proliferate in response to growth factors [4,6,9,16,19].

Once tumor cells invade the BBB to establish a brain metastasis, endothelial cells form a blood-tumor barrier (BTB). Almost nothing is known of the BTB in the human or in model systems [9,19]. One hallmark of brain metastases is the edema that surrounds the tumor, an effect possibly caused by altered permeability of tumor-associated endothelial cells that permits greater leakage of fluid into the tumor. An improved understanding of the interactions between tumor and epithelial cells could assist in the development of new therapeutic approaches [9].

The brain parenchyma is populated by astrocytes, which can synthesize a host of biologically interesting proteins including multiple interleukins (e.g., ILs 1, 3, and 6), interferon gamma (INF-γ), tumor necrosis factor (TNF-α), and a variety of other growth factors such as TGF-β, IGF-1, platelet-derived growth factor (PDGF)-1, and other cytokines [9,19]. Astrocytes can also serve as antigen-presenting cells for immune responses [9,19]. Although glial cells have traditionally been thought to provide structural support for neurons, we now know that they also actively support neuronal physiology and influence brain and BBB integrity. Therefore cytokines from the brain microenvironment may provide part of the "soil" in which the metastatic "seed" grows [9,13–19].

A. Molecular Mechanisms: Metastasis Suppressors

Analogous to the identification of tumor suppressor genes such as *RB*, Steeg et al. hypothesized that metastasis might also involve loss of gene functions that strive to maintain the normal, differentiated state [8,20]. Since the identification in 1988 of this first metastasis suppressor gene, *NM-23*, additional genes have been identified whose expression is reduced in tumors with high metastatic potential [8,20]. At least 10 such genes have been identified and confirmed to function as metastasis suppressors in some, but not all, epithelial cancers (Table 1) [4,6,8,20]. Interestingly, most of these proteins do not appear to be involved actively in the early stages of metastasis such as intravasation or in angiogenesis. Rather, most appear to act at the final stage of tumor cell colonization and expansion within the metastatic site. Mkk4 appears to influence the apoptotic response to stress; Mkk4, Rkip, and Nm23 all have direct effects on the MAP kinase pathway; and BRMs1 interacts with several histone deacetylases complexes, which can influence transcriptional activation. Nm23 also appears to play a role in the differentiation of epithelial cells. Epigenetic changes such as promoter hypermethylation, rather than gene mutation or deletion, appear to be the main mechanism by which the influence of these metastasis suppressors is lost [8,20]. Finally, early work suggests that some standard chemotherapeutic agents may work, at least in part, through modulation of metastasis suppressor expression or function [8,20]. Several reviews cover this topic thoroughly [4,8,20].

Table 1 Representative Metastasis Suppressor Genes

Gene	Cancer/Metastatic Tumor	Function(s) of Protein	OMIM Number	Chromosome Location
NM23	Breast, colon, melanoma	A histidine kinase. Nm23 phosphorylates KSR and can lead to decreased ERK 1/2 activation. Appears to play a role in normal development and differentiation.	156490	17q21.3
MKK4	Breast, ovarian, prostate	A mitogen-activated protein kinase kinase (MAPKK) that phosphorylates p38 and Jun (JNK) kinases.	601335	17p11.2
BRMS1	Breast, melanoma	Functions in gap-junction communication.	606259	11q13.1-q13.2
KiSS1	Breast, melanoma	A G-protein coupled receptor ligand. Also known as *metastin*.	603286	1q32
KAI1 (CD82)	Bladder, breast, lung, pancreas, prostate	Interacts with beta-catenin-reptin and histone deacetylases. May desensitize EGFR activity. Also known as *kangai*.	600623	11p11.2
CD44	Breast, colon, lung, melanoma, prostate	An integral cell membrane glycoprotein that affects cell adhesion. Decreased expression due in part to hypermethylation.	107269	11pter-p13
CRSP3	Melanoma	A transcriptional co-activator that may work through the enhancer binding factor Sp1.	605042	
RHOGDI2	Bladder, breast, colon, kidney, liver, lung, prostate	Regulates function of Rho and Rac, GTP-binding proteins of the Ras superfamily.		11p11.2
VDUP1	Melanoma	A differentiation factor via thoreduxin inhibition.	606599	1q21
PTEN/MMAC1	Breast, colon, endometrial, germ cell, kidney, lung, melanoma, thyroid	A homologue of cytoskeletal tension, leading to invasion and metastasis through interaction with actin filaments at focal adhesions.	601728	10q23.31

EGFR, Epidermal growth factor receptor; *ERK*, extracellular signal-regulated kinase; *JNK*, Jun-terminal kinase; *KSR*, kinase suppressor of Ras; *OMIM*, online Mendelian inheritance in man identification number.
See www.ncbi.nlm.nih.gov for detailed information about OMIM and references for these genes and their protein products and potential functions.

B. Clinical Pathophysiology: The Her-2 Connection in Breast Cancer Metastasis to the CNS

Her-2 is a member of the epidermal growth factor (EGF) receptor superfamily [9,11]. It is overexpressed in approximately 20%–30% of breast carcinomas via gene amplification, and its overexpression correlates with poor patient survival. Her-2 dimerizes with other members of the EGF receptor superfamily and receptor activation by ligand initiates numerous signaling pathways that control or influence diverse aspects of growth and differentiation. Trastuzumab is a recombinant, humanized monoclonal antibody to Her-2 that improves survival in metastatic breast cancer patients when used alone or in combination with cytotoxic chemotherapy [9,11].

Many potential molecular mechanisms have been suggested to mediate the aggressive behavior of Her-2–positive breast cancers [9,11,21–24]. For example, increased activation of Her-2 signaling enhances cell proliferation, survival, apoptosis resistance, migration, and invasion. Bendell et al. retrospectively studied 122 women who were treated with trastuzumab alone or in combination with chemotherapy. Based on a median follow-up of 23 months, 34% of patients were diagnosed with CNS metastases, well above historical rates; 50% had responded to trastuzumab systemically. In a second report by Clayton et al., 25% of 93 patients with metastatic breast cancer patients developed brain metastases over a median follow-up period of 10.8 months from the initiation of trastuzumab therapy. Of 23 patients developing CNS metastases, 78% had stable disease at other sites while on trastuzumab therapy. The CNS was the first site of symptomatic disease progression in 82% of patients and the only site of disease progression in nearly 70% of patients. Both studies report frequencies of brain metastases above those reported historically for all breast cancer patients. Furthermore, metastases generally occurred in patients receiving trastuzumab who had stable systemic disease. Finally, Miller et al. screened 155 women with metastatic breast cancer, but no symptomatic CNS metastases, with magnetic resonance imaging (MRI) before entry into several molecularly based, antiangiogenic clinical trials. Nearly 15% of the women screened had occult brain metastases, and Her-2 overexpression by the primary tumor was predictive of occult brain metastases. Survival among patients with occult brain metastases was shorter than that of patients without CNS disease but was similar to the survival of patients with symptomatic brain metastases.

The causes of these trends are unknown [9,11,21–24]. One theory suggests that Her-2 overexpression endows tumor cells with increased metastatic aggressiveness to sites such as the lungs and may similarly augment metastatic propensity to the CNS. Second, by enhancing patient survival, trastuzumab may permit brain metastases to develop and become symptomatic. Third, trastuzumab is effective against systemic metastases but is likely to be relatively ineffective against CNS metastases due to its poor penetration of the BTB. Taken together, the data regarding Her-2-postive breast cancer that is metastatic to the CNS support Paget's conclusion in his 1889 treatise: "The best work in the pathology of cancer now is done by those who... are studying the seed. They are like scientific botanists, and he who turns over the records of the cases of cancer is only a ploughman [sic], but his observations of the properties of the soil might also be useful."

C. Pathophysiology and Clinical Symptoms

Diagnosis of brain metastasis is based on patient symptoms, signs, and neuroimaging [25]. Common symptoms of a parenchymal brain metastasis include headache or alterations in cognition, mental status, and behavior. Frequent signs that generally reflect the location of the tumor and the influence of peritumoral cerebral edema—which reflects the incompetence of the BTB and the process of tumor-induced neovascularization—include nausea and vomiting, seizures, deficits in sensation, motor function, speech, and/or vision. Lesions in the cerebellum and brainstem, which are less common than those in the cerebral hemispheres, can cause ataxia, cranial neuropathies, and upper motor neuron dysfunction as well as additional signs and symptoms related to hydrocephalus, such as headache, memory loss, or behavioral problems. Contrast-enhanced neuroimaging (i.e., computed tomography [CT] or MRI) is the mainstay of diagnostic evaluation. Ancillary studies such as lumbar puncture may be indicated in some situations in which symptoms and signs such as headache, cranial neuropathy, or alterations in cognition or neuroimaging suggest leptomeningeal carcinomatosis rather than a parenchymal mass.

Solid epithelial cancer involving the CNS tends to be a late complication of progressive metastatic disease [9,25]. Of all brain metastatic patients, those with controlled extracranial tumor, age less than 65 years, and a favorable general performance (Karnofsky Performance Status ≥ 70) fare better; older patients with a KPS < 70 fare less well. Patients with solitary metastases and with a longer disease-free interval also tend to fare better.

Treatment strategies have been reviewed in several monographs and reviews [25–29]. Many of the randomized studies cited pertain to patients with brain metastases from multiple-cancer histologies, without separate stratification for histological subtype. Corticosteroids can reduce peritumoral edema and provide symptomatic relief. Chemotherapy has not generally been useful in the treatment of most epithelial cancers that metastasize to the brain due to the limitations on drug delivery imposed by the BBB or BTB, although actual penetration of various agents has not been studied well and deserves greater consideration in some patients. Whole brain radiation can provide a median survival of 4–5 months, which can be further extended by stereotactic radiosurgery. Several nonrandomized studies have suggested that stereotactic radiosurgery may provide nearly equivalent outcomes compared with surgery followed by whole brain irradiation. Surgery tends to reduce symptoms quickly and prolong life significantly, with concurrent increases in quality of life. Multiple metastases (up to three) can be removed surgically with a risk similar to that of a single lesion and can provide similar benefits. At present, adjuvant radiotherapy follows surgical resection because this combined approach has been shown in general to prolong median survival significantly: about 12 months depending on the factors noted previously. A growing body of evidence suggests that surgery may be useful in select patients with recurrent brain metastases.

Mean survival from diagnosis of a brain metastasis varies among studies but ranges from 2 to 16 months, depending on involvement of the CNS, the extent of the extracranial metastatic disease, and the treatment applied. The mean 1-year survival is estimated to be approximately 20% [25–29]. The ability to control extracranial (systemic) disease has traditionally been the main limiting factor. However, as systemic therapies improve, control of extracranial disease may become a less predictive factor. As previously noted, this point is strengthened by studies of Her-2 positive patients with breast cancer treated with trastuzumab [11,21–24]. In the study reported by Bendell et al., the median survival of patients with metastatic breast cancer treated with trastuzumab was 13 months, and nearly half of all patients died as a result of progressive CNS disease.

VI. Conclusion

Brain metastases from systemic cancer are a common and important problem. Given continued success in obtaining locoregional control of many solid epithelial cancers, it is likely that the incidence and burden from these tumors will increase. Novel and innovative therapies, perhaps based on some of the pathophysiological mechanisms discussed in this chapter, are needed.

References

1. CBTRUS, Central Brain Tumor Registry of the United States. www.cbtrus.org.
2. Kleihues, P., and Cavenee, W. (2000). *Pathology and genetics of tumors of the nervous system. World Health Organization Classification of Tumours.* IARC Press, Lyon, France.
3. Weiss, L. (2000). Metastasis of cancer: a conceptual history from antiquity to the 1990s. *Cancer Metastasis Rev.* **19**, 193–400.
4. Fidler, I. J. (2003). The pathogenesis of cancer metastasis: the "seed and soil" hypothesis revisited. *Nature Rev. Cancer* **3**, 453–458.
5. Paget, S. (1889). The distribution of secondary growths in cancer of the breast. *Lancet* **1**, 571–573.
6. Chambers, A. F., Groom, A.C., and MacDonald, I. C. (2002). Dissemination and growth of cancer cells in metastatic sites. *Nature Rev. Cancer* **2**, 563–572.
7. Hwang, T.-L., Close, T. P., Grego, J. M., Brannon, W. L., and Gonzales, F. (1996). Predilection of brain metastasis in gray and white matter junction and vascular border zones. *Cancer* **77**, 1551–1555.
8. Steeg, P. S. (2003). Metastasis suppressors alter the signal transduction of cancer cells. *Nature Rev. Cancer* **3**, 55–63.
9. Weil, R. J., Palmieri, D. C., Bronder, J. L. Stark, A. M., and Steeg, P. S. (2005). Breast cancer metastasis to the central nervous system. *Am. J. Pathol.* **167**, 913–920.
10. Hida, K., and Klagsbrun, M. (2005). A new perspective on tumor endothelial cells: unexpected chromsome and centrosome abnormalities. *Cancer Research* **65**, 2507–2510.
11. Lin, N., Bellon, J., and Winer, E. (2004). CNS metastases in breast cancer. *J. Clin. Oncol.* **22**, 3608–3617.
12. Zhang, R., Fidler, I., and Price, J. (1991). Relative malignant potential of human breast carcinoma cell lines established from pleural effusions and a brain metastasis. *Invasion Metastasis* **11**, 204–215.
13. Brinckerhoff, C. E., and Matrisian, L. M. (2002). Matrix metalloproteinases: a tail of a frog that became a prince. *Nat. Rev. Mol. Cell Biol.* **3**, 207–214.
14. Zlotnick, A. (2004). Chemokines in neoplastic progression. *Semin. Cancer Biol.* **14**, 181–185.
15. Monsky, W., Carreira, C. M., Tsuzuki, Y., Gohongi, T., Fukumura, D., and Jain, R. (2002). Role of host microenvironment in angiogenesis and microvascular functions in human breast cancer xenografts: mammary fat pad versus cranial tumors. *Clin. Cancer Res.* **8**, 1008–1013.
16. Vernon, A. E., and LaBonne, C. (2004). Tumor metastasis: a new twist on epithelial-mesenchymal transitions. *Curr. Biol.* **14**, 719–721.
17. Dolle, L., Adriaenssens, E., El Yazidi-Belkoura, I., Le Bourhis, X., Nurcombe, V., and Hondermarck, H. (2004). Nerve growth factor receptors and signaling in breast cancer. *Curr. Cancer Drug Targets* **4**, 463–470.
18. Decock, J., Paridaens, R., and Cufer, T. (2005). Proteases and metastasis: clinical relevance nowadays? *Curr. Opin. Oncol.* **17**, 545–550.
19. Hawkins, B. T., and Davis, T. P. (2005). The blood-brain/neurovascular unit health and disease. *Pharmacol. Rev.* **57**, 173–185.
20. Steeg, P. S. (2004). Perspectives on classic article: metastasis suppressor genes. *J. Natl. Cancer Inst.* **96**, E4 (review).
21. Bendell, J., Domchek, S., Burstein, H., Harris, L., Younger, J., Kuter, I., Bunnell, C., Rue, M., Gelman, R., and Winer, E. (2003). Central nervous system metastases in women who receive trastuzumab-based therapy for metastatic breast carcinoma. *Cancer* **97**, 2972–2977.
22. Lai, R., Dang, C. T., Malkin, M. G., and Abrey, L. E. (2004). The risk of central nervous system metastases after trastuzumab therapy in patients with breast carcinoma. *Cancer* **101**, 810–816.
23. Clayton, A. J., Danson, S., Jolly, S., Ryder, W. D., Burt, P. A., Stewart, A. L., Wilkinson, P. M., Welch, R. S., Magee, B., Wilson, G., Howell, A., and Wardley, A. M. (2004). Incidence of cerebral metastases in patients treated with trastuzumab for metastatic breast cancer. *Br. J. Cancer* **91**, 639–643.
24. Miller, K. D., Weathers, T., Hanley, L., Timmerman, R., Dickler, M., Shen, J., and Sledge, G. W., Jr. (2003). Occult central nervous system involvement in patients with metastatic breast cancer: prevalence, predictive factors and impact on overall survival. *Ann. Oncol.* **2003**, 1072–1077.
25. Shaffrey, M. E., Mut, M., Asher, A. L., Burri, S. H., Chahlavi, A., Chang, S. M., Farace, E., Fiveash, J. B., Lang, F. F., Lopes, M. B., Markert, J. M., Schiff, D., Siomin, V., Tatter, S. B., and Vogelbaum, M. A. (2004). Brain metastases. *Curr. Probl. Surg.* **41**, 665–741.

26. Peereboom, D. M. (2005). Chemotherapy in brain metastases. *Neurosurgery* **57(S4)**, 54–65.
27. Mehta, M.P., and Khuntia, D. (2005). Current strategies in whole-brain radiation therapy for brain metastases. *Neurosurgery* **57(S4)**, 33–44.
28. McDermott, M. W., and Sneed, P. K. (2005). Radiosurgery in metastatic brain cancer. *Neurosurgery* **57(S4)**, 45–53.
29. Lesniak, M., and Brem, H. (2004). Targeted therapy for brain tumours. *Nature Rev. Drug Discov.* **3**, 499–508.

35

Meningioma

Katrin Lamszus, MD
Christian Hagel, MD
Manfred Westphal, MD

Keywords: *angiogenesis, animal models, classification, clinical presentation, hormones, meningioma, molecular genetics*

I. History
II. Incidence and Etiology
III. Meningioma Morphology and Subtypes
IV. Clinical Aspects
V. Pathogenesis and Molecular Mechanisms
VI. Special Forms of Meningioma
VII. Pathophysiology
VIII. Cell Lines and Animal Models
IX. Future Directions
    References

I. History

Meningiomas are tumors that arise from the arachnoidal cap cells. The term *meningioma* was first introduced by Cushing and Eisenhardt in 1922. Before that, meningiomas were called by a variety of names including "fungus of the dura mater," "psammoma," "endothelioma," and "mesothelioma" among others [1,2]. Meningiomas present with a variety of histological patterns. In 1938 Cushing and Eisenhardt distinguished 9 main types and 20 subtypes of meningioma. To simplify matters, a working classification in 1941 identified only four major types: syncytial, fibroblastic, transitional, and angioblastic and recognized also mixed patterns. The initial World Health Organization (WHO) classification in 1979 distinguished six different subtypes corresponding with grade I as well as to papillary and anaplastic meningioma, which both correspond with either grade II or III. The 1993 WHO classification identified 11 subtypes of benign meningiomas in addition to papillary, atypical, and anaplastic meningiomas. The current WHO

Table 1 WHO Classification of Meningiomas

WHO Grade I: Low Risk of Recurrence and Aggressive Growth
Meningothelial meningioma
    Fibrous (fibroblastic) meningioma
    Transitional (mixed) meningioma
    Psammomatous meningioma
    Angiomatous meningioma
    Microcystic meningioma
    Secretory meningioma
    Lymphoplasmacyte-rich meningioma
    Metaplastic meningioma

WHO Grade II: Greater Likelihood of Recurrence and/or Aggressive Growth
    Atypical meningioma
    Clear cell meningioma (intracranial)
    Chordoid meningioma

WHO Grade III: High Proliferation Index and/or Brain Invasion of Any Subtype or Grade
    Rhabdoid meningioma
    Papillary meningioma
    Anaplastic (malignant) meningioma

classification, published in 2000 [3], includes nine different subtypes of benign meningiomas (WHO grade I), three different types of grade II meningiomas, and three different types of grade III meningiomas (Table 1).

The constant change in nomenclature demonstrates that meningioma typing is a matter of permanent dispute rather than unequivocal criteria. In the most recent revision of the WHO classification, several substantial changes were introduced. These changes are based mainly on a series of studies from the Mayo Clinic in which histopathological parameters were correlated with clinical prognostic parameters [4]. Whereas the 1993 WHO classification had ill-defined borders between benign, atypical, and anaplastic meningiomas, grading criteria are now much more stringent and objective. Futhermore, for the first time the revised WHO classification of 2000 includes genetic observations, in addition to pathological and immunohistochemical findings. Genetic analyses have made some important contributions to our understanding of meningiomas; for example, the recognition of hemangiopericytomas as a distinct entity.

II. Incidence and Etiology

Meningiomas constitute approximately 20% of all primary intracranial tumors with an annual incidence of about 6 cases per 100,000 people. The highest incidence is between the sixth and seventh decade of life (Fig. 1). The mean patient age at tumor operation in the neuropathological series in Hamburg (1984–2005) was 55 years for females and 54 years for males (see Fig. 1). Meningiomas are significantly more common in females; in particular, among middle-aged patients the female:male incidence ratio is greater than 2:1.

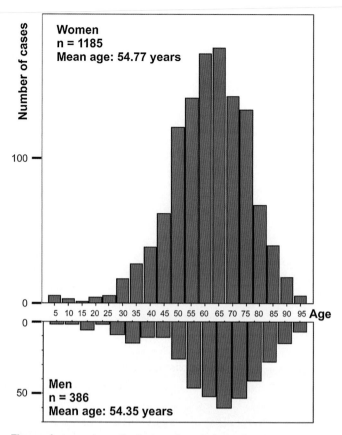

Figure 1 Age and sex distribution of meningiomas based on 1571 cases treated in the Department of Neurosurgery, University Hospital, Hamburg-Eppendorf, Germany.

Several endogenous and exogenous factors predispose a patient to meningioma development. Most patients with neurofibromatosis type 2 (NF2) develop meningiomas during the course of life. In addition, several non-NF2 families with an increased susceptibility to meningiomas have been reported. The most important exogenous association is with ionizing radiation. Following irradiation, meningiomas usually develop, with a latency period of several decades. The majority of cases were immigrants to Israel who received low-dose irradiation for tinea capitis during the 1950s. Also, survivors of atomic bomb explosions in Hiroshima and Nagasaki have an increased incidence of meningioma. In addition, evidence suggests that even exposure to full-mouth series of dental x-rays may carry an increased risk of meningioma development [5].

III. Meningioma Morphology and Subtypes

A. WHO Classification

The majority of meningiomas (80–90%) are benign and classified as WHO grade I (MI). Nine distinct subtypes belong to this category (see Table 1) (Fig. 2A–I). Of the various subtypes, meningothelial (see Fig. 2A), fibrous (see Fig. 2B), and transitional (see Fig. 2C) meningiomas are by far the most common (Fig. 3). Occasional mitoses and pleomorphic nuclei are tolerable features within meningiomas grade I and do not connote more aggressive behavior.

In various clinical series, between 5% and 15% of meningiomas are classified as atypical, corresponding with WHO grade II (MII). The introduction of far more stringent and objective criteria to distinguish between atypical and benign meningiomas is a major achievement of the WHO classification of 2000 compared with the

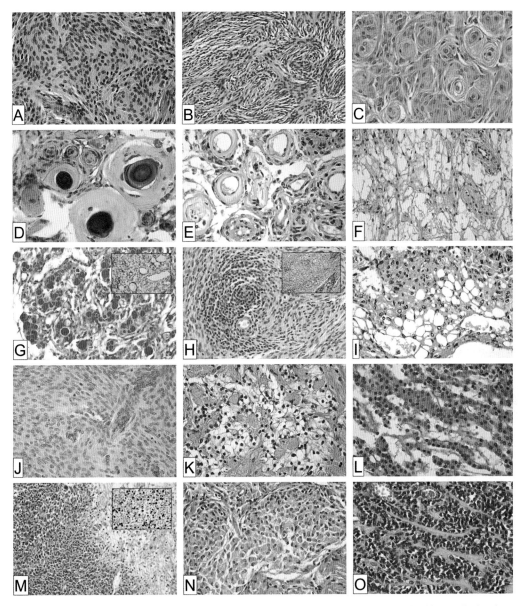

Figure 2 Histological features of meningioma. **A**, Meningothelial meningioma composed of uniform cells that form lobules. **B**, Fibrous meningioma formed by spindle-shaped cells. **C**, Transitional meningioma with numerous whorls. **D**, Psammomatous meningioma with lamellar calcifications (psammomas). **E**, Angiomatous meningioma containing numerous fibrotic blood vessels. **F**, Microcystic meningioma characterized by large intracellular cysts. **G**, Secretory meningioma with periodic acid-Shiff (PAS)–positive pseudopsammoma bodies surrounded by tumor cells immunoreactive for carcino-embryonic antigen (*inset*). **H**, Lymphoplasmacyte-rich meningioma with lymphocytes and plasma cells immunoreactive for leukocyte common antigen (*inset*). **I**, Metaplastic meningioma with lipomatous differentiation. **J**, Atypical meningioma with increased mitotic activity. **K**, Clear cell meningioma composed of cells with clear glycogen-rich cytoplasm. **L**, Chordoid meningioma with a trabecular cell pattern. **M**, Anaplastic meningioma with high cell density, pleomorphism, necrosis, and high proliferative activity (*inset: MIB-1 labeling*). **N**, Rhabdoid meningioma composed of large eosinophilic tumor cells with eccentric nuclei. **O**, Papillary meningioma with perivascular pseudopapillae.

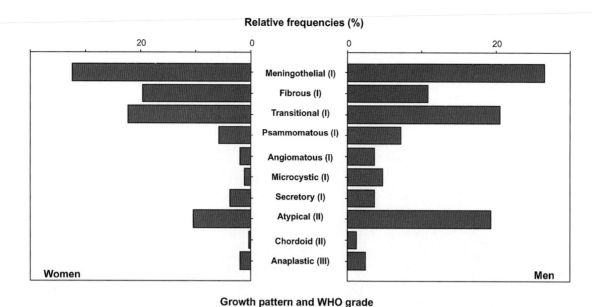

Figure 3 Distribution of meningioma subtypes classified according to the WHO classification of 2000 in the Institute of Neuropathology, Hamburg, Germany, from 2001–2005. N, 351 cases.

previous classification of 1993. Diagnostic criteria for atypical meningiomas are now defined as either the presence of increased mitotic activity (defined as 4 or more mitoses per 10 high-power fields of $0.16\,\text{mm}^2$) or at least three of the following features: increased cellularity, small cells with high nucleus:cytoplasm ratios, prominent nucleoli, uninterrupted patternless or sheetlike growth, and necroses (see Fig. 2J). In addition, clear cell (see Fig. 2K) and chordoid (see Fig. 2L) meningiomas are also classified as WHO grade II.

Anaplastic (malignant) meningiomas (MIII) are rare and constitute between 1% and 3% of meningiomas. They display histological features of frank malignancy far in excess of the abnormalities present in atypical meningiomas (see Fig. 2M). Such features include either a high mitotic index (20 or more mitoses per 10 high-power fields) or an obviously malignant cytology (e.g., resemblance to sarcoma, carcinoma, or melanoma). Due to their highly aggressive behavior, the rare variants of rhabdoid (see Fig. 2N) or papillary (see Fig. 2O) meningioma also belong to the category of WHO grade III meningiomas.

Notably, neither brain invasion nor proliferation indices such as the MIB-1 labeling index were incorporated into the current WHO grading criteria. Some authors prefer to assign brain-invasive meningiomas to WHO grade II because these meningiomas tend to recur and behave similar to atypical meningiomas [6]. However, invasion is a feature more pertinent to staging than malignancy grade, and molecular genetic studies have failed to show alterations typical of high-grade meningiomas in structurally benign meningiomas with brain invasion. Proliferation indices were not included in grading criteria because it is impossible to determine reliable cut-off levels for different grades due to high variations among institutions and individual researchers who quantify and interpret the stains for proliferating cells. Nevertheless, proliferative activity and the presence or absence of brain invasion provide valuable additional prognostic information, and the working group who devised the current WHO classification recommended the addition of phrases such as "with brain invasion" or "with high proliferative activity" to diagnoses whenever appropriate.

B. Immunohistochemistry

A variety of antigens expressed by meningiomas can help distinguish these tumors from other neoplasias with similar morphology. However, none of these antigens are exclusively typical of meningiomas. The most specific marker is epithelial membrane antigen (EMA), which is expressed by the vast majority of meningiomas, although less consistently so in the malignant variants. All meningiomas stain for vimentin; however, the discriminatory value of this staining is poor because vimentin is also expressed by most hemangiopericytomas and sarcomas. The most helpful marker to distinguish hemangiopericytomas from meningiomas is CD34, which is typically expressed by hemangiopericytomas but absent in meningiomas. Secretory meningiomas usually display strong immunoreactivity for carcino-embryonic antigen (CEA) in the pseudopsammoma bodies and for cytokeratins in cells adjacent to these bodies (see Fig. 2G). Other membrane and intercellular junction–associated molecules such as desmoplakin, E-cadherin, or connexin have been stained for research purposes but have not yet been used in routine pathological work.

IV. Clinical Aspects

A. Clinical Presentation

The majority of meningiomas are attached to the dura mater and arise within the intracranial cavity, the spinal canal, or (rarely) the orbit (Fig. 4). Some meningiomas develop as primarily intraosseous tumors, especially those related to the sphenoid wing. The clinical presentation depends upon the tumor location. Meningiomas are usually slow-growing tumors, and they can sometimes become very large in size before the patient experiences symptoms. The most frequent symptom is seizure, which occurs in approximately 40% of meningioma patients. Other signs are neurological deficits due to brain, cranial nerve, or spinal compression; symptoms of raised intracranial pressure due to tumor size and edema; ventricular obstruction and hydrocephalus; obstruction of the dural venous sinuses; ataxia in the case of posterior fossa meningiomas; and others [7,8].

B. Neuroradiology

Meningioma is diagnosed by contrast-enhanced computed tomography (CT), magnetic resonance imaging (MRI) scan, or both. CT is particularly useful to delineate bone involvement, calcification, and hyperostosis. The typical appearance on CT scans is that of a well-demarcated, dural-based mass that enhances brightly and homogenously with contrast and frequently may be calcified. However, meningiomas that grow *en plaque* along the dura, without ever forming a larger mass, are difficult to detect on CT scans. In these cases, MRI is mandatory. On unenhanced T1-weighted MRI images, meningiomas typically appear isointense or hypointense to grey matter. On unenhanced proton-density or T2-weighted images, they are typically isointense or hyperintense. MRI is also valuable in the evaluation of the extent of peritumoral edema, which appears as hyperintensity on T2-weighted MRI. Extensive edema is typically associated with the secretory variant of meningioma (Fig. 5) but can also occur in other subtypes. Angiography has only a very limited place at present and mainly is used to assess the patency of infiltrated sinuses and the collateral venous situation. In rare instances, a highly vascularized basal meningioma of the posterior fossa or tentorium is selected for preoperative embolization via meningeal feeders.

C. Prognosis

The most important factors that determine the likelihood of meningioma recurrence are the extent of tumor resection and histological grade. Extent of surgical resection is still classified according to a five-step scheme devised by Simpson in 1957, ranging from grade 1 (complete resection) to grade 5 (decompression only) [9]. The location in which meningiomas arise is critical for tumor resectability and prognosis. Whereas benign meningiomas of the convexity are usually curable by surgical resection, skull base tumors often have an unfavorable outcome because they cannot be removed completely. In particular, meningiomas that arise in the petroclival region, the cavernous sinus, or the orbit often display slow but relentless growth, leading to widespread invasion and bone destruction. These tumors often remain histologically benign over many years but also can eventually progress to a higher histological grade.

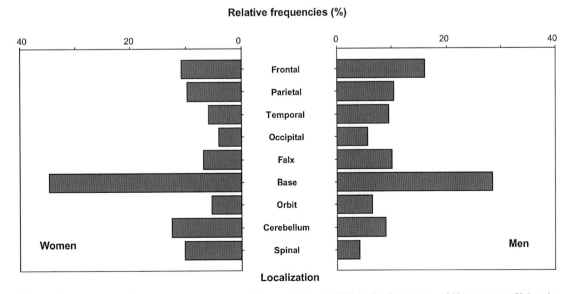

Figure 4 Localization of meningiomas operated on between 1984 and 2005 in the Department of Neurosurgery, University Hospital Hamburg-Eppendorf, Germany.

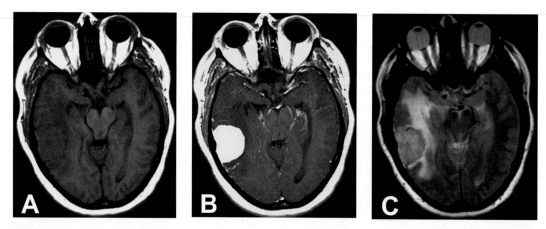

Figure 5 Axial magnetic resonance images of a 47-year-old female patient with the typical features of a secretory meningioma. **A**, T1 weighted images without gadolinium. **B**, T1 weighted images with gadolinium, showing a strong homogeneous enhancement. **C**, TurboSpin Echo (TSE) sequence showing the pronounced collateral edema, which, in addition to the mass effect of the tumor, causes an additional midline shift that results in partial obliteration of the basal cisterns and beginning midline shift.

Generally, WHO grade I meningiomas carry a relatively low risk of recurrence (7–20%) and aggressive behavior. The risk of recurrence for atypical meningiomas is 30–40%; for anaplastic meningiomas, 50–80%. Malignant meningiomas are usually fatal within less than 2 years after diagnosis. In addition to infiltrative and destructive local growth, these tumors can metastasize to extracerebral locations, most commonly the lungs and bones. Interestingly, skull base meningiomas are more frequently benign than tumors of the convexity ($p < 0.001$ in our institutional series). However, despite their more favorable histology, skull base tumors tend to carry a worse prognosis than convexity meningiomas because they are often only incompletely resectable.

D. Treatment

Most meningiomas require treatment. In asymptomatic meningiomas in elderly people, which are found incidentally, it is common practice to "wait and see" and only plan therapy when progression has been documented. The primary therapy for a meningioma is resection. However, a paradigm shift has occurred in respect to radicality. In cases in which radiographically complete resection results in unacceptable morbidity, tumor residuals now are left behind and referred to stereotactic radiotherapy (either single-dose radiosurgery or a standardized fractionated regimen). Meningiomas that are restricted to the cavernous sinus are now primarily treated by such modalities. Conventional radiotherapy has been reported to be of clinical benefit 1) after incomplete resection, 2) after recurrence, and 3) when tumor histology shows atypia or anaplasia. However, none of these studies were randomized, controlled, or prospective, and few had long-term follow-ups [8]. Chemotherapy has yielded disappointing results for meningiomas. Temozolomide, which is effective against malignant gliomas, had no effect against recurrent meningiomas in a phase II study [5]. Hydroxyurea is a powerful inhibitor of meningioma cell proliferation *in vitro* and can inhibit meningioma growth in nude mice. However, clinical studies of patients with nonresectable, recurrent, and/or residual meningiomas are controversial. Although some investigators reported rare cases with tumor regression or modest long-term tumor control, others found hydroxyurea to be of no benefit in the treatment of patients with meningioma [5,7].

V. Pathogenesis and Molecular Mechanisms

A. Genetics

1. Chromosome 22

Early cytogenetic studies discovered monosomy of chromosome 22 in as many as 70% of meningiomas [10,11]. This was the first characteristic cytogenetic alteration associated with a solid neoplasm. Molecular genetic studies identified loss of heterozygosity (LOH) at polymorphic markers on 22q in 40–70% of all meningiomas [11]. Subsequently, mutations in the *NF2* gene were detected in as many as 60% of sporadic meningiomas and were found to be typically associated with LOH 22q.

The *NF2* gene product is termed *merlin* (or *schwannomin*). Strongly reduced or absent immunoreactivity for merlin can be found in the majority of meningiomas and is typically associated with LOH 22q. Overexpression of merlin in both *NF2*-negative and *NF2*-positive meningioma cells inhibits their proliferation *in vitro*, suggesting that merlin is a negative regulator of tumor growth. Furthermore, loss of merlin in mouse embryonic fibroblasts was found to be associated with defects in cell growth and motility. However, the mechanism by which merlin inhibits tumor growth and acts as a tumor suppressor protein is poorly understood. Merlin is a member of the 4.1 family of structural proteins and shares sequence similarity with other members of this family, specifically ezrin, radixin, and moesin (ERM proteins). ERM proteins can bind to actin and spectrin. In addition, merlin can interact with several interacting proteins including a sodium-hydrogen exchange regulatory factor (NHE-RF), hepatocyte growth factor–regulated tyrosine kinase substrate (HRS), schwannomin interacting protein-1 (SCHIP-1), paxillin, and other ERM proteins, as well as with the transmembrane-signaling proteins β1-integrin and CD44. One possible mechanism for protein 4.1 growth regulation involves the engagement of protein 4.1 tumor suppressors at the cell surface in association with transmembrane proteins such as CD44 and propagation of the growth-arrest signal through downstream interacting effectors such as HRS [12].

Several groups reported that the frequency of *NF2* inactivation differs among the three major groups of WHO grade I meningiomas. Fibroblastic and transitional meningiomas carry mutations in 70–80% of cases, a frequency that is similar to those in atypical and anaplastic variants. In contrast, meningothelial meningiomas harbor mutations in only 25% of cases. In addition, a close correlation between LOH 22q and the fibroblastic variant was found. Correspondingly, reduced merlin immunoreactivity was detected in the majority of fibroblastic and transitional meningiomas, but rarely in meningothelial ones. These observations suggest that the genetic origin of meningothelial tumors is largely independent of *NF2* gene inactivation. In addition, the similar frequency of *NF2* mutations in atypical and anaplastic as well as in fibroblastic and transitional meningiomas suggests that *NF2* mutations are not responsible for the progression to higher-grade meningiomas. Instead, *NF2* mutation appears to be an early event that is already involved in the formation of most benign meningiomas (Fig. 6).

Several findings suggest that other tumor suppressor genes on chromosome arm 22q could also be involved in meningioma pathogenesis: 1) the frequency of LOH 22q exceeds that of *NF2* mutations in meningiomas, and molecular genetic studies discovered interstitial deletions on 22q that did not include the *NF2* locus in some tumors; 2) several non-*NF2* families with multiple meningiomas in different members have been described; 3) in a patient with multiple meningiomas no *NF2* mutation but monosomy for chromosome 22 was found.

Genes on 22q that were screened in the context of meningiomas include *ADTB1* (β-*adaptin, BAM22*), *RRP22*, and *GAR22*, all of which map to the 22q12.2 region in relatively close proximity to the *NF2* gene. However, no mutations in these genes have been reported thus far, although 12% of

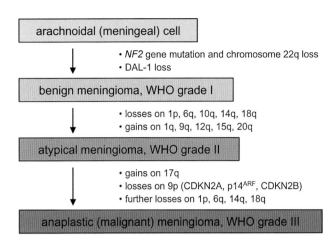

Figure 6 Hypothetic model of genetic alterations associated with the formation of benign meningiomas and progression toward atypia and anaplasia. Only those alterations that were found in more than 30% of tumors of a specific grade in the majority of studies are listed as contributing to the development of a particular grade. Alterations that increase in frequency by at least 20% (averaged over several studies) in tumors of the next higher histological grade, despite already contributing to the previous grade, are listed more than once.

sporadic meningiomas lacked *ADTB1* transcripts. Several other candidate genes map outside the 22q12 region. *MN1* at 22q12.2 was found to be disrupted in a single meningioma; however, it acts as an oncogenic transcription coactivator rather as a tumor suppressor. The *hSNF5/INI1* (*SMARCB1*) gene at 22q11.23 is frequently mutated in atypical teratoid/rhabdoid tumors. An identical missense mutation in this gene was identified in 4 of 126 meningiomas, 3 of which also contained an *NF2* mutation. Expression of the *CLTCL1/CLH-22* gene at 22q11.21 was found to be absent in 80% of meningiomas analyzed, including cases both with and without chromosome 22 deletions. However, mutations in this gene have not been reported, and the mechanism of its loss of expression is unclear. The *LARGE* gene at 22q12.3 is another interesting candidate; however, it is one of the largest human genes, and mutation or expression analyses have not been published to date.

2. Chromosome 18

The 4.1B (DAL-1) protein shares significant homology with merlin and also has tumor suppressor properties. The *4.1B* gene maps to chromosomal region 18p11.3. Similar to merlin, 4.1B also interacts with CD44 and βII-spectrin but not with HRS or SCHIP-1. It further binds to 14-3-3 and to a transmembrane protein, tumor suppressor in lung cancer-1 (TSLC1) [12]. The 14-3-3 family of proteins are important regulators of cell survival and apoptosis, and TSLC1 is important for cell growth as well as adhesion and motility so that different pathways may contribute to the tumor suppressive effects of 4.1B.

Immunoreactivity for 4.1B is lost in 76% of sporadic meningiomas, a frequency similar to that reported for loss of merlin. Lack of the 4.1B protein is not significantly more frequent in anaplastic meningiomas (87%) than in benign and atypical ones (70–76%), suggesting that, similar to loss of merlin, it represents an early event in meningioma tumorigenesis [12]. LOH at 18p11.3 was detected in 71% of investigated cases. However, despite multiple attempts, mutations could not be identified within the nondeleted allele. Therefore epigenetic silencing (e.g., by methylation) might be operative in 4.1B-deficient tumors.

Combined loss of 4.1B and merlin was detected in 58% of investigated cases, suggesting that both are not part of the same growth regulatory pathway in which inactivation of either member causes the same effect. A tendency toward more frequent combined 4.1B and merlin loss was observed in anaplastic meningiomas (70%) compared with atypical (60%) and benign (50%) meningiomas, suggesting that although both alterations are considered early changes, simultaneous loss of both proteins may provide a selective growth advantage.

Comparative genomic hybridization (CGH) analyses frequently identified losses of chromosomal material from the long arm of chromosome 18. These were associated with increasing histological grade and therefore are presumed to be associated with meningioma progression (see Fig. 6). Losses on 18q were detected in 67% of WHO grade III and 40% of WHO grade II but only 13% of WHO grade I meningiomas. The tumor suppressor genes *MADH2*, *MADH4*, *APM-1*, and *DCC*, which are located at 18q21 were analyzed for mutations and expression levels in 37 meningiomas. However, only one *APM-1* mutation was found, and transcripts for all four genes were detectable in all tumors; therefore alterations of these genes do not seem to play a major role in meningiomas. CGH analyses suggest that a putative meningioma progression–associated tumor suppressor gene may be located distally to the 18q21 region, between 18q22 and qter.

3. Chromosome 1

Deletions on the short arm of chromosome 1 are the second most common chromosomal abnormality in meningiomas. The frequency of 1p deletions increases with tumor grade and was observed in 13–26% of benign, 40–76% of atypical, and 70–100% of anaplastic meningiomas [13]. In addition, 1p deletions were reported to be associated with tumor progression in several individual cases of recurrent meningiomas. These findings implicate loss of genomic information from 1p in meningioma progression rather than tumor formation (see Fig. 6).

The prototypical erythrocyte protein 4.1 gene (*4.1R*) at 1p36 is another member of the 4.1 family that is relevant to meningiomas [12]. Defects in protein 4.1R were shown to be responsible for a form of hereditary elliptocytosis. Protein 4.1R also interacts with spectrin and CD44, similar to merlin and protein 4.1B. Loss of protein 4.1R expression was found in 2 meningioma cell lines and in 6 of 15 sporadic meningiomas. In addition, overexpression of protein 4.1R resulted in reduced meningioma cell proliferation *in vitro*, suggesting that protein 4.1R functions as another tumor suppressor in the pathogenesis of meningioma.

Other genes on 1p that were screened for alterations in meningiomas include *CDKN2C* ($p18^{INK4c}$, located at 1p32), *TP73* (1p36.32), *RAD45L* (1p32), and *ALPL* (1p36.1-p34). However, in three studies of more than 100 meningiomas, only one mutation in the *CDKN2C* gene and one homozygous deletion were found. No loss of *CDKN2C* transcripts and no aberrant methylation were discovered, which suggests that this gene is rarely altered in meningiomas. Similarly, no *RAD54L* mutations were detected in a series of 29 meningiomas; however, an association between a silent polymorphism at nucleotide 2290 and a higher frequency of meningioma development was discovered. The *ALPL* gene encodes alkaline phosphatase. Loss of alkaline phosphatase (ALP) enzymatic activity was reported to be strongly associated with loss of 1p in meningiomas, which led to the speculation that *ALPL* might have tumor suppressor function. However, alterations of the *ALPL* gene have not been documented, and functional evidence of tumor suppressor properties of ALP is lacking. Another candidate gene is *TP73*. However, only one mutation was found in more than 50 meningiomas analyzed, which argues against a significant role of *TP73* inactivation in meningioma progression. Interestingly, p73 protein expression was reported to increase with tumor grade, suggesting that it might have a dominant oncogenic function rather than a tumor suppressor function.

Deletion mapping studies suggest that more than one tumor suppressor gene on 1p may be involved in meningioma progression. At least two distinct regions of common deletion could be identified on 1p: one was initially mapped to 1p32, the other to 1pter-p34. However, sequence information from the Human Genome Project indicates that mapping data, previously assembled mainly on the basis of recombination events, require revision in several instances. According to current databases (genome.cse.ucsc.edu/cgi-bin/ and ncbi.nlm.nih.gov/mapview/maps.cgi), many of the microsatellite markers previously used for deletion analyses meanwhile have been mapped to other chromosomal regions than initially thought. For example, the commonly deleted region that was mapped to 1p32 would now map to 1p34-p32 [13]. An extensive reanalysis of the published studies would be necessary to obtain reliable information regarding cluster regions of deletion on 1p. Naturally, this confusion is not restricted to meningiomas; deletions on 1p also are very common in many other tumor types, including oligodendrogliomas, neuroblastomas, and many epithelial cancers.

4. Chromosome 14

Cytogenetic studies identified loss of chromosome 14 as the third most common chromosomal abnormality following aberrations of chromosomes 22 and 1 [11]. Deletions on chromosome arm 14q appear to be associated with meningioma progression (see Fig. 6). LOH and fluorescence *in situ* hybridization (FISH) analyses detected deletions on 14q in as many as 31% of MI cases, 40–57% of MII, and 55–100% of MIII [13]. Evidence suggests that deletions on 14q are more common in meningiomas that subsequently recur.

No specific tumor suppressor gene relevant to meningiomas has yet been identified on chromosome 14. Deletion mapping studies identified several commonly deleted regions on 14q; however, as mentioned in the discussion of 1p, these findings need to be interpreted cautiously because currently available sequence information often places the previously used markers in a different order. Reported cluster regions of deletion on the long arm of chromosome 14 include 14q24.3-q32.3, 14q22-q24, 14q21, and others. The most obvious conclusion from this diversity is that no consistent commonly deleted region has been identified yet on 14q, and no particular tumor suppressor gene has emerged thus far as prime candidate for a meningioma progression gene.

5. Chromosome 10

Deletions on chromosome 10 are nearly as common in meningiomas as are deletions on 14q. Likewise, they seem to be a relatively late event associated with meningioma progression rather than formation (see Fig. 6). LOH and CGH analyses have identified deletions predominantly on the long arm of chromosome 10. Frequencies of 10q deletions in several large studies were between 5% and 12% for MI cases, between 29% and 40% for MII, and between 40% and 58% for MIII [13]. Correlative analysis revealed an unfavorable prognostic significance for LOH at microsatellites mapping to 10p14 and 10q26.3, which predicted higher tumor grade, and of microsatellites at 10q26.12 and 10q26.3, which predicted shorter survival and/or faster recurrence.

No specific tumor suppressor gene on chromosome 10 has yet been found to be inactivated in a substantial fraction of meningiomas. The *PTEN* gene at 10q23.3 was analyzed in a large number of cases; however, mutations were only detected in two cases of MIII, and homozygous deletions were also absent. Likewise, no homozygous deletions were detected in the *DMBT1* gene at 10q26.11-q26.12.

Mapping studies have defined various distinctive, commonly deleted regions on chromosome 10, including at least three regions on 10q and at least one on 10p. Again, these studies require reanalysis on the basis of currently available sequence information. Given the relatively large number and broad distribution of commonly deleted regions identified, it is to be expected that a variety of candidate tumor suppressor regions, rather than a single cluster, would emerge from such an analysis.

6. Chromosome 9

Loss of genetic material on chromosome 9 occurs frequently in malignant meningiomas but rarely in benign or atypical ones. In particular, the short arm of chromosome 9 has received attention because it contains the tumor suppressor genes *CDKN2A* ($p16^{INK4a}$/*MTS1*), $p14^{ARF}$, and *CDKN2B* ($p15^{INK4b}$/*MTS2*) at 9p21, which are inactivated at a high rate in a variety of human tumors. CGH and microsatellite analyses revealed losses on 9p in 38% of MIII cases, 18% of MII, and 5% of MI [13]. Using FISH analysis, the frequencies of detected deletions were about two- to threefold higher, most likely due to the fact that FISH analysis on tumor sections also detects alterations that are restricted only to certain tumor areas or subsets of cells.

Homozygous deletions of *CDKN2A*, $p14^{ARF}$, and *CDKN2B* were found in 36% of MIII cases and 3% of MII, but never in MI. In 2 of 13 MIII cases investigated, somatic point mutations in *CDKN2A* and $p14^{ARF}$ were found. In addition, five tumors without homozygous loss or mutation lacked transcripts for at least one of the three genes (5% of MI cases, 9% of MII, and 8% of MIII). Thus the majority of MIII displays alterations of *CDKN2A*, $p14^{ARF}$, and *CDKN2B*, whereas these are infrequent in MI. Another study showed that 36% of nonmalignant but 71% of malignant meningiomas lacked p14 protein, and only about 26% of MI and MII but 57% of MIII cases lacked both p16 and p15 proteins. Furthermore, *CDKN2A* alterations appear to be of prognostic relevance because MIII patients with *CDKN2A* deletions had a shorter survival. Taken together, these studies demonstrate that both the pRB pathway (p15 and p16) and the p53 pathway (p14) are disrupted in the majority of cases of MIII [13].

7. Chromosome 17

The tumor suppressor gene *TP53* on chromosome arm 17p is one of the most frequently mutated genes in human neoplasias. However, several relatively large studies reported either no meningiomas or only single meningiomas with sporadic *TP53* mutations [13]. *TP53* mutation can lead to increased stability of the p53 protein, which then becomes detectable by immunohistochemistry. Several studies demonstrated an association between p53 accumulation in meningiomas and malignancy grade, although other studies contradict this finding. Given the absence of p53 mutations in meningiomas, the significance of p53 accumulation in meningiomas is unclear.

CGH analyses suggest that the long arm of chromosome 17 may contain one or more genes relevant to meningiomas. High-level amplification on 17q was detected in

42% of MIII cases but in almost no cases of nonmalignant meningiomas. Subsequently, copy-number increases at microsatellite loci on 17q were identified in 61% of MIII cases but in only 21% of MII and 14% of MI. However, high levels of amplification were present in only a small minority of cases. Amplification patterns defined several distinct common regions of increased allele copy number, one of which contained the putative proto-oncogene *PS6K* at 17q23. Whereas a FISH analysis identified *PS6K* amplification in 3 of 22 MIII cases, a real-time polymerase chain reaction (PCR) study found only low-level copy-number increases in 13 of 44 meningiomas (4 of 19 MII and 9 of 18 MIII); instead, high-level amplification at adjacent loci in 2 MIII cases were detected in the latter study. Thus *PS6K* is unlikely to be the major target gene for high-level amplifications in malignant meningiomas, although its amplification in tumor cell subpopulations (as detected by FISH) may have some relevance to progression.

8. Other Chromosomes

Numerous additional chromosomal alterations have been identified in meningiomas, although less consistently than those described previously. Relatively frequent losses occur on 3p, 6q, X, and Y, as do gains on 1q, 9q, 12q, 15q, and 20q [13]. Few studies have looked for specific gene alterations on these chromosome arms. Some studies reported that the *CDK4* and *MDM2* genes on 12q are rarely amplified in meningiomas. No other specific gene alterations have yet been identified on the less commonly altered chromosome arms.

9. Telomerase

Telomeres consist of stretches of repetitive DNA sequences at the ends of chromosomes. These are progressively shortened with each mitosis until cells become senescent at a critical minimal length. Telomerase is a reverse transcriptase that stabilizes telomere length. It is not active in most normal adult tissues but is active in germ-line and most embryonic cells as well as in many cancer types.

Telomerase activity was detected in 100% of MIII cases and in 58–92% of MII, but only in 3–21% of MI [13]. It was described to be associated with poor clinical outcome in benign meningiomas, suggesting a value as a prognostic marker [13]. Telomerase contains an RNA subunit, telomerase RNA (hTR), which contains a template sequence for telomeric repeat synthesis, in addition to a reverse transcriptase subunit, telomerase reverse transcriptase (hTERT). Generally, hTERT expression correlates best with telomerase activity, whereas hTR can be present even when enzyme activity is absent. The detection rate for hTERT expression in meningiomas is even higher than that for telomerase activity, and all telomerase-positive tumors were found to express hTERT, but not vice versa. In addition, some tumors that subsequently recurred expressed hTERT in the absence of detectable telomerase activity. These observations suggest that expression of hTERT might be an even more sensitive marker for an aggressive clinical course than telomerase activity.

VI. Special Forms of Meningioma

A. Radiation-Induced Meningiomas

Ionizing radiation to the skull is the most important risk factor for meningioma development. Most of the knowledge of this association stems from the observation of immigrants to Israel who had received low-dose cranial irradiation for tinea capitis between 1948 and 1960. Typically, these patients had been treated during childhood and developed meningiomas with a latency period of 20–40 years. Radiation-induced meningiomas (RIM) are often more aggressive (corresponding with MII or MIII) than sporadic meningiomas and are often multiple in number.

Some studies analyzed the genetic mechanisms involved in RIM development. *NF2* mutations and LOH22q are relatively rare in these cases, occurring in less than 25% of RIM compared with more than 50% in sporadic control cases. Mutations in the *TP53* and *PTEN* genes were found to be restricted to single cases, and no mutations were detected in the *HRAS*, *KRAS*, and *NRAS* genes. Allelic losses on 1p were slightly more common in RIM than in sporadic cases, and CGH analysis identified losses on 7p, which are uncommon in sporadic meningiomas, in 67% of RIM. Thus putative tumor suppressor genes on 1p and 7p may play a role in RIM development, although the available series are too small to draw reliable conclusions.

B. Multiple Meningiomas

Multiple meningiomas are spatially separate, independently arising tumors. They occur in 1–8% of meningioma patients. Multiple meningiomas are particularly frequent in NF2 patients and in the rare non-NF2 families with a hereditary predisposition to meningioma development. Several studies addressed the question of whether multiple meningiomas are due to metastatic meningeal seeding of a primary tumor or representative of *de novo* formation of separate tumors. Identical *NF2* mutations and patterns of X chromosome inactivation indicate that the majority of multiple and recurrent meningiomas are of clonal origin, so multiplicity most likely represents subarachnoid spread in these cases. Alternatively, *NF2* mosaicism could be responsible for some cases. Interestingly, *NF2* mutations and clonal origin were more frequently detected in patients with larger numbers of tumors than in patients with only two tumors, which is compatible with either mosaicism or with an increased ability of seeding via the CSF caused by a mutant *NF2* gene.

One study focused on differences between multiple meningiomas in sporadic versus familial non-NF2 cases. All detected *NF2* mutations occurred in tumors from patients with no affected relatives, suggesting that *NF2* inactivation is frequently involved in multiple sporadic meningiomas but is rare in multiple meningioma kindreds. In line with this finding, linkage analysis of a multiple meningioma family showed no segregation with the *NF2* locus. In one multiple meningioma kindred, merlin immunoreactivity of the tumors cells was detected, indicating that the *NF2* gene was not inactivated. Interestingly, most non-NF2 meningioma family members develop meningothelial meningiomas, which is in line with the observation that this variant appears to arise independently of *NF2* inactivation.

C. Meningiomas in NF2 and Pediatric Meningiomas

Most NF2 patients develop meningiomas. NF2-associated meningiomas differ from their sporadic counterparts in several respects: 1) they usually manifest several decades earlier in life; 2) they are often multiple; and 3) they more frequently belong to the fibroblastic variant. NF2-associated meningiomas display deletions on 22q in almost 100% of cases, a frequency that is significantly higher than in sporadic meningiomas. Additional genetic alterations such as deletions on 1p, 6q, 9p, 10q, 14q, and 18q occur at a similar frequency as in sporadic tumors [14].

Most studies found that the frequency of atypical or malignant meningiomas was not increased among NF2-associated tumors. The MIB-1 labeling index in NF2-associated meningiomas was slightly higher in one study, but a more detailed analysis of a larger number of cases revealed no difference compared with sporadic meningiomas. The experience in our institution is that NF2 patients rarely develop malignant meningiomas and that the mortality in these patients is usually due to other causes.

Pediatric meningiomas present most commonly in the second decade of life but may occur at any age, including infancy. These tumors show a higher frequency of large tumor size, high-grade histology, aggressive behavior, and aggressive variants in particular clear cell and papillary meningiomas. Meningiomas in patients younger than 18 years of age were observed in 18 of 1155 cases in the Hamburg series, including 8 WHO grade I meningiomas, 4 atypical meningiomas, and 6 anaplastic meningiomas. In addition, pediatric meningiomas are more often found in unusual locations, including the lateral ventricles, posterior fossa, and spinal epidural regions. A pediatric meningioma can be the first manifestation of NF2. Another predisposing factors is prior irradiation. Genetically, a high incidence of *NF2* mutation and *4.1B* inactivation, as well as of deletions on 1p and 14q, is found in pediatric meningiomas [15].

VII. Pathophysiology

A. Angiogenesis, Edema, and Growth Factors

The blood supply of meningiomas is predominantly derived from meningeal vessels that originate in the external carotid circulation. In approximately 60% of patients, additional supply from cerebral-pial blood vessels is present. The vascularity of meningiomas varies, ranging from sparsely vascularized tumors to highly vascular angiomatous variants. Meningiomas further display variable degrees of peritumoral brain edema, ranging from absent to life-threatening conditions. In particular, secretory meningiomas are often associated with extensive brain edema (see Fig. 5). In contrast to gliomas, no obvious association exists between edema and histological grade in meningiomas.

A key regulator of angiogenesis and edema formation in tumors is vascular endothelial growth factor-A (VEGF-A), which has also been termed *vascular permeability factor*. Meningioma cells express the VEGF-A mRNA. Several studies found VEGF-A levels in meningiomas to correlate with the extent of peritumoral edema [13]. In addition, meningiomas with striking VEGF-A expression were found to usually receive some blood supply through cerebral-pial arteries, and the presence of cerebral-pial blood supply also appears to correlate with the extent of brain edema.

Two small studies suggested that VEGF-A mRNA expression may correlate with the degree of vascularization in meningiomas. However, when VEGF-A protein levels were measured in 69 meningiomas, no association with microvessel density was found [16]. Nor was there a correlation between microvessel density and histological grade, which is in contrast to gliomas. However, VEGF-A levels were increased tenfold in MIII and twofold in MII compared with MI. Moreover, protein extracts of meningiomas that contained VEGF-A induced the formation of capillary-like endothelial tubes *in vitro*. Another study reported a correlation between VEGF-A protein expression and recurrence of MI. Taken together, these studies suggest that VEGF-A is probably involved in vascular remodeling and angiogenesis in meningiomas; however, this is not reflected in a net increase in vessel number with increasing histological grade. Because malignant meningiomas supposedly have a higher oxygen and metabolic demand than benign ones, VEGF-A might facilitate adaptation by modulating vascular permeability.

Several other growth factors, including VEGF-B, placenta growth factor, scatter factor/hepatocyte growth factor, fibroblast growth factor-2, epidermal growth factor

(EGF), platelet-derived growth factor (PDGF), and insulin-like growth factor (IGF) have also been studied in meningiomas [12,17,18]. Autocrine loops between these factors and their receptors may contribute to meningioma growth. For example, PDGF-BB and its receptor PDGFR-β are overexpressed in meningiomas. The EGF receptor (EGFR) is almost universally expressed in meningiomas but not in normal or reactive meningothelial cells, and the EGF ligands transforming growth factor-alpha (TGF-α) and EGF are also expressed by the tumor cells. In contrast to glioblastomas, the EGFR gene is not amplified in meningiomas, which suggests other mechanisms for protein overexpression. A high ratio between IGF-II and its receptor, IGF binding protein-2 (IGFBP-2), appears to be associated with meningioma progression. However, it has been difficult to establish an association between histological grade and most other growth factors.

B. Hormones

Several factors suggest a role of sex hormones in meningioma development and growth: 1) meningiomas are more than twice as frequent in females than males; 2) accelerated meningioma growth has been observed during pregnancy and during the luteal phase of the menstrual cycle; and 3) meningiomas seem to occur at a slightly increased rate in breast cancer patients.

Estrogen receptors were first discovered on meningiomas in 1979. Numerous subsequent studies, most of which were conducted in the early 1990s using immunohistochemistry, have looked for both estrogen and progesterone receptors [7,17]. These studies demonstrated that estrogen receptors are only expressed in approximately 10% of meningiomas and only at low levels [17–19]. In addition, the effect of the antiestrogen agent tamoxifen has been clinically disappointing [7].

Progesterone and androgen receptors are present in approximately two-thirds of meningiomas. Both are more frequently expressed in females than males; approximately 80% of meningiomas in females are progesterone receptor positive compared with only 40% in males. Progesterone receptor expression is inversely associated with histological grade and is a favorable prognostic factor. Proliferation of meningioma cells can be inhibited by the progesterone receptor antagonist mifepristone (RU486) *in vitro*. However, mifepristone showed no benefit compared with placebo in a phase III double-blind randomized study [7]. Antiandrogens also have antiproliferative effects on meningioma cells in culture, but their clinical usefulness has not yet been investigated.

Meningiomas also express nonsteroid hormone receptors, including growth hormone, and somatostatin. Some experimental *in vitro* or *in vivo* studies demonstrated antiproliferative effects of the growth hormone receptor antagonist pegvisomant and the somatostatin agonist octreotide. However, full-scale clinical trials with these substances have not been reported.

C. Neurotransmitters

In an experimental model, fetal exposure to the irreversible dopamine receptor blocking agent 6-OH-dopamine causes complete ablation of the meninges in rats [20]. Consequently, dopamine receptors could also be demonstrated on human meningiomas; however, a therapeutic exploitation had disappointing results. Bromocriptine was used in exploratory small studies that unfortunately were unconvincing, so no large-scale trials have followed.

VIII. Cell Lines and Animal Models

Only few human meningioma cell lines have been established; these were exclusively derived from malignant meningiomas. When injected subcutaneously or intracranially into nude mice, these cell lines form tumors *in vivo*. However, it is difficult to implant intracranial dural-based tumors in such a way that they accurately model the human condition. An alternative to use of meningioma cell lines is the injection of meningioma cells from patients after short-term culture over only one or two passages. When these cells are mixed with matrigel (an extracellular matrix solution that contains basement membrane proteins and growth factors and enhances tumorigenicity), they form subcutaneous tumors in nude mice. Another alternative is to implant meningioma cells into the renal capsules of nude mice.

In addition to the various xenograft models, genetic models also have been developed that facilitate the study of meningioma growth in animals. Using Cre recombinase technology, the NF2 gene was specifically inactivated in arachnoidal cells, which resulted in the formation of intracranial meningothelial hyperplasia and meningiomas in approximately 30% of the mice [21].

IX. Future Directions

The completion of the Human Genome Project, together with the introduction of increasingly more sophisticated molecular research techniques, facilitate genome-wide studies at a higher resolution than ever before. A major advancement in this context is matrix-CGH (array-CGH) technology, which has an approximately 100-fold higher resolution than conventional CGH analysis and is able to narrow down chromosomal gains or losses to 75,000 bp segments. Another expanding field is that of cDNA microarrays, which have so far been applied to meningiomas in only a few studies. Gene expression profiling revealed characteristic profiles for different meningioma grades and thus can provide prognostic information. Patterns of gene overexpression or underexpression may also lead to discovery of unknown tumor suppressor genes or oncogenes, respectively. However, for some genes that lack expression in a substantial fraction of meningiomas, no structural gene alterations can be identified. Inactivation of these genes apparently does not follow the classic tumor suppressor gene paradigm. Instead, epigenetic inactivation by aberrant methylation or suppression of transcription by other mechanisms probably accounts for the absence of gene expression. Array technology was expanded to methylation analyses, so high-throughput screening may generate comprehensive profiles of methylated genes. The extent to which aberrant methylation can explain inactivation of genes relevant to meningioma formation or progression remains one of the major challenges in the field. In addition, further elucidation of the stepwise molecular alterations that contribute to meningioma formation and progression can be expected from the further refinement of genetic animals models.

References

1. Kepes, J. J. (1982). "Meningiomas." Masson Publishing, New York.
2. Lantos, P. L., VandenBerg, S. R., and Kleihues, P. (1997). Tumours of the nervous system. *In:* "Greenfield's Neuropathology" (D. I. Graham and P. L. Lantos, eds.), vol. 2, pp. 583–879. Arnold, London.
3. Louis, D. N., Scheithauer, B. W., Budka, H., von Deimling, A., and Kepes, J. J. (2000). Meningiomas. *In:* "Pathology and Genetics of Tumours of the Nervous System" (P. Kleihues, and W. K. Cavenee, eds.), pp. 176–184. IARC Press, Lyon, France.
4. Kleihues, P., Louis, D. N., Scheithauer, B. W., Rorke, L. B., Reifenberger, G., Burger, P. C., and Cavenee, W. K. (2002). The WHO classification of tumors of the nervous system. *J. Neuropathol. Exp. Neurol.* **61**, 215–225.
5. Lusis, E., and Gutmann, D. H. (2004). Meningioma: an update. *Curr. Opin. Neurol.* **17**, 687–692.
6. Perry, A., Scheithauer, B. W., Stafford, S. L., Lohse, C. M., and Wollan, P. C. (1999). "Malignancy" in meningiomas: a clinicopathologic study of 116 patients, with grading implications. *Cancer* **85**, 2046–2056.
7. Drummond, K. J., Zhu, J. J., and Black, P. M. (2004). Meningiomas: updating basic science, management, and outcome. *Neurologist* **10**, 113–130.
8. Whittle, I. R., Smith, C., Navoo, P., and Collie, D. (2004). Meningiomas. *Lancet* **363**, 1535–1543.
9. Simpson, D. (1957). The recurrence of intracranial meningiomas after surgical treatment. *J. Neurochem.* **20**, 22–39.
10. Collins, V. P., Nordenskjold, M., and Dumanski, J. P. (1990). The molecular genetics of meningiomas. *Brain Pathol.* **1**, 19–24.
11. Zang, K. D. (2001). Meningioma: a cytogenetic model of a complex benign human tumor, including data on 394 karyotyped cases. *Cytogenet. Cell Genet.* **93**, 207–220.
12. Perry, A., Gutmann, D.H., and Reifenberger, G. (2004). Molecular pathogenesis of meningiomas. *J. Neurooncol.* **70**, 183–202.
13. Lamszus, K. (2004). Meningioma pathology, genetics, and biology. *J. Neuropathol. Exp. Neurol.* **63**, 275–286.
14. Lamszus, K., Vahldiek, F., Mautner, V. F., Schichor, C., Tonn, J., Stavrou, D., Fillbrandt, R., Westphal, M., and Kluwe, L. (2000). Allelic losses in neurofibromatosis 2-associated meningiomas. *J. Neuropathol. Exp. Neurol.* **59**, 504–512.
15. Perry, A., and Dehner, L. P. (2003). Meningeal tumors of childhood and infancy. An update and literature review. *Brain Pathol.* **13**, 386–408.
16. Lamszus, K., Lengler, U., Schmidt, N. O., Stavrou, D., Ergun, S., and Westphal, M. (2000). Vascular endothelial growth factor, hepatocyte growth factor/scatter factor, basic fibroblast growth factor, and placenta growth factor in human meningiomas and their relation to angiogenesis and malignancy. *Neurosurgery* **46**, 938–947.
17. Black, P., Carroll, R., and Zhang, J. (1996). The molecular biology of hormone and growth factor receptors in meningiomas. *Acta. Neurochir. Suppl.* **65**, 50–53.
18. Sanson, M., and Cornu, P. (2000). Biology of meningiomas. *Acta. Neurochir. (Wien.)* **142**, 493–505.
19. McCutcheon, I. E. (1996). The biology of meningiomas. *J. Neurooncol.* **29**, 207–216.
20. Sievers, J., Pehlemann, F. W., Baumgarten, H. G., and Berry, M. (1985). Selective destruction of meningeal cells by 6-hydroxydopamine: a tool to study meningeal-neuroepithelial interaction in brain development. *Dev. Biol.* **110**, 127–135.
21. Kalamarides, M., Niwa-Kawakita, M., Leblois, H., Abramowski, V., Perricaudet, M., Janin, A., Thomas, G., Gutmann, D. H., and Giovannini, M. (2002). Nf2 gene inactivation in arachnoidal cells is rate-limiting for meningioma development in the mouse. *Genes Dev.* **16**, 1060–1065.

36

Primary Central Nervous System Lymphoma

Maciej Mrugala, MD, PhD, MPH
Anne Newcomer
Tracy Batchelor, MD, MPH

Keywords: *B cell, cancer, central nervous system, human immunodeficiency virus, lymphoma*

I. Introduction
II. Epidemiology
III. Pathology and Biology
IV. Clinical Features of PCNSL
V. Treatment Options for PCNSL
VI. Summary
    References

I. Introduction

Primary central nervous system lymphoma (PCNSL) is a rare and aggressive form of extranodal non-Hodgkin's lymphoma (NHL) involving the brain, leptomeninges, or eyes. PCNSL typically remains confined to the nervous system [1]. As a unique subtype of extranodal NHL, PCNSL requires its own diagnostic algorithm and management plan. Unlike other forms of NHL, relatively little is known about the biology of PCNSL. Although insight into the biology of all forms of NHL may be improved with studies of extraneural lymphomas, it is becoming increasingly apparent that PCNSL is associated with a unique set of biological, clinical, and therapeutic features. However, a challenge for researchers interested in PCNSL is the relative paucity of archival or fresh frozen tissue available for research purposes. The disease is commonly diagnosed by stereotactic brain biopsy, and the tissue is typically consumed by the diagnostic process.

II. Epidemiology

Although the disease was considered extremely rare several decades ago, the incidence of PCNSL increased nearly threefold between 1973 and 1984 [2]. However, data suggest that the rates of newly diagnosed PCNSL may be stabilizing or declining slightly (Fig. 1) [3,4]. PCNSL accounted for 3.1% of all primary brain tumors diagnosed in the United States from 1998 to 2002 [5]. The demographics of immunocompetent and immunocompromised patients with PCNSL are varied. Immunocompetent patients who develop PCNSL are predominantly older adults with a slight male dominance [1]. Although PCNSL tends to affect males more than females in both groups, this difference is more pronounced in the immunocompromised population (male:female ratio of 7.38:1) than in the immunocompetent population (1.35:1). The incidence of PCNSL peaks at the age of 57 years in immunocompetent populations, compared with a peak range of 31–35 years in immunocompromised patients [6,7].

A. Immunological Risk Factors

Congenital or acquired immunodeficiency is the only established risk factor for PCNSL. The human immunodeficiency virus (HIV) pandemic is the primary factor responsible for the increase in the incidence of PCNSL. In fact, persons infected with HIV have a 3600-fold higher risk of PCNSL compared with the general population [8]. HIV-related PCNSL is associated with a reduction in the number of circulating CD4+ cells, a factor linked to patient survival [9]. With the advent of highly active antiretroviral therapy (HAART), the proportion of HIV-infected persons with CD4+ cell counts less than $50\,cell/mm^3$ has declined, and this has correlated with a decline in the incidence of PCNSL.

An increased risk of lymphoproliferative diseases such as PCNSL has been seen in organ transplant recipients who receive immunosuppressive drugs. However, the increased risk of PCNSL in allograft recipients was observed mainly in patients who received older, azathioprine-based immunosuppressive regimens that are not commonly used today and is less common in patients treated with cyclosporine-based therapy [10].

The Epstein-Barr virus (EBV), a lymphotropic virus involved in malignant B-cell transformation [11], is involved in the pathogenesis of PCNSL in immunocompromised patients. PCNSL in immunocompromised patients is strongly associated with EBV infection [12]. In the immunocompromised state, chronic immune stimulation by EBV may lead to B-cell proliferation and, ultimately, malignant transformation [13]. Cerebrospinal fluid (CSF) EBV titers

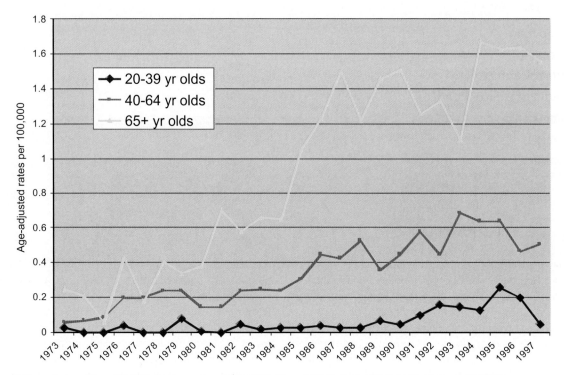

Figure 1 Age-adjusted PCNSL incidence rates, 1973–1977. (From O'Neill, B. P. [2004]. Epidemiology of PCNSL. *In:* "Lymphoma of the Nervous System" [T. T. Batchelor, ed.], pp. 43–51. Butterworth-Heinemann, Boston.)

are higher in acquired immunodeficiency syndrome (AIDS) patients with PCNSL compared with AIDS patients with systemic lymphomas, suggesting a distinct pathogenesis for PCNSL in this patient population. Meeker and colleagues demonstrated that tumor tissue in all PCNSL cases was EBV+ [14], and the EBV CSF titer is widely used as a diagnostic test in the immunocompromised patient population [11]. Conversely, EBV does not appear to be involved in the pathogenesis of PCNSL in immunocompetent patients. Other viruses, namely human herpes simplex type 6 (HHV6) and type 8 (HHV8) and simian virus 40 (SV40), have been implicated in the pathogenesis of PCNSL, although definitive evidence regarding the causative role of these viruses is lacking [15].

Higher frequencies of autoimmune and immunomodulating diseases, as well as prior malignancies, have been observed in patients who develop PCNSL. The nature of these diseases implicates immune dysregulation as a risk factor for PCNSL [1].

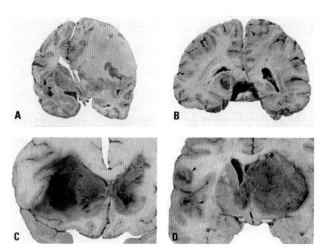

Figure 2 Gross pathology specimens of patients with PCNSL. **A**, Large necrotizing B-cell lymphoma in an HIV-infected infant. **B**, B-cell lymphoma in the medial temporal and occipital lobes. **C** and **D**, PCNSL involving basal ganglia with extension to cortex (*arrows*). (From Kleihues, P. and Cavenee, W. K., eds. [2000]. Tumours of the nervous system. World Health Organization Classification of Tumours, ed X. International Agency for Research on Cancer. 2000, IARC Press, Lyon, France.

B. Genetic Risk Factors

Homocysteine and folate metabolism is closely linked to DNA methylation and contribute substantially to preservation of DNA integrity. Consequently, genetic polymorphisms that influence homocysteine and folate metabolism are associated with different types of cancer including NHL, acute leukemias, and colorectal cancer. Methionine synthase (MS) is an important enzyme in folate metabolism, and the MS missense dimorphism c.2756A > G(D919G) is associated with a lower risk of developing colorectal cancer and systemic NHL. Linnebank et al. analyzed 31 cases of PCNSL and concluded that the MS c.2756A > G(Gallele) may also have a protective function against PCNSL [16].

III. Pathology and Biology

The pathological characterization of PCNSL and tissue-based biological studies of PCNSL have been limited by several factors. First, the diagnosis is typically achieved with a stereotactic biopsy and most of the specimen is used in the diagnostic process. Thus there is a paucity of tissue available for research purposes. Additionally, many patients receive corticosteroid treatment before tissue sampling. Corticosteroids promote apoptosis of malignant lymphoid cells, leading to changes in tumor architecture. This often prevents pathologists from making an accurate histopathological diagnosis and limits the amount of fresh, "untreated" tumor tissue for molecular analysis.

Autopsy examinations of patients with PCNSL have typically demonstrated poorly circumscribed, deep-seated masses in the periventricular regions [1] (Fig. 2). PCNSL diffusely infiltrates brain parenchyma in an angiocentric pattern (Fig. 3). The tumor is composed of immunoblasts or centroblasts with a predilection for blood vessels, resulting in lymphoid clustering around small cerebral vessels. Reactive T-cell infiltrates can also be present to varying degrees, making it difficult for the pathologist to discriminate between PCNSL and a reactive process. Relatively high percentages of activated cytotoxic T lymphocytes (CTLs) have been detected in CNS lymphomas as compared with lymphoma in other sites; yet, the anti-tumor effect of these CTLs may be impaired by an escape mechanism common to aggressive B-cell lymphomas of the brain. Specifically, for CTLs to recognize and attack tumor cells, human lymphocyte antigen (HLA) class I and HLA class II molecules must be expressed by these cells. When expression of these two molecules is lost, an efficient anti-tumor response cannot be mounted. A recent study of HLA I and HLA II expression suggests that loss of HLA expression provides a strong growth advantage for lymphoma cells and is much more common in B-cell lymphomas of the central nervous system (CNS) than in extraneural lymphomas [17].

A. Molecular Pathogenesis of Diffuse Large B-Cell Lymphoma

Diffuse large B-cell lymphoma (DLBCL) is the most common type of systemic lymphoma, comprising approximately 30% of all NHL cases worldwide. DLBCL encompasses a diverse group of aggressive B-cell lymphomas [19]. These tumors arise from B cells arrested at the

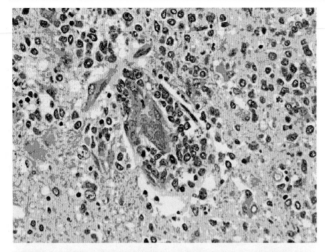

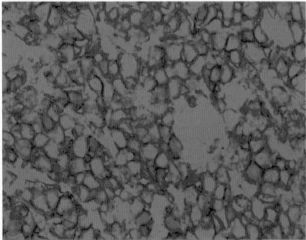

Figure 3 Microscopic section from a tumor specimen in a patient with PCNSL. *Top*, The tumor cells are pleomorphic with large nuclei and a coarse chromatin pattern. Tumor cells are clustered around a cerebral blood vessel in a pattern typical of PCNSL. (Hematoxylin and eosin stain.) *Bottom*, Most of the tumor cells stain positively (brown) with the CD20 (pan-B-cell) immunostain, demonstrating that this is a B-cell malignancy.

certain differentiation stage when neoplastic transformation took place [20]. Based on gene expression profiling, two distinct subgroups of DLBCL have been described: 1) germinal center (GC)-like, and 2) non–GC-like. These two subgroups originate from distinct B cell counterparts and employ distinct mechanisms of malignant transformation [19]. Notably, the clinical outcome for GC-like DLBCL is more favorable compared with that of non–GC-like DLBCL (Fig. 4).

Normal GC B cells have been implicated as the likely origin for GC-like DLBCL, whereas in non–GC-like DLBCL, the activated B cell-like (ABC-like) subtype appears to derive from a post-GC stage such as memory B cells or plasma cells [21,22]. GC-like DLBCL tumors are marked by characteristic genetic variations that distinguish them from non–GC-like DLBCL tumors. Somatic mutations in the variable (V) region of immunoglobulin (Ig) genes are commonly used as markers of passage through the GC because normal pregerminal center lymphocytes harbor unmutated Ig genes. Also, the presence of intraclonal heterogeneity in Ig gene mutations is regarded as a marker of ongoing somatic mutations that occur almost exclusively in the GC and thus is a marker of GC origin [19,23].

DLBCL accounts for 85% of PCNSL cases. The other 15% consists of poorly characterized, low-grade lymphomas, Burkitt's lymphomas, and T-cell lymphomas (Fig. 5). In contrast to extraneural DLBCL, PCNSL is poorly understood in terms of its precise histogenetic origin and molecular pathogenesis. However, recent immunohistochemical studies and gene profiling experiments suggest that PCNSL is derived from GC B cells. Although there is no evidence of GC formation in the brain, the high levels of accumulated somatic mutations, ongoing hypermutation, and *BCL-6* gene mutations—observations consistent with transition through the germinal center—are observed in tumor tissue from immunocompetent PCNSL patients. These results strongly suggest a GC origin for PCNSL [20,24,25].

Preliminary gene expression profiling studies indicate that PCNSL is a distinct biological subtype of large cell lymphoma. In one study involving 35 frozen tumor sections from PCNSL patients, approximately 100 genes had at least a twofold level of differential expression between PCNSL, nodal lymphomas of similar histology, and nonneoplastic brains. The gene expression signature of PCNSL includes genes involved in B-cell differentiation, proliferation, apoptosis, and cytokine signaling [27].

B. Immunophenotype

Braaten et al. [18] investigated the histogenetic origin of PCNSL with respect to stage of B-cell differentiation and possible prognostic markers. Within a cohort of 33 immunocompetent patients with PCNSL, 79% of tumors expressed bcl-6, a protein primarily found in B lymphocytes in the GC stage of development (Fig. 6). A zinc-finger transcriptional repressor encoded by the *BCL-6* gene, bcl-6 promotes cell proliferation and blocks differentiation. Although expression of bcl-6, which is highly associated with the GC, supports the theory that the cell of origin in PCNSL is related to the GC stage of B-cell differentiation, only 19% of tumor samples in this study displayed a definite GC-like phenotype, as denoted by coexpression of two GC antigens, bcl-6 and CD10, combined with an absence of the post-GC antigens, vs38c and CD138. However, 10 additional bcl-6 positive cases were CD10 negative but lacked CD44 and vs38c as well, possibly suggestive of late GC differentiation. Combining these two categories, 50% of the PCNSL cases in these series had either a definite or possible GC-like phenotype. All but one tumor expressed

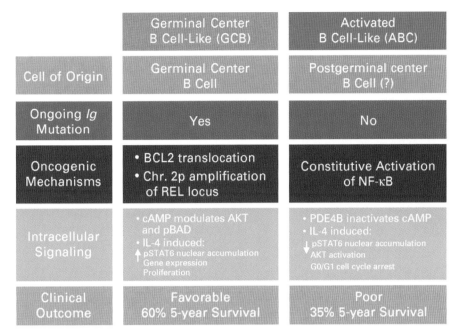

Figure 4 Molecular, pathogenic, and clinical features distinguishing GC–like and activated B-cell–like diffuse large B-cell lymphoma. Translocation involving the *BCL-2* gene and the amplification of the *c-rel* locus on chromosome 2p have been detected exclusively in GC-like DLBCL. NF-κB (nuclear factor B) family members activate transcription and mediate proliferation, apoptosiscell survival. High expression of NF-κB target genes has been observed in ABC-like DLBCL but not in GC-like DLBCL cell lines. PDE4B inactivates cAMP, an intracellular second messenger that modulates several signaling pathways and induces cell cycle arrest and apoptosis of B cells. PDE4B is highly expressed in ABC-like DLBCL. Both GC-like and ABC-like DLBCL cell lines exhibit distinct responsiveness and intracellular signaling in response to IL-4 stimulation. The response to stimulation is different: in ABC-like DLBCL cell lines, IL-4 phosphorylated AKT does not induce a sustained increase in nuclear phosphorylated STAT6 (signal transducer and activator of transcription 6), but in GC-like DLBCL cell lines, IL-4 activates STAT6 intracellular signaling and mildly increases cell proliferation. (From Lossos, I. S. [2005]. Molecular pathogenesis of diffuse large B-cell lymphoma. *J. Clin. Oncol.* **23**, 6351–6357.)

MUM-1, a transcriptional factor involved in differentiation from GC to plasma cells and a generally accepted marker of the ABC-like phenotype. These findings suggest a late-GC or post-GC origin for the PCNSL cases studied. Furthermore, seven tumors were bcl-6-/CD10-, consistent with a non-GC-like immunophenotype. True plasmacytic differentiation, as evidenced by membrane staining for CD138, was not seen, and cytoplasmic staining for vs38c was distinctly rare. Taken together, these immunophenotypic findings are suggestive of a GC or possibly late GC stage of B-cell differentiation for most cases of PCNSL (Table 1). In the same study, 93% of evaluable tumors expressed bcl-2, denoted by strong membranous staining. BCL-2 protein expression prevents cellular apoptosis and is down-regulated by normal GC cells. Its significance in PCNSL has not been determined.

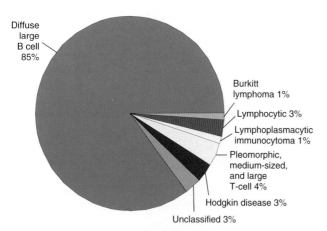

Figure 5 Classification of 72 immunocompromised patients with PCNSL using the Revised European and American Lymphoma (REAL) system. (From Camilleri-Broet, S., Martin, A., et al. [1998]. Primary central nervous system lymphomas in 72 immunocompetent patients: pathologic findings and clinical correlations. *Am. J. Clin. Pathol.* **110**, 607–612.)

Table 1 Immunohistochemistry for GC-Related Antigens

Antigen	Positive Incidence (%)
bcl-6[1]	26/33 (79%)
CD10[2]	6/32 (19%)
MUM1[3]	31/32 (97%)
CD44[4]	11/31 (35%)
vs38c[5]	4/33 (12%)
CD138[6]	0/32 (0%)
BCL-2[7]	27/29 (93%)

GC, Germinal center.
[1] The bcl-6 protein is a zinc-finger transcriptional repressor encoded by the *BCL-6* gene.
[2] CD10 is a marker that consistently reflects GC origin in both reactive lymphoid tissue and lymphomas.
[3] MUM1 is a late GC/early post-GC marker.
[4] CD44 is expressed most strongly by post-GC mantle zone cells.
[5,6] The presence of vs38c and CD138 denote plasmacytic and/or post-GC differentiation.
[7] BCL-2 protein expression prevents cellular apoptosis and is down-regulated by normal GC cells.
From Braaten, K. M., Betensky, R. A., deLeval, L., et al. (2003). BCL-6 expression predicts improved survival in patients with primary central nervous system lymphoma. *Clin. Cancer Res.* **9**, 1063–1104.

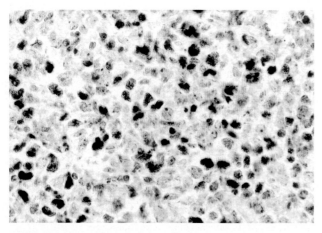

Figure 6 Primary CNS lymphoma cells with nuclear staining for bcl-6 (immunoperoxidase technique). (From Braaten, K. M., Betensky, R. A., deLeval, L., et al. [2003]. BCL-6 expression predicts improved survival in patients with primary central nervous system lymphoma. *Clin. Cancer Res.* **9**, 1063–1104.)

C. Other Genetic Alterations

Montesinos-Rongen et al. investigated the possible role of aberrant somatic hypermutation in growth regulatory genes in the development of PCNSL. These investigators observed extremely high mutation frequencies for *IgH* genes in PCNSL. They also found that PCNSL exhibited aberrant somatic hypermutation with involvement of 4 proto-oncogenes: *PAX5*, *PIM1*, *c-MYC*, and *RhoH/TTF*. These genes have a role in B-cell development and differentiation as well as in the regulation of proliferation and apoptosis [26]. Although mutation frequencies reported for *IgH* genes are much higher (60-fold) than mutation frequencies for *PAX5*, *PIM1*, *c-MYC*, and *RhoH/TTF*, the involvement of these four genes in the pathogenesis of PCNSL is potentially significant. Moreover, individual comparison of these four genes in PCNSL and extraneural DLBCL revealed that their mean mutation frequencies were two- to fivefold higher in PCNSL [26]. Such high mutation frequencies may arise during a prolonged interaction of the tumor cell (or its precursor) in the GC microenvironment [25].

The $p16^{INK4a}$ gene is frequently inactivated by either homozygous deletion (40–50%) or 5'-CpG hypermethylation (15–30%) in PCNSL patients [31]. Inactivation of $p14^{ARF}$ and $p16^{INK4a}$ genes by either homozygous deletion or promoter hypermethylation may represent an important step in the molecular pathogenesis of PCNSL. The $p14^{ARF}$ gene, for example, normally induces growth arrest and stabilizes p53 protein in the cell nucleus. Its deletion has been reported in glioblastomas and systemic NHL. Both $p14^{ARF}$ and $p16^{INK4a}$ genes are frequently co-deleted in human neoplasms; moreover, mice lacking the murine homologue of $p14^{ARF}$ develop a variety of tumors, including lymphomas, sarcomas, and gliomas [32–34]. In contrast, mutations in the *TP53* gene have been observed in only a small proportion of PCNSL specimens. Figure 7 summarizes major genetic events that are potentially important in the pathogenesis of diffuse B-cell lymphomas, including PCNSL.

Specific chromosomal translocations found frequently in PCNSL tumor specimens involve the *IgH* gene and the *BCL-6* gene. Such translocations are thought to originate by recombination mechanisms following double-strand DNA breaks; these breaks may be generated during class switch recombinations and somatic hypermutation of the *Ig* genes. Montesinos-Rongen et al. have proposed that the resulting juxtaposition of some oncogenes with regulatory elements may lead to oncogenetic activation. *BCL-6* translocations in PCNSL appear similar to translocations found in other extranodal forms of DLBCL and are thought to be pathogenetically relevant as well [28].

In addition to the genetic alterations discussed previously, there are differences in the expression of chemokines, chemokine receptors, cytokines, and cytokine receptors in extranodal lymphomas such as PCNSL. Compared with other extranodal lymphomas, PCNSL exhibited significantly higher expression of Th1-type cytokine IL-2 and Th2-type cytokine IL-13 [29]. It was initially believed that the B-cell-attracting chemokine 1 (CXCL13 [BCA-1]) was exclusively expressed in malignant lymphocytes and vascular endothelium within the CNS; however, further studies demonstrated that BCA-1 expression was not specific for CNS lymphomas and can also be found in extraneural lymphomas [29,30].

D. Biomarkers of Prognosis

Lymphomas thought to be derived from GC cells, such as follicular lymphoma, express *BCL-6*, whereas lymphomas derived from naïve B cells, like chronic lymphocytic leukemia or mantle cell lymphoma, do not express *BCL-6* [37,38]. Consistent with other studies demonstrating a favorable prognostic effect of *BCL-6* expression in patients with extraneural DLBCL, Braaten et al. showed that overexpression of *BCL-6* in PCNSL patients was associated with improved survival (101 months) compared with patients with tumors that did not express *BCL-6* (14.7 months) (Fig. 8) [18].

Analysis of chromosomal imbalances by comparative genomic hybridization has demonstrated frequent chromosome 6q deletions in PCNSL (60%) [39]. The incidence of chromosome 6q deletions is higher in PCNSL than in extraneural lymphomas. Patients with LOH on 6q have a shorter survival compared with PCNSL patients without LOH on 6q (Fig. 9). Further study, which consisted of deletion mapping of chromosome 6q in PCNSL, disclosed a region suspected to harbor a lymphoma-related tumor suppressor gene (6q22-23) [40]. This locus is known to contain the

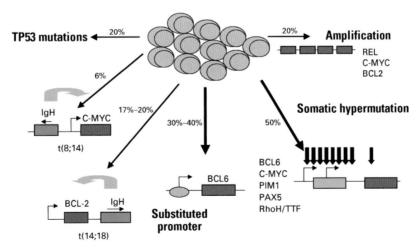

Figure 7 Model for diffuse large B-cell lymphoma (DLBCL) pathogenesis. Only the most common genetic lesions and their frequencies are shown. Genetic lesions more specific for PCNSL include somatic hypermutations as well as chromosomal translocations, especially involving the *IgH* gene and the *BCL-6* gene. *TP53* mutations are rare in PCNSL. (From Lossos, I. S. [2005]. Molecular pathogenesis of diffuse large B-cell lymphoma. *J. Clin. Oncol.* **23**, 6351–6357.)

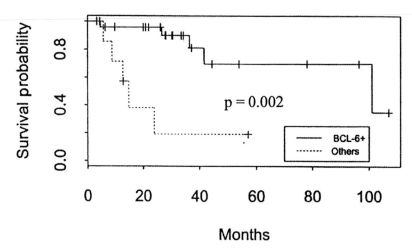

Figure 8 Kaplan-Meier estimates for the distributions of time to death by *BCL-6* status. Overexpression of *BCL-6* was associated with improved survival (101 months) compared with patients with tumors that did not express *BCL-6* (14.7 months). (From Braaten, K. M., Betensky, R. A., deLeval, L., et al. [2003]. BCL-6 expression predicts improved survival in patients with primary central nervous system lymphoma. *Clin. Cancer Res.* **9**, 1063–1104.)

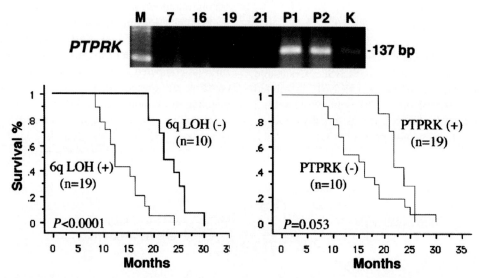

Figure 9 A, Representative raw data of RT-PCR as shown: *P1* and *P2*, normal prostate tissue; *K*, normal kidney tissue; *M*, size marker. Numbers represent cases of PCNSL (7, 16, 19, and 21). High-level expression of transcript was present in prostate tissues, and low-level expression was detected in kidney tissue. Complete loss of *PTPRK* expression is shown in PCNSL cases. **B**, Overall survival in PCNSL patients as a function of LOH on 6q (*left graph*) and PTPRK expression (*right graph*). (From Nakamura, M., Kishi, M., Sakaki, T., et al. [2003]. Novel tumor suppressor loci on 6q22-23 in primary central nervous system lymphomas. *Cancer Res.* **63**, 737–741.)

PTPRK gene in humans. Products of this gene are important for regulation of cell contact and adhesion [40]. Earlier death was observed in patients with PCNSL showing loss of PTPRK expression as compared with those maintaining PTPRK expression; however, this correlation was not statistically significant [40] (see Fig. 9).

In preliminary gene expression studies in PCNSL, Rubenstein et al. have demonstrated that differential expression of genes may distinguish between patients with long- and short-term survival [27]. Ongoing gene expression profiling studies may yield other markers with specific clinical relevance.

IV. Clinical Features of PCNSL

A. Clinical Presentation

Patients with PCNSL typically present with neurological signs and symptoms rather than systemic B symptoms such as fever, weight loss, and night sweats. In a study that focused on the clinical presentations of 248 immunocompetent patients with PCNSL, Bataille and colleagues reported that 70% had focal neurological deficits, 43% had neuropsychiatric symptoms, 33% had increased intracranial pressure, 14% had seizures, and 4% had ocular symptoms [41]. In the 41% of PCNSL patients with leptomeningeal involvement at the time of diagnosis, most showed no clinical signs of leptomeningeal disease. Those patients with ocular involvement at the time of diagnosis generally had bilateral ocular disease and complained of floaters, blurred vision, diminished visual acuity, and painful red eyes [42]. In immunocompromised patients, mental status changes and seizures are observed more frequently as presenting symptoms.

B. Diagnostic Evaluation

PCNSL may involve any part of the craniospinal axis as well as the eyes. At the time of diagnosis, patients should have a contrast-enhanced cranial magnetic resonance imaging (MRI) study, and, if a lumbar puncture can be performed safely, CSF should be obtained for cytological evaluation, flow cytometry, and polymerase chain reaction (PCR) for *IgH* gene studies. All patients should have a complete ophthalmology evaluation that includes a slit-lamp examination. It is also recommended that patients have contrast-enhanced computerized tomography (CT) scans of the chest, abdomen, and pelvis, as well as a bone marrow biopsy and aspirate because 3.9–12.5% of patients with "primary" CNS lymphoma are found to have occult extraneural disease on further evaluation [43–45]. Although positron emission tomography (PET) imaging has proven to be a useful modality for the staging of many cancers, its role in screening for occult systemic disease in patients with presumed PCNSL has yet to be defined. Both clinical and ultrasound testicular examination should also be considered in men with PCNSL. Routine blood testing including HIV serology, a complete blood cell count, basic electrolyte analysis, and serum lactate dehydrogenase (LDH) level is indicated as part of the initial evaluation [46]. The International PCNSL Collaborative Group (IPCG) has published guidelines for standardized baseline evaluation in patients with newly diagnosed PCNSL (Table 2) [47].

1. Neuroimaging

Contrast-enhanced cranial MRI is the optimal imaging modality for assessing patients with PCNSL. In patients for whom MRI is contraindicated, contrast-enhanced cranial CT scans are recommended. PCNSL is often isodense to hyperdense on CT images and isointense to hypointense on T2-weighted MRI, a finding that is attributed to its high cell density and scant cytoplasm (Fig. 10). On post-contrast CT or MRI, there is typically a homogeneous pattern of enhancement in immunocompetent PCNSL patients. In an MRI series of 100 patients with newly diagnosed

Table 2 International Primary CNS Lymphoma Collaborative Group (IPCG) Guidelines for Baseline Evaluation for Clinical Trials

Pathology	Clinical	Laboratory	Extent-of-Disease Evaluation (Staging)
Centralized review of pathology	Complete medical and neurological examination	HIV serology	Contrast-enhanced cranial MRI[3]
Immunophenotyping	Dilated eye examination including slit lamp evaluation	Serum LDH level	CT of chest, abdomen, and pelvis
	Record prognostic factors (age, performance status)	CSF cytology, flow cytometry, IgH PCR	Bone marrow biopsy with aspirate
	Serial evaluation of cognitive function[1]	24-hour urine collection for creatinine clearance[2]	Testicular ultrasound in elderly males

CSF, Cerebrospinal fluid; *CT*, computed tomography; *HIV*, human immunodeficiency virus; *IgH*, immunoglobin H; *LDH*, lactate dehydrogenase; *MRI*, magnetic resonance imaging; *PCR*, polymerase chain reaction.

[1] Mini-mental status examination is used commonly, although improved instruments are under development.
[2] For patients who will receive high-dose methotrexate.
[3] Contrast-enhanced cranial CT should be obtained in patients who have a contraindication for MRI (e.g., pacemaker) or who cannot tolerate MRI (e.g., claustrophobia).
Adapted from Abrey, L. E., Batchelor, T. T., Ferreri, A., et al. (2005). Report of an international workshop to standardize baseline evaluation and response criteria for primary CNS lymphoma. *J. Clin. Oncol.* **23**, 5034–5043.

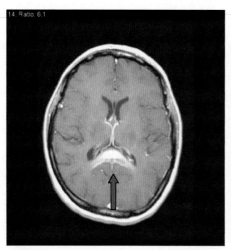 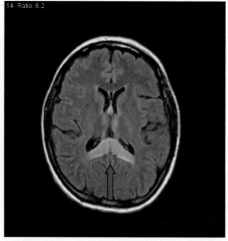

Figure 10 MRI radiographs of a patient with PCNSL. Typical corpus callosum involvement (*arrows*) seen both on post-contrast images (*left*) and FLAIR sequences (*right*).

PCNSL, lesions were solitary in 65% of patients and multifocal in 35% of patients. Lesions were located in the hemispheres (38%), thalamus/basal ganglia (16%), corpus callosum (14%), ventricular region (12%), and cerebellum (9%). Contrast enhancement was strong in 85% of patients, moderate in 10%, and absent in 1% [48]. Enhancement along the Virchow-Robin spaces, although not constant, is a highly specific feature [49].

In immunocompromised patients, multifocal disease and basal ganglia involvement are seen more frequently. The signal characteristics on both CT and MRI are more variable. PCNSL lesions often appear hypodense on CT and hyperintense on T2-weighted MRI. Contrast enhancement is more heterogeneous, and ring-enhancing lesions are more common. The variability in the radiographic appearance of AIDS-related PCNSL makes it difficult to distinguish between lymphoma and other AIDS-related masses, particularly toxoplasmosis. Adjunctive imaging techniques such as thallium 201 single photon-computed tomography (SPECT), PET, magnetic resonance spectroscopy (MRS), and MR-perfusion imaging may be helpful in making these distinctions [46].

2. Cerebrospinal Fluid

Lumbar puncture with CSF sampling should be performed at the time of initial assessment in patients with suspected or confirmed PCNSL, although increased intracranial pressure may be a relative contraindication to lumbar puncture. CSF evaluation should include cell counts, protein and glucose levels, cytology, flow cytometry, immunoglobulin heavy chain gene rearrangement studies, and, in immunocompromised patients, EBV antibody titers. Increases in the white blood cell count ($>7 \text{cells/mm}^3$) and high protein concentrations (median concentration = 69 mg/dL) are often present, as well as low CSF glucose levels relative to serum glucose levels. CSF cytology may be abnormal, typically showing clumped pleomorphic cells with enlarged nuclei and coarse chromatin in 26% to 31% of PCNSL patients [46]. However, two-thirds of PCNSL patients who developed positive CSF cytologic findings had negative results on an initial examination, suggesting that serial CSF samples would result in increased diagnostic sensitivity. In another study of 96 patients with newly diagnosed PCNSL, the initial CSF cytologic studies were positive in only 15% of cases [50]. Therefore most patients require stereotactic biopsy of the brain lesion for definitive diagnosis. Patients who manifest serious or worsening clinical symptoms should immediately undergo stereotactic biopsy rather than risk any diagnostic delay associated with serial lumbar punctures.

3. Ocular Examination

As mentioned previously, ophthalmological evaluation should be performed in all patients during the staging process. In addition to a fundoscopic examination, a slit lamp should be used to visualize the vitreous fluid. CNS lymphoma also may be associated with retinal deposits of tumor. Cytological evaluation of the vitreal fluid may be required for definitive diagnosis of intraocular lymphoma (Fig. 11).

4. Prognostic Scoring

PCNSL is a form of extranodal NHL and is classified as stage IE disease according to the Ann Arbor staging system, in which stage I represents localized lymphoma and E denotes extranodal disease. However, the prognostic significance of the Ann Arbor staging system does not apply to PCNSL. In a review of a large historical patient database, the International Extranodal Lymphoma Study Group (IELSG) reported that the following parameters were associated with a poor prognosis in PCNSL patients: age

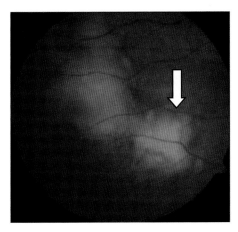 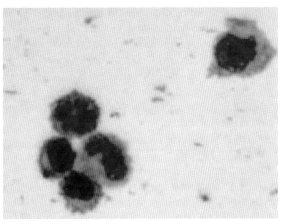

Figure 11 Ocular lymphoma. **A**, Yellow subretinal infiltrates in primary intraocular lymphoma. **B**, Abnormal lymphocytes in the vitreous fluid with basophilic cytoplasm and large irregular nuclei. (Adapted with permission from Chan, C. C. and Wallace, D. J. [2004]. Intraocular lymphoma: update on diagnosis and management. *Cancer Control* **11**, 285–295.)

older than 60 years; performance status greater than 1 on the Eastern Cooperative Oncology Group (ECOG) performance status scale; elevated serum LDH; high CSF protein concentration; and tumor location within the deep regions of the brain (i.e., periventricular regions, basal ganglia, brainstem, and/or cerebellum). Patients with 0 or 1, 2 or 3, or 4 or 5 of these adverse risk factors had 2-year overall survival rates of 80%, 48%, and 15%, respectively (Fig. 12) [51].

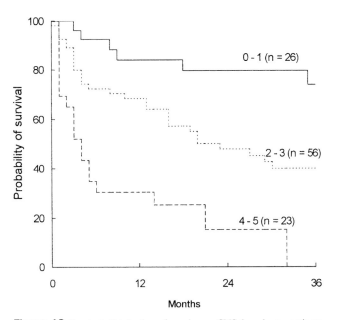

Figure 12 Survival distributions for primary CNS lymphoma patients grouped according to their International Extranodal Lymphoma Study Group prognostic scores. Patients with 0 to 1, 2 to 3, or 4 to 5 unfavorable features have a 2-year survival of $80\% \pm 8\%$, $48\% \pm 7\%$, and $15\% \pm 7\%$, respectively ($P < 0.00001$). (From Ferreri, A. J. M., Abrey, L. E., Blay, J., et al. [2003]. Summary statement on primary central nervous system lymphomas from the Eighth International Conference on Malignant Lymphoma, Lugano, Switzerland, June 12–15, 2002. *J. Clin. Oncol.* **21**, 2407–2414.)

V. Treatment Options for PCNSL

PCNSL, unlike most other primary malignant brain tumors, is responsive to radiation and chemotherapy and is considered a potentially curable brain cancer. However, it is important to note that choice of treatment, response to treatment, and adverse treatment effects differ according to the patient's immune status. For immunocompetent patients with newly diagnosed PCNSL, treatment options include any combination of corticosteroids, whole brain radiation therapy (WBRT), and chemotherapy. For immunocompromised patients, due to the higher likelihood of toxicity from chemotherapy, the standard of care has generally consisted of WBRT alone. However, antiviral and immunomodulatory treatments have recently shown some promise in this population. Due to the fact that PCNSL is a rare cancer, there have been few multicenter, randomized treatment trials; virtually all the treatment recommendations for PCNSL patients are based on smaller, uncontrolled studies. As many as to 40% of patients with PCNSL will experience a significant initial response to corticosteroids alone [52]. However, most patients relapse quickly when corticosteroids are used as the sole therapeutic modality. Because of their lymphotoxic properties and the subsequent difficulty in establishing a definitive histopathological diagnosis, corticosteroids should not be administered before tumor biopsy when at all possible.

Special Consideration for Treatment of PCNSL in the Immunocompromised Patient

The etiologies of mass lesions in immunocompromised patients, particularly those with HIV infection, can be variable. They include opportunistic infections with parasites, fungi, and bacteria, as well as neoplasms and cerebrovascular diseases. Specific guidelines were created by the

American Academy of Neurology on how to approach an immunocompromised patient with a brain mass [53]. The optimal treatment of immunocompromised patients remains an area of research. HIV-infected PCNSL patients generally have CD4+ cell counts less than 50 cells/mm^3. If these patients have a multidrug-resistant strain of HIV, treatment may be palliative with corticosteroids and WBRT. Donahue et al. reported that the median survival of HIV+ PCNSL patients treated with WBRT alone was 2.1 months [54]. There have been reports of HIV-infected PCNSL patients achieving complete responses to methotrexate-based chemotherapy [55]. Recent reports suggest that the natural history of PCNSL in immunocompromised patients may be altered indirectly by restoration of immune competence with HAART and directly by antiviral targeting of EBV [12,56,57]. Raez et al. sought to exploit both of these therapeutic pathways in a novel approach employing parenteral zidovudine, ganciclovir, and interleukin-2 (IL-2). Of five patients treated with this regimen, three achieved a partial response and one experienced a complete response that lasted 6 years [57]. Therapy with high-dose zidovudine, ganciclovir, and IL-2 has been reported to be successful in a small number of patients in other studies as well [58,59]. The description of long-term remission in a patient with post-transplant EBV-associated PCNSL who received EBV-specific therapy suggests that the anti-EBV mechanism may mediate some of the antilymphoma effect [12].

In patients with post-transplant lymphoproliferative diseases (PTLD) of the CNS, several different treatment approaches have been used, but no consensus exists. Treatment approaches include reduction of immunosuppression, antiviral agents, rituximab, chemotherapy, and radiation. Nozzoli and colleagues described a case of PTLD of the CNS successfully treated with rituximab, cidofovir, and intrathecal methotrexate [60].

A. Surgery

A stereotactic biopsy is warranted for almost all patients with suspected PCNSL. Partial or gross total resections confer no additional survival benefit in the PCNSL patient population because of the multifocal pattern of tumor growth and the deep location of the tumors. Moreover, PCNSL may affect the entire neuraxis, including the eyes and leptomeninges. Survival after stereotactic biopsy alone is 1 to 4 months [61,62]. Thus the surgical approach of choice is an early stereotactic biopsy to establish the diagnosis.

B. Radiation

Because of the multifocal nature of PCNSL, radiation must be administered to the entire brain. The use of WBRT as the sole treatment modality for PCNSL was associated with an overall response rate (ORR) of 90% but a median survival of only 12.2 months in one study. Moreover, PCNSL relapse occurred in 61% of patients within the irradiated field [63], indicating that WBRT alone not does produce durable remissions for most patients. WBRT is associated with an increased likelihood of treatment-related, delayed neurotoxicity, especially in patients older than 60 years. Because of limited efficacy as a single therapeutic modality and because of the risk of delayed neurotoxicity, radiation is often deferred in newly diagnosed, elderly patients.

C. Chemotherapy

Methotrexate, a folate antagonist that interrupts DNA biosynthesis, is the most effective and most widely used drug for PCNSL. Although high-dose methotrexate (HD-MTX) is the most effective drug against PCNSL, outcome-determining variables related to its administration schedule have not been well defined. The impact on toxicity and outcome of the area under the curve (AUC_{MTX}), dose intensity (DI_{MTX}), and infusion rate (IR_{MTX}) of MTX and plasma creatinine clearance (CL_{crea}) was investigated in a retrospective series of 45 PCNSL patients treated with three different HD-MTX-based combinations [64]. Slow creatinine clearance ($CL_{crea} \leq 85$ ml min^{-1}) and high AUC_{MTX} (>1100 mmol $\cdot$ h $\cdot$ l^{-1}) were two factors associated with significant improvement in survival, independent of age, DI_{MTX}, IR_{MTX}, and other therapeutic variables. However, slow CL_{crea} was also associated with severe toxicity by multivariate analysis. Based on their observations in this latter study, the authors recommended that future studies of methotrexate-based chemotherapy include a methotrexate dose ≥ 3 g/m^2 as a 4–6 hour infusion given every 3–4 weeks.

Some patients, however, do not respond to therapy with MTX or develop resistance to MTX over time. Mechanisms of MTX resistance may involve changes in the uptake or polyglutamation of MTX, hydrolysis of MTX polyglutamates, or interaction of MTX with dihydrofolate reductase [65]. The major route for cellular uptake of MTX involves the reduced folate carrier (RFC), a bidirectional anion transporter with high affinity for reduced folates and antifolates but low affinity for folic acid. In vitro models of transport mechanisms have shown that down-regulation of *RFC* activity results in transport-mediated MTX resistance. Lack of *RFC* expression in human tumor cell lines has been attributed to the methylation of a CpG island in the *RFC* gene promoter region [66]. Ferreri et al. investigated the prognostic role of CpG island methylation of the *RFC* gene promoter in immunocompetent PCNSL patients [67]. Forty specimens were analyzed for the status of *RFC* promoter methylation. It was concluded that *RFC* methylation was more common in PCNSL compared with systemic DLBCL and was associated with a lower complete response (CR) rate to MTX-based chemotherapy.

Combined Modality Therapy Combined modality therapy, which consists of WBRT plus chemotherapy, was developed with the goal of improving response rates and survival in patients with PCNSL. Early studies included chemotherapy regimens that were effective for other types of NHL but produced poor results in patients with PCNSL, mainly because the chemotherapeutic drugs had insufficient brain penetration.

Methotrexate-based chemotherapy combined with WBRT is associated with radiographic responses in more than 50% of patients with PCNSL and a 2-year survival of 43–73% [68]. One methotrexate-based, combined-modality regimen consists of intravenous methotrexate ($2.5\,g/m^2$), intravenous vincristine, oral procarbazine, and intraventricular methotrexate (12 mg), followed by WBRT (45 Gy) and post-WBRT high-dose cytarabine. In a multicenter study of 98 patients with PCNSL, this regimen was associated with an ORR of 94% (58% CRs and 36% PRs), a median progression-free survival (PFS) of 24 month, and an overall survival (OS) of 36.9 months. There were 12 cases of neurotoxicity (15%) and 8 deaths [69].

In another multicenter study of methotrexate-based, combined modality therapy, 52 PCNSL patients received MBVP, consisting of methotrexate ($3\,g/m^2$, days 1 and 15), teniposide ($100\,mg/m^2$, days 2 and 3), carmustine ($100\,mg/m^2$, day 4), methylprednisolone ($60\,mg/m^2$, days 1 to 5), and intrathecal methotrexate (15 mg) plus cytarabine (40 mg), followed by WBRT (40 Gy) The ORR was 81% and the median OS was 46 months. Treatment-related deaths occurred in 10% of the study population [70].

In an earlier study of 226 patients with PCNSL, the combination of methotrexate and WBRT was associated with a high risk of delayed neurotoxicity, especially in patients older than 60 years [71]. After a median follow-up of 76 months, median overall survival was 16 months. Age greater than 60 years, poor performance status, CSF protein level greater than 0.6 g/L, involvement of the corpus callosum or subcortical gray structures, detectable lymphoma cells in CSF, and increased serum LDH all correlated significantly with shorter survival. Although treatment with methotrexate and WBRT was associated with longer survival ($P=0.01$), the projected incidence of treatment-induced late neurotoxicity was 26% at 6 years. Median survival after the diagnosis of delayed neurotoxicity was 12 months. In multivariate analysis, treatment with radiotherapy followed by chemotherapy was the only parameter that correlated with the development of late neurotoxicity (relative risk, 11.5; $P = 0.0007$).

Chemotherapy Alone The high incidence of neurotoxicity following radiation-containing treatment regimens, especially in patients older than 60 years, has led to an interest in the use of chemotherapy alone in patients with PCNSL.

Complete radiographic responses occur in 30–100% of patients when methotrexate at a dose of $8\,g/m^2$ is administered alone and in 70–94% of patients when methotrexate is used as part of a multiagent regimen [50]. Durable responses are possible, although most patients eventually relapse. In a multicenter phase 2 study, single-agent therapy with high-dose methotrexate ($8\,g/m^2$) resulted in a CR rate of 52%, PFS of 12.8 months, and an OS of 55.4 months, with only minimal toxicity.

Methotrexate-based multiagent chemotherapy regimens without radiation were assessed in a phase 2 study in which 65 patients received high-dose methotrexate, cytarabine, vincristine, ifosfamide, cyclophosphamide, and intraventricular methotrexate, cytarabine, and prednisolone. The investigators reported an ORR of 71%, with 61% CRs and 10% partial responses (PRs), a median time to progression of 21 months, and a median OS of 50 months. However, this intensive regimen was associated with 6 treatment-related deaths (9%) and 12 cases of Ommaya reservoir infection (19%) [72]. The incidence of adverse events is higher in multiagent regimens that include methotrexate than when methotrexate is used as a single agent [69,73].

High-dose chemotherapy followed by autologous stem cell rescue has also been investigated in patients with newly diagnosed PCNSL. In one study, 28 patients received induction chemotherapy with high-dose intravenous methotrexate ($3.5\,g/m^2$) and cytarabine ($3\,g/m^2$ daily for 2 days). Neuroimaging detected chemosensitive disease after induction in 14 patients, who went on to receive high-dose therapy with BEAM regimen (carmustine, etoposide, cytarabine, and melphalan), followed by autologous stem cell rescue. The median event-free survival was just 9.3 months, with 8 patients experiencing progression at a median of 2.3 months after stem cell infusion [74]. The poor response to induction therapy and the significant number of patients who relapsed soon after stem cell rescue indicate that this approach must be reassessed, especially because less toxic and potentially more effective treatment regimens are available [68]. Additional therapies used in treatment of PCNSL are listed in Table 3.

Salvage Therapy Few prospective large-scale studies of salvage therapy have been conducted in patients with relapsed or refractory PCNSL, and no consensus exists on which salvage therapy to use in these patients (see Table 3) [75]. Individuals with disease that is refractory to first-line therapy are less likely to achieve durable remissions with subsequent therapies [69]. However, in a study of 22 PCNSL patients who relapsed after prior CR to methotrexate, the ORR after a second course of methotrexate was 91% and the median OS from the time of relapse was 61.9 months [76]. Another study of 27 patients who were initially treated with chemotherapy alone reported

Table 3 Salvage Therapy for Relapsed or Refractory PCNSL

Treatment	Number of Subjects	ORR	OS (months)
WBRT [93]	27	20/27	10.9
Methotrexate* [76]	22	20/22	61.9
PCV** [92]	7	6/7	16+
Temozolomide [91]	23	6/23	3.5+
Temozolomide [90] and Rituximab	15	8/15	14
Topotecan [89]	15	6/15	32
Intensive chemotherapy and stem cell rescue [77]	10	10/10	24+

References appear in brackets following the name of each therapy. ORR, Overall response rate; OS, overall survival; PCV, procarbazine, CCNU (lomustine), vincristine; WBRT, whole brain radiation therapy.
*In patients who experience relapse after prior complete response to methotrexate.
**Procarbazine, CCNU lomustine, vincristine.

that WBRT (36 Gy) was an effective salvage therapy at the time of relapse, with an ORR of 74% (37% CR and 37% PR) and a median OS of 10.9 months. High-dose chemotherapy with autologous stem cell rescue may also be effective as salvage therapy. In one small study, all 10 patients who achieved a radiographic response after high-dose chemotherapy received an autologous stem cell transplant. Median survival from the time of relapse was greater than 24 months [77], although patients experienced significant treatment-related toxicity. Other regimens have been tested as salvage therapy, including temozolomide, temozolomide plus rituximab, combination of procarbazine, CCNU and vincristine (PCV), and topotecan. The ORR and OS associated with each regimen are summarized in Table 3.

D. Neurotoxicity

Neurotoxicity is a potentially devastating complication that can occur in response to antineoplastic therapies including WBRT and chemotherapy. Although the pathogenesis of radiation-induced cognitive injury is still an active area of investigation, it appears to involve the loss of neural precursor cells from the subgranular zone of the hippocampal dentate gyrus. Impaired neurogenesis, associated with a significant inflammatory response, was observed in young adult male C67BL mice after whole brain irradiation [78]. Blockage of this "neuroinflammation" by indomethacin has been observed to restore and augment neurogenesis after cranial radiation in another animal model [79]. These preclinical observations offer insight into the underlying pathogenesis of radiation-induced neurotoxicity and potential methods to block or reverse this process.

Neurotoxicity may manifest in a variety of ways, including impaired cognition, ataxia, and incontinence, and is often associated with a significant decline in the quality of life. Elderly patients with PCNSL are especially vulnerable to treatment-related neurotoxicity. In one study, 100% of patients older than 60 years who were treated with WBRT-containing regimens for PCNSL experienced clinical neurotoxicity manifesting as dementia, ataxia, and incontinence, with a median time to onset of 13.2 months [80]. By contrast, only 30% of patients younger than 60 years developed neurotoxicity over 96 months of follow-up. Because of this increased risk of neurotoxicity in elderly patients, many authorities recommend deferral of WBRT in individuals older than 60 years with newly diagnosed PCNSL.

Patients affected by neurotoxicity often demonstrate MRI signal abnormalities in the periventricular white matter of both hemispheres, central and cortical atrophy, and ventricular enlargement (Fig. 13). Autopsy studies in patients with neurotoxicity reveal myelin and axonal loss, gliosis, pallor, spongiosis, and rarefaction of the cerebral white matter. Tissue necrosis and atherosclerosis of large cerebral vessels have also been identified, suggesting vascular injury and subsequent tissue infarction and necrosis as potential mechanisms of WBRT-related neurotoxicity [81].

Among the different therapeutic strategies for PCNSL, WBRT followed by methotrexate-based chemotherapy seems to produce the highest incidence of neurotoxicity, followed by WBRT alone and then chemotherapy alone [82]. In fact, chemotherapy alone has been associated with improvement or stabilization of cognitive function in several studies, and more than half of the patients treated only with chemotherapy in two prospective studies had normal cognitive function on follow-up neuropsychological testing [83,84]. Fliessbach et al. performed neuropsychological testing on 23 PCNSL patients treated with methotrexate-based chemotherapy without WBRT. Although the authors did report MTX-induced white matter changes on MRI scans of the study subjects, these radiographic changes did not correlate with cognitive impairment. Another comparative study in PCNSL survivors identified that patients treated with WBRT with or without chemotherapy displayed more pronounced cognitive dysfunction than patients treated with MTX-based chemotherapy alone [82].

There is no effective therapy once neurotoxicity develops, and patients diagnosed with this condition often die of neurotoxicity-related complications in the absence of recurrent or active lymphoma. CSF shunting may be partially beneficial in a subset of patients.

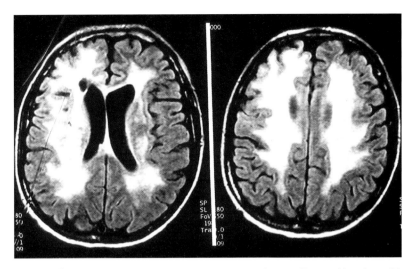

Figure 13 Neurotoxicity. Cranial FLAIR MR images from a 67-year-old patient with PCNSL, 1 year after achieving a complete response to WBRT. The patient developed memory failure, gait ataxia, and incontinence in the postradiation setting. There is increased signal throughout the cerebral white matter, cortical atrophy, and ventricular enlargement.

VI. Summary

PCNSL, a rare form of extranodal NHL, is typically a diffuse large B-cell lymphoma that is confined to the nervous system and eyes. The diagnosis of PCNSL is supported by CT and MRI studies, as well as CSF testing, but is ultimately confirmed on the basis of stereotactic biopsy in most patients. The incidence of HIV-associated PCNSL has declined in the modern era of HAART. However, the prognosis for PCNSL in this immunocompromised population remains poor with standard WBRT. The promising preliminary data using EBV-targeted antiviral therapy deserves further investigation. Although PCNSL in the immunocompetent host is a potentially curable primary brain tumor, current treatment regimens are achieving long-term remissions in only a small fraction of patients and more effective treatment regimens are desperately needed. Methotrexate-based, multiagent chemotherapy is currently the treatment of choice, especially in the elderly patient population, which is the group at highest risk of developing radiation-related neurotoxicity. The optimal role and timing of WBRT in the management of newly diagnosed PCNSL patients have yet to be established. The risk and impact of treatment-related neurotoxicity cannot be overemphasized. The development of neurotoxicity has a negative impact on quality of life, emotional status, motor function, employment, and cognition. Moreover, patients who develop this complication often die from neurotoxicity-related complications without evidence of active lymphoma. Thus an important objective is to minimize the risk of neurotoxicity by deferring WBRT in patients over 60 years of age.

A better biological understanding of PCNSL will enable the development of customized treatment approaches for this patient population. Until recently, biological studies of PCNSL were lacking, mainly due to the rare frequency of the disease and the paucity of available research specimens. However, multicenter collaboration has resulted in the first gene expression profile study, and other studies of biomarkers of prognosis have been published. Although there is a large gap between our understanding of other forms of extranodal B-cell lymphomas and PCNSL, it is anticipated that this gap will close in the coming years as multicenter collaboration, tissue preservation methods, and molecular techniques are improved and refined.

References

1. Pfannl, R., Ferry, J. A., and Harris, N. L. (2004). Pathology of primary central nervous system lymphoma and related conditions. In: "Lymphoma of the Nervous System" (T. T. Batchelor, ed.), pp. 29–41. Butterworth-Heinemann, Boston.
2. Eby, N. L., Grufferman, S., Flannelly, C. M., Schold, S. C. Jr., Vogel, F. S., and Burger, P. C. (1988). Increasing incidence of primary brain lymphoma in the U.S. Cancer **62**, 2461–2465.
3. O'Neill, B. P. (2004). Epidemiology of PCNSL. In: "Lymphoma of the Nervous System" (T. T. Batchelor, ed.), pp. 43–51. Butterworth-Heinemann, Boston.
4. Kadan-Lottick, N. S., Skluzarek, M. C., and Gurney, J. G. (2002). Decreasing incidence rates of primary central nervous system lymphoma. Cancer **95**, 193–202.
5. Central Brain Tumor Registry of the United States (CBTRUS). (2005). Statistical report: primary brain tumors in the United States, 1998–2002. CBTRUS, Chicago.

6. Fine, H. A. and Mayer, R. J. (1993). Primary central nervous system lymphoma. *Ann. Intern. Med.* **119**, 1093–1104.
7. Schabet, M. (1999). Epidemiology of primary CNS lymphoma. *J. Neurooncol.* **43**, 199–201.
8. Cote, T. R., Manns, A., Hardy, C. R., Yellin, F. J., and Hartge, P. (1996). Epidemiology of brain lymphoma among people with or without acquired immunodeficiency syndrome. *J. Natl. Cancer Inst.* **88**, 675–679.
9. Bashir, R., Chamberlain, M., Ruby, E., and Hochberg, F. H. (1996). T-cell infiltration of primary CNS lymphoma. *Neurology* **46**, 440–444.
10. Ferry, J. and Harris, N. (1994). Pathology of post-transplant lymphoproliferative disorders. *In:* "Pathology and Rejection Diagnosis in Solid Organ Transplantation" (K. Solez et al., eds.), pp. 277–301. Marcel Dekker: New York.
11. Rubin, N. and Chamberlain, M. C. (2004). Primary central nervous system lymphoma in acquired immune deficiency syndrome. *In:* "Lymphoma of the Nervous System" (T. T. Batchelor, ed.), pp. 97–103. Butterworth-Heinemann, Boston.
12. Roychowdhury, S., Peng, R., Baiocchi, R., Bhatt, D., Vourganti, S., Grecula, J., Gupta, N., Eisenbeis, C. F., Nuovo, G. J., Yang, W., Schmalbrock, P., Ferketich, A., Moeschberger, M., Porcu, P., Barth, R. F., and Caligiuri, M. A. (2003). Experimental treatment of Epstein-Barr virus–associated primary central nervous system lymphoma. *Cancer Res.* **63**, 965–971.
13. Tanner, J. E. and Alfieri, C. (2001). The Epstein-Barr virus and post-transplant lymphoproliferative disease: interplay of immunosuppression, EBV, and the immune system in disease pathogenesis. *Transpl. Infect. Dis.* **3**, 60–69.
14. Meeker, T. C., Shiramizu, B., Kaplan, L., Herndier, B., Sanchez, H., Grimaldi, J. C., Baumgartner, J., Rachlin, J., Feigal, E., and Rosenblum, M., et al. (1991). Evidence for molecular subtypes of HIV-associated lymphoma: division into peripheral monoclonal, polyclonal and central nervous system lymphoma. *AIDS* **5**, 669–674.
15. Montesinos-Rongen, M., Besleaga, R., Heinsohn, S., Siebert, R., Kabisch, H., Wiestler, O. D., and Deckert, M. (2004). Absence of Simian Virus 40 DNA sequences in primary central nervous system lymphoma in HIV-negative patients. *Virchows Arch.* **444**, 436–438.
16. Linnebank, M., Schmidt, S., Kolsch, H., Linnebank, A., Heun, R., Schmidt-Wolf, I. G., Glasmacher, A., Fliessbach, K., Klockgether, T., Schlegel, U., and Pels, H. (2004). The methionine synthase polymorphism D919G alters susceptibility to primary central nervous system lymphoma. *Br. J. Cancer* **90**, 1969–1971.
17. Riemersma, S. A., Oudejans, J. J., Vonk, M. J., Dreef, E. J., Prins, F. A., Jansen, P. M., Vermeer, M. H., Blok, P., Kibbelaar, R. E., Muris, J. J. F., Schuuring, E. M. D., and Kluin, P. M. (2005). High numbers of tumour-infiltrating activated cytotoxic T lymphocytes, and frequent loss of HLA class I and II expression, are features of aggressive B cell lymphomas of the brain and testis. *J. Pathol.* **206**, 328–336.
18. Braaten, K. M., Betensky, R. A., deLeval, L., Okada, Y., Hochberg, F. H., Louis, D. N., Harris, N. L., and Batchelor, T. T. (2003). BCL-6 expression predicts improved survival in patients with primary central nervous system lymphoma. *Clin. Cancer Res.* **9**, 1063–1104.
19. Lossos, I. S. (2005). Molecular pathogenesis of diffuse large B-cell lymphoma. *J. Clin. Oncol.* **23**, 6351–6357.
20. Thompsett, A. R., Ellison, D. W., Stevenson, F. K., and Zhu, D. (1999). V(H) gene sequences from primary central nervous system lymphomas indicate derivation from highly mutated germinal center B cells with ongoing mutational activity. *Blood* **94**, 1738–1746.
21. Alizadeh, A. A., Eisen, M. B., Davis, R. E., Ma, C., Lossos, I. S., Rosenwald, A., Boldrick, J. C., Sabet, H., Tran, T., Yu, X., Powell, J. I., Yang, L., Marti, G. E., Moore, T., Hudson, J. Jr., Lu, L., Lewis, D. B., Tibshirani, R., Sherlock, G., Chan, W. C., Greiner, T. C., Weisenburger, D. D., Armitage, J. O., Warnke, R., Levy, R., Wilson, W., Grever, M. R., Byrd, J. C., Botstein, D., Brown, P. O., and Staudt, L. M. (2000). Distinct types of diffuse large B-cell lymphoma identified by gene expression profiling. *Nature* **403**, 503–511.
22. Rosenwald, A., Wright, G., Chan, W. C., Connors, J. M., Campo, E., Fisher, R. I., Gascoyne, R. D., Muller-Hermelink, H. K., Smeland, E. B., Giltnane, J. M., Hurt, E. M., Zhao, H., Averett, L., Yang, L., Wilson, W. H., Jaffe, E. S., Simon, R., Klausner, R. D., Powell, J., Duffey, P. L., Longo, D. L., Greiner, T. C., Weisenburger, D. D., Sanger, W. G., Dave, B. J., Lynch, J. C., Vose, J., Armitage, J. O., Montserrat, E., Lopez-Guillermo, A., Grogan, T. M., Miller, T. P., LeBlanc, M., Ott, G., Kvaloy, S., Delabie, J., Holte, H., Krajci, P., Stokke, T., Staudt, L. M., and Lymphoma/Leukemia Molecular Profiling Project. (2002). The use of molecular profiling to predict survival after chemotherapy for diffuse large B-cell lymphoma. *N. Engl. J. Med.* **346**, 1937–1947.
23. Lossos, I. S., Alizadeh, A. A., Eisen, M. B., Chan, W. C., Brown, P. O., Botstein, D., Staudt, L. M., and Levy, R. (2000). Ongoing immunoglobulin somatic mutation in germinal center B cell–like but not in activated B cell–like diffuse large cell lymphomas. *Proc. Natl. Acad. Sci. USA* **97**, 10209–10213.
24. Larocca, L. M., Capello, D., Rinelli, A., Nori, S., Antinori, A., Gloghini, A., Cingolani, A., Migliazza, A., Saglio, G., Cammilleri-Broet, S., Raphael, M., Carbone, A., and Gaidano, G. (1998). The molecular and phenotypic profile of primary central nervous system lymphoma identifies distinct categories of the disease and is consistent with histogenetic derivation from germinal center-related B cells. *Blood* **92**, 1011–1019.
25. Montesinos-Rongen, M., Kueppers, R., Schleuter, D., Spieker, T., Van Roost, D., Schaller, C., Reifenberger, G., Wiestler, O. D., and Deckert-Schluter, M. (1999). Primary central nervous system lymphomas are derived from germinal-center B-cells and show a preferential usage of the V4-34 gene segment. *Am. J. Pathol.* **155**, 2077–2086.
26. Montesinos-Rongen, M., Roost, D. V., Schaller, C., Wiestler, O. D., and Deckert, M. (2004). Primary diffuse large B-cell lymphomas of the central nervous system are targeted by aberrant somatic hypermutation. *Blood* **103**, 1869–1875.
27. Rubenstein, J. L., Shen, A., Fridlyand, J., Jain, A. J., Aldape, K., Batchelor, T., Berger, M. S., McDermott, M. W., Prados, M. D., Khodabakhsh, D., Treseler, P. A., Haqq, C. M., and Shuman, M. A. (2003). Gene expression profile analysis of primary CNS lymphoma: class distinction and outcome prediction. Presented at the Society for Neuro-Oncology Eighth Annual Meeting, Keystone, CO.
28. Montesinos-Rongen, M., Zuhlke-Jenisch, R., Gesk, S., Martin-Subero, J. I., Schaller, C., Van Roost, D., Wiestler, O. D., Deckert, M., and Siebert, R. (2002). Interphase cytogenetic analysis of lymphoma-associated chromosomal breakpoints in primary diffuse large B-cell lymphomas of the central nervous system. *J. Neuropathol. Exp. Neurol.* **61**, 926–933.
29. Fujii, A., Ohshima, K., Hamasaki, M., Utsunomiya, H., Okazaki, M., Kagami, Y., Seto, M., and Kikuchi, M. (2004). Differential expression of chemokines, chemokine receptors, cytokines and cytokine receptors in diffuse large B cell malignant lymphoma. *Int. J. Oncol.* **24**, 529–538.
30. Smith, J. R., Braziel, R. M., Paoletti, S., Lipp, M., Uguccioni, M., and Rosenbaum, J. T. (2003). Expression of B-cell-attracting chemokine 1 (CXCL13) by malignant lymphocytes and vascular endothelium in primary central nervous system lymphoma. *Blood* **101**, 815–821.
31. Nakamura, M., Sakaki, T., Hashimoto, H., Nakase, H., Ishida, E., Shimada, K., and Konishi, N. (2001). Frequent alterations of the p14(ARF) and p16(INK4a) genes in primary central nervous system lymphomas. *Cancer Res.* **61**, 6335–6339.
32. Kamijo, T., Bodner, S., van de Kamp, E., Randle, D. H., and Sherr, C. J. (1999). Tumor spectrum in ARF-deficient mice. *Cancer Res.* **59**, 2217–2222.

33. Nakamura, M., Watanabe, T., Klangby, U., Asker, C., Wiman, K., Yonekawa, Y., Kleihues, P., and Ohgaki, H. (2001). p14(ARF) deletion and methylation in genetic pathways to glioblastoma. *Brain Pathol.* **11**, 159–168.
34. Ichimura, K., Schmidt, E., Goike, H. M., and Collins, V. P. (1996). Human glioblastomas with no alterations of the CDKN2A (p16INK4A, MTS1) and CDK4 genes have frequent mutations of the retinoblastoma gene. *Oncogene* **13**, 1065–1072.
35. Dent, A. L., Shaffer, A. L., Yu, X., Allman, D., and Staudt, L. M. (1997). Control of inflammation, cytokine expression, and germinal center formation by BCL-6. *Science* **276**, 589–592.
36. Chang, C. C., Ye, B. H., Chaganti, R. S. K., and Dalla-Favera, R. (1996). BCL-6, a POZ/zinc-finger protein, is a sequence-specific transcriptional repressor. *Proc. Natl. Acad. Sci. USA* **93**, 6947–6952.
37. Falini, B., Fizzotti, M., Liso, A., Bigerna, B., Marafioti, T., Gambacorta, M., Pacini, R., Alunni, C., Natali-Tanci, L., Ugolini, B., Sebastiani, C., Cattoretti, G., Pileri, S., Dalla-Favera, R., and Stein, H. (1997). Bcl-6 protein expression in normal and neoplastic lymphoid tissues. *Ann. Oncol.* **8(2)**, S101–S104.
38. Dogan, A., Bagdi, E., Munson, P., and Isaacson, P. G. (2000). CD10 and BCL6 expression in paraffin sections of normal lymphoid tissue and B-cell neoplasms. *Am. J. Surg. Pathol.* **24**, 846–852.
39. Weber, T., Weber, R. G., Kaulich, K., Actor, B., Meyer-Puttlitz, B., Lampel, S., Buschges, R., Weigel, R., Deckert-Schluter, M., Schmiedek, P., Reifenberger, G., and Lichter, P. (2000). Characteristic chromosomal imbalances in primary central nervous system lymphomas of the diffuse large B-cell type. *Brain Pathol.* **10**, 73–84.
40. Nakamura, M., Kishi, M., Sakaki, T., Hashimoto, H., Nakase, H., Shimada, K., Ishida, E., and Konishi, N. (2003). Novel tumor suppressor loci on 6q22-23 in primary central nervous system lymphomas. *Cancer Res.* **63**, 737–741.
41. Bataille, B., Delwail, V., and Menet, E., Vandermarcq, P., Ingrand, P., Wager, M., Guy, G., and Lapierre, F. (2000). Primary intracerebral malignant lymphoma: a report of 248 cases. *J. Neurosurg.* **92**, 261–266.
42. Henson, J. W. and Batchelor, T. T. (2004). Intraocular lymphoma. In: "Lymphoma of the Nervous System" (T. T. Batchelor, ed.), pp. 183–188. Butterworth-Heinemann, Boston.
43. O'Neill, B. P., Dinapoli, R. P., Kurtin, P. U., and Habermann, T. M. (1995). Occult systemic non-Hodgkin's lymphoma in patients initially diagnosed as primary central nervous system lymphoma: how much staging is enough? *J. Neurooncol.* **25**, 67–71.
44. Ferreri, A. J. M., Reni, M., Zoldan, M. C., Terreni, M. R., and Villa, E. (1996). Importance of complete staging in non-Hodgkin's lymphoma presenting as a cerebral mass lesion. *Cancer* **77**, 827–833.
45. Loeffler, J. S., Ervin, T. J., Mauch, P., Skarin, A., Weinstein, H. J, Canellos, G., and Cassady, J. R. (1985). Primary lymphomas of the central nervous system: patterns of failure and factors that influence survival. *J. Clin. Oncol.* **3**, 490–494.
46. Fitzsimmons, A. L., Upchurch, K. U., and Batchelor, T. T. (2005). Clinical features and diagnosis of primary central nervous system lymphoma. *Hematol. Oncol. Clin. North Am.* **19**, 689–703.
47. Abrey, L. E., Batchelor, T. T., Ferreri, A., Gospodarowicz, M., Pulczynski, E. J., Zucca, E., Smith, J. R., Korfel, A., Soussain, C., DeAngelis, L. M., Neuwelt, E. A., O'Neill, B. P., Thiel, E., Shenkier, T., Graus, F., van den Bent, M., Seymour, J. F., Poortmans, P., Armitage, J. O., Cavalli, F., and International Primary CNS Lymphoma Collaborative Group. (2005). Report of an international workshop to standardize baseline evaluation and response criteria for primary CNS lymphoma. *J. Clin. Oncol.* **23**, 5034–5043.
48. Kuker, W., Nagele, T., Korfel, A., Heckl, S., Thiel, E., Bamberg, M., Weller, M., and Herrlinger, U. (2005). Primary central nervous system lymphomas (PCNSL): MRI features at presentation in 100 patients. *J. Neurooncol.* **72**, 169–177.
49. Atlas, S. W., Lavi, E., and Fisher, P. G. (2002). Intraaxial brain tumors. In: "Magnetic Resonance Imaging of the Brain and Spine" (S. W. Atlas, ed.), pp. 565–693. Lippincott, Williams & Wilkins, Philadelphia.
50. Balmaceda, C., Gaynor, J. J., Sun, M., Gluck, J. T., and DeAngelis, L. M. (1995). Leptomeningeal tumor in primary central nervous system lymphoma: recognition, significance, and implications. *Ann. Neurol.* **38**, 202–209.
51. Ferreri, A. J., Blay, J.-Y., Reni, M., Pasini, F., Spina, M., Ambrosetti, A., Calderoni, A., Rossi, A., Vavassori, V., Conconi, A., Devizzi, L., Berger, F., Ponzoni, M., Borisch, B., Tinguely, M., Cerati, M., Milani, M., Orvieto, E., Sanchez, J., Chevreau, C., Dell'Oro. S., Zucca, E., and Cavalli, F. (2003). Prognostic scoring system for primary CNS lymphomas: the international extranodal lymphoma study group experience. *J. Clin. Oncol.* **21**, 266–272.
52. DeAngelis, L. M., Yahalom, J., Heinemann, M.-H., Cirrincione, C., Thaler, H. T., and Krol, G. (1990). Primary CNS lymphoma: combined treatment with chemotherapy and radiotherapy. *Neurology* **40**, 80–86.
53. Quality Standards Subcommittee of the American Academy of Neurology (1998). *Evaluation and management of intracranial mass lesions in AIDS*. American Academy of Neurology, Minneapolis.
54. Donahue, B., Sullivan, J. W., and Cooper, J. S. (1995). Additional experience with empiric radiotherapy for presumed human immunodeficiency virus-associated primary central nervous system lymphoma. *Cancer* **76**, 163–166.
55. Jacomet, C., Girard, P. M., Lebrette, M. G., Farese, V. L., Monfort, L., and Rozenbaum, W. (1997). Intravenous methotrexate for primary central nervous system non-Hodgkin's lymphoma in AIDS. *AIDS* **11**, 1725–1730.
56. Skiest, D. J., and Crosby, C. (2003). Survival is prolonged by highly active antiretroviral therapy in AIDS patients with primary central nervous system lymphoma. *AIDS* **17**, 1787–1793.
57. Raez, L., Cabral, L., Cai, J. P., Landy, H., Sfakianakis, G., Byrne, G. E. Jr., Hurley, J., Scerpella, E., Jayaweera, D., and Harrington, W. J. Jr. (1999). Treatment of AIDS-related primary central nervous system lymphoma with zidovudine, ganciclovir, and interleukin 2. *AIDS Res. Human Retrovirus* **15**, 713–719.
58. Kurokawa, M., Ghosh, S. K., Ramos, J. C., Mian, A. M., Toomey, N. L., Cabral, L., Whitby, D., Barber, G. N., Dittmer, D. P., and Harrington, W. J. Jr. (2005). Azidothymidine inhibits NF-kappaB and induces Epstein-Barr virus gene expression in Burkitt lymphoma. *Blood* **106**, 235–240.
59. Wu, W., Rochford, R., Toomey, L., Harrington, W. J., and Feuer, G. (2005). Inhibition of HHV-8/KSHV infected primary effusion lymphomas in NOD/SCID mice by azidothymidine and interferon-alpha. *Leuk. Res.* **29**, 545–555.
60. Nozzoli, C., Bartolozzi, B., Guidi, S., Orsi, A., Vannucchi, A. M., Leoni, F., and Bosi, A. (2006). Epstein-Barr virus–associated post-transplant lymphoproliferative disease with central nervous system involvement after unrelated allogeneic hematopoietic stem cell transplantation. *Leuk. Lymphoma* **47**, 167–169.
61. Henry, J. M., Heffner, R. R., Dillard, S. H., Earle, K. M., and Davis, R. L. (1974). Primary malignant lymphomas of the central nervous system. *Cancer* **34**, 1293–1302.
62. Murray, K., Kun, L., and Cox, J. (1986). Primary malignant lymphoma of the central nervous system. *J. Neurosurg.* **65**, 600–607.
63. Nelson, D. F., Martz, K. L., Bonner, H., Nelson, J. S., Newall, J., Kerman, H. D., Thomson, J. W., and Murray, K. J. (1992). Non-Hodgkin's lymphoma of the brain: can high-dose, large-volume radiation therapy improve survival? Report on a prospective trial by the Radiation Therapy Oncology Group (RTOG): RTOG 8315. *Int. J. Radiat. Oncol. Biol. Phys.* **23**, 9–17.
64. Ferreri, A. J., Guerra, E., Regazzi, M., Pasini, F., Ambrosetti, A., Pivnik, A., Gubkin, A., Calderoni, A., Spina, M., Brandes, A., Ferrarese, F., Rognone, A., Govi, S., Dell'Oro, S., Locatelli, M., Villa, E., and Reni, M. (2004). Area under the curve of methotrexate and creatinine clearance are outcome-determining factors in primary CNS lymphomas. *Br. J. Cancer* **90**, 353–358.

65. Gorlick, R., Goker, E., Trippett, T., Waltham, M., Banerjee, D., and Bertino, J. R. (1996). Intrinsic and acquired resistance to methotrexate in acute leukemia. *N. Engl. J. Med.* **335**, 1041–1048.
66. Worm, J., Kirkin, A. F., Dzhandzhugazyan, K. N., and Guldberg, P. (2001). Methylation-dependent silencing of the reduced folate carrier gene in inherently methotrexate-resistant human breast cancer cells. *J. Biol. Chem.* **276**, 39990–40000.
67. Ferreri, A. J., Dell'Oro, S., Capello, D., Ponzoni, M., Iuzzolino, P., Rossi, D., Pasini, F., Ambrosetti, A., Orvieto, E., Ferrarese, F., Arrigoni, G., Foppoli, M., Reni, M., and Gaidano, G. (2004). Aberrant methylation in the promoter region of the reduced folate carrier gene is a potential mechanism of resistance to methotrexate in primary central nervous system lymphomas. *Br. J. Haematol.* **126**, 657–664.
68. Ferreri, A. J. M., Abrey, L. E., Blay, J.-Y., Borisch, B., Hochman, J., Neuwelt, E. A., Yahalom, J., Zucca, E., Cavalli, F., Armitage, J., and Batchelor, T. (2003). Management of primary central nervous system lymphoma: a summary statement from the 8th International Conference on Malignant Lymphoma. *J. Clin. Oncol.* **21**, 2407–2414.
69. DeAngelis, L. M., Seiferheld, W., Schold, S. C., Fisher, B., Schultz, C. J., and Radiation Therapy Oncology Group Study 93–10. (2002). Radiation Therapy Oncology Group Study 93–10. Combination chemotherapy and radiotherapy for primary central nervous system lymphoma. *J. Clin. Oncol.* **20**, 4643–4648.
70. Poortmans, P. M., Kluin-Nelemans, H. X., Haaxma-Reiche, H., Van't Veer, M., Hansen, M., Soubeyran, P., Taphoorn, M., Thomas, J., Van den Bent, M., Fickers, M., Van Imhoff, G., Rozewicz, C., Teodorovic, I., van Glabbeke, M., and European Organization for Research and Treatment of Cancer Lymphoma Group. (2003). High-dose methotrexate-based chemotherapy followed by consolidating radiotherapy in non–AIDS-related primary central nervous system lymphoma: European Organization for Research and Treatment of Cancer Lymphoma Group Phase II Trial 20962. *J. Clin. Oncol.* **21**, 4483–4488.
71. Blay, J.-Y., Conroy, T., Chevreau, C., Thyss, A., Quesnel, N., Eghbali, H., Bouabdallah, R., Coiffier, B., Wagner, J. P., Le Mevel, A., Dramais-Marcel, D., Baumelou, E., Chauvin, F., and Biron, P. (1998). High-dose methotrexate for the treatment of primary cerebral lymphomas: analysis of survival and late neurological toxicity in a retrospective series. *J. Clin. Oncol.* **16**, 864–871.
72. Pels, H., Schmidt-Wolf, I. G., Glasmacher, A., Schulz, H., Engert, A., Diehl, V., Zellner, A., Schackert, G., Reichmann, H., Kroschinsky, F., Vogt-Schaden, M., Egerer, G., Bode, U., Schaller, C., Deckert, M., Fimmers, R., Helmstaedter, C., Atasoy, A., Klockgether, T., and Schlegel, U. (2003). Primary central nervous system lymphoma: results of a pilot and phase II study of systemic and intraventricular chemotherapy with deferred radiotherapy. *J. Clin. Oncol.* **21**, 4489–4495.
73. Batchelor, T. T., Carson, K., O'Neill, A., Grossman, S. A., Alavi, J., New, P., Hochberg, F., and Priet, R. (2003). NABTT CNS Consortium. The treatment of primary central nervous system lymphoma (PCNSL) with methotrexate and deferred radiotherapy: NABTT 96-07. *J. Clin. Oncol.* **21**, 1044–1049.
74. Abrey, L. E., Moskowitz, C. H., Mason, W. P., Abrey, L. E., Moskowitz, C. H., Mason, W. P., Crump, M., Stewart, D., Forsyth, P., Paleologos, N., Correa, D. D., Anderson, N. D., Caron, D., Zelenetz, A., Nimer, S. D., and DeAngelis, L. M. (2003). Intensive methotrexate and cytarabine followed by high-dose chemotherapy with autologous stem-cell rescue in patients with newly diagnosed primary CNS lymphoma: an intent to treat analysis. *J. Clin. Oncol.* **21**, 4151–4156.
75. Plotkin, S. R., and Batchelor, T. T. (2001). Primary nervous system lymphoma. *Lancet Oncol.* **2**, 354–365.
76. Plotkin, S. R., Betensky, R. A., Hochberg, F. H., Grossman, S. A., Lesser, G. J., Vabors, L. B., Chon, B., and Batchelor, T. T. (2004). Treatment of central nervous system lymphoma with sequential courses of high-dose methotrexate. *Clin. Cancer Res.* **10**, 5643–5646.
77. Soussain, C., Suzan, F., Hoang-Xuan K., Cassoux, N., Levy, V., Azar, N., Belanger, C., Achour, E., Ribrag, V., Gerber, S., Delattre, J. Y., and Leblond, V. (2001). Results of intensive chemotherapy followed by hematopoietic stem-cell rescue in 22 patients with refractory or recurrent primary CNS lymphoma or intraocular lymphoma. *J. Clin. Oncol.* **19**, 742–749.
78. Mizumatsu, S., Monje, M. L., Morhardt, D. R., Rola, R., Palmer, T. D., and Fike, J. R. (2003). Extreme sensitivity of adult neurogenesis to low doses of X-irradiation. *Cancer Res.* **63**, 4021–4027.
79. Monje, M.L.(2003). Inflammatory blockade restores adult hippocampal neurogenesis. *Science* **302**, 1760–1765.
80. Abrey, L. E., DeAngelis, L. M., and Yahalom, J. (1998). Long-term survival in primary CNS lymphoma. *J. Clin. Oncol.* **16**, 859–863.
81. Lai, R., Abrey, L. E., Rosenblum, M. K., and DeAngelis, L. M. (2004). Treatment-induced leukoencephalopathy in primary CNS lymphoma: a clinical and autopsy study. *Neurology* **62**, 451–456.
82. Correa, D. D., DeAngelis, L. M., Shi, W., Thaler, H., Glass, A., and Abrey, L. E. (2004). Cognitive functions in survivors of primary central nervous system lymphomas. *Neurology* **62**, 548–555.
83. Fliessbach, K., Helmstaedter, C., Urbach, H., Althaus, A., Pels, H., Linnebank, M., Juergens, A., Glasmacher, A., Schmidt-Wolf, I. G., Klockgether, T., and Schlegel, U. (2005). Neuropsychological outcome after chemotherapy for primary CNS lymphoma: a prospective study. *Neurology* **64**, 1184–1188.
84. Neuwelt, E. A., Guastadisegni, P. E., Varallyay, P., and Doolittle, N. D. (2005). Imaging changes and cognitive outcome in primary CNS lymphoma after enhanced chemotherapy delivery. *Am. J. Neuroradiol.* **26**, 258–265.
85. Kleihues, P. and Cavenee, W. K., eds. (2000). Tumours of the nervous system. World Health Organization Classification of Tumours, ed X[CN25]. International Agency for Research on Cancer. 2000, IARC Press, Lyon, France.
86. Camilleri-Broet, S., Martin, A., Moreau, A., Angonin, R., Henin, D., Gontier, M. F., Rousselet, M. C., Caulet-Maugendre, S., Cuilliere, P., Lefrancq, T., Mokhtari, K., Morcos, M., Broet, P., Kujas, M., Hauw, J. J., Desablens, B., and Raphael, M. (1998). Primary central nervous system lymphomas in 72 immunocompetent patients: pathologic findings and clinical correlations. *Am. J. Clin. Pathol.* **110**, 607–612.
87. Chan, C. C., and Wallace, D. J. (2004). Intraocular lymphoma: update on diagnosis and management. *Cancer Control* **11**, 285–295.
88. Ferreri, A. J. M., Abrey, L. E., Blay, J., Borisch, B., Hochman, J., Neuwelt, E. A., Yahalom, J., Zucca, E., Cavalli, F., Armitage, J., and Batchelor, T. (2003). Summary statement on primary central nervous system lymphomas from the Eighth International Conference on Malignant Lymphoma, Lugano, Switzerland, June 12–15, 2002. *J. Clin. Oncol.* **21**, 2407–2414.
89. Voloschin, A., Wen, P., Hochberg, F., and Batchelor, T. (2004). Topotecan as salvage therapy for refractory or relapsed primary central nervous system lymphoma: final report. *Neurology* **62**, A478.
90. Enting, R. H., Demopoulos, A., DeAngelis, L. M., and Abrey, L. E. (2004). Salvage therapy for primary CNS lymphoma with a combination of temozolomide and rituximab. *Neurology* **63**, 901–903.
91. Reni, M., Mason, W., Zaja, F., Perry, J., Franceschi, E., Bernardi, D., Dell'Oro, S., Stelitano, C., Candela, M., Abbadessa, A., Pace, A., Bordonaro, R., Latte, G., Villa, E., and Ferreri, A. J. (2004). Salvage chemotherapy with temozolomide in primary CNS lymphomas: preliminary results of a phase II trial. *Eur. J. Cancer* **40**, 1682–1688.
92. Herrlinger, U., Brugger, W., Bamberg, M., Kuker, W., Dichgans, J., and Weller, M. (2000). PCV salvage chemotherapy for recurrent primary CNS lymphoma. *Neurology* **54**, 1707–1708.
93. Nguyen, P. L., Chakravarti, A., Finkelstein, D. M., Hochberg, F. H., Batchelor, T. T., and Loeffler, J. S. (2005). Results of whole-brain radiation as salvage of methotrexate failure for immunocompetent patients with primary CNS lymphoma. *J. Clin. Oncol.* **23**, 1507–1513.

37

Neurofibromatosis 1

Linda Piersall, MS
David H. Gutmann, MD, PhD

Keywords: *café-au-lait spot, glioma, Lisch nodule, malignant peripheral nerve sheath tumor, neurofibroma, neurofibromatosis, neurofibromin, NF1, RAS*

I. History
II. Nomenclature
III. Etiology
IV. Pathogenesis
V. Natural History
VI. Structural Detail
VII. Biochemistry and Molecular Mechanisms
VIII. Symptoms in Relation to Pathophysiology
References

I. History

The first published clinical descriptions of neurofibromatosis type 1 (NF1) date to 1768, and drawings of persons with features suggestive of NF1 appear even earlier. A stone icon from 300 BCE depicts an individual with cutaneous, tumorous lesions on the torso. Joseph Merrick, who was famously known as the "Elephant Man," was born in England in 1862 and subsequently showcased as a circus attraction because of his significant disfiguring tumors and bony abnormalities. Although Merrick's story brought a great deal of attention to NF1, it is now widely thought that he suffered not from NF1 but from a rare condition known as Proteus syndrome.

Recognition of NF1 as a clinical entity began with the German pathologist Friedrich Daniel von Recklinghausen, who published his famous monograph on NF1 in 1882. Because von Recklinghausen was credited with being the first to recognize that neurofibromas in NF1 arose from cells in the myelin sheath, the disease is often called von Recklinghausen disease. In the early 1900s, NF1 was often lumped together with other neurocutaneous syndromes, including neurofibromatosis type 2 (NF2). It would not be until 1956 that the genetic basis of NF1 was proposed by Frank Crowe, William Schull, and James Neel.

In the 1970s, it became clear that there were two distinct types of neurofibromatosis and that individuals affected with NF1 required multidisciplinary care in subspecialty clinics [1]. The diagnostic criteria for NF1 and NF2 were formally established as a result of a National Institutes of Health (NIH) Consensus Development Conference, held in 1987 [2]. The establishment of these criteria allowed the diagnosis to be rendered confidently by physicians and paved the way to the collection of families necessary for the identification of the *NF1* gene. This international effort resulted in the positional cloning of the *NF1* gene in 1990 in addition to the identification of the *NF1* protein, neurofibromin [3].

II. Nomenclature

The neurofibromatoses are divided into two types. Although both type 1 and type 2 NF predispose to benign and malignant tumors, they are clinically and genetically distinct disorders. Individuals with NF1 have characteristic cutaneous manifestations, iris hamartomas, and tumors that primarily involve the peripheral nervous system and occasionally the central nervous system (CNS) (e.g., optic gliomas). As will be discussed in greater detail in this chapter, individuals with NF1 also have involvement of many organ systems and may exhibit bony abnormalities, learning disabilities, and vascular problems. In contrast, persons with NF2 develop cataracts, hearing loss, and tumors predominantly of the CNS, most notably the vestibular schwannoma [2]. Clinical studies have suggested that other forms of neurofibromatosis might exist, including spinal neurofibromatosis, schwannomatosis, and segmental NF. These forms are not discussed in this chapter.

Although it is occasionally still referred to as "von Recklinghausen disease" or "peripheral neurofibromatosis," *NF1* is the preferred nomenclature in the medical literature. The term *neurofibroma*, from which the disorder derives its name, was first coined by von Recklinghausen to refer to the tumors that involve both the nerve ("neuro") and fibrous tissue ("fibroma") [1]. NF1 is also frequently classified with tuberous sclerosis and von Hippel-Lindau disease as one of the phacomatoses, a group of disorders that present with hamartomas of the central and peripheral nervous system, eye, and skin [4].

III. Etiology

NF1 is one of the most common single gene disorders. NF1 affects approximately 1 in every 3000 individuals and is more prevalent than cystic fibrosis, Huntington's disease, and Tay-Sachs disease combined. It exhibits no gender, racial, or ethnic bias and has been reported in every culture and country [5,6]. NF1 is inherited in an autosomal-dominant manner, in which half of all individuals are the first individual in the family with the disorder as a result of a new or *de novo* germline mutation. The rest of affected individuals inherit NF1 from an affected parent [7].

NF1 results from a mutation in the *NF1* gene such that affected individuals start life with one functional copy and one nonfunctional (mutant) copy of the *NF1* gene in every cell in the body. Although all people who harbor a mutated *NF1* gene carry the diagnosis of NF1 (i.e., complete penetrance), the clinical features in any given individual can be quite variable, even within families with the same *NF1* gene mutation (i.e., variable expressivity).

IV. Pathogenesis

NF1 is a protean disorder that can result in a wide variety of cutaneous, ocular, peripheral nerve, and orthopedic manifestations. Individuals must exhibit at least two features from the list of seven diagnostic criteria to be diagnosed with NF1 [2]. These diagnostic criteria are shown in Table 1.

A hallmark of NF1 is the presence of multiple café-au-lait macules, a reference to the characteristic "coffee with milk" color of the spots (Fig. 1A). They are typically the most common and earliest manifestation of NF1 in affected individuals. Children are frequently born with café-au-lait spots and may continue to develop new lesions over the first 2 to 5 years of life [2]. Café-au-lait spots associated with NF1 are usually ovoid and have smooth borders, unlike the irregular café-au-lait spots that are seen in other genetic conditions such as McCune-Albright or Russell-Silver syndrome [4]. In individuals with NF1, the café-au-lait spots are homogeneous in pigmentation, although they may darken with sun exposure or fade in adulthood [1]. To count as a diagnostic feature, there must be six or more spots, each measuring 1.5 cm or larger in postpubertal individuals and 0.5 cm or larger in children before the onset of puberty [2,4].

The diagnosis of NF1 can be difficult to confirm in an infant or a young child because café-au-lait spots may be the only clinical feature present for several years. Although one or two café-au-lait spots are common in individuals without NF1, there should be a high index of suspicion for NF1 in any child presenting with multiple café-au-lait spots [2,4]. Children with multiple café-au-lait spots frequently develop a second feature of NF1 later in childhood, typically by age 8, such that a diagnosis can definitively be made at that time [4]. Thus although NF1 is fully penetrant and all individuals who inherit a mutated copy of the *NF1* gene will develop features of NF1 [3], a diagnosis cannot always be readily established in a young child because the development of additional features is age dependent. Most

Table 1 NIH Consensus Diagnostic Criteria for Neurofibromatosis 1 (NF1)

The individual should have two or more of the following:
- Six or more café-au-lait spots
  1.5 cm or larger in postpubertal individuals
  0.5 cm or larger in prepubertal individuals
- Two or more neurofibromas, or one or more plexiform neurofibromas
- Axillary or inguinal freckling
- Optic glioma
- Two or more Lisch nodules
- Distinctive bony lesion
  Sphenoid bone dysplasia or tibial dysplasia/pseudarthrosis
- A first-degree relative with NF1

NF1, Neurofibromatosis 1; *NIH*, National Institutes of Health.

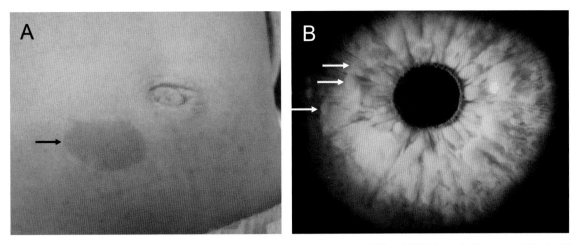

Figure 1 Pigmentary features in NF1. **A**, Typical café-au-lait macule in a young child with NF1 (*arrow*). (Courtesy of Dr. David Gutmann, Washington University School of Medicine, St. Louis.) **B**, Lisch nodules in an adult with NF1 (*arrows*). (Courtesy of Dr. Gregg Lueder, St. Louis Children's Hospital, St. Louis.)

clinicians who care for children with NF1 will follow children with only multiple café-au-lait macules in the same manner they follow children with NF1 based on a high index of suspicion, including assessments of bone growth and annual ophthalmological examinations.

Skinfold freckling in non–sun-exposed areas is the second most common feature of NF1 in children. These freckles are found in the axillary and inguinal regions and begin to appear between 3 and 5 years of age. Approximately 80% of children with NF1 have skinfold freckling by the age of 6, and 90% of affected adults have axillary freckling [2]. Adults may also have freckling under the chin, in redundant skinfolds, and under the breasts in women [2].

Lisch nodules are hamartomatous lesions on the iris. These slightly pigmented, raised lesions are often difficult to visualize at the bedside and are best seen by an experienced ophthalmologist using a slit lamp (see Fig. 1B). These lesions are considered pathognomonic for NF1 and have no clinical significance other than to aid in the diagnosis. The nodules are seen in virtually all adults with NF1 but in only 50% of children by the age of 5 years [2].

Neurofibromas, for which the disorder is named, are discrete cutaneous or subcutaneous growths that arise from a single nerve fascicle. They may be seen or palpated as discrete masses growing on or just beneath the skin (Fig. 2A). The appearance of neurofibromas can be quite variable, with some appearing as flattened lesions and others assuming a more pedunculated form [2,4]. Neurofibromas can grow anywhere on the body, including the scalp, nipples, and soles of feet. Neurofibromas typically begin to appear just before the onset of puberty and increase in number and size during adolescence and adulthood. An accelerated increase in both the rate of new tumor development and the growth of existing neurofibromas is commonly reported in pregnant women [4]. Neurofibromas can also develop on visceral and spinal nerve roots. The spinal neurofibromas may involve multiple nerves and appear on magnetic resonance imaging (MRI) as dumbbell-shaped intraforaminal spinal tumors [2].

Although these discrete neurofibromas pose no risk for malignancy, they may press on other anatomical structures as they grow, causing pain or other medical problems. They can occasionally cause nerve impingement,

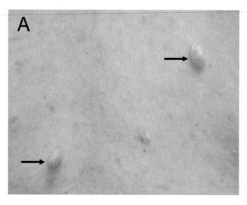 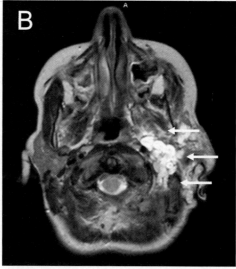

Figure 2 Neurofibromas in NF1. **A**, Typical cutaneous dermal neurofibromas in an adult with NF1 (*arrows*). **B**, Facial and neck plexiform neurofibroma in a child with NF1. *Arrows* on the MRI scan demonstrate the multilobulated tumor.

resulting in weakness, pain, paresthesias, or autonomic dysfunction [2,8]. Most commonly, neurofibromas may be tender when irritated and cause significant cosmetic problems for affected individuals. In most cases, cutaneous and subcutaneous neurofibromas can be safely resected, but they may regrow.

Histological examination of neurofibromas demonstrates that these tumors are composed of a variety of different cell types, including Schwann cells, mast cells, perineurial cells, and fibroblasts [8]. It is the Schwann cell that represents the neoplastic cell in the neurofibroma, whereas the other cellular elements may contribute to tumor formation and continued growth.

In contrast to the discrete neurofibromas, the plexiform neurofibroma is a diffuse multinodular tumor that runs along the length of a peripheral nerve. These more diffuse neurofibromas can involve multiple nerve fascicles, nerve branches, and plexi (see Fig. 2B) [9,10]. Plexiform neurofibromas can grow as superficial masses or develop as internal tumors affecting the face, neck, limbs, and torso. Clinically, these growths have a nonhomogeneous texture and may be associated with overlying skin hyperpigmentation, fine hair growth, skin thickening, and increased vascular markings [9,10]. Plexiform neurofibromas are detected in one-third of patients with NF1 and are thought to be slow-growing congenital or early developmental lesions. In keeping with their congenital origin, the majority of plexiform neurofibromas are initially found during childhood [9,10]. Despite its early development, a plexiform neurofibroma may go undetected if it is not visible and does not cause any clinical symptoms. Often these benign tumors do not result in clinical sequela, and they are not typically painful except when hit by a direct force or blow [9,10].

However, plexiform neurofibromas can sometimes cause significant disfigurement by distorting the surrounding soft tissues or resulting in abnormal bone growth. In this regard, plexiform neurofibromas can stimulate long bone growth to result in a leg length discrepancy or can disrupt normal bone development to cause orbital bone dysplasia.

Surgical resection is difficult because of the rich vascular network supporting the plexiform neurofibroma, the involvement of healthy tissue, and the risk of morbidity associated with surgery [9,10]. The greatest concern associated with a plexiform neurofibroma is the risk for malignant transformation of a previously benign tumor into a malignant peripheral nerve sheath tumor (MPNST). The lifetime risk of malignant transformation is 10% [9–11]. Although MRI scans are useful in determining the location and extent of a plexiform neurofibroma, they are not useful in differentiating between a benign tumor and one that has undergone malignant transformation. There is an emerging use of positron emission tomography (PET) to distinguish benign tumors from MPNSTs. Although it is not currently used in routine practice, it is likely that PET will become an adjuvant test in the evaluation of a patient with NF1 and a suspected MPNST. At present, true cut biopsy or surgical excision remains the only way to determine whether a plexiform neurofibroma has undergone malignant transformation [9,10].

The second most common tumor in children with NF1 is the optic pathway glioma (OPG). OPG is present in 15–20% of NF1 patients and may involve the optic nerve (unilateral or bilateral), chiasm, hypothalamus, and occasionally the optic tracts [12,13]. OPG is a tumor almost exclusively of childhood, with 4.5 years representing the mean age of diagnosis. Although they are classified as benign World Health Organization grade I tumors (pilocytic astrocytoma), optic gliomas can result in visual impairment and hypothalamic dysfunction. In contrast to sporadic OPG in children without NF1, half of all OPGs in children with NF1 remain clinically asymptomatic and do not require treatment [12–14]. Histologically, NF1-associated OPGs are identical to the more clinically aggressive pilocytic astrocytomas that occur in individuals without NF1 [12,13].

Although the period of greatest risk for OPG is younger than 6 years and late progression is uncommon, individuals presenting with clinical progression of OPG after the age of 10 years have been reported in as many as 30% of children with NF1 and OPG. Children with a symptomatic OPG may present with decreased visual acuity, decreased visual fields, abnormal pupillary function, decreased color vision, optic atrophy, precocious puberty, or proptosis [12–14]. In contrast, an asymptomatic OPG is often discovered as an incidental finding on MRI. Because of the inability of young children to reliably report changes in their vision, all children with NF1 require annual ophthalmological evaluations [2]. Children with abnormal ophthalmological examinations are evaluated by brain MRI. OPG may appear as a thickened or tortuous optic nerve, sometimes exhibiting robust gadolinium contrast enhancement on MRI (Fig. 3A). In addition, OPG may also involve the optic chiasm, postchiasmatic optic nerve tracks, and hypothalamus. In the latter case, hypothalamic involvement may result in precocious puberty. The development of precocious puberty in a child with NF1 should also prompt brain MRI.

The use of routine screening MRI to detect OPG was historically a source of heated clinical debate. However, screening MRI does not lead to an earlier detection of symptomatic gliomas, does not enhance treatment options for the symptomatic glioma, and is not warranted in the management of children with NF1 [1,12].

Less commonly, children with NF1 may harbor pilocytic astrocytoma involving the brainstem. These NF1-associated brainstem gliomas often remain asymptomatic and pose less morbidity and mortality than brainstem gliomas in the general population [14]. Additionally, other gliomas (excluding OPG) rarely occur throughout life, including higher-grade malignant astrocytomas [14].

In addition to tumors, children with NF1 are also at an increased risk for specific orthopedic concerns including tibial dysplasia, sphenoid wing dysplasia, and dystrophic scoliosis. Tibial dysplasia with or without pseudarthroses is seen in approximately 5% of individuals with NF1. Affected children present with congenital bowing, typically of the long bones of the arms or legs, which can progress to fracture and the development of a "false" joint or pseudarthrosis. Children with suspected long bone dysplasia should be referred to an orthopedic specialist for evaluation and treatment [1,4]. Another NF1-associated congenital bony abnormality is sphenoid wing dysplasia [4]. In affected individuals, the sphenoid wing can be incompletely formed or is absent, resulting in proptosis or a sunken eye. Referral to a craniofacial expert is often required.

Approximately 10–20% of individuals with NF1 develop scoliosis, which tends to present earlier in these individuals than in the general population. In most cases, scoliosis is mild and is managed similarly to idiopathic scoliosis in the general population [1,2,4]. Although dystrophic scoliosis does not frequently occur, it can progress quickly during childhood, often requiring multiple complicated surgical

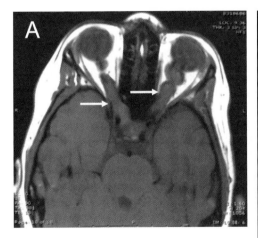
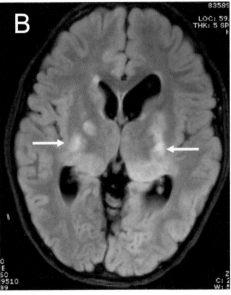

Figure 3 Brain abnormalities in NF1. **A**, Bilateral optic nerve gliomas in a young child with NF1 (*arrows*). **B**, Bilateral T2-hyperintense lesions in the basal ganglia (*arrows*) are often seen on brain MRI of young children with NF1. The hyperintensities are often referred to as *unidentified bright objects* or *UBOs*.

interventions. Dystrophic features may also include vertebral scalloping, wedging of vertebral bodies, dysplastic pedicles, and a paravertebral soft tissue mass.

MPNSTs typically arise from a preexisting plexiform neurofibroma [9,11,14] but may also develop in an individual with NF1 in whom a preexisting plexiform neurofibroma was not clinically obvious. These malignant tumors are highly aggressive, almost universally fatal, and frequently metastatic cancers. Generally, MPNSTs develop in 10% of NF1 patients, usually in the third to fourth decade of life [11,14]. In contrast, MPNSTs occur infrequently in the general population in the sixth or seventh decade of life. Unlike sporadic MPNST, individuals with NF1 may develop more than one MPNST. In the setting of a known plexiform neurofibroma, the development of unexplained unremitting pain, sudden growth, a change in tumor texture (e.g., soft to hard), or an associated neurological deficit in the area of an existing plexiform neurofibroma should raise concern for malignant transformation [9,11,14]. The mainstay of treatment is wide surgical resection followed by local radiation; however, tumor recurrence and metastasis are common. MPNSTs do not respond well to currently available chemotherapy or radiation, and the 5-year survival rate following the diagnosis of an MPNST in patients with NF1 is poor [11,14].

Learning disabilities, which can manifest as both language-based and nonverbal neurocognitive dysfunction, are seen in 30–60% of all individuals with NF1 [6,15]. Although verbal, performance, and full-scale intelligence quotients (IQs) in individuals with NF1 are typically lower than in their unaffected siblings [6,15], frank mental retardation is uncommon. The specific learning problems in children with NF1 are similar to those encountered in the general population, or those people who do not have NF. In this regard, early identification of learning disabilities and the development of an aggressive, individualized education plan are essential. In addition to specific learning disabilities, children with NF1 are reported to have an increased incidence of hypotonia, coordination difficulties, and behaviors associated with attention-deficit hyperactivity disorder (ADHD) [4,6].

Some studies have suggested that specific MRI findings, referred to as *unidentified bright spots* (UBOs), may correlate with lower IQs in children with NF1 [6,15]. UBOs appear as hyperintense lesions on T2 weighted MRI scans and are most commonly located in the basal ganglia, optic tracts, brainstem, and cerebellum (see Fig. 3B). These UBOs do not enhance with gadolinium contrast administration and often disappear as children enter early adulthood. UBOs are seen in roughly half of all children with NF1. Given the benign nature of these MRI findings, their neuropathological correlates are unclear. It is unlikely that they are hamartomas.

Because NF1 is an incredibly heterogeneous disorder, other findings associated with NF1 include macrocephaly, seizures, short stature relative to family members, leg length discrepancy, renal artery stenosis leading to hypertension in adults, congenital heart disease, cerebrovascular stenosis resulting in moya moya syndrome, and stroke [2,4,16]. Other malignancies have also been reported with increased frequency, including pheochromocytoma and leukemia [14].

Occasionally an individual presents with café-au-lait spots and neurofibromas localized to one area or side of the body. These patients do not have the more common generalized form of NF1 and may be difficult to diagnose using the established diagnostic criteria. These individuals are referred to as having *segmental* or *mosaic NF1*. It is believed that this form of NF1 arises during development as a result of an *NF1* gene mutation in a small subset of embryonic tissues. The timing of the mutational event is reflected in the extent of the tissue expressing the *NF1* mutation [3]. Depending on the distribution of the NF1 features, the risk of genetic transmission can be variable. However, a child born to an individual with segmental NF1 will have the generalized form and not segmental NF1.

V. Natural History

Although NF1 demonstrates complete penetrance, the development of individual features associated with NF1 is age dependent. As previously mentioned, NF1 displays intrafamilial and interfamilial variable expressivity [2,3,5]. In this regard, individuals with identical *NF1* gene mutations can have markedly different clinical features. The fact that the specific *NF1* gene mutation does not account for the clinical phenotype suggests that modifying genes present in the genome contribute to the variation in expression in NF1 [3]. Recent studies in mice have begun to identify these modifier loci.

The clinical care of individuals with NF1 must take into account the age of the patient. Children with NF1 are typically noted to harbor multiple café-au-lait spots either at birth or shortly thereafter and may continue to develop more café-au-lait spots during the first few years of life. In contrast, skinfold freckling and Lisch nodules are usually not present in infancy but typically begin to develop by the age of 6 years [2,3,4]. Consequently, most individuals will fulfill at least two of the diagnostic criteria by late childhood or early adolescence. During early childhood, attention must be paid to the developing skeleton to exclude the presence of tibial dysplasia and sphenoid wing dysplasia. Similarly, careful inspection of young children for asymmetrical and soft tissue masses is key to the detection of a plexiform neurofibroma. Similarly, the ages of greatest risk for OPG are between 2 and 6 years, and all children should be evaluated by an experienced pediatric ophthalmologist during the first decade of life. By the end of the first decade of life, scoliosis may be detected [4]. With puberty and the onset of adolescence comes the initial growth of neurofibromas, which may continue to develop throughout adulthood. Although the number of neurofibromas that will develop is unpredictable, individuals with NF1 tend to develop more neurofibromas as they age [2,8]. Lastly, malignant transformation of an existing plexiform neurofibroma into an MPNST typically does not occur until the second or third decade of life [9,11]. Few studies have formally evaluated the lifespan of persons with NF1, however, it is generally thought that the average life expectancy in individuals with NF1 is somewhat reduced compared with the general population (61.6 years vs 70 years) due to a higher rate of malignancy [1,6].

VI. Structural Detail

The *NF1* gene maps to chromosome 17q11.2 and spans approximately 350 kilobases of genomic DNA. Genetic studies have demonstrated that the new mutation rate for the *NF1* gene is tenfold higher than most genes, which likely explains the frequency of *de novo* germline mutations in the *NF1* gene [17]. The majority of these de novo mutations occurs on the paternal chromosome and may reflect the greater likelihood of developing a new mutation in developing sperm rather than ova [3].

Mutations are distributed throughout the *NF1* gene. The majority of mutations are novel; only 10 recurring mutations are reported in the medical literature, each accounting for approximately 2.9% of all identified germline mutations [7,17]. Large deletions, microdeletions, missense mutations, nonsense mutations, insertions, and splicing mutations have all been reported. Most mutations create a premature stop codon or frameshift, resulting in a premature truncation of the encoded protein [7,17]. There are no obvious genotype-phenotype correlations associated with these reported mutations [17], suggesting that the type of *NF1* gene mutation does not determine the clinical features that will manifest in a child or adult. The only exception to this lack of a genotype-phenotype correlation is the group of patients with large genomic deletions that typically encompass the entire *NF1* gene as well as flanking loci on chromosome 17q11.2. Individuals with this microdeletion syndrome have a higher likelihood of developing neurofibromas at an earlier age, a lower IQ (including mental retardation), characteristic facial dysmorphism (Noonan's facies), and a higher lifetime risk for MPNST [7,17].

Genetic testing is available for NF1 and can detect the causative *NF1* gene mutation more than 90% of the time. However, the identification of a mutation by clinically available DNA testing does not provide prognostic information that would aid in clinical management. Conversely, the inability to identify a mutation does not exclude the diagnosis of NF1 and should not alter the medical management of an individual who already meets clinical diagnostic criteria or in whom the physician has a high level of suspicion for NF1 [17]. Lastly, it is important that all individuals who undergo genetic testing for NF1 receive genetic counseling.

The *NF1* gene encodes a large 220–250 kDa protein comprised of 2818 amino acids, called *neurofibromin* (see Fig. 4A). Although *NF1* transcripts can be detected in

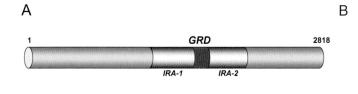

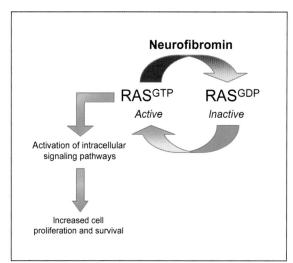

Figure 4 **A**, Predicted protein structure of neurofibromin. Neurofibromin is composed of 2818 amino acids. A small portion of neurofibromin shares structural and functional similarity with other proteins that inactivate RAS, called GTPase activating proteins (GAPs). The GAP-related domain (GRD) is flanked by two domains shared with yeast GAP molecules (IRA-1, IRA-2). **B**, Neurofibromin functions in part as a negative regulator of the RAS proto-oncogene. As a RAS-GAP, neurofibromin accelerates the conversion of RAS from its active, GTP-bound form to its inactive, GDP-bound form. Loss of neurofibromin expression results in increased RAS-GTP, activation of intracellular signaling pathways, and increased cell proliferation and survival.

many tissues, neurofibromin is expressed in neurons, astrocytes, oligodendrocytes, Schwann cells, adrenal medullary cells, white blood cells, and gonadal tissue in adults [18]. Sequence analysis of the predicted amino acid sequence of neurofibromin revealed a small 360 amino acid region that shares sequence and functional homology with the catalytic domain of guanosine triphosphatase (GTPase) activating proteins (GAP molecules) [18]. Flanking this GAP-related domain (GRD) are sequences shared with yeast GAP molecules, termed *IRA domains*. Their functional significance is unknown. The remaining 90% of neurofibromin contains no structural motifs or domains that provide additional clues to neurofibromin function.

VII. Biochemistry and Molecular Mechanisms

The *NF1* gene is a tumor suppressor gene that negatively regulates cell growth. Classic tumor suppressor genes are defined as genes inactivated in human cancers whose protein products normally function to inhibit cell growth. The elevated risk of cancer development in individuals with tumor suppressor gene disorders (tumor predisposition syndromes) can be explained on the molecular level by the two-hit hypothesis, originally proposed by Alfred Knudson for retinoblastoma (Fig. 5). Individuals who do not have NF1 have two functional copies of the *NF1* gene and a significantly lower incidence of tumor formation. However, in the case of NF1, individuals are born with one mutated (nonfunctional; "first hit") and one nonmutated (functional) copy of the *NF1* gene. Later, an acquired somatic mutation ("second hit") in the previously functioning *NF1* gene leads to complete inactivation of both copies of the *NF1* gene, loss of neurofibromin expression, and an increased growth advantage to that cell [19].

GAP molecules, including neurofibromin, act as negative regulators of RAS. In its active state, RAS is bound to GTP and provides a growth-promoting signal to many cell types. Early studies of activated RAS demonstrated that it could transform cells in vitro and in vivo, suggesting that constitutively active RAS functioned as an oncogene. In its inactive state, RAS is bound to guanosine diphosphate (GDP) [18]. GAP molecules regulate RAS by accelerating the hydrolysis of the active, GTP-bound form of RAS to the inactive, GDP-bound form of RAS (see Fig. 4B), and by doing so, turn off RAS activity, reduce RAS-mediated mitogenic signaling, and inhibit cell proliferation. Neurofibromin has been shown to exhibit RAS-GAP activity both *in vitro* and *in vivo* [18]. Loss of *NF1* gene expression has been shown to result in absent neurofibromin RAS-GAP function, increased RAS activity, and increased cell proliferation. Studies in mammalian cells have shown that replacement of the RAS-GAP domain of neurofibromin is

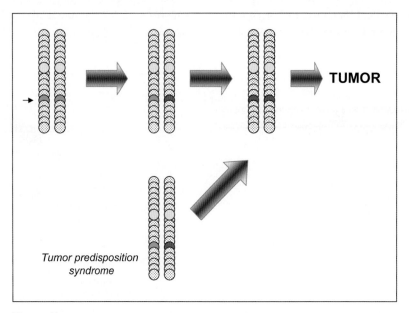

Figure 5 Knudson two-hit hypothesis. Individuals without NF1 begin life with two functional copies of the *NF1* gene (*green*). In this case, tumor formation requires inactivation of both copies of the *NF1* gene. Individuals with NF1 start life with one functional copy and one mutated (nonfunctional [*red*]) copy of the *NF1* gene. In this case, tumor formation only requires one additional genetic event: the loss of the one remaining functional *NF1* allele. For this reason, individuals with tumor predisposition syndromes such as NF1 have an increased likelihood of tumor formation.

sufficient to restore normal RAS activity and inhibit *NF1*-deficient cell growth in vitro [19].

The RAS-GAP domain represents only 10% of the coding region of the *NF1* gene [19], raising the intriguing possibility that neurofibromin has other functions unrelated to its RAS-GAP activity. Neurofibromin has also been shown to positively regulate cyclic adenosine monophosphate (cAMP) generation in mammalian nervous system cells [20]. Further studies are required to determine the role of cAMP regulation in neurofibromin growth regulation, tumor formation, and brain function.

VIII. Symptoms in Relation to Pathophysiology

Since the discovery of the *NF1* gene, a tremendous amount of progress has been made in understanding the pathogenesis of the tumors associated with NF1. The RAS-GAP function of neurofibromin has provided a framework to understand the molecular pathogenesis of NF1-associated tumor formation, yet little is known about the relationship between the function of neurofibromin and other clinical manifestations of NF1, such as learning disabilities and bony growth.

In keeping with the proposed role of neurofibromin as a RAS-GAP, *NF1* gene inactivation, neurofibromin loss, and increased RAS activity have been reported in a variety of NF1-associated tumors, including astrocytomas, leukemias, pheochromocytoma, neurofibromas, and MPNSTs [19]. In most of these tumor types, hyperactivation of RAS appears to be the driving force behind tumorigenesis.

In the case of the neurofibroma, the neoplastic cell lacking neurofibromin expression in the tumor is the Schwann cell [20]. In addition to Schwann cells, other cells in the tumor may contribute to tumor formation, such as mast cells and fibroblasts. To gain insights into the molecular pathogenesis of human neurofibroma formation, Parada and colleagues developed mice in which the *Nf1* gene could be selectively inactivated in specific cell types [21]. Selective inactivation of the *Nf1* gene in Schwann cell precursors did not result in tumor formation, although Schwann cell hyperplasia was observed in these mice. These exciting findings demonstrated that loss of neurofibromin was not sufficient in itself for tumor formation and suggested that the $NF1+/-$ cells (one functional and one nonfunctional *NF1* gene) in the surrounding peripheral nerve environment (e.g., mast cells and fibroblasts) might play a role in tumor initiation. To model the human condition, Parada and colleagues then generated $Nf1+/-$ mice that lacked neurofibromin in Schwann cell precursors. Similar to humans with NF1, these mice had one functional and one nonfunctional *NF1* gene in every cell in their body but only lacked neurofibromin expression in select cells. These mice developed tumors similar to human plexiform neurofibromas, firmly establishing the need for *NF1* heterozygous cells ($NF1+/-$) for neurofibroma tumor development (Fig. 6) [20]. Based on

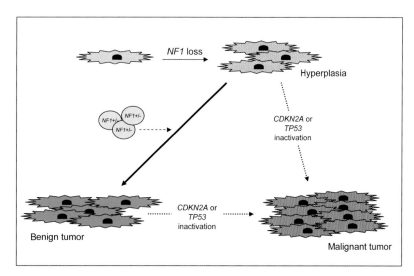

Figure 6 Evolution of NF1-associated tumors. In the peripheral and central nervous system, biallelic *NF1* inactivation in Schwann cells or astrocytes results in increased cell proliferation and hyperplasia in vitro and in vivo but not tumor formation. Neurofibroma development requires the presence of $NF1+/-$ cells (e.g., $NF1+/-$ mast cells), which are recruited to the site of the developing neurofibroma and have been hypothesized to promote tumor formation. Similarly, factors produced by the $NF1+/-$ brain environment (e.g., $NF1+/-$ microglia) are required for the formation of low-grade tumors. The formation of malignant cancers (malignant nerve sheath tumors or high-grade gliomas) is associated with *CDKN2A* or *TP53* inactivation in $NF1-/-$ Schwann cells.

these studies and others, it has been proposed that $NF1+/-$ mast cells are attracted to the site of neurofibroma formation by factors released by $NF1$-deficient Schwann cells. These recruited mast cells in turn secrete additional factors that promote Schwann cell proliferation and transformation.

In the brain, loss of neurofibromin expression and increased RAS activity are seen in the astrocytes of human NF1-associated OPG. Interestingly, sporadic OPG does not demonstrate mutations in the $NF1$ gene or loss of neurofibromin expression, suggesting that NF1-associated and sporadic OPG are histologically identical but genetically distinct. Studies aimed at defining the genetic changes associated with NF1-associated astrocytoma formation have shown that these tumors do not have the typical genetic changes found in sporadic low-grade glioma [18,20].

Mouse modeling of NF1-associated optic glioma has yielded important insights into the molecular and cellular determinants of tumor formation. Similar to mouse NF1-associated plexiform neurofibroma pathogenesis, inactivation of the $Nf1$ gene in astrocytes is insufficient for glioma formation. However, $Nf1+/-$ mice lacking neurofibromin in astrocytes develop optic nerve glioma. These mice with $Nf1$ optic nerve glioma exhibit increased astrocyte proliferation in the optic nerve and chiasm by 3 weeks of age and develop discrete low-grade tumors by 2 months of age. Similar to children with NF1, gliomas are restricted to the optic nerve and chiasm, and progression to high-grade glioma has not been observed. Moreover, these tumors exhibit maximal proliferation in young mice, with relatively little proliferative activity after 4 months of age. Analysis of the optic nerves of these mice at 3 weeks of age, before tumor formation, demonstrates the presence of infiltrating microglia and new blood vessel formation, suggesting that microglia may play a role in tumorigenesis in the brain similar to mast cells in the genesis of neurofibromas in the peripheral nervous system [20].

In addition to providing insights into the pathogenesis of glioma and neurofibroma development in individuals with NF1, these mouse models provide unique opportunities to identify and evaluate potential new therapies for NF1-associated tumors [18,20]. We have exploited the $Nf1$ optic glioma mouse model to correlate MRI findings with tumor proliferation. Studies using small-animal MRI of these tumors demonstrate gadolinium contrast enhancement, similar to their human counterparts, but have begun to yield improved imaging methods to distinguish between growing and quiescent optic glioma. In addition, we have identified several new targets for potential NF1 tumor therapeutic drug design. Recent studies in our laboratory have shown that NF1-associated mouse and human tumors exhibit hyperactivation of the mTOR signaling pathway, which can be reversed by treatment with rapamycin, a drug already in clinical use for other tumors. The availability of suitable mouse models opens the door to the development of tailored therapies for NF1-associated tumors.

Loss of neurofibromin is seen in both sporadic and NF1-associated MPNST but is not sufficient for MPNST development. Accumulation of additional somatic mutations in the Schwann cell is likely required, including p53, p27, and $CDKN2A$ locus inactivation [20]. Additionally, epidermal growth factor receptor (EGFR), which is not expressed in normal Schwann cells, shows increased expression in MPNST. Activation of EGFR results in activation of RAS as well as other mitogenic or survival pathways, suggesting that EGFR expression may provide an additional growth advantage relevant to MPNST formation [14].

MPNST formation can be modeled in mice that harbor germline mutations in both the $Nf1$ and p53 genes. These mice develop aggressive sarcoma-like tumors that are histologically similar to MPNSTs in individuals with NF1. The role of EGFR amplification, p27 loss, or $CDKN2A$ inactivation in the formation of MPNST in mice is unclear [20].

Interestingly, mice with one functional and one nonfunctional copy of the $Nf1$ gene exhibit learning disabilities. These $Nf1+/-$ mice perform poorly on tests of spatial memory and learning. Recent studies have suggested that this learning problem reflects abnormal gamma-aminobutyric acid (GABA)–mediated inhibition in the hippocampus as a result of RAS hyperactivation. $Nf1+/-$ mice do not demonstrate UBOs on brain MRI.

In summary, NF1 is a complex, protean disorder characterized by the development of pigmentary abnormalities, benign and malignant tumors, cognitive dysfunction, and other clinical features. It is highly recommended that individuals with NF1 receive care in the setting of a multidisciplinary NF clinical program, which can provide up-to-date information and clinical treatments. The availability of a team of specialists focused on NF1 provides the optimal context for the medical management of this complicated hereditary disease.

References

1. Korf, B. R., and Rubenstein, A. E. (2005). "Neurofibromatosis: A Handbook for Patients, Families, and Health Care Professionals." Thieme Medical, New York.
2. Gutmann, D. H., Aylsworth, A., Carey, J. C., Korf, B., Marks, J., Pyeritz, R. E., Rubenstein, A., and Viskochil, D. (1997). The diagnostic evaluation and multidisciplinary management of neurofibromatosis 1 and neurofibromatosis 2. *J. Am. Med. Assoc.* **278**, 51–57.
3. Viskochil, D. V. (2002). Genetics of neurofibromatosis 1 and the *NF1* gene. *J. Child Neurol.* **17**, 562–570.
4. Korf, B. R. (2002). Clinical features and pathobiology of neurofibromatosis 1. *J. Child Neurol.* **17**, 573–577.
5. Carey, J. C. and Viskochil, D. V. (1999). Neurofibromatosis type 1: a model condition for the study of the molecular basis of variable expressivity in human disorders. *Am. J. Med. Genet.* **89**, 7–13.
6. Kayl, A. E., and Moore, B. D. III. (2000). Behavioral phenotype of neurofibromatosis, type 1. *Ment. Retard. Dev. Disabil. Res. Rev.* **6**, 117–124.

7. Thomson, S. A., Fishbein, L., and Wallace, M. R. (2002). *NF1* mutations and molecular testing. *J. Child Neurol.* **17**, 555–561.
8. Rosser, T., and Packer, J. R. (2002). Neurofibromas in children with neurofibromatosis 1, *J. Child Neurol.* **17**, 585–594.
9. Korf, B. R. (1999). Plexiform neurofibromas. *Am. J. Med. Genet.* **89**, 31–37.
10. Packer, R. J., Gutmann, D. H., Rubenstein, A., Viskochil, D., Zimmerman, R. A., Vezina, G., Small, J., and Korf, B. (2002). Plexiform neurofibromas in NF1: toward biologic-based therapy. *Neurology* **58**, 1461–1470.
11. Ferner, R. E., and Gutmann, D. H. (2002). International consensus statement on malignant peripheral nerve sheath tumors in neurofibromatosis 1. *Can. Res.* **62**, 1573–1577.
12. Listernick, R., Louis, D. N., Packer, R. J., Guttman, D. H. (1997). Optic pathway gliomas in children with neurofibromatosis 1: consensus statement from the NF1 Optic Pathway Glioma Task Force. *Ann. Neurol.* **41**, 143–149.
13. Listernick, R., Charrow, J., and Gutmann, D. H. (1999). Intracranial gliomas in neurofibromatosis type 1. *Am. J. Med. Genet.* **89**, 38–44.
14. Korf, B. R. (2000). Malignancy in neurofibromatosis type 1. *Oncologist* **5**, 477–485.
15. North, K. N., Riccardi, V., Samango-Sprouse, C., Ferner, R., Moore, B., Legius, E., Ratner, N., and Denckla, M. B. (1997). Cognitive function and academic performance in neurofibromatosis 1: consensus statement from the NF1 Cognitive Disorders Task Force. *Neurology* **48**, 1121–1127.
16. Friedman, J. M., Arbiser, J., Epstein, J. A., Gutmann, D. H, Huot, S. J., Lin, A. E., McManus, B., and Korf, B. R. (2002). Cardiovascular disease in neurofibromatosis 1: report of the NF1 Cardiovascular Task Force. *Genet. Med.* **4**, 105–111.
17. Messiaen, L. M., Callens, T., Mortier, G., Beysen, D., Vandenbroucke, I., Van Roy, N., Speleman, F., and Paepe, A. D. (2000). Exhaustive mutation analysis of the *NF1* gene allows identification of 95% of mutations and reveals a high frequency of unusual splicing defects. *Hum. Mut.* **15**, 541–555.
18. Dasgupta, B., and Gutmann, D. H. (2003). Neurofibromatosis 1: closing the GAP between mice and men. *Curr. Op. Gen. Devel.* **13**, 20–27.
19. Gutmann, D. H. (2001). The neurofibromatoses: when less is more. *Hum. Mol. Gen.* **10**, 747–755.
20. Rubin, J. B., and Gutmann, D. H. (2005). Neurofibromatosis 1: A model for nervous system tumor formation? *Nature Rev. Cancer.* **(5)7**, 557–564.
21. Zhu, Y., Ghosh, P., Charnay, P., Burns, D. K., and Parada, L. F. (2002). Neurofibromas in NF1: Schwann cell origin and role of tumor environment. *Science* **296**, 920–922.

38

Medulloblastoma and Primitive Neuroectodermal Tumors

Said Elshihabi, MD
James T. Rutka, MD, PhD

Keywords: *medulloblastoma, p53, pediatric brain tumor, primitive neuroectodermal cell, sonic hedgehog*

I. Historical Concepts and Definitions
II. Neuroembryogenesis of the Cerebellum: Clues to the Origin of Medulloblastoma
III. Histopathology of Medulloblastoma
IV. Molecular Genetics of Medulloblastoma
V. Medulloblastoma and Inherited Cancer Syndromes
VI. Developmental Signaling Pathways and Medulloblastoma
VII. Molecular Profiling and Patient Outcome in Medulloblastoma
VIII. Treatment of Medulloblastoma: The Promise of New Pharmacotherapeutics
IX. Summary
References

I. Historical Concepts and Definitions

The term *medulloblastoma* was introduced by Harvey Cushing and Percival Bailey in June 1924 as a result of their exhaustive review of more than 400 gliomas. Their now-historic classification of this disease entity revealed it to be a tumor composed of cells with numerous mitotic figures, small round nuclei, and scanty cytoplasm. They proposed that its genesis was from undifferentiated embryonal cell rests in the roof and ependymal lining of the fourth ventricle. Today, medulloblastoma is classified as a malignant and invasive embryonal tumor of the cerebellum, corresponding to a World Health Organization (WHO) grade of IV. Medulloblastoma is included within the group of tumors known as *primitive neuroectodermal tumors* (PNETs) as first described by Hart and Earle in 1973 and elaborated by Rorke et al. in 1983 [1]. Whereas PNETs may be found in diverse locations throughout the central nervous system (CNS), the term *medulloblastoma* is reserved for those PNETs arising from the cerebellum. Interestingly, gene expression studies have shown that medulloblastoma is molecularly distinct from the supratentorial PNET [2].

II. Neuroembryogenesis of the Cerebellum: Clues to the Origin of Medulloblastoma

During neuroembryogenesis, the external granule cell layer of the cerebellum proliferates in response to the soluble ligand, sonic hedgehog (SHH), secreted by Purkinje cells (Fig. 1). Following proliferation, the external granule cells migrate past the Purkinje cells and take up residence in the internal granule cell layer. They then become synaptically competent by sending cell processes and axonal connections into the molecular layer. It is thought that perturbations in the SHH signaling pathway as it affects external granule cells may represent the "first hit" that predisposes an embryo to medulloblastoma formation. A second hit occurring in postnatal life that mutates or alters the remaining allele of a gene in the SHH pathway may be sufficient to explain the origin of at least some medulloblastomas. The members at the SHH pathway that are known to contribute to the pathogenesis of medulloblastoma are discussed in Section VI.

III. Histopathology of Medulloblastoma

The majority of medulloblastomas (85%) arise from the cerebellar vermis. In the minority of cases, they arise laterally from the cerebellar hemispheres; this is particularly true of older children and adults. Medulloblastomas are pinkish gray to purple masses that reveal a highly cellular neoplasm on light microscopy. Mitoses, apoptotic bodies, and differentiation along glial, neuronal, and/or melanotic lineages are commonly seen. The basic histological variants usually can be discerned using standard histopathological stains. The "classic" medulloblastoma subtype is the most

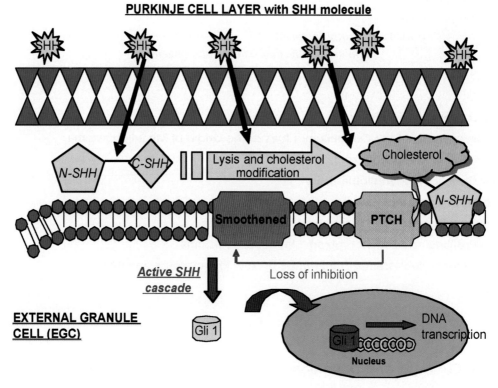

Figure 1 SHH developmental signaling pathway in relationship to medulloblastoma formation. SHH, secreted by Purkinje cells, is processed into an active form (N-SHH) and binds to PTCH, thereby releasing its tonic inhibition on Smoothened. Smoothened can then activate a polymeric protein complex comprised of Fused, Suppressor of Fused, Costal-2, and Gli1, such that the Gli transcription factors can enter the nucleus to transactivate downstream target genes, leading to cellular proliferation.

common variant and is seen in approximately 70% of cases. Figure 2A-C depicts the typical histological appearance of a medulloblastoma and some of its more common subtypes. The desmoplastic variant, found in approximately 10–20% of cases, contains paucicellular islands of well-differentiated cells surrounded by reticulin and collagen, as well as neighboring regions of poorly differentiated medulloblastoma cells. It has been reported that this group of medulloblastomas carries a more favorable prognosis [3]. Another subtype with a favorable prognosis is the medulloblastoma with extensive nodularity and neuronal components. The large cell/anaplastic variant carries the worst prognosis and has a greater percentage of poorly differentiated, anaplastic cellular regions. This subtype of medulloblastoma has been reported in as many as 20% of cases.

Medulloblastoma must be differentiated from another childhood tumor arising from the cerebellum: the atypical teratoid/rhabdoid tumor (ATRT). ATRT is an aggressive, malignant tumor that looks similar to medulloblastoma by light microscopy but can be distinguished with advanced histochemical and molecular marker techniques. ATRTs demonstrate losses in the *hSNF5/INI1* gene [4].

IV. Molecular Genetics of Medulloblastoma

It has been known for quite some time that sporadic medulloblastoma is characterized by a number of nonrandom cytogenetic abnormalities affecting chromosomes 5, 6q, 16q, and 17p. One of the most interesting cytogenetic abnormalities is the presence of isochromosome 17q. We have used advanced cytogenetic techniques such as fluorescence *in situ* hybridization (FISH), spectral karyotyping (SKY), and comparative genomic hybridization (CGH) to characterize a panel of childhood medulloblastomas [5]. In addition to the chromosomal

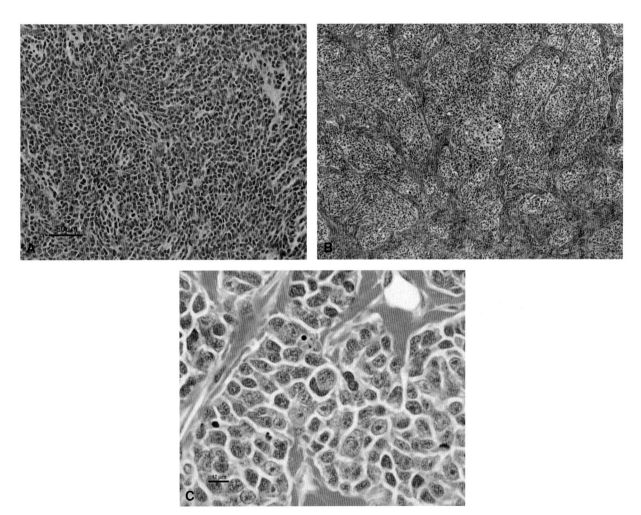

Figure 2 History of medulloblastoma. **A**, Medium-powered view of a classic medulloblastoma with typical small, round blue cells with high nuclear:cytoplasmic ratios. **B**, Low-powered view of a typical desmoplastic medulloblastoma showing nests of cells separated by reticulin network. Desmoplastic medulloblastoma is common in patients with Gorlin's syndrome, adult medulloblastoma, and patients with mutations in *Suppressor of Fused*. **C**, High-powered view of an anaplastic, large-cell medulloblastoma with frequent mitoses.

alterations listed earlier, we showed that frequent genetic alterations in medulloblastoma include rearrangement of chromosome 7, amplification of chromosome 2q, and loss of chromosome 10.

Although the molecular genetic alterations of medulloblastoma are becoming better understood, the genetic alterations associated with supratentorial PNET are less well defined. As a rule, the genetic alterations in pinealoblastomas, hemispheric PNETs, and retinoblastoma do not correlate with those found in the routine medulloblastomas. Loss of heterozygosity (LOH) on chromosome 17, which is routinely found in many medulloblastomas, is not typically associated with its supratentorial counterparts.

V. Medulloblastoma and Inherited Cancer Syndromes

The occurrence of medulloblastoma in patients with inherited cancer syndromes offers a unique opportunity to increase our understanding of the pathogenesis of this tumor. Although rare in comparison with sporadic medulloblastoma, medulloblastomas arising in families with well-characterized germline defects in specific genes have provided clues to the molecular pathways at play in these tumors. For medulloblastoma, the prototypical example is Gorlin's syndrome.

Gorlin's syndrome, also known as *nevoid basal cell carcinoma syndrome* (NBCCS), is caused by a germline mutation in a gene called *PTCH*, which is localized to chromosome 9q22. Patients with Gorlin's syndrome develop a number of developmental anomalies such as jaw cysts, macrocephaly, rib abnormalities, and dural calcifications. However, they are also predisposed to developing cancers, of which basal cell carcinoma and medulloblastoma are the most common. Interestingly, *PTCH* is a member of the SHH signaling pathway, a pathway that is frequently the target of inactivating mutations in several of its members (Section VI).

Another inherited cancer syndrome that is associated with medulloblastoma is Turcot's syndrome, in which patients have a germline mutation in the adenomatous polyposis coli (*APC*) gene located on chromosome 5q21. They develop hundreds to thousands of colonic polyps but are also predisposed to brain tumors, of which glioma and medulloblastoma are the most common types. The relative risk of developing a cerebellar medulloblastoma in patients with this gene mutation is 92 ($P < 0.001$) [6]. The inactivation of the *APC* gene in the tumors of patients with Turcot's syndrome leads to aberrant signaling in the WNT signaling pathway with excess production of beta-catenin (Section VI).

The role of the p53 tumor suppressor gene in the pathogenesis of medulloblastomas has been controversial. Some studies have suggested that patients with Li-Fraumeni syndrome, in which there is an inherited germline mutation of p53, are predisposed to medulloblastoma formation. However, examinations of sporadic medulloblastomas have failed to reveal any significant mutations of p53. Furthermore, p53 knock-out mice do not commonly develop medulloblastomas or PNETs. Interestingly, there are some studies showing that supratentorial PNETs, but not medulloblastomas, demonstrated p53 mutations in about half the cases [7]. This is further evidence that cerebellar medulloblastoma and supratentorial PNET may have unique molecular signatures. For medulloblastoma, inactivation of the p53 is not sufficient to result in tumor formation. However, when *ptch* heterozygous mice are crossed with p53 null mice, 100% of the newborn mice develop medulloblastomas by 12 weeks of age [8]. The acceleration of medulloblastomas in these mice may be a consequence of increased genomic instability associated with loss of p53 function that may enhance the rate of acquisition of secondary mutations.

The different molecular genetic alterations and defects in developmental signaling pathways that are associated with the different histopathological subtypes of medulloblastoma are shown in Figure 3.

VI. Developmental Signaling Pathways and Medulloblastoma

As mentioned previously, two key developmental signaling pathways now known to be associated with medulloblastoma are the SHH and WNT signaling pathways [9]. The SHH protein is a strong mitogen for granule cell precursors in the external granule cell layer, the supposed cell layer of origin for medulloblastomas. SHH is secreted by the Purkinje cells and binds to the PATCHED (PTCH) receptor on external granule cells (see Fig. 1). After the SHH ligand is bound to the PTCH receptor, its tonic inhibition of another multipass cell membrane protein, Smoothened, is released, which in turn activates downstream signaling elements through modifications of a polymeric protein complex comprised of Fused, Suppressor of Fused, Costal2, and Gli. Therefore inactivation of *PTCH* as occurs with mutations in the gene in Gorlin's syndrome will be functionally similar to SHH protein excess. When the SHH signaling cascade is active, Gli activity is stimulated. Once initiated, and with the assistance of their transcriptional coactivator CBP, Gli enters the nucleus and activates transcription. When the SHH pathway is inactive, PTCH inhibits Smoothened activity, and other Gli downstream elements are inactivated via ubiquitin-mediated degradation and/or phosphorylation by protein kinase A (PKA).

The clinical significance of this pathway in human medulloblastoma is evident when one considers the mutations that have already been identified in tumors in several of

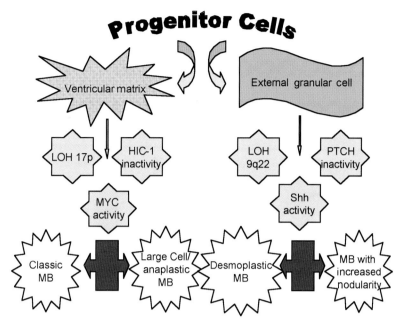

Figure 3 Molecular genetic and developmental signaling pathway alterations leading to medulloblastoma. Evidence is mounting to suggest that the different histopathological subtypes of medulloblastoma arise from different genetic disturbances, as illustrated here.

the members of this pathway including *PTCH, Smoothened,* and *Suppressor of Fused.* Mutations in these genes may account for approximately 15% of all medulloblastoma tumors. The mutations found in *Suppressor of Fused* are of interest as they have been found not only in the medulloblastoma tumor but also in the germline of patients [10]. Additional experimental evidence in support of the SHH pathway causing medulloblastoma is provided by a study of genetically engineered PTCH heterogynous knock-out mice in which 30% of affected mice developed tumors that were histopathologically identical to human medulloblastomas [11]. Although mutations of this pathway have been preferentially discovered in the desmoplastic subtype, some groups have also reported them in classic-type medulloblastomas [12]. The value of identifying the SHH pathway as a critical pathway in medulloblastoma formation is that its members can now be selectively targeted as a means to inhibit the growth of this tumor (Section VIII).

It has been suggested that the SHH pathway does not work in isolation and that cooperative signaling pathways are required. For example, the combined activation of SHH pathway and insulin-like growth factor (IGF) nearly tripled the frequency of tumor formation [13,14].

As mentioned in the previous section, another example of a molecular pathway involved in the uncontrolled growth of medulloblastoma progenitor cells is the WNT signaling pathway. The APC protein functions by regulating the binding of WNT to its receptor, the seven-transmembrane protein Frizzled, with an eventual goal of up-regulating beta-catenin expression. After the membrane receptor Frizzled becomes activated, the signal is then carried through to Disheveled and onward to a multiprotein complex containing APC, Axin, and beta-catenin. In the absence of WNT signaling, some of these proteins down-regulate the activity of beta-catenin by phosphorylation of its amino-terminal residues and subsequent ubiquitin-proteasome degradation. If, however, the WNT cascade continues and beta-catenin enters the nucleus, it will transcribe and activate other growth regulating genes, some of which are oncogenes, including *c-MYC*, and *Cyclin D*. Interestingly, the large cell or anaplastic medulloblastoma variant has been shown to have a high incidence of *c-MYC* gene amplification, a risk factor now associated with a poor prognosis. Figure 4 illustrates the basic elements of the WNT signaling pathway.

VII. Molecular Profiling and Patient Outcome in Medulloblastoma

Prognosis in medulloblastoma can now be predicted using clinical parameters, histological subtype, and molecular profiling. As previously mentioned, the progression of the disease in patients with the large cell/anaplastic subtype is worse than for those with the classic or desmoplastic variants. Historically, the natural course of the disease in infants and young children is worse than in older children or adults. It has also been noted that this more aggressive course is in relation to an increased prevalence of metastases at time of diagnosis [15].

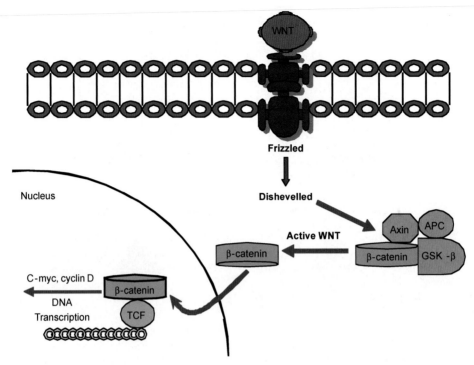

Figure 4 Wingless (WNT) signaling pathway. WNT binds to Frizzled at the plasma membrane, which in turn activates Dishevelled. Dishevelled then reacts with a polymeric protein complex comprised of Axin, APC, GSK-3β, and β-catenin. β-catenin is then released from this complex to enter the nucleus to transactivate downstream target genes such as c-myc and cyclin D to stimulate the cell cycle.

In a study by Lamont et al., patients with medulloblastomas were stratified according to histopathological and molecular cytogenetic aberrations [16]. *In situ* hybridization techniques were used to detect abnormalities of chromosome 17, losses of 9q22 and 10q24, and amplification of the *MYC-C* and *MYC-N* oncogenes. These data, combined with the tumor cytoarchitecture, predicted that certain groups of patients would have a worse prognosis. Large cell/anaplastic tumors with a loss of 17p13.3 and over-amplification of the *MYC* oncogene exhibited a statistically significant reduction in long-term survival compared with those tumors without these features.

Ray et al. determined the contributions of biological and clinical predictors in the overall survival of 119 patients with medulloblastoma [7]. The immunohistochemical markers that were studied included ErbB2, Ki-67, p53, MYC, TrkC, and platelet-derived growth factor receptor (PDGFR)-alpha, as well as TUNEL assay for apoptosis. Although prior studies correlated single prognostic markers with long-term outcome, this was the first study to examine multiple markers on the same set of tumors, which was accomplished using tissue microarray technology. The clinical markers used in this study included age at onset, extent of tumor resection, and use of radiation therapy and chemotherapy. Among the clinical features evaluated, only metastatic disease at presentation correlated with a poorer prognosis, and age at presentation and extent of resection were only significant in univariate analysis. With respect to the immunohistochemical arrays, p53 immunopositivity was the only marker predictive of a poor outcome in both univariate and multivariate analyses. TrkC reactivity was the only factor that approached statistical significance for predicting a good outcome in this patient population. Interestingly, in a study by Grotzer et al., TrkC expression was also examined in a large number of PNETs, including medulloblastoma, and was found to be the only factor that predicted a favorable patient outcome [17].

In another study to determine the importance of p53 expression as a prognostic indicator of medulloblastomas, intense staining for p53 with an antibody that recognized both the wild-type and mutant form of the protein was the only variable correlated with a decreased disease-free survival rate in the adult population [3].

In attempts to determine how pediatric medulloblastomas differ from those occurring in adults, Sarkar et al. evaluated the histopathology and clinical behavior of more than 180 medulloblastomas [18]. The childhood tumors were most often of the classic subtype originating from the vermian region, whereas older patients tended to have tumors of desmoplastic origin in a lateral location. Comparisons of various labeling indices such as MIB-1, apoptotic index, p53, and Bcl-2 expression between these two groups revealed medulloblastomas in the adult population to be less aggressive, with lower growth rate parameters than those in the childhood population.

VIII. Treatment of Medulloblastoma: The Promise of New Pharmacotherapeutics

The standard of care for most patients with medulloblastomas begins with an attempted gross total resection of the tumor, followed by adjuvant chemotherapy and radiation therapy to the involved neuraxis, depending on clinical risk factors [15]. Although refinement in irradiation techniques has allowed for reported survival rates of up to 60%, the deleterious side effects of radiation have led to advances in the pharmacological treatment of this disease. Radiation is typically avoided in children younger than 3 years. At the outset, patients must be stratified into either high- or standard-risk categories. Patients with more than $1.5\,cm^2$ postoperative tumor residual, age 3 years or younger, or the presence of metastases are deemed high risk. Various chemotherapeutic agents have been used to date, and efforts are underway to standardize the best chemotherapeutic protocol for all patients.

A trial by Chi et al. assessed the feasibility and response to induction chemotherapy intensified with high-dose methotrexate for children in the high-risk category [19]. Following maximal surgical resection, 21 patients were given five cycles of vincristine, cisplatin, etoposide, cyclophosphamide/mesna, and methotrexate with leucovorin rescue. The 3-year overall survival was 60%, and it was proposed that the addition of methotrexate in this high-risk group was beneficial and tolerable [18].

In a bold study by Rutkowski et al., intraventricular methotrexate and a similar chemotherapeutic regimen were used without adjuvant radiation therapy in the high-risk group [20]. They monitored the patients with neuropsychological tests and imaging studies to rule out leukoencephalopathy, which may be a toxic effect of methotrexate. Overall, 43 patients were studied, and at the completion of the chemotherapy, 14 patients had residual disease and received further chemotherapy, radiation therapy, or both. Of this patient subset, 10 patients died due to disease progression. The estimated 5-year progression-free and survival rates were 58% and 66%, respectively. However, those children without metastases or postoperative tumor residual had progression-free and overall survival rates of 82% and 93%, respectively. The main pharmacological side effect noted from this intense chemotherapeutic regimen was myelotoxicity, with patients requiring a median of two transfusions during therapy. With the exception of two patients that reportedly developed seizures following intraventricular methotrexate, the remainder of the patients tolerated this portion of the treatment. Moreover, 20 of 31 patients that did not have macroscopic metastases were spared radiation therapy and the serious cognitive sequelae that can arise from this therapy [20].

The future of pharmacotherapeutic treatment for medulloblastoma is optimistic, based on the targeted inhibition of the developmental signaling pathways described in this chapter. For example, Taipale et al. showed that cyclopamine could reverse the effects of oncogenic mutations in *Patched* and *Smoothened* in an experimental model of medulloblastoma [21]. In another study, Romer et al. used a small molecular inhibitor of the SHH pathway to significantly inhibit the growth of medulloblastomas in *ptch* heterozygous mice [22]. This inhibitor blocked the function of Smoothened and resulted in suppression of several downstream SHH pathway elements. The investigators showed that this novel drug inhibited cell proliferation, increased apoptosis, and prolonged medulloblastoma-free survival. Although experimental at this phase, such studies lend hope to the notion that such a therapeutic option will be available in the treatment of human medulloblastomas in the near future [23].

IX. Summary

The essential elements in the developmental biology of medulloblastoma formation are becoming increasingly well recognized. Despite its aggressive nature and malignant behavior, medulloblastomas are now associated with increased disease-free survival rates with the advancement of chemotherapeutic trials and techniques in molecular biology. The ultimate goal of curing patients with this malignant brain tumor will likely be accomplished by continued progress in these fields.

Acknowledgments

This work was supported by grants from the Ontario Cancer Research Network, the Pediatric Brain Tumor Foundation, and the National Cancer Institute of Canada.

References

1. Hart, M. N., and Earle, K. M. (1973). Primitive neuroectodermal tumors of the brain in children. *Cancer* **32**, 890–897.
2. Pomeroy, S. L., Tamayo, P., Gaasenbeek, M., Sturla, L. M., Angelo, M., McLaughlin, M. E., Kim, J. Y., Goumnerova, L. C., Black, P. M., Lau, C., Allen, J. C., Zagzag, D., Olson, J. M., Curran, T., Wetmore, C., Biegel, J. A., Poggio, T., Mukherjee, S., Rifkin, R., Califano, A., Stolovitzky, G., Louis, D. N., Mesirov, J. P., Lander, E. S., and Golub, T. R. (2002). Prediction of central nervous system embryonal tumour outcome based on gene expression. *Nature* **415**, 436–442.
3. Provias, J. P., and Becker, L. E. (1996). Cellular and molecular pathology of medulloblastoma. *J. Neuro-Oncol.* **29**, 35–43.

4. Sevenet, N., Lellouch-Tubiana, A., Schofield, D., Hoang-Xuan, K., Gessler, M., Birnbaum, D., Jeanpierre, C., Jouvet, A., and Delattre, O. (1999). Spectrum of hSNF5/INI1 somatic mutations in human cancer and genotype-phenotype correlations. *Hum. Mol. Genet.* **8**, 2359–2368.
5. Bayani, J., Zielenska, M., Marrano, P., Kwan Ng, Y., Taylor, M. D., Jay, V., Rutka, J. T., and Squire, J. A. (2000). Molecular cytogenetic analysis of medulloblastomas and supratentorial primitive neuroectodermal tumors by using conventional banding, comparative genomic hybridization, and spectral karyotyping. *J. Neurosurg.* **93**, 437–448.
6. Hamilton, S. R., Liu, B., Parsons, R. E., Papadopoulos, N., Jen, J., Powell, S. M., Krush, A. J., Berk, T., Cohen, Z., Tetu, B., et al. (1995). The molecular basis of Turcot's syndrome. *N. Engl. J. Med.* **332**, 839–847.
7. Ray, A., Ho, M., Ma, J., Parkes, R. K., Mainprize, T. G., Ueda, S., McLaughlin, J., Bouffet, E., Rutka, J. T., and Hawkins, C. E. (2004). A clinicobiological model predicting survival in medulloblastoma. *Clin. Cancer Res.* **10**, 7613–7620.
8. Wetmore, C. J., and Curran, T. (2000). Accelerated central nervous system tumorigenesis in mice with both Patched and p53 genes. *Proc. Am. Assoc. Cancer Res.* **41**, 228.
9. Gilbertson, R. J. (2004). Medulloblastoma: signaling a change in treatment. *Lancet Oncol.* **5**, 209–218.
10. Taylor, M. D., Liu, L., Raffel, C., Hui, C. C., Mainprize, T. G., Zhang, X., Agatep, R., Chiappa, S., Gao, L., Lowrance, A., Hao, A., Goldstein, A. M., Stavrou, T., Scherer, S. W., Dura, W. T., Wainwright, B., Squire, J. A., Rutka, J. T., and Hogg, D. (2002). Mutations in SUFU predispose to medulloblastoma. *Nat. Genet.* **31**, 306–310.
11. Zurawel, R. H., Allen, C., Wechsler-Reya, R., Scott, M. P., and Raffel, C. (2000). Evidence that haploinsufficiency of Ptch leads to medulloblastoma in mice. *Genes Chromosomes Cancer* **28**, 77–81.
12. Schofield, D., West, D. C., Anthony, D. C., Marshal, R., and Sklar, J. (1995). Correlation of loss of heterozygosity at chromosome 9q with histological subtype in medulloblastomas. *Am. J. Pathol.* **146**, 472–480.
13. Rao, G., Pedone, C. A., Valle, L. D., Reiss, K., Holland, E. C., and Fults, D. W. (2004). Sonic hedgehog and insulin-like growth factor signaling synergize to induce medulloblastoma formation from nestin-expressing neural progenitors in mice. *Oncogene* **23**, 6156–6162.
14. Weiner, H. L., Bakst, R., Hurlbert, M. S., Ruggiero, J., Ahn, E., Lee, W. S., Stephen, D., Zagzag, D., Joyner, A. L., and Turnbull, D. H. (2002). Induction of medulloblastomas in mice by sonic hedgehog, independent of Gli1. *Cancer Res.* **62**, 6385–6389.
15. Rutka, J. T., and Hoffman, H. J. (1996). Medulloblastoma: a historical perspective and overview. *J. Neuro-Oncol.* **29**, 1–7.
16. Lamont J. M., McManamy, C. S., Pearson, A. D., Clifford, S. C., and Ellison, D. W. (2004). Combined histopathological and molecular cytogenetic stratification of medulloblastoma patients. *Clin. Cancer Res.* **10**, 5482–5493.
17. Grotzer, M. A., Janss, A. J., Fung, K., Biegel, J. A., Sutton, L. N., Rorke, L. B., Zhao, H., Cnaan, A., Phillips, P. C., Lee, V. M., and Trojanowski, J. Q. (2000). TrkC expression predicts good clinical outcome in primitive neuroectodermal brain tumors. *J. Clin. Oncol.* **18**, 1027–1035.
18. Sarkar, C., Pramanik, P., Karak, A. K., Mukhopadhyay, P., Sharma, M. C., Singh, V. P., Mehta, V. S. (2002). Are childhood and adult medulloblastomas different? A comparative study of clinicopathological features, proliferation index and apoptotic index. *J. Neuro-Oncol.* **59**, 49–61.
19. Chi, S. N., Gardner, S. L., Levy, A. S., Knopp, E. A., Miller, D. C., Wisoff, J. H., Weiner, H. L., and Finlay, J. L. (2004). Feasibility and response to induction chemotherapy intensified with high-dose methotrexate for young children with newly diagnosed high-risk disseminated medulloblastoma. *J. Clin. Oncol.* **22**, 4881–4887.
20. Rutkowski, S., Bode, U., Deinlein, F., Ottensmeier, H., Warmuth-Metz, M., Soerensen, N., Graf, N., Emser, A., Pietsch, T., Wolff, J. E., Kortmann, R. D., and Kuehl, J. (2005). Treatment of early childhood medulloblastoma by postoperative chemotherapy alone. *N. Engl. J. Med.* **352**, 978–986.
21. Taipale, J., Chen, J. K., Cooper, M. K.,Wang, G., Mann, R. K., Milenkovic, L., Scott, M. P., and Beachy, P. A. (2000). Effects of oncogenic mutations in Smoothened and Patched can be reversed by cyclopamine. *Nature* **406**, 944–945.
22. Romer, J. T., Kimura, H., Magdaleno, S., Sasai, K., Fuller, C., Baines, H., Connelly, M., Stewart, C. F., Gould, S., Rubin, L. L., and Curran, T. (2004). Suppression of the SHH pathway using a small molecule inhibitor eliminates medulloblastoma in Ptc 1(+/−) p53(−/−) mice. *Cancer Cell* **6**, 229–240.
23. Romer, J., and Curran, T. (2005). Targeting medulloblastoma: small-molecule inhibitors of the Sonic Hedgehog pathway as potential cancer therapeutics. *Cancer Res.* **65**, 49.

39

Glioma

Anders I. Persson, PhD
QiWen Fan, PhD
Joanna J. Phillips, PhD
William A. Weiss, PhD

Keywords: *brain, cancer, glioma, progenitor, tumor, stem cell*

I. Introduction
II. Nomenclature
III. Locations and Symptoms
IV. Epidemiology
V. Etiology
VI. Histopathology
VII. Genetic Alterations
VIII. Cell of Origin
IX. Signaling Pathways
X. Pharmacology
XI. Future Therapeutic Interventions
References

I. Introduction

Decades of research on infiltrating gliomas still leave important questions regarding the etiology, cellular origin, and role of different cell types within glial malignancies. Despite surgical resection, radiotherapy, and chemotherapy, the median survival for high-grade glioma remains 1–3 years. Advances in the understanding of the normal development in the central nervous system (CNS) and in the pathogenesis of glioma, specifically focusing on cells of origin and critical signaling pathways, will lead to a better understanding of the biology of these tumors and more effective treatments for patients with glioma. In this chapter, we provide the reader with an overview on gliomas, including pharmacological treatments and findings regarding cell of origin in relation to cellular heterogeneity in gliomas.

II. Nomenclature

Gliomas, including astrocytomas, oligodendrogliomas, mixed oligoastrocytomas, and ependymomas, are classified based on their histological and immunohistological features [1]. The World Health Organization (WHO) classification divides gliomas into grades based on histopathological features of anaplasia, including nuclear atypia, mitotic activity, microvascular proliferation, and/or necrosis. Pilocytic astrocytomas, corresponding to WHO grade I, have clinical, pathological, and molecular features that are distinct from the infiltrating gliomas, which include the diffusely infiltrating astrocytomas, oligodendrogliomas, and mixed oligoastrocytomas. Pilocytic astrocytomas (WHO grade I), diffuse astrocytomas (WHO grade II), oligodendrogliomas (WHO grade II), and oligoastrocytomas (WHO grade II) are all classified as low-grade gliomas. Anaplastic astrocytomas (WHO grade III), anaplastic oligodendrogliomas (WHO grade III), anaplastic oligoastrocytoma (WHO grade III), and glioblastoma multiforme (WHO grade IV) are classified as high-grade gliomas. Glioblastomas themselves can be further classified as primary or secondary. Primary glioblastoma present *de novo*, whereas secondary glioblastomas arise from a low-grade precursor lesion through a sequential series of genetic events. Oligodendrogliomas differ from astrocytomas in their characteristic morphology and by the observation of a subset of anaplastic tumors (marked by deletion of 1p and 19q) that are responsive to alkylating chemotherapeutic agents. Ependymal tumors and other, less common types of gliomas of different grades are not discussed in this chapter (for a broader review on gliomas, see reference 1).

III. Locations and Symptoms

All low-grade gliomas can be found throughout the CNS, although they occur more frequently in some areas than in others. Depending on the tumor location, signs such as seizures, macrocephaly, headache, increased intracranial pressure, visual deficits, and lack of coordination may be present in patients. In contrast to other low-grade gliomas, pilocytic astrocytomas are not typically localized to cortical areas but are more often found in other areas, including the optic nerve and cerebellum; therefore, seizures are uncommon. All high-grade gliomas develop throughout the CNS, with a preference for the cerebral hemispheres. In a comparison between low-grade gliomas and glioblastoma, the low-grade gliomas were more commonly found in supplementary motor area (27.3% vs 10.8%) and within the insula (25% vs 10.8%) [2]. Due to the more aggressive character of high-grade gliomas, signs are similar to low-grade astrocytomas but with a shorter history.

IV. Epidemiology

Investigations of the inherited and environmental contributions to glioma have not provided any fundamental insights that can explain the occurrence of this disease [3]. In Western countries, the incidence of brain tumor in adults is approximately 4 to 7 per 100,000 persons per year in males and 3 to 5 per 100,000 in females. Primary brain tumors represent approximately 2% of all cancers in the United States. Pilocytic astrocytoma is the most common glioma in children and usually occurs during the first and second decades of life. All other gliomas are frequently found after 30 years of age (Table 1). Secondary glioblastomas often occur in younger adults (aged 30–45 years), whereas primary glioblastomas are more frequent in older adults (mean age, 55 years). Some studies suggest that brain tumors are more common in Caucasians than in people of African or Asian descent. However, a higher incidence rate of brain tumors in developed, industrial countries might reflect differences in socioeconomic status and diagnostic ascertainment, rather than a true increase in genetic or environmental susceptibility.

V. Etiology

Studies of carcinogenic risks have assessed more than 900 exposures classified as carcinogens in humans. For only nine of these exposures have there been reports of possible or weak associations with nervous system tumors in humans. Substances of aromatic nature (petroleum products), metals (lead, arsenic, and mercury) or reactive

Table 1 Most Common Gliomas as a Function of Age

Type	Grade	Average Age in Years
Pilocytic astrocytoma	I	<20
Diffuse astrocytoma	II	30–40
Oligodendroglioma	II	35–45
Mixed oligoastrocytoma	II	35–45
Anaplastic astrocytoma	III	~40
Anaplastic oligodendroglioma	III	~50
Anaplastic oligoastrocytoma	III	40–50
Secondary glioblastoma multiforme	IV	<45
Primary glioblastoma multiforme	IV	~55

dipolar substances (vinyl chloride) have all been associated with glioma risk (for lead: odds ratio 2:1, 95% confidence interval [CI] 1.1–4.0) [3]. Because these families of molecules alter developmental processes, it is perhaps not surprising that the incidence of glial tumors and primitive neuroectodermal tumors was increased in children with parents working in agriculture (odds ratio 1:3, 95% CI 1.0–1.8). The same was true for children with male parents working as electricians (odds ratio 1:1, 95% CI 0.9–1.5), drivers (odds ratio 1:3, 95% CI 1.0–1.7), or mechanics (odds ratio 1:5, 95% CI 1.0–2.3). Children with female parents in the textile industry had a significantly higher incidence of these brain tumors (odds ratio 1:7, 95% CI 1.1–2.7) [4]. Parents working in chemical industry were at higher risk of having children with astroglial tumors (father's odds ratio 2:1, 95% CI 1.1–3.9 and mother's odds ratio 3:3, 95% CI 1.4–7.7). Physicians and firefighters represent other occupations linked to higher glioma risk (physician's odds ratio 4:7, 95% CI 0.5–42.7, and firefighter's odds ratio 2:7, 95% CI 0.3–26.1).

Diets high in N-nitroso compounds, which are present in cured meat, cooked ham, processed pork, and fried bacon, were associated with elevated glioma risk (relative risk [RR] of 1.48, 95% CI 1.20–1.83) in adults [3]. An inverse association has been noted for frequent intake of fruits, fresh vegetables, and vitamin C. The developmental aspect is important because several studies found that high consumption of meats by women during pregnancy was associated with elevated risk of astrocytic gliomas (RR of 1.68, 95% CI 1.30–2.17) in children. Ingestion of meats may lead to increased oxidative stress, which has been implicated as a contributing factor in tumorigenesis.

Leukemia patients who received therapeutic intracranial irradiation had an increased risk of brain tumors. This risk was most evident in children younger than 5 or 6 years who were treated for acute lymphoblastic leukemia. Some of these children developed glioma as early as 7–9 years after gamma-irradiation. The number of brain tumors among patients who had received cranial radiation was nearly 27 times greater than expected, whereas no such tumors were seen after chemotherapy. The developing brain contains a significant number of proliferating progenitor cells. Studies in rodents have shown that g-irradiation causes acute loss of neural progenitors [5]. The neural progenitors reappeared after 1 week but may be more susceptible to transformation into tumorigenic cells due to damage to deoxyribonucleic acid (DNA). Registers of the 80,160 survivors of the Hiroshima and Nagasaki atomic bombings also demonstrated a dose-related increase in CNS tumors (RR per sievert dose 1.2, 95% CI 0.6–2.1). Risk of glioma development was also increased but not statistically significant (RR per sievert dose 0.6, 95% CI 0.2–2.0). No clear correlation has been found between exposure to electromagnetic fields or cellular phones and CNS tumors.

VI. Histopathology

A. Low-Grade Gliomas

Pilocytic astrocytomas are slow-growing, generally well-circumscribed tumors. Histologically, these tumors are of low to moderate cellularity and show a biphasic architecture, with compact areas composed of bipolar (piloid) tumor cells alternating with loose, microcystic areas composed of multipolar tumor cells. Other common features include Rosenthal fibers, eosinophilic granular bodies, and low mitotic activity (Fig. 1A). The bipolar tumor cells exhibit strong immunoreactivity for glial fibrillary acidic protein (GFAP).

Diffuse astrocytomas exhibit marked cellular differentiation and are characterized by slow growth. These tumors infiltrate neighboring brain structures and over time can undergo malignant progression. Histologically, tumors demonstrate a moderate increase in cellularity, mild to moderate nuclear atypia, and little mitotic activity. Tumors are composed of well-differentiated neoplastic astrocytes that can have a fibrillary, gemistocytic, or protoplasmic appearance. In biopsy samples, the presence of bare neoplastic nuclei in a moderately hypercellular background often suggests the diagnosis (see Fig. 1B). The most common variant, fibrillary astrocytoma, is composed of multipolar neoplastic GFAP-positive astrocytes with scant cytoplasm and fine cell processes that form a fiber-rich glial matrix. Tumor cells have few mitoses and show neither vascular proliferation nor necrosis.

Oligodendrogliomas are well-differentiated, diffusely infiltrating tumors composed of monomorphic neoplastic cells with round, relatively uniform nuclei. Typically the tumors have well-delineated borders, may appear gelatinous, and frequently contain calcifications. A characteristic artifact of major diagnostic relevance is the perinuclear cytoplasmic clearing seen on standard tissue preparation (see Fig. 1C). The isomorphic nuclei reveal a speckled, dense chromatin pattern and are slightly larger than the nuclei of oligodendrocytes. The proliferative index is generally low, and a dense network of branching capillaries often appears in the background.

By far the most common mixed gliomas are the oligoastrocytomas. The diagnosis of oligoastrocytoma requires the identification of two distinct populations of neoplastic cells types that resemble the neoplastic cells in oligodendroglioma and diffuse astrocytoma. Both biphasic and intermingled variants are recognized. Overall, oligoastrocytomas are moderately cellular and exhibit no or low mitotic activity.

B. High-Grade Gliomas

Relative to the diffuse astrocytoma, anaplastic astrocytomas show increased cell density, increased nuclear atypia, and significant mitotic activity (see Fig. 1D). Many tumor cells express GFAP. Anaplastic oligodendrogliomas

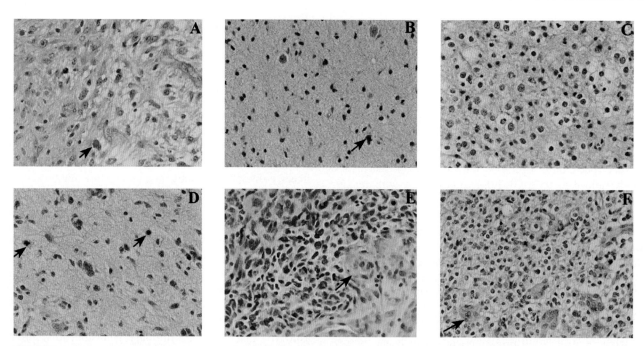

Figure 1 Common glial neoplasms of the central nervous system (hemotoxylin & eosin stain). **A**, Pilocytic astrocytomas often have a biphasic appearance with loose cystic areas and compact fascicular areas. Eosinophilic granular bodies are a common finding (*arrow*). **B**, Diffuse astrocytomas, WHO grade II, insidiously infiltrate brain tissue and typically exhibit mild cytologic atypia with mild nuclear pleomorphism and hyperchromasia (*arrow*). **C**, Oligodendroglioma, WHO grade II, is composed of neoplastic cells with round, relatively uniform nuclei and a fine capillary network. **D**, Anaplastic astrocytomas, WHO grade III, demonstrate increased cellularity, increased cytologic atypia, and significant mitotic activity (*arrows*). **E**, Glioblastoma, WHO grade IV, often exhibits prominent cytologic atypia, mitotic activity, and well-developed microvascular proliferation (*arrow*). **F**, Anaplastic oligodendroglioma, WHO grade III, demonstrates increased cytologic atypia, increased cellularity, and high mitotic activity. Microvascular proliferation (*arrow*) and necrosis may also be present.

retain certain histological features of oligodendroglioma but also demonstrate anaplastic features such as increased cellularity, nuclear atypia, increased mitotic activity, and sometimes microvascular proliferation and necrosis (see Fig. 1E). Anaplastic mixed oligoastrocytomas also show histological features of malignancy including increased cellularity, nuclear atypia, cellular pleomorphism, and high mitotic activity. The anaplastic features can be present in the oligodendroglioma-like components, astrocytoma-like components, or both.

Glioblastoma is a densely cellular malignant neoplasm composed of poorly differentiated neoplastic astrocytes. The neoplastic cells exhibit prominent cytologic and nuclear atypia and high mitotic activity. The presence of well-developed microvascular proliferation, necrosis, or both is essential for the diagnosis (see Fig. 1F). Necrosis within a glioblastoma can consist of large areas of ischemic necrosis or multiple, small foci of pseudopalisading necrosis. Tumors can show extensive cellular heterogeneity, including multinucleated cells; undifferentiated, round tumor cells; bipolar fusiform cells; and more differentiated, neoplastic, GFAP-positive astrocytic-like cells.

If the tumor is dominated by multinucleated cells, it is considered *giant cell glioblastoma*. Vascular proliferation is often found in necrotic areas and the peripheral zone of the tumor. GFAP staining is typical but may be patchy because more astrocytic-like cells express GFAP whereas small, highly proliferative and undifferentiated tumor cells are negative.

VII. Genetic Alterations

Cancer development is characterized by the accumulation of mutations in distinct molecular pathways that regulate fundamental aspects of proliferation and homeostasis. Gain of function is acquired in genes that positively regulate proliferation; loss of function mutations are acquired in genes that negatively regulate proliferation (Table 2).

The genetic alterations of gliomas involve three main signaling pathways: *RB1* (retinoblastoma 1 gene) dysregulation, *p53* pathway abnormalities, and aberrant signaling through receptor tyrosine kinases and downstream components (reviewed in [6]). In low-grade gliomas, two genetic alterations are frequently observed: *p53* inactivation (associated with astrocytomas) and the loss of 1p and 19q chromosome arms in oligodendrogliomas. High-grade gliomas show deletion of *CDKN2A* ($p16^{INK4A}/p14^{ARF}$) on chromosome 9p21, inactivation of *RB1* on chromosome 13q, and amplification of *CDK4* on chromosome

Table 2 Common Genetic and Epigenetic Alterations in Glioma

Gene	Typical Alteration	Protein Functions	Locations of Common Alterations
Tumor suppressor genes			
TP53	Mutation	Regulator of apoptosis, cell cycle progression, DNA repair	Diffuse astrocytomas, anaplastic astrocytomas, glioblastomas (secondary > primary)
Rb1	Mutation hypermethylation	Cell cycle regulation	Glioblastomas
CDKN2A	Homozygous deletion Hypermethylation	Inhibitor of CDK4/6	Glioblastomas, anaplastic astrocytomas, anaplastic oligodendrogliomas
$P14^{ARF}$	Homozygous deletion Hypermethylation	Inhibitor of MDM2	Glioblastomas
PTEN	Mutation, hypermethylation	Lipid phosphatase, negative regulator of PI3K	Anaplastic astrocytomas, anaplastic oligodendrogliomas, glioblastomas
Oncogenes			
EGFR	Amplification and overexpression Genomic rearrangement	Tyrosine kinase growth factor receptor	Diffuse and anaplastic astrocytomas, glioblastomas
PDGFR	Amplification and overexpression	Tyrosine kinase growth factor receptor	Glioblastomas
CDK4	Amplification and overexpression	Cyclin-dependent kinase, promoter of G1/S phase progression	Glioblastomas
MDM2	Amplification and overexpression	Inhibitor of p53 function	Glioblastomas

12q. Primary glioblastoma show frequent amplification of *EGFR*, inactivation of *PTEN*, deletion of 10q, and loss of *CDKN2A*. Secondary glioblastoma progresses over time from low- to high-grade tumor and usually shows mutations in *p53* and *RB* and amplification of *CDK4*. Some of the most common genetic and epigenetic changes in gliomas are presented in Table 39.2. These genes are downstream effectors of growth factor signaling and are discussed later in this chapter (Fig. 2).

Epigenetic changes that alter acetylation of histones and methylation of CpG islands within DNA sequences influence gene expression and contribute to malignant progression in glioma. Dividing glioma cells depend on correct acetylation of core histone proteins during replication, when cells have a looser chromatin structure. DNA is packed by histones into nucleosomes, which are composed of 147 base pairs of DNA and core histone proteins. The histone acetyltransferases and histone deacetylases regulate acetylation and deacetylation of the conserved lysine residues present in the amino terminal tails of the core histones. Acetylation of the core histones has been shown to weaken the histone-DNA interactions and consequently increase DNA accessibility. However, pretreatment with histone deacetylase inhibitors of cultured glioblastoma cells increased the efficiency of anticancer drugs that target DNA [7].

Several tumor suppressor genes in gliomas are hypermethylated (see Table 2). DNA methylation involves transfer of a methyl group to cytosine in a CpG dinucleotide by DNA methyltransferases that create or maintain methylation patterns. Methylation of CpG islands located in the 5' promoter region of the genes occurs frequently in glioblastomas, is also observed in mouse models for glioma, and

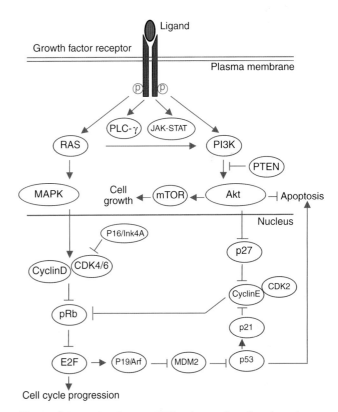

Figure 2 Interactions between RTK pathways, the cell cycle, and apoptosis. RTKs are activated in gliomas by growth factors, such as EGF and PDGF, that in turn signal through PI3-K/Akt, RAS/MAPK, PLCγ, and JAK-STAT pathways. Ras and PI3-K engage the cell cycle machinery and apoptosis at different levels. Most identified genetic alterations result in either abnormal activation of signal transduction pathways downstream of growth factor receptors or disruption of cell cycle arrest pathways.

results in gene silencing [8]. Epigenetic silencing of the O6-methylguanine-DNA methyltransferase (MGMT) gene by promoter methylation has been associated with longer overall survival in glioblastoma patients treated with radiation and the alkylating agent temozolomide [9]. Epigenetic changes in gliomas have also been observed for the tumor suppressor genes *PTEN*, *Rb1*, $p14^{ARF}$, $p15^{INK4B}$, and $p16^{INK4A}$ [10].

VIII. Cell of Origin

Similar to normal stem cells, tumor stem cells are rare cells that also have the ability to perpetuate themselves through self-renewal and to differentiate into the various cell types that drive tumorigenesis. The clear identification of stem cells in hematopoietic malignancies has led investigators to attempt comparable isolation of tissue-specific stem and progenitor cells in solid tumors, including breast, prostate, pancreatic, and brain tumors. These cells presumably comprise a small minority of the cells in different cancers and are hypothesized to be tumor-initiating cells.

Gliomas have previously been thought to arise from dedifferentiation of a mature brain cell in response to genetic alterations. An alternative hypothesis suggested that these tumors arise from transformation of a resident immature brain cell. Since the 1990s, it has been commonly accepted that the adult mammalian hippocampus and subventricular zone (SVZ) contain neural stem cells (NSCs) [11]. This was a rediscovery based on original findings in the 1960s. The adult human brain has since been shown to contain multipotent NSCs with extensive capacity for self-renewal and the ability to differentiate into neurons and glia. It was recently shown that radial glia give rise to adult NSCs in the SVZ. These NSCs express nestin and GFAP, two proteins that

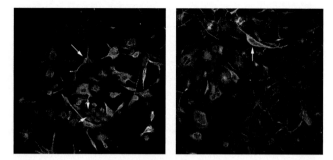

Figure 3 Cellular heterogeneity in primary cultures of human glioblastoma. The cultures present populations of cells with varying morphology and expression of markers. Some elongated cells (*long white arrows*) that are positive for nestin (*green*) also express GFAP (*red*). Small round cells (*short white arrows*) are often negative for these markers but express high levels of Sox2 (*blue*).

are found in subpopulations of tumorigenic cells isolated from human gliomas (Fig. 3). A subpopulation of the nestin-positive NSCs and isolated glioma cells also express the cell surface marker CD133. The CD133 negative cells comprise the majority of the tumor and therefore largely comprise the characteristic cells on which diagnostic criteria are based (Fig. 4A).

Both CD133 immunoreactive SVZ NSCs and their progenitors may be the targets for transformation, leading to glioma (see Fig. 4B). The longevity of the NSCs targets them for the accumulation of genetic mutations. Chromosome loss occurs frequently in SVZ NSCs as well as neurons and glia. Several recent studies have identified a subpopulation of cells in brain tumors that express immunohistochemical markers often found on NSCs [12].

Do NSCs or immature glial precursor cells represent the cellular origin of gliomas? The cancer stem cell hypothesis suggests that not all tumor cells have the same ability to proliferate and maintain the growth of the tumor. In

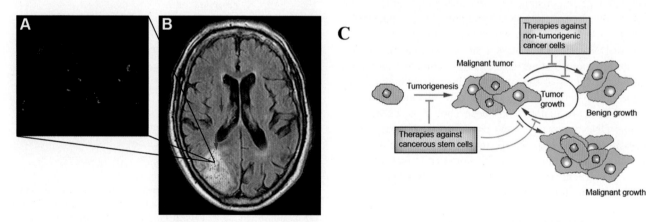

Figure 4 **A**, Only a few percent of tumor cells (nuclear counterstain in *blue*) isolated from glioblastomas express the CD133 cell surface protein (*green*). **B**, It has been hypothesized that the progeny of transformed SVZ CD133-positive NCSs (*red arrows*) migrate away from the ventricles and form gliomas. **C**, Recent studies suggest that only CD133-positive cancerous stem cells (*gray*) are able to maintain tumor growth. These cells generate more differentiated tumor cells (*green*) with limited self-renewal capacity. Therefore future treatments should target both populations of cells. It is most important to target the CD133-positive cancerous stem cells that can give rise to new tumors.

fact, gliomas are heterogenous and are comprised of cells expressing phenotypes of more than one neural lineage (see Fig. 3). Neoplastic cells in gliomas that express the cell surface marker CD133 have been identified as cancerous stem cells due to their increased self-renewal capacity, ability to form tumor spheres in culture, and ability to generate cellular phenotypes expressing both neuronal and glial markers [12,13]. Even more important, separation of CD133-positive and CD133-negative cells demonstrated that only the CD133-positive subpopulation supported tumor growth when xenotransplanted into mice [14].

Moreover, signalling pathways that mediate self-renewal and proliferation of NSCs have also been shown to induce brain tumors in mice. Using viruses that infected only subpopulations of nestin- or GFAP-positive cells, researchers have shown that expression of platelet-derived growth factor (PDGF), epidermal growth factor receptor (EGFR), or the downstream targets K-Ras or Akt can induce glioma in mice. Recent data suggest that chronic PDGF receptor activation in glial progenitors gives rise to oligodendroglioma, whereas chronic activation of Ras and Akt gives rise to astrocytoma.

IX. Signaling Pathways

A. Receptor Tyrosine Kinase Signaling in Glioma

Primary glioblastomas frequently amplify and/or overexpress oncogenes such as EGFR and PDGFR. In addition, mutations that commonly occur in glioma impact the *p53, RB1,* and phosphatidylinositol 3-kinase (PI3-K) signaling pathways. Mutations such as amplification of EGFR, gain of function in PIK3CA, or loss of PTEN all activate the PI3-K/Akt signaling pathway, downstream targets of which include mTOR and the FOXO proteins. The PDGFR is also frequently overexpressed, resulting in activation of many of the same pathways as EGFR and the constitutively activated *EGFRvIII* (also known as $\Delta EGFR$ or del2-7EGFR). Therefore EGFR, *EGFRvIII*, PDGFR, and the RAS/mitogen-activated protein kinase (MAPK) and PI3-K/mTOR pathways present attractive targets for glioblastoma therapy (see Fig. 2) [15].

1. PI3-K, Akt and PTEN Signaling

Activation of PI3-K can occur from numerous agonists and receptors, including PDGF, EGFR, fibroblast growth factor receptor (FGFR), insulin-like growth factor-1 (IGFR-1), vascular endothelial growth factor receptor (VEGFR), interleukin (IL) receptors, interferon (IFN) receptors, integrin receptors, and the Ras pathway. The PI3-Ks can be grouped into three classes based on substrate preference and sequence homology. The substrate preference of the Class I PI3Ks is the lipid phosphatidylinositol 4,5-bisphosphate PIP_2, which is phosphorylated by PI3-K to phosphatidylinositol 3,4,5 trisphosphate PIP_3. Other targets of Class I PI3-Ks include Rac, $p70^{s6k}$, and certain isoforms of protein kinase C (PKC). Class I PI3-Ks are heterodimers of approximately 200 kDa, composed of a regulatory subunit (p55 or p85) and a catalytic subunit (p110). The regulatory subunit, in particular p85, contains several different regions including Src homology domains (SH3 domain and two SH2 domains). The p110 subunit has $\alpha, \beta, \delta,$ and γ isoforms; each contains a kinase domain and a Ras interaction site. In addition, the $\alpha, \beta,$ and δ isoforms have an interaction site for the p85 subunit, which stabilizes p110. Class 1 PI3-Ks can be further subdivided into A and B groups. The class 1A p110 isoforms $\alpha, \beta,$ and δ are able to active PDK-1 and AKT. Class 1B PI3-Ks contain the γ isoform of p110 and cannot interact with Akt. Class II PI3-Ks are all 170-210-kDa monomeric proteins that preferentially phosphorylate PtdIns and PtdIns(4)P. Class III PI3-Ks only phosphorylate PtdIns. The best evidence supporting a role for class I PI3-Ks in glioma comes from recent studies in which this family of proteins was sequenced in a number of patients with glioma. Approximately 5–25% of patients showed activating mutation in p110a, with no mutations observed in other p110 molecules.

AKT (PKB) is a 56-kDa serine/threonine protein kinase with three human isoforms (AKT1-3) and two phosphorylation sites. Once AKT is associated with the membrane and bound to PIP_3, a conformational change occurs that allows phosphorylation at the threonine 308-position of the catalytic domain by protein-dependent kinase-1 (PDK-1). A second phosphorylation event is then mediated by PDK-2. PDK-2 phosphorylates the serine 473-position of the hydrophobic carboxy-terminal tail. Phosphorylation of both sites is required for full activation of AKT. Activated AKT phosphorylates numerous downstream effectors, including glycogen synthase kinase (GSK)-3α and GSK-3β, mTOR, IkB, the pro-apoptotic molecules caspase 3 and 9, and the forkhead transcription factors FOXO1, FOXO3, and FOXO4, among others. One signal responsible for cell survival is mediated through BAD, a proapoptotic member of the Bcl-2 family. Akt phosphorylates BAD on serine 136, creating a binding site for the adaptor protein 14-3-3. Once BAD is bound by 14-3-3, it is unable to heterodimerize with and inhibit the pro-survival proteins Bcl-2 or $Bcl-X_L$.

Akt is also involved modifying cell cycle function and enhanced cell proliferation. Akt phosphorylates and inhibits the CDK inhibitors $p21^{WAF1/CIP1}$ and $p27^{KIP1}$. Phosphorylation of mdm2 by Akt promotes the degradation of *p53* and induction of E2F transcriptional activity, leading to enhanced cell cycle activity at the G_1/S interface. In addition, Akt is able to modify GSK3β-mediated phosphorylation and degradation of cyclin D1.

PTEN mutation and methylation are frequent in high-grade astrocytomas and are responsible for the abnormally high levels of activity in the PI3-K/Akt signaling pathway. The *PTEN* gene, which consists of nine exons, is located on chromosome 10q23.3 and encodes a 403-amino acid cytoplasmic protein that contains two domains in the *N*-terminus, a region with lipid phosphatase activity that dephosphorylates PIP_3 to PIP_2, and a region that interacts with cellular cytoskeleton. PTEN can dephosphorylate tyrosine-, serine-, and threonine-phosphorylated peptides, although no physiologic protein substrates have been identified. In normal cells with a functioning *PTEN* gene, PIP_3 and activated Akt are maintained at low levels. In tumors and *PTEN*-deficient tumor cell lines, basal levels of PIP_3 and phospho-Akt [16] are high, and multiple proteins are activated to promote cellular survival.

2. mTOR Signaling

mTOR (also known as FRAP, RAFT, or RAPT) is a recently characterized signal transduction mediator that is intimately linked with the PI3-K/Akt signaling pathway and regulates protein synthesis and cell growth. mTOR is activated in response to growth factor signals through the PI3-K/Akt pathway. Stimulation of PI3-K leads to activation of Akt, with subsequent phosphorylation of mTOR on ser-2448. In addition, phospho-Akt is able to phosphorylate the Tsc1/Tsc2 complex on the Thr-1462 of Tsc2 that is inhibitory to mTOR [16]. Recent studies demonstrate that inhibitors of mTOR activate Akt signaling, suggesting that the clinical efficacy of mTOR inhibitors may be augmented by concurrent blockade of Akt.

3. RAS/RAF/MEK/ERK Signaling

The small guanosine-triphosphate (GTP)–binding proteins include several important mediators of the growth factor receptor pathway, including RAS, RHO, RAC, and CDC42 family members. These proteins promote cell cycle progression, invasion, motility, and angiogenesis. Unlike many systemic cancers, the infiltrating gliomas typically do not express mutant RAS but may exhibit increased levels of all three major RAS proteins (N-RAS, H-RAS, and Ki-RAS), resulting in increased levels of MAPK [17,18].

B. Hypoxic Regulation of Angiogenesis in Glioblastoma

When grade III astrocytomas progress to grade IV glioblastomas, tumor biology changes dramatically. Actively proliferating tumor cells outgrow their existing blood supply and progressively become hypoxic and necrotic. A key transcription factor mediating such responses is hypoxia inducible factor (HIF-1), a heterodimer that consists of a HIF-1α subunit and a constitutively expressed HIF-1β. In low oxygen conditions, HIF-1α binds to hypoxia-response elements (HREs), thereby inducing the expression of numerous hypoxia-responsive genes involved in tumor angiogenesis and invasion, cell survival, and glucose metabolism.

Angiogenesis is a complex and tightly controlled physiological process that involves growth and maintenance of blood vessels within tissues and organs. A delicate equilibrium exists between opposing proangiogenic (e.g., vascular endothelial growth factor [VEGF], transforming growth factor-b [TGF-β], PDGF, IL-8, and FGF) and antiangiogenic factors (e.g., thrombospondin TSP-1, endostatin, and angiostatin). Glioblastomas are among the most highly vascularized tumors with elevated levels of numerous proangiogenic factors. Hypoxia is one of the most potent stimulators of VEGF expression (Fig. 5). In addition to hypoxic regulation, VEGF expression and glioma angiogenesis are also affected by the genetic status of the tumor cells. EGFR activation leads to an up-regulation of VEGF by increasing the levels of the hypoxia inducible factor (HIF1-α) via a PI3-K dependent pathway that is distinct from signals induced by hypoxia. Mutation or loss of *PTEN* function is also characteristic of these tumors and activates the PI3-K pathway.

C. Mechanisms of Glioma Cell Migration and Invasion

Cellular migration is dependent on adhesion to the surrounding matrix. Extracellular matrix (ECM) proteins (fibronectin, laminin, collagen, vitronectin, and tenascin) in the brain are localized to the perivascular space. These proteins interact with adhesion molecules such as integrins and cadherins. Altered levels of the ECM proteins in the tumor cells and surrounding tissue contribute to tumor cell migration.

One of the most important hallmarks of malignant gliomas is their invasive behavior. Glioma migration and invasion are influenced by overexpression of matrix metalloproteinases (MMPs) in glioma. Invasion of tumor cells is regulated by several proteins. Rho proteins are proteins that cycle between GTP- and GDP-bound states and therefore have a multitude of functions, including effects on the migratory behavior of glioma cells. Focal adhesion kinase (FAK) and proline-rich tyrosine kinase (PYK2) are nonreceptor tyrosine kinases. Inhibition or overexpression of these proteins has been shown to alter migration of glioma cells. Some studies suggest that radiotherapy reduces glioma invasion in xenografts and cultures. However, clinical trials to date have not demonstrated reduced invasion of tumor cells in patients.

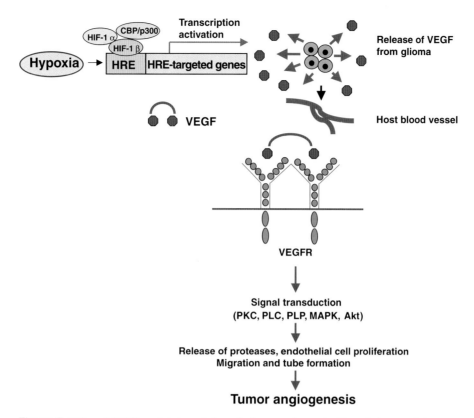

Figure 5 HIF- and VEGF-mediated regulation of glioma angiogenesis. Hypoxia is one of the most potent stimulators of VEGF expression and acts in part through transcription factors of the HIF family. VEGF acts to bind to either the homodimers or heterodimers of the VEGFR and subsequently activates intrinsic tyrosine kinase in the intracellular domain followed by autophosphorylization of the receptor. The tyrosine-phosphorylated receptor is reorganized by cytoplasmatic signaling molecules. The activated receptor signals through the transduction cascade and promotes cellular responses. Activated endothelial cells release proteases, proliferate, and migrate towards the glioma.

X. Pharmacology

The standard of care for gliomas includes surgery followed by radiation and chemotherapy. Chemotherapy has typically used nonspecific cytotoxic approaches that generally act via damaging DNA. Alkylating agents such as temozolomide and nitrosoureas have been more commonly used in recent research trials. The purpose of this section is to highlight therapies that target signaling pathways discussed earlier in this review; this section is not intended to be a comprehensive review of all targets and therapeutics currently under development.

A. Targeted Molecular Therapy

Responses to pharmacological treatment vary considerably even within a classified type of glioma. This is not surprising because these tumors are found in different areas of the CNS, contain high cellular heterogeneity, and also show variations in genetic alterations. Glioblastomas are highly suitable for targeted molecular therapy because they have a set of defined molecular lesions and signaling pathway disruptions that present clear targets. However, the ability of drugs to cross the blood-brain barrier (BBB) and the low incidence of glioma in relation to more common adult neoplasms present relative barriers to glioma-specific drug development. The initial studies of targeted molecular therapies for malignant gliomas have focused on the inhibition of tyrosine kinases such as EGFR, PDGFR, and the mTOR, mainly due to availability of agents that target these pathways.

B. Receptor Tyrosine Kinase Inhibitors

The EGFR is an attractive therapeutic target in glioblastoma. The *EGFR* gene is amplified in 40–50% and overexpressed in more than 60% of glioblastomas. Approximately 40% of tumors with EGFR amplification have gene rearrangements in EGFR that result in the constitutive activation. Overactivity of the EGFR pathway results in cell proliferation; increased tumor invasiveness, motility, and angiogenesis; inhibition of apoptosis; and is associated with resistance to treatment with radiation therapy and chemotherapy. Several small-molecule tyrosine kinase inhibitors of the EGFR are being evaluated in malignant gliomas [19].

Gefitinib (ZD1839, Iressa) and erlotinib (OSI-774, Tarceva) are small molecule inhibitors of the EGFR. The use of gefitinib alone in patients with glioblastoma has been disappointing: the 6-month median progression-free survival (PFS) was 13.2% [19]. Erlotinib showed slightly encouraging efficacy in a phase II trial. The PFS for glioblastoma was 12 weeks, with all patients progressing at 24 weeks [19].

PDGFR is also an attractive target for therapy because PDGFR signaling promotes glioblastoma proliferation and survival. Imatinib is an ATP competitive inhibitor of Abl that also has a high level of activity against PDGFR. Imatinib has already dramatically impacted the survival of chronic myeloid leukemia patients. Imatinib also inhibits the growth of glioblastoma xenografts *in vivo*; however, clinical trials with this agent have not documented activity against glioma.

C. Combination Therapy

An obvious issue that could explain the failure of EGFR inhibitors in patients relates to the ability of these inhibitors to shut down signaling through PI3-K. If *PTEN* is lost, or if PI3-K shows gain-of-function mutations, then inhibition of EGFR will not lead to inhibition of PI3-K. We have found in our laboratory that gefitinib and LY294002 showed combinatorial efficacy on growth of xenografts derived from *PTEN*-mutant U87MG glioblastoma cells transduced with *EGFRvIII*. This experiment demonstrates that biologically based therapies targeting EGFR and PI3-K effectively cooperate in the preclinical treatment of gliomas [20] and argues that a similar approach should be applied in patients with glioma.

The PI3-Ks are lipid kinases. The basic chemical understanding of inhibitors of lipid kinases lags about a decade behind the basic chemical understanding of protein kinases structure and function. Existing inhibitors of the PI3-Ks, wortmannin and LY294002, have demonstrated activity in preclinical testing but are not applicable to the clinical arena because of unfavorable pharmacological characteristics, including problems with instability and insolubility. Another concern is that the first-generation PI3-K inhibitors have such a broad inhibitory spectrum against the entire class of PI3-Ks, as well as more distant PI3-K-like kinases (e.g., ATM). This lack of specificity translates into a poor therapeutic index and unacceptable clinical toxicity. In order to translate our preclinical observations to test inhibitors of PI3-K in combination with inhibitors of EGFR, new drugs targeting PI3-K need to be developed. The isoform-selective inhibitors of p110 show particular promise in this regard [16] because these should effectively block specific isoforms of PI3-K active in tumors while sparing those isoforms critical to growth and development of normal cells.

D. Inhibition of the RAS/MAPK Pathway

Constitutive Ras activation in glioblastoma arises primary from EGFR and PDGFR signaling, suggesting that patients whose tumors overexpress these receptors may derive benefit. The synthetic farnesyltransferase inhibitors SCH66336 and R115777 demonstrate promising results in preclinical models, including glioblastoma cell lines, although it is unclear whether these agents actually impact Ras. In early clinical trials, these drugs have been effective in some patients with a variety of cancers, particularly when used in combination with conventional cytotoxic agents.

E. mTOR Inhibitors

Whereas direct PI3-K inhibitors are generally toxic, there are several well-tolerated inhibitors of mTOR, including sirolimus (Rapamycin), temsirolimus (CCI-779), everolimus (RAD001), and AP23573. These mTOR inhibitors inhibit glioblastoma cell proliferation in culture and intracerebral xenografts and are undergoing clinical testing [19]. However, if these drugs block mTOR and feedback to activate Akt, then these drugs will likely be effective only in combination with PI3-K or Akt inhibitors, neither of which are currently available for use in patients.

XI. Future Therapeutic Interventions

Due to the heterogeneity of gliomas, successful future treatment will most likely include multimodal therapy. For example, recent studies have shown that treatment of glioblastoma patients using radiotherapy in combination with temozolomide increases the median survival by a few months compared with radiotherapy alone. Furthermore, the benefit from temozolomide in glioblastoma patients has been positively correlated with epigenetic silencing of the MGMT promoter through methylation. Maximal potentiation by the MGMT inhibitor O^6-benzylguanine requires complete and prolonged suppression of MGMT. This supports the use of O^6-benzylguanine as a pretreatment to achieve full benefit of alkylating agents, particularly temozolomide, in the chemotherapy of gliomas.

The role of drug delivery in the treatment of CNS neoplasms is a crucial consideration in the development of any agent. The delivery of all substances into the brain is tightly regulated by the BBB. Lipophilic substances can penetrate the BBB whereas large or charged molecules cannot. Alternative approaches involve the use of intracranial catheters (termed *convection-enhanced delivery*), specialized vehicles (e.g., liposomes) or stem cells, or compounds to improve BBB penetration.

Another approach relies on the ability of monoclonal antibodies to find targets on tumor cells. As important components of the immune system, antibodies can potentially initiate immune responses, including complement-mediated or antibody-dependent cytotoxicity. Antibodies may also block protein-protein interactions, including growth-factor and integrin-ligand binding. Internalizing antibodies have been labelled with radioisotopes or toxins to kill tumor cells. Thus far, antibodies directed against the EGFR have shown promising results in preclinical and clinical studies, although delivery of antibodies across the BBB remains an active issue. In other clinical trials, ligands specific for cell surface receptors on glioma cells have been fused to toxins. The advantage of this approach is that ligands are usually very specific and can be used in low concentrations, reducing unwanted side effects.

NSCs have been shown to target glioma cells that are transplanted into murine brains. In these experiments, NSCs were distributed around tumor cells and reduced tumor growth. Due to the infiltrative nature of gliomas, the migration of NSCs might represent an alternative approach to target tumor cells dispersed throughout the brain. In most studies, NSCs have been used as vectors expressing interleukins or enzymes that are know to reduce tumor growth. However, even endogenous NSCs have been shown to "home in" on tumor cells and reduce tumor growth. This migratory capacity of NSCs is not restricted to tumors. In a similar way, NSCs also migrate toward injured and ischemic parts of the brain. The NSCs seem to migrate toward a gradient of stromal cell–derived factor-1 (SDF-1) because they express the receptor chemokine receptor 4 (CXCR4), and the effect is inhibited using an antagonist against this receptor. The understanding of the cellular heterogeneity of gliomas further suggest that future interventions will be based on combined effects from different therapies.

Although some of the treatments mentioned here reduce tumor growth, the majority of glioma patients develop recurrent and lethal tumors. These observations suggest that our current therapeutic approaches do not affect a substantial reservoir of neoplastic tumor cells and that this reservoir is sufficient to reestablish the malignancy.

Both NSCs and glial progenitors are found in brain areas including the SVZ, hippocampus, and subcortical white matter. Glial progenitors that are unable to generate neuronal progeny have been isolated throughout the neuraxis. Abnormal activation of the sonic hedgehog (SHH), WNT, NOTCH-1, and EGFR pathways in NSCs and glial progenitors may initiate hyperplasia by increasing proliferation, altering differentiation, or changing the proportion of cells undergoing asymmetrical division. Alternatively, differentiated adult glia may undergo neoplastic transformation in response to carcinogens through reactivation or overexpression of these pathways. Glioblastoma is a multiforme tumor and is comprised of diverse cell types. The stem cell hypothesis suggests that this diversity results, in part, from the differentiation of a relatively homogeneous progenitor population. If tumor stem cells represent the ideal therapeutic target in glioma, then preclinical and clinical interventions should be tested on this population specifically, in addition to, or instead of the entire tumor as a whole (see Fig. 4C).

Existing data suggest that SHH signaling plays a significant role in initiating and sustaining medulloblastoma growth. Gli, a transcription factor family activated by SHH, is expressed in both low- and high-grade gliomas, suggesting a possible role in this cancer. While SHH is involved in generation of the ventral telencephalic cells during neural development, the sequential signaling of WNT and fibroblast growth factor specifies cells of dorsal telencephalic character. Overexpression of human Dkk-1, a gene encoding a WNT antagonist, sensitizes glioblastoma cells to apoptosis following alkylation damage to DNA, suggesting a protective role for the WNT pathway in glioblastoma cells.

Glioblastoma progenitor cells may be isolated based on excluding Hoechst dye (side-population). The more tumorigenic cells in the side-population express high levels of both β-catenin, a transcription factor downstream of WNT, and Notch-1. Oncogenic Ras also activates Notch-1 signaling and is necessary to maintain the neoplastic phenotype of fibroblasts. Notch-1 therefore represents a potential therapeutic target for various cancers including gliomas. In the SVZ of the adult rodent, EGF-responsive, transit-amplifying progenitors constitute a large population of migratory, rapidly dividing cells. Stable transfection of the *EGFRvIII* variant into neural progenitor cells resulted in highly migratory and fast-dividing cells that infiltrated different brain regions when transplanted into the rodent brain. These and other pathways shared between NSCs and glioma cells represent potential therapeutic targets.

Research should develop markers to identify glioma stem cells and other subpopulations of tumor cells and to assess the impact of preclinical and clinical interventions on this population specifically, rather than on the entire tumor as a whole (see Fig. 4C). Identification of differentially expressed targets for signaling pathways or cell-surface antigens on cancerous stem cells and the more differentiated tumor cells in gliomas may enable the development of therapies that not only inhibit tumor growth but also reduce the recurrence of tumor formation.

Acknowledgments

We thank Eric Burton for his critical review of the manuscript. A.I.P. was supported by grants from the Medical Faculty of Göteborg University, Hjärnfonden, the

Swedish Society of Medicine, the Swedish Society for Medical Research, the Lars Hiertas Minne, the Sahlgrenska University Hospital, and the Assar Gabrielsson and Edit Jacobsson Foundations. W.A.W acknowledges research support from the Goldhirsh, Sandler, and Waxman Foundations; Thrasher and Burroughs Wellcome Funds; and the Brain Tumor Society.

References

1. Kleihues, P., and Webster, K. C., eds. (1997). "Pathology and Genetics—Tumours of the Nervous System." IARC Press, Lyon, France.
2. Duffau, H., and Capelle, L. (2004). Preferential brain locations of low-grade gliomas. *Cancer* **100**, 2622–2626.
3. Ohgaki, H. and Kleihues, P. (2005). Epidemiology and etiology of gliomas. *Acta. Neuropathol. (Berl.)* **109**, 93–108.
4. Cordier, S., Mandereau, L., Preston-Martin, S., Little, J., Lubin, F., Mueller, B., Holly, E., Filippini, G., Peris-Bonet, R., McCredie, M., Choi, N. W., and Arsla, A. (2001). Parental occupations and childhood brain tumors: results of an international case-control study. *Cancer Causes Control* **12**, 865–874.
5. Fukuda, H., Fukuda, A., Zhu, C., Korhonen, L., Swanpalmer, J., Hertzman, S., Leist, M., Lannering, B., Lindholm, D., Bjork-Eriksson, T., Marky, I., and Blomgren, K. (2004). Irradiation-induced progenitor cell death in the developing brain is resistant to erythropoietin treatment and caspase inhibition. *Cell Death Differentiation* **11**, 1166–1178.
6. Sanson, M., Thillet, J., and Hoang-Xuan, K. (2004). Molecular changes in gliomas. *Curr. Opin. Oncol.* **16**, 607–613.
7. Kim, M. S., Kim, E. M., Kim, N. J., Chang, K. A., Choi, Y., Ahn, K. W., Lee, J. H., Kim, S., Park, C. H., and Suh, Y. H. (2003). Inhibition of histone deacetylase increases cytotoxicity to anticancer drugs targeting DNA. *Cancer Res.* **63**, 7291–7300.
8. Ching, T. T., Ching, T. T., Maunakea, A. K., Jun, P., Hong, C., Zardo, G., Pinkel, D., Albertson, D. G., Fridlyand, J., Mao, J. H., Shchors, K., Weiss, W. A., and Costello, J. F. (2005). Epigenome analyses using BAC microarrays identify evolutionary conservation of tissue-specific methylation of SHANK3. *Nat. Genet.* **37**, 645–651.
9. Hegi, M. E., Diserens, A. C., Gorlia, T., Hamou, M. F., de Tribolet, N., Weller, M., Kros, J. M., Hainfellner, J. A., Mason, W., Mariani, L., Bromberg, J. E., Hau, P., Mirimanoff, R. O., Cairncross, J. G., Janzer, R. C., and Stupp, R. (2005). MGMT gene silencing and benefit from temozolomide in glioblastoma. *N. Engl. J. Med.* **352**, 997–1003.
10. Feinberg, A. P. (2004). The epigenetics of cancer etiology. *Semin. Cancer Biol.* **14**, 427–432.
11. Eriksson, P. S., Perfilieva, E., Bjork-Eriksson, T., Alborn, A. M., Nordborg, C., Peterson, D. A., and Gage, F. H. (1998). Neurogenesis in the adult human hippocampus. *Nat. Med.* **4**, 1313–1317.
12. Hemmati, H. D., Nakano, I., Lazareff, J. A., Masterman-Smith, M., Geschwind, D. H., Bronner-Fraser, M., and Kornblum, H. I. (2003). Cancerous stem cells can arise from pediatric brain tumors. *Proc. Natl. Acad. Sci. USA* **100**, 15178–15183.
13. Galli, R., Binda, E., Orfanelli, U., Cipelletti, B., Gritti, A., De Vitis, S., Fiocco, R., Foroni, C., Dimeco, F., and Vescovi, A. (2004). Isolation and characterization of tumorigenic, stem-like neural precursors from human glioblastoma. *Cancer Res.* **64**, 7011–7021.
14. Singh, S. K., Hawkins, C., Clarke, I. D., Squire, J. A., Bayani, J., Hide, T., Henkelman, R. M., Cusimano, M. D., and Dirks, P. B. (2004). Identification of human brain tumour initiating cells. *Nature* **432**, 396–401.
15. Mischel, P. S. and Cloughesy, T. F. (2003). Targeted molecular therapy of glioblastoma. *Brain Pathol.* **13**, 52–61.
16. Newton, H. B. (2004). Molecular neuro-oncology and development of targeted therapeutic strategies for brain tumors. Part 2: PI3K/Akt/PTEN, mTOR, SHH/PTCH and angiogenesis. *Expert Rev. Anticancer Ther.* **4**, 105–128.
17. Kapoor, G. S. and O'Rourke, D. M. (2003). Receptor tyrosine kinase signaling in gliomagenesis: pathobiology and therapeutic approaches. *Cancer Biol. Ther.* **2**, 330–342.
18. Sridhar, S. S., Hedley, D., and Siu, L. L. (2005). Raf kinase as a target for anticancer therapeutics. *Mol. Cancer Ther.* **4**, 677–685.
19. Drappatz, J., and Wen, P. Y. (2004). Non-cytotoxic drugs as potential treatments for gliomas. *Curr. Opin. Neurol.* **17**, 663–673.
20. Fan, Q. W., Specht, K. M., Zhang, C., Goldenberg, D. D., Shokat, K. M., and Weiss, W. A. (2003). Combinatorial efficacy achieved through two-point blockade within a signaling pathway—a chemical genetic approach. *Cancer Res.* **63**, 8930–8938.

40

Bacterial and Fungal Infections of the Nervous System

Thomas P. Bleck, MD, FCCM

Keywords: *arteritis*, Aspergillus fumigatus, *epidural abscess*, Escherichia coli, *herniation*, Listeria monocytogenes, *meningitis*, Mycobacterium leprae, Mycobacterium tuberculosis, *septic venous thrombosis, spinal cord compression*, Streptococcus agalactiae, Staphylococcus aureus, Staphylococcus epidermidis, Streptococcus pneumoniae, *subdural empyema, vasculitis, ventriculitis, Virchow-Robin space*

I. History and Nomenclature
II. Etiology
III. Pathogenesis
IV. Relevant Structural Detail
V. Pathophysiology
VI. Pharmacology
VII. Signs and Symptoms
VIII. Natural History
References

I. History and Nomenclature

Central nervous system (CNS) infections were well known in the ancient world, and descriptions of meningitis date to the 16th century. Leprosy, the prototypical bacterial infection of the peripheral nervous system, was also known to the ancients, although there remains considerable debate regarding the identification of that condition with present-day Hansen's disease. The major syndromes of neurological infection were delineated during the heyday of descriptive pathology in the 19th century. The study of host defense in protection against and response to infection dates to the same period. Quincke's invention of the lumbar puncture needle in 1891 and the development of computed tomography (CT) and magnetic resonance imaging (MRI) in the 20th century round out our current understanding.

The comprehension of nervous system infections first requires knowledge of anatomy because the consequences of infection vary both with the anatomical spaces involved and the functions of the tissues located within them [1]. Although a few organisms are notable for their ability to cross tissue planes, most infections initially manifest in one or two spaces.

Epidural infections are typically excluded from either direct extension into or immunological effects compromising CNS structures [2]. The manifestations of these infections are primarily due to compression of adjacent tissues. Inside the cranium, signs and symptoms correspond with the elevation of intracranial pressure produced by added volume of alien tissue (e.g., headache and altered consciousness) and with the compression of the brain beneath the abscess. Spinal epidural abscesses cause local pain, then nerve root compression, and finally spinal cord compression. Subdural empyema, in contrast, often incites a substantial inflammatory response in the underlying brain or spinal cord and thereby causes findings seemingly out of proportion to the volume of inflammatory exudates produced [3]. Although either the dura mater or the pia mater may become infected, meningitis commonly refers to infection in the subarachnoid space [4]. Rarely, infection is confined to the lining tissues themselves, producing pachymeningitis or arachnoiditis. When the infection is predominantly within the ventricular system, ventriculitis or ependymitis may develop. The brain or spinal cord themselves may become infected, initially in the form of cerebritis or myelitis, which then evolve into a parenchymal abscess [5]. The blood vessels may also become infected, producing either an arteritis (also termed vasculitis). When veins or venous sinuses are involved, the pathophysiological consequences are typically due to thrombosis, so the terms septic venous thrombosis or sinus thrombosis are often employed [6]. Although not technically parts of the nervous system, infections of the cranium or spinal column (osteomyelitis) or the bony sinuses (sinusitis) are important antecedents of CNS infection [7]. Although most infections of peripheral nerve are viral, a disorder such as leprosy is considered a neuritis.

II. Etiology

Bacteria and fungi, by virtue of their intrinsic properties and the host responses they incite, tend to produce specific clinical syndromes when they infect the nervous system. Because the host's defenses play such a critical part in shaping the signs and symptoms, abnormalities of host response may cause different syndromes in different patients infected with the same organism. The major infecting organisms are grouped by their syndromes and host characteristics in Table 1.

III. Pathogenesis

The infecting organism must first gain entry into the target tissue. The common routes of nervous system infection include 1) hematogenous spread via the arterial blood and 2) direct extension from another site of infection, such as an infected bone or sinus [8]. The extreme case of direct extension occurs when trauma has produced a direct communication between the outside world and the CNS. Once the organism has invaded, it must elude or subvert the local host defenses in order to survive and reproduce. These local host defense mechanisms may then invoke a more systemic response. These defensive responses are intended to clear the infection, but their effects are often deleterious to the nervous system tissue itself; many of the signs and symptoms of infection are consequences of the inflammatory response and its aftermath.

IV. Relevant Structural Detail

The linings of the CNS provide protection from external forces and invading organisms. Because of their structures, these linings also produce distinct syndromes when infections occur; as they develop, these infections give rise to distinct syndromes that are often more consistent with their location than the particular organism involved (Fig. 1).

V. Pathophysiology

The infecting organism may produce symptoms and signs by interfering with the function of the nervous system tissue being invaded or compressed, but many findings reflect the inflammatory response produced by the host in response to the infection. Figure 2 presents a summary of steps in the pathogenesis of infection, using bacterial meningitis as a paradigm.

In bacterial meningitis, the structural barriers imposed by the dura and the arachnoid provide substantial protection from bacterial invasion. However, infection by direct infection, usually from the skull (including the sinuses) does occur, presumably because the inoculum of the infecting organism is too large for the extradural defenses to contain.

Table 1 Common Organisms and Syndromes of Central Nervous System Infection

Organism	Syndrome	Typical Patient Age	Common Host Characteristics
Bacteria			
Streptococcus pneumoniae	Acute bacterial meningitis	Adult	Normal; may be impaired
Neisseria meningitidis	Acute bacterial meningitis	Adolescent	Normal
Hemophilus influenzae type B	Acute bacterial meningitis	Child	Normal
Listeria monocytogenes [14]	Acute bacterial meningitis	Infant	Normal, associated with maternal infection
	Brain abscess (typically involving the pons, termed rhombencephalitis)	Any	Normal
	Acute or subacute bacterial meningitis	Older adult	Often immunocompromised
Staphylococcus aureus (coagulase-positive staphylococcus)	Acute bacterial meningitis	Any	Anatomical or surgical defect in skull or meninges
	Brain abscess	Any	Hematogenous dissemination (e.g., endocarditis)
	Epidural abscess	Any	Direct extension from osteomyelitis
Staphylococcus epidermidis (coagulase-negative staphylococcus)	Subacute bacterial meningitis	After neurosurgical procedures	
Streptococcus agalactiae (group B streptococcus)	Acute or subacute bacterial meningitis	Newborns and older adults	Colonization during delivery; gastrointestinal source of bacteremia
Escherichia coli	Acute bacterial meningitis	Newborns or patients with anatomical or surgical defects	
Other gram-negative rods	Meningitis	Any	After neurosurgical procedures
	Brain abscess	Any	After neurosurgical procedures
Streptococcus milleri group	Brain abscess	Any	
Bacteroides species and other anaerobes	Brain abscess	Any	
Bacillus anthracis [15]	Meningitis	Any	
Treponema pallidum [16]	Meningitis (secondary syphilis; later, meningovascular syphilis)	Any	Accelerated course in HIV patients
	Encephalitis (general paresis)	Older adults	
	Parenchymal neurosyphilis	Older adults	
	Tabes dorsalis	Older adults	
	Congenital neurosyphilis	Newborns	
Borrelia burgdorferi	Meningitis	Any	
	Peripheral neuropathy	Any	
Rickettsii (e.g., Rickettsia rickettsii, Rocky Mountain spotted fever)	Encephalitis	Any	
Higher bacteria			
Nocardia	Brain abscess	Any	Cell-mediated immunity defects
Actinomyces	Brain abscess	Any	
Mycobacteria			
Mycobacterium tuberculosis [17]	Meningitis	More severe in children	
	Brain abscess (tuberculoma)	Any	
	Epidural abscess (Pott's disease)	Any	
Mycobacterium avium complex	Meningitis	Any	HIV patients most commonly affected
	Encephalitis	Any	HIV patients most commonly affected
Mycobacterium leprae	Peripheral neuropathy	Any	
Fungi [18]			
Aspergillus spp. [19]	Brain abscess	Any	Granulocytopenic patients
	Meningitis	Any	
Cryptococcus neoformans [20]	Meningitis	Ay	Patients with HIV infection and others with cell-mediated immune defects (especially steroids)
	Brain abscess	Any	Patients with HIV infection and others with cell-mediated immune defects (especially steroids)

(Continued)

Table 1 (Continued)

Organism	Syndrome	Typical Patient Age	Common Host Characteristics
Coccidiodes immitis	Meningitis	Any	Geographically limited
	Brain abscess		
Candida albicans (and other species)	Meningitis	Any	
	Brain abscess	Any	
Rhizopus, Mucor, and related fungi	Meningitis with vasculitits	Any (usually adults)	Acidotic patients

HIV, Human immunodeficiency virus.
References are indicated by bracketed numbers.

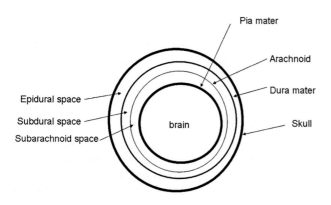

Figure 1 Anatomy of central nervous system infection.

Whether the organisms reach the subarachnoid space by traversing these membranes or by hematogenous spread, the subarachnoid space itself provides a relatively favorable location for bacterial or fungal replication. The subarachnoid space and the brain are often considered immunologically "privileged" sites because they are extensively protected by anatomical barriers, but at the same time they partially exclude many of the cells and substances (e.g., complement); when infection does occur, they are in part immunologically deprived [9].

The major impediments to the movement of both microorganisms and inflammatory substances from the systemic circulation into the CNS are the blood-brain and blood–cerebrospinal fluid (CSF) barriers. Anatomically, the barrier resides in the investment of cerebral capillaries (which themselves have tight junctions) by astrocytic foot processes that completely envelop the cerebral microvasculature. These barriers normally regulate the transit of molecules from the bloodstream into the brain. However, inflammation and trauma can damage the barrier, making it much less selective. Matrix metalloproteinase-9 is probably a major mediator of this effect. This is not entirely deleterious, as it allows many antimicrobial agents to enter the CNS in higher concentrations when an infection is present. Endogenous substances, such as complement, also cross in higher concentrations but generally less than those

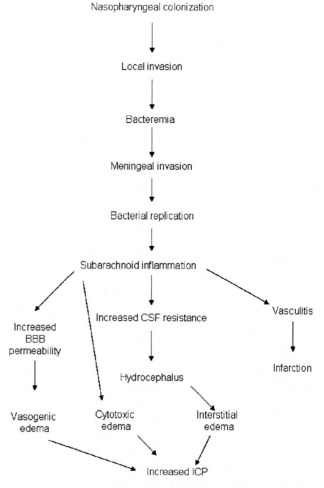

Figure 2 Pathogenetic sequence of central nervous system infection. (Adapted from Leib S. L. and Täuber, M. G. [2004]. Pathogenesis and pathophysiology of bacterial infections; and from Roos, K. L., Tunkel, A. R., and Scheld, W. M. [2004]. Acute bacterial meningitis. *Both in*: "Infections of the Central Nervous System" [W. M. Scheld et al., eds.], ed. 3, pp. 331–422. LWW, New York.)

of the blood. Antibodies needed to combat infection also cross, but as the inflammatory response continues, local plasma cells will produce considerably more immunoglobulin locally [10].

The cellular components of the immune response are less dependent on changes in blood-brain and blood-CSF barrier function than the humoral components. Microglia, the resident macrophages of the CNS, are predominantly derived from precursors in the bone marrow and easily pass in and out of these otherwise protected spaces. When inflamed, the CNS releases activated complement components, cytokines (e.g., tumor necrosis factor, interleukin [IL]-1β, IL-6, IL-8, and IL-10), and chemokines, which trigger the chemotaxis of neutrophils and other inflammatory cells. The initial response to infection causes neutrophils in the small vessels to begin rolling along the endothelial surface due to the effect of selectins; these cells are then attached to the vessel wall by integrins and soon migrate through the vessel wall under the influence of IL-8. The initial neutrophilic response to most infections is followed by other components of the macrophage/monocyte system and the eventual arrival of lymphocytes, some of which will become plasma cells and produce antibody locally.

Because the CNS lacks lymphatics, the debris of inflammation exits the parenchyma via the perivascular space (Virchow-Robin space). This space is actually a continuation of the subarachnoid space along penetrating blood vessels. Histological evidence of inflammatory infiltrate in this space is one of the pathological criteria of many types of infection [10]. The accumulation of this debris in the subarachnoid space may explain the occurrence of cerebral arterial vasculitis and cerebral venous thrombosis as common complications of diseases such as bacterial meningitis.

Neuronal damage is mediated at least in part by the effects of nitric oxide (NO) and excitatory amino acids (EAA). NO is released by both inflammatory cells and neurons in the course of their normal activities, but when inflammatory cells undergo an oxidative burst to kill microorganisms, the excessive NO produced can trigger apoptosis in neurons, particularly those of the hippocampus [11]. The involvement of microglia via toll-like receptors in this process indicates that the innate immune system of the brain also participates in this potentially deleterious process [12]. Understanding of the role of EAAs continues to evolve; blocking their effects may stop seizures but not apoptosis [14].

Foreign bodies in the CNS provide sites at which bacteria may multiply with some protection from immune surveillance and attack. The bacteria involved are often less pathogenic than those that usually cause meningitis (e.g., infection of ventriculoperitoneal shunt tubing often involves *Staphylococcus epidermidis* (coagulase-negative staphylococci), which produce a disease that is initially much less severe than *Staphylococcus aureus*. However, the presence of the foreign body also makes both the host's immune attack and antibiotic treatment incompletely effective; foreign bodies must often be removed to cure such an infection.

VI. Pharmacology

The definitive treatment of CNS infection depends on correct identification and antimicrobial treatment of the infecting organism, relief of the excessive pressure or mass effect exerted by the organism, and modulation of the host's immune response to allow clearance of the organism while minimizing the deleterious consequences of excessive inflammation.

The specifics of antimicrobial treatment are beyond the scope of this chapter, but a few general comments are appropriate. First, the use of antibiotics alone is usually inadequate in the treatment of abscesses. Therefore cranial and spinal epidural abscesses and subdural empyemas almost always require surgical or CT-guided drainage in addition to antibiotics for resolution. The situation with brain abscesses is more problematic. At the point in time when the abscess appears to have developed a capsule when viewed by CT or MRI, the tissue is usually still in the histological state of cerebritis rather than having a necrotic center that could be drained. Obtaining a sample of the infected tissue for microbiological studies as soon as possible is important regardless of the age of the abscess, but it may be necessary to drain or resect the abscess later when it has become more mature. Second, one must choose antimicrobial agents that will achieve a useful concentration in the infected tissue. Epidural abscesses, for example, are outside of the CNS and can be treated with antibiotics that do not penetrate the brain. Clindamycin does not achieve useful concentrations in the CSF and should not be used to treat meningitis. Third, potent drugs such as the third-generation cephalosporins are effective in sterilizing the CSF of many organisms that cause meningitis, but the rapid release of bacterial cell wall components into the CSF prompts a profound inflammatory response, which causes some of the complications of meningitis, such as seizures and sensorineural hearing loss. Pretreatment with corticosteroids decreases this inflammatory response and appears to lower the incidence of some of these complications.

The emergence of penicillin and, to a lesser extent, cephalosporin resistance in pneumococci led to several changes in the empirical treatment of meningitis. Thus empirical therapy for acute bacterial meningitis now includes a third-generation cephalosporin (e.g., ceftriaxone) as well as vancomycin. Unfortunately, corticosteroid administration decreases the penetration of vancomycin into the CSF; this has limited enthusiasm for the use of steroids prior to antibiotic administration [13].

VII. Signs and Symptoms

The signs and symptoms of CNS infections can be divided into those related to local tissue dysfunction (negative phenomena in Hughlings Jackson's nosology), those due to inflammation and irritation (positive phenomena), those due to vascular compromise, and those due to mass effect. Because these types of findings rarely occur in isolation, it is more useful to consider them together as they affect the brain or spinal cord at different locations. The findings attributable to local tissue dysfunction and inflammation often give important clues related to the location of the infection, and the time course of their development often illuminates the etiology of the disorder.

The findings of acute bacterial meningitis reflect both inflammation of the meninges (causing a stiff neck, and often Kernig's and Brudzinski's signs, which betray irritation of the dura when it is stretched) and increased intracranial pressure (e.g., diffuse headache, nausea, vomiting, and alteration or loss of consciousness). Similar disturbances characterize subdural empyema, with the addition of signs of focal tissue dysfunction (e.g., weakness or seizures) reflecting inflammation in the cortex beneath the empyema. In contrast, cranial epidural abscesses typically present with localized headache (probably reflecting the cranial osteomyelitis at onset and later stretching of the dura at the base of the brain, which is the major pain-sensitive structure within the cranium) followed by signs of increased intracranial pressure. Subtentorial abscesses are not associated with seizures but are more likely to cause hydrocephalus by obstruction of the ventricular system. Some types of meningitis (typically the chronic meningitides, such as tuberculous meningitis or coccidioidal meningitis) also cause hydrocephalus, either by obstructing the foramina of Luschka and Magendie or by interfering with CSF resorption at the arachnoid granulations. Rarely, one sees cranial nerve disorders related to cranial osteomyelitis, as in the syndrome of malignant otitis observed in diabetic patients with *Pseudomonas aeruginosa* infection causing the jugular foramen syndrome (paresis of cranial nerves IX, X, and XI).

Spinal epidural abscess is a common condition with a very characteristic progression of findings. Early recognition of this condition is vital to preservation of spinal cord function. The initial symptom is localized back pain, usually occurring as a consequence of spinal osteomyelitis. As the infection progresses, the patient develops radicular pain and weakness, reflecting compression of the nerve roots. If the infection is in the cervical or lumbosacral regions, weakness is easy to detect; as a lower motor neuron disorder, it is associated with diminished muscle tone and hypoactive or absent reflexes without signs of upper motor neuron dysfunction (e.g., Babinski's sign). Because the findings reflect local compression, they are confined to the nerve roots actually being compressed. When the lesion is in the thoracic region, weakness is much less apparent and is not easily elicited in the course of the usual physical examination. However, electromyography confirms the presence of a lower motor neuron lesion. Although electromyography would not be performed if an epidural abscess were suspected, spinal epidural abscess must be considered in the differential diagnosis of lower motor neuron lesions detected by this technique. When the volume of the abscess increases to the point of spinal cord compression, the patient develops the typical long tract findings associated with this problem (e.g., upper motor neuron weakness below the level of the lesion; sensory loss below the level of the lesion; bowel and bladder dysfunction, acutely manifested as urinary retention and loss of anal sphincter tone) and autonomic dysfunction appropriate to the level of the lesion. Early suspicion of the possibility of an epidural abscess is important because the preservation or restoration of spinal cord function depends on rapid decompression (measured in hours, not days).

Some types of CNS infection also produce prominent systemic signs and symptoms of infection (e.g., fever and hypotension). These are sometimes consequences of a bacteremia, as occurs in many cases of bacterial meningitis, but they may also reflect the systemic inflammatory response syndrome elicited by cytokine release into the systemic circulation (commonly seen with subdural empyema). However, brain abscesses and epidural abscesses are associated with fever in only about half of cases, and even when fever is present, it is rarely high or spiking as may be seen with undrained abscesses in other organs.

Compression of the brain or spinal cord parenchyma causes distortion of tissues, resulting in herniation out of the normal location of the involved structures. The panoply of symptoms related to herniation is beyond the scope of this chapter; the classical syndromes and their pathophysiology have been discussed elsewhere [16]. Two points require emphasis here. First, the movement of the brain or spinal cord parenchyma can also distort blood vessels, leading to arterial or venous distribution infarctions that may be distant to the site of herniation and thus cause clinical findings not expected from the location of the infections. Second, interventions that increase pressure gradients across areas of focal infection, causing tissue swelling, may prompt herniation (e.g., performing a lumbar puncture in a patient with a posterior fossa abscess).

VIII. Natural History

In the pre-antibiotic era, the natural history of most CNS infections was dismal. Although patients would rarely recover spontaneously from bacterial meningitis, the mortality often exceeded 90%. Epidural abscesses of the cranium or spine could be drained surgically and thus had some possibility of improvement, and brain abscesses could sometimes be treated by resection.

The prognosis of these conditions changed drastically with the availability of antimicrobial agents. The mortality of bacterial meningitis has fallen to about 30%. This percentage has remained relatively constant over the past six decades, which probably reflects the coincidence of more effective drugs with the development of antibiotic resistance among bacteria and the increasing number of immunocompromised patients. Data for the other, less numerous, infections are less clear but generally show a substantial benefit of antibiotic treatment.

Survival is not the only measure of outcome, and the survivors of CNS infections frequently have serious neurological compromise. Seizures and focal neurological deficits are common following the otherwise successful treatment of brain abscesses and subdural empyemas. Patients who have recovered from bacterial meningitis, including tuberculous meningitis, may have chronic problems related to vascular complications. Cranial nerve deficits are also common. Meninigitis in children may result in mental retardation; one must be careful in these patients to exclude hearing loss as the cause of or a contributor to a decline in school performance after meningitis because eighth cranial nerve problems are very frequent in this population.

The natural history of spinal cord compression due to epidural abscess highlights the potential for rapid deterioration and the need for alacrity in diagnosis and treatment. When the condition is diagnosed and treated at the stage of local pain, complete recovery is typical. If intervention is delayed until radicular symptoms are present, the spinal cord is spared but the nerve roots may not recover. When long tract findings appear, permanent disability may develop within hours; this constitutes a medical and surgical emergency that must be managed immediately.

References

1. Bleck, T. P. and Greenlee, J. E. (2000). Approach to the patient with central nervous system infection. *In:* "Principles and Practice of Infectious Diseases" (G. M. Mandell, J. E. Bennett, and R. Dolin, eds.), ed. 5, pp. 950–959. Churchill Livingstone, New York.
2. Bleck, T. P. and Greenlee, J. E. (2000). Epidural abscess. *In:* "Principles and Practice of Infectious Diseases" (G. M. Mandell, J. E. Bennett, and R. Dolin, eds.), ed. 5, pp. 1031–1034. Churchill Livingstone, New York.
3. Bleck, T. P. and Greenlee, J. E. (2000). Subdural empyema. *In:* "Principles and Practice of Infectious Diseases" (G. M. Mandell, J. E. Bennett, and R. Dolin, eds.), ed. 5, pp. 1028–1031. Churchill Livingstone, New York.
4. Roos, K. L., Tunkel, A. R., and Scheld, W. M. (2004). Acute bacterial meningitis. *In:* "Infections of the Central Nervous System" (W. M. Scheld, R. J. Whitley, and C. M. Marra, eds.), ed. 3, pp. 347–422. LWW, New York.
5. Kastenbauer, S., Pfister, H.-W., Wispelwey, B., et al. (2004). Brain abscess. Acute bacterial meningitis. *In:* "Infections of the Central Nervous System" (W. M. Scheld, R. J. Whitley, and C. M. Marra, eds.), ed. 3, pp. 479–508. LWW, New York.
6. Bleck, T. P. and Greenlee, J. E. (2000). Suppurative intracranial phlebitis. *In:* "Principles and Practice of Infectious Diseases" (G. M. Mandell, J. E. Bennett, and R. Dolin, eds.), ed. 5, pp. 1034–1036. Churchill Livingstone, New York.
7. Gullipalli, D. and Bleck, T. P. (2002). Bacterial infections of the central nervous system. *In:* "Diseases of the Nervous System: Clinical Neuroscience and Therapeutic Principles" (A. K. Asbury, G. M. McKhann, W. I. McDonald, P. J. Goadsby, and J. C. McArthur, eds.), ed. 3, pp. 1728–1744. Saunders, Philadelphia.
8. Leib S. L. and Täuber, M. G. (2004). Pathogenesis and pathophysiology of bacterial infections. *In:* "Infections of the Central Nervous System" (W. M. Scheld, R. J. Whitley, and C. M. Marra, eds.), ed. 3, pp. 331–346. LWW, New York.
9. Lucas, S. M., Rothwell, N. J., and Gibson, R. M. (2006). The role of inflammation in CNS injury and disease. *Br. J. Pharmacol.* **147(Suppl 1)**, S232–S240.
10. Pfister, H.-W. and Black, T. P. (2003). Bacterial infections. *In:* "Neurological Disorders: Course and Treatment" (T. Brandt, L. R. Caplan, J. Dichgans, H. C. Diener, and C. Kennard, eds.), ed. 2, pp. 529–544. Academic Press, San Diego.
11. Roberts, M., Carmichael, A., and Martin, P. (2004). Cerebral vasculitis caused by *Aspergillus* species in an immunocompetent adult. *Infection* **32**, 360–363.
12. Yamaguchi, A., Tamatani, M., Matsuzaki, H., Namikawa, K., Kiyama, H., Vitek, M. P., Mitsuda, N., and Tohyama, M. (2001). Akt activation protects hippocampal neurons from apoptosis by inhibiting transcriptional activity of p53. *J. Biol. Chem.* **276**, 5256–5264.
13. Iliev, A. I., Stringaris, A. K., Nau, R., and Neumann, H. (2004). Neuronal injury mediated via stimulation of microglial toll-like receptor-9 (TLR9). *FASEB J.* **18**, 412–414.
14. Kolarova, A., Ringer, R., Täuber, M. G., and Leib, S. L. (2003). Blockade of NMDA receptor subtype NR2B prevents seizures but not apoptosis of dentate gyrus neurons in bacterial meningitis in infant rats. *BMC Neurosci.* **4**, 21.
15. Bleck, T. P. (2002). Therapy for bacterial infections. *In:* "Current Therapy in Neurologic Disease" (R. T. Johnson and J. W. Griffen, eds.), ed. 6, pp. 151–153. Mosby, Chicago.
16. Blech, T. P. (2002). Diagnosis and management of the comatose patient in the intensive care unit. *In:* "Critical Care Medicine: Perioperative Management" (M. J. Murray, D. B. Coursin, R. G. Pearl, and D. S. Prough, eds.), ed. 2, pp. 244–252. LWW, Philadelphia.
17. Gerner-Smidt, P., Ethelberg, S., Schiellerup, P., Christensen, J. J., Engberg, J., Fussing, V., Jensen, A., Jensen, C., Petersen, A. M., and Brunn, B. G. (2005). Invasive listeriosis in Denmark 1994–2003: a review of 299 cases with special emphasis on risk factors for mortality. *Clin. Microbiol. Infect.* **11**, 618–624.
18. Sejvar, J. J., Tenover, F. C., and Stevens, D. S. (2005). Management of anthrax meningitis. *Lancet Infect. Dis.* **5**, 287–295.
19. Marra, C. (2004). Neurosyphilis. *In:* "Infections of the Central Nervous System" (W. M. Scheld, R. J. Whitley, and C. M. Marra, eds.), ed. 3, pp. 649–658. LWW, New York.
20. Almeida A. (2005). Tuberculosis of the spine and spinal cord. *Eur. J. Radiol.* **55**, 193–201.
21. Perfect, J. R. (2004). Fungal meningitis. *In:* "Infections of the Central Nervous System" (W. M. Scheld, R. J. Whitley, and C. M. Marra, eds.), ed. 3, pp. 691–712. LWW, New York.
22. Singh, N. (2005). Invasive aspergillosis in organ transplant recipients: new issues in epidemiologic characteristics, diagnosis, and management. *Med. Mycol.* **43(Suppl 1)**, S267–S270.
23. Lee, S. C., Dickson, D. W., and Casadevall, A. (1996). Pathology of cryptococcal meningoencephalitis: analysis of 27 patients with pathogenetic implications. *Hum. Pathol.* **27**, 839–847.

41

Parasitic Infections

Ana-Claire Meyer, MD
Gretchen L. Birbeck, MD, MPH

Keywords: Acanthamoeba, *amebic brain abscess*, Angiostrongylus cantonensis, *aseptic meningitis*, *Balamuthia*, Baylisascaris procyonis, bilharzias, *cerebral hydatid disease*, *cerebral malaria*, *cerebral schistosomiasis*, *chorioretinitis*, Echinococcus, Entamoeba histolytica, *eosinophilic meningitis*, Gnathostoma spinigerum, *granulomatous amebic encephalitis*, *human African trypanosomiasis*, *malaria*, Naegleria fowleri, *neural larva migrans*, *neurocysticercosis*, Plasmodium falciparum, *primary amoebic meningoencephalitis*, raccoon roundworm, rat lungworm, *schistosomiasis*, schistosoma mansoni, *sleeping sickness*, Strongyloides stercoralis, Taenia solium, *tick paralysis*, Toxocara canis, Toxocara cati, *toxoplasmosis*, Trichinella spiralis, Trypanosoma brucei gambiense, Trypanosoma brucei rhodesiense, *trypanosomiasis, vasculitis*

I. Introduction
II. Protozoans
III. Helminths: Cestodes
IV. Helminths: Nematodes
V. Helminths: Trematodes
   References

I. Introduction

Parasitic infections are widespread among the animal kingdom and have adapted to and co-evolved with *Homo sapiens* for millions of years. Helminth parasites were first described in Egyptian papyri between 3000–400 BCE and later in the writings of Greek and Arabic physicians. Humans harbor more than 70 species of protozoa and 300 species of helminth worms, although only a small proportion of these are responsible for the bulk of human disease [1].

Parasites are distributed worldwide, although the tropical latitudes are disproportionately affected. Many human parasites have animal reservoirs or are transmitted via insect, copepod, or snail vectors, and proximity to and contact with these populations and their excreta are important risk factors for infection. Thus many helminthic infections have almost vanished with improvements in housing and sanitation. The burden of disease lies in poor or politically unstable regions in which diseases such as malaria and trypanosomiasis have increased in incidence over the past few decades.

The reasons for the resurgence of human parasitic diseases are varied. Population growth in both rural and urban areas without corresponding improvements in the public health infrastructure, housing, sanitation, and water supply leads to increased prevalence of vectors and pathogens. Increased trade and travel brings people from nonendemic areas into contact with parasites such as malaria. Deforestation, urbanization, creation of dams, irrigation, clearing of wetlands, and other changes in land use have had both positive and negative effects on the populations of vectors such as mosquitoes, ticks, copepods, and snails. Global warming lengthens the growth season of many vectors and may extend their area of distribution [2].

Neurological manifestations of parasitic infections are less common than systemic ones, but some carry severe morbidities and high mortality (Table 1). For example, human African trypanosomiasis, cerebral malaria, and amebic encephalitides very high mortality rates, and survivors often have devastating neurological deficits. Also, neurocysticercosis is the leading cause of acquired epilepsy worldwide. Despite their public health importance and often devastating effects, relatively little is known about the pathophysiology and molecular mechanisms of these infections.

The pathogenesis of parasitic diseases is complex. In general, parasites can cause pathology by mechanical disruption as they migrate through or displace tissues. Others secrete toxic substances, but the predominant pathological mechanism is via the human immune response to infection. Eggs and degenerating larvae often induce granuloma formation, which can cause fibrosis or mass effect on other tissues if excessive. Protozoans and helminths alike induce disordered inflammatory responses resulting in meningitis, encephalitis, or localized inflammatory responses.

As parasites are constantly exposed to the immune system, they have developed many strategies to evade and, more importantly, to modulate the immune response. Protozoans, for example, can evade the immune system via antigenic variation of their surface proteins or by sequestration in granulomas, cysts, or intracellular locations (Fig. 1) [3]. On the other hand, the larger helminths must live in exposed extracellular locations such as lymphatics, the bloodstream,

Table 1 Human Parasitic Infections of the Nervous System

Protozoans	*Micronema deletrix*
Malaria	*Dracunculus medinensis*
Plasmodium falciparum	*Dipetalonema perstans*
Human African trypanosomiasis	*Onchocerca volvulus*
Trypanosoma brucei gambiense	*Loa Loa*
T. b. rhodesiense	Trematodes
Amoebiasis	Schistosomes
Acanthamoeba	*Schistosoma mansoni*
Balamuthia	*S. japonicum*
Naegleria fowleri	*S. hematobium*
Entamoeba histolytica	Paragonimus
Toxoplasmosis	*Paragonimus westermani*
Toxoplasma gondii	*P. mexicanus*
Helminths	*P. miyazakii*
Cestodes	Ectoparasites
Neurocysteircercosis	Tick paralysis
Taenia solium	North America
Echinococcosis	*Dermacentor andersoni*
Echinococcus granulosus	*Dermacentor variabilis*
Others	*Amblyomma maculatum*
Diphyllobothrium latum	*Amblyomma americanum*
Spirometra spp.	*Ixodes scapularis*
Taenia multiceps	Russia
Nematodes	*Ornithodoros laborensis*
Strongyloidiasis	South Africa
Strongyloides stercoralis	*Otobius megnini*
Angiostrongyliasis	*Ixodes rubicundus*
Angiostrongylus cantonensis	*Rhipipcephalus simus*
Gnathostomiasis	Mexico
Gnathostoma spinigerum	*Ixodes tancitarus*
Toxocariasis	Phillipines
Toxocara cani	*Amblyomma cyprium*
T. cati	*aeratipes*
Trichinosis	Eastern Australia
Trichinella spiralis	*Ixodes holocyclus*
Other nematodes	*Ixodes cornuatus*
Baylisascaris procyonis	
Wuchereria bancrofti	

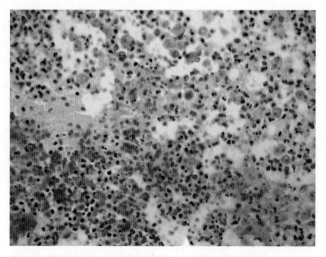

Figure 1 Encysted *Acanthamoeba* in the cerebellum. (Image courtesy of E. Tessa Hedley-Whyte, MD, Director of Neuropathology, Professor of Pathology, Massachusetts General Hospital, Boston, Massachusetts.)

within tissues, or the gastrointestinal tract. Rather than evade the immune system, helminths use different strategies to modulate and suppress the immune response. Surface glycoproteins modulate the immune response. Some species can secrete cytokine homologs and protease inhibitors that stimulate regulatory (suppressor) T cells, induce anergy in T cells, or suppress macrophages [3]. It is often the death of the helminth while still within the body that generates the pathology of the disease, as in the clinical worsening of patients with neurocysticercosis after the initiation of treatment.

Effective treatments are available for only some parasitic infections, and despite sustained research efforts, no vaccines are available. Research is hampered by lack of funding, and control efforts are fragmented because of political instability. Sadly, most of these diseases would be cured with safe drinking water and food and improved housing, but these continue to be unavailable to the majority of the world's population.

II. Protozoans

A. Cerebral Malaria

1. Brief History

Malaria infects 300–500 million people per year and results in more than 1 million deaths, most of which occur among African children [4]. The poor are disproportionately affected: 60% of malaria deaths worldwide occur in the poorest 20% of the population. Recent estimates suggest that 48% of the global population, or 3 billion people, are exposed to malaria and are at risk of infection, despite the fact that control efforts have reduced the area of malaria risk from 53% to 27% of the earth's land surface [5]. The extent of morbidity is difficult to estimate, but malaria causes severe maternal anemia in pregnant women, low birth weight, reduced attendance at school and work, and a drain on the healthcare system [4]. Approximately 90% of all malaria cases occur in Africa, where infection results in the loss of an estimated $12 billion (U.S. dollars) in economic revenue annually [6].

Malaria control has deteriorated in the past decade for a variety of reasons including climactic instability with droughts and floods, global warming, increased travel, human immunodeficiency virus (HIV), drug resistance, insecticide resistance, and civil disturbances that result in the collapse of control programs and crowding of refugees from nonendemic areas. Current prevention efforts target prophylactic antimalarial medications in travelers and during pregnancy, vaccine development, vector control with insect traps, widespread insecticide treatments, repellents, and insecticide-treated nets [5].

2. Etiology

Four protozoan species (*Plasmodium falciparum, Plasmodium vivax, Plasmodium malariae*, and *Plasmodium ovale*) cause malaria and are transmitted by the bite of several species of female anopheline mosquitoes. *P. falciparum* causes all cases of severe malaria and cerebral malaria and most of the anemia seen in endemic areas. Estimates of the incidence of severe malaria are highly variable depending on the study site. Severe malaria occurs primarily in patients from nonendemic regions ("nonimmune"), pregnant women, and children [4].

The malarial life cycle is complex, with multiple stages within the mosquito as well as multiple stages within the human, including both intracellular and extracellular phases (Fig. 2).

3. Pathogenesis

In cerebral malaria, pathological studies reveal a slate-gray brain with petechial hemorrhages. Capillaries and venules are packed with parasitized red blood cells (Fig. 3). Immunohistochemical staining suggests endothelial activation. Inflammatory cells, platelets, and immune complex deposition are occasionally seen [7].

In vitro studies showed that infected red blood cells contain a membrane protein named *P. falciparum* erythrocyte membrane protein-1 (PfEMP-1). PfEMP-1 is a well-recognized virulence factor, and this family of proteins enables antigenic variation and interacts with endothelial ligands on capillary walls such as intracellular adhesion molecule-1. This interaction is thought to initiate a complex series of signaling events leading to sequestration of red blood cells in vital organs such as the brain, liver, and placenta [8]. Parasites that cause severe malaria express a different subset of these proteins than those that cause uncomplicated infections [4]. This led to the hypothesis that cerebral malaria was the result of mechanical obstruction and hypoperfusion, but the striking lack of strokes as a part of this syndrome suggests another etiology. Further investigations regarding the role of the blood brain barrier in pathogenesis have revealed little conclusive evidence for breakdown of the blood-brain barrier (BBB) in cerebral malaria. Primate and murine models suggest an immunopathological etiology, but these models differ from human pathology in that they do not show the classic sequestration of red blood cells in the capillaries in the brain [8].

In endemic regions, repeated exposure to infection results in clinical immunity that limits the inflammatory response to infection, as well as antiparasite immunity that results in parasite death or inhibition of replication. Interestingly, cerebral malaria occurs primarily in children who have already acquired a significant degree of antimalarial immunity as evidenced by their low parasite densities and resistance to severe anemia. This suggests that cerebral malaria

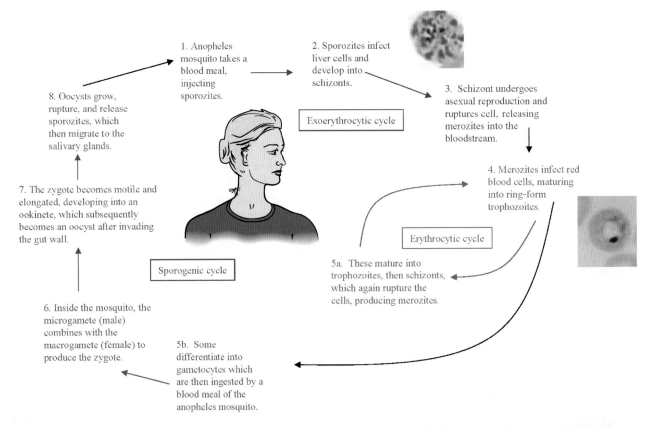

Figure 2 Life cycle of *P. falciparum*. (Adapted from DPDx: CDC's Web site for parasitology identification; http://www.dpd.cdc.gov/dpdx.)

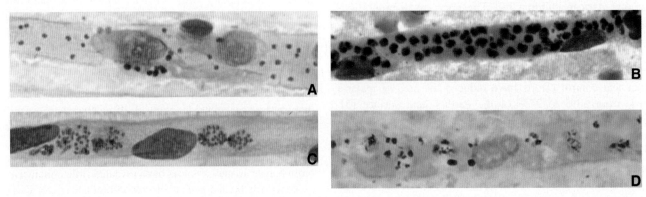

Figure 3 Red blood cells infected by malaria sequestered in cerebral capillaries. (Images courtesy of Dan A. Milner, Jr., MD, The Blantyre Malaria Project and Brigham & Women's Hospital, Boston, Massachusetts.)

results from a dysregulated immune response rather than direct damage from the infection. Prior infections serve to prime the immune system. However, adults from nonendemic regions also develop severe malaria and cerebral malaria with death rates higher than in children. Here a cross-priming effect has been hypothesized as experiments show that malaria-responsive T-helper 1 cells cross-react with antigens including *Toxoplasma gondii*, tetanus toxoid, adenovirus, and mycobacterial, fungal, and streptococcal antigens [9]. The pathogenesis of cerebral malaria remains incompletely understood.

4. Pathophysiology

In a nonimmune host, fever is almost always present and is accompanied by nonspecific symptoms such as chills, malaise, headache, myalgias, cough, and gastrointestinal symptoms. Splenomegaly is frequently found on exam. Headache, arthralgias, and chills are common complaints in children and are accompanied by anemia, splenomegaly, and hepatomegaly. In endemic areas, as many as 80% of people with parasitemia will be completely asymptomatic [6].

The neurological manifestations of *P. falciparum* infection include headache, agitation, psychosis, seizures, and impaired consciousness. Cerebral malaria is the most severe neurological complication and is characterized by a diffuse encephalopathy, often with brainstem or other focal neurological signs. The definition of cerebral malaria suggested by the World Health Organization is unarousable coma in the presence of asexual parasitemia with the exclusion of other causes of encephalopathy such as post-ictal state or hypoglycemia [8].

In cerebral malaria, adults usually become unconscious after several days of nonspecific feverish symptoms, although the progression has been as rapid as a few hours and deaths have been reported within 24 hours of symptom onset. Coma may develop following a generalized seizure and typically persists for 24–72 hours. Meningeal signs are rare. Disorders of conjugate gaze, retinal hemorrhages, convergent spasm, ocular bobbing, horizontal and/or vertical nystagmus, and sixth nerve palsies are occasionally observed. Bruxism and a brisk jaw jerk may be present. Pout reflexes can be seen, although other primitive reflexes are typically absent. Motor examination reveals symmetrical upper motor neuron dysfunction and may be accompanied by opisthotonos, decorticate, or decerebrate posturing. Occasionally, generalized extensor spasms may be accompanied by upward deviation of the eyes, pouting, and periods of stertorous breathing resembling dyskinesia [5]. Magnetic resonance imaging (MRI) findings are minimal, although increased brain volume without edema and decreased cerebral blood flow have been observed. Cerebral edema is usually only present during agonal stages [7].

In children, severe falciparum malaria presents with impaired consciousness, metabolic acidosis, respiratory distress, or severe anemia. Children have a higher incidence of seizures. Cerebral malaria in children presents with a 1- to 4-day history of fever and focal motor or generalized tonic clonic convulsions, followed by coma. Seizures, and in particular, prolonged seizures are associated with a poor outcome. Children typically regain consciousness after 48–72 hours with treatment. Most deaths occur within 24 hours of starting treatment and are accompanied by brainstem signs consistent with transtentorial herniation. Opening cerebrospinal fluid (CSF) pressures are usually elevated, and invasive monitoring has shown elevated intracranial pressures, thought to be secondary to increased cerebral blood volume secondary to sequestration. Computed tomography (CT) scan shows cerebral edema [7].

Several postinfectious syndromes have been described, including cerebellar ataxia and a postmalaria neurological syndrome consisting of an acute confusional state, acute psychosis, generalized convulsions, and tremor [7].

5. Natural History

The prognosis for cerebral malaria is poor. Adults usually die from noncerebral complications such as pulmonary edema and renal failure. Mortality is near 20% [7]. Among children, mortality ranges from 15–30%, and approximately 10–30% of children have neurological deficits at discharge. Some deficits (e.g., cerebellar ataxia) are transient, but others (e.g., hemiparesis or cortical blindness) may improve but not completely resolve. Recent studies suggest that as many as 25% of surviving children have persistent, long-term cognitive deficits. Increased rates of epilepsy are also seen [8].

B. Human African Trypanosomiasis (Sleeping Sickness)

1. Brief History

Although initially described in 1803, it was not until 1894 that investigators discovered the link between the fly disease of hunters, the disease of "negro lethargy," and *nagana*, a disease of cattle. Shortly thereafter, their relationship to the tsetse fly was described [1]. Human African trypanosomiasis (HAT) occurs in 36 countries in sub-Saharan Africa. Approximately 10 million km^2 are infested by the tsetse fly vector, an area equivalent to one-third of the land mass of Africa. The widespread decimation caused by the disease has left large areas of land uninhabited [10].

Large epidemics in the early 20th century lead to aggressive control programs, and by the 1950s HAT was almost brought under control. However, in recent years this disease has reemerged. War in countries such as Angola, the Republic of Congo, and Sudan has led to population migrations and disruption in control efforts, which in turn resulted in increasing incidence of this devastating disease. For example, in the Republic of Congo during the late 1980s, only 10,000 cases per year were reported. However, in the early 1990s, aid was interrupted because of political conflicts, and the annual incidence rose to 30,000 new cases per year. Other factors contributing to the reemergence of HAT include inadequate financing of treatment, increasing drug resistance, changes in climate and vegetation, and migration of animal reservoirs [10].

The potential for HAT to cause large epidemics and its negative impact on economic development increases the public health importance of this disease despite moderate incidence rates. Further, HAT ranks third among African parasitic diseases when its impact is measured in disability adjusted life-years. Currently an estimated 60 million people worldwide are at risk, with an annual incidence of 300,000 cases. Because less than 10% of people at risk are under active surveillance, the true incidence is likely underestimated [11].

2. Etiology

HAT is caused by the extracellular protozoan *Trypanosoma brucei rhodesiense* (eastern and southern Africa) and *Trypanosoma brucei gambiense* (west and central Africa). *T. b. rhodesiense* is closely related to *T. b. brucei*, which only affects animals. The human infectivity of *T. b. rhodesiense* is conferred by a serum resistance–associated (SRA) gene product that makes the parasite resistant to lytic activity in human serum. *T. b. gambiense* is distinct genetically and biochemically from the other two species. Its mechanisms of infection and resistance to lysis are unknown [12].

Infected feral animals or domestic animals such as cattle act as reservoirs for the human disease. The trypanosome is taken up by a blood meal of *Glossina* female tsetse fly from a human or animal and undergoes a complex cycle of morphological and biochemical development. Flies become infectious about 21 days after feeding on an infected host [11]. The tsetse fly again takes a blood meal and injects the trypanosomes via its saliva into the skin of the human. The parasite is taken up by local lymph nodes and enters the bloodstream (Fig. 4) [11,13].

3. Pathogenesis

During the hemolytic stage of the disease, trypanosomes can be identified on thin and thick blood smears. *T. b. rhodesiense* is more reliably found given its relatively stable parasitemia, whereas *T. b. gambiense* is less reliably found as its clinical course is characterized by intermittent waves of parasitemia. Although the opsonized trypanosomes are easily cleared, trypanosomes have developed an elegant method to evade the immune system. The cell membrane of each trypanosome is composed of a single variant surface glycoprotein (VSG). The parasite is able to switch VSG molecules at a rate of 1–2% per generation, and the genome has a repertoire of more than 1000 VSG [11,12]. By varying the antigen presented on its cell membrane, the trypanosome evades detection by the immune system, thus allowing the waves of parasitemia that characterize the slow progression of disease in *T. b. gambiense*.

The VSG and its anchor to the cell membrane, glycoinositol phosphate (GIP), elicit both T-cell and independent B-cell responses and also act as potent macrophage activators. The antibody response to VSG stimulates polyclonal B-cell activation and generates autoantibodies and immune complex disease. The inflammatory response likely results in the nonspecific constitutional symptoms associated with the initial stages of the disease [13].

Trypanosomes circulate in the bloodstream, lymphatics, and CSF but are not thought to cross the BBB [12]. Neither human nor mouse models of disease have shown trypanosomes in brain parenchyma except in areas thought to be outside the BBB, such as dorsal root ganglia and circumventricular organs, including the median eminence, posterior pituitary, and possibly the hypothalamus. These sites in the central nervous system (CNS) show the earliest localization of parasites, induction of proinflammatory cytokines, and activation of microglia. Involvement of the dorsal root ganglia likely leads to the sensory hyperesthesia commonly seen in late stages of the disease, and involvement of the hypothalamus likely results in the clinical alterations in circadian rhythms and sleep architecture [13,14].

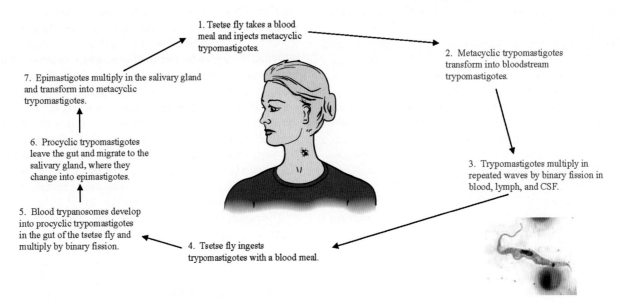

Figure 4 Life cycle of human African trypanosomiasis. (Adapted from DPDx: CDC's Web site for parasitology identification; http://www.dpd.cdc.gov/dpdx.)

Pathological specimens from autopsy cases and mouse models of the late stages show a meningoencephalitis with cellular proliferation in the leptomeninges and diffuse white matter infiltration consisting of lymphocytes, plasma cells, and macrophages. Perivascular cuffing is prominent, and the adjacent parenchyma contains activated astrocytes and macrophages, as well as the pathognomonic morular or Mott cells, which are modified plasma cells with eosinophilic inclusions containing immunoglobulin (Ig)-M [11].

The pathology in the CNS is thought to result from dysregulated inflammatory immune responses to antigenic stimulation rather than direct effects of the trypanosome. In short, HAT stimulates proinflammatory cytokines via the VSG and GIP, resulting in damage to the CNS, but also stimulates antiinflammatory cytokines such as IL-10 that decrease the ability of the immune system to clear infection. Interestingly, in the late stages of disease, levels of proinflammatory cytokines such as interferon (IFN)-gamma and tumor necrosis factor (TNF)-alpha are decreased, although initially both correlate with disease severity [12].

4. Pathophysiology

a. Initial Stage A chancre often forms at the site of the bite approximately 5–15 days postinoculation. The chancre is thought to represent a local inflammatory response to trypanosomes and Glossina products and subsides after about 4 weeks. It is most commonly found in patients who originate from nonendemic areas [12].

b. Early or Hemolytic Stage One to three weeks after the bite, patients present with 1–7 days of fever and generalized lymphadenopathy. Other symptoms are nonspecific and include general malaise, headache, arthralgia, weight loss, and weakness. Physical examination often shows marked generalized lymphadenopathy, hepatomegaly, and splenomegaly, and laboratory evaluation shows anemia [12]. During this period, many organs are infected including the spleen, liver, skin, cardiovascular system, endocrine system, and eyes. In some cases, disseminated intravascular coagulation and multiple hemorrhages may develop. In severely ill patients with *T. b. rhodesiense* infections, death from myocarditis may occur [11].

c. Late or Meningoencephalitic Stage The onset of the late stage of infection is insidious, and the clinical spectrum is broad. Typical clinical features include psychiatric/cognitive changes, motor and sensory abnormalities, and sleep abnormalities. Psychiatric and cognitive symptoms include irritability, lassitude, headache, personality changes, and frank psychosis with violence and/or hallucinations. Motor findings include limb tremors, fasciculations of the tongue and limbs, hypertonia, pyramidal weakness, choreiform and athetoid movements, dysarthria, cerebellar ataxia, and polyneuritis. Pout and palmomental reflexes may be present. Sensory symptoms are characterized by a painful hyperesthesia and Kerandel's sign. Kerandel's sign consists of a delayed deep hyperesthesia after a slight blow to a bony projection [12].

Lumbar puncture during late phases shows a lymphocytic pleocytosis, elevated protein, elevated intrathecal IgM, and trypanosome deoxyribonucleic acid (DNA) via polymerase chain reaction (PCR) [11]. In addition, specific IgG/IgM and polyclonal IgM responses are found frequently in CSF [12]. MRI findings are nonspecific, showing diffuse asymmetrical white matter abnormalities, diffuse hyperintensities in the basal ganglia, and ventricular enlargement.

The onset of sleep changes heralds the terminal stages of HAT. Sleep abnormalities are characterized by spontaneous uncontrollable urges to sleep and reversal of the normal sleep/wake cycle. Sleep monitoring shows that the disease process is not characterized by hypersomnia but instead by loss of circadian rhythms resulting in diurnal somnolence and nocturnal insomnia. The diurnal rhythm of the sleep/wake cycle disappears progressively. During late stages, rapid eye movement (REM) sleep occurs throughout the entire sleep/wake cycle; latency to slow wave sleep and REM sleep is decreased; and sleep onset REM occurs frequently, often without any intermediary non-REM sleep. Alterations in circadian fluctuations of cortisol, prolactin, growth hormone, and plasma renin activity have been described, although melatonin seems unaffected. Specific studies of the suprachiasmatic nucleus (SCN) in infected rats showed abnormalities in its responsiveness to light day cycles and light pulses, as well as alterations in intrinsic SCN firing with decreased spontaneous postsynaptic excitatory activity [11,14].

d. Post-Treatment Reactive Encephalopathy (PTRE) When the trivalent arsenic melarsoprol is used to treat the late stages of HAT, approximately 10% of patients develop a posttreatment reactive encephalopathy consisting of seizures, coma, cerebral edema, and fever. PTRE itself has a 50% mortality. Pathological specimens of PTRE in humans and in mouse models show exacerbation of the inflammatory meningoencephalitis seen in untreated disease. The cause of PTRE is not known, although hypotheses include subcurative chemotherapy, abnormal immune responses to killed parasites, immune complex deposition, arsenic toxicity, and other autoimmune mechanisms [11].

5. Natural History

Wide variation occurs both in the presentation and progression of disease. Asymptomatic carriers can be found with both species. In general, *T. b. gambiense* presents as a chronic infection with progression over months to the late stage, which itself can last from months to years. The clinical course of *T. b. rhodesiense* infection is much more rapid, progressing to the late stage over several months and often leading to death within 3 months. However, several populations in Malawi have been found to have mild infections and chronic disease evolution even with *T. b. rhodesiense* [12]. The varied clinical course results from a combination of parasitic virulence and host resistance. If untreated, the disease progresses to the final stages in which patients ultimately develop seizures, severe somnolence, double incontinence, cerebral edema, coma, systemic organ failure, and death [11,12].

C. Amoebiasis

1. Brief History

Globally, amoeba are among the most widely distributed of the parasitic organisms, although infections of the CNS are relatively rare and are of three types: amebic granulomatous encephalitis is caused by *Acanthamoeba* species, primary amoebic meningoencephalitis is caused by *Naegleria fowleri*, and brain abscesses are rare complication of *Entamoeba histolytica* infections.

2. Amebic Granulomatous Encephalitis

Amebic granulomatous encephalitis is caused by *Acanthamoeba* species and rarely by *Balamuthia mandrillaris*. *Acanthamoebae* are distributed worldwide and have been isolated from soil, dust, and air, as well as natural, treated, and sea water. They are found in swimming pools, sewage, sediment, air conditioning units, tap water, bottled water, drinking water treatment plants, dental treatment units, hospitals and dialysis units, eyewash stations, contact lenses, and contaminants in bacterial, yeast, and mammalian cell cultures. *Acanthamoebae* have been isolated from fish, amphibians, reptiles, mammals, and the nasal mucosa of asymptomatic humans. Despite frequent exposure, infection is rare and typically affects only immunocompromised patients. Amebic keratitis is the exception and is primarily seen in immunocompetent patients [16–18].

a. Etiology The life cycle consists of a dormant cyst and an actively feeding and dividing trophozoite. The cyst forms under adverse environmental conditions such as food deprivation, desiccation, and changes in temperature and pH. Cysts have been shown to retain infectivity for up to 24 years, although their virulence appears to decrease over time. The trophozoite feeds on bacteria, algae, and yeast. Uptake of food can occur by pseudopod formation and phagocytosis, or by food cup formation and ingestion of particulate matter. Locomotion is sluggish and involves the formation of a hyaline pseudopodium [16–18].

b. Pathogenesis The portal of infection can be through the nasal passages, lungs, cornea, or skin lesions. The incubation period is unknown, but at least several weeks or possibly months are thought to pass before clinical signs become apparent. The infection is spread to the CNS hematogenously or directly by extension through the nasal epithelium [16–18].

Pathological findings vary according to host immune status. In immunosuppressed patients, findings include severe hemorrhagic necrosis, fibrin thrombi, and inflammation. The cerebral hemispheres show moderate to severe edema. Multifocal lesions are present in the midbrain, brainstem, corpus callosum, and cerebellum. A chronic inflammatory exudate can be found over the cortex and consists of polymorphonuclear leukocytes and mononuclear cells. Severe angiitis with perivascular cuffing by lymphocytes can also be observed. Numerous trophozoites can be identified within the tissue (Fig. 5) [16–18]. By contrast, well-developed granulomas are found in immunocompetent individuals. Amebic forms are fewer and may include cyst forms. Rarely localized mass lesions have been described.

The extensive tissue damage seen in immunosuppressed patients results partly from the phagocytic processes and cytotoxic substances such as proteases and elastases released by the amoeba. The adherence factors that allow invasion in the host appear to directly affect virulence. However, amoeba also stimulate antibody formation and activate complement. Neutrophils, macrophages, and microglial cells migrate to areas of infection, but only the latter two are able to phagocytose and destroy amoeba [16–18].

c. Pathophysiology *Acanthamoeba* spp. produce a number of clinical diseases, including amebic granulomatous encephalitis; amebic keratitis; and nasopharyngeal, cutaneous, and disseminated infections. Although some cases have been reported in immunocompetent children, most occur in the immunocompromised, such as patients with human immunodeficiency virus (HIV)/acquired immune deficiency syndrome (AIDS), with organ transplants, those who are taking corticosteroids, or those who are otherwise debilitated [16–18].

Symptoms include headache, confusion, nausea, vomiting, fever, lethargy, stiff neck, focal neurological deficits, or signs of increased intracranial pressure. CSF findings show pleocytosis with abundant lymphocytes and neutrophils, low glucose, and high protein. Wet preparation of the CSF reveals the organisms that are often confused with macrophages. Imaging can show enhancing or nonenhancing lesions resembling abscesses or brain tumors [16–18].

d. Natural History There are few reports of successful treatment, although attempts with multidrug regimens with trimethoprim-sulfamethoxazole, ketoconazole, and rifampin have had some success in immunocompetent patients. For rare localized lesions, surgical excision has been successful [16–18].

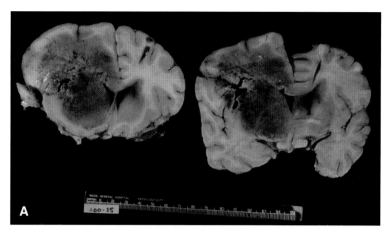
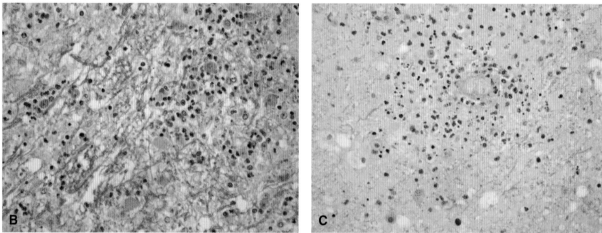

Figure 5 A, Gross pathology of granulomatous amebic encephalitis shows extensive severe hemorrhagic necrosis in bilateral frontal lobes. B, Microscopic pathology shows numerous amoeba and extensive inflammation and edema in the medulla of the same patient. C, Many amoeba and severe angiitis leading to a necrotic blood vessel is seen in the cingulate gyrus of the same patient. (Images courtesy of E. Tessa Hedley-Whyte, MD, Director of Neuropathology, Professor of Pathology, Massachusetts General Hospital, Boston, Massachusetts.)

3. Primary Amebic Meningoencephalitis

a. Etiology *Naegleria fowleri* is the causal agent of primary meningoencephalitis. *Naegleria* species are amoebo-flagellates, found worldwide in soil and water, and are not as ubiquitous as *Acanthamoeba*. *Naegleria* grows best at somewhat elevated temperatures, and this amoeba has been isolated from warm freshwater bodies including manmade lakes and ponds, hot springs, and thermally polluted streams and rivers. They are more sensitive to environmental conditions, such as drying and pH extremes, and cannot survive in seawater [15].

Most infections have occurred as a result of people swimming or diving in manmade bodies of water, disturbed natural habitats, or areas in which soil and unchlorinated/unfiltered water are in contact. A few cases have been reported after bathing in hot springs or washing with contaminated water. Young, healthy individuals are most commonly affected [15].

The *Naegleria* life cycle includes an invasive reproductive trophozoite (amoeboid form), a transient, motile, flagellate, pear-shaped form that appears when the amoeba is placed in water, and a double-walled cystic stage that can survive in lower temperatures while maintaining infectivity. The amoeba feeds on bacteria and other organic matter [15].

b. Pathogenesis *N. fowleri* have been isolated from the nasal mucosa of healthy asymptomatic children and from the nasal passages of patients with primary amebic meningoencephalitis. The trophozoite and cysts are able to penetrate the mucosal epithelial layer and migrate along the olfactory nerve tracts, crossing the cribiform plate and entering the brain. In children the cribiform plate is more porous than in adults, which presumably puts them at higher risk for infection [15].

The frontal and olfactory lobes of the brain are initially involved, but the infection rapidly spreads to the base of the brain, brainstem, and cerebellum. In addition, amoeba can be found in the subarachnoid space of the meninges as well as in perivascular areas. Imaging during the course of the disease often reveals meningeal and basilar enhancement as well as hypodense areas suggestive of areas of cerebral infarction. Severe cerebral edema is a late finding and suggests poor outcome.

Pathological examination reveals an edematous cortex with numerous petechial hemorrhages. The olfactory bulbs and tracts are hemorrhagic and necrotic. Grossly congested vessels are found in the pia and arachnoid matter as well as the cerebral hemispheres. Microscopic examination reveals that the gray matter is widely infiltrated with fibrinous purulent exudates apparent along the leptomeninges. The exudate contains neutrophils, monocytes, and lymphocytes as well as small numbers of amoebae. However, amoebae are present in far greater numbers in the perivascular spaces [15].

c. Pathophysiology Symptoms develop days after exposure and include severe headache, nausea, vomiting, fever, and behavioral abnormalities. The meningoencephalitis follows a fulminant course with rapid progression to coma. The infection is usually fatal within 1–2 weeks [15].

d. Natural History No more than a dozen survivors have been reported of an estimated 200 cases, despite aggressive therapy including intrathecal and intravenous amphotericin B, miconazole, and rifampin [15].

4. Amebic Brain Abscess

Entamoeba histolytica is the second leading cause of death from parasitic disease worldwide. Ingestion of the quadrinucleate cyst of *E. histolytica* from fecally contaminated food or water leads to intestinal amoebiasis. The amoebae then invade through the intestinal mucosa, causing amoebic colitis and bloody diarrhea. They can invade through the portal circulation and cause liver abscesses. Liver abscesses are rarely accompanied by the development of brain abscesses (< 0.1% of liver abscess cases). In these cases, more than half of patients have rapidly progressive disease that results in death, although outcomes have improved with use of metronidazole [19].

D. Toxoplasmosis

1. Brief History

Toxoplasmosis is caused by the widely distributed obligate intracellular protozoan parasite *Toxoplasma gondii*. Over 500 million of the world's population have serum antibodies to *T. gondii*, although most infected adults are asymptomatic. Reactivation of latent disease in immunocompromised patients results in disseminated toxoplasmosis and toxoplasmic encephalitis. These conditions are still rare but are increasing in incidence with the rise of the HIV/AIDS pandemic [20].

Congenital toxoplasmosis is found more commonly and has a birth prevalence from 1–10% to per 10,000 live births. Maternal infection in the first or second trimester may lead to severe congenital toxoplasmosis and intrauterine death, whereas late maternal infections usually result in normal appearing newborns. About 85% of newborns with congenital toxoplasmosis have subclinical infection [20].

2. Etiology

T. gondii is a remarkably clonal species and only three clonal lineages have been identified that differ in virulence and epidemiology. Types I and II have been recorded in congenital disease, type II in patients with AIDS, and type III in strains isolated from animals [20]. Factors including inoculum size, virulence of the organism, genetic background of the host, sex, and immunological status affect the course of infection in humans and animal models [20].

The definitive hosts of *T. gondii* are members of the cat family. Sexual replication of the parasite occurs in the intestine, resulting in the production of large numbers of oocysts in the feces. Sporulation takes 7–21 days, after which the oocysts become infective when ingested by mammals and develop into tachyzoites. The tachyzoite is the rapidly multiplying asexual form of the parasite and is able to invade all nucleated cells. Replication leads to death of the host cell, as well as neighboring cells, and incites a strong inflammatory response. This immune response stimulates the production of cysts containing bradyzoites, which then are undetectable by the immune system. The bradyzoites slowly multiply within the cysts and can be re-released and transform into tachyzoites, causing recrudescence of infection. Cysts can form in brain, skeletal, and heart muscle and are infective if consumed [20].

Infection is transmitted by contact with cat litter or the ingestion or handling of undercooked or raw meat that contains tissue cysts, especially pork and lamb. Ingestion of water or food contaminated with oocysts also results in infections. Recently, transmission via transplanted organs has been documented. Maternal-fetal transmission occurs when the mother develops her primary infection during gestation. Prior maternal infection, however, poses little or no risk to the fetus [20].

3. Pathogenesis

After oral ingestion, the oocyst actively invades intestinal epithelial cells or is phagocytosed, resulting in the release of tachyzoites. Various proteins including host laminin, parasite laminin, lectins, and surface proteins have been implicated in the mediation of preliminary attachment. Two specialized organelles, rhoptries and micronemes, appear to assist the parasite in entering the host cell. Once it has entered the intracellular space, *T. gondii* induces the formation of a parasitophorous vacuole, which uses secreted parasite proteins to exclude host proteins that would promote phagosome maturation and lysosome fusion. In that vacuole, *T. gondii* then replicates in a geometrical fashion and finally bursts from its dying host cell [21]. An intense inflammatory response is thus generated and within 2 weeks after infection, IgG, IgM, IgA, and IgE antibodies can be detected.

Infection induces a strong and persisting T-helper-1 response characterized by production of proinflammatory cytokines such as interleukin (IL)-12, IFN-gamma, and TNF-alpha. Together these protect the host against rapid replication of tachyzoites. In murine models, dendritic cells are the main activators of the T-helper-1 response, although granulocytes also contribute. The activated macrophage is the primary effector cell, with assistance from CD4+ and CD8+ T lymphocytes [21].

Murine models suggest that IFN-gamma plays a major role in the resistance to infection with *T. gondii*. It is produced by a variety of cells in response to *T. gondii* infection and stimulates cytotoxic effector cells including CD4+ and CD8+ T lymphocytes, macrophages, microglia, and astrocytes. These in turn directly interfere with tachyzoite replication and clear the infection. IFN-gamma also stimulates the production of the enzyme IDO (indolamine 2,3 dioxygenase), which depletes intracellular tryptophan pools and decreases the availability of intracellular iron, thus slowing the intracellular replication of tachyzoites. Finally, IFN-gamma stimulates the production of IGTP, a recently identified family of six IFN-responsive genes. Their gene products localize to the endoplasmic reticulum and appear to interfere with the processing or trafficking of immunologically relevant proteins [3]. In summary, IFN-gamma is the central mediator of resistance to toxoplasmic encephalitis [22].

The pathology of toxoplasmosis differs depending on the site of infection and immune status of the patient. In active CNS lesions, multiple enlarging foci of necrosis and microscopic microglial nodules are observed. Necrotic areas frequently calcify. In patients with AIDS, multiple abscesses are observed. Both cerebral hemispheres are involved, although *T. gondii* has a predilection for the basal ganglia. Cysts and tachyzoites are seen, but only the latter is pathognomonic of infection. Tachyzoites are observed within and in proximity to necrotic areas, as well as in glial nodules, perivascular regions, and unaffected regions of the brain [21]. In infants, periaqueductal and periventricular vasculitis and necrosis develop, often leading to obstruction of the aqueduct of Sylvius or foramen of Monro, resulting in obstructive hydrocephalus [21].

Toxoplasmic lymphadenitis in immunocompetent patients is characterized by reactive follicular hyperplasia, irregular clusters of epithelioid histiocytes that blur the margins of the germinal centers, and focal distension of sinuses with monocytoid cells. Giant cells, granulomas, and parasites are notably absent [21]. Acute chorioretinitis in immunocompetent patients is accompanied by severe inflammation and necrosis. Granulomatous inflammation of the choroid is frequently observed and is thought to be secondary to the necrotizing retinitis. Rarely, tachyzoites and cysts can be observed in the retina. Recurrent chorioretinitis frequently occurs secondary to either a hypersensitivity reaction or cyst rupture and reinfection [21].

Additionally, biopsy-proven polymyositis and myocarditis have been documented in both immunocompetent individuals, as well as those taking corticosteroids [21].

4. Pathophysiology

The clinical presentation of toxoplasmosis is quite varied. Primary infection in children, pregnant women, and other adults is usually asymptomatic. Approximately 10% of primary infections can cause a self-limited and nonspecific illness. The usual clinical manifestation is a nontender, nonsuppurative isolated cervical or occipital lymphadenopathy lasting 4–6 weeks. A chronic fluctuating lymphadenopathy that lasts for months occurs occasionally. Rarely, immunocompetent individuals can develop myocarditis, polymyositis, pneumonitis, hepatitis, or encephalitis [20]. Toxoplasmic chorioretinitis can be seen as a result of acute infection or reactivation of congenital disease. Findings include white focal lesions with an overlying and intense vitreal inflammatory reaction leading to a "headlight in the fog" appearance. Recurrent lesions are typically found at the borders of chorioretinal scars [20].

In immunocompromised patients, the disease can be life threatening and is almost always reactivation of a latent infection. The CNS is the most commonly affected site. Toxoplasmic encephalitis begins with a subacute and gradual increase in symptoms over several weeks, often developing into a more acute presentation of a confusional state with or without focal neurological signs. Meningeal signs are rare; however, constitutional symptoms such as fever and malaise may be present. Toxoplasmic encephalitis can be associated with a chorioretinitis, pneumonitis or multiorgan involvement leading to acute respiratory failure and hemodynamic abnormalities similar to septic shock [20]. In patients with HIV or AIDS, abscesses may form that are difficult to distinguish on imaging from primary CNS lymphoma (Fig. 6).

In congenital disease, abnormalities observed in utero include intracranial calcifications, ventricular dilation, hepatic enlargement, ascites, and increased placental thickness. Neonatal manifestations include hydrocephalus, microcephaly, intracranial calcifications, chorioretinitis, strabismus, blindness, seizures, psychomotor or mental retardation, petechiae due to thrombocytopenia, or anemia.

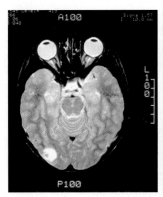
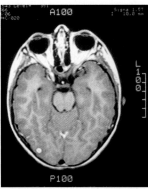

Figure 6 Thick-walled enhancing toxoplasmic abscess is seen in adjacent sections of the right parietal lobe on contrast-enhanced MRI scan. (Image courtesy of Michael J. Potchen, Clinical Director, Michigan State University Department of Radiology, East Lansing, Michigan.)

The classic triad of chorioretinitis, hydrocephalus, and cerebral calcifications is rare [20]. Grossly and by imaging, these brain deformities resemble those caused by other TORCH infections.

5. Natural History

Toxoplasmic lymphadenitis and chorioretinitis in immunocompetent patients typically resolve on their own without treatment. However, if the retinal lesions are in close proximity to the fovea or there is a severe inflammatory response, treatment is typically initiated. Treatment of pregnant women with evidence of acute infection during pregnancy decreases the incidence of neonatal toxoplasmosis, and the infant should be treated throughout the first year of life [20].

In immunocompromised patients, however, intensive multidrug treatment with pyrimethamine, sulfadiazine, and folinic acid is standard for induction. Clindamycin can be used instead of sulfadiazine in sulfadiazine-intolerant patients. After the induction phase, maintenance with half-dose should be continued for the life of the patient, or until the CD4 count is greater than 200 and the viral load is well controlled in patients with HIV or AIDS [20].

III. Helminths: Cestodes

A. Neurocysticercosis

1. Brief History

Neurocysticercosis has a worldwide distribution and is endemic in most of Latin America, Africa, and Asia. Immigration of carriers has led to increasing incidence in Europe and North America. In endemic regions, 30–60% of pigs and 10–25% of humans are seropositive, although fewer show clinical signs of the disease. Further, neurocysticercosis is the most common cause of acquired epilepsy worldwide [23]. Epilepsy rates in Latin America and Africa are three to six times greater than the United States and are thought to result largely from increased rates of neurocysticercosis. For example, surveys in neurological centers in Mexico and Peru show that 30–50% of individuals with epilepsy show evidence of neurocysticosis, compared with 5% of individuals with other neurological symptoms [23,24].

2. Etiology

In humans, *Taenia solium* infections result in two clinical syndromes: taeniasis and cysticercosis. Taeniasis, or intestinal infection with the adult tapeworm, is a benign disease that often does not come to medical attention and results from ingestion of the tapeworm larvae (i.e., cysticerci) in uncooked or poorly cooked meat. Symptoms (if present) are mild and include abdominal pain, distension, diarrhea, and nausea. Patients often do not notice passage of tapeworm segments in the stool because these segments are smaller and less motile than those of other tapeworms. Poor sensitivity of stool microscopy makes diagnosis difficult [23].

Human cysticercosis represents infection with the larval stage of *T. solium* and can be acquired by egg ingestion via fecal-oral contamination from tapeworm carriers or possibly by internal autoinfection via regurgitation of proglottids into the stomach. Although autoinfection has not been definitely proven, nearly 15% of patients with neurocysticercosis harbor a tapeworm at the time of diagnosis, and as many as 25% have a history of a tapeworm. Infection of vegetarians and persons who keep a kosher diet have occasionally resulted from having a carrier in the household [23].

After ingestion of the eggs, the embryos or oncospheres are liberated by the action of gastric acids and actively cross the bowel wall, enter the bloodstream, and are carried to other tissues including the subcutaneous tissues, muscles, heart, brain, and eyes [23,24].

3. Pathogenesis

In neurocysticercosis, the brain parenchyma is most likely seeded via hematogenous dissemination, and the ventricular system and subarachnoid space are likely infected via the choroid plexus. Cysts are uniformly round or oval vesicles, varying in size from a few millimeters to 1–2 centimeters. The most common location is in the cerebral hemisphere at the gray-white junction. Cysts may also be found in the cerebellum, ventricles, brainstem, subarachnoid space, basal cisterns, and extramedullary or intramedullary spinal cord. The number can range from a single cyst to more than one thousand cysts. Intraventricular cysts tend to be free floating and are found more frequently in dependent positions such as the fourth ventricle. The rare racemose variant of neurocysticercosis consists of a large lobulated vesicle without a scolex (larvae), which develops at the base of the brain or in the ventricles. This form is associated with high mortality because it grows quickly, filling the basal cisterns or ventricles, resulting in obstructive hydrocephalus [23,24].

The cysts develop over time at individual rates. Intracerebral cysts are traditionally classified into four groups. The initial or vesicular stage is characterized by noncystic embryos that lodge at the gray-white junction, basal ganglia, or cerebellum and that elicit little response from the surrounding tissue. As the embryo develops into cystic larvae, edema in the surrounding tissue develops over several weeks. Cysts are thin walled (typically 1–2 cm in diameter) with a mural nodule that represents the invaginated scolex [24].

Over the next several years, the colloid vesicular stage develops and is characterized by degeneration of the cyst, increasing immune response, thickening of cyst wall, and increasingly proteinaceous fluid. On imaging, the cyst wall may enhance. This is followed by the granular nodular stage in which the capsule thickens further and absorption of the cyst fluid begins. Homogenous enhancement with surrounding edema is seen. Finally, as the dead larvae calcify, the nodular calcified stage develops and multiple calcified nodules can be seen at sites of previous cysts [24].

4. Pathophysiology

Cysticerci can be found in the subcutaneous tissues, muscles, heart, and eye. Intraocular cysts are found most frequently in the vitreous humor, although they may also present in the anterior chamber, conjunctiva, extraocular muscles, or retro-ocular space. Visual changes are the result of retinal damage, chronic uveitis, or optic nerve compression, depending on the site of the lesion [23]. Involvement of the CNS is seen in 60–90% of patients and takes many forms: parenchymal brain lesions often resulting in seizures, intraventricular lesions resulting in increased intracranial pressure or hydrocephalus, giant cysts resulting in mass effect, massive infection with acute encephalitis and strokes, and spinal cord lesions [23].

Parenchymal lesions and the resulting seizures are the most common presentation of neurocysticercosis (Fig. 7). Seizures are seen in 50–80% of patients with parenchymal disease. Cysticercosis is thought to cause seizures via a variety of mechanisms. Seizures occur early in disease during periods of intense inflammation associated with viable or degenerating cysts. Later in disease, seizures recur intermittently during episodes of perilesional edema that are poorly characterized. Seizures may occur late in disease as a result of infarcts, encephalomalacia, and gliosis from areas of prior cysts. Finally, the calcified quiescent lesions are also potential epileptogenic foci [25].

Intracranial hypertension and hydrocephalus result from cysts in the cerebral ventricles or basal cisterns, which block the circulation of CSF by mechanical obstruction,

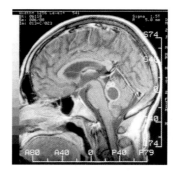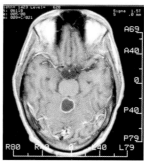

Figure 7 Thin-walled, ring-enhancing cyst of neurocysticercosis seen in the cerebellum on axial and sagittal views of contrast-enhanced MRI scan. (Image courtesy of Michael J. Potchen, Clinical Director, Michigan State University Department of Radiology, East Lansing, Michigan.)

ependymal inflammation, or residual fibrosis. Acute intermittent hydrocephalus (Bruns' syndrome) may result from a mobile intraventricular cyst [23,24]. Giant cysts result in symptoms that are similar to any mass lesion with focal neurological deficits and intracranial hypertension. Stroke is seen primarily in subarachnoid neurocysticosis and consists mostly of deep lacunar infarcts resulting from endarteritis of small penetrating arteries in the brainstem, internal capsule, or subcortical white matter. Occlusion of large vessels and hemorrhagic strokes are rare [23].

In children and teenagers, an acute encephalitic picture can present and is thought to result from a pronounced inflammatory response to a massive infection. This is often associated with marked intracranial hypertension [23]. Spinal cord lesions are rare (1% of cases). Presentation is usually that of spinal cord compression. Lesions in the spinal cord are usually found in the subarachnoid space, although intramedullary cysts have been described [23].

5. Natural History

The natural history of neurocysticercosis has yet to be fully studied. Intraparenchymal neurocysticercosis typically follows a favorable course, with degeneration of the parasites and eventual calcification of lesions. The phenomenon whereby calcified lesions activate with increased perilesional edema many years after initial infection is not well understood, although it often results in intermittent seizures. Some hypothesize that the edema causes the seizures, whereas others postulate that the edema results from the seizure. Both during active infection and the long term, seizures represent the major cause of morbidity from intraparenchymal lesions. Cysticerci in the subarachnoid space and ventricles have increased morbidity and mortality as compared with parenchymal lesions because they can cause acute obstructive hydrocephalus or intracranial hypertension [23,24].

The timing of treatment and determination of which syndromes to treat is the subject of much debate. It is clear that treatment with antiparasitic drugs causes a worsening of symptoms secondary to the inflammatory reaction of the degenerating cysts. Evidence suggests that antiparasitic treatment and steroids should be given to those with moderate infections and viable cysts, subarachnoid cysts, giant cysts, or racemose cysts. Intraventricular cysts, intracranial hypertension, hydrocephalus, and spinal cord lesions require surgical treatment. It remains unclear whether there is benefit in treating patients with massive infections with viable cysts and in patients with many degenerating cysts, given the risk for clinical decline. Finally, treatment seems unnecessary in a patient with only one or a few viable cysts, given the overall favorable course. Treatment with antiparasitic medications should not be given for calcified lesions [23,24].

B. Other Cestodes

See Table 2.

IV. Helminths: Nematodes

A. Strongyloidiasis

1. Etiology

The nematode *Strongyloides stercoralis* is estimated to infect 30–100 million people worldwide and is endemic in Africa, Asia, Southeast Asia, and Central and South America. Decades ago, it was also found in temperate climates such as rural Appalachia. Because this can be a chronic asymptomatic infection enduring decades, a history of travel to or residence in endemic countries, even decades before, is a risk for infection [35].

2. Pathogenesis

Infective larvae penetrate intact human skin, usually via bare feet. The life cycle is more complex than most nematodes and includes both free-living and parasitic cycles, as well as the ability to multiply and autoinfect the host [35].

3. Pathophysiology

Acute infection can often cause a local dermatological reaction that can last as long as several weeks. Pulmonary symptoms such as cough and tracheal irritation occur several days postinfection as larvae migrate through the lungs, followed several weeks later by nonspecific gastrointestinal symptoms including diarrhea, constipation, anorexia, and abdominal pains. Larvae are detectable in the stool after 3–4 weeks [35].

Chronic infection is common with intermittent non-specific gastrointestinal symptoms, larva currens (a rapidly migratory and pruritic linear eruption due to migration of larvae), and peripheral eosinophilia. Gram-negative bacteremia and bacterial infections can be associated with migrating larvae [35].

Disseminated hyperinfection with uncontrolled endogenous worm replication occurs in approximately 1.5–2.5% of patients, although estimates are probably inaccurate because of the large number of undiagnosed infections. Risk factors for disseminated disease include immunosuppressive therapy, transplantation, hematological malignant disease, and human T-lymphotropic virus-1 infection. Glucocorticoids are frequently implicated in disseminated infection. Interestingly, cyclosporine is not associated with disseminated strongyloides infection and in fact seems to have some antihelminthic properties. Other predisposing factors include malnutrition, diabetes mellitus, chronic renal failure, and chronic alcohol consumption. Rare reports have been made in patients with HIV or AIDS [36].

The onset of disseminated infection is varied and can be acute or insidious, with or without fever, usually with no specific constitutional symptoms and often with a suppressed eosinophil count. Infection is usually found in the gastrointestinal tract resulting in ileus or small bowel obstruction, protein-losing enteropathy, or gastrointestinal bleeding. Pulmonary manifestations include cough, choking, hemoptysis, acute respiratory distress syndrome, and respiratory collapse [36].

The CNS may also be involved; larvae have been found in CSF, meningeal vessels, the dura, and in epidural, subdural, and subarachnoid spaces. The most typical presentation is that of an aseptic meningitis with meningeal signs and symptoms and with CSF showing a pleocytosis, elevated protein, and normal glucose. Co-infection with enteric gram-negative rods is also found, as is hyponatremia [35,36].

4. Natural History

After an acute infection, most people develop a relatively asymptomatic chronic infection. Treatment with antihelminthics such as thiabendazole results in clearance of the infection, although refractory infections have been reported in immunosuppressed patients. Disseminated infections develop rarely but carry a 75–80% mortality even with treatment [35,36].

B. Other Nematodes

A variety of nematodes cause various clinical syndromes (see Table 2). Isolated case reports of fatal meningoencephalitis have been reported for *Lagochilascaris minor*, *Micronema deletrix*, *Meningonema peruzzi*, and *Baylisascaris procyonis*. Also, isolated reports of extradural spinal cord compression have been reported as a complication of *Dracunculus* infection. *Oncocerca volvulus* and *Loaloa* involve the eye, often resulting in blindness.

Table 2 Rare Helminthic Infections

Organism	Distribution and Epidemiology	Pathogenesis	Pathophysiology	Natural History
Cestodes				
Echinococcosis (hydatid disease) *Echinococcus granulosum*	Worldwide distribution: endemic in the Middle East, Australia, New Zealand, South America, Turkey, north and east Africa, China, and south and central Europe. Age range 1–75 years old, bulk of infections between ages 4 and 15 years; populations at risk include farm laborers and animal herders [26]. Humans are an accidental host of *Echinococcus* species, tapeworms whose definitive hosts are dogs, wolves or other wild carnivores that are infected by ingestion of meat contaminated with hydatid cysts with viable protoscoleces.	In accidental hosts such as humans, the oncosphere larva is released from the egg and migrates passively through lymphatics or blood vessels to the liver, lung, or other organs. About 5 days after ingestion, the larvae develop into hydatid cysts that induce a granulomatous host reaction. Most cysts are between 1–15 cm in diameter, although larger cysts >20 cm have been reported. Most cysts are univesicular, but smaller daughter cysts can form within the larger mother cysts. More than 90% of cysts occur in the liver, lungs, or both. Occasional reports are made of cysts in the kidney, spleen, peritoneal cavity, skin, and muscles. Very rarely (<1% each) are cysts found in the heart, brain, vertebral column, or ovaries [26,27].	Incubation period likely months to years. Symptoms are due to cyst rupture or mass effect. Systemic immune response possible with cyst leak or rupture. Cerebral cysts may be single or multiple, uni- or multiloculated and thin or thick walled; lesions commonly seen in the MCA territory, frequently involving the parietal lobes. Clinical symptoms include headache, vomiting, seizures, visual disturbances, papilledema, and other focal neurological findings. The cysts grow slowly and often remain asymptomatic until they are quite large [26].	60% asymptomatic; case fatality of 22%.
Diphyllobothrium latum Broad or fish tapeworm	Distribution includes Arctic North America, Northern Europe, Eastern Siberia, Japan, and Argentina. Intermediate hosts: copepods and freshwater fish such as pike, burbot, perch, ruff, and turbot.	Disease becomes evident when passing immature proglottids. It is unclear whether they produce sparganosis in humans.	Disease has various manifestations including marked B_{12} deficiency and diphyllobothriasis with weight loss, diarrhea, paresthesias, impairment of skin sensitivity, gait disturbance, and incoordination.	Not well described
Spirometra species	Distributed widely in eastern and western hemispheres. Found in small intestine of cats and dogs. Humans acquire infection by ingesting contaminated drinking water or food containing larval stage or raw or undercooked flesh containing spargana (second larval stage).	Sparganosis: larvae migrate into subcutaneous or connective tissue of muscles, abdomen, legs, or under the peritoneum or pleura and cause severe inflammatory reactions. Symptoms include urticaria, painful edema, and irregular subcutaneous nodules.	In CNS, can produce severe damage because of size and mobility of larvae, including secondary hemorrhages.	Not well described.
Taenia multiceps	Distribution is primarily in Africa, with a few cases described in Europe and North and South America. Intermediate hosts: herbivores and omnivores. Definitive host: canidae.	An abdominal tapeworm in which humans are an accidental host.	Larvae produce space-occupying lesions and frequently invade the brain producing lethal lesions.	Not well described.

(Continued)

Table 2 (Continued)

Organism	Distribution and Epidemiology	Pathogenesis	Pathophysiology	Natural History
Nematodes				
Angiostronglyiasis *Angiostrongylus cantonensis*	Worldwide distribution but found primarily in Southeast Asia and throughout the Pacific basin (Hawaii, Japan, Indonesia, the Philippines, and Papua New Guinea). Adult worms live and lay eggs in the pulmonary arteries of rats. First-stage larvae hatch and migrate via the trachea and gastrointestinal tract to the feces. Snails and slugs that feed on rodent excrement serve as an intermediary host. Transmission is via the consumption of raw snails, vegetables contaminated with mollusk slime, or carrier hosts (land crabs or freshwater shrimp).	*Angiostrongylus* is neurotropic, and larvae are transported via the bloodstream to the CNS, where they burrow into neural tissue. The migration of larvae leaves minute tracks with few hemorrhages [29]. The clinical symptoms and findings of the disease are secondary to an inflammatory response to invasion of the larvae into the CNS. The inflammatory response is able to kill the larvae but results in an eosinophilic meningitis [29].	Eosinophilic meningitis with or without focal neurological symptoms; paresthesias of the extremities, trunk, or face are a distinctive finding and can persist for weeks to months after other symptoms have resolved; complications include increased intracranial pressure and persistent headaches [29].	Most patients recover completely in 3–6 weeks with only supportive treatment. Heavy infections can lead to a chronic form of the disease, resulting in persistent neurological deficits that last for months. The mortality rate is <1% [29].
Gnathostomiasis *Gnathostoma spinigerum*	Endemic in Southeast Asia, including Thailand, Korea, Japan, and southern China, also identified in Mexico and South America. Transmission occurs via the consumption of raw or inadequately cooked freshwater fish, poultry, or pork. Less frequently, the disease is acquired by drinking water containing infected copepods [29].	Adult worms are present in the stomach of dogs and cats, their definitive hosts. The eggs are released in the feces. Ten to twelve days after reaching water, the eggs hatch and release first-stage larvae. These infect the intermediate host, a small freshwater crustacean of the genus *Cyclops*, and mature into second-stage larval forms. The third stage matures in the second intermediate host: freshwater fish, frogs, snakes, chickens, or pigs. In humans, the third-stage larvae penetrate the gastric wall and migrate throughout the body. The typical presentation is with subcutaneous migratory swellings, but occasionally the larvae reach the nervous system by migrating along the nerve roots. [29].	The clinical symptoms result from mechanical damage from migrating larvae, which leave necrotic tracks in the brain and spinal cord. Subarachnoid, intracerebral and intraventricular hemorrhages are not uncommon. The migration along nerve roots causes radiculitis associated with severe nerve root pain. Paralysis of one or more extremities and urinary retention are common, as is involvement of the cranial nerves. Myelitis and an eosinophilic meningoencephalitis may also develop. CSF is typically xanthrochromic or bloody with an eosinophilic pleocytosis [29].	Antihelminthic agents are not effective and may worsen symptoms due to inflammation surrounding the dying worm. Therefore treatment is primarily supportive. Corticosteroids are sometimes used to relieve symptoms. Permanent focal neurological deficits are common, with 38% of patients in one study demonstrating sequelae of infection. Mortality is due primarily to cerebral hemorrhages and occurs in 7–25% of patients [29].

Toxocariasis *Toxocara canis* *Toxocara cati*	Definitive hosts: the domestic dog and cat. Worldwide distribution. Ingestion of eggs leads to infection in humans, an aberrant host. The life cycle cannot be completed in humans and instead the larvae remain in the juvenile stages and wander through the body for months or even years [32,33].	Most *Toxocara* larvae localize to the liver, lung, or other viscera. Eosinophilic granulomas develop in response to dead and dying larvae, especially in the lungs, liver, eye, and brain. The *Toxocara* are able to shed a variety of excretory-secretory proteins that consist of antigenic mucins as they migrate, and they are thought to vary the secreted mucins in the hopes of evading the immune system [32,33].	Ocular and CNS involvement are rare; eosinophilic meningitis and meningoencephalitis, meningomyelitis, extramedullary space-occupying lesions, vasculitis, and seizures. Ocular larva migrans results in retinal damage and retinal granulomas [34]. Children < 5 years old are most commonly affected.	Visceral involvement usually runs a benign, self-limited course but can be treated with albendazole. Case reports of involvement of the CNS show excellent response to treatment with minimal neurological sequelae [32,33].
Trichinosis *Trichinella spiralis*	Widely distributed and can be found in arctic, temperate, and tropical zones. Most mammals are susceptible to infection, although humans seem particularly susceptible to developing clinical disease after ingestion of undercooked or raw meat contaminated with the larvae. Pigs, horses, sheep, goats, and cattle and wild game have been implicated. More than 20,000 cases have occurred in Europe from 1991–2000; continued surveillance of the food supply worldwide has been a public health priority for some time and is credited with the control of this formerly epidemic disease [37,38].	All stages of the life cycle occur in individual mammalian hosts and can be transmitted by the ingestion of skeletal muscle containing infected larvae. Adult worms persist in the gut for only a few weeks before being expelled, although in immunocompromised patients, they may survive for much longer. Meanwhile, the larvae are able to penetrate the submucosa and are carried via the bloodstream to various organs including myocardium, brain, lungs, retina, lymph nodes, pancreas, and CSF. Only those that reach skeletal muscle survive and encyst after 21–30 days. The larvae then become infective and remain inside a modified "nurse muscle cell" for many years, without causing disease [37,38]. Pathophysiology of CNS infection is not well understood but is thought to result from damage due to wandering larvae that cause a vasculitis and granulomatous inflammatory reaction. Alternatively, neural cells may be damaged by eosinophil degranulation products resulting in a variant of hypereosinophilic syndrome [37,38].	**Enteral phase**: may be asymptomatic; or prolonged diarrhea, nausea, anorexia, malaise, fever, and type I hypersensitivity reaction usually lasting 2–7 days but may persist for weeks. **Parenteral phase** (trichinellotic syndrome): inflammatory and allergic response to muscle invasion. Incubation period ranges from 7–30 days. Symptoms include fever, nausea, diarrhea, myalgias, headaches, and periorbital edema; muscles are stiff, hard, and edematous; symptoms improve over 2–3 weeks [37,38]. CNS involvement has been estimated at 6–24% of cases. Manifestations vary widely and consist of a combination of encephalopathy and focal neurological findings, which can appear cortical, cerebellar, myelopathic, or radicular in distribution. In a review of 55 patients, meningoencephalitic signs were the most common (96%), followed by focal symptoms (73%), and delirium (71%) [37,38].	Severity of disease is affected by the size of the inoculum, sex, age, ethnic group, and immune status of the host. Most infected individuals recover without treatment. The goal of treatment with antihelminthic medications is to minimize skeletal muscle invasion and reduce muscle damage. Treatment with mebendazole and corticosteroids usually results in marked if not complete resolution of symptoms [37,38].

CNS, Central nervous system; *CSF*, cerebrospinal fluid; *MCA*, middle cerebral artery.

V. Helminths: Trematodes

A. Schistosomiasis

1. Brief History

Theodor Bilharz described the parasitic infection that is now known as *schistosomiasis* (formerly called "bilharzia") in Cairo in 1851. As of 2000, more than 200 million people in 74 countries have the disease. Of these, 120 million are symptomatic, and 20 million have severe disease. Travelers to endemic countries are also at risk [40].

2. Etiology

Schistosomiasis is caused by a sexually dimorphic parasitic trematode that resides primarily inside the bloodstream of the mesenteric or pelvic veins. Five species of schistosomes cause human disease: *Schistosoma haematobium, Schistosoma mansoni, Schistosoma intercalatum, Schistosoma japonicum*, and *Schistosoma mekongi* [40]. *S. haematobium, S. mansoni*, and *S. intercalatum* are found in sub-Saharan Africa. *S. mansoni* is additionally found in Brazil, Venezuela, and the Caribbean nations. *S. Japonicum* has largely been eliminated from Japan but is found in China, Indonesia, and the Philippines. *S. mekongi* infection can be found in Cambodia and Laos along the Mekong River [40].

Schistosome infections follow direct contact with fresh water that harbors the free-swimming larval forms of the parasite, called cercariae. Cercariae penetrate the skin and shed their bifurcated tails, and the resulting schistosomula enter capillaries on their way to the lungs. From the lung vessels they reach the systemic circulation, finally settling into the portal or pelvic venous system. There the worms mature and copulate, attaching themselves in pairs to the mesenteric veins or vesicular venous plexus. Egg production begins 4–6 weeks after infection and continues for the life of the worm, usually about 3–5 years but sometimes much longer. The cycle is completed when the hatched eggs pass from the lumen of the blood vessel into the intestinal or bladder mucosa and are passed out of the body in urine or feces. The eggs hatch once they reach fresh water. The larvae, called miracidia, are able to seek and infect the species-specific host snails in which they undergo a complex life cycle, ultimately resulting in the release of cercariae into bodies of fresh water (Fig. 8) [40].

3. Pathogenesis

Embolization of eggs laid in the mesenteric or vesicular veins to the CNS occurs via two separate mechanisms. Pulmonary arteriovenous shunts and portopulmonary anastomoses via the azygous vein allow eggs to reach the CNS through the arterial system. Retrograde venous flow into the Batson vertebral epidural venous plexus connects the portal venous system and vena cava with the spinal cord and cerebral veins. This permits embolization of eggs via the venous system. Also, ectopic adult worms can migrate to sites close to the CNS and lay eggs in close proximity to each other.

For unknown reasons, the eggs of *S. japonicum* cause brain lesions, whereas those of *S. mansoni* and *S. hematobium* typically affect the lower spinal cord. Once deposited in tissue, the eggs secrete antigenic products that generate a periovular granulomatous reaction. Once large numbers of eggs are deposited, the intense granulomatous reaction causes damage via inflammation, vascular damage, and mass effect [41].

4. Pathophysiology

Rarely, a maculopapular eruption may develop several hours after exposure and up to 1 week later at the site of penetration by the cercariae. Acute schistosomiasis or Katayama fever can be seen in areas with high transmission rates and in persons first infected as adults. It develops 14–84 days after exposure. The symptoms are thought to be secondary to immune complex formation in reaction to egg deposition into host tissues. Common symptoms at presentation include fever, headache, generalized myalgias, right upper quadrant pain, and bloody diarrhea. This can be accompanied by respiratory symptoms in as many as 70% of those infected with *S. mansoni*. On exam, tender hepatomegaly is present, and splenomegaly is seen in about 30% of cases. Aseptic meningitis is rare. Laboratory abnormalities show marked eosinophilia and positive serologies [40].

Chronic schistosomiasis results from the host's vigorous granulomatous reaction to schistosome eggs, which can lead to chronic fibroobstructive disease, primarily of the liver, intestines, or genitourinary tract, depending on the infective species. Granulomas have been found in many types of tissue including the skin, lung, brain, adrenal glands, and skeletal muscle. *S. mansoni, S. japonicum* and *S. mekongi* cause mostly intestinal and hepatic lesions leading to periportal hepatic fibrosis and portal venous hypertension. Clinical manifestations include nonbloody or guaiac-positive diarrhea, abdominal pain, hepatomegaly, splenomegaly, and variceal bleeding. *S. haematobium* causes primarily urinary tract pathology, including hematuria, dysuria, chronic cystitis, and obstructive uropathy. Recurrent infections are associated with higher risk of squamous cell carcinoma of the bladder. In children, severe infection affects growth, nutrition, activity, and school performance [40].

Neurological involvement may be symptomatic or asymptomatic. Frequently eggs are deposited in the brain and spinal cord without symptoms. However, symptomatic neurological involvement comes in two forms: cerebral schistosomiasis caused by *S. japonicum*, and a myeloradiculopathy caused by *S. mansoni* and *S. hematobium* [41].

Cerebral schistosomiasis has been recorded in 2–4% of individuals infected with *S. japonicum*. The clinical manifestations include an acute encephalitis or encephalomyelitis

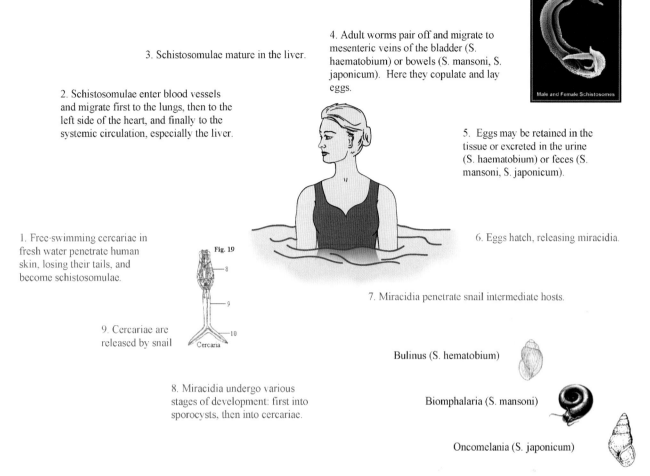

Figure 8 Life cycle of schistosomiasis. (Adapted from DPDx: CDC's Web site for parasitology identification; http://www.dpd.cdc.gov/dpdx.)

concurrent with or immediately after the systemic manifestations of the acute phase of infection. The encephalitis is transient, lasting several days or weeks, and treatment usually results in a complete recovery. Rarely, cerebral schistosomiasis presents as a slowly expanding mass lesion resulting in focal neurological symptoms, increased intracranial pressure, or seizures [41].

Myeloradiculopathy has been reported more rarely, with approximately 500 cases described since the initial description of this manifestation of schistosomiasis in 1930. Clinical symptoms usually begin with lumbar or lower extremity pains, often radicular in nature. They are followed within hours or days by a transverse myelitis including muscular weakness, sensory impairment, and autonomic and bladder dysfunction, which usually localize to the lower spinal cord, the cauda equina, or both. Hallmarks include rapid progression, disease ascending through the levels, and variable severity of disease, including permanent paraplegia [41].

There is no definitive way to diagnose cerebral or spinal involvement, but typical lesions accompanied by eggs in the stool or antibodies in the CSF are strongly suggestive. CSF typically shows increased protein and lymphocytosis. Eosinophils may or may not be present [41].

5. Natural History

Treatment is with praziquantel and is quite efficacious when given early even in single-dose regimens, although the preferred regimen has 2–3 doses total. In cases involving the CNS, steroids are also used [40].

B. Other Trematodes

Cerebral paragonimiasis is a rare complication of infection with *Paragonimus* spp. and may produce mass lesions, meningitis, and seizures.

References

1. Cox, F. E. (2002). History of human parasitology. *Clin. Microbiol. Rev.* **15**, 595–612.
2. Sutherst, R. W. (2004). Global change and human vulnerability to vector-borne disease. *Clin. Microbiol. Rev.* **17**, 136–173.
3. Maizels, R. M., Balic, A., Gomez-Escobar, N., Nair, M., Taylor, M. D., and Allen, J. E. (2004). Helminth parasites–masters of regulation. *Immunol. Rev.* **201**, 89–116.
4. Greenwood, B. M., Bojang, K., Whitty, C. J., and Targett, G. A. (2005). Malaria. *Lancet* **365**, 1487–1495.
5. Hay, S. I., Guerra, C. A., Tatem, A. J., Noor, A. M., and Snow, R. W. (2004). The global distribution and population at risk of malaria: past, present, and future. *Lancet Infect. Dis.* **4**, 326–336.
6. Suh, K. N., Kain, K. C., and Keystone, J. S. (2004). Malaria. *CMAJ* **170**, 1693–1702, 2004.
7. Newton, C. R., and Warrell, D. A. (1998). Neurological manifestations of Falciparum malaria. *Ann. Neurol.* **43**, 695–702.
8. Gitau, E. N., and Newton, C. R. (2005). Blood-brain barrier in falciparum malaria. *Trop. Med. Int. Health* **10**, 285–292.
9. Artavanis-Tskonas, K., Tongren, J. E., and Riley, E. M. (2003). The war between the malaria parasite and the immune system: immunity, immunoregulation and immunopathology. *Clin. Exp. Immunol.* **133**, 145-152.
10. Cattand, P. (2001). L'epidemiologyie de la trypansomiase humaine africaine: une histoire mulifactorielle complexe. *Med. Trop.* **61**, 313–322.
11. Kennedy, P. G. E. (2004). Human African trypanosomiasis of the CNS: current issues and challenges, *J. Clin. Investigat.* **113**, 496–504.
12. Sternberg, J. M. (2004). Human African trypanosomiasis: clinical presentation and immune response, *Parasite Immunol.* **26**, 469–476.
13. Mhlanga, J. D. M., Bentivoglio, M., and Kristensson, K. (1997). Neurobiology of cerebral malaria and African sleeping sickness. *Brain Res. Bull.* **44**, 579–589.
14. Lundkvist, G. B., Kristensson, K., and Bentivoglio M. (2004). Why trypanosomes cause sleeping sickness. *Physiology* **12**, 198–206.
15. Barnett, N. D., Kaplan, A. M., Hopkin, R. J., Saubolle, M. A., and Rudinsky, M. F. (1996). Primary amoebic meningoencephalitis with *Naegleria fowleri*: clinical review. *Ped. Neurol.* **15**, 230–234.
16. Marciano-Cabral, F., and Cabral, G. (2003). *Acanthamoeba* spp. as agents of disease in humans. *Clin. Microbiol. Rev.* **16**, 273–307.
17. Khan, N. A. (2003). Pathogenesis of Acanthamoeba infections. *Microb. Path.* **34**, 277–285.
18. Schuster, F. L. and Visvesvara, G. S. (2004). Free-living amoebae as opportunistic and non-opportunistic pathogens of humans and animals. *Int. J. Parasitol.* **34**, 1001–1027.
19. Stanley, S. (2003). Amoebiasis. *Lancet* **361**, 1025–1034.
20. Montoya, J. G. and Liesenfeld, O. (2004). Toxoplasmosis. *Lancet* **363**, 1965–1976.
21. Bhopale, G. M. (2003). Pathogenesis of toxoplasmosis. *Comp. Immunol. Microbiol. Infect. Dis.* **26**, 213–222.
22. Suzuki, Y. (2002). Immunopathogenesis of cerebral toxoplasmosis. *J. Infect. Dis.* **186(Suppl 2)**, S234–S240.
23. Garcia, H. H., Gonzalez, A. E., Evans, C. A., Gilman, R. H., and Cysticercosis Working Group in Peru. (2003). Taenia solium cystercercosis. *Lancet* **361**, 547–556.
24. Hawk, M. W., Shahlaie, K., Kim, K. D., and Theis, J. H. (2005). Neurocystercercosis: a review. *Surg. Neurol.* **63**, 123–132.
25. Nash, T. E., Del Brutto, O. H., Butman, J. A., Corona, T., Delgado-Escueta, A., Duron, R. M., Evans, C. A., Gilman, R. H., Gonzalez, A. E., Loeb, J. A., Medina, M. T., Pietsch-Escueta, S., Pretell, E. J., Takayanagui, O. M., Theodore, W., Tsang, V. C., and Garcia, H. H. (2004). Calcific neurocystircercosis and epileptogenesis. *Neurology* **62**, 1934–1938.
26. Raether, W., and Hanel, H. (2003). Epidemiology, clinical manifestations and diagnosis of zoonotic cestode infections: an update. *Parsitol. Rev.* **91**, 412–438.
27. Eckert, J., and Deplazes, P. (2004). Biological, epidemiological and clinical aspects of echinococcosis, a zoonosis of increasing concern. *Clin. Microbiol. Rev.* **17**, 107–135.
28. Bukte, Y., Kemaloglu, S., Nazaroglu, H., Ozkan, U., Ceviz, A., and Simsek, M. (2004). Cerebral hydatid disease: CT and MR imaging findings. *Swiss Med. Wkly.* 134, 459–467.
29. Lo Re, V., Glucman, S. J. (2003). Eosinophilic meningitis. *Am. J. Med.* **114**, 217–223.
30. Wise, M. E., Sorvillo, F. J., Shafir, S. C., et al. (2005). Severe and fatal central nervous system disease in humans caused by *Baylisascaris procyonis*, the common roundworm of raccoons: a review of the current literature. *Microbes Infection* 7, 317–323.
31. Murray, W. J., and Kazacos, K. R. (2004). Raccoon roundworm encephalitis. *Clin. Infect. Dis.* **39**, 1484–1492.
32. Vidal, J. E., Sztajnbok, J., and Seguro, A. C. (2003). Eosinophilic meningoencephalitis due to *Toxocara canis*: case report and review of the literature. *Am. J. Trop. Med. Hyg.* **69**, 341–343.
33. Despommier, D. (2003). Toxocariasis: clinical aspects, epidemiology, medical ecology, and molecular aspects. *Clin. Microbiol. Rev.* **16**, 265–272.
34. Moreira-Silva, S. F., Rodrigues, M. G., Pimenta, J. L. et al. (2004). Toxocariasis of the central nervous system: with report of two cases. *Rev. Soc. Brasil. Med. Trop.* **37**, 169–174.
35. Lim, S., Katz, K., Krajden, S., et al. (2004). Complicated and fatal *Strongyloides* infection in Canadians: risk factors, diagnosis and management. *CMAJ* **171**, 479–484.
36. Keiser, P. B., and Nutman, T. B. (2004). *Strongyloides stercoralis* in the immunocompromised population. *Clin. Microbiol. Rev.* **17**, 208–217.
37. Fourestie, V., Douceron, H., Brugieres, P., et al. (1993). Neurotrichinosis: a cerebrovascular disease associated with myocardial injury and hypereosinophilia. *Brain* 116, 603–616.
38. Bruschi, F., and Murrell, K. D. (2002). New aspects of human trichinellosis: the impact of new *Trichinella* species. *Postgrad. Med. J.* **78**, 15–22.
39. Gelal, F., Kumral, E., Vidinli, B. D., et al. (2005). Diffusion-weighted and conventional MR imaging in neurotrichinosis. *Acta. Radiologica* **46**, 196–199.
40. Ross, A. G. P., Bartley, P. B., Sleigh, A. C., Olds, G. R., Li, Y., Williams, G. M., and McManus, D. P. (2002). Schistosomiasis. *New Engl. J. Med.* 346, 1212–1220.
41. Ferrari, T. C. (2004). Involvement of the central nervous system in the schistosomiasis. *Mem. Inst. Oswaldo Cruz* **99(Suppl 1)**, 59–62.

42

Prion Diseases

Adriano Aguzzi, MD, PhD, DVM, hc, FRCP, FRCPath
Frank L. Heppner, MD
Ingo M. Westner, MD
Markus Glatzel, MD

Keywords: bovine spongiform encephalopathy, Creutzfeldt-Jakob disease, neuroinvasion, prion, PrP^C, PrP^{Sc}, prion pathogenesis, prion therapeutics, transmissible spongiform encephalopathies

I. History of Prion Diseases
II. The "Protein-Only" Hypothesis (What Is a Prion?)
III. Prion Diseases of Animals and Humans
IV. Animal Models of Prion Diseases
V. Peripheral Prion Pathogenesis
VI. Therapy
   References

I. History of Prion Diseases

Prion diseases have been known for almost two centuries (Table 1). Scrapie, the prototypical prion disease affecting sheep and goats, had been a concern since the 18th century. The crucial breakthrough was achieved in the 1930s by the experimental transmission of scrapie to goats. Next, Carleton Gajdusek demonstrated that kuru, a disease affecting the aboriginal people of Papua New Guinea, was a transmissible spongiform encephalopathy. Interestingly, the first attempts at transmitting kuru to primates failed for the same reason that experimental transmission of scrapie among sheep had failed for decades: the incubation time of the disease was longer than the investigators' persistence. Following a suggestion by William Hadlow that kuru resembled scrapie and hence might exhibit a very long incubation time, Gajdusek achieved transmission of kuru to chimps and, shortly thereafter, transmission of Creutzfeldt-Jakob disease (CJD), today's most prevalent human prion disease. Therefore prion diseases are also called *transmissible spongiform encephalopathies* (TSEs), a term that emphasizes their infectious nature [1].

Table 1 History of Prion Research

Mid-18th century	Earliest description of scrapie recorded
1898	Neuronal vacuolation discovered in brains from sheep with scrapie.
1918	Contagious spread of scrapie in natural conditions suspected.
1920	First cases of CJD described by Creutzfeldt and Jakob.
1937	Epidemic of scrapie in Scotland following administration of formalin-treated louping ill vaccine prepared from sheep brain.
1939	Experimental transmission of scrapie reported.
1955–1957	Kuru discovered among Fore people of Papua New Guinea.
1959	Similarities between kuru and scrapie noted.
1961	Multiple strains of scrapie agent described.
1961	Scrapie transmitted to mice.
1963	Transmission of kuru to chimpanzees reported.
1966	Scrapie agent found to be highly resistant to ionizing radiation and ultraviolet light.
1967	First enunciation of the protein-only hypothesis.
1968	CJD transmitted to chimpanzees.
	Description of gene affecting scrapie incubation period in mice (*Sinc*).
1974	First documented case of iatrogenic prion transmission (corneal graft).
1980	Protease resistant, highly hydrophobic protein discovered in hamster brain fractions highly enriched for scrapie infectivity.
1982	Prion concept enunciated, states that infectious agent is principally proteinaceous.
1985	Gene encoding PrP^C cloned.
1986	PrP^C and PrP^{Sc} isoforms shown to be encoded by same host gene.
1987	Linkage demonstrated between PrP and scrapie incubation period in mice.
	First report of BSE in cattle.
1989	Mutation in PrP linked to GSS syndrome.
	Importance of isologous PrP^C/PrP^{Sc} interactions established.
1992	Ablation of *Prnp* by gene targeting in mice.
1993	$Prnp^{o/o}$ mice are resistant to scrapie inoculation.
	Structural differences between PrP^C and PrP^{Sc} isoforms noted.
1994	Cell-free conversion of PrP^C to protease-resistant PrP.
1996	New variant of CJD identified.
	BSE prion strain carries a distinct glycotype signature.
	First NMR structure of core murine PrP^C solved.
1997	Experimental evidence that vCJD is caused by the BSE agent.
	B-lymphocytes are necessary for peripheral prion pathogenesis.
1998	Formal demonstration that postulated genes controlling incubation period in mice are congruent with PrP.
1999	Discovery of the PrP^C homologue.
2000	Temporary depletion of lymphoid FDCs impairs prion replication.
	Experimental transmission of BSE in sheep by blood transfusion.
2001	Involvement of complement system in prion replication. Proof of principle of anti-prion immunization in transgenic mice.
2003	Transgenic expression of soluble dimeric PrP binds PrP^{Sc} and inhibits prion replication.
2004–2005	Prions synthesized or amplified in cell-free systems.

BSE, Bovine spongiform encephalopathy; *CJD*, Creutzfeld-Jakob disease; *FDC*, follicular dendritic cell, *GSS*, Gerstmann-Sträussler-Scheinker (syndrome).

When Stanley Prusiner started his first attempts to tackle the cause of TSEs [2], this group of diseases was not of public interest. However, bovine spongiform encephalopathy (BSE) was recognized a few years later [1] and considerably changed the public perception of prion diseases. CJD in humans was, and fortunately continues to be, exceedingly rare: its incidence is typically $1/10^6$ inhabitants/year but has been reported to reach $3/10^6$ inhabitants/year in Switzerland [3]. However, several aspects of CJD epidemiology continue to be enigmatic, and a screen for recognized or hypothetical risk factors for CJD has not exposed any causal factors to date. Several scenarios may account for the increase in incidence, including improved reporting, iatrogenic transmission, and transmission of a prion zoonosis.

Although only less than 1% of all reported cases of CJD can be traced to a defined infectious source, the identification of BSE and its subsequent epizootic spread has highlighted prion-contaminated meat-and-bone meal as an efficient vector for bovine prion diseases [4]. Infectious prions do not completely lose their infectious potential even after extensive autoclaving [5]. When transmitted to primates, BSE produces a pathology strikingly similar to that of variant Creutzfeldt-Jakob disease (vCJD). BSE is most likely also transmissible to humans, and strong circumstantial evidence indicates that BSE is the cause of vCJD [6], which has resulted in more than 140 deaths in the United Kingdom [7] and a much smaller number in some other countries.

It is remarkable that some of the questions formulated in the 19th century are still unanswered. For example, it is still not known whether prion diseases in animals are predominantly of genetic or infectious nature. If the latter is true, how do prions spread within flocks? The emerging epizootic of chronic wasting disease in North American cervids [8] drastically emphasizes the general importance of these issues.

II. The "Protein-Only" Hypothesis (What Is a Prion?)

The most widely accepted hypothesis on the nature of the infectious agent causing TSEs (which was termed *prion* by Stanley B. Prusiner) predicates that it consists essentially of PrP^{Sc}, an abnormally folded, protease-resistant, beta-sheet rich isoform of a normal cellular protein, PrP^C. According to this theory, the prion does not contain any informational nucleic acids, and its infectivity propagates simply by recruitment and "autocatalytic" conformational conversion of cellular prion protein into disease-associated PrP^{Sc} [2].

A large body of experimental and epidemiological evidence is compatible with the protein-only hypothesis, and stringently designed experiments have failed to disprove it. It would go well beyond the scope of this chapter to review all efforts that have been undertaken to this effect. Perhaps most impressively, knock-out mice carrying a homozygous deletion of the *Prnp* gene that encodes PrP^C, fail to develop disease upon inoculation with infectious brain homogenate, nor do their brains carry prion infectivity. Reintroduction of *Prnp* by transgenesis—even in a shortened, redacted form—restores infectibility and prion replication in $Prnp^{o/o}$ mice [9]. In addition, all familial cases of human TSEs are characterized by *PRNP* mutations [10]. Recent studies essentially settle the score as to the nature of the infectious agent and further support the protein-only hypothesis by demonstrating that prions may be synthesized in cell-free systems [11].

III. Prion Diseases of Animals and Humans

A. Prion Diseases of Animals

Prion diseases in animals were recognized long before their human counterparts. A prion disease occurring in sheep and goats, termed *scrapie*, was the first mentioned in 1738 [1]. Since then a number of clinically diverse entities have been described in various mammals. All of these diseases share features characteristic of prion diseases, including their transmissibility to adequate hosts. Three diseases deserve to be mentioned in detail: scrapie, BSE, and chronic wasting disease.

1. Scrapie of Sheep and Goats

Scrapie of sheep and goats is endemic in most European countries and the Americas, but it is not known to exist in Australia or New Zealand [1]. Most clinical cases of scrapie occur in sheep between 2 and 5 years of age, and unlike other prion diseases, there is evidence that scrapie can be transmitted both vertically, from ewe to lamb before or shortly after birth, and horizontally, from sheep to sheep either directly or via the environment. The genetic makeup (i.e., certain polymorphisms within the PrP gene) has a strong influence on whether an animal will develop clinical scrapie. There is no direct evidence that the infectious agent of scrapie may be transmitted to humans [12]. Nevertheless, the fact that sheep may have been exposed to BSE prions has opened the possibility that these animals could, in theory, constitute a risk to humans. This has prompted several programs aimed at eliminating scrapie from sheep flocks in the United Kingdom.

2. Chronic Wasting Disease of Deer and Elk

Chronic wasting disease (CWD) of deer and elk was initially reported in the late 1960s in research facilities in Colorado. The first cases of CWD in wild ranging animals were recognized in 1981. Since then, CWD has been detected in Colorado, Wyoming, New Mexico, Wisconsin, Illinois, Nebraska, and Utah in the United States and Saskatchewan in Canada. Additionally, CWD has been found in captive deer and elk in a number of states in North America and in South Korea. Similar to scrapie, the disease spreads horizontally within affected herds; yet, the efficiency of horizontal transmission seems to be very high in this disease entity. Clinical signs of CWD are remarkably subtle and nonspecific, characterized by lethargy, weight loss, flaccid hypotonic facial muscles, polydipsia/polyuria, excessive salivation, and behavioral changes such as loss of fear of humans. Analogous to scrapie, PrP^{CWD} is abundant in secondary lymphoid tissues and can be detected in the tonsils, Peyer's patches, and ileocecal lymph nodes. The capacity for CWD transmission to other species is clearly an area of great concern. Unfortunately, little is known about the risk of other wildlife species, domestic ruminants, or humans contracting the disease, yet preliminary evidence suggests that zoonotic transmission of CWD to humans is very unlikely [13].

3. Bovine Spongiform Encephalopathy

In 1986 an epidemic of a previously unknown disease appeared in cattle in Great Britain: BSE, or "mad cow" disease. BSE was shown to be a prion disease by demonstrating protease-resistant PrP in brains of ill cattle. The BSE epidemic, which has caused more than 180,000 cases to date, reached its peak in 1992 in the U.K. and has become a rare disease in British and European cattle. The BSE epidemic may have resulted from the feeding of scrapie-containing sheep meat-and-bone meal to cattle, and there is strong evidence that the epidemic was amplified by feeding rendered bovine meat-and-bone meal to young calves. Clinical signs of BSE are nonspecific, with initial signs of weight loss and restlessness eventually evolving to ataxia and frequent falls of affected animals. BSE has been experimentally transmitted to a number of different animal species, and transmission of BSE prions has almost certainly resulted in a novel form of human prion disease (vCJD, see Section III.B.6) [4,14].

B. Human Prion Diseases

Human prion diseases manifest as sporadic, genetic, and acquired disorders and are referred to as sporadic CJD (sCJD), familial or genetic CJD (fCJD), and iatrogenic (iCJD) or variant CJD (vCJD) (Table 2).

1. Clinical Diagnosis of Human Prion Diseases

The diagnosis of human prion diseases is based on the appraisal of clinical signs and symptoms and a number of auxiliary examinations. For a long time, electroencephalography was the method of choice in order to substantiate the diagnosis of a human prion disease. Because the overall sensitivity of this method is limited, the usefulness of this investigation has been questioned [15]. An alternative auxiliary test that is able to confirm the clinical suspicion of a human prion disease is the elevation of markers of neuronal injury in the cerebrospinal fluid (CSF). Several of these markers have been monitored in CSF of patients suffering from human prion disease. The most popular of these surrogate markers is the 14.3.3 protein. Because elevation of this protein is also reported in a range of non–prion-related diseases such as encephalitis, cerebral infarction, and paraneoplastic neurological disorders, satisfactory sensitivity and specificity can only be achieved in selected cohorts. Due to these drawbacks, this test cannot be recommended as a screening test for human prion diseases. Recent advances in neuroimaging, especially in magnetic resonance imaging (MRI), may lead to the establishment of specific patterns for human prion diseases. For vCJD, the "pulvinar sign," a high T2 MRI signal in the posterior thalamus, seems to be relatively unique and is present in about 75% of patients with vCJD. For sporadic CJD, fluid-attenuated inversion recovery and diffusion-weighted MRI sequences are associated with relatively high sensitivity and specificity and may represent a relatively noninvasive method to corroborate the diagnosis of a human prion disease [15].

Pathological and biochemical examination of specimens removed bioptically is only possible if adequate biosafety measures are ensured and can only be recommended to exclude the diagnosis of diseases in which therapeutic options are available. Until recently, it has been thought that PrP^{Sc} may only be detected in central nervous system (CNS) tissue of patients with prion disease. Recently, it has become obvious that PrP^{Sc} may be detected in lymphoid tissue of vCJD patients and in the olfactory mucosa and muscle tissue of sCJD patients [16]. The coming years will determine whether any of these methods might facilitate the diagnosis of human prion diseases.

2. Molecular Diagnosis of Human Prion Diseases

Molecular diagnosis of human prion diseases relies on the combination of genetic, biochemical, and neuropathological investigations in conjunction with the clinical data.

a. Genetic Investigations Sequencing of *PRNP* enables the exclusion of genetically caused CJD. In addition, this investigation provides information on the status on a polymorphism on codon 129. There is compelling evidence from studies of genetically modified mice and from clinical studies of patients with human prion diseases that homozygosity for methionine on codon 129 constitutes a risk factor for the development of prion disease. Methionine homozygotes are clearly overrepresented among sCJD patients. Furthermore, all individuals affected by vCJD are codon 129 methionine homozygotes. In addition to constituting a risk factor for the development of prion diseases, this polymorphism has a considerable effect on the clinical, biochemical, and neuropathological presentation of prion-diseased individuals [17].

b. Biochemical Investigations The biochemical characterization of PrP^{Sc} by western blotting, following limited digestion by proteinase K, provides important information on the predominant type of PrP^{Sc} that is produced within the CNS. This information, together with clinical and genetic data, helps to determine the type of human prion disease (see Section III.B.7).

c. Histological Investigations Routine neuropathological investigations include sampling of defined regions within

Table 2 Human Prion Diseases: Clinical, Diagnostic, and Neuropathological Features

	Sporadic CJD (sCJD)	Inherited CJD			Acquired CJD	
		Genetic CJD	GSS	FFI	Variant CJD	Iatrogenic CJD
Clinical features						
Age at onset (years)	60–70	50–60	50–60	50 (20–63)	26 (12–74)	*
Disease duration (mean)	6 months (1–35 months)	6 months (2–41 months)	5–6 years (3 months–13 years)	13–15 months	14 months (6–24 months)	Similar to sCJD
Leading clinical symptoms	Progressive dementia and neurological signs (e.g., myoclonus, cerebellar ataxia, visual problems, extrapyramidal symptoms)	Clinical symptoms similar to sCJD	Cerebellar dysfunction (e.g., ataxia, nystagmus, dysarthria)	Insomnia, autonomic dysfunction	Early psychiatric symptoms (e.g., depression, anxiety, social withdrawal), dysaesthesia, later neurological deficits and cognitive decline	Clinical symptoms similar to sCJD
Diagnostic tests						
CSF 14-3-3	Positive in >90%	Positive in >90%	Usually negative	Negative	Positive in 50%	Positive in 77%
EEG	PSWC 60–70%	PSWC 7%	Nonspecific alterations	Nonspecific alterations	Nonspecific alterations, no PSWC	Similar to sCJD
MRI	Brain atrophy, hyperintensities in basal ganglia and/or cortical 67%	Similar to sCJD	Normal or nonspecific cerebral or cerebellar atrophy	Normal or nonspecific cerebral or cerebellar atrophy	Hyperintensities in the posterior thalamus ("Pulvinar sign") 78%	Similar to sCJD
Biopsy	(Brain), muscle	–	–	–	(Brain), tonsils	–
Genetics: codon 129	MM 70%, MV 14%, VV 16%	–	M (on the mutated allele)	M (on the mutated allele)	MM 100%	MM 57%, MV 20%, VV 23%
Genetics: *PRNP* mutation	NO	(Over 25 disease-associated mutations, such as E200K)	P102L (plus 7 less-common mutations)	D178N	NO	NO
Postmortem neuropathological examination						
Histopathological features	Spongiform changes, neuronal loss, astrogliosis, PrP-deposition (various patterns)	Similar to sCJD	Spongiform changes, neuronal loss, astrogliosis, PrP-deposition (multicentric plaques)	Involvement of thalami	Spongiform changes, neuronal loss, astrogliosis, PrP-deposition (florid plaques)	Similar to sCJD
Biochemical tests	PrP^{Sc} typing (WB)	Same as sCJD	Same as sCJD	Same as sCJD	Same as sCJD	Same as sCJD

CSF, Cerebrospinal fluid; EEG, electroencephalogram; FFI, fatal familial insomnia; GSS, Gerstmann-Sträussler-Scheinker (syndrome); MRI, magnetic resonance imaging; NO, not observed; PSWC, periodic sharp wave complexes.
* Age at onset depending on iatrogenic exposure: incubation period 1–30 years.

the CNS and immunohistochemical demonstration of PrP. Special emphasis is placed on the classification of the distinct deposition pattern of PrP in various regions of the CNS, such as the cerebellum and the thalamus [18].

3. Sporadic Creutzfeldt-Jakob Disease

sCJD is a rapidly progressive dementia, usually leading to death within 12 months of disease onset [19]. Initial symptoms include cognitive deficits, sleep disturbances, and behavioral abnormalities. As the disease progresses, other clinical features such as extrapyramidal and pyramidal symptoms, ataxia, and visual disturbances become obvious, and the patient usually develops myoclonus [19]. Terminally sCJD-affected patients fall into a state of akinetic mutism prior to death. Unlike other dementing diseases such as cerebrovascular dementia, Alzheimer's, and Parkinson's disease, whose incidence rises exponentially with age, the peak incidence of sCJD is between 55 and 65 years of age.

According to the previously mentioned criteria, sCJD can be subdivided into several groups with distinct genetic, biochemical, neuropathological, and clinical features. The typical rapidly progressing form of sCJD shows homozygosity for methionine on codon 129 and, upon western blotting, a PrP^{Sc} type with a relatively long (thus slower-migrating) unglycosylated PrP^{Sc} fragment of approximately 21kDa. Clinically, atypical sCJD patients very often show heterozygosity of codon 129 and, upon western blotting, a shorter (thus faster-migrating) unglycosylated PrP^{Sc} fragment of approximately 19kDa [20].

4. Inherited Human Prion Diseases

This group of conditions can be subdivided into three phenotypes: fCJD, Gerstmann-Straussler-Scheinker (GSS) syndrome, and fatal familial insomnia. The mode of inheritance in each disease, which co-segregates with mutations in *PRNP*, is autosomal dominant [19].

fCJD is often not associated with distinctive clinical features and may only be diagnosed on sequencing of *PRNP*. Penetrance of *PRNP* mutations is usually high, although the existence of healthy octogenarian carriers of certain mutations clearly argues in favor of the existence of non–*PRNP*-related disease modifiers.

GSS syndrome is characterized by a slowly progressive cerebellar ataxia, beginning in the fifth or sixth decade, accompanied by cognitive decline. In contrast to other inherited human prion diseases, GSS syndrome shows unique neuropathological features consisting of widespread, multicentric PrP plaques. Although various *PRNP* mutations have been described for phenotypes of GSS syndrome, the P102L and the G131V mutations are most commonly found.

Fatal familial insomnia presents with a profound disruption of the normal sleep-wake cycle, insomnia, and sympathetic hyperactivity. The clinicopathological features of fatal familial insomnia segregate with the D178N mutation only when combined with methionine homozygosity at codon 129.

5. Acquired Human Prion Diseases: Iatrogenic Creutzfeldt-Jakob Disease

iCJD is caused by prion exposure of individuals during medical procedures, such as implantation of human dura mater, corneal graft implantation, or parenteral treatment with human cadaveric pituitary extracts. iCJD is rare, with less than 300 published cases. The majority of cases were caused by implantation of dura mater and injection of pituitary growth hormone.

The site of prion inoculation seems to dictate the incubation time until onset of prion disease–related symptoms. Direct intracerebral exposure to prions and implantation of prion-contaminated dura, for example, are associated with short incubation periods (16–28 months), whereas peripheral extraneural exposure to prions results in long incubation times ranging from 5 to 30 years [21]. Furthermore, there is evidence that the route of prion exposure influences the clinical presentation. Dura mater or growth hormone–related cases of iCJD present with a predominantly ataxic phenotype, whereas cases in which prions were directly introduced in the CNS present with dementia as the initial symptom.

6. Variant Creutzfeldt-Jakob Disease

vCJD, a relatively new member of the group of human prion diseases, was first reported in 1996 [21]. In the past years, biochemical, neuropathological, and transmission studies have substantiated the concern that vCJD represents transmission of BSE prions to humans [6,23]. From 1996 to 2001 the incidence of vCJD in the United Kingdom rose annually, evoking fears of a large upcoming epidemic. Since 2001, however, the incidence of vCJD in the United Kingdom appears to be stabilizing, and only a small number of countries other than the United Kingdom have seen isolated cases of vCJD. Although predictions of the future of the vCJD epidemic are still marred by imprecision, there is growing evidence that the total number of vCJD victims will be limited.

The fact that vCJD carries a distinct clinicopathological profile has facilitated the formulation of diagnostic criteria. In contrast to sCJD, most vCJD victims are very young (median age at death 29 years). In addition, sCJD and vCJD have different initial features and illness durations. About 60% of vCJD patients present with psychiatric symptoms and the median illness duration is 14 months, whereas sCJD rarely presents with early psychiatric symptoms and the disease course is much more rapid (median 5 months). Neuropathologically, the CNS of patients with vCJD shows widespread PrP plaques, some of which are encircled by vacuoles and are termed *florid* plaques (Fig. 1).

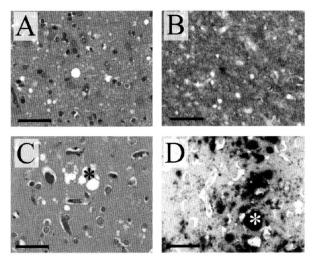

Figure 1 Histological features of prion diseases. Histological findings in brain tissue of sCJD (**A** and **B**) and vCJD (**C** and **D**) patients, showing astrogliosis and widespread spongiform changes. PrP depositions are synaptic (**A** and **B**) and in the form of florid plaques (*asterisk*, **C** and **D**). **A** and **C** are hematoxylin-eosin stains; **B** and **D** are immunohistochemical stains for PrP (scale bar = 50 μm).

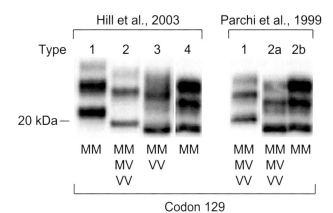

Figure 2 Western blot analysis of PrP^{Sc}. PrP^{Sc} types according to proposed schemes [20]. Both discriminate PrP^{Sc} types based on the mobility of the unglycosylated band of PrP^{Sc} and information on the relative abundance of the signal intensity produced by the di-, mono-, and unglycosylated forms of PrP^{Sc}. Whereas [20] differentiate four principal PrP^{Sc} types (1–4), [38] propose 3 PrP^{Sc} types (1, 2a, and 2b).

7. Biochemical and Structural Properties of Prions

Structural studies have shown that PrP^C has a high alpha-helix content (about 40%) and little beta-sheet (3%), whereas PrP^{Sc} contains more beta-sheets (about 40%) and less alpha-helical structures (about 30%). Although the three-dimensional structure of PrP^C has been resolved by nuclear magnetic resonance spectroscopy, similar analyses have not been possible for PrP^{Sc} [24].

The basis of biochemical characterization of PrP^{Sc} resides in the relative resistance of PrP^{Sc} toward proteolytic degradation. Whereas PrP^C is entirely digested by proteinase K, identical treatment leads to the removal of a variable number of N-terminal amino acids in the case of PrP^{Sc}. This results in the appearance of three distinct bands, corresponding with the di-, mono-, and unglycosylated forms of PrP^{Sc}, upon western blotting.

The molecular classification of PrP^{Sc} takes two factors into account: the first is the size and thus mobility of the unglycosylated band of PrP^{Sc} on polyacrylamide gel electrophoresis, and the second is the relative abundance of the signal intensity produced by the di-, mono-, and unglycosylated forms of PrP^{Sc}. The resulting information is then used to establish the "type" of PrP^{Sc} that may be classified according to proposed schemes (Fig. 2) [20]. Depending on the exact conditions under which the protease digestion and the western blotting procedure is performed, between three and six different PrP^{Sc} types can be distinguished [20,25] (see Fig. 2). Distinct PrP^{Sc} types are thought to represent the molecular correlate of distinct prion strains. The fact that the PrP^{Sc} type found in patients suffering from vCJD is similar to the PrP^{Sc} type present in BSE-diseased cattle has been purported as one of the main arguments supporting the theory that BSE prions are responsible for the vCJD epidemic in humans.

It may appear difficult to understand how a glycotype ratio can be propagated with any fidelity during prion replication. Although this question is essentially unanswered, experiments with yeast prions indicate that this can incontrovertibly occur in a synthetic prion replication system. This phenomenon may be related to the quaternary structure of prion aggregates.

IV. Animal Models of Prion Diseases

Animal models have contributed many important insights into the field of prion biology, including the understanding of the molecular basis of the species barrier for prions as well as the structure-activity relationship of the PrP gene with regard to scrapie susceptibility. In addition, animal models have been instrumental in understanding the pathogenesis and defining new therapeutic concepts of prion diseases.

Although transmission of prions within a species is easily achievable, transmission from one species to another is usually accompanied by a prolongation of the incubation period in the first passage and incomplete penetrance of the disease. Subsequent passage in the same species occurs with high frequency and shortened incubation times. This so-called *species barrier* can be overcome by introducing into the recipient host PrP transgenes derived from the prion donor. Initial transgenetic studies had shown that the species barrier between mice and Syrian hamsters (SHa) for the transmission of prions can be abolished by expression of a SHaPrP transgene in mice. However, transgenic mice expressing human PrP^C, or Tg(HuPrP), failed to develop

CNS dysfunction when inoculated with human prions [26]. To clarify this apparent discrepancy, mice expressing a chimeric human/mouse (Mo) PrP transgene, designated Tg(MHu2M), were generated and were highly susceptible to human prions, suggesting that Tg(HuPrP) mice have difficulty converting Hu PrP^C into PrP^{Sc} [26]. Although MoPrP and HuPrP differ at 28 residues, only 9 or perhaps fewer amino acids in the region between codons 96 and 167 feature the species barrier in the transmission of human prions into mice.

When $MoPrP^C$ was removed by gene ablation in Tg(HuPrP) mice, Tg(HuPrP) mice were rendered susceptible to human prions, suggesting that $MoPrP^C$ inhibited the conversion of $HuPrP^C$ into PrP^{Sc} in Tg(HuPrP) mice [9]. Whereas earlier studies argued that PrP^C forms a complex with PrP^{Sc} during the formation of nascent PrP^{Sc}, these later findings suggested that PrP^C also binds to at least one additional macromolecule during the conversion process, which has been termed *protein X*. Although no physical evidence for protein X has been found to date and exceptions to the genetic observations that had sparked the invention of protein X have been observed, at present it is still possible that protein X is indeed a prion replication cofactor. However, it is equally possible that it is nonexistent, that it exists but is not a protein, or that multiple host-encoded factors (proteins X, Y, and Z) distinct from PrP may bind to PrP^{Sc} [27] and influence prion replication in vivo through a variety of direct and indirect mechanisms.

Ablation of the PrP gene ($Prnp^{o/o}$) affected neither development nor behavior of the mice, and $Prnp^{o/o}$ mice have remained healthy for more than 2 years. $Prnp^{o/o}$ mice are resistant to prions and do not propagate scrapie infectivity [9]. Because the absence of PrP^C expression does not provoke disease, it is likely that prion diseases result not from an inhibition of PrP^C function due to PrP^{Sc} but rather from interference with some as yet undefined cellular process. Mice heterozygous ($Prnp^{+/o}$) for ablation of the PrP gene had prolonged incubation times when inoculated with mouse prions. Therefore the amount of available PrP^C is rate limiting for the development of disease, a finding with significant implications for prion therapy. These findings are in agreement with studies of Tg(SHaPrP) mice in which increased SHaPrP expression was accompanied by diminished incubation times and reinforce the idea that the concentration of PrP^C in brain is rate limiting for its conversion into PrP^{Sc} [9].

To model GSS syndrome in mice, the codon 102 point mutation was introduced into the MoPrP gene, and Tg(MoPrP-P101L)H mice were created that expressed high (H) levels of the mutant transgene product. Two lines of Tg(MoPrP-P101L)H mice spontaneously developed CNS degeneration, characterized by clinical and neuropathological signs indistinguishable from experimental murine scrapie. Surprisingly, transmission of brain homogenate from Tg(MoPrP-P101L)H mice to healthy mice expressing lower levels of the same transgene, but not to wild-type mice, induced spongiform encephalopathy [27]. Considering that overexpression of wild-type PrP by itself from a cosmid PrP transgene causes a spontaneous disease phenotype in transgenic mice, the 101 Pro-> Leu mutation (equivalent to codon 102 in the human PrP gene) was introduced into the endogenous murine PrP gene. Mice either homozygous or heterozygous for the codon 101 mutation remained healthy for more than 650 days without showing any signs of spontaneous CNS disease and without abnormal pathology [28]. This shows that the presence of the mutated PrP at physiological levels is not sufficient to cause spontaneous neurodegeneration in mice.

V. Peripheral Prion Pathogenesis

The fastest and most efficient method for inducing spongiform encephalopathy in the laboratory is intracerebral inoculation of brain homogenate. Inoculation of 1,000,000 ID_{50} infectious units (defined as the amount of infectivity that will induce TSE with 50% likelihood in a given host) will yield disease in approximately half a year; a remarkably strict inverse relationship can be observed between the logarithm of the inoculated dose and the incubation time [2].

However, the previous situation does not correspond with what typically happens in the field. In these instances, acquisition of prion infectivity through any of several peripheral routes is the rule. Therefore the peripheral route of prion uptake merits discussion.

Prions can find their way through the body to the brain of their host and colonize various extracerebral organs including the lymphoreticular system and skeletal muscles. However, histopathological changes have not been identified in organs other than in the CNS. The process by which prions travel through the body to the CNS is termed *neuroinvasion* and comprises two phases [30]. A primary phase of prion accumulation and replication in organs of the lymphoreticular system (LRS) is followed by a secondary phase, which is initiated once the agent has gained access to peripheral nerves. It may be argued that this model oversimplifies the complex mechanisms of prion neuroinvasion, and indeed there are exceptions: Some studies have shown that prion neuroinvasion along peripheral nerves can occur independently of lymphoid prion replication, and not all lymphoid organs are always colonized by prions. However, we maintain that the vast majority of experimental data can be explained on the basis of the two-step model of prion neuroinvasion.

Under normal circumstances, the immune cells responsible for extraneural prion replication are confined to lymphoid organs. However, a variety of pathological conditions, most notably chronic inflammatory states, may lead to extravasation of activated B-lymphocytes to sites of pathology. As such ectopically located B-lymphocytes express lymphotoxin-α and -β, they may lead to maturation of stromal cells with the characteristics of follicular dendritic cells, which are the primary sites of prion replication in lymphoid organs. Indeed, it emerged that scrapie infection of mice with chronic inflammatory conditions broadens the tropism of prions and leads to the formation of ectopic prion reservoirs at the sites of inflammation [31].

VI. Therapy

A. Prionostatic Approaches

To date, no antiprion therapeutical leads have proven to be useful in clinical settings. One of the possible problems derives from the fact that most antiprion compounds were identified in cell culture assays, where chronically prion-infected neuroblastoma cells are "cured" of their prion burden. A multitude of substances appears to possess such prion-curing properties, including compounds as diverse as Congo red, amphotericin B, anthracyclines, sulfated polyanions, porphyrins, branched polyamines, "beta-sheet breakers", quinacrine, and the spice curcumin (tumeric). Disappointingly, none of these compounds thus far have proved to be effective for actual therapy of prion-diseased animals or humans [32].

Based on earlier experiments [33], our laboratory recently showed that transgenic expression of an immunoglobulin Fcγ domain fused to PrPC (PrP-Fc2) results in a drastic delay in the onset of prion disease upon peripheral and intracerebral prion challenge [32].

Another approach generated great expectations of turning into a potential prion immunotherapy: chronic administration of Cytidyl-guanyl oligodeoxynucleotides (CpG-ODN), which binds Toll-like receptor 9 (TLR9) and stimulates innate immune responses, was shown to delay prion disease in scrapie-infected mice. A follow-up study initiated by our laboratory revealed that chronic CpG-ODN administration resulted in severe immunotoxicity and germinal center destruction [34]. As lymphoid germinal centers are crucial to peripheral prion pathogenesis [35], a destruction of the lymphoarchitecture by CpG-ODN may explain the CpG-ODN-mediated antiprion properties. Because the use of CpG-ODN was also found to result in multifocal liver necrosis and hemorrhagic ascites [36], feasibility of the usefulness of CpG-ODN for postexposure prion prophylaxis has to be scrutinized thoroughly.

B. Immunotherapeutic Approaches against Prions

Prions are known to take advantage of immune and lymphoreticular cells to gain access to the peripheral and then the central nervous systems [32]. Conversely, humoral immune responses to the prion protein were shown to antagonize prion infection. This is true even when such responses are directed against PrPC and do not selectively target PrPSc: anti-PrP antisera have been shown to reduce the titer of infectious hamster brain homogenates in vitro. Moreover, anti-PrP antibodies prevented formation of PrPSc in a cell-free system. Also, monoclonal antibodies and F(ab) fragments recognizing PrP repressed de novo scrapie infection and abolished prion infectivity from chronically scrapie-infected neuroblastoma cells [32].

Because many cells in neural and extraneural compartments express PrPC, straightforward evaluation of the efficacy of anti-PrP immune responses in vivo was thwarted by host tolerance to PrPC. To bypass PrPC tolerance, we transgenetically expressed anti-PrP antibodies in mice and found that intraperitoneal inoculation of prions resulted in protection from scrapie pathogenesis [39]. Notably, co-expression of anti-PrPC antibodies and PrPC at physiological levels did not induce obvious autoimmune side effects—a finding of some relevance when contemplating the possibility of anti-prion vaccination. Enthusiasm for anti-prion immunoprophylaxis was even more stronger after White and colleagues demonstrated that passive transfer of anti-PrP antibodies into intraperitoneally prion inoculated wild-type mice delayed the onset of scrapie significantly [32].

The prionostatic efficacy of anti-PrP antibodies in the previously mentioned in vivo studies, however, was confined to extraneural compartments: Transgenic 6H4μ mice expressing anti-PrP-specific immunoglobulin (Ig)-M molecules were not protected when prions were administered intracerebrally, and passive transfer of PrP-specific IgG immunoglobulins was inefficient in intracerebrally prion inoculated mice, most likely due to the limited influx of immunoglobulins into the CNS. In principle, such limitations can be circumvented by means of active immunization, which results in the generation of stable and persistent titers.

However, one major obstacle to devising effective regimens of active immunization is host tolerance to endogenous PrPC that inhibits a host-derived, anti-PrP antibody immune response. Active immunization attempts thus far have resulted in the induction of at best meager anti-PrP titers and limited biological efficacy, emphasizing the need for alternative strategies [36]. Conversely, administration of antibodies generated in Prnp-ablated animals ("passive immunization"), although feasible and effective, suffers from the intrinsic problem of poor diffusion from vessels into tissues, particularly the central nervous tissue. Worryingly, intracerebellar or intrahippocampal injection of monoclonal anti-PrP holoantibodies induced neurotoxicity, whereas injection of monovalent F(ab)1 fragments prepared from the same monoclonal antibodies was innocuous, suggesting that antibody-induced cross-linking of PrPC may be toxic [37].

To obviate these problems, a recent cell culture study explored the antiprion potential of recombinant single chain antibody fragments (scFv) derived from monoclonal anti-PrP antibody 6H4. When co-cultured with cells secreting anti-PrP scFv, chronically prion-infected neuroblastoma cells ceased to produce PrPSc, even if antibody-producing cells were physically separated from target cells in transwell cultures [40]. It is important to note that the monovalent anti-PrP scFvs cannot cross-link PrPC and should be safer than bivalent full-fledged antibodies that may confer neurotoxicity [37]. The delivery of scFvs by genetic transduction in vivo may also allow for sustained production of scFvs at predefined sites for prolonged periods of time. The latter is difficult to obtain by direct systemic administration of monoclonal antibodies, which are very large molecules and therefore may present limited bioavailability, and by injection of recombinant scFvs, which suffer from extremely short half-lives in blood. However, future studies will need to address the feasibility of this approach.

References

1. Aguzzi, A., and Polymenidou, M. (2004). Mammalian prion biology. One century of evolving concepts. *Cell* **116**, 313–327.
2. Prusiner, S. B. (1998). Prions. *Proc. Natl. Acad. Sci. USA* **95**, 13363–13383.
3. Glatzel, M., Ott, P. M., Linder, T., Gebbers, J. O., Gmur, A., Wust, W., Huber, G., Moch, H., Podvinec, M., Stamm, B., and Aguzzi, A. (2003). Human prion diseases: epidemiology and integrated risk assessment. *Lancet Neurol.* **2**, 757–763.
4. Weissmann, C., and Aguzzi, A. (1997). Bovine spongiform encephalopathy and early onset variant Creutzfeldt-Jakob disease. *Curr. Opin. Neurobiol.* **7**, 695–700.
5. Taylor, D. M. (2000). Inactivation of transmissible degenerative encephalopathy agents: a review. *Vet. J.* **159**, 10–17.
6. Aguzzi, A., and Weissmann, C. (1996). Spongiform encephalopathies: a suspicious signature. *Nature* **383**, 666–667.
7. CJD Surveillance Unit, University of Edinburgh, Edinburgh, U.K. Accessed online October 2005 at www.cjd.ed.ac.uk/figures.htm.
8. Williams, E. S., and Young, S. (1980). Chronic wasting disease of captive mule deer: a spongiform encephalopathy. *J. Wildl. Dis.* **16**, 89–98.
9. Weissmann, C., and Bueler, H. (2004). A mouse to remember. *Cell* **116(Suppl 2)**, S111–S115.
10. Prusiner, S. B., Scott, M. R., DeArmond, S. J., and Cohen, F. E. (1998). Prion protein biology. *Cell* **93**, 337–348.
11. Castilla, J., Saa, P., Hetz, C., and Soto, C. (2005). In vitro generation of infectious scrapie prions. *Cell* **121**, 195–206.
12. Scott, M. R., Peretz, D., Nguyen, H. O., Dearmond, S. J., and Prusiner, S. B. (2005). Transmission barriers for bovine, ovine, and human prions in transgenic mice. *J. Virol.* **79**, 5259–5271.
13. Williams, E. S., and Miller, M. W. (2002). Chronic wasting disease in deer and elk in North America. *Rev. Sci. Tech.* **21**, 305–316.
14. Taylor, D. M. (1996). Bovine spongiform encephalopathy—the beginning of the end? *Br. Vet. J.* **152**, 501–518.
15. Zerr, I., and Poser, S. (2002). Clinical diagnosis and differential diagnosis of CJD and vCJD. With special emphasis on laboratory tests. *APMIS* **110**, 88–98.
16. Glatzel, M., Abela, E., Maissen, M., Aguzzi, A. (2003). Extraneural pathologic prion protein in sporadic Creutzfeldt-Jakob disease. *N. Engl. J. Med.* **349**, 1812–1820.
17. Pocchiari, M., Puopolo, M., Croes, E. A., Budka, H., Gelpi, E., Collins, S., Lewis, V., Sutcliffe, T., Guilivi, A., Delasnerie-Laupretre, N., Brandel, J. P., Alperovitch, A., Zerr, I., Poser, S., Kretzschmar, H. A., Ladogana, A., Rietvald, I., Mitrova, E., Martinez-Martin, P., de Pedro-Cuesta, J., Glatzel, M., Aguzzi, A., Cooper, S., Mackenzie, J., van Duijn, C. M., and Will, R. G. (2004). Predictors of survival in sporadic Creutzfeldt-Jakob disease and other human transmissible spongiform encephalopathies. *Brain* **127**, 2348–2359.
18. Budka, H., Aguzzi, A., Brown, P., Budka, H., Aguzzi, A., Brown, P., Brucher, J. M., Bugiani, O., Gullotta, F., Haltia, M., Hauw, J. J., Ironside, J. W., Jellinger, K., et al. (1995). Neuropathological diagnostic criteria for Creutzfeldt-Jakob disease (CJD) and other human spongiform encephalopathies (prion diseases). *Brain Pathol.* **5**, 459–466.
19. Gambetti, P., Kong, Q., Zou, W., Parchi, P., and Chen, S. G. (2003). Sporadic and familial CJD: classification and characterisation. *Br. Med. Bull.* **66**, 213–239.
20. Hill, A. F., Joiner, S., Wadsworth, J. D., Sidle, K. C., Bell, J. E., Budka, H., Ironside, J. W., and Collinge, J. (2003). Molecular classification of sporadic Creutzfeldt-Jakob disease. *Brain* **126**, 1333–1346.
21. Collins, P. S., Lawson, V. A. and Masters, P. C. (2004). Transmissible spongiform encephalopathies. *Lancet* **363**, 51–61.
22. Will, R. G., Ironside, J. W., Zeidler, M., Ironside, J. W., Zeidler, M., Cousens, S. N., Estibeiro, K., Alperovitch, A., Poser, S., Pocchiari, M., Hofman, A., and Smith, P. G. (1996). A new variant of Creutzfeldt-Jakob disease in the UK. *Lancet* **347**, 921–925.
23. Aguzzi, A. (1996). Between cows and monkeys. *Nature* **381**, 734.
24. Riesner, D. (2003). Biochemistry and structure of PrP(C) and PrP(Sc). *Br. Med. Bull.* **66**, 21–33.
25. Notari, S., Capellari, S., Giese, A., Westner, I., Baruzzi, A., Ghetti, B., Gambetti, P., Kretzschmar, H. A., and Parchi, P. (2004). Effects of different experimental conditions on the PrPSc core generated by protease digestion: implications for strain typing and molecular classification of CJD. *J. Biol. Chem.* **279**, 16797–16804.
26. Telling, G. C., Scott, M., Hsiao, K. K., Telling, G. C., Scott, M., Hsiao, K. K., Foster, D., Yang, S. L., Torchia, M., Sidle, K. C., Collinge, J., DeArmond, S. J., and Prusiner, S. B. (1994). Transmission of Creutzfeldt-Jakob disease from humans to transgenic mice expressing chimeric human-mouse prion protein. *Proc. Natl. Acad. Sci. USA* **91**, 9936–9940.
27. Maissen, M., Roeckl, C., Glatzel, M., Goldmann, W., and Aguzzi, A. (2001). Plasminogen binds to disease-associated prion protein of multiple species. *Lancet* **357**, 2026–2028.
28. Telling, G. C., Haga, T., Torchia, M., Tremblay, P., DeArmond, S. J., and Prusiner, S. B. (1996). Interactions between wild-type and mutant prion proteins modulate neurodegeneration transgenic mice. *Genes Develop.* **10**, 1736–1750.
29. Moore, R. C., and Melton, D. W. (1997). Transgenic analysis of prion diseases. *Mol. Hum. Reprod.* **3**, 529–544.
30. Aguzzi, A., and Heikenwalder, M. (2005). Prions, cytokines, and chemokines: a meeting in lymphoid organs. *Immunity* **22**, 145–154.
31. Heikenwalder, M., Zeller, N., Seeger, H., Prinz, M., Klohn, P. C., Schwarz, P., Ruddle, N. H., Weissmann, C., and Aguzzi, A. (2005). Chronic lymphocytic inflammation specifies the organ tropism of prions. *Science* **307**, 1107–1110.
32. Weissmann, C., and Aguzzi, A. (2005). Approaches to therapy of prion diseases. *Annu. Rev. Med.* **56**, 321–344.
33. Telling, G. C., Scott, M., Mastrianni, J., Gabizon, R., Torchia, M., Cohen, F. E., DeArmond, S. J., and Prusiner, S. B. (1995). Prion propagation in mice expressing human and chimeric PrP transgenes implicates the interaction of cellular PrP with another protein. *Cell* **83**, 79–90.

34. Heikenwalder, M., Polymenidou, M., Junt, T., Sigurdson, C., Wagner, H., Akira, S., Zinkernagel, R., and Aguzzi A. (2004). Lymphoid follicle destruction and immunosuppression after repeated CpG oligodeoxynucleotide administration. *Nat. Med.* **10**, 187–192.
35. Aguzzi, A. (2003). Prions and the immune system: a journey through gut, spleen, and nerves. *Adv. Immunol.* **81**, 123–171.
36. Aguzzi, A., and Sigurdson, C. J. (2004). Antiprion immunotherapy: to suppress or to stimulate? *Nat. Rev. Immunol.* **4**, 725–736.
37. Solforosi, L., Criado, J. R., McGavern, D. B., Wirz, S., Sanchez-Alavez, M., Sugama, S., DeGiorgio, L. A., Volpe, B. T., Wiseman, E., Abalos, G., Masliah, E., Gilden, D., Oldstone, M. B., Conti, B., and Williamson, R. A. (2004). Cross-linking cellular prion protein triggers neuronal apoptosis in vivo. *Science* **303**, 1514–1516.
38. Parchi, P., Giese, A., Capellari, S., Brown, P., Schulz-Schaeffer, W., Windl, O., Zerr, I., Budka, H., Kopp, N., Piccardo, P., Poser, S., Rojiani, A., Streichemberger, N., Julien, J., Vital, C., Ghetti, B., Gambetti, P., and Kretzschmar, H. (1999). Classification of sporadic Creutzfeldt-Jakob disease based on molecular and phenotypic analysis of 300 subjects. *Ann. Neurol.* **46**, 224–233.
39. Heppner, F. L., Musahi, C., Arrighi, I., Klein, M. A., Rülicke, T., Oesch, B., Zinkernagel, R. M., Kalinke, U., and Aguzzi, A. (2001). Prevention of scrapie pathogenesis by transgenic expression of anti-prion protein antibodies. *Science* **294**, 178–182.
40. Donofrio, G., Heppner, F. L., Polymenidou, M., Musahi, C., and Aguzzi, A. (2005). Paracrine inhibition of prion propagation by anti-PrP single-chain Fv miniantibodies. *J. Virol.* **79**, 8330–8338.

43

Central Nervous System Viral Infections: Clinical Aspects and Pathogenic Mechanisms

Christopher Power, MD, FRCP (C)
Farshid Noorbakhsh, MD, PhD

Keywords: *herpes simplex, human immunodeficiency virus, neuroinvasion, neurosusceptibility, neurotropism, neurovirulence*

I. Introduction
II. Neuropathogenesis of CNS Viral Infections
III. Diagnostic Considerations
IV. Acute CNS Viral Infection: Herpes Simplex Encephalitis
V. Chronic CNS Viral Infection: Neurocognitive Syndromes in HIV Infection
VI. Future Perspectives
    References

I. Introduction

In this chapter, we review the general aspects of human central nervous system (CNS) viral infections, together with basic pathogenic mechanisms and diagnostic tools. Because there are more than 30 recognized viruses or distinct strains thereof that cause human neurological disease [1], coupled with limited space availability, we have concentrated on two prototypical CNS viral infections, herpes simplex virus encephalitis and human immunodeficiency virus dementia, which highlight many of the concepts that underlie the clinical presentation [2], diagnosis, and treatment of these diverse groups of neurological disorders. Over the past 20 years the fields of virology and neuroscience have experienced important developments, ranging from fundamental molecular and epidemiological advances to unprecedented epidemics in humans. Several factors have contributed to the increasing emergence of neurological infections, the least of which is the exponential rise in the global human population size, which is now 6.2 billion but expected to reach 10 billion by 2050. Indeed, as the population rises, there is increased pollution and trafficking of humans and animals and their products across international borders, together with new food production practices [3]. The human immunodeficiency virus (HIV) together with acquired immunodeficiency syndrome (AIDS) epidemic represent a prime example of several newly recognized neurological infectious disorders, including primary HIV-induced neurovirological syndromes such as HIV-associated dementia and distal sensory polyneuropathy or secondary/opportunistic neurovirological disorders including the JC virus–related progressive multifocal leukoencephalopathy or herpes zoster, which causes myelitis and multifocal leukoencephalopathy [4]. HIV/AIDS also reflects changing social and economic milieus that have resulted in the emergence of diseases previously unknown in humans, probably contracted by humans by eating contaminated foods. However, other mechanisms of zoonotic infection have recently been shown to cause new neurological infections in humans, including Nipah virus transmission from pigs and West Nile virus from mosquitoes transported from warmer climates [5]. Increased travel among humans and the global exportation of animals have provided ripe opportunities for transcontinental spread of neurological infectious agents. Fifty years ago, it would have taken months for livestock to be shipped across oceans, during which time sick animals would have been culled, whereas today's shipments are done in hours. Social developments over the past 2 decades have also contributed to the appearance of new diseases, including increased global urbanization and changing trends in the workforce. Similarly, with increased population size, poverty, and war, select regions of the world that were already beset with other diseases are now faced with the ever increasing HIV/AIDS epidemic and its accompanying opportunistic infections of the brain, with few resources to cope with the associated economic and social burdens.

Neurological infections represent unique infections because they are evolutionary cul-de-sacs for most pathogens, frequently rendering the host unable to support further pathogen replication and spread or, worse still, dead. Within the CNS, three principal clinical manifestations of direct viral infections are recognized, including 1) *meningitis* involving infection and inflammation of the meninges (e.g., nuchal rigidity, confusion, headache, focal cranial neuropathies); 2) *encephalitis* that can affect specific regions of the brain (e.g., altered awareness, seizures, movement and behavioral disorders); and finally 3) *myelitis,* which may have effects on any region of the spinal cord (e.g., back pain, limb weakness, bowel and bladder dysfunction). Moreover, the temporal sequence of symptoms and signs varies widely but is frequently categorized based on the rapidity (acute vs chronic) and persistence of clinical features (Tables 1 and 2). These disorders usually overlap, with simultaneous infection of the meninges, brain, and spinal cord parenchyma leading to heterogenous clinical presentations; moreover, they are often accompanied by systemic symptoms and signs, including fever, rash, and myalgias. Several systemic viral infections are associated with a *postinfectious meningoencephalomyelitis*, which can be difficult to distinguish from viral infections concentrated in the meninges and CNS parenchyma. Indeed, this syndrome is thought to reflect an autoimmune process arising in response to select viral antigens [6].

Unlike systemic infections, infectious pathogens in the CNS pose a more ominous predicament for clinicians because of the brain's enhanced vulnerability due to its limited host defenses, which are largely dependent on innate immunity in many diseases, including HIV and rabies virus infections [7]. In addition, delivery of therapeutics to the brain remains a major obstacle, even when suitable drugs are available. Newer treatment strategies may also yield new infections, including the proposed use of stem cells from human embryos or animal organs in xenotransplantation. The use of molecular tools for infectious disease epidemiology was barely in its infancy a decade ago; yet, today it is the cornerstone of controlling disease outbreaks and monitoring ongoing epidemics. Indeed, a greater understanding of host susceptibility to infection has arisen through the identification of specific genotypes associated with disease including single nucleotide polymorphisms. Neurological diagnostics and therapeutics have also evolved with the advent of improved neuroimaging, highly sensitive and specific molecular tools, elaborate neurocognitive testing, and the wider availability of specific drugs for neurological infections.

II. Neuropathogenesis of CNS Viral Infections

The brain is particularly vulnerable to damage caused by infectious and immune diseases because of its circumscribed and interdependent structures in conjunction with a lack of resident lymphoid cells [8]. Furthermore, the CNS is comprised of multiple groups of topographically differentiated cell types that exhibit very specific properties. Among neurons, many different subpopulations exist and are defined by electrophysiological properties, predominantly expressed protein or neurotransmitters, morphology, and location in

Table 1 Viruses Causing Acute CNS Disease

Virus	Family	Nucleic Acid	CNS Disease	Cell Tropism	Nonhuman Host	Route of Entry to CNS	Treatment
Herpes simplex viruses-1 and -2	Herpesviridae	DNA	Meningo-encephalitis	Neuron	ND	Intraneuronal	Acyclovir
Rabies virus	Rhabdoviridae	RNA	Encephalomyelitis	Neuron	Carnivores	Intraneuronal	Post-exposure prophylaxis: RIG and vaccine
West Nile virus	Flaviviridae	RNA	Meningo-encephalomyelitis	Neuron	Birds, horses	Hematogenous	IVIG, interferon-α
Nipah virus	Paramyxoviridae	RNA	Encephalitis	Neuron, endothelia	Pigs, fruit bats	Hematogenous	Ribavirin
Equine encephalitis viruses	Togaviridae	RNA	Meningitis, encephalitis	Neuron, astrocyte	Horses	Hematogenous	Supportive care
Mumps virus	Paramyxoviridae	RNA	Meningitis, encephalitis, myelitis	Neuron, ependymal cell	ND	Hematogenous	IVIG
Rubella virus	Togaviridae	RNA	Meningitis, encephalitis	Not specified	ND	Hematogenous	Plasmapheresis
Coxsackie virus, echovirus	Picornaviridae	RNA	Meningitis, meningo-encephalitis, myelitis	Neuron	ND	Hematogenous	Pleconaril
Polio virus	Picornaviridae	RNA	Meningitis, myelitis	Neuron	ND	Hematogenous	Pleconaril
California encephalitis virus	Bunyaviridae	RNA	Meningitis, encephalitis	Neuron	Small mammals	Hematogenous	Ribavirin

CNS, Central nervous system; *DNA*, deoxyribonucleic acid; *IVIG*, intravenous immunoglobulin; *ND*, not determined, *RIG*, rabies immune globulin, *RNA*, ribonucleic acid.

the CNS. Likewise, glia including astrocytes, oligodendrocytes, and monocytoid cells such as parenchymal microglia, dendritic cells, and perivascular macrophages also display marked heterogeneity in their phenotypes. Ependymal and endothelial cells are also heterogenous but can act as barriers to the brain parenchyma. Indeed, each of these cell types may be influenced by blood-derived cells such as activated lymphocytes trafficking through the CNS as part of normal immune surveillance [9]. The individual neural cell types are permissive to infection by specific viruses (Fig. 43.1), giving rise to distinct clinical syndromes. A critical and complex structure within most of the CNS is the blood-brain barrier (BBB), which is composed of endothelial cells connected by tight junctions, surrounded by a basement membrane and abutting processes of astrocytes, macrophages, and neurons (see Fig. 43.1). The BBB is unique in its capacity to regulate the transport of soluble factors from blood into the brain. CNS infections may depend on the permissiveness of CNS endothelial cells to virus infection with subsequent release of virus into the brain following transcytosis of virions across endothelial cells (Fig. 43.2). Viruses can also traverse the BBB by infecting trafficking cells (monocytes as well as lymphocytes), termed the *Trojan horse model* of viral infection of the CNS, as suggested for HIV (Fig. 43.3). Other mechanisms by which different viruses enter the CNS include crossing the blood–cerebrospinal fluid barrier (B-CSF-B) at the level of the choroid plexus (see Fig. 43.2). Finally, retrograde axonal transport of a virus from the periphery following initial exposure represents a classic mode of CNS infection, arising from a bite in muscle, as suggested for rabies infection.

Many viruses enter the CNS soon after systemic infection, termed *neuroinvasion*, although infection of resident neural cells is not a requisite outcome [1]. Inherently *neurotropic* viruses are able to infect and replicate in CNS cells without necessarily causing neurological disease. In fact, many neurotropic viruses engage specific host cell membrane proteins as receptors that mediate cell entry. *Neurovirulent* viruses cause neurological disease and thus are both neuroinvasive and neurotropic by definition. The chief determinants of neurovirulence include the virus's intrinsic properties, defined by its molecular composition and heterogeneity, replicative capacity, and the *neurosusceptibility* of the infected individual (i.e., age, general level of health, underlying genetic attributes). For example, the

Table 2 Viruses Causing Chronic CNS Disease

Virus	Family	Nucleic Acid	CNS Disease	Cell Tropism	Nonhuman Host	Route of Entry to CNS	Treatment
Human immunodeficiency virus	Retroviridae	RNA	Encephalitis, meningitis, myelitis	Microglia, macrophage, astrocyte	ND	Hematogenous	HAART and neuroprotective agents
Human T-cell leukemia viruses-1 and -2	Retroviridae	RNA	Myelitis	Astrocyte, leukocyte	ND	Hematogenous	Zidovudine, lamivudine, glucocorticoids
JC virus	Polyomaviridae	DNA	Progressive multifocal leukoencephalopathy	Oligodendrocyte, astrocyte	ND	Hematogenous	Interferon-α, cidofovir
Varicella-zoster virus	Herpesviridae	DNA	Leukoencephalitis, cerebellitis, meningitis, myelitis	Neurons, satellite cell	ND	Hematogenous	Acyclovir
Measles virus	Paramyxoviridae	RNA	Encephalitis; SSPE	Neuron	ND	Hematogenous	Ribavirin, interferon-α, isoprinosine
Cytomegalovirus	Herpesviridae	DNA	Encephalitis	Neuron, ependymal cell, oligodendrocyte, monocytoid cell, endothelia	ND	Hematogenous	Foscarnet, ganciclovir
Epstein-Barr virus	Herpesviridae	DNA	Encephalitis, meningitis, myelitis	Infiltrating mononuclear cells	ND	Hematogenous	Acyclovir, ganciclovir (?)

CNS, Central nervous system; DNA, deoxyribonucleic acid; ND, not determined, RNA, ribonucleic acid; SSPE, subacute sclerosing panencephalitis.

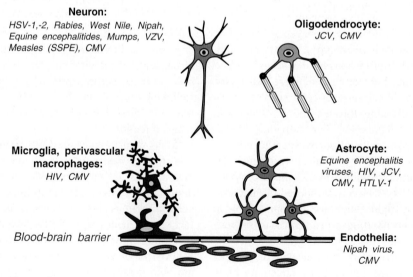

Figure 1 Neurotropism of different viruses in the central nervous system (CNS). The brain is protected from circulating pathogens and toxins by the blood-brain barrier. Viruses infect specific cell types within the CNS depending on the specific properties of the virus together with individual cell membrane proteins expressed on permissive cell types.

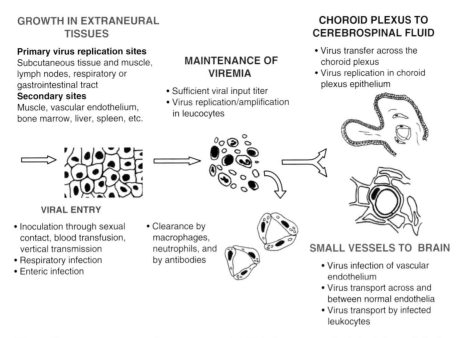

Figure 2 Infection of the central nervous system (CNS) by hematogenously derived viruses. Infection of the CNS often occurs in a sequential manner involving the initial infection of extraneural tissues and resulting in a marked viremia that permits eventual seeding of the CNS through direct infection of CNS endothelia or via infected cells traversing the blood-brain barrier. (Adapted from Johnson, R. T. [1998]. "Viral Infections of the Nervous System," ed. 2. Lippincott-Raven, Philadelphia.)

contemporaneous level of immunosuppression caused by HIV-1 is an integral determinant of the development of CNS disease. The disease pattern characterizing CNS viral infections often consists of acute systemic infection, which can elicit an eventual immune response, but in rapid succession results in CNS neuroinvasion (see Fig. 43.2). Alternatively, there is a long period of asymptomatic infection with the virus subsequently reactivated, causing CNS disorders and hastening the host's morbidity or mortality.

Neurotropism is determined in part by the individual cell type's permissiveness to viral entry and replication and the specific strain of the infecting virus. For example, certain strains of West Nile virus replicate more efficiently in neurons and have been reported to influence the severity of the neurological lesions and symptoms that develop following infection of the nervous system [10,11]. Likewise, herpes simplex virus (HSV)-1 replicates chiefly in neurons, where it lies dormant for many years before being re-activated to cause disease, as evidenced by acute encephalitis (see later discussion). Conversely, cells infected by HIV-1 in the brain are primarily microglia, perivascular macrophages, and to a lesser extent, astrocytes [12–15]. JC virus, which causes progressive multifocal leukoencephalopathy, productively infects oligodendrocytes and astrocytes only, underscoring the unique neurotropic properties of each of these viruses.

As mentioned previously, entry and infection of the CNS are not the sole determinants of neurological disease or *neurovirulence*. The inflammatory responses elicited by the infected cells as well as the activation and dysregulation of bystander cells (often microglia and astrocytes) are considered key factors in viral CNS disease development [16–18]. Viral neuropathogenesis is characterized by both direct and indirect activation of immune responses in the CNS with resultant neural cell damage and death [16]. In viral infections of the CNS, the initial activation of *innate immune responses* in the CNS manifests itself as activation and infiltration of monocytoid cells, NK cells, and neutrophils with concomitant upregulation of cytokines, chemokines, complement, free radicals, and proteases [19]. These inflammatory molecules are usually derived from activated or infected cells including infected cells including microglia/macrophages, astrocytes, and infiltrating lymphocytes. The increase in proinflammatory molecules following infection may recruit additional inflammatory leucocytes into the nervous system [20,21], while both cytokines and chemokines, through interactions with their cognate receptors present on astrocytes and neurons, also have toxic effects on these cell types or result in the release of molecules with neurotoxic actions [22–25]. Moreover, infection by several viruses or mere exposure to their gene products results in the release of other neurotoxic molecules by microglia and astrocytes [16–18]. Other molecular pathways also mediate neurovirulence, including elevated nitric oxide metabolites (i.e., peroxynitrite), altered tryptophan metabolism, and activation of

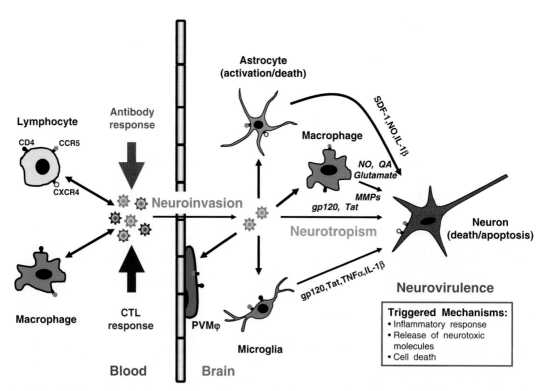

Figure 3 Neuroinvasion, neurotropism, and neurovirulence during HIV-1 infection of the central nervous system (CNS). In the periphery or blood, the immune response (either cellular or humoral) and the availability of cells for infection, such as lymphocytes and macrophages and cells in other viral reservoirs, determine viral evolution towards increased neuroinvasion (CNS entry), neurotropism (infection of microglia, macrophages, neurons, and astrocytes), and ultimately neurovirulence (neuronal damage or death). Viral diversity, represented by the color variation in the virions, results from differing selection pressures in the periphery. After neuroinvasion, either through direct transport of virions or trafficking of infected cells across the blood-brain barrier, the virus in turn drives neurovirulence following infection. Viral replication in the perivascular monocytoid cells (PVMϕ) and microglia (neurotropism), together with the incomplete infection of astrocytes, result in induction of the release of neurotoxins: nitric oxide (NO), quinolinic acid (QA), glutamate, matrix metalloproteinases (MMP), viral proteins (gp120, Tat), and neuroinflammatory responses (TNF-α, IL-1β, MMPs), culminating in cell death among neurons and astrocytes. (Adapted from van Marle, G., and Power, C. [2005]. Human immunodeficiency virus type 1 genetic diversity in the nervous system: evolutionary epiphenomenon or disease determinant? *J. Neurovirol.* **11**, 107–128.)

arachidonic acid metabolism [26], whereas upregulation of cell cycle regulators and deregulation of differentiation factors may also impair neural cell survival and function [27–29]. Ultimately, each of these mechanisms is driven by the presence of virus in the brain and stochastic events defined by the interaction [2] between a particular viral protein and a host cellular pathway. In addition to the resident innate immune cells in the CNS, cells mediating *adaptive immunity* frequently participate in the pathogenesis of CNS viral infections. B and T lymphocytes are involved in multiple CNS infections, infiltrating the CNS from the circulation and subsequently proliferating in response to the presence of specific antigens [19]. There is a large literature showing that activated T cells can be detected in the CNS during specific viral infections, exerting both neuroprotective effects, as observed by infiltrating CD4+ Th1 T lymphocytes, as well as the ability of cytotoxic T lymphocytes (CTLs) to facilitate clearing virus-infected cells [30] or neurotoxic effects in which otherwise healthy bystander cells (neurons) are attacked by misdirected CTLs, resulting in cellular injury and death [31]. Likewise, the other major component of the adaptive immune system, B cell–producing antibodies generated during CNS viral infections, can wield beneficial effects by neutralizing viruses, thereby abrogating infection or targeting neural cell antigens, which may exacerbate neurological disease, possibly through molecular mimicry [19].

The individual host's *neurosusceptibility* is an important disease determinant in CNS viral infections, similar to other infectious systemic diseases in which age and genetic polymorphisms confer vulnerability to neurological disease [32–34]. For example, West Nile virus encephalitis is rare in children, whereas older adults and patients with immunosuppression (with the exception of HIV/AIDS patients) are at greater risk for infection [10]. Host genetic studies in humans have identified individual genetic polymorphisms that are associated with the onset of AIDS or its progression, including the development of HIV-1 neurological disease. The host neurosusceptibility genes identified in relation to human T-cell lymphoma virus (HTLV)-1 infection include specific human lymphocyte antigen (HLA) haplotypes [35]. Indeed, it is likely that an increasing number of polymorphisms in host-immune genes will be identified as risk factors for viral-induced neurological diseases in the future.

III. Diagnostic Considerations

In the past decade the precision and accuracy of neurological diagnosis has improved markedly, largely due to the wider availability of large-scale epidemiological studies, molecular diagnostic tools, and high-quality neuroimaging (reviewed in [36]). Aside from conventional studies in blood and CSF, including cell counts and differential analyses, serology, total protein levels, glucose, oligoclonal bands and routine culturing methods, the field of neurodiagnostics has moved ahead with the advent of molecular tools such as polymerase chain reaction (PCR), which permits the rapid amplification, detection, and quantification of viral genomes in different body fluids or tissues. Moreover, in situ hybridization of deoxyribonucleic acid (DNA) or ribonucleic acid (RNA) has also proven to be useful in the detection of some viruses. Although variability exists among different laboratories, these tools are used widely with comparative success. Newer neuroimaging techniques including magnetic resonance (MR) spectroscopy and angiography, together with refined positron emission tomography (PET) using new isotopically labeled molecules and MR imaging that involves fluid-attenuated inversion recovery (FLAIR) or diffusion-weighted imaging (DWI) with new contrast agents, await greater exploitation in their use for CNS viral infection diagnosis and prognosis. Nonetheless, older diagnostic methods such quantitative serology and (stereotaxic) brain biopsy continue to have substantial value, particularly for patients in whom the diagnosis is unclear [37]. It is important to have strong laboratory resources including microbiology and pathology in the event that these latter techniques are required, especially if specialized methods such as electron microscopy are anticipated. Indeed, it is often valuable to discuss the patient's potential diagnoses in advance of the procedure or sample submission to ensure that maximal information is gleaned from the diagnostic test, serial serology, or brain biopsy. Future diagnostic and prognostic tools include the use of DNA, RNA, or protein arrays on nanochips or multiplex PCR, which permits simultaneous detection of several infectious agents, together with higher-resolution neuroimaging techniques.

IV. Acute CNS Viral Infection: Herpes Simplex Encephalitis

A. Epidemiology

Herpes simplex encephalitis (HSE) is recognized as the most common cause of acute sporadic encephalitis in children older than 6 months of age and in adults in the United States [39]. The incidence varies from 1 in 250,000 to 1 in 500,000 individuals [39]; however, the actual incidence is likely underestimated considering the number of undiagnosed cases. In contrast to some other virus-induced encephalitides (such as the arbovirus encephalitides), the occurrence of HSE does not show any seasonal or sexual predilections [40]. HSE can occur at any age, with a higher frequency in patients younger than 20 years of age or older than 50 years [38,41]. In the absence of therapy, disease prognosis is poor; the mortality rate is greater than 70% with substantial neurological morbidity in the survivors [42]. Treatment decreases the mortality rate to 30% percent, with an accompanying high rate of permanent neurological sequelae that necessitate long-term care and support [42].

More than 90% of HSE cases in immunocompetent hosts are due to HSV-1 infection, with the remainder due to HSV-2 [43]. Approximately one-third of cases of HSV-1 encephalitis result from primary infection with the virus, and the remaining cases are due to the reactivation of a latent infection [40,44]. In contrast, HSV-2 encephalitis is believed to be chiefly due to primary infection [43]. HSE does not occur at a higher frequency in immunocompromised hosts, although the disease course might be atypical in immunocompromised patients (see later discussion).

B. Pathogenesis

Mucocutaneous surfaces are the site of primary infection by the closely related DNA viruses, HSV-1 and HSV-2 [45]. After replication in the site of infection, envelope-less viral capsid is transported to the peripheral sensory ganglia by the retrograde axonal flow [46], where it can establish a latent infection. Reactivation of latent infection leads to viral replication and shedding at mucocutaneous surfaces. Although both viruses can cause clinically and anatomically indistinguishable disease, HSV-1 chiefly infects trigeminal nerve ganglia with its reactivation causing orofacial lesions of herpes labialis, whereas HSV-2 mainly infects sacral dorsal root ganglia with its reactivation causing genital herpes [45]. It remains uncertain how the virus gains access to the CNS. Reactivation of the virus in the trigeminal ganglion may be followed by the spread of virus to frontal and temporal lobes through trigeminal nerve axons [47,48]. In primary infections, the virus might enter the CNS through the olfactory bulb and spread through the olfactory pathways to anterior and middle cranial fossae [49,50]. HSE begins with nonspecific changes including the congestion of cortical and subcortical white matter capillaries and small vessels, followed by perivascular cuffing and glial nodules in the second and third weeks of infection [51]. Full-blown HSE is characterized by diffuse perivascular mononuclear cell infiltration and gliosis with ensuing widespread hemorrhagic necrosis [39,52]. Inflammation and hemorrhagic necrosis are most prominent in the temporal lobes in adult patients with HSE, unlike the more diffuse distribution in neonatal cases [39,53].

C. Clinical Features

Although HSV-induced infection of the brain is usually defined by severe focal necrotizing encephalitis, introduction of PCR analysis of CSF as the principal diagnostic tool has also disclosed the occurrence of relatively mild encephalitis with spontaneous resolution in some cases [54]. Nonspecific signs and symptoms of meningoencephalitis including altered consciousness (97%), fever (90%), and headache (81%) are followed by personality and behavioral changes (85%), focal neurological signs (e.g., hemiparesis, dysphasia) (35%), and seizures (50%) [39,55]. In some cases, disease might have an explosive onset, with the altered consciousness developing in a matter of hours and usually accompanied by fever. In immunocompromised patients, HSE may exhibit a less severe but progressive disease course associated with meningitis and fewer focal neurological deficits [56,57].

D. Diagnosis

Given the high mortality and morbidity rates associated with HSE and the role of antiviral therapy in determining the prognosis of disease, immediate diagnosis is imperative. PCR analysis of CSF to detect viral DNA is the diagnostic method of choice for HSE, with approximately 96% sensitivity and 99% specificity [58,59]. However, negative PCR results should be cautiously interpreted in the context of compelling clinical manifestations, as false negative cases can happen due to early testing (i.e., in the first few days after the onset of disease), presence of PCR inhibitors (e.g., hemoglobin in a bloody CSF sample), or poor PCR sensitivity because of technical limitations [58,60]. Quantitative real-time PCR to measure viral DNA copy numbers in CSF has been used to assess disease severity or monitor the success of antiviral therapy. Viral DNA load has been reported to be correlated with disease severity and drops dramatically with antiviral therapy [61–63]. PCR requires corroborative CSF examinations and neuroimaging. CSF is abnormal in more than 95% of HSE cases, showing nonspecific changes including lymphocytic pleocytosis (10–500 cell/mm^3), elevated protein levels (60–700 mg/dL), and mild hypoglycorrhachia (30–40 mg/dL) in some cases [38,40]. Red blood cells (10–500 cells/mm^3) are detectable as a result of the necrotizing pathology of disease, but they are not always present nor is their presence specific to HSE [38,39,55]. Serological examination of CSF can be performed to detect the intrathecal production of anti-HSV antibodies. Because the passive leakage of antibodies from serum can be the source of CSF antibodies, titers should be corrected using CSF and serum albumin levels [64,65]. CSF cultures are rarely positive for virus in adult cases of HSE but may be used as an adjunctive test in neonatal cases [66].

Magnetic resonance imaging (MRI) shows abnormalities early in the course of the disease in the majority of HSE cases. Damage to the BBB with ensuing edema and temporal lobe abnormalities are detectable on MRI. MRI might also show a lesion of high signal intensity on T2-weighted and flair images at temporal and frontal lobes [67,68].

Electroencephalography (EEG) provides limited diagnostic information but may reveal focal findings, which are characterized by spike and slow-wave activity and, in some patients, periodic lateralizing epileptiform discharge (PLEDs) arising from the temporal lobe [69]. Although PLEDs can also be associated with other forms of focal brain injury (e.g., stroke, intracranial tumor or abscess), their occurrence in the context of relevant clinical manifestations can suggest the diagnosis of HSE. Although EEG changes are frequently detectable in HSE, their resolution is not closely associated with clinical recovery [70].

Before the advent of PCR-based techniques, brain biopsy and culture to detect virus were the "gold standard" for HSE diagnosis. With the availability of PCR analysis of CSF, brain biopsy, an invasive procedure that requires stereotaxic needle aspiration or open craniotomy, is rarely performed today for the diagnosis of HSE. If performed, brain histopathology should be examined by an experienced neuropathologist, together with immunofluorescent staining for viral antigens and culture [66].

E. Treatment

HSE is one of the first viral infections to have been successfully treated with antiviral chemotherapy [71,72]. The current standard therapy for HSE is 10–15 mg/kg intravenous acyclovir every 8 hours for 14–21 days [41]. The increased period of treatment to 2–3 weeks has been recommended after the occurrence of some relapses following 10-day therapeutic courses of acyclovir [73]. Although it is not clearly understood how long the viral DNA persists in the CSF after treatment, additional 1–2 week treatment is recommended upon detection of viral DNA in CSF reanalysis [74]. Acyclovir has proved to have a better efficiency and safety profile compared with vidarabine, another available treatment [71,75]. Different factors, including patient age, duration of encephalitis, seizures, and level of consciousness on presentation, affect treatment outcome and prognosis [71]. Additional therapeutic measures include intubation and mechanical ventilation, anticonvulsant therapies, and monitoring of the elevated intracranial pressure [41].

V. Chronic CNS Viral Infection: Neurocognitive Syndromes in HIV Infection

A. Epidemiology

Unlike acute viral infections of the CNS, the retrovirus HIV-1 usually manifests its neuropathogenic effects long after initial infection, with ensuing cognitive, motor, and behavioral dysfunction among affected patients (reviewed in [76]). HIV-associated dementia (HAD) represents a constellation of progressive symptoms and signs that usually begin once an individual's blood CD4 T-cell counts dips below 200 cells/μl of blood; not surprisingly, HAD is an AIDS-defining illness. With the availability of highly active antiretroviral therapy (HAART), HAD is now presenting with CD4 T^+ lymphocyte levels above 200. Most importantly, the diagnosis of HAD heralds a worsened mean survival prognosis, with or without HAART [77]. Before the era of HAART, the annual incidence of HAD was 53%, although the overall prevalence was only 6% [77], which was likely a consequence of the high mortality rate after HAD onset because survival time after diagnosis was only 5.1 months [5,78]. With the advent of HAART, the incidence of HAD has fallen to less than 10%, and survival time after diagnosis has leapt to 38.5 months, resulting in overall higher prevalence rates. Risk factors for HAD include low CD4 levels, high viral loads in CSF or plasma, anemia, extremes of age, and intravenous drug use.

It has also been postulated that minor cognitive and motor deficit (MCMD) is a risk factor for progression to HAD [79,80]. MCMD is a syndrome that exhibits many clinical aspects of HAD, although its signs and symptoms are less severe. Because MCMD has been identified in patients with higher CD4 counts, there is some suggestion that it may be a precursor to HAD [81]. MCMD may affect as many as 30% of HIV/AIDS patients in North American clinics [43]. Nonetheless, co-morbidities, including chronic drug abuse, head injury, and other risk factors for neurocognitive impairments, may contribute to the diagnosis of MCMD. Among children with HIV infection, the risk of frank HIV encephalopathy or merely neurodevelopmental delay is high, approaching 50% in some studies [82].

B. Pathogenesis

At the time of primary HIV infection, circulating HIV-infected (and activated) leukocytes including macrophages and lymphocytes adhere to endothelia within the neural compartment and subsequently enter the brain parenchyma [83]. On entry, HIV establishes infection of perivascular macrophages, resident microglia, and to a lesser extent, astrocytes; HIV infection of neurons is minimal or nonexistent, as evidenced by the comparative absence of virus-encoded RNA, DNA (through reverse transcription and integration), or protein detected in these cell types by different techniques [34]. HIV infects macrophages and microglia by binding to the CD4 molecule acting as the primary receptor together with the chemokine receptors (CCR5 and CXCR4) as co-receptors for infection. HIV-1 exerts its neuropathogenic effects through two principal mechanisms, including stimulation of neuroimmune cells within the CNS or peripheral nerve system to produce host proinflammatory molecules such as cytokines, chemokines, prostaglandins, redox reactants, excitotoxic amino acids or derivatives thereof, and proteases, which damage neurons and the proximate astrocytes that support them. An alternative mechanism by which neural cells are injured is through direct (neurotoxic) interactions between HIV-encoded proteins including gp120, gp41, Tat, Vpr, and potential target cells, including neurons or astrocytes. In fact, these mechanisms overlap in a complementary manner because perivascular macrophages and microglia are the chief sources of both the host neuropathogenic molecules and the secreted neurotoxic viral proteins, whereas some of the host molecules activate viral replication, many of the virus-encoded proteins can also activate neuropathogenic host gene responses.

The CNS pathological hallmarks of HIV infection include multinucleated giant cells, diffuse white matter pallor, perivascular cuffs composed of monocytes and lymphocytes, microglial nodules, and the presence of HIV-1 antigens [84–86]. The diffuse white matter pallor shows preserved myelin protein expression but concurrent deposition of serum proteins in white matter, implying that altered permeability of the BBB rather than frank demyelination underlies diffuse myelin pallor [87,88]. Neuronal and astrocyte injury and death are defined by dendritic "pruning," together with cell death involving both necrotic and apoptotic mechanisms depending on the effector molecule and the selective vulnerability of the target cell. For example, neurons in the basal ganglia represent a highly vulnerable population, partly because of the high density of microglia in this region of the CNS but also because of the intrinsic properties of this group of neurons.

A limited correlation exists between the clinical entity HAD and the pathological entity HIV encephalitis, defined by the presence of multinucleated giant cells or viral antigens [84,89,90]. There may be a correlation between HIV antigen load and HAD [91], although other studies have suggested that macrophage and microglia presence and activation in the basal ganglia is a better predictive marker for HIV-associated dementia [92]. Although neuronal injury and death in the frontal cortex and deep gray matter occur in the brains of patients with AIDS [93–95], the degree of neuronal apoptosis is not correlated with the severity of HAD. Recent studies also imply that the astrocyte death is also associated with rapidly progressive HAD [96].

C. Clinical Features

Because HIV preferentially affects the basal ganglia and deep white matter, HAD predictably presents with many of the features of a subcortical dementia. Although HAD can show remarkable diversity in its clinical phenotype [97–99], the classic triad of *neurocognitive impairments, behavioral disturbances*, and *motor abnormalities* represent the defining features of HAD (reviewed in [100]). Patients frequently complain of an advancing process of mental slowing and find they are no longer as "quick" as they once were. They initially report minor forgetfulness, especially the inability to remember simple items such as phone numbers and names, although memory function declines steadily. Concentration is also impaired and is one of the features of HAD that most disturbs patients. Patients often lose the ability to read books or focus on television. These symptoms are usually supported by family members' and caregivers' reports, but employment history can also be a useful indicator of neurocognitive dysfunction. Patients with HAD frequently display marked apathy and social withdrawal. These symptoms preclude patients from being part of the work force and are a significant cause for concern in family members and care providers. Indeed, HAD can be mistaken for depression, but when asked about his or her mood, the patient will deny depressive symptoms despite exhibiting features of depression. Other behavioral manifestations can include marked irritability, mental inflexibility, and decreased sex drive. HAD can also be accompanied by psychosis or mania, seizures, or specific movement disorders, although these features are not as common [101,102]. As HAD progresses, patients begin to experience generalized weakness, imbalance, and ataxic gait. They can also display slowing of motor function. On physical examination, there is usually a diffuse increase in tone, tremor, and hyperreflexia together with interrupted smooth-pursuit eye movements. Parkinsonian signs are common, including a masked face and slow stooped gait. Frontal release signs and myoclonus are occasionally present, particularly in advanced stages of disease [103]. It is also worth considering that patients with marked immunosuppression who are antiretroviral therapy naïve may experience an exaggerated immune response, termed the immune reconstitution inflammatory syndrome (IRIS), after HAART introduction [104]. Indeed, IRIS can occasionally manifest as transient cognitive dysfunction, along with signs and symptoms of acute meningoencephalitis, although preexisting neurological complications of AIDS may also be exacerbated with IRIS [105,106].

D. Diagnosis

It is important to note that HAD is not detected by the Mini-Mental Status Exam unless the patient is severely demented, likely because HAD is a subcortical dementia. If there is a suspicion that an HIV-seropositive patient is suffering from HAD, more useful screening tools are applicable, including the HIV Dementia Scale [107], the Mental Alteration Test [108], the Executive Interview [109], and the HIV Dementia Assessment [110]. A widely accepted tool for clinical staging of HAD is the Memorial Sloan-Kettering Scale, which provides a qualitative measure of dementia severity, allowing the physician to track progression of the dementia over time [111].

Radiological features accompanying HAD include cerebral and basal ganglia atrophy and diffuse periventricular white matter hyperintensities on MRI T2 weighted images [112,113]. Unfortunately, it is difficult to correlate the presence of these radiological changes with the presence of HAD because HIV/AIDS patients without dementia also display these changes on neuroimaging [114]. MR spectroscopy studies show diminished N-acetyl aspartate levels in the brain, implying neuronal injury or loss [115]. Other critical investigations include CSF analyses, chiefly to exclude opportunistic processes and also to assess the levels of viral replication in the neural compartment. A high viral load in CSF with a concomitant low viral load in plasma in a patient without another obvious cause for neurocognitive impairment raises the possibility of drug-resistant virus localized to the CNS and necessitates a change in antiretroviral regimen, together with evaluation of drug resistance profile in CSF. In addition to neuroimaging and CSF analyses, detailed neuropsychological assessment is an invaluable tool in confirming the diagnosis of HAD and also in facilitating evaluation of the response to therapy in patients with HAD or MCMD.

E. Treatment

The course of the dementia is variable, with an abrupt decline in function over weeks among some individuals, whereas others display a protracted course over several years culminating in death [116]. The most effective treatment for HAD is to treat the underlying cause. HAART routinely consists of two nucleoside analogue reverse transcriptase inhibitors (NRTIs) and either a potent protease inhibitor (PI) or a non-nucleoside analogue reverse transcriptase inhibitor (NNRTIs). Clinical trials have shown that neuropsychological testing improves in HIV/AIDS patients treated with two NRTIs and the NNRTI nevirapine [117], as well as with two NRTIs and a PI [118,119]. Specific antiretroviral drugs have higher permeability to the BBB than do others, including zidovudine (AZT) (NRTI), stavudine (d4T) (NRTI), abacavir (NRTI), nevirapine (NNRTI), and indinavir (PI) [103]. The resulting high CSF levels may act to decrease viral load in the CNS. If a particular patient is having symptoms of mania or psychosis, it is best to avoid use of efavirenz (NNRTI), as this particular NNRTI may cause hallucinations, vivid dreams, and behavioral changes, all of which may exacerbate an individual's existing symptoms. Occasionally the addition of methylphenidate and amantadine as adjunct therapies may alleviate some symptoms of psychomotor retardation, thus increasing the quality of life [120].

Unfortunately, HAART shows limited efficacy in reversing HAD, making it essential for clinicians to consider other treatments with potential neuroprotective properties. Several agents have been investigated in the past, although few have had significant beneficial effects. Selegiline may have an antiapoptotic effect and slow the progression of HAD [81,121]. Memantine has been shown to block neurotoxicity induced by the HIV viral proteins Tat and gp120 [122]. A Phase II multicenter trial to test the efficacy of memantine in alleviating symptoms of HAD is currently ongoing. Prinomastat is a matrix metalloproteinase inhibitor that has been shown to inhibit HIV Tat-associated neurotoxicity and may be a potential neuroprotective agent in HAD [123]. Human growth hormone has been shown to be neuroprotective and may also be a component of HAD treatment in the future [124]. CPI-1189 blocks the effect of tumor necrosis factor-α (TNF-α) but was not beneficial in clinical trials [125]. Antioxidants such as OPC-14117, which is structurally similar to vitamin E, also had no effect in clinical trials [126]. In addition, the effect in clinical trials [126]. In addition, the L-type Ca2+ blocker nimodipine had no effect on HAD, although it shows some promise for HIV-associated peripheral neuropathies [127].

VI. Future Perspectives

As the human population continues to grow and with greater international travel and transport of goods; evolving social, economic and scientific milieus; changes in climate and geographical (e.g., mining, deforestation, dams) conditions; and pervasive global poverty, the potential for new viral infections of the CNS to emerge is immense. Indeed, Nipah virus was thought to be limited to Southeast Asia but is now found in India, and related viruses have been identified in Australia. Likewise, the potential emergence of new influenza virus strains, some with neurovirulent properties, represent serious threats to human and animal health [128]. At the same time, there is a growing interest in the pathogenic roles of resident brain flora; for example, the human herpes virus-6 and HSV-1 have been implicated in the pathogenesis of multiple sclerosis and Alzheimer's disease, respectively [129]. The stellar rise in availability of new therapeutics for neurological disease has been a boon for clinicians in the past 2 decades. New anticonvulsants with both antiepileptic and analgesic properties have improved the quality of life and health outcomes for patients affected with these disorders. Similarly, immunotherapies including intravenous immunoglobulin, plasmapheresis, and immunosuppressive drugs have had substantial impact neurological disease morbidity and mortality. Nonetheless, growing concerns about drug-resistant virus strains haunt clinics and present serious therapeutic limitations for common infections such as HIV [130] and HSV [131]. Indeed, new therapies are also on the horizon: RNA-interference strategies and small (antiviral) molecules are leading the way together with the new uses for established drugs as neuroprotective agents, as recently suggested for several antibiotics [132,133].

Acknowledgments

The authors thank for Sherry Sweeney for assistance with manuscript preparation and members of the Neurovirology Laboratory for helpful discussions. C. P. holds a Canada Research Chair (T1) in Neurological Infection and Immunity.

References

1. Johnson, R. T. (1998). "Viral Infections of the Nervous System" ed. 2. Lippincott-Raven, Philadelphia.
2. Kanki, P. J., Hamel, D. J., Sankale, J. L., Hsieh, C., Thior, I., Barin, F., Woodcock, S. A., Gueye-Ndiaye, A., Zhang, E. Montano, M., Siby, T. Marlink, R., N. D. I., Essex, M. E., and M. B. S. (1999). Human immunodeficiency virus type 1 subtypes differ in disease progression. *J. Infect. Dis.* **179**, 68–73.
3. Wilson, M. E. (2005). Ecological disturbances and emerging infections: travel, dams, shipment of goods, and vectors. *In:* "Emerging Neurological Infections" (C. Power and R. T. Johnson, eds.), pp. 35–58. Taylor & Francis, Boca Raton, FL.
4. Brew, B. J. (2001). "HIV Neurology," pp. 132–175. Oxford University Press, Oxford.
5. Portegies, P., de Gans, J., and Lange, J. M. (1989). Declining incidence of AIDS dementia complex after introduction of zidovudine treatment. *Brit. Med. J.* **299**, 819–821.
6. Coyle, P. K. (2000). Postinfectious encephalomyelitis. *In:* "Infectious Diseases of the Nervous System" (L. E. David et al., eds.), pp. 43–108. Butterworth-Heinemann, City.
7. Cassady, K. A., and Whitley, R. J. (2004). Pathogenesis and pathophysiology of viral infections of the central nervous system. *In:* "Infections of the Central Nervous System" (W. M. Scheld et al., eds.), ed. 3, pp. 57–74. Lippincott, Williams & Wilkins, City.

8. Barker, R., and Wharton, S. (2001). Infection. In: "Brain Damage, Brain Repair" (J. W. Fawcett et al., eds.), pp. 57–78. Oxford University Press, City.
9. Nath, A., Galey, D., and Berger, J. R. (2003). Neuropathogenesis of viral infections. In: "Clinical Neurovirology" (A. Nath, et al., eds.), pp. 21–34. Marcel Dekker, City.
10. Johnson, R. T. (2002). West Nile virus in the US and abroad. Curr. Clin. Top. Infect. Dis. **22**, 52–60.
11. Komar, N. (2003). West Nile virus: epidemiology and ecology in North America. Adv. Virus Res. **61**, 185–234.
12. Bagasra, O., Lavi, E., Bobroski, L., Khalili, K., Pestaner, J. P., Tawadros, R., and Pomerantz, R. J. (1996). Cellular reservoirs of HIV-1 in the central nervous system of infected individuals: identification by the combination of in situ polymerase chain reaction and immunohistochemistry. AIDS **10**, 573–585.
13. Torres-Munoz, J., Stockton, P., Tacoronte, N., B. Roberts, B., Maronpot, R. R., and Petito, C. K. (2001). Detection of HIV-1 gene sequences in hippocampal neurons isolated from postmortem AIDS brains by laser capture microdissection. J. Neuropathol. Exp. Neurol. **60**, 885–892.
14. Nuovo, G. J., Gallery, F., MacConnell, P., and Braun, A. (1994). In situ detection of polymerase chain reaction-amplified HIV-1 nucleic acids and tumor necrosis factor-alpha RNA in the central nervous system. Am. J. Pathol. **144**, 659–666.
15. Trillo-Pazos, G., Diamanturos, A., Rislove, L., Menza, T., Chao, W., Belem, P., Sadiq, S., Morgello, S., Sharer, L., and D. J. Volsky, D. J. (2003). Detection of HIV-1 DNA in microglia/macrophages, astrocytes and neurons isolated from brain tissue with HIV-1 encephalitis by laser capture microdissection. Brain Pathol. **13**, 144–154.
16. Kaul, M., Garden, G. A., and Lipton, S. A. (2001). Pathways to neuronal injury and apoptosis in HIV-associated dementia. Nature **410**, 988–994.
17. Wesselingh, S. L., and Thompson, K. A. (2001). Immunopathogenesis of HIV-associated dementia. Curr. Opin. Neurol. **14**, 375–379.
18. Mollace, V., Nottet, H. S., Clayette, P., Turco, M. C., Muscoli, C., Salvemini, D., and Perno, C. F. (2001). Oxidative stress and neuroAIDS: triggers, modulators and novel antioxidants. Trends Neurosci. **24**, 411–416.
19. Talbot, P. J., Jacomy, H., and Gruslin, H. (2005). Principles of immune-virus interactions in the nervous system. In: "Clinical Neuroimmunology" (J. Antel, et al., eds.), ed. 2, pp. 115–128. Oxford University Press, City.
20. Lane, J. H., Sasseville, V. G., Smith, M. O., Vogel, P., Pauley, D. R., Heyes, M. P., and Lackner, A. A. (1996). Neuroinvasion by simian immunodeficiency virus coincides with increased numbers of perivascular macrophages/microglia and intrathecal immune activation. J. Neurovirol. **2**, 423–432.
21. Sasseville, V. G., Smith, M. M., Mackay, C. R., Pauley, D. R., Mansfield, K. G., Ringler, D. J., and Lackner, A. A. (1996). Chemokine expression in simian immunodeficiency virus-induced AIDS encephalitis. Am. J. Pathol. **149**, 1459–1467.
22. Gabuzda, D., and Wang, J. (2000). Chemokine receptors and mechanisms of cell death in HIV neuropathogenesis. J. Neurovirol. **6(Suppl 1)**, S24–S32.
23. Koirala, T. R., Sharma, S., Morikawa, S., Ishida, T. (2000). Expression of CXCR4 in the brain of feline immunodeficiency virus infected cat. Indian J. Pathol. Microbiol. **43**, 285–290.
24. Zheng, J., Ghorpade, A., Niemann, D., Cotter, R. L., Thylin, M. R., Epstein, L., Swartz, J. M., Shepard, R. B., Liu, X., Nukuna, A., and Gendelman. H. E. (1999). Lymphotropic virions affect chemokine receptor-mediated neural signaling and apoptosis: implications for human immunodeficiency virus type 1-associated dementia. J. Virol. **73**, 8256–8267.
25. Zheng, J., Thylin, M. R., Ghorpade, A., Xiong, H., Persidsky, Y., Cotter, R., Niemann, D., Che, M., Zeng, Y. C., Gelbard, H. A., Shepard, R. B., Swartz, J. M., and Gendelman, H. E. (1999). Intracellular CXCR4 signaling, neuronal apoptosis and neuropathogenic mechanisms of HIV-1-associated dementia. J. Neuroimmunol. **98**, 185–200.
26. Garden, G. A. (2002). Microglia in human immunodeficiency virus-associated neurodegeneration. Glia **40**, 240–251.
27. Jordan-Sciutto, K. L., Wang, G., Murphy-Corb, M., and Wiley, C. A. (2000). Induction of cell-cycle regulators in simian immunodeficiency virus encephalitis. Am. J. Pathol. **157**, 497–507.
28. Peruzzi, F., Gordon, J., Darbinian, N., and Amini, S. (2002). Tat-induced deregulation of neuronal differentiation and survival by nerve growth factor pathway. J. Neurovirol. **8**, 91–96.
29. van Marle, G., Antony, J. M., Silva, C., Sullivan, A., and Power, C. (2005). Aberrant cortical neurogenesis in a pediatric neuroAIDS model: neurotrophic effects of growth hormone. AIDS **19**, 1781–1791.
30. Irani, D. N., and Griffin, D. E. (2001). Regulation of T cell responses during central nervous system viral infections. Adv. Virol. **56**, 175–198.
31. Giuliani, F., Goodyer, C. G., Antel, J. P., and Yong, V. W. (2003). Vulnerability of human neurons to T cell-mediated cytotoxicity. J. Immunol. **171**, 368–379.
32. O'Brien, S. J., Nelson, G. W., Winkler, C. A., and Smith, M. W. (2000). Polygenic and multifactorial disease gene association in man: lessons from AIDS. Annu. Rev. Genet. **34**, 563–591.
33. Patrick, M. K., Johnston, J. B., and Power, C. (2002). Lentiviral neuropathogenesis: comparative neuroinvasion, neurotropism, neurovirulence and host neurosusceptibility. J. Virol. **76**, 7923–7931.
34. van Marle, G., and Power, C. (2005). Human immunodeficiency virus type 1 genetic diversity in the nervous system: evolutionary epiphenomenon or disease determinant? J. Neurovirol. **11**, 107–128.
35. Sabouri, A. H., Saito, M., Usuku, K., Bajestan, S. N., Mahmoudi, M., Forughipour, M., Sabouri, Z., Abbaspour, Z., Goharjoo, M. E., Khayami, E., Hasani, A., Izumo, S., Arimura, K., Farid, R., and Osame, M. (2005). Differences in viral and host genetic risk factors for development of human T-cell lymphotropic virus type 1 (HTLV-1)–associated myelopathy/tropical spastic paraparesis between Iranian and Japanese HTLV-1-infected individuals. J. Gen. Virol. **86**, 773–781.
36. Booss, J., and Esiri, M. M. (2003). Viral encephalitis in humans. ASM Press, 1–20.
37. Cohen, B. A., and Levy, R. M. (2003). The role of the brain biopsy in the diagnosis of CNS viral infections. In: "Clinical Neurovirology" (A. Nath, et al., eds.), pp. 35–42. Marcel Dekker, City.
38. Tyler, K. L. (2004). Herpes simplex virus infections of the central nervous system: encephalitis and meningitis, including Mollaret's. Herpes **11(Suppl 2)**, 57A–64A.
39. Whitley, R. J., and Kimberlin, D. W. (2005). Herpes simplex encephalitis: children and adolescents. Semin. Pediatr. Infect. Dis. **16**, 17–23.
40. Whitley, R. J., Soong, S. J., Linneman, C. Jr., Liu, C., Pazin, G., and Alford, C. A. (1982). Herpes simplex encephalitis. Clinical Assessment. JAMA **247**, 317–320.
41. Schmutzhard, E. (2001). Viral infections of the CNS with special emphasis on herpes simplex infections. J. Neurol. **248**, 469–477.
42. Whitley, R., Arvin, A., Prober, C., Burchett, S., Corey, L., Powell, D. Plotkin, S., Starr, S., Alford, C., Connor, J., et al. (1991). A controlled trial comparing vidarabine with acyclovir in neonatal herpes simplex virus infection. Infectious Diseases Collaborative Antiviral Study Group. N. Engl. J. Med. **324**, 444–449.
43. Aurelius, E., Johansson, B., Skoldenberg, B., and Forsgren, M. (1993). Encephalitis in immunocompetent patients due to herpes simplex virus type 1 or 2 as determined by type-specific polymerase chain reaction and antibody assays of cerebrospinal fluid. J. Med. Virol. **39**, 179–186.
44. Nahmias, A. J., Whitley, R. J., and Visintine, A. N. (1982). Herpes simplex virus encephalitis: laboratory evaluations and their diagnostic significance. J. Infect. Dis. **145**, 829–836.
45. Stevens, J. G. (1989). Human herpesviruses: a consideration of the latent state. Microbiol. Rev. **53**, 318–332.

46. Cook, M. L., and Stevens, J. G. (1973). Pathogenesis of herpetic neuritis and ganglionitis in mice: evidence for intra-axonal transport of infection. *Infect. Immun.* **7**, 272–288.
47. Davis, L. E., and Johnson, R. T. (1979). An explanation for the localization of herpes simplex encephalitis? *Ann. Neurol.* **5**, 2–5.
48. Steiner, I., Spivack, J. G., O'Boyle, D. R. II, Lavi, E., and Fraser, N. W. (1988). Latent herpes simplex virus type 1 transcription in human trigeminal ganglia. *J. Virol.* **62**, 3493–3496.
49. Barnett, E. M., Cassell, M. D., and Perlman, S. (1993). Two neurotropic viruses, herpes simplex virus type 1 and mouse hepatitis virus, spread along different neural pathways from the main olfactory bulb. *Neuroscience* **57**, 1007–1025.
50. Liedtke, W., Opalka, B., Zimmermann, C. W., and Lignetz, E. (1993). Age distribution of latent herpes simplex virus 1 and varicella-zoster virus genome in human nervous tissue. *J. Neurol. Sci.* **116**, 6–11.
51. Booss, J., and Kim, J. H. (1984). Biopsy histopathology in herpes simplex encephalitis and in encephalitis of undefined etiology. *Yale J. Biol. Med.* **57**, 751–755.
52. Garcia, J. H., Colon, L. E., Whitley, R. J., and Wilmes, F. J. (1984). Diagnosis of viral encephalitis by brain biopsy. *Semin. Diagn. Pathol.* **1**, 71–81.
53. Kimberlin, D. (2004). Herpes simplex virus, meningitis and encephalitis in neonates. *Herpes* **11(Suppl 2)**, 65A–76A.
54. Domingues, R. B., Tsanaclis, A. M., Pannuti, C. S., Mayo, M. S., and Lakeman, F. D. (1997). Evaluation of the range of clinical presentations of herpes simplex encephalitis by using polymerase chain reaction assay of cerebrospinal fluid samples. *Clin. Infect. Dis.* **25**, 86–91.
55. Whitley, R. J., and Gnann, J. W. (2002). Viral encephalitis: familiar infections and emerging pathogens. *Lancet* **359**, 507–513.
56. Grover, D., Newsholme, W., Brink, N., Manji, H., and Miller, R. (2004). Herpes simplex virus infection of the central nervous system in human immunodeficiency virus-type 1-infected patients. *Int. J. STD AIDS* **15**, 597–600.
57. Schiff, D., and Rosenblum, M. K. (1998). Herpes simplex encephalitis (HSE) and the immunocompromised: a clinical and autopsy study of HSE in the settings of cancer and human immunodeficiency virus-type 1 infection. *Hum. Pathol.* **29**, 215–222.
58. Lakeman, F. D., and Whitley, R. J. (1995). Diagnosis of herpes simplex encephalitis: application of polymerase chain reaction to cerebrospinal fluid from brain-biopsied patients and correlation with disease. National Institute of Allergy and Infectious Diseases Collaborative Antiviral Study Group. *J. Infect. Dis.* **171**, 857–863.
59. Tebas, P., Nease, R. F., and Storch, G. A. (1998). Use of the polymerase chain reaction in the diagnosis of herpes simplex encephalitis: a decision analysis model. *Am. J. Med.* **105**, 287–295.
60. Puchhammer-Stockl, E., Presterl, E., Croy, C., Aberle, S., Popow-Kraupp, T., Kundi, M., Hofmann, H., Wenninger, U., and Godl, I. (2001). Screening for possible failure of herpes simplex virus PCR in cerebrospinal fluid for the diagnosis of herpes simplex encephalitis. *J. Med. Virol.* **64**, 531–536.
61. Ando, Y., Kimura, H., Miwata, H., Kudo, T., Shibata, M., and Morishima, T. (1993). Quantitative analysis of herpes simplex virus DNA in cerebrospinal fluid of children with herpes simplex encephalitis. *J. Med. Virol.* **41**, 170–173.
62. Domingues, R. B., Lakeman, F. D., Mayo, M. S., and Whitley, R. J. (1998). Application of competitive PCR to cerebrospinal fluid samples from patients with herpes simplex encephalitis. *J. Clin. Microbiol.* **36**, 2229–2234.
63. Wildemann, B., Ehrhart, K., Storch-Hagenlocher, B., Meyding-Lamade, U., Steinvorth, S., Hacke, W., and Haas, J. (1997). Quantitation of herpes simplex virus type 1 DNA in cells of cerebrospinal fluid of patients with herpes simplex virus encephalitis. *Neurology* **48**, 1341–1346.
64. Klapper, P. E., Laing, I., and Longson, M. (1981). Rapid non-invasive diagnosis of herpes encephalitis. *Lancet* **2**, 607–609.
65. Vandvik, B., Skoldenberg, B., Forsgren, M., Stiernstedt, G., Jeansson, S., and Norrby, E. (1985). Long-term persistence of intrathecal virus-specific antibody responses after herpes simplex virus encephalitis. *J. Neurol.* **231**, 307–312.
66. Boivin, G. (2004). Diagnosis of herpesvirus infections of the central nervous system. *Herpes* **11(Suppl 2)**, 48A–56A.
67. Domingues, R. B., Fink, M. C., Tsanaclis, A. M., de Castro, C. C., Cerri, G. G., Mayo, M. S., and Lakeman, F. D. (1998). Diagnosis of herpes simplex encephalitis by magnetic resonance imaging and polymerase chain reaction assay of cerebrospinal fluid. *J. Neurol. Sci.* **157**, 148–153.
68. McCabe, K., Tyler, K., and Tanabe, J. (2003). Diffusion-weighted MRI abnormalities as a clue to the diagnosis of herpes simplex encephalitis. *Neurology* **61**, 1015–1016.
69. Garcia-Morales, I., Garcia, M. T., Galan-Davila, L., Gomez-Escalonilla, C., Saiz-Diaz, R., Martinez-Salio, A., de la Pena, P., and Tejerina, J. A. (2002). Periodic lateralized epileptiform discharges: etiology, clinical aspects, seizures, and evolution in 130 patients. *J. Clin. Neurophysiol.* **19**, 172–177.
70. Misra, U. K., and Kalita, J. (1998). Neurophysiological studies in herpes simplex encephalitis. *Electromyogr. Clin. Neurophysiol.* **38**, 177–182.
71. Whitley, R. J., Alford, C. A., Hirsch, M. S., Schooley, R. T., Luby, J. P., Aoki, F. Y., Hanley, D., Nahmias, A. J., and Soong, S. J. (1986). Vidarabine versus acyclovir therapy in herpes simplex encephalitis. *N. Engl. J. Med.* **314**, 144–149.
72. Longson, M., Klapper, P. E., and Cleator, G. M. 1983. The treatment of herpes encephalitis. *J. Infect.* **6**, 15–16.
73. VanLandingham, K. E., Marsteller, H. B., Ross, G. W., Hayden, F. G. (1988). Relapse of herpes simplex encephalitis after conventional acyclovir therapy. *JAMA* **259**, 1051–1053.
74. Cinque, P., Cleator, G. M., Weber, T., Monteyne, P., Sindic, C. J., and van Loon, A. M. (1996). The role of laboratory investigation in the diagnosis and management of patients with suspected herpes simplex encephalitis: a consensus report. The EU Concerted Action on Virus Meningitis and Encephalitis. *J. Neurol. Neurosurg. Psychiatry* **61**, 339–345.
75. Skoldenberg, B., Forsgren, M., Alestig, K., Bergstrom, T., Burman, L., Dahlqvist, E., Forkman, A., Fryden, A., Lovgren, K., Norlin, K., and et al. (1984). Acyclovir versus vidarabine in herpes simplex encephalitis. Randomised multicentre study in consecutive Swedish patients. *Lancet* **2**, 707–711.
76. Gonzalez-Scarano, F., and Martin-Garcia, J. (2005). The neuropathogenesis of AIDS. *Nat. Rev. Immunol.* **5**, 69–81.
77. McArthur, J. C., Hoover, D. R., Bacellar, H., Cornblath, D. R., Selnes, O. A., Ostrow, D., Johnson, R. T., Phair, J., and Polk, B. F. (1993). Dementia in AIDS patients: incidence and risk factors. Multicenter AIDS Cohort Study. *Neurology* **43**, 2245–2252.
78. Dore, G. J., McDonald, A., Li, Y., Kaldor, J. M., and Brew, B. J. (2003). Marked improvement in survival following AIDS dementia complex in the era of highly active antiretroviral therapy. *AIDS* **17**, 1539–1545.
79. Ellis, R. J., Hsia, K., Spector, S. A., Nelson, J. A., Heaton, R. K., Wallace, M. R., Abramson, I., Atkinson, J. H., Grant, I., and McCutchan, J. A. (1997). Cerebrospinal fluid human immunodeficiency virus type 1 RNA levels are elevated in neurocognitively impaired individuals with acquired immunodeficiency syndrome. HIV Neurobehavioral Research Center Group. *Ann. Neurol.* **42**, 679–688.
80. Ellis, R. J., Deutsch, R., Heaton, R. K., Marcotte, T. D., McCutchan, J. A., Nelson, J. A., Abramson, I., Thal, L. J., Atkinson, J. H., Wallace, M. R., and Grant, I. (1997). Neurocognitive impairment is an independent risk factor for death in HIV infection. San Diego HIV Neurobehavioral Research Center Group. *Arch. Neurol.* **54**, 416–424.

81. Dana Consortium on the Therapy of HIV Dementia and Related Cognitive Disorders. (1998). A randomized, double-blind, placebo-controlled trial of deprenyl and thioctic acid in human immunodeficiency virus-associated cognitive impairment [comments]. *Neurology* **50**, 645–651.
82. Tardieu, M., and Boutet, A. (2002). HIV-1 and the central nervous system. *Curr. Top. Microbiol. Immunol.* **265**, 183–195.
83. Power, C., and Johnson, R. T. (2001). Neurovirological and neuroimmunological aspects of HIV infection. *Adv. Virus Res.* **56**, 579–624.
84. Navia, B. A., Cho, E. S., Petito, C. K., and Price, R. W. (1986). The AIDS dementia complex: II. Neuropathology. *Ann. Neurol.* **19**, 525–535.
85. Sharer, L. R. (1992). Pathology of HIV-1 infection of the central nervous system. A review. *J. Neuropathol. Exp. Neurol.* **51**, 3–11.
86. Vinters, H. V., and Anders, K. H. (1990). "Neuropathology of AIDS." CRC Press, Boca Raton, FL.
87. Petito, C. K., and Cash, K. S. (1992). Blood-brain barrier abnormalities in the acquired immunodeficiency syndrome: immunohistochemical localization of serum proteins in postmortem brain. *Ann. Neurol.* **32**, 658–666.
88. Power, C., Kong, P. A., Crawford, T. O., Wesselingh, S., Glass, J. D., McArthur, J. C., and Trapp, B. D. (1993). Cerebral white matter changes in acquired immunodeficiency syndrome dementia: alterations of the blood-brain barrier. *Ann. Neurol.* **34**, 339–350.
89. Glass, M., Faull, R. L., Bullock, J. Y., et al. (1996). Loss of A1 adenosine receptors in human temporal lobe epilepsy. *Brain Res.* **710**, 56–68.
90. Sharer, L. R., Epstein, L. G., Cho, E. S., et al. (1986). Pathologic features of AIDS encephalopathy in children: evidence for LAV/HTLV-III infection of brain. *Hum. Pathol.* **17**, 271–284.
91. Achim, C. L., Heyes, M. P., and Wiley, C. A. (1993). Quantitation of human immunodeficiency virus, immune activation factors, and quinolinic acid in AIDS brains. *J. Clin. Invest.* **91**, 2769–2775.
92. Glass, J. D., Fedor, H., Wesselingh, S. L., McArthur, J. C. (1995). Immunocytochemical quantitation of human immunodeficiency virus in the brain: correlations with dementia. *Ann. Neurol.* **38**, 755–762.
93. Everall, I. P., Luthert, P. J., and Lantos, P. L. (1991). Neuronal loss in the frontal cortex in HIV infection. *Lancet* **337**, 1119–1121.
94. Masliah, E., Ge, N., Achim, C. L., Hansen, L. A., and Wiley, C. A. (1992). Selective neuronal vulnerability in HIV encephalitis. *J. Neuropathol. Exp. Neurol.* **51**, 585–593.
95. Masliah, E., Ge, N., Morey, M., DeTeresa, R., Terry, R. D., and Wiley, C. A. (1992). Cortical dendritic pathology in human immunodeficiency virus encephalitis. *Lab. Invest.* **66**, 285–291.
96. Thompson, K. A., McArthur, J. C., and Wesselingh, S. L. (2001). Correlation between neurological progression and astrocyte apoptosis in HIV-associated dementia. *Ann. Neurol.* **49**, 745–752.
97. Maher, J., Choudhri, S., Halliday, W., Power, C., and Nath, A. (1997). AIDS dementia complex with generalized myoclonus. *Mov. Disord.* **12**, 593–597.
98. Mirsattari, S. M., Power, C., and Nath, A. (1998). Parkinsonism with HIV infection. *Mov. Disord.* **13**, 684–689.
99. Navia, B. A., Jordan, B. D., and Price, R. W. (1986). The AIDS dementia complex: I. Clinical features. *Ann. Neurol.* **19**, 517–524.
100. McArthur, J. C., Brew, B. J., and Nath, A. (2005). Neurological complications of HIV infection. *Lancet Neurol.* **4**, 543–555.
101. Alciati, A., Fusi, A., D'Arminio Monforte, A., Coen, M., Ferri, A., and Mellado, C. (2001). New-onset delusions and hallucinations in patients infected with HIV. *J Psychiatry Neurosci.* **26**, 229–234.
102. Koutsilieri, E., Scheller, C., Sopper, S., ter Meulen, V., and Riederer, P. (2002). Psychiatric complications in human immunodeficiency virus infection. *J. Neurovirol.* **8(Suppl 2)**, 129–133.
103. Nath, A., and Berger, J. (2004). HIV dementia. *Curr. Treat Options Neurol.* **6**, 139–151.
104. Shelburne, S. A. III and Hamill, R. J. (2003). The immune reconstitution inflammatory syndrome. *AIDS Rev.* **5**, 67–79.
105. Miller, R. F., Isaacson, P. G., Hall-Craggs, M., Lucas, S. F. Gray, F., Scaravilli, F. and An, S. F. (2004). Cerebral CD8+ lymphocytosis in HIV-1 infected patients with immune restoration induced by HAART. *Acta. Neuropathol. (Berl.)* **108**, 17–23.
106. Vendrely, A., Bienvenu, B., Gasnault, J., Thiebault, J. B., Salmon, D., and Gray, F. (2005). Fulminant inflammatory leukoencephalopathy associated with HAART-induced immune restoration in AIDS-related progressive multifocal leukoencephalopathy. *Acta Neuropathol (Berl).* **109**, 449–455.
107. Power, C., Selnes, O. A., Grim, J. A., and McArthur, J. C. (1995). HIV Dementia Scale: a rapid screening test. *J. AIDS Hum. Retrovirol.* **8**, 273–278.
108. Jones, B. N., Teng, E. L., Folstein, M. F., and Harrison, K. S. (1993). A new bedside test of cognition for patients with HIV infection. *Ann. Intern. Med.* **119**, 1001–1004.
109. Berghuis, J. P., Uldall, K. K., and Lalonde, B. (1999). Validity of two scales in identifying HIV-associated dementia. *J. AIDS* **21**, 134–140.
110. Grassi, M., Perin, P. C., Borella, M., and Mangoni, A. (1999). Assessment of cognitive function in asymptomatic HIV-positive subjects. *Eur. Neurol.* **42**, 225–229.
111. Price, R. W., and Brew, B. J. (1988). The AIDS dementia complex. *J. Infect. Dis.* **158**, 1079–1083.
112. Dal Pan, G. J., McArthur, J. H., Aylward, E., Selnes, O. A., Nance-Sproson, T. E., Kumar, A. J., Mellits, E. D., and McArthur, J. C. (1992). Patterns of cerebral atrophy in HIV-1-infected individuals: results of a quantitative MRI analysis. *Neurology* **42**, 2125–2130.
113. Simpson, D. M., and Tagliati, M. (1994). Neurologic manifestations of HIV infection [published erratum appears in Ann. Intern. Med. 1995, **122**, 317]. *Ann. Intern. Med.* **121**, 769–785.
114. McArthur, J. C., Cohen, B. A., Farzedegan, H., Cornblath, D. R., Selnes, O. A., Ostrow, D., Johnson, R. T., Phair, J., and Polk, B. F. (1988). Cerebrospinal fluid abnormalities in homosexual men with and without neuropsychiatric findings. *Ann. Neurol.* **23**, S34–S37.
115. Chang, L., Ernst, T., Leonido-Yee, M., Walot, I., and Singer, E. (1999). Cerebral metabolite abnormalities correlate with clinical severity of HIV-1 cognitive motor complex. *Neurology* **52**, 100–108.
116. Power, C., McArthur, J. C., Johnson, R. T., Griffin, D. E., Glass, J. D., Perryman, S., and Chesebro, B. (1994). Demented and nondemented patients with AIDS differ in brain-derived human immunodeficiency virus type 1 envelope sequences. *J. Virol.* **68**, 4643–4649.
117. Price, R. W., Yiannoutos, T., Clifford, D. B., Zaborski, L., Tselis, A., Sidtis, J. J., Cohen, B., Hall, C. D., Erice, A., and Henry, K. (1999). Neurological outcomes in late HIV infection: adverse impact of neurological survival and protection effect of antiretroviral therapy. *AIDS* **13**, 1677–1685.
118. Sacktor, N. C., Lyles, R. H., Skolasky, R. L., Anderson, D. E., McArthur, J. C., McFarlane, G., Selnes, O. A., Becker, J. T., Cohen, B., Wesch, J., and Miller, E. N. (1999). Combination antiretroviral therapy improves psychomotor speed performance in HIV-seropositive homosexual men. Multicenter AIDS Cohort Study (MACS). *Neurology* **52**, 1640–1647.
119. von Giesen, H. J., Koller, H., Theisen, A., and Arendt, G. (2002). Therapeutic effects of nonnucleoside reverse transcriptase inhibitors on the central nervous system in HIV-1-infected patients. *J. AIDS* **29**, 363–367.
120. Brown, G. R. (1995). The use of methylphenidate for cognitive decline associated with HIV disease. *Int. J. Psychiatry Med.* **25**, 21–37.
121. Sacktor, N. C., Skolasky, R. L., Lyles, R. H., Esposito, D., Selnes, O. A., and McArthur, J. C. (2000). Improvement in HIV-associated motor slowing after antiretroviral therapy including protease inhibitors. *J. Neurovirol.* **6**, 84–88.
122. Nath, A., Haughey, N. J., Jones, M., Anderson, C. J., Bell, J. E., and Geiger, J. D. (2000). Synergistic neurotoxicity by human immunodeficiency virus proteins Tat and gp120: protection by memantine. *Ann. Neurol.* **47**, 186–194.

123. Zhang, K., McQuibban, G. A., Silva, C., Butler, G. S., Johnston, J. B., Holden, J., Clark-Lewis, I., Overall, C. M., and Power, C. (2003). HIV-induced metalloproteinase processing of the chemokine stromal cell derived factor-1 causes neurodegeneration. *Nat. Neurosci.* **6**, 1064–1071.
124. Silva, C., Zhang, K., Tsutsui, S., Holden, J. K., Gill, M. J., and Power, C. (2003). Growth hormone prevents human immunodeficiency virus-induced neuronal p53 expression. *Ann. Neurol.* **54**, 605–614.
125. Clifford, D. B. (2000). Human immunodeficiency virus-associated dementia. *Arch. Neurol.* **57**, 321–324.
126. The Dana Consortium on the Therapy of HIV Dementia and Related Cognitive Disorders. (1997). Safety and tolerability of the antioxidant OPC-14117 in HIV-associated cognitive impairment. *Neurology* **49**, 142–146.
127. Navia, B. A., Dafni, U., Simpson, D., Tucker, T., Singer, E., McArthur, J. C., Yiannoutsos, C., Zaborski, L., and Lipton. S. A. (1998). A phase I/II trial of nimodipine for HIV-related neurologic complications. *Neurology* **51**, 221–228.
128. Lee, C. W., Suarez, D. L., Tumpey, T. M., Sung, H. W., Kwon, Y. K., Lee, Y. J., Choi, J. G., Joh, S. J., Kim, M. C., Lee, E. K., Park, J. M., Lu, X., Katz, J. M., Spackman, E., Swayne, D. E., and Kim, J. H. (2005). Characterization of highly pathogenic H5N1 avian influenza A viruses isolated from South Korea. *J. Virol.* **79**, 3692–3702.
129. Mayne, M., and Johnston, J. B. (2005). Latent and activated brain flora: human herpesvirus, endogenous retroviruses, coronaviruses, and chlamydia and their role in neurological disease. *In:* "Emerging Neurological Infections" (C. Power and R. T. Johnson, eds.), pp. 363–396. Taylor & Francis, Boca Raton, FL.
130. Loy, C. T., Tomlinson, S., and Brew, B. J. (2005). HIV-related neurological disease in the era of HAART. *In:* "Emerging Neurological Infections" (C. Power, et al., eds.), pp. 427–471. Taylor & Francis, Boca Raton, FL.
131. Tenser, R. B. (2005). Herpes simplex virus drug resistance-HSV thymidine kinase mutants. *In:* "Emerging Neurological Infections" (C. Power, et al., eds.), pp. 397–414. Taylor & Francis, Boca Raton, FL.
132. Yong, V. W., Wells, J., Giuliani, F., Casha, S., Power, C., Metz, L. M., and Tabira, T. Related articles. The promise of minocycline in neurology. *Lancet Neurol.* **3**, 744–751.
133. Rothstein, J. D., Patel, S., Regan, M. R., Haenggeli, C., Huang, Y. H., Bergles, D. E., Jin, L., Dykes Hoberg, M., Vidensky, S., Chung, D. S., Toan, S. V., Bruijn, L. I., Su, Z. Z., Gupta, P., and Fisher, P. B. (2005). Beta-lactam antibiotics offer neuroprotection by increasing glutamate transporter expression. *Nature* **433**, 73–77.
134. Petersen, L. R., Marfin, A. A., and Gubler, D. J. (2003). West Nile virus. *JAMA* **290**, 524–528.

44

Spinal Muscular Atrophy

Charlotte J. Sumner, MD
Kenneth H. Fischbeck, MD

Keywords: *motor neuron, spinal muscular atrophy, survival motor neuron*

I. Introduction
II. Spinal Muscular Atrophy
III. Genetic Basis of SMA
IV. SMN Gene Expression
V. Splicing of SMN Transcripts
VI. SMN Protein
VII. Animal Models of SMA
VIII. Therapeutic Strategies for SMA
IX. Remaining Questions
References

I. Introduction

Spinal muscular atrophy (SMA) is a currently untreatable, autosomal recessive motor neuron disease. With an incidence of approximately 1 in 10,000 live births and a carrier frequency of 1 in 50, SMA is the leading inherited cause of infant mortality. Over the past decade, research efforts have led to the discovery of the genetic basis of SMA, an understanding of some of the functions of the survival motor neuron (SMN) protein, and the development of SMA animal models that closely mimic the human disease. Although more work is needed to fully understand the mechanism of SMA disease pathogenesis, promising targets for SMA therapeutics have been identified and early clinical trials of compounds directed to these targets are ongoing in SMA patients.

II. Spinal Muscular Atrophy

A. Clinical Features

SMA was first described in the 1890s by Guido Werdnig of the University of Vienna and Johann Hoffman of

Heidelberg University. The cardinal signs of SMA in all patients are muscle weakness and atrophy due to motor neuron loss. The pattern of weakness is symmetrical and proximal, with the legs more affected than the arms and the arms more affected than the facial muscles and diaphragm. It has long been recognized that SMA disease severity is widely heterogeneous, prompting debate (before the causative gene was identified) as to whether this represented a single entity or multiple different diseases. SMA is currently classified into three types: types I, II, and III based on international consensus (Table 1). This classification provides clinical utility, although it is recognized that the boundaries between types are arbitrary and the disease actually has a continuous range of severity. Patients with type I SMA (Werdnig-Hoffman disease) have severe, generalized muscle weakness and hypotonia at birth or within the first 6 months of life. Patients never achieve the ability to sit, and death usually occurs from respiratory insufficiency within the first 2 years if ventilatory support is not provided. Patients with type II SMA have onset after 6 months of age and can sit but never walk unaided. Prognosis in this group is largely dependent on the degree of respiratory involvement, and survival for decades is expected with aggressive respiratory management. Patients with type III SMA (Kugelberg-Welander disease) usually have their first symptoms between 18 months of age and early childhood. They are able to stand and walk but often become wheelchair dependent during youth or adulthood. Life expectancy is not usually reduced in this group. There are also well-described cases in which patients have onset in late childhood or adulthood (as late as the fifth or sixth decade). Some investigators have classified these cases as SMA type IV.

B. Disease Course

The disease course in SMA is distinct from other degenerative motor neuron diseases such as amyotrophic lateral sclerosis (ALS) and spinal and bulbar muscular atrophy (Kennedy's disease). Rather than an inexorably progressive disease course after onset, patients with SMA tend to have the greatest rate of loss of muscle power at disease onset. This may manifest as a clear loss of strength in some infants; in others, there may be only the absence of normal motor milestone gains with early growth and development. After this early phase, residual muscle strength often stabilizes for many years in surviving patients. In some patients, subsequent deterioration in functional abilities may result from the secondary effects of scoliosis and contractures rather than changes in muscle strength. This disease course, as well as pathological observations of immature muscle in severe forms of SMA (see Section C), has led to the concept that SMA may be, at least in part, a developmental disease.

C. Pathological Features

The predominant feature of autopsy studies of patients with SMA is loss of motor neurons in the ventral horn of the spinal cord and in brainstem motor nuclei (reviewed in [1]). Remaining motor neurons may appear normal, or they may show swelling of the perikaryon and loss of Nissl substance or occasionally may appear atrophic. Associated mild ventral horn gliosis is present. Upper motor neurons and the corticospinal tracts are preserved; however, ballooned neurons have also been described within Clarke's nucleus of the spinal cord, in the ventrolateral region of the thalamus, and within the dorsal root ganglia in type I SMA autopsies, indicating that other neuronal types may be involved. Ultrastructural studies of ballooned motor neurons in SMA have demonstrated a loose accumulation of intermediate filaments at the periphery of the cell and an accumulation of mitochondria, vesicles, and lysosomes at the center. It has not been definitively established from these autopsy studies whether axonal degeneration precedes neuronal cell body loss.

Muscle biopsies in patients with type III SMA reveal changes typical of chronic neurogenic atrophy, with small angular fibers, grouped atrophy, and fiber type grouping. In contrast, SMA type I and type II muscle biopsies often show large groups of small, rounded atrophic fibers that can involve the whole fascicle. Rather than grouped atrophic fibers of the same type, fibers are often in a normal checkerboard pattern of interspersed types I and II fibers. Only scattered hypertrophic type I fibers are present. In SMA type I muscle, in particular, many of the small muscle fibers have an immature appearance, with central nuclei. Ultrastructural analysis also indicates an appearance similar to developing myotubes.

III. Genetic Basis of SMA

The SMA disease gene was mapped by linkage analysis to a complex region of chromosome 5q in 1990 [2,3]. This region contains a large inverted duplication and consequently at least four genes that are present in telomeric and centromeric copies: survival motor neuron gene *(SMN)*, neuronal apoptosis inhibitor protein gene *(NAIP)*, basal transcription factor subunit p44 gene, and a gene encoding

Table 1 SMA Disease Classification

SMA Type	Age of Onset (Months)	Motor Milestones	Age of Death (Years)
I	<6	Never sit	<2
II	<18	Sit but never stand	>2
III	>18	Stand	Adult

a protein of unknown function, H4F5t. In 1995 Lefebvre et al. reported that homozygous mutations in the telomeric copy of the *SMN* gene *(SMN1)* cause SMA [4]. These mutations are most often deletion mutations involving at least exons 6 through 8 but can also be nonsense, frameshift, or missense mutations. Although 5% of normal individuals lack the centromeric copy of *SMN (SMN2)*, all patients with SMA retain the *SMN2* gene. *SMN1* and *SMN2* differ by only five nucleotides; however, a translationally silent C→T transition located within an exonic splicing region of *SMN2* leads to frequent exon 7 skipping during transcription of *SMN2* [5,6]. Consequently, whereas *SMN1* produces full-length transcripts, the majority of transcripts that arise from *SMN2* lack exon 7 (Fig. 1). The transcripts lacking exon 7 encode a truncated protein that has impaired ability to oligomerize and associate with its binding partners and may be rapidly degraded [7,8]. *SMN2* is therefore unable to compensate completely for the loss of *SMN1*, and SMA results from a deficiency of full-length SMN protein.

The SMA chromosomal region is unstable, and gene conversion events of *SMN1* to *SMN2* mean that the *SMN2* gene copy number varies in different individuals. The *SMN2* copy number in SMA patients has a very important modifying effect on disease severity. Most patients with SMA type I have one or two *SMN2* copies, most patients with type II have three *SMN2* copies, and most patients with type III have three or four *SMN2* copies. It has also been demonstrated in cell lines isolated from patients that increased levels of full-length SMN2 transcript and SMN protein correlate with increased copy number of the *SMN2* genes and decreased disease severity. Thus the *SMN2* copy number may explain much of the phenotypic variation in SMA; however, rare families have been reported in which markedly different degrees of disease severity are present in siblings with the same *SMN2* copy number. Efforts are ongoing to identify other disease-modifying factors in these families. The *SMN* neighboring genes, NAIP and H4F5t, have been shown to be more often deleted in type I SMA patients, but it is not clear that these are legitimate disease-modifying genes.

IV. SMN Gene Expression

The human *SMN1* and *SMN2* gene promoters are nearly identical in sequence and activity [9,10]. The principal transcriptional initiation site for the *SMN* genes is located 163 base pairs upstream of the translation initiation site. A second transcriptional initiation site has been mapped to 246 base pairs upstream and appears to be used during fetal development [11]. An approximately 150 base-pair region upstream of the translation initiation site contains the sequences necessary for minimal promoter activity [12], although regulatory sequences relevant to *SMN* gene expression may be present as far as 4.6 kb upstream of the transcription initiation site. *SMN* expression varies in different tissue types and likely decreases with development from embryonic to early postnatal stages. SMN promoter activity has also been demonstrated to decrease with cellular differentiation. It has been shown that the *SMN* promoter contains binding sites for and binds the cAMP-response element binding protein (CREB), the Sp family of proteins, and the interferon regulatory factor (IRF-1), all of which can modulate the promoter activity [12–14]. In addition, it has been demonstrated that the *SMN* promoter is associated with the

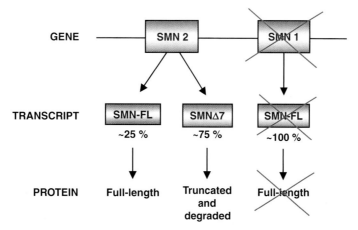

Figure 1 In normal individuals, most full-length SMN transcript and protein arises from the *SMN1* gene. Patients with SMA have homozygous mutations in the *SMN1* gene but retain at least one copy of the *SMN2* gene. During transcription of *SMN2*, exon 7 is frequently skipped. Consequently, the majority of transcripts that arise from *SMN2* lack exon 7 and code for truncated protein that is rapidly degraded; a minority of transcripts include exon 7 and code for full-length protein. SMA results from a deficiency of full-length SMN protein.

histone deacetylase (HDAC) 1 and 2 proteins, which likely modulate the histone acetylation state at the *SMN* promoter, thereby playing a role in determining the activity of the promoter during development [15]. More work is needed to dissect the deoxyribonucleic acid (DNA)–binding proteins and epigenetic modifications that are most critical in regulating *SMN* gene expression.

V. Splicing of SMN Transcripts

The molecular mechanisms that direct splicing of the *SMN* gene transcripts have been investigated in detail. Full-length SMN transcript is encoded by nine exons (1, 2a, 2b, and 3–8). Exons 1–7 are translated into SMN protein. The majority of transcripts that arise from the *SMN1* gene are full length, with a very small minority of transcripts lacking exon 7. In contrast, transcripts arising from *SMN2* often lack exon 7. Two models have been proposed to explain the inhibitory effect of the C→T transition in *SMN2* on exon 7 inclusion [16–18]. According to the exonic splice enhancer model, *SMN1* exon 7 contains a heptamer sequence motif that recruits the splicing factor SF2/ASF and promotes exon 7 inclusion. When this motif is interrupted by the C→T (C→U in messenger ribonucleic acid [mRNA]) transition that is present in SMN2 transcripts, the SF2/ASF factor is not recruited, the 3′ splice site is not recognized, and exon 7 is excluded [17]. The silencer model, in contrast, proposes that the C→U transition creates an exon silencer element that interacts with hnRNP A1 and represses exon 7 inclusion [18]. It has been shown that exon 7 skipping can be partially overcome by a complex of splicing factors that binds an AG-rich exonic splicing enhancer region downstream of the heptamer motif. Htra2-b1 binds directly and specifically to this nucleic acid recognition sequence and facilitates exon 7 inclusion [19]. The function of Htra2-b1 is enhanced by interaction with hnRNP-G [20]. Srp30c also binds Htra2-b1 and may further stabilize the complex [21]. It is likely that other exonic and intronic splice enhancer and silencer motifs play roles in SMN transcript splicing. It has been estimated that in some tissues, 75% of transcripts arising from *SMN2* lack exon 7 and 25% are full length; however, the relative quantity of alternatively spliced versus full-length SMN transcripts in different tissue types and, most importantly, in motor neurons remains unknown.

VI. SMN Protein

SMN is a ubiquitously expressed protein with a molecular weight of 38 kDa. It has been highly conserved through evolution, and orthologues have been identified in a wide range of organisms, including nematode *(Caenorhabditis elegans)*, fly *(Drosophila melanogaster)*, and mouse *(mus musculus)*. In mammals, SMN expression is most abundant in brain, spinal cord, and muscle, with lower expression levels in lymphocytes and fibroblasts. SMN is present in both the cytoplasm and nucleus. In the nucleus, it is concentrated in punctate structures called "gems" that overlap with or are closely apposed to Cajal bodies [22]. Cajal bodies contain high levels of factors involved in transcription and processing of many types of nuclear RNAs. Cajal bodies and gems co-localize in some cell lines and tissues and are distinct in others. Gem number in cell lines or tissues from SMA patients correlates inversely with disease severity, with type I patients showing few or no gems [23].

The SMN protein has several identified motifs, including a lysine-rich basic region encoded by exon 2, a Tudor motif (important in RNA processing) encoded by exon 3, a polyproline region encoded by exons 4 and 5, and a region enriched in tyrosine and glycine (Y-G) residues encoded by exon 6 (Fig. 2). A number of different proteins

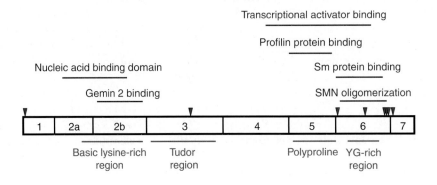

Figure 2 The SMN full-length transcript codes for a protein that contains several sequence motifs (pictured in *green*). Many different proteins and nucleic acid have been shown to bind the SMN protein at different regions (pictured in *blue*).

have been shown to bind SMN at specific regions. Missense mutations have been identified in several of these functional regions, suggesting that each of these interactions may be functionally important.

SMN has been shown to oligomerize and form a stable multiprotein complex with at least six other proteins: gemins 2–7. This SMN complex interacts with several other proteins, including the Sm and Sm-like proteins, RNA helicase A, fibrillarin, GAR1, and the ribonuclear proteins (RNP) hnRNP U, hnRNP Q, and hnRNP R. The common feature of these proteins is a domain rich in arginine (R) and glycine (G) residues that are essential for the SMN interaction. Modification of specific arginine residues within these RG-rich domains by symmetrical dimethylation enhances the affinity of several of these proteins for SMN. This modification is mediated by an arginine methyltransferase complex that contains JBP1/PRMT5, pIC1n, and MEP50.

Many of the SMN interacting proteins are components of various RNP complexes that are involved in distinct aspects of RNA processing. The SMN complex may therefore play a role in diverse aspects of RNA metabolism, including pre-RNA splicing, transcription, and metabolism of ribosomal RNAs. Presently, the best-characterized function of the SMN complex is regulation of the assembly of a specific class of RNA-protein complexes, the uridine-rich small nuclear ribonuclear proteins (snRNPs) (reviewed in [24]). The snRNPs are a critical component of the spliceosome; a large RNA-protein complex that catalyzes pre-mRNA splicing. The biogenesis of snRNPs involves a complex series of steps (Fig. 3). First, a small nuclear RNA (snRNA) is transcribed by RNA polymerase II and rapidly exported to the cytoplasm. In the cell cytoplasm, the SMN complex mediates the ATP-dependent assembly of a ring of Sm proteins (B or B', D1, D2, D3, E, F, and G) around an Sm site on the correct snRNA molecule. To mediate this assembly, SMN interacts directly with Sm proteins and the snRNA and provides specificity to the assembly process. After proper Sm core assembly, there is 5'-cap hypermethylation and 3'-end processing of the snRNA. The complex is then imported to the nucleus to process newly transcribed pre-mRNA into mature RNA. Recent work shows that a decrease in SMN protein levels in patient-derived cells correlates with defects in snRNP assembly [25]. This could mean that deficits in RNA metabolism are directly deleterious to motor neurons; however, the mode by which this deficit is selective for motor neurons remains unclear.

In addition to its known housekeeping activities, SMN may have other unique functions in motor neurons. The SMN protein has been shown to co-localize with cytoskeletal proteins in the dendrites and axons of motor neurons. SMN can also form granules that are actively transported and associated with β-actin mRNA in neuronal processes and growth cones [26]. This localization may be facilitated by a five amino acid motif encoded by exon 7 (QNQKE) of SMN. In other work, it has been shown that in neuronal processes SMN binds the protein hnRNPR, which in turn binds to the 3'-untranslated region of β-actin mRNA [27]. This interaction is required for the efficient transport of β-actin mRNA to growth cones of motor neurons. Motor neurons isolated from mice deficient in SMN showed shortened axons and small growth cones. These growth cones also are deficient in β-actin mRNA and protein. β-actin mRNA and protein localization in the growth cone is known to be necessary for axonal outgrowth, as the actin cytoskeleton is the driving force for growth cone mobility. Actin dynamics are regulated by a series of actin-binding proteins. One of these proteins, profilin, has been shown to bind the proline-rich region of SMN and co-localize with SMN in neuronal processes [28]. These studies raise the possibility that SMN is important in the transport of mRNPs and mRNAs in the motor neuron axon and growth cone and that this function is critical for normal motor neuron outgrowth and maintenance.

VII. Animal Models of SMA

The *SMN2* gene is unique to humans; most other organisms possess a single copy of the *SMN* gene. Nematode, fly, and mouse models with no functional SMN protein have a uniformly early embryonic lethal phenotype, limiting their utility for further study. Nonetheless, fly, zebrafish, and mouse models of SMA have now been engineered that are providing important insights into the pathophysiology of SMA.

A. Drosophila

Invertebrate models can be useful in modeling human neurogenetic disorders because many genes are well conserved. A SMA Drosophila model was established when several missense mutations in the Drosophila *smn* gene were identified that resulted in elimination of the ability of the resulting protein to self-associate [29]. The resulting mutant embryos survived to the late larval stage, but only because of maternal contribution of wild-type SMN in these early life stages. Before death, these larvae developed severe motor abnormalities. The phenotype was associated with disorganization of the neuromuscular junction, including impaired clustering of post-synaptic neurotransmitter receptors. This phenotype could be rescued by expression of wild-type SMN protein in both motor neuron and muscle, but not in either tissue alone. This model suggests that SMN deficiency causes defects of both the presynaptic and postsynaptic sides of the neuromuscular junction.

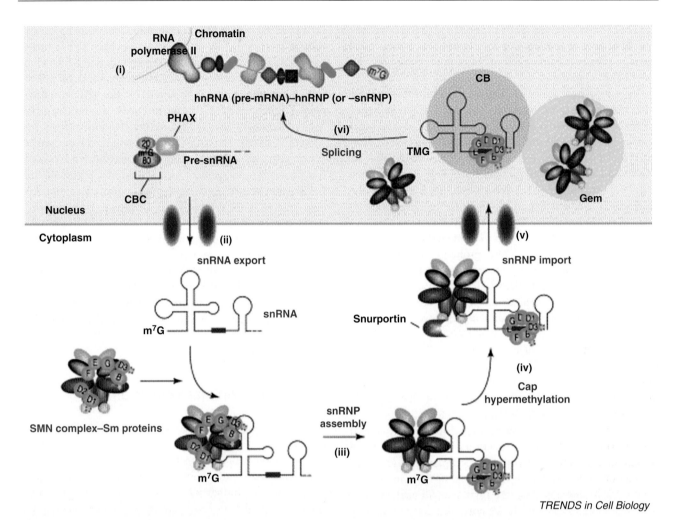

Figure 3 The role of the survival of motor neurons (SMN) complex in the assembly of spliceosomal small nuclear RNAs (snRNAs). (i), The major Sm-site-containing snRNAs—U1, U2, U4, and U5—and the minor spliceosomal snRNAs are transcribed in the nucleus by RNA polymerase II. The primary transcripts contain a monomethylated m7GpppG (m7G) cap structure at the 5'-end and are (ii) rapidly exported to the cytoplasm with the 20-kDa and 80-kDa nuclear cap-binding complex (CBC) proteins and the export adapter PHAX. (iii), The formation of the Sm core is required for the subsequent (iv) hypermethylation of the m7G cap of these snRNAs to convert it to a 2,2,7-trimethyl guanosine (m3G or TMG) and for 3'-end maturation (18 and 62). A properly assembled Sm core and the TMG cap structure are prerequisites for small nuclear ribonucleoprotein (snRNP) import into the nucleus (v). In the nucleus, the newly imported snRNPs are initially concentrated in Cajal bodies (CBs), from which they transit to pre-mRNAs for splicing (vi). The SMN complexes in the nucleus are found throughout the nucleoplasm but are particularly concentrated in gems, the "twins" of the snRNP-rich CBs. (From Yong, J., Wan, L., and Dreyfuss, G. [2004]. Why do cells need an assembly machine for RNA-protein complexes? *Trends Cell Biol.* **14**, 226–232.)

B. Zebrafish

A zebrafish model of SMA was created by using morpholino antisense oligonucleotide "knock-down" technology to decrease the level of endogenous SMN protein [30]. Morphilinos were injected at the one- to four-cell stage, which resulted in an approximately 60% decrease in SMN protein throughout the embryo. These fish showed defects in motor neuron axonal outgrowth and pathfinding during development, without abnormalities of other neuronal types or in muscle. Further knock-down caused a worsening of this phenotype. This model suggests that SMN functions in motor axon development and that early developmental defects of the axon could lead to subsequent motor neuron cell loss.

C. Mouse

In 1997 the murine homolog of the *SMN* gene (*Smn*) was knocked out. Null mice showed massive cell death at an early embryonic stage. Heterozygous *Smn* (+/−) mice showed an approximately 50% reduction in spinal motor neuron number by 6 months of age, but these mice never developed overt weakness, nor did they show a reduction in lifespan. To overcome the embryonic lethality of

Smn (−/−) mice and create mouse models of SMA with a clear behavioral phenotype, two different strategies were pursued. Frugier et al. made a conditional knock-out of exon 7 of the murine *Smn* gene using the Cre-LoxP system [31]. Mice carrying exon 7 flanked by LoxP sites were crossed with mice expressing Cre recombinase in different tissues. When the recombinase was expressed under the control of the neuron-specific enolase promoter, only truncated SMN was expressed in neuronal cells. These mice displayed a SMA phenotype with rapid motor deterioration and death at 4 weeks of age. Further characterization of these mice revealed a dramatic loss of motor axons and neurogenic atrophy of muscle, with only a mild reduction in motor neuron cell body number. In addition, there was aberrant cytoskeletal organization at the neuromuscular junction characterized by abnormal accumulation of neurofilaments in terminal axons associated with a defect in axonal sprouting. These data suggest that loss of motor neuron cell bodies in SMA results from a dying back axonopathy. This group has subsequently deleted SMN exon 7, specifically in skeletal muscle [8]. These mutant mice showed dystrophic changes in muscle, which led to muscle paralysis and death. These results, similar to those in Drosophila, may support the idea that there is primary involvement of muscle in SMA.

The other strategy that was used to generate SMA mice was to express the human *SMN2* gene in the *Smn* knock-out background. This was done by two different groups and each showed that this resulted in a SMA phenotype, the severity of which was dependent on *SMN2* copy number [32,33]. In this way, these mice closely modeled the human disease. Specifically, one group showed that mice expressing one copy of *SMN2* died within 12 hours of birth and mice with two *SMN2* copies survived to 6 days and showed reduced number of motor neurons by day 5 [33]. Mice expressing 8 to 16 copies of *SMN2* were completely rescued from the SMA phenotype. These results dramatically confirm the modifying effect of the *SMN2* gene. This group has subsequently generated two further mouse models on the *SMN2*+/+; *Smn*−/− background. The first is a line of mice that also express a *SMN* transgene containing an A2G missense mutation that has been described in SMA patients [34]. Heterozygous A2G; *SMN2*+/+; *Smn*−/− mice have a mild phenotype characterized by weakness starting at 3 weeks of age and survival to approximately 1 year. Interestingly, the A2G transgene alone is unable to rescue *Smn*−/− embryonic lethality, indicating that the A2G SMN protein likely retains function only by interacting with the normal SMN protein. A second line of mice express high levels of a SMN transcript lacking exon 7 (SMNΔ7) on the *SMN2*+/+; *Smn*−/− background [35]. Lifespan on these mice is extended from approximately 5 days to 13 days, indicating that the SMNΔ7 transcript and protein is not deleterious, as had been previously reported in an *in vitro* study, but rather serves a protective role, perhaps through complexing with the full-length protein.

The SMA mouse models that have been developed to date have confirmed the unique susceptibility of motor neurons to SMN deficiency and have verified that motor neuron degeneration can be prevented by increased dosage of *SMN2*. This has pointed to activation of the *SMN2* gene as a potential therapeutic strategy in SMA patients. The currently available lines of mice that carry the human *SMN2* transgene are now being used to test potential therapeutics (see Section VIII), but these studies remain challenging because mice with severe phenotypes die within 1 to 2 weeks of age, limiting the methods and duration of therapeutic delivery. Similarly, the A2G mice are not ideal for drug studies because of their mild phenotype. Because of these difficulties, efforts are ongoing to develop mice with intermediate disease severity phenotypes that can be more easily used for therapeutic trails. In addition, there is an effort to develop an inducible mouse model in which SMN can be activated during different stages of development. This may help to answer the question of when SMN replacement therapy must be delivered to patients in order to achieve a benefit.

VIII. Therapeutic Strategies for SMA

As a result of progress in understanding the genetic basis and pathophysiology of SMA, potential approaches to the treatment of SMA have emerged. Based on the data from transgenic mouse models, one might predict that increased SMN protein levels would correct SMA before symptoms occur. In symptomatic SMA patients, increased SMN levels might help preserve remaining motor neuron function. Several strategies are being actively investigated to increase full-length SMN protein levels, including activating *SMN2* gene expression, preventing *SMN2* exon 7 skipping, and stabilizing SMN protein (Fig. 4). Other SMA therapeutic strategies being pursued are the identification of drugs that might provide neuroprotection to motor neurons with low SMN levels, replacement of the *SMN1* gene using gene therapy, and replacement of motor neurons using embryonic stem (ES) cells.

A. SMN2 Gene Activation: Histone Deacetylase Inhibitors and Other Drugs

One class of drugs that has been investigated for the ability to activate the *SMN2* gene is the histone deacetylase (HDAC) inhibitors. Control of the acetylation state of histones is an important mechanism regulating gene expression. The basic unit of chromatin, the nucleosome, consists of approximately 150 base pairs of DNA wrapped

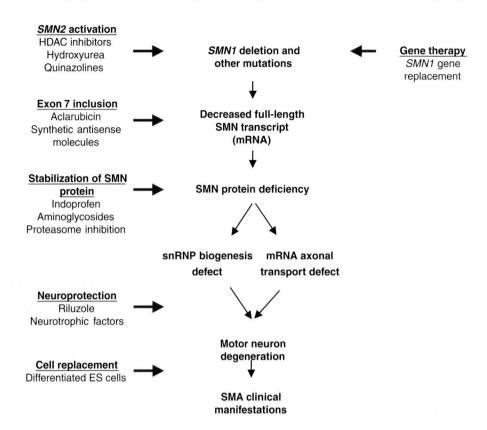

Figure 4 Strategies for therapeutic intervention in SMA.

around an octamer of core histone proteins. The NH_2-terminus of the histone (or "histone tail") protrudes from the nucleosome, and lysine residues within these tails can undergo various post-translational modifications, including acetylation. When histone tails are acetylated, this region of chromatin takes on a more relaxed chromatin structure and is more transcriptionally active due to increased accessibility of DNA to the transcriptional machinery. The level of histone acetylation is determined by the balance of activities of acetyltransferases (HATs), which acetylate histones, and HDACs, which deacetylate histones. HDAC inhibitors are compounds that can modulate the expression of certain genes by inhibiting deacetylases. In 2001 Chang and colleagues demonstrated that sodium butyrate, one of the earliest-discovered HDAC inhibitors, increases full-length SMN2 transcript levels and protein levels in lymphoblastoid cell lines derived from type I SMA patients. In addition, when sodium butyrate was administered to the mothers of SMA transgenic mice, it improved survival in their offspring [36]. Unfortunately, this drug has a short half-life, making it unsuitable for human use. Subsequent studies have shown that the SMN2 promoter can be activated and full-length SMN mRNA and protein levels increased in SMA patient-derived cells by several other HDAC inhibitors, including phenylbutyrate, valproic acid, and suberoyl anilide hydroxamic acid (SAHA) [15,36–39].

Two of these drugs, phenylbutyrate and valproic acid, have been in clinical use for many years for other indications and have well-established pharmacokinetic and safety profiles. Consequently, these drugs have been used in early clinical trials in SMA patients, although they have not yet been shown to be effective in SMA animal models. Clinical trials of phenylbutyrate and valproic acid are currently ongoing or planned in the United States, Italy, and Germany.

Another drug that has been proposed to activate the SMN2 gene is hydroxyurea. In the past, this compound has been shown to activate the fetal globin gene and thus is used to treat patients with sickle cell disease and thalassemias. Hydroxyurea has been reported to increase the amount of full-length SMN transcript and protein in patient-derived lymphoblastoid cell lines and is currently being studied in clinical trials of SMA patients in Taiwan and the United States.

Recently, a cell-based high-throughput screen of 550,000 compounds was performed to identify drugs that activate the SMN2 promoter in a motor neuron–like cell line, NSC34 [40]. Seventeen distinct compounds belonging to nine different scaffolds were identified as activators of SMN2 promoter activity. Two of these scaffolds were confirmed to increase full-length SMN transcript and protein in patient-derived fibroblast cells lines. Of the two, an indole compound and the quinazoline compounds, the quinazoline scaffold is more attractive because it showed the highest potency and minimal cellular toxicity and is amenable to chemical modification. The quinazoline scaffold is currently undergoing a medicinal chemistry effort to improve its drug properties. These compounds do not appear to act as HDAC inhibitors.

B. Exon 7 Inclusion in SMN2 Transcripts

Another strategy that could be successful in increasing SMN protein levels is to promote exon 7 inclusion in *SMN2* derived transcripts. After a screening of patient fibroblasts, Andreassi et al. demonstrated that the chemotherapeutic drug aclarubicin stimulated exon 7 inclusion and consequently increased SMN protein levels [41]. Unfortunately the toxicity profile of this drug prohibits its long-term use in SMA patients. A nonpharmacological strategy to enhance exon 7 inclusion is the use of synthetic antisense oligonucleotides that bind to *SMN2*-derived transcripts and promote exon 7 inclusion during splicing. Cartegni and Krainer have used small chimeric molecules consisting of an antisense moiety that recognizes and hybridizes to exon 7 sequence covalently linked to a peptide; these molecules mimic the function of SR proteins (a family of highly conserved splicing factors) [42]. Skordis et al. have used oligoribonucleotides that are complementary to exon 7 and contain exonic splicing enhancer motifs to provide trans-acting enhancers 43]. Both groups have demonstrated that such strategies can work *in vitro* to increase exon 7 inclusion.

C. Stabilizing SMN Protein

Related strategies for SMA treatment are to increase translation of *SMN2*-derived protein or to stabilize the protein that arises from the *SMN2* gene. Lunn et al. performed a cell-based high-throughput screen of approximately 47,000 compounds to identify drugs that increase *SMN2*-derived protein but not *SMN1*-derived protein [44]. Indoprofen selectively increased *SMN2*-derived protein and also increased endogenous SMN protein in patient-derived fibroblasts. This drug does not act at the transcriptional level, but rather appears to act at the translational level, perhaps by increasing the efficiency of translation of *SMN2*-derived transcripts. In other work, Wolstencroft et al. recently showed that aminoglycosides increase SMN protein levels and gem counts in patient-derived fibroblasts [45]. Aminoglycosides are known to alter translation by promoting read-through of stop codons. The authors speculate that these drugs enable read-through of the initial stop codon in exon 8 of SMN2 transcripts. This results in a SMN protein with a slightly elongated C-terminus, which promotes its stability. Unfortunately aminoglycosides have poor central nervous penetration. In order to apply this approach to SMA patients, new compounds will need to be identified that retain this activity and penetrate the blood-brain barrier. It has also been demonstrated that the SMN protein is degraded by the ubiquitin-proteasome system and that drugs inhibiting this pathway increase SMN protein levels in patient-derived fibroblasts [46]. Further understanding of how the SMN protein is degraded may lead to other SMA therapeutic targets.

D. Neuroprotection

Another important goal in SMA therapeutics research is to identify ways to protect SMN-deficient motor neurons from degeneration. Unlike efforts to identify compounds that increase *SMN2* expression or exon 7 inclusion, there have not yet been extensive high-throughput screens to identify compounds that might be neuroprotective because an *in vitro* motor neuronal model of SMA has been lacking to date. Currently, efforts are underway to develop such a model using ES cells differentiated into motor neurons.

Riluzole has been shown to be modestly effective in the treatment of patients with ALS, perhaps by limiting excitotoxicity induced by excessive glutamate. This drug has been shown to have beneficial effects in a SMA mouse model [47], and clinical trials are in progress in SMA patients in the United States and Europe. Another preclinical study showed that SMA mice showed a beneficial response to the neurotrophic factor, cardiotrophin-1 (CT-1) [48]. CT-1 was delivered by intramuscular injection in an adenoviral vector. Even at low doses, CT-1 was able to improve survival and delay the motor defect in SMA mice with targeted deletion of exon 7 in neurons. Neurotrophic factors deserve further study as potential neuroprotective drugs in SMA.

E. Gene Therapy and Cell Replacement

A strategy that could theoretically lead to a cure for SMA is the replacement of the *SMN1* gene using a gene therapy approach. Unfortunately, development of gene therapy in general has suffered setbacks because of the technical difficulties of efficient gene delivery to target tissues as well as problems related to random insertion of the therapeutic gene into the host DNA. Nonetheless, a recent study in SMA mice showed promising early results. Azzouz et al. showed that after injection of the *SMN1* gene packaged in a lentivector gene transfer system into muscle, the *SMN1* gene was retrogradely transported to and expressed in spinal motor neurons [49]. This led to a modest increase in the lifespan of SMA mice by approximately 3–5 days. Another strategy that might eventually play a role in SMA treatment is cell replacement. ES cells are pluripotent cells that can be differentiated into motor neurons. In a recent study, differentiated ES cells were transplanted into the spinal cord of rats with Sindbus virus–induced motor neuron injury [50]. The cells survived and produced axons that were able to grow into the ventral root.

IX. Remaining Questions

Great progress has been made in SMA research in the past 10 years. Although this has pointed to targets for therapeutics in SMA, important questions remain. Will SMN replacement work in SMA patients? If so, in what cells and tissues will SMN replacement be required? Is delivery to motor neurons sufficient, or must SMN also be delivered to muscle? How late in the disease process can SMN replacement rescue the phenotype? Does data suggesting that SMN has a developmental role in proper motor neuron outgrowth indicate that SMN must be delivered very early? Can the specific susceptibility of motor neurons to SMN deficiency be identified? Does motor neuron degeneration result directly from snRNP assembly deficiency, or does SMN have a unique function in motor neurons? Answers to these questions will likely lead to the identification of new targets for SMA therapeutics and, ultimately, effective treatment for the disease.

References

1. Ince, S. W. (2003). Pathology of motor neuron disorders. *In:* "Motor Neuron Disorders" (P. J. Strong, et al., eds.), pp. 17–49. Butterworth Heinemann, Philadelphia.
2. Brzustowicz, L. M., Lehner, T., Castilla, L. H., Penchaszadeh, G. K., Wilhelmsen, K. C., Daniels, R., Davies, K. E., Leppert, M., Ziter, F., Wood, D. et al. (1990). Genetic mapping of chronic childhood-onset spinal muscular atrophy to chromosome 5q11.2-13.3. *Nature* **344**, 540–541.
3. Melki, J., Sheth, P., Abdelhak, S., Burlet, P., Bachelot, M. F., Lathrop, M. G., Frezal, J., and Munnich, A. (1990). Mapping of acute (type I) spinal muscular atrophy to chromosome 5q12-q14. The French Spinal Muscular Atrophy Investigators. *Lancet* **336**, 271–273.
4. Lefebvre, S., Burglen, L., Reboullet, S., Clermont, O., Burlet, P., Viollet, L., Benichou, B., Cruaud, C., Millasseau, P., Zeviani, M. et al. (1995). Identification and characterization of a spinal muscular atrophy-determining gene. *Cell* **80**, 155–165.
5. Lorson, C. L., Hahnen, E., Androphy, E.J., and Wirth, B. (1999). A single nucleotide in the SMN gene regulates splicing and is responsible for spinal muscular atrophy. *Proc. Natl. Acad. Sci. U S A* **96**, 6307–6311.
6. Monani, U. R., Lorson, C. L., Parsons, D. W., Prior, T. W., Androphy, E. J., Burghes, A. H., and McPherson, J. D. (1999). A single nucleotide difference that alters splicing patterns distinguishes the SMA gene SMN1 from the copy gene SMN2. *Hum. Mol. Genet.*, **8**, 1177–1183.
7. Lorson, C. L., Strasswimmer, J., Yao, J. M., Baleja, J. D., Hahnen, E., Wirth, B., Le, T., Burghes, A. H., and Androphy, E. J. (1998). SMN oligomerization defect correlates with spinal muscular atrophy severity. *Nat. Genet.* **19**, 63–66.
8. Cifuentes-Diaz, C., Frugier, T., Tiziano, F. D., Lacene, E., Roblot, N., Joshi, V., Moreau, M. H., and Melki, J. (2001). Deletion of murine SMN exon 7 directed to skeletal muscle leads to severe muscular dystrophy. *J. Cell. Biol.* **152**, 1107–1114.
9. Echaniz-Laguna, A., Miniou, P., Bartholdi, D., and Melki, J. (1999). The promoters of the survival motor neuron gene (SMN) and its copy (SMNc) share common regulatory elements. *Am. J. Hum. Genet.* **64**, 1365–1370.
10. Monani, U. R., McPherson, J. D., and Burghes, A. H. (1999). Promoter analysis of the human centromeric and telomeric survival motor neuron genes (SMNC and SMNT). *Biochim. Biophys. Acta* **1445**, 330–336.
11. Germain-Desprez, D., Brun, T., Rochette, C., Semionov, A., Rouget, R., and Simard, L. R. (2001). The SMN genes are subject to transcriptional regulation during cellular differentiation. *Gene* **279**, 109-117.
12. Rouget, R., Vigneault, F., Codio, C., Rochette, C., Paradis, I., Drouin, R. and Simard, L. R. (2005). Characterization of the survival motor neuron (SMN) promoter provides evidence for complex combinatorial regulation in undifferentiated and differentiated P19 cells. *Biochem. J.* **385**, 433–443.
13. Baron-Delage, S., Abadie, A., Echaniz-Laguna, A., Melki, J., and Beretta, L. (2000). Interferons and IRF-1 induce expression of the survival motor neuron (SMN) genes. *Mol. Med.* **6**, 957–968.
14. Majumder, S., Varadharaj, S., Ghoshal, K., Monani, U., Burghes, A. H., and Jacob, S. T. (2004). Identification of a novel cyclic AMP-response element (CRE-II) and the role of CREB-1 in the cAMP-induced expression of the survival motor neuron (SMN) gene. *J. Biol. Chem.* **279**, 14803–14811.
15. Kernochan, L. E., Russo, M. L., Woodling, N. S., Huynh, T. N., Avila, A. M., Fischbeck, K. H., and Sumner, C. J. (2005). The role of histone acetylation in SMN gene expression. *Hum. Mol. Genet.* **14**, 1171–1182.
16. Lorson, C. L., and Androphy, E. J. (2000). An exonic enhancer is required for inclusion of an essential exon in the SMA-determining gene SMN. *Hum. Mol. Genet.* **9**, 259–265.
17. Cartegni, L., and Krainer, A. R. (2002). Disruption of an SF2/ASF-dependent exonic splicing enhancer in SMN2 causes spinal muscular atrophy in the absence of SMN1. *Nat. Genet.* **30**, 377–384.
18. Kashima, T., and Manley, J. L. (2003). A negative element in SMN2 exon 7 inhibits splicing in spinal muscular atrophy. *Nat. Genet.* **34**, 460–463.
19. Hofmann, Y., Lorson, C. L., Stamm, S., Androphy, E. J.. and Wirth, B. (2000). Htra2-beta 1 stimulates an exonic splicing enhancer and can restore full-length SMN expression to survival motor neuron 2 (SMN2). *Proc. Natl. Acad. Sci. U S A* **97**, 9618–9623.
20. Hofmann, Y., and Wirth, B. (2002). hnRNP-G promotes exon 7 inclusion of survival motor neuron (SMN) via direct interaction with Htra2-beta1. *Hum. Mol. Genet.* **11**, 2037–2049.
21. Young, P. J., DiDonato, C. J., Hu, D., Kothary, R., Androphy, E. J., and Lorson, C. L. (2002). SRp30c-dependent stimulation of survival motor neuron (SMN) exon 7 inclusion is facilitated by a direct interaction with hTra2 beta 1. *Hum. Mol. Genet.* **11**, 5779–5787.
22. Liu, Q., and Dreyfuss, G. (1996). A novel nuclear structure containing the survival of motor neurons protein. *Embo. J.* **15**, 3555–3565.
23. Patrizi, A. L., Tiziano, F., Zappata, S., Donati, M. A., Neri, G., and Brahe, C. (1999). SMN protein analysis in fibroblast, amniocyte and CVS cultures from spinal muscular atrophy patients and its relevance for diagnosis. *Eur. J. Hum. Genet.* **7**, 301–309.
24. Yong, J., Wan, L., and Dreyfuss, G. (2004). Why do cells need an assembly machine for RNA-protein complexes? *Trends Cell. Biol.* **14**, 226–232.
25. Wan, L., Battle, D. J., Yong, J., Gubitz, A. K., Kolb, S. J., Wang, J., and Dreyfuss, G. (2005). The Survival of Motor Neurons Protein Determines the Capacity for snRNP Assembly: Biochemical Deficiency in Spinal Muscular Atrophy. *Mol. Cell. Biol.* **25**, 5543–5551.
26. Zhang, H. L., Pan, F., Hong, D., Shenoy, S. M., Singer, R. H., and Bassell, G. J. (2003). Active transport of the survival motor neuron

protein and the role of exon-7 in cytoplasmic localization. *J. Neurosci.* **23**, 6627–6637.

27. Rossoll, W., Jablonka, S., Andreassi, C., Kroning, A. K., Karle, K., Monani, U. R., and Sendtner, M. (2003). Smn, the spinal muscular atrophy-determining gene product, modulates axon growth and localization of beta-actin mRNA in growth cones of motoneurons. *J. Cell. Biol.* **163**, 801–812.

28. Sharma, A., Lambrechts, A., Hao, L. T., Le, T. T., Sewry, C. A., Ampe, C., Burghes, A. H., and Morris, G. E. (2005). A role for complexes of survival of motor neurons (SMN) protein with gemins and profilin in neurite-like cytoplasmic extensions of cultured nerve cells. *Exp. Cell. Res.* in press.

29. Chan, Y. B., Miguel-Aliaga, I., Franks, C., Thomas, N., Trulzsch, B., Sattelle, D. B., Davies, K. E., and van den Heuvel, M. (2003). Neuromuscular defects in a Drosophila survival motor neuron gene mutant. *Hum. Mol. Genet.* **12**, 1367–1376.

30. McWhorter, M. L., Monani, U. R., Burghes, A. H., and Beattie, C. E. (2003). Knockdown of the survival motor neuron (Smn) protein in zebrafish causes defects in motor axon outgrowth and pathfinding. *J. Cell. Biol.* **162**, 919–931.

31. Frugier, T., Tiziano, F. D., Cifuentes-Diaz, C., Miniou, P., Roblot, N., Dierich, A., Le Meur, M., and Melki, J. (2000). Nuclear targeting defect of SMN lacking the C-terminus in a mouse model of spinal muscular atrophy. *Hum. Mol. Genet.* **9**, 849–858.

32. Hsieh-Li, H. M., Chang, J. G., Jong, Y. J., Wu, M. H., Wang, N. M., Tsai, C. H., and Li, H. (2000). A mouse model for spinal muscular atrophy. *Nat. Genet.* **24**, 66–70.

33. Monani, U. R., Sendtner, M., Coovert, D. D., Parsons, D. W., Andreassi, C., Le, T. T., Jablonka, S., Schrank, B., Rossol, W., Prior, T. W. et al. (2000). The human centromeric survival motor neuron gene (SMN2) rescues embryonic lethality in Smn(−/−) mice and results in a mouse with spinal muscular atrophy. *Hum. Mol. Genet.* **9**, 333–339.

34. Monani, U. R., Pastore, M. T., Gavrilina, T. O., Jablonka, S., Le, T. T., Andreassi, C., DiCocco, J. M., Lorson, C., Androphy, E. J., Sendtner, M. et al. (2003). A transgene carrying an A2G missense mutation in the SMN gene modulates phenotypic severity in mice with severe (type I) spinal muscular atrophy. *J. Cell. Biol.* **160**, 41–52.

35. Le, T. T., Pham, L. T., Butchbach, M. E., Zhang, H. L., Monani, U. R., Coovert, D. D., Gavrilina, T. O., Xing, L., Bassell, G. J., and Burghes, A. H. (2005). SMNDelta7, the major product of the centromeric survival motor neuron (SMN2) gene, extends survival in mice with spinal muscular atrophy and associates with full-length SMN. *Hum. Mol. Genet.* **14**, 845–857.

36. Chang, J. G., Hsieh-Li, H. M., Jong, Y. J., Wang, N. M., Tsai, C. H., and Li, H. (2001). Treatment of spinal muscular atrophy by sodium butyrate. *Proc. Natl. Acad. Sci. U S A* **98**, 9808–9813.

37. Sumner, C. J., Huynh, T. N., Markowitz, J. A., Perhac, J. S., Hill, B., Coovert, D. D., Schussler, K., Chen, X., Jarecki, J., Burghes, A. H. et al. (2003). Valproic acid increases SMN levels in spinal muscular atrophy patient cells. *Ann. Neurol.* **54**, 647–654.

38. Brichta, L., Hofmann, Y., Hahnen, E., Siebzehnrubl, F. A., Raschke, H., Blumcke, I., Eyupoglu, I. Y., and Wirth, B. (2003). Valproic acid increases the SMN2 protein level: a well-known drug as a potential therapy for spinal muscular atrophy. *Hum. Mol. Genet.* **12**, 2481–2489.

39. Andreassi, C., Angelozzi, C., Tiziano, F. D., Vitali, T., De Vincenzi, E., Boninsegna, A., Villanova, M., Bertini, E., Pini, A., Neri, G. et al. (2004). Phenylbutyrate increases SMN expression in vitro: relevance for treatment of spinal muscular atrophy. *Eur. J. Hum. Genet.* **12**, 59–65.

40. Jarecki, J., Chen, X., Bernardino, A., Coovert, D. D., Whitney, M., Burghes, A., Stack, J., and Pollok, B. A. (2005). Diverse small-molecule modulators of SMN expression found by high-throughput compound screening: early leads towards a therapeutic for spinal muscular atrophy. *Hum. Mol. Genet.* **14**, 2003–2018.

41. Andreassi, C., Jarecki, J., Zhou, J., Coovert, D. D., Monani, U. R., Chen, X., Whitney, M., Pollok, B., Zhang, M., Androphy, E. et al. (2001). Aclarubicin treatment restores SMN levels to cells derived from type I spinal muscular atrophy patients. *Hum. Mol. Genet.* **10**, 2841–2849.

42. Cartegni, L., and Krainer, A. R. (2003). Correction of disease-associated exon skipping by synthetic exon-specific activators. *Nat. Struct. Biol.* **10**, 120–125.

43. Skordis, L. A., Dunckley, M. G., Yue, B., Eperon, I. C., and Muntoni, F. (2003). Bifunctional antisense oligonucleotides provide a trans-acting splicing enhancer that stimulates SMN2 gene expression in patient fibroblasts. *Proc. Natl. Acad. Sci. U S A* **100**, 4114–4119.

44. Lunn, M. R., Root, D. E., Martino, A. M., Flaherty, S. P., Kelley, B. P., Coovert, D. D., Burghes, A. H., Man, N. T., Morris, G. E., Zhou, J. et al. (2004). Indoprofen upregulates the survival motor neuron protein through a cyclooxygenase-independent mechanism. *Chem. Biol.* **11**, 1489–1493.

45. Wolstencroft, E. C., Mattis, V., Bajer, A. A., Young, P. J., and Lorson, C. L. (2005). A non-sequence-specific requirement for SMN protein activity: the role of aminoglycosides in inducing elevated SMN protein levels. *Hum. Mol. Genet.* **14**, 1199–1210.

46. Chang, H. C., Hung, W. C., Chuang, Y. J., and Jong, Y. J. (2004). Degradation of survival motor neuron (SMN) protein is mediated via the ubiquitin/proteasome pathway. *Neurochem. Int.* **45**, 1107–1112.

47. Haddad, H., Cifuentes-Diaz, C., Miroglio, A., Roblot, N., Joshi, V., and Melki, J. (2003). Riluzole attenuates spinal muscular atrophy disease progression in a mouse model. *Muscle Nerve* **28**, 432–437.

48. Lesbordes, J. C., Cifuentes-Diaz, C., Miroglio, A., Joshi, V., Bordet, T., Kahn, A., and Melki, J. (2003). Therapeutic benefits of cardiotrophin-1 gene transfer in a mouse model of spinal muscular atrophy. *Hum. Mol. Genet.* **12**, 1233–1239.

49. Azzouz, M., Le, T., Ralph, G. S., Walmsley, L., Monani, U. R., Lee, D. C., Wilkes, F., Mitrophanous, K. A., Kingsman, S. M., Burghes, A. H. et al. (2004). Lentivector-mediated SMN replacement in a mouse model of spinal muscular atrophy. *J. Clin. Invest.* **114**, 1726–1731.

50. Harper, J. M., Krishnan, C., Darman, J. S., Deshpande, D. M., Peck, S., Shats, I., Backovic, S., Rothstein, J. D., and Kerr, D. A. (2004). Axonal growth of embryonic stem cell-derived motoneurons in vitro and in motoneuron-injured adult rats. *Proc. Natl. Acad. Sci. U S A* **101**, 7123–7128.

45

Amyotrophic Lateral Sclerosis–Like Syndromes Associated with Malignancy

Zachary Simmons, MD

Keywords: *amyotrophic lateral sclerosis, cancer, malignancy, motor neuron disease, paraneoplastic syndrome*

I. Brief History and Nomenclature
II. Etiology
III. Pathogenesis, Pathophysiology, and Relevant Structural Detail
IV. Pharmacology, Biochemistry, and Molecular Mechanisms
V. Explanation of Symptoms in Relation to Pathophysiology
VI. Natural History
   References

I. Brief History and Nomenclature

Amyotrophic lateral sclerosis (ALS) is a progressive disorder of unknown cause characterized primarily by loss of upper and lower motor neurons. This includes the large pyramidal motor neurons of the cerebral cortex, brainstem motor nuclei, and anterior horn cells of the spinal cord. Because there are no specific biological markers or radiographic findings that define the disease, diagnostic criteria have been established based on the neurological examination and electrodiagnostic studies, supplemented by neuroimaging and laboratory studies to exclude disorders that may mimic ALS, and by neuropathological studies that may support or exclude the diagnosis. Patients with ALS demonstrate evidence of upper motor neuron (UMN) and lower motor neuron (LMN) dysfunction, with spread of symptoms or signs over time.

It is possible for a patient to develop a clinical syndrome that is characterized by some or all of these features and occurs in the setting of a known malignancy, or one in which a malignancy is identified as being present concurrently with, or shortly after the appearance of, the neurological syndrome. These ALS-like syndromes may or may not meet all criteria for ALS and may be associated with other neurological symptoms and signs outside the motor system. The term *motor neuron disease* (MND) may be used to describe some such patients. Although sometimes used as a synonym for ALS, MND often is a broader term used to describe all patients who have features of motor neuron degeneration, whether focal or generalized; sporadic, familial, or infectious; UMN only or LMN only, or a combination of UMN and LMN. The concurrent occurrence of ALS-like syndromes (MND) and malignancies has been noted for more than 50 years [1–3].

II. Etiology

The observed association of ALS-like syndromes and malignancies in some patients has led to speculation that malignancies sometimes may be a cause of ALS. However, the frequency of occurrence of malignancies in patients with ALS is not high, suggesting that the concurrent presence of ALS and malignancy is statistically a chance association. For example, only 1 of 80 autopsy cases of ALS demonstrated such an association in one series [2]. Consistent with this, several epidemiological studies have not found an increased incidence of cancer in patients with ALS/MND [4,5]. A large review of 3029 patients with MND found cancer in only 61 patients [4].

Although it appears that gross epidemiological data do not support a relationship between ALS and malignancies, it is possible that the association is meaningful in some individual cases. In this regard, case reports may be of value. The possibility of malignancies playing a role in causing ALS-like syndromes is suggested by two observations of patients with ALS-like syndromes and malignancies: 1) such patients may have clinical features that are atypical for idiopathic ALS; and 2) some of these patients demonstrate neurological improvement with treatment of their malignancies.

Although some patients with an ALS-like syndrome and a malignancy may be indistinguishable from patients with ALS and no malignancy [5,6], the possibility of an underlying malignancy in a patient with an ALS-like syndrome should be considered when there are unexpected systemic features or when the neurological syndrome is atypical. Brain noted this as early as 1965, and this observation has been made frequently since then [2–4,7]. Examples appear in Table 1.

Reports of responsiveness of the neurological syndrome to treatment of the malignancy are mixed, but there are cases in the literature of patients whose neurological syndromes clearly improved with treatment of their malignancies. Rosenfeld and Posner [4] summarized eight such reports. Some illustrative examples are presented in Table 1.

Table 1 Patients with ALS-Like Syndromes and Malignancies: Atypical Clinical Features and Improvement with Treatment of the Malignancy

Atypical Feature	Underlying Malignancy	Response of Neurological Syndrome to Treatment of Malignancy
Chest pain, clubbing of fingers and toes	Adenocarcinoma of the lung	Lobectomy resulted in partial improvement
Hemoptysis	Large cell carcinoma of the lung	Resection of cancer resulted in complete remission
Sensory loss, mental status changes	Renal cell carcinoma	Nephrectomy resulted in marked improvement, but then developed metastatic disease
Cough, early-onset dyspnea	Squamous cell carcinoma of the bronchus	Improvement after surgery and radiation therapy
Oromandibular dyskinesia, sedimentation rate of 121–125, platelet count of 690,000	Renal cell carcinoma	Normalization of strength after nephrectomy
Nystagmus, facial palsy, abnormal cervical magnetic resonance imaging	Ductal adenocarcinoma of the breast	Partial improvement after surgery

Case reports such as these suggest that at least some of the time, there may be a causal relationship between a malignancy and an ALS-like syndrome. In several groups of patients, the relationship between an ALS-like syndrome and malignancy has been reasonably well explored (Table 2), leading in some cases to a better understanding of possible pathogenetic mechanisms. Illustrative descriptions of patients in whom such associations were present are provided in the following discussion.

A. Anti-Hu Antibodies

Three patients with an ALS-like syndrome and anti-Hu antibodies were described in one series. All had atrophy, fasciculations, and a LMN pattern of weakness. Two also had brisk reflexes. EMG in all cases demonstrated

Table 2 ALS-Like Syndromes Associated with Malignancies

- ALS-like syndromes and paraneoplastic autoantibodies
  - Anti-Hu antibodies
  - Anti-Purkinje cell antibodies
  - Other anti-neuronal antibodies
- ALS-like syndromes and lymphoma, paraproteinemia, and other lymphoproliferative disorders
- Upper motor neuron syndrome and breast cancer

fibrillation potentials and positive sharp waves. All had some additional neurological symptoms and signs, including sensory deficits, mental status changes, cerebellar symptoms, seizures, autonomic dysfunction, and cranial neuropathies. Two had small-cell lung cancer, and one had prostate cancer. All had anti-Hu antibodies [5]. Another patient [6] has been described with progressive weakness and atrophy of the upper extremities, which eventually generalized. Reflexes were decreased to absent, although there was a unilateral Babinski sign. Needle electromyography (EMG) demonstrated denervation in upper and lower extremity and paraspinal muscles. Anti-Hu antibody titers were elevated.

B. Anti-Purkinje Cell Antibodies

A woman experienced progressive weakness and atrophy of all limbs, brisk reflexes, and upgoing plantar responses bilaterally. Sensation was normal. Nerve conduction studies demonstrated revealed low amplitude or absent motor responses and slightly decreased upper extremity sensory nerve amplitudes. Needle EMG demonstrated positive sharp waves and fibrillation potentials diffusely. Ovarian adenocarcinoma with peritoneal metastases was identified. She was found to have type 1 anti-Purkinje cell antibodies (anti-Yo antibodies, or PCA1 antibodies) [8]. In another report, a patient with MND was found to have renal cell carcinoma and anti-Purkinje cell antibodies that differed from PCA1/anti-Yo antibodies [9].

C. Other Antineuronal Antibodies

A patient was described with LMN weakness affecting the left upper extremity, progressing to quadriparesis without sensory abnormalities and with absent reflexes. She had transient nystagmus. She eventually also developed a right facial palsy. Needle EMG revealed fibrillation potentials and positive sharp waves in affected muscles with normal nerve conduction studies. A ductal adenocarcinoma of the breast with metastases to lymph nodes was discovered after the neurological presentation. She was found to have antineuronal antibodies reacting with axonal initial segments and nodes of Ranvier [10].

D. ALS-Like Syndromes and Lymphoma, Paraproteinemia, and Other Lymphoproliferative Disorders

The first patients described with lymphoma and an ALS-like syndrome demonstrated a LMN disorder but no UMN dysfunction. In 1963 Rowland and Schneck [1] described two female patients with lower motor neuron syndromes who had Hodgkin's lymphoma. A group of 10 such patients was later published [11]. These individuals were noted to have a lower motor neuron syndrome associated with Hodgkin's disease or non-Hodgkin's lymphoma. All demonstrated subacute, painless, progressive, often asymmetric weakness without significant sensory involvement. The authors termed this a *subacute motor neuronopathy*. Subsequent papers reported UMN as well as LMN abnormalities in patients with lymphoma and other lymphoproliferative disorders. One study found definite or probable UMN signs in 8 of 9 lymphoma patients [12]. A broader study reviewed 26 patients with MND and lymphoproliferative disorders, including multiple myeloma, Waldenström's macroglobulinemia, chronic lymphocytic leukemia, follicular cell carcinoma, and Hodgkin's disease, only 3 of whom had exclusively LMN findings. When cases from the literature were added to reach a total of 56 patients with MND and lymphoproliferative disorders, the authors again found that UMN findings were common; 55% of their patients had both UMN and LMN findings clinically [13].

E. Upper Motor Neuron Syndrome and Breast Cancer

Several years ago, a report described 5 patients with breast cancer and a predominant UMN syndrome. Sensory exams were normal. Cranial nerves were normal except for spastic dysarthria in 4 patients and mild dysphagia in 3. The needle EMG was normal in 3, one of whom then went on to develop rapidly progressive LMN dysfunction and death. The needle EMG in 2 patients revealed some mild LMN involvement [5]. One separate report of a UMN syndrome and breast cancer describes a woman with a progressive UMN disorder consistent with primary lateral sclerosis who was found 6 years after the onset of her neurological syndrome to have breast cancer on a routine screening mammogram [6].

III. Pathogenesis, Pathophysiology, and Relevant Structural Detail

ALS-like syndromes have been described in patients with paraneoplastic syndromes associated with several different types of antineuronal autoantibodies. Paraneoplastic syndromes have been defined as "... rare neurological syndromes that occur with increased frequency in patients with cancer and are etiologically linked to the cancer but not due to metastases or invasion by tumor" [4]. In paraneoplastic neurological disorders, certain antigens expressed in a cancer are also found in the central nervous system (CNS). The immune system recognizes the antigens in the cancer as foreign, resulting in an immune response. CNS structures expressing the antigen are then damaged, resulting in neurological symptoms [15,16]. The best-studied syndromes are those associated with anti-Hu or anti-Purkinje cell antibodies.

A. ALS-Like Syndromes and Anti-Hu Antibodies

Anti-Hu antibodies (type 1 antineuronal nuclear antibodies, or ANNA1) are antineuronal autoantibodies that appear to be synthesized in the CNS and also systemically. They react with protein antigens of 35-40 kD molecular mass expressed in nuclei or (to a lesser extent) cytoplasm of neurons and some cancer cells, primarily small-cell lung cancer but also other malignancies, including neuroblastoma, non–small cell lung cancer, breast cancer, and prostate cancer [6,7,17]. The binding of anti-Hu antibodies to neurons is widespread, including the cerebral hemispheres, limbic system, cerebellum, brainstem, spinal cord, motor neurons, dorsal root ganglia, nerve roots, and autonomic ganglia. There is loss of Betz cells in the motor cortex with secondary corticospinal tract degeneration and loss of anterior horn cells, in addition to extensive inflammatory infiltrates of mononuclear cells, including T and B lymphocytes and plasma cells. Autopsy studies of patients with anti-Hu antibodies and ALS-like syndromes also demonstrate widespread findings outside the motor system, including neuronal loss and lymphocytic infiltration in dorsal root ganglia, degeneration and macrophage infiltration in posterior columns of the spinal cord, loss of Purkinje cells and gliosis in the cerebellum, and abnormalities in the brainstem [5,7].

B. ALS-Like Syndromes and Anti-Purkinje Cell Antibodies

Type 1 anti-Purkinje cell antibodies (anti-Yo antibodies, or PCA1 antibodies) are polyclonal antibodies that bind to the cytoplasm of Purkinje cells in the cerebellum and to other cytoplasm-rich neurons in the central and peripheral nervous systems (e.g., brain, spinal cord, peripheral nerves). The antigenic target is the 50-60kD cdr2 protein. Patients demonstrate a severe, diffuse loss of Purkinje cells in the cerebellar cortex. Many are found on autopsy to have a more diffuse encephalomyelitis, as is common for the paraneoplastic syndromes in general. Virtually all patients with anti-Yo antibodies have cancer of the breast, ovary, or female genital tract. Occasionally these antibodies occur in patients with other adenocarcinomas [17].

C. ALS-Like Syndromes and Other Antineuronal Antibodies

In the single case previously described of a patient with antineuronal antibodies reacting with axonal initial segments and nodes of Ranvier, the authors identified the target of the autoantibodies as 2 isoforms of βIV spectrin. They also identified antibodies directed against one or more surface antigens enriched at axonal initial segments, which could be the direct cause of the LMN syndrome. The authors speculated that the mechanism is humoral autoimmunity to multiple antigens that are part of the same macromolecular complex and that the surface neuronal antigen recognized by the patient's autoantibodies may be a transmembrane protein associated with βIV spectrin [18].

D. ALS-Like Syndromes and Lymphoma, Paraproteinemia, and Other Lymphoproliferative Disorders

The LMN syndrome (subacute motor neuronopathy) associated with lymphoma has been characterized as a "remote effect of the lymphoma." However, the pathogenetic mechanism remains unknown. It has been speculated that radiation therapy or an opportunistic viral infection might play a role [11]. Others have questioned the radiation theory based on the presence of a similar neurological syndrome in patients with lymphoma who have not received radiation therapy [12]. Although most patients with ALS/MND do not have a paraprotein or a lymphoproliferative disorder, both appear to be more common in individuals with ALS than in the general population [13]; lymphoproliferative diseases have been estimated to occur in 2.5–5% of patients with MND, compared with a frequency of 0.5% or less in the general population. A paraproteinemia was found in 9% of patients with MND in one large series, which is higher than that expected in the general population [19]. Interestingly, some patients with MDN have a lymphoproliferative disorder by bone marrow examination, even when a paraprotein is absent, although it is not known whether the frequency of this finding exceeds that of the general population. These associations remain intriguing, but it remains uncertain what pathophysiological role, if any, these paraproteins and lymphoproliferative disorders play in producing a motor neuron disease.

The relationship between lymphoproliferative disorders and ALS-like syndromes may be different for patients with only LMN syndromes than for patients with combined UMN and LMN disease. Autopsy studies in some series of patients with lymphoproliferative disorders and ALS-like syndromes have demonstrated findings that are typical for ALS or that differ minimally from typical idiopathic ALS [12,13]. However, autopsy studies of patients with LMN syndromes and lymphoma have shown neuron loss outside the motor system, including posterior horns, Clarke's column, the intermediolateral cell column, and commissure nuclei. Demyelination and gliosis of spinal white matter has been noted, as well as patchy areas of segmental demyelination of spinal roots and brachial and lumbar plexuses [11].

E. Upper Motor Neuron Syndrome and Breast Cancer

In the series of patients with UMN syndromes and breast cancer, the authors postulate that there is a relationship between breast cancer and UMN syndromes (primary lateral sclerosis) because of the relatively large number of patients seen at their center in whom the two disorders were present concurrently [5]. They were sufficiently concerned about a causal relationship between the two that they recommend that women with a UMN syndrome be evaluated for the possibility of breast cancer. However, it remains unknown whether there is truly a causal relationship between breast cancer and a UMN syndrome. No pathophysiological mechanism linking the two has been identified. The relationship between cancer and a UMN syndrome is even less certain in the other case report cited previously because of the long time interval between the onset of the neurological syndrome and the identification of the cancer [14].

IV. Pharmacology, Biochemistry, and Molecular Mechanisms

Pharmacology, biochemistry, and molecular mechanisms have not been identified in ALS-like syndromes associated with malignancy, beyond the descriptions provided previously.

V. Explanation of Symptoms in Relation to Pathophysiology

A. ALS-Like Syndromes and Anti-Hu Antibodies

When anti-Hu antibodies produce an ALS-like syndrome, the clinical picture usually differs from classical ALS due to the widespread binding of the anti-Hu antibodies, which produces multifocal nervous system involvement. The areas of the CNS to which the anti-Hu antibodies bind determine the neurological symptoms and signs [5,7]. The most common clinical picture is a subacute sensory neuronopathy due to involvement of dorsal root ganglia. However, 20–30% of patients have a sensory neuronopathy as a minor component. These patients present with a variety of neurological manifestations, including cerebellar degeneration, limbic encephalitis, autonomic dysfunction, extrapyramidal disorders, and most relevant for this discussion, motor neuron disease. In a series of 71 patients, 45% of those with anti-Hu antibodies had weakness, but only 20% (14 patients) exhibited predominantly a motor neuron syndrome, characterized by weakness (often asymmetrical), fasciculations, and muscle atrophy, with or without UMN findings. Importantly, although these patients had what could be broadly classified as an ALS-like syndrome, none of the 14 patients in this series had the pure motor neuron syndrome of classical ALS [7]. Similarly, all three of the patients described in a smaller series demonstrated symptoms outside of the motor system [5]. Rarely, a patient with anti-Hu antibodies may have a pure ALS-like syndrome, presumably due to preferential binding of the anti-Hu antibodies to motor neurons [6].

B. ALS-Like Syndromes and Anti-Purkinje Cell Antibodies

In most patients with anti-Yo antibodies, the antibodies target the Purkinje cells of the cerebellum, resulting in a subacute cerebellar syndrome with dysarthria, ataxia, nystagmus, ocular dysmetria, and opsoclonus. But like other paraneoplastic autoantibodies, the anti-Purkinje cell antibodies are not specific for one particular neuron, so often there are also more widespread findings of encephalomyelitis such as lethargy, cognitive dysfunction, bulbar dysfunction, corticospinal tract findings, sensory neuronopathy, and limb weakness as the antibodies target limbic system neurons and sensory neurons [8]. Rarely, an ALS-like syndrome, such as the case described earlier in this paper, may result. There is no postmortem data available for a patient with anti-Purkinje cell antibodies and a paraneoplastic syndrome, but presumably the clinical symptoms result from preferential targeting by the autoantibodies of the upper and lower motor neurons.

C. ALS-Like Syndromes and Other Antineuronal Antibodies

In the single case described earlier of a patient with antineuronal antibodies reacting with βIV spectrin, the binding of the antineuronal antibodies, as in many paraneoplastic syndromes, appeared to be widespread, affecting the brainstem and spinal cord, producing a paraneoplastic encephalomyelitis [18].

D. ALS-Like Syndromes and Lymphoma, Paraproteinemia, and Other Lymphoproliferative Disorders

Because the pathophysiology of ALS-like syndromes associated with lymphoma, paraproteinemia, and other lymphoproliferative disorders remains unknown, the origin of the neurological symptoms remains unexplained. The relatively large number of cases of a LMN syndrome (subacute motor neuronopathy) associated with lymphoma certainly suggests a possible causal relationship, but the mechanism remains unclear. The authors have characterized this as a "remote effect of the lymphoma" [11], but a specific autoantibody or other biomarker has not been identified. Similarly, it is not known what role, if any, other lymphoproliferative disorders or paraproteins may play in producing an ALS-like syndrome.

E. Upper Motor Neuron Syndrome and Breast Cancer

Because the pathophysiology of UMN syndromes associated with breast cancer remains unknown, the origin of the neurological symptoms remain unexplained. The relatively large number of cases of the disorders occurring concomitantly in one series suggests a possible causal relationship, but the mechanism remains unclear [5].

F. General Considerations Regarding Symptoms and Pathophysiology

There is general agreement that ALS-like syndromes usually are not associated with malignancy and that paraneoplastic autoantibodies are rare in patients with MND. Therefore if a patient is seen with an ALS-like syndrome, it is unlikely that it is a paraneoplastic disorder. An evaluation for an underlying malignancy in such a patient usually is not undertaken unless certain features are present (Table 3) [3–5,15]. Note that these guidelines inevitably will miss some ALS-like syndromes that are associated with malignancy, particularly those cases presenting as "typical" ALS. However, because the association of an ALS-like syndrome with malignancy is rare, the number of patients with such an association who do not fit within the guidelines provided here is very small.

Table 3 Features in ALS-Like Syndromes Suggestive of an Underlying Malignancy

Clinical Feature	Possible Underlying Malignancy
Lower motor neuron signs only	Lymphoma
Neurological abnormalities outside of the motor system (encephalomyelopathy)	Paraneoplastic disease associated with antineuronal autoantibodies, most commonly lung, breast, or ovarian cancer
Clinical or laboratory evidence of a lymphoproliferative disease (adenopathy, abnormal chest radiograph, high erythrocyte sedimentation rate [ESR], anemia, leukocytosis)	Lymphoproliferative disorder
Upper motor neuron signs only	Breast cancer
Very young age of onset	Nonspecific—a frequent recommendation in the literature, but without a clear evidence basis

With regard to those patients with an ALS-like syndrome due to paraneoplastic autoantibodies, it has been emphasized that the variety of neurological manifestations associated with most paraneoplastic autoantibodies is broad and not limited to the typical syndromes most familiar to neurologists, such as sensory neuronopathy with anti-Hu antibodies and small-cell lung cancer, or cerebellar ataxia with anti-Yo antibodies and breast or ovarian cancer. Thus, the antoantibody profile predicts the underlying neoplasm, not the specific neurological syndrome, and it has been recommended that patients suspected of having a paraneoplastic syndrome be tested for a broad range of autoantibodies [20]. This is consistent with the finding that ALS-like syndromes may occur in patients with several types of antineuronal antibodies, as described previously.

VI. Natural History

A. ALS-Like Syndromes and Paraneoplastic Autoantibodies, Including Anti-Hu and Anti-Purkinje Cell Antibodies

A common clinical feature of paraneoplastic neurological disorders is that symptoms usually develop rapidly over a few days to weeks, then stabilize to leave the patient with severe deficits, which rarely improve much with treatment of the tumor or with immunomodulating therapy. Patients with ALS-like syndromes associated with anti-Hu and anti-Purkinje cell antibodies have been found to follow the expected clinical course for paraneoplastic neurologic syndromes [5–9,17]. It has been speculated that there is early destruction of neurons in the CNS in these disorders such that immunotherapy, even if effective, would not change the neurological syndrome [15,16]. The patient with antibodies to βIV spectrin demonstrated progressive neurological impairment until the time of surgery for her cancer, after which she demonstrated partial improvement [18].

B. ALS-Like Syndromes and Lymphoma, Paraproteinemia, and Other Lymphoproliferative Disorders

The presence of an ALS-like disorder with only LMN manifestations appears to carry a better prognosis that a disorder with combined LMN and UMN manifestations. In a series of 10 patients with "subacute motor neuronopathy" (LMN findings only) and lymphoma, the course of the neurological deficits appeared to be independent of the course of the lymphoma. Some patients had slow progression of their neurological deficits, but most (7 of 10) improved partially or completely in a gradual fashion, months to years after onset [11]. In contrast, other patients with lymphoma and a combination of UMN and LMN findings usually demonstrated progressive neurological dysfunction [12]. When patients with a broader range of lymphoproliferative disorders were reviewed, the only two who demonstrated improvement or stabilization were those whose ALS-like syndrome was characterized by LMN disease only, whereas all other patients showed progression of their neurological disease [13].

C. Upper Motor Neuron Syndrome and Breast Cancer

In the series of patients with a UMN syndrome and breast cancer, the neurological disease was progressive in all five patients [5].

References

1. Rowland, L. P., and Schneck, S. (1963). Neuromuscular disorders associated with malignant neoplastic disease. *J. Chron. Dis.* **16**, 777–795.
2. Rowland, L. P. (1997). Paraneoplastic primary lateral sclerosis and amyotrophic lateral sclerosis. *Ann. Neurol.* **41**, 703–705.
3. Younger, D. S. (2000). Motor neuron disease and malignancy. *Muscle Nerve* **23**, 658–660.
4. Rosenfeld, M. R., and Posner, J. B. (1991). Paraneoplastic motor neuron disease. *Adv. Neurol.* **56**, 445–459.
5. Forsyth, P. A., Dalmau, J., Graus, F., Cwik, V., Rosenblum, M. K., and Posner, J. B. (1997). Motor neuron syndromes in cancer patients. *Ann. Neurol.* **41**, 722–730.
6. Verma, A., Berger, J. R., Snodgrass, S., and Petito, C. (1996). Motor neuron disease: a paraneoplastic process associated with anti-Hu antibody and small-cell lung carcinoma. *Ann. Neurol.* **40**, 112–116.
7. Dalmau, J., Graus, F., Rosenblum, M. K., and Posner, J. B. (1992). Anti-Hu-associated paraneoplastic encephalomyelitis/sensory neuronopathy: a clinical study of 71 patients. *Medicine* **71**, 59–72.
8. Khwaja, S., Sripathi, N., Ahman, B. K., and Lennon, V. A. (1998). Paraneoplastic motor neuron disease with type 1 Purkinje cell antibodies. *Muscle Nerve* **21**, 943–945.
9. Vianello, M., Vitaliani, R., Pezzani, R., Nicolao, P., Betterle, C., Keir, G., Thompson, E. J., Tavolato, B., Scaravilli, F., and Giometto, B. (2004). The spectrum of antineuronal autoantibodies in a series of neurological patients. *J. Neurol. Sci.* **220**, 29–36.
10. Ferracci, F., Fassetta, G., Butler, M. H., Floyd, S., Solimena, M., and De Camilli, P. (1999). A novel antineuronal antibody in a motor neuron syndrome associated with breast cancer. *Neurology* **53**, 852–855.
11. Schold, S. C., Eun-Sook, C., Somasundaram, M., and Posner, J. B. (1979). Subacute motor neuronopathy: a remote effect of lymphoma. *Ann. Neurol.* **5**, 271–287.
12. Younger, D. S., Rowland, L. P., Latov, N., Hays, A. P., Lange, D. J., Sherman, W., Inghirami, G., Pesce, M. A., Knowles, D. M., Powers, J., et al. (1991). Lymphoma, motor neuron disease, and amyotrophic lateral sclerosis. *Ann. Neurol.* **29**, 78–86.
13. Gordon, P. H., Rowland, L. P., Younger, D. S., Sherman, W. H., Hays, A. P., Louis, E. D., Lange, D. J., Trojaborg, W., Lovelace, R. E., Murphy, P. L., and Latov, N. (1997). Lymphoproliferative disorders and motor neuron disease: an update. *Neurology* **48**, 1671–1678.
14. Corcia, P., Honnorat, J., Guennoc, A. M., de Toffol, B., and Autret, A. (2000). Primary lateral sclerosis with breast cancer, a potential paraneoplastic neurological syndrome. *Rev. Neurol.* **156**, 1020–1022.
15. Posner, J. B. (2003). Immunology of paraneoplastic syndromes: overview. *Ann. NY Acad. Sci.* **998**, 178–186.
16. Bataller, L., and Dalmau, J. O. (2004). Paraneoplastic disorders of the central nervous system: update on diagnostic criteria and treatment. *Sem. Neurol.* **24**, 461–471.
17. Dropcho, E. J. (2002). Remote neurologic manifestations of cancer. *Neurol. Clin.* **20**, 85–122.
18. Berghs, S., Ferracci, F., Maksimova, E., Gleason, S., Leszczynski, N., Butler, M., De Camilli, P., and Solimena, M. (2001). Autoimmunity to beta IV spectrin in paraneoplastic lower motor neuron syndrome. *PNAS* **98**, 6945–6950.
19. Younger, D. S., Rowland, L. P., Latov, N., Sherman, W., Pesce, M., Lange, D. J., Trojaborg, W., Miller, J. R., Lovelace, R. E., Hays, A. P., et al. (1990). Motor neuron disease and amyotrophic lateral sclerosis: relation of high CSF protein content to paraproteinemia and clinical syndromes. *Neurology* **40**, 595–599.
20. Pittock, S. J., Kryzer, T. J., and Lennon, V. A. (2004). Paraneoplastic antibodies coexist and predict cancer, not neurological syndrome. *Ann. Neurol.* **56**, 715–719.

46

Amyotrophic Lateral Sclerosis: Idiopathic and Inherited

Nicholas J. Maragakis, MD
Jeffrey D. Rothstein, MD, PhD

Keywords: *ALS2, apoptosis, axon transport, astroglia, COX2, dynactin, gene therapy, glutamate, glutamate transport, IGF1, mitochondria, motor neuron, muscle atrophy, neurodegeneration, neurofilament, senataxin, SOD1, VEGF*

I. History and Nomenclature
II. Epidemiology
III. Etiology and Pathogenesis
IV. Pathophysiology
V. Symptoms in Relation to Pathophysiology and Natural History
VI. Structural Detail
VII. Molecular Mechanisms
VIII. Pharmacology and Treatment
References

I. History and Nomenclature

Although commonly known as Lou Gehrig's disease in the United States, after the famous New York Yankees baseball player who acquired the disease in the 1930s, amyotrophic lateral sclerosis (ALS) was described in 1874 by Charcot [1]. *Amyotrophy* comes from the Greek word meaning "without muscle nourishment," and asymmetric muscle atrophy can be one of the most obvious signs of the disease. The term "lateral sclerosis" refers to the appearance of the lateral columns of the spinal cord in pathology specimens. Following the loss of motor neurons in the cortex, the descending lateral corticospinal tracts of the spinal cord subsequently degenerate. These axon tracts are replaced by gliotic tissue that has the gross appearance of being hardened or "sclerotic."

II. Epidemiology

ALS is the most common form of adult motor neuron disease. It is an uncommon but not a particularly rare disease, with an incidence of 1–3/100,000 individuals [2] and has a male:female ratio of 1:4 to 2:5 [3]. The mean duration of disease from onset to death or ventilator dependence is 2–5 years, although a significant percentage (19%–39%) survive 5 years and a smaller percentage, 8%–22%, survive 10 years without ventilator use.

Factors suggested as predictors of survival include age at onset, gender, clinical presentation (bulbar vs spinal), and rate of disease progression. Age at onset appears to be a powerful predictor of disease duration, with younger patients surviving longer. Also, progression to death is faster among patients with bulbar onset [4,5].

The majority (>90%) of patients present with sporadic disease, although an important subset of patients inherit the disease, typically in an autosomal dominant pattern. Although ALS is historically considered a neuromuscular disease, current theories based on pathophysiology and natural history suggest that it is a neurodegenerative disease. This designation is more appropriate and places ALS in the category of other neurodegenerative diseases including Parkinson's disease, Alzheimer's disease, and Huntington's disease. Although each of these disorders has unique clinical presentations and pathology, there is a significant overlap in pathophysiological and clinical features.

III. Etiology and Pathogenesis

Numerous mechanisms have been proposed as possibly leading to motor neuron death. Neuroepidemiological studies have analyzed toxic exposures, heavy metal exposure, occupation, physical activities, electrocution injuries, smoking, and alcohol use, among other mechanisms [6–8]. However, age and family history remain the most well-established risk factors for the disease. Infectious etiologies have historically received attention, especially given that another motor neuron disease, polio, is caused by an enterovirus. Human immunodeficiency virus (HIV) has produced an "ALS-like" disease with some reversibility [9], and enteroviruses have been reported in ALS tissue [10]. Other reports, however, have not found a relationship [11], and the theory of viral infection as an etiology for ALS remains controversial. These viruses have not produced ALS-like disease in animals or in vitro models. Ultimately, the link between the presence of virus in "sick" patients and the pathogenesis of disease remains unclear. The relationship between chronic Lyme disease and ALS has been particularly prominent on Internet forums. However, the majority of proposed etiologies are based on anecdotal cases, making the assessment of possible causative links difficult to interpret.

A. Inherited ALS

1. Superoxide Dismutase Mutations

A key discovery was made in 1993 by Rosen and colleagues when they identified a mutation in superoxide dismutase (SOD1) that lead to an autosomal dominant form of the disease. SOD1 is a highly abundant, free radical, scavenging enzyme that forms a major component of the intracellular defense mechanisms employed by most cells to guard against free radical species produced during cellular metabolism. Mutations in the SOD1 gene are estimated to account for 15%–20% of all inherited forms of the disease, and although this mutation probably only represents 1%–2% of all ALS cases, it has spawned intense investigation into the mechanisms underlying sporadic ALS. More than 110 different mutations of this 153 amino acid protein have been described.* ALS from SOD1 mutations are inherited in an autosomal dominant fashion with a high penetrance of the disease between generations.

Subsequently, several transgenic mouse models of the mutant form of human SOD1 (G93A [glycine substituted to alanine at position 93], G37R, G85R) resulted in the development of an important animal model of the familial disease [12]. All the mutant SOD1 mice develop a slowly progressive hindlimb paralysis with early death. This was accompanied by motor neuron loss, astrogliosis, and in most of the models, cytosolic SOD1 inclusions [13,14]. Notably, the pathology is present in neural and nonneural cells in the central nervous system (CNS). This first genetic-based model has formed the cornerstone for the study of ALS biology and potential therapeutics.

2. ALS2: Alsin Mutations

The identification of the SOD1 mutation spurred the hope that other genes would soon be revealed. Among the most fruitful of these studies has been the identification of a gene named *ALS2* that has been mapped to chromosome 2q33. Found in a small number of families, the disease is recessively inherited. The full *ALS2* gene encodes a 1675 amino acid 184-kDa protein (alsin) that is derived from 34 exons that span 80 kb, but a shorter, alternatively spliced form that is 396 amino acids also exists. Affected individuals carry either a single or double base deletion that generates a homozygous frameshift, which results in a loss of function mutant.

---

*See www.alsod.org.

The mechanism by which this loss of *ALS2* leads to motor neuron death in disease is still to be elucidated, but sequence analysis of the protein reveals that the amino terminal contains the motifs characterizing several guanine-nucleotide exchange factor domains (GEFs). Functionally individual GEFs are responsible for the recycling of specific G-protein from their guanosine diphosphate (GDP)-bound states to the guanosine triphosphate (GTP) state. The specific G-protein partner for ALS2, a putative GEF, is still undetermined but it shows high levels of homology to RCC1 (a regulator of chromosome condensation), a well-studied GEF that acts on RAN, a G-protein required for nuclear import and export. At the carboxy-terminal portion of the protein, a pleckstrin homology domain and a DBL homology domain have been identified. A mutation found in one of the families that spares the shorter polypeptide but eliminates the full-length protein resulted in a distinct form of motor neuron disease termed *primary lateral sclerosis* (PLS), in which pathology is restricted only to the upper motor neurons [15,16].

3. ALS4: Senataxin Mutations

ALS4 is a rare juvenile form of ALS inherited in an autosomal dominant form. It is characterized by distal muscle weakness and atrophy, normal sensation, and pyramidal signs. Individuals affected with ALS4 usually have onset of symptoms at age <25 years, a slow rate of progression, and a normal life span. Missense mutations in Senataxin (a deoxyribonucleic acid [DNA]/ribonucleic acid [RNA] helicase) were found in this gene, suggesting a potential role for dysfunction in DNA helicase activity or RNA processing in the development of this disorder [17].

B. Other Genetic Forms of ALS

A locus related to a recessively inherited form of a juvenile-onset ALS-like disorder has been found in seven Tunisian families. The locus lies at 15q15-22, but the gene affected has yet to be determined. In this case, affected individuals initially suffer atrophy and weakness in the hands and feet followed by a later onset of upper motor neuron involvement.

An autosomal dominant form of ALS associated with frontotemporal dementia has been mapped to chromosome 9q21-q22 [18]. This dementia has been characterized by socially inappropriate, impulsive behavior and deterioration in the ability to perform routine daily tasks. These patients also have the corticospinal and lower motor neuron features seen in ALS. Other familial forms of the disease await further characterization and genetic linkage. Additional studies may also focus on potential disease-related genes identifiable in sporadic ALS cases.

IV. Pathophysiology

Although ALS is categorized as a motor neuron disease, the lower motor neurons of the spinal cord are not the only ones affected (as occurs in polio). ALS is a neurodegenerative disorder of both the upper motor neurons of the motor cortex and the lower motor neurons in the anterior horn of the spinal cord. This combination results in unique clinical features of the disease. However, extensive pathological studies in human sporadic and familial tissue, as well as transgenic models of the SOD1 mutations, clearly document widespread neural degeneration of nonmotor neurons in spinal cord and motor cortex. In addition, histological and functional abnormalities of nonneuronal cells, such as astroglia, have been described. The alterations of these other neurons and glia, as described in Section VI, certainly contribute to disease pathophysiology.

V. Symptoms in Relation to Pathophysiology and Natural History

The diagnosis of the disease is made by the presence of upper and lower motor neuron findings with progression of the disease over a 12-month period. Formal criteria for diagnosis as possible, probable, and definite ALS using El-Escorial World Federation of Neurology (WFN) have been established [19]. These criteria are used to aid in uniformity of diagnosis and enrollment of patients in clinical trials. The diagnosis of ALS is sometimes delayed because of the heterogeneous presentation of the disease with either prominent lower motor neuron signs and symptoms or, conversely, primarily upper motor neuron signs and symptoms. Indeed, in two motor neuron disease variants, clinical features are largely limited to either lower motor neurons (progressive spinal muscular atrophy [PSMA]) or upper motor neurons (PLS). These motor neuron diseases are considered distinct entities and in general have a better prognosis than patients diagnosed with ALS.

ALS is also often categorized as spinal or bulbar onset. The spinal form of the disease may present with lower motor neuron symptoms including weakness and muscle atrophy. These symptoms may have an insidious onset and progress over time. The spinal form may affect one limb asymmetrically, with subsequent spread to adjacent limbs, and eventually affect muscles of respiration. Upper motor neuron signs and symptoms in the spinal form included hyperreflexia, extensor plantar responses, and limb spasticity.

Bulbar ALS initially and prominently affects muscles of speech, swallowing, and respiration. Bifacial weakness, dysarthria, and tongue atrophy may be prominent signs. Dyspnea is often an early symptom with reduced breath sounds in the lung bases and reduced diaphragmatic excursion on exam. Pulmonary function testing may reveal a reduced forced vital capacity (FVC) consistent with diaphragmatic weakness, which may predate the onset of clinically significant dyspnea. Upper motor neuron symptoms may be subtle but include a pseudobulbar affect with emotional incontinence.

Typically in ALS, sensation and autonomic function are spared, and other diagnoses should be sought if these systems are prominently involved.

Traditionally, ALS patients were told that cognition was spared in the disease. However, emerging data suggest that the coexistence of dementia is higher than previously thought. A particular association has been made between ALS and frontotemporal lobar dementia (FTLD). In one study of 36 patients with FTLD, 5 of them met criteria for definite ALS and another 13 met criteria for possible ALS [20]. In an extension of this observation, nearly half of ALS patients had some evidence of frontal executive deficits [21]. These findings highlight the heterogeneity of ALS and reinforce its classification as a neurodegenerative disease rather than a neuromuscular disease only.

VI. Structural Detail

Gross pathological specimens in ALS show selective atrophy of the motor cortex, atrophy of the anterior spinal nerve roots, grayness of the lateral columns of the spinal cord, and muscle atrophy [19]. Microscopically, the most prominent pathological features of ALS are the loss of motor neurons from the motor cortex, motor nuclei of the brainstem, and anterior horn of the spinal cord. It should be emphasized, however, that pathology is not limited to these regions. Neuronal loss is also seen in the surrounding cortices (premotor, sensory, and temporal) [22]. Pathologically, the muscle shows denervation atrophy with small angular fibers, fiber type grouping, and grouped atrophy. Most often, ALS spares the peripheral nervous system, Onufrowicz nucleus, and cranial nerves III, IV, and VI. Microscopically, cellular changes may include axonal spheroids [23], Bunina bodies (eosinophilic refractile granules in the perikarya of neurons) [24], Lewy body–like hyaline inclusions [25], and ubiquitinated cytoplasmic inclusions. Astrogliosis is always seen in ALS tissue, suggesting that pathological changes are not limited to motor neurons. Microglia activation is also widely seen in affected tissues. The significance of altered astrocyte and microglial biology in the development and propagation of the disease is under investigation. Recent transgenic animal studies strongly suggest that these nonneuronal cells contribute to cellular degeneration [26].

VII. Molecular Mechanisms

The first insights into the possible causes of the disease came with the use of modern genetics and the identification of mutations in the gene for Cu/Zn superoxide dismutase (SOD1), followed by the discovery of a second gene, *ALS2*, that codes for a protein termed *alsin*. More recently, a mutation in the gene coding for the transporter protein dynactin was found in a family with motor neuron disease. This protein is required for retrograde transport of vesicles and organelles along microtubules [27]. Other genes implicated as possible contributors to the disease include those encoding structural proteins such as neurofilaments and growth factors such as vascular endothelial cell growth factor (VEGF). These genetic factors, taken with the biochemical and pathological observations made in ALS postmortem tissue, have allowed several intriguing hypotheses for disease mechanisms to be formulated.

A. Cu/Zn Superoxide Dismutase (SOD1) Mutations and Toxicity in Motor Neurons

SOD1 is a free radical scavenging enzyme that forms a major component of the intracellular defense mechanisms employed by a variety of cells to guard against free radical species produced during cellular metabolism. Ubiquitously expressed and functioning as a homodimer, SOD1 catalyzes the conversion of superoxide to hydrogen peroxide and oxygen in two asymmetrical steps using an essential copper atom in the active site of the enzyme.

To date, more than 110 mutations have been reported; this tally continues to grow as more patients are analyzed. However, the mechanism by which these mutant proteins induce ALS has not been fully elucidated. A loss of enzymatic antioxidant function was proposed to lead to an increase in free radical–mediated injury by superoxide. This was rapidly dismissed because studies in a transgenic mouse expressing familial ALS-linked mutant SOD1^{G93A} (i.e., glycine substituted to alanine at position 93) developed progressive motor neuron disease despite the presence of elevated SOD1 activity [28]. The development of other transgenic mice expressing a number of different SOD1 mutants all manifest with motor neuron disease, regardless of increased or unchanged wild-type SOD1 activity [13,14,29,30].

Moreover, transgenic mice in which the functional SOD1 gene has been deleted altogether (SOD1 knock-out mice) live to adulthood without developing symptoms of motor neuron disease [29]. In light of these findings, current theories suggest that SOD1 mutations cause disease by acquiring a toxic gain of function. These gains of function are purported to stem from alterations in the polypeptide's tertiary structure and the availability of the enzymatic active site, the catalytic copper ion's access to alternative substrates, or a change in the protein's binding affinity for zinc. In addition, release of copper into the cytosol from SOD1 may allow it to catalyze harmful oxidative reactions. Such alterations give rise to a number of proposed pathways by which SOD1 toxicity may be mediated. These include aberrant oxidative chemistry of misfolded or aggregated SOD1 subunits, possibly leading to more chronic misfolding of abundant SOD1 and inhibition of proteosome degradation by misfolded SOD1 mutants [31,32]. Proposed aberrant substrates include peroxynitrite and hydrogen peroxide. The spontaneous reaction of superoxide with nitric oxide yields peroxynitrite, which SOD1 can use for tyrosine nitration of proteins. Studies have reported increased levels of free nitrotyrosine in the spinal cords of both FALS and SALS patients, but to date, no specific nitration targets have been identified [33].

In the case of hydrogen peroxide, there is the potential to produce the highly reactive hydroxyl radical. A normal reaction cycle releases hydrogen peroxide and an oxidized form of the enzyme. The use of peroxide as a substrate by the enzyme in a reduced form generates the hydroxyl radical, which can initiate a cascade of peroxidation. *In vivo* it is unclear whether higher peroxidation will occur by such a mechanism because elevated products were only found in SOD1^{G93A} [34,35] mice and in no other transgenic mouse models at any stage of disease [36].

Another alternative to the peroxynitrite hypothesis involves zinc-depleted wild-type and mutant SOD1 protein. The reduction in bound zinc results in a rapid reduction of SOD1 to the Cu$^+$ form, which in turn converts oxygen to superoxide. As previously described, this can react with nitric oxide to form peroxynitrite and promote protein nitration. Unfortunately this appears unlikely to occur in vivo because the wild type is as toxic as the mutant protein and many mutants do not show diminished zinc binding.

The hypotheses proposed previously assume a role for copper in the generation of the oxidative damage induced by mutant SOD1. A correlation between the degree of copper loading in SOD1 and the resulting toxicity has been determined using the discovery of a copper chaperone for SOD1 (CCS) [37]. Initially isolated in yeast, a mammalian CCS protein has been identified that serves to specifically load copper into both wild-type human and mutant SOD1 subunits [38,39]. The absence of CCS results in the absolute loss of copper loading onto mutant SOD1^{G37R} as measured through *in vivo* labeling with radioactive copper. This mutant remains fully active [40]. In the presence of CCS, it robustly loads copper. It has been demonstrated in SOD1^{G37R}, SOD1^{G85R}, and SOD1^{G93A} mutant mice that markedly reduced copper loading appears to have no effect on disease onset, progression, or pathology. This finding suggests that CCS-dependent copper loading has no bearing on mutant generated toxicity and offers evidence against copper mediated toxicity. These studies also show that 10%–20% residual SOD1 activity can be detected *in vitro* in tissue extracts from both CCS-null SOD1^{G93A} and SOD1^{G37R} mice. The finding that toxicity is insensitive to the absence of CCS contradicts an underlying oxidative mechanism.

Protein aggregation is a common pathological feature in a wide variety of neurodegenerative disorders. The hypothesis that protein aggregates are a cause of toxicity in ALS was founded on the discovery of prominent, cytoplasmic intracellular inclusions, in both neurons and astrocytes surrounding them, in the SOD1 transgenic mouse models employed. Importantly, their presence coincides with the onset of clinical disease and continues to increase during progression of disease. The toxicity of aggregates may be attributable to a number of intrinsic properties of the inclusion such as aberrant oxidative chemistry and loss of protein function through co-aggregation or extrinsic factors such as reduced chaperone activity or reduced proteosome activity. However, not all mutant SOD1 mice develop prominent cytoplasmic inclusions, and similar inclusions are not widely seen in the CNS of ALS patients; thus aberrant biology of inclusions and mutant SOD1 toxicity is not understood. Furthermore, in some neurodegenerative diseases (e.g., spinocerebellar ataxia), cellular inclusions have been proposed to act in a protective capacity and are not necessarily toxic.

Protein-folding chaperones or heat shock proteins (HSP) have the ability to prevent heat denaturation of the cellular proteins to which they associate. They are also involved in the folding and assembly of other polypeptides. In SOD1^{G93A} mice, the levels of chaperone activity are reduced by approximately 25% compared with control mice. Further studies using an in vitro model of ALS showed that simultaneous elevation of the protein-folding chaperone HSP70 (heat shock protein, 70 kDa) could ameliorate both aggregates and acute toxicity [41,42]. Arimoclomol, a coinducer of heat shock proteins, was also found to be neuroprotective in a very small number of SOD1^{G93A} mice, suggesting heat shock proteins as components of SOD1-induced neurotoxicity.

B. Apoptotic Factors in ALS

Apoptosis is the active process and energy-requiring mechanism of programmed cell death. It involves a variety of pathways that interact and regulate each other through the actions of various factors (genetic regulation, death receptors and pro- or anti-apoptotic proteins), which eventually leads to the controlled death of cells. A role for apoptotic cell death in ALS was first highlighted in a group of ALS patients with the demonstration that DNA fragmentation and increased immunoreactivity for the proapoptotic protein Bax occurred in selectively vulnerable regions of the CNS [43]. The case for an apoptotic mechanism in motor neuron death was strengthened by transfection studies that showed the proapoptotic effects of mutant SOD1 on cultured neurons. These neurons displayed prominent hallmarks of apoptotic death (i.e., DNA fragmentation, caspase activation and altered expression of the antiapoptotic protein Bcl-2) in transgenic mice carrying a familial SOD1 mutation [42,44]. Similarly, Bcl-2 family members have been examined in the transgenic SOD1^{G93A} mouse model of ALS. Expression of the anti-apoptotic proteins, Bcl-2 and Bcl-xL, and proapoptotic proteins, Bad and Bax, were found to be similar in both asymptomatic SOD1^{G93A} and normal mice. However, with the onset of the disease in the SOD1^{G93A} mice, a decrease in the expression of Bcl-2 and Bcl-xL was noted with an increase in the expression levels of Bad and Bax [45]. In conjunction with these findings, the overexpression of Bcl-2 in SOD1^{G93A} mice resulted in a slowing of disease onset. The survival time was improved by 3–4 weeks [46].

Other apoptotic factors now known to be activated by mutant SOD1 and playing a critical role in execution of cell death pathways are the caspases, a group of cysteine proteases with specificity for aspartate. Intracellularly, these proteins exist as procaspases and lack enzymatic activity until activated. In both motor neurons and astrocytes, activation of caspase 3 plays a central role in the cell death mediated by mutant SOD1 at the time of earliest onset in three mouse models, the SOD1^{G93A}, SOD1^{G37R} and SOD1^{G85R} mice [47–49]. In the case of the SOD1^{G93A} mutant, cytochrome c is released from mitochondria, followed by the activation of caspase 9, which is believed to be an effector for the subsequent activation of caspases 3 and 7 [50]. Interestingly, the activation of another caspase, caspase 1, appears to precede both neuronal death and the phenotypic expression of ALS in SOD1 mutants by many months. This is in stark contrast to apoptosis observed during development, where cell death occurs rapidly after initial caspase activation. This temporal activation of caspase 1 and 3 has been shown to occur within the same neurons in vitro but has yet to be confirmed fully in mouse models.

The use of caspase inhibitors in mutant mice has been shown to increase mutant SOD1^{G93A} mouse survival. Intrathecal administration of zVAD-fmk (N-benzylocarbonyl-Val-Ala-Asp-fluoromethylketone), a tetrapeptide pan-caspase inhibitor, prolonged survival in SOD1^{G93A} mice by 25% of their lifespan [48]. The inability of anti-apoptotic drugs or genetic manipulations to stop disease may suggest that these cell death pathways are partial contributors but not the primary insult of SOD1-mediated neurotoxicity.

C. Intermediate Filaments and ALS

Neurofilament (NF) proteins represent the majority of cytoskeletal proteins that are present in motor neurons. These proteins play a significant role in determining the shape of cells, caliber of axonal projections, and maintenance of axonal transport. Three distinct neurofilament protein subunits exist, differing in molecular weight: NF-heavy, NF-medium, and NF-light. Structurally the NF-light subunit forms the core of the neurofilament around which the two larger subunits associate and contribute to the side arm projections radiating from the filament. Assembly of the filaments occurs in the motor neuron cell body, where they are then transported down the axon [51]. Abnormal synthesis of filament units and the accumulation of these proteins in the cell body and proximal axons of motor neurons is a hallmark pathological feature of the disease, observed in both familial and sporadic cases of ALS in patients as well as in SOD1 mutant mice. Currently, it is unclear whether accumulation of neurofilaments occurs as a result of axonal transport blockade or whether (conversely) the build-up of protein leads to secondary impairment of axonal transport. Excessive phosphorylation of neurofilaments [52–54] has been suggested as a factor affecting axonal transport, with some studies demonstrating an increase in perikaryal expression of phosphorylated filaments in ALS cases, but this finding has been refuted and remains inconclusive [55]. Immunoreactivity of antibodies to neurofilament epitopes has also been shown in ubiquitinated inclusions with compact or Lewy body–like morphology in residual motor neurons in ALS cases [56]. Under normal physiological conditions, the covalent addition of ubiquitin to cellular proteins usually marks them for degradation by an ATP-dependent, non-lysosomal proteolytic system. In several cases of SOD1-related FALS, the detection of both nonphosphorylated and phosphorylated neurofilaments in dramatic hyaline conglomerate inclusions has been revealed within the perikarya and axons of motor neurons [56,57].

In a manner similar to SOD1, the identification of genetic mutations and the use of transgenic animals have provided evidence that highlights the role for neurofilaments in both the maintenance of healthy motor neurons and the pathogenesis of the various forms of ALS. Characterization of mRNA levels of all three neurofilaments genes in the motor neurons of several ALS patients reveals a reduction in NF-light mRNA, with no significant changes in both NF-medium and NF-heavy mRNA. In a number of ALS cases, small in-frame insertion and deletion mutations in the NF-H subunit have been elucidated [58,59]. Experimentally, the knock-out of NF-light in transgenic mice has provoked the accumulation of NF-medium and NF-heavy in motor neurons, a finding that correlates to the pathology seen in both familial and sporadic ALS tissue. Other transgenic lines in which NF-light or NF-heavy subunits have been overexpressed or the NF-light gene has been mutated have all been shown to develop motor neuron pathology, although none develop a clinical phenotype truly similar to human ALS [60–62].

Conversely, in SOD1 mutant mouse models, the overexpression of NF-heavy [63] or the elimination of NF-light

[64] expression has been shown to protect against motor neuron disease, suggesting that neurofilaments may serve as a target for the toxic effects of mutant SOD1. Evidence to support this theory comes from the transfection of motor neuron cell lines with mutant SOD1. NF-light chains accumulate within the cell, a feature not seen with the introduction of wild-type SOD1. Oxidative stress models in neuronal culture systems have also shown that NF proteins are more susceptible to damage caused by free radicals, with the NF-light subunit being more prone to nitrosylation than other CNS proteins. In contrast, the accumulation of neurofilaments may serve to protect motor neurons against SOD1 toxicity through their ability to buffer intracellular calcium, a major factor in the mechanisms of oxidative damage and excitotoxicity.

D. Defects in Axonal Transport

It has been postulated that impaired axonal transport may play a role in motor neuron disease. Support for these hypotheses was the transgenic mutant SOD1 mouse in which deficits in slow axonal transport occurred early in the disease course [65]. A link to human motor neuron disease was then made following the identification of a point mutation in the p150 subunit of dynactin. This protein complex is required for dynein-mediated retrograde transport of vesicles and organelles along microtubules. Clinically this was manifested in adult patients with vocal fold paralysis, progressive facial weakness, and weakness in the hands. Distal lower limb weakness occurred later in the course of disease [27]. A potential relationship between dynactin mutations was later made in ALS patients, suggesting that allelic variants in the dynactin gene may confer a genomic risk factor for the development of ALS [66].

Mutations in the dynein protein were found to cause a progressive motor neuron disorder in mice [67], and overexpression of the protein dynamitin (part of the dynein-dynactin complex of microtubule transport) resulted in the development of late-onset motor neuron disease in a transgenic mouse model [68]. Subsequent studies of the dynein-dynactin complex with mutant SOD1 demonstrated that an inhibition of microtubule transport from muscle to cell body occurred [69]. Taken together, these data potentially redefine motor neuron disease and suggest an expansion of the biology related to motor neuron degeneration outside the cell body. These studies also have implications for the discovery and development of new therapeutics targeted to heretofore unexplored aspect of motor neuron biology.

E. Excitotoxicity and Glutamate Transporters

The death of neurons resulting from glutamate excitotoxicity is believed to play a part in the etiology of a number of neurodegenerative diseases. In the case of ALS, glutamate-mediated neurotoxicity was first suggested as mechanism of motor neuron death with the discovery of increased levels of glutamate in the cerebrospinal fluid of 40% of sporadic ALS patients, confirmed in larger follow-up studies a decade later [70]. This finding led to the belief that abnormalities in glutamate metabolism may be responsible for motor neuron degeneration and implicated the glutamate transporters in the pathogenesis of the disease. Although several glutamate transporter subtypes exist, EAAT2, the primary astrocyte glutamate transporter, has been shown to be responsible for approximately 95% of glutamate transport in some brain regions [71]. Studies of the EAAT2 transporter in sporadic ALS tissue showed that in about two-thirds of patients, a marked loss of up to 95% of astroglial EAAT2 protein and activity in affected areas was observed [72]. The mechanism for loss of EAAT2 protein is not certain and could involve transcriptional and/or translational processes. One of the causal mechanisms for such losses is thought to arise not from mutations in the EAAT2 gene but as a result of selective errors during splicing of EAAT2 mRNA. Several aberrantly spliced mutant EAAT2 glutamate transporter proteins have been detected, with two prominent products being retained. The first involves a transcript in which exon 9 is skipped; the second results from the retention of intron 7. In both cases, nonfunctional truncated products are translated, with the intron 7 variant profoundly inhibiting normal EAAT2 mediated glutamate transport. Aberrant RNA has been shown to be abundant in neuropathologically affected areas of ALS patients [73]. The production of truncated EAAT2 protein by aberrant RNA splicing shows that truncated mutants have less ability to transport glutamate and may lead to the retention of normal EAAT2 protein within the cytoplasm. This may be due to disrupted trafficking of normal EAAT2 to the cell membrane and the formation of protein aggregates comprised of a mixture of truncated and normal EAAT2 proteins. However, some of the splice variants have been found in normal tissue; the exact pathophysiological implications of this splicing abnormality is not certain. Interestingly, other transcription and splicing abnormalities in ALS, including glutamate receptors, have also been reported. A related inherited motor neuron disease, spinal muscular atrophy, is the result of a defect in the splicing protein complex.

Another mechanism by which glutamate transporter dysfunction may contribute to motor neuron toxicity involves the inactivation of EAAT2 protein by mutant SOD1 [74]. SOD1 mutants, but not wild-type SOD1, were found to catalyze the inactivation of EAAT2 in the presence of elevated concentrations of hydrogen peroxide. This led to a rise in synaptic glutamate and contributed to glutamate neurotoxicity. This process could provide a link between the familial and sporadic forms of ALS. The involvement of EAAT2 points to a role for astrocytic dysfunction, rather than inherent motor neuron dysfunction, as a component in motor neuron death and a factor in the pathogenesis of ALS. Overexpression of EAAT2 in a transgenic mouse resulted in a twofold increase in spinal cord glutamate transport. When crossed with the SOD1^{G93A} mouse, the resulting mice had a delay in grip strength decline and motor neuron loss but, ultimately, not in survival time [75]. Investigators who searched for potential modulators of glutamate transport using a screening of FDA-approved compounds demonstrated that β-lactam antibiotics increased EAAT2, both in vitro and in vivo. Treatment of SOD1^{G93A} mice with one of these compounds, ceftriaxone, similarly resulted in an increase in EAAT2 expression and an improved outcome in the mouse model [76]. This neuroprotective effect appeared independently of ceftriaxone's antibiotic activity and suggests an additional method for modulating glutamate transporter expression.

AMPA receptors are the primary glutamate receptors on motor neurons. Investigators overexpressed the AMPA subunit GluB-(N) (a subunit with a particularly high permeability to calcium) in a transgenic mouse model. When crossed with the SOD1^{G93A} mouse, the resultant offspring had a more rapid decline in motor performance and a shortened lifespan [77]. Interestingly, the AMPA antagonist NBQX was neuroprotective in this model [78]. However, a number of studies have not shown a correlation between the GluR2 subunit (part of the AMPA receptor with high calcium permeability) and motor neuron pathology in mouse models [79] or human ALS spinal cord [80]. These data suggest that although glutamate neurotransmission may not play a central role in the initiation of disease, it does appear to participate in disease progression, based on the findings of glutamate receptor activation, transporter dysfunction, and finally, neuroprotection in patients by the anti-glutamate drug riluzole.

F. Growth Factors

The discovery of a putative role in ALS for VEGF stems from the genetic manipulation of the control mechanism responsible for the expression of inducible VEGF gene in mice. Physiologically, VEGF is a critical factor that controls the growth and permeability of blood vessels. Under conditions of hypoxia, VEGF can maintain and restore vascular perfusion of normal tissues as well as stimulate the growth of new blood vessels. The induction of VEGF in such situations is governed through transcription factors that react to low oxygen tension. Studies in a transgenic mouse model in which the VEGF gene had the specific hypoxia-response element deleted reported that although the mice maintained normal baseline levels of VEGF expression, there was a severe deficit in the ability to induce VEGF during bouts of hypoxia.

Interestingly in a subset of these altered mice surviving through early development, profound and gradually increasing motor deficits were observed between 5 to 7 months. They progressed to display all the hallmark features of ALS: accumulation of neurofilaments in motor neurons, degeneration of motor axons, and the characteristic denervation-induced muscle atrophy. The effects of VEGF on motor neurons could in part be through a direct action on these cells, serving as a neurotropic or neuroprotective factor. *In vitro* studies have demonstrated that VEGF can support the survival of primary motor neurons and protect against cell death induced by hypoxia or serum deprivation [81]. Alternatively, it may act indirectly by regulating the blood supply to motor neurons, which consume high levels of energy to sustain a high rate of electrical firing. In some measure it appears that both mechanisms could play a part in the maintenance of motor neurons.

In ALS, VEGF was found to potentially be a modifier of the disease, with patients who were homozygous for some VEGF haplotypes showing an increased risk of developing the disease [82]. Furthermore, when a VEGF mutant mouse was crossed with the SOD1^{G93A} mouse, the course of the disease was accelerated, thus suggesting a potential neuroprotective role for VEGF. With these theories in mind, investigators showed that the delivery of VEGF to muscles with a lentiviral vector with subsequent retrograde transport to the spinal cord resulted in a significant prolongation in survival of the SOD1^{G93A} mouse [83]. Intracerebroventricular delivery of VEGF into an SOD1^{G93A} rat model also resulted in a delay in hindlimb paralysis and prolonged survival [84]. The translation of these preclinical studies to human clinical trials is anticipated.

G. Mitochondrial Dysfunction

Mitochondrial dysfunction is thought to play a potential role in a host of neurological and nonneurological diseases. Although substantial evidence for mitochondrial dysfunction in human ALS tissues is lacking, suggestions that mitochondria may be involved in motor neuron degeneration have come from ALS animal models. In the G93A mouse and rat models of ALS, vacuoles believed to be mitochondrial in origin have been observed [13,85,86]. However, these vacuoles are not observed in mouse models of G85R where the disease is more prolonged and where even very low levels of the protein are toxic. The SOD2 isoform is localized to the mitochondrial matrix. A partial deficiency of this enzyme appeared to exacerbate the disease in transgenic SOD1 mice [87]. A selective recruitment of mutant SOD1 to spinal cord mitochondria, but not to mitochondria in unaffected tissue, was observed and may account for the prominence of pathology observed in the spinal cord of this animal model [88]. In support of mitochondrial energy dysfunction in propagating the disease, creatine (a compound that may enhance energy storage capacity) was found effective in slowing the progression of disease in a SOD1^{G93A} mouse model [89]. Although these data are intriguing, mitochondrial pathophysiology awaits further study to establish the role of mitochondrial dysfunction in ALS pathogenesis. Creatine therapy in sporadic ALS patients has not proven effective in altering disease progression.

H. Neuroinflammation

Although ALS tissues lack the robust inflammatory responses observed in autoimmune disorders of the CNS such as multiple sclerosis, microglial activation has been commonly observed in ALS postmortem spinal cord specimens. Microglia are immune cells of the CNS and serve as mediators of neuroinflammation [90]. The relevance of neuroinflammation to the development or propagation of ALS has been studied in mouse models of ALS in which the elaboration of proinflammatory molecules such as cyclooxygenase 2(COX2), TNF-α, and interleukin 1-B was observed [91–93]. Evidence for modulation of these neuroinflammatory properties was supported by three trials of minocycline in mouse models of motor neuron disease. Minocycline is known to block microglial activation. Administration of minocycline to transgenic mutant SOD1 mice resulted in a reduction in microglial activation and prolonged survival [94–96]. Celecoxib, a COX2 inhibitor, has also been shown to be effective in prolonging survival in the SOD1^{G93A} mouse [97], along with other COX2 inhibitors.

VIII. Pharmacology and Treatment

Attempts to understand the molecular mechanisms of ALS pathogenesis have resulted in the development of a number of potentially unique pharmacological and novel technological treatments for ALS.

A. Modulators of Glutamate Neurotoxicity

In support of a potential role for glutamate excitotoxicity in the propagation of ALS is the efficacy of riluzole in improving outcome. Riluzole is currently the only FDA-approved drug for treating ALS. Riluzole was studied because of its ability to modulate glutamatergic neurotransmission. It appears to have several mechanisms of action on the glutamatergic system including the inhibition of glutamic acid release, blockade of amino acid receptors, and inhibition of voltage-dependent sodium channels on dendrites and cell bodies [98].

Riluzole's efficacy was established in two important ALS clinical trials. In the first trial, a more robust effect in survival was seen in patients with bulbar-onset disease. The significance of this finding is not well understood, but the effect was clear. In the riluzole-treated group, 74% of patients were alive at 12 months compared with 58% in the placebo-treated group [99].

The second human clinical trial was much larger, with 959 ALS patients treated for 18 months. The most efficacious dose of riluzole was also determined in this study. This study showed that at 18 months, survival rates were 50.4% for patients treated with a placebo and 56.8% for those treated with 100 mg/day riluzole. Adjustment for baseline prognostic factors showed a 35% decreased risk of death with the 100 mg dose compared with the placebo [100]. The consistent results in these human clinical trials have led to the use of riluzole as a standard-of-care in the pharmacological treatment of ALS. More recent follow-up studies have continued to demonstrate the drug's efficacy, especially in the setting of a multidisciplinary ALS clinic [101].

Conversely, the antiglutamatergic compounds topiramate, lamotrigine [102], dextromethorphan [103], and gabapentin [104] have not proven beneficial in human clinical trials, although in some trials, inappropriately low drug doses were used.

B. Antioxidants

Oxidative damage has been proposed as a central mechanism in neurodegenerative diseases. This has been supported by evidence of aberrant substrates in ALS mouse models as well as in human ALS tissue [105]. Vitamin E, an antioxidant, delayed the progression of the disease from a mild to a more severe state but failed to increase survival in a human clinical trial [106]. Epidemiologically, however, subjects who took vitamin E were less prone to develop ALS [107].

Other compounds with antioxidant activity, including N-acetyl-L-cysteine (NAC), did not demonstrate an effect on survival or progression of the disease [108].

C. Energy Metabolism

Creatine and phosphocreatine constitute an intricate cellular energy buffering and transport system that connects sites of energy production (mitochondria) with sites of energy consumption. Creatine administration stabilizes the mitochondrial creatine kinase and inhibits opening of the mitochondrial transition pore. When administered orally to mutant SOD1 mice, an effect in prolonging survival was seen [109], but two human clinical trials of creatine in ALS did not demonstrate any effect [110,111].

D. Antiinflammatory Agents

Prostaglandins are synthesized from arachidonic acid by cyclooxygenase-2 (COX2). Studies have demonstrated that prostaglandins induce glutamate release from astrocytes, thereby contributing to glutamate neurotoxicity [112]. Therefore COX2 inhibition may reduce this release and afford neuroprotection.

Celecoxib, a COX2-inhibitor, has been demonstrated to have an effect on motor neuron survival in a spinal cord culture model of ALS [113]. Further study of this compound in SOD1^{G93A} mice, as well as several other COX2 inhibitors, also demonstrated a modest prolongation of survival [114]. In a human clinical trial, however, celecoxib was not effective.

Lysine acetylsalicylate (LAS), a soluble salt of aspirin, delayed the appearance of motor deficits in SOD1^{G93A} mice. This beneficial effect of treatment was maintained up to 18 weeks of age, until just before onset of end-stage disease. However, neither the onset of paralysis nor end-stage disease were improved by the ALS treatment [115].

E. Neurotrophins

Neurotrophins have been extensively studied in preclinical animal models of ALS and repeatedly showed benefits in numerous in vitro studies in the early 1990s. In light of these findings, clinical trials of several of these compounds were undertaken. Trials of neurotrophic factors such as brain-derived neurotrophic factor (BDNF), glial-derived neurotrophic factor (GDNF), and ciliary neurotrophic factor (CNTF) failed to show an improvement in survival and had a significant number of side effects, most notably debilitating anorexia, nausea, and vomiting [116,117]. Administration of insulin-like growth factor (IGF-1) had a beneficial effect in one trial [118], but a subsequent follow-up study failed to support a significant impact on survival [119]. A follow-up trial of recombinant IGF-1 is now underway. The failure of these proteins to alter disease may have been due in part to their poor penetration into the CNS.

F. Anti-Retrovirals and Protease Inhibitors

In an uncontrolled study, patients with HIV and ALS who were treated with zidovudine and other retroviral agents resulted in improvement of symptoms and perhaps slowing of the progression of ALS. The unusual rapid extension of the disease course and the positive response to antiretroviral therapy suggest that ALS syndrome and HIV infection may be etiologically related. HIV-1 might cause an ALS-like disorder by several mechanisms: via neuronal infection, secretion of toxic viral substance, induction of the immune system to secrete cytokines, or induction of an autoimmune disease [120]. Based on these data, a double-blind, placebo-controlled trial of the protease inhibitor indinavir was undertaken. Side effects including nephrolithiasis and gastrointestinal upset were prominent, and there was no effect of this compound on disease progression [121].

G. Immunomodulatory Therapies

An immune-mediated hypothesis as an etiology of ALS has been proposed for nearly 3 decades. Although no clear causal relationship has been established, intriguing findings continue to fuel potential immunomodulatory therapies for the disease [122]. Unfortunately, immunotherapies such as total lymphoid irradiation [123], cyclophosphamide [124], and cyclosporine [125] have not proven beneficial in human trials.

H. New Technologies

New technologies such as stem cell transplantation, viral vector gene delivery, and siRNA techniques for molecular knock-down of proteins have received abundant media attention.

For stem cell transplantation studies, targets would include either the replacement of motor neurons, the transplant of nonneuronal cells thought to be involved in ALS pathogenesis such as astrocytes, or both. In vitro studies of motor neuron cell death have suggested a role for glial precursor transplantation in motor neuron protection through the delivery of glutamate transporters, as well as other potential neuroprotective properties of healthy astroglia [126]. In a virally induced model of motor neuron disease, transplantation of embryoid-body derived cells resulted in improvements in hindlimb function potentially related to the secretion of neurotrophic factors [127], with other studies showing the potential of motor neuron precursors to develop into mature motor neurons. Transplantation of bone marrow–derived cells and umbilical cord cells into the SOD1^{G93A} mouse model has shown evidence of engraftment and potential responses [128,129].

The delivery of compounds with potential neuroprotective effects by viral-vector delivery has shown promising results in preclinical animal models. The delivery of IGF-1 by adeno-associated virus injection into relevant muscles resulted in retrograde transport of IGF-1 into motor neurons and subsequent improvement in survival in the SOD1^{G93A} mouse [130]. The delivery of the neurotrophic factor VEGF by lentiviral vector delivery yielded similar results, suggesting that these techniques will offer additional approaches to ALS therapeutics [83].

A gene-specific silencing technique has emerged as an attractive approach to treating SOD1-mediated motor neuron disease. Investigators used a lentiviral vector to deliver interfering RNA (RNAi) to spinal motor neurons of the SOD1^{G93A} mouse via injection into the muscle. The RNAi was then retrogradely transported into the spinal cord and resulted in a down-regulation of SOD1 and an impressive prolongation of motor strength and survival [131]. A similar lentiviral delivery method was also used via direct intraspinal injection of RNAi, with similar results [132]. This approach may be of clinical use in a small but important group of patients with familial ALS and mutations in the SOD1 gene.

These methods may offer new approaches to the traditional pharmacotherapeutical approach to treating neurodegenerative diseases and are fertile grounds for further study.

References

1. Charcot, J. M. (1874). De la sclerose laterale amyotrophique. *Prog. Med.* **23**, 235–237; **24**, 341–342; **29**, 453–455.
2. Yoshida, S., Mulder, D. W., Kurland, L. T., Chu, C. P., and Okazaki, H. (1986). Follow-up study on amyotrophic lateral sclerosis in Rochester, Minn., 1925 through 1984. *Neuroepidemiology* **5**, 61–70.
3. Mitsumoto H. (1998). "Amyotrophic Lateral Sclerosis," pp. 19–21. 1998. F. A. Davis, Philadelphia.
4. Eisen, A., Schulzer, M., MacNeil, M., Pant, B., and Mak, E. (1993). Duration of amyotrophic lateral sclerosis is age dependent. *Muscle Nerve* **16**, 27–32.
5. Haverkamp, L. J., Appel, V., and Appel, S. H. (1995). Natural history of amyotrophic lateral sclerosis in a database population. Validation of a scoring system and a model for survival prediction. *Brain* **118**, 707–719.
6. Armon, C., Kurland, L. T., Daube, J. R., and O'Brien, P. C. (1991). Epidemiologic correlates of sporadic amyotrophic lateral sclerosis. *Neurology* **41**, 1077–1084.
7. Armon, C. (2003). An evidence-based medicine approach to the evaluation of the role of exogenous risk factors in sporadic amyotrophic lateral sclerosis. *Neuroepidemiology* **22**, 217–228.
8. Felmus, M. T., Patten, B. M., and Swanke, L. (1976). Antecedent events in amyotrophic lateral sclerosis. *Neurology* **26**, 167–172.
9. MacGowan, D. J., Scelsa, S. N., and Waldron, M. (2001). An ALS-like syndrome with new HIV infection and complete response to antiretroviral therapy. *Neurology* **57**, 1094–1097.
10. Berger, M. M., Kopp, N., Vital, C., Redl, B., Aymard, M., and Lina, B. (2000). Detection and cellular localization of enterovirus RNA sequences in spinal cord of patients with ALS. *Neurology* **54**, 20–25.
11. Nix, W. A., Berger, M. M., Oberste, M. S., Brooks, B. R., McKenna-Yasek, D. M., Brown, Jr., R. H., Roos, R. P., and Pallansch, M. A. (2004). Failure to detect enterovirus in the spinal cord of ALS patients using a sensitive RT-PCR method. *Neurology* **62**, 1372–1377.
12. Gurney, M. E., Pu, H., Chiu, A. Y., Dal Canto, M. C., Polchow, C. Y., Alexander, D. D., Caliendo, J., Hentati, A., Kwon, Y. W., Deng, H. X., et al. (1994). Motor neuron degeneration in mice that express a human Cu,Zn superoxide dismutase mutation. *Science* **264**, 1772–1775.
13. Wong, P. C., Pardo, C. A., Borchelt, D. R., Lee, M. K., Copeland, N. G., Jenkins, N. A., Sisodia, S. S., Cleveland, D. W., and Price, D. L. (1995). An adverse property of a familial ALS-linked SOD1 mutation causes motor neuron disease characterized by vacuolar degeneration of mitochondria. *Neuron* **14**, 1105–1116.
14. Bruijn, L. I., Houseweart, M. K., Kato, S., Anderson, K. L., Anderson, S. D., Ohama, E., Reaume, A. G., Scott, R. W., and Cleveland, D. W. (1998). Aggregation and motor neuron toxicity of an ALS-linked SOD1 mutant independent from wild-type SOD1. *Science* **281**, 1851–1854.
15. Hadano, S., Hand, C. K., Osuga, H., Yanagisawa, Y., Otomo, A., Devon, R. S., Miyamoto, N., Showguchi-Miyata, J., Okada, Y., Singaraja, R., Figlewicz, D. A., Kwiatkowski, T., Hosler, B. A., Sagie, T., Skaug, J., Nasir, J., Brown, R. H. Jr, Scherer, S. W., Rouleau, G. A., Hayden, M. R., and Ikeda, J. E. (2001). A gene encoding a putative GTPase regulator is mutated in familial amyotrophic lateral sclerosis 2. *Nat. Genet.* **29**, 166–173.
16. Yang, Y., Hentati, A., Deng, H. X., Dabbagh, O., Sasaki, T., Hirano, M., Hung, W. Y., Ouahchi, K., Yan, J., Azim, A. C., Cole, N., Gascon, G., Yagmour, A., Ben-Hamida, M., Pericak-Vance, M., Hentati, F., and Siddique, T. (2001). The gene encoding alsin, a protein with three guanine-nucleotide exchange factor domains, is mutated in a form of recessive amyotrophic lateral sclerosis. *Nat. Genet.* **29**, 160–165.
17. Chen, Y. Z., Bennett, C. L., Huynh, H. M., Blair, I. P., Puls, I., Irobi, J., Dierick, I., Abel, A., Kennerson, M. L., Rabin, B. A., Nicholson, G. A., Auer-Grumbach, M., Wagner, K., De Jonghe, P., Griffin, J. W., Fischbeck, K. H., Timmerman, V., Cornblath, D. R., and Chance, P. F. (2004). DNA/RNA helicase gene mutations in a form of juvenile amyotrophic lateral sclerosis (ALS4). *Am. J. Hum. Genet.* **74**, 1128–1135.
18. Hosler, B. A., Siddique, T., Sapp, P. C., Sailor, W., Huang, M. C., Hossain, A., Daube, J. R., Nance, M., Fan, C., Kaplan, J., Hung, W. Y., McKenna-Yasek, D., Haines, J. L., Pericak-Vance, M. A., Horvitz, H. R., and Brown, R. H. Jr. (2000). Linkage of familial amyotrophic lateral sclerosis with frontotemporal dementia to chromosome 9q21-q22. *JAMA* **284**, 1664–1669.
19. Brooks, B. R. (1994). El Escorial World Federation of Neurology criteria for the diagnosis of amyotrophic lateral sclerosis. Subcommittee on Motor Neuron Diseases/Amyotrophic Lateral Sclerosis of the World Federation of Neurology Research Group on Neuromuscular Diseases and the El Escorial "Clinical limits of amyotrophic lateral sclerosis" workshop contributors. *J. Neurol. Sci.* **124 (Suppl)**, 96–107.
20. Lomen-Hoerth, C., Anderson, T., and Miller, B. (2002). The overlap of amyotrophic lateral sclerosis and frontotemporal dementia. *Neurology* **59**, 1077–1079.
21. Lomen-Hoerth, C., Murphy, J., Langmore, S., et al. (2003). Are amyotrophic lateral sclerosis patients cognitively normal? *Neurology* **60**, 1094–1097.
22. Kiernan, J. A., and Hudson, A. J. (1991). Changes in sizes of cortical and lower motor neurons in amyotrophic lateral sclerosis. *Brain* **114**, 843–853.
23. Carpenter, S. (1968). Proximal axonal enlargement in motor neuron disease. *Neurology* **18**, 841–851.
24. Bunina, T. L. (1962). [On intracellular inclusions in familial amyotrophic lateral sclerosis.]. *Zh. Nevropatol. Psikhiatr. Im S. S. Korsakova* **62**, 1293–1299.
25. Hirano, A., Kurland, L. T., and Sayre, G. P. (1967). Familial amyotrophic lateral sclerosis. A subgroup characterized by posterior and spinocerebellar tract involvement and hyaline inclusions in the anterior horn cells. *Arch. Neurol.* **16**, 232–243.
26. Clement, A. M., Nguyen, M. D., Roberts, E. A., Garcia, M. L., Boillee, S., Rule, M., McMahon, A. P., Doucette, W., Siwek, D., Ferrante, R. J., Brown, R. H. Jr, Julien, J. P., Goldstein, L. S., and Cleveland, D. W. (2003). Wild-type nonneuronal cells extend survival of SOD1 mutant motor neurons in ALS mice. *Science* **302**, 113–117.
27. Puls, I., Jonnakuty, C., LaMonte, B. H., Holzbaur, E. L., Tokito, M., Mann, E., Floeter, M. K., Bidus, K., Drayna, D., Oh, S. J., Brown, R. H. Jr, Ludlow, C. L., and Fischbeck, K. H. (2003). Mutant dynactin in motor neuron disease. *Nat. Genet.* **33**, 455–456.

28. Gurney, M. E., Pu, H., Chiu, A. Y. Dal Canto, M. C., Polchow, C. Y., Alexander, D. D., Caliendo, J., Hentati, A., Kwon, Y. W., Deng, H. X., et al. (1994). Motor neuron degeneration in mice that express a human Cu,Zn superoxide dismutase mutation [see comments]. *Science* **264**, 1772–1775.
29. Reaume, A. G., Elliott, J. L., Hoffman, E. K., Kowall, N. W., Ferrante, R. J., Siwek, D. F., Wilcox, H. M., Flood, D. G., Beal, M. F., Brown, R. H. Jr, Scott, R. W., and Snider, W. D. (1996). Motor neurons in Cu/Zn superoxide dismutase-deficient mice develop normally but exhibit enhanced cell death after axonal injury. *Nat.Genet.* **13**, 43–47.
30. Ripps, M. E., Huntley, G. W., Hof, P. R., Morrison, J. H., and Gordon, J. W. (1995). Transgenic mice expressing an altered murine superoxide dismutase gene provide an animal model of amyotrophic lateral sclerosis. *Proc. Natl. Acad. Sci. USA* **92**, 689–693.
31. Johnston, J. A., Dalton, M. J., Gurney, M. E., and Kopito, R. R. (2000). Formation of high molecular weight complexes of mutant Cu,Zn-superoxide dismutase in a mouse model for familial amyotrophic lateral sclerosis. *Proc. Natl. Acad. Sci. USA* **97**, 12571–12576.
32. Shinder, G. A., Lacourse, M. C., Minotti, S., and Durham, H. D. (2001). Mutant Cu/Zn-superoxide dismutase proteins have altered solubility and interact with heat shock/stress proteins in models of amyotrophic lateral sclerosis. *J. Biol. Chem.* **276**, 12791–12796.
33. Beal, M. F., Ferrante, R. J., Browne, S. E., Matthews, R. T., Kowall, N. W., and Brown, R. H., Jr. (1997). Increased 3-nitrotyrosine in both sporadic and familial amyotrophic lateral sclerosis. *Ann. Neurol.* **42**, 644–654.
34. Andrus, P. K., Fleck, T. J., Gurney, M. E., and Hall, E. D. (1998). Protein oxidative damage in a transgenic mouse model of familial amyotrophic lateral sclerosis. *J. Neurochem.* **71**, 2041–2048.
35. Hall, E. D., Andrus, P. K., Oostveen, J. A., and Gurney, M. E. (1998). Relationship of oxygen radical-induced lipid peroxidative damage to disease onset and progression in a transgenic model of familial ALS. *J. Neurosci. Res.* **53**, 66–77.
36. Bruijn, L. I., Beal, M. F., Becher, M. W., Schulz, J. B., Wong, P. C., Price, D. L., and Cleveland, D. W. (1997). Elevated free levels, but not protein-bound nitrotyrosine or hydroxylradicals, throughout amyotrophic lateral sclerosis (ALS)-like disease implicate tyrosine nitration as an aberrant property of one familial ALS-linked superoxide dismutase 1 mutant. *Proc. Natl. Acad. Sci.* **94**, 7606–7611.
37. Culotta, V. C., Klomp, L. W., Strain, J., Casareno, R. L., Krems, B., and Gitlin, J. D. (1997). The copper chaperone for superoxide dismutase. *J. Biol. Chem.* **272**, 23469–23472.
38. Corson, L. B., Strain, J. J., Culotta,V. C., and Cleveland, D. W. (1998). Chaperone-facilitated copper binding is a property common to several classes of familial amyotrophic lateral sclerosis-linked superoxide dismutase mutants. *Proc. Natl. Acad. Sci. USA* **95**, 6361–6366.
39. Wong, P. C., Waggoner, D., Subramaniam, J. R., Tessarollo, L., Bartnikas, T. B., Culotta, V. C., Price, D. L., Rothstein, J., and Gitlin, J. D. (2000). Copper chaperone for superoxide dismutase is essential to activate mammalian Cu/Zn superoxide dismutase. *Proc. Natl. Acad. Sci. USA* **97**, 2886–2891.
40. Borchelt, D. R., Lee, M. K., Slunt, H. S., Guarnieri, M., Xu, Z. S., Wong, P. C., Brown, R. H. Jr, Price, D. L., Sisodia, S. S., and Cleveland, D. W. (1994). Superoxide dismutase 1 with mutations linked to familial amyotrophic lateral sclerosis possesses significant activity. *Proc. Natl. Acad. Sci. USA* **91**, 8292–8296.
41. Bruening, W., Roy, J., Giasson, B., Figlewicz, D. A., Mushynski, W. E., and Durham, H. D. (1999). Up-regulation of protein chaperones preserves viability of cells expressing toxic Cu/Zn-superoxide dismutase mutants associated with amyotrophic lateral sclerosis. *J. Neurochem.* **72**, 693–699.
42. Durham, H. D., Roy, J., Dong, L., and Figlewicz, D. A. (1997). Aggregation of mutant Cu/Zn superoxide dismutase proteins in a culture model of ALS. *J. Neuropathol. Exp. Neurol.* **56**, 523–530.
43. Martin, L. J. (1999). Neuronal death in amyotrophic lateral sclerosis is apoptosis: Possible contribution of a programmed cell death mechanism. *J. Neuropathol. Exp. Neurol.* **58**, 459–471.
44. Spooren, W. P., and Hengerer, B. (2000). DNA laddering and caspase 3-like activity in the spinal cord of a mouse model of familial amyotrophic lateral sclerosis. *Cell Mol. Biol.* **46**, 63–69.
45. Vukosavic, S., Dubois-Dauphin, M., Romero, N., and Przedborski, S. (1999). Bax and Bcl-2 interaction in a transgenic mouse model of familial amyotrophic lateral sclerosis. *J. Neurochem.* **73**, 2460–2468.
46. Kostic, V., Jackson-Lewis, V., de Bilbao, F., Dubois-Dauphin, M., and Przedborski, S. (1997). Bcl-2: prolonging life in a transgenic mouse model of familial amyotrophic lateral sclerosis. *Science* **277**, 559–562.
47. Vukosavic, S., Stefanis, L., Jackson-Lewis, V., Dubois-Dauphin, M., and Przedborski, S. (2000). Delaying caspase activation by Bcl-2: A clue to disease retardation in a transgenic mouse model of amyotrophic lateral sclerosis. *J. Neurosci.* **20**, 9119–9125.
48. Li, M., Ona, V. O., Guegan, C., Jackson-Lewis, V., Andrews, L. J., Olszewski, A. J., Stieg, P. E., Lee, J. P., Przedborski, S., and Friedlander, R. M. (2000). Functional role of caspase-1 and caspase-3 in an ALS transgenic mouse model. *Science* **288**, 335–339.
49. Pasinelli, P., Houseweart, M. K., Brown, R. H. Jr., and Cleveland, D. W. (2000). Caspase-1 and -3 are sequentially activated in motor neuron death in Cu,Zn superoxide dismutase-mediated familial amyotrophic lateral sclerosis. *Proc. Natl. Acad. Sci. USA* **97**, 13901–13906.
50. Guegan, C., Vila, M., Rosoklija, G., Hays, A. P., and Przedborski, S. (2001). Recruitment of the mitochondrial-dependent apoptotic pathway in amyotrophic lateral sclerosis. *J. Neurosci.* **21**, 6569–6576.
51. Nixon, R. A., and Shea, T. B. (1992). Dynamics of neuronal intermediate filaments: a developmental perspective. *Cell Motil. Cytoskeleton* **22**, 81–91.
52. Manetto, V., Sternberger, N. H., Perry, G., Sternberger, L. A., and Gambetti, P. (1988). Phosphorylation of neurofilaments is altered in amyotrophic lateral sclerosis. *J. Neuropathol. Exp. Neurol.* **47**, 642–653.
53. Munoz, D. G., Greene, C., Perl, D. P., and Selkoe, D. J. (1988). Accumulation of phosphorylated neurofilaments in anterior horn motoneurons of amyotrophic lateral sclerosis patients. *J. Neuropathol. Exp. Neurol.* **47**, 9–18.
54. Sobue, G., Hashizume, Y., Yasuda, T., Mukai, E., Kumagai, T., Mitsuma, T., and Trojanowski, J. Q. (1990). Phosphorylated high molecular weight neurofilament protein in lower motor neurons in amyotrophic lateral sclerosis and other neurodegenerative diseases involving ventral horn cells. *Acta. Neuropathol. (Berl.)* **79**, 402–408.
55. Leigh, P. N., Dodson, A., Swash, M., Brion, J. P., and Anderton, B. H. (1989). Cytoskeletal abnormalities in motor neuron disease. An immunocytochemical study. *Brain* **112**, 521–535.
56. Ince, P. G., Tomkins, J., Slade, J. Y., Thatcher, N. M., and Shaw, P. J. (1998). Amyotrophic lateral sclerosis associated with genetic abnormalities in the gene encoding Cu/Zn superoxide dismutase: Molecular pathology of five new cases, and comparison with previous reports and 73 sporadic cases of ALS. *J. Neuropathol. Exp. Neurol.* **57**, 895–904.
57. Rouleau, G. A., Clark, A. W., Rooke, K., Pramatarova, A., Krizus, A., Suchowersky, O., Julien, J. P., and Figlewicz, D. (1996). SOD1 mutation is associated with accumulation of neurofilaments in amyotrophic lateral sclerosis. *Ann. Neurol.* **39**, 128–131.
58. Figlewicz, D. A., Krizus, A., Martinoli, M. G., Meininger, V., Dib, M., Rouleau, G. A., and Julien, J. P. (1994). Variants of the heavy neurofilament subunit are associated with the development of amyotrophic lateral sclerosis. *Hum. Mol. Genet.* **3**, 1757–1761.
59. al-Chalabi, A., Powell, J. F., and Leigh, P. N. (1995). Neurofilaments, free radicals, excitotoxins, and amyotrophic lateral sclerosis. [Review]. *Muscle Nerve* **18**, 540–545.

60. Cote, F., Collard, J. F., and Julien, J. P. (1993). Progressive neuronopathy in transgenic mice expressing the human neurofilament heavy gene: a mouse model of amyotrophic lateral sclerosis. *Cell* **73**, 35–46.
61. Xu, Z., Cork, L. C., Griffin, J. W., and Cleveland, D. W. (1993). Increased expression of neurofilament subunit NF-L produces morphological alterations that resemble the pathology of human motor neuron disease. *Cell* **73**, 23–33.
62. Lee, M. K., Marszalek, J. R., and Cleveland, D. W. (1994). A mutant neurofilament subunit causes massive, selective motor neuron death: implications for the pathogenesis of human motor neuron disease. *Neuron* **13**, 975–988.
63. Couillard-Despres, S., Zhu, Q., Wong, P. C., Price, D. L., Cleveland, D. W., and Julien, J. P. (1998). Protective effect of neurofilament heavy gene overexpression in motor neuron disease induced by mutant superoxide dismutase. *Proc. Natl. Acad. Sci. USA* **95**, 9626–9630.
64. Williamson, T. L., Bruijn, L. I., Zhu, Q., Anderson, K. L., Anderson, S. D., Julien, J. P., and Cleveland, D. W. (1998). Absence of neurofilaments reduces the selective vulnerability of motor neurons and slows disease caused by a familial amyotrophic lateral sclerosis-linked superoxide dismutase 1 mutant. *Proc. Natl. Acad. Sci. USA* **95**, 9631–9636.
65. Williamson, T. L., and Cleveland, D. W. (1999). Slowing of axonal transport is a very early event in the toxicity of ALS-linked SOD1 mutants to motor neurons. *Nat. Neurosci* **2**, 50–56.
66. Munch, C., Sedlmeier, R., Meyer, T., Homberg, V., Sperfeld, A. D., Kurt, A., Prudlo, J., Peraus, G., Hanemann, C. O., Stumm, G., and Ludolph, A. C. (2004). Point mutations of the p150 subunit of dynactin (DCTN1) gene in ALS. *Neurology* **63**, 724–726.
67. Hafezparast, M., Klocke, R., Ruhrberg, C., Marquardt, A., Ahmad-Annuar, A., Bowen, S., Lalli, G., Witherden, A. S., Hummerich, H., Nicholson, S., Morgan, P. J., Oozageer, R., Priestley, J. V., Averill, S., King, V. R., Ball, S., Peters, J., Toda, T., Yamamoto, A., Hiraoka, Y., Augustin, M., Korthaus, D., Wattler, S., Wabnitz, P., Dickneite, C., Lampel, S., Boehme, F., Peraus, G., Popp, A., Rudelius, M., Schlegel, J., Fuchs, H., Hrabe de Angelis, M., Schiavo, G., Shima, D. T., Russ, A. P., Stumm, G., Martin, J. E., and Fisher, E. M. (2003). Mutations in dynein link motor neuron degeneration to defects in retrograde transport. *Science* **300**, 808–812.
68. LaMonte, B. H., Wallace, K. E., Holloway, B. A., Shelly, S. S., Ascano, J., Tokito, M., Van Winkle, T., Howland, D. S., and Holzbaur, E. L. (2002). Disruption of dynein/dynactin inhibits axonal transport in motor neurons causing late-onset progressive degeneration. *Neuron* **34**, 715–727.
69. Ligon, L. A., LaMonte, B. H., Wallace, K. E., Weber, N., and Kalb, R. G. (2005). Mutant superoxide dismutase disrupts cytoplasmic dynein in motor neurons. *Neuroreport* **16**, 533–536.
70. Spreux-Varoquaux, O., Bensimon, G., Lacomblez, L., Salachas, F., Pradat, P. F., Le Forestier, N., Marouan, A., Dib, M., and Meininger, V. (2002). Glutamate levels in cerebrospinal fluid in amyotrophic lateral sclerosis: a reappraisal using a new HPLC method with coulometric detection in a large cohort of patients. *J. Neurol. Sci.* **193**, 73–78.
71. Tanaka, K., Watase, K., Manabe, T., Yamada, K., Watanabe, M., Takahashi, K., Iwama, H., Nishikawa, T., Ichihara, N., Kikuchi, T., Okuyama, S., Kawashima, N., Hori, S., Takimoto, M., and Wada, K. (1997). Epilepsy and exacerbation of brain injury in mice lacking the glutamate transporter GLT-1. *Science* **276**, 1699–1702.
72. Bristol, L. A., and Rothstein, J. D. (1996). Glutamate transporter gene expression in amyotrophic lateral sclerosis motor cortex. *Ann. Neurol.* **39**, 676–679.
73. Lin, C. L., Bristol, L. A., Jin, L., Dykes-Hoberg, M., Crawford, T., Clawson, L., and Rothstein, J. D. (1998). Aberrant RNA processing in a neurodegenerative disease: The cause for absent EAAT2 a glutamate transporter, in amyotrophic lateral sclerosis. *Neuron* **20**, 589–602.
74. Trotti, D., Rolfs, A., Danbolt, N. C., Brown, R. H., Jr., and Hediger, M. A. (1999). SOD1 mutants linked to amyotrophic lateral sclerosis selectively inactivate a glial glutamate transporter. *Nat. Neurosci.* **2**, 427–433.
75. Guo, H., Lai, L., Butchbach, M. E., Stockinger, M. P., Shan, X., Bishop, G. A., and Lin, C. L. (2003). Increased expression of the glial glutamate transporter EAAT2 modulates excitotoxicity and delays the onset but not the outcome of ALS in mice. *Hum. Mol. Genet.* **12**, 2519–2532.
76. Rothstein, J. D., Patel, S., Regan, M. R., Haenggeli, C., Huang, Y. H., Bergles, D. E., Jin, L., Dykes, Hoberg, M., Vidensky, S., Chung, D. S., Toan, S. V., Bruijn, L. I., Su, Z. Z., Gupta, P., and Fisher, P. B. (2005). Beta-lactam antibiotics offer neuroprotection by increasing glutamate transporter expression. *Nature* **433**, 73–77.
77. Kuner, R., Groom, A. J., Bresink, I., Kornau, H. C., Stefovska, V., Muller, G., Hartmann, B., Tschauner, K., Waibel, S., Ludolph, A. C., Ikonomidou, C., Seeburg, P. H., and Turski, L. (2005). Late-onset motoneuron disease caused by a functionally modified AMPA receptor subunit. *Proc. Natl. Acad. Sci. USA* **102**, 5826–5831.
78. Van Damme, P., Leyssen, M., Callewaert, G., Robberecht, W., and Van Den, B. L. (2003). The AMPA receptor antagonist NBQX prolongs survival in a transgenic mouse model of amyotrophic lateral sclerosis. *Neurosci. Lett.* **343**, 81–84.
79. Morrison, B. M., Janssen, W. G., Gordon, J. W., and Morrison, J. H. (1998). Light and electron microscopic distribution of the AMPA receptor subunit, GluR2, in the spinal cord of control and G86R mutant superoxide dismutase transgenic mice. *J. Comp Neurol.* **395**, 523–534.
80. Kawahara, Y., Kwak, S., Sun, H., Ito, K., Hashida, H., Aizawa, H., Jeong, S. Y., and Kanazawa, I. (2003). Human spinal motoneurons express low relative abundance of GluR2 mRNA: an implication for excitotoxicity in ALS. *J. Neurochem.* **85**, 680–689.
81. Oosthuyse, B., Moons, L., Storkebaum, E., Beck, H., Nuyens, D., Brusselmans, K., Van Dorpe, J., Hellings, P., Gorselink, M., Heymans, S., Theilmeier, G., Dewerchin, M., Laudenbach, V., Vermylen, P., Raat, H., Acker, T., Vleminckx, V., Van Den Bosch, L., Cashman, N., Fujisawa, H., Drost, M. R., Sciot, R., Bruynickx, F., Hicklin, D. J., Ince, C., Gressens, P., Lupu, F., Plate, K. H., Robberecht, W., Herbert, J. M., Collen, D., and Carmeliet, P. (2001). Deletion of the hypoxia-response element in the vascular endothelial growth factor promoter causes motor neuron degeneration. *Nat. Genet.* **28**, 131–138.
82. Lambrechts, D., Storkebaum, E., Morimoto, M., Del-Favero, J., Desmet, F., Marklund, S. L., Wyns, S., Thijs, V., Andersson, J., van Marion, I., Al-Chalabi, A., Bornes, S., Musson, R., Hansen, V., Beckman, L., Adolfsson, R., Pall, H. S., Prats, H., Vermeire, S., Rutgeerts, P., Katayama, S., Awata, T., Leigh, N., Lang-Lazdunski, L., Dewerchin, M., Shaw, C., Moons, L., Vlietinck, R., Morrison, K. E., Robberecht, W., Van Broeckhoven, C., Collen, D., Andersen, P. M., and Carmeliet, P. (2003). VEGF is a modifier of amyotrophic lateral sclerosis in mice and humans and protects motoneurons against ischemic death. *Nat. Genet.* **34**, 383–394.
83. Azzouz, M., Ralph, G. S., Storkebaum, E., Walmsley, L. E., Mitrophanous, K. A., Kingsman, S. M., Carmeliet, P., and Mazarakis, N. D. (2004). VEGF delivery with retrogradely transported lentivector prolongs survival in a mouse ALS model. *Nature* **429**, 413–417.
84. Storkebaum, E., Lambrechts, D., Dewerchin, M., Moreno-Murciano, M. P., Appelmans, S., Oh, H., Van Damme, P., Rutten, B., Man, W. Y., De Mol, M., Wyns, S., Manka, D., Vermeulen, K., Van Den Bosch, L., Mertens, N., Schmitz, C., Robberecht, W., Conway, E. M., Collen, D., Moons, L., and Carmeliet, P. (2005). Treatment of motoneuron degeneration by intracerebroventricular delivery of VEGF in a rat model of ALS. *Nat. Neurosci.* **8**, 85–92.

85. Dal Canto, M. C., and Gurney, M. E. (1995). Neuropathological changes in two lines of mice carrying a transgene for mutant human Cu,ZN SOD, and in mice overexpressing wild type human SOD: a model of familial amyotrophic lateral sclerosis (FALS). *Brain Res.* **676**, 25–40.
86. Kong, J. M., and Xu, Z. S. (1998). Massive mitochondrial degeneration in motor neurons triggers the onset of amyotrophic lateral sclerosis in mice expressing a mutant SOD1. *J. Neurosci.* **18**, 3241–3250.
87. Andreassen, O. A., Ferrante, R. J., Klivenyi, P., Klein, A. M., Shinobu, L. A., Epstein, C. J., and Beal, M. F. (2000). Partial deficiency of manganese superoxide dismutase exacerbates a transgenic mouse model of amyotrophic lateral sclerosis. *Ann. Neurol.* **47**, 447–455.
88. Liu, J., Lillo, C., Jonsson, P. A., Vande Velde, C., Ward, C. M., Miller, T. M., Subramaniam, J. R., Rothstein, J. D., Marklund, S., Andersen, P. M., Brannstrom, T., Gredal, O., Wong, P. C., Williams, D. S., and Cleveland, D. W. (2004). Toxicity of familial ALS-linked SOD1 mutants from selective recruitment to spinal mitochondria. *Neuron* **43**, 5–17.
89. Klivenyi, P., Ferrante, R. J., Matthews, R. T., Bogdanov, M. B., Klein, A. M., Andreassen, O. A., Mueller, G., Wermer, M., Kaddurah-Daouk, R., and Beal, M. F. (1999). Neuroprotective effects of creatine in a transgenic animal model of amyotrophic lateral sclerosis. *Nature Med.* **5**, 347–350.
90. Kreutzberg, G. W. (1996). Microglia: a sensor for pathological events in the CNS. *Trends Neurosci.* **19**, 312–318.
91. Alexianu, M. E., Kozovska, M., and Appel, S. H. (2001). Immune reactivity in a mouse model of familial ALS correlates with disease progression. *Neurology* **57**, 1282–1289.
92. Elliott, J. L. (2001). Cytokine upregulation in a murine model of familial amyotrophic lateral sclerosis. *Brain Res. Mol. Brain Res.* **95**, 172–178.
93. Nguyen, M. D., Julien, J. P., and Rivest, S. (2001). Induction of proinflammatory molecules in mice with amyotrophic lateral sclerosis: no requirement for proapoptotic interleukin-1beta in neurodegeneration. *Ann. Neurol.* **50**, 630–639.
94. Kriz, J., Nguyen, M. D., and Julien, J. P. (2002). Minocycline slows disease progression in a mouse model of amyotrophic lateral sclerosis. *Neurobiol. Dis.* **10**, 268–278.
95. Van Den, B. L., Tilkin, P., Lemmens, G., and Robberecht, W. (2002). Minocycline delays disease onset and mortality in a transgenic model of ALS. *Neuroreport* **13**, 1067–1070.
96. Zhu, S., Stavrovskaya I. G., Drozda, M., Kim, B. Y., Ona, V., Li, M., Sarang, S., Liu, A. S., Hartley, D. M., Wu du, C., Gullans, S., Ferrante, R. J., Przedborski, S., Kristal, B. S., and Friedlander, R. M. (2002). Minocycline inhibits cytochrome c release and delays progression of amyotrophic lateral sclerosis in mice. *Nature* **417**, 74–78.
97. Drachman, D. B., Frank, K., Dykes-Hoberg, M., Teismann, P., Almer, G., Przedborski, S., and Rothstein, J. D. (2002). Cyclooxygenase 2 inhibition protects motor neurons and prolongs survival in a transgenic mouse model of ALS. *Ann. Neurol.* **52**, 771–778.
98. Doble, A. (1996). The pharmacology and mechanism of action of riluzole. *Neurology* **47**, S233–S241.
99. Bensimon, G., Lacomblez, L., Meininger, V., and The ALS/Riluzole Study Group. (1994). A controlled trial of riluzole in amyotrophic lateral sclerosis. *N. Engl. J. Med.* **330**, 585–591.
100. Lacomblez, L., Bensimon, G., Leigh, P. N., Guillet, P., Powe, L., Durrleman, S., Delumeau, J. C., and Meininger, V. (1996). A confirmatory dose-ranging study of riluzole in ALS. ALS/Riluzole Study Group-II. *Neurology* **47**, S242–S250.
101. Traynor, B. J., Alexander, M., Corr, B., Frost, E., and Hardiman, O. (2003). Effect of a multidisciplinary amyotrophic lateral sclerosis (ALS) clinic on ALS survival: a population based study, 1996–2000. *J. Neurol. Neurosurg. Psychiatry* **74**, 1258–1261.
102. Eisen, A., Stewart, H., Schulzer, M., and Cameron, D. (1993). Antiglutamate therapy in amyotrophic lateral sclerosis: a trial using lamotrigine. *Can. J. Neurol. Sci.* **20**, 297–301.
103. Gredal, O., Werdelin, L., Bak, S., Christensen, P. B., Boysen, G., Kristensen, M. O., Jespersen, J. H., Regeur, L., Hinge, H. H., and Jensen, T. S. (1997). A clinical trial of dextromethorphan in amyotrophic lateral sclerosis. *Acta Neurol. Scand.* **96**, 8–13.
104. Miller, R. G., Moore, D. H. II, Gelinas, D. F., Dronsky, V., Mendoza, M., Barohn, R. J., Bryan, W., Ravits, J., Yuen, E., Neville, H., Ringel, S., Bromberg, M., Petajan, J., Amato, A. A., Jackson, C., Johnson, W., Mandler, R., Bosch, P., Smith, B., Graves, M., Ross, M., Sorenson, E. J., Kelkar, P., Parry, G., Olney, R., and Western ALS Study Group. (2001). Phase III randomized trial of gabapentin in patients with amyotrophic lateral sclerosis. *Neurology* **56**, 843–848.
105. Cleveland, D. W., and Rothstein, J. D. (2001). From Charcot to Lou Gehrig: deciphering selective motor neuron death in ALS. *Nat. Rev. Neurosci.* **2**, 806–819.
106. Desnuelle, C., Dib, M., Garrel, C., Favier, A. (2001). A double-blind, placebo-controlled randomized clinical trial of alpha-tocopherol (vitamin E) in the treatment of amyotrophic lateral sclerosis. ALS Riluzole-Tocopherol Study Group. *Amyotroph. Lateral. Scler. Other Motor Neuron Disord.* **2**, 9–18.
107. Ascherio, A., Weisskopf, M. G., O'Reilly, E. J., Jacobs, E. J., McCullough, M. L., Calle, E. E., Cudkowicz, M., and Thun, M. (2005). Vitamin E intake and risk of amyotrophic lateral sclerosis. *Ann. Neurol.* **57**, 104–110.
108. Louwerse, E. S., Weverling, G. J., Bossuyt, P. M., Meyjes, F. E., and de Jong, J. M. (1995). Randomized, double-blind, controlled trial of acetylcysteine in amyotrophic lateral sclerosis. *Arch. Neurol.* **52**, 559–564.
109. Klivenyi, P., Ferrante, R. J., Matthews, R. T., Bogdanov, M. B., Klein, A. M., Andreassen, O. A., Mueller, G., Wermer, M., Kaddurah-Daouk, R., and Beal, M. F. (1999). Neuroprotective effects of creatine in a transgenic animal model of amyotrophic lateral sclerosis. *Nat. Med.* **5**, 347–350.
110. Shefner, J. M., Cudkowicz, M. E., Schoenfeld, D., Conrad, T., Taft, J., Chilton, M., Urbinelli, L., Qureshi, M., Zhang, H., Pestronk, A., Caress, J., Donofrio, P., Sorenson, E., Bradley, W., Lomen-Hoerth, C., Pioro, E., Rezania, K., Ross, M., Pascuzzi, R., Heiman-Patterson, T., Tandan, R., Mitsumoto, H., Rothstein, J., Smith-Palmer, T., MacDonald, D., Burke, D., and NEALS Consortium. (2004). A clinical trial of creatine in ALS. *Neurology* **63**, 1656–1661.
111. Groeneveld, G. J., Veldink, J. H., van der Tweel, I., Kalmijn, S., Beijer, C., de Visser, M., Wokke, J. H., Franssen, H., and van den Berg, L. H. (2003). A randomized sequential trial of creatine in amyotrophic lateral sclerosis. *Ann. Neurol.* **53**, 437–445.
112. Bezzi, P., Carmignoto, G., Pasti, L., Kalmijn, S., Beijer, C., de Visser, M., Wokke, J. H., Franssen, H., and van den Berg, L. H. (1998). Prostaglandins stimulate calcium-dependent glutamate release in astrocytes. *Nature* **391**, 281–285.
113. Drachman, D. B., and Rothstein, J. D. (2000). Inhibition of cyclooxygenase-2 protects motor neurons in an organotypic model of amyotrophic lateral sclerosis. *Ann. Neurol.* **48**, 792–795.
114. Drachman, D. B., Frank, K., Dykes-Hoberg, M., Teismann, P., Almer, G., Przedborski, S., and Rothstein, J. D. (2002). Cyclooxygenase 2 inhibition protects motor neurons and prolongs survival in a transgenic mouse model of ALS. *Ann. Neurol.* **52**, 771–778.
115. Barneoud, P., and Curet, O. (1999). Beneficial effects of lysine acetylsalicylate, a soluble salt of aspirin, on motor performance in a transgenic model of amyotrophic lateral sclerosis. *Exp. Neurol.* **155**, 243–251.

116. Kasarskis, E. J., Shefner, J. M., Miller, R., et al. (1999). A controlled trial of recombinant methionyl human BDNF in ALS. *Neurology* **52**, 1427–1433.
117. Miller, R. G., Petajan, J. H., Bryan, W. W., Armon, C., Barohn, R. J., Goodpasture, J. C., Hoagland, R. J., Parry, G. J., Ross, M. A., and Stromatt, S. C. (1996). A placebo-controlled trial of recombinant human ciliary neurotrophic (rhCNTF) factor in amyotrophic lateral sclerosis. rhCNTF ALS Study Group. *Ann. Neurol.* **39**, 256–260.
118. Lai, E. C., Felice, K. J., Festoff, B. W., Gawel, M. J., Gelinas, D. F., Kratz, R., Murphy, M. F., Natter, H. M., Norris, F. H., and Rudnicki, S. A. (1997). Effect of recombinant human insulin-like growth factor-I on progression of ALS. A placebo-controlled study. The North America ALS/IGF-I Study Group. *Neurology* **49**, 1621–1630.
119. Borasio, G. D., Robberecht, W., Leigh, P. N., Emile, J., Guiloff, R. J., Jerusalem, F., Silani, V., Vos, P. E., Wokke, J. H., and Dobbins, T. (1998). A placebo-controlled trial of insulin-like growth factor-I in amyotrophic lateral sclerosis. European ALS/IGF-I Study Group. *Neurology* **51**, 583–586.
120. Moulignier, A., Moulonguet, A., Pialoux, G., and Rozenbaum, W. (2001). Reversible ALS-like disorder in HIV infection. *Neurology* **57**, 995–1001.
121. Scelsa, S. N., MacGowan, D. J., Mitsumoto, H., Imperato, T., LeValley, A. J., Liu, M. H., DelBene, M., and Kim, M. Y. (2005). A pilot, double-blind, placebo-controlled trial of indinavir in patients with ALS. *Neurology* **64**, 1298–1300.
122. Appel, S. H., Smith, R. G., Engelhardt, J. I., and Stefani, E. (1994). Evidence for autoimmunity in amyotrophic lateral sclerosis. [Review]. *J. Neurol. Sci.* **124 (Suppl)**, 14–19.
123. Drachman, D. B., Chaudhry, V., Cornblath, D., Kuncl, R. W., Pestronk, A., Clawson, L., Mellits, E. D., Quaskey, S., Quinn, T., Calkins, A., et al. (1994). Trial of immunosuppression in amyotrophic lateral sclerosis using total lymphoid irradiation. *Ann. Neurol.* **35**, 142–150.
124. Smith, S. A., Miller, R. G., Murphy, J. R., and Ringel, S. P. (1994). Treatment of ALS with high dose pulse cyclophosphamide. *J. Neurol. Sci.* **124(Suppl)**, 84–87.
125. Appel, S. H., Stewart, S. S., Appel, V., Harati, Y., Mietlowski, W., Weiss, W., and Belendiuk, G. W. (1988). A double-blind study of the effectiveness of cyclosporine in amyotrophic lateral sclerosis. *Arch. Neurol.* **45**, 381–386.
126. Maragakis, N. J., Rao, M. S., Llado, J., Wong, V., Xue, H., Pardo, A., Herring, J., Kerr, D., Coccia, C., and Rothstein, J. D. (2005). Glial restricted precursors protect against chronic glutamate neurotoxicity of motor neurons in vitro. *Glia* **50**, 145–159.
127. Kerr, D. A., Llado, J., Shamblott, M. J., Maragakis, N. J., Irani, D. N., Crawford, T. O., Krishnan, C., Dike, S., Gearhart, J. D., and Rothstein, J. D. (2003). Human embryonic germ cell derivatives facilitate motor recovery of rats with diffuse motor neuron injury. *J. Neurosci.* **23**, 5131–5140.
128. Garbuzova-Davis, S., Willing, A. E., Zigova, T., Saporta, S., Justen, E. B., Lane, J. C., Hudson, J. E., Chen, N., Davis, C. D., and Sanberg, P. R. (2003). Intravenous administration of human umbilical cord blood cells in a mouse model of amyotrophic lateral sclerosis: distribution, migration, and differentiation. *J. Hematother. Stem Cell Res.* **12**, 255–270.
129. Comi, G. P., Bordoni, A., Salani, S., Franceschina, L., Sciacco, M., Prelle, A., Fortunato, F., Zeviani, M., Napoli, L., Bresolin, N., Moggio, M., Ausenda, C. D., Taanman, J. W., and Scarlato, G. (1998). Cytochrome c oxidase subunit I microdeletion in a patient with motor neuron disease. *Ann. Neurol.* **43**, 110–116.
130. Kaspar, B. K., Llado, J., Sherkat, N., Rothstein, J. D., and Gage, F. H. (2003). Retrograde viral delivery of IGF-1 prolongs survival in a mouse ALS model. *Science* **301**, 839–842.
131. Ralph, G. S., Radcliffe, P. A., Day, D. M., Carthy, J. M., Leroux, M. A., Lee, D. C., Wong, L. F., Bilsland, L. G., Greensmith, L., Kingsman, S. M., Mitrophanous, K. A., Mazarakis, N. D., and Azzouz, M. (2005). Silencing mutant SOD1 using RNAi protects against neurodegeneration and extends survival in an ALS model. *Nat. Med.* **11**, 429–433.
132. Raoul, C., Abbas-Terki, T., Bensadoun, J. C., Guillot, S., Haase, G., Szulc, J., Henderson, C. E., and Aebischer, P. (2005). Lentiviral-mediated silencing of SOD1 through RNA interference retards disease onset and progression in a mouse model of ALS. *Nat. Med.* **11**, 423–428.

47

Hereditary Spastic Paraplegia and Primary Lateral Sclerosis

Philip A. Wilkinson, BSc, MBBS, MRCP
Michael Swash, MD, FRCP, FRCPath

Keywords: *familial spastic paraparesis, Strümpell-Lorrain syndrome*

I. Definitions
II. Historical Aspects
III. Epidemiology
IV. Clinical Features of HSP
V. Neuropathology of HSP
VI. Molecular Genetics of HSP
VII. Nosology of PLS
VIII. Common Molecular Mechanisms Underlying HSP and PLS
References

I. Definitions

The term *hereditary spastic paraplegia* (HSP) defines a clinically and genetically heterogeneous group of inherited neurological disorders characterized by progressive symmetrical spasticity and weakness of the lower limbs. Additional features are often present. The phenotype is classified as *pure* if symptoms and signs are confined to progressive spastic paraparesis, with occasional posterior column or bladder involvement, or *complicated* if major additional neurological or other features are present. Inheritance may be autosomal dominant (AD), autosomal recessive (AR), or X-linked for both pure and complicated forms.

Primary lateral sclerosis (PLS) is a variant of amyotrophic lateral sclerosis (ALS), in which the predominant feature is progressive spastic weakness involving spinal regions and usually also pseudobulbar features [1]. The disorder is almost always sporadic, and most cases are of adult onset. A number of associated features may develop, including frontal lobe dementia, urgency of micturition, and lower motor neuron (LMN) involvement. Progression to ALS may develop, but only after several years of pure upper motor neuron (UMN) disorder.

II. Historical Aspects

The first description of HSP was made by Seeligmüller in 1876, who reported four siblings with progressive spasticity and weakness of the lower limbs. Four years later, Strümpell, after whom the disease is often named, reported two brothers with this condition. Strümpell also referred to the father as being "a little lame," the first suggestion of AD inheritance of HSP and probably an example of the variable disease expression that is common even within the same family.

In the early part of the 20th century, the condition became widely recognized with numerous cases described in different communities. Reports began to appear of families in which additional neurological or other clinical features were inherited in association with spastic paraparesis, although there was much debate as to whether these represented separate disease entities. It was not until the pioneering work of Harding in the 1980s that formal criteria for the classification of HSP cases were proposed. Harding recognized the occurrence of pure and complicated cases and noted different patterns of phenotype and progression according to the age of onset. She also recognized early-onset (before age 35 years) and late-onset cases (after age 35 years). With the advent of genetic testing, this classification has begun to be superseded, although the causation of phenotypic variation remains a problem, presumably indicating modifier genetic or environmental factors.

PLS was also described in the 19th century, probably first by Erb in 1875, but for a long period the diagnosis was in eclipse, until clinical diagnostic criteria were suggested by Pringle and colleagues in 1992. Since then, with the reassurance of magnetic resonance imaging to exclude multiple sclerosis and other cervical and brain disorders, the clinical diagnosis of PLS has become more secure.

III. Epidemiology

HSP has been reported worldwide, although its prevalence varies. The population prevalence ranges from 0.1 to 9.6 per 100,000 people. In part, this may reflect methodology, especially when secondary cases were sought by examination of all at-risk relatives. Many affected individuals are presymptomatic or do not seek medical attention, leading to underestimation of disease incidence.

AD inheritance occurs in about 80% of HSP cases. The majority of families show the pure phenotype. AR HSP is observed more frequently in societies in which consanguineous marriages are commonplace; this may account for a significant proportion of apparently sporadic cases. Complicated phenotypes predominate in AR HSP families; for example, a number of distinct clinical syndromes have been recognized in isolated communities, such as the Old Order Amish in the United States.

PLS is a rare sporadic disorder, but the phenotype has also been reported in some families with ALS associated with SOD-1 mutations.

IV. Clinical Features of HSP

The index clinical feature in HSP is progressive, symmetrical, spastic paraparesis. The disease may present at any time in life. It may even cause delayed motor milestones in early childhood or occur in the later decades. The onset is insidious, with a variable but linear rate of progression. The spectrum of resulting disability ranges from entirely asymptomatic (10%–20% of cases) to wheelchair-dependent or, in the most extreme cases, bedridden individuals. The most common presenting symptoms are stiffness in the legs, difficulty walking or running with frequent stumbling or falls, premature wear on the shoes, or the observation by others of an abnormal gait. Urgency and frequency of micturition are also frequently reported, although urinary symptoms tend to present later in the course of the disease. Urinary incontinence may eventually become a problem, and this is frequently worsened by problems with mobility.

Neurological examination in patients with pure HSP reveals symmetrical spastic paraparesis with increased tone, pyramidal weakness, hyperreflexia, and extensor plantar responses. Characteristically, spasticity is very marked and the degree of weakness slight. Pes cavus and other foot deformities are commonly reported, particularly in early-onset cases. Abnormalities of posterior column sensation are frequent, with varying degrees of diminished vibration sense and proprioception. Minor upper limb incoordination has also been described in some patients, although it is unclear if this is due to sensory or cerebellar ataxia. Mild muscle wasting may also occur in long-standing, uncomplicated cases; this most often reflects disuse atrophy in immobile patients. However, subtle cognitive defects are increasingly being recognized in patients with pure HSP phenotypes.

A. Complicated HSP

Unlike pure forms of HSP, complicated HSP is most commonly inherited in an AR fashion. Most X-linked cases, although rare, also present with complicated phenotypes. A number of distinct clinical syndromes have been described (Table 1). However, classification of these syndromes is not comprehensive because many families do not fall into such clear categories. A number of syndromes have also been included, not generally recognized as forms of HSP, in which spasticity is a minor or variable feature.

Phenotypic variability has long been recognized both between and within HSP pedigrees, indicating that mutations in the same gene may result in different clinical phenotypes. Genetic testing for known mutations in HSP and other disorders has led to recognition that some syndromes ascribed to HSP represent atypical presentations of other diseases such as Friedreich's ataxia and presenilin mutations in Alzheimer's disease. Eventually, genetic understanding of the basis of HSP will inform the clinical classification, but at present, clinical classification remains a useful tool for grouping families for research purposes.

V. Neuropathology of HSP

Autopsy findings in patients with HSP are rarely reported. Early studies were typically of patients with long-standing, uncomplicated HSP phenotypes and were focused on changes within the spinal cord. The common pathological feature appears to be dying-back degeneration of the longest motor and sensory axons of the central nervous system (CNS) (i.e., corticospinal tracts, fasciculus gracilis, and spinocerebellar tracts) (Fig. 1), so that degeneration is maximum in the terminal portion of these axons. In addition, there may be mild loss of anterior horn cells. Additional cortical pathology, including neuronal loss with tau-immunoreactive neurofibrillary tangles and balloon cells, has been described in a small number of HSP patients

Table 1 Complicated HSP Phenotypes

Type	Other Clinical Features	Inheritance/Gene
CRASH syndrome	Agenesis of the *c*orpus callosum, mental *r*etardation, *a*dducted thumbs, *s*pastic paraparesis, and *h*ydrocephalus	X-linked/L1CAM
MASA syndrome	*M*ental retardation, *a*phasia, *s*pasticity, and *a*dducted thumbs	X-linked/L1CAM
Silver syndrome	Distal amyotrophy	AD/SPG17
Kjellin syndrome	Mental retardation, amyotrophy, and macular dystrophy	AR/SPG15
Troyer syndrome	Dysarthria, amyotrophy, and skeletal abnormalities in Old Order Amish population	AR/SPG20 (Spartin)
Mast syndrome	Dementia, cerebellar signs, and extrapyramidal features in Old Order Amish population	AR/SPG21 (Maspardin)
Sjögren-Larson syndrome	Ichthyosis and mental retardation	AR/Fatty aldehyde dehydrogenase
Charlevoix-Saguenay syndrome	Spastic ataxia, neuropathy, dysarthria, and prominent myelinated nerve fibers on funduscopy (Quebec)	AR/Sacsin
With hyperekplexia	Single family with GLRA1 mutation in which hyperekplexia and spastic paraparesis cosegregate	AD/GLRA1
With dementia	Variable degrees of cognitive decline	AD/AR (including presenilin 1, spastin)
With cataracts and gastroesophageal reflux	Plus distal amyotrophy	AD/SPG9
With cerebellar signs	Dysarthria, nystagmus, and ataxia	AD/AR
With neuropathy	Motor and sensory neuropathies described	AD/AR
With optic atrophy	Decreased visual acuity	AD/AR
With disordered skin pigmentation	Hypo and hyperpigmented skin lesion	AD/AR (including SPG24)
With thin corpus callosum	Associated with cognitive decline (Japan)	AR (including SPG11)
With epilepsy	Absence, myoclonic, partial and generalized seizures all described	AD/AR
With extrapyramidal features	Choreoathetosis, dystonia and rigidity all described	AD/AR
With syndactyly	Syndactylia in upper limbs	AD

AD, Autosomal dominant; *AR*, autosomal recessive; *HSP*, hereditary spastic paraplegia.

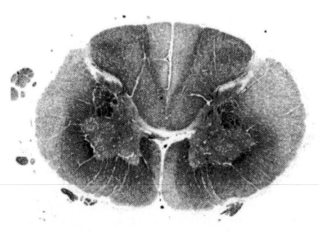

Figure 1 Pathology of HSP: note pallor of crossed and uncrossed corticospinal tracts, and to a lesser extent, of the posterior columns.

with an associated dementia, although the significance of these changes remains uncertain.

VI. Molecular Genetics of HSP

The pathophysiology underlying the various forms of HSP is beginning to be understood. To date, 28 HSP loci have been mapped and pathogenic mutations identified in 11 different genes (Table 2). These developments have helped establish a number of common, overlapping molecular mechanisms associated with abnormal development or degeneration of the corticospinal pathways in HSP and other neurodegenerative disorders [2].

A. X-linked HSP

X-linked forms of HSP demonstrate both clinical and genetic heterogeneity. Mutations in two genes (L1CAM and PLP) have been identified in association with X-linked HSP, and at least two further genetic loci may exist. Although an X-linked mode of inheritance cannot be definitively excluded in families lacking male-to-male transmission, overall they appear to be relatively rare. The phenotypes associated with X-linked HSP are mostly complicated HSP, and several distinct clinical syndromes are recognized.

1. SPG1 (L1CAM)

Mutations in the gene that encodes the neural cell adhesion molecule L1 (L1CAM) were the first to be described in patients with complicated X-linked forms of HSP. L1CAM mutations may cause a range of clinical presentations including varying degrees of hydrocephalus, mental retardation, lower limb spasticity, and flexion-adduction abnormalities of the thumbs. Enlargement of the cerebral ventricles with abnormal development of the corpus callosum and corticospinal tracts may occur. The term *CRASH syndrome* (*c*orpus callosum hypoplasia, *r*etardation, *a*dducted thumbs, *s*pastic paraparesis and *h*ydrocephalus) has been adopted for L1CAM mutations with variations of this constellation of features.

L1 is a transmembrane glycoprotein that is part of the superfamily of immunoglobulin (Ig)-related cell adhesion molecules. It is composed of six Ig-like domains and five fibronectin type-III repeats in the extracellular region, a single transmembrane domain, and a short cytoplasmic tail. L1 therefore acts not only as an adhesion molecule but also functions as a receptor and plays a critical role in growth and guidance of axons in the developing nervous system. More than 100 different L1 mutations have been described. There is a striking correlation between mutations in the L1CAM gene and the severity of the disease. Mutations that produce truncations in the extracellular domain of the L1 protein are more likely to produce severe hydrocephalus, grave mental retardation, or early death than are point mutations in the extracellular domain or mutations affecting only the cytoplasmic domain of the protein. Although less severe than extracellular truncations, point mutations in the extracellular domain generally produce more severe neurological problems than mutations in the cytoplasmic domain, which tend to result in hydrocephalus alone.

2. SPG2 (PLP)

SPG2 maps to Xq22 and results from mutations in PLP and its splice variant DM20 (produced by alternative splicing of the same gene). Both PLP and DM20 are major protein components of myelin in the CNS. Pelizaeus-Merzbacher disease (PMD), an infantile-onset progressive leukodystrophy characterized by neonatal hypotonia, nystagmus, psychomotor retardation, spasticity, dystonia, and ataxia, is also caused by mutations or duplication of the this gene. The mechanisms underlying genotype-phenotype correlations between these different presentations associated with PLP mutations remains incompletely understood. An association between the trafficking capability of mutant PLP and the severity of the disease has been postulated. Disruption of PLP mediated axonal-glial interactions is thought to underlie axonal degeneration identified in both patients and animal models. An increased dose of PLP due to gene duplication or mutations results in structurally altered translated PLP, which is thought to engulf the secretory pathway and compromise oligodendrocyte function, leading to the PMD phenotype. Conversely, mutations allowing PLP and DM20 to traverse the secretory pathway and reach the cell surface generally result in milder phenotypes. However, there may be phenotypic variation even within families, indicating that additional genetic and possibly environmental factors act as disease modifiers.

Table 2 Genetic Classification of HSP

SPG Locus	Chromosomal Location	Inheritance	Gene Product	Phenotype
SPG1	Xq28	X-linked	L1CAM	Complicated
SPG2	Xq22	X-linked	PLP/DM20	Pure and complicated
SPG3	14q11-q21	AD	Atlastin	Pure
SPG4	2p22-p21	AD	Spastin	Pure and complicated
SPG5	8q11.1-q21.2	AR	–	Pure
SPG6	15q11.1	AD	NIPA1	Pure
SPG7	16q24.3	AR	Paraplegin	Pure and complicated
SPG8	8q24	AD	–	Pure
SPG9	10q23.3-q24.2	AD	–	Complicated
SPG10	12q13	AD	KIF5A	Pure
SPG11	15q13-q15	AR	–	Pure and complicated
SPG12	19q13	AD	–	Pure
SPG13	2q24-q34	AD	HSP60	Pure
SPG14	3q27-q28	AR	–	Complicated
SPG15	14q22-q24	AR	–	Complicated
SPG16	Xq11.2	X-linked	–	Pure and complicated
SPG17	11q12-q14	AD	BSCL2	Complicated
SPG18	Reserved	–	–	–
SPG19	9q33-q34	AD	–	Pure
SPG20	13q12.3	AR	Spartin	Complicated
SPG21	15q22.3	AR	Maspardin	Complicated
SPG22	Reserved	–	–	–
SPG23	1q24-q32	AR	–	Complicated
SPG24	13q14	AR	–	Pure
SPG25	6q23-q24.1	AR	–	Complicated
SPG26	12p11.1-q14	AR	–	Complicated
SPG27	10q22.1-q24.1	AR	–	Pure
SPG28	14q21.3-q22.3	AR	–	Pure
SPG29	1p31.1-p21.1	AD	–	Complicated
SPG30	2q37.3	AR	–	Complicated

AD, Autosomal dominant; *AR*, autosomal recessive; *HSP*, hereditary spastic paraplegia.

B. Autosomal Recessive HSP

AR HSP is relatively rare outside societies in which consanguineous marriages are frequent, although it is possible that a significant proportion of seemingly sporadic cases may actually be AR in nature. The phenotype is more frequently a complicated form of HSP that includes a number of distinct clinical syndromes.

1. SPG7 (Paraplegin)

SPG7 was the first AR HSP gene to be characterized [3]. The protein product, paraplegin, is composed of 795 amino acids and forms part of a protein complex at the inner mitochondrial membrane. It shares amino acid sequence homology with a group of yeast mitochondrial metalloproteases. These proteins are members of the AAA protein superfamily (ATPase associated with diverse cellular activities), which are found widely in both prokaryotic and eukaryotic cells and play an important role in a variety of cellular activities including cell division, transcription, organelle biogenesis, vesicle transport, and enzyme assembly. Yeast mitochondrial ATPases are known to possess both proteolytic and chaperone-like activities at the inner mitochondrial membrane, where they are involved in the assembly and degradation of proteins in the respiratory chain complex.

SPG7 mutations have been identified in patients with both pure and complicated HSP phenotypes. Muscle biopsies from severely affected individuals have shown mitochondrial oxidative phosphorylation defects including ragged-red fibers and cytochrome oxidase (COX)–negative and succinate dehydrogenase (SDH)–positive fibers. Decreased activity of complex 1 activity in the mitochondrial respiratory chain has been identified in cultured cells from patients with SPG7 mutations and may play a significant role in the pathogenesis of this form of HSP through an increased sensitivity to oxidative stress.

2. SPG20 (Spartin) and SPG21 (Maspardin)

Troyer syndrome and Mast syndrome are complicated forms of AR HSP originally described in the Old Order Amish population. The cardinal features of Troyer syndrome are of a progressive spastic paraparesis associated with pseudobulbar palsy, distal amyotrophy, mild developmental delay, and short stature. The Troyer syndrome locus has been mapped to a small region on chromosome 13q12.3 (SPG20), where a single base deletion (1110delA) was found in exon 4 of a gene encoding a novel protein, spartin [4]. The precise role of spartin remains unclear, but the identification of a conserved functional domain, also present in spastin and molecules involved in endosomal trafficking, the MIT (Microtubule-Interacting and Trafficking) domain, suggests a possible role in microtubule dynamics and cellular trafficking.

Mast syndrome is characterized by a later-onset spastic paraparesis associated with progressive cognitive decline. Cerebellar and extrapyramidal features may also occur in advanced cases. Homozygosity mapping in Amish families localized Mast syndrome to chromosome 15q22.31 (SPG21) [5]. Sequencing of the three overlapping transcripts from within this region identified a single base insertion (601insA) in 33kDa acid-cluster protein gene (ACP33). This frameshift mutation results in the truncation of the protein product maspardin. Maspardin localizes to endosomal and trans-Golgi transport vesicles and is predicted to play a role in protein transport and sorting. To date, confirmed cases of Troyer syndrome and Mast syndrome are confined to the Old Order Amish population.

C. Autosomal Dominant HSP

HSP is most commonly inherited as an AD trait. The phenotype is usually pure, often with a later age of onset than recessive and X-linked forms. However, phenotypic variability and incomplete penetrance are frequently encountered.

1. SPG4 (Spastin)

Mutations in the SPG4 gene, encoding the protein spastin, account for 20%–40% of all cases of AD HSP [6]. More than 100 different pathogenic mutations thus far have been identified. Missense, nonsense, and splice-site mutations, along with deletions and insertions, have been described. The phenotype is usually pure, although complicated forms have been reported and mild cognitive impairment in the later stages is increasingly recognized. However, no clear correlation between genotype and phenotype has been established.

Like paraplegin, spastin is also a member of the AAA family of proteins, with a conserved domain containing ATPase motifs. The N-terminal region also contains the MIT domain found in spartin and other microtubule-interacting proteins. Spastin binds to microtubules in an ATP-dependent fashion and regulates microtubule function. Therefore, defective neuronal trafficking due to disruption of microtubule stability has been proposed as the underlying pathogenic mechanism in HSP associated with spastin mutations.

2. SPG3 (Atlastin)

Mutations in the SPG3 gene encoding the protein atlastin are the second most common cause of AD HSP, accounting for approximately 10% of all cases [7]. The phenotype is typically an early-onset pure HSP with relatively slow disease progression and less disability than many of the other forms of HSP. Atlastin is related to the dynamin family of GTPases, which are known to play a role in a wide variety of vesicle trafficking events. In addition, this group of proteins has been implicated in the maintenance and distribution of mitochondria and is associated with cytoskeletal elements such as actin and microtubules. The precise mechanism by which atlastin mutations result in HSP remains unclear. However, recent studies have confirmed that atlastin is localized predominantly in pyramidal neurons in the cerebral cortex and in the hippocampus. In cultured cortical neurons, atlastin co-localizes with markers of the Golgi apparatus, and membrane fractionation and protease protection assays have confirmed that atlastin is an integral membrane protein with two transmembrane domains. This suggests that atlastin is predominantly involved in Golgi membrane dynamics or vesicle trafficking and that disruption of this process in the adult CNS underlies the neurodegenerative process.

3. SPG6 (NIPA1)

The SPG6 locus was initially mapped to the centromeric region of chromosome 15q. Deletions in this part of chromosome 15 are associated with Prader-Willi syndrome or Angelman syndrome, conditions characterized by genetic imprinting. Subsequently, mutations have been identified in SPG6 linked families in a nonimprinted gene from within this region, NIPA1 [8]. The precise function of the NIPA1 protein remains unclear, although predictions based on the amino acid sequence and hydrophobicity analysis suggest it is an integral membrane protein that acts as either a receptor or transporter. The fact that patients with Prader-Willi syndrome or Angelman syndrome do not develop spastic paraparesis suggests that mutations in NIPA1 exert a pathogenic effect through a gain-of-function mechanism.

4. SPG10 (KIF5A) and SPG13 (Hsp60)

A pathogenic missense mutation (A767G) in the motor domain of the neuronal kinesin heavy chain gene KIF5A has been identified in a single family linked to the SPG10 locus [9]. The microtubule-associated proteins of the kinesin and dynein families act as molecular motors that transport intracellular material along microtubules: kinesins in the anterograde direction and dyneins in the retrograde direction. This missense mutation in KIF5A has been predicted to disrupt stimulation of the motor ATPase by microtubule binding. Therefore although this form of HSP is rare, this mutation supports defective axonal transport in the pathogenesis of HSP [10]. A missense mutation in the gene encoding the mitochondrial chaperonin Hsp60 has also been identified as the cause of a pure form of AD HSP in a family linked to the SPG13 locus. This member of the heat-shock protein group forms part of a multimeric chaperone complex and adds supportive evidence for mitochondrial dysfunction in the pathogenesis of certain forms of HSP.

5. SPG17 (BSCL2)

Heterozygous mutations in the BSCL2 gene are associated with Silver syndrome, a form of HSP in which there is prominent distal amyotrophy in addition to spastic paraparesis [11]. However, phenotypic variability is marked, with some patients having isolated distal amyotrophy (a clinical picture that was originally termed "distal HMN V") and with a high degree of incomplete penetrance within families. Null mutations in BSCL2, which encodes the protein seipin, were previously shown to be associated with AR Berardinelli-Seip congenital lipodystrophy. Seipin is a transmembrane protein that is localized to the endoplasmic reticulum. The precise function of seipin remains unknown, although its homology to midasin, an AAA domain–containing nuclear protein that is involved in RNA transport, might explain how mutant forms of seipin can cause such clinically distinct syndromes.

VII. Nosology of PLS

Since its first description, the nosology of PLS has been problematic. Initially inseparable from other causes of progressive spastic quadriplegia (e.g., cervical spondylotic myelopathy, and especially multiple sclerosis), the diagnosis only began to be recognized in the modern era after Pringle et al. suggested diagnostic criteria [12] (Table 3). The diagnosis has been increasingly recognized, and PLS has again been classified as a syndrome within the group of motor neuron diseases, as a degenerative disorder involving primarily the motor system [13]. Many cases presenting with the PLS syndrome, perhaps with very LMN involvement, evolve into ALS after a period of 3–5 years, although in other cases, the syndrome remains as a limited UMN disorder for a much longer time. It is assumed that the pathology of ALS resembles that of classical Charcot ALS, although there are very few pathological studies in PLS. Pathology resembling that of frontotemporal dementia, with τ-positive, α synuclein–negative neuronal inclusions, has also been noted. Bunina bodies and ubiquitin-positive skeins have been reported in spinal motor neurons, strongly suggesting that PLS is related to classical ALS. Therefore, it is probably appropriate to regard PLS as the UMN end of the combined UMN and LMN degeneration, with the more widespread cortical involvement, especially frontal lobe degeneration, that characterizes motor neuron disease itself.

Because both HSP and PLS may show a complex phenotype with LMN features and bladder symptoms, clinical overlap exists between the two groups of disorders. However, inherited forms of PLS have not been described except in the context of families with familial ALS with variable phenotype, and at present, PLS is best regarded as a sporadic disorder.

Table 3 Diagnostic Criteria for PLS

Clinical
Insidious onset of progressive spastic paresis, usually beginning in legs, but often also involving bulbar musculature and upper limbs
Adult onset, usually in fifth decade
Sporadic disease
Course >3 years
Clinically isolated UMN dysfunction
Symmetrical signs

Additional suggestive features
Preserved bladder function
Abnormal cortical-evoked motor latencies with normal CMAPs
Focal atrophy of precentral gyrus on MRI
Decreased glucose consumption in precentral region on PET

Laboratory studies
All normal

CMAP, Compound motor action potential; *MRI*, magnetic resonance imaging, *PET*, positron emission tomography; *PLS*, primary lateral sclerosis; *UMN*, upper motor neuron.

From Pringle, C. E., Hudson, A. J., Munoz, D. G., et al. (1992). Primary lateral sclerosis: clinical features, neuropathology and diagnostic criteria. *Brain* **115**, 495–520.

VIII. Common Molecular Mechanisms Underlying HSP and PLS

Advances in the identification of causative genes have increased our understanding of a number of overlapping molecular mechanisms underlying the various forms of HSP. Disruption of the corticospinal pathways, manifesting in a spastic paraparesis, may result from either abnormal initial development or subsequent neurodegeneration (Fig. 2).

With mutations in the L1CAM and PLP genes, the failure of appropriate neuronal migration during development or subsequent myelination results in infantile-onset disease that usually affects multiple systems. The role of mitochondrial

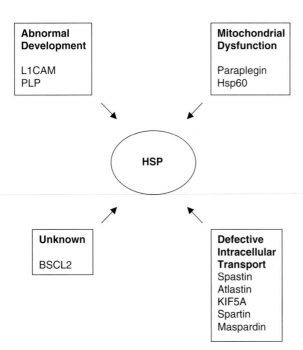

Figure 2 Pathological mechanisms associated with HSP genes.

respiratory chain dysfunction in neurodegenerative disorders is well established, and mutations in both paraplegin and Hsp60 provide supportive evidence for this mechanism in HSP. Although they may also affect multiple systems and manifest as complicated phenotypes, the terminal ends of the longest axons seem to be particularly vulnerable to defects in energy metabolism.

Defective cellular trafficking and transport appears to be a major factor underlying neurodegeneration in HSP and ALS and therefore also in PLS [14]. Spastin, atlastin, KIF5A, spartin, and maspardin all have proposed functions in various pathways associated with intracellular transport processes, particularly those involving microtubule dynamics. The ubiquitinated neuronal inclusions found in ALS and PLS are also recognized as interfering with axonal transport, suggesting the possibility of a final common pathway of metabolic disorder leading to neuronal and axonal dysfunction in susceptible long neurons in the motor system. The fact that these are highly energy-dependent processes may also explain in part the mechanism associated with mitochondrial respiratory chain dysfunction.

At present, there are no disease-modifying treatments for patients with HSP, and no clinical trials are available regarding the possible effect of riluzole (the only medication known to have any effect in slowing progression of ALS) in PLS. All these syndromes show progressive disability.

References

1. Swash, M., Desai, J., and Misra, V. P. (1999). What is primary lateral sclerosis? *J. Neurol. Sci.* **170**, 5–10.
2. Reid, E. (2003). Science in motion: common molecular pathological themes emerge in the hereditary spastic paraplegias. *J. Med. Genet.* **40**, 81–86.
3. Casari, G., De Fusco, M., Ciarmatori, S., Zeviani, M., Mora, M., Fernandez, P., De Michele, G., Filla, A., Cocozza, S., Marconi, R., Durr, A., Fontaine, B., and Ballabio, A. (1998). Spastic paraplegia and OXPHOS impairment caused by mutations in paraplegin, a nuclear-encoded mitochondrial metalloprotease. *Cell* **93**, 973–983.
4. Patel, H., Cross, H., Proukakis, C., Hershberger, R., Bork, P., Ciccarelli, F. D., Patton, M. A., McKusick, V. A., and Crosby, A. H. (2002). SPG20 is mutated in Troyer syndrome, a hereditary spastic paraplegia. *Nat. Genet.* **31**, 347–348.
5. Simpson, M. A., Cross, H., Proukakis, C., Pryde, A., Hershberger, R., Chatonnet, A., Patton, M. A., and Crosby, A. H. (2003). Maspardin is mutated in Mast syndrome, a complicated form of hereditary spastic paraplegia associated with dementia. *Am. J. Hum. Genet.* **73**, 1147–1156.
6. Hazan, J., Fonknechten, N., Mavel, D., Paternotte, C., Samson, D., Artiguenave, F., Davoine, C. S., Cruaud, C., Durr, A., Wincker, P., Brottier, P., Cattolico, L., Barbe, V., Burgunder, J. M., Prud'homme, J. F., Brice, A., Fontaine, B., Heilig, B., and Weissenbach, J. (1999). Spastin, a new AAA protein is altered in the most frequent form of autosomal dominant spastic paraplegia. *Nat. Genet.* **23**, 296–303.
7. Zhao, X., Alvarado, D., Rainier, S., Lemons, R., Hedera, P., Weber, C. H., Tukel, T., Apak, M., Heiman-Patterson, T., Ming, L., Bui, M., and Fink, J. K. (2001). Mutations in a novel GTPase cause autosomal dominant hereditary spastic paraplegia. *Nat. Genet.* **29**, 326–331.
8. Rainier, S., Chai, J. H., Tokarz, D., Nicholls, R. D., and Fink, J. K. (2003). NIPA1 gene mutations cause autosomal dominant hereditary spastic paraplegia (SPG6). *Am. J. Hum. Genet.* **73**, 967–971.
9. Reid, E., Kloos, M., Ashley-Koch, A., Hughes, L., Bevan, S., Svenson, I. K., Graham, F. L., Gaskell, P. C., Dearlove, A., Pericak-Vance, M. A., Rubinsztein, D. C., and Marchuk, D. A. (2002). A kinesin heavy chain (KIF5A) mutation in hereditary spastic paraplegia (SPG10). *Am. J. Hum. Genet.* **71**, 1189–1194.
10. Hansen, J. J., Durr, A., Cournu-Rebeix, I., Georgopoulos, C., Ang, D., Nielsen, M. N., Davoine, C. S., Brice, A., Fontaine, B., Gregersen, N., and Bross, P. (2002). Hereditary spastic paraplegia SPG13 is associated with a mutation in the gene encoding the mitochondrial chaperonin Hsp60. *Am. J. Hum. Genet.* **70**, 1328–1332.
11. Windpassinger, C., Auer-Grumbach, M., Irobi, J., Patel, H., Petek, E., Horl, G., Malli, R., Reed, J. A., Dierick, I., Verpoorten, N., Warner, T. T., Proukakis, C., Van den Bergh, P., Verellen, C., Van Maldergem, L., Merlini, L., De Jonghe, P., Timmerman, V., Crosby, A. H., and Wagner, K. (2004). Heterozygous missense mutations in BSCL2 are associated with distal hereditary motor neuropathy and Silver syndrome. *Nat. Genet.* **36**, 271–276.
12. Pringle, C. E., Hudson, A. J., Munoz, D. G., Kiernan, J. A., Brown, W. F., and Ebers, G. C. (1992). Primary lateral sclerosis: clinical features, neuropathology and diagnostic criteria. *Brain* **115**, 495–520.
13. Le Forestier, N., Maisonobe, T., Piquard, A., Rivaud, S., Crevier-Buchman, L., Salachas, F., Pradat, P. F., Lacomblez, L., and Meininger, V. (2001). Does primary lateral sclerosis exist? *Brain* **124**, 1989–1999.
14. Crosby, A. H., and Proukakis, C. (2002). Is the transportation highway the right road for hereditary spastic paraplegia? *Am. J. Hum. Genet.* **71**, 1009–1016.

48

Poliomyelitis

Eric J. Sorenson, MD

Keywords: *electrophysiology, epidemiology, polio, post-polio fatigue, post-polio syndrome*

I. Background
II. Epidemiology
III. Pathophysiology
IV. Treatment
V. Summary
   References

I. Background

Epidemics of paralytic poliomyelitis occurred regularly before the introduction of the polio vaccination. Such epidemics were most problematic in the developed world. In undeveloped countries, cases were generally sporadic with few epidemics. This difference was largely the result of public health practices in the developed world. In undeveloped countries, polio was transmitted via contaminated drinking water and typically infected infants at a young age. In this young age group, few subjects developed the paralytic form of the disease and most recovered without sequelae but with lifelong immunity. In the developed world with a protected water supply, inoculation occurred later in life when the probability of the paralytic form of the disease was higher. In addition, without the continuous exposure, conditions were favorable for epidemics.

Epidemics typically occurred in the late summer. Poliovirus, the causative agent of paralytic poliomyelitis, is an enterovirus spread by the oral route. The principal infection associated with the poliovirus is enteritis with the prodromal illness of fever, headache, arthralgia, vomiting, and diarrhea lasting 3–4 days. About half of the patients do not develop paralytic manifestations. In the remaining, a biphasic course evolves. As the initial enteritis subsides, the paralysis begins. Severe back and limb pain, headache, and meningismus develop, accompanied by severe and disabling muscle spasms. Paralysis tends to occur in a patchy, multifocal distribution. Weakness of individual muscles comes on rapidly over days and typically reaches a maximum within 1 week. The virus has a specific trophism for the motor neurons, resulting in motor neuron death. Virtually any of the skeletal muscles, including bulbar, limb, and respiratory muscles, can be affected. Exceptions to this are the extraocular eye muscles and sphincter muscles, which are generally spared. In a very small minority of cases, paralytic poliomyelitis is accompanied by encephalitis that usually resolves without sequelae within a few days. Recovery depends on collateral sprouting and reinnervation of muscles by surviving motor axons. The paralysis improves over many months to years. Permanent deficits occur in about 50% of survivors. In these subjects there is a prominent loss of motor neurons such that the surviving motor neurons are incapable of fully reinnervating the denervated muscle groups. The acute mortality of 10%–20% is usually due to severe encephalitis or involvement of respiratory and bulbar muscles. Residual weakness is managed by physical therapy, bracing, and orthopedic procedures to transfer tendons or stabilize joints. Acute paralytic poliomyelitis was virtually eliminated from the developed world in dramatic fashion with the introduction of the Salk vaccine in 1955.

The Salk vaccine was the first polio vaccine that was available for mass production and widespread human immunization. Its success was dependent on the identification of all virulent strains (three) and the development of an in vitro culture system. Inactivation of the cultured poliovirus with formalin resulted in the first mass-produced polio vaccine for use in humans [1]. Within 2 years there was a dramatic reduction in cases of poliomyelitis.

Despite the success of the Salk vaccine, the vaccine had its limitations. The vaccine had to be injected to be effective and immunity tended to diminish with time. As a result, Sabin introduced the attenuated oral vaccine in 1960. The advantages of the live-attenuated vaccine are that it can be administered orally and offers a more sustained immunological response. Unfortunately, live-attenuated virus may cause polio in very rare cases. The question of which vaccine is superior remains a public health debate. The World Health Organization has now eradicated wild-type polio from all but four countries limited to central Africa. It is hoped that if mass vaccination programs are allowed to continue in central Africa, eradication there will be complete within a few more years.

Despite the eradication of acute poliomyelitis, there remains a large population of patients with significant motor deficits who were infected before the onset of the vaccination programs. Later in life, these people frequently note the constellation of fatigue, pain, and weakness. In the past it has been reported that these patients may also have progressive motor deficits after many years of stability. The combination of muscle aching or joint pain, fatigue, and a perception of progressive weakness has been named the *post-polio syndrome*.

Progressive neurological deficits in patients with prior poliomyelitis have been reported for more than 100 years [2–6]. The etiology of the syndrome has been debated. It has been suggested that it is the result of attritional loss of anterior horn cells with aging [2]. Others have proposed poor collateral reinnervation from terminal axonal sprouts in massively reinnervated muscles as the main causative factor [7–9]. An inflammatory response within affected muscles has been shown in some instances [7]. More recently, it has been shown that denervation continues later in life and that enlarged motor units demonstrate instability [9]. Another group has shown that the number of motor units drops with time [10]. Although this syndrome appears prevalent among the polio survivors, the exact etiology remains elusive.

II. Epidemiology

It has been estimated that there are nearly 250,000 survivors of paralytic poliomyelitis [11]. Undoubtedly, this number is much larger worldwide. Among these survivors the syndrome of progressive weakness, fatigue, and pain occurring after years of stability has been commonly reported (i.e., post-polio syndrome). Although these complaints are common among polio survivors, the clinical entity of the post-polio syndrome is controversial.

A number of large, population-based, cross-sectional studies have been completed to date. These studies have documented the prevalence of these symptoms in various European and U.S. populations. In one of the largest studies, 2392 Norwegian polio survivors were surveyed. The researchers received responses from 1449 subjects (61%). In those health surveys, 85% of subjects complained of increasing weakness in muscles previously affected and 58% had increased weakness in previously unaffected muscles [12]. Fatigue and pain was common in that study, occurring in 80% and 58% of subjects, respectively. Approximately 4% used walkers and 30% used either a manual or electric wheelchair. These self-reported rates were nearly twice the rates reported earlier in their course. The researchers concluded that more than 50% of polio survivors have symptoms suggestive of a progressive post-polio syndrome with an increasing need for adaptive equipment and rehabilitative services. In another Norwegian study, progressive symptoms were identified in 81% of the participating subjects [13]. The late symptomatic decline was attributed to polio-related deficits in slightly more than 50% of the subjects.

A population-based cross-sectional survey in the Netherlands identified similar trends. Ivanyi surveyed 350 randomly chosen subjects from 1784 cases of acute poliomyelitis in the Netherlands with a 74% response rate. Of the respondents, 58% indicated progressive weakness after years of stability [14]. The study population was similar to that in the Norwegian study, with the exception of age. The mean age in the Netherlands study was 44.6 years of age, or nearly a decade younger than in the Norwegian study [12,14].

In the United States, a smaller study of 50 subjects in Olmsted County, Minnesota, identified a prevalence rate of 64% for progressive symptoms after years of stability [15]. In a retrospective analysis, the only risk factor for identifying those subjects at risk was the magnitude of their residual deficit following the acute poliomyelitis [15].

In summary, a high prevalence rate for the symptoms of progressive weakness, fatigue, and pain has been reproducibly identified by independent authors in differing populations. This has ranged from a prevalence of 58% (in the youngest group) to 85% of the subjects with a remote history of paralytic poliomyelitis (Fig. 1). Although a high prevalence rate of late progressive symptoms has been convincingly identified, these rates are largely self-reported and retrospective. The natural history of polio survivors has been studied prospectively in only a small number of studies.

The Netherlands population was studied prospectively over 6 years by Nollet [16]. This prospective cohort study compared health status of polio survivors without late progressive symptoms with those with these symptoms. Over the 6-year period, there was no difference between the two groups with respect to their measures of physical performance or perceived health status. They found no apparent decline in function of either group, including measures of strength and physical performance. This is despite the fact that more than 58% of these subjects subjectively reported progressive loss of function.

The Olmsted County (Minnesota) population was similarly followed for a 5-year period. As in the Netherlands population, no decline in strength or functional performance could be demonstrated [17]. Again, this is despite the fact that more than 60% of these subjects self-reported symptoms of progressive weakness over this same time period. These studies have been criticized for having too short a follow-up period with insufficient power to identify the decline in strength and performance.

Recently, a 15-year follow-up has been reported from the Olmsted County population. Over that timeframe, there

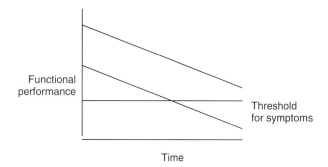

Figure 2 Graph depicting two subjects with an equal magnitude of decline. One subject begins the time period with a poorer functional performance and becomes symptomatic in the follow-up period; the other remains above the symptomatic threshold, thereby remaining asymptomatic.

was a modest decline in strength and functional performance for all polio subjects, independent of the symptom status [18]. Over this same timeframe, fewer than 5% of the subjects acquired new adaptive equipment or gait aids or made any new modifications to their homes [19]. This again was in contrast to the fact that more than two-thirds of the subjects complained of progressive weakness and functional decline. In the prospective Olmsted County cohort study, the only independent risk factor predictive of these symptoms was the magnitude of residual deficit following the poliomyelitis. A similar trend was noted in the Netherlands population; however, this association did not reach significance over the 6-year period of follow-up in that cohort [16].

There is an apparent discrepancy between the self-reported prevalence rates of progressive weakness and the prospective cohort studies demonstrating stability over time. The reasons for this discrepancy remain unclear. It has been speculated that this difference is largely due to the loss of functional reserve capacity in the more severely affected polio subjects (Fig. 2). For example, subjects with prominent loss of strength in their legs may notice a modest decline in strength, whereas the same decline in a nonaffected person may go unnoticed. Whether polio survivors decline over time differently than normal aging subjects remains a major unanswered question.

III. Pathophysiology

Careful study of subjects with a remote history of paralytic poliomyelitis has determined a number of interesting observations. Electrophysiology testing has demonstrated very large motor unit potentials in clinically affected muscles. However, there is also a high prevalence of neurogenic involvement in clinically unaffected limbs [10]. In a small minority of subjects with a remote history of poliomyelitis, fibrillation potentials can be identified on

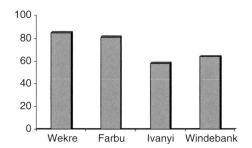

Figure 1 Prevalence rates of progressive weakness (expressed as a percentage) in the large population-based cross-sectional studies performed to date.

muscle electromyography [20]. These fibrillation potentials correlate with the degree of weakness; the greater the muscle weakness, the greater the likelihood of fibrillation potentials. Fibrillation potentials are unlikely to be identified in subjects with only mild weakness. Quantitative measures including macro-electromyography (EMG) and single-fiber EMG have demonstrated massively reinnervated motor unit potentials and increased jitter occur commonly in previously affected muscles [8,9,21,22]. Longitudinal studies have also demonstrated a loss of motor units in late survivors of polio [10,18].

Muscle pathology demonstrates chronic denervation with fiber type grouping. Additionally reported is the presence of small, angulated fibers consistent with ongoing denervation [9]. A minority of muscle biopsies may demonstrate interstitial inflammation [23] or vacuolated inclusions similar to inclusion body myositis [24]. Unfortunately, although these findings have been reproducibly identified in subjects with prior poliomyelitis, they did not distinguish between subjects with and without symptoms of late progressive weakness, pain, and fatigue, nor have any of these factors been associated with the risk to develop these symptoms. The reasons why some develop the features of the post-polio syndrome and others do not remain controversial. Three principal theories are behind the etiology of the post-polio syndrome: 1) massively reinnervated motor units are unable to maintain the axonal sprouts, and a dying-back phenomenon occurs that results in progressive weakness; 2) there is attritional loss of previously affected anterior horn cells later in life; and 3) physiological changes of aging are superimposed on a remote neurological deficit.

Based on the EMG findings, several authors have proposed that the loss of axonal sprouts in these large motor unit potentials later in life causes the remote deterioration in polio survivors. Evidence supports the theory that massively reinnervated motor units late in life become unstable with subsequent dying-back of the axonal sprouts. It has been demonstrated that polio survivors later in life do have motor unit instability [22].

Additionally, it has been demonstrated by macro-EMG that large motor units can diminish in size in polio survivors, suggestive of a loss of these axonal sprouts [26]. However, there was no association between motor unit instability and neuromuscular function, nor was there an association of these parameters with change in strength over a 7-year period [21,26]. Additionally, no differences have been demonstrated between those polio survivors with and without the post-polio syndrome.

Attritional loss of previously affected motor neurons has been proposed for many years [2]. The physiological loss of motor neurons with aging is well known. McComas demonstrated a loss of motor unit potentials in both normal aging and a cohort of polio survivors over a 3-year period. The rate of motor unit loss was greater in the polio group than in the normal controls [10]. This loss of motor units in the polio population was confirmed in a population-based cohort study [18]. However, despite the loss of motor unit potentials in the polio population on the whole, no differences have been demonstrated between those polio survivors with and without the post-polio syndrome [18].

Nearly all the prospective polio studies have identified either stability of function or only modest declines in function over long periods of time, even in those subjects with subjective symptoms of progressive weakness, pain, and fatigue [14,16–18]. These findings raise doubts as to the validity of the subjective complaints in this population and suggest that the post-polio syndrome may merely be physiological aging on a remote static deficit. These prospective studies have confirmed that even with symptoms, the long-term prognosis for the polio survivors remains good, with years of relatively stable neuromuscular function.

IV. Treatment

Successful treatment depends on an accurate diagnosis. As with any patient, a careful history and examination needs to be completed to ascertain the true cause of the underlying complaints. These subjects normally develop the common comorbidities of aging. Furthermore, the weakened limbs appear to be at risk for accelerated degenerative changes. This is secondary to either years of maladaptive use or to overuse syndromes. In one population study, two-thirds of the polio survivors complaining of late weakness, fatigue, or pain were found to have an alternative musculoskeletal diagnosis as an explanation for their pain [17]. Many of these alternative diagnoses were treatable, with the subjects' symptoms relieved. Degenerative joint disease involving muscle groups clinically affected by the polio frequently occurs. It is relatively common for subjects not to perceive the full extent of their strength limitations. Subjects frequently report that regions or limbs are unaffected, only to discover after a thorough examination that asymptomatic weakness is present [15]. This highlights the importance of correct diagnosis. These symptoms should not simply be attributed to the post-polio syndrome without a thorough and complete evaluation of alternative diagnoses.

A. Fatigue

Once secondary causes have been excluded, treatment should be approached in terms of symptomatic management. A number of drug agents have been tried without success in managing fatigue associated with the post-polio syndrome. Fatigue is frequently the most prominent symptom. It has been proposed that fatigue is a result of impaired neuromuscular transmission seen in aging polio survivors. Pyridostigmine is an anticholinesterase inhibitor. Because this drug impedes the degradation of acetylcholine, it prolongs the postsynaptic muscle end-plate potential, thus improving neuromuscular transmission. This mechanism is the basis of its effectiveness in the treatment of myasthenia gravis, for which it has been used for many years; in addition, it has been suggested anecdotally in post-polio fatigue. In uncontrolled, open-label reports, pyridostigmine was reportedly beneficial to these patients [27,28]. This led to two placebo-controlled, double-blind trials of pyridostigmine treatment of post-polio syndrome [29,30]. These clinical trials failed to identify any improvement of fatigue in patients. Other clinical trials tested prednisone and amantadine. These studies did not demonstrate any effect on fatigue or strength [31]. Other agents that have been anecdotally suggested include amitriptyline, fluoxetine, and bromocriptine. However, none of these have gained widespread acceptance or use in this population, and no randomized clinical trial evidence is available to justify their use. To date, no pharmacological agent has been identified as improving the fatigue and weakness that are the hallmarks of the post-polio syndrome.

B. Pain

Pain is a poorly characterized but frequently occurring symptom among polio survivors. Pain syndromes include frequent muscle cramps, myofascial pain, and arthralgia. The myofascial pain syndrome is clinically similar to that of fibromyalgia. Treatment success depends on the ability of the clinician to accurately identify the correct pain syndrome. No randomized clinical trials have tested analgesic medications in this population. Treatment relies on experience with other similar pain syndromes.

For frequent muscle cramps, the most effective treatment is often a regular structured stretching program [32]. Membrane-stabilizing medications such as anticonvulsants or quinine may also be tried. The role of additional physical modalities such as heat therapy or electrical and galvanic stimulation remains unclear and untested.

For the myofascial pain syndrome, the treatment most often suggested includes mood-stabilizing medication in combination with biofeedback, muscle relaxation therapy, or both. Although widely used in the fibromyalgia syndrome, the effectiveness of this treatment approach in post-polio syndrome remains untested. However, this approach is safe and not unreasonable.

Arthralgia is often secondary to advanced or early degenerative joint changes and occurs more frequently in this population given the mechanical disadvantage that the joints have supported since the acute paralysis. In these circumstances, arthralgia should be addressed with proper analgesic medications. Treatment typically begins with acetaminophen and escalates to stronger analgesia as necessary for pain control. Joint replacement is controversial. Case reports present conflicting results but suggest that joint replacement may be effective for pain relief in some instances [33]. The largest case series suggested that the success of the arthroplasty depended on the degree of weakness present before surgery [34]. The greater the presurgical weakness, the less satisfactory the outcome. In selected cases, joint arthroplasty should be considered an option for the management of the subject's pain.

In other cases, joint stability and function may be improved with the use of orthotics and bracing. The weight of the orthotic must be considered. Frequently, heavy braces are not practical for limbs that are already weakened as the braces may actually worsen the functional state of the limb. Because polio survivors have individualized deficits, they frequently have to identify a creative orthotist who is willing to test a variety of designs before finding one that is satisfactory. Additionally, lifestyle modification may be advisable to limit the ongoing stress to the affected joints.

C. Weakness

Of the symptoms commonly associated with the post-polio syndrome, the most feared by subjects is progressive weakness. Often their biggest concern is prognosis for ambulation, independent living, and employment. Longitudinal studies have demonstrated that the long-term prognosis is good, even in polio survivors symptomatic from their progressive weakness [18,19]. The majority of patients are not required to adopt new orthotics or gait aids (including wheelchairs) or to make additional modifications to their home or work environment [19].

Classically it has been suggested that rest and energy conservation are important in management of post-polio weakness. However, a number of studies have now demonstrated that submaximal or nonfatiguing strengthening exercises can be beneficial to strength and endurance without inducing overuse injuries [35–37]. Currently it is recommended that polio survivors maintain a submaximal strengthening program under the guidance of an experienced physical therapist to avoid overuse and damage to secondary muscles, tendons, or ligaments. To date, however, there is no evidence that such an exercise program alters the long-term outcome of these subjects. As subjects curtail their activities and become more sedentary, obesity may become a problem. Obesity further complicates their condition by increasing the strain on the joints, ligaments, and muscles. If obesity develops, a downward cycle often arises with increasing weight leading to less activity, which causes further weight gain.

Weakness, in combination with osteoarthritis, often impairs the subject's gait. New gait problems are present in the majority of polio survivors as they age [38]. Only a minority, however, require the new use of additional gait aids [19]. Often polio survivors are reluctant to use additional adaptive equipment that would simplify their activities of daily living. Frequently they are concerned about becoming dependent on the adaptive equipment and fear losing their independence.

In summary, the weakness associated with the post-polio syndrome is best managed with the assistance of an experienced physiatrist or physical therapist. A satisfactory outcome requires a cooperative effort, and frequently a bit of ingenuity, between the patient and the caregiver.

D. Bulbar Dysfunction

Although it is uncommon, some polio survivors complain of worsening dyspnea and dysphagia later in life. This phenomenon is felt to be analogous to the progressive limb weakness that is more commonly symptomatic. Rarely does the dysphagia lead to aspiration or problems in malnutrition [39]. Dysphagia evaluation and treatment should include an experienced speech pathologist. Often conservative diet modifications and body positioning are sufficient to alleviate the symptoms.

Respiratory complaints are often multifactorial. Contributors include diaphragmatic weakness related to the denervation of the diaphragm and scoliosis causing restriction of chest wall motion. Deteriorating respiratory function is much more likely to occur in subjects who required respiratory support during the acute paralytic phase [40]. Overnight oximetry can be monitored. If nocturnal desaturations are detected, subjects may benefit from noninvasive intermittent positive airway pressure (Bi-PAP) treatments. If scoliosis is a prominent contributor, then spine bracing should be considered.

V. Summary

The post-polio syndrome, which includes progressive fatigue, pain, and weakness, is present late in life in a majority of survivors of paralytic poliomyelitis. Although it is a common diagnosis, the epidemiological studies suggest that the rate of decline is modest in the majority of patients and carries a benign prognosis in most. The pathophysiology of the post-polio syndrome remains debated, but leading theories include dying-back of axonal sprouts in massively reinnervated motor units, attritional loss of affected anterior horn cells and physiological aging on a fixed neurological deficit. Treatment of the post-polio syndrome is limited to symptomatic care. Fatigue has been unresponsive to multiple medication trials and responds best to energy conservation. Pain requires a thorough evaluation with treatment directed at specific etiologies. Management of accelerated degenerative joint disease is controversial.

Judicious use of orthotics and bracing is most prudent. Arthroplasty may be indicated in select circumstances. Weakness is best addressed with the assistance of an experienced rehabilitative team including submaximal, nonfatiguing strengthening exercises. The prevention of obesity is important. Late bulbar dysfunction is uncommon but typically presents with dysphagia and dyspnea when it does occur. Dysphagia can frequently be addressed conservatively by a knowledgeable speech pathologist. Dyspnea is often multifactorial with a combination of diaphragm denervation and chest restriction secondary to scoliosis. Noninvasive positive airway pressure ventilation may be indicated in cases with nocturnal desaturations as evident by overnight oximetry. Overall, the prognosis of the post-polio syndrome is good. Most patients can expect years of neuromuscular stability, and only a minority require additional lifestyle modifications later in life.

References

1. Salk, J. (1953). Studies in human subjects on active immunization against poliomyelitis. I. A preliminary report of experiments in progress. *JAMA* **151**, 1081–1098.
2. Mulder, D. W., Rosenbaum, R. A., and Layton, D. D. Jr. (1972). Late progression of poliomyelitis or forme fruste amyotrophic lateral sclerosis? *Mayo Clin. Proc.* **47**, 756–761.
3. Jubelt, B., and Cashman, N. R. (1987). Neurological manifestations of the post-polio syndrome. *Crit. Rev. Neurobiol.* **3**, 199–220.
4. Wiechers, D. O. (1987). Late effects of polio: historical perspectives. *Birth Defects* **23**, 1–11.
5. Raymond, M. (1875). Paralysis essentielle de l'enfance, atrophie musculaire consecutive. *Comptes Rendus Soc. Biol.* **27**, 158–165.
6. Vulpian, E. F. A. (1879). Paralysie atrohique de l'enfance, ayany dans son evolution procede par poussees successives. Surcharge adipeuse dans lower extremities regions envahies l'atrophie. *Clin. Med. Hopital Charite* 778–784.
7. Dalakas, M. C. (1988). Morphologic changes in the muscles of patients with postpoliomyelitis neuromuscular symptoms. *Neurology* **38**, 99–104.
8. Cashman, N. R., Maselli, R., Wollman, R. L., Roos, R., Simon, R., and Antel, J. P. (1987). Late denervation in patients with antecedent paralytic poliomyelitis. *N. Engl. J. Med.* **317**, 7–12.
9. Grimby, G., Stalberg, E., Sandberg, A., and Stibrant Sunnerhagen, K. (1998). An 8-year longitudinal study of muscle strength, muscle fiber size, and dynamic electromyogram in individuals with late polio. *Muscle Nerve* **21**, 1428–1437.
10. McComas, A. J., Quartly, C., and Griggs, R. C. (1997). Early and late losses of motor units after poliomyelitis. *Brain* **120**, 1415–1421.
11. U.S. Department of Health and Human Services. (1981). Prevalence of selected impairments, United States: 1977. *Vital Health Stat.* **10**, 15.
12. Wekre, L. L., Stanghelle, J. K., Lobben, B., and Oyhaugen, S. (1998). The Norwegian Polio Study 1994: a nation-wide survey of problems in long-standing poliomyelitis. *Spinal Cord* **36**, 280–284.
13. Farbu, E., Rekand, T., and Gilhus, N. E. (2003). Post-polio syndrome and total health status in a prospective hospital study. *Eur. J. Neurol.* **10**, 407–413.
14. Ivanyi, B., Nollet, F., Redekop, W. K., de Haan, R., Wohlgemuth, M., van Wijngaarden, J. K., and de Visser, M. (1999). Late onset polio sequelae: disabilities and handicaps in a population-based cohort of

the 1956 poliomyelitis outbreak in the Netherlands. *Arch. Phys. Med. Rehabil.* **80**, 687–690.
15. Windebank, A. J., Litchy, W. J., Daube, J. R., Kurland, L. T., Codd, M. B., and Iverson, R. (1991). Late effects of paralytic poliomyelitis in Olmsted County, Minnesota. *Neurology* **41**, 501–507.
16. Nollet, F., Beelen, A., Twisk, J. W., Lankhorst, G. J., and De Visser, M. (2003). Perceived health and physical functioning in postpoliomyelitis syndrome: a 6-year prospective follow-up study. *Arch. Phys. Med. Rehabil.* **84**, 1048–1056.
17. Windebank, A. J., Litchy, W. J., Daube, J. R., and Iverson, R. A. (1996). Lack of progression of neurological deficit in survivors of paralytic polio: a 5-year prospective population-based study. *Neurology* **46**, 80–84.
18. Sorenson, E. J., Daube, J. R., and Windebank, A. J. (2005). A 15-year follow-up of neuromuscular function in patients with prior poliomyelitis. *Neurology* **64**, 1070–1072.
19. Sorenson, E. J., and Windebank, A. J. (2005). Incidence of adaptive equipment use in subjects with a remote history of paralytic poliomyelitis. *Neurology* **65**, 963.
20. Ghavanini, M. R., and Ghavanini, A. A. (1998). Fibrillation potentials and positive sharp waves in patients with antecedent paralytic poliomyelitis. *EMG Clin. Neurophysiol.* **38**, 455–458.
21. Rodriquez, A. A., Agre, J. C., and Franke, T. M. (1997). Electromyographic and neuromuscular variables in unstable postpolio subjects, stable postpolio subjects and control subjects. *Arch. Phys. Med. Rehabil.* **78**, 986–991.
22. Wiechers, D. O., and Hubbell, S. L. (1981). Late changes in the motor unit after acute poliomyelitis. *Muscle Nerve* **4**, 524–528.
23. Dalakas, M. C. (1986). New neuromuscular symptoms in patients with old poliomyelitis: a 3-year follow-up study. *Eur. Neurol.* **25**, 381–387.
24. Semino-Mora, C., and Dalakas, M. C. (1998). Rimmed vacuoles with beta-amyloid and ubiquitinated filamentous deposits in the muscles of patients with long-standing denervation (postpoliomyelitis muscular atrophy): similarities with inclusion body myositis. *Hum. Pathol.* **29**, 1128–1133.
25. Ivanyi, B., Ongerboer de Visser, B. W., Nelemans, P. J., and de Visser, M. (1994). Macro EMG follow-up study in post-poliomyelitis patients. *J. Neurol.* **242**, 37–40.
26. Rodriquez, A. A., Agre, J. C., Harmon, R. L., Franke, T. M., Swiggum, E. R., and Curt, J. T. (1995). Electromyographic and neuromuscular variables in post-polio subjects. *Arch. Phys. Med. Rehabil.* **76**, 989–993.
27. Trojan, D. A., Gendron, D., and Cashman, N. R. (1993). Anticholinesterase-responsive neuromuscular junction transmission defects in post-poliomyelitis fatigue. *J. Neuro. Sci.* **114**, 170–177.
28. Trojan, D. A., and Cashman, N. R. (1995). An open trial of pyridostigmine in post-poliomyelitis syndrome. *Can. J. Neurol. Sci.* **22**, 223–227.
29. Trojan, D. A., Collet, J.-P., Shapiro, S., Jubelt, B., Miller, R. G., Agre, J. C., Munsat, T. L., Hollander, D., Tandan, R., Granger, C., Robinson, A., Finch, L., Ducruet, T., and Cashman, N. R. (1999). A multicenter, randomized, double-blinded trial of pyridostigmine in postpolio syndrome. *Neurology* **53**, 1225–1233.
30. Horemans, H. D., Nollet, F., Beelen, A., Drost, G., Stegeman, D. F., Zwarts, M. J., Bussmann, J. B., de Visser, M., and Lankhorst, G. J. (2003). Pyridostigmine in postpolio syndrome: no decline in fatigue and limited functional improvement. *J. Neurol. Neurosurg. Psychiatry* **74**, 1655–1661.
31. Dalakas, M. C., Bartfeld, H., and Kurland, L. T. (1995). The postpolio syndrome: advances in the pathogenesis and treatment. *Ann. NY Acad. Sci.* **753**, 1–411.
32. Kottke, F. J. (1990). Therapeutic exercises to maintain mobility. In "Krusen's Handbook of Physical Medicine and Rehabilitation" (F. J. Kottke, et al., eds.). Saunders, Philadelphia.
33. Evangelista, G. T., and Zuckerman, J. D. (2003). Total knee arthroplasty in a patient with quadriceps paralysis secondary to poliomyelitis: a case report. *Am. J. Orthop.* **32**, 593–597.
34. Giori, N. J., and Lewallen, D. G. (2002). Total knee arthroplasty in limbs affected by poliomyelitis. *J. Bone Joint Surg. Am.* **84**, 1157–1161.
35. Ernstoff, B., Wetterqvist, H., Kvist, H., and Grimby, G. (1996). The effects of endurance training on individuals with post-poliomyelitis. *Arch. Phys. Med. Rehabil.* **77**, 843–848.
36. Agre, J. C., Rodriquez, A. A., and Franke, T. M. (1997). Strength, endurance, and work capacity after muscle strengthening exercise in postpolio subjects. *Arch. Phys. Med. Rehabil.* **78**, 681–686.
37. Agre, J. C., and Rodriquez, A. A. (1997). Muscular function in late polio and the role of exercise in post-polio patients. *Neurorehab* **8**, 107–118.
38. Agre, J. C. (1995). The role of exercise in the patient with post-polio syndrome. *Ann. NY Acad. Sci.* **753**, 321–334.
39. Sonies, B. C., and Dalakas, M. C. (1995). Progression of oral-motor and swallowing symptoms in the post-polio syndrome. *Ann. NY Acad. Sci.* **753**, 87–95.
40. Bach, J. R. (1995). Management of post-polio respiratory sequelae. *Ann. NY Acad. Sci.* **753**, 96–102.

49

Spinobulbar Muscular Atrophy (Kennedy's Disease)

Jeffrey D. Zajac, MBBS, PhD, FRACP
Karen J. Greenland, BSc, (Hons) PhD

Keywords: *androgen receptor, motor neuron disease, polyglutamine repeat expansion, spinal and bulbar muscular atrophy, testosterone*

I. Introduction
II. Clinical Features
III. Pathological Features
IV. Kennedy's Disease as a Polyglutamine Disease: Molecular Pathogenesis
V. Female Carriers of Kennedy's Disease
VI. Pharmacology: The Role of Ligand in the Pathogenesis
VII. Genetically Modified Animal Models
VIII. Symptoms in Relation to Pathophysiology
IX. Potential for Therapy
X. Conclusions
   References

I. Introduction

Spinobulbar muscular atrophy (SBMA), also known as Kennedy's disease, is an X-linked adult-onset form of motor neuron disease caused by a trinucleotide CAG repeat expansion in the first exon of the androgen receptor gene. This illness, although rare, is one of the better-characterized CAG repeat diseases because the function of the mutant gene, the androgen receptor, is known. The androgen receptor, a steroid hormone–dependent transcription factor, is the protein responsible for mediating the effects of the male sex hormone testosterone. SBMA was first described by Kennedy et al. in 1968 [1], and the androgen receptor as the genetic basis was identified in 1991 by La Spada et al. [2], who mapped the gene to the long arm of the X-chromosome (Xq11-q12). SBMA was the first disease identified that is caused by an expanded triplet nucleotide repeat. Clinical aspects and mechanisms underlying disease pathogenesis have been previously reviewed [3–7]. In this chapter, we describe the neurobiological basis for the features of this disease.

II. Clinical Features

Kennedy's disease displays characteristic X-linked genetics, with onset of symptoms in males usually between 30 and 50 years. It has a variable onset usually after the age of 30. Muscle cramps are very common and may precede muscle weakness by as long as 30 years. Most affected patients develop weakness and atrophy of the pelvic and shoulder girdle muscles, with weakness occurring first in the lower limbs. Bulbar involvement results in weakness and wasting of the tongue, jaw, and facial muscles. Fasciculation of affected muscles is common and is particularly noticeable in the tongue and chin. Tremor is also common. Deep tendon reflexes are usually absent, and there is no evidence of upper motor neuron involvement. As a result of the bulbar involvement, difficulty in swallowing is a frequent symptom. Dysarthria is common in the later stages. Intermittent choking is a common problem, and aspiration pneumonia can occur in the terminal stages of the disease. Sensation can be abnormal but is usually asymptomatic, detected only on more active testing.

Because the mutation in the androgen receptor results in the partial loss of function [8], signs of mild androgen insensitivity are common. These include gynecomastia, testicular atrophy, and reduced fertility. Gynecomastia, enlargement of the breast, is present in more than 50% of affected males. The degree of breast enlargement is often slight to moderate but is noticeable if looked for specifically. Testicular volumes may be normal but are commonly significantly reduced. Creatine kinase (CK) is elevated as a result of muscle atrophy. Testosterone levels have been reported to be slightly low, normal, or elevated [3,9].

Prevalence of Kennedy's disease is estimated to be approximately 1 in 40,000, although several studies have indicated that it is an underdiagnosed condition. The most common misdiagnosis is amyotrophic lateral sclerosis (ALS). Misdiagnosis may be commonly attributed to the significant overlap in clinical features with other motor neuron diseases, as well as variability in expression of disease symptoms and ignorance of the condition. Definitive diagnosis based on genetic analysis is possible and straightforward, given that the causative mutation is known. Assessment of the CAG triplet length is now a routine procedure in molecular genetic diagnostic laboratories. Endocrine features relating to loss of androgen receptor activity, such as reduced testicular function and gynecomastia, sets Kennedy's disease apart from other motor neuron diseases. The gynecomastia, although mild, is a strong feature suggestive of Kennedy's disease. Similarly the presence of a family history of males with muscle weakness is a clue. Kennedy's disease is a much milder illness than ALS. Patients with this diagnosis may live 30 or 40 years. Progression of symptoms and signs is relatively slow. After 20 years, patients often require a walking stick or wheelchair. Swallowing and breathing become more problematic, and the terminal episode is often aspiration pneumonia. It should be noted that a significant number of patients are still relatively well 20 or 30 years after diagnosis.

III. Pathological Features

A. Neurons

Many studies have described the pathological changes in SBMA, and autopsies have been recorded by a number of research groups. The major clinical manifestations of Kennedy's disease result from atrophy and loss of lower motor neurons, especially anterior horn cells of the spinal cord and of the cranial nerves of the brainstem. The main pathological changes are substantial loss of bulbar and spinal motor neurons. There is a marked reduction of the absolute number of anterior horn motor neurons in all segments. Those cells remaining are atrophic. In some cases, these areas are completely devoid of neurons. In contrast, sensory neurons are well preserved, although some loss of sensory motor neurons has also been described. The nerve cells in Clarke's cell columns and the posterior horns are well preserved. The white matter is well conserved. The anterior spinal nerve roots contain a decreased number of fibers compared with the posterior nerve roots. The distribution of large and small myelinated lateral corticospinal tract fibers are the same in SBMA and normal controls. Atrophy of the ventral horns is observed through all spinal segments. Large, medium, and small neurons in the intermediate zone of the ventral horn are depleted. Inclusion bodies are sometimes seen in affected neurons. There is minimal astroglial reaction. Electron microscopic studies of rough endoplasmic reticulum show disaggregation of the polyribosomes in the endoplasmic reticulum of anterior horn cells. In one report, Nagashima et al. [10] describes two cases. The report describes degeneration of cranial and spinal motor nuclei with marked neuronal loss, particularly involving cranial nerves 5, 7, 11, and 12, as well as the anterior horn of the spinal cord. In contrast, neurons in the motor cortex and upper pyramidal tracts were well preserved, as were cranial nuclei of cranial nerves 3, 4, 6, 8, and 10. Upper motor neurons are spared, and the disease is usually symmetrical.

B. Muscle

Histological assessment of skeletal muscle biopsies typically shows progressive neuropathic atrophy of muscle fibers, initially small group atrophy among normal-sized fibers and later large group atrophy. This is replaced by adipose tissue. Scant endomysial connective tissue may follow, but fibrosis is rare. These progressive neuropathological changes include atrophy of muscle fibers. Increases in the number of clumps of pyknotic nuclei are apparent. Hypertrophy of remaining muscle fibers is quite common with a detectable increase in the number of nuclei. Later, some of these enlarged fibers undergo splitting and fragmentation. Enzyme histochemical analysis shows atrophy of all three fiber types (1, 2a, and 2b). Collateral reinnervation can occur after some time, with numbers of the same fibers forming fiber groups that lead to type grouping.

We have described a postmortem case of a 72-year-old man [11], in which gross examination showed wasting of the anterior nerve rootlets of the spinal cord and severe loss of anterior horn neurons in sections of the spinal cord (Fig. 1).

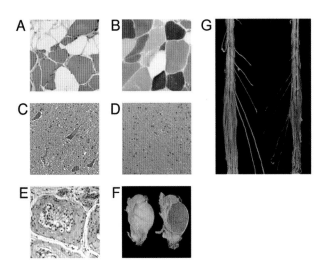

Figure 1 Post mortem findings in SBMA. (**A**) Quadriceps femoris muscle, showing replacement of muscle fibers by fat, small angulated fibers and increased number of internal nuclei. (**B**) Fiber typing in quadriceps femoris muscle, showing angulated, denervated fibers of type I, type 2A, and type 2B fiber types. (**C**) Anterior horn region of cervical spinal cord from control subject (age-matched), showing large motor neuron cell bodies. (**D**) Anterior horn region of cervical spinal cord from SBMA subject, showing marked loss of motor neurons. (**E**) Testis, showing atrophy of seminiferous tubules, thickened basal lamina, and no spermatogenesis. Original magnification 200× for (**A**) and (**B**), 100× for (**C**), (**D**), (**E**). Sections stained with hematoxylin and eosin (**A**), (**C**), (**D**), (**E**), or for ATPase activity (**B**). (**F**) Testis showing atrophy. (**G**) Ventral view of distal spinal cord showing atrophy of the anterior nerve roots compared with an age and sex matched control patient.

IV. Kennedy's Disease as a Polyglutamine Disease: Molecular Pathogenesis

The molecular basis of Kennedy's disease is an expansion of CAG trinucleotide repeats within the first exon of the androgen receptor gene. This polymorphic repeat region normally numbers between 11 and 35; however, expansion to more than 40 repeats renders the translated gene product, the androgen receptor, toxic to selective sets of vulnerable neurons. Patients with Kennedy's disease may have between 40 and 70 repeats. There is no overlap between the numbers of repeats in Kennedy's disease and the normal population. Increased triplet length is associated with decreased age of onset. Severity may also be worse, but this is controversial. This CAG repeat region encodes a polyglutamine (polyQ) tract within the N-terminal, or transactivation domain, of the androgen receptor (Fig. 2). The human androgen receptor gene consists of eight exons and a 2.7-kb open reading frame, encoding a 919 amino acid protein with a molecular mass of approximately 110 kDa. It is the receptor for the male sex hormones, testosterone and dihydrotestosterone (DHT). Functionally, it is a ligand-dependent transcription factor that plays a central role in the masculinization of the XY fetus, as well as the development and maintenance of secondary male sexual characteristics. Although mild partial loss of receptor activity is a feature of Kennedy's disease, resulting in decreased sexual functioning, testicular atrophy, and a decline in fertility, this is not considered a loss-of-function mutation because sexual development in affected individuals is normal. Furthermore, XY individuals with complete loss of androgen receptor function, termed *complete androgen insensitivity syndrome* (cAIS), do not have a male phenotype but develop as females. These individuals with no androgen receptor activity have no neuropathy and no features of Kennedy's disease.

Androgen receptors are expressed in the normal brain and spinal cord, but their role in neural tissue is poorly understood, although there is some evidence that androgens have a neuroprotective or neurotrophic role in certain populations of neurons, including motor neurons [12]. The polyQ expansion is believed to confer a gain of toxic property that leads to a gradual demise in functioning of the affected cell and eventual cell death.

Kennedy's disease is one of a group of at least 9 inherited neurodegenerative diseases caused by a polyQ expansion. Other diseases sharing the same genetic basis include Huntington's disease, dentatorubral and pallidoluysian atrophy (DRPLA), and several forms of

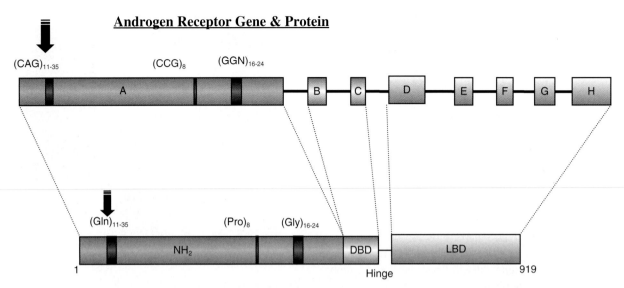

Figure 2 Structure of the androgen receptor gene (**A**) and protein (**B**). The androgen receptor is a steroid-dependent transcription factor consisting of four predominant domains: an N-terminal transactivation domain, a deoxyribonucleic acid (DNA)–binding domain *(DBD)*, a hinge region, and a ligand-binding domain *(LBD)*. The polyQ *(Gln)* tract resides in the N-terminal transactivation domain of the receptor. This domain contains two other repeat regions, a polyproline *(Pro)* repeat and a glycine *(Gln)* repeat region. The polyQ region is heterogenous in number within the normal population. Expansion of this region to >40 glutamines causes Kennedy's disease.

spinocerebellar ataxia (SCA1, SCA2, SCA3/Machado-Joseph Disease, SCA6, SCA7, SCA17), all of which have an autosomal dominant mode of inheritance. The X-linked pattern of inheritance of Kennedy's disease is therefore unique among CAG triplet diseases. Although they share the same causative mutation type, the affected proteins for each of the polyQ-expansion diseases are functionally unrelated. A significant unifying feature of these diseases is that the predominant cells affected are in the central nervous system. Despite significant advances being made in recent years, the basis of selective neuronal vulnerability remains poorly understood. This family of diseases shares certain features of pathogenesis. However, disease-specific differences, such as the neuronal subsets that are affected, are likely to be based on protein-specific characteristics of the affected proteins such as levels and cell specificity of expression and interaction with other intracellular factors [13].

The molecular impact of the polyQ expansion is heterogenous in nature. Evidence to date indicates that pathogenesis is multifaceted, with cellular dysfunction induced by the polyQ expansion modulating a number of intracellular functions and pathways, including disruption of proteolytic processing and degradation, as well as transcriptional regulation and axonal transport. Although not all aspects are discussed here, many of the mechanisms believed to contribute to the pathogenesis of Kennedy's disease are represented schematically in Figure 3. Predominant theories of polyQ-induced neurodegeneration include nuclear accumulation of polyQ-expanded proteins and proteolytic cleavage and aggregation of the mutant proteins [14]. As with other polyQ expansion diseases and a number of other neurodegenerative diseases, the presence of intracellular aggregates is a pathological hallmark. These aggregates contain sequestered mutant proteins as well as a number of other intracellular proteins, such as transcription factors and components of the cellular machinery necessary for protein degradation, and tend to accumulate in nuclear inclusions (NIs). These NIs are evident not only in spinal and brainstem motor neurons, the predominant targets of Kennedy's disease–induced neurodegeneration, but also in a number of visceral organs such as the scrotal skin, dermis, kidneys, heart, and testes [15]. Recent evidence has demonstrated that diffuse nuclear accumulation of mutant protein is more widespread and extensive than NIs [16] and also is thought to contribute to pathophysiology. The role of aggregates and NIs in Kennedy's disease–induced neurodegeneration remains a point of controversy within the scientific and medical research communities. Specifically, there is debate regarding whether aggregates play a causative or pathogenic role or whether they are produced in the cell as secondary knock-on or protective effects in response to other cellular insults imparted by the polyQ tract. Inhibition of axonal transport by protein fragments containing expanded triplets can occur in the absence of inclusions. Notwithstanding, studies into the content and impact of these aggregates have provided essential clues in unraveling mechanisms involved in disease progression. Although a number of reports have dissociated the presence of inclusions from toxicity in both animal models and humans, they are significant as a pathological marker.

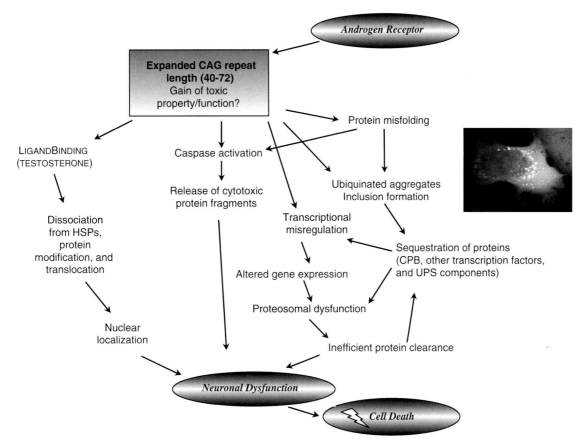

Figure 3 Schematic diagram of some of the pathogenic mechanisms leading to neurodegeneration in Kennedy's disease. Although the majority of mechanisms have been described for polyQ-induced neurodegeneration in general, such as proteolytic cleavage and protein aggregation, the role of ligand-binding in the induction of pathogenesis is at this stage a specific mechanism relating only to Kennedy's disease. Shown is a cell transfected with a Kennedy's disease mutant human androgen receptor linked to green-fluorescent protein expression vector, with expression driven by a cytomegalovirus CMV promoter. Significant aggregation of the mutant protein is evident. In cells transfected with normal androgen receptor (not shown), the protein is uniformly seen throughout the cytoplasm.

Evidence suggests that the proteolytic cleavage of mutant polyQ-expanded proteins is a key step in disease pathogenicity. Indeed, research in animal models indicates that truncated forms of the mutated polyQ-expanded proteins, formed through proteolytic cleavage, confer greater pathogenicity than the full-length protein. Aggregated proteins in NIs are detectable using only antibodies directed against N-terminal epitopes, not with C-terminal epitopes. The activation of caspase 3, a cysteine protease involved in apoptotic death, plays a central role in the cleavage of mutant androgen receptor, a step that is critical in mediating cellular toxicity [17]. PolyQ-induced intracellular aggregates have been found not only to consist of mutant androgen receptor fragments but also to co-localize with components of the ubiquitin-proteasome system (UPS). Evidence also shows that intracellular aggregates inhibit proteasome function, thereby leading to cellular dysfunction due to the disruption of normal degradation processes within the cell. This appears to be a common pathological feature among neurodegenerative disorders [18].

Aggregation and accumulation of mutant proteins is thought to contribute to the disruption of an array of fundamental cellular functions. Intranuclear aggregates have been found to sequester transcription factors such as cAMP-response element binding protein (CBP), causing transcriptional dysregulation of responsive genes [19]. Inhibition of histone acetylation activity induced by polyQ mutant protein has also been identified as contributing factor in transcriptional dysregulation, and the use of inhibitors of histone deacetylase (HDAC) inhibits motor impairment in disease models [20]. Genes with altered expression in Kennedy's disease models include CBP gene targets such as the trophic factor vascular endothelial growth factor (VEGF), the dysregulation of which may be associated with pathogenesis [21]. Indeed, rescue of Kennedy's disease–induced neuronal death by VEGF treatment has been demonstrated in a transgenic mouse model [21]. The downstream consequences of the polyQ-induced intracellular dysfunction include inhibition of fast axonal transport, an effect that was found to be exacerbated in the presence of truncated proteins [22]. Such insights provide essential clues as to the cellular impact of polyQ-induced neural cell dysfunction.

Certain intracellular signal transduction pathways have also been implicated as playing a significant role in polyQ-induced toxicity. One such pathway demonstrated to be central to this process is the p44/42 MAP kinase pathway, the inhibition of which rescues cellular demise [23]. The toxic insult of the polyQ protein in neuronal cells appears to be closely reliant on disruption of fundamental cellular functions such as protein degradation and the activation of specific signaling pathways involved in cell death. How their roles could play a part in the development of therapeutic approaches is yet to be realized. Collectively, cellular dysfunction and death may be the result of a number of interconnected cellular insults. Rescue of cell functioning may therefore be achievable by exploiting the sites of known dysfunction using a number of different pharmacological approaches.

V. Female Carriers of Kennedy's Disease

Heterozygous female carriers of the Kennedy's disease mutation may develop subclinical manifestations of motor neuron dysfunction. Signs of chronic denervation have been reported in some carriers, and signs of bulbar motor neuron symptoms may develop later in life [24]. Muscle biopsies from heterozygous females have shown moderate type 2 fiber predominance and an increased spread of fiber diameters [25]. Mild sensory nerve involvement has also been found [26]. Many theories as to why females do not develop Kennedy's disease have been proposed. The absence of features of Kennedy's disease in female carriers is not caused by the subtle decrease in androgen receptor signal transduction function. As noted earlier, complete deletion of the sole copy of the AR in XY individuals does not cause neuronal degeneration. Thus the mutation is positive or a toxic gain of function. However, females with one mutant gene and one normal gene have only a very mild form of the disease. The explanation of this apparent paradox was initially attributed to mutant specific lyonization, or X-chromosome inactivation, with a lower dosage affect contributing to the lack of disease. However, recent evidence in animal models indicates a central role for ligand-binding in disease pathogenesis (see Section VI). Therefore it is likely that females escape development of the disease due to the low circulating levels of testosterone compared with males. This is further supported by the absence of a significant disease phenotype in homozygous females [27], despite the presence of mild phenotypic manifestations.

VI. Pharmacology: The Role of Ligand in the Pathogenesis

Perhaps one of the most important developments in field of Kennedy's disease research in recent years has been the characterization of the contributory role of ligand binding to the androgen receptor, in disease pathogenesis (reviewed in [6,7]). The fact that the causative protein, the androgen receptor, is able to bind ligand—thereby altering conformation, phosphorylation state, and cellular distribution—sets Kennedy's disease apart from other polyQ expansion diseases. The androgen receptor in its unbound state is confined to an inactive complex in the cell cytoplasm, bound with chaperone molecules known as *heat shock proteins* (HSPs). Binding of the cognate ligands, testosterone, or the more potent DHT, induces conformational changes, dissociation from HSPs, receptor dimerization, and translocation into the nuclear compartment of the cell, the intracellular site where polyQ-induced toxicity is triggered [28]. Multiple studies in animal and cell models support a central role for ligand stimulation of the androgen receptor as a key event in triggering pathogenic processes leading to the development of Kennedy's disease.

Although the predominant pathogenic trigger may be induction of nuclear translocation, stimulation by androgen binding contributes to toxicity via a number of other mechanisms. Ligand binding induces dissociation from the stable conformation in which the androgen receptor is bound to HSPs, rendering the androgen receptor more susceptible to proteolytic cleavage. Several models have demonstrated that overexpression of HSPs significantly reduces aggregate formation and polyQ-induced cell death, most likely due to facilitated re-folding of mutant protein. Evidence also suggests that mutant protein degradation is also enhanced by HSP overexpression, leading to a decline on accumulation of the toxic protein within the cell. HSPs therefore appear to play a role in protecting the cell against polyQ-induced toxicity, an effect that has been demonstrated across a number of different polyQ-expansion diseases. In the absence of ligand stimulation, the mutant protein remains in the inactive complex in association with HSPs, protecting the protein from cleavage and therefore transformation into a more toxic form, as well as from nuclear translocation. The mechanisms through which androgen binding to the androgen receptor increases toxicity of the mutant protein also appear to be modulated via phosphorylation [23]. Evidence suggests that ligand-stimulated phosphorylation on the mutant protein enhances cytotoxicity. Phosphorylation appears to render the androgen receptor protein more susceptible to cleavage by caspase 3 [23], resulting in the release of the more toxic truncated form of the polyQ-mutant protein.

The endocrine function of the androgen receptor is modified in Kennedy's disease. The binding affinity of the androgen receptor is significantly decreased in cultured genital skin fibroblasts from patients with Kennedy's disease [8]. There is also a decrease in androgen receptor messenger RNA (mRNA) expression with increasing triplet number [29], as well as diminished transactivational activity.

VII. Genetically Modified Animal Models

The toxic effect of the CAG expansion has been demonstrated in multiple animal models of many of the CAG triplet diseases, including mice and flies. Kennedy's disease has been replicated in a number of mouse models. These studies have afforded important insights into the significance of ligand binding in disease pathogenesis in recent years, including the prevention of nuclear accumulation of mutant androgen receptor protein and disease severity via androgen reduction/ablation [30]. The role of androgens in disease pathogenesis also provides a molecular explanation for the lack of disease expression in heterozygous or homozygous female carriers, although subclinical manifestations may be evident in some cases [27]. The role of ligands in disease development opens treatment options based on hormonal manipulation in symptomatic Kennedy's disease sufferers, particularly the suppression of androgens. Such approaches have been demonstrated to be efficacious in the treatment of Kennedy's disease in transgenic mouse models. The endocrine role in Kennedy's disease was first demonstrated by Katsuno and colleagues [31] in a transgenic mouse model expressing the full-length androgen receptor containing either 24 (normal) or 97 (expanded) CAGs, with expression driven by a cytomegalovirus enhancer and β-actin promoter. A disease phenotype was observed only in the mice expressing mutant androgen receptor, and symptoms reflected Kennedy's disease in humans. Indications for the role of androgens were given by gender-related differences in phenotypes, the same as for humans. Hormonal manipulations confirmed the role of androgens, with androgen ablation achieving rescue of disease symptoms in male mice and androgen treatment of females exacerbating the disease phenotype. Examination of the cellular distribution of androgen receptor supported the essential role of nuclear localization in mediating the phenotypic effects observed, with acceleration of disease symptoms being associated with an increase in mutant protein in the nuclear compartment. Androgen ablation can be achieved via surgical castration or via the use of luteinizing hormone–releasing hormone (LHRH) agonists such as leuprorelin, both of which have achieved rescue from disease pathogenesis in animal models. The essential role of androgens in disease development has now been demonstrated in a number of different animal and cell culture models of Kennedy's disease. The question of whether rescue from neurodegenerative processes through hormonal manipulation can be achieved in humans is yet to be answered, although clinical trials are currently underway.

VIII. Symptoms in Relation to Pathophysiology

Kennedy's disease initially presents with very mild symptoms and signs. For example, patients often notice leg cramps. These relatively nonspecific features take on significance in genetically susceptible individuals. Examination of such patients often reveals decreased or absent reflexes in the lower limbs. No pathological information is available on individuals in the early stages of Kennedy's disease. This phase may last 10 to 30 years. Muscle strength gradually decreases, affecting the shoulder girdle, limb girdle, bulbar, and facial muscles. Fasciculation becomes obvious at this stage. Pathophysiological information is only available in late-stage disease, by which time the full gamut of clinical signs and symptoms is usually present. It remains unclear whether changes in the androgen receptor have a direct effect on muscle function. There is no direct evidence for this in humans. Subtle changes in the binding affinity, decreased transactivational activity, and reduced expression of the androgen receptor are the likely explanations for the endocrine features such as gynecomastia, decreased testicular size, and reduced fertility. As with the impairment of sexual function, testosterone levels may be low, reflecting testicular atrophy, or high, reflecting partial androgen insensitivity.

IX. Potential for Therapy

The current knowledge of the neurobiology of Kennedy's disease presents many potential approaches to treatment. Decreasing expression of toxic proteins and inhibiting proteolytic processing or nuclear transport are potential targets. Inhibiting aggregation may (or may not) be useful. Finally, specific approaches are available in the treatment of Kennedy's disease because the mutation is in the androgen receptor. Reduction in the levels of circulating androgens is feasible and may prove effective. As yet, there is no proven therapy for Kennedy's disease. Given the rarity of this illness, placebo-controlled trials will be difficult to organize. It seems likely for the immediate future that small observational studies or open trials will provide the only data available. As discussed previously, direct evidence from androgen ablation studies in animals suggest that androgen ablation in human subjects could be beneficial. Initial treatment of Kennedy's disease involved androgen replacement. The theoretical basis for this was that the partial loss of androgen receptor functioning was causing the neurological deficits. Another rationale was based on evidence of androgen-induced neuroprotective and trophic effects in a number of nerve injury and cell culture models [6,7]. Androgens are also known to exert potent anabolic effects on muscle. Despite this, there was no direct evidence supporting the use of androgens for Kennedy's treatment and the practice was never widely implemented. Androgen ablation therapy is widely used in the treatment of prostate cancer in men—a field from which the development of therapies for Kennedy's disease can greatly benefit. Gonadotrophin agonists are widely available and safe. Small numbers of patients with Kennedy's disease are currently being treated with these agents in the absence of significant human data. Theoretically, testosterone levels may increase at the onset of therapy. In prostate cancer, this flare is covered with androgen receptor blockade using drugs such as flutamide. However, in animal models of Kennedy's disease, this drug does not ablate the pathological effect of the mutation. The problem with the use of the antagonist flutamide is that, like androgens, flutamide promotes androgen receptor nuclear translocation, a mechanism demonstrated to be central to mediating polyQ-induced neurodegeneration in Kennedy's disease. Clearly, a reduction in androgens in affected males is not without potential undesirable side effects, including osteoporosis, muscle wasting, loss of vitality, and sexual problems. Such approaches are also not recommended in males still wishing to reproduce. This central role for androgen binding and nuclear translocation of the mutant androgen receptor not only links protein function inextricably with disease pathogenesis but also provides avenues for slowing or preventing disease prevention. Given the safety of gonadotrophin agonists, their use could be considered in individual patients.

X. Conclusions

Kennedy's disease or SBMA is a rare form of lower motor neuron degeneration. Its importance is twofold. First, it serves as a model for the other polyglutamine expansion diseases because the function of the gene is known. However, the mechanism of the toxic affect remains unclear. Secondly, compared with other forms of motor neuron disease, Kennedy's disease is very slowly progressive. The much worse prognosis of ALS compared with Kennedy's disease highlights the importance of careful diagnosis and genetic testing in appropriate individuals. The next step is to use the similarities between the CAG triplet diseases to identify the underlying pathways. It is likely that there is a common underlying pathogenic mechanism modified by cell specificity of the mutant gene and its function and expression. An understanding of this will allow the development of effective therapy. In this respect, perhaps Kennedy's disease will lead the way. Diagnosis is readily available with genetic testing, and potential therapies derived from animal models may soon be available.

References

1. Kennedy, W. R., Alter, M., and Sung, J. H. (1968). Progressive proximal spinal and bulbar muscular atrophy of late onset. A sex-linked recessive trait. *Neurology* **18**, 671–680.
2. La Spada, A. R., Wilson, E. M., Lubahn, D. B., Harding, A. E., and Fischbeck, K. H. (1991). Androgen receptor gene mutations in X-linked spinal and bulbar muscular atrophy. *Nature* **352**, 77–79.
3. Zajac, J. D., and MacLean, H. E. (1998). Kennedy's disease: clinical aspects. In: "Genetic Instabilities and Hereditary Neurological Diseases" (R. D. Wells, S. T. Warren, and M. Sarmiento, eds.), pp. 87–100. Academic Press, San Diego.
4. Walcott, J. L., and Merry, D. E. (2002). Trinucleotide repeat disease. The androgen receptor in spinal and bulbar muscular atrophy. *Vitam. Horm.* **65**, 127–147.
5. Greenland, K. J., and Zajac, J. D. (2004). Kennedy's disease: pathogenesis and clinical approaches. *Intern. Med. J.* **34**, 279–286.
6. Katsuno, M., Adachi, H., Tanaka, F., and Sobue, G. (2004). Spinal and bulbar muscular atrophy: ligand-dependent pathogenesis and therapeutic perspectives. *J. Mol. Med.* **82**, 298–307.
7. Greenland, K. J., Zajac, J. D., and Warne, G. L. (2004). The role of androgens in the pathogenesis of Kennedy's disease. *Curr. Top. Ster. Res.* **4**, 85–92.
8. MacLean, H. E., Choi, W. T., Rekaris, G., Warne, G. L., and Zajac, J. D. (1995). Abnormal androgen receptor binding affinity in subjects with Kennedy's disease (spinal and bulbar muscular atrophy). *J. Clin. Endocrinol. Metab.* **80**, 508–516.
9. Dejager, S., Bry-Gauillard, H., Bruckert, E., Eymard, B., Salachas, F., LeGuern, E., Tardieu, S., Chadarevian, R., Giral, P., and Turpin, G. (2002). A comprehensive endocrine description of Kennedy's disease revealing androgen insensitivity linked to CAG repeat length. *J. Clin. Endocrinol. Metab.* **87**, 3893–3901.
10. Nagashima, T., Seko, K., Hirose, K., Mannen, T., Yoshimura, S., Arima, R., Nagashima, K., and Morimatsu, Y. (1988). Familial bulbospinal muscular atrophy associated with testicular atrophy and sensory neuropathy (Kennedy-Alter-Sung syndrome). Autopsy case report of two brothers. *J. Neurol. Sci.* **87**, 141–152.
11. Maclean, H. E., Gonzales, M., Greenland, K. J., Warne, G. L., and Zajac, J. D. (2005). Age-dependent differences in androgen binding affinity in a family with spinal and bulbar muscular atrophy. *Neurol. Res.* **27**, 548–551.
12. Perez, J., and Kelley, D. B. (1996). Trophic effects of androgen: receptor expression and the survival of laryngeal motor neurons after axotomy. *J. Neurosci.* **16**, 6625–6633.
13. La Spada, A. R., and Taylor, J. P. (2003). Polyglutamines placed into context. *Neuron* **38**, 681–684.
14. Walsh, R., Storey, E., Stefani, D., Kelly, L., and Turnbull, V. (2005). The roles of proteolysis and nuclear localisation in the toxicity of the polyglutamine diseases. A review. *Neurotox. Res.* **7**, 43–57.
15. Li, M., Nakagomi, Y., Kobayashi, Y., Merry, D. E., Tanaka, F., Doyu, M., Mitsuma, T., Hashizume, Y., Fischbeck, K. H., and Sobue, G.

(1998). Nonneural nuclear inclusions of androgen receptor protein in spinal and bulbar muscular atrophy. *Am. J. Pathol.* **153**, 695–701.

16. Adachi, H., Katsuno, M., Minamiyama, M., Waza, M., Sang, C., Nakagomi, Y., Kobayashi, Y., Tanaka, F., Doyu, M., Inukai, A., Yoshida, M., Hashizume, Y., and Sobue, G. (2005). Widespread nuclear and cytoplasmic accumulation of mutant androgen receptor in SBMA patients. *Brain* **128**, 659–670.

17. Ellerby, L. M., Hackam, A. S., Propp, S. S., Ellerby, H. M., Rabizadeh, S., Cashman, N. R., Trifiro, M. A., Pinsky, L., Wellington, C. L., Salvesen, G. S., Hayden, M. R., and Bredesen, D. E. (1999). Kennedy's disease: caspase cleavage of the androgen receptor is a crucial event in cytotoxicity. *J. Neurochem.* **72**, 185–195.

18. Bence, N. F., Sampat, R. M., and Kopito, R. R. (2001). Impairment of the ubiquitin-proteasome system by protein aggregation. *Science* **292**, 1552–1555.

19. McCampbell, A., Taylor, J. P., Taye, A. A. Robitschek, J., Li, M., Walcott, J., Merry, D., Chai, Y., Paulson, H., Sobue, G., and Fischbeck, K. H. (2000). CREB-binding protein sequestration by expanded polyglutamine. *Hum. Mol. Genet.* **9**, 2197–2202.

20. Minamiyama, M., Katsuno, M., Adachi, H., Waza, M., Sang, C., Kobayashi, Y., Tanaka, F., Doyu, M., Inukai, A., and Sobue, G. (2004). Sodium butyrate ameliorates phenotypic expression in a transgenic mouse model of spinal and bulbar muscular atrophy. *Hum. Mol. Genet.* **13**, 1183–1192.

21. Sopher, B. L., Thomas, P. S. Jr., LaFevre-Bernt, M. A., Holm, I. E., Wilke, S. A., Ware, C. B., Jin, L. W., Libby, R. T., Ellerby, L. M., and La Spada, A. R. (2004). Androgen receptor YAC transgenic mice recapitulate SBMA motor neuronopathy and implicate VEGF164 in the motor neuron degeneration. *Neuron* **41**, 687–699.

22. Szebenyi, G., Morfini, G. A., Babcock, A., Gould, M., Selkoe, K., Stenoien, D. L., Young, M., Faber, P. W., MacDonald, M. E., McPhaul, M. J., and Brady, S. T. (2003). Neuropathogenic forms of huntingtin and androgen receptor inhibit fast axonal transport. *Neuron* **40**, 41–52.

23. LaFevre-Bernt, M. A., and Ellerby, L. M. (2003). Kennedy's disease. Phosphorylation of the polyglutamine-expanded form of androgen receptor regulates its cleavage by caspase-3 and enhances cell death. *J. Biol. Chem.* **278**, 34918–34924.

24. Mariotti, C., Castellotti, B., Pareyson, D., Testa, D., Eoli, M., Antozzi, C., Silani, V., Marconi, R., Tezzon, F., Siciliano, G., Marchini, C., Gellera, C., and Donato, S. D. (2000). Phenotypic manifestations associated with CAG-repeat expansion in the androgen receptor gene in male patients and heterozygous females: a clinical and molecular study of 30 families. *Neuromuscul. Disord.* **10**, 391–397.

25. Schoenen, J., Delwaide, P. J., Legros, J. J., and Franchimont, P. (1979). Hereditary motor neuron disease: the proximal, adult, sex-linked form (or Kennedy disease). Clinical and neuroendocrinologic observations. *J. Neurol. Sci.* **41**, 343–357.

26. Guidetti, D., Vescovini, E., Motti, L., Ghidoni, E., Gemignani, F., Marbini, A., Patrosso, M. C., Ferlini, A., and Solime, F. (1996). X-linked bulbar and spinal muscular atrophy, or Kennedy disease: clinical, neurophysiological, neuropathological, neuropsychological and molecular study of a large family. *J. Neurol. Sci.* **135**, 140–148.

27. Schmidt, B. J., Greenberg, C. R., Allingham-Hawkins, D. J., and Spriggs, E. L. (2002). Expression of X-linked bulbospinal muscular atrophy (Kennedy disease) in two homozygous women. *Neurology* **59**, 770–772.

28. Yang, W., Dunlap, J. R., Andrews, R. B., and Wetzel, R. (2002). Aggregated polyglutamine peptides delivered to nuclei are toxic to mammalian cells. *Hum. Mol. Genet.* **11**, 2905–2917.

29. Nakamura, M., Mita, S., Matuura, T., Nagashima, K., Tanaka, H., Ando, M., and Uchino, M. (1997). The reduction of androgen receptor mRNA in motoneurons of X-linked spinal and bulbar muscular atrophy. *J. Neurol. Sci.* **150**, 161–165.

30. Katsuno, M., Adachi, H., Doyu, M., Minamiyama, M., Sang, C., Kobayashi, Y., Inukai, A., and Sobue, G. (2003). Leuprorelin rescues polyglutamine-dependent phenotypes in a transgenic mouse model of spinal and bulbar muscular atrophy. *Nat. Med.* **9**, 768–773.

31. Katsuno, M., Adachi, H., Kume, A., Li, M., Nakagomi, Y., Niwa, H., Sang, C., Kobayashi, Y., Doyu, M., and Sobue, G. (2002). Testosterone reduction prevents phenotypic expression in a transgenic mouse model of spinal and bulbar muscular atrophy. *Neuron* **35**, 843–854.

50

Neurobiology of Genetic Mental Retardation

Walter E. Kaufmann, MD
John C. Carter
Irena Bukelis
David N. Lieberman, MD, PhD

Keywords: *Down syndrome, fragile X syndrome, mental retardation, Rett syndrome, synapses, X-linked mental retardation*

I. Introduction
II. Dendritic and Synaptic Abnormalities: Fundamental Features of Genetic MR
III. Down Syndrome
IV. X-Linked MR Disorders
V. Other Genetic MR Disorders
VI. Conclusion
   References

I. Introduction

Mental retardation (MR) is one of the most common and severe developmental neurological syndromes, affecting 1%–3% of individuals in the general population. MR is typically described as a relatively nonprogressive primary impairment of cognitive function and adaptive behavior. The definition of MR depends on whether medical, educational, legal, or other aspects are emphasized; however, there is widespread agreement in classifying individuals with an intelligence quotient (IQ) lower than 70 at age 5 years or older (when cognitive performance

---

We are grateful to the families who have participated in research projects at the Center for Genetic Disorders of Cognition & Behavior; their partnership is critical for advancing the field. We also thank Dr. Michael Johnston for his continuous support of our research and the Center.

is more stable) as having MR. Severe MR, corresponding with an IQ below 40–50, is frequently associated with genetic abnormalities, comprising 25%–50% of all forms of MR. Although environmental factors appear to play an important role in the etiology of MR in mildly affected individuals, it is unquestionable that even these exogenous agents act through disturbances in genetic developmental programs [1,2]. Due to space limitations, this review will not discuss metabolic and degenerative disorders of childhood, which may resemble MR during the course of their evolution but usually evolve in the setting of a normally developed brain. Among the known "primary" genetic abnormalities associated with MR, a disproportionately high percentage involves genes on the X chromosome (6–8 times more frequent than expected for the proportion of X chromosome genes in the human genome) [3]. Altogether, these observations suggest that important clues about the pathogenesis of MR will emerge from the study of key genes involved in brain development, particularly those located on the X chromosome. Whereas the focus of this review is the neurobiology of MR, clinical and neurobiological evidence has suggested considerable overlap between the presentations of MR and autistic spectrum disorders. Importantly, some of the literature on neuroanatomical changes in MR is based on cases that would be labeled as autism under current diagnostic standards [2]. Consequently, some of the principles and features described in the following sections could be applicable to individuals with autism spectrum disorders.

II. Dendritic and Synaptic Abnormalities: Fundamental Feature of Genetic MR

Since the application in the early 1970s of dendritic labeling techniques (e.g., Golgi impregnations) to postmortem brain samples, it has become evident that reductions in dendritic branching and abnormalities of dendritic spine morphology are the most consistent anatomical features in genetic and environmental conditions associated with MR (Table 1, Fig. 1). More extensive anatomical derangements such as cortical malformations secondary to disturbed neuronal proliferation, migration, or both are commonly associated with "stagnated" cognitive and motor development and therefore may resemble the early global developmental delay seen in patients with MR. However, patients' reduced life expectancy due to severe brain dysfunction and associated non–central nervous system (CNS) malformations precludes their diagnostic classification among the genetic MR syndromes [2,4]. Although genetically identified causes of MR are more prevalent in syndromic compared with nonsyndromic forms of MR, considerable progress has been made in identifying single-gene causes

Table 1 Neocortical Cytoarchitectonic and Dendritic Abnormalities in Genetic Disorders Associated with MR*

Disorder	Laminar Disturbance	Increased Packing Density	Reduced Dendritic Length	Spine Dysgenesis
Down syndrome	Y	N	Y	Y
Fragile X syndrome	N	N	N	Y
Neurofibromatosis-1	Y (focal)	N	?	?
Tuberous sclerosis	Y (focal)	Y (focal)	Y (focal)	Y (focal)
Williams syndrome	Y	Y	?	?
Rett syndrome	N	Y	Y	Y
Phenylketonuria	N	Y	Y	Y
Patau syndrome	Y	N	Y	Y
Rubinstein-Taybi syndrome	N	Y	?	?

MR, Mental retardation; *N*, no; *Y*, yes; ?, unknown or questionable.
*The conditions have been listed according to estimated incidence.
Adapted from Kaufmann, W. E., and Moser, H. W. (2000). Dendritic anomalies in disorders associated with mental retardation. *Cereb. Cortex* **10**, 981–991.

Figure 1 Dendritic spine abnormalities in Patau syndrome and Down syndrome (DS). Drawings are from Golgi preparations depicting comparable segments of apical dendrites from layer V pyramidal neurons (motor cortex). **A–E**, Different developmental stages in normal subjects (5th gestational month, 7th gestational month, neonatal period, 2nd postnatal month, 8th postnatal month, respectively). **F**, A neonate with 13–15 trisomy. **G**, An 18-month-old infant with DS (trisomy 21). Note the progressive increase in spine density, associated with a reduction in spine length, during normal development. Spines in Patau syndrome are not only sparse but also longer than expected for a neonate. On the other hand, the infant with DS had shorter and thinner rather than long spines. (Reprinted with permission of Marin-Padilla, M. [1972]. Structural abnormalities of the cerebral cortex in human chromosomal aberrations. A Golgi study. *Brain. Res.* **44**, 625–629.)

of X-linked nonsyndromic forms of MR. The relevance of dendritic and other synaptic abnormalities to MR has been underscored by their reproduction in animal models of these genetic disorders and by the demonstration that many genes mutated in MR are involved directly or indirectly in synaptic function [3]. Comparative analyses of dendritic profiles of patients affected by genetic MR (e.g., fragile X syndrome [FXS]) and their related animal models (e.g., *FMR1* knock-out mouse) have also provided valuable insights into the nature and evolution of synaptic anomalies in these disorders.

III. Down Syndrome

Down syndrome (DS) is the most common genetic cause of MR, occurring in an estimated 1:1000 live births [1,6]. The DS phenotype is linked in most cases to chromosome 21 trisomy and involves characteristic aberrations of physical and neurological growth. Physical abnormalities include cardiac and gastrointestinal malformations. Neurological abnormalities include decreased brain size, immature gyral patterns, volumetric reductions in selective brain regions (e.g., frontal cortex), delayed myelination of cortical fibers, and abnormal neocortical lamination [2,4,6,7]. For a list of common phenotypic features of DS, see Table 2. As in other MR disorders, and in correspondence with changes in dendritic length, abnormalities in cortical cytoarchitecture are also present (e.g., decrease in neuronal density in granular layers of the neocortex) [2,8]. In addition, poor laminar distinction in the superior temporal gyrus and aberrations of neuronal morphology and orientation have been noted [4,8]. The profile of cortical dendritic abnormalities in DS suggests a postnatal process, characterized by initial normal development followed by progressive neurodegeneration. Several studies have reported normal numbers of dendrites and normal dendritic morphology in newborns with DS, with some reports of greater than normal dendritic arborization in DS infants younger than 6 months, in contrast to marked reductions in length, branching, and spine density in early childhood (see Fig. 1) [2,8]. Cross-sectional studies of adults with DS have revealed degenerative neuronal abnormalities with respect to dendritic length and morphology [8]. The course and nature of these dendritic abnormalities appear unique to DS and, in some reports, are related to the cognitive profile of DS [2,8].

The study of the molecular bases of neuronal pathology in DS has been aided by the recent sequencing of the human genome and the application of data derived from animal models. As DS involves triplication of an entire chromosome that is estimated to contain upward of 300 genes [8], the number of related gene products is immense, and thus dendritic/synaptic abnormalities in this condition may be the result of altered expression levels of a variety of proteins coded for by genes on chromosome 21 (Table 3). Dendritic development is a dynamic process involving changing expression profiles of dendritic proteins across the course of development, often in a time-dependent manner. In the case of DS, molecular analyses indicate an accumulation of several cytoskeletal proteins, most likely due to dendritic involution [9]. Recent genetic and molecular studies have implicated a number of chromosome 21 gene products. DYRK1A, a serine-threonine kinase, controls dendrite growth and differentiation and is a modulator for cAMP response element-binding protein (CREB), which is involved in neuronal differentiation and plasticity. In mouse models haplo-insufficient for DYRK1A, cortical layer III pyramidal neurons show decreased soma size, dendritic arborization, and spine distribution compared with wild-type mice [8]. Additionally, Down syndrome cell adhesion molecule (DSCAM), an inhibitor of synaptogenesis and outgrowth of neurites, is elevated in DS brains [10]. Drebrin, an actin-binding protein co-localized with actin in dendrites and filopodia in cortex, has been shown to be decreased in DS brains during the second trimester of prenatal life. Similarly, levels of α-SNAP and SNAP-25, two synaptosomal proteins associated with drebrin, have also been reported to be lower in prenatal DS cortex during the second trimester [10]. Initial studies of channel and receptor function in the (partial trisomic) Ts65Dn mouse model have demonstrated abnormalities in action potential duration; rates of depolarization and repolarization; and sodium, potassium, and calcium channel kinetics, as well as unusual distribution patterns of these channels [10]. Further evidence from the Ts65Dn mouse suggests that hippocampal electrophysiology involving protein kinases A and C and the phosphoinositide pathway may be aberrant in DS, such that long-term potentiation (LTP), long-term depression (LTD), and related learning and memory paradigms are affected [10].

Table 2 Phenotypic Features of Down Syndrome (Adapted from [6])

Physical Features	Neuro-behavioral Features
Epicanthal folds	Mental retardation
Flat nasal bridge	Alzheimer-type dementia
Open mouth	
Short stature	Autistic-like features
Broad hands	Depression
	Attention-deficit hyperactivity disorder
Brachycephaly	Obsessive-compulsive disorder
Brachydactyly	Anxiety
Fifth finger clinodactyly	Aggressive-disruptive behaviors
Cataracts	
Wide 1–2 toe gap	Hypotonia
Lax ligaments	Seizures
Mitral valve prolapse	
Duodenal atresia	
Hypo/hyper-thyroidism	
Cervical myelopathy	

Table 3 Genes on Chromosome 21 That Are Possibly Involved in Brain Development and Degeneration in Down Syndrome

Symbol	Name	Possible Effect in Down Syndrome
SIM2	Single-minded homolog 2 (Drosophila)	Brain development: required for synchronized cell division and establishment of proper cell lineage
DYRK1A	Dual-specificity tyrosine- (Y)-phosphorylation regulated kinase 1A	Brain development: expressed during neuroblast proliferation and believed to be an important homolog in regulation of cell-cycle kinetics during cell division
GART	Phosphoribosylglycinamide formyltransferase Phosphoribosylglycinamide synthetase Phosphoribosylaminoimidazole synthetase	Brain development: expressed during prenatal development of the cerebellum
PCP4	Purkinje cell protein 4	Brain development: function unknown but found exclusively in the brain and most abundantly in the cerebellum
DSCAM	Down syndrome cell adhesion molecule	Brain development and possible candidate gene for congenital heart disease: expressed in all regions of the brain and believed to have a role in axonal outgrowth during development of the nervous system
GRIK1	Glutamate receptor, ionotropic, kainate 1	Neuronal loss: function unknown, found in the cortex in fetal and early postnatal life and in adult primates, most concentrated in pyramidal cells in the cortex
APP	Amyloid beta (A4) precursor protein (protease nexin-II, Alzheimer's disease)	Alzheimer's-type neuropathy: seems to be involved in plasticity, neurite outgrowth, and neuroprotection
S100B	S100 calcium-binding protein, beta (neural)	Alzheimer's-type neuropathy; stimulates glial proliferation
SOD1	Superoxide dismutase 1, soluble (amyotrophic lateral sclerosis, adult)	Accelerated aging? Scavenges free superoxide molecules in the cell and might accelerate aging by producing hydrogen peroxide and oxygen

Adapted from Roizen, N. J., and Patterson, D. (2003). Down's syndrome. *Lancet* **361**, 1281–1289.

Taken together, the molecular, electrophysiological, and genetic data from DS and its animal models point to a predominantly functional (as opposed to structural) underlying neural abnormality. Although this information has not yet been applied toward improving the diagnostic classification of this relatively heterogeneous disorder, treatments aimed at remedying the neurological pathobiology in DS are beginning to emerge. Operating under the premise that homeostatic and metabolic processes are disrupted in DS [6], many dietary supplements including vitamins and minerals have been proposed to increase cognitive function, although such treatments have been shown to be largely ineffective [6,11]. Pharmacological treatments have focused primarily on improving cognition in DS, mainly by targeting cholinergic and glutamatergic neurotransmission, although none of the currently available pharmacological interventions have consistently proven effective in this regard. (*Note:* The type and magnitude of cholinergic abnormalities in children with DS is still controversial.) Nootropic drugs such as the cyclic GABA-derivative piracetam, although initially showing promise in animal studies, have since shown poor results in DS [6,8]. The association between DS and Alzheimer's disease, including the accumulation of beta-amyloid (Aβ) plaques, subsequent cholinergic presynaptic deficit, the development of dementia in both conditions, and the efficacy of cholinesterase inhibitors in targeting AD-related symptoms, has spurred research focusing on the effects of these drugs in DS. Nonetheless, drugs targeting the cholinergic system, including muscarinic agonists, acetylcholine analogs, and acetylcholinesterase inhibitors, specifically the cholinesterase inhibitor donepezil, have shown mixed results, with some studies reporting no significant clinical benefit and others reporting increases in socialization, expressive language, and adaptive behavior domains in subjects with and without dementia [8]. Hormone therapies, including steroid, thyroid, and growth hormone treatments, have shown little promise, although new research on estrogen administration in the Ts65Dn mouse suggests that women with DS and dementia may benefit from such therapeutic regimens with respect to cognitive function and cholinergic phenotype [8]. Finally, in response to cognitive deficits and related LTP and LTD, drugs designed to target the glutamatergic system are emerging as a promising area of new research. Alternative, nonpharmacological therapies such as environmental/social enrichment have shown mixed results, both in humans and Ts65Dn mice [2]. Environmental enrichment and related interventions in the trisomic mouse have produced measurable improvements in cognitive function (e.g., performance in a Morris water maze), as well as increases in dendritic complexity and number of spines in female mice, although these results were strongly gender specific: Environmental enrichment actually worsened Morris water maze performance in male Ts65Dn mice [8]. In children with DS, educational interventions have demonstrated efficacy in improving performance on measures of cognitive function, although these effects appear minor and transient [6,8].

IV. X-Linked MR Disorders

Among genetic disorders associated with MR, the group in which the primary defect involves genes on the X chromosome, termed *X-linked MR (XLMR)*, deserves special consideration because it comprises 20%–25% of cases of MR, with a high proportion of inheritance. XLMR includes a relatively large number of conditions in which the only clinical manifestations are MR-related cognitive and behavioral abnormalities (nonsyndromic XLMR), as well as a few disorders in which physical or metabolic features are present (syndromic XLMR) [3]. Although mutated genes have been identified in only about 10% of families with XLMR at present, this represents a comparatively high percentage in the context of MR in general. Of particular interest is the group of genes linked to nonsyndromic XLMR because many of these genes are involved in the Rho-GTPase signaling pathway. Examples of these are oligophrenin-1, a RhoA-GAP (GTPase-activating protein) that interacts with several postsynaptic density proteins; α-PIX, a guanine nucleotide exchange factor (GEF) for Rac and Cdc42; and PAK 3, a RhoA effector protein. Since the Rho signaling system is involved in key processes leading to dendritic and axonal formation, such as neuronal and spine morphogenesis polarity, neurite outgrowth and synapse formation (Table 4) [12], it is reasonable to speculate that these XLMR conditions are primary synaptic disorders. Because different components of the Rho system have opposite effects on dendritic development (e.g., Rac and Cdc42 promote, activated RhoA inhibits), these XLMR disorders appear to represent imbalances in synaptic formation and plasticity. Novel therapeutic

Table 4 Genes Involved in Nonsyndromic and Syndromic XLMR, with Emphasis on Rho Signaling*

Gene	Locus	Protein Name(s)	Function(s)	Effects of Mutation(s)	MRX Family[a]	MIM Number[b]
Nonsyndromic MR (MRX) genes						
ARHGEF6	Xq26	αPIX, Cool-2	GEF for Rac/Cdc42; activator of PAKs	Exon skipping, 28 amino acid deletion in CH domain, loss-of-function owing to translocation	MRX46	300267
FMR2	Xq28	FMR2	Transcriptional activator? (AF-4 like)	Truncation, loss-of-expression		309548
GDI1	Xq28	αGDI, RABGDIA	GDI for Rab3a,b; neurotransmitter release and vesicle trafficking	Decrease and loss-of-function, truncation and loss of expression	MRX41,tam R; MRX48	300104
IL1RAPL	Xq22	ILRAPL	IL signaling?	Deletions, truncation and decreased expression	MRX34	300206
OPHN1	Xq12	Oligophrenin-1	Rho family GAP inhibitor of RhoA?	Frame-shift and decreased expression, loss-of-function	MRX60	300127
PAK3	Xq22	PAK3, β-PAK	Ser/Thr protein kinase Cdc42/Rac1 effector	Truncation, loss-of-kinase activity, mutation in GTPase-binding domain	MRX30; MRX47	300142
TM4SF2	Xq11	Tetraspanin 2	Integrin-mediated signaling?	Truncation, mutation in extracellular loop, translocation and decreased expression		300096
Syndromic MR genes involved in Rho signaling						
FGDY	Xp11	FGD1	GEF for Cdc42	Truncation, mutations in PH and GEF domain	Faciogenital dysplasia Aarskog-Scott syndrome	305400
LIMK1	7q11	LIM kinase-1	Tyr kinase downstream of Rac/Cdc42 and PAK; phosphorylates and inactivates cofilin	Deletion of multiple genes, including elastin gene	Williams syndrome (probably involved)	601329

ARHGEF, Rho guanine nucleotide exchange factor; *Cool-2*, cloned out of library 2; *FGD*, faciogenital dysplasia; *FMR*, fragile site mental retardation; *GDI*, guanine dissociation inhibitor; *GEF*, guanine nucleotide exchange factor; *IL*, interleukin; *IL1RAPL*, IL-1 receptor accessory protein like; *LIMK*, LIM domain containing kinase; *MR*, mental retardation; *PAK*, p21-associated kinase; *PH*, pleckstirin homology; *PIX*, PAK-interacting exchange factor; *RABGDIA*, Rab GDP-dissociation inhibitor a; *TM4SF2*, transmembrane-4 superfamily 2; *XLMR*, X-linked mental retardation.

[a] As listed in the XLMR Update web site (xlmr.interfree.it/home.html).
[b] Entry numbers for Online Mendelian Inheritance in Man database; for further information and references, see www.ncbi.nlm.nih.gov/entrez/query.fcgi?db=OMIM.
*This table also includes a gene on chromosome 7, *LIMK1*, which is also implicated in Rho signaling.
Adapted from Ramakers, G. J. A. (2002). Rho proteins, mental retardation and the cellular basis of cognition. *Trends Neurosci.* **25,** 191–199.

approaches targeting GTPase-related disorders, including perturbations of the balance of the Rho system, may lead to specific treatments for this group of XLMR conditions. Furthermore, recent approaches for the identification of mutations involving genes on the X chromosome (e.g., customary microarray technology), in conjunction with the characteristic pedigrees of families affected by X-linked disorders, hold the promise of increasing the number of identified XLMR conditions. The following two sections review important neurobiological aspects of the two most common XLMR syndromes: fragile X syndrome (FXS) and Rett syndrome (RTT).

A. Fragile X Syndrome

Fragile X syndrome (FXS) is the most prevalent form of inherited MR, affecting 1:4000 males and 1:6000 females. The disorder is linked to the expansion of a CGG polymorphism in the (5'UTR) regulatory region of *FMR1*. When the normal ~30 CGG repeats increase to >200 (full mutation), *FMR1* promoter hypermethylation, *FMR1* transcriptional silencing, and FXS phenotype occur. Intermediate level expansions (60–200 CGG repeats), which are termed *premutation*, are not associated with FXS [13]. In addition to mild to moderate MR, FXS is characterized by dysmorphic features, connective tissue abnormalities (e.g., lax joints) and other non-CNS phenotypical anomalies (e.g., macroorchidism after puberty). However, variable cognitive and language impairments and associated neurobehavioral problems that include attentional difficulties, hyperactivity, anxiety, and autistic disorder constitute the major medical and educational concerns for patients with FXS (Table 5) [13]. As expected in an X-linked condition, the characteristic features of FXS are more prominent in affected males.

Neuroimaging studies have shown that in FXS, the brain is larger in general (i.e., most cortical regions, caudate, hippocampus) although of normal configuration. Nonetheless, certain areas such as the temporal neocortex and the posterior cerebellar vermis are decreased in both absolute and relative size [7]. Postmortem analyses of the cerebral cortex in males with FXS demonstrate that the most consistent anatomical abnormality is an aberrant conformation of dendritic spines, which are longer and tortuous (Fig. 2) [2]. Similar profiles have been observed in mice lacking the *FMR1* product, the fragile X mental retardation protein (FMRP) [2]. This spine configuration resembles that of early stages of normal dendritic development (see Fig. 1A–D), which suggests that in FXS a maturational arrest of dendritic spine formation occurs. The combination of studies of patient samples, animal models, and in vitro systems has led to an explosion of knowledge about the neurobiology of FXS, in particular the dendritic and synaptic abnormalities in FXS [2,15,16]. FMRP is an

Table 5 Phenotypic Features of Fragile X Syndrome (Adapted from [13])

Physical Features	Neuro-behavioral features
Large ears	Mild to moderate mental retardation
Thick nasal bridge	Language delay, predominantly expressive
Prominent jaw	Rapid/burst-like speech
High-arched/narrow palate	Attentional-organizational dysfunction
Pale blue irides	Visuospatial impairment
Strabismus	Hyperactivity
Pectus excavatum	Autistic-like features
Kyphoscoliosis	Aggressive behavior
Lax joints	Hyperarousal
	Anxiety, particularly social
Single palmar crease	Stereotypic/perseverative behavior
Flat feet	
Cutis laxa	Hypotonia
	Nystagmus
Mitral valve prolapse	Seizures
Macroorchidism	

Figure 2 Examples of typical spine morphologies on Golgi-impregnated dendrites from (**A**) a person afflicted by fragile X syndrome and (**B**) an unaffected control. Dendrites are at extremes of density differences and not intended to depict the norm. (Reprinted with permission of Irwin, S. A., Galvez, R., and Greenough, W. T. [2000]. Dendritic spine structural anomalies in fragile-X mental retardation syndrome. *Cereb. Cortex* **10**, 1038–1044.)

RNA-binding protein that associates with polyribosomes and is involved in the transport and translational regulation (mainly inhibition) of a selected subset of transcripts at synaptic sites [16]. FMRP appears also to be linked to the microRNA pathway [17]. Among the mRNAs regulated by FMRP are those coding for key cytoskeletal (e.g., MAP1-B) and synaptic (e.g., Arc) constituents. It is postulated that FMRP's role is to regulate protein synthesis in response to synaptic activity, in this way enabling the

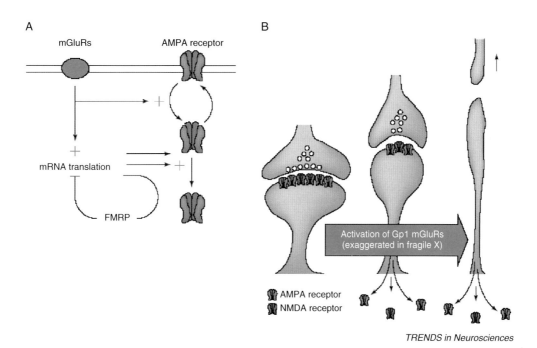

Figure 3 Models of protein synthesis-dependent, functional and structural consequences of group 1 (Gp1) metabotropic glutamate receptor (mGluR) activation at hippocampal synapses, and the role of FMRP. **A**, Model to account for exaggerated mGluR long-term depression (mGluR-LTD) in the *FMR1* knock-out mouse, based on the assumption that FMRP is synthesized in response to mGluR activation and functions as a translational repressor. **B**, Model relating the net loss of synaptic AMPA and NMDA receptors and elongation of dendritic spines observed following Gp1 mGluR activation in cultured hippocampal neurons. (Reprinted with permission of Bear, M. F., Huber, K. M., and Warren, S. T. [2004]. The mGluR theory of fragile X mental retardation. *Trends Neurosci.* **27**, 370–377.)

adaptation of the postsynaptic site in particular to multiple influences during development and plasticity [16]. Of importance for understanding the mechanisms of action of FMRP deficiency is the consistent observation of LTD enhancement in the hippocampus of the *FMR1* knock-out mouse (other synaptic abnormalities are less well characterized). This group 1 metabotropic glutamate receptor (mGluR) – and protein synthesis–dependent type of LTD has been linked to a variety of neurobehavioral manifestations of FXS, as well as to the unique appearance of dendritic spines in the mouse model (Fig. 3) [15]. Exaggerated group 1 mGluR activity would lead to a reduction in α-amino-5-hydroxy-3-methyl-4-isoxazole propionic acid (AMPA) and NMDA subtype of glutamate receptors on the postsynaptic surface that, in conjunction with an imbalance in key postsynaptic proteins, would result in a morphological and functionally immature/deficient synapse unable to deal with subtle changes in synaptic activity. Another potential mechanism for dendritic changes in FXS is the interaction between FMRP and the cytoplasmic interacting proteins CYFIP1 and CYFIP2, which bind activated Rac1. This provides a pathogenetic link between FXS and the Rho system–dependent abnormalities (e.g., actin cytoskeleton remodeling) in nonsyndromic XLMRs, as described in the preceding section.

The knowledge of specific targets of FMRP in the CNS, as well as of defective synaptic processes, offers the opportunity to improve current diagnostic and therapeutic approaches to FXS. Despite the single-gene nature of FXS, at present it is not possible to anticipate which patient with *FMR1* full mutation will develop a severe neurobehavioral phenotype. Molecular profiling of patients with FXS, through the measurement in peripheral samples of relevant biomarkers, may help in diagnostic classification, prognosis, and subject selection for treatment. In addition, pharmacological interventions aiming at restoring a balance in glutamate receptor activity (some are currently under development), such as decreasing group 1 mGluR activity and enhancing AMPA receptor function, would provide a neurobiologically based amelioration of FXS manifestations.

B. Rett Syndrome

Rett syndrome (RTT) is an XLMR condition that affects predominantly females, with an incidence of 1:10,000–15,000 births. RTT is a severe disorder that is lethal in most male cases and the second-leading cause of global developmental delay and severe MR in females [18,19]. The majority of RTT cases are associated with mutations in the coding region of *MeCP2*, a gene located on Xq28, which encodes the transcriptional repressor methyl-CpG–binding protein 2 (MeCP2) [18–20]. RTT is a dynamic condition; apparently normal development, from birth until 6–18 months of age, is followed by deceleration of head growth and failure to attain new milestones. At this "regressive" stage, progressive loss of language and motor skills characteristically develops. Additional manifestations include respiratory irregularities, impaired social interaction (i.e., "autistic-like" features), seizures, and the stereotypical hand-wringing movements that characterize the disorder (Tables 6 and 7) [18]. Typically, after age 3–4 years, the neurological manifestations tend to stabilize or improve (e.g., language) for a period of years. By adolescence, a slow motor decline that lasts for decades becomes apparent and in most cases leads to wheelchair dependency [18]. Other non-CNS clinical features, such as disturbed gastrointestinal motility and abnormal autonomic vascular regulation, appear to be related to peripheral neuronal dysfunction. The described "classic" RTT phenotype may not be fully developed in a variety of atypical forms of RTT, including rare male cases [18,20]. Mutations in *MeCP2* and brain tissue changes in MeCP2 expression have been associated with other phenotypes, such as nonspecific MR, Angelman syndrome, and autism spectrum disorder [18,20].

The neuroanatomical phenotype of RTT is characterized by a marked brain hypoplasia, with slightly greater involvement of the cerebral gray matter [19–21]. At the microscopic level, there is a generalized increase in cell packing density and marked decrease in the size of the neuronal soma and dendritic tree. Abnormal dendritic spine profiles have also been described in RTT postmortem cortical samples [2,19,20]. Although neuroimaging and postmortem anatomical studies indicate more severe changes in areas such as the frontal cortex, subcortical cholinergic neuronal groups, and monoaminergic nuclei of the brainstem (e.g., substantia nigra hypopigmentation), severe neuronal pathology is widespread in RTT [19–21]. The previously mentioned neuronal structural features, as well as immunochemical data showing reductions in cytoskeletal proteins linked to neuronal differentiation (Fig. 4) [9], suggest that the fundamental cellular phenotype in RTT is a disturbance in neuronal maturation. As in FXS, some of the key cellular and neurological features of the disorder have been reproduced in mouse models lacking MeCP2 expression [19,20]. The latter provides further evidence for RTT's classification as a disorder of synaptic development. Neurotransmitter abnormalities associated with either the early stages (i.e., reduction in monoamines) or the period of regression (i.e., increase in glutamate levels and NMDA receptor binding) of RTT have not yet been reported in the recently developed animal models of the disorder.

Table 6 Diagnostic Criteria for Rett Syndrome

Manifestations/Age	Comments
Infant apparently normal initially	Pre/perinatal period as well as first 6 months of life or longer
Head circumference stagnation 3 months–4 years	Normal at birth, then a decelerating growth rate
Purposeful hand skill loss 9 months–2.5 years	Communicative dysfunction, social withdrawal, mental deficiency, loss of speech/babbling
Classical stereotypic hand movements after 1–3 years	Hand washing/wringing or clapping/tapping
Gait/posture dyspraxia 2–4 years	Gait "ataxia"/more or less jerky truncal "ataxia"

From Hagberg, B. (2002). Clinical manifestations and stages of Rett syndrome. *Ment. Retard. Dev. Disabil. Res. Rev.* **8**, 61–65.

Table 7 Clinical Stages of Classical Rett Syndrome

Original Staging System	Later Additions
Stage I: early-onset stagnation	
Onset age: 6 months–1.5 years	Onset from 5 months of age
Developmental progress delayed	Early postural delay
Developmental pattern still not significantly abnormal	Dissociated development "Bottom shufflers"
Duration: weeks to months	
Stage II: developmental regression	
Onset age: 1–3 or 4 years	Loss of acquired skills: fine finger, babble/words, active playing
Loss of acquired skills/communications	
Mental deficiency appears	Occasionally "in another world"
Duration: weeks to months, possibly 1 year	Eye contact preserved
	Breathing problems still modest Seizures in only 15%
Stage III: pseudostationary period	
Onset: after passing stage II	"Wake-up" period
Some communication restitution	Prominent hand apraxia/dyspraxia
Apparently preserved ambulant ability	
Inapparent, slow neuromotor regression	
Duration: years to decades	
Stage IV: late motor deterioration	*Subgrouping introduced:*
Onset: when stage III ambulation ceases	Stage IVA: previous walkers, now non-ambulant
Complete wheelchair dependency	Stage IVB: never ambulant
Severe disability: wasting and distal distortion	
Duration: decades	

From Hagberg, B. (2002). Clinical manifestations and stages of Rett syndrome. *Ment. Retard. Dev. Disabil. Res. Rev.* **8**, 61–65.

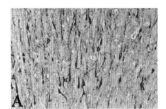
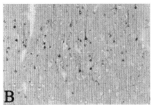

Figure 4 Immunohistochemical pattern of MAP-2 in motor cortex. **A**, In control subjects, perikaryal and (predominantly) dendritic staining is seen throughout layer V. **B**, In Rett syndrome, there is a marked reduction in MAP-2 staining involving both somas and dendrites in the same lamina. Note also the smaller cell bodies in RS. (**A** and **B**, Original magnification ×140.) (Reprinted with permission of Kaufmann, W. E., MacDonald, S. M., and Altamura, C. [2000]. Dendritic cytoskeletal protein expression in mental retardation: an immunohistochemical study of the neocortex in Rett syndrome. *Cereb. Cortex* **10**, 992–1004.)

MeCP2 binds methylated deoxyribonucleic acid (DNA) sequences through its methyl-binding domain and subsequently recruits histone deacetylases via interactions between its transcriptional repressor domain and co-repressors (e.g., Sin3A). This process, which alters chromatin architecture from an active to inactive state, may not be the only mechanism of action of MeCP2 [19]. Regardless, as expected from MeCP2 deficit, increases in levels of acetylated histones have been demonstrated in a mouse model of *MeCP2* nonsense (i.e., protein truncation) mutation [20]. In contrast to FXS, the search for genes transcriptionally silenced by MeCP2 has been a frustrating endeavor. Brain-derived neurotrophic factor (BDNF) is one of the few recognized targets of MeCP2, which also supports a role for MeCP2 in synaptic plasticity [20]. The recent demonstration of two *MeCP2* transcripts, with differential tissue distribution, and MeCP2 immunoreactivities of different molecular weight and intracellular localization suggest that MeCP2 regulation is a complex phenomenon with unique temporal and spatial dimensions [20]. Available data on MeCP2's pattern of expression suggest that it plays a role in the maintenance of neuronal differentiation and in other synaptic activity–dependent processes. In the CNS, MeCP2 expression is almost exclusively neuronal and begins when the differentiated neuroblast phenotype is established and synaptogenesis begins (Fig. 5) [19,20]. In contrast with other transcriptional regulators, MeCP2 levels in several brain regions (e.g., cerebral cortex) increase steadily until the adult stage [19,20]. Neuronal activation and neural circuit manipulation lead to changes in MeCP2 levels and possibly in the protein's conformation, as suggested by the de-repression of the BDNF promoter secondary to MeCP2 phosphorylation in stimulated neurons. Altogether, these data support the hypothesis that MeCP2 plays a critical role in synaptic plasticity, both during development and in the processes of learning and memory in the adult brain (Table 8) [20]. Consequently, MeCP2 deficit during synaptic formation and stabilization, after the establishment of the neuronal phenotype, would lead to the marked reductions of dendritic arborization seen in RTT.

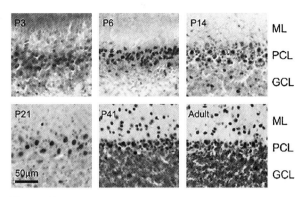

Figure 5 Ontogeny of MeCP2 staining in the rat cerebellum. MeCP2 expression patterns change dramatically with age; early postnatal ages exhibit significant staining only in Purkinje cells and Golgi cells within the granule cell layer *(GCL)*. Granule cells exhibited minimal MeCP2 staining up through the first 3 postnatal weeks but were darkly stained by postnatal day *(P)* 41. *ML*, Molecular layer; *PCL*, Purkinje cell layer. Reprinted with permission of Mullaney, B. C., Johnston, M. V., and Blue, M. E. [2004]. Developmental expression of methyl-CpG binding protein 2 is dynamically regulated in the rodent brain. *Neuroscience* **123**, 939–949.)

Table 8 Evidence Supporting the Role of MeCP2 in Synaptic Plasticity

ERK1-like N-terminal motif (MeCP2B)
Extranuclear localization (MeCP2A, MeCP2B)
Postsynaptic localization
Membrane depolarization–dependent BDNF silencing
100-kd MeCP2-like band in synaptic fractions
Neuronal group–specific pattern of expression
High-expressing neuronal subtype (postnatally)
Levels regulated by deafferentation (olfactory system)
Induction by hippocampal kindling

From Kaufmann, W. E., Johnston, M. V., and Blue, M. E. (2005). MeCP2 expression and function during brain development: implications for Rett syndrome's pathogenesis and clinical evolution. *Brain Dev.* **27**, S77–S87.

Although identification of MeCP2 targets is still in its early stages, preliminary work on histone acetylation and methylation profiles indicate that these and other indices of MeCP2 function in peripheral cells hold the promise of improving the limited predictive value of mutation patterns in RTT [20]. Alternative or complementary diagnostic classifications to the existing ones, which are based on type of mutation and clinical profile, could be of great importance in the near future because new treatments are emerging. As in DS, recently available modulators of cholinergic and glutamatergic activity are being considered as the first group of neurobiologically based pharmacological interventions for RTT. Insight into the genetic defect also offers the possibility of modifying defective DNA methylation binding and histone deacetylation; clinical trials with methyl group donors such as folate are currently in progress [23].

V. Other Genetic MR Disorders

There is indirect evidence that dendritic/synaptic abnormalities similar to those reported in DS and XLMR exist in other genetic MR conditions [1,2]. On the basis of studies of RTT and idiopathic autism, increased neuronal density by cytoarchitectonic analyses, which can be interpreted as a likely reduction in length and complexity of dendritic trees, has been demonstrated in Williams syndrome (WS) and Rubinstein-Taybi syndrome (RTS) (see Table 1) [2]. WS is caused by a submicroscopic deletion on chromosome 7q11.23, which includes the *elastin* gene, the *HPC-1/syntaxin 1A* (*STX1A*) gene that codes for a protein involved in the docking of synaptic vesicles, and the gene coding for the LIM kinase-1 (LIMK1) that is linked to the Rho signaling system (see Table 3). Patients with WS show a distinctive cognitive and social phenotype; although they demonstrate relatively preserved language and face processing abilities, WS subjects are typically impaired in visual-spatial domains. In addition, they are hypersociable, with engaging personalities and excessive sociability with strangers [24]. In correspondence with their profile of strengths and weaknesses, morphometric neuroimaging studies have shown relative reductions in parietooccipital regions and enlargement of frontal temporal structures involved in emotion and face processing [25]. Neuropathological data on WS are limited; however, generalized reductions in cortical columnar organization and increases in cell packing density, as well as abnormal neuronal orientation that is more severe in posterior regions, have been reported [2]. Despite the lack of direct evidence of postsynaptic abnormalities in WS, defective LIMK1 function may lead to dendritic spine dysgenesis and enhanced LTP, as shown in *LIMK1* knockout mice [26]. LIMK1 is downstream of Rac/Cdc42 and PAK, which phosphorylates and inactivates cofilin, thus influencing actin cytoskeletal dynamics (i.e., increasing turnover) and dendritic morphology [12,26]. Interestingly, levels of phospho-cofilin levels are reduced in *FMR1* knockout mice, suggesting that WS and FXS may share dendritic spine abnormalities.

In addition to MR, RTS is characterized by selective deficits in expressive language and by maladaptive behavior. Initially described somatic/skeletal abnormalities include short stature, facial dysmorphia, and broad thumbs and first toes [2]. The genetic defect in RTS (16p13.3) has been reported to involve CREB-binding protein (CBP), a protein that is recruited by CREB to bind DNA and activates the basal transcription factor–enzyme complex [27]. Limited neuroimaging and neuropathological investigations have shown an association between RTS and several CNS malformations, such as agenesis of the corpus callosum. Nonetheless, the most comprehensive neuropathological evaluation of an adult RTS brain reported mild reduction in brain weight and overall preserved cortical architecture but decreased neuronal size and marked increase in cell packing density [2]. In correspondence to CBP's critical role in neuronal function both as a transcriptional activator and a histone acetyltransferase, mice that modeled a haploinsufficiency form of RTS show decreased chromatin acetylation and impairment in some forms of long-term memory and the late phase of hippocampal LTP [28]. Of mechanistic and therapeutic importance, the memory and LTP deficits can be ameliorated by enhancing the expression of CREB-dependent genes and by inhibiting histone deacetylase activity [28].

VI. Conclusion

There is considerable evidence that a disturbance of neuronal differentiation and, to a lesser extent, of other CNS developmental processes underlies most genetic forms of MR [2,4]. Furthermore, morphological and molecular data indicate that impaired neuronal and synaptic formation in MR represents a defect in signaling pathways involved in synaptic plasticity [29]. Because there is an economy of genetic programs responsible for neuronal adaptation to environmental changes, most key molecules involved in the modeling of synapses during development are also implicated in cognition and complex behavior in the adult brain [27,29]. This molecular "double jeopardy" in genetic MR results in an abnormally developed brain that also lacks the "tools" for responding properly or optimally to environmental stimuli. While this is an apparently negative situation, it also represents an opportunity because modifying the fundamental molecular or neurotransmitter disturbances in adulthood can lead to improvement of cognition and behavior, as demonstrated in animal models [15,28]. This means that there is a lifelong possibility of ameliorating the neurobehavioral phenotype in genetic MR, although the ultimate goal is intervening before synaptic development is complete. Considering that environmental causes of MR (e.g., lead poisoning) could also affect signaling programs disturbed in genetic MR [29], therapeutic approaches for genetic disorders may have a broader application. The postsynaptic compartment, which comprises neurotransmitter receptors (e.g., glutamate receptors), cytoskeletal elements (e.g., actin), signaling proteins (e.g., kinases), and anchoring molecules that "connect" the former constituents (e.g., Homer), seems to be particularly affected in MR. Therefore the dendritic abnormalities in patients with MR appear to be the structural correlate of less "adaptable" postsynaptic sites.

Although neurobiologically based treatments for genetic MR are emerging, several major challenges remain in the field. First, there is a need to understand how changes in dendritic morphology or postsynaptic components translate into functional (e.g., neurotransmitter) alterations in synaptic transmission. The extent to which presynaptic abnormalities are also present in MR should also be better characterized; the current emphasis on dendritic pathology is in part the consequence of methodological advantages of studying postsynaptic elements. Although knowledge of synaptic anomalies has led to novel pharmacological strategies, there is the possibility of using neurobiological information to plan more effective nonpharmacological interventions. For instance, the role of environmental stimulation in genetic MR treatment is still unclear. Neurobiologically supported pharmacological enhancement of different therapies is also a potential alternative. Despite all these issues, the future of the clinical management of MR is quite promising because it will be able to take advantage of innovations in genetics, developmental neurobiology, neuroimaging, and other relevant fields.

Acknowledgments

Dr. Walter Kaufmann is supported by NIH grants MH067092, HD24448, and HD24061 and by the FRAXA Research Foundation and the International Rett Syndrome Association. We are grateful to the families who have participated in research projects at our Center; their partnership is critical for advancing the field. We also thank Dr. Michael Johnston for his continuous support of our research and the Center.

References

1. Kaufmann, W. E. (1996). Mental retardation and learning disabilities: a neuropathologic differentiation. *In:* "Developmental Disabilities in Infancy and Childhood" (A. J. Capute, et al., eds.), vol. 2, pp. 49–70. Paul H. Brookes, Baltimore.
2. Kaufmann, W. E., and Moser, H. W. (2000). Dendritic anomalies in disorders associated with mental retardation. *Cereb. Cortex* **10**, 981–991.
3. Inlow, J. K., and Restifo, L. L. (2004). Molecular and comparative genetics of mental retardation. *Genetics* **166**, 835–881.
4. Kaufmann, W. E. (2003). Cortical histogenesis. *In:* "Encyclopedia of the Neurological Sciences" (M. J. Aminoff and R. B. Daroff, eds.), vol 1, pp. 777–784. Academic Press, San Diego.
5. Marin-Padilla, M. (1972). Structural abnormalities of the cerebral cortex in human chromosomal aberrations. A Golgi study. *Brain. Res.* **44**, 625–629.
6. Roizen, N. J., and Patterson, D. (2003). Down's syndrome. *Lancet* **361**, 1281–1289.
7. Kates, W. R., Folley, B. S., Lanham, D. C., Capone, G. T., and Kaufmann, W. E. (2002). Cerebral growth in Fragile X syndrome: review and comparison with Down syndrome. *Microsc. Res. Tech.* **57**, 159–167.
8. Benavides-Piccione, R., Ballesteros-Yanez, I., de Lagran, M. M., Elston, G., Estivill, X., Fillat, C., Defelipe, J., and Dierssen, M. (2004). On dendrites in Down syndrome and DS murine models: a spiny way to learn. *Prog. Neurobiol.* **74**, 111–126.
9. Kaufmann, W. E., MacDonald, S. M., and Altamura, C. (2000). Dendritic cytoskeletal protein expression in mental retardation: an immunohistochemical study of the neocortex in Rett syndrome. *Cereb. Cortex* **10**, 992–1004.
10. Galdzicki, Z., Siarey, R., Pearce, R., Stoll, J., and Rapoport, S. I. (2001). On the cause of mental retardation in Down syndrome: extrapolation from full and segmental trisomy 16 mouse models. *Brain Res. Rev.* **35**, 115–145.
11. Salman, M. S. (2002). Systematic review of the effect of therapeutic dietary supplements and drugs on cognitive function in subjects with Down syndrome. *Eur. J. Paediatr. Neurol.* **6**, 213–219.
12. Ramakers, G. J. A. (2002). Rho proteins, mental retardation and the cellular basis of cognition. *Trends Neurosci.* **25**, 191–199.
13. Kaufmann, W. E., and Reiss, A. L. (1999). Molecular and cellular genetics of Fragile X syndrome. *Am. J. Med. Genet.* **88**, 11–24.
14. Irwin, S. A., Galvez, R., and Greenough, W. T. (2000). Dendritic spine structural anomalies in fragile-X mental retardation syndrome. *Cereb. Cortex* **10**, 1038–1044.
15. Bear, M. F., Huber, K. M., and Warren, S. T. (2004). The mGluR theory of fragile X mental retardation. *Trends Neurosci.* **27**, 370–377.
16. Bagni, C., and Greenough, W. T. (2005). From mRNP trafficking to spine dysmorphogenesis: the roots of fragile X syndrome. *Nat. Rev. Neurosci.* **6**, 376–387.
17. Jin, P., Alisch, R. S., and Warren, S. T. (2004). RNA and microRNAs in fragile X mental retardation. *Nat. Cell Biol.* **6**, 1048–1053.
18. Hagberg, B. (2002). Clinical manifestations and stages of Rett syndrome. *Ment. Retard. Dev. Disabil. Res. Rev.* **8**, 61–65.
19. Akbarian, S. (2003). The neurobiology of Rett syndrome. *Neuroscientist* **9**, 57–63.
20. Kaufmann, W. E., Johnston, M. V., and Blue, M. E. (2005). MeCP2 expression and function during brain development: implications for Rett syndrome's pathogenesis and clinical evolution. *Brain Dev.* **27**, S77–S87.
21. Naidu, S., Kaufmann, W. E., Abrams, M. T., Pearlson, G. D., Lanham, D. C., Fredericksen, K. A., Barker, P. B., Horska, A., Golay, X., Mori, S., Wong, D. F., Yablonski, M., Moser, H. W., and Johnston, M. V. (2001). Neuroimaging studies in Rett syndrome. *Brain Dev.* **23**, S62–S71.
22. Mullaney, B. C., Johnston, M. V., and Blue, M. E. (2004). Developmental expression of methyl-CpG binding protein 2 is dynamically regulated in the rodent brain. *Neuroscience* **123**, 939–949.
23. Percy, A. K. (2002). Clinical trials and treatment prospects. *Ment. Retard. Dev. Disabil. Res. Rev.* **8**, 106–111.
24. Bellugi, U., Lichtenberger, L., Mills, D., Galaburda, A., and Korenberg, J. R. (1999). Bridging cognition, the brain and molecular genetics: evidence from Williams syndrome. *Trends Neurosci.* **22**, 197–207.
25. Reiss, A. L., Eckert, M. A., Rose, F. E., Karchemskiy, A., Kesler, S., Chang, M., Reynolds, M. F., Kwon, H., and Galaburda, A. (2004). An experiment of nature: brain anatomy parallels cognition and behavior in Williams syndrome. *J. Neurosci.* **24**, 5009–5015.
26. Meng, Y., Zhang, Y., Tregoubov, V., Janus, C., Cruz, L., Jackson, M., Lu, W. Y., MacDonald, J. F., Wang, J. Y., Falls, D. L., and Jia, Z. (2002). Abnormal spine morphology and enhanced LTP in LIMK-1 knockout mice. *Neuron* **35**, 121–133.
27. Kaufmann, W. E., and Worley, P. F. (1999). The role of early neural activity in regulating immediate early gene expression in the cerebral cortex. *Ment. Retard. Dev. Disabil. Res. Rev.* **5**, 41–50.
28. Alarcon, J. M., Malleret, G., Touzani, K., Vronskaya, S., Ishii, S., Kandel, E. R., and Barco, A. (2004). Chromatin acetylation, memory, and LTP are impaired in CBP+/- mice: a model for the cognitive deficit in Rubinstein-Taybi syndrome and its amelioration. *Neuron* **42**, 947–959.
29. Johnston, M. V. (2004). Clinical disorders of brain plasticity. *Brain Dev.* **26**, 73–80.

51

Cerebral Palsy

Alexander H. Hoon, Jr., MD, MPH

Keywords: *cerebral palsy, diffusion tensor imaging, magnetic resonance imaging, neuroimaging, periventricular leukomalacia*

I. Brief History and Nomenclature
II. Etiology
III. Pathogenesis
IV. Pathophysiology
V. Relevant Structural Detail
VI. Pharmacology, Biochemistry, and Molecular Mechanisms
VII. Symptoms in Relation to Pathophysiology
VIII. Natural History
References

I. Brief History and Nomenclature

Cerebral palsy (CP) describes a group of motor-impairment syndromes secondary to a wide range of genetic and acquired disorders of early brain development. This definition allows inclusion of new genetic, metabolic, and other causes of CP that may be discovered in the future and serves as an ongoing reminder to clinicians to remain vigilant to the diagnostic challenges in this group of children. Correct diagnosis has important ramifications for treatment, prognosis, and recurrence risk.

Although there is some variability in etiological antecedents among developing and developed countries, prevalence is approximately 2:1000 children in most areas, making it the most common chronic neurological motor disorder of childhood [1]. Although comprehensive longitudinal studies are limited, the majority of children grow into adulthood, actively participating in societal life but facing specific, recognized challenges.

Historically there has been ongoing debate in regard to timing and etiology, in terms of both definitions and treatment. The causal factors leading to CP have been debated over the past 150 years, ever since Sir William Little linked CP to perinatal events in the 1860s. In 1897 Freud offered a differing view, stating, "Difficult birth in certain cases is merely a symptom of deeper effects that influence the development of the fetus" [2]. Currently, CP is most commonly linked to prenatal factors in full-term infants and to a combination of prenatal and perinatal factors in preterm infants [3]. Finally, although there have been major improvements in obstetrical and neonatal care in developed nations, which are associated with decreases in perinatal mortality, there have been no significant overall changes in the prevalence of CP.

Despite a broad range of widely applied and accepted medical and rehabilitative interventions for children with CP, there is often variability in use and inconsistency in outcome. For example, children born prematurely with spastic diplegia and good antigravity strength are readily offered selective dorsal rhizotomy in some centers and orthopedic surgery in others. This variability in treatment interventions, resulting from differing etiologies and classifications, lack of quantitative measurement tools, and insufficient controlled trials, is currently being addressed [4]. The World Health Organization's publication "The International Classification of Functioning, Disability and Health" describes health for all people including those with cerebral palsy. It is intended to assist in the determination of measurement tools, appropriate goals and decision making, and meaningful outcomes [5].

The nomenclature used to describe children with CP has neurological, topographical, and functional components (Table 1) [6]. On the basis of neurological examination, children may be classified into spastic, extrapyramidal (dystonic, athetotic, choreic, rigid, hypotonic, and ataxic subtypes), and mixed phenotype categories. The topographical terminology of leg and arm involvement applied to spastic CP includes diplegia (primarily lower extremities), quadriplegia (significant four-limb involvement) and hemiplegia (arm dominant or leg dominant). Children with extrapyramidal phenotypes commonly have significant limitations in both arm and leg function. Functional classification is characterized by degree of mobility, from wheelchair use to community mobility (Table 2).

Positive language empowers people with disabilities. In written and spoken use, the term *child with cerebral palsy* (rather than "cerebral-palsied child") promotes self-esteem and societal acceptance. Although a variety of other phrases including "physically challenged," "differently abled," "child with motor encephalopathy," and "child with static encephalopathy" may be used, *child with cerebral palsy* is a preferred term because it emphasizes the person first and reflects respect for the individuality and dignity of the child, adolescent, or adult.

Table 1 Definition of Cerebral Palsy

Cerebral palsy (CP) describes a group of disorders of the development of movement and posture, causing activity limitation, that are attributed to nonprogressive disturbances that occurred in the developing fetal or infant brain. The motor disorders of cerebral palsy are often accompanied by disturbances of sensation, cognition, communication, perception, and/or behavior, and/or by a seizure disorder.

From Bax, M., Goldstein, M., Rosenbaum, P., Leviton, A., Paneth, N., Dan, B., Jacobsson, B., and Damiano, D.; Executive Committee for the Definition of Cerebral Palsy. (2005). Proposed definition and classification of cerebral palsy. *Dev. Med. Child. Neurol.* **47**, 571–576.

Table 2 Components of Cerebral Palsy Classification

1. **Motor abnormalities**
   A. *Nature and typology of the motor disorder:* the observed tonal abnormalities assessed on examination (e.g., hypertonia or hypotonia) as well as the diagnosed movement disorders present, such as spasticity, ataxia, dystonia, or athetosis.
   B. *Functional motor abilities:* the extent to which the individual is limited in his or her motor function in all body areas, including oromotor and speech function.

2. **Associated impairments**
   The presence or absence of associated non-motor neurodevelopmental or sensory problems, such as seizures, hearing, or vision impairments, or attentional, behavioural, communicative, and/or cognitive deficits, and the extent to which impairments interact in individuals with cerebral palsy.

3. **Anatomic and radiological findings**
   A. *Anatomic distribution:* the parts of the body (e.g., limbs, trunk, or bulbar region) affected by motor impairments or limitations.
   B. *Radiological findings:* the neuroanatomical findings on computed tomography or magnetic resonance imaging, such as ventricular enlargement, white matter loss, or brain anomaly.

4. **Causation and timing**
   Whether there is a clearly identified cause, as is usually the case with postnatal cerebral palsy (e.g., meningitis or head injury) or when brain malformations are present, and the presumed timeframe during which the injury occurred, if known.

From Bax, M., Goldstein, M., Rosenbaum, P., Leviton, A., Paneth, N., Dan, B., Jacobsson, B., and Damiano, D.; Executive Committee for the Definition of Cerebral Palsy. (2005). Proposed definition and classification of cerebral palsy. *Dev. Med. Child. Neurol.* **47**, 571–576.

II. Etiology

Although an etiology can be established for some children with CP, such as LIS1 mutations leading to a lissencephalic cortex and quadriplegia, precise causal links are more difficult to establish for common causes such as periventricular leukomalacia (PVL). When considered in the context of the wide range of genetic and acquired (prenatal, perinatal, and postnatal) risk factors and disorders that lead to or are associated with CP, etiology can be considered from these two interrelated perspectives: 1) recognized disorders that lead to CP; and 2) risk factors that may operate in isolation or in combination to produce cerebral palsy phenotypes (Table 3).

Brain magnetic resonance imaging (MRI) has been of great benefit in the determination of cause and shows evidence of brain dysgenesis or injury in 70%–90% of affected children. Based on an understanding of normal brain development, careful interpretation shows patterns of selective vulnerability in brain structures, characteristic of the nature of the insult, gestational timing, and severity [7,8]. In specific circumstances, other imaging techniques may be of diagnostic benefit. For example, cranial ultrasound—an inexpensive, noninvasive, repeatable modality—has utility both in diagnosis and management of the high-risk neonate. MR spectroscopy and diffusion-weighted and diffusion-tensor imaging are of benefit in appropriate clinical settings.

Using neuroimaging as the primary diagnostic tool, children with CP can be broadly grouped into 4 etiological groups: 1) disorders of early brain formation; 2) injury associated with prematurity (PVL); 3) neonatal encephalopathies; and 4) a heterogeneous group of postnatal disorders [9]. Although regional epidemiological data vary, approximately 30% are associated with brain malformations, 40% with prematurity, 20% with neonatal encephalopathies, and 10% with postnatal causes.

The etiology of developmental brain malformations is related to both recognized and suspected genetic disorders and to acquired disorders such as in utero infection with cytomegalovirus [10]. Genetic factors range from chromosomal aneuploidy to abnormalities in genetic imprinting. For example, lissencephaly is related to at least five recognized gene defects, including LIS1, 14-3-3epsilon, DCX, RELN, and ARX [11].

In children born preterm with CP, recognized risk factors include hypoxemia-ischemia and indirect infection and direct infection, with the primary area of vulnerability in deep white matter [12]. Although early studies focused on intraventricular or intraparenchymal hemorrhage (Grade III–IV IVH), with advances in perinatal and neonatal care, the incidences of IVH as well as cystic periventricular leukomalacia (cPVL) have decreased significantly. Currently, the most common form of white matter injury is termed *non-cystic PVL* [13].

A wide range of disorders and risk factors, many beginning antenatally, lead to neonatal encephalopathy (seizures, decreases in state, tone, oromotor function, and respiratory drive in the term or near-term infant) and often progress to various CP phenotypes. These include fetal and maternal infections, placental disorders, maternal thyroid disorders, nutritional deficiencies, brain malformations, genetic disorders such as Prader Willi syndrome, and myotonic dystrophy as well as a family history of epilepsy [14].

As the susceptibility of white matter to hypoxic and infectious insults decreases toward term, cortical and deep gray neurons become more vulnerable to injury associated with hypoxic ischemic encephalopathy (HIE) and increased serum bilirubin (kernicterus). "Acute, total" asphyxial events often lead to injury in the putamen and ventrolateral nucleus of the thalamus; whereas "partial, prolonged" hypoxic-ischemic insults may result in multicystic encephalomalacia. The globus pallidus and subthalamic nucleus are vulnerable to bilirubin toxicity [15].

Table 3 Risk Factors for Cerebral Palsy

Preexisting factors
Autoimmune and coagulation disorders
History of infertility
Intellectual impairment
Long menstrual cycles
Maternal age (<20 or >35 years of age)
Maternal epilepsy
Maternal diabetes
Maternal thyroid abnormalities
Previous pregnancy loss or neonatal death
Prior child with cerebral palsy
Short or long interpregnancy interval

Pregnancy and fetal factors
Abnormal fetal presentation
Abruption
Congenital infection
Congenital malformation
Genetic disorders
Infertility treatment and in vitro fertilization
In utero growth restriction
Male gender
Multiple gestation
Perinatal asphyxia
Perinatal infection
Placental abnormalities
Preclampsia
Prematurity or low birth weight
Prolonged rupture of membranes

Neonatal factors, morbidities, and interventions
Hyperbilirubinemia
Perinatal stroke
Pneumothorax
Postnatal steroids
Prolonged mechanical ventilation

Postnatal childhood factors
Accidental and nonaccidental trauma
Central nervous system infections
Genetic/metabolic disease

Adapted from Jacobsson, B., and Hagberg, G. (2004). Antenatal risk factors for cerebral palsy. *Best. Pract. Res. Clin. Obstet. Gynaecol.* **18**, 425–436; and from Stashinko E. E., and Kammann, H. (1996). Epidemiology. *In:* "Developmental Disabilities in Infancy and Childhood" (P. Accardo, ed.), Paul H. Brookes Publishing Co., Baltimore.

In those children with normal prenatal and perinatal courses who develop CP, there are a large number of associated genetic and acquired disorders. On MRI, characteristic findings in the basal ganglia may be seen in a number of genetic disorders, including mitochondrial encephalopathies, pantothenate kinase–associated neurodegeneration, methylmalonic acidemia, juvenile Huntington disease, and glutaric aciduria type 1. Although immunizations have reduced the incidence of some types of bacterial meningitis leading to CP, other bacterial forms of meningitis and encephalitis, as well as central nervous system (CNS) involvement in human immunodeficiency virus (HIV), are important causes. Accidental and nonaccidental trauma are also associated with CP.

Children with hemiplegic CP have distinctive etiological antecedents and clinical profiles [16]. Most originate in the prenatal period, with smaller numbers associated with neonatal cerebral infarction and stroke during early childhood. Prenatal etiologies include malformations such as schizencephaly, unilateral polymicrogyria, or hemimegalencephaly; injury to periventricular white matter; and cortical-subcortical strokes. Although the dysgenesis or injury can be described using neuroimaging, pathogenetic mechanisms are frequently unknown and may be the result of genetic and/or environmental factors.

In children with CP, etiological understanding promotes treatment, prognosis, and understanding of recurrence risk. A small but growing number of genetic disorders, including dopa-responsive dystonia, glutaric aciduria type 1, and methylmalonic acidemia, may significantly improve or be ameliorated with appropriate medical therapy. Furthermore, a specific diagnosis can reduce or eliminate feelings of parental guilt that may be present.

III. Pathogenesis

As *cerebral palsy* refers to the motor impairment secondary to a wide range of genetic and acquired disorders, pathogenesis is as varied as etiology. For some genetic etiologies such as lissencephaly, pathogenesis is fairly well established. In others such as holoprosencephaly, risk factors are being investigated [17]. Furthermore, there can be varied clinical expression with specific gene mutation, determined partially by the type of gene abnormality. For example, within the homeobox gene ARX (aristaless-related homeobox), clinical manifestations can range from autism to dystonia, and brain malformations from lissencephaly to midbrain anomalies based on whether the mutations are truncating or insertion/missense; manifestations are further modulated on the basis of X chromosome inactivation [18].

An important mechanism of injury in both preterm and term infants is excitotoxicity, the excessive stimulation of excitatory synapses, leading to downstream cascades of intracellular events and resulting in cell death by apoptosis, necrosis, or both [19]. Glutamate is the most common excitatory neurotransmitter in the human brain and the one most commonly implicated in excitotoxic brain injury.

In preterm infants, PVL has three characteristics: 1) it is closely related to selective vulnerability of preoligodendrocytes; 2) it occurs in areas of vascular end zones or border zones; and 3) activated microglia are present, supporting the concept that ischemia/reperfusion plays an important role in pathogenesis [12]. Immature oligodendroglia express AMPA receptors, which may provide a link between neuronal activity and myelination. Excessive glutamate release may result in oligodendroglial cell death. Furthermore, glutamate release may result in microglia activation, with subsequent release of soluble factors including cytokines, nitric oxide, and free radicals, also leading to cell death.

In term infants with HIE and excitotoxic brain injury, encephalopathy occurs after a latent period of 6–36 hours and correlates with the levels of glutamate and aspartate in cerebrospinal fluid (CSF). Release of these excitatory neurotransmitters leads to calcium influx into neurons and triggers cell death by several mechanisms, including mitochondrial dysfunction and activation of cell death pathways via caspases and PARP-1 (poly ADP-ribose polymerase-1). MRI and positron emission tomography (PET) studies confirm activation and injury in parts of the brain with active glutamatergic circuits.

IV. Pathophysiology

A key concept in pathophysiology is that differing insults can lead to similar CP phenotypes. For example, spastic diplegia is most commonly associated with PVL but may also occur secondary to brain malformations, hydrocephalus, HIV infection, dopa-responsive dystonia, or arginase deficiency. How these differing etiological insults lead to similar clinical presentations is under active investigation.

V. Relevant Structural Detail

Neuroimaging techniques, most notably MRI, now permit exquisite in vivo identification of structural abnormalities in 70%–90% of children with CP. For example, in children with the dystonic form of extrapyramidal CP, MRI often shows hyperintense signal, atrophy, or both in various deep gray structures. A number of genetic disorders, including mitochondrial encephalopathies and pantothenate kinase–associated neurodegeneration (PKAN), target the globus pallidus, whereas hypoxia injures the putamen and ventrolateral nucleus of the thalamus, resulting in dystonia.

However, the correlation between patterns of injury and clinical phenotype does not always hold, as kernicterus also targets the globus pallidus but leads to chorea.

A second useful example is PVL, which was felt to result primarily from injury to descending motor pathways. Neuropathological studies have shown lesions in corticospinal, thalamocortical, optic radiation, superior occipitofrontal, and superior longitudinal pathways. In vivo identification of injury in specific white matter pathways is below the resolution of conventional MR techniques. However, diffusion tensor imaging has demonstrated injury to both sensory and motor pathways in children with PVL, leading to the hypothesis that in some children, CP is a disorder of sensorimotor function [20].

VI. Pharmacology, Biochemistry, and Molecular Mechanisms

Neurological signs may be amenable to pharmacological intervention based on neuroanatomical location of injury. Antispasticity medications, including benzodiazepines, baclofen, tizanidine, and dantrolene, may be used to reduce spastic hypertonicity in selected children. Beneficial effects on tone reduction must be weighed against unwanted cognitive side effects. Although these medications reduce tone, currently there are no double-masked, randomized controlled trials showing that these medications improve motor capabilities.

In syndromes with athetosis or dystonia, up-regulating dopamine with levodopa/carbidopa or down-regulating acetylcholine with trihexyphenidyl may have beneficial effects on function. In children with chorea or hemiballismus, down-regulating dopamine (reserpine, tetrabenazine) or increasing GABA flux (benzodiazepines, anticonvulsants) may be beneficial.

For children with CP secondary to genetic-metabolic disorders such as glutaric aciduria type 1 or methylmalonic academia in which the biochemistry is understood, early identification, prevention of metabolic stress, and mitochondrial co-factor therapy may prevent further injury. For those with primary mitochondrial disorders, a similar approach is recommended.

VII. Symptoms in Relation to Pathophysiology

CP is a clinical diagnosis made on the basis of significant delay in gross and/or fine motor function, with characteristic findings of tone, posture, and movement on examination. Although the neurological abnormalities in CP include loss of selective motor control, agonist/antagonist muscle imbalance, impaired balance/coordination, and sensory deficits, neurological examination is often the basis for diagnosis, classification, and medical treatment.

Children may have predominantly spastic, rigid, dystonic, or mixed hypertonicity. Spastic CP syndromes include diplegia, quadriplegia, and hemiplegia and are most commonly associated with injury to descending white matter pathways, as seen in PVL. In contrast, rigid hypertonicity is bidirectional and is often associated with diffuse brain injury on MRI. Dystonic hypertonicity, characterized by co-contraction of agonist-antagonist muscles and associated with twisting and repetitive movements, usually occurs during voluntary movement or with voluntary maintenance of a body posture. It is often associated with disorders of the basal ganglia and thalamus. Ataxic CP syndromes, a heterogeneous group of individually rare disorders that are often genetically mediated, frequently have associated cerebellar abnormalities on imaging.

VIII. Natural History

The variability in motor phenotypes, associated cognitive and neurobehavioral impairments, family structure and support, and personal characteristics of the affected child make determination of natural history difficult. Currently, most children with CP survive into adulthood, although the death rate is higher than in unaffected children. Mortality is closely related to the degrees of intellectual impairment and motor disability and is often associated with respiratory problems. Co-morbid problems in adulthood include pain, orthopedic deformities, bowel/bladder dysfunction, malnutrition, seizures, reflux and esophagitis, and osteoporosis. Mental health problems are common.

Natural history can be positively affected by improvements in delivering clinical services, enhancing self-management, improving decision support for providers, developing clinical information systems, and enhancing the organization of healthcare delivery. Children with CP develop to their full potential when treatment programs optimize motor capabilities, minimize orthopedic deformities, and address associated impairments.

References

1. Costeff, H. (2004). Estimated frequency of genetic and nongenetic causes of congenital idiopathic cerebral palsy in west Sweden. *Ann. Hum. Genet.* **68**, 515–520.
2. Freud, S. (1968). "Infantile Cerebral Paralysis." University of Miami Press, Coral Gables, FL.
3. Kuban, K. C. K., and Leviton, A. (1994). Cerebral palsy. *New Engl. J. Med.* **330**, 188–195.
4. Butler, C., and Campbell, S. (2000). Evidence of the effects of intrathecal baclofen for spastic and dystonic cerebral palsy (Annotation). *Dev. Med. Child. Neurol.* **42**, 634–645.

5. World Health Organization. (2001). International classification of functioning, disability and health. World Health Organization, Geneva.
6. Bax, M., Goldstein, M., Rosenbaum, P., Leviton, A., Paneth, N., Dan, B., Jacobsson, B., and Damiano, D.; Executive Committee for the Definition of Cerebral Palsy. (2005). Proposed definition and classification of cerebral palsy, *Dev. Med. Child. Neurol.* **47**, 571–576.
7. Jacobsson, B., and Hagberg, G. (2004). Antenatal risk factors for cerebral palsy. *Best. Pract. Res. Clin. Obstet. Gynaecol.* **18**, 425–436.
8. Hoon, A. H. Jr., Reinhardt, E. M., Kelley, R. I., Breiter, S. N., Morton, D. H., Naidu, S. B., and Johnston, M. V. (1997). Brain magnetic resonance imaging in suspected extrapyramidal cerebral palsy: observations in distinguishing genetic-metabolic from acquired causes. *J. Pediatr.* **131**, 240–245.
9. Hoon, A. H., Belsito, K. M., and Nagae-Poetscher, L. M. (2003). Neuroimaging in spasticity and movement disorders. *J. Child. Neurol.* **18**, S25–S39.
10. van der Knaap, M. S., Vermeulen, G., Barkhof, F., Hart, A. A., Loeber, J. G., and Weel, J. F. (2004). Pattern of white matter abnormalities at MR imaging: use of polymerase chain reaction testing of Guthrie cards to link pattern with congenital cytomegalovirus infection. *Radiology* **230**, 529–536.
11. Kato, M., and Dobyns, W. B. (2003). Lissencephaly and the molecular basis of neuronal migration. *Hum. Mol. Genet.* **12**, R89–R96.
12. Volpe, J. J. (2003). Cerebral white matter injury of the premature infant–more common than you think. *Pediatrics* **112**, 176–180.
13. Wilson-Costello, D., Friedman, H., Minich, N., Fanaroff, A. A., and Hack, M. (2005). Improved survival rates with increased neurodevelopmental disability for extremely low birth weight infants in the 1990's, *Pediatrics* **115**, 997–1003.
14. Johnston, M. V. (2003). MRI for neonatal encephalopathy in full term infants, *Lancet* **361**, 713–714.
15. Johnston, M. V., Hoon, A. H. (2000). Possible mechanisms in infants for selective basal ganglia damage from asphyxia, kernicterus, or mitochondrial encephalopathies, *J. Child. Neurol.* **15**, 588–591.
16. Takanashi, J., Barkovich, A. J., and Ferriero, D. M. (2003). Widening spectrum of congenital hemiplegia. *Neurology* **61**, 531–533.
17. Stashinko, E. E., Clegg, N. J., Kammann, H. A., Sweet, V. T., Delgado, M. R., Hahn, J. S., and Levey, E. B. (2004). A retrospective survey of perinatal risk factors of 104 living children with holoprosencephaly. *Am. J. Med. Gen.* **128**, 114–119.
18. Sherr, E. (2003). The ARX study (epilepsy, mental retardation, autism and cerebral malformations): one gene leads to many phenotypes. *Curr. Opin. Pediatr.* **15**, 567–571.
19. Johnston, M. V. (2005). Excitotoxicity in perinatal brain injury. *Brain Pathol.* **15**, 234–240.
20. Hoon, A. H., Lawrie, W. T. Jr., Melhem, E. R., Reinhardt, E. M., Van Zijl, P. C., Solaiyappan, M., Jiang, H., Johnston, M. V., and Mori, S. (2002). Diffusion tensor imaging of periventricular leukomalacia shows affected sensory cortex white matter pathways. *Neurology* **59**, 752–756.

52

Autism

Martha R. Herbert, MD, PhD

Keywords: *autism, excitation:inhibition ratio, macrocephaly, neurobehavioral disorder, neuroinflammation, social-emotional impairment, underconnectivity*

I. Brief History and Nomenclature
II. Etiology
III. Pathogenesis
IV. Pathophysiology
V. Brain Structure
VI. Pharmacology, Biochemistry, and Molecular Mechanisms
VII. Explanation of Symptoms in Relation to Pathophysiology
VIII. Natural History
    References

Although autism is a neurobehavioral syndrome defined exclusively by behavioral criteria, abnormalities have been identified at many other levels in individuals and cohorts who meet these criteria, including genetics, neurochemistry and metabolism, immunology, neuroanatomy, and electrophysiology, and in multiple domains including sensory-motor and social-emotional functioning as well as perception and attention. However, no findings to date have been consistent, predictive, or diagnostic. It appears instead that the population of individuals who meet rigorous behavioral criteria for autism may be quite heterogeneous at every level, even including aspects of behavioral manifestations. Given the specificity with which the behavioral features of the syndrome can be defined, such heterogeneity poses a challenge: How do the different biological underpinnings converge on a consistent behavioral syndrome? This question cannot yet be answered, but posing it provides a useful lens through which to view the range of scientific findings in autism.

I. Brief History and Nomenclature

Autism was first described by Leo Kanner in 1943 in a group of 11 children. Kanner noted that children tended to be aloof and to have odd behavior but a good rote memory and skill with objects; he also noted oversensitivity to stimuli and either mutism or language that did not have communicative intent and observed that some of the children had large heads. The term *autism* had been coined early in the 20th century by Bleuler, but Kanner's was the first clinical description. Within 1 year Hans Asperger published a report of a study of four children with "autistic psychopathy," although this paper did not receive attention in the English-speaking world until nearly 4 decades later. In earlier years, autism was not clearly distinguished from schizophrenia and psychosis. Over several decades the definition of autism has evolved; from the initial notion of "early infantile autism" there developed a series of modifications, including Michael Rutter's 1978 distinction between autism, infantile autism, and childhood autism, and the distinction between autistic disorder and pervasive developmental disorder (PDD) not otherwise specified in the *Diagnostic and Statistical Manual of Mental Disorders (DSM)-III-R* in 1987. Today, autism is defined according to DSM-IV criteria (See Table I) [1], and the current autism classification system recognizes five disorders under PDD, including Rett syndrome, childhood disintegrative disorder (CDD), autistic disorder, Asperger's syndrome, and pervasive developmental disorder not otherwise specified (PDD-NOS). Rett syndrome and CDD involve progressive deterioration not seen in the other three disorders, and although the origin of CDD remains obscure, the MECP2 gene has been associated with Rett syndrome. Although many mutations are involved, some individuals with mutations do not develop the disorder, and mutations have been found in mentally retarded males and in children with autism who do not have Rett syndrome.

II. Etiology

Some of the earliest autism studies identified frequent patterns of environmental and occupational exposure in the small numbers of family members of autistic children whom they discussed, but this theme did not persist in the literature. For several decades, psychogenic theories of autism were commonly held, but these have been totally discredited and autism is now considered to be a biologically based disorder. The publication of a twin study by Folstein and Rutter in 1977, which identified a substantially higher monozygotic than dizygotic concordance, inaugurated a search for the genetic roots of autism. Subsequent twin studies, which have varied somewhat in their methodologies, have found monozygotic concordance to be 36–91%, whereas dizygotic concordance has been in the range of 0–10%. Monozygotic concordance is higher (around 90%) if only one sibling needs to meet full criteria for autism (with the other being in the "broad autistic spectrum"); if both members of the twin pair are required to meet full criteria for autism, the monozygotic concordance is about 60% [2]. Linkage findings on chromosomes 7q, 2q, and 15q have been most consistently replicated, whereas for the most part associations from candidate gene studies have not been replicated. Particularly promising candidate genes have included GABA receptor and serotonin transporter genes, as well as Engrailed 2, Neuroligin, MeCP2, WNT2, and BDNF. Suggestive immunogenetic findings include abnormalities in the major histocompatibility complex, with a particularly strong association for the null allele of C4B in the class III region [3].

The role of environment and epigenetic mechanisms in the etiology of autism arises for a number of reasons. Twin studies show less than 100% concordance between monozygotic twins, whereas the dizygotic concordance in some studies is higher than might be expected for a purely genetic disorder. Rates of autism being reported have gone up from less than 3 per 10,000 in the 1970s and 1980s to 30–60 per 10,000 more recently; these increases may be due in part to increased awareness and altered diagnostic categories; however, the reported rates have continued to increase even in more recent years while the diagnostic categories have remained stable and while there has not been a concomitant decrease in the rates of other developmental disorders with which autistic children could have been diagnosed in earlier years when there was less autism awareness [4]. Regarding candidate environmental factors, various chemical, biological, and infectious agents have been considered, but no single agent has emerged as either a clear or unique trigger or cause [5]. Given the likely heterogeneity of relevant environmental influences, large populations will be needed for these studies, and there are considerable methodological challenges in the study of emergent synergistic effects of interactions among multiple xenobiotics to which individuals are co-exposed in low but chronic doses. In addition to xenobiotic chemicals, a variety of congenital viral infections have been implicated in autism, including rubella, herpes simplex, enteroviruses, and human immunodeficiency virus (HIV), as well as syphilis [6]. There may be a significant role for interindividual variability in genetic vulnerability to environmental factors. The systematic study of environmental influences in autism is just beginning, and two large research projects sponsored by the National Institute of Environmental Health Sciences are underway to pursue this at various levels [5].

III. Pathogenesis

Pathogenic mechanisms of autism have not been identified; they are likely to be heterogeneous and some may be nonspecific. Broadly speaking, autism can be divided into two categories: symptomatic and idiopathic. Autism

Table 1 DSM-IV Criteria for Pervasive Developmental Disorders

299.00 Autistic Disorder
A. A total of six (or more) items from (1), (2), and (3), with at least two from (1), and one each from (2) and (3):
  (1) qualitative impairment in social interaction, as manifested by at least two of the following:
    (a) marked impairment in the use of multiple nonverbal behaviors, such as eye-to-eye gaze, facial expression, body postures, and gestures to regulate social interaction
    (b) failure to develop peer relationships appropriate to developmental level
    (c) a lack of spontaneous seeking to share enjoyment, interests, or achievements with other people (e.g., by a lack of showing, bringing, or pointing out objects of interest)
    (d) lack of social or emotional reciprocity
  (2) qualitative impairments in communication, as manifested by at least one of the following:
    (a) delay in, or total lack of, the development of spoken language (not accompanied by an attempt to compensate through alternative modes of communication such as gesture or mime)
    (b) in individuals with adequate speech, marked impairment in the ability to initiate or sustain a conversation with others
    (c) stereotyped and repetitive use of language or idiosyncratic language
    (d) lack of varied, spontaneous make-believe play or social imitative play appropriate to developmental level
  (3) restricted, repetitive, and stereotyped patterns of behavior, interests, and activities as manifested by at least one of the following:
    (a) encompassing preoccupation with one or more stereotyped and restricted patterns of interest that is abnormal either in intensity or focus
    (b) apparently inflexible adherence to specific, nonfunctional routines or rituals
    (c) stereotyped and repetitive motor mannerisms (e.g., hand or finger flapping or twisting or complex whole-body movements)
    (d) persistent preoccupation with parts of objects
B. Delays or abnormal functioning in at least one of the following areas, with onset before age 3 years: (1) social interaction, (2) language as used in social communication, or (3) symbolic or imaginative play.
C. The disturbance is not better accounted for by Rett's disorder or childhood disintegrative disorder.

299.80 Pervasive Developmental Disorder, Not Otherwise Specified
This category should be used when there is a severe and pervasive impairment in the development of reciprocal social interaction or verbal and nonverbal communication skills, or when stereotyped behavior, interests, and activities are present, but the criteria are not met for a specific pervasive developmental disorder, schizophrenia, schizotypal personality disorder, or avoidant personality disorder. For example, this category includes "atypical autism"– presentations that do not meet the criteria for autistic disorder because of late age of onset, atypical symptomatology, or subthreshold symptomatology, or all of these.

299.80 Asperger's Disorder
A. Qualitative impairment in social interaction, as manifested by at least two of the following:
  (1) marked impairment in the use of multiple nonverbal behaviors, such as eye-to-eye gaze, facial expression, body postures, and gestures to regulate social interaction
  (2) failure to develop peer relationships appropriate to developmental level
  (3) a lack of spontaneous seeking to share enjoyment, interests, or achievements with other people (e.g., by a lack of showing, bringing, or pointing out objects of interest to other people)
  (4) lack of social or emotional reciprocity
B. Restricted, repetitive, and stereotyped patterns of behavior, interests, and activities, as manifested by at least one of the following:
  (1) encompassing preoccupation with one or more stereotyped and restricted patterns of interest that is abnormal either in intensity or focus
  (2) apparently inflexible adherence to specific, nonfunctional routines or rituals
  (3) stereotyped and repetitive motor mannerisms (e.g., hand or finger flapping or twisting, or complex whole-body movements)
  (4) persistent preoccupation with parts of objects
C. The disturbance causes clinically significant impairment in social, occupational, or other important areas of functioning.
D. There is no clinically significant general delay in language (e.g., single words used by age 2 years, communicative phrases used by age 3 years).
E. There is no clinically significant delay in cognitive development or in the development of age-appropriate self-help skills, adaptive behavior (other than in social interaction), and curiosity about the environment in childhood.
F. Criteria are not met for another specific pervasive developmental disorder or schizophrenia.

299.80 Rett's Disorder
A. All of the following:
  (1) apparently normal prenatal and perinatal development
  (2) apparently normal psychomotor development through the first 5 months after birth
  (3) normal head circumference at birth
B. Onset of all of the following after the period of normal development:
  (1) deceleration of head growth between ages 5 and 48 months
  (2) loss of previously acquired purposeful hand skills between ages 5 and 30 months with the subsequent development of stereotyped hand movements (i.e., hand-wringing or hand washing)
  (3) loss of social engagement early in the course (although often social interaction develops later)
  (4) appearance of poorly coordinated gait or trunk movements
  (5) severely impaired expressive and receptive language development with severe psychomotor retardation

(Continued)

Table 1 (Continued)

299.10 Childhood Disintegrative Disorder

A. Apparently normal development for at least the first 2 years after birth as manifested by the presence of age-appropriate verbal and nonverbal communication, social relationships, play, and adaptive behavior.
B. Clinically significant loss of previously acquired skills (before age 10 years) in at least two of the following areas:
   (1) expressive or receptive language
   (2) social skills or adaptive behavior
   (3) bowel or bladder control
   (4) play
   (5) motor skills
C. Abnormalities of functioning in at least two of the following areas:
   (1) qualitative impairment in social interaction (e.g., impairment in nonverbal behaviors, failure to develop peer relationships, lack of social or emotional reciprocity)
   (2) qualitative impairments in communication (e.g., delay or lack of spoken language, inability to initiate or sustain a conversation, stereotyped and repetitive use of language, lack of varied make-believe play)
   (3) restricted, repetitive, and stereotyped patterns of behavior, interests, and activities, including motor stereotypies and mannerisms
D. The disturbance is not better accounted for by another specific pervasive developmental disorder or by schizophrenia.

is described as "symptomatic" if the individual is diagnosed with a condition known to be associated with an incidence of autism greater than the population background rate. "Symptomatic" autism might include a history of in utero infection, for example. A subset of "symptomatic" is "syndromic," which includes such genetically based conditions or diseases as tuberous sclerosis, fragile X syndrome, tuberous sclerosis, Angelman syndrome, duplication of 15q11-q13, Down syndrome, San Filippo syndrome, MECP2-related disorders, phenylketonuria, Smith-Magenis syndrome, Smith-Lemli-Opitz syndrome, 22q13 deletion, and adenylosuccinate lyase deficiency [7]. Within the domain of symptomatic autism are theories about pathogenic mechanisms related to some of the disorders associated with autism, although it is not known why the syndrome of autistic behaviors only occurs in some but not in all such affected individuals. In tuberous sclerosis, autism has not been causally linked to any one factor, whether electroencephalogram (EEG) abnormalities, infantile spasms, cognitive impairment, or location of tubers; and although there is an idea that the autism is some effect of the underlying genetic abnormality itself, this has not been established [8]. In fragile X syndrome, dendritic spine abnormalities and abnormal activation of metabotropic glutamate receptors (see later discussion) appear to play a role in the neurobiological mechanisms underlying the autism that commonly develops in affected individuals [9,10]. "Idiopathic" autism currently may account for 85–90% of autism cases and includes cases in which no etiology or associated disease state has been identified and the underlying pathogenic mechanism is unknown. Presumably with the progress of autism research, subgroups of "idiopathic" cases will receive biomedical diagnoses and shift to the "symptomatic" category.

Another distinction in autism is between regressive and nonregressive autism. Roughly 30% of children diagnosed with autism have a largely or entirely symptom-free interval of between 1 and 2 years' duration, after which they lose speech and social reciprocity and became much more absorbed in repetitive and ritualistic behaviors [11]. The electrophysiological features of autistic regression are not well characterized because the clinical regression may be initially subtle and may not be evaluated diagnostically for a substantial time interval after its onset, but it may be more associated with subclinical abnormalities than with frank seizures. Although autism is sometimes associated with a variety of physical symptoms such as gastrointestinal disease or recurrent infection, these are not reliably associated with regression [12]. Retrospective analysis of videotapes has shown that in some cases the regression is preceded by a period of gradual decline, whereas in other cases it is fairly abrupt. Prospective studies are in progress of younger siblings of children diagnosed with autism, who are at substantially increased risk (2–8%, which is much higher than the prevalence rate in the general population) [2]. Early signs in children who are later diagnosed with autism include atypical patterns of visual attention, reduced orienting and responsiveness to both social and nonsocial stimuli, excessive fixation on nonsocial features of the visual environment, early delays in imitation and language skills, less vocalization, poor eye contact, increased irritability, intense responses to sensory stimulation that may be accompanied by signs of distress, as well as few initiations and less responsiveness to attempts to engage their attention [13]. Failure to develop pointing and joint attention is noted. Retrospective analysis of videotapes of movement patterns of infants later diagnosed with autism also show abnormal movement patterns well before the first birthday [14]. Although some prospective magnetic resonance imaging (MRI) neuroimaging data are becoming available, the measures obtained on these children have been for the most part behavioral, with no candidate biomarkers included in the research designs of the first several studies. Thus with the exception of brain size, we do not yet have measures of what is changing in association with the early years of autism, with or without regression, at the various nonbehavioral levels that have been implicated as manifesting abnormalities in autism.

Nevertheless, based on various findings there are a number of intriguing models for autism pathogenesis. These can be reviewed in order of the time during development implicated by each model. Individuals who are affected by a series of genetic disorders (including Moebius syndrome, CHARGE association, and Goldenhar syndrome) and teratological factors (including thalidomide embryopathy, valproate embryopathy, and misoprostol) involving early embryogenic defects frequently manifest autistic behaviors; the timing associated with these disorders has raised the possibility of involvement in autism of the brainstem or of genes associated with early embryogenesis [15]. In utero exposure to maternal antibodies or neurotoxins later in gestation, as well as in utero or perinatal infection, also appear to be associated with the development of autism [6]. A number of genes associated with synaptic modulation or maintenance (e.g., MeCP2, Neuroligin-1, and UBE3A) have been associated with autism [16] and are suggested to be relevant given that the postnatal symptom onset suggests autism is related to disturbances of synaptic maturation, connectivity, or stabilization. Abnormal cortical columns related to early in utero formation or possibly to postnatal sculpting of these columns [17,18] have been proposed to underlie the abnormal inhibition and connectivity identified at other levels of autism research (see later discussion). The postnatal pattern of brain enlargement (also see later discussion) may selectively affect higher-order association areas, particularly the frontal lobe, where brain development processes proceed for a more protracted period [18].

Belmonte [19] has reviewed electrophysiological and functional MRI studies and notes common patterns among the findings, which appear to indicate that autistic individuals experience hyperarousal in neuroendocrine, neurochemical, hormonal, behavioral, and cardiovascular domains. In the brain this appears to involve a heightened degree of activation in earlier stimulus-driven processing but a less than typical degree of activation in higher-order processing centers. These abnormalities appear to involve a disturbance in modulation of perceptual filtering, with an impairment of early filtering but a suppression of irrelevant stimuli in later stages of processing, where such filtering is less appropriate and less efficient. Also noted are attentional deficits including problems with rapid shifting of attention between sensory modalities, spatial locations, and features of objects. Difficulties in parsing of transient sound differences have been identified in some individuals with autism that, when present, may contribute to language impairment [20].

The downstream consequences of this type of abnormal processing are multiple and in particular include an abnormal organization of response to stimuli that hinders higher-level integrative processing [19], which in turn could epigenetically affect the development of experience-dependent brain areas. Thus the abnormality of the trajectory of brain and behavior development can potentially amplify over time [21]. Hyperexcitability in early processing along with impaired attention and higher-order processing may yield deficits ranging from early movement abnormalities [14] to poorly developed processing of complex social stimuli such as faces and nuanced emotion. It has been proposed that the failure to develop pointing and joint attention may be in some cases or in part a consequence of processing-related motor organization difficulties that impede the development of the physical coordination necessary to carry out the hand and eye movements necessary for these activities. Neuroanatomically altered brain development may yield widespread anatomical asymmetries [21], miswiring of associated hemispheric specialization, and abnormal trajectory of serotonin synthesis (which in turn may be implicated in the abnormal trajectory of autistic brain volume development, given the role of serotonin in modulating developmental processes) [22], as well as a late (i.e., during adolescence) amplification of right hemispheric language dominance [23]. The avoidance of higher-order information processing may yield less network integration and greater local processing [24]. Not all of the consequences of this altered processing style may be "deficits," however; the tendency to local processing and "weak central coherence" may also be associated with domains of heightened skill and talent found in some autistic individuals [25].

IV. Pathophysiology

Pathophysiological investigations in autism have proceeded at multiple levels. Cognitive models proposed to date to characterize the alterations in cognitive processing that might underlie the observed autistic behavioral profile include an impairment in "theory of mind" (i.e., the ability to infer the mental state of others), executive dysfunction, "weak central coherence" (i.e., an orientation toward parts rather than wholes) [25], and impaired complex processing [26]. A number of functional neuroimaging studies have been designed to study the neural circuits involved in these functions. Inferring mental states of others elicited less frontal and amygdalar activation. In a task demanding explicit appraisal of emotions, autistic individuals did not activate the bilateral fusiform gyrus, while when implicitly appraising emotional facial expressions, they did not activate the left amygdala or left cerebellum. In a task assessing the relative strength of configural and segmented processing in embedded figures—which autistic individuals do quite well because cognitively they are not distracted by "central coherence" or gestalt perception and thus see parts more easily—both control and autistic groups activated the middle and inferior temporal gyrus, the supramarginal gyrus and precuneus, the inferior frontal gyrus, and the middle occipital gyrus. However, the controls showed more activation in the right dorsolateral prefrontal cortex and bilateral parietal cortex. In face processing, a task for which psychological studies have demonstrated that autistic individuals scan faces differently and spend less time scanning the eyes, the ventral temporal fusiform face area (FFA) failed to activate in several studies of face processing [27], although some subsequent studies showed robust activation in this area [28]. Substantial interindividual variability in patterns of activation has also been demonstrated among autistic subjects who did not activate the FFA [29].

Functional neuroimaging research can also be interpreted from the vantage point of network perturbances by highlighting the atypical connections between levels of the processing hierarchy that are implicit across many of the findings [19]. A model formulated to link the phenomenon of impaired higher-order or complex processing [26] with underlying brain networks is "underconnectivity" [24], which has been proposed as a core mechanism yielding the autism profile. This model is based on functional MRI observations of modest but widespread and consistent reduction in the degree of synchrony of coactivation between associated brain regions, which is a diffuse rather than localized phenomenon (Fig. 1). Thus although there is circuit abnormality, there is also a broader phenomenon of reduced covariance. The consequences of this underconnectivity include altered network properties and a tendency to local, anatomically contiguous rather than distributed network formation. A similar phenomenon of reduced covariance phenomenon was previously reported in resting glucose metabolism positron emission tomography (PET) studies by Horwitz [30].

A common accompaniment of autism is epilepsy, which develops in approximately one-third of individuals with autism. EEG abnormalities are quite common in autism, although the documented proportion of individuals has affected varies widely depending on study design. Interictal epileptiform abnormalities are more likely to be focal than primarily generalized. Atypical sleep architecture is also common and may include longer sleep latency, more frequent nocturnal awakenings, lower sleep efficiency, increased duration of stage 1 sleep, decreased non–rapid eye movement (REM) sleep and slow-wave sleep, fewer stage 2 EEG sleep spindles, and a lower number of rapid eye movements during REM sleep than in nonautistic individuals [31]. Autism may follow infantile spasms, which occur mainly during the first year of life. Continuous spike-wave during slow-wave sleep, which may be associated with language regression, may also be associated with both autism and cognitive decline. Seizures and epilepsy are more common among children with autistic spectrum

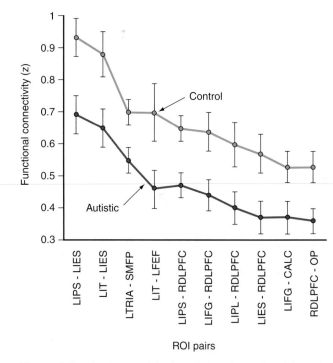

Figure 1 Functional connectivity for autistic and control participants in the 10 ROI pairs, with a reliable ($P < 0.05$) difference between autistic and control participants (presented in descending order of mean connectivity). The pattern of functional connectivities among these 10 ROI pairs is very similar for the two groups ($r = 0.98$). Error bars represent the standard error of the mean. *L*, Left; *R*, right; *CALC*, calcarine fissure; *DLPFC*, dorsolateral prefrontal cortex; *FEF*, frontal eye field; *IES*, inferior extrastriate; *IFG*, inferior frontal gyrus; *IPL*, inferior parietal lobe; *IPS*, intraparietal sulcus; *IT*, inferior temporal; *OP*, occipital pole; *SMFP*, superior medial frontal paracingulate; *TRIA*, triangularis. *Source*: (From Just, M. A., Cherkassky, V. L., Keller, T. A., et al. [2004]. Cortical activation and synchronization during sentence comprehension in high-functioning autism: evidence of underconnectivity. *Brain* **127**, 1811–1821.)

disorder who experience language regression, especially those who experience language regression after the age of 2 years, and particularly in autistic children with mental retardation and motor abnormalities [32].

Autonomic nervous system disorders have also been documented in autism. Sleep disorders are found in a large majority of children (up to 80%) [33], and the frequently encountered gastrointestinal symptoms such as chronic constipation or diarrhea may also have a major autonomic component. Additional autonomic abnormalities that have been reported include abnormal skin conductance, blunted autonomic arousal to social stimuli, and increased tonic electrodermal activity [34]. Autonomic hyperarousal appears to be a common feature of autism, although hypoarousal has also been seen.

Neuroinflammation with activated microglia and astroglia has been identified in the cerebral cortex and white matter, and particularly in the cerebellum, of a small series of postmortem autistic brain specimens ranging in age from 5 to 44 years. In addition, inflammatory cytokines and chemokine profiles were identified in these brains and in the cerebrospinal fluid (CSF) of living persons with autism, with macrophage chemoattractant protein (MCP)-1 and tumor growth factor (TGF)-β1, derived from neuroglia, as the most prevalent cytokines [35]. The immune activation identified in this study was mediated by innate but not adaptive immunity. Many of the cytokines identified in this study also act as growth and differentiation factors during brain development. Microglial and astroglial changes can contribute to significant neuronal and synaptic changes [34].

V. Brain Structure

Both regional and widespread differences from normal brains have been identified. Neuropathological studies have been limited due to the small number of postmortem autistic samples available, and their interpretation has been limited by variable and often poorly documented histories; confounding comorbidities including metabolic diseases, mental retardation, and seizures; and variable postmortem intervals. Nevertheless, these studies have produced a number of provocative findings that have sparked further investigations. Observations have included abnormalities in the limbic system, with small and tightly packed cells, as well as in the cerebellum with a reduced number of Purkinje cells. Scattered signs of cortical dysgenesis have been observed, although not consistently. Brainstem and olivary nucleus changes have been noted in several studies. Several areas (including limbic and olivary nucleus) showed larger, darker cells in younger subjects compared with smaller, paler cells in older subjects [36]. Heavy brain weight has been noted frequently, more in younger than in older brains [37]. Digital analysis of minicolumn structure in the superior and middle frontal gyrus, middle temporal gyrus, temporal-parietal auditory area, and frontal lobe has shown that minicolumn width, peripheral neurophil space, and compactness were reduced in subjects with autism compared with controls, and it has been suggested that a depletion of GABAergic inhibitory neurons is particularly implicated in this phenomenon.

Neuroimaging studies have the advantage over neuropathology of recruiting prospectively and allowing the accumulation of a cohort for study that is much more carefully characterized and consistent in features. The most consistently replicated brain structure finding in autism is a tendency toward larger brains. Younger brains have been large, but the increase in volume compared with controls does not persist past late puberty or early adolescence. Macrocephaly (i.e., head circumference above the 97th percentile) has been found in 20% (10–30% in various studies) of autistic individuals, while most have head circumferences above the 50th percentile. Brain size at birth (determined by retrospective study of head circumference) is average or even below average but then increases strikingly during the first 2–3 years, beginning during the first postnatal month or two, with the growth sharply diminishing thereafter [37] (Fig. 2). In the youngest subjects imaged to date, both white and gray matter are increased, but the white matter increase is greater. In mid-childhood, the white matter volume increase is discernible, whereas the gray matter increase is less dramatic, and the white matter volume increase is most prominent in the outer (radiate) zone and the frontal and especially prefrontal lobes—all areas that myelinate late [21]. The corpus callosum has been smaller or no different than controls, in spite of larger brain volume in every sample studied to date, although the regional distribution of this volume decrease differed among studies and among individuals in some studies. The tissue composition of these volume changes is not known at this time because neuropathological investigations of white matter to determine whether the increase is related to myelin, axons, microglial cells, vasculature, or something else are just beginning. Abnormal cortical minicolumn architecture might contribute to an altered number of axons and altered connectivity patterns. Diffusion tensor imaging studies are just beginning to appear and have not yet addressed this question, while MR spectroscopy in one case has shown a reduction of N-acetylaspartate in the cerebral cortex, suggesting a lower rather than an increased density of neurons [38,39].

Early region-of-interest approaches to volumetric imaging analysis yielded findings of a large subgroup with a small midline cerebellar vermis area and another subgroup in whom this area is larger. A smaller number of studies

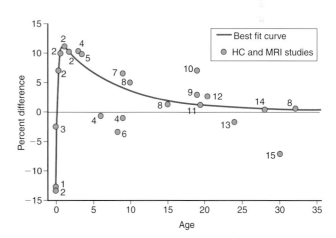

Figure 2 Head circumference (HC) and magnetic resonance imaging (MRI) percent difference are illustrated as percent different from control, by age. Each point on the graph represents a published study of head circumference or brain volume as measured by MRI (the specific studies are listed in Table 1 of the figure source [37]), and is plotted by the mean age of the study subjects. The best fitted curve shows that the most rapid rates of increased deviation from normal brain size in autism occurs within the first year of life, and that the greatest rates of decrease in deviation from normal occur during middle and late childhood.

have measured overall cerebellar volume, and all have found it to be larger. Findings in many other parts of the brain have been variable, both within and among studies as well across ages. The amygdala and hippocampus have been measured as larger, smaller, and no different from controls. Only a few reports on thalamus and basal ganglia are available. A modest number of brainstem measurements have been variable. Cingulate gyrus has been reported to be both smaller and hypometabolic in the same subjects. Asymmetry reversals have been identified in inferior frontal gyrus, particularly in autistic subjects with documented language impairment, but asymmetry changes relative to controls have also been found quite widely distributed throughout the cerebral cortex, most strikingly in higher-order association cortical areas; highly similar alterations were also identified in nonautistic children with developmental language disorder. A morphometric profile of gray and white matter structure volumes throughout the brain in school-aged, high-functioning autistic boys found nonuniform scaling of volume changes across regions: In the setting of total brain volume that was modestly larger, one set of regions were absolutely but not proportionally larger after adjusting for overall size increase, another set was absolutely the same as controls but proportionally smaller, and a third (comprised only of cerebral white matter) was both relatively and absolutely larger than controls. The potential contribution of spectroscopic, diffusion tensor, and other imaging modalities to the pool of data, which might help understand the tissue character of regional brain changes in autism, has not yet been realized [38].

Nearly a dozen studies of brain perfusion in autism have been performed, mostly using SPECT neuroimaging. Regarding localization of perfusion changes, one finds substantial variability across findings: temporal, parietal and frontal, anterior cingulate, and right hemisphere hypoperfusion have been noted. But if one reads these studies with an eye for underlying biological disease mechanisms, then one notes that low perfusion has been documented in at least eight SPECT studies as well as in some PET studies, with lack of decrement in perfusion compared with controls being rare and hyperperfusion virtually never being reported (in contrast to glucose metabolism imaging studies, in which inter-study variability has included findings of increase as well as decrease in that measure) [38].

VI. Pharmacology, Biochemistry, and Molecular Mechanisms

Abnormalities related to a variety of neurotransmitters have been identified in autistic cohorts [40]. Serotonin abnormalities have been documented most extensively, and serotonin-enhancing drugs are frequently helpful in alleviating symptoms in autism. More than two dozen studies have found that approximately 30% of children with autism have about a 50% increase in serotonin blood levels (serotonin is mostly in platelets). Most of the 5-HT is absorbed by platelets after being manufactured in the gut, and the connection between peripheral and central serotonin is uncertain. A study by Chugani et al. of serotonin synthesis capacity using α-[11C]methyl-L-tryptophan and PET showed reduced serotonin synthesis and an atypical developmental trajectory in the quantity synthesized in brains of autistic children, and a later study also showed atypical asymmetries and in this measure. When tryptophan (a serotonin precursor) is depleted, which can happen for a variety of reasons including increased use of tryptophan in the competing inflammation-associated kynurenine pathway, autistic symptoms worsen [3], whereas drugs enhancing serotonin reduce aggression and obsessive-compulsive rituals in many autistic individuals [40,41]. A number of serotonin-related genes have been evaluated as candidate genes for autism, and animals in which serotonin was altered in development show "autistic-like" behaviors as well as changes in brain regions that are also affected in autism. Because autism is not diagnosed until after the first few years of life, we do not have data on serotonin levels during in utero and early development, but serotonin plays a number of important roles in early brain development as a developmental signal, and disturbances in serotonin metabolism may be related not only to behavioral problems, as previously mentioned, but also to epilepsies. A positive relationship was observed between elevated serotonin levels and the major histocompatibility complex (MHC) types previously reported to be associated with autism [42].

A model of an increased excitation:inhibition ratio has been proposed as the basis for some cases of autism [43]. Such a change would be consistent with multiple aspects of the autism phenotype, including hyperexcitability of early cortical processing, anxiety, sleep disturbances and seizures, as well as alteration of perception and attention [44]. The model suggests that this increased ratio (which could be due to an increase in the numerator [excitation], a decrease in the denominator [inhibition], or both) can be caused by combinatorial effects of genetic and environmental variables that impinge on a given neural system [43]. Factors associated with autism that could contribute to a situation consistent with this model are numerous. GABAergic signaling may be reduced or glutamatergic activity increased through gene-based GABA or glutamate receptor abnormalities, abnormalities in GABAergic cortical interneurons, altered modulation of GABAergic inhibition, or increased glutamatergic activity related to excitotoxicity secondary to neuroinflammation or upregulation of glutamate due to polychlorinated biphenyls (PCB) exposure [43]. Mutations in the membrane-associated proteins neuroligin and β-neurexin, which modulate the ratio of inhibitory and excitatory synapses on central nervous system (CNS) neurons, may also contribute to the autism phenotype by altering the excitation:inhibition ratio [45]. The absence of fragile X mental retardation protein (FMRP) leads to overactive or inappropriate Gp1 metabotropic glutamate receptor signaling and in animal models is associated with seizures; repetitive and anxious behaviors; sleep disturbance; gut dysmotility; an increased density of long, thin dendritic spines; and loss of motor coordination, which are key features of fragile X syndrome; these also include multiple "secondary" features associated with autism [9]. Altered interneuron development could change the number or activity of GABAergic neurons or their processes, modifying the excitation/inhibition ratio [46]. Reelin, also implicated in autism by some genetic and neuropathological studies, is expressed by GABAergic interneurons; the reelin signaling pathway can be disturbed by mutations or selective hypermethylation of the Reln gene promoter, as well as following various prenatal or postnatal insults [47].

Neuropeptide abnormalities are also implicated in autism. Abnormalities were identified in archived neonatal blood spot samples, with neonatal concentrations of vasoactive intestinal peptide (VIP), calcitonin gene-related peptide (CGRP), brain-derived neurotrophic factor (BDNF), and neurotrophin 4/5 (NT4/5) found to be higher in children with either autism and mental retardation than they were in either control children or children with cerebral palsy; concentrations of these substances were similar in autistic subgroups, including those with and without regression [48]. Oxytocin and vasopressin have been of interest because although these neuropeptides have been difficult to localize in the human brain, animal models support the importance of central actions of oxytocin and vasopressin in the regulation of various features of social behavior, including social recognition and social bonding, that are relevant to autism. No data on CSF oxytocin are available, but plasma oxytocin is reduced in autistic children [49]. Of note, early suggestions that pitocin induction of delivery could be associated with a higher risk for autism have not been supported by subsequent studies.

Numerous immune system abnormalities have been identified in individuals with autism, although there are substantial inconsistencies among studies. These abnormalities include autoantibodies (particularly to CNS proteins) [50] and deficits in immune cell subsets, cytokine abnormalities, impaired responses to viral infections, and prolonged and recurrent infections [3], as well as vulnerability factors including family history of autoimmune disease and genetic variants associated with autoimmunity [50]. Although enhanced autoimmunity has been noted, autism does not meet criteria for being an autoimmune disease, and the presence of autoantibodies to any particular protein is not necessarily either causal or predictive of an effect on that protein or system; moreover, autoantibodies differ considerably among autistic individuals [50]. Cytokine abnormalities may also be involved in altering aspects of brain development or ongoing functions such as circadian rhythms, emotional responses, or attention, but given the large number of cytokines that have an individual developmental time course and response to infectious and immune perturbation, there may be no single pathognomonic cytokine abnormality in autism [6].

Abnormal folate metabolism and low glutathione, found in other neurological disorders such as Alzheimer's, Parkinson's disease, schizophrenia, and Down syndrome, have recently been reported in autism. In one study, children with autism had significantly lower baseline plasma concentrations of methionine, S-adenosylmethionine (SAM), homocysteine, cystathionine, cysteine, and total glutathione and significantly higher concentrations of S-adenosyl-homocysteine (SAH), adenosine, and oxidized glutathione. [51]. This metabolic profile is consistent with impaired capacity for methylation (significantly lower ratio of SAM to SAH) and increased oxidative stress (significantly lower redox ratio of reduced glutathione to oxidized glutathione) in children with autism. Methylation impairment may impact neurotransmitter synthesis, metabolism of xenobiotic substances, and DNA regulation.

Only a modest amount of data implicates acetylcholine and catecholamines in autism. Neuropathological investigations have identified reduced gene expression of the α4-β2 nicotinic receptor in the cerebral cortex, while posttranscriptional abnormalities of both this and the α7 subtype are apparent in the cerebellum. The findings point to dendritic and/or synaptic nicotinic receptor abnormalities that may relate to disruptions in cerebral circuitry development. Reduced dopamine hydroxylase and increased norepinephrine in serum from autistic children and their parents suggest possible involvement of these neurotransmitters [44].

VII. Explanation of Symptoms in Relation to Pathophysiology

The linkages between autism symptoms and pathophysiology are only beginning to be clarified in detail. No regional brain changes have been identified that are consistently present. Thus although social and emotional deficits may in some cases be attributable to disturbances in limbic system or ventral-temporal face processing areas, these brain abnormalities are not present in every autistic individual based on investigative methods available to date. Nor is cerebellar vermis hypoplasia or loss of Purkinje cells reliably present. The relatively more pronounced problems with higher-order associational functions could be due to a relatively greater impact on later developing brain areas, particularly frontal lobe; but it could also be due to widely distributed, fairly pervasive neurochemical, synaptic, dendritic, interneuron, or white matter changes that preferentially target functions dependent on a high degree of interregional coordination. It is possible that a reduction in connectivity sufficient to lead to impaired higher-order processing and an increase in local processing might be a consequence of multiple pathophysiological processes, which differ among individuals. This might be one of the reasons autism is so heterogeneous and difficult to pin down either genetically or biologically: impaired complex processing and reduced connectivity could well be final common pathways rather than phenomena that are biologically specific.

It may be that studying comorbidities, even though they are not strictly speaking part of "autism" as behaviorally defined, can contribute to identifying mechanisms and distinguishing subgroups. Seizures may be a consequence of a variety of pathophysiological abnormalities reviewed previously, including increased excitation:inhibition ratios, serotonin abnormalities, or altered interneuron development. Anxiety, sleep disturbances, and repetitive behaviors may be due to similar factors or to immunochemical dysregulation. Gastrointestinal issues, when present, may indicate up-regulated glutamate [9] or immune/inflammatory problems [52] and may also be related to regional brain abnormalities, given that chronic inflammatory bowel disease has shown to activate periventricular gray, hypothalamic/visceral thalamic stress axes and cortical domains, and septal/preoptic/amygdala, brain areas in which abnormalities have been identified often in autism [53]. However, the study of the prevalence and covariance of core and comorbid symptoms in autism is only beginning.

VIII. Natural History

Although Rett syndrome and CDD are degenerative, the remaining disorders listed in the DSM-IV autism spectrum are not degenerative and affected individuals can have normal lifespans. Causes of death that are more frequent in autism include accidents, drowning, and epilepsy. The first two of these are a consequence of limitations in ability to assess danger, motor control problems, and other issues devolving from the underlying autism.

However, postnatal changes in the brain are beginning to suggest that the disorder may involve factors that change over the lifespan. Biological markers in autism are dynamic at multiple levels, including both neuroanatomy (e.g., brain volumes) and neurochemistry (e.g., alterations in serotonin metabolism). It is also the case that some autistic individuals improve or recover, either spontaneously or as a result of intervention. Study of such cases may also be of some utility in identifying subgroups capable of such improvements and discerning whether such individuals comprise subgroups whose autism is due to distinctive pathophysiological mechanisms.

Given confluent advances on multiple fronts in autism research, our integrated understanding of the mechanisms driving this disorder is likely to increase substantially in the next few years. Some of these mechanisms may lead to the identification of treatment targets that may help improve the condition of at least some proportion of individuals with autism.

References

1. American Psychiatric Association. (1994). "Diagnostic and Statistical Manual of Mental Disorders, ed. 4 (DSM-IV)." Washington, D.C., The Association.
2. Santangelo, S. L., and Tsatsanis, K. (2005). What is known about autism: genes, brain, and behavior. *Am. J. Pharmacogenomics* **5**, 71–92.
3. Ashwood, P., and Van de Water, J. (2004). A review of autism and the immune response. *Clin. Dev. Immunol.* **11**, 165–174.
4. Newschaffer, C. J., Falb, M. D., and Gurney, J. G. (2005). National autism prevalence trends from United States special education data. *Pediatrics* **115**, e277–e282.
5. Lawler, C. P., Croen, L. A., Grether, J. K., Van de Water, J. (2004). Identifying environmental contributions to autism: provocative clues and false leads. *Ment. Retard. Dev. Disabil. Res. Rev.* **10**, 292–302.
6. Hornig, M., and Lipkin, W. I. (2001). Infectious and immune factors in the pathogenesis of neurodevelopmental disorders: epidemiology, hypotheses, and animal models. *Ment. Retard. Dev. Disabil. Res. Rev.* **7**, 200–210.
7. Cohen, D., Pichard, N., Tordjman, S., Baumann, C., Burglen, L., Excoffier, E., Lazar, G., Mazet, P., Pinquier, C., Verloes, A., Heron, D. (2005). Specific genetic disorders and autism: clinical contribution towards their identification. *J. Autism Dev. Disord.* **35**, 103–116.
8. Curatolo, P., Porfirio, M. C., Manzi, B., Seri, S. (2004). Autism in tuberous sclerosis. *Eur. J. Paediatr. Neurol.* **8**, 327–332.
9. Bear, M. F., Huber, K. M., and Warren, S. T. (2004). The mGluR theory of fragile X mental retardation. *Trends Neurosci.* **27**, 370–377.
10. Dong, W. K., and Greenough, W. T. (2004). Plasticity of nonneuronal brain tissue: roles in developmental disorders. *Ment. Retard. Dev. Disabil. Res. Rev.* **10**, 85–90.
11. Luyster, R., Richler, J., Risi, S., Hsu, W. L., Dawson, G., Bernier, R., Dunn, M., Hepburn, S., Hyman, S. L., McMahon, W. M., Goudie-Nice, J., Minshew, N., Rogers, S., Sigman, M., Spence, M. A., Goldberg, W. A., Tager-Flusberg, H., Volkmar, F. R., and Lord, C. (2005). Early regression in social communication in autism spectrum disorders: a CPEA Study. *Dev. Neuropsychol.* **27**, 311–336.
12. Richler, J., Luyster, R., Risi, S., Hsu, W., Dawson, G., Bernier, R., Dunn, M., Hepburn, S., Hyman, S., McMahon, W., Goudie-Nice, J., Minshew, N., Rogers, S., Sigman, M., Spence, M., Goldgerg, W., Tager-Flusberg, H., Volkmar, F., and Lord, C. (2006). Is there a 'regressive phenotype' of autisms spectrum disorder associated with the measles-mumps-rubella vaccine? A CPEA study. *J. Aut. Deve. Disord.* **36**, 299–316.
13. Zwaigenbaum, L., Bryson, S., Rogers, T., Roberts, W., Brian, J., and Szatmari, P. (2005). Behavioral manifestations of autism in the first year of life. *Int. J. Dev. Neurosci.* **23**, 143–152.
14. Teitelbaum, O., Benton, T., Shah, P. K., Prince, A., Kelly, J. L., and Teitelbaum, P. (2004). Eshkol-Wachman movement notation in diagnosis: the early detection of Asperger's syndrome. *Proc. Natl. Acad. Sci. USA* **101**, 11909–11914.

15. Miller, M. T., Stromland, K., Ventura, L., Johansson, M., Bandim, J. M., and Gillberg, C. (2005). Autism associated with conditions characterized by developmental errors in early embryogenesis: a mini review. *Int. J. Dev. Neurosci.* **23**, 201–219.
16. Zoghbi, H. Y. (2003). Postnatal neurodevelopmental disorders: meeting at the synapse? *Science* **302**, 826–830.
17. Casanova, M. F., Buxhoeveden, D., and Gomez, J. (2003). Disruption in the inhibitory architecture of the cell minicolumn: implications for autisim. *Neuroscientist* **9**, 496–507.
18. Courchesne, E., and Pierce, K. (2005). Brain overgrowth in autism during a critical time in development: implications for frontal pyramidal neuron and interneuron development and connectivity. *Int. J. Dev. Neurosci.* **23**, 153–170.
19. Belmonte, M. K., Cook, E. H., Anderson, G. M. Rubenstein, J. L., Greenough, W. T., Beckel-Mitchener, A., Courchesne, E., Boulanger, L. M., Powell, S. B., Levitt, P. R., Perry, E. K., Jiang, Y. H., DeLorey, T. M., and Tierney, E. (2004). Autism as a disorder of neural information processing: directions for research and targets for therapy. *Mol. Psychiatry* **9**, 646–663.
20. Oram Cardy, J. E., Flagg, E. J., Roberts, W., and Roberts, T. P. (2005). Delayed mismatch field for speech and non-speech sounds in children with autism. *Neuroreport* **16**, 521–525.
21. Herbert, M. R., Ziegler, D. A., Deutsch, C. K., O'Brien, L. M., Kennedy, D. N., Filipek, P. A., Bakardjiev, A. I., Hodgson, J., Takeoka, M., Makris, N., and Caviness, V. S. Jr (2005). Brain asymmetries in autism and developmental language disorder: a nested whole-brain analysis. *Brain* **128**, 213–226.
22. Chandana, S. R., Behen, M. E., Juhasz, C., Muzik, O., Rothermel, R. D., Mangner, T. J., Chakraborty, P. K., Chugani, H. T., and Chugani, D. C. (2005). Significance of abnormalities in developmental trajectory and asymmetry of cortical serotonin synthesis in autism. *Int. J. Dev. Neurosci.* **23**, 171–182.
23. Flagg, E., Oram Cardy, J., Roberts, W., and Roberts, T. (2005). Language lateralization development in children with autism: insights from the late field magnetoencephalogram. *Neurosci. Lett.* **386**, 82–87.
24. Just, M. A., Cherkassky, V. L., Keller, T. A., and Minshew, N. J. (2004). Cortical activation and synchronization during sentence comprehension in high-functioning autism: evidence of underconnectivity. *Brain* **127**, 1811–1821.
25. Hill, E. L., and Frith, U. (2003). Understanding autism: insights from mind and brain. *Philos. Trans. R. Soc. Lond. B. Biol. Sci.* **358**, 281–289.
26. Minshew, J., Goldstein, G., and Siegel, D. (1997). Neuropsychologic functioning in autism: profile of a complex informational processing disorder. *J. Int. Neuropsychol. Soc.* **3**, 303–316.
27. Schultz, R. T. (2005). Developmental deficits in social perception in autism: the role of the amygdala and fusiform face area. *Int. J. Dev. Neurosci.* **23**, 125–141.
28. Dalton, K. M., Nacewicz, B. M., Johnstone, T., Schaefer, H. S., Gernsbacher, M. A., Goldsmith, H. H., Alexander, A. L., and Davidson, R. J. (2005). Gaze fixation and the neural circuitry of face processing in autism. *Nat. Neurosci.* **8**, 519–526.
29. Pierce, K., Muller, R. A., Ambrose, J., Allen, G., and Courchesne, E. (2001). Face processing occurs outside the fusiform 'face area' in autism: evidence from functional MRI. *Brain* **124**, 2059–2073.
30. Horwitz, B., Rumsey, J. M., Grady, C. L., and Rapoport, S. I. (1988). The cerebral metabolic landscape in autism. Intercorrelations of regional glucose utilization. *Arch. Neurol.* **45**, 749–755.
31. Limoges, E., Mottron, L., Bolduc, C., Berthiaume, C., and Godbout, R. (2005). Atypical sleep architecture and the autism phenotype. *Brain* **128**, 1049–1061.
32. Trevathan, E. (2004). Seizures and epilepsy among children with language regression and autistic spectrum disorders. *J. Child. Neurol.* **19(Suppl 1)**, S49–S57.
33. Malow, B. A. (2004). Sleep disorders, epilepsy, and autism. *Ment. Retard. Dev. Disabil. Res. Rev.* **10**, 122–125.
34. Zimmerman, A. W., Connors, S. L., and Pardo, C. A. (2006). Neuroimmunology and neurotransmitters in autism. *In:* "Autism: A Neurobiological Disorder of Early Brain Development" pp. 141–159 (R. Tuchman, et al., eds.). MacKeith Press, London.
35. Vargas, D. L., Nascimbene, C., Krishnan, C., Zimmerman, A. W., and Pardo, C. A. (2005). Neuroglial activation and neuroinflammation in the brain of patients with autism. *Ann. Neurol.* **57**, 67–81.
36. Palmen, S. J., van Engeland, H., Hof, P. R., and Schmitz, C. (2004). Neuropathological findings in autism. *Brain* **127**, 2572–2583.
37. Redcay, E., and Courchesne, E. (2005). When is the brain enlarged in autism? A meta-analysis of all brain size reports. *Biol. Psychiatry* **58**, 1–9.
38. Herbert, M. R., and Caviness, V. S. (2006). Neuroanatomy, and neuroradiology. *In:* "Autism: A Neurobiological Disorder of Early Brain Development" pp. 115–140 (R. Tuchman, et al., eds.). MacKeith Press, London.
39. Cody, H., Pelphrey, K., and Piven, J. (2002). Structural and functional magnetic resonance imaging of autism. *Int. J. Dev. Neurosci.* **20**, 421–438.
40. Lam, K. S., Aman, M. G., and Arnold, L. E. (2006). Neurochemical correlates of autistic disorder: A review of the literature. *Res. Dev. Disabil.* **27**, 254–289.
41. Chugani, D. C. (2004). Serotonin in autism and pediatric epilepsies. *Ment. Retard. Dev. Disabil. Res. Rev.* **10**, 112–116.
42. Warren, R. P., and Singh, V. K. (1996). Elevated serotonin levels in autism: association with the major histocompatibility complex. *Neuropsychobiology* **34**, 72–75.
43. Rubenstein, J. L., and Merzenich, M. M. (2003). Model of autism: increased ratio of excitation/inhibition in key neural systems. *Genes Brain Behav.* **2**, 255–267.
44. Polleux, F., and Lauder, J. M. (2004). Toward a developmental neurobiology of autism. *Ment. Retard. Dev. Disabil. Res. Rev.* **10**, 303–317.
45. Cline, H. (2005). Synaptogenesis: a balancing act between excitation and inhibition. *Curr. Biol.* **15**, R203–R205.
46. Levitt, P., Eagleson, K. L., and Powell, E. M. (2004). Regulation of neocortical interneuron development and the implications for neurodevelopmental disorders. *Trends Neurosci.* **27**, 400–406.
47. Fatemi, S. H. (2005). Reelin glycoprotein: structure, biology and roles in health and disease. *Mol. Psychiatry* **10**, 251–257.
48. Nelson, K. B., Grether, J. K., Croen, L. A., Dambrosia, J. M., Dickens, B. F., Jelliffe, L. L., Hansen, R. L., and Phillips, T. M. (2001). Neuropeptides and neurotrophins in neonatal blood of children with autism or mental retardation. *Ann. Neurol.* **49**, 597–606.
49. Winslow, J. T., and Insel, T. R. (2002). The social deficits of the oxytocin knockout mouse. *Neuropeptides* **36**, 221–229.
50. Ashwood, P., and Van de Water, J. (2004). Is autism an autoimmune disease? *Autoimmun. Rev.* **3**, 557–562.
51. James, S. J., Cutler, P., Melnyk, S., Jernigan, S., Janak, L., Gaylor, D. W., and Neubrander, J. A. (2004). Metabolic biomarkers of increased oxidative stress and impaired methylation capacity in children with autism. *Am. J. Clin. Nutr.* **80**, 1611–1617.
52. Jyonouchi, H., Geng, L., Ruby, A., and Zimmerman-Bier, B. (2005). Dysregulated innate immune responses in young children with autism spectrum disorders: their relationship to gastrointestinal symptoms and dietary intervention. *Neuropsychobiology* **51**, 77–85.
53. Welch, M. G., Welch-Horan, T. B., Anwar, M., Anwar, N., Ludwig, R. J., and Ruggiero, D. A. (2005). Brain effects of chronic IBD in areas abnormal in autism and treatment by single neuropeptides secretin and oxytocin. *J. Mol. Neurosci.* **25**, 259–274.

53

Neurobiology of Dyslexia

Sheryl L. Rimrodt, MD
Laurie E. Cutting, PhD

Keywords: *dyslexia, neuroimaging, reading disability*

I. Definition and History
II. Natural History
III. Behavioral Models of Dyslexia
IV. Neurobiology of Dyslexia
V. Relating Neurobiology to Reading Behavior
VI. Summary
References

I. Definition and History

Dyslexia is a developmental disorder of the neurological system that results in a relatively selective impairment of an individual's ability to learn to read. It is described as an unexpected difficulty in learning to read in an individual who otherwise possesses the intelligence, motivation, and schooling considered necessary for accurate and fluent reading [1].

Dyslexia was originally called "congenital word-blindness" and was first described in the medical literature in 1896 by W. Pringle Morgan; his paper was a response to James Hinshelwood's earlier descriptions of "acquired word-blindness." Hinshelwood, an ophthalmologist, also encountered several cases of congenital word-blindness and discussed these at length in a 1917 monograph. He formulated the problem as a "congenital deficiency of the visual memory of words" [2]. His observations included a familial tendency (which has been corroborated by many authors) and a greater frequency among males than females (which has been challenged by recent research) [1]. Based on the lesions found in cases of acquired word-blindness with similar clinical characteristics, he postulated that congenital word-blindness was the result of faulty development resulting in lower functional activity in the left angular and supramarginal gyri [2]. Hinshelwood's early hypothesis has been supported by modern neuroimaging evidence that demonstrates the importance of the left temporoparietal region to reading [3].

Another pioneer in the field of developmental reading impairment was Samuel Orton, a neurologist whose work included examinations of children with reading, writing, and speech problems. He expanded the concept of developmental reading disorder to "a graded series including all degrees of severity of the handicap" [4] that was therefore more common than Hinshelwood's congenital word-blindness. He also recognized the presence of other frequently comorbid developmental disorders including dysgraphia, speech delay, stuttering, receptive language deficits, and motor coordination deficits [4,5]. He also observed "strephosymbolia," or written letter, sound, and syllable reversals (e.g., confusing "b" for "d" or "saw" for "was"), in children with poor reading. Strephosymbolia has sometimes been misinterpreted as evidence of a visual perceptual abnormality in dyslexia, but Orton himself attributed this symptom to a lack of unilateral left hemisphere cerebral dominance for language [4]. Recent work continues to suggest that poor readers use more bilateral cortex during reading than do nonimpaired readers, although the extent of differences between these groups in anatomical symmetry of brain regions that are important for reading remains unclear.

II. Natural History

Estimates of the prevalence of dyslexia among school-aged children range from 5% to 17% depending on the population studied and the criteria used to establish the diagnosis [1]. Male predominance for dyslexia has generally been noted by investigators and educators since the early descriptions by Hinshelwood and Orton; however, this finding was not supported by a recent large population study that found equal prevalence of the disorder among males and females [1]. Longitudinal studies of reading "growth" have demonstrated that dyslexia does not represent a transient developmental lag but rather a disorder in which affected individuals continue to demonstrate characteristic deficits even as reading skill improves [1,6]; compensatory physiological mechanisms continue to be present in adults with a history of developmental reading impairment [7]. Although there is little debate about the neurological basis for the disorder because it is well established, evidence also suggests that environmental influences (e.g., type and amount of reading instruction) can affect the outcome for affected individuals [8].

Specific language impairment (SLI) is closely associated with dyslexia. Among a cohort of children with SLI, 77% did poorly on tests of single-word reading and 98% did poorly on tests of reading comprehension [9]. The association is frequent enough that it has been hypothesized the two diagnoses are not completely distinct entities [9].

Children with dyslexia may also have other specific learning disabilities in mathematics or written expression, but the most common comorbid condition with dyslexia is attention deficit hyperactivity disorder (ADHD). It is estimated that 20–25% of children with dyslexia have ADHD and that anywhere from 10–50% of children with ADHD also have dyslexia [10]. This is important when considering treatment for dyslexia. Specific and systematic educational intervention is the core of successful treatment for dyslexia, and there is no recognized pharmacological agent to treat this disorder. However, in appropriate cases, adjunctive treatment of comorbid ADHD, including pharmacotherapy with stimulant medications, may have a positive effect on a student's ability to benefit from an educational intervention.

Regarding etiology, the prevalence of dyslexia is increased among individuals with certain genetic syndromes (e.g., neurofibromatosis or Klinefelter's syndrome) and among children with past medical histories that include prematurity, in utero drug exposure, congenital infection, hydrocephalus, or other early neurological insults (e.g., neonatal stroke) [11]. However, most individuals with dyslexia do not have any of these risk factors.

There is significant support for dyslexia being an inherited disorder. Among adults with a known history of dyslexia, it is estimated that between 23% and 65% have a child with dyslexia. Conversely, 27–49% of children with dyslexia have an affected parent. The prevalence of dyslexia among siblings of affected children is 40%, with an 84–100% concordance among monozygotic twins [1]. However, the genetic mechanisms of this heritability have been shown to be complex and heterogeneous. To date, linkage analyses have reported the presence of genes associated with dyslexia on chromosomes 1, 2, 3, 6, 7, 15, and 18 [12,13].

III. Behavioral Models of Dyslexia

Reading can be described as the act of deciphering and comprehending written language. By the time most children begin learning to read, they have already learned to understand and produce oral language proficiently. Nevertheless, the task of understanding and producing language in written form generally does not occur spontaneously but rather requires direct instruction. The core difference between written and spoken language is the additional step of making the connection between speech sounds and their abstract visual representations (i.e., the alphabetic principle) [14]. Converging lines of evidence indicate that this step requires an explicit awareness of and ability to manipulate (e.g., separate and re-blend) the speech sounds that comprise words in a manner more explicit than is necessary to master oral language [14]. The broad term that encompasses these skills is *phonological awareness*.

The importance of phonological skill to reading is demonstrated when someone tries to read an unfamiliar word or a pseudoword (a pronounceable string of letters that does not form a real word). Such a task requires an individual to access knowledge of the speech sounds associated with the letters and to blend the sounds together according to the rules for combining sounds (i.e., phonology) without the benefit of any rote memory of the word; individuals who demonstrate deficits in phonological skill show particular difficulty with this aspect of reading. The relationship between phonological skill and reading is also demonstrated by evidence that consistently links dyslexia to phonological deficits and improved reading to interventions that include phonological awareness training [1,8,9,14].

In addition to phonological awareness, another skill that is a significant predictor of reading skill is the ability to rapidly name a limited set of digits, letters, pictures, or colors, which are typically presented in a 50–60 item array (i.e., rapid naming task) [15]. In particular, in languages with a transparent correspondence between sounds and the symbols that represent them (e.g., Italian), it is impaired rapid naming rather than phonological awareness that has been more consistently linked to dyslexia [15].

Other theories about the causes of impaired reading also have been proposed but primarily serve as explanations of additional variability in reading skill that is not accounted for by the phonological deficit theory [16]. Other models have focused on findings relevant to subsets of individuals with auditory or visual perception deficits [16,17] or those who lack an automatic quality of reading due to cerebellar dysfunction [17].

IV. Neurobiology of Dyslexia

Reading is a later-acquired skill that is built on the foundation of speech and language systems. Therefore it is not surprising that lesion studies of adults with acquired reading impairments demonstrate a left-hemisphere dominance for reading [18]. According to this model, the process of reading text begins in the primary and secondary visual cortices of the occipital lobes. The phonological process of relating letters and letter combinations to speech sounds and the semantic process of relating meaning to the perceived words are subserved by the left temporoparietal region. This region includes the superior temporal gyrus including Wernicke's area, the angular gyrus, and the supramarginal gyrus. Left inferior frontal gyrus (Broca's area), linked bidirectionally to Wernicke's area, subserves articulation as well as some types of syntax processing (i.e., the rules for connecting words into sentences). This model has informed much of the work in postmortem pathology studies as well as anatomical and functional neuroimaging studies of dyslexia.

Comparisons of postmortem morphology among individuals with and without a history of dyslexia have suggested several areas of difference between the groups. More ectopic and dysplastic cortical tissue has been found in individuals with a history of dyslexia, including in the region of the perisylvian fissure [18]. There has also been evidence of an absence of the usual left greater than right asymmetry of the planum temporale (superior temporal gyrus region) in dyslexia [18]. These differences in anatomical symmetry have been proposed as possible behavioral mechanisms for a lack of left cerebral dominance in dyslexia [19]. At a cytopathological level, examination of thalamic magnocellular layers has demonstrated smaller cells in less well-organized layers in subjects with dyslexia than in nonimpaired readers. However, the difficulty in obtaining pathology material and the inability to select for specific characteristics (e.g., age, sex, type and degree of reading impairment, or comorbid diagnoses) among donors has limited the ability to generalize these findings.

Advances in neuroimaging techniques now allow *in vivo* morphometric studies of neuroanatomy using magnetic resonance imaging (MRI). This allows investigators to select volunteers with specific characteristics (e.g., age, sex, reading ability, and comorbidity), with the additional option of obtaining repeated measures on a subject for a longitudinal assessment of morphological changes. To date, morphometric MRI studies have not consistently supported the differences in anatomical asymmetry suggested by the postmortem pathology findings [19]; differences among studies with regard to technique (e.g., how to define a region such as the planum temporale) and subject selection may still limit the ability to generalize these findings. An emerging approach to assessing asymmetry is the use of diffusion tensor imaging (a relatively new anatomical MRI technique) to examine organization of white matter tracts in human volunteers. In addition to preliminary evidence suggesting that left perisylvian white matter tracts are less well-organized in adults with dyslexia as compared with nonimpaired adult readers, a correlation has been demonstrated between reading ability and the degree of organization of the white matter in this region [20].

In addition to expanding the possibilities for anatomical neuroimaging, recent advances also enable investigators to examine physiological parameters associated with reading behavior in subjects without brain injuries. This is done using *functional* neuroimaging techniques—including positron emission tomography (PET), evoked response potentials (ERP), and more recently, functional magnetic resonance imaging (fMRI) and magnetoencephalography (MEG)—that link specific behavioral tasks to surrogate measures of local neural activity and are uniquely suited to investigate the neurological correlates of a complex human task such as reading [21]. The functional neuroimaging techniques have been the primary source of new information about the neurobiology of reading and dyslexia in recent years.

The behavioral measures used in functional neuroimaging are usually simple tasks, carefully designed to tap a particular skill (e.g., reading pseudowords) or an implicit neural process (e.g., perception of a visual stimulus). The type of brain imaging information varies depending on the technique. fMRI and PET both approximate local neural activity based on the regional hemodynamic response: fMRI measures local changes in blood oxygenation, and PET measures local changes in blood flow. fMRI is the more recent innovation and has the distinct safety advantage of not exposing the subject to any radiation. Both techniques can provide excellent spatial localization (within a few millimeters), but neither provides information about the timing of local neural activity because the hemodynamic responses that are measured are much slower (on the order of several seconds) than neural activity (on the order of milliseconds). ERP and MEG are both excellent for tracking the temporal progression of neural activity although spatial localization is less precise. They are based on local changes in either electrical currents or magnetic dipole moments, respectively, produced by the summed electrical activity of a local population of cells. The importance of a left-hemisphere network of cortical regions (similar to those identified by lesion studies) to phonological skill, reading, and dyslexia has been supported by the spatial and temporal information obtained from functional imaging studies. However, the findings from these studies and the published interpretation of these data should not be accepted uncritically. Differences in technique, both at the level of image acquisition and in the details of the behavioral task, can limit the accuracy of extrapolation between studies. Differences in theoretical approach to reading frequently result in differences in interpretation of the same finding. The neurobiological findings presented in this chapter are more widely accepted than some others but are by no means unequivocal truths; they are also primarily interpreted through the "lens" of the phonological deficit model.

PET and fMRI imaging studies have demonstrated both posterior and anterior left hemisphere–dominant locations of cortical activation associated with reading [3] that roughly correspond with the regions implicated by lesion studies; results of these studies are generally discussed within the context of three broadly defined brain regions. The first region, the occipitotemporal (OT) region, includes the middle temporal gyrus, the inferior temporal gyrus, and the OT junction. The OT junction is implicated in rapid recognition of written symbols as words (orthographic processing): the posterior portion is most responsive to general visual stimuli, with gradually increasing specificity for linguistic stimuli (e.g., letters) in the more anterior portions. Some authors have suggested that this OT junction accesses a lexicon of familiar whole words, whereas others suggest that this region facilitates rapid orthographic to phonologic mapping. In either case, this area has been shown to be positively correlated with reading skill across a variety of age groups and diagnoses [3]. Based on this information, the OT junction may be most accurately described as a brain region associated with skilled reading [3].

The second region consistently activated during functional imaging studies of reading includes temporal and temporoparietal areas, primarily the posterior superior temporal gyrus (Wernicke's area), the angular gyrus, and the superior marginal gyrus. The activity of this region appears to correlate with demand for phonological processing [3]. This temporal/temporoparietal region and the OT junction region are often referred to as the *posterior regions*.

The third (anterior) region is the left inferior frontal gyrus (IFG). In general, the IFG (Broca's area) appears to work closely with the temporoparietal region to access phonological pronunciations to decode unfamiliar words [3]; this is analogous to models based on lesion studies that demonstrate a cooperative relationship between Broca's area and Wernicke's area. Some evidence suggests specialization within the IFG, such that the anterior portion is associated with recalling the meaning of closed-class words while the posterior portion participates more in phonological processing [3].

The time sequence information obtained from ERP and MEG studies of reading complements the lesion studies model and the spatial information from fMRI and PET. There is evidence of activity in the occipital primary visual cortex at 100 milliseconds that does not discriminate between stimuli with or without linguistic content (i.e., is not specific for words). At 150 milliseconds, there is evidence of occipitotemporal area activation that is more positively correlated with letter strings than other visual stimuli. At 200–400 milliseconds, there is activation of the left superior temporal gyrus that shows greater response to pseudowords than real words [22].

V. Relating Neurobiology to Reading Behavior

Multiple studies have demonstrated a relative underactivation of left temporoparietal and OT junction areas in individuals with dyslexia during phonological tasks. These differences persist across different languages and in studies of both adults and children [3]. The between-group differences emerge as the tasks progress from nonphonological (e.g., matching line orientations) to phonological (e.g., nonword rhyming), further supporting the role of phonology in dyslexia.

fMRI and PET studies of phonological processing have demonstrated differences in activation in the left IFG and right temporoparietal and OT junction regions between individuals with dyslexia and nonimpaired readers [3]. It has also been shown that there is less correlation between activity in the left OT junction region and the left IFG in dyslexic readers than in nonimpaired readers [3]. These differences in activation have been interpreted as a possible compensatory response to support or bypass the phonological impairment.

ERP and MEG studies also show differences between dyslexic and nonimpaired individuals during word reading tasks. Some studies have found that nonimpaired readers demonstrate a pronounced activation at 150 milliseconds in OT junction that is not present in dyslexic readers [22]. Dyslexic readers demonstrate a weaker signal and a delayed onset (400 milliseconds) of the left superior temporal gyrus activation, compared with 200 milliseconds in nonimpaired readers; dyslexic readers also show a greater engagement of the homologous right temporoparietal region and greater activation in the 200–400 millisecond timeframe in the left IFG than that seen in nonimpaired readers [3,22]. These data echo the fMRI and PET findings suggesting less activation of left temporoparietal and OT junction regions and differences in activation of left IFG and right temporoparietal regions in individuals with dyslexia.

As noted previously, educational intervention is the treatment of choice for dyslexia. There is also a renewed commitment to Orton's principles of explicit instruction in the alphabetic principle, based on a recent summary of the extant literature on reading intervention [23]. Beyond the behavioral response of improved reading skill after phonological intervention [8], there has also been demonstration of a long-term physiological response to intervention. Children with dyslexia successfully treated with a phonologically based educational intervention showed fMRI changes toward the pattern observed in children with typical reading development [3].

VI. Summary

Functional neuroimaging studies support long-held hypotheses that implicate both specific left-hemisphere language-related brain regions and an atypical pattern of right- versus left-hemisphere activity in the pathophysiology of dyslexia. Compared with typical readers, individuals with dyslexia show relatively decreased neural activity in posterior brain regions implicated in phonological processing and rapid recognition of linguistic stimuli and skilled reading. Dyslexia is also associated with differences in neural activity in the left inferior frontal gyrus and in homologous right temporoparietal and right occipitotemporal junction areas, possibly to compensate for relative underactivity in left posterior regions associated with phonological processing. There is also evidence that although typical reading is associated with correlated neural activity between these two posterior regions and the left inferior frontal gyrus, this correlation is less strong in individuals with dyslexia. Educational interventions, particularly those involving phonological training, are associated with improved reading skill and with changes in functional neuroimaging activation patterns toward those found in more skilled readers.

References

1. Shaywitz, S. E., and Shaywitz, B. A. (2003). The science of reading and dyslexia. *J. AAPOS* **7**, 158–166.
2. Hinshelwood, J. (1917). "Congenital Word-Blindness." H. K. Lewis, London.
3. Sandak, R., Mencl, W. E., Frost, F. J., and Pugh, K. R. (2004). The neurobiological basis of skilled and impaired reading: recent findings and new directions. *Sci. Studies Read.* **8**, 273–292.
4. Orton, S. T. (1937). "Reading, Writing and Speech Problems in Children." W. W. Norton, New York.
5. Doris, J. L. (1998). Dyslexia: the evolution of a concept. *In:* "Specific Reading Disability: A View of the Spectrum" (B. K. Shapiro, et al., eds.), pp. 3–20. York Press, Timonium, Maryland.
6. Shaywitz, S. E., Fletcher, J. M., Holahan, J. M., Shneider, A. E., Marchione, K. E., Stuebing, K. K., Francis, D. J., Pugh, K. R., and Shaywitz, B. A. (1999). Persistence of dyslexia: the Connecticut Longitudinal Study at adolescence. *Pediatrics* **104**, 1351–1359.
7. Shaywitz, S. E., Shaywitz, B. A., Fulbright, R. K., Skudlarski, P., Mencl, W. E., Constable, R. T., Pugh, K. R., Holahan, J. M., Marchione, K. E., Fletcher, J. M., Lyon, G. R., and Gore, J. C. (2003). Neural systems for compensation and persistence: young adult outcome of childhood reading disability. *Biol. Psychiatry* **54**, 25–33.
8. Torgesen, J. K., Wagner, R. K., Rashotte, C. A., Lindamood, P., Rose, E., Conway, T., and Garvan, C. (1999). Preventing reading failure in young children with phonological processing disabilities: group and individual responses to instruction. *J. Ed. Psych.* **91**, 579–593.
9. Bishop, D. V., and Snowling, M. J. (2004). Developmental dyslexia and specific language impairment: same or different? *Psychol. Bull.* **130**, 858–886.
10. Beitchman, J. H., and Young, A. R. (1997). Learning disorders with a special emphasis on reading disorders: a review of the past 10 years. *J. Am. Acad. Child. Adolesc. Psychiatry* **36**, 1020–1032.
11. Ewen, J. B., and Shapiro, B. K. (In press). Specific learning disabilities. *In:* "Developmental Disabilities in Infancy and Childhood" (P. J. Accardo, ed.). Paul H. Brookes, Baltimore.
12. Demonet, J. F., Taylor, M. J., and Chaix, Y. (2004). Developmental dyslexia. *Lancet* **363**, 1451–1460.
13. Kaminen, N., Hannula-Jouppi, K., Kestila, M., Lahermo, P., Muller, K., Kaaranen, M., Myllyluoma, B., Voutilainen, A., Lyytinen, H., Nopola-Hemmi, J., and Kere, J. (2003). A genome scan for developmental dyslexia confirms linkage to chromosome 2p11 and suggests a new locus on 7q32. *J. Med. Genet.* **40**, 340–345.
14. Liberman, I. Y., Shankweiler, D., and Liberman, A. M. (1989). The alphabetic principle and learning to read. *In:* "Phonology and Reading Disability: Solving the Reading Puzzle" (D. Shankweiler, et al., eds.), pp. 1–33. University of Michigan Press, Ann Arbor.
15. Misra, M., Katzir, T., Wolf, M., and Poldrack, R. A. (2004). Neural systems for rapid automatized naming in skilled readers: unraveling the RAN-reading relationship. *Sci. Studies Read.* **8**, 241–256.

16. Ramus, F. (2003). Developmental dyslexia: specific phonological deficit or general sensorimotor dysfunction? *Curr. Opin. Neurobiol.* **13**, 212–218.
17. Beaton, A. A. (2002). Dyslexia and the cerebellar deficit hypothesis. *Cortex* **38**, 479–490.
18. Rumsey, J. M., and Eden, G. (1998). Functional neuroimaging of developmental dyslexia: regional cerebral blood flow in dyslexic men. *In:* "Specific Reading Disability: A View of the Spectrum" (B. K. Shapiro, et al., eds.), pp. 35–62. York Press, Timonium, Maryland.
19. Eckert, M. (2004). Neuroanatomical markers for dyslexia: a review of dyslexia structural imaging studies. *Neuroscientist* **10**, 362–371.
20. Nagy, Z., Westerberg, H., and Klingberg, T. (2004). Maturation of white matter is associated with the development of cognitive functions during childhood. *J. Cogn. Neurosci.* **16**, 1227–1233.
21. Demonet, J. F., Thierry, G., and Cardebat, D. (2005). Renewal of the neurophysiology of language: functional neuroimaging. *Physiol. Rev.* **85**, 49–95.
22. Salmelin, R., and Helenius, P. (2004). Functional neuroanatomy of impaired reading in dyslexia. *Sci. Studies Read.* **8**, 257–272.
23. National Reading Panel (U.S.) and National Institute for Literacy (U.S.). (2001). Reading: know what works. National Institute for Literacy and U.S. Dept. of Education, Washington, D.C.

54

Neonatal Brain Injuries

Steven P. Miller, MD, MAS
Donna M. Ferriero, MD, MS

Keywords: *diffusion tensor imaging, excitotoxicity, hypoxia, inflammation, ischemia, magnetic resonance imaging, newborn, oxidative stress, periventricular leukomalacia, spectroscopy, stroke, white matter*

I. Nomenclature
II. Etiology and Pathogenesis
III. Pathophysiology: Relevant Structural Detail and Molecular Mechanisms
IV. Explanation of Symptoms in Relation to Pathophysiology and Natural History
V. Summary
References

I. Nomenclature

Neonatal brain injuries represent a group of common yet heterogeneous disorders that result in long-term neurodevelopmental deficits. Neonatal brain injury is recognized clinically by a characteristic encephalopathy that evolves over days in the newborn period [1]. This clinical syndrome includes a lack of alertness, poor tone, abnormal reflex function, poor feeding, compromised respiratory status, and seizures [2]. Neonatal encephalopathy, the most common clinical syndrome to result from neonatal brain injury, occurs in as many as 6 in 1000 live term births [1] and is a major cause of neurodevelopmental disability in term infants [1]. Other forms of neonatal brain injury are also common; stroke in the term newborn has an incidence as high as 1 in 4000 live births [3]. More than 95% of infants with neonatal stroke survive to adulthood and many have some form of motor or cognitive disability.

This chapter focuses on the etiology and pathophysiology of nontraumatic brain injury in the term and preterm newborn with specific attention to ways in which recent advances in neuroimaging and developmental biology have contributed to our understanding of the basic mechanisms of neonatal brain injury. We also address the important clinical syndromes resulting from neonatal brain injury. Advanced neuroimaging modalities, such as magnetic resonance imaging (MRI), magnetic resonance spectroscopy (MRS), and diffusion weighted imaging (DWI), have shown patterns of brain injury that are unique to the

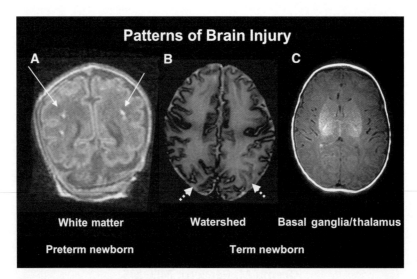

Figure 1 The predominant patterns of injury in the preterm and term newborn. **A**, White matter injury pattern in the preterm newborn: coronal spoiled gradient echo volumetric image demonstrating several foci of T1 hyperintensity *(arrow)* in the absence of marked T2 hypointensity. **B**, Watershed injury pattern in the term newborn: axial T2 weighted image at the level of the corona radiata, above the body of the lateral ventricles, demonstrating the characteristic T2 hyperintensity of the cortex and white matter in the posterior vascular watershed regions *(dashed arrows)*. **C**, Basal nuclei injury pattern in the term newborn: axial T1 weighted image at the level of the foramen of Monroe, demonstrating marked T1 hyperintensity of the basal ganglia and thalamus.

immature brain after a hypoxic-ischemic insult and that depend on the severity and duration of the insult [4,5]. The pattern of brain injury also depends on the maturity of the brain at the time the insult occurs, with distinct patterns observed in the preterm and term newborn [4,5] (Fig. 1). In addition to defining injury patterns, the ability to safely image the newborn brain serially has allowed us to confirm that neonatal brain injury evolves over days, if not weeks [6]. These clinical investigations are consistent with the prolonged temporal evolution of brain injury in neonatal animal models [6]. Understanding this time course will allow the implementation of therapeutic interventions instituted not only hours, but also days, after newborn insults, with different interventions needed as the injury evolves [6].

II. Etiology and Pathogenesis

It is increasingly recognized that neonatal brain injury is a heterogeneous condition with multiple etiologies. It is clear, however, that most neonatal brain injury is metabolic, whether from transient ischemia-reperfusion or defects in inherited metabolic pathways expressed soon after birth. Hypoxic-ischemic encephalopathy certainly accounts for a substantial percentage of neonatal brain injuries, yet many cases of neonatal encephalopathy have no documented hypoxic-ischemic insult [1,6]. Metabolic abnormalities (including acute bilirubin encephalopathy, electrolyte disturbances, and inborn errors of metabolism), infection, trauma, and malformations of cerebral development can also result in neonatal encephalopathy. Because many etiologies of neonatal encephalopathy have specific therapies, a critical part of clinical management is to determine the underlying *etiology* through careful history taking and neurological examination with laboratory and neuroimaging studies. This chapter focuses on hypoxic-ischemic damage to the neonatal brain.

The concept that neonatal brain injury is almost always secondary to acquired insults such as "birth asphyxia" is being modified by careful epidemiological studies [7]. The studies are beginning to elucidate the antenatal, perinatal, and postnatal factors that underlie the vulnerability of the newborn brain, as well as the mechanisms that contribute to potential resilience and recovery. In a prior study of *risk factors* for neonatal encephalopathy, 69% of cases had antepartum risk factors such as maternal hypothyroidism or preeclampsia, 5% had only intrapartum risks such as cord prolapse or abruptio placentae, and 24% had both antepartum and intrapartum risks [7]. Whether neonatal encephalopathy is primarily related to brain injury occurring in the antepartum or intrapartum period continues to be controversial [5,7]. Although many risk factors are clearly prenatal [7], recent evidence from prospective cohorts of neonatal encephalopathy using MRI demonstrates that the brain injury actually happens at or near the time of birth [4,5].

III. Pathophysiology: Relevant Structural Detail and Molecular Mechanisms

A. Patterns of Brain Injury

Advanced neuroimaging techniques, such as MRI, can now be applied in the human newborn to better understand the heterogeneity of brain injury associated with neonatal encephalopathy. In a primate model of term neonatal brain injury, the specific regional distribution of injury was associated with different durations and severities of ischemia: partial asphyxia caused cerebral white matter injury, and acute and profound asphyxia produced selective injury to the basal ganglia and thalamus [1]. A comparable regional vulnerability is observed in the term newborn, with two major patterns of injury detectable by MRI. The *watershed* pattern predominantly involves the white matter, particularly in the vascular watershed, and extends to cortical gray matter when severe. The basal ganglia/thalamus predominant pattern characteristically involves the deep gray nuclei (basal ganglia and thalamus) and perirolandic cortex, extending to the total cortex when severe [4].

Focal infarctions of the brain are an additional but underrecognized pattern of injury in the term newborn that occur in the setting of perinatal stroke, either arterial or venous. Although perinatal strokes occur in 1 in 4000 live births [3], the pathogenesis of this condition remains poorly understood. Many cases of perinatal stroke are associated with infection and cardiac and blood disorders [3]. Recent data suggest that focal infarctions are also associated with neonatal encephalopathy in the term newborn and hypoxic-ischemic injury [5]. Consistent with observations of childhood stroke, many newborns with stroke have multiple risk factors for brain injury, including intrapartum complications.

The premature newborn characteristically sustains injury to the white matter [1]. This pattern of injury evolves over the first weeks of life and is often seen in the setting of cardio-respiratory instability. When severe, this injury results in cystic periventricular leukomalacia (PVL), regions of coagulation necrosis and liquefaction in the periventricular white matter that are usually small, multiple, bilateral, fairly symmetrical, and well detected by ultrasonography [1]. It has also been recognized that these multifocal areas of coagulation necrosis are accompanied by a more diffuse abnormality of the cerebral white matter [1]. In studies using MRI to determine the spectrum of white matter injury in premature newborns, cystic PVL was distinctly uncommon, whereas noncystic white matter injury was very common and not accurately detected by ultrasound [8,9]. Premature newborns are also at particular risk of intraventricular hemorrhage (IVH) and periventricular hemorrhagic infarction. IVH, originating from the subependymal germinal matrix, is described based on the extent of bleeding into the lateral ventricles (Grades 1 to 3) [1]. Periventricular hemorrhagic infarction, previously referred to as "Grade 4" intraventricular hemorrhage, is characterized by hemorrhagic necrosis in the white matter adjacent to the ventricles that is usually unilateral and asymmetrical. Periventricular hemorrhagic infarction should be considered a form of venous stroke.

Because these different patterns of injury may be associated with different antenatal and perinatal risk factors, MRI can also be used to define more homogenous groups of newborns when studying the antecedents of neonatal encephalopathy and responses to new therapeutic interventions.

B. Biochemistry and Molecular Mechanisms

Neonatal brain injury involves a complex set of interrelated biochemical and molecular pathways including oxidative stress, excitotoxicity, inflammation, and genetic effects. Underlying these mechanisms is a selective vulnerability of specific cell types at specific maturational stages. Together these factors underlie the regional vulnerability of the developing brain, including the characteristic patterns of injury, as well as the prolonged temporal evolution of brain injury in the human newborn.

1. Selective Cellular Vulnerability

At different gestational ages in the newborn, characteristic patterns of injury are observed after acute hypoxic-ischemic insults. These areas of regional susceptibility suggest that specific cell types are selectively vulnerable at specific developmental stages in both humans and animal models. "Selective vulnerability" of neurons has been recognized for decades in the mature brain in a variety of disease states such as Parkinson's and Huntington's diseases [6,10]. Likewise, neuronal and glial cells in the immature nervous system appear to be selectively vulnerable to an ischemic insult when it occurs early in gestation [6,10]. It is now known that pre-oligodendrocytes and oligodendrocyte progenitor cells are more vulnerable than mature oligodendrocytes to acute ischemic injury [6,10]. In addition, another population of selectively vulnerable cells exists in the premature brain, the subplate neuron. These cells play a critical role in normal visual thalamocortical development and functional maturation [6,10]. The vulnerability of these cells based on maturational stage may explain why premature newborns are the population at highest risk of white matter injury and visual impairment. In contrast, the neonate born at term gestation is particularly at risk of injury to specific neurons in the deep gray nuclei and perirolandic cortex. The propensity for the neurons in the deep gray nuclei to be vulnerable is reflected in the patterns of injury observed on brain MRI in neonates who have suffered hypoxic-ischemia and who later develop severe neurodevelopmental sequelae [4]. The mechanisms responsible for this regional and cell type selective vulnerability are not entirely known, but oxidative stress and excitotoxicity are two of the primary mechanisms proposed [10].

2. Oxidative Stress

The developing brain faces unique challenges from oxidative stress, and its responses to injury and protective mechanisms are different from those of the mature nervous system. The neurons and glia in the neonatal brain are particularly vulnerable to oxidative damage because of the high concentration of unsaturated fatty acids, high rate of oxygen consumption, low concentrations of antioxidants, and availability of "free" redox-active iron (reviewed in [11]). Normally, reactive oxygen species (ROS) are produced at low concentrations as part of normal metabolism. Cells are protected from the deleterious effects of ROS via several enzymatic and nonenzymatic antioxidant pathways (Fig. 2). Enzymatic antioxidants include superoxide dismutase (SOD), which exists as cytoplasmic Cu,Zn-SOD (SOD1), mitochondrial Mn-SOD (SOD2), and extracellular-SOD (SOD3). SOD dismutates superoxide anion (O_2^-), an oxygen free radical produced after hypoxic-ischemic injury as a result of mitochondrial dysfunction, to hydrogen peroxide (H_2O_2). H_2O_2 is itself a toxic ROS and must be further detoxified to water and oxygen. This is accomplished enzymatically via mitochondrial glutathione peroxidase (GPx) or cytoplasmic catalase. GPx converts glutathione (GSH) to oxidized glutathione (GSSG), and GSH is then reformed via glutathione reductase (GR) at the expense of NADPH. Thus adequate cellular energy is essential for continued antioxidant protection.

Several lines of evidence suggest that activities and the responses of these enzymes to oxidative stress are age dependent. For instance, GPx levels are low early in gestation and then gradually increase to reach their maximal levels during adulthood. In fact, a marked decrease in GPx activity in the developing rat brain may occur during the first postnatal week. GPx activity remains constant from time 0 through 144 hours after hypoxia-ischemia in adult rats, but it falls dramatically at 2 and 24 hours after hypoxia-ischemia in neonatal mice. Furthermore, SOD1 overexpression results in marked neuroprotection in adult rats after hypoxia-ischemia, whereas SOD1 overexpression in the neonatal animal exacerbates brain injury. A possible explanation for the variable effect during different stages of development is that SOD1 transgenic adult mice show an adaptive rise in catalase activity, whereas neonatal SOD1 transgenic brains do not show an adaptive increase in either GPx or catalase. In fact, the brains of postnatal day 7 (p7) mice that overexpress SOD1 have a significantly higher H_2O_2 concentration at 24 hours after hypoxia-ischemia as compared with wild-type P7 mice. Several lines of evidence suggest that H_2O_2 is an important mediator of cell death. For instance, H_2O_2 has been shown to cause apoptosis in cultured oligodendroglia and neurons. In addition, studies show that degeneration in Down syndrome (DS) neurons, which overexpress SOD1 because of a third copy of the gene, occurs through apoptosis and can be prevented by treatment with free-radical scavengers or catalase [12]. The consequent imbalance in antioxidant enzyme activities allows for greater H_2O_2 production after hypoxia-ischemia and thus greater injury. In animals overexpressing Gpx, brain injury is ameliorated compared with their wild-type littermates, providing further evidence of the importance of H_2O_2 in the generation of hypoxic-ischemic brain injury [13–15].

In the very immature brain, oligodendrocyte progenitor cells and preoligodendrocytes are selectively vulnerable to antioxidant depletion or exposure to exogenous free radicals [16]. Mature oligodendrocytes, in comparison, are highly resistant to oxidative stress in part due to differences in expression levels of antioxidant enzymes and proteins involved in programmed cell death. These particular oligodendrocyte maturational characteristics may explain why white matter is often injured selectively in the premature newborn brain.

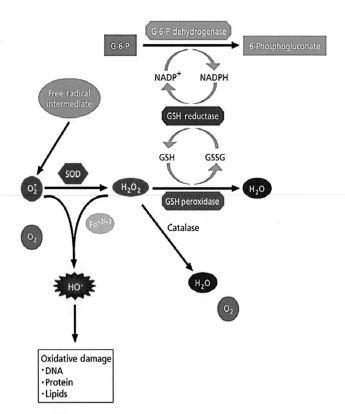

Figure 2 Oxidative stress. Mechanisms of oxidative damage and antioxidant pathways. Oxygen free radicals are generated when the enzymatic defenses within the cell are overwhelmed due to developmentally regulated deficiencies in times of energy crises.

3. Excitotoxicity

Excitotoxicity refers to excessive activation of glutamatergic pathways leading to cell death. In the newborn brain, evidence abounds for excitotoxic cell death in many cell types. For example, oligodendrocyte progenitor cells express non-*N*-methyl-*D*-aspartate (NMDA) glutamate receptors, including α–amino-3-hydoxy-5-methyl-4-isoxazolepropionic acid (AMPA) and kainate receptors. Blockade of these glutamate receptors protects against hypoxic-ischemic white matter injury in the immature rodent. The selective vulnerability of oligodendrocyte progenitor cells is due in part to aberrant activation of the calcium permeable AMPA/kainate receptor. The neuronal nitric oxide synthase–containing (nNOS) interneurons throughout the immature central nervous system (CNS) generate nitric oxide (NO) and are resistant to both acute and chronic oxidative stress (reviewed in [10]). The production of NO depends on the coupling and activation of the NMDA receptor and the nNOS enzyme through postsynaptic density proteins that allow for calcium entry and activation of the calcium-calmodulin dependent nNOS. Interruption of the PSD complex can prevent ischemic cell death [17], and in immature animals both PSD93 and PSD95 appear to be necessary for this to occur [18]. Therefore when NO is produced in excessive amounts during periods of oxidative stress in vulnerable brain regions such as the basal ganglia, it becomes a potent mediator of free radical damage. However, because NO-producing neurons themselves are resistant to both hypoxia-ischemia and NMDA-mediated excitotoxicity, and only in the immature brain, these cells become important mediators of the maturational response to ischemic injury. In regions where the immature NMDA receptor is expressed, the basal ganglia nNOS neurons are especially abundant. Elimination of nNOS neurons and disruption of the postsynaptic density complex that links NMDA to nNOS result in reduced ischemic injury [10,17]. Therefore both excitotoxicity and oxidative stress, and their interaction, appear to mediate neonatal brain damage from hypoxia-ischemia. However, mechanisms of selective vulnerability must be considered within the context of normal cell and system development because differences in receptor and enzyme system ontogeny contribute to the patterns of early brain injury.

4. Inflammation

Experimental evidence suggests that inflammatory signals from proinflammatory cytokines, prostaglandins, and lipopolysaccharide trigger microglia to further release cytokines and oxygen free radicals (reviewed in [19]). These inflammatory signals may have a wide range of effects on neuronal and glial structure and function [19]. Chorioamnionitis has been associated with poor neurological outcome in the preterm and term infant. However, the identification of maternal infection as a cause for neonatal brain injury is problematic given the limited descriptors used to define chorioamnionitis and the frequent lack of placental histology to support this diagnosis. Although white matter injury in the premature newborn has been associated with chorioamnionitis, the association of brain injury with elevated levels of proinflammatory cytokines is more variable. This variability may relate directly to the tissues being sampled (e.g., amniotic fluid, cord blood, or neonatal blood). Despite these conflicting data and methodological issues, the role of the *fetal* systemic and CNS inflammatory response is critical to understanding the genesis of brain injury in the newborn.

In a recent large multicenter cohort, postnatal infection in low-birth-weight newborns was associated with impaired neurodevelopmental and growth outcomes in childhood [20]. Elevations of proinflammatory cytokines in the amniotic fluid, umbilical cord blood, and cerebrospinal fluid in premature newborns have been associated with white matter injury [19]. However, the association of elevated proinflammatory cytokines in premature infants with adverse neurodevelopmental outcomes is less consistent. This discrepancy may be due, at least in part, to the deleterious effects of other noninflammatory conditions that affect premature infants. The relationship between infection, proinflammatory cytokines, white matter injury, and adverse neurodevelopmental outcome is complex. The mechanisms by which postnatal infection and other proinflammatory states may lead to adverse neurodevelopmental outcomes need to be clarified in future investigations.

5. Genetic Effects

Clinical observation tells us that not every baby with a similar "insult" will have the same injury observed on MRI, nor the same neurodevelopmental outcome. The variability is also observed in animal models [6] and appears to depend on the genetic background. Evidence suggests that certain polymorphisms may increase the risk for many complex diseases. Recently, polymorphisms of the Apolipoprotein E genes, which regulate cholesterol and fatty acid metabolism, have been linked to neurodevelopmental performance at 2 years of age [21]. Despite these observations, genetic "susceptibility" factors for neonatal brain injury have not yet been clearly identified. Large population-based studies are needed to investigate genetic risk factors for neonatal brain injury.

C. Potential Therapies

As expected from these laboratory experiments, the immature brain responds differently to therapies than does the mature brain. Therapies designed to ameliorate brain injury in the adult may even accentuate developmental cell death programs. In the rodent brain, pharmacological blockade of NMDA receptors or potentiation of $GABA_A$ receptors triggers widespread apoptosis [22]. Despite this, several drugs are effective in large animal models, even when used *after* the insult. Allopurinol, desferoxamine, and 3-iminobiotin, drugs that interrupt free radical injury, ameliorate brain injury in a variety of models [6]. Pharmacologically administered erythropoietin may also prove to be neuroprotective in the newborn [23]. Exciting early results from several studies indicate that moderate hypothermia is safe in the high-risk newborn, and in at least one randomized clinical trial, moderately encephalopathic newborns (determined by amplitude integrated electroencephalography [EEG] cerebral function monitor [CFM] criteria) had better neurodevelopmental outcomes at 18 months compared with the normothermic group [24]. A key question that remains is how best to combine therapies aimed at various phases of the injury cascade to optimize protection and repair.

IV. Explanation of Symptoms in Relation to Pathophysiology and Natural History

A. Clinical Syndromes

The clinical signs and symptoms of neonatal brain injury depend on the severity, timing, and duration of the insult. Just as the newborn's gestational age is a critical determinant of the regional distribution of brain injury, the newborn's maturity needs to be taken into consideration when evaluating the clinical syndrome. Neonatal encephalopathy and seizures are the most overt manifestation of the severity of brain injury in the newborn period. Equally important are the "silent" clinical syndromes, which have subtle manifestations that may not be recognized on clinical examination. For example, stroke and perinatal white matter injury are often not recognized in the neonatal period. The clinical syndromes are related to the pathophysiology of neonatal brain injury, and their progression reflects the natural history of injury in the developing brain.

B. Neonatal Encephalopathy

Neonatal encephalopathy is strongly associated with neurodevelopmental disability in term infants: 15–20% of affected infants die during the newborn period, and an additional 25% sustain permanent neurological deficits [1]. The terms *hypoxic-ischemic encephalopathy* and *birth asphyxia* have been used to describe this clinical state, but they are misleading because hypoxia-ischemia or asphyxia are not *documented* in many cases of neonatal encephalopathy.

From studies of *risk factors* for neonatal encephalopathy, many cases have antepartum risks such as maternal hypotension, infertility treatment, and thyroid disease, and some have both antepartum and intrapartum indicators. A minority of cases have only intrapartum risk factors such as maternal fever, difficult delivery involving forceps, breech extraction, cord prolapse, or abruptio placentae [7]. Postnatal complications such as severe respiratory distress and sepsis are present in fewer than 10% of cases of neonatal encephalopathy in the *term* infant. Although antenatal risk factors may be present, infants with neonatal encephalopathy studied prospectively with MRI are more often diagnosed with brain injury occurring at or near the time of birth [4,5].

The severity of neonatal encephalopathy depends on both timing and duration of the insult. The major components of the neonatal encephalopathy syndrome include alertness, tone, respiratory status, reflexes, feeding, and seizures. Symptoms usually evolve over days, making *serial* detailed examinations important. The infant's gestational age is relevant to the interpretation of the symptoms. In the first hours after a severe ischemic insult, the neonate exhibits a depressed level of consciousness, often accompanied by periodic breathing with apnea or bradycardia. At this time, cranial nerve function is likely spared, with intact pupillary responses and eye movements if the injury is not severe. Involvement of cortical regions is manifested by hypotonia with decreased movement or as seizure activity seen in the severely affected infants very soon after the insult. Over the first days of life, there may be a transient increase in the level of alertness, but other signs of neurological improvement do not accompany this change. Refractory seizures are seen during this period, accompanied by apneic episodes, shrill cry, and jitteriness. Weakness in the proximal limbs and exaggeration of Moro and muscle stretch reflexes are observed. Eventually, level of consciousness deteriorates with respiratory arrest and other signs of brainstem dysfunction. Surviving newborns may have an improvement in level of consciousness; hypotonia and weakness in the proximal limbs, face, and bulbar musculature persist. The use of an encephalopathy score as a bedside tool can standardize the approach to the newborn with brain injury and help in the selection of neonates who require therapeutic intervention [2] (Table 1).

The neurodevelopmental outcome following neonatal encephalopathy is variable and may include deficits of motor, visual and cognitive functions. The severity and duration of the encephalopathy as well as the severity of brain lesions on MRI are predictive of neurodevelopmental outcome following neonatal encephalopathy in the term newborn. Although the risk of an abnormal neurodevelopmental outcome increases with the severity of the injury, the pattern of brain injury on neuroimaging also conveys important prognostic information. Abnormal signal intensity in the posterior limb of the internal capsule on MRI and injury predominantly in the basal ganglia and thalamus

Table 1 The Encephalopathy Score

Sign	Score = 0	Score = 1
Feeding	Normal	Gavage feeds, gastrostomy tube or not tolerating oral feeds
Alertness	Alert	Irritable, poorly responsive, or comatose
Tone	Normal	Hypotonia or hypertonia
Respiratory status	Normal	Respiratory distress (need for CPAP or mechanical ventilation)
Reflexes	Normal	Hyperreflexia, hyporeflexia, or absent reflexes
Seizure	None	Suspected or confirmed clinical seizure

CPAP, Continuous positive airway pressure.
Newborn infants were scored daily for the first 3 days of life, and the maximal score was used for analysis. The ES was assigned only on days that the subject was not sedated or paralyzed. From Miller, S. P., Latal, B., Clark, H., et al. (2004). Clinical signs predict 30-month neurodevelopmental outcome after neonatal encephalopathy. *Am. J. Obstet. Gynecol.* **190**, 93–99.

are associated with severely impaired motor and cognitive outcomes [4,25]. Cognitive deficits may result from the cerebral cortical injury that frequently accompanies basal nuclei injury on MRI. In contrast, the watershed pattern of injury is most commonly associated with cognitive impairments that are not accompanied by major motor deficits [4]. The cognitive deficits associated with watershed injury may not be apparent in early infancy and often need more prolonged follow-up for detection. It should be stressed that abnormal outcome following neonatal encephalopathy is not limited to cerebral palsy *and* often requires follow-up beyond 1 year of age to be detected [4]. Postnatal factors such as socioeconomic status also have important effects on neurodevelopmental outcome following brain injury in the term and preterm newborn.

C. Neonatal Seizures

Seizure activity in neonates is often subtle and difficult to distinguish on clinical examination alone. The most common manifestation of seizures in the newborn are ocular movements such as tonic horizontal deviation of the eyes, sustained eye opening or blinking, orolingual movements such as tongue or lip smacking or sucking, rowing or bicycling movements of the extremities, or recurrent apnea. Focal clonic seizures are seen often in the setting of arterial or venous infarction. Bedside monitoring devices such as the amplitude-integrated EEG (aEEG) [26] aid in the evaluation of newborns for seizure activity but often require validation with a standard EEG.

As with neonatal encephalopathy, it must be recognized that many etiologies for seizures in the newborn exist, including metabolic disorders and malformations of cerebral development. Imaging the brain with MRI can be very helpful in distinguishing between hypoxic ischemic and other forms of metabolic or genetic disease [27]. MRI also provides information regarding traumatic and infectious etiologies in conjunction with laboratory testing. Reversible causes of seizures such as hypoglycemia, hypocalcemia, hyponatremia, hypoxemia, acidosis, and hyperbilirubinemia should be diagnosed because they are frequently associated with an underlying disorder and may exist in the setting of a global hypoxic-ischemic insult (Fig. 3). Additionally, the timing of neonatal seizures often provides important information regarding their etiology. Lumbar puncture is a critical tool for diagnosing certain genetic disorders, such as the pediatric neurotransmitter diseases and glucose transporter defects, in addition to its essential role in excluding infectious etiologies (e.g., meningitis or encephalitis).

The neurodevelopmental prognosis of neonatal seizures appears to depend on their etiology. In the setting of neonatal encephalopathy, seizures are associated with an

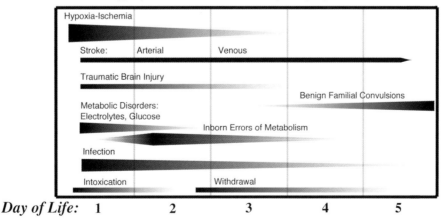

Figure 3 Common etiologies of neonatal seizures. The most common etiologies of neonatal seizures are plotted by their most common day(s) of presentation.

increased risk of adverse neurodevelopmental outcome [2]. Yet neonatal seizures do not always imply a poor neurodevelopmental prognosis, particularly in the setting of transient metabolic disturbances or benign epilepsy syndromes such as benign familial neonatal convulsions. Although there is some evidence from experimental models and clinical investigation that seizures themselves may be harmful to the developing brain [1], it is still unclear how aggressively neonatal seizures should be treated, for how long, and with what medication. Furthermore, seizures in the newborn period are difficult to treat effectively with the medications most commonly used in clinical practice such as phenobarbital and phenytoin. Clinical trials are desperately needed to determine how to optimally treat neonatal seizures and improve neurodevelopmental outcome.

D. Neonatal Stroke

Stroke occurs in 1 in 4000 live births [3]. In many affected newborns the diagnosis is only made retrospectively because they appear healthy in the immediate newborn period [3]. Neonatal strokes are often arterial in origin and ischemic in nature, although at least 30% are due to sinovenous thrombosis [28]. Prepartum factors such as preeclampsia and intrauterine growth restriction have been implicated in the etiology of neonatal stroke [29]. Other risk factors include coagulation abnormalities (protein C or S and antithrombin III deficiencies, and elevated lipoprotein-a and homocysteine) as well as certain genetic polymorphisms (factor V G1691A, factor II G 20210A, methylenetetrahydrofolate reductase C677T) [3]. These disorders may be particularly common in newborns with venous stroke from cerebral venous thrombosis where multiple risk factors, including perinatal hypoxia-ischemia, are frequently present. Fortunately, the risk of recurrence for neonatal stroke is extremely low (<5%) and is associated with intercurrent illnesses or complications of systemic disorders such as congenital heart disease.

IVH that occurs commonly in very premature newborns should not be confused with in utero hemorrhagic stroke, as IVH is not associated with poor neurodevelopmental outcome in the absence of parenchymal brain injury or hydrocephalus [30]. The age of the baby at the time of the stroke influences the pattern of brain injury on MRI and the clinical findings later in life. Although hemiplegic cerebral palsy is a common sequela of strokes occurring late in gestation, some children present later in life with more subtle findings (e.g., mild dystonia and cognitive disabilities). In children presenting later in life, neuroimaging with MRI often does not distinguish the precise timing of the injury. The development and application of more accurate in utero measures of brain injury, such as fetal MRI, may help elucidate the timing and etiology of these conditions.

E. Subtle Neonatal Syndromes

Subtle brain injury syndromes, such as subclinical stroke in the term newborn or white matter injury in the preterm newborn, may take months to identify. Advances in brain monitoring and neuroimaging in the intensive care nursery have increased our ability to identify early brain injury in high-risk newborns. These procedures include head ultrasound and aEEG. MRI has been the definitive test for identifying clinically silent injury such as white matter injury and brain volume loss but is not yet available for sick newborns in many hospitals. MRI-compatible incubators for transporting sick newborns are expensive and still commercially available in a limited fashion.

F. Insights from Neuroimaging

MRI is increasingly recognized as a valuable tool for determining the timing and etiology of neonatal brain injury. MRI used shortly after birth can also elucidate the timing and evolution of the injury [4,5,31]. Because the first step in caring for an encephalopathic newborn is establishing the etiology of the brain injury, MRI should be performed initially to narrow the differential diagnosis that may include, in addition to hypoxic-ischemic encephalopathy, an underlying metabolic, genetic, vascular, or inflammatory disease requiring intervention.

1. Advanced MR Techniques

Advanced MR techniques, such as diffusion and spectroscopic imaging, allow us to objectively determine the progression of brain injury in the newborn. DWI is a MR technique that is sensitive to alteration in free water diffusion. Diffusion tensor imaging (DTI) uses special MR techniques to quantify both the amount of water motion reflected in the apparent diffusion coefficient of water (ADC, or average diffusivity) and the directionality of this motion, reflected in the fractional anisotropy. DTI must be considered in the context of brain maturation, as there are dramatic changes in these diffusion parameters with brain development. With increasing gestational age, ADC decreases in gray and white matter, and fractional anisotropy increases in white matter, at a time before myelination is evident on conventional MR imaging [32,33]. The developmental changes in these diffusion parameters also differ in specific brain regions that appear homogenous on conventional MRI. DWI and DTI also provide sensitive *in vivo* measures of regional brain injury [31,34]. In addition, DTI has been used in the human newborn following brain injury to detect abnormalities of microstructural brain development remote from the sites of primary injury in areas of the brain that are normal on conventional MRI [32,33,35].

MRS provides an *in vivo* quantitative measure of brain biochemistry from a given region of the brain. MRS can be done serially to provide a unique window on the change in brain metabolism following brain injury in the newborn. Of the compounds measured by MRS at long echo times, N-acetylaspartate (NAA) and lactate are the most useful in assessing brain injury because abnormalities in these compounds are highly predictive of neurodevelopmental outcome following neonatal brain injury [27,36]. NAA is an acetylated amino acid found in high concentrations in the neurons of the CNS [27,36]. Metabolite levels measured by MRS change dramatically with brain development, with an increase in NAA related to advancing cerebral maturity [27,36]. Relevant to neonatal encephalopathy, NAA decreases with cerebral injury or impaired cerebral metabolism [27,36]. Lactate is normally produced in the brain by astrocytes and used as fuel by neurons to replenish energy stores via oxidative phosphorylation [37]. Elevation of lactate reflects an ongoing disturbance in cerebral energy substrate delivery and oxidative metabolism such as that secondary to hypoxia-ischemia. Myo-inositol is another metabolite measured via proton MRS and is best measured at short echo times. Myo-inositol is one of the major osmolytes present in the brain and is elevated following hypoxic-ischemic brain injury in the term newborn, reflecting increased brain osmolality after ischemia [38].

2. Progression of Neonatal Brain Injury

DTI also allows earlier detection of brain injury than standard MRI techniques in neonates because brain regions with acutely injured cells have decreased local water diffusion, even within 24 hours of the insult [31,34]. It is critical to interpret diffusion imaging studies in the newborn in the context of the time between injury and the acquisition of the diffusion data [31,34]. Data in human and animal models demonstrate that reductions in free water diffusion reach their nadir by 2–4 days after perinatal injury [31]. Thus MR diffusion images obtained before the nadir, as in the first day following an injury, may not show the full extent of injury. Importantly, diffusion abnormalities persist for 7–8 days in the newborn before returning toward normal values (pseudonormalization) and ultimately reflect increased diffusion [31,34]. Data from MRS studies in human newborns mirror this prolonged time course of injury progression. In the first 24 hours following brain injury in the term newborn, lactate increases, followed by a decrease in NAA in the 3 days postinjury [27,34,36]. As short–echo time MRS is more challenging to quantify and not as widely available as long–echo time MRS, less is known about the dynamic changes in myo-inositol in term neonatal encephalopathy. The prolonged progression of neonatal brain injury days after the insult is consistent with the mechanisms of cell injury that persist for days following neonatal hypoxic-ischemic brain injury and the findings of delayed neurodegeneration in animal models with this condition [6].

Diffusion imaging can be used to identify periventricular white matter injury in its earliest state in the preterm newborn [32,33]. These early and persistent white matter abnormalities are associated with impaired cerebral development in cortical gray matter and in widespread white matter regions [32,33,35]. Additionally, diffuse white matter atrophy, ventriculomegaly, immature gyral development, and enlarged subarachnoid space are often observed when newborns delivered extremely preterm (<26 weeks gestation) are scanned with MRI at term-equivalent age [8].

Together these data suggest an evolution of neonatal brain injury in newborns *after* birth and that the opportunity to intervene to prevent or ameliorate brain injury may extend over *days* in the term newborn and possibly over *weeks* in the preterm newborn. Furthermore, these advanced MR techniques offer a dynamic surrogate measure of brain injury in the newborn that can be safely used to determine the short-term effects of novel intervention strategies.

In summary, MRI can be used to identify patterns of brain injury that are predictive of particular neurodevelopmental syndromes later in childhood [4,25]. Because the neonatal brain is still myelinating and has a greater water content than the mature brain, not only does injury appear differently, the time course of evolution of injury differs from that seen in the adult brain [4,5,25]. Advanced MR imaging techniques, such as diffusion tensor imaging and MR spectroscopic imaging, can be used to define the injury at its inception and to predict neurodevelopmental outcome to determine which infant in the newborn period is at highest risk of adverse neurological sequelae. Because recent data suggest that a prolonged window of opportunity to test interventions may exist, selecting neonates who should receive therapy becomes a major priority. The combination of clinical signs of neonatal encephalopathy and the pattern and severity of injury on neuroimaging should improve our ability to evaluate and implement novel strategies of cerebral protection.

V. Summary

The past decade has seen tremendous advances in our understanding of neonatal brain injury, including the basic mechanisms of neonatal brain injury and the use of advanced neuroimaging techniques. There remains a pressing need to establish the mechanistic link between antenatal and perinatal risk factors and the etiology of brain injury in the newborn. This is critical to the prevention of acquired neonatal brain injury and may be achieved with the development and application of more accurate in utero measures of brain injury, such as fetal MRI. The early recognition of newborns at risk using advanced neuroimaging modalities, combined with rational intervention and therapeutics, may result in the prevention or reduction of lifelong disabilities such as cerebral palsy, epilepsy, and behavioral and learning disorders. These disabilities continue to affect the surviving baby, the family, and society. A better understanding of the mechanisms that underlie the vulnerability of the developing CNS to injury may provide opportunities to intervene therapeutically and influence outcome for all ages.

References

1. Volpe, J. (2001). "Neurology of the Newborn," ed 4. Saunders, Philadelphia.
2. Miller, S. P., Latal, B., Clark, H., Barnwell, A., Glidden, D., Barkovich, A. J., Ferriero, D. M., and Partridge, J. C. (2004). Clinical signs predict 30-month neurodevelopmental outcome after neonatal encephalopathy. *Am. J. Obstet. Gynecol.* **190**, 93–99.
3. Nelson, K. B., and Lynch, J. K. (2004). Stroke in newborn infants. *Lancet Neurol.* **3**, 150–158.
4. Miller, S. P., Ramaswamy, V., Michelson, D., Barkovich, A. J., Holshouser, B., Wycliffe, N., Glidden, D. V., Deming, D., Partridge, J. C., Wu, Y. W., Ashwal, S., and Ferriero, D. M. (2005). Patterns of brain injury in term neonatal encephalopathy. *J. Pediatr.* **146**, 453–460.
5. Cowan, F., Rutherford, M., Groenendaal, F., Eken, P., Mercuri, E., Bydder, G. M., Meiners, L. C., Dubowitz, L. M., and de Vries, L. S. (2003). Origin and timing of brain lesions in term infants with neonatal encephalopathy. *Lancet* **361**, 736–742.
6. Ferriero, D. M. (2004). Neonatal brain injury. *N. Engl. J. Med.* **351**, 1985–1995.
7. Badawi, N., Kurinczuk, J. J., Keogh, J. M., Alessandri, L. M., O'Sullivan, F., Burton, P. R., Pemberton, P. J., and Stanley, F. J. (1998). Intrapartum risk factors for newborn encephalopathy: the Western Australian case-control study. *B. M. J.* **317**, 1554–1558.
8. Inder, T. E., Wells, S. J., Mogridge, N. B., Spencer, C., and Volpe, J. J. (2003). Defining the nature of the cerebral abnormalities in the premature infant: a qualitative magnetic resonance imaging study. *J. Pediatr.* **143**, 171–179.
9. Miller, S. P., Ferriero, D. M., Leonard, C., Piecuch, R., Glidden, D. V., Partridge, J. C., Perez, M., Mukherjee, P., Vigneron, D. B., and Barkovich, A. J. (2005). Early brain injury in premature newborns detected with MRI is associated with adverse early neurodevelopmental outcome. *J. Pediatr.* **147**, 609–616.
10. McQuillen, P. S., and Ferriero, D. M. (2004). Selective vulnerability in the developing central nervous system. *Pediatr. Neurol.* **30**, 227–235.
11. Ferriero, D. M. (2001). Oxidant mechanisms in neonatal hypoxia-ischemia. *Dev. Neurosci.* **23**, 198–202.
12. Busciglio, J., and Yankner, B. A. (1995). Apoptosis and increased generation of reactive oxygen species in Down's syndrome neurons in vitro. *Nature* **378**, 776–779.
13. Sheldon, R. A., Jiang, X., Francisco, C., Christen, S., Vexler, Z. S., Tauber, M. G., and Ferriero, D. M. (2004). Manipulation of antioxidant pathways in neonatal murine brain. *Pediatr Res.* **56**, 656–662.
14. McLean, C., and Ferriero, D. (2004). Mechanisms of hypoxic-ischemic injury in the term infant. *Semin. Perinatol.* **28**, 425–432.
15. McLean, C. W., Mirotchnitchenko, O., Claus, C. P., Noble-Haeussslein, L. J., and Ferriero, D. M. (2005). Overexpression of glutathione peroxidase protects immature murine neurons from oxidative stress. *Dev. Neurosci.* **27**, 169–175.
16. Baud, O., Greene, A. E., Li, J., Wang, H., Volpe, J. J., and Rosenberg, P. A. (2004). Glutathione peroxidase-catalase cooperativity is required for resistance to hydrogen peroxide by mature rat oligodendrocytes. *J. Neurosci.* **24**, 1531–1540.
17. Aarts, M., Liu, Y., Liu, L., Besshoh, S., Arundine, M., Gurd, J. W. Wang, Y. T., Salter, M. W., and Tymianski, M. (2002). Treatment of ischemic brain damage by perturbing NMDA receptor-PSD-95 protein interactions. *Science* **298**, 846–850.
18. Jiang, X., Mu, D., Sheldon, R. A., Glidden, D. V., and Ferriero, D. M. (2003). Neonatal hypoxia-ischemia differentially upregulates MAGUKs and associated proteins in PSD-93-deficient mouse brain. *Stroke* **34**, 2958–2963.
19. Hagberg, H., and Mallard, C. (2005). Effect of inflammation on central nervous system development and vulnerability. *Curr. Opin. Neurol.* **18**, 117–123.
20. Stoll, B. J., Hansen, N. I., Adams-Chapman, I., Fanaroff, A. A., Hintz, S. R., Vohr, B., Higgins, R. D., and National Institute of Child Health and Human Development Neonatal Research Network. (2004). Neurodevelopmental and growth impairment among extremely low-birth-weight infants with neonatal infection. *JAMA* **292**, 2357–2365.
21. Wright, R. O., Hu, H., Silverman, E. K., Tsaih, S. W., Schwartz, J., Bellinger, D., Palazuelos, E., Weiss, S. T., and Hernandez-Avila, M. (2003). Apolipoprotein E genotype predicts 24-month bayley scales infant development score. *Pediatr. Res.* **54**, 819–825.
22. Ikonomidou, C., Bosch, F., Miksa, M., Bittigau, P., Vockler, J., Dikranian, K., Tenkova, T. I., Stefovska, V., Turski, L., and Olney, J. W. (1999). Blockade of NMDA receptors and apoptotic neurodegeneration in the developing brain. *Science* **283**, 70–74.
23. Sola, A., Wen, T. C., Hamrick, S. E., and Ferriero, D. M. (2005). Potential for protection and repair following injury to the developing brain: a role for erythropoietin? *Pediatr. Res.* **57**, 110R–117R.
24. Gluckman, P. D., Wyatt, J. S., Azzopardi, D., Ballard, R., Edwards, A. D., Ferriero, D. M., Polin, R. A., Robertson, C. M., Thoresen, M., Whitelaw, A., and Gunn, A. J. (2005). Selective head cooling with mild systemic hypothermia after neonatal encephalopathy: multicentre randomised trial. *Lancet* **365**, 663–670.
25. Rutherford, M. A., Pennock, J. M., Counsell, S. J., Mercuri, E., Cowan, F. M., Dubowitz, L. M., and Edwards, A. D. (1998). Abnormal magnetic resonance signal in the internal capsule predicts poor neurodevelopmental outcome in infants with hypoxic-ischemic encephalopathy. *Pediatrics* **102**, 323–328.
26. Toet, M. C., van der Meij, W., de Vries, L. S., Uiterwaal, C. S., and van Huffelen, K. C. (2002). Comparison between simultaneously recorded amplitude integrated electroencephalogram (cerebral function monitor) and standard electroencephalogram in neonates. *Pediatrics* **109**, 772–779.
27. Barkovich, A. J. (2000). "Pediatric Neuroimaging," ed. 3. Lippincott, Williams & Wilkins, Philadelphia.
28. deVeber, G., and Andrew, M. (2001). Cerebral sinovenous thrombosis in children. *New Engl. J. Med.* **345**, 417–423.
29. Wu, Y. W., March, W. M., Croen, L. A., Grether, J. K., Escobar, G. J., and Newman, T. B. (2004). Perinatal stroke in children with motor impairment: a population-based study. *Pediatrics* **114**, 612–619.
30. De Vries, L. S., Van Haastert, I. L., and Rademaker, K. J., Koopman, C., Groenendaal, F. (2004). Ultrasound abnormalities preceding cerebral palsy in high-risk preterm infants. *J. Pediatr.* **144**, 815–820.
31. McKinstry, R. C., Miller, J. H., Snyder, A. Z., Mathur, A., Schefft, G. L., Almli, C. R., Shimony, J. S., Shiran, S. I., and Neil, J. J. A prospective, longitudinal diffusion tensor imaging study of brain injury in newborns. *Neurology* **59**, 824–833.
32. Neil, J. J., and Inder, T. E. (2004). Imaging perinatal brain injury in premature infants. *Semin. Perinatol.* **28**, 433–443.
33. Huppi, P. S., and Inder, T. E. (2001). Magnetic resonance techniques in the evaluation of the perinatal brain: recent advances and future directions. *Semin. Neonatol.* **6**, 195–210.

34. Barkovich, A. J., Westmark, K. D., Bedi, H. S., Partridge, J. C., Ferriero, D. M., and Vigneron, D. B. (2001). Proton spectroscopy and diffusion imaging on the first day of life after perinatal asphyxia: preliminary report. *Am. J. Neuroradiol.* **22**, 1786–1794.
35. Miller, S. P., Vigneron, D. B., Henry, R. G., Bohland, M. A., Ceppi-Cozzio, C., Hoffman, C., Newton, N., Partridge, J. C., Ferriero, D. M., and Barkovich, A. J. (2002). Serial quantitative diffusion tensor MRI of the premature brain: Development in newborns with and without injury. *J. MRI* **16**, 621–632.
36. Novotny, E., Ashwal, S., and Shevell, M. (1998). Proton magnetic resonance spectroscopy: an emerging technology in pediatric neurology research. *Ped. Res.* **44**, 1–10.
37. Pellerin, L., and Magistretti, P. J. (2004). Neuroscience. Let there be (NADH) light. *Science* **305**, 50–52.
38. Robertson, N. J., Lewis, R. H., Cowan, F. M., Allsop, J. M., Counsell, S. J., Edwards, A. D., and Cox, I. J. (2001). Early increases in brain myo-inositol measured by proton magnetic resonance spectroscopy in term infants with neonatal encephalopathy. *Pediatr. Res.* **50**, 692–700.

55

Spina Bifida

Stephen L. Kinsman, MD

Keywords: *Chiari malformation, hydrocephalus, hydromyelia, infantile brainstem syndrome, lipomyelomeningocele, meningomyelocele, myelodysplasia, myelomeningocele, neural tube defect, spina bifida, spinal dysraphism, syrinx, tethered spinal cord*

I. Brief History and Nomenclature
II. Etiology
III. Pathogenesis
IV. Pathophysiology
V. Relevant Structural Details
VI. Pharmacology, Biochemistry, and Molecular Mechanisms
VII. Natural History
   References

I. Brief History and Nomenclature

The term *spina bifida* refers to a group of malformations of the central nervous system (CNS) that always includes a spinal cord malformation. This author does not include isolated defects in the formation of posterior vertebral bodies within the group of disorders known as spina bifida. These vertebral body defects, if restricted to one or two vertebrae, are very common and are only rarely associated with underlying spinal cord malformations. The only exception is when cutaneous lesions overlie the bony defect. Two or more cutaneous lesions in the midline are highly associated with underlying spinal cord malformations and tethered spinal cord (TSC).

Most of the conditions considered as spina bifida are felt to result from abnormal neural tube fusion, closure, or both; therefore they are also known as *neural tube defects* (NTDs). The most common type of NTD, myelomeningocele (MMC), is a form of spina bifida associated with outpouching of the spinal cord and its coverings through an open defect of the posterior vertebral arches and overlying tissues. This form of spina bifida also involves much of the remaining neuraxis with the frequent presence of hydrocephalus of the brain, Chiari malformation of the posterior fossa and its contents, and involvement of the cervical spinal cord.

Early investigators of these conditions viewed these lesions as destructive in nature, although a few saw the possibility of abnormal embryogenesis as the cause. In fact, the term *spinal dysraphism*, another term often used to describe these conditions, derives from notions of abnormal embryogenesis.

Much early experimental embryology was done on chick embryos, laying the groundwork for our understanding of the various tissue planes involved in CNS development (i.e., ectoderm, mesoderm, and endoderm). The inductive phenomenon required for the orderly creation of the neural plate, then the neural folds, and finally the neural tube continues to be elucidated [1]. *Dysraphism* refers to this process and its aberrations.

Table 1 outlines the various types of spinal dysraphism. Open lesions can range from severe forms of anencephaly and craniorachischisis to small sacral spinal bifida aperta or cystica. Closed lesions range from the complex diastematomyelia to asymptomatic MMCs.

The introduction of cerebrospinal fluid (CSF) ventricular shunting in the late 1950s changed the natural history of this condition, both in terms of marked decrease in mortality and also significant improvements in function. More recent concepts of the TSC syndrome and Chiari malformation pathophysiology have led to increased surgical intervention in MMC. Rapid evolution of clinical management in a setting where natural history has been inadequately studied and understood has led to a situation in which there is some confusion as to the magnitude of added benefit these interventions contribute. However, clinical consensus is that these interventions do indeed contribute significantly to the health and well-being of individuals with these conditions.

Table 1 Various Forms of Spinal Dysraphism

Open types
Cranial
    Anencephaly
Spinal (myelomeningocele)
    Aperta
    Cystica
    Craniorachischisis
Closed types
Posterior vertebral body fusion defect without spinal cord malformation
    (so-called *spina bifida occulta*)
Forms of occult spinal dysraphism
    Frequently with overlying cutaneous and/or subcutaneous
        abnormalities
        Lipomyelomeningocele and lipomeningocele
        Diastematomyelia and other split cord malformations
        Dermal sinus tracts
    Often without overlying cutaneous and/or subcutaneous abnormalities
        Spinal epidermoids and dermoids
        Thickened and/or fatty filum terminale

It is important that surgeons and other health professions with substantial experience with these conditions carefully consider their indications in individuals.

The biggest advance in this disabling neurological condition has been in primary prevention. Epidemiological studies pointed to a possible nutritional factor in the pathogenesis of neural tube defects [2]. Reduction of recurrence risks using folic acid supplementation introduced a few months before conception led to a 70% reduction in the incidence of neural tube defects. Initial trials were done in high-risk families with high recurrence risks. Later trials generalized this finding to wider populations [2].

Advances in uterine wall clamping led to a resurgence in fetal surgery in the 1990s. One surgery with initial promising reports was the closure of fetal MMC. The initial rationale for closing the open cele during the fetal period was the concept that exposure of the placode and nerve roots to amniotic fluid causes a secondary damage leading to increased disability in this condition. This rationale was supported by studies in animal models. It should be noted, however, that to date there has not been a demonstration of improved sensory, motor, or bladder/bowel function in those who have undergone fetal surgery for MMC [3]. On the other hand, surgeons performing these procedures noted a decrease in the occurrence and severity of hydrocephalus in those who underwent fetal MMC closure. Further investigation has led to the notion that the core improvement is in the severity and appearance of the Chiari malformation, most likely due to expansion of the fetal posterior fossa. This improvement leads to better CSF flow and lessening of hydrocephalus. The risk-benefit ratio of this procedure remains controversial, including the risk of premature birth, and currently the procedure is only sanctioned as part of a controlled trial at multiple centers within the United States.

More recently, the pace of discovery has quickened in several areas of inquiry into the pathogenesis of NTDs; these areas are beginning to converge. The development of molecular genetic studies in mouse, improvements in gene cloning, and the ability to disturb gene function via knock-out technology have led to an increase of mouse models of the neural tube defects. The study of *Drosophila* and *Caenorhabditis elegans* developmental genetics has led to a large number of new candidate genes important in the formation and closure of the neural tube [4,5]. Finally, unraveling of the molecular pathways critical to the normal development of the CNS has also increased our appreciation for the numbers of potential genes involved in this complex and highly regulated process [6,7].

II. Etiology

The etiology of MMC is most often multifactorial and includes both genetic and environmental factors [2]. There is a familial tendency, with a recurrence risk for neural tube defects among siblings of 5% after the first affected child. The risk increases to 10% after two affected children [8]. As noted previously, an association between folic acid intake and the risk of neural tube defects has been demonstrated. The multicenter Medical Research Council Vitamin Study confirmed a protective effect of periconceptual folic acid intake in preventing NTDs in more than 70% of at-risk pregnancies [9]. A more recent prospective study proved that the protective effects of folic acid extend to all pregnancies, not just those at risk. Some cases appear to be explained by a thermolabile variant of 5,10-methylenetetrahydrofolate reductase that causes high plasma homocysteine levels [5].

Other observed associations of an increased risk of NTDs are with valproic acid intake during early pregnancy, maternal hyperthermia, diabetes mellitus, obesity, Meckel-Gruber syndrome, Dandy-Walker syndrome, and trisomy 13. A paper describing an association between MMC and Waardenburg syndrome (type 3) in patients with interstitial deletions of 2q35 and the PAX3 gene suggested a digenic inheritance as a possible cause of some NTDs [10]. Recent mouse models of PAX3 mutants will hopefully elucidate potential mechanisms.

Geographical and temporal variations in the birth prevalence rates of NTDs are found [2,11]. The highest prevalence of NTDs has been recorded in Ireland (4 per 1000 live births) and in China (up to 10 per 1000 live births). The lowest prevalence occurs in the United States (0.5 per 1000 live births), continental Europe (1 per 1000 live births), and Japan (1 per 1000 live births).

The human studies into the preventative role of folic acid in reducing NTDs opened the question of folic acid's mechanism of action. Researchers began to test whether folic acid would reduce the incidence of NTDs in the various mouse mutants know to have a high incidence of NTDs [4]. Similar to the human situation, folic acid prevents NTD occurrence in several mouse models, including homozygous mice for a *Cart1* null mutation, *Cited2* knock-out mice, and the *splotch* mouse [4]. Also similarly, not all mutants are folic acid responsive, including cranial NTD in ephrin-A5 knock-out mice and spinal NTD in the curly-tail mutant mouse [4]. This has led investigators to look for other molecules that might also prevent NTDs, either independently of folic acid or synergistically. Another molecule found to have promise is inositol or myo-inositol [4].

III. Pathogenesis

The spinal cord is formed during embryogenesis by two processes: 1) neurulation and 2) canalization. Errors in either process can lead to a NTD, although most recent studies suggest that problems of neurulation are most important. Neurulation occurs at the third and fourth week of gestation when the flat neural plate forms the cylindrical neural tube. Completion of neurulation includes closure of both the anterior neuropore of the developing brain (at 21 days of gestation) and the posterior neuropore of the distal spinal cord (at 28 days of gestation). Caudal to this point of posterior closure (which is now believed to be at the lower sacral spinal cord level), the cauda equina and nerve roots form by clustering and cavitation of a group of cells, a process called *secondary neurulation*. Thus neural tube defects can be classified into 2 groups: 1) neurulation defects (upper neural tube defects) and 2) canalization defects (lower neural tube defects). The two types were believed by some to be different malformations with different etiologies, but more recent evidence suggests upper and lower spinal NTDs may be similar [2].

The biology of the formation of the neural folds and then their fusion is being unraveled [6]. Two major processes involved are apical constriction and convergent extension. Apical constriction is the process by which the midline cells of the neural plate become wedge like and contribute the shape changes necessary to convert the neural plate into neural folds. Convergent extension is the process by which the ventral midline cells of the neural tube rearrange in a longer, narrower array, thus providing shape changes critical to the formation of the spinal cord. Abnormalities in convergent extension can keep the nascent neural folds apart and thus prevent fusion of the neural folds.

Increasingly, the developmental processes that play out during ontogeny are found to be driven by molecular cascades and signaling pathways. These molecular systems of communication are then used to regulate various cellular processes needed for development. Some of these pathways underlie a basic cellular mechanism believed to be important in neural tube formation and closure, planar cell polarity (PCP) [7]. Evidence is increasing that PCP plays a role in convergence extension and that alterations in this pathway can lead to NTDs in mouse models.

The future holds great promise for a greater understanding of the causes of NTDs as each of these parallel lines of discovery continues to converge. A continued explosion of candidate genes is anticipated. Better clinical studies will improve our ability to test these candidate genes, as well as potential gene-environmental interactions. Although there is great redundancy in these molecular pathways, evidence suggests the possibility of double heterozygosity as a potential mechanism for the etiology of NTDs in humans [7].

IV. Pathophysiology

This section focuses on the pathophysiology of the complications of MMC. As MMC affects all parts of the neuraxis, each area is discussed briefly.

A. Chiari Malformation

The pathogenesis of symptoms and signs referable to Chiari malformation is multifactorial, including a relationship to hydrocephalus, vascular compromise due to compression, direct neuronal and axonal distortion, and congenital neural malformation [12]. These abnormal forces appear more important beyond infancy. In newborns with brainstem symptoms, it is less clear that posterior fossa decompression improves function and outcome.

B. Hydrocephalus

The causes of hydrocephalus relate to any pathophysiological process capable of altering CSF production, circulation, or absorption in such a way as to allow an increased accumulation of CSF in a brain that is unable to compensate for the resulting increase in CSF volume [13]. Much of hydrocephalus in MMC relates to CSF outflow problems, most likely secondary to Chiari malformation and other posterior fossa–related abnormalities.

C. Tethered Spinal Cord

During fetal life, the spinal cord extends to the sacral (coccygeal) area. As fetal development progresses, there is greater growth of the vertebral column than the spinal cord distally, which results in ascension of the spinal cord to the adult level of between the L1-L2 vertebral interspace by a few weeks before term.

When reviewing the etiology of TSC, one needs to separate the etiology of the situation causing the tethering from the causes for the neurological deterioration that is an integral part of the syndrome. What follows is a review of why tethering (or associated pathologies) is believed to lead to first reversible and then irreversible neurological deterioration.

Integral to the notion of cord tethering is a mechanical stretch of the distally tethered spinal cord [14]. This stretch secondarily leads to the following: 1) further stretch with growth and movement; 2) injury to spinal cord axons and nerve roots due to stretch or shear forces; and 3) cord or nerve ischemia as a result of vascular stretching or compromise.

However, there is evidence that not all problems and situations associated with TSC fit this pathophysiology. First, Hoffman's large series of cases found the majority of children with frank neurological deterioration to be around 6 years of age [15]. This is well before the major growth spurt of early adolescence, the time when maximal stretch on the tethered cord due to growth should occur. Second, many cases show atrophy of spinal cord up into the thoracic spinal cord, an area likely to be outside the area of influence exerted by the mechanical stretch of the distally tethered cord. Both of these findings suggest that other mechanisms are at least partially at work in causing the clinical manifestations of TSC.

Associated pathologies are likely to also play a role in pathogenesis as well, including: 1) dermoid sinuses or cysts resulting in chemical irritation; 2) compression from mass lesions such as lipomas and arachnoid cysts; 3) cord ischemia leading to the release of excitotoxic neurotransmitters and a cycle of delayed neuronal degeneration; 4) undetected past episodes of "subclinical" hydro- or syringomyelia with spinal cord distention and possibly resultant spinal canal compression, leaving an atrophic spinal cord with little evidence of the previous pathological process; and 5) potential "water-hammer" effects of external back trauma, transmitting traumatic forces to the underlying spinal cord.

D. Syrinx Formation

Many individuals with MMC also have one or more syrinxes. The cause of this complication is believed to be partly embryological and partly secondary to complications of CSF flow, CSF absorption, or both, over long periods of time. Much like in adult syringomyelia, expansion of the CSF-filled cavity leads to dysfunction in the surrounding cell types and fiber tracts. In MMC, most of these syrinxes exist within the central canal; sometimes they involve the entire length of the spinal cord and occasionally extend up into the brainstem. Resolution of the fluid accumulation may require aggressive treatment of hydrocephalus, Chiari malformation decompression, and sometimes direct syrinx decompression.

E. Neurogenic Bladder and Bowel

Abnormal innervation to the sphincters of the bladder outlet and anorectal apparatus leads to dysregulation of normal bladder and bowel physiology [16]. The obvious consequence of this is varying degrees of difficulty with continence and efficient evacuation of urine and stool. Less well known is the effect of this dysregulation on bladder pressure and function. Many individuals with neurogenic bladder end up with high-pressure bladders that are poorly compliant to filling with urine. This can lead to bladder wall hypertrophy, vesicoureteral reflux, and kidney damage. Advances in the management of neurogenic bladder, including regular and early use of anticholinergic agents, clean intermittent catheterization, and bladder surgeries, have led to significant decreases in these problems. Neurogenic bowel in MMC is dominated by lower motor neuron–type problems in the anus and pelvic floor musculature, leading to both leakage and poor evacuation of stool.

Fecal incontinence and fecal retention with constipation and/or obstipation are the most common complications. Management is directed at maintaining stool consistency and timed evacuation using such aids as suppositories, enemas, or both.

F. Orthopedic Deformities

Over time, the various neuromuscular imbalances seen in MMC lead to common orthopedic complications [11]. These include scoliosis, hip dislocations, pelvic obliquity, foot deformities, and contractures of the hips, knees, and ankles.

V. Relevant Structural Details

The structural issues related to spinal dysraphism relate to defining the nature of the primary lesion as well as the extent and severity of associated malformations. As discussed previously, the presence of open- vs closed-type lesions is pivotal in determining the presence of MMC. Most closed lesions are forms of occult spinal dysraphism. Most open lesions, whether cystic or flat, are MMC (Fig. 1). The complications of hydrocephalus, Chiari malformation, disorders of neuronal migration, and degrees of agenesis of the corpus callosum are uncommon in occult spinal dysraphism and almost uniformly found in MMC. TSC pathophysiology is often present in all forms of spinal dysraphism, whether open or closed, and should be actively considered as a cause of functional deterioration.

Most cases of occult spinal dysraphism are sacral in location, although diastematomyelia and dermal sinuses can

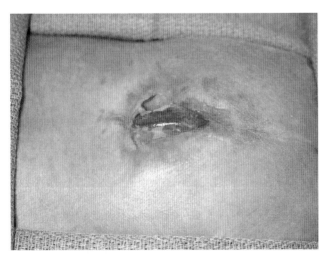

Figure 1 Example of an open spina bifida in a newborn with myelomeningocele. Note the neural placode in the center of the defect. Rostral is to the left. (Courtesy of Arthur DiPatri, Chicago, IL.)

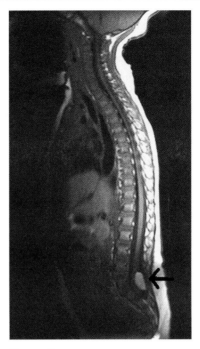

Figure 2 Sagittal magnetic resonance image of a child with lipomyelomeningocele. Note the high signal lesion in the posterior sacral region (*arrow*), which represents the abnormal accumulation of fat dorsal to the tethered spinal cord. Also note the posterior displacement of the spinal cord and lack of a distinct conus medullaris higher, at L1–L2.

occur at higher levels rostral to the primary dysraphic lesion (Fig. 2). Open spinal dysraphisms are distributed up and down the spinal axis. Roughly one-third of these lesions are sacral, one-third lumbar, and one-third thoracolumbar. Cervical lesions are less common. Cases of anencephaly (involvement of the rostral neural tube) and craniorachischisis (involvement of the entire neuraxis) are not compatible with life.

It is important to recognize that all levels of the neuraxis can be involved in MMC. Clinical deterioration requires investigation of all possible causes of altered pathophysiology because multiple mechanisms can cause similar symptoms. Decompensation of hydrocephalus needs to always be considered, even in individuals believed to have a working ventricular shunt. Many individuals with MMC have a syrinx above the cele that can become symptomatic at any time. Some of these individuals present with atrophy in the hand muscles. Both MMC and cases of occult spinal dysraphism can develop involvement of all four extremities due to TSC.

Chiari malformation is a complex malformation [12]. Structural components include a small posterior fossa, herniation of the cerebellum including the vermis and/or tonsils, herniation and/or downward herniation or tethering of the brainstem into the cervical spinal canal, and hypoplasia of

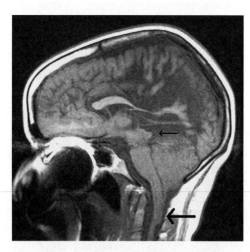

Figure 3 Sagittal view of a Chiari II malformation in a child with myelomeningocele. Note the herniation of the cerebellar vermis and brainstem (*thick arrow*). Also seen is the beaking of the tectum (*thin arrow*). Note the absence of a normal-shaped fourth ventricle.

the tentorium cerebelli (Fig. 3). An associated abnormality of the brainstem is the so-called *beaking of the tectum*, the severity of which has been recently associated with eye movement problems.

Hydrocephalus is a critical co-morbidity in MMC. The main pathophysiological determinant of hydrocephalus is accumulation of CSF under pressure [13]. Therefore expansion of the ventricular system is a structural marker of hydrocephalus severity. However, several confounding variables are present. Most importantly, much of the hydrocephalus occurs in utero and in the neonatal period, when skull compliance is high and may lead to stretching of the brain without concomitant injury. Also, some ventriculomegaly may be due to white matter hypoplasia rather than injury and atrophy and therefore may not correlate as well with the hydrocephalic process. This implies a degree of aberrant CNS connectivity and consequent effects on CNS function and development independent of hydrocephalus and its consequences. Also, sometimes scarring of the ventricular lining and surrounding brain occurs and decreases brain compliance. In this situation, high intracranial pressures can be seen with normal-sized or even slit-like ventricles.

VI. Pharmacology, Biochemistry, and Molecular Mechanisms

The dosage of folic acid supplementation required to optimize the prevention of this disabling birth defect remains controversial. Also, the administration of the folic acid remains a subject of controversy both in the United States and worldwide. The U.S. Public Health Service has made some recommendations regarding folic acid supplementation [17]. Others advocate for the fortification of food as a preferred approach to supplementation. Recent studies showing fortification to be more effective than vitamin use suggest that more work needs to be done to determine optimal approaches and their attendant public policies [2].

Table 2 outlines the sensory motor deficits seen in MMC and other dysraphisms. In general, the higher the anatomical lesion present, the more severe the sensorimotor impairment. However, it should be noted that exceptions exist. Some individuals with high lesions have little deficit. On the other hand, some individuals with small sacral lesions have greater deficits, particularly motor ones. Much of this appears due to brain- and brainstem–associated lesions and occasionally to the severity of hydrocephalus and its complications.

The degree of complete or incomplete paraplegia of the muscle groups below the level of the spinal lesion determines the subsequent motor impairment. The degree of motor impairment determined by the spinal level directly relates to the potential for functional ambulation. However, it has been described that individuals with similar muscle paresis can exhibit different ambulatory function for a variety of reasons, including deformities, cognition/motivation, and balance.

Symptoms and signs of increased intracranial pressure including headache, nausea, vomiting, lethargy, irritability, and personality change can occur at any time in the life of an individual with MMC. It is important to note that this

Table 2 Functional Levels in Myelomeningocele

Lesion Level	Motor Impairment	Ambulatory Function
S2 or below	Possible normal function	Can be normal
	Weakness of toes	May require AFO
	Foot deformities	May have Trendelenburg gait
S1	Weakness of plantar flexion	Usually requires AFO
L4–L5	Variable weakness of ankle dorsiflexion	Usually requires AFO
		Trendelenburg gait
L3–L4	Variable weakness of hamstrings and quadriceps	Often requires KAFOs
		May have a crouch gait
L1–L2	No movement at knees or below	Requires HKAFO or RGOs to ambulate
		Later uses wheelchair
T12 and above	No movement at hips or below	Usually uses wheelchair for mobility
		May ambulate when young with RGO or HKAFO

AFO, Ankle-foot-orthosis; *HKAFO*, hip-knee-ankle-foot-orthosis; *KAFO*, knee-ankle-foot-orthosis; *RGO*, reciprocating gait orthosis.

is true in both those with or without ventricular shunts in place. Those without a shunt usually present in the first few months of life with obvious increased intracranial pressure. However, some older, un-shunted individuals (including adults) with MMC eventually develop the need for a shunt to relieve intracranial pressure. Also, individuals with shunts can present with problems related to shunt malfunction at any time. Just because a shunt has not clinically malfunctioned for many years does not mean that it is now working effectively or that the shunt is not needed at all. Those with a ventricular shunt in place should always be viewed as being at risk for increased intracranial pressure from shunt malfunction or inadequacy.

The Chiari malformation is a common problem in patients with spina bifida, particularly MMC, where it occurs in more than 90% of individuals. The presence and degree of hydrocephalus and ventriculomegaly are particularly important clinical matters in such cases because hydrocephalus often worsens a symptomatic Chiari malformation. Also, in MMC an infantile presentation of Chiari malformation is seen in 10–20% of patients. Because of its severity, along with the potentially poor prognosis of this infantile presentation, the infant brainstem syndrome of MMC is considered separately.

For children, adolescents, and adults with MMC, clinical manifestations are similar to those of Chiari malformation in those without MMC. Sleep apnea (both the central and obstructive types) appears to be common in this group. Some cases of obstructive sleep apnea appear to be related to abnormal control of pharyngeal airway patency in sleep. Symptoms from accompanying syrinx or hydromyelia are also common, especially those of progressive upper extremity dysfunction, scoliosis, or both.

Symptoms of brainstem dysfunction such as stridor, apnea, and dysphagia, are seen in 10–20% of newborns and infants with MMC and can be ominous signs [18]. In some series, up to one-half of these children died. Persistent crying can be a presenting sign. Some authors have referred to this group of symptoms in the newborn with MMC as the *infantile brainstem syndrome*.

Serial assessment of the neurological level of impairment (motor and sensory) to detect deterioration from such potentially reversible problems as ventricular shunt malfunction, TSC, hydromyelia, and spinal cord compression from lumbar stenosis or arachnoid cysts is required. Other neurological changes of potential importance include decrease of muscle strength or change in tone and/or reflexes. Also, assessment of upper extremity function (in particular, decreases in handgrip strength, changes in reflexes, evidence of hand atrophy, pain or paresthesia in the hands or arms, or onset of upper extremity weakness) should raise the suspicion of a symptomatic syrinx or Chiari malformation.

Treatment of urological abnormalities ensures health and is a high management priority [11]. Individuals who are at high risk for renal deterioration need to be identified. They include individuals with high-pressure bladders, vesicoureteric abnormalities leading to reflux and hydronephrosis, and recurrent urinary tract infections. It is important to emphasize that urological status is rarely static in people with spinal dysraphism; individuals can convert from the low- to the high-risk group. Frequent monitoring and aggressive management centered on clean intermittent catheterization are needed to prevent such complications as hypertension and renal failure.

Another major pathophysiological problem is urinary incontinence. The management of incontinence has advanced considerably over the past 2 decades, particularly with the acceptance of clean intermittent catheterization and advances in bladder augmentation surgery. A rational management plan must be based on adequate information about voiding function. This usually requires a cystometrogram to determine bladder tone and pressure, outlet function, and the coordination or degree of synergy or dyssynergy between the two. Management of urinary continence often includes the use of medications such as oxybutynin for bladder relaxation, sympathomimetics such as ephedrine to promote bladder outlet contraction, and imipramine to do both [11]. Doses must be titrated and side effects monitored. Finally, issues of sexual function need to be considered as the patients approach adolescence.

Serial monitoring of these symptoms is critical in the management of individuals with spinal dysraphism. Many of the problems encountered are either preventable or reversible (at least in part) if they are detected early and the proper intervention(s) are instituted.

VII. Natural History

The natural history of this condition has received little focused attention despite many efforts to intervene both surgically and medically over many decades. The notion of MMC as a complex and disabling condition is clearly true. However, data that quantify and analyze the dimensions and magnitude of this complexity are scarce.

Patients with MMC continue to have a higher than normal mortality rate throughout the lifespan [19]. There is clearly a peak in the newborn period, most likely related to those cases with multiple malformations, chromosomal abnormalities, and multiple severe impairments, particularly in brainstem function. Before aggressive urological management, death from renal failure was common, and renal dysfunction remains an important problem that requires ongoing management. Other common contributors to mortality include infection, pressure ulcers, shunt malfunction and infection, latex allergy, and pulmonary dysfunction with or without sleep apnea.

Although the introduction of CSF shunting markedly improved cognitive function in individuals with MMC, cognitive impairment remains an important cause of disability. Some degree of relationship exists between lesion level and severity of cognitive impairments [20].

Motor function and impairment of ambulation is clearly a major feature of spina bifida. Functional capacity to ambulate is correlated with level of motor function, as described in Table 2.

The natural history of the spinal dysraphisms is one of increased disability with age [19]. Part of this is due to the effects of growth and weight changes on body mass and center of gravity. This often leads to a loss of ambulatory ability over time. Disuse atrophy and deconditioning often occur after the multiple surgeries undergone by these individuals, particularly during late childhood, adolescence, and adulthood. These problems often lead to a decline in function that is difficult to overcome without intensive rehabilitation in my experience. Cognitive, mental health, and environmental factors also contribute to the level of overall disability. The converse of this is the ability of some individuals to function at higher than expected levels due to strengths in these factors as well as successful accommodations.

References

1. Sadler, T. W. (2005). Embryology of neural tube development. *Am. J. Med. Genet. C. Semin. Med. Genet.* **135**, 2–8.
2. Mitchell, L. E. (2005). Epidemiology of neural tube defects. *Am. J. Med. Genet. C. Semin. Med. Genet.* **135**, 88–94.
3. Tubbs, R. S., Chambers, M. R., Smyth, M. D., Bartolucci, A. A., Bruner, J. P., Tulipan, N., and Oakes, W. J. (2003). Late gestational intrauterine myelomeningocele repair does not improve lower extremity function. *Pediatr. Neurosurg.* **38**, 128–132.
4. Greene, N. D., and Copp, A. J. (2005). Mouse models of neural tube defects: investigating preventive mechanisms. *Am. J. Med. Genet. C. Semin. Med. Genet.* **135**, 31–41.
5. Boyles, A. L., Hammock, P., and Speer, M. C. (2005). Candidate gene analysis in human neural tube defects. *Am. J. Med. Genet. C. Semin. Med. Genet.* **135**, 9–23.
6. Wallingford, J. B. (2005). Neural tube closure and neural tube defects: studies in animal models reveal known knowns and known unknowns. *Am. J. Med. Genet. C. Semin. Med. Genet.* **135**, 59–68.
7. Doudney, K., and Stanier, P. (2005). Epithelial cell polarity genes are required for neural tube closure. *Am. J. Med. Genet. C. Semin. Med. Genet.* **135**, 42–47.
8. Milhan, S. (1962). Increased incidence of anencephalus and spina bifida in siblings of affected cases. *Science* **138**, 593.
9. EUROCAT Working Group. (1991). Prevalence of neural tube defects in 20 regions of Europe and the impact of prenatal diagnosis, 1980–1986. *J. Epidemiol. Comm. Health* **45**, 52–58.
10. Lynch, S. A. (2005). Non-multifactorial neural tube defects. *Am. J. Med. Genet. C. Semin. Med. Genet.* **135**, 69–76.
11. Chauvel, P. J., and Kinsman, S. L. (1996). Spina bifida and hydrocephalus. *In:* "Developmental Disabilities in Infancy and Childhood" (A. J. Capute, et al., eds.), pp. 179–188. Paul H. Brookes, Baltimore.
12. McLone, D. G., and Dias, M. S. (2003). The Chiari II malformation: cause and impact. *Childs Nerv. Syst.* **19**, 540–550.
13. Rekate, H. L. (2001). Hydrocephalus classification and pathophysiology. *In:* "Pediatric Neurosurgery: Surgery of the Developing Nervous System" (D. G. McLone, ed.), pp. 457–474. Saunders, Philadelphia.
14. Yamada, S., Knerium, D. S., Mandybur, G. M., Schultz, R. L., and Yamada, B. S. (2004). Pathophysiology of tethered cord syndrome and other complex factors. *Neurol. Res.* **26**, 722–726.
15. Hoffman, H. J., Hendrick, E. B., and Humphreys, R. P. (1976). The tethered spinal cord: its protean manifestations, diagnosis and surgical correction. *Childs Brain* **2**, 145–155.
16. Snodgrass, W. T., and Adams, R. (2004). Initial urologic management of myelomeningocele. *Urol. Clin. North Am.* **31**, 427–434.
17. (1992). Recommendations for the use of folic acid to reduce the number of cases of spina bifida and other neural tube defects. *MMWR Recomm. Rep.* **41**, 1–7.
18. Charney, E. B., Rorke, L. B., Sutton, L. N., and Schut, L. (1987). Management of Chiari II complications in infants with myelomeningocele. *J. Pediatr.* **111**, 364–371.
19. Bowman, R. M., McLone, D. G., Grant, J. A., Tomita, T., and Ito, J. A. (2001). Spina bifida outcome: a 25-year prospective. *Pediatr. Neurosurg.* **34**, 114–120.
20. Fletcher, J. M., Copeland, K., Frederick, J. A., Blaser, S. E., Kramer, L. A., Northrup, H., Hannay, H. J., Brandt, M. E., Francis, D. J., Villarreal, G., Drake, J. M., Laurent, J. P., Townsend, I., Inwood, S., Boudousquie, A., and Dennis, M. (2005). Spinal lesion level in spina bifida: a source of neural and cognitive heterogeneity. *J. Neurosurg.* **102**, 268–279.

56

Circuits to Synapses: The Pathophysiology of Tourette Syndrome

Harvey S. Singer, MD
Kendra Harris, MSc

Keywords: *dopamine, neurophysiology, tics, Tourette syndrome*

I. Introduction
II. Brief History and Nomenclature
III. Natural History
IV. Relevant Structural Detail and Neurophysiology
V. Location of the Primary Dysfunction
VI. Neurochemical Basis for TS
VII. Autoimmunity as a Mechanism for TS
VIII. Summary
   References

I. Introduction

Tourette syndrome (TS) is a childhood-onset neuropsychiatrical disorder characterized by chronic motor and vocal tics that are frequently accompanied by coexisting problems such as obsessive-compulsive disorder (OCD) and attention deficit hyperactivity disorder (ADHD). The precise underlying pathophysiological mechanism(s) for tics is unknown. Neuroanatomical and neurophysiological studies suggest an abnormality involving cortico-striatal-thalamo-cortical (CSTC) circuits, but the site of primary dysfunction is controversial. Many investigators have focused on the striatal component, perhaps influenced by knowledge of associations between basal ganglia dysfunction and movement disorders. Others, however, suggest an abnormality in prefrontal cortex. Neurotransmitters involved in CSTC circuits have been hypothesized as the primary pathological factor, with recent evidence suggesting a dopaminergic tonic-phasic release abnormality. In addition, autoimmunity has been proposed as a mechanism in a small subset of TS patients.

This chapter begins with an overview of clinical issues pertaining to tic disorders and then discusses the pathobiology of TS. In the latter, CSTC circuits are used a framework for the discussion of current pathophysiological hypotheses. A notable implication of a circuit hypothesis is that a lesion in one part of the circuit (e.g., prefrontal cortex) can produce signs and symptoms similar to that in another component (e.g., striatum). Hence, analogous to the situation at a racetrack, a poor outcome (abnormal behaviors) can result from difficulties arising anywhere within the circuit (frontal lobe, striatum, globus pallidum, or thalamus). Therefore a key question is whether the site of the primary abnormality underlying this disorder can be localized.

II. Brief History and Nomenclature

Tourette syndrome is named after the French physician Georges Gilles de la Tourette, who in 1885 reported nine patients with chronic tic disorders characterized by a combination of involuntary motor movements and spontaneous shouting and cursing, an affliction he called *maladie des tics* [1]. Although many of the clinical characteristics described in his early reports are valid, the diagnostic features have been redefined (Table 1) and coexisting problems clarified. TS occurs worldwide, with increasing evidence of common features in all cultures and races. The prevalence of TS is currently estimated at about 1–10 per 1000 children and adolescents (reviewed in [2]). It is an inherited disorder, but the precise pattern of transmission and the identification of the gene(s) have remained elusive.

Tics, the hallmark of TS, are easily observed but broadly defined as sudden, rapid, repetitive movements or vocalizations. They occur in a nonrhythmic, stereotyped fashion that serve no purpose, occur for no apparent reason, and usually present in clusters or bouts. Tics typically consist of simple or coordinated gestures or utterances that mimic fragments of normal behavior; characterized by Charcot as "caricatures of natural acts" [1]. Motor tics are often classified according to the type of movement. Simple motor tics are brief rapid movements that usually involve a single muscle group (e.g., eye blink, head jerk, and shoulder shrug). Complex motor tics are abrupt movements that involve either a cluster of simple movements or a more coordinated sequence of movements (e.g., trunk bending or gyrating, jumping, smelling, making obscene gestures [copropraxia], and imitating others' gestures [echopraxia]). Simple phonic tics include sounds and noises such as sniffing, throat clearing, grunting, barking, blowing, and making sucking sounds. Complex phonic tics are linguistically meaningful utterances and verbalizations and include syllables, phrases, echolalia (repetition of others' words), palilalia (repetition of one's own words), and coprolalia (the shouting of obscenities, profanities, or otherwise socially inappropriate words or phrases). Coprolalia, one of the most distressing and recognizable symptoms of TS, occurs in only about 10% of patients. Tics are more common in males than in females (about 3:1), have a higher incidence in winter than spring, and are more common in children attending special education programs than in those attending regular classes (reviewed in [2,3]).

Characteristics of tics include exacerbation by anxiety, excitement, anger, or fatigue and reduction by absorbing activities or sleep. Tics can be briefly suppressed, but this is often associated with a build-up of inner tension. Finally, tics may be preceded by an urge, impulse, tension, pressure, itch, or tingle (premonitory urge) that is often localized to a discrete anatomical region. This tic-prompting sensation occurs more frequently in adults than children.

III. Natural History

The onset of tics is usually before age 11 years, typically beginning between 3 and 8 years of age [4]. Motor tics often begin with transient periods of intense eye blinking or other facial tics. Phonic tics typically follow the onset of motor tics by several years. The course of tics is one of waxing and waning numbers, frequency, intensity, and complexity. The precise mechanism underlying the fluctuation of tics remains unknown but may involve environmental (e.g., psychosocial stress, adversity, or infection) and biological factors. Although the long-term course of TS can be quite variable, most studies suggest that tics improve in late adolescence or early adulthood. Early tic severity is not a good predictor of later tic severity and individuals with only tics are less impaired than those with coexisting problems such as ADHD, OCD, anxiety, depression, self-injurious behaviors, and episodic rage (for more on natural history, see [3]).

Table 1 Diagnostic Criteria for Tourette Syndrome

Onset: by age 21
Motor tics: multiple
Vocal tics: at least one
Course: gradual; add and subtract
Duration: more than 1 year
Medications: not due to use of tic-provoking substances (e.g., stimulants)
Medical conditions: not associated with other diseases (e.g., Huntington's chorea, postviral encephalitis)
Witnessed: tics observed by a knowledgeable individual

IV. Relevant Structural Detail and Neurophysiology

A. CSTC Circuits

A simplified schematic diagram of CSTC circuits highlighting the three major components (cortex, striatum, and thalamus) is presented in Figure 1.

1. Striatum and Its Cortical Afferents

The striatum, long considered the major pathophysiological site in TS, has three major anatomical subdivisions: the caudate, putamen, and ventral "limbic" striatum (nucleus accumbens, portions of the olfactory tubercle, and the ventral medial aspect of the caudate and putamen). Striatal neurons are divided into two main subtypes: 1) medium-sized spiny neurons (MSSNs), so named because of their size and the presence of numerous dendritic spines, which provide direct projections to structures outside of the striatum; and 2) aspiny interneurons, which are large aspiny cholinergic neurons; medium-sized aspiny neurons expressing somatostatin, neuropeptide Y, and nitric oxide or calretinin; and medium-sized aspiny GABAergic neurons expressing parvalbumin, a calcium-binding protein.

The cerebral cortex provides the striatum with massive excitatory glutamatergic projections that make asymmetrical synaptic contact on the head of dendritic spines of MSSNs. Two different types of cortical projecting neurons have been proposed: 1) pyramidal tract (PT)–type neurons that innervate MSSNs, which in turn contribute to the "indirect" efferent striatal pathway neurons; and 2) intratelencephalically projecting (IT)–type neurons, which preferentially contact MSSNs that are part of the "direct" striatal efferent system [5]. Different types of cortical inputs help to topographically organize striatal sectors and impose a functional subdivision of the striatum into a dorsolateral "somatosensory," an intermediate (or central) "associative," and a ventromedial "limbic" (emotional-motivational) sector.

2. Cortico-Striatal Pathways

Five distinct but integrated parallel circuits that connect cortex to striatum have been described in primates (reviewed in [6]). Although presented as distinct pathways, there is evidence that these circuits may be more integrated than previously thought. The *motor* circuit, a potential site for generation of tics, originates primarily from the supplementary motor cortex and projects to the putamen in a somatotopic distribution. The *oculomotor* circuit, a potential site for ocular tics, begins principally in the frontal eye fields and connects to the central region of the caudate. The *dorsolateral pre-frontal (cognitive)* circuit links Brodmann's areas 9 and 10 with the dorsolateral head of the caudate and appears to be involved in executive function and motor planning. Dysfunction of this pathway could lead to attentional difficulties. The *lateral orbitofrontal (personality)* circuit originates in the inferolateral prefrontal cortex and projects to the ventromedial caudate. Orbitofrontal injury is associated with OCD, personality changes, disinhibition, irritability, and mania. Finally, the *anterior cingulate (limbic)* circuit arises in the anterior cingulate gyrus and projects to the ventral striatum (olfactory tubercle, nucleus accumbens, and ventral medial aspect of the caudate and putamen), which receives additional input from the amygdala, hippocampus, and entorhinal and perirhinal cortices. Mutism, apathy, and OCD are associated with this circuit.

3. Striato-Thalamic Pathways

The simplified striato-thalamic projections in Figure 1 show two separate striatal output pathways ("direct" and "indirect") projecting from the striatum to the globus pallidus interna (GPi)/substantia nigra pars reticulata (SNpr) and a single pathway from the GPi/SNpr to the thalamus. The direct pathway transmits striatal information monosynaptically to the GPi/SNpr, whereas the indirect system conveys information to these same regions via a disynaptic relay from globus pallidus externa (GPe) to the subthalamic nucleus (STN). Outputs are somatotopically organized, with the head and eyes represented in the SNpr and the rest of the body in the GPi. The parallel direct and indirect pathways have opposing effects on GABAergic GPi/SNpr output neurons (the direct pathway inhibits and the indirect pathway stimulates) and in turn produce a reverse effect on thalamocortical (ventral anterior [VA]/ventral lateral [VL]) neurons. The two general pathways are thus associated with the facilitation (direct) or suppression (indirect) of thalamic outputs, which influence movements and cognitive processes. For example, disinhibition of excitatory neurons in the thalamus results in hyperexcitability of cortical motor areas and possibly the release of tics.

Although the aforementioned two-pathway striatal output system appears relatively straightforward, anatomical data suggests a more complex circuitry, such as direct GPe and STN input to the thalamus [7]. Additionally, although not illustrated in Figure 1, GPi/SNpr neurons target a variety of functionally diverse systems with projections to different regions of the thalamus (ventral motor, mediodorsal, and the centromedian-parafascicular thalamic complex), the pedunculopontine tegmental nucleus, superior colliculus, and habenula [7].

A second organizational system for the striatum has also been described [8]. In brief, based on its intrinsic anatomical organization, the striatum is subdivided into two neurochemically defined compartments, which are variably designated as striosomes and extrastriosomal matrix, patches and

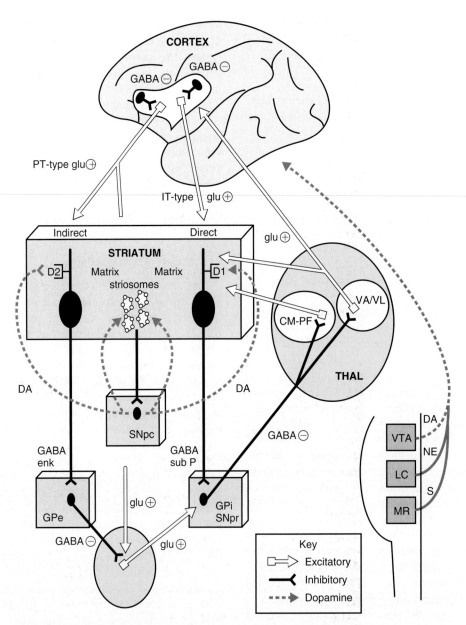

Figure 1 Cortico-striatal-thalamo-cortical circuits. Illustration of PT-type cortical neurons projecting to the striatal *indirect* pathway neurons, which in turn transmit information monosynaptically to the GPi/SNpr. IT-type cortical neurons preferentially contact the striatal *direct* pathway neurons, which in turn connect disynaptically to the GPi via the GPe and STN. The *striosomal* pathway transmits information to the SNpc, which in turn exerts dopaminergic influence back to the striatum. GPi/SNpr projections synapse on a variety of thalamic nuclei, which in turn project to the striatum, cortex, or both. *CM-PF*, Centromedian-parafascicular complex; *DA*, dopamine; *enk*, enkephalins; *glu*, glutamate; *GPe*, globus pallidus externa; *GPi*, globus pallidus interna; *IT-type neurons*, intratelencephalically projecting–type neurons; *LC*, locus coeruleus; *MR*, median raphe; *NE*, norepinephrine; *PT-type neurons*, pyramidal tract–type neurons; *S*, serotonin; *SNpc*, substantia nigra pars compacta; *SNpr*, substantia nigra pars reticulata; *STN*, subthalamic nucleus; *subP*, substance P; *THAL*, thalamus; *VA/VL*, ventral anterior/ventral lateral nuclei; *VTA*, ventral tegemental area.

matrix, or islands and matrix. Cells within the matrix (matrisomes) receive innervation from sensorimotor and association areas of the neocortex and represent the site of origin of both direct and indirect pathways. Smaller striosomal areas receive input from the orbitofrontal, anterior cingulate, and posterior medial prefrontal cortex and project to the substantia nigra pars compacta (SNpc) and its immediate area.

4. Thalamo-Cortical and Thalamo-Striatal Pathways

In the past, discussions of thalamic innervation often focused on hypotheses of excess glutamatergic projection outputs from the VA/VL thalamic complex to motor-related frontal cortical areas. More recently, however, the successful reduction of tics with deep brain stimulation of intralaminar thalamic nuclei [9] has led many investigators to reexamine the highly specific thalamo-striatal system. Although other thalamo-striatal projections exist, the major source appears to be the caudal intralaminar nuclei, which include the centromedian (CM)-parafascicular (PF) nuclear complex. As shown in Figure 2, highly topographical and specific projections from the CM-PF complex influence widespread striatal regions that are involved in the processing of functionally segregated information [10]. Studies have also shown that CM-PF thalamic inputs to the striatum innervate both MSSNs and interneurons, have a greater projection to direct output neurons, form synapses on MSSN dendritic shafts, and are not in close proximity

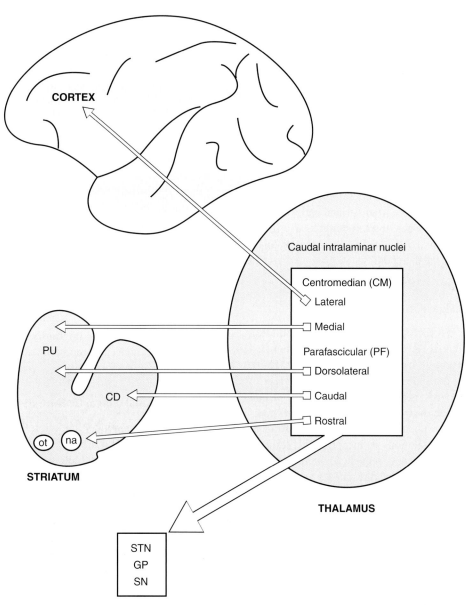

Figure 2 Caudal intralaminar thalamo-striatal pathways. The caudal intralaminar nuclei (CM-PF complex) of the thalamus and highly topographical glutamatergic projections to the striatum and cortex. *CD*, Caudate; *GP*, globus pallidus; *na*, nucleus accumbens; *ot*, olfactory tubercle; *PU*, putamen; *SN*, substantia nigra; *STN*, subthalamic nucleus.

to dopaminergic afferents (located on the neck of dendritic spines) [10]. Similarly, VA/VL nuclei also send topographically and functionally organized projections to the entire rostrocaudal extent of the dorsal striatum.

V. Location of the Primary Dysfunction

Pathophysiological hypotheses for TS have been formulated based on studies of neuroanatomy, neurophysiology, and neurochemistry. To date, overwhelming evidence suggests involvement of components of CSTC circuits in TS. The challenge remaining is to identify the primary site of the abnormality. In this section, we review the specific data (clinical assessment, neurophysiological evaluations, magnetic resonance imaging [MRI] studies, and functional imaging of glucose metabolism and blood flow) that support involvement of cortical and striatal regions of the CSTC circuits.

A. Evidence Supporting Frontal Lobe Dysfunction in TS

The frontal cortex, the site of the primary motor cortex as well as the origin of five major cortico-striatal pathways that influence both motor and behavioral systems, is a primary candidate for an essential role in TS. Specific evidence supporting a fronto-striatal hypothesis is presented in the following sections.

1. Neuropsychological Testing

Comprehensive psychological evaluations in children with TS, with or without ADHD, have repeatedly identified the presence of executive dysfunction (reviewed in [11]). Executive function, attributed to the frontal lobe, refers to self-regulating behaviors necessary to select and sustain actions and guide behavior within the context of goals or rules (e.g., initiation, planning, organization, shifting of thought or attention, inhibition of inappropriate thought or behavior, and sustained and sequenced behavior). TS patients display accurate but slowed responses on a number of measures of executive function, particularly on those that are timed (e.g., Letter Word Fluency Test) and those requiring verbal fluency and an efficient memory search. In contrast, TS patients do not demonstrate impairment on the Semantic Word Fluency Test, which identifies deficits in the posterior language system. These findings have led to the hypothesis that patients with TS have a linguistic executive dysfunction involving fronto-striatal circuits. Other psychological studies supporting this hypothesis include findings that demonstrate disorganized language formulation, output, concreteness, and speech dysfluencies.

2. Motor Testing

Voluntary movement abnormalities, which were studied by use of several different paradigms, have been suggestive of fronto-striatal dysfunction involving the motor or dorsolateral prefrontal cortex (reviewed in [6]). Serial choice reaction time studies using a button-pushing procedure identified increased movement sequencing deficits, without progressive slowing, as the level of advanced information was reduced. Unimanual/bimanual rhythmical tasks with handheld objects showed inaccuracies in precision grip, which were associated with reduced activation of secondary motor areas on functional MRI (fMRI) indicative of movement organization difficulties. Finally, a lateralized fronto-striatal deficit affecting visuospatial attention has been hypothesized based on results that show an abnormally right-biased line bisection task.

3. Neurophysiological Studies

Polysomnographic studies in patients with TS have shown markedly altered sleep quality and difficulties with initiating and maintaining sleep (reviewed in [12]). Based on a positive correlation between variables of sleep disturbance and the severity of TS, some investigators have suggested a disorder of hyperarousal. Prepulse inhibition of the startle reflex, a measure of inhibitory sensorimotor gating, is deficient in TS. Studies of visual attention on tactile performance, a vibrotactile choice reaction time task, and visual modalities have also been used to support the theory of alterations within CSTC circuits. Similarly, measures of event-related brain potentials (small voltage fluctuations recorded from the scalp that vary as a function of stimulus perception or in conjunction with cognitive processes) suggest altered inhibitory processes, sustaining difficulties, and abnormal modulation of circuits involving motor excitation or inhibition.

4. Transcranial Magnetic Stimulation (TMS)

Transcranial magnetic stimulation (TMS) has been used to study cortical excitability in TS patients [13]. Although results are somewhat variable among TS studies, all show either a reduced pre-pulse inhibition, also known as *intracortical inhibition* (the ratio of amplitude of motor action potential generated by a suprathreshold stimulus to that after a conditioning paradigm that uses a subthreshold stimulus followed by a standard suprathreshold stimulus), a shortened cortical silent period (the period of electrical silence after the TMS-evoked motor-evoked potential in a voluntarily contracted muscle), or both. These findings suggest that tics originate from a primarily subcortical disorder that affects the motor cortex through disinhibited afferent signals, from impaired inhibition directly at the level of the motor cortex, or both.

5. Oculomotor Paradigms

The control of various types of ocular saccades has been attributed to different cerebral areas; for example, the frontal eye field and posterior parietal cortex are associated with initiating saccades; the dorsolateral prefrontal cortex, supplementary eye field, and cingulate cortex are involved with more volitional and cognitive aspects of saccade control; and all the aforementioned areas interact with the basal ganglia [14]. Therefore findings of delays in initiation of a motor response to a visual stimulus in TS, as evidenced by prolonged latency on prosaccades, are believed to be representative of abnormalities in circuits that involve motor/premotor cortices (including the frontal eye field).

6. Corpus Callosum Abnormalities

MRI measurements of the corpus callosum in males with TS have consistently shown abnormalities, although results have been conflicting. In one study, the rostral area, which carries interhemispherical connections between frontal lobes, was significantly increased in size. In contrast, a cross-sectional case control study of 158 subjects with TS and 121 healthy comparisons found that younger individuals with TS had a smaller overall corpus callosum size [15]. Size correlated positively with tics and the volume of premotor regions and negatively with the volume of prefrontal cortices. The authors suggest that a smaller corpus callosum in TS may help modulate the severity of tics by reducing excitation across the cerebral hemispheres, thereby reducing the net inhibition and enhancing excitation of the prefrontal regions that help control tic behaviors. This hypothesis is consistent with suggestions that cortical hyperexcitability in TS may be the result of reduced activity of cortical interneurons.

7. Volumetric MRI Studies

Direct evidence for pathophysiological involvement of the frontal cortex is provided by volumetric studies (reviewed in [2,6,16]). One investigation has shown larger dorsolateral prefrontal regions in children with TS but significantly smaller volumes in adults with the disorder [17]. These differences in prefrontal volume size correlated inversely with the severity of tic symptoms. Age also correlated inversely in the orbitofrontal cortex, and the size of this prefrontal region was also inversely proportional to maximal tic severity. These data have suggested that volume reductions may prevent compensation for tic symptoms. The authors also speculate that TS children with relatively small prefrontal volumes are more likely to have enduring symptoms. Other studies have reported a larger percentage of white matter in the right frontal lobe or volumetric decreases in the left deep frontal white matter of TS patients compared with controls. Changes in white matter, especially deep white matter, suggest that abnormalities are present in long association and projection fiber bundles.

8. Functional Magnetic Resonance Imaging

Several fMRI studies have supported alterations within frontal cortical regions (reviewed in [2,6,16]; see also [18]). One study, in which images acquired during periods of voluntary tic suppression were compared with those during spontaneous expression of tics, suggested that tic suppression involves activation of the prefrontal cortex. Event-related positron emission tomography (PET) techniques have revealed correlations between tics and cortical brain activity in many regions, including the dorsolateral-rostral prefrontal cortex. A standard motor task paradigm, performed to determine whether an abnormal organization of motor functions could be detected in TS patients, showed an increased area of cerebral activation in both sensorimotor cortex and supplementary motor areas as compared with healthy subjects. Finally, an fMRI study using usual and unusual self-paced voluntary movements showed no differences between tasks in TS subjects, whereas controls had significantly increased premotor and SMA activation only during the unusual task. Hence TS patients had strong activation of both areas during the usual and unusual tasks, suggesting to the authors that activation in TS patients may reflect the involvement of a greater area of cerebral cortex to perform a voluntary motor task, possibly secondary to an additional effort required to suppress tic activity [18].

9. Glucose Metabolism and Blood Flow Studies

PET and SPECT studies have identified abnormalities within several cortical areas (reviewed in [16,19]). Examination by PET following injection of [^{18}F]2-fluoro-2-deoxyglucose has shown decreased activity in the frontal, cingulate, and insular cortices and hypometabolism of the orbitofrontal cortex in TS patients. An innovative functional neuroimaging study of tics using event-related [^{15}O]H$_2$O PET combined with time-synchronized videotaping showed that numerous brain regions contribute to brain activity involved in the generation of tics, including the medial and lateral premotor cortices, anterior cingulate cortex, dorsolateral-rostral prefrontal cortex, inferior parietal cortex, putamen, caudate, primary motor cortex, supplementary motor and premotor cortex, Broca's area, superior temporal gyrus, insula, and claustrum. The particular region that accounts for the initiation, rather than execution, of diverse motor and vocal behaviors remains unknown. Other investigators have emphasized hyperactivity of motor cortex and supplementary motor area or changes in cingulate activity in TS.

B. Evidence Supporting Striatal Dysfunction in TS

1. Neuropathology

Although routine postmortem studies have failed to identify a specific focus abnormality, two single-case reports have contained detailed neuropathological examinations [20]. In one report, findings in a 42-year-old male showed an increased packing density of neurons in the caudate and putamen, closely resembling that of an infant (i.e., "arrested development" of the striatum). Neuropathological and radiographic investigations in individuals with secondary tics ("acquired Tourette") also suggest associations with structural abnormalities (reviewed in [6]). For example, in a postmortem study of encephalitis lethargica, individuals with acquired tics differed from those without tics by the presence of an array of small focal lesions in the central gray matter that extended into the midbrain tegmentum. Tourette-like symptoms have also appeared in association with a variety of chronic movement disorders, such as torsion dystonia, Huntington's disease, pantothenate kinase–associated neurodegeneration, and neuroacanthocytosis.

2. Volumetric MRI

Despite the importance of imaging as circumstantial evidence for basal ganglia involvement in TS, results have been inconsistent (reviewed in [16]). Several small structural studies have been reported that show either the caudate or the lenticular nuclei are abnormal in volume or asymmetry compared with control subjects. For example, one study in children showed significant differences in the symmetry of the putamen; that is, TS patients had a right-sided predominance whereas controls had a left-sided predominance. In contrast, one study in adults showed decreased volumes of the left lenticular nucleus whereas another study showed no differences in basal ganglia volumes. In a larger study that excluded subjects with prior exposure to neuroleptics, caudate nucleus volumes were significantly smaller in children and adults with TS; the lenticular nucleus was also smaller in individuals with comorbid OCD. Finally, a study of monozygotic twins who were discordant for tic severity found smaller caudate volumes in the more severely affected twin.

3. Functional MRI

An fMRI study of tic suppression in TS suggests that active tic suppression is associated with activation of the basal ganglia in addition to orbitofrontal, limbic-related cortical areas (reviewed in [16]). In particular, this study found that prefrontal activity correlated with activity in the right caudate nucleus and that increases in the caudate signal in turn were associated with decreases elsewhere in the basal ganglia. This finding has led investigators to suggest that inputs to or outputs from the caudate are responsible for the inability to suppress tics.

4. Glucose Metabolism and Blood Flow Studies

The most consistent PET and single photon emission computed tomography (SPECT) studies in TS have involved the basal ganglia [16,19]. Two studies have demonstrated decreased glucose utilization in the ventral striatum. Changes in the caudate nucleus have been reported to be either up- or down-regulated in TS. Some data suggest that perfusion deficits may be more prominent on the left side of the basal ganglia; decreased metabolism of the left caudate has been found to correlate with tic severity in different centers. During voluntary tic suppression, increased blood flow has been seen in the right caudate. Both fluorodeoxyglucose-positron emission tomography (FDG-PET) and fMRI studies have reported hypometabolism of the putamen, although none of the results correlated with severity. From a comparative point of view, it is interesting that Parkinson's disease shows putaminal hypermetabolism.

5. Ventral Striatum

Several investigators have emphasized the involvement of the ventral striatum in TS. In studies assessing functional coupling of regional cerebral metabolic rates for glucose, connectivity of the ventral striatum differentiated TS from control groups. Monoaminergic hyperinnervation has also been described in the right ventral striatal region of TS patients. As previously described, this region is supplied by the limbic circuit and is considered the area involved in coupling natural rewards with the expression of behavior. The ventral striatum is also involved with the formation of habit memories and is a regulator of stereotyped behaviors. Based on suggested roles for the dorsal striatum (e.g., causing repetitive behaviors) and the ventral striatum (e.g., involved in various aspects of reward-based learning processes), it has been postulated that tics might result from an imbalance between areas that are involved in habit formation [21].

C. Evidence Supporting Thalamic Dysfunction in TS

Recently identified thalamo-striatal circuitry and the beneficial response to neurosurgical procedures reinforce the functional importance of thalamic regions. Results of case studies have shown that high-frequency stimulation of the median and rostral intralaminar thalamic nuclei can produce a significant reduction of tics. Other surgical procedures used to treat patients with medically intractable tics have included thalamotomy (inner part of the ventral oral nucleus and median and rostral intralaminar nuclei), infrathalamic lesioning or stimulation, and a novel combination of cingulotomy and infrathalamic lesioning. The improvement of tics determined by lesion or stimulation of the thalamic nuclei is believed to be mediated directly by the effect of these midline thalamic nuclei on tonically active striatal neurons or indirectly on broadly distributed cortical systems and their cortico-striatal projections.

D. Other Potential Neuroanatomical Regions Involved in TS

Brain regions including the midbrain, brainstem, and other areas that affect the circuits described previously have also been hypothesized to play a role in TS. Although largely ignored for more than 20 years [22], the midbrain tegmentum has received some renewed interest based on the resolution of tics following treatment of an abnormality (caused by thiamine deficiency) to this area and the results of voxel-based morphometry showing increased gray matter mainly in the left mesencephalon.

VI. Neurochemical Basis for TS

Neurochemical hypotheses tend to be based on extrapolations from clinical trials evaluating the response to specific medications; from CSF, blood, and urine studies in relatively small numbers of patients; from neurochemical assays of a few postmortem brain tissues; and from PET/SPECT studies (reviewed in [6]). Most neurotransmitters involved in CSTC circuitry, including the dopaminergic (DA), glutamatergic, GABAergic, serotonergic, cholinergic, noradrenergic, and opioid systems, have been implicated. Which, if any, of these proposals represents the primary pathological factor has yet to be definitively determined. Although our bias is that the dopaminergic system plays the dominant role, the multiple systems involved in the production of complex actions make it possible, indeed probable, that imbalances exist within several transmitter systems. Because space for this chapter is limited, we focus on evidence for a dopamine abnormality in TS, specifically on the proposed role of a DA tonic-phasic release abnormality.

A. Dopaminergic Pathways in the CSTC Circuits

Dopaminergic input from the substantia nigra pars compacta (SNpc) influences striatal activity via several mechanisms, including: 1) postsynaptic D1 and D2 receptors on MSSNs (inhibitory D2 receptors are present on indirect pathway neurons, whereas activating D1 receptors are located on direct pathway neurons); 2) postsynaptic DA receptors on striatal cholinergic interneurons;

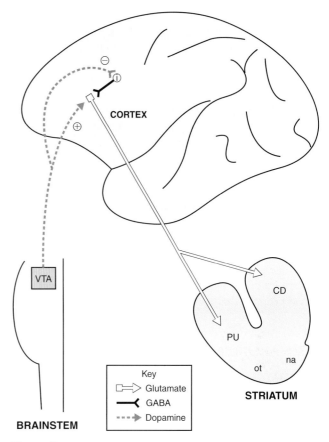

Figure 3 Prefrontal dopamine innervation. Dopaminergic projections from the ventral tegmental area (*VTA*) to the prefrontal cortex inhibit cortical interneurons (*i*) and excite cortico-striatal projection neurons. See text for more discussion. *CD*, Caudate; *na*, nucleus accumbens; *PU*, putamen.

and 3) presynaptic DA receptors on presynaptic glutamatergic cortico-striatal terminals. As is evident in Figure 3, dopaminergic fibers, arising from the ventral tegmental area, innervate prefrontal regions. Dopamine receptors are present on both pyramidal neurons (stimulating) and interneurons (inhibiting) in prefrontal cortex. Although the concentrations of dopamine and its receptors are lower in the frontal cortex than in the striatum, the influence of dopamine in the cortex is well established.

B. Dopaminergic Abnormality in TS

The possibility of a dopaminergic abnormality in TS continues to receive strong consideration because of the therapeutic response to neuroleptics, preliminary data from postmortem studies, and a variety of nuclear imaging protocols (reviewed in [6]). Excess nigrostriatal dopaminergic activity, whether via supersensitive dopamine receptors, dopamine hyperinnervation, or abnormal presynaptic terminal function, would be expected to cause a significant hyperkinetic effect. Activation of D1 receptors is excitatory to the movement-releasing direct pathway, whereas activation of D2 receptors inhibits the indirect pathway, which in turn inhibits movement. Hence the result of either action is the disinhibition of excitatory neurons in the thalamus that in turn could cause hyperexcitability or disinhibition of cortical motor areas and the release of tics. In addition to short-term effects, DA can modulate cortico-striatal transmission by the mechanism of long-term depression or potentiation. This modulation either strengthens or weakens the efficacy of cortico-striatal synapses and thus can mediate reinforcement of specific discharge patterns. DA-induced fluctuating abnormalities in the resting potentials of striatal neurons have been hypothesized to influence tic waxing and waning and might explain the lack of identifiable abnormality in DA transmission [23]. Finally, in addition to dopamine's important role in the prefrontal cortex and striatal two-pathway system, it also has influential effects on the striosomal circuit and in the ventral striatum.

C. Dopaminergic Tonic-Phasic Model for TS

Numerous studies have attempted to clarify whether the dopaminergic abnormality in TS is associated with supersensitive postsynaptic receptors, hyperinnervation, a presynaptic abnormality, or elevated intrasynaptic release (reviewed in [6]). With some inconsistency, PET/SPECT studies of the striatum have shown a slight increase in the number of dopamine receptors [24], greater binding to dopamine transporters (DAT) [25,26], and an increased release of dopamine following amphetamine stimulation [27]. Similarly, biochemical evaluation of postmortem prefrontal cortex from TS patients shows elevated levels of dopamine and increased concentrations of dopaminergic receptors and transporters [28]. These data in turn have led to a unifying hypothesis, described in Figure 4, which involves the tonic-phasic release of dopamine.

Tonic dopamine, which exists extracellularly in low concentration, determines long-term homeostasis including regulation of postsynaptic receptors. The tonic (or basal) level of dopamine is defined primarily as an extrasynaptic measure and is calculated by use of microdialysis and electrophysiological measurements. Dopamine autoreceptors (D_2 and D_3 subtypes) are proposed as regulators of tonic dopamine control. In contrast, phasic dopamine is spike-dependent dopamine released into the synaptic cleft. With sufficient stimulation or when an uptake blocker is given in high concentrations, phasic dopamine can escape this region. A surrogate measure of phasic dopamine is the intrasynaptic dopamine release induced by the use of stimulant challenges, such as with amphetamine.

One component of the tonic-phasic hypothesis involves a reduction in tonic dopamine levels, which would explain up-regulation of dopamine receptors. Several potential

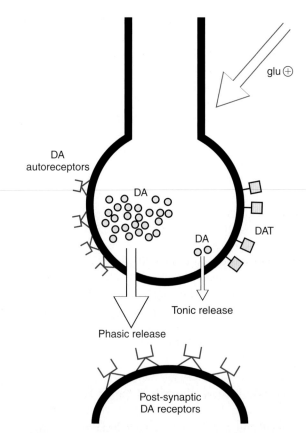

Figure 4 Tonic-phasic release of DA. DA is released in both a tonic and phasic fashion from the axon terminal. One pathophysiological hypothesis for TS, discussed in the text, involves an abnormal phasic release of DA.

explanations for a lower tonic dopamine level include an increased activity of the dopamine transporter (DAT), diminished cortical afferent input, or decreased phasic overflow from the synaptic cleft to the extracellular space. Based on a significant increase in the number of presynaptic dopamine transporter carrier sites in postmortem striatum [25] and frontal cortex [28], as well as increased SPECT/PET binding in pediatric and adult imaging protocols [26], we believe the primary defect is most likely due to an overactive dopamine transporter system. This situation would create reduced levels of extracellular dopamine, higher concentrations of dopamine in the axon terminal, increased stimulus-dependent dopamine release, and autoreceptor supersensitivity at the presynaptic site. It is also possible that the hyperresponsive spike-dependent (phasic) dopaminergic system in TS is secondary to a primary alteration of other afferents projecting to dopaminergic neurons. Nevertheless, irrespective of the primary mechanism, a phasic abnormality has been documented in [^{11}C]raclopride PET studies of TS patients [27]. This physiological hypothesis is further supported by several clinical findings: 1) the exacerbation of tics by stimulant medications could be secondary to enhanced dopamine release from the axon terminal; 2) environmental stimuli, such as stress, anxiety, and medications, which are well known to exacerbate tics, have been shown to increase phasic bursts of dopamine; and 3) tic suppression with very low doses of neuroleptics could occur because a reduced amount of tonic dopamine is available for competition at the postsynaptic receptor. Although the dopaminergic tonic-phasic model hypothesis could exist in either the cortex or striatum, we favor a prefrontal dopaminergic abnormality. Additional studies are currently in progress to confirm that dopaminergic abnormalities identified in Brodmann's area 9 [28] are also present in other prefrontal regions supplied by the ventral tegmental area.

VII. Autoimmunity as a Mechanism for TS

The suggestion of an autoimmune etiology for TS has gained some notoriety due to a reported link between tics and/or obsessive-compulsive symptoms and group A β-hemolytic streptococcal (GABHS) infection. The existence of this etiology, which is labeled as pediatric autoimmune neuropsychiatric disorders associated with streptococcal infection (acronym PANDAS), is controversial [29,32]. On the basis of a model proposed for Sydenham's chorea (SC), it is hypothesized that the underlying pathology in PANDAS involves an immune-mediated mechanism with molecular mimicry. Antibodies produced against GABHS are believed to cross-react with neuronal tissue in specific brain regions (i.e., become antineuronal antibodies) and in turn result in tic and/or behavioral symptoms. A single study that examined the response of patients with PANDAS to immunomodulatory therapy supports this hypothesis.

Antineuronal antibodies have been assessed in patients with PANDAS with variable results. Investigators using enzyme-linked immunosorbent assay (ELISA) and Western immunoblotting methodologies to evaluate sera from 40 children with movement disorders associated with streptococcal infections (including 16 with PANDAS) suggested that this cohort could be readily differentiated from a variety of disease controls [33]. ELISA results were significantly elevated in this cohort compared with those in control populations. Further, by use of a colorimetric assay with frozen postmortem basal ganglia tissue as the epitope, immunoblotting detected limited reactivity in controls but significantly more bands (at 60, 45, and 40 kDa) in poststreptococcal patients. In contrast, other researchers using ELISA assays with several human postmortem epitopes including supernatant, pellets, and synaptosomal fractions from fresh caudate, putamen, and globus pallidus were unable to distinguish 15 PANDAS subjects from controls [34]. Immunoblotting assays in this cohort, with the use of supernatant and pellet fractions from the caudate and putamen and an electrochemiluminescent method, showed that no molecular weight bands had measurable activity exclusive to either PANDAS or controls for any of the antigenic tissues. In a larger study from our laboratory, we compared antineuronal antibody profiles in subjects with TS (n=48) and PANDAS (n=46) to those of age-matched controls (n=43) [35]. Results confirmed that ELISA optical density readings were similar among the groups. Immunoblotting showed complex staining patterns in both patient and control groups, with no differences in any tissue. Thus the number of bands with reactivity peaks at molecular weights 98, 60, 45, and 40 kDa, and total area under ScanPack-derived peaks were not different. Additionally, we sought to determine the presence of serum antibodies against several commercially obtained specific antigens including pyruvate kinase M1, α- and γ-enolase, and aldolase C; antigens proposed to be the purified epitopes identified in poststreptococcal patients at MW 60, 45 (a doublet), and 40 kDa, respectively. Again, our investigation identified no difference in the frequency of serum antibodies. To determine whether detected antibodies cross-react with streptococcal proteins, several assays were repeated after preabsorption with homogenized GABHS. Our failure to alter immunoreactivity by the prior exposure of sera to streptococcal antigens further suggests a lack of association between antibodies against these putative antigens and a preceding GABHS infection.

Because the presence of autoantibodies in the serum of TS patients does not imply causation (autoantibodies are also found in controls), animal models have been developed to study whether serum (IgG) can induce stereotypies that may be analogous to human tics in rodents. Results from three studies using this approach were inconsistent, and further investigation indicated that methodological variables have a significant impact on findings [36]. Although convincing evidence for an immunologic etiology of TS is lacking, longitudinal studies correlating antineuronal antibodies with clinical symptoms and streptococcal markers are currently in progress.

VIII. Summary

A growing body of evidence indicates that an abnormality in CSTC circuits and their neurotransmitter systems are likely to be associated with tics and coexisting problems in TS. Early evidence for a primarily striatal defect has been matched by expanding investigations that point to abnormalities in the frontal cortex. Neurochemically, evidence continues to support involvement of the dopaminergic system, but other neurotransmitters within CSTC circuits may also be involved. Although assigning a role to the immune system in the etiology of TS remains an intriguing hypothesis, convincing evidence supporting an immune-mediated process is not available. We look forward to continued work characterizing the precise neuropathological abnormality in TS.

References

1. Kushner, H. I. (1999). "A Cursing Brain? The Histories of Tourette Syndrome." Harvard University Press, Cambridge, MA.
2. Singer, H. S. (2005). Tourette's syndrome: from behaviour to biology. Lancet Neurol. 4, 149–159.
3. Kurlan, R. (ed.). (2005). "Handbook of Tourette's Syndrome and Related Tic and Behavioral Disorders," ed. 2. Marcel Dekker, New York.
4. Leckman, J. F. (2003). Phenomenology of tics and natural history of tic disorders. Brain Dev. **25(Suppl 1)**, S24–S28.
5. Lei, W., Jiao, Y., Del Mar, N., and Reiner, A. (2004). Evidence for differential cortical input to direct pathway versus indirect pathway striatal projection neurons in rats. J. Neurosci. **24**, 8289–8299.
6. Singer, H. S., and Minzer, K. (2003). Neurobiology of Tourette syndrome: concepts of neuroanatomical localization and neurochemical abnormalities. Brain Dev. **25(Suppl l)**, S70–S84.
7. Smith, Y., Bevan, M. D., Shink, E., and Bolam, J. P. (1998). Microcircuitry of the direct and indirect pathways of the basal ganglia. Neuroscience **86**, 353–387.
8. Graybiel, A. M., Canales, J. J., and Capper-Loup, C. (2000). Levodopa-induced dyskinesias and dopamine-dependent stereotypies: a new hypothesis. Trends Neurosci. **23(Suppl 10)**, S71–S77.
9. Krauss, J. K., Pohle, T., Weigel, R., and Burgunder, J. M. (2002). Deep brain stimulation of the centre median-parafascicular complex in patients with movement disorders. J. Neurol. Neurosurg. Psychiatry **72**, 546–548.
10. Smith, Y., Raju, D. V., Pare, J., and Sidebe, M. (2004). The thalamostriatal system: a highly specific network of the basal ganglia circuitry. Trends Neurosci. **27**, 520–527.
11. Channon, S., Pratt, P., and Robertson, M. M. (2003). Executive function, memory, and learning in Tourette's syndrome. Neuropsychology **17**, 247–254.
12. Rothenberger, A., Kostanecka, T., Kinkelbur, J., Cohrs, S., Woerner, W., and Hajak, G. (2001). Sleep and Tourette syndrome. Adv. Neurol. **85**, 245–259.
13. Moll, G. H., Wischer, S., Heinrich, H., Tergau, F., Paulus, W., and Rothenberger, A. (1999). Deficient motor control in children with tic disorder: evidence from transcranial magnetic stimulation. Neurosci. Lett. **272**, 37–40.
14. Sweeney, J. A., Takarae, Y., Macmillan, C., Luna, B., and Minshew, N. J. (2004). Eye movements in neurodevelopmental disorders. Curr. Opin. Neurol. **17**, 37–42.
15. Plessen, K. J., Wentzel-Larsen, T., Hugdahl, K., Feineigle, P., Klein, J., Staib, L. H., Leckman, J. F., Bansal, R., and Peterson, B. S. (2004). Altered interhemispheric connectivity in individuals with Tourette's disorder. Am. J. Psychiatry **161**, 2028–2037.
16. Gerard, E., and Peterson, B. S. (2003). Developmental processes and brain imaging studies in Tourette syndrome. J. Psychosomatic Res. **55**, 13–22.
17. Peterson, B. S., Staib, L., Scahill, L., Zhang, H., Anderson, C., Leckman, J. F., Cohen, D. J., Gore, J. C., Albert, J., and Webster, R. (2001). Regional brain and ventricular volumes in Tourette syndrome. Arch. Gen. Psychiatry **58**, 427–440.
18. Fattapposta, F., Restuccia, R., Colonnese, C., Labruna, L., Garreffa, G., and Bianco, F. (2005). Gilles de la Tourette syndrome and voluntary movement: a functional MRI study. Psychiatry Res. **138**, 269–272.
19. Adams, J., Troiano, A., and Calne, D. (2004). Functional imaging in Tourette's syndrome. J. Neural Transm. **11**, 1495–1506.
20. Richardson, E. P. Jr. (1982). Neuropathological studies of Tourette syndrome. Adv. Neurol. **35**, 83–87.
21. Groenewegen, H. J., van den Heuvel, O. A., Cath, D. C., Voorn, P., and Veltman, D. J. (2003). Does an imbalance between the dorsal and ventral striatopallidal systems play a role in Tourette's syndrome? A neuronal circuit approach. Brain Dev. **25(Suppl 1)**, S3–S14.
22. Devinsky, O. (1983). Neuroanatomy of Gilles de al Tourette's syndrome. Possible midbrain involvement. Arch. Neurol. **40**, 508–514.
23. Mink, J. W. (2001). Basal ganglia dysfunction in Tourette's syndrome: a new hypothesis. Pedriatr. Neurol. **25**, 190–198.
24. Wong, D. F., Singer, H. S., Brandt, J., Shaya, E., Chen, C., Brown, J., Kimball, A. W., Gjedde, A., Dannals, R. F., Ravert, H. T., Wilson, P. D., and Wagner, H. N. Jr. (1997). D2-like dopamine receptor density in Tourette syndrome measured by PET. J. Nucl. Med. **38**, 1243–1247.
25. Singer, H. S., Hahn, I. H., and Moran, T. H. (1991). Abnormal dopamine uptake sites in postmortem striatum from patients with Tourette's syndrome. Ann. Neurol. **30**, 558–562.
26. Serra-Mestres, J., Ring, H. A., Costa, D. C., Gacinovic, S., Walker, Z., Lees, A. J., Robertson, M. M., and Trimble, M. R. (2004). Dopamine transporter binding in Gilles de la Tourette syndrome: a [^{123}I]FP-CIT/SPECT study. Acta. Psychiatr. Scand. **109**, 140–146.
27. Singer, H. S., Szymanski, S., Giuliano, J., Yokoi, F., Dogan, A. S., Brasic, J. R., Zhou, Y., Grace, A. A., and Wong, D. F. (2002). Elevated intrasynaptic dopamine release in Tourette's syndrome measured by PET. Am. J. Psychiatry **159**, 1329–1336.
28. Minzer, K., Lee, O., Hong, J. J., and Singer, H. S. (2004). Increased prefrontal D2 protein in Tourette syndrome: a postmortem analysis of frontal cortex and striatum. J. Neurol. Sci. **219**, 55–61.
29. Swedo, S. E., Leonard, H. L., Garvey, M., Mittleman, B., Allen, A. J., Perlmutter, S., Lougee, L., Dow, S., Zamkoff, J., and Dubbert, B. K. (1998). Pediatric autoimmune neuropsychiatric disorders associated with streptococcal infections: clinical description of the first 50 cases. Am. J. Psychiatry **155**, 264–271.
30. Swedo, S. E., Leonard, H. L., and Rapoport, J. L. (2004). The pediatric autoimmune neuropsychiatric disorders associated with streptococcal infection (PANDAS) subgroup: separating fact from fiction. Pediatrics **113**, 907–911.
31. Kurlan, R., and Kaplan, E. L. (2004). The pediatric autoimmune neuropsychiatric disorders associated with streptococcal infection (PANDAS) etiology for tics and obsessive-compulsive symptoms: hypothesis or entity? Practical considerations for the clinician. Pediatrics **113**, 883–886.
32. Singer, H. S., and Loiselle, C. (2003). PANDAS: a commentary. J. Psychosom. Res. **55**, 31–39.
33. Church, A. J., Dale, R. C., and Giovannoni, G. (2004). Anti-basal ganglia antibodies: a possible diagnostic utility in idiopathic movement disorders? Arch. Dis. Child **89**, 611–614.
34. Singer, H. S., Loiselle, C. R., Lee, O., Minzer, K., Swedo, S., and Grus, F. H. (2004). Anti-basal ganglia antibodies in PANDAS. Mov Disord. **19**, 406–415.
35. Singer, H. S., Hong, J. J., Yoon, D. Y., and Williams, P. N. (2005). Serum autoantibodies do not differentiate PANDAS and Tourette syndrome from controls. Neurology **65**, 1701–1707.
36. Singer, H. S., Mink, J. W., Loiselle, C. R., Burke, K. A., Ruchkina, I., Morshed, S., Parveen, S., Leckman, J. F., Hallett, J. J., and Lombroso, P. J. (2005). Microinfusion of antineuronal antibodies into rodent striatum: failure to differentiate between elevated and low titers. J. Neuroimmunol. **163**, 8–14.

57

Attention-Deficit Hyperactivity Disorder

Paul H. Lipkin, MD
Stewart Mostofsky, MD

Keywords: *attention-deficit hyperactivity disorder, basal ganglia, cerebellum, dopamine, executive function, frontal lobe, neurobiology, norepinephrine, pathophysiology, pharmacology, response inhibition*

I. History and Nomenclature
II. Etiology
III. Pathogenesis and Pathophysiology
IV. Relevant Neurological Findings
V. Pharmacology, Biochemistry, and Molecular Mechanisms
VI. Natural History
   References

Attention-deficit hyperactivity disorder (ADHD) is the most common developmental disorder of childhood, affecting approximately 3–9% of schoolchildren [1,2]. ADHD can affect a person's ability to function at school, at home, and in social settings and is one of the more common reasons for referral to pediatric neurologists, neurodevelopmental pediatricians, and child psychiatrists. It is also important that adult clinicians be able to recognize and manage the disorder; approximately 50% of children with ADHD continue to have symptoms into adulthood, with prevalence estimates in adults ranging from 1–6%.

I. History and Nomenclature

Medical interest in attention problems and hyperactivity can be traced back as early as the mid-19th century, when a German physician, Heinrich Hoffman, wrote two poems featuring boys with behaviors now equated with ADHD. In "The Story of Fidgety Phillip," we hear of a boy who "wriggles and giggles, and then, I declare swings backwards and forwards and tilts up his chair." Hoffman's poem "The Story of Johnny Head-in-the-Air" begins with the following description of Johnny: "As he trudg'd along to school, it was always Johnny's-rule, to be looking at the sky, and the clouds that floated by; but what just before him lay, in his way, Johnny never thought about." However, it was George Still, an English physician, who is credited with generating modern medical interest in the behaviors now referred to as ADHD. In his address to the Royal College of Physicians, he suggested that "the occurrence of defective moral control" represented a "morbid condition in children" and a "mental pathology," frequently seen in children with intellectual impairment and without "any particular gross lesion of the brain." Further interest and greater suggestion of an organic neurological etiology emerged with the influenza epidemic of 1917–1918, when children with encephalitis were noted to have similar symptomatology. In 1937 Charles Bradley identified a pharmacological means of treating this behavioral syndrome through the use of an amphetamine (Benzedrine) in children, thus ushering in the medical management of children with behavior disorders.

The next several decades saw further development of the nosology of this syndrome. The group of behavior disorders marked by hyperactivity, impulsivity, and emotional lability came to be known variously as minimal brain damage, minimal brain dysfunction, minimal cerebral dysfunction, hyperkinetic syndrome, hyperactive child syndrome, and hyperkinetic reaction of childhood and were recognized as frequently being accompanied by learning and perceptual dysfunction, with key research and writing provided by Gessel and Amatruda, Strauss and Werner, Clements, and others. A major change in conceptual framework and nomenclature occurred in 1980 with publication of the third edition of the Diagnostic and Statistical Manual (DSM) of the American Psychiatric Association (APA). Rather than focusing on issues of organicity and hyperactive behavior, the APA renamed the disorder attention-deficit disorder, reframing the disorder as a behavioral syndrome of which the primary deficit was in attention rather than motor activity, with inclusion of an inattentive subtype, in which manifestations of hyperactive and impulsive behavior are absent or mild.

This terminology has continued to the present day, modified as *attention-deficit hyperactivity disorder*. Currently, the *Diagnostic and Statistical Manual of Mental Disorders*, fourth edition, revised (DSM IV-R) [1] includes three subtypes: "predominantly inattentive," "predominantly hyperactive/impulsive," and a combined type (Table 1). By definition, symptoms must occur in more than one setting (home, school/occupational, or social) and must be present before 7 years of age.

Two other important historical landmarks specifically regarding medical treatment of ADHD must be noted. The pharmacological management of this disorder began with the use of the amphetamines benzedrine in 1937, and later, dexedrine. But it was with the development of methylphenidate in 1955 and its later marketing that the frequency of medical treatment of symptoms later defined by ADHD rapidly expanded. Then, in 1983, atomoxetine, a selective norepinephrine reuptake inhibitor, was developed as a new medication with a different action, with its first trials in adults with ADHD in 1998 and children in 2001. These treatments have remained the mainstay of medical management through the present day.

II. Etiology

Etiological investigations of ADHD suggest a strong genetic contribution, accounting for as much as 85% of the risk. Multifactorial patterns of inheritance are seen with a complex genetic network, frequently linked to catecholaminergic (particularly dopaminergic) neurotransmitter systems [4]. Family studies show parents with ADHD having greater than a 50% chance of having an affected child. Conversely, 25% of children with ADHD will have an affected parent. Further support to the heritability of this disorder is revealed in twin studies in which a concordance of 80% has been reported in monozygotic twins and approximately 50% in dizygotic, with high concordance for the inattentive and combined subtypes.

The loci for the dopamine postsynaptic D4 receptor (DRD4) and the DAT transporter system (DAT1) appear to be important contributors to the genetics of this disorder based on molecular genetic studies, further supporting the importance of the dopamine system in this disorder [4,5]. Some polymorphisms of the 480-bp allele of the dopamine transporter gene, DAT, have been associated with ADHD, as has the 7-repeat allele of DRD4. The genes for the D5 receptor, the serotonin transporter, the serotonin 1B receptor, and synaptosomal-associated protein 25 (SNAP-25) may also be playing a role. New techniques including genome scanning and the candidate gene approach have had mixed results. Genome-wide scanning has been unsuccessful in identifying specific gene regions implicated in ADHD, although regions 16p13 and 17p11 may contain genes related to this disorder. A new approach to identifying the genetic roots of ADHD is being pursued through identification of endophenotypes, biologically based phenotypes of the disorder based on neuropsychological variables that are independent of the core symptoms of the disorder [6].

As further evidence for a genetic contribution, ADHD symptoms are commonly associated with known chromosomal and genetic syndromes [7]. Most males with fragile X syndrome (gene map locus Xq27.3) have hyperactive behavior. Similar problems are seen in other common chromosomal disorders, including Down syndrome (trisomy 21), velocardiofacial syndrome (gene map locus 22q11.2), and Williams syndrome (gene map locus 7q11.2), as well as rarer anomalies. Other common genetic disorders

Table 1 Diagnostic Criteria for Attention-Deficit Hyperactivity Disorder

A. Either (1) or (2)
  (1) Inattention: six (or more) of the following symptoms of inattention have persisted for at least 6 months to a degree that is maladaptive and inconsistent with developmental level:
    (a) often fails to give close attention to details or makes careless mistakes in schoolwork, work, or other activities
    (b) often has difficulty sustaining attention in tasks or play activities
    (c) often does not seem to listen when spoken to directly
    (d) often does not follow through on instructions and fails to finish schoolwork, chores, or duties in the workplace (not due to oppositional behavior or failure to understand directions)
    (e) often has difficulties organizing tasks and activities
    (f) often avoids, dislikes, or is reluctant to engage in tasks that require sustained mental effort (such as schoolwork or homework)
    (g) often loses things necessary for tasks or activities (e.g., school assignments, pencils, books, or tools)
    (h) is often easily distracted by extraneous stimuli
    (i) is often forgetful in daily activities
  (2) Hyperactivity-impulsivity: six (or more) of the following symptoms of hyperactivity-impulsivity have persisted for at least 6 months to a degree that is maladaptive and inconsistent with developmental level:
  Hyperactivity
    (a) often fidgets with hands or feet or squirms in seat
    (b) often leaves seat in classroom or in other situations in which remaining seated is expected
    (c) often runs about or climbs excessively in situations in which it is inappropriate (in adolescents or adults, may be limited to subjective feelings of restlessness)
    (d) often has difficulty playing or engaging in leisure activities quietly
    (e) is often "on the go" or often acts as if "driven by a motor"
    (f) often talks excessively
  Impulsivity
    (g) often blurts out answers to questions before the questions have been completed
    (h) often has difficulty awaiting turn
    (i) often interrupts or intrudes on others (e.g., butts into conversations or games)
B. Some hyperactive-impulsive or inattentive symptoms that caused impairment were present before age 7 years.
C. Some impairment from the symptoms is present in two or more settings (e.g., at school [or work] and at home).
D. There must be clear evidence of clinically significant impairment in social, academic, or occupational functioning.
E. The symptoms do not occur exclusively during the course of a pervasive developmental disorder, schizophrenia, or other psychotic disorder, and are not better accounted for by another mental disorder (e.g., mood disorder, anxiety disorder, dissociative disorder, or a personality disorder).

Code based on type:
Attention-deficit hyperactivity disorder, combined type: if both criteria A1 and A2 are met for the past 6 months
Attention-deficit hyperactivity disorder, predominantly inattentive type: if criterion A1 is met but criterion A2 is not met for the past 6 months
Attention-deficit hyperactivity disorder, predominantly hyperactive-impulsive type: if criterion A2 is met but criterion A1 is not met for the past 6 months
Coding note: For individuals (especially adolescents and adults) who currently have symptoms that no longer meet full criteria, "in partial remission" should be specified.

From American Psychiatric Association Task Force on DSM-IV. (2000). "Diagnostic and Statistical Manual of Mental Disorders: DSM-IV-TR," ed. 4. American Psychiatric Association, Washington, D.C.

with known gene loci frequently present with hyperactive behavior, including phakomatoses such as neurofibromatosis type 1 metabolic disorders such as phenylketonuria, homocystinuria, and maple syrup urine disease, storage disorders such as Hurler syndrome, and thyroid disorders. As many as 90% of children with Tourette syndrome have associated ADHD.

Other biological and environmental risk factors have also been linked with ADHD [2,8,9]. Prenatal exposure to nicotine through maternal cigarette smoking increases the risk of ADHD two- to threefold in children. Animal studies support the nicotine link, with hyperactivity seen in exposed offspring. Influence of this exposure on the dopaminergic system and the proliferation of nicotine receptors have been implicated. Furthermore, both animal and human studies extend the relationship between nicotine exposure and ADHD into adolescence and adulthood, with nicotine decreasing the DAT density and causing downregulation of receptor function.

Alcohol exposure through maternal consumption has similarly been linked to an increased risk of ADHD [7,9]. It has been directly implicated as a teratogen in fetal alcohol syndrome, in which ADHD symptomatology has been described. As with smoking, an association with the disorder is also seen when examining maternal exposure through birth cohorts.

Both animal and human studies report on the sensitivity of the dopamine system to hypoxia and ischemia [10]. Similarly, injuries to the frontal-subcortical circuits, including cortical lesions associated with stroke and trauma and shearing white matter injuries associated with trauma, have been shown to result in ADHD symptoms. Evidence of such mechanisms for the development of ADHD is also borne out in the perinatal outcome literature. In large population samples, low-birth-weight status and frank intrauterine brain injury are seen to increase the risk for ADHD. At the same time, long-term follow-up studies of preterm infants without cerebral injury and with known intraventricular hemorrhage and periventricular leukomalacia commonly report ADHD as a significant sequela of these perinatal events. ADHD is also a known consequence of other forms of pediatric brain injury, such as cerebral palsy, hypoxic-ischemic encephalopathy, and traumatic brain injury. Adverse environmental factors have been associated with symptoms of ADHD. These may include infectious pathogens, as with children with infectious diseases such as human immunodeficiency virus (HIV) or viral encephalitides who may display hyperactivity, impulsivity, and inattention [10]. Autoimmunity has also been implicated in the pediatric autoimmune neuropsychiatric disorders associated with *Streptococcus* infections (PANDAS), in which tics, obsessive-compulsive symptoms, recurring and remitting illness temporally related to streptococcal infections, and associated adventitious movements, hyperactivity and emotional lability may be seen [11]. Prolonged exposure to the environmental toxin of lead has been associated with impaired cognition primarily but may also manifest itself with ADHD symptoms [7].

Social environmental factors are frequently noted as contributors to ADHD, particularly adverse family variables [8,10]. Although these may not be primary in causality, they are frequently noted as modifiers of function of the ADHD child. Parental mental illness, chronic family discord, and poor family cohesion have all been implicated. Television exposure in young children younger than age 3 has also been associated with attentional problems in a large American longitudinal survey of children [12]. However a similar association was not established in a Danish cohort, and the direction of the causality of these findings are unknown [13]. If such an association is confirmed, important questions will be raised about the effects of early visual and auditory experience (e.g., television, music, or video games) on attention, cognition, and related neurotransmitters.

III. Pathogenesis and Pathophysiology

The current nomenclature stresses abnormalities of "attention" as being central to ADHD. Neurological models of attention emphasize the importance of posterior parietal structures in sensory attentional systems that are responsible for stimulus detection, vigilance, disengagement, and shifting and the importance of frontal and interconnected subcortical structures (forming a frontal network) in motor intentional systems involved in initiating and sustaining a response to a stimulus, inhibiting inappropriate responses, and stopping a movement when it is no longer required. Most researchers suggest that ADHD is fundamentally the result of dysfunction within frontal intentional networks [14]. In this model, the core symptoms of ADHD are thought to be secondary to abnormal selection of motor response to stimuli (i.e., difficulty in preparing the response to, rather than attending to, stimuli). The result is unresponsiveness to stimuli that should lead to action and defective inhibition of responses to those that should not, with the latter contributing to impulsive and hyperactive behavior [15]. Neuropsychological findings of impairment on response inhibition tasks, both skeletomotor and oculomotor, lend support to this hypothesis. Although deficits other than poor inhibitory control have been reported, findings from neuropsychological studies of individuals with ADHD support an emphasis on response inhibition. As noted by Durston in a review of the biological bases of ADHD [16], "while it is yet unresolved whether poor inhibitory control is the only cognitive deficit in ADHD, resulting in the deficits observed in other domains, the evidence from neuropsychological studies does support the notion that poor inhibitory control is central to ADHD."

IV. Relevant Neurological Findings

Findings from functional imaging, lesion, and electrophysiology studies implicate frontal-subcortical circuits in response inhibition. At its most basic level, premotor circuits, including those originating in the supplementary motor area, are likely critical for selection of motor responses, including inhibition, with prefrontal circuits important for processing of cognitive and socioemotional information necessary to guide response selection.

Abnormalities within frontal regions and interconnected subcortical regions (i.e., the basal ganglia and cerebellum) are consistent findings in anatomical magnetic resonance imaging (MRI) studies of ADHD, particularly in boys with the disorder [16,17]. Most anatomical imaging studies of ADHD have found overall reductions in total brain and total cerebral volume. Decreases in frontal regional volumes are the most consistent finding, although abnormalities in more posterior regions have been reported. Whether the frontal abnormalities are localized to a specific frontal/subcortical region (or circuit), and if so, which region(s) and what level (cortical or subcortical), remains unclear.

Current neurological models of frontal lobe structure and function have their basis in a well-described series of parallel frontal-subcortical circuits. Original models identified five circuits, two of which are related to motor function, originating in skeletomotor and oculomotor regions of the cortex; the other three, originating in dorsolateral prefrontal, anterior cingulate and orbitofrontal cortices, are thought to be crucial in cognitive ("executive") and socioemotional control. Neuroanatomical studies in primates have helped to further delineate these circuits and have lent support to the original proposed model in which frontal projections to the basal ganglia and cerebellum form a series of frontal-striatal-thalamo-frontal and frontal-cerebello (dentato)-frontal circuits.

Findings from neuroimaging studies suggest that abnormalities associated with ADHD may not be confined to a specific frontal circuit. At the cerebral cortical level, decreased volume within several frontal lobe regions has been reported. Initial studies of cerebral cortical volumes in ADHD used an approach that relied on callosal landmarks to subdivide ("parcellate") the cerebrum in a gridlike pattern. Findings from these studies revealed decreased volumes in multiple regions, including right "anterior frontal," bilateral "anterior inferior" (frontal), and right "anterior superior" (frontal) regions. More recently, we used a semi-automated approach that relied on gyral/sulcal landmarks to parcellate the frontal lobe into functionally relevant volumes based on models of neuroanatomical circuits as previously discussed [18]; findings once again revealed abnormalities in multiple circuits with decreased volumes in both prefrontal and premotor regions. The findings suggest that the clinical picture of ADHD encompasses dysfunctions that are attributable to anomalous development of both premotor and prefrontal circuits.

Studies since have focused on examining single specific prefrontal regions, with findings of one group reporting smaller right dorsolateral prefrontal volume and another smaller left orbitofrontal volumes. It will be important for future studies to include a comprehensive examination of specific prefrontal (e.g., dorsolateral prefrontal, orbitofrontal, anterior cingulate, lateral prefrontal) and premotor (e.g., supplementary motor area, frontal eye fields, ventral premotor) regions within the same patient population; this will provide a means of determining which circuits are affected in ADHD and whether variations in clinical presentation (e.g., inattentive vs combined types) are associated with different neuroanatomical findings.

At the subcortical level, abnormalities within the basal ganglia have been reported in both the caudate, which receives projections from the prefrontal cortices, and the putamen, which receives projections from the motor and premotor cortices [16,17]. Within the caudate, investigators have reported volume and asymmetry differences, although the findings have not been consistent across studies. Studies of the putamen have yielded similarly inconsistent findings. Acquired lesions of the putamen have been associated with a higher incidence of secondary ADHD, although studies of putamen volume have generally found no significant differences. Finally, volume of the globus pallidus, which receives projections from the caudate and putamen, was found to be decreased in two studies of ADHD, although they differed as to whether volume reduction was greater on the right or left side.

Cerebellar abnormalities have been reported in children with ADHD, including two separate studies revealing decreased size of the posterior-inferior vermis and others revealing decreased overall size of the cerebellum [16,17]. Findings of cerebellar projections (via the dentate nucleus) to prefrontal association areas (e.g., Brodmann's areas 9 and 46) provide a basis for its contribution to cognitive and behavioral impairments associated with ADHD, and although the cerebellum has been traditionally viewed as being principally involved in motor coordination, evidence from the past 2 decades increasingly shows that the cerebellum is more broadly involved in coordination and control of more complex aspects of behavior and cognition, including executive functions relevant to ADHD.

Findings from neurological and neuropsychological studies of individuals with ADHD also implicate abnormalities within frontal circuits and suggest involvement of both premotor and prefrontal systems [14]. Subtle abnormalities on motor examination (e.g., slow speed, variable response time, and excessive overflow movements) suggest dysfunction of motor/premotor circuits in ADHD. These motor manifestations received greater emphasis in previous diagnostic constructs, in particular, "minimal brain dysfunction"; nevertheless, studies of ADHD have continued to reveal that the disorder is associated with motor impairment and that the incidence of developmental coordination disorder is markedly increased in ADHD. As evidence of impaired prefrontal function, individuals with ADHD often have difficulties with executive functioning linked to prefrontal function, including deficits in working memory and planning, organizing, and generating strategies for future actions (see [15,19] for further discussion of executive dysfunction and its relation to the pathophysiology of ADHD).

The association between clinical impairments and neuroanatomical findings in ADHD has been investigated by examining brain-behavior correlations and using functional imaging. Studies of the association between measures of clinical symptoms (quantified using rating scales) and anatomical measures have found greater severity to be correlated with smaller frontal, temporal, caudate, and cerebellar volumes. Results from studies examining correlations between performance on behavioral tasks and anatomical MRI measurements in individuals with ADHD are somewhat conflicting. One group found a significant correlation between right anterior cortical volume and performance on a measure of response inhibition (auditory Go/No-go task) in boys with ADHD but not in controls. In another study, investigators did not find significant correlation between anatomical MRI measures and a cognitive measure of response inhibition (Stroop interference task), although they did report correlation between both right and left anterior superior white matter and measures of externalizing behavior on a parent rating scale.

Tasks used during functional MRI in ADHD have ranged from those involving basic motor execution (e.g., repetitive finger sequencing) to those involving more complex cognitive/behavioral control functions, such as response inhibition and working memory [20]. Differences in activation have been observed within multiple frontal-subcortical circuits, including prefrontal and motor cortices, the cerebellum, and basal ganglia; findings also include differences in activation in posterior regions of the cerebral cortex.

V. Pharmacology, Biochemistry, and Molecular Mechanisms

Catecholaminergic medications have been the mainstay of pharmacological management of ADHD since the introduction of the amphetamines in 1937. The stimulants, amphetamine and methylphenidate, have been recognized as raising extracellular dopamine (DA) and norepinephrine (NE) in the frontal cortex in both animal and human studies, resulting in a decrease in the symptoms associated with this disorder [5,21,22]. Three mechanisms have been identified. Recent investigation specifically links these medications to their effects on the dopamine transporter (DAT) as well as the norepinephrine transporter (NET), with each blocking the reuptake of DA and NE. The amphetamines have also been found to enhance their release at the synaptic cleft. Both also decrease the catabolism of these transporters via inhibition of monoamine oxidase. The importance of the NE system has been highlighted with the development of the newer agent, atomoxetine, which selectively affects the NET by blocking reuptake of this neurotransmitter. Animal studies demonstrate that methylphenidate and amphetamine each exert their effects on the striatum, prefrontal cortex, and nuclear accumbens, whereas atomoxetine affects the prefrontal cortex alone. The α-agonists clonidine and guanfacine are used less commonly, usually as adjuncts in the treatment of ADHD. They appear to exert their primary effect on the NE system by stimulating pre- and postsynaptic α-2 receptors, with involvement of the prefrontal cortex observed in monkeys (Table 2).

Although each medication has specific sites of action in the brain and on different neurotransmitters in the catecholamine system, clinical advantages of choosing specific DAT or NET selective compounds have not been established. Instead, the choice of medication for treating ADHD usually is centered on clinical efficacy, onset of action, and duration of effect being sought (Table 3).

Through hundreds of clinical trials extending across several decades, the stimulant medications have been found to be highly effective [23], with response rates estimated at 75–80%. In affecting the dopaminergic or noradrenergic systems, they modulate attention, alertness, vigilance, and arousal. This appears to occur through enhancement of inhibitory control systems and potentiation of delays between stimuli and responses, thereby reducing impulsive and off-task behavior. Both methylphenidate and the amphetamines have onset of action within 30–60 minutes postingestion. Both are available in short-acting or long-acting forms. The short-acting compounds reach peak clinical effect within 1–2 hours and have a total duration of effect of approximately 4 hours. They are available in liquid, chewable, and standard tablet forms. The longer-acting sustained-release preparations, provided in capsular form, have similar times of onset (30–60 minutes), with duration of effect ranging from 8 (amphetamine salt, methylphenidate) to a maximum of 12 hours (OROS-methylphenidate). New delivery systems involving topical administration via skin patch are currently being tested.

Atomoxetine, the selective NE reuptake inhibitor, is now available for the treatment of ADHD [22]. It is ingested in capsule form and has a serum concentration peak at 1–3 hours. It is metabolized in the liver and has a plasma half-life of 5 hours, with excretion in the urine. However, as many as 10% of users are slow metabolizers, with the half-life extending up to 24 hours. Although the initial trials involved dosing at 12-hour intervals, current recommendations are that it should be used once daily. Due to the longer half-life, clinical effects may not be observed for 1 week or longer after initiation.

Limited clinical trials have been performed with the α-adrenergic agonists clonidine and guanfacine and antidepressants, including bupropion and the tricyclic antidepressants (TCA) [21]. Efficacy of the α-agonists has been demonstrated in treating hyperactivity and impulsivity, as well as the common sleep problems observed in ADHD. Bupropion and TCAs may play a role in DA and NE reuptake inhibition. Trials of these antidepressants similarly have shown benefit in children and adults with ADHD, but their use has been limited by side effects and toxicity associated with these medications. Comparative trials with stimulant medications have established the stimulants to be superior. The selective serotonin reuptake inhibitors have not been shown to be effective in treatment of ADHD, consistent with the concept of ADHD being a disorder of DA and NE pathways.

VI. Natural History

ADHD has been primarily conceptualized as a phenomenon of school-aged children, with most descriptions of its symptomatology focused on behaviors occurring among children of this age in the school setting, in the process of learning, at play, or at home. It is most

Table 2 Neurotransmitters Associated with Attention-Deficit Hyperactivity Disorder

Neurotransmitter	Function	Area of Brain	Medication	Genes
Dopamine (DA)	Movement, allows expression of cortically initiated voluntary movement	Substantia nigra, basal ganglia, prefrontal and cingulate cortices, nucleus accumbens	Stimulants	DAT1, DRD4
			Dextroamphetamine: increases DA release	DAT1, DRD2, DRD5
				DA-B-hydroxylase gene
			Methylphenidate: blocks DA transporters	DRD4
Norepinephrine (NE)	Role in arousal and attention	Neurons originate in the brainstem (locus coeruleus)	Atomoxetine: blockade of the presynaptic NE transporter	No specific evidence for linkage of α1c, 2A, and 2C or NET1 genes
	Possibly important in anxiety		α-2 noradrenergic agonists: stimulate pre- and postsynaptic receptors	ADRA2A, ADRA2C, COMT, NET, DBH, PNMT
			Tricyclic antidepressants: action on NE reuptake	
			Stimulants: dextroamphetamine blocks reuptake of NE and increase release	
Serotonin	Coordination of brain functions: sleep, impulse control and personality, mood and psychosis, etc.	Brainstem nuclei: dorsal and median raphe, supply forebrain		HTT, HTR1A, TPH

commonly diagnosed during this period, with the peak period of medication treatment occurring at approximately age 10–11 years. The DSM IV-R subtypes differ clinically in terms of age of onset, with the hyperactive/impulsive presenting earlier than the combined subtype and much earlier than the inattentive subtype. Furthermore, it is important to recognize that the forms can change over the lifespan; not uncommonly, one individual can present with the hyperactive/impulsive type as a preschooler, develop the full syndrome until middle school, and manifest the inattentive type thereafter.

The current criteria require that some symptoms must be present before age 7 years. A growing literature verifies the common presentation of ADHD in the preschool years, with symptoms of hyperactivity and impulsivity often presenting before age 4 and described as early as infancy. Its validity as a diagnosis in the preschool period has been debated, with investigators often preferring the term "preschool hyperactivity" or "hard-to-manage preschoolers" to "ADHD" during this time period [24]. When one looks at symptoms of preschool children, problems similar to those in school-aged children are seen, including inattention, hyperactivity, and impulsivity. However, children with primary inattention are less likely to be identified in this age period, despite their known risk for later academic problems. Related problems with compliance and antisocial behavior are also frequently described in those with the hyperactive or combined subtypes, and significant overlap between symptoms of ADHD and oppositional defiant disorder (ODD) is recognized in preschoolers. Neuropsychological deficits can also be identified with testing in preschool children. Problems are seen in working memory, planning, inhibition, attention flexibility, and delay sensitivity. Although the literature on medication efficacy for preschool hyperactivity is quite limited, increasing numbers of preschool children are being treated with stimulant medications. Psychotherapeutic behavioral interventions also have a role in the preschool population, although preferred techniques and their benefit, particularly as compared with medication management, remains unclear.

It is during the school-age period, from ages 6–11 years, that most children with ADHD are identified and treated [8,24]. Children with the combined and hyperactive subtypes generally are identified earlier than those with the inattentive subtype, often in response to disruptive behaviors in the classroom as they enter formal structured education. In contrast, children with the inattentive subtype present later with problems primarily in academic achievement and are less likely to experience social concerns. Although the criteria state that symptoms must cause impairment before age 7 years, these children may in fact begin to show impairment at a later age, as academic demands and expectations rise. It is during the elementary school–age period that other associated conditions also emerge. Problems in school behavior and academic achievement frequently uncover associated learning disorders, which affect approximately one-third of children with ADHD, with rates that are highest for reading disability (dyslexia) and other language-related

Table 3 Medications Used to Treat ADHD

Medication	Dose*	Daily Schedule	Common Side Effects
Stimulants			
Amphetamine	0.3–1.5 mg/kg/day		*Common to all stimulants:*
Short-acting (dexedrine tablets)		Twice or three times	Insomnia
			Appetite suppression
Intermediate-acting (Adderall, dexedrine spansules)		Once or twice	Tic exacerbation
			Depression, anxiety
Long-acting (Adderall-XR)		Once	*Preparation-specific:*
			Rebound phenomena (more common with short-acting preparations)
Methylphenidate	0.5–2.0 mg/kg/day		
Short-acting (Ritalin, Metadate, Focalin)		Twice to four times	
Long-acting (Concerta, Ritalin LA, Metadate CD, Focalin XR)		Once	
Selective norepinephrine reuptake inhibitors			
Atomoxetine (Strattera)	0.5–1.4 mg/kg/day	Once or twice	Abdominal pain, nausea/vomiting, constipation, dry mouth, appetite suppression, liver toxicity (rare)
Antidepressants			
*Tricyclic antidepressants (TCAs)*** (e.g., imipramine, nortriptyline)	2.0-5.0 mg/kg/day for imipramine 1.0–3.0 mg/kg/day for nortriptyline	Once or twice	Dry mouth, constipation, weight change, EKG changes
Bupropion	1.0–6.0 mg/kg/day		Irritability, insomnia
Short-acting (Wellbutrin)		Three times	Lower seizure threshold
Long-acting (Wellbutrin SR)		Once	Contraindicated in bulimia
Antihypertensives			
Clonidine (Catapres)	3–10 μg/kg/day	Twice to four times	Sedation (less with guanfacine)
			Dry mouth
			Depression
Guanfacine (Tenex)	30–100 μg/kg/day	Twice	Hypotension (including orthostatic) and associated symptoms of lightheadedness, dizziness

* Recommended doses by weight (mg/kg/day) serve only as a guide. Optimal dose varies among patients, and several medications have maximal doses recommended by the manufacturer. For all medications, slow titration is recommended to achieve a dose that provides maximal benefit with minimal side effects.
** Monitoring of blood levels can be useful in guiding TCA dosing.

disorders but also significant for mathematics disability. Aggression and antisocial behaviors occur frequently in these children, with as many as one-half meeting the criteria for ODD and one-third or more meeting criteria for the more severe conduct disorder (CD). Those children with a history of aggressive behavior appear to be at greatest risk for these disruptive behavior disorders.

Adolescence frequently presents new challenges for those with ADHD. The clinical symptoms of hyperactivity, inattention, and impulsivity generally improve, along with a greater demand for autonomy, resulting in a decrease in the rates of medication use. When compared with peers, the majority of affected teenagers continue to display impairments in these areas and meet criteria for ADHD. Adolescents undergo many social and physical changes. In facing these new demands, problems may result, with anxiety disorders, mood problems, or frank major depression occurring in as many as 25% of teenagers with ADHD, particularly those with an associated history of anxiety. Adolescents with ADHD also face higher risks for cigarette smoking, alcohol use, and illegal substance use and are more likely than peers to experience driving accidents and traffic violations. However, a significant reduction of risk in these behaviors occurs with medication treatment during adolescence.

Although ADHD was originally conceived of as a developmental disorder occurring in childhood, it is now apparent that symptoms often persist into adulthood [25].

Approximately one-half to two-thirds of individuals diagnosed with ADHD in childhood continue to have symptoms into adult life. It is not uncommon for patients to first present during adolescence or even adulthood, in part due to the demands for behavior control and self-organization required as an individual moves from primary and secondary school to college and then to "real-world" situations such as job-related activities and raising a family. Adults with ADHD are more likely to have problems in the workplace and in sustaining employment. Issues frequently arise related to concentration, memory, and impulsivity, with resulting disorganization, interpersonal conflict, and time-related problems. Difficulties with relationships at home and in the community are more common as well, with increased frequency of marital conflicts. As with adolescents, substance abuse is more common in adults with ADHD who are untreated. Finally, genetic factors can result in affected adults having children with ADHD and associated parent-child problems. Current literature suggests continuing benefit from treatment of ADHD in affected adults, leading to reductions in symptoms and improved functioning.

References

1. American Psychiatric Association Task Force on DSM-IV. (2000). "Diagnostic and Statistical Manual of Mental Disorders: DSM-IV-TR," ed. 4. American Psychiatric Association, Washington, D.C.
2. Rowland, A. S., Lesesne, C. A., and Abramowitz, A. J. (2002). The epidemiology of attention-deficit/hyperactivity disorder (ADHD): a public health view. *Ment. Retard. Dev. Disabil. Res. Rev.* **8**, 162–170.
3. Hoffmann, H. (1995). "Struwwelpeter" (in English translation). Dover, New York.
4. Faraone, S. V., Perlis, R. H., Doyle, A. E., Smoller, J. W., Goralnick, J. J., Holmgren, M. A., and Sklar, P. (2005). Molecular genetics of attention-deficit/hyperactivity disorder. *Biol. Psychiatry* **57**, 1313–1323.
5. Solanto, M. V. (2002). Dopamine dysfunction in AD/HD: integrating clinical and basic neuroscience research. *Behav. Brain Res.* **130**, 65–71.
6. Doyle, A. E., Willcutt, E. G., Seidman, L. J., Biederman, J., Chouinard, V. A., Silva, J., and Faraone, S. V. (2005). Attention-deficit/hyperactivity disorder endophenotypes. *Biol. Psychiatry* **57**, 1324–1335.
7. Pearl, P. L., Weiss, R. E., and Stein, M. A. (2001). Medical mimics. Medical and neurological conditions simulating ADHD. *Ann. NY Acad. Sci.* **931**, 97–112.
8. Biederman, J. (2005). Attention-deficit/hyperactivity disorder: a selective overview. *Biol. Psychiatry* **57**, 1215–1220.
9. Spencer, T. J., Biederman, J., Wilens, T. E., and Faraone, S. V. (2002). Overview and neurobiology of attention-deficit/hyperactivity disorder. *J. Clin. Psychiatry* **63(Suppl)**, 3–9.
10. Voeller, K. K. (2004). Attention-deficit hyperactivity disorder (ADHD). *J. Child Neurol.* **19**, 798–814.
11. Swedo, S. E., and Grant, P. J. (2005). Annotation: PANDAS: a model for human autoimmune disease. *J. Child Psychol. Psychiatry* **46**, 227–234.
12. Christakis, D. A., Zimmerman, F. J., DiGiuseppe, D. L., and McCarty, C. A. (2004). Early television exposure and subsequent attentional problems in children. *Pediatrics* **113**, 708–713.
13. Obel, C., Henriksen, T. B., Dalsgaard, S., Linnet, K. M., Skajaa, E., Thomsen, P. H., and Olsen, J. (2004). Does children's watching of television cause attention problems? Retesting the hypothesis in a Danish cohort. *Pediatrics* **114**, 1372–1373; author reply, 1373–1374.
14. Denckla, M. (2003). ADHD: topic update. *Brain Dev.* **25**, 383–389.
15. Barkley, R. A. (1997). Behavioral inhibition, sustained attention, and executive functions: constructing a unifying theory of ADHD. *Psych. Bull.* **121**, 65–94.
16. Durston, S. (2003). A review of the biological bases of ADHD: what have we learned from imaging studies? *Ment. Retard. Dev. Disabil. Res. Rev.* **9**, 184–195.
17. Seidman, L. J., Valera, E. M., and Makris, N. (2005). Structural brain imaging of attention-deficit/hyperactivity disorder. *Bio. Psychiatry* **57**, 1263–1272.
18. Mostofsky, S. H., Cooper, K. L., Kates, W. R., Denckla, M. B., and Kaufmann, W. E. (2002). Smaller prefrontal and premotor volumes in boys with ADHD. *Biol. Psychiatry* **52**, 785–794.
19. Denckla, M. B. (1996). A theory and model of executive function: a neuropsychological perspective. *In:* "Attention, Memory, and Executive Function" (G. R. Lyon and N. A. Krasneger, eds.), pp. 263–278. Paul H. Brookes Publishing Co., Baltimore.
20. Bush, G., Valera, E. M., and Seidman, L. J. (2005). Functional neuroimaging of attention-deficit/hyperactivity disorder: a review and suggested future directions. *Biol. Psychiatry* **57**, 1273–1284.
21. Pliszka, S. R. (2005). The neuropsychopharmacology of attention-deficit/hyperactivity disorder. *Biol. Psychiatry* **57**, 1385–1390.
22. Madras, B. K., Miller, G. M., and Fischman, A. J. (2005). The dopamine transporter and attention-deficit/hyperactivity disorder. *Biol. Psychiatry* **57**, 1397–1409.
23. Wilens, T. E., Biederman, J., and Spencer, T. J. (2002). Attention deficit/hyperactivity disorder across the lifespan. *Annu. Rev. Med.* **53**, 113–131.
24. Sonuga-Barke, E. J., Auerbach, J., Campbell, S. B., Daley, D., and Thompson. M. (2005). Varieties of preschool hyperactivity: multiple pathways from risk to disorder. *Dev. Sci.* **8**, 141–150.
25. Wender, P. H., Wolf, L. E., and Wasserstein, J. (2001). Adults with ADHD. An overview. *Ann. NY Acad. Sci.* **931**, 1–16.

58

Congenital Hydrocephalus

Stephen L. Kinsman, MD

Keywords: *brain malformations, cerebrospinal fluid, hydrocephalus, increased intracranial pressure, ventriculomegaly*

I. Brief History and Nomenclature
II. Etiology
III. Pathogenesis
IV. Pathophysiology
V. Relevant Structural Details
VI. Pharmacology, Biochemistry, and Molecular Mechanisms
VII. Natural History
References

I. Brief History and Nomenclature

In the early 1700s, Vesalius was the first to accurately recognize hydrocephalus as an accumulation of fluid within the cerebral ventricles. During the remainder of the 18th century, Morgagni and others described the neuroanatomical and pathological causes of hydrocephalus. In the next century, the physiology of cerebrospinal fluid (CSF) circulation began to be elucidated. Magendie is credited with developing the concept of an active bulk flow of CSF.

The first half of the 20th century was a time of rapid increases in our understanding of the clinical and radiographic aspects of hydrocephalus. Of particular importance is the contribution of Dandy with the introduction of pneumoencephalography in 1918. The work of Russell is important with regards to the addition of systematic pathological studies of the causes of hydrocephalus.

The past 3 decades have provided us with better, less invasive techniques for the diagnosis of hydrocephalus with the developments of computed tomography, cranial sonography, and magnetic resonance imaging (MRI). MRI has been particularly important because of its ability to accurately image hindbrain structures as well as to identify areas of CSF flow. Finally, improvements in CSF pressure monitoring and analysis have added to our ability to determine the presence of decompensated CSF accumulation, although to date this is used more frequently for adult hydrocephalus than for pediatric forms [1].

All of these developments have led to modern definitions of hydrocephalus. Rekate has developed an acceptable definition: "an accumulation of excessive volumes of CSF within the intracranial compartment with higher pressures than would occur in a steady state and which may be treated with a drainage procedure before it becomes intractable" [2]. The notion of early hydrocephalus versus late or intractable hydrocephalus is particularly important when determining the urgency of intervention and possibly prognosis.

The modern era of hydrocephalus began with the development of the valve-regulated shunt system by Nulsen and Spitz in the 1950s. Since that time, valve-regulated shunts have become the standard of care. Recently, there has been resurgence in the use of endoscopic third ventriculostomy as a method of treatment.

A few studies have reported effective medical treatments of certain forms of infantile hydrocephalus using isosorbide or furosemide and acetazolamide. However, study results have not been consistent, and these treatments have not received widespread acceptance. Also, some investigators have suggested medical approaches to provide at least temporary treatment of the cellular-damaging effects of the late decompensated (or intractable) phase of hydrocephalus [3].

A review of nomenclature is useful for this discussion. Increased accumulation of CSF suggests an alteration in CSF flow at some level. It is first important to identify the site of the blockade to CSF circulation, absorption, or both. Terms originating from the era of pneumoencephalography, such as "internal" versus "external" or "obstructive" versus "nonobstructive" hydrocephalus, are confusing and outdated. External hydrocephalus is sometimes mistaken with subdural hygromas, and all hydrocephalus is to some degree caused by obstruction to normal CSF circulation (thus it would all be termed "internal hydrocephalus"). The terminology that best describes CSF blockade is *communicating* versus *noncommunicating* hydrocephalus. Noncommunicating hydrocephalus denotes a blockade of CSF pathways at or proximal to the outlet foramina of the fourth ventricle (foramina of Luschka and Magendie). Communicating hydrocephalus denotes a blockade distal to this point, in the basal subarachnoid cisterns, in the subarachnoid spaces over the brain surface, or within the arachnoid granulations (specialized units at which CSF is absorbed back into the circulation) [2].

II. Etiology

The causes of hydrocephalus relate to any pathophysiological process capable of altering CSF production, circulation, or absorption in such a way as to allow an increased accumulation of CSF in a brain that is unable to compensate for the resulting increase in CSF volume and, in some circumstances, decreased CSF flow. Causes of hydrocephalus include congenital malformations of the central nervous system (CNS), infections, hemorrhage, trauma, teratogens, and tumors. Increasingly, we are seeing the power of investigational convergence, brought about by progress in the study of genetic animal models and human genetic disease, in advancing our understanding of the etiology of hydrocephalus. Analysis of human cases of X-linked hydrocephalus, particularly in multiplex families, has led to an understanding of the role of L1CAM, a molecule now felt to be more of a neural recognition molecule than a cell adhesion molecule, in producing a hydrocephalic state. Many cases of congenital hydrocephalus with adducted thumbs are accounted for by a mutation in the L1CAM gene [4].

With regard to understanding the clinical etiologies of hydrocephalus, distinguishing between nongenetic and genetic causes of hydrocephalus helps in organizing the differential diagnosis of this condition. Among genetic causes, further separation into syndromic and nonsyndromic etiologies is also helpful [5]. Certain metabolic disorders, such as Hurler's syndrome and nonketotic hyperglycinemia (NKH), also can be associated with hydrocephalus. Conditions associated with skeletal anomalies such as VACTERL (vertebral anomalies, anal atresia, cardiac defect, tracheoesophageal fistula, renal abnormalities, and limb abnormalities) and skullbase anomalies are also often associated with hydrocephalus. Finally, abnormalities of the development of the posterior fossa are also seen in association with hydrocephalus; this is especially true of the Dandy-Walker malformation.

The etiologies of several mouse models of congenital hydrocephalus have recently been unraveled via molecular analysis. Recently, the *Hydin* gene, which when mutated is responsible for causing hydrocephalus in the mouse, has been cloned [6]. The product of this novel gene with unknown function is expressed in the ciliated ependymal cell layer that lines the lateral, third, and fourth ventricles and may interact with the cytoskeleton. Its role in altering CSF dynamics remains unknown. The hydrocephalus mouse model *hyh* has been shown to be due to a mutation in the gene encoding α-SNAP. α-SNAP plays a key role in a wide variety of membrane fusion events in eukaryotic cells, including the regulated exocytosis of neurotransmitters [7]. The *hyh* mutant points to potential roles of α-SNAP in embryogenesis and brain development [8]. The next step is to link the human homologues of these genes to conditions associated with clinical hydrocephalus. This approach has been very successful with many other genetic conditions.

III. Pathogenesis

Pathogenesis of hydrocephalus should consider both biomechanical and fluid dynamic factors. Equally important is the brain's built-in compensatory mechanisms as it responds to alterations in the previously mentioned processes. Finally, once compensatory capacity has been exhausted, hydrocephalus begins to cause injury to the CNS, mostly in the form of periventricular axonal cellular damage [3].

The pathogenesis of hydrocephalus encompasses the anatomical (gross and microscopic), biochemical, and physiological abnormalities that can result in an increased accumulation of CSF. As stated previously, anatomical, biochemical, and physiological processes also attempt to accommodate (compensate for) the resulting CSF accumulation. These processes include changes in skull shape, lymphatic drainage within the dura, and changes in brain tissue composition.

A. Basic Physiology and Anatomy of CSF Circulation

The normal circulation of CSF requires a balance between the formation of CSF and its absorption and the unimpeded flow of CSF through its normal travel routes. Most of the formation of CSF (at least 80%) takes place in the choroid plexus [2]. This complex tissue is found in the lateral, third, and fourth ventricles, with some of the tissue extending via the foramina of Luschka into the cerebellopontine angles. The bulk of this tissue resides within the lateral ventricles. The cellular composition of the tissue includes endothelial cells that form capillaries and specialized cuboidal epithelium. Specialized tight junctions that are present at the apical sides of the epithelial cells provide the blood-CSF barrier. This barrier provides for the chemical integrity of the CSF. CSF is formed by the ultrafiltration of plasma through choroidal capillary epithelium as well as through uptake of the ultrafiltrate and secretion of CSF (by an active metabolic process) into the ventricular system. It should be noted that several enzymes are believed to be important in the process of CSF formation, with sodium-potassium-adenosine triphosphate pumps and carbonic anhydrase being particularly prominent. It appears that the other 20% of CSF production, so-called *nonchoroidal CSF*, is formed from brain extracellular fluid. Brain extracellular space accounts for about 15% of brain volume under normal conditions [2]. With the exception of the choroids plexus papilloma, there is scant evidence that CSF overproduction/secretion is an important factor in the pathogenesis of hydrocephalus. Under normal circumstances, CSF is produced at a rate of 0.3–0.35 mL/minute, and production equals absorption. CSF is absorbed into the sagittal sinus through the arachnoid villi. The process is passive and therefore requires no energy.

B. Pathophysiology of Abnormal CSF Circulation

It is important to note that the causes of hydrocephalus are extremely varied. Etiology plays a critical role in determining such factors as the age of onset, presence and degree of blockade to CSF circulation, degree and course of ventricular dilatation, brain and cranial compliance, and type of associated intracranial pathology. It should be emphasized at this juncture that the term "congenital hydrocephalus" is deceiving because presentation of hydrocephalus caused by a congenital malformation or process has been reported to present clinically at any age. Therefore age of onset does not always lead immediately to easy determination of the etiology. However, most so-called "congenital hydrocephalus" becomes symptomatic in the fetal, newborn, infant, or childhood intervals.

Classification of hydrocephalus starts with identification of the site of CSF circulation blockage, the probable or precise etiology, the state of progression, and the dynamic status of the disorder (i.e., progressive or arrested). Not all ventriculomegaly, even if progressive, is hydrocephalus because it is sometimes difficult to determine whether a case of progressive ventriculomegaly is due to hydrocephalus or atrophy. Certain patterns of ventriculomegaly suggest that hydrocephalus is likely present, in particular the size and shape of the temporal horns and the shape of the third ventricle on midsagittal MRI [9].

The main branch point of many classification systems for the etiology of hydrocephalus is a separation between congenital cases and those due to other causes, such as inflammatory or neoplastic. However, it would be better to abandon the term "congenital hydrocephalus" when discussing etiology. The issue is really whether the cause of the hydrocephalus is a malformation (such as Dandy-Walker malformation), a genetic process (e.g., X-linked or autosomal recessive types), inflammation of either the leptomeninges or ventricular lining, chemical irritation (e.g., subarachnoid or ventricular blood), or neoplasm. Also important in the development of hydrocephalus is the viscoelastic response of the brain to the biomechanical effects of hydrocephalus, so-called *brain turgor* [2]. Differences in the brain's response to the accumulation of CSF and stretch of brain tissues and their effects on cerebral blood volume all influence how the abnormal accumulation of CSF leads to clinical symptoms and signs in an individual with hydrocephalus. One can also hypothesize that individual differences in this process may account at least in part for the likelihood and possibly severity of brain injury seen in hydrocephalus.

Studies to date do not address the issue of effects of pathophysiological processes on altering compensatory mechanisms that could lead to the development and/or maintenance of hydrocephalus in various situations. Clinical experience supports the concept that "the ability to recruit alternative pathways of CSF absorption appears to differ among individuals" [2]. One important difference is an individual's ability to alter central venous pressure (and possibly volume). Another important factor is brain turgor.

IV. Pathophysiology

Hydrocephalus is a condition that involves multiple intracranial compartments, is dynamic in nature, is variable in its onset and severity for each etiology, includes both biomechanical and fluid dynamic aspects, and is in need of novel treatment strategies. This section focuses on the measurement of abnormal intracranial pressure (ICP) and CSF flow. Increased understanding and detection of the various abnormal processes in a dynamic and potentially more longitudinal manner will enhance our understanding and treatment of hydrocephalus. In cases of communicating hydrocephalus, traditional measurements of ICP have been made by lumbar puncture and consist of static pressure measurements. A level of >20 centimeters of H_2O has been considered abnormally elevated. Others consider a mean ICP of 15 mm Hg on continuous ICP monitoring to be abnormal [1]. The detection of hydrocephalus in newborns and infants is more related to the detection of the clinical symptoms and/or signs of increased CSF pressure. In this age group, increasing head circumference is particularly important. Widening of the cranial sutures and fullness of the fontanelles are also important signs. Neuroimaging then provides evidence of abnormal accumulation of CSF (see later discussion), sometimes with evidence of increased periventricular accumulation of fluid (Fig. 1).

The condition known as *normal-pressure hydrocephalus* denotes abnormal CSF accumulation with significant CNS dysfunction but no increase in intracranial pressure. This condition can occur in children. A decrease in brain turgor occurring at the same time as an increased resistance to CSF outflow into the spinal subarachnoid space is the best current explanation of this condition's pathophysiology [2]. Others note that a better term for this clinical situation might be *chronic hydrocephalus*, which separates the term from an instrument-based definition [10]. Importantly, the notion that hydrocephalus can even exist as a low-pressure state has received some attention [11].

One unique pathophysiological state, unique to shunted hydrocephalus, has been called the *slit ventricle syndrome*. This is considered a state of long-term shunt overdrainage and brain stiffness. These individuals, who have small or slit-sized ventricles, lose their ability to compensate for small changes in CSF volume and respond to small increases in volume with large increases in ICP. Often this condition can be treated with shunt revision, but sometimes other treatment is required, such as furosemide, acetazolamide, corticosteroids, antimigrainous treatments, and subtemporal cranial vault decompression (Fig. 2). This condition is considered by some to represent two quite different clinical problems, the aforementioned one and one in which slit ventricles result from shunt overdrainage.

It is important to recognize that some individuals with congenital hydrocephalus go undetected for many years or are in fact asymptomatic for many years. Symptoms then begin in adulthood. Some authors have called this condition *long-standing ventriculomegaly* (LOVA), described as a unique form of hydrocephalus that develops during childhood and manifests symptoms during adulthood. Some of these individuals become symptomatic with symptoms

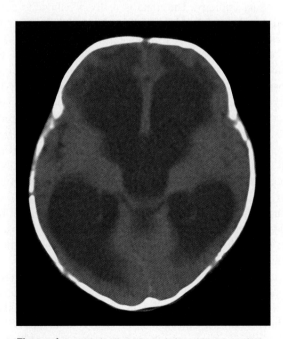

Figure 1 Acute infantile hydrocephalus with periventricular fluid accumulation.

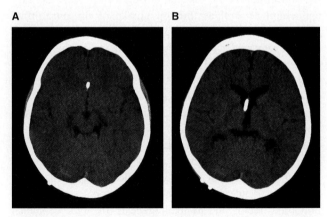

Figure 2 A, Computed tomography (CT) scan of a child with shunted hydrocephalus and slit ventricles. Child had multiple episodes of headache and altered consciousness. **B**, After a diagnosis of slit ventricle syndrome was made, the shunt was revised and the child became asymptomatic. This follow-up CT scan shows that the lateral ventricles have expanded to a more normal size.

and signs of increased ICP, whereas others present more a more normal pressure hydrocephalus (NPH)-like picture of dementia/mental retardation, gait disturbance, and urinary incontinence [12].

V. Relevant Structural Details

The first step in the workup of infantile hydrocephalus is to determine the distribution of abnormal CSF accumulation. Cases of external hydrocephalus are either familial, postprematurity, traumatic (usually nonaccidental), or metabolic (in particular, glutaric aciduria). In internal hydrocephalus, the distribution of abnormal CSF accumulation helps differentiate the site of blockade to CSF flow. The site of the blockade is often helpful in narrowing the search for causes of the hydrocephalus. However, it is often the case that a given pathological process demonstrates elements of both communicating and noncommunicating hydrocephalus, although one type usually predominates. The final step in proper diagnosis is to determine the etiology of the blockade.

It is important to reemphasize that not all ventriculomegaly is the result of a hydrocephalic process. Brain atrophy also leads to ventriculomegaly, so-called *hydrocephalus ex vacuo*. Important differences between true hydrocephalus and the *ex vacuo* form include the degree of temporal horn dilatation relative to sulcal prominence, the angle of frontal horn displacement, and alterations in the shape of the third ventricle [9].

Fetal ventriculomegaly poses some special circumstances in this regard. Most ventriculomegaly identified in the fetal period is static in nature. Brain architecture is not mature enough to distinguish between loss of brain tissue volume and tissue stretch from abnormal CSF accumulation. Serial sonographic measurement of the ventricles is required to distinguish these situations [13]. Optimal treatment of this situation remains unproven, and fetal surgery for this condition remains for the most part unsuccessful.

A few comments are also in order with regard to the relationship between malformations of the posterior fossa and congenital hydrocephalus. Dandy-Walker syndrome (DWS), which is a triad of agenesis or hypoplasia of the cerebellar vermis, cystic dilatation of the fourth ventricle, and supratentorial hydrocephalus, accounts for 1–4% of childhood hydrocephalus. Most of these patients have normal ventricles at birth and that nearly 80% develop symptomatic hydrocephalus by 1 year of age [14].

VI. Pharmacology, Biochemistry, and Molecular Mechanisms

The molecular mechanisms of congenital hydrocephalus are beginning to be unraveled. As noted previously, several mouse models of hydrocephalus have been created via molecular cloning. These discoveries suggest that the molecular function of the ependymal lining plays a critical role in CSF dynamics. Experimental mutation of the axonemal dynein heavy chain gene *Mdnah5* also effects ependymal cell function and leads to hydrocephalus [15]. The *Mdnah5* is specifically expressed in ependymal cells and is essential for ultrastructural and functional integrity of ependymal cilia. In these *Mdnah5*-mutant mice, lack of ependymal flow causes closure of the aqueduct and subsequent formation of triventricular hydrocephalus during early postnatal brain development. These findings suggest that the higher incidence of aqueduct stenosis and hydrocephalus formation in patients with ciliary defects proves the relevance of this novel mechanism in humans.

The mouse model of hydrocephalus *hyh* has been found to be due to a mutation in the gene *Mf1*, a winged helix/forkhead transcription factor [16]. The mechanisms by which this altered transcription factor causes hydrocephalus remain to be elucidated. Alteration of *Msx1*, a regulatory gene involved in epithelio-mesenchymal interactions during organogenesis, leads to mice that lack a subcommissural organ, with resultant collapse of the cerebral aqueduct and the development of hydrocephalus [17]. Finally, increased expression of transforming growth factor (TGF)-β, as seen in subarachnoid hemorrhage, leads to hydrocephalus by a proposed mechanism of alteration in the composition of the brain's extracellular matrix [18].

VII. Natural History

The advent of the valve-regulated shunt system has radically changed the prognosis for individuals with most forms of hydrocephalus. In the U.S., virtually all children with hydrocephalus, with the possible exception of some with massive in utero–acquired hydrocephalus, receive intervention. Laurence and Coates's series of 182 untreated individuals gives us some idea of the historical prognosis for untreated cases [19]. In this series, only 20% of infants reached adulthood. Many of the deaths occurred in the first 1.5 years of life. Of those with untreated hydrocephalus who lived, 60% had intellectual impairment, 25% were "completely ineducable," and 25% had "cocktail party syndrome." Some patients had intelligence quotient (IQ) scores in the normal range but often with significant disparities between verbal and performance scores.

Also of concern was the finding that "some cases with compensated hydrocephalus" had sudden death or a downhill spiral of deterioration. There is little question that for most individuals, untreated hydrocephalus is bad for health, intellect, and function. How well do patients do with treatment? Studies show that in general, a postshunting mantle of about 3 cm or more is important for normal intellectual development [20]. Preshunt scans do not appear predictive of ultimate function in most cases. The most important prognostic factors are the cause of hydrocephalus, duration of the condition before shunting, and presence of associated brain malformations or injury.

A study of the efficacy of various shunt types in contributing to shunt failure rates in the first year post-shunt placement is our best analysis of the symptoms and signs presenting in childhood hydrocephalus [21]. Table 1 summarizes these findings. This study also gives us some sense of the frequency of etiologies presenting with childhood hydrocephalus, as summarized in Table 2. This study was a multicenter trial of shunt placement, not an epidemiological study. However, epidemiological studies show similar incidences. Most interestingly, this study concludes that clearly more research is needed "into the complexities of CSF and brain fluid dynamics and biomechanics." As noted previously, technologies now exist to better study CSF flow and composition, alterations in brain and ventricular architecture over the course of hydrocephalus and its successful (and unsuccessful) treatment, and imaging of brain water content as well as brain tissue compromise, such as the magnetic resonance property anisotropy.

A key question remains: Is there a critical point during the development of hydrocephalus at which irreversible damage occurs? When children with infection and related complications are excluded from the sample, studies yielded normal IQ scores within groups of uncomplicated, shunted hydrocephalic children. Longitudinal study of the effects of hydrocephalus on cognitive development has been difficult. When studies examine a more select and appropriate group of these "uncomplicated, shunted hydrocephalic children," some findings do seem to emerge. As Wills summarizes in a 1993 review:

> Overall, the consensus of these findings appears to be that it is the presence of brain anomalies, cytoarchitectonic defects, or a history of trauma or infection rather that the presence and extent of hydrocephalus per se that accounts for most cognitive deficits. The "pure" effect of hydrocephalus, if such exists, seems to impair visuospatial and visuomotor performance specifically and may depend on selective compression or posterior brain regions. [22]

A quantitative MRI-based study by Fletcher and colleagues adds quantitative confirmation of this assertion [23].

Table 1 Presenting Symptoms and Signs of Childhood Hydrocephalus

Signs or Symptoms	Percentage of Cases
Presenting symptoms	
Irritability	26.6
Delayed developmental milestones	19.8
Nausea or vomiting	19.0
Headache	17.5
Lethargy	17.5
New seizures or change in seizures	6.6
Diplopia	5.8
Worsening school performance	4.2
Fever	2.6
Presenting signs	
Increased head circumference	81.3
Bulging fontanelle	70.6
Delayed developmental milestones	20.9
Loss of upward gaze	15.8
Decreased level of consciousness	12.6
Other focal neurological deficits	12.4
Papilledema	12.0
Sixth nerve(s) palsy	4.6
Hemiparesis	3.8
Nuchal rigidity	1.3

From Drake, J. M., Kestle, J. R., Milner, R., et al. (1998). Randomized trial of cerebrospinal fluid shunt valve design in pediatric hydrocephalus. *Neurosurgery* **43**, 294–305.

Table 2 Hydrocephalus Etiology*

Cause	Percentage of Cases
Intraventricular hemorrhage	24.1
Myelomeningocele	21.1
Tumor	9.0
Aqueduct stenosis	7.0
Cerebrospinal fluid infection	5.2
Head injury	1.2
Other	11.3
Unknown	11.0
Two or more causes	8.7

*Excluding Dandy-Walker malformation.

From Drake, J. M., Kestle, J. R., Milner, R., et al. (1998). Randomized trial of cerebrospinal fluid shunt valve design in pediatric hydrocephalus. *Neurosurgery* **43**, 294–305.

References

1. Eide, P. K. (2005). Assessment of childhood intracranial pressure recordings using a new method of processing intracranial pressure signals. *Pediatr. Neurosurg.* **41**, 122–130.
2. Rekate, H. L. (2001). Hydrocephalus classification and pathophysiology. In: "Pediatric Neurosurgery: Surgery of the Developing Nervous System" (D. G. McLone, ed.), pp. 457–474. Saunders, Philadelphia.
3. Del Bigio, M. R. (2004). Cellular damage and prevention in childhood hydrocephalus. *Brain Pathol.* **14**, 317–324.
4. Finckh, U., Schroder, J. Ressler, B., Veske, A, and Gal, A. (2002). Spectrum and detection rate of L1CAM mutations in isolated and familial cases with clinically suspected L1-disease. *Am. J. Med. Genet.* **92**, 40–46.
5. Schrander-Stumpel, C., and Fryns, J. P. (1998). Congenital hydrocephalus: nosology and guidelines for clinical approach and genetic counselling. *Eur. J. Pediatr.* **157**, 355–362.

6. Davy, B. E., and Robinson, M. L. (2003). Congenital hydrocephalus in hy3 mice is caused by a frameshift mutation in Hydin, a large novel gene. *Hum. Mol. Genet.* **12**, 1163–1170.
7. Hong, H. K., Chakravarti, A., and Takahashi, J. S. (2004). The gene for soluble N-ethylmaleimide sensitive factor attachment protein alpha is mutated in hydrocephaly with hop gait (hyh) mice. *Proc. Natl. Acad. Sci. USA* **101**, 1748–1753.
8. Chae, T. H., Kim, S., Marz, K. E., Hanson, P. I., and Walsh, C. A. (2004). The hyh mutation uncovers roles for alpha Snap in apical protein localization and control of neural cell fate. *Nat. Genet.* **36(3)**, 264–270.
9. Barkovich, A. J. (2000). Hydrocephalus. *In:* "Pediatric Neuroimaging," pp. 581–620. Lippincott, Williams & Wilkins, Philadelphia.
10. Bret, P., Guyotat, J., and Chazal, J. (2002). Is normal pressure hydrocephalus a valid concept in 2002? A reappraisal in five questions and proposal for a new designation of the syndrome as "chronic hydrocephalus." *J. Neurol. Neurosurg. Psychiatry* **73**, 9–12.
11. Pang, D., and Altschuler, E. (1994). Low-pressure hydrocephalic state and viscoelastic alterations in the brain. *Neurosurgery* **35**, 643–655; discussion 655–656.
12. Oi, S., Shimoda, M., Shibata, M., Togo, K., Shinoda, M., Tsugane, R., Sato, O. (2000). Pathophysiology of long-standing overt ventriculomegaly in adults. *J. Neurosurg.* **92**, 933–940.
13. Volpe, J. J. (2001). Neural tube formation and prosencephalic development. *In:* "Neurology of the Newborn," pp. 3–44. W. B. Saunders Company, Philadelphia.
14. Arai, H., and Sato, K. (2001). Dandy-Walker syndrome. *In:* "Pediatric Neurosurgery: Surgery of the Developing Nervous System" (D. G. McLone, ed.), pp. 483–488. Saunders, Philadelphia.
15. Ibanez-Tallon, I., Pagenstecher, A., Fliegauf, M., Olbrich, H., Kispert, A., Ketelsen, U. P., North, A., Heintz, N., and Omran, H. (2004). Dysfunction of axonemal dynein heavy chain Mdnah5 inhibits ependymal flow and reveals a novel mechanism for hydrocephalus formation. *Hum. Mol. Genet.* **13**, 2133–2141.
16. Hong, H. K., Lass, J. H., and Chakravarti, A. (1999). Pleiotropic skeletal and ocular phenotypes of the mouse mutation congenital hydrocephalus (ch/Mf1) arise from a winged helix/forkhead transcription factor gene. *Hum. Mol. Genet.* **8**, 625–637.
17. Fernandez-Llebrez, P., Grondona, J. M., Perez, J., Lopez-Aranda. M. F., Estivill-Torrus, G., Llebrez-Zayas, P. F., Soriano, E., Ramos, C., Lallemand, Y., Bach, A., and Robert, B. (2004). Msx1-deficient mice fail to form prosomere 1 derivatives, subcommissural organ, and posterior commissure and develop hydrocephalus. *J. Neuropathol. Exp. Neurol.* **63**, 574–586.
18. Crews, L., Wyss-Coray, T., and Masliah, E. (2004). Insights into the pathogenesis of hydrocephalus from transgenic and experimental animal models. *Brain Pathol.* **14**, 312–316.
19. Laurence, K. M., and Coates, S. (1962). The natural history of hydrocephalus. Detailed analysis of 182 unoperated cases. *Arch. Dis. Child* **37**, 345–362.
20. Nulsen, F. E., and Rekate, H. L. (1982). Results of treatment for hydrocephalus as a guide to future treatment. *In:* "Pediatric Neurosurgery: Surgery of the Developing Nervous System" (R. L. McLaurin, ed.), pp. 229–241. Grune & Stratton, New York.
21. Drake, J. M., Kestle, J. R., Milner, R., Cinalli, G., Boop, F., Piatt, J. Jr., Haines, S., Schiff, S. J., Cochrane, D. D., Steinbok, P., and MacNeil, N. (1998). Randomized trial of cerebrospinal fluid shunt valve design in pediatric hydrocephalus. *Neurosurgery* **43**, 294–303; discussion 303–305.
22. Wills, K. E. (1993). Neuropsychological functioning in children with spina bifida and/or hydrocephalus. *J. Clin. Child Psychol.* **22**, 247–265.
23. Fletcher, J. M., McCauley, S. R., Brandt, M. E., Bohan, T. P., Kramer, L. A., Francis, D. J., Thorstad, K., and Brookshire, B. L. (1996). Regional brain tissue composition in children with hydrocephalus. Relationships with cognitive development. *Arch. Neurol.* **53**, 549–557.

59

Neurological Manifestations of Hematological Disease

Babette B. Weksler, MD

Keywords: *ADAMTS13, hyperviscosity, M-protein associated neuropathy, myeloproliferative disease, platelet activation, sickle cell vasculopathy, vitamin B_{12} deficiency*

I. Vitamin B_{12} Deficiency
II. Paraproteinemias: Syndromes Associated with a Monoclonal Immunoglobulin
III. Hemoglobinopathies
IV. Acute Porphyrias
V. Thrombotic Thrombocytopenic Purpura
VI. Myeloproliferative Syndromes
References

A wide variety of blood diseases affect the central nervous system (CNS), including vitamin B_{12} deficiency, paraproteinemias, hemoglobinopathies, porphyrias, thrombotic thrombocytopenic purpura, and myeloproliferative syndromes. The neurological manifestations of these conditions are discussed in this chapter, with the exception of leukemias and lymphomas and neurological complications caused by their treatments. These subjects are discussed elsewhere in this textbook.

I. Vitamin B_{12} Deficiency

Neurological or psychiatric symptoms occur in about 40% of patients with cobalamin (vitamin B_{12}) deficiency, in association with progressive damage to the spinal cord, peripheral nerves, and cerebrum [1]. Cobalamin is a vitamin that serves as a cofactor for one-carbon transfers during DNA synthesis and in myelin synthesis, and cobalamin deficiency results in specific demyelination of the posterior and lateral columns of the spinal cord ("combined systems disease"), producing a characteristic sensory neuropathy that usually begins as loss of vibratory and position sense in the feet and legs, progresses to a sensory ataxia that later affects the upper extremities and the corticospinal tracts, and finally results in a spastic paraparesis. Visual changes, optic atrophy, and dementia ("megaloblastic madness") may also occur. The blood manifestations of cobalamin deficiency consist of a slowly progressive macrocytic anemia accompanied by leukopenia and thrombocytopenia with megaloblastic maturation in the bone marrow of all three hematopoietic lineages. Although macrocytic anemia and neuropathy of cobalamin deficiency often occur in parallel, either may dominate the clinical course. Particularly in older adults, neurological symptoms may occur earlier than marked anemia, or the anemia may be masked by concomitant folate therapy.

The most common cause of cobalamin deficiency is autoimmune chronic atrophic gastritis because a protein cofactor secreted by the gastric parietal cells (intrinsic factor) is necessary for the active transport of cobalamin across the distal ileal mucosa, the specific absorption site for this vitamin. Vitamin B_{12} is the only vitamin that requires a specific transport system for uptake. Classical pernicious anemia results from this autoimmune process. Humans do not produce cobalamin and must acquire the vitamin from food, mainly from animal protein. Because the liver normally stores at least 3 years' worth of cobalamin, the onset of the deficiency can be very gradual. Additional causes of cobalamin deficiency include gastrectomy, severe sprue, ileal disease or ileectomy, or extremely strict vegetarianism. A rare inherited lack of ileal receptors for the vitamin B_{12} intrinsic factor complex, Imerslund syndrome, may also cause cobalamin deficiency. Folic acid deficiency produces the same blood picture of megaloblastic anemia as vitamin B_{12} deficiency but does not cause neurological disease (see later discussion). There has been some interest in whether mild cognitive impairment due to Alzheimer's disease or dementia might respond to the combination of cobalamin and folic acid, but controlled trials of the two vitamins for up to 12 weeks failed to have any effect.

Vitamin B_{12} is a cofactor in the conversion of homocysteine to methionine by the enzyme methionine synthase, a reaction required for the regeneration of reduced folic acid, which is itself an essential one-carbon donor of reduced methyl groups for the synthesis of thymine from uracil, a key step in the synthesis of deoxyribonucleic acid (DNA). A deficiency of either cobalamin or folic acid slows DNA synthesis, resulting in an underproduction of all cells that normally have a rapid turnover, such as those in the gut mucosa and skin and hematopoietic cells in the bone marrow. The result of this impediment in DNA synthesis is the appearance of giant cells that are unable to divide in many tissues, notably the bone marrow, blood, skin, mucosal surfaces, and intestine. The peripheral smear shows well-hemoglobinized macroovalocytes with an elevated mean cell volume (MCV), hypersegmented neutrophils, and giant platelets, all reflecting the delayed cellular division resulting from impaired DNA synthesis. This is a hemolytic anemia with ineffective hematopoiesis that typically produces a hypercellular megaloblastic bone marrow. In early cobalamin deficiency, the anemia may be very mild but the MCV is often elevated.

In addition to its key role in DNA synthesis, cobalamin is an essential cofactor in the CNS in a completely independent biochemical reaction, the conversion of methylmalonyl-CoA to succinyl-CoA, a fatty acid precursor needed for normal myelin synthesis. Although folic acid can bypass cobalamin deficiency to correct the hematopoietic defect in DNA synthesis, it cannot substitute for cobalamin's role in the CNS. Thus folic acid deficiency produces a macrocytic anemia but no neurological dysfunction, and conversely, folic acid can partially correct a macrocytic anemic due to cobalamin deficiency but cannot correct the neurological dysfunction caused by lack of cobalamin.

The severity of neurological symptoms correlates with the length of time that vitamin B_{12} deficiency has been present and may also correlate with age, being greater in older patients. It is not clearly related to the serum vitamin B_{12} level. In a large series of patients (n = 143) with cobalamin deficiency involving the CNS, pernicious anemia was the most common underlying cause, and neurological complaints (paresthesias or ataxia) were the most common initial symptoms [1]. The median duration of symptoms before the specific diagnosis and treatment of vitamin B_{12} was 4 months. The range of neurological symptoms varied from diminished vibratory sense and proprioception to ataxia, hypoesthesia, muscle weakness, abnormal deep tendon reflexes, spasticity, incontinence, orthostatic hypotension, dementia, and psychosis. In 27% of these patients the initial hematocrit was normal, and in 23% the MCV was normal. In nonanemic patients in whom diagnosis was delayed, neurological progression often occurred without anemia. Which patients will have predominantly anemia or will show the neuropathy of cobalamin deficiency is not at all clear. Recent studies suggest that measuring serum total serum cysteine or glutathione levels may help to predict patients at risk, in that individuals with pure pernicious anemia with neurological symptoms had higher cysteine levels than those without neurological symptoms. There appears to be no correlation between mutations in methyltetrahydrofolate reductase (MTHFR), a common polymorphism that may lead to increased blood homocysteine levels, and the occurrence of neurological symptoms in cobalamin deficiency, but larger studies are needed. In patients presenting with abnormal vibration sense, diminished proprioception and abnormally brisk reflexes, a diagnosis of cobalamin neuropathy is made with a low serum cobalamin level, and evidence of megaloblastic hematopoiesis. Serum methylmalonic acid and homocysteine elevations are also sensitive indicators. The presence of intrinsic factor antibodies is a presumptive sign of the specific diagnosis, but these antibodies are much more common than vitamin B_{12} deficiency.

The treatment is parenteral vitamin B_{12}, usually 1 mg weekly intramuscularly for 1 month and then 1 mg monthly for life. A therapeutic trial may be required to confirm the diagnosis. Improvement in neurological signs may occur in the first 3 months with further improvement over a longer time period, but the reversibility of the neurological dysfunction depends on the duration of symptoms, and major disability may persist in advanced cases. In the large series mentioned previously, neurological recovery was complete in 47% of patients and the severity of neurological disability improved in 91%. Although folic acid therapy corrects the macrocytic anemia, it neither corrects the neurological dysfunction nor precipitates neurological symptoms. However, by improving only the anemia in patients with cobalamin deficiency, folic acid therapy may permit the neurological damage to progress. Several studies have indicated that large daily doses of oral cobalamin may, by a mass effect, also correct the vitamin B_{12} deficiency.

II. Paraproteinemias: Syndromes Associated with a Monoclonal Immunoglobulin

Neurological abnormalities associated with the presence of a monoclonal immunoglobulin (Ig) (M-protein) are found in a range of diseases, from overt malignancies such as multiple myeloma, Waldenström's macroglobulinemia, and lymphomas to monoclonal globulin of uncertain significance (MGUS), cryoglobulinemias, primary amyloidosis, and the POEMS syndrome [2]. The incidence of M-proteins increases with age, such that 1% of persons >50 years have a monoclonal Ig, 3% of those >70, and >10% of those >80. About 3% of persons with a monoclonal Ig have a neuropathy, and the neuropathy may be the presenting symptom of the disorder, especially in MGUS and amyloidosis. Conversely, about 5% of all patients with polyneuropathy will manifest an M-protein, such that the prevalence of M protein–associated neuropathy in the general population older than 50 years is roughly 80 per 100,000, which makes it a major cause of neuropathy in older persons. Although in some cases the M-protein has demonstrable anti-nerve activity, often there is no direct correlation between the properties of the monoclonal antibody and the neurological damage [3].

A. Multiple Myeloma

Multiple myeloma represents a clonal malignancy of plasma cells in which the bone marrow contains >10% monoclonal plasma cells; the M-protein they produce is usually detectable in the serum (>3 grams/dL), urine, or both. Commonly, the urine contains Bence-Jones protein, which consists of the monoclonal light chains of the particular M-protein. Myeloma is one of the more common hematological malignancies, with an incidence of about 3 cases per 100,000 people. The incidence of the different M-protein types follows the distribution of normal immunoglobulins, with IgG-κ being most frequent, then IgG-λ, then IgA and rarely, IgD or IgE. In "light chain disease," only the κ or λ light chain, not the entire Ig molecule, is secreted by the malignant clone, and in rare cases the myeloma is nonsecretory and no protein product is made. Myeloma frequently presents with the clinical triad of fatigue, back pain, and proteinuria, and as it progresses, it manifests osteopenic or lytic bone lesions, renal dysfunction, cytopenias, hypogammaglobulinemia (of normal immunoglobulins) leading to increased risk of infections, and hypercalcemia. Bone lesions are osteosclerotic in only 5% of cases, but these patients have an increased incidence of neuropathies (see later discussion as well as discussion of POEMS syndrome).

The most frequent cause of neurological abnormalities in myeloma is radiculopathy representing spinal cord or nerve root compression due to collapse of vertebra because of severe osteopenia, lytic bone lesions, or extradural plasmacytoma. Impending cord compression is a medical emergency, requiring immediate initiation of steroids and local irradiation. Plasmacytomas at the base of the skull may cause cranial nerve palsies. Leptomeningeal involvement by the malignant plasma cells or intraparenchymal CNS plasmacytomas are rare complications. In some cases of myeloma, monoclonal light chains precipitate in tissues to form amyloid deposits that damage many organs as well as nerves. Light chain amyloidosis, either associated with myeloma or as primary amyloidosis (see later discussion), causes 30–40% of M protein–associated neuropathies. Hypercalcemia, which accompanies extensive myeloma bone disease, may cause mental status changes or obtundation.

The neuropathies associated with multiple myeloma often respond to treatment of the underlying disease. However, light chain amyloidosis that causes neuropathy is extremely resistant to treatment. The overall treatment of myeloma remains palliative rather than curative, although newer drug treatments such as Revlimid and Velcade, used as supplements to the more classical melphalan and prednisone, are resulting in longer remissions and better disease control. Tandem bone marrow transplantation to permit use of high-dose chemotherapy also provides more disease-free time for patients but has not been generally curative.

B. Waldenström's Macroglobulinopathy

In Waldenström's macroglobulinopathy (WM), which is a clonal B-cell malignancy, a pentameric monoclonal

IgM is produced, and the clinical picture and neurological manifestations differ considerably from those in myeloma [4]. WM is quite rare, accounting for only 2% of hematological malignancies and occurring mostly in Caucasians. Familial clustering of B-cell proliferative disorders and autoimmune syndromes is commonly associated with WM. There is no specific genetic marker for WM, the IgH class-switch rearrangements common in myeloma are absent, and the responsible cell is lymphoplasmacytoid in appearance and appears to be a post–germinal center B cell.

WM is a disease of the elderly, presenting with nonspecific symptoms such as anemia and purpura. Lymphoplasmacytic infiltration of the bone marrow occurs, often accompanied by hepatosplenomegaly, lytic bone lesions, or extensive Bence-Jones proteinuria are uncommon. The monoclonal IgM is predominantly κ in light chain type and remains intravascular because of its high molecular weight; at high concentrations it increases plasma viscosity, causing a hyperviscosity syndrome with fatigue, headache, blurred vision, and mental status changes, even coma. The monoclonal IgM may have antibody activity causing typical anti–myelin-associated glycoprotein (MAG) sensimotor neuropathy; may be a cold agglutinin causing hemolytic anemia, Raynaud's phenomenon, and acrocyanosis; or may be a cryoglobulin causing sensorimotor neuropathy in combination with mononeuropathies, including motor neuropathy. Because the monoclonal protein is mainly intravascular, treatment includes plasmapheresis as well as cytotoxic anti–B cell therapy.

C. Monoclonal Gammopathy of Undetermined Significance

In contrast to patients with multiple myeloma, persons with MGUS are generally in good health and do not have back pain, fatigue, anemia, bone disease, or renal dysfunction. However, their blood shows a monoclonal Ig of 3 grams/dL or less, without suppression of normal Ig. A small amount of Bence-Jones protein may be present in the urine. The average incidence is 1:1000 persons > 50 years old. In some individuals, MGUS persists for many years without changes, but in about 1% per year, it progresses to multiple myeloma or lymphoma. The neurological manifestations of MGUS are quite different from those in myeloma and are typically demyelinating peripheral neuropathy with some axonal degeneration [3]. These neuropathies are mainly sensory and symmetrical, progress slowly, and do not affect cranial nerves. Neuropathy occurs in 5–28% of patients with MGUS, mostly of the IgM-type. Conversely, about 10% of patients who present with sensorimotor neuropathy of unknown cause have an M-protein. Both the clinical picture and the response to treatment differ according to the Ig heavy chain class of the M-protein.

Neuropathies associated with IgM MGUS tend to be more sensory than motor, manifest tremor and ataxia, display demyelination on electrophysiological tests, and respond poorly to plasma exchange or intravenous immunoglobulin (IV Ig). There is no correlation between the size of the M-protein peak and the severity of the neuropathy. In half of patients with IgM gammopathy and neuropathy, IgM binds to MAG, the major glycoprotein of peripheral myelin. Patients with IgM anti-MAG M-proteins are more likely to have mainly distal acquired demyelinating symmetrical (DADS) chronic neuropathy, with mainly sensory features and a slow progression over many years. High titers of the anti-MAG implicate worse prognosis. Pathological studies of nerve biopsies show segmental demyelination with widening of the myelin lamellae, deposition of IgM M-protein, and deposition of complement. The prognosis is relatively good, and treatment to reduce M-protein can produce clinical improvement. The drugs Fludarabine or Rituximab have been reported to be effective in small series. In other cases the IgM binds chondroitin sulfate, gangliosides, or glycolipids, all components of peripheral nerve. Anti-sulfatide M-proteins are the second most common antineural IgM proteins associated with neuropathy and cause sensorimotor symptoms and small fiber painful neuropathy.

In contrast, neuropathies associated with IgA MGUS develop very slowly, tend to occur in males, are sensory but often with autonomic manifestations, and are improved by IV Ig, plasma exchange, and/or steroids. IgG M-proteins are less commonly associated with neuropathy, but individual cases have been described. There are no clinically distinguishing features. Evidence that IgA or IgG M-proteins deposit in neural tissue is limited to a very few cases. The neuropathy associated with IgG MGUS may show less demyelination and less prolongation of distal latency than in IgM MGUS.

In animal models, injection with purified IgM or IgA M-proteins taken from MGUS patients with neuropathy has resulted in demyelination. Injection of Fab fragments of the M-proteins also produced demyelination. These M-proteins do not necessarily show anti-MAG activity.

The relationship between M-proteins and chronic inflammatory demyelinating polyradiculopathy (CIDP) is not clear at present, although patients with CIDP and M-protein have been reported to have a milder course, fewer relapses, less weakness, and less severe sensory changes that those with classical CIDP without MGUS. Whereas MGUS tends to increase with age and the associated neuropathy is more sensory than motor, CDIP occurs at any age, motor symptoms predominate, and the course is more relapsing. CDIP is more responsive to intravenous Ig therapy than MGUS-associated neuropathies.

D. Primary Amyloidosis

Deposition of amyloid protein fibrils, consisting of the insoluble, variable portions of Ig light chains in tissues including the peripheral nerves, skin, liver, kidney, heart, tongue, bone marrow, and spleen, is characteristic of the rare syndrome of primary amyloidosis (incidence <1 per 100,000) and is usually a disease of the elderly [4]. Sensory (later, sensorimotor) peripheral neuropathy starting in the lower extremities and autonomic symptoms including hypotension, syncope, or impotence are common and progressive, with cardiomyopathy and nephrotic syndrome frequently developing; cardiac or renal failure are frequent causes of death. Diagnosis of amyloidosis is made by biopsy of subcutaneous abdominal fat, nerves, or bone marrow with demonstration of characteristic homogeneous, Congo red–staining perivascular amyloid deposits. In sural nerve biopsies, endoneural and perineural amyloid is present, and loss of small myelinated and unmyelinated fibers and axonal degeneration occur. Carpal tunnel syndrome is common (25%). An M-protein may be found in the urine, but plasma cell infiltration of the bone marrow is usually <10%. In patients with primary amyloidosis, hemostasis is impaired with marked skin purpura and severe bleeding on trauma or after surgical procedures that may be difficult to control; in addition, acquired deficiency of factor X, which can be sequestered by amyloid infiltrates, may lead to life-threatening bleeding. The acquired factor X deficiency is almost always associated with λ light chain–type amyloid. Median survival in primary amyloidosis is about 1 year, and response to alkylating agents and or prednisone is very poor. Bone marrow transplantation may prolong survival and improve neuropathy, but data are limited to a very few cases.

E. POEMS Syndrome and Osteosclerotic Myeloma

POEMS syndrome comprises *p*eripheral neuropathy, *o*rganomegaly, *e*ndocrinopathy, a *m*onoclonal plasma cell disorder, and *s*kin change [5]. In 95% of cases, either osteosclerotic bone lesions or Castleman's syndrome (angiofollicular lymph node hyperplasia) occur. Conversely, as many as 50% of patients with osteosclerotic myeloma have peripheral neuropathy. Visual changes and headache are common symptoms, and optic disc edema is observed on retinal examination. Other common features of POEMS include edema, papilledema, effusions and ascites, and thrombocytosis. Anemia does not occur, and about 20% of patients show erythrocytosis. The M-protein is almost always of λ type. Typically, the bone marrow contains <5% plasma cells, and hypercalcemia or renal insufficiency are rare, which distinguishes POEMS syndrome from multiple myeloma. The peak incidence is among ages 50–60 years. The prognosis is more favorable than that of multiple myeloma, with about a fourfold better survival.

Hepatosplenomegaly is present in 50% of patients. Diabetes mellitus and hypothyroidism are increased in incidence, and gonadal dysfunction is common. Pulmonary hypertension, restrictive lung disease, and vascular thrombosis (both arterial and venous) are common. Abnormal pulmonary function tests most commonly show restriction due to neuromuscular weakness and impaired diffusion capacity. Increased levels of circulating cytokines, especially vascular endothelial growth factor (VEGF), interleukin (IL)-1β, tumor necrosis factor (TNF)-α, and IL-6 have been detected, as well as increased levels of tissue metalloproteases.

The polyneuropathy of POEMS, which is a major feature of the syndrome, is mainly motor and symmetrical, starting with distal paresthesias in the feet, progressing proximally with nerve conduction abnormalities and motor symptoms, and occasionally with pain. Nerve biopsies show axonal degeneration, demyelination, and endoneurial edema but no deposition of amyloid or M-protein. Nerve conduction velocities are slowed, with prolonged distal latencies and abnormal muscle action potentials. The differential diagnosis of POEMS includes CIDP, primary amyloidosis, cryoglobulinemia, and MGUS-associated neuropathy. The presence of osteosclerotic lesions is an important distinguishing criterion; they may be solitary or multiple, and their treatment with radiation can reverse the neuropathy, although more slowly than reversal of the other clinical manifestations. In contrast to CIDP, POEMS patients do not respond to plasmapheresis or IV Ig.

III. Hemoglobinopathies

A. Sickle Cell Disease

In the past decade, understanding of the frequency and pathogenesis of neurological complications of the sickle hemoglobinopathies has been clearly elucidated. Both occlusive and hemorrhagic stroke are common in these disorders, particularly in children. Improved appreciation of the pathogenesis of stroke risk and cognitive developmental problems has completely changed the management of pediatric sickle cell disease in recent years.

Sickle cell diseases are a group of hemoglobinopathies including sickle cell anemia (Hbb SS), sickle cell-beta thalassemia syndromes, and sickle cell-hemoglobin C disease. In sickle cell anemia, a single homozygous mutation in the beta globin chain of hemoglobin at position 7 results in substitution of glutamic acid by valine (Hb SS, $\alpha 2\beta^S\beta^S$), which alters surface charge of the molecule, favors oxygen free radical generation, permits aggregation of the beta chains. and decreases their solubility. In sickle cell-beta zero thalassemia, heterozygosity for β^S is accompanied by a beta-chain thalassemic mutation that greatly reduces (β+thalassemia) or abolishes (β⁰thalassemia) the synthesis of the other, normal beta chain allele. In sickle cell Hb-C disease, there is double heterozygosity for HbS and Hgb C (another beta chain mutation also at position 6, substituting lysine for glutamic acid, $\alpha 2\beta^S\beta^C$). The most severe clinical syndromes are in sickle cell anemia and sickle-beta zero thalassemia. Hb SC is less severe with milder anemia and less vascular damage. Sickle cell trait, inheritance of one β^S and one normal β^A gene ($\alpha^2\beta^A\beta^S$), is usually an asymptomatic state that produces no anemia and almost no rheological abnormalities or other symptoms. The presence of high quantities of Hb F, fetal hemoglobin, interferes with the self-association of Hb SS molecules and also ameliorates the clinical disease.

The mutation-producing Hb S changes the surface charge on the hemoglobin molecule to favor tight intracellular stacking of the hemoglobin molecules into liquid crystals of decreased solubility that distort and rigidify the shape of the red blood cells and damage their membranes, resulting in hemolysis and favoring obstruction in the microcirculation, especially when blood is desaturated as in veins and organs where blood flow is slow. In addition, the sickled red cells are easily hemolyzed, become prothrombotic, and, in turn, activate and damage the vascular endothelium, promoting thrombosis. Furthermore, the hemoglobin released from the red cells antagonizes vascular nitric oxide, a vasodilator normally produced by healthy endothelium, resulting in vasoconstriction, platelet activation, inflammation, and release of growth factors that produce vessel wall proliferation. Particularly in the brain, this pathophysiological sequence produces the combination of microvascular occlusion and larger arterial stenoses.

Children with Hb SS have a high incidence (7%) of clinical stroke and a higher incidence (17%) of clinically silent stroke that reflects the progressive acquisition of arterial stenoses within the cerebral circulation [6]. These strokes have a strong predilection for the watershed areas between major arterial distribution beds, reflecting the pathophysiological combination of arterial stenoses in medium to large intracerebral arteries and the microvascular obstruction of small venules and capillaries. In addition, neovascular proliferations in the form of "moya-moya" capillary lesions that display augmented vascular fragility accompany these other changes and bleed easily. Transcranial Doppler (TCD) ultrasonography is a sensitive technique for detecting the abnormal increases in blood flow that signal stenosed intracranial arteries, whereas magnetic resonance imaging (MRI) confirms infarction. Mean velocity in the middle cerebral or internal carotid arteries of >220 centimeters/second on one examination or of >200 centimeters/second on three examinations is considered abnormally high and is indicative of high stroke risk. These noninvasive techniques have shown that the first areas of the brain to be affected are subcortical, with later appearance of abnormalities in the deep white matter. Strokes are more likely to occur in the frontal lobes, as well as in the watershed areas, in children with sickle cell disease.

In addition to having a high incidence of first stroke, children with sickle cell disease who demonstrate arterial stenotic lesions are at a very high risk of developing stroke recurrence (70%) unless treated by chronic transfusion/exchange transfusion therapy to bring the percentage of HbS below 50% and preferably below 30%. About 90% of recurrent strokes in such children can be prevented. This improved outcome has been documented in several controlled clinical trials in which the incidence of recurrent stroke in transfused children has been markedly diminished. Moreover, with transfusion to maintain a substantially reduced percentage of HbS, the stenoses may reverse and the abnormal cerebral blood flow may return to a normal pattern. An important recent trial (STOP II) has, unfortunately, clearly demonstrated that when transfusions are discontinued even after the cerebral blood flow patterns normalize, the incidence of recurrent stroke rises again sharply [7]. Reversion to abnormal cranial artery flow velocities or appearance of new stroke was observed after a mean of 4.5 ± 2.6 months without transfusion therapy, and almost half of the patients in whom transfusion was halted had a primary event within 10 months. Trials are currently underway to determine whether the addition of hydroxyurea therapy to raise the percentage of protective Hb F can reduce transfusion requirements in children with sickle cell disease at high risk of stroke. Iron chelation therapy is customarily given to prevent iron overload and its pathological complications in children who receive chronic transfusions.

The neuropsychological consequences of sickle cell brain pathology include learning disabilities, impaired attention and concentration, poor spatial-visual skills and, consequently, poor academic achievement and social skills [8]. Because much brain pathology in sickle cell disease is silent in terms of overt neurological dysfunction, an appreciation of "silent" lesions using TCD ultrasound permits early detection, and most likely, prevention of these consequences as well. Moreover, TCD can detect an expanding arterial lesion early enough that transfusion therapy may be able to reverse the lesion. Recent studies also document early hearing loss and hearing impairment in sickle cell patients of all ages, especially females.

In adults with sickle cell disease, there is also an increased risk of retinopathy, which represents the summation of repeated microvascular occlusions and local tissue hypoxia with consequent neovascularization. Retinopathy, associated with retinal hemorrhages, is also increased in severe anemia of any type (Hb < 8 grams/dL) or severe thrombocytopenia (platelets < 50,000/uL). The incidence of retinopathy as documented in a recent study was 23% in patients with either and 38% in those with both anemia and thrombocytopenia [9].

In iron deficiency anemia in children, a number of additional neurological disorders are common, including developmental delay, stroke, breath holding, pseudotumor cerebri, and cranial nerve palsies [10].

B. Thalassemia

In the thalassemia syndromes, mutations or deletions in the alpha or beta globin chains of the hemoglobin genes impair the efficiency of hemoglobin production, leading to a variety of clinical syndromes that are characterized by hypochromic microcytic anemias varying from minimal (one alpha or beta gene deletion in thalassemia trait) to lethal in utero (four alpha gene deletions in hydrops fetalis). Neurological complications are limited to the most severe thalassemias, such as beta thalassemia major, which is a very severe anemia that requires chronic transfusions for survival after infancy, or thalassemia intermedia, a severe anemia in which transfusion is usually not required. Because of the marked inefficiency of red cell production, there is extreme compensatory expansion of bone marrow within long bones, skull, and vertebrae in addition to extramedullary hematopoiesis that may present as soft tissue paravertebral masses detectable by MRI or other imaging techniques. These masses can cause spinal cord and nerve root compression, presenting as back pain and lower extremity weakness. Radiotherapy of the masses, which are highly radiosensitive, and, more recently, hydroxyurea therapy can shrink the masses and produce remission of the symptoms in a noninvasive manner. In addition to this mechanism, thalassemia is a hypercoagulable state, and patients with thalassemia syndromes who carry coincident thrombophilic genes, have even higher risk for vascular thrombosis.

Delayed puberty and hypogonadism, demonstrating pituitary dysfunction, are common in patients with thalassemic syndromes and may represent, in part, pituitary damage due to iron overload as a result of abnormally increased iron absorption, which is characteristic of thalassemia, plus transfusion-related iron storage.

IV. Acute Porphyrias

Hepatic porphyrias, whether acute or inducible, result from inherited deficiencies of several enzymes in the heme synthesis pathway. These deficiencies lead to recurrent attacks of neurovisceral symptoms, typically abdominal pain but also neuropathy, muscle weakness, and CNS dysfunction [11]. Four acute porphyrias cause neurovisceral symptoms; they are listed in order of the sequence of the responsible enzyme in heme synthesis: aminolevulinic acid (ALA) dehydratase deficiency porphyria (very rare); acute intermittent porphyria, involving deficiency in porphobilinogen (PBG) deaminase; hereditary coproporphyria, involving deficiency of coproporphyrinogen III oxidase; and variegate porphyria, involving deficiency of protoporphyrinogen oxidase. These are all autosomal dominant disorders, and many different mutations of each gene have been recorded. With the exception of ALA dehydratase deficiency, which needs to be <5% of normal to cause clinical disease, levels of 50% (heterozygous state) of the key enzymes (the heterozygous state) in the other acute porphyries are common in persons with clinical symptoms. Coproporphyria and variegate porphyria have cutaneous manifestations as well. The incidence of these porphyrias is about 5 per 100,000 people, but about 80% of affected individuals remain asymptomatic. Several other porphyrias exist, but these do not have neurological manifestations.

Colicky abdominal pain is the most common symptom of acute porphyria and lasts for hours to days, frequently accompanied by tachycardia and hypertension, which indicate a sympathomimetic autonomic neuropathy. Severe constipation is also common. Proximal motor peripheral neuropathy, usually in the upper rather than lower extremities, is a less frequent occurrence. Sensory loss over the trunk is common, and neuropsychiatric manifestations of anxiety, depression, disorientation, hallucinations, and paranoia may appear during attacks. Hyponatremia resulting from vomiting and hypotonic fluid replacement, or from excessive antidiuretic hormone secretion that may occur despite decreases in plasma and blood volume, is also common. Cranial nerve signs, seizures, coma, respiratory and bulbar paralysis, and even sudden death due to arrhythmias have been reported.

Disease pathogenesis relates to increased levels of porphyrin intermediates. Attacks of acute porphyrias are rare before puberty and may be triggered by stress, starvation, alcohol, smoking, certain drugs, infections, or menses (in some women, attacks are cyclical, correlating with the luteal phase of the menstrual cycle, and are controlled by gonadotrophin-releasing hormone analogs or low-dose estrogen). Among drugs that trigger attacks are barbiturates, hydantoins, rifampin, and progestins. Endogenous steroid hormone cycling may also cause attacks. The known triggers often increase demand for hepatic heme (e.g., induction of cytochromes P450) or induce the up-regulation of ALA synthase, the first enzyme in heme biosynthesis. If this occurs, the inherited partial enzyme deficiencies later in the heme synthetic pathways become rate limiting, leading to the accumulation of intermediates, which appear to have the neurotoxic effects listed previously.

Because the enzyme deficiencies are partial, residual enzymatic activity suffices to maintain adequate hepatic heme synthesis, such that anemia is not part of these syndromes.

Diagnosis is made by documenting increased porphobilinogen (PBG) in fresh urine during the acute attacks [12]. The traditional Watson-Schwartz qualitative test has been supplanted by the commercially available Trace PBG kit (Trace American/Trace Diagnostics, Louisville, Colo.) that is semi-quantitative and shows marked urinary PBG elevation in all porphyrias with the exception of ALA dehydratase deficiency, the rarest type. If the PBG level is increased, additional tests will define the precise disorder: decreased erythrocyte PBG deaminase levels detect AIP; urine and fecal porphyrin levels are elevated in coproporphyria; and increased plasma porphyrin levels and fluorescence pattern detect variegate porphyria. Management of acute attacks include removal of inciting factors; administration of large amounts of glucose (which suppresses ALA synthetase production), usually intravenously; use of sodium to correct hyponatremia; and prompt administration of hemin for severe attacks (Panhematin, Ovation Pharmaceuticals, Deerfield, Ill.; 3–5 milligrams/kilogram/day for 3–5 days). Treatment of hypertension with cautious use of beta blockers, analgesia with morphine, and additional phenothiazine to control anxiety and nausea are useful adjuncts to control symptoms.

V. Thrombotic Thrombocytopenic Purpura

Thrombotic thrombocytopenic purpura (TTP), a syndrome involving thrombocytopenia, microangiopathic hemolytic anemia, fever, neurological abnormalities, and renal dysfunction, is a rare but treatable acute illness that is characterized by arteriolar and capillary platelet-fibrin thrombi present in many organs including the brain [13]. It was first described in 1925 in a teenage girl who died of stroke and heart failure. About 60% of TTP patients present with neurological symptoms, most often altered mental status but also with stroke, seizures, bizarre behavior, and paresthesias. The incidence of TTP is 3–4 cases/million patient years, with a peak in the third decade of life; it occurs more often among women than men (3:2) and is markedly increased among patients with human immunodeficiency virus (HIV) and acquired immune deficiency syndrome (AIDS). Since treatment by plasma exchange was begun in the 1970s, TTP's prognosis has changed from 95% fatal to usually curable, and the full pentad of symptoms is less common, with diagnosis now frequently made from the combination of thrombocytopenia and microangiopathic hemolytic anemia that appear without an apparent inciting cause.

Fluctuating neurological abnormalities are present at onset in the majority of patients and eventually develop in 90%. These include focal and generalized seizures, waxing and waning neurological deficits, paresis, aphasia, paresthesias, visual abnormalities, altered mental status, and encephalopathy. Abnormalities on computed tomography (CT) scan, MRI, or electroencephalogram (EEG) are infrequent, and permanent neurological damage is unusual in patients who recover. During the acute phase, medical imaging techniques may document signs of cerebral edema or posterior leukoencephalopathy syndrome, which probably reflect the microvascular nature of the pathology. One recent study showed normal neuroimaging in the acute phase in 87% of patients [14]. The presence of hemorrhages or infarctions on imaging studies likely indicates more severe cases. The EEG shows diffuse slowing in patients with generalized seizures but is generally normal and uncorrelated with clinical status; however, EEG may be helpful in documenting nonconvulsive status epilepticus in those patients with persistent mental status abnormalities. Laboratory tests show the presence of schistocytes and thrombocytopenia on the peripheral blood smear, elevation in bilirubin and creatinine, and a marked elevation in serum lactic dehydrogenase (LDH) (released from the damaged erythrocytes). The LDH is an excellent biological marker for following the course of the disease. There are usually no indications of disseminated intravascular coagulation, and Coombs' test is negative.

The observation that patients with TTP either have an immune-mediated acquired deficiency of a normal plasma metalloprotease, ADAMTS13, or (rarely) an inherited deficiency of this enzyme helps to explain the pathophysiology of the microvascular platelet-fibrin thrombosis that is the hallmark of the syndrome [15]. ADAMTS13 is a zinc metalloprotease that normally cleaves ultra-large polymers of von Willebrand factor (vWF), freshly released from the vascular endothelium, to mature multimeric vWF. The acquired form of the disease results from development of inhibitory antibodies to ADAMTS13 that prevent such cleavage of vWF. Ultra-large vWF polymers have an elongated molecular shape that rapidly binds and activates circulating platelets, creating platelet thrombi that favor intravascular fibrin formation and microvascular occlusion. In contrast, mature vWF that has been cleaved from the ultralarge forms by ADAMTS13 is globular and only supports platelet activation at sites of injury or high shear, thus promoting normal hemostasis without causing thrombosis. A defect in ADAMTS13 function develops clinical importance when endothelium is injured; for example, after some infections, with use of certain drugs such as ticlopidine or mitomycin C, or in states of immune dysregulation such as postpartum or in HIV/AIDS, when ultra-large vWF polymers are released but cannot be cleaved and microvascular thrombosis is initiated in certain organs such as the brain, heart, skin, or kidney. Formation of intravascular fibrin strands "guillotine" red blood cells passing through the microvasculature, leading to the appearance of schistocytes, to hemolysis, and to production of thrombocytopenia from the consumption of platelets. Secondary hemorrhage may occur. The microvascular nature of the vascular thrombi accounts for the waxing and waning of neurological symptoms, as some of the thrombi are intermittently cleared by normal fibrinolytic mechanisms. This pathogenesis indicates why successful treatment comprises plasmapheresis (to remove the anti-ADAMTS13 antibodies) and plasma exchange (to replenish ADAMTS13), treatments that may have to be continued for months. In difficult cases, recent evidence suggests that rituximab, which destroys the B-lymphocytes responsible for producing the pathological antibodies, is useful therapy. In the case of inherited ADAMTS13 deficiency, TTP presents early in life and does not respond to plasmapheresis (because no antibodies are involved) but is controlled by infusion of small amount of normal plasma.

VI. Myeloproliferative Syndromes

Myeloproliferative syndromes include polycythemia vera (PV), essential thrombocythemia (ET), idiopathic myelofibrosis (IMF), and chronic myelocytic leukemia (CML). Neurological complication of these primary bone marrow diseases, which are characterized by increased, unregulated hematopoietic cell production of mature cells, relate mainly to hyperviscosity in *polycythemia vera* and microvascular occlusion by platelets in *essential thrombocythemia*, with these patients experiencing a mixture of thrombotic and hemorrhagic sequelae. Hematopoietic progenitors from patients with myeloproliferative diseases appear hypersensitive to cytokines such as erythropoietin or thrombopoietin, and recently an acquired gain-of-function mutation in the negative regulatory region of the JAK2 gene (V617F) was identified in 95% of patients with PV and 50% of patients with ET or IMF [16]. JAK2, via activation of Stat 3, is important for signaling by the hematopoietic growth factors erythropoietin and thrombopoietin, and this JAK2 mutation increases the sensitivity of target cells to these growth factors. Patients with PV and ET who show the V617F mutation have more thrombotic and hemorrhagic complications of their disease than those who are negative for the mutation.

In PV, arterial thrombosis dominates venous problems and affects the large cerebral arteries [17]. A key characteristic of PV is an increased red cell mass in the presence of normal oxygen saturation. Because whole blood viscosity rises sharply with >45% hematocrits, studies have documented decreased cerebral blood flow in PV patients with elevated hematocrits. In addition, arterial rheology favors thrombosis when the hematocrit is high: in flowing blood, red cells exhibit axial flow; that is, they stream toward the center of the column of arterial blood, pushing the platelets and plasma closer to the arterial wall where their activation via increased shear forces occurs. This effect is exaggerated at high hematocrits and promotes thrombosis, as does the decrease in overall flow velocity, because activated plasmatic clotting factors are less efficiently cleared. Furthermore, in PV patients treated with phlebotomy until iron deficiency is induced, the resulting hypochromic, microcytic red cells are less flexible than normal, worsening the flow problems.

In ET, not only is the platelet count high, but the platelet mass is markedly increased due to the large number of giant platelets and megakaryocyte fragments that circulate. It is estimated that about 25% of ET patients have episodes of neurological impairment that are not explained by usual cardiovascular risk factors. No differences in age, platelet count, hemoglobin, or leukocyte counts were noted between those ET patients with neurological episodes and those without, but the great majority of ET patients with neurological symptoms were female. The most common neurological complication was cerebrovascular ischemia, with headache, transient ischemic attacks, and dizziness also being common [18]. One neurovascular complication peculiar to ET is *erythromelalgia*, a syndrome of microvascular thrombosis characterized by erythema, vascular congestion, warmth, and severe burning pain of the hands and feet due to microvascular platelet aggregates that form in the digits. This may progress to acrocyanotic ischemia and digital gangrene. Autonomic responses, especially sweating, are often abnormal with signs of distal small-fiber denervation and loss of sympathetic vasoconstrictor responses clinically resembling reflex sympathetic dystrophy, with the exception that pain in erythromelalgia is relieved by cold. Fortunately, erythromelalgia is extremely responsive to aspirin treatment, which presumably decreases platelet aggregation and relieves symptoms of pain and abnormal vasomotor responses.

Although both thrombosis and hemorrhage occur in PV and ET, thromboses predominate. In the ECLAP study, age >60 years and a prior thrombotic event strongly predicted increased risk of recurrent thrombosis, and conventional risk factors for cardiovascular disease further enhanced risk [19]. It is clear that platelet activation is enhanced in myeloproliferative disorders, especially platelet procoagulant activity and the formation of platelet microparticles, expression of activation markers on the platelet surface, and deficiency of lipoxygenase that may favor formation of thromboxane. These platelet changes may explain why cytoreductive therapy to decrease platelet number is less effective in reducing thrombotic events than is administration of antiplatelet therapy such as aspirin. Platelet activation may also explain why control of the red cell count by phlebotomy alone in PV is not associated with a decrease in thrombotic events. Data also exist for neutrophil activation, neutrophil-platelet interaction, and endothelial activation in PV and ET, all of which promote thrombotic events. Because elevated neutrophil and platelet counts are common in PV (and sometimes in ET), decreasing the levels of all three lines of blood cells by use of hydroxyurea or interferon appears to have the added benefit of decreasing thrombotic risk [20].

Thus primary hematological disorders have a great variety of neurological manifestations of differing pathophysiologies. Because of recent advances in understanding the particular disease mechanisms in each hematological disorder, a more rational approach to the diagnosis and management of the particular neurological syndromes involved is now possible.

References

1. Healton, E. B., Savage, D. G., Brust, J. C., Garrett, T. J., and Lindenbaum, J. (1991). Neurologic aspects of cobalamin deficiency. *Medicine* **70**, 229–245.
2. Drappatz, J., and Batchelor, T. (2004). Neurologic complications of plasma cell disorders. *Clin. Lymphoma* **5**, 163–171.
3. Nobile-Orazio, E., and Carpo, M. (2001). Neuropathy and monoclonal gammopathy. *Curr. Opin. Neurol.* **14**, 615–620.
4. Gertz, M. A., Merlini, G., and Treon, S. P. (2004). Amyloidosis and Waldenstrom's macroglobulinemia. *Hematology 2004*, 257–282.
5. Dispenzieri, A., Kyle, R. A., Lacy, M. Q., Rajkumar, S. V., Therneau, T. M., Larson, D. R., Greipp, P. R., Witzig, E., Basu, R., Suarez, G. A., Fonseca, R., Lust, J. A., and Gertz, M. A. (2003). POEMS syndrome: definitions and long-term outcome. *Blood* **101**, 2496–2506.
6. Adams, R. J., McKie, V. C., Nichaols, F. T., Files, B., Vichinsky, E., Pegelow, C., Abboud, M., Gallagher, D., Kutlar, A., Nichols, F. T., Bonds, D. R., Brambilla, D. (1998). Prevention of a first stroke by transfusions in children with sickle cell anemia and abnormal results on transcranial Doppler ultrasonography. *New Engl. J. Med.* **339**, 5–11.
7. Adams, R. J., Brambilla, D., and the Optimizing Primary Stroke Prevention in Sickle Cell Anemia (STOP 2) Trial Investigators. (2005). Discontinuing prophylactic transfusions used to prevent stroke in sickle cell disease. *New Engl. J. Med.* **353**, 2769–2778.
8. Kral, M. C., Brown, R. T., and Hynd, G. W. (2001). Neuropsychological aspects of pediatric sickle cell disease. *Neuropsychol. Rev.* **11**, 179–196.
9. Carraro, M. C., Rossetti, L., and Gerli, G. C. (2001). Prevalence of retinopathy in patients with anemia or thrombocytopenia. *Eur. J. Haematol.* **67**, 238–244.
10. Yager, J. Y., and Hartfield, D. S. (2002). Neurologic manifestations of iron deficiency in childhood. *Ped. Neurol.* **27**, 85–92.
11. Anderson, K. E., Bloomer, J. R., and Bonkovsky, H. L. (2005). Recommendations for the diagnosis and treatment of acute porphyrias. *Ann. Int. Med.* **142**, 439–450.
12. Bonkovsky, H. L. (2005). Neurovisceral porphyrias: what a hematologist needs to know. *Hematology 2005*, 24–30.
13. Tsai, H.-M. (2006). Current concepts in thrombotic thrombocytopenic purpura. *Ann. Rev. Med.* **57**, 419–436.
14. Meloni, G., Proia, A., Antonini, G., De Lena, C., Guerrisi, V., Capria, S., Trisolini, S. M., Ferrazza, G., Sideri, G., and Mandelli, F. (2001). Thrombotic thrombocytopenic purpura: prospective neurologic, neuroimaging and neurophysiologic evaluation. *Haematologica* **86**, 1194–1199.
15. Coppo, P., Bengoufa, D., Veyradier, A., Wolf, M., Bussel, A., Millot, G. A., Malot, S., Heshmati, F., Mira, J. P., Boulanger, E., Galicier, L., Durey-Dragon, M. A., Fremeaux-Bacchi, V., Ramakers, M., Pruna, A., Bordessoule, D., Gouilleux, V., Scrobohaci, M. L., Vernant, J. P., Moreau, D., Azoulay, E., Schlemmer, B., Guillevin, L., Lassoued, K.; Reseau d'Etude des Microangiopathies Thrombotiques de l'Adulte. (2004). Severe ADAMTS13 deficiency in adult idiopathic thrombotic microangiopathies defines a subset of patients characterized by various autoimmune manifestations, lower platelet count and mild renal involvement. *Medicine* **83**, 233–244.
16. Kralovics, R., Passamonti, F., Buser, A. S., Teo, S. S., Tiedt, R., Passweg, J. R., Tichelli, A., Cazzola, M., and Skoda, R. C. (2005). A gain-of-function mutation of JAK2 in myeloproliferative disorders. *New Engl. J. Med.* **352**, 1779–1790.
17. Marchioli, R., Finazzi, G., Landolfi, R., Kutti, J., Gisslinger, H., Patrono, C., Marilus, R., Villegas, A., Tognoni, G., and Barbui, T. (2005). Vascular and neoplastic risk in a large cohort of patients with polythemia vera. *J. Clin. Oncol.* **23**, 2224–2232.
18. Kesler, A., Ellis, M. H., Manor, Y., Gadoth, N., and Lishner, M. (2000). Neurological complications of essential thrombocytosis (ET). *Acta. Neurol. Scand.* **102**, 299–302.
19. Tarach, J. S., and Nowicka-Tarach, B. M. (2000). Myeloproliferative disorders—neurologic complications. *Med. Sci. Monit.* **6**, 421–425.
20. Harrison, C. N. (2005). Platelets and thrombosis in myeloproliferative diseases. *Hematology 2005*, 409–415.

60

Neurological Manifestations of Renal Disease

Ajay K. Singh, MB, MRCP
Radhika Bhatia, MD

Keywords: *kidney disease, renal failure, toxic-metabolic syndrome, uremia*

I. Introduction
II. Neurological Syndromes in Chronic Kidney Disease
III. Neurological Complications of Dialysis
IV. Renal Complications of Neurological Syndromes
V. Neurological Manifestations of Electrolyte Imbalances
VI. Drug Toxicities
   References

I. Introduction

In the U.S., chronic kidney disease (CKD) affects 5–10% of the general population (Table 1). It is estimated that nearly 20 million Americans have some degree of chronic kidney disease as defined by an estimated glomerular filtration rate (GFR) of less than 60 milliliters/minute or evidence of kidney damage by imaging study, biopsy, biochemical testing, or urine tests with an estimated GFR more than 60 milliliters/minute [1–3]. There are currently 1,065,000 people on hemodialysis worldwide. Ninety percent of them live in North America, Japan, and Europe [3,4]. Although the life expectancy of patients with end-stage renal disease has improved since the introduction of dialysis in the 1960s [5,6], it is still far shorter that of the general population. Therefore there is a need for early diagnosis, prevention, and treatment of kidney disease to prevent complications that may ensue and necessitate renal replacement therapy [3,4,7].

Central nervous system (CNS) abnormalities are common in patients with kidney disease. Patients with end-stage renal disease (ESRD) manifest a variety of neurological disorders. Symptoms range from mild altered sensorium and cognitive dysfunction to tremors, coma, and death. These syndromes can occur during dialysis or persist even despite dialysis. This chapter outlines the important neurological complications of

Table 1 Stages and Prevalence of Chronic Kidney Disease

Stage	Description	GFR	Prevalence* Number (Thousands)	Percentage
1	Kidney damage with normal or ↑ GFR	≥90	5900	3.3
2	Kidney damage with mild ↓ GFR	60–89	5300	3.0
3	Moderate ↓ GFR	30–59	7600	4.3
4	Severe ↓ GFR	15–29	400	0.2
5	Kidney failure	<15 (or dialysis)	300	0.1

GFR, Glomerular filtration rate.
*Data on prevalence obtained from NKF K/DOQI Clinical Practice Guidelines for Chronic Kidney Disease, which were estimated using data from NHANES III (1988–1994) and USRDS (1998).

Table 2 Neurological Syndromes in Renal Insufficiency

Syndromes in chronic kidney disease patients
- Uremic encephalopathy
- Uremic polyneuropathy
- Uremic myopathy
- Autonomic neuropathy
- Depression
- Intellectual dysfunction

Syndromes in chronic dialysis patients
- Dialysis-dependent encephalopathy
- Dialysis disequilibrium syndrome
- Dialysis dementia
- Stroke
- Sexual dysfunction
- Peripheral neuropathy
- Seizure disorder

CKD and dialysis therapy (Table 2). The more common CNS disorders that occur both during and despite renal replacement therapy have been emphasized. In addition, the chapter presents a brief outline of renal complications that can occur in specific neurological and multisystemic disorders as well as drug toxicities in relation to neurological and kidney diseases.

II. Neurological Syndromes in Chronic Kidney Disease

A. Uremic Encephalopathy

Uremic encephalopathy is an acute or subacute organic brain syndrome that manifests with CNS signs and symptoms from inadequate kidney function. Uremic encephalopathy can occur in acute or chronic renal failure (ARF or CRF). Patients are at especially high risk when GFR is less than 10% of normal [1,3]. The clinical manifestations are summarized in Figure 1. The initial presentation includes lassitude, fatigue, a sluggish response to dialysis, insomnia, itching, nausea, vomiting, slurred speech, and cognitive decline (Table 3). Other salient characteristics are restless leg syndrome, asterixis, confusion, cranial nerve signs, dysarthria and nystagmus, sensory and motor involvement, weakness, fasciculations, hyperreflexia, and unsustained clonus. Nervous system dysfunction may include cognitive, neuromuscular, autonomic, and sensory impairment; the severity of CNS dysfunction progression is more marked in acute renal failure compared with chronic renal failure [6,8–10]. The earliest and most reliable sign is sensorial clouding.

The etiopathogenesis of uremic encephalopathy is complex. The syndrome reflects the accumulation of toxins and metabolites in the cerebrospinal fluid (CSF) that normally would be excreted by the kidney. Studies indicate an imbalance in neurotransmitter amino acids within the brain. Mild uremic encephalopathy appears to be associated with an imbalance of glycine levels (increasing) over glutamine and GABA levels (decreasing). Other abnormalities include perturbations in brain dopamine and serotonin metabolism that may contribute to sensorial clouding. In more advanced uremia, low levels of GABA and high levels of glycine, dopamine, and serotonin may precipitate myoclonus, seizures, and more severe changes in mental status. In addition to these changes, it is possible that inadequate endocrine function of the kidney also may contribute. For example, deficiency of erythropoietin may be important because of the presence of erythropoietin receptors in the brain, and erythropoietin crosses the blood-brain barrier (BBB). Impaired oxygen utilization by the brain occurs in uremic encephalopathy and is only partially reversed by correction of anemia [8–11]. Parathyroid hormone (PTH) has been implicated as a uremic toxin and has been shown to be responsible for certain adverse effects on the CNS [12]. Both parathyroidectomy and suppression of PTH cause a remarkable improvement in CNS signs and symptoms; therefore the symptoms may be attributed to both PTH and a high brain calcium content.

The electroencephalogram (EEG), assessment of evoked potentials (EPs), and formal cognitive testing are the troika of tests that are commonly used in working up patients

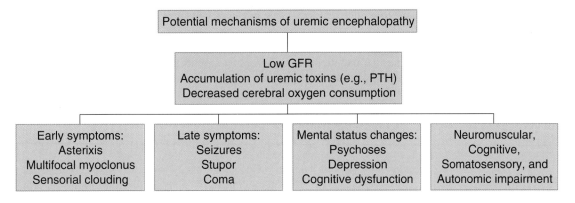

Figure 1 Uremic encephalopathy.

Table 3 Nonneurological Manifestations of Uremia

Gastrointestinal
Cheilitis
Stomatitis
Glossitis
Halitosis
Metallic taste
Anorexia
Nausea
Vomiting
Dyspepsia
Weight loss

Cardiovascular
Pericardial effusion/tamponade
Pericarditis
Hypertension
Volume overload and congestive heart failure (CHF)

Hematological
Anemia (secondary to erythropoietin and/or iron deficiency)
Easy bruising and bleeding (secondary to a platelet function disorder)

Pulmonary
Volume overload with dyspnea
Metastatic calcifications
Pleural effusions

Endocrine
Osteodystrophy (secondary to activated vitamin D and perturbations in calcium/phosphorous metabolism)
Amenorrhea
Loss of libido
Impotence

Skin
Pallor
Bruising

suspected of uremic encephalopathy. A lumbar puncture usually is not performed because no specific abnormalities of uremia manifest in analysis of CSF fluid. Typical EEG findings in patients with uremic encephalopathy include 1) slowing and loss of alpha frequency waves, 2) disorganization, and 3) intermittent bursts of theta and delta waves with slow background activity. EPs (i.e., EEG signals that occur at a reproducible time after the brain receives a sensory stimulus [visual, auditory, somatosensory]) demonstrate prolongation of the latency of the cortical visual-evoked response. Notably, although the auditory-evoked responses are generally not altered in uremia, delays in the cortical potential of the somatosensory-evoked response frequently are observed. Several cognitive function tests are available to evaluate uremic encephalopathy: 1) the trail-making test, which measures psychomotor speed; 2) the continuous memory test, which measures short-term recognition; and 3) the choice reaction time test, which measures simple decision making. Alterations in choice reaction time appear to correlate best with uremic encephalopathy.

B. Uremic Polyneuropathy

Peripheral neuropathy is the most common neurological sign of CRF and is present in about 65% of ESRD patients at the time of starting dialysis. Uremic peripheral neuropathy usually manifests as a distal, symmetric, mixed polyneuropathy and an associated secondary demyelinating process in the posterior columns of the spinal cord. The lower limbs are more involved than the upper limbs. Sensory changes follow the typical glove and stocking distribution. Common signs and symptoms seen are the restless-leg and burning-feet syndromes [3,12,13]. Burning feet syndrome used to be a common symptom of hemodialysis due to the removal of thiamine by the dialyzers. This is now prevented by thiamine replacement in dialysis patients and the syndrome is not as common now. The incidence is now less than 10%. Other signs of nerve dysfunction include loss of deep tendon reflexes and pain, light touch, and vibration sensations [14,15].

The pathophysiology of uremic polyneuropathy has been studied widely but no single toxin has been implicated; rather, the chronic accumulation over years of several toxins is likely. The neuropathy is of a primary axonal degeneration type with secondary segmental demyelination and may be due to a metabolic failure in the perikaryon.

Because of the improvement of neuropathy symptoms in patients on dialysis, it is hypothesized that the metabolites involved in the pathogenesis are dialyzable metabolites or "middle molecules." Supporting this notion are 1) the lower incidence of neuropathy in patients on peritoneal dialysis and 2) the improvement of symptoms after renal transplantation [1,16].

III. Neurological Complications of Dialysis

A. Dialysis-Dependent Encephalopathy

The newly coined term *chronic dialysis-dependent encephalopathy* (CDDE) denotes a CNS disorder occurring in patients despite adequate dialysis (Fig. 2). CDDE encompasses the combination of an underlying organic mental disorder with a psychiatric disorder occurring in a patient receiving dialysis. Dialysis-dependent encephalopathy usually presents in patients after more than 10 years of dialysis. Neurological symptoms comprise decreased intellectual capacity (without evidence of stroke), impaired cognition, depression, decreased physical activity capacity, suicidal tendencies, sexual dysfunction, and psychoses. Brain pathology and neuroimaging are generally nonspecific. Generalized neuronal degeneration with variable anatomical location may be present, as well as focal glial degeneration (2%) and intracerebral hemorrhages with necrotic foci (10%) [3,17].

Three main biochemical processes have been implicated in the pathophysiology of CDDE: 1) apoptosis, 2) oxidative stress, and 3) inflammation. Apoptosis can be initiated by the generation of free radicals, excessive glutamate, hypoxia, and an elevated level of calcium ions. Diabetes mellitus can predispose to brain cell death via increased apoptosis. The accumulation of advanced glycosylation end products (AGEs) initiates oxidative stress and damage by generation of reactive oxygen species. The inflammatory processes involved in the pathogenesis of encephalopathy include an elevated C-reactive protein (CRP), homocysteine, occult infection, and an occult inflammatory nidus. Homocysteine, in turn may be an important factor predisposing uremic patients to atherosclerosis through enhanced oxidative stress.

B. Dialysis Disequilibrium Syndrome

One of the many disorders that may arise as a result of hemodialysis is dialysis disequilibrium syndrome (DDS), first described in 1962. It is thought to arise as a consequence of cerebral edema and is seen in patients who have just been initiated on dialytic therapy. The risk is increased in patients with a markedly elevated blood urea nitrogen levels (BUN > 175 milligrams/dl) [7,15]. Other risk factors include an older or pediatric age group, metabolic acidosis, and preexisting CNS disorders.

Symptoms are variable and can range from mild dizziness, muscle cramping, and restlessness to tremors, seizures, confusion, coma, and possibly death. DDS was previously associated with a rapid and intense initiation of dialysis, but its incidence has now decreased due to a more gradual and earlier initiation of dialysis. DDS is seen exclusively in patients on hemodialysis and is not seen in peritoneal dialysis patients. Glenn et al. found that of 180 patients on dialysis, 7% had dialysis-associated seizures; with one exception, all of this subset were on hemodialysis. Other studies found that the syndrome of dialysis disequilibrium occurred in patients in the first four of their initial hemodialysis treatments. Sporadic cases of death due to this disease have occurred [13,18,19].

The symptoms of DDS occur due to a rapid correction of plasma osmolality that causes cerebral edema without enough time for internal removal of idiogenic osmoles. There are currently two theories to explain this cerebral edema in the pathogenesis of DDS: 1) a reverse osmotic shift and 2) intracerebral acidosis [20,21]. Rapid dialysis results in the removal of urea, resulting in a reduction in BUN,

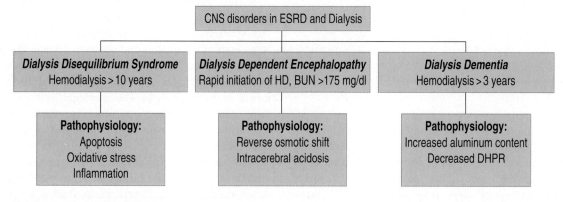

Figure 2 Neurological complications of dialysis.

thereby lowering plasma osmolality. This in turn creates an osmotic gradient and movement of water into the cells and results in cerebral edema in the brain. The loss of water in the extracellular space causes extracellular volume depletion and ensuing hypotension. Thus, when there is not enough time for urea to equilibrate itself during rapid hemodialysis, BUN is reduced and urea acts as an effective osmole and promotes water movement into the brain.

The second and more recent theory of pathogenesis involves intracerebral acidosis. A fall in cerebral intracellular pH occurs due to unknown mechanisms and is hypothesized to be of importance because a *reverse osmotic shift* cannot account for the rapid cerebral edema in DDS [19]. An increased production of organic acids and the displacement of sodium and potassium with hydrogen ions cause an increase in osmolality, promoting cerebral edema, increased intracranial pressure, and encephalopathy [20–22].

C. Dialysis Dementia

Dialysis dementia (DD) is a progressive, often fatal disease that is observed almost exclusively in patients on chronic hemodialysis for a period of more than 3 years. DD is also thought to occur as a component of a multisystemic disorder that includes anemia, osteomalacia, and a proximal myopathy. DD, which was first described in several reports between 1970 and 1973, is linked to a high level of brain aluminum in patients on chronic hemodialysis [23,24]. The initial presenting signs include dysarthria, apraxia, and speech problems. Eventually, symptoms evolve to include personality changes, psychosis, seizures, dementia, and death within 6 months of onset. EEG changes precede the symptoms by 6 months.

The epidemic form of DD has been linked to aluminum and occurred mainly in dialysis units that did not use reverse osmosis or deionization techniques for water purification. Patients thus exposed to high levels of aluminum during hemodialysis developed symptoms of dementia and osteomalacia. Today, DD has been minimized by improved water purification techniques [25,26].

The aluminum content is three times greater in patients with DD than those on hemodialysis without dementia. High levels of aluminum in patients with kidney disease are due to both an increased gastrointestinal absorption and a decreased renal capacity for elimination [22]. Another possible source is aluminum contained in the oral phosphate binders commonly given to uremic patients [27–29]. Altman et al. found that a high aluminum content results in a decreased availability of dihydropteridine reductase (DHPR), an enzyme essential for the synthesis of important neurotransmitters such as acetylcholine and tyrosine [23]. DHPR activity was inversely related to the serum aluminum concentration, and aluminum inhibited DHPR activity in red blood cells. This effect was reversed by aluminum chelation.

In a different study, Altman and associates found a significant correlation between erythrocyte DHPR activity and psychomotor performance. Even at a high blood aluminum level, most of it is bound to transferrin and therefore cannot bind to transferrin receptors in the brain [26]. A low transferrin binding capacity may also play a role in the pathogenesis of DD.

D. Stroke in Dialysis Patients

Cerebrovascular diseases are a common cause of morbidity and mortality in hemodialysis patients. In the United States, the incidence of cardiovascular disease is higher in dialysis patients than in the normal population [2–4]. The most frequent causes of death are myocardial infarction, stroke, and infection. The incidence of stroke in renal transplant recipients is 8%. Risk factors include hypertension, smoking, and diabetes mellitus in ESRD patients [2,4]. High cholesterol and obesity are also common risk factors. Naganuma et al. reported the presence of a silent cerebral infarction to be a new risk factor for vascular events in hemodialysis patients [30].

The pathogenesis of cardiovascular disease involves chronic inflammatory cells, cytokines, and acute phase reactants. AGEs play an important role in ESRD patients with diabetes, whereas CRP is an important risk factor of disease, especially in females.

In dialysis patients, the interaction between blood and dialyzers leads to activation of mononuclear cells [2,3,31,32]. Specifically, the cytokines interleukin (IL)-1β, tumor necrosis factor (TNF)-α, and IL-6 are thought to be of importance in dialysis-related mortality.

The presence of renal failure, even with a normal GFR, may result in a higher risk of myocardial infarction and death. Cardiovascular and cerebrovascular diseases lead to cerebral ischemia and the activation of free radicals, N-methyl-D-aspartate (NMDA), and apoptosis [31,32]. Brain cell destruction in stroke may be mediated through anoxic injury, hence the increased production of nitric oxide (NO) and free radical formation. Activation of NMDA through glutamate can lead to apoptosis; this programmed cell death is another mode of brain injury in stroke patients [2–4].

E. Seizure Disorders in Dialysis Patients

Seizures may occur in approximately 7% of patients undergoing hemodialysis, especially in severely uremic patients who require acute dialysis. Seizure activity tends to occur either during or immediately after dialysis [33,34].

The causes of seizures include several etiologies discussed elsewhere in this chapter, such as uremic encephalopathy, dialysis disequilibrium syndrome, drug toxicity, cerebrovascular disease, dialysis dementia, and electrolyte disorders. CNS dysfunction can occur due to any of these factors alone or in combination but is less likely to occur in patients who are not in acute renal failure [20]. The threshold for seizure activity is lower in uremic patients and in patients with electrolyte imbalances, and aggressive dialysis adds additional insult to the CNS that can provoke seizure activity.

Uremic patients usually have generalized seizures and symptoms. The tendency toward recurrent seizures usually abates after initiation of renal replacement therapy. Individual seizures can be prevented by recognition and correction of predisposing factors such as extreme uremia and severe electrolyte abnormalities [20,33].

IV. Renal Complications of Neurological Syndromes

A. Thrombotic Thrombocytopenic Purpura

Thrombotic thrombocytopenic purpura (TTP) is a multisystemic disorder, first described in 1925, that is characterized by thrombotic microangiopathy in certain target organs. TTP occurs mainly in young females and is more common in African-Americans [35]. Primary disease can be idiopathic (classical), familial, or chronic relapsing, whereas secondary disease is preceded by pregnancy, systemic disorders, carcinomas, or medications.

Common symptoms of TTP include fever, bleeding, and CNS manifestations in 50% of cases. The dominant feature of the syndrome is brain involvement, which occurs in 90% of patients. Neurological symptoms include headache, focal neurological deficits, aphasia, confusion, hemiparesis, motor instability, and seizures. Coma occurs in 10%. Renal involvement is common and includes hematuria, proteinuria, and azotemia. Other organ system involvement is manifested by hemorrhages in the form of gastrointestinal bleeding, epistaxis, menorrhagia, and purpura [35–37].

Recent observations in the pathogenesis of TTP show a high circulating von Willebrand's factor (vWF) during the acute phase of classic TTP. vWF reacts with platelets and promotes platelet aggregation [35]. vWF-cleaving protease may be deficient in TTP patients and is thought to be associated with an immunoglobulin (Ig)-G protease inhibitor autoantibody.

B. Lupus Nephritis

Lupus nephritis is a disease process that may be limited to the kidney or can be a part of a multisystemic disorder such as systemic lupus erythematosus (SLE). It is unusual for lupus nephritis to occur in isolation because SLE reflects a systemic autoimmune process. Approximately 25% of patients with SLE develop renal manifestations, which include microscopic hematuria, proteinuria, and cellular casts. Of all connective tissue disorders, SLE has the highest incidence of CNS complications, occurring in 30–75% of patients. Neurological symptoms include seizures and psychoses, depression, mood changes, dementia, and altered consciousness [38,39].

C. Behçet's Syndrome

Behçet's syndrome is a multisystemic disorder involving the oral and genital mucosa. Renal involvement is rare but when present includes either a rapidly progressing crescentic glomerulonephritis or mesangial IgA deposition [40]. Arterial and venous thromboses are common. Gastrointestinal involvement may include nausea and diarrhea, and CNS involvement may occur as a result of venous and arterial thromboses [40,41]. Renal and peripheral nervous system involvement have a lower prevalence in Behçet's syndrome as compared with the other vasculitides.

V. Neurological Manifestations of Electrolyte Imbalances

Electrolyte disorders associated with sodium and osmolality produce CNS depression and encephalopathy as their major clinical manifestations, whereas disorders of potassium result in peripheral nervous system (PNS) depression and muscle weakness. Calcium and magnesium imbalances can produce both types of symptoms and usually manifest as tetany and seizures [42–45].

A. Hyponatremia

Patients with hyponatremia usually are asymptomatic until serum sodium is less than 125 millimoles/liter. At a cellular level, fluid shifts, cellular swelling, and alterations in the content of intracellular molecules occur. CNS symptoms include confusion, weakness, and seizures. Brain swelling with brainstem compression leading to respiratory arrest is an important cause of death in patients with severe hyponatremic encephalopathy. Cerebral edema occurs due to osmotic shifts of water into the brain owing to a decrease in the effective plasma osmolality.

Osmotic demyelination or central pontine myelinosis (CPM) can occur as a result of rapid correction of sodium in chronic hyponatremia. Symptoms of CPM include cognitive, behavioral, and movement disorders that may not be apparent for days after correction of hyponatremia [44,45]. Acute hyponatremia is more safely reversible and carries a lower risk of osmotic demyelination, whereas the risk is higher in chronic hyponatremia associated with a rapid correction.

B. Hypernatremia

Hypernatremia is defined as a serum sodium level greater than 145 millimoles/liter. CNS symptoms begin to appear above a level of 160 millimoles/liter and include weakness, confusion, lassitude, and a progression to coma. The pathophysiology involves a net water loss or a hypertonic sodium gain. An initial movement of CSF into the interstitial areas of the brain and an increase in intracellular electrolyte levels protect the brain from the effects of high sodium. As the hypernatremia progresses, there is an increased volume loss at the cellular level that can lead to rupture of cerebral blood vessels with severe neurological consequences [45,46].

C. Disorders of Potassium Metabolism

Hyperkalemia is defined as a plasma potassium level of more than 5 millimoles/liter. High potassium levels affect the PNS and the cardiovascular system. Symptoms include neuromuscular defects, weakness, and cardiac conduction defects. As the potassium levels increase, cell membrane excitability increases, causing a decrease in cellular sodium transport that leads to a decrease in membrane excitability and ensuing neuromuscular weakness. Fasting hyperkalemia may occur in dialysis patients due to a decrease in plasma insulin [46,47] causing a shift in potassium from the intracellular to the extracellular compartments, which in dialysis patients cannot be excreted, resulting in a fasting hyperkalemia.

Hypokalemia occurs when the plasma potassium falls below 3.5 millimoles/liter. Mild hypokalemia can be asymptomatic. As the potassium level falls below 3 millimoles/liter, PNS symptoms begin to manifest and weakness and muscle pain can occur, especially in the lower limbs [46–48]. At a level below 2.5 millimoles/liter, paralysis and rhabdomyolysis may occur.

VI. Drug Toxicities

Many medications may precipitate and result in neurological symptoms, particularly seizures. Some of these are outlined in this section.

A. Erythropoietin

Erythropoietin (EPO) is commonly used to treat anemia in CKD patients. Administration of this drug may cause a rapid rise in blood pressure, resulting in hypertensive encephalopathy and seizure activity. Hypertensive encephalopathy occurs in 2–17% of hypertensives and is associated with the use of a higher-than-recommended dosage. Seizure activity is rare in normotensive patients receiving EPO. Symptoms of EPO toxicity include a prodrome of headache, nausea, and visual disturbances, the so-called *posterior reversible leukoencephalopathy syndrome* (PRES) [49]. The pathogenesis may involve hyperperfusion of the cerebral blood vessels and a breakdown of cerebral autoregulation. When given via the intravenous route, endothelin-1 is stimulated more readily and is a potent vasoconstrictor, which may be the mediator of the resultant side effects. The subcutaneous route of administration is used to prevent this [49,50].

B. Lithium

Lithium is used for the treatment of patients with bipolar disorder (manic-depressive disorder). Chronic lithium ingestion has been associated with several different forms of renal injury. Polyuria occurs in 20% of patients due to an accumulation of lithium in the collecting tubules, which interferes with the ability of anti-diuretic hormone (ADH) to increase water permeability. Renal tubular acidosis (RTA) is another effect of lithium on the kidney cells. This occurs due to a tubular defect in the distal nephron, which impairs the kidney's ability to acidify urine, and is manifested as a distal RTA (type 1).

Infrequently, lithium has been associated with nephrotic syndrome. Most cases are due to minimal change disease but cases of focal glomerulosclerosis have also been described. Other effects of lithium on the kidney, which are rare and controversial, include hypercalcemia and chronic interstitial nephritis.

C. Acyclovir

Acyclovir is widely used in the treatment of herpes virus infections, especially herpes simplex virus (HSV) and varicella-zoster virus infections. It is usually well tolerated in patients; however, neurological and renal toxicities exist. In the kidney, insoluble acyclovir crystals may accumulate and precipitate, leading to acute renal failure. This is usually a complication of intravenous therapy and can be avoided with prior hydration and a slower drug infusion rate.

Neurological toxicity is rare but may occur particularly in patients with renal failure. Symptoms include agitation, tremors, delirium, and hallucinations. Severe neurotoxicity has been described in patients requiring hemodialysis; its symptoms include delirium and coma. Severe toxicity carries a greater risk in patients being treated with peritoneal dialysis as compared with those on hemodialysis [51,52].

D. Lead

The toxic lead level in the United States is determined by the Centers for Disease Control (CDC) based on the concentration of lead in the blood. From 1970 to 1991, the toxic level was gradually decreased from 60 micrograms/dL (2.90 micromoles/liter) to 10 micrograms/dL. Compared with adults, children younger than 6 years of age are more susceptible to lead toxicity resulting in lead encephalopathy because of 1) an incomplete BBB permitting the entry of lead into the developing CNS; 2) a greater prevalence of iron deficiency, which is associated with lead poisoning; and 3) an increased absorption of lead from the gastrointestinal tract [53]. The neurological symptoms of lead toxicity range from a developmental delay and a loss of milestones to severe encephalopathy. Other symptoms include cognitive and behavioral problems, hearing loss, peripheral neuropathy, and a decreased nerve conduction velocity.

At a molecular level, lead can disrupt signal transduction primarily by the activation of protein kinase C, causing a competition with magnesium binding, and secondarily by inhibition of cyclic nucleotide hydrolysis via phosphodiesterases. Lead also can cause uncoupling of the mitochondrial oxidative phosphorylation cycle in the CNS [53,54].

Renal effects of lead poisoning can be characterized by chronic interstitial nephritis. Chronic lead exposure results in lead nephropathy, proximal tubular function damage, aminoaciduria, glycosuria, and hyperphosphaturia. Common etiologies include occupational lead exposure (lead-containing batteries or lead-containing aerosols in the workplace), moonshine whiskey ingestion, and, in children, the ingestion of lead-containing paint.

A chronic and low-level lead exposure in the general population may cause a moderate decline in GFR. The incidence of a low-level lead toxicity may be higher in patients with renal failure because an underlying renal disease may accelerate lead retention.

E. Other Drugs Associated with Seizure Activity

A number of drugs need to be titrated, monitored, or altogether avoided in patients with renal disease and while on dialysis. Penicillin and cephalosporins may precipitate seizures at high doses in patients on hemodialysis. Meperidine is associated with the accumulation of the toxic metabolic product normeperidine and causes neuromuscular instability and seizures. Phenothiazines such as metoclopramide decrease the seizure threshold and predispose a person to seizure activity. Some antiepileptic drugs such as phenobarbital and primidone are dializable drugs and are removed by dialysis. They need to be titrated to reach therapeutic levels. Other drugs that may result in seizures include amantadine, acyclovir, L-dopa, lithium, ethanol, and intravenous contrast at high doses [55,56].

References

1. Burn, D. J., and Bates, B. (1998). Neurology and the kidney. *J. Neurol. Neurosurg. Psychiatry* **65**, 810–821.
2. U. S. Renal Data System. (2000). Renal data system annual report, 1996–1998. National Institutes of Health, National Institute of Diabetes and Digestive and Kidney Diseases, Bethesda, Md.
3. Arieff, A. I. (2004). Neurological complications of renal insufficiency. *In:* "Breener and Rector's The Kidney" (B. M. Brenner, et al., eds.), vol. 7, pp. 2227–2253. Saunders, Philadelphia.
4. Fraser, C. L., and Arieff, A. I. (2001). Nervous system manifestations of renal failure. *In:* "Diseases of the Kidney and Urinary Tract" (R. W. Scrier, et al., eds.), ed. 7, pp. 2769–2794. Lippincott, Williams & Wilkins, Philadelphia.
5. Brouns, R., and De Deyn, P. P. (2004). Neurological complications in renal failure: a review. *Clin. Neurol. Neurosurg.* **107**, 1–16.
6. Lockwood, A. H. (1989). Neurologic complications of renal disease. *Neurol. Clin.* **7**, 617–627 [review].
7. Raskin, N. H., and Roberts, J. K. (2005). Renal disease. *In:* "Merritt's Neurology" (L. P. Rowland, ed.), ed. 11, pp. 1080–1083. Lippincott, Williams & Wilkins, Philadelphia.
8. Agodoa, L. Y., and Eggers, P. W. (1995). Renal replacement therapy in the United States: data from the United States Renal Data System. *Am. J. Kidney Dis.* **25**, 119–133.
9. Locke, S. J., Merrill, J. P., and Tyler, H. R. (1961). Neurologic complications of acute uremia. *Arch. Intern. Med.* **108**, 519–530.
10. Bolton, C. F., and Young, G. B. (1990). "Neurological Complications of Renal Disease." Butterworth, Boston.
11. Cogan, M. G., Covey, C. M., Arieff, A. I., and Wisniewski A. (1978). Central nervous system manifestations of hyperparathyroidism. *Am. J. Med.* **65**, 963–970.
12. Ponticelli, C., and Campise, M. R. (2005). Neurological complications in kidney transplant recipients. *J. Nephrol.* **18**, 521–528.
13. Fraser, C. L., and Arieff, A. I. (1988). Nervous system complications in uremia. *Ann. Intern. Med.* **109**, 143–153 [review].
14. Palmer, B. F., and Henrich, W. L. (2005). Uremic polyneuropathy. *In:* "UpToDate" (B. D. Rose, ed.). UpToDate, Waltham, Mass.
15. Kovalik, E. C. (2005). Endocrine and neurologic manifestations of kidney disease. *In:* "Primer on Kidney Diseases" (A. Greenburg, ed.), vol. 4, pp. 529–536. Elsevier, New York.
16. Raskin, N. H., and Fishman, R. A. (1976). Neurologic disorders in renal failure. *New Engl. J. Med.* **294**, 143–148 [review].
17. Levy, N. B., and Cohen, L. M. (2001). Central and peripheral nervous systems in uremia. *In:* "Textbook of Nephrology" (S. G. Massry, et al., eds.), ed. 4, pp. 1279–1282. Lippincott Williams & Wilkins, Philadelphia.
18. Mailloux, L. U. (2005). Dialysis disequilibrium syndrome. *In:* "UpToDate" (B. D. Rose, ed.). UpToDate, Waltham, Mass.
19. Arieff, A. I. (1994). Dialysis disequilibrium syndrome: current concepts on pathogenesis and prevention. *Kidney Int.* **45**, 629–635.
20. Glenn, C. M., and Astley, S. J. (1992). Dialysis-associated seizures in children and adolescents. *Pediatr. Nephrol.* **6**, 182–186.
21. Vanholder, R., De Smet, R., Glorieux, G., Argiles, A., Baurmeister, U., Brunet, P., Clark, W., Cohen, G., De Deyn, P. P., Deppisch, R., Descamps-Latscha, B., Henle, T., Jorres, A., Lemke, H. D., Massy, Z. A., Passlick-Deetjen, J., Rodriguez, M., Stegmayr, B., Stenvinkel, P., Tetta, C., Wanner, C., Zidek, W.; European Uremic Toxin Work Group (EUTox). (2003). Review on uremic toxins: classification, concentration, and interindividual variability. *Kidney Int.* **63**, 1934–1943 [review].
22. Greca, G., Dettori, P., and Biasoli, S. (1980). Brain density studies in dialysis. *Lancet* **2**, 582.
23. Altmann, P., Dhanesha, U., Hamon, C., Cunningham, J., Blair, J., and Marsh, F. (1989). Disturbance of cerebral function by

aluminum in hemodialysis patients without overt aluminum toxicity. *Lancet* **2**, 7–12.
24. Davison, A. M., Walker, G. S., Oli, H., Lewins, and A. M. (1982). Water supply aluminum concentration, dialysis dementia, and effect of reverse osmosis water treatment. *Lancet* **2**, 785–787.
25. Tzamaloukas, A. H., and Agaba, E. I. (2004). Neurological manifestations of uraemia and chronic dialysis. *Niger. J. Med.* **13**, 98–105 [review].
26. Altmann, P., Al-Salihi, F., Butter, K., Cutler, P., Blair, J., Leeming, R., Cunningham, J., and Marsh, F. (1987). Serum aluminum levels and erythrocyte dihydropteridine reductase activity in patients on hemodialysis. *New Engl. J. Med.* **317**, 80–84.
27. Gault, P. M., Allen, K. R., and Newton, K. E. (2005). Plasma aluminium: a redundant test for patients on dialysis? *Ann. Clin. Biochem.* **42**, 51–54.
28. Mahurkar, S. D., Salta, R., Smith, E. C., Dhar, S. K., Meyers, L. Jr., and Dunea, G. (1973). Dialysis dementia. *Lancet* **1**, 1412–1415.
29. Flendrig, J. A., Kruis, H., and Das, H. A. (1976). Aluminum and dialysis dementia. *Lancet* **1**, 1235.
30. Naganuma, T., and Uchida, J. (2005). Silent cerebral infarction predicts vascular events in hemodialysis patients. *Kidney Int.* **67**, 2434–2439.
31. Mahooney, C. A., and Arieff, A. I. (1982). Uremic encephalopathies: clinical, biochemical, and experimental features. *Am. J. Kidney Dis.* **2**, 324–336 [review].
32. Held, P. J., Brunner, F., Odaka, F., Garcia, J. R., Port, F. K., and Gaylin, D. S. (1990). Five-year survival for end-stage renal disease patients in the United States, Europe, and Japan, 1982 to 1987. *Am. J. Kidney Dis.* **15**, 451–457.
33. Barri, Y. M., and Golper, T. A. (2005). Seizures in patients undergoing hemodialysis. *In:* "UpToDate" (B. D. Rose, ed.). UpToDate, Waltham, Mass.
34. Benna, P., Lacquianiti, F., Triolo, G., Ferrero, P., and Bergamasco, B. (1981). Acute neurologic complications of hemodialysis. Study of 14,000 hemodialyses in 103 patients with chronic renal failure. *Ital. J. Neurol. Sci.* **2**, 53–57.
35. Rose, B. D., and George, J. N. (2005). Causes of thrombotic thrombocytopenic purpura-hemolytic uremic syndrome in adults. *In:* "UpToDate" (B. D. Rose, ed.). UpToDate, Waltham, Mass.
36. Ono, T., Minuro, J., Madoiwa, S., Soejima, K., Kashiwakura, Y., Ishiwata, A., Takano, K., Ohmori, T., and Sakata, Y. (2006). Severe secondary deficiency of von Willebrand factor-cleaving protease (ADAMTS13) in patients with sepsis-induced disseminated intravascular coagulation: its correlation with development of renal failure. *Blood* **107**, 528–534.
37. Hollenbeck, M., Kutkuhn, B., and Aul, C. (1998). Haemolytic-uraemic syndrome and thrombotic-thrombocytopenic purpura in adults: clinical findings and prognostic factors for death and end-stage renal disease. *Nephrol. Dial. Transplant.* **13**, 76–81.
38. Contreras, G., Pardo, V., Cely, C., Borja, E., Hurtado, A., De La Cuesta, C., Iqbal, K., Lenz, O., Asif, A., Nahar, N., Leclerq, B., Leon, C., Schulman, I., Ramirez-Seijas, F., Paredes, A., Cepero, A., Khan, T., Pachon, F., Tozman, E., Barreto, G., Hoffman, D., Almeida Suarez, M., Busse, J. C., Esquenazi, M., Esquenazi, A., Garcia Mayol, L., and Garcia Estrada, H. (2005). Factors associated with poor outcomes in patients with lupus nephritis. *Lupus* **14**, 890–895.

39. Jennekens, F. G., and Kater, L. (2002). The central nervous system in systemic lupus erythematosus. Part 2. Pathogenetic mechanisms of clinical syndromes: a literature investigation. *Rheumatology* (Oxford) **41**, 619–630 [review].
40. Yurdakul, S., Hamryudan, V., and Yazici, H. (2004). Behçet syndrome. *Curr. Opin. Rheumatol.* **16**, 38–42 [review].
41. O'Duffy, J. D. (1993). "Behçet's syndrome. Primer on the Rheumatic Diseases," ed. 10, p. 206. Arthritis Foundation, Atlanta.
42. Krishnan, A. V. (2005). Altered motor nerve excitability in end-stage kidney disease. *Brain* **128**, 2164–2174.
43. Riggs, J. E. (1989). Neurologic manifestations of fluid and electrolyte disturbances. *Neurol. Clin.* **7**, 509–523.
44. Lampl, C., and Yazdi, K. (2002). Central pontine myelinolysis. *Eur. Neurol.* **47**, 3–10 [review].
45. Kurtz, I., and Nguyen, M. K. (2005). Evolving concepts in the quantitative analysis of the determinants of the plasma water sodium concentration and the pathophysiology and treatment of the dysnatremias. *Kidney Int.* **68**, 1982–1993 [review].
46. Berne, R. M., and Levy, M. N. (1981). "Cardiovascular Physiology," ed. 4, pp. 7–17. Mosby, St Louis.
47. Rose, B. D. (2005). Clinical manifestations and treatment of hyperkalemia. *In:* "UpToDate" (B. D. Rose, ed.). UpToDate, Waltham, Mass.
48. Stanton, B. A. (1989). Renal potassium transport: morphological and functional adaptations. *Am. J. Physiol.* **257**, R989–R997 [review].
49. Beccari, M. (1994). Seizures in dialysis patients treated with recombinant erythropoietin. Review of the literature and guidelines for prevention. *Int. J. Artif. Organs* **17**, 5–13 [review].
50. Strippoli, G. F., Craig, J. C., Manno, C., and Schena, F. P. (2004). Hemoglobin targets for the anemia of chronic kidney disease: a meta-analysis of randomized, controlled trials. *J. Am. Soc. Nephrol.* **15**, 3154–3165.
51. Davenport, A., Goel, S., and Mackenzie, J. C. (1992). Neurotoxicity of acyclovir in patients with end-stage renal failure treated with continuous ambulatory peritoneal dialysis. *Am. J. Kidney Dis.* **20**, 647–649.
52. Hellden, A., Odar-Cederlof, I., and Diener, P. (2003). High serum concentrations of the acyclovir main metabolite 9-carboxymethoxymethylguanine in renal failure patients with acyclovir-related neuropsychiatric side effects: an observational study. *Nephrol. Dial. Transplant.* **18**, 1135–1141.
53. Rose, B. D. (2005). Lead nephropathy. *In:* "UpToDate" (B. D. Rose, ed.). UpToDate, Waltham, Mass.
54. Muntner, P., He, J., Vupputuri, S., Coresh, J., and Batuman, V. (2003). Blood lead and chronic kidney disease in the general United States population: results from NHANES III. *Kidney Int.* **63**, 1044–1050.
55. Hassan, H., Bastani, B., and Gellens, M. (2000). Successful treatment of normeperidine neurotoxicity by hemodialysis. *Am. J. Kidney Dis.* **35**, 146–149.
56. Porto, I., John, E. G., and Heilliczer, J. (1997). Removal of phenobarbital during continuous cycling peritoneal dialysis in a child. *Pharmacotherapy* **17**, 832–835.

61

Diabetes and Endocrine Disorders

Rodica Pop-Busui, MD, PhD
Zachary London, MD
Aaron Kellogg, BS

Keywords: *adrenal insufficiency, diabetes mellitus, diabetic neuropathies, disorders of calcium metabolism, glucocorticoid excess states, growth hormone excess, thyroid disorders*

I. Diabetes Mellitus
II. Diabetic Neuropathies
III. Other Endocrine Disorders
  References

I. Diabetes Mellitus

The incidence of diabetes and its complications is increasing at a staggering rate. Currently, the World Health Organization (WHO) reports an overall prevalence of 130 million, but by 2025 it is estimated that 300 million individuals will have diabetes mellitus.

II. Diabetic Neuropathies

According to the San Antonio Consensus Conference, *diabetic neuropathy* (DN) is defined as a demonstrable disorder in the somatic and/or autonomic parts of the peripheral nervous system, either clinically evident or subclinical, that occurs in the setting of diabetes mellitus without other causes for peripheral neuropathy [1]. Confirmation can be established with electrophysiology, quantitative sensory, and autonomic function testing. It is to be noted that other forms of neuropathy, including uremia, vitamin B_{12} deficiency, hypothyroidism, and chronic inflammatory demyelinating polyneuropathy may occur more frequently in patients with diabetes and should be ruled out.

DNs are among the most common long-term complications of diabetes. Considering the epidemic explosion of diabetes, mainly of type 2 diabetes in the U.S. and other developed and developing countries, DNs represent a significant cause of morbidity and mortality.

A. Brief History and Nomenclature

The presence of a diabetic sensorimotor polyneuropathy in patients with diabetes has been described as early as 1885. Between 1885 and 1890, Althaus, Leyden, and Bruns described various neurological deficits complicating diabetes such as proximal diabetic, truncal, median, and ulnar neuropathies. Leyden reported that the various manifestations associated with diabetic polyneuropathy appear to include three main clinical subtypes (painful, ataxic, and paralytic) that were associated with the presence of peripheral nerve degeneration on autopsies studies of these patients.

DNs are a heterogeneous group of disorders and present with a wide range of abnormalities. Although the clinical spectrum of DNs syndromes is quite broad, the currently accepted classification is shown in Table 1.

Distal symmetrical peripheral neuropathy (DPN) is a significant cause of pain, suffering, and disability and is the leading cause of nontraumatic lower limb amputations, accounting for about 85,000 amputations/year in the U.S. Although predominantly sensory in nature, DPN is associated with distal motor abnormalities and with autonomic dysfunction.

Diabetic autonomic neuropathy (DAN) may affect virtually any sympathetic or parasympathetic function. Manifestations include abnormalities of heart rate control and vascular dynamics, pupillary dysfunction, gastroparesis, constipation, cystopathy, impotence, abnormal sweating, and reduced hypoglycemia awareness. Cardiovascular autonomic neuropathy (CAN) is the most prominent focus of autonomic dysfunction because of the life-threatening consequences of this complication, such as sudden cardiac death and silent myocardial ischemia. Multiple observational studies have consistently documented an increased risk of mortality in subjects with CAN.

Focal and multifocal neuropathies have usually a sudden onset and can occur as a result of the involvement of the median (5.8%), ulnar (2.1%), radial (0.6%), or common peroneal and cranial nerves (involving primarily cranial nerves III, IV, VI, and VII).

An accurate assessment of the true prevalence of diabetic neuropathy is difficult because the criteria for diagnosis vary and because epidemiological studies of DN are limited by the still-large numbers of people with diabetes who remain undiagnosed. However, in landmark studies of DN, such as the Rochester Diabetic Neuropathy Study and Pirart's 25-year prospective study, almost two-thirds of people with diabetes experienced symptoms and/or clinical signs of neuropathy [2].

B. Etiology and Pathogenesis

1. Hyperglycemia

The contribution of hyperglycemia to the pathogenesis of microvascular complications, including neuropathy, in both type 1 [3] and type 2 diabetic subjects [4] is now beyond dispute. The Diabetes Control and Complications Trial (DCCT) has provided the first strong evidence for the importance of hyperglycemia, insulin deficiency, or both in the pathogenesis of diabetic neuropathy [3]. Intensive insulin therapy, aimed at reducing blood glucose levels to as close as possible to the normal range in subjects with type 1 diabetes and no neuropathy, reduced the prevalence of clinical diabetic neuropathy, confirmed by abnormal nerve conduction or autonomic function, by 60% after 5 years.

Pirart's 25-year prospective study of 4400 unselected type 2 diabetic patients provides an additional convincing epidemiological link between the duration and severity of hyperglycemia and the presence of clinical diabetic neuropathy. *Neuropathy*, defined as loss of ankle reflexes combined with diminished vibratory sensation in the presence or absence of "more dramatic polyneuropathy," was present in 12% of patients at the time of diabetes diagnosis. The cumulative prevalence of neuropathy increased linearly with duration of diabetes to nearly 50% after 25 years [5] but did not differ substantially as a function of age at time of diagnosis.

A close relationship between the severity and/or duration of hyperglycemia with the development and progression of diabetes complications is also supported by the results of the Epidemiology of Diabetes Interventions and Complications (EDIC) study. The EDIC study follow-up demonstrated that the differences in retinal, renal, and cardiovascular outcomes between the intensive and conventional treatment groups observed at the end of the DCCT have persisted and even increased for as long as 12 years, despite the narrowing of glycemic differences that appeared to explain the majority of the treatment differences during the DCCT. These prolonged salutary effects of intensive therapy and the prolonged deleterious effects of conventional therapy were termed *imprinting* or *metabolic memory*. Although these studies implicate glucose- or insulin-related metabolic factors as important pathogenetic elements in the DNs disease process, it is not yet known whether imprinting or metabolic memory apply to development of DNs. Although

Table 1 Classification of Diabetic Neuropathies

Distal symmetrical peripheral neuropathy
Diabetic autonomic neuropathy
Focal and multifocal neuropathy
   Ulnar
   Radial
   Common peroneal
   Cranial

the DCCT conclusively demonstrated that intensive insulin therapy reduced neuropathy in type 1 diabetic subjects, 25% of DCCT patients with an average HbA1C of 7.2% developed new nerve conduction abnormalities, and 40% of patients showed signs of progression in both primary prevention and secondary intervention cohorts over the 6.5-year follow-up. Furthermore, the VA Cooperative Study did not observe a difference in the prevalence of DPN or DAN in patients with type 2 diabetes after 2 years of tight glycemic control (HbA1C of 7.3%). It was also shown, in another small cohort of patients with type 2 diabetes, that although a targeted, intensive intervention of hyperglycemia in concert with multiple other cardiovascular disease (CVD) risk factors significantly reduced the prevalence of autonomic neuropathy, it was without a significant effect on the peripheral neuropathy [6].

These and other smaller trials suggest that either the achieved glycemic control may have been inadequate to prevent the development and progression of neuropathy or that other factors, such as hypertension and hyperlipidemia, contribute to its pathogenesis. In UKPDS, for instance, the greatest benefits were observed in subjects who achieved intensive glycemic and hypertension control [7].

A recent report showed that other vascular risk factors may accelerate the effects of hyperglycemia on nerve dysfunctions in patients with type 1 diabetes and that even slight improvements in lipids are associated with a significantly lower risk of DPN [8].

Therefore despite a close epidemiological association between clinical neuropathy and the duration and severity of hyperglycemia in diabetic populations, the onset of clinically overt diabetic neuropathy in an individual patient does not appear to be a predictable event that necessarily reflects concurrent metabolic control, nor does it follow inexorably from prolonged and severe hyperglycemia. This somewhat loose clinical association can be understood in terms of the indolent and occult nature of the underlying subclinical nerve damage but also may indicate the presence of other independent pathogenetic variables such as genetic, medical (e.g., hypertension and hyperlipidemia), nutritional, toxic (e.g., alcohol), and mechanical (e.g., entrapment and compression) factors that may influence the appearance of clinical signs and symptoms in individual patients.

C. Pathophysiology and Molecular Mechanisms

As previously described, although mounting evidence provides support for a microvascular basis of DNs, their pathophysiology is still poorly understood. Multiple pathogenetic factors have been widely considered in experimental diabetes, including increased oxidative and nitrosative stress, redox imbalance secondary to enhanced aldose reductase (AR) activity, nonenzymatic glycation of structural nerve proteins, impaired protein kinase C (PKC) activity, impaired nitric oxide (NO) synthesis and endothelial dysfunction, alterations in cyclooxygenase (COX) activity with subsequent perturbations in prostaglandin (PG) metabolism, direct hypoxia and ischemia of nerve trunks and ganglia, deficiencies in the neurotrophic support of neurons, and deficiencies in C-peptide DPN [5,9].

1. Oxidative Stress

There is convincing experimental and clinical evidence that the generation of reactive oxygen species (ROS) is increased in both types of diabetes and that the onset of diabetes complications is closely associated with oxidative stress. Hyperglycemia is reported to induce oxidative stress through multiple pathways such as redox imbalances secondary to enhanced AR activity, increased advanced glycation end products (AGE), increase in PKC (especially β-isoform), prostaglandin imbalances, and mitochondrial overproduction of superoxide [9,10]. All these pathways converge in producing oxidative stress in the peripheral nerves, dorsal root, sympathetic ganglia, and the vasculature of the peripheral nervous system, thus contributing to nerve blood flow and nerve conduction deficits, impaired neurotrophic support, changes in signal transduction and metabolism, and morphological abnormalities characteristic of DNs [9] (Fig. 1).

Although an overwhelming body of data supports the role of oxidative stress in the pathogenesis of DN and demonstrates the protective effects of antioxidants in animal DN models, the general lack of a significant benefit observed with a number of antioxidants in human trials is disappointing. It has been suggested that single-nucleotide polymorphisms of the genes for mitochondrial (SOD2) and extracellular (SOD3) superoxide dismutases may confer an increased risk for the development of neuropathy, which might partially explain the failure of the human trials [11]. Several large multicenter, randomized, double-blind placebo trials in Europe and North America have demonstrated some benefits on neuropathic symptoms and electrophysiological testing with the powerful antioxidant α-lipoic acid (ALA), which scavenges ROS and regenerates glutathione. Therefore studies of DN using earlier therapeutic interventions and longer-term assessment are warranted.

2. Nitrosative Stress

An increase of nitrotyrosine and impairment of antioxidant defense systems typically occur to a similar extent in diabetic patients with or without complications. Oxidized proteins have been reported to be significantly higher in diabetic patients with complications than in those without complications [12]. This could suggest that NO-mediated nitrosylation of proteins may be less significant in producing complications than other free radicals. Nitration of proteins

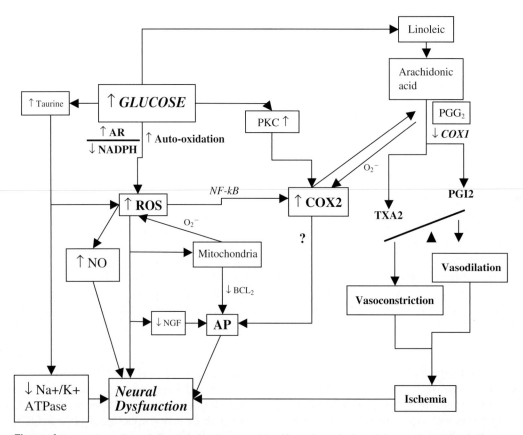

Figure 1 Proposed paradigm of glucose-induced neurotoxicity. Hyperglycemia, through increased mitochondrial overproduction of superoxide, auto-oxidation and activation of aldose-reductase (AR) pathway, with secondary NADPH and taurine depletion, generates reactive oxygen species (HO^-, O_2^-) (ROS) and increased oxidative stress. Alternatively, glucose-induced increase of PKC (protein kinase C) activity induces increased O_2^- production, further promoting increased oxidative stress. Oxidative stress also induces impaired nitric oxide (NO) synthesis and endothelial dysfunction. Increased ROS, via modulation of NF-kB (nuclear factor kappa B), and PKC signaling induce COX-2 expression, which regulates the conversion of arachidonate to vasoconstricting and proinflammatory PG (prostaglandins). This precipitates the imbalance in TXA2 (thromboxane)/PGI2 (prostacyclin) ratio and therefore favors vasoconstriction and ischemia. Reciprocally, COX-2 up-regulation increases the rate of PGG2 to PGH2 conversion and ROS generation, further exacerbating oxidative stress. ROS may promote apoptosis (AP) by diminishing the NGF neurotrophic support and by inducing mitochondrial dysfunction. All these contribute to neural dysfunction.

can lead to rapid proteasomal degradation and therefore can be removed from the cell and resynthesized. Carbonylated proteins and peptides are also inactivated by oxidative stress. Measurements of protein carbonyls are highly sensitive and are detected in plasma of both type 1 and type 2 diabetic patients even without complications [9].

3. Aldose Reductase Pathway

Growing evidence indicates that the AR pathway plays a key role in peripheral nerve dysfunction. Two distinct (but not mutually exclusive) biochemical consequences of increased metabolic flux of glucose through the AR pathway have been invoked to explain the short-term and long-term deleterious effects of hyperglycemia on diabetic nerve and other complication-prone tissues. These are the osmoregulatory consequences of sorbitol accumulation with compensatory depletion of taurine and myo-inositol (the "osmotic" or "compatible osmolyte" hypothesis) and the effects of sorbitol pathway on the NADH/NAD and NADPH/NADP redox couples (the "redox" hypothesis). Glucose flux through AR and sorbitol dehydrogenase oxidizes the NADPH/NADP and reduces the NADH/NAD redox couple, thereby disturbing the adenine nucleotide–linked reaction that functions near the redox equilibrium. Oxidation of the NADPH/NADP couple has been involved in the depletion of reduced GSH, thus increasing the susceptibility to oxidative stress in the peripheral nerve and endoneurial vasculature. Studies in transgenic mouse models overexpressing the human AR have shown that diabetic mice overexpressing human AR develop severe functional and structural abnormalities in the peripheral nerves, such as decreased nerve conduction velocity (NCV), significant reduction of GSH, and myelinated fiber atrophy. These results are similar to those found in rats with streptozotocin (STZ)-induced diabetes and in BioBreeding (BB) rats that spontaneously develop insulin-deficient diabetes, suggesting that AR is a key contributor to oxidative stress and peripheral nerve dysfunction under diabetic conditions [5,9].

Studies with specific aldose reductase inhibitors (ARIs) in experimental diabetic neuropathy have demonstrated that ARIs prevent, reverse, or moderate various defects in NCV and ameliorate nerve fiber damage and loss. However, in humans, similar to trials with antioxidants, the situation is not clear; most of the trials demonstrate either a lack of efficacy or unacceptable side effects. A meta-analysis of 19 randomized controlled trials of ARIs, which tested various ARIs for a median of 24 weeks, demonstrated a small but statistically significant reduction in decline of median (0.66 m/s, 95% CI 0.18–1.14 m/s) and peroneal (0.53 m/s, 0.02–1.04) motor NCV without benefit in sensory nerves [11]. Considering the recent evidence of polymorphisms in the AR promoter region leading to a highly significant decrease in the frequency of the $Z+2$ allele and of differences in the set point for AR activity in various diabetic patients, possible reasons for the marginal benefit of the human trials may be related to the lack of a targeted approach that identified those most genetically susceptible to alterations in AR activity and therefore those most likely to benefit from AR inhibition [11]. However, the new and potent ARI fidarestat (Sanwa Kagaku KenKyusho, Nagoya, Japan) remains in clinical trials, as current studies are showing therapeutic benefit both in STZ rats and in diabeticpatients. It is also becoming clear that AR inhibition may be insufficient in its own right to achieve the desirable degree of metabolic enhancement in patients with a multitude of biochemical abnormalities. Combinations of therapy with ARIs and antioxidants may become critical if the relentless progress of DN is to be abated.

4. Endothelial Dysfunction

The endothelium-derived relaxing factor NO mediates vasodilatation by activating soluble guanylate cyclases that regulate ion channels, macrophage cytotoxicity, and neurotransmission, and it is highly susceptible to oxidative stress in diabetes. NO synthase (NOS), the enzyme that catalyzes the conversion of L-arginine to citrulline and NO at the expense of NADPH, is critically situated in endothelial cells, vascular smooth muscle cells, and sympathetic ganglia. An impaired synthesis of NO has been linked to polyol pathway activation through an NADPH-mediated mechanism (via the redox hypothesis), to alterations in PK activation and calcium levels (via the osmotic hypothesis), and to increased formation of AGE. Recent studies have shown that activation of poly(ADP-ribose) polymerase (PARP) within the vasculature is an important factor in the pathogenesis of endothelial dysfunction. In animal studies, PARP inhibitors have demonstrated beneficial effects in preventing neurovascular dysfunction.

5. PKC Activation

The contributory role of PKC in diabetic neuropathy still requires clarification. In particular, PKC-ß has been proposed to play a major role in DN through its effects on vascular blood flow and microvascular disease rather than directly in neuronal cells. A variety of diabetes-induced effects on 1,2-diacylglycerol (DAG) levels and PKC activity in sciatic nerve have been reported. Initial studies suggested that hyperglycemia, through diminished phosphatidylinositol (PI) turnover and subsequent reduced availability of DAG, induces a decrease in neuronal PKC activity. Diminished PKC activity reduces the phosphorylation of Na^+-K^+-ATPase, leading to a decrease in nerve conduction [5]. Nevertheless, even in nerves of diabetic animals, a fall in DAG levels and a consistent pattern of change in PKC activity has not been observed. More recently it has been shown that increased PKC activity within the vasa nervorum mediates increased O_2^- production and reduced NO availability, thus contributing to endothelial dysfunction. Increased PKC activity also is involved in the activation of the COX pathway by glucose. Inhibition of PKC-β in various studies of diabetic rats appears to correct reduced nerve blood flow and NCV. Once again, however, a phase II clinical trial in humans demonstrated only modest benefit in diabetic patients with neuropathy despite beneficial effects on retinopathy and nephropathy [13].

6. Advanced AGE

It has been shown that the glycation process is enhanced in peripheral nerves in both diabetic patients and animals. In the process of nonenzymatic glycation, reducing sugars such as glucose, fructose, or galactose initially react with free amino groups of proteins, lipids, or nucleic acids to form the early reversible products such as Schiff bases and ketamines or Amadori adducts. In addition, autooxidative glycosylation takes place through the Wolff pathway, forming glycoxidation products. Over time these products undergo chemical rearrangement, dehydration, and fragmentation reactions and cross-linking to form irreversible AGE [9]. AGE act on specific receptors (RAGEs), inducing monocytes and endothelial cells to increase the production of cytokines and adhesion molecules. Glycation has also recently been shown to have an effect on matrix metalloproteinases (MMPs), in particular MMP-2, which degrades type IV collagen, tissue inhibitors of MMPs (TIMP)-1 and -2, and transforming growth factor-ß (TGF-β) [11]. Studies of various AGE inhibitors, such as the nucleophilic compounds pyridoxamine, tenilsetam, 2,3-diaminophenazone, or aminoguanidine, have shown that these changes can be prevented in experimental diabetes, but no trial data are currently available for human DN.

D. Other Mechanisms

1. COX-2 Activation

Various pathways linking hyperglycemia to ROS generation and oxidative stress have been already discussed. Increased oxidative stress results in nuclear factor (NF)-β activation, COX-2 mRNA induction, and COX-2 protein expression. COX-2 up-regulation leads to an altered PG profile with an increased production of vasoconstricting PGH_2, TXA, and $PGF_{2\alpha}$ and reduction in vasodilatory prostacyclin (PGI_2), thus favoring vasoconstriction and ischemia. Reciprocally, COX-2 up-regulation increases the rate of PGG_2 to PGH_2 conversion and ROS generation, further exacerbating oxidative stress (see Fig. 1) [9]. Studies of different COX inhibitors in diabetic rats have shown inconsistent effects on neurovascular and functional deficits, reflecting differences in the relative degree of inhibition of the two COX pathways. We have reported that COX-2 selective inhibition, but not nonselective COX inhibition, prevented deficits of NCV and NBF in experimental diabetes in rats. Also, we have recently found that COX-2 gene-deficient STZ-D mice demonstrate a differential protection against biochemical and functional markers of experimental neuropathy; the extent of this protection appears to be dependent on the degree of COX-2 gene deficiency.

2. C-Peptide Deficiency

Recent evidence has also shown that perturbations in C-peptide signaling appear to play a role in the pathogenesis of diabetic complications, including diabetic neuropathy [14]. Indirect evidence supporting this notion was obtained from the DCCT trial, in which optimal glucose control did not prevent the progression of diabetic neuropathy in 40% of patients. Impaired insulin and C-peptide signaling results in abnormal regulations of proteins and genes that are not involved in the glucose regulatory process, such as neurotrophic genes IGF-I, IGF-II, IGF-I-receptor, and the insulin receptor itself, as well as effects on Na(+)/K(+)-ATPase activity, endothelial NO synthase, and deoxyribonucleic acid (DNA) binding of transcription factors and modulation of apoptotic phenomena [14]. These direct effects of insulin signaling, optimized by C-peptide, are likely to account for emerging differences in underlying mechanisms that have been identified among both experimental and human type 1 and type 2 diabetic neuropathy.

3. Immune Mechanisms

Studies suggest that sera from type 1 diabetic patients with neuropathy contain an autoimmune immunoglobulin that induces complement-independent, calcium-dependent apoptosis in neuronal cells. The expression of these cytotoxic factors has been related to the severity of neuropathy and the type of neuronal cell killed. Therefore it has been suggested that such toxic factors may contribute to DN by acting in concert with hyperglycemia to damage sensory and/or autonomic neurons [11].

4. Apoptosis

Apoptosis has been identified as the mechanism of cell death in a number of human neurodegenerative disorders and has been implicated in the development of chronic diabetic complications, including neuropathy. Neurons in the retina have been shown to display apoptosis-associated DNA fragmentation in experimental and human diabetes, and sera from diabetic subjects have been demonstrated to induce apoptosis in cultured neuroblastoma cells. However, data reported in sensory neurons are controversial. Some authors reported that sensory neurons are susceptible to the induction of apoptosis during hyperglycemia and that apoptosis contributes to sensory and autonomic neuron dysfunction in STZ-D rats and in acutely hyperglycemic nondiabetic rats [15,16]. More recently, other authors have shown that sensory neurons demonstrate a remarkable resistance apoptosis, despite their expression of caspase-3, a molecule often considered to be inexorably linked to apoptosis [17]. A parallel massive loss of axons in models or in humans shortly after the onset of diabetes has not been described.

5. Growth Factors

Neurotrophic growth factors modulate neuronal outgrowth during development, sustain innervation, and promote local regeneration thereafter.

Nerve growth factor (NGF) is relatively specific for sensory and autonomic neurons and their axonal processes. NGF production and gene expression in autonomic and sensory neuronal targets varies reciprocally with the density of innervation by NGF-sensitive neurons. Peripherally synthesized NGF is retrogradely transported by the axon from target organs to neuronal cell bodies for normal maintenance and regeneration of the peripheral nervous system. Ganglionic NGF is required for signal transduction, neurotransmitter synthesis, protein phosphorylation, methylation, and gene expression of *ras*-like proteins in sympathetic and sensory neurons. Studies in diabetic patients with sensory fiber dysfunction demonstrated contradictory results with respect to both skin NGF expression and the effects of recombinant human NGF in phase II clinical trials.

Insulin-like growth factor (IGF)-1 has been shown to have a protective effect via PI 3-kinase in preventing glucose-mediated apoptosis in both cell culture and animal models, and a deficiency in circulating IGF-1 levels as a consequence of hyperglycemia has been described in STZ-diabetic and the BB/Wor rat. However, IGF-1 and IGF-1 receptor messenger ribonucleic acid (mRNA) levels have not been shown to differ in the sural nerve of diabetic patients compared with control subjects [11].

VEGF is another growth factor with direct effects on neurons and glial cells; it stimulates their growth, survival, and axonal outgrowth. Studies in different animal models have demonstrated restoration of nerve blood flow as well as both large- and small-fiber dysfunction 4 weeks after intramuscular gene transfer of plasmid DNA encoding VEGF-1 or VEGF-2.

E. Pathological Structural Changes

Histological studies of autopsy and nerve biopsy material from patients with diabetic neuropathy have identified hallmark deficits involving peripheral nerve axons, including a progressive damage to and loss of large and small myelinated and unmyelinated nerve fibers. The most commonly described lesions are Wallerian degeneration, segmental and paranodal demyelination, and proliferation of endoneurial connective tissue including thickening and reduplication of the basement membranes of nerve fibers, endoneurial blood vessels, and the perineurium. The proximal-to-distal increase in morphological abnormalities and the topographic and temporal distribution of neurological signs and symptoms in the distal symmetrical polyneuropathy of diabetes suggest a primary axonopathy, preferentially involving longer myelinated axons. However, defects in Schwann cells, perineurial cells, and endoneurial vascular elements, such as basement membrane thickening and reduplication, endothelial cell swelling and proliferation, and platelet aggregation resulting in vessel occlusion, have been also described. Diabetic polyneuropathy has been also characterized by defective axonal regeneration, which has been reported to be both positively and negatively correlated with the degree of nerve fiber loss.

A quantitative increase in these vascular abnormalities in association with focal loss of myelinated fibers in older diabetic subjects has been interpreted as suggestive of hypoxic or ischemic damage to nerve fibers in diabetic subjects with focal neuropathies.

A characteristic neurophysiological defect in diabetes is slowing of motor and sensory NCV, which can be attributed to several types of physiological and anatomical abnormalities. Maximal NCV primarily reflects the integrity of the largest and most rapidly conducting myelinated nerve fibers, and it is only modestly decreased with the selective loss of the largest myelinated fibers because smaller fibers conduct only slightly more slowly. On the other hand, conduction is markedly slowed with widespread demyelination. Patients with long-standing, established diabetes exhibit consistent but mild evidence of motor and sensory conduction slowing, whereas patients with clinically overt diabetic neuropathy have slightly more severe electrophysiological abnormalities than those with subclinical neuropathy.

The neuropathology of diabetic autonomic neuropathy has been less well studied, in part because biopsy material is not readily available. A few autopsy studies have demonstrated axonal degenerative changes and fiber loss in the paravertebral sympathetic chain, vagus nerve, esophageal and splanchnic nerves, and intrinsic nerves of the bladder, as well as swelling and vacuolization of autonomic ganglionic neurons. Direct assessment of cardiac sympathetic integrity has become possible with the introduction of radiolabeled analogues of norepinephrine, which are actively taken up by the sympathetic nerve terminals of the heart, thus providing valuable information on the integrity of the sympathetic nerve terminals. Quantitative scintigraphic assessment of the pattern of sympathetic innervation of the human heart is possible with either [^{123}I] metaiodobenzylguanidine (MIBG) or [^{11}C] hydroxyephedrine (HED), which may offer greater sensitivity to detect more subtle degrees of DAN than is possible by reflex testing. In cross-sectional studies, for example, deficits of LV [^{123}I] MIBG and [^{11}C] HED retention have been identified in diabetic subjects without abnormalities on cardiovascular reflex testing; these deficits have been reported even in newly diagnosed diabetes [18–20].

The relationship of metabolic control to progression of abnormalities of autonomic function in both type 1 and type 2 diabetes has been demonstrated in the DCCT and STENO 2 trials in which, as mentioned previously, intensive therapy was able to slow the progression and development of abnormal autonomic tests [6,21] and confirmed by quantitative scintigraphic imaging techniques such as HED-PET (positron emission tomography) and MIBG SPECT (single photon emission computed tomography).

III. Other Endocrine Disorders

A. Glucocorticoid Excess States

Glucocorticoid excess states, such as seen in Cushing's disease, ectopic adrenocorticotropic (ACTH) production, cortisol-producing adrenal adenomas and carcinomas, or chronic iatrogenic steroid therapy, can contribute to a variety of neurological manifestations including myopathy, psychiatric changes, spinal cord compression, and impaired visual acuity. In his original case description, Cushing noted severe proximal weakness and atrophy, features that have since been reported in 50–80% of affected patients [22].

Steroid myopathy comprises both proximal weakness and myalgias that involve the lower more than the upper extremities but spare the cranial nerve-innervated muscles. Although the precise mechanisms by which glucocorticoids produce steroid myopathy are not known, impairment of muscle protein and carbohydrate metabolism appears to play an important role. Glucocorticoids induce muscle protein catabolism by promoting protein degradation and inhibiting protein synthesis in a dose-dependent fashion, decreasing the rate of amino acid uptake and stimulating the oxygenation of branched-chain amino acids and myosin degradation. With regard to the muscle carbohydrate metabolism, glucocorticoids induce an insulin-resistant state by antagonizing the action of insulin after insulin binds to its surface receptors, a process that likely involves modulation of gene expression. Experimental models of glucocorticoid excess have also demonstrated an increase in muscle glycogen, which may result from reduced glycogen phosphorylase activity and increased muscle glycogen synthetase activity [22].

The classic histopathological finding in steroid myopathy is selective atrophy of type II (fast-twitch) muscle fibers, especially type IIb fibers [23]. Skeletal muscle fibers contain a high concentration of glucocorticoid receptors, and type I muscle fibers actually have a higher density of receptors than type II muscle fibers. However, type I and type IIa fibers seem to be relatively resistant to glucocorticoid-induced atrophy. One explanation for this is that they have a higher capacity to compensate for impaired glycolytic activity via conversion to oxidative metabolism than do type IIb fibers. Furthermore, the effects of glucocorticoids on protein catabolism appear to be more pronounced in type II muscle fibers than in type I muscle fibers, leading to more profound negative protein balance [22]. Elevated levels of ACTH may have myopathic actions distinct from those of glucocorticoids. Patients with Cushing's disease who are treated with adrenalectomy have high ACTH levels, proximal weakness, and wasting. Excessive ACTH decreases the quantal content of the end-plate potential, thereby impairing neuromuscular transmission. Unlike glucocorticoids, ACTH administration to rabbits causes focal necrosis of muscle fibers and extensive deposition of lipids [22].

Myelopathy also occurs in patients with glucocorticoid excess due to epidural fat accumulation and direct compression of the spinal cord and surrounding neural structures.

Psychiatric disorders such as agitated depression, lethargy, and psychosis are quite common in Cushing's syndrome regardless of cause. Memory and cognitive function may also be affected. Patients with Cushing's syndrome have decreased cerebrospinal fluid (CSF) levels of somatostatin-like immunoreactivity (SLI), ACTH, beta-endorphin (BE), and norepinephrine. They also have significant slowing of brain-wave electrical activity on electroencephalography. The previously mentioned changes appear to be related to the behavioral disturbances seen in glucocorticoid excess states, particularly those relating to mood and cognition [24].

B. Adrenal Insufficiency

Well-recognized neurological manifestations of adrenal insufficiency include myopathy and exercise intolerance (present in 25–50% of patients), seizures, and psychiatric abnormalities.

Myopathy was first mentioned by Addison in his original description of adrenal insufficiency; he noted generalized muscle weakness, cramping, and fatigue as part of the constellation of symptoms. Muscle enzymes and electromyography (EMG) usually are within normal limits, and muscle biopsies usually are unremarkable with the exception of increased glycogen content.

Potential mechanisms that may contribute to the development of myopathy in adrenal insufficiency are impairment of muscle carbohydrate metabolism and an increased affinity of the sarcolemmal insulin receptor, leading to enhanced sensitivity to insulin. Muscle glycogen stores are depleted, and epinephrine-induced glycogen phosphorylase is suppressed in both skeletal muscle and liver. The extreme exercise intolerance seen in patients with Addison's disease may also be due, in part, to exercise-induced hypotension. Mineralocorticoid deficiency leads to decreased membrane Na-K-ATPase activity and diminished β-adrenergic stimulation of the Na-K pump. Intracellular potassium is depleted, and patients become hyperkalemic. Glucocorticoids also have a permissive effect on adrenergic sensitivity. The combination of sodium wasting by the kidneys and the loss of adrenergic vasoconstriction can result in hyponatremia and hypovolemia. This leads to exercise-induced hypotension and markedly diminished work capacity [22].

Psychiatric symptoms, ranging from depression and irritability to an acute confusional state or even coma, are reported in 20–70% of patients with Addison' disease [23]. It is interesting to note that despite their divergent physiological mechanisms, Addison's disease and Cushing's disease may have similar psychiatric manifestations. It is unclear whether the cognitive and behavioral manifestations are due to adrenal cortical deficiency itself or to the compensatory increases in corticotropin and corticotropin-releasing factor. There is evidence that the latter compounds do appear to affect attentional and affective processes to some degree.

Seizures are an uncommon complication of Addison's disease and may be intractable. Hypoglycemia secondary to glucocorticoid deficiency is the most likely cause, although electrolyte disturbances secondary to mineralocorticoid deficiency may be a contributing factor.

C. Thyroid Disorders

1. Hyperthyroidism

Thyrotoxicosis can cause a variety of neurological sequelae, including myopathy, periodic paralysis, agitation, seizures, tremor, ophthalmopathy, and changes in tendon reflexes.

The **myopathy** of hyperthyroidism was originally described by Graves and Von Basedow. Weakness affects proximal muscles more than distal muscles and tends to spare the bulbar and sphincter muscles. Weakness is often out of proportion to the degree of atrophy. Respiratory insufficiency requiring ventilation has been reported as a rare complication of thyrotoxic myopathy.

The pathogenesis of thyrotoxic myopathy is likely multifactorial. One of the primary effects of thyroid hormone is to increase glucose uptake and glycolytic activity and stimulate glycogenolysis in skeletal muscle. High circulating levels of active thyroid hormone produce insulin resistance, with fasting hyperglycemia and glucose intolerance despite increased insulin levels [25]. This insulin-resistance state, in addition to the associated accelerated muscle protein degradation and decreased protein synthesis, result in muscle glycogen depletion and reduced concentrations of adenosine triphosphate (ATP) and creatine phosphate. The excess of thyroid hormones increases the basal metabolic rate and increases skeletal muscle heat production through thyroid-induced activation of muscle Na-K-ATPase. Animal studies have suggested that 85% of the increased energy use in thyrotoxic muscles is due to increases in Na-K-ATPase [25].

In addition, hyperthyroidism causes brisk tendon reflexes by increasing the velocity of muscle contraction and shortening relaxation time. The shortening of twitch duration is the result of a shift in the expression of myosin heavy and light chains in slow-twitch muscle toward that characteristic of fast-twitch muscle. Elevated levels of thyroid hormone also lead to an increased rate of calcium uptake by the sarcoplasmic reticulum and a shift in the calcium sensitivity of the contractile proteins so that tension develops at lower calcium concentrations [22].

Thyrotoxic periodic paralysis (TPP) is another neurological manifestation of hyperthyroidism characterized by episodes of muscle weakness lasting minutes to days. Attacks can be triggered by muscle cooling, rest after exercise, or a carbohydrate challenge. Although the exact mechanism of TPP is unclear, a sarcolemmal depolarization with subsequent inactivation of sodium channels and loss of membrane excitability has been described. In addition, the thyrotoxicosis-induced increase in the number of muscle Na-K pumps may explain the hypokalemia seen during attacks [22].

Impairments in cognitive function are generally milder than those seen in hypothyroidism. Nervousness, irritability, restlessness, and fatigue are most commonly described. In addition, an exaggerated essential tremor in the hands, eyelid, or tongue that may progress to chorea in some patients and new-onset seizures may be seen. The metabolic changes that underlie these disorders are poorly understood, but they may be associated with up-regulation of β-adrenergic receptors.

D. Hypothyroidism

Hypothyroidism may have a variety of neurological sequelae, including myopathy, neuropathy, a myasthenic syndrome, and a wide spectrum of cognitive and behavioral changes.

The **myopathy of hypothyroidism** is characterized by proximal weakness, fatigue, myalgia, slowed movements, and slowed reflexes. In children with cretinism, enlarged muscles are commonly seen, giving patients the classic "infant Hercules" appearance. Myoedema, a localized knot of contracting muscle, can be induced by percussion or manual irritation of the muscle. Hypothyroidism produces myopathy, partly by decreasing adrenergic sensitivity. In the absence of physiological levels of active thyroid hormone, the number of β-adrenergic receptors on muscle cells is reduced, leading to diminished glycogenolysis. Energy metabolism is also impaired by reductions in gluconeogenesis and reduced oxidative and glycolytic capacity. The decrease in mitochondrial oxidation capacity leads to reduced oxygen consumption and a reduced basal metabolic rate. Changes in protein metabolism may also contribute to the pathogenesis of hypothyroid myopathy. Protein synthesis and protein degradation are both reduced in hypothyroidism, with a balance resulting in a net protein catabolism. Diminished protein synthesis may be due to reduced DNA transcription. Reduced protein degradation is associated with impaired lysosomal protease activity [22].

Sluggish contraction and delayed relaxation of muscles in hypothyroidism are related to changes in fiber type distribution. The distribution of myosin, lactic acid dehydrogenase (LDH), and myofibrillar ATPase are altered from a pattern characteristic of fast-twitch fibers to one more characteristic of slow-twitch fibers, which may contribute to the slowing of both muscle contraction and relaxation. The active form of thyroid hormone increases the transcription of calcium ATPase in the sarcoplasmic reticulum of skeletal and cardiac muscle, which is involved in the clearance of cytoplasmatic calcium after muscle contraction [26]. Therefore an impaired calcium clearance in hypothyroidism results in prolongation of muscle relaxation, which explains the myoedema and the delayed tendon relaxation.

Paresthesias and mononeuropathies are commonly reported in hypothyroidism, whereas polyneuropathy is rare even in areas of the world where myxedemic cretinism is endemic.

Carpal tunnel syndrome, most likely due to a reduced space within the flexor retinaculum, is the most common mononeuropathy associated with hypothyroidism. Deposits of mucopolysaccharide protein complexes in nerves, tendons, synovial sheaths, and other soft tissues of the carpal tunnel, together with joint effusion and synovial thickening, have been described in the thyroid-deficient state; this explains the symptoms [27]. These factors remit readily with hormone replacement, obviating the need for surgery.

The existence of diffuse sensorimotor polyneuropathy directly related to hypothyroidism has been accepted only recently. The neuropathy of hypothyroidism affects sensory function more than motor function and may affect the upper extremities more than the lower extremities. Evidence suggests dysfunction of both the Schwann cells and axons. The exact metabolic changes that cause hypothyroid neuropathy are not fully understood. It is known that thyroid hormone is important in mediating microtubule assembly. Slow axonal transport has been found to be decreased in the sciatic nerves of hypothyroid rats, a finding consistent with microtubule dysfunction [27]. Patients improve with hormone replacement, suggesting a degree of plasticity associated with whatever metabolic changes underlie the pathophysiology of hypothyroid neuropathy.

A myasthenic syndrome consisting of fatigable limb girdle and trunk weakness and bulbar dysfunction has been described in hypothyroid patients. It has been hypothesized that weakness is due to a desensitization block at the postsynaptic membrane or a conduction block in the motor nerve terminals [27].

Mild hypothyroidism can cause changes in cognition and behavior such as fatigue, apathy, inattention, and slowed response times. A floridly psychotic state, "myxedema madness," can occur in 5% of myxedematous patients [26]. Hypothyroidism is considered one of the reversible causes of dementia, although it has not been conclusively established that myxedematous patients meet strict criteria for dementia or that hormone replacement truly improves cognitive function. The physiological mechanisms underlying the cognitive and psychiatric disturbances in hypothyroidism are incompletely understood, but the effects of thyroid hormone on the brain may be related to its action on neurotransmitters. In animal models, an increase in dopamine and tyrosine hydroxylase activity and a decrease in serotonin synthesis are observed.

E. Growth Hormone Excess

The neurological manifestations of growth hormone (GH) excess include myopathy, neuropathy, myelopathy, and sleep apnea. About 50% of patients with acromegaly have proximal weakness and decreased exercise tolerance, although clinically these patients present increased muscle bulk.

GH excess may cause muscle weakness by multiple mechanisms: increased fatty acid oxidation with subsequent inhibition of oxidative and glycolytic carbohydrate metabolism [22]; impaired glycogenolysis and insulin resistance with subsequent decreased glucose use; increased protein synthesis and decreased protein breakdown; decreased surface membrane excitability and impaired myofibrillar contractile ability, inducing tetanic tendons and reduced twitch in the hypertrophied muscles; and alterations in vasculature and impaired muscle blood flow. A decreased energy level and easy fatigue may also be secondary to unrecognized sleep apnea, a common finding in acromegaly that is most likely related to increased upper airway tissue, although central apnea has also been described [23].

Median mononeuropathy and polyneuropathy have both been described in patients with acromegaly. Predominant pathology findings are increased edema or hypertrophic changes within the nerve and, more rarely, synovial edema, hyperplasia of ligaments and tendons, and proliferation of cartilage and bone [23,27]. Polyneuropathy associated with acromegaly has also been described, although fewer cases have been published. It has been postulated that the high incidence of diabetes mellitus among patients with acromegaly may be responsible, although in one series many of the affected patients had normal or minimally impaired glucose tolerance. Histological studies have shown a reduction in the number of myelinated fibers and segmental demyelination with "onion bulb" formation [27]. Hypertrophic changes within the endoneurium and perineurium have been noted and may predispose nerves to pressure and trauma.

In addition, acromegaly can place patients at risk for compressive myelopathy by causing enlargement of vertebral bodies and extradural soft tissue.

F. Disorders of Calcium Metabolism

1. Hyperparathyroidism and Osteomalacia

Proximal weakness and wasting occur in primary hyperparathyroidism or hyperparathyroidism secondary to renal disease. Nearly half of all patients with osteomalacia develop proximal weakness and myalgias, sometimes even before the characteristic bony changes are evident. All these muscle disorders appear to result from elevations of parathyroid hormone (PTH) and impaired activation of vitamin D. PTH stimulates protein degradation in skeletal muscle by activating intracellular proteases. It also decreases mitochondrial oxygen consumption and the activity of mitochondrial ATPase. Furthermore, PTH increases cAMP-dependent phosphorylation of troponin-1, diminishing the calcium sensitivity of the calcium-binding subunit and thus the whole contractile apparatus [22,23]. An uncommon neurological manifestation of hyperparathyroidism is spinal cord or root compression induced by collapse of decalcified vertebrae. Because vitamin D stimulates calcium absorption in the gut, bone resorption, and renal reabsorption of phosphate, one of the consequences of the electrolyte disturbances caused by vitamin D deficiency is impairment of excitation-contraction coupling. The calcium uptake and storage capacity of mitochondria and sarcoplasmic reticulum are impaired, and myofibrillar ATPase activity is diminished. Protein synthesis is also decreased, which could partially explain the atrophy seen in patients with vitamin D deficiency.

As many as 50% of patients with hyperparathyroidism suffer from psychiatric symptoms ranging from mental slowness or depression to mania, psychosis, and acute confusional state. These symptoms are generally but not invariably related to both the degree of hypercalcemia and the rapidity of change in electrolyte concentration. The negative magnesium balance in hyperparathyroidism may also contribute to delirium and psychosis in these patients.

2. Hypoparathyroidism and Pseudohypoparathyroidism

Hypoparathyroidism can cause perioral and distal paresthesias, carpopedal spasm, and diffuse muscle cramping [25]. The muscle disorder most commonly associated with hypoparathyroidism is tetany, although true myopathy does occur rarely. Nerve fibers are rendered hyperexcitable by hypocalcemia and hypomagnesemia. Voltage-sensitive channels in the nerve are controlled by the transmembrane electric field. Extracellular cations such as calcium can cluster around the surface of membrane phospholipids, reducing the surface charge and raising the threshold for depolarization. When serum calcium is decreased, the normal amount of surface charge screening is reduced, which reduces the transmembrane potential just as in depolarization [22]. In normocalcemic individuals, tapping the facial nerve produces a subthreshold depolarization of nerve fibers. In the setting of hypocalcemia, the depolarization may be sufficient to generate action potentials in the nerve fibers, inducing latent tetany (Chvostek's sign). Ischemia may also lowers the depolarization threshold of nerve fibers; in hypocalcemic patients, tetany therefore can be induced by occluding venous return in the arm (Trousseau's sign) [25].

A cognitive impairment manifested as psychosis, depression, and anxiety has been also described in hypoparathyroidism and does not appear to be associated exclusively with hypocalcemia.

Basal ganglia calcifications are a common incidental finding in hypoparathyroidism. Although cases of chorea in association with hypoparathyroidism have been reported, there does not appear to be a consistent correlation between the presence of basal ganglia calcifications and the occurrence of movement disorder [23].

References

1. American Diabetes Association and American Academy of Neurology. (1988). Consensus statement: report and recommendations of the San Antonio conference on diabetic neuropathy. *Diabetes Care* **11**, 592–597.
2. Dyck, P. J., Kratz, K. M., Karnes, J. L., Litchy, W. J., Klein, R., Pach, J. M., Wilson, D. M., O'Brien, P. C., Melton, L. J., and Service, F. J. (1993). The prevalence by staged severity of various types of diabetic neuropathy, retinopathy, and nephropathy in a population-based cohort: the Rochester Diabetic Neuropathy Study. *Neurology* **43**, 817–824 [published erratum appears in **43**, 2345].
3. Diabetes Control and Complications Trial Research Group. (1993). The effect of intensive treatment of diabetes on the development and progression of long-term complications in insulin-dependent diabetes mellitus. *New Engl. J. Med.* **329**, 977–986 [see comments].
4. U. K. Prospective Diabetes Study (UKPDS) Group. (1998). Intensive blood-glucose control with sulphonylureas or insulin compared with conventional treatment and risk of complications in patients with type 2 diabetes (UKPDS 33). *Lancet* **352**, 837–853 [see comments].
5. Stevens, M. J., Pop-Busui, R., Greene, D. A., Obrosova, I. G., and Feldman, E. L. (2002). Pathogenesis of diabetic neuropathy. *In:* "Ellenberg and Rifkin's Diabetes Mellitus" (D. Porte, et al, eds.), pp. 741–770. Appleton & Lange, Stamford, Conn.
6. Gaede, P., Vedel, P., Larsen, N., Jensen, G. V., Parving, H. H., and Pedersen, O. (2003). Multifactorial intervention and cardiovascular disease in patients with type 2 diabetes. *New Engl. J. Med.* **348**, 383–393.
7. Adler, A. I., Stratton, I. M., Neil, H. A., Yudkin, J. S., Matthews, D. R., Cull, C. A., Wright, A. D., Turner, R. C., and Holman, R. R. (2000). Association of systolic blood pressure with macrovascular and microvascular complications of type 2 diabetes (UKPDS 36): prospective observational study. *B. Med. J.* **321**, 412–419.
8. Tesfaye, S., Chaturvedi, N., Eaton, S. E., Ward, J. D., Manes, C., Ionescu-Tirgoviste, C., Witte, D. R., Fuller, J. H. (2005). Vascular risk factors and diabetic neuropathy. *New Engl. J. Med.* **352**, 341–350.
9. Pop-Busui, R., Sima, A., and Stevens, M. J. (2006). Diabetic neuropathy and oxidative stress. *Diabetes, Metabolism Research Reviews* **22**, 257–273.
10. Brownlee, M. (2003). A radical explanation for glucose-induced beta cell dysfunction. *J. Clin. Invest.* **112**, 1788–1790.
11. Boulton, A. J., Malik, R. A., Arezzo, J. C., and Sosenko, J. M. (2004). Diabetic somatic neuropathies. *Diabetes Care* **27**, 1458–1486.
12. Low, P. A., Ward, K., Schmelzer, J. D., and Brimijoin, S. (1985). Ischemic conduction failure and energy metabolism in experimental diabetic neuropathy. *Am. J. Physiol.* **248**, E457–E462.
13. Boulton, A. J., Vinik, A. I., Arezzo, J. C., Bril, V., Feldman, E. L., Freeman, R., Malik, R. A., Maser, R. E., Sosenko, J. M., and Ziegler, D. (2005). Diabetic neuropathies: a statement by the American Diabetes Association. *Diabetes Care* **28**, 956–962.
14. Sima, A. A., Kamiya, H., and Li, Z. G. (2004). Insulin, C-peptide, hyperglycemia, and central nervous system complications in diabetes. *Eur. J. Pharmacol.* **490**, 187–197.
15. Vincent, A. M., Brownlee, M., and Russell, J. W. (2002). Oxidative stress and programmed cell death in diabetic neuropathy. *Ann. NY Acad. Sci.* **959**, 368–383.
16. Russell, J. W., Sullivan, K. A., Windebank, A. J., Herrmann, D. N., and Feldman, E. L. (1999). Neurons undergo apoptosis in animal and cell culture models of diabetes. *Neurobiol. Dis.* **6**, 347–363.
17. Cheng, C., and Zochodne, D. W. (2003). Sensory neurons with activated caspase-3 survive long-term experimental diabetes. *Diabetes* **53**, 2363–2371.
18. Stevens, M. J., Raffel, D. M., Allman, K. C., et al. (1999). Regression and progression of cardiac sympathetic dysinnervation complicating diabetes: an assessment by C-11 hydroxyephedrine and positron emission tomography. *Metabolism* **48**, 92–101.
19. Ziegler, D., Weise, F., Langen, K. J., Schwaiger, M., and Wieland, D. M. (1998). Effect of glycaemic control on myocardial sympathetic innervation assessed by [^{123}I] metaiodobenzylguanidine scintigraphy: a 4-year prospective study in IDDM patients. *Diabetologia* **41**, 443–451.
20. Pop-Busui, R., Kirkwood, I., Schmid, H., Marinescu, V., Schroeder, J., Larkin, D., Raffel, D., Stevens, M. J. (2004). Sympathetic dysfunction in type I diabetes: implications for cardiovascular injury. *JACC* **44**, 2368–2374.
21. DCCT Study Group. (1995). Effect of intensive diabetes treatment on nerve conduction in the Diabetes Control and Complications Trial. *Ann. Neurol.* **38**, 869–880.
22. Anagnos, A., Ruff, R. L., and Kaminski, H. J. (1997). Endocrine neuromyopathies. *Neurol. Clin.* **15**, 673–696.
23. Kaminski, H. J., and Ruff, R. L. (1989). Neurologic complications of endocrine diseases. *Neurol. Clin.* **7**, 489–508.

24. Wolkowitz, O. M. (1994). Prospective controlled studies of the behavioral and biological effects of exogenous corticosteroids. *Psychoneuroendocrinology* **19**, 233–255.
25. Ubogu, W., Ruff, R. L., and Kaminski, H. J. (2004). Endocrine myopathies. *In:* "Myology" (A. G. Engel, et al., eds.), pp. 1713–1738. McGraw-Hill, New York.
26. Gelb, D. Hypothyroidism, S. G., Ed. Medlink Neurology, www.medlink.com, San Diego.
27. Pollard, J. D. (2005). Neuropathy in disease of the thyroid and pituitary glands. *In:* "Peripheral Neuropathy" (P. Dyck, et al., eds.), pp. 2039–2049. Saunders, Philadelphia.

62

Mechanisms and Consequences of Central Nervous System Hypoxia

Meredith L. Turetz, MD
Ronald G. Crystal, MD

Keywords: *cerebral blood flow, cerebral hypoxia, cerebral oxygen supply, clinical consequences, neurobiology of hypoxia, ventilatory control*

I. Cerebral Energy Metabolism
II. Cerebral Oxygen Supply
III. Cerebral Blood Flow
IV. Ventilatory Control
V. Oxygen-Sensing Mechanisms
VI. Neurobiology of Hypoxic-Ischemic Injury
VII. Clinical Consequences of Hypoxia
VIII. Conclusions
References

The brain has high metabolic demands and depends on a continuous source of oxygenated blood that contains glucose, the exclusive substrate required for central nervous system (CNS) energy metabolism [1]. Oxygen is vital in this process because oxidative phosphorylation is the major contributor of adenosine triphosphate (ATP), the energy source needed to maintain cellular function [2]. Cerebral hypoxia, the lack of adequate oxygen supply to brain tissues, can have consequences ranging from reversible behavioral changes to coma and death depending on the severity and duration of the insult [3]. Delivery of oxygen to the brain depends on the respiratory and cardiovascular systems, red cell oxygen–carrying capacity, and the ability of the tissues to use oxygen. By far, the two major causes of brain hypoxia are 1) hypoxemia, an inadequate partial pressure of oxygen in the arteries; and 2) hypoperfusion, or inadequate blood flow to the brain [3].

I. Cerebral Energy Metabolism

When glucose enters the cell, it is converted to pyruvate by an anaerobic mechanism in the cytoplasm, creating two molecules of ATP per molecule of glucose [4]. Alone, this would not result in enough energy to power the cells of the brain. However, under aerobic conditions, pyruvate can enter the mitochondrial matrix and can be metabolized in the citric acid cycle to generate nicotinamide adenine dinucleotide and flavin adenine dinucleotide (FADH2), which are used in the electron transport chain to generate an additional 36 molecules of ATP via oxidative phosphorylation. In total, the glycolytic and aerobic pathways of metabolism generate 38 molecules of ATP per molecule of glucose.

Different cells have different energy requirements, and therefore different areas of the CNS may have different cerebral metabolic rates for oxygen. For example, neurons are more metabolically active than glia and astrocytes and require a larger amount of energy to maintain ionic gradients across the cell membranes and maintain cell integrity [3]. Thus gray matter, with its higher concentration of neurons, has a higher oxygen consumption than white matter [5].

II. Cerebral Oxygen Supply

The body needs to obtain an adequate amount of oxygen from the atmosphere and deliver it to the cerebral tissues. Oxygen inspired from the atmosphere diffuses into the alveoli of the lungs based on an oxygen pressure differential. Oxygen continues to diffuse down this differential across the alveolar-capillary membrane to the blood, where it is transported by reversible binding to hemoglobin (1.34 milliliters of oxygen per gram of fully saturated hemoglobin) and dissolved in plasma ($0.0031 \times PaO_2$) [1]. The total amount of oxygen in the arterial blood is the arterial oxygen content, and the total amount in the venous blood is the venous oxygen content. The difference between the two—the arteriovenous O_2 difference—is a measure of the extraction of oxygen by tissues [1]. Oxygen continues to diffuse along its pressure differential by extraction from the blood into the tissues and finally to the mitochondria, the site of oxidative phosphorylation. Across the normal adult human brain, jugular oxygen venous saturation can serve as a marker of oxygen supply and consumption. Assuming that the cerebral artery oxygen saturation entering the brain is 100% and the normal jugular oxygen venous saturation is 55–70%, in the normal state, the brain extracts 25–50% of the oxygen it is supplied [1].

Importantly, the brain is unable to store a significant amount of glucose or oxygen. Therefore the brain depends on the constant delivery of oxygen, determined by the product of the cardiac output and the arterial oxygen content. Hypoxemia and hypoperfusion are the main causes of brain hypoxia (Table 1). Less commonly, pure anemia or disturbed intracellular metabolism is responsible for CNS hypoxia.

Cerebral hypoxia is most commonly due to a combination of both hypoxemia and hypoperfusion. It is unlikely that pure arterial hypoxemia ever causes CNS tissue injury and death because, at very severe levels of hypoxemia, cardiovascular depression and hypotension (and thus hypoperfusion of the CNS) usually occur.

A. Hypoxemia

CNS hypoxia resulting from hypoxemia occurs when either delivery of oxygen from the environment to the alveoli or from the alveoli to the pulmonary circulation is inadequate [6]. At sea level, the normal expected alveolar oxygen tension is 100 mm Hg. The alveolar-arterial oxygen difference (referred to as the *A-a gradient*) is a measure of how effectively air from the alveoli is transferred to the pulmonary capillaries.

Table 1 Causes of Hypoxia

Mechanism	Examples
Hypoxemic hypoxia	
Low alveolar PO_2	High altitude
Hypoventilation	CNS depressants, muscle weakness, chest wall disease
Right to left shunt	Parenchymal lung disease, cardiac shunt, AVM
Ventilation-perfusion mismatch	Obstructive lung disease, parenchymal lung disease
Impaired red cell transit time	Interstitial lung disease, emphysema
Hypoperfusion hypoxia	Cardiac arrest, shock (hypovolemic, cardiogenic, distributive, obstructive), local hypoperfusion (e.g., stroke, cerebral vascular disease)
Anemic hypoxia	Decreased hemoglobin, carbon monoxide poisoning, methemoglobinemia
Histotoxic hypoxia	Cyanide poisoning, hydrogen sulfide

AVM, Arteriovenous malformation; *CNS*, central nervous system.

The first mechanism that can cause hypoxemia is reduced inspired oxygen tension. This most commonly occurs at high altitudes where the decrease in atmospheric pressure directly decreases the alveolar oxygen tension, as can be seen by the alveolar gas equation. In this case, the A-a gradient remains normal.

Hypoventilation, the second potential mechanism of hypoxemia, is a state in which ventilation is inadequate to perform gas exchange. The pressure of carbon dioxide in the blood increases while the pressure of oxygen decreases. This causes a reduced alveolar oxygen tension in the setting of a normal A-a gradient. Pure hypoventilation is caused by conditions that depress the CNS (e.g., drug overdose or medullary infarcts), abnormal respiratory neuromuscular function (e.g., Guillain-Barré syndrome or amyotrophic lateral sclerosis), states of muscle weakness (e.g., muscular dystrophies), and diseases of the chest wall (e.g., kyphoscoliosis) [7].

The third mechanism, right to left shunting, is a form of hypoxemia that is difficult to correct with supplemental oxygen. In this state, venous blood bypasses the alveoli and remains unoxygenated, leading to both hypoxemia and an elevated A-a gradient. Parenchymal lung diseases such as pneumonia and acute respiratory distress syndrome, as well as arteriovenous malformations, are responsible for shunting.

The fourth mechanism, ventilation-perfusion mismatch, also causes hypoxemia with an elevated A-a gradient. Imbalances in the relationship between ventilation and blood flow are commonly associated with obstructive lung disease and chronic parenchymal lung diseases. Hypoxemia in this setting can usually be improved with supplemental oxygen.

The fifth potential mechanism of hypoxemia is impairment in red cell transit time across the alveoli. In the past, this concept referred to diffusion impairment, with the idea that thickened alveolar-capillary membranes (such as from interstitial lung disease or pulmonary edema) impair the diffusion of oxygen across the membrane. However, it is now recognized that what was thought to be "diffusion block" is actually an inability of oxygen to fill up sites on the hemoglobin in red blood cells that are passing too quickly across the alveoli. Normally, the red blood cells within the capillary have 0.75 seconds of exposure to the oxygen coming from the alveoli, and equilibration normally happens at 0.25 seconds. During exercise, there is increased blood flow and red cells spend less time in contact with the oxygen from the alveoli. With a variety of lung disorders, if there is a loss of capillaries, the blood flow from the right ventricle is found to transit the pulmonary circulation more rapidly than normal, resulting in decreased red cell saturation with oxygen.

B. Hypoperfusion

CNS hypoxia can also result from hypoperfusion, or inadequate supply of blood to the tissues. In this setting, the alveolar and arterial pressures of oxygen and oxygen content are normal, but delivery of oxygen to tissues is reduced. Local hypoperfusion, or ischemia, of a particular vascular bed (e.g., in the setting of a cerebral vascular accident) can cause hypoxia to specific cerebral tissues. Alternatively, the many causes of circulatory collapse (including hypovolemic shock, cardiogenic shock, distributive shock, and obstructive shock) lead to systemic hypoperfusion of tissues. On a chronic basis, hypoperfusion can result from blockage of CNS blood flow from cerebral vascular disease.

C. Anemia

In the setting of severe anemia, the marked decrease in hemoglobin concentration causes decreased arterial oxygen content despite a normal alveolar and arterial PO_2 [8]. If the body is unable to compensate by increasing cardiac output, tissue hypoxia can result. Functional anemia is caused by inability of oxygen to bind to hemoglobin. For example, hemoglobin has a higher affinity for carbon monoxide than for oxygen, accounting for the cerebral hypoxia in carbon monoxide poisoning. In methemoglobinemia, the iron in hemoglobin changes from ferrous to ferric form; this prevents both binding of new oxygen and unloading of already bound oxygen, again leading to functional anemia.

D. Histotoxic Hypoxia

Histotoxic hypoxia refers to tissue hypoxia due to disruption of the cellular metabolic processes that use oxygen to produce energy [8]. For example, cyanide interferes with cytochrome oxidase, disrupting the electron transport chain and inhibiting aerobic metabolism. Overdoses of sodium nitroprusside, an antihypertensive that forms cyanide as a byproduct, can be life threatening. Hydrogen sulfide, which is found in volcanic gases and petroleum deposits, acts similarly to cyanide.

III. Cerebral Blood Flow

The high energy demands of the brain require a constant cerebral blood flow that brings nutrients such as glucose and oxygen to the cells. The brain comprises only 2–3% of the total body weight of a human adult but requires a cerebral blood flow of 45–60 mL/100 grams/minute, representing nearly 20% of the total resting cardiac output [9]. The critical ischemic threshold for cerebral blood flow in humans is about 18 mL/100 grams/minute, the level at which neurons change from aerobic to anaerobic metabolism [4]. As cerebral blood flow drops to 10 mL/100 grams per minute, neuron membrane integrity is disrupted, which can lead to irreversible damage and functional consequences. Of note, reductions in arterial oxygen tension (PaO_2) with preserved cerebral blood flow rates can also lead to functional impairment. For example, in an acute setting, a PaO_2 of 30 mm Hg will cause lack of consciousness. Recent advances in cerebral monitoring and the ability to measure brain tissue oxygen tension suggest that at a critical level of 15 to 20 mm Hg, the likelihood of cell death and tissue infarction increases.

Through changes in cerebrovascular resistance, the brain is able to match energy consumption with cerebral blood flow to provide a constant source of energy based on metabolic demands, both through cerebral autoregulation and local metabolic control [9]. Cerebral blood flow is determined by the ratio of the perfusion gradient (cerebral perfusion pressure, the difference between the mean arterial pressure and intracranial pressure) to cerebrovascular resistance. As a protection from damage at the lower or higher ends of cerebral perfusion pressure, the brain provides constant flow mediated by changes in resistance of vessels, which constrict or dilate within a wide range of mean arterial blood pressures (60–150 mm Hg). However, outside of this range, or in the setting of hypoperfusion or traumatic brain injury, autoregulation is lost and perfusion then becomes passive and pressure dependent. Activation of the sympathetic nervous system can shift the lower and upper limits of autoregulation upward, as can chronic hypertension.

Local metabolic factors are also essential in maintaining normal tissue oxygenation. Cerebral blood flow is extremely sensitive to increases in arterial carbon dioxide tension ($PaCO_2$) [9]. Hypercapnia, mediated by changes in hydrogen ion concentrations and pH, leads to dilation of the cerebral vessels and increased perfusion. Conversely, hypocarbia leads to vasoconstriction and deceased flow. A reduction in arterial PaO_2 also results in an increase in cerebral blood via vasodilation, likely due to both a direct effect on cerebral vessels and indirect effects, such as the release of vasoactive mediators including hydrogen ions, potassium, adenosine, and nitric oxide.

IV. Ventilatory Control

With careful control of ventilation, PaO_2 and $PaCO_2$ are maintained within tight limits. Chemoreceptors detect changes in partial pressures of gas as well as changes in acid-base status and relay this information to the brainstem, which houses the respiratory centers [10]. The brainstem then coordinates information and sends impulses to the muscles of respiration, including the diaphragm, intercostal muscles, abdominals, and accessory muscles. To some degree, the cortex can override the involuntary function of the brainstem and produce voluntary hyperventilation, although voluntary hypoventilation is more difficult to produce.

Both central and peripheral chemoreceptors respond to changes in gas tension and hydrogen ion concentration [10]. The central chemoreceptors, located in the medulla, respond to changes in hydrogen concentration of the surrounding extracellular fluid. As blood $PaCO_2$ rises, CO_2 diffuses from the cerebral vessels through the blood-brain barrier (BBB) into the cerebrospinal fluid (CSF), freeing up H^+ ions that stimulate chemoreceptors and cause ventilation. The cerebral vasodilation that occurs with increased $PaCO_2$ helps with this diffusion into the CSF. With prolonged aberrations in CSF pH, bicarbonate (HCO_3^-) is transported across the BBB to compensate for the acidosis. This corrects the CSF pH to near-normal before the kidneys are able to compensate and correct the blood pH. The ventilatory response to CO_2 is reduced with age, sleep, and genetic factors.

Peripheral chemoreceptors include the carotid bodies and aortic bodies that respond to decreased PaO_2 and pH and increased $PaCO_2$ by increasing ventilation [10]. In the absence of these receptors, severe hypoxemia depresses ventilation. Normally, the response to CO_2 is the most important factor in controlling ventilation. However, the hypoxemic drive to breathe becomes increasingly important in individuals with severe lung disease and chronic CO_2 retention. In these individuals, the pH of the brain returns to near-normal values and the hypercapnic drive to breathe is lost. If they are given too much supplemental oxygen, the hypoxemic drive for ventilation is suppressed, leading to hypoventilation with worsening hypercapnia and potential CO_2 narcosis.

V. Oxygen-Sensing Mechanisms

In addition to vasodilation of the cerebral vasculature and oxygen sensing via central and peripheral chemoreceptors, changes occur at the cellular level to maintain physiological oxygen levels. Hypoxia-inducible factor 1 (HIF-1) is a transcription factor that has a major role in changing gene expression in the setting of hypoxia in the brain, heart, and other organs. The HIF-1 complex is a heterodimer composed of HIF-1α and HIF-1β, members of the basic helix-loop-helix proteins of the PAS family. HIF-1β, also known as *aryl hydrocarbon receptor nuclear translocator*, is constitutively expressed in all cells. HIF-1α is made continuously in hypoxic cells, but in the presence of oxygen it is hydroxylated by prolyl-4-hydroxylases, which allows ubiquitination and targeting for proteosomal degradation. In the setting of hypoxia, the prolyl-hydroxylases are inhibited, and HIF-1α is phosphorylated and dimerizes with HIF-1β, binds to cofactor p300/CBP (CREB, or cyclic-AMP-response element binding protein), and activates hypoxia response elements in HIF target genes, which allows transcription of these genes [11].

The HIF-1 system regulates the transcription of genes involved in oxygen homeostasis. More than 60 probable HIF target genes are involved in vasomotor control, angiogenesis, erythropoiesis, and energy metabolism, including nitric oxide synthase, vascular endothelial growth factor (VEGF), and glucose transporters 1 and 3, among others. Additionally, HIF-1 is necessary for embryological development of the brain. Mouse embryos that lack HIF-1 do not survive, and mice that are HIF-1 deficient can have a variety of abnormalities including hydrocephalus and decreased neural cells and memory. HIF-1 is also involved in preconditioning, which is a brief period of hypoxia that protects against a more severe injury hours to days later. When rats are exposed to hypoxia for several hours, levels of HIF-1 and target genes such as VEGF, erythropoietin, and glucose transporter 1 increase in the brain, which protects against stroke and ischemia a day later [11,12].

VI. Neurobiology of Hypoxic-Ischemic Injury

At rest, about 40% of cerebral energy is used to maintain ionic gradients and membrane potential; even more is used during activity [2]. The ATP-dependent Na^+/K^+ pump consumes the largest portion of energy, with other mechanisms such as the Na^+/Cl^- ion pump and Na^+/Ca^{++} exchanger using less energy. The Na^+/K^+ pump actively extrudes $3Na^+$ to the extracellular space and maintains $2K^+$ intracellularly, creating a resting membrane potential of approximately -60 to $70\,mV$, with the interior of the cell negative relative to the exterior. When this potential becomes less negative, depolarization occurs and an action potential is transmitted down the neuronal axon, with release of neurotransmitters into the synaptic cleft and subsequent diffusion into the postsynaptic membrane. The neurotransmitters bind with receptors and, depending on whether they are excitatory (glutamate and aspartate) or inhibitory (gamma-amino butyric acid or glycine), cause either depolarization or hyperpolarization of the responding neuron.

Within seconds of hypoxia, loss of consciousness occurs, reflecting neuronal depression [2]. However, this is not initially associated with an immediate decrease in cerebral energy levels. Rather, the cells try to conserve energy to match energy supply with demand by shutting down processes such as spontaneous electrical activity. Animal models of the hippocampal neurons, an area involved in memory and learning that is particularly susceptible to hypoxic injury, suggest that cells are initially protected by membrane hyperpolarization and blockage of excitatory potentials [13]. There is increased K^+ conductance and efflux out of the cell, possibly mediated by a small rise in intracellular Ca^{++} or fall in ATP levels. This leads to hyperpolarization of the neuron and a cessation of neuronal firing. The neuroinhibitory nucleoside adenosine, the breakdown product of ATP, is released during times of hypoxia and seems to be involved in metabolic depression both by increasing conductance of K^+ and inhibiting neurotransmitter release at excitatory synapses. If there is hypoxia but adequate glucose supply, the cells can continue to carry on anaerobic metabolism and supply the minimal amount of ATP required to power the Na^+/K^+ pump, which is needed for cell survival.

As hypoxia continues and glucose is diminished, ATP becomes completely depleted, and the Na^+/K^+ pump is no longer able to function. There is an efflux of K^+ into the extracellular space, with resultant loss of the resting membrane potential, depolarization of the cell, and an influx of Na^+ and Ca^{++} into the cell. These ionic changes lead to osmolarity shifts, and water follows into the cells, contributing to neuronal edema, injury, and death.

The exact mechanism of cell death associated with CNS hypoxia is still unclear [14]. There is likely a neurotoxic cascade mediated by hundreds of different mediators of cellular injury. However, Ca^{++} is likely one of the key effectors of hypoxic-ischemic injury leading to structural and functional damage of cells. Calcium influx into the cell is mediated by voltage-gated ion channels as well as by the process of excitotoxicity. Depolarization causes the release of glutamate, an excitatory neurotransmitter, into the synaptic cleft; this activates N-methyl-D-aspartate (NMDA) receptors and allows increased amounts of Ca^{++} into cells. Excess intracellular Ca^{++} activates Ca^{++}-sensitive enzymes, including nitric oxide synthase, leading to the generation of nitric oxide, which is toxic on its own and also combines with other superoxide ions to form more toxic reactive oxygen species (ROS). Calcium triggers various lipases and proteases, which are involved in a proinflammatory cascade, causing membrane and microvascular injury, endonuclease release, and damage to deoxyribonucleic acid (DNA), as well as further free radical production. Mitochondria are damaged by both the overload of Ca^{++} and by free radicals, causing an inability to generate ATP and maintain ionic gradients, thus perpetuating a cycle of membrane depolarization and NMDA receptor channel opening. Damaged mitochondria also generate more ROS. Myriad events are involved in these cascades, and other potential mediators include acidosis, ions such as Na^+ and K^+, other excitatory amino acids, and apoptotic enzymes such as capsases.

Whether the actual cause of injury and death to cells induced by CNS hypoxia is by way of necrosis or apoptosis is still a matter of debate. Recent thinking suggests that instead of two dichotomous pathways, cells may die along an apoptosis-necrosis continuum [14]. Whereas necrosis suggests pathological cell death due to an outside insult to the cell that destroys membrane and function, apoptosis involves the turning on of a genetic program that kills cells. It is believed that damaged mitochondria release cytochrome-c, which activates capsases and carries out apoptosis pathways. Whether cells die due to apoptosis or necrosis may depend on several factors. More intense excitotoxicity and mitochondrial dysfunction is more likely to cause necrosis. The age and maturity of the CNS are likely involved because apoptosis seems more prevalent in the immature brain, with adult cells rarely exhibiting a purely apoptotic morphology. Necrosis likely plays a larger role in NMDA receptor subtypes.

VII. Clinical Consequences of Hypoxia

The clinical sequelae of cerebral hypoxia depend on the severity, duration, and rate of onset of the insult. Often hypoxia and ischemia are combined, although one typically predominates and it is often difficult to distinguish between the two because severe hypoxia causes myocardial depression and circulatory collapse.

A. Acute Hypoxia

In acute cerebral hypoxia, cognitive function declines at a PaO_2 of 40–50 mm Hg with typical symptoms of inattentiveness, poor judgment, and motor incoordination [15,16]. If the acute hypoxia is not severe enough to cause loss of consciousness, then there is no permanent damage. Consciousness is typically lost at a PaO_2 of less than 30 mm Hg within 6 to 8 seconds if there is complete cessation of cerebral circulation and a few seconds later if there is continued circulation in the absence of oxygen. Continued hypoxia results in a loss of brainstem reflexes and loss of motor responses. Loss of consciousness, which reflects neuronal depression, is not associated with an immediate decrease in ATP levels. Instead, a slowing of metabolic processes occurs in an attempt to conserve energy. After 2 minutes of anoxia, neuronal ATP stores are lost and further neurobiological sequelae and cell injury and death occur. Different CNS cell populations are selectively vulnerable to hypoxia, possibly as a function of their metabolic rates. The neurons of the hippocampus, parieto-occipital lobe, and cerebellum are particularly sensitive. As hypoxia becomes more severe, damage extends to the entire cerebral cortex and deep nuclei and ultimately to the brainstem.

Recovery of function depends on the speed with which oxygen is restored to the brain. Hypothermia prolongs the period of tolerable hypoxia by slowing the cerebral metabolic rate for oxygen. With immediate restoration of oxygen delivery to the CNS, consciousness typically returns within seconds to minutes and usually without clinical consequences. If anoxia has lasted for a couple of minutes, confusion or stupor may occur on return of consciousness and irreversible damage may begin after 2 to 4 minutes. If the individual survives, longer periods of anoxia usually lead to permanent sequelae such as memory deficits, myoclonic and generalized seizures, extrapyramidal disorders, and cerebellar ataxia. More prolonged global anoxia leads to coma. If brainstem reflexes are preserved, then partial recovery is possible, although usually with diffuse cerebral damage. If reflexes are not preserved, the likelihood of brain death and mortality is high. Prognosis for anoxic encephalopathy depends on the severity and duration of hypoxia as well as on underlying diseases. The main goal is to restore oxygenation and treat the underlying cause of the hypoxia.

B. Chronic Hypoxia

Both pulmonary and extrapulmonary disorders can cause chronic hypoxia. The major classes of pulmonary disorders associated with chronic hypoxia are chronic obstructive lung disease (COPD), fibrotic/intersitial lung disease, and the sleep apnea syndromes. Extrapulmonary disorders include neuromuscular diseases, such as myasthenia gravis and muscular dystrophy, as well as neurological disorders that disrupt normal ventilatory patterns (e.g., stroke). Clinical symptoms of chronic hypoxia include headaches, impaired cognition, and motor disturbances [15]. Many individuals with chronic hypoxia also have hypercapnia, and it may be difficult to determine which is the cause of clinical symptoms. It is often difficult to study the effects of pure hypoxia because severe hypoxia causes cardiovascular depression and thus hypoperfusion. Two populations have been studied that potentially represent "pure" hypoxia: individuals with COPD and individuals exposed to high altitudes. Studies in these populations have shown that even in the absence of hypoxia severe enough to cause loss of consciousness, chronic hypoxia may have long-lasting neurobehavioral consequences.

C. Chronic Obstructive Lung Disease

Cognitive impairment has been documented in patients with COPD. For example, Grant et al. [17] combined data from the Nocturnal Oxygen Therapy Trial and the Intermittent Positive Pressure Breathing Trial to study the neurobehavioral responses in 302 patients with COPD. They found that neurobehavioral deficits increased from 27% in those with mild hypoxemia (PaO_2 68 ± 6 mm Hg) to 61% among those with severe hypoxemia (PaO_2 44 ± 4 mm Hg). In addition to hypoxemia, older age and lower educational level also correlated with severity of impairment. Of note, the $PaCO_2$ of these groups increased from a mean of 35 to

46 mm Hg and neither group was acidemic, suggesting that hypercapnia was not responsible for the changes. Furthermore, neurocognitive functioning seems to improve with continuous oxygen treatment. The mechanism of these deficits does not seem to be due to cell death but may be due to changes in the synthesis of neurotransmitters.

D. High-Altitude Sickness

Ascent to higher altitudes is accompanied by a fall in barometric pressures with a resultant fall in the partial pressure of oxygen and hypoxemia. At sea level, the barometric pressure is 760 mm Hg and the inspired partial pressure of oxygen is 149 mm Hg, with an alveolar partial pressure of O_2 of 100 mm Hg. In contrast, at the top of Mt. Everest, at 29,028 feet, the barometric pressure is 253 mm Hg and the inspired partial pressure of oxygen is 43 mm Hg. The neurological manifestations of high-altitude exposure and subsequent hypoxemia (estimated arterial PaO_2 of 28 mm Hg on the summit of Mt. Everest) include acute mountain sickness and high-altitude cerebral edema, as well as chronic altitude sickness. Acute mountain sickness is characterized by headache, nausea and vomiting, difficulty in concentrating, and insomnia [18,19]. These symptoms are worsened at night during the normal hypoventilation of sleep and also are worsened with sedatives or alcohol. Symptoms occur within hours to days of rapid ascent from sea level to 8000–10,000 feet and are quickly reversible with descent. High-altitude cerebral edema is characterized by severe headache, hallucinations, ataxia, and cerebral edema [19]. Coma can occur in severe cases. The proposed pathogenesis is that hypoxemia induces cerebral vasodilation, which causes vasogenic brain edema and elevated intracranial pressure. Supplemental oxygen and descent to lower altitudes is mandatory.

Often these neurological sequelae can be prevented with slow acclimatization to altitudes. Interestingly, Hornbein et al. [20] studied mountaineers before and after ascent as well as subjects undergoing simulation in a high-altitude chamber. He found a slight decrease in visual and verbal long-term memory as well as mild aphasic defects and decreased speed of finger tapping after altitude exposure. Those who had a higher ventilatory response to hypoxia, and hence the best oxygenation, actually had poorer outcomes. These investigators postulated that hypocarbia induced by the increased ventilation caused cerebral vasoconstriction and that although the higher ventilatory response may have maintained oxygen delivery to muscles, it resulted in worsening hypoxia to the brain.

E. Adaptation to High Altitudes

Acclimatization refers to adaptation to high altitudes and occurs primarily through hyperventilation as well as increased erythropoiesis. Carotid body chemoreceptors sense decreased inspired PO_2 and stimulate increased alveolar ventilation to augment the alveolar PO_2. Hyperventilation occurs fairly quickly; in contrast, the process of increasing erythrocyte concentration and oxygen-carrying capacities takes several days and is not completed for weeks. High-altitude adaptations are important for those who usually reside at sea level and travel to these environments for recreation and work, as well as for several populations of highlanders who live in these environments [19].

Different groups living at high altitude adapt in different ways to hypoxia exposure. For example, a study of populations living in the Tibetan and Andean plateaus found that the Tibetans had higher resting ventilation and higher hypoxic ventilatory response but that the Andean population had higher hemoglobin concentrations and higher oxygen saturations [21]. This suggests that in addition to acquired adaptations, genetic adaptations to hypoxia may exist; these genetic adaptations may differ among populations. The specific genetic influences are not yet defined, and research involving specific factors, such as HIF-1 and its multiple target genes, is underway.

VIII. Conclusions

Cerebral hypoxia, typically due to respiratory or cardiovascular causes, deprives the brain of the ability to produce a sufficient amount of energy to sustain its high metabolic demands. Continued hypoxia leads to cell injury and death by a variety of neurotoxic cascades that remain to be fully elucidated. The challenge will be to discover if there is a way to intervene in these pathways to prevent the cellular destruction and clinical consequences.

Acknowledgments

We thank N. Mohamed for help in preparing this manuscript. These studies were supported in part by the Will Rogers Memorial Fund, Los Angeles.

References

1. De Georgia, M. A., and Deogaonkar, A. (2005). Multimodal monitoring in the neurological intensive care unit. *Neurologist* 11, 45–54.
2. Morris, J. C., and Ferrendelli, J. A. (1990). Metabolic encephalopathy. In: "Neurobiology of Disease" (A. L. Pearlman, et al., eds.), pp. 356–379. Oxford University Press, City.
3. Hornbein, T. F. (1997). Hypoxia and the brain. In: "The Lung" (R. G. Crystal, et al., eds.), pp. 1981–1987. Lippincott–Raven, Philadelphia.
4. Zauner, A., Daugherty, W. P., Bullock, M. R., and Warner, D. S. (2002). Brain oxygenation and energy metabolism. Part I. Biological function and pathophysiology. *Neurosurgery* 51, 289–301.

5. Erecinska, M., and Silver, I. A. (2001). Tissue oxygen tension and brain sensitivity to hypoxia. *Respir. Physiol.* **128**, 263–276.
6. J. F. Nunn, ed. (1993). "Nunn's Applied Respiratory Physiology." Butterworth-Heinemann, Oxford.
7. Weinberger, S. E., Schwartzstein, R. M., and Weiss, J. W. (1989). Hypercapnia. *New Engl. J. Med.* **321**, 1223–1231.
8. M. G. Levitzky, ed. (1995). "Pulmonary Physiology." McGraw-Hill, New York.
9. Markus, H. S. (2004). Cerebral perfusion and stroke. *J. Neurol. Neuro surg. Psychiatry* **75**, 353–361.
10. West, J. B. (2000). Control of ventilation: how gas exchange is regulated. *In:* "Respiratory Physiology: the Essentials." (P. J. Kelly, ed.), pp. 103–116. Lippincott, Willams & Wilkins, Baltimore.
11. Sharp, F. R., and Bernaudin, M. (2004). HIF-1 and oxygen sensing in the brain. *Nat. Rev. Neurosci.* **5**, 437–448.
12. Acker, T., and Acker, H. (2004). Cellular oxygen sensing need in CNS function: physiological and pathological implications. *J. Exp. Biol.* **207**, 3171–3188.
13. Nieber, K., Eschke, D., and Brand, A. (1999). Brain hypoxia: effects of ATP and adenosine. *In:* "Progress in Brain Research, vol 120, Nucleotides and Their Receptors in the Nervous System" (B. Illes, et al., eds.), pp. 287–297. Elsevier Science, Amsterdam; New York.
14. Martin, L. J., Al-Abdulla, N. A., Brambrink, A. M., Kirsch, J. R., Sieber, F. E., and Portera-Cailliau, C. (1998). Neurodegeneration in excitotoxicity, global cerebral ischemia, and target deprivation: a perspective on the contributions of apoptosis and necrosis. *Brain Res. Bull.* **46**, 281–309.
15. Kirsch, D. B., and Jozefowicz, R. F. (2002). Neurologic complications of respiratory disease. *Neurol. Clin.* **20**, 247–264.
16. Ropper, A. H., and Brown, R. H. (2005). The acquired metabolic disorders of the nervous system. *In:* "Adams and Victor's Principles of Neurology" (A. H. Ropper, et al., eds.), pp. 959–982. McGraw-Hill, New York.
17. Grant, I., Prigatano, G. P., Heaton, R. K., McSweeny, A. J., Wright, E. C., and Adams, K. M. (1987). Progressive neuropsychologic impairment and hypoxemia. Relationship in chronic obstructive pulmonary disease. *Arch. Gen. Psychiatry* **44**, 999–1006.
18. Hackett, P. H., and Roach, R. C. (2001). High-altitude illness. *New Eng. J. Med* **345**, 107–114.
19. West, J. B. (2004). The physiologic basis of high-altitude diseases. *Ann. Intern. Med.* **141**, 789–800.
20. Hornbein, T. F., Townes, B. D., Schoene, R. B., Sutton, J. R., and Houston, C. S. (1989). The cost to the central nervous system of climbing to extremely high altitude. *New Engl. J. Med.* **321**, 1714–1719.
21. Beall, C. M. (2000). Tibetan and Andean patterns of adaptation to high-altitude hypoxia. *Hum. Biol.* **72**, 201–228.

63

Neurological Manifestations of Gastrointestinal and Hepatic Diseases

Brian P. Bosworth, MD
Brian R. Landzberg, MD

Keywords: acute intermittent porphyria cholestasis, celiac disease, hemochromatosis, hepatic encephalopathy, inflammatory bowel disease, neurological manifestations of gastrointestinal and hepatic diseases, vitamin deficiencies, Whipple's disease, Wilson's disease

I. Introduction
II. Diseases of the Alimentary Tract
III. Hepatic Diseases
IV. Vitamin and Mineral Deficiencies
V. Conclusion
   References

I. Introduction

Not all gastrointestinal and hepatic maladies present with symptoms easily attributable to the digestive tract and liver, such as malabsorption, diarrhea, or jaundice. Rather, many such diseases have protean manifestations including rheumatological (e.g., Crohn's disease with arthritis), dermatological (e.g., celiac disease with dermatitis herpetiformis), and most importantly for this chapter, neurological.

Neurological symptoms in these conditions (e.g., pernicious anemia and B_{12} deficiency) may precede, coincide with, develop after, or occur in the complete absence of gastrointestinal or hepatic symptoms. Therefore it is critical for the physician to be aware of the neurological issues likely to arise in treating digestive disease patients and of the possibility of making a diagnosis of a gastrointestinal

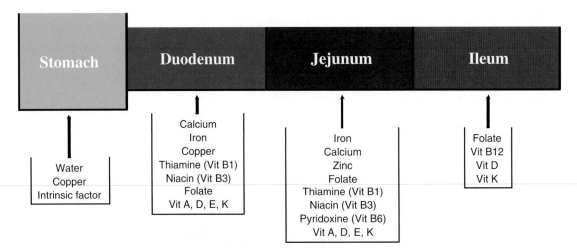

Figure 1 Sites of nutritional absorption in the gastrointestinal tract. Courtesy of Drs. Jonathan Waitman and Scott Shikora, New York, NY.

or hepatic illness in the neurological patient. Studying the scientific basis of the overlap between digestive and nervous system diseases may also lend insight into the primary etiology of idiopathic neurological disease in general.

The chapter is divided into two sections. The larger first section discusses those digestive diseases with direct effects on the central and peripheral nervous systems (CNS and PNS), with emphasis on celiac disease. A smaller second section reviews the neurological sequelae of several vitamin deficiencies that may be nonspecifically associated with a number of malabsorptive diseases (Fig. 1).

The chapter cannot hope to accomplish an exhaustive discussion of all the scientific literature regarding digestive disease and its impact on the nervous system. Rather, it seeks to highlight a handful of common or prototypical disease entities in this category, presenting a brief clinical context and our best current understanding of the science underlying the connections. The list of references will also prove useful to those who wish to explore specific neurological manifestations in more depth. Multiple congenital defects and inborn errors of metabolism are related to the liver and alimentary tract and may, for example, result in hyperammonemia or hyperbilirubinemia and present with neurological sequelae in infants and children. However, we will confine the scope of this chapter to gastrointestinal and hepatic conditions with neurological manifestations that present in adulthood.

II. Diseases of the Alimentary Tract

A. Celiac Disease

1. Background

Celiac disease (also known as celiac sprue, nontropical sprue, or gluten-sensitive enteropathy) is an autoimmune, inflammatory disease of the small intestine, triggered by dietary exposure to proteins found in many common grains. The disease is quite common, affecting between 1 in 200 and 1 in 150 individuals in the United States. However, it frequently goes undiagnosed because patients presenting with the typical syndrome of malabsorptive diarrhea reflect only the tip of the celiac iceberg and are vastly outnumbered by those with irritable bowel syndrome, isolated iron or folate deficiency, osteoporosis, arthritis, infertility, skin conditions such as dermatitis herpetiformis, associated autoimmune thyroid disease, lymphoma, and a host of neurological manifestations.

Pathogenesis begins with exposure to the dietary protein gluten (gliadin and glutenin, found in wheat, rye, and barley) in genetically susceptible individuals. Gluten is degraded in the intestinal lumen to gliadin peptides and amino acids. A 33-amino acid residue of gluten is resistant to digestion by gastric and pancreatic amylases. This polypeptide is the substrate for tissue transglutaminase (tTG), a ubiquitous calcium-dependent enzyme in the lamina propria that can deamidate glutamine residues into glutamate. This allows for avid binding in the HLA-DQ2 (DQA1*05/DQB1*02) or HLA-DQ8 (DQA1*0301/DQB1*0302), which are the only haplotypes that allow celiac disease to occur. A T-cell mediated inflammatory response then ensues, causing destruction of the intestinal villi. Villous atrophy leads to the classical symptoms of malabsorption, chronic diarrhea, and iron-deficiency anemia.

The gold standard for diagnosis is intestinal biopsy demonstrating villous atrophy, but serological tests of the targets of the autoimmune response (anti-tTG, anti-endomysial [EMA], and anti-gliadin antibodies) have a very high sensitivity and specificity for celiac disease and are easily available to the internist or neurologist [1,2]. EMA and its surrogate antigen marker, anti-tTG, are probably the most useful because gliadin antibodies, particularly the immunoglobulin (Ig)-G form, are nonspecific. When interpreting these tests, one needs to keep in mind that a significant percentage of celiac patients will be IgA deficient and therefore unable to generate these IgA markers; therefore the total IgA level should also be determined. A second caveat is that if the patient has self-imposed a gluten-free diet, serological markers and duodenal biopsies may revert to normal. Genetic testing for the haplotypes DQ2 and DQ8 is also widely available that, when negative, effectively rule out the diagnosis regardless of diet. Positive genetic testing, however, is very unhelpful because these haplotypes may be seen in one-third of the general population and are felt to be necessary but not sufficient for the diagnosis.

In addition to villous atrophy, an intraepithelial lymphocytosis occurs with a CD8 predominance. The mechanisms responsible for this are unknown, but the T-cell mediated immune responses are directed against damaged epithelial cells that express IFN-γ-induced molecules (MIC and HLA-E), which are recognized by the natural killer cell receptors NKG2D and CD94.

At present, the only treatment for celiac disease is strict adherence to a gluten-free diet that, when meticulous, is effective in resolving the intestinal symptoms, serological markers, and endoscopic and histological findings in the vast majority of cases.

The neurological sequelae of celiac disease, present in roughly 10% of patients, are widely varied and may involve both the CNS and PNS, including syndromes of cerebellar and myoclonic ataxia, encephalopathy and dementia (occasionally associated with bilateral occipital calcifications), seizures, CNS vasculitis and progressive multifocal leukoencephalopathy, peripheral neuropathy, and myopathy. These neurological syndromes are not necessarily associated with the intestinal symptoms and in fact may be the only presenting complaints. To make matters more difficult, many patients have neurological symptoms similar to those associated with celiac disease (see later discussion) and are positive for the HLA haplotypes and serologies of celiac disease but have normal duodenal biopsies [3]. Such patients cannot be said to have celiac disease per se (although some experts suggest that the endoscopic biopsies may be falsely negative due to sampling error) but may enjoy an abatement of their neurological symptoms on a gluten-free diet. Others may have celiac disease proven by duodenal biopsy but neurological symptoms that do not improve on a gluten-free diet, suggesting either a noncausal association or noncompliance with diet.

2. Cerebellar Ataxia (Gluten Ataxia)

Cooke and Smith first reported a series of patients with ataxia, peripheral neuropathy, and celiac disease [4]. In subsequent case series, cerebellar involvement was reported. In patients with cerebellar ataxia of unknown cause, biopsy examination demonstrated celiac disease in 2–41% of patients. The prevalence of gluten sensitivity in hereditary ataxia has been reported to be higher than in sporadic ataxias, suggesting that that the relationship between the two entities is quite complex but not necessarily causal. Cerebellar atrophy may be detected by magnetic resonance imaging (MRI) in patients with gluten ataxia, and postmortem findings have included atrophy, gliosis, Purkinje cell loss, and degeneration of the posterior columns of the spinal cord. Milder degrees of associated cerebral atrophy have also been reported.

It remains controversial whether gluten sensitivity without intestinal involvement should be considered the cause of cerebellar degeneration in ataxia of otherwise idiopathic cases. Hadjivassiliou et al. studied 224 patients with ataxia, which they subdivided into idiopathic forms, hereditary forms, or multisystem atrophy and then compared with normal controls. Antigliadin antibodies were detected in 41% of the idiopathic group, 14% of the hereditary ataxia group, 15% of the multisystem atrophy group, and 12% of controls [5], confirming that this antibody is the least specific of the celiac serologies. The study also showed that antigliadin antibodies and sera from most patients with presumed gluten ataxia patients reacted with Purkinje cells, but this finding remains controversial.

The effect of gluten-free diet on ataxia has been variable, and only a few reports show efficacy of immunomodulation with intravenous Ig. Although the majority of celiac-related neurological syndromes are not related to nutritional deficiency, vitamin E deficiency has been postulated to be an independent factor in some patients with cerebellar ataxia, and this testing for vitamin E deficiency or empiric supplementation should be done in all such patients [6,7].

3. Epilepsy

A high prevalence of epilepsy has been reported in patients with celiac disease compared with normal controls (3.5%–5.5%). Conversely, a higher prevalence of celiac disease was noted among patients with epilepsy compared with the general population (0.8%–2.5%). Numerous reports characterize cerebral calcification at the epileptiform center. These calcified lesions have been described more commonly in the occipital cortex (associated with transient blindness) [8,9]. Although the majority of patients reported complex partial seizures referable to the occipital or temporal lobes, secondarily generalized seizures, other types of seizures, and episodic headaches also were described.

The etiology of the calcified lesions in celiac disease is similar but not identical to Sturge-Weber syndrome (SWS) (characterized by port wine stain or nevus flammeus). Unlike in SWS, the occipital calcifications of celiac disease are bilateral, lack contrast enhancement on MRI, and can extend into the subcortical region on computed tomography (CT) scans. The calcifications consist of patchy pial angiomas, fibrosed veins, and large jagged microcalcifications. Whereas in SWS, the calcification has been attributed to blood stasis in the angiomata, chronic ischemia, and altered metabolism, the calcifications in celiac patients may be related to low serum folate.

The effect of a gluten-free diet on epilepsy control is variable. There seems to be a decrease in the use of antiepileptic medication use but not a complete resolution of the seizures. Surgical intervention has been recommended for intractable seizures localized to the occipital or temporal lobes [10].

4. Neuropathy and Myopathy

Gluten sensitivity has been frequently implicated as a cause of neuropathy, with as many as 49% of celiac disease patients experiencing some form of neuropathy. Most commonly, a chronic distal and symmetrical sensory neuropathy occurs. However, pure motor neuropathy, mononeuritis multiplex, Guillain-Barré–like syndrome, and autonomic neuropathy also have been reported [11,12].

Chin et al. report screening 400 neuropathy patients with antigliadin and anti-tTG antibodies, finding 20 (5%) with positive serologies and subsequent biopsy confirmation of celiac disease. However, 14 of those 20 patients carried a previous diagnosis of celiac disease. Nerve biopsies and electrodiagnostic testing usually found primary sensory axonopathy. Anti-ganglioside antibodies were present in 13 of those 20 patients. These anti-ganglioside antibodies bind to the Schwann cell surface, nodes of Ranvier, and axons in the peripheral nerves [13] and commonly are associated with autoimmune peripheral neuropathies—specifically, IgG antibodies to GM and GD1b in Guillain-Barré, anti-GQ1b and GT1b in Miller Fisher syndrome, and IgM antibodies to GM in multifocal motor neuropathy. In at least one patient with mononeuritis multiplex, biopsy found evidence for neuronal vasculitis that subsequently was steroid responsive [12].

Myopathy has been reported less frequently than neuropathy in association with celiac disease and is less well understood and probably more multifactorial. Vacuolar myopathy was observed on muscle biopsy in a patient with hypokalemia and hypocalcemia, which resolved promptly with correction of electrolytes. A syndrome of proximal muscle weakness and atrophy without creatine kinase (CK) or electromyogram (EMG) anomalies has also been observed. These patients demonstrated noninflammatory type II fiber atrophy on muscle biopsy [14,15]. Although some of these patients showed signs of osteomalacia, suggesting a possible role for vitamin D or other nutritional deficiency, many did not. Another subset of myopathy patients demonstrated a classical polymyositis picture by clinical and pathological criteria. These patients generally did not exhibit signs of nutritional deficiency and often improved with a gluten-free diet, although some did require immunomodulatory therapy [16].

The effect of gluten-free diet on peripheral neuropathy and other neuromuscular disorders associated with celiac disease is unclear. Only a few reports show that it is efficacious, whereas others show progression of the neuropathy despite well-documented intestinal healing [17].

5. Other Neurological Entities

A few studies show an association between celiac disease and headache (migraine and tension headache) [18–20]. Headache appeared to improve with a gluten-free diet. Depression has been reported as a common complication of celiac disease; it is described in as many as one-third of patients. The mechanism remains unclear but may be associated with hypothyroidism, nutritional deficiencies (such as vitamin B_6), or other linked autoimmune diseases. Decreased levels of tryptophan have been found in celiac patients, resulting in decreased cerebrospinal fluid (CSF) levels of serotonin, dopamine, and norepinephrine metabolites. Rapid improvement of depression has been reported and frequently observed in clinical practice with adherence to a gluten-free diet.

Given that there are at least reports of improvement in the various neurological conditions associated with celiac disease on implementation of a gluten-free diet, a causal relationship between the two seems quite likely. Although more studies on the pathology of CNS diseases associated with celiac disease are necessary, screening for celiac disease should be strongly considered in many idiopathic neurological conditions, especially cerebellar ataxia or sensory axonopathy, particularly when other clinical or laboratory signs of celiac disease or its associated conditions (e.g., diarrhea, iron or folate deficiency, dermatitis herpetiformis, autoimmune thyroid disease, or insulin-dependent diabetes mellitus) are present [10,13].

B. Inflammatory Bowel Disease

Inflammatory bowel disease (IBD) reflects a chronic autoimmune inflammation of the alimentary tract that requires the entry of bacterial antigens into the gut lamina propria of the genetically predisposed host, antigen presentation by enterocytes and macrophages, and an ultimate imbalance between the proinflammatory and antiinflammatory mediators that regulate gut immune function. Diagnosis rests on typical endoscopic, radiographic, and histological findings, and treatment consists of various immunomodulating agents, probiotic therapy, and surgery. Traditionally, IBD has been divided into two distinct entities: ulcerative colitis (UC) and Crohn's disease (CD). An innovative and nuanced view presents IBD as an immunoinflammatory spectrum of chronic and recurring disease of the intestines.

This newly gained perspective holds the promise of moving treatment in a more proactive direction—toward targeting molecules, rather than treating symptoms. In addition, IBD also has a genetic component, with a familial linkage and associations with the NOD2 gene.

Neurological manifestations of IBD have been long recognized; however, the incidence of such concomitant diseases ranges from <1% to 35% among patients with IBD. These manifestations may precede or follow a diagnosis of IBD [21]. Elsehety and Bertorini report that neurological and neuropsychiatric complications were evident in 84 of 253 patients with Crohn's disease, an incidence of 33.2%. They computed a direct relationship between the two in 19.3% of patients and determined that a variety of neurological and neuropsychologic events, such as seizure disorder, peripheral neuropathy, and myopathy, were also associated with CD [22]. Anxiety and depression were also prevalent.

Seizure disorder ranks high in prevalence among the neurological manifestations of IBD. Although a variety of seizure types have been reported, both the electroencephalogram (EEG) and brain MRI of these patients are usually normal. Some patients require daily prophylactic antiepileptic medications, and others develop isolated seizure activity at the time of a flare of IBD. A vasculopathy is likely the underlying cause of these events.

Myelopathy has been commonly reported as well. Patients may develop slowly progressive spastic paraparesis, increased tendon reflexes, and bilateral extensor responses. Lossos reports that in his series, no patient had a spinal sensory level and none had lower motor neuron signs or urinary incontinence [23]. In isolated cases, an elevated CSF protein level and oligoclonal bands can be identified; otherwise there are no positive tests pointing to a specific etiology of the neurological dysfunction.

Multiple sclerosis also has been identified as being associated with IBD, but no clear mechanism of causality has been defined. It remains unclear whether there is simply a genetic linkage between the two or a shared pathogenesis or whether the circulating cytokines as a result of inflammatory activity in IBD are responsible for demyelination. It is interesting to note, however, that natalizumab was effective in both multiple sclerosis and Crohn's disease [24–27]. Natalizumab is a humanized IgG4 monoclonal antibody that blocks the adhesion and subsequent migration of leukocytes into the gut by binding α_4 integrin. Integrins are adhesion molecules that stabilize interactions between cells and their environment. They also act as cellular sensors and signaling molecules. All integrins are composed of noncovalently linked α and β chains that dimerize to confer activity. Natalizumab affects both the $\alpha_4\beta_1$ integrin (VLA4 or CD49d–CD29) and the $\alpha_4\beta_7$ integrin. It likely has therapeutic effects via its ability to block $\alpha_4\beta_1$ and $\alpha_4\beta_7$ binding to their respective endothelial counter-receptors, vascular cell adhesion molecule 1 (VCAM-1) and mucosal addressin cell adhesion molecule 1 (MAdCAM-1). These molecular interactions are required for lymphocytes to enter the CNS (mediated by $\alpha_4\beta_1$ and VCAM-1) and the intestine (mediated by $\alpha_4\beta_7$ and MAdCAM-1) [28]. Although the studies were quite promising in both CD and multiple sclerosis, the reports of several cases of progressive multifocal leukoencephalopathy led to its withdrawal from the market by the Food and Drug Administration (FDA) [15,29–34].

Patients with IBD are hypercoagulable and therefore predisposed to cerebral vein thromboses. Proteins C and S and antithrombin III deficiency, as well as reactive thrombocytosis and hyperfibrinogenemia, have been seen frequently and are likely contributory in these patients. Associated cerebral vasculitis has also been rarely reported.

PNS associations with IBD include a peripheral neuropathy typically seen in ulcerative colitis, with examination and electrodiagnostic findings consistent with acute inflammatory demyelinating polyradiculoneuropathy, mononeuritis multiplex, or brachial plexopathy [35]. In patients with CD, as opposed to UC, PNS manifestations more typically take the form of axonal motor and sensory neuropathy, not unlike that seen in celiac disease, another disorder of small bowel. The Lossos and Elsehety reviews also found inflammatory myopathy in 3 of 638 and 4 of 263 CD patients, respectively [21,22].

C. Whipple's Disease

George Hoyt Whipple described the clinical and pathological state of a 36-year-old missionary who had experienced a 5-year unknown illness characterized by "gradual loss of weight and strength, stools consisting chiefly of neutral fat and fatty acids, indefinite abdominal signs, and a peculiar multiple arthritis" [36]. He called the unknown disorder "intestinal lipodystrophy" because of the frothy "fat" deposits and vacuoles in the cytoplasm of macrophages and enlarged mesenteric lymph nodes. He also identified a rod-shaped organism but did not consider this a possible etiology of the condition. This organism has since been named *Tropheryma whippelii*. Polymerase chain reaction (PCR) of the bacterial 16S ribosomal ribonucleic acid (RNA) gene has allowed the phylogenetic classification of the agent as actinomycetes. Fewer than 1000 cases have been reported in the literature. The disease may occur at any age but is most commonly seen in the fourth through sixth decades of life and has a male predominance. The diagnosis of Whipple's disease is based on a high index of suspicion leading to duodenal or proximal jejunal biopsy followed by PCR for the organism. Light microscopy can demonstrate periodic acid-Schiff (PAS)–positive, mucopolysaccharide-filled macrophages, and electron microscopy confirms the extracellular bacilli.

Clinical features of Whipple's disease are quite varied, with a gastrointestinal tract predominance but manifestations in other organ systems as well. The classic triad consists of weight loss, chronic diarrhea, and arthralgia. When this triad is also associated with fever and peripheral lymphadenopathy, the clinical suspicion for Whipple's disease should be quite high. The diarrhea is generally watery, and there may be fat malabsorption as well, leading to weight loss and cachexia. Furthermore, protein-losing enteropathy caused by lymphatic hypertension is very common and is associated with severe hypoalbuminemia. Of the many extraintestinal manifestations, arthropathy is most commonly seen, affecting 80–90% of patients. The joint symptoms often precede the diagnosis by several years and consist of chronic, symmetrical, migratory, nondestructive seronegative joint disease [32].

The neurological complications of Whipple's disease represent the greatest risk for long-term disability and are present in 10–40% of cases [38]. However, CSF PCR for *T. whippleii* was positive in 67% of patients with Whipple's disease who did not have neurological symptoms [39]. The symptoms are quite varied and include mental abnormalities (e.g., dementia, depression, cognitive alterations, memory loss, confusion, behavioral abnormalities, and personality change); movement disorders (e.g., both supranuclear and intranuclear ophthalmoplegia, oculomasticatory myorhythmia, oculofacial-skeletal myorhythmia, myoclonus, and ataxia); hypothalamic symptoms (e.g., sleeping abnormalities, polydipsia, and hyperphagia); and others (e.g., epilepsy, cerebellar dysfunction, and dysphasia) [37].

The term *oculomasticatory myorhythmia* (OMM) refers to pendular vergence oscillations, which occur synchronously with slow rhythmic mouth and palatal movements. Oculofacial-skeletal myorhythmias (OFSM) are slow pendular vergence oscillations, which occur synchronously with rhythmic movements of the mouth, face, and extremities and persist during sleep. Although quite specific for Whipple's disease, these signs are not sensitive for the diagnosis, with only 20% of the reported cases demonstrating OMM or OFSM. The clinical triad of dementia, ophthalmoplegia, and myoclonus occurs in 10% of patients, but two of the three signs occur in 41% [40]. Generally, the onset and outcome of the neurological sequelae are insidious and progressive.

There appears to be a defective cell-mediated immune function with a proliferation of mature CD8 T-cells with a reduced response to phytohemagglutinin, concanavalin A, and other mitogens [37]. Furthermore, there is a deregulated Th1/Th2 response with decreased macrophage IL-12 production.

Chloramphenicol was the first successful therapy reported in Whipple's disease, and since then a variety of antibiotics and schedules have been successfully used. The most commonly used series of medications associated with clinical success and low risk for relapse is the combination of penicillin G (1.2 million units/day intramuscularly) and streptomycin (1 gram/day intramuscularly) and/or a third-generation cephalosporin for 2 weeks followed by the administration of trimethoprim-sulfamethoxazole (160 milligrams/800 milligrams, twice daily) for at least 1 year.

D. Campylobacter-Associated Guillain-Barré Syndrome

In 30–40% of individuals afflicted with Guillain-Barré syndrome (GBS), an antecedent infection with *Campylobacter jejuni* can be identified [41]. *Campylobacter* is a frequent cause of infectious diarrhea, and 1 in 1058 cases of *Campylobacter* infection leads to GBS. Particular strains of *Campylobacter*, types O:19 and O:41, result in GBS in up to 1 in 158 cases. The lipo-oligosaccharide from the *Campylobacter* bacterial wall contains ganglioside-like structures, and its injection into rabbits induces a neuropathy that resembles acute motor axonal neuropathy. This is likely a form of molecular mimicry that leads to development of anti-ganglioside antibodies [42]. Antibodies to GM1, GM1b, GD1a, and GalNac-GD1a are particularly implicated in acute motor, as well as motor and sensory, axonal neuropathy. The Fisher syndrome subtype is especially associated with antibodies to GQ1b, and similar cross-reactivity with ganglioside structures in the wall of *Campylobacter* has been discovered. Anti-GQ1b antibodies have been shown to damage the motor nerve terminal in vitro by a complement-mediated mechanism [43].

Campylobacter-associated GBS is generally of greater severity and eventually requires ventilatory support. This subset of patients is also more likely to have EMG evidence of axonal degeneration and may be associated with demyelination. Furthermore, a subgroup of these patients are rapid progressors and become quadriplegic within 48 hours of symptom onset.

E. Iron Overload

1. Background

Iron (Fe) serves as a metal cofactor for many proteins and enzymes and is essential for certain biological functions in the form of heme for oxygen transport (e.g., hemoglobin) and electron transport (e.g., mitochondrial proteins), as well as in the form of non-heme for deoxyribonucleic acid (DNA) synthesis. Although it is quite a versatile element, given its ability to exist in both an oxidized and a reduced state, the acquisition of iron by the body is rather difficult and complex.

Heme is taken up intact via specific high-affinity heme-binding sites in the mucosal brush border of the proximal duodenum, and it is the greater source of dietary Fe. After entering the intestinal epithelial cells, the Fe is released from heme by heme oxygenase. A divalentmethyltransferase (Nramp2/DMT1) transports inorganic Fe^{2+} from the intestine. All other cells in the body take up iron that is bound to transferrin. This ligand binds to transferring receptor complexes on the cell surface and is then internalized via endocytosis. The iron at this stage is in an oxidized Fe^{3+} state but needs to be reduced to be transported through the endosomal membrane via an internal Nramp2/DMT1.

The turnover of transferrin Fe is approximately 30 milligrams/day, and generally 80% of this Fe is transported to the bone marrow for hemoglobin synthesis. Senescent erythrocytes are phagocytosed by the reticuloendothelial system in the liver, where the free Fe is liberated from the heme and returned to the circulation. Because of the high degree of Fe conservation by the body, only about 1 milligram of dietary Fe is absorbed daily and is very tightly controlled by the bowel mucosa. In conditions such as hereditary hemochromatosis, this system is dysregulated, leading to large amounts of daily iron absorption from the small bowel.

The blood-brain barrier (BBB) does not allow for passage of large molecules such as transferrin, yet transferrin receptors are present on the endothelial cells of the cerebral vasculature. Maximal brain Fe uptake occurs during rapid brain growth, but Fe uptake continues throughout life. However, this is a highly regulated process, and brain iron overload is rare, even in patients with hemochromatosis [44].

2. Neuroferritinopathy

Ferritin is a protein that functions as a means of soluble iron storage. An autosomal dominant disease of the ferritin gene has been described in patients with iron accumulation in various parts of the brain, particularly in the basal ganglia. The patients generally present between the ages of 40 and 55 years with involuntary movements. Other symptoms of extrapyramidal dysfunction include choreoathetosis, dystonia, spasticity, and rigidity. The patients did not have other signs or symptoms of iron overload (e.g., diabetes, abnormal liver function tests, and such). However, brain histochemistry revealed accumulation of iron and ferritin in the globus pallidus, as well as inclusions in the forebrain, cerebellum microglia, and oligodendroglia.

3. Friedreich's Ataxia

Friedreich's ataxia is the most common inherited ataxia and is an autosomal recessive neurodegenerative disease. Generally, onset is before age 25, and clinical symptoms include progressive limb and gait ataxia, absent lower extremity deep tendon reflexes, dysarthria, areflexia, sensory loss, and cardiomyopathy. The defective gene is *FRDA*, a nuclear gene that codes for a mitochondrial protein known as *frataxin*. Frataxin is highly expressed in tissues rich in mitochondria, such as heart, liver, and skeletal muscle, and the mutant alleles in Friedreich's ataxia cause a severe reduction in frataxin levels. Individuals affected with Friedreich's ataxia develop myocardial fibrosis and degeneration of the cardiac muscle, associated with iron deposits in myocardial cells. Although these patients do not have the generalized iron overload found in patients with hereditary hemochromatosis, there is recent evidence that iron chelators, particularly lipophilic chelators, may be capable of crossing the BBB, a property lacking in the more commonly used chelator desferrioxamine.

III. Hepatic Diseases

A. Hepatic Encephalopathy

Hepatic encephalopathy (HE) is a frequent and occasionally refractory complication of cirrhosis in which brain function slows due to increased levels of ammonia and other toxins. HE represents a reversible decrease in neurological function caused by liver disease and occurs most notably in subjects with portal hypertension and shunting of blood away from the liver. Onset of HE in the setting of chronic liver disease is often insidious and is characterized by subtle and sometimes intermittent changes in memory, personality, concentration, and reaction times.

Patients with HE present with a wide spectrum of clinical findings ranging from merely abnormal psychometric testing to coma. Neurological abnormalities become more apparent with progression and are commonly graded on a numerical scale reflecting increasing degrees of neurological dysfunction. The presence of abnormal psychometric testing in the absence of symptoms has been classified as minimal HE. Stage I HE is characterized by the presence of disturbed sleep pattern, slight inattentiveness, and subtle personality changes. Stage II HE is characterized by the presence of a characteristic irregular flapping tremor, termed *asterixis*.

The development of HE at any stage in a cirrhotic patient has important implications. Patients with only minimal HE have impaired daily functioning, are at risk for the subsequent development of clinical HE, and have decreased survival. One retrospective study estimated a cumulative survival time from the first HE episode of 42% at 1 year and 23% at 3 years [45].

Glucose utilization is increased in brains of acutely hyperammonemic animals, and an ammonia-induced stimulation of glycolysis has been demonstrated in brain sections. Furthermore, ammonia activates brain phosphofructokinase, a rate-limiting enzyme in glycolysis. Additionally, activities of several glycolytic enzymes (phosphofructokinase, aldolase, glyceraldehydes-3-phosphate dehydrogenase, enolase, and pyruvate kinase) are elevated in the cerebral cortex, cerebellum, and brainstem of rats treated with ammonium [46]. Ammonia interrupts the malate-aspartate shuttle and stimulates astrocytic glycolysis [47].

Multiple pathogenic factors have been incriminated in the development of HE. Current theories focus on portal-systemic shunting of increased neurotoxin levels arising from gastrointestinal flora, as well as increased ammonia levels of either gut bacterial or renal origin [48,49]. Gastroparesis and intestinal motility dysfunction due to altered smooth muscle contractility and autonomic nervous system dysfunction may contribute to development of HE by leading to bacterial overgrowth and colonic inertia. Both conditions are common in cirrhosis, with the frequency and intensity of intestinal dysmotility increasing as the liver disease progresses.

Treatment options currently available in the United States for HE have been limited to nonabsorbable disaccharides (lactulose or, more recently, lactitol) and the nonabsorbable antibiotic, neomycin, but these drugs have several important limitations. Specifically, lactulose is frequently poorly tolerated, and many patients are noncompliant with its use. Also, neomycin is both ototoxic and nephrotoxic, which is especially problematic because HE is frequently reported in patients with renal insufficiency.

Rifaximin is a newly available, orally administered rifamycin derivative that acts on the B-subunit of DNA-dependent RNA polymerase in susceptible microorganisms and thus has a broad spectrum of activity. The addition of a pyridoimidazole ring to the rifampin molecule makes rifaximin nonabsorbable, which results in no known drug interactions, minimal systemic absorption ($<0.4\%$), and a tolerability profile comparable to placebo. Rifaximin, 200 milligrams taken 3 times a day (TID) for 3 days, was approved by the FDA for the treatment of travelers' diarrhea (TD) in May 2004 and is currently marketed for adults and children, 12 years or older, in the United States by Salix Pharmaceuticals, Inc., Morrisville, NC. Rifaximin also has been marketed since 1985 by Alfa-Wasserman (Bologna, Italy) for treatment of acute bacterial diarrhea, HE, and small-bowel bacterial overgrowth. Rifaximin is currently licensed in 17 countries, including in Europe (Italy and Spain), Mexico, South America, and Asia.

B. Acute Intermittent Porphyria

Acute intermittent porphyria (AIP) is an autosomal dominant metabolic disease of the liver that results from the partial deficiency of porphobilinogen deaminase (PBGD) and is characterized by life-threatening, acute neurological attacks. The heterozygous form of the disease affects 1 in 20,000 individuals and is the most common hepatic porphyria. Rarely, cases of homozygous individuals have been reported. PBGD is an enzyme in heme biosynthesis and catalyzes the polymerization of porphobilinogen (PBG). The PBGD gene is located on chromosome 11q24, and although approximately 200 mutations have been described, the clinical manifestations are uniform [50].

Affected heterozygotes have half-normal hydroxymethylbilane synthase (HMBS) activity and accumulate the porphyrin precursors d-aminolevulinic acid (ALA) and PBG. The acute neurological attacks are precipitated by metabolic, hormonal, and environmental factors that induce hepatic 5-aminolevulinate synthase (ALAS1) activity. Some of these factors include alcohol, infections, fasting, and endogenous changes in the sex-hormone balance. As ALAS1 activity increases, porphyrin precursor levels of ALA and PBG increase. The half-normal hepatic HMBS activity is insufficient to prevent the pathological precursor accumulation that leads to the neurological symptoms [51].

These neurological symptoms during acute attacks are characterized by severe abdominal pain, vomiting, constipation, hypertension, tachycardia, and bladder dysfunction, presumably due to an autonomic neuropathy. Motor weakness and sensory involvement like results from a peripheral axonal neuropathy because no central cause has been established. The peripheral neuropathy is manifested as pain in the extremities but may progress to a severe motor neuropathy. Cranial nerve palsies, epileptic seizures, respiratory insufficiency, abnormal sphincter function, hallucinations, and mental changes may also be present. Given the protean manifestations of the disease, patients may present first to the neurologist, psychiatrist, gastroenterologist, or surgeon.

Although the pathogenesis of the neurological manifestations is poorly understood, there are two leading hypotheses: 1) heme or hemoprotein deficiency in nerve cells and 2) neurotoxicity of ALA, PBG, or other porphyrin precursors. In particular, ALA is a γ-aminobutyric acid (GABA), glutamate, and aspartate analogue and thus may have neurotoxic properties via its interaction with GABA receptors and/or inhibition of glutamate uptake.

C. Pruritus of Cholestasis

Cholestasis is seen with many hepatobiliary disorders that produce a disruption of intrahepatic biliary trafficking or an extrahepatic biliary obstruction, particularly in primary biliary cirrhosis (PBC) and primary sclerosing cholangitis (PSC). Pruritus is defined as an unpleasant sensation that elicits the scratch reflex. It is a particularly problematic consequence of cholestasis and can range in severity from mild to moderate (in which sleep is disturbed) to extreme (in which the lifestyle of the patient is completely disrupted).

No high-quality studies of the pathophysiology of the pruritus of cholestasis are available. Although the pathogenesis of pruritus in cholestasis is unknown, various hypotheses have been proposed, including bile acid accumulation in the skin and an increase in the concentration or activity of endogenous opioids. The medullary dorsal horn in monkeys and rats has been identified as an area in which the injection of opioids leads to a scratching behavior [52].

Elevated levels and deposition of bile acids in the skin of patients with cholestatic liver disease act as pruritogens. This theory has credence because bile acids have been recovered from the surface of the skin of affected patients. Furthermore, the injection of exogenous bile acids in the skin lead to increased pruritus in both human and animal subjects. However, several observations are not consistent with a primary role for bile acids as the cause of pruritus: pruritus occasionally subsides, despite ongoing cholestasis and persistently elevated bile acids in the plasma; many patients with cholestasis and elevated bile acid levels do not have pruritus; and there does not appear to be any correlation between the presence or severity of pruritus and concentrations of bile acids in skin of patients with chronic cholestasis.

There is increasing evidence, however, that the pruritus of cholestasis is more related to endogenous opioids. It is known that opioid receptor agonists, such as morphine, have been associated with pruritus. Treating cholestatic patients with opiate antagonists such as naloxone greatly improves their pruritus. Other evidence to support a role for opioids include increased plasma concentration of some of the opioid peptides in cholestatic patients; down-regulation of central opioid receptors in the brain of rats with cholestasis; and administration of naloxone, which prevents heightened scratching in rats injected with the plasma extracts of cholestatic patients [53].

Additionally, the serotonin system may play a role in the pruritus of cholestasis. Although no specific data implicates an altered state of serotonergic transmission in cholestasis, an increased central opioidergic tone can result in increased serotonergic tone. Preliminary studies show that 5-HT3 antagonists provide some relief for patients with pruritus of cholestasis.

The cannabinoid system may also play a role in the sensation and mediation of pruritus. Dronabinol, a cannabinoid receptor agonist, has been reported to alleviate the pruritus in cholestatic patients. This system may modulate nociception, and pruritus is certainly a nociceptive stimulus [54].

D. Wilson's Disease

Wilson's disease is an autosomal recessive inherited disorder of copper metabolism resulting in pathological accumulation of copper in many organs and tissues. It has a prevalence of approximately 1 in 30,000 live births in most populations. Copper is an essential dietary nutrient and facilitates electron transfer reactions when incorporated into specific cuproproteins. These cuproproteins are needed for such diverse processes as mitochondrial respiration, melanin biosynthesis, dopamine metabolism, iron homeostasis, antioxidant defense, connective tissue formation, and peptide amidation.

Approximately 60% of dietary copper, or 0.6 to 3 milligrams/day, is absorbed in the upper small bowel, binds to circulating albumin, and is taken up by various tissues. In normal individuals, copper is incorporated in the liver into apoceruloplasmin, forming ceruloplasmin. Ceruloplasmin is secreted into plasma, where it represents more than 90% of the circulating copper [55]. An impairment in biliary excretion of copper leads to the accumulation in the liver. Over time the liver is progressively damaged and eventually becomes cirrhotic. The hepatic injury is believed to be caused by excess copper that leads to mitochondrial damage with alteration of lipid oxidation, resulting in marked hepatic steatosis.

Once cirrhosis occurs and the capacity of the liver to store copper is exhausted, the excess copper is released into the circulation and accumulates in and damages other tissues. However, its toxicity is most pronounced in the CNS. The incorporation of copper into ceruloplasmin is also impaired. This accounts for the decreased serum ceruloplasmin concentrations present in most patients with Wilson's disease.

Wilson's disease presents as a variety of conditions but most commonly as liver disease and/or neuropsychiatric disturbances. Neurological disorders are present in up to 35% of patients with Wilson's disease. The Kayser-Fleischer ring is a hallmark of the disease but is not universally present. It is found in more than 95% of patients with neurological symptoms but only in 50–60% of those without neurological stigmata. The rings themselves are pigmented granular deposits of copper in Descemet's membrane in the cornea, close to the endothelial surface. Because copper is primarily complexed with sulfur, the deposits have a characteristic color. The rings are usually most pronounced at the inferior and superior poles of the cornea.

Neurological disorders are present in up to 35% of patients with Wilson's disease and usually develop in the second and third decades of life but may also present at a later time. Signs include a Parkinsonian-like tremor, rigidity, clumsy gait, slurred speech, inappropriate and uncontrollable grinning (risus sardonicus), and drooling. Bradykinesia, rigidity, cognitive impairment, and an organic mood syndrome were associated with dilatation of the third ventricle by MRI. Ataxia and tremor were associated with focal thalamic lesions. Dyskinesia, dysarthria, and an organic personality syndrome were associated with focal lesions in the putamen and pallidum, often termed *hepatolenticular degeneration*. Approximately one-third of patients initially present with psychiatric abnormalities. Symptoms range from subtle personality changes and deteriorating performance at school to overt depression, labile mood, paranoia, catatonia, and frank psychosis.

Treatment for Wilson's disease has progressed from intramuscular administration of the chelator British anti-lewisite (BAL or dimercaptopropanol) to the more easily administered oral chelator, penicillamine. This compound promotes urinary excretion of copper. However, worsening of neurological symptoms has been reported in 10–50% of those treated with penicillamine during the initial phase of treatment. The numerous side effects necessitate discontinuation of the drug in 20–30% of patients. These side effects include sensitivity reactions with fever and cutaneous eruptions, neutropenia, nephrotoxicity, and development of a lupus-like syndrome [23].

Because of these reactions, a newer chelator, trientine, was developed. As with penicillamine, trientine promotes copper excretion by the kidneys. It has few side effects, but over-treatment resulting in copper deficiency can be associated with a reversible sideroblastic anemia. Tetrathiomolybdate has been studied more recently and appears to act differently than other anti-copper drugs. It forms a very stable tripartite complex with copper and protein. In this manner, it complexes dietary copper with food protein, rendering that copper unabsorbable. Furthermore, it binds copper in salivary, gastric, and intestinal secretions, thus placing the patient in an immediate negative copper balance. It has been shown to improve neurological manifestations of Wilson's disease.

Other treatment strategies include zinc, which interferes with the uptake of copper from the gastrointestinal tract; antioxidants, such as vitamin E; and avoidance of foods with high concentrations of copper (e.g., shellfish, nuts, chocolate, and mushrooms).

IV. Vitamin and Mineral Deficiencies

The digestive diseases previously discussed produce direct effects on the nervous system or had associations with neurological syndromes with an unclear degree of causality. In the next section, we review the effects of several common vitamin deficiencies on the nervous system. These may stem nonspecifically from a variety of digestive diseases. Disease entities associated with fat malabsorption—either due to insufficiency of the exocrine pancreas, villous atrophy, or short bowel syndrome due to extensive small bowel resection—tend to result in deficiency in the fat-soluble vitamins A, D, E, and K. Diseases that affect the quality or effective quantity (resection or bypass) of the proximal small bowel (i.e., duodenum and proximal jejunum) tend to result in iron, folate, and calcium deficiency (celiac sprue), whereas gastrectomy (due to intrinsic factor) and diseases of the ileum tend to be associated with B_{12} deficiency (CD). Another interesting example is tropical sprue (TS), a likely infectious chronic malabsorptive diarrheal disease (of declining frequency) acquired in certain areas in the Indian subcontinent and Southeast Asia. TS affects the entire small bowel and therefore is often associated with both B_{12} and folate deficiency, with reported subacute combined degeneration, nutritional neuropathies, and myopathy [56,57]. Unlike the autoimmune celiac sprue, neurological manifestations are almost entirely nutritional in origin.

A. Vitamin A

Vitamin A is a subclass of a family of fat-soluble compounds referred to as *retinoic acids*. These consist of four isoprenoid units joined in a head-to-tail fashion. Vitamin A occurs in three forms: retinols, beta-carotenes, and carotenoids. Retinol (preformed vitamin A) is the most active form and is mostly found in animal sources of food. Beta-carotene (provitamin A) is the plant source of retinol from which mammals make two-thirds of their vitamin A. Carotenoids, the largest of the vitamin A subclasses, contain multiple conjugated double bonds and exist in a free alcohol or in a fatty acyl-ester form.

Vitamin A deficiency, although rarely seen in the U.S., is the third most common nutritional deficiency worldwide. The neurological complication most commonly associated with vitamin A deficiency is night blindness. Other manifestations include complete blindness, xerophthalmia, and abnormalities in corneal and conjunctival development.

Two types of retinal photoreceptor cells are involved in the visual process. While the cone cells are responsible for the absorption of light and color vision in bright light, the rod cells detect motion and are responsible for night vision. In the rod cells of the retina, all-trans-retinol is converted to 11-cis-retinol, which then combines with a membrane-bound protein, termed *opsin*, to yield rhodopsin. (A similar type of reaction occurs in the cone cells of the retina to produce iodopsin.) The light-activated transformation of these complexes leads to a cascade of hyperpolarization of rod cell membrane, thus enabling the transmission of light stimuli to the CNS.

B. Vitamin B_1

Thiamine (vitamin B_1) serves as a coenzyme, along with nicotinamide adenine dinucleotide (NAD) and coenzyme A (derived from vitamin B_5, or pantothenic acid), in the oxidative decarboxylation reaction mediated by pyruvate dehydrogenase. Thiamine also plays a role in the initiation of nerve impulse propagation, independent of its coenzyme functions.

Beriberi is the prototypical disease associated with thiamine deficiency. In infants, the disease generally becomes clinically apparent between 2 and 3 months of age and is characterized by cardiomegaly, tachycardia, cyanosis, dyspnea, vomiting, and a loud piercing cry. In adults, the disease is separated into "dry" and "wet" beriberi. In dry beriberi, a symmetrical peripheral neuropathy develops, characterized by sensory as well as motor impairments. It generally affects the distal extremities. Wet beriberi has both a neuropathic component and a cardiac component that resembles infant beriberi.

In chronic alcoholics with thiamine deficiency, Wernicke-Korsakoff syndrome may be found. Wernicke's disease is a combination of nystagmus, ophthalmoplegia, ataxia, and confusion. Korsakoff's psychosis is characterized by impaired short-term memory and confabulation with otherwise grossly intact cognition. Because not all patients with thiamine deficiency are affected by this malady, there may be a genetic predisposition or an impact of the alcohol itself. Because of the fear of inducing a Wernicke's encephalopathy in a potentially thiamine-deficient patient by prompting the oxidative decarboxylation of carbohydrates, alcoholics are generally not given dextrose without concomitant thiamine supplementation.

Leigh's syndrome is a subacute necrotizing encephalomyopathy that can be seen in thiamine deficiency. It is a sporadic mitochondrial disorder with a subacute neurological course, developmental regression, ataxia, and hypotonia with subsequent respiratory and brainstem dysfunction.

C. Vitamin B_3

Niacin (vitamin B_3) contains both nicotinic acid and nicotinamide and is the principal component of NAD and NAD phosphate (NADP). Many enzymatic reactions depend on NAD and NAPD as a means of electron transport and oxidation/reduction reactions.

Pellagra is the classic niacin deficiency, characterized by a photosensitive pigment dermatitis in sunexposed areas. It is most common in alcoholics who are generally nutritionally deficient. Other findings include a red tongue and nonspecific symptoms such as nausea, vomiting, and diarrhea. Neurological symptoms include insomnia, anxiety, disorientation, delusions, dementia, and encephalopathy.

D. Vitamin B_6

Pyridoxine (vitamin B_6) is involved in the transamination of amino acids, gluconeogenesis, neurotransmitter synthesis, and immune function, among many other biological processes. It is commonly found in the human diet, and deficiencies are quite rare. However, a relative deficiency could influence memory function and might contribute to age-associated cognitive impairment and dementia. In older adults with normal vitamin B_6 levels, supplementation does benefit either mood or cognition.

E. Vitamin B_{12}

Pernicious anemia (PA) is the most common cause of vitamin B_{12} deficiency and is associated with chronic atrophic gastritis. As many as 1.9% of individuals older than 60 years have undiagnosed PA, and symptoms can present before the onset of actual anemia. Chronic atrophic gastritis is characterized by loss of gastric mucosal folds and thinning of the gastric mucosa. It can be autoimmune in nature (with antiparietal cell antibodies associated with the hypochlorhydria of the stomach and high serum gastrin levels) or nonautoimmune (usually associated with *Helicobacter pylori* and low serum gastrin concentrations). In addition to PA, B_{12} deficiency can be seen in patients with gastric resection, resection or disease of terminal ileum such as CD, and some parasitic infestations such as *Diphyllobothrium latum*.

Gastric acid production is important for the absorption of vitamin B_{12}, which must be first complexed to intrinsic factor to be absorbed in the terminal ileum. Intrinsic factor is secreted in the stomach by the parietal cells as an inactive glycoprotein. In the setting of an acidic gastric milieu, intrinsic factor is cleaved from its R-protein and avidly binds vitamin B_{12}. Anti–intrinsic factor antibodies can prevent the formation of the vitamin B_{12}–intrinsic factor complex.

Vitamin B_{12} deficiency may cause peripheral neuropathy and lesions in the posterior and lateral columns of the spinal cord (a subacute combined degeneration) and in the cerebrum. These lesions progress from demyelination to axonal degeneration. Patients have loss of vibration sense and proprioception and may have a sensory ataxia with a positive Romberg's sign, as well as limb weakness and spasticity. The most frequent manifestations, however, are paresthesias and numbness. The cerebral symptoms range from mild personality defects and memory loss to frank psychosis. These neurological complications are not universally reversible after replacement therapy.

F. Vitamin E

Vitamin E, also known as *tocopherol*, is a fat-soluble vitamin whose principal action is to act as a free-radical scavenger and protect polyunsaturated fatty acids from oxidation. It is quite abundant in foods, and thus a deficiency state is rare. Neuromuscular disorders associated with vitamin E deficiency are mostly of the neuropathic and myopathic type. The major manifestations of vitamin E deficiency are skeletal myopathy, spinocerebellar ataxia, and pigmented retinopathy.

However, certain genetic disorders lead to congenital vitamin E deficiency. A defect in the alpha-tocopherol transfer protein (A-TTP) may lead to a condition termed *ataxia with vitamin E deficiency* (AVED). This autosomal recessive disease is caused by mutations in the A-TTP gene on chromosome 8q13.1. It may present as a slowly progressive ataxia syndrome with neuropathy resembling Friedreich's ataxia. High doses of vitamin E typically lead to neurological improvement, although recovery may be slow and incomplete. Heterozygotes are phenotypically normal but have serum vitamin E concentrations that are 25% lower than normal.

G. Minerals: Zinc

Zinc was not recognized as a distinct and essential mineral until the twentieth century. Zinc is present in all organs, but 90% is stored in bone and skeletal muscle. It is primarily absorbed in the jejunum and, to a lesser extent, in the stomach and large bowel. During digestion, dietary zinc is released and forms complexes with different ligands: amino acids, phosphates, organic acids, and histidines. Zinc-ligand complexes are then absorbed through the intestinal mucosa by both an active and a passive process. Once absorbed, the portal circulation carries zinc to the liver. There is an intricate homeostatic control of zinc absorption, regulated by metallothionein, and this may be impaired in pancreatic disease or insufficiency. Pancreatic enzymes are necessary for release of dietary zinc, and pancreatic juices may contain zinc-complexing ligands. The major route of zinc excretion is via the gastrointestinal tract, and urinary excretion of zinc depends on the serum levels of the ion and drops appropriately in response to low dietary intake. Zinc also has a critical role in the structure of cell membranes, and its deficiency leads to increased susceptibility of the phospholipid cell membrane to free-radical damage and oxidative changes [58].

Mild zinc deficiency impairs growth velocity, whereas severe depletion of zinc leads to growth retardation. Principal neurological manifestations of zinc deficiency include dysgeusia, night blindness, and night blindness with other clinical manifestations, including delayed sexual maturation, impotence, hypogonadism, oligospermia, alopecia, immune dysfunction, impaired wound healing, and various skin lesions. A possible mortality benefit in adults was described in a randomized controlled trial, in which the primary focus was evaluation of zinc supplementation in ocular disorders [59].

V. Conclusion

As we have discussed, digestive diseases may have protean neurological manifestations. An awareness of these manifestations is critical when performing neurological consultation in patients with or without a known history of gastrointestinal or hepatic disease because the neurological symptoms may dominate or precede digestive symptoms. Further study of the connections between the digestive and nervous systems, which is clearly required, may also lend insight into and better understanding of the etiology of idiopathic neurological diseases in general.

References

1. Alaedini, A., and Green, P. H. (2005). Narrative review: celiac disease: understanding a complex autoimmune disorder. *Ann. Intern. Med.* **142**, 289–298.
2. Green, P. H. (2005). The many faces of celiac disease: clinical presentation of celiac disease in the adult population. *Gastroenterology* **128(Suppl 1)**, S74–S78.
3. Hadjivassiliou, M., Grunewald, R. A., Chatopadhyay, A. K., Davies-Jones, G. A., Gibson, A., Jarratt, J. A., Kandler, R. H., Lobo, A., Powell, T., and Smith, C. M. (1998). Clinical, radiological, neurophysiological, and neuropathological characteristics of gluten ataxia. *Lancet* **352**, 1582–1585.
4. Cooke, W. T., and Smith, W. T. (1966). Neurological disorders associated with adult coeliac disease. *Brain* **89**, 683–722.
5. Hadjivassiliou, M., Grunewald, R. A., Sharrack, B., Sanders, D., Lobo, A., Williamson, C., Woodroofe, N., Wood, N., and Davies-Jones, A. (2003). Gluten ataxia in perspective: epidemiology, genetic susceptibility and clinical characteristics. *Brain* **126**, 685–691.
6. Ackerman, Z., Eliashiv, S., Reches, A., and Zimmerman, J. (1989). Neurological manifestations in celiac disease and vitamin E deficiency. *J. Clin. Gastroenterol.* **11**, 603–605.
7. Mauro, A., Orsi, L., Mortara, P., Costa, P., and Schiffer D. (1991). Cerebellar syndrome in adult celiac disease with vitamin E deficiency. *Acta. Neurol. Scand.* **84**, 167–170.
8. Gobbi, G., Ambrosetto, P., Zaniboni, M. G., Lambertini, A., Ambrosioni, G., and Tassinari, C. A. (1992). Celiac disease, posterior cerebral calcifications and epilepsy. *Brain Dev.* **14**, 23–29.
9. Gobbi, G., Bouquet, F., Greco, L., Lambertini, A., Tassinari, C. A., Ventura, A., and Zaniboni, M. G. (1992). Coeliac disease, epilepsy, and cerebral calcifications. The Italian Working Group on Coeliac Disease and Epilepsy. *Lancet* **340**, 439–443.
10. Bushara, K. O. (2005). Neurologic presentation of celiac disease. *Gastroenterology* **128(Suppl 1)**, S92–S97.
11. Kaplan, J. G., Pack, D., Horoupian, D., DeSouza, T., Brin, M., and Schaumburg, H. (1988). Distal axonopathy associated with chronic gluten enteropathy: a treatable disorder. *Neurology* **38**, 642–645.
12. Kelkar, P., Ross, M. A., and Murray, J. (1996). Mononeuropathy multiplex associated with celiac sprue. *Muscle Nerve* **19**, 234–236.
13. Chin, R. L., and Latov, N. (2005). Peripheral neuropathy and celiac dsease. *Curr. Treat. Options Neurol.* **7**, 43–48.
14. Hall, W. H. (1968). Proximal muscle atrophy in adult celiac disease. *Am. J. Dig. Dis.* **13**, 697–704.
15. Hardoff, D., Sharf, B., and Berger, A. (1980). Myopathy as a presentation of coeliac disease. *Dev. Med. Child Neurol.* **22**, 781–783.
16. Henriksson, K. G., Hallert, C., Norrby, K., and Walan, A. (1982). Polymyositis and adult coeliac disease. *Acta. Neurol. Scand.* **65**, 301–319.
17. Green, P. H., Alaedini, A., Sander, H. W., Brannagan, T. H. 3rd, Latov, N., and Chin, R. L. (2005). Mechanisms underlying celiac disease and its neurologic manifestations. *Cell Mol. Life Sci.* **62**, 791–799.
18. Cicarelli, G., Della Rocca, G., Amboni, M., Ciacci, C., Mazzacca, G., Filla, A., and Barone, P. (2003). Clinical and neurological adnormalities in adult celiac disease. *Neurol. Sci.* **24**, 311–317.
19. D'Amico, D., Rigamonti, A., Spina, L., Bianchi-Marzoli, S., Vecchi, M., and Bussone, G. (2005). Migraine, celiac disease, and cerebral calcifications: a new case. *Headache* **45**, 1263–1267.

20. Morello, F., Ronzani, G., and Cappellari, F. (2003). Migraine, cortical blindness, multiple cerebral infarctions and hypocoagulopathy in celiac disease. *Neurol. Sci.* **24**, 85–89.
21. Lossos, A., River, Y., Eliakim, A., and Steiner, I. (1995). Neurologic aspects of inflammatory bowel disease. *Neurology* **45**, 416–421.
22. Elsehety, A., and Bertorini, T. E. (1997). Neurologic and neuropsychiatric complications of Crohn's disease. *South. Med. J.* **90**, 606–610.
23. Schilsky, M. L. (2001). Treatment of Wilson's disease: what are the relative roles of penicillamine, trientine, and zinc supplementation? *Curr. Gastroenterol. Rep.* **3**, 54–59.
24. Ghosh, S., Goldin, E., Gordon, F. H., Malchow, H. A., Rask-Madsen, J., Rutgeerts, P., Vyhnalek, P., Zadorova, Z., Palmer, T., Donoghue, S., Natalizumab Pan-European Study Group. (2003). Natalizumab for active Crohn's disease. *New Engl. J. Med.* **348**, 24–32.
25. Lew, E. A., and Stoffel, E. M. Natalizumab for active Crohn's disease. *New Engl. J. Med.* **348**, 1599; author reply, 1599.
26. Miller, D. H., Khan, O. A., Sheremata, W. A., Blumhardt, L. D., Rice, G. P., Libonati, M. A., Willmer-Hulme, A. J., Dalton, C. M., Miszkiel, K. A., O'Connor, P. W.; International Natalizumab Multiple Sclerosis Trial Group. (2003). A controlled trial of natalizumab for relapsing multiple sclerosis. *New Engl. J. Med.* **348**, 15–23.
27. Sandborn, W. J., Colombel, J. F., Enns, R., Feagan, B. G., Hanauer, S. B., Lawrance, I. C., Panaccione, R., Sanders, M., Schreiber, S., Targan, S., van Deventer, S., Goldblum, R., Despain, D., Hogge, G. S., Rutgeerts, P., International Efficacy of Natalizumab as Active Crohn's Therapy (ENACT-1) Trial Group; Evaluation of Natalizumab as Continuous Therapy (ENACT-2) Trial Group. (2005). Natalizumab induction and maintenance therapy for Crohn's disease. *New Engl. J. Med.* **353**, 1912–1925.
28. von Andrian, U. H., and Engelhardt, B. (2003). Alpha4 integrins as therapeutic targets in autoimmune disease. *New Engl. J. Med.* **348**, 68–72.
29. Alvarez-Cermeno, J. C., Masjuan, J., and Villar, L. M. (2005). Progressive multifocal leukoencephalopathy, natalizumab, and multiple sclerosis. *New Engl. J. Med.* **353**, 1744–1746; author reply, 1744–1746.
30. Berger, T., and Deisenhammer, F. (2005). Progressive multifocal leukoencephalopathy, natalizumab, and multiple sclerosis. *New Eng. J. Med.* **353**, 1744–1746; author reply, 1744–1746.
31. Drazen, J. M. (2005). Patients at risk. *New Engl. J. Med.* **353**, 417.
32. Kleinschmidt-DeMasters, B. K., and Tyler, K. L. (2005). Progressive multifocal leukoencephalopathy complicating treatment with natalizumab and interferon beta-1a for multiple sclerosis. *New Engl. J. Med.* **353**, 369–374.
33. Langer-Gould, A., Atlas, S. W., Green, A. J., Bollen, A. W., and Pelletier, D. (2005). Progressive multifocal leukoencephalopathy in a patient treated with natalizumab. *New Engl. J. Med.* **353**, 375–381.
34. Van Assche, G., Van Ranst, M., Sciot, R., Dubois, B., Vermeire, S., Noman, M., Verbeeck, J., Geboes, K., Robberecht, W., and Rutgeerts, P. (2005). Progressive multifocal leukoencephalopathy after natalizumab therapy for Crohn's disease. *New Engl. J. Med.* **353**, 362–368.
35. Skeen, M. B. (2002). Neurologic manifestations of gastrointestinal disease. *Neurol. Clin.* **20**, 195–225, vii.
36. Whipple, G. (1907). A hitherto undescribed disease characterized anatomically by deposits of fat and fatty acids in the intestinal and mesenteric lymphatic tissues. *Bull. Johns Hopkins Hosp.* **18**, 382–391.
37. Bai, J. C., Mazure, R. M., Vazquez, H., Niveloni, S. I., Smecuol, E., Pedreira, S., and Maurino, E. (2004). Whipple's disease. *Clin. Gastroenterol. Hepatol.* **2**, 849–860.
38. Adams, M., Rhyner, P. A., Day, J., DeArmond, S., and Smuckler, E. A. (1987). Whipple's disease confined to the central nervous system. *Ann. Neurol.* **21**, 104–108.
39. von Herbay, A., Ditton, H. J., Schuhmacher, F., and Maiwald, M. (1997). Whipple's disease: staging and monitoring by cytology and polymerase chain reaction analysis of cerebrospinal fluid. *Gastroenterology* **113**, 434–441.
40. Marth, T., and Raoult, D. (2003). Whipple's disease. *Lancet* **361**, 239–246.
41. Nachamkin, I., Allos, B. M., and Ho, T. (1998). *Campylobacter* species and Guillain-Barré syndrome. *Clin. Microbiol. Rev.* **11**, 555–567.
42. von Wulffen, H., Hartard, C., and Scharein, E. (1994). Seroreactivity to *Campylobacter jejuni* and gangliosides in patients with Guillain-Barré syndrome. *J. Infect. Dis.* **170**, 828–833.
43. Hughes, R. A., and Cornblath, D. R. (2005). Guillain-Barré syndrome. *Lancet* **366**, 1653–1666.
44. Ponka, P. (2004). Hereditary causes of disturbed iron homeostasis in the central nervous system. *Ann. NY Acad. Sci.* **1012**, 267–281.
45. Hartmann, I. J., Groeneweg, M. Quero, J. C., Beijeman, S. J., de Man, R. A., Hop, W. C., and Schalm, S. W. (2000). The prognostic significance of subclinical hepatic encephalopathy. *Am. J. Gastroenterol.* **95**, 2029–2034.
46. Ratnakumari, L., and Murthy, C. R. (1993). Response of rat cerebral glycolytic enzymes to hyperammonemic states. *Neurosci. Lett.* **161**, 37–40.
47. Kala, G., and Hertz, L. (2005). Ammonia effects on pyruvate/lactate production in astrocytes—interaction with glutamate. *Neurochem. Int.* **47**, 4–12.
48. Groeneweg, M., Moerland, W., Quero, J. C., Hop, W. C., Krabbe, P. F., and Schalm, S. W. (2000). Screening of subclinical hepatic encephalopathy. *J. Hepatol.* **32**, 748–753.
49. Groeneweg, M., Quero, J. C., De Bruijn, I., Hartmann, I. J., Essink-bot, M. L., Hop, W. C., and Schalm, S. W. (1998). Subclinical hepatic encephalopathy impairs daily functioning. *Hepatology* **28**, 45–49.
50. Kauppinen, R., and von und zu Fraunberg, M. (2002). Molecular and biochemical studies of acute intermittent porphyria in 196 patients and their families. *Clin. Chem.* **48**, 1891–1900.
51. Solis, C., Martinez-Bermejo, A., Naidich, T. P., Kaufmann, W. E., Astrin, K. H., Bishop, D. F., and Desnick, R. J. (2004). Acute intermittent porphyria: studies of the severe homozygous dominant disease provides insights into the neurologic attacks in acute porphyrias. *Arch. Neurol.* **61**, 1764–1770.
52. Mela, M., Mancuso, A., and Burroughs, A. K. (2003). Review article: pruritus in cholestatic and other liver diseases. *Aliment Pharmacol. Ther.* **17**, 857–870.
53. Bergasa, N. V. (2003). Pruritus and fatigue in primary biliary cirrhosis. *Clin. Liver Dis.* **7**, 879–900.
54. Bergasa, N. V. (2004). An approach to the management of the pruritus of cholestasis. *Clin. Liver Dis.* **8**, 55–66, vi.
55. Brewer, G. J., and Askari, F. K. (2005). Wilson's disease: clinical management and therapy. *J. Hepatol.* **42(Suppl)**, S13–S21.
56. Connor, B., and Landzberg, B. (2004). Persistent travelers' diarrhea. *In:* "Travel Medicine" (Keystone, J. et al., eds.), pp. 502–515. Mosby, St. Louis.
57. Iyer, G. V., Taori, G. M., Kapadia, C. R., Mathan, V. I, and Baker, S. J. (1973). Neurologic manifestations in tropical sprue. A clinical and electrodiagnostic study. *Neurology* **23**, 959–966.
58. King, J., and Keen, C. L. (2000). Zinc. *In:* "Modern Nutrition in Health and Disease" (M. Shils, et al., eds.), pp. 223–235. Lippincott, Philadelphia.
59. Clemons, T. E., Kurinij, N., and Sperduto, R. D. (2004). Associations of mortality with ocular disorders and an intervention of high-dose antioxidants and zinc in the Age-Related Eye Disease Study: AREDS Report No. 13. *Arch. Ophthalmol.* **122**, 716–726.

64

Neurosarcoidosis

John J. Caronna, MD

Keywords: *central nervous system, cranial neuropathies, involvement, meningitis, myopathy, peripheral neuropathy, sarcoidosis*

I. Neurosarcoidosis
II. Neurosarcoidosis: Sites of Involvement
III. Conclusion
   References

I. Neurosarcoidosis

Sarcoidosis is a multisystem granulomatous disease of unknown etiology that most commonly affects young adults. Most patients who have sarcoidosis are asymptomatic, and the disease often is detected on a routine chest radiograph. Presenting symptoms include fatigue, fever, weight loss, cough, shortness of breath, and arthritis. The lungs are affected most often, but the skin, eyes, central and peripheral nervous systems, heart, kidneys, bones, joints, and muscles also may be affected. When sarcoidosis affects the central nervous system (CNS) or peripheral nervous system (PNS), it is referred to as neurosarcoidosis [1]. Sarcoidosis of the CNS is uncommon.

The peripheral neuropathies are well recognized as a part of systemic sarcoidosis, especially in patients with clinical manifestations of sarcoidosis elsewhere, such as uveoparotid syndrome. Sarcoidosis of the CNS is an uncommon but severe, and sometimes life-threatening, manifestation of the disease. The overall incidence of neuromuscular involvement in patients with systemic sarcoidosis is 5% [2–4]. Autopsy studies, however, suggest that the diagnosis is made antemortem in only 50% of patients with neurosarcoidosis [5]. Sarcoidosis isolated to the CNS, that is, without evidence of systemic involvement, is rare.

A. Incidence

In the United States, the prevalence of sarcoid varies widely but generally is more common in African Americans, and the prevalence is much higher in the southeastern United States than the rest of the country. In New York City, the prevalence is 30 per 100,000. In the Puerto Rican population, a retrospective study found prevalence as high as 175 per 100,000. In Los Angeles, Mexican immigrants constituted 7% of the observed cases compared with 82% of African Americans. In Europe, the incidence of sarcoid has been reported to be from 0.04 per 100,000 in Spain to 64 per 100,000 in Sweden. The prevalence in London was found to be 27 per 100,000 for patients born in the United Kingdom and 200 per 100,000 for those born in the West Indies [6]. Sarcoidosis is extremely rare in China and Southeast Asia. There have been a few reports from Africa and from Japan. Sarcoidosis is slightly more common in women than in men [7]. The age at onset of neurological sarcoidosis has ranged from 3 months to the eighth decade but commonly occurs in adults aged 25 to 50 years [8]. The mean age at diagnosis is higher for patients with neurosarcoidosis than for those with systemic sarcoidosis as a presenting symptom. The mean age at diagnosis is higher in women than in men. Sarcoid disease in childhood has a different clinical manifestation than in adulthood and is characterized by cutaneous nodules, arthritis, and uveitis. The rate of ocular involvement in children has been reported at 100% in the presence of iritis and/or anterior vitreitis. Children older than 8 years have a clinical picture similar to adults.

The neuromuscular manifestations of sarcoidosis are polymorphic and can develop at different times in the course of the systemic illness. Neurological signs are the first indication of sarcoidosis in about one third of patients. The mode of onset of neuromuscular symptoms is variable but overall is usually subacute or chronic. The myopathy of sarcoidosis, however, is slowly progressive in all cases. There are racial differences in the frequency of the neurological manifestations of sarcoidosis: whites have a lower frequency of cranial nerve palsies than do blacks. Whites also have a higher frequency of peripheral neuropathy and myopathy but a lower incidence of CNS involvement and meningitis than blacks.

B. Cause

The cause of sarcoidosis is unknown [9]. Current evidence suggests that active sarcoid disease results from an exaggerated cellular immune response to either foreign or self-antigens. The T helper cells proliferate, resulting in an exaggerated response that leads to pathology. The T helper lymphocytes undergo differentiation to a T helper 1-type cell under the influence of interleukin 4 (IL4) and the costimulator CD28. The T helper 1 cell induces IL2 and interferon-γ (IFN-γ) on the macrophages, followed by a cascade of chemotactic factors that promote formation of granulomas. IFN-γ increases the expression of major histocompatibility complex II on macrophages and activated macrophage receptors carry an Fc receptor of immunoglobulin G (IgG) that potentiates their phagocytotic function. The result is tissue destruction and granuloma formation.

Several hypotheses have been proposed to explain the mechanism of the exaggerated immune response. One hypothesis is that a persistent antigen, either foreign or self, triggers the T helper cell response. Another theory is that the response of the suppressor arm of the immune response is inadequate, preventing T helper cells from shutting down. A third possibility is that an inherited or acquired genetic difference in response genes leads to the exaggerated response.

In addition to the cellular immune response, active sarcoidosis is characterized by hyperglobulinemia with antibodies against several infectious agents, including mycobacteria and tuberculosis, as well as IgM anti-T cell antibodies. The presence of these antibodies may be due to a nonspecific polyclonal stimulation of B cells by activated T cells at the site, and it is not thought that these antibodies play a role in disease pathogenesis [4,5].

C. Pathology

The histology of neurosarcoidosis is characterized by the formulation of granulomas in the CNS [10]. The lesion consists of a collection of lymphocytes and mononuclear phagocytes with frequent giant cells surrounding a noncaseating, nonnecrotizing epithelioid cell granuloma. Individual granulomas are visible to the naked eye and resemble the lesions of miliary tuberculosis. Sarcoidosis of the CNS, therefore, was at first difficult to distinguish from tuberculosis. Although the individual granulomas of neurosarcoid can coalesce to form large masses, in distinction to tuberculosis, caseation does not occur. However, necrosis has been reported in sarcoid lesions; therefore, the distinction from the caseating granuloma of tuberculosis is not always clear.

In the leptomeninges, granulomas are thickly scattered and the arachnoid is opaque and infiltrated by lymphocytes. A dense adhesive arachnoiditis can occur around the spinal cord. The sites of predilection for sarcoid granulomas are the basal cisterns in the optic chiasm, but sarcoid meningitis can develop anywhere over the surface of the brain or spinal cord and be either diffusely or sharply localized. Granulomas also are common in the ependymal lining of the ventricles and the choroid plexus [11].

In the parenchyma of the brain, sarcoid granulomas can occur anywhere within the brain, spinal cord, or optic nerves. They tend to cluster densely in periventricular areas or in the Virchow-Robin spaces. Granulomatous masses up to several centimeters in diameter can form over the convexity of the brain or within the ventricles or basal cisterns, or they may compress the spinal cord. The region of the hypothalamus and the pituitary gland are common sites for tumor formation in neurosarcoidosis.

Perivascular granulomas, particularly in the meninges, may invade the arterial walls, producing occlusion of blood vessels and multiple cerebral infarcts. Veins may be similarly affected. The arterial lesions are mainly in small vessels and resemble those of giant cell angiitis [12].

The pathology of sarcoidosis in the PNS is less well established, and some cases of sarcoid polyneuritis have not been associated with infiltration of nerve trunks on biopsy or at autopsy. The pathology of cranial nerve palsies has not been established because such patients rarely come to autopsy or surgery [13,14]. Likewise, the pathogenesis of transient symptoms involving either the CNS or the PNS has not been established.

In systemic sarcoidosis, granulomas can be present in skeletal muscle. However, the incidence of symptomatic involvement of muscles in sarcoidosis is low [15–17].

II. Neurosarcoidosis: Sites of Involvement

In 1948, Colover [18] collected the published information on 115 cases of neurosarcoidosis and added three cases of his own. Subsequently there have been numerous reports of neurosarcoidosis. In 1990, Chapelon and associates [19] reported 35 patients with neurosarcoidosis whose diagnoses had been proven histologically by biopsy of various tissues. Of the 35 patients, 6, or 17%, had neurological symptoms as the only clinical manifestation of sarcoidosis, despite an average follow-up period of more than 100 months. It may be, however, that the corticosteroid treatment that these patients received prevented the identification of extraneurological lesions.

The extraneurological manifestations of neurosarcoidosis in this series of patients are listed in Table 1, and the CNS manifestations of neurosarcoidosis in these patients are presented in Table 2.

The clinical presentation and sites of involvement have been categorized by Matthews [1] as described in the next sections.

Table 1 Extraneurologic Manifestations in 35 Patients

Manifestation	%
Constitutional symptoms (18:35)	51
Fatigue	37
Fever > 38° C	23
Weight loss > 2 kg	20
Intrathoracic lesions	71
Skin nodules	29
Peripheral lymph nodes	26
Splenomegaly	6
Parotids	17
Cardiac symptoms	14
Hepatomegaly	9
Uveitis	9
Exophthalmos	6
Other (e.g., Sicca syndrome)	26

*From Chapelon, C., Ziza, J. M., Piette, J. C., Levy, Y., Raquin, G., Wechsler, B., Bitker, M. O., Bletry, O., Laplane, D., Bousser, M. D., and Godeau, P. (1990). Neurosarcoidosis: signs, course and treatment in 35 confirmed cases. *Medicine* **69**, 261–274.

Table 2 CNS Manifestations in Six Patients with Isolated Neurosarcoidosis

Manifestation	n
Seizures	5
Babinski's response	5
Psychiatric symptoms	3
Diabetes insipidus	3
Headache	2
Ataxia	2
Amenorrhea	2
Vertigo	1
Sphincter dysfunction	1
Optic nerve involvement	1
Galactorrhea	1

*From Chapelon, C., Ziza, J. M., Piette, J. C., Levy, Y., Raquin, G., Wechsler, B., Bitker, M. O., Bletry, O., Laplane, D., Bousser, M. D., and Godeau, P. (1990). Neurosarcoidosis: signs, course and treatment in 35 confirmed cases. *Medicine* **69**, 261–274.

A. Skull and Vertebrae

Asymptomatic lytic lesions of the skull in sarcoid are discovered accidentally and usually do not extend intracranially, although subdural sarcoid infiltration mimicking a subdural hematoma has been reported. Lytic or sclerotic lesions of the spinal column may cause pain, pathological fracture, and even paraplegia.

Sarcoidosis of bone is usually seen in the presence of chronic disease in several organ systems. There is about a 5% frequency of radiologic evidence of bone involvement in patients with sarcoid. The roentographic manifestations vary with the region of the skeleton affected. Generalized osteopenia has been observed, and characteristically the small bones of the hands, and to a lesser extent those of the feet and the wrists, are involved. Lesions of the skull and vertebral bodies are less common. The lesions have a characteristic "punched out" appearance. Less typically, there is a localized or generalized osteosclerosis. In rare cases, there is a widespread increase in skeletal radiodensity resembling the changes of Paget's disease, fluorosis, or skeletal metastases such as those from prostate carcinoma.

B. Cranial Nerve Palsies

CN I: Anosmia has been reported in a number of patients with neurosarcoidosis, but in some patients it may be due to a direct involvement of the nasal mucosa rather than an intracranial cause [20].

CN II: The optic nerve may be involved by sarcoidosis in various ways: The nerve head may swell during an attack of optic neuritis. Increased intracranial pressure can cause papilledema; the optic nerve may atrophy due to intracranial compression, and the nerve may be invaded by sarcoid granulomas. The signs of optic nerve or chiasmal involvement in sarcoidosis occur in the appropriate clinical contexts and therefore do not present any distinctive syndrome.

CN III, IV, VI: External ophthalmoplegia due to raised intracranial pressure, invasion of the orbit, or intracranial disease has been reported. Diplopia due to partial third nerve palsy has been reported in the presence of sarcoid lesions of the midbrain. Unilateral and sixth nerve palsies have been reported as presenting symptoms and chronic sarcoid meningoencephalitis, internal ophthalmoplegia, or abnormalities of the pupil have been reported in some cases of neurosarcoidosis, but these abnormalities were possibly related to iridocyclitis.

CN V: Paresthesias of facial numbness have been reported in cases of sarcoid cranial polyneuritis.

CN VII: Paralysis of the seventh cranial nerve is the most common single sign of sarcoidosis of the CNS and occurs at some stage of the disease in more than a third of cases. The facial palsy of neurosarcoidosis is indistinguishable from idiopathic Bell's palsy. Bilateral palsy is common in neurosarcoidosis, but the two sides are not affected simultaneously. It has been stated that, in facial palsy due to sarcoidosis, taste often is lost but hyperacusis is rare; the opposite is true in patients with idiopathic Bell's palsy. In cases of sarcoid with parotid gland involvement, the enlarged parotid gland causes compression of the facial nerve. In these cases, the facial palsy is not accompanied by any disorder of hearing or taste.

CN VIII: Deafness or vestibular function is commonly involved in neurosarcoidosis. In one series of 19 cases, the usual presentation in patients with neurosarcoid cranial polyneuritis was of progressive hearing loss combined with the absence or reduction of coloric vestibular reactions. In a provocative paper [21], Drake suggested that the traditional explanation for Beethoven's deafness, namely, otosclerosis, is wrong and that sarcoidosis could explain many features of Beethoven's medical history, particularly his deafness. Sarcoidosis, although not identified in the nineteenth century, could explain his progressive deafness and many of his gastrointestinal, rheumatic, and thoracic symptoms [14].

CN IX–XII: The lower cranial nerves frequently are involved in sarcoid cranial polyneuritis, and isolated bulbar palsy has been reported.

C. Neurosarcoid of the Peripheral Nervous System

Sarcoidosis of the spinal nerves, nerve roots, and peripheral nerves can take the form of a single nerve palsy, multiple nerve palsies, or a symmetrical polyneuropathy [22]. The characteristic feature of this minor nerve involvement by sarcoid is loss of sensation, which is often accompanied by pain over large areas of the trunk. Mononeuritis multiplex can occur without cranial neuropathy, and patients have been reported with neuropathy with one or both median nerves, phrenic nerve paralysis, and loss of sensation in the distribution of thoracic or abdominal nerves. A symmetrical polyneuropathy that resembles the Guillain-Barré syndrome may occur but is relatively uncommon

D. Central Nervous System Neurosarcoidosis

CNS involvement in patients with sarcoidosis is indicated by the presence of features such as seizures, ataxia, cognitive disturbances, endocrine changes, and abnormal findings on neurological examination such as hemianopsia and extensor plantar responses. Lesions include mass lesions that mimic metastatic tumors, hydrocephalus due to obstruction of subarachnoid pathways for reabsorption of spinal fluid, meningoencephalitis, and hypothalamic or pituitary gland involvement. The most intracranial sites of involvement by sarcoidosis are the third ventricular region, the hypothalamus, and the pituitary gland. Patients who have pituitary and hypothalamic involvement often have presenting symptoms of diabetes insipidus, such as polyuria, polydipsia, and hyperglycemia.

Sarcoid granulomas occur in the pituitary gland, and female patients with sarcoidosis can have galactorrhea and amenorrhea.

Sarcoidosis affecting the brain can appear as a single or multiple space-occupying lesion(s) that is/are indistinguishable clinically from intracranial tumors. Neurosarcoidosis of the meninges causes hydrocephalus by obstructing outlet of cerebrospinal fluid (CSF) from the fourth ventricle without signs of focal brain damage. The course can be chronic or acute. The aqueduct of Silvius may be partially obstructed by a granulomatous mass lesion in the tegument of the midbrain or by multiple granulomas in the parenchyma. Sarcoid masses also can occur within the third ventricle. Patients with cerebral sarcoid may have hydrocephalus and increased intracranial pressure without focal neurological signs.

1. Spinal Cord

The spinal cord can be involved by intra- or extramedullary granulomas and can be the site of multiple parenchymatous granulomas. It can be involved by adhesive arachnoiditis or damaged by ischemic infarction secondary to sarcoid-induced vasculitis. The clinical presentation of sarcoidosis affecting the spinal cord is one of progressive paraplegia.

E. Sarcoid Myopathy

The incidence of symptomatic involvement of the muscles in sarcoidosis is low. The nature of the pathological process by which myopathic symptoms are produced is controversial. In some cases the granulomas are large and may have a local mass effect to compress or displace muscle fibers. In other cases there is evidence of a granulomatous polymyositis; namely, there is muscle fiber atrophy around sarcoid granulomas but no evidence of disease in the remainder of the muscle. The most common presentation of sarcoid myopathy is a slowly progressive, symmetrical, mainly proximal weakness and wasting of limb girdle muscles in middle-aged or elderly women. In some cases, dysphagia may be a prominent or presenting symptom; in others, the myopathy appears as an acute myositis with the rapid onset of fever, arthralgias, severe muscle pain, and tenderness, but without weakness.

III. Conclusion

Sarcoidosis is a multisystem granulomatous disease of unknown etiology associated with several immunological abnormalities. Neurological manifestations in sarcoidosis are uncommon, and isolated neurosarcoidosis is rare. Therefore, each patient with sarcoidosis or with unexplained neurological signs and symptoms requires a thorough and accurate diagnostic evaluation because, left untreated, the extraneurological and neurological impairments caused by sarcoidosis can be severe.

References

1. Matthews, W. B. (1979). Neurosarcoidosis. *In:* "Handbook of Clinical Neurology" (P. J. Vinken and G. W. Bruyn, eds.), vol. 38, pp. 521–542. Elsevier/North-Holland Biomedical Press, Amsterdam.
2. Delane, P. (1977). Neurologic manifestations of sarcoidosis. *Ann. Intern. Med.* **87**, 336–345.
3. Newman, L. S., Rose, C. S., and Maier, L. A. (1997). Sarcoidosis. *N. Engl. J. Med.* **336**, 1224–1234.
4. Gullapalli, D., and Philips, L. H. (2002). Neurologic manifestations of sarcoidosis. *Neurol. Clin.* **20**, 59–83.
5. Iwai, K., Tachibana, T., Takemura, T., Matsui, Y., Kitaichi, M., and Kawabata, Y. (1993). Pathological studies on sarcoidosis autopsy: I. Epidemiological features of 320 cases in Japan. *Acta Pathol. Jpn.* **43**, 372–376.
6. Zajicek, J. P., Scolding, N. J., Foster, O., Rovaris, M., Evanson, J., Moseley, I. F., Scadding, J. W., Thompson, E. J., Chamoun, V., Miller, D. J., McDonald, W. I., and Mitchell, D. (1999). Central nervous system sarcoidosis: diagnosis and management. *Q. J. Med.* **92**, 103–117.
7. Wiederholt, W. C., and Siekert, R. C. (1965). Neurological manifestations of sarcoidosis. *Neurology* **15**, 1147–1154.
8. Herring, A. B., and Urich, H. (1969). Sarcoidosis of the central nervous system. *J. Neurol. Sci.* **9**, 405–422.
9. Barnard, J., and Newman, L. S. (2001). Sarcoidosis: immunology, rheumatic involvement and therapeutics. *Curr. Op. Rheumatol.* **13**, 84–91.
10. Said, G., Lacroix, D., Plante-Bordeneuve, V., Le Page L, Pico F, Presles O, Senant J, Remy P, Rondepierre P, Mallecourt, J. (2002). Nerve granulomas and vasculitis in sarcoid peripheral neuropathy: a clinicopathological study of 11 patients. *Brain* **125**, 264–275.
11. Nowak, D. A., and Widenka, D. C. (2001). Neurosarcoidosis: a review of its intracranial manifestation. *J. Neurol.* **248**, 363–372.
12. Reske-Nielsen, E., and Hermsen, A. (1962). Periangiitis and panangiitis as a manifestation of sarcoidosis of the brain. *J. Nerv. Ment. Dis.* **135**, 399–412.
13. Quinones-Hinogosa, A., Chang, E. F., Khan, S. A., McDermott, M. W. (2003). Isolated trigeminal nerve sarcoid granuloma mimicking trigeminal schwannoma: case report. *Neurosurgery* **52**, 700–705.
14. Tobias, S., Prayson, R. A., and Lee, J. H. (2002). Necrotizing neurosarcoidosis of the cranial base resembling enplaque sphenoid wing meningioma: case report. *Neurosurgery* **51**, 1290–1294.
15. Silverstein, A., and Siltzback, L. E. (1969). Muscle involvement in sarcoidosis: asymptomatic myositis and myopathy. *Arch. Neurol.* **21**, 235–241.
16. Thorpe-Thorpe, C. (1972). Muscle weakness due to sarcoid myopathy. *Neurology* **22**, 917–928.
17. Berger, C., Sommer, C., and Meinck, H. M. (2002). Isolated sarcoid myopathy. *Muscle Nerve* **26**, 553–556.
18. Colover, J. (1948). Sarcoidosis with involvement of the nervous system. *Brain* **71**, 451–475.
19. Chapelon, C., Ziza, J. M., Piette, J. C., Levy, Y., Raquin, G., Wechsler, B., Bitker, M. O., Bletry, O., Laplane, D., Bousser, M. D., and Godeau, P. (1990). Neurosarcoidosis: signs, course and treatment in 35 confirmed cases. *Medicine* **69**, 261–274.
20. Scott, T. F. (1993). Neurosarcoidosis: progress and clinical aspects. *Neurology* **43**, 8–12.
21. Drake, M. E., Jr. (1994). Deafness, dysesthesia, depression, diarrhea, dropsy, and death: the case for sarcoidosis in Ludwig van Beethoven. *Neurology* **44**, 562–565.
22. Zuniga, G., Ropper, A. H., and Frank, J. (1991). Sarcoid peripheral neuropathy. *Neurology* **41**, 1558–1563.

65

REM Sleep Behavior Disorder

Carlos H. Schenck, MD
Mark W. Mahowald, MD

Keywords: *dreaming, narcolepsy-cataplexy, parasomnia, parkinsonism, REM sleep behavior disorder*

I. Brief History and Nomenclature
II. Etiology
III. Pathogenesis
IV. Pathophysiology
V. Pharmacology, Biochemistry, and Molecular Mechanisms
VI. Explanation of Symptoms in Relation to Pathophysiology
VII. Natural History
    References

I. Brief History and Nomenclature

Rapid eye movement (REM) sleep behavior disorder (RBD) is a multifaceted motor, behavioral, and experiential disorder of REM sleep in which the affected person (usually a middle-aged or older male) typically enacts in his sleep the stereotypically altered dreams that have intruded on him, featuring confrontation, aggression, and violence, but lacks any daytime tendency for increased aggressiveness [1]. The vigorous and violent behaviors of RBD commonly result in injury, which at times can be severe and even life-threatening. The core electromyographic (EMG) abnormalities of RBD consist of intermittent loss of the usual skeletal muscle atonia of REM sleep (REM atonia), with increased muscle tone and/or excessive phasic muscle twitching during REM sleep [1,2], as shown in Figure 1. Periodic limb movements and nonperiodic twitching during nonrapid eye movement (NREM) sleep are common, indicating generalized REM and NREM sleep motor dyscontrol in RBD [1–3]. There is increased slow-wave sleep (SWS) for age and increased electroencephalogram (EEG) delta power across sleep in RBD [1,4]. During wakefulness,

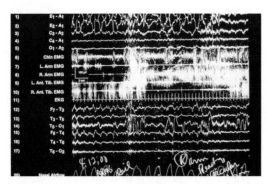

Figure 1 Polygraphic Correlates of Nocturnal Dream-Enacting Behavior. Rapid eye movement (REM) sleep contains dense, high-voltage REM activity (1–2) and an activated electroencephalogram (3–5, 12–17). The electrocardiogram (11) has a constant rate of 64 per minute, despite vigorous limb movements. Chin (submental) electromyographic (EMG) tone is augmented with phasic accentuations (6). Arms (7–8) and legs (9–10) show aperiodic bursts of intense EMG twitching, which accompany gross behavior noted by the technician. This sequence culminates in a spontaneous awakening, when the patient reports a dream in which his movements match his behavior in the sleep lab as observed by the sleep technologist.

Table 1 Diagnostic Criteria for RBD[a]

A. Presence of REM sleep without atonia: the EMG finding of excessive amounts of sustained or intermittent elevation of submental EMG tone or excessive phasic submental or (upper or lower) limb EMG twitching

B. At least one of the following is present:
   i. Sleep related injurious, potentially injurious, or disruptive behaviors by history
   ii. Abnormal REM sleep behaviors documented during polysomnographic monitoring

C. There is no EEG epileptiform activity during REM sleep unless RBD can be clearly distinguished from any concurrent REM sleep-related seizure disorder

D. The sleep disturbance is not better explained by another sleep disorder, medical or neurological disorder, mental disorder, medication use, or substance use disorder

[a]From American Academy of Sleep Medicine. (2005). "International classification of sleep disorders: diagnostic and coding manual," 2nd edition. American Academy of Sleep Medicine, Westchester, IL.

olfactory dysfunction [5], cortical EEG slowing [6,7], visuospatial constructional dysfunction, and visuospatial memory impairment on neuropsychological testing [8] have been documented in idiopathic and symptomatic RBD. Therefore, RBD is a global disorder of REM sleep, NREM sleep, and wakefulness and is strategically situated at an important crossroads of the neurosciences and clinical medicine [1].

RBD was first recognized as a distinct category of parasomnia in 1985 and was named in 1987 [1]. In 1990, RBD was included within the official International Classification of Sleep Disorders (ICSD), with diagnostic criteria being established [1]. In the revision of the sleep nosology, ICSD-2, RBD is the only 1 of 12 primary parasomnias that requires polysomnographic documentation for its diagnosis. Table 1 contains the diagnostic criteria. RBD in humans is usually a chronic condition that is either idiopathic or symptomatic of a neurological disorder, most commonly a parkinsonian neurodegenerative disorder, narcolepsy, or stroke [1,9].

RBD occurs naturally across mammalian species, and there is an experimental animal model of RBD induced by dorsal pontine tegmental lesions in cats [1]. There is a close correspondence among the categories of behaviors released during REM sleep in humans and in lesioned cats with RBD: minimal limb twitching and jerking, orientation, exploration, and attack behavior. (Talking, shouting, singing, and laughing during REM sleep can also occur in human RBD.) Also, the categories of behaviors not found in REM sleep in humans with RBD are those categories not found in lesioned cats with RBD: sexual, feeding, or grooming behaviors; micturition; and defecation. Despite the prominent tonic and phasic motor abnormalities during REM sleep, all other features of REM sleep remain intact in RBD, such as latency to REM sleep, REM sleep as a percentage of total sleep time, the number of REM sleep periods, and the customary REM sleep–NREM sleep cycling.

The nomenclature of RBD includes these most commonly used terms: acute RBD, chronic RBD, idiopathic RBD, symptomatic RBD, and subclinical RBD.

II. Etiology

RBD can be an acute or chronic disorder [2]. Acute RBD is primarily found in drug withdrawal or drug intoxication states and is generally a reversible condition [2]. Chronic RBD can be an idiopathic or symptomatic disorder that generally persists without any eventual remission and therefore requires ongoing pharmacotherapy. In three large series of RBD patients, involving 96, 93, and 52 patients, respectively, idiopathic RBD was the initial diagnosis after clinical and polysomnographic evaluations in 42%, 43%, and 25% of the three series [1,9]. Central nervous system (CNS) disorders causally related to RBD onset were diagnosed in 48%, 57%, and 75% of patients in the three series. RBD has been linked with most categories of CNS disorders [2], as listed in Table 2. In an autopsy study of 15 men with RBD and a neurodegenerative disorder, the pathological diagnosis was Lewy body disease in 12 men and multiple system atrophy (MSA) in 3 men [10]. These findings suggest that when associated with dementia and/or parkinsonism RBD usually predicts an underlying synucleinopathy. Finally, all of the disorders listed in Table 2 can also manifest as subclinical RBD, in which the polysomnographic markers of RBD are present but there is no clinical

Table 2 Neurological Disorders Associated with Chronic RBD

Category	Etiology
Toxic-Metabolic	Tricyclic antidepressants
	Monoamine oxidase inhibitors
	Serotonin reuptake inhibitors
	Venlafaxine
	Mirtazapine
	Selegiline
Vascular	Subarachnoid hemorrhage
	Vasculitis
	Ischemic
Tumor	Acoustic neuroma
	Pontine neoplasm
Infectious, postinfectious, autoimmune, degenerative	Guillain-Barré syndrome
	Multiple sclerosis
	Fatal familial insomnia
	Amyotrophic lateral sclerosis
	Potassium channel antibody-associated limbic encephalitis
	Normal pressure hydrocephalus
	Olivopontocerebellar degeneration
	Dementia with Lewy bodies
	Alzheimer's disease–Lewy body variant
	Corticobasal degeneration
	Parkinson's disease
	Multiple system atrophy
	Shy-Drager syndrome
	Progressive supranuclear palsy
Developmental, congenital, familial	Narcolepsy
	Tourette's syndrome
	Group A xeroderma pigmentosum
	Mitochondrial encephalomyopathy

parasomnia correlate. It is currently unknown to what extent patients with subclinical RBD will eventually develop frank clinical RBD.

Recent immunocytochemical analyses have revealed that Parkinson's disease (PD), dementia with Lewy bodies (DLB), and MSA share the similarity of α-synuclein-positive intracellular inclusions, and these disorders are now considered collectively as the *synucleinopathies*. There is growing evidence that RBD may be a sensitive and specific clinical marker for the synucleinopathies. For example, in a 7-year follow-up of a previously published series [11], 65% of men older than 50 years who were initially diagnosed with idiopathic RBD eventually developed a parkinsonian disorder at a mean interval of 13 years [12]. Brain imaging studies in idiopathic RBD using positron emission tomography (PET) and single-photon emission computed tomography (SPECT) have found a decreased striatal dopaminergic innervation and a reduced presynaptic striatal dopamine transporter binding in RBD, similar to what has been found with early PD [1,12].

One study detected a profound impairment of olfactory function in idiopathic RBD and in RBD associated with PD and with narcolepsy [5]. These new findings correlated with the neuropathological staging of PD (stages 1–3) proposed by Heiko Braak [5]: In stage 1, the anterior olfactory nucleus or olfactory bulb is affected, as is the dorsal motor nucleus of the glossopharyngeal and vagal nerves. In stage 2, additional lesions consistently remain confined to the medulla oblongata and pontine tegmentum, which are critical areas for generating RBD. In stage 3, there are midbrain lesions, in particular degeneration of dopaminergic neurons in the substantia nigra pars compacta. Therefore, idiopathic RBD patients with olfactory impairment might have stage 2 preclinical α-synucleinopathy.

Cortical CNS activity and reactivity is impaired in idiopathic RBD, during both wakefulness and sleep. Compared with controls, RBD patients have a higher theta power in the frontal, temporal, and occipital regions, with a lower beta power in the occipital region during wakefulness [6]. The entire mean EEG power spectrum in the occipital region appears to be shifted toward slower frequencies compared with that of control subjects. There are similarities between the topographic distribution of the EEG slowing in patients with idiopathic RBD and the pattern of perfusional and metabolic impairment observed in DLB and PD. Furthermore, EEG slowing documented during wakefulness in nondemented patients with PD is strongly related to the presence of RBD [7]. Only patients with PD-RBD had a slowing of the EEG and of the dominant occipital frequency. These findings lend support to a common etiopathogenesis between these conditions. Evidence of altered EEG reactivity in RBD is demonstrated by reduced EEG arousal responses associated with periodic limb movements of NREM sleep compared with controls [3].

In a study of 28 idiopathic RBD patients and 28 controls, the idiopathic RBD patients spent significantly more time in SWS compared with controls, and spectral analyses showed significantly increased all-night delta power in idiopathic RBD patients compared with controls [4]. The authors commented that the dopaminergic system has a strong negative interaction with the adenosine system. Adenosine administration increases SWS, and its increase in cholinergic basal forebrain increases delta spectral power during SWS. Considering the colocalization of adenosine A2 and dopamine D2 in the striatum, and the antagonistic effects of adenosine and dopamine, it is possible that the greater SWS and delta power in RBD patients resulted from increased adenosinergic activity, a consequence of decreased dopaminergic activity.

A neuropsychological study of idiopathic RBD patients found impairment, compared with controls, in visuospatial construction scores and in visuospatial learning scores that suggested a link with similar impairment found in DLB [8]. This finding, together with other abnormal CNS findings in idiopathic RBD, supports the growing belief that "cryptogenic" RBD is more accurate than "idiopathic" RBD [8].

Approximately 90% of patients with MSA have RBD [1], and 58% of patients with PD have "REM without atonia," including 33% with frank RBD [13]. Therefore, synucleinopathies are strongly interconnected with RBD. Nevertheless, in a recent series, 6 of 10 cases of parkinsonism with *parkin* mutations (Park2), that is, parkinsonism without synucleinopathy, had RBD [14]. Also, a recent study has shown that RBD and subclinical RBD are just as common in progressive supranuclear palsy (a tauopathy) as in PD (a synucleinopathy) [15]. Therefore, the location of the brain lesion in a neurodegenerative disorder appears to be more important than the type of lesion for induction of RBD.

A close association of RBD with narcolepsy-cataplexy and their treatments has been described in children, adolescents, and adults [1]. RBD can thus be regarded as an additional REM abnormality within the narcolepsy syndrome. This calls attention to the following inverse forms of motor dyscontrol in narcolepsy-RBD: episodic loss of muscle tone—cataplexy—in wakefulness and episodic increase in muscle tone and excessive muscle twitching in REM sleep.

III. Pathogenesis

Clinical or subclinical RBD can be induced reversibly or irreversibly by pharmacological, immunological, physiological, and/or anatomical interference with the structure or functioning of neuronal centers and/or connecting pathways subserving the two basic REM sleep motor circuits, in turn subserving REM atonia and phasic motor-behavioral activity and possibly dream-generating circuits [2].

RBD has several recognized variants, including the parasomnia overlap disorder, status dissociatus, and agrypnia excitata [2].

A. Parasomnia Overlap Disorder

A subgroup of RBD patients has been identified with polysomnographic-documented overlapping parasomnias consisting of sleepwalking, sleep terrors, and RBD [1]. These cases demonstrate motor-behavioral dyscontrol extending across NREM and REM sleep, in addition to the usual findings in RBD of increased periodic limb movements and nonperiodic limb movements during NREM sleep. In a series of 33 patients, the mean age of parasomnia onset was 15 years (range: 1–66) and 70% (n = 23) were males. An idiopathic subgroup (n = 22) had a significantly earlier mean age of parasomnia onset (9 years; range: 1–28) than a symptomatic subgroup (n = 11) (27 years; range: 5–66). In the symptomatic subgroup, 6 had parasomnia that began with neurological disorders (congenital Möbius syndrome, narcolepsy, multiple sclerosis, brain tumor [and treatment], brain trauma, indeterminate disorder [exaggerated startle response/atypical cataplexy]); 1 that began with nocturnal paroxysmal atrial fibrillation; 1 that began with posttraumatic stress disorder or major depression; 1 that began with chronic ethanol or amphetamine abuse and withdrawal; and 2 that began with mixed disorders (schizophrenia, brain trauma, substance abuse).

A family has been reported in which three adult first-degree relations were documented to have RBD, sleepwalking, sleep terrors, narcolepsy, periodic limb movements, and nonperiodic limb movements of NREM sleep in various combinations, thus revealing a broad spectrum of intrafamilial REM and NREM sleep motor dyscontrol [2]. An experimental animal model crucial for understanding parasomnia overlap disorder—in conjunction with the pontine-lesioned animal RBD model—consists of a cat model involving kainic acid injection into the midbrain reticular core [1]. Concurrent with an activated, wakeful EEG, there is an immediate onset of hallucinatory defense and attack behavior identical to the REM sleep behaviors elicited in pontine-lesioned cats.

B. Status Dissociatus

Status dissociatus [16] is the most extreme form of RBD and appears to represent the complete breakdown of state-determining boundaries. Clinically, these patients appear to be either awake or asleep; however, their "sleep" is atypical, characterized by frequent muscle twitching, vocalization, and reports of dreamlike mentation on spontaneous or forced awakening. Polygraphically, there are no features of either conventional REM or NREM sleep; rather, there is the simultaneous admixture of elements of wakefulness, REM sleep, and NREM sleep. "Sleep" is often perceived as "normal" and restorative, despite the nearly continuous motor and verbal behaviors and absence of polysomnographic-defined REM or NREM sleep. Conditions associated with status dissociatus include protracted withdrawal from alcohol abuse, narcolepsy, olivopontocerebellar degeneration, and prior open-heart surgery. An AIDS-related case with prominent brainstem involvement has been reported [2]. Similar signs and symptoms can be seen in "fatal familial insomnia," a prion disease associated with preferential thalamic degeneration.

C. Agrypnia Excitata

The recently described condition of agrypnia excitata is characterized by generalized overactivity associated with loss of SWS; oneirism (dream-enacting); wakeful oneirism (the inability to initiate and maintain sleep with wakeful dreaming); and by marked motor and autonomic sympathetic activation seen in such diverse conditions as delirium tremens, Morvan's fibrillary chorea, and fatal familial insomnia [2]. Oneiric dementia is likely a related condition.

RBD and narcolepsy are prototypic examples of state dissociation, underscoring the concept that the three states of being—wakefulness, NREM sleep, and REM sleep—are not mutually exclusive but rather may occur in various combinations, resulting in fascinating clinical experiences and behaviors. Clearly, states of being are not necessarily global brain phenomena. Sleep or wakefulness occurring asynchronously in "bits and pieces" of the brain is a useful concept for understanding this diverse and peculiar set of nocturnal behaviors.

IV. Pathophysiology

Animal research has demonstrated two separate motor systems involved in normal REM sleep: the tonic motor system, which generates the REM atonia, and the phasic motor system, including locomotor activity, which is predominantly suppressed during REM sleep [1]. The generators for both the tonic and the phasic events of REM sleep are located in the brainstem. REM atonia results from an active inhibitory pathway that originates in the perilocus coeruleus region in the pons, which sends an excitatory signal to the nucleus reticularis magnocellularis in the medulla via the lateral tegmentoreticular tract. The nucleus reticularis magnocellularis, in turn, actively inhibits spinal α-motoneurons via the ventrolateral reticulospinal tract [1].

In the cat model of RBD, the appearance of each behavioral category is dependent on the location and size of the pontine tegmental lesions [17]: (1) a minimal syndrome of generalized limb or truncal twitching and jerking, which can intermittently become prominent and violent; (2) orienting and exploratory behaviors, involving staring, head raising, head turning, grasping, and searching; (3) stalking imaginary prey and episodic attack behavior; and (4) locomotion. These animal experiments have revealed that loss of REM atonia is alone insufficient to generate RBD. There must presumably also be (or instead be) disinhibition of motor pattern generators in the mesencephalic locomotor region to result in phasic motor overactivation with behavioral release during REM sleep.

The pathophysiology of the characteristic dream disturbance in RBD may be explained by the "activation-synthesis" model of dream formation, which postulates that during REM sleep brainstem generators phasically activate motor, perceptual, affective, and cognitive circuits whose rostral flow is synthesized into dreams within the forebrain [2]. With RBD, intensified activity and/or biased activation of particular circuits originating in the brainstem would induce corresponding changes in dream process and content that would emerge isomorphically with the simultaneous acting-out of those dreams. In other words, this model would predict a common pathophysiological basis for the generation and the simultaneous physical display of altered dream activity.

V. Pharmacology, Biochemistry, and Molecular Mechanisms

RBD can be induced or aggravated by various psychotropic and other medications that alter CNS serotonergic, monoaminergic, and cholinergic neurotransmission, including serotonin-specific reuptake inhibitors, venlafaxine, mirtazapine, selegiline, tricyclic antidepressants, monoamine oxidase inhibitors, anticholinergics, and β-blockers [2]. Also, caffeine excess, including chocolate excess, can cause or aggravate RBD [2].

Striatal monoaminergic deficits, in a PET imaging study, have been found to be strongly correlated with the extent of loss of REM atonia in RBD associated with MSA [18]. This finding reflects degeneration of dopaminergic neurons in the substantia nigra and suggests that decreased nigrostriatal dopaminergic projections may contribute to RBD in MSA. Furthermore, in a related SPECT study, increased muscle activity during REM sleep correlated with a significant decrease of striatal presynaptic dopamine transporters across a descending severity gradient involving PD, RBD, and subclinical RBD [19].

The only published autopsy case of idiopathic RBD involved an 84-year-old man with a 20-year history of RBD confirmed by polysomnography [20]. At postmortem examination, histopathology revealed Lewy body disease with marked decrease of pigmented neurons in the locus coeruleus and substantia nigra. These findings provided the first documented evidence of a loss of brainstem monoaminergic neurons in clinically idiopathic RBD. Furthermore, they suggested that Lewy body disease might represent the true basis for "idiopathic" RBD in the elderly.

Finally, it is now more evident that the emergence of RBD results from lesion localization related to any underlying neurological disorder rather than from the molecular processes of the neurological disorders themselves. This would support the concept of a "strategic hit" (anatomical) basis, rather than a molecular mechanism basis, for triggering RBD and would explain how an array of etiologically different CNS disorders could trigger RBD.

However, a strategic-hit basis for RBD would also include physiological and pharmacological "hits" that are medication or immune-system related and would ultimately affect molecular mechanisms. Therefore, understanding the molecular mechanisms underlying medication-induced and immune-induced RBD is an important research area.

VI. Explanation of Symptoms in Relation to Pathophysiology

As already stated, loss of REM atonia alone is not sufficient to cause RBD [1,17]. There must also presumably be disinhibition of motor pattern–locomotor generators in the brainstem (perhaps originating in multiple higher centers, including limbic system and forebrain) for the various behaviors of RBD to emerge. Lesion studies in the experimental cat model have shown that the location of the brainstem lesion determines which category of behavior is released in REM sleep [17]. One of the most fascinating questions involving RBD concerns the pathophysiology of the stereotypic dream changes and associated attempts at dream enactment. There is a close correspondence between dream action and observed or tape-recorded behaviors representing dream enactment. There is typically a tandem relationship among the onset of the behavioral and dream disturbances of RBD, their relapse (temporary or sustained), and their shared responsivity to treatment. These observations suggest that there is a common underlying pathophysiology for the dream and behavioral (dream-enacting) disturbances of RBD. Finally, the signs and symptoms of RBD, including the experiential components, can be placed within a broad spectrum of dissociated states of sleep and wakefulness [2,16].

VII. Natural History

Apart from acute RBD that usually emerges during drug-induced states, toxic-metabolic states, and drug withdrawal states, RBD is a chronic disorder that rarely has spontaneous remissions. RBD is often the harbinger of a parkinsonian or other neurodegenerative disorder and can therefore be regarded as a sentinel or "cryptogenic" disorder. A currently unresolved question concerns the natural history of medication-induced RBD, both when the offending medication is still being administered (out of clinical necessity) and after the medication has been discontinued. Also unknown is whether medication-induced RBD, females with RBD, children and adults with parasomnia overlap disorder, or patients with mixed RBD-narcolepsy are at increased risk for eventually developing parkinsonism or another tardive neurological disorder.

References

1. Schenck, C. H., and Mahowald, M. W. (2002). REM sleep behavior disorder: clinical, developmental, and neuroscience perspectives 16 years after its formal identification in SLEEP. *Sleep* **25**, 120–138.
2. Mahowald, M. W., and Schenck, C. H. (2005). REM Sleep Parasomnias. *In:* "Principles and Practice of Sleep Medicine" (M. H. Kryger, T. Roth, W. C. Dement, eds.), 4th edition, pp. 897–916. Elsevier Saunders, Philadelphia.
3. Fantini, M. L., Michaud, M., Gosselin, N., Lavigne, G., and Montplaisir, J. (2002). Periodic leg movements in REM sleep behavior disorder and related autonomic and EEG activation. *Neurology* **59**, 1889–1894.
4. Massicotte-Marquez, J., Carrier, J., Decary, A., Mathieu, A., Vendette, M., Petit, D., and Montplaisir, J. (2005). Slow-wave sleep and delta power in rapid eye movement sleep behavior disorder. *Ann. Neurol.* **57**, 277–282.
5. Stiasny-Kolster, K., Doerr, Y., Moller, J. C., Hoffken, H., Behr, T. M., Oertel, W. H., and Mayer, G. (2005). Combination of "idiopathic" REM sleep behaviour disorder and olfactory dysfunction as possible indicator for alpha-synucleinopathy demonstrated by dopamine transporter FP-CIT-SPECT. *Brain* **128**, 126–137.
6. Fantini, M. L., Gagnon, J. F., Petit, D., Rompre, S., Decary, A., Carrier, J., and Montplaisir, J. (2003). Slowing of electroencephalogram in rapid eye movement sleep behavior disorder. *Ann. Neurol.* **53**, 774–780.
7. Gagnon, J. F., Fantini, M. L., Bedard, M. A., Petit, D., Carrier, J., Rompre, S., Decary, A., Panisset, M., and Montplaisir, J. (2004). Association between waking EEG slowing and REM sleep behavior disorder in PD without dementia. *Neurology* **62**, 401–406.
8. Ferini-Strambi, L., Di Gioia, M. S., Castronovo, V., Oldani, A., Zucconi, M., and Cappa, S. F. (2004). Neuropsychological assessment in idiopathic REM sleep behavior disorder (RBD): does the idiopathic form of RBD really exist? *Neurology* **62**, 41–45.
9. Olson, E., Boeve, B., and Silber, M. (2000). Rapid eye movement sleep behavior disorder: demographic, clinical, and laboratory findings in 93 cases. *Brain* **123**, 331–339.
10. Boeve, B. F., Silber, M. H., Parisi, J. E., Dickson, D. W., Ferman, T. J., Benarroch, E. E., Schmeichel, A. M., Smith, G. E., Petersen, R. C., Ahlskog, J. E., Matsumoto, J. Y., Knopman, D. S., Schenck, C. H., and Mahowald, M. W. (2003). Synucleinopathy pathology and REM sleep behavior disorder plus dementia or parkinsonism. *Neurology* **61**, 40–45.
11. Schenck, C. H., Bundlie, S. R., and Mahowald, M. W. (1996). Delayed emergence of a parkinsonian disorder in 38% of 29 older males initially diagnosed with idiopathic REM sleep behavior disorder. *Neurology* **46**, 388–393.
12. Fantini, M. L., Ferini-Strambi, L., and Montplaisir, J. (2005). Idiopathic REM sleep behavior disorder: toward a better nosologic definition. *Neurology* **64**, 780–786.
13. Gagnon, J. F., Bedard, M. A., Fantini, M. L., Petit, D., Panisset, M., Rompre, S., Carrier, J., and Montplaisir, J. (2002). REM sleep behavior disorder and REM sleep without atonia in Parkinson's disease. *Neurology* **59**, 585–589.
14. Kumru, H., Santamaria, J., Tolosa, E., Valldeoriola, F., Munoz, E., Marti, M. J., and Iranzo, A. (2004). Rapid eye movement sleep behavior disorder in parkinsonism with *PARKIN* mutations. *Ann. Neurol.* **56**, 599–603.
15. Arnulf, I., Merino-Andreu, M., Bloch, F., Konofal, E., Vidailhet, M., Cochen, V., Derenne, J. P., and Agid, Y. (2005). REM sleep behavior disorder and REM sleep without atonia in patients with progressive supranuclear palsy. *Sleep* **28**, 349–354.
16. Mahowald, M. W., and Schenck, C. H. (1991). Status dissociatus: a perspective on states of being. *Sleep* **14**, 69–79.
17. Hendricks, J., Morrison, A., and Mann, G. (1982). Different behaviors during paradoxical sleep without atonia depend on pontine lesion site. *Brain Res.* **239**, 81–105.
18. Gilman, S., Koeppe, R. A., Chervin, R. D., Consens, F. B., Little, R., An, H., Junck, L., and Heumann, M. (2003). REM sleep behavior disorder is related to striatal monoaminergic deficit in MSA [multiple system atrophy]. *Neurology* **61**, 29–34.
19. Eisensehr, I., Linke, R., Tatsch, K., Kharraz, B., Gildehaus, J. F., Wetter, C. T., Trenkwalder, C., Schwarz, J., and Noachtar, S. (2003). Increased muscle activity during rapid eye movement sleep correlates with decrease of striatal dopamine transporters. IPT and IBZM SPECT imaging in subclinical and clinically manifest idiopathic REM sleep behavior disorder, Parkinson's disease, and controls. *Sleep* **26**, 507–512.
20. Uchiyama, M., Isse, K., Tanaka, K., Yokota, N., Hamamoto, M., Aida, S., Ito, Y., Yoshimura, M., and Okawa, M. (1995). Incidental Lewy body disease in a patient with REM sleep behavior disorder. *Neurology* **45**, 709–712.

66

Neurobiology of Narcolepsy and Hypersomnia

Emmanuel Mignot, MD, PhD
Jamie M. Zeitzer, PhD

Keywords: *autoimmune, cataplexy, histamine, human leukocyte antigen, hypersomnia, hypocretin, orexin, narcolepsy, sleep, sleep disorder, wakefulness*

I. Narcolepsy-Cataplexy
II. Narcolepsy without Cataplexy
III. Idiopathic Hypersomnia
   References

Narcolepsy and idiopathic hypersomnia are common causes of excessive daytime sleepiness in the general population. In sleep disorder clinics, these pathologies form the fourth most frequently diagnosed group of pathologies, after obstructive sleep apnea, insomnia, and restless legs syndrome. Exact population prevalence figures are established only for classic narcolepsy with cataplexy (0.05% in the United States and western Europe) [1,2]. This number is likely to increase as narcolepsy, especially narcolepsy without cataplexy, and hypersomnia are increasingly diagnosed, largely as a consequence of the rapid growth of sleep medicine usage and the associated testing of an increased number of subjects through the multiple sleep latency test (MSLT). A diagnosis of narcolepsy or idiopathic hypersomnia can lead to long-term therapies with amphetamine stimulants, antidepressants, and hypnotics; yet, little is known about the pathophysiology of most of these disorders. The one exception is narcolepsy-cataplexy, a human leukocyte antigen (HLA)-associated disorder typically caused by the destruction of most hypocretin-producing cells, a small population of 50,000–100,000 hypothalamic neurons [2,3].

The recognition of hypocretin deficiency as the cause of most narcolepsy-cataplexy cases and the growth of sleep medicine usage has led to a redefinition of these disease entities in the revised International Classification of Sleep Disorders (Table 1) [4]. Most notably, narcolepsy-cataplexy, a disorder tightly associated with

HLA-DQB1*0602 and hypocretin deficiency (demonstrated by low cerebrospinal fluid hypocretin-1 levels), and narcolepsy without cataplexy (5%–30% with hypocretin deficiency) are now separate diagnostic entities. Narcolepsy without cataplexy is now defined as excessive daytime sleepiness and multiple sleep-onset rapid eye movement periods (SOREMPs) on the MSLT. Ancillary REM-related symptoms, such as sleep paralysis and hypnagogic hallucinations, are no longer considered critical for this diagnosis, as the prevalence of these symptoms in the general population is high and not strongly associated with HLA-DQB1*0602.

Another change in the classification scheme is the separation of idiopathic hypersomnia into two subtypes [4]. The first subtype of idiopathic hypersomnia is associated with long sleep times. It is a rare disorder of unknown etiology, characterized by excess amounts of unrefreshing nocturnal sleep (≥ 10 h/day) and subsequent severe subjective sleepiness. The disorder was historically described as idiopathic or central nervous system (CNS) hypersomnia proper [5,6]. The second subtype, idiopathic hypersomnia without long sleep time, was created to accommodate a growing patient population with excessive daytime sleepiness, short mean sleep latency on the MSLT, but one or no SOREMP and normal nocturnal sleep amounts. Of note, in this case the term *hypersomnia* may be a misnomer, as total amounts of daily sleep are not increased.

Other hypersomnias include periodic hypersomnias such as Kleine-Levin syndrome, a rare neurological disorder with recurrent episodes of sleepiness and derealization and occasionally associated with hyperphagia, aggressivity, and hypersexuality of unknown etiology [7]. This disorder affects adolescents and typically disappears 2–10 years after onset. A viral or autoimmune etiology has been suggested based on a recently reported association with HLA-DQB1*02. As this group of disorders is exceptional, it will not be further discussed. Most of this chapter will focus on narcolepsy, the only one of these disorders with a well-established neuropathology.

I. Narcolepsy-Cataplexy

The classic narcolepsy "tetrad" includes daytime sleepiness, cataplexy, sleep paralysis, and hypnagogic hallucinations [1,2,8]. Cataplexy, a sudden loss of muscle tone, triggered by strong emotions, is almost pathognomonic. Other symptoms include disrupted nighttime sleep and automatic behaviors. Polysomnographic studies indicate short REM latency (a SOREMP in 50% of cases), fragmented sleep with low sleep efficiency, and, not uncommonly, increased periodic leg movements during sleep. The MSLT indicates multiple SOREMPs and a mean sleep latency less than 8 minutes in approximately 85% of the cases.

In narcolepsy-cataplexy, the disorder typically starts around adolescence, occasionally abruptly or more insidiously. It is often impossible to pinpoint a specific trigger, although precipitating events as various as a bee sting, infection, head trauma, or sudden psychological stress are often retrospectively reported. Daytime sleepiness is most often the first symptom to appear, followed by cataplexy, most often within a couple of years. Naps are often refreshing, but the rested feeling evaporates within a few hours after awakening. The disorder is not typically characterized by increased sleep amounts but by an inability to stay awake or asleep for prolonged amounts of time.

Table 1 International Classification of Sleep Disorders (ICSD): Definitions and Pathophysiology of Narcolepsy and Hypersomnias

Condition	Diagnostic criteria	Pathophysiology
Narcolepsy with cataplexy	Presence of definite cataplexy (and usually abnormal MSLT results)	90% with low CSF Hcrt-1 and HLA-DQB1*0602 positivity
Narcolepsy without cataplexy	MSLT: mean sleep latency ≤ 8 min, ≥ 2 SOREMPs; no or doubtful cataplexy	Unknown, probably heterogeneous; 5%–30% with low CSF Hcrt-1, 40% HLA-DQB1*0602 positive
Secondary narcolepsy	As above, but due to other known medical conditions (e.g., neurological)	With or without Hcrt deficiency
Idiopathic hypersomnia, prolonged sleep	MSLT: short mean sleep latency, <2 SOREMPs; long (≥ 10 h) unrefreshing nocturnal sleep	Unknown, probable heterogeneous etiology
Idiopathic hypersomnia, normal sleep length	MSLT: short mean sleep latency, <2 SOREMPs; normal nightly sleep amounts (<10 h)	Unknown, probable heterogeneous etiology
Periodic hypersomnia (includes Kleine-Levin syndrome)	Recurrent (>1×/yr) sleepiness (lasting 2–28 days), normal function between occurrences	Unknown, possible autoimmune or viral etiology

Note: CSF, (lumbar sac) cerebrospinal fluid; Hcrt, hypocretin; HLA, human leukocyte antigen; MSLT, multiple sleep latency test; SOREMP, sleep-onset rapid eye movement period.

A. Genetic Predisposition to Narcolepsy-Cataplexy

The occurrence of narcolepsy-cataplexy involves both genetic predisposition and environmental triggers [2,9]. Approximately 75% of reported monozygotic twins are discordant for narcolepsy, suggesting the importance of environmental factors. Multiplex families are rare, but a 10- to 40-fold increase in relative risk is reported in first-degree relatives.

One of the major genetic susceptibility factors for narcolepsy-cataplexy is the HLA-DQB1 locus (Fig. 1). Almost all patients with typical cataplexy carry HLA-DQB1*0602, an HLA subtype found in 12% of Japanese, 25% of Caucasians, and 38% of African Americans. HLA-DQB1*0602 is always associated with DQA1*0102, another HLA allele encoded by the nearby DQA1 gene, located less than 20 kb telomeric to DQB1 (see Fig. 1). The DQB1 association is especially tight in subjects with hypocretin deficiency, as only four non–HLA-DQA1*0102–DQB1*0602-negative narcolepsy subjects with low or undetectable cerebrospinal fluid (CSF) hypocretin-1 have been identified to date (out of several hundred subjects with CSF testing). The HLA association in narcolepsy is primary and not due to linkage disequilibrium with other loci [2,10]. Association studies across ethnic groups and the study of additional genetic markers in the region all indicate a primary DQ effect (Table 2). It is mostly due to DQA1*0102–DQB1*0602, but it is not simply a dominant effect. Indeed, DQB1*0602 homozygotes have a two- to fourfold higher risk than DQB1*0602/X heterozygotes, and specific DQB1*0602 heterozygotes are either at increased (e.g., DQB1*0602/DQB1*0301) or decreased (e.g., DQB1*0602/DQB1*0601 or DQB1*0602/DQB1*0501) risk [10]. Additional smaller effects have also been suggested for specific DRB1 alleles, another HLA locus located in the same region (see Fig. 1). The complexity of the HLA allele association in narcolepsy [10] mirrors that reported in various autoimmune diseases, such as type I diabetes.

Genetic factors other than HLA are likely to be involved in the genetic susceptibility to narcolepsy-cataplexy, but these are not as well established or replicated. They are also likely to contribute far less to overall genetic risk than HLA-DQ. Studies in multiplex families suggest linkage to 4p13-q31 and 21q11.2 and possible associations with tumor necrosis factor-α (TNF-α), catechol-O-methyltransferase, and TNF receptor-2 polymorphisms [9]. Notably absent in this list is the preprohypocretin gene itself and its two receptors, hypocretin receptor-1 (HCRTR1) and hypocretin receptor-2 (HCRTR2). Indeed, only a single mutation in any of these three loci has been identified. In this case, an HLA-DQB1*0602-negative boy with hypocretin deficiency had an unusually early onset of narcolepsy-cataplexy at 6 months of age [11]. This boy had a potential dominant mutation in the signal peptide of the preprohypocretin gene

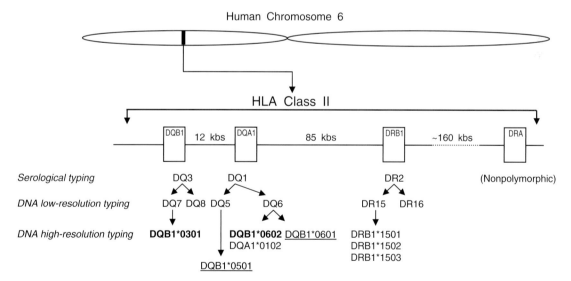

Figure 1 Major HLA Subtypes Involved in Narcolepsy Susceptibility. Subtypes in bold increase susceptibility, and underlined subtypes reduce predisposition to narcolepsy. Human leukocyte antigen (HLA) class II DR and DQ genes are heterodimers encoded by two genes each, one generating an α-chain, the other a β-chain. All of these genes are located within a small genetic distance, leading to extremely high linkage disequilibrium. DRA is nonpolymorphic and does not contribute significantly to HLA diversity, in contrast with DQA1, DQB1, and DRB1, which have several hundred possible alleles. The most important genetic factor in narcolepsy is HLA-DQB1*0602, a subtype of DQ1 (DQ6). This subtype is most often found in linkage disequilibrium with DRB1*1501, in the context of the DRB1*1501–DQA1*0102–DQB1*0602 haplotype. The effects of subtypes other than DQB1*0602 are far less important quantitatively.

Table 2 Genotype Distribution of DQB1 Alleles in Narcolepsy versus Controls in Three Ethnic Groups

DQB1 Alleles	Caucasian		African American		Japanese	
	Narcolepsy (n = 238)	Control (n = 146)	Narcolepsy (n = 77)	Control (n = 243)	Narcolepsy (n = 105)	Control (n = 698)
0602/0602	18.1%[b]	2.1%	23.4%[b]	3.7%	15.2%[b]	0.7%
0602/0301	20.6%[b]	2.7%	15.6%[b]	3.3%	19.0%[b]	0.9%
0602/other DQB1	45.8%[b]	17.1%	48.1%[b]	23.0%	56.0%[b]	7.7%
0602/0501	5.0%	1.4%	6.5%[a]	1.2%	1.9%	0.7%
0602/0601	0%	0%	0%	0%	7.6%[b]	2.1%
Non0602/non0602	10.5%[b]	76.7%	6.5%[b]	68.7%	0%[a]	87.8%

Allele groupings are ranked from the most predisposing to most protective combinations. The highest risk is carried by subjects homozygous for DQB1*0602, followed by DQB1*0602/0301 heterozygotes. DQB1*0501, a subtype common in many ethnic groups, reduces susceptibility even in the presence of DQB1*0602. DQB1*0601, a predominantly Asian subtype, is also relatively protective in the presence of DQB1*0602. Non-DQB1*0602 combinations confer a high protective effect for narcolepsy-cataplexy.
[a] $p < 0.05$.
[b] $p < 0.005$.

that produces impaired hypocretin trafficking and, presumably, cell death.

B. Narcolepsy-Cataplexy Is Closely Associated with Hypocretin Deficiency

The role of hypocretin in narcolepsy was first demonstrated in familial canine narcolepsy, a disease caused by mutations in the HCRTR2 gene, and through the study of a mouse preprohypocretin gene knockout model. This was followed by the demonstration of low CSF hypocretin-1 in seven of nine subjects with narcolepsy-cataplexy, a result that has now been extended to several hundred patients [2,3]. Most recently, postmortem studies have shown a selective loss of 50,000–100,000 posterior hypothalamic neurons that produce the neuropeptide hypocretin (orexin) [11,12]. The HLA association and associated hypocretin deficiency is particularly strong (>90%) in patients with definite cataplexy [2]. This association lends support to the hypothesis that narcolepsy is an autoimmune disorder.

C. Neurobiology of the Hypocretin Systems and Relevance to Narcolepsy

The hypocretin peptides (also called orexins) are encoded by the preprohypocretin gene. This gene encodes two related amidated polypeptides, hypocretin-1 and hypocretin-2, with high sequence homology [13]. Hypocretin-1 (orexin-A), a 33-amino acid peptide, has two disulfide bounds and is remarkably stable in biological fluids. It has an approximately equal affinity for the two known hypocretin receptors, HCRTR1 and HCRTR2. Hypocretin-2 (orexin-B), a smaller peptide of 29 amino acids, has a 10-fold higher affinity for HCRTR2 [13].

Hypocretin neurons have tight functional interactions with cholinergic and monoaminergic systems regulating sleep (Fig. 2). These cell groups are enriched in hypocretin receptors, with a complementary distribution of the two receptors at the anatomical level. HCRTR1 is particularly abundant in the adrenergic locus coeruleus, whereas HCRTR2 is highly expressed on histaminergic cells of the tuberomammillary nucleus [13,14]. Other regions of interest include the hippocampus, paraventricular thalamus, and ventromedial hypothalamic nucleus for HCRTR1, and the cortex, septal nuclei, paraventricular thalamus, paraventricular and arcuate hypothalamic nuclei, and pontine gray area for HCRTR2. In almost all cases, these two receptors have excitatory effects, although there is suggestion of promiscuous coupling of these receptors to multiple G proteins with occasional inhibitory effects [13,14].

It has been hypothesized that the loss of excitatory input to monoaminergic cell groups mediates sleepiness and short REM sleep in narcolepsy [1,2,13,14]. Indeed, activity of most monoaminergic cell groups—for example, locus coeruleus, raphe, and tuberomammillary nuclei—but not of the dopamine systems in the ventral tegmental area and substantia nigra are depressed during sleep and almost silent during REM sleep. Narcolepsy is typically treated with drugs that indirectly stimulate monoaminergic transmission (e.g., amphetamine-like compounds and antidepressants) [2,8], which is also in line with this hypothesis.

Interestingly, hypocretin effects on wakefulness after intracerebroventricular injections are largely blocked or attenuated by monoaminergic blockers, most notably antihistaminergic compounds, suggesting a primary role for downstream histaminergic systems. An important role of histamine is also suggested by recent findings indicating significantly decreased CSF histamine levels in narcolepsy-cataplexy, a result contrasting with more variable changes in other monoamine and metabolites [15].

Without hypocretin, narcoleptic patients have sleepiness, inappropriate REM paralysis during wakefulness (cataplexy, sleep paralysis), REM dreaming before falling asleep

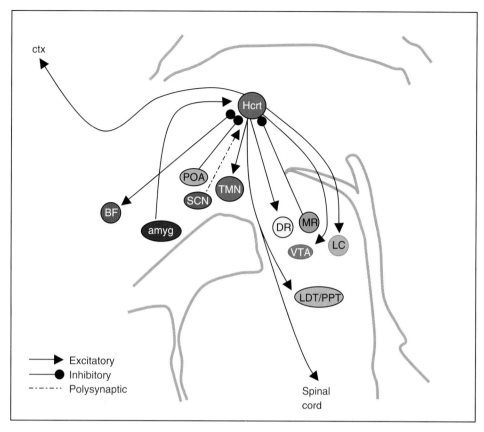

Figure 2 Neuroanatomical Connectivity of Hypocretin Neurons with Areas Involved in Control of Sleep and Wakefulness. All connections are monosynaptic except the innervation of the hypocretin neurons from the suprachiasmatic nucleus (SCN), which is likely via the retrochiasmatic area or the dorsomedial hypothalamus. The reciprocal connections between basal forebrain (BF) and hypocretin neurons are likely important for the stabilization of sleep and wake states. *Note:* amyg, amygdala; BF, (cholinergic) basal forebrain; ctx, cortex; DR, (serotonergic) dorsal raphe; Hcrt, hypocretin; LC, (noradrenergic) locus coeruleus; LDT/PPT, (cholinergic) laterodorsal tegmentum/pedunculopontine tegmentum; MR, (serotonergic) median raphe; POA, (sleep-active) preoptic area; SCN, suprachiasmatic nucleus (circadian pacemaker); TMN, (histaminergic) tuberomammillary nucleus; VTA, (dopaminergic) ventral tegmental area.

(hypnagogic hallucinations), and disorganized nighttime sleep. SOREMPs, rapid transitions into REM sleep (REM latency less than 20 minutes), can be observed during nocturnal sleep and while napping. Finally, narcolepsy is often associated with an increased body mass index and obesity, a possible consequence of sleepiness and of decreased energy expenditure.

Little is known regarding the normal function of hypocretins. Hypocretin neurons integrate metabolic and sleep- and wake-related inputs [13,16–18]. The relative contribution of these factors may vary across species. In rodents, these cells have receptors for leptin and ghrelin and are sensitive to glucose changes, likely facilitating wakefulness in reaction to changes in food availability. The activity of hypocretin cells is also tightly coupled with locomotor activity in rodents, an effect less evident in wake-consolidated species such as the squirrel monkey [18]. Hypocretin activity is also driven by the circadian clock and peaks at the end of the active (wake) period. Finally, hypocretin activity is increased by even brief sleep deprivation, suggesting a role for the hypocretin system in consolidating wakefulness in the face of a mounting sleep debt. This role of hypocretin may explain the inability of narcolepsy patients to stay awake for periods typically longer than a few hours, because sleep debt is increasing. As expected for a wake-promoting signal, these cells are more active during the wake period, decrease their activity during non-REM sleep, and are least active during REM sleep. The sleep stage-specific activity is consistent with a role in promoting wakefulness and inhibiting REM sleep.

D. Is Narcolepsy an Autoimmune Disorder?

The tight HLA association and the involvement of both genetic and nongenetic factors in the pathophysiology of narcolepsy suggest an autoimmune basis for the disorder. This hypothesis is consistent with the peripubertal onset of the disease but not with the even sex ratio found in the disorder. The recent discovery of hypocretin-producing cells as a potential target has given further credence to the autoimmune hypothesis. HLA-DQ is expressed on the surface of B cells, macrophages (including microglia), and activated T cells. It is a heterodimer composed of an α-chain and a β-chain, encoded by the polymorphic DQA1 and DQB1 genes, two genes in tight linkage disequilibrium (see Fig. 1). HLA-DQ binds peptide fragments and can present the resulting complex to other cells of the immune system, as recognized by the T cell receptor complex.

How HLA molecules in general and HLA-DQ in particular predispose to autoimmune disorders is not well understood. Predisposition probably involves the ability of some HLA alleles to bind and consequently present specific autoantigens to the rest of the immune system. In other HLA-associated diseases, however, antibodies directed toward the potential target (e.g., islet cell antibodies in type I diabetes) or T cell clones with primary responsiveness to target antigens (e.g., against myelin basic protein in multiple sclerosis) can be detected. In narcolepsy, however, all attempts at demonstrating T cell reactivity against hypocretin peptides or at detecting autoantibodies directed at hypocretin-producing cells have failed. A recent report in mice suggests a possible functional antibody with indirect effects on cholinergic systems after passive transfer, but this finding requires replication [2].

These generally negative results do not necessarily exclude the possibility of an autoimmune mechanism, but they do raise the possibility of other, more complex neuroimmune mechanisms. Another possibility is an infectious agent with particular tropism toward hypocretin-producing cells, although HLA association in infectious diseases are usually not tight or primarily to a single allele. Interest in the autoimmune arena has been recently rekindled by case reports that intravenous immunoglobulin may reduce the development of narcolepsy severity if administered within a year of onset [19].

II. Narcolepsy without Cataplexy

The importance of cataplexy in the diagnosis of narcolepsy has long been suggested by clinical studies. The finding that almost all subjects with definite cataplexy (defined as recurrent episodes of muscle paralysis triggered, at least sometimes, by typical emotions such as laughing or joking) have HLA-DQB1*0602 positivity strongly suggests etiological homogeneity in these subjects. This was further confirmed by the finding that almost all HLA-positive patients with cataplexy have low CSF hypocretin-1. In contrast, HLA-DQB1*0602 is typically found in 40% of patients without cataplexy or with doubtful or atypical cataplexy (rare events or only events triggered by unusual emotions). This is matched by a much lower percentage of patients without cataplexy who have low CSF hypocretin-1 (5%–30% of patients, depending on the specific case series) [2]. Some of these subjects (approximately one-third) are children with recent onset (≤4 years) who are likely to develop cataplexy within a few years.

These current results are consistent with two possible, nonexclusive pathophysiological models for narcolepsy without cataplexy. In the first model, narcolepsy with and without cataplexy are part of the same disease continuum, with similar pathophysiological effects on hypocretin transmission. Narcolepsy with cataplexy is a generally more severe form of the disease, most commonly associated with an almost complete destruction of the hypocretin system. HLA-DQB1*0602 may be a severity factor more likely to predispose to complete hypocretin destruction. Only in rare cases with cataplexy would hypocretin cell loss be partial but sufficient to produce normal levels of hypocretin and a narcolepsy phenotype. In these cases, compensation by the remaining hypocretin cells likely maintains CSF hypocretin-1 at normal levels. Projections with more effects on CSF hypocretin-1 but of less functional importance for the phenotype (e.g., hypocretin afferents to the spinal cord) also may be spared more in patients with cataplexy and normal CSF levels. The observation that rare HLA-DQB1*0602-negative cases with cataplexy, patients who have generally normal CSF hypocretin-1, may share an HLA genetic susceptibility with HLA-DQB1*0602-positive cases beyond DQB1*0602—for example, both groups have increased DQB1*0301 frequency—also supports this hypothesis [10].

In narcolepsy without cataplexy, cell loss would be less pronounced. The severity of the disease would be generally lower and the cell loss would be insufficient to result in low CSF hypocretin-1. The HLA-DQB1*0602 association with narcolepsy without cataplexy is weaker. Only in rare cases might neuroanatomical destruction be pronounced, resulting in low CSF hypocretin-1 but still sparing a select hypocretin cell subpopulation projecting to cataplexy triggering pathways. In favor of this model is the report by Thannickal et al., reporting 14% remaining hypocretin cells in a subject without cataplexy versus 4.4%–9.4% remaining cells in five other subjects with cataplexy [12]. Lesion studies in rats also indicate a 50% decrease in CSF hypocretin-1 with a 77% destruction of hypocretin cells, suggesting some degree of compensation.

HLA-DQB1*0602 is slightly increased in frequency in subjects without cataplexy but with normal CSF hypocretin-1, which is also consistent with this hypothesis. This finding, however, remains to be confirmed in a population-based sample, as HLA positivity may have been increased in clinical samples due to a bias in inclusion (some clinicians use HLA typing to confirm the diagnosis of narcolepsy without cataplexy). The model also predicts that a large number of cases of narcolepsy without cataplexy may exist in the general population but have a milder phenotype, consistent with our observation of slightly decreased REM latency in normal population subjects with HLA-DQB1*0602. Similar milder forms of the disease or long latent periods have been suggested for other autoimmune diseases such as DQ2-associated celiac disease, B27-associated spondylarthropathies, and DR3/DR4-associated type I diabetes. In type I diabetes, for example, a larger number of individuals, especially relatives of affected probands, may be positive for islet cell antibodies without ever developing the disease.

In the second model, the cause of cases without low CSF hypocretin does not involve partial hypocretin cell loss. Other systems downstream of hypocretin itself, for example, hypocretin receptors, histamine, or neuroanatomical systems unconnected with hypocretin neurobiology, may be involved. Of note, the two models may not be entirely exclusive. A partial hypocretin cell loss may, for example, not be sufficient in and of itself to produce symptoms, but it could lead to narcolepsy when associated with additional defects downstream. A similar model, albeit speculative, has been proposed in some children with obese-type diabetes where both type I and type II may coexist. In these cases, subjects with partially reduced islet cell numbers may be asymptomatic when lean but develop insulin resistance if obesity develops and the remaining islet population is unable to produce enough insulin to keep up with increased tissue demand.

To distinguish among these models, neuropathological studies of cases of narcolepsy without cataplexy are urgently needed. It is also imperative to study narcolepsy without cataplexy not only in clinical samples but also in the general population, as clinical cases may represent a more severe and selected subpopulation. We believe that the most likely explanation will be a combination of these models, including both disease heterogeneity with respect to hypocretin neuropathology and severity gradients for hypocretin cell loss as discussed previously. The extent of this overlap remains to be defined.

Two important questions remain unanswered in the area of narcolepsy without cataplexy. The first pertains to the importance of ancillary symptoms such as sleep paralysis and hypnagogic hallucinations, symptoms that have been deleted from current diagnostic criteria. Even in subjects with cataplexy, these symptoms are only reported by 40%–60%. In addition, population-based studies have found a high prevalence of these symptoms in the general population (up to several percent on a weekly or monthly basis) and in patients with obstructive sleep apnea. In the general population, significant correlations of these symptoms are found with both sleepiness (of many causes) and other neuropsychiatric symptoms (e.g., depression and anxiety) but not with DQB1*0602 positivity, suggesting a lack of specificity. Importantly, however, when present in narcoleptic patients, these symptoms can be bothersome and intense, so larger differences may be present if severe occurrence or an association with a frightening sensation is added.

A second important question in this area is the real, population-based prevalence of narcolepsy without cataplexy (defined as unexplained sleepiness plus MSLT mean sleep latency of ≤ 8 min and ≥ 2 SOREMPs). In sleep disorder clinics, only 20%–40% of patients diagnosed with narcolepsy do not have cataplexy. The only population- and registry-based study of narcolepsy without cataplexy indicates a prevalence of approximately 0.02%, a similar value to that reported for cases with cataplexy [20]. Surprisingly for a test used clinically to diagnose narcolepsy, there are no large-scale normative data using the MSLT. Preliminary data by our group and others suggest that a small percentage ($\sim 2\%$–4%) of the population have unexplained SOREMPs and short mean sleep latency on the MSLT. It is thus likely that more subjects, many with no or minimal sleep complaints, will be identified and diagnosed as having narcolepsy without cataplexy as the population is increasingly screened for sleep disorders. Whether a small or large portion of these subjects have a mild degree of hypocretin dysfunction (possibly related pathophysiologically to narcolepsy-cataplexy) is unknown.

III. Idiopathic Hypersomnia

Idiopathic hypersomnia disorders are poorly defined conditions characterized by excessive daytime sleepiness unexplained by narcolepsy (no REM abnormalities during the MSLT), insufficient sleep (no chronic sleep deprivation), periodic leg movements during sleep, or sleep apnea. It is an evolving clinical concept in search of a proper clinical and pathophysiological definition and most likely represents a heterogeneous group of disorders and etiologies.

The concept of idiopathic hypersomnia was first proposed by Dr. Bedrich Roth, a Czech neurologist, to describe patients with long episodes of nocturnal sleep, sleep drunkenness, and unrefreshing naps [6]. A monosymptomatic form with isolated sleepiness and a polysymptomatic form including the other symptoms were distinguished. The polysymptomatic form of the disorder is rare and currently labeled "idiopathic hypersomnia with long sleep time." Nocturnal sleep polysomnography indicates increased total sleep time, short sleep latency, and normal "harmonious" sleep-stage distribution. Some patients have autonomic nervous system disturbances (cold hands and feet, orthostatic hypotension) or a history of viral illness around the disease onset (typically around adolescence) [5]. MSLT sleep latency is short but often not as short as it occurs in narcolepsy and not in proportion with subjective sleepiness. Sleep paralysis and hypnagogic hallucinations may be present. A familial tendency for the disorder is noted in some cases. There is no HLA-DQB1*0602 association, although increased frequency of HLA-Cw2 and DR11 has been reported by one group [21]. Idiopathic hypersomnia with prolonged sleep time must be differentiated from depression with hypersomnia and excessive daytime sleepiness in association with various neurological disorders, for example, posttraumatic hypersomnia. It is also important to exclude mild sleep-disordered breathing.

More recently, however, the concept of idiopathic hypersomnia has evolved to include a growing population of patients with unexplained sleepiness, as documented by a short mean sleep latency on the MSLT, and normal nocturnal sleep [4]. Problematically, however, current population prevalence studies suggest that a large portion of the population may have a short MSLT sleep latency. It is critical to exclude mild degrees of sleep-disordered breathing or insufficient sleep in these patients.

Little pathophysiological research has been conducted in this area. Spectral analysis of electroencephalogram during sleep suggests decreased sleep pressure as measured during the first two nocturnal sleep cycles of idiopathic hypersomnia subjects. CSF hypocretin-1 levels are normal in all subjects tested to date [2,3]. Neurochemical studies of CSF monoamine metabolites have suggested decreased monoaminergic metabolites, in particular noradrenergic metabolites [5]. Most recently, however, we and others have found decreased CSF histamine levels in a subset of idiopathic hypersomnia and narcolepsy patients with normal hypocretin levels [15]. This result is interesting considering that histamine levels were not decreased in patients with sleepiness secondary to sleep apnea [15]. Whether this decrease is a passive reflection of daytime sleepiness or is causally involved will require further investigation.

References

1. Dauvilliers, Y., Billiard, M., and Montplaisir, J. (2003). Clinical aspects and pathophysiology of narcolepsy. *Clin. Neurophysiol.* **114**, 2000–2017.
2. Mignot, E. (2005). Narcolepsy: pharmacology, pathophysiology, and genetics. *In:* "Principles and Practice of Sleep Medicine" (M. H. Kryger, T. Roth, and W. C. Dement, eds.), 4th edition, pp. 761–779. Elsevier Saunders, Philadelphia.
3. Mignot, E., Lammers, G. J., Ripley, B., Okun, M., Nevsimalova, S., Overeem, S., Vankova, J., Black, J., Harsh, J., Bassetti, C., Schrader, H., and Nishino, S. (2002). The role of cerebrospinal fluid hypocretin measurement in the diagnosis of narcolepsy and other hypersomnias. *Arch. Neurol.* **59**, 1553–1562.
4. American Academy of Sleep Medicine. (2005). "International Classification of Sleep Disorders: Diagnostic and Coding Manual." 2nd edition. American Academy of Sleep Medicine, Westchester, IL.
5. Billiard, M., and Dauvilliers, Y. (2001). Idiopathic hypersomnia. *Sleep Med. Rev.* **5**, 349–358.
6. Roth, B. (1980). "Narcolepsy and Hypersomnia." Karger, Basel, Switzerland.
7. Arnulf, I., Zeitzer, J. M., File, J., Farber, N., and Mignot, E. (2005). Kleine-Levin syndrome: a meta-analysis and review of 186 cases in the literature. *Brain* **128**, 2763–2776.
8. Lammers, G. J., and Overeem, S. (2003). Pharmacological management of narcolepsy. *Expert Opin. Pharmacother.* **4**, 1739–1746.
9. Dauvilliers, Y., Maret, S., and Tafti, M. (2005). Genetics of normal and pathological sleep in humans. *Sleep Med. Rev.* **9**, 91–100.
10. Mignot, E., Lin, L., Rogers, W., Honda, Y., Qiu, X., Lin, X., Okun, M., Hohjoh, H., Miki, T., Hsu, S., Leffell, M., Grumet, F., Fernandez-Vina, M., Honda, M., and Risch, N. (2001). Complex HLA-DR and -DQ interactions confer risk of narcolepsy-cataplexy in three ethnic groups. *Am. J. Hum. Genet.* **68**, 686–699.
11. Peyron, C., Faraco, J., Rogers, W., Ripley, B., Overeem, S., Charnay, Y., Nevsimalova, S., Aldrich, M., Reynolds, D., Albin, R., Li, R., Hungs, M., Pedrazzoli, M., Padigaru, M., Kucherlapati, M., Fan, J., Maki, R., Lammers, G. J., Bouras, C., Kucherlapati, R., Nishino, S., and Mignot, E. (2000). A mutation in a case of early onset narcolepsy and a generalized absence of hypocretin peptides in human narcoleptic brains. *Nat. Med.* **6**, 991–997.
12. Thannickal, T. C., Moore, R. Y., Nienhuis, R., Ramanathan, L., Gulyani, S., Aldrich, M., Cornford, M., and Siegel, J. M. (2000). Reduced number of hypocretin neurons in human narcolepsy. *Neuron* **27**, 469–474.
13. Taheri, S., Zeitzer, J. M., and Mignot, E. (2002). The role of hypocretins (orexins) in sleep regulation and narcolepsy. *Annu. Rev. Neurosci.* **25**, 283–313.
14. Mignot, E., Taheri, S., and Nishino, S. (2002). Sleeping with the hypothalamus: emerging therapeutic targets for sleep disorders. *Nat. Neurosci.* **5 (Suppl)**, 1071–1075.
15. Kanbayashi, T., Kodama, T., Kondo, H., Satoh, S., Miyazaki, N., Kuroda, K., Abe, M., Nishino, S., Inoue, Y., and Shimizu, T. (2004). CSF histamine and noradrenaline contents in narcolepsy and other sleep disorders. *Sleep* **27**, A236.
16. Sakurai, T., Nagata, R., Yamanaka, A., Kawamura, H., Tsujino, N., Muraki, Y., Kageyama, H., Kunita, S., Takahashi, S., Goto, K., Koyama, Y., Shioda, S., and Yanagisawa, M. (2005). Input of orexin/hypocretin neurons revealed by a genetically encoded tracer in mice. *Neuron* **46**, 297–308.
17. Yamanaka, A., Beuckmann, C. T., Willie, J. T., Hara, J., Tsujino, N., Mieda, M., Tominaga, M., Yagami, K., Sugiyama, F., Goto, K., Yanagisawa, M., and Sakurai T. (2003). Hypothalamic orexin neurons regulate arousal according to energy balance in mice. *Neuron* **38**, 701–713.
18. Zeitzer, J. M., Buckmaster, C. L., Parker, K. J., Hauck, C. M., Lyons, D. M., and Mignot, E. (2003). Circadian and homeostatic regulation of hypocretin in a primate model: implications for the consolidation of wakefulness. *J. Neurosci.* **23**, 3555–3560.
19. Dauvilliers, Y., Carlander, B., Rivier, F., Touchon, J., and Tafti, M. (2004). Successful management of cataplexy with intravenous immunoglobulins at narcolepsy onset. *Ann. Neurol.* **56**, 905–908.
20. Silber, M. H., Krahn, L. E., Olson, E. J., and Pankratz, V. S. (2002). The epidemiology of narcolepsy in Olmsted County, Minnesota: a population-based study. *Sleep* **15**, 197–202.
21. Poirier, G., Montplaisir, J., Decary, F., Momege, D., and Lebrun, A. (1986). HLA antigens in narcolepsy and idiopathic central nervous system hypersomnolence. *Sleep* **9**, 153–158.

67

Restless Legs Syndrome and Periodic Limb Movements in Sleep

Richard P. Allen, PhD

Keywords: *dopamine, iron, iron metabolism, periodic leg movements, restless legs syndrome, striatum, substantia nigra*

I. Natural History of Restless Legs Syndrome
II. Etiology of Restless Legs Syndrome
III. Pathophysiology and Molecular Mechanisms of Restless Legs Syndrome
IV. Explanation of Restless Legs Syndrome Symptoms in Relation to Pathophysiology
V. Periodic Limb Movements
VI. Conclusion
References

A strange thing happened on the way to understanding periodic leg movements in sleep (PLMS, as defined later). The striking PLMS patterns of repetitive, short (but sometimes large) activations of anterior tibialis on an electromyography often accompanied by brief electroencephalogram (EEG) arousal certainly seemed to be a potent source of sleep disruption. A periodic limb movement disorder characterized by PLMS and sleep-wake complaints was defined to recognize the presumed sleep-disrupting effects of PLMS and was reported to be a common cause of insomnia and daytime sleepiness. But appearances are deceptive.

Subsequent studies demonstrated that these events do not, by themselves, relate to insomnia or excessive daytime sleepiness. Rather, they provide an important motor sign of sensorimotor dysfunction probably related to the abnormalities in subcortical central nervous system (CNS) dopaminergic systems [1]. Thus PLMS are more important as markers of disease and its severity than as disrupters of sleep. Periodic limb movement disorder is rare if it exists. These periodic movements are particularly significant for restless legs syndrome (RLS), occurring in 80%–90% of RLS patients recorded for one night and perhaps nearly all RLS patients with repeated nights of evaluations. RLS thus provides an important venue for evaluating these periodic movements. RLS is a disorder of the transition states between wake and sleep and includes these periodic leg movements during quiet yet awake resting (periodic leg movements of wake, or PLMW), as well as sleep (PLMS). Thus RLS is a major sleep disorder with PLMS and PLMW representing a motor sign of this disorder.

RLS itself has long been recognized in the medical literature, starting with the writings of the famous seventeenth-century London physician Thomas Willis. It has also been largely ignored. Ekbom systematically described the disorder and its conditions in the middle of the last century [2], but only at the dawn of the twenty-first century do we have a clear diagnostic definition of the disorder, as developed by a consensus of experts, the international RLS study group [3]. This group established the definition of RLS that follows:

Periodic leg movements (PLMs): A series of at least four leg movements (on an activity meter) or an anterior tibialis electromyographic burst lasting 0.5–10 seconds (older criteria limit each burst to 5 seconds) with onsets separated by 5–90 seconds. These occur during sleep (PLMS) or while lying or sitting up awake with legs horizontal (PLMW).

Restless legs syndrome (RLS): A sensorimotor disorder often severely affecting sleep defined by four essential diagnostic criteria: a strong urge to move the legs (focal akathisia), usually accompanied by a strange feeling in the leg; episodes are precipitated by rest with inactivity (quiescegenic); the urges and any accompanying feelings are relieved by movement or arousal; and it is worse in the evening or night than in the morning.

In this chapter, we focus first on RLS, where there have been remarkable gains in our understanding of the disorder. We then consider the limited knowledge we have about PLMD, the disorder attributed to PLMS.

I. Natural History of Restless Legs Syndrome

RLS can start at any age but most commonly starts in adult life or late adolescence. When it starts early in life, before age 45, it often has a slow, insidious development of symptoms, with daily symptoms commonly occurring between ages 50 and 65. Late-onset RLS usually progresses faster, with daily symptoms occurring in less then 5 years and in some cases almost immediately with the start of the disorder. Not all RLS patients progress to daily symptoms; some will have only infrequent symptoms with little or no sign of progression with age. Once RLS starts it usually appears to persist, although a small percentage of patients report a lasting spontaneous remission from symptoms. Many patients report brief periods with reduced RLS symptoms.

Some studies have suggested that RLS improves for those older than 70, but this has not been adequately documented and may be an artifact of the limited sample sizes and communication problems for older adults in these studies. The more severely affected early-onset RLS patient, however, shows a continuing slow progression of symptoms into older adult life with no indication of improvement with age.

II. Etiology of Restless Legs Syndrome

A. Environmental Factors

1. Systemic Iron Deficiency

RLS occurs both as a primary disorder and secondarily to three other medical conditions. When RLS appears secondary to these conditions, the RLS symptoms resolve or improve when the causal condition resolves or improves. Studies from Europe and the Americas show that RLS occurs in 20%–60% of patients with end-stage renal disease (ESRD), in 15%–27% of pregnant women, and in about 40% of patients with iron deficiency. The occurrence in these conditions significantly exceeds the prevalence of 7%–10% of European and American adults [4]. In most cases, a successful kidney transplant, the end of a pregnancy, or the correction of the iron deficiency produces complete and lasting freedom from all RLS symptoms. For ESRD, the occurrence of RLS decreases survival time independent of premature termination of treatment and the return of RLS symptoms after transplant provides an early sign of transplant failure. These three medical conditions causing RLS share one factor: they all compromise iron status [5]. Patients with RLS secondary to ESRD when treated aggressively with intravenous (IV) iron and erythropoietin show reduction in RLS and, in one blinded study, short-term improvement of the RLS not shown in a placebo-treated group [6]. The risk of pregnant women developing RLS is greater if they have lower iron status before or during pregnancy. Thus the secondary RLS in these three conditions results from the compromised iron status. This led Allen and Earley to hypothesize that all conditions that compromise iron status increase the risk of RLS [5]. To date, this hypothesis has proven correct whenever it has been tested. Frequent blood donations, gastric surgery, rheumatoid arthritis compared with osteoarthritis, and low-density lipoprotein apheresis—all of which significantly reduce iron status—have been associated with significantly increased prevalence of RLS. Thus any compromise in systemic iron status is a major environmental factor producing secondary RLS. In fact, this is the only known confirmed etiological factor for RLS. It deserves note that not all patients with compromised iron status develop RLS, indicating significant involvement of either genetic or other environmental factors.

2. Other Environmental Factors

Population-based surveys for RLS have been used to suggest other possible environmental factors increasing the risk of RLS. Unfortunately, many of these studies did not use the full diagnostic criteria for defining RLS and most failed to use any validated questionnaire to select RLS patients. Moreover, it is hard to know from these surveys whether or not the findings result from RLS or, instead, increase the risk of RLS. For example, the consistent finding of decreased alcohol use in RLS patients [7] likely reflects the adverse effect of alcohol on RLS symptoms rather than indicates a factor contributing to development of RLS. Decreased systemic iron status, which would be expected to be a significant environmental factor, at least for secondary RLS, has been confirmed in one of two population-based surveys examining iron status [8].

Isolated case reports relate RLS occurring after use of some serotonin-specific reuptake inhibitors and diphenhydramine (an antihistaminergic agent with muscarinic effects), but there has been no consistent documentation of these effects. Dopamine (DA) antagonists are known to produce a significant general akathisia, but, unlike RLS, this involves the whole body and is less restricted to the legs. The similarity, however, raises the possibility that alterations of the dopaminergic system might produce RLS. In favor of this hypothesis, dopaminergic agonists and levodopa certainly provide effective treatment for RLS (discussed later).

B. Brain Iron Insufficiency

Earley and Allen extended the iron deficiency hypothesis by suggesting that brain iron could be compromised despite normal systemic iron status in a significant portion (if not most) of primary RLS cases. Three different types of studies support this hypothesis. First, several studies have documented low brain iron in RLS patients with normal systemic iron status. RLS patients compared with controls showed in two independent cerebrospinal fluid (CSF) studies increased transferrin and decreased ferritin [9,10]. Reduced brain iron is also suggested by two magnetic resonance imaging (MRI) studies reporting significantly reduced ferritin iron content in the substantia nigra. In Figure 1 the MRI images from a normal subject show well-marked areas with decreased intensity (white), indicating the iron-rich parts of the substantia nigra and red nucleus. However, the image from an age-matched, severely affected RLS patient shows the same areas with greatly decreased intensity, indicating markedly less iron [11]. These reports of decreased

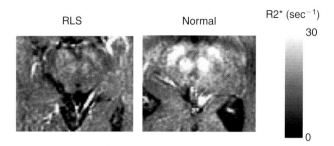

Figure 1 MRI R2* Images in a 70-Year-Old RLS Patient and a 71-Year-Old Control Subject. Much lower R2* relaxation rates indicating less iron density are apparent in the restless legs syndrome (RLS) case in both red nucleus and substantia nigra. Allen, R. P., Barker, P. B., Wehrl, F., Song, H. K., and Earley, C. J. (2001). MRI measurement of brain iron in patients with restless legs syndrome. *Neurology* **56(2)**, 263–265.

iron in the substantia nigra of RLS patients have subsequently been confirmed by a transcranial ultrasound study reporting reduced hyperechogenicity signal in the substantia nigra area [12]. Autopsy evaluation of the nigra from RLS patients compared with controls also showed decreased iron, decreased heavy-chain ferritin, and increased transferrin both for sections from the nigra and for neuromelanin cells obtained by microlaser capture [13,14]. Examples presented in Figure 2 show these changes for normal compared with RLS tissue, including one panel showing the particularly striking decrease in heavy-chain ferritin for the RLS compared with the control tissues. These autopsy results, however, were documented only in RLS patients whose symptoms started before age 45. This may be important given apparently different pathophysiology for early-versus late-onset RLS (discussed later).

A second type of study used IV iron to treat RLS in patients with normal systemic iron status. This approach was first tried in 1952 by Nordlander, who, with repeated IV iron doses, showed RLS remission in 21 of 22 cases [15]. Earley and associates reported that 1000 mg of iron dextran produced complete relief from all RLS symptoms for 2 months to 2 years in 6 of 10 patients [16]. The objective measure of PLMs while asleep supported the subjective report of symptom relief, indicating this result was unlikely to be a placebo effect. Enhancing peripheral iron status seems likely to correct brain iron deficiency noted for RLS patients and thereby to reduce the RLS symptoms.

A third type of study correlated CSF ferritin content in adult RLS patients with self-reported age of onset for RLS. A relatively strong correlation (r = 0.65) was found in cases with early-onset RLS (before age 45), indicating that the degree of brain iron deficiency observed in adult life reflects the severity of underlying pathology, leading to expression of symptoms at an earlier age. The brain iron deficiency hypothesis may thus be more relevant for early-onset cases, and indeed in the most recent studies,

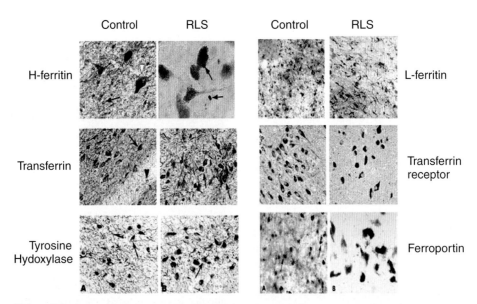

Figure 2 Substantia Nigra Sections with Blue Staining of Proteins. Some examples of proteins are noted by arrows; the white arrow is for heavy-chain ferritin (H-ferritin) in the neuromelanin cell. Rearranged figures from Connor, J. R., Boyer, P. J., Menzies, S. L., Dellinger, B., Allen, R. P., Ondo, W. G., and Earley, C. J. (2003). Neuropathological examination suggests impaired brain iron acquisition in restless legs syndrome. *Neurology* **61**, 304–309.

MRI and nocturnal CSF ferritin abnormalities occurred only in the early-onset form of RLS; highly variable MRI and CSF results were observed in late-onset RLS, suggesting multiple etiologies. It is important to note that this was not a post-hoc demarcation of patients by age at onset; rather, prior studies had already defined the two age-of-onset phenotypes for RLS [5]. The CNS iron studies used these previously defined age-of-onset phenotypes. These three types of studies provide strong support for the hypothesis that brain iron insufficiency causes most cases, if not all, of early-onset primary RLS. The brain iron insufficiency itself seems likely to involve some genetic predisposition.

C. Genetic Factors

RLS can start at any age, but as noted previously the age of onset clearly defines two different phenotypes. Early-onset RLS (onset before age 45) generally has a rather insidious course and a slowly progressive worsening of symptoms. RLS tends to run in families. A family history study found a relative four- to sevenfold increased risk of RLS in first-degree relatives of an early-onset RLS patient but only a twofold increased risk of RLS in first-degree relatives of late-onset RLS cases. Two subsequent segregation analyses have similarly shown that (1) the general population distribution of RLS age of onset is better matched to two distributions than one and (2) patients in the earlier-onset group are more likely to have a genetic basis, and the familial pattern of those in the latter-onset group is not consistent with single or complex genetic transmission alone. The critical ages of onset separating the two populations differ somewhat for each of these three studies with a range from 26 to 45, but the concept holds for all of the studies [17–19]. Thus RLS etiology appears to have both a genetic and a strong environmental component, with the genetic component being more important for early- than for late-onset RLS. The late-onset RLS may have a largely environmental etiology with little significant genetic component.

The genetic contribution to RLS appears to be complex, probably interacting with environmental factors. As expected in this situation, the exploration of the possible genetic basis for RLS has proven difficult. Although no specific gene has been identified as associated with increased risk of RLS, studies have documented linkage on three different chromosomes: 14q in one large family from Tyrol [20], 9p24-22 for data from 15 families from the United States [21], and 12q (D12S1044 to D12S78) for several large French Canadian families [22]. A 12q linkage signal was also found in two sets of German families [23,24] and among RLS patients in Iceland with a somewhat reduced range [25]. All the linkage peaks have been found in only small subsets of RLS families and did not replicate across all RLS families, suggesting heterogeneity. The replication of a linkage signal on 12q is, however, encouraging. It is likely that these genetic studies would need a more precise and uniform definition of the RLS phenotype to be successful and comparable. Studying age of onset is a step in that direction, but perhaps some assessment of iron status would help.

In summary, there is only one known etiology for RLS, namely iron insufficiency both systemically for secondary RLS and possibly primarily in the brain for early-onset primary RLS. Linkage findings within a relatively small region of chromosome 12 and on chromosome 9 raise the possibility of discovering RLS-predisposing polymorphisms.

III. Pathophysiology and Molecular Mechanisms of Restless Legs Syndrome

A. Dopaminergic Abnormalities

The serendipitous discovery of dramatic relief of RLS symptoms from relatively low doses of levodopa has led to what is now a generally accepted hypothesis: that DA abnormalities are in the final pathway producing RLS symptoms. A number of studies, described in the next sections, have explored this possibility.

1. Brain Imaging

Striatal imaging studies have produced mixed results, providing limited support for a dopaminergic abnormality in RLS [26]. One positron emission tomography (PET) study using raclopride showed reduced striatal D2 binding. Two fluorodopa studies similarly found reduced uptake in the striatum. Two of three single-photon emission computed tomography (SPECT) studies, however, failed to show any striatal abnormalities, although the third study found decreased binding similar to that from the PET study. None of the SPECT studies showed any abnormalities in binding for the DA transporter. The D2-binding differences, when significantly abnormal, are only about 10% reduced for RLS compared with matched controls. The D2 down-regulation occurred in both the putamen and the caudate, although somewhat more consistently in the putamen. Considering the known regulation of D2 receptors by DA release, the down-regulation could be primary or secondary to an increased DA activity or could be the result of both phenomena. All imaging studies reported to date were done in the daytime, when RLS symptoms are minor or even absent. It is tempting to consider that evening or night imaging (when the RLS symptoms are more prominent) might have shown more pronounced abnormalities. At this point, however, the imaging studies do not clearly confirm a dopaminergic abnormality.

2. Autopsy Evaluations

Autopsy studies of RLS tissue have found no indications for DA cell loss, but surprisingly they have reported for both the putamen and the substantia nigra increased tyrosine hydroxylase (TH) and an even greater increase in phosphorylated TH, along with actually increased DA content [27]. Thus the results of the imaging studies noted previously may reflect increased DA availability rather than simply decreased receptors.

3. Cerebrospinal Fluid Studies

Although two CSF studies failed to find any abnormalities in DA metabolites or DA-related proteins [5], a more recent CSF study reported two interesting DA-related abnormalities [28]. In that study, CSF was analyzed from samples taken at 10 A.M. from one set of patients (with matched controls) and at 10 P.M. from a second set. The circadian pattern of tetrahydrobiopterin showed normal levels at 10 P.M. but abnormally increased levels at 10 A.M. Given CSF circulation time, the 10 A.M. increase would be consistent with increased brain DA production at a period about 6–7 hours before this sample was taken, about 3–4 A.M., a time when RLS symptoms typically decrease. Both of these samples also had significant increases in 3-O-methyldopa. The increase in this metabolite occurs when levodopa is not decarboxylated into DA by action of the aromatic L-amino acid decarboxylase (AAD) but rather is catabolized through the catechol-O-methyltransferase into 3-O-methyldopa. This result indicates either a failure to have adequate AAD activity or an overload in the production of levodopa exceeding the capacity of the AAD metabolic path. The latter would be compatible with autopsy results and the increased biopterin reported in the morning CSF samples.

4. Hormonal Studies

In RLS patients, both growth hormone and prolactin show an abnormal response to a levodopa challenge given in the late evening but not when the same challenge is given in the morning. These results were highly significant for growth hormone but may be partly confounded by better sleep after levodopa (as the result of decreased symptomatology with treatment) [29]. The abnormal prolactin response is more specific to DA regulation, but the results were only significant at one of multiple time points evaluated and only marginally significant when the area under the curve was analyzed. A separate study showed a remarkably large release of prolactin in RLS patients following a daytime dopaminergic challenge [10], but it did not include control subjects, making the results somewhat uncertain [30]. Although these studies taken together indicate abnormal functioning of the tuberoinfundibular system, replication is needed before we can be confident of this abnormality.

5. A11 System

There has been considerable interest in the A11 system in RLS. This DA system has its cell bodies in the hypothalamus but sends processes descending the entire spinal column and ascending into the thalamus. This system is involved in nociception, and some animal models of RLS have been proposed based on the effects of A11 lesions producing increased motor activity [31] and perhaps some changes in pain thresholds. D3 receptor knockout mice have been reported to have increased activity levels and facilitation of some spinal reflexes similar to those seen in RLS [32]. Because D3 receptors are also located on A11 cells, this particular change could involve this A11 spinal system. However, none of these behavioral patterns have been shown to match the basic clinical symptoms of RLS, such as the circadian pattern of symptoms, nor has cell loss been reported in any of the RLS autopsy results. Nonetheless, the location and function of the A11 system makes it an attractive putative system for abnormal DA functioning in RLS.

6. Summary

Some DA abnormalities have been found in RLS. However, they appear to be pervasive, involving multiple systems; do not involve cell loss but rather involve receptor dysfunction; and, surprisingly, involve increased DA production. Thus RLS, unlike Parkinson's disease, appears to result from a hyperdopaminergic state with receptor dysfunction.

B. Iron Metabolism Abnormalities

1. Cellular Iron Metabolism

The autopsy studies not only confirm the brain iron deficiency in early-onset RLS but also provide some limited indication of the nature of the iron metabolism disorder in RLS [13,14]. Ferritin is a large, complex globular protein formed as a spherical shell composed of 24 subunits (peptide chains) with a center core that can contain up to 4500 atoms of iron. Each peptide chain is made up of one of two types of ferritin, heavy chain (H-ferritin) or light chain (L-ferritin). The ratio of these chains in a ferritin protein differs by cell type, with H-ferritin being relatively more common in the brain than in the liver. RLS produces a dramatic reduction in H-ferritin in the neuromelanin cells of the substantia nigra and, apparently, generally in all substantia nigra cells, particularly the oligodendrocytes. In contrast, L-ferritin is not reduced, but outside of the neuromelanin cells it appears to be distributed more in cells with considerable processes, such as astrocytes or microglial cells, rather than the normal predominance in oligodendrocytes that have more limited processes. (See the upper right panel in Fig. 2.) In addition, ferroportin, which is involved in iron export from the cell, is significantly decreased, and transferrin, which is involved in providing iron for the cell, is dramatically increased in the substantia nigra neuromelanin cells of the early-onset RLS brain. The H-ferritin, ferroportin, and transferrin results would be expected for iron-deficient cells. However, these cells also show decreased DMT1, a protein that serves to export iron from the endosomes into the intracellular labile iron pool. Although somewhat of a puzzle, this may still reflect abnormally low iron status of the cell. More of a problem is the significantly decreased transferrin receptor (TfR) found in these RLS neuromelanin cells. The TfR provides the major avenue for bringing iron into the cell; consequently, iron-deficient cells increase this protein. This abnormality may be one of the causes of the cells' iron deficiency. Fig. 2 shows the staining for the iron-related proteins of H- and L-ferritin, transferrin, and TfR in the 4 upper panels as marked. Two iron regulatory proteins (IRP1 and IRP2) posttranscriptionally regulate TfR, and these IRPs are increased with cellular iron deficiency. In the RLS substantia nigra neuromelanin cells, IRP2 is significantly increased as expected; in contrast, IRP1 is significantly decreased. IRP1 exists in an active form with a messenger ribonucleic acid (mRNA)-binding or an inactive form with aconitase activity. With iron deficiency, IRP1 normally shifts into the mRNA-binding form. Both IRP forms are significantly decreased in these RLS cells. The decrease in the active form of IRP1 could lead to reduced post-transcriptional stabilization of TfR and thus the observed TfR decrease occuring despite iron deficiency. This IRP1 abnormality may represent a fundamental source of the brain iron metabolism problems in RLS.

Thus at the cellular level there are two abnormalities in iron metabolism. First, an abnormal cellular distribution of L-ferritin raises the issue of problems in appropriate delivery of iron to the cells after it has crossed the blood-brain barrier. Second, the reduced IRP1 levels and activity and the reduced TfR deprive the cell of access to needed iron. The degree to which the IRP1 and TfR abnormalities pervade other cells in the brain and the body remains to be established, but at least for the critical DA-producing cells in the substantia nigra these abnormalities would lead to the observed cellular iron deficiency. The environmental and genetic factors producing these abnormalities remain to be established.

2. Blood-Brain Barrier Transport

Iron has to be actively transported across the blood-brain barrier, and the activity of this process may be partially reflected in the almost linear relation between serum and CSF ferritin in healthy, normal adults. RLS patients also show this linear relationship, but the increase in CSF ferritin for higher levels of serum ferritin is much less than normal [9,10]. In particular, for the RLS patients, the projection of this linear relation indicates normal CSF serum values would occur only when serum ferritin levels reach 300–600 μg/L, levels which are above normal and approaching those considered potentially toxic. Although these between-subjects correlations fail to establish causal relations, they nonetheless suggest abnormally reduced transport of iron across the blood-brain barrier. This could explain the common RLS situation of brain iron deficiency occurring with normal systemic iron status (status in the body outside of the brain), but this remains to be more thoroughly evaluated.

3. Peripheral or Systemic Iron Metabolism

Although most RLS patients show normal systemic levels of iron, as indicated by any of the usual serum iron measures (e.g., ferritin, percentage transferrin saturation, and total iron-binding capacity), there are some indications of abnormalities in systemic iron metabolism. The most striking finding is the reported rapid iron loss after an IV iron treatment of some RLS patients. A small group of patients given an IV dose of 1000 mg of iron dextran all experienced, as expected, a significant increase in serum ferritin but unexpectedly showed a decline in ferritin occurring over several weeks at a rate 3 to 14 times faster than expected for normal iron metabolism. Six of 10 patients showed complete remission of their RLS symptoms after the single large dose of IV iron, but for 5 of these patients, symptoms returned over the next year. Additional IV iron doses of 450 mg of iron gluconate were then given when RLS symptoms returned, provided serum ferritin levels were below 300 µg/L. The rapid ferritin loss observed with the initial treatment persisted but decreased with successive treatments for 4 of the 5 patients; accordingly, the duration of the treatment benefit increased. Two of the 5 subjects reached essentially normal levels of ferritin decrease after the fifth treatment for the first patient and the third for the second [33]. This remarkable correction of the rate of ferritin loss suggests that the initial rapid decline was not an artifact of the large dose but rather reflected abnormal systemic iron metabolism. Serum ferritin levels are generally considered one of the best measures of body iron stores, particularly with relation to iron status in the bone marrow. The abnormally rapid decrease of ferritin may therefore indicate abnormal systemic loss of iron in RLS, but it remains unclear where the iron is going. There may be some critical body stores, isolated from the feedback regulating ferritin, that need to be replenished for RLS patients, or the iron may be excreted in urine or stools. Presumably, if there is an abnormal rate of excretion, it will correct, at least partly, with repeated challenge to the system from exposure to large iron doses. These concepts need to be accepted cautiously, given the small number of patients treated and the lack of any normal control, but overall these studies are consistent with abnormal systemic iron metabolism in RLS related to allocating iron to body stores, excretion, or both. It deserves note that the L-ferritin results from the autopsy data described previously also indicate abnormal cellular allocation of iron in the brain after it has passed the blood-brain barrier.

4. Summary

All relevant studies to date have documented that abnormal iron metabolism would produce a state of brain iron deficiency in at least early-onset RLS. The abnormality appears to be pervasive, involving various cellular mechanisms in the brain, transport across the blood-brain barrier, and cellular allocation of iron both systemically and in the brain.

IV. Explanation of Restless Legs Syndrome Symptoms in Relation to Pathophysiology

A. Iron-Dopamine Connection

In RLS, two pathophysiologies must combine to produce the observed symptoms. One is the "wiring" problems indicated principally by the abnormalities noted previously in the dopaminergic systems, and the other is abnormal iron metabolism. The question then becomes, How does iron deficiency affect the dopaminergic system? Several animal and *in vitro* studies demonstrate the effects of iron deficiency on DA, and one series of studies suggests a putative mechanism linking iron and DA status.

1. Iron-Deprived Rodents

Studies of rats and mice that were iron-deprived after weaning into early adult life have shown two remarkable results. First, the activity levels appear to be increased during the rest period. These behavioral and sleep findings are consistent with the circadian pattern of RLS symptoms, with symptoms becoming worse in the evening and night. Second, these animals show brain DA abnormalities characterized by decreased density of D2 receptors, impaired DA transporter functioning, increased TH activity, increased intracellular DA, and increased extracellular DA. The amount of decrease in the D2 receptors in the striatum is tightly linked to the brain iron status, with greater decrease for lower iron levels. Both TH and phosphorylated TH (pTH) and the ratio pTH:TH were increased in the iron-deficient animals. The TH activity is closely linked to DA production, so these increases in TH would be consistent with the increased intra- and extracellular levels of DA resulting from increased DA production. The extracellular DA was measured over the end of the lights on (inactive) period and the first part of the lights off (active) period and showed iron deprivation, both increased DA levels overall and, more significantly, a major increase in the amplitude of the normal rise in DA during the first part of the active cycle [34]. Thus iron deprivation produces not only increased DA production with increased extracellular DA but also increased amplitude of the normal circadian pattern (Fig. 3).

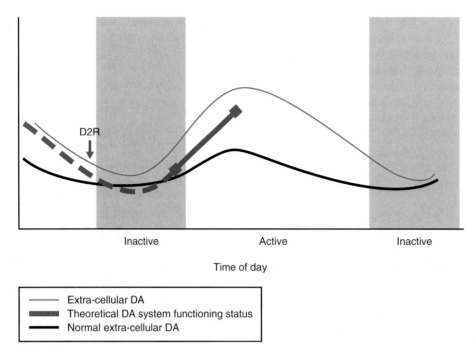

Figure 3 Putative Effects of Iron Deficiency on Functioning of Dopaminergic Systems. The red line below the white line is the time for restless legs syndrome (RLS) symptom occurrence. Note the asymmetrical pattern with rapid rise at the start of the active period. *Note:* DA, dopamine.

2. *In vitro* Studies

Iron chelation of PC12 cells (rat adrenal pheochromocytoma cells) using deferoxamine produced increases in pTH with increasing degree of chelation. At the most extreme level of chelation there was also an increase in TH. The ratio pTH:TH also significantly increased with iron chelation. DA content in the cells also increased.

3. Putative Mechanism Linking Iron and Dopamine Status

One cell adhesion molecule, Thy-1, was observed to be sensitive to iron status and was therefore selected as a putative link between iron and DA. Thy-1 was found to be reduced with iron chelation of PC12 cells, reduced for iron-deprived rodents, and reduced in the substantia nigra neuromelanin cells from postmortem RLS tissue [35]. Thy-1 is abundant on the cell surface of mature neurons and in the synaptic vesicles. Its functions include supporting regulation of vesicular release of neurotransmitters. The reduction of Thy-1 with iron deficiency may lead to alterations in synaptic status disrupting normal feedback regulation of DA production and thereby producing the increased DA production seen in RLS. This explains the iron-deprived animal results better than the *in vitro* studies do.

4. Summary

Iron deficiency produced *in vitro* by iron chelation or *in vivo* by dietary iron deprivation increased DA production. Thus the iron-deficiency produces the hyperdopaminergic state shown from preceding autopsy and CSF analyses. Animal studies indicate a marked increase in the amplitude of the normal circadian variation in extracellular DA during iron deficiency. The *in vivo* results are primarily from the striatum, but some limited studies of other DA brain areas indicate fairly similar results in some of the areas. Thus a pervasive iron deficiency produces the type of pervasive DA abnormality consistent with that seen in RLS. Thy-1 provides a possible mechanism linking iron deficiency with DA dysfunction.

B. Dopamine System Relationship to Restless Legs Syndrome Symptoms

1. Circadian Pattern

RLS has a remarkably pronounced circadian pattern of symptoms characterized by a gradual increase in symptom expression across the afternoon into the nighttime, with symptoms generally at their worst near the body core temperature minimum. Mornings appear to be relatively "protected" with few, if any, symptoms occurring even in fairly severe patients. Thus as RLS becomes worse, the symptoms spread into the early evening and late afternoon but rarely occur in the morning at the end of the sleep period. This pattern is not dependent upon sleep per se, although loss of sleep exacerbates RLS symptoms. The changes observed for the DA system (Fig. 3) would account for this symptom pattern. One view of the model suggested in Figure 3 would be that the postsynaptic receptor function fails to follow the large fluctuation in extracellular DA release. This would produce reduced DA function during the evening and at night, leading to RLS symptoms. In contrast, a hyperdopaminergic function in the morning would confer protection with few, if any, RLS symptoms. Dopaminergic agonists given in the evening would correct for the lack of DA during the evening and night but might also contribute to the instability of DA function during the daytime.

2. Dopamine Sensory and Motor Regulation

The primary symptom in RLS is an urge to move the legs (a focal akathisia). This symptom has the elements of a compulsive behavior, including craving and expectancy for an event (the leg movement). Three conditions with a similar compulsory nature come to mind: obsessive-compulsive disorder, addiction, and urgency of action in Tourette's syndrome. It is notable that these three disorders are all known to involve the DA system. Recently, addiction has also been related to the hypocretin (orexin) system, but this has not been evaluated for either obsessive-compulsive behavior or Tourette's. Thus the dopaminergic system seems likely to be involved in RLS. The akathisia focused on the legs and often associated with abnormal sensations in the legs could result from the abnormal functioning of either the sensorimotor integration in the nigrostriatal system or the nociception of the A11 system. The functioning of both of these systems is likely to be impaired with iron deficiency, and either one or both provide the types of sensory regulation that could produce the RLS symptoms.

The DA system is traditionally viewed as supporting activation, and most stimulants work by increasing DA production. Thus the increased DA production in the morning is likely to be activating. Similarly, an appropriate level of decreased dopaminergic activity may contribute to the ability to inhibit movement and sensation, allowing the organism to remain at rest, thereby facilitating the onset of sleep. We unfortunately know little about the neurobiology of this "predormitum" state preceding sleep. Some mammals and humans appear to have such a quiet state that facilitates the onset of sleep but also occurs without sleep onset. The postsynaptic activity in the striatum during this predormitum state when modulated by a relative deficient DA tone may produce inappropriate activation, setting the stage for an expected neural event that normally would not occur. Thus a relative nocturnal compared with morning DA deficiency in the RLS patient could lead to anticipation or a demand for movement, stimulation, or both. Activity or any alertness would provide different stimulation, ending the neurobiological characteristics of the predormitum state.

C. Other Neurotransmitter Systems and Restless Legs Syndrome Daytime Arousal

The RLS studies have focused primarily on the DA system largely because of the treatment benefit from levodopa and the dopaminergic agonists, but it seems likely that the pervasive iron metabolic disorder found in RLS will affect other neurotransmitter systems. Two have already been suggested. CSF hypocretin-1/orexin-A appears to be abnormally elevated in samples obtained around 10:30 P.M. but not for samples from early evening (5–6 P.M.), and as noted previously this system is involved in addiction. Histamine also appears to have some role in RLS because antihistamines that cross the blood-brain barrier can precipitate RLS symptoms more dramatically than even DA antagonists. This would suggest excessive histamine activation in the day, perhaps working with or responding to increased daytime hypocretin activity, producing increased CNS arousal. The hyperdopaminergic state in the morning would similarly support increased daytime arousal.

One of the RLS symptoms rarely mentioned is the lack of sleepiness in the daytime despite poor sleep at night. When severe, RLS produces a profound chronic loss of sleep, possibly worse than that of any other sleep disorder, yet there is no profound frank sleepiness in the daytime. RLS patients do not generally fall asleep at stoplights or when talking on the phone. They report fatigue and trouble concentrating and may have some mild executive function loss that occurs with sleep loss, but they appear to be, if anything, overstimulated in the daytime. Removing that stimulation by a centrally active antihistamine appears to release or engender the RLS symptoms. It deserves note that the GABAergic medications, although certainly as sedating as antihistamines, do not produce the profound symptom exacerbation observed with the sedating antihistamines [36].

V. Periodic Limb Movements

A. Etiology and Pathogenesis

PLMS have been found to occur in a range of disorders, including narcolepsy, Parkinson's disease, REM behavior disorder, iron deficiency, RLS, and attention deficit disorder; they also occur with aging [1]. The disorders with which they are associated predominantly involve DA pathology, and it has been suggested that any disruption of the sensorimotor dopaminergic systems, including the A11 system, will produce PLMS. As an individual ages, there is a decrease in these DA cells, which may account for the aging effect. The one puzzling phenomenon is the apparent development of some PLMS following a complete high cord transection. The movements after a cord transection are, however, less pronounced than the PLMS with the other conditions [37].

Recent studies have associated PLMS with several autonomic changes often preceding the events and continuing after the event. These changes include cardiac acceleration starting before the movement, which is then followed by a deceleration below resting level; EEG delta power increases before the movement, followed by faster EEG after the movement; and even transient blood pressure increases occurring with the movements [38]. These autonomic changes may have more clinical significance then any sleep disruption from the movement itself. The occurrence of the PLMs have also been linked with the A2 and A3 phase of the EEG cycling alternating pattern observed in sleep [39].

B. Pathophysiology

Studies of the spinal flexor reflex show that, during sleep, patients with PLMS have a lower threshold and a greater spatial spread for these reflexes [40]. This indicates increased spinal cord excitability or, conversely, a failure of the subcortical inhibition of the spinal sensorimotor system. It seems likely that this inhibition involves both the striatal and, more directly, the A11 DA systems. Thus disorders disrupting the spinal inhibitory function of these systems, including cord transection, will produce PLMS. In this regard PLMS are a motor sign of the subcortical dysfunction and provide an excellent marker for the severity of the dysfunction, but they prove somewhat less useful for diagnosis given the range of alternatives that have to be considered. It deserves note that during sleep of normal controls without PLM the threshold for the spinal flexor is slightly increased compared with waking, in contrast to the decreased threshold for patients with PLMS [40]. Thus normally there is, if anything, increased inhibitory control over the spinal system during sleep or decreased excitability that fails to function adequately for patients with RLS, producing the opposite results of increased spinal excitability.

C. Summary

Despite the multiple causes for the pathology producing PLMS, these movements remain an important, easily measured marker of the functioning of subcortical DA systems. As a motor sign they provide both an objective basis for confirming suspected diagnoses and an objective marker of the degree of the dysfunction. This is particularly useful for RLS.

VI. Conclusion

A model underlying the pathophysiology of RLS is emerging. Early-onset RLS (starting before age 45) appears to result mostly from a pervasive iron metabolism abnormality producing brain iron insufficiency. The impaired iron status produces a hyperdopaminergic state with an exaggerated circadian pattern of DA release. The iron deficiency probably also disrupts other neurotransmitter systems, such as hypocretin (orexin) and histamine.

Late-onset RLS (starting after age 45) appears to have somewhat more diverse causes. Abnormal iron metabolism is also probably involved in some cases. Nonetheless, patients appear to have DA abnormalities similar to those seen in early-onset RLS cases.

All RLS cases, whether early or late onset, share some elements of dopaminergic dysfunction. The DA abnormality also produces PLMS and even PLMs in resting wakefulness. Whether the dopaminergic dysfunction is sufficient to produce the urge to move remains unclear. The possible involvement of other systems such as hypocretin (orexin) and histamine needs to be better evaluated.

References

1. Montplaisir, J., Michaud, M., Denesle, R., and Gosselin, A. (2000). Periodic leg movements are not more prevalent in insomnia or hypersomnia but are specifically associated with sleep disorders involving a dopaminergic impairment. *Sleep Med.* **1(2)**, 163–167.
2. Ekbom, K. A. (1945). "Restless Legs." Ivar Haeggströms, Stockholm.
3. Allen, R. P., Picchietti, D., Hening, W. A., Trenkwalder, C., Walters, A. S., Montplaisir, J., Restless Legs Syndrome Diagnosis and Epidemiology workshop at the National Institutes of Health, International Restless Legs Syndrome Study Group. (2003). Restless legs syndrome: diagnostic criteria, special considerations, and epidemiology; a report from the restless legs syndrome diagnosis and epidemiology workshop at the National Institutes of Health. *Sleep Med.* **4(2)**, 101–119.
4. Allen, R. P., Walters, A. S., Montplaisir, J., Hening, W., Myers, A., Bell, T. J., and Ferini-Strambi, L. (2005). Restless legs syndrome prevalence and impact: REST general population study. *Arch. Intern. Med.* **165(11)**, 1286–1292.
5. Allen, R. P., and Earley, C. J. (2001). Restless legs syndrome: a review of clinical and pathophysiologic features. *J. Clin. Neurophysiol.* **18(2)**, 128–147.
6. Sloand, J. A., Shelly, M. A., Feigin, A., Bernstein, P., and Monk, R. D. (2004). A double-blind, placebo-controlled trial of intravenous iron dextran therapy in patients with ESRD and restless legs syndrome. *Am. J. Kidney Dis.* **43(4)**, 663–670.
7. Phillips, B., Young, T., Finn, L., Asher, K., Hening, W. A., and Purvis, C. (2000). Epidemiology of restless legs symptoms in adults. *Arch. Intern. Med.* **160(14)**, 2137–2141.
8. Hogl, B., Kiechl, S., Willeit, J., Saletu, M., Frauscher, B., Seppi, K., Muller, J., Rungger, G., Gasperi, A., Wenning, G., and Poewe, W. (2005). Restless legs syndrome: a community-based study of prevalence, severity, and risk factors. *Neurology* **64(11)**, 1920–1924.
9. Mizuno, S., Mihara, T., Miyaoka, T., Inagaki, T., and Horiguchi, J. (2005). CSF iron, ferritin and transferrin levels in restless legs syndrome. *J. Sleep Res.* **14(1)**, 43–47.
10. Earley, C. J., Connor, J. R., Beard, J. L., Malecki, E. A., Epstein, D. K., and Allen, R. P. (2000). Abnormalities in CSF concentrations of ferritin and transferrin in restless legs syndrome. *Neurology* **54(8)**, 1698–1700.

11. Allen, R. P., Barker, P. B., Wehrl, F., Song, H. K., and Earley, C. J. (2001). MRI measurement of brain iron in patients with restless legs syndrome. *Neurology* **56(2)**, 263–265.
12. Schmidauer, C., Sojer, M., Seppi, K., Stockner, H., Hogl, B., Biedermann, B., Brandauer, E., Peralta, C. M., Wenning, G. K., and Poewe, W. (2005). Transcranial ultrasound shows nigral hypoechogenicity in restless legs syndrome. *Ann. Neurol.* **58(4)**, 630–634.
13. Connor, J. R., Boyer, P. J., Menzies, S. L., Dellinger, B., Allen, R. P., Ondo, W. G., and Earley, C. J. (2003). Neuropathological examination suggests impaired brain iron acquisition in restless legs syndrome. *Neurology* **61**, 304–309.
14. Connor, J. R., Wang, X. S., Patton, S. M., Menzies, S. L., Troncoso, J. C., Earley, C. J., and Allen, R. P. (2004). Decreased transferrin receptor expression by neuromelanin cells in restless legs syndrome. *Neurology* **62(9)**, 1563–1567.
15. Nordlander, N. B. (1953). Therapy in restless legs. *Acta Med. Scand.* **145**, 453–457.
16. Earley, C. J., Heckler, D., and Allen, R. P. (2004). The treatment of restless legs syndrome with intravenous iron dextran. *Sleep Med.* **5(3)**, 231–235.
17. Allen, R. P., La Buda, M. C., Becker, P., and Earley, C. J. (2002). Family history study of the restless legs syndrome. *Sleep Med.* **3(Suppl)**, S3–S7.
18. Hening, W. A., Mathias, R. A., Allen, R. P., Washburn, M., Lesage, S., Wilson, A. F., and Earley, C. J. (2005). A segregation analysis of restless leg syndrome families. Abstract. *Neurology* **64(Suppl)**, 140.
19. Winkelmann, J., Muller-Myhsok, B., Wittchen, H. U., Hock, B., Prager, M., Pfister, H., Strohle, A., Eisensehr, I., Dichgans, M., Gasser, T., and Trenkwalder, C. (2002). Complex segregation analysis of restless legs syndrome provides evidence for an autosomal dominant mode of inheritance in early age at onset families. *Ann. Neurol.* **52(3)**, 297–302.
20. Bonati, M. T., Ferini-Strambi, L., Aridon, P., Oldani, A, Zucconi, M., and Casari, G. (2003). Autosomal dominant restless legs syndrome maps on chromosome 14q. *Brain* **126(Pt 6)**, 1485–1492.
21. Chen, S., Ondo, W. G., Rao, S., Li, L., Chen, Q., and Wang, Q. (2004). Genomewide linkage scan identifies a novel susceptibility locus for restless legs syndrome on chromosome 9p. *Am. J. Hum. Genet.* **74(5)**, 876–885.
22. Desautels, A., Turecki, G., Montplaisir, J., Sequeira, A., Verner, A., and Rouleau, G. A. (2001). Identification of a major susceptibility locus for restless legs syndrome on chromosome 12q. *Am. J. Hum. Genet.* **69(6)**, 1266–1270.
23. Kock, N., Culjkovic, B., Maniak, S., Schilling, K., Muller, B., Zuhlke, C., Ozelius, L., Klein, C., Pramstaller, P. P., and Kramer, P. L. (2002). Mode of inheritance and susceptibility locus for restless legs syndrome, on chromosome 12q. *Am. J. Hum. Genet.* **71(1)**, 205–208; discussion 208.
24. Winkelmann, J., Lichtner, P., Putz, B., Trenkwalder, C., Hauk, S., Meitinger, T., Strom, T., and Muller-Myhsok, B. (2006). Evidence for further genetic locus heterogeneity and confirmation of RLS-1 in restless legs syndrome. *Mov. Disord.* **21(1)**, 28–33.
25. Rye, D. B. (2005). "Genetics of RLS." World Association of Sleep Medicine, Berlin, Germany.
26. Allen, R. (2004). Dopamine and iron in the pathophysiology of restless legs syndrome (RLS). *Sleep Med.* **5(4)**, 385–391.
27. Connor, J. R. (2005). "Iron and RLS." World Association of Sleep Medicine, Berlin, Germany.
28. Earley, C. J., Hyland, K., and Allen, R. P. (2006). Diurnal changes in CSF dopaminergic measures in restless legs syndrome. *Sleep Med.* **7(3)**, 263–268.
29. Garcia-Borreguero, D., Larrosa, O., Granizo, J. J., de la Llave, Y., and Hening, W. A. (2004). Circadian variation in neuroendocrine response to L-dopa in patients with restless legs syndrome. *Sleep* **27(4)**, 669–673.
30. Winkelmann, J., Schadrack, J., Wetter, T. C., Zieglgansberger, W., and Trenkwalder, C. (2001). Opioid and dopamine antagonist drug challenges in untreated restless legs syndrome patients. *Sleep Med.* **2(1)**, 57–61.
31. Ondo, W. G., He, Y., Rajasekaran, S., and Le, W. D. (2000). Clinical correlates of 6-hydroxydopamine injections into A11 dopaminergic neurons in rats: a possible model for restless legs syndrome. *Mov. Disord.* **15(1)**, 154–158.
32. Clemens, S., and Hochman, S. (2004). Conversion of the modulatory actions of dopamine on spinal reflexes from depression to facilitation in D3 receptor knock-out mice. *J. Neurosci.* **24(50)**, 11,337–11,345.
33. Earley, C. J., Heckler, D., and Allen, R. P. (2005). Repeated IV doses of iron provides effective supplemental treatment of restless legs syndrome. *Sleep Med.* **6(4)**, 301–305.
34. Chen, Q., Beard, J. L., and Jones, B. C. (1995). Abnormal rat brain monoamine metabolism in iron deficiency anemia. *J. Nutr. Biochem.* **6**, 486–493.
35. Wang, X., Wiesinger, J., Beard, J., Felt, B., Menzies, S., Earley, C., Allen, R., and Connor, J. (2004). Thy-1 expression in the brain is affected by iron and is decreased in Restless Legs Syndrome. *J. Neuro. Sci.* **220(1–2)**, 59–66.
36. Allen, R. P., Lesage, S., and Earley, C. J. (2005). Antihistamines and benzodiazepines exacerbate restless legs syndrome (RLS) symptoms. *Sleep* **28**, A279.
37. de Mello, M. T., Poyares, D. L., and Tufik, S. (1999). Treatment of periodic leg movements with a dopaminergic agonist in subjects with total spinal cord lesions. *Spinal Cord* **37(9)**, 634–637.
38. Sforza, E., Pichot, V., Barthelemy, J. C., Sforza, E., Pichot, V., Barthelemy, J. C., Haba-Rubio, J., and Roche, F. (2005). Cardiovascular variability during periodic leg movements: a spectral analysis approach. *Clin. Neurophysiol.* **116(5)**, 1096–1104.
39. Parrino, L., Boselli, M., Buccino, G. P., Spaggiari, M. C., Di Giovanni, G., and Terzano, M. G. (1996). The cyclic alternating pattern plays a gate-control on periodic limb movements during non-rapid eye movement sleep. *J. Clin. Neurophysiol.* **13**, 314–323.
40. Bara-Jimenez, W., Aksu, M., Graham, B., Sato, S., and Hallett, M. (2000). Periodic limb movements in sleep: state-dependent excitability of the spinal flexor reflex. *Neurology* **54(8)**, 1609–1616.

68

Neurobiology of Insomnia

Michael L. Perlis, PhD
Wilfred R. Pigeon, PhD
Sean Patrick Andrews Drummond, PhD

Keywords: *ascending reticular activating system, homeostasis, inhibition of wakefulness, insomnia, neurobiology, ventrolateral preoptic area*

I. Introduction
II. Definition of Insomnia
III. Theoretical Perspectives on Insomnia
IV. Brief Review of the Neurobiology of Sleep and Wakefulness
V. Neurobiology of Sleep and Wakefulness: Implications for Insomnia
VI. Neurophysiologic, Neuroendocrine, and Neuroimaging Measures of Insomnia
VII. Call for an Integrative Perspective on Insomnia
References

I. Introduction

Although there has been significant progress with respect to the delineation of the neurobiology of sleep and wakefulness (e.g., see references 1 and 2), the circadian aspects of sleep (e.g., see references 3–7), the nonrapid eye movement–rapid eye movement (NREM–REM) oscillation (e.g., see references 8–14), and sleep-related phasic events (e.g., see references 15–17), substantially less is known about the neurobiological abnormalities that give rise to sleep disorders in general and insomnia in particular. The lack of findings with respect to insomnia is directly attributable to several factors, including the persistent point of view that insomnia is strictly a psychological or behavioral disorder, the absence of an animal model of chronic insomnia with which to conduct proper neuroscientific investigation, and the fundamental lack of agreement about what insomnia is and how it should be defined for research purposes.

This chapter has six sections. In the first section, we review issues that pertain to the definition of insomnia, giving special consideration to how the definition of insomnia may influence the attempt to define the neurobiological aspects of the disorder. In the second section, we provide information about two models of insomnia that attempt to more precisely specify "what insomnia is" and in so doing may focus the effort to delineate the neurobiological abnormalities that occur with insomnia. In the third section, we briefly review the regulation of sleep and wakefulness. In the fourth section, we proffer some speculations regarding how insomnia may occur as an aberration of the normal regulation of sleep and wakefulness. In the fifth section, we review and discuss the neurophysiological, neuroimaging, and neuroendocrine findings regarding insomnia. Finally, in the sixth section, we provide a framework for the consideration of insomnia as a disorder that not only has multiple determinants but also can be described in terms of multiple parallel processes. Our hope is that this framework encourages investigators from throughout the sleep research community to adopt an integrative and multidisciplinary approach to the problem of insomnia.

II. Definition of Insomnia

The three most common classification systems for insomnia can be found in the American Psychiatric Association's *Diagnostic and Statistical Manual of Mental Disorders*, fourth edition, text revision (DSM-IV-TR) [18]; the American Academy of Sleep Medicine's *International Classification of Sleep Disorders Manual*, revised edition (ICSD-II) [19]; and in an article by a task force of the American Academy of Sleep Medicine, which provides for research diagnostic criteria [20]. The general definition for insomnia, as framed within the ICSD-II, is provided in Figure 1. Each system allows for a disease or disorder definition of insomnia (as opposed to characterizing insomnia as a symptom) and each has the following description at its core:

> The individual reports one or more of the following sleep-related complaints:
> 1. Difficulty initiating sleep
> 2. Difficulty maintaining sleep
> 3. Waking up too early
> 4. Sleep that is chronically nonrestorative or poor in quality
>
> These symptoms occur frequently and chronically (e.g., more than 3 days per week for longer than 3 months).

The difficulty with the core definition is that it is overly inclusive and excessively nondescript. With respect to the former, there are two issues that are debatable. First, the

ICSD-II General Criteria for Insomnia

A. A complaint of difficulty initiating sleep, difficulty maintaining sleep, waking up too early, or sleep that is chronically nonrestorative or poor in quality. In children, the sleep difficulty is often reported by the caretaker and may consist of observed bedtime resistance or inability to sleep independently.

B. The above sleep difficulty occurs despite adequate opportunities and circumstances to sleep.

C. At least one of the following forms of daytime impairment related to the nighttime sleep difficulty is reported by patient:
  1. Fatigue or malaise
  2. Attention, concentration, or memory impairment
  3. Social or vocational dysfunction or poor school performance
  4. Mood disturbance or irritability
  5. Daytime sleepiness
  6. Motivation, energy, or initiative reduction
  7. Proneness for errors or accidents at work or while driving
  8. Tension, headaches, or gastrointestinal symptoms in response to sleep loss
  9. Concerns or worries about sleep

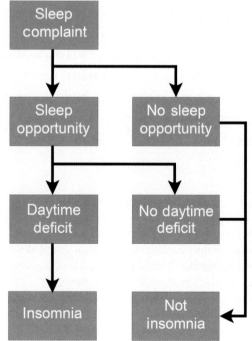

Figure 1 ICSD-II General Criteria for Insomnia.

inclusion of sleep that is "nonrestorative or poor in quality" as a form of insomnia, irrespective of whether the individual has concomitant problems with sleep initiation and/or maintenance, blurs the diagnostic distinction between the disorders of initiating and maintaining sleep and the disorders of excessive somnolence. The lack of diagnostic specificity, in turn, can be expected to substantially complicate the search for, and identification of, the neuronal systems that, when functioning abnormally, produce these different outcomes. Second, the core definition of insomnia all but ignores the various phenotypic expressions of insomnia. That is, initial, middle, late, and mixed insomnia are not presented as subtypes of a larger disorder but instead delineated as more or less the same problem. Thus it is assumed that there is a common underlying neurobiological abnormality, which, alone or interacting with unspecified factors, gives rise to all four types of insomnia. It is equally likely, however, that there are different control mechanisms that interact to produce sleep initiation, maintenance, and termination. If this is the case, the search for a single neurobiological defect may be ill advised. Accordingly, it may be more productive to allow for the possibility that various forms of insomnia are unique. This may facilitate the effort to explain how and why the various insomnia subtypes first occur, may occur chronically, and/or may vary within the individual over time.

Finally, there is the issue that the definitions used are nondescript. That is, it is assumed that the presenting symptom and the illness state are one and the same. The presenting symptom is that sleep cannot be initiated and/or maintained. The illness state, therefore, is thought to be rooted in the dysregulation of the neuronal systems related to sleep initiation or maintenance. Although this is plausible, it is possible to conceive of insomnia differently. For example, the inability to fall or stay sleep may occur in association with the "failure to inhibit wakefulness" [21,22]. This may occur in association with a variety of functional abnormalities, each with unique neurobiological underpinnings, such as the following:

- The inability to inhibit sensory flow
- The failure to engage "dysattentiveness"
- An increase in information processing at or around sleep onset and during sleep
- The failure to engage the normal mesograde amnesia of sleep
- Conditioned central nervous system or cortical arousal (which may be permissive of all the preceding)

In short, as the effort to describe the neurobiology of insomnia goes forward, it will be useful to adopt definitions that are not only increasingly specific with respect to presentation type, illness severity, and chronicity but also specific with respect to function.

III. Theoretical Perspectives on Insomnia

Although there are certainly more than two models of insomnia [23], we have focused on the two that are most inclusive (i.e., include hyperarousal, homeostatic, cognitive, and behavioral components) and/or have direct implications for the neurobiology of insomnia.

A. Neurocognitive Model

This model is based on the behavioral perspective that chronic insomnia occurs acutely in association with predisposing and precipitating factors and chronically in association with perpetuating factors (e.g., extension of sleep opportunity) [24]. The model extends the behavioral perspective by explicitly allowing for the possibility that conditioned arousal may act as a perpetuating factor. The concept of arousal is expressed in terms of somatic, cognitive, and cortical arousal. Somatic arousal corresponds to measures of metabolic rate, cognitive arousal typically refers to mental constructs such as worry rumination, and cortical arousal refers to the level of cortical activation (but may also include all of the central nervous system). Cortical arousal, it is hypothesized, occurs as a result of classical conditioning and allows for abnormal levels of sensory and information processing, as well as long-term memory formation. These phenomena, in turn, are directly linked to sleep continuity disturbance and/or sleep state misperception (i.e., paradoxical insomnia).

Specifically, *enhanced sensory processing* around sleep onset and during NREM sleep is thought to make the individual particularly vulnerable to perturbation by environmental stimuli (and/or interoception or nocioception), and these events directly interfere with sleep initiation and/or maintenance. *Enhanced information processing* during NREM sleep may blur the phenomenological distinction between sleep and wakefulness. That is, one cue for "knowing" that one is asleep is the lack of awareness for events occurring during sleep. Enhanced information processing may therefore account for the tendency in insomnia to judge polysomnographic sleep as wakefulness 25–31. Finally, *enhanced long-term memory* (i.e., the attenuation of the normal mesograde amnesia of sleep) around sleep onset and during NREM sleep may interfere with the subjective experience of sleep initiation and duration. Normally, subjects cannot recall information from periods immediately before sleep 32–35, during sleep 36–38, or during brief arousals from sleep [39,40]. An enhanced ability to encode and retrieve information in insomnia would be expected to influence judgments about sleep latency, wakefulness after sleep onset, and sleep duration. A schematic representation of the neurocognitive model is provided in Figure 2.

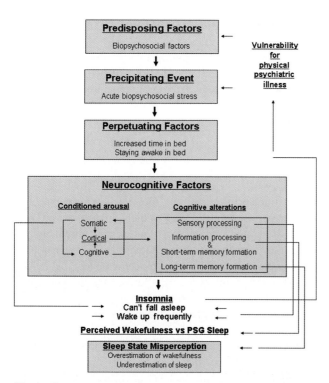

Figure 2 Neurocognitive Model. This version of the neurocognitive model is adapted from the schematic recently published in the fourth edition of the *Principles and Practice of Sleep Medicine*. The representation differs from those previously published in that (1) the model explicitly incorporates the Spielman three-factor model, (2) long-term memory formation is explicitly linked to the phenomenon of sleep state misperception, and (3) sleep state misperception itself is allowed to be related to either the overestimation of wakefulness and/or the underestimation of sleep.

The relative strengths of the neurocognitive model is that it defines hyperarousal, it embraces the notion that cognitive arousal and cortical arousal are related phenomena, and it specifies how cognitive or cortical arousal may interfere with sleep initiation and maintenance through heightened sensory and information processing. The last of these is particularly important for the neurobiological study of insomnia because it suggests that structures other than the "on-and-off" switches for sleep and wakefulness may be altered in insomnia.

B. Psychobiological Inhibition Model

Although the majority of insomnia models conceptualize insomnia as a disorder of hyperarousal, Espie [21,22] has proposed an important alternative point of view, suggesting that insomnia occurs, at least initially, in association with the failure to inhibit wakefulness. This perspective suggests that insomnia may arise as a natural response to stress. That is, when the organism is under threat, it is adaptive to sustain wakefulness until the threat is resolved. This adaptive response, however, is thought to evolve into a chronic condition as a result of two cognitive phenomena. First, when individuals are unable to sleep, attention is drawn to an otherwise automatic process. The very processes of attending, in turn, prevent perceptual and behavioral disengagement and sleep initiation. Second, when subjects are unable to sleep, effort is expended "trying" to fall asleep and this effort, like enhanced attention, only extends wakefulness. This increased attention and intention result in sustained wakefulness, undermine what is normally an automatic process, and set the stage for additional cognitive, behavioral, and perhaps neurobiological changes.

The relative strengths of the psychobiological inhibition model are that it defines how acute insomnia may become chronic, it specifies how this change may occur in association with cognitive factors, and, most importantly, it conceptualizes the inability to fall asleep in terms of the failure to inhibit wakefulness, thus potentially implicating different neuronal mechanisms than those associated with sleep initiation and termination.

IV. Brief Review of the Neurobiology of Sleep and Wakefulness

Based on the early work of Von Economo [41] and Moruzzi [42–44], it has become well established that cortical arousal is regulated by the ascending reticular activating system (ARAS). This system originates in the brainstem and has two major branches.

One branch originates from a network of neurons including the pedunculopontine and the laterodorsal tegmental nuclei, inputs into the thalamus, and activates the thalamic relays that densely innervate the cortex. This system is primarily cholinergic, and the source neurons fire maximally during wakefulness and REM sleep [1,2,45,46].

The other branch originates from a series of neurons including the locus coeruleus, the dorsal and medial raphe, and the tuberomammillary cells, and bypasses the thalamus to innervate neurons in the lateral hypothalamic area, in the basal forebrain, and throughout the cortex. The ascending aspect of this system is noradrenergic, serotoninergic, and histaminergic. The output from the lateral hypothalamus may be augmented through orexin, and the end target neurons contain acetylcholine or γ-aminobutyric acid. Neurons within the monoaminergic aspect of this system fire maximally during wakefulness, fire more slowly during NREM sleep, and are relatively silent during REM sleep. Neurons within the orexinergic aspect of this system fire maximally during both wakefulness and REM sleep [1,2,45,46].

Unclear from this characterization is how this system is permissive of, or actively promotes, sleep. To accomplish this it is necessary to posit that there is either a gating system or a related descending system that exerts influence over the structures that initiate cortical arousal. In the case of the cholinergic branch of the ARAS, there is substantial evidence to suggest that the reticular nucleus of the thalamus blocks afferent flow and thereby permits cortical synchronization.

In the case of the monoaminergic branch of the ARAS, investigators (during the 1980s and 1990s) found a candidate mechanism for what might serve as the switch for a "descending dearousal system"; the switch being the ventrolateral preoptic area (VLPO) [45,46]. The VLPO, which is maximally active during sleep, has major *outputs* to most, if not all, hypothalamic and brainstem components of the monoaminergic aspect of the ARAS and contains inhibitory neurotransmitters (i.e., galanin and γ-aminobutyric acid). Thus the VLPO appears to be uniquely positioned to function as an off switch. This putative function was confirmed by Saper and colleagues, who have shown that lesions within this region reduce NREM and REM sleep by more than 50% [45,46].

The Saper group has also demonstrated that the VLPO has major *inputs* from the hypothalamic and brainstem components of the monoaminergic aspect of the ARAS and that the VLPO is strongly inhibited by noradrenaline and serotonin. The existence of such inputs and neurotransmitter effects suggests that the VLPO not only inhibits wakefulness but also is inhibited by wakefulness. Saper and colleagues have likened this reciprocal relationship between the VLPO and the ARAS to the functioning of a "flip-flop circuit" [45,46]. This electrical engineering–based analogy provides a framework for conceptualizing how the wake-promoting and sleep-promoting halves of the circuit may strongly inhibit the other and in so doing create a bistable feedback loop that reinforces wakefulness and sleep and prevents intermediate states. This elegant conceptualization, however, does not explain how sleep is initiated and terminated; it only explains how sleep and wakefulness tend to occur in a consolidated fashion. To accomplish this, there must also be a system that impinges on the circuit and allows for homeostasis and possibly allostasis.

In the case of sleep/wake homeostasis, there must be a process that represents the accumulation of wakefulness and/or sleep that can act to "trip the switch." The concept of sleep/wake homeostasis (and its interaction with the circadian system) has been described theoretically and tested empirically by Borbely and colleagues [47–50]. In this model, the accumulation of wakefulness is represented by "Process S" and is measured in terms of the relationship between the duration of wakefulness and the discharge of slow-wave activity during NREM sleep. To date, the neurobiological structures that comprise the "sleep homeostat" are unknown. One candidate for a process that may represent the duration of wakefulness is the accumulation of adenosine within the basal forebrain. Experimental work with this hypothesis has shown that adenosine levels rise in proportion to the duration of wakefulness and that, when injected into the basal forebrain, adenosine induces sleep and promotes activity within the VLPO. In the case of sleep/wake allostasis, it has been proposed that orexin neurons within the posterior half of lateral hypothalamus reinforce wakefulness (promote sustained wakefulness) and thereby act as a "finger" on the flip-flop switch that prevents unwanted transitions into sleep.

V. Neurobiology of Sleep and Wakefulness: Implications for Insomnia

The preceding description of the normal regulation of sleep and wakefulness suggests that insomnia may occur in association with one of several neurobiological abnormalities. First, the switch itself may be malfunctioning. Saper and colleagues describe this as follows:

> Mathematical models show that when either side of a flip-flop neural circuit is weakened, homeostatic forces cause the switch to ride closer to its transition point during both states. As a result, there is an increase in transitions, both during the wake and the sleep periods, regardless of which side is weakened. This is certainly seen in animals with VLPO lesions, which fall asleep about twice as often as normal animals, wake up much more often during their sleep cycle and, on the whole, only sleep for about one-quarter as long per bout—in other words, they wake up and are unable to fall back asleep during the sleep cycle, but also are chronically tired, falling asleep briefly and fitfully during the wake cycle. [45]

This description appears to characterize not so much primary insomnia but rather sleep as it occurs in neonates or infants, and insomnia as it occurs in the elderly (i.e., polyphasic sleep with middle and/or late insomnia) and/or in patients with narcolepsy. Thus it would seem that an inherently "faulty switch" is not likely to be the neurobiological defect that occurs with primary insomnia.

Second, chronic activation of the monoaminergic branch of the ARAS might lead to some form of desensitization and/or a compensatory down-regulation, which results in insufficient "force" to trip the switch, which tends to favor the wake (on) position (i.e., there is a failure to inhibit wakefulness and/or substantially more wakefulness is required to flip the switch to the sleep position). In this instance, we might expect decreased activation within the nuclei that input to the VLPO (e.g., the locus coeruleus, the dorsal and/or medial raphe, and/or the tuberomammillary cells). From a neuroendocrine point of view, however, we might expect to see continued evidence of hyperarousal in parallel with the neurobiological down-regulation; that is, patients with chronic insomnia would exhibit hypercortisolemia, excessive secretion of the monoamines, and/or even hypocretin (orexin), despite diminished central nervous system activity. Evidence for some of these possibilities, which are presaged by the psychobiological inhibition model, is reviewed in the section on neuroendocrine and neuroimaging measures of insomnia.

Finally, it is possible that the neurobiological abnormalities that occur with insomnia occur either predominately within (or simultaneously within) the "cholinergic" branch of the ARAS and appear as altered functioning within the thalamus and basal forebrain and across the cortex. For example, we might expect the following:

- Reduced activity during wakefulness within the adenosinergic regions of the basal forebrain
- Overall decreased cortical arousal during wakefulness
- Increased activity during the sleep period within the thalamic nuclei related to sensory processing, and reduced activity within the sensory gating nuclei (i.e., the reticular nucleus)
- Overall increased cortical arousal during sleep

The evidence for these possibilities, which are presaged by the neurocognitive model, is also reviewed in the next section.

VI. Neurophysiologic, Neuroendocrine, and Neuroimaging Measures of Insomnia

A. Neurophysiologic Measures of Insomnia

To date, several studies have shown that patients with primary insomnia exhibit more cortical arousal than either good sleepers or patients with insomnia secondary to major depression [51–57]. Specifically, these studies show that patients with primary insomnia exhibit more high-frequency electroencephalogram activity (beta and gamma frequencies) at sleep onset and during NREM sleep. There is also evidence that (1) patients with sleep state misperception (i.e., paradoxical insomnia) exhibit more beta electroencephalogram activity than good sleepers or patients with primary insomnia [57], and (2) beta activity is negatively associated with the perception of sleep quality [58,59] and is positively associated with the degree of subjective-objective discrepancy [56]. Taken together, these data suggest that cortical arousal may occur uniquely in association with primary insomnia (versus secondary insomnia) and that this form of arousal may be associated with the tendency toward sleep state misperception.

Although the data acquired from this measurement strategy appear to strongly support the idea that cortical arousal may be a biomarker for insomnia (theoretically appealing to the extent that the increased occurrence of beta and gamma activity is thought to be permissive of increased sensory and information processing), the lack of replication across larger-scale contemporary investigations (e.g., see reference 60) and unpublished studies (e.g., see references 61 and 62) suggests that this approach has some limitations. In our hands, the occurrence of beta and gamma activity not only varies with trait considerations (diagnostic category) but also appears to be mediated or moderated by a variety of factors, including first-night effects (e.g., see references 63), prior sleep debt, degree of circadian dysrhythmia, type of insomnia, technical considerations, and extent to which there is environmental noise.

B. Neuroendocrine Measures of Insomnia

Activation of the hypothalamic-pituitary-adrenal (HPA) axis may provide further evidence that insomnia involves, or results from, chronic activation of the stress response system. Other neuroendocrine measures, including norepinephrine and melatonin, have been examined as potential correlates of insomnia.

1. Urinary Measures

An early study of urinary-free 11-hydroxycorticosteroids (11-OHCS) in young-adult good and poor sleepers found that the mean 24-hour rate of 11-OHCS excretion over 3 days was significantly higher in the poor sleepers [64]. A subsequent study of urinary cortisol and epinephrine in middle-aged good and poor sleepers found no significant differences, although poor sleepers showed a trend toward higher urinary cortisol and epinephrine secretion [6]. More recently, Vgontzas and colleagues [65,66] collected 24-hour urine specimens for urinary-free cortisol, catecholamines (DHPG and DOPAC), and growth hormone and correlated these measures with polysomnographic measures of sleep continuity and sleep architecture in subjects with primary insomnia. Urinary-free cortisol levels were positively correlated with total wake time, and the catecholamine measures were positively correlated with percent stage 1 sleep and with wake-after-sleep-onset time. Although not statistically significant, norepinephrine levels tended to correlate positively with percent stage 1 wake after sleep onset and negatively with percent slow-wave sleep. These data suggest that HPA axis and sympathetic nervous system activity is associated with objective sleep disturbance.

2. Plasma Measures

Plasma measures of adrenocorticotropic hormone (ACTH) and cortisol have also been compared among patients with primary insomnia and matched good sleepers. In one study, insomnia patients had significantly higher mean levels of ACTH and cortisol over 24 hours, with the largest group differences observed in the evening and first half of the night [65,66]. Patients with a high degree of sleep disturbance (sleep efficiency <70%) secreted higher amounts of cortisol than patients with less sleep disturbance. In contrast to these findings, a recent study of patients with primary insomnia and age- and gender-matched good sleepers found no differences in the mean amplitude or area under the curve for cortisol secretion over a 14-hour period (7 P.M. to 9 A.M.) [67].

Some of the variability of neuroendocrine findings in insomnia may be explained by intrusion of wakefulness into the measured sleep period. This is a particular concern for studies using urinary measures, which integrate biological activity over a long period. This possibility is important when considering causality, that is, whether increased HPA activity leads to insomnia or whether insomnia leads to increased HPA activity.

Finally, although the findings from various studies are not entirely consistent, the elevations in ACTH and cortisol before and during sleep in insomnia patients may help shed light on the intimate association between insomnia and major depression, which is also associated with HPA axis activation. Specifically, not only is insomnia a risk factor for major depression [68–76], but it is a prodromal symptom [77] and a ubiquitous [78,79] and persistent symptom of the condition as well. The common link may be that acute stress leads to activation of the HPA axis and insomnia and that chronic insomnia, in turn, leads to persistent activation of the HPA axis.

C. Neuroimaging Measures of Insomnia

Functional neuroimaging methods such as single-photon emission computed tomography (SPECT) and positron emission tomography (PET) may be used to identify regional brain blood flow or metabolic activity associated with particular tasks or states. Functional imaging techniques have been used to identify regional brain metabolic changes associated with sleep and sleep stages (for a review, see the work of Maquet and colleagues in references 80 and 81 and/or that of Nofzinger in reference 82) and have recently been applied to the study of insomnia. To date, two studies have been undertaken, one using TC-99HMPAO SPECT and one using fluorodeoxyglucose PET. In the SPECT study, imaging was conducted around the sleep onset interval for patients with primary insomnia and good sleeper subjects. Contrary to expectation, patients with insomnia exhibited a consistent pattern of hypoperfusion across eight preselected regions of interest, with the most prominent effect observed in the basal ganglia [83]. The frontal medial, occipital, and parietal cortices also showed significant decreases in blood flow compared with good sleepers. In the PET study, imaging data were acquired from patients with chronic insomnia and control subjects for an interval during wakefulness and during consolidated NREM. Patients with insomnia exhibited increased global cerebral glucose metabolism during wakefulness and NREM sleep [84]. In addition, it was found that patients with insomnia exhibited smaller declines in relative glucose metabolism from wakefulness to sleep in wake-promoting regions, including the ARAS, hypothalamus, and thalamus. A smaller decrease was also observed in areas associated with cognition and emotion, including the amygdala, hippocampus, and insular cortex, and in the anterior cingulate and medial prefrontal cortices.

Although results from these studies appear to be inconsistent, numerous methodological differences may help explain differences in the findings. For instance, the SPECT study, with its short time resolution, may have captured a more transient phenomenon that occurs when subjects with chronic and severe insomnia first achieve persistent sleep. The PET study, with its longer time resolution, may have captured a more stable phenomenon that occurs throughout NREM sleep in subjects with moderately chronic and severe insomnia. In addition to the temporal resolution issues, the PET study used a sample of insomnia patients who did not show objective sleep continuity disturbances in the laboratory, whereas the SPECT study included patients with objective sleep continuity disturbances. Thus the samples may have differed with respect to the type of insomnia, the degree of partial sleep deprivation, and the degree of sleep state misperception. Although further studies are needed, these preliminary investigations clearly demonstrate the feasibility of using functional neuroimaging methods in the study of insomnia, and suggest that insomnia complaints may have a basis in altered brain activity. For additional information on how imaging may be informative regarding the neurobiology of insomnia, refer to an article by Drummond and colleagues, published in 2004 [85].

VII. Call for an Integrative Perspective on Insomnia

In the introduction we noted that neuroscience research on insomnia has been limited by, among other things, the persistent and pervasive point of view that insomnia is strictly a psychological or behavioral disorder (and that "it" does not have a neurobiological cause, concomitant, or consequence). Although it is likely that cognitive and behavioral factors moderate, and may even mediate, the occurrence and intensity of insomnia, it is also abundantly clear that sleep and sleeplessness are "of the brain, and by the brain" [86]. Despite the likelihood that chronic insomnia occurs as an interaction of mind and body, many investigators appear to persist in taking one or the

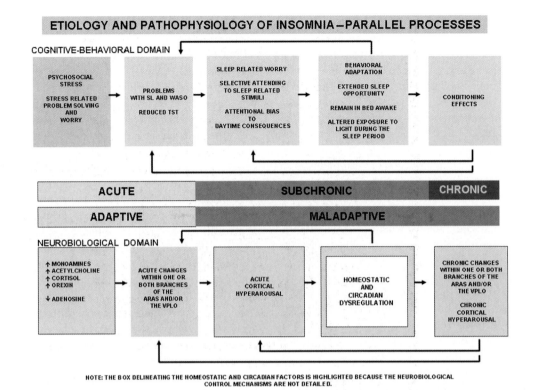

Figure 3 Etiology and Pathophysiology of Insomnia: Parallel Processes.

other perspective. Perhaps what is needed is an integrative perspective that allows for parallel processes. Figure 3 is intended to provide for such a representation, which might, in turn, encourage a collaborative approach to the delineation of both the mind and the body aspects of insomnia.

In closing, it should be noted that in June 2005 the National Institutes of Health convened a state-of-the-science conference on the topic of insomnia. At this conference, a panel of advisers was charged with assessing what is and is not known about insomnia and what types of research will be required to move the field forward. The full text of this document may be found at http://consensus.nih.gov.

With respect to neurobiology, the panel concluded that

The neural mechanisms underlying chronic insomnia are poorly understood. Studies aiming to identify neural mechanisms should use animal models and *in vivo* neural imaging approaches in people with insomnia and in individuals with normal sleep.

As we have reviewed, imaging work in insomnia is under way, but limited. Similarly, there are at least two nascent efforts to develop animal models of insomnia, one with Drosophila and one in rodents. Both animal models are in their preliminary stages of development and have been reported on only as meeting presentations [87,88]. This said, both of these experimental approaches, as highlighted by the panel, are likely to provide important insights into the neurobiology of insomnia.

References

1. Jones, B. E. (1993). The organization of central cholinergic systems and their functional importance in sleep-waking states. *Prog. Brain Res.* **98**, 61–71.
2. Jones, B. E. (1978). Toward an understanding of the basic mechanisms of the sleep-waking cycle. *Behav. Brain Sci.* **1(3)**, 495.
3. Turek, F. W. (1985). Circadian neural rhythms in mammals. *Annu. Rev. Physiol.* **47**, 49–64.
4. Turek, F. W. (2004). Circadian rhythms: from the bench to the bedside and falling asleep. *Sleep* **27(8)**, 1600–1602.
5. Turek, F. W. (2000). Introduction to chronobiology: sleep and the circadian clock. *In:* "Principles and Practice of Sleep Medicine" (M. H. Kryger, T. Roth, and W. C. Dement, eds.), pp. 319–320. W. B. Saunders: Philadelphia.
6. Turek, F. W. (1983). Neurobiology of circadian-rhythms in mammals. *Bioscience* **33(7)**, 439–444.
7. Czeisler, C. A., and Khalsa, B. S. (2000). The human circadian timing system and sleep-wake regulation. *In:* "Principles and Practice of Sleep Medicine" (M. H. Kryger, T. Roth, and W. C. Dement, eds.), pp. 353–376. W. B. Saunders: Philadelphia.
8. McCarley, R. W., and Hobson, J. A. (1975). Neuronal excitability modulation over the sleep cycle: a structural and mathematical model. *Science* **189(4196)**, 58–60.
9. Hobson, J. A., McCarley, R. W., and Wyzinski, P. W. (1975). Sleep cycle oscillation: reciprocal discharge by two brainstem neuronal groups. *Science* **189(4196)**, 55–58.
10. McCarley, R. W. (2004). Mechanisms and models of REM sleep control. *Arch. Ital. Biol.* **142(4)**, 429–467.

11. McCarley, R. W., and Massaquoi, S. G. (1992). Neurobiological structure of the revised limit cycle reciprocal interaction model of REM cycle control. *J. Sleep Res.* **1**(2), 132–137.
12. McCarley, R. W., and Massaquoi, S. G. (1986). Further discussion of a model of the REM sleep oscillator. *Am. J. Physiol.* **251**(6 Pt 2), R1033–R1036.
13. Massaquoi, S. G., and McCarley, R. W. (1992). Extension of the limit cycle reciprocal interaction model of REM cycle control: an integrated sleep control model. *J. Sleep Res.* **1**, 138–143.
14. McCarley, R. W., and Massaquoi, S. G. (1986). A limit cycle mathematical model of the REM sleep oscillator system. *Am. J. Physiol.* **251**, R1011–R1029.
15. Amzica, F., and Steriade, M. (1998). Cellular substrates and laminar profile of sleep K-complex. *Neuroscience* **82**(3), 671–686.
16. Steriade, M., Wyzinski, P., and Oakson, G. (1971). Activities in synaptic pathway between the motor cortex and ventrolateral thalamus underlying EEG spindle waves. *Int. J. Neurol.* **8**, 211–229.
17. Steriade, M., McCormick, D. A., and Sejnowski, T. J. (1993). Thalamocortical oscillations in the sleeping and aroused brain. *Science* **262**(5134), 679–685.
18. American Psychiatric Association. (1994). "Diagnostic and Statistical Manual of Mental Disorders," 4th edition, text revision. APA, Washington, D.C.
19. American Sleep Disorders Association. (1997). "The International Classification of Sleep Disorders: Diagnostic and Coding Manual," revised edition. ASDA, Rochester, MN.
20. Edinger, J. D., Bonnet, M. H., Bootzin, R. R., Doghramji, K., Dorsey, C. M., Espie, C. A., Jamieson, A.O., McCall, W. V., Morin, C. M., Stepanski, E. J., and American Academy of Sleep Medicine Work Group. (2004). Derivation of research diagnostic criteria for insomnia: report of an American Academy of Sleep Medicine Work Group. *Sleep* **27**(8), 1567–1596.
21. Espie, C. A., Broomfield, N. M., MacMahon, K. M. A., Macphee, L. M., and Taylor, L. M. (2006). The attention-intention-effort pathway in the development of psychophysiologic insomnia: an invited theoretical review. *Sleep Med. Rev.* **10**, 214–245.
22. Espie, C. A. (2002). Insomnia: conceptual issues in the development, persistence, and treatment of sleep disorder in adults. *Annu. Rev. Psychol.* **53**, 215–243.
23. Perlis, M. L., Smith, M. T., and Pigeon, W. R. (2005). Etiology and Pathophysiology of Insomnia. *In:* "Principle and Practice of Sleep Medicine" (M. H. Kryger, T. Roth, and W. C. Dement, eds.), pp. 714–725. Elsevier Saunders: Philadelphia.
24. Spielman, A., Caruso, L., and Glovinsky, P. (1987). A behavioral perspective on insomnia treatment. *Psychiatr. Clin. North Am.* **10**(4), 541–553.
25. Borkovec, T., Lane, T., and van Oot, P. (1981). Phenomenology of sleep among insomniacs and good sleepers: wakefulness experience when cortically asleep. *J. Abnorm. Psychol.* **90**, 607–609.
26. Coates, T. J., Killen, J., Silberman, S., Marchini, J., Hamilton, S., and Thoresen, C. (1983). Cognitive activity, sleep disturbance, and stage specific differences between recorded and reported sleep. *Psychophysiology* **20**, 243.
27. Coates, T. J., Killen, J., George, J., Marchini, E., Hamilton, S., and Thoresen, C. (1982). Estimating sleep parameters: a multitrait-multimethod analysis. *J. Consult. Clin. Psychol.* **50**, 345–352.
28. Mendelson, W., James, S., Garnett, D., Sack, D., and Rosenthal, N. (1986). A psychophysiological study of insomnia. *Psychiatry Res.* **19**, 267–284.
29. Mendelson, W., Martin, J., Stephens, H., Giesen, H., and James, S. (1988). Effects of flurazapam on sleep, arousal threshold, and perception of being asleep. *Psychopharmacology (Berl.)* **95**, 258–262.
30. Engle-Friedman, M., Baker, E., and Bootzin, R. (1985). Reports of wakefulness during EEG identified stages of sleep. *Sleep Res.* **14**, 152.
31. Mercer, J. D., Bootzin, R. R., and Lack, L. C. (2002). Insomniacs' perception of wake instead of sleep. *Sleep* **25**(5), 564–571.
32. Wyatt, J. K., Bootzin, R. R., Allen, J. J., and Anthony, J. L. (1997). Mesograde amnesia during the sleep onset transition: replication and electrophysiological correlates. *Sleep* **20**(7), 512–522.
33. Wyatt, J. K., Bootzin, R. R., Anthony, J. L., and Bazant, S. (1994). Sleep onset is associated with retrograde and anterograde amnesia. *Sleep* **17**(6), 502–511.
34. Guilleminault, C., and Dement, W. (1977). Amnesia and disorders of excessive daytime sleepiness. *In:* "Neurobiology of Sleep and Memory" (R. Drucker-Colin and J. McGaugh, eds.), pp. 439–456. New York: Academic Press.
35. Portnoff, G., Baekeland, F., Goodenough, D. R., Karacan, I., and Shapiro, A. (1966). Retention of verbal materials perceived immediately prior to onset of non-REM sleep. *Percept. Mot. Skills* **22**, 751–758.
36. Wood, J., Bootzin, R., Kihlstrom, J., and Schachter, D. (1992). Implicit and explicit memory for verbal information presented during sleep. *Psychol. Sci.* **3**(4), 236–239.
37. Bootzin, R., Fleming, G., Perlis, M. L., Wyatt, J., and Schachter, D. (1991). Short and long term memory for stimuli presented during sleep. *Sleep Res.* **20**(A), 258.
38. Koukkou, M., and Lehmann, D. (1968). EEG and memory storage experiments with humans. *Electroencephalogr. Clin. Neurophysiol.* **25**, 455–462.
39. Goodenough, D., Sapan, J., and Cohen, H. (1971). Some experiments concerning the effects of sleep on memory. *Psychophysiology* **8**, 749–762.
40. Bonnet, M. (1983). Memory for events occurring during arousal from sleep. *Psychophysiology* **20**, 81–87.
41. Von Economo, C. (1949). Sleep as a problem of localization. *J. Nerv. Ment. Dis.* **71**, 249–259.
42. Moruzzi, G. (1965). Sleep mechanisms: summary statement. *Prog. Brain Res.* **18**, 241–243.
43. Moruzzi, G. (1964). Reticular influences on the EEG. *Electroencephalogr. Clin. Neurophysiol.* **16**, 2–17.
44. Moruzzi, G., and Magoun, H. W. (1995). Brain stem reticular formation and activation of the EEG. 1949. *J. Neuropsychiatry Clin. Neurosci.* **7**(2), 251–267.
45. Saper, C. B., Scammell, T. E., and Lu, J. (2005). Hypothalamic regulation of sleep and circadian rhythms. *Nature* **437**(7063), 1257–1263.
46. Saper, C. B., Chou, T. C., and Scammell, T. E. (2001). The sleep switch: hypothalamic control of sleep and wakefulness. *Trends Neurosci.* **24**(12), 726–731.
47. Borbely, A. A., and Achermann, P. (1999). Sleep homeostasis and models of sleep regulation. *J. Biol. Rhythms* **14**(6), 557–568.
48. Borbely, A. A. (1998). Processes underlying sleep regulation. *Horm. Res.* **49**(3–4), 114–117.
49. Borbely, A. A., and Achermann, P. (2000). Sleep homeostasis and models of sleep regulation. *In:* "Principles and Practice of Sleep Medicine" (M. H. Kryger, T. Roth, and W. C. Dement, eds.), 3rd edition, pp. 377–390. W. B. Saunders: Philadelphia.
50. Borbely, A. A. (1982). A two process model of sleep regulation. *Hum. Neurobiol.* **1**(3), 195–204.
51. Freedman, R. (1986). EEG power in sleep onset insomnia. *Electroencephalogr. Clin. Neurophysiol.* **63**, 408–413.
52. Merica, H., and Gaillard, J. M. (1992). The EEG of the sleep onset period in insomnia: a discriminant analysis. *Physiol. Behav.* **52**(2), 199–204.
53. Merica, H., Blois, R., and Gaillard, J. M. (1998). Spectral characteristics of sleep EEG in chronic insomnia. *Eur. J. Neurosci.* **10**, 1826–1834.
54. Lamarche, C. H., and Ogilvie, R. D. (1997). Electrophysiological changes during the sleep onset period of psychophysiological insomniacs, psychiatric insomniacs, and normal sleepers. *Sleep* **20**(9), 724–733.

55. Jacobs, G. D., Benson, H., and Friedman, R. (1993). Home-based central nervous system assessment of a multifactor behavioral intervention for chronic sleep-onset insomnia. *Behav. Ther.* **24**, 159–174.
56. Perlis, M. L., Smith, M. T., Orff, H. J., Andrews, P. J., and Giles, D. E. (2001). Beta/Gamma EEG activity in patients with primary and secondary insomnia and good sleeper controls. *Sleep* **24(1)**, 110–117.
57. Krystal, A. D., Edinger, J. D., Wohlgemuth, W. K., and Marsh, G. R. (2002). NREM sleep EEG frequency spectral correlates of sleep complaints in primary insomnia subtypes. *Sleep* **25(6)**, 630–640.
58. Hall, M., Buysse, D. J., Nowell, P. D., Nofzinger, E. A., Houck, P., Reynolds, C. F., III, and Kupfer, D. J. (2000). Symptoms of stress and depression as correlates of sleep in primary insomnia. *Psychosom. Med.* **62(2)**, 227–230.
59. Nofzinger, E. A., Price, J. C., Meltzer, C. C., Buysee, D. J., Villemagne, V. L., Miewald, J. M., et al. Sembrat, R.C., Steppe, D. A., and Kupfer, D. J. (2000). Towards a neurobiology of dysfunctional arousal in depression: the relationship between beta EEG power and regional cerebral glucose metabolism during NREM sleep. *Psychiatry Res.* **98(2)**, 71–91.
60. Bastien, C. H., LeBlanc, M., Carrier, J., and Morin, C. M. (2003). Sleep EEG power spectra, insomnia, and chronic use of benzodiazepines. *Sleep* **26(3)**, 313–317.
61. Buysse, D. (2005). Failure to replicate that patients with insomnia exhibit increased Beta/Gamma activity as compared to good sleepers. Personal communication.
62. Perlis, M. L. (2005). Failure to replicate that patients with insomnia exhibit increased Beta/Gamma activity as compared to good sleepers. Unpublished work.
63. Curcio, G., Ferrara, M., Piergianni, A., Fratello, F., and De Gennaro, L. (2004). Paradoxes of the first-night effect: a quantitative analysis of anteroposterior EEG topography. *Clin. Neurophysiol.* **115(5)**, 1178–1188.
64. Johns, M. W. (1971). Relationship between sleep habits, adrenocortical activity and personality. *Psychosom. Med.* **33(6)**, 499–508.
65. Vgontzas, A. N., Tsigos, C., Bixler, E. O., Stratakis, C. A., Zachman, K., Kales, A., Vela-Bueno, A., and Chrousos, G. P. (1998). Chronic insomnia and activity of the stress system: a preliminary study. *J. Psychosom. Res.* **45(1 Spec No)**, 21–31.
66. Vgontzas, A. N., Bixler, E. O., Lin, H. M., Prolo, P., Mastorakos, G., Vela-Bueno, A., Kales, A., and Chrousos, G. P. (2001). Chronic insomnia is associated with nyctohemeral activation of the hypothalamic-pituitary-adrenal axis: clinical implications. *J. Clin. Endocrinol. Metab.* **86(8)**, 3787–3794.
67. Riemann, D., Klein, T., Rodenbeck, A., Feige, B., Horny, A., Hummel, R., Weske, G., Al-Shajlawi, A., and Voderholzer, U. et al. (2002). Nocturnal cortisol and melatonin secretion in primary insomnia. *Psychiatry Res.* **113(1–2)**, 17–27.
68. Ford, D. E., and Kamerow, D. B. (1989). Epidemiologic study of sleep disturbances and psychiatric disorders. An opportunity for prevention? [see comments]. *J. Am. Med. Assoc.* **262(11)**, 1479–1484.
69. Dryman, A., and Eaton, W. W. (1991). Affective symptoms associated with the onset of major depression in the community: findings from the US National Institute of Mental Health Epidemiologic Catchment Area Program. *Acta Psychiatr. Scand.* **84(1)**, 1–5.
70. Breslau, N., Roth, T., Rosenthal, L., and Andreski, P. (1996). Sleep disturbance and psychiatric disorders: a longitudinal epidemiological study of young adults. *Biol. Psychiatry* **39(6)**, 411–418.
71. Chang, P. P., Ford, D. E., Mead, L. A., Cooper-Patrick, L., and Klag, M. J. (1997). Insomnia in young men and subsequent depression: the Johns Hopkins Precursors Study. *Am. J. Epidemiol.* **146(2)**, 105–114.
72. Livingston, G., Blizard, B., and Mann, A. (1993). Does sleep disturbance predict depression in elderly people? A study in inner London [see comments]. *Br. J. Gen. Pract.* **43(376)**, 445–448.
73. Mallon, L., Broman, J. E., and Hetta, J. (2000). Relationship between insomnia, depression, and mortality: a 12-year follow-up of older adults in the community. *Int. Psychogeriatr.* **12(3)**, 295–306.
74. Roberts, R. E., Shema, S. J., Kaplan, G. A., and Strawbridge, W. J. (2000). Sleep complaints and depression in an aging cohort: a prospective perspective [in process citation]. *Am. J. Psychiatry* **157(1)**, 81–88.
75. Vollrath, M., Wicki, W., and Angst, J. (1989). The Zurich study. VIII. Insomnia: association with depression, anxiety, somatic syndromes, and course of insomnia. *Eur. Arch. Psychiatry Neurol. Sci.* **239(2)**, 113–124.
76. Weissman, M. M., Greenwald, S., Nino-Murcia, G., and Dement, W. C. (1997). The morbidity of insomnia uncomplicated by psychiatric disorders. *Gen. Hosp. Psychiatry* **19(4)**, 245–250.
77. Perlis, M. L., Giles, D. E., Buysse, D. J., Tu, X., and Kupfer, D. J. (1997). Self-reported sleep disturbance as a prodromal symptom in recurrent depression. *J. Affect Disord.* **42(2,3)**, 209–212.
78. Perlis, M. L., Giles, D. E., Buysse, D. J., Thase, M. E., Tu, X., and Kupfer, D. J. (1997). Which depressive symptoms are related to which sleep electroencephalographic variables? *Biol. Psychiatry* **42(10)**, 904–913.
79. Thase, M. E. (1999). Antidepressant treatment of the depressed patient with insomnia. *J. Clin. Psychiatry* **60(Suppl 17)**, 28–31.
80. Maquet, P., and Franck, G. (1989). Cerebral glucose metabolism during the sleep-wake cycle in man as measured by positron emission tomography. *In:* "Sleep '88" (J. Horne, ed.), pp. 76–78. Gustav Fischer, Stuttgart, Germany.
81. Maquet, P., and Phillips, C. (1998). Functional brain imaging of human sleep. *J. Sleep Res.* **7(Suppl 1)**, 42–47.
82. Nofzinger, E. A. (2005). Neuroimaging and sleep medicine. *Sleep Med. Rev.* **9(3)**, 157–172.
83. Smith, M. T., Perlis, M. L., Chengazi, V. U., Pennington, J., Soeffing, J., Ryan, J. M., and Giles, D. E. (2002). Neuroimaging of NREM sleep in primary insomnia: a Tc-99-HMPAO single photon emission computed tomography study. *Sleep* **25(3)**, 325–335.
84. Nofzinger E. A., Buysee, D. J., Germain, A., Price, J. C., Miewald, J. M., and Kupfer, D. J. (2004). Functional imaging evidence for hyperarousal in insomnia. *Am. J. Psychiatry.* **161(11)**, 2126-2128.
85. Drummond, S. P., Smith, M. T., Orff, H. J., Chengazi, V., and Perlis, M. L. (2004). Functional imaging of the sleeping brain: review of findings and implications for the study of insomnia. *Sleep Med. Rev.* **8(3)**, 227–242.
86. Hobson, J. A. (2005). Sleep is of the brain, by the brain and for the brain. *Nature* **437(7063)**, 1254–1256.
87. Cano, G., and Saper CB. (2005). Mechanisms underlying stress-induced insomnia. Abstract. 35rd Annual Meeting of the Society for Neuroscience, Washington, D.C.
88. Seugnet, L., Israel, S., Toledo, R., and Shaw, P. J. (2004). Evaluating a *Drosophila* model of insomnia. Abstract. *Sleep* **27(A859)**.

69

Organic Chemicals

John L. O'Donoghue, VMD, PhD, DABT

Keywords: *mechanisms of action, neuropathology, organic chemicals, pathogenesis*

I. History and Nomenclature
II. Epidemiology
III. Risk Factors
IV. Pathogenesis: Etiological Agents and Factors Affecting Mode of Action
V. Neuropathology of Organic-Chemical-Induced Neurotoxicity
VI. Pathophysiology
VII. Biochemistry and Molecular Mechanisms
VIII. Explanation of Signs and Symptoms in Relation to Pathophysiology
IX. Natural History
X. Management of Organic-Chemical-Induced Neurotoxicity
   References

I. History and Nomenclature

Organic chemistry is the branch of science that deals with chemicals made from carbon. Chemicals classified as organic are minimally composed of carbon and hydrogen. Prior to the nineteeth century, the term *organic* was used to identify a subclass of chemicals because initially it was thought that such chemicals could only be formed with the aid of a vital force existing in living cells. In 1827, this view began to change when Friedrich Wöhler was able to synthesize urea, an organic substance, from ammonium cyanate, an inorganic substance. By 1850, the view that a vital force was required for the creation of chemicals such as acetic acid was no longer prevalent, as a number of examples of synthetic chemicals had been published. Organic chemicals that induce neurotoxicity may be naturally occurring or synthetic. Those that are naturally occurring products of plants or animals are referred to as *toxins*, and those that are synthetic are referred to as *toxicants*. Beginning in the early 1850s and accelerating since the 1940s, the growth of organic chemistry has been rapid and strong. Consequently, the American Chemical Society's Chemical Abstract Service (CAS) reported that by September 2005 it had listed more than 26.6 million substances; of these, almost 9 million were available commercially and approximately 239,000 were inventoried or regulated substances.

Although these numbers include inorganic and organic substances, the majority of these substances are organic chemicals. The continuing growth of organic chemistry can be seen by the number of new CAS-listed chemicals, which increases by about 4000 new substances each day. With the advent of nanotechnology, organic chemicals with physicochemical properties that are different from existing chemicals are being introduced into commerce and medicine. The ability of some of these nanoparticulates to enter the brain through the olfactory pathway is likely to affect how the neurotoxicology of organic substances is assessed.

Organic chemicals are broadly divided into those that are aliphatic (carbon atoms are linked in straight chains) and those that are aromatic (carbon atoms are linked in a cyclic structure). Heterocycles are cyclic organic molecules with a least one heteroatom (e.g., oxygen, nitrogen, phosphorus, or sulfur) in the ring. Polymers can be organic or inorganic chemicals depending on their atomic makeup. Organic chemicals can be further subdivided depending on the substituent groups attached to them. For example, alcohols, ketones, aldehydes, and acids make up groups of chemicals with varying degrees of oxygenation. Similarly, halogens, sulfur, and other atoms contribute to defining a chemical's structure and how it does or does not interact with the nervous system. Well-recognized man-made polymers include plastics. Essential organic biopolymers include proteins, nucleic acids, and polysaccharides.

Ionization of an organic chemical can play a role in its potential for inducing neurotoxicity, as ionized chemicals typically do not cross the blood-brain barrier. The potential for ionization of organic chemicals can be altered by their exposure to mineral acids or bases (e.g., hydrochloric acid), creating salts of various chemicals. This is commonly done with pharmaceutical preparations because it makes them more water soluble and increases their uptake and urinary excretion. Ingestion of organic chemicals can also lead to the formation of salts when the chemicals are exposed to the hydrochloric acid normally present in gastric secretions. Organic chemicals that enter the body through the skin or lungs are not ionized in this manner. Therefore, the potential for chemicals to readily pass through the blood-brain barrier may differ depending upon route of exposure.

Although the history of neurotoxicity due to exposure to naturally occurring animal and plant toxins and minerals extends back to antiquity, the neurotoxicity associated with exposure to synthetic chemicals only began in the late 1800s and early 1900s, when the industrial production and refining of organic chemicals led to their increasingly widespread use and substitution for inorganic chemicals and plant- and animal-derived materials. The early use of organic chemicals in large-scale manufacturing operations was often accompanied by a lack of adequate ventilation and the absence of personal protective devices. By 1925, Alice Hamilton of the Harvard Medical School had collected enough case reports and records to publish a book highlighting that chemical exposure in the workplace was resulting in significant morbidity and mortality [1]. Although much of the information she presented deals with the toxicity associated with inorganic and natural substances, a range of effects, including many descriptions of neurotoxicity due to synthetic organic substances, are also described. For example, she included reports detailing the effects of inhalation or dermal absorption of methyl alcohol, including headache, dizziness, nausea, trembling, weakness, unsteady gait, impaired vision, and blindness resulting from optic nerve damage. The availability of chlorinated organic solvents in the early 1900s led to their use as substitutes for simple alcohols, benzene, and other fat-soluble substances. The introduction of these new solvents into commerce led to significant workplace exposures, including to trichloroethylene. As early as 1916, W. Plessner described the first cases of trigeminal sensory neuropathy in three men who had used a trichloroethylene solvent for only one full day or less in the manufacture of munitions [1]. Hamilton clearly described other situations in which synthetic organic substances caused significant neurotoxicity in workers, including those in which methyl bromide and 2,4-dinitrophenol were used [1].

Organic chemicals used for a number of purposes have been identified as neurotoxic either in animal studies or in case reports and epidemiological studies following human exposures. The exposure profiles of organic substances with neurotoxic properties include uses of industrial chemicals (acrylamide and n-hexane), pharmaceuticals (disulfiram and valproic acid), over-the-counter drugs (hexachlorophene), vitamins (pyridoxine, or vitamin B_6, and retinoic acids, or vitamin A), pesticides (lindane, methyl bromide, and organophosphates), environmental pollutants (methylmercury, and tetraethyl lead), fuel (gasoline), consumer products (pyrithione and toluene), cosmetics (acetyl ethyl tetramethyl tetralin), chemical warfare agents (sarin), food contaminants (tri-o-tolyl phosphate, or TOCP, and Spanish toxic oil, beverages (ethanol and methanol), and illicit drugs (1-methyl-4-phenyl-1,2,3,6-tetrahydropyridine, or MPTP, and methamphetamine).

II. Epidemiology

The ability to make a diagnosis of organic-chemical-induced neurotoxicity depends on the ability to show a dose relationship between a clinical situation and an exposure. The mere presence of a chemical in the body may be indicative of an exposure in certain cases, but with the ever-increasing sensitivity of analytical chemical methods, minute quantities of various chemicals are often reported. For example, 2,5-hexanedione is considered the proximate causative agent in the peripheral neuropathies induced by exposure to n-hexane and methyl *n*-butyl ketone. Its presence in human urine has been used as a marker for exposure to these chemicals [2]. However, the presence of 2,5-hexanedione has been recognized in "normal" human urine; therefore, a biological exposure limit recommended by the American Conference of Governmental Industrial Hygienists to control occupational exposures has been set at 5 mg of 2,5-hexanedione per gram of creatinine [3]. Urinary levels below this concentration are potentially derived from diet and ordinary exposures that occur during daily life, as well as occupational exposures. Although the neurotoxicity of many industrial, commercial, and pharmaceutical chemicals have been described and categorized, the size and growth of the chemicals listed by the CAS would strongly suggest that additional chemicals with neurotoxic properties have yet to be described.

The prevalence and incidence of disease due to exposure to neurotoxic organic chemicals have not been quantified by the National Institute of Neurological Disorders and Stroke or others, although it has been recognized that neurotoxicity may play a role in the incidence of certain relatively common conditions, such as Parkinson's disease. Neurodegenerative diseases (Parkinson's disease, Alzheimer's disease, and motoneuron disease) are among the most common diseases often hypothesized to be associated with environmental and occupational exposures of adults to organic chemicals. Park and associates [4] found that of approximately 2,614,346 deaths that occurred from 1992 to 1998 in 22 states, neurodegenerative diseases caused or contributed to 112,805 deaths (4.3%). In this study, groups associated with statistically significant higher rates of neurodegenerative diseases included individuals with exposure to solvents, hairdressers, and farmers. However, the ratios attached to these associations were generally small. If the estimates of the association between solvent exposure and neurodegenerative diseases are unbiased, they imply that approximately 10% of the deaths caused or affected by neurodegenerative diseases may be of individuals in occupations with the highest solvent exposures.

Because of the relatively high rate of ethanol ingestion in pregnant women (10.1% according to a 2002 Communicable Disease Center report), prenatal alcohol exposure is likely to be a leading cause of organic-chemical-induced neurotoxicity, with an estimated incidence of 10 per 1000 U.S. births for fetal alcohol syndrome, alcohol-related neurodevelopmental disorders, and alcohol-related birth defects. Neurotoxicity is a common undesirable side effect of many therapeutic agents and can limit the extent of treatment. Drug-induced neurotoxicity can also result in transgeneration effects, such as the neurodevelopmental deficits seen in children whose mothers were treated with valproic acid for seizure disorders.

Many of the early epidemiological studies of outbreaks of neurotoxicity due to food or beverage contaminants or occupational exposures were descriptive or investigational studies directed at identifying causation in well-recognized disease outbreaks. The earliest of these described a 1930 outbreak that was referred to as the Ginger-Jake syndrome and involved sensorimotor deficits, including signs and symptoms of numbness, tingling, weakness, and paralysis. These deficits resulted from axonal degeneration in the central nervous system (CNS) and peripheral nervous system (PNS). Individuals affected by "Jake paralysis" had consumed an extract of Jamaican ginger that had been adulterated with Lyndol, an oily material that is used in lacquers and varnishes and that contains TOCP. Estimates of the number of people affected by Jake paralysis are in the range of 20,000–100,000. A clinically similar epidemic of paralysis of initially unknown cause occurred in Meknes, Morocco, where an attack rate of 300 new cases of acute peripheral neuritis was being reported daily in September 1959. Over time, approximately 10,000 cases of peripheral nerve damage were reported from this single incident. Eventually, it was discovered that an aviation lubricating oil containing 3% TOCP had been sold into the local cooking oil market, resulting in organophosphate-induced axonopathy. Following the epidemiological investigation of a series of cases of sensorimotor neuropathy related to exposure to methyl *n*-butyl ketone, Allen [5] proposed a series of criteria that would be useful for investigating suspected neurotoxicants (Table 1).

Early epidemiological studies of neurotoxicity involved situations in which frank clinical disease was clearly prevalent, whereas more recent studies have often involved circumstances in which the health outcome of concern cannot be easily discerned from the normal background of neurological disease or impairment. Of the various epidemiological study designs available, the cross-sectional study has become the observational design most commonly used for neurotoxicological investigations. The methods used to detect effects in large populations of people often depend upon questionnaires or on manual or computer-administered neurobehavioral batteries. Less often, quantitative sensory testing, tremor analysis, electrophysiology, neuroimaging, and noninvasive testing methods are used for population studies. When studies are designed to be prospective or longitudinal, it may be possible to obtain reliable measures of environmental exposures or absorbed dose through analysis of urinary excretion rates or breath analysis. However, in retrospective studies exposure measurements are often not available. If the study groups are collected from a range of work sites or industries, assessment of exposures is even more problematic. In place of exposure analysis, job analyses conducted by individuals familiar with industrial and occupational exposures allocate test subjects to a range of potential exposure levels for calculation of dose–response effects. The World Health Organization has published criteria documents reviewing best practices for investigating chemicals or situations involving neurotoxicity [6,7].

Table 1 Criteria for Investigating Suspect Neurotoxicants

Criterion	Applicable Generalizations
Background disease incidence	The incidence of the disease under investigation must exceed the incidence of unrelated neurological disease in the population under study.
Consistency of clinical effects	The clinical pattern of the outbreak must be consistent with the pattern of the suspected agent (if known) or should be consistent with the well-recognized pattern of well-defined cases as the study progresses.
Dose–incidence relationship	There must be a relationship between the attack rate and the indexes of exposure to the suspected agent.
Dose–severity relationship	The severity of the disorder should be related to the exposure to the suspected agent, taking into account other susceptibility factors.
Analytical verification of exposure	The presence of the suspected agent in the exposure environment should be verified by analytical chemistry.
Confounding variables	In the presence of multiple suspected toxicants, epidemiological approaches should be used to isolate confounding exposures.
Prevention of new cases	Elimination of the suspect agent should result in no new cases.
Recovery from reversible effects	If the disorder is reversible, affected individuals should continue to improve after elimination of the suspect agent.
Experimental reproducibility of clinical manifestations	The neurotoxic effect should be reproducible in animal studies.
Experimental reproduction of morphological aspects	The histopathology of the experimental disorder should conform to that seen in human cases.

III. Risk Factors

Given that an organic chemical has the ability to interact with the CNS and PNS in a potentially adverse manner, important risk factors include those circumstances that allow significant exposures to occur. Table 2 summarizes a series of general scenarios associated with outbreaks of neurotoxicity in humans. A common observation in reports of occupation-related neurotoxicity is that the diseases are associated with poor personal or poor industrial hygiene practices. These include inadequate ventilation; open vats or containers that allow organic chemicals to escape into the environment; dipping hands and arms into chemical containers; cleaning equipment and clothing with workplace solvents; the absence of or incorrect use of personal protective equipment; and smoking tobacco, eating, or drinking in the workplace. In some situations, the spouses of exposed individuals have been affected when contaminated work clothes have been worn and/or cleaned at home. A related risk factor is the inadequate, inappropriate, or confusing labeling of containers, which has led to widespread food chain exposure to neurotoxic chemicals. In some cases, famine or criminal activity have resulted in contamination of the food supply with neurotoxic substances. Consequently, relatively large numbers of people consuming neurotoxic plants (e.g., *Lathyrus sativus*) or unknowingly consuming beverages (e.g., Ginger Jake Paralysis due to TOCP) or food stuffs (e.g., cooking oil containing TOCP) that have been adulterated with industrial chemicals have developed severe neurotoxicities. In abusive situations, children and young adults often intentionally sniff high concentrations of volatile organic substances (e.g., glues, solvents, and paints) to alter mood states (see the chapter on drug abuse). Less common are dietary fads that have led to high exposures to dietary factors with neurotoxic properties. Other, less common exposure routes are listed in Table 2.

Coexposure to other nonneurotoxic organic substances can be a risk factor for the development of neurotoxicity from organic substances (e.g., coexposure to methyl ethyl ketone increases the neurotoxicity of *n*-hexane and methyl *n*-butyl ketone even though methyl ethyl ketone alone does not produce peripheral neuropathy). Many organic substances undergo metabolic transformations following absorption. Although many of these transformations are mechanisms of detoxification, some of them activate an otherwise nontoxic material (e.g., *n*-hexane and a number of organophosphates that produce peripheral neuropathy require metabolic activation). Because these transformations are commonly inducible and the enzymatic processes involved can interact with multiple chemicals, coexposure to certain chemicals can lead to higher rates of activation of neurotoxic substances, lower rates of detoxification, or competition for binding sites on enzymes. The results of these interactions can enhance or inhibit neurotoxicity.

Pharmacogenetics related to organic-chemical-induced neurotoxicity is a little explored area. However, it is clear that genetic polymorphisms affect neurotoxicity and may help explain some situations in which highly variable or idiosyncratic human responses to organic chemicals are observed. This is perhaps more evident in neurotoxicities involving pharmaceuticals, where large numbers of people are intentionally exposed to organic chemicals. A genetic polymorphism affecting acetylation of drugs is a well-recognized explanation for the greater

Table 2 Risk Factors for Human Neurotoxicity Due to Organic Chemicals

Factor	Summary	Examples
Poor personal or industrial hygiene	Poor ventilation; lack of personal protective equipment; eating, drinking, or smoking at work site; cleaning work clothes at home	n-Hexane, methyl n-butyl ketone, chlordecone
Product mixing or labeling errors	Improper or confusing labeling or inadequate ingredient storage, leading to use of incorrect ingredients or ingredients at toxic levels	Hexachlorophene baby powder toxicity
Criminal activity	"Recycling" industrial oils for human consumption, substitution of cheaper toxic ingredients for more expensive nontoxic ingredients	Spanish toxic oil poisonings, Ginger Jake paralysis, Moroccan toxic oil paralysis, methanol sold in place of ethanol
Drug or chemical abuse or addiction	Recreational use of household products, industrial chemicals, medications, illicit drugs	Nitrous oxide, methyl n-butyl ketone, glue solvents, parkinsonian-like disease due to MPTP
Environmental contamination	Contamination of ambient environment with natural or synthetic chemicals	Shellfish toxin, methylmercury
Attempted suicide	Neurological sequelae of failed suicide attempts	Vacor rat poison
Dietary fads	Unusually high consumption of food substances or supplements	Pyridoxine ingestion, vitamin A
Famine or inadequate food supply	Consumption of toxic plants	Neurolathyrism
Side effects of therapy	Induction of neurotoxicity during therapeutic treatment	Antibiotic-induced deafness, vincristine neuropathy

risk for peripheral neuropathy induced with isoniazid in individuals of Caucasian ancestry. Acetylation reactions that lead to detoxification occur more slowly in Caucasians than in individuals with different racial backgrounds.

Although differences in biotransformation rates, because of sex, diet, and aging, are known to occur in humans and have been reported to affect the neurotoxicity of organic chemicals in laboratory animals, they generally have not shown significant effects on human neurotoxicity.

For neurotoxicants that damage the developing nervous system, the stage of brain development and the absence of an effective blood-brain barrier are significant risk factors determining the degree to which the nervous system is damaged and the location of injury.

IV. Pathogenesis: Etiological Agents and Factors Affecting Mode of Action

Organic chemicals can produce many effects on the nervous system [8,9]. Table 3 includes a list of nervous system functions affected by a variety of organic chemicals. Underlying these functional changes are cellular and subcellular changes that mediate the clinical and pathological appearance of the neurotoxicity.

The pathogenesis of neurotoxicity induced by exposure to organic chemicals depends on the specific chemical involved, the dose rate, and the developmental status of the exposed individual. Within classes or groupings of similar chemicals, the pathogenesis may vary significantly depending on the number and spacing of substituent groups. For example, among dicarbonyls, commonly referred to as diketones, the gamma spacing between the carbonyls, such as in 2,5-hexanedione, is critical for development of peripheral neuropathy. If the carbonyls are one carbon closer together, as in 2,4-pentanedione, foci of necrosis are seen in the brainstem. If the carbonyls are one carbon further apart, as in 2,6-hexanedione, no neurotoxicity is observed. Similarly, 2-mercaptopropionic acid and 3-mercaptopropionic acid, although similar in structure, produce different forms of neurotoxicity. Table 4 provides a simple grouping of various chemical structures and compares and contrasts examples of neurotoxic effects of the chemicals within that group.

Although chemical structure is an important determinant of what type of neurotoxicity occurs, the exposure or dose rate can also change the pathogenesis of neurotoxicity. Acrylamide is an example of a substance studied for years as a prototype for chemicals causing peripheral neuropathy. However, the clinical and morphological appearance of acrylamide-induced neurotoxicity in humans and laboratory animals varies depending on the dose rate. In laboratory animals given acrylamide at a high dose rate (≥ 30 mg/kg/day), ataxia and weakness predominate in the clinical picture and cerebellar Purkinje's cells and nerve terminals degenerate in the CNS and PNS. However, at a lower dose rate (10 mg/kg/day), nerve terminals still degenerate and peripheral nerves show evidence of degeneration, but Purkinje cell lesions are no longer seen.

Table 3 Examples of Nervous System Functions Affected by Organic Chemicals

Function	Effect	Chemical
Global	Irritability	Chlordecone
	Apathy or lethargy	2-Methoxyethanol
	Attention	Benzene
	Hallucinations or delusions	Methyl bromide, cocaine
	Anxiety	Amphetamines
	Mania	Corticosteroids
Cranial nerve	Smell	Styrene
	Sight	Methanol, clioquinol
	Color perception	Toluene, styrene
	Taste	Terbinafine
	Hearing	Neomycin, kanamycin
	Equilibrium	Streptomycin
	Trigeminal nerve function	Trichloroethylene, dichloroacetylene
Somatosensory	Altered touch, pressure, or pain	Allyl chloride
	Proprioception	Acrylamide
Motor	Muscular weakness	n-Hexane
	Paralysis	TOCP
	Spasticity	β-N-Oxalyamino-L-alanine
	Rigidity	MPTP
	Tremor	Chlordecone
	Dystonia	Phenothiazine
	Incoordination	Methylmercury
	Hyperactivity	2,4-Dinitrophenol
	Myoclonus	Methyl bromide
	Fasciculations	Anticholinesterases
	Cramps	Statins
	Seizures	Lindane
Autonomic	Sweating	Acrylamide
	Temperature control	Chlordane
	Gastrointestinal function	Methyl chloride
	Appetite	Dinitrobenzenes (o-, m-, p-)
	Blood pressure and cardiovascular function	Vacor
	Bladder control	DMAP
	Sexual function	DMAP, β-chloroprene

Note: DMAP, dimethylaminopropionitrile; MPTP, 1-methyl-4-phenyl-1,2,3,6-tetrahydropyridine; TOCP, tri-o-tolyl phosphate.

The stage of development of the organism (human or experimental animal) can have a profound effect on the pathogenesis of neurotoxicity. One of the best examples is from exposure to ethanol [10]. In adults, ethanol consumption causes inebriation that, except in individuals who heavily consume ethanol, results in mild reversible effects. However, as noted previously, ethanol is a serious public health problem. The pathogenesis of fetal neurotoxicity due to ethanol is complex because it differs depending on the developmental stage of the fetus and cell type involved. In addition, it may involve the toxicity of ethanol or its metabolites. The nervous system is most susceptible to ethanol intoxication during synaptogenesis, when neurite elaboration occurs, synapses form, and interneuronal signaling begins. For humans this occurs in the last trimester and into the first few years of life. The most readily observable effect of ethanol exposure in the developing fetal brain is neuronal cell death, which occurs by necrosis and apoptosis. The metabolism of alcohol forms reactive oxygen species and depletes antioxidants, two activities that can cause cell necrosis. Widespread cell death occurs when ethanol produces activation of BAX-dependent caspase-3, which is sufficient to trigger apoptosis within 6 hours of exposure in animal models [11]. Cell death affects a number of processes because the loss of neurons reduces production, migration, and differentiation of neuronal cell lines that underlie diminished or lost structures and functions later in life. Ethanol exposure also triggers premature maturation of precursor cells to astrocytes, an event that interferes with the migration of neuronal cells to their normal location in the brain. Ethanol further interferes with growth factors such as insulin-like growth factor, which is necessary for neuronal cell maintenance. Developing neurotransmitter systems, including the serotonin system and N-methyl-D-aspartate receptor function, are altered by exposure to ethanol. All of these interactions between ethanol and

Table 4 Chemical Structure and Neurotoxicity in Studies with Humans or Experimental Animals

General Chemical Structure	Examples of Neurotoxic Effects	Examples of Specific Chemical
Organometals	Cerebellar degeneration	Methylmercury
	Myelinopathy	Trimethyl tin, triethyl tin
	Encephalopathy	Tetraethyl lead
Nitriles	Vestibular degeneration	3,3-Iminodipropionitrile, allyl nitrile, cis-crotononitrile, 2-butenenitrile
	Inferior olivary nucleus degeneration	2,4-Hexanedinitrile, trans-crotononitrile
	Astrocytoma	Acrylonitrile
Organophosphorus compounds	Cholinesterase inhibition	Majority to varying degree
	Peripheral neuropathy	TOCP: the prototypical chemical for this class
Organosulfur compounds	Peripheral neuropathy	Disulfiram, tetramethylthiuram disulfide, pyridinethione salts
	Cerebellar granule cell degeneration	Thiophene, 2-mercaptopropionic acid
	Seizures	3-Mercaptopropionic acid
Aliphatic chemicals containing oxygen	Peripheral neuropathy	n-Hexane, methyl n-butyl ketone, 2-hexanol, 2,5-hexanedione, ethylene oxide
	Personality changes	2-Methoxyethanol
	Optic nerve degeneration	Methanol
	Fetotoxicity affecting the nervous system	Ethanol
Aliphatic chemicals containing halogens	Cerebellar degeneration	Methyl chloride, ethyl chloride, 2-chloropropionic acid, 2-bromopropionic acid, 2-bromobutyric acid, 1-bromopropane
	Encephalopathy	Methyl bromide, methyl iodide
	Astrocytoma	Vinyl chloride
	Trigeminal neuropathy	Trichloroethylene, dichloroacetylene
	Peripheral neuropathy	Allyl chloride
	Myelinopathy	Dichloroacetic acid
Simple aromatic hydrocarbons	Encephalopathy	Toluene
Halogenated aromatics	Myelinopathy	Hexachlorophene
	Myotonia	2,4-dichlorophenoxyacetic acid
	Tremors	Chlordecone

the developing fetal nervous system depend on life-stage changes that occur in immature neurons and glial cells; these neurons and cells are not seen in these cell populations in mature animals. Ethanol presents one of the clearest situations in which the neurotoxicity observed in the adult and the developing brain differs greatly.

As exemplified by the effects of ethanol on the developing nervous system, multiple cell types may be involved in the pathogenesis of organic-chemical-induced neurotoxicity. This aspect of neurotoxicity is often overlooked; because of their complexity, reductionist thinking has often been necessary in studying these neurotoxicities. In addition to affecting multiple cell types, organic chemicals may have to affect different cell types in a certain sequence for neurotoxicity to become evident. MPTP is an example of a heterocyclic compound that has induced a parkinsonian-like neurotoxicity in drug abusers who inadvertently used an illicit drug contaminated with it [12]. The clinical disease induced by MPTP has many of the characteristics of idiopathic Parkinson's disease. Nigrostriatal degeneration induced by MPTP is the consequence of a complex series of events that leads to accumulation of 1-methyl-4-phenylpyridinium (MPP^+) in dopaminergic neurons. MPTP's lipophilicity makes it capable of passing through the blood-brain barrier; nevertheless, MPTP alone is not neurotoxic, and critical neurotoxic metabolites of MPTP formed in other tissues are blocked by the blood-brain barrier. The neurotoxicity of MPTP depends on its metabolic activation, which occurs by its oxidation to 1-methyl-4-phenyl-2,3-dihydropyridium and then to MPP^+, primarily in astrocytes that catalyze the first critical metabolic step by the enzyme monoamine oxidase type B. The MPP^+ formed in astrocytes is subsequently released into the extracellular space, from where it is taken up and accumulated intracellularly to toxic concentrations in neuromelanin containing dopaminergic neurons. Once taken up by neurons, MPP^+ accumulates in mitochondria, where it inhibits the flow of electrons through the respiratory chain.

This causes the death of nigrostriatal cells and the depletion of dopamine in the extrapyramidal system, resulting in parkinsonian signs and symptoms. The toxicity of MPTP would not occur except for the critical interplay between the astrocyte metabolism and the dopaminergic system uptake and accumulation of MPP^+.

The final pathways by which many, if not all, organic chemicals produce neurotoxicity in adults and in the developing nervous system involve essentially the same processes (e.g., necrosis, apoptosis, interference with intracellular movement, and interference with oxidative metabolism). What makes the neurotoxicities distinctive are the triggers that induce these processes and the regional differences in anatomy, physiology, and pharmacokinetics that allow the triggers to be expressed in different cell populations in the CNS and PNS.

V. Neuropathology of Organic-Chemical-Induced Neurotoxicity

Except in some noteworthy situations, such as Minamata disease due to methylmercury exposure, and in individual case reports, pathology specimens are often not available for analysis from human cases of organic-chemical-induced neurotoxicity. Therefore, much of what is known about the neuropathology of these diseases is derived from studies with experimental animals. This applies particularly to the early stages of neurotoxic diseases; tissues from experimental animals can be collected over time to create a time-sequenced analysis of neurotoxic events. In the majority of organic-chemical-induced neurotoxicities, the brain, spinal cord, and peripheral nerves will appear relatively unremarkable from a macroscopic perspective. Acute neurotoxicities that interfere with receptor binding (e.g., organophosphates and carbamates) may not produce morphological lesions unless the receptor binding is sustained. A small number of materials (e.g., acetyl ethyl tetramethyl tetralin, o- and m-diethylbenzene) form chromogenic metabolites that react with CNS tissues to form a blue coloration. Industrial dyes that do not form protein complexes in the blood, as most dyes do, can accumulate in CNS and PNS tissues to a concentration that can cause macroscopic tissue discoloration. However, these situations are rare and most organic chemicals, even highly colored ones, do not discolor the nervous system, presumably because they do not pass through the blood-brain and blood-nerve barriers at concentrations high enough to induce color changes. Changes in the size of the nervous system are also uncommon in most neurotoxicities except (1) in cases of chronic encephalopathy in which the cerebral or cerebellar cortices undergo diffuse atrophy, (2) in certain neurotoxicities in which localized or regional cortical atrophy occurs (e.g., atrophy of the calcarine cortex with methyl mercury-induced encephalopathy), (3) with chemicals that induce edema (e.g., hexachlorophene), (4) with chemicals that induce hydrocephalus (e.g., cuprizone), and (5) in neurotoxicities in which significant cell death is induced during brain development (e.g., ethanol, methylmercury, and valproic acid).

The distribution of lesions within the CNS and PNS, and the histological appearance of lesions, will vary from chemical to chemical and over time or by dose rate for the same chemical. Changes in distribution and appearance of lesions due to chemical structure are commonly understood; it is well recognized that for a chemical to induce neurotoxicity directly (not through effects on another organ system) it must be able to be transported to and interact with a target site in the nervous system. Different cell types and regions of the nervous system have differing vulnerabilities to neurotoxic organic chemicals depending on their metabolic capability, their relationship to the blood-brain and blood-nerve barriers, receptor content, blood flow, and other factors. What is less appreciated is that over time the appearance of lesions may change or that the dose rate may affect the location and appearance of lesions, as discussed previously for acrylamide: Purkinje's cell damage occurs at the highest dose level, a terminalopathy occurs at lower dose levels, and axonal degeneration is more prominent as time progresses [13].

The most common types of neurotoxicity in adults are the toxic neuropathies. In these conditions, the initial appearance of lesions differs according to whether the chemical affects the proximal or distal axon, the axonal terminals, the neuronal cell body, the myelin sheath, or the myelinating cells (Schwann cells or oligodendroglia). Over time, many of these conditions take on a similar appearance, as axons undergo fragmentation and wallerian-like degeneration and myelin sheaths undergo fragmentation and phagocytosis.

Large axonal swellings involving the proximal segments of large caliber axons in the CNS and PNS, similar in appearance to those seen in amyotrophic lateral sclerosis, are a hallmark of the neurotoxicity induced in animals given materials similar to β,β'-iminodipropionitrile. Ultrastructurally, these swellings are filled with masses of poorly aligned neurofilaments, which are associated with slowing of axonal transport and atrophy or, in some fibers, degeneration of the axon distal to the swellings. More commonly, toxic neuropathies involving axons are seen morphologically as affecting the distal axon or nerve terminus. Degeneration of the distal axon can take on a number of morphologies depending on the stage of the intoxication, but it is not unusual to see a predominant morphology early in the intoxication. Thus the hexacarbon neuropathies (e.g., n-hexane, methyl n-butyl ketone, and 2,5-hexanedione) are associated with "giant" axonal swellings typically filled with masses of neurofilaments. The axonopathy associated with exposure to acrylamide includes axonal profiles filled with tubulovesicular structures. Although these differences are commonly reported, the morphologies induced by acrylamide and the hexacarbons overlap, with swellings occurring in acrylamide neuropathies and tubulovesicular-filled structures appearing in profiles of axons affected by the hexacarbons. There has been a considerable amount of work completed recently to understand the anatomical site of the earliest lesions associated with the distal axonopathies. Earlier reports identified the earliest lesions as being located in the distal or preterminal axon, but more recent reports have found significant degeneration of axonal terminals before the onset of degeneration in the distal or preterminal axon [13]. This difference in the distribution of early lesions is important because it may indicate that the organic chemicals inducing these neuropathies may affect membrane processes at the axon terminus in addition to their well-recognized effects on axonal transport and neuronal metabolism. These effects may be critical for the development of neuropathy. Many of the toxic neuropathies are regarded solely as peripheral neuropathies; however, detailed morphological investigations have shown that axons and their terminal endings in the CNS are affected morphologically at the same time as those in the PNS, even though the predominant clinical picture is one of a peripheral neuropathy. Although the literature has often indicated that toxic neuropathies affect primarily the large-diameter, myelinated motor and sensory axons, unmyelinated axons of the peripheral nerves and the autonomic nervous system are also often affected by these intoxications. In some distal axonopathies (e.g., 2-*t*-butylazo-2-hydroxy-5-methylhexane), affected individuals have had significant impairment of pain sensation and bladder and sexual functions, suggesting advanced involvement of small axonal fibers in the PNS and autonomic nervous system.

Myelin degeneration secondary to axonal degeneration is commonly observed in toxic neuropathies predominantly viewed as axonopathies. However, myelin and myelinating cells can also be the primary morphological targets for some organic chemicals by induction of cytotoxicity or disruption of the myelin sheath. Examples of organic chemicals inducing toxicity to Schwann cells or oligodendroglia are primarily limited to natural toxins such as diphtheria toxin or experimental models (e.g., ethidium bromide) that interfere with protein synthesis. Examples of organic chemicals altering the structure of myelin sheaths are more common than those exhibiting cytotoxicity; however, some materials regarded as myelin toxicants may eventually be found to induce their effects by impairing cellular function rather than directly affecting myelin. Myelin damage in humans has been associated with exposure to hexachlorophene and triethyltin, both of which induce the accumulation of fluid within the myelin sheath. Significant accumulation of fluid in the myelin sheath results in the appearance of widespread vacuolization or status spongiosis in the brain, which causes swelling of the brain and secondary changes such as optic nerve compression. In experimental animals, the swelling of the myelin sheath can be associated with hydrocephalus in some situations (e.g., cuprizone intoxication), presumably because of stenosis of the aqueduct of Sylvius. Stenosis is caused by the compressive force exerted by the expanded myelin sheaths.

Organic chemicals can also induce toxic neuropathies that appear morphologically to be direct effects on neuronal cell bodies. In some cases, neurons are vulnerable because of their location outside of the blood-brain barrier, as exemplified by the experimental toxicity of doxorubicin, an anthracycline used to treat cancer that induces neuronal necrosis in the dorsal root ganglia, trigeminal ganglia, and other unprotected sites. Organic chemicals can induce cell death that morphologically appears to be necrosis or apoptosis, or both morphologies may be associated with exposure to the same substance. Necrotic neurons are associated with acute cell death and appear to have shrunken nuclei without fragmentation and swollen or lysed cell bodies. Apoptotic neurons typically undergo delayed cell death without cell swelling. Intermediate morphologies have also been described. 3-Nitropropionic acid, an inhibitor of succinate dehydrogenase that reduces intracellular adenosine 5'-triphosphate (ATP), has been reported experimentally to induce activation of N-methyl-D-aspartate receptors. In the presence of moderate levels of glutamate, this triggers excitotoxic neuron necrosis after 3–8 hours [14]. Apoptosis is also induced in 3-nitropropionic acid–exposed cells after 24–48 hours; however, the mechanism involved is not mediated by receptors or altered by the presence of glutamate [14]. The loss of neurons, whether through necrosis or through apoptosis, can be specific. This is exemplified by allylnitrile, which causes degeneration of the auditory, vestibular, and olfactory sensory cells but not other parts of the CNS. On the other hand, neuronal cell death due to necrosis, apoptosis, or both during development of the brain can lead to widespread neuron loss, resulting in effects such as fetal alcohol syndrome or an autism-like syndrome induced in animals exposed to valproic acid [15].

VI. Pathophysiology

Although much work has been completed in using model chemicals to understand the pathophysiological processes associated with organic-chemical-induced neurotoxicities, a coherent picture of the critical steps needed to induce toxicity has yet to emerge. For some chemicals, the effect of an individual chemical on a physiological process has been studied in depth; however, this information often does not provide a clear definition of a mechanism of action for the neurotoxicity. This may be because chemicals often have multiple pathophysiological effects and it is the interaction among these effects or the summation of effects that leads to cell degeneration or malfunction.

For organic chemicals to reach many of the neural structures that are potential target tissues for interaction, they must be able to pass through the blood-brain and blood-nerve barriers in sufficient quantity to be toxic or to interact with targets not surrounded by these barriers. Lipid-soluble chemicals and other nonionized chemicals can pass through these permeability barriers, which are made up of blood vessels without fenestrations and glial or perineurial and endoneurial supporting cell processes. Neural structures unprotected by permeability barriers include the nerve terminus of sensory and motor fibers, particularly such structures as the peripheral terminus of the olfactory nerve, which provides a link directly into the CNS, autonomic ganglia, dorsal root ganglia, and area postrema. The blood-brain barrier can be compromised by concurrent viral infections and is incomplete during development of the nervous system. Mannitol is an organic chemical that has been used clinically to osmotically increase the permeability of the blood-brain barrier, but in general there are few organic chemicals recognized for their ability to compromise the blood-brain barrier that do not also cause vascular damage.

The cells of the nervous system highly depend on energy to maintain their structure and function. Adequate blood supplies of oxygen and glucose, which generate ATP, are necessary. There are many examples of acute or peracute neurotoxicity induced by exposure to asphyxiant gases, inorganic substances (e.g., carbon dioxide, cyanide, or cyanide-generating substances), methemoglobin formers, or plant-derived substances (e.g., 3-nitropropionic acid) that interfere with oxidative metabolism in the nervous system. However, relatively few synthetic organic substances (e.g., 2,4-dinitrophenol and related substances that interfere with mitochondrial respiration) have been reported to directly produce their effects in humans by interfering with oxidative metabolism.

Subchronic or chronic interference with oxidative metabolism following exposure to organic substances has been more commonly implicated in neurotoxicities involving degenerative processes. Interference with oxidative metabolism by 2,5-hexanedione has been postulated to be one of the mechanisms by which *n*-hexane and methyl *n*-butyl ketone exposures produce peripheral neuropathy. *In vitro* studies with 2,5-hexanedione have shown it to inhibit glyceraldehyde-3-phosphate dehydrogenase, enolase, and phosphofructokinase activities and to reduce ATP production in rat brain mitochondria. Inhibition of glycolysis interferes with fast axonal transport, which is an energy-dependent process shown to be impaired by exposure to 2,5-hexanedione.

In older individuals, age-related declines in oxidative metabolism or oxygenation due to impaired circulation may play a role in neurodegenerative processes. Impaired oxidative metabolism is also capable of potentiating the excitotoxic effects of endogenous glutamate.

The effects of organic chemicals on axonal transport systems, including fast anterograde and retrograde transport and slow transport, have been studied with prototypic neurotoxic chemicals for close to 50 years [16]. However, with certain exceptions the pathophysiological role of organic-chemical-induced interference with axonal transport in the induction of neurotoxicity remains hypothetical. Vincristine, a drug used in cancer treatment, binds to tubulin subunits and disrupts the integrity of microtubules, which interferes with both anterograde and retrograde fast axonal transport and is followed by axonal degeneration. Taxol binds to β-tubulin-stabilizing microtubular structures and in high doses induces axonal degeneration. Other organic chemicals, such as 2,5-hexanedione, are known to bind to and cross-link cytoskeletal proteins, which is presumably one way that 2,5-hexanedione interferes with axoplasmic transport. However, it is not clear whether this effect or the effects of acrylamide on the transport system are responsible for axonal degeneration. More than 40 years of studies on the effects of acrylamide on axonal transport have resulted in a large amount of data showing that a single exposure to acrylamide can transiently reduce axonal transport within just a few hours of exposure. Repeated exposure to acrylamide eventually results in axonal degeneration, which has been hypothesized to result from covalent binding of acrylamide to the sulfhydryl groups on the protein kinesin. This provides the force necessary for fast anterograde axonal transport when it interacts with microtubules [16]. Although it has been suggested that acrylamide may affect retrograde transport by interacting with dynein, the motor protein associated with retrograde transport, there is little support for this hypothesis. Acrylamide, 2,5-hexanedione, and similar organic chemicals are reactive *in vivo* and bind to many sulfur-containing proteins, making it difficult to show conclusively that binding to kinesin or any other individual protein is causally related to the induction of axonal degeneration through interference with axonal transport mechanisms.

VII. Biochemistry and Molecular Mechanisms

The primary site of metabolism for organic substances is the liver where biotransformation reactions (Phase I, functionalization, and Phase II, conjugation) detoxify and speed the excretion of xenobiotics. During development, maternal hepatic metabolism is an important determinant of chemical exposure to the fetal brain. In some cases, Phase I biotransformation reactions bioactivate neurotoxicants. As an example, the P450-facilitated oxidation of n-hexane to 2-hexanol, methyl n-butyl ketone, and subsequently 2,5-hexanedione is necessary for the expression of n-hexane-induced axonal degeneration. In contrast, oxidation of n-hexane to 1-hexanol or 3-hexanol does not result in axonopathy. Differences in individual susceptibility to neurotoxic substances can result from age-related expression of detoxification enzyme activity or genetic polymorphisms. One of the best-known metabolic susceptibilities to neurotoxicants induces prolonged inhibition of acetylcholinesterase by certain organophosphate substances in individuals with reduced detoxification capacity, caused by low levels or low catalytic capacity of paraoxonase.

The induction of neurotoxicity by organic chemicals can be enhanced or potentiated in certain situations by coexposure to nonneurotoxic substances; hence, coexposure to methyl ethyl ketone and methyl n-butyl ketone causes clinical signs and symptoms to appear more quickly than exposure only to methyl n-butyl ketone. The induction of neurotoxicity can also be inhibited by coexposure to other organic substances. For example, coexposure to toluene and n-hexane increases the time to onset of n-hexane-induced axonopathy. In some situations, an organic chemical can inhibit the induction of neurotoxicity if it is given before exposure to the neurotoxicant and can promote neurotoxicity if it is given after exposure to a neurotoxicant. Phenylmethanesulfonyl fluoride (PMSF) and methamidophos are weak inducers of organophosphate-induced delayed neuropathy (OPIDN). When these chemicals are given at dose levels that do not result in OPIDN in experimental animals, they protect against the development of OPIDN; however, if they are administered after exposure to an OPIDN-inducing compound, they promote the induction of OPIDN.

Although hepatic metabolism is the primary location for biotransformation reactions, localized intracerebral metabolism is key to the induction of some neurotoxicities. The metabolism of MPTP to 1-methyl-4-phenylpyridinium or MPP^+ in the liver is considered a detoxification reaction and does not result in nigrostriatal degeneration because MPP^+ formed in the liver does not cross the blood-nerve barrier. However, when MPP^+ is formed from MPTP in the brain in association with dopaminergic neurons, it does result in nigrostriatal degeneration. Likewise, although ethanol is metabolized in the maternal liver, its metabolism in the developing brain of the fetus is hypothesized to induce fetal alcohol syndrome.

A number of the organic chemicals that induce neurotoxicity are reactive substances or are metabolized to reactive substances. The nervous system may be partially protected from exposure to some of these substances, as they may circulate in blood bound to serum proteins, hemoglobin, or both and are thereby partially excluded from the nervous system by the blood-brain and blood-nerve barriers. Once in contact with neural tissues, reactive organic chemicals may interact with nucleophiles in these tissues. Doxorubicin is an example of a chemical that can interact with deoxyribonucleic acid (DNA) and intercalate itself into the DNA structure. Because DNA repair is poor in neurons, the reaction of doxorubicin and DNA is essentially irreversible, leading to interference with ribonucleic acid synthesis and eventually protein synthesis that results in the loss of sensory ganglion cells in experimental animals. Methylmercury is a substance that binds strongly to the sulfur of thiol groups present in many proteins. This binding can reduce substances that protect against intracellular oxidation because of its depletion of glutathione, interference with protein metabolism, or inhibition or inactivation of important enzyme systems. The adduction of cysteine residues on soluble N-ethylmaleimide-sensitive fusion attachment receptor (SNARE) proteins by acrylamide or an acrylamide metabolite is postulated to inhibit membrane fusion in the terminus of PNS axons and trigger the onset of axonal degeneration [13]. Other substances, such as the organophosphates, bind to receptor acetylcholinesterase, inactivating it and preventing or slowing the hydrolysis of acetylcholine. This results in signs and symptoms of cholinergic toxicity. Certain organophosphates bind to neuropathy target esterase (a poorly characterized neuronal esterase) and, by a yet-to-be-discovered mechanism, trigger delayed axonopathies.

Other neurotoxicities occur through biochemical mechanisms that enhance or inhibit vitamin function. The simplest of these are the hypervitaminoses induced by overexposure to the vitamin, such as the increase in cerebrospinal fluid pressure induced by vitamin A. Others interfere with vitamin function (e.g., Vacor effects on niacin) and produce a toxic axonopathy that affects the PNS and the autonomic nervous system because of a vitamin deficiency. The most serious of these neurotoxicities are those that interfere with gene expression in the developing brain. Retinoic acid is the endogenous signal that initiates the homeobox (Hox) cascade in developing CNS tissue. If retinoic acid levels are too high, too many Hox gene products are produced; conversely, if retinoic acid levels are too low, the Hox genes do not produce enough of their gene products. In either situation, the nervous system can become malformed [17].

Organic chemicals can also chelate essential ions (e.g., sodium, potassium, and calcium) responsible for nerve action potential conduction; many of these are plant- or animal-derived toxins discussed elsewhere. Chelation of other essential ions, such as zinc and copper, by substances such as dithiocarbamates has led to neurotoxicity in experimental animals. In other cases, substitution of one ion for another can lead to neurotoxicity; thus the replacement of the chloride ion by ionic forms of methyl bromide or methyl iodide can produce an acute encephalopathy in humans.

Interference with transmitter systems, such as that induced by the nigrostriatal toxicity of MPTP or the release of glutamate in the developing nervous system by ethanol, can have both short-term or lifetime effects on the structure and function of the nervous system.

VIII. Explanation of Signs and Symptoms in Relation to Pathophysiology

The neurotoxicities induced by exposure to organic chemicals are a highly heterogeneous group of disorders with variable clinical and pathological phenotypes depending on the chemical involved, the exposure received, the age of the affected individual, and the part of the nervous system affected.

Organic chemicals that produce toxic axonopathies result in a constellation of signs and symptoms that depends on which type of axon (e.g., motor vs sensory and large diameter vs small diameter) is predominantly or first affected. Much of what is known about these types of intoxication has come from studying model chemicals. In the toxic axonopathies induced by organic chemicals, changes in fast, slow, and retrograde axoplasmic flow and oxidative metabolism have been clearly demonstrated; however, relating these changes to the onset of intoxication and symptoms has been controversial. In humans under occupational conditions, substances such as acrylamide most commonly induce a symmetrical, distal sensorimotor neuropathy. Common symptoms include numbness and tingling of extremities, weakness, unsteady gait, and excessive sweating of the hands and feet. These symptoms can be related to cerebellar Purkinje cell degeneration, loss of large-diameter axons in the PNS, and degeneration in dorsal root ganglion neurons, Pacinian corpuscles, muscle stretch receptors, and muscle spindle receptors as observed in experimental animals. Whether the degenerative changes observed in these tissues result from changes in axoplasmic flow and oxidative metabolism in the affected cells or they result from deficits in the ability of synaptic vesicles, transport vesicles, and their cognate membranes is an unresolved issue. Acrylamide, the hexacarbons (e.g., n-hexane and methyl n-butyl ketone), and organophosphate chemicals principally induce symptoms of motor and sensory neuropathy. Organic substances (e.g., 2-t-butylazo-2-hydroxy-5-methylhexane) produce prominent symptoms of cranial nerve dysfunction (loss of peripheral, color, and night vision), CNS dysfunction (memory deficits and reduced ability to perform calculations), and autonomic dysfunction (urinary incontinence). Experimental replication of these intoxications in animals has shown axonal degeneration in the cranial nerves, CNS, and PNS consistent with the symptoms observed.

Organic chemicals that induce myelin damage have variable symptom complexes depending on the underlying processes responsible for myelin degeneration. Acute demyelination of peripheral nerve axons is an uncommon expression of organic-chemical-induced neurotoxicity in humans and experimental animals, but it can occur with exposure to diphtheria toxin. Such exposure produces clinical symptoms similar to Guillain-Barré syndrome, including a predominantly motor neuropathy with loss of reflexes and signs of cranial nerve dysfunction. The relatively rapid loss of function and rapid recovery in mild cases parallels myelin damage and remyelination, which affect nerve conduction velocities. The myelin segments of remyelinated axons typically have shorter internodal lengths than those formed during development of the nervous system. Although the presence of more frequent nodes should delay the conduction of action potentials, this is not typically observed functionally. More commonly seen in experimental situations (e.g., cuprizone and acetyl ethyl tetramethyl tetralin) is myelin degeneration that follows a subchronic time course and is accompanied by axonal damage. This damage changes nerve function because of a combination of pathophysiological mechanisms resulting from myelin damage and interference in axonal function and integrity.

Organic chemicals such as hexachlorophene present an entirely different clinical picture. In children intoxicated by dermal application of hexachlorophene, signs and symptoms of increased intracranial pressure, seizures, paresis, alterations in consciousness, and deaths have been reported. The clinical manifestations of intoxication are the result of the accumulation of fluid within myelin sheaths of the CNS, particularly in the cerebrum and cerebellum. This accumulation of fluid causes vacuolar degeneration or spongiosis and leads to increased intracranial pressure, which results in the signs and symptoms observed. Although hexachlorophene binds to myelin and interferes with mitochondrial oxidative metabolism, causing signs of hyperthermia, the mechanism by which hexachlorophene induces fluid accumulation in the brain is not known.

Organic substances that affect the neuronal cell bodies result in a range of symptoms depending on which cells are intoxicated. In MPTP, for example, the pathophysiological changes accompanying the induction of parkinsonian-like symptoms shortly after exposure are well described and closely parallel what is known about spontaneously occurring Parkinson's disease. However, the pathophysiological processes involved in the progression of nigrostriatal degeneration in the absence of continued exposure to MPTP are unknown [18]. Although the effects of organic chemicals on neurons of the adult nervous system can be severe, more widespread effects are seen in the developing nervous system. One of the most common and significant organic-chemical-induced neurotoxicities affecting neurons is fetal alcohol syndrome and associated developmental disabilities. Signs associated with fetal alcohol syndrome include craniofacial abnormalities and behavioral abnormalities, including irritability, hyperactivity, poor fine motor control, mental retardation, and poor impulse control. Although it is difficult to assign exact behavioral abnormalities with specific neuroanatomic effects, the basal ganglia, corpus callosum, cerebellum, and hippocampus have been reported to be affected by prenatal exposure to ethanol. In the developing brain, neurons undergo a series of processes including multiplication, growth, differentiation, and migration that must occur during specific periods if an anatomically and functionally normal brain is to form. Exposure of the developing nervous system has been shown to induce neuronal cell death by several mechanisms that can cause necrosis, apoptosis, or both. The consequences of cell death during the developmental period of the brain result from the severity of the exposure to ethanol and the timing of the exposure during neuronal developmental process. Hence the spectrum and severity of clinical outcomes in an exposed child can be quite broad and last for the affected individual's life span, as there are no current therapeutic processes that can reverse developmental abnormalities such as fetal alcohol syndrome and its associated clinical conditions.

IX. Natural History

The natural history of organic-chemical-induced neurotoxicity is highly variable depending on the individual chemical, the dose received, the frequency of exposure, the medium through which the exposure occurs (i.e., food contamination, drug administration, drug abuse, or workplace use), and the age of the exposed individual. The time course associated with neurotoxicity is often highly variable. For materials that have pharmacological effects (e.g., acetylcholinesterase inhibitors), the onset and recovery from clinical effects often depend on blood levels attained and biochemical clearance mechanisms.

The relationship between exposure concentration or dose and exposure time is often complex. Fritz Haber is credited with developing the concept often referred to as Haber's law, which states the product of the concentration (C) of a substance and the length of the exposure time (T) results in a fixed level of effect for a given end point (i.e., $C \times T = k$). Although some neurotoxic substances (e.g., acrylamide) appear to follow Haber's law, there are a number of instances in which Haber's law appears to either overestimate or underestimate effects. In these instances, other relationships among dose, time, and effect need to be considered. For example, high concentrations (4000–8000 parts per million for 6 hours) of 1,1,2-trichloroethylene produce midfrequency hearing loss in rats; however, as the exposure concentration is lowered and exposure time is increased, hearing loss does not occur as a fixed product of $C \times T$. Thus the predicted effects of 1,1,2-trichloroethylene are overestimated at lower concentrations and underestimated at higher concentrations [19]. Similarly, ethylene oxide exposure of pregnant rats results in greater effects when exposures occur at high concentrations for short periods rather than lower concentrations for long periods [20]. Consequently, studies conducted at high concentrations for short periods may overestimate risk, and those conducted at lower concentrations for longer periods may underestimate risk when attempts are made to extrapolate one exposure scenario to the other. Because it is often necessary to make such extrapolations between animal data and human exposure conditions to understand the hazards and make risk management decisions, the common lack of simple relationships recommends the use of large safety margins when the potential for human neurotoxicity exists based on animal toxicity information.

For some substances that cause a delayed peripheral neuropathy (e.g., TOCP and related substances), the relationship between exposure and effect is complicated because there is typically about 3 weeks following exposure in which the affected individual or test animal appears normal before the onset of the signs of peripheral neuropathy. During the latency period, there is no outward sign or symptom of the impending development of sensorimotor deficits.

Of potentially greater concern is the progression of neurodegenerative disease in individuals exposed to clinically subtoxic doses of organic chemicals capable of inducing neurotoxicity. This is best exemplified by the experience of Langston and colleagues with MPTP. Langston and colleagues originally identified the onset of a parkinsonian-like condition among intravenous drug users who used a synthetic heroin containing MPTP. Of an estimated 400 individuals who used the drug containing MPTP, 7 developed a frank parkinsonian syndrome. Some of the individuals exposed to MPTP were available for repeat examination using fluorodopa positron emission tomography (FD-PET) to assess their nigrostriatal dopaminergic function [18]. Of the 10 individuals (mean age 32.7 years) initially examined, 5 were clinically normal at the time of their first examination and 5 had limited signs of parkinsonism. When reexamined 7 years later, 5 had new clinical deficits and FD-PET indicated that nigrostriatal function had declined at the rate of 2.3% per year, which was significantly faster aging than that observed for control subjects. Overall, the experience of MPTP-intoxicated individuals supports the hypothesis that parkinsonism may be induced by short-term exposure to toxic agents and suggests that other neurodegenerative diseases may occur after a long latency period following exposure to neurotoxicants at dose levels that do not produce frank neurotoxicity initially. MPTP-induced parkinsonism and spontaneous parkinsonism have similar natural histories, except for the age of patients and their drug histories, suggesting that identification of other substances that might result in delayed neurodegenerative diseases is a difficult task.

X. Management of Organic-Chemical-Induced Neurotoxicity

Treatment and management of organic-chemical-induced neurotoxicity in affected individuals depends on the specific chemical involved and the underlying mechanism by which toxicity occurs. Those toxicities that involve interference with pharmacological processes require supportive therapies to allow the patient to survive the acute phase of the intoxication. Those toxicities that involve structural damage to the nervous system of adults have variable recovery depending on the anatomical site involved in the toxicity and the extent of damage to the site. Intoxication involving axonal structures, particularly in the PNS, may have prolonged recovery periods, as regeneration of peripheral nerve axons occurs at the rate of 1–2 mm/day. Intoxications involving neuronal cell bodies can lead to the permanent loss of neurons and their axons. Variable degrees of recovery may occur if other neural structures are able to provide support through axonal sprouting or other reparative processes. In cases in which the developing nervous system is intoxicated, the effects on nervous system function can lead to the need for special education and lifetime support.

The primary management process of handling organic-chemical-induced neurotoxicity is through the identification of materials that present a hazard to the adult or developing nervous system; the identification of dose–response relationships so that effect and no-effect levels for neurotoxicity and safe levels of exposure with adequate margins of safety can be set; and the dissemination of information including hazard warnings, labels, material safety data sheets, and personal protection equipment recommendations. Methods for testing for neurotoxicity and conducting hazard, dose–response, and risk assessments have been published by national and international organizations [6,7,21,22]. Based on properly conducted risk assessments, risk management decisions can be made to eliminate, reduce, or control the identified risk. In some cases (e.g., methyl *n*-butyl ketone), materials with identified hazards are no longer manufactured. Agencies such as the U.S. Environmental Protection Agency have promulgated new regulations, such as a significant new use rule, that prevent the reintroduction of materials for manufacture or importation without prior notification to regulatory authorities. In other cases (e.g., alcoholic beverages, pharmaceuticals, and pesticides), product packages must carry warnings about identified hazards so that individuals and corporations can assess the risk associated with the use of a particular material.

References

1. Hamilton, A. (1925). "Industrial Poisons in the United States." Macmillan, New York.
2. Perbellini, L., Mozzo, P., Brugnone, F., and Zedde, A. (1986). Physiologico-mathematical model for studying human exposure to organic solvents: kinetics of blood/tissue *n*-hexane concentrations and of 2,5-hexanedione in urine. *Br. J. Ind. Med.* **43**, 760–768.
3. American Conference of Governmental Industrial Hygienists. (2000). "2000 TLVs and BEIs: Threshold Limit Values for Chemical Substances and Physical Agents and Biological Exposure Indices." American Conference of Governmental Industrial Hygienists, Cincinnati.
4. Park, R. M., Schulte, P. A., Bowman, J. D., Walker, J. T., Bondy, S. C., Yost, M. G., Touchstone, J. A., and Dosemeci, M. (2005). Potential occupational risks for neurodegenerative diseases. *Am. J. Indust. Med.* **48**, 63–77.
5. Allen, N. (1980). Identification of methyl *n*-butyl ketone as the causative agent. *In:* "Experimental and Clinical Neurotoxicology" (P. S. Spencer and H. H. Schaumburg, eds.), pp. 834–845. Williams and Wilkins, Baltimore.
6. World Health Organization. (1986). "Environmental Health Criteria 60: Principles and Methods for the Assessment of Neurotoxicity Associated with Exposure to Chemicals." WHO, Geneva.
7. World Health Organization. (2001). "Environmental Health Criteria 223: Neurotoxicity Risk Assessment for Human Health; Principles and Approaches." WHO, Geneva.
8. O'Donoghue, J. L. (1985). "Neurotoxicity of Industrial and Commercial Chemicals," vol. I and II. CRC Press, Boca Raton, FL.
9. Spencer, P. S., and Schaumburg, H. H. (2000). "Experimental and Clinical Neurotoxicology," 2nd edition. Oxford University Press, New York.
10. Welch-Carre, E. (2005). The neurodevelopmental consequences of prenatal alcohol exposure. *Adv. Neonatal Care* **5**, 217–229.
11. Nowoslawski, L., Klocke, B. J., and Roth, K. A. (2005). Molecular regulation of acute ethanol-induced neuron apoptosis. *J. Neuropathol. Exp. Neurol.* **64**, 490–497.
12. Langston, J. W., Ballard, P. A., Tetrud, J. W., and Irwin, I. (1983). Chronic parkinsonism in humans due to a product of meperidine analog synthesis. *Science* **219**, 979–980.
13. LoPachin, R. M. (2004). The changing view of acrylamide neurotoxicity. *Neurotoxicology* **25**, 617–630.
14. Pang, Z., and Geddes, J. W. (1997). Mechanisms of cell death induced by the mitochondrial toxin 3-nitropropionic acid: acute excitotoxic necrosis and delayed apoptosis. *J. Neurosciences* **17**, 3064–3073.
15. Ingram, J. L., Peckham, S. M., Tisdale, B., and Rodier, P. M. (2000). Prenatal exposure of rats to valproic acid reproduces the cerebellar anomalies associated with autism. *Neurotox. Teratol.* **22**, 319–324.
16. Sickles, D. W., Stone, J. D., and Friedman, M. A. (2002). Fast axonal transport: a site of acrylamide neurotoxicity? *Neurotoxicology* **23**, 223–251.
17. Rodier, P. M. (2004). Environmental causes of central nervous system maldevelopment. *Pediatrics* **113**, 1076–1082.
18. Vingerhoets, F. J. G., Snow, B. J., Tetrud, J. W., Langston, J. W., Schulzer, M., and Caine, D. B. (2004). Positron emission tomographic evidence for progression of human MPTP-induced dopaminergic lesions. *Ann. Neurol.* **36**, 765–770.
19. Crofton, K. M., and Zhao, X. (1997). The ototoxicity of trichloroethylene: extrapolation and relevance of high-concentration, short-duration animal exposure data. *Fund. Appl. Toxicol.* **38**, 101–106.
20. Weller, E., Long, N., Smith, A., Williams, P., Ravi, S., Gill, J., Henessey, R., Skomik, W., Brain, J., Kimmel, G., Holmes, L., and Ryan, L. (1999). Dose-rate effects of ethylene oxide exposure on developmental toxicity. *Toxicol. Sci.* **50**, 259–270.
21. U.S. Environmental Protection Agency. (1998). "OPPTS Harmonized Test Guidelines Series 870 Health Effects Test Guidelines." EPA, Washington, D.C.
22. U.S. Environmental Protection Agency. (1998). Guidelines for neurotoxicity risk assessment. *Fed. Reg.* **63**, 26926–26954.

70

Metals

Luigi Manzo, MD

Keywords: *arsenic, cisplatin, bismuth, developmental neurotoxicity, lead, lithium, manganese, mercury, metals, methylmercury, thallium, zinc deficiency*

I. Introduction
II. Metals Causing Nervous System Disease
III. Pathophysiology
IV. Targets
V. Biochemical and Molecular Mechanisms
VI. Nature of Neurotoxic Syndromes
VII. Age-Related Variables
References

I. Introduction

The nervous system is a susceptible target for a variety of toxic metals. Metal-induced neurotoxicity has been described in humans in connection with environmental pollution (lead, methylmercury), occupational exposure (elemental mercury, lead, manganese), use of medicinal agents (lithium, cisplatin, bismuth, gold salts), and accidental ingestion of pesticides and rodenticides containing arsenic or thallium salts.

Neurobiology of metals has received widespread scientific attention, and several monographs and books dealing with this topic have been published in the past decade [1–3]. This chapter provides a summary of the metals and metallic compounds that can be classified as genuine neurotoxicants because of their documented etiological role in human disease (Table 1). The list includes agents that have no known biological role (e.g., lead, mercury, and thallium) and those metals, such as manganese, that are essential for life but can be toxic when absorbed in excess amounts. Significant exposure to these agents is known to produce a consistent pattern of neurotoxic changes in most individuals, as shown by epidemiological or clinical studies. The induction of a particular response is usually related to the absorbed dose and the presence of a critical metal concentration at the target site in the nervous system.

Chronic exposure to certain neurotoxic metals has been implicated in the pathophysiology of certain neurodegenerative disorders, such as Alzheimer's disease, Parkinson's disease, and amyotrophic lateral sclerosis. However, data are still controversial [4,5]. Lead, aluminum, and manganese are often (but not invariably) found in excess in autopsy

Table 1 Metals and Human Nervous System Disease

Genuine neurotoxicants
    Lead, organic and inorganic
    Mercury, organic and inorganic
    Lithium
    Manganese
    Thallium
    Tin compounds, organic
    Arsenic
    Cisplatin
    Bismuth
    Gold salts
Metals accumulated in brain in neurodegenerative disorders
    Lead
    Mercury
    Manganese
    Aluminum
    Copper
Metal deficiencies associated with human disease
    Copper
    Zinc
    Manganese

nervous tissue samples obtained from victims of these disorders, but there are uncertainties regarding dose–response relationships, molecular mechanisms, and extrapolation to humans of results obtained in laboratory animals.

Nervous system disease can also be induced by metal deficiency. Zinc, copper, and manganese are examples of beneficial trace elements whose deficiency may lead to neurological disorders. Metal deficiency can result from inadequate dietary intake, effects of drugs, and other situations in which body losses of the metal are increased. Zinc deficiency accompanied by neurological alterations may occur as a consequence of hemodialysis, intravenous amino acid infusion, and use of excessive doses of diuretics. Heavy alcohol consumption also induces zinc deficiency, which has been implicated as a possible etiological factor in fetal alcohol syndrome, a complex pattern of neurobehavioral disorders affecting the developing organism due to alcohol abuse during pregnancy. Depletion of copper can be observed in patients with peripheral neuropathy caused by disulfiram and carbon disulfide.

II. Metals Causing Nervous System Disease

A. Lead and Mercury

In the last decades, neurological hazards associated with chronic low-level exposure to lead and mercury have emerged into a public health problem.

Hazardous exposure to lead may be widespread and pervasive. Toxicologically relevant sources include dust, soil, paint chips, food, and drinking water. By the 1950s, many cases of severe encephalopathy were diagnosed, resulting from the ingestion of lead-containing paint chips flaking off the surface of walls and furniture in inner-city homes [2]. Drinking water may be the dominant source of lead where water is supplied through lead-containing pipes and plumbing or where lead has been used for internal plumbing.

In adults, occupational exposure to lead has been associated with poisoning manifested by peripheral and autonomic nervous system changes, often accompanied by hematological disorders. In children, neurotoxicity of lead is even more pronounced. Subtle cognitive and behavioral deficits have been reported in young infants with blood levels in the range from 10 to 20 µg/dL.

Mercury neurotoxicity in the working environment has been known for a long time, notably the Mad Hatter syndrome resulting from exposure to inorganic mercury compounds and erethism from elemental mercury vapor in mining and in the chloralkali industry. Mercurial erethism is characterized by severe behavioral and personality changes, increased excitability, loss of memory, and insomnia, which may develop into depression.

Outside occupational settings, the best-known neurological diseases caused by mercury are the major incidents that occurred in the Japanese fishing village of Minamata and in rural Iraq. In Japan, the industrial discharge of mercury into an ocean bay in the 1950s led to a mass health disaster, with thousands of subjects affected with severe brain damage. The neurotoxic agent was methylmercury. It was found that inorganic mercury is methylated by organisms in fresh and marine water and concentrated as methylmercury through the aquatic food chain in the tissues of fish and marine mammals.

The devastating effects of organic mercury in the nervous system have been further documented in the victims of the outbreak of poisoning that occurred in Iraq (1971–1972), when consumption of bread baked from grain treated with alkylmercury-containing fungicides led to the hospitalization of more than 6000 people and caused 400 deaths. All signs and symptoms arose from damage to the nervous system. Toxic insult was remarkably selective, predominantly involving focal areas of the granule cells of the cerebellum and the neurons in the visual cortex.

B. Lithium

Lithium is widely used as a treatment for psychiatric disorders. It is generally acknowledged to be a safe drug if adequate precautions are observed to monitor serum lithium levels. Intoxication occurs as a result of lithium accumulation to serum levels above 2 mEq/L (the therapeutic concentration is about 0.5–1.2 mEq/L).

Lithium neurotoxicity encompasses a spectrum of signs and symptoms referable to the central and peripheral nervous systems [6]. Mental status changes are the most common manifestations of acute poisoning. An encephalopathy (disorientation, poor memory, incoherence) may appear several hours to several days after an overdose and can progress to paralysis, stupor, seizures, and coma. Toxic effects observed during maintenance therapy include loss of appetite, vomiting, polydipsia, polyuria, incontinence, visual disturbances, extrapyramidal symptoms, ataxic tremor, proximal muscle weakness, and fasciculation.

There is an increased risk of lithium neurotoxicity in patients taking the drug with carbamazepine or neuroleptics (haloperidol, phenothiazines).

C. Manganese

Manganese is unique among the metals discussed in this chapter because its deficiency and its excess both cause neurotoxicity.

Manganese exposure can induce a neurological syndrome that resembles Parkinson's disease. A dopamine deficiency occurs in both Parkinson's disease and chronic manganese poisoning, suggesting that the mechanism underlying manganese neurotoxicity is related to functional abnormalities of the extrapyramidal system. Manganese-induced parkinsonism can be differentiated from idiopathic Parkinson's disease given the predilection of manganese to accumulate in, and damage, the pallidum and striatum. The clinical syndrome, response to levodopa, imaging studies with magnetic resonance imaging and positron emission tomography, and pathological features help distinguish these two conditions and permit the correct diagnosis to be established [7].

Neurotoxicity has also been reported in association with manganese deficiency. Incoordination, impaired orientation, and other behavioral deficits can be observed in animals maintained on manganese-deficient diets. Rat pups born to manganese-deficient mothers show pronounced ataxia, loss of equilibrium, and postural abnormalities. These effects have been associated with biochemical changes involving biogenic amines and loss of integrity of the maculae of the inner ear [1].

D. Thallium

Legal restrictions in industrialized countries have substantially reduced the use of thallium-containing rodenticides and, consequently, the incidence of thallium poisoning. However, thallium toxicosis is still reported worldwide and remains one of the most common acute diseases from all toxic metals [8]. The lethal dose of thallium ranges from 0.5 to 1.0 g after a single ingestion. In acute poisoning, neurotoxicity is usually a distal, predominantly sensory neuropathy with some involvement of the central nervous system. The neuropathological substrate is nonspecific degeneration of distal regions of axons. Axonal degeneration commences distally and slowly proceeds toward the neuronal cell body, resulting in symmetrical, distal clinical signs in the legs and arms. Clinical features correlate with the neuropathological changes and include gradual insidious onset, with initial findings usually involving stocking-glove sensory and motor loss of the extremities. Recovery may take months or years.

E. Aluminum

During the last decades, aluminum neurotoxicity has received widespread attention, chiefly as the result of studies implicating aluminum as an etiological factor in two separate human diseases: dialysis encephalopathy and Alzheimer's disease [4].

Dialysis encephalopathy (dialysis dementia) has been described in patients who received long-term hemodialysis for chronic renal failure. The syndrome is characterized initially by intermittent speech difficulties and changes in the electroencephalogram consisting of a slowing of the dominant rhythm, increased low-voltage theta activity, periodic bursts of high-voltage delta activity, and occasional spike and wave complexes. These changes usually precede by several months the onset of clinical manifestations, including dyspraxia, tremor, motor incoordination, myoclonus, seizures, loss of memory, and personality changes. Patients dying from dialysis encephalopathy have greater concentrations of aluminum in brain gray matter and other tissues than patients dying from other causes. Aluminum-containing phosphate-binding gels increase gut aluminum absorption. These agents have been considered a likely source of aluminum accumulation in body tissues, in addition to the dialysis water itself because aluminum compounds are commonly used in water purification.

Elevated aluminum levels found in the nucleus of tangle-bearing neurons from patients with Alzheimer's disease and certain neuropathological features, such as the presence of neurofibrillary changes in the neuronal perikaryon, observed in experimental aluminum-induced encephalopathy have suggested a possible involvement of aluminum in Alzheimer's disease. However, the exact role, if any, of this metal in human disease remains unclear. The accumulation of aluminum in the brain of patients with dialysis encephalopathy is greater than that in the brain of patients with Alzheimer's disease. The localization of aluminum within the neurons is different, and the mechanism of action also is presumably different.

Aluminum is poorly absorbed from the gastrointestinal tract. However, concomitant intake of low–molecular-weight organic ligands, like the common dietary constituents citrate and maltolate, greatly enhances intestinal absorption of the metal. Orange juice, which contains citric acid, increases severalfold the absorption of aluminum from an aluminum-containing antacid [4].

F. Other Metals

Arsenic has been used in agriculture as an insecticide, herbicide, and growth stimulant. It has also been used as an ant and rat poison. Today, it is available in insecticides containing arsenic trioxide. Arsenic can also be found as a homeopathic remedy in some health food stores. The classical presentation of acute arsenic poisoning is described as dysphagia associated with a metallic taste. This is followed by nausea, vomiting, diarrhea, and cardiovascular collapse. Seizures and coma may occur after a massive ingestion, followed by an encephalopathy days later. The encephalopathy may result from cerebral edema followed by microhemorrhages and areas of necrosis. A demyelinating neuropathy resembling Guillain-Barré syndrome, with areflexia and cranial nerve dysfunction, may appear within 3 weeks of ingestion. Peripheral neuropathy accompanied by dermal changes is the most common complication seen in patients with subacute and chronic arsenic poisoning.

Cisplatin and gold salts can also cause peripheral neuropathy. High doses of cisplatin, such as those used in cancer chemotherapy, can induce a sensory neuronopathy syndrome in which the initial biochemical and morphological changes occur in the neuron cell body. Clinical manifestations are restricted to segments innervated by affected nerve cell bodies. There is rapid onset following exposure with diffuse sensory loss and ataxia with no signs of central nervous system disease. Recovery is variable depending on whether the insult caused the death of the nerve cell body, with permanent loss of axons, or is less severe, with only slight impairment of some cells that can therefore reconstitute their axons. Cisplatin neurotoxicity is potentiated by Taxol. The interaction has been demonstrated in patients treated simultaneously or sequentially with these anticancer drugs [2].

A clinical picture reminiscent of Guillain-Barré syndrome has also been described in rheumatoid arthritis patients treated with gold salts preparations, such as sodium aurothiomalate and aurothioglucose. Cranial neuropathy due to gold salts may also occur, either in association with polyneuropathy or in isolation.

Cases of bismuth intoxication have resulted from the therapeutic use of over-the-counter mixtures containing bismuth subsalicylate. Bismuth neurotoxicity typically appears subacutely and gradually after prolonged oral administration, with mental changes (memory loss, insomnia, confusion, excitation, delirium, psychosis), ataxia, dysarthria, tremors, myoclonus, and seizures. The syndrome is usually reversible over several weeks or months when bismuth intake is stopped.

Organic tin compounds have a number of applications as polymer stabilizers, biocides, antioxidants in rubber products, and surface disinfectants. The commercial use of these agents has increased steadily over the past several decades.

Although metallic tin has a low degree of toxicity, some organotin derivatives, mainly trialkyltins, are highly toxic. The most neurotoxic alkyltin compounds are trimethyl tin and triethyl tin.

III. Pathophysiology

A. Metabolism and Structure-Activity Relationship

Metals can induce different patterns of neurotoxicity, depending on the chemical form absorbed. Chemical, physical, or biological processes may convert neurotoxic metals to bioavailable forms. The ultimate form of the metal absorbed in the body may originate from chemical transformations occurring in the environment, as shown with mercury, or from man-made processes involving metals of industrial and commercial importance (e.g., tin and platinum). Because toxicity varies with the chemical state of the metal, transformation processes substantially control the level of the human health hazard [9].

Organic tin compounds provide an example of how the neurotoxicity profile may change depending even on small differences in the chemical form (Table 2). Triethyl tin causes a pervasive metabolic derangement with a generalized brain edema involving the white matter, which is related to its ability to bind to myelin with high affinity; trimethyl tin produces neuronal cell death in discrete brain regions, most notably the limbic system, neocortex, and sensory areas.

Similarly, the patterns of effects induced by inorganic forms of lead and mercury in the nervous system differ considerably from those caused by exposure to organic species such as triethyl lead and methylmercury. In general, organic metal derivatives are considerably more toxic than the corresponding inorganic forms. Neurotoxicity is induced by organic tin, gold, and platinum compounds but not by the metal itself. On the other hand, inorganic arsenic is more toxic than organic arsenic derivatives.

Biotransformation processes are of considerable importance for metals of environmental interest. Inorganic forms of mercury can be biomethylated in the environment to generate the more toxic methylmercury. This phenomenon is an important public health concern since the alkyl species are lipophilic and have a greater tendency to accumulate in the aquatic food chain and in mammalian tissues compared with the inorganic compounds [9].

Table 2 Neurotoxic Organotin Compounds

Triethyl tin

Pathophysiology	Disturbance of cerebral energy metabolism, pervasive metabolic derangement with distinct myelinopathy, and generalized brain edema involving the white matter.
Signs and symptoms of toxicity	Generalized depression of the central nervous system and autonomic nervous system alterations with decreases in blood pressure and heart rate, accompanied by vasodilation and slower respiration. Decrease in body temperature and metabolic rate, hypoglycemia, slowing of the electroencephalographic rhythm, general depression of behavioral processes with concomitant sensory impairments.

Trimethyl tin

Pathophysiology	Selective disruption of specific neuronal populations and neurotransmitter systems. Degeneration of neurons in the hippocampus, brainstem, and primary sensory structures. Alterations in the regional levels of γ-aminobutyric acid and dopamine. Neuronal cell death in discrete brain regions, most notably the limbic system, neocortex, and sensory areas. Alterations in visual, auditory, and somatosensory processing.
Signs and symptoms of toxicity	Seizures, hyperreactivity, and impairment of learning and memory processes. Behavioral alterations including hyperactivity and cognitive impairment.

B. Vulnerability

Metals can virtually affect any part of the neuraxis, with a variable degree of toxic potency (Table 3). They can produce widely divergent clinical outcomes depending on the agent and chemical species involved, patterns of exposure (acute or chronic), nutritional state, coexisting pathological findings, and abilities of the affected structures to repair damage. Dramatic differences are also related to the age and maturity of the exposed organism. For example, inorganic lead induces a peripheral neuropathy in occupationally exposed adults and an acute encephalopathy with subtle behavioral disorders in children.

Metal neurotoxicity can be expressed as neuropathology, altered neurochemistry and electrophysiology, or behavioral and psychiatric symptoms. Functional changes may

Table 3 Vulnerability to Neurotoxic Metals

Part of the Neuraxis	Metal
Cerebellum	Mercury
Basal ganglia	Manganese
Hippocampus	Trimethyl tin, organic lead
Peripheral nervous system	Thallium, arsenic, cisplatin
Autonomic nervous system	Thallium, lead

Table 4 Special Sense Alterations in Metal-Related Disease

Sense	Metal
Olfaction	Cadmium oxide
Taste perception	Arsenic, zinc deficiency
Vision	Thallium, methylmercury, lithium
Auditory	Cisplatin, trimethyl tin, methylmercury

develop even in the absence of any recognizable structural damage, especially after chronic low-level exposure. Sensory systems are typically involved in metal neurotoxicity (Table 4). Distorted taste has been reported in patients treated with lithium and gold salts. Thallium can induce visual disorders secondary to optic nerve damage. Treatment with cisplatin may cause ototoxicity, characterized by tinnitus and hearing loss, resulting from a direct effect of the metal on hair cells. Hearing impairment and deafness can also occur after developmental exposure to methylmercury [10].

Metal-induced neurological disorders are usually classified in terms of the anatomical site and the neural cell populations most susceptible to each agent, or according to the clinical presentation. However, there are practical limitations with both of these methods. For most neurotoxic metals, the primary locus of cellular damage has not yet been identified or the initial insult is biochemical, so two or more contemporaneous primary lesions may occur.

IV. Targets

A. Neurons and Their Processes

Axonal transport processes represent important targets for certain toxic metals. Lead, manganese, and inorganic mercury accumulate in neurons after retrograde axonal transport from the periphery to neuronal somata. Studies in rats have demonstrated that cadmium is transported from the olfactory primary sensory neurons to the olfactory bulb. Retrograde transport of cadmium from the olfactory epithelium to the bulb could be a mechanism by which cadmium induces anosmia in humans.

Thallium and arsenic are examples of metals that perturb axonal transport mechanisms and interrupt normal intra-axonal flow of materials along central and peripheral nervous system fibers. These processes may have an important role in the mechanisms underlying thallium- and arsenic-induced peripheral neuropathy, as illustrated later.

Axonal transport is also affected in methylmercury and cisplatin toxicity, but the peripheral nerve alterations caused by these metals seem to involve somal synthesis and processing rather than distal axons [2].

B. Neuroglia and Myelin

Astrocytes possess a potent array of protective systems. They take up glutamate from the synaptic gap, thus playing a crucial role in protecting neural cells against excitotoxic damage. Astrocytes also remove exogenous neurotoxicants from the extracellular space, as shown with methylmercury. These cells commonly respond to toxic exposure by changes in phenotype with up-regulation of a large number of molecules, including the metal-binding proteins—metallothioneins—that control the protective systems. Astrocytic functions also include secretion of neurotrophic factors; control of extracellular pH, uptake, and metabolism of neurotransmitters; and preservation of the normal relationships between neurons and glia. Recent studies have indicated that astrocytes are important targets for metal-induced neurotoxic injury [11,12]. Lead and methylmercury can induce damage to the nervous system by disrupting either the relationship between neurons and glia or the specific astrocyte function.

Myelinating cells represent another susceptible target for toxicants. Several agents can disrupt the synthesis, transport, and insertion of materials required to maintain cellular and myelin integrity. Demyelination at multiple sites in the nervous system has been reported in patients as persistent sequelae of lithium intoxication [6]. Triethyl tin induces a diffuse encephalopathy, characterized by intramyelinic edema without apparent damage or loss of the myelin-forming cells.

Microglial cells represent the third major population of glial cells susceptible to toxic agents. These cells are quite sensitive to even minor disturbances in nervous system homeostasis. In pathological conditions associated with chemical injury, microglial cells are readily activated, showing changes that involve cell morphology, cell number, cell surface receptor expression, and production of growth factors and cytokines. Reactive microgliosis is a common event that can be regarded as a cellular effort to initiate reparative measures in the damaged brain.

C. Nonneural Targets

Few agents exclusively damage nerve tissue because most biochemical mechanisms are diffuse and molecular targets are present in a range of cell types. Clinically relevant examples of adverse effects caused by neurotoxic metals in nonneural tissues include mercury-induced nephrotoxicity, blood alterations occurring in lead poisoning, and alopecia associated with thallium toxicity. The occurrence of symptoms involving nonneural organs may be of considerable diagnostic importance. The focus on neurotoxicity should not diminish recognition of the other adverse effects.

In neurotoxic events, the brain and the rest of the body may interact in an interdependent manner involving both neural and nonneural systems. Response to certain neurotoxicants may occur in the form of complex interactions among neurobehavioral, endocrine, and immune processes. Neurotransmitters can modulate the neuroendocrine system, and chemicals that alter neurotransmitter function can produce specific neuroendocrine effects even at moderate doses. This has been shown to occur with manganese. Studies involving workers who were occupationally exposed to manganese have indicated changes in the tuberoinfundibular dopaminergic system, with pituitary function alterations, abnormal circulating levels of prolactin, and neurobehavioral abnormalities [1].

Thyroid function disorders accompanied by neurobehavioral changes have been reported in animals exposed to lead and organic mercury. These agents are known to cause brain damage during periods of thyroid-dependent brain development, and several data suggest that some developmental effects of lead and mercury could be mediated through the thyroid system. Thyroid hormones regulate neuronal proliferation, migration, and differentiation in the developing brain and play a regulatory role in the maturation of cholinergic and dopaminergic systems, both of which are necessary for normal cytoskeletal assembly and stability in neural cells of the developing organism.

V. Biochemical and Molecular Mechanisms

Most neurotoxic metals bind rapidly and with high affinity to most ligands present in living cells, including such common ligands as sulfhydryl, phosphate, amino, and carboxyl. Consequently, they have the potential to induce multiple diffuse changes with inhibition of enzymes, disruption of cell membranes and ion channel function, damage to structural proteins, and derangement involving the genetic code of nucleic acids [13].

A. Metal Ions and Trace Elements

There are metals (lead, mercury, lithium, and thallium) whose neurotoxicity is associated with alterations of specific cations (potassium, calcium, zinc, and others) known to serve crucial roles in neurotransmitter function, initiation of action potentials, axonal transport, and other key functions in neurons and glial cells.

Element–element interactions may be relevant to metal neurotoxicity. Lead toxicity is mitigated by calcium supplementation, and lead may displace zinc, iron, and copper from metal-containing macromolecules of physiological importance, inducing deficiency syndromes of these essential elements. Disruption of zinc metabolism may be of particular importance.

With copper, zinc serves in the metalloenzyme superoxide dismutase, which is thought to protect neurons from damage by superoxide radicals. Zinc also acts as a negative modulator of the N-methyl-D-aspartate receptor channel and a modulator of fast axonal transport. Effects of zinc deficiency include impaired growth and maturation of nerve cells and neuropsychological disorders. Certain manifestations associated with zinc deficiency, such as hyperactivity, emotionality, and impaired cognition, resemble those observed in children with lead poisoning.

Inorganic lead causes an inhibition of the zinc-dependent ∂-aminolevulinic acid dehydratase, a key enzyme in heme biosynthesis. This may result in elevated circulating levels of ∂-aminolevulinic acid. This mechanism may be implicated in some of the neurotoxic effects of lead [1]. ∂-Aminolevulinic acid was shown to inhibit Na,K-ATPase in nerve tissue and compete with the γ-aminobutyric acid in synaptic reuptake and binding processes involving this inhibitory neurotransmitter.

Hyperzincuria accompanied by neurological changes (abnormal dark adaptation, taste abnormalities, emotional disorders, and lethargy) has been reported during long-term total parenteral nutrition and in patients treated with the chelating agent D-penicillamine for Wilson's disease.

Alterations of iron homeostasis may have important influences on metal neurotoxicity because of the widespread nature or iron deficiency. Iron deficiency during growth has been shown to enhance intestinal absorption of lead, possibly exacerbating the special susceptibility of children to lead poisoning. Other data suggest that iron deficiency may enhance manganese toxicity. Anemic miners appear to be more susceptible to the toxic effects of manganese exposure [1].

Other element–element interactions have been investigated as possible causative factors in the etiology of a clinical variant of amyotrophic lateral sclerosis. This variant was often associated with parkinsonism and/or Alzheimer-type dementia that used to be common in the western Pacific (Guam disease). It has been suggested that chronic nutritional deficiencies of calcium and magnesium, combined with a basic defect in mineral metabolism, induces a form of secondary hyperparathyroidism that leads to increased gastrointestinal absorption of aluminum and deposition of high levels of aluminum, calcium, and silicon as aluminosilicates or hydroxyapatites in neurons [5].

B. Neurotransmitter and Second Messenger Systems

Metal neurotoxicity may result from an imbalance in neurotransmission. Mechanisms involve neurotransmitter synthesis, transport, presynaptic release and reuptake, interaction between neurotransmitters and postsynaptic receptors, or removal of neurotransmitters from the synaptic gap. Alteration of neurotransmitter systems may in turn affect ligand-gated ion channels and signal transduction cascades involving adenylate cyclase and phosphoinositide systems coupled to intracellular effectors.

Several data suggest that neurotransmitter systems are critically implicated in manganese poisoning. Manganese was shown to alter the levels and metabolism of dopamine, norepinephrine, and serotonin and modify the sensitivity of some neurotransmitter receptors [1]. Cholinergic systems can be affected by lithium and methylmercury at low doses [1,14].

Other neurotoxic events promoted by mercury, lead, and manganese are mediated by amino acid neurotransmitters [11]. Excitotoxicity is an aberrant condition that occurs in a variety of disorders associated with metal exposure. Excitotoxicity describes the supraphysiological stimulation of glutamate receptor subtypes, such as the N-methyl-D-aspartate receptor. Excessive activation of these receptors results in intracellular Ca^{2+} overload that, coupled with failure of intracellular Ca^{2+} sequestration mechanisms, activates Ca^{2+}-dependent processes that amplify neuronal injury and induce cell death. Metal-related excitotoxicity induces cell death by either apoptosis or necrosis depending on the intensity and type of insult [15].

Glutamate released from nerve terminals is removed from the synaptic gap by an astroglial transport system. Metals that perturb the glutamate transport system may impair the uptake process and increase concentration or residence time of synaptic glutamate, thus promoting excitotoxicity and neuronal degeneration. Methylmercury was shown to preferentially accumulate in astrocytes and impair the ability of astrocytes to take up excitatory amino acids and maintain the composition of the extracellular fluid [11]. The consequent elevation of glutamate levels in the extracellular space may trigger or accelerate processes of excitotoxic neurodegeneration. These changes are exacerbated by concomitant disruption of cellular energy metabolism, causing uncontrolled efflux of glutamate from nerve terminals and astrocytes.

Neurotoxic metals can interact at low concentrations with G proteins and other macromolecules coupling a large number of receptors for extracellular mediators that activate second messenger systems in the cell [16]. Interactions with second messenger systems in nerve cells have been documented with mercury (both organic and inorganic), lithium, aluminum, manganese, and organic tin compounds. Modification of intracellular cell signaling may alter cell response to hormones, neurotransmitters, and growth factors and thus have important physiological implications.

C. Energy Deficits

Metal toxicity may directly depend on an effect on intermediary metabolism. Mitochondria can accumulate certain metals. Exposure to lead, mercury, thallium, and organotins has been shown to inhibit oxidative phosphorylation both *in vitro* and *in vivo* [1]. Consequences include altered calcium ion transport and storage within the cell. As already indicated, lead can disrupt heme synthesis, causing loss of heme-containing enzymes. This process may affect mitochondrial function and derange energy metabolism.

Peripheral neuropathy induced by arsenic and thallium poisoning is an example of nerve injury associated with alterations of energy metabolism. Proposed mechanisms of arsenic-induced peripheral nerve damage include substitution of pentavalent arsenic for phosphate in the oxidation of glyceraldehyde-3-phosphate to form diphosphoglycerate and reaction of trivalent arsenic with cellular thiols such as lipoic acid, required in the terminal step of the pyruvate dehydrogenase complex. Thallium compounds form insoluble salts with riboflavin, thereby inhibiting flavin-dependent oxidative degradation of pyruvate and electron transport in mitochondria. These changes are thought to occur in initial steps of thallium-induced axonopathy.

There are typical features in the pathology of energy-dependent toxic neuropathies. Sensory fibers are more likely to show changes than motor ones. The earliest changes are always distal, and spinal tracts are generally not affected except for the dorsal columns [2].

D. Oxidative Stress

Several data support a role for reactive oxygen species (ROSs) in cellular responses to neurotoxic metals, such as methylmercury, manganese, and aluminum. Early generation of ROSs may be crucial in the cascade of events leading to neural cell damage, namely, membrane alteration, increased permeability of the blood-brain barrier, altered tubulin formation, inhibition of mitochondrial respiration, and perturbation of synaptic transmitter and ion function.

Enhanced tissue ROSs generation usually requires high doses of the offending agent and therefore may not represent the primary cause of injury in neurotoxic events occurring at low-level exposure. However, manganese exposure was shown to induce oxidative stress in brain tissue as the result of manganese-catalyzed auto-oxidation of dopamine and interference with mitochondrial respiration [1].

Other factors, such as elevated concentrations of polyunsaturated fatty acids, predispose patients to oxidative brain damage and may be implicated in methylmercury neurotoxicity. Methylmercury can stimulate cytosolic phospholipase A2 in phospholipid membranes and increase the generation of arachidonic acid, which then promotes the release of glutamate and excitotoxic events. It is also believed that the homolytic breakdown of methylmercury produces alkyl radicals and free radicals, which can activate lipid peroxidation, leading to cellular injury.

The role of glutathione (GSH) in protecting neural and glial cells from oxidative stress is well established. High concentrations of neurotoxic metals have been shown to cause depletion of GSH contents or derange intracellular GSH homeostasis in brain tissue.

Radical scavengers such as estrogens, testosterone, and vitamin E (α-tocopherol) provide some degree of protection against oxidative stress-induced nervous system damage. 17β-Estradiol protects cerebellar granule cells from methylmercury-induced apoptotic cell death. Antioxidant protection by estrogens also occurs in astrocytes. Neuroprotective estrogens can be formed in glial cells from precursor androgens through the enzyme aromatase, whose expression is induced by different forms of brain injury.

VI. Nature of Neurotoxic Syndromes

A. Susceptibility Factors

Neurotoxic metals readily distribute to all tissues in the body, and the brain has concentrations usually no higher than those of other tissues. Thus the selective action presented by these agents in the nervous system is surprising, as it contrasts with the general profile of their chemical properties.

The selectivity of neurotoxic effects and the mechanism by which a specific cell type is more susceptible to toxic insult compared with other cells have been explained only in a few cases. Studies have indicated that protein synthesis is similarly inhibited by methylmercury in cerebellar granule cells (which is a target cell usually damaged in methylmercury poisoning) and in nontarget cells, such as cells from the cerebrum and Purkinje cells (which are usually spared in methylmercury poisoning) [12]. However, in the granule cells, in contrast to nontarget cells, no recovery takes place, suggesting that susceptibility and selective action may occur because certain cells cannot repair initial damage.

One explanation for the high susceptibility of the immature brain to methylmercury is that this agent affects processes unique to the developing nervous system, namely, cell migration and cell division [14]. An important mechanism underlying selective damage involves cytoskeletal elements. Methylmercury avidly reacts with tubulin monomers, leading to rapid depolymerization of brain microtubules essential for cell division.

Other factors modulating susceptibility to neurotoxic damage implicate physiological barriers and the transport of metals into the nervous tissue. Even the most potent agent may fail to trigger a neurotoxic response if access to neural tissue is denied. Certain neurotoxic chemicals can be sequestered in, or their effects can be blocked by, nonneural tissues. The absorbed lead can be temporarily sequestered in bone. Binding of lead and methylmercury to macromolecules in blood, liver, and kidney may prevent or mitigate neurotoxic attack. Metallothioneins form high-affinity thiolate clusters with several metals. They are likely to reduce the ability of methylmercury to react with functional biomolecules, keeping it in a relatively nontoxic form within neurons and astrocytes.

The blood-brain barrier plays a major role in restricting passage of neurotoxic chemicals from blood into the brain and maintaining the microenvironment of the nerve tissue [17]. Organic lead (triethyl lead), because of its lipophilic properties, is able to penetrate the blood-brain barrier and thus is more selectively neurotoxic than inorganic lead.

The integrity of the blood-brain barrier may be damaged by disease states (infection or high blood pressure) and toxic conditions (lead or triethyl tin poisoning), which promote leakage of blood-borne agents into the brain and render the individual more susceptible to injury. The blood-brain barrier is incompletely developed at birth, thus predisposing the infant brain to neurotoxins that normally would be excluded in adults.

B. Nature of Response

A fundamental event that mediates cellular response to chemical insult in the nervous system is now reasonably well understood. It involves excessive extracellular accumulation of glutamate, free-cytosolic Ca^{2+} overload, and neural cell death. When such cellular insults occur, neurons mobilize a disparate array of cellular defenses and adaptations in the attempt to decrease the likelihood of neuron death or to decrease the harm to neighboring neurons. Examples of defense mechanisms that are induced by toxic metals include decreased neuronal excitability, decreased extracellular glutamate accumulation, reduced cytosolic calcium mobilization, induced proteins of the heat shock family, and activated antioxidant processes; in addition, a neuron can be biased toward apoptotic rather than necrotic death [15].

In metal neurotoxicity, the dose level may be critical in determining the pattern of response and the severity may largely depend on frequency and duration of exposure. A single massive dose of arsenic causes seizures, coma, and encephalopathy; repetitive or continuous exposures to lower doses usually produce alterations confined to peripheral nerves.

In some instances, exposure to marginally elevated levels of neurotoxic metals may result in an "asymptomatic" nervous system disease characterized by modest decline in performance, which may go unnoticed by the individual. This has often been observed in workers with subclinical toxic neuropathy who deny any disabilities, despite the presence of mild sensory dysfunction [2]. The nature of response also depends on functional reserve. The nervous system appears to possess greater neuronal connectivity than is necessary for the maintenance of apparently normal function. Therefore, minor neurotoxic changes may be unrecognized.

On the other hand, neurological disease may erroneously be ascribed to metal exposure. This misconception often happens in environmental medicine when the presence of a potentially neurotoxic agent is recognized simultaneously with an atypical or inadequately diagnosed naturally occurring condition. An example illustrating the condition is the association of exposure to manganese with cases of atypical parkinsonism.

C. Progress or Propagation of Insult and Recovery

Metal-induced neurotoxicity may appear and disappear rapidly, or it may evolve slowly over days or weeks and regress over months or years in the presence or absence of visible structural damage. Neurotoxic illness may improve gradually after exposure to the agent ceases. However, after significant neuronal loss or irreparable damage to the brain or spinal cord, recovery may be absent or incomplete. A complex set of responses occasionally occurs, seemingly unrelated to the immediate relevant toxic effect of the agent.

Chronic exposure to certain metals may be followed by a latent period of days or weeks before structural and functional changes become evident in the nervous system. This is required either to achieve a sufficiently high concentration of the agent or to develop adequate tissue damage for functional effects to surface. Examples of this phenomenon include thallium-induced polyneuropathy and the latency to onset of symptoms in mercury intoxication. Monkeys receiving a low dose of methylmercury for the first 7 years of life were shown to develop no signs of poisoning until 13 years of age, that is, after a latency of 6 years [10]. Latency periods as long as 15 years were reported after the Minamata outbreak [18].

A recent report of an individual case of methylmercury poisoning clearly indicates that short-term exposure can lead to neurotoxic insult possessing a long period of clinical "silence" before clinical expression [18]. In cases of manganism, active neuronal degeneration was observed in the substantia nigra of patients more than a decade after they had been intoxicated by occupational exposure even for a relatively short period.

How the neurotoxic disorder can progress even after cessation of exposure to the culpable agent remains to be elucidated. Age-related loss of neurons, an established biological phenomenon, was proposed to represent a mechanism by which subclinical damage from a neurotoxic event occurring early in life could spread progressively with the advanced age [18]. In theory, an initial neurotoxic insult might "erode" the capacity of physiological defense systems with alterations that become evident later in life. However, most, if not all, defense processes in the body are efficiently protected by homeostatic mechanisms that should allow recovery after a single chemical insult.

VII. Age-Related Variables

A. Developmental Neurotoxicity

The developing nervous system is a susceptible target for toxic metals, especially lead [19] and mercury [20]. The response to chemical injury during development is dictated largely by the stage of fetal growth. There are critical periods of organ susceptibility, so the timing of exposure during prenatal development is of considerable importance in determining the effects on the developing brain.

Several mechanisms have been proposed to explain the vulnerability of the developing brain to metal toxicity. The immature brain is especially prone to damage by impaired energy supply. Metals that cause energy deficits in brain cells, such as lead and organic mercury, may severely retard the maturation of the brain.

In developing neurons, even submicromolar concentrations of methylmercury induce intracellular Ca^{2+} overload, leading to inappropriate activation of degradative enzymes (phospholipases, proteases, and endonucleases), and cause fragmentation of microtubules, an event preceding apoptotic cell death. The integrity of microtubule function is critical for normal brain development in relation to essential processes such as axodendritic transport, extension of neurites, and migration of postmitotic neurons to form the cortical layers of the cerebrum and cerebellum.

Methylmercury alters the cholinergic modulation of neurotrophic processes necessary for neuronal cell integrity, growth, migration, and patterning during embryonic development. Developmental toxicity also involves glutamate-dependent mechanisms. Glutamate receptors are expressed on the majority of cell types in the developing brain. They mediate a continuous bidirectional signaling between neurons and glia. This form of communication modulates cell proliferation and differentiation, as well as synaptic efficacy. As previously illustrated, methylmercury preferentially accumulates in astrocytes and potently inhibits astrocytic glutamate uptake [14]. Consequences include dysregulation of excitatory amino acid homeostasis, astrocyte swelling, and excitotoxic neuronal damage.

Children are more susceptible than adults to lead toxicity because of their more effective gastrointestinal absorption, which leads to higher lead intake. The developing organism may suffer an extra risk of lead toxicity due to an immature blood-brain barrier.

B. Metal Neurotoxicity and Aging

Aging is generally associated with an increased risk for some forms of neurotoxicity, an observation that stems in part from reduced hepatic and renal circulation, decreased biotransformation capacity, and diminished renal and biliary excretion of xenobiotics. Vulnerability to neurotoxicants late in life also results from loss of plasticity and reduced capacity to compensate for impairment, as well as from the postmature decline in the efficiency of homeostatic mechanisms. In the aging brain, neurotoxicants may hasten the progressive erosion of function observed with certain abilities and may produce greater effects because the aging nervous system has already undergone a reduction in its ability to withstand toxic challenges. A reduced ability of the nervous tissue to handle free radicals and oxidant damage may also play a role as a condition that enhances vulnerability of the aged brain to metal-induced injury.

References

1. Bondy, S. C., and Prasad, K. N. (eds.). (1998). "Metal Neurotoxicity." CRC Press, Boca Raton.
2. Spencer, P. S., and Schaumburg, H. H. (eds.). (2000). "Experimental and Clinical Neurotoxicology." Oxford University Press, New York.
3. Yasui, M., Strong, M. J., Ota, K., and Verity, M. A. (eds.). (1996). "Mineral and Metal Neurotoxicology." CRC Press, Boca Raton.
4. Flaten, T. P. (1996). Neurotoxic metals in the environment: some general aspects. In: "Mineral and Metal Neurotoxicology" (M. Yasui, M. J. Strong, K. Ota, and M. A. Verity, eds.), pp. 17–25. CRC Press, Boca Raton.
5. Candura, S. M., Manzo, L., and Costa, L. G. (1998). Role of occupational neurotoxicants in psychiatric and neurodegenerative disorders. In: "Occupational Neurotoxicology" (L. G. Costa and L. Manzo, eds.), pp. 131–167. CRC Press, Boca Raton.
6. Adityanjee, Munshi, K. R., and Thampy, A. (2005). The syndrome of irreversible lithium-effectuated neurotoxicity. Clin. Neuropharmacol. **28**, 38–49.
7. Olanow, C. W. (2004). Manganese-induced parkinsonism and Parkinson's disease. Ann. N. Y. Acad. Sci. **1012**, 209–223.
8. Manzo, L. (2000). Thallium. In: "Experimental and Clinical Neurotoxicology" (P. S. Spencer and H. H. Schaumburg, eds.), pp. 1168–1177. Oxford University Press, New York.
9. Dopp, E., Hartman, L. M., Florea, A. M., Rettenmeier, A. W., and Hirner, A. V. (2004). Environmental distribution, analysis, and toxicity of organometa(loid) compounds. Crit. Rev. Toxicol. **34**, 301–333.
10. Rice, D. (1996). Evidence of delayed neurotoxicity produced by methylmercury. Neurotoxicology **17**, 583–596.
11. Fitsanakis, V. A., and Aschner, M. (2005). The importance of glutamate, glycine, and gamma-aminobutyric acid transport and regulation in manganese, mercury and lead neurotoxicity. Toxicol. Appl. Pharmacol. **204**, 343–354.
12. Costa, L. G., Aschner, M., Vitalone, A., Syversen, T., and Soldin, O. P. (2004). Developmental neuropathology of environmental agents. Annu. Rev. Pharmacol. Toxicol. **44**, 87–110.
13. Manzo, L. (2001). Cellular response to injury by neurotoxicants. In: "Clinics in Occupational and Environmental Medicine" (M. L. Bleeker, ed.), pp. 463–488. Saunders, Philadelphia.
14. Castoldi, A. F., Coccini, T., and Manzo, L. (2003). Neurotoxic and molecular effects of methylmercury in humans. Rev. Environ. Health **18**, 19–31.

15. Nicotera, P., Leist, M., and Manzo, L. (1999). Neuronal cell death: a demise with different shapes. *Trends Pharmacol. Sci.* **20**, 46–51.
16. Rossi, A., Manzo, L., Orrenius, S., Vahter, M., and Nicotera, P. (1991). Modifications of cell signalling in the cytotoxicity of metals. *Pharmacol. Toxicol.* **68**, 424–429.
17. Zheng, W., Aschner, M., and Ghersi-Egea, J. F. (2003). Brain barrier systems: a new frontier in metal neurotoxicological research. *Toxicol. Appl. Pharmacol.* **192**, 1–11.
18. Weiss, B., Clarkson, T. W., and Simon, W. (2002). Silent latency periods in methylmercury poisoning and in neurodegenerative disease. *Environ. Health Perspect.* **110 (Suppl 5)**, 851–854.
19. Lidsky, T. I., and Schneider, J. S. (2006). Adverse effects of childhood lead poisoning: the clinical neuropsychological perspective. *Environ. Res.* **100**, 284–293.
20. Davidson, P. W., Myers, G. J., and Weiss, B. (2004). Mercury exposure and child development. *Pediatrics* **113 (4 Suppl)**, 1023–1029.

71

Neurobiology of Drug Addiction

Timothy P. Condon, PhD
Curtis W. Balmer, PhD

Keywords: *addiction, impulsivity, incentive salience, motive circuit, relapse, reward*

I. Drug Addiction versus Dependence
II. Traditional Conceptions of Drug Addiction: The Hedonic Model
III. Dopamine as the Hedonic Signal
IV. Incentive Salience Model of Drug Addiction
V. Aberrant Learning Model of Drug Addiction
VI. Drug Addiction Involves Multiple Neural Circuits, Transmitter Systems, and Processes
VII. Drug Addiction Reflects the Interaction of Neurobiological, Genetic, and Environmental Factors
VIII. Stress and Vulnerability to Drug Abuse
References

Drug addiction is a chronic relapsing disorder characterized by compulsive drug seeking and taking that occurs at the expense of important activities and despite adverse consequences, including severely impaired social relationships and occupational function, medical illness, and incarceration. Licit and illicit drug abuse and addiction remain prevalent in the United States and around the world, and they exact an enormous toll on individuals and societies. According to the 2004 National Survey on Drug Use and Health (NSDUH), of Americans aged 12 years or older, 19.1 million (7.9%) had used an illicit drug during the month before the survey, 121 million (50.3%) reported being current drinkers of alcohol, and 70.3 million reported use of a tobacco product in the past month [1]. In 2004, according to the same report, 22.5 million Americans aged 12 or older, or 9.4% of the population, reported substance dependence or abuse in the past year. The overall costs of drug abuse in the United States in 2002, including health- and crime-related costs and losses in potential productivity, were estimated at nearly $181 billion [2]. As of 2005, the cost of tobacco use alone, including medical expenditures and losses in productivity, was estimated to be in excess of $155 billion per year [3]. Furthermore, the economic costs of alcohol abuse in 1998—the most recent year for which estimates are available—are estimated to have been nearly $185 billion [4]. As staggering as these numbers are, they provide only a limited perspective on an even larger and more devastating tragedy—the human toll this disease exacts on those struggling with addiction, and the families, friends, neighbors, and communities who care for and suffer with them.

I. Drug Addiction versus Dependence

Although often used interchangeably, the terms *drug addiction* and *drug dependence* refer to related but distinct adaptational phenomena. Dependence refers to adaptations to repeated drug exposure that manifest as tolerance and/or withdrawal symptoms (e.g., dysphoria and somatic symptoms) when drug taking ceases [5,6]. Most drugs have the potential to produce such dependence. Indeed, traditional conceptions of drug addiction credited the avoidance of such symptoms as a primary force driving addicts' uncontrolled drug administration [7]. Although clearly a dimension of the addicts' drug experience, dependence reflects only one of several forms of adaptations to repeated drug exposure and does not account fully for the compulsive drive for drugs of abuse that defines addictive behavior [5].

II. Traditional Conceptions of Drug Addiction: The Hedonic Model

Contemporary conceptions of drug addiction have typically emphasized the role of positive and negative affective states (e.g., euphoria and dysphoria) in driving compulsive drug seeking and taking. In this view, initial drug use is motivated and perpetuated by the desire to experience and reexperience the acute pleasurable effects produced by drugs of abuse. Homeostatic neuroadaptive responses to repeated drug exposure, however, produce tolerance, withdrawal, or both, which manifest as decreases in the pleasure experienced with the same amount of drug and increasingly unpleasant withdrawal symptoms when drug taking ceases. Thus it has been generally assumed that the transition from casual, controlled drug use to compulsive drug taking reflects an initial pursuit of the euphoria or pleasure that drugs of abuse produce (positive affective state) and an avoidance or alleviation of aversive withdrawal symptoms (negative affective state) experienced when drug taking is suspended. One statement of this pleasure-withdrawal theory of addiction, the opponent-process theory, posits that the subjective experience of the drug user reflects the summation of two opposing processes: an *a-process* that is activated in brain reward circuits by acute drug administration and is directly responsible for the pleasure experienced when taking the drug and a *b-process* regulatory mechanism that is activated by the a-process and acts to restore homeostasis in the brain. Initially, the a-process dominates the subjective drug experience and the user experiences primarily euphoria (positive affective state) and few aversive effects (negative affective state). With repeated drug use, however, the b-process, which gives rise to the unpleasant sensations of withdrawal, increases in magnitude and duration (reflecting the development of drug tolerance) and comes to dominate the drug experience. Thus addicts take drugs primarily to avoid the unpleasant sensations of withdrawal rather than to recapture the high experienced with initial drug use [7].

III. Dopamine as the Hedonic Signal

Pleasure-withdrawal explanations of addiction are guided by the central assumption that the neural pathways and transmitter systems that subserve natural reward and reinforcement also mediate addiction. The experience of natural rewards such as food and sex is accompanied by increases in extracellular dopamine concentrations in the mesolimbic pathway, one of four major dopaminergic pathways in the brain and a key component of the brain's system of reward and reinforcement. Originating in the ventral tegmental area (VTA)—a dopamine-rich nucleus in the ventral midbrain—the mesolimbic pathway projects to several brain regions including the amygdala, bed nucleus of the stria terminalis, lateral septal area, and lateral hypothalamus (Fig. 1). The mesolimbic dopaminergic axons, however, project primarily to the nucleus accumbens (NAc)—a component of the limbic system located in the ventral striatum and itself a target of action of many drugs of abuse. Mesolimbic dopaminergic projections to the NAc comprise the key component of the brain reward circuitry.

Like natural rewards, most, if not all, drugs of abuse increase extracellular dopamine in the NAc and, therefore, like natural rewards, can shape behavior. This increase in synaptic dopamine, particularly within the shell region of the NAc, is necessary for addictive drugs to produce

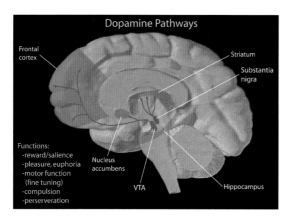

Figure 1 Dopaminergic transmission from the ventral tegmental area (VTA) to the nucleus accumbens and between other limbic and cortical structures is critical to the development and perpetuation of addiction, as well as relapse, to drug taking.

their rewarding effects. Evidence linking mesolimbic activation with the "high" produced by drugs but not with drug craving suggested that mesolimbic dopamine was acting as a hedonic signal directly mediating the pleasurable sensations associated with acute drug administration. Within this context, compulsive drug taking was thought to be driven by an ongoing need to maintain elevated mesolimbic dopamine concentrations so as to reexperience associated positive affective states (e.g., euphoria) and avoid the negative affective states (e.g., dysphoria) produced by dopamine depletion.

IV. Incentive Salience Model of Drug Addiction

The pursuit of drug-induced euphoria and the avoidance of associated withdrawal symptoms likely contribute to the establishment and perpetuation of drug addiction. Ongoing research, however, suggests that drug-related positive and negative affective states (e.g., euphoria and dysphoria) do not fuel all or even most of the compulsive drive for drugs [8]. Instead, it has been hypothesized that addiction reflects the attribution of excessive incentive salience, or the motivational state of "wanting," to drugs and associated stimuli [7]. Thus incentive salience models posit that the transition to addiction reflects the development of pathological *motivational* states (i.e., excessive drug wanting) rather than an aberrant pursuit and avoidance of particular drug-induced *affective* states. This emerging emphasis on drug wanting represents a significant departure from traditional hedonic interpretations of addiction that highlight drug liking (and the disliking of withdrawal symptoms) as the primary force driving compulsive drug use. In the context of incentive salience, drug liking, while no doubt a significant component of the addict's subjective drug experience, is by itself not sufficient to motivate the uncontrolled administration of drugs. This hypothesis further implies that wanting and liking are distinct and independent experiences and neurobiological processes and that a drug's addictive potential is not necessarily a function of its particular euphoria-inducing properties [8].

Incentive salience models of addiction reflect evolving interpretations of the role of mesolimbic dopamine efflux in the development of compulsive drug use and relapse. Rather than directly mediating the hedonic or pleasurable effects of rewards, repeated elevations in extracellular mesolimbic dopamine concentrations (in the NAc in particular) are interpreted as the neural mechanism by which previously neutral stimuli acquire heightened salience and become potent motivational incentives that elicit a state of wanting for associated behaviors and goals. This attachment of excessive wanting to drugs and related stimuli is believed to reflect drug-induced "sensitization" of "NAc-related circuitry" [7]. This sensitization is thought to involve enduring changes in dopamine D1 receptor function and alterations in neuronal structure and in dopamine, glutamate, γ-aminobutyric acid, and several other neurotransmitter systems (Fig. 2). In keeping with the observation that only a small percentage of individuals who ever use potentially addictive drugs transition to uncontrolled, compulsive drug taking, the incentive salience model posits that vulnerability to neural sensitization exhibits individual variation and is influenced by several genetic, hormonal, behavioral, and environmental

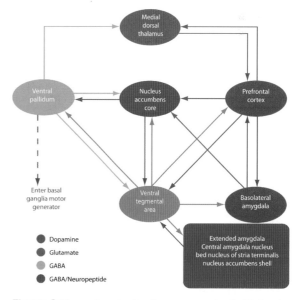

Figure 2 The motive circuit reflects progress in elucidating the brain functions and neurotransmitter systems involved in drug addiction. *Note*: GABA, γ-aminobutyric acid. *Source*: Redrawn, with permission, from Kalivas, P. W., and Volkow, N. D. (2005). The neural basis of addiction: a pathology of motivation and choice. *Am. J. Psychiatry* **162(8)**, 1403–1413.

factors. In addition, it is hypothesized that neural sensitization induced by one drug may facilitate sensitization to other drugs, as well as hypersensitivity to stress. Likewise, previous stress may increase vulnerability to drug-induced neural sensitization.

V. Aberrant Learning Model of Drug Addiction

Other investigators have posited that drug addiction is a pathological form of learned behavior mediated by the same neural substrates that underlie learning and memory associated with natural rewards such as food and sex [7,9]. In this view, the transition from controlled to compulsive drug use reflects the formation and long-term retention of unusually potent and durable associations between drug use and related stimuli. It is hypothesized that drug-related cues acquire heightened motivational significance that exceeds that of natural rewards (presumably by eliciting greater and more reliable dopamine release) and that drug-related cues activate both an overwhelming state of wanting and highly automatized behavioral responses that effectively facilitate drug procurement and use.

As with the previously presented conceptions of drug addiction, learning-based theories cast mesolimbic dopamine release in a central role. Rather than mediating the hedonic effects of drugs of abuse or assigning incentive salience, however, aberrant learning models of addiction interpret dopamine release in the VTA–NAc circuit as critical to two types of reward-related learning thought to be "hijacked" by addictive drugs. First, it is hypothesized that mesolimbic dopamine efflux promotes *stimulus-reward* learning in which stimuli associated with a particular event (drug use, in this instance) acquire varying degrees of predictive significance depending on how strongly they predict the delivery or acquisition of a specific reward. Second, it is posited that mesolimbic dopamine efflux is involved in *stimulus-action* learning by which behavioral responses to stimuli are refined so that the specific action sequences that most effectively increase the probability of obtaining the desired reward or goal are retained and used. Thus within the context of aberrant learning, mesolimbic dopamine release is thought to function in identifying and valuing drug cues and in shaping, directing, and activating relevant behavioral responses.

Learning reflects the acquisition, storage, and appropriate recall of relevant detailed information. Thus learning-based theories of addiction must provide a neurobiological means of recording and retaining the details of drug-associated experience and coordinating behavioral responses to drug cues. Although mesolimbic dopamine, as outlined previously, figures centrally in aberrant learning models of addiction, the architecture of the midbrain dopamine system has suggested to at least some investigators that it is not directly involved in recording and retaining detailed information associated with drug use [9]. Instead, midbrain dopamine is thought to be involved in coordinating the interaction among functionally distinct brain regions that record and store reward-related information and subserve associative long-term memory processes [9].

VI. Drug Addiction Involves Multiple Neural Circuits, Transmitter Systems, and Processes

Although much remains to be learned, there is growing evidence that addiction involves multiple brain regions, circuits, and transmitter systems. Accordingly, some investigators have proposed a more holistic interpretation that conceives of addiction as a pathological behavioral adaptation reflecting drug-induced disruptions in neural substrates mediating several processes, including reward, salience attribution, conditioned learning, motivation, decision-making, impulse inhibition, and various forms of memory [10].

A. Initiation of Drug Addiction: Dopamine and the Motive Circuit

Adaptive behavior is mediated at least partly by an integrated set of neural circuits referred to collectively as the motive circuit (see Fig. 2). The motive circuit interconnects several functionally diverse brain regions and translates motivationally relevant events (e.g., natural rewards such as eating and sex) into specific behavioral responses directed at obtaining these rewards again. Sufficiently novel or salient stimuli elicit the release of dopamine from the VTA throughout the motive circuit. Dopamine release is necessary for the reinforcing effects of rewards and facilitates neuroplastic changes underlying the formation of strong associations between a specific salient event and the related environmental stimuli. Through their association with the event, these stimuli acquire heightened salience and are able to independently evoke dopamine release from the VTA. In doing so, these stimuli alert the system to the pending appearance of the desired event or goal. Dopamine release is also thought to facilitate the learning of directed behavioral responses to specific salient events [10].

It appears that drugs of abuse initiate addiction partly by hijacking the natural processes of novelty detection, salience attribution, and conditioned learning. The magnitude and duration of drug-induced dopamine release (recall that most drugs of abuse elicit dopamine release from the VTA) far exceed those of natural rewards and produce an exaggerated, aberrant response in the motive circuit. Drug taking is interpreted by the system as a highly salient event: related environmental stimuli become strongly associated with drug administration, and drug procurement and taking behaviors are powerfully learned. As with natural rewards, these stimuli acquire heightened salience and dopamine-releasing effects of their own and thereby serve as cues that activate pathological drug seeking and taking behaviors.

B. Drug Addiction and the Prefrontal Cortex: Decision-Making and Impulse Inhibition

Traditionally defined as a disorder of the limbic reward system, several lines of evidence suggest that addiction also involves the disruption of several executive functions mediated by prefrontal cortical components of the motive circuit (Figs. 2 and 3). Addicts' uncontrolled, compulsive pursuit of drugs in the face of almost certain severe consequences is indicative of impaired decision-making and a reduced capacity to inhibit innate and conditioned motivational drives or impulses. These internal drive states are thought to be modulated by active inhibitory processes that allow slower, more deliberate cognitive functions to dominate decision-making and behavior [11].

The proper modulation of response inhibition and decision-making depends critically on discrete regions of the prefrontal cortex, including the orbitofrontal cortex [11,12]. Indeed, individuals with lesions of the OFC exhibit impaired decision-making, characterized by marked insensitivity to future positive and negative consequences of a course of action [13] and a diminished ability to reverse stimulus-reinforcement associations in response to changing contingencies [14]. In other words, individuals with impaired OFC function engage in maladaptive behaviors that persist even when their long-term consequences become adverse and severe.

Similar patterns of diminished inhibitory restraint and maladaptive decision-making have been observed empirically in alcoholics and illicit drug abusers, suggesting that drug seeking and administration are associated with functional deficits in prefrontal cortical circuitry [15–17]. Indeed, neuroimaging studies have consistently shown that activity in the OFC is altered in drug addicts [18]. In the absence of drug administration or related stimuli, OFC activity is decreased relative to nonaddicted control subjects. When presented with drug-related stimuli, however, addicts exhibit increased OFC activation. Collectively these results strongly suggest that addiction involves disruption of prefrontal cortical circuitry that participates in mediating inhibitory control and decision-making [7,8,11,18]. This interpretation is supported by preclinical studies reporting evidence of drug-induced changes in neuronal morphology in the prefrontal cortex (PFC).

C. Drug Addiction and Excitatory Transmission

Theories of drug addiction have traditionally emphasized the role of mesolimbic dopamine as the key neurotransmitter in all major aspects of addiction. More recent findings, however, suggest that although dopamine is clearly important in initiating addiction, it is enduring pathological adaptations in excitatory transmission that underlie the transition to compulsive drug taking and persistent vulnerability to relapse [10,19].

Adaptive behavioral learning (e.g., the formation associations between motivationally relevant events and related stimuli) is mediated partly by modifications in excitatory transmission within the motive circuit, including alterations in presynaptic glutamate release and changes in postsynaptic dendritic morphology that alter responsiveness to released glutamate (see Fig. 2) [10,20]. Repeated exposure to drugs of abuse is thought to induce pathological forms of plasticity in excitatory transmission that incorporate the PFC and associated glutamatergic projections into the circuitry mediating addiction. Drug-induced modifications in excitatory projections from the PFC to the NAc appear to be particularly important in mediating drug seeking, as well as drug craving and relapse.

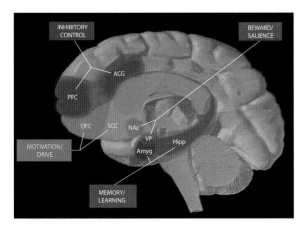

Figure 3 Addiction reflects a disruption in the normal balance of multiple brain functions, including reward, salience, memory, learning, motivation, drive, and inhibitory control. *Note*: ACG, anterior cingulate gyrus; Amyg, amygdala; Hipp, hippocampus; NAc, nucleus accumbens; OFC, orbitofrontal cortex; PFC, prefrontal cortex; SCC, subcallosal cortex; VP, ventral pallidum.

VII. Drug Addiction Reflects the Interaction of Neurobiological, Genetic, and Environmental Factors

The phenomenon of drug addiction is even more complex because not all individuals who initiate use of substances with addictive potential progress to compulsive, uncontrolled drug taking [21]. This differential response to repeated drug exposure suggests individuals vary, perhaps widely, in their vulnerability to addiction [21]. Indeed, progress in drug abuse research over the past 30 years has revealed that drug taking per se is not sufficient to produce addiction. Rather, the transition from controlled to compulsive drug taking reflects the complex interaction of neurobiological, genetic, and environmental factors. This interaction modulates vulnerability to addiction, protecting some individuals and placing others at significant risk.

Animal and human studies indicate that the effect of genes and environment on the risk of addiction is substantial [22,23]. It is estimated that 30%–60% of individual susceptibility to addiction is attributable to genetic factors [22–24]. Although researchers have begun the arduous but necessary task of identifying allelic variants of human genes that influence vulnerability [22], the contribution of any single gene is thought to be small [21]. Instead, it is believed that most genetic predisposition to addiction arises from the interaction of multiple genes [24]. Several environmental factors, including socioeconomic status, quality of parenting, family environment, and drug availability, are also recognized as contributing substantially to an individual's susceptibility to addiction [22,23].

How genetic and environmental factors act alone and in concert to modulate individual responses to drugs is not yet fully understood. Nonetheless, accumulating evidence suggests that genes and environment influence vulnerability in a variety of ways through several mechanisms. For example, some genetic variants may have a broader, generalized effect and predispose addiction to multiple classes of drugs, and others increase susceptibility to only a single drug class [23,24]. Furthermore, genes and environment both appear to influence specific, yet different, stages of drug use and therefore affect the progression from initial drug exposure, to abuse, to addiction in distinct ways [23].

Genetic and environmental factors exert their influence on addiction partly by altering or disrupting relevant cellular and molecular events [22,23]. For example, some human genes with alleles that influence vulnerability to addiction have been found to alter pharmacokinetic properties of drugs, including drug metabolism [22,23]. Similarly, polymorphisms identified in genes associated with elevated risk of addiction may increase vulnerability by disrupting normal receptor function and subsequent drug effects [22,23].

Certain environmental factors may also influence specific cellular events relevant to drug use and addiction. For example, studies in nonhuman primates suggest that social status may influence the expression in the brain of at least one class of dopamine receptor (D2) and thereby modulate the propensity to self-administer cocaine [25]. This data is bolstered by separate animal studies showing that experimentally altering dopamine D2 receptor levels in the NAc influences alcohol consumption [26]. Exactly how changes in social status may result in alterations in drug receptor expression is unclear. It has been suggested, however, that resultant stress may be the common currency by which many environmental factors influence cellular and molecular events contributing to individual vulnerability to addiction [22].

More indirectly, genetic and environmental factors may modulate vulnerability to addiction through their contribution to the development of personality traits that bias toward or against drug use and addiction. For example, impulsivity, risk-taking, and novelty-seeking are associated with increased propensity to experiment with drugs of abuse and increased risk of addiction and may specifically influence the transition to an addicted state [23].

Although it is important to understand as much as possible how specific genetic and environmental factors independently affect vulnerability to addiction, it is evident that these factors interact in complex ways to influence one another's effects [23]. Specific environmental events may be key in determining the extent to which a genetic predisposition to addiction is expressed. Likewise, genetic factors may exacerbate or attenuate drug-associated behaviors, attitudes, and traits facilitated by the environment. Thus the sum of these factors is most important in determining individual susceptibility to drug addiction.

Drug abuse and addiction are also strongly associated with mental illness [22]. Although not yet well understood, the relationship between psychiatric and addictive diseases is clearly complex and their high comorbidity is likely the result of several interacting factors [22]. For instance, mental illness and drug addiction probably share some common neurobiological, genetic, and environmental underpinnings; thus factors that elevate risk for one may increase vulnerability to the other [22]. Furthermore, efforts to ameliorate their symptoms through self-medication with licit and illicit substances likely accounts in part for increased rates of drug abuse and addiction among individuals with mental illness [22,27]. Conversely, chronic drug use may induce changes in brain chemistry, structure, and function that increase the risk of psychiatric pathology [22].

Although this discussion of genes and environment has focused primarily on their role in increasing vulnerability to addiction, it is evident that these factors can also protect against drug use and addiction. For example, the consumption of alcohol by individuals deficient in one or more alleles of acetaldehyde dehydrogenase (a deficiency is particularly prevalent among Asian populations) produces an intense flushing response that is highly aversive and thus protective against alcoholism [28]. There is also evidence that polymorphisms in the CYP2A6 gene, which decrease the rate of nicotine inactivation, may reduce the risk of smoking among some populations [29].

VIII. Stress and Vulnerability to Drug Abuse

Stress is generally believed to play an important role in increasing susceptibility to substance abuse and addiction. Accordingly, it figures prominently in most major theories of drug addiction [30–32]. Relevant studies in humans, although relatively few in number and inconclusive as yet, suggest that multiple types of stress (e.g., early physical and sexual abuse and inadequate social and parenting support) are associated with an increased propensity to abuse drugs [31]. This association is strongly supported by numerous animal studies demonstrating that stress—whether acute or chronic, physical or psychological—enhances the acquisition of drug self-administration [30,32]. This appears to be true whether the stress is experienced at maturity or during pre- or postnatal development.

Several factors, including intensity, duration, and predictability, influence the extent to which various stressors facilitate drug taking behaviors [30]. For example, a noxious stimulus (e.g., electric footshock) of sufficient intensity to evoke a physiological stress response increases the acquisition of drug self-administration most dramatically when its application is unpredictable.

How the experience of stress renders the individual more prone to drug use is not clear. It is thought that this effect likely results from stress-induced increases in the activity of brain motivation and reward systems [30,32]. Because these are the same systems that mediate the reinforcing effects of substances of abuse, it has been proposed that modification of these neurobiological substrates would enhance responsiveness to the motivational and reinforcing properties of drugs.

Reaction to stress is regulated by the hypothalamic-pituitary-adrenal (HPA) axis—a major component of the neuroendocrine system comprising parts of the hypothalamus, the anterior lobe of the pituitary gland, and the adrenal cortices. The release of corticotropin-releasing factor (CRF) from the hypothalamus is influenced by stress and stimulates the release of adrenocorticotropic hormone (ACTH) from the anterior pituitary into the general circulation. ACTH in turn stimulates the cortex of the adrenal glands to increase the biosynthesis and secretion of glucocorticoids including cortisol (corticosterone in rats).

Like drugs of abuse, stress has been shown to increase activity in brain motivation and reward systems, including increased dopamine release in the NAc [30,32]. Elevated circulating glucocorticoid levels have also been associated with increased mesolimbic dopaminergic neurotransmission and, thus, have been proposed to be the link between stress and increased propensity to use drugs [30,32]. Indeed, experimentally increasing dopamine release in the NAc increases drug self-administration [30,32]. Furthermore, elevating glucocorticoids to stress-induced levels increases both the release of dopamine in the NAc and the propensity to self-administer at least some types of drugs of abuse [30,32]. These effects are reversed when stress-induced release of glucocorticoids is blocked by a variety of experimental means [30–32]. Interestingly, animal studies suggest that significant stress experienced early in life may also induce enduring changes in CRF-HPA function such that the subject experiences increased sensitivity to stressors throughout life [31]. Thus stress may not only affect neural systems mediating motivation and reward important in drug abuse and addiction but also fundamentally alter the function of the stress response system itself.

A. Stress and Relapse

Stress is also a major precipitant of relapse to substance abuse [33,34]. Arguably the most troublesome and problematic dimension of addiction, vulnerability to relapse persists long after detoxification, rendering addicts highly susceptible to renewed drug taking.

Accordingly, elucidating the neural substrates underlying this susceptibility is a major focus of addiction research. Ethical considerations prohibit directly studying the neurobiological mechanisms mediating relapse in humans. However, using animal models approximating key features of drug relapse and vulnerability in humans, investigators have begun to identify specific neural circuits and transmitters mediating the reinstatement of drug self-administration following exposure to a variety of precipitating factors [33,34]. This work has revealed that stress, drug priming (i.e., reexposing the drug-free animal to the substance it was trained to self-administer), and drug-associated environmental stimuli or cues (e.g., lights or tones paired with drug administration) robustly reinstate drug self-administration in animals in which this behavior was established and subsequently extinguished [33,34]. These same factors are strongly associated with relapse in humans.

Once thought to be highly dissociable, recent findings suggest that there is overlap among the neural substrates mediating reinstatement induced by stress, drug priming, and cues [32,34,35]. For example, stress-associated CRF-induced activation of the HPA axis appears to also contribute to the reinstatement of cocaine seeking following exposure to drug cues [32,35]. How stress so potently induces relapse is not clear. However, animal studies indicate that stress-induced reinstatement directly involves brain CRF and norepinephrine, as well as glutamatergic mesocorticolimbic projections [8,31].

Elucidating how vulnerability to relapse varies and is regulated following withdrawal is critical to both understanding the phenomenon of addiction in all its dimensions and improving the long-term treatment of substance abuse. Animal studies suggest that drug craving, rather than being greatest immediately following withdrawal, as might intuitively seem to be the case, may incubate—or increase steadily over several weeks or months to elevated levels that persist for an extended period before decreasing [34,36]. It has been suggested that the incubation of cocaine craving reported in cocaine-dependent lab animals reflects long-lasting drug-induced changes in glutamate receptor regulation observed in several brain regions, including the VTA, NAc, and amygdala [34]. Changes in brain-derived neurotrophic factor—a type of growth factor involved in neuronal survival and synaptic plasticity—and an associated signaling pathway are also thought to be involved [34].

These findings and interpretations would seem to suggest that neuroadaptations induced by repeated drug exposure are critical mediators of the vulnerability to relapse. And, indeed, that may be the case. However, much remains to be learned before the contribution of drug-induced adaptations to relapse in humans can be defined with certainty [36].

References

1. Substance Abuse and Mental Health Services Administration. (2005). "Overview of Findings from the 2004 National Survey on Drug Use and Health" (Office of Applied Studies, NSDUH Series H-27, DHHS Publication No. SMA 05-4061). Rockville, MD. (2005).
2. Office of National Drug Policy (2004). "The Economic Costs of Drug Abuse in the United States: 1992–2002." Executive Office of the President (Publication No. 207303), Washington, D. C.
3. Centers for Disease Control and Prevention. (2005). National Center for Chronic Disease Prevention and Health Promotion "Targeting Tobacco Use: The Nation's Leading Cause of Death, 2005." Centers for Disease Control and Prevention, Atlanta.
4. Harwood, H. (2000). "Updating Estimates of the Economic Costs of Alcohol Abuse in the United States: Estimates, Update Methods, and Data Report." Prepared by the Lewin Group for the National Institute on Alcohol Abuse and Alcoholism, Rockville, MD.
5. Chao, J., and Nestler, E. J. (2004). Molecular neurobiology of drug addiction. *Annu. Rev. Med.* **55**, 113–132.
6. Koob, G. F., and Le Moal, M. (2005). "Neurobiology of Addiction," 1st ed. Academic Press, Boston.
7. Robinson, T. E., and Berridge, K. C. (2003). Addiction. *Annu. Rev. Psychol.* **54**, 25–53.
8. Adinoff, B. (2004). Neurobiologic processes in drug reward and addiction. *Harv. Rev. Psychiatry* **12(6)**, 305–320.
9. Hyman, S. E. (2005). Addiction: a disease of learning and memory. *Am. J. Psychiatry* **162(8)**, 1414–1422.
10. Kalivas, P. W., and Volkow, N. D. (2005). The neural basis of addiction: a pathology of motivation and choice. *Am. J. Psychiatry* **162(8)**, 1403–1413.
11. Jentsch, J. D., and Taylor, J. R. (1999). Impulsivity resulting from frontostriatal dysfunction in drug abuse: implications for the control of behavior by reward-related stimuli. *Psychopharmacology (Berl.)* **146(4)**, 373–390.
12. Bechara, A., Damasio, H., and Damasio, A. R. (2000). Emotion, decision making and the orbitofrontal cortex. *Cereb. Cortex* **10(3)**, 295–307.
13. Davis, C., Levitan, R. D., Muglia, P., Bewell, C., and Kennedy, J. L. (2004). Decision-making deficits and overeating: a risk model for obesity. *Obes. Res.* **12(6)**, 929–935.
14. Berlin, H. A., Rolls, E. T., and Kischka, U. (2004). Impulsivity, time perception, emotion and reinforcement sensitivity in patients with orbitofrontal cortex lesions. *Brain* **127(Pt 5)**, 1108–1126.
15. Morgan, M. J. (1998). Recreational use of "ecstasy" (MDMA) is associated with elevated impulsivity. *Neuropsychopharmacology* **19(4)**, 252–264.
16. Bechara, A., Dolan, S., Denburg, N., Hindes, A., Anderson, S. W., and Nathan, P. E. (2001). Decision-making deficits, linked to a dysfunctional ventromedial prefrontal cortex, revealed in alcohol and stimulant abusers. *Neuropsychologia* **39(4)**, 376–389.
17. Semple, S. J., Zians, J., Grant, I., and Patterson, T. L. (2005). Impulsivity and methamphetamine use. *J. Subst. Abuse Treat.* **29(2)**, 85–93.
18. Volkow, N. D., Fowler, J. S., and Wang, G. J. (2004). The addicted human brain viewed in the light of imaging studies: brain circuits and treatment strategies. *Neuropharmacology* **47(Suppl 1)**, 3–13.
19. Jones, S., and Bonci, A. (2005). Synaptic plasticity and drug addiction. *Curr. Opin. Pharmacol.* **5(1)**, 20–25.
20. Kalivas, P. W. (2004). Recent understanding in the mechanisms of addiction. *Curr. Psychiatry Rep.* **6(5)**, 347–351.
21. Hiroi, N., and Agatsuma, S. (2005). Genetic susceptibility to substance dependence. *Mol. Psychiatry* **10(4)**, 336–344.
22. Volkow, N. D., and Li, T. K. (2004). Drug addiction: the neurobiology of behaviour gone awry. *Nat. Rev. Neurosci.* **5(12)**, 963–970.
23. Kreek, M. J., Nielsen, D. A., Butelman, E. R., and LaForge, K. S. (2005). Genetic influences on impulsivity, risk taking, stress responsivity and vulnerability to drug abuse and addiction. *Nat. Neurosci.* **8(11)**, 1450–1457.
24. Uhl, G. R., and Grow, R. W. (2004). The burden of complex genetics in brain disorders. *Arch. Gen. Psychiatry* **61(3)**, 223–229.
25. Nader, M. A., and Czoty, P. W. (2005). PET imaging of dopamine D2 receptors in monkey models of cocaine abuse: genetic predisposition versus environmental modulation. *Am. J. Psychiatry* **162(8)**, 1473–1482.
26. Thanos, P. K., Rivera, S. N., Weaver, K., Grandy, D. K., Rubinstein, M., Umegaki, H., Wang, G. J., Hitzemann, R., and Volkow, N. D. (2005). Dopamine D2R DNA transfer in dopamine D2 receptor-deficient mice: effects on ethanol drinking. *Life Sci.* **77(2)**, 130–139.
27. Dani, J. A., and Harris, R. A. (2005). Nicotine addiction and comorbidity with alcohol abuse and mental illness. *Nat. Neurosci.* **8(11)**, 1465–1470.
28. Assanangkornchai, S., Noi-pha, K., Saunders, J. B., and Ratanachaiyavong, S. (2003). Aldehyde dehydrogenase 2 genotypes, alcohol flushing symptoms and drinking patterns in Thai men. *Psychiatry Res.* **118(1)**, 9–17.
29. Malaiyandi, V., Sellers, E. M., and Tyndale, R. F. (2005). Implications of CYP2A6 genetic variation for smoking behaviors and nicotine dependence. *Clin. Pharmacol. Ther.* **77(3)**, 145–158.
30. Piazza, P. V., and Le Moal, M. (1998). The role of stress in drug self-administration. *Trends Pharmacol. Sci.* **19(2)**, 67–74.
31. Sinha, R. (2001). How does stress increase risk of drug abuse and relapse? *Psychopharmacology (Berl.)* **158(4)**, 343–359.
32. Goeders, N. E. (2003). The impact of stress on addiction. *Eur. Neuropsychopharmacol.* **13(6)**, 435–441.
33. Shalev, U., Grimm, J. W., and Shaham, Y. (2002). Neurobiology of relapse to heroin and cocaine seeking: a review. *Pharmacol. Rev.* **54(1)**, 1–42.

34. Bossert, J. M., Ghitza, U. E., Lu, L., Epstein, D. H., and Shaham, Y. (2005). Neurobiology of relapse to heroin and cocaine seeking: an update and clinical implications. *Eur. J. Pharmacol.* **526(1–3)**, 36–50.
35. Weiss, F. (2005). Neurobiology of craving, conditioned reward and relapse. *Curr. Opin. Pharmacol.* **5(1)**, 9–19.
36. Shaham, Y., and Hope, B. T. (2005). The role of neuroadaptations in relapse to drug seeking. *Nat. Neurosci.* **8(11)**, 1437–1439.

72

Assessment of Neurobiological Diseases with Magnetic Resonance Spectroscopy

Jeffry R. Alger, PhD

Keywords: *brain, brain disease, magnetic resonance spectroscopic imaging, magnetic resonance spectroscopy, metabolite*

I. Introduction
II. History, Nomenclature, and Basic Technological Concepts
III. Biochemistry and the Interpretation of Magnetic Resonance Spectra
IV. Pathophysiology of Neurological Diseases of the Brain and Magnetic Resonance Spectroscopy
V. Summary
References

I. Introduction

Magnetic resonance spectroscopy (MRS) detects electromagnetic signals produced by atomic nuclei within molecules. It can be used to obtain *in situ* concentration measures for certain chemicals present in living tissues and has been used to study the human brain to a greater degree than any other organ system. This chapter introduces the technological concepts on which MRS is based and the basic biochemical concepts necessary for interpreting MRS studies of the brain. It concludes with an overview of the clinical use of MRS for the assessment of neurobiological diseases.

II. History, Nomenclature, and Basic Technological Concepts

MRS is essentially identical to nuclear magnetic resonance (NMR) spectroscopy, which has been an important technique in chemistry and physics for more than 50 years. The term *magnetic resonance spectroscopy* has been used for biomedical and clinical applications instead of *nuclear magnetic resonance spectroscopy* since the 1980s. Terminology that eliminates *nuclear* reflects that MRS uses neither radioactive nuclei nor ionizing radiation.

In MRS, magnetic resonance signals are detected from chemical compounds other than water, with the goal of evaluating *in vivo* biochemistry. In contrast, the magnetic resonance signals used to generate magnetic resonance imaging (MRI) images of anatomy are most often produced by water. MRS and MRI share the same signal detection technology and, accordingly, MRS studies are performed using the same equipment used for MRI.

MRS can be understood at a basic level through consideration of each of the component terms (i.e., magnetic, resonance, and spectroscopy). MRS is based on the fact that certain atomic nuclei behave as spinning bar magnets. These "nuclear magnets" produce magnetic fields that can interact with an externally applied magnetic field. It is through this *magnetic* interaction that energy can be exchanged between the nuclear magnetic fields and the externally applied field. Energy exchange of this sort is known as *resonance*. In physics and chemistry, the term *spectroscopy* is used to describe the study of the specific oscillation frequency at which an energy exchange interaction occurs. The oscillation frequency typically detected in MRS is similar to that of the electromagnetic waves used in radio and television communication.

MRS requires that living tissue be placed in a strong time-invariant (static), externally applied magnetic field (the MRI magnet). Human MRS typically uses a magnetic field strength of 1.5 tesla (T) or higher. MRS signal strength generally improves as the static magnetic field strength is increased, so it is desirable to use the strongest available magnetic field. A second externally applied magnetic field, whose strength oscillates as radio waves do (e.g., 63 megahertz, or MHz, when a 1.5 T magnet is used), is also necessary. It is through this oscillating magnetic field that the specific interaction frequencies can be detected (i.e., the spectroscopy). The term *radio frequency (RF) field* is often used to denote this oscillating magnetic field. The device that produces the RF field is the RF coil, which must be placed around or near the body part being examined. Therefore, the procedure for acquiring brain MRS data is to place the subject's head in the MRI magnet and within an RF coil. The RF coil is turned on (i.e., pulsed) briefly to create an RF field that interacts with the magnetic fields produced by a broad range of types of atomic nuclei in the brain. Once the RF pulse is turned off, the magnetic interaction between the nuclear magnetism and the static magnetic field produces distinct electromagnetic signals in the RF coil. These signals are subjected to analog and digital processing, resulting in a magnetic resonance spectrum, which is a two-dimensional plot of frequency on the horizontal axis versus intensity of resonance interaction (i.e., signal strength) on the vertical axis (Fig. 1). Many variations of this basic procedure are available to meet specific needs. The RF pulse is usually turned on and off more than once during the data acquisition, and the term *pulse sequence* conveys the idea that the data are generated by a complicated series of pulses designed to meet specific experimental needs. These experimental needs include such issues as defining the anatomic location from which signals are detected (i.e., spatial localization) and simplifying (i.e., editing) the spectrum so that only a few key signals can be detected with clarity and, therefore, measured with the greatest possible degree of accuracy.

Special hardware known as gradient coils are also located within the MRI instrument, and these can be turned on and off at the appropriate times during the pulse sequence. These coils produce spatially inhomogeneous magnetic fields that accentuate or diminish the magnetic field produced by the MRI magnet, resulting in a composite magnetic field strength that depends on location within the brain. This spatially inhomogeneous magnetic field configuration is used in conjunction with RF pulses to spatially localize the source of the detected MRS signal to a particular brain region. Two complementary methodologies are available for attaining anatomic localization of the detected MRS signals [1]. In localized single-volume MRS, a conventional image is first acquired with MRI. This is used to identify a location of interest, which is typically defined as a rectilinear voxel (Fig. 2). The magnetic resonance spectrum is then acquired from only this location using a pulse sequence designed to optimally detect signals produced by the defined region and to suppress, to the greatest degree possible, signals that arise from other regions. The alternative method is magnetic resonance spectroscopic imaging (MRSI). In MRSI studies (Fig. 3), MRS signals are simultaneously acquired from a grid containing a large number of rectilinear voxels that include the tissue of interest. MRSI is also sometimes referred to as multivoxel MRS, spectroscopic imaging (SI), or chemical shift imaging (CSI).

The random movement of electrically charged particles within living tissues produces random magnetic fields detected as noise in the magnetic resonance signal detection system. This noise appears as random fluctuations throughout the magnetic resonance spectrum. (See spectra shown in Figs. 1 and 2.) For a single data acquisition, the MRS signal intensities generated by living tissues are generally similar in magnitude to this noise. The signal-to-noise ratio (SNR) can be enhanced by repeating the signal detection procedure multiple times and averaging the results. The spectra shown in Figures 1 and 2 are actually the result of averaging 64 (see Fig. 1) and 128 (see Fig. 2) acquisitions so that the signals can be readily identified above the noise. Signal averaging is productive because the desired signal is identical on each repetition, whereas the noise is random

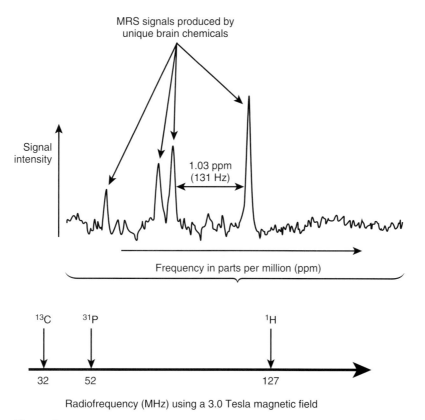

Figure 1 Signal Frequency Measurements in MRS. The frequency of a magnetic resonance spectroscopy (MRS) signal identifies the signal-producing nucleus. When a magnetic field strength of 3.0 T is used, all proton (^{1}H) nuclei produce signals in the immediate vicinity of 127 megahertz (MHz). Carbon (^{13}C) and phosphorus (^{31}P) nuclei produce all of their signals in the frequency bands centered around 32 and 52 MHz, respectively. Small perturbations of the magnetic field strength due to circulation of the molecular electrons lead to frequency differences about a million times smaller. The ^{1}H magnetic resonance spectrum shown was taken from the brain of a normal subject. Two characteristic ^{1}H-MRS signals are separated by 131 hertz (Hz). This separation occurs because the ^{1}H nuclei that produce these two signals are parts of different molecules and have different electronic configurations surrounding them. It is customary to express the smaller frequencies differences of "chemical shifts" in parts per million (ppm) relative to the nuclear frequency. For example, the two signals are separated by 1.03 ppm (= 131 Hz/127 MHz).

and, therefore, differs for each repetition. The disadvantage of signal averaging is that it is time consuming. The need for signal averaging is one reason MRS studies require more time than many other types of MRI studies. To enhance the SNR without extensive signal averaging, MRS measurements must be taken from a larger tissue volume. This is because the noise arises from the entire body part that is within the RF coil, and the signal arises from only within the defined volume of interest. The relatively unfavorable SNR characteristics of MRS result in MRS volume resolution that is relatively poor compared with many other MRI techniques.

Only certain atomic nuclei (isotopes) of biological significance (e.g., ^{1}H, ^{31}P, ^{13}C, ^{7}Li, and ^{19}F) have suitable magnetic properties and are therefore capable of producing MRS signals. The MRS signals produced by these isotopes are easily distinguished from one another by their different characteristic frequencies (see Fig. 1). Moreover, the frequency of an MRS signal produced by a particular isotope (e.g., ^{1}H) depends on the molecular framework surrounding it, because the electrons that form the chemical bonds surrounding a nucleus circulate in ways that tend to alter the magnetic field at the nucleus to a small but significant extent and the MRS signal frequency is proportional to the exact magnetic field strength at the nucleus. Hence, molecule-dependent alterations in the MRS signal frequency exist. This allows the identification of specific resonance signals from individual nuclei within individual molecules (see Fig. 2). The different nuclei of biological significance have different attributes and practical limitations with respect to MRS detection. The proton (^{1}H) produces the strongest, most easily detected MRS signal and

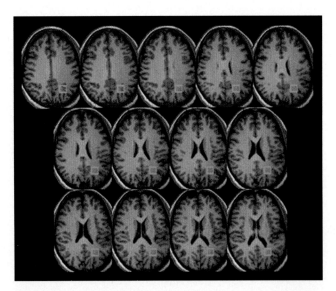
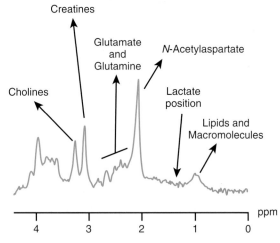

Figure 2 Localized Single-Volume ^{1}H-MRS of a Normal Human Brain. A green square superimposed on the series of T1-weighted magnetic resonance imaging (MRI) slices (left) identifies the brain volume that produced the magnetic resonance spectrum shown on the right. The use of a series of approximately 1-mm-thick MRI slices conveys that the localized magnetic resonance spectroscopy (MRS) volume has a finite thickness along the caudal/cranial dimension. The localized volume has dimensions of $15 \times 15 \times 15$ mm (= 3.4 cm^3). The MRS data were obtained using a 3.0-T MRI unit with time of repetition = 3000 msec, time to echo = 30 msec, and 128 signal averages. The MRS data acquisition time was 6.4 minutes. These parameters are typical for clinical MRS studies of brain. Various key MRS signals are identified. The frequency axis is calibrated based on a signal produced by tetramethylsilane that appears at 0 parts per million by convention. Tetramethylsilane is not present in brain, so a secondary reference signal produced by one of the brain signals is usually used. Note that it is customary to present spectra so that higher frequencies (more parts per million) are to the *left* of lower frequencies.

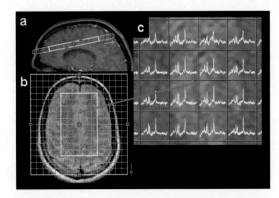

Figure 3 Typical Proton Magnetic Resonance Spectroscopic Imaging (^{1}H-MRSI) Study. Spectra are simultaneously collected from a grid of locations (shown in yellow/green) in one or more "slabs" prescribed from a magnetic resonance imaging study (**a** and **b**). In addition, a large localized volume (white rectangle in **b**) is selected using a localized volume technique analogous to that shown in Fig. 2. The purpose of this volume selection is to suppress the acquisition of interfering outer volume lipid signals (Fig. 4). Each grid location within the selected volume produces a spectrum as shown in **c**.

is, therefore, most often used for routine MRS studies of the human brain. Moreover, ^{1}H-MRS conveniently uses the same signal detection system used for MRI, because the ^{1}H signal of water is used to create MRI images. Phosphorus (^{31}P) produces the second most intense set of MRS signals.

^{31}P-MRS has been the basis for some clinical MRS examinations, but the technical limitations associated with detection of ^{31}P signals compared with ^{1}H signals have led to far more common uses of ^{1}H-MRS in recent years. Other atomic nuclei, particularly ^{13}C, a nonradioactive isotope of carbon, are of research interest but are not presently used in routine clinical MRS studies.

MRS has a strict requirement for a spatially homogeneous static magnetic field. The applied static magnetic field intensity must vary by less than approximately 100 parts per billion over the intended sampling volume. Anatomic features, such as bone and air-filled sinuses, that lie near the tissue of interest can distort the shape and intensity of the applied magnetic field to an unacceptable extent and thereby introduce problems with the detection of MRS signals from certain brain regions. Therefore, it is often challenging to obtain high-quality MRS data from the inferior and mesial portions of the frontal and temporal lobes. Similarly, the presence of certain magnetic materials, such as hemoglobin degradation products in hemorrhages or the paramagnetic contrast agent that remains in brain lesions following contrast-enhanced MRI procedures, can be a problem. In addition, the study of patients who had surgeries before their MRS examinations can sometimes be problematic as a result of magnetic materials left during the surgical procedure.

Magnetic resonance instruments can provide a highly accurate measure of signal frequency but are limited in their ability to provide exact measurements of signal amplitude (the area under the signal) calibrated against a defined universal standard. Hence, MRS permits ready conclusions about whether a particular molecule's signal is present above the noise level, but tissue concentration measurements are less precise. Furthermore, measures of signal intensity must be quoted with reference to some calibration signal that is also present in the spectrum being analyzed or that can be acquired from some other MRS or MRI measurement. Therefore, MRS results are often reported as the ratio of two signals, although more recent studies have emphasized more absolute approaches to quantification of the concentration of signal-producing compounds (Fig. 4).

Special problems related to signal overlap apply to ^{1}H-MRS. The tissue water signal is usually at least 10,000-fold stronger than the signals produced by tissue chemicals, and it overlaps with some chemical signals, rendering them hard to identify and measure. Therefore, the pulse sequences used in ^{1}H-MRS usually have a component designed to suppress water signal relative to tissue chemical signals. Another signal overlap problem is evident in the spectrum shown in Fig. 2. There is considerable overlap of various brain chemical signals, some of which merge into broad signals that make the baseline difficult to define accurately. Baseline signals such as these introduce systematic errors into signal strength quantification. Figure 5 demonstrates the most common approach to spectral simplification. All pulse sequences have an inherent time to echo (TE) which is defined analogously with the TE used in T2-weighted MRI. MRS signals attenuate and become T2-weighted as TE is lengthened, just as the signals in MRI do. Some of the complicating broad baseline signals actually attenuate favorably relative to the stronger, sharper signals when larger TE values are used. Accordingly, many MRS studies use these larger TE values to obtain a more readily quantifiable spectrum. A third special problem of ^{1}H-MRS is that the fatty acyl components of certain lipid molecules, such as the triglycerides in adipocytes, produce strong signals that may overlap with the desired MRS signals (Fig. 6). Interestingly, other lipid molecules, such as the phospholipids and cholesterol esters in biological membranes, do not produce large, interfering MRS signals because of their tight packing structure. This means that ^{1}H-MRS chemical signals can only be detected from tissues that do not contain a substantial amount of triglyceride. One reason ^{1}H-MRS applications have developed more rapidly for the brain than

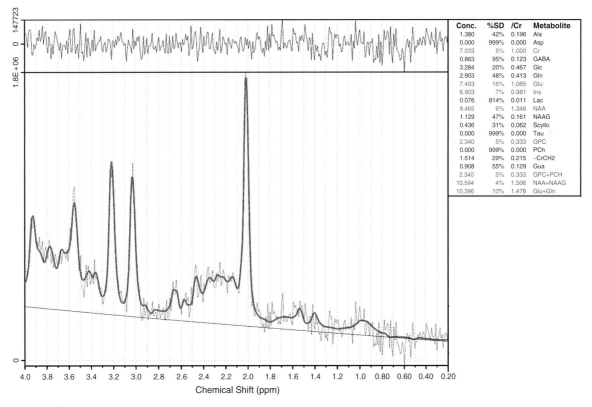

Figure 4 ^{1}H-MRS Signal Quantification. The molecular tissue concentrations are proportional to the area under their magnetic resonance spectroscopy (MRS) signals. The graph shows the spectral data (black) and a synthesized model spectrum that fits the data (red). The upper trace shows the difference between the model and the data. This trace will show random variation around zero when the model fits the data. The model provides estimates of the areas under each signal, which are used to derive relative and absolute tissue concentrations (table upper right). Commercial LCModel software (www.s-provencher.com/pages/lcmodel.shtml) was used for the signal processing shown in this figure.

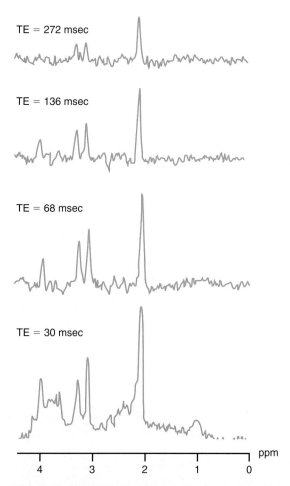

Figure 5 Increasing TE Simplifies the Brain Proton Magnetic Resonance Spectrum. Proton magnetic resonance spectroscopy data were acquired as described in Figure 2 except that time to echo (TE) was varied. The number of averages at each TE was 64 (acquisition time for each TE was 3.2 minutes). As the TE is increased, the smaller signals disappear, producing a less complicated spectrum with fewer signals and a better-behaved baseline. An unfavorable aspect of using longer TE to simplify the spectrum is decrease in the signal-to-noise ratio.

for other body parts is that the lipid components of the brain tissue do not interfere with the detection the brain chemical signals. This is true even for the lipid-rich white matter regions. However, for brain studies, there is a definite need to suppress the interference created by triglyceride signals arising from the overlying muscle and fatty tissues, as shown by Figure 6. Localization procedures that suppress signals from tissues that lie outside the brain are needed, and this is the primary reason for use of volume localization of the type illustrated in Figure 3 in many ^{1}H-MRSI applications.

More detailed presentations of the technological underpinnings of brain MRS may be found in recent chapters by Maudsley [1] and Barker [2] and in the first three chapters of the monograph by Danielsen and Ross [3].

III. Biochemistry and the Interpretation of Magnetic Resonance Spectra

MRS cannot detect signals generated by molecules that have molecular weights greater than a few thousand daltons or by smaller molecules bound to macromolecular arrays (e.g., proteins, membranes, or nucleic acid polymers). Only the small, relatively mobile molecules present within brain tissue are detectable [4,5]. Typically, such tissue-associated small molecules are metabolites involved in intermediary metabolism. Accordingly, MRS is often said to detect tissue metabolites, and from this comes the (somewhat imprecise) generalization that MRS "measures" metabolism. In this light, MRS is often discussed in association with certain nuclear medicine imaging procedures designed to measure metabolic rates through the use of radioactive tracer molecules. However, radiotracer imaging technologies and MRS do not sense metabolism in the same manner. MRS detects the signals produced by certain metabolites that are inherently present in the tissue. With appropriate calibration, such measures can be used to obtain the tissue metabolite concentration. This is inherently different from measuring the rate of a metabolic pathway, as is done by radiotracer imaging. A convenient summary of the actual capabilities of MRS versus radiotracer imaging is that MRS measures *metabolites* whereas nuclear imaging measures *metabolism*. From this point of view, MRS offers a complementary view of metabolism.

MRS has an inherently low sensitivity compared with many destructive techniques (e.g., molecular assay of biopsy material) for detecting molecules in tissue. Indeed, only a few of the most heavily concentrated tissue molecules are readily detected with MRS [5]. For ^{1}H-MRS signal detection, the standard rule of thumb is that there must be at least 1.0 micromole of the molecules of interest within the volume of interest.

Space limitations prevent more than the following cursory summary of the biochemical and metabolic knowledge related to each MRS-detectable metabolite. For further information, consult the review by Imamura [5].

A. N-Acetylaspartate

N-acetylaspartate (NAA) produces a sharp, intense signal at 2.0 parts per million (ppm) in the ^{1}H spectrum of brain tissue (see Fig. 2). The NAA molecule is produced only in normal neurons [5,6]. Astrocytes and other prominent differentiated central nervous system cell types do not contain NAA. NAA signal reduction is observed in many neurological diseases, and this observation is usually attributed to neuronal cell death. Brain cancers provide an extreme example of a condition in which the NAA signal decreases because normal brain tissue is replaced by a mass

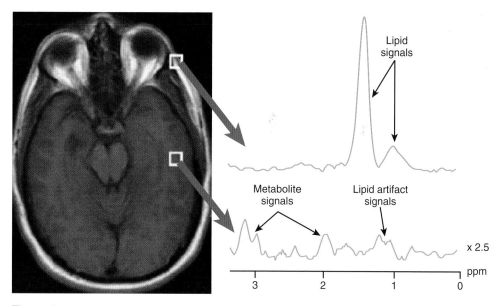

Figure 6 Lipid Signal Interference in Brain Proton Magnetic Resonance Spectroscopy (^{1}H-MRS). Anatomic structures that contain a substantial amount of triglyceride (more intense signal on T1-weighted magnetic resonance imaging, left) produce spectra with strong characteristic signals on the right side of the spectrum (upper spectrum). These lipid signals are near the N-acetylaspartate and lactate signal positions and are large compared with brain metabolite signals (lower spectrum). Care must be taken to suppress detection of such interfering "outer volume" lipid signals to avoid the presence of lipid signal artifacts in brain spectra that can have signal strengths similar to brain metabolite signals.

devoid of neurons. A few studies suggest that NAA signal loss may be reversible in some conditions, such as remitting multiple sclerosis. This observation, with the knowledge that an enzyme associated with the neuronal mitochondria synthesizes NAA, had led to the possibility that NAA levels detected with MRS may also be reporting more subtle alterations in neuronal functional integrity.

B. Cholines

The cholines signal appears at 3.2 ppm in the ^{1}H spectrum of the brain and many other tissues. The ^{1}H-MRS signals produced by choline (Cho), phosphocholine (PCho), and glycerophosphocholine (GPC) cannot be resolved from one another. Hence, the signal seen at approximately 3.2 ppm in a typical *in vivo* ^{1}H-MRS study is generated by the combination of all three molecules and is often referred to as the *total cholines signal* or just the *cholines signal*. Acetylcholine undoubtedly also contributes to the cholines signal, but its level is so much lower than the total of these three "metabolic" cholines that it is not a significant contributor. All three of the total cholines contributors (Cho, PCho, and GPC) participate in the metabolism of the phospholipids that form structural elements of biological membranes [7]. Two of these compounds (Cho and PCho) are phospholipid anabolites; GPC is a phospholipid catabolite. The inability to resolve the signals produced by the three unique choline-containing metabolites leads to uncertainty about the underlying cause for disease-associated changes in the cholines signal. Without being able to resolve PCho, Cho, and GPC signals, it is not possible to know whether alteration of anabolic or catabolic activity is responsible for the choline signal changes that may be associated with disease. This uncertainty has led some authors to use terminology like "increased membrane turnover" and "altered membrane or phospholipid metabolism" as means of acknowledging uncertainty about whether enhanced anabolic or catabolic phospholipid metabolism is responsible.

Prominent abnormalities in the cholines signal strength occur in the context of cancer, which typically shows a profound increase in the intensity of the cholines signal [5,7,8]. Figure 7 illustrates a pronounced cholines signal increase evident in a case of oligodendroglioma. Accordingly, cholines signal increases associated with neoplasia are usually attributed to the altered phospholipid metabolism that accompanies proliferation of cancer cells. Similarly, the more subtle cholines signal increases sometimes apparent in a variety of neurological diseases including multiple sclerosis, infections, and epilepsy are usually attributed to the cell proliferation associated with the brain's inflammatory response to these diseases.

C. Lactate

In the brain, glucose is the sole source of cellular energy. Under normal conditions it is catabolized by an enzyme

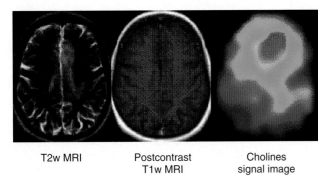

Figure 7 ^{1}H-MRSI of an Oligodendroglioma. T2-weighted and postcontrast T1-weighted magnetic resonance imaging (MRI) of a patient with an oligodendroglioma are shown on the left (grayscale). The color image (right) is the "cholines spectroscopic image," which conveys the spatial distribution of the cholines proton magnetic resonance spectroscopy (^{1}H-MRS) signal strength from the slices shown on MRI. Red hues represent more intense choline signals. Blue hues represent less intense signals. The figure illustrates the typical elevation of cholines signal (in comparison to surrounding brain) that frequently occurs in brain tumors. Among brain tumors, oligodendrogliomas produce some of the most profound cholines signal elevations. Oligodendrogliomas also illustrate that cholines signal elevation occurs in tumors that do not contrast enhance. The cholines spectroscopic image was obtained by using a time to echo of 272 msec with an ^{1}H magnetic resonance spectroscopic imaging (MRSI) technique that is tolerant to outer volume lipid signal artifact and can collect spectra almost to the brain surface. The choline signal area in each MRSI grid element was measured and converted to a color. The heavy spatial smoothing used to produce the cholines spectroscopic image is a result of the low resolution of the MRSI grid compared with MRI resolution.

pathway that includes anaerobic and aerobic steps. The pathway catabolizes glucose to pyruvate without oxygen; the further catabolism of pyruvate to carbon dioxide and water requires oxygen. Both the aerobic and anaerobic pathways yield energy; however, the aerobic portion of the pathway does so more efficiently than the anaerobic portion. If the tissue is deprived of oxygen, optimal utilization of the aerobic pathway is not possible and pyruvate levels rise. Pyruvate quickly interconverts with lactate; therefore, lactate accumulates in tissues that do not have a sufficient supply of oxygen to support their energy needs. Alternatively, disease processes may lead to deranged mitochondrial function and the loss of feedback regulation between the aerobic and the anaerobic pathways that cause lactate production even in the presence of adequate oxygenation. Once the cellular lactate level exceeds about 1 μmol per gram, the lactate formed as a result of abnormal glucose metabolism can be detected with ^{1}H-MRS as a signal at 1.33 ppm (see Fig. 2). Such lactate signal elevations are typically seen in acute stroke [9,10] and in some brain tumors [5,7,8].

D. Amino Acid Neurotransmitters

The amino acid neurotransmitters glutamate, glutamine, and γ-aminobutyric acid (GABA) can be detected with ^{1}H-MRS (see Fig. 2). A recent review by Novotny and colleagues [11] summarizes the current observational capabilities and their potential clinical uses. Two general observations are relevant to this body of investigation: (1) The signals produced by these amino acid neurotransmitters are quite complex. Their signal intensities are low with respect to other ^{1}H-MRS signals, such as NAA or total cholines, and they overlap. As a result, more advanced MRS techniques such as ^{13}C metabolic enrichment, magnetic fields stronger than 1.5 T, and editing pulse sequences are usually needed to quantify the absolute levels of these amino acid neurotransmitters with ^{1}H-MRS. (2) ^{1}H-MRS measures the total tissue concentration of these compounds, and such measurements are not necessarily reflective of either metabolic activity or neurotransmission rates. Accordingly, MRS techniques discussed in the review by Novotny and colleagues [11] that use ^{13}C enrichment of metabolic pathways to measure the synthesis and turnover rates of specific neurotransmitters probably represent areas of future work.

IV. Pathophysiology of Neurological Diseases of the Brain and Magnetic Resonance Spectroscopy

MRS can be performed using commercially available MRI instruments with specialized pulse sequences and the software needed for signal processing and display of the spectroscopic information. Several MRS accessories that operate within commercial MRI products have been cleared for marketing by the U.S. Food and Drug Administration since the early 1990s. Most academic radiology departments and many private radiology practices throughout the world are, therefore, capable of performing brain MRS studies for clinical purposes. However, the medical insurance industry typically does not reimburse for clinical MRS studies because there is not sufficient scientific evidence in support of a conclusion that clinical use of MRS alters clinical outcomes [12]. Therefore, neurologists and neurosurgeons may request that MRS studies be performed for their patients outside of research protocols, but such studies are not likely to be reimbursed.

For MRS to be useful in the evaluation of a particular disease, the disease must alter the concentration of one of the tissue metabolites that can be detected and quantified with MRS. More often than not, a clear understanding of how a particular disease alters the biochemistry of MRS-detectable metabolites is not available and an empirical approach to understanding the MRS features of brain disease has been employed. Indeed, much of the altered biochemistry that is exploited in clinical MRS studies was "discovered" by doing exploratory clinical MRS studies. For instance, before the availability of ^{1}H-MRS, the NAA metabolism in the brain was largely unexplored [6]. Thousands of MRS studies covering all major categories of human brain disease have been published since the technique became available. ^{1}H-MRS has been used in most of these studies, and ^{1}H is now essentially the only magnetic nucleus used for routine clinical MRS work. More extensive consideration of the literature related to clinical use of ^{1}H-MRS than can be given here is available in a recent monograph [3], which also contains numerous examples of how an experienced research group has applied early ^{1}H-MRS techniques to specific diagnostic problems in clinical neurology.

Most MRS studies of brain disease show some combination of three changes: loss of NAA signal, interpreted to signify loss of functional neurons; elevation of lactate, signifying lack of tissue oxygen or infiltration of anaerobic cells such as macrophages; and an elevated choline signal, thought to reflect acceleration of certain metabolic pathways involving components of biological membranes. These changes reflect general pathophysiological processes more often than disease-specific phenomena.

The use of MRS as a diagnostic tool is limited by the absence of a quantitative summary of the normal range of metabolite MRS signal levels as a function of brain location. A variety of different MRS pulse sequences are now in use. Different centers prefer to use different pulse sequence parameters (e.g., TE and time of repetition), and there has been substantial evolution in MRS pulse sequence technology. These between-center differences have resulted in the need for each center to acquire its own normative data each time new MRS technology is implemented. In addition, currently available MRS techniques tend to be inefficient with regard to assaying metabolites throughout the entire brain. Many centers use only single-voxel ^{1}H-MRS, and the commonly used ^{1}H-MRSI techniques typically can acquire data from only a few slices of brain tissue. Because of these technological limitations, there are substantial logistical problems associated with collection of normative MRS data from the entire brain in large populations of normal subjects. Hence, MRS diagnoses in individual patients tend to rely on the comparison to a locally available database of normal MRS studies or on the experience of a radiologist or other physician who has expertise in MRS.

A. Neoplasia

Clinical MRS has been used more often for the evaluation of brain cancer than for any other purpose. Recent reviews of this MRS application are available [7,8,13,14]. Much of the enthusiasm for MRS studies of brain cancer undoubtedly occurs because many cancers produce profound, easily recognizable MRS signal changes (e.g., see Fig. 7). For some brain cancers, the NAA signal decreases and the choline signal increases are much greater than 100% relative to normal brain tissue. The MRS signal changes in many other brain diseases tend to be far more subtle. However, clinical factors have played a role as well. Brain cancer represents a serious health problem with limited therapeutic alternatives. There are dangers and difficulties associated with neurosurgical procedures and with cranial radiation therapy. Furthermore, chemotherapeutic treatments for brain cancer are often ineffective. These clinical factors suggest the need for better noninvasive diagnostic procedures that enable a more exact definition of regional pathology, that provide a measure of therapeutic response (or failure), and that can detect tumor regrowth following therapy. ^{1}H-MRS is believed to offer significant opportunities in each of these areas [7]. As a result, in many advanced cancer centers, brain cancer patients now undergo ^{1}H-MRS studies as part of their routine assessments.

Considerable effort has been expended toward using ^{1}H-MRS technology for defining histological tumor type and grade before biopsy, with the eventual goal a reduced reliance on surgical biopsies for diagnostic purposes [12,14]. Early ^{1}H-MRS studies of cerebral neoplasia suggested that the MRS signal patterns may identify tumor type and grade. Subsequent studies have generally supported this concept, although there are yet to be clearly defined diagnostic standards for typing or grading brain tumors of individual patients with ^{1}H-MRS. Different grades and types of intracranial neoplastic mass lesions have statistically significant differences in mean spectroscopic patterns, but there is also considerable between-subject variability that limits application on a case-by-case basis. Nevertheless, present knowledge permits use of the ^{1}H-MRS signal patterns to suggest the most likely pathology. In some cases, it may be possible to rule out certain types of disease. It is clear that spectrum pattern analysis systems now being developed will likely provide the most efficient approach to using ^{1}H-MRS for preoperative diagnosis, in preference to the evaluation of a single spectroscopic signal [14].

Other uses of ^{1}H-MRS in the context of brain cancer have also been promoted [15]. In particular, ^{1}H-MRSI appears to provide a means of targeting biopsies and can be helpful in defining the spatial extent of neoplastic invasion before surgery or radiation therapy. Here the low spatial resolution of ^{1}H-MRSI has been a limiting factor, although with the appropriate techniques assessments can be made with a spatial resolution of approximately 0.2 cm^3. The significant MRS pattern differences between normal and neoplastic tissues indicate that ^{1}H-MRSI can augment conventional MRI in the delineation of lesion extent.

There have been exploratory studies in a number of areas related to treatment of brain tumors. For instance, studies have demonstrated that progressive or recurrent glioma produce large changes in the cholines signal. Accordingly, many neuro-oncologists hope to use ^{1}H-MRS as a means of distinguishing recurrent and progressing tumor from radiation necrosis for defining when to initiate chemotherapy. In many centers, fluorodeoxyglucose positron emission tomography (FDG-PET) is regarded as the best available method for differentiating recurrent brain tumors from radiation necrosis. There is only one small study that directly compares the clinical value of ^{1}H-MRSI to FDG-PET for differentiating recurrence from necrosis [16]. This study concluded that ^{1}H-MRSI was somewhat more valuable than FDG-PET. However, the authors also recognized that the available ^{1}H-MRSI technique is limited in its ability to evaluate the full three-dimensional extent of brain tumors and the tissues that surround them. It is, furthermore, important to note that newer PET tracers, such as carbon-11-methionine, probably offer improvements over FDG-PET for the identification of recurrent tumors within a previously radiated brain field [17].

B. Stroke

MRS studies of stroke have focused on detecting the decrease of the NAA signal and the increase of the lactate signal that accompanies stroke. Each of these signal changes has been documented in human cases of stroke [9,10]. Initial efforts with human patients in this area were made in the early 1990s and were targeted at developing MRS as a tool for identifying and, perhaps, imaging the ischemic penumbra. The hypothetical basis is that brain regions that exhibit NAA signal preservation simultaneously with lactate signal increase are not yet irreversibly damaged but are not receiving adequate blood flow and oxygen. The interest in using MRS for identification of the ischemic penumbra in human patients has largely waned as a result of the introduction of diffusion and perfusion MRI techniques, which accomplish the same purpose and are more easily performed than MRS. However, MRS continues to be used in studies of experimental brain ischemia [18]. These studies emphasize the importance MRSI techniques that can efficiently image NAA and lactate signals over wide brain areas. Therefore, it is likely that interest in using MRS in human stroke will be renewed as faster, more efficient MRSI techniques become available for human imaging.

MRS may also have a role to play in the evaluation of chronic cerebrovascular disease. Brain NAA and lactate signal measurements in patients who have carotid artery disease correlate with other clinical measures of disease severity and may even eventually predict which patients would benefit from specific treatments.

C. Epilepsy

A large body of investigational work has focused on use of ^{1}H-MRS in the evaluation of temporal lobe epilepsy [19,20]. A specific goal has been to evaluate whether ^{1}H-MRS signal patterns aid in the diagnosis of which temporal lobe is the source of epileptic activity. Studies indicate that NAA signal decrease is a marker of hippocampal sclerosis and can be helpful for this purpose. Studies of changes in other ^{1}H-MRS signals, particularly the elevation of the cholines signal, are now under way. Knowlton and colleagues [21] compared FDG-PET and ^{1}H-MRSI to ictal electroencephalographic lateralization in patients with temporal lobe epilepsy. FDG-PET showed a greater concordance (87%) with ictal electroencephalographic lateralization than did ^{1}H-MRSI (61%). However, the same group published a subsequent study [22] that used more advanced ^{1}H-MRSI techniques and signal processing without FDG-PET comparison. This later study showed that ^{1}H-MRSI concordance with an electroencephalogram can be as high as 85%, which compares quite favorably with FDG-PET. As a result, ^{1}H-MRS is used at some epilepsy centers in conjunction with other techniques as part of the presurgical planning routine. It is important to emphasize, however, that use of ^{1}H-MRS for identification of seizure-generating foci outside of the temporal lobe is not supported by scientific evidence. There is not sufficient knowledge about the normal range of variation in ^{1}H-MRS signal patterns throughout the brain; therefore, it is difficult to conclusively identify abnormal signal patterns at arbitrary brain locations in individual patients.

Advanced ^{1}H-MRS procedures are capable of measuring brain GABA levels. This has led to the potential for using ^{1}H-MRS to quantitatively assess pharmacological treatments for epilepsy that are targeted at GABA metabolism or function [23].

D. Multiple Sclerosis

The use of MRS in the evaluation of multiple sclerosis (MS) has been reviewed by de Stefano [24]. Decreased NAA signal is regularly observed in acute MS plaques, implying that more axonal damage occurs in such lesions than had been previously thought. However, it has also been shown that the NAA signal recovers partially in demyelinating lesions in a manner that correlates with clinical recovery. This observation was among the first that suggested the NAA signal may act as a biomarker of neuronal function, as well as neuronal density. More recent efforts have focused on using ^{1}H-MRS to demonstrate that the "normal appearing" white and gray matter are metabolically abnormal in patients who suffer from MS. This MRS evidence has generally supported the shift toward thinking of MS as a disease of the whole brain rather than a selective disease of white matter.

E. Other Neurological Diseases

Space limitations do not permit a discussion of the numerous MRS observations that have been made in many other neurological diseases, including Alzheimer's disease and other dementias, parkinsonism, motoneuron disease, human immunodeficiency virus (HIV) encephalopathy, and metabolic disorders. As a starting point for development of an understanding of this body of work, see a recent review by Lin and colleagues [12].

This chapter has focused on brain diseases because MRS studies have, in general, not been possible in the spinal cord. Except for cervical cord segments, the cord dimensions are small relative to the typical MRS volume requirements. In addition to this, magnetic field inhomogeneity arising from the vertebral bodies, the concentrated triglyceride in the vertebral bone marrow, and respiratory motion are all limiting factors. However, it is possible that recent technical advancements in performing MRI studies of the thoracic cavity may eventually be used to facilitate MRS studies of the thoracic cord elements.

V. Summary

MRS provides a means of measuring the concentrations of the more abundant tissue metabolites in the brain using MRI technology. MRS measurements are made with regional specificity without exposure to ionizing radiation. Past trends suggest an increased use of MRS for addressing clinical management problems in neurology but also to further our understanding of the unique biochemistry and physiology of the nervous system. Two attributes of MRS provide compelling support for this assertion. First, MRS clearly provides information beyond that provided by other forms of MRI. MRI provides a vivid anatomic depiction of the brain and some physiological information (e.g., perfusion or oxygenation), and MRS provides complementary biochemical information. Therefore, the ease with which MRS can be incorporated into an MRI study protocol guarantees its increased use even though the biochemical information provided by MRS is not yet fully understood in many cases. Second, MRS provides one of the few means of assessing aspects of tissue biochemistry without exposure to radioisotopes or to ionizing radiation. MRS can be performed in an individual patient at a frequency only limited by finances and logistics. This suggests that MRS may become a management tool of considerable significance because an individual patient can undergo almost unlimited MRS surveillance during and following a variety of treatments.

Finally, it is most important to emphasize that MRS is a "moving technology." There has been sustained technological progress in its clinical implementation during the past 20 years. It is clear that this technological progress will continue and this will lead to a further acceleration in the use of MRS for the clinical assessment of neurological problems.

References

1. Maudsley, A. A. (2002). Magnetic resonance spectroscopic imaging. *In:* "Brain Mapping: The Methods" (A. W. Toga and J. C. Mazziotta, eds.), 2nd edition, pp. 351–378. Academic Press, San Diego.
2. Barker, P. B. (2004). Fundamentals of MR spectroscopy. *In:* "Clinical MR Neuroimaging: Diffusion, Perfusion and Spectroscopy" (J. H. Gillard, A. D. Waldman, and P. B. Barker, eds.). Cambridge University Press, Cambridge.
3. Danielsen, E. R., and Ross, B. D. (1999). "Magnetic Resonance Spectroscopy Diagnosis of Neurological Disease." Marcel Dekker, New York.
4. Alger, J. R. (2004). MRS of the brain. *In:* "Encyclopedia of Neuroscience" (G. Adelman and B. A. Smith, eds.). Elsevier Science, Amsterdam.
5. Imamura, K. (2003). Proton MR spectroscopy of the brain with a focus on chemical issues. *Magn. Reson. Med.* **2**, 117–132.
6. Demougeot, C., Marie, C., Giroud, M., and Beley, A. (2004). N-acetylaspartate: a literature review of animal research on ischemia. *J. Neurochem.* **90**, 776–783.
7. Alger, J. R. (2005). Magnetic resonance spectroscopy and cancer. *In:* "New Techniques in Oncologic Imaging" (A. Padhani and P. Choyke, eds.). Taylor & Francis CRC Press, New York.
8. Alger, J. R. (2004). Magnetic resonance spectroscopy of brain tumors in adults. *In:* "Clinical MR Neuroimaging: Diffusion, Perfusion and Spectroscopy." (J. H. Gillard, A. D. Waldman, and P. B. Barker, eds.). Cambridge University Press, Cambridge.
9. Graham, G. D., Kalvach, P., Blamire, A. M., Brass, L. M., Fayad, P. B., and Prichard, J. W. (1995). Clinical correlates of proton magnetic resonance spectroscopy findings after acute cerebral infarction. *Stroke* **26**, 225–229.
10. Barker, P. B. (1997). Metabolism: magnetic resonance spectroscopy and spectroscopic imaging. *In:* "Primer on Cerebrovascular Diseases" (K. M. A. Welch, L. R. Caplan, D. J. Reis, B. K. Siesjo, and B. Weir, eds.), pp. 650–660. Academic Press, San Diego.
11. Novotny, E. J., Fulbright, R. K., Perl, P. L., Fulbright, R. K., Pearl, P. L., Gibson, K. M., and Rothman, D. L. (2003). Magnetic resonance spectroscopy of neurotransmitters in human brain. *Ann. Neurol.* **54(Suppl 6)**, S25–S31.
12. Lin, A., Ross, B. D., Harris, K., and Wong, W. (2005). Efficacy of proton magnetic resonance spectroscopy in neurological diagnosis and neurotherapeutic decision making. *NeuroRx* **2**, 197–214.
13. Law, M. (2004). MR spectroscopy of brain tumors. *Top. Magn. Reson. Imaging* **15**, 291–313.
14. Howe, F. A., and Opstad, K. S. (2003). ^{1}HMR spectroscopy of brain tumours and masses. *NMR Biomed.* **16**, 123–131.
15. Nelson, S. J. (2003). Multivoxel magnetic resonance spectroscopy of brain tumors. *Mol. Cancer Ther.* **2**, 497–507.
16. Lichy, M. P., Henze, M., Plathow, C., Bachert, P., Kauczor, H. U., and Schlemmer, H. P. (2004). Metabolic imaging to follow stereotactic radiation of gliomas: the role of ^{1}H-MR spectroscopy in comparison to FDG-PET and IMT-SPECT. *Rofo* **176**, 1114–1121.
17. Jacobs, A. H., Kracht, L. W., Gossmann, A., Ruger, M. A., Thomas, A. V., and Thiel, A., Herholz, K. (2005). Imaging in neurooncology. *NeuroRx* **2**, 333–347.
18. Weber, R., Ramos-Cabrer, P., and Hoehn, M. (2006). Present status of magnetic resonance imaging and spectroscopy in animal stroke models. *J. Cereb. Blood Flow Metab.* **26(5)**, 591–604.
19. Hammen, T., Stefan, H., Eberhardt, K. E., W-Huk, B. H., and Tomandl, B. F. (2003). Clinical applications of ^{1}H-MR spectroscopy in the evaluation of epilepsies: what do pathological spectra stand for with regard to current results and what answers do they give to common clinical questions concerning the treatment of epilepsies? *Acta Neurol. Scand.* **108**, 223–238.

20. Briellmann, R. S., Wellard, R. M., and Jackson, G. D. (2005). Seizure-associated abnormalities in epilepsy: evidence from MR imaging. *Epilepsia* **46**, 760–766.
21. Knowlton, R. C., Laxer, K. D., Ende, G., Hawkins, R. A., Wong, S. T., Matson, G. B., Rowley, H. A., Fein, G., and Weiner, M. W. (1997). Presurgical multi-modality neuroimaging in electroencephalographic lateralized temporal lobe epilepsy. *Ann. Neurol.* **42**, 829–837.
22. Capizzano, A. A., Vermathen, P., Laxer, K. D., Ende, G. R., Norman, D., Rowley, H., Matson, G. B., Maudsley, A. A., Segal, M. R., Weiner, M. W. (2001). Temporal lobe epilepsy: qualitative reading of ^{1}H-MR spectroscopic images for presurgical evaluation. *Radiology* **22(4)**, 625–631.
23. Petroff, O. A. (2002). GABA and glutamate in the human brain. *Neuroscientist* **8**, 562–573.
24. De Stefano, N., Bartolozzi, M. L., Guidi, L., Stromillo, M. L., and Federico, A. (2005). Magnetic resonance spectroscopy as a measure of brain damage in multiple sclerosis. *J. Neurol. Sci.* **233**, 203–208.

73

Magnetic Resonance Imaging

John A. Detre, MD

Keywords: *brain, diffusion, magnetic resonance imaging, molecular imaging, perfusion, spinal cord, relaxation*

I. History of Magnetic Resonance Imaging
II. Magnetic Resonance Imaging Hardware
III. Basic Principles of Magnetic Resonance Imaging
IV. Image Analysis
V. Basic Clinical Magnetic Resonance Imaging
VI. Perfusion and Diffusion Magnetic Resonance Imaging
VII. Other Sources of Image Contrast in Proton Magnetic Resonance Imaging
VIII. Exogenous Contrast Agents
IX. Imaging Nuclei Other Than Protons
X. Conclusion
 References

High-quality imaging of the central nervous system (CNS) is more challenging than in other organs because the brain and spinal cord are soft tissue encased in dense bony structures. Although even routine x-ray can provide diagnostic information about organs in the torso, only limited inferences can be drawn about the brain from skull x-rays, primarily based on asymmetries or shifts in the location of a few possibly calcified structures. Computed tomography (CT) revolutionized the assessment of the CNS, providing sufficient dynamic range to visualize brain tissue within the skull—albeit with careful windowing with respect to bone, which is much denser than nervous tissue, dominates the signal, and often degrades image quality in the brainstem and spinal cord where the ratio of tissue to bone is reduced.

CT scanning remains an important modality for rapid assessment of the CNS (e.g., head trauma), though tissue contrast in CT is limited to changes in tissue density and enhancement following exogenous contrast administration.

Magnetic resonance imaging (MRI) currently is the most versatile and informative imaging modality for the CNS. There are a variety of reasons for this. First, MRI is detected as radio frequency (RF) signals that can penetrate the skull and spinal column without degradation. Second, the hardware used to generate MRI scans allows tremendous flexibility in imaging orientation and spatial resolution. Finally, there are numerous sources of image contrast that contribute to the characterization of structure, function, and lesions of the CNS. Some of these capabilities are covered in chapters on functional MRI and magnetic resonance spectroscopy (MRS) in this text.

I. History of Magnetic Resonance Imaging

Clinical MRI is the result of an extraordinary number of scientific and engineering advances [1]. The first successful nuclear magnetic resonance (NMR) spectroscopy experiments were independently demonstrated in the 1945 by Felix Bloch and Edward Purcell, who shared the Nobel Prize in Physics in 1952 for the finding. For the next few decades, NMR experiments were mainly used for chemical and physical analysis of small samples that could be fit into small-bore NMR spectrometers. High-resolution NMR continued to evolve into a powerful modality for detailed chemical analysis of molecules, but NMR imaging of spatially resolved signals developed in a different direction. Spatial encoding in NMR is accomplished through the use of magnetic field gradients, which can introduce spatial variations in the main magnetic field. The concept of generating images with NMR arose from Paul Lauterbur's 1972 idea of applying field gradients in all three dimensions, using back-projection methods borrowed from CT scanning to generate images. This inspired development of Fourier transform reconstruction by Richard Ernst in 1974, which is the predominant approach used today. Around that time, wide-bore NMR systems capable of imaging living animals and human limbs were available, and larger magnets capable of accommodating a human body were being considered. Peter Mansfield reported the first *in vivo* image of human anatomy in 1977, a cross-sectional image through a finger. The potential diagnostic value of changes in NMR relaxation was suggested by Raymond Damadian and others and further motivated the development of MRI for clinical use. *Nuclear magnetic resonance* imaging was renamed *magnetic resonance imaging* to avoid the undesirable connotations of the word *nuclear* among the lay public. In 2003, Lauterbur and Mansfield shared the Nobel Prize in Medicine for MRI.

II. Magnetic Resonance Imaging Hardware

Several component systems are combined to create a modern MRI scanner, illustrated in Figure 1. The major components are the magnet, the gradient system, the RF system, and one or more computers to control these systems and process data.

The main magnetic field is typically generated by a wide-bore superconducting solenoid magnet. Superconducting materials capable of maintaining a useful magnetic field must be maintained at liquid helium temperature. The coils are immersed within a Dewar containing liquid helium that is either refrigerated or cooled within a second Dewar containing liquid nitrogen. The main magnet can be "quenched" in the case of an emergency or accidentally if a ferromagnetic object affects the structure. During this process the liquid gases cooling the magnet are rapidly evaporated and must be vented away to prevent harm to people in the vicinity. Some low-field "open" MRI systems continue to use permanent magnets for the main magnetic field, and electromagnets have been used for prototype intraoperative MRI systems that can be switched on and off. Large-bore magnets used for MRI create strong fringe fields that must be considered in siting these instruments. Advances both in magnet technology and in passive and active shielding of the main magnetic field have allowed increasingly high field-strength systems to be deployed.

Magnetic field gradient coils are placed inside the main magnet and are used to produce spatial variations in magnetic field strength. Most systems use gradients in the X, Y, and Z directions, which can be combined to produce a gradient in any Cartesian direction. These gradient coils are driven by powerful amplifiers under computer control. Another set of weaker shim gradients is used to fine-tune the homogeneity of the main magnetic field. Signals are transmitted and received from the sample by the RF system, which is similar to that used in radio and television. Most MRI scanners include a large "body" RF coil and smaller local coils for focused studies. For MRI of the brain, a volume coil is usually placed around the head. An improved signal-to-noise ratio can be obtained using arrays of even smaller surface coils. These multicoil arrays require multiple RF channels, which is an emerging technology in commercial MRI systems. The best sensitivity for imaging the brain and spine is now achieved using arrays of 4–16 coils working in tandem to receive signals, following excitation with the body coil. RF excitation also uses a powerful amplifier operating in the megahertz range, and care must be taken to avoid excessive heating of the sample. In the United States, the Food and Drug Administration specifies acceptable limits for field strength, RF deposition, and gradient switching rates in clinical MRI.

Images are created using pulse sequences that specify RF and gradient events with microsecond accuracy. RF signals

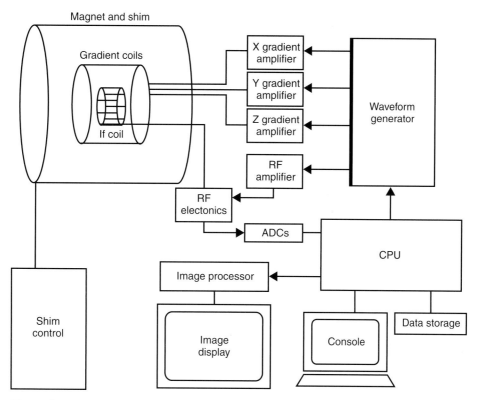

Figure 1 Schematic Diagram Indicating the Components of a Magnetic Resonance Imaging Scanner. *Note:* ADCs, analog-to-digital converters; CPU, central processing unit; RF, radio frequency. *Source:* Figure courtesy of Dr. Stuart Clare, Oxford University.

received from the sample are digitized and converted to images using reconstruction algorithms. These processes are controlled by several general-purpose and custom computers. Software for clinical MRI systems allows definition of complex imaging protocols, graphical interfaces for prescription of imaging slices and volumes, and sophisticated postprocessing.

III. Basic Principles of Magnetic Resonance Imaging

MRI provides enormous flexibility in image resolution, orientation, and source of contrast. These are obtained through variations in the *pulse sequence* used by the hardware to generate the signal and in the approaches used to reconstruct images from the raw data. Only a brief overview of key concepts in MRI physics can be provided here. Certain biologically relevant nuclei have an intrinsic angular momentum, termed *spin*, that gives rise to NMR signals in the presence of a magnetic field. Of these, a smaller number can produce NMR signals amenable to imaging, in that they may be present in high enough concentration, and generate signals strong enough and long-lived enough to be spatially encoded. Most MRI is based on the proton (^{1}H) signal in water (H_2O), which is present in high concentrations (approximately 110 molar) in living tissues. Some imaging has also been carried out using phosphorus (^{31}P) and sodium (^{23}Na) signals and with exogenously administered compounds containing the stable isotopes of fluorine (^{19}F), hydrogen (^{2}H), and oxygen (^{17}O). The signal strength varies for each nucleus and is quantitatively related to the number of nuclei in the imaging volume, resulting in a trade-off between the signal-to-noise ratio and the spatial resolution. Signal strength is also proportional to the strength of the main magnetic field. Although there are numerous challenges to implementing MRI at ever higher field strengths, the benefit is a higher signal-to-noise ratio, which can be used to reduce scan times, improve spatial resolution, or detect signals from nuclei with a weak signal or at low concentrations. Whereas a decade ago 1.5 tesla was considered a high-field MRI scanner, many institutions currently have experimental MRI systems operating at 7 tesla.

While spins (protons, in the case of clinical MRI) align with the main magnetic field, application of RF energy at the resonance frequency creates a second magnetic field that rotates these spins into the transverse plane where an RF signal is emitted at the resonance frequency. RF excitation pulses can be characterized by the degree to which excited spins are rotated from the direction of the main magnetic field. A 90-degree pulse produces the maximum transverse signal; a 180-degree pulse produces no signal but inverts the excited spins. Relaxation times T1 and T2 are the time constants for an exponential process by which the NMR signal decays, and regional variations in the molecular environment of water in normal or pathological tissue result in regional variations in T1 and T2 for water protons that give rise to image contrast in routine T1-weighted and T2-weighted MRI scanning. T1 reflects the time it takes to return to the resting spin orientation with respect to the main magnetic field. When spins are excited by RF energy and flipped into the transverse plane, they are initially in phase. T2 reflects the time it takes for this phase coherence to be disrupted by interactions with the molecular environment and is much shorter than T1 for ^{1}H-MRI in biological tissues. T1 and T2 are increased in proportion to the concentration of free water in tissues, and factors that disrupt local field homogeneity can shorten T1 and T2.

Spatial encoding of tissue signals is accomplished by pulsing the magnetic field gradients to generate a spatially varying resonance frequency or phase. To provide sufficient time to apply spatial encoding gradient pulses, most MRI scans are acquired using pulse sequences that induce an echo signal. This is accomplished either by applying a 180-degree pulse (spin echo) or by reversing the polarity of the frequency encoding gradient (gradient echo). An "echo" signal is then observed following a time equal to the time between the initial excitation and the 180-degree pulse or gradient inversion. The time to echo (TE) is equal to twice this duration. The strength of the signal obtained at TE increases with T2. T1-weighted imaging is obtained by reducing the time between phase-encoding steps with respect to T1. The time between phase-encoding steps is the time of repetition (TR). Regions or lesions with short T1 provide increased signal intensity because spins are able to relax more during TR than regions with longer T1. Two-dimensional fast Fourier transformation is used to covert from the time domain signal acquired to spatially encoded images. For T2-weighted imaging, relatively long TE are used to increase the conspicuity of regions or lesion with long T2. Spin echo imaging can eliminate signal loss due to local variations in magnetic field homogeneity, but gradient echo imaging cannot. For example, local magnetic susceptibility effects due to methemoglobin in hemorrhages produce prominent areas of signal loss in gradient echo imaging. A pulse sequence diagram including the basic components of a spin echo imaging sequence is illustrated in Figure 2.

IV. Image Analysis

MRI affords numerous sources of image contrast within a single imaging modality. For various contrasts obtained during a single imaging session, the images are inherently coregistered in space. In radiological applications, images are typically analyzed by visual inspection. However, in basic and clinical neuroscience applications, quantitative morphometry using semiautomated algorithms is usually employed. Segmentation into compartments of interest can be carried out on one set of images and applied to another. For example, segmentation into gray and white matter compartments can be carried out on T1-weighted MRI, which provides excellent gray-white contrast, and then applied to images acquired with perfusion contrast. Specific brain regions or lesions may also be segmented manually or semiautomatically for quantitative volumetry or to assess for other regional effects. Assessment of lesion volume provides a surrogate marker of disease progression. Such "region of interest" analyses allow data reduction into scalar values for comparisons across groups of subjects or time. A region of interest may also characterize whole brain effects, such as the measurement of progressive atrophy based on the ratio between gray matter and cerebrospinal fluid. Figure 3 illustrates segmentation into gray matter, white matter, and cerebrospinal fluid compartments based on T1 contrast.

More elaborate procedures are required to compare volumetric data across subjects. This typically entails spatial normalization of the data into a standard neuroanatomic space. Procedures to accomplish this reliably are still evolving and must be capable of handling intersubject variability in gyral structure and, for clinical applications, the presence of focal lesions. The assessment of structural effects in volumetric imaging has been termed *voxel-based morphometry* (VBM) and provides a potentially powerful approach to exploring structure–function relationships. In VBM, statistical methods are used to probe spatially normalized volumetric imaging data from one or more cohorts to identify specific voxels that correlate with a particular phenotype [2]. Analogous to genotype–phenotype assessments, VBM methods must account for multiple comparisons, with up to several hundred thousand voxels being

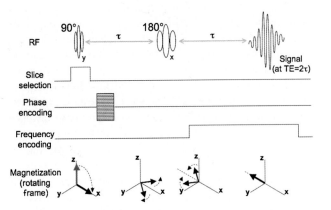

Figure 2 Schematic Diagram Showing the Basic Radio Frequency and Gradient Events Leading to Magnetic Resonance Imaging Signal in a Two-Dimensional Spin Echo Imaging Sequence. *Note*: RF, radio frequency; TE, time to echo.

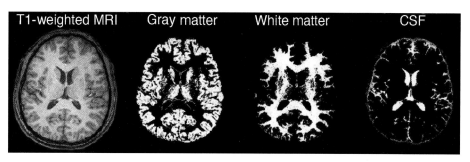

Figure 3 Automated Segmentation of T1-Weighted Structural Magnetic Resonance Imaging (MRI), Gray Matter, White Matter, and Cerebrospinal Fluid (CSF) Regions using SPM2 Software (www.fil.ion.ucl.ac.uk/spm).

assessed, but they do not require an a priori hypothesis about which brain regions might be affected.

V. Basic Clinical Magnetic Resonance Imaging

The early applications of MRI to clinical neuroscience mainly focused on T1-weighted (short TR), T2-weighted (long TE), and spin density (long TR, short TE) imaging. With just these sources of image contrast, significant advantages over CT scanning were observed, mainly because of improved resolution and particularly lack of bone artifact in the brain, brainstem, and spinal cord. A steady stream of improvements has greatly extended the diagnostic utility of MRI over the past two decades, with no evidence yet of nearing its full potential. Major enhancements included the development of intravenous MRI contrast agents for detecting blood-brain barrier breakdown, fast scanning techniques to shorten the duration of clinical studies, the use of cerebrospinal fluid suppression in T2-weighted imaging to increase conspicuity of lesions near the ventricles and subarachnoid spaces, and the development of magnetic resonance angiography (MRA) and magnetic resonance venography (MRV) to assess vascular patency. Increased field strength and improved hardware are also allowing ever-smaller lesions to be reliably identified, with submillimeter resolution now routinely possible.

Several additional sources of contrast have also been characterized, some of which are discussed later. At present, MRI, MRA, or MRV with contrast represents the definitive neuroimaging examination for most CNS disorders. Use of myelography has been virtually eliminated by MRI. CT scanning is now carried out primarily in hyperacute situations or in patients for whom MRI is contraindicated due to a pacemaker, severe obesity, or claustrophobia, though the advent of faster and higher resolution CT scanners has provided improved spatial and temporal resolution for that modality and contrast-based CT angiography and dynamic CT perfusion imaging is increasingly used in the evaluation of cerebrovascular disease. Although the sensitivity of MRA is now approaching that of catheter angiography for examining large vessels such as the carotid arteries, there remain limitations with resolution for small intracranial vessels and difficulties in visualizing turbulent flow distal to stenoses. Catheter angiography remains the most sensitive and specific approach for vascular imaging, except in cases such as carotid dissection where soft tissue visualization of the vessel wall is also desired. MRI can also be used for soft tissue characterization of carotid plaque [3]. For details on the use of various MRI contrasts in clinical diagnosis, refer to a neuroradiological MRI text (e.g., see reference 4).

Although the greatest effect of MRI in basic and clinical neuroscience has been for visualization of the brain and spinal cord, the high spatial resolution and soft tissue contrast in MRI has also provided new opportunities for visualization of the peripheral nervous system [5]. Thus far, most MRI applications in peripheral nerves have been for the clinical assessment of compression, inflammation, or trauma in nerve roots, plexi, and larger peripheral nerves. Muscle imaging is useful in documenting the progression of myopathy, and the distribution muscle signal changes due to denervation can be used for localizing neuropathies in a manner analogous to electromyography.

VI. Perfusion and Diffusion Magnetic Resonance Imaging

Perfusion is a fundamental physiological parameter, and disorders of perfusion account for a major portion of health morbidity. The ability to measure cerebral perfusion is useful not only in cerebrovascular diseases but also in brain tumors, degenerative diseases, and epilepsy, where changes in regional perfusion occur secondarily. The most widely used method for assessing perfusion with MRI uses dynamic scanning to quantify regional signal changes following intravenous administration of an MRI contrast agent such as gadolinium [6]. As the contrast agent passes through the vasculature, a transient reduction in image intensity is observed in susceptibility-weighted (gradient echo) imaging. A variety of hemodynamic parameters can be derived from the time course of this effect, including mean transit time (MTT), cerebral blood volume (CBV), and cerebral blood flow (CBF). All these variables can be inferred using the central volume principle (CBF = CBV / MTT). Because of its ease of measurement, in many clinical applications the time to peak (TTP) susceptibility effect has been used to quantify dynamic susceptibility contrast perfusion MRI. Accurate calculation of CBF from dynamic susceptibility contrast perfusion MRI requires consideration of the arterial input function to the brain to eliminate the effects of temporal smearing of the contrast bolus in the heart and lungs. In the absence of this, the approach is often termed *perfusion-weighted imaging* (PWI). An alternative approach to measuring CBF with MRI uses magnetically labeled arterial blood water as a diffusible flow tracer. This approach is analogous to CBF measurements with $^{15}O-H_2O$ and positron emission tomography and provides quantitative CBF more directly [7]. However, this approach is technically challenging and is only beginning to be considered for clinical applications [8].

For decades it has been recognized that diffusion effects can contribute to signal loss in spin echoes and that this approach can be used to quantify accurately diffusion coefficients in samples. This effect can be enhanced by adding large, symmetrical field gradient pulses before and after the 180-degree pulse in a spin echo sequence to dramatically increase the spatial dependence of the echo amplitude. Thus spins spatially displaced during the milliseconds between the first and the second gradient pulses will not regain full phase coherence at the TE, and their signal will be reduced.

More recently, diffusion weighting has been used to produce contrast in biomedical imaging [9]. In biological tissues, diffusion is restricted due to structural components. In addition, an apparent diffusion coefficient (ADC) of water representing a weighted sum of diffusion compartments in each voxel can be calculated based on MRI intensities with and without the strong diffusion-encoding gradients. In diffusion-weighted imaging (DWI), the addition of strong gradients to a spin echo imaging sequence produces a reduction in signal intensity in regions with rapid diffusion. In a calculated ADC image, these regions appear bright. Although most clinical applications of diffusion MRI have used DWI, it must be noted that there is also T2-weighting in DWI and accurate differentiation of T2 changes from diffusion changes requires an ADC image. The major clinical application of diffusion MRI has been in stroke, where shifts in water compartmentalization within brain parenchyma occur within minutes of ischemia as a result of loss of ion homeostasis and a reduction in water diffusivity is observed with this early cytotoxic injury [10]. DWI can rapidly and accurately visualize the extent and distribution of ischemic changes in acute stroke. In addition, as stroke evolves and cells eventually lyse, regional diffusivity increases so that DWI can also readily distinguish between acute and chronic infarcts that may have a similar appearance on T2-weighted MRI. Furthermore, a combination of DWI and PWI can be used to operationally define an ischemic penumbra at risk for infarction, and growing evidence suggests that consideration of mismatches between DWI and PWI lesions in acute stroke may be useful in triaging patients for thrombolytic therapy [11]. In centers where hyperacute MRI is used routinely, a carefully prescribed imaging protocol can provide perfusion, diffusion, MRA, T2-weighted MRI, and gradient echo imaging in approximately 15 minutes. Hyperacute MRI is typically carried immediately on presentation to the emergency department to avoid delaying therapy, but it may also have a role in expanding the current 3-hour window for intravenous thrombolysis and for providing insight into mechanisms of stroke. An example of hyperacute MRI data is shown in Figure 4.

Water diffusion in gray matter is reasonably isotropic, meaning that water movement occurs equally in all directions. However, in white matter water diffusion is anisotropic and is facilitated along white matter tracts. As noted previously, diffusion encoding in MRI is accentuated through the use of strong field gradient pulses. In regions with anisotropic diffusion, differing signal attenuation will be observed depending on which gradient is used for sensitization. Diffusion MRI data acquired with varying gradients can be combined to create a diffusion tensor image, in which each voxel is represented not only by an ADC but also by one or more tensors (multidirectional vectors) that include information about the directionality of water diffusion. Such data can be used to generate images of fiber tracts based on algorithms that integrate diffusion tensors from adjoining

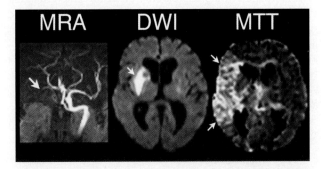

Figure 4 Magnetic Resonance Imaging in Hyperacute Stroke Obtained from a Patient with Left Hemiparesis. Magnetic resonance angiography (MRA) shows right-middle cerebral artery occlusion. Diffusion-weighted imaging (DWI) shows hyperintensity in a lenticulostriate distribution, indicating cytotoxic edema and suggesting infarction. The mean transit time (MTT) shows prolonged transit time in the entire middle cerebral artery distribution. This mismatch between the region showing altered hemodynamics and the region showing infarction has been operationally defined as a "penumbral" region potentially amenable to therapy. *Source:* Images courtesy of Dr. Gottfried Schlaug, Harvard University.

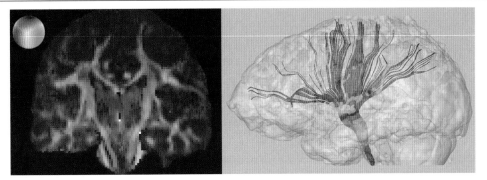

Figure 5 Diffusion Tensor Imaging Data. *Left:* Fiber orientation map, color coded for directionality based on diffusion tensors. Inset shows color scheme, with blue indicating superior/inferior projections, red indicating left/right projections, and green indicating anterior/posterior projections. *Right:* Diffusion tensor tractography showing long corticoefferent fibers projecting through corona radiata, internal capsule, cerebral peduncle, and brainstem. Corticopontine fibers are shown in red-purple. Corticospinal fibers are shown in blue. *Source:* Left image courtesy of Dr. Carlo Pierpoli, National Institute of Neurological Diseases and Stroke. Right image courtesy of Dr. Susumu Mori, Johns Hopkins University.

voxels [12] or to assess myelination as manifested by fractional anisotropy [13]. This is an extremely active area of research and development. An example of diffusion tensor imaging is shown in Figure 5.

VII. Other Sources of Image Contrast in Proton Magnetic Resonance Imaging

Images acquired using a gradient echo pulse sequence do not refocus signals that have been dephased during the TE by a local susceptibility gradient. As noted previously, gradient echo imaging can visualize microhemorrhages based on the susceptibility effects of methemoglobin deposits. It has also been used to assess iron deposition in degenerative diseases [14]. Such local susceptibility effects shorten T2*, a parameter that includes both T2 relaxation and dephasing due to local susceptibility. A quantitative comparison of spin and gradient echo signal changes can be used to quantify a parameter T2' that specifically relates to susceptibility effects. Dynamic changes in susceptibility due to changes in the concentration of deoxyhemoglobin can be used to monitor regional brain activation and are the basis for blood oxygenation level-dependent (BOLD) contrast used for functional MRI. Over the past decade this has become a major neuroscience application of MRI and is discussed in a separate chapter.

Magnetic resonance methods are also capable of measuring molecular exchange processes through an effect termed *magnetization transfer*. In magnetization transfer imaging, the exchange rate of protons between water and macromolecules provides image contrast [15] and is altered in white matter disorders such as multiple sclerosis and diffuse axonal injury. To detect magnetization transfer, RF is applied away from the resonance frequency where it selectively excites macromolecules, which have a broad resonance line shape spanning several kilohertz. The resulting saturation of the macromolecular protein pool is then transferred to the on-resonance water signal through chemical exchange of protons, producing a reduction in signal intensity proportional to the exchange rate.

VIII. Exogenous Contrast Agents

The most widely used exogenous contrast agents for MRI are the paramagnetic metal lanthanide gadolinium (Gd^{3+}), manganese (Mn^{2+}) and iron nanoparticles. Gd^{3+} alone is neurotoxic but can be safely administered as a chelate that encapsulates the ion. These agents alter the signal in ^{1}H-MRI by shortening water relaxation through fast exchange. Gd^{3+} and Mn^{2+} primarily shorten T1, and iron dextran microparticles, also termed *superparamagnetic iron oxide* (SPIO), shorten T2 even at low concentrations. These contrast agents do not readily cross the blood-brain barrier under normal circumstances. The major clinical use of Gd^{3+} in neuroradiology is to detect blood-brain barrier breakdown in tumors or inflammatory lesions or to assess hemodynamics in PWI as described previously. However, in experimental systems where blood-brain barrier permeability limitations can be overcome, there are growing applications of exogenous MRI contrast agents as markers of cellular and molecular processes in the CNS. Because of its similarity to calcium, Mn^{2+} is taken up by neurons in an activity-dependent manner and can be used to monitor integrated neural activity or to label neural tracts. SPIO particles can be taken up by stem cells or lymphocytes and can be used for cell tracking. Functional and molecular contrast agents have been demonstrated that dynamically alter the relaxivity of a contrast agent, using chemical moieties that affect water access based on changes in concentrations of calcium, pH, or pO_2 or protein binding [16].

IX. Imaging Nuclei Other Than Protons

Although most MRI is based on the ^{1}H signal from water, several other nuclei have the potential for contributing biologically relevant information. The ^{31}P-NMR spectrum includes resonances attributable to the high-energy phosphates phosphocreatine and adenosine triphosphate; however, their signals are weak and difficult to image. ^{17}O is a stable isotope of oxygen that can be detected directly or indirectly through its effects on the ^{1}H resonance. The latter approach allows ^{17}O to be detected with the sensitivity of ^{1}H-MRI. Exogenous administration of ^{17}O$_2$ allows direct measurement of oxygen consumption based on its conversion to H$_2$^{17}O in oxidative metabolism. Sodium is present in reasonably high concentration in biological tissues, and endogenous ^{23}Na produces a weak NMR signal suitable for imaging. The *in vivo* ^{23}Na signal appears to vary, depending on whether it is intracellular or extracellular, and applications in the assessment of stroke and brain tumor evolution where alterations in ion pumping occur are being assessed. Fluorine also produces a relatively strong NMR signal that can generate low-resolution images. Although there is minimal ^{19}F in biological tissues, many pharmaceuticals are fluorinated and ^{19}F-MRI has been used to determine their distribution. The considerable experience derived from making ligands with ^{18}F for positron emission tomography studies may facilitate the development of ^{19}F-MRI probes. Hyperpolarization uses a laser-based spin-exchange optical approach to generate magnetization effects in noble gases that are several orders of magnitude greater than those that can be achieved with conventional RF excitation. Hyperpolarized gases can generate MRI signals and have primarily been used as exogenous tracers for imaging air-filled cavities such as the lungs, but they could also be used to measure CBF.

X. Conclusion

MRI currently provides the most versatile modality for imaging in the nervous system. Multiple sources of structural and functional image contrast are available, and new sources of contrast are still being characterized. Continued improvements in MRI hardware are increasing sensitivity and reducing scan times, and developments in image processing are providing greater ease and accuracy in data interpretation. MRI has become the imaging procedure of choice for most clinical applications in clinical neuroscience. In clinical research, MRI-derived parameters such as lesion load or gray matter volume provide readily quantifiable biomarkers for disease progression and therapeutic response. In basic research, MRI provides a critical link between humans and animal models. Because MRI provides this information noninvasively and without ionizing radiation, data can also be acquired from normal subjects and children without significant risk.

Acknowledgments

The author gratefully acknowledges the assistance of Dr. Felix Wehrli in reviewing this chapter.

References

1. Wehrli, F. W. (1995). From NMR diffraction and zeugmatography to modern imaging and beyond. *Prog. Nucl. Mag. Res. Sp.* **28**, 87–135.
2. Ashburner, J., and Friston, K. J. (2000). Voxel-based morphometry: the methods. *Neuroimage* **11(6 Pt 1)**, 805–821.
3. Gillard, J. H. (2003). Imaging of carotid artery disease: from luminology to function? *Neuroradiology* **45**, 671–680.
4. Atlas, S. W. (2002). "Magnetic Resonance Imaging of the Brain and Spine," 3rd edition. Lippincott Williams & Wilkins, Philadelphia.
5. Filler, A. G., Maravilla, K. R., and Tsuruda, J. S. (2004). MR neurography and muscle MR imaging for image diagnosis of disorders affecting the peripheral nerves and musculature. *Neurol. Clin.* **22(3)**, 643–682, vi–vii.
6. Ostergaard, L. (2004). Cerebral perfusion imaging by bolus tracking. *Top. Magn. Reson. Imaging* **15(1)**, 3–9.
7. Detre, J. A., and Wang, J. (2002). Technical aspects and utility of fMRI using BOLD and ASL. *Clin. Neurophysiol.* **113**, 621–634.
8. Calamante, F., Thomas, D. L., Pell, G. S., Wiersma, J., and Turner, R. (1999). Measuring cerebral blood flow using magnetic resonance imaging techniques. *J. Cereb. Blood Flow Metab.* **19(7)**, 701–735.
9. Le Bihan, D., Breton, E., Lallemand, D., Grenier, P., Cabanis, E., and Laval-Jeantet, M. (1986). MR imaging of intravoxel incoherent motions: application to diffusion and perfusion in neurologic disorders. *Radiology* **161**, 401–407.
10. Neumann-Haefelin, T., Moseley, M. E., and Albers, G. W. (2000). New magnetic resonance imaging methods for cerebrovascular disease: emerging clinical applications. *Ann. Neurol.* **47(5)**, 559–570.
11. Albers, G. W. (2001). Advances in intravenous thrombolytic therapy for treatment of acute stroke. *Neurology* **57(5 Suppl 2)**, S77–81.
12. Mori, S., and van Zijl, P. C. (2002). Fiber tracking: principles and strategies; a technical review. *NMR Biomed.* **15**, 468–480.
13. Neil, J., Miller, J., Mukherjee, P., and Huppi, P. S. (2002). Diffusion tensor imaging of normal and injured developing human brain: a technical review. *NMR Biomed.* **15(7–8)**, 543–552.
14. Vymazal, J., Righini, A., Brooks, R. A., Canesi, M., Mariani, C., Leonardi, M., and Pezzoli, G. (1999). T1 and T2 in the brain of healthy subjects, patients with Parkinson disease, and patients with multiple system atrophy: relation to iron content. *Radiology* **211(2)**, 489–495.
15. Van Buchem, M. A., and Tofts, P. S. (2000). Magnetization transfer imaging. *Neuroimaging Clin. N. Am.* **10(4)**, 771–788, ix.
16. Jasanoff, A. (2005). Functional MRI using molecular imaging agents. *Trends Neurosci.* **28(3)**, 120–126.

74

Neurovascular Computed Tomography Angiography*

Stuart R. Pomerantz, MD
Michael H. Lev, MD

Keywords: *aneurysm, computed tomography angiography, computed tomography perfusion, computed tomography scanning, multidetector row computed tomography, neurovascular disease, neurovascular imaging, stroke*

I. Introduction
II. Fundamentals
III. Specific Neurovascular Clinical Scenarios
IV. Conclusions and Future Directions
   References

I. Introduction

Remarkable advances in computed tomography (CT) scanner technology over the last decade have enabled computed tomography angiography (CTA) to become the first-line imaging study for most neurovascular applications, most notably those of acute stroke and subarachnoid hemorrhage. Indeed, at many centers, CTA has replaced catheter arteriography as the "gold standard" for a growing number of clinical indications.

The increasing clinical impact of CTA over the past decade can largely be attributed to the development of helical scanners with increasing numbers of detector rows. By acquiring numerous image slices simultaneously from a single, rotating x-ray source, these multidetector row computed tomography (MDCT) scanners enable acquisition of CTA with greatly increased speed and quality. Given the dramatic increase in image slices per study, as well as the challenges of depicting the complex neurovasculature, three-dimensional (3D) postprocessing has become essential for diagnostic review of CTA. Powerful tools for image manipulation are used in stand-alone 3D workstations or are increasingly incorporated into CT scanner consoles and

---

*Portions of this manuscript appear with permission from Lev, M. H., and Gonzalez, R. G. (2002). CT angiography and CT perfusion imaging. *In:* "Brain Mapping: The Methods" (J. C. Mazziotta and A. W. Toga, eds.), 2nd edition, pp. 427–484. Academic Press, San Diego.

diagnostic review workstations (picture archive and communication system). CTA acquisition parameters and standardized 3D views are tailored to best demonstrate the relevant pathology for various CTA indications.

In this chapter, we first discuss some fundamental technical aspects underlying successful CTA, including the requisite advances in MDCT capability, radiation and image quality considerations, and strategies for the safe and effective use of contrast agent. Subsequently, we survey specific neurovascular disease entities, highlighting the critical role CTA can play in their diagnostic evaluation. Pearls and pitfalls of CTA image acquisition, 3D postprocessing, and interpretation are illustrated.

II. Fundamentals

A. Evolution of Multidetector Row Computed Tomography

The great CT advance in the early 1990s was from step-and-shoot axial imaging to helical CT in which an unbroken stream of data was acquired as an x-ray source-and-detector combination spun continuously around the patient [1]. This advance first enabled clinically useful volumetric imaging. The subsequent introduction of scanners with progressively more detector rows then enabled significantly larger coverage in a shorter amount of time and now true isotropic resolution (image voxels of comparable size in all three dimensions) [2]. The latest 16- and 64-slice CT units can scan entire vascular territories in 15 to 30 seconds, well within the time of dynamic administration of a single bolus of intravenous contrast agent [3,4]. These capabilities are especially advantageous for the challenging demands of new applications in cardiac imaging and neurovascular CTA.

With MDCT, each channel of the detector array generates a separate helix of imaging data. The reconstruction algorithms that merge these data channels to create MDCT image slices are complex and relate to the particular detector configuration [5]. Depending on the clinical indication, detector rows of equal or varying thickness are activated in particular combinations to produce a certain number of image slices of a given thickness and resolution. Thus a 4-slice scanner (i.e., able to acquire 4 image slices simultaneously) actually may have up to 16 detector rows. Hence the confusion of coexisting terminologies, with multidetector row CT and multislice CT both correct descriptors for modern CT scanners [1]. The total length of the detector array along the long axis of the patient direction (z-axis) in 16-slice scanners is typically 20 mm. The latest MDCT scanners can acquire up to 64 image slices simultaneously, with most manufacturers using a 40-mm-long array of 64 equally sized detector rows.

B. Radiation and Image Quality Considerations

The typical challenge when CT scanning is to maximize length of coverage, resolution, and signal-to-noise ratio yet minimize slice thickness, contrast dose, and radiation to the patient. The effect of various parameter changes on radiation dose can be assessed by referring to the dose length product and CT dose index, whose display on current scanner consoles is a mandated international safety standard. Increasing table speed and table increment per gantry rotation (pitch factors) decreases scan duration and lowers the total administered radiation dose. However, higher pitch settings can decrease image quality and cause particular artifacts (Fig. 1). The increased number of detector rows available with each new CT generation has minimized the inherent conflict between these goals, though protocol design remains challenging as new opportunities become available.

C. Intravenous Contrast Issues

1. Contrast Agent Safety

Nonionic CT contrast agents have been shown to be generally safe, even in the setting of cerebral ischemia. In an animal model of middle cerebral artery (MCA) stroke, no significant neuronal toxicity from nonionic contrast agents was observed, even to already ischemic neurons [6,7].

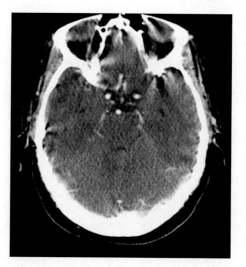

Figure 1 Windmill Artifact. Linear streaks radiate from the peripheral osseous interfaces, most notably at the petrous bones, and are most conspicuous when scrolling through an image stack. This artifact is specific to multidetector row computed tomography scanners and arises as an increasing number of detector rows intersect the scan plane at higher pitch settings.

However, some patients are at higher risk for contrast-induced nephropathy (CIN) at baseline, especially those who have diabetes, preexisting renal dysfunction, or both. Factors such as serum creatinine level and creatinine clearance help determine whether iodinated contrast can be safely administered. Creatinine clearance is an easily calculated estimate of the glomerular filtration rate. It is recognized as more accurate than serum creatinine for assessing renal function, as it takes into account the patient's body weight and gender [8].

Multiple strategies to reduce the risk of CIN are available. Because nephrotoxicity from contrast media is dose dependent [9,10], CTA protocols are designed to use the least amount of contrast possible. The use of denser contrast agents and innovative strategies enabled by the latest generation of power injectors may increasingly result in reduced contrast loads [11]. Recently, Aspelin and associates demonstrated the potential benefit of using an iso-osmolar agent, iodixanol, in patients with diabetes and borderline renal function. This agent, however, is more expensive and more viscous at room temperature than standard nonionic agents [12]. Adequate pre- and postprocedure hydration is considered by most experts to be the most important factor in preventing CIN.

For those patients allergic to iodinated contrast media, premedication with antihistamines and steroids can blunt the anaphylactoid response. In the setting of an acute stroke, where there is insufficient time to complete a course of steroid administration, a gadolinium magnetic resonance contrast agent may be used as a clinically effective alternative for CT [13]. It should be noted that though the scans are usually diagnostic (Fig. 2), peak vessel opacification is much less than with iodinated contrast agents, even at gadolinium doses several times higher than typically used for magnetic resonance imaging (MRI). Because gadolinium at these concentrations may theoretically be even more nephrotoxic than iodinated contrast, gadolinium should be used with caution when the contraindication to iodinated contrast administration is renal insufficiency rather than allergy.

2. Optimizing Contrast Injection Strategies

CTA and especially computed tomography perfusion (CTP) studies benefit from rapid contrast bolus infusion rates. Thus an antecubital intravenous catheter with at least a 20-gauge bore is required to perform CTA or CTP, though a larger bore, such as 18- or 16-gauge, is preferable [14]. The catheter optimally is placed in the right arm because of the more direct course into the central circulation and the higher likelihood of venous stenoses in the left brachiocephalic vein from cardiac pacemaker leads. Undiluted contrast retained in the brachiocephalic veins during scanning can obscure the arteries arising from the aortic arch and can even reflux high into the venous collateral system of the neck (Fig. 3A).

Successful CTA depends on achieving a prolonged, uniformly high level of contrast density in the target vasculature throughout the period of scan acquisition (Fig. 4). Synchronizing scan initiation and duration to contrast circulation time has become a critical challenge with the much faster 16- and 64-slice scanners. Physiological and pharmacokinetic modeling of contrast enhancement has clarified key parameters affecting contrast bolus geometry in CTA.

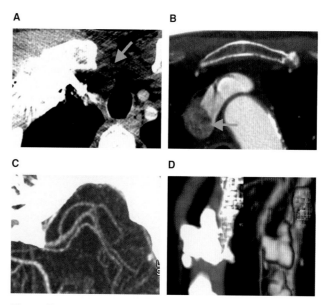

Figure 3 Pitfalls in Computed Tomography Angiography Imaging and Postprocessing. (**A**, arrow) Streak artifact from undiluted inflowing contrast can obscure proximal large vessel segments. (**B**, arrow) Pseudothrombus appearance from incomplete contrast mixing of unopacified and opacified blood can occur in the superior vena cava. (**C**) Pseudobeading artifact can occur because of superimposition of background noise on poorly opacified vessels on maximum intensity projection (MIP) views. (**D**) Thick atherosclerotic calcification overlying lumen can preclude residual luminal diameter measurement on MIP and volume-rendered views.

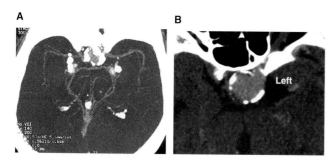

Figure 2 Gadolinium Computed Tomography Angiography Performed in a Patient with Severe Iodinated Contrast Allergy. (**A**) Despite suboptimal vascular opacification relative to iodinated contrast, (**B**) the diagnosis of a 1.9-cm left posterior communicating artery aneurysm is still clear.

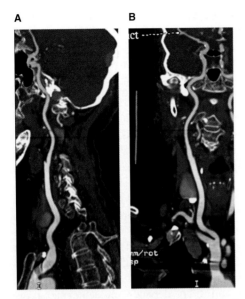

Figure 4 Computed Tomography Angiography of the Neck Performed with Saline Chaser. **(A, B)** A 90-cc injection of high-density contrast material (370 mg of iodine per milliliter) followed by 40 cc of saline through a dual-barrel power injector results in homogeneous opacification from the aortic arch to the circle of Willis on curved-reformatted images. Homogenous opacification is crucial for the success of semiautomated post-processing techniques (see Fig. 12).

These include *patient-specific parameters* such as cardiac output and blood pool size and *injection parameters* such as rate, volume, duration, and concentration of contrast administration [11,15]. Techniques to determine the appropriate scan delay after initiation of injection, such as bolus tracking or test injections, are continuing to evolve. Novel strategies, such as varying the rate of contrast administration during the course of injection or using a saline chaser, help achieve more homogenous peak opacification levels and potentially use less contrast for a given scan duration [16–18]. Table 1 summarizes the advantages and disadvantages of these various strategies to optimize contrast administration for CTA.

A number of CTA imaging and postprocessing pitfalls are encountered. Many relate to variability of vessel lumen opacification (see Fig. 3A–C) or the presence of overlapping structures (see Fig. 3D). Table 2 highlights some of the common CTA imaging pitfalls discussed throughout the text.

III. Specific Neurovascular Clinical Scenarios

A. Stroke

According to the National Stroke Association web site, "stroke is our nation's third leading cause of death, killing nearly 160,000 Americans every year." Death or severe disability can be prevented or diminished if thrombolytic treatment is administered shortly after onset of embolic stroke. By helping to rapidly identify intracranial thrombus, vascular stenosis, and parenchymal ischemia, CTA, with CTP analysis, has become a critical part of the early management of such patients in many institutions.

Table 1 Strategies for Optimizing Contrast Enhancement

	Advantages	Disadvantages
Higher-density contrast medium	• Better opacification • Cost savings because price is usually by volume and less is required per scan	• Higher contrast quantity in lower volume restricts coverage length
Higher injection rate	• Greater peak enhancement	• Larger intravenous access • Vascular injury
Higher injection volume	• Greater peak enhancement	• Higher iodine load
Iso-osmolar contrast	• Less nephrotoxic	• Costlier • Increased viscosity
Gadolinium	• CTA agent in patients with severe iodine allergy	• Relatively poor CT attenuation • Potentially nephrotoxicity at CTA doses
Saline chaser	• Prolonged enhancement plateau or less contrast media • Decreased streak artifact at origin of large vessels	• Requires specialized dual-head CT power injectors
Fixed contrast prep delay	• Simple • Straightforward for large technologist pool in large multisite centers	• Bolus may arrive too early or late with new rapid scanners (16- or 64-slice CT) • Lengthens delay with atrial fibrillation
Bolus tracking	• Synchronize scan with peak contrast enhancement by accounting for specific patient physiology	• Increased setup time and complexity • Sensitive to patient motion if small ROI target
Test bolus	• Synchronize scan with peak contrast enhancement by accounting for specific patient physiology	• Extra complex step • Extra 10–15 ml of contrast

Note: CTA, computed tomography angiography; CT, computed tomography; ROI, region of interest.

Table 2 Neurovascular CTA Protocols

Protocol	Indication	Variation
"Acute" stroke CTA	• Stroke onset <9–12 hours prior	• CTA head and neck with cine-CTP
"Subacute" stroke CTA	• Stroke onset >9–12 hours prior	• CTA head and neck without cine-CTP
Neck CTA	• Cervical steno-occlusive disease	• Subacute stroke protocol with coverage only from circle of Willis to aortic arch
	• Trauma	
	• Dissection	• Optional delayed series if apparent ICA occlusion to distinguish true occlusion from hairline lumen
	• Subarachnoid bleed	
Aneurysm CTA	• Subarachnoid hemorrhage	• CTA head, thinnest possible slices for fine detail
	• Confirm or follow unruptured aneurysm	• Higher contrast injection rate (4–5 cc/s)
	• Evaluate arteriovenous malformation or fistula	• Optimal pitch (0.562)
		• Thinner slice reconstruction (0.625 mm) section
Vasospasm CTA	• Postaneurysmal rupture vasospasm	• CTA head with cine-CTI
CTV	• Dural and cavernous sinus thrombosis	• CTA head protocol with prolonged 40-second scan delay after start of contrast administration
	• Surgical planning	• Subtraction of skull bones to reveal dural venous sinuses

Note: CTA, computed tomography angiography; CTP, computed tomography perfusion; CTV, computed tomography venography; ICA, internal carotid artery.

A combined CTA-CTP study in patients suspected of having embolic stroke facilitates their diagnosis and, importantly, the triage to appropriate therapy. MRI techniques such as diffusion-weighted imaging (DWI) and perfusion-weighted imaging (PWI) have also revolutionized stroke imaging; DWI clearly delineates infarcted brain tissue within minutes, and PWI defines areas of cerebral hypoperfusion [19]. The choice of CT or magnetic resonance techniques may depend on which modality is more readily available. At least one study in acute stroke patients has demonstrated equal accuracy of CTA and magnetic resonance angiography (MRA) for vessel delineation and close to equal accuracy of CTA source image analysis compared with DWI in determining infarct volumes [20]. When both are available, the modalities are complementary, with CTA providing excellent assessment of proximal and collateral vessel status and DWI or PWI providing conspicuous identification of the infarct core and whole brain perfusion analysis. In this section, we focus on the rationale and efficacy of a CTA or CTP imaging approach for acute stroke and how such studies are acquired and analyzed.

1. "Time Is Brain": Triage Tool for Acute Stroke

Following stroke onset, the National Institute of Neurological Disorders and Stroke multicenter trial suggests an absolute 3-hour window for the administration of *intravenous* thrombolytics. Beyond 3 hours, the risk of severe intraparenchymal hemorrhage exceeds the risk of treatment [21]. The results of the prolyse in acute cerebral thromboembolism (PROACT) II trial also demonstrated a benefit of *intra-arterial* thrombolytic therapy administered within 6 hours of symptom onset for the treatment of proximal large-vessel stroke [22]. With the recent advent of the desmoteplase in acute ischemic stroke trial results, this window may soon be extended to as long as 9 hours [23]. Mechanical thrombectomy and clot-retrieval devices are also under evaluation. These have the potential benefits of reduced time to reperfusion and lower hemorrhagic risk compared with pharmacological thrombolysis [24].

Typically, patients with acute ischemic stroke symptoms undergo an unenhanced head CT scan as their first imaging test to determine whether contraindications to thrombolytic treatment exist. Such contraindications include hemorrhage (an absolute contraindication) and a "large" parenchymal hypodensity. The hypodensity corresponds to already infarcted tissue. "Large" is defined as greater than one-third the vascular territory [21,25] (Fig. 5). However, findings of early infarction on unenhanced CT are often subtle and may go undetected, even by experienced physicians [26] (Figs. 5 and 6). Also, unenhanced CT scanning alone, although of some value in predicting patients most likely to be *harmed by* thrombolysis, is of little value in predicting patients most likely to *benefit from* thrombolysis, specifically those with proximal large-vessel vascular occlusions.

Because unenhanced CT and clinical exam alone are limited in their ability to detect large-vessel thrombus, CTA has become the first-line diagnostic test for patients with signs and symptoms of acute stroke at many institutions. Because of the narrow window available to initiate thrombolytic agent, rapid triage is crucial. Therefore, the rationale in the evaluation of an acute stroke patient is to identify as quickly as possible those patients who may benefit from intravenous or intra-arterial thrombolysis or other acute stroke treatments. Importantly, CTA excludes from treatment patients with stroke mimics, such as transient ischemic

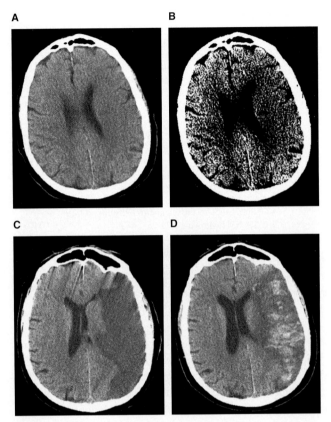

Figure 5 Evolution of Large Left MCA Infarct. (**A**) On initial unenhanced computed tomography (CT), subtle infarct is hard to detect with standard window width and level settings. (**B**) It becomes more conspicuous using narrow "stroke" settings. (**C**) Large territorial infarct is more conspicuous after 24 hours, confirming that more than one-third of middle cerebral artery (MCA) territory evolved. Because of the size of infarct on presentation CT, the patient was considered at too high a risk for hemorrhagic conversion to receive thrombolysis. (**D**) Hemorrhagic conversion occurred spontaneously at 2 weeks.

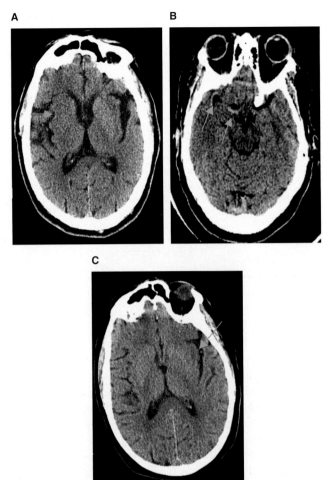

Figure 6 Subtle Early Signs of Ischemia on Noncontrast Head Computed Tomography in Different Patients. (**A**, arrow) Obscuration of lentiform nucleus, (**B**, arrows) hyperdense middle cerebral artery sign, and (**C**, arrow) loss of insular ribbon.

attack, complex migraine, and seizure. These patients will not benefit from, and may be harmed by, such therapies.

Because helical CT scanners are less expensive and more readily available at most hospital emergency departments than are MRI scanners, performing CTA or CTP can be a quick and natural extension of the unenhanced head CT exam—an exam routinely obtained as part of the prethrombolysis workup at most institutions [27–29]. The addition of a CT angiographic study seldom adds more than 10 minutes to the scanning time of a conventional CT examination. Required postprocessing can typically be performed in minutes, during which time the patient could be prepared for thrombolysis, should the decision to proceed with treatment be made [27,30].

Given the risks and expense associated with performing intra-arterial thrombolysis, CTA has great value in positively identifying patients with true proximal vessel occlusion [27,30]. With further research, CTA with CTP may also help identify those patients who, because of adequate collateral circulation supplying reversibly ischemic tissue, may be candidates for intravenous or intra-arterial thrombolysis beyond the currently accepted time windows [31,32].

2. Detection and Prognosis

Multiple studies have confirmed the ability of CTA to reliably detect large-vessel intravascular clot with an accuracy approaching 99% [27,33–35] (Fig. 7A). CTA has also been shown to be useful for the evaluation of collateral circulation distal to an occlusion, as well as for improving the conspicuity of acute cerebral ischemia [36,37]. CTA has been shown to have higher sensitivity and less interoperator variability than MRA for intracranial steno-occlusive disease and was found superior even to digital subtraction angiography (DSA) for detecting posterior-circulation involvement when slow flow is present [38] (Fig. 8). Because CTA through the neck is included in the CTA evaluation of patients with acute stroke, an embolic source

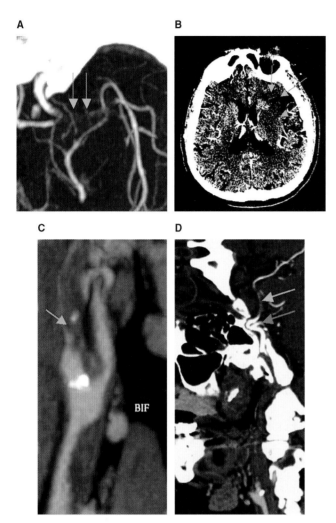

Figure 7 CTA of the Head and Neck and CTA Source Image Analysis. (**A**, arrows) Left proximal middle cerebral artery (MCA) segmental occlusion is demonstrated on head computed tomography angiography (CTA). (**B**, arrows) Hypoperfused and possibly ischemic brain tissue is most conspicuous on CTA source images as a hypoattenuated region. (**C**, arrow) Maximum intensity projection of the carotid bifurcation (BIF) demonstrates source of embolic infarct with clot occluding the proximal left internal carotid artery (LICA). (**D**) Curved-reformatted view demonstrates absence of flow through the occluded cervical internal carotid artery, reconstitution of flow within the cavernous or supraclinoid carotid segments (red arrow), and again more distal obstruction of proximal MCA segment (blue arrow).

of infarct, such as thrombus or occlusion at the carotid bifurcation, can often be identified (see Fig. 7C–D).

CTA can also be used for risk stratification and prognosis. Among patients treated with intravenous tissue plasminogen activator, those who had patent vasculature or only occult distal occlusion on pretreatment CTA had a better prognosis characterized by fewer hemorrhages, a better score on the National Institutes of Health stroke scale, and better early improvement [39]. Risk stratification can also be accomplished by evaluating the parenchyma on CTA

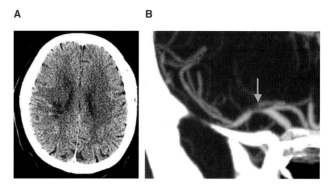

Figure 8 CTA of Intracranial Vascular Stenosis. (**A**) Numerous chronic right-sided lacunar infarcts suggest proximal right middle cerebral artery (MCA) disease involving lateral lenticulostriate branches. (**B**) Severe stenosis in proximal right MCA is confirmed with computed tomography angiography (CTA).

source images. Assuming an approximate steady-state level of contrast enhancement during scan acquisition, the source images from the CTA dataset can be considered whole-brain perfused blood volume images, generally referred to as *CTA source images*. Hypodensity on such images helps facilitate detection of subtle parenchymal ischemic changes associated with distal embolic occlusions [28] and can be used for risk stratification (see Fig. 7B). A recent retrospective study from our institution has positively correlated the degree of parenchymal hypoattenuation on initial CTA source images with the likelihood of hemorrhagic transformation and poor clinical outcome after intra-arterial reperfusion therapy [40].

Following CTA source images, a dedicated CTP study acquisition can be acquired using *first-pass cine slab technique*. Following rapid bolus infusion of contrast material (~7 cc per second), image slices are acquired once per second over a period of 45–60 seconds. This period is sufficient to track the first pass of the contrast bolus through the intracranial vasculature without recirculation effects. More coverage with a second cine-CTP slab may be accomplished if renal function and total contrast dose limitations permit a second bolus injection. Following acquisition, cine-CTP images are processed into quantitative maps of cerebral blood flow (CBF), cerebral blood volume (CBV), and mean transit time (MTT). These maps help outline the region of "ischemic penumbra," which is understood as the abnormally perfused tissue surrounding a core of infarcted tissue. The penumbra, though potentially viable, is felt to be at risk for imminent infarction and may benefit from thrombolysis (Fig. 9). Alternatively, CTP maps may indicate that all or most of the abnormally perfused tissue is already inevitably progressing to infarction. In this situation, initiating thrombolytic treatment could significantly worsen the outcome by precipitating intracranial hemorrhage. First-pass CTP is currently limited in the extent of coverage that can be obtained during a single bolus injection of contrast. This

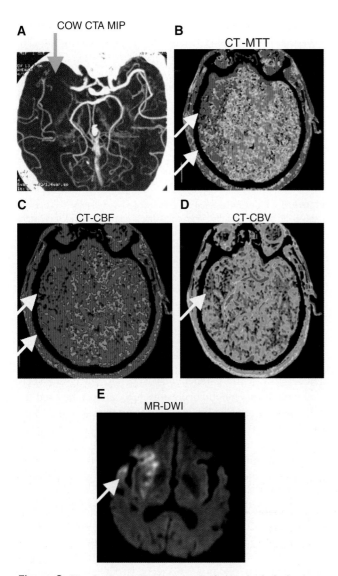

Figure 9 Cine-Computed Tomography Perfusion Analysis in Acute Stroke. (**A**, arrow) Middle cerebral artery stem occlusion is clearly visualized on computed tomography angiography (CTA). Mismatch is present between (**B** and **C**, arrows) large territorial perfusion abnormality on the mean transit time (MTT) and cerebral blood flow (CBF) maps and (**D**, arrow) small area of decreased perfusion on the cerebral blood volume (CBV) map. (**E**, arrow) Final infarct on diffusion-weighted imaging (DWI) after successful thrombolysis is confined to area of abnormality on the CBV map, suggesting successful salvage by thrombolysis of larger penumbra. *Note:* MIP, maximum intensity projection; MR, magnetic resonance.

Table 3 CTA Imaging Pitfalls

Streak artifact from hyperdense, undiluted contrast in inflowing veins
Pseudoclot in large veins from inhomogeneous contrast mixing
  poor visualization of arteries in skull base on 3D views because of overlying bone or cavernous sinus opacification
Pseudobeading artifact from poor contrast opacification
Obscuration of lumen by heavily calcified vessels on MIP and VR images
Overestimation of calcific vessel stenosis because of incorrect window and level settings
Difficulty in detecting short-segment vessel occlusion on thin source images
Streak artifact from aneurysm clip or embolization material
MIP "exclusion" artifact (see Fig. 11)

Note: MIP, maximum intensity projection; VR, volume-rendered.

[28]. Depending on the MDCT scanner generation, high-resolution CTA coverage of the complete neurovascular system, from aortic arch to the vertex, can be performed in 15–35 seconds. Table 3 summarizes the protocol components for acute stroke CTA and the variations on that protocol used for other specific neurovascular indications, as discussed later in the chapter. It should be noted that protocol parameters are included only as guidelines and may need to be varied for specific clinical considerations.

Because short-segment thromboembolic occlusions and small aneurysms can be virtually undetectable when scrolling through the thin axial source images, reformatted views that display vessels in an "angiographic" manner are an essential component of CTA diagnosis (Fig. 10). At Massachusetts General Hospital, we use dedicated technologists in a 3D-imaging lab to create standard views of each vascular segment in a timely fashion. However, for emergent triage, angiographic maximum intensity projection (MIP) views can be semiautomatically reformatted at the scanner console in less than 1 additional minute after scan acquisition. They allow confident diagnosis of circle of Willis proximal branch vessel occlusion even before the patient has been removed from the scanner. Because MIP images are of arbitrary thickness, they can be used to quickly exclude overlapping bone from the angiographic views. A pitfall to be aware of is a tortuous vessel segment that loops out of the MIP slab plane, mimicking an occluded segment (Fig. 11).

B. Neck

1. Chronic Carotid Artery Steno-Occlusive Disease

C. Miller Fisher of the Massachusetts General Hospital first elucidated the relationship between extracranial artery disease and embolic stroke in 1951 [41]. Severe atheromatous narrowing of the proximal internal carotid artery

limitation is sometimes clinically restricting but should be less so with the 40-mm maximum collimation available for each CTP slab on 64-slice scanners.

The acute stroke protocol is a multisequence scan consisting of the following components: (1) routine unenhanced head CT, (2) CTA of the head and neck, and (3) an optional single- or double-slab cine-CTP study

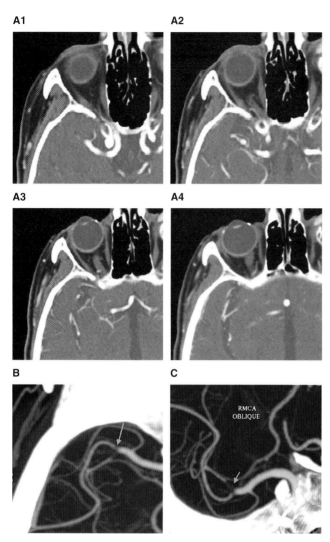

Figure 10 Importance of MIP Images for Detection of Small Segmental Occlusions. Occlusive thrombus at right middle cerebral artery (RMCA) bifurcation is (**A**1–4) difficult to detect on sequential axial source computed tomography angiography images though (**B, C** arrows) conspicuous on maximum intensity projection (MIP) images. Good opacification of vessels distal to occlusion by collateral flow is noted.

(ICA) predisposes the patient to the formation of platelet-fibrin emboli in regions of turbulent flow; these emboli can be the source of transient ischemic attack or stroke. Two large randomized trials, the North American Symptomatic Carotid Endarterectomy Trial (NASCET) and the European Carotid Surgery Trial, have proven that surgical management will significantly reduce the morbidity and mortality of symptomatic patients with severe stenosis (70%–99%) when compared with medical management [42,43]. Though the accepted gold standard for evaluation of carotid artery stenosis is DSA, noninvasive imaging modalities, including MRA, ultrasound, and CTA, are increasingly in the forefront. This trend partially reflects the increased risks associated with DSA (1% risk of a major stroke and small risk

of death [44,45]) but also that noninvasive methods demonstrate a high sensitivity and specificity for determining the degree of stenosis, especially when part of a multimodality approach. Whereas CTA has become the test of choice for emergent neurovascular indications, ultrasonography, which is completely noninvasive and less expensive, still has a primary role in outpatient screening of chronic steno-occlusive disease. CTA is an important problem-solving tool when MRI cannot be performed or the results of ultrasonography and MRA are equivocal [46,47]. CTA is often requested before carotid endarterectomy or stenting to confirm the presence and degree of a suspected stenosis and to evaluate the aortic arch origins of the large vessels. CTA can preoperatively distinguish important anatomic variants, such as vascular loops, calcifications, aneurysms, and adjacent osseous structures [48,49]. DSA is generally reserved for those few cases with complex multivessel disease, for which detailed knowledge of collateral flow patterns is necessary to make rational treatment decisions [46,47,50,51].

CTA measurements of residual luminal diameter have compared favorably with those of DSA, unenhanced MRA, and ultrasonography [50,52–58]. However, the CTA user must pay careful attention to appropriate window and level display settings. Beam-hardening artifact from heavy circumferential calcifications can result in overestimation of the degree of stenosis in both axial and longitudinal views. For measurement of small lumen diameters, even without calcified plaque, the accuracy of CTA measurement is limited by the pixel size—typically 0.4 to 0.5 mm (assuming a 20- to 25-cm field of view and a 512×512 pixel imaging matrix). In serial examinations, differences in window settings, level settings, and contrast density from one CTA study to another can produce large differences in the measured lumen size [55,59,60]. Software solutions for semiautomated detection of the cross-sectional area and true luminal diameters ($D_{min}, D_{max}, D_{mean}$), orthogonal to a computer-generated centerline, are now available and promise great utility for reducing interobserver variability and increasing postprocessing efficiency [61] (Fig. 12). Dense contrast opacification throughout the entire vascular segment being evaluated is a prerequisite for success with such semiautomated techniques.

Degree of carotid stenosis can be expressed in terms of percent stenosis, residual lumen area, or residual lumen diameter. Percent stenosis, the ratio of maximal luminal narrowing to the normal ICA distal to the bulb, was the severity index in NASCET and is the most commonly used measure by practitioners in North America. However, the reference diameter of the distal ICA typically ranges from 5 to 8 mm, which can significantly alter the calculated percent stenosis. At Massachusetts General Hospital, we report the degree of vascular stenosis based on residual lumen diameter using 1.5 mm as the cutoff for hemodynamically significant stenosis (Table 4). A residual lumen diameter of

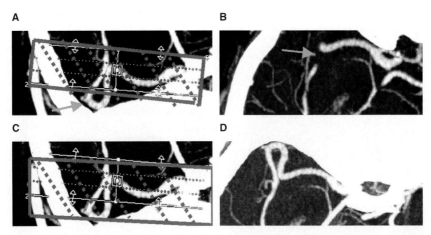

Figure 11 MIP "Exclusion" Artifact. (**A**, arrow) A tortuous vascular loop dips below the maximum intensity projection (MIP) slab, (**B**, arrow) giving rise to artifactual segmental occlusion on an oblique axial MIP. (**C**) When the MIP slab is thickened, (**D**) the tortuous loop is included in the reformatted view.

1.5 mm correlates approximately to ultrasound peak systolic velocity of more than 250 cm/sec and NASCET measurement of 70% stenosis [62].

Plaque characteristics such as ulceration, amount of calcification, thin fibrous cap, lipid core, and hemorrhage have been evaluated as predictors of stroke risk. These features can sometimes be identified on CTA, but not consistently [49,63].

Before performing CTA of the neck, it is important to review prior noninvasive ultrasound, magnetic resonance, CT, and angiographic studies to determine the questions that must be answered. Our CTA neck protocol is similar to the stroke CTA protocol except coverage extends only from the circle of Willis through the aortic arch. An optional *delayed* series helps detect the slow opacification of a hairline residual lumen and thus distinguish it from complete occlusion. The distinction is critically important: those with a hairline residual lumen are still at risk for embolic stroke and thus candidates for carotid endarterectomy or stenting, whereas those with complete occlusion are usually treated medically [64]. *Curved-reformatted* (CR) images, prepared in our 3D lab, facilitate the survey of long, tortuous vascular segments in the neck and cavernous sinus region for rapid identification of arterial stenoses and occlusion (see Figs. 3, 7D).

2. Dissection, Trauma, and Fibromuscular Dysplasia

Internal carotid and vertebral artery dissections are among the more common causes of stroke in young adults. They typically have a good prognosis, though anticoagulation with oral warfarin is administered for several months to prevent thromboembolic complications. Dissections tend to occur in regions of vascular tethering, such as in the distal extracranial ICA at the skull base.

Though DSA has long been the reference standard for diagnosis of dissection, CTA is advantageous in that it is minimally invasive and can provide high-resolution images of the both the arterial lumen and the arterial wall, the locus of pathology in dissection. Whereas DSA often only demonstrates nonspecific irregular stenoses or occlusion in the setting of dissection, CTA can demonstrate the intramural hematoma, in addition to other signs including intimal flap, double-lumen appearance, and pseudoaneurysm [65] (Fig. 13A–B). High sensitivity and specificity for the CTA diagnosis of extracranial internal carotid dissection and vertebral artery dissection (see Fig. 13C–D) have been demonstrated [65,66].

CTA is also highly sensitive and practical for screening and evaluating patients at suspected risk of blunt and penetrating traumatic cerebrovascular injury [67,68]. The ascending course of the vertebral artery through the transverse foramina of C6 to C3 vertebrae and over the posterior arch of the atlas places it at increased risk for injury in the setting of fracture or dislocation spinal injury (see Fig. 13E–G).

Fibromuscular dysplasia is a noninflammatory idiopathic process predominantly affecting the renal and internal carotid arteries in young to middle-aged women. CTA of the neck can depict the classic "string of beads" appearance associated with fibromuscular dysplasia and other sequelae, including dissection and macroaneurysms [69] (Fig. 14).

C. Subarachnoid Hemorrhage

Subarachnoid hemorrhage from a ruptured saccular aneurysm accounts for 6%–8% of all strokes; in North America, the incidence of subarachnoid hemorrhage is approximately 11–12 per 100,000 people [70]. Outcome for patients with subarachnoid hemorrhage remains poor, with

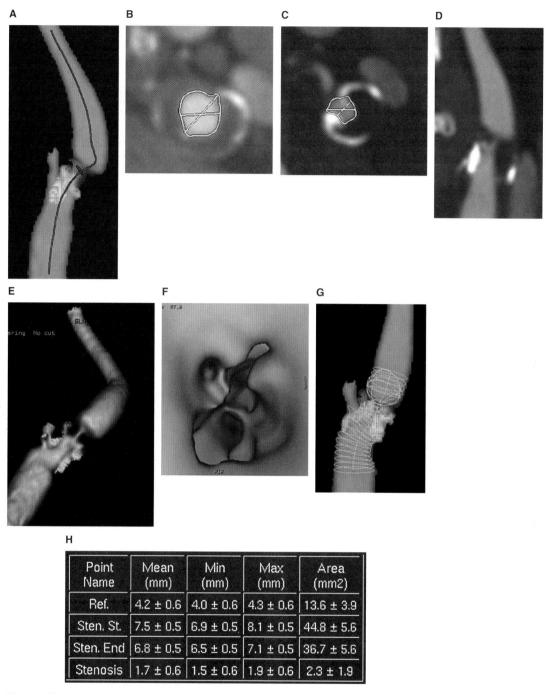

Figure 12 Semiautomated Carotid Stenosis Evaluation Software. (**A**) A vascular centerline is automatically generated. The cross-sectional area and true luminal diameters (D_{min}, D_{max}, D_{mean}), orthogonal to the centerline, are then calculated. Software must distinguish between (**B**) luminal contrast opacification and soft plaque and (**C**) mural calcification. Once the centerline model has been segmented, multiple postprocessing options are available, including (**D**) maximum intensity projection, (**E**) volume rendering, and (**F**) endoluminal navigation. Analysis options include (**G**) volume of stenosis and (**H**) automated generation of residual luminal diameters and percentage stenoses for surgical decision-making.

overall mortality rates of 25% and significant morbidity among approximately 50% of survivors [70]. The risk of aneurysm rupture is related to size: for aneurysms less than 10 mm the risk is estimated at 0.05% per year, and for those more than 25 mm the risk is as high as 6.0% per year [71]. These risks are up to 11 times higher if there has been previous subarachnoid hemorrhage. Posterior communicating artery and vertebrobasilar junction aneurysms may

Table 4 Carotid Stenosis Grading System

Minimal Residual Lumen Diameter	Degree of Stenosis
2.0–2.5 mm	Moderate stenosis
<1.5–2.0 mm	Moderate-to-severe stenosis
<1.0–1.5 mm[a]	Several stenosis
<1.0 mm	Critical stenosis

[a] <1.5 mm is considered "hemodynamically significant."

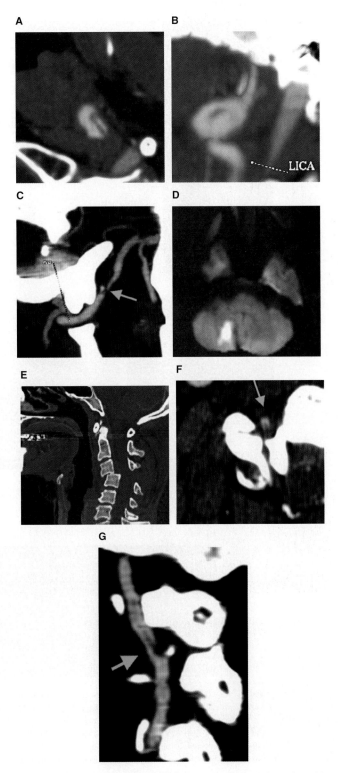

Figure 13 Arterial Dissection. (**A**) Spontaneous carotid dissection with flap and (**B**) pseudoaneurysm formation; (**C**, arrow) vertebral artery dissection with (**D**) associated posterior inferior cerebellar artery territory infarct on diffusion-weighted magnetic resonance imaging; (**E**) traumatic vertebral dissection due to bilateral facet dislocations; (**F**) dissection flap seen on axial source image and (**G**) oblique sagittal maximum intensity projection. *Note:* LICA, left internal carotid artery.

also be at increased risk of rupture into the subarachnoid space, whereas small intracavernous aneurysms are generally not treated [71]. Multiple aneurysms, or those with daughter sacs, also have a higher risk for rupture [71].

Traditionally, the workup of patients with nontraumatic subarachnoid hemorrhage has required DSA, the gold standard study for evaluation of cerebral aneurysms. However, the remarkable advances in MDCT image acquisition and 3D postprocessing over the past 8 years have brought CTA evaluation of cerebral aneurysms sufficiently close to DSA to enable a major shift in approach. At our institution, the neurosurgery staff uses the results of CTA in place of DSA as the *only* diagnostic and pretreatment planning study for patients with *ruptured and unruptured* cerebral aneurysms. A prospective study at Massachusetts General Hospital involving 223 patients validated decision-making based on CTA alone as both safe and effective in most patients [72]. Other institutions are experiencing similar success with this approach [73,74]. Though MRA is still used as a screening method for patients at increased risk of cerebral aneurysms (e.g., based on family history or polycystic kidney disease), CTA is superior in detection of smaller aneurysms (<3 mm) and in characterizing the precise relationship between aneurysms and adjacent osseous structures (Fig. 15).

When CTA fails to identify a cause for subarachnoid hemorrhage, DSA is generally performed and even repeated if necessary to identify small ruptured aneurysms. However, when a characteristically perimesencephalic pattern of subarachnoid hemorrhage is present on CT, the probability that the cause is a ruptured vertebrobasilar aneurysm is low (approximately 4%). Given the added sensitivity of a negative CTA in this situation, some have argued that the risks of DSA outweigh the extremely small residual chance of a missed aneurysm in this patient subgroup [75,76].

Aneurysm detection with CTA is not significantly compromised by the presence of subarachnoid hemorrhage, which, although dense on CT images, is much less attenuating than iodinated contrast material [77,78] (Fig. 16). The ability of the neuroradiologist to make a confident diagnosis of a ruptured aneurysm immediately at the CT console from overlapping MIP images is of great benefit to patients requiring emergent neurosurgical or endovascular treatment. Our 3D-imaging technologists also prepare a

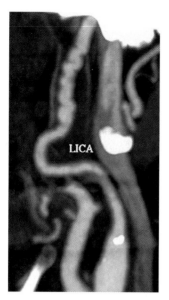

Figure 14 Fibromuscular Dysplasia. Beaded narrowing in distal cervical left internal carotid artery (LICA) consistent with fibromuscular dysplasia.

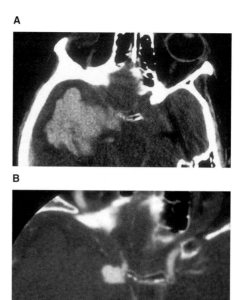

Figure 16 Conspicuity of Opacified Aneurysm Sac despite Surrounding Hyperdense Hematoma. (**A**) Dense subarachnoid hemorrhage seen on noncontrast head computed tomography is not (**B**) a contraindication to computed tomography angiography, because the attenuation of contrast is much greater than that of extravasated blood.

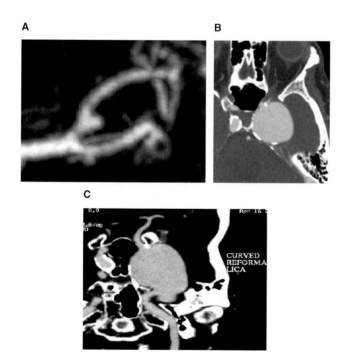

Figure 15 CTA Detection and Characterization of Cerebral Aneurysms. (**A**) Small aneurysm (<3 mm) at the left middle cerebral artery bifurcation is well depicted on a computed tomography angiography (CTA) maximum intensity projection image. The relationship of a giant cavernous carotid aneurysm to surrounding skull base structures is well depicted on (**B**) source and (**C**) curved-reformatted views. *Note:* LICA, left internal carotid artery.

standard set of reformatted angiographic views highlighting the origins and bifurcations of every major intracranial vessel (the most common sites of aneurysm formation) in at least two projections. These additional views are particularly helpful for detecting small or multifocal aneurysms. Color 3D volume-rendered images optimally display branch vessel morphology and the relationship of aneurysms to osseous landmarks; thus they are used primarily for surgical planning (Fig. 17). In addition to reviewing static 3D images, neurosurgeons can dynamically interact with the CTA model at a 3D workstation, gaining invaluable vascular insights before operating for a variety of neurosurgical indications (Fig. 18).

D. Vasospasm

Vasospasm can begin as early as 3 days after the onset of a subarachnoid hemorrhage. It typically peaks within 6–8 days and resolves within 2 weeks. In this vulnerable period, vasospasm often results in delayed ischemia and, more than rebleeding, remains the most significant cause of death and disability following aneurysm rupture [70]. "Angiographic" spasm, the visually apparent reduction in vessel caliber that can be detected by transcranial Doppler ultrasonography, CTA, or DSA, should be distinguished from "clinical" spasm, the syndrome of confusion and decreased level of

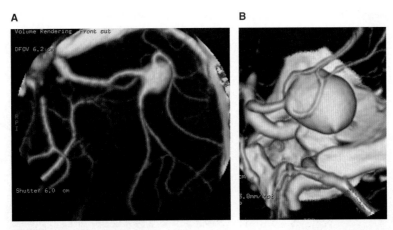

Figure 17 Relationship of Branch Vessels to Aneurysm Sac, Best Demonstrated on Volume-Rendered Views. **(A)** Major M2 branch vessels arise directly off of the middle cerebral artery bifurcation aneurysm sac, which complicates neurosurgical clipping. **(B)** In a different patient, A2 branches of the anterior cerebral artery appear to drape over the aneurysm dome of a large anterior communicating artery aneurysm, suggesting the possibility of adherence. Note focal peaked contour abnormality on the side of the aneurysm, which is likely the site of recent rupture.

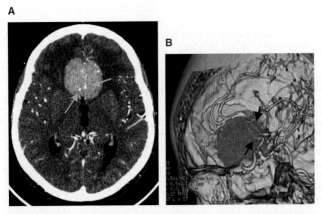

Figure 18 Surgical Planning for Olfactory Groove Meningioma Resection. Relationship of tumors to A2 segments of the anterior cerebral arteries and adjacent orbitofrontal branches. **(A)** Encasement of the A2 anterior cerebral artery segments is best appreciated on axial source images (blue arrows) with **(B)** the volume-rendered view providing the surgical "craniotomy" perspective.

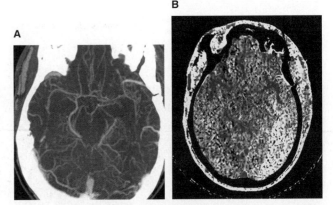

Figure 19 Vasospasm with Perfusion Abnormality on CTP. **(A)** Widespread, predominantly left-sided, irregular vessel narrowing is shown on computed tomography angiography, secondary to post-subarachnoid hemorrhage vasospasm. **(B)** Mean transit time elevation on computed tomography perfusion (CTP) identifies the most severely effected region, which helps plan catheter-based therapy.

consciousness associated with reduced blood flow to the brain parenchyma. Angiographic spasm occurs in approximately 50% of patients following aneurysmal subarachnoid hemorrhage; clinical spasm occurs in about 30%. Roughly half of clinical spasm cases result in infarction [70].

CTA has the potential to detect angiographic vasospasm, whereas CTP aids diagnosis of clinical vasospasm (Fig. 19). In a study that compared CTA MIP images with DSA, it was concluded that CTA was highly accurate for the detection of vasospasm in proximal branches of the circle of Willis and only slightly less accurate in more distal vessels [79]. CTA has also been shown to provide complementary information to transcranial Doppler imaging for assessing vertebrobasilar vasospasm [80]. Magnetic resonance, CT, and SPECT blood flow studies have revealed a correlation among angiographic spasm, reduction in CBF, and clinically symptomatic spasm [81,82]. In the future, measures of cerebrovascular reserve could prove to be of value in the triage of subarachnoid hemorrhage patients between medical and endovascular treatments.

E. Vasculitis

Though CTA has high sensitivity for intracranial steno-occlusive disease [38], it does not currently have significant clinical utility for diagnosing primary CNS vasculitis. This is not surprising given the insensitivity of even DSA for pathology predominantly affecting arterioles and venules less than 300 μm in diameter [83]. However, inflammation of the large and medium-sized blood vessels, as seen with Takayasu's arteritis and the use of chemotherapeutic agents such as gemcitabine, can be visualized with CTA of the aortic arch and cervical vasculature [84,85] (Fig. 20).

F. Computed Tomography Venography

Computed tomography venography (CTV) is a variation of our head CTA protocol that optimizes visualization of the intracranial venous anatomy. The most common indication is to evaluate dural and cavernous sinus thrombosis, although it is increasingly recognized for its utility in surgical planning.

1. Cerebral Venous Thrombosis

Unlike arterial stroke, thrombosis of the cerebral veins and sinuses typically affects young adults and children. The presentation is variable, including headaches, seizures, and coma, though more than 80% of all patients have a good neurological outcome. Sequelae arise both from the local effects of cerebral vein thrombosis, including localized brain edema, venous infarction, and hemorrhage, and from the more global effects of intracranial hypertension due to occlusion of the major sinuses [86]. Common causes include prothrombotic conditions such as pregnancy and oral contraceptive use, infection such as otitis and meningitis, and head injury. Dehydration is a particular risk factor for cerebral venous thrombosis in the pediatric age group. Iatrogenic causes include jugular catheterization, neurosurgery, and even lumbar puncture [86].

Both CTV and magnetic resonance venography are highly reliable in the workup of venous sinus thrombosis [87] (Fig. 21). Advantages of CTV over magnetic resonance techniques include better delineation of cerebral venous and dural sinus anatomy, lower cost, less vulnerability to motion artifact, and easier patient monitoring. CTV can also better demonstrate adjacent soft tissue and calcific structures. This is beneficial when evaluating extrinsic lesions, such as a meningioma or calvarial metastasis compressing or invading the dural venous sinuses [88,89]. CTV has the further advantage of not depending on flow like time-of-flight and phase-contrast magnetic resonance venography techniques. This enables visualization of smaller diameter veins, reduces confusion from loss of signal from in-plane flow, and helps identify common variants such as a hypoplastic transverse sinus [87].

The utility of CTV in children has also been demonstrated. This is noteworthy given the higher risk in this

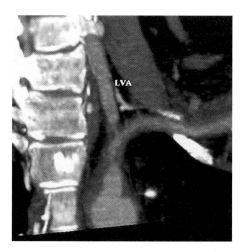

Figure 20 Takayasu's Arteritis. Large and medium-sized vessel stenoses are characteristic of this disease, as depicted here with severe narrowing at the origin of the left vertebral artery (LVA).

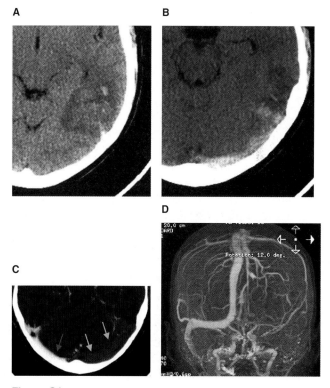

Figure 21 Venous Sinus Thrombosis. (**A**) Unenhanced head computed tomography demonstrates a hemorrhagic venous infarction pattern, (**B**) adjacent to hyperdense thrombus in left transverse sinus. (**C**) The axial computed tomography venography source image demonstrates the filling defect in the left transverse sinus (blue arrows) compared with normal opacification of the right transverse sinus (red arrow). (**D**) The posterior oblique maximum intensity projection view following bone subtraction displays the extent of left transverse sinus thrombosis.

population and the need to scan with reduced contrast volumes and injection rates [86,90].

2. Venous Anatomy for Surgical Planning

CTV can also greatly aid preoperative planning, particularly for skull base surgery. It readily documents surgically relevant anatomic variants such as those found in the superficial middle cerebral and basal veins [91–93]. With aneurysm surgery, preoperative assessment of venous anatomy can help assess the risks of postoperative complications from sacrificing critical veins, identify bridging veins that limit retraction of adjacent brain tissue, and identify adherent veins that may restrict aneurysmal dissection [94,95]. Fortunately, given the short arteriovenous transit time in the cerebral circulation, this venous anatomy is often apparent on examinations performed with the aneurysm CTA protocol without need for a separate CTV.

IV. Conclusions and Future Directions

Rapid advances in CT technology, particularly the rapid evolution of MDCT, have brought CTA to the forefront in the emergent management of the most important neurovascular indications. Advances in our understanding of contrast enhancement physiology and in the capabilities and convenience of 3D postprocessing have enabled CTA to be an even more effective tool. In patients presenting with acute stroke, CTA is highly accurate for identifying the location of intracranial vascular occlusion. CTA, in combination with CTP, also provides information on the state of the parenchyma, increasing its value in the rapid triage of acute stroke patients to appropriate therapy, including intravenous and intra-arterial thrombolytic treatment. By including coverage through the neck in the acute stroke CTA evaluation, plaque burden at the carotid bifurcations (important potential sources of emboli) can be determined. CTA is also the test of choice for dissection and traumatic vascular injury in the neck region. In the diagnostic evaluation of patients who have subarachnoid hemorrhage, CTA in many institutions has essentially replaced the more invasive and resource-intensive conventional angiographic procedure. It has become the primary, and often the only, procedure performed before neurosurgical or endovascular repair of most ruptured and unruptured aneurysms. Though postprocessing is more time-consuming, CTV is highly effective for evaluation of venous and cavernous sinus thrombosis.

Scanners with further dramatic increases in the number of detector rows, such as a 256-slice prototype, are not far off. It may soon be possible to routinely extend head-and-neck CTA exams to include evaluation of the left side of the heart, thus identifying mural thrombus as a potential cause of a patient's infarct. The need for acute stroke patients with atrial fibrillation to undergo transesophageal echocardiogram, which carries the rare but life-threatening risk of esophageal rupture [96], may be obviated [97]. The use of area detectors, large enough to cover entire organs in one axial sweep, may become practical in the next few years. The ability to repeatedly image through the entire brain in a short period will allow dynamic volumetric acquisition of angiographic information during arterial, capillary, and venous phases of contrast enhancement, further narrowing the advantages of conventional catheter angiography. The development of new applications including functional and whole-brain volume perfusion studies may follow [1,98].

Acknowledgments

The authors' sincere thanks go to Erin K. Murphy, ScB, for her help in preparing this manuscript; Javier Romero, MD, Hemali Desai, MD, and Shams Sheikh, MD, for providing background materials; R. Gilberto Gonzalez, MD, for advancing CTA as a clinical tool at Massachusetts General Hospital; and Gordon Harris, PhD, for his leadership of the 3D Visualization Laboratory at Massachusetts General Hospital.

References

1. Flohr, T., Ohnesorge, B., and Schaller, S. (2004). Design, technique, and future perspective of multislice CT scanners. *In:* "Multislice CT" (M. F. Reiser, M. Takahashi, M. Modic, and C. Becker, eds.), pp. 3–16. Springer-Verlag, Berlin.
2. Hu, H., He, H. D., Foley, W. D., and Fox, S. H. (2000). Four multidetector-row helical CT: image quality and volume coverage speed. *Radiology* **215**, 55–62.
3. Fox, S. H., Tanenbaum, L. N., Ackelsberg, S., He, H. D., Hsieh, J., and Hu, H. (1998). Future directions in CT technology. *Neuroimaging Clin. N. Am.* **8**, 497–513.
4. Rydberg, J., Buckwalter, K. A., Caldemeyer, K. S., Phillips, M. D., Conces, D. J. Jr, Aisen, A. M., Persohn, S. A., and Kopecky, K. K. (2000). Multisection CT: scanning techniques and clinical applications. *Radiographics* **20**, 1787–1806.
5. Taguchi, K., and Aradate, H. (1998). Algorithm for image reconstruction in multi-slice helical CT. *Med. Phys.* **25**, 550–561.
6. Kendell, B., and Pullicono, P. (1980). Intravascular contrast injection in ischemic lesions: II. Effect on prognosis. *Neuroradiology* **19**, 241–243.
7. Doerfler, A., Engelhorn, T., von Kummer, R., Weber, J., Knauth, M., Heiland, S., Sartor, K., and Forsting, M. (1998). Are iodinated contrast agents detrimental in acute cerebral ischemia? An experimental study in rats. *Radiology* **206**, 211–217.
8. Bettmann, M. A. (2004). Frequently asked questions: iodinated contrast agents. *Radiographics* **24(Suppl 1)**, S3–S10.
9. Morcos, S. K. (1998). Contrast media-induced nephrotoxicity: questions and answers. *Br. J. Radiol.* **71**, 357–365.
10. Morcos, S. K., Thomsen, H. S., and Webb, J. A. (1999). Contrast-media-induced nephrotoxicity: a consensus report. Contrast Media Safety Committee, European Society of Urogenital Radiology (ESUR). *Eur. Radiol.* **9**, 1602–1613.

11. Cademartiri, F., van der Lugt, A., Luccichenti, G., Pavone, P., and Krestin, G. P. (2002). Parameters affecting bolus geometry in CTA: a review. *J. Comput. Assist. Tomogr.* **26**, 598–607.
12. Aspelin, P., Aubry, P., Fransson, S. G., Strasser, R., Willenbrock, R., and Berg, K. J. (2003). Nephrotoxic effects in high-risk patients undergoing angiography. *N. Engl. J. Med.* **348**, 491–499.
13. Henson, J. W., Nogueira, R. G., Covarrubias, D. J., Gonzalez, R. G., and Lev, M. H. (2004). Gadolinium-enhanced CT angiography of the circle of Willis and neck. *AJNR Am. J. Neuroradiol.* **25**, 969–972.
14. Herts, B. R., O'Malley, C. M., Wirth, S. L., Lieber, M. L., and Pohlman, B. (2001). Power injection of contrast media using central venous catheters: feasibility, safety, and efficacy. *AJR Am. J. Roentgenol.* **176**, 447–453.
15. Brink, J. A. (2003). Use of high concentration contrast media (HCCM): principles and rationale; body CT. *Eur. J. Radiol.* **45(Suppl 1)**, S53–S58.
16. Fleischmann, D., and Hittmair, K. (1999). Mathematical analysis of arterial enhancement and optimization of bolus geometry for CT angiography using the discrete Fourier transform. *J. Comput. Assist. Tomogr.* **23**, 474–484.
17. Bae, K. T., Tran, H. Q., and Heiken, J. P. (2004). Uniform vascular contrast enhancement and reduced contrast medium volume achieved by using exponentially decelerated contrast material injection method. *Radiology* **231**, 732–736.
18. Schoellnast, H., Tillich, M., Deutschmann, M. J., Deutschmann, H. A., Schaffler, G. J., and Portugaller, H. R. (2004). Aortoiliac enhancement during computed tomography angiography with reduced contrast material dose and saline solution flush: influence on magnitude and uniformity of the contrast column. *Invest. Radiol.* **39**, 20–26.
19. Schellinger, P. D., Fiebach, J. B., and Hacke, W. (2003). Imaging-based decision making in thrombolytic therapy for ischemic stroke: present status. *Stroke* **34**, 575–583.
20. Schramm, P., Schellinger, P. D., Fiebach, J. B., Heiland, S., Jansen, O., Knauth, M., Hacke, W., and Sartor, K. (2002). Comparison of CT and CT angiography source images with diffusion-weighted imaging in patients with acute stroke within 6 hours after onset. *Stroke* **33**, 2426–2432.
21. National Institute of Neurological Disorders and Stroke rt-PA Stroke Study Group. (1995). Tissue plasminogen activator for acute ischemic stroke. *N. Engl. J. Med.* **333**, 1581–1587.
22. Furlan, A., Higashida, R., Wechsler, L., Gent, M., Rowley, H., Kase, C., Pessin, M., Ahuja, A., Callahan, F., Clark, W. M., Silver, F., and Rivera, F. (1999). Intra-arterial prourokinase for acute ischemic stroke. The PROACT II study: a randomized controlled trial. Prolyse in Acute Cerebral Thromboembolism. *J. Am. Med. Assoc.* **282**, 2003–2011.
23. Hacke, W., Albers, G., Al-Rawi, Y., Bogousslavsky, J., Davalos, A., Eliasziw, M., Fischer, M., Furlan, A., Kaste, M., Lees, K. R., Soehngen, M., Warach, S., and the DIAS Study Group. (2005). The desmoteplase in acute ischemic stroke trial (DIAS): a phase II MRI-based 9-hour window acute stroke thrombolysis trial with intravenous desmoteplase. *Stroke* **36(1)**, 66–73.
24. Versnick, E. J., Do, H. M., Albers, G. W., Tong, D. C., and Marks, M. P. (2005). Mechanical thrombectomy for acute stroke. *AJNR Am. J. Neuroradiol.* **26**, 875–879.
25. Von Kummer, R., Allen, K. L., Holle, R., Bozzao, L., Bastianello, S., Manelfe, C., Bluhmki, E., Ringleb, P., Meier, D. H., and Hacke, W. (1997). Acute stroke: usefulness of early CT findings before thrombolytic therapy [see comments]. *Radiology* **205**, 327–333.
26. Lev, M. H., Farkas, J., Gemmete, J. J., Hossain, S. T., Hunter, G. J., Koroshetz, W. J., and Gonzalez, R. G. (1999). Acute stroke: improved nonenhanced CT detection; benefits of soft-copy interpretation by using variable window width and center level settings. *Radiology* **213**, 150–155.
27. Lev, M. H., Farkas, J., Rodriguez, V. R., Schwamm, L. H., Hunter, G. J., Putman, C. M., Rordorf, G. A., Buonanno, F. S., Budzik, R., Koroshetz, W. J., and Gonzalez, R. G. (2001). CT angiography in the rapid triage of patients with hyperacute stroke to intra-arterial thrombolysis: accuracy in the detection of large vessel thrombus. *J. Comput. Assist. Tomogr.* **25**, 520–528.
28. Lev, M. H., and Nichols, S. J. (2000). Computed tomographic angiography and computed tomographic perfusion imaging of hyperacute stroke. *Top. Magn. Reson. Imaging* **11**:273–287.
29. Lev, M. H., Segal, A. Z., Farkas, J., Hossain, S. T., Putman, C., Hunter, G. J., Budzik, R., Harris, G. J., Buonanno, F. S., Ezzeddine, M. A., Chang, Y., Koroshetz, W. J., Gonzalez, R. G., and Schwamm, L. H. (2001). Utility of perfusion-weighted CT imaging in acute middle cerebral artery stroke treated with intra-arterial thrombolysis: prediction of final infarct volume and clinical outcome. *Stroke* **32**, 2021–2028.
30. Koroshetz, W. J., and Gonzales, R. G. (1999). Imaging stroke in progress: magnetic resonance advances but computed tomography is poised for counterattack. *Ann. Neurol.* **46**, 556–558.
31. Warach, S. (2001). New imaging strategies for patient selection for thrombolytic and neuroprotective therapies. *Neurology* **57**, S48–S52.
32. Lee, K. H., Lee, S. J., Cho, S. J., Na, D. G., Byun, H. S., Kim, Y. B., Song, H. J., Jin, I. S., and Chung, C. S. (2000). Usefulness of triphasic perfusion computed tomography for intravenous thrombolysis with tissue-type plasminogen activator in acute ischemic stroke. *Arch. Neurol.* **57**, 1000–1008.
33. Knauth, M., von Kummer, R., Jansen, O., Hahnel, S., Dorfler, A., and Sartor, K. (1997). Potential of CT angiography in acute ischemic stroke [see comments]. *AJNR Am. J. Neuroradiol.* **18**, 1001–1010.
34. Shrier, D., Tanaka, H., Numaguchi, Y., Konno, S., Patel, U., and Shibata, D. (1997). CT angiography in the evaluation of acute stroke. *AJNR Am. J. Neuroradiol.* **18**, 1011–1020.
35. Wildermuth, S., Knauth, M., Brandt, T., Winter, R., Sartor, K., and Hacke, W. (1998). Role of CT angiography in patient selection for thrombolytic therapy in acute hemispheric stroke. *Stroke* **29**, 935–938.
36. Ponzo, J., Hunter, G., Hamburg, L., et al. (1998). Evaluation of collateral circulation in acute stroke patients using CT angiography. Proceedings of the 23rd International Conference on Stroke and Cerebral Circulation, Orlando, FL.
37. Barest, G., Hunter, G., Hamberg, L., et al. (1997). Dynamic contrast-enhanced helical CT improves conspicuity of acute cerebral ischemia. Proceedings of the 83nd Scientific Assembly and Annual Meeting of the Radiological Society of North America, (Hot topics), Chicago, IL.
38. Bash, S., Villablanca, J. P., Jahan, R., Duckwiler, G., Tillis, M., Kidwell, C., Saver, J., and Sayre, J. (2005). Intracranial vascular stenosis and occlusive disease: evaluation with CT angiography, MR angiography, and digital subtraction angiography. *AJNR Am. J. Neuroradiol.* **26**, 1012–1021.
39. Sims, J. R., Rordorf, G., and Smith, E. E., Koroshetz, W. J., Lev, M. H., Buonanno, F., and Schwamm, L. H. (2005). Arterial occlusion revealed by CT angiography predicts NIH stroke score and acute outcomes after IV tPA treatment. *AJNR Am. J. Neuroradiol.* **26**, 246–251.
40. Schwamm, L. H., Rosenthal, E. S., Swap, C. J., Rosand, J., Rordorf, G., Buonanno, F. S., Vangel, M. G., Koroshetz, W. J., and Lev, M. H. (2005). Hypoattenuation on CT angiographic source images predicts risk of intracerebral hemorrhage and outcome after intra-arterial reperfusion therapy. *AJNR Am. J. Neuroradiol.* **26**, 1798–1803.
41. Fisher, C. M. (1951). Occlusion of the internal carotid artery. *Arch. Neurol. Psychiat.* **65**, 346–377.
42. North American Symptomatic Carotid Endarterectomy Trial Collaborators. (1991). Beneficial effect of carotid endarterectomy in symptomatic patients with high-grade carotid stenosis. *N. Engl. J. Med.* **325**, 445–453.
43. European Carotid Surgery Trialists' Collaborative Group. (1991). MRC European carotid surgery trial: interim results for symptomatic

patients with severe (70–99%) or with mild (0–29%) carotid stenosis. *Lancet* **337**, 1235–1243.
44. Davies, K. N., and Humphrey, P. R. (1993). Complications of cerebral angiography in patients with symptomatic carotid territory ischaemia screened by carotid ultrasound. *J. Neurol. Neurosurg. Psychiatry* **56**, 967–972.
45. Bendszus, M., Koltzenburg, M., Burger, R., Warmuth-Metz, M., Hofmann, E., and Solymosi, L. (1999). Silent embolism in diagnostic cerebral angiography and neurointerventional procedures: a prospective study. *Lancet* **354**, 1594–1597.
46. Ackerman, R. (1995). Neurovascular non-invasive evaluation. Chapter 50. *In:* "Radiology: Diagnosis, Imaging, Intervention" (J. Taveras and J. Ferrucci, eds.). Lippincott Williams & Wilkins, Philadelphia.
47. Ackerman, R., Candia, M., and May, Z. (1999). Technical advances and clinical progress in carotid diagnosis. *AJNR Am. J. Neuroradiol.* **20**, 187–189.
48. Dillon, E., Van Leeuween, M., Fernandez, M. A., Eikelboom, B., and Mali, W. (1993). CT angiography: application to the evaluation of carotid artery stenosis. *Radiology* **189**, 211–219.
49. Oliver, T. B., Lammie, G. A., Wright, A. R., Wardlaw, J., Patel, S. G., Peek, R., Ruckley, C. V., and Collie, D. A. (1999). Atherosclerotic plaque at the carotid bifurcation: CT angiographic appearance with histopathologic correlation. *AJNR Am. J. Neuroradiol.* **20**, 897–901.
50. Lev, M., Ackerman, R., Chehade, R., et al. (1996). The clinical utility of spiral computed tomographic angiography in the evaluation of carotid artery disease: review of our first 50 patients. Proceedings of the 21st International Conference on Stroke and Cerebral Circulation, San Antonio, TX.
51. Bluemke, D. A., and Chambers, T. P. (1995). Spiral CT angiography: an alternative to conventional angiography. *Radiology* **195**, 317–319.
52. Berg, M., Zhang, Z., Ikonen, A., Sipola, P., Kalviainen, R., Manninen, H., and Vanninen, R. (2005). Multi-detector row CT angiography in the assessment of carotid artery disease in symptomatic patients: comparison with rotational angiography and digital subtraction angiography. *AJNR Am. J. Neuroradiol.* **26**, 1022–1034.
53. Josephson, S. A., Bryant, S. O., Mak, H. K., Johnston, S. C., Dillon, W. P., and Smith, W. S. (2004). Evaluation of carotid stenosis using CT angiography in the initial evaluation of stroke and TIA. *Neurology* **63**, 457–460.
54. Lev, M., Ackerman, R., Rabinov, J., et al. (1997). Hairline residual lumen or total occlusion of the internal carotid artery? The clinical utility of spiral CT angiography. Proceedings of the 35th Annual Meeting of the American Society of Neuroradiology, Toronto, Canada.
55. Lev, M., Ackerman, R., Lustrin, E., and Brown, J. (1995). A procedure for accurate spiral CT angiographic measurement of lumenal diameter. Proceedings of the 81st Scientific Assembly and Annual Meeting of the Radiological Society of North America, Chicago, IL.
56. Link, J., Brossmann, J., Grabener, M., Mueller-Huelsbeck, S., Steffens, J. C., Brinkmann, G., and Heller, M. (1996). Spiral CT angiography and selective digital subtraction angiography of internal carotid artery stenosis. *AJNR Am. J. Neuroradiol.* **17**, 89–94.
57. Schwartz, R., Tice, H., Hooten, S., Hsu, L., and Stieg, P. (1994). Evaluation of cerebral aneurysms with helical CT: correlation with conventional angiography and MR angiography. *Radiology* **192**, 717–722.
58. Anderson, G. B., Ashforth, R., Steinke, D. E., Ferdinandy, R., and Findlay, J. M. (2000). CT angiography for the detection and characterization of carotid artery bifurcation disease. *Stroke* **31**, 2168–2174.
59. Dix, J., Evans, A., Kallmes, D., Sobel, A., and Phillips, C. (1997). Accuracy and precision of CT angiography in a model of the carotid artery bifurcation. *AJNR Am. J. Neuroradiol.* **18**, 409–415.
60. Liu, Y., Hopper, K. D., Mauger, D. T., and Addis, K. A. (2000). CT angiographic measurement of the carotid artery: optimizing visualization by manipulating window and level settings and contrast material attenuation. *Radiology* **217**, 494–500.
61. Zhang, Z., Berg, M. H., Ikonen, A. E., Vanninen, R. L., and Manninen, H. I. (2004). Carotid artery stenosis: reproducibility of automated 3D CT angiography analysis method. *Eur. Radiol.* **14**, 665–672.
62. Suwanwela, N., Can, U., Furie, K. L., Southern, J. F., Macdonald, N. R., Ogilvy, C. S., Hansen, C. J., Buonanno, F. S., Abbott, W. M., Koroshetz, W. J., and Kistler, J. P. (1996). Carotid Doppler ultrasound criteria for internal carotid artery stenosis based on residual lumen diameter calculated from en bloc carotid endarterectomy specimens. *Stroke* **27**, 1965–1969.
63. Walker, L. J., Ismail, A., McMeekin, W., Lambert, D., Mendelow, A. D., and Birchall, D. (2002). Computed tomography angiography for the evaluation of carotid atherosclerotic plaque: correlation with histopathology of endarterectomy specimens. *Stroke* **33**, 977–981.
64. Lev, M. H., Romero, J. M., Goodman, D. N., Bagga, R., Kim, H. Y., Clerk, N. A., Ackerman, R. H., and Gonzalez, R. G. (2003). Total occlusion versus hairline residual lumen of the internal carotid arteries: accuracy of single section helical CT angiography. *AJNR Am. J. Neuroradiol.* **24**, 1123–1129.
65. Chen, C. J., Tseng, Y. C., Lee, T. H., Hsu, H. L., and See, L. C. (2004). Multisection CT angiography compared with catheter angiography in diagnosing vertebral artery dissection. *AJNR Am. J. Neuroradiol.* **25**, 769–774.
66. Leclerc, X., Godefroy, O., Salhi, A., Lucas, C., Leys, D., and Pruvo, J. P. (1996). Helical CT for the diagnosis of extracranial internal carotid artery dissection. *Stroke* **27**, 461–466.
67. Berne, J. D., Norwood, S. H., McAuley, C. E., and Villareal, D. H. (2004). Helical computed tomographic angiography: an excellent screening test for blunt cerebrovascular injury. *J. Trauma* **57**, 11–17, discussion 17–19.
68. Munera, F., Soto, J. A., Palacio, D., Velez, S. M., and Medina, E. (2000). Diagnosis of arterial injuries caused by penetrating trauma to the neck: comparison of helical CT angiography and conventional angiography. *Radiology* **216**, 356–362.
69. Slovut, D. P., and Olin, J. W. (2004). Fibromuscular dysplasia. *N. Engl. J. Med.* **350**, 1862–1871.
70. Mayberg, M. R., Batjer, H. H., Dacey, R., Diringer, M., Haley, E. C., Heros, R. C., Sternau, L. L., Torner, J., Adams, H. P. Jr, Feinberg, W., et al. (1994). Guidelines for the management of aneurysmal subarachnoid hemorrhage. A statement for healthcare professionals from a special writing group of the Stroke Council, American Heart Association. *Stroke* **25**, 2315–2328.
71. Investigators ISoUIA. (1998). Unruptured intracranial aneurysms: risk of rupture and risks of surgical intervention. International Study of Unruptured Intracranial Aneurysms Investigators. *N. Engl. J. Med.* **339**, 1725–1733.
72. Hoh, B. L., Cheung, A. C., Rabinov, J. D., Pryor, J. C., Carter, B. S., and Ogilvy, C. S. (2004). Results of a prospective protocol of computed tomographic angiography in place of catheter angiography as the only diagnostic and pretreatment planning study for cerebral aneurysms by a combined neurovascular team. *Neurosurgery* **54**, 1329–1340, discussion 1340–1342.
73. Boet, R., Poon, W. S., Lam, J. M., and Yu, S. C. (2003). The surgical treatment of intracranial aneurysms based on computer tomographic angiography alone: streamlining the acute management of symptomatic aneurysms. *Acta Neurochir. (Wien)* **145**, 101–105, discussion 105.
74. Gonzalez-Darder, J. M., Pesudo-Martinez, J. V., and Feliu-Tatay, R. A. (2001). Microsurgical management of cerebral aneurysms based in CT angiography with three-dimensional reconstruction (3D-CTA) and without preoperative cerebral angiography. *Acta Neurochir. (Wien)* **143**, 673–679.

75. Ruigrok, Y. M., Rinkel, G. J., Buskens, E., Velthuis, B. K., and van Gijn, J. (2000). Perimesencephalic hemorrhage and CT angiography: a decision analysis. *Stroke* **31**, 2976–2983.
76. Velthuis, B. K., Rinkel, G. J., Ramos, L. M., Witkamp, T. D., and van Leeuwen, M. S. (1999). Perimesencephalic hemorrhage: exclusion of vertebrobasilar aneurysms with CT angiography. *Stroke* **30**, 1103–1109.
77. Alberico, R. A., Patel, M., Casey, S., Jacobs, B., Maguire, W., and Decker, R. (1995). Evaluation of the circle of Willis with three-dimensional CT angiography in patients with suspected intracranial aneurysms. *AJNR Am. J. Neuroradiol.* **16**, 1571–1578, discussion 1579–1580.
78. Vieco, P. T., Shuman, W. P., Alsofrom, G. F., and Gross, C. E. (1995). Detection of circle of Willis aneurysms in patients with acute subarachnoid hemorrhage: a comparison of CT angiography and digital subtraction angiography. *AJR Am. J. Roentgenol.* **165**, 425–430.
79. Anderson, G. B., Ashforth, R., Steinke, D. E., and Findlay, J. M. (2000). CT angiography for the detection of cerebral vasospasm in patients with acute subarachnoid hemorrhage. *AJNR Am. J. Neuroradiol.* **21**, 1011–1015.
80. Goldsher, D., Shreiber, R., Shik, V., Tavor, Y., and Soustiel, J. F. (2004). Role of multisection CT angiography in the evaluation of vertebrobasilar vasospasm in patients with subarachnoid hemorrhage. *AJNR Am. J. Neuroradiol.* **25**, 1493–1498.
81. Ohkuma, H., Manabe, H., Tanaka, M., and Suzuki, S. (2000). Impact of cerebral microcirculatory changes on cerebral blood flow during cerebral vasospasm after aneurysmal subarachnoid hemorrhage. *Stroke* **31**, 1621–1627.
82. Nabavi, D. G., LeBlanc, L. M., Baxter, B., Lee, D. H., Fox, A. J., Lownie, S. P., Ferguson, G. G., Craen, R. A., Gelb, A. W., and Lee, T. Y. (2001). Monitoring cerebral perfusion after subarachnoid hemorrhage using CT. *Neuroradiology* **43**, 7–16.
83. Kadkhodayan, Y., Alreshaid, A., Moran, C. J., Cross, D. T., 3rd, Powers, W. J., and Derdeyn, C. P. (2004). Primary angiitis of the central nervous system at conventional angiography. *Radiology* **233**, 878–882.
84. Scatarige, J. C., Urban, B. A., Hellmann, D. B., and Fishman, E. K. (2001). Three-dimensional volume-rendering CT angiography in vasculitis: spectrum of disease and clinical utility. *J. Comput. Assist. Tomogr.* **25**, 598–603.
85. Bendix, N., Glodny, B., Bernathova, M., and Bodner, G. (2005). Sonography and CT Vasculitis During Gemcitabine Therapy. *AJR Am. J. Roentgenol.* **184**, S14–S15.
86. Stam, J. (2005). Thrombosis of the cerebral veins and sinuses. *N. Engl. J. Med.* **352**, 1791–1798.
87. Ozsvath, R. R., Casey, S. O., Lustrin, E. S., Alberico, R. A., Hassankhani, A., and Patel, M. (1997). Cerebral venography: comparison of CT and MR projection venography. *AJR Am. J. Roentgenol.* **169**, 1699–1707.
88. Eskey, C. J., Lev, M. H., Tatter, S. B., and Gonzalez, R. G. (1998). Cerebral CT venography in surgical planning for a tentorial meningioma. *J. Comput. Assist. Tomogr.* **22**, 530–532.
89. Casey, S. O., Alberico, R. A., Patel, M., Jimenez, J. M., Ozsvath, R. R., Maguire, W. M., and Taylor, M. L. (1996). Cerebral CT venography. *Radiology* **198**, 163–170.
90. Alberico, R., Barnes, P., Robertson, R., and Burrows, P. (1999). Helical CT angiography: dynamic cerebrovascular imaging in children. *AJNR Am. J. Neuroradiol.* **20**, 328–334.
91. Suzuki, Y., and Matsumoto, K. (2000). Variations of the superficial middle cerebral vein: classification using three-dimensional CT angiography. *AJNR Am. J. Neuroradiol.* **21**, 932–938.
92. Hoffmann, O., Klingebiel, R., Braun, J. S., Katchanov, J., and Valdueza, J. M. (2002). Diagnostic pitfall: atypical cerebral venous drainage via the vertebral venous system. *AJNR Am. J. Neuroradiol.* **23**, 408–411.
93. Suzuki, Y., Ikeda, H., Shimadu, M., Ikeda, Y., and Matsumoto, K. (2001). Variations of the basal vein: identification using three-dimensional CT angiography. *AJNR Am. J. Neuroradiol.* **22**, 670–676.
94. Kaminogo, M., Hayashi, H., Ishimaru, H., Morikawa, M., Kitagawa, N., Matsuo, Y., Hayashi, K., Yoshioka, T., and Shibata, S. (2002). Depicting cerebral veins by three-dimensional CT angiography before surgical clipping of aneurysms. *AJNR Am. J. Neuroradiol.* **23**, 85–91.
95. Suzuki, Y., Nakajima, M., Ikeda, H., Ikeda, Y., and Abe, T. (2004). Preoperative evaluation of the venous system for potential interference in the clipping of cerebral aneurysm. *Surg. Neurol.* **61**, 357–364, discussion 364.
96. Shapira, M. Y., Hirshberg, B., Agid, R., Zuckerman, E., and Caraco, Y. (1999). Esophageal perforation after transesophageal echocardiogram. *Echocardiography* **16**, 151–154.
97. Zaidat, O. S. J., Gilkeson, R., et al. (2002). The utility of ultrafast cardiac cycle gated and contrasted spiral computerized axial chest tomography in evaluation of patients with acute stroke and comparison with transesophageal echocardiography. Proceedings of the 27th International Stroke Conference of the American Stroke Association, San Antonio, TX.
98. Freiherr, G. (2001). CT vendors press quest for more speed, slices. *Diagn. Imaging Vol. Tech. Advis.* 25–26.

75

PET Imaging in Parkinson's Disease and Other Neurodegenerative Disorders

Vijay Dhawan, PhD
David Eidelberg, MD

Keywords: *differential diagnosis, dopamine cell implant, glial cell line-derived neurotrophic factor, Parkinson's disease, positron emission tomography, statistical parametric mapping*

I. Introduction
II. Diagnosis of Parkinson's Disease
III. Treatment
IV. Cognitive Impairment: New Imaging Approaches
V. PET Imaging and Clinical Diagnosis
VI. Conclusion
   References

I. Introduction

Many exciting new developments have recently taken place in the application of positron emission tomography (PET) imaging tools for the study of neurological disorders. With the aging of the North American and European populations, significant emphasis has been placed on the use of these new tools in the study of neurodegenerative diseases and their treatment. The primary focus of these developments has been to expand the horizon of *what* and *how* to image in these diseases. A secondary focus has been to advance the methodology used in the analysis of imaging data. Multimodality imaging, an integrative approach that merges structural (computed tomography and magnetic resonance imaging) and functional (PET, single-photon emission computed tomography (SPECT), functional magnetic resonance imaging, and optical imaging) techniques has been critical to advancing the scope of data acquisition. Similarly, a hierarchical approach integrating quantitative parameter estimation, voxel-based statistical tools, and spatial covariance pattern analysis has greatly broadened the interpretation of imaging data in the study of disease processes.

Parkinson's disease (PD) is one of the best-studied neurodegenerative illnesses from a functional imaging standpoint. In this review, we summarize recent advances in the use of new imaging approaches for clinical investigation in PD and related disorders. We also briefly discuss the application of these imaging strategies in the study of other neurodegenerative diseases in which cognitive impairment is a dominant feature. Several reviews in this area have appeared recently [1,2].

The original application of PET in PD was to detect presynaptic nigrostriatal dopamine dysfunction as a means of diagnosing this disorder [1]. It was understood that injury to the substantia nigra caused defects in the storage and release of dopamine in the striatum. Thus fluorine-18 fluorodopa (FDOPA) was developed as a PET tracer to quantify this process. Indeed, FDOPA-PET clearly demonstrated a decrease in tracer uptake in the posterior putamen in PD patients even at early disease stages. However, similar decreases were found in atypical neurodegenerative parkinsonian syndromes such as striatonigral degeneration or multiple system atrophy, progressive supranuclear palsy, and corticobasal ganglionic degeneration [3].

PET assessment of regional glucose metabolism with ^{18}F fluorodeoxyglucose (FDG) has provided an alternative approach to differential diagnosis of parkinsonism. FDG-PET studies have demonstrated a consistent pattern of basal ganglia and thalamus hypermetabolism in patients with classical PD [4]. This finding, in the company of decreased striatal uptake of presynaptic dopaminergic tracers, is compatible with a diagnosis of clinical PD. In addition, pattern analysis of FDG-PET data can provide an accurate means of diagnosing atypical forms of parkinsonism [5]. Similarly, spatial covariance analysis can be used to detect small metabolic signals characteristic of PD and other movement disorders [6]. Recent studies have found that such disease-related patterns evolve with progressive disease and can be modulated by successful antiparkinsonian therapy [7,8].

II. Diagnosis of Parkinson's Disease

A. Dopaminergic Imaging

1. Presynaptic Dopaminergic Function

a. Fluorodopa Uptake, Metabolism, and Storage PET studies using FDOPA measure the uptake of this tracer and its conversion to ^{18}F fluorodopamine by the enzyme dopa decarboxylase (DDC) in the striatal dopaminergic nerve terminals. In the plasma, FDOPA is metabolized by catechol-*O*-methyl-transferase (COMT) to 3-*O*-methyl-fluorodopa [1]. In PET experiments, peripheral DDC can be blocked by the administration of the DDC inhibitor carbidopa before tracer administration. FDOPA transport across the blood-brain barrier follows the same channel as the large neutral amino acid.

FDOPA-PET data can be analyzed either by using simple noninvasively derived target-to-background ratios, obtained by dividing striatal count rates by those in a nonspecific uptake brain region such as the occipital cortex (striatal-to-occipital ratio), or by using multiple time graphical analysis using plasma or brain input functions. In the latter, the gradient of the linear regression of the data, described as the net influx constant (K_i), reflects the rate of FDOPA decarboxylation and storage. Compartmental models have also been developed to estimate the specific kinetic rate constant (k_3) for striatal DDC activity.

Many studies have shown that FDOPA-PET yields quantitative parameters that can discriminate PD patients from healthy control subjects and correlate with independent disease severity measures. More importantly, it has been shown that putamen FDOPA uptake correlates with nigral dopamine cell counts measured in postmortem specimens [1]; similar correlations have been found in experimental animal models of parkinsonism.

b. Dopamine Transporter: Synaptic Reuptake The development of radiotracers that bind to the striatal dopamine transporter (DAT) has led to another means for imaging the nigrostriatal dopaminergic system with PET or SPECT. The most extensively studied agents in this category are the cocaine analogues, such as 2β-carbomethyl-3β-(4-iodophenyl) tropane (β-CIT) and its fluoroalkyl esters [2] (Fig. 1). The use of DAT tracers has several potential advantages over FDOPA: (1) Although dopaminergic neurons decline in normal aging, striatal FDOPA uptake appears to be maintained by the up-regulation of DDC activity. Therefore, FDOPA-PET may be relatively insensitive to age-related decrements in presynaptic dopaminergic function. By contrast, the DAT-binding agents are more sensitive to age-related dopaminergic attrition. Nevertheless, this sensitivity may require the introduction of age corrections in longitudinal studies of disease progression in PD. (2) The transport of 3-*O*-methyl-fluorodopa

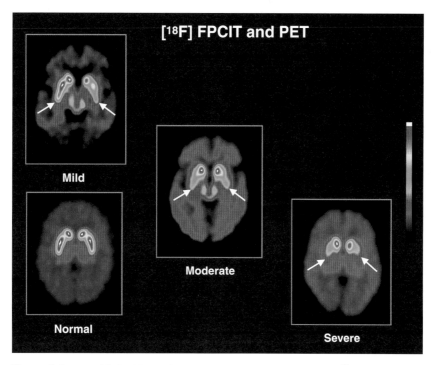

Figure 1 Images of Striatal Dopamine Transporter Binding Obtained Using ^{18}F-fluoropropyl-β-CIT (^{18}F-FPCIT) and Positron Emission Tomography (PET) [30]. Scans from patients with mild, moderate, and severe Parkinson's disease exhibit progressive declines in radiotracer uptake in the posterior putamen. Early disease is associated with marked asymmetry, which becomes bilateral with advancing motor symptoms. Caudate uptake is relatively preserved until later stages of disease. A scan from a normal volunteer (lower left) is included for comparison.

across the blood-brain barrier can affect quantification of DDC activity with FDOPA-PET. (3) The signal-to-noise ratio is potentially higher for DAT imaging than for FDOPA-PET.

Despite these advantages, recent studies have raised the possibility that DAT binding may be altered by concurrent levodopa therapy [9]. More research will be needed to determine how these findings will influence the use of this tracer in future neuroprotection trials.

c. Vesicular Monoamine Transporter The presynaptic vesicular monoamine transporter (VMAT) is involved in the packaging and transport of monoamines to storage vesicles located in nerve terminals. Radioactive ligands that bind to VMAT sites such as carbon-11 dihydrotetrabenazine (DTBZ) can be used as a reliable measure of monoaminergic and nerve terminal density. Indeed, VMAT binding is reduced in the putamen of PD patients [10,11]. VMAT binding does not appear to be affected by concurrent antiparkinsonian dopaminergic therapy, giving it potential advantage over other dopaminergic tracers. However, this method has a comparably low signal-to-noise ratio and may not be specific for dopaminergic terminals.

2. Postsynaptic Dopaminergic Function

PET studies using ligands that bind selectively to D_1 and D_2 receptors in the striatum can be used to measure changes in postsynaptic dopaminergic function occurring with disease progression and pharmacotherapy.

Radioligands such as with ^{11}C raclopride (RAC) and ^{11}C N-methylspiperone can provide sensitive measures of local D_2 receptor density. RAC-PET studies have demonstrated a decrement of dopamine D_2 receptor binding with normal aging of approximately 0.6% per year [12]. The postsynaptic response to nigrostriatal deafferentation in PD is likely to differ from that related to normal aging. It has been suggested that the loss of dopaminergic nerve terminals, in association with changes in postsynaptic dopamine receptors, underlies the motor complications that occur in the course of treatment [13]. A relative increase of striatal dopamine D_2 receptor binding has been reported in early, untreated parkinsonian patients, particularly in the putamen contralateral to the more affected body side [12]. However, initial dopamine D_2 receptor up-regulation in the putamen may reverse with disease progression; binding values in the normal range or lower may be encountered in advanced PD patients [12,14]. Because striatal RAC and FDOPA-PET changes are associated throughout the disease course, it is likely that dopamine D_2 receptor changes result from the decline in presynaptic dopaminergic drive.

RAC, although a widely used D_2 tracer, is easily displaced by endogenous dopamine, whereas N-methylspiperone has a higher affinity for D_2 receptors and is not as easily displaced. Indeed, an increase or decrease in RAC binding may reflect endogenous dopamine levels instead of D_2 receptor density [11]. As a measure of endogenous dopamine, RAC-PET imaging has provided important new information regarding the dynamic functions of the intact nigrostriatal dopamine terminals in PD and other movement disorders.

The ultimate utility of D_1 neuroreceptor quantification with PET in the study of PD remains unknown. It is likely that quantification of D_1 and D_2 striatal neuroreceptors in individual patients will provide a better understanding of the mechanism of the levodopa response, including the potentiation of dyskinesias.

B. Differential Diagnosis of Idiopathic Parkinson's Disease and Atypical Parkinsonism

Atypical neurodegenerative parkinsonian syndromes such as striatonigral degeneration, multiple system atrophy, progressive supranuclear palsy, and corticobasal ganglionic degeneration are often difficult to distinguish from PD on clinical grounds. These diseases may have different imaging features that can be used in differential diagnosis. In patients with early stage idiopathic PD, FDOPA uptake is relatively preserved in the caudate and anterior putamen. By contrast, patients with atypical parkinsonism can exhibit decrements in tracer uptake in these structures and the posterior putamen. However, this differential dopaminergic topography is often insufficient to discriminate individual patients with idiopathic PD from atypical subjects at early disease stages [15].

Recently, FDG-PET imaging has been employed to identify characteristic patterns of regional glucose metabolism in patients with idiopathic PD, as well as in cohorts with variant forms of parkinsonism. Single-case statistical parametric mapping techniques can be used to categorize individual PET scans based on characteristic disease templates [5] (Fig. 2). This pattern recognition approach correctly diagnosed 92.4% of the subjects in accord with clinical assessments made 2 years after PET imaging by blinded movement disorders specialists. Thus FDG-PET performed at the time of initial referral for parkinsonism accurately predicted the clinical diagnosis of individual patients made at subsequent follow-up.

Correct diagnosis can be used to optimize treatment and to reduce the risks attendant to potentially ineffective treatment strategies. Accurate differential diagnosis is also of particular importance in the conduct of treatment trials in parkinsonism. Inadvertent inclusion of atypical patients into pharmacological trials for PD is likely to reduce statistical power by increasing the heterogeneity of the treatment cohorts. Accurate differential diagnosis in patients with parkinsonism supported by the use of imaging techniques such as FDG-PET may improve the power of future clinical trials by identifying potentially nonresponsive participants before randomization.

C. Metabolic Brain Networks in Parkinson's Disease: Spatial Covariance Patterns

Although the primary pathological abnormality in PD responsible for motor dysfunction is confined to the substantia nigra, the degeneration of dopaminergic nigral projections to the striatum results in widespread alterations in the functional activity of the basal ganglia and its output to ventral thalamus and brainstem. Specifically, the functional organization of the basal ganglia predicts that the loss of inhibitory dopaminergic input to the striatum causes increased inhibitory output from the putamen to the external globus pallidus, diminished inhibitory output from the external globus pallidus to the subthalamic nucleus, and functional overactivity of the subthalamic nucleus and internal globus pallidus. These changes decrease output from the ventrolateral thalamus to the motor cortex [16] and are accompanied by alterations in regional cerebral glucose metabolism and blood flow [4].

In most degenerative neurological conditions, regional cerebral blood flow and metabolism are generally coupled and relate to synaptic activity within a given brain region [17]. However, measurements of local rates of metabolism are insufficient to describe the spatially distributed neural networks associated with a neurodegenerative process or the modulation of these networks during treatment. We have modeled such systems using a spatial covariance approach based on principal components analysis [18]. In this approach, principal components analysis is applied to combined samples of patient and control scans. Patterns with significantly different expression in the disease group are considered "disease related" [6]. These metabolic patterns reflect covarying regional increases or decreases in brain function relative to a baseline defined by the normal population.

In PD, the topography of disease-related covariance patterns corresponds well to the functional architecture of the neural pathways involved in this disorder [17,18]. Parkinson's disease–related metabolic patterns (PDRPs) are characterized by relative pallidothalamic and pontine hypermetabolism covarying with reductions in cortical motor and sensory association regions. This topography is highly reproducible across patient populations and tomographs [6]. Network analysis can also be used to quantify the expression of these patterns in individual subjects on a scan-by-scan basis [19]. Prospective computation of PDRP scores can discriminate PD patients from normal controls and patients with atypical parkinsonism with approximately

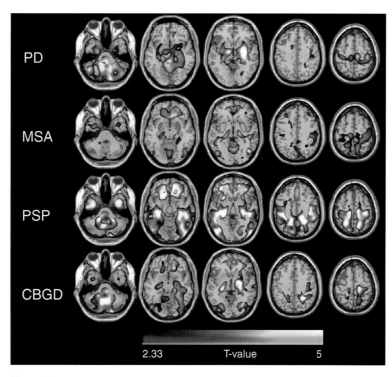

Figure 2 Templates of Abnormal Glucose Metabolism. Patients have classical Parkinson's disease (PD), multiple system atrophy (MSA), progressive supranuclear palsy (PSP), and corticobasal ganglionic degeneration (CBGD) identified by group analysis using statistical parametric mapping. These templates can be used with fluorodeoxyglucose positron emission tomography for purposes of differential diagnosis of single cases [5]. Increased glucose metabolism is indicated by "hot" colors and decrease metabolism by "winter" colors. For image analysis, the hemispheres opposite the body sides more affected clinically appeared on the right. PD is characterized by hypermetabolism of the putamen and globus pallidus associated with metabolic increases in the thalamus and cerebellum and metabolic decreases in the parietal cortex. MSA is characterized by metabolic reductions in the putamen, cerebellum, and brainstem. PSP is characterized by metabolic decreases in the brainstem and midfrontal regions. CBGD is characterized by asymmetrical metabolism in the cortex and basal ganglia with relative reductions contralateral to the side most affected clinically. Hypometabolism of the brainstem and/or frontal areas can also be seen in this disorder.

85% sensitivity. In contrast to neurochemical dopaminergic markers such as FDOPA or β-CIT, PDRP expression has the attribute of increasing with disease progression. This may provide greater sensitivity in the detection of longitudinal changes in brain function, critically important in the assessment of potential disease-modifying agents.

Recent clinical trials of potential neuroprotective agents in PD have employed dopaminergic imaging with FDOPA and β-CIT SPECT to determine whether the nigrostriatal degenerative process is slowed by treatment. In the course of the studies, it was found that 10%–15% of subjects with parkinsonism had normal radiotracer uptake [9]. Similarly, in a recently conducted retrospective study of patients with parkinsonism referred for FDOPA-PET, we found that 16.2% had normal dopaminergic scans. PDRP expression was quantified in FDG-PET scans conducted at or near the time of dopaminergic imaging [5]. In these subjects, pattern expression was normal, indicating that these individuals were not likely to have had classical PD. Indeed, at long-term clinical follow-up, these subjects were found to have secondary forms of parkinsonism including dopa-responsive dystonia, a syndrome occasionally confused with PD in young subjects [19]. Thus PDRP expression may be useful in differentiating PD from atypical variant conditions, including secondary forms of parkinsonism in which nigrostriatal dopaminergic terminals are relatively preserved.

III. Treatment

A. Role of Imaging in Therapy Assessment for Parkinson's Disease

Reliable *in vivo* markers of neuronal activity are needed to assess therapeutic outcome. Available clinical scales are relatively insensitive and inherently variable. By contrast,

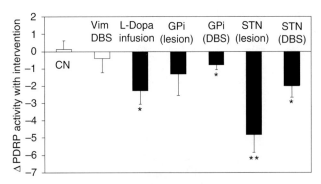

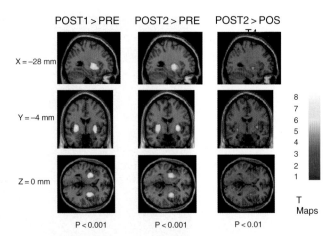

Figure 3 Bar Graph Illustrating Relative Changes in the Expression of the Parkinson's Disease–Related Metabolic Covariance Pattern (Δ PDRP, see [6]).[a] Changes occurred during antiparkinsonian therapy (filled bars) with levodopa infusion (L-Dopa), unilateral ventral pallidotomy, pallidal, and subthalamic nucleus (STN) deep brain stimulation (DBS), and subthalamotomy. For the surgical interventions, Δ PDRP reflects changes in network activity in the operated hemispheres. With levodopa infusion, the PDRP changes were averaged across hemispheres. The control data (open bars) represent values in the *unoperated* contralateral hemispheres (CN) of the surgical patients scanned in the nonmedicated state. (From Eckert, T., Eidelberg, D., Neuroimaging and therapeutics in movement disorders. Neuro Rx(r) 2: 361–371. Copyright © 2005, American Society for Experimental NeuroTherapeutics. All rights reserved.)

[a] Asterisks represent p values with respect to the untreated condition (paired student's t test): *, $p < 0.01$; **, $p < 0.005$. *Note:* GPi, globus pallidus internus; Vim, ventralis intermedius.

Figure 4 Mapping of FDOPA Uptake at Baseline (PRE) and at 1 (POST1) and 2 (POST2) Years after Embryonic Dopamine Cell Implantation [23]. Data from 19 transplant recipients are compared by paired tests using statistical parametric mapping. Transplantation results in significantly increased FDOPA uptake in the bilateral posterior putamen at both postoperative (POST) times. Comparison of POST2 and POST1 shows significant bilateral increase in the most posterior part of ventral putamen, reflecting continued innervation 2 years after implantation. (From Ma et al., PET and embryonic dopamine cell transplantation in Parkinson's disease. In: Bioimaging in neurodegeneration. P. A. Broderick, D. N. Rabni, and E. H. Kolodny, eds. New Jersey: Humana Press; 2005: 45–58.)

quantitative functional brain imaging markers may be suitable as outcome measures to assess new symptomatic therapies. We have found that FDG-PET may be a useful tool in predicting optimal candidates for certain surgical interventions [7]. Moreover, effective antiparkinsonian therapy is associated with significant reductions in PDRP expression. PDRP modulation has been described with levodopa infusion, as well as with either lesioning or deep brain stimulation of the pallidum and of the subthalamic nucleus [6] (Fig. 3). The assessment of PDRP activity in FDG-PET scans conducted at baseline and during treatment may prove useful as a means of assessing the efficacy of novel agents and stereotaxic interventions such as subthalamic gene therapy [20].

1. Dopamine Cell Implantation

The implantation of dopamine-producing cells has potential promise in the treatment of advanced PD. FDOPA-PET has been valuable in determining graft viability in the context of several blinded trials of embryonic nigral cell implantation [21,22]. Voxel-based approaches have been implemented to assess transplant effects in treated versus placebo groups [23], as well as to track graft activity during "open-label" follow-up (Fig. 4). Although changes in FDOPA uptake are only minimally correlated with clinical outcome following transplantation, this imaging approach has been critical in examining the basis for transplant-related dyskinesias. In five transplant recipients with this complication, FDOPA uptake was elevated in discrete regions of the putamen relative to controls without dyskinesia [24]. In addition to the posterodorsal zone in which a prominent reduction in uptake was present at baseline, the dyskinesia group displayed a relative increase ventrally, in which preoperative dopaminergic input was relatively preserved. However, such focal abnormalities may be a universal feature of cell transplantation procedures for PD [22].

Further technological development is required to optimize dopamine cell implantation as a therapeutic option in advanced parkinsonism. Stem cell implantation can be considered in future studies once the mechanism of graft-related dyskinesias has been better elucidated in experimental animal models of parkinsonism. Likewise, further research is needed to better understand the disconnection between changes in FDOPA-PET signal and clinical outcome in other new interventions such as intraputaminal glial cell line-derived neurotrophic factor. Adjunctive FDG-PET techniques may be useful in quantifying the network changes that occur with treatment and their relationship to dopaminergic function and clinical outcome.

IV. Cognitive Impairment: New Imaging Approaches

Two novel radioligands are being developed for the clinical investigation of Alzheimer's disease and related disorders. 2-(1-{6-[(2-[^{18}F]Fluroethyl)(methyl)amino]-2-naphthyl}ethylidene)malononitrile (^{18}FDDNP) can be used to target neurofibrillary tangles and β-amyloid senile plaques [25]. Likewise, N-methyl-[^{11}C]2-(4'-methylaminophenyl)-6-hydroxybenzothiazole (^{11}C-PIB) may be useful because of its high specificity for β-amyloid [26]. These PET compounds may also be applicable to the study of cognitive dysfunction in PD.

Using a novel approach, Koeppe and associates [27] suggest that a single-scan DTBZ can diagnose and differentiate among Alzheimer's disease, frontotemporal dementia, and dementia with Lewy bodies. They used a bolus-plus-infusion protocol with dynamic ^{11}C-DTBZ imaging to generate images of K_1 and distribution volume. These parametric images may provide similar information to that obtained with two independent PET scans with FDG and ^{11}C-DTBZ. However, PD subjects with associated Alzheimer's disease pathology cannot be discriminated with DTBZ.

In a review article, Gilman [28] discusses the use of ^{11}C-DTBZ and N-[^{11}C]methylpiperidin-4-yl propionate (a tracer for acetylcholinesterase activity) in rapid eye movement sleep behavior disorder in PD, dementia with Lewy bodies, and multiple system atrophy. New radiotracers that target α-synuclein may be useful for imaging glial cytoplasmic inclusions in cases of multiple system atrophy and other parkinsonian syndromes. Of note, ^{11}C (R)-PK11195 (1-[2-chlorophenyl]-N-methyl-N-[1-methylpropyl]-3-isoquinolone carboxamide) PET has recently been used to localize *in vivo* microglial activation in patients with multiple system atrophy [29]. Increased ^{11}C (R)-PK11195 binding was primarily found in the dorsolateral prefrontal cortex, putamen, pallidum, pons, and substantia nigra, reflecting the known distribution of neuropathological changes in multiple system atrophy.

V. PET Imaging and Clinical Diagnosis

A. Parkinson's Disease

1. *Clinical indication:* Questionable parkinsonism with or without tremors. Is it essential tremor or PD?

Scan: ^{18}F-FDOPA-PET scan to rule out presynaptic dopaminergic dysfunction. (This compound has been the subject of the largest number of publications.) Other tracers that can be substituted for FDOPA include dopamine reuptake transporter ligands such as ^{18}F-fluoropropyl-β-CIT (^{18}FPCIT-PET), iodine-123 β-CIT, FPCIT, or technetium-99m (SPECT). ^{11}C-DTBZ for VMATs can be considered if available.

2. *Clinical indication:* Parkinsonian features with rapid progression or limited response to dopaminergic medication. Is it an atypical parkinsonian variant?

Scan: ^{18}F-FDG-PET scan for differential diagnosis. FDG has distinctive patterns for multiple system atrophy, corticobasal ganglionic degeneration, and progressive supranuclear palsy. (Blood flow tracers that may be substituted for FDG-PET include ^{99m}Tc-HMPAO and ^{99m}Tc-ECD SPECT, although pattern analysis has not been validated using these imaging modalities.)

B. Huntington's Disease

Clinical indication: Emotional, cognitive, and motor disturbances in individuals at risk for Huntington's disease.

Scan: ^{11}C-RAC-PET scan to rule out postsynaptic defect and perform longitudinal studies. Other D_2 receptor imaging agents include ^{11}C-methylspiperone (PET) and ^{123}I-IBZM (SPECT). These may be applicable in Huntington's disease. FDG-PET studies showing reductions in metabolism in caudate and putamen may also be helpful in supporting this diagnosis.

C. Alzheimer's Disease and Other Dementias

Clinical indication: Mild cognitive impairment. Is it a neurodegenerative dementia or a secondary problem?

Scan: FDG-PET can be useful in supporting a diagnosis of Alzheimer's disease if parietotemporal and parieto-occipital metabolic decrements are present. Decreased FDG uptake in the frontal and temporal regions (especially anteriorly) is consistent with frontotemporal dementia. SPECT imaging of blood flow (^{99m}Tc HMPAO and ^{99m}Tc ECD) can potentially substitute for FDG-PET. Other investigational PET tracers that may be used to identify regions with abnormal tau protein, amyloid deposits, or both include ^{18}FDDNP and ^{11}C-PIB.

VI. Conclusion

Modern data analytical approaches have opened new possibilities for automated differential diagnosis in PD. Similarly, these methods have allowed high throughput image analysis to facilitate multicenter clinical trials using PET biomarkers. In addition, new radiotracer techniques have been developed to quantify neurochemical deficits associated with neurodegenerative processes. A major contribution of this research has been a combined approach using both *in vivo* neurochemical measurements and network analytical strategies to investigate the relationships between localized neuronal attrition and expression of disease-related spatial covariance patterns. These complementary PET techniques may greatly advance the understanding of the pathophysiology of PD and the functional changes that occur with successful therapy.

Acknowledgments

This review was supported by NIH NS RO1 32368, 35069, and 37564, as well as generous grants from the Parkinson's Disease Foundation and the American Parkinson's Disease Association.

References

1. Dhawan, V., and Eidelberg, D. (2003). PET imaging in Parkinson's disease. *Adv. Clin. Neurosci.* **13**, 251–276.
2. Trost, M., Dhawan, V., Feigin, A., and Eidelberg, D. (2005). PET/SPECT. *In:* "Neurodegenerative Diseases: Neurobiology, Pathogenesis, and Therapeutics" (M. F. Beal, A. Lang, and A. Ludolph, eds.), pp. 290–300. Cambridge University Press, Cambridge.
3. Eckert, T., and Eidelberg, D. (2004). The role of functional neuroimaging in the differential diagnosis of idiopathic Parkinson's disease and multiple system atrophy. *Clin. Auton. Res.* **14**, 84–91.
4. Eidelberg, D., Edwards, C., Mentis, M., Dhawan, V., and Moeller, J. R. (2000). Movement disorders: Parkinson's disease. *In:* "Brain Mapping: The Disorders" (J. C. Mazziotta, A. W. Toga, and R. Frackowiak, eds.), pp. 241–261. Academic Press, San Diego.
5. Eckert, T., Barnes, A., Dhawan, V., Frucht, S., Gordon, M. F., Feigin, A. S., and Eidelberg, D. (2005). FDG PET in the differential diagnosis of parkinsonian disorders. *NeuroImage* **26**, 912–921.
6. Eckert, T., and Eidelberg, D. (2005). Neuroimaging and therapeutics in movement disorders. *NeuroRx.* **2**, 361–371.
7. Carbon, M., Edwards, C., and Eidelberg, D. (2003). Functional brain imaging in Parkinson's disease. *Adv. Neurol.* **91**, 175–181.
8. Asanuma, K., Dhawan, V., Carbon, M., and Eidelberg, D. (2004). Assessment of disease progression in parkinsonism. *J. Neurol.* **251(Suppl 7)**, 4–8.
9. Fahn, S., Oakes, D., Shoulson, I., Kieburtz, K., Rudolph, A., Lang, A., Olanow, C. W., Tanner, C., Marek, K., and the Parkinson Study Group. (2004). Levodopa and the progression of Parkinson's disease. *N. Engl. J. Med.* **351**, 2498–2508.
10. Frey, K. A., Koeppe, R. A., Kilbourn, M. R., Vander Borght, T. M., Albin, R. L., Gilman, S., and Kuhl, D. E. (1996). Presynaptic monoaminergic vesicles in Parkinson's disease and normal aging. *Ann. Neurol.* **40**, 873–874.
11. Stoessl, A., and Ruth, T. (1998). Neuroreceptor imaging: new developments in PET and SPECT imaging of neuroreceptive binding (including dopamine transporters, vesicle transporters, and postsynaptic receptor sites). *Curr. Opin. Neurol.* **11**, 327–333.
12. Antonini, A., Schwarz, J., Oertel, W. H., Pogarell, O., and Leenders, K. L. (1997). Long-term changes of striatal dopamine D_2 receptors in patients with Parkinson's disease: a study with positron emission tomography and [^{11}C] raclopride. *Mov. Disord.* **12**, 33–38.
13. Stoessl, A., and de la Fuente-Fernandez, R. (2003). Dopamine receptors in Parkinson's disease: imaging studies. *Adv. Neurol.* **91**, 65–71.
14. Brooks, D. J., Ibanez, V., Sawle, G. V., Playford, E. D., Quinn, N., Mathias, C. J., Lees, A. J., Marsden, C. D., Bannister, R., and Frackowiak, R. S. (1992). Striatal D_2 receptor status in patients with Parkinson's disease, striatonigral degeneration, and progressive supranuclear palsy, measured with ^{11}C-raclopride and positron emission tomography. *Ann. Neurol.* **31**, 184–192.
15. Eidelberg, D., Moeller, J. R., Ishikawa, T., Dhawan, V., Spetsieris, P., Chaly, T., Belakhlef, A., Mandel, F., Przedborski, S., and Fahn, S. (1995). Early differential diagnosis of Parkinson's disease with ^{18}F-fluorodeoxyglucose and positron emission tomography. *Neurology* **45**, 1995–2004.
16. Wichmann, T., and DeLong, M. R. (1998). Models of basal ganglia function and pathophysiology of movement disorders. *Neurosurg. Clin. N. Am.* **9**, 223–236.
17. Eidelberg, D., Moeller, J. R., Kazumata, K., Antonini, A., Sterio, D., Dhawan, V., Spetsieris, P., Alterman, R., Kelly, P. J., Dogali, M., Fazzini, E., and Beric, A. (1997). Metabolic correlates of pallidal neuronal activity in Parkinson's disease. *Brain* **120**, 1315–1324.
18. Eidelberg, D., Moeller, J. R., Dhawan, V., Spetsieris, P., Takikawa, S., Ishikawa, T., Chaly, T., Robeson, W., Margouleff, D., Przedborski, S., et al. (1994). The metabolic topography of parkinsonism. *J. Cereb. Blood Flow Metab.* **14**, 783–801.
19. Asanuma, K., Ma, Y., Huang, C., Carbon-Correll, M., Edwards, C., Raymond, D., Bressman, S. B., Moeller, J. R., and Eidelberg, D. (2005). The metabolic pathology of dopa-responsive dystonia. *Ann. Neurol.* **57**, 596–600.
20. During, M. J., Kaplitt, M. G., Stern, M. B., and Eidelberg, D. (2001). Subthalamic GAD gene transfer in Parkinson disease patients who are candidates for deep brain stimulation. *Hum. Gene Ther.* **12**, 1589–1591.
21. Freed, C. R., Greene, P. E., Breeze, R. E., Tsai, W. Y., DuMouchel, W., Kao, R., Dillon, S., Winfield, H., Culver, S., Trojanowski, J. Q., Eidelberg, D., and Fahn, S. (2001). Transplantation of embryonic dopamine neurons for severe Parkinson's disease. *N. Engl. J. Med.* **344**, 710–719.
22. Olanow, C. W., Goetz, C. G., Kordower, J. H., Stoessl, A. J., Sossi, V., Brin, M. F., Shannon, K. M., Nauert, G. M., Perl, D. P., Godbold, J., and Freeman, T. B. (2003). A double-blind controlled trial of bilateral fetal nigral transplantation in Parkinson's disease. *Ann. Neurol.* **54**, 403–414.
23. Ma, Y., Dhawan, V., Freed, C., Fahn, S., and Eidelberg, D. (2005). Positron Emission Tomography and embryonic dopamine cell transplantation in Parkinson's disease. *In:* "Bioimaging in Neurodegeneration" (P. A. Broderick, D. N. Rabni, and E. H. Kolodny, eds.), pp. 45–58. Humana Press, Totowa, NJ.
24. Ma, Y., Feigin, A., Dhawan, V., Fukuda, M., Shi, Q., Greene, P., Breeze, R., Fahn, S., and Freed, C., Eidelberg D. (2002). Dyskinesia after fetal cell transplantation for parkinsonism: a PET study. *Ann. Neurol.* **52**, 628–634.
25. Agdeppa, E. D., Kepe, V., Liu, J., Small, G. W., Huang, S. C., Petric, A., Satyamurthy, N., and Barrio, J. R. (2003). 2-Dialkylamino-6-acylmalononitrile substituted naphthalenes (DDNP analogs): novel diagnostic and therapeutic tools in Alzheimer's disease. *Mol. Imaging Biol.* **5**, 404–417.
26. Klunk, W. E., Engler, H., Nordberg, A., Wang, Y., Blomqvist, G., Holt, D. P., Bergstrom, M., Savitcheva, I., Huang, G. F., Estrada, S., Ausen, B., Debnath, M. L., Barletta, J., Price, J. C., Sandell, J., Lopresti, B. J., Wall, A., Koivisto, P., Antoni, G., Mathis, C. A., and Langstrom, B. (2004). Imaging brain amyloid in Alzheimer's disease with Pittsburgh Compound-B. *Ann. Neurol.* **55**, 306–319.
27. Koeppe, R. A., Gilman, S., Joshi, A., Liu, S., Little, R., Junck, L., Heumann, M., Frey, K. A., and Albin, R. L. (2005). ^{11}C-DTBZ and ^{18}F-FDG PET measures in differentiating dementias. *J. Nucl. Med.* **46**, 936–944.
28. Gilman, S. (2005). Functional imaging with positron emission tomography in multiple system atrophy. *J. Neural Transm.* **112**, 1647–1655.
29. Brooks, D. J. (2005). Positron emission tomography and single-photon emission computed tomography in central nervous system drug development. *NeuroRx* **2**, 226–236.
30. Kazumata, K., Dhawan, V., Chaly, T., Antonini, A., Margouleff, C., Belakhlef, A., Neumeyer, J., and Eidelberg, D. (1998). Dopamine transporter imaging with fluorine-18-FPCIT and PET. *J. Nucl. Med.* **39**, 1521–1530.

76

Single-Photon Emission Computed Tomography

Joseph C. Masdeu, MD, PhD

Keywords: *brain disease, brain perfusion, functional neuroimaging, receptors, single-photon emission computed tomography*

I. Brief History and Method
II. SPECT in Diseases of the Brain
   References

I. Brief History and Method

Single-photon emission computed tomography (SPECT) was introduced in the early 1980s as an instrument for the evaluation of regional cerebral perfusion and receptor density studies [1]. For the performance of SPECT, a flow tracer or a receptor-binding substance is tagged with a radionuclide and injected intravenously (Fig. 1). The flow tracer is assumed to accumulate in different areas of the brain proportionally to the rate of delivery of nutrients to that volume of brain tissue [2]. SPECT images are generated using gamma cameras or ring-type imaging systems that record photons emitted by the tracer trapped in the brain (Fig. 2). SPECT results in better image quality than two-dimensional or planar imaging because focal sources of activity are not superimposed on one another. As a result, the contrast between the target and the background (the signal-to-noise ratio) is greatly increased. Depending on the type of imaging system and tracer used, the resolution ranges from 14–17 mm full width at half maximum (FWHM) for single-head gamma cameras, now seldom used for brain imaging, to 8–10 mm FWHM for three- and four-head camera systems and to 7–8 mm FWHM for special-purpose ring-type imaging systems. In general, system cost is directly proportional to the number and complexity of camera heads or crystals.

Scanning time in SPECT depends on the imaging system, the type of radiopharmaceutical, and the quality of image desired. High-resolution images of the whole brain can be obtained with current technology in about 20 to 30 minutes. The volume imaging capacity of most SPECT systems permits reconstruction at any angle—including the axial, coronal, and sagittal planes—or at the same angle of imaging obtained with computed tomography (CT) or magnetic resonance imaging (MRI) to facilitate image comparisons.

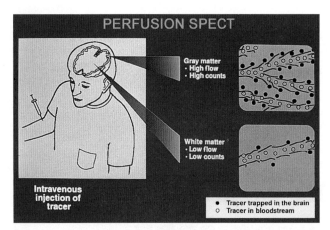

Figure 1 Injection of a SPECT Perfusion Agent. These highly lipid-soluble agents tend to penetrate the blood-brain barrier and remain trapped in the perivascular space for several hours, enough to allow imaging. Brain areas with high perfusion, such as the cerebral cortex, have higher counts than those with lower perfusion, such as the white matter.

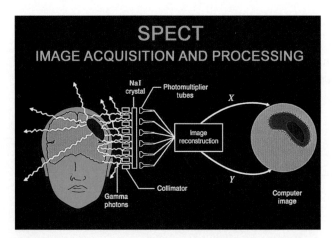

Figure 2 Distribution of a Radionuclide in the Brain Imaged by Quantification of the Photons That Interact with Sodium Iodine (NaI) Detectors. Careful collimation is essential for image quality. Mathematical algorithms similar to the ones used by other computed tomographic techniques, including Fourier transform, are used to reconstruct the distribution of the isotope in the brain.

A number of commercially available and experimental radiopharmaceuticals have been applied to SPECT studies of cerebral perfusion. The radiotracer is assumed to accumulate in different areas of the brain proportionally to the rate of delivery of nutrients to that volume of brain tissue and is described in units of ml/min/100 g. Studies in animals and humans have demonstrated that under properly controlled conditions SPECT data obtained with perfusion agents approximates perfusion closely enough to be meaningful in clinical and research studies [2]. Furthermore, most routine clinical applications of brain perfusion SPECT do not require quantitation of regional cerebral blood flow (rCBF) and rely exclusively on the generation of images that reflect tracer uptake and retention only. Thus areas of abnormal activity are said to be hyper- or hypoperfused compared with a reference set, often the average cerebellar activity, when this structure is unlikely to be affected by the disease under study, or the average cortical activity in the same anatomical slice.

The U.S. Food and Drug Administration (FDA) has approved three brain perfusion radiopharmaceuticals for clinical use. The oldest one, iodine-123 isopropyliodoamphetamine (IMP, Spectamine), distributes proportionally to rCBF over a range of flows but may be decreased with low plasma pH as in cerebral ischemia or acidosis. Brain activity remains relatively constant from 20 to at least 60 minutes after injection. Not commercially available in the United States, it is widely used in Japan. Technetium-99m hexamethylpropyleneamine oxime (HMPAO, Ceretec), a lipid-soluble macrocyclic amine, is available for routine clinical use [2]. Brain uptake is rapid and reaches its maximum within 10 minutes. Radiotracer distribution remains constant for many hours after injection. A third radiopharmaceutical, ^{99m}Tc ethyl cysteinate dimer (ECD, Neurolite), has a rapid blood clearance, resulting in high brain-to-soft tissue activity ratios early and with less exposure to radiation [2]. The Tc-radiolabeled compounds are stable for about 6 hours, facilitating their use for the study of episodic phenomena, such as seizures.

The inert gas xenon-133 has also been used to study rCBF. ^{133}Xe SPECT is performed after inhalation of the gas and is based on clearance techniques that relate the change in radiotracer activity over time to blood flow [2]. The principal advantage over other tracers that remain in the brain is that rCBF can be measured quantitatively and repeatedly without arterial sampling. ^{133}Xe does have major limitations, including poor spatial resolution and the need for specialized instrumentation.

In addition to their use in determining perfusion, radiotracers can be used to determine biochemical interactions such as receptor binding. Postmortem studies have reported a severe depletion of cocaine recognition sites associated with the dopamine transporter (DAT) system in the striatum of patients with Parkinson's disease (PD). Several analogues of cocaine have been investigated to develop DAT selective radioligands for SPECT imaging, of which [^{123}I] 2β-carbomethyl-3β-(4-iodophenyl) tropane (β-CIT) and ^{123}I-Ioflupane (^{123}I-fluoropropyl-β-CIT) are the most widely used. The main advantage of ^{123}I-Ioflupane is that images can be acquired from 3 to 6 hours after injection, compared with the 18 to 24 hours required for [^{123}I] β-CIT [3]. A D_2 receptor marker (^{123}I-iodobenzamide) is also commercially available in Europe. Other ^{123}I-labeled ligands have been developed for imaging the cholinergic, noradrenergic, and GABAergic receptor systems.

Brain perfusion SPECT is a safe procedure. The whole-body effective dose equivalent received from the administration of ^{99m}Tc HMPAO is 0.7 roentgen equivalent man (rem) per 20 millicurie dose. This effective dose equivalent value is similar to that received during a radionuclide bone scan, is 1.5 times that received from a CT of the abdomen and pelvis, and is 43% of the average annual background radiation in the United States. Most state-of-the-art imaging systems are designed to reduce head motion and patient discomfort. Most clinical applications do not require arterial sampling.

One of the major reasons for the interest in SPECT is that it represents a less expensive technique to do functional neuroimaging. The older and more accurate modality for functional neuroimaging is positron emission tomography (PET). Unlike PET, SPECT cannot measure regional cerebral metabolism, but it provides a qualitative estimate of rCBF, which in many neurological disorders is tightly coupled with brain metabolism. Thus SPECT provides functional information not available by conventional CT or MRI at a cost similar to that of CT.

II. SPECT in Diseases of the Brain

The interest in SPECT has spawned a rich literature in the past few years. Unfortunately, many of the reports comprise only a few patients; large, well-controlled studies are rare. In addition, most studies have used as a standard the clinical diagnosis, lacking pathological confirmation. Another caveat has to do with the variable quality of the techniques used for clinical SPECT, which depends more on the operator than the CT or MRI. The reported results have usually been obtained at well-established nuclear medicine services and may not be generalizable to all institutions. Given the complexity of information derived from functional neuroimaging, the interpretation of results often requires a close collaboration between nuclear medicine physicians and clinicians. Therefore, the practical application of SPECT to clinical work varies widely among institutions, depending on the interests of nuclear medicine physicians and, more often, of neurologists or psychiatrists who refer patients for SPECT. Of the applications reviewed in this chapter, perhaps the most useful and widespread is ictal SPECT for epilepsy. But there are institutions, particularly in Europe, at which SPECT is used routinely in the management of patients with acute stroke, particularly when aggressive therapies, such as hemicraniectomy for massive hemispherical swelling, are carried out in select cases.

A. Stroke

With the increasing availability of perfusion CT and diffusion and perfusion MRI, is SPECT still a useful tool to study perfusion, an obviously important variable in stroke and stroke-prone patients? As interesting as this question may be, we only answer it partially in this review. Leaving aside for a moment the real-life availability of these techniques, there are no controlled studies comparing their usefulness and cost in the different settings relevant to clinical cerebrovascular disease, namely, (1) stroke prediction in the presymptomatic subject; (2) stroke prediction in the patient at risk; (3) diagnosis of an acute ischemic event; and (4) prognosis of the acute event [4]. Moreover, it is likely that such studies will never be carried out because the field of stroke imaging is evolving quickly and the real applicability of the different modalities used to study stroke depends on the availability of each modality at the clinical setting. Therefore, this discussion focuses on the potential of SPECT to answer clinically relevant questions, such as the following: (1) In the patient at risk of stroke, what is the mechanism and likelihood of suffering a stroke? (2) In the patient with a strokelike syndrome, what is the mechanism, how should it be treated, and what is the prognosis for recovery?

1. Risk of Stroke in Subjects at Risk

a. Assessment of Cerebral Perfusion with Large-Vessel Stenosis Vasodilation caused by hypercapnia (inhalation of 5% CO_2) or by acetazolamide injection (1 g intravenously 15 minutes before radionuclide injection) can be used to assess the vascular reserve of brain regions supplied by a stenotic or occluded artery. By using two isotopes, a Tc agent for a baseline blood flow measurement and an iodine agent for a postacetazolamide blood flow measurement, the separate distributions can be measured using two windows on some modern three-headed scanners [2]. Acetazolamide stress brain-perfusion SPECT has been found useful as a complementary method in determining selective carotid shunting during carotid endarterectomy. Shunts were necessary for 8 of 8 patients with a severely reduced vascular reserve compared with 4 of 67 who had a less severe reduction [4]. Patients who have had cerebral infarction are also at risk for additional ischemic lesions. In a group of patients with watershed hemispherical infarcts ipsilateral to a stenotic artery, those with white matter infarcts (Fig. 3) had worse reserve than those with cortical infarcts [4]. The second group may have compensated between the time of the infarction and the time of the SPECT or suffered infarction not on a hemodynamic but on an embolic basis.

b. Subarachnoid Hemorrhage Ischemia from vasospasm is a major cause of morbidity and mortality following subarachnoid hemorrhage. Given the ability of SPECT to detect surface ischemia, this technique, when combined with transcranial Doppler, is a sensitive screening tool for the early detection of vasospasm and delayed ischemic deficits in an entire hemisphere or an isolated cortical branch. Acetazolamide SPECT has been used within the first 18 days

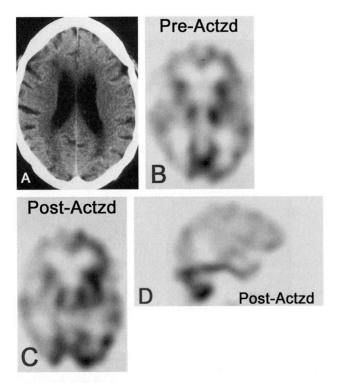

Figure 3 Use of Acetazolamide Challenge to Assess Vascular Reserve and Differentiate Ischemia from Diaschisis. A 51-year-old had transient left arm weakness. (**A**) A computed tomography scan showed hypodense areas in the watershed of the left hemispherical white matter. Duplex ultrasound revealed occlusion of the right internal carotid artery at the bifurcation. (**B**) hexamethylpropyleneamine oxime single-photon emission computed tomography (SPECT) before acetazolamide showed hypoperfusion of the right middle cerebral artery (MCA) territory and the thalamus (usually supplied by the posterior circulation). (**C** and **D**) Two days later, SPECT was repeated after oral loading with acetazolamide. (**C**) The difference between the hemispheres became more obvious. Perfusion of the previously hypoperfused right thalamus became normal (diaschisis), but not in the cortical distribution of the right MCA. (**D**) Note the sharp difference between the ischemic territory of the MCA and the normal cortex supplied by the posterior cerebral artery, from the posterior circulation. *Note:* Actzd, acetazolamide.

after subarachnoid hemorrhage to predict which patients are likely to develop cerebral infarction [4]. Early and extensive reduction in cerebral vasodilatory capacity correlated with the development of cerebral infarction due to vasospasm following subarachnoid hemorrhage. As vasospasm can be treated with angioplasty, the application of techniques for early and accurate diagnosis of dangerous vasospasm is of obvious importance.

2. Acute Strokelike Syndrome

For the diagnosis of stroke, time is of the essence. As evidence mounts that the therapeutic time window for acute stroke is narrow and that delaying effective therapy results in more tissue damage, the pressure is on to diagnose the cause of stroke as quickly as possible. Because thrombolysis is effective in acute stroke and intracerebral hemorrhage is its main contraindication, CT has become the standard imaging modality for acute stroke. MRI with echo-planar imaging is replacing CT at many institutions, because it adds information on tissue damage and potentially reversible perfusion defects. It also provides vessel visualization. Many see SPECT as an interesting technique but one whose potential benefits would be offset by the attendant delay in diagnosis and treatment. Although SPECT was first approved by the FDA for the study of cerebral ischemia, this technique is not used for acute stroke in the majority of U.S. hospitals. However, there are a number of active stroke centers around the world using perfusion SPECT for the evaluation of acute stroke [4].

SPECT may have a role in helping diagnose nonischemic neurological deficits and in predicting the need for thrombolysis, separating the patients who would not benefit from it either (1) because by the time the procedure is ready there is no perfusion deficit to be corrected by thrombolysis or (2) because the infarct is large and the ischemia is profound, with attendant capillary necrosis and a high risk for bleeding. In patients with large infarcts, SPECT may help determine which patients may benefit from holohemispheric decompression.

a. Nonischemic Neurological Deficits Occasionally, the clinical situation arises of differentiating ischemia from epilepsy as the cause of a sudden neurological deficit, particularly in cases with a prolonged focal discharge, when the onset was not witnessed and the patient is unable to give a reliable account. Ischemia will cause an area of hypoperfusion on SPECT, whereas epileptic phenomena are often manifested by hyperperfusion, although postictally there is hypoperfusion (Fig. 4). Differentiating ischemia from focal epileptic phenomena is critical, as thrombolysis and other techniques to treat acute stroke are now available. Increased perfusion of the hemisphere contralateral to the affected limbs has also been reported with infantile alternating hemiplegia, transient aphasia, or neurosarcoidosis [4].

b. Prediction of Transient or Mild Ischemic Attack In patients with acute ischemic neurological symptoms studied with ^{99m}Tc-ECD SPECT, those who had no perfusion deficit detectable by visual inspection of the SPECT scan despite clinical symptoms were symptom free after 7 days [4]. In the semiquantitative SPECT analysis, these patients had abnormal count densities in the affected region (activity less than 90% but more than 70% compared with the contralateral side). All patients with subsequent infarction (n = 59) had values less than 70%. Performance of the procedure in the first few hours seems critical for the usefulness of the test from the patient management point of view and because the findings are less clear when the study is delayed. The apparently normal perfusion of a necrotic area has been called *nonnutritional reperfusion* [4]. This pattern on serial SPECT (hypoperfusion, normal perfusion, hypoperfusion) results

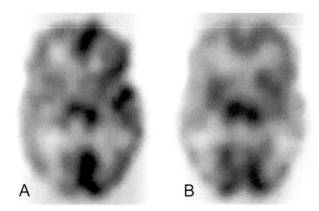

Figure 4 Peri-ictal Hemiplegia. (**A**) Axial single-photon emission computed tomography (SPECT) showing increased perfusion of the left hemisphere obtained in a 58-year-old woman 26 hours after she was found in the street with a right-sided hemiplegia, global aphasia, and a gaze deviation to the left. Clinical findings persisted at the time of this SPECT. Computed tomography was negative. An electroencephalogram obtained 12 hours before the SPECT showed marked slowing over the left hemisphere but no epileptiform activity. Improvement ensued over a 1-week period, when the patient could give a history of having had a seizure disorder since her teens. (**B**) Follow-up SPECT after the patient had improved showed symmetrical perfusion of the hemispheres.

most likely from perfusion being restored after neuronal damage has already occurred. Inflammatory tissue at the site of infarction causes arteriolar dilation in the days and weeks following the infarct, with the resultant "normal" perfusion pattern. As the necrosed tissue is reabsorbed and inflammation abates, regional perfusion decreases and the area again becomes abnormal on SPECT. This mechanism would also explain why failure to show an area of necrosis may happen more readily with HMPAO, a better marker of perfusion, than with ECD, which is poorly retained by areas of necrosis [4].

Lesions in the cortex or deep gray nuclei are more likely to cause SPECT defects than purely white matter lesions, so a normal SPECT study may also predict a lacunar stroke [4].

c. Acute Ischemic Neurological Deficit After answering the critical question of whether the stroke is ischemic or hemorrhagic, by the use of CT or gradient echo MRI, the next step is to predict the likely usefulness of intravenous or, where permitted by protocol, intra-arterial thrombolysis. In some centers, decompression of a massively swollen hemisphere due to a large middle cerebral artery (MCA) infarct has been found to have a positive outcome if the procedure is performed before additional pressure damage occurs. It is therefore important to predict early who will develop severe hemispherical swelling. Using ECD SPECT in the first 6 hours after stroke, Barthel and co-workers were able to determine which patients would develop massive MCA-territory necrosis, with hemispherical herniation [5]. These patients have a high risk of hemorrhage following thrombolysis and could be helped by early decompressive hemicraniectomy. Complete MCA infarctions were predicted with significantly higher accuracy with early SPECT (area under receiver operating characteristic curve [AUC] index 0.91) compared with early CT (AUC index 0.77) and clinical parameters (AUC index 0.73, p < 0.05). Furthermore, the predictive value increased when the findings on CT, clinical examination, and SPECT were considered [5]. Other studies have found SPECT to add predictive value to the clinical score on admission [4]. In summary, a patient with a normal ECD SPECT study performed within 3 hours of stroke onset will most likely recover spontaneously and does not require thrombolysis. A patient with a dense deficit in the entire MCA distribution has a high risk of hemorrhage with thrombolysis and, depending on age and other factors, should be considered for decompressive hemicraniectomy. The patients most likely to benefit from thrombolysis are the ones with less massive lesions [4].

A mismatch between the area acutely affected, as seen on diffusion-weighted imaging, and an MRI or SPECT perfusion study predicts shortly after the acute event which infarcts will become larger in the hours and days following the stroke. The infarct grows to match the area of severely decreased perfusion (between 12 and 20 ml per 100 g every minute) [4]. The area of mismatch is the ischemic penumbra, which often becomes necrotic as the process evolves. Using only ^{99m}Tc-ECD SPECT, the dynamic study approximates a perfusion map of the region, whereas decreased uptake in the static study shows the nonviable area of the brain, albeit with somewhat poor spatial resolution [4]. This information is important because patients with a mismatch may benefit from neuroprotective therapies. Ideally, a study on acute neuroprotection should include only these patients.

Because serial SPECT studies can be conducted about 30 minutes after one another using an initial smaller dose of the radioisotope and then a larger dose, after a few minutes using two different isotopes (^{99m}Tc and ^{123}I), or after about 24 hours using the same isotope and dose, SPECT has been used to evaluate the effect of different therapeutic strategies on brain perfusion (Fig. 5) [2]. Among them are intra-arterial urokinase, intravenous recombinant tissue plasminogen activator, streptokinase, and rheopheresis [4]. Strategies that seek to open an occluded vessel can be particularly well evaluated by perfusion SPECT, a much less invasive procedure than arterial angiography or iodine-contrast CT and a less costly procedure than perfusion MRI. For intra-arterial thrombolysis, ischemic tissue with perfusion greater than 55% of cerebellar flow may be salvageable, even with treatment initiated 6 hours after onset of symptoms [4]. Ischemic tissue with perfusion greater than 35% of cerebellar flow still may be salvageable with early treatment (less than 5 hours). Ischemic tissue with perfusion less than 35% of cerebellar flow may be at risk for hemorrhage with thrombolysis [4].

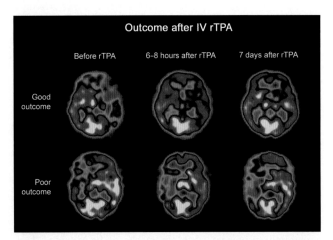

Figure 5 Repeated SPECT with Technetium-99m Ethyl Cysteinate Dimer in Two Patients. Patients (top and bottom rows) at baseline before intravenous (IV) recombinant tissue plasminogen activator (r-tPA) (left column), 6 to 8 hours after IV r-tPA (middle column), and 7 days after IV r-tPA (right column). The patient whose SPECT is displayed at the top row had a good outcome, unlike the one at the bottom. Reprinted, with permission, from Berrouschot, J., Barthel, H., Hesse, S., Knapp, W. H., Schneider, D., and von Kummer, R. (2000). Reperfusion and metabolic recovery of brain tissue and clinical outcome after ischemic stroke and thrombolytic therapy. *Stroke* **31**, 1545–1551.

SPECT has also been used to evaluate the effect on cerebral perfusion of a number of vasodilators, including pentoxifylline, olprinone, and nimodipine, and of vasoconstrictors, including cocaine [4]. In this regard, xenon CT allows for the quantitative determination of rCBF, which can also be obtained with xenon SPECT, whereas SPECT with the more widely available radionuclides only provides a relative estimate [4].

B. Neoplasms

Tracers labeled with thallium-201 have been used to attempt to quantify the malignancy grades of gliomas and to differentiate radiation necrosis from tumor recurrence. Higher malignancy tumors have a greater radionuclide uptake. As the uptake is higher with higher tumor perfusion, a mismatch is sought such that a high uptake on Tl SPECT more reliably points to tumor recurrence when perfusion SPECT shows a low tumor perfusion. Dual-isotope SPECT with ^{201}Tl to label the tumor and ^{99m}Tc HMPAO to evaluate perfusion has given reliable results in small series [6]. More recently, ^{123}I-α-methyltyrosine has been used to differentiate radiation necrosis from the recurrence of malignant astrocytomas, brain metastases, primitive neuroendocrine tumors, clivus chordomas, ependymomas, pituitary tumors, and anaplastic meningiomas. False negatives are common with small tumors (<13 mm in diameter) because they test the limits of the resolution of clinical SPECT units [7].

C. Infectious Diseases

1. Human Immunodeficiency Virus Encephalopathy

In human immunodeficiency virus (HIV) encephalopathy SPECT shows decreased cortical uptake, often with focal defects that give the cortex a moth-eaten appearance [8]. The central gray nuclei generally have multifocal defects. Perfusion of the hemispherical white matter is also decreased. In some instances, decreased uptake in the periventricular regions mimics ventricular dilation, out of proportion to the dilation appreciated on CT or MRI. The pathogenesis of perfusion changes in HIV encephalopathy is still unclear. Microvascular changes, perhaps induced by cytokines secreted by infected macrophages, may result in decreased perfusion. An association exists between the severity of ocular microangiopathy, measured by conjunctival sludge and the number of retinal cotton-wool spots, and the severity of cerebral hypoperfusion [9]. By damaging the metabolic machinery of neurons, cytokines may also cause regional neuronal hypometabolism, leading to a decreased oxygen demand and therefore decreased regional perfusion. SPECT is particularly useful in instances of psychosis, mild attentional impairment, or depression in HIV-positive individuals with normal CT or MRI. The characteristic pattern described previously would favor HIV encephalopathy rather than reactive psychosis or depression. However, the SPECT findings in HIV encephalopathy are not pathognomonic. Similar perfusion changes can be observed in chronic cocaine users, patients with the chronic fatigue syndrome, or those with mild head trauma [9]. Because the perfusion pattern is not specific, it has to be evaluated in the context of the clinical presentation and with the information provided by MRI or CT.

2. Herpes Simplex Encephalitis

Early treatment of herpes simplex encephalitis with acyclovir is particularly rewarding in patients who have a milder form of the disorder because they are likely to recover more fully. Unfortunately, the diagnosis in these patients may be delayed because they present with psychiatric syndromes or seizures and have negative MRI and CT. The cerebrospinal fluid may also be initially negative. SPECT changes may help in these cases [9]. An abnormally high accumulation of radiotracer in the affected temporal lobe may be present even at an early stage, when MRI is normal.

D. Head Trauma

Early SPECT reveals perfusion abnormalities more extensive than anatomic changes seen on CT early after trauma, even in mild head injury [10]. The lesions on

SPECT of mild head trauma are of two types: (1) sharply circumscribed areas of hypoperfusion, with borders showing relative hyperperfusion; and (2) more diffuse areas of hypoperfusion, involving the occipitotemporal regions. The first type probably corresponds to contusions, with areas of cortical ischemia. The second type may be related to smaller, widespread contusions; to axonal injury of the long white matter tracts running anteroposteriorly, such as the visual radiations; or to both [10].

Few studies have looked at the neuropathological substrate of the changes observed on perfusion SPECT [10]. By studying small-vessel ultrastructure and mapping rCBF at different times within the first 3 weeks of head injury, a zone of ischemic brain was found in areas of hypoperfusion, which persisted for weeks or months. Ischemic areas had astrocytic swelling and microvascular compression, seen on electron microscopy. Focal zones of hyperemia were also present in 42% of patients within the first 2 weeks of injury. Early focal hyperemia appeared only within apparently normal tissue as judged by late MRI or CT.

The extent of acute SPECT changes seems to correlate with the clinical severity of the posttraumatic syndrome [11]. As MRI may depict lesions not visible on CT, and vice versa, the combination of the two procedures is a more powerful predictor than either alone.

E. Seizure Disorders

Seizures are associated with dramatic increases in cerebral blood flow, localized in partial seizures and global during generalized seizures, reflected on perfusion SPECT [12]. A SPECT study showing increased regional cerebral perfusion ictally in the same region that shows decreased regional cerebral perfusion interictally provides strong evidence for the epileptogenic nature of the lesion. In complex partial seizures, the seizure focus can be identified in 71% to 93% of ictal SPECT studies with a positive predictive value of 95%. In secondarily generalized epilepsy, SPECT may show increased cerebral blood flow locally despite a clinical picture suggesting a nonfocal onset. SPECT changes may also be useful in differentiating an epileptic disorder from a psychogenic one (pseudoseizure).

As Tc-labeled compounds are bound on first pass through the brain, where they remain trapped for several hours, electroencephalogram (EEG)-guided ictal injections are used to obtain a snapshot of ictal perfusion [2]. The SPECT scan is performed hours later when the patient has recovered from the seizure and is cooperative. To be accurately recorded with SPECT, the seizure should be at least 5 seconds in duration, and the time from seizure onset to injection of the SPECT tracer should ideally be less than 45 seconds [12]. Longer delays may cause the tracer to be picked up by regions of seizure propagation, rather than the focus of origin. The ictal scan is analyzed for regions of increased and decreased perfusion by subtracting the interictal perfusion scan to create a difference image. The computer-aided subtraction ictal SPECT is then co-registered to the patient's MRI—the entire procedure is abbreviated as SISCOM—to facilitate interpretation and increase diagnostic and prognostic accuracy [12]. A SISCOM study requires hospitalization and long-term EEG monitoring, usually longer in the case of adults, with more infrequent seizures. As the imaging compound should be available at the time of seizure onset, only ECD or stable HMPAO can be used. Both compounds are stable for about 6 hours after reconstitution. If the patient has no seizures in this period, the compound has to be discarded.

SPECT may define the site of seizure onset but lacks the specificity of MRI to determine the nature of the lesion causing epilepsy. Patient management depends on the nature of the lesion. For instance, the treatment of a medial temporal glioma differs from that of mesial temporal sclerosis. Therefore, MRI is the first-line noninvasive imaging modality in focal epilepsy or generalized epilepsy with a focal origin. Quantitative MRI has a sensitivity of 80%–90% for the lateralization of temporal lobe epilepsy and is helpful in the detection of cortical abnormalities in children with intractable epilepsy. However, there are instances in which ictal SPECT has identified lesions not detected by MRI. Depth electrocorticography and intraoperative electrocorticography, though accurate in many forms of epilepsy, are both highly invasive. In cortical developmental disorders, these EEG techniques often fail to localize the epileptogenic area. PET is useful in focal epilepsy by showing hypometabolism of the affected areas. One study comparing interictal PET and ictal SPECT found similar performance of the two techniques [13].

Currently, SPECT in epilepsy is used mainly in the presurgical evaluation of patients and, rarely, in the evaluation of patients suspected of having pseudoseizures. When the ictal EEG bears such artifacts as to render it useless and the patient's behavior is puzzling, a focal SPECT abnormality that correlates well with the ictal behavior would favor the diagnosis of true seizures. In the presurgical evaluation of patients with focal epilepsy, SISCOM is useful when the MRI is normal or when the localization of seizure origin by MRI and EEG are at odds [12]. It may also be helpful in patients with multilobar pathology. SISCOM may be used to identify a "target" for placement of intracranial EEG electrodes. Localization with SISCOM may obviate the need for intracranial EEG recordings in selected patients [12]. For example, patients with normal MRI and seizures of temporal lobe origin may not require chronic intracranial EEG monitoring if the extracranial ictal EEG pattern and peri-ictal SPECT studies are concordant.

Other applications are at an investigational stage. SPECT has been used to map the area perfused with barbiturate during Wada's test in the course of the evaluation for temporal lobe epilepsy surgery. The accuracy of SPECT localization of a seizure focus can be expected to improve further with the use of new radiopharmaceuticals directed at specific neurotransmitter- or antiepileptic medication-binding sites.

F. Alzheimer's Disease

In Alzheimer's disease (AD) SPECT shows decreased perfusion in the association cortex of the parietal lobe and the posterior temporal regions [14]. Frontal association cortex is predominantly affected in some cases, but usually it is not involved until late in the course of the disease. The occipital lobes are less involved, and the paracentral cortex is spared. Using a statistical factorial system to compare regional perfusion with SPECT in AD and controls, Johnson [15] could prove that regional perfusion was decreased in the AD group in the following regions: parietotemporal cortex, hippocampus, anterior and posterior cingulum, and dorsomedial and anterior nucleus of the thalamus. This pattern had a sensitivity of 86% and a specificity of 80%. Presymptomatically, those who developed AD showed a decreased perfusion in the hippocampus, anterior and posterior cingulate gyrus, and dorsomedial and anterior nucleus of the thalamus, all of them structures with an important role in memory and attention. This finding had a sensitivity of 78% and a specificity of 71% [15]. This and other clinical studies suffer from the lack of neuropathological confirmation of the diagnosis. In a group of 70 patients with dementia and 14 controls, all with autopsy, Jagust and colleagues [16] compared the diagnostic accuracy of the clinical criteria without and with the help of SPECT. The clinical diagnosis of *probable AD* was associated with a probability of 84% of a neuropathological diagnosis of AD. A positive SPECT increased the probability of a diagnosis of AD to 92%, whereas a negative SPECT lowered that figure to 70%. SPECT was most useful when the clinical diagnosis was of *possible AD*, with a probability of a diagnosis of AD of 67% without SPECT, of 84% with a positive SPECT, and of 52% with a negative SPECT [16]. The average score on the Mini-Mental test of the patients in this study was 13, indicating that they were suffering from serious dementia. However, it is interesting that the group in which SPECT supported most the diagnosis was that of possible AD, which logically includes those patients at an earlier stage.

Whether SPECT can help identify at risk individuals was studied in a family with a presenilin-1 gene mutation [17]. Of this family, 23 people did not have the mutation, 18 had it but were cognitively intact, and 16 had already shown signs of cognitive impairment. Compared with those who did not have the mutation, those who did have it but did not have cognitive impairment had decreased perfusion in the hippocampus, anterior and posterior cingulum, and parietal and frontal association cortex. This pattern on perfusion SPECT could separate correctly 86% of gene carriers and controls, indicating abnormalities in cerebral perfusion even in asymptomatic individuals with a presenilin-1 gene mutation.

SPECT is also useful to help in the differential diagnosis of dementia, particularly in the early stages, when the cognitive findings can be ascribed to psychogenic disorders [14]. Patients with Pick's disease or other frontotemporal dementias have prefrontal and anterior temporal uptake below the control range. Those with diffuse Lewy-body disease have lowered mesial occipital perfusion. Standard perfusion SPECT does not separate well vascular dementia from AD. Acetazolamide causes an increase of cerebral perfusion due to vasodilation in areas of the brain with an intact vascular reserve but not where the arterioles are already fully dilated, such as in ischemic areas or in areas with small-vessel arteriopathies. After the intravenous administration of 1 g of acetazolamide, perfusion increases in the originally hypoperfused parietotemporal areas of AD patients, whereas it remains the same or decreases in patients with cerebrovascular disease.

G. Basal Ganglia Disorders

SPECT has been used to aid in the differential diagnosis of the parkinsonian syndromes, particularly by imaging the striatal DAT, and thereby providing an estimate of the involvement of the nigrostriatal pathway [3]. It has also been used to test potential neuroprotective therapies and the effect on neurodegeneration of medications used for PD, although the results obtained with SPECT and their interpretation are controversial [18].

Early in the disease it may be difficult to classify some patients with a parkinsonian syndrome. In typical PD there is a loss of presynaptic dopaminergic terminals in the striatum; therefore, there is a decrease of striatal uptake when the patient is studied with radioligands for DAT, such as $[^{123}I]$ β-CIT or ^{123}I-Ioflupane. Patients who have normal studies are unlikely to have typical PD. In the ELLDOPA trial, 21 out of 135 cases (16%) felt clinically to have PD had normal ^{123}I-β CIT SPECT [3]. These subjects have now been followed for up to 6 years, and both their clinical syndromes and their imaging findings are unchanged, not having progressed as in classical PD. Imaging of the cardiac sympathetic system reveals denervation in PD patients, even in early disease stages, and may differentiate between patients with multiple system atrophy and those with PD [19]. In Huntington's chorea the caudate nuclei appear hypoperfused on SPECT, even in the early stages when there is no structural evidence of caudate atrophy on CT or MRI. Combined with genetic information, imaging studies of patients at risk may be helpful in monitoring disease progression in early symptomatic or presymptomatic stages. Disease progression also must be monitored to test the effectiveness of new therapies aimed at halting neurodegeneration.

H. Diagnosis of Death on Neurological Criteria

Cerebral-perfusion imaging with SPECT agents, particularly ^{99m}Tc compounds, has been used for the diagnosis of brain death in difficult instances [20]. Absent cerebral perfusion is clearly shown on SPECT, but tomography is generally not needed; the anterior and two lateral planar views are sufficient. This technique has not been compared with conventional angiography, but it is more convenient than angiography when the diagnosis of death based on neurological criteria requires an arterial perfusion study.

Acknowledgments

This review was supported by the UTE Fundación para la Investigación Médica Aplicada (Foundation for Applied Medical Research), Pamplona, Spain.

References

1. Catafau, A. M. (2001). Brain SPECT in clinical practice. Part I: perfusion. *J. Nucl. Med.* **42**, 259–271.
2. Devous, M. D. (2005). Single-photon emission computed tomography in neurotherapeutics. *NeuroRx* **2**, 237–249.
3. Catafau, A. M., and Tolosa, E. (2004). Impact of dopamine transporter SPECT using ^{123}I-Ioflupane on diagnosis and management of patients with clinically uncertain Parkinsonian syndromes. *Mov. Disord.* **19**, 1175–1182.
4. Masdeu, J. C. (2003). Imaging of stroke with SPECT. *In:* "Imaging Cerebrovascular Disease" (V. Babikian, L. Wechsler, and R. T. Higashida, eds.), pp. 131–144. Butterworth-Heinemann, Boston.
5. Barthel, H., Hesse, S., Dannenberg, C., Rossler, A., Schneider, D., Knapp, W. H., Dietrich, J., and Berrouschot, J. (2001). Prospective value of perfusion and x-ray attenuation imaging with single-photon emission and transmission computed tomography in acute cerebral ischemia. *Stroke* **32**, 1588–1597.
6. Vallejos, V., Balana, C., Fraile, M., Roussos, Y., Capellades, J., Cuadras, P., Ballester, R., Ley, A., Arellano, A., and Rosell, R. (2002). Use of ^{201}Tl SPECT imaging to assess the response to therapy in patients with high grade gliomas. *J. Neurooncol.* **59**, 81–90.
7. Plotkin, M., Amthauer, H., Eisenacher, J., Wurm, R., Michel, R., Wust, P., Stockhammer, F., Rottgen, R., Gutberlet, M., Ruf, J., and Felix, R. (2005). Value of ^{123}I-IMT SPECT for diagnosis of recurrent non-astrocytic intracranial tumours. *Neuroradiology* **47**, 18–26.
8. Tucker, K. A., Robertson, K. R., Lin, W., Smith, J. K., An, H., Chen, Y., Aylward, S. R., and Hall, C. D. (2004). Neuroimaging in human immunodeficiency virus infection. *J. Neuroimmunol.* **157**, 153–162.
9. Masdeu, J. C., Van Heertum, R. L., and Abdel-Dayem, H. (1995). Viral infections of the brain. *J. Neuroimaging* **5(Suppl 1)**, S40–S44.
10. Masdeu, J. C., Abdel-Dayem, H., and Van Heertum, R. L. (1995). Head trauma: use of SPECT. *J. Neuroimaging* **5(Suppl 1)**, S53–S57.
11. Lee, B., and Newberg, A. (2005). Neuroimaging in traumatic brain imaging. *NeuroRx* **2**, 372–383.
12. Cascino, G. D., So, E. L., Buchhalter, J. R., and Mullan, B. P. (2004). The current place of single-photon emission computed tomography in epilepsy evaluations. *Neuroimaging Clin. N. Am.* **14**, 553–561, X.
13. Ho, S. S., Berkovic, S. F., Berlangieri, S. U., Newton, M. R., Egan, G. F., Tochon-Danguy, H. J., and McKay, W. J. (1995). Comparison of ictal SPECT and interictal PET in the presurgical evaluation of temporal lobe epilepsy. *Ann. Neurol.* **37**, 738–745.
14. Masdeu, J. C., Zubieta, J. L., and Arbizu, J. (2005). Neuroimaging as a marker of the onset and progression of Alzheimer's disease. *J. Neurol. Sci.* **236**, 55–64.
15. Johnson, K. A., Jones, K., Holman, B. L., Becker, J. A., Spiers, P. A., Stalin, A., and Albert, M. S. (1998). Preclinical prediction of Alzheimer's disease using SPECT. *Neurology* **50**, 1563–1571.
16. Jagust, W., Thisted, R., Devous, M. D., Sr., Van Heertum, R., Mayberg, H., Jobst, K., Smith, A. D., and Borys, N. (2001). SPECT perfusion imaging in the diagnosis of Alzheimer's disease: a clinical-pathologic study. *Neurology* **56**, 950–956.
17. Johnson, K. A., Lopera, F., Jones, K., Becker, A., Sperling, R., Hilson, J., Londono, J., Siegert, I., Arcos, M., Moreno, S., Madrigal, L., Ossa, J., Pineda, N., Ardila, A., Roselli, M., Albert, M. S., Kosik, K. S., and Rios, A. (2001). Presenilin-1-associated abnormalities in regional cerebral perfusion. *Neurology* **56**, 1545–1551.
18. Ravina, B., Eidelberg, D., Ahlskog, J. E., Albin, R. L., Brooks, D. J., Carbon, M., Dhawan, V., Feigin, A., Fahn, S., Guttman, M., Gwinn-Hardy, K., McFarland, H., Innis, R., Katz, R. G., Kieburtz, K., Kish, S. J., Lange, N., Langston, J. W., Marek, K., Morin, L., Moy, C., Murphy, D., Oertel, W. H., Oliver, G., Palesch, Y., Powers, W., Seibyl, J., Sethi, K. D., Shults, C. W., Sheehy, P., Stoessl, A. J., and Holloway, R. (2005). The role of radiotracer imaging in Parkinson disease. *Neurology* **64**, 208–215.
19. Eckert, T., and Eidelberg, D. (2005). Neuroimaging and therapeutics in movement disorders. *NeuroRx* **2**, 361–371.
20. Facco, E., Zucchetta, P., Munari, M., Baratto, F., Behr, A. U., Gregianin, M., Gerunda, A., Bui, F., Saladini, M., and Giron, G. (1998). ^{99m}Tc-HMPAO SPECT in the diagnosis of brain death. *Intensive Care Med.* **24**, 911–917.
21. Berrouschot, J., Barthel, H., Hesse, S., Knapp, W. H., Schneider, D., von and Kummer, R. (2000). Reperfusion and metabolic recovery of brain tissue and clinical outcome after ischemic stroke and thrombolytic therapy. *Stroke* **31**, 1545–1551.

77

Functional Magnetic Resonance Imaging

Gereon R. Fink, MD, PhD

Keywords: *aphasia, apraxia, blood oxygenation level-dependent effect, dementia, functional neuroimaging, genotyping, neglect, neuronavigation, neuropharmacology, phenotyping, recovery of function*

I. Overview and Methods
II. Functional Magnetic Resonance Imaging of Cognition and Cognitive Deficits
III. Clinical Application in Neuronavigation
IV. Monitoring Recovery of Function
V. Phenotyping and Genotyping
VI. Conclusion
   References

I. Overview and Methods

This chapter reviews the key contributions of functional magnetic resonance imaging (fMRI) to our understanding of human cognition and its various disorders. Following a brief introduction on fMRI methodology, the application of fMRI to elucidate the neural mechanisms underlying some core aspects of human cognition (language, praxis, spatial cognition, memory) is illustrated. The convergence (and divergence) of functional imaging and neurological data from patients with lesions of the central nervous system is then reviewed. Special emphasis is placed on how this informs our knowledge of the pathophysiology of disorders of cognition and neurological disease before the use of robust fMRI paradigms in presurgical patient evaluation and the planning of neurosurgical approaches to, for example, brain tumors (i.e., neuronavigation) is illustrated. Thereafter, the application of fMRI in developing novel approaches to the treatment of patients with neurological and neuropsychological deficits is highlighted. The chapter demonstrates the powerful potential of applying fMRI in both a clinical and a more research-oriented environment and its promising perspective to deliver new insights into neurological diseases and their treatment in the years to come. The chapter finishes by highlighting two particularly promising areas: human pharmacological fMRI and studies that link genotypic and phenotypic information. These intriguing new research avenues allow mapping of modulatory effects of pharmacological agents on large-scale neural systems supporting cognitive functions. The data enable inferences with reference to pharmacodynamics, specific neurotransmitters supporting specific cognitive operations, and most recently, changes in neurophysiological drug effects associated with genetic variations.

Functional imaging of the human brain today is largely using fMRI because 1.5 or 3 tesla (T) MRI scanners are universally available. Furthermore, the techniques for determining which parts of the brain are activated (e.g., when moving a finger or making a decision) are easy to implement in a standard clinical environment. Human "brain mapping" is achieved by setting up the MRI scanner in a way that it detects changes in the regional distribution of blood flow to tissue with higher oxygen demand due to increased neural activity. There are different methods to achieve this; the most common technique uses the blood oxygenation level-dependent (BOLD) effect [1], which can be measured if the MRI scanner has been equipped with echo-planar imaging capability. The technique relies on physiological contrast effects; that is, no contrast agent is administered. The latter is possible because magnetic resonance image intensity results from various tissue contrast mechanisms (e.g., proton density, T1 and T2 relaxation rates, in-flow in blood plasma protons). Importantly, some of these tissue contrast mechanisms show functional sensitivity and can, therefore, be used to noninvasively detect functional changes in the human brain.

The BOLD effect occurs because the microvascular magnetic resonance signal on T2- and T2*-weighted images is heavily influenced by the blood oxygenation state: the rate of loss of proton spin phase coherence is a measure of T2, and the local magnetic field homogeneity (T2*) is modulated by the presence or absence of deoxyhemoglobin. Iron in blood hemoglobin is used as a magnetic susceptibility-induced T2*-shortening intravascular contrast agent, which serves as a local indicator of functional activation. Oxygenated blood contains oxygenated hemoglobin, which is diamagnetic and has a small magnetic susceptibility effect. Deoxyhemoglobin is significantly more paramagnetic and thereby changes the local magnetic field, B0. The local T2* critical in fMRI is then determined by the balance of deoxygenated to oxygenated hemoglobin in a given voxel of brain tissue. By increasing the flow of oxygenated blood (as a result of increased neural activity), an increase in local intravoxel T2* occurs, which leads to an increased image intensity. All increases in oxygenated arterially delivered blood result in a relatively longer regional T2* and an increased image intensity. Importantly, this BOLD effect does not measure tissue perfusion or regional cerebral blood flow; rather, it indirectly detects changes that characterize increased local oxygen demand and brain activity [2].

The size of the BOLD effect depends on the task performed and typically varies between approximately 0.3% and 3%. It takes the BOLD effect about 4–6 seconds from stimulus or event onset until it reaches its maximum (i.e., the maximum oxygenation in response to the increased local energy demand). This sluggish response directly reflects hemodynamic characteristics, and it takes the signal another 20–30 seconds to return to baseline. This so-called hemodynamic response function (HRF) is a function not only of blood oxygenation but also of blood flow and blood volume; hence, differences have been described across both brain regions and individuals.

Numerous approaches to image processing and data analysis are available. Basic image processing and statistical analysis include (1) the removal of motion artifacts (i.e., correction for head motion across the time series by image realignment—for some studies even brainstem motion related to the cardiac cycle needs to be corrected using cardiac gating), (2) normalization of the individual structural and functional magnetic resonance images into a standard anatomic reference frame (originally provided by Talairach and Tournoux [3]) to compensate for individual differences in brain size and shape and thereby enable between-subject comparison or stereotactic localization before a neurosurgical intervention, and (3) smoothing to conform to the assumption of multivariate Gaussian distribution on which many data analysis packages are based (e.g., Statistical Parametric Mapping, www.fil.ion.ucl.ac.uk, Wellcome Department of Imaging Neuroscience, London). Thus preprocessed local changes in the BOLD signal are then assessed for their significance; thereafter, the localization of significant changes can be obtained by relating the local maxima to the standard stereotactic atlas [3], anatomic probability maps [4], or superimposition of the significant activation(s) on the structural magnetic resonance image transformed into the same standard stereotactic space.

To isolate the neural mechanisms underlying specific cognitive processes, different study designs have been developed. For example, cognitive subtraction designs are based on the assumption that task A differs from task B by only one cognitive variable. Subtracting B from A then isolates the cognitive variable of interest and its associated neural processes. Because the isolation of one variable requires a control task (task B), which is often difficult to find, two factorial designs allow assessment of not only the main factors but also any interactions among them. In contrast, a conjunction analysis enables assessment of neural activity common to a variety of tasks. The latter is of particular interest if, for example, a specific cognitive function is assumed to be part of many cognitive processes (e.g., executive processes) [5].

Another important issue is whether to assess a cognitive state or a specific process well defined in time. Although, at least in principle, fMRI is limited by the sluggishness of the HRF, echo-planar imaging nowadays is readily capable of imaging the entire brain volume every 1 to 2 seconds; at this sampling density MRI can accurately follow the time course of brain activation. Event-related fMRI uses these advances and overcomes the problem of the sluggish HRF by "jittering" the magnetic resonance scans: the event of interest and the time series of the magnetic resonance scans are dispersed from each other and from other events or cognitive processes of no interest by introducing variable delays (e.g., between the stimulus onset and the response). Thus intimately linked cognitive processes, like motor intention and preparation, can be separated [6]. Some study designs and image analysis packages even allow an online analysis of the data, enabling "real-time" fMRI.

II. Functional Magnetic Resonance Imaging of Cognition and Cognitive Deficits

Through fMRI we have gained a much more sophisticated understanding of the neural mechanisms underlying normal higher-level motor control, language processing, imagery, memory, visual and spatial processing, and executive or supervisory functions. Before the advent of functional imaging, our knowledge of these higher-order functions was based on electrophysiological data (obtained in both macaque and man), human brain surgery, and lesion studies. fMRI has enabled groundbreaking research in both basic and clinically oriented neurosciences. It is beyond the scope of this chapter to review the literature of even the core aspects of cognition (and the deficits resulting from injury). Rather, the following paragraphs highlight some key components of human cognition that are related to neurological diseases. The interested reader is referred to reviews dealing specifically with the respective topics.

A. Language and Aphasia

Language was the first mental faculty to be studied systematically in terms of its anatomical substrate. From 1825 onward, Jean-Baptiste Bouillaud published many papers on impaired speech after frontal lobe damage in explicit support of Franz Joseph Gall's localization of "l'organe du langage articulé," but he never recognized the significance of the laterality of the lesion. Paul Pierre Broca eventually became convinced that unilateral lesions of the left third frontal convolution typically gave rise to loss of "the memory of the procedure that is employed to articulate language," and equivalent lesions of the right hemisphere did not. Shortly thereafter, Karl Wernicke observed the association between injury to left temporal cortex and fluent, paraphasic speech, combined with impairment of language comprehension. More than 150 years of research into the neural organization of language based on this anatomoclinical method have stressed the central importance of two regions and their interconnections: Broca's area in inferior frontal cortex and Wernicke's area in the posterior superior temporal region. As a result of this research, the cortical organization of language has long been considered largely modular [7], with a typically rather large anatomical substrate of each module (e.g., Brodmann's areas 44 and 45 or 22).

fMRI and structural MRI have shown, however, that the story is far more complex. For example, small lesions restricted to Broca's area produce few permanent deficits. The range of language deficits subsumed under the label of Broca's aphasia usually implicates a larger lesion extending back along the Sylvian fissure. These deficits include problems with articulation, naming, fluent sentence production, morphology, syntax, and the comprehension of some complex syntactic structures. The range of these impairments makes it difficult to assume that all underlying functions are indiscriminately computed in one large area. Rather, different lesions around the inferior frontal gyrus correlate with different aphasic symptoms within the overall syndrome of Broca's aphasia [8]. In line with this concept, many functional imaging studies of language have highlighted the need to define which language function is being assessed (e.g., grammar, semantics, phonology, syntax, prosody, comprehension, articulation, reading, or listening). As a result of these studies, in recent years the concept of "language" and its associated neural bases have changed dramatically: comprehension or production of even the simplest sentence is now widely accepted to involve the coordinated effort of various mechanisms, including both specific linguistic processes and non-language-specific "general purpose" processes shared by a variety of cognitive tasks. As a direct consequence of this variety of cognitive processes serving various aspects of language processing, the studies that attempt to locate component processes of language to specific brain regions are hampered by the poor correlation between levels of performance on a task and areas of brain damage. Unfortunately, this limitation of the anatomoclinical method (i.e., lesion-deficit approach) applies at least partly to the functional neuroimaging method. Accordingly, more recent approaches to functional imaging of aphasic stroke patients concentrate on assessing the potential for a better understanding of the neural mechanisms that sustain language recovery.

For example, recent studies of speech production in aphasic patients suggest that recovery from aphasia may depend on slowly evolving activation changes in the left hemisphere. In contrast, right hemisphere activation changes are interpreted as callosal disinhibition that does not reflect recovery of function. This interpretation is based on the observation that right hemisphere activation changes occur early after stroke in areas homologous to the lesion and that they do not correlate with recovery of language function. Contrary to this, the few studies that have investigated auditory speech comprehension seem to suggest that unlike speech production, recovery of speech comprehension may depend on both left and right temporal cortex activations [9]. Although more studies of patients with aphasic stroke are needed, which include more focal lesions and larger patient numbers, the preceding studies highlight the potential to apply functional imaging to gain deeper insights into the neural mechanisms underlying recovery of function, which depend on the task, lesion site, and time elapsed from the insult. Furthermore, they illustrate the need to differentiate between neural reorganization that is and neural reorganization that is not associated with recovery of function.

B. Praxis and Apraxia

One of the most intriguing findings over the last decade related to higher motor cognition was the description of a *mirror neuron system* in macaques, the existence of which in man is now undisputed. The neurons originally observed in monkey premotor cortex (area F5) were shown to discharge both when the monkey performs a specific hand action (e.g., grasping an object) and when it observes another individual performing the same action. This has led to the hypothesis that the mirror neuron system subserves the capacity to recognize or execute actions. The mirror neuron system is not restricted to ventral premotor cortex; it is part of a parietal-premotor circuit.

An important part of this circuit is the cortex lying next to and in the intraparietal sulcus (IPS). The macaque IPS consists of highly specialized functional entities (or modules, i.e., the anterior intraparietal area, ventral intraparietal area, medial intraparietal area, lateral intraparietal area, and caudal intraparietal area), and in humans putative equivalents have been isolated using fMRI [10]. Despite some interspecies differences, possibly reflecting evolutional changes of the dorsal visual stream and the inferior parietal lobule, the functional significance of the IPS for visuomotor tasks comprising target selection for arm and eye movements, object manipulation, and visuospatial attention for action in space is undisputed and is evidenced by lesions to this region resulting in such deficits as apraxia, neglect, or Balint's syndrome. For example, lesions centered on the left human anterior IPS are known to impair object grasping, with reaching being less impaired [11]. Other deficits observed following anterior parietal cortex damage include pure tactile apraxia (i.e., impaired active tactile-shape recognition without tactile agnosia) and tactile agnosia (i.e., a loss of tactile-shape representations). These deficits suggest that the anterior intraparietal area and its neighboring areas are involved in the creation of "pragmatic object representations" in which intrinsic object properties such as shape, size, and orientation are coded to allow the selection of the appropriate object-related movement.

Typically, lesions of inferior parietal cortex go beyond inferior anterior parietal cortex and extend into the supramarginal gyrus and the angular gyrus, resulting in apraxia. Neurological studies of patients suffering from apraxia strongly imply a left hemisphere bias for skillful object use. fMRI has helped to elucidate the neural mechanisms underlying both normal and impaired skillful object use. Acquiring skillful object use involves a variety of cognitive processes. For example, exquisite timing in response to spatial triggers must be achieved. Using fMRI, the neural location where temporal and spatial information are combined has been assigned to the left supramarginal gyrus (Fig. 1). Another important key process is the correct assignment of an action to a presented object (a specific perceptual trigger). For instance, when a patient with apraxia is presented an object and asked to use it (or pantomime its use), the patient must have learned the appropriate movement and must be able to associate it with its eliciting condition. Functional imaging has been able to demonstrate that producing a range of skilled actions triggered by objects (controlled for perceptual, motor, semantic, and lexical effects) activates left inferior parietal cortex and thereby provides an explanation for why patients with lesions including left parietal cortex suffer from ideational apraxia (as assessed by impaired object use and pantomiming to visually presented objects) [12].

C. Spatial Cognition and Neglect

Clinical pathologies of spatial attention, including visual extinction, simultanagnosia, and unilateral neglect imply frontal, temporal, and particularly parietal cortices in spatial cognition. Consistent with this, human functional neuroimaging highlights the importance of posterior parietal cortex for attending to spatial locations (space-based attention) or features of visual stimuli (object-based attention). More importantly, however, fMRI data extend the information provided by lesion studies: where neuropsychological studies are limited by large lesion sizes and remote (diaschitic) effects, fMRI allows a more detailed specification of the areas involved in the cognitive operations underlying some core deficits. For example, in concurrence with electrophysiological single-unit recordings in macaques, fMRI studies indicate that parietal and frontal cortices mediate the covert and overt allocation of attention. Cortical activation occurs primarily in the areas around the IPS when a location is attended before target presentation, whereas the right temporoparietal junction is activated when a target is detected, particularly at an unattended location [13] (i.e., when a stimulus is salient and unexpected), thus working as a circuit breaker for the dorsal (where) pathway, and directing attention to a salient event. Defective awareness of

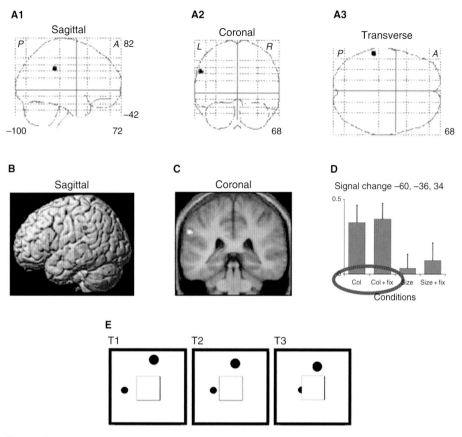

Figure 1 Cortical Activations Associated with the Collision Task. In the collision task, healthy subjects had to combine temporal and spatial information, indicating by simple button presses whether two moving objects (i.e., dots of the same size or different sizes) would or would not collide with each other. This task was contrasted with a control task that employed the same stimuli and identical motor responses but in which the size of the two moving objects had to be compared. To assess putative differential eye-movement effects, both tasks were performed with and without central fixation (i.e., under free vision). Analysis of the fMRI data (for the group of subjects) employing a random-effects model showed for the main effect of collision (i.e., collision, both with and without fixation, is greater than size, both with and without fixation) only one significant activation ($p < 0.05$, corrected), which was located in the left supramarginal gyrus, (**A**) shown as through projections onto representations of standard stereotactic space. (**A1**) sagittal view, (**A2**) coronal view, (**A3**) transverse view. (**B**) The activation is projected onto a three-dimensional surface rendering of the standard normalized brain provided by Statistical Parametric Mapping and (**C**) onto a coronal section ($y = -36$ mm) through the group's normalized mean anatomic magnetic resonance. (**D**) The relative signal change associated with the four conditions at the voxel of local maximum signal change ($x = -60$ mm, $y = -36$ mm, and $z = 34$ mm, coordinates in standard stereotactic space) is plotted. (**E**) Examples of the visual displays for the collision task ("Do the two dots collide behind the inner square?") and size task (for control, "Do the two dots have the same size?"), showing the two moving dots at three different times (T1, T2, T3) of a trial. The dots move in straight lines and at a constant speed from the outer square to the opaque inner square, behind which they disappear. *Note:* col, collision judgments, no fixation; col+fix, collision judgments with fixation; size, size judgments, no fixation; size+fix, size judgments with fixation. Adapted, with permission, from Figs. 1 and 2 of Assmus, A., Marshall, J. C., Ritzl, A., Noth, J., Zilles, K., and Fink, G. R. (2003). Left inferior parietal cortex integrates time and space during collision judgments. *Neuroimage* **20**, S82–S88.

contralesional sensory input, namely, the inability to detect and to report events in the contralesional space, is a key feature of unilateral neglect.

However, right posterior parietal cortex is involved not only in target detection and orienting but also in spatial judgments. Using variants of the line bisection task (where a subject is asked to bisect a line) and the landmark task (where a subject is asked to judge whether a line has been correctly prebisected or whether the mark has been misplaced, i.e., left or right of the true center), it has been repeatedly demonstrated that visuospatial judgments activate right parietal cortex along the IPS (Fig. 2).

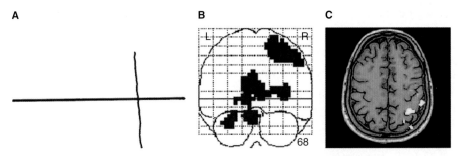

Figure 2 Neural Activations Associated with Visuospatial Judgments. (**A**) Example of a stimulus used for the visuospatial judgment task ("Is the line correctly prebisected?"). (**B**) All significant activations (p < 0.05, corrected, for the group of subjects studied) associated with the visuospatial judgment task (relative to a control task in which subjects judged whether or not there was a bisection mark) shown as a through projection onto a coronal representation of standard stereotactic space. (**C**) The local maximum of activation in right posterior parietal cortex is shown for a single subject projected on the same subject's anatomic magnetic resonance. Adapted, with permission, from Figs. 1 and 2 of Fink, G. R., Marshall, J. C., Shah, N. J., Weiss, P. H., Halligan, P. W., Grosse-Ruyken, M., Ziemons, K., Zilles, K., and Freund, H. J. (2000). Line bisection judgments implicate right parietal cortex and cerebellum as assessed by fMRI. *Neurology* **54**, 1324–1331.

As mentioned previously, spatial positions can be referenced to the subject's body or to objects in the environment. Furthermore, spatial positions can be coded in eye-, head-, or body-centered coordinates. Thus egocentrically referenced and allocentrically referenced sensory information needs to be represented and integrated. In both macaques and humans, posterior parietal cortex contributes to the computation of these different spatial reference frames. For example, fMRI studies demonstrate increased neural activity in a predominantly right hemispherical network of frontal and parietal areas when subjects perform tasks that involve the computation of the subjective midsagittal plane or judgments of the location of stimuli with respect to either their body or another object (for a review, see reference 14).

The loss of conscious awareness of left space in unilateral spatial neglect implies that inferior parietal cortex, the temporoparietal junction, and their connections contribute to spatial experience. Consistent with this, interference from distractors that appear close to a target in either time (backward masking) or space (flanker masking) engages the cortex around IPS. Such interference may contribute to extinction (loss of conscious awareness of a stimulus previously reported when a more ipsilesional stimulus is presented simultaneously) and neglect when damage to inferior parietal cortex reduces the processing capacity of the attentional system. Likewise, in normal subjects, attentional modulation of perceptual processes in early visual or auditory processing areas resulting from top-down influences of a frontoparietal network has been demonstrated. The same network is implicated when attention is directed to, for example, global or local aspects of hierarchically organized visual stimuli or to the color, shape, or velocity of an object.

D. Memory and Amnestic Deficits

Memory and its associated neural mechanisms have long been of interest to neuroscientists, as well as clinicians, because memory processes underlie many fundamental cognitive operations: memory is the ability of a living organism to retain and use acquired information. This can occur both implicitly (e.g., in the motor domain) and explicitly (e.g., in the episodic autobiographical domain). Memory processes also span encoding (i.e., the acquisition of information), consolidation (i.e., formation of a stable memory trace), and retrieval thereof. Despite all efforts, to date the precise neural processes underlying these memory processes remain to be further elucidated. In particular, it remains to be established where and how memories are consolidated and retained. Most researchers would agree that there is some sort of an engram (i.e., the stored representation of the information in the brain) but that the engram is not located in a single brain region; rather, multiple features distributed across different brain regions are bound together by the hippocampus and related structures in the medial temporal lobe. Here, functional imaging has had an enormous impact: fMRI studies of working memory, semantic memory, episodic memory, and priming have significantly enhanced our understanding not only of the different brain structures involved in the various types of memories but also of the different cognitive processes involved and their relationship to issues of executive function and control [15].

That memory is not a unitary process is also reflected in the many ways memories can go awry. For example, memories can fail in the explicit domain while implicit learning is maintained. Likewise, memories can become distorted. Each of these types of failures is associated with different pathologies, all of which are amenable to functional imaging of the neural processes underlying either success or failure of the cognitive process of interest. fMRI can even be used to predict the success (or failure) of encoding and retrieving the contents of memories with which a subject is processing.

Recently, the need to investigate the cognitive and neural mechanisms of aging-associated changes has led to the emergence of a new discipline in neuroscience: the cognitive neuroscience of aging. The issues addressed by this new discipline are intimately linked to neurodegenerative disorders such as organic dementia. This is an endemic syndrome that affects up to 20% of the population depending on age and diagnostic criteria. Although easy to diagnose in its full-blown clinical stage, first-time memory complaints are often nonspecific. Furthermore, although Alzheimer's disease (AD) is the most common form of dementia, no specific objective marker indicating AD is currently known. Yet the advances in the field of neuroprotective agents and symptomatic treatment for AD call for a diagnosis at the earliest possible time. Here, fMRI offers a huge potential to classify patients with memory complaints even at an early stage: fMRI can be used to assess patients with subjective memory complaints and to differentiate patients with AD from patients with mild cognitive impairments and controls. Furthermore, fMRI allows differentiation of patients with AD from patients with major depressive disorder and depressive pseudodementia, thus offering an objective neural marker (Fig. 3) for diagnostic classification. This also facilitates drug trials with evaluation of specific treatments. At present, the differential diagnosis for the individual patient still relies on converging evidence from more than one source of information. However, the availability of a method that allows the functional assessment of the integrity of the hippocampus and the medial temporal lobe structures at early stages of the disease, where hippocampal damage may still be incipient, is a major step forward toward the use of fMRI for clinical applications rather than only as a research tool.

III. Clinical Application in Neuronavigation

Despite the groundbreaking applications of fMRI in basic and clinical neuroscience, from a public health perspective it is of foremost importance to demonstrate that fMRI can be applied clinically. One domain where functional imaging has already reached clinical application in a routine way is neuronavigation. The tools available to guide neurosurgeons when planning their surgical approach have traditionally been limited—an issue of foremost importance when trying to spare eloquent cortex. The method of electrocorticography, made famous by Otfrid Foerster and Wilder Penfield, is still used today, for example, in epilepsy surgery. The technique, however, has severe limitations (e.g., most of the cortex is buried in sulci) and obvious risks for the patient (e.g., the complexity of cognitive functions contrasts with the limited number of cognitive operations that can be studied during open skull neurosurgery; semipermanent surface electrodes have surprisingly little or even ill-determined localization to offer). The continued use of Wada's test, a method in which amobarbital (Amytal) is injected into one of the two carotid arteries and inferences are made from the temporary loss of brain functions subsequent to the injection, speaks to how limited the information is that is gained from electrocorticography (or related surface electroencephalogram recordings).

Here, fMRI offers a significant advance in localizing specific brain functions. For example, fMRI is more accurate than Wada's test in assessing language dominance. Numerous studies have compared fMRI with Wada's test (for both language and memory); they have shown a correlation of about 90% but, more importantly, some of these studies have been able to demonstrate a significantly better correlation between presurgical fMRI and postsurgical outcome. Likewise, determinations of areas of hand functions or other motor activity are a routine component of

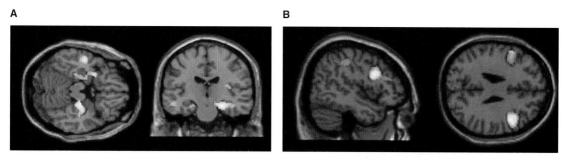

Figure 3 Objective Neural Marker for AD. (A) Compared with healthy age matched controls, mildly affected patients with Alzheimer's disease (AD) (mini mental state examination score of approximately 26) show a significant reduction of the hippocampal activation associated with an episodic memory task (abstract pattern recognition). The figure shows the results of the group comparison of neural activity associated with pattern recall (i.e., control subjects show higher activation in the hippocampus than AD patients) superimposed on T1-weighted images in standardized space. (B) In contrast, brain regions not yet or less affected by the neurodegenerative process, like prefrontal cortex, may show increased neural activation associated with pattern recall (i.e., AD patients show higher activation of prefrontal cortex than age-matched controls). Adapted, with permission, from Fig. 2 of Grön, G., Bittner, D., Schmitz, B., Wunderlich, A. P., and Riepe, M. W. (2002). Subjective memory complaints: Objective neural markers in patients with Alzheimer's disease and major depressive disorder. *Ann. Neurol.* **51**, 491–498.

presurgical evaluation using fMRI. Therefore, current data support fMRI as a valid alternative to the invasive intracarotid amobarbital procedure [16]. The obvious advantage of using fMRI rather than using Wada's test presurgically lies in the huge potential of fMRI to assess subtle cognitive functions, thereby redefining the concept of "eloquent" cortex.

A. Cerebral Localization of Function versus Integration of Function

It is important to stress that the preceding examples of attempts to localize the neural mechanisms underlying a particular cognitive operation are limited because general conceptual problems arise when reducing the approach to higher cognition to pure "localizationalism." Although the false part of Gall's doctrine—reading competence and character from bumps of the skull—rapidly became a joke, phrenology in the sense of deploying the anatomoclinical method rapidly became established, making it fashionable to claim that mental faculties have a distinct neural substrate in different parts of the brain. This framework led, and unfortunately still leads, some behavioral neurologists to show how discrete brain lesions may provoke discrete cognitive deficits.

However, solid evidence nowadays suggests that skepticism about the entire paradigm is advisable: functional specialization of brain regions is not such a fixed property as previously supposed. For example, Broca's area (Brodmann's areas 44 and 45) is involved in natural language syntactic processing and coordinated articulatory movements, musical syntax processing, and rhythm perception, processes that all can be subsumed as related to language functions. However, activation of the same area has also been ascribed to imagining movement trajectories or local visual search, functions difficult to assign to the same single cognitive component. Thus a more reasonable conjecture is that Brodmann's areas 44 and 45 receive functional significance by their particular interaction within different neural networks and that exactly how the functional specialization is determined depends on its anatomic and functional connections. Accordingly, functional specialization is only meaningful in the context of functional integration, and vice versa [17]. This functional integration can best be assessed by fMRI and analysis of effective connectivity or electrophysiological methods that allow the assessment of synchronization of neural activity across distributed neuronal populations.

IV. Monitoring Recovery of Function

fMRI can also be used to monitor recovery of function, for example, following lesion-induced neurological and neuropsychological deficits. Using fMRI, spontaneous recovery, training-induced plasticity, and the effects of pharmacological treatment have been investigated. For example, the influence of a single dose of fluoxetine (a selective serotonin reuptake inhibitor) on the motor activation of lacunar stroke patients was investigated behaviorally and using fMRI in the early phase of recovery with a prospective, double-blind, crossover, placebo-controlled study on eight patients with pure motor hemiparesis [18]. Each patient underwent two fMRI examinations: one with fluoxetine and one with placebo. The first was performed 2 weeks after stroke onset and the second a week later. During the two fMRI examinations, patients performed an active motor task with the affected hand (and a passive one conducted by the examiner with the same hand). Motor performances were evaluated under placebo and under fluoxetine immediately before the fMRI studies to investigate the behavioral effect of fluoxetine. Under fluoxetine, during the active motor task, increased activation in the ipsilesional primary motor cortex was found. Moreover, fluoxetine significantly improved motor skills of the affected side. The data exemplify the potential of pharmacological fMRI to show the effect of drugs that modulate task-associated neural activations in patients with neurological or neuropsychological deficits.

Instead of a drug, behavioral training can be used to induce recovery of function and associated cortical plasticity. As mentioned previously, deficits of attention are observed subsequent to stroke in approximately 50% of all patients. These deficits span both deficits of spatial cognition (neglect, see Sect. II.C) and alertness (attention intensity). Alertness is known to draw on a right hemispherical frontoparietal-thalamic-brainstem network, which overlaps in the parietal cortex with the neural network subserving spatial cognition. Based on the hypothesis that behavioral alertness training may enhance the right hemisphere network for spatial attention, the effects of a 3-week computerized alertness training program on stroke patients with chronic (>3 months) visuospatial hemineglect were investigated prospectively by neuropsychological testing and fMRI. Following the alertness training, the group showed improved alertness and significant improvement in the performance of a neglect test battery beyond any improvement during a 3-week baseline phase. Improvements in the neglect tasks were associated with increased neural activity in right frontal cortex, right anterior cingulate cortex, right precuneus, right cuneus, and right angular gyrus, areas previously implicated in spatial attention and alertness. The data show that a behavioral training can improve performance both in the task trained and in another related function and that these behavioral effects are associated with reactivation in areas associated with these functions (but spared by the lesion itself) [19].

The latter, however, imposes a methodological problem: increased task performance can be associated with differences in neural activations [20]. The critical observation from such studies is a differential pattern of activation in patients relative to controls or within the same patients scanned on subsequent occasions (prerecovery, postrecovery). Regional neural underactivity is interpretable only when the patients make normal behavioral responses. In this context, a failure to activate a component region of the network normally involved in the task then implies that this region was not necessary for task performance. In contrast, overactivity indicates either cognitive or neuronal reorganization. Neural reorganization is indicated only if the patient performs the task using the same set of cognitive operations as normal subjects. Cognitive reorganization can be demonstrated if the same activation pattern is elicited by normal subjects when they are coerced into using the same cognitive implementation as the patient (i.e., when a task can be performed with different strategies). Accordingly, the interpretation of neuroimaging studies of neurologically or psychologically impaired patients depends on both intact task performance and detailed task analysis. When these criteria are met, patient studies can be used to identify (1) necessary and sufficient brain systems, (2) dysfunction at sites distant to damage, (3) perilesional activation, and (4) compensation either at a neural level when preexisting cognitive strategies are reinstantiated, using duplicated neuronal systems (degeneracy), or at a cognitive level when alternative cognitive strategies (and their corresponding brain systems) are adopted [20].

Considering these methodological constraints, fMRI allows new perspectives for a hypothesis-driven, pathophysiology-based development of novel treatment strategies (behavioral, pharmacological, minimally invasive techniques).

V. Phenotyping and Genotyping

Finally, much can be gained from combining phenotypic (using fMRI as one variable of interest) and genotypic information. Mattay and colleagues showed in normal, healthy subjects that individual variations in the brain response to amphetamines may vary depending on the catechol-O-methyltransferase (COMT) val 158-met genotype [21]. It has long been known that the clinical effects of amphetamines are variable and hence hard to predict: although some subjects show positive effects on mood and cognition, other individuals may show negative responses. Mattay and colleagues tested the hypothesis that such differential individual effects of amphetamines may result from individual differences in cortical efficiency related to variations in monoaminergic function and monoamine system genes [21]. To test this hypothesis, the effect of a functional polymorphism (val 158-met) in the COMT gene, which has been shown to modulate prefrontal dopamine in animals and prefrontal cortical function in humans, on the modulatory actions of amphetamine on human prefrontal cortex was explored. Val/val genotype subjects, val/met genotype subjects, and met/met genotype subjects were investigated using fMRI, an N-back working memory task, and both placebo and amphetamine were administered before fMRI. Amphetamine enhanced the efficiency of prefrontal cortex function as measured with fMRI during the working memory task in homozygous subjects with the val/val genotype and high COMT activity (who presumably have relatively less prefrontal synaptic dopamine) at all levels of task difficulty. In contrast, homozygous subjects with the met/met genotype and low COMT activity showed no drug effect on cortical efficiency at a low to moderate working memory load but reduced cortical efficiency at a high working memory load. These data are consistent with the notion of an inverted-U function of dopaminergic response in prefrontal cortex. The results also imply that individuals with the met/met COMT genotype are at risk for an adverse response to amphetamine. The study illustrates the potential of applying functional neuroimaging in pharmacogenomics and highlights the need to take into account both behavioral measures and neurophysiological measures of brain function.

VI. Conclusion

The preceding examples show how fMRI can provide deeper insights into the functional or anatomic brain networks that mediate cognitive functions and how pharmacological approaches can be used to understand better the neurochemistry underlying cognition and deficits thereof. They also highlight the potential for the development of novel pharmacological approaches for recovery of function following a stroke. The combination of genotypic and phenotypic information promises to be particularly helpful in understanding the relationship between behavior and neurophysiological measures of brain function and is likely to contribute significantly to neuroscience in general and medicine in particular: only by understanding why some subjects respond to a stimulus (e.g., a behavioral training) or a drug and others do not will effective cognitive and pharmacological treatment become available.

References

1. Ogawa, S., Lee, T. M., Nayak, A. S., (1990). Oxygenation-sensitive contrast in magnetic resonance imaging of rodent brain at high magnetic fields. *Magn. Res. Med.* **14**, 68–78.
2. Logothetis, N. K., Pauls, J., Augath, M., Trinath, T., and Oeltermann, A. (2001). Neurophysiological investigation of the basis of the fMRI signal. *Nature* **412**, 150–157.

3. Talairach, J., and Tournoux, P. (1988). "Co-Planar Stereotactic Atlas of the Human Brain." Thieme Verlag, Stuttgart.
4. Eickhoff, S. B., Stephan, K. E., Mohlberg, H., Grefkes, C., Fink, G. R., Amunts, K., and Zilles, K. (2005). A new SPM toolbox for combining probabilistic cytoarchitectonic maps and functional imaging data. *Neuroimage* **25**, 1325–1335.
5. Price, C. J., and Friston, K. J. (1997). Cognitive conjunction: a new approach to brain activation experiments. *Neuroimage* **5**, 261–270.
6. Toni, I., Thoennissen, D., and Zilles, K. (2001). Movement preparation and motor intention. *Neuroimage* **14**, S110–S117.
7. Bookheimer, S. (2002). Functional MRI of language: new approaches to understanding the cortical organization of semantic processing. *Annu. Rev. Neurosci.* **25**, 151–188.
8. Alexander, M. P., Naeser, M. A., and Palumbo, C. (1990). Broca's area aphasias: aphasia after lesions including the frontal operculum. *Neurology* **40**, 353–362.
9. Price, C. J., and Crinion, J. (2005). The latest on functional imaging studies of aphasic stroke. *Curr. Opin. Neurol.* **18**, 429–434.
10. Grefkes, C., and Fink, G. R. (2005). The functional organization of the intraparietal sulcus in humans and monkeys. *J. Anat.* **207**, 3–17.
11. Binkofski, F., Dohle, C., Posse, S., Stephan, K. M., Hefter, H., Seitz, R. J., Freund, and H. J. (1998). Human anterior intraparietal area subserves prehension. A combined lesion and fMRI study. *Neurology* **50**, 811–815.
12. Rumiati, R. I., Weiss, P. H., Shallice, T., Ottoboni, G., Noth, J., Zilles, K., and Fink, G. R. (2004). Neural basis of pantomiming the use of visually presented objects. *Neuroimage* **21**, 1224–1231.
13. Corbetta, M., and Shulman, G. L. (2002). Control of goal-directed and stimulus-driven attention in the brain. *Nat. Rev. Neurosci.* **3**, 201–215.
14. Halligan, P. W., Fink, G. R., Marshall, J. C., and Vallar, G. (2003). Spatial cognition: evidence from visual neglect. *Trends Cogn. Sci.* **7**, 125–133.
15. Wagner, A. D., Bunge, S. A., and Badre, D. (2004). Cognitive control, semantic memory, and priming: contributions from prefrontal cortex. *In:* "The Cognitive Neurosciences III" (M. S. Gazzaniga, ed.), pp. 709–726. MIT Press, Cambridge, MA.
16. Kloppel, S., and Buchel, C. (2005). Alternatives to the Wada test: a critical view of functional magnetic resonance imaging in preoperative use. *Curr. Opin. Neurol.* **18**, 418–423.
17. Friston, K. J. (2002). Beyond phrenology: what can neuroimaging tell us about distributed circuitry? *Annu. Rev. Neurosci.* **25**, 221–250.
18. Pariente, J., Loubinoux, I., Carel, C., Albucher, J. F., Leger, A., Manelfe, C., Rascol, O., and Chollet, F. (2001). Fluoxetine modulates motor performance and cerebral activation of patients recovering from stroke. *Ann. Neurol.* **50**, 718–729.
19. Thimm, M., Fink, G. R., Kust, J., Karbe, H., and Sturm, W. (2006). Impact of alertness training on spatial neglect: a behavioural and fMRI study. *Neuropsychologia* **44(7)**, 1230–1246.
20. Price, C. J., and Friston, K. J. (1999). Scanning patients with tasks they can perform. *Hum. Brain Mapp.* **8**, 102–108.
21. Mattay, V. S., Goldberg, T. E., Fera, F., Hariri, A. R., Tessitore, A., Egan, M. F., Kolachana, B., Callicott, J. H., and Weinberger, D. R. (2003). Catechol-*O*-methyltransferase val158-met genotype and individual variation in the brain response to amphetamine. *Proc. Natl. Acad. Sci. USA* **100**, 6186–6191.

78

Impaired Glucose Regulation and Neuropathy

James W. Russell, MD, MS, FRCP
A. G. Smith, MD
J. R. Singleton, MD

Keywords: *apoptosis, diabetes, impaired glucose tolerance, oxidative stress, peripheral neuropathy, protein kinases*

I. Introduction
II. Spectrum of Impaired Glucose Regulation
III. Pathophysiology of Neuropathy Related to Impaired Glucose Regulation
IV. Symptoms in Relation to Pathophysiology
V. Natural History
VI. Conclusion
    References

I. Introduction

Type 2 diabetes mellitus (T2DM) affects nearly 20 million people in the United States, and prediabetes or impaired glucose tolerance (IGT) affects a considerably larger, but unknown, population group. The development of peripheral neuropathy is the most common complication of T2DM and occurs in up to 50% of subjects based on current epidemiological data [1], although such studies did not use more sensitive measures of neuropathy such as intraepidermal nerve fiber density and may have underestimated the point prevalence of neuropathy in diabetic subjects. Several observational studies indicate that IGT and impaired fasting glucose (IFG) are also associated with neuropathy (for review, see reference 2). This chapter addresses the spectrum of impaired glucose regulation (IGR), the pathophysiology of IGR and neuropathy, and the clinical presentation's relationship to the physiology of the neuropathy.

II. Spectrum of Impaired Glucose Regulation

Definitions for each of these disorders are based on fasting venous glucose, or venous glucose values following an oral glucose tolerance test (OGTT) based on the criteria of the Expert Committee on the Diagnosis and Classification of Diabetes Mellitus [3]. Investigations of the natural history of IGT reveal a dynamic and reversible state. The Diabetes Prevention Program study randomized 3244 patients with IGT to treatment with placebo, metformin, or intensive diet and exercise counseling. Nearly 30% of 1082 subjects receiving placebo progressed from IGT to T2DM in 3 years, but during this same period 25% reverted to postprandial normoglycemia [4]. Similar results were obtained in patients from a large study in China, the Da Qing study [5], and in elderly Finnish diabetic patients. In the Finnish study, at the end of a 3-year follow-up period of nearly 100 patients more than one-third of the people with baseline normal glucose tolerance had progressed to IGT or T2DM [6]. Almost one-third of those with baseline IGT had reverted to normal glucose tolerance, half had persisting IGT, and one-fifth had progressed to T2DM. Almost one-fifth of the subjects with baseline T2DM had reverted to IGT, and only 1 had reverted to normal glucose tolerance. Overall, most patients progress toward greater glycemic dysregulation, but this progress appears to be slow. Unmonitored patients probably experience many years of occult insulin resistance and postprandial hyperglycemia before developing typical symptoms of diabetes (Fig. 1). Thus, although blood glucose values are used to define IFG, IGT, and T2DM, these definitions are artificial because they fail to recognize that IGR is a dynamic surrogate marker for an underlying metabolic disturbance associated with the metabolic syndrome and that the glucose level fluctuates depending on changes in insulin resistance.

Although both IFG and OGTT correlate with insulin resistance, corresponding to a level of hyperglycemia greater than which acute phase insulin secretion is lost in response to a glucose load, several studies have shown that IFG underestimates the prevalence of IGR in population studies. Furthermore, use of the American Diabetes Association–defined IFG, rather than IGT with an OGTT, underestimated the hyperglycemic risk for cardiovascular disease. Thus, the fasting venous plasma glucose is probably too insensitive to accurately determine IGT, and it is important to recognize that the OGTT is more sensitive in diagnosing IGT and diabetes than the fasting glucose or hemoglobin A1c.

III. Pathophysiology of Neuropathy Related to Impaired Glucose Regulation

A. Glycemic Control

The diabetes control and complications trial established clear links among impaired glycemic control, neuropathy, and retinopathy. This study prospectively followed 1441 insulin-dependent participants with type 1 diabetes mellitus (T1DM) for a mean of 6.5 years to assess the effect of intensive insulin therapy on the development of diabetic complications (reviewed in reference 7). Patients were divided into a primary-prevention group and a secondary-intervention group and treated with intensive or conventional insulin therapy. In the secondary-intervention cohort, intensive insulin therapy reduced the appearance of clinical neuropathy by 60% over a 5-year follow-up. The results for patients who had neither retinopathy nor significant albuminuria at the start of the study (primary-prevention cohort) were even more impressive. In this group, intensive therapy reduced the appearance of neuropathy by 69%, compared with only 10% with conventional therapy, indicating that early optimal glycemic control can prevent the development of neuropathy before the development of retinopathy and microvascular injury. Furthermore, it is clear from this data that any increase in glucose above normal is associated with an increased risk of end-organ injury, including neuropathy. The results of the diabetes control and complications trial and similar studies suggests that early impaired glycemic control is associated with peripheral neuropathy and may be the primary pathology at presentation before the development of other end-organ injury such as retinopathy or nephropathy [7]. Although diabetics may have neuropathy at presentation, intervention in subjects with IGT may prevent some, if not all, cases of neuropathy.

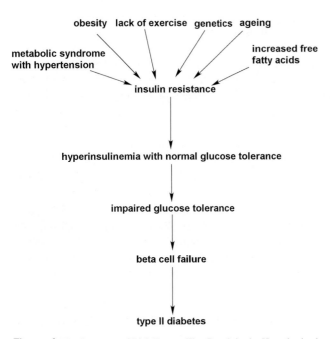

Figure 1 The Spectrum of Risk Factors That Result in the Hypothesized Progression from Metabolic Syndrome to Impaired Glucose Tolerance to Diabetes.

B. Diet and Exercise

In one of the largest studies of subjects with IGT, 577 subjects were randomized either to a control group or to one of three active treatment groups: diet only, exercise only, or diet plus exercise [5]. Follow-up evaluation examinations were conducted at 2-year intervals over a 6-year period to identify subjects who developed T2DM. The cumulative incidence of T2DM at 6 years was 67.7% (95% confidence interval, or CI, 59.8–75.2%) in the control group compared with 43.8% (95% CI, 35.5–52.3%) in the diet group, 41.1% (95% CI, 33.4–49.4%) in the exercise group, and 46.0% (95% CI, 37.3–54.7%) in the diet-plus-exercise group. The relative decrease in progression to diabetes was similar in the active treatment groups after stratifying subjects (according to body mass index, or BMI) as lean (BMI $\leq 25\,kg/m^2$), or overweight (BMI $> 25\,kg/m^2$). The diet, exercise, and diet-plus-exercise interventions were associated with 31% ($p < 0.03$), 46% ($p < 0.0005$), and 42% ($p < 0.005$) reductions in risk of developing diabetes, respectively [5]. Similarly, the Diabetes Prevention Program study randomly assigned 3234 nondiabetic people with elevated fasting and postload plasma glucose concentrations to one of three groups: placebo, metformin (850 mg twice daily), or a lifestyle-modification program with the goals of at least a 7% weight loss and at least 150 minutes of physical activity per week [4]. The average follow-up period was 2.8 years. The incidence of diabetes was 11.0 cases per 100 person-years in the placebo group, 7.8 in the metformin group, and 4.8 in the lifestyle group. The lifestyle intervention reduced the incidence of diabetes by 58% (95% CI, 48–66%), and metformin reduced it by 31% (95% CI, 17–43%), as compared with placebo. The conclusion of this study was that the lifestyle intervention was significantly more effective than metformin. Furthermore, to prevent one case of diabetes during 3 years, 6.9 people would have to participate in the lifestyle-intervention program and 13.9 would have to receive metformin. Older patients showed a greater relative response to the diet-and-exercise program, whereas the difference between diet-and-exercise benefit and metformin was less in younger patients. In the Finnish Diabetes Prevention Study [8], 522 middle-aged, overweight subjects with IGT were randomized to either usual care or an intensive lifestyle-intervention group. The subjects in the intervention group received individualized dietary counseling from a nutritionist, underwent circuit-type resistance training sessions, and were advised to increase overall physical activity. The intensive lifestyle intervention produced sustained improvement in diet, physical activity, and metabolic factors. Diet and exercise interventions are likely to affect several pathways that control insulin resistance, glucose handling, and mitochondrial function in muscle as described in the following sections [7].

C. Changes in Redox Potential

One potential mechanism of injury to the peripheral nervous system (PNS) is by oxidative stress [9]. In the diabetic rat, levels of oxidative stress and reduced antioxidant defense parallel neuropathy; blocking oxidative stress in the diabetic animal restores normal blood flow and sciatic and saphenous nerve conduction velocities. Increasing evidence indicates that a change in the inner mitochondrial membrane potential is associated with induction of reactive oxygen species (ROS) [9]. Increased metabolic mitochondrial flux due to high glucose increases formation of ROS, including peroxynitrite, superoxide (O_2^-), and hydroxyl radicals (Fig. 2). Generation of ROS is associated with mitochondrial permeability transition, opening of the adenine nucleotide translocase- or voltage-dependent anion channel, and mitochondrial swelling that disrupts the integrity of the outer membrane [10]. In addition, there are membrane lipid peroxidation and degradation of deoxyribonucleic acid (DNA), all of which are associated with neuronal or axonal injury [9]. Consistent with the concept that increased oxidative injury is associated with loss of mitochondrial integrity, swelling of mitochondria has been confirmed in dorsal root ganglion (DRG) neurons that contain ballooned mitochondria with disrupted cristae [9]. Regulation of increased electron flux through the mitochondrial electron transfer chain associated with increased glycemic load, or stabilization of the inner mitochondrial membrane potential, prevents mitochondrial disruption and neuroaxonal injury.

D. Role of Nitric Oxide and Nitrosative Injury in Diabetic Neuropathy

Nitric oxide (NO) is ubiquitous in the nervous system and plays an important role in the regulation of many normal biological functions, including neural signaling in the myenteric plexus of the intestine and in peripheral resistance vessels (reviewed in reference 11). However, under conditions of oxidative stress NO may act as a stimulus for O_2^- generation [11]. Evidence of increased production of reactive nitrogen species, NO, and peroxynitrite, coupled with evidence of programmed cell death (PCD), implicates nitrosative injury in models of diabetic neuropathy [11]. Therefore, diabetic DRG neuronal injury may involve the formation of peroxynitrite in the presence of increased concentrations of O_2^- and NO [11].

NO is formed by activation of NO synthase, which catalyzes the oxidation of L-arginine to NO and citrulline [11]. Neuronal NO is the primary constitutively active isoform in neurons. In general, NO-induced toxicity depends on the degree of local generation of NO and/or O_2^-. In the neuronal mitochondrion, peroxynitrite can inactivate electron transfer complexes I and II

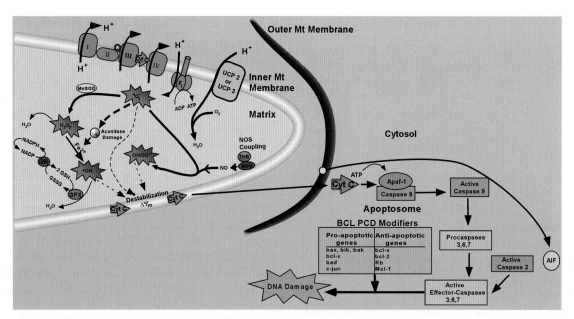

Figure 2 Mechanisms for Oxidative Injury in the Peripheral Nervous System. Under diabetic conditions, incomplete electron transfer through complexes I, II, III, and IV generates superoxides (O_2^-) that facilitate the production of other radical species, for example, hydrogen peroxide (H_2O_2), or hydroxyl radicals (OH^-). Manganese superoxide dismutase (MnSOD) converts O_2^- to H_2O_2, which is then converted to water though either catalase or glutathione peroxidase (GPX). Reduced glutathione (GSH) is regenerated from glutathione disulfide (GSSG) through glutathione reductase (GR) and nicotinamide adenine dinucleotide phosphate (NADPH). Nitric oxide synthase (NOS) with its cofactor tetrahydrobiopterin (THB) generates nitric oxide (NO), which couples with O_2^- to create peroxynitrite ($ONOO^-$). O_2^-, $ONOO^-$, and OH^- destabilize the inner mitochondrial membrane, with resultant release of cytochrome C (Cyt C) from the inner mitochondrial membrane space into the cytosol. Cyt C is released and combines with cell death pathway components to form the apoptosome complex, consisting of caspase-9 and apoptosis protease-activating factor 1 (Apaf-1). The formation of this complex leads to cleavage of caspase-9 and downstream activation of effector caspase-3, caspase-6, and caspase-7. The effector caspases damage structural proteins and lead to deoxyribonucleic acid (DNA) damage and cell death. Apoptosis-inducing factor (AIF) is released from the mitochondrion with induction of apoptosis and translocates to the nucleus, causing DNA fragmentation. Bcl genes that regulate apoptosis (both activators and inhibitors) are listed. Uncoupling of oxidative phosphorylation in the neuronal mitochondrion by uncoupling proteins (UCP 2 or UCP 3) can restore the proton equilibrium under diabetic conditions and prevent the generation of reactive oxygen species and programmed cell death (PCD). *Note:* ADP, adenosine diphosphate; ATP, adenosine triphosphate; Mt, mitochondrial.

and adenosinetriphosphatase, reversibly nitrosylate- or irreversibly nitrate-critical proteins and enzymes including manganese superoxide dismutase, cytochrome C, aconitase, and the voltage-dependent anion channel (reviewed in reference 7). In combination, these effects of NO increase oxidative stress and nitrosative stress through the overproduction of S-nitrosylated proteins.

It is not certain what controls the balance between neuroprotective and neurotoxic effects of NO. Understanding this issue is important in comprehending how inhibition or activation of NO synthase affects the severity of diabetic complications. In general, the toxicity of NO depends on the local concentration of NO produced and/or the presence of O_2^- [11]. For example, a low level of NO produced with activation of endothelial NO synthase results in normal vasodilation in the peripheral vasculature and helps prevent peripheral nerve ischemia in diabetic neuropathy. In the presence of excess O_2^-, the constitutive production of NO in endothelial or neuronal cells may favor the formation of peroxynitrite. NO is relatively unstable *in vivo*. However, NO can bind thiol-containing proteins in the cytosol and mitochondria through a process called S-nitrosylation, thereby sustaining its effect. S-nitrosylation can alter the structure and activity of multiple proteins. For example, NO-mediated S-nitrosylation may modify post-translationally transcription factors such as nuclear factor-κβ (NF-κβ), thus sensitizing susceptible cells to injury, or may inactivate caspases, thus preventing cell death. Nitration results in more permanent inactivation of proteins. For example, nitrated cytochrome C reduces state 4 respiration, and this leads to failure of mitochondrial function [11].

E. Uncoupling Proteins and Oxidative Injury

Integral mitochondrial proteins, or the uncoupling proteins (UCPs) (e.g., UCP2 and UCP3), may regulate ROS generation during oxidative phosphorylation by facilitating proton leak across the mitochondrial membrane, thus limiting a high inner mitochondrial membrane potential and reducing the generation of electron-rich intermediates capable of generating O_2^- radicals. Because UCPs are associated with the regulation of energy metabolism, they may play an important role in diabetes and diabetic complications (reviewed in reference 7). The UCPs are inner mitochondrial membrane proteins that can dissipate the proton-electrochemical gradient as heat; that is, they uncouple mitochondrial electron transfer from oxidative phosphorylation and regulate proton conductance [7]. UCP1 is expressed primarily in brown adipose tissue, whereas UCP2 and in particular UCP3 are expressed in the PNS [10]. UCPs share considerable amino acid sequence homology to one another, suggesting that they all possess an uncoupling function. UCP1 is a classic uncoupling protein; however, UCP2 and UCP3 may have important roles other than their uncoupling activity. In the UCP3 knockout mouse, there is no compensatory up-regulation of other UCPs, yet they display normal thermoregulation—even though UCPs are absent in muscle, a major site of thermogenesis. This implies an alternate role for UCP3, namely, control of proton transport in the mitochondrion, and has important implications for diabetic complications. Because UCPs increase nonphosphorylating (uncoupled) mitochondrial respiration, one potential role for the UCPs is regulation of ROS production by reducing O_2^-. The corollary to this is that O_2^- increases proton conductance through effects on UCP1–3, thereby reducing the inner mitochondrial membrane potential and generation of further ROS [10]. When UCP levels are reduced as they are in diabetic neurons, then the inner mitochondrial membrane potential is abnormally high and there is increased backpressure on the inner mitochondrial membrane proton pumps, resulting in an increased half-life of electrons in the electron transfer chain. This enhances ubiquinone (coenzyme Q) mediated interaction of electrons with free O_2 to produce O_2^- and other ROS. Evidence for the role of UCP2 or UCP3 in stabilizing the inner mitochondrial membrane potential and preventing the generation of ROS is provided by two separate observations. First, overexpression of UCPs in peripheral sensory neurons stabilizes the inner mitochondrial membrane potential and prevents oxidative injury. Second, in the UCP3 knockout mouse there is an increase in production of O_2^- in mitochondria during state 4 respiration.

Further evidence points to the important role of UCPs in diabetes and diabetic complications. The UCP2 gene has been mapped to loci associated with obesity and hyperinsulinemia and has led to investigations into the role of this UCP in weight regulation and energy balance (reviewed in references 7 and 10). It has been shown that UCP2 may be increased in pancreatic β-cells in the prediabetic state and that this relates to impaired glucose-induced insulin secretion. One mechanism for increased UCP2 in prediabetes is the presence of a polymorphism in the UCP2 promoter that leads to increased expression of the gene. Although UCP2 overexpression in β-cells results in hyperglycemia, reduced expression of UCP3 is observed in muscle in T2DM, in dorsal root ganglia from streptozotocin-induced diabetic rats [10], and in Zucker diabetic fatty rats. In contrast, overexpression of UCPs reduces oxidative stress and induction of downstream PCD pathways in DRG neurons. Thus, UCPs in neurons may help prevent neuronal oxidative injury, and therapeutic regimens designed to up-regulate UCPs may enhance this ability to prevent neuronal injury.

F. Apoptosis and Neuropathy

Pathological changes consistent with apoptosis have been described in the PNS in models of diabetic neuropathy. In diabetic dorsal root ganglia the following changes have been observed: condensation of chromatin and shrinkage of the nucleus and cell cytoplasm [12]. These apoptotic changes occur side by side with healthy cells, which is consistent with single-cell deletion, a feature of apoptosis. The changes are evenly distributed throughout the DRG neurons and are observed in dorsal root Schwann cells. Frequent large vacuoles are observed and are most prominent in cells showing early apoptotic changes, such as mild chromatin aggregation with neuronal and cytoplasmic shrinkage. In contrast, these changes are not observed in control animals.

There is controversy as to whether classic apoptosis occurs in the PNS and if loss of DRG neurons by apoptosis is responsible for the neuropathic deficits observed in both animals and humans. Most studies to date indicate activation of caspases in DRG neurons both *in vitro* and *in vivo* (reviewed in reference 13), and there is evidence of neuronal nuclear DNA fragmentation. Most studies indicate some loss of DRG neurons, and in particular some studies indicate a statistically significant loss of large DRG neurons, although the total loss of DRG neurons may not be statistically significant. Even though there is evidence of DRG neuronal loss, the number of DRG neurons showing evidence of caspase-3 cleavage or TUNEL staining appears to be greater than the measured loss of neurons. This may occur because activation of caspases does not invariably result in neuronal death or because there is an intrinsic capacity for repair within the neuron resulting from DNA repair or by activation of neurotrophic protective signaling pathways [13]. PCD is a balance between caspase activation and blocking by inhibitors of apoptosis. One possible repair mechanism would be by elevated expression of the DNA repair enzyme poly(ADP-ribose) polymerase (PARP). However, activation of PARP-1 may itself lead to cell death. When the DNA damage is moderate, PARP-1 participates in the DNA repair process, but in the case of massive DNA injury, elevated PARP-1 activation leads to rapid and massive nicotinamide adenine dinucleotide($^+$) or adenosine triphosphate (ATP) consumption and cell death by necrosis [13]. Corresponding to this evidence of apoptosis are changes in the intrinsic pathway of PCD. In affected

DRG neurons, there is a decrease in Bcl-2 levels and translocation of cytochrome C from the mitochondrion to the cytoplasm. In hyperglycemic conditions there is early serine phosphorylation of Bcl-2 followed by a reduction in Bcl-2 expression and loss of the inner mitochondrial membrane potential [13].

Less is known about regulation of apoptosis in the autonomic nervous system. In rat superior cervical ganglion (SCG) cultures, there is evidence of glucose-induced apoptosis in SCG neurons, although they are considerably less sensitive to glucose toxicity than DRG neurons treated with the same high glucose conditions [13]. Apoptosis in SCG neurons is also coupled with degeneration of SCG neurites consistent with degeneration of autonomic fibers. In agreement with these findings, acute streptozotocin diabetes is associated with evidence of PCD in a small number of autonomic neurons and activation of the apoptotic cascade occurs relatively early in diabetic autonomic neuropathy. However, significant neuron loss in the autonomic nervous system in chronic diabetes is lacking, indicating that apoptotic neuronal cell death alone is unlikely to account for the severity of autonomic neuropathy observed in T1DM.

Loss of autonomic neurons is more severe in animal models of T1DM compared with T2DM and, when coupled with the ability of specific growth factors to prevent neuronal PCD, suggests a protective role for insulin and insulin-like growth factor-I (IGF-I). The failure of these protective systems may account for the severity of autonomic dysfunction in T1DM [7]. Models of T1DM develop marked hyperglycemia, a deficiency in both circulating insulin and IGF-I, and neuroaxonal dystrophy in nerve terminals in the prevertebral sympathetic ganglia and the distal portions of noradrenergic ileal mesenteric nerves. In contrast, animal models of T2DM, despite developing severe hyperglycemia comparable to that in the streptozotocin- and biobreeding/Worcester-diabetic rat models, do not develop neuroaxonal dystrophy. This may be because animal models with T2DM have significant hyperinsulinemia and normal levels of plasma IGF-I that protect autonomic neurons.

Apoptosis also occurs in Schwann cells in models of diabetic neuropathy and in human diabetic neuropathy [7]. Schwann cells obtained from the dorsal root of diabetic animals and in human nerves exhibit chromatin clumping and disruption of the myelin surrounding atrophic axons. Schwann-like satellite cells from corresponding diabetic dorsal root ganglia show severe chromatin clumping and perikaryal vacuolation consistent with mitochondrial ballooning and disruption of the internal mitochondrial cristae structure. Further evidence of apoptosis signaling in Schwann cells is supported by the observation that the antiapoptotic Bcl family protein, Bcl-xL, is expressed in normal Schwann cells but is not significantly increased or decreased under high glucose conditions. However, overexpression of Bcl-xL protects Schwann cells from apoptosis *in vitro*.

G. Signaling Pathways Associated with Glucose-Induced Apoptosis

Apoptosis is regulated by a variety of important cellular signaling events. The mitogen-activated protein kinases (MAPKs) are a group of serine or threonine kinases that partly help regulate both cell survival and apoptosis. MAPKs are in turn activated by dual-specificity MAPKs that include c-Jun-NH2-terminal kinases (JNKs) or stress-activated protein kinase (SAPK), p38 MAPK, and extracellular signal-related kinases (ERKs), for example, MEK1 and MEK2. Diabetes activates all three groups of MAPKs in sensory ganglia of the PNS. Under high glucose conditions, elevated glucose can lead to sustained phosphorylation of p38 and to MEK1 cleavage, an event blocked by caspase inhibitors. Inhibition of ERKs and p38, stress-responsive members of the MAPK family, prevents nerve damage. Antioxidants and aldose reductase inhibitors that improve neuronal function in diabetic rats are also able to inhibit activation of ERKs and p38 in dorsal root ganglia and increase activation of JNKs. The role of JNKs in diabetes is complex. In streptozocin-induced diabetic rat sensory neurons, high glucose-induced JNK activation prevents cell death, but in contrast activation of p38 may induce neuropathy. It is likely that activation of JNK because of a combination of raised glucose and oxidative stress protects DRG neurons from glycemic damage.

p53 may also be required for glucose-induced apoptosis in that phosphorylated p53 activates JNKs while O-glycosylation of p53 leads to Bax activation. Furthermore, glucose transport and high glucose induction of Bax and apoptosis is impaired in p53-null mice compared with wild-type animals. This suggests that p53 is required for apoptosis. Further support for the role of p53 in diabetic complications is provided by evidence that apoptotic DNA damage is associated with p53-dependent activation of Bax-dependent cytochrome C release and cleavage of caspase-9, caspase-2, and caspase-3 [13].

Another important pathway, phosphatidylinositol 3-kinase (PI3K), is implicated in regulation of apoptosis (reviewed in reference 13). In high glucose–treated neurons, the PI3K pathway is the primary pathway regulating Bcl-2 and Bcl-xL, changes in the inner mitochondrial membrane potential, and apoptosis [7]. Stimulation of the PI3K/Akt pathway phosphorylates the following Akt effectors: the survival transcription factor cyclic–adenosine monophosphate response element-binding (CREB) protein and the proapoptotic effector proteins glycogen synthase kinase-3β and forkhead. Increased accumulation of phospho-Akt in DRG neuronal nuclei, increased CREB-mediated transcription, and nuclear exclusion of forkhead are associated with DRG neuronal survival. Similar changes have been observed with insulin treatment. Insulin stabilizes the inner mitochondrial membrane potential by activation of the PI3K signaling pathway and phosphorylation of Akt and CREB.

Activation of protein kinase C (PKC) and corresponding increases in diacylglycerol (DAG) levels are associated with hyperglycemia. Increased glycolytic pathway flux or aldose reductase pathway activity promotes *de novo* DAG synthesis by glycerol-3-phosphate following increased levels of intracellular glyceraldehyde-3-phosphate. Chronically elevated DAG, in turn, increases PKC activity. Activation of PKC isoforms (α, β, δ, ε, ξ) are reported in some tissues prone to diabetic complications. Hyperglycemia-induced oxidative stress may also mediate the adverse effects of PKC-β isoforms by activating the DAG-PKC pathway, because treatment with antioxidants may prevent glucose-induced cellular injury and inhibit DAG-PKC activation. Activation of PKC promotes vasoconstriction and ischemia, increased permeability, NO dysregulation, and increased leukocyte adhesion, further inducing diabetic neuropathy. There is further evidence of an important pathogenetic interaction between the PKC and the aldose reductase pathway. Using transgenic mice that overexpress human aldose reductase, PKC activity in dorsal root ganglia was markedly reduced in diabetic transgenic mice and the changes were associated with reduced expression of a membrane PKC-α isoform translocated to cytosol [14]. The high-affinity PKC-β inhibitor ruboxistaurin mesylate has been evaluated for treatment of human diabetic neuropathy in a randomized, Phase II, double-blind, placebo-controlled parallel-group trial comparing 32 mg/day or 64 mg/day of ruboxistaurin with placebo for 1 year. The primary end point was the change in the vibration detection threshold, and the secondary end point measures included effects of ruboxistaurin versus placebo on the Neuropathy Total Symptom Score-6 and other clinical and electrodiagnostic measures of neuropathy. In this study, the primary and secondary efficacy measures did not differ among treatment groups. However, a subgroup of patients with less severe diabetic neuropathy showed a trend toward improvement in the primary and secondary efficacy measures [15].

H. Metabotropic Glutamate Receptors and Oxidative Injury

Metabotropic glutamate receptors (mGluRs) may also regulate CREB signaling intermediates and prevent neuronal cellular injury [16]. The mGluRs are a subfamily of glutamate receptors that are G protein coupled and linked to second messenger systems [16]. The glutamate carboxypeptidase II (GCPII) inhibitor 2-(phosphonomethyl)pentanedioic acid (2-PMPA) is protective against glucose-induced PCD and neurite degeneration in DRG neurons in a cell culture model of diabetic neuropathy [16], likely by activating mGluR3. Preclinical data indicates that GCPII inhibitors ameliorate diabetic neuropathy in animal models [7,9]. Furthermore, the direct mGluR3 agonist aminopyrrolidine dicarboxylate prevents induction of ROS. Together these findings are consistent with an emerging concept that mGluRs may protect against cellular injury by regulating oxidative stress in models of diabetic neuropathy.

I. Advanced Glycation End-Products

Increases in tissue glucose levels induce the generation of advanced glycation end-products (AGEs) in peripheral nervous tissue. Glucose in the tissue is metabolized to 3-deoxyglucosone, methylglyoxal, and N^{ε}-(carboxymethyl)lysine (CML). Furthermore, activation of the polyol pathway causes overproduction of fructose through sorbitol. Fructose is further converted to fructose-3-phosphate and 3-deoxyglucosone. Binding of these AGE ligands to the receptor for advanced glycation end-products (RAGEs) has been associated with several diabetic complications. Is the accumulation of AGEs in diabetic tissues a cause of the observed pathology or a consequence of an alternate pathological mechanism, or are both likely to be true? The answer to this question has not been resolved for diabetic neuropathy, although recent evidence indicates that there is an increase in AGE and RAGE accumulation in the serum, perineurium, and epi- and endoneurial microvessels of nerves from patients with IGT and that this is associated with an increase in expression of the transcription factor NF-$\kappa\beta$ [17]. In the human diabetic peripheral nerve, immunoelectron microscopy indicates that CML is present in vascular endothelial cells, pericytes, and basement membrane, as well as axons and Schwann cells. Modification of neurofilament and tubulin with AGEs may interfere with the axonal transport and lead to atrophy and degeneration of nerve fibers. Glycation of P0 protein with AGEs may also lead to demyelination of nerve fibers. AGEs can reduce inducible NO synthase and generation of NO, reducing nerve blood flow and increasing microvascular hypoxia. In this setting, activation of RAGEs would also increase the generation of reactive intracellular or microvascular oxygen species, whereas inhibition of O_2^- formation prevents glucose-induced formation of AGEs and activation of PKC [9].

J. Growth Factors and Diabetes

In animal models of diabetes, levels of several growth factors may be reduced and treatment with growth factors may improve neuropathy. Reduced levels of nerve growth factor (NGF), brain-derived neurotrophic factor (BDNF), and IGF-I have been observed in models of diabetes. In contrast, administration of these factors protects against diabetic neuropathy in animals (reviewed in reference 9). NGF primarily supports the survival of small-fiber sensory neurons associated with pain and temperature sensation and would be expected to benefit diabetic patients with early diabetic neuropathy. A reduction in trophic support increases oxidative injury, whereas an increase in growth factor support may prevent neuronal oxidative stress by increasing antioxidant defense. An example is the ability of growth factors to increase intracellular concentrations of GSH and catalase in neurons [7]. Furthermore, NGF can regulate the inner mitochondrial membrane potential and prevent mitochondrial injury in Schwann cells, probably through regulation of the p75 neurotrophin receptor [18]. In a similar fashion, IGF-I prevents glucose-induced down-regulation of Bcl-2, inner mitochondrial membrane depolarization, and mitochondrial swelling [7]. BDNF may improve glycemic control and, at least in animal models, may reduce the severity of diabetic neuropathy [9]. BDNF enhances neurite outgrowth, supports DRG neuronal survival, is up-regulated in injured peripheral nerves, and promotes axonal regeneration. Systemic administration of BDNF also decreases nonfasted blood glucose in obese, non–insulin-dependent diabetic C57BLKS-Leprdb/leprdb (db/db) mice, and this effect can persist for weeks after cessation of BDNF treatment [7].

IGF-I activity is reduced in both clinical and experimental diabetes. Sensory neurons and supporting Schwann cells from diabetic rodents express lowered amounts of IGF-I and its receptor and are susceptible to glucose-induced injury [12]. In sciatic nerve from diabetic animals, there is a decrease in serum IGF-I and a reduction in IGF-I messenger ribonucleic acid. These changes in IGF-I are coupled with evidence that administration of IGF-I reduces DRG neuronal injury, ameliorates neuropathy in diabetic rats, and promotes nerve regeneration (reviewed in reference 9). In humans, IGF-I and IGF-I receptor levels are decreased in diabetic patients with neuropathy compared with diabetics without neuropathy [9]. Taken together, these changes indicate that IGF-I plays a critical role in the development of diabetic neuropathy.

However, promising results in animal models of diabetes and in experimental diabetic neuropathy have not translated to clinical studies in which no measurable benefit was obtained [7,9]. Although NGF was well tolerated at the doses selected for this study, it is likely that the study failed to achieve concentrations found to be therapeutic in prior animal studies. Similarly, a Phase II trial of human recombinant BDNF showed a small improvement in cold perception but not intraepidermal nerve fiber density [19]. Despite the potential for IGF-I treatment for human diabetic neuropathy, human clinical trials have not been completed because of a concern for potential IGF-I toxicity.

IV. Symptoms in Relation to Pathophysiology

There is a wide clinical spectrum of diabetic neuropathy (Table 1) (reviewed in reference 20).

A. Diabetic Symmetrical Sensory Polyneuropathy

Diabetic symmetrical sensory polyneuropathy is the most common presentation in diabetic neuropathy. Initially, patients with diabetic peripheral neuropathy experience sensory loss in the toes and feet due to length-dependent dysfunction of predominantly myelinated fibers. Small myelinated and unmyelinated fibers convey sensations of light touch, pain, and temperature. Large fiber loss decreases vibratory sensation and joint position sense. This distal "dying back" type of neuropathy is consistent with a metabolic disturbance in the PNS, coupled with failure of key signaling systems, and DRG neuronal injury.

B. "Small Fiber" Neuropathy

"Small fiber" neuropathy is characterized by superficial burning pain in the feet. Clinical findings include allodynia, deep aching pain, sympathetic vasomotor changes, pallor alternating with rubor, cyanosis, and mottling. Examination usually reveals normal strength and mild impairment of sensation distally with reflexes often being normal. In the largest prospective series, 81% of neuropathy patients with IGT had exclusively sensory complaints and 92% recognized neuropathic pain as a dominant symptom of their neuropathy [2]. In small-fiber neuropathy there is preferential involvement of the small unmyelinated nerve fibers that mediate pain, temperature sensation, and autonomic function. Direct quantitation of unmyelinated intraepidermal nerve fibers from skin biopsies shows similar fiber loss and altered morphology in patients who have neuropathy associated with IGT and early diabetes [2]. Thus, there is good evidence that abnormal small-fiber function, as seen in painful diabetic neuropathy, is an early finding in IGR.

Table 1 Classification of Diabetic Neuropathy

Distal symmetrical sensorimotor polyneuropathy
"Small fiber" neuropathy
Autonomic neuropathy
Amyotrophy
Polyradiculopathy
Cranial mononeuropathies
Limb mononeuropathies
Mononeuropathy multiplex

C. Diabetic Autonomic Neuropathy

Diabetic autonomic neuropathy often exists with somatic neuronal involvement. Autonomic changes include cardiac parasympathetic impairment, orthostatic hypotension, abnormal sweat patterns, gastric motor abnormalities, and erectile dysfunction. The symptoms depend on which specific autonomic component is affected (Table 2). Clinical autonomic failure increases with the length of the illness and the age of the patient. The severity of autonomic neuropathy varies between T1DM and T2DM and may reflect the relative level of insulin and IGF-I as described in Sect. III. Signs of dysfunction are present in approximately 16%–20% of all subjects and up to 75% of newly diagnosed subjects with T1DM at the time of first diagnosis. In patients with T2DM, autonomic parasympathetic changes occur in 20% at 5 years and 65% at 10 years. Sympathetic changes are present in 7% at 5 years and 24% at 10 years. However, standardized reflex tests are more sensitive at detecting early deficits of parasympathetic compared with sympathetic function and may partly explain the discrepancy between parasympathetic and sympathetic diabetic autonomic neuropathy. A similar pattern of autonomic neuropathy is seen in IGT patients [21].

D. Other Types of Diabetic Neuropathy and Radiculopathy

Motor weakness is less common in diabetic polyneuropathy, and, if present, is a late manifestation of the disorder. When motor weakness is present because of the diabetic state, the pattern parallels that of sensory loss. Mild weakness begins distally in toe flexors and extensors and progresses to affect more proximal muscles.

Table 2 Clinical Features of Autonomic Nervous System Failure

Distal sympathetic failure
- Difficulty keeping feet warm
- Early increased sweating
- Late anhidrosis
- Associated neuropathic pain
- Peripheral edema
- Rubor
- Acrocyanosis

Adrenergic failure
- Lightheadedness
- Tachycardia
- Pallor
- Blurred vision
- Stress incontinence
- Tremulousness
- Feeling clammy
- Nausea

The reason for the more severe sensory compared with motor susceptibility to diabetic injury is unknown. A small subset of patients experience more significant focal weakness due to nerve compression. Focal forms of diabetic neuropathy include involvement of single or multiple peripheral nerves (mononeuropathy and mononeuropathy multiplex). With the exception of peripheral nerve mononeuropathies, these focal forms of diabetic neuropathy are relatively uncommon. Diabetic lumbosacral radiculoplexus neuropathy is an unusual diabetic complication associated with asymmetrical weakness of the lower extremities and often with pain. The degree of weakness is often severe, resulting in considerable disability with slow recovery. The disorder is thought to be caused by an inflammatory ischemic vasculitis, possibly resulting from activation of key immune and cell death pathways.

V. Natural History

In an unselected population with IGT the prevalence is 11.2% for ages 50–59, peaking at 14.2% for ages 60–75, and declining in individuals older than 75. Investigations of the natural history of IGT reveal a dynamic and reversible state. Overall, most patients progress toward greater glycemic dysregulation, but this progress appears slow. Unmonitored patients probably experience many years of occult insulin resistance and postprandial hyperglycemia before developing typical symptoms of diabetes. During this time, the metabolic disturbance will increase the chance of developing neuropathy. This neuropathy may remain subclinical for several years, and distal nerve degeneration may be reversible early in the disorder, similar to the observed reversibility in IGT. At a certain point in advanced diabetes, the degree of neuropathy is severe enough that reversibility is unlikely. However, in IGR the full natural history of neuropathy from inception is unknown.

VI. Conclusion

IGR constitutes a spectrum of impaired glucose and metabolic regulation that can result in neuropathy. There are several pathways of injury in the diabetic PNS that include metabolic dysregulation induced by metabolic syndrome, oxidative stress, failure of NO regulation, activation of AGEs, and dysfunction of certain key signaling pathways, including MAPK, PI3K, PKC, and NF-κβ. Potentially, oxidative stress can directly injure both DRG neurons and peripheral nerve axons. Modulation of the NO system may have detrimental effects on endothelial function and neuronal survival. ROS alter mitochondrial function, protein, and DNA structure; interfere with signaling pathways; and deplete antioxidant defenses. AGE formation and

ROS, are activated by and, in turn, regulate key signal transduction pathways.

Acknowledgments

The authors acknowledge Denice Janus for secretarial support and Alexander Kaminski for technical assistance with Figure 2. This work was supported in part by the National Institutes of Health NS42056, NS40458, NS40458 and DK064814 (AGS and JRS), MO1-RR00042, and DK02-016; the Juvenile Diabetes Research Foundation Center for the Study of Complications in Diabetes; and the Office of Research Development (Medical Research Service), Department of Veterans Affairs (JWR).

References

1. Dyck, P. J., Kratz, K. M., Karnes, J. L., Litchy, W. J., Klein, R., and Pach, J. M., Wilson, D. M., O'Brien, P. C., Melton, L. J. 3rd, and Service, F. J. (1993). The prevalence by staged severity of various types of diabetic neuropathy, retinopathy, and nephropathy in a population-based cohort: the Rochester Diabetic Neuropathy Study. *Neurology* **43**, 817–824.
2. Singleton, J. R., Smith, A. G., Russell, J. W., and Feldman, E. L. (2003). Microvascular complications of impaired glucose tolerance. *Diabetes* **52**, 2867–2876.
3. Genuth, S., Alberti, K. G., Bennett, P., Buse, J., Defronzo, R., Kahn, R., Kitzmiller, J., Knowler, W. C., Lebovitz, H., Lernmark, A., Nathan, D., Palmer, J., Rizza, R., Saudek, C., Shaw, J., Steffes, M., Stern, M., Tuomilehto, J., Zimmet, P.; and the Expert Committee on the Diagnosis and Classification of Diabetes Mellitus. (2003). Follow-up report on the diagnosis of diabetes mellitus. *Diabetes Care* **26(11)**, 3160–3167.
4. Knowler, W. C., Barrett-Connor, E., Fowler, S. E., Hamman, R. F., Lachin, J. M., Walker, E. A., Nathan, D. M.; and the Diabetes Prevention Program Research Group. (2002). Reduction in the incidence of type 2 diabetes with lifestyle intervention or metformin. *N. Engl. J. Med.* **346**, 393–403.
5. Pan, X. R., Li, G. W., Hu, Y. H., Wang, J. X., Yang, W. Y., An, Z. X., Hu, Z. X., Lin, J., Xiao, J. Z., Cao, H. B., Liu, P. A., Jiang, X. G., Jiang, Y. Y., Wang, J. P., Zheng, H., Zhang, H., Bennett, P. H., and Howard, B. V. (1997). Effects of diet and exercise in preventing NIDDM in people with impaired glucose tolerance: the Da Qing IGT and Diabetes Study. *Diabetes Care* **20(4)**, 537–544.
6. Hiltunen, L., Laara, E., and Keinanen-Kiukaanniemi, S. (1999). Changes in glucose tolerance during three years' follow-up in an elderly population. *Public Health* **113(4)**, 181–184.
7. Russell, J. W., and Kaminsky, A. J. (2005). Oxidative injury in diabetic neuropathy. *In:* "Nutrition and Diabetes: Pathophysiology and Management" (E. Opara, ed.), pp. 381–397. Taylor & Francis, Boca Raton, FL.
8. Lindstrom, J., Louheranta, A., Mannelin, M., Rastas, M., Salminen, V., Eriksson, J., Uusitupa, M., Tuomilehto, J.; and the Finnish Diabetes Prevention Study Group. (2003). The Finnish Diabetes Prevention Study (DPS): lifestyle intervention and 3-year results on diet and physical activity. *Diabetes Care* **26(12)**, 3230–3236.
9. Cowell, R. M., and Russell, J. W. (2004). Peripheral neuropathy and the Schwann cell. *In:* "Neuroglia" (H. Kettenmann and B. R. Ransom, eds.), pp. 573–585. Oxford University Press, New York.
10. Vincent, A. M., Olzmann, J. A., Brownlee, M., Sivitz, W. I., and Russell, J. W. (2004). Uncoupling proteins prevent glucose-induced neuronal oxidative stress and programmed cell death. *Diabetes* **53**, 726–734.
11. Cowell, R. M., and Russell, J. W. (2004). Nitrosative injury and antioxidant therapy in the management of diabetic neuropathy. *J. Investig. Med.* **52(1)**, 33–44.
12. Russell, J. W., Sullivan, K. A., Windebank, A. J., Herrmann, D. N., and Feldman, E. L. (1999). Neurons undergo apoptosis in animal and cell culture models of diabetes. *Neurobiol. Dis.* **6**, 347–363.
13. Russell, J. W., Cowell, R. M., and Feldman, E. L. (2006). Neuronal and Schwann cell death in diabetic neuropathy. *In:* "The Clinical Management of Diabetic Neuropathy" (A. Veves and R. Malik, eds.). Humana Press, Totowa, NJ.
14. Uehara, K., Yamagishi, S., Otsuki, S., Chin, S., and Yagihashi, S. (2004). Effects of polyol pathway hyperactivity on protein kinase C activity, nociceptive peptide expression, and neuronal structure in dorsal root ganglia in diabetic mice. *Diabetes* **53(12)**, 3239–3247.
15. Vinik, A. I., Bril, V., Kempler, P., Litchy, W. J., Tesfaye, S., Price, K. L., Bastyr, E. J. 3rd; and the MBBQ Study Group. (2005). Treatment of symptomatic diabetic peripheral neuropathy with the protein kinase C beta-inhibitor ruboxistaurin mesylate during a 1-year, randomized, placebo-controlled, double-blind clinical trial. *Clin. Ther.* **27(8)**, 1164–1180.
16. Berent-Spillson, A., Robinson, A., Golovoy, D., Slusher, B., Rojas, C., and Russell, J. W. (2004). Protection against glucose-induced neuronal death by NAAG and GCP II inhibition is regulated by mGluR3. *J. Neurochem.* **89**, 90–99.
17. Haslbeck, K. M., Schleicher, E., Bierhaus, A., Nawroth, P., Haslbeck, M., Neundorfer, B., and Heuss, D. (2005). The AGE/RAGE/NF-κB pathway may contribute to the pathogenesis of polyneuropathy in impaired glucose tolerance (IGT). *Exp. Clin. Endocrinol. Diabetes* **113(5)**, 288–291.
18. Vincent, A. M., Brownlee, M., and Russell, J. W. (2002). Oxidative stress and programmed cell death in diabetic neuropathy. *Ann. N. Y. Acad. Sci.* **959**, 368–383.
19. Wellmer, A., Misra, V. P., Sharief, M. K., Kopelman, P. G., and Anand, P. (2001). A double-blind placebo-controlled clinical trial of recombinant human brain-derived neurotrophic factor (rhBDNF) in diabetic polyneuropathy. *J. Peripher. Nerv. Syst.* **6(4)**, 204–210.
20. Feldman, E. L., Stevens, M. J., Russell, J. W., and Greene, D. A. (2003). Somatosensory neuropathy. *In:* "Ellenberg and Rifkin's Diabetes Mellitus" (D. Porte Jr., R. S. Sherwin, and H. Rifkin, eds.), pp. 771–788. McGraw-Hill Professional, New York.
21. Russell, J. W., Peltier, A., Sheikh, K., Howard, J., Goldstein, J., Smith, A. G., Feldman, E. L., and Singleton, J. R. (2005). Autonomic dysfunction in subjects with impaired glucose tolerance. *J. Peripher. Nerv. Syst.* **10**, 78.

79

Acquired Inflammatory Demyelinating and Axonal Neuropathies

James W. Teener, MD
James W. Albers, MD, PhD

Keywords: *acute inflammatory demyelinating polyneuropathy, acute motor axonal neuropathy, chronic inflammatory demyelinating polyneuropathy, Fisher syndrome, monoclonal gammopathy, multifocal acquired demyelinating sensory and motor neuropathy, multifocal motor neuropathy*

I. Acute Inflammatory Neuropathies
II. Chronic Inflammatory Neuropathies
III. Chronic Inflammatory Demyelinating Polyneuropathy and Closely Related Neuropathies
IV. Neuropathies Associated with Monoclonal Gammopathies
V. Treatment
VI. Summary
   References

Although diabetes is the leading cause of polyneuropathy worldwide, the inflammatory neuropathies are a particularly important group of disorders because many of them are among the most treatable neuropathies encountered in clinical practice. Treatment of inflammatory neuropathies has advanced as the understanding of the immunopathology of this group of disorders has grown.

The family of inflammatory neuropathies can be subdivided in several ways. Perhaps the most fundamental is to separate acute from chronic processes. This division, although somewhat arbitrary, is likely relevant to pathophysiology because the acute inflammatory neuropathies are typically self-limited. In contrast, the chronic inflammatory neuropathies are characterized by persistent autoimmune

Table 1 Acquired Inflammatory Demyelinating and Axonal Neuropathies

I. Acute inflammatory neuropathies
   A. Primarily demyelinating
      1. AIDP
      2. FS
   B. Primarily axonal
      1. AMSAN
      2. AMAN
II. Chronic inflammatory neuropathies
   A. Classic CIDP
   B. CIDP variants
      1. DADS neuropathy
      2. MMN
      3. MADSAM
   C. Neuropathies associated with monoclonal gammopathy
      1. IgM related
      2. Anti-MAG
      3. CANOMAD
      4. IgG and IgA related
      5. CIDP
      6. POEMS

AIDP, Acute inflammatory demyelinating polyneuropathy; AMAN, acute motor axonal neuropathy; AMSAN, acute motor and sensory axonal neuropathy, CANOMAD, chronic ataxic neuropathy, ophthalmoplegia, IgM paraprotein, cold agglutinins, and disialosyl antibodies; CIDP, chronic inflammatory demyelinating polyneuropathy; DADS, distal acquired demyelinating symmetrical; FS, Fisher syndrome; Ig (A, G, and M), immunoglobulin (A, G, and M); MADSAM, multifocal acquired demyelinating sensory and motor neuropathy; MAG, myelin-associated glycoprotein; MMN, multifocal motor neuropathy; POEMS, polyneuropathy, organomegaly, endocrinopathy, monoclonal gammopathy, and skin changes.

activity. Further subdivision can be made based on the primary site of immunological attack, that is, the axon itself or the myelin sheath. Finally, inflammatory neuropathy may be a feature of a more widespread disorder and may be characterized as such. Table 1 lists the inflammatory neuropathies discussed in this chapter.

I. Acute Inflammatory Neuropathies

The acute inflammatory neuropathies fall under the recently expanded classification of the Guillain-Barré syndrome (GBS). In the Western world, acute inflammatory demyelinating polyneuropathy (AIDP) accounts for most GBS. The Fisher syndrome (FS) in its purest form is restricted to ophthalmoparesis, areflexia, and ataxia, but these clinical features may be accompanied by features of AIDP, in which case the moniker "atypical FS" is sometimes applied. Two subtypes of GBS involve a primary immune attack on peripheral nerve axons. These are termed *acute motor axonal neuropathy* (AMAN) and *acute motor and sensory axonal neuropathy* (AMSAN) depending on whether the syndrome is restricted to motor function, as in the former, or involves both sensory and motor function, as in the latter.

A. Clinical Features

1. AIDP and AMSAN

Although AIDP and AMSAN differ substantially in terms of pathology, their clinical presentations are indistinguishable. The classic description of GBS is of ascending paralysis, but sensory symptoms such as numbness and paresthesias and even deep aching pain may predominate in the earliest stage of both AIDP and AMSAN. Although weakness typically begins distally and "ascends" to involve more proximal muscles, any muscle may be affected. Early loss of tendon stretch reflexes is typical. Nerve conduction studies often allow differentiation of demyelinating from axonal subtypes. Extensive pathological study may also allow this distinction to be made, but this is rarely performed and in most cases the diagnosis is based on clinical and electrodiagnostic criteria.

Electrodiagnostic studies play several roles in the evaluation of the patient with acute weakness. Alternative diagnoses such as myopathy, myasthenia gravis, botulism, and others may be excluded. The presence of demyelination may be confirmed in cases of AIDP. Extensive axonal loss, as suggested by low or absent sensory and motor response amplitudes and by abnormal spontaneous activity on needle electromyography, correlates with poor prognosis.

Weakness progresses over several weeks, reaching maximal severity within 4 weeks in 90% of patients. Related conditions such as chronic inflammatory demyelinating polyradiculoneuropathy must be considered if progression is prolonged. The rate of progression and ultimate severity vary widely. In the mildest forms of AIDP, patients have limited weakness that begins to improve within days. In its most virulent form, AIDP causes rapidly progressive weakness and respiratory failure. This possible outcome represents a true neurological emergency and serves as justification for close early observation of patients suspected of having GBS. Along with respiratory failure, which may develop in up to 40% of GBS patients, autonomic failure may lead to severe complications requiring intensive medical treatment.

After several weeks of progression, most patients achieve a plateau, often lasting days to weeks. Recovery begins following the plateau, and its course is highly variable. In the best scenario recovery is rapid, and a few patients have been observed to move from ventilator dependence to walking within 1 week. GBS mortality has been markedly reduced by modern intensive care unit practice. Current mortality rates of 1.25 to 2.5% compare favorably to the 13% rate observed as recently as the early 1980s [1]. The profound reduction in mortality has unfortunately not been mirrored by a similar reduction in morbidity. Recovery is often slow and incomplete, with approximately 15% of GBS patients having permanent limitations in daily activities [2]. Initial reports of prognosis in AMSAN suggested a particularly poor prognosis, but more recent reports suggest that some cases have an outcome similar to that of AIDP [3].

2. AMAN

AMAN is characterized by acute progressive weakness without sensory impairment. The axonal forms of GBS, particularly AMAN, are more common in China, Japan, India, Korea, and Mexico than in the United States and Europe. Electrodiagnostic studies in AMAN patients reveal low or absent motor responses with preserved sensory responses. No evidence of demyelination such as conduction block or slowing is present [4]. Despite the electrodiagnostic findings consistent with severe axonal involvement, the prognosis does not appear to differ from that of AIDP [3]. Possible pathophysiological explanations for this paradox include axonal conduction failure caused by antibody attack without true axonal loss and axonal pathology confined to the nerve terminal region [5]. These will be discussed in later sections of this chapter.

B. Pathology

1. AIDP

Despite our long time recognition of GBS, relatively few extensive studies of its pathology are available. The main reported features include lymphocytic inflammation, demyelination, endoneurial edema, and some degree of axonal loss [6]. The findings are most prominent in the spinal roots and the nerve terminals. This distribution may reflect the less robust blood-nerve barrier in these regions.

Lymphocytes are present in cuffs around epineurial vessels and within the endoneurial space. The number of lymphocytes typically appears sparse, but their pathological importance is clear, as discussed in later sections. Both CD4 and CD8 lymphocytes are present. Macrophages also are present, sometimes in abundance even in the relative absence of T cells. Macrophages are likely involved in stripping and phagocytosis of compact myelin, and they are found in widened nodes of Ranvier [7]. The most prominent demyelination is often found in paranodal regions where macrophages cluster.

Complement deposition is an additional important feature of the immune response in GBS. In AIDP, activated complement is deposited on the abaxonal (outermost) Schwann cell plasmalemma. Complement activation products are present in both plasma and cerebrospinal fluid in AIDP patients.

In most cases of AIDP there is some degree of axonal degeneration [6]. The mechanism of axonal loss is incompletely understood. Postulated mechanisms include specific immune attack on axons, "bystander effects" of severe inflammation, and increased endoneurial fluid pressure resulting in ischemic nerve damage.

2. AMAN and AMSAN

Electrodiagnostic study of the axonal forms of GBS provides evidence of axonal loss but no demyelination. Pathological study of nerve biopsy and autopsy specimens confirms the absence of demyelination in AMAN and AMSAN. Lymphocytic infiltration also is sparse in comparison with most cases of AIDP [4,8]. AMAN and AMSAN differ pathologically primarily in the absence or presence of sensory nerve involvement. In AMSAN, extensive sensory nerve fiber degeneration may be present, but no more than minimal Wallerian degeneration of sensory axons is present in AMAN [9].

Early pathological change in the axonal forms of GBS consists of lengthening of nodal gaps at the nodes of Ranvier. Activation products of complement are present in the nodal region. The complement components localize to the adaxonal (innermost) Schwann cell plasmalemma; in contrast, in AIDP complement localizes to the abaxonal membrane. Complement provides a chemotactic signal for macrophage invasion of the periaxonal space. The presence of macrophages, as well as antibody deposition, has been postulated to cause conduction failure in the absence of extensive axonal pathology. This mechanism could account for the unexpected rapid recovery noted among some patients with axonal forms of GBS, particularly AMAN, as such recovery is not possible in the setting of severe axonal degeneration. Another possible explanation for rapid recovery is Wallerian degeneration localized solely to the distal nerve terminals, allowing recovery through terminal sprouting and axonal regrowth over a short distance [5]. In more severe cases of AMAN, there is severe axonal degeneration beginning at the ventral root exit zone with degeneration of the entire distal axon.

3. Fisher Syndrome

Because FS is relatively mild and nonfatal, the pathological findings are not completely understood. The cases that have come to autopsy have been somewhat atypical ones in which the features of FS were present with paralysis. In these cases, the electrodiagnostic and pathological features were those of AIDP. Bickerstaff's brainstem encephalitis is a disorder that may be closely related to FS. Ophthalmoparesis is shared in these disorders, but Bickerstaff's encephalitis also includes enhancing posterior fossa lesions demonstrable with magnetic resonance imaging. In both disorders, antibodies to ganglioside GQ1b are often measurable in serum [10]. The role of these antibodies, and other antiganglioside antibodies, is discussed later.

C. Pathogenesis

1. General Concepts

Substantial recent progress in understanding the pathogenesis of GBS has been fueled by recognition of molecular mimicry as the leading putative mechanism in many cases of GBS [11]. The term *molecular mimicry* as used here refers to antigenic determinants shared between infectious agents and peripheral nerve components. If the immune response mounted against an infectious agent is directed toward an epitope shared with a peripheral nerve, immune-mediated damage to the peripheral nervous system may follow. Despite growing evidence in support of molecular mimicry, many questions remain. The concept of self-injury implies a loss of tolerance to self-antigen. The mechanism underlying this break in tolerance is unknown. It is unknown why only a small minority of patients infected with organisms known to be capable of producing molecular mimicry develops GBS. The factors influencing the longevity of the autoimmune attack also are not understood. Finally, our understanding of the interaction of host and organism factors in producing GBS remains incomplete.

A major breakthrough in understanding the pathogenesis of GBS occurred with recognition of the role of antecedent infection with *Campylobacter jejuni* in the development of AMAN [12]. *C. jejuni* infection has been identified preceding AMSAN, FS, and AIDP as well. *C. jejuni*, a microaerophilic, gram-negative, nonflagellate rod is a leading cause of gastroenteritis worldwide. Multiple serotypes are recognized, and some have been particularly linked with the development of GBS. The heat-stable (HS or O) serotype HS/O:19 is overrepresented in GBS patients in Japan and China, being found in up to 81% of Japanese patients. This is in stark contrast to its 2% rate of isolation among all patients with *Campylobacter*-related gastroenteritis in Japan. Similarly, the HS/O:41 strain of *C. jejuni* is a rare cause of gastroenteritis but is commonly associated with GBS in South Africa. Based on data from the United States, the risk of developing GBS following *C. jejuni* infection is 1 in 1058.

In much of the world, handling and ingesting chicken is likely the leading cause of infection. Other routes of infection, such as waterborne infection, may be predominant in some areas. Infection may be asymptomatic, although an acute self-limited gastroenteritis with watery or bloody diarrhea is common. The organism is typically no longer present in the stool at the time of development of GBS, so serological studies must be employed to demonstrate antecedent infection in GBS patients.

Investigation of the role of *C. jejuni* in GBS, as well as other research, has focused attention on glycolipids, particularly gangliosides, as important immune targets in both acute and chronic neuropathies. Antiglycolipid antibodies to numerous glycolipids have been identified in GBS patients (Table 2).

Table 2 Glycolipids to Which Antibodies Have Been Reported in GBS

GM1	GM1(NeuGc)	GM1b
GalNAc-GM1b	GD1a	GalNAc-GD1a
GD1b-o-acetyl GD1b	GD3	GT1a
GT1b	GQ1b	GQ1bα
LM1	Galactocerebroside	SGPG

Note: SGPG; sulfoglucuronyl paragloboside.

Gangliosides are a large family of glycosphingolipids characterized by sialic acid linked to an oligosaccharide core. They occur in many body tissues but are highly concentrated in the nervous system. Different gangliosides predominate in different locations in peripheral nerves and in different fiber types. This has led to efforts to correlate clinical phenomena, such as the specific attack on motor axons in AMAN, with the presence of antibodies directed against gangliosides known to be highly represented in appropriate locations or fiber types, such as the motor axon. Such studies have been only partially successful. For example, the ganglioside GM1 is present in ventral roots, and anti-GM1 antibodies are commonly identified in patients with purely or predominately motor neuropathies, such as AMAN and multifocal motor neuropathy. However, GM1 is also present at many nervous system sites that remain unaffected, suggesting that the simple distribution of gangliosides is not sufficient to fully account for the observed clinical pathology. Besides simply the presence of the ganglioside in peripheral nerves, several other factors may influence pathogenicity of an antiganglioside antibody. The ganglioside may be folded within a membrane, making it inaccessible to the antibody. Also, the blood-nerve barrier may prevent binding to gangliosides, and the breech of this barrier may be an important pathological step in the development of immune-mediated neuropathy. Ganglioside function may also play a role in pathogenesis of neuropathy. Gangliosides have been demonstrated to have calcium-binding properties, signal transduction roles, and even receptor function. Immune-mediated interference in these normal functions could be a contributing factor in the development of neuropathy.

The pathogenic role of antiganglioside antibodies is a point of considerable debate. Several possible relationships between the antibodies and the inflammatory neuropathies have been suggested. A few investigators maintain that ganglioside antibodies are irrelevant in the development of GBS and are present only as secondary markers of disease. Other authorities believe that there are fine, as-yet-undiscovered differences among antibodies that account for the clinical phenotype. A possible example of this phenomenon is the known propensity for some GM1 antibodies to be highly specific, whereas others cross-react with other gangliosides. Another possible explanation for differing neuropathy phenotypes developing in relation to antiganglioside antibodies recognizes the complex biology of gangliosides. Ganglioside epitopes exist both in myelin and in axolemma membranes in differing concentrations or configurations, creating the possibility of preferential antibody binding under different circumstances in different individuals. Thus one individual may be susceptible to axonal damage on development of an antiganglioside antibody, and another individual may experience demyelination with development of the same antiganglioside antibody. This concept is supported by the epidemiological patterns of antiganglioside antibodies and prevalent forms of GBS, which vary markedly among geographic regions. The association between particular antiganglioside antibodies and various clinical syndromes is discussed further in later sections devoted to each of the clinical subtypes of GBS.

One trigger for the development of antiganglioside antibodies is infection with *C. jejuni*. Such infection is the most commonly recognized antecedent infection in the axonal forms of GBS and may precede AIDP and FS. Much research has focused on the role of molecular mimicry in the development of antiganglioside antibodies and GBS following *C. jejuni* infection. The lipopolysaccharide and lipo-oligosaccharide molecules in the bacterial cell wall are the leading focus of attention. In *C. jejuni* strains isolated from GBS patients, homology of lipo-oligosaccharide and lipopolysaccharide regions with GM1, GD1a, GalNAc-GD1a, GM1b, GT1a, GD3, GM2, and GQ1b gangliosides has been reported [11]. Epidemiological studies of *C. jejuni* strains have revealed that ganglioside-like moieties are far more common on strains that are associated with the development of GBS than on strains that are only diarrhea related. There is also evidence that *C. jejuni* infection may affect autoimmunity other than by triggering antiganglioside antibody production. The lipo-oligosaccharide of *C. jejuni* may interfere with the natural oral tolerance mechanism. Lipo-oligosaccharide has been shown to block the typical oral tolerance, which develops on oral feeding of antigen in experimental allergic neuritis (EAN). If such a mechanism were active in humans, it could contribute to the development of GBS in patients infected with *C. jejuni*.

Although much recent research has focused on the role of humoral immune processes in GBS, cell-mediated immunity clearly plays a critical role [13]. Lymphocytic infiltration of nerves is a prominent pathological finding in AIDP. EAN, the primary experimental model of GBS, is largely a cell-mediated disorder. Activated T cells may play multiple roles in the pathogenesis of GBS. T cells may help B cells engineer antibodies by releasing proliferation signals such as interleukins 4, 5, and 6. T cells activate macrophages by releasing proinflammatory cytokines tumor necrosis factor-α (TNF-α) and interferon-γ. In EAN, T cells inflict cytotoxic damage on Schwann cells, and they may play a similar role in GBS. T cells also have the potential to modulate and even stop acute inflammatory processes through suppressor functions. CD4 Th2 cells secrete the down-regulatory cytokines interleukins 4 and 10, as well as transforming growth factor-β. $CD8^+$ suppressor cells may play a similar role. In EAN, it is likely that T cell apoptosis is involved in terminating the inflammatory response.

Macrophages also play multiple roles in the pathogenesis of GBS. Their presence in nerves may derive from invasion of circulating monocytes or represent endoneurial macrophages. They present antigens, engage in phagocytosis, and release proinflammatory cytokines and other toxic mediators. The inflammatory cytokines include interleukin-1 and TNF-α. The toxic mediators include proteases, complement components, oxygen radicals, and nitric oxide eicosanoids.

2. Pathogenesis of Specific Guillain-Barré Syndrome Subtypes

a. Fisher Syndrome FS represents only about 5% of all GBS cases, and a small number of FS cases follow *C. jejuni* infection. However, the syndrome is of considerable interest because the strongest association between antiganglioside antibodies and a GBS subtype occurs with FS. Nearly 90% of FS patients have anti-GQ1b antibodies [13]. The anti-GQ1b antibodies are able to fix complement. Because of the recognized close relationship between FS and anti-GQ1b antibodies, the antibodies have been extensively studied. In FS, the anti-GQ1b antibodies typically cross-react with ganglioside GT1a, and in many cases cross-reactivity with gangliosides such as GD3, GD1b, and GT1b can be demonstrated. This cross-reactivity is correlated with the clinical phenotype. Oropharyngeal weakness is associated with GT1a reactivity. GQ1b reactivity is associated with ophthalmoplegia in FS and when it occurs in AIDP.

The clinical phenotype, particularly ophthalmoplegia, may also be related to the relative abundance of GQ1b in extraocular motor cranial nerves, particularly at nodes of Ranvier. The reduction in muscle stretch reflexes may be related to binding to peripheral proprioceptive components, including dorsal root ganglion neurons, muscle spindles, and intrafusal muscle fibers, which are labeled by antibodies present in many FS patients.

Infections are common triggers for FS. Upper respiratory infections and *C. jejuni* gastroenteritis have been documented preceding FS. *C. jejuni* isolates from FS provide important evidence for molecular mimicry. Both GD3- and GT1a-like structures have been identified in *C. jejuni* isolates. Although structural studies have not demonstrated any apparent structural similarity with GQ1b, antibody-binding assays have revealed GQ1b cross-reactive moieties in *C. jejuni* lipopolysaccharide.

Studies of the pathogenic effects of antiganglioside antibodies have been particularly important in FS because of the relative lack of autopsy data for this disorder. The available autopsy data is from cases with atypical FS, including more widespread clinical involvement than just ophthalmoplegia, ataxia, and areflexia. In these cases the predominant pathology is demyelination indistinguishable from AIDP. The major pathological effects of GQ1b and related antibodies are at the neuromuscular junction [14]. In experimental preparations, anti-GQ1b and related antibodies cause initial massive quantal release of acetylcholine followed by failure of neuromuscular transmission. These effects are complement dependent. The conduction failure is followed by degeneration of motor nerve terminals. Other studies have suggested both pre- and postsynaptic effects at the neuromuscular junction, raising the possibility of multiple pathological effects of anti-GQ1b antibodies.

b. AMAN The nodes of Ranvier on motor axons appear to be the site of early pathology in AMAN. A highly selective antibody-mediated attack on the axon is likely given the presence of IgG and complement deposits on the axolemma, along with macrophage recruitment. The macrophages insert processes into the node and enter the periaxonal space of the internode. This may result in transaction of the axon with distal Wallerian degeneration, but the rapid recovery of strength in some AMAN patients suggests alternative pathological processes. Antibody binding at the nodes of Ranvier and the effects of macrophages could produce physiological axonal conduction block without axonal degeneration. Alternatively, the axonal transaction may be limited to the distal nerve terminals, allowing rapid axonal regrowth and early clinical recovery.

Numerous reports have demonstrated the presence of anti-GM1 IgG antibodies in AMAN patients. Certainly not all patients with AMAN have GM1 antibodies, and GM1 antibodies have been identified in patients with other forms of GBS. Despite this discrepancy, it is likely that GM1 antibodies play a direct pathological role in at least some AMAN cases. Antibodies to ganglioside GD1a also have been identified in many Chinese AMAN patients.

c. AIDP AIDP has been considered a T cell–mediated disorder [13]. Lymphocytic infiltration of nerves is common. However, the recognition of the likely importance of antibodies in the pathogenesis of other GBS subtypes has fueled a search for antibody-mediated pathology in AIDP. Antibodies against gangliosides, basal lamina components, and other proteins have been reported in relatively few AIDP cases. The lack of a single antigen has cast doubt on the pathogenic role of antibodies in AIDP. However, there appear to be definite correlations between the presence of antiganglioside antibodies and the clinical features of AIDP. For example, GM1 antibodies are present in approximately 15%–20% of AIDP cases and correlate with prominent motor involvement and particularly severe disease, often with inexcitable nerves. Studies suggest prolonged recovery in GM1-seropositive AIDP patients. In AIDP patients with serological evidence of recent cytomegalovirus infection, antibodies to GM2 are detected about 50% of the time. These patients tend to have more prominent sensory symptoms than do AIDP patients without GM2 antibodies. Antibodies have been detected against several other gangliosides, including GM1b, GalNAc-GD1a, and LM1 in AIDP patients. Host factors play an unexplained role in the development of antibodies. In a group of Dutch AIDP patients, immunoglobulin M (IgM) antiganglioside antibodies were common, whereas in Japanese AIDP patients, IgG antibodies predominated.

Antibodies to a variety of whole nerve or myelin proteins other than gangliosides have also been reported and may play a pathogenetic role in some AIDP cases. Using an extremely sensitive C1 fixation and transfer assay, Koski found circulating IgM antibodies to peripheral nerve myelin in more than 90% of AIDP cases [15]. The antigen is uncertain, although a neutral glycolipid resembling the Forssman hapten is a possible target. Other studies have not confirmed this result, although somewhat less sensitive assays were used. Antibodies to cardiolipin, sulfoglucuronyl paragloboside, and galactocerebroside have also been identified in AIDP patients.

II. Chronic Inflammatory Neuropathies

The division of the entire spectrum of inflammatory neuropathies into acute and chronic disorders is somewhat arbitrary but highlights a fundamental pathological difference. Although GBS is a monophasic, self-limited disorder, the chronic inflammatory neuropathies display persistent autoimmune attack on peripheral nerve elements. It thus becomes important to identify chronic inflammatory neuropathies because treatment is typically necessary to halt their progression. Thankfully, many of the chronic inflammatory neuropathies respond nicely to immune-modulating therapy.

III. Chronic Inflammatory Demyelinating Polyneuropathy and Closely Related Neuropathies

The presenting symptom of classic chronic inflammatory demyelinating polyneuropathy (CIDP) is relatively symmetrical, proximal, and distal weakness progressing over more than 2 months [16]. Impaired sensation is typical, and tendon reflexes are reduced or absent. Cerebrospinal fluid protein levels are increased. Electrodiagnostic studies and nerve biopsy provide evidence of demyelination, although there is considerable controversy over the relative importance of these confirmatory tests. Several groups have proposed diagnostic criteria for CIDP. It should be noted that these are primarily intended as tools for research and as such have significant limitations in clinical use. The criteria tend to be quite specific but may not be sufficiently sensitive to identify CIDP in clinical practice. Given that CIDP often responds well to immunomodulatory therapy, this is a critical limitation of the research criteria. Related forms of chronic inflammatory neuropathies have similar clinical presentations, but some have unique presentations that have lead to their classification as separate clinical entities. In many cases, the clinical differences reflect known pathogenetic differences among the chronic inflammatory neuropathies.

Chronic inflammatory demyelinating neuropathies with distinct clinical presentations include *distal acquired demyelinating symmetrical (DADS) neuropathy*, *multifocal motor neuropathy* (MMN), and *multifocal acquired demyelinating sensory and motor neuropathy* (MADSAM or the Lewis-Sumner syndrome) [17].

DADS neuropathy occurs with increased frequency among older men. It is characterized by distal sensory loss with mild weakness and prominent gait disturbance. Nearly two-thirds of DADS neuropathy patients have paraproteinemia, often of the IgM class. When associated with an IgM paraproteinemia, this condition is often resistant to treatment.

MMN is an important clinical entity because it represents a treatable mimic of amyotrophic lateral sclerosis. Sensory signs and symptoms are absent. Weakness is asymmetrical, often more prominent in the arms. In many patients, partial motor conduction block can be demonstrated on nerve conduction studies. In some patients, antiganglioside antibodies, particularly GM1 antibodies, are present. MMN is resistant to treatment with corticosteroids and plasma exchange (PEx) but typically improves following treatment with intravenous infusion of human immunoglobulin (IVIG) or cyclophosphamide [17].

MADSAM is sometimes linked with MMN because of its asymmetry and frequent conduction block. However, the presence of clinical and electrodiagnostic sensory involvement clearly separates the disorders. Like classic CIDP, cerebrospinal fluid protein levels are typically elevated. Ganglioside antibodies have been detected in some MADSAM patients. MADSAM patients typically improve with treatment with IVIG and cyclophosphamide, as do patients with MMN, but corticosteroids and PEx are also sometime effective [17].

Other CIDP subgroups have been proposed, including axonal forms. However, only small numbers of patients in these groups have been reported. The clinical and pathological features of these groups are incompletely understood and are beyond the scope of this chapter.

CIDP may occur in association with concurrent diseases, most notably monoclonal gammopathy of undetermined significance, lymphoma, melanoma, diabetes, Sjögren's syndrome, inflammatory bowel disease, and infection with hepatitis C and the human immunodeficiency virus. The role of the concurrent disease in the development of CIDP is often uncertain and likely differs among the concurrent diseases. For example, there is considerable evidence that in certain subtypes of paraproteinemia the monoclonal protein has direct pathological effects on peripheral nerve. In others, the paraprotein cannot be shown to be reactive with peripheral nerve epitopes. The relationship to melanoma is intriguing because of the shared embryological lineage of melanocytes and Schwann cells, both derivatives of neural crest tissues. Shared antigens are likely between melanoma cells and Schwann cells, raising the possibility of antibodies or other immune attack directed at the melanoma cross-reacting with peripheral nerve [18].

A. Pathology

Nerve biopsy in CIDP patients may reveal characteristic features of demyelination, but it is not uncommon to have minimal or nonspecific pathological findings. CIDP often involves motor more than sensory function, and the pathology may be widely scattered or primarily involve proximal nerve segments. Thus biopsy of a distal sensory nerve has obvious limitations. This problem is compounded if the biopsy is not performed properly or the nerve specimen is mishandled. Classic features of demyelination include large fibers devoid of myelin, thinly myelinated large fibers indicative of remyelination, and onion bulb formation. The specialized technique of teasing nerve fibers enhances the detection of demyelination. Focal areas of demyelination are readily apparent on teased fiber preparations. Some degree of axonal loss is noted in about one-third of biopsies otherwise typical of CIDP. The presence of endoneurial inflammatory cells is relatively uncommon, occurring in 10%–15% of biopsies. If seen, the inflammatory infiltrate is usually rather scant. Special stains typically identify these cells as T cells and macrophages. Evidence of phagocytosis by macrophages may be seen. Endoneurial edema is periodically identified as well.

B. Immunopathogenesis

There is evidence supporting both a prominent cell-mediated immune response and a humoral immune response in CIDP [19]. Activated T cells can be identified in nerve specimens from CIDP patients [20]. The activation process for T cells is complex but has been studied in some detail in CIDP patients. The initial steps of T cell activation include (1) presentation of immunogenic material with major histocompatibility complex molecules by antigen-presenting cells, (2) interaction of specific molecules on antigen-presenting cells with T cell receptors, (3) adhesion and extravasation of activated T cells, and (4) local amplification of the immune response within nerves [21]. The antigenic target of activated T cells in CIDP remains unknown. It has been shown that gamma delta T cells, which are capable of recognizing ganglioside (nonprotein) antigens, are present in more than 60% of CIDP nerve biopsy specimens. There is evidence that endoneurial macrophages may act as antigen-presenting cells in CIDP. They express the major histocompatibility complex-like molecules CD1a and CD1b.

Breech of the blood-nerve barrier is necessary for the full immune cascade to occur, except at limited sites in the nerve root entry zone and nerve terminals. Lymphocytes migrate across the blood-nerve barrier by interacting with adhesion molecules on endothelial cells. Proinflammatory cytokines like TNF-α and matrix metalloproteinases (MMPs) are implicated in breakdown of the blood-nerve barrier. MMP-2 and MMP-9 are up-regulated in nerve biopsy specimens from CIDP patients. Further evidence of breakdown of the blood-nerve barrier is the down-regulation of tight junction proteins claudin-5 and zonula occludens-1 in sural nerve biopsy specimens from CIDP patients. A breach of the blood-nerve barrier also allows elements of the humoral immune system access to neural antigens.

Additional recruitment of immune cells across the blood-nerve barrier is mediated by chemoattractant cytokines called chemokines. Chemokines also activate macrophages, induce nitric oxide synthesis, and activate naïve T cells. A variety of chemokines are elevated in the cerebrospinal fluid of CIDP patients. A chemokine receptor, CXCR-3, has been reported to be up-regulated in inflammatory neuropathies.

Cell-mediated nerve injury and demyelination occurs by multiple mechanisms. The main effector cells in immune-mediated demyelination are macrophages and $CD8^+$ T cells. Macrophages are apparently able to slip cellular processes into intact myelin and strip it from the underlying axon. Activated T cells and macrophages also secrete cytotoxic substances such as TNF-α [19].

Many of these fundamental immune processes do not differ substantially between acute and chronic inflammatory neuropathies. The lack of termination of the immune response in CIDP is the main feature separating the disorders. Factors influencing this lack of termination are poorly understood. Schwann cells may contribute to cessation of the immune response by expression of FasL, a molecule that induces apoptosis in activated T cells. Impairment of expression of FasL by Schwann cells could theoretically result in a persistent immune response [18].

It has long been recognized that antibodies play a role in the pathogenesis of CIDP. Complement and immunoglobulin deposition on myelinated nerve fibers in CIDP patients has been demonstrated. Passive transfer experiments have shown that purified immunoglobulin from CIDP patients causes demyelination and conduction block in rat nerves.

As in the acute inflammatory neuropathies, gangliosides and other glycolipids are likely target antigens in CIDP. The similarity continues in that a few CIDP patients are known to have had an antecedent *C. jejuni* infection. Antibodies against a nonmyelin Schwann cell epitope may be important in some CIDP patients.

Axon loss is common in CIDP and its magnitude likely determines the long-term prognosis in CIDP patients. The pathogenic mechanism of axonal loss is unknown. Toxic cytokines such as TNF-α and other cytotoxins such as nitric oxide and metalloproteinases have been suggested as enhancers of axonal destruction.

IV. Neuropathies Associated with Monoclonal Gammopathies

A monoclonal gammopathy is present in nearly 10% of patients with peripheral neuropathy of otherwise uncertain cause but only in 1%–5% in the general population, increasing with age [22]. Monoclonal gammopathies occur in association with a variety of malignant processes, including lymphoma, amyloidosis, Waldenström's macroglobulinemia, myeloma, and Castleman disease. However, much of the time no apparent malignant process is present. In this setting, the term *monoclonal gammopathy of undetermined significance* (MGUS) is applied, accounting for the known tendency for malignant transformation to occur at a rate of up to 1% per year. Inflammatory neuropathies may occur in association with MGUS. In some cases, a direct causative role for the monoclonal antibody is likely, but in many cases the neuropathy is indistinguishable from CIDP and the pathogenic relationship is less certain. Occasionally, an axonal neuropathy without evidence of inflammation is found in patients with MGUS. It is likely that this represents a chance occurrence and that the monoclonal gammopathy is not playing a direct pathogenic role.

A. Immunoglobulin M Monoclonal Gammopathy

Although neuropathy is recognized to occur in relationship with IgG, IgA, and IgM monoclonal gammopathies, the relationship between the gammopathy and the pathogenesis of the underlying neuropathy is best understood for IgM paraproteinemia. More than 50% of patients with a paraproteinemic neuropathy have an IgM gammopathy, but only 11%–27% of gammopathies in general are of the IgM class. Paraproteins of the IgM class are the most likely to be known to recognize antigens in the peripheral nervous system. The best-studied example is anti-myelin-associated glycoprotein (anti-MAG) IgM-related neuropathy [23].

Large studies of patients with measurable IgM antibodies directed against MAG have demonstrated a typical phenotype. Most patients have predominantly sensory symptoms and signs, with electrophysiological evidence of prominent conduction slowing in both sensory and motor nerves. Distal weakness may occur, and a profound sensory ataxia is common. Rarely, a patient may have severe motor involvement. The anti-MAG antibody is often detected as an IgM monoclonal gammopathy but may be present in the absence of a detectable paraproteinemia.

Sural nerve biopsy among patients with anti-MAG neuropathy typically reveals segmental demyelination, endoneurial IgM deposits, and no inflammatory cell infiltration. Ultrastructural studies reveal a characteristic widening of the outer myelin lamellae, as well as widening of myelin lamellae at the Schmidt-Lanterman incisures and paranodal loops. A 23-nm spacing develops between the outer leaflets of the Schwann cell plasma membrane in the location expected for the intraperiod line. This pathology is seen in scattered fibers in anti-MAG neuropathy, as well as in paraproteinemic neuropathies with antigen specificity other than MAG.

Transfer experiments provide considerable evidence that the anti-MAG antibodies can produce the pathological hallmarks of the neuropathy. Despite this, the specific mechanism of demyelination or other nerve damage remains uncertain. Complement fixation appears not to play a role, given the absence of cellular infiltrates and the lack of demonstrated complement binding. One theory holds that anti-MAG antibodies interfere with remyelination through their effects on MAG turnover.

IgM antibodies directed against the NeuAc(α2–8) NeuAc(α2–3) Gal-type disialylated gangliosides (including GD1b, GD3, GT1b, and GQ1b) have been found to be associated with a neuropathy that causes a particularly prominent sensory ataxia. These antibodies are typically identified in the setting of an IgM paraproteinemia. The antibody has cold-agglutinating properties in some patients. In addition to the sensory ataxia, clinical features of this neuropathy include areflexia and little limb motor involvement. Oculomotor and other bulbar muscles may sometimes be involved. When ocular involvement is prominent, the acronym CANOMAD (chronic ataxic neuropathy, ophthalmoplegia, IgM paraprotein, cold agglutinins, and disialosyl antibodies) has been applied to this neuropathy [17]. Experimental studies on these antibodies have included passive transfer experiments in which a patient's IgM antibodies induced cell death in rat dorsal root ganglion neurons. Other studies have indicated preferential immunostaining of dorsal horn laminae, but not motor neurons, by antibodies directed to these gangliosides.

Ganglioside antibodies in motor syndromes were first described in patients with IgM paraproteinemia and motor neuron disease. The antibodies were found to have specificity to GM1. Monoclonal anti-GM1 antibodies have been identified in patients with motor neuropathies without paraproteinemia. MMN, as described previously, occurs both with and without measurable GM1 antibodies. Depending on the detection method used, anti-GM1 antibodies are detected in 25%–50% of patients with MMN. Anti-GM2 antibodies are also associated with chronic, motor-predominant neuropathies. However, similar antibodies have been detected in control subjects, so the pathogenetic relationship remains uncertain.

B. Immunoglobulin G and A Monoclonal Gammopathies

IgG and IgA paraprotein-related chronic inflammatory neuropathies occur in patients with myeloma or MGUS. Neuropathy is particularly associated with the rare variant of myeloma, osteosclerotic myeloma. This form of myeloma accounts for only 3% of all myeloma, but 50% of patients with osteosclerotic myeloma have neuropathy. The presence of neuropathy often leads to the initial detection of osteosclerotic myeloma. The typical neuropathy associated with osteosclerotic myeloma is distal and symmetrical and involves both sensory and motor fibers, although the neuropathy may involve proximal muscles and be indistinguishable from typical CIDP. Both pathological studies and nerve conduction studies reveal evidence of acquired demyelination and often some degree of superimposed axonal loss. The paraprotein is IgA or IgG, almost always with lambda light chains. The pathogenesis of neuropathy with osteosclerotic myeloma is uncertain. Other neuropathic conditions to consider in myeloma patients include compression neuropathy related to bony involvement or nerve infiltration by plasma cells. Typical CIDP is occasionally seen in multiple myeloma patients.

A defined syndrome is also associated with osteosclerotic myeloma. The POEMS syndrome, characterized by polyneuropathy, organomegaly, endocrinopathy, monoclonal gammopathy, and skin changes, occurs with nonmalignant IgG or IgA gammopathy, osteolytic myeloma, extramedullary plasmacytoma, and lymphatic hyperplasia. Few patients display all components of full-blown POEMS syndrome, but many patients with osteosclerotic myeloma and a demyelinating neuropathy will have at least one of the other systemic manifestations of POEMS [17].

V. Treatment

Inflammatory neuropathies are likely the most treatable of all generalized neuropathic disorders. Treatment trials involving large numbers of patients have confirmed the efficacy of various immunomodulatory treatments in both acute and chronic inflammatory neuropathies.

For GBS, the largest trials have focused on AIDP. Both PEx and IVIG have been shown to be beneficial. The time to recovery is shortened, and it is likely that both therapies prevent progression to more severe impairment. A head-to-head comparison trial found equal benefit with IVIG or PEx, and the use of IVIG following PEx provided no additional benefit. A smaller study suggests that patients with GM1 antibodies and antecedent *C. jejuni* infection respond better to IVIG than to PEx. Interestingly, corticosteroids have been shown to be ineffective when used as the sole immune-modulating treatment. A factor in the care of GBS patients as important as immune-modulating therapy is care in a modern intensive care unit. A drop in GBS-related mortality from 15%–20% to 1.23%–2.5% has been realized with the advent of dedicated critical care units.

As might be assumed from the heterogeneity of the group of chronic inflammatory neuropathies, no single treatment is effective in all patients. CIDP has been most extensively studied. Most patients respond to treatment with a relatively high dose of corticosteroids, typically prednisone at 60–100 mg per day initially. Response rates of up to 95% are reported for this regimen. A slow taper is begun when a clinical response is obtained. The taper may take up to 18 months before an acceptable level of daily or alternate-day prednisone dosing is reached. Many patients do not tolerate or have contraindications to such prolonged corticosteroid treatment. The use of IVIG and PEx is supported by several clinical trials as well. IVIG is used more than PEx, particularly for chronic therapy. It is typical for patients to be treated with periodic IVIG infusions as a long-term therapy, although there is little trial data to fully support this practice. Pulsed cyclophosphamide has been advocated for use in rapidly progressive disease and may be useful for other patients who fail to respond to or do not tolerate other regimens. Long-term immunosuppression with azathioprine or mycophenolate is employed by many neuromuscular specialists, but this practice is based on small case series and clinical experience, without the evidence-based support from large, randomized clinical trials.

VI. Summary

The inflammatory demyelinating and axonal neuropathies may be viewed as a heterogeneous collection of disorders that share the common feature of autoimmunity underlying their pathogenesis. A few of the subtypes of inflammatory neuropathies have well-recognized antibodies that clearly react with nerve antigens, suggesting a plausible pathogenic relationship. Among other inflammatory neuropathies antibodies may be identified, but their role in pathogenesis remains obscure. In others, cell-mediated immunity may be the primary pathogenic mechanism. Some disorders seem to be triggered by an antecedent infection, but in most the inciting event is unknown. It remains uncertain why in some forms of neuropathy inflammation is self-limited, whereas at other times a long-lasting autoimmune attack persists. Despite the large number of unanswered questions regarding inflammatory neuropathies, as a group they make up the most treatable of the generalized neuropathic disorders. It is likely that even better treatments will follow an improved understanding of the pathogenesis of the inflammatory neuropathies.

References

1. McKhann, G. M., Griffin, J. W., Cornblath, D. R., Mellits, E. D., Fisher, R. S., and Quaskey, S. A. (1988). Plasmapheresis and Guillain-Barré syndrome: analysis of prognostic factors and the effect of plasmapheresis. *Ann. Neurol.* **23(4)**, 347–353.
2. Fletcher, D. D., Lawn, N. D., Wolter, T. D., and Wijdicks, E. F. (2000). Long-term outcome in patients with Guillain-Barré syndrome requiring mechanical ventilation. *Neurology* **54(12)**, 2311–2315.
3. Ho, T. W., Li, C. Y., Cornblath, D. R., Gao, C. Y., Asbury, A. K., Griffin, J. W., and McKhann, G. M. (1997). Patterns of recovery in the Guillain-Barré syndromes. *Neurology* **48(3)**, 695–700.
4. McKhann, G. M., Cornblath, D. R., Griffin, J. W., Ho, T. W., Li, C. Y., Jiang, Z., Wu, H. S., Zhaori, G., Liu, Y., Jou, L. P., et al. (1993). Acute motor axonal neuropathy: a frequent cause of acute flaccid paralysis in China. *Ann. Neurol.* **33(4)**, 333–342.
5. Ho, T. W., Hsieh, S. T., Nachamkin, I., Willison, H. J., Sheikh, K., Kiehlbauch, J., Flanigan, K., McArthur, J. C., Cornblath, D. R., McKhann, G. M., and Griffin, J. W. (1997). Motor nerve terminal degeneration provides a potential mechanism for rapid recovery in acute motor axonal neuropathy after *Campylobacter* infection. *Neurology* **48(3)**, 717–724.
6. Asbury, A. K., Arnason, B. G., and Adams, R. D. (1969). The inflammatory lesion in idiopathic polyneuritis: its role in pathogenesis. *Medicine* **48(3)**, 173–215.
7. Brechenmacher, C., Vital, C., Deminiere, C., Laurentjoye, L., Castaing, Y., Gbikpi-Benissan, G., Cardinaud, J. P., and Favarel-Garrigues, J. P. (1987). Guillain-Barré syndrome: an ultrastructural study of peripheral nerve in 65 patients. *Clin. Neuropathol.* **6(1)**, 19–24.
8. Feasby, T. E., Hahn, A. F., Brown, W. F., Bolton, C. F., Gilbert, J. J., and Koopman, W. J. (1993). Severe axonal degeneration in acute Guillain-Barré syndrome: evidence of two different mechanisms? *J. Neurol. Sci.* **116(2)**, 185–192.
9. Griffin, J. W., Li, C. Y., Ho, T. W., Tian, M., Gao, C. Y., Xue, P., Mishu, B., Cornblath, D. R., Macko, C., McKhann, G. M., and Asbury, A. K. (1996). Pathology of the motor-sensory axonal Guillain-Barré syndrome. *Ann. Neurol.* **39(1)**, 17–28.
10. Odaka, M., Yuki, N., and Hirata, K. (2001). Anti-GQ1b IgG antibody syndrome: clinical and immunological range. *J. Neurol. Neurosurg. Psychiatry* **70(1)**, 50–55.
11. Willison, H. J., and Yuki, N. (2002). Peripheral neuropathies and anti-glycolipid antibodies. *Brain* **125**, 2591–2625.
12. Yuki, N. (2001). Infectious origins of, and molecular mimicry in, Guillain-Barré and Fisher syndrome. *Lancet Infectious Diseases* **1**, 29–37.

13. Hughes, R. A. C., Hadden, R. D. M., Gregson, N. A., and Smith, K. J. (1999). Pathogenesis of Guillain-Barré syndrome. *J. Neuroimmunol.* **100**, 74–97.
14. Willison, H. J., and O'Hanlon, G. M. (1999). The immunopathogenesis of Miller-Fisher syndrome. *J. Neuroimmunol.* **100**, 3–12.
15. Koski, C. L., Humphrey, R., and Shin, M. L. (1985). Anti-peripheral myelin antibody in patients with demyelinating neuropathy: qualitative and kinetic determination of serum antibody by complement component 1 fixation. *Proc. Nat. Acad. Sci. USA* **82**, 905–909.
16. Emilia-Romagna Study Group on Clinical and Epidemiological Problems in Neurology. (1998). Guillain-Barré syndrome variants in Emilia-Romagna, Italy, 1992–1993: incidence, clinical features, and prognosis. *J. Neurol. Neurosurg. Psychiatry* **65**, 218.
17. Lewis, R. A. (2005). Chronic inflammatory demyelinating polyneuropathy and other immune-mediated demyelinating neuropathies. *Semin. Neurol.* **25(2)**, 217–218.
18. Kiesser, B. C., Keifer, R., Gold, R., Hemmer, B., Willison, H. J., and Hartung, H. P. (2004). Advances in understanding and treatment of immune mediated disorders of the peripheral nervous system. *Muscle Nerve* **30**, 131–156.
19. Koller, H., Schroeter, B. C., Kieseier, B. C., and Hartung, H. P. (2005). Chronic inflammatory demyelinating polyneuropathy: update on pathogenesis, diagnostic criteria and therapy. *Curr. Opin. Neurol.* **18**, 273–278.
20. Rezania, K., Gundogdu, B., and Soliven, B. (2004). Pathogenesis of chronic inflammatory demyelinating polyradiculoneuropathy. *Front. Biosci.* **9**, 939–945.
21. Kieseier, B. C., Dalakas, M. C., and Hartung, H. P. (2002). Immune mechanisms in chronic inflammatory demyelinating neuropathy. *Neurology* **59(12 Suppl 6)**, S7–S12.
22. Nobile-Orazio, E., and Carpo, M. (2001). Neuropathy and monoclonal gammopathy. *Curr. Opin. Neurol.* **14**, 615–620.
23. Nobile-Orazio, E. (2004). IgM paraproteinaemic neuropathies. *Curr. Opin. Neurol.* **17**, 599–605.

80

Toxic and Drug-Induced Neuropathies

Guido Cavaletti, MD

Keywords: *drugs, chemotherapy, medications, peripheral neuropathy, toxic*

I. Antimicrobial Agents
II. Nucleoside Reverse Transcriptase Inhibitors
III. Cardiovascular Drugs
IV. Statins
V. Disulfiram
VI. Colchicine
VII. Chloroquine
VIII. Gold Salts
IX. Leflunomide
X. Tacrolimus (FK506)
XI. Interferons
XII. Antiepileptic Drugs
XIII. Antineoplastic Drugs
XIV. New Generation Antineoplastic Drugs
   References

Peripheral neuropathy is a side effect of several commonly used drugs [1–4]. The correct diagnosis of medication-induced neuropathy can be difficult, and a high index of suspicion is required to identify a medication as the cause of peripheral neuropathy. In this case, the diagnosis may only be made after a detailed and negative screening of other, more common causes of neuropathy. This is an important issue, because a delay in recognizing a toxic neuropathy may lead to an unfavorable course and incomplete recovery after drug withdrawal or the beginning of treatment. However, the temporal association between the use of a drug and the onset of peripheral neuropathy does not necessarily indicate a causal relationship. Criteria have been proposed to clarify this issue. An association with a medication is considered likely when a strong dose–response relationship is present, clinical manifestations are consistent and occur rapidly after drug administration is initiated, amelioration is obtained after drug withdrawal, and similar pathological alterations can be observed in human biopsies and animal models [3]. Nevertheless, these criteria are not always applicable in clinical practice.

An accurate patient history is the essential first step in any case of possible drug-induced neuropathy, because its onset and course are variable, depending on the drugs and early symptoms, and signs can be subtle. In most cases, nerve damage is chronic and ensues after lengthy exposure, but the onset may be acute or subacute. Symptoms and signs generally tend to disappear after drug withdrawal, although worsening can occur despite drug discontinuation, thus raising doubts as to the correct diagnosis. Depending on the site and type of neurotoxic action, the clinical symptoms and signs of drug-induced neuropathy, as well as its course, may vary. In most cases, the clinical presentation is that of length-dependent, distal, symmetrical polyneuropathy. Axonal degeneration is the most common type of pathological change observed after neurotoxic drug exposure, and demyelination is much rarer. However, dorsal root ganglion (DRG) neuronopathy can also occur because the peripheral nervous system is not protected by an efficient blood–nerve barrier at this site. In this case, symptoms and signs may be "length independent." When dorsal root ganglia are the main target of the neurotoxic action, sensory symptoms and signs are largely predominant and axonal and/or myelin damage may be associated with sensory and motor impairment. Accordingly, demyelinating neuropathies tend to recover more easily and more quickly than in cases with severe axonal damage. In the case of severe DRG neuronopathy, permanent impairment is not uncommon.

Another clinically relevant issue is that drug-induced peripheral neuropathies are not equally severe in all exposed patients, suggesting that specific factors may predispose patients to this side effect of therapy [5]. Among the known factors, decreased drug metabolism or clearance may occur, particularly in elderly patients, and preexisting neuropathies (even asymptomatic) may be associated with increased neurotoxicity of several compounds. However, it is likely that still unknown factors are also a basis for the variation in individual susceptibility to drug-induced peripheral neurotoxicity. Identifying these factors is one of the targets of ongoing pharmacogenomic studies.

Neurophysiological examination is generally effective to better characterize the type and site of damage, to monitor the treatment, and to follow up the course of the neuropathy. Nerve biopsy is rarely required for the diagnosis, particularly when known neurotoxic drugs are involved. However, sural nerve biopsy may be indicated when new agents are involved because, in most cases, sensory impairment is relevant. The use of less invasive pathological methods, such as skin biopsy, to estimate intraepidermal nerve fiber density has been suggested. Skin biopsy could be particularly useful for monitoring treatments with an expected high incidence of sensory symptoms, and early changes should be demonstrated, such as during antineoplastic chemotherapy. This approach is minimally invasive and potentially sensitive; consequently, this method has been used in other neuropathies (e.g., diabetic neuropathy). However, validation of the use of skin biopsies in toxic neuropathies has not yet been performed; therefore, at the moment, its use should be considered only as a supportive experimental procedure.

The investigation into toxic neuropathies has shown that preclinical *in vitro* and *in vivo* models have already proven useful and will be so in the future—provided that they reliably reflect the clinical manifestations described in patients, a requirement that cannot always be met despite extensive efforts.

Table 1 lists compounds that have been associated with iatrogenic peripheral neurotoxicity (although, in some cases, the demonstration of a true causal effect is debatable). This chapter focuses on the most clinically relevant drug-induced

Table 1 Summary of Medications Reported as Neurotoxic to the Peripheral Nervous System

Predominant Pathology	Axonal	Axonal	Demyelinating	Mixed
Predominant clinical presentation	Sensory	Sensory + motor		
	Almitrine	Chloramphenicol	Perhexiline	Amiodarone
	Bortezomib	Colchicine	Procainamide	Chloroquine
	Doxorubicin	Dapsone	Zimeldine	FK506 (tacrolimus)
	Ethambutol	Disulfiram		Gold salts
	Ethionamide	Epothilones		Suramin
	Etoposide (VP-16)	Leflunomide		
	Gemcitabine	Lithium		
	Glutethimide	Methyl bromide		
	Hydralazine	Metronidazole		
	Ifosfamide	Nitrofurantoin		
	Interferon-α	Taxanes		
	Isoniazid	Vinca alkaloids		
	Misonidazole			
	NRTIs			
	Phenytoin			
	Platinum analogues			
	Propafenone			
	Pyridoxine			
	Statins			
	Thalidomide			

Note: NRTIs, nucleoside reverse transcriptase inhibitors.

neuropathies, paying particular attention to cases in which this side effect is dose limiting and requires drug withdrawal or dose reduction, such as during antineoplastic treatment.

I. Antimicrobial Agents

Several antimicrobial drugs have been reported as causing peripheral neuropathy, although clear evidence of these compounds having a causal role has been demonstrated only for some of the drugs and only after long-term administration [3].

Chloramphenicol, metronidazole, and nitrofurantoin have all been described as causing axonal, length-dependent, sensorimotor neuropathies with a close correlation with chronic use (e.g., for long-term prophylaxis of recurrent urinary tract infections or treatment of inflammatory bowel diseases) and high-dose schedules. Optic neuropathy can precede, follow, or be associated with peripheral neuropathy during chloramphenicol treatment. With all of these drugs, however, the incidence of reported cases is low compared with the number of patients exposed to them, and the availability of safer and effective alternative drugs will further reduce the future occurrence of neuropathy.

More detailed information is available for medications used to treat tuberculosis, such as ethambutol, ethionamide, and isoniazid. Although ethambutol and ethionamide neuropathies are rare, mild, predominantly sensory, and result from axonal damage through an unknown mechanism, isoniazid neuropathy is more common and can be severe, particularly when doses of 300 mg/day or more are used. Optic neuropathy and central nervous system involvement (ataxia, seizures), as well as myalgias and rhabdomyolysis, have also been reported with high doses of isoniazid. Peripheral neuropathy affects around 1% of the treated patients, with a higher risk for people with a slow rate of drug metabolism (slow acetylators), concomitant malnutrition, or alcohol abuse and high-dose schedules. Distal, symmetrical sensory symptoms and signs are predominant over motor impairment, and they are secondary to isoniazid-induced pyridoxine depletion. Prophylactic administration of pyridoxine (25–50 mg/day) prevents the onset of isoniazid neuropathy and allows the planned treatment to be completed in most cases. Pyridoxine administration is also effective in treating isoniazid neuropathy once symptoms and signs have ensued and even if axonal damage is already present.

However, higher-than-recommended doses of pyridoxine can be neurotoxic to the peripheral nerves and they can make the neuropathy worse. The interaction between isoniazid treatment and pyridoxine (which is important for protein, carbohydrate, and fatty acid metabolism and for sphingomyelin synthesis) is not completely understood, and no correlation has been demonstrated between the severity of the neuropathy and the pyridoxine levels. Pure axonal sensory neuropathy is the most common type of pyridoxine-induced damage, but in the most severe cases, because of marked pyridoxine overdose, DRG neuron damage with poor recovery may occur.

Dapsone is used mainly to treat leprosy and dermatitis herpetiformis, but in recent years it has also been used as a second-line drug for rheumatoid arthritis and for the prophylaxis of opportunistic infections in immunocompromised patients. It acts by facilitating the conversion of myeloperoxidase into an inactive form and by inhibiting neutrophil adherence to vascular endothelium. The risk of toxic effects on the peripheral nervous system is higher in slow acetylators and with the long-term administration of high doses (i.e., more than 300 mg/day). Neuropathy is predominantly motor, also severely affecting the arms with a symmetrical, distal distribution, and coasting is common. The neuropathy is primarily axonal, but the mechanism of action is unknown.

II. Nucleoside Reverse Transcriptase Inhibitors

Nucleoside reverse transcriptase inhibitors (NRTIs) are a key therapy in human immunodeficiency virus (HIV) infections, and they can cause both axonal degeneration and myelin damage associated with mitochondrial dysfunction.

Drugs such as zalcitabine, didanosine, stavudine, fialuridine, and lamivudine (but not zidovudine, which induces myopathy) are associated with largely sensory distal, symmetrical polyneuropathy with acute or subacute onset that may be dose limiting [3,6]. Zalcitabine seems to be the most neurotoxic drug within the NRTI family. A typical clinical feature of NRTI neuropathy is pain, a symptom that may make it difficult to differentially diagnose NRTI neuropathy from HIV painful neuropathy (which occurs in at least one-third of HIV patients at the beginning of their history), although the latter is generally more insidious in its onset. An accurate differential diagnosis between the two clinical conditions is essential but often difficult, because NRTI painful neuropathy can be dose limiting. A clue to discriminating between the two conditions may be represented by serum lactate measurement, because high levels of lactate are often associated with NRTI neuropathy but not with HIV neuropathy. The incidence of NRTI neuropathy is high (it occurs in up to 40% of the treated patients), particularly when high doses of the drugs are administered, and recovery may require months, even after drug withdrawal. Neuropathological studies have demonstrated axonal damage with typical enlargement and vacuolation in mitochondria, associated with electron-dense structures and myelin degeneration.

A mechanism for NRTI neurotoxicity has recently been suggested and is strictly related to NRTIs' pharmacological properties. These compounds have been designed to be phosphorylated and then to compete with natural substrates for the HIV reverse transcriptase. Unfortunately, they react also with γ-deoxyribonucleic acid (γ-DNA) polymerase, which is an enzyme required for mitochondrial DNA replication. This interference reduces mitochondrial DNA and mitochondrial DNA-encoded enzymes, and it subsequently depletes neuronal energy during NRTI treatment [6]. An indirect confirmation of mitochondrial dysfunction during NRTI administration is represented by the analysis of the most common side effects associated with NRTI-induced peripheral neuropathy—that is, myopathy, pancreatitis, lactic acidosis, cardiomyopathy, and bone marrow suppression, which closely resemble the clinical features of mitochondrial diseases. Reduced acetyl-L-carnitine levels have been demonstrated in symptomatic patients as further evidence of mitochondrial damage, and the effectiveness of the exogenous administration of this substance in the treatment of NRTI-induced neuropathy has been reported.

III. Cardiovascular Drugs

Peripheral neuropathies induced by cardiovascular drugs have occasionally been reported [3], but consistent peripheral damage has been described only for chronic amiodarone administration.

Although it is an old drug, amiodarone is still a cornerstone of antiarrhythmic treatment. Neurological toxicity of amiodarone is not limited to the peripheral nerves (here an incidence of more than 5% has been reported in chronically treated patients) but includes optic neuropathy, myopathy, and central nervous system symptoms with ataxia and tremors.

Peripheral neuropathy induced by chronic amiodarone treatment was initially recognized in 1974. Subsequent studies have revealed that the drug accumulates within peripheral nerves (concentration may be 80 times higher than in serum) where it causes segmental myelin damage and axonopathy. The mechanism of amiodarone toxicity is unknown. A prominent feature of amiodarone neuropathy is represented by abundant lysosomal lamellar inclusions, which are observed not only in peripheral nerves (within Schwann cells, fibroblasts, and endothelial and perineurial cells) but also in autonomic, dorsal root ganglia, and myenteric plexus neurons and nonneural cells.

Clinical symptoms, generally occurring after chronic use of more than 400 mg/day, are those of length-dependent, symmetrical sensorimotor neuropathy with recovery after amiodarone treatment discontinuation. The long half-life of amiodarone can result in a delayed recovery. Although autonomic symptoms are rare, orthostatic hypotension has been occasionally associated with amiodarone neuropathy. Altered response in diabetic patients, suggesting an enhancement of a preexisting autonomic neuropathy, has also been reported [7].

Hydralazine, used to treat arterial hypertension, can cause sensory axonal neuropathy, particularly in slow acetylators. Pyridoxine depletion has been suggested as the mechanism of toxic action.

IV. Statins

Various agents are available for the treatment of hypercholesterolemia, but in recent years the most commonly used drugs are the 3-hydroxy-3-methylglutaryl coenzyme A (HMG-CoA) reductase inhibitors, commonly known as *statins*. They are better tolerated than similar drugs, although cases of severe muscular damage have occurred and careful monitoring is required. Cases of peripheral neuropathy have also been reported. Although the incidence of statin-induced neuropathies is probably low, the potential number of subjects who are candidates for chronic treatment is high, thus making it necessary to consider this potential side effect [8]. The clinical features of statin-induced neuropathy have not yet been clearly defined: in the cases reported so far, the most common symptoms are distal sensory and motor impairment indicating symmetrical polyneuropathy. Complete recovery generally occurs after drug withdrawal. Neurophysiological studies indicate axonopathy, but in some patients fasciculations associated with proximal and distal weakness (suggesting motor neuron damage) have been observed and incomplete recovery may occur. Concomitant evidence of subclinical muscle damage with electromyographic examination is common even in asymptomatic patients and is generally associated with a slight increase in serum creatine kinase levels at rest or following mild physical exercise. HMG-CoA reductase inhibitor interference with mitochondrial activity through the blocking of ubiquinone and dolichol has been postulated as the main mechanism of neuronal toxicity of these compounds, but interference with axonal transport and mitochondrial dysfunction have also been suggested [9].

V. Disulfiram

The use of disulfiram in the treatment of alcohol addiction has been associated with the onset of toxic peripheral neuropathy, although it is often difficult to separate the features of the damage induced by the drug from the underlying alcohol-induced changes.

However, it has been demonstrated that chronic high-dose administration (i.e., more than 200–300 mg/day) of disulfiram can cause neurofilament swelling within axons, causing distal sensorimotor polyneuropathy. Motor signs are generally rather mild and limited to the most distal segments of the limbs. However, severe or fulminant cases of peripheral neuropathy have rarely been associated with disulfiram use. The mechanism of disulfiram toxicity is not clear, although it might be related to the action of one of its metabolites, carbon disulfide, which is a known neurotoxin and causes neurofilamentous distal axonopathy in animal models.

VI. Colchicine

Colchicine is a natural compound obtained from the tuber of *Gloriosa superba*. It was introduced into clinical practice more than 200 years ago and is still used to treat spontaneous gout and cyclosporine A-induced gout in organ transplant recipients. The major neurological side effect of colchicine is toxic, vacuolar myopathy with lysosomal accumulation, but peripheral neuropathy may also occur, probably secondarily to impaired microtubular assembly. The risk of developing peripheral neuropathy is increased by concomitant renal impairment, a rather common complication of gout, and by chronic administration of the drug, which is generally required. Overall, the incidence of colchicine neuropathy is low, and even in symptomatic patients the severity of the distal sensorimotor impairment is mild. Axonal damage involving both myelinated and unmyelinated fibers is evident in the affected patients, and recovery is generally complete after drug withdrawal. However, intracytoplasmatic inclusions in endothelial, perineurial, and Schwann cells, as well as segmental demyelination, have occasionally been reported in sural nerve studies. At the neurophysiological level, the typical changes resulting from axonopathy are generally associated with abnormal spontaneous activity more evident in the proximal muscles at an electromyographic examination, thereby giving rise to the condition known as *colchicine myoneuropathy*.

VII. Chloroquine

Chloroquine causes a toxic myopathy but also produces a toxic neuropathy. Chloroquine is an antimalarial drug also used for the treatment of severe rheumatoid arthritis, systemic lupus erythematosus, and other dermatological diseases. The clinical features of chloroquine-induced neuropathy are those of symmetrical sensorimotor polyneuropathy. Both axonal damage and primary demyelination have been reported. A prominent pathological feature of chloroquine neuropathy is ultrastructural intracytoplasmatic lamellar inclusions, resembling those described in amiodarone neuropathy, within Schwann cells and in perineurial and endothelial cells but not in axons. Only long-term treatment has been associated with chloroquine neuropathy, followed by complete recovery after drug withdrawal.

VIII. Gold Salts

Gold salts, which are now used only occasionally to treat severe rheumatoid arthritis, are a rare cause of peripheral neuropathy described for the first time more than 70 years ago. Although the incidence of gold salt neuropathy is low, it must always be considered in differential diagnoses, particularly when high cumulative doses of gold have been administered. However, peripheral neuropathy has been reported with cumulative doses of gold ranging from 30 to 2600 mg. Sodium aurothiomalate is the most common preparation associated with neuropathy. The onset is generally slow and secondary to axonal damage, but cases with sudden motor weakness (sometimes asymmetrical) and prominent segmental demyelination have also been reported. In both chronic and acute neuropathies, motor signs (weakness, muscle cramps and atrophy, myokymia) predominate over sensory impairment and recovery is invariably slow. The neurophysiological examination is often normal or a mild reduction in nerve conduction velocity may be demonstrated, but denervation in affected muscles is not uncommon. The presence of elevated cerebrospinal fluid levels in some patients has raised the possibility of radicular involvement. The pathogenesis of gold-induced neuropathy is unknown, although the different clinical presentations suggest that the medication may act with different mechanisms.

IX. Leflunomide

Leflunomide is a new immunosuppressive prodrug used for the treatment of rheumatoid arthritis, which can induce the onset of axonal sensorimotor neuropathy. Pain can sometimes be a prominent clinical feature. The clinical onset is usually slow, generally after 4–6 months of treatment, with a slow recovery after drug withdrawal. The mechanism of peripheral neurotoxicity is still unknown [8].

X. Tacrolimus (FK506)

Tacrolimus is a potent immune-suppressant drug acting on T cell proliferation after specific binding to a 12-kDa protein resulting in the inhibition of calcineurin activity and interleukin-12 synthesis. It is used particularly in the prevention of rejection response after organ transplantation. Tacrolimus is a macrolide antibiotic produced by *Streptomyces tsukubaensis* and is often associated with central nervous system toxicity (reversible leukoencephalopathy, seizures, tremors). Peripheral neuropathy due to tacrolimus administration occurs in about 0.5% of patients after 2 to 10 weeks of treatment and is represented by a predominantly sensory impairment with milder motor dysfunction. At the pathological level, both axonopathy and segmental demyelination have been reported in tacrolimus-treated patients [1]. In these patients, whose clinical course closely resembles chronic inflammatory demyelinating polyradiculoneuropathy, it has been suggested that the altered function of T cells induced by the drug administration may be the underlying mechanism of action, and intravenous immunoglobulins or plasmapheresis treatments have been reported as being effective.

Despite these observations, tacrolimus has also been evaluated as a neuroprotectant drug in NRTI and acrylamide peripheral neurotoxicity animal models [8].

XI. Interferons

Interferons are used to treat several medical conditions ranging from hepatitis C and hematological malignancies (α-interferon) to multiple sclerosis (β-interferon). β-Interferon is generally considered devoid of any neurotoxic effect on the peripheral nervous system, leading to the proposal that it could be used as a possible treatment for cases of chronic demyelinating polyradiculoneuropathy refractory to conventional treatment. However, patients with multiple sclerosis who developed polyneuropathy, or had exacerbation of previously subclinical neuropathy, during treatment with β-interferon have been recently reported [10]. In some of these patients the neuropathy improved after discontinuation of treatment, and in two patients it relapsed on rechallenge. Peripheral neuropathies attributed to α-interferon administration have been reported more often, although the pathogenesis is unknown. In most instances, α-interferon has been used to treat chronic viral hepatitis (a condition that may itself be a cause of peripheral neuropathy), and the reported side effect is the onset or worsening of predominantly sensory axonal polyneuropathy. Although in most cases the peripheral nerve impairment was dose dependent, some cases have been reported after low-dose α-interferon administration. Similarly, asymmetrical neuropathies, brachial plexopathies, or mononeuropathies have occasionally been associated with the use of α-interferon. The incidence of α-interferon neuropathy is definitely low, probably no more than 0.1% even after chronic treatment, as suggested by epidemiological studies in viral hepatitis patients.

XII. Antiepileptic Drugs

In recent years several antiepileptic drugs have been reported as being neurotoxic, although in most cases a definite demonstration of this effect at the pathological level has not been reported and the overall results are conflicting. In several cases neurophysiological changes, represented by a reduction in sensory and motor conduction velocity, occurred after mono- or polytherapy, generally in the absence of any symptoms being reported by the affected subjects. Neurological examination in these cases occasionally revealed deep tendon reflex impairment, but in most cases it was normal. Recent epidemiological results have demonstrated a high incidence of these subclinical abnormalities after single-agent administration (in decreasing order, in long-term phenytoin-, carbamazepine- and phenobarbital-treated patients but not after sodium valproate treatment) [11], although previous studies performed with slightly different methods failed to show a significant correlation between antiepileptic drug administration and peripheral neuropathy. Most data on single-agent-induced neuropathy regards phenytoin, a drug that can cause an acute and reversible reduction in nerve conduction velocity at high serum levels, as demonstrated in preclinical models and in healthy volunteers. Phenytoin neuropathy, which has been described mostly on a neurophysiological basis in cohort studies on epileptic patients and which has never been reported as being a severe side effect of the treatment, has been reproduced in a rat model. The pathological changes observed in nerve conduction velocity were attributed to protein kinase C activation secondary to phenytoin-induced depletion of reduced glutathione [12]. Low folate levels have been advocated as a cofactor in phenytoin-induced neuropathy in humans. A few sural nerve biopsies have been performed in antiepileptic drug-treated patients, revealing inconsistent changes in the different cases ranging from axonopathy to demyelination and raising some concern about the true correlation with drug administration. Both phenobarbital and carbamazepine have been reported as being neurotoxic either when intoxication occurred or in animal models, but no consistent results have been reported suggesting that they can cause peripheral neuropathy, even after long-term administration. The current opinion is that the clinical relevance of antiepileptic drug neuropathy is minimal when plasma levels are maintained within the normal therapeutic range.

XIII. Antineoplastic Drugs

Among the various neurotoxic drugs, antineoplastic agents represent one of the major clinical problems given their widespread use, the potential severity of their toxicity, and the absence (in several cases) of alternative drugs with the same effectiveness [2,13,14]. Given the use of high-intensity schedules, the peripheral neurotoxicity of antineoplastic agents often represents one of their dose-limiting side effects in the treatment of cancer. Moreover, even when antineoplastic agents' peripheral neurotoxicity is not a dose-limiting side effect, its onset may severely affect the quality of life of cancer patients and cause chronic discomfort. This is a serious problem for all patients, but it can be particularly important for those subjects with the best outcome in terms of anticancer treatment leading to long survival. On clinical grounds, several classes of effective drugs induce sensory and/or motor impairment during chemotherapy or even after treatment withdrawal, depending on the target and site of the neurotoxic action. Different neuroprotective strategies have been tested, but although encouraging results have been reported with several compounds in animal models, no clear-cut evidence of a clinically relevant effect of these compounds has been obtained in humans [8,15,16]. Finally, although rarely considered, environmental exposure of hospital personnel exposed to antineoplastic agents should always be considered when peripheral neuropathy of unknown origin ensues in individuals at risk.

For all these reasons, this part of the chapter reviews in detail the features of chemotherapy-induced peripheral neurotoxicity resulting from the administration of the most widely used neurotoxic drugs. Moreover, the earliest data regarding the neurotoxicity of some new classes of promising antineoplastic agents, such as epothilones and proteasome inhibitors, is discussed.

A. Platinum-Based Drugs

Since the introduction of platinum-based chemotherapy, particularly cisplatin (which was available for its first clinical trials in 1972), the outcome of treatment of many solid tumors has changed. The main platinum compounds in cancer chemotherapy are cisplatin, carboplatin, and oxaliplatin, but several new agents are under investigation and peripheral neurotoxicity should be considered as one of their potential side effects [17]. The platinum drugs in use share some structural similarities, although they have marked differences in their therapeutic use, pharmacokinetics, and adverse effect profiles.

Platinum drugs are virtually unable to cross the blood-brain barrier that protects the central nervous system, but they have easy access to the peripheral nervous system, particularly at the DRG level. Platinum drugs appear to affect primarily the sensory neurons in the dorsal root ganglia where no blood-brain barrier is present (Fig. 1). However, there are significant pharmacokinetic differences among cisplatin, carboplatin, and oxaliplatin, and these differences are probably at the root of the different clinical features of their peripheral neurotoxicity. Cisplatin is the most highly protein bound (>90%), followed by oxaliplatin (85%) and carboplatin (24–50%), but a peculiarity of oxaliplatin is that, after its intravenous administration, about 33% of the dose is bound to erythrocytes, thus making it difficult to clearly correlate the dose administered with the neurotoxicity.

The precise mechanisms of the cytotoxic and neurotoxic actions of the platinum compounds have not yet been fully elucidated. Several interstrand and intrastrand cross-links in DNA, including two adjacent guanine or two adjacent guanine–adenine bases, can be observed following cisplatin exposure. However, although it is clear that the antineoplastic activity of platinum drugs is largely caused by platinum–DNA interaction, and platinum–DNA adducts have been reported in DRG neurons in experimental models, other mechanisms are probably involved in platinum drugs' neurotoxicity, including interaction with mitochondrial DNA, intracytoplasmatic protein binding, and ion channel interaction.

Cisplatin is the most severely neurotoxic among the platinum drugs. Cisplatin-induced neuropathy is sensory and is predominantly characterized by symptoms of large myelinated fiber damage, such as numbness and tingling, paresthesias of the upper and lower extremities, reduced vibration and position sense perception, reduced deep tendon reflexes, and incoordination with gait disturbance. Pain and thermal impairment are rare. Occasionally, Lhermitte's sign is reported by patients, suggesting the involvement of the centripetal branch of the DRG neuron axon in the spinal cord. Ototoxicity is another common side effect of cisplatin. The first symptoms are often observed after a cumulative dose of 300–500 mg/m^2 of cisplatin. It has been reported that risk factors for more severe neurotoxicity include diabetes mellitus, alcohol consumption, and inherited neuropathies, all conditions that by themselves induce peripheral nerve damage. Advanced age has not been identified as an independent risk factor when there is no comorbidity. However, most related data are anecdotal and are not based on large-scale observations.

After completion of cisplatin chemotherapy, only a small proportion of the patients have significant neurotoxic symptoms, whereas 3–5 months later the proportion is definitely higher. This phenomenon (called *coasting*) is clinically relevant, because it makes it difficult to assess the severity of the DRG neuron damage during cisplatin administration and it prevents correct decisions from being made. Resolution or amelioration of symptoms occurs in most of the patients over the following months (although abnormal neurological examination is often permanent), and in patients with mild signs of cisplatin-related neuropathy, retreatment with platinum drugs is generally feasible after several months.

Conventional dosages of carboplatin have been associated with a lower risk of peripheral neuropathy than cisplatin. Although they are generally less severe, qualitatively the symptoms of carboplatin peripheral neuropathy

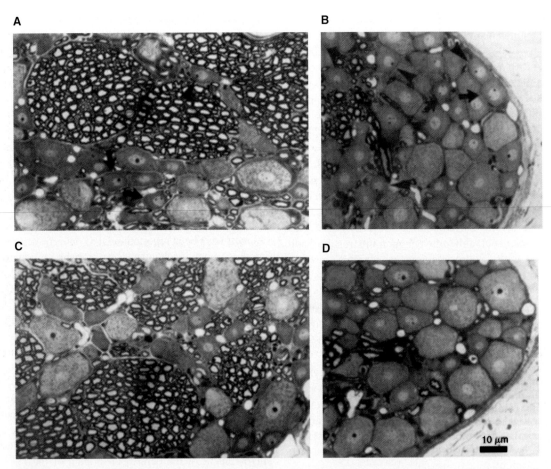

Figure 1 (**A, B**) Light micrographs of dorsal root ganglia in high-dose carboplatin-treated rats and (**C, D**) in controls. Carboplatin-treated neurons have an increased number of eccentric nucleoli (arrowheads) or are multinucleolated (arrows). Reprinted, with permission, from Cavaletti, G., Fabbrica, D., Minoia, C., Frattola, L., and Tredici, G. (1998). Carboplatin toxic effects on the peripheral nervous system of the rat. *Ann. Oncol.* **9**, 443–447.

are the same as those observed with cisplatin. When high-dose regimens (megadose schedules) have been tested to achieve a better antineoplastic response, carboplatin peripheral neurotoxicity has become clinically relevant, with symptoms and signs identical in their severity to those observed after cisplatin administration.

The features of oxaliplatin neurotoxicity are rather different from those of cisplatin and carboplatin. Besides chronic sensory neurotoxicity, in about 90% of patients oxaliplatin treatment is associated with a peculiar acute neurosensory toxicity predominantly affecting the fingers, toes, pharyngolaryngeal tract, and perioral and oral regions, which is generally induced or aggravated by exposure to cold (e.g., drinking a cold beverage). Such symptoms, which sometimes need to be treated with different antiepileptic agents, may occur 30–60 minutes from the beginning or shortly after each course of oxaliplatin. Acute neurotoxicity is generally mild in severity; it disappears within a few hours or days and does not require oxaliplatin treatment withdrawal. Some patients may also develop muscle cramps or spasms, but myalgias are rare. The acute neurotoxic effects of oxaliplatin result from drug-related inhibition of voltage-gated sodium currents. It has been suggested that oxalate ions, which are released during the complex oxaliplatin metabolism, might be responsible for the inhibitory effects on the voltage-gated sodium channels because of their chelating activity.

In addition to the acute neurotoxic symptoms caused by oxaliplatin, about 15%–20% of patients treated with this agent develop a moderate neuropathy, particularly after cumulative doses of 600–800 mg/m^2. The symptoms of chronic neuropathy include non–cold-related dysesthesias, paresthesias, superficial and deep sensory loss with sensory ataxia, and functional impairment that persist between treatment cycles. Although the coasting phenomenon often occurs, most of these symptoms usually disappear months after oxaliplatin withdrawal.

Neurophysiological studies in platinum-drug-treated patients show reduction in the amplitude of the sensory potentials with minimal changes in the sensory nerve conduction velocity. Pathological examination of sural nerve biopsies has demonstrated axonal degeneration without any evidence of primary demyelination. Although human autopsies have shown an accumulation of platinum in the dorsal root ganglia, clear data on the DRG neuron pathological changes are available only in rodent models, demonstrating predominant nucleolar changes with segregation of its granular and fibrillar components (Fig. 2).

B. Antitubulin Drugs

The members of the large family of antitubulin drugs are among the most promising agents for the current and future treatment of cancer, because tubulin is deeply involved in mitosis and cellular replication. Not only are they already part of well-established chemotherapy schedules, but several new drugs are currently in the preclinical phase or are in the first or second phases of clinical development. Two families of antitubulin agents, acting with different mechanisms of action, are commonly used to treat cancer with a range of indications: the taxanes and the vinca alkaloids.

1. Taxanes

Paclitaxel (recognized as an antiproliferative agent in the 1960s during a U.S. National Cancer Institute–supported screening project involving more than 30,000 natural products but only available for the first clinical trials in 1983 because of solubility problems) and docetaxel are two effective drugs belonging to the *taxane* family [18]. Current data suggest that these two drugs have similar efficacy, but regimens based on either of them have shown potentially important differences in their toxicity profiles. This issue is particularly important, because several chemotherapy schedules are based on combinations of taxanes and platinum drugs (generally cisplatin or carboplatin), thus raising concerns about the possible interaction in terms of their peripheral neurotoxicity.

A complete and effective chemical synthesis of taxanes is not yet available because they are large and complex molecules, and this has raised some concerns about their large-scale use. Paclitaxel is a natural drug derived from the stem bark of the Pacific yew tree *(Taxus brevifolia)*, but its concentration is low, so a large number of trees have to be used. On the contrary, docetaxel is a semisynthetic taxane derived from the needles of the European yew tree *(Taxus baccata)*.

Both components are highly bound (about 90%) to proteins and minimally excreted in the urine, with metabolism, biliary excretion, and tissue distribution responsible for the total body clearance. Taxanes are virtually unable to cross the blood-brain barrier, the main limitations being their large molecular weight and the presence of P-glycoprotein, which acts as an extremely efficient drug efflux pump.

The mechanism of the antineoplastic action of taxanes is based on interference with the mitotic spindle and subsequent inactivation because of increased aggregation of intracellular microtubules. In this way, taxanes induce the arrest of the cell cycle at the G2/M phase and the initiation of the apoptotic processes. However, there is increasing evidence that G2/M arrest induced by taxanes is not the only mechanism through which the drug commits cells to apoptosis. Changes in intracellular signaling pathways have been demonstrated in a variety of human cells of nonneuronal origin undergoing paclitaxel-induced apoptosis. Like the antineoplastic action, the mechanism of taxane neurotoxicity is only partially understood. The toxicity of these compounds on postmitotic neurons cannot be

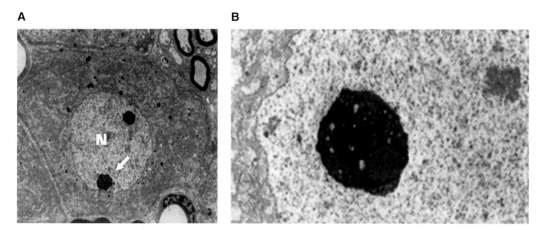

Figure 2 (**A**) Electron micrographs of dorsal root ganglion specimens in cisplatin-treated rats showing multiple eccentric nucleoli (arrows) and (**B**) segregation of the granular and fibrillar components of the nucleolus. *Note:* N, nucleus. Reprinted, with permission, from Tredici, G., Tredici, S., Fabbrica, D., Minoia, C., and Cavaletti, G. (1998). Experimental cisplatin neuronopathy in rats and the effect of retinoic acid administration. *J. Neurooncol.* **36**, 31–40.

explained by the mechanism of antireplicative action previously described. An impairment of axonal transport because of altered structure and functioning secondary to microtubular aggregation seems reasonable, but an effect on DRG neurons and Schwann cells has also been suggested. The few pathological data in humans are limited to sural nerve biopsies, demonstrating unspecified axonal changes, and in animal models axonal microtubular aggregates have been reported (Fig. 3).

Paclitaxel is more neurotoxic than docetaxel, but for the latter drug neurotoxicity may be dose limiting. The clinical features of the peripheral neurotoxicity induced by taxanes are qualitatively identical, and they are mostly represented by distal, symmetrical hypoesthesia in the upper and lower extremities with a length-dependent distribution. In most cases, all sensory modalities are affected and deep tendon reflex loss is an early feature of taxanes' peripheral neurotoxicity. Motor signs and symptoms can occur during treatment with paclitaxel or docetaxel, although only rarely is motor impairment a clinically relevant feature. Distal neuropathic pain may ensue during taxane treatment, and myalgias are a typical symptom in the absence of other evidence of muscular damage. The signs of taxanes' peripheral neurotoxicity tend to be reversible, but in a few cases persistence of sensory impairment is observed and incomplete recovery can occur.

Nerve conduction studies in patients treated with taxanes show a reduction in sensory (and more rarely motor) potential amplitudes, with a mild reduction in sensory and motor conduction velocity.

2. Vinca Alkaloids

The family of vinca alkaloids includes two natural agents, vincristine and vinblastine, and several semisynthetic drugs, such as vindesine and vinorelbine. Vinca alkaloids act by arresting cell mitosis after binding with intracellular tubulin, but the effect is the opposite of that described for taxanes. The exposure to vinca alkaloids induces disassembly of the normal microtubular array. The first molecule of this family to be used as an anticancer agent was vincristine, a drug still in use because of its effectiveness despite the availability of less neurotoxic derivatives (e.g., vinorelbine and vindesine) [19].

Like taxanes, vinca alkaloids are unable to cross the intact blood-brain barrier, but no data point to the dorsal root ganglia as their neurotoxicity target. Most patients treated with vincristine develop dose-dependent and potentially treatment-limiting neurotoxicity, with the clinical features of sensorimotor peripheral neuropathy caused by inhibition of anterograde and retrograde axonal transport secondary to microtubule damage. Sensory signs and symptoms are predominant in most patients, and distal symmetrical hypoesthesia and dysesthesia involve all sensory modalities (i.e., deep and superficial), sometimes with a painful component. Muscular cramps can occur, in association with reduced strength in the distal muscles, in the most severely affected patients. Severe impairment of motor function has occasionally been described in children or in patients with preexisting hereditary neuropathies. Clinically relevant autonomic nervous system involvement, which is rare with other antineoplastic agents, is observed in about one-third of the subjects exposed to vincristine who have orthostatic hypotension, urinary bladder dysfunction, and erectile impotence. The most severe and dose-limiting autonomic side effect is the occurrence of constipation because of a paralytic ileus or megacolon. The toxic signs induced by vincristine are reversible in most cases, but long-lasting impairment or incomplete recovery is common.

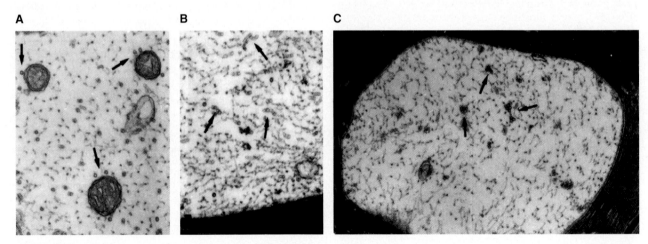

Figure 3 (A) Electron micrographs from the peripheral nerves of paclitaxel-treated rats: microtubules show an increased tendency to surround mitochondria (arrows) and (B) for the formation of linear aggregates (arrows). (C) Neurotubular grouping is more evident after polyethyleneimine staining (arrows). Reprinted, with permission, from Cavaletti, G., Tredici, G., Braga, M., and Tazzari, S. (1995). Experimental peripheral neuropathy induced in adult rats by repeated intraperitoneal administration of Taxol. *Exp. Neurol.* **133**, 64–72.

Nerve conduction studies show decreased distal motor and sensory nerve action potentials with less prominent changes in nerve conduction velocity, and electromyography shows denervation in distal muscles. These findings are consistent with the largely predominant axonal damage described in sural nerve biopsies.

C. Suramin

Suramin was introduced into clinical treatment in 1920 as an antiparasitic agent. More recently it has been demonstrated that suramin is a potent reverse transcriptase inhibitor (but it is not effective against HIV infection). It has also been recognized as the first member of a new family of agents able to interfere with cancer cell growth, probably through an action on cell growth factor receptors. This effect of suramin, which has been demonstrated at least for the epidermal growth factor, platelet-derived growth factor, nerve growth factor, and insulin-like growth factor, is likely to be a mechanism at the basis of its neurotoxicity [1].

Tissue distribution of suramin is generally poor, and virtually no detectable amount of the drug, which is a highly charged molecule, is present in the central nervous system.

From the clinical point of view, suramin is not yet widely used as an anticancer drug, but it is potentially an interesting agent because it is one of the few antineoplastic compounds effective in hormone refractory prostate cancer. Its mechanism of peripheral neurotoxicity is still unclear, although some evidence indicates that the drug can block metabotropic (P2Y) and ionotropic (P2X) P2 purine receptors present on the DRG neurons (P2Y) and on the Schwann cells (P2X) and involved in Ca^{++} homeostasis. Although it is possible that suramin may be neurotoxic through a perturbation of Ca^{++}-dependent intracellular pathways, no conclusive data are available to confirm this hypothesis.

The clinical features of suramin-induced peripheral neurotoxicity are those of unspecified axonal sensorimotor neuropathy in most cases. However, a proportion of patients develop potentially severe demyelinating proximal and distal polyradiculoneuropathy, affecting both sensory and motor fibers, with the clinical, neurophysiological, and pathological features of Guillain-Barré syndrome (including the presence of lymphocytic infiltration observed in sural nerve biopsy studies). The risk of developing suramin-induced peripheral neurotoxicity is closely correlated with the peak plasma levels (about 40% of patients with plasma levels higher than 350 μg/ml of suramin develop sensorimotor polyneuropathy), which should, therefore, be carefully monitored during treatment.

D. Thalidomide

Thalidomide was first introduced in the 1950s as a sedative, but because it was found to be the cause of severe birth defects when it was used as a hypnotic drug in pregnant women the drug was no longer used for this purpose. Recently thalidomide has been described as being effective in the treatment of some severe cutaneous diseases and of refractory myeloma, and its use has been allowed again under strict monitoring. Peripheral neurotoxicity is one of the most severe side effects that can be observed [20,21].

Thalidomide has poor solubility in water, and its bioavailability after oral administration is not completely known. Its absorption is slow, and the exact metabolic route and fate of thalidomide are unknown in humans. The mean half-life of thalidomide is about 5–7 hours, but the drug is metabolized into a large number of active metabolites and their role in thalidomide neurotoxicity is still unknown.

Given all these uncertainties, it is not surprising that, although thalidomide sensory neurotoxicity was first recognized in the early 1960s, the features of this dose-limiting side effect of the drug are still not completely understood and several important issues have yet to be resolved [8].

Although several clinical studies have investigated thalidomide-induced sensory neurotoxicity, a consensus has not yet been reached even on some key aspects, such as the site of action, the dose dependency, the incidence, the correlation between clinical and neurophysiological results, and the course of the neuropathy once it has ensued.

The reasons for the discrepancies observed in the literature might be due at least partly to the evaluation of small series of patients, often exposed to a limited dose range of thalidomide and examined with different methods. It has been demonstrated that thalidomide-induced peripheral neurotoxicity is dose dependent when relatively high doses are administered (particularly in the case of the treatment of myeloma), and this dose effect is not evident when thalidomide is administered at lower doses (i.e., for the treatment of dermatological or rheumatological diseases). The clinical features in most patients are those of length-dependent sensory neuropathy, mainly involving tactile and thermal modalities, with a reduction or disappearance of deep tendon reflexes and distal dysesthesia. DRG neuron damage may occur at least in some cases, as demonstrated by clinical and neuroradiological findings. Occasionally, mild distal motor impairment has been reported, but the neurological impairment is generally not clinically relevant.

Thalidomide neurotoxicity is generally assessed during treatment in humans through serial sensory nerve examinations. According to the scientific literature, thalidomide should be withdrawn or the dose should be modified if the amplitude of the sensory potentials is reduced by more than 50%. Even in severe cases, sensory nerve conduction velocities change only slightly. Motor nerve conduction changes may occasionally occur, but they are generally of no clinical relevance. It is noteworthy that clinical signs of thalidomide sensory neurotoxicity may also occur in the presence of "normal" (i.e., sensory action potential amplitudes are not reduced) neurophysiological results. This is particularly true with low-dose thalidomide and may represent early DRG damage, as seen with cisplatin treatment.

E. Doxorubicin

Doxorubicin is the lead compound of the anthracycline family. The main dose-limiting side effect of this class of compounds is cardiotoxicity, leading to heart failure in the most severe cases. Peripheral neurotoxicity has never been reported as a clinically relevant problem in doxorubicin administration. However, animal studies have clearly demonstrated that its chronic administration can cause a severe DRG neuronopathy. This finding raises some concern about the possible neurotoxicity of doxorubicin because of a cumulative effect once administered in combination schedules with other anticancer drugs [22], particularly if they damage the DRG neurons.

F. Misonidazole

Misonidazole is a drug with a chemical similarity to metronidazole used to enhance the effectiveness of radiation therapy. A significant proportion of the exposed patients (around one-third of the subjects) may develop a cumulative, dose-related, sensory, and painful distal neuropathy. The mechanism of neurotoxicity is unknown, but it has been demonstrated that misonidazole administration causes large-diameter fiber axonal degeneration with secondary demyelination.

XIV. New Generation Antineoplastic Drugs

A. Epothilones

As partially discussed earlier, a range of compounds interact with the tubulin system. In recent years, a number of such substances have been isolated, particularly from marine animals and from microbial agents. Among these natural antitubulin substances, epothilone (a substance isolated in 1987 from the cellulolytic myxobacterium *Sorangium cellulosum*) has aroused considerable excitement, being extremely active as a microtubule stabilizer with an effect similar to (but more potent than) that observed with taxanes. Since its discovery, it has been recognized that the effect of epothilone is superior to the effect of paclitaxel or docetaxel. Moreover, epothilone is effective against taxane-resistant tumor cells. Several epothilone analogues are now available for preclinical and clinical investigation.

Given the common mechanism of action, it is not surprising that epothilones, like taxanes, can induce dose-limiting peripheral neurotoxicity. However, the data available so far do not make it possible to have a clear picture of the clinical presentation of epothilone-induced nerve impairment [8]. Using the most active schedules reported, it seems that the incidence of clinically relevant peripheral neuropathy is increased (up to two-thirds of patients exposed to BMS-247550, an effective epothilone B analogue [23], had sensory symptoms and signs). Although no conclusive data are available on the time course of epothilone-induced neuropathy or on its site of action, it is likely that this side effect is a problem to be faced in the clinical use of this promising class of antineoplastic drugs.

B. Proteasome Inhibitors

The ubiquitin-proteasome pathway is the major proteolytic system in the cytosol and nucleus of all eukaryotic cells and is involved in a large number of processes. Among other activities, it is responsible for many cellular regulatory mechanisms, including cell cycle control through proteasome degradation of cyclins and cyclin-dependent kinases. Another important function is the selective removal of mutant, damaged, and misfolded proteins. The key role of the proteasome in normal and pathological cellular activities makes this structure prone to be a target of chemotherapy. The aberrant regulation of cell cycle proteins can result in accelerated and uncontrolled cell division leading to tumorigenesis and the growth and spread of cancer: the hypothesis is that selective blocking of the proteasome may counteract these pathological events. Several studies have shown that proteasome inhibitors can induce apoptotic cell death in cancer cell lines at doses comparatively nontoxic to untransformed cells.

The 26S proteasome is an adenosine 5'-triphosphate-dependent, multifunctional proteolytic complex that is the target of bortezomib (formerly known as PS-341), the first proteasome inhibitor available for clinical investigation.

Although it is not clear how and where bortezomib administration can affect the peripheral nervous system, at least one-third of patients under treatment have had evidence of clinically relevant sensory peripheral neuropathy [24]. The clinical and neurophysiological data available do not allow any clear conclusions to be drawn as to the site and mechanism of neurotoxic action of bortezomib [8], although it inhibits nuclear factor-κβ activation, a key molecule for sensory neuron survival. In some cases treatment withdrawal resulted in moderate clinical improvement, but the low number of patients studied should be taken into consideration, as should the fact that they were generally treated with bortezomib in the context of second- or third-line chemotherapy schedules. Further studies are essential to clarify the important issue of the clinical relevance of proteasome inhibitors' peripheral neurotoxicity and to define the mechanisms of this toxic effect, because it potentially may be a dose-limiting side effect in the future clinical use of these substances.

References

1. Peltier, A. C., and Russell, J. W. (2002). Recent advances in drug-induced neuropathies. *Curr. Opin. Neurol.* **15**, 633–638.
2. Quasthoff, S., and Hartung, H. P. (2002). Chemotherapy-induced peripheral neuropathy. *J. Neurol.* **249**, 9–17.
3. Pratt, R. W., and Weimer, L. H. (2005). Medication and toxin-induced peripheral neuropathy. *Semin. Neurol.* **25**, 204–216.
4. Weimer, L. H. (2003). Medication-induced peripheral neuropathy. *Curr. Neurol. Neurosci. Rep.* **3**, 86–92.
5. Chaudhry, V., Chaudhry, M., Crawford, T. O., Simmons-O'Brien, E., and Griffin, J. W. (2003). Toxic neuropathy in patients with pre-existing neuropathy. *Neurology* **60**, 337–340.
6. Dalakas, M. C. (2001). Peripheral neuropathy and antiretroviral drugs. *J. Peripher. Nerv. Syst.* **6**, 14–20.
7. Iervasi, G., Clerico, A., Bonini, R., Nannipieri, M., Manfredi, C., Sabatino, L., Biagini, A., and Donato, L. (1998). Effect of antiarrhythmic therapy with intravenous loading dose of amiodarone: evidence for an altered response in diabetic patients. *Int. J. Clin. Pharmacol. Res.* **18**, 109–120.
8. Umapathi, T., and Chaudhry, V. (2005). Toxic neuropathy. *Curr. Opin. Neurol.* **18**, 574–580.
9. Baker, S. K., and Tarnopolsky, M. A. (2005). Statin-associated neuromyotoxicity. *Drugs Today (Barc.)* **41**, 267–293.
10. Ekstein, D., Linetsky, E., Abramsky, O., and Karussis, D. (2005). Polyneuropathy associated with interferon beta treatment in patients with multiple sclerosis. *Neurology* **65**, 456–458.
11. Bono, A., Beghi, E., Bogliun, G., Cavaletti, G., Curtò, N., Marzorati, L., and Frattola, L. (1993). Antiepileptic drugs and peripheral nerve function: a multicenter screening investigation of 141 patients with chronic treatment. Collaborative Group for the Study of Epilepsy. *Epilepsia* **34**, 323–331.
12. Raya, A., Gallego, J., Bosch-Morell, F., Roma, J., and Romero, F. J. (1995). Phenytoin-induced glutathione depletion in rat peripheral nerve. *Free Radic. Biol. Med.* **19**, 665–667.
13. Windebank, A. J. (1999). Chemotherapeutic neuropathy. *Curr. Opin. Neurol.* **12**, 565–571.
14. Hilkens, P. H., and ven den Bent, M. J. (1997). Chemotherapy-induced peripheral neuropathy. *J. Peripher. Nerv. Syst.* **2**, 350–361.
15. Verstappen, C. C., Heimans, J. J., Hoekman, K., and Postma, T. J. (2003). Neurotoxic complications of chemotherapy in patients with cancer: clinical signs and optimal management. *Drugs* **63**, 1549–1563.
16. Cavaletti, G., and Zanna, C. (2002). Current status and future prospects for the treatment of chemotherapy-induced peripheral neurotoxicity. *Eur. J. Cancer* **38**, 1832–1837.
17. Christian, M. C. (1992). The current status of new platinum analogs. *Semin. Oncol.* **19**, 720–733.
18. Katsumata, N. (2003). Docetaxel: an alternative taxane in ovarian cancer. *Br. J. Cancer* **89(Suppl 3)**, S9–S15.
19. Verstappen, C. C., Koeppen, S., Heimans, J. J., Huijgens, P. C., Scheulen, M. E., Strumberg, D., Kiburg, B., and Postma, T. J. (2005). Dose-related vincristine-induced peripheral neuropathy with unexpected off-therapy worsening. *Neurology* **64**, 1076–1077.
20. Pichardo, D., Singhal, S., Mehta, J., and Rosen, S. (2003). Recent developments and future directions in the treatment of multiple myeloma. *Cancer Biother. Radiopharm.* **18**, 497–511.
21. Cavaletti, G., Beronio, A., Reni, L., Ghiglione, E., Schenone, A., Briani, C., Zara, G., Cocito, D., Isoardo, G., Ciaramitaro, P., Plasmati, R., Pastorelli, F., Frigo, M., Piatti, M., and Carpo, M. (2004). Thalidomide sensory neurotoxicity: a clinical and neurophysiologic study. *Neurology* **62**, 2291–2293.
22. Gehl, J., Boesgaard, M., Paaske, T., Vittrup Jensen, B., and Dombernowsky, P. (1996). Combined doxorubicin and paclitaxel in advanced breast cancer: effective and cardiotoxic. *Ann. Oncol.* **7**, 687–693.
23. Eng, C., Kindler, H. L., Nattam, S., Ansari, R. H., Kasza, K., Wade-Oliver, K., and Vokes, E. E. (2004). A Phase II trial of the epothilone B analog, BMS–247550, in patients with previously treated advanced colorectal cancer. *Ann. Oncol.* **15**, 928–932.
24. Kondagunta, G. V., Drucker, B., Schwartz, L., Bacik, J., Marion, S., Russo, P., Mazumdar, M., and Motzer, R. J. (2004). Phase II trial of bortezomib for patients with advanced renal cell carcinoma. *J. Clin. Oncol.* **22**, 3720–3725.

81

Inherited Peripheral Neuropathies

Angelo Schenone, MD
Lucilla Nobbio, PhD

Keywords: *axonal, Charcot-Marie-Tooth disease, demyelinating, inherited neuropathies, Schwann cell*

I. Introduction
II. Classification
III. Different Types of Hereditary Neuropathies: Clinical, Genetic, and Pathological Features and Pathomechanisms
IV. Hereditary Motor Neuropathies
V. Hereditary Sensory and Autonomic Neuropathies
VI. Hereditary Brachial Plexus Neuropathy
VII. Rare Forms of Hereditary Peripheral Neuropathies
References

I. Introduction

Hereditary peripheral neuropathies affect approximately 1 in 2500 people and are among the most common inherited diseases of the nervous system. Based on clinical, neurophysiological, and neuropathological features, hereditary neuropathies have been classified into hereditary motor and sensory neuropathies (HMSNs), hereditary sensory and autonomic neuropathies (HSANs), and hereditary motor neuronopathies (HMNs) [1]. HMSNs are also known as Charcot-Marie-Tooth disease (CMT) and related neuropathies, named after the authors who in 1886 independently described a syndrome characterized by a slowly progressive, distal sensorimotor process beginning in childhood [2,3]. Clinically, CMT neuropathies are characterized by onset in childhood or early adulthood, distal weakness, sensory loss, foot deformities (pes cavus and hammertoes), and absent reflexes. However, patients with an early onset and a severe course or with a mild disorder have been described. On the basis of neurophysiological properties and neuropathology, CMT has been divided into primary demyelinating and primary axonal types. Most CMT neuropathies are demyelinating, although up to one-third appear to be primary axonal disorders. The primary demyelinating neuropathies include Charcot-Marie-Tooth disease type 1 (CMT1), Dejerine-Sottas disease (DSD), congenital hypomyelinating neuropathy (CHN), and hereditary neuropathy with liability to pressure palsies (HNPP). The primary axonal neuropathies have been classified as CMT2. Most CMT neuropathies, of both the demyelinating and the axonal type, show autosomal dominant inheritance, although X-linked dominant (CMTX) and autosomal recessive forms (CMT4) have been also described. Apparent sporadic cases exist, because dominantly inherited disorders may begin as a new mutation in a given patient. The advancements of molecular genetics in the last 15 years allowed important discoveries in the field of hereditary neuropathies. At least 35 genes are known to cause inherited neuropathies, and more than 50 distinct loci have been identified. The inherited peripheral neuropathies mutation database (www.molgen.ua.ac.be/CMTMutations/default.cfm) provides a comprehensive and updated list of all known mutations. Molecular and cell biology played an important role in understanding how particular mutations cause disease. A deeper knowledge of the underlying pathogenic mechanisms will allow the development of therapeutic strategies for CMT neuropathies, as is happening for the 1A type of CMT (see later discussion).

II. Classification

HMSNs were further divided by Dyck and colleagues into different types (I to VII) based on clinical, neurophysiological, neuropathological, and genetic criteria [1]. However, HMSNs are now best known and classified as CMT and related neuropathies (Table 1) [4]. This group also includes HNPP, which was previously classified, along with the hereditary neuralgic amyotrophy (HNA), among the hereditary recurrent focal neuropathies [5]. HSANs and HMNs retain their original classification (Table 2) [1]. The CMT neuropathies may be further subdivided into a group with motor nerve conduction velocity (NCV) below 38 m/s (CMT1) and a group with motor NCV above 38 m/s (CMT2) [6]. In general terms, this classification based on electrophysiological findings is still valid, although molecular genetic studies allowed further definition of demyelinating CMT (described later).

In this review, we base the classification on the most recent advancements of molecular genetic studies, which allowed the identification of specific genetic defects in numerous hereditary neuropathies. CMT and related neuropathies will be divided into CMT1 if the patient has an autosomal dominant demyelinating neuropathy, CMT2 if the neuropathy is dominantly inherited and axonal, CMTX if the patient has an X-linked neuropathy, and CMT4 if the neuropathy is recessive. Cases of CMT1, CMT2, and CMT4 will be further subdivided based on differences in genetic abnormalities or linkage studies [4]. DSD and CHN will be also included in the classification. DSD and CHN may be considered variants of the same group of demyelinating CMT because they are both severely disabling hereditary neuropathies beginning in infancy. Originally they were defined as autosomal recessive, but molecular genetic studies have clarified that DSD and CHN patients may have autosomal dominant mutations in various genes. Finally, as some families show NCVs that make it difficult to assign the label CMT1 or CMT2 and display typical autosomal dominant inheritance, the term *dominant intermediate Charcot-Marie-Tooth disease* (DI-CMT) was recently introduced [4].

As for HSANs and HMNs, the original classification will be maintained and further refined using molecular genetic studies. HNA and giant axonal neuropathy will be classified as separate entities, although they share some similarities with HNPP and axonal CMT, respectively.

III. Different Types of Hereditary Neuropathies: Clinical, Genetic, and Pathological Features and Pathomechanisms

A. Charcot-Marie-Tooth Disease Type 1

CMT1 patients show slowly progressive distal muscle weakness and wasting, initially affecting the small foot and peroneal muscles. Later in the disease course, weakness and wasting of the small hand muscles may appear. Sensory loss is variable and affects both large (vibration and proprioception) and small (pain and temperature) fiber modalities. Deep tendon reflexes are almost invariably absent in CMT1 patients. Most patients show foot deformities, like high arches and hammertoes (Fig. 1). The course of CMT1 is usually benign, and most patients remain ambulatory throughout life, never becoming wheelchair dependent. Additional features, including postural tremor (referred as Roussy-Levy syndrome) and muscle cramps, may occur. Occasional patients develop a severe phenotype in infancy, and others develop minimal disability during their life. The presence of mutations in genes other than *peripheral myelin protein-22 (PMP22)* may account for this clinical variability [7]. Slowing of NCV below 38 m/s is part of the definition of CMT1. However, a variable degree of axonal impairment as suggested by progressive reduction in the amplitude of motor and sensory action potential (MAP and SAP, respectively) is also present in CMT1 and is responsible for the severity of clinical impairment. Often, SAPs

Table 1 Hereditary Motor and Sensory Neuropathies: Classification

Class	Inheritance	Chromosome	Gene
CMT1			
CMT1A	AD	17p11.2	*PMP22* duplication or point mutation
CMT1B	AD	1q21-23	*MPZ*
CMT1C	AD	16p13.1-12.3	*LITAF/SIMPLE*
CMT1D	AD	10q21-22	*EGR2*
CMT1F	AD	8p21	*NEFL*
DSD			
	AD	1q21-23	*MPZ*
	AD	17p11.2	*PMP22* point mutation
	AD	10q21-22	*EGR2*
	AR	Unknown	Unknown
CHN			
	AD	17p11.2	*PMP22* point mutation
	AD or AR	10q21-22	*EGR2*
	AD or AR	1q22-23	*MPZ*
HNPP			
	AD	17p11.2	*PMP22* deletion or point mutation
CMTX			
CMTX1	XL	Xq13-22	*GJB1* (connexin 32)
CMTX2	XL	Xp22.2	Unknown
CMTX3	XL	Xq26-q28	Unknown
CMTX4	XL	Xq24-q26	Unknown
CMT2			
CMT2A1	AD	1p35-36	*MFN2*
CMT2A2	AD	1p35-36	*KIF1Bβ*
CMT2B	AD	3q13-22	*RAB 7*
CMT2C	AD	12q23-24	Unknown
CMT2D	AD	7p15	*GARS*
CMT2E	AD	8p21	*NEFL*
CMT2F	AD	7q11-21	*HSBP1 (HSP27)*
CMT2G	AD	12q12-q13.3	Unknown
CMT2I and J	AD	1q22-23	*MPZ*
CMT2L	AD	12q24	*HSBP8 (HSP22)*
CMT4			
CMT4A (demyelinating)	AR	8q13	*GDAP1*
CMT4B1	AR	11q23	*MTMR2*
CMT4B2	AR	11p15	*MTMR13/SBF2*
CMT4C	AR	5q32	*KIAA1985*
CMT4C1 (axonal)	AR	1q21.2-q21.3	*LMNA*
CMT4C2 (axonal)	AR	8q21.3	Unknown
CMT4C3 (axonal)	AR	19q13.3	Unknown
CMT4C4 (axonal)	AR	8q21	*GDAP1*
CMT4D (HMSN-L)	AR	8q24	*NDRG1*
CMT4E (see CHN)	AR	10q21-22	*EGR2*
CMT4F	AR	19q13.13-13.2	*PRX*
HMSNR	AR	10q23.2	Unknown
DI-CMT			
DI-CMTA	AD	10q24.1-q25.1	Unknown
DI-CMTB	AD	19p12-p13.2	*Dynamin-2*
DI-CMTC	AD	1p34-p35	Unknown
DI-CMTD (see also CMT1B)	AD	1q22-q23	*MPZ*

Note: CHN, congenital hypomyelinating neuropathy; CMT1, Charcot-Marie-Tooth disease type 1; CMT2, Charcot-Marie-Tooth disease type 2; CMT4, Charcot-Marie-Tooth disease type 4; CMTX, Charcot-Marie-Tooth disease type X-linked dominant; DI-CMT, dominant intermediate Charcot-Marie-Tooth disease; DSD, Dejerine-Sottas disease; HNPP, hereditary neuropathy with liability to pressure palsies.

Table 2 Hereditary Motor Neuronopathies (Distal) and Hereditary Sensory and Autonomic Neuropathies: Classification

Class	Inheritance	Chromosome	Gene
dHMN			
dHMN I	AD	7q11-21	HSBP1 (HSP27)
dHMN II	AD	12q24	HSBP8 (HSP22)
dHMN III	AD	Unknown	Unknown
dHMN IV	AR	11q13	Unknown
dHMN Va	AD	7p15	GARS
dHMN Vb	AD	11q12-14	BSLC2
dHMN VI	AR	11q13-q23	IGHMBP2
dHMN VIIa	AD	2q14	Unknown
dHMN VIIb	AD	2q13	Unknown
dHMN-pyramidal/ASL4	AD	9q34	SETX
dHMN-Jerash type	AR	9p21.1-p12	Unknown
HSAN			
HSAN1	AD	9q22	SPTLC1
HSAN2	AR	12p13.33	HSN2
HSAN3	AR	9q31-q33	IKBKAP
HSAN4	AR	1q21-q22	NTRK1 (TrkA)
HSAN5	AR	1q21-q22/ 1p11.2-p13.2	NTRK1/NGFB

Note: dHMN, distal hereditary motor neuropathy; HSAN, hereditary sensory and autonomic neuropathy.

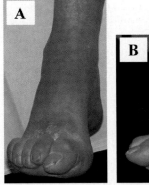
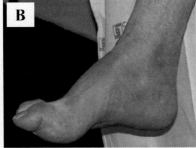

Figure 1 A 50-Year-Old CMT1A Patient. Front (**A**) and lateral (**B**) views of a pes cavus with hammertoes. Foot deformities, although common for Charcot-Marie-Tooth disease (CMT) patients, may be lacking in some families. Moreover, similar abnormalities may be observed in other inherited disorders of the peripheral and central nervous system.

are so severely reduced that they are not even recordable. Chronic denervation is usually observed by electromyography (EMG).

In 1991 the first specific genetic mutation causing a distinct form of CMT1 was described [8]. Since then, several other mutations in genes expressed by myelinating Schwann cells have been found.

1. Charcot-Marie-Tooth Disease Type 1A

CMT1A is the most typical type of CMT1. It is caused by a 1.5 megabase tandem duplication on chromosome 17p11.2, containing the gene encoding PMP22 [8]. Approximately 60% of CMT1 patients have this duplication. Almost all *de novo* CMT1A duplications are caused by an unequal crossing-over event during spermatogenesis [9]. Both sides of the CMT1A duplication are flanked by highly homologous proximal and distal tandem repeat sequences. Misalignment of the CMT1A repeat sequences during meiosis causes the CMT1A duplication [9]. One of the two resulting haploid spermatocytes contains two copies of the *PMP22* gene (duplication), and the other one contains one copy of it (deletion). In this way, the genotype of HNPP (see Sect. III.E) also may be explained. Several pieces of evidence confirm the role of the *PMP22* gene and of its duplication in causing CMT1A [9]: (1) naturally occurring mouse models of CMT1 (Trembler and TremblerJ) are caused by missense mutations in the *PMP22* gene; (2) transgenic mice and rats bearing extra copies of the *PMP22* gene develop a CMT1A-like neuropathy; (3) some patients with missense mutations in the *PMP22* gene develop a similar phenotype; and (4) patients homozygous for the duplication are usually more severely affected than heterozygous ones.

PMP22 is overexpressed in CMT1A nerve biopsies [9], suggesting that an increased level of PMP22 is the most likely pathogenic mechanism for CMT1A. PMP22 is a 22-kDa glycoprotein found in the compact region of peripheral myelin, where it accounts for only 2%–5% of total myelin proteins. The function of PMP22 is not completely known, but it probably has dual roles: as a growth-arrest protein in fibroblasts and Schwann cells and as a structural adhesive component of the myelin sheath [9]. The mechanisms by which an increased PMP22 dose leads to CMT1A are still unclear. Overexpression of PMP22 may affect Schwann cell maturation and differentiation, leading to altered myelin formation and subsequent demyelination [9]. A role for immune-mediated demyelination has been also proposed in an animal model of CMT1A, as for connexin 32 (Cx32) and myelin protein zero (MPZ) mutations [10].

Neurophysiological and neuropathological studies suggest that axonal damage plays a relevant role in CMT1A neuropathy [11]. The pathomechanisms of axonal impairment in CMT1A are unknown. A selective deficit of the ciliary neurotrophic factor has been found in animal models of CMT1A and in human CMT1A nerves [12]. The ciliary neurotrophic factor is one of the most important neurotrophic factors produced by myelinating Schwann cells, and its deficiency may contribute to development of axonal atrophy in CMT1A.

Morphological studies in sural nerves from patients with CMT duplications show rather stereotyped abnormalities [13]. Compared with control nerves, there is an increase in transverse fascicular area. The density of myelinated

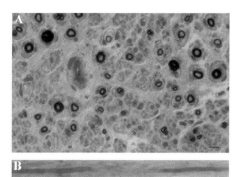

Figure 2 Sural Nerve Biopsy from a CMT1A Patient. (A) The density of myelinated fibers is reduced. Onion bulbs made up of concentric layers of Schwann cell cytoplasm surrounding thinly or normally myelinated fibers may be seen. Semithin section, toluidine blue. Bar 10 μm. (B) Segmental demyelination is often observed in teased fibers preparation. Osmium tetroxide.

fibers is reduced, and most of the remaining myelinated fibers are surrounded by "onion bulbs," made up of concentric layers of Schwann cell cytoplasm, that wrap around thinly or near-normally myelinated axons (Fig. 2). Consistent with axonal loss, there is a reduction in myelinated fibers density, ranging from moderate to severe. As expected in a demyelinating disease, unmyelinated fiber density appears normal or only mildly reduced. Some onion bulbs may contain several myelinated or unmyelinated axons, suggesting that the axons are regenerating nerve sprouts.

The occasional presence of large axons devoid of a myelin sheath, sometimes surrounded by an onion bulb formation, confirms the demyelinating nature of the pathological process. This is even more evident in teased fibers preparation, where most fibers show segmental demyelination and/or remyelination (see Fig. 2). Morphometric studies show that larger fibers are more affected than smaller ones. The mean internodal length is less than in normal nerves, demonstrating that remyelination is also prominent in nerve fibers of CMT1A patients. The presence of myelinated fibers with reduced axon diameter compared with myelin thickness again suggests the presence of an associated axonal impairment The relevance of axonal atrophy in CMT1A nerves has been confirmed in xenografts of sural nerves from CMT1A patients into nude mice sciatic nerves [14]. In fact, within the CMT1A xenografts but not the control ones, distal axonal loss and atrophy of the nude mice axons were observed.

Neuropathies caused by PMP22 missense mutations are relatively rare and often severer than those caused by duplication of the gene. The pathology of these cases is also severer, though genotype–phenotype correlations are not possible. Finally, inflammatory changes may be observed in sural nerve biopsies of CMT1A patients [13]. This observation suggests a concomitant inflammatory process and may relate to previous reports of steroid-responsive inherited neuropathies [1]. However, some evidence in animal models suggests that a macrophage-mediated process may underlie demyelination in the early stages of CMT1A [10]. Recently, in animal models of CMT1A therapeutic approaches using onapristone, a progesterone antagonists, and vitamin C have been successfully used [15]. Clinical trials with vitamin C in human CMT1A are now under way (Pareyson, D., personal communication).

2. Charcot-Marie-Tooth Disease Type 1B

CMT1B occurs less often than CMT1A, accounting for 4%–5% of all CMT1 cases, and is caused by various mutations in the gene encoding MPZ on chromosome 1. MPZ is the most abundant (nearly 50%) myelin protein in peripheral nerves and is a member of the immunoglobulin superfamily. It has a single transmembrane domain and is necessary for the adhesion of concentric myelin wraps in the internode.

Patients with CMT1B show a variety of phenotypes, from severe forms with CHN presenting symptoms *in utero* and DSD beginning in infancy to milder CMT2-like cases beginning in adulthood [16]. Many neuropathies caused by *MPZ* gene mutations tend to cluster with early onset, with presenting symptoms before 1 year of age, or late onset neuropathies, with presenting symptoms well into adulthood [17]. The mechanism by which different mutations in the *MPZ* gene cause distinct phenotypes is unclear. The type and location of the mutation on the *MPZ* coding region may determine the severity of the neuropathy. Recently, some nonsense and frameshift mutations associated with a mild phenotype were related to a haploinsufficiency caused by a nonsense-mediated decay of the mutated messenger ribonucleic acid (RNA) [18]. Mutations that escape this mechanism of messenger RNA degradation encode apparent dominant-negative or gain-of-function proteins responsible for a severer phenotype. Intracellular accumulation of the mutant protein within the endoplasmic reticulum leading to apoptosis has been found in case of MPZ-truncated proteins that escaped the nonsense-mediated decay and are associated with severer phenotypes. Interestingly, treatment with curcumin, a chemical compound that can rescue misfolded proteins, relieves the toxic effect of mutant-aggregation apoptosis [19].

In typical CMT1B cases, the sural nerve biopsy shows a variable loss of myelinated fibers, onion bulbs made up of concentric layers of Schwann cell cytoplasm surrounding thinly myelinated fibers and segmental demyelination in teased fibers [13]. Focal thickenings of the myelin sheath have been observed in CMT1B patients. In other families, ultrastructural examination shows uncompacted myelin in several fibers, thus suggesting two divergent neuropathological phenotypes in CMT1B, the first dominated by myelin thickenings and the second by loosening of myelin lamellae [20]. Morphometric studies show, like in CMT1A, a preferential loss of larger fibers. Total transverse fascicular area is not as enlarged as in typical CMT1A. Most remaining fibers, outside of the myelin thickenings, have high g-ratios, thus confirming the severity of the demyelinating process [20]. Axonal degeneration is the prominent feature of nerve biopsies from patients with late onset disease, and demyelination is less evident. Most biopsies show marked loss of myelinated fibers of all calibers, with numerous clusters of regenerating axons [13].

3. Charcot-Marie-Tooth Disease Type 1C

CMT1C is caused by mutations in the *lipopolysaccharide-induced tumor necrosis factor (LITAF)* gene, also known as *SIMPLE:* small integral membrane protein of lysosome/late endosome [21]. The function of the *LITAF/SIMPLE* gene is unknown, as are the mechanisms underlying CMT1C. These patients develop distal muscle weakness and atrophy, sensory loss, and slow nerve conduction velocities in the range of 20–25 m/s [21]. The neuropathological abnormalities are suggestive of a hypertrophic demyelinating neuropathy, like CMT1A. In particular, myelin loss and onion bulbs similar to those observed in CMT1A have been reported in one large family with CMT1C [21].

4. Charcot-Marie-Tooth Disease Type 1D

Mutations in the *early growth response 2 (EGR2,* also called *krox20)* gene, a transcription factor involved in the regulation of myelination, cause CMT1D [4]. Variable phenotypes are associated with *EGR2* gene mutations. Most patients are severely affected, as it may be observed in DSD or CHN, but milder cases with an adult onset have been described [16].

Patients with CMT1D display neuropathological changes ranging from severe to relatively mild loss of myelinated fibers. Onion bulbs, although present, are not as prominent as in CMT1A. Fibers showing a reduced axon diameter compared with myelin thickness have also been observed in a CMT1D family, consistent with axonal atrophy in addition to demyelination [13].

5. Charcot-Marie-Tooth Disease Type 1F

A particular type of autosomal dominant demyelinating neuropathy may be also caused by mutations in the *neurofilament light chain (NEFL)* gene [4]. These patients are normally severely affected and may show signs of axonal damage. Families exclusively showing axonal features have been classified as having CMT2E (see Sect. III.G.5).

B. Dejerine-Sottas Disease

Although DSD is probably not a distinct clinical and genetic entity, the term is useful to describe a hypertrophic demyelinating polyneuropathy with early onset and a severe disabling clinical course, whose genotype may be caused by point mutations in different genes *(PMP22, MPZ, EGR2,* and *periaxin)*. DSD is normally caused by autosomal dominant or dominant *de novo* mutations. Rarely, a homozygous mutation either in the *PMP22* or in the *MPZ* gene has been described [4]. Motor NCV is markedly lowered, usually below 12 m/s, and sensory NCV is usually not recordable because of the extremely low amplitude of SAPs. Nerve biopsy shows a profound loss of myelinated fibers and a diffuse formation of onion bulbs [13]. Occasionally, onion bulbs around demyelinated axons are made up only of concentric layers of basal lamina.

C. Congenital Hypomyelinating Neuropathy

CHN has an early onset in infancy and is characterized by hypotonia, areflexia, distal muscle weakness, and slow NCV. In some extreme cases, premature death may occur. Other patients, however, show a slow and steady improvement after years. Milder cases may be difficult to distinguish from DSD. Sural nerve biopsies may help in differentiating between these two diseases: in CHN patients, most axons are devoid of myelin or only surrounded by a thin myelin sheath, well-organized onion bulbs are rare, and active myelin breakdown is absent. Therefore, a primary defect of myelin formation may be postulated. In CHN families, point mutations in the *MPZ, PMP22,* and *EGR2* genes have been found [4].

D. Charcot-Marie-Tooth Disease Type X-Linked Dominant

In CMTX the inheritance pattern has classic X-linked features, like the absence of male-to-male transmission. CMTX is the second most common form of CMT, accounting for 10%–16% of cases. A severer clinical course and a significant lowering of motor NCV in males (between 18 and 40 m/s) compared with females (between 25 and 61 m/s) characterize CMTX. The most common type of CMTX (CMTX1) is caused by missense mutations in the *Cx32 kd* gene, also known as gap junction-β1, located on the X chromosome [4]. Cx32 is a major component of noncompact myelin at the nodes of Ranvier and Schmidt-Lanterman incisures. This protein presumably plays a role in the gap junction formation, permitting the passage of small molecules and ions between adjacent loops of the paranode or incisures. More than 250 mutations of Cx32 have been identified. Onset of symptoms in CMTX patients is usually in late teenage years or young adulthood. Clinical abnormalities are usually slowly progressive, limited to the distal legs and hands. Wasting of calf muscles is often more pronounced in CMTX than in CMT1A patients. It is now known that this neuropathy is caused by primary axonal damage with possible secondary demyelination and that the severity of the neuropathy is related to the degree of axonal loss [16]. In fact, the neuropathology of CMTX differs from that of CMT1A and CMT1B because it is mainly characterized by axonal changes, with atrophy and clusters of axonal regeneration and minimal signs of de-remyelination [13]. In agreement with the distribution of Cx32 in noncompact myelin, electron microscopy analysis reveals unusual findings at the nodal and paranodal regions. Widening of the Schmidt-Lanterman incisures and of the nodes of Ranvier has been described, along with a separation of the myelin sheath from the axon that leaves a clear periaxonal space appearing either empty or filled with vesicular material whose significance is unknown [22].

Extremely rare forms of CMTX are CMTX4, also know as Cowchock syndrome, linked to chromosome Xq24-q26.1 [23] and two other types mapped to chromosome Xq26-q28 (CMTX3) and to chromosome Xp22.2 (CMTX2) [4].

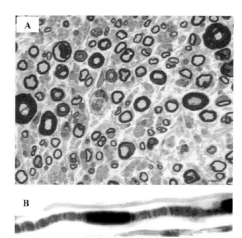

Figure 3 Sural Nerve Biopsy from an HNPP Patient. (**A**) The density of myelinated fibers is normal. Several fibers show thickenings of the myelin sheath compared with axonal diameter (tomacula). Semithin section, toluidine blue. Bar 10 μm. (**B**) Tomacula are easily seen in the teased fibers preparation. Osmium tetroxide.

E. Hereditary Neuropathy with Liability to Pressure Palsies

HNPP is an autosomal dominant demyelinating neuropathy caused by a deletion of the 17p11.2 chromosomal region, which is duplicated in CMT1A. [24]. HNPP is characterized by recurrent episodes of peripheral nerve palsies caused by mechanical compression or trauma of the nerve trunks. The symptoms normally improve over days or months. Between episodes, a patient's neurological examination is often normal or only minimally abnormal. Electrophysiological studies have shown nerve conduction slowing that is more evident at common entrapment sites, prolonged distal latencies, and sometimes conduction blocks. Some variability in clinical and neurophysiological phenotype has been observed, from a CMT1A-like, progressive neuropathy to recurrent, painless brachial plexus palsies. The examination of sural nerve biopsies discloses a variable number of "sausage like" thickenings of the myelin sheath (tomacula) with a near-normal number of myelinated fibers (Fig. 3). Typical tomacula are found in both internodal and paranodal regions and are approximately 40 and 250 μm in length [5]. The analysis of longitudinal sections and teased fiber preparations is sometimes needed to identify these peculiar myelin abnormalities. At the electron microscopic level, tomacula look like redundant loops of myelin with irregularly folded lamellae, which are also enormously increased in number. Rarely, HNPP sural nerves show a severe reduction of myelinated fibers and a high number of onion bulbs. These unusual patients, although carrying the 17p11.2 deletion, have been reported to develop a clinical and neurophysiological phenotype indistinguishable from CMT1. Some cases of HNPP may be also caused by point mutations of the PMP22 gene, often leading to insertion of a stop codon or a frameshift [16].

Because only one copy of the *PMP22* gene is present in HNPP patients and PMP22 underexpression at both the messenger RNA and the protein levels have been demonstrated, a gene dosage mechanism has been proposed as the most likely pathogenic factor for the disease. How down-regulation of PMP22 leads to HNPP is, however, still unknown. The development of PMP22-deficient transgenic mice, which reproduce the human HNPP neuropathological phenotype, is a further demonstration that a gene dosage mechanism is causative for the disease [25].

F. Dominant Intermediate Charcot-Marie-Tooth Disease

Some dominant CMT families showing NCV between 25 m/s and normal and absence of X-linked inheritance are considered intermediate types of CMT and have been classified as DI-CMT. Three chromosomal loci have been linked to DI-CMT: 10q24.1-q25 (DI-CMTA), 19p12-p13.2 (DI-CMTB), and 1p34-p35 (DI-CMTC). Recently, mutations in the pleckstrin homology domain of *dynamin-2* gene were found in DI-CMTB [26]. Patients with DI-CMT show a typical CMT phenotype with a fairly early onset, progression up to 50 years of age, and then clinical stabilization. Histological examination of peripheral nerves show combined axonal and demyelinating features [26].

Given the broad range of NCV changes, it may be difficult to include in the DI-CMT group apparently sporadic cases or patients coming from small families. Before making this diagnosis, accurate studies to exclude acquired neuropathies like the chronic inflammatory demyelinating neuropathy must be carried out.

G. Charcot-Marie-Tooth Disease Type 2

The phenotype of CMT2 is similar to that of CMT1, with distal weakness and atrophy, acral sensory loss, and foot deformities. Therefore, a distinction between the two forms based only on clinical features is impossible. However, CMT2 patients have a wider age range of symptom onset and degree of disability than those with CMT1. Neurophysiological studies discriminate the two types of CMT. CMT2 is the axonal form and displays NCVs above 38 m/s; typically, compound motor action potential (CMAP) and SAP amplitudes are reduced and needle EMG shows evidence of active denervation and partial reinnervation. CMT2 is dominantly inherited and includes approximately 30% of autosomal dominant CMT. The discovery of the guanosine triphosphatase mitofusin 2 gene (MFN2) mutations in most CMT2 patients (CMT2A) is important because we finally have a mutated gene to test in CMT2 families, after years of difficulties resulting from the lack of known mutations [27]. However, according to a recent classification, based on genetic features nine additional subtypes of CMT2 may be identified (A to L) [4].

The neuropathological phenotype of most CMT2 patients is similar. In sural nerve biopsies, the number of myelinated fibers is only mildly decreased, particularly for large-diameter fibers. Occasionally, clusters of small regenerating fibers or small onion bulbs may be seen. Irregular foldings of the myelin sheath have been rarely described, particularly in biopsies of patients with *MPZ* and *NEFL* gene mutations [28,29]. Teased fibers preparations have often shown fibers with short internodes, but there has been no evidence of clear segmental demyelination.

1. Charcot-Marie-Tooth Disease Type 2A

CMT2A shows a typical CMT phenotype. The locus of CMT2A is at chromosome 1p36, and mutations in the *MFN2* gene have been found in most cases of CMT2A [27]. Approximately 20% of CMT2 patients are expected to carry mutations in *MFN2*. Significant clinical variability with an age-dependent penetrance effect has been observed in CMT2A [30]. Some authors include in CMT2A a family showing mutations in the kinesin motor protein 1B (KIF1B), although it is unclear whether these patients also had mutations in the *MFN2* gene, which is located near the KIF1B gene [16]. MFN2 is a mitochondrial membrane protein that participates in mitochondrial fusion in mammalian cells.

It may trigger mitochondrial energization by regulating expression of the oxidative phosphorylation system [31]. How this may lead to CMT2A is still unknown.

2. Charcot-Marie-Tooth Disease Type 2B

CMT2B, because of mutations in the *small guanosine triphosphatase late endosomal protein RAB7* on chromosome 3q21 [16], is a predominantly sensory dominant neuropathy complicated by foot ulcers and amputations. Therefore, there is debate about whether cases should be considered under pure sensory neuropathies.

3. Charcot-Marie-Tooth Disease Type 2C

CMT2C is a rare sensorimotor axonal neuropathy characterized by diaphragmatic paresis of vocal cords and respiratory compromise due to intercostal muscle weakness. Recently, a link was established to chromosome 12 in families with CMT2C, suggesting that it is, at least, a genetically distinct disorder [32].

4. Charcot-Marie-Tooth Disease Type 2D

CMT2D is characterized by sensorimotor symptoms and signs predominantly in the upper limbs. CMT2D maps to chromosome 7p14 and results from mutations in the *glycyl transfer RNA synthetase (GARS)* gene [4]. However, families have been described that share this genotype but show a pure motor syndrome classified as distal hereditary motor neuropathy type V (dHMN V) or distal spinal muscular atrophy. The GARS protein is ubiquitously expressed and has an enzymatic activity necessary for the esterification of glycine to its cognate transfer RNAs. The exact mechanism by which mutations in this protein lead to CMT2D or dHMN Va remains unknown.

5. Charcot-Marie-Tooth Disease Type 2E

CMT2E shows a typical CMT2 phenotype and is caused by mutations in the *NEFL* gene [4]. Because the NEFL protein is an important constituent of the neurofilaments used in axonal transport systems, and neurofilaments phosphorylation is known to be abnormal in demyelinating forms of CMT, CMT2E may provide important clues into mechanisms of axonal damage not only in CMT2 but also in CMT1. Sural nerve biopsies have revealed a relatively high frequency of thin onion bulbs, which would be consistent with a demyelinating process secondary to the axonal impairment. The recent observation of giant axons in the sural nerve biopsy of a family with a Pro22Ser mutation of the *NEFL* gene further supports this possibility [33].

6. Additional Types of Charcot-Marie-Tooth Disease Type 2

Kuhlenbäumer and colleagues [4] included in their classification of CMT2 subtypes F through L. All these types are characterized by mutations in different genes, the phenotype being similar in all families. Only CMT2I and J show peculiar clinical features characterized by late onset with predominant sensory involvement (CMT2I) and late onset with deafness and pupillar abnormalities (CMT2J). Both CMT2I and J are caused by mutations in the *MPZ* gene. The CMT2F and L types are caused by mutations in genes coding for the *small heat shock proteins (sHSP) 27 and 22*, respectively. Mutations in the *sHSP27* gene have been also found in families with a clinical picture of dHMN (see Sect. IV). CMT2G has been linked to chromosome 12q12-q13.3.

H. Charcot-Marie-Tooth Disease Type 4

The autosomal recessive forms of CMT1 have been classically defined as CMT4. However, because axonal types were recently included in the CMT4 type, we think that this term should be referred to as autosomal recessive CMT irrespective of the demyelinating or axonal nature. CMT4 forms are rare and have heterogeneous phenotypes. Usually CMT4 neuropathies are severer than the autosomal dominantly inherited ones. They may have systemic symptoms, such as cataracts and deafness. Multiple subtypes of CMT4 may be identified based on molecular genetic tests. Some types also show peculiar neuropathological features.

1. Charcot-Marie-Tooth Disease Type 4A

CMT4A was first identified in Tunisian families and is linked to chromosome 8q13-q21.1. We now know that CMT4A is caused by mutations in *ganglioside-induced differentiation associated protein 1 (GDAP1)*, a novel protein whose action is probably related to the mitochondrial function [34]. CMT4A is an early onset, predominantly motor neuropathy with mild sensory symptoms and usually slow NCV (average of 30 m/s). However, some patients with neurophysiological features suggesting prominent axonal damage have been described. Also, neuropathological studies showed abnormalities supportive of a primary demyelinating process in some CMT4A patients and of a mainly axonal process in others [13]. Although the number of mutations described so far in the *GDAP1* gene is too low to permit reliable genotype–phenotype correlations, CMT4A, like CMTX, combines axonal and demyelinating changes. GDAP1 is expressed at mitochondria, which is consistent with this interpretation [34].

2. Charcot-Marie-Tooth Disease Type 4B

CMT4B1 is a severely disabling recessive neuropathy characterized by a unique pathological feature: the presence of focally folded myelin sheaths in nerve biopsies (myelin outfoldings) (Fig. 4). The genetic locus is on chromosome 11q23 and encodes a gene called *myotubularin-related protein-2 (MTMR2)* [4]. Patients show distal and proximal weakness, foot deformities, scoliosis, peculiar voice changes, and onset in early childhood. In the second or third decade of life, they become wheelchair dependent. Motor conduction velocities are severely reduced with temporal dispersion, CMAPs are reduced, and SAPs are often absent. Segmental demyelination is also found in nerve biopsies.

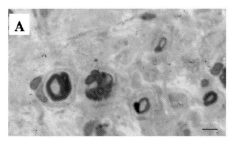
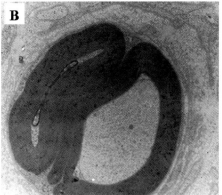

Figure 4 Sural Nerve Biopsy from a CMT4B Patient. (**A**) The density of myelinated fibers is severely reduced from normal. Myelin outfoldings may be seen on most remaining fibers. Semithin section, toluidine blue. Bar 10 μm. (**B**) Electron microscopy shows that myelin outfoldings are characterized by complex outfolding of the myelin sheath within the Schwann cell cytoplasm. Lead citrate and uranyl acetate, 20,000 times magnification.

Linkage to a distinct locus on chromosome 11p15 in a Turkish family with CMT1B was recently identified (CMT4B2). Mutations were then identified in a novel gene, named *SET binding factor 2 (SBF2)*, that lies within the interval on chromosome 11p15 and segregates with the neuropathy in affected families [4]. SBF2 is a member of the pseudophosphatase branch of myotubularins with striking homology to the *MTMR2* gene. MTMR2-null mice, which reproduce the CMT4B phenotype, were recently developed [35]. However, as MTMR2 is expressed in myelinating and nonmyelinating Schwann cells, or in neurons, it is unclear whether loss of this protein in Schwann cells, neurons, or both causes the neuropathy in transgenic mice. Recent studies with specific ablation of MTMR2 in either Schwann cells or motoneurons showed that only inactivation of the gene in Schwann cells is sufficient to produce experimental CMT4B1 neuropathy [36].

3. Charcot-Marie-Tooth Disease Type 4C

CMT4C is an early onset, severely demyelinating sensorimotor neuropathy, caused by homozygous or compound heterozygous mutations in the previously uncharacterized *KIAA1985* gene [4]. Scoliosis is an early sign and may precede weakness or sensory loss. The *KIAA1985* gene is strongly expressed in neural tissue and encodes for a protein of unknown function. Motor NCV in CMT4C is severely reduced, and the sural nerve biopsy shows a predominantly demyelinating neuropathy with a severe loss of large myelinated fibers ($>8\,\mu m$), abnormally thin myelin sheaths in the remaining fibers, and extensive onion bulb formation [37]. Onion bulbs made up of concentric, basal membrane layers surrounding a demyelinated axon are often seen (basal lamina onion bulbs), along with supernumerary Schwann cells processes connecting isolated unmyelinated axons [13].

a. Charcot-Marie-Tooth Disease Type 4C1 CMT4C1 is an autosomal recessive sensorimotor neuropathy caused by mutations in the *lamin A/C nuclear envelope protein (LMNA)* gene. CMT4C1 has been described in Moroccan families. Motor NCV is within normal limits, needle EMG shows signs of chronic denervation, and nerve biopsies display the typical features of CMT2 [38].

b. Charcot-Marie-Tooth Disease Type 4C2 CMT4C2 is an autosomal recessive, axonal, sensorimotor neuropathy described in a Tunisian family, and linked to chromosome 8q21.11, near the *GDAP1* gene. Thus CMT4A and CMT4C2 may be allelic disorders. Clinically, patients show distal weakness and wasting, acral sensory symptoms, and brisk tendon reflexes [38].

c. Charcot-Marie-Tooth Disease Type 4C3 Only a family from Costa Rica was described as having an early onset, autosomal recessive, sensorimotor, axonal neuropathy linked to chromosome 19q13.3 [38].

4. Charcot-Marie-Tooth Disease Type 4D

CMT4D is also known as HMSN of the Lom type, because it is an autosomal, recessive HMSN first described in Bulgarian Roma gypsies living close to the town of Lom. After linkage to chromosome 8q24, mutations in the *N-myc downstream-regulated gene (NDRG1)* have been described in CMT4D patients [4]. The neuropathy begins as distal muscle wasting and weakness, sensory loss, foot and hand deformities, and loss of deep tendon reflexes. Deafness and abnormalities in auditory-evoked potentials are invariably present. NCVs are severely reduced in younger patients and unobtainable after 15 years of age. Sural nerve biopsy findings in CMT4D are suggestive of a severe demyelinating process, with profound loss of myelinated fibers and prominent onion bulb formation [13].

5. Charcot-Marie-Tooth Disease Type 4F

CMT4F is caused by mutations in the *periaxin (PRX)* gene on chromosome 19. It has been described in a Lebanese family in which affected subjects show slightly delayed motor milestones, unsteady gait, distal muscle wasting and weakness, acral sensory loss, scoliosis, and foot deformities [39]. The PRX gene is expressed in Schwann cells and encodes L- and S-periaxin. L-periaxin is a constituent of the destroglycan-dystrophin-related protein-2 complex linking the Schwann cell cytoskeleton to the extracellular matrix. Nerve conduction studies are markedly slowed, and sural nerve biopsies show a severe loss of myelinated fibers accompanied by prominent onion bulb formation and, sometimes, hypermyelination [39]. The paranodal region shows incomplete myelination and separation of multiple terminal myelin loops from the axon, suggesting that PRX may play an important role in mediating Schwann cell–axon adhesion at the node of Ranvier [39]. Interestingly, PRX-null mice develop neuropathological features strikingly similar to the human pathology, supporting the view that studies in this animal model may lead to a better understanding of the human disease [40].

I. Hereditary Motor and Sensory Neuropathy-Russe

Hereditary motor and sensory neuropathy-Russe (HMSNR) also known as CMT of the Russe type, is linked to chromosome 10q23.2, a small-interval telomeric to the *EGR2* gene [38]. Patients develop primarily severe sensory loss, although motor conduction velocities are also moderately reduced (average is 32 m/s).

IV. Hereditary Motor Neuronopathies

HMNs are characterized by the degeneration of lower motoneurons. When distal distribution of clinical symptoms occurs, they have been classified as dHMN or spinal CMT. The term *spinal muscular atrophy* is instead reserved to the motor neuronopathies with primary proximal involvement and will not be discussed further in this review.

dHMNs comprise about 10% of all HMNs and have been tentatively classified into seven subtypes based on clinical manifestations, age at onset, and mode of inheritance. Genetic loci and specific gene mutations for several subtypes of dHMN have now been identified.

dHMN I shows a clinical presentation similar to CMT, but sensory loss is typically absent. dHMN I is inherited as an autosomal dominant trait, and recently mutations in sHSP27 (or HSBP1) were found in affected subjects [4]. Mutations in sHSP27 have been found also in CMT2F. The presence or absence of sensory symptoms should therefore be the only feature distinguishing these two hereditary neuropathies. The prognosis of dHMN is generally good, and progression is slow.

dHMN II is an autosomal dominant motor neuronopathy, which has been linked in a large Belgian family to chromosome 12q24.3. Distal motor symptoms (weakness of the extensor muscles of the great toe and foot) start in the second or third decade of life and often progress to complete leg paralysis in few years. Occasional patients have been described with decreased vibratory sensation. NCVs are normal with needle EMG showing chronic denervation. Mutations in the *sHSP22* gene were recently described in dHMN II [4].

sHSPs are induced by stress, like elevated temperature. Their function is not well known but is thought to involve signal transduction pathways, protection from apoptosis, and stabilization of cytoskeletal systems. Mutations in different HSPs also cause myopathies and congenital cataracts. sHSP22 and sHSP27 interact with each other; therefore, dHMN I, dHMN II, and CMT2F may share common disease mechanisms. What these mechanisms are, however, remains to be elucidated [41]. As both sHSP22 and sHSP27 have chaperone-like properties, they may participate in processing proteins for intracellular trafficking [41]. However, they have also been shown to inhibit apoptosis and to stabilize cytoskeletal systems. In this regard, it is noteworthy that mutations in sHSP27 disrupt neurofilament assembly [41]. Mutations in the *NEFL* gene cause CMT2E. Taken together, these results suggest that the sHSPs may play critical roles in regulating or maintaining the axonal cytoskeleton and axonal transport. Up-regulation of sHSP27 is also necessary for the survival of injured motoneurons or sensory neurons, and sHSP27 is overexpressed in motoneurons of the SOD1 mouse model of familial amyotrophic lateral sclerosis. Therefore, sHSP-mediated pathways may also prove to play important roles in many neurodegenerative disorders [41].

dHMN III is an early onset, autosomal recessive motor neuronopathy, which clinically resembles dHMN I and II. There are no studies available that confirm the linkage of dHMN III to a single chromosomal locus [41].

dHMN IV is an autosomal recessive motor neuronopathy but is severer and extends to the proximal leg muscles. Linkage to chromosome 11q13 has been demonstrated in a Lebanese family with dHMN IV [41].

dHMN V may be further divided into two autosomal dominant subtypes, Va and Vb, based on genetic features [41]. Clinically they are characterized by muscle weakness and wasting predominantly involving the upper limbs, the lower ones being involved only later and at a lower degree in the disease course. In dHMN Vb pyramidal signs in the lower limbs have been also described. dHMN Va has been localized to chromosome 7p and is caused by mutations in the *GARS* gene on chromosome 7p15 [41]. dHMN Vb is caused by mutations in the *Berardinelli-Seip congenital lipodystrophy 2 (BSCL2)* gene [42]. As with dHMN Va, mutations in the *BSCL2* gene have been related to a continuum of clinical phenotypes ranging from typical dHMN V (pure motor neuronopathy predominantly involving upper limbs), to Silver syndrome (amyotrophy of the small hand muscles and spasticity of the lower extremities), to CMT2 (distal weakness and wasting of the lower and upper limbs and sensory abnormalities) [41]. Asymptomatic or minimally affected subjects have been also described [41].

dHMN VI, also known as spinal muscular atrophy with respiratory distress, is a severe and often lethal congenital motor neuronopathy characterized mainly by diaphragmatic paralysis with eventration and predominant involvement of upper limbs [41]. It is caused by mutations in *the immunoglobulin μ-binding protein 2 (IGHMBP2)* gene.

dHMN VII may be distinguished as two subtypes based on clinical and genetic features [41]. dHMN VIIa is characterized by weakness and wasting of the small hand muscles, subsequently involving the distal muscles of the legs, and by vocal cord paralysis. It is similar to CMT2C, which also shows sensory signs and symptoms. dHMN VIIb, mapped to chromosome 2q14, shows additional facial weakness. Mutations in the *dynactin* gene have been found [41]. Binding assays show decreased binding of the mutant protein to microtubules, suggesting that dysfunction of dynactin-mediated transport caused the motoneuron degeneration [41].

dHMN pyramidal/amyotrophic lateral sclerosis-4 is an autosomal dominant, distal motor syndrome with pyramidal signs mapping to chromosome 9q34. The clinical course of these patients is much milder than that of typical amyotrophic lateral sclerosis. Pyramidal/amyotrophic lateral sclerosis 4 is characterized by a clinical onset between 6 months and 21 years of age, a slow progression, and a normal life span. Patients develop distal muscle weakness and atrophy and pyramidal signs, but they have normal sensation. The disorder is caused by missense mutations in the *senataxin (SETX)* gene [41]. The function of SETX is unknown, but the gene contains a DNA/RNA helicase domain with strong homology to human RENT1 and IGHMBP2, two genes encoding proteins involved in RNA processing. It is hypothesized that mutations in SETX may cause neuronal degeneration through dysfunction of the helicase activity or other steps in RNA processing in motoneurons [41].

dHMN of the Jerash type is a recessive dHMN found in a large Jordanian family with a locus mapped to chromosome 9p21.1-p12 [41]. Patients show symmetrical distal atrophy, weakness of the lower limbs, and gait instability before the age of 10 and a few years later develop wasting and weakness of hand muscles. Pes cavus is present in all affected subjects. Sensory loss is absent [41].

V. Hereditary Sensory and Autonomic Neuropathies

HSANs are clinically characterized by predominant peripheral sensory and autonomic disturbances. HSANs have been classified into five types according to clinical, neurophysiological, and neuropathological features and to mode of inheritance [43]. The neuropathological findings, represented by a prevailing degeneration and a loss of fibers in the sensory nerves, posterior columns, posterior roots, and spinal ganglia, mirror the clinical picture [43].

A. Hereditary Sensory and Autonomic Neuropathy Type 1

In patients affected by HSAN1, mutations in the *serine-palmitoyltransferase-1 (SPTLC1)* gene were recently found [44]. The clinical picture is characterized by a variable combination of spontaneous pain in the lower limbs, foot complication ranging from acral ulcers to amputations, loss of pain and thermal sensations with a distal distribution, reduced sweating, and decreased deep reflexes. Tactile sensation and muscle strength may be normal or only mildly affected. However, patients with significant distal weakness and wasting have been reported [45]. In this case, CMT2B may be considered in the differential diagnosis. Normally, in CMT2B patients all sensory modalities are equally affected, motor signs are severer, and spontaneous lancinating pain is absent. In typical HSAN1 patients, the neurophysiological examination suggests an axonal sensory neuropathy with only mild involvement of motor conduction. The sural nerve biopsy shows a diffuse loss of large and small myelinated fibers. Unmyelinated fibers are not markedly decreased. The function of SPTLC1 as an enzyme of sphingolipid synthesis and the elevated concentrations in lymphoblasts of HSAN1 patients of glucosylceramide, a molecule inducing cell apoptosis, may suggest pathogenic mechanisms involving neuronal death in sensory ganglia [44].

B. Hereditary Sensory and Autonomic Neuropathy Type 2

HSAN2 is an autosomal recessive, congenital sensory neuropathy mapping to chromosome 12p13.33 and caused by mutations in a novel gene *(HSN2)* of unknown function [44]. In *HSAN2* sensory loss affects all modalities of sensation and involves lower and upper limbs, as well as the trunk [44]. Impairment of sweating is prominent but without other autonomic symptoms like postural hypotension or impotence in the male. SAPs are typically absent, whereas motor NCV and CMAPs are normal or only minimally affected. The sural nerve biopsy shows an almost complete absence of myelinated fibers with a relative preservation of unmyelinated fibers. Nerve fascicles also show a decrease in transverse fascicular area [13].

C. Hereditary Sensory and Autonomic Neuropathy Type 3

HSAN3, also known as familial dysautonomia or Riley-Day syndrome, is an autosomal recessive disease affecting Ashkenazi Jews, recently related to mutations in the *inhibitor-of-κ-light-polypeptide (IKBKAP)* gene [44]. Interestingly, genetic tests allowed the identification of an *IKBKAP* gene mutation in a patient with typical HSAN3 but no Ashkenazi Jewish ancestry [46].

Clinically, HSAN3 is characterized by a history of poor feeding, repeated episodes of vomiting, severe autonomic dysfunction (defective lacrimation and temperature control, excessive perspiration, tachycardia, hypertension or postural hypotension), insensitivity to pain, and areflexia. Nerve conduction studies reveal reduced SAPs; motor NCV may be also slightly affected. The sural nerve biopsy shows an approximately normal density of myelinated fibers and a striking loss of unmyelinated fibers [43]. A marked reduction of autonomic and spinal ganglia volume and a decreased density of sympathetic and sensory neurons were also observed in HSAN3. These findings may suggest an arrest of embryogenesis in the neural crest [43]. However, although the function of the protein coded by the *IKBKAP* gene in the transcription control is known, the pathomechanisms of HSAN3 remain to be elucidated [44].

D. Hereditary Sensory and Autonomic Neuropathy Types 4 and 5

These two rare and clinically similar disorders of the sensory system are caused by mutations in genes related to neurotrophic factors, namely the *neurotrophin receptor tyrosine kinase-1 (NTRK1)* gene [44], which encodes for the nerve growth factor (NGF) high-affinity receptor TRK1 and the *NGF-β (NGFB)* gene [44], which encodes for NGFβ itself. Congenital insensitivity to pain and anhidrosis are the hallmarks of the two diseases. Mental retardation may be present in HSAN4 but not in HSAN5. Autonomic dysfunction is more evident in HSAN4 than in HSAN5. The neuropathological picture is somehow different in these two diseases. The sural nerve biopsy, in the previously described cases of HSAN4, showed a nearly normal density of myelinated fibers and a virtual absence of unmyelinated ones [43]. Instead, in HSAN5, a selective decrease of small myelinated and of unmyelinated fibers was found in the sural nerve biopsy [43]. However, too few patients have been described with these disorders and *NTRK1* or *NGFB* gene mutations to correctly describe genotype–phenotype correlations from the neuropathological point of view. Finding mutations in genes involved in the NGF pathway may explain why these patients have a loss of small myelinated and of unmyelinated fibers, which are sustained by NGF in their embryonic development.

VI. Hereditary Brachial Plexus Neuropathy

Hereditary brachial plexus neuropathy (HBPN) is a rare autosomal dominant disorder characterized by episodes of acute pain, weakness, and sensory loss in the upper extremities. Attacks may be preceded by infections, surgery, or immunization [5]. Recovery usually begins to occur several weeks to months after the onset of symptoms. Subsequent attacks may occur in the same or opposite arm. Minor dysmorphic features, including short stature, hypotelorisms, epicanthal folds, and cleft palate, are common associated findings [5]. Neurophysiology studies show active axonal degeneration at needle EMG, during attacks, and usually normal motor and sensory NCVs. A reduced amplitude of SAPs and CMAPs in the affected limbs is often found. After mapping HBPN to chromosome 17q24-25, mutations in the gene *septin 9 (SEPT9)* in six families with HNA linked to this chromosomal region were found [47]. Septins are implicated in formation of the cytoskeleton, cell division, and tumorigenesis; however, the mechanisms by which mutations in this gene lead to HNA are still unknown.

VII. Rare Forms of Hereditary Peripheral Neuropathies

Kuhlenbäumer and colleagues [4] included in the classification of hereditary neuropathies some rare diseases characterized by symptoms and signs of peripheral nerve damage associated to multisystem involvement: (1) giant axonal neuropathy, (2) agenesis of the corpus callosum with peripheral neuropathy or Anderman syndrome, (3) CHN with central dysmyelination and intestinal pseudo-obstruction or Waardenburg-Hirschsprung disease, (4) hereditary peripheral neuropathy and deafness, (5) minifascicular peripheral neuropathy with partial gonadal dysgenesis, (6) Fabry disease, (7) Tangier disease, and (8) Navajo neuropathy.

A detailed description of these disorders is beyond the scope of this review.

Acknowledgments

This research was financially supported by Telethon GGP02169 and GUP04002 (to A.S.).

References

1. Dyck, P. J., Chance, P., Lebo, R., and Carney, A. J. (1993). Hereditary motor and sensory neuropathies. *In*: "Peripheral Neuropathy" (P. J. Dyck, P. K. Thomas, J. W. Grillin, P. A. Low, and J. F. Poduslo, eds.), 3rd edition, pp. 1094–1136. W. B. Saunders, Philadelphia.
2. Charcot, J. M., and Marie, P. (1886). Sur une forme particuliere d'atrophie musculaire progressive, souvent familiale, debutant pas les pieds et les jambes et atteignant plus trad les mains. *Rev. Med.* **6**, 97–138.
3. Tooth, H. H. (1886). "The Peroneal Type of Progressive Muscular Atrophy." H. K. Lewis, London.
4. Kuhlenbäumer, G. (2005). Overview of the classification and genetics of hereditary peripheral neuropathies and rare unclassified forms. *In*: "Hereditary Peripheral Neuropathies" (G. Kuhlenbäumer, F. Stögbauer, E. B. Ringelstein, and P. Young, eds.), pp. 73–91. Springer, Steinkopff Darmstadt.
5. Windebank, A. J. (1993). Inherited recurrent focal neuropathies. *In:* "Peripheral Neuropathy" (P. J. Dyck, P. K. Thomas, J. W. Grillin, P. A. Low, and J. F. Poduslo, eds.), 3rd edition, pp. 1137–1148. W. B. Saunders, Philadelphia.
6. Harding, A. E., and Thomas, P. K. (1980). The clinical features of hereditary motor and sensory neuropathy types I and II. *Brain* **103(2)**, 259–280.
7. Hodapp, J. A., Carter, G. T., Lipe, H. P., Michelson, S. J., Kraft, G. H., and Bird, T. D. (2006). Double trouble in hereditary neuropathy: concomitant mutations in the PMP-22 gene and another gene produce novel phenotypes. *Arch. Neurol.* **63(1)**, 112–117.
8. Lupski, J. R., de Oca-Luna, R. M., Slaugenhaupt, S., Pentao, L., Guzzetta, V., Trask, B. J., Saucedo-Cardenas, O., Barker, D. F., Killian, J. M., Garcia, C. A., Chakravarti, A., and Patel, P. I. (1991). DNA duplication associated with Charcot-Marie-Tooth disease type 1A. *Cell* **66(2)**, 219–232.
9. Schenone, A., and Mancardi, G. L. (1999). Molecular basis of inherited neuropathies. *Curr. Opin. Neurol.* **12**, 603–616.
10. Kobsar, I., Hasenpusch-Theil, K., Wessig, C., Muller, H. W., and Martini, R. (2005). Evidence for macrophage-mediated myelin disruption in an animal model for Charcot-Marie-Tooth neuropathy type 1A. *J. Neurosci. Res.* **81(6)**, 857–864.
11. Hanemann, C. O., and Gabreels-Festen, A. A. (2002). Secondary axonal atrophy and neurological dysfunction in demyelinating neuropathies. *Curr. Opin. Neurol.* **15**, 611–615.
12. Vigo, T., Nobbio, L., Hummelen, P. V., Abbruzzese, M., Mancardi, G., Verpoorten, N., Verhoeven, K., Sereda, M. W., Nave, K. A., Timmerman, V., and Schenone, A. (2005). Experimental Charcot-Marie-Tooth type 1A: a cDNA microarrays analysis. *Mol. Cell. Neurosci.* **28(4)**, 703–714.
13. Schenone, A. (2005). Principles of pathology and nerve biopsy. *In:* "Hereditary Peripheral Neuropathies" (G. Kuhlenbäumer, F. Stögbauer, E. B. Ringelstein, and P. Young, eds.), pp. 41–70. Springer, Steinkopff Darmstadt.

14. Sahenk, Z., Chen, L., and Mendell, J. R. (1999). Effects of PMP22 duplication and deletions on the axonal cytoskeleton. *Ann. Neurol.* **45**, 16–24.
15. Grandis, M., and Shy, M. E. (2005). Current therapy for Charcot-Marie-Tooth disease. *Curr. Treat. Options Neurol.* **7**, 23–31.
16. Shy, M. E. (2004). Charcot-Marie-Tooth disease: an update. *Curr. Opin. Neurol.* **17(5)**, 579–585.
17. Shy, M. E., Jani, A., Krajewski, K., Grandis, M., Lewis, R. A., Li, J., Shy, R. R., Balsamo, J., Lilien, J., Garbern, J. Y., and Kamholz, J. (2004). Phenotypic clustering in MPZ mutations. *Brain* **127**, 371–384.
18. Inoue, K., Khajavi, M., Ohyama, T., Hirabayashi, S., Wilson, J., Reggin, J. D., Mancias, P., Butler, I. J., Wilkinson, M. F., Wegner, M., and Lupski, J. R. (2004). Molecular mechanism for distinct neurological phenotypes conveyed by allelic truncating mutations. *Nat. Genet.* **36(4)**, 361–369.
19. Khajavi, M., Inoue, K., Wiszniewski, W., Ohyama, T., Snipes, G. J., and Lupski, J. R. (2005). Curcumin treatment abrogates endoplasmic reticulum retention and aggregation-induced apoptosis associated with neuropathy-causing myelin protein zero-truncating mutants. *Am. J. Hum. Genet.* **77(5)**, 841–850.
20. Gabreels-Festen, A. A., Hoogendijk, J. E., Meijerink, P. H., Gabreels, F. J., Bolhuis, P. A., van Beersum, S., Kulkens, T., Nelis, E., Jennekens, F. G., de Visser, M., van Engelen, B. G., Van Broeckhoven, C., and Mariman, E. C. (1996) Two divergent types of nerve pathology in patients with different P0 mutations in Charcot-Marie-Tooth disease. *Neurology* **47**, 761–765.
21. Street, V. A., Bennett, C. L., Goldy, J. D., Shirk, A. J., Kleopa, K. A., Tempel, B. L., Lipe, H. P., Scherer, S. S., Bird, T. D., and Chance, P. F. (2003). Mutation of a putative protein degradation gene LITAF/SIMPLE in Charcot-Marie-Tooth disease 1C. *Neurology* **60(1)**, 22–26.
22. Hahn, A. F., Ainsworth, P. J., Bolton, C. F., Bilbao, J. M., and Vallat, J. M. (2001). Pathological findings in the X-linked form of Charcot-Marie-Tooth disease: a morphometric and ultrastructural analysis. *Acta Neuropathol.* **101**, 129–139.
23. Cowchock, F. S., Duckett, S. W., Streletz, L. J., Graziani, L. J., and Jackson, L. G. (1985). X-linked motor-sensory neuropathy type-II with deafness and mental retardation: a new disorder. *Am. J. Med. Genet.* **20**, 307–315.
24. Chance, P. F., Alderson, M. K., Leppig, K. A., Lensch, M. W., Matsunami, N., Smith, B., Swanson, P. D., Odelberg, S. J., Disteche, C. M., and Bird, T. D. (1993). DNA deletion associated with hereditary neuropathy with liability to pressure palsies. *Cell* **72(1)**, 143–151.
25. Adlkofer, K., Martini, R., Aguzzi, A., Zielasek, J., Toyka, K. V., and Suter, U. (1995). Hypermyelination and demyelinating peripheral neuropathy in PMP22-deficient mice. *Nat. Genet.* **11(3)**, 274–280.
26. Zuchner, S., Noureddine, M., Kennerson, M., Verhoeven, K., Claeys, K., De Jonghe, P., Merory, J., Oliveira, S. A., Speer, M. C., Stenger, J. E., Walizada, G., Zhu, D., Pericak-Vance, M. A., Nicholson, G., Timmerman, V., and Vance, J. M. (2005). Mutations in the pleckstrin homology domain of dynamin 2 cause dominant intermediate Charcot-Marie-Tooth disease. *Nat. Genet.* **37(3)**, 289–294.
27. Reilly, M. M. (2005). Axonal Charcot-Marie-Tooth disease: the fog is slowly lifting! *Neurology* **65(2)**, 186–187.
28. De Jonghe, P., Timmerman, V., Ceuterick, C., Nelis, E., De Vriendt, E., Lofgren, A., Vercruyssen, A., Verellen, C., Van Maldergem, L., Martin, J. J., and Van Broeckhoven, C. (1999). The Thr124Met mutation in the peripheral myelin protein zero (MPZ) gene is associated with a clinically distinct Charcot-Marie-Tooth phenotype. *Brain* **122**, 281–290.
29. Jordanova, A., De Jonghe, P., Boerkoel, C. F., Takashima, H., De Vriendt, E., Ceuterick, C., Martin, J. J., Butler, I. J., Mancias, P., Papasozomenos, S. Ch., Terespolsky, D., Potocki, L., Brown, C. W., Shy, M., Rita, D. A., Tournev, I., Kremensky, I., Lupski, J. R., and Timmerman, V. (2003). Mutations in the neurofilament light chain gene (NEFL) cause early onset severe Charcot-Marie-Tooth disease. *Brain* **126**, 590–597.
30. Lawson, V. H., Graham, B. V., and Flanigan, K. M. (2005). Clinical and electrophysiologic features of CMT2A with mutations in the mitofusin 2 gene. *Neurology* **65(2)**, 197–204.
31. Pich, S., Bach, D., Briones, P., Liesa, M., Camps, M., Testar, X., Palacin, M., and Zorzano, A. (2005). The Charcot-Marie-Tooth type 2A gene product, Mfn2, up-regulates fuel oxidation through expression of OXPHOS system. *Hum. Mol. Genet.* **14(11)**, 1405–1415.
32. Klein, C. J., Cunningham, J. M., Atkinson, E. J., Schaid, D. J., Hebbring, S. J., and Anderson, S. A., Klein, D. M., Dyck, P. J., Litchy, W. J., Thibodeau, S. N., and Dyck, P. J. (2003). The gene for HMSN2C maps 12q23-24: a region of neuromuscular disorders. *Neurology* **60**, 1151–1156.
33. Fabrizi, G. M., Cavallaro, T., Angiari, C., Bertolasi, L., Cabrini, I., Ferrarini, M., and Rizzuto, N. (2004). Giant axons and neurofilament accumulation in Charcot-Marie-Tooth disease type 2E. *Neurology* **62**, 1429–1431.
34. Niemann, A., Ruegg, M., La Padula, V., Schenone, A., and Suter, U. (2005). Ganglioside-induced differentiation associated protein 1 is a regulator of the mitochondrial network: new implications for Charcot-Marie-Tooth disease. *J. Cell Biol.* **170(7)**, 1067–1078.
35. Bolino, A., Bolis, A., Previstali, S. C., Dina, G., Bussini, S., Dati, G., Amadio, S., Del Carro, U., Mruk, D. D., Feltri, M. L., Cheng, C. Y., Quattrini, A., and Wrabetz, L. (2004). Disruption of MTMR2 produces CMT4B1-like neuropathy with myelin outfoldings and impaired spermatogenesis. *J. Cell Biol.* **167**, 711–721.
36. Bolis, A., Coviello, S., Bussini, S., Dina, G., Pardini, C., Previtali, S. C., Malaguti, M., Morana, P., Del Carro, U., Feltri, M. L., Quattrini, A., Wrabetz, L., and Bolino, A. (2005). Loss of MTMR2 phosphatase in Schwann cells but not in motor neurons causes Charcot-Marie-Tooth type 4B1 neuropathy with myelin outfoldings. *Neurobiol. Dis.* **25(37)**, 8567–8577.
37. Kessali, M., Zemmouri, R., Guilbot, A., Maisonobe, T., Brice, A., LeGuern, E., Grid, D. (1997). A clinical, electrophysiologic, neuropathologic, and genetic study of two large Algerian families with an autosomal recessive demyelinating form of Charcot-Marie-Tooth disease. *Neurology* **48**, 867–873.
38. Nelis, E., De Jonghe, P., and Timmerman, V. (2005). Charcot-Marie-Tooth type 1 (CMT1) and hereditary neuropathy with liability to pressure palsy (HNPP). *In:* "Hereditary Peripheral Neuropathies" (G. Kuhlenbäumer, F. Stögbauer, E. B. Ringelstein, and P. Young, eds.), pp. 92–120. Springer, Steinkopff Darmstadt.
39. Takashima, H., Boerkoel, C. F., De Jonghe, P., Ceuterick, C., Martin, J. J., Voit, T., Schroder, J. M., Williams, A., Brophy, P. J., Timmerman, V., and Lupski, J. R. (2002). Periaxin mutations cause a broad spectrum of demyelinating neuropathies. *Ann. Neurol.* **51**, 709–715.
40. Gillespie, C. S., Sherman, D. L., Fleetwood-Walker, S. M., Cottrell, D. F., Tait, S., Garry, E. M., Wallace, V. C., Ure, J., Griffiths, I. R., Smith, A., and Brophy, P. J. (2000). Peripheral demyelination and neuropathic pain behavior in periaxin-deficient mice. *Neuron* **26**, 523–531.
41. Stögbauer, F., and Kuhlenbäumer, G. (2005). Distal hereditary motor neuropathies. *In:* "Hereditary Peripheral Neuropathies" (G. Kuhlenbäumer, F. Stögbauer, E. B. Ringelstein, and P. Young, eds.), pp. 146–156. Springer, Steinkopff Darmstadt.
42. Auer-Grumbach, M., Schlotter-Weigel, B., Lochmüller, H., Strobl-Wildemann, G., Auer-Grumbach, P., Fischer, R., Offenbacher, H., Zwick, E. B., Robl, T., Hartl, G., Hartung, H. P., Wagner, K., Windpassinger, C., and the Austrian Peripheral Neuropathy Study Group. (2005). Phenotypes of the N88S Berardinelli-Seip congenital lipodystrophy 2 mutation. *Ann. Neurol.* **57**, 415–424.

43. Dyck, P. J., Chance, P., Lebo, R., and Carney, J. A. (1993). Neuronal atrophy and degeneration predominantly affecting peripheral sensory and autonomic neurons. *In:* "Peripheral Neuropathy" (P. J. Dyck, P. K. Thomas, J. W. Grillin, P. A. Low, and J. F. Poduslo, eds.), 3rd edition, pp. 1065–1093. W. B. Saunders, Philadelphia.
44. De Jonghe, P., and Kuhlenbäumer, G. (2005). Hereditary sensory neuropathies (HSANs). *In:* "Hereditary Peripheral Neuropathies" (G. Kuhlenbäumer, F. Stögbauer, E. B. Ringelstein, and P. Young, eds.), pp. 157–169. Springer, Steinkopff Darmstadt.
45. Houlden, H., King, R., Blake, J., Groves, M., Love, S., Woodward, C., Hammans, S., Nicoll, J., Lennox, G., O'Donovan, D. G., Gabriel, C., Thomas, P. K., and Reilly, M. (2006). Clinical, pathological and genetic characterization of hereditary sensory and autonomic neuropathy type I (HSANI). *Brain* **129(Pt. 2)**, 411–425.
46. Leyne, M., Mull, J., Gill, S. P., Cuajungco, M. P., Oddoux, C., Blumenfeld, A., Maayan, C., Gusella, J. F., Axelrod, F. B., and Slaugenhaupt, S. A. (2003). Identification of the first non-Jewish mutation in familial *Dysautonomia. Am. J. Med. Genet. A.* **118(4)**, 305–308.
47. Kuhlenbäumer, G., Hannibal, M. C., Nelis, E., Schirmacher, A., Verpoorten, N., Meuleman, J., Watts, G. D., De Vriendt, E., Young, P., Stogbauer, F., Halfter, H., Irobi, J., Goossens, D., Del-Favero, J., Betz, B. G., Hor, H., Kurlemann, G., Bird, T. D., Airaksinen, E., Mononen, T., Serradell, A. P., Prats, J. M., Van Broeckhoven, C., De Jonghe, P., Timmerman, V., Ringelstein, E. B., and Chance, P. F. (2005). Mutations in SEPT9 cause hereditary neuralgic amyotrophy. *Nat. Genet.* **37(10)**, 1044–1046.

82

Neurological Manifestations of Vasculitis

Safwan S. Jaradeh, MD

Keywords: *central nervous system, peripheral nervous system, vasculitis*

I. Introduction
II. History and Nomenclature
III. Pathogenesis
IV. Primary Vasculitides
V. Secondary Vasculitides
VI. Central and Peripheral Nervous System Manifestations
References

I. Introduction

Vasculitis is a pathological process characterized by inflammation of blood vessels. This inflammation produces narrowing and eventually occlusion of blood vessels. The symptoms and signs of vasculitis reflect the particular organs or tissues involved. There is ischemia or infarction from the narrowing or occlusion of blood vessels, but there is also an inflammatory component damaging the surrounding tissue. Vasculitis may affect blood vessels of one or of a range of sizes. When vasculitis involves small to medium-sized arteries, it may affect epineurial vessels in the vasa nervorum and produce peripheral neuropathies. When large vessels are primarily involved, there is involvement of the central nervous system (CNS) with relative sparing of epineurial vessels in the peripheral nervous system (PNS). Vasculitis may affect a single or multiple organs. In some individuals, vasculitis remains confined to the PNS or CNS, producing a chronic, progressive, or relapsing neurological impairment. Vasculitis may also be the primary disease process or may complicate another disease, but the neurological manifestations of vasculitis are usually similar whether it is primary or secondary. They often mimic the manifestations of other nonvasculitic neurological conditions, and these diseases should be considered in the differential diagnosis of vasculitis.

II. History and Nomenclature

In 1990, the American College of Rheumatology (ACR) developed their classification of vasculitis based on data obtained from 1000 patients with vasculitis evaluated at 48 medical centers [1]. These criteria classified 678 of these patients into seven specific forms of vasculitis: polyarteritis nodosa (PAN), Churg-Strauss syndrome (CSS), Wegener's granulomatosis (WG), hypersensitivity vasculitis, Henoch-Schonlein purpura (HSP), giant cell (temporal) arteritis (GCA), and Takayasu's arteritis. In addition, 141 patients were classified into vasculitis secondary to a connective tissue disease, 52 with Kawasaki disease, and 129 with vasculitis without one of the seven specific forms of vasculitis or a definite connective tissue disease. Because these criteria were not appropriate for diagnosing new patients with suspected vasculitis, and because of some overlap in nomenclature and the lack of distinction among vasculitis and other nonvasculitic diseases, an international consensus conference (ICC) was held in Chapel Hill, North Carolina, in 1994 to construct standardized nomenclature for the vasculitides [2]. The classification and nomenclature proposed by the ICC include the original seven vasculitides recognized by the ACR but define some of them differently. The ICC classification also defines three other types of systemic vasculitis. The names and definitions for these 10 types of vasculitis were based on the size and histopathology of the involved vessels. According to this system, temporal arteritis and Takayasu's arteritis primarily involve large vessels; classic PAN and Kawasaki disease affect small to medium-sized arteries; and WG, CSS, HSP, microscopic polyangiitis (MPA), cutaneous leukocytoclastic angiitis, and essential cryoglobulinemic vasculitis predominantly involve the microvasculature. The ICC classification provided straightforward pathological definitions and was the first to recognize MPA as a separate diagnostic entity. However, it relied on histological features, which limits the clinical utility when appropriate histology is inconclusive. Thus two classification systems for systemic vasculitis are in common usage today. Although many terms are identical, the 1990 ACR criteria do not include MPA and tend to diagnose WG and, to a lesser extent, PAN more often than the ICC criteria do [3] (Fig. 1).

III. Pathogenesis

The pathogenesis of vasculitis is complex and is the result of various autoimmune reactions, both humoral and cell mediated. There are multiple triggering events or antigens leading to various immunological and histological responses. This complicated autoimmune process activates circulating leukocytes, endothelial cells, and various antigen-presenting cells in and around blood vessel

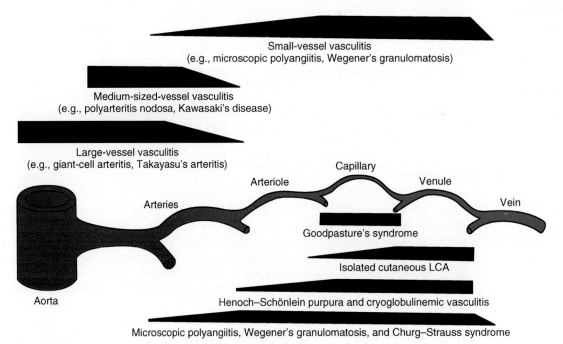

Figure 1 Preferred Sites of Vascular Involvement by Selected Vasculitides. The widths of the trapezoids indicate the frequencies of involvement of various portions of the vasculature. LCA denotes leukocytoclastic angiitis. Reproduced with permission from Jennette, J. C., and Falk, R. J. (1997). Small vessel vasculitis. *N. Engl. J. Med.* **337**, 1512–1523.

walls. The secretion of various inflammatory regulators and cytokines and the activation of cellular adhesion molecules lead to the recruitment of additional effector cells (neutrophils, eosinophils, monocytes, and macrophages), and extravascular migration of leukocytes. Subsequently, there is damage to blood vessel walls by degradative enzymes, cytotoxic cytokine release, formation of free radicals, perforin or complement complex-mediated cytolysis, and apoptosis [4–7]. This destructive cascade exposes tissue factors that stimulate thrombosis, vessel wall fibrosis, and angiogenesis. As the inflammatory activity begin to slow, the subsequent histological repair results in a thick, scarred, and narrow or occluded vessel.

Recently, a greater understanding of the role of cytokines has lead to therapeutic implications. The cytokines are a diverse group of soluble polypeptides that are produced by immune cells and that function as growth factors and hormones of the immune and hematopoietic systems [8–10]. They include interleukins (IL), chemokines, interferons, and colony-stimulating factors. They act on many different cell types and may interact with one another. Major proinflammatory cytokines are IL1, IL8, interferon-γ (IFN-γ), and tumor necrosis factor (TNF)-α. Primary anti-inflammatory cytokines are IL4, IL10, and IL13. Cytokines distinguish subsets of CD4 T helper (Th) lymphocytes. Th1 cells secrete IFN-γ, TNF-α, IL2, and IL12, and Th2 cells secrete IL4, IL5, IL10, and IL13. Th2 cells stimulate B cell and eosinophil proliferation and reduce IFN-γ production by Th1 cells, thereby promoting humoral and allergic responses. Th1 cells promote cellular immune responses by activating macrophages and T cell proliferation.

Table 1 outlines a classification for vasculitis. The primary systemic vasculitides are classified mainly as proposed by the ICC. The systemic vasculitides can be subdivided into three groups based on the size of blood vessel involvement. Large vessels include the aorta and its major branches. Medium-sized vessels are main arteries [11]. Small vessels include arterioles, venules, and capillaries. There is significant overlap among different vasculitides, and vasculitis that affects small and medium-sized vessels is considered a small-vessel vasculitis [12] (see Fig. 1). Because neither classification for systemic vasculitis includes nonsystemic vasculitis limited to the nervous system, Kissel and Mendell [13] proposed adding the category of localized vasculitis to that of primary vasculitis.

In general, the vasculitides are uncommon diseases. Temporal arteritis and the predominantly cutaneous vasculitides are the most common, followed by rheumatoid vasculitis, WG, and MPA. However, their protean neurological manifestations make them an important consideration in the differential diagnosis of many undetermined neurological conditions.

This review focuses on the primary vasculitides.

Table 1 Classification of Vasculitis

I. Primary vasculitis
  A. Systemic vasculitis
    1. Large-vessel vasculitis
      a. Giant cell (temporal) arteritis
      b. Takayasu's arteritis
    2. Medium-sized-vessel vasculitis
      a. Polyarteritis nodosa
      b. Kawasaki disease
    3. Small-vessel vasculitis (with or without medium-sized-vessel involvement)
      a. Pauci-immune small-vessel vasculitis
        i. Wegener's granulomatosis
        ii. Churg-Strauss syndrome
        iii. Microscopic polyangiitis
      b. Immune complex small-vessel vasculitis (with or without medium-sized-vessel involvement)
        i. Henoch-Schönlein purpura
        ii. Essential cryoglobulinemic vasculitis
        iii. Cutaneous leukocytoclastic vasculitis
  B. Localized vasculitis
    1. Isolated angiitis of the central nervous system
    2. Nonsystemic vasculitic neuropathy
II. Secondary vasculitis
  A. Connective tissue disease
    1. Systemic lupus erythematosus
    2. Rheumatoid arthritis
    3. Sjögren's syndrome
    4. Scleroderma
    5. Mixed connective tissue disease
    6. Behçet's syndrome
  B. Infection
    1. Varicella zoster virus
    2. Spirochetes
    3. Rickettsiae
    4. Fungi
    5. Bacterial and mycobacterial meningitis
    6. Human immunodeficiency virus
  C. Neoplasm
  D. Substance abuse

IV. Primary Vasculitides

A. Systemic Vasculitis

1. Large-Vessel Vasculitis

GCA is the most common systemic vasculitis, accounting for 21.4% of all vasculitides included in the 1990 ACR series [14]. Temporal arteritis was first described in 1937 by B. T. Horton and colleagues, and the presence of granulomatous giant cell inflammation in the vessels was further characterized one year later by G. H. Jennings and colleagues. The disease is primarily an autoimmune process in reaction to elastin, which explains its predilection for extradural vessels. The condition affects mainly the ophthalmic, posterior ciliary, superficial temporal, occipital, facial, and internal maxillary arteries. Clinically, patients are older than

50 years and often have presenting symptoms of a new-onset headache [15]. About 50% of patients have symptomatic polymyalgia rheumatica, with aching and morning stiffness of the neck, torso, shoulders, upper arms, hips, or proximal thighs. Claudication of the mastication muscles with chewing is uncommon but highly diagnostic. Amaurosis fugax, permanent visual loss, and ischemic infarction of extraocular muscles or nerves are more specific, are more ominous than headache, and occur in approximately 20% of patients. Peripheral neuropathy is less common in GCA than in many of the small- and medium-sized-vessel vasculitides, although in one series of consecutive patients with GCA 14% had a peripheral neuropathy and half of those had a single or multiple mononeuropathies. Transient ischemic attacks, strokes, ischemic cervical myelopathy, and ischemic encephalopathy are rare complications. Limb claudication is uncommon. Occasionally, the condition becomes generalized with involvement of the aorta, its branches, and other large arteries. Signs include prominent and tender temporal arteries, sometimes with decreased temporal artery pulse and increase scalp sensitivity. Laboratory evaluation reveals an elevated erythrocyte sedimentation rate (ESR) greater than 50 mm/h. The diagnosis is confirmed by a temporal artery biopsy that reveals granulomatous inflammation with multinucleated giant cells. It should be done urgently, and treatment with corticosteroids should be started immediately.

Takayasu's arteritis results from granulomatous inflammation of the aorta or its major extracranial branches and is difficult to distinguish pathologically from GCA. Clinically, patients with Takayasu's arteritis are younger than age 40 and present with claudication of a limb. Most patients are female. Signs include a bruit over the aorta or a subclavian artery, a decrease in peripheral arterial pulses in one or more limbs, and a difference greater than 10 mm Hg in systolic blood pressure between the arms. If suspected clinically, arteriography is the diagnostic test of choice before biopsy. The younger age of onset helps distinguish between Takayasu's arteritis and GCA [16]. Also, in Takayasu's arteritis there is generally marked, full-thickness cicatrization including the intima, which is usually spared in GCA. Takayasu's arteritis rarely produces CNS or PNS manifestations. Neurological sequelae usually occur late in the disease, when gradual obliteration of the vessels leads to ischemia. Manifestations resemble those of temporal arteritis, with headache, amaurosis fugax, monocular blindness, orthostatic dizziness, syncope, stroke, and corneal opacification.

2. Medium-Sized-Vessel Vasculitis

PAN is the prototype of necrotizing arteritis, a group of systemic disorders that also includes CSS, MPA syndrome, Kawasaki disease, and overlap syndrome. PAN is one of the two systemic vasculitides defined differently in the two major classifications. By the 1990 ACR criteria, PAN is the second most common form of systemic vasculitis after GCA, with a frequency of 12%. The presence of 3 or more of the following 10 features supports the diagnosis of PAN: weight loss of 4 kg or more, livedo reticularis, testicular pain or tenderness, myalgias, mononeuropathy or polyneuropathy, diastolic blood pressure greater than 90 mm Hg, elevated blood urea nitrogen or serum creatinine levels, presence of hepatitis B reactants in serum, arteriographic abnormality, and presence of granulocyte or mixed leukocyte infiltrate in an arterial wall on biopsy. However, the ICC criteria defined PAN as a syndrome in which there is necrotizing inflammation of medium or small arteries without glomerulonephritis or vasculitis of arterioles, capillaries, or venules. By excluding PAN in cases with small-vessel involvement and by adding the category of MPA, the frequency of PAN becomes lower. The affected vessels exhibit inflammation and fibrinoid necrosis, but small aneurysmal dilatations may occur, hence the term *nodosa*. Because of involvement of the vasa nervorum, peripheral neuropathy is the primary presenting symptom in more than one-third of patients and most have peripheral neuropathy by the end of the first year of symptoms [17,18] (Fig. 2). CNS manifestations are less common and are of two types. The first type includes stroke and cerebral hemorrhage. The second type manifests as a diffuse encephalopathy, with or without seizures. Retinal infarcts may also occur. ESR is often elevated.

Kawasaki disease is chiefly a pediatric disease and is characterized by viral exanthema that leads to necrotizing arteritis. Children usually have presenting symptoms of acute fever, cervical lymphadenitis, mucocutaneous edema, erythema of the mouth, lips, and palms, and desquamation of the skin of the fingertips. The arteritis involves large, medium, and small arteries. The coronary arteries are most

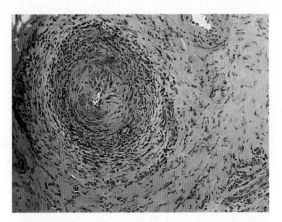

Figure 2 Necrotizing Arteritis in a Medium Epineurial Artery in a Superficial Peroneal Nerve Biopsy Specimen from a Patient with Polyarteritis Nodosa. The muscularis contains areas of fibrinoid necrosis. Hematoxylin and eosin, 200 times magnification.

commonly affected. The condition is usually self-limited, but about 25% of affected children are left with aneurysmal damage to the coronary arteries. Intravenous γ-globulin, in conjunction with aspirin, is effective in reducing the inflammation and preventing coronary artery abnormalities if administered within the first 10 days of illness. Acute encephalopathy and cerebral infarction have been described, but neurological involvement is rare [19].

3. Small-Vessel Vasculitis

WG is also defined differently in the two major classification systems. By the ACR criteria, WG accounted for 9% of all vasculitides [20]. The diagnostic criteria are both sensitive and specific and include two or more of the following four criteria: abnormal findings on chest radiograph (nodules, cavities, or fixed infiltrates), oral ulcers or nasal discharge, abnormal urinary sediment (red cell casts or more than five red blood cells per high power field), and granulomatous inflammation on biopsy. The ICC criteria define WG as a syndrome with granulomatous inflammation involving the respiratory tract and necrotizing vasculitis affecting small and medium vessels (e.g., capillaries, venules, arterioles, and arteries). And although necrotizing glomerulonephritis is common, it is not essential to the diagnosis of WG (Fig. 3).

WG is one of three primary systemic vasculitides strongly associated with antineutrophil cytoplasmic antibodies (ANCAs). ANCAs assay by direct immunofluorescence reveals two patterns of neutrophilic staining: cytoplasmic staining (c-ANCA) and perinuclear staining (p-ANCA). More specific testing by enzyme-linked immunosorbent assays shows two patterns of reactivity: antibodies to a lysosomal proteinase active in killing bacteria, proteinase 3, present in most cases with a c-ANCA pattern, and antibodies to myeloperoxidase, an enzyme that generates oxygen radicals, present in most cases with a p-ANCA pattern. The presence of anti–proteinase 3 antibody is highly specific for WG, whereas antimyeloperoxidase antibodies are characteristic of other vasculitides, such as PAN and MPA. As stated earlier, the diagnostic criteria applied lead to variation in frequency. In one series of 24 patients with primary vasculitis [3], 15 of 24 patients (63%) had WG with the ACR criteria but only 5 (21%) had WG with the ICC criteria. Discrepant cases were diagnosed as MPA, a diagnosis not recognized in the 1990 ACR criteria.

The disease is more common in men. In WG, the vasculitis affects very small vessels and CNS involvement is not common. In one large series, 13 of 324 patients with WG had an associated cerebrovascular event [21]. Other CNS complications include a cranial neuropathy, seizures, or encephalopathy. There are rare reports of basilar artery occlusion associated with WG. Exceptionally, there are instances of direct extension of the granulomatous process from the sinuses to the orbits and basal meninges, resulting in visual and oculomotor disturbances. On the other hand, WG often leads to PNS involvement. Peripheral neuropathy occurs in up to 44% of patients [21,22].

CSS is one of the least common forms of primary systemic vasculitis. The ACR criteria are also sensitive and specific. They include four or more of the following six criteria: asthma, peripheral eosinophilia exceeding 10% on differential white blood cell count, mononeuropathy (single or multiple) or polyneuropathy, nonfixed pulmonary infiltrates on chest x-ray, paranasal sinus abnormality, and biopsy containing a blood vessel with extravascular eosinophils. The ICC criteria defined CSS as a syndrome with: (1) eosinophil and granulomatous inflammation of the respiratory tract, (2) necrotizing vasculitis affecting small to medium vessels, and (3) asthma and eosinophilia. CSS is also associated with ANCAs. Peripheral neuropathy develops in more than 50% of patients, whereas cranial neuropathy and stroke each occur in 5% [23].

MPA is a syndrome with necrotizing vasculitis that affects small vessels. There are few or no immune deposits. Necrotizing glomerulonephritis and pulmonary capillaritis are common. MPA is also associated with the presence of ANCAs. MPA was not recognized in the ACR classification. Peripheral neuropathy is the main neurological complication and occurs in up to 36% of patients [24] (Fig. 4).

HSP is as common as WG (9% of vasculitis), but it rarely affects the nervous system. The diagnostic criteria include at least two of the following: age less than or equal to 20 years at disease onset, palpable purpura, acute abdominal pain, and biopsy showing granulocytes in the walls of small arterioles or venules. The ICC defined HSP as a small-vessel vasculitis with immunoglobulin A (IgA) immune deposits that typically involves skin, gut, and glomeruli. It is commonly associated with arthralgias or arthritis. HSP rarely affects the nervous system, but there are cases complicated by seizures, encephalopathy, or neuropathy [25–27].

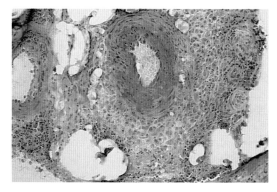

Figure 3 Vasculitis with Mononuclear Infiltrates and Epithelioid Cells in a Patient with Wegener's Granulomatosis. Trichrome, 200 times magnification.

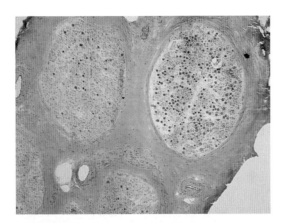

Figure 4 Nonnecrotizing Arteritis in a Small Epineurial Artery in a Sural Nerve Biopsy Specimen from a Patient with Antineutrophil Cytoplasmic Antibody-Associated Microscopic Polyangiitis. There is striking multifocal axonal loss among various fascicles. Trichrome, 200 times magnification.

Essential cryoglobulinemic vasculitis involves small vessels, but with cryoglobulin immune deposits that affect the same vessels. Cryoglobulins are serum proteins that precipitate at a low temperature and dissolve at higher temperatures. Type I has monoclonal immunoglobulins, type II has a combination of polyclonal and monoclonal immunoglobulins, and type III has polyclonal immunoglobulins. Types II and III are designated mixed essential cryoglobulins when there is no associated lymphoproliferative disorder. The level of circulating cryoglobulins in the serum may fluctuate. Most cases are associated with hepatitis C infection [28,29], and these cases actually represent a secondary vasculitis. Nonetheless, hepatitis C serology should be obtained when cryoglobulins are detected. However, because hepatitis C viral messenger ribonucleic acid (mRNA) may be concentrated up to 1000 times within the cryoprecipitate, the polymerase chain reaction assay for viral mRNA may need to be performed on the cryoprecipitate if the initial screen is negative. Other laboratory findings that support a diagnosis of cryoglobulinemic vasculitis include low levels of complement C4 and the presence of an IgM rheumatoid factor without antinuclear antibodies.

Patients with cryoglobulinemic vasculitis commonly have presenting symptoms of peripheral neuropathy, but stroke is rare. The neuropathy often takes the form of mononeuritis multiplex, because epineurial vasculitis may induce fascicular ischemia and multifocal axon loss. Renal involvement may coexist [30,31].

A prospective randomized, controlled trial found that IFN-α2a led to marked improvement in 60% of patients with cryoglobulinemia associated with hepatitis C virus infection [32]. However, treatment of cryoglobulinemia with negative hepatitis C serology is more debatable, although one small study of two patients also showed a response to IFN-α2a [33]. In patients with neurological complications, plasmapheresis may be used as short-term treatment to remove circulating cryoglobulins and to prevent their deposition in tissues. Immunosuppressive therapy with prednisone and cytotoxic drugs, such as cyclophosphamide, usually is initiated to suppress cryoglobulin production. Recently, rituximab, an anti-CD20 monoclonal antibody that depletes B cells, showed promise in cases that are hepatitis C seropositive and seronegative [35]. However, hepatitis C viral mRNA levels may increase in some treated patients, and the level must be monitored during treatment with rituximab.

Cutaneous leukocytoclastic vasculitis is a dermatological angiitis without other system involvement. Still, the ICC included this diagnosis in their nomenclature for systemic vasculitis. It is likely a form of hypersensitivity vasculitis. Nervous system involvement by definition excludes the diagnosis of cutaneous leukocytoclastic vasculitis.

B. Localized Vasculitis

1. Isolated Angiitis of the Central Nervous System

In 1959, Cravioto and Feigin reported two new patients and reviewed six other cases with a noninfectious granulomatous angiitis of the nervous system [35]. All patients had headaches and multifocal CNS signs. Seizures and stupor were common. Cerebrospinal fluid was normal in one patient. It had elevated protein in four and mononuclear pleocytosis in four. Autopsy on all patients revealed a granulomatous angiitis of the CNS (exclusive in six, predominant in two). In this condition, there is a necrotizing process involving segments of leptomeningeal and blood vessels that measure 200–500 µm in diameter. The vessel walls are infiltrated by lymphocytes, plasma cells, epithelioid cells, and giant cells. Granulomas may be found in brain tissue adjacent to inflamed vessels. Few cases are caused by the varicella-zoster virus as documented by polymerase chain reaction testing. ESR is increased in 65% of these patients, and the white blood cell count is elevated in about 50%. Other serological tests are negative.

The prognosis was poor until a series of four patients with isolated angiitis of the CNS showed a better response to a combination of cyclophosphamide and prednisone compared with prednisone alone [36]. If the diagnosis is suspected, magnetic resonance angiography or cerebral angiography may be helpful in illustrating arterial irregularities consistent with this diagnosis [37]. However, brain and meningeal biopsy is the diagnostic procedure of choice before starting treatment [36–39].

2. Nonsystemic Vasculitic Neuropathy

In 1985, Kissel and colleagues found 4 cases of necrotizing vasculitis without systemic involvement out of

16 patients with vasculitic neuropathy [40]. Two years later, Dyck and colleagues [41] described 20 new cases and coined the term *nonsystemic vasculitic neuropathy*. Among these 20 patients, 12 had mononeuritis multiplex, 4 had an asymmetrical polyneuropathy, and 4 had a distal symmetrical polyneuropathy. Subsequent series have shown that about one-third of vasculitic neuropathies are caused by nonsystemic vasculitis limited to the peripheral nerve and muscle [42]. Patients have a mean age of 51 years (range 23 to 74 years), and neuropathic symptoms often begin acutely or subacutely. Constitutional symptoms are absent or minimal. Laboratory evaluation for systemic involvement is negative, although 2 patients had an ESR of more than 50 mm/h. The prognosis is more benign than with systemic vasculitis or CNS angiitis, and aggressive treatment with cyclophosphamide is not necessary. The condition usually responds to oral corticosteroids, but some patients may require chronic maintenance on low-dose immunosuppressive therapy, in which case agents such as azathioprine may be used.

V. Secondary Vasculitides

Vasculitides secondary to defined connective tissue diseases are common and account for 14% of the ACR cases of vasculitis [1]. The rheumatological syndromes that can be associated with CNS vasculitis include systemic lupus erythematosus, rheumatoid arthritis, mixed connective-tissue disease, Behçet's syndrome, Sjögren's syndrome, progressive systemic sclerosis, and Reiter's syndrome.

Systemic lupus erythematosus is an autoimmune disorder with dermal, joint, renal, cardiac, and hematological manifestations. The dominant immune mechanism is that of immune complex deposition, and small vessels are preferentially involved. In addition to microinfarctions, microhemorrhages may occur in the subarachnoid space or subcortical white matter. The CNS is far more affected than the PNS. Neuropsychiatric symptoms may be the initial presentation in a small percentage of patients, but they subsequently develop in more than a third of patients during the course of the disease [43]. Other CNS manifestations include optic neuropathy, transverse myelitis, headache, stroke, pseudotumor cerebri, myelopathy, chorea, dementia, and affective disorders. Changes in the mental status and elevated level of antinuclear antibodies should suggest the diagnosis. PNS involvement is less common and, as with other vasculitides, patients may develop a distal symmetrical polyneuropathy, mononeuropathy multiplex, or polyradiculoneuropathy. More specific tests include double-stranded deoxyribonucleic acid and anti-Smith antibodies. The sensitivity of these markers ranges from 25 to 75%. In addition to vasculitis, systemic lupus erythematosus may lead to a nonvasculitic cerebral angiopathy caused by circulating IgG and IgM antiphospholipid antibodies. A stroke syndrome associated with circulating lupus anticoagulants and antiphospholipid antibodies can also occur.

Rheumatoid arthritis is a granulomatous disease that involves multiple organs, including the nervous system. The rheumatoid lesions consist of infiltrates of T lymphocytes and plasma cells and scattered areas of fibrinoid necrosis. During flare-ups, neutrophils also accumulate and destroy the tissue by releasing lytic enzymes and producing oxygen free radicals [5]. The lesions form nodules in serous membranes but may form sheetlike plaques in the nervous system. The main complications include rheumatic pachymeningitis, encephalopathy, seizures, cerebral hemorrhage, and myelopathy. The spinal cord may also be compressed because of erosive skeletal manifestations leading to vertebral collapse, odontoid or atlantoaxial subluxation, and spinal stenosis from extradural pannus formation. Peripheral neuropathy affects less than 10% of patients, is mild, and tends to appear in chronic disease. Median neuropathy at the wrist is often caused by compression secondary to joint deformity and synovial thickening.

About 20% of patients with *Sjögren's syndrome* have CNS symptoms that may mimic multiple sclerosis [44], but stroke, hemorrhage, seizures, aseptic meningoencephalitis, and transverse myelitis have all been reported. Peripheral neuropathy also occurs, and most of these cases are caused by vasculitic neuropathy involving mainly the epineurial arterioles and venules [45]. The neuropathy can be either sensory or sensorimotor. Myositis may also occur. When skin is involved, urticaria, erythematous macules, and purpura appear. The syndrome is associated with keratoconjunctivitis sicca, and antineuronal antibodies such as extractable RNA proteins Ro or Sjögren syndrome-A, and intranuclear RNA-associated antigen La or Sjögren syndrome-B are usually present.

In *scleroderma* or systemic sclerosis, there is widespread microangiopathy and diffuse tissue fibrosis affecting skin, heart, lungs, and gastrointestinal tract. Neurological manifestations include headache, encephalopathy, seizures, and myositis and often follow renal involvement and malignant hypertension. One variant is the CREST syndrome (Raynaud's phenomenon, esophageal dysmotility, sclerodactyly, and telangiectasia). Trigeminal nerve involvement is well described, but peripheral neuropathy is rare [42,44]. Some patients have autoantibodies against centromere, SCL-70 (scleroderma) or topoisomerase, RNA-polymerase III determinants, and the HLA-DQB1 haplotype.

In *mixed connective tissue disease*, there are features of systemic lupus erythematosus, scleroderma, and polymyositis. Neurological manifestations occur in up to 10% of patients and include headache, aseptic meningitis, optic neuropathy, encephalopathy, seizures, ataxia, transverse myelitis, and inflammatory neuropathy.

Behçet's syndrome is a chronic, relapsing inflammatory vasculitic syndrome endemic in the Far East, Middle East, and Mediterranean basin. HLA-B51 is the main allele associated with this disease. The international diagnostic criteria require the presence of recurrent oral ulcerations and at least two of the following: recurrent genital ulcers, skin lesions (erythema nodosum, pseudofolliculitis, papulopustular lesions, or acneiform nodules), ocular lesions (anterior uveitis, posterior uveitis, or retinal vasculitis) and a positive pathergy test [46,47]. The neurological involvement in Behçet's syndrome is often caused by inflammation of small veins with focal or multifocal CNS parenchymal involvement, but a less common form is caused by cerebral venous sinus thrombosis and has a better prognosis. One prospective study found neurological complications in 13% of men and 6% of women [47]. Headaches, subacute brainstem syndrome, and intracranial hypertension from venous sinus thrombosis are the most common CNS manifestations, but hemispherical and myelopathic presentations are also known. The Cerebrospinal fluid (CSF) is often abnormal, with increased protein and lymphocytic pleocytosis. PNS involvement is uncommon and appears as a mononeuritis multiplex, a polyneuropathy, or a polyradiculoneuritis [46].

The *antiphospholipid-antibody syndrome* is characterized by the presence of arterial and venous thrombosis, thrombocytopenia, fetal loss in women, and the presence of antiphospholipid antibodies, lupus anticoagulants, or both [48]. There are increased titers of anticardiolipin antibodies, mainly of the IgG type, and a false-positive serological test for syphilis may be present. These antibodies are not specific to any particular disorder. The antibodies may enhance platelet aggregation by direct platelet phospholipid binding or by interfering with endothelial cell function, but the exact pathogenic mechanisms are not fully known. Cerebral or ocular ischemia occurs most often in patients with other risk factors for stroke [49]. Cerebral angiography is abnormal in two-thirds of cases and shows occlusion of extracranial or intracranial arteries, but vasculitis is usually not seen.

Vasculitis can also complicate certain *infections*, especially hepatitis B and C, varicella-zoster, cytomegalovirus infections, and several bacterial, fungal, and parasitic infections [50].

In *paraneoplastic vasculitis* [51,52], the cancer is usually of the hematological type, and hairy-cell leukemia and malignant histiocytosis have been associated with polyangiitis. Another malignancy that mimics primary cerebral vasculitis is intravascular lymphomatosis [53]. This condition is a variant of non-Hodgkin's lymphoma in which the neoplastic cells accumulate in small vessels and may cause multiple infarctions of the brain. Only small- and medium-sized arteries are involved, which explains the low yield of cerebral angiography. Another unusual form of lymphoma associated with vasculitis is that of lymphomatoid granulomatosis, an angiocentric and angiodestructive lymphoproliferative disease in which the lungs are usually affected as well. On magnetic resonance imaging (MRI) there are multiple punctate or linear-enhancing lesions, found primarily in the perivascular spaces, or multiple periventricular and deep white-matter abnormalities; multiple infarctions are rare [54]. If an underlying cancer is suspected, a computed tomography scan of the thorax, abdomen, and pelvis should be done, and assays for α-fetoprotein, prostate-specific antigen, and carcinoembryonic antigen may be indicated.

Vasculitis may be secondary to *substance abuse*, especially of amphetamines and cocaine [55]. There is debate about whether sympathomimetic drugs cause a vasculitis or a vasculopathy because pathological studies have found various lesions ranging from narrow vascular lumens to a frank necrotizing angiitis. The syndrome is more common with amphetamines than cocaine and may be caused by the drug itself or to the vehicle used to mix the drug, such as talc. The pathogenesis is usually that of an immune complex small-vessel angiitis. Other mechanisms account for the ischemic stroke caused by sympathomimetic drugs, including acute vasospasm and increased platelet aggregation. If this syndrome is suspected, MRI and cerebral angiography are helpful. If the CSF shows pleocytosis, the possibility of a CNS vasculitis is strong and the temporary use of corticosteroids may be considered, but further immunosuppressive therapy is not needed. Brain biopsy is not recommended in these patients. Cerebral angiography should also be considered when sympathomimetic drugs lead to a hemorrhagic stroke in a young adult (younger than 45 years of age) because these agents may unmask a preexisting arteriovenous malformation or aneurysm. There are numerous cases of ANCA-positive vasculitis in patients taking antithyroid medication. Almost all cases followed the administration of propylthiouracil, with only two cases occurring after receiving methimazole. There are usually crescentic glomerulonephritis and a lupus-like syndrome, with positive tests for antinuclear antibodies and leukopenia.

VI. Central and Peripheral Nervous System Manifestations

A. Central Nervous System Manifestations of Vasculitis

CNS involvement in systemic vasculitis is uncommon relative to PNS involvement. In isolated CNS angiitis, headaches and altered mental status are often the presenting symptoms. Focal neurological symptoms are less common and occur after several weeks of nonfocal changes. Focal motor abnormalities are more common than seizures or language impairment. Myelopathy may occur in up to 15% of patients [36].

Unless immunotherapy is instituted, there is progressive deterioration to stupor and death over weeks or months. Fever, anorexia, and weight loss are common in most vasculitides. When a patient's presenting symptoms are a combination of headache, focal signs, seizures, and encephalopathy, vasculitis with CNS involvement should be considered if other etiologies are not found. The involvement of other tissues or organs such as skin, lungs, kidneys, or peripheral nerves is a useful clue to the diagnosis. The evaluation should include complete blood count, ESR, antinuclear antibody, rheumatoid factor, ANCA, extractable nuclear antibody panel, complement levels, cryoglobulins, hepatitis B and C serology, urine for protein and microscopic analysis, and chemistry tests of liver and renal function. Chest x-ray and other pertinent imaging studies should be obtained. With CNS involvement, MRI of the brain and CSF analysis should be done. As mentioned earlier, the CSF is often abnormal in CNS vasculitis and shows mononuclear pleocytosis in 80% of patients, with more than 100 cells/mm^3 in 60% of 40 patients in one series [36]. In the same series, CSF protein was increased in 92% of patients and exceeded 100 mg/dL in 72% of them. CSF glucose is seldom decreased. The absence of headache and normal CSF studies should make the clinician question the diagnosis of CNS vasculitis, but rare exceptions occur [38]. MRI of the brain is sensitive in CNS vasculitis, often revealing multiple lesions [56], but the findings often lack specificity or the MRI may be normal [57].

Cerebral angiography may be useful to the diagnosis when it reveals multifocal segmental stenosis. However, the sensitivity of cerebral arteriography is lower than that of MRI, and in one series of 14 biopsy proven cases, 9 had normal results [38]. If the diagnosis remains in doubt, brain biopsy is necessary for confirmation. The biopsy ideally targets the affected tissue, but if the MRI abnormalities are diffuse, a biopsy of the meninges and nondominant temporal tip are usually done.

Once a primary vasculitis with CNS involvement is confirmed, treatment with a combination of prednisone and cyclophosphamide is often necessary. In isolated CNS angiitis, recent reviews suggest adding cyclophosphamide to prednisone using individualized clinical judgment [39].

B. Peripheral Nervous System Manifestations of Vasculitis

Mononeuropathy multiplex is the prototypical presentation in vasculitic neuropathy. It is usually subacute and painful. The initial symptom is often a deep aching pain in the proximal limb affected by vasculitis. Paresthesiae in the distribution of the affected nerve usually occur several hours to days after the proximal pain. Numbness and weakness typically follow the positive sensory symptoms. Although this pattern is highly suggestive, vasculitic neuropathy appears as often with symmetrical or asymmetrical length-dependent polyneuropathy. This is often because of the cumulative effect of multifocal nerve ischemia and inflammation, leading to partially or fully confluent patterns [42].

Among systemic vasculitides, PAN has the highest incidence of mononeuropathy multiplex. Because some nerves are more vulnerable to ischemia, the distribution of nerve infarction is not random. In one series of 94 cases of vasculitic neuropathy, the peroneal nerve or division was involved in 76%, the ulnar nerve in 28%, the tibial nerve in 11%, and the median nerve in 9% [58]. In general, the peroneal nerve is most often affected in the lower limb and the ulnar nerve is most commonly involved in the upper limb. The peroneal nerve has a watershed zone of relatively poor perfusion in the midthigh, and the ulnar nerve has a watershed zone in the mid–upper arm. On neurological examination, localized impairment of pain and temperature sensation dominates that of large fiber sensory modalities [59]. Localized weakness is also common. Hyporeflexia or areflexia is usually distal.

When vasculitic neuropathy is suspected, electrodiagnostic studies are important to assess possible asymmetry, which may also help in selecting suitable muscle and nerve for biopsy. Because the process leads most often to axonal degeneration in a multifocal pattern (Fig. 4), nerve conduction studies typically reveal low amplitudes of compound nerve action potentials, sensory more than motor. The distribution may not depend on length. Conduction velocities are usually normal or minimally slowed. On needle EMG, there is usually multifocal active partial denervation. The EMG abnormalities also identify acute partial denervation in proximal muscles. The systemic evaluation is similar to that mentioned earlier in CNS vasculitis. Biopsy of an affected muscle and cutaneous nerve should be arranged to establish the diagnosis.

Once vasculitic neuropathy is confirmed, the treatment depends on its type. Nonsystemic vasculitic neuropathy often responds to prednisone followed occasionally by azathioprine, but cyclophosphamide is not necessary. PAN usually requires a combination of prednisone and cyclophosphamide because of its high mortality. However, some authors do not recommend the routine addition of cyclophosphamide after one large prospective trial showed that remission and survival rates were similar for those treated with steroids and plasma exchange, with or without cyclophosphamide [60]. Most clinicians reserve the use of cyclophosphamide for cases in which histology documents the presence of arterial necrosis or systemic involvement is critical (lung or kidney disease). A recent randomized trial in patients with generalized vasculitis and ANCA antibodies found that the withdrawal of cyclophosphamide and the substitution of azathioprine after 3 months of remission did not increase the rate of relapse while it reduced the cumulative toxicity from exposure to cyclophosphamide [61].

References

1. Bloch, D. A., Michel, B. A., Hunder, G. G., McShane, D. J., Arend, W. P., Calabrese, L. H., Edworthy, S. M., Fauci, A. S., Fries, J. F., Leavitt, R. Y., et al. (1990). The American College of Rheumatology 1990 criteria for the classification of vasculitis: patients and methods. *Arthritis Rheum.* **33(8)**, 1068–1073.
2. Jennette, J. C., Falk, R. J., Andrassy, K., Bacon, P. A., Churg, J., Gross, W. L., Hagen, E. C., Hoffman, G. S., Hunder, G. G., Kallenberg, C. G. et al. (1994). Nomenclature of systemic vasculitides: proposal of an international consensus conference. *Arthritis Rheum.* **37(2)**, 187–192.
3. Bruce, I. N., and Bell, A. L. (1997). A comparison of two nomenclature systems for primary systemic vasculitis. *Br. J. Rheumatol.* **36(4)**, 453–458.
4. Crow, M. K. (2001). Structure and function of monocytes/macrophages and other antigen-presenting cells. In: "Arthritis and Allied Condition: A Textbook of Rheumatology" (W. J. Koopman, ed.), 14th edition, pp. 317–336. Lippincott Williams & Wilkins, Philadelphia.
5. Suzuki, K. (2001). Neutrophil functions of patients with vasculitis related to myeloperoxidase-specific anti-neutrophil antibody. *Int. J. Hematol.* **74**, 134–143.
6. Goronzy, J. J. (2002). T cells in vascular disease. In: "Inflammatory Diseases of Blood Vessels" (G. S. Hoffman and C. M. Weyand, eds.), pp. 57–68. Marcel Dekker, New York.
7. Russell, J. H., and Ley, T. J. (2002). Lymphocyte-mediated cytotoxicity. *Annu. Rev. Immunol.* **20**, 323–370.
8. Csernok, E., and Gross, W. L. (2002). Cytokines and vascular inflammation. In: "Inflammatory Diseases of Blood Vessels" (G. S. Hoffman and C. M. Weyand, eds.), pp. 97–112. Marcel Dekker, New York.
9. Lotz, M. K. (2001). Cytokines and their receptors. In: "Arthritis and Allied Condition: A Textbook of Rheumatology" (W. J. Koopman, ed.), 14th edition, pp. 437–477. Lippincott Williams & Wilkins, Philadelphia.
10. Borish, L. C., and Steinke, J. W. (2003). Cytokines and chemokines. *J. Allergy Clin. Immunol.* **111**, S460–S475.
11. Weyand, C. M., and Goronzy, J. J. (2003). Medium- and large-vessel vasculitis. *N. Engl. J. Med.* **349**, 160–169.
12. Jennette, J. C., and Falk, R. J. (1997). Small vessel vasculitis. *N. Engl. J. Med.* **337**, 1512–1523.
13. Kissel, J. T., and Mendell, J. R. (1992). Vasculitic neuropathy. *Neurol. Clin.* **10(3)**, 761–781.
14. Hunder, G. G., Bloch, D. A., Michel, B. A., Stevens, M. B., Arend, W. P., Calabrese, L. H., Edworthy, S. M., Fauci, A. S., Leavitt, R. Y., Lie, J. T., et al. (1990). The American College of Rheumatology 1990 criteria for the classification of giant cell arteritis. *Arthritis Rheum.* **33(8)**, 1122–1128.
15. Caselli, R. J., and Hunder, G. G. (1997). Giant cell (temporal) arteritis. *Neurol. Clin.* **15(4)**, 893–902.
16. Michel, B. A., Arend, W. P., and Hunder, G. G. (1996). Clinical differentiation between giant cell (temporal) arteritis and Takayasu's arteritis. *J. Rheumatol.* **23(1)**, 106–111.
17. Bouche, P., Leger, J. M., Travers, M. A., Cathala, H. P., and Castaigne, P. (1986). Peripheral neuropathy in systemic vasculitis: clinical and electrophysiologic study of 22 patients. *Neurology* **36(12)**, 1598–1602.
18. Guillevin, L., Le Thi Huong, D., Godeau, P., Jais, P., and Wechsler, B. (1988). Clinical findings and prognosis of polyarteritis nodosa and Churg-Strauss angiitis: a study in 165 patients. *Br. J. Rheumatol.* **27(4)**, 258–264.
19. Tabarki, B., Mahdhaoui, A., Selmi, H., Yacoub, M., and Essoussi, A. S. (2001). Kawasaki disease with predominant central nervous system involvement. *Pediatr. Neurol.* **25(3)**, 239–241.
20. Leavitt, R. Y., Fauci, A. S., Bloch, D. A., Hunder, G. G., Arend, W. P., Calabrese, L. H., Fries, J. F., Lie, J. T., Lightfoot, R. W. Jr, et al. (1990). The American College of Rheumatology 1990 criteria for the classification of Wegener's granulomatosis. *Arthritis Rheum.* **33(8)**, 1101–1107.
21. Nishino, H., Rubino, F. A., DeRemee, R. A., Swanson, J. W., and Parisi, J. E. (1993). Neurological involvement in Wegener's granulomatosis: an analysis of 324 consecutive patients at the Mayo Clinic. *Ann. Neurol.* **33**, 4–9.
22. De Groot, K., Schmidt, D. K., Arlt, A. C., Gross, W. L., and Reinhold-Keller, E. (2001). Standardized neurologic evaluations of 128 patients with Wegener granulomatosis. *Arch. Neurol.* **58(8)**, 1215–1221.
23. Sehgal, M., Swanson, J. W., DeRemee, R. A., and Colby, T. V. (1995). Neurologic manifestations of Churg-Strauss syndrome. *Mayo Clin. Proc.* **70(4)**, 337–341.
24. Lhote, F., Cohen, P., Genereau, T., Gayraud, M., and Guillevin, L. (1996). Microscopic polyangiitis: clinical aspects and treatment. *Ann. Med. Interne (Paris)* **147(3)**, 165–177.
25. Chen, C. L., Chiou, Y. H., Wu, C. Y., Lai, P. H., and Chung, H. M. (2000). Cerebral vasculitis in Henoch-Schonlein purpura: a case report with sequential magnetic resonance imaging changes and treated with plasmapheresis alone. *Pediatr. Nephrol.* **15(3–4)**, 276–278.
26. Sokol, D. K., McIntyre, J. A., Short, R. A., Gutt, J., Wagenknecht, D. R., Biller, J., and Garg, B. (2000). Henoch-Schonlein purpura and stroke: antiphosphatidylethanolamine antibody in CSF and serum. *Neurology* **55(9)**, 1379–1381.
27. Bulun, A., Topaloglu, R., Duzova, A., Saatci, I., Besbas, N., and Bakkaloglu, A. (2001). Ataxia and peripheral neuropathy: rare manifestations in Henoch-Schonlein purpura. *Pediatr. Nephrol.* **16(12)**, 1139–1141.
28. Dammacco, F., Sansonno, D., Piccoli, C., Tucci, F. A., and Racanelli, V. (2001). The cryoglobulins: an overview. *Eur. J. Clin. Invest.* **31(7)**, 628–638.
29. Ferri, C., Zignego, A. L., and Pileri, S. A. (2002). Cryoglobulins. *J. Clin. Pathol.* **55(1)**, 4–13.
30. Apartis, E., Leger, J. M., Musset, L., Gugenheim, M., Cacoub, P., Lyon-Caen, O., Pierrot-Deseilligny, C., Hauw, J. J., and Bouche, P. (1996). Peripheral neuropathy associated with essential mixed cryoglobulinaemia: a role for hepatitis C virus infection? *J. Neurol. Neurosurg. Psychiatry* **60(6)**, 661–666.
31. Origgi, L., Vanoli, M., Carbone, A., Grasso, M., and Scorza, R. (1998). Central nervous system involvement in patients with HCV-related cryoglobulinemia. *Am. J. Med. Sci.* **315(3)**, 208–210.
32. Misiani, R., Bellavita, P., Fenili, D., Vicari, O., Marchesi, D., Sironi, P. L., Zilio, P., Vernocchi, A., Massazza, M., Vendramin, G., et al. (1994). Interferon α-2a therapy in cryoglobulinemia associated with hepatitis C virus. *N. Engl. J. Med.* **330**, 751–756.
33. Casato, M., Lagana, B., Pucillo, L. P., and Quinti, I. (1998). Interferon for hepatitis C virus-negative type II mixed cryoglobulinemia. *N. Engl. J. Med.* **338**, 1386–1387.
34. Zaja, F., De Vita, S., Mazzaro, C., Sacco, S., Damiani, D., De Marchi, G., Michelutti, A., Baccarani, M., Fanin, R., and Ferraccioli, G. (2003). Efficacy and safety of rituximab in type II mixed cryoglobulinemia. *Blood* **101**, 3827–3834.
35. Carvioto, H., and Feigin, I. (1959). Noninfectious granulomatous angiitis with a predilection for the nervous system. *Neurology* **9**, 599–609.
36. Vollmer, T. L., Guarnaccia, J., Harrington, W., Pacia, S. V., and Petroff, O. A. (1993). Idiopathic granulomatous angiitis of the central nervous system: diagnostic challenges. *Arch. Neurol.* **50(9)**, 925–930.
37. Moore, P. M., and Richardson, B. (1998). Neurology of the vasculitides and connective tissue diseases. *J. Neurol. Neurosurg. Psychiatry* **65**, 10–22.

38. Alrawi, A., Trobe, J. D., Blaivas, M., and Musch, D. C. (1999). Brain biopsy in primary angiitis of the central nervous system. *Neurology* **53(4)**, 858–860.
39. Younger, D. S., Calabrese, L. H., and Hays, A. P. (1997). Granulomatous angiitis of the nervous system. *Neurol. Clin.* **15(4)**, 821–834.
40. Kissel, J. T., Slivka, A. P., Warmolts, J. R., and Mendell, J. R. (1985). The clinical spectrum of necrotizing angiopathy of the peripheral nervous system. *Ann. Neurol.* **18(2)**, 251–257.
41. Dyck, P. J., Benstead, T. J., Conn, D. L., Stevens, J. C., Windebank, A. J., and Low, P. A. (1987). Nonsystemic vasculitic neuropathy. *Brain* **110(Pt 4)**, 843–853.
42. Olney, R. K. (1998). Neuropathies associated with connective tissue disease. *Semin. Neurol.* **18(1)**, 63–72.
43. Brey, R. L., Holliday, S. L., Saklad, A. R., Navarrete, M. G., Hermosillo-Romo, D., Stallworth, C. L., Valdez, C. R., Escalante, A., del Rincon, I., Gronseth, G., Rhine, C. B., Padilla, P., and McGlasson, D. (2002). Neuropsychiatric syndromes in lupus: prevalence using standardized definitions. *Neurology* **58(8)**, 1214–1220.
44. Nadeau, S. E. (2002). Neurologic manifestations of connective tissue disease. *Neurol. Clin.* **20(1)**, 151–178.
45. Mellgren, S. I., Conn, D. L., Stevens, J. C., and Dyck, P. J. (1989). Peripheral neuropathy in primary Sjögren's syndrome. *Neurology* **39**, 390–394.
46. Siva, A., Altintas, A., and Saip, S. (2004). Behçet's syndrome and the nervous system. *Curr. Opin. Neurol.* **17(3)**, 347–357.
47. Kural-Seyahi, E., Fresko, I., Seyahi, N., Ozyazgan, Y., Mat, C., Hamuryudan, V., Yurdakul, S., and Yazici, H. (2003). The long-term mortality and morbidity of Behçet syndrome: a 2-decade outcome survey of 387 patients followed at a dedicated center. *Medicine (Baltimore)* **82**, 60–76.
48. Levine, S. R., Brey, R. L., Sawaya, K. L., Salowich-Palm, L., Kokkinos, J., Kostrzema, B., Perry, M., Havstad, S., and Carey, J. (1995). Recurrent stroke and thrombo-occlusive events in the antiphospholipid syndrome. *Ann. Neurol.* **38**, 119–124.
49. Brey, R. L., and Escalante, A. (1998). Neurological manifestations of antiphospholipid antibody syndrome. *Lupus* **7(Suppl 2)**, S67–S74.
50. Gerber, O., Roque, C., and Coyle, P. K. (1997). Vasculitis owing to infection. *Neurol. Clin.* **15(4)**, 903–925.
51. Greer, J. M., Longley, S., Edwards, N. L., Elfenbein, G. J., and Panush, R. S. (1988). Vasculitis associated with malignancy: experience with 13 patients and literature review. *Medicine (Baltimore)* **67**, 220–230.
52. Oh, S. J. (1997). Paraneoplastic vasculitis of the peripheral nervous system. *Neurol. Clin.* **15(4)**, 849–863.
53. Glass, J., Hochberg, F. H., and Miller, D. C. (1993). Intravascular lymphomatosis: a systemic disease with neurologic manifestations. *Cancer* **71**, 3156–3164.
54. Tateishi, U., Terae, S., Ogata, A., Sawamura, Y., Suzuki, Y., Abe, S., and Miyasaka, K. (2001). MR imaging of the brain in lymphomatoid granulomatosis. *AJNR Am. J. Neuroradiol.* **22**, 1283–1290.
55. Brust, J. C. (1997). Vasculitis owing to substance abuse. *Neurol. Clin.* **15(4)**, 945–957.
56. Pomper, M. G., Miller, T. J., Stone, J. H., Tidmore, W. C., and Hellmann, D. B. (1999). CNS vasculitis in autoimmune disease: MR imaging findings and correlation with angiography. *AJNR Am. J. Neuroradiol.* **20(1)**, 75–85.
57. Wasserman, B. A., Stone, J. H., Hellmann, D. B., and Pomper, M. G. (2001). Reliability of normal findings on MR imaging for excluding the diagnosis of vasculitis of the central nervous system. *AJR Am. J. Roentgenol.* **177(2)**, 455–459.
58. Said, G., Lacroix-Ciaudo, C., Fujimura, H., et al. (1988). The peripheral neuropathy of necrotizing arteritis: a clinicopathological study. *Ann. Neurol.* **23(5)**, 461–465.
59. Hellmann, D. B., Laing, T. J., Petri, M., Whiting-O'Keefe, Q., and Parry, G. J. (1988). Mononeuritis multiplex: the yield of evaluations for occult rheumatic diseases. *Medicine (Baltimore)* **67(3)**, 145–153.
60. Guillevin, L., Lhote, F., Cohen, P., Jarrousse, B., Lortholary, O., Genereau, T., Leon, A., and Bussel, A. (1995). Corticosteroids plus pulse cyclophosphamide and plasma exchanges versus corticosteroids plus pulse cyclophosphamide alone in the treatment of polyarteritis nodosa and Churg-Strauss syndrome patients with factors predicting poor prognosis: a prospective, randomized trial in sixty-two patients. *Arthritis Rheum.* **38(11)**, 1638–1645.
61. Jayne, D., Rasmussen, N., Andrassy, K., Bacon, P., Tervaert, J. W., Dadoniene, J., Ekstrand, A., Gaskin, G., Gregorini, G., de Groot, K., Gross, W., Hagen, E. C., Mirapeix, E., Pettersson, E., Siegert, C., Sinico, A., Tesar, V., Westman, K., Pusey, C.; European Vasculitis Study Group. (2003). A randomized trial of maintenance therapy for vasculitis associated with antineutrophil cytoplasmic autoantibodies. *N. Engl. J. Med.* **349(1)**, 36–44.

83

Neuropathies Associated with Infections

Paul Twydell, DO
David N. Herrmann, MBBCh

Keywords: *diphtheria, hepatitis C, human immunodeficiency virus infection, leprosy, Lyme disease, pathogenesis, peripheral neuropathy, West Nile virus*

I. Introduction
II. Peripheral Neuropathy Syndromes Related to Bacterial, Mycobacterial, and Spirochetal Infections
III. Peripheral Neuropathy Syndromes Related to Viral Infections
IV. Conclusion
   References

I. Introduction

A variety of peripheral neuropathies occur during the course of human infections or as parainfectious or postinfectious syndromes. Direct cytopathic effects of the infectious agent, peripheral neurotoxicity mediated by dysregulated host immune responses to the infectious agent, and neurotoxicity from antimicrobial therapies are shared mechanisms underlying infection-related neuropathies. However, for most neuropathy syndromes, including those described in this chapter, our understanding of pathogenesis remains fragmentary. This chapter focuses on peripheral neuropathies occurring during infection and reviews in detail the pathogenesis of selected infection-related neuropathies, some of historical and others of contemporary clinical interest.

II. Peripheral Neuropathy Syndromes Related to Bacterial, Mycobacterial, and Spirochetal Infections

A. Lyme Disease

Lyme disease is a multisystem disorder caused by three species of the tick-borne *(Ixodes dammini, Ixodes rinus, Ixodes scapularis)* spirochete: *Borrelia burgdorferi* (Bb) sensu lato, Bb sensu stricto (United States and Europe), *Borrelia garinii*, and *Borrelia afzelii* (Europe) [1].

Lyme disease was first recognized in the United States in 1975, and Bb was described as the causative organism in 1982. However, the clinical manifestations of the disorder had long been described in Europe as a tick-borne meningopolyneuritis [1]. Peripheral nervous system (PNS) manifestations of Lyme disease occur with both acute Bb infection and late-stage Lyme disease (months to years after initial infection) [1,2]. Early disseminated Bb infection is associated with a painful meningopolyradiculoneuropathy and cranial neuropathies (typically bifacial palsy) (Garin-Bujadoux-Bannwarth syndrome) [1,2]. Late disseminated Lyme disease is primarily associated with a chronic predominantly sensory or sensorimotor (symmetrical or asymmetrical) axonal polyneuropathy [1,2]. Nearly 40% of patients with late Lyme borreliosis as a presenting symptom will have peripheral nerve involvement [1,2]. Less clear associations exist between Lyme disease and brachial plexopathies, carpal tunnel syndrome, and a Guillain-Barré syndrome-like illness.

In the United States, transmission is most likely to occur in the late summer or early fall in the mid- to northeast Atlantic coast, upper Midwest, and occasionally California and requires about 36 hours of tick attachment. Once the *Ixodes* tick ingests host blood, the spirochete begins to proliferate in its gut. Subsequent dissemination in the tick and transfer to the host leads to infection. Within 1 month of infection, a centrifugally expanding erythema migrans rash will develop. This is followed by acute disseminated disease in which a flulike syndrome occurs and multiple systems become involved.

1. Pathology

Early- and late-stage neuropathies disclose similar pathological changes, although inflammation may be more prominent in early-stage cases. Pathologically, there is multifocal axon degeneration and loss of myelinated and unmyelinated fibers, with epineurial, perivascular, and/or transmural inflammation (lymphocytes, plasma cells, histiocytes) involving small arterioles [3]. Small epineurial vessel thrombosis and recanalization may occur, but fibrinoid necrosis is not seen. Perineuritis is a feature in some cases [3]. Electrophysiological findings are consistent with the preceding pathological findings and indicate a primarily axon loss mechanism [2].

2. Pathogenesis

The pathogenesis of PNS Lyme disease is uncertain [1,4–6]. Given the pathological similarities between early and late PNS manifestations of Lyme disease, a common pathogenesis has been suggested. Bb has been considered to mediate neural damage through several mechanisms.

a. Direct Cytopathic Effects of Borrelia burgdorferi *Infection* PNS manifestations of Lyme disease in general respond well to antibiotic therapy, which implicates persistent Bb infection in their pathogenesis. Bb can bind *in vitro* to Schwann cells and galactocerebroside moieties and is cytotoxic toward neonatal brain cells and oligodendrocytes [1]. However, in contrast to other tissues, Bb has seldom been isolated from the PNS in human Lyme disease, although one recent report described the presence of Bb (on polymerase chain reaction) in a patient with a chronic sensorimotor polyneuropathy [6]. Although it has been suggested that local infection may be important in the pathogenesis of acute meningoradiculitis associated with early disseminated Lyme infection, the failure to isolate Bb from PNS tissue in most pathological studies argues against direct cytopathic effects of Bb as a dominant mechanism of PNS Lyme disease [1].

b. Cytokine-Mediated Neurotoxicity Bb is known to stimulate T cells, be mitogenic for B cells, and stimulate production of several proinflammatory cytokines, including tumor necrosis factor-α (TNF-α) and interleukin 6 (IL6), which can mediate neural and Schwann cell damage [1,5]. Bb lipoproteins are believed to play an important role in cytokine dysregulation. Lipoprotein-mediated stimulation of TNF-α (but not IL6) appears to be mediated through a mitogen-activated protein kinase–dependent pathway [5].

Pathological studies are lacking to support the hypothesis of cytokine-mediated axonal or peripheral neuronal damage in Lyme disease; however, cytokines are emerging as a relevant effector pathway in several infection-related neuropathies, including human immunodeficiency virus (HIV) associated distal symmetrical polyneuropathy.

c. Borrelia burgdorferi *Induction of Microvasculitis with PNS Axon Loss Secondary to Ischemia* Although several mechanisms may contribute to PNS Lyme disease, the pathological features of multifocal axon loss, epineurial perivascular and transmural inflammatory infiltrates, and epineurial vessel thrombosis and recanalization suggest an immune-mediated microvasculitic mechanism, whereby axon degeneration or loss is mediated partly by ischemia [4,6]. However, definite pathological evidence of vasculitis (transmural inflammation) is not present in many cases. Thus the question of whether a microvasculitis is a universal pathway to neuropathy in Lyme disease is controversial. Recent detailed immunopathological investigations in a patient with a sensorimotor polyneuropathy that developed 4 years after a tick bite demonstrated presence of Bb deoxyribonucleic acid (DNA) by polymerase chain reaction in the sural nerve and the presence of perivascular, perineurial, and endoneurial mononuclear cell infiltrates [6]. Terminal complement deposition (membrane attack complex) deposition was widespread in epineurial blood vessels, and activated (CD68$^+$) macrophages surrounded and invaded epineurial vessel walls. Occasional CD4$^+$ T lymphocytes were observed in a perivascular distribution [6]. HLA DR was widely expressed on Schwann cells and macrophages. No immunoglobulin deposition or anti–peripheral nerve antibodies in serum were identified [6]. These features suggest an immune-mediated perivasculitic mechanism for this polyneuropathy in tertiary Lyme disease, possibly triggered and sustained by the presence of Bb in peripheral nerves.

d. Borrelia burgdorferi Induction of Autoantibodies That Cross-React with the PNS (Molecular Mimicry) The possibility of immune-mediated PNS involvement related to molecular mimicry has been hypothesized [4]. Studies indicate that antibodies to the flagellin and Osp A epitopes on Bb cross-react with peripheral nerve and dorsal root ganglion (DRG) cells [4]. Immunoglobulin M (IgM) and IgG autoantibodies to terminal gangliosides with Gal(B11-3) GalNac also occur in some patients with Bb infection [4]. Studies demonstrating lymphocyte reactivity toward cross-reactive neural epitopes in the serum of patients with PNS Lyme disease and reproduction of PNS Lyme disease in an animal model are lacking to further support molecular mimicry as a relevant mechanism [4].

3. Prognosis

The meningoradiculoneuropathy associated with early-stage Lyme disease is self-limited with excellent recovery. Most patients with late-stage PNS Lyme disease improve or make a good recovery with appropriate antimicrobial therapy [2].

B. Leprosy

Leprosy is one of the "oldest" human infectious diseases, with biblical descriptions and significant historical and social influence [7]. Caused by the obligate intracellular acid-fast bacillus, *Mycobacterium leprae*, it is a leading cause of peripheral neuropathy in the developing world. Gerard Henrik Armauer Hansen identified *M. leprae* as the causative organism in 1873 [7]. With the development of multidrug therapy and bacille Calmette-Guerin vaccination, the prevalence of leprosy has declined considerably; however, about 500,000 new cases still occur each year [7]. There is a wide spectrum of disease (Table 1) ranging from relatively mild tuberculoid leprosy (TT), where host cell–mediated immunity (CMI) toward *M. leprae* is preserved and mycobacterial organisms are few (paucibacillary leprosy), to devastating lepromatous leprosy (LL), which is associated with poor host CMI, and overwhelming infection (multibacillary leprosy) [7].

Anesthetic skin patches and multifocal palpable enlargement of nerves are a hallmark of leprosy. There is a predilection for colder areas of the body, including the ear lobes, chin, knees, and dorsum of the hands and feet. TT manifests with a limited number of discreet asymmetrical, anesthetic skin lesions, with involvement of epidermal and dermal nerve fibers [7]. There is initial impairment of temperature and pain sensation and then loss of pressure sensation. Peripheral nerves proximal to anesthetic skin patches may be involved and enlarged.

Borderline leprosy produces more numerous, anesthetic skin patches with more ill-defined edges. Nerve trunks are affected and enlarged, and the patient has a progressive mononeuropathy multiplex (MM). LL has presenting symptoms of numerous, nonanesthetic skin lesions and a more slowly evolving and relatively symmetrical sensorimotor polyneuropathy that characteristically spares the palms of the hands and soles of the feet and often deep tendon reflexes. A minority of leprosy patients develop neuritic leprosy in which there are no skin lesions but an MM. The absence of skin lesions may delay diagnosis [7].

1. Pathogenesis of Leprosy-Related Neuropathies

M. leprae is thought to be transmitted primarily through nasal droplet infection or contact with infected soil, with humans and armadillos as known reservoirs. There is no

Table 1 Ridley-Jopling Classification of Leprosy[a]

Form of Leprosy		Neuropathy Features	Histopathology
Tuberculoid (TT)	Excellent *M. leprae*–specific CMI	Asymmetrical nerve enlargement proximal to skin lesions	Granuloma formation disrupting nerve structure; rare bacilli
Borderline tuberculoid		Shared features of TT and borderline	Shared features of TT and borderline
Borderline		Early involvement of multiple nerves	Spectrum of features of TT and LL with variable bacilli
Borderline lepromatous		Shared features of borderline and LL	Shared features of borderline and LL
Lepromatous (LL)	Poor *M. leprae*–specific CMI	Late distal symmetrical neuropathy	Foamy cell infiltration of the perineurium; many bacilli seen

[a] Arrow denotes decreasing *M. leprae*–specific CMI.
Note: CMI, cell-mediated immunity.

agreed on route of entry into the PNS; however, direct binding to dermal nerve Schwann cells in cool body areas in TT and direct perineurial invasion through hematogenous dissemination in LL are the most likely routes of entry [7].

a. Host Immunity and Clinical Manifestations of Leprosy Severity of disease is related to the hosts' immune response, with more than 90% of infected individuals lacking clinical manifestations. The Ridley-Jopling classification of leprosy (see Table 1) is based on this immune response. Infected individuals express varying degrees of disease based on their genetic susceptibilities. Cytokine expression, natural resistance–associated macrophage protein 1, HLA genes, toll-like receptors (TTRs), interferon-γ (IFN-γ) expressivity, and leukocyte immunoglobulin receptor expression have all been implicated in disease resistance or expression [7]. Individuals expressing a T helper 1 (Th1) cell response to the mycobacterial infection are able to mount a stronger cellular immune response than those with primarily T helper 2 (Th2) cells whose macrophages are unable to combat the invading bacilli. TT, where disease-specific CMI is more effective, is thus associated with a Th1 profile of cytokine expression (IL2, IFN-γ, IL12, and IL18) and TNF-α. LL is associated with a Th2 response (IL4, IL5, IL10, and reduced TNF-α) [7]. The TTR is involved in upregulating antigen-presenting molecules and cytokine expression. Reduced expression of the TTR leads to more disseminated forms of leprosy [7]. Increases in T cell production of IFN-α provokes killing of intracellular *M. leprae* by infected macrophages [7]. The leukocyte immunoglobulin receptor interferes with the TTR and alters interleukin production, promoting mycobacterial proliferation [7]. IL10 promoter single-nucleotide polymorphisms have also been associated with risk of development of either multibacillary or paucibacillary forms of leprosy. The profile of host cytokine production in response to *M. leprae* infection thus conditions whether the host will remain asymptomatic or develop pauci- or multibacillary leprosy.

b. Mechanisms of Cutaneous Sensory Loss in Leprosy Anesthetic and hypesthetic skin patches are an early feature of several forms of leprosy-related neuropathies. The early prominence of cutaneous sensory loss has been attributed the predilection of *M. leprae* for a cooler environment and the Schwann cells of unmyelinated nerve fibers [8]. Immunohistochemical studies across the leprosy spectrum have shown reduced nerve growth factor immunoreactivity in the basal layer of the epidermis and a loss of intraepidermal nerve fibers compared with controls in both affected and clinically unaffected mirror image skin [8]. Moreover, studies have shown decreased numbers of tyrosine kinase A and nerve growth factor–dependent NaV1.8-positive nerve fibers in the subepidermis in leprosy skin [8]. It has been suggested that a neurotrophic defect occurs early in leprosy and may explain decreases in sensation before detectable loss of unmyelinated dermal nerve fibers and lack of local pain, despite intense inflammation in some forms of cutaneous leprosy [8].

c. Pathogenesis of Neuropathy in Multibacillary (Lepromatous) Leprosy PNS damage in LL is mediated primarily by proliferation of *M. leprae* within Schwann cells in the dermis and peripheral nerves, within the context of a defective host cell–mediated immune response to the organism [8].

M. leprae binds to Schwann cells through an organism-specific cell wall antigen, the phenolic glycolipid-1 complex that attaches specifically to laminin-2, a component of the extracellular basal lamina [7]. This complex then binds to α-dystroglycan, which is associated with the plasma membrane [7]. *M. leprae* is able to invade non-myelin-producing Schwann cells (e.g., of unmyelinated dermal nerve fibers), which have been suggested to serve as reservoirs of infection [7]. *M. leprae* binds to myelin-producing Schwann cells, in a similar manner, but is not able to invade them. Once bound to myelinating Schwann cells, signal transduction because of dystrophin-related protein-2 disrupts myelination, leading to demyelination of nerves [7]. Once these myelin-producing Schwann cells are converted to non-myelin-producing Schwann cells, the organism is able to invade them and proliferate, causing further myelin destruction, disruption of Schwann cell–axon interactions, and neural degeneration [7]. Additional pathogenic factors in PNS leprosy are believed to include release of cytokines from macrophages (e.g., IL1) and growth factors (e.g., fibroblast growth factor) with progressive fibrosis of the endoneurium and perineurium, which might impede axon regeneration.

These pathogenic factors contribute to the pathology of LL. The fascicular structure of the nerve (epineurium/perineurium/endoneurium) is relatively preserved in LL compared with in TT. There is initial infection of the epineurium and perineurium and uneven fascicular involvement. *M. leprae* is widespread within macrophages (foamy cells), fibroblasts, and endothelial cells. The perineurium becomes hypertrophic and assumes an onion skin appearance with edema, infiltration of foamy macrophages, and perineurial cell proliferation. These features explain the palpable nerve enlargement in leprosy. The foamy macrophages of LL and borderline lepromatous leprosy show immunoreactivity for transforming growth factor-β (TGF-β) [9]. TGF-β is believed to be a macrophage suppressor that may contribute to ineffective *M. leprae* clearance and relate to the preponderance of immature forms of macrophages in LL lesions. This contrasts with TT (see Sect. II.B.1.d), where macrophages do not express TGF-β and where macrophages differentiate and proliferate into epithelioid cells [9].

Proliferation of *M. leprae* within Schwann cells in the endoneurium produces segmental demyelination [7]. As the disease progresses, inflammation increases and axon loss occurs with progressive neural fibrosis. Giant cells, granulomas, and true vasculitis are not features of LL. The pathology of LL is thus mediated by *M. leprae* proliferation within various cells types, including Schwann cells, demyelination, disruption of axon–Schwann cell interactions and fibrosis.

d. Pathogenesis and Pathology of Tuberculoid Leprosy
The neural pathology of TT is primarily mediated by the host cell–mediated immune (hypersensitivity) response to *M. leprae* antigens in the skin and peripheral nerves [7]. In paucibacillary TT, *M. leprae* is not evident on Fite staining of skin and nerve biopsy specimens. *M. leprae* organisms are evident in Schwann cells on electron microscopy, and *M. leprae* antigens are present in the dermis and peripheral nerve using immunohistochemical techniques. In TT, macrophages effectively inactivate *M. leprae* organisms, and an intense antigen-specific T cell response with Th1 cytokine production occurs [7]. Macrophages develop epithelioid features, and Langerhans and foreign body–type giant cells develop with granuloma formation [7]. The neuropathy of TT is primarily an axon loss neuropathy, mediated partly as a bystander effect of the intense inflammatory reaction and granuloma formation. In contrast to LL, involved nerves show marked disruption of their fascicular arrangement. Involvement of the nerve is multifocal, with an overwhelming inflammatory infiltrate comprising granulomas with epithelioid histiocytes, multinucleated giant cells, lymphocytes, and plasma cells. There is marked inflammatory thickening of the perineurium. As the disease evolves, nerve architecture is destroyed, leaving a fibrotic residue with complete axon loss. Beyond bystander effects, there is evidence that macrophages associated with granuloma formation in leprosy express inducible nitric oxide synthase and nitrotyrosine [10]. Nitrotyrosine immunoreactivity has been observed within dermal neurofilament aggregates and on *M. leprae* cell walls at the ultrastructural level [10]. These data implicate peroxynitrite and nitrosative stress in the neural damage. Other macrophage products, including proteases, may contribute to axonal damage in TT.

e. Pathogenesis and Pathology of Borderline Forms of Leprosy Neuropathy In borderline leprosy the cardinal pathological features are the presence of plentiful epithelioid histiocytes and *M. leprae* bacilli but no foamy cells (LL) and no giant cells or granulomas (TT). Pathology is usually accentuated in the perineurium and subperineurium with proliferation of epithelioid histiocytes. Initial involvement of Schwann cells and demyelination occur as in LL, followed by axon loss. Borderline lepromatous leprosy and borderline tuberculoid leprosy forms show pathological features that are intermediate between those of borderline leprosy and those of the polar forms (LL and TT) of leprosy.

2. Course and Prognosis of Leprosy-Related Neuropathies

LL neuropathies stabilize or improve with therapy, although the degree of improvement depends on the pretreatment severity of the neuropathy. In TT, loss of skin sensation may be permanent. Immune-mediated leprosy reactions can also occur during the course of the disease and therapy as changes in host immunity occur. Leprosy reactions can produce acute worsening of leprosy-related neuropathies.

Type 1 reactions or *M. leprae* reactions occur mostly in borderline leprosy and represent a change in the host immune response [7]. They are subdivided into reversal and downgrading reactions, with reversal reactions occurring as a result of treatment and downgrading reactions representing a progression of disease. They are characterized by a worsening of symptoms coinciding with new skin and nerve lesions. They also are characterized by a delayed hypersensitivity response with CD4$^+$ T lymphocyte and macrophage infiltration and increased levels of IL2 and IFN-γ [7]. Caseous necrosis and microabscess formation in affected nerves can occur in tuberculoid and lepromatous forms, respectively [7].

Type 2 reactions or erythema nodosum reactions occur mainly in LL, are associated with widespread destruction of *M. leprae* organisms, and are characterized by painful erythematous papules, acute worsening of neuropathy, fatigue, fever, lymphadenopathy, and glomerulonephritis. Increased TNF-α expression and immune complex deposition are thought to be responsible [7].

Finally, therapeutic interventions for leprosy may themselves lead to a peripheral neuropathy. Dapsone produces a dose-related, motor-predominant, distal, axonal polyneuropathy with a predilection for the upper extremities [7]. Dose reduction or drug cessation may reverse the neuropathy; however, some individuals will progress for a period after the drug is discontinued. Thalidomide, used to treat erythema nodosum reactions, causes a dose-related, distal, symmetrical, primarily sensory axon loss polyneuropathy that progresses until the drug is withdrawn with partial recovery [7].

C. Diphtheria

Diphtheria is caused by infection with *Corynebacterium diphtheriae*, a gram-positive, aerobic rod and release of exotoxin (diphtheria toxin) [11,12]. Diphtheria is today a relatively rare disorder as a result of advances in public health and dedicated vaccination programs. Although it has been known about for centuries, it was first accurately described by Pierre-Fidele Bretonneau in the first quarter of the nineteenth century, when epidemics had broken out in France, Norway, and Denmark. Edwin Klebs and Friedrich

Löeffler isolated the bacillus in 1883. The neuropathy of diphtheria, however, may not have been described until 1888 when two colleagues of Louis Pasteur, Emile Roux and Alexandre Emile John Yersin, demonstrated that the exotoxin was responsible for widespread effects in rabbits, guinea pigs, and pigeons. Ninety percent of the world's reported cases of diphtheria arose from an epidemic within the former Soviet Union between 1990 and 1995 [11,12]. During that time, more than 125,000 people were infected, resulting in more than 4000 deaths.

1. Clinical Characteristics of Diphtheria Polyneuropathy

Polyneuropathy develops in about 20% of patients with diphtheria. Its occurrence is associated with the severity of pharyngeal infection and may lead to bulbar and diaphragmatic paralysis, necessitating mechanical ventilation; a painful polyradiculoneuropathy; and autonomic and circulatory dysfunction [11,12]. It is characterized by a biphasic course, with initial effects of local infection and later systemic effects. Cranial nerve involvement, particularly cranial nerves IX and X, occurs within the first 2 weeks, followed by progressive ventilatory failure (in 20% of cases), quadriparesis, and eventually sensory disturbances [11,12].

2. Pathogenesis

C. diphtheriae is harbored in the nasopharyngeal pathways of humans and transmitted by respiratory droplets or human-to-human contact. Its exotoxin is a 58.3 KD protein with two functional components. The B portion binds to cells using the heparin-binding epidermal growth factor (diphtheria toxin receptor), and the A portion is responsible for its toxic effects. The exotoxin inhibits protein production in affected cells by catalyzing a covalent attachment of an adenosine diphosphate–ribose moiety of nicotinamide adenine dinucleotide to elongation factor-2 [13]. Diphtheritic neuropathy is characterized pathologically by a noninflammatory segmental and paranodal demyelination of nerve roots, dorsal root and autonomic ganglia, and peripheral nerves with phagocytic macrophages infiltration and Schwann cell proliferation. *In vitro* studies using sciatic nerves from chick embryos have suggested that diphtheria toxin enters Schwann cells and induces segmental demyelination, not by degradation of myelin proteins but by inhibition of the synthesis of proteolipid and basic myelin proteins [14]. This disrupts the normal turnover of myelin proteins.

3. Prognosis of Diphtheria Polyneuropathy

The course of diphtheritic neuropathy can be devastating, with about 15% mortality and 40% of survivors unable to return to manual labor by 1 year. In contrast to Guillain-Barré syndrome, diphtheria neuropathy evolves more slowly (over a median of 49 days in one study) and the time to onset of recovery is slower (a median of 73 days) [11]. Diphtheria toxin antitoxin lessens the severity of the disorder and reduces mortality rates, but only if administered within the first 48 hours after onset of neuropathic symptoms [11].

III. Peripheral Neuropathy Syndromes Related to Viral Infections

A. HIV Infection-Associated Distal Symmetrical Polyneuropathy

HIV infection has been associated with various forms of peripheral neuropathy (Table 2) [15]. Some occur early (CD4$^+$ lymphocyte counts ≥ 500 cells/mm^3), for example, acute inflammatory demyelinating polyneuropathy; may

Table 2 Peripheral Nervous System Disorders in Human Immunodeficiency Virus Infection[a]

Stage of HIV	Syndrome
Early	Acute inflammatory demyelinating polyneuropathy
(≥ 500 CD4$^+$ cells/µl)	Chronic inflammatory demyelinating polyneuropathy
	Brachial plexopathy
	Mononeuropathies (e.g., facial neuropathy)
Moderately advanced	Chronic inflammatory demyelinating polyneuropathy
(200–500 CD4$^+$ cells/µl)	Diffuse infiltrative lymphocytosis syndrome
	Syphilitic polyradiculopathy
	Hepatitis C distal neuropathy
	HTLV-1-related neuropathy
Advanced	Distal sensory polyneuropathy
(<200 CD4$^+$ cells/µl)	Autonomic neuropathy
	Cytomegalovirus-related mononeuropathy multiplex
	HIV-related lumbosacral polyradiculopathy
	Cytomegalovirus-related polyradiculopathy
	Polyradiculopathy due to other infectious or neoplastic meningitides
Any stage	Antiretroviral therapy–related neuropathies
	Other medication-related neuropathies
	Mononeuropathy multiplex

[a] Modified from Brew, B. J. (2003). The peripheral nerve complications of human immunodeficiency virus (HIV) infection. *Muscle Nerve* **28**, 542–552.
Note: HIV, human immunodeficiency virus; HTLV-1, human T lymphotropic virus type 1.

coincide with HIV seroconversion; are immune mediated; and respond to standard immunotherapies. Others develop primarily with advanced HIV infection and an immunocompromised state and are facilitated by HIV replication, neurotoxicity from antiretroviral therapies, and coinfection with opportunistic pathogens (Table 2) [15]. Some syndromes (e.g., MM) may occur at any stage of HIV infection; however, the pathogenesis depends on the stage of infection. Thus MM with HIV seroconversion is typically immune mediated and often self-limited; with moderate immunosuppression, it may be associated with PNS vasculitis and during advanced HIV infection with cytomegalovirus coinfection and necrotizing vasculitis [15]. MM in HIV infection may also be associated with cryoglobulinemia (CG), hepatitis B or C coinfection, and/or systemic arteritis (polyarteritis nodosa). Antigen-antibody immune complex formation and deposition may lead to vasculitis in some of these instances.

Here we only review in detail current understanding of the pathogenesis of HIV-associated distal symmetrical polyneuropathy (HIV-DSP), the most common form of peripheral neuropathy with HIV infection [15,16]. HIV-associated symptomatic distal symmetrical polyneuropathy (HIV-SDSP) occurred in 30–35% of patients with acquired immunodeficiency syndrome (AIDS) before the advent of highly active antiretroviral therapy (HAART) in 1996 [16]. Since the introduction of HAART, the incidence of SDSP appears to have decreased. However, its prevalence has increased, coinciding with increased patient survival [16,17].

HIV-DSP is regarded as a late manifestation of HIV infection, usually occurring when CD4$^+$ cell counts drop lower than 200 cells/mm^3 [15]. It is a symmetrical and often painful, distal, predominantly small-fiber sensory axonal polyneuropathy [15,16]. HIV-DSP should be distinguished from an acute toxic neuropathy due to dideoxynucleoside (DDN) antiretroviral agents (zalcitabine, didanosine, and stavudine) [15–17]. Acute toxic neuropathy occurs in the setting of the recent initiation of a DDN agent and improves with discontinuation of the offending agent. However, some patients may "coast" for 6–8 weeks after drug withdrawal. The incidence of DDN-caused acute toxic neuropathy is associated with higher doses (in common use in the early 1990s), advanced immunosuppression, and concomitant pharmacological interventions (e.g., hydroxyurea) [15–17]. Mitochondrial toxicity is the likely mechanism underlying DDN neuropathies. DDNs inhibit DNA-polymerase-γ and mitochondrial DNA synthesis [17].

1. Epidemiology of HIV-Associated Distal Symmetrical Polyneuropathy

Risk factors for development of SDSP in the pre-HAART era included a history of AIDS-defining illnesses, poorer physical function scores, advanced immunosuppression, higher HIV plasma viral load, older age, and underlying comorbidities (e.g., diabetes mellitus) [15–17].

The relationship between antiretroviral therapy and HIV-DSP is complex [16,17]. In the pre-HAART era, the incidence of SDSP appeared to be strongly associated with DDN use [16,17]. Since introduction of HAART, and with changing patient selection and lower dosing of DDNs, the effect of antiretroviral therapy on incidence of SDSP is less clear. At least one large prospective cohort (the North East AIDS Dementia Cohort) has found that use of DDN within 6 months of study entry was not a risk factor for SDSP [16]. Although DDNs are toxic to DRG neurons in culture, they have generally failed to produce SDSP in animal models. DDNs may unmask an asymptomatic neuropathy in a susceptible HIV-infected host rather than cause *de novo* peripheral neurotoxicity [16,17].

Protease inhibitors, an integral component of HAART, are not known to be neurotoxic but do induce the metabolic syndrome, with impaired glucose tolerance or diabetes mellitus. The effect of this drug-induced metabolic syndrome on the occurrence and progression of HIV-SDSP is unclear. Alcohol and intravenous drug abuse and hepatitis C virus (HCV) coinfection are also potential risk factors for SDSP that require further investigation [15–17].

2. Pathology of HIV-Associated Distal Symmetrical Polyneuropathy

Pathological findings are distally dominant and involve predominantly sensory pathways [16]. Dorsal root ganglia show a mild loss of neurons [16]. Axonal degeneration of the central and peripheral projections of DRG neurons occurs with a proximal–distal severity gradient [16]. Sural nerve biopsies disclose an axon loss picture, with a greater loss of unmyelinated than of myelinated nerve fibers. Perivascular mononuclear inflammatory cell collections may be seen in the epineurium or endoneurium; however, vasculitis is not a feature of HIV-DSP. Punch skin biopsies demonstrate a loss of unmyelinated epidermal C fibers in most patients. In mild cases, epidermal nerve fiber density may remain within a normal range, but epidermal and subepidermal nerve fibers show morphological abnormalities [16]. The predominant pathological involvement of unmyelinated sensory fibers may explain the presentation of SDSP as a painful, predominantly sensory polyneuropathy.

3. Pathogenesis of HIV-Associated Distal Symmetrical Polyneuropathy

The pathogenesis of HIV-DSP is uncertain. Current evidence supports dysregulated immune activation and cytokine-mediated neurotoxicity triggered by HIV infection or HIV viral proteins (e.g., the GP120 envelope protein) rather than direct viral neuronal infection and neurotoxicity [16–18].

HIV-1 has been isolated from peripheral nerve tissue; however, studies have generally failed to demonstrate productive HIV infection of peripheral neuronal tissue, arguing against a hypothesis of direct HIV neurotoxicity [16–19]. In contrast, HIV mRNA and protein has been observed in epidermal Langerhans cells, satellite cells in the dorsal root ganglia, and macrophages in peripheral nerve and dorsal root ganglia [19]. HIV-1 isolates from peripheral nerve tissue in HIV-SDSP have shown a predominance of macrophage tropic, CCR5 chemokine receptor–dependent strains [19]. In support of a hypothesis of dysregulated immune activation, macrophage and T lymphocyte (CD8 predominance) infiltration, and up-regulation of cytokines (in particular, TNF-α, interferon-γ, IL1, and IL6) has been observed at all levels of the PNS in DSP, with HIV-infected individuals without DSP showing lower levels of proinflammatory cytokines and macrophage activation [16,18,19]. The HIV envelope protein GP120, which may be released by HIV-1-infected glial cells, is neurotoxic to DRG neurons in cell culture models [16,19,20]. Two pathways of GP120 DRG neurotoxicity have been delineated in these culture models. First, GP120 may bind to chemokine receptors on Schwann cells with release of RANTES (regulated on activation, normally T expressed, and presumably secreted), a chemokine molecule. This triggers TNF-α-mediated neurotoxicity using caspase-3-dependent pathways [16,20]. Second, GP120 has been shown to induce apoptosis of neonatal rat DRG neurons, likely by binding to the CXCR4 chemokine receptor on dorsal root ganglia, and inducing activation of the C-Jun N-terminal kinase pathway [20]. It is uncertain to what extent each of these pathways contributes *in vivo* to neurotoxicity in HIV-SDSP [20]. Cerebrospinal fluid levels of the cytokine macrophage colony-stimulating factor have also been associated with active HIV replication and with development of SDSP [16].

Elevated titers of IgG antisulfatide antibodies have recently been reported to occur in about 36% of patients with HIV-SDSP but not in those without DSP. This raises the hypothesis that autoantibody-mediated peripheral nerve injury may also contribute to HIV-SDSP in some patients; however, the significance of sulfatide antibodies in HIV-DSP is uncertain and may yet represent an epiphenomenon [21].

Peripheral nerve injury through the preceding pathways is suggested to produce neuropathic pain through sensitization of nociceptors, alterations in protein transcription (e.g., voltage-gated sodium channels), and central sensitization [16].

4. Prognosis

HIV-SDSP is a chronic, slowly progressive neuropathy. Symptoms of neuropathic pain can be variably controlled with symptomatic agents; however, it has yet to be determined whether virologically successful HAART can arrest or partially reverse HIV-SDSP.

B. Hepatitis C Virus–Related Peripheral Neuropathies

Peripheral neuropathy is a relatively common accompaniment of chronic HCV infection, a single-stranded RNA virus belonging to the *Flaviviridae* family [22,23].

Peripheral neuropathies associated with HCV infection are largely related to mixed type II or III CG, which occurs in about 50% of HCV patients. HCV has an overall prevalence of peripheral neuropathy of about 9%, with a prevalence in patients with concomitant mixed CG of about 30% [22]. In one study, 78% of HCV patients with peripheral neuropathy had associated CG [22]. Peripheral neuropathy is uncommon but does occur in association with chronic HCV without CG or other etiological explanation [22]. There have been rare reports of neuropathy in HCV infection related to polyarteritis nodosa. Secondary Sjögren's syndrome has increasingly been associated with chronic HCV infection and may be a cause of various peripheral neuropathy syndromes.

1. Clinical Syndromes of Hepatitis C Virus–Related Neuropathies

The most common presentation, irrespective of the presence of CG, is a distal sensory or sensorimotor axonal polyneuropathy. Some patients have MM in the limbs, cranial neuropathies, or a combination of the two. There is an uncertain association between HCV and acquired demyelinative neuropathies (e.g., chronic inflammatory demyelinating polyneuropathy) [22]. If CG is present, Raynaud's phenomenon, purpura, digital necrosis, arthralgias, hepatosplenomegaly, and renal involvement are variable accompaniments. Peripheral neuropathy associated with HCV infection is, in general, severer when CG is present [22,23].

2. Pathology and Pathogenesis of Hepatitis C Virus Neuropathy Syndromes

Sural nerve biopsies from patients with HCV, with and without CG, show similar pathological features consistent with an angiopathic ischemic mechanism [22]. Approximately 50% of HCV-associated neuropathies show evidence of epineurial vasculitis involving small- and medium-sized arterioles [22,24]. Axon loss is multifocal, with intrafascicular and interfascicular variation in the distribution of axon loss. Both of these features support an ischemic mechanism of axonal degeneration [22].

Understanding of how HCV infection leads to cryoprecitable immune complex formation is emerging [23,24]. Most patients with HCV and CG have type II mixed CG, where the CG are composed of a monoclonal IgM component with rheumatoid factor (RF) activity and polyclonal IgG. HCV stimulates a polyclonal activation of B cells ($CD5^+$) and production of RF, with later development of a monoclonal IgM with RF activity. The RF IgM has been shown to form a complex with the Fc portion of polyclonal IgG with anti-HCV core-binding properties [23]. In this way, immune complexes (cryoprecipitates) are triggered by HCV infection. These deposit in the walls of vasa nervorum, bind some complement components (C1q), and activate the complement pathway. There are consequent inflammatory mural infiltrates, fibrinoid necrosis, vessel occlusion with multifocal infarction of nerve fascicles, and axonal degeneration. Aside from induction of vasculitis in small- and medium-sized epineurial arterioles, CG may precipitate within the lumen of epineurial arterioles and produce vessel occlusion [23]. The demonstration of such endothelial deposits of CG has been relatively uncommon in the reported literature, suggesting that this is a secondary mechanism. Studies of nerve biopsies from patients with CG, with and without neuropathy, have suggested that damage to endoneurial capillaries occurs preferentially in cases associated with neuropathy. Thus an endoneurial microangiopathy may be a factor in development of peripheral neuropathy.

Mechanisms whereby PNS vasculitis or angiopathy might occur in the absence of CG are less clear. HCV RNA has been detected in a proportion of nerve biopsy specimens from patients with peripheral neuropathy, both with and without CG, when sensitive techniques (nested reverse transcription-polymerase chain reaction) have been employed [24]. Although this raises the possibility of a direct role for HCV in axonal injury, the dominant pathology (as in cases with CG) is one of epineurial vasculitis and multifocal axon loss, which argues in favor of an immune-mediated angiopathy and ischemic axonal injury. It is uncertain how HCV might trigger vasculitis in the absence of CG, but it has been suggested that HCV could act by (1) HCV-mediated activation of complement, (2) formation of HCV–HCV antibody immune complexes with deposition in vasa nervorum, or (3) activation of natural killer cells directed against viral proteins [22].

3. Prognosis of Hepatitis C Virus–Associated Neuropathies

Primary treatment of HCV with IFN-α has been reported to produce short-term stabilization or improvement in CG neuropathies and is now a primary form of therapy [23]. Plasma exchange, cyclophosphamide, corticosteroids, and rituximab have also produced stabilization or improvement in HCV-associated vasculitic neuropathies [23]. The long-term prognosis of HCV-associated neuropathies is uncertain.

Increased use of IFN-α therapy has also lead to reports of worsening neuropathy [25]. Of the reported cases, typical worsening was not dose dependent and usually occurred between 2 and 28 weeks after initiation of therapy. Although the precise mechanism is unclear, several types of neuropathies have been reported, including MM, predominantly sensory polyneuropathies, acute autonomic and sensory neuropathies, chronic inflammatory demyelinating polyneuropathy, and cranial neuropathies. Evidence for necrotizing vasculitis has been reported in some cases. Corticosteroids, plasmapheresis, and cyclosporine have all been reported to stabilize the worsening neuropathy.

C. West Nile Virus–Associated Flaccid Paralysis

West Nile virus (WNV) is a mosquito-borne single-stranded RNA flavivirus initially isolated in 1937 and is one of the most widely distributed flaviviruses. Since 1999, the virus has been recognized in North America following an epizootic infection among birds and horses and an epidemic of meningitis and encephalitis in humans [26,27]. Since outbreaks in Romania in 1996 and the United States since 1999, epidemiological studies have shown that 1 in 150 patients will develop neurological disease, mostly encephalitis or aseptic meningitis. WNV also causes a "poliomyelitis-like" acute flaccid paresis of one or more limbs, which is differentiated from Guillain-Barré syndrome by its typical occurrence during the febrile portion of the illness and often asymmetrical weakness. A lymphocytic pleocytosis is found in the cerebrospinal fluid with moderately elevated protein and normal glucose levels. Electrophysiological studies are most consistent with acute, predominantly lower motoneuron dysfunction [26,27].

1. Pathology and Pathogenesis of West Nile Virus Flaccid Paralysis

The precise mechanisms of WNV flaccid paralysis remain uncertain. Human autopsy reports in WNV flaccid paralysis have demonstrated a loss of anterior horn cells, perivascular inflammatory cell collections consisting predominantly of $CD8^+$ lymphocytes and $CD68^+$ macrophages, focal microglial nodules, and gliosis. Anterior horn cell loss is focal within the spinal cord, which explains the asymmetrical, segmental distribution of weakness [26,27].

A mouse model of WNV-associated flaccid paralysis has revealed both frequent pyknotic neurons (particularly among anterior horn cells) and lymphocyte infiltration in the spinal cord [28]. Anterior horn cell neurons show frequent TUNEL positivity, suggesting apoptotic cell death [28]. In contrast, WNV-infected mice without flaccid paralysis show normal spinal cord neuronal morphology and scant inflammation [28]. Immunohistochemical studies have revealed that paralyzed WNV-infected mice have positive immunostaining for WNV antigens (consistent with viral infection) in about 35% of their spinal neurons as opposed to about 5% of neurons in nonparalyzed mice [28]. These animal and human autopsy data suggest two mechanisms of anterior horn cell loss and myelitis in WNV: direct neuronal toxicity of WNV infection and immune-mediated neuronal injury [28]. WNV infection has been shown to increase major histocompatibility I molecule expression, and pathological studies show that CD8+ cells predominate in inflammatory infiltrates in WNV; thus it has been hypothesized that infected neurons may be subject to antigen-restricted killing by cytotoxic T lymphocytes [28].

An embryonic stem cell–neuronal culture model has indicated that WNV can infect and trigger apoptosis of such a culture (but not undifferentiated embryonic stem cells) in the absence of inflammation, supporting a hypothesis of direct viral neurotoxicity [28]. These data do not, however, address the relative contribution of immune-mediated neurotoxicity in WNV flaccid paralysis.

Although most patients with WNV flaccid paralysis show limited improvement in strength with time (consistent with a severe loss of anterior horn cells), a subgroup of WNV-infected individuals develops an asymmetrical flaccid paralysis with acute denervation electrophysiologically but with rapid and marked improvement within weeks to months [27]. In one such subject, muscle biopsy has disclosed, along with signs of acute denervation, the presence of prominent CD8+ lymphocyte collections within intramuscular nerves. It has been suggested that an immune-mediated motor nerve terminal neuropathy may contribute to the flaccid paralysis, in some patients, and explain prompt improvement in such cases [27].

WNV infection is common, yet symptomatic infection is uncommon. With predisposing factors of age, and immunocompromised states, it appears that organism and host genetic factors may confer susceptibility. Two lineages of WNV have been described; lineage 1 is more virulent. Mashimo and colleagues have demonstrated a genetic locus responsible for determining susceptibility or resistance to WNV in mice on chromosome 5, where genes encoding for certain interferon-inducible proteins reside [29].

2. Prognosis

WNV infection is fatal in 3–15% of patients who develop severe infections, with higher death rates in those older than 65 years. Incomplete and limited recovery of muscle strength is the rule and occurs largely within the first 6 months after onset of weakness. Occasional patients will make a near-full recovery [27]. Motor unit number estimates of functioning motor units may be a predictor of the ultimate extent of motor recovery [27].

IV. Conclusion

This chapter reviewed the present, albeit incomplete, understanding of the pathogenesis of several infection-related neuropathies, most of current clinical importance but others of historical and mechanistic relevance. Summarized in Tables 3 and 4 are mechanisms of several additional infection-related disorders of nerve terminals and peripheral nerve that are beyond the scope of this review.

Table 3 Mechanisms of Other Infection-Related Disorders of the Nerve Terminal and Peripheral Nerve (Bacterial and Parasitic Organisms)

Syndrome	Presentation	Pathogenesis
Bacterial		
Brucellosis (*Brucella abortus, melitensis, suis*)	Cranial nerve palsies; acute polyradiculoneuropathy	Uncertain pathogenesis; persistent infection, response to antimicrobials
Botulism (*Clostridium botulinum*)	Diffuse weakness, autonomic dysfunction	Toxin-mediated blockade of presynaptic acetylcholine release at the neuromuscular junction and sympathetic and parasympathetic nerve terminals
Parasitic		
Chagas disease (*Trypanosoma cruzi*)	Autonomic > sensorimotor neuropathy	Uncertain; molecular mimicry leading to immune cross-reactivity between the parasite and the enteric neurons; elevated antisulfatide, asialoganglioside, and monosialoganglioside antibodies; development of autoreactive cytotoxic T cells and delayed hypersensitivity T cells
Tick paralysis (*Dermacentor andersoni/variabilis, Ixodes holocyclus*)	Acute ascending weakness resembling Guillain-Barré syndrome	*Ixodes:* Toxin-mediated presynaptic neuromuscular transmission blockade at motor nerve terminals; *Dermacentor:* Toxin-mediated blockade of axonal sodium channels

Table 4 Mechanisms of Other Infection-Related Disorders of the Nerve Terminal and Peripheral Nerve (Viral Organisms)

Syndrome	Presentation	Pathogenesis
Polio picornaviridae (enterovirus)	Acute febrile flaccid paralysis	Infection of AHC through PVR; T cell–mediated inflammation, destruction of AHC
Herpes zoster (*Herpesviridae*)	Painful sensory > motor radiculoneuropathy; transverse myelitis	Latent infection in neurons of cranial nerve, dorsal root ganglia, and autonomic ganglia; reactivation in setting of decreased Zoster-specific immunity; infection and inflammation of ganglia that may spread to plexus, nerve roots, and spinal cord; disease manifestations due to neuronal infection, inflammation, and vasculitis
Cytomegalovirus (*Herpesviridae*)	Progressive polyradiculopathy; mononeuropathy multiplex	Cytomegalovirus infection of nerve roots, spinal ganglia, dorsal root ganglia, and peripheral nerve in setting of HIV infection or other immunosuppression; PNS vasculitis in some cases
HTLV-1 (retrovirus)	Mild distal symmetrical or asymmetrical sensory or sensorimotor, primarily axonal neuropathy; rare reports of sensory neuronopathy	Nerve biopsy shows axonal degeneration, regeneration, and occasional features of remyelination; no evidence of direct infection of peripheral nerves (in contrast to HAM). HAM has an immunological basis, but the pathogenesis of HTLV-1-associated neuropathy is uncertain. HTLV-1-associated Sjögren's syndrome may be related to neuropathy in some cases.
Paralytic rabies (rhabdovirus)	Acute flaccid paralysis, may simulate Guillain-Barré syndrome with either demyelinative or axonal electrophysiology	Virus accesses nerve at NMJ, retrograde axonal transport of virus and infection of dorsal root ganglia and AHC; disimmune and direct viral neurotoxicity hypotheses exist for peripheral neuropathy of paralytic rabies, but the pathogenesis is uncertain.

AHC, Anterior horn cell; HAM, HTLV-1-associated myelopathy; HIV, human immunodeficiency virus; HTLV-1, human T lymphotropic virus type 1; NMJ, neuromuscular junction; PNS, peripheral nervous system; PVR, polio virus receptor.

References

1. Sigal, L. H. (1997). Immunologic mechanisms in Lyme neuroborreliosis: the potential role of autoimmunity and molecular mimicry. *Semin. Neurol.* **17**, 63–68.
2. Logigian, E. L., and Steere, A. C. (1992). Clinical and electrophysiologic findings in chronic neuropathy of Lyme disease. *Neurology* **42**, 303–311.
3. Meier, C., Grahmann, F., Engelhardt, A., and Dumas, M. (1989). Peripheral nerve disorders in Lyme-Borreliosis. Nerve biopsy studies from eight cases. *Acta Neuropathol. (Berl.)* **79**, 271–278.
4. Alaedini, A., and Latov, N. (2005). Antibodies against OspA epitopes of Borrelia burgdorferi cross-react with neural tissue. *J. Neuroimmunol.* **159**, 192–195.
5. Ramesh, G., and Philipp, M. T. (2005). Pathogenesis of Lyme neuroborreliosis: mitogen-activated protein kinases Erk1, Erk2, and p38 in the response of astrocytes to Borrelia burgdorferi lipoproteins. *Neurosci. Lett.* **384**, 112–116.
6. Maimone, D., Villanova, M., Stanta, G., Bonin, S., Malandrini, A., Guazzi, G. C., and Annunziata, P. (1997). Detection of Borrelia burgdorferi DNA and complement membrane attack complex deposits in the sural nerve of a patient with chronic polyneuropathy and tertiary Lyme disease. *Muscle Nerve* **20**, 969–975.
7. Ooi, W. W., and Srinivasan, J. (2004). Leprosy and the peripheral nervous system: basic and clinical aspects. *Muscle Nerve* **30**, 393–409.
8. Facer, P., Mann, D., Mathur, R., Pandya, S., Ladiwala, U., Singhal, B., Hongo, J., Sinicropi, D. V., Terenghi, G., and Anand, P. (2000). Do nerve growth factor–related mechanisms contribute to loss of cutaneous nociception in leprosy? *Pain* **85**, 231–238.
9. Goulart, I. M., Figueiredo, F., Coimbra, T., and Foss, N, T. (1996). Detection of transforming growth factor-β1 in dermal lesions of different clinical forms of leprosy. *Am. J. Pathol.* **148**, 911–917.
10. Schon, T., Hernandez-Pando, R., Baquera-Heredia, J., Negesse, Y., Becerril-Villanueva, L. E., Eon-Contreras, J. C., Sundqvist, T., and Britton, S. (2004). Nitrotyrosine localization to dermal nerves in borderline leprosy. *Br. J. Dermatol.* **150**, 570–574.
11. Logina, I., and Donaghy, M. (1999). Diphtheritic polyneuropathy: a clinical study and comparison with Guillain-Barré syndrome. *J. Neurol. Neurosurg. Psychiatry* **67**, 433–438.
12. Piradov, M. A., Pirogov, V. N., Popova, L. M., and Avdunina, I. A. (2001). Diphtheritic polyneuropathy: clinical analysis of severe forms. *Arch. Neurol.* **58**, 1438–1442.
13. Papini, E., Santucci, A., Schiavo, G., Domenighini, M., Neri, P., Rappuoli, R., and Montecucco, C. (1991). Tyrosine 65 is photolabeled by 8-azidoadenine and 8-azidoadenosine at the NAD binding site of diphtheria toxin. *J. Biol. Chem.* **266**(4), 2494–2498.
14. Pleasure, D. E., Feldmann, B., and Prockop, D. J. (1973). Diphtheria toxin inhibits the synthesis of myelin proteolipid and basic proteins by peripheral nerve *in vitro*. *J. Neurochem.* **20**, 81–90.
15. Brew, B. J. (2003). The peripheral nerve complications of human immunodeficiency virus (HIV) infection. *Muscle Nerve* **28**, 542–552.
16. Herrmann, D. N., and Schifitto, G. (2005). HIV: associated neuropathies and myelopathy *In:* "Principles of Neurological Infectious Disease" (K. Roos, ed.). McGraw Hill, New York.
17. Cherry, C. L., McArthur, J. C., Hoy, J. F., and Wesselingh, S. L. (2003). Nucleoside analogues and neuropathy in the era of HAART. *J. Clin. Virol.* **26**, 195–207.

18. Tyor, W. R., Wesselingh, S. L., Griffin, J. W., McArthur, J. C., and Griffin, D. E. (1995). Unifying hypothesis for the pathogenesis of HIV-associated dementia complex, vacuolar myelopathy, and sensory neuropathy. *J. Acquir. Immune. Defic. Syndr. Hum. Retrovirol.* **9**, 379–388.
19. Jones, G., Zhu, Y., Silva, C., Tsutsui, S., Tsutsui, S., Pardo, C. A., Keppler, O. T., McArthur, J. C., and Power, C. (2005). Peripheral nerve-derived HIV-1 is predominantly CCR5-dependent and causes neuronal degeneration and neuroinflammation. *Virology* **334**, 178–193.
20. Bodner, A., Toth, P. T., and Miller, R. J. (2004). Activation of c-Jun N-terminal kinase mediates gp120IIIB- and nucleoside analogue-induced sensory neuron toxicity. *Exp. Neurol.* **188**, 246–253.
21. Lopate, G., Pestronk, A., Evans, S., Li, L., and Clifford, D. (2005). Anti-sulfatide antibodies in HIV-infected individuals with sensory neuropathy. *Neurology* **64**, 1632–1634.
22. Nemni, R., Sanvito, L., Quattrini, A., Santuccio, G., Camerlingo, M., and Canal, N. (2003). Peripheral neuropathy in hepatitis C virus infection with and without cryoglobulinaemia. *J. Neurol. Neurosurg. Psychiatry* **74**, 1267–1271.
23. Sansonno, D., and Dammacco, F. (2005). Hepatitis C virus, cryoglobulinaemia, and vasculitis: immune complex relations. *Lancet Infect. Dis.* **5**, 227–236.
24. Authier, F. J., Bassez, G., Payan, C., Guillevin, L., Pawlotsky, J. M., Degos, J. D., Gherardi, R. K., Belec L. (2003). Detection of genomic viral RNA in nerve and muscle of patients with HCV neuropathy. *Neurology* **60**, 808–812.
25. Boonyapisit, K., and Katirji, B. (2002). Severe exacerbation of hepatitis C–associated vasculitic neuropathy following treatment with interferon alpha: a case report and literature review. *Muscle Nerve* **25**, 909–913.
26. Li, J., Loeb, J. A., Shy, M. E., Shah, A. K., Tselis, A. C., Kupski, W. J., and Lewis, R. A. (2003). Asymmetric flaccid paralysis: a neuromuscular presentation of West Nile virus infection. *Ann. Neurol.* **53**, 703–710.
27. Cao, N. J., Ranganathan, C., Kupsky, W. J., and Li, J. (2005). Recovery and prognosticators of paralysis in West Nile virus infection. *J. Neurol. Sci.* **236**, 73–80.
28. Shrestha, B., Gottlieb, D., and Diamond, M. S. (2003). Infection and injury of neurons by West Nile encephalitis virus. *J. Virol.* **77**, 13203–13213.
29. Mashimo, T., Lucas, M., Simon-Chazottes, D., Frenkiel, M. P., Montagutelli, X., Ceccaldi, P. E., Deubel, V., Guenet, J. L., and Despres, P. (2002). A nonsense mutation in the gene encoding 2'-5'-oligoadenylate synthetase/L1 isoform is associated with West Nile virus susceptibility in laboratory mice. *Proc. Natl. Acad. Sci. USA* **99**, 11311–11316.

84

Muscular Dystrophies

Leland E. Lim, MD, PhD
Charles A. Thornton, MD
Thomas A. Rando, MD, PhD

Keywords: *dystrophin, muscular dystrophy, myotonic dystrophy, nuclear membrane, sarcoglycan, sarcomere*

I. Introduction
II. Duchenne's and Becker's Muscular Dystrophies
III. Limb-Girdle Muscular Dystrophies
IV. Congenital Muscular Dystrophies
V. Myotonic Dystrophy
VI. Facioscapulohumeral Muscular Dystrophy
VII. Conclusion
    References

I. Introduction

The muscular dystrophies include a large number of hereditary, degenerative diseases of skeletal muscle that differ in clinical presentation, mode of inheritance, and genetic basis. They share certain pathologic features, including muscle fiber necrosis, acute and chronic changes associated with regeneration and inflammation, and end-stage changes including prominent fibrosis and replacement of muscle by adipose tissue. Major progress has been made in identifying the genetic basis for most types of muscular dystrophy. However, the pathogenetic cascades leading to changes in muscle tissues remain to be elucidated even for the most common forms of the dystrophies. One major area of research is to identify commonalities and divergences in pathogenesis. This chapter summarizes recent developments in our understanding of these pathogenetic mechanisms.

We focus in this chapter on the most common forms of muscular dystrophy, Duchenne's and Becker's muscular dystrophies in childhood and myotonic muscular dystrophy in adulthood, and on the most informative biochemical abnormalities, such as those associated with defects in nuclear membrane proteins and with defects in protein glycosylation. We also review representative members of different groups of dystrophies based on clinical presentation (such as the limb-girdle muscular dystrophies) and dystrophies that may result from unique genetic or pathogenetic mechanisms (such as facioscapulohumeral muscular dystrophy).

The classification of muscular dystrophy initially was based on the clinical pattern of weakness, mode of inheritance, and age of onset (Table 1). With the discovery of genes responsible for most of these diseases, there has been a gradual shift to a classification based on genes affected and function of the respective proteins. In Table 1 we provide a summary that reflects both methods of classification. The subsequent sections do not follow a specific classification scheme but rather address the diseases in which the most progress has been made in understanding the underlying pathobiology.

Table 1 The Muscular Dystrophies

Muscular dystrophy	Inheritance	Gene location	Protein product
Duchenne/Becker	XR	Xp21	Dystrophin
Limb-girdle			
LGMD1A	AD	5q31	Myotilin
LGMD1B	AD	1q21	Lamin A/C
LGMD1C	AD	3p25	Caveolin-3
LGMD1D	AD	6q23	Unknown
LGMD1E	AD	7q	Unknown
LGMD2A	AR	15q15	Calpain-3
LGMD2B	AR	2p13	Dysferlin
LGMD2C	AR	13q12	γ-Sarcoglycan
LGMD2D	AR	17q12-21	α-Sarcoglycan
LGMD2E	AR	4q12	β-Sarcoglycan
LGMD2F	AR	5q33	δ-Sarcoglycan
LGMD2G	AR	17q12	Telethonin
LGMD2H	AR	9q31-34	TRIM32 ubiquitin ligase
LGMD2I	AR	19q13	FKRP
LGMD2J (and tibial)	AR	2q31	Titin
Congenital			
Merosin-deficient	AR	6q2	Laminin-α2
Fukuyama	AR	9q31-33	Fukutin
Muscle-eye-brain	AR	1p3	POMGnT1
Walker-Warburg syndrome	AR	9q34	POMT1
MDC1C	AR	19q1	FKRP
MDC1D	AR	22q12	LARGE glycosyltransferase
Ullrich disease	AR	21q22	COL6A2
	AR	2q37	COL6A3
Emery-Dreifuss, X-linked	XR	Xq28	Emerin
Emery-Dreifuss, dominant	AD	1q21	Lamin A/C
Facioscapulohumeral	AD	4q35	Unknown
Myotonic dystrophy type 1	AD	19q13	Myotonin
Myotonic dystrophy type 2	AD	3q21	Zinc finger protein 9

Note: COL6A2, collagen VI-α2 subunit; COL6A3, collagen VI-α3 subunit; FKRP, fukutin-related protein; LGMD, limb-girdle muscular dystrophy; POMGnT12, protein O-mannose β-1, 2-N-acetylglucosaminyltransferase; POMT1, protein O-mannosyltransferase 1.

II. Duchenne's and Becker's Muscular Dystrophies

Duchenne's muscular dystrophy (DMD) and its milder allelic variant Becker's muscular dystrophy (BMD) are X-linked recessive disorders. DMD is the most common muscular dystrophy, affecting approximately 1 in 3000 male births [1]. Affected children are normal at birth. Early signs of DMD include delay in achieving motor milestones and pseudohypertrophy of the calves. Muscle weakness, typically affecting the proximal muscles at first, becomes evident by late childhood, manifested by a waddling gait. By age 12, most DMD patients are confined to a wheelchair. Death occurs typically in the third decade of life, usually because of involvement of the muscles of respiration or the heart. BMD, in contrast, is characterized by slower progression of symptoms, though most individuals eventually become wheelchair bound as well. Death in BMD is also caused by respiratory compromise or cardiomyopathy, which is more common in BMD than in DMD [1].

DMD and BMD are caused by mutations in the dystrophin gene at the chromosomal location Xp21. The dystrophin gene is among the largest known genes, spanning 2.4 megabases (Mb) and encoding a 14-kb transcript composed of 79 exons. The protein product of this locus, dystrophin, is a large subsacrolemmal protein with a molecular weight of 427 kilodaltons (kDa) [2]. Most disease-causing mutations are deletions or duplications. Those that produce a shift of the translational reading frame and thus result in a complete absence of dystrophin cause DMD. Other mutations, such as point mutations, cause DMD if they result in a complete absence of the protein. Deletions that preserve the reading frame and thus result in an internally deleted protein cause BMD, as do mutations that result in reduced levels of dystrophin expression.

Dystrophin has four major functional domains [3]. The N-terminal domain binds filamentous actin (F-actin) and has homology to other actin-binding proteins, such as β-spectrin. The rod domain consists of 24 spectrin-like repeats interrupted by four "hinge" regions. Contained within this domain is another binding site for F-actin [4]. The cysteine-rich region toward the carboxyl terminus (C-terminus) of dystrophin is the binding site for β-dystroglycan, a transmembrane glycoprotein. Through its interaction with the dystroglycan complex, dystrophin localizes to the sarcolemma. The C-terminal domain of dystrophin is the binding site for the syntrophins and dystrobrevin. Dystrophin, the dystroglycan complex, the sarcoglycan complex (discussed in more detail later), and their associated proteins form a functional unit referred to as the dystrophin-glycoprotein complex (DGC) (Fig. 1) [5]. The DGC, in turn, is known to be associated with many other proteins, including several involved in various intracellular signaling pathways [6].

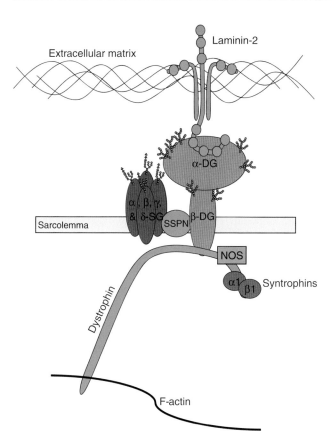

Figure 1 The Dystrophin-Glycoprotein Complex. Dystrophin binds F-actin of the cytoskeleton and is associated with a transmembrane complex that consists of the dystroglycans, sarcoglycans, and sarcospan. α-Dystroglycan binds laminin-2 (among other proteins) of the extracellular matrix. In addition, proteins with roles in cell signaling, such as NOS, are also associated with the DGC. *Note:* DG, dystroglycan; NOS, nitric oxide synthase; SG, sarcoglycan; SSPN, sarcospan.

In the absence of dystrophin, all other components of the DGC are either absent or mislocalized [7]. Multiple hypotheses about how the disruption of the DGC causes muscular dystrophy have been proposed. The disruption of the DGC disrupts the link between the cytoskeleton and the extracellular matrix (ECM), as mediated by the binding of dystrophin to actin and the binding of dystroglycan to ECM proteins such as laminin and perlecan. The disruption of the DGC also interferes with normal signal transduction processes that typically associate with and may modulate the DGC [6]. Dystrophin-deficient muscle is susceptible to mechanical stresses and to metabolic challenges such as oxidative stress and hypoxia. The result of these susceptibilities is not a failure of muscle development but rather a gradual degenerative process in which individual fibers or groups of fibers undergo focal and segmental necrosis. Although the precise pathogenetic mechanisms remain to be elucidated, among the most prevalent cellular changes that ultimately lead to necrosis are disruption of plasma membrane integrity and increase of intracellular calcium. The absence of nitric oxide synthase (NOS), an important signaling protein associated with the DGC, may contribute to the increased susceptibility and fiber degeneration. Dystrophic mice that overexpress NOS have reduced dystrophic pathology [8], which suggests the importance of NOS to myofiber survival. Over time, the cumulative effects of chronic degeneration and regeneration lead to the characteristic dystrophic phenotype seen in later stages of DMD [1].

III. Limb-Girdle Muscular Dystrophies

The limb-girdle muscular dystrophies (LGMDs) represent a genetically heterogeneous classification of autosomally transmitted muscular dystrophies (see Table 1). Previously, these dystrophies were classified according to their similarity to DMD or BMD; early literature refers to these conditions as either Duchenne-like or Becker-like autosomal recessive or dominant muscular dystrophies. Clinically, LGMD primarily has presenting symptoms of weakness of the shoulder and pelvic girdle muscles with relative sparing of the distal musculature. Onset is typically during the second or third decade of life, but this can vary widely. The severity and rate of progression can also be quite variable, ranging from mild exercise intolerance and normal life expectancy to rapidly progressive weakness and death in early adulthood.

The LGMDs are divided into two subgroups. The autosomal dominant forms are categorized under the subgroup LGMD1. There are six known autosomal dominant LGMDs, designated LGMD1A–F. The autosomal recessive forms are categorized under the subgroup LGMD2, of which ten have been described (LGMD2A–J). Thus far, the genes responsible for these muscular dystrophies have been identified in all but three of the dominant forms. The genes implicated in LGMDs are involved in a variety of cellular processes and structures. In the following sections, we consider groups of disorders based on shared biochemical or presumed pathogenetic features.

A. LCMD2C–F: "Sarcoglycanopathies"

The sarcoglycans (α-, β-, γ-, and δ-sarcoglycan) are integral components of the DGC (see Fig. 1). All sarcoglycans share similar structures, consisting of a small cytosolic N-terminal domain, a single transmembrane domain, and a large C-terminal extracellular domain that contains at least one site for N-linked glycosylation and cysteine residues that may contribute to secondary structure. They range in molecular weight from 35 to 50 kDa [9]. Studies *in vitro* have demonstrated that the sarcoglycan complex is formed in the endoplasmic reticulum before its translocation to the sarcolemma [10]. The sarcoglycans do not interact directly with dystrophin. To date, no clear extracellular ligands for any of the sarcoglycans have been identified, nor have any sarcoglycans been shown to mediate signal transduction process, although both have been suggested [6]. As mentioned previously, the sarcoglycans are absent in DMD.

Clinically, the sarcoglycanopathies are quite heterogeneous. Mutations in different regions of a sarcoglycan protein may cause dystrophies of differing severity, but in some cases identical mutations have been shown to result in different phenotypes. Whether an individual will follow a rapidly progressive or a more indolent course cannot be predicted by genetic analysis. In addition to skeletal muscle disease, sarcoglycan mutations may also cause cardiomyopathy, especially with mutations in γ- and δ-sarcoglycan [11]. Again, the likelihood of cardiomyopathy is difficult to predict; in some families, one sibling may develop a muscular dystrophy whereas another may develop cardiomyopathy [12–14].

Sarcoglycan mutations are the basis of LGMD2C–F (see Table 1). Immunohistochemical analysis reveals that dystrophin and the dystroglycans are present at normal levels at the sarcolemma but that the sarcoglycans are absent. Thus a mutation in one sarcoglycan results in the deficiency of all four components, presumably because of disruption of the intracellular assembly of the sarcoglycan complex. The absence of the sarcoglycans causes similar pathological changes to those seen in DMD, with disruption of membrane integrity and increases in cytosolic calcium levels leading to myofiber necrosis, suggesting the possibility of shared pathogenetic mechanisms between dystrophin deficiency and sarcoglycan deficiency, perhaps mediated by the latter in each case.

B. LGMD1C and 2B: Other Membrane Protein Defects (Caveolin-3 and Dysferlin)

Mutations in other sarcolemmal proteins have been shown to cause other forms of muscular dystrophy. Unlike the sarcoglycans, whose normal functions remain largely unknown, two proteins—caveolin-3 (mutations of which causes LGMD1C) [15] and dysferlin (mutations of which causes LGMD2B) [16]—have well-documented roles in sarcolemmal function.

Caveolin-3 is the muscle-specific isoform of the caveolin protein family. The caveolins are the principal component of caveolae, which are small plasma membrane invaginations that play important roles in cell signaling and transverse tubule formation during myogenesis. Although caveolin-3 appears to be associated with the DGC, mutations that cause LGMD1C do not appear to exert their effects through dysfunction of caveolin-3 through dysfunction of the DGC [17]. Ultrastructural analysis of muscle from LGMD1C patients reveals a decreased number of caveolae at the sarcolemma. Mutations in caveolin-3 disrupt the normal signaling in pathways that use this protein as a scaffold, and it may be through those altered signaling pathways, several of which are associated with cell survival and cell death pathways, that the pathogenetic processes are mediated [18]. However, caveolae are complex structures, and caveolin-3 mutations have been implicated in muscle disorders in addition to LGMD, including rippling muscle disease, distal myopathy, and hyperCKemia [12,13].

Dysferlin gets its name from its homology to the *Caenorhabditis elegans* fertility factor fer-1. Mutations in fer-1 cause infertility because of an inability of intracellular organelles to fuse with the plasma membrane. In mammalian muscle, dysferlin is localized both to sarcolemma and to intracellular organelles, suggesting a role of the ferlin family of proteins in membrane trafficking. Furthermore, intracellular dysferlin is increased and sarcolemmal dysferlin is decreased in DMD and other forms of LGMD. This leads to the hypothesis that dysferlin plays a role in sarcolemmal repair in response to muscle damage associated with the muscular dystrophies. Indeed, membrane repair is abnormal in dysferlin-null mice [19]. Interestingly, dysferlin is ubiquitously expressed, but mutations only appear to cause muscle pathology. This may be because muscle is more dependent on dysferlin-mediated membrane repair process than other tissues. Like caveolin-3 mutations, dysferlin mutations can causes more than one distinct phenotype. In addition to LGMD, dysferlin mutations cause a distal myopathy called Miyoshi myopathy. In fact, identical dysferlin mutations may cause either of these clinical entities in different individuals, even within a single pedigree.

C. LGMD1A, 2A, 2G, and 2H: Defects in Sarcomeric Proteins and Sarcomere Organization

Mutations in a large number of sarcomeric proteins cause a variety of congenital myopathies and cardiomyopathies. Mutations in three sarcomeric proteins cause distinct forms of LGMD. In general, these cases are rare and only a few families in distinct populations have been identified. Nevertheless, these clusters of patients indicate that dysfunction of the sarcomere is another mechanism of pathogenesis of LGMD.

Mutations in myotilin, which localizes to the Z-line of the sarcomere, cause the autosomal dominant LGMD1A [20]. Myotilin has been shown to bind α-tropomyosin and appears to be involved in bundling F-actin during myofibrilogenesis. Histopathologically, LGMD1A muscle contains rimmed vacuoles and may resemble nemaline myopathy.

Titin is a large protein with a molecular weight of 4200 kDa. In addition to actin and myosin, titin is the third myofilament of the sarcomere, extending from the Z-disc to the M-line. It is believed to be involved in the assembly of the sarcomere and interacts directly with several other sarcomeric components. It also has intrinsic kinase activity. Mutations in titin cause tibial muscular dystrophy, a late-onset autosomal dominant disorder affecting the anterior compartment of the lower extremities. In rare cases of homozygous expression of titin mutations, patients develop LGMD2J. These dystrophies are found primarily in the Finnish population [13,21].

Telethonin mutations cause LGMD2G, which has been found in four Brazilian families [22]. Like myotilin, telethonin is a component of the Z-line and is a substrate for titin. On biopsy, rimmed vacuoles may be present, but this is not as consistent a finding as in LGMD1A. Ultrastructural examination of affected muscle does not reveal any evidence of disrupted sarcomeric structure.

Finally, muscular dystrophies due to mutations in genes encoding cytoplasmic enzymes may relate, in some cases, to altered sarcomeres. Mutations in both calpain-3 (LGMD2A) [23] and TRIM32 (LGMD2H) [24] result in disruption of sarcomere structure, in both cases perhaps by interfering with normal myofibril remodeling processes.

D. LGMDs Due to Defects in Nuclear Membrane Proteins

The nuclear envelope comprises the outer and inner nuclear membranes, nuclear pore complexes, and the nuclear lamina. The inner membrane contains several integral protein components that mediate its interaction with the lamina. The pore complexes allow bidirectional transport of material between the nucleoplasm and the cytoplasm. The nuclear lamina is a meshwork of intermediate filaments that underlies the inner nuclear membrane. In addition to providing structure and support to the nuclear envelope, the lamina plays a role in the reconstitution of the nucleus after mitosis. The lamina binds chromatin and is involved in DNA replication and mRNA synthesis [25,26].

Over the past several years, muscular dystrophies due to dysfunction of the nuclear envelope have been identified. X-linked Emery-Dreifuss muscular dystrophy (X-EDMD) is caused by mutations in emerin, a 34-kDa integral protein of the inner nuclear membrane. EDMD is characterized by progressive muscle weakness and wasting; joint contractures involving the neck, elbows, and Achilles tendon; and cardiomyopathy with conduction block. Immunohistochemical and ultrastructural analysis of EDMD tissue reveals absence of emerin from the inner nuclear membrane and disorganization of the nuclear lamina.

Emerin interacts with several components of the nucleus and nuclear envelope. It binds the lamins, the intermediate filaments of the lamina, thus anchoring the lamina to the inner membrane. It interacts with nuclear F-actin and is thought to play a role in the organization and stabilization of the nuclear cytoskeleton. It also binds certain transcription regulatory elements, especially transcription repressors. Despite expanding understanding of the role of emerin, the precise mechanism by which its absence causes muscular dystrophy is not clear [25].

Mutations in the alternatively spliced lamin A/C gene have been implicated in at least five disorders, including two muscular dystrophies. Lamin A/C mutations cause autosomal dominant EDMD and the autosomal dominant LGMD1B. The clinical features of autosomal dominant EDMD are similar to those of X-EDMD, with muscle wasting and contractures as prominent features. In contrast, LGMD1B is characterized by relatively mild, proximal weakness and rare contractures. As in EDMD, LGMD1B patients develop cardiomyopathy and conduction abnormalities [12,26].

IV. Congenital Muscular Dystrophies

The congenital muscular dystrophies (CMDs) are also a genetically and clinically heterogeneous classification of muscular dystrophy. These disorders are characterized by weakness evident at birth or within the first few months of life. Children born with CMD are often described as "floppy babies." Unlike DMD and the LGMDs, brain developmental abnormalities and mental retardation are common in CMDs. Recent discoveries about the molecular basis of many CMDs have shed light on this phenomenon and have revealed a novel mechanism of muscle pathogenesis, as described in the following sections.

A. Merosin-Deficient Congenital Muscular Dystrophy

The most common form of CMD is caused by mutations in merosin (laminin-2). As in all CMDs, infants with merosin-deficient CMD (MDC1A) are hypotonic, with weakness evident in all muscle groups. The immediate postnatal course is often complicated by poor feeding effort and respiratory compromise. Contractures may be present at birth. In addition to muscle weakness, a significant proportion of affected children have evidence of central nervous system involvement, such as cognitive dysfunction or epilepsy.

The laminins are heterotrimers, composed of a heavy α-chain and two lighter β- and γ-chains. The laminins are expressed in basement membranes; laminin-2, which consists of the α2-, β1-, and γ1-chains, is the predominant isoform in skeletal muscle. Laminin-2 is an important ligand for α-dystroglycan, a key component of the DGC, and certain integrins. Thus laminin is an important mediator of signaling from the ECM through transmembrane proteins for skeletal muscle cells.

In MDC1A, mutations in the laminin-α2-chain drastically reduce merosin in the ECM, thus disrupting the cytoskeleton-dystrophin-dystroglycan-ECM axis [27]. Histologically, skeletal muscle from MDC1A patients shows changes typical of muscular dystrophy, such as evidence of degeneration and regeneration, inflammatory cell infiltrates, and fatty change. Immunohistochemical analysis reveals absent staining of merosin and normal levels of dystrophin and DGC components. Studies of animal models of MDC1A have revealed that disruption of myofiber membrane integrity is less prominent than in animal models of DMD or sarcoglycanopathies, suggesting distinct pathogenetic mechanisms [28].

In addition to these skeletal muscle abnormalities, MDC1A patients have evidence of brain defects. White matter abnormalities on magnetic resonance imaging are commonly seen in affected individuals. Because laminin-2 is a component of the basal lamina of cerebral blood vessels, a breakdown in the blood-brain barrier may be the cause of this abnormal signal intensity. Other abnormalities include occipital agyria and cerebellar hypoplasia, reflecting the importance of this ECM protein in aspects of brain development, such as neuronal migration.

B. Ullrich Disease

A subset of CMD characterized by proximal contractures, distal joint laxity, and normal intelligence was first recognized in 1930. Called Ullrich disease or scleroatonic muscular dystrophy, it has been identified in only a few dozen patients since it was recognized as a distinct clinical entity. Mutations in the α2 and α3 subunits of collagen VI, a major component of the ECM in muscle, have been identified in patients with Ullrich disease. This represents the second ECM protein implicated in muscular dystrophy. Unlike merosin, collagen VI does not interact directly with elements of the DGC. The mechanism of pathogenesis may be related to alterations in ECM composition, which is important for myotube formation and maturation during development [29,30].

C. Disorders of Glycosylation

As mentioned previously, α-dystroglycan is a receptor for laminin and is an integral component of the DGC. In addition, dystroglycan binds other ECM components, such as perlecan and agrin. Dystroglycan is ubiquitously expressed and is involved in basement membrane formation in a variety of tissues. Homozygous dystroglycan-null mice have early, severe developmental abnormalities that result in embryonic lethality. As this is almost certainly true across mammalian species, this is likely why no muscular dystrophy or other disorder due to mutations in dystroglycan has been identified in humans [31].

α-Dystroglycan is heavily glycosylated. The N- and O-linked carbohydrate residues attached to the core polypeptide account for more than half of dystroglycan's molecular weight. In recent years, certain forms of CMDs have been shown to be caused by mutations in proteins with either known or putative roles in glycosylation. In each case, the glycosylation of α-dystroglycan is affected, usually manifested by a decrease in its apparent molecular weight. This alteration presumably disrupts the functional role that the dystroglycan complex plays in the transmission of physical and biochemical signals between the ECM and the intracellular milieu.

In addition to causing muscle disease, mutations in these glycosyltransferases lead to characteristic eye and brain abnormalities, underscoring the role of dystroglycan in the development of multiple tissues and organs. Fukuyama congenital muscular dystrophy (FCMD) is the second most common muscular dystrophy in Japan. In addition to muscle weakness, affected individuals universally have some degree of mental retardation, usually severe. Seizures occur in more than 50%, and retinal dysplasia occurs in up to 90% of patients. Histopathological examination of the brains of FCMD patients reveals cortical dysplasia and evidence of a neuronal migration defect. The protein product of the FCMD locus is fukutin. Affected individuals have a common ancestral allele that contains a retrotransposon element in the 3′-untranslated region of the mRNA, resulting in diminished levels of functional fukutin. FCMD is caused by either a second mutation in the nonancestral allele or the homozygous expression of the founder allele. The precise function of fukutin has not been determined, but sequence analysis reveals homology to several glycosyltransferases. A role for fukutin in glycosylation is supported by the observation that α-dystroglycan is incompletely glycosylated in FCMD patients [32,33].

Walker-Warburg syndrome is characterized by lissencephaly, agenesis of the corpus callosum, pontocerebellar hypoplasia, and structural abnormalities of the eye. This condition is caused by mutations in the protein O-mannosyltransferase 1 gene (POMT1), which catalyzes the addition of mannose to a serine or threonine residue. Similarly, mutations in the protein O-mannose β-1,2-N-acetylglucosaminyltransferase gene (POMGnT1) cause muscle-eye-brain (MEB) disease, which is found primarily among isolated populations in Finland. The clinical phenotype of MEB disease is similar to that of Walker-Warburg syndrome. Like FCMD, both conditions are associated with neuronal migration defects and abnormal glycosylation of α-dystroglycan [32,33].

Other CMDs due to impaired protein glycosylation have been identified. Mutations in LARGE, a putative glycosyltransferase mutated in the myodystrophy (myd) mouse, have now been shown to cause a form of human CMD (MDC1D) [34]. These patients have severe mental retardation in addition to muscular dystrophy, but thus far identified structural brain abnormalities are comparatively mild. As in FCMD, Walker-Warburg syndrome, and MEB disease, dystroglycan glycosylation is affected. Mutations in the fukutin-related protein (FKRP) gene cause a severe CMD but do not affect brain structure or function; patients with this form of CMD (MDC1C) are cognitively intact. FKRP gene mutations also result in reduced molecular weight of dystroglycan and a secondary reduction in laminin-α2 [35]. Interestingly, FKRP has also been identified as the gene responsible for LGMD2I. In most LGMD2I patients, a point mutation is present that results in leucine-to-isoleucine change. This substitution of a structurally similar amino acid may decrease but not eliminate the function of the FKRP gene, thus resulting in the milder phenotype [36].

V. Myotonic Dystrophy

A. Myotonic Dystrophy Type 1

Myotonic dystrophy (or dystrophia myotonica) type 1 (DM1) is the most common inherited disease of skeletal muscle in adults (estimated frequency of 1 per 7500). The classical form of the disease is dominantly inherited and is characterized by myotonia and muscle wasting that begins in the second to fifth decade. Classical DM1 is a multisystem degenerative disease that also affects smooth muscle, the specialized conduction tissue of the heart, select populations of neurons, the lens, the testes, and other organs. However, when the onset of disease occurs *in utero*, DM1 manifests an entirely different set of clinical signs that reflect abnormal development of skeletal muscle and brain, often with the presenting symptom of respiratory failure in newborns. Affected infants who survive the neonatal period have mental retardation and later develop the degenerative features of DM1. By contrast, the onset of DM1 in some individuals is delayed until after age 50 and consists only of cataracts, with or without mild muscle weakness.

The range of phenotypes in affected members of a single family is extremely broad. Also, there is a striking tendency for symptoms to occur at an earlier age and with increased severity in successive generations, a phenomenon known as *anticipation*. The genetic basis for anticipation in DM1 was revealed when the mutation was discovered in 1992. DM1 results from an expansion of CTG repeats in the dystrophia myotonica protein kinase (DMPK) gene [37]. The number of CTG repeats at this locus is variable in the general population, ranging from 4 to 37, whereas individuals with DM1 have more than 50. Two aspects of the mutation account for anticipation. First, the expanded CTG repeat is unstable and shows a tendency to undergo further expansion when transmitted from one generation to the next. Second, increasing length of the expanded repeat is associated with earlier age of symptom onset and greater disease severity. Individuals with classical DM1 usually have several hundred CTG repeats, whereas individuals with congenital DM1 generally have expansions of more than a thousand repeats.

The instability of the DM1 mutation is apparent not only in the dynamics of intergenerational transmission and anticipation but also in the somatic cells of an individual over time. In proliferating cells, somatic instability reflects error-prone replication of the expanded CTG tract. Somatic expansion may also occur in nondividing cells by mechanisms that involve DNA repair. The result is that CTG expansions in muscle, heart, and brain tissue of adults with classic DM1 are 1000–4000 repeats in length, severalfold larger than the expansions measured in circulating blood cells of the same individual.

The expanded repeat is located in the 3′-untranslated region of the DMPK gene. In this position, the mutation does not interrupt the DMPK coding region or lead to the production of mutant protein. Complete absence of this protein kinase in mice causes only minimal impairment of cardiac and skeletal muscle, the tissues that express the gene most highly. Thus evidence does not suggest that this kinase, whose biochemical function is unknown, plays a major role in the pathogenesis of DM1. The expanded repeat, however, does affect the amount of DMPK protein synthesized because it blocks the nuclear export of the mutant mRNA. Mutant transcripts are retained in the nucleus in discrete foci [38]. Studies clearly show that the nuclear retention of this mRNA is triggered by the expanded CUG repeat because insertion of this element in other transcripts causes a similar block of nuclear export and formation of nuclear RNA foci.

There is now considerable evidence that DM1 results from an RNA-mediated disease process in which mutant DMPK transcripts have an adverse effect on nuclear function. For example, in transgenic mouse models, expression of CUG expansion RNA in skeletal muscle results in myotonia and morphological changes in muscle that resemble DM1 [39]. This effect is associated with the accumulation of mutant RNA in nuclear foci, and it can occur when the CUG expansion is placed in the DMPK mRNA or when it is inserted into the 3′-untranslated region of an unrelated mRNA. Thus DM1 is the first example of an inherited disease in which the mRNA transcribed from a mutant gene has a direct pathogenic effect, independent of the protein it encodes.

B. Myotonic Dystrophy Type 2

Genetic testing for the DM1 mutation revealed a subset of patients, some of whom were considered to have "atypical myotonic dystrophy," who did not have an expanded CTG repeat or any other mutation at the DM1 locus. This led to the recognition of myotonic dystrophy type 2 (DM2). DM2 tends to be less severe than DM1, and the earliest signs are in proximal rather than distal muscles. Unlike DM1, DM2 does not appear to cause congenital or juvenile-onset disease. However, the myotonia, muscle histopathology, testicular atrophy, cataracts, and magnetic resonance imaging signal changes in cerebral white matter are similar.

DM2 is caused by an expanded CCTG tetramer repeat in the zinc-finger protein 9 (ZNF9) gene, which encodes a nucleic acid–binding protein [40]. The expanded CCTG repeat lies in the first intron of ZNF9. This intronic expansion does not appear to disrupt expression of ZNF9. Furthermore, transcripts from the mutant ZNF9 allele form nuclear foci of expanded repeat RNA, similar to those observed in DM1. This implies that the shared features of DM1 and DM2 reflect a common RNA-mediated disease process.

C. Common Pathogenetic Mechanisms of the Myotonic Dystrophies

Both types of DM are associated with abnormal regulation of alternative splicing. This was first observed for cardiac troponin T; it has been subsequently demonstrated for more than 10 other genes [41]. The defect in RNA processing observed in DM is specific, in the sense that it is restricted to a particular set of exons that are alternatively spliced. This implies a biochemical lesion that involves particular factors that regulate alternative splicing, as opposed to a general disruption of the splicing machinery. Furthermore, effects on alternative splicing have been linked to specific symptoms of DM. The clearest example concerns the chloride channel, ClC-1. In a transgenic mouse model of DM1, the myotonia induced by expanded CUG RNA is caused by reduced ClC-1 protein and loss of ClC-1 activity [42]. Examination of the ClC-1 mRNA in the transgenic mice, and subsequently in patients with DM1 and DM2, showed several abnormalities of alternative splicing. In every example examined to date, the abnormality of alternative splicing in DM1 consists of the inappropriate expression in mature skeletal muscle of splice isoforms that normally are expressed in neonatal muscle. Thus it appears that many of the biochemical derangements in DM skeletal muscle result from failure to maintain a set of physiological alternative splicing transitions that normally occur soon after birth.

A current focus of research in DM is to determine how the accumulation of CUG or CCUG expansion RNA in nuclear foci leads to misregulated alternative splicing of pre-mRNA. Muscleblind (MBNL) proteins are a family of nuclear proteins that bind to expanded CUG or CCUG RNA *in vitro* and in DM1 and DM2 cells [43]. Of three mammalian genes in the MBNL family, MBNL1 is expressed most highly in adult muscle. In DM1 cells, this protein binds to repeat expansion RNA and becomes depleted from the nucleoplasm. MBNL1 has been shown to function as a splicing factor, and it regulates alternative splicing for some of the transcripts misregulated in DM1. Furthermore, disruption of the MBNL1 gene in mice leads to myotonia, cataracts, and misregulated alternative splicing that are similar to those in DM1 and DM2 [44]. Thus it seems likely that sequestration of MBNL1 in nuclear foci of repeat expansion RNA is a key step in the pathogenesis of both types of DM. Current efforts are directed to identifying other proteins whose activity is affected by repeat expansion RNA.

VI. Facioscapulohumeral Muscular Dystrophy

Facioscapulohumeral muscular dystrophy (FSHD) is the third most common form of muscular dystrophy (estimated frequency of 1 per 20,000). This autosomal dominant disorder is characterized by progressive weakness of the facial, periscapular, and upper arm muscles, which is often asymmetrical. A subset of patients, particularly those with early onset or severe disease, also develop vasculopathy of the retinal vessels or sensorineural hearing loss.

Regions near the ends of human chromosomes are enriched for repetitive sequences and devoid of genes. One class of repetitive DNA in the subtelomere is the D4Z4 repeat. This 3.3-kb element is found in the subtelomeric regions of chromosomes 4 and 10, where it forms large arrays of head-to-tail repeats. Repetitive DNA in subtelomeric regions is inherently prone to genetic rearrangement, but this is particularly true for the D4Z4 repeat. The normal subtelomeric array on chromosome 4 has 11–100 D4Z4 repeats. FSHD results from deletions within the array that leave only 1–10 D4Z4 repeats [45]. Short residual arrays containing only 1–4 D4Z4 repeats, though uncommon, are associated with earlier onset and more severe disease [46]. Most patients with FSHD, however, have 5–10 residual D4Z4 repeats, and within this range the manifestations are quite variable, ranging from absence of symptoms (nonpenetrance) to marked impairment. Up to 30% of FSHD results from *de novo* mutations. Therefore, a negative family history is not evidence against the diagnosis.

Despite discovery of the 4q genetic rearrangement in 1992, the general nature of the disease process in FSHD, and the identity of the responsible gene or genes, remains obscure. The deletion usually is confined within the D4Z4 array and does not extend into the flanking DNA. Furthermore, it does not appear that any part of the D4Z4 repeat is transcribed. The deletion, therefore, does not appear to have a direct effect on any gene. Recent investigations have focused on the possibility that changes in chromatin structure because of contraction of the repeat array affect the function of neighboring genes. Evidence suggests that the normal D4Z4 repeat is embedded in heterochromatin, a condensed chromosomal conformation that precludes access of transcription factors and leads to repression of gene expression. Contraction of the repeat may cause a transition to a more accessible, euchromatin conformation, allowing increased expression of flanking genes. Consistent with this hypothesis, contraction of the D4Z4 repeat is associated with reduced methylation of DNA in the repeat (a marker of inactive chromatin) and reduced binding of a transcriptional repressor complex [46]. However, efforts thus far have not identified any genes in the close vicinity of 4q whose expression is consistently affected in FSHD.

VII. Conclusion

As illustrated here, the muscular dystrophies represent a vast assortment of disorders marked by heterogeneity in many aspects. This heterogeneity encompasses genetics, clinical presentation, and pathogenesis. With regard to the latter, most of what is understood about the mechanisms of muscular dystrophy is largely speculative; in many cases the precise roles of the affected proteins have not been definitively proven.

Though the causes of these disorders are scattered among a variety of muscle functions, what unifies them is a dystrophic muscle phenotype and a lack of effective treatment for any of them. A considerable amount of effort has been devoted to developing treatments for DMD, particularly in the field of gene replacement therapy. Although this technique may eventually prove successful, the diverse etiologies leading to a common clinical phenotype suggest that studies of pathophysiology may reveal therapeutic approaches that can be used to treat dystrophies as a whole, regardless of the particular gene defect. Corticosteroid use, for example, results in a multitude of changes in gene expression and can improve the clinical course of DMD. Techniques that interfere with calcium-dependent degenerative pathways may be helpful in dystrophies associated with impaired sarcolemma integrity. Up-regulation of factors that promote satellite cell proliferation and myogenesis could be useful across the entire spectrum of muscular dystrophies.

Underlying the development of novel therapeutic techniques is the need for better knowledge of the basic physiologic mechanisms that cause diseases. Research over the last twenty years has yielded a wealth of information about the muscular dystrophies, but it also highlights the tremendous amount of work that remains before these disorders can truly be understood.

References

1. Emery, A. E. H. (1993). "Duchenne Muscular Dystrophy." Oxford University Press, New York.
2. Hoffman, E. P., Brown, R. H., Jr., and Kunkel, L. M. (1987). Dystrophin: the protein product of the Duchenne muscular dystrophy locus. *Cell* **51**, 919–928.
3. Koenig, M., Monaco, A. P., and Kunkel, L. M. (1988). The complete sequence of dystrophin predicts a rod-shaped cytoskeletal protein. *Cell* **53**, 219–226.
4. Rybakova, I. N., Amann, K. J., and Ervasti, J. M. (1996). A new model for the interaction of dystrophin with F-actin. *J. Cell Biol.* **135**, 661–672.
5. Ervasti, J. M., and Campbell, K. P. (1993). Dystrophin and the membrane skeleton. *Curr. Opin. Cell Biol.* **5**, 82–87.
6. Rando, T. A. (2001). The dystrophin-glycoprotein complex, cellular signaling, and the regulation of cell survival in the muscular dystrophies. *Muscle Nerve* **24**, 1575–1594.
7. Matsumura, K., and Campbell, K. P. (1994). Dystrophin-glycoprotein complex: its role in the molecular pathogenesis of muscular dystrophies. *Muscle Nerve* **17**, 2–15.
8. Wehling, M., Spencer, M. J., and Tidball, J. G. (2001). A nitric oxide synthase transgene ameliorates muscular dystrophy in mdx mice. *J. Cell Biol.* **155**, 123–131.
9. Lim, L. E., and Campbell, K. P. (1998). The sarcoglycan complex in limb-girdle muscular dystrophy. *Curr. Opin. Neurol.* **11**, 443–452.
10. Holt, K. H., and Campbell, K. P. (1998). Assembly of the sarcoglycan complex: insights for muscular dystrophy. *J. Biol. Chem.* **273**, 34,667–34,670.
11. Hack, A. A., Groh, M. E., and McNally, E. M. (2000). Sarcoglycans in muscular dystrophy. *Microsc. Res. Tech.* **48**, 167–180.
12. Kirschner, J., and Bonnemann, C. G. (2004). The congenital and limb-girdle muscular dystrophies: sharpening the focus, blurring the boundaries. *Arch. Neurol.* **61**, 189–199.
13. Laval, S. H., and Bushby, K. M. (2004). Limb-girdle muscular dystrophies: from genetics to molecular pathology. *Neuropathol. Appl. Neurobiol.* **30**, 91–105.
14. Piccolo, F., Moore, S. A., Mathews, K. D., and Campbell, K. P. (2002). Limb-girdle muscular dystrophies. *Adv. Neurol.* **88**, 273–291.
15. Minetti, C., Sotgia, F., Bruno, C., Scartezzini, P., Broda, P., Bado, M., Masetti, E., Mazzocco, M., Egeo, A., Donati, M. A., Volonte, D., Galbiati, F., Cordone, G., Bricarelli, F. D., Lisanti, M. P., and Zara, F. (1998). Mutations in the caveolin-3 gene cause autosomal dominant limb-girdle muscular dystrophy. *Nat. Genet.* **18**, 365–368.
16. Liu, J., Aoki, M., Illa, I., Wu, C., Fardeau, M., Angelini, C., Serrano, C., Urtizberea, J. A., Hentati, F., Hamida, M. B., Bohlega, S., Culper, E. J., Amato, A. A., Bossie, K., Oeltjen, J., Bejaoui, K., McKenna-Yasek, D., Hosler, B. A., Schurr, E., Arahata, K., de Jong, P. J., and Brown, R. H. Jr. (1998). Dysferlin, a novel skeletal muscle gene, is mutated in Miyoshi myopathy and limb girdle muscular dystrophy. *Nat. Genet.* **20**, 31–36.
17. Crosbie, R. H., Yamada, H., Venzke, D. P., Lisanti, M. P., and Campbell, K. P. (1998). Caveolin-3 is not an integral component of the dystrophin glycoprotein complex. *FEBS Lett.* **427**, 279–282.
18. Smythe, G. M., Eby, J. C., Disatnik, M. H., and Rando, T. A. (2003). A caveolin-3 mutant that causes limb girdle muscular dystrophy type 1C disrupts Src localization and activity and induces apoptosis in skeletal myotubes. *J. Cell Sci.* **116**, 4739–4749.
19. Bansal, D., Miyake, K., Vogel, S. S., Groh, S., Chen, C. C., Williamson, R., McNeil, P. L., and Campbell, K. P. (2003). Defective membrane repair in dysferlin-deficient muscular dystrophy. *Nature* **423**, 168–172.
20. Hauser, M. A., Horrigan, S. K., Salmikangas, P., Torian, U. M., Viles, K. D., Dancel, R., Tim, R. W., Taivainen, A., Bartoloni, L., Gilchrist, J. M., Stajich, J. M., Gaskell, P. C., Gilbert, J. R., Vance, J. M.,

Pericak-Vance, M. A., Carpen, O., Westbrook, C. A., and Speer, M. C. (2000). Myotilin is mutated in limb girdle muscular dystrophy 1A. *Hum. Mol. Genet.* **9**, 2141–2147.

21. Hackman, P., Vihola, A., Haravuori, H., Marchand, S., Sarparanta, J., De Seze, J., Labeit, S., Witt, C., Peltonen, L., Richard, I., and Udd, B. (2002). Tibial muscular dystrophy is a titinopathy caused by mutations in TTN, the gene encoding the giant skeletal-muscle protein titin. *Am. J. Hum. Genet.* **71**, 492–500.

22. Moreira, E. S., Wiltshire, T. J., Faulkner, G., Nilforoushan, A., Vainzof, M., Suzuki, O. T., Valle, G., Reeves, R., Zatz, M., Passos-Bueno, M. R., and Jenne, D. E. (2000). Limb-girdle muscular dystrophy type 2G is caused by mutations in the gene encoding the sarcomeric protein telethonin. *Nat. Genet.* **24**, 163–166.

23. Richard, I., Broux, O., Allamand, V., Fougerousse, F., Chiannilkulchai, N., Bourg, N., Brenguier, L., Devaud, C., Pasturaud, P., Roudaut, C., et al. (1995). Mutations in the proteolytic enzyme calpain 3 cause limb-girdle muscular dystrophy type 2A. *Cell* **81**, 27–40.

24. Frosk, P., Weiler, T., Nylen, E., Sudha, T., Greenberg, C. R., Morgan, K., Fujiwara, T. M., and Wrogemann, K. (2002). Limb-girdle muscular dystrophy type 2H associated with mutation in TRIM32, a putative E3-ubiquitin-ligase gene. *Am. J. Hum. Genet.* **70**, 663–672.

25. Bengtsson, L., and Wilson, K. L. (2004). Multiple and surprising new functions for emerin, a nuclear membrane protein. *Curr. Opin. Cell Biol.* **16**, 73–79.

26. Nagano, A., and Arahata, K. (2000). Nuclear envelope proteins and associated diseases. *Curr. Opin. Neurol.* **13**, 533–539.

27. Wewer, U. M., and Engvall, E. (1996). Merosin/laminin-2 and muscular dystrophy. *Neuromuscul. Disord.* **6**, 409–418.

28. Straub, V., Rafael, J. A., Chamberlain, J. S., and Campbell, K. P. (1997). Animal models for muscular dystrophy show different patterns of sarcolemmal disruption. *J. Cell Biol.* **139**, 375–385.

29. Camacho, V. O., Bertini, E., Zhang, R. Z., Petrini, S., Minosse, C., Sabatelli, P., Giusti, B., Chu, M. L., and Pepe, G. (2001). Ullrich scleroatonic muscular dystrophy is caused by recessive mutations in collagen type VI. *Proc. Natl. Acad. Sci. USA* **98**, 7516–7521.

30. Demir, E., Sabatelli, P., Allamand, V., Ferreiro, A., Moghadaszadeh, B., Makrelouf, M., Topaloglu, H., Echenne, B., Merlini, L., and Guicheney, P. (2002). Mutations in COL6A3 cause severe and mild phenotypes of Ullrich congenital muscular dystrophy. *Am. J. Hum. Genet.* **70**, 1446–1458.

31. Cohn, R. D. (2005). Dystroglycan: important player in skeletal muscle and beyond. *Neuromuscul. Disord.* **15**, 207–217.

32. Grewal, P. K., and Hewitt, J. E. (2003). Glycosylation defects: a new mechanism for muscular dystrophy? *Hum. Mol. Genet.* **12**, R259–R264.

33. Martin-Rendon, E., and Blake, D. J. (2003). Protein glycosylation in disease: new insights into the congenital muscular dystrophies. *Trends Pharmacol. Sci.* **24**, 178–183.

34. Longman, C., Brockington, M., Torelli, S., Jimenez-Mallebrera, C., Kennedy, C., Khalil, N., Feng, L., Saran, R. K., Voit, T., Merlini, L., Sewry, C. A., Brown, S. C., and Muntoni, F. (2003). Mutations in the human LARGE gene cause MDC1D, a novel form of congenital muscular dystrophy with severe mental retardation and abnormal glycosylation of alpha-dystroglycan. *Hum. Mol. Genet.* **12**, 853–861.

35. Brockington, M., Blake, D. J., Prandini, P., Brown, S. C., Torelli, S., Benson, M. A., Ponting, C. P., Estournet, B., Romero, N. B., Mercuri, E., Voit, T., Sewry, C. A., Guicheney, P., and Muntoni, F. (2001). Mutations in the fukutin-related protein gene (FKRP) cause a form of congenital muscular dystrophy with secondary laminin alpha2 deficiency and abnormal glycosylation of alpha-dystroglycan. *Am. J. Hum. Genet.* **69**, 1198–1209.

36. Brockington, M., Yuva, Y., Prandini, P., Brown, S. C., Torelli, S., Benson, M. A., Herrmann, R., Anderson, L. V., Bashir, R., Burgunder, J. M., Fallet, S., Romero, N., Fardeau, M., Straub, V., Storey, G., Pollitt, C., Richard, I., Sewry, C. A., Bushby, K., Voit, T., Blake, D. J., and Muntoni, F. (2001). Mutations in the fukutin-related protein gene (FKRP) identify limb girdle muscular dystrophy 2I as a milder allelic variant of congenital muscular dystrophy MDC1C. *Hum. Mol. Genet.* **10**, 2851–2859.

37. Brook, J. D., McCurrach, M. E., Harley, H. G., Buckler, A. J., Church, D., Aburatani, H., Hunter, K., Stanton, V. P., Thirion, J. P., Hudson, T., et al. (1992). Molecular basis of myotonic dystrophy: expansion of a trinucleotide (CTG) repeat at the 3′ end of a transcript encoding a protein kinase family member. *Cell* **68**, 799–808.

38. Taneja, K. L., McCurrach, M., Schalling, M., Housman, D., and Singer, R. H. (1995). Foci of trinucleotide repeat transcripts in nuclei of myotonic dystrophy cells and tissues. *J. Cell Biol.* **128**, 995–1002.

39. Day, J. W., and Ranum, L. P. (2005). RNA pathogenesis of the myotonic dystrophies. *Neuromuscul. Disord.* **15**, 5–16.

40. Liquori, C. L., Ricker, K., Moseley, M. L., Jacobsen, J. F., Kress, W., Naylor, S. L., Day, J. W., and Ranum, L. P. (2001). Myotonic dystrophy type 2 caused by a CCTG expansion in intron 1 of ZNF9. *Science* **293**, 864–867.

41. Philips, A. V., Timchenko, L. T., and Cooper, T. A. (1998). Disruption of splicing regulated by a CUG-binding protein in myotonic dystrophy. *Science* **280**, 737–741.

42. Mankodi, A., Takahashi, M. P., Jiang, H., Beck, C. L., Bowers, W. J., Moxley, R. T., Cannon, S. C., and Thornton, C. A. (2002). Expanded CUG repeats trigger aberrant splicing of ClC-1 chloride channel pre-mRNA and hyperexcitability of skeletal muscle in myotonic dystrophy. *Mol. Cell* **10**, 35–44.

43. Miller, J. W., Urbinati, C. R., Teng-Umnuay, P., Stenberg, M. G., Byrne, B. J., Thornton, C. A., and Swanson, M. S. (2000). Recruitment of human muscleblind proteins to (CUG)(n) expansions associated with myotonic dystrophy. *EMBO J.* **19**, 4439–4448.

44. Kanadia, R. N., Johnstone, K. A., Mankodi, A., Lungu, C., Thornton, C. A., Esson, D., Timmers, A. M., Hauswirth, W. W., and Swanson, M. S. (2003). A muscleblind knockout model for myotonic dystrophy. *Science* **302**, 1978–1980.

45. Van Deutekom, J. C., Wijmenga, C., van Tienhoven, E. A., Gruter, A. M., Hewitt, J. E., Padberg, G. W., van Ommen, G. J., Hofker, M. H., and Frants, R. R. (1993). FSHD associated DNA rearrangements are due to deletions of integral copies of a 3.2 KB tandemly repeated unit. *Hum. Mol. Genet.* **2**, 2037–2042.

46. Van der Maarel, S. M., and Frants, R. R. (2005). The D4Z4 repeat-mediated pathogenesis of facioscapulohumeral muscular dystrophy. *Am. J. Hum. Genet.* **76**, 375–386.

85

Myasthenia Gravis and Myasthenic Syndromes

Angela Vincent, MA (MB, BS, MSc Lond.), FRCPath, FMedSci

Keywords: *autoantibodies, autoimmune, Lambert-Eaton myasthenic syndrome, mechanisms, myasthenia gravis, neuromuscular junction, neuromyotonia*

I. Neuromuscular Junction
II. Autoimmune Neuromuscular Junction Disorders
III. Congenital Myasthenic Syndromes
VI. What Determines the Phenotypic Variability in Neuromuscular Junction Disorders?
References

I. Neuromuscular Junction

The neuromuscular junction represents a synapse between the motor nerve terminal and the surface of the muscle fiber, but the synaptic and postsynaptic elements are somewhat different from those in a central synapse [1]. The unmyelinated motor nerve terminals are separated by a 500-ampere synaptic cleft from the postsynaptic muscle membrane (Fig. 1). The synaptic cleft contains a basal lamina that includes collagen IV, laminins, fibronectin, and perlecan. Acetylcholine esterase (AChE) is anchored by ColQ, a collagen-like molecule. Agrin and neuregulins, secreted from the nerve terminal, concentrate in the basal lamina. Interactions among these and other basal lamina proteins and membrane proteins, such as voltage-gated calcium channels (VGCCs) presynaptically and the dystroglycans postsynaptically, are being identified and the importance of the structure and function of the basal lamina is being recognized [2].

The postsynaptic membrane at the neuromuscular junction forms a series of folds that are particularly deep in human muscle. Acetylcholine receptors (AChRs) are found at the top one-third of these folds, whereas the voltage-gated sodium channels are anchored at the bottom of the folds (diagrammatically illustrated in Fig. 2). The development of the neuromuscular junction is an area of active research but is outside the scope of this review. However, it is evident that several key proteins, for example, muscle-specific kinase (MuSK), are essential for normal neuromuscular junction formation and these are of some relevance to the diseases to be discussed (see Fig. 2).

Neuromuscular transmission depends on motor nerve terminal depolarization by the nerve action potential and subsequent calcium influx through VGCCs located in active zones along the terminal membrane where it abuts the muscle. Acetylcholine (ACh) is released in packets or

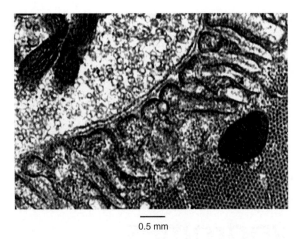

Figure 1 The Neuromuscular Junction and Acetylcholine Receptors. Electron-microscopic image of the human neuromuscular junction. The nerve terminal contains mitochondria and many synaptic vesicles. The basal lamina can be seen as a thin line between the pre- and the postsynaptic membranes. The tops of the postsynaptic folds are electron dense because of the high density of acetylcholine receptors. The folds are long, and below them the muscle fibrils can be seen in cross-section. Courtesy of Prof. Clarke Slater, University of Newcastle.

trode as an end plate potential (EPP). Miniature end plate potentials (MEPPs) are the much smaller depolarizations that occur when a single packet of ACh is spontaneously released. When the EPP reaches a critical firing threshold, voltage-gated sodium channels open and an action potential is initiated. AChRs close spontaneously, ACh dissociates from AChRs, and ACh is hydrolyzed by AChE, thus limiting the duration of the response. Meanwhile, the presynaptic calcium channels close spontaneously, the membrane depolarization is reversed by efflux of potassium through voltage-gated potassium channels (VGKCs), and the resting state is restored. The sodium/potassium adenosine triphosphatase restores the membrane potential. The extent to which the EPP exceeds what is necessary to initiate the action potential is usually called the *safety factor for neuromuscular transmission*.

The neuromuscular junction is vulnerable to a variety of disease largely because it has no blood-brain barrier and is accessible to circulating factors. For instance, in various parts of the world, muscle paralysis or hyperactivity (fasciculations and cramps) are commonly caused by envenomation (e.g., from snakes, spiders, and scorpions), and botulism is still a major problem in some developing countries. Poisoning by environmental insecticides or self-administered insecticides is common, and there are a variety of plant extracts that interfere with neuromuscular transmission. In the Western world, however, the most common disorders are caused by autoantibodies

quanta, and initially most of it succeeds in avoiding hydrolysis by AChE and binds instead to AChRs on the postsynaptic membrane. This leads to the opening of AChR-associated ion channels and depolarization of the motor end plate, which can be measured by an intracellular elec-

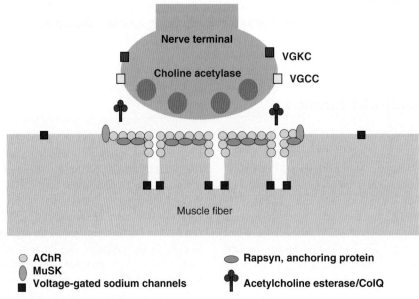

Figure 2 Diagrammatic Representation of the Neuromuscular Junction. Included are the ion channels, MuSK, and enzymes (choline acetylase, acetylcholine esterase) essential for normal neuromuscular junction function. Antibodies specific for the acetylcholine receptor (AChR), muscle-specific kinase (MuSK), voltage-gated calcium channel (VGCC), and voltage-gated potassium channel (VGKC) are pathogenic in different autoimmune diseases. Mutations in choline acetyltransferase, ColQ, AChR, MuSK, rapsyn, and the sodium channels underlie different congenital myasthenic syndromes.

to molecules on the postsynaptic membrane that cause autoimmune myasthenia gravis (MG). There are also presynaptic autoimmune diseases caused by antibodies to VGCCs and VGKCs, respectively (Lambert-Eaton myasthenic syndrome, neuromyotonia) and rare genetic forms of myasthenia caused mainly by mutations in the genes encoding the AChR or rapsyn. This review concentrates on the mechanisms at the neuromuscular junction and does not discuss in any detail the immunological aspects of these conditions.

II. Autoimmune Neuromuscular Junction Disorders

There are now four well-defined disorders in which autoantibodies are thought to be pathogenic. Their identification has depended on a variety of approaches, often including an element of serendipity and deductive reasoning.

A. Myasthenia Gravis

1. History

MG is a condition that has been recognized since Thomas Willis's probable first description in 1672. Traditionally, it first presents itself in young adult or adult life as muscle weakness and fatigue, which can be generalized or limited in extent. Typically the muscles of eye movement are involved, causing double vision; ptosis may result from weakness of eye-lid elevation; and there may be weakness on any task that requires repetitive movements. Involvement of the facial and bulbar muscles can result in loss of smile, poor speech, and choking. Respiratory muscle involvement can be life-threatening.

It was suggested from the beginning of the twentieth century that this weakness in MG originates at the neuromuscular junction and that "autotoxic" circulating factors might be involved. Studies in the 1950s showed that the MEPPs are substantially reduced in amplitude in the patients' muscle biopsies, and it was rightly deduced that either the amount of ACh in each packet was reduced or the postsynaptic sensitivity to ACh was reduced. At that time the existence of "receptors" for ACh was still hypothetical, and there was no direct way of measuring them. But it was shown that a snake toxin, α-bungarotoxin, from the Taiwan banded krait, bound irreversibly to the neuromuscular junction. Further studies indicated that it was probably binding to "receptors" for ACh. Since then, radioactively, peroxidase- and fluorescently labeled bungarotoxin has been used to localize, identify, and tag AChRs from a variety of sources. One of the first practical uses was to show that the number of AChRs was reduced at the neuromuscular junctions of MG patients [3].

The electric organs of eels and rays are made up of stacks of flat cylindrical plates whose ventral surface is studded with AChRs. The electric organs are derived from branchial arch myotubes and are related, therefore, to muscle. AChRs were purified from detergent extracts of these tissues using the slightly lower–affinity Krait neurotoxins, such as α-cobratoxin, for affinity chromatography. AChRs were first purified from the electric eel by Patrick and Lindstrom [4], and the subunit structure was subsequently worked out by a number of investigators. This led to the first N-terminal protein sequences of the AChR subunits and eventually to cloning and expression of all five of them. We now know that the AChR is an oligomeric membrane protein consisting of five subunits (see Fig. 2): α^2, β, γ, and δ in embryonic or denervated muscle (fetal form), and α^2, β, δ, and ε at the adult end plate (adult form). A possibly serendipitous finding was that rabbits, or other species, immunized with purified fish AChR developed muscle weakness that responded to cholinesterase inhibitors [4], paralleling the clinical response of MG patients to cholinesterase inhibitors (first shown by Mary Walker in 1934) and providing an experimental autoimmune model of MG (EAMG). As a result, Lindstrom and colleagues began to look for AChR antibodies in patients with MG. These antibodies were first systematically measured by immunoprecipitation of iodine-125 (^{125}I) α-bungarotoxin–labeled AChR extracted from ischemic human muscle, which consists mainly of the fetal form (because ischemia causes denervation); most groups now use a mixture of adult and fetal AChRs extracted from muscle-like cell lines. AChR antibody titers are highly variable among MG patients, ranging from 0 to more than 1000 nm/L, and the titers do not correlate well with clinical severity among individuals. However, passive transfer of purified immunoglobulin G (IgG) from MG patients to mice resulted in reduced MEPPs and reduced AChR numbers (like those in the patients' muscles), confirming that the antibodies were pathogenic [5]. Moreover, despite the variability among patients, the level of antibody within an individual correlates well with clinical scores after plasma exchange, thymectomy, and/or immunosuppressive treatments, providing supportive evidence that the disease is caused by these, and perhaps other, antibodies. Recent reviews discuss the early studies on electric fish [6], the use of snake venom neurotoxins [7], and the history of myasthenia research [8].

2. Etiology

Despite several decades of research, the etiology of MG is unknown. Various hypotheses have been put forward at different times, including theories regarding molecular mimicry between the AChR and bacterial or viral antigens, but none have been proven. However, it is clear that at least some patients have genetic susceptibility to development of MG and that in some this relates to polymorphic variants in the AChR genes, as well as in the major histocompatibility genes [9].

The antibodies are heterogeneous and bind to several sites on the AChR. Even within an individual, they can be of both kappa and lambda light chain and of different IgG subclasses, although IgG1 and IgG3 subclasses predominate in most patients. A large number of studies indicate that the antibodies, at least those measured in the radioimmunoprecipitation assays, are highly specific for the intact human AChR and do not bind well to denatured or recombinant subunits. In many patients a proportion of the anti-AChR antibody competes with monoclonal antibodies directed toward the main immunogenic regions (MIRs) on each of the α subunits (Fig. 3). This is the dominant epitope defined by monoclonal antibodies derived from animals immunized with purified AChR. Most MIR monoclonal antibodies discriminate between the native AChR, to which they bind strongly, and the denatured or recombinant AChR subunits, to which they bind weakly or not at all. Nevertheless, mapping studies with synthetic peptides and site-directed mutagenesis show that the sequence α67–76 makes an important contribution towards the MIR (for a review, see reference 10). However, other studies indicate that some patients do not have MIR antibodies and some have antibodies that compete with monoclonal antibodies directed against other epitopes on the α or remaining subunits. One of these is specific to fetal AChR and is on the γ subunit.

Overall, the high affinity (around 10^{-10} molar) and specificity of MG anti-AChRs indicate that the MG antibodies are induced by some form of the native AChR molecule. The possibility remains that an initiating event in MG may induce a low-affinity antibody that subsequently matures and expands, through the processes of somatic mutation and determinant spreading, to form the heterogeneous, high-affinity antibodies that we recognize *in vitro*. These issues are discussed in more detail elsewhere [11].

3. Pathogenic Mechanisms

The physiological and subsequent bungarotoxin-binding studies showed that the main defect in MG was a loss of AChRs from the neuromuscular junction; indeed, there was a correlation between MEPP amplitudes and α-bungarotoxin binding to the neuromuscular junction in individual patients. However, the situation is not so simple, as will be reviewed here.

Several authors clearly demonstrated cross-linking of AChRs by divalent antibodies. This process was studied first in cell lines expressing AChRs. The ^{125}I-bungarotoxin-labeled AChRs are normally internalized and degraded with release of ^{125}I into the medium. In cultured cells this process has a half-life of 10–16 hours, but in the presence of AChR antibodies the half-life falls to less than 10 hours. Similarly, at the intact neuromuscular junction, experiments using *in vivo* labeling of AChRs with injected ^{125}I-bungarotoxin, showed that the increased rate of internalization and degradation of AChRs reduces the half-life for AChRs from about 10 days to less than 5 days. If AChRs were pretreated with cold bungarotoxin, to block-specific binding of ^{125}I-bungarotoxin, the appearance of new ^{125}I-bungarotoxin-binding sites (AChRs) could be determined. Interestingly, the antibodies increased the synthesis rate of AChRs, as well as the degradation rate. Since then, analysis of muscles from MG patients and from AChR-immunized mice has confirmed an increase in the expression of AChR subunits.

Antibodies to the ACh or α-bungarotoxin-binding site are found in many MG patients, but they often compose only a small proportion of the total AChR antibody population. Direct inhibition of ion channel function, therefore, is relatively uncommon, although may contribute to the pathogenetic mechanisms in some patients. Further details can be found in Drachman [12].

Engel and his colleagues [13] studied in detail the immunopathology of the neuromuscular junction in MG, as well as its animal model, EAMG, in rats. They found

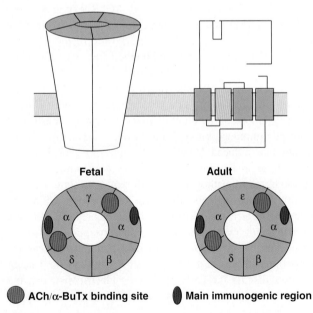

Figure 3 The AChR Is a Pentameric Membrane Protein That Occurs in an Adult and Fetal Isoform. Acetylcholine (ACh) and α-bungarotoxin bind to sites on the interfaces between the α and the adjacent subunits. Many of the antibodies in myasthenia bind to a main immunogenic region on the two α subunits. Null- or low-expressor mutations in the ε subunit are the most common cause of congenital myasthenic syndromes, but single nucleotide mutations in each of the subunits can cause kinetic defects in the acetylcholine receptor (AChR) channel function.

IgG localized to the neuromuscular junction postsynaptically. The distribution of IgG tended to correspond to the distribution of AChRs as determined by peroxidase-bungarotoxin labeling. There were deposits of complement components C3 and C9 and of the membrane attack complex. The synaptic cleft was widened and contained debris that also immunostained for IgG and complement. Moreover, the extent of staining for the membrane attack complex of complement was roughly inversely related to the number and density of AChRs, suggesting that the terminal complement membrane attack complex is responsible for at least some of the damage and AChR loss. Similar changes were found in rats immunized with electric fish AChR, although when pertussis toxin was used as an additional adjuvant, an acute stage occurred within 10 days, during which there was a cellular infiltration of the neuromuscular junction.

Complement-mediated damage to the neuromuscular junction is probably the most important mechanism by which the AChR antibodies cause failure of neuromuscular transmission in humans. This is not only because the complement attack leads to loss of postsynaptic membrane, and release of membrane debris that may stimulate the immune response (although this has never been proven), but also because it leads to loss of the postsynaptic folds. At the bottom of these folds, the voltage-gated sodium channels are located (see Fig. 2). These are essential for the generation of the action potential that will lead to muscle contraction. The antibodies, by reducing the postsynaptic folds, lead to loss of the sodium channels and, therefore, an increased threshold for generation of the action potential, as demonstrated by Ruff and Lennon [14]. The amount of current that is needed to be injected postsynaptically (simulating an EPP) to initiate an action potential was increased at MG neuromuscular junctions, presumably because the loss of postsynaptic folds had reduced the number of voltage-gated sodium channels.

Like most biological systems, however, the neuromuscular junction probably tries to compensate for the loss of AChRs that causes interference with neuromuscular transmission by increasing AChR synthesis, as mentioned previously. Interestingly, the presynaptic motor nerve also seems to recognize the impaired neuromuscular transmission. Plomp and colleagues [15] showed that in MG muscle biopsies there was an overall increase in the number of ACh packets released and the increase was inversely associated with the size of the MEPPs at each end plate. Thus, retrograde signaling, from the postsynaptic to the presynaptic component, can lead to compensatory changes in ACh release. How this occurs and whether it involves modulation of existing nerve terminal proteins or synthesis of new proteins, either in the motoneuron cell body or at the motor nerve terminal, is still unknown.

Overall, whatever the combination of mechanisms involved, the antibodies cause a reduction in the EPP so that it does not reach the critical firing threshold—either initially or during repeated effort (when the release of ACh naturally decreases a little). As a result of these changes, there is weakness at rest, increasing fatigue, or both during sustained efforts. The disease can remain localized (e.g., ocular MG), involve mainly specific muscle groups (e.g., bulbar MG), or spread to include most muscles (generalized MG).

4. Subtypes of Myasthenia

a. Generalized Acetylcholine Receptor The patients with generalized MG and AChR antibodies can be further subdivided on the basis of their thymic pathology and age at onset: young patients, often female, with thymic hyperplasia; older patients of both sexes with normal or involuted thymus glands; and patients of both sexes with thymoma. However, there is no evidence that the AChR antibody characteristics or pathogenic mechanisms differ among these three groups (see reference 11), suggesting that the antibodies are the common final pathway that can be reached by a number of different etiological routes. The most important practical consequence of this division is the recognition that younger patients benefit from thymectomy whereas older patients probably do not.

This is because the thymus is a source of the AChR antibody in the younger patients. Normally the thymus does not contain many B cells or antibody-producing plasma cells, but it contains AChRs on myoid cells that lie sparsely in the thymic medulla. These muscle-like cells were first identified by German anatomists in the nineteenth century and have since been shown to express a variety of muscle antigens both *in situ* and after culture *in vitro*. In early-onset MG, the thymus often contains germinal centers similar to those found in lymph nodes. The germinal centers contain T and B cells specific for the AChR [16], and the myoid cells are sometimes found at the edge of the germinal centers, suggesting that their AChR may stimulate the immune response. Removal of the thymus results in a moderate fall in the AChR antibody with clinical improvement in most patients, although most also require long-term immunosuppressive treatments. In those patients with a thymoma, the role of the thymus is less clear. Thymectomy generally does not lead to clinical improvement, and most patients require immunosuppressive treatments. The thymic tumor can recur, and these patients often develop other autoimmune diseases (see reference 11).

b. Ocular Myasthenia Gravis Some patients with MG never develop generalized symptoms but continue to have marked ocular muscle weakness. There is considerable interest in the neurophysiology of the ocular muscles, although their size and inaccessibility make them difficult to study in detail. Earlier studies showed that the extraocular muscles consisted of twitch- and tonic-type fibers, some of which have multiple neuromuscular junctions. The motor unit sizes are smaller than in limb muscles, and the firing frequencies are high. Normally the muscles are resistant to fatigue with high blood flow, mitochondrial content, and metabolic rate [17]. At one time it was thought that the ocular muscles contained mainly fetal-type AChR rather than adult-type AChR, but this is not the case; adult AChR is expressed at higher concentration than fetal, and use of adult muscle AChR in the radioimmunoprecipitation assays improves the diagnosis of patients with ocular MG. Nevertheless, there may be some fibers in which neuromuscular transmission depends on fetal AChR.

Regardless of these considerations, the ocular muscles are susceptible to circulating factors. During the onset of envenomation or botulinum toxin poisoning, ocular symptoms are common. This may be because the safety factor is low in the twitch fibers and the tonic fibers do not generate action potentials but depend on the EPP to activate contraction. Thus any reduction in EPP may be quickly translated into muscle weakness. In addition, recent studies show that the neuromuscular junctions have low expression of the complement regulators now known to be expressed at higher concentration in other muscles [18] and this could make them more vulnerable to complement-mediated damage in MG. Perhaps it is relevant that in Lambert-Eaton myasthenic syndrome (see Sect. V) which is not complement mediated, ocular weakness is uncommon.

c. Neonatal Myasthenia Gravis and Arthrogryposis Multiplex Congenita A proportion of babies born to MG mothers have transient respiratory and feeding difficulties because of transplacental transfer of maternal anti-AChR. In a small number of cases, mothers with MG have babies with arthrogryposis multiplex congenita, a relatively common condition that involves fixed joint contractures associated with inadequate development of the lungs and is thought to arise as the result of lack of fetal movement *in utero*. A few instances are known in which the mother had a number of consecutive pregnancies affected by this condition before a diagnosis of MG was made. One mother remains symptom free. These women have AChR antibodies that completely inhibit the function of fetal AChR but do not necessarily affect adult AChR function; this explains the fetal paralysis and development of deformities [19]. An animal model of this condition was induced by injecting the maternal plasmas into pregnant mice; the offspring were found to be paralyzed and to exhibit fixed joint contractures. Maternal antibodies to other antigens are beginning to be considered as a possible cause of other developmental disorders.

B. Muscle-Specific Kinase Antibody-Associated Myasthenia Gravis

About 10% to 15% of all MG patients with generalized symptoms do not have detectable AChR antibodies by current laboratory methods. They have muscle weakness and fatigue similar to those of patients with AChR antibody–positive generalized MG, but bulbar symptoms are often marked. Because they respond well to plasma exchange and their immunoglobulins passively transfer a defect in neuromuscular transmission to mice, they appear to have an antibody-mediated disease. In some patients without AChR antibodies there appears to be a non-IgG, IgM-containing, plasma fraction that reduces ACh-induced currents in whole cell clamped TE671 (muscle-like) cells. Electrophysiological observations point to enhanced desensitization of the AChR, possibly by binding of low-affinity IgM antibodies to the AChR itself (see also reference 20). In a proportion of the patients without evidence of AChR antibodies, there are antibodies to MuSK.

MuSK was an attractive candidate antigen because, although it plays a crucial role during development of the neuromuscular junction (see reference 1), like the AChR it is a transmembrane protein localized to the neuromuscular junction in mature muscle. The antibodies to MuSK were first identified by an enzyme-linked immunosorbent assay in up to 70% of AChR antibody–negative MG patients [21], but a number of recent studies suggest that the incidence is variable throughout the world and in some countries there are few or no MuSK antibody–positive patients (see reference 22). Moreover, some patients report preceding infections that might be involved in the initiation of their disease. Together these features suggest that environmental agents might be important in the etiology.

MuSK antibodies are mainly IgG4, contrasting with IgG1 and IgG3 in AChR-MG, and it is not yet clear how the MuSK antibodies cause a defect in neuromuscular transmission (discussed in reference 22). IgG4 does not activate complement efficiently, and neither AChR loss nor complement deposition is found in the limb muscles of most MuSK-MG patients [23]. One difficulty with the studies on MuSK-MG patients is that limb muscles may not show abnormal *in vivo* electrophysiology [24], suggesting that these are not the best muscles in which to investigate pathogenic mechanisms. Indeed, antisense inhibition of MuSK in mature rat muscle did result in disassembly of the neuromuscular junctions with dispersal of AChRs [25], indicating that MuSK plays a critical role during adult life.

One aspect of MuSK-MG that needs to be considered is that it is relatively commonly associated with marked bulbar and facial weakness (see reference 20). Magnetic resonance imaging shows muscle atrophy in tongue and facial muscles in up to 50% of patients, and facial muscles show typical electrophysiological evidence of neuromuscular junction defects [26]. It is likely that the antibodies interfere with some aspect of MuSK signaling, and there is preliminary evidence that they up-regulate a ring-finger ligase, MURF-1, that is involved in induction of muscle atrophy. Animal studies are required to determine the role of

MuSK in adult muscle and to look for differences between facial and limb neuromuscular junctions.

C. Lambert-Eaton Myasthenic Syndrome

1. History

In 1957 Edward H. Lambert and Lee McKendre Eaton [27] described a myasthenic syndrome that was electrophysiologically distinct from MG. On stimulation of a motor nerve *in vivo*, the compound muscle action potential (CMAP) amplitude following supramaximal nerve stimulation was small and became smaller at low rates of stimulation (<10 per second). But during stimulation at higher rates, or following a few seconds of voluntary contraction, the amplitude increased substantially. These findings contrasted with those in MG, where the CMAP is not usually reduced initially and fails to show an increment following high-frequency stimulation. Lambert-Eaton myasthenic syndrome (LEMS) is more common in males than in females. The distribution of weakness is different from that in MG, usually affecting the legs first with relatively little ocular muscle involvement. Reflexes are absent or depressed but may become stronger after voluntary contraction. Autonomic symptoms (dry mouth, constipation, impotence) are present in many patients, suggesting that the target antigen may be common to certain autonomic systems [28].

2. Etiology

LEMS is associated with small cell lung cancer in about 50% of patients, and the evidence strongly suggests that the immune response is primarily directed against the tumor in these cases. Thus, LEMS is a member of the expanding family of paraneoplastic autoimmune conditions. However, about 50% of patients never develop a tumor and have an acquired autoimmune disease of unknown etiology. These patients often have other autoimmune disorders, such as thyroid disease, vitiligo, pernicious anemia, celiac disease, and juvenile onset diabetes mellitus. There is also an association with the major histocompatibility gene B8, particularly in the patients without cancer.

3. Pathophysiology

In vitro recordings from patients' muscle biopsies revealed normal MEPPs but small EPPs, indicating that the quantal content (the number of packets of ACh released per nerve impulse) was reduced [29]. The EPPs increased during repetitive stimulation and in response to increases in extracellular calcium concentration. It seems that increasing the buildup of intracellular calcium concentration, either by increasing extracellular calcium or by high-frequency stimulation, increases ACh release and improves neuromuscular transmission. These results pointed to a possible defect in the presynaptic VGCCs.

Freeze fracture electron microscopic studies of motor nerve terminals in healthy human muscle biopsy samples reveal, as in other species, double parallel rows of intramembranous particles (each about 10–12 nm in diameter) associated with the presynaptic active zones. In LEMS, there is a reduction in the number of active zone particles and in the number of particles per active zone [29]. The particles are thought to be VGCCs.

The paradigms established for the demonstration that MG is an antibody-mediated disease were followed in LEMS. Plasma exchange was found to lead to clinical improvement associated with an increase in CMAP amplitudes. In addition, most patients improve clinically when given long-term immunosuppressive drugs or intravenous immunoglobulin therapy. Daily injection of LEMS plasmas, or IgG fractions, into mice reproduced the principal neurophysiological changes of LEMS in the recipients [29].

Quantitative freeze fracture electron microscopy in mice injected with LEMS IgG showed reductions in the number of active zones that were similar to those seen at the nerve terminals of LEMS patients. The clustering of the active zone particles was preceded by a reduction in the distance between particles, suggesting that the divalent antibodies were acting by cross-linking adjacent particles. IgG was demonstrated at the presynaptic active zones of mice that had received multiple intraperitoneal doses of LEMS IgG. Importantly, divalent F(ab)$_2$ IgG molecules, but not monovalent F(ab)s, were able to cause the neurophysiological changes, indicating that divalency of the antibody is essential. F(ab)2 IgGs do not fix complement; therefore, complement activation is not required. These observations implicate cross-linking and internalization as the main mechanisms for VGCC loss in LEMS; why the antibodies do not activate complement is not yet clear, as the IgG subclasses have not been identified.

All of these observations implicate VGCCs as the main target in LEMS. VGCC subtypes are transmembrane proteins comprising $\alpha 1$, β, and $\alpha 2/\delta$ subunits. The $\alpha 1$ subunits, of which there are eight subtypes, contain the Ca^{2+}-conducting channel and are the principal determinant of the functional properties of the particular subtype. The cone snail–derived toxin ω-conotoxin MVIIC reduces the EPP at the mouse neuromuscular junction, indicating that $\alpha 1A$ (P/Q-type) VGCC mediates ACh release, and mice lacking these VGCC channels have altered ACh release [30]. Solubilized VGCCs prelabeled with ^{125}I-ω-conotoxin MVIIC can be immunoprecipitated by LEMS IgG. The binding sites of the anti-VGCC antibodies have not been identified, but their specificity for the $\alpha 1A$ subtype was shown by applying the LEMS IgG to cultured human embryonic kidney-293 cells expressing the different types of VGCC. After overnight incubation, the LEMS IgG substantially reduced calcium currents through VGCCs expressing the α1A subtype but not through those expressing the other subtypes [31]. In cultured cerebellar granular neurons, there was a decrease in α1A channels and an up-regulation of other subtypes. Similarly, in mice injected with LEMS IgG, an up-regulation of N-type VGCCs was found, illustrating, again, the ability of the neuromuscular junction to compensate for disease-induced changes [32].

Autonomic dysfunction is common in LEMS, suggesting that LEMS IgG may interfere with neurotransmission at autonomic synapses. Multiple subtypes of VGCC are involved in the release of neurotransmitter at mouse bladder muscle and the vas deferens, including the P, Q, N, and R types. Mice injected with LEMS IgG show reduced muscle tension generated at high frequency because of reduced activity through P-type VGCCs with relatively little change in N-type VGCCs [28]. However, there was no obvious clinical effect in the mice, and it would be interesting to know more about the role of these channels in normal autonomic functions.

D. Acquired Neuromyotonia

1. History

Neuromyotonia (NMT), or Isaacs's syndrome, is a syndrome of spontaneous and continuous muscle fiber contraction. The clinical features include muscle stiffness, cramps, myokymia (visible undulation of the muscle), pseudomyotonia (slow relaxation after contraction) and weakness, most prominent in the limbs and trunk. Increased sweating is common. Myokymia characteristically continues during sleep and even during general anesthesia, indicating its peripheral origin. The diagnosis largely rests on a combination of clinical and electromyographic findings of spontaneous motor unit discharges that occur in distinctive doublets, triplets, or longer runs. These neuromyotonic discharges have a high intraburst frequency and usually occur at irregular intervals of 1 to 30 seconds. Some patients have sensory symptoms, and central nervous system symptoms such as insomnia, hallucinations, delusions, and personality change are not infrequent. Cramp fasciculation syndrome, formerly thought to be a different disorder, may represent part of the same spectrum [33,34]. These conditions usually begin between 25 and 60 years.

2. Etiology

As in the other diseases considered in this chapter, there is no clear information regarding the etiology of NMT. However, there are anecdotal reports of the condition occurring in association with infections, and some patients seem to have a monophasic illness that recovers spontaneously within 1 to 2 years. These observations suggest that there may be specific conditions, probably infections, that can predispose to development of these antibodies. It may be relevant that, like in MuSK-MG, most of the antibodies are IgG4 [L. Clover, A. Vincent, unpublished observations].

3. Pathogenic Mechanisms in Neuromyotonia

Although muscle biopsies have not been studied in NMT, *in vivo* electrophysiology indicates that the muscle hyperactivity is caused by hyperexcitability of the motor nerves, generated mainly distally, perhaps at the neuromuscular junction itself, but in some cases more proximally. The underlying basis is not clear, but evidence suggests that a loss of VGKCs is likely to be involved. In some patients there may be an associated inflammatory demyelinating polyneuropathy, and the pathogenic mechanisms may be different in these cases. There are also genetic disorders that cause similar symptoms [34].

NMT may be associated with other autoimmune diseases or other autoantibodies, suggesting an autoimmune basis, and cerebrospinal fluid analysis may show an increase in IgG or "oligoclonal bands," suggesting an immune response to neuronal antigens. Like MG, NMT can be associated with a thymoma [34]. Moreover, plasma exchange results in a clinical improvement and a reduction in the frequency of the abnormal muscle discharges on electromyography. Electrophysiological evidence of changes in motor nerve terminal activity can be passively transferred to mice by injections of plasma or IgG from patients with NMT. In these experiments, the phrenic nerve diaphragm preparations from animals treated with NMT IgG showed increased β-tubocurarine resistance and a moderate increase in the quantal content of the EPP (i.e., there were more ACh quanta released per nerve impulse). Moreover, purified IgG injected into mice led to prolonged sensory nerve action potentials and to repetitive action potentials in dorsal root ganglion cultures. Crucially, the results were similar to those obtained with low concentrations of the VGKC blockers, 4-aminopyridine or 3,4-diaminopyridine.

The most likely target for the autoantibodies, therefore, was a VGKC. Aminopyridines or specific toxins such as α-dendrotoxin that block certain VGKCs cause the axon membrane to become more excitable and to generate repetitive or spontaneous bursts of action potentials. A functional VGKC consists of four transmembrane α subunits that combine as homomultimeric and heteromultimeric tetramers. A large number of VGKC α subunits have been identified. VGKCs of the subtypes Kv1.1 and 1.2 are highly expressed in the peripheral nervous system at nodes of Ranvier. The subtype expressed at the motor nerve terminal, however, is not known, and α-dendrotoxin has relatively small effects on neuromuscular transmission *in vitro*.

Antibodies to VGKCs can be detected in about 40% of NMT patients by immunoprecipitation of ^{125}I-α-dendrotoxin-labeled VGKCs extracted from human frontal cortex. α-Dendrotoxin binds to Kv1.1, 1.2, and 1.6, but attempts to define which of these subunits is the main target of the antibodies have not been very successful. Antibodies to Kv1.2 probably dominate in most patients, but antibodies to Kv1.1 or Kv1.6 may be important in some sera [35]. The antibodies, or IgG fractions, can reduce potassium channel currents in a neuroblastoma cell line, NB-1, and this does not require complement but does require divalent F(ab)2 antibodies [36]. So the immunological mechanisms are likely to be similar to those in LEMS. For a recent review of these mechanisms, see reference 37.

III. Congenital Myasthenic Syndromes

In contrast to the autoimmune conditions discussed previously, the congenital myasthenic syndromes are rare inherited disorders that result from mutations in different key proteins at the neuromuscular junction. The main disorders identified so far are summarized in Table 2, and details can be found in the work of Engel and colleagues [38]. Most are first found neonatally or in early childhood with ptosis, poor suck and feeding problems, respiratory distress (particularly during infections), and delayed milestones. However, some congenital myasthenic syndrome patients do not show symptoms until adolescence or young adulthood. Even within a family the time of presentation can differ.

In the AChR deficiency syndrome, which is caused by mutations in the AChR genes (usually ε) or in the gene encoding rapsyn that anchors the AChR (see Fig. 2), the main defect is a loss of AChRs from the postsynaptic membrane at the neuromuscular junction. The electrophysiology, both *in vivo* and *in vitro*, is similar to that of MG. But the disease does not respond to immunotherapies, and there are no antibodies to the AChR. Babies do not die at birth and many patients survive well into adult life because the γ subunit of the AChR (see Fig. 3) continues to be synthesized in human muscle and can persist at the neuromuscular junction.

Choline acetylase is the enzyme that synthesizes ACh in the presynaptic nerve terminal. Deficiency of this enzyme is usually associated with heteroallelic mutations in the gene encoding the enzyme that reduce function. Typically the patients have normal neuromuscular transmission at rest, with normal EPPs, but during repetitive firing the amount of ACh in each quanta decreases, resulting in a parallel decrease in the EPP until it becomes subthreshold.

There are two conditions that, in contrast to the preceding ones, probably act through excitotoxic mechanisms. In AChE deficiency, the genetic defect is in the gene encoding ColQ, the collagen-like molecule that anchors AChE. The condition is recessive and leads to complete loss of AChE at the neuromuscular junction. As a result, the action of ACh on AChRs is prolonged, with repetitive binding and prolonged time course of the EPP. This can lead to a repetitive activation of the action potential, with a "double response" to a single stimulus on *in vivo* electromyography. The patients usually show symptoms in early childhood. The disease is progressive with marked muscle wasting and disability. The dominant, slow channel syndrome is similar in that the duration of individual AChR openings is prolonged, resulting in prolonged EPPs and repetitive muscle activity. In this case, however, the defect is in one of the AChR subunit genes. The prolonged bursts of ACh openings can be because of either slowing of AChR channel closure or increase in the affinity of the ACh for the AChR so that the neurotransmitter does not dissociate from the receptor, which reopens repetitively in prolonged bursts. In both of these conditions, it is likely that the prolonged AChR

Table 1 Antibodies and Mechanisms of Action at the Neuromuscular Junction

	MG	MuSK-MG	LEMS	NMT
Target	AChR	MuSK	α1A VGCC	VGKC
Principal subclass of antibody	IgG1, IgG3	IgG4	Not known (possibly IgG4)	IgG4
Principal mechanisms of action at NMJ	Complement-mediated destruction of folds (includes loss of AChR and sodium channels), increased turnover, direct block of function	Not clear yet	Increased turnover	Increased turnover
Compensatory mechanisms identified	Increased presynaptic ACh release, increased postsynaptic AChR synthesis	Not clear yet	Involvement of other VGCCs in release process	Probable up-regulation of other VGKCs

Note: AChR, acetylcholine receptor; Ig, immunoglobulin; LEMS, Lambert-Eaton myasthenic syndrome; MG, myasthenia gravis; MuSK, muscle-specific kinase; NMJ, neuromuscular junction; NMT, neuromyotonia; VGCC, voltage-gated calcium channel; VGKC, voltage-gated potassium channels.

Table 2 Congenital Myasthenic Syndromes

	Genetic Target	Presentation	Prognosis	Mechanism
AChR deficiency	AChR, mainly ε subunit	Within first 2 years of life	Good, may improve with age	Loss of adult AChR expression, incorporation of fetal AChR
AChR deficiency	Rapsyn	*In utero*, with arthrogryposis, in infancy with episodes of weakness during infections, or in adult life	Generally good	Loss of AChR at NMJ
Choline acetylase deficiency	Enzyme that synthesizes ACh	Usually early childhood with episodic apnea	Good as long as treated during apneic attacks	Reduced ACh synthesis during repetitive activity
AChR fast channel syndrome	Various AChR subunits	Any time from early childhood to adult life	Generally good	Reduced cation flux through shortened channel openings
AChR slow channel syndrome	Various AChR subunits	Any time from early childhood to adult life	Poor; progressive weakness and wasting common	Excess cation flux through prolonged channel openings
AChE deficiency	ColQ that anchors AChE at NMJ	Usually early childhood	Poor; progressive weakness and wasting	Probably similar to slow channel syndrome

Note: ACh, acetylcholine; AChE, acetylcholine esterase; AChR, acetylcholine receptor; ColQ, NMJ, neuromuscular junction.

activations lead to excessive influx of calcium through the AChR (which is not selective for sodium) and an "end plate myopathy." Little is known about the mechanisms involved, but transgenic mouse models reproducing many of the features of the human disease are now available [39]. There is also a fast channel syndrome, in which the AChR ion channel closes too quickly.

IV. What Determines the Phenotypic Variability in Neuromuscular Junction Disorders?

In all conditions discussed here, including the genetic conditions, there is considerable phenotypic variability among patients. In the autoimmune conditions particularly there can be marked variability from day to day. How can one explain these features? Because neuromuscular transmission depends on the EPP reaching a threshold, it is an all-or-none process (except in some extraocular muscles) in each fiber. Anything that increases the EPPs sufficiently for threshold to be reached in a number of different fibers, previously inactive, may result in successful transmission. Conversely, if the EPP is just greater than the threshold, although weakness will not be present, any further decrease will lead to transmission failure.

In the autoimmune conditions, presumably some variability is in the nature of the autoantibodies themselves, although these are not likely to be responsible for daily variations. In addition, other features of the immune response, such as complement activity, could modify the pathogenic mechanisms. It now appears that there are complement regulators at the neuromuscular junction, and their concentration may be important in determining the extent of complement-mediated damage. But it is also important to appreciate the geometry of the neuromuscular junction. The neuromuscular synaptic gap can be thought of as a tiny disklike space (30 μm diameter, 0.05 μm depth) with, on one side, a high density of AChRs (total number approximately 2×10^7). For these AChRs to be saturated by antibodies, a considerable flow of extracellular fluid through the synaptic space is required. Meanwhile, new AChRs are being synthesized, probably at an increased rate. The balance between antibody-mediated degradation and synthesis will be crucial, and this may depend partly on blood flow into the muscle and the delivery of antibodies (as well as the antibody level and affinity). The natural turnover of AChRs may also be variable and not only related to AChR synthesis. A series of studies on EAMG in rats found that older animals were not susceptible to induction of AChR loss. One determinant of this resistance is the amount of rapsyn at the neuromuscular junction. Rapsyn is an intracellular protein that anchors AChRs (see Fig. 2), and AChR loss in rats induced by injection of monoclonal antibody to AChR can be prevented by overexpression of rapsyn [40]. This means that differences in rapsyn expression in different muscles or in different individuals might influence susceptibility to antibody-mediated degradation. Another factor that might differ among individuals would be the ability of the presynaptic nerve terminal to up-regulate ACh release to compensate for impaired postsynaptic activity. In LEMS and NMT similar considerations apply, but in this situation the evidence suggests that compensatory increases in the expression of other VGCCs or VGKCs probably take place.

In the genetic disorders, there are the different questions of why patients with null alleles in the same subunit can be clinically diverse and how the phenotype can differ even within a family with the same genetic defect. The extent to which the AChR γ subunit is expressed and can be incorporated into α_2-, β-, and δ-containing AChRs may determine the degree of weakness experienced by patients with AChR ε subunit mutations. In AChE deficiency and the slow channel syndrome, intracellular mechanisms that determine the extent of excitotoxic damage are likely to be important. If more of the determinants of the phenotypic variability and compensatory mechanisms in all these conditions can be identified, it should help define new approaches to treatments.

References

1. Burden, S. J. (2002). Building the vertebrate neuromuscular synapse. *J. Neurobiol.* **53**, 501–511.
2. Patton, B. L. (2003). Basal lamina and the organization of neuromuscular synapses. *J. Neurocytol.* **32**, 883–903.
3. Fambrough, D. M., Drachman, D. B., and Satyamurti, S. (1973). Neuromuscular junction in myasthenia gravis: decreased acetylcholine receptors. *Science* **19**, 293–295.
4. Patrick, J., and Lindstrom, J. (1973). Autoimmune response to acetylcholine receptor. *Science* **25**, 871–872.
5. Toyka, K. V., Drachman, D. B., Griffin, D. E., Pestronk, A., Winkelstein, J. A., Fishbeck, K. H., and Kao, I. (1977). Myasthenia gravis. Study of humoral immune mechanisms by passive transfer to mice. *N. Engl. J. Med.* **296**, 125–131.
6. Keesey, J. (2005). How electric fish became sources of acetylcholine receptor. *J. Hist. Neurosci.* **14(2)**, 149–164.
7. Chu, N. S. (2005). Contribution of a snake venom toxin to myasthenia gravis: the discovery of alpha-bungarotoxin in Taiwan. *J. Hist. Neurosci.* **14(2)**, 138–148.
8. Vincent, A. (2002). Unravelling the pathogenesis of myasthenia gravis. *Nat. Rev. Immunol.* **2(10)**, 797–804.
9. Garchon, H. J. (2003). Genetics of autoimmune myasthenia gravis, a model for antibody-mediated autoimmunity in man. *J. Autoimmun.* **21(2)**, 105–110.
10. Tzartos, S. J., Barkas, T., Cung, M. T., Mamalaki, A., Marraud, M., Orlewski, P., Papanastasiou, D., Sakarellos, C., Sakarellos-Daitsiotis, M., Tsantili, P., and Tsikaris, V. (1998). Anatomy of the antigenic structure of a large membrane autoantigen, the muscle-type nicotinic acetylcholine receptor. *Immunol. Rev.* **163**, 89–120.
11. Vincent, A., Willcox, N., Hill, M., Curnow, J., MacLennan, C., and Beeson, D. (1998). Determinant spreading and immune responses to acetylcholine receptors in myasthenia gravis. *Immunol. Rev.* **164**, 157–168.
12. Drachman, D. B. (1994). Myasthenia gravis. *N. Engl. J. Med.* **330**, 1797–1810.
13. Engel, A. G. (1984). Myasthenia gravis and myasthenic syndromes. *Ann. Neurol.* **16**, 519–534.
14. Ruff, R. L., and Lennon, V. A. (1998). End-plate voltage-gated sodium channels are lost in clinical and experimental myasthenia gravis. *Ann. Neurol.* **43**, 370–379.
15. Plomp, J. J., Van Kempen, G. T., De Baets, M. B., Graus, Y. M., Kuks, J. B., and Molenaar, P. C. (1995). Acetylcholine release in myasthenia gravis: regulation at single end-plate level. *Ann. Neurol.* **37**, 627–636.
16. Roxanis, I., Micklem, K., McConville, J., Newsom-Davis, J., and Willcox, N. (2002). Thymic myoid cells and germinal center formation in myasthenia gravis; possible roles in pathogenesis. *J. Neuroimmunol.* **125**, 185–197.
17. Yu Wai Man, C. Y., Chinnery, P. F., and Griffiths, P. G. (2005). Extraocular muscles have fundamentally distinct properties that make them selectively vulnerable to certain disorders. *Neuromuscul. Disord.* **15(1)**, 17–23.
18. Kaminski, H. J., Li, Z., Richmonds, C., Lin, F., and Medof, M. E. (2004). Complement regulators in extraocular muscle and experimental autoimmune myasthenia gravis. *Exp. Neurol.* **189**, 333–342.
19. Jacobson, L., Polizzi, A., and Vincent, A. (1998). An animal model of maternal antibody-mediated arthrogryposis multiplex congenita (AMC). *Ann. N. Y. Acad. Sci.* **841**, 565–567.
20. Vincent, A., Bowen, J., Newsom-Davis, J., and McConville, J. (2003). Seronegative generalised myasthenia gravis: clinical features, antibodies, and their targets. *Lancet Neurol.* **2**, 99–106.
21. Hoch, W., McConville, J., Helms, S., Newsom-Davis, J., Melms, A., and Vincent, A. (2001). Auto-antibodies to the receptor tyrosine kinase MuSK in patients with myasthenia gravis without acetylcholine receptor antibodies. *Nat. Med.* **7**, 365–368.
22. Vincent, A., and Leite, M. I. (2005). Neuromuscular junction autoimmune disease: MuSK antibodies and treatments for myasthenia gravis. *Curr. Opin. Neurol.* **18(5)**, 519–525.
23. Shiraishi, H., Motomura, M., Yoshimura, T., Fukudome, T., Fukuda, T., Nakao, Y., Tsujihata, M., Vincent, A., and Eguchi, K. (2005). Acetylcholine receptors loss and postsynaptic damage in MuSK antibody-positive myasthenia gravis. *Ann. Neurol.* **57**, 289–293.
24. Farrugia, M. E., Kennett, R. P., Newsom-Davis, J., Hilton-Jones, D., and Vincent, A. (2006). Single-fiber electromyography in limb and facial muscles in muscle-specific kinase antibody and acetylcholine receptor antibody myasthenia gravis. *Muscle Nerve* **33**, 568–570.
25. Kong, X. C., Barzaghi, P., and Ruegg, M. A. (2004). Inhibition of synapse assembly in mammalian muscle in vivo by RNA interference. *EMBO Rep.* **5**, 183–188.
26. Farrugia, M. E., Robson, M. D., Clover, L., Anslow, P., Newsom-Davis, J., Kennett, R., Hilton-Jones, D., Matthews, P. M., and Vincent, A. (2006). MRI and clinical studies of facial and bulbar muscle involvement in MuSK antibody-associated myasthenia gravis. *Brain* **129**, 1481–1492.
27. Elmqvist, D., and Lambert, E. H. (1968). Detailed analysis of neuromuscular transmission in a patient with the myasthenic syndrome sometimes associated with bronchogenic carcinoma. *Mayo Clin. Proc.* **43**, 689–713.
28. Waterman, S. A., Lang, B., and Newsom-Davis, J. (1997). Effect of Lambert-Eaton myasthenic syndrome antibodies on autonomic neurons in the mouse. *Ann. Neurol.* **42**, 147–156.
29. Engel, A. G. (1991). Review of evidence for loss of motor nerve terminal calcium channels in Lambert-Eaton myasthenic syndrome. *Ann. N. Y. Acad. Sci.* **635**, 246–258.
30. Urbano, F. J., Piedras-Renteria, E. S., Jun, K., Shin, H. S., Uchitel, O. D., and Tsien, R. W. (2003). Altered properties of quantal neurotransmitter release at endplates of mice lacking P/Q-type Ca^{2+} channels. *Proc. Natl. Acad. Sci. USA* **100**, 3491–3496.

31. Pinto, A., Gillard, S., Moss, F., Whyte, K., Brust, P., Williams, M., Stauderman, K., Harpold, M., Lang, B., Newsom-Davis, J., Bleakman, D., Lodge, D., and Boot, J. (1998). Human autoantibodies specific for the alpha1A calcium channel subunit reduce both P-type and Q-type calcium currents in cerebellar neurons. *Proc. Natl. Acad. Sci. USA* **95**, 8328–8333.
32. Giovannini, F., Sher, E., Webster, R., Boot, J., and Lang, B. (2002). Calcium channel subtypes contributing to acetylcholine release from normal, 4-aminopyridine-treated and myasthenic syndrome auto-antibodies-affected neuromuscular junctions. *Br. J. Pharmacol.* **136(8)**, 1135–1145.
33. Newsom-Davis, J., Buckley, C., Clover, L., Hart, I., Maddison, P., Tuzum, E., and Vincent, A. (2003). Autoimmune disorders of neuronal potassium channels. *Ann. N. Y. Acad. Sci.* **998**, 202–210.
34. Hart, I. K., Maddison, P., Newsom-Davis, J., Vincent, A., and Mills, K. R. (2002). Phenotypic variants of autoimmune peripheral nerve hyperexcitability. *Brain* **125**, 1887–1895.
35. Kleopa, K. A., Elman, L. B., Lang, B., Vincent, A., and Scherer, S. S. (2006). Neuromyotonia and limbic encephalitis sera target mature Shaker-type K+ channels: subunit specificity correlates with clinical manifestations. *Brain* **129**, 1570–1584.
36. Tomimitsu, H., Arimura, K., Nagado, T., Watanabe, O., Otsuka, R., Kurono, A., Sonoda, Y., Osame, M., and Kameyama, M. (2004). Mechanism of action of voltage-gated K^+ channel antibodies in acquired neuromyotonia. *Ann. Neurol.* **56(3)**, 440–444.
37. Arimura, K., Sonoda, Y., Watanabe, O., Nagado, T., Kurono, A., Tomimitsu, H., Otsuka, R., Kameyama, M., and Osame, M. (2002). Isaacs' syndrome as a potassium channelopathy of the nerve. *Muscle Nerve* **Suppl 11**, S55–S58. Review.
38. Engel, A. G., Ohno, K., and Sine, S. M. (2003). Sleuthing molecular targets for neurological diseases at the neuromuscular junction. *Nat. Rev. Neurosci.* **4**, 339–352.
39. Vohra, B. P., Groshong, J. S., Maselli, R. A., Verity, M. A., Wollmann, R. L., and Gomez, C. M. (2004). Focal caspase activation underlies the endplate myopathy in slow-channel syndrome. *Ann. Neurol.* **55**, 347–352.
40. Losen, M., Stassen, M. H. W., Martinez-Martinez, P., Machiels, B. N., Duimel, H., Fredrik, P., Veldman, H., Wokke, J. H. J., Spans, F., Vincent, A., and De Baets, M. G. (2005). Increased expression of rapsyn in muscles prevents acetylcholine receptor loss in experimental autoimmune myasthenia gravis. *Brain* **128(Pt. 10)**, 2327–2337.

86

Metabolic Myopathies

Michio Hirano, MD

Keywords: *adenosine triphosphate, coenzyme Q10, fatty acid, glycogen, metabolic, mitochondria, mitochondrial deoxyribonucleic acid, mutation, myopathy, oxidative phosphorylation, respiratory chain*

I. Brief History and Nomenclature
II. Etiology
III. Pathogenesis
IV. Pathophysiology
V. Pharmacology, Biochemistry, and Molecular Mechanisms
VI. Explanation of Symptoms in Relation to Pathophysiology
VII. Natural History
   References

I. Brief History and Nomenclature

Metabolic myopathies comprise a broad group of disorders caused by defects in the biochemical pathways that produce adenosine triphosphate (ATP) [1]. It is not surprising that defects of ATP synthesis cause diseases of muscle, because this tissue consumes large amounts of energy.

The concept of metabolic myopathy originated in 1951 with Brian McArdle's description of a patient with ischemic exercise-induced painful muscle contractures without electrical activity by electromyography [2]. McArdle noted that the patient's venous lactate and pyruvate did not rise after exercise; therefore, he correctly deduced that the disorder was caused by a defect in glycogen conversion to lactate. Nine years later, deficiency of myophosphorylase, a muscle-specific enzyme that catalyzes the first step of glycogen breakdown, was identified as the cause of McArdle disease. In 1962, Rolf Luft reported the first mitochondrial myopathy patient—a young woman with severe hypermetabolism and normal thyroid function [3]. Ultrastructural analyses revealed massive proliferation of mitochondria, and biochemical studies revealed loose coupling between oxygen consumption and phosphorylation (ATP synthesis) in isolated mitochondria from muscle. In 1973, Salvatore DiMauro identified the first myopathy caused by a defect in fatty acid metabolism in two brothers with recurrent myoglobinuria and deficiency of carnitine palmitoyl-transferase (CPT) [4]. At that time, activities of CPT I and

CPT II could not be distinguished, but subsequently this disease was linked to CPT II deficiency. These three seminal papers described the first defects of glycogen metabolism, mitochondrial oxidative phosphorylation, and fatty acid metabolism; dysfunction of these three biochemical pathways is responsible for most metabolic myopathies.

II. Etiology

Metabolic myopathies are genetic disorders, predominantly inherited in an autosomal recessive manner and typically caused by severe loss of function of an enzyme. Rare metabolic myopathies are caused by autosomal dominant mutations, which can produce toxic gain of function or haploinsufficiency (partial loss of function). X-linked recessive metabolic myopathies affect male patients but may occasionally affect women because of skewed lyonization. An astonishingly large number of maternally inherited myopathies are caused by mutations in mitochondrial deoxyribonucleic acid (mtDNA). Finally, some sporadic cases of metabolic myopathy are caused by spontaneous mutations that arise either in germline cells or in early embryogenesis.

III. Pathogenesis

Numerous causes of metabolic myopathy exist because of the many biochemical reactions required to produce cellular energy, mainly in the form of ATP. Skeletal muscle generates ATP from three major sources: fatty acids, glycogen, and high-energy phosphate compounds such as phosphocreatine. The type and duration of muscle activity dictate the relative proportions of energy derived from these three sources. At rest, fatty acid oxidation accounts for the bulk of ATP production. During the first several minutes of moderate exercise, high-energy phosphate compounds regenerate ATP from adenosine diphosphate. After 5–10 minutes of exercise, glycogen becomes the major energy source, and after longer periods of exertion, fatty acids are the predominant sources of ATP. As a consequence, defects of glycogen breakdown, for example, myophosphorylase deficiency (McArdle disease), are evident after relatively short intervals of moderate to intense exercise, such as walking uphill or sprinting short distances. In contrast, disorders of fatty acid metabolism, such as CPT II deficiency, are more likely to produce symptoms after prolonged periods of exercise.

IV. Pathophysiology

To describe the numerous metabolic steps involved in the breakdown of fatty acid and glycogen to form ATP is beyond the scope of this chapter; refer to other sources for further details [5]. Nevertheless, a brief overview of the major biochemical pathways may be helpful to place the various metabolic myopathies into a broader perspective. These processes include lipid metabolism, glycogen metabolism, the Krebs or tricarboxylic acid cycle, and oxidative phosphorylation.

Most of the body's lipids are stored in adipocytes, which release free fatty acids that are then incorporated into muscle and other cells. Short-chain 4-carbon and medium-chain 8-carbon fatty acids freely cross the mitochondrial membranes, and long-chain 14-carbon and very long-chain 24-carbon fatty acids must be transported into mitochondria by CPT I and CPT II. Within the mitochondrial matrix, fatty acids undergo β-oxidation to form acetyl coenzyme A (acetyl-CoA) which, in turn, is processed by the Krebs or tricarboxylic acid cycle (Fig. 1). Therefore, metabolic disorders of lipid metabolism can be caused by defects of fatty acid transport or β-oxidation.

In contrast to lipids, most of the body's glycogen is contained in liver and muscle. Liver stores of glycogen are mainly responsible for maintaining blood glucose levels, and muscle glycogen is mainly used for internal consumption to produce ATP. Glucose is osmotically active; therefore, high concentrations of glucose cannot be stored. However, large amounts of the relatively insoluble glycogen can be sequestered in cells. Glycogen constitutes about 1% of the total muscle mass and is maintained within a narrow range of concentrations by a balance between glycogen synthesis and glycogen degradation. Defects in either process can cause clinical disorders. Glycogen synthesis involves several enzymes and is controlled mainly by glycogen synthetase, which adds glucosyl residues from uridine diphosphate glucose to the end of glycogen molecules. Every 8–12 residues, branches of glycogen are added to the stems to produce highly branched glycogen molecules. Breakdown of glycogen is primarily regulated by phosphorylase, which liberates glucose-1-phosphate from glycogen (Fig. 2). Glucose is degraded by a series of reactions called glycolysis to form pyruvate, which enters the mitochondria, where it is converted by pyruvate dehydrogenase complex to acetyl-CoA.

Within the mitochondrial matrix (the innermost compartment of mitochondria), acetyl-CoA, derived from both glycolysis and β-oxidation, enters the Krebs cycle to produce reduced nicotinamide adenine dinucleotide and reduced flavin adenine dinucleotide, which transfer reducing equivalents (electrons) to the mitochondrial respiratory chain that leads to ATP synthesis through oxidative phosphorylation (see Fig. 1). The transfer of electrons through complexes I–IV of the respiratory chain is coupled to transfer of protons pumping from the mitochondrial matrix into the intermembrane space, resulting in an electrochemical gradient across the inner mitochondrial membrane. This

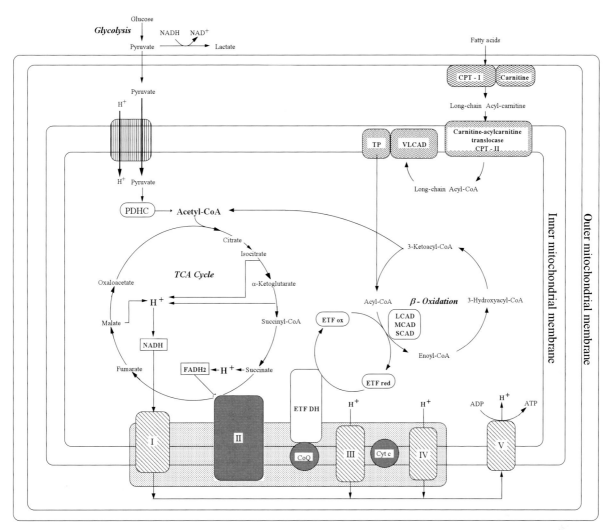

Figure 1 Schematic Representation of Mitochondrial Metabolism. *Note:* ADP, adenosine diphosphate; ATP, adenosine triphosphate; CoA, coenzyme A; CoQ, coenzyme Q; CPT, carnitine palmitoyltransferase; Cyt *c*, cytochrome *c*; ETF, electron transfer flavoprotein; $FADH_2$, flavin adenine dinucleotide; LCAD, long-chain acetyl-coenzyme A dehydrogenase; MCAD, medium-chain acetyl-coenzyme A dehydrogenase; NADH, nicotinamide adenine dinucleotide; PDHC, pyruvate dehydrogenase complex; SCAD, short-chain acetyl-coenzyme A dehydrogenase; TCA, tricarboxylic acid; TP, trifunctional protein; VLCAD, very long-chain acetyl-coenzyme A dehydrogenase. Roman numerals refer to mitochondrial respiratory chain enzymes.

proton gradient is used by complex V of the respiratory chain to produce ATP.

Because energy metabolism requires numerous enzymatic reactions and transporting functions, it is not surprising that many biochemical defects can cause metabolic myopathies.

V. Pharmacology, Biochemistry, and Molecular Mechanisms

In this section, the main biochemical and clinical features of several metabolic myopathies are described according to etiologies.

A. Defects of Glycogen Metabolism

During moderate- to high-intensity exercise (70%–80% of maximum aerobic capacity), glycogen is the major source of stored energy for ATP production. When glycogen metabolism is defective, the supply of substrate for energy production cannot meet the demands of intense exercise. The resulting symptoms include exercise intolerance, muscle pain, and myoglobinuria. Other organ systems depend on glycogen metabolism; therefore, liver, kidney, heart, and brain can also be affected. Fasting hypoglycemia is another manifestation.

Of the 12 glycogenoses, 10 have been associated with neuromuscular syndromes (see Fig. 2) [5,6]. The two forms without neuromuscular involvement are type I, caused by

1. Generalized Glycogen Storage Diseases

Glycogen storage disease type II, AMD, is distinct from the other forms of glycogen metabolism defects in that the enzyme is located in lysosomes rather than in the cytosol. As a consequence, in AMD glycogen accumulates mainly within membrane-bound vacuoles in muscle. Clinically, three forms have been defined by ages at onset: infantile, childhood, and adult AMD. Infantile AMD (Pompe's disease) begins in the first weeks or months of life with diffuse hypotonia and weakness including respiratory muscles; macroglossia is common. There is marked cardiomegaly and mild hepatomegaly. The disease is invariably fatal before 2 years of age because of pulmonary and cardiac failure. Childhood and adult AMD are primarily disorders of skeletal muscle manifesting as slowly progressive weakness. The childhood form can resemble Duchenne muscular dystrophy. Respiratory muscle weakness out of proportion to the limb weakness can be a clinical clue to the diagnosis.

Debrancher deficiency, type III, is clinically characterized by childhood onset, liver dysfunction with hepatomegaly, growth failure, fasting hypoglycemia, and occasionally hypoglycemic seizures. Symptoms tend to spontaneously resolve around puberty; however, cirrhosis and hepatic failure may develop. Type IIIa affects both muscle and liver, whereas type IIIb spares muscle. The myopathy primarily affects distal leg and intrinsic hand muscles and is slowly progressive and rarely incapacitating. Normally, after muscle phosphorylase has shortened the peripheral chains of glycogen to about 4-glucosyl units, producing a partially digested polysaccharide called *phosphorylase-limit dextrin;* debrancher enzyme removes the residual oligosaccharide "twigs."

Branching enzyme deficiency (Andersen's disease, type IV), is typically a severe, rapidly progressive disease of infancy clinically dominated by liver dysfunction with hepatosplenomegaly, cirrhosis, and death by 4 years of age because of hepatic failure or gastrointestinal bleeding [7]. In addition, branching enzyme deficiency can begin with neuromuscular dysfunction in infants, children, or adults manifesting as myopathy, cardiomyopathy, or both. In infants, muscle wasting, hypotonia, and contractures are prominent. In Askenazi Jewish patients with adult polyglucosan body disease, brancher enzyme deficiency has been identified in leukocytes. Adult polyglucosan body disease is characterized by late-onset (fifth or sixth decade) progressive upper and lower motor neuron degeneration, sensory neuropathy, bladder and bowel incontinence, and in some patients, dementia. Branching enzyme catalyzes the final step of glycogen synthesis by adding short new glucosyl branches (about 7 glucosyl units) to linear peripheral chains of glycogen.

PBK deficiency (type VIII) has been associated with four distinct phenotypes based on the mode of inheritance and tissue involvement: (1) liver disease, typically a benign

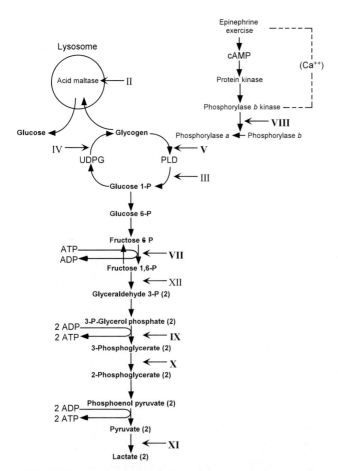

Figure 2 Scheme of Glycogen Metabolism and Glycolysis. Roman numerals denote muscle glycogenoses caused by defects in the following enzymes: II, acid maltase; III, debrancher; IV, brancher; V, myophosphorylase; VII, phosphofructokinase; VIII, phosphorylase *b* kinase; IX, phosphoglycerate kinase; X, phosphoglycerate mutase; XI, lactate dehydrogenase; and XII, aldolase. Bold numerals designate glycogenoses associated with exercise intolerance, pain, and myoglobinuria. Non-bold numerals correspond to glycogenoses causing weakness. *Note:* ADP, adenosine diphosphate; ATP, adenosine triphosphate; cAMP, cyclic adenosine monophosphate; PLD, phosphorylase-limit dextrin; UDPG; uridine diphosphate glucose.

glucose-6-phosphatase deficiency and causing liver and kidney dysfunction, and type VI, caused by hepatic phosphorylase deficiency and affecting liver and erythrocytes. Four types of glycogen storage disease have exclusively presenting skeletal muscle symptoms: type V, muscle phosphorylase deficiency; type VII, phosphofructokinase deficiency; type X, muscle phosphoglycerate mutase deficiency; and type XI, muscle lactate dehydrogenase deficiency. Type II, acid maltase deficiency (AMD), and type VIII, phosphorylase *b* kinase (PBK) deficiency, have multiple clinical phenotypes and can have myopathy or multiorgan disease. The other glycogen storage diseases, including aldolase deficiency, are generalized disorders with multiple tissue involvement.

condition of infancy or childhood with hepatomegaly, growth retardation, delayed motor development, and fasting hypoglycemia and usually inherited as an X-linked trait; (2) liver and muscle disease with a static myopathy inherited as an autosomal-recessive trait; (3) myopathy alone, also inherited in an X-linked recessive pattern; and (4) fatal infantile cardiomyopathy. In patients with myopathy, serum creatine kinase (CK) is variably increased. PBK is composed of four subunits (α, β, δ, and γ) and acts on two enzymes, glycogen synthetase and phosphorylase. Specifically, PBK converts phosphorylase from the less active *b* form to the more active *a* form while converting glycogen synthetase from a more active dephosphorylated form to a less active phosphorylated form.

Aldolase deficiency (type XII) was identified in a 4-year-old boy with exercise intolerance, mild weakness, developmental delay, hemolysis, and repeated episodes of rhabdomyolysis during febrile illnesses. The muscle biopsy did not reveal excess glycogen by histochemistry; however, it showed an isolated severe defect in aldolase activity by biochemical analysis (4% of normal). A homozygous point mutation was identified.

2. Glycogen Storage Diseases Primarily Affecting Skeletal Muscle

Myophosphorylase deficiency (McArdle disease, type V), muscle phosphofructokinase deficiency (Tarui's disease, type VII), phosphoglycerate mutase deficiency (type X), and lactate dehydrogenase A deficiency (type XI) have similar presenting myopathic features: exercise intolerance with increased fatigue, myalgia, and painful muscle contractions relieved by rest. Symptoms are more likely to occur during intense isometric exercise (lifting heavy weights) or during less intense but sustained dynamic exercise (walking up stairs). Patients with McArdle disease experience a "second wind" phenomenon. If these individuals slow or rest briefly at the onset of symptoms, they can resume exercising at the original pace. Patients often experience acute muscle necrosis and myoglobinuria after exercise, which, if severe, can cause acute renal failure. Some patients develop fixed weakness with time. Phosphofructokinase deficiency can also cause a hemolytic anemia, jaundice, and gouty arthritis, which can be helpful in making the diagnosis. PGK deficiency can be clinically asymptomatic or may manifest as hemolytic anemia, seizures, and mental retardation with or without myopathy. Isolated myopathy with intolerance of vigorous exercise, cramps, and myoglobinuria has been reported in a few patients with PGK deficiency. Mutations in the genes encoding the enzymes have been identified in all four of these disorders.

B. Defects of Lipid Metabolism

Lipids are the most important and efficient fuel source in the body. Fatty acids are particularly vital during periods of fasting, because liver glycogen stores are depleted within a few hours of a meal. The fatty acids serve three major functions: (1) partial oxidation of fatty acids in the liver produces ketones, which are an important auxiliary fuel for almost all tissues and especially the brain; (2) fatty acids constitute a major source of energy for cardiac and skeletal muscle, particularly during rest and during prolonged exercise; and (3) ATP produced from fatty acid oxidation provides energy for gluconeogenesis and ureagenesis.

1. Carnitine Deficiency

L-carnitine is a vital molecule for the transport of long-chain fatty acid into mitochondria. Other physiological functions of L-carnitine include buffering of the acyl-CoA/CoASH ratio, scavenging of potentially toxic acyl-CoA, and oxidation of branched-chain amino acids. About 75% of L-carnitine is derived from dietary sources, and the rest is synthesized in liver and kidney. Most total body carnitine (95%) is stored in muscle.

Primary deficiencies of L-carnitine manifest in three phenotypic forms: (1) dilated cardiomyopathy, (2) myopathy with lipid storage, and (3) hypoketotic hypoglycemia with recurrent encephalopathy [8,9]. Patients often show overlapping phenotypes. The age at onset of symptoms ranges from 1 month to 7 years with a mean of about 2 years. The dilated cardiomyopathy is progressive and ultimately fatal unless treated with L-carnitine supplementation. The myopathic form is the least common phenotype and is usually associated with a cardiomyopathy, encephalopathy, or both. The myopathy begins with motor delay, hypotonia, or slowly progressive proximal limb weakness. Acute metabolic encephalopathy is associated with hypoketotic hypoglycemia, usually in younger infants. Typically, the episodes are triggered by intercurrent illnesses and stress complicated by recurrent vomiting and decreased oral intake. Persistent central nervous system signs can develop because of severe hypoglycemic encephalopathy and cardiac or respiratory arrest.

Laboratory studies can reveal low serum carnitine concentrations (usually less than 10% of normal), and the acylcarnitine ester fraction is proportionally decreased. The diagnosis can be confirmed by documenting decreased carnitine uptake in cultured skin fibroblasts. Primary carnitine deficiency responds dramatically to oral L-carnitine therapy (in children daily doses of 100–200 mg/kg of body weight in divided doses or in adults 1000–3000 mg daily in divided doses). Although primary carnitine deficiencies are important to identify, secondary carnitine deficiencies are more common and are caused by a variety of underlying defects. Causes of secondary carnitine deficiencies include defects of β-oxidation, malnutrition states, excessive carnitine loss (i.e., renal Fanconi syndrome), and valproic acid therapy, which causes iatrogenic carnitine excretion.

2. Carnitine Palmitoyltransferase Deficiency

Mitochondria contain two CPTs vital in the transport of long-chain fatty acids into mitochondria. CPT I is located in the inner aspect of the outer mitochondrial membrane, and CPT II is bound to the inner mitochondrial membrane.

CPT I deficiency begins in infancy with attacks of fasting-induced life-threatening hypoketotic hypoglycemia. The hypoglycemic episodes manifest as lethargy, coma, and seizures and may cause death. They may also cause psychomotor developmental delay, hemiplegia, or generalized epilepsy. Myopathy is not a typical manifestation, but it has been reported. The diagnosis is confirmed by demonstrating decreased CPT I activity in cultured fibroblasts, leukocytes, or hepatocytes; however, CPT I activity is normal in skeletal muscle, accounting for the absence of clinical myopathy. Two tissue-specific isoforms of CPT I exist: liver and muscle. Only mutations in the liver isoform have been reported.

CPT II deficiency, in contrast to CPT I deficiency, has variable clinical manifestations. Three forms of CPT II deficiency have been described: infantile, late infantile, and adult. The early-infantile phenotype is rare and occurs at birth with severe hypoketotic hypoglycemia and generalized steatosis, and it usually causes death within a few days. Multiple organ malformations are often present. The late-infantile hepatomuscular form is clinically similar to CPT I deficiency with acute, episodic fasting hypoglycemia and hypoketosis, lethargy, coma, and death. Seizures, hepatomegaly, cardiomegaly, arrhythmias, and pancreatitis have been described. In both infantile forms, CPT II activity is less than 10% of normal.

The adult form of CPT II deficiency is a common cause of exercise-induced myoglobinuria. The disorder typically begins in young adulthood with complaints of muscle pain and pigmenturia after prolonged exercise. Severe bouts of rhabdomyolysis with myoglobinuria can cause acute renal failure. As a presenting symptom, some infants have acute muscle breakdown induced by fever. Adult patients may also have acute rhabdomyolysis precipitated by fever or other stress. Lipid storage cardiomyopathy has also been reported. CPT activity is less than 30% of normal. Mutations in the CPT II gene have been identified.

3. β-Oxidation Defects

The breakdown of fatty acids in mitochondria requires two distinct but linked systems: the inner mitochondrial membrane portion, which metabolizes long-chain acyl-CoA, and the matrix β-oxidation spiral, which acts on medium- and short-chain acyl-CoA (SCAD).

a. β-Oxidation Defects of the Inner Mitochondrial Membrane System Very long-chain acyl-CoA dehydrogenase (VLCAD) deficiency begins in infancy with hypoketotic hypoglycemia, x hepatic steatosis, cardiomyopathy, and elevated plasma levels of long-chain acylcarnitines. Metabolic acidosis, dicarboxylic aciduria, and increased serum CK with myoglobinuria have also been noted. Mutations in the VLCAD gene have been identified.

Trifunctional protein is a multienzyme complex with three identified functions: long-chain 3-hydroxy-acyl-CoA dehydrogenase (LCHAD), enoyl-CoA hydratase, and 3-ketoacyl-CoA thiolase. Most patients with trifunctional protein defects have isolated LCHAD deficiency, and a small number of individual have had combined defects of all three enzymes. The clinical feature of LCHAD deficiency include infantile onset; Reye-like episodes; hypoketotic hypoglycemia with hepatic dysfunction; progressive myopathy, recurrent myoglobinuria, or both; cardiomyopathy; and sudden infant death syndrome.

b. β-Oxidation Defects of the Mitochondrial Matrix System The mitochondrial β-oxidation matrix system shortens the fatty acid backbone of acyl-CoA by two carbon fragments during each turn through the β-oxidation spiral. In this process, acetyl-CoA is produced and becomes a substrate for the Krebs cycle. In addition, electron transfer flavoprotein is reduced; reduced electron transfer flavoprotein provides reducing equivalents to the oxidative-phosphorylation pathway. Human diseases can be caused by defects in several steps of this mitochondrial pathway.

Defects of long-chain, medium-chain, and short-chain acyl-CoA dehydrogenases begin in infancy. Medium-chain acyl-CoA dehydrogenase (MCAD) is the most common defect of β-oxidation. MCAD typically begins in the first 2 years of life with fasting intolerance, nausea, vomiting, hypoketotic hypoglycemia, lethargy, and coma; however, clinical expression is variable and some patients are asymptomatic. MCAD activity in most tissues (fibroblasts, lymphocytes, and liver) is low: 2%–20% of normal. Early diagnosis and treatment can lead to good long-term outcome. Dietary therapy is aimed at avoidance of fasting and provision of adequate caloric intake. LCAD and SCAD deficiency have been documented in only a few patients.

Multiple acyl-CoA dehydrogenase (MAD) deficiency or glutaric aciduria type II is a clinical syndrome characterized by metabolic acidosis, hypoketotic hypoglycemia, a characteristic odor, and early death. Three distinct clinical presentations of MAD deficiency exist: a severe neonatal form with congenital abnormalities; a severe neonatal form without congenital abnormalities; and a mild, later-onset form. The biochemical abnormality is reduced activity for various acyl-CoA dehydrogenases, resulting in urinary excretion of large amounts of numerous organic acids. The three defects that lead to MAD deficiency are electron transfer flavoprotein deficiency, electron transfer flavoprotein-coenzyme Q-oxidoreductase deficiency, and riboflavin (B_2) responsive MAD. In patients with the riboflavin-responsive MAD, vitamin B_2 supplementation (100 mg daily) can produce dramatic clinical improvements.

C. Mitochondrial Respiratory Chain Defects

Mitochondria are unique mammalian organelles because they possess their own genetic material. MtDNA is a small (16.5 kilobases) circular molecule encoding 22 transfer RNAs (tRNA), 13 polypeptides, and 2 ribosomal RNAs. The mtDNA-encoded polypeptides are functionally important because they are subunits of the respiratory or electron transport chain. However, most mitochondrial proteins are encoded in the nuclear DNA (nDNA). Thus mitochondria are the products of two genomes and defects in either genome can cause mitochondrial dysfunction [5,10]. To date, most respiratory chain defects that have been characterized at the molecular genetic level are caused by mtDNA mutations; however, the number of identified nDNA mutations is growing rapidly.

An important principle of mtDNA genetics is *heteroplasmy*. Each mitochondrion contains 2–10 copies of mtDNA and, in turn, each cell contains multiple mitochondria. Alterations of mtDNA may be present in some of the mtDNA molecules of a cell (heteroplasmy) or in all of the molecules (homoplasmy). As a consequence of heteroplasmy, the proportion of mtDNAs harboring a deleterious mtDNA mutation can vary widely among patients and in different cells of an individual. An individual who harbors a large proportion of mutant mtDNA will be more severely affected by the mitochondrial dysfunction than a person with a low percentage of the same mutation; therefore, the spectrum of clinical severity among patients with a given mitochondrial mutation is wide.

A second factor that can influence the expression of an mtDNA mutation is the *tissue distribution* of that mutation. The best example of tissue distribution variation involves large-scale mtDNA deletions. Infants with a high proportion of deleted mtDNA in their blood can develop Pearson's syndrome, consisting of sideroblastic anemia often accompanied by exocrine pancreatic dysfunction. Presumably, these infants have a high proportion of deleted mtDNA in the bone marrow stem cells. Some children survive the anemia with blood transfusions and subsequently recover because the stem cells with a high proportion of deleted mtDNA are under a negative selection bias. Later in life, however, those children may develop the multisystem mitochondrial disorder, Kearns-Sayre syndrome (KSS), characterized by ophthalmoplegia, pigmentary retinopathy, and cardiac conduction block. Thus the variable tissue distribution broadens the clinical spectrum of pathogenic mtDNA mutations.

The third factor that determines clinical manifestations of an mtDNA mutation is tissue *threshold effect*. Cells with high metabolic activities are severely and adversely affected by mtDNA mutations; therefore, these disorders tend to affect disproportionately brain and muscle (encephalomyopathies). The threshold effect is well illustrated by the mtDNA T8993G mutation in the ATP synthase-6 gene. When the mutation is present at 70%–90% heteroplasmy, patients develop neuropathy, ataxia, and retinitis pigmentosa, but when present at higher levels, infants or young children develop maternally inherited Leigh disease, a devastating multisystem disease that mainly affects the brain and causes subacute necrotizing encephalopathy.

A fourth unusual characteristic of mtDNA is *maternal inheritance*. During the formation of the zygote, the mtDNA is derived exclusively from the oocyte. Thus mtDNA is transmitted vertically in a non-Mendelian fashion from the mother to both male and female progeny. This inheritance pattern is important to recognize in determining whether a family is likely to harbor an mtDNA mutation. A caveat to this principle is that maternal relatives who have a lower percentage of an mtDNA mutation may have fewer symptoms (oligosymptomatic) than the proband or they may even be asymptomatic. Therefore, in taking the family history, it is important to ask about the presence of subtle symptoms or signs in maternally related family members who might be oligosymptomatic.

These peculiar features of "mitochondrial genetics" contribute to the clinical complexity of human mitochondrial disorders. Variable heteroplasmy of mtDNA mutations produces an extensive range of disease severity, and tissue distribution and tissue threshold of mtDNA mutations explain the frequent but variable involvement of multiple organ systems. Although most mtDNA mutations are heteroplasmic at varying levels throughout the body, pathogenic homoplasmic mtDNA mutations have also been identified and are sometimes found predominantly or exclusively in skeletal muscle causing myopathy.

In addition to mtDNA mutations, nDNA defects can cause mitochondrial dysfunction. In fact, nDNA encodes most of the mitochondrial components. Mutations of nDNA have been identified in genes encoding polypeptide subunits of the respiratory chain enzymes or ancillary factors required for the assembly of these multisubunit complexes. Finally, a third group of genetic mitochondrial disorders includes defects of intergenomic communication caused by mutations of nDNA genes regulating replication and expression of the mitochondrial genome.

1. Mitochondrial Deoxyribonucleic Acid Mutations

The mitochondrial encephalomyopathies comprise a heterogeneous group of multisystem disorders. The discovery of distinct mtDNA mutations has demonstrated that, in general, clinical phenotypes have specific genotypes; however, some patients do not fit into any clinical syndrome or have an atypical presentation for a particular mtDNA mutation. Because clinicians are confronted with patients, the clinical classification of the mitochondrial disorders has pragmatic significance in guiding diagnostic evaluation and directing therapy.

Rowland and colleagues defined KSS by the obligate triad of ophthalmoplegia, pigmentary retinopathy, and onset before age 20, accompanied by at least one of the following features: cardiac conduction block, ataxia, and cerebrospinal fluid protein greater than 100. The existence of KSS is supported by the more than 150 patients with these characteristics that have been reported. About 90% of the KSS patients have a single large-scale rearrangement of mtDNA, which may consist of deletions, duplications, or both. Typically, KSS patients are sporadic, because the mtDNA rearrangements seem to originate in oogenesis or early in embryogenesis.

In contrast to KSS, myoclonus epilepsy ragged-red fibers (MERRF) syndrome is maternally inherited. The disease is clinically defined by seizures, myoclonus, ataxia, and ragged-red fibers in the muscle biopsy. Other common clinical manifestations associated with MERRF are hearing loss, dementia, peripheral neuropathy, short stature, exercise intolerance, lipomas, and lactic acidosis. Although most MERRF patients have a history of affected maternally related family members, not all may have the full syndrome. An A-to-G transition mutation at nucleotide 8344 (A8344G) of the mtDNA tRNA^{Lys} gene has been found in about 90% of MERRF patients tested. Within families of an MERRF proband, oligosymptomatic and asymptomatic members harbor the same mtDNA mutation, but the phenotype is presumably attenuated by heteroplasmy and tissue distribution of the mtDNA mutation.

Mitochondrial encephalomyopathy, lactic acidosis, and strokelike episodes (MELAS) syndrome is another maternally inherited disorder whose defining clinical features include (1) strokelike episodes at a young age (typically before 40 years); (2) encephalopathy manifested by seizures, dementia, or both; and (3) mitochondrial dysfunction with lactic acidosis and ragged-red fibers [11]. In addition, to secure the diagnosis, at least two of the following clinical features should be present: normal early development, recurrent headaches, or recurrent vomiting. Other commonly encountered manifestations include myopathic weakness, exercise intolerance, myoclonus, ataxia, short stature, and hearing loss. It is uncommon for more than one family member to have the full MELAS syndrome. In most pedigrees, there is only one MELAS patient, whereas other relatives in the maternal lineage remain oligosymptomatic or asymptomatic. A mutation in the tRNA$^{Leu(UUR)}$ gene (A3243G) has been identified in about 80% of the MELAS patients.

In addition to these three phenotypes, many other clinical syndromes have been associated with defects of oxidative-phosphorylation. Despite the complexity and the heterogeneity of mitochondrial disorders, there are several clinical themes that are common. First, the disorders tend to affect children and young adults. Second, the syndromes are often multisystemic and usually affect brain, skeletal muscle, or both. Third, maternal inheritance is pathognomonic of mtDNA point mutations, and patients with single large-scale rearrangements tend to be sporadic. Fourth, there is great variability of phenotypic expression in families with mtDNA point mutations. Fifth, most pathogenic mtDNA mutations affect tRNA genes and therefore affect mitochondrial protein synthesis.

Nonetheless, as with most "rules" of clinical medicine, exceptions exist. In contrast to the prototypical maternally inherited multisystem mitochondrial diseases with heteroplasmic mtDNA mutations, a syndrome of sporadic exercise intolerance with homoplasmic or heteroplasmic mtDNA mutations in skeletal muscle has been described in about 20 patients. Interestingly, these patients harbor mutations in polypeptide coding rather than in tRNA genes.

2. Nuclear Deoxyribonucleic Acid Mutations

The first nDNA causing a mitochondrial respiratory chain defect was described in 1995 in a pair of siblings with Leigh disease who were found to have compound heterozygous mutations in the flavoprotein subunit of complex II [12]. Since then, autosomal recessive mutations encoding structural subunits of complexes I and II have been identified in a number of patients, mainly with Leigh disease [13]. Curiously, sequencing of nDNA genes encoding complexes III, IV (cytochrome *c* oxidase, or COX), and V subunits have not revealed mutations; however, defects have been identified in genes encoding factors required for the assembly of these multisubunit complexes. Mutations in the nuclear-encoded ancillary protein, BCS1L, have been identified in infants and young children with complex III deficiency and a multisystem disease designated by the acronym GRACILE (growth retardation, aminoaciduria, cholestasis, iron overload, lactic acidosis, and early death). By contrast, complex IV deficiency has been found to be caused by mutation in six genes encoding COX assembly factors: SURF1, SCO1, SCO2, COX10, COX15, and LRPPRC. SURF1 and LRPPRC mutations have been associated with Leigh disease, whereas patients with SCO2 mutations have had Leigh disease–like features with a severe hypertrophic cardiomyopathy. The one patient with a homozygous mutation in COX10 had a multisystem disorder characterized by encephalopathy, myopathy, and renal tubulopathy. Mutations in ATP12, an assembly factor for complex V, have been found to cause a rapidly fatal mitochondrial encephalomyopathy. These mutations are likely to be representative of a much larger pool of nuclear mutations causing dysfunction of the mitochondrial respiratory chain.

3. Defects of Intergenomic Communication

A growing number of autosomal diseases have been associated with secondary mtDNA alterations or defects of mitochondrial protein synthesis [14]. These disorders are thought to disrupt pathways that regulate the integrity and quantity of mtDNA. Autosomal dominant progressive external ophthalmoplegia with multiple mtDNA deletions was the first of these disorders to be identified. A second example, mtDNA depletion syndrome, was originally reported in infants with severe hepatopathy or myopathy. Mitochondrial neurogastrointestinal encephalomyopathy (MNGIE) was the first disease of intergenomic communication to be molecularly defined [15]. This autosomal recessive disorder is associated with depletion, multiple deletions, and point mutations of mtDNA and is caused by defects in the gene encoding thymidine phosphorylase, a cytosolic enzyme thought to contribute to the regulation of nucleotide pools in the mitochondria.

Additional nDNA mutations in genes encoding other factors involved in mitochondrial nucleotide metabolism have been identified [16]. Autosomal dominant progressive external ophthalmoplegia with multiple mtDNA deletions in muscle has been associated with mutations in the ANT1 gene, encoding the muscle and heart isoform of adenine nucleotide translocator; the Twinkle gene, encoding a mitochondrial helicase; and the POLG gene, encoding mitochondrial polymerase γ. POLG mutations also cause autosomal recessive progressive external ophthalmoplegia, a multisystem syndrome called SANDO (sensory ataxia, neuropathy, dysphagia, and ophthalmoplegia), a MNGIE-like phenotype, parkinsonism, and Alpers syndrome, an autosomal recessive disorder of childhood with hepatopathy, cerebral gray matter degeneration, and depletion of mtDNA. Three other genes encoding enzymes involved in mitochondrial nucleotide metabolism have been identified in young patients with mtDNA depletion. The myopathic form has been associated with mutations in the TK2 gene, which encodes the mitochondrial thymidine kinase; a hepatocerebral form is caused by mutations in the dGK gene, encoding deoxyguanosine kinase; and a severe encephalomyopathy is caused by mutation in SUCLA2, whose gene product interacts with the mitochondrial nucleotide diphosphate kinase.

In addition to affecting mtDNA, intergenomic communication defects may impair mitochondrial protein synthesis. A pair of siblings with consanguineous parents died in infancy of hepatocerebral dysfunction. They had a homozygous mutation in the EFG1 gene, which encodes an elongation factor required for mitochondrial protein synthesis. In addition, in an infant with dysmorphic features, hypotonia, limb edema, elevated liver transaminases, and lactic acidosis leading to death, a homozygous mutation was identified in the mitochondrial ribosomal protein subunit 16 gene (MRPS16).

D. Defects of the Lipid Milieu

The mitochondrial respiratory chain enzymes are embedded in the predominantly lipid inner membrane. Cardiolipin is an acidic lipoprotein that is abundant in the inner mitochondrial membrane and is required for normal respiratory chain enzyme activity. Synthesis of cardiolipin is decreased in Barth syndrome, an X-linked recessive disease characterized by cardiomyopathy, mitochondrial myopathy, and cyclic neutropenia. Barth syndrome is caused by mutations in the G4.5 gene, which encodes an acyltransferase required for cardiolipin synthesis. Another vital component of the mitochondrial inner membrane is coenzyme Q10 (CoQ10), which transfers electrons to complex III from complexes I and II. In addition, CoQ10 is a potent antioxidant and membrane stabilizer. Deficiency of CoQ10 has been associated with three major phenotypes: (1) a myopathic form with recurrent myoglobinuria, encephalopathy, and ragged-red fibers in muscle; (2) a predominantly ataxic form with marked cerebellar atrophy and other neurological features; and (3) a severe infantile encephalomyopathy with renal disease. All three forms of primary CoQ10 deficiency can be diagnosed by measurement of CoQ10 in skeletal muscle, and all forms respond to CoQ10 supplementation.

VI. Explanation of Symptoms in Relation to Pathophysiology

Impaired ATP production can cause progressive skeletal muscle weakness. Alternatively, metabolic myopathies can show presenting symptoms of exercise intolerance with recurrent, acute, and reversible episodes of muscle dysfunction that manifest as muscle cramps and breakdown of muscle (rhabdomyolysis), which, in turn, allows CK and myoglobin to spill into the patient's serum. When there is massive muscle breakdown, myoglobin passes through the kidneys and is visible as dark urine (pigmenturia). In some metabolic myopathies, both progressive weakness and episodic muscle dysfunction coexist.

VII. Natural History

The natural history of metabolic myopathies varies according to the specific etiology and severity of the biochemical defect. Many of these diseases have been well characterized at the biochemical level, and molecular genetics studies are rapidly revealing their causes. Expanding knowledge in this field will enhance our scientific understanding and may eventually translate into therapies for patients with these often devastating diseases.

References

1. DiMauro, S., Hays, A. P., and Tsujino, S. (2004). Nonlysosomal glycogenoses. *In:* "Myology" (A. G. Engel and C. Franzini-Armstrong, eds.), pp. 1535–1558. McGraw-Hill, New York.
2. McArdle, B. (1951). Myopathy due to a defect in muscle glycogen breakdown. *Clin. Sci.* **10**, 13–33.
3. Luft, R., Ikkos, D., Palmieri, G., Ernster, L., and Afzelius, B. (1962). A case of severe hypermetabolism of nonthyroid origin with a defect in the maintenance of mitochondrial respiratory control: a correlated clinical, biochemical, and morphological study. *J. Clin. Invest.* **41**, 1776–1804.
4. DiMauro, S., and DiMauro-Melis, P. M. (1973). Muscle carnitine palmityltransferase deficiency and myoglobinuria. *Science* **182**, 929–931.
5. DiMauro, S., and Hirano, M. (2005). Mitochondrial encephalomyopathies: an update. *Neuromuscul. Disord.* **15**, 276–286.
6. DiMauro, S., Servidei, S., and Tsujino, S. (2003). Glycogen storage diseases. *In:* "The Molecular and Genetic Basis of Neurologic and Psychiatric Disease" (R. N. Rosenberg et al., eds.), pp. 583–589. Butterworth-Heinemann, Philadelphia.
7. Bruno, C., van Diggelen, O. P., Cassandrini, D., Gimpelev, M., Giuffre, B., Donati, M. A., Introvini, P., Alegria, A., Assereto, S., Morandi, L., Mora, M., Tonoli, E., Mascelli, S., Traverso, M., Pasquini, E., Bado, M., Vilarinho, L., van Noort, G., Mosca, F., DiMauro, S., Zara, F., and Minetti C. (2004). Clinical and genetic heterogeneity of branching enzyme deficiency (glycogenosis type IV). *Neurology* **63**, 1053–1058.
8. Di Donato, S., and Taroni, F. (2003). Disorders of lipid metabolism. *In:* "The Molecular and Genetic Basis of Neurologic and Psychiatric Disease" (R. N. Rosenberg et al., eds.), pp. 591–601. Butterworth-Heinemann, Philadelphia.
9. Tein, I. (2003). Carnitine transport: pathophysiology and metabolism of known molecular defects. *J. Inherit. Metab. Dis.* **26**, 147–169.
10. DiMauro, S., and Schon, E. A. (2003). Mitochondrial respiratory-chain diseases. *N. Engl. J. Med.* **348**, 2656–2668.
11. Hirano, M., and Pavlakis, S. (1994). Mitochondrial myopathy, encephalopathy, lactic acidosis, and strokelike episodes (MELAS): current concepts. *J. Child Neurol.* **9**, 4–13.
12. Bourgeron, T., Rustin, P., Chretien, D., Birch-Machin, M., Bourgeois, M., Viegas-Pequignot, E., Munnich, A., and Rotig, A. (1995). Mutation of a nuclear succinate dehydrogenase gene results in mitochondrial respiratory chain deficiency. *Nature Genet.* **11**, 144–149.
13. Ugalde, C., Vogel, R., Huijbens, R., Van Den Heuvel, B., Smeitink, J., and Nijtmans, L. (2004). Human mitochondrial complex I assembles through the combination of evolutionary conserved modules: a framework to interpret complex I deficiencies. *Hum. Mol. Genet.* **13**, 2461–2472.
14. Hirano, M., Martì, R., Vilà, M. R., and Nishigaki, Y. (2004). MtDNA maintenance and stability genes: MNGIE and mtDNA depletion syndromes. *In:* "Mitochondrial Function and Biogenetics" (C. Köhler and M. F. Bauer, eds.), pp. 177–200. Springer-Verlag, Berlin.
15. Hirano, M., Nishigaki, Y., and Martì, R. (2004). Mitochondrial neurogastrointestinal encephalomyopathy (MNGIE): a disease of two genomes. *Neurologist* **10**, 8–17.
16. Spinazzola, A., and Zeviani, M. (2005). Disorders of nuclear-mitochondrial intergenomic signaling. *Gene* **354**, 162–168.

87

Immunobiology of Autoimmune Inflammatory Myopathies

Marinos C. Dalakas, MD

Keywords: *inclusion body myositis, major histocompatibility complex class I–associated signaling pathways, polymyositis, stress response, T cell–mediated cytotoxicity*

I. Introduction
II. Epidemiology and Immunogenetics
III. Clinicohistological Features
IV. Immunopathogenesis
V. Treatment: Present and Future
References

I. Introduction

The autoimmune inflammatory myopathies constitute a heterogeneous group of subacute, chronic, and rarely acute acquired diseases of skeletal muscle that have in common the presence of moderate to severe muscle weakness and inflammation in the muscle. The diseases are clinically important because they represent the largest group of acquired and potentially treatable myopathies. On the basis of clinical, demographic, histological, and immunopathological studies, the most common and clearly defined inflammatory myopathies, occurring in isolation or in association with other systemic disorders or viral infections, include polymyositis (PM), dermatomyositis (DM), and sporadic inclusion body myositis (IBM) [1–5]. This review describes progress in understanding of these disorders with emphasis on pathology, immunopathogenesis, and mechanisms of myofiber cell loss.

II. Epidemiology and Immunogenetics

DM affects both children and adults, and it affects women more often than men. PM is seen after the second decade of life, and it is rare in childhood; IBM is more frequent in men over the age of 50. The exact incidence of PM and DM is unknown; estimates ranging from 0.6 to 1 per 100,000 may not be reliable because few of these studies had taken into consideration the distinction between PM and IBM. In all age groups, DM is more common than PM; IBM is the most common in patients older than 50 years. The prevalence of IBM worldwide is also unknown. In Australia the reported prevalence of IBM is 9.3 per 1 million with an age adjusted prevalence of 35.3 per 1 million over the age of 50. A prevalence of 4.3 per 1 million has been reported in the Netherlands [1–5].

Genetic factors may play a role based on rare familial occurrences and association with certain HLA genes, especially DRB1 *0301 alleles for PM and IBM. The HLA-DQA1 0501 allele may confer a genetic risk for juvenile DM, and the tumor necrosis factor 308A polymorphism may contribute to the photosensitivity of DM. HLA-DR3 has a strong association with PM in patients who have Jo-1 antibodies.

In IBM patients there appears to be a strong association with the ancestral haplotype marked by DR3 and C4A*Q0, pointing to a susceptibility locus within the region close to major histocompatibility complex class II (MHC-II). An increased frequency of $DR\beta_1 0301$ and $DQ\beta_1 0201$ alleles associated with DR and DQ haplotypes was also documented in up to 75% of IBM patients. In a series of 52 IBM patients, 33% had other autoimmune disorders, such as multiple sclerosis, autoimmune thyroid disease, rheumatoid arthritis, or Sjögren's syndrome. Furthermore, the B8-DR3-DR52-DQ2 haplotype was found in 67% of the patients, regardless of whether they had another autoimmune disease. This frequency is similar to that seen in other autoimmune disorders, such as myasthenia gravis or myasthenic syndrome.

III. Clinicohistological Features

A. Dermatomyositis

DM occurs in both children and adults. It is a distinct clinical entity identified by a characteristic rash accompanying, or more often preceding, muscle weakness. The skin manifestations include a heliotrope rash (blue-purple discoloration) on the upper eyelids that is often associated with edema. Gottron papules are raised, violaceous, scaly lesions on the extensor surface of the metacarpophalangeal and interphalangeal joints. The erythematous rash also occurs at the knees, elbows, malleoli, neck, anterior chest (often in a V sign), back, and shoulders (in a "shawl" sign). The rash may be exacerbated after exposure to the sun, and it may be pruritic. When chronic, the rash becomes scaly with a shiny appearance. Dilated capillary loops at the base of the fingernails with irregular, thickened, and distorted cuticles are also characteristic. The lateral and palmar areas of the fingers may become rough with cracked, horizontal lines, resembling those common on mechanic's hands.

The weakness varies from mild to severe, leading to quadriparesis. At times the muscle strength appears normal, hence the term *DM sine myositis* or *amyopathic DM*. When muscle biopsy is performed in such cases, however, significant perivascular and perimysial inflammation can be seen. Whether subclinical myopathy is present in all cases of amyopathic DM or there are also cases limited to the skin is unclear. Amyopathic and myopathic DM are probably part of the spectrum of DM affecting skin and muscle to a varying degree. In addition to the primary myopathy, a number of extramuscular manifestations or coexisting conditions may be present, including the following:

1. *Systemic symptoms*, such as fever, malaise, weight loss, arthralgia, and Raynaud's phenomenon, especially when DM is associated with a connective tissue disorder
2. *Joint contractures*, especially in children
3. *Dysphagia and gastrointestinal symptoms* resulting from involvement of the oropharyngeal striated muscles and upper esophagus
4. *Cardiac disturbances*, including atrioventricular conduction defects, tachyarrhythmias, dilated cardiomyopathy, and low ejection fraction
5. *Pulmonary dysfunction* caused by primary weakness of the thoracic muscles, interstitial lung disease, or drug-induced pneumonitis (e.g., from methotrexate)
6. *Arthralgias, synovitis*, or *deforming arthropathy* of the interphalangeal joints with subluxation in some patients with anti-Jo-1 antibodies (described later)
7. *Subcutaneous calcifications*, sometimes extruding on the skin and causing ulcerations and infections, primarily in children

DM is also associated with an increase frequency of malignancies. Ovarian, gastrointestinal, lung, and breast cancers and non-Hodgkin lymphomas are the most common malignancies requiring continuous vigilance for early recognition, especially in older people and during the first 3 years of disease onset. DM also can occur as part of an "overlap" syndrome with systemic sclerosis and mixed connective tissue disease. Signs such as sclerotic thickening of the dermis, contractures, esophageal hypomotility, microangiopathy, and calcium deposits are present in both DM and scleroderma [1].

DM is easily suspected because of the characteristic skin rash, the myopathic muscle weakness, and the elevation of creatine kinase (CK), which can be mild or high. The diagnosis is confirmed with a muscle biopsy. In DM, the muscle fibers undergo phagocytosis and necrosis, often in groups (microinfarcts) involving a portion of a muscle fasciculus or at the periphery of the fascicle, resulting in perifascicular atrophy. Perifascicular atrophy, characterized by 2–10 layers of atrophic fibers at the periphery of the fascicles, is diagnostic of DM, even in the absence of inflammation. The inflammation is predominantly perivascular or in the interfascicular septae and around rather than within the fascicles. The intramuscular blood vessels show endothelial hyperplasia with tubuloreticular profiles, fibrin thrombi, and obliteration of capillaries. The active skin lesions also show perivascular inflammation with $CD4^+$ cells in the dermis. In chronic stages there is dilatation of superficial capillaries.

B. Polymyositis

PM is best defined as a subacute myopathy that evolves over weeks to months, affects adults but rarely children, and has a presenting symptom of weakness of the proximal muscles resulting in limitation of daily activities. Unlike DM, in which the rash secures early recognition, the actual onset of PM cannot be easily determined. PM mimics many myopathies and remains a diagnosis of exclusion. It is a subacute inflammatory myopathy affecting adults who do not have any of the following: rash, involvement of the extraocular and facial muscles, family history of a neuromuscular disease, history of exposure to myotoxic drugs or toxins, endocrinopathy, neurogenic disease, muscular dystrophy, biochemical muscle disorder (deficiency of a muscle enzyme), or IBM as excluded by muscle biopsy analysis (see Sect. III.C). It is increasingly recognized that PM, occurring in isolation, is a rare and rather overdiagnosed disorder; it is more often seen with a systemic autoimmune or connective tissue disease or with known viral or bacterial infections. The most common myopathy misdiagnosed as PM is IBM; this disease is often suspected in retrospect when a patient with presumed PM has not responded to therapy. Especially in men older than age 50, a PM-like disease should be considered as IBM until proved otherwise. Other myopathies misdiagnosed as PM include toxic and endocrine myopathies, DM sine dermatitis, and certain dystrophies. In a patient diagnosed as PM, the diagnosis should be challenged and reconsidered if the myopathic symptoms have developed slowly, especially before the age of 16, or if the symptoms consist predominantly of myalgia and fatigue rather than muscle weakness [1–5].

In patients with active disease the CK is elevated and the electromyograph shows an irritative myopathy with spontaneous activity and myopathic units or voluntary activation. The diagnosis is confirmed with a muscle biopsy that shows multifocal lymphocytic infiltrates surrounding and invading healthy muscle fibers. The inflammation is *primary*, a term coined to indicate that lymphocytes ($CD8^+$ cells) invade histologically healthy, MHC-I-expressing muscle fibers. I refer to this lesion as the *CD8/MHC-I complex*, as described later. In chronic stages, connective tissue is increased and may react with alkaline phosphatase. When, in addition to primary inflammation, there are vacuolated muscle fibers with basophilic granular deposits around the edges (rimmed vacuoles) and congophilic amyloid deposits within or next to the vacuoles, the diagnosis of IBM is likely.

C. Inclusion Body Myositis

In patients at least 50 years old, IBM is the most common of the inflammatory myopathies. It is often misdiagnosed as PM and suspected only retrospectively when a patient with presumed PM does not respond to therapy. Weakness and atrophy of the distal muscles, especially foot extensors and deep finger flexors, occur in almost all cases of IBM and may be a clue to early diagnosis. Some patients fall because their knees collapse from early quadriceps weakness. Others have a presenting symptom of weakness in the small muscles of the hands, especially finger flexors; they also complain of inability to hold certain objects, such as golf clubs, or perform certain tasks, such as turning keys or tying knots. On occasion, the weakness and accompanying atrophy can be asymmetrical and selectively involve the quadriceps, iliopsoas, triceps, biceps, and finger flexors, resembling a lower motoneuron disease. Dysphagia is common, occurring in up to 60% of IBM patients, and may lead to episodes of choking. Facial muscle weakness is also common and at times prominent. Sensory examination is generally normal; some patients have mildly diminished vibratory sensation at the ankles that presumably is age related. The distal weakness does not represent motoneuron or peripheral nerve involvement but results from the myopathic process affecting distal muscles as confirmed by macroelectromyographic analysis. The diagnosis is always made by characteristic findings on the muscle biopsy, as discussed later in this section. Disease progression is slow but steady, and most patients require an assistive device such as cane, walker, or wheelchair within several years of onset [1–5].

In at least 20% of cases, IBM is associated with systemic autoimmune or connective tissue diseases. Familial aggregation has also been noted in coaffected siblings with typical IBM; such cases have been designated *familial inflammatory IBM*. This disorder is distinct from *hereditary inclusion body myopathy* (h-IBM), which describes a heterogeneous group of recessive and less often dominant, inherited syndromes. The h-IBMs are noninflammatory myopathies with clinical profiles distinct from IBM. A subset of h-IBM that spares the quadriceps muscles has emerged as a distinct entity. This disorder, originally described in Iranian Jews and now seen in many ethnic groups, is linked to chromosome 9p1 and results from mutations in the uridine diphosphate-N-acetylglucosamine 2-epimerase/N-acetylmannosamine kinase (GNE) gene.

The diagnosis of IBM is confirmed with a muscle biopsy that shows the following: (1) intense endomysial inflammation with T cells invading MHC-I-expressing muscle fibers, in a pattern similar to (but often severer than) that seen in PM, that are almost never vacuolated or necrotic but rather appear to be healthy; (2) vacuolated fibers, not surrounded or invaded by T cells, that contain basophilic granular deposits distributed around the edge of slitlike vacuoles (rimmed vacuoles); (3) loss of fibers, replaced by fat tissue, increased connective tissue, hypertrophic fibers, and angulated or round fibers, scattered or in small groups; (4) eosinophilic cytoplasmic inclusion; (5) abnormal mitochondria characterized by the presence of ragged-red fibers or cytochrome-oxidase-negative fibers; (6) tiny congophilic amyloid deposits within or next to the vacuoles, best visualized by Texas-red fluorescent optics; and (7) characteristic filamentous inclusions seen by electron microscopy in the vicinity of the rimmed vacuoles. Although demonstration of the filaments by electron microscopy was previously thought to be essential for the diagnosis of IBM, this is not necessary if all other characteristic light-microscopic features, including amyloid deposits, are present [1–5].

It has become clear that IBM is a complex disease. The observation that the vacuolated muscle fibers are rarely invaded by T cells yet the apparently intact fibers are those characterized by primary inflammation and autoaggressive T cells has led my colleagues and I to propose that in IBM two processes occur in parallel: the primary immune process described later and a degenerative process characterized by vacuolization and amyloid-related degenerative molecules. The factors supporting the autoimmune pathogenesis of IBM include the following: (1) association with other autoimmune diseases and autoantibodies similar in frequency to those seen in classic autoimmune disorders such as MG; (2) occurrence of disease with the same sporadic phenotype in family members of the same generation (familial inflammatory IBM), as seen with other autoimmune disorders; (3) strong association in up to 70% of patients with HLA class I and II antigens with the autoimmune-prone HLA-B8-DR3 ancestral haplotype (irrespective of the presence of coexisting autoimmune diseases); (4) association with common variable immunodeficiency; (5) association with paraproteinemias in up to 22.8%, much beyond the frequency found in age-matched controls (2%), suggesting disturbed immunoregulation; (6) autoinvasive T cells that are antigen driven and form an immunological synapse, as described later; (7) frequent association with retroviruses as noted later; and (8) strong up-regulation of cytokines, chemokines, and their receptors at the protein, messenger ribonucleic acid (mRNA), and gene level (see Sect. IV.C).

The degenerative process of IBM is supported by the presence of vacuoles (almost always in fibers noninvaded by T cells) and deposits of β-amyloid within the vacuolated muscle fibers, which immunoreact for amyloid precursor protein (APP), chymotrypsin, presenilin-1, apolipoprotein E, and phosphorylated tau. It is unclear, however, whether these deposits directly contribute to disease pathogenesis or are secondary phenomena. Identical vacuoles are observed in several myopathies that lack inflammation, such as h-IBM due to GNE mutations, facioscapulohumeral muscular dystrophy, X-linked Emery-Dreifuss muscular dystrophy, rigid-spine syndrome, and myofibrillar myopathies; they were even observed in chronic neurogenic conditions such as postpolio. Similarly, the intracellular accumulation of AβPP, phosphorylated tau, presenilin-1, apolipoprotein E, γ-tubulin, clusterin, α-synuclein, gelsolin, oxidative stress proteins, and the catalytic core of the proteasome have been also found to be expressed in IBM and other vacuolar myopathies, leading to the conclusion that the accumulations are not unique to IBM. Although it remains unclear whether they are the primary or the secondary process in IBM, the accumulation of these proteins may be fundamental in understanding the ongoing stressor mechanisms that lead to muscle fiber degeneration, as discussed later.

IV. Immunopathogenesis

A. Autoantibodies

Autoantibodies directed against nuclear or cytoplasmic antigens, ribonucleoproteins involved in protein synthesis (antisynthetase antibodies), or translational transport (anti–signal-recognition particle antibodies), are found in approximately 20% of patients with DM or PM. These antibodies are useful clinical markers because of their frequent association with interstitial lung disease. The antibodies against histidyl-t-RNA synthetase, called anti-Jo-1, account for 80% of all antisynthetases. Anti-Jo-1 also seems to confer specificity for identifying a small subset of patients with the combination of myositis, nonerosive arthritis, "mechanic's hands," and Raynaud's phenomenon. The antisynthetase syndrome has a strong association with the DR3, DRW52, and DQA1*0501 HLA haplotypes. The significance of these antibodies remains unclear because they are not tissue or disease-subset specific, occur in less than 25% of PM or DM, are not pathogenic, and can be seen with equal frequency in patients with interstitial lung disease who do not have myositis. Other autoantibodies are anti-Mi-2, found in DM and PM; anti-PM-Scl, found in DM associated with scleroderma; and anti-KL6, associated with interstitial lung disease.

B. Signaling Pathways and Immunobiology of the Inflammatory Process

1. Dermatomyositis

The primary antigenic target in DM is the vascular endothelium of the endomysial capillaries. The disease begins when putative antibodies directed against endothelial cells activate complement C_3 that forms C_{3b} and C_{4b} fragments leading to the formation of the C5b-9 membranolytic attack complex (MAC), the lytic component of the complement pathway (Fig. 1). MAC, C_{3b}, and C_{4b} are detected early in the patients' serum and deposited on capillaries before inflammatory or structural changes are seen in the muscle. Sequentially, the complement-mediated alterations begin with swollen endothelial cells followed by vacuolization, necrosis of capillaries, perivascular inflammation, ischemia, and muscle fiber destruction. The characteristic perifascicular atrophy appears to be a reflection of the endofascicular hypoperfusion. Finally, there is marked reduction in the number of capillaries per muscle fiber with compensatory dilatation of the lumen of the remaining capillaries. The release of cytokines and chemokines related to complement activation up-regulates vascular cell adhesion molecule-1 and intercellular adhesion molecule-1 on the endothelial cells. These molecules serve as ligands for the integrins very late activation antigen-4 and leukocyte function-associated antigen-1, expressed on T cells and facilitate their exit through the blood vessel wall to the perimysial and endomysial spaces (see Fig. 1). Immunophenotypic analysis of the lymphocytic infiltrates demonstrates B cells, CD4$^+$ cells, and plasmacytoid and dendritic cells in the perimysial and perivascular regions, supporting the view that a humoral-mediated mechanism plays the major role in the disease. Other molecules up-regulated in DM include transforming growth factor-β (TGF-β) and, in the perifascicular regions, the cathepsins and signal transducer and activator of transcription-1, probably triggered by interferon-γ (IFN-γ). Using gene expression arrays, a number of adhesion molecules, cytokine genes, and chemokine genes are up-regulated in the muscles of DM patients. Most notable among those genes are the Kallmann syndrome-1 adhesion molecule, which is up-regulated *in vitro* by TGF-β and may

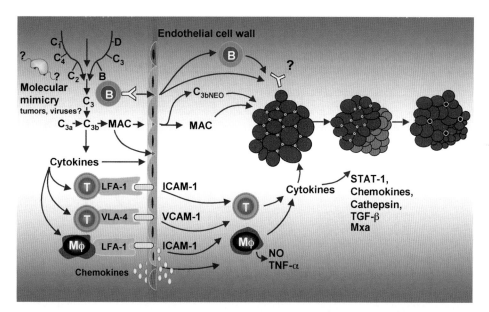

Figure 1 Sequence of Immunopathological Changes in Dermatomyositis. The disease probably begins with activation of complement and formation of C_3 through the classic or alternative complement pathway by putative antibodies against endothelial cells. Activated C_3 leads to formation of C_{3b}, C_{3bNEO}, and the membranolytic attack complex (MAC), which is deposited in and around the endothelial cell wall of the endomysial capillaries. Deposition of the MAC leads to destruction and a reduced number of capillaries, with ischemia or microinfarcts most prominent in the periphery of the fascicle. Finally, a reduced number of capillaries with a dilated diameter remain and perifascicular atrophy ensues. Complement-fixing antibodies, B cells, CD4$^+$ T cells, and macrophages traffic into the muscle. The migration of cells from the circulation is facilitated by vascular cell adhesion molecule-1 (VCAM-1) and intercellular adhesion molecule-1 (ICAM-1), whose expression on the endothelial cells is up-regulated by the released cytokines. T cells and macrophages bind to VCAM-1 and ICAM-1 through their integrins, very late activation antigen-4 (VLA-4) and leukocyte function-associated antigen-1 (LFA-1), and transgress to the muscle through the endothelial cell wall. *Note:* Mφ, macrophage; NO, nitric oxide; STAT-1, signal transducer and activator of transcription-1; TGF-α, transforming growth factor-α; TNF-α, tumor necrosis factor-α; Mxa, myxovirus resistance protein.

have a role in inducing fibrosis, and the myxovirus resistance protein [6,7]. This protein is also induced by IFN, and in DM it is predominantly immunolocalized in the perifascicular regions [6]. The cellular source of the abundant IFN-α or IFN-β is probably the large number of plasmacytoid dendritic cells, suggesting that in DM the innate immune response is involved, a pattern similar to that observed in systemic lupus erythematosus. Gene expression profiling in the muscles of children genetically susceptible to DM has also shown IFN-inducible genes, implying virus-driven autoimmune dysregulation [8]. However, no viruses have been amplified from the muscles of these patients.

2. Polymyositis and Inclusion Body Myositis

PM and IBM are T cell–mediated diseases in which CD8+ cells invade MHC-I antigen-expressing muscle fibers. The immune components associated with this process are similar in both PM and IBM, despite poor response to immunotherapies of the latter, as described later [1–5,7,9]. These components are discussed here and illustrated in Figure 2. Thus sensitized autoinvasive cytotoxic CD8+ T cells form an immunological synapse with muscle fibers expressing MHC-I antigen.

a. Major Histocompatibility Complex Expression In PM and IBM, the CD8+ cells, surround and invade healthy but MHC-I antigen-expressing, nonnecrotic muscle fibers. Muscle fibers normally do not express detectable amount of MHC-I or MHC-II antigens. In all inflammatory myopathies, however, widespread overexpression of MHC-I, and occasionally MHC-II, is seen even in areas remote from the inflammation. In other chronic myopathies or dystrophies, the muscle fibers do not express MHC-I antigen, or the costimulatory molecules described later, in

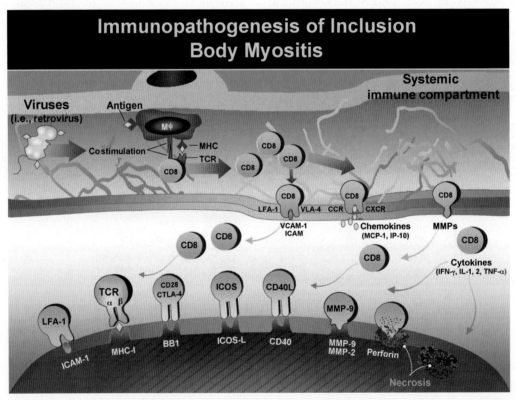

Figure 2 Immunopathogenesis of Inclusion Body Myositis. Binding of leukocyte function-associated antigen-1 (LFA-1) expressed by CD8+ T cells and intercellular adhesion molecule-1 (ICAM-1) expressed on the muscle fiber initiates the formation of an immunological synapse between the muscle fiber and the antigen-receiving T cell receptor (TCR) of the CD8+ T cells. Stimulation is supported and enhanced by the engagement of BB1, CD40, and additional costimulatory molecules with their CD28 ligand, cytotoxic T lymphocyte–associated protein-4 (CTLA-4), and CD40 ligand. Matrix metalloproteinases (MMPs) facilitate the migration of T cells and their attachment to the muscle surface. Muscle fiber necrosis occurs through the perforin granules released by the autoaggressive T cells. A direct myocytotoxic effect exerted by the released interferon-γ (IFN-γ), interleukin-1 (IL-1), or tumor necrosis factor (TNF) might also play a role. Death of the muscle fiber is mediated by a form of necrosis rather than apoptosis, presumably because of the counterbalancing effect or protection by antiapoptotic molecules. *Note:* ICOS-L, inducible (T cell) costimulator-ligand; MHC, major histocompatibility complex; Mϕ, macrophage; TGF, transforming growth factor; VCAM, vascular cell adhesion molecule; VLA, very late activation antigen.

a ubiquitous and consistent pattern. Furthermore, in dystrophies the occasional autoinvasive T cells are not antigen driven because they are clonally diverse, do not form the immunological synapse described later, and do not express the specific activation markers of cytotoxicity. In human myotubes, MHC molecules are up-regulated by IFN-γ and play a major role as a stress factor in the myofiber, as discussed later [1–5,7,9].

b. Activated Cytotoxic T Cells In PM and IBM the T cells are activated, as evidenced by their expression of memory and activation markers CD45RO and intercellular adhesion molecule-1, as well as MHC-I and inducible costimulator (ICOS) on their surface. Furthermore, the ICOS-positive autoinvasive CD8$^+$ T cells are cytotoxic, as indicated by their overexpression of perforin and granzyme granules at the protein and mRNA level. Release of these granules induces muscle fiber necrosis (see Fig. 2). Thus the perforin pathway seems to be the major cytotoxic effector mechanism. In contrast, the Fas–Fas-L-dependent apoptotic process is not functionally involved, despite expression of the Fas antigen on muscle fibers and of Fas-L on the autoinvasive CD8$^+$ cells. The coexpression of the antiapoptotic molecules Bcl-2, FLICE-like inhibitory protein, and human inhibitor of apoptosis-like protein may confer resistance of muscle to Fas-mediated apoptosis [1–5,7,9].

c. Rearrangement of the T Cell Receptor Gene of the Endomysial T Cells The T cells recognize an antigen through the T cell receptors (TCRs), a heterodimer of two α- and β-chains encoded by the V (variable), D (diversity), J (joining), and C (constant) regions of the TCR gene. The part of the TCR that recognizes an antigen is the CDR3 region, which is encoded by the V-J and V-D-J segments of the TCR gene. If the endomysial T cells are selectively recruited by a specific autoantigen, the use of the V and J genes of the TCR should be restricted and the amino acid sequence in their CDR3 region should be conserved. In patients with PM and IBM, but not in those with DM or dystrophies, only certain T cells that use specific TCR-α and TCR-β families are recruited into muscle from the circulation. Cloning and sequencing of the amplified endomysial or autoinvasive TCR gene families has demonstrated a restricted use of the Jβ gene with conserved amino acid sequence in the CDR3 region, indicating that these cells are specifically selected and clonally expanded *in situ*. Furthermore, these specific T cell clones express perforin, indicating their cytotoxic potential against the muscle. Immunocytochemistry combined with polymerase chain reaction and sequencing of the most prominent TCR families has shown that the autoinvasive, but not the perivascular, CD8$^+$ endomysial cells are the clonally expanded cells. Using a novel technique of CDR3 spectratyping combined with molecular laser-assisted microdissection and single-cell polymerase chain reaction, it was confirmed that the autoinvasive CD8 cells are clonally expanded and these clones can be traced to the circulation [10]. Sequential muscle biopsy specimens obtained during a 19- to 22-month period in three patients with IBM has further shown a persistent clonal restriction of the same Vβ families among the autoinvasive CD8$^+$ cells. These cells not only exhibited restricted usage of certain Vβ gene families but also had a conserved amino acid sequence homology in the complementary CDR3-determining region. Of interest, in the CDR3 region only a small number of amino acids was found, suggesting that the MHC-I-expressing muscle fibers present a limited number of antigenic peptides to the autoinvasive CD8$^+$ T cells during the course of the disease. The observation that in IBM identical T cell clones with restricted amino acid sequence in the CDR3 region belong to autoinvasive T cells and persist for several years even in different muscles has been confirmed in several studies. Collectively, in PM and IBM there is overwhelming evidence of clonal restriction of the autoinvasive endomysial T cells, which are specifically recruited to the muscle and appear to expand *in situ*, probably driven by the same antigen or antigens [2,4,5,7,9,10].

d. Presence of Costimulatory Molecules For the activation of T cells and antigen recognition, the presence of costimulatory molecules and their counterreceptors is fundamental. Autoinvasive CD8$^+$ T cells are primed to receive specific antigenic peptides presented by the MHC-I molecule expressed on the muscle fibers. For antigen presentation, however, these MHC-I-positive fibers should also express the B7 family of costimulatory molecules (B7-1, B7-2, BB1, or ICOS-L), and the autoinvasive CD8$^+$ T cells express the counterreceptors CD28, cytotoxic T lymphocyte–associated protein-4 (CTLA-4), or ICOS. Several studies have confirmed that in PM and IBM muscles the BB1 (CD80) protein is expressed on MHC-I-positive muscle fibers that contact the CD28 or CTLA-4 ligands on the autoinvasive CD8$^+$ T cells. The concept of the immunological synapse between CD8$^+$ cells and muscle fibers was reinforced by studies of ICOS-ligand (ICOS-L), another B7-family molecule associated with memory T cells. In PM and IBM, muscle fibers express ICOS-L, and their autoinvasive T cells express ICOS, supporting the view that muscle fibers can behave like antigen-presenting cells. Furthermore, the ICOS-positive T cells are cytotoxic, expressing perforin granules. Because ICOS-L is a functional molecule up-regulated by tumor necrosis factor-α, in PM and IBM the ICOS–ICOS-L interactions are fundamental in facilitating clonal expansion and costimulation of the effector functions of memory T cells [2,4,5,7,9,10].

Although the muscle fibers can behave like antigen-presenting cells, information about the hematopoietic dendritic cells, the most effective cells for antigen presentation, is slowly emerging. Within the endomysial infiltrates of PM and DM there are rare mature dendritic cells but abundant immature dendritic cells expressing the CCL2$^-$/CCR6 chemokine receptor complex. However, the microenvironment necessary for antigen presentation by dendritic cells, T cell interaction, and signaling in muscle is still unclear.

C. T Cell Transmigration, Cytokine Signaling, Chemokines, and Chemokine Receptors

Cytokines and chemokines are essential in enhancing the activation of T cells or induction of costimulatory molecules. Polymerase chain reaction studies of muscle tissue from PM, DM, and IBM have confirmed a varying degree of amplification of mRNA of the various cytokines, including interleukins 1 and 2, tumor necrosis factor-α, IFN-γ, TGF-β, granulocyte–macrophage colony-stimulating factor, and interleukins 6 and 10. Some of these cytokines can also exert a direct cytotoxic effect on the muscle fibers *in vivo* and on human myotubes *in vitro* [2,4,7,9,11].

Chemokines, a class of small cytokines, are known to participate in the recruitment and activation of leucocytes at the site of inflammation. Among them, macrophage chemoattractant protein-1 and macrophage inflammatory protein-1α are strongly up-regulated in IBM muscles at the protein and mRNA level. The IFN-γ-inducible chemokines Mig and IP-10 and the Mig receptor CXCR-3 are also strongly expressed on the muscle fibers and on a subset of autoinvasive CD8$^+$ cells. Because Mig and IP-10 are produced by myotubes on IFN-γ stimulation, they could facilitate the recruitment of activated T cells to the muscle and contribute to self-sustaining nature of endomysial inflammation as commonly seen in IBM. The presence of macrophage chemoattractant protein-1 and macrophage inflammatory protein-1α in the extracellular matrix and the possibility of *in situ* synthesis may have an effect in promoting tissue fibrosis in the late stages of IBM. Cytokines and chemokines also activate adhesion molecules such as vascular cell adhesion molecule, intercellular adhesion molecule-1, very late activation antigen-4, and matrix metalloproteinases (MMPs), a family of calcium-dependent zinc endopeptidases involved in the remodeling of the extracellular matrix. These actions facilitate the transmigration of lymphocytes toward the muscle fiber. Among the MMPs, MMP-9 and MMP-2 are up-regulated on the nonnecrotic and MHC-I-expressing muscle fibers of patients with IBM. Furthermore, MMP-2 immunostains the autoinvasive CD8$^+$ T cells. Of interest, gene expression profiling has shown that up-regulation of chemokine and cytokine genes is much higher in the muscles of patients with IBM compared with muscles of patients with DM.

D. Viruses as Possible Triggering Factors

Several viruses, including coxsackieviruses, influenza, paramyxoviruses, mumps, cytomegalovirus, and Epstein-Barr virus, have been indirectly associated with chronic and acute myositis. For the coxsackieviruses, an autoimmune myositis triggered by molecular mimicry has been proposed because of structural homology between the histidyl-transfer RNA synthetase that is the target of the Jo-1 antibody and a coat protein of encephalomyocarditis virus. Sensitive polymerase chain reaction studies, however, have repeatedly failed to confirm the presence of such viruses in muscle biopsies from these patients [12].

The best evidence of a viral connection in PM and IBM is with the retroviruses. Monkeys infected with simian immunodeficiency virus and humans infected with human immunodeficiency virus (HIV) and human T cell lymphotropic virus (HTLV-1) develop PM or IBM [4,13]. In humans infected with HIV or HTLV-1, an isolated inflammatory myopathy may occur as the initial manifestation of the retroviral infection or myositis may develop later in the disease course. It appears that IBM is also seen with low frequency in humans with HIV and HTLV-1 infection. This association is more than a coincidence because in retroviral-positive patients the disease always appears before the age of 50 but several years after the first manifestations of the retroviral infection. At least seven HIV- or HTLV-1-positive patients with IBM have been reported, and my colleagues and I have seen six more cases in the last 2 years, suggesting that the disease is now more often recognized in HIV-positive patients who live longer and harbor the virus for several years. The clinical phenotype and muscle histology in HIV-IBM patients are identical to those of retroviral-negative IBM. The predominant endomysial cells are CD8$^+$, cytotoxic T cells that, along with macrophages, invade or surround MHC-I antigen-expressing nonnecrotic muscle fibers. Using *in situ* hybridization, polymerase chain reaction, immunocytochemistry, and electron microscopy, viral antigens could not be detected within the muscle fibers but only in occasional endomysial macrophages. My colleagues and I have interpreted these observations to suggest that in HIV-IBM and HTLV-1 IBM there is no evidence of the virus or viral replication within the muscle fibers. HIV-IBM or HTLV-1 IBM is caused by a clonally driven subpopulation of activated T cells that invade MHC-I-expressing muscle fibers in a pattern similar to retroviral-negative IBM, as described previously. These findings suggest that the chronic retroviral infection and immune recognition are sufficient to trigger the inflammatory process that leads to IBM. In the retroviral-positive patients, the retroviral-positive endomysial macrophages may facilitate the autoimmune process by secreting cytokines, chemokines, or nitric oxide, which up-regulate MHC-I or costimulatory molecules and enhance myocytotoxicity.

E. Major Histocompatibility Complex Class I Antigen Associated with Endoplasmic Reticulum Stress in the Myofiber

In PM and IBM, MHC-I antigens are expressed in all muscle fibers regardless of whether they are invaded by T cells or contain vacuoles; in contrast, the vacuolated fibers always express MHC-I antigens but are almost never surrounded by T cells. Such strong up-regulation of MHC-I is emerging as an important stress cell factor [14].

The assembly and folding of MHC-I molecules occurs in the endoplasmic reticulum (ER) and begins with the association of a heavy chain glycoprotein with β2-microglobulin that forms an unstable heterodimeric complex, which matures only when bound to an antigenic peptide [7,15,16] (Fig. 3). Antigenic peptides are generated in the cytosol by immunoproteasomes that contain two MHC-encoded subunits, large multifunctional proteases 2 and 7 (LMP2 and LMP7), and are transported to the ER by transporter-associated protein (TAP). A system of chaperone proteins (including calnexin, calreticulin, GRP94, GRP78, and ERp72) and two antigen-processing proteins (TAP and tapasin) ensure the proper maturation of MHC-I antigens. The ER maintains quality control by processing, folding, and exporting the MHC-I loaded with antigen. If the complex of glycoprotein, β2-microglobulin, calnexin, calreticulum, GR57, TAP, and tapasin—collectively called the MHC-I loading complex—does not bind to suitable

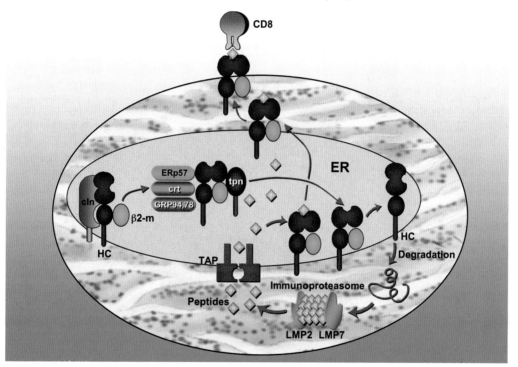

Assembly of MHC-I Peptide Complexes in the ER of the Muscle Fiber for Antigen Presentation and Translocation of Misfolded MHC-1 Heavy Chain Glycoproteins in the Cytoplasm

Figure 3 Proposed Mechanisms in the Pathogenesis of Polymyositis and Inclusion Body Myositis. Viral triggers lead to clonal expansion of CD8[+] T cells and T cell–mediated cytotoxicity through the perforin pathway. The released cytokines result in at least a threefold increase in the expression of major histocompatibility complex class I (MHC-I) molecules, which cannot exit the endoplasmic reticulum (ER) because they cannot be conformationally assembled with antigenic peptide. As a result, there is an increase in the MHC-peptide loading complex and ER stress response, which leads to the activation of nuclear factor-κB (NF-κB) and the accumulation of MHC misfolded glycoproteins that include amyloid-related proteins. Activation of NF-κB leads to the transcription of genes encoding inflammatory mediators, such as cytokines and chemokines, which further stimulates the MHC/CD8 complex and results in self-sustaining inflammatory response. The amyloid-related misfolded proteins could also enhance the expression of cytokines, closing the loop between inflammatory and degeneration-associated molecules [15]. *Note:* β2-m, β2-microglobulin; cln, calnexin; crt, calreticulin; HC, heavy chain; IFN, interferon-1β; IL-1β, interleukin; LMP, large multifunctional protease; TAP, transporter-associated protein; TNF, tumor necrosis factor; tpn, tapasin.

antigens, the heavy chain glycoprotein is misfolded and removed from the ER to the cytosol for degradation [15].

In PM and IBM, the muscle fibers are overloaded by MHC-I molecules, which lead to ER stress response and further protein misfolding, presumably because the antigenic peptides cannot undergo proper conformational change to bind to the MHC-I loading complexes. This is supported by the following findings on the muscle biopsies of patients with PM and IBM [7,16]: (1) enhanced immunoproteasome activity, evidenced by up-regulation of LMP2 and LMP7 with increased expression of MHC-I antigens; (2) enhanced MHC-I assembly activity, evidenced by increased protein and mRNA expression of MHC-I antigens, GRP78, GRP94, ERp72, calnexin, and calreticulin; (3) misfolding of glycoproteins, evidenced by the accumulation of amyloid-related molecules and aggreasomes with ER chaperone proteins and MHC-I molecules; and (4) cell stress response, evidenced by activation of nuclear factor-κB (a means by which the cells protect themselves from ER stress), and upstream up-regulation of αB-crystalline (a major stress protein) in the intact fibers and its colocalization with MHC-I molecules [14–19]. Such stressor effects are seen not only in PM and IBM but also in MHC-I antigen transgenic mice, suggesting that overexpression of MHC-I antigens alone might be sufficient to induce ER stress.

Accordingly, as shown in Figure 4, PM or IBM might be triggered by viral infection, leading to clonal expansion of T cells and T cell–mediated, MHC-I antigen–restricted cytotoxicity through the perforin pathway. The resultant ER stress response, which leads to NFκB activation and increased expression of inflammatory mediators, further stimulates the MHC-I/CD8 complex, inducing

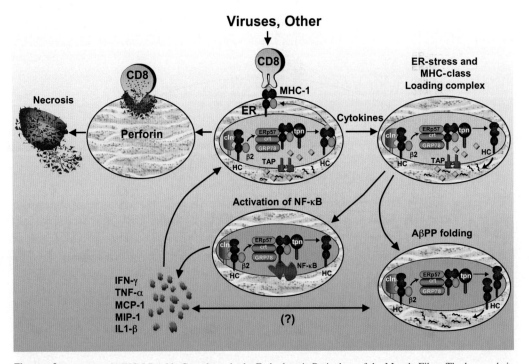

Figure 4 Assembly of MHC-I Peptide Complexes in the Endoplasmic Reticulum of the Muscle Fiber. The heavy chain of major histocompatibility complex class I (MHC-I) molecules associates with the chaperone protein calnexin; binds with β2-microglobulin (β2-m); is incorporated into a series of chaperone proteins such as ERp57, GRP78, GRP94, calreticulin, transporter-associated protein (TAP), and tapasin; and forms the MHC-I peptide loading complex. Antigenic peptides (derived from degraded proteins in the cytosol and synthesized by immunoproteasomes that contain two MHC-encoded subunits, large multifunctional protease-2 and -7) are transported to the endoplasmic reticulum (ER) by TAP and bind with the MHC-I-peptide loading complex. High-affinity antigenic peptides induce a conformational change in MHC-I molecules, promoting their release from the loading complex. In the muscle, the assembled MHC-I peptides move to the cell surface of the muscle fiber, which acts as an antigen-presenting cell, for recognition by CD8+ T cells. If suitable antigenic peptides are absent, or cannot go through the proper conformational change to bind to the MHC-I complex, the MHC-I-peptide loading complex becomes unstable and the heavy chain of MHC-I molecules are misfolded and removed from the ER to the cytosol for degradation [7,15]. *Note:* cln, calnexin; crt, calreticulin; HC, heavy chain; LMP, large multifunctional protease; tpn, tapasin.

a self-sustaining inflammatory response. In IBM, upregulation of cytokine and chemokine mRNA is correlated with increased mRNA for AβPP, microtubule-associated protein tau, and ubiquitin, supporting the previously proposed interrelationship between proinflammatory and degenerative molecules [16]. Consistent with the preceding is the observation that cytokines, such as interleukin-1β (IL-1β), TGF-β, and MMPs, share common antigenic determinants and colocalize with MHC-I and AβPP [16]. Because AβPP enhances IL-1β production and IL-1β up-regulates AβPP expression (IL-1β ↔ AβPP ↔ IL-1β ↔ inflammation), an interaction between amyloid and inflammatory mediators has been proposed [5].

V. Treatment: Present and Future

A. Polymyositis and Dermatomyositis

The goal of therapy in inflammatory myopathies is to improve activities of daily living by increasing muscle strength and improving extramuscular manifestations, such as skin rash, dysphagia, dyspnea, arthralgia, and fever. Unfortunately, there are only a handful of controlled clinical trials, conducted mostly in DM and IBM [4,5]. Overall, DM responds better than PM (especially if the presentation is acute and the treatment starts early), whereas IBM shows minimal, transient, or no response. The agents in the following sections are used in the treatment of PM and DM.

1. Glucocorticoids

Although controlled trials have not been performed, oral prednisone is the initial treatment of choice. The effectiveness and side effects of this therapy determine the future need for stronger immunosuppressive drugs [4,5,7,20]. High-dose prednisone, at least 1 mg/kg/day, is initiated as early in the disease as possible. After an initial period of 3 to 4 weeks, prednisone is tapered slowly over a period of 10 weeks to 1 mg/kg every other day. Then, if there is evidence of efficacy and no serious side effects, the dosage is further reduced by 5 or 10 mg every 3 to 4 weeks until the lowest possible dose that controls the disease is reached. The efficacy of prednisone is determined by an increase in muscle strength and activities of daily living, which almost always occurs by the third month of therapy. A feeling of increased energy or a reduction of the CK level without a concomitant increase in muscle strength is not a reliable sign of improvement. If prednisone provides no objective benefit after about 3 months of high-dose therapy, the disease is probably unresponsive to the drug and tapering should be accelerated while the next-in-line immunosuppressive drug is started. Although controlled trials have not been performed, almost all patients with true PM or DM respond to glucocorticoids to some degree and for some period of time; in general, DM responds better than PM.

2. Combined Steroid with Immunosuppressive Therapy

In severe cases, combined initial therapy of steroids with another immunosuppressant, such as methotrexate or azathioprine, may be associated with a lower relapse rate or better long-term outcome.

In other cases, a second immunosuppressive drug is necessary to sustain or enhance steroid-induced improvement.

Drug selection is largely empirical, with choices based on personal experiences, relative efficacy, and safety. The following agents are commonly used: (1) Azathioprine is well tolerated, has few side effects, and may be as effective for long-term therapy as other drugs. (2) Methotrexate has a faster onset of action than azathioprine. A rare side effect is methotrexate pneumonitis, which can be difficult to distinguish from the interstitial lung disease of the primary myopathy associated with Jo-1 antibodies (described earlier). (3) Cyclophosphamide has limited success and significant toxicity. (4) Chlorambucil has variable results. (5) Cyclosporine has inconsistent and mild benefit. (6) Mycophenolate mofetil has been tried recently.

3. Immunomodulating Procedures

In a double-blind study of patients with refractory DM, intravenous immunoglobulin (IVIg) improved not only the strength and rash but also the underlying immunopathology. The benefit can be impressive, but in most cases it is short lived (≤ 8 weeks); repeated infusions every 6 to 8 weeks are often required to maintain improvement. A dose of 2 g/kg divided over 2 to 5 days per course is recommended. A controlled double-blind study in PM is not completed (probably because of the rarity of the disease), but uncontrolled observations suggest that IVIg is beneficial in most PM patients. Neither plasmapheresis nor leukapheresis appear to be effective in PM or DM.

B. Inclusion Body Myositis

Despite the previously described primary immune factors, IBM remains resistant to most immunotherapies, prompting the contention that it could be a degenerative rather than an autoimmune disease. However IBM is not the only immune disease unresponsive to such therapies. Primary progressive multiple sclerosis is a classic example in which immune and degenerative features coexist and the disease is resistant to therapies. Although the common immunotherapeutic agents, such as corticosteroids, azathioprine, methotrexate, cyclosporine, and cyclophosphamide, are generally ineffective in patients with IBM, some patients have responded to immunomodulating therapies to a certain degree or for short periods. In two double-blind studies of IVIg in IBM, minimal benefit in up to 30% of the patients was found. The strength gains, however, were not of sufficient magnitude to justify routine use of this drug. A second controlled trial combining IVIg with prednisone was ineffective in 36 IBM patients. IVIg has also improved the dysphagia of IBM in both a controlled study and an uncontrolled series, suggesting that more specific or potent agents may be promising.

References

1. Dalakas, M. C. (1991). Polymyositis, dermatomyositis and inclusion-body myositis. *N. Engl. J. Med.* **325**, 1487–1498.
2. Engel, A. G., and Hohlfeld, R. (2004). The polymyositis and dermatomyositis syndromes. *In:* "Myology" (A. G. Engel and C. Franzini-Armstrong, eds.), 3rd edition, pp. 1321–1366. McGraw-Hill, New York.
3. Mastaglia, F. L., and Phillips, B. A. (2002). Idiopathic inflammatory myopathies: epidemiology, classification and diagnostic criteria. *Rheum. Dis. Clin. N. Am.* **28**, 723–741.
4. Dalakas, M. C., and Hohlfeld, R. (2003). Polymyositis and dermatomyositis. *Lancet* **362**, 971–982.
5. Dalakas, M. C. (2004). Inflammatory disorders of muscle: progress in polymyositis, dermatomyositis and inclusion body myositis. *Curr. Opin. Neurol.* **17**, 561–567.
6. Greenberg, S. A., Pinkus, J. L., Pinkus, G. S., Burleson, T., Sanoudou, D., Tawil, R., Barohn, R. J., Saperstein, D. S., Briemberg, H. R., Ericsson, M., Park, P., and Amato A. A. (2005). Interferon-alpha/beta-mediated innate immune mechanisms in dermatomyositis. *Ann. Neurol.* **57(5)**, 664–678.
7. Dalakas, M. C. (2006). Signaling pathways and immunobiology of inflammatory myopathies. *Nature Clin. Pract. Rheumol.* **2**, 219–227.
8. Tezak, Z., Hoffman, E. P., Lutz, J. L., Fedczyna, T. O., Stephan, D., Bremer, E. G., Krasnoselska-Riz, I., Kumar, A., and Pachman, L. M. (2002). Gene expression profiling in DQA1*0501+ children with untreated dermatomyositis: a novel model of pathogenesis. *J. Immunol.* **168**, 4154–4163.
9. Wiendl, H., Hohlfeld, R., and Kieseier, B. C. (2005). Immunobiology of muscle: advances in understanding an immunological microenvironment. *Trends Immunol.* **26**, 373–380.
10. Hofbauer, M., Wiesener, S., Babbe, H., Roers, A., Wekerle, H., Dornmair, K., Hohlfeld, R., and Goebels, N. (2003). Clonal tracking of autoaggressive T cells in polymyositis by combining laser microdissection, single-cell PCR and CDR3 spectratype analysis. *Proc. Natl. Acad. Sci. USA* **100**, 4090–4095.
11. Figarella-Branger, D., Civate, M., Bartoli, C., and Pellissier, J. F. (2003). Cytokines, chemokines, and cell adhesion molecules in inflammatory myopathies. *Muscle Nerve* **28(6)**, 659–682.
12. Leff, R. L., Love, L. A., Miller, F. W., Greenberg, S. J., Klein, E. A., Dalakas, M. C., and Plotz, P. H. (1992). Viruses in the idiopathic inflammatory myopathies: absence of candidate viral genomes in muscle. *Lancet* **339**, 1192–1195.
13. Illa, I., Nath, A., and Dalakas, M. C. (1991). Immunocytochemical and virological characteristics of HIV-associated inflammatory myopathies: similarities with seronegative polymyositis. *Ann. Neurol.* **29**, 474–481.
14. Nagaraju, K., Casciola-Rosen, L., Lundberg, I., Rawat, R., Cutting, S., Thapliyal, R., Chang, J., Dwivedi, S., Mitsak, M., Chen, Y. W., Plotz, P., Rosen, A., Hoffman, E., and Raben, N. (2005). Activation of the endoplasmic reticulum stress response in autoimmune myositis: potential role in muscle fiber damage and dysfunction. *Arthritis Rheum.* **52**, 1824–1835.
15. Grandea, A. G., 3rd, and Van Kaer, L. (2001). Tapasin: an ER chaperone that controls MHC class I assembly with peptide. *Trends Immunol.* **22**, 194–199.
16. Dalakas, M. C. (In press). Sporadic inclusion body myositis: diagnosis, pathogenesis and therapeutic strategies. *Nature Clin. Pract. Neurol.*
17. Ferrer, I., Martin, B., and Castano, J. G. (2004). Proteasomal expression, induction of immunoproteasome subunits, and local MHC class I presentation in myofibrillar myopathy and inclusion body myositis. *J. Neuropathol. Exp. Neurol.* **63**, 484–498.
18. Vattemi, G., Engel, W. K., McFerrin, J., and Askanas, V. (2004). Endoplasmic reticulum stress and unfolded protein response in inclusion body myositis muscle. *Am. J. Pathol.* **164**, 1–7.
19. Fratta, P., Engel, W. K., McFerrin, J., Davies, K. J., Lin, S. W., and Askanas, V. (2005). Proteasome inhibition and aggresome formation in sporadic inclusion-body myositis and in amyloid-{beta} precursor protein-overexpressing cultured human muscle fibers. *Am. J. Pathol.* **167(2)**, 517–526.
20. Hohlfeld, R., and Dalakas, M. C. (2003). Basic principles of immunotherapy in neurological diseases. *Semin. Neurol.* **23**, 121–132.

88

Central Autonomic Network

Peter Novak, MD, PhD

Keywords: *autonomic nervous system, brainstem, central autonomic network, hypothalamus*

I. Introduction
II. Functional Anatomy
III. Neurotransmitters and Neuromodulators
References

I. Introduction

The central autonomic network (CAN) consists of several reciprocally interconnected structures at different levels of the brain [1–6]. The main centers are located in the cortex, hypothalamus, brainstem, and spine. The main function of the CAN is control of the tonic, reflex, and adaptive autonomic functions. In addition, the CAN modulates a number of other brain functions, including emotional, attentional, behavioral, endocrine, respiratory, vestibular, sexual, and pain responses. The CAN affects essentially all organs through direct innervations by the parasympathetic and sympathetic nerves.

An important aspect of the CAN is the concept of integration centers. These centers functionally and anatomically interface autonomic centers with other, nonautonomic centers. For example, the periventricular nucleus of the hypothalamus, which regulates production of the steroid hormones as a response to the stress, has reciprocal connections to autonomic centers, thus establishing integration between the hormonal stress responses and autonomic activity. The CAN is a neuronal substrate for autonomic reflexes and characteristic "autonomic patterns" aiming to preserve homeostasis [6]. Another typical feature of the CAN is its redundancy and resistance to a single lesion. It can be difficult to locate a CAN lesion on the basis of symptoms. Usually, a bilateral lesion is needed to produce a symptomatic effect. Furthermore, age-related changes may have an impact on the manifestation of a particular lesion within the CAN.

Many of the autonomic structures are defined by their function rather than by distinct cytoarchitectonic landmarks. A typical example is the brainstem, where the reticular formation consists of a loose network of neurons distributed through the brainstem, forming more or less compact clusters of neurons that share common functions.

The past decade witnessed an explosive increase in information regarding the CAN. Many aspects of the autonomic afferent pathways have been clarified. Numerous new neurotransmitters and receptors have been described, in many cases with potent effect on autonomic functions. Although this review summarizes current anatomical, physiological, and biochemical knowledge, it is by no means exhaustive. Interested readers should consult the included references.

II. Functional Anatomy

The CAN consists of reciprocally interconnected structures with integration centers at the cortex, hypothalamus, brainstem, and spine (Figs. 1 and 2). The roles of the cerebellum and thalamus in the context of the CAN remain to be established.

A. Cerebral Cortex (Medial Prefrontal Cortex, Insula, and Anterior Cingulate)

The medial prefrontal cortex (MPC) is a premotor area that projects to the rostral ventrolateral medulla, a major source of cardiovascular sympathoexcitatory neurons [7]. The pathways are probably multisynaptic and include relay at the parabrachial nuclei and the nucleus tractus solitarius. Activation of the MPC results in inhibition of the sympathetic drive with bradycardia and hypotension. The insula is located beneath the frontal, parietal, and temporal opercula. The insula integrates sensory, noxious, endocrine, and autonomic responses. Right anterior cingulate cortex activity correlates with peripheral changes in the blood pressure, and it was proposed that the cingulate might be involved in the integration of peripheral cardiovascular changes with cognitive activity, motor preparedness, and emotional states.

B. Basal Forebrain (Amygdala and Bed Nucleus of the Stria Terminalis)

The central nucleus of amygdala and the bed nucleus of the stria terminalis integrate autonomic, neuroendocrine, and behavioral responses to emotions. Amygdala projects to the hypothalamus and autonomic brainstem nuclei, including the nucleus of the solitary tract, the parabrachial nucleus, and the rostral ventromedial medulla. The bed nucleus of the stria terminalis has reciprocal connections with many brainstem autonomic centers, particularly with the parabrachial nucleus.

C. Hypothalamus

The hypothalamus is a main center for neuroendocrine regulations. The hypothalamus regulates autonomic activity through the endocrine path via the pituitary gland and by direct connections with the autonomic centers [8]. The majority of the autonomic inputs to the hypothalamus arise in the nucleus of the solitary tract using glutamate, catecholamines, or peptides as major neurotransmitters. Some of the afferent autonomic fibers from the nucleus of the solitary tract are relied through the ventrolateral medulla, parabrachial nuclei, or locus coeruleus before they reach the hypothalamus. The majority of the catecholaminergic terminals in the hypothalamus originate in the ventrolateral medulla and in the locus coeruleus. Although the brainstem noradrenergic fibers supply the whole hypothalamus, the autonomic control is concentrated mainly in the paraventricular nucleus and to a lesser degree in the arcuate nucleus, lateral hypothalamic area, posterior periventricular area, and zona incerta.

D. Paraventricular Nucleus

The paraventricular nucleus (PVN) integrates autonomic responses with stress, baroregulation, thermoregulation, and energy balance. The PVN produces a number of hormones active in neuroendocrine regulations, including corticotropin-releasing hormone, vasopressin, oxytocin, thyrotropin-releasing hormone, and somatostatin. The PVN is also the principal source of descending autonomic pathways. The PVN sends direct projections to the parasympathetic preganglionic neurons (the salivatory nuclei, the dorsal nucleus of vagus, the nucleus ambiguous) and sympathetic preganglionic neurons in the intermediolateral cell column of the spine. These descending fibers are peptidergic, using oxytocin and vasopressin as major transmitters, but the fibers are positive also for corticotropin-releasing hormone, enkephalin, dynorphins, thyrotropin-releasing hormone, somatostatin, vasoactive intestinal polypeptide, and angiotensin II. The glutaminergic tier of the descending autonomic paraventricular fibers target the brainstem catecholaminergic autonomic centers, including the A1 group at the caudal ventrolateral medulla and A5 group at the rostral ventrolateral medulla.

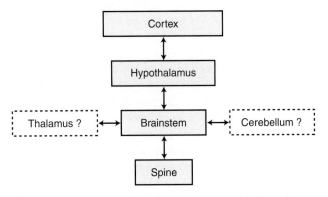

Figure 1 Location of the Main Integrative Centers of the Central Autonomic Network.

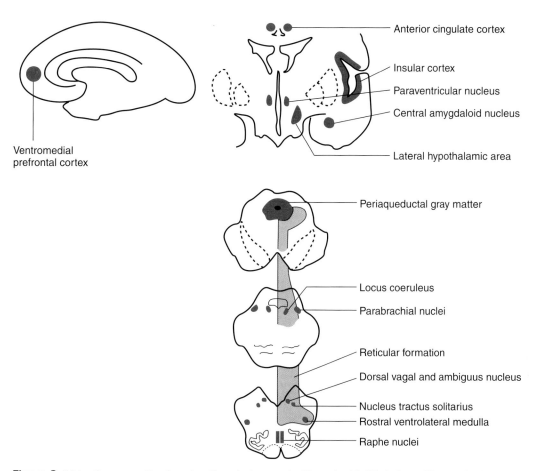

Figure 2 Main Structures Forming the Central Autonomic Network. Modified from Westmoreland, B. F., Benarroch, E. E., Daube, J. R., et al. (1994). "Medical Neurosciences: An Approach to Anatomy, Pathology, and Physiology by Systems and Levels." 3rd ed. Boston, Little, Brown and Co.

1. Stress Response Modulation

The corticotropin-releasing-hormone-producing cells in the PVN integrate the endocrine and autonomic responses to stress. The PVN monitors the homeostasis via chemoreceptors, circulating hormones, and local hypothalamic circuits and by using visceral input through the ascending autonomic tracts. Sensory information from the thorax and abdomen is provided by the vagal and glossopharyngeal afferents. The sensory input is relayed at the nucleus of the solitary tract, dorsal nucleus of vagus, and dorsal and ventral lateral medulla. These brainstem nuclei directly project to the corticotropin-releasing-hormone-producing PVN neurons. Although the specific mechanisms are poorly understood, the PVN is capable of integrating a variety of chemical and neural signals in a secretory output to the adrenal cortex and simultaneously activating the autonomic structures via descendent autonomic projections. The PVN also receives multiple projections from the limbic system, thus enabling the integration of emotional and autonomic responses. Furthermore, inflammation that is associated with release of endotoxin and proinflammatory cytokines can directly activate autonomic centers, including the nucleus tractus solitarius and the rostral ventrolateral medulla neurons, with the net effect of increased activation of the corticotropin-releasing hormone neurons in the PVN.

2. Baroregulation

Vasopressin-related baroregulation should be distinguished from the baroreceptor reflex. Whereas baroregulation is controlled mainly at the PVN, the baroreceptor reflex is regulated at the brainstem level. Hyperosmolality, hypovolemia, and low systemic pressure promote release of vasopressin in the PVN and induce water resorption in the kidneys. Osmolality sensing is carried out by widely dispersed osmoreceptor cells. The afferent portion of the baroreceptor reflex also provides baroregulatory information to the PVN through a relay at the nucleus tractus solitarius that sends adrenergic projections to the PVN.

3. Thermoregulation

Thermoregulation is an important part of homeostasis. Warm- and cold-sensitive neurons are located predominantly in the medial preoptic region. Preoptic warming induces heat loss through vasoconstriction, sweating, increased respiration, and inhibition of the uncoupling protein-1 in brown adipose tissue, a mediator of thermogenesis. The thermoregulatory information from the preoptic region is relayed to the autonomic structures through the PVN and via direct projections to the periaqueductal gray matter [9]. The latter controls thermogenesis by modulating skin vasomotor responses and heat production in the brown adipose tissue. The efferents from the periaqueductal gray matter to the preganglionic neurons supplying the skin vasomotor fibers as well as the sudomotor pathways controlling sweat production are not well understood.

4. Energy Balance

The PVN and the arcuate nucleus (ARN) integrate energy balance, including food intake, with the autonomic system [10,11]. The complex endocrine control, which involves both gut hormones and hypothalamic peptides, remains to be elucidated. The circulating signals of energy balance converge at the arcuate nucleus, followed by an interplay between pro-appetite and anti-appetite circuits. Pro-opiomelanocortin and cocaine- and amphetamine-regulated transcript pathways inhibit the food intake, whereas the neuropeptide Y and agouti-related peptide circuits stimulate food intake. These pathways are modulated by, for example, insulin (which suppresses food intake) and leptin (which suppresses food intake and activates the nucleus of the solitary tract), whereas cholecystokinin and ghrelin have an orexigenic effect (Table 1).

The ARN receives afferents from several brainstem autonomic centers, including the parabrachial nuclei, the nucleus of the solitary tract, and the ventrolateral medulla, and it projects to the PVN and directly to the preganglionic parasympathetic and sympathetic neurons, establishing integration between the energy balance–autonomic functions.

E. Thalamus

The role of the thalamus in control of autonomic functions is incompletely understood. Several thalamic nuclei are interconnected to the CAN (Fig. 3). For example, the paraventricular nucleus projects to the medial prefrontal cortex, whereas the mediodorsal nucleus is connected to several limbic areas with autonomic influences. The role of the thalamus in autonomic regulation can be inferred from fatal familial insomnia (FFI). FFI is a rare disorder associated with autonomic (predominantly sympathetic) overactivity, severe sleep disorder, and altered neuroendocrine regulation [12]. Advanced-stage FFI results in a disappearance of circadian autonomic and neuroendocrine rhythms. FFI selectively damages the dorsomedial and the anteroventral nuclei of the thalamus while sparing the hypothalamus and brainstem autonomic areas. These thalamic nuclei may modulate autonomic functions, although the specific mechanisms are unknown.

F. Cerebellum

The role of the cerebellum in the context of the CAN is less clear. Earlier studies showed that stimulation of the paleocerebellum can induce alterations in arterial blood pressure, heart rate, piloerection, pupils (dilatation), and the urinary and gastrointestinal tracts [13].

G. Brainstem

Brainstem autonomic centers represent important sensory-motor integrative centers and relay stations between the hypothalamus and preganglionic autonomic neurons. Although there are several discrete autonomic structures (the locus coeruleus, nucleus of the solitary tract, dorsal nucleus of vagus, and nucleus ambiguous), most of the autonomic centers are part of the reticular formation, without having distinct cytoarchitectonic landmarks.

H. Periaqueductal Gray Matter

The periaqueductal gray matter (PGM) is a complex structure that coordinates the antinociceptive, behavioral, and autonomic reactions to stress and injury. The PGM receives input from the autonomic cortex, amygdala, and hypothalamus and projects to the ventrolateral medulla. The activity at the PGM is mediated by glutamate, acetylcholine, serotonin, and opioid peptides. Stimulation of the dorsolateral portion of the PGM produces hypertension, tachycardia, and aggressive behavior consistent with a "fight or flight" reaction, whereas stimulation of the ventrolateral PGM produces an opposite response, including hypotension, bradycardia, and passive behavior [14].

I. Locus Coeruleus

The locus coeruleus (LC) is a major source of norepinephrine in the brain. The LC receives input from the medial prefrontal cortex, nucleus paragigantocellularis in the ventrolateral medulla, nucleus prepositus hypoglossi, and lateral hypothalamus. Axons of the LC project widely into the whole brain, including the preganglionic sympathetic neurons in the intermediolateral column. The rostral ventrolateral medulla activates the LC.

J. Nucleus Tractus Solitarius

The nucleus tractus solitarius (NTS), located at the upper medulla, is the first relay of the visceral input. NTS

Table 1 Overview of Neurotransmitters and Neuropeptides Involved in the Function of the Central Autonomic Network

Neurotransmitters/ Neuropeptides	Source	Receptor	Location of Receptors	Effect
Norepinephrine	RVLM, CVLM-A1, RF, LC, NTS-A2, A5	α1, α2, β1, β3,	Ubiquitous, vascular smooth muscles	Mediates visceral sensory information, hypothermia; main excitatory neurotransmitter in sympathetic system
Epinephrine	RVLM – C1 NTS – C2, C3	α1, α2, β1, β2,	VLM, IML	Hypothermia
Acetylcholine	RF, NTS ARNM, IML	Nicotinic Muscarinic (m1-5)	Ubiquitous,	Hyperthermia; stimulates colonic motility; vasodilator
CCK		CCK-A, CCK-B	NTS, ARN	Anorexigenic; stimulates colonic motility
Vasopressin	SO, PVN	V1-V3	RVLM, CVLM, IML, NTS	Fluid and electrolyte balance, sympathoexcitation
Oxytocin	SO, PVN	OXTR	Amygdala, NTS	Fluid and electrolyte balance, sympathoexcitation
Substance P	RVLM, CVLM	NK1	Ubiquitous LC, NTS	Increases cardiorespiratory reflexes
Somatostatin	NTS	STTR1-5	PGM, NTS, LC	Hyperthermia
Neurotensin	NTS, PVN, PGM	NTR1-3	Hypothalamus	Hyperthermia
Enkephalin	CVLM, PVN, NTS	Opioid receptors	Ubiquitous	Sympathoinhibition
Neurokinin A, B	Ubiquitous	NK2, NK3	Ubiquitous	Central control of the blood pressure
Nitric oxide	NTS, PVN, IML			Sympathoinhibition
VIP	PVN, SCN	VPAC1-2	Ubiquitous	Sympathoexcitation
NPY	Ubiquitous ARN	Y1-Y5	Ubiquitous	Mediates visceral sensory information; orexigenic; hypothermia; stimulates parasympathetic system
Galanin	LC, RVLM, PVN	GALR1-3	Ubiquitous	Orexigenic; suppresses effect of norepinephrine and acetylcholine
5-HT	Raphe nucleus, CVLM DMV, IML	5-HT1-7	Hypothalamus LC, RF	Hyperthermia; stimulates ventilation; sympathoexcitation or inhibition
CGRP-α, CGRP-β	PBN	CGRPR1-3	Ubiquitous	Modulates sympathetic activity; vasodilator
Angiotensin II	PVN, circumventricular organ	AT1, AT2	NTS, VRLM	Hyperthermia, sympathoexcitation
Orexins A, B (hypocretins)	Lateral hypothalamus	OX1, OX2	ARN	Increase food intake; sympathoexcitation
Adenosine	NTS	A1, A2a, A2b, A3	Vessels, NTS	Mediates baroreflex
CRF	PVN	CRF-R1, CRF-R2	LC, dorsal raphe nuclei	Stress response
Activins	Ovary, testis	I-IV	Ubiquitous	Mediate visceral sensory information
GLP-1	NTS, CVLM, intestine	GLR-1R	Ubiquitous	Mediates visceral sensory information; sympathoexcitatory; inhibits food intake
Carbon monoxide	NTS			
ATP	Ubiquitotous	P2X-Y	Ubiquitous	Cotransmitter; modulates NTS-mediated cardiovascular control
Cannabis	Hypothalamus	CB1, CB2	Brain immune cells	Hypo/hyperthermia, dose-dependent
Leptin	Adipocytes	OB-R	ARN, NTS	Sympathoexcitation, anorexigenic
Ghrelin	PVN, ARN, stomach	GHS-R	ARN	Modulates GI and cardiovascular functions; orexigenic
Endothelin	PVN		Ubiquitous	Vasoconstrictor

CCK, Cholecystokinin; VIP, vasoactive intestinal peptide; NPY, neuropeptide Y; 5-HT, serotonin; CGRP, calcitonin gene-related peptide; PVN, paraventricular nucleus of the hypothalamus; SO, supraoptic nucleus of hypothalamus; CRF, corticotropin-releasing factor; RVLM, rostral ventrolateral medulla; CVLM, caudal ventrolateral medulla; RF, reticular formation; DMV, dorsal motor nucleus of vagus; IML, intermediolateral cell column; ARNM, arcuate nucleus of the medulla; ARN, arcuate nucleus of the hypothalamus; GLP, glucagon-like peptide; LC, locus coeruleus; PBN, parabrachial nucleus; NTS, nucleus of the solitary tract; PGM, periaqueductal gray matter; SCN, suprachiasmatic nucleus.

receives sensory input via the trigeminal, facial, vagal, and glossopharyngeal nerves—also called the "cranial parasympathetic afferent system"—and via spinal afferents—the "sympathetic afferent system" [6]. A number of neuropeptides have been localized in the NTS. The NTS is also the first relay of the baroreflex and baroregulatory circuits. Bilateral lesion of the NTS produces hypertension or labile blood pressure.

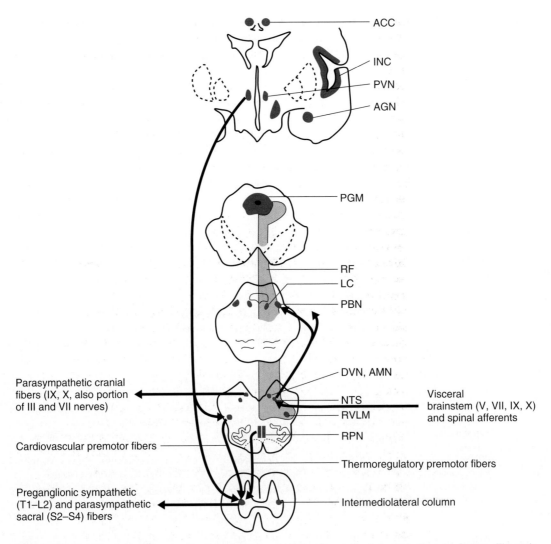

Figure 3 Main Connections Between the Central Autonomic Network and the Autonomic Nervous System. The main efferent (left) and afferent (right) pathways are drawn. Most of the central autonomic network structures are reciprocally interconnected, forming multiple feedback loops. For simplicity, only unilateral connections are shown, although most of the projections are bilateral. For definitions of abbreviations, see Figure 2 and Table 1.

K. Dorsal Nucleus of Vagus and Nucleus Ambiguus

The majority of the parasympathetic preganglionic neurons originate in the dorsal nucleus of vagus and the nucleus ambiguous. Both nuclei innervate the heart and gastrointestinal tract.

L. Reticular Formation

The reticular formation (RF), phylogenetically one of the oldest portions of the brain, consists of reciprocally interconnected neurons forming a network ("reticulum" = netlike structure) that spreads throughout the brainstem. RF modulates most of the brain functions, including arousal, sleep, attention, respiration, muscle tone, and autonomic functions. Most studied autonomic centers are located in the parabrachial nuclei and the ventrolateral and ventromedial medulla, including the raphe nuclei.

M. Parabrachial Nuclei

The parabrachial nuclei (PBN), located in the dorsolateral tegmentum of the pons, relay visceral inputs from the nucleus tractus solitarius to the forebrain, including the hypothalamus and amygdala. The PBN have direct projections to the rostral ventrolateral medulla, thus forming an additional inhibitory feedback loop (the nucleus tractus solitarius–PBN–rostral ventrolateral medulla) in the baroreceptor reflex circuits. The main transmitter in the PBN is glutamate.

N. Rostral Ventrolateral Medulla and Caudal Ventrolateral Medulla

The ventrolateral medulla contains catecholaminergic (A1, A5, and C1 cell groups) and noncatecholaminergic neurons that regulate cardiovascular activity [15,16]. Functionally, the ventrolateral medulla has been divided into the rostral (RVLM) and caudal (CVLM) portions. The RVLM and the closely related parapyramidal region (PPr) contain premotor sympathoexcitatory neurons that directly project to the intermediolateral cell column, providing major input to the sympathetic preganglionic neurons. The spinally projecting RVLM neurons are glutamatergic, noradrenergic (A5 group), and adrenergic (C1 group). The A5 noradrenergic cell group receives input from the parabrachial nucleus and also has reciprocal connections with the periventricular nucleus of the hypothalamus and the nucleus tractus solitarius. Integrity of the RVLM is necessary for maintenance of the basal sympathetic vasomotor tone and mean arterial pressure. The caudal ventrolateral medulla (CVLM) contains A1 noradrenergic neurons. The CLVM has reciprocal connections with the RVLM and the periventricular nucleus of the hypothalamus, but the CLVM does not project to the intermediolateral column. The main effect of CVLM activation is depression of the mean arterial pressure.

O. Medullary Raphe Nuclei

The medullary raphe nuclei (MRN) contain both serotoninergic and nonserotoninergic neurons. The MRN have extensive projection from the hypothalamus and project to many brainstem autonomic nuclei. The rostral MRN receive input from the thermosensitive medial preoptic region of the hypothalamus and project widely to the brainstem autonomic centers. The rostral MRN is a part of the thermoregulatory circuit and control thermogenesis by sympathetic output to the adipose tissue. Recently it was shown that the rostral MRN contain the sympathetic premotor thermoregulatory neurons [17]. The MRN inhibit or excite the RVLM premotor sympathoexcitatory neurons.

P. Neuronal Substrates of the Arterial Baroreceptor Reflex

The baroreceptor reflex (BR) controls the heart rate and arterial pressure on a beat-to-beat basis [2]. Several brainstem structures are involved in the BR circuit, with multiple input, output, and feedback loops. High arterial pressure activates the baroreceptors in the carotid sinus and aortic arch. Primary afferents from baroreceptors excite second-order neurons in the NTS via glutamate. The NTS neurons activate the CVLM neurons, and then the CVLM GABAergic neurons inhibit the RVLM sympathoexcitatory premotor neurons, resulting in withdrawal of the sympathetic drive and vasodilatation. The NTS also activates the preganglionic parasympathetic neurons in the dorsal nucleus of vagus and the nucleus ambiguus that is associated with bradycardia.

III. Neurotransmitters and Neuromodulators

The principal neurotransmitters in the CAN are excitatory amino acid glutamate and probably aspartate, whereas γ-aminobutyric acid is the main inhibitory neurotransmitter [4,5]. The responses elicited by these classic transmitters are fast, short, and mediated predominantly thorough specific ion-channel receptors. There is also a ubiquitous presence of catecholamines [18] and acetylcholine in many autonomic centers. A number of additional (especially peptidergic) neurotransmitters in the CAN have been described and have a profound effect on autonomic functions (Table 1). Many of these transmitters colocalize with the principal transmitters and hence may play a modulatory function by fine-tuning the autonomic responses. Most of the neuromodulators act through G protein–coupled specific receptors [19,20].

A. Norepinephrine

Norepinephrine (NE) is a primary excitatory neurotransmitter of the CAN. NE is a catecholamine that is synthesized by a series of enzymatic steps from tyrosine. The NE neurons are located in the locus ceruleus, subcoeruleus, ventrolateral pons, and ventrolateral and dorsomedial medulla. The NE system broadly projects to the whole brain, subserving a variety of functions, including regulation of cardiovascular, gastrointestinal, urinary, and respiration functions, sleep, dreaming, and motivation. The effect of catecholamines is mediated by distinct G protein–coupled adrenergic receptors. The action of NE is terminated by its reuptake by NE transporter on the presynaptic membrane.

B. Acetylcholine

Acetylcholine (ACh) is synthesized from choline and acetyl coenzyme A by choline acetyltransferase. ACh acts on muscarinic and nicotinic receptors. Five muscarinic (m1-5) G protein–coupled receptors have been described. Nicotinic receptors are fast ligand-gated ion channels. Activation of the nicotinic receptors is excitatory, whereas activation of the muscarinic receptors can be either excitatory or inhibitory. Muscarinic receptors activate sympathetic premotor neurons in the rostral ventrolateral medulla and modulate the baroreceptor reflex in the nucleus tractus solitarius.

C. Serotonin

Serotonin (5-hydroxytryptamine, or 5-HT) is formed by the hydroxylation and decarboxylation of tryptophan. Most of the serotonin receptors are G protein–coupled (5-HT1, -2, -4), whereas 5-HT3 receptors are ion channel receptors. The main sources of the serotonin are the raphe nuclei that project widely to respiratory and autonomic nuclei at the brainstem. Serotonin potentiates the parasympathetic outflow.

D. Vasopressin and Oxytocin

Vasopressin and oxytocin are nine-amino-acid peptides that are major transmitters in the descending autonomic hypothalamic pathways. They activate the sympathoexcitatory neurons at the rostral ventrolateral medulla and preganglionic sympathetic neurons in the intermediolateral column.

E. Neuropeptide Y

Neuropeptide Y (NPY) is the most abundant neuropeptide in the brain. It is a small peptide composed of 36 amino acid residues. NPY is highly concentrated in the amygdala, the arcuate nucleus of the hypothalamus, the periaqueductal gray matter, and the locus coeruleus. NPY facilitates action of norepinephrine, thus affecting a variety of autonomic functions. NPY mediates energy balance and food intake.

F. Angiotensin II

Angiotensin II is an octapeptide that can play an important role in the signaling of the CAN. Angiotensin receptors are expressed in the nucleus tractus solitarius and the rostral and caudal ventrolateral medulla. Angiotensin activates sympathoexcitatory neurons in the rostral ventrolateral medulla.

G. Tachykinins (Substance P, Neurokinin A and B)

The tachykinins, widely distributed in the brain, are a family of peptide neurotransmitters that share the c-terminal sequence. Substance P is an 11-amino-acid peptide and neurokinin A and B are 10-amino-acid polypeptides. They bind to specific receptors (NK1–NK3): substance P, neurokinin A, and neurokinin B have preferential affinity for NK1, NK2, and NK3, respectively. The tachykinins have diverse actions, including facilitating synaptic transmission, synaptic plasticity, and development of the central nervous system. Tachykinins also cause vasodilatation.

H. Orexins

The orexins (hypocretins) are a family of peptides (33 amino acid, Orexin A, 28 amino acid, and Orexin B) linking arousal, appetite, and neuroendocrine-autonomic control. The orexins are produced at the lateral hypothalamus and stimulate food intake. The orexinogergic neurons project into the brainstem cholinergic and monoaminergic nuclei, including the reticular formation, locus coeruleus, nucleus tractus solitarius, and dorsal nucleus of vagus. Orexins increase firing of the noradrenergic neurons and facilitate the pressor responses.

I. Leptin

Leptin is a 167-amino-acid protein produced at the adipocytes that reduces food intake by acting on the arcuate neurons of the hypothalamus. Leptin has a sympathoexcitatory effect, mediated most likely via stimulation of the paraventricular and arcuate nucleus of the hypothalamus.

J. Adenosine

Adenosine is a purine nucleoside that modulates a variety of physiological functions, involving cardiovascular brainstem centers. Adenosine facilitates glutamate release from baroreceptor afferents. Four adenosine receptors have been described, and some of them have opposite functions. A1 receptors mediate vasodilatation and slow down the heart rate, and A2A-B receptors have an anti-inflammatory effect. The physiological function of A3 receptors remains unclear. Adenosine induces apoptosis via nonreceptor adenosine signaling.

K. Nitric Oxide

Nitric oxide (NO) is a gaseous compound that functions as a diffusible neurotransmitter. Depending on its central neural target, nitric oxide may either activate or inhibit the cardiovascular system.

L. Carbon Monoxide

Carbon monoxide (CO) is a putative gaseous neurotransmitter that has an effect similar to that of adenosine. CO and adenosine promote release of each other in the nucleus of the solitary tract.

M. Adenosine Triphosphate

Adenosine triphosphate is a nucleotide that in addition to being a major source of energy also acts as a cotransmitter of norepinephrine.

N. Ghrelin

Ghrelin is a 28-amino-acid peptide produced in the stomach, placenta, kidney, pituitary, and hypothalamus. Ghrelin stimulates secretion of growth hormone, activates the feeding center in the hypothalamus, stimulates gastric emptying, and increases cardiac output.

References

1. Loewy, A. D., and Spyer, K. M., eds. (1990). "Central Regulation of Autonomic Functions." Oxford University Press, New York.
2. Eckberg, D. L., and Sleight, P. (1992). "Human Baroreflexes in Health and Disease" Oxford University Press, Oxford.
3. Saper, C. B. (1995). The central autonomic system. *In:* "The Rat Nervous System" (G. Paxinos, ed.), pp. 155–185. Academic Press, San Diego.
4. Low, P., ed. (1997). "Clinical Autonomic Disorders." Lippincott-Raven, Philadelphia.
5. Benarroch, E. E. (1997). "Central Autonomic Network: Functional Organization and Clinical Correlations." Futura Publishing, Armonk, New York.
6. Saper, C. B. (2002). The central autonomic nervous system: conscious visceral perception and autonomic pattern generation. *Annu. Rev. Neurosci.* **25**, 433–469.
7. Loewy, A. D. (1991). Forebrain nuclei involved in autonomic control. *Prog. Brain Res.* **87**, 253–268.
8. Palkovits, M. (1999). Interconnections between the neuroendocrine hypothalamus and the central autonomic system. Geoffrey Harris Memorial Lecture, Kitakyushu, Japan, October 1998. *Front Neuroendocrinol.* **20**, 270–295.
9. Yoshida, K., Konishi, M., Nagashima, K., Saper, C. B., and Kanosue, K. (2005). Fos activation in hypothalamic neurons during cold or warm exposure: projections to periaqueductal gray matter. *Neuroscience.* **133**, 1039–1046.
10. Shirasaka, T., Takasaki, M., and Kannan, H. (2003). Cardiovascular effects of leptin and orexins. *Am. J. Physiol. Regul. Integr. Comp. Physiol.* **284**, R639–R651.
11. Wynne, K., Stanley, S., McGowan, B., and Bloom, S. (2005). Appetite control. *J. Endocrinol.* **184**, 291–318.
12. Benarroch, E. E., and Stotz-Potter, E. H. (1998). Dysautonomia in fatal familial insomnia as an indicator of the potential role of the thalamus in autonomic control. *Brain Pathol.* **8**, 527–530.
13. Dietrichs, E., and Haines, D. E. (2002). Possible pathways for cerebellar modulation of autonomic responses: micturition. *Scand. J. Urol. Nephrol. Suppl.* **210**, 16–20.
14. Behbehani, M. M. (1995). Functional characteristics of the midbrain periaqueductal gray. *Prog. Neurobiol.* **46**, 575–605.
15. Morrison, S. F. (2001). Differential control of sympathetic outflow. *Am. J. Physiol. Regul. Integr. Comp. Physiol.* **281**, R683–R698.
16. Van Bockstaele, E. J., and Aston-Jones, G. (1995). Integration in the ventral medulla and coordination of sympathetic, pain and arousal functions. *Clin. Exp. Hypertens.* **17**, 153–165.
17. Nakamura, K., Matsumura, K., Kobayashi, S., and Kaneko, T. (2005). Sympathetic premotor neurons mediating thermoregulatory functions. *Neurosci. Res.* **51**, 1–8.
18. Grace, A. A., Gerfen, C. R., and Aston-Jones, G. (1998). Catecholamines in the central nervous system: overview. *Adv. Pharmacol.* **42**, 655–670.
19. Quirion, R., Björklund, A., and Hökfelt, T., eds. (2002). "Peptide Receptors, Part I." Elsevier, New York.
20. Quirion, R., Björklund, A., and Hökfelt, T., eds. (2002). "Peptide Receptors, Part II." Elsevier, New York.

89

Autonomic Neuropathies

Steven Vernino, MD, PhD
Phillip A. Low, MD

Keywords: *amyloid, autoimmune, Composite Autonomic Severity Scale, diabetic, distal small fiber neuropathy, ganglionic antibody*

I. Introduction
II. History and Nomenclature
III. Etiology
IV. Pathogenesis and Pathophysiology
V. Structural Basis
VI. Pharmacology, Biochemistry, Molecular Mechanisms
VII. Pathophysiological Basis of Symptoms
VIII. Natural History
References

I. Introduction

Many peripheral neuropathies affect autonomic fibers in isolation or in addition to somatic sensory and motor fibers. When involvement of autonomic fibers is prominent, the term *autonomic neuropathy* is used. The most commonly encountered autonomic neuropathies are listed in Table 1. In diabetes, for instance, involvement of autonomic nerve fibers is common and sometimes leads to disabling symptomatology. Other neuropathies, inherited or acquired, specifically target autonomic fibers. Our current understanding of the autonomic neuropathies has improved due to better understanding of the molecular basis of the inherited autonomic neuropathies, development of objective and noninvasive tests of autonomic function, and recognition of an immunological basis for many cases of subacute acquired autonomic neuropathy (the autoimmune autonomic neuropathies).

II. History and Nomenclature

Autonomic neuropathy can be caused by certain drugs and toxins or in association with systemic disorders that affect the nerves more diffusely (e.g., diabetes amyloidosis).

Table 1 Autonomic Neuropathies

Autonomic Neuropathies with Acute/Subacute Onset
1. Autoimmune autonomic neuropathy
2. Paraneoplastic autonomic neuropathy
3. Guillain-Barré syndrome
4. Botulism
5. Porphyric neuropathies
6. Drug-induced autonomic neuropathies
7. Toxic autonomic neuropathies

Chronic Autonomic Neuropathies
1. Diabetic autonomic neuropathies
2. Hereditary sensory and autonomic neuropathies types I–V
3. Amyloid neuropathy
4. Distal small fiber neuropathy
5. Fabry disease
6. Postural tachycardia syndrome (POTS) (some cases)

Diabetic autonomic neuropathy can involve distal autonomic nerves, typically in conjunction with involvement of somatic C fibers (distal small-fiber neuropathy, or DSFN), autonomic innervation of the heart and blood vessels (cardiovascular autonomic neuropathy, or CAN), and/or involvement of the autonomic nerves controlling bowel motility (diabetic enteric neuropathy). It has been proposed that the presence and severity of CAN are independent risk factors for mortality [1]. Autonomic nerves can also be affected selectively in inherited neuropathies. These fall in the category of hereditary sensory and autonomic neuropathy (HSAN) to distinguish them from the hereditary motor and sensory neuropathies (HMSN, also known as Charcot-Marie-Tooth disease).

Once toxic and metabolic causes are excluded, many cases of acquired acute or subacute autonomic failure may be attributed to autoimmunity targeting the autonomic nerves and/or ganglia. Pure acute dysautonomia was first described as a discrete clinical entity by Young et al. in 1969 [2]. This disorder was characterized by a subacute-onset, monophasic course with recovery, sympathetic and parasympathetic failure, and no significant evidence of somatic peripheral neuropathy. Subsequent studies in larger patient groups revealed that this syndrome often follows a viral prodrome and can be associated with elevated levels of cerebrospinal fluid (CSF) protein [3,4]. In addition to the time course of the symptoms, subacute autonomic neuropathy may occur in association with known autoimmune neuromuscular disorders, such as myasthenia gravis, Lambert-Eaton myasthenic syndrome, or acquired neuromyotonia, or it may be associated with occult cancer (paraneoplastic autonomic neuropathy). Initially, this disorder was considered to be an autonomic variant of Guillain-Barré syndrome (GBS). However, damage to small, unmyelinated autonomic fibers would not be expected in an inflammatory demyelinating neuropathy, and the autonomic symptoms in GBS are usually related to autonomic instability rather than severe diffuse autonomic failure. On the basis of clinical features and recent laboratory studies, the term *autoimmune autonomic neuropathy* (AAN) has been proposed as a more appropriate description for idiopathic immune-mediated peripheral autonomic neuropathy and helps distinguish this disorder from GBS. Paraneoplastic autonomic neuropathy is a form of immune-mediated autonomic neuropathy that occurs as a remote complication of cancer and likely has a different pathophysiology than nonparaneoplastic AAN.

III. Etiology

There are many causes of autonomic neuropathy (Table 1). The acute/subacute autonomic neuropathies are generally autoimmune, due to metabolic disorders, toxins, or drugs. AAN can occur in isolation, where about half the cases are associated with neuronal acetylcholine receptor (ganglionic AChR) antibody. Other cases are associated with cancer, especially small-cell lung carcinoma (SCLC), thymoma, and lymphoma.

GBS is an acute/subacute immune-mediated demyelinating neuropathy. Autonomic instability is common and autonomic problems of some degree, particularly involving cardiovascular and gastrointestinal function, are found in two-thirds of patients and may be a life-threatening complication of the disease [5].

Autonomic involvement is an integral component of diabetic neuropathy. It can be subacute, as in diabetic radiculoplexus neuropathy, diabetic cranial neuropathy, or diabetic gastroparesis, or it can be chronic. Vagal involvement occurs early, as does distal sudomotor loss [6].

Botulism causes an acute autonomic neuropathy that follows the ingestion of food contaminated by *Clostridium botulinum*. The patient develops an acute cholinergic neuropathy with widespread anhidrosis, sicca syndrome, ileus, and urinary retention [6].

The hepatic porphyries are autosomal dominant disorders of heme biosynthesis. The clinical features are those of an acute peripheral and autonomic neuropathy, skin symptoms, and central nervous system manifestations. Autonomic involvement is common in acute intermittent porphyria, and dysautonomia may also occur in variegate porphyria. Autonomic hyperactivity with persistent tachycardia may precede the onset of somatic neuropathy. Abdominal pain, nausea, vomiting, obstinate constipation, bladder distension, and sweating abnormalities are also present. Sudden death during an acute attack has been related to cardiac arrhythmias.

Drugs that can cause autonomic neuropathy include cisplatin, vincristine, amiodarone, and perhexiline. Toxins that cause autonomic neuropathy are the heavy metals, hexacarbons, and acrylamide. The neuropathies caused by toxins and drugs typically are acute axonal neuropathies.

IV. Pathogenesis and Pathophysiology

A. Autoimmune Autonomic Neuropathies

AAN and paraneoplastic autonomic neuropathies are neuroimmunological disorders. These disorders occur when the host immune system becomes sensitized to self antigens in the nervous system. The most persuasive evidence for autoimmunity in these conditions is the presence of antibodies in the serum and CSF that specifically bind to neuronal autoantigens. In some disorders, these antibodies have a direct role in pathogenesis, whereas in other disorders, the antibodies are nonpathogenic markers of a cell-mediated attack on neurons. The neurological autoimmune response can be initiated in several ways, but in many cases, the initial immunological stimulus remains unknown.

The clinical features of AAN (also known as acute panautonomic neuropathy, idiopathic autonomic neuropathy, or acute pandysautonomia) reflect involvement of parasympathetic, sympathetic, and enteric nervous systems. Less common patterns are those of selective cholinergic failure, selective adrenergic neuropathy, or isolated gastrointestinal dysmotility. The acute or subacute onset of autonomic failure occurs with relative or complete sparing of somatic nerve fibers. Neuropathic symptoms, such as tingling in the distal extremities, occur in approximately 25% of patients, but these symptoms are not accompanied by objective signs or electrophysiological evidence of somatic neuropathy. A presumed antecedent viral infection may be reported in about 60% of cases, with flu-like illness or upper respiratory infection being the most frequent association [3]. Specific preceding viral infections have been reported, most commonly Epstein-Barr virus infection. Presumably, these viral infections stimulate the immune system either nonspecifically (in patients predisposed to autoimmunity) or specifically (through molecular mimicry by presenting antigens that are similar to those found in the autonomic nervous system). The observation that some patients respond to intravenous immunoglobulin or plasma exchange suggests that this disorder may be caused by pathogenic autoantibodies.

About 50% of patients with the typical clinical features of AAN have high titers of autoantibodies directed against the ganglionic acetylcholine receptor (AChR) [7]. This receptor is a pentameric transmembrane complex consisting of two AChR α3 subunits in combination with AChR β subunits. The α3-type ganglionic AChR mediates fast synaptic transmission in all peripheral autonomic ganglia and is homologous but genetically and immunologically distinct from the AChR at the neuromuscular junction. Serum ganglionic AChR antibody levels in AAN cases correlate with the severity of autonomic neuropathy clinically and in laboratory testing [7]. Although the finding of high levels of ganglionic AChR antibody is specific for the diagnosis of AAN, a negative antibody test does not rule out the diagnosis. Ganglionic AChR antibodies can also be found in patients with lung cancer–related and thymoma-related autonomic neuropathy, so a positive test does not exclude a paraneoplastic cause [7]. Clinically, patients with ganglionic AChR antibodies more often have a subacute onset and generally show more prominent cholinergic dysautonomia (sicca complex, pupillary abnormalities, and gastrointestinal tract symptoms), compared with patients with seronegative autonomic neuropathy [8].

The definition of AAN as an antibody-mediated disorder and a better understanding of the pathophysiology of the disorder come from experimental animal models of autoimmune autonomic neuropathy. A critical step is demonstration that the specific antibodies reproduce the cardinal features of the disease in animals. Antibodies can be obtained from affected individuals and administered to the animal (passive transfer) or induced in the animal by immunization with the antigen of interest (active immunization). In the ideal setting (as with myasthenia gravis and experimental autoimmune myasthenia gravis), both passive transfer and active immunization reproduce the disease.

Experimental autoimmune autonomic neuropathy (EAAN) can be produced by immunizing rabbits with a recombinant α3 AChR subunit fusion protein in adjuvant [9]. Within a few weeks of immunization, rabbits begin to produce ganglionic AChR antibodies and develop chronic severe generalized dysautonomia with prominent "cholinergic" failure. Rabbits producing high levels of ganglionic AChR antibody develop gastrointestinal dysmotility, severe parasympathetic dysfunction, and reduced levels of plasma catecholamine. As in patients with AAN, the severity of autonomic disturbances is greater in rabbits with higher antibody levels. Neurons in autonomic ganglia are intact but show a loss of surface ganglionic AChR [10]. Electrophysiological studies of mesenteric ganglia isolated from EAAN rabbits demonstrate a failure of ganglionic synaptic transmission consistent with the loss of postsynaptic receptors [9].

Autoimmune autonomic neuropathy can also be induced transiently in mice by passive transfer of IgG from affected rabbits or humans [11]. The ability to transfer the disease with antibody alone is persuasive evidence that AAN is primarily caused by the effects of ganglionic AChR antibodies.

The pathogenesis of autonomic neuropathy in the context of a systemic malignancy may be somewhat different. These syndromes likely represent a complication of naturally occurring tumor immunity that is beneficial to limiting tumor metastasis and growth. Under certain conditions, the anti-tumor immune response can lead to a specific anti-neuronal autoimmune reaction [12]. Some tumors aberrantly express antigens that are usually restricted to neurons. Many of these antigens are constituents of peripheral autonomic neurons. For example, the Hu proteins, a family of neuronal nuclear RNA-binding proteins, can be found in SCLC cells and in most neurons, including those in autonomic ganglia.

Some SCLC cells express neuronal ion channels including ganglionic AChR [9]. See the chapter on paraneoplastic pathophysiology in this text for a more detailed discussion of that condition

Autonomic neuropathy is a common paraneoplastic syndrome. The autoantibody most commonly associated with paraneoplastic autonomic neuropathy is anti-Hu, also known as anti-neuronal nuclear type 1 (ANNA-1). SCLC is found in more than 80% of patients who are seropositive for ANNA-1 [13]. Among patients with ANNA-1 or collapsin response mediator protein-5 (CRMP-5, another common paraneoplastic antibody), up to 30% have autonomic neuropathy. The most common manifestation of paraneoplastic dysautonomia is gastrointestinal dysmotility, and enteric neuropathy can sometimes be the sole manifestation of a paraneoplastic syndrome [13]. Pathologically, paraneoplastic dysmotility has been associated with an inflammatory destructive process affecting myenteric ganglia of the gut. In postmortem or surgical samples of the esophagus, stomach bowel, and colon, every area shows abnormalities in the myenteric plexus, with reduction in neurons and axons and lymphocytoplasmic infiltration [14]. Some investigators have demonstrated that IgG extracted from the serum of patients with paraneoplastic enteric neuropathy can induce cell death in cultured myenteric neurons, but the majority of evidence suggests that the enteric ganglionitis is caused primarily by cell-mediated immunity rather than a direct antibody effect. By inference, more-diffuse paraneoplastic autonomic neuropathy is probably due to cytotoxic lymphocytic infiltration and destruction of neurons in peripheral autonomic ganglia more generally and possibly also in central autonomic nuclei.

B. Diabetic Autonomic Neuropathy

DAN is unique in the multiplicity of targets of oxidative injury, demonstrated in experimental and human diabetic neuropathy [15]. There is involvement of neural microvessels and macrovessels, autonomic and sensory ganglia, axons and Schwann cells, and the interstitium [16]. Many distinct forms of diabetic neuropathy and diabetic autonomic neuropathy are described. This recognition provides some explanation for the great success in preventing diabetic neuropathy and lack of success in treating established DAN.

Peripheral autonomic neuropathy (distal anhidrosis and vasomotor instability) accompanies the diabetic sensorimotor peripheral neuropathy and represents damage to the long axons of both small and large nerve fibers. More "proximal" autonomic failure (such as CAN and enteric neuropathy) may also be length-dependent in that involvement concerns, in major part, distal ends of the long vagus and splanchnic-mesenteric nerves.

C. Inherited Autonomic Neuropathy

Some of the inherited autonomic neuropathies are due to specific mutations (Table 2). HSAN I is reported to be due to a mutation of the serine palmitoyltransferase gene (*SPTLC1*) located on chromosome 9q22.9,23. Serine palmitoyltransferase (SPT) is thought to be the rate-limiting enzyme in the synthesis of sphingolipids, including ceramide and sphingomyelin. Because ceramide is important in regulation of programmed cell death in a number of tissue types, including differentiating neuronal cells, this mutation may be pathogenetically important in the production of neuropathy. There are also some families with the HSAN I phenotype and associated distal weakness with linkage to human chromosome 3q, indicating some genetic heterogeneity in the HSAN I group.

The HSAN III gene was previously mapped to a 0.5-cM region on chromosome 9q31. Recently, the minimal candidate region was sequenced and cloned and its five genes were characterized [17]. One of these, *IKBKAP*, harbors two mutations that can cause HSAN III. The major haplotype

Table 2 Hereditary Sensory Autonomic Neuropathy (HSAN)

Disease Type	Onset	Inheritance	Neurons (Axons)			Sudomotor	Loci	Gene
			Aα	Aδ	C			
1[a]	2+ decade	AD	+	++	++	LS+	9q22.1	SPTLC1
2	C	AR	++	++	+	G	Unknown	Unknown
3	C	AR	++	++	++	G	9q31	IKBKAP
4	C	AR	N	N	++	G	1q21	trKA
5	C	AR	N	++	?N	G	?1q21[b]	?trKA[b]

[a] See text for other autosomal dominant forms without genetic characterization.
[b] A single patient reported.

Note: C, congenital; AD, autosomal dominant; AR, autosomal recessive; N, normal; +, affected; ++, severely affected; G, generalized; LS+, lumbosacral plus.

mutation is located in the donor splice site of intron 20. This mutation can result in skipping of exon 20 in the mRNA of patients with HSAN III. The mutation associated with the minor haplotype in four patients is a missense (R696P) mutation in exon 19, which is predicted to disrupt a potential phosphorylation site. Almost all cases of HSAN III are caused by an unusual splice defect. The cause of HSAN IV is mutation of the *NTRK1* gene coding for the neurotrophic tyrosine kinase receptor type 1.

D. Amyloid Autonomic Neuropathy

Amyloid neuropathy results from the accumulation and deposition of insoluble protein aggregates (amyloid) in blood vessels and nerves. Sporadic primary amyloidosis is invariably associated with peripheral neuropathy and autonomic failure. Cardiomyopathy due to amyloid deposition in cardiac muscle may contribute to the cardiovascular autonomic symptoms as well. Inherited amyloid neuropathy is due to mutations of three proteins: TTR, apolipoprotein A1, and gelsolin.

Formerly known as familial amyloid polyneuropathy type I, TTR-related FAP was originally described by Andrade in northern Portugal. It is dominantly inherited and is the most common type of FAP and has also been described by researchers in Brazil, Japan, Sweden, and elsewhere. More than 80 point mutations of the TTR gene have been reported. The TTR mutation (Val30Met) was the first described and is the most common. Age at onset and penetrance of the disease vary significantly. There is variability depending on the population at risk. In Portugal, the penetrance for TTR amyloidosis is high and the progression is rapid. The opposite occurs in Sweden, with low penetrance and slow progression of the disease.

The clinical manifestations of apolipoprotein A1 FAP are similar to those of TTR FAP except for early renal involvement and high incidence of duodenal ulcers. Autonomic involvement is not as florid as in TTR FAP. The amyloid fibrils consist of a variant of apolipoprotein A1.

Originally described in Finland and formerly known as FAP IV, hereditary gelsolin amyloidosis is a systemic disorder caused by a point mutation in the gelsolin gene on chromosome 9. A substitution of asparagine for aspartic acid at residue 187 (654G-A) is the most common gelsolin mutation. Other substitutions have been reported (654G-T). These mutations lead to abnormal proteolysis of gelsolin, resulting in an amyloidogenic fragment [18]. Gelsolin amyloidosis is characterized by ocular manifestations. Corneal opacity due to amyloid infiltration, referred to as lattice corneal dystrophy, is a cardinal clinical feature. Reduction in heart rate variability in most patients with orthostatic hypotension (OH) is commonly present [19].

V. Structural Basis

Involvement of autonomic nerve fibers can occur at the preganglionic, ganglionic, or postganglionic level to cause a loss of function. Depending on the end organ, the subject might develop neurogenic bladder, ileus, loss of sweating, loss of pupillary control, or OH.

Biopsy of a peripheral nerve, such as a sural nerve biopsy, allows a detailed qualitative and quantitative examination of all fiber populations and blood vessels. There are certain characteristic diagnostic findings that can be made when selective immunolabels are used. For instance, amyloid deposits are readily identified, perivascular round-cell inflammation can be seen, and the distribution of myelinated and unmyelinated fibers can be determined. In patients with autonomic neuropathies, there is typically a disproportionate loss of unmyelinated fibers.

The importance to the splanchnic-mesenteric bed in the maintenance of postural normotension is supported by the close relationship between pathological changes affecting the greater splanchnic nerve and OH. In a study of patients with diabetic vs alcoholic neuropathy of similar severity, there was good agreement between denervation of this preganglionic nerve and OH [6]. The fiber density was significantly reduced in the greater splanchnic nerve of diabetics and not in alcoholics. Disordered blood pressure (BP) control in diabetes correlated with the pathological abnormalities in the sympathetic nervous system.

Skin biopsy with labeling of intraepidermal nerve fibers is a relatively simple approach to sample the skin and examine somatic unmyelinated fibers. In a prospective study of somatic and autonomic C-fiber function, we evaluated 11 healthy control subjects and 38 patients with different clinical patterns of neuropathy [20]. Intraepidermal nerve fiber density was used to evaluate distal somatic C fibers in patients with distal small-fiber neuropathy (DSFN), nonspecific peripheral neuropathy, diabetic neuropathy, neuropathic postural tachycardia syndrome (POTS), and AAN. Both a quantitative sudomotor axon reflex test and skin norepinephrine content measurement were done at the biopsy site to assess distal autonomic C-fiber function. Postganglionic sudomotor, adrenergic, and cardiovagal functions were evaluated by autonomic reflex testing and quantified with the Composite Autonomic Severity Scale (CASS). Skin norepinephrine concentration was significantly related to CASS. Diabetic neuropathy was associated with somatic and autonomic C-fiber impairment with good agreement. POTS was associated with selective distal autonomic deficit. DSFN had combined distal somatic and C-fiber impairment. AAN showed combined and selective distal and generalized autonomic C-fiber impairment. The somatic neuropathies had C-fiber impairment affecting both populations to varying degrees.

VI. Pharmacology, Biochemistry, Molecular Mechanisms

The physiology and pharmacology of BP control is typically impaired in the autonomic neuropathies [6]. The baroreflex is important in maintaining postural normotension. High and low pressure baroreceptors from the thorax and neck (carotid sinus and aortic arch) [21] send afferents via the IX and X cranial nerves to the nucleus of the tractus solitarius. From this nucleus, a polysynaptic cardiovagal pathway travels to the nucleus ambiguus and dorsal motor nucleus of the vagus and, thence, via the vagus nerve to the sinoatrial node to control heart rate. Sympathetic function is regulated by the rostroventrolateral nucleus of medulla, which projects to the intermediolateral column of the thoracic spinal cord, which, in turn, provides sympathetic innervation to the heart and arterioles [22]. Autonomic neuropathy can involve the afferent or efferent limbs or both of the baroreflex. The lesion is typically postganglionic in the autonomic neuropathies. Postganglionic denervation disconnects resistance vessels. Upregulation of α_1-adrenoreceptors occurs, and the denervated arteriole responds to its normal neurotransmitter (norepinephrine) with an excessive response (denervation supersensitivity).

The splanchnic-mesenteric capacitance bed is a large-volume, low-resistance system of great importance in the maintenance of postural normotension in humans. It constitutes 25% to 30% of the total blood volume [23]. Unlike muscle veins, splanchnic veins have an abundance of smooth muscle and a rich sympathetic innervation. The mesenteric capacitance bed is markedly responsive to both arterial and venous baroreflexes. Venoconstriction is mediated by α-adrenergic receptors [24]. The nerve supply to the mesenteric bed is mostly from preganglionic axon in the greater splanchnic nerve, with cell bodies in the intermediolateral column (mainly T4 to T9) that synapse in the celiac ganglion, from whence postganglionic adrenergic fibers supply effector cells. Abnormalities in the splanchnic autonomic outflow have been found in human diabetic neuropathy, indicating that preganglionic fibers can be affected [25].

VII. Pathophysiological Basis of Symptoms

Denervation of end organs leads to loss of function. Hence, patients with generalized autonomic failure will have widespread anhidrosis, OH, and neurogenic bladder and bowel. Additionally, in postganglionic adrenergic denervation, there can be an upregulation of adrenergic receptors on denervated arterioles (denervation supersensitivity).

Vasomotor tone and heart rate are controlled by the baroreflex. Baroreflexes are often referred to as "buffer" nerves because they maintain a constant BP in all positions. Baroreflex failure results in the triad of OH, supine hypertension, and loss of diurnal variation in BP. In normal subjects, BP is lower at night than during the day, whereas the converse occurs in baroreflex failure [26]. Autonomic neuropathies can affect the baroreflex in its afferent or efferent limbs.

The splanchnic-mesenteric capacitance bed is also critical in maintaining postural normotension in humans. Splanchnic veins have an abundance of smooth muscle and a rich sympathetic innervation. The mesenteric capacitance bed is markedly responsive to both arterial and venous baroreflexes. Following a meal, there is a 200% to 300% increase in mesenteric blood flow, and this can greatly worsen OH [27].

Autonomic ganglion involvement occurs in a number of autonomic neuropathies, including diabetic and AAN. Because the baroreflexes are relayed through autonomic ganglia, such lesions interrupt baroreflexes and OH can occur. Ganglionic involvement is often apparent on the thermoregulatory sweat test, where regional areas of anhidrosis are seen [28].

The distal ends of unmyelinated C fibers are commonly affected in the neuropathies. Both somatic and autonomic C fibers are involved. Partially injured C fibers are prone to be spontaneously active. Spontaneous firing of somatic C fibers is presumed to be responsible for the distal burning, stabbing, and lancinating pain. Similar activity of autonomic C fibers occurs. Adrenergic overactivity results in vasomotor changes. Patients complain of coldness and blanching of the feet, sometimes with episodes of blueness or redness. Some patients develop the full clinical picture of erythromelalgia [29]. Loss of function can result in a vasodilated warm extremity. Similarly, sudomotor dysfunction can be manifested as acral hyperhidrosis followed by anhidrosis [30].

VIII. Natural History

The natural history of the autonomic neuropathies depends on the specific pathogenesis and pathophysiology. For instance, the autoimmune autonomic neuropathies have a much better prognosis than diabetic neuropathy. In the case of GBS, the target is myelin, which is capable of regeneration, and autonomic nerve involvement is typically partial and subject to recovery. Typically, autonomic neuropathy is GBS improves in concert with improvement in motor and sensory nerve function. Long-term autonomic sequelae are not typical. With AAN, the illness is often monophasic. Prognosis is variable, but spontaneous improvement can occur. Autonomic improvement occurs in concert with a fall in ganglionic antibody titer [7]. In contrast, diabetic autonomic neuropathy is characterized by inexorable progression of autonomic failure. In this autonomic neuropathy, there is structural damage to the triad of microvessels (and macrovessels), neurons, and Schwann cells, so that little is reversible by the time established autonomic neuropathy occurs.

The natural history of autonomic distal neuropathies due to toxins and drugs is more favorable. The brunt of the pathology is manifested at the distal ends of the longest nerve fibers. Although the cell body can be affected, its involvement is usually partial and capable of recovery. Following withdrawal of the toxin or cessation of the drug, slow recovery is expected. The natural history of positive somatic and autonomic symptomatology is complex. Recovery is typical but not invariable. Part of the incomplete recovery relates to the incompleteness of fiber regeneration. However, there can be functional alterations. For instance, with nerve injury to sympathetic sudomotor fibers, there can be a switch from sympathetic cholinergic to adrenergic innervation [31], which has been proposed as the basis for protopathic pain.

Acknowledgments

This work was supported in part by grants NS 32352 (Vernino; Low), NS 44233 (Low), and NS 43364 (Low), from the National Institutes of Health, Bethesda, Maryland; grant MO1 R00585 from the Mayo General Clinical Research Center, Rochester, Minnesota, and Mayo Funds (Low).

References

1. Rathmann, W., Ziegler, D., Jahnke, M., Haastert, B., and Gries, F. A. (1993). Mortality in diabetic patients with cardiovascular autonomic neuropathy. *Diabet. Med.* **10**, 820–824.
2. Young, R. R., Asbury, A. K., Corbett, J. L., and Adams, R. D. (1975). Pure pan-dysautonomia with recovery. Description and discussion of diagnostic criteria. *Brain* **98**, 613–636.
3. Suarez, G. A., Fealey, R. D., Camilleri, M., and Low, P. A. (1994). Idiopathic autonomic neuropathy: Clinical, neurophysiologic, and follow-up studies on 27 patients. *Neurology* **44**, 1675–1682.
4. Hart, R. G., and Kanter, M. C. (1990). Acute autonomic neuropathy: two cases and a clinical review. *Arch. Intern. Med.* **150**, 2373–2376.
5. Zochodne, D. W., Ward, K. K., and Low, P. A. (1988). Guanethidine adrenergic neuropathy: an animal model of selective autonomic neuropathy. *Brain Res.* **461**, 10–16.
6. Low, P. A., Vernino, S., and Suarez, G. A. (2003). Autonomic dysfunction in peripheral nerve disease. *Muscle Nerve* **27**, 646–661.
7. Vernino, S., Low, P. A., Fealey, R. D., Stewart, J. D., Farrugia, G., and Lennon, V. A. (2000). Autoantibodies to ganglionic acetylcholine receptors in autoimmune autonomic neuropathies. *N. Engl. J. Med.* **343**, 847–855.
8. Sandroni, P., Vernino, S., Klein, C. M., Lennon, V. A., Benrud-Larson, L., Sletten, D., and Low, P. A. (2004). Idiopathic autonomic neuropathy: comparison of cases seropositive and seronegative for ganglionic acetylcholine receptor antibody. *Arch. Neurol.* **61**, 44–48.
9. Lennon, V. A., Ermilov, L. G., Szurszewski, J. H., and Vernino, S. (2003). Immunization with neuronal nicotinic acetylcholine receptor induces neurological autoimmune disease. *J. Clin. Invest.* **111**, 907–913.
10. Vernino, S., Low, P. A., and Lennon, V. A. (2003). Experimental autoimmune autonomic neuropathy. *J. Neurophysiol.* **90**, 2053–2059.
11. Vernino, S., Ermilov, L. G., Sha, L., Szurszewski, J. H., Low, P. A., and Lennon, V. A. (2004). Passive transfer of autoimmune autonomic neuropathy to mice. *J. Neurosci.* **24**, 7037–7042.
12. Roberts, W. K., and Darnell, R. B. (2004). Neuroimmunology of the paraneoplastic neurological degenerations. *Curr. Opin. Immunol.* **16**, 616–622.
13. Lucchinetti, C. F., Kimmel, D. W., and Lennon, V. A. (1998). Paraneoplastic and oncologic profiles of patients seropositive for type 1 antineuronal nuclear autoantibodies. *Neurology* **50**, 652–657.
14. Jun, S., Dimyan, M., Jones, K. D., and Ladabaum, U. (2005). Obstipation as a paraneoplastic presentation of small cell lung cancer: case report and literature review. *Neurogastroenterol. Motil.* **17**, 16–22.
15. Low, P. A. (2003). Oxidative stress: an integrative view. In: "Textbook of Diabetic Neuropathy" (F. A. Gries, N. E. Cameron, P. A. Low, and D. Ziegler, eds). pp. 123–128. Thieme, New York.
16. Dyck, P. J., and Giannini, C. (1996). Pathologic alterations in the diabetic neuropathies of humans: a review. *J. Neuropathol. Exp. Neurol.* **55**, 1181–1193.
17. Slaugenhaupt, S. A., Blumenfeld, A., Gill, S. P., Leyne, M., Mull, J., Cuajungco, M. P., Liebert, C. B., Chadwick, B., Idelson, M., Reznik, L., Robbins, C., Makalowska, I., Brownstein, M., Krappmann, D., Scheidereit, C., Maayan, C., Axelrod, F. B., and Gusella, J. F. (2001). Tissue-specific expression of a splicing mutation in the IKBKAP gene causes familial dysautonomia. *Am. J. Hum. Genet.* **68**, 598–605.
18. Levy, E., Haltia, M., Fernandez-Madrid, I., Koivunen, O., Ghiso, J., Prelli, F., and Frangione, B. (1990). Mutation in gelsolin gene in Finnish hereditary amyloidosis. *J. Exp. Med.* **172**, 1865–1867.
19. Kiuru, S., Matikainen, E., Kupari, M., Haltia, M., and Palo, J. (1994). Autonomic nervous system and cardiac involvement in familial amyloidosis, Finnish type (FAF). *J. Neurol. Sci.* **126**, 40–48.
20. Singer, W., Spies, J. M., Hauer, P., McArthur, J. C., Griffin, J. W., and Low, P. A. (2000). Epidermal nerve fiber density in the neuropathic type of the postural tachycardia syndrome. *Neurology* **54 (Suppl 3)**, A226.
21. Korner, P. I., West, M. J., Shaw, J., and Uther, J. B. (1974). "Steady-state" properties of the baroreceptor-heart rate reflex in essential hypertension in man. *Clin. Exp. Pharmacol. Physiol.* **1**, 65–76.
22. Joyner, M. J., and Shepherd, J. T. (1997). Autonomic regulation of circulation. In: "Clinical Autonomic Disorders: Evaluation and Management" (P. A. Low, ed). pp. 61–71. Lippincott-Raven, Philadelphia.
23. Rowell, L. B. (1973). Regulation of splanchnic blood flow in man. *Physiologist.* **16**, 127–142.
24. Thirlwell, M. P., and Zsoter, T. T. (1972). The effect of propranolol and atropine on venomotor reflexes in man. Venous reflexes: effect of propranolol and atropine. *J. Med.* **3**, 65–72.
25. Low, P. A., Walsh, J. C., Huang, C. Y., and McLeod, J. G. (1975). The sympathetic nervous system in diabetic neuropathy. A clinical and pathological study. *Brain* **98**, 341–356.
26. Carvalho, M. J., van Den Meiracker, A. H., Boomsma, F., Lima, M., Freitas, J., Veld, A. J., and Falcao De Freitas, A. (2000). Diurnal blood pressure variation in progressive autonomic failure. *Hypertension* **35**, 892–897.
27. Fujimura, J., Camilleri, M., Low, P. A., Novak, V., Novak, P., and Opfer-Gehrking, T. L. (1997). Effect of perturbations and a meal on superior mesenteric artery flow in patients with orthostatic hypotension. *J. Auton. Nerv. Syst.* **67**, 15–23.

28. Fealey, R. D., Low, P. A., and Thomas, J. E. (1989). Thermoregulatory sweating abnormalities in diabetes mellitus. *Mayo Clin. Proc.* **64**, 617–628.
29. Davis, M. D., Sandroni, P., Rooke, T. W., and Low, P. A. (2003). Erythromelalgia: vasculopathy, neuropathy, or both? A prospective study of vascular and neurophysiologic studies in erythromelalgia. *Arch. Dermatol.* **139**, 1337–1343.
30. Stewart, J. D., Low, P. A., and Fealey, R. D. (1992). Distal small fiber neuropathy: results of tests of sweating and autonomic cardiovascular reflexes. *Muscle Nerve* **15**, 661–665.
31. Chemali, K. R., Gorodeski, R., and Chelimsky, T. C. (2001). Alpha-adrenergic supersensitivity of the sudomotor nerve in complex regional pain syndrome. *Ann. Neurol.* **49**, 453–459.

90

Thermoregulation and Its Disorders

Robert D. Fealey, MD

Keywords: anhidrosis, brown adipose tissue, central autonomic network, fever, human thermoregulation, hyperhidrosis, hyperthermia, hypothalamic set-point, hypothermia, neuroendocrine integration, pyrogen, shivering, sweating, vasoregulation

I. Chapter Overview
II. Human Thermoregulation
III. Pyrogens and Fever
IV. Effect of Aging on Human Thermoregulation
V. Some Disorders of Thermoregulation
VI. Thermoregulatory Failure Due to Degenerative Disorders
References

I. Chapter Overview

Thermoregulation is a splendid example of the integrative functioning of the human hypothalamus. Autonomic neural circuitry, adaptive behavioral responses, endocrine, and other hypothalamic regulatory functions are recruited and blended automatically to accomplish thermal balance necessary for the close regulation of body temperature. This "set-point" core temperature is maintained in humans by effector mechanisms that include vasodilation and sweat secretion (which powerfully protect us from hyperthermia) and shivering and vasoconstriction, which in adults are the primary and less robust ways we have of defending against hypothermia.

This chapter illustrates the hierarchy of structures and processes involved in thermoregulation and then discusses some disorders in which components of thermoregulation fail or become excessive. Examples of hyperthermia, hypothermia, hyperhidrosis and anhidrosis, and cutaneous vasomotor dysfunction are discussed. When possible, an illustrative clinical example depicting the underlying neurobiological explanation of the disorder is given.

II. Human Thermoregulation

Humans are homeotherms who regulate body temperature by adjusting the rates of heat production and heat loss. Metabolic activity in visceral organs and muscle continuously produce body heat that is sensed, regulated, and balanced by several heat-loss mechanisms to dynamically produce a relatively constant core temperature.

Human thermoregulatory mechanisms include behavioral adjustments to environmental temperature and, most important, autonomic nervous system effector responses. The latter include sweating and cutaneous vasomotor responses for heat dissipation and shivering and nonshivering thermogenesis and cutaneous vasomotor adjustments for heat production and heat conservation respectively.

Thermoregulation can be thought of in terms of afferent and efferent pathways conveying thermal and nonthermal signals to and from an integrative center in a subcortical autonomic network (predominantly the hypothalamus), which maintains core temperature at a set-point. Figure 1 depicts the principal components of this thermoregulatory system [1].

The medial preoptic (MPO) region of the hypothalamus acts as an adjustable thermostat sensitive to body core and skin temperature input as well as signals from nonthermal homeostatic regulatory systems such as neuroendocrine peptide/hormonal secretory units [2,3]. The MPO region contains warm-sensitive neurons that initiate responses leading to heat dissipation (skin vasodilatation and sweating) and inhibit cold-sensitive neurons that trigger heat production via shivering or brown fat metabolism and through skin vasoconstriction.

Heating warm-sensitive neurons facilitates ionic channel prepotentials, which increase spike generation and firing rate. These cells receive inhibitory input from temperature-insensitive neurons, further enhancing thermal discrimination. Warm-sensitive neurons are the most important effector output of the MPO region and directly or indirectly control most thermoregulatory responses.

Cold-sensitive neurons receive inhibitory inputs from the warm-sensitive neurons and tonic excitatory inputs from temperature-insensitive MPO interneurons. They begin to fire when temperature decreases below a certain level.

Most of the ascending pathways relaying inputs from the skin and spinal thermoreceptors to the MPO region go to warm-sensitive neurons and allow these neurons to integrate central and peripheral thermal information. These pathways include the spinoparabrachio-hypothalamic and spino-hypothalamic tracts. The dendrites of MPO neurons are oriented mediolaterally, perpendicular to the third ventricle, which allows them to collect sensory signals arriving from the medial forebrain bundle laterally and the periventricular system medially. This structural organization at the cellular level provides the functional connectivity to "fine tune" core temperature modulation to afferent influences.

MPO neurons are targeted by many thermoreactive chemical substances. Those that inhibit warm-sensitive neurons and increase set-point temperature when administered into the MPO region include serotonin, acetylcholine, histamine, prostaglandins, endorphins, and Mu opioid agonists. Substances that decrease set-point temperature include dopamine, norepinephrine, thyrotropin-releasing hormone (TRH), neurotensin, and arginine vasopressin. TRH, when administered intravenously, may increase activity of warm-sensitive neurons, thus lowering the set-point temperature, which in turn produces a generalized sweating response. This action of TRH may require dopamine release in the MPO region, and an interesting finding is its marked attenuation in patients with Parkinson's disease.

Progesterone decreases the firing rate of warm-sensitive neurons and increases that of cold-sensitive neurons, elevating "set-point" and core temperature, whereas estradiol has the opposite effects. These hormonal-hypothalamic interactions explain the variations of core temperature that occur during the menstrual cycle.

Pyrogens such as prostaglandin E_2 decrease firing of the warm-sensitive neurons, resulting in suppression of heat-loss responses and disinhibition of the cold-sensitive neurons. This heat production and retention causes an elevated set-point temperature—in other words, a fever.

The most important effector mechanisms for thermoregulation in adult humans depend on the sympathetic outflow to the skin, which stimulates sweat glands via M_3 muscarinic cholinergic receptors and affects skin blood flow via activation or withdrawal of sympathetic vasoconstrictor activity and activation of skin vasodilator nerves, probably through nitric oxide (NO)–related mechanisms. The adult human has about 3 million sweat glands, each one capable of producing 10 nl of sweat per minute to be evaporated on the skin surface. For the entire body this amounts to 1.8 L of sweat/hour and gives us the capacity of dissipating 1000 kcal of heat, or about the amount of heat energy produced by an athlete vigorously exercising steadily for 1 hour [4]. In hot environments, skin arteriovenous anastomoses open and superficial veins dilate. In the cold these anastomoses close and blood is shunted to the deeper veins, the venae comitantes.

Body temperature shows diurnal variation, lower in the early morning and higher in the evening. Similar diurnal variation in the cutaneous blood flow response to heat stress has been observed.

There is evidence of plasticity in thermoregulation. For example, repeated exposure to a hot environment, sufficient

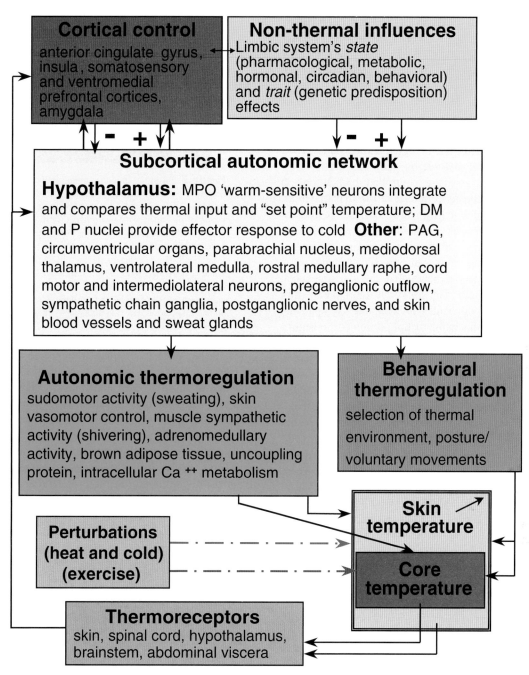

Figure 1 Anatomical and physiological schema of human thermoregulatory sweating. DM, Dorsomedial; MPO, medial preoptic; P, posterior; PAG, periaqueductal gray. (Modified with permission from Thermoregulatory Failure. [2000]. *In:* "Handbook of Clinical Neurology" [O. Appenzeller, ed.], Vol. 75, Series 31: The Autonomic Nervous System, Part II: Dysfunctions. p. 54. Elsevier Science B. V., Amsterdam.)

to raise core (brain) temperature, can lead to heat acclimatation. This adaptation involves a decreased sodium concentration in sweat and a reduced inhibitory effect of elevated plasma osmolarity on thermoregulatory responses to heat stress. Many animal studies show alterations in neuroeffector tissue (for example, the expression of uncoupling protein 1 in mitochondria of brown adipocytes) that are strongly induced by environmental temperature exposure.

The primary responses to cooling in humans are vasoconstriction, which decreases heat loss through the skin, and shivering, which increases muscle metabolic activity and heat production. In animals and human infants, another important mechanism of thermogenesis is known as 'nonshivering thermogenesis,' which involves sympathetic nerve activation of β3 adrenoreceptors in brown adipose tissue (BAT) [5]. This property of BAT is mainly attributable to uncoupling protein (UCP-1), expressed by the brown adipocyte mitochondria. This protein dissipates the hydrogen ion gradient across the inner mitochondrial membrane and uncouples mitochondrial respiration from phosphorylation. This energy is liberated as heat in BAT. The molecular neurobiology of this response is under intense scrutiny at present and involves cAMP-dependent and protein kinase A–dependent activation of a protein kinase that promotes the UCP-1 gene expression via the orchestrated activity of multiple transcription factors.

The responses to cold are controlled by neurons in the dorsomedial and posterior nuclei of the hypothalamus. These neurons are inhibited by warm-sensitive neurons of the MPO and become active when there is a decrease in firing of these neurons (and thence an increase in firing of cold-sensitive neurons) on exposure to cold stimuli.

Experimental studies indicate that the periaqueductal gray (PAG) is an important relay of thermoregulatory signals from the hypothalamus to medullary sympathetic control centers. Neurons in the rostral portion of the ventrolateral PAG receive excitatory warm signals from the MPO region and are involved in active skin vasodilatation responses, whereas neurons in the caudal PAG receive a cold signal from the dorsomedial and posterior hypothalamic nuclei and mediate responses to cold. These cold-induced responses likely involve a relay in the rostral medullary raphe, which, via projections to the intermediolateral (IML) cell column, activates skin vasoconstrictor neurons, and increases sympathetic outflow to BAT. These raphe neurons receive an inhibitory signal from warm-sensitive neurons of the MPO region. Projections to the lateral medullary reticular formation and via tectospinal and rubrospinal tracts to spinal motor neurons activate the shivering response [2].

There is a dynamic link between skin thermal afferent input and central thermoregulation effector responses, as was recently demonstrated in a remarkable study by Egan et al., utilizing positron-emission tomographic (PET) activation imaging of subjects clad in a water-perfused suit [6]. This group has shown human brain regions that respond to changes in skin temperature in the somatosensory cortex, insula, anterior cingulate, thalamus, and hypothalamus, with evidence that the hypothalamic response codes for the direction of temperature change (Fig. 2).

The ventral-anterior medial hypothalamus appeared activated with warming of the skin. This produced forearm vasodilation and sweating. As somatosensory cortical areas showed activation in a similar manner with heat and cold, they mainly seem to reflect registration of the location of the thermal sensation. The cingulate activation may reflect the affective component of the sensory experience involved in the behavioral response to the stimulus.

This study and one by Nagai [7], using functional MRI to image brain regions active during stimuli effecting skin

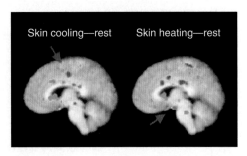

Figure 2 Positron emission tomographic (PET) scan depiction of the thermoregulatory central autonomic network. Regional cerebral blood flow changes show activation of the anterior cingulate gyrus (arrow on left) with heating and cooling and of the anterior hypothalamus (right arrow) with heating of skin. (Modified with permission from Egan, G. F., Johnson, J., Farrell, M., McAllen, R., Zamarripa, F., McKinley, M. J., Lancaster, J., Denton, D., and Fox, P. T. [2005]. Cortical, thalamic, and hypothalamic responses to cooling and warming the skin in awake humans: a positron-emission tomography study. *Proc. Natl. Acad. Sci. USA.* **102**, 5262–5267.)

conductance (primarily demonstrating emotional sweating control regions) have provided exciting visual mapping evidence of those cortical and limbic system regions that are involved in thermoregulation. Such brain regions are critically important to producing the behavioral adjustments necessary for satisfactory thermoregulation.

Thermal-skin autonomic interactions are organized at the spinal cord level as well. Lamina I neurons are thermoreceptive and provide monosynaptic ipsilateral and contralateral projections to the sympathetic preganglionic neurons and may provide the substrate for the somatosympathetic reflex depicted in Figure 3, where segmental skin temperature

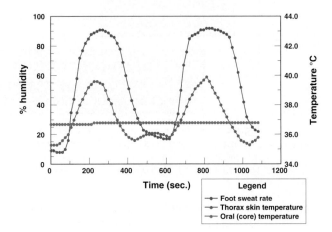

Figure 3 Somatosympathetic reflex regulation of sweating. Note the exquisitely close regulation of sweat rate (on foot) during skin heating (of trunk) while core temperature (oral) remains constant. (Modified with permission from Thermoregulatory Failure. [2000]. *In:* "Handbook of Clinical Neurology" [O. Appenzeller, ed.], Vol. 75, Series 31: The Autonomic Nervous System, Part II: Dysfunctions. p. 57. Elsevier Science B. V., Amsterdam.)

elevation causes widespread activation of sweating, with core temperature remaining constant.

Many of us have utilized a corollary of this reflex by brief cooling of a limb or the face by a cool breeze or splash of water, for example, to stop a generalized sweating response. This reflex is particularly active when core temperature is near the threshold of generalized sweat gland activation, and it is evidence of an important thermoregulatory role for skin temperature.

III. Pyrogens and Fever

The activity of MPO warm-sensitive neurons, derived from the integrated effects of their synaptic input and circulating thermoactive substances, determines the set-point temperature. During fever, a regulated and elevated set-point temperature, exogenous bacterial pyrogens stimulate phagocytic leukocytes to produce endogenous pyrogen (interleukin 1 [IL-1] and IL-6, tumor necrosis factor [TNF], and γ-interferon). Circumventricular organs (CVOs) in the area postrema, subfornical area, and organum vasculosum laminae terminalis (OVLT) may provide access points to neurons that respond to circulating pyrogens and, through their efferent projections, activate central pathways, which release prostaglandin E2 (PGE2) in the ventromedial MPO and the parvicellular subnuclei of the paraventricular nucleus of the hypothalamus (PVN).

Recently it has been postulated that the cytokine-like property of Toll-like receptor (TLR) signal transduction provides an explanation by which any microbial product can cause fever by engaging its specific TLR on the vascular network supplying the thermoregulatory center in the anterior hypothalamus. Animal research has shown remarkable evidence of nuclear signal transcriptional activation in glial cells in CVO and transcriptional down-regulation of proteins involved in PGE2 inactivation to facilitate elevated levels of PGE2 in the brain. Others have shown changes in electrical activity of neurons, induction of other transcription factors causing modifications in gene expression, and localized release of secondary signal molecules, which interact with neural structures inside the blood-brain barrier.

Cytokines that act not only as circulating hormones are also intrinsic modulators in the brain. Neurons immunoreactive for IL-1 are present in the PVN and arcuate nuclei of the human hypothalamus, and IL-1 immunoreactive fibers innervate the MPO, ventromedial hypothalamus, PVN, and lateral hypothalamic area. Signals that stimulate IL-1 production in the brain include humoral factors, such as circulating IL-6, and activation of peripheral C-fibers and vagal afferents. These various mechanisms cause the activity of warm-sensitive neurons to decrease and that of cold-sensitive neurons to increase, raising the set-point temperature. During this time we feel chilled, anorexic, and sleepy, become vasoconstricted and dry, and may shiver as heat-producing and heat-conservation changes are occurring so as to raise body temperature to the new set-point.

Simultaneously, antipyretic mechanisms are activated (increasing, for example, arginine vasopressin, alpha- and gamma-melanocyte stimulating hormones, adrenocorticotropic hormone [ACTH], and glucocorticoid levels in the hypothalamus), inhibiting cyclo-oxygenase enzyme and PGE2 production to limit core temperature rise.

The increasing phase of fever is often associated with a markedly increased heart rate and cardiac output. Defervescence is also often accompanied by tachycardia. Thus, cardiovascular complications are common throughout the febrile course and constitute a major clinical consequence of fever.

Besides the MPO region, thermoreceptors are also found in the skin, medulla, midbrain, spinal cord, and abdominal viscera; each has varying influences on the regulation of temperature. Afferent pathways for thermoregulation enter the cord via the dorsal root ganglion and ascend in the spinothalamic tract as well as in multisynaptic fibers in the lateral spinal cord, projecting to the reticular formation of the brainstem and eventually to the MPO region of the hypothalamus. The vagus nerve and the splanchnic nerves may transmit thermal and nonthermal abdominal visceral sensation to central pathways as well. Recently, experimentally induced hyperthermia has been shown to produce a pronounced increase in the baseline activity and excitability vagal C-fibers probably involving activation of transient, temperature-sensitive ion channel potential vanilloid (TRPV) receptors, expressed on the C-fiber endings.

It is helpful to present a macroscopic example of how fever elevates core set-point temperature at which heat-dissipating (sweating) activity occurs. Figure 4 shows such a situation in a patient with newly diagnosed peripheral neuropathy due to leprosy. The distribution of the neuropathy is usually well demonstrated by thermoregulatory sweat–testing the subject and documenting the areas of anhidrosis. However, during the first sweat test attempt, the patient was febrile and no sweating was produced even after elevating core temperature to 38.0°C. It seemed improbable that the entire cutaneous innervation of the body was involved by leprosy; thus, we retested the patient and she was afebrile. The lower, normal set-point temperature for sweat activation was reached during the test this time, allowing the characteristic pattern of anhidrosis to be visualized.

IV. Effect of Aging on Human Thermoregulation

As might be expected, aging tends to attenuate our ability to thermoregulate. The following age-related observations

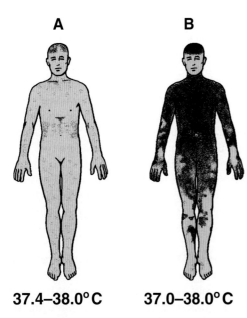

Figure 4 Effect of fever on initiation of thermoregulatory sweating. A patient with lepromatous neuropathy was febrile during her initial thermoregulatory sweat test (TST) **(A)** and failed to sweat even after core temperature increased to 38.0°C. When tested in the afebrile state, sweating commenced at a lower core temperature, revealing the anticipated pattern of neuropathic involvement **(B)** (sweating in dark shaded area).

from many investigators over the past 3 decades support this notion:

1. loss of IML cell column autonomic neurons
2. decreased ability to augment individual sweat gland secretory rate
3. regional decreased sweat output, especially in summertime, causing a significant impediment to maintenance of body temperature with passive heating
4. diminished reactivity of cutaneous vascular tone
5. declines in cold-induced metabolic heat production via skeletal muscle shivering due to reduced muscle mass
6. attenuated BAT cold-induced nonshivering thermogenic capacity due to declines in adiposities, in GDP binding, and in UCP levels in brown fat
7. decreased insulative properties of body fat.

These observations undoubtedly contribute to the vulnerability of the elderly to accidental hypothermia/hyperthermic injury. However, these observations in the elderly can be quite variable and dependent on the genetic predisposition, gender, and conditioning of the individual.

There are several later-life perturbations of thermoregulation that are most common in perimenopausal and postmenopausal women. The most common is the hot flash, which initially may respond to estrogen and progesterone replacement. The serotonin reuptake inhibitors paroxetine and duloxetine have been recently reported to be helpful. These drugs may act to inhibit warm-sensitive neurons and thus elevate the set-point temperature at which vasodilatation (the hot flash) occurs and decrease hot flash frequency. A number of women in the postmenopausal state develop regional hyperhidrosis of the upper body, especially the head, without provocation. The condition is of multifactorial cause with hormonal-hypothalamic feedback factors predominating. In some women the condition begins with stopping of hormone replacement therapy (HRT). Thermoregulatory sweat response to heat is most often normal. Treatment measures range from blocking peripheral M3 cholinergic receptors to using pharmacologic strategies to elevate the set-point temperature.

V. Some Disorders of Thermoregulation

Thermoregulation's complex neuroanatomy and integration with neurobehavioral, immunological, and endocrine-metabolic systems explain the many disorders of thermoregulation that may occur. The following sections are therefore an abbreviated account of some disorders that illustrate the neurobiology of the altered thermoregulation particularly well.

A. Malignant Hyperthermia

Malignant hyperthermia (MH) is a life-threatening and frequently fatal disorder triggered by commonly used anesthetics. MH susceptibility is a genetically determined predisposition to the development of MH. MH is a rare, inherited (often autosomal dominant with variable penetrance) disorder. Mutations in the ryanodine receptor type 1 (RYR1) gene are the major cause of MH susceptibility. Patients with a muscle disorder known as central core disease also have genetic alterations in the ryanodine receptor and can develop MH.

RNA and DNA extracted from muscle tissue or blood lymphocytes can be used for known genetic defects. In one reported technique, the entire RYR1 coding region is amplified in 57 overlapping fragments and subjected to denaturing high-performance liquid chromatography analysis, followed by direct nucleotide sequencing to characterize RYR1 alterations. Denaturing high-performance liquid chromatography analysis of RNA samples extracted from the biopsied skeletal muscle, followed by DNA sequencing, is a highly efficient methodology for RYR1 mutation detection. This approach allows increasing the rate of mutation detection to 70% and identifying mutations in the entire RYR1 coding region. Mutations tend to occur at three "hot spots" (domains) in the receptor molecule. Figure 5 depicts the complex receptor structure and shows the numbered

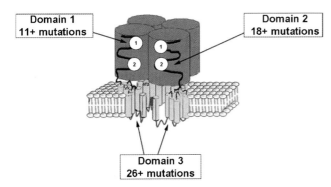

Figure 5 Ryanodine Receptor. The diagram depicts R1Y1 receptor structure and shows three domains and approximate number of genetic mutations reported to be associated with malignant hyperthermia and/or central core disease. (Modified with permission from Treves, S. et al. "Ryanodine receptor 1 mutations, dysregulation of calcium homeostasis and neuromuscular disorders." *Neuromuscul. Disord.*, 2005. V. 15(9–10): Figure 1 Page 580)

domains where many of the described genetic mutations have been identified [8].

The ryanodine (RYR1) receptor is a calcium-release channel of the sarcoplasmic reticulum (SR). MH reactions are often triggered by halothane, succinylcholine, isoflurane, desflurane, sevoflurane, and enflurane. MH reactions may be caused by a prolonged opening of calcium channels with an excessive release of calcium ions from the SR and a consequent increase in heat, water, CO_2, and lactic acid production and in O_2 and adenosine triphosphate (ATP) utilization in the muscle fiber cytoplasm. The molecular mechanisms underlying these disorders in humans are under intense study. Most recently this has been shown, in patients with central core disease activation of the ryanodine receptor, that myotubes release IL-6, a potent circulating endogenous pyrogen. This effect was dependent on de novo protein synthesis and could be blocked by dantrolene, thus explaining the latter's specific beneficial effect in this disorder.

A related disorder of thermoregulatory failure is the neuroleptic malignant syndrome (NMS), an uncommon but increasingly recognized and potentially fatal disorder of thermoregulation. A recent hypothesis on this disorder by Gurrera [9] suggests that NMS results when tonic inhibitory inputs from higher central nervous system centers are disrupted, allowing autonomous but dysregulated sympathetic nervous system hyperactivity. Many of the clinical features of the disorder—hyperthermia, increased muscle tone and tremor, sweating, tachycardia, and labile blood pressure—can be attributed to sympathoadrenal hyperactivity. There is evidence of a trait susceptibility [10], that is, a genetic predisposition to more extreme sympathetic nervous system activation and/or dysfunction in response to emotional or psychological stress, which when coupled with acute psychic distress or dopamine receptor antagonism (due to neuroleptic drugs) gives rise to the NMS syndrome.

B. Episodic Hypothermia

Spontaneous periodic hypothermia, or Shapiro's syndrome, is an uncommon condition that likely has multiple pathophysiological mechanisms. One hypothesis, based on the observation of excessive sweating in spite of very low core temperature, postulates a resetting of the hypothalamic set-point to a lower temperature. The variety of reported treatments for this rare disorder, however, indicates multiple thermoregulatory neurotransmitter disturbances can produce the hypothermia. The hypothermia of Wernicke's encephalopathy due to thiamine deficiency has been attributed to damage of the posterior hypothalamus. Recent MRI evidence [11] suggests other thermoregulatory relay nuclei (for example, the PAG matter and medial periventricular thalamic nuclei) are also implicated.

Decreased cellular heat production, which occurs in hypothyroidism, was long thought to be the main cause of hypothermia. New research in animals, however, shows that the hypothermia may be regulated (a lower set-point temperature), maintained by behavioral thermoregulation environmental selection. Such studies have led to more recent experiments with neuropeptide-transmitter analogs (like neurotensin-77) that produce a lower core temperature, utilizing the integrative neuroendocrine function of the hypothalamus instead of forced external cooling and blockade of heat defense mechanisms [12,13]. In one animal model, this regulated hypothermia was associated with less oxidative stress than forced external cooling during experimental ischemia trials, an observation that may influence future therapy for brain injury.

C. Anhidrosis and Hyperhidrosis: Chronic Idiopathic Anhidrosis

Thermoregulatory dysfunction can occasionally be caused by isolated failure of sympathetic sudomotor activity. Low and colleagues suggested the name *chronic idiopathic anhidrosis* (CIA) for this condition. The main clinical features include symptoms of heat intolerance; patients become hot, flushed, dyspneic, lightheaded, and weak when the ambient temperature is high or when exercising [14]. Reviews of this entity, however, have emphasized the heterogeneous features of this condition [15], and we have encountered distinct subtypes, two of which are discussed next. Not uncommonly, patients first note asymmetric hyperhidrosis from a body segment that is still capable of sweating. This type of excessive sweating is known as compensatory and/or perilesional hyperhidrosis.

1. Segmental Type of CIA

The patient shown in Figure 6 was aware of his asymmetric sweating for several years before his 1982 sweat test. He had moderate exercise and heat intolerance. From 1983 to 1985 he noted gradual loss of left leg and right arm sweating and increased heat intolerance. There was progressive sweat loss on the thermoregulatory sweat test (from 68% to 93% anhidrosis).

Interestingly, the areas of progressive loss (the left leg and foot, for example) initially showed preservation of the quantitative sudomotor axon reflex response, strong evidence that the lesion was affecting a site proximal to the sympathetic chain ganglion cells (i.e., synaptic or preganglionic axons or IML cell column of the cord). After several more years the peripheral sweat response was lost in the left foot.

Isolated failure of sweating can be caused by altered humoral and/or cell-mediated autoimmune responses. Nakazato has described cases of humorally mediated acquired anhidrosis that are acute to subacute in onset. He has found elevated levels of IgE antibodies in a significant number of cases, and some have responded dramatically to corticosteroids. Skin biopsies have not shown inflammation of sweat glands.

Yet a third type of CIA is presented in Figure 7. Here skin biopsy, does show perieccrine infiltration with

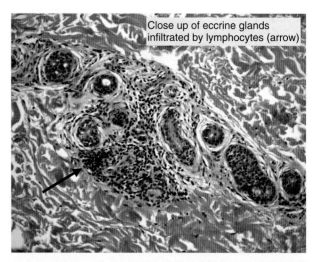

Figure 7 Perieccrine lymphocytic infiltration in chronic idiopathic anhidrosis. High-power microscopic image of a hematoxylin and eosin–stained thick-skin biopsy specimen, showing multiple sweat gland secretory coils and exuberant perieccrine lymphocytic infiltration.

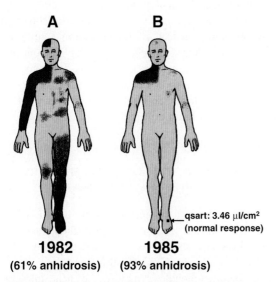

Figure 6 Chronic idiopathic anhidrosis (CIA). Serial thermoregulatory sweat tests in a patient having CIA, showing progressive widespread thermoregulatory failure associated with heat intolerance. Note the preserved QSART (quantitative sudomotor axon reflex test) response in the anhidrotic left foot (**B**), indicating a preganglionic or synaptic ganglionic lesion in this case (sweating in dark-shaded areas). (Modified with permission from Thermoregulatory Failure. [2000]. *In:* "Handbook of Clinical Neurology" [O. Appenzeller, ed.], Vol. 75, Series 31: The Autonomic Nervous System, Part II: Dysfunctions. p. 66. Elsevier Science B. V., Amsterdam.)

small lymphocytes that presumably is the cause of anhidrosis [From Fealey, R. D., unpublished observations]. The antigenic target of the CD3+ lymphocytes is unclear but possibly is the M3 acetylcholine receptor.

The direct demonstration of cellular and humoral mediated immunopathology has provided new insights into the pathophysiology and treatment of CIA. Examining skin biopsy specimens with electron and laser scanning confocal fluorescence microscopy [16] to discern the ultrastructural changes disrupting the neuroglandular unit and determining cell surface markers of infiltrating lymphocytes are among current and proposed methods to understand the neurobiology of this disorder, of conditions such as Fabry's disease [17], and of small fiber neuropathies in general [18].

D. Autoimmune Autonomic Neuropathy

Another interesting disorder causing acquired anhidrosis has been described by Vernino and colleagues [19,20]. These patients develop subacute autonomic failure and high titers of alpha-3 AChR ganglionic antibodies. Presumably, cholinergic synaptic transmission is affected by the antibodies in some of these cases. An association with occult malignancy may occasionally occur. Antibody titers may correlate with disease severity and measures of autonomic failure such as the percentage of body surface anhidrosis (Fig. 8).

An animal model of the disorder has been developed (known as EAAN, for experimental, autoimmune, autonomic neuropathy), and alpha-3 ganglionic AChR

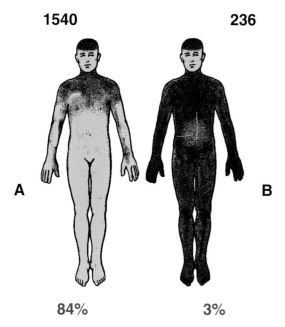

Figure 8 Autoimmune autonomic neuropathy. One of the originally reported cases, showing the correlation between alpha-3 ganglionic acetylcholine receptor antibody titer (top, in black) with the percentage of anhidrosis on sweat testing (bottom, in red). Axon reflex sweat response was present in anhidrotic areas on thermoregulatory sweat testing, suggesting a preganglionic or synaptic lesion (left). Patient had a spontaneous recovery (thermoregulatory sweat test, on right; sweating in purple shade).

antibodies have been shown to cause failure of synaptic transmission between preganglionic nerve terminal and postganglionic cell body (Fig. 9) [21].

VI. Thermoregulatory Failure Due to Degenerative Disorders

One degenerative disorder that best illustrates the neurobiology of thermoregulatory failure is multiple system atrophy. This disorder is due to loss of specific neuronal populations with accumulation of alpha-synuclein in cells. Autonomic failure often results from loss of the IML cell column autonomic neurons of the spinal cord. This lesion produces loss of thermoregulatory sweating mediated by pathways projecting from the hypothalamus to the IML cell column and from the IML cell column to the sympathetic chain. The sympathetic chain ganglion cell is often intact early on, as can readily be demonstrated by combined testing of thermoregulatory and axon reflex sweating [22].

In Figure 10, note the widespread anhidrosis with a sweat level on the trunk (higher on the right); quantitative axon reflex sweat output from anhidrotic lower-extremity effectors (terminal axons and sweat glands) is normal. This pattern of thermoregulatory and quantitative sudomotor axon reflex sweat tests is highly indicative of a preganglionic (probably ILC) lesion. Recently, techniques evaluating the degree of sympathetic denervation of the heart have been used to distinguish between multiple-system atrophy (MSA) and Parkinson's disease [23].

We are beginning to understand the molecular neurobiology of MSA. In the past few years, neuropathological analyses of the brains of patients dying of MSA have revealed widespread accumulation of phosphorylated alpha-synuclein in both glial and neuronal cells, a distinct

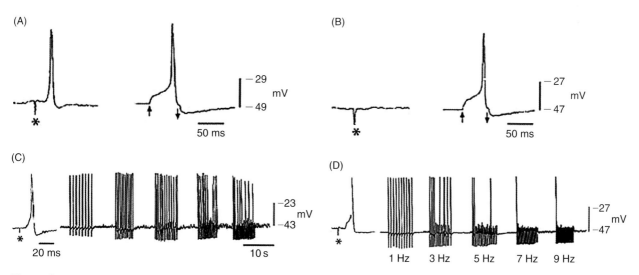

Figure 9 Experimental autoimmune autonomic neuropathy (EAAN). **A and B**, inferior mesenteric ganglion neuron action potential to preganglionic (colonic nerve) and direct electrical stimulation. (**A**) Control animal and (**B**) an animal with severe EAAN; note loss of the colonic nerve stimulation response. Supramaximal stimulation of lumbar colonic nerve in control rabbit (**C**) and a rabbit with relatively mild EAAN (**D**). Synaptic transmission failed more frequently and at lower rates of stimulation in neurons from rabbits with EAAN. (Modified with permission from Lennon, V. A., Ermilov, L. G., Szurszewski, J. H., and Vernino, S. (2003). Immunization with neuronal nicotinic acetylcholine receptor induces neurological autoimmune disease. *J. Clin. Invest.* **111**, 907–913.)

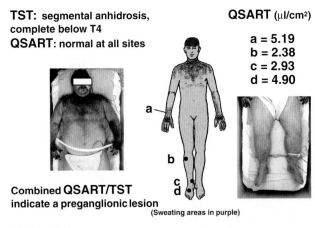

Figure 10 Sporadic multiple system atrophy. Thermoregulatory sweat test (TST; photos and central figure) and quantitative axon reflex test (QSART) responses (volume output at sites **a**, **b**, **c**, and **d** shown). Normal QSART sweat volumes in areas of total anhidrosis with segmental pattern on TST are highly suggestive of a preganglionic lesion, likely in the intermediolateral cell column.

pathological feature in patients suffering from MSA [24]. The recent generation of transgenic mouse models of oligodendroglial alpha-synucleinopathy has enabled investigation of how alpha-synuclein causally contributes to MSA neuropathology. Moreover, human disease–specific pathological modifications of alpha-synuclein were recapitulated in transgenic mice, including insolubility phosphorylation at serine-129 and ubiquitination. Thus, the transgenic mice will be useful tools to assess cellular risk factors such as protein folding stress, protein kinase hyperactivity, and failure of the ubiquitin-proteasome system. Understanding these molecular alterations may be the key to developing treatment strategies that block or reverse such changes before accumulation of alpha-synuclein causes cellular disruption and death [25]. It is not too unreasonable to speculate that brain mapping of alpha-synuclein aggregates in vivo via scintigraphy will become a powerful diagnostic tool in MSA.

References

1. Fealey, R. D. (2000). Thermoregulatory failure. *In:* "The Autonomic Nervous System II" (O. Appenzeller, ed), pp. 53–84. Elsevier, Amsterdam.
2. Benarroch, E. E. (2006). "Basic Neurosciences with Clinical Applications," pp. 758–761. Butterworth Heinemann Elsevier, Philadelphia.
3. Boulant, J. A. (2000). Role of the preoptic-anterior hypothalamus in thermoregulation and fever. *Clin. Infect. Dis.* **31 Suppl 5**, S157–S161.
4. Sato, K. (1997). Normal and abnormal sweat gland function. *In:* "Clinical Autonomic Disorders" (P. A. Low, ed), 2nd edition, pp. 97–108. Lippincott-Raven, Philadelphia.
5. Sell, H., Deshaies, Y., and Richard, D. (2004). The brown adipocyte: update on its metabolic role. *Int. J. Biochem. Cell Biol.* **36**, 2098–2104.
6. Egan, G. F., Johnson, J., Farrell, M., McAllen, R., Zamarripa, F., McKinley, M. J., Lancaster, J., Denton, D., and Fox, P. T. (2005). Cortical, thalamic, and hypothalamic responses to cooling and warming the skin in awake humans: a positron-emission tomography study. *Proc. Natl. Acad. Sci. USA.* **102**, 5262–5267.
7. Nagai, Y., Critchley, H. D., Featherstone, E., Trimble, M. R., and Dolan, R. J. (2004). Activity in ventromedial prefrontal cortex covaries with sympathetic skin conductance level: a physiological account of a "default mode" of brain function. *Neuroimage.* **22**, 243–251.
8. Treves, S., Anderson, A. A., Ducreux, S., Divet, A., Bleunven, C., Grasso, C., Paesante, S., and Zorato, F. (2005). Ryanodine receptor 1 mutations, dysregulation of calcium homeostasis and neuromuscular disorders. *Neuromuscul. Disord.* **15**, 577–587.
9. Kishida, I., Kawanishi, C., Furuno, T., Kato, D., Ishigami, T., and Kosaka, K. (2004). Association in Japanese patients between neuroleptic malignant syndrome and functional polymorphisms of the dopamine D(2) receptor gene. *Mol. Psychiatry* **9**, 293–298.
10. Nolli, M., Barbieri, A., Pinna, C., Pasetto, A., and Nicosia, F. (2005). Wernicke's encephalopathy in a malnourished surgical patient: clinical features and magnetic resonance imaging. *Acta Anaesthesiol. Scand.* **49**, 1566–1570.
11. Katz, L. M., Young, A. S., Frank, J. E., Wang, Y., and Park, K. (2004). Regulated hypothermia reduces brain oxidative stress after hypoxic-ischemia. *Brain Res.* **1017**, 85–91.
12. Gordon, C. J., McMahon, B., Richelson, E., Padnos, B., and Katz, L. (2003). Neurotensin analog NT77 induces regulated hypothermia in the rat. *Life Sci.* **73**, 2611–2623.
13. RDF, Perieccrine infiltrates in CIA, in Powerpoint. (2005).
14. Low, P. A., Fealey, R. D., Sheps, S. G., Su, W. P., Troutman, J. C., and Kuntz, N. L. (1985). Chronic idiopathic anhidrosis. *Annals of Neurology* **18**, 344–348.
15. Nakazato, Y., Tamura, N., Ohkuma, A., Yoshimaru, K., and Shimazu, K. (2004). Idiopathic pure sudomotor failure: anhidrosis due to deficits in cholinergic transmission. *Neurology* **63**, 1476–1480.
16. Kennedy, W. R., Wendelschafer-Crabb, G., and Brelje, T. C. (1994). Innervation and vasculature of human sweat glands: an immunohistochemistry-laser scanning confocal fluorescence microscopy study. *J. Neurosci.* **14**, 6825–6833.
17. Banerjee, T. K. (2004). Fabry disease with special reference to neurological manifestations. *Eur. Rev. Med. Pharmacol. Sci.* **8**, 275–281.
18. Lauria, G., Cornblath, D. R., Johansson, O., McArthur, J. C., Mellgren, S. I., Nolano, M., Rosenberg, N., Sommer, C.; European Federation of Neurological Societies. (2005). EFNS guidelines on the use of skin biopsy in the diagnosis of peripheral neuropathy. *Eur. J. Neurol.* **12**, 747–758.
19. Vernino, S., Adamski, J., Kryzer, T. J., Fealey, R. D., and Lennon, V. A. (1998). Neuronal nicotinic ACh receptor antibody in subacute autonomic neuropathy and cancer-related syndromes. *Neurology.* **50**, 1806–1813.
20. Vernino, S., Low, P. A., Fealey, R. D., Stewart, J. D., Farrugia, G., and Lennon, V. A. (2000). Autoantibodies to ganglionic acetylcholine receptors in autoimmune autonomic neuropathies. *N. Engl. J. Med.* **343**, 847–855.
21. Lennon, V. A., Ermilov, L. G., Szurszewski, J. H., and Vernino, S. (2003). Immunization with neuronal nicotinic acetylcholine receptor induces neurological autoimmune disease. *J. Clin. Invest.* **111**, 907–913.
22. Fealey, R. (2001). Use of the thermoregulatory sweat test in the evaluation of patients with autoimmune autonomic neuropathy and early multiple system atrophy syndromes. *Japanese Jour. of Perspiration Research* **8**, 37–40.
23. Goldstein, D. S., Holmes, C. S., Dendi, R., Bruce, S. R, and Li, S. T. (2002). Orthostatic hypotension from sympathetic denervation in Parkinson's disease. *Neurology.* **58**, 1247–1255.
24. Goedert, M. (2004). Alpha-synuclein and neurodegeneration. *In:* "Primer on the Autonomic Nervous System" (D. Robertson, ed.), 2nd edition, pp. 204–207. Elsevier Academic Press, San Diego, California.
25. Wenning, G. K., and Jellinger, K. A. (2005). The role of alpha-synuclein in the pathogenesis of multiple system atrophy. *Acta Neuropathol. (Berl.)* **109**, 129–140.

91

Control of Blood Pressure—Normal and Abnormal

Michael J. Joyner, MD
William G. Schrage, PhD
John H. Eisenach, MD

Keywords: *baroreflexes, fainting, vasovagal syncope*

I. History and Nomenclature
II. Key Elements of Short-Term Blood Pressure Regulation in Humans
III. The Physiological Problem of Upright Posture in Humans
IV. Failure of Blood Pressure Regulation during Orthostatic Stress
V. Ideas about Physiological "Causes" of Common Fainting
VI. Who Faints, How Often, and When?
VII. Summary
References

I. History and Nomenclature

Arterial blood pressure is perhaps the key regulated variable in the cardiovascular system [1–3]. This concept is especially true in the short term on a beat-to-beat, or moment-to-moment, basis. Because blood flow (and oxygen delivery) to vital organs such as the brain and heart is under significant autoregulatory control, this means that flow to these organs is generally adequate as long as mean arterial pressure stays within a modest range (usually about 60–140 mm Hg). In this context, the brain and heart are usually considered the "vital" organs because ischemia and/or severe hypoxia can do irreversible damage to these organs faster than almost any other, and because severe brain or heart dysfunction leads rapidly to permanent disability or death.

When blood flow (or oxygen delivery) to the brain is inadequate, there can be loss of consciousness. When this occurs in a standing human, the loss of consciousness is accompanied by a faint ("syncope"; however, an emotional faint can occur while supine). There are a collection of symptoms and syndromes that can generally be described as a family of disorders know as "neurogenic orthostatic hypotension" [4–9]. One teleological line of thinking argues that the "purpose" of a faint is to lower the head relative to heart level so that relative perfusion pressure of the brain is increased. Thus, fainting when blood pressure falls during orthostatic stress can be seen as a protective mechanism that is activated when blood pressure regulation fails. With these introductory comments as background, this chapter will

- review key elements of short-term blood pressure regulation in humans;
- discuss the physiological "problem" (more or less unique to humans) that is presented by the upright posture;
- discuss how physiological regulation of blood pressure can fail during orthostatic stress in humans;
- highlight several ideas about the "cause" of common fainting in humans;
- discuss who faints, how often, and when; and
- discuss how knowledge of the physiological responses to orthostatic stress can be used to diagnose and treat patients with orthostatic symptoms.

II. Key Elements of Short-Term Blood Pressure Regulation in Humans

Mean arterial pressure (MAP) is the product of cardiac output (CO) and total peripheral resistance (TPR):

$$\mathrm{MAP} = \mathrm{CO} \times \mathrm{TPR}$$

TPR is a calculated variable and only MAP and CO can be measured. Whereas measuring arterial pressure is straightforward and can be done cheaply and noninvasively (with a simple blood pressure cuff, by one person with minimal training), measuring cardiac output or even obtaining a reasonably accurate, noninvasive estimate takes significant equipment and technical skill. Implicit in the earlier equation is the idea that MAP might be regulated by changing either CO or TPR (also called vascular resistance) [1–3]. If there is an acute fall in blood pressure, physiological responses that tend to maintain or improve CO occur, and the blood vessels are constricted so that vascular resistance rises. If blood pressure is raised acutely, generally opposite directional changes occur. How does this happen?

There are sensory afferents located throughout the cardiovascular system that respond to mechanical events associated with the cardiac cycle. In general there are two main groups of mechanoreceptors that play an essential role in the beat-to-beat regulation of arterial pressure so that MAP does not swing wildly with postural changes and during activities of daily living.

The carotid sinus and aortic arch possess the so-called arterial baroreceptors (these areas also possess chemoreceptors) that are innervated by cranial nerves IX and X, respectively (Fig. 1). Mechanosensitive afferents in these areas respond to changes in arterial pressure (i.e., stretch) and evoke reflex changes in heart rate and vascular resistance when there are changes in blood pressure. The arterial baroreceptors are stimulated when blood pressure is higher, with some afferents appearing more sensitive to static distention and others to phasic deformation (pulse pressure). When stimulated, the baroreceptors send signals to the brainstem cardiovascular centers that inhibit sympathetic outflow and stimulate cardiac vagal traffic, leading to vasodilation and a slower heart rate. When arterial pressure falls, afferent traffic from the baroreceptors falls; sympathetic outflow is no longer inhibited and vagal outflow is no longer stimulated. Thus, both heart rate and sympathetic vasoconstrictor tone increase. Figure 2 is an individual record of the heart rate and muscle sympathetic nerve responses to changes in arterial pressure evoked by sequential boluses of vasodilating and constricting drugs in a volunteer subject.

The thoracic cavity, great veins, and cardiac chambers are also innervated by mechanosensitive (and chemosensitive) afferents [2,3,10,11]. At least some of these afferents sense mechanical events related to cardiac filling, and in general, when active, these afferents are sympathoinhibitory. This means that when central blood volume is high, sympathetic outflow is reduced. In general, the cardiopulmonary afferents do not play a prominent role in the regulation of heart rate, but information from them can act centrally and modify the heart rate responses to arterial baroreceptor loading and unloading. Cardiopulmonary afferents can also modulate release of fluid-regulating hormones from the hypothalamus. When central blood volume is high the afferents are stimulated, and this inhibits the activation of fluid-retaining-hormone release and other mechanisms that conserve body fluids.

For many years it was assumed that the cardiopulmonary afferents served as an early warning system so that small reductions in central blood volume subtly deactivated the cardiopulmonary afferents and evoked increases in sympathetic outflow before the arterial baroreceptors sensed a fall in arterial pressure (Fig. 3). However, this view has been challenged by studies in humans showing that small changes in central blood volume affect mechanical events that are likely sensed by the arterial receptors (for discussion, see [11]). This highlights the difficulty of studying blood pressure regulation in humans. First, for anatomical reasons, in humans it is difficult to isolate all but the carotid receptors for selective stimulation. Second, any reflex responses

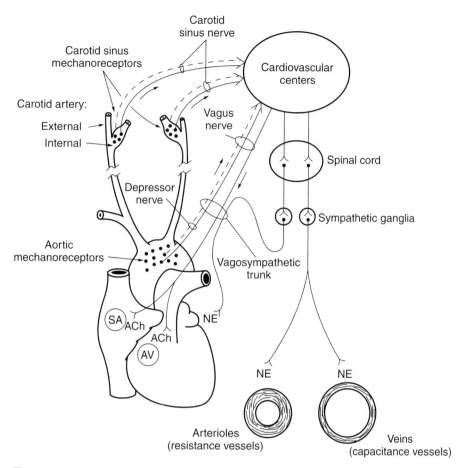

Figure 1 Schematic of key blood pressure–regulating systems in humans. This figure shows the carotid and aortic mechanoreceptors that respond to mechanical deformation in the great vessels associated with changes in arterial pressure. Information from the carotid receptors travels via the carotid sinus nerve to the brainstem cardiovascular centers. Information from the aortic mechanoreceptors travels via the vagus nerve. Together these receptors sense events related to changes in arterial pressure and evoke changes in heart rate and sympathetic outflow that help regulate arterial blood pressure. Not shown in this figure are the so-called cardiopulmonary receptors that are part of an extensive vagal afferent system that innervates the cardiac chambers, great veins, and other areas in the thorax. (Reprinted, with permission, from Shepherd, J. T., and Vanhoutte, P. M. (1979). "The Human Cardiovascular System: Facts and Concepts." Raven Press, New York.)

evoked by "selective" activation of one afferent pool evoke changes in systemic hemodynamic variables that are sensed by the other afferent pools, which then (in turn) evoke additional compensatory responses that make it difficult to interpret the overall behavior of the system.

In summary, arterial and perhaps cardiopulmonary receptors play a key role in the short-term regulation of arterial pressure in humans. Stimulation of these receptors by stretch associated with increased arterial pressure or central blood volume inhibits sympathetic outflow to blood vessels and stimulates vagal outflow to the heart. When blood pressure falls, there is less baroreceptor afferent activity and therefore more sympathetic outflow to vessels and withdrawal of vagal tone to the heart; responses that tend to maintain or increase arterial pressure.

III. The Physiological Problem of Upright Posture in Humans

Standing up presents two basic mechanical challenges to blood pressure regulation that are more or less unique to humans ([1] and Fig. 4). First, the head is above heart level, and this reduces the effective perfusion pressure of the brain. In an average human the brain is about 30–40 cm above the aortic root and there is a ~ 0.8-mm Hg reduction in pressure per cm. This means that if blood pressure at the aortic root is 120/80 (MAP, ~ 93) it would be ~ 25 mm Hg lower at the brain. Second, on standing, there is a shift in blood volume of 5–10 mL/kg from the central circulation (thorax) to the veins in the lower abdomen and dependent legs, which reduces venous return and cardiac filling.

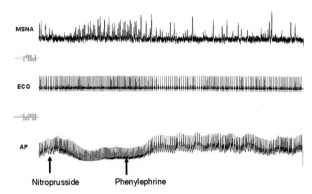

Figure 2 This figure demonstrates the concepts outlined in the text and Figure 1. It is an individual record of muscle sympathetic nerve activity (MSNA), heart rate (ECG), and arterial pressure. As part of the modified Oxford technique, changes in blood pressure are evoked by systemic boluses of nitroprusside (100 μg), followed 1 min later by the vasoconstrictor phenylephrine (150 μg). This evokes a predictable fall in blood pressure, followed by a rise in blood pressure. When blood pressure falls, there is a reflex increase in muscle sympathetic nerve activity and a speeding of the heart rate. After the phenylephrine bolus, blood pressure rises and sympathetic traffic is inhibited and heart rate falls. (Unpublished figure courtesy of Dr. Nisha Charkoudian.)

In patients without a functioning autonomic nervous system due to conditions such as pure autonomic failure, blood pressure falls rapidly on assumption of the upright posture (Fig. 5). This happens because there are no reflex increases in sympathetic vasoconstrictor outflow and heart rate when these patients stand up. (Notably, these same patients can have marked hypertension when supine.)

There are several basic physiological responses to the challenges noted earlier. First, the upright posture is associated with roughly a doubling of sympathetic vasoconstrictor outflow. Second, this is accompanied by a rise in heart rate of 10–15 beats per minute [1,12]. Together, these responses keep mean arterial pressure close to the value observed while supine. However, cardiac output is reduced by about 20%, diastolic blood pressure increases, and there is a fall in pulse pressure. Additionally, small movements of the legs in conjunction with the actions of the one-way venous valves work together as part of a "muscle pump" to improve venous return and drive at least some of the blood volume in the dependent legs toward the central circulation. Figure 4 reviews the physiological responses that permit humans to stand for long periods.

IV. Failure of Blood Pressure Regulation during Orthostatic Stress

There are numerous ways that blood pressure regulation can fail during orthostatic stress and lead to syncope. A number of pathophysiological conditions, including certain forms of valvular heart disease, cardiac arrhythmias, and diseases of the autonomic nervous system, can all disrupt or limit the normal adaptations to orthostatic stress. Orthostatic intolerance and syncope can also be a side effect of drugs, systemic diseases such as diabetes (peripheral neuropathy), and destruction of key neural structures by any number of means. However, by far the most common form of syncope is simple fainting (vasovagal) in otherwise healthy people [4,5,8,9].

A simple vasovagal faint has several phases. First there is assumption of the upright posture (or imposition of orthostatic stress by lower body negative pressure, or LBNP) and the expected increases in heart rate and sympathetic vasoconstrictor outflow, along with the expected narrowing of pulse pressure. Second, as the orthostatic stress continues, it is likely that there is progressive venous pooling that further limits cardiac filling and leads to more dramatic increases in sympathetic outflow and heart rate. Third, the subject suddenly feels cold and sweaty; there is a sensation of mild nausea, and sometimes there is a feeling of "impending doom." These symptoms are frequently called "presyncopal," and if the orthostatic stress is terminated there is no loss of consciousness and the faint is avoided. Fourth and finally, concurrent with or usually a few seconds following the onset of presyncopal symptoms, there is frequently a deep breath, sudden fall in blood pressure, loss of consciousness, and a faint that occurs simultaneously with dramatic bradycardia and profound sympathetic silence. If the subject is upright there is a loss of postural tone as he or she "goes to the ground."

On the basis of the earlier description, a typical orthostatic faint has been termed vasovagal syncope, neurocardiogenic syncope, and more broadly neurogenic orthostatic hypotension. Whatever the name, the obvious question, then, is why the expected and seemingly adaptive compensatory responses (vasoconstriction and tachycardia) to decreasing venous return and blood pressure suddenly give way to vasodilation and bradycardia. What regulatory systems are responsible that might explain these seemingly paradoxical responses?

V. Ideas about Physiological "Causes" of Common Fainting

The most obvious idea about what causes common vasovagal or neurocardiogenic syncope is that bradycardia is somehow the culprit. Bradycardia was noted as a feature of common fainting and experimental administration of atropine to stop it was tried in 1932, by Sir Thomas Lewis, with little effect [13]. Subsequently, vasovagal syncope has been seen in heart transplantation patients, and pacing in patients with recurrent syncope is largely ineffective [14,15]. Thus, blood pressure can fall and syncope can occur when the bradycardia is prevented.

Control of Blood Pressure

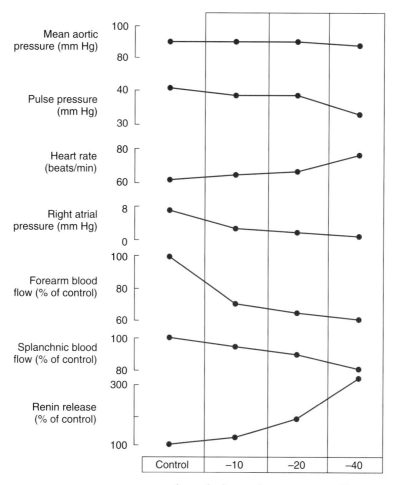

Figure 3 Integrated blood pressure and hemodynamic responses to graded venous pooling evoked by lower body suction. Throughout the period of suction up to 40 mm Hg (roughly similar to standing up), there is little change in mean aortic (arterial) pressure. During lower levels of suction there is no change in pulse pressure, but as suction increases, pulse pressure falls. Heart rate typically does not increase dramatically until there is a fall in pulse pressure. This observation has been used to argue that arterial baroreceptors are the main afferent system regulating the orthostatic heart rate responses in humans. In contrast to changes in arterial pressure and heart rate, right atrial pressure falls immediately with mild levels of venous pooling, and this evokes marked vasoconstriction, as noted by the reduction in forearm blood flow. In the classical view, these data argue that changes in central venous pressure sensed by so-called cardiopulmonary receptors play an important role in regulating sympathetic outflow during orthostatic stress. It is of interest that there is little change in splanchnic blood flow until more severe levels of venous pooling and that there is also a graded increase in renin release. (Reprinted, with permission, from [2,3,10].)

The previous brief discussion tells us that the "vagal" portion of the term "vasovagal" comes from the bradycardia and perhaps nausea, but what about the "vaso" element of this descriptive name? After the advent of direct recordings of muscle sympathetic nerve activity in humans, several serendipitous individual records of nerve traffic were recorded during a faint ([12] and Fig. 6). As described earlier, the end of the presyncopal phase of the physiological responses to orthostatic stress was marked by sudden and dramatic sympathetic silence. This sympathetic withdrawal led to a total loss of peripheral vascular tone, and together with the reduced cardiac output associated with the limited venous return (and perhaps bradycardia), it was responsible for the marked and sudden fall in blood pressure seen during the vasovagal faint. The idea that peripheral vasodilation is a major contributing factor to a faint was noted more than 200 years ago during clinical observations of blood letting [16–18]. It is also likely that sympathetic withdrawal, in conjunction with reduced venous return, explains why fainting is still seen in patients after heart transplantation

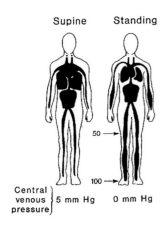

- ↓ CBV
- ↓ Arterial pulse pressure
- Mechanoreceptor unloading
- ↑ Sympathetic outflow
- ↓ Vagal tone
- Veno-arteriolar reflexes
- Activation of renin angiotensin system

Figure 4 Schematic representation of the mechanical effects of going from supine to standing. With the assumption of the upright posture, central venous pressure falls from about 5 mm Hg to 0. There is also marked venous pooling in the dependent extremities, especially the legs. This leads to a reduction in central blood volume (CBV), decrease in arterial pulse pressure, and arterial cardiopulmonary mechanoreceptor unloading. These effects evoke a reflex increase in sympathetic outflow and a reduction in vagal tone. There is also activation of the so-called venoarterial reflexes, which also act locally to evoke vasoconstriction. The renin-angiotensin system is also stimulated to promote retention of fluid. (Reprinted, with permission, from [1,2].)

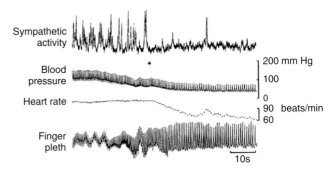

Figure 6 Individual record of sympathetic traffic, arterial blood pressure, heart rate, and finger blood flow (finger pleth) in a subject during a tilt-induced faint. Sympathetic activity is high, with nearly one sympathetic burst per heart beat. It drops suddenly at the same time that there is an accelerated fall in blood pressure and marked bradycardia, which occurs over a few beats (star). Associated with this is a marked rise in finger blood flow, indicative of peripheral vasodilation. (Reprinted, with permission, from Wallin, B. G., and Sundlöf, G. (1982). Sympathetic outflow to muscles during vasovagal syncope. *J. Auton. Nerv. Syst.* **6**, 287–291.)

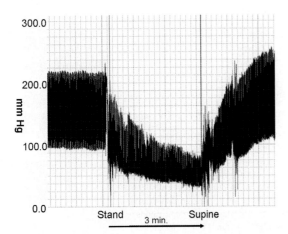

Figure 5 Effects of standing up on blood pressure in a patient with pure autonomic failure (an autoimmune disease which destroys the sympathetic nervous system). While supine, blood pressure is relatively high. Immediately on assumption of the upright posture, blood pressure falls, there is no reflex vasoconstriction or tachycardia, and over the course of several minutes mean pressure drops below the autoregulatory range and the patient feels light-headed. This series of events reverses immediately on standing. It is of interest to compare these changes in blood pressure to the stable blood pressure values seen during orthostatic stress in a normal subject (see Fig. 3). (Unpublished data from authors.)

and with pacing [14,15]. The venous pooling and sympathetic silence alone are clearly enough to cause a faint.

Another important factor that might contribute to fainting in at least some subjects is a sudden surge in epinephrine just before the dramatic fall in blood pressure [19]. The idea is that in subjects who are especially sensitive to fainting, there is a sudden increase in circulating epinephrine levels. This increase in epinephrine evokes β_2-adrenergic receptor–mediated vasodilation and contributes to the fall in pressure as presyncope becomes syncope. This concept is attractive for several reasons. First, there is evidence of sympathoadrenal imbalance in subjects who are especially susceptible to fainting [19]. These subjects' catecholamine responses to orthostatic stress are marked by what might be described as excessive epinephrine levels. Second, in some studies and at least anecdotally in some patients, β-adrenergic blockade can limit the tendency to faint. Third, emerging evidence suggests that common genetic variants of the β-adrenergic receptor differ in their vasodilator responsiveness to catecholamines, and perhaps individuals with selected variants are especially susceptible to fainting [20]. Finally, there is also genetic variation in the catecholamine synthetic pathways, and some subjects might be especially prone to release large amounts of epinephrine during sympathoexcitatory stress [21].

The ideas discussed earlier about what causes common fainting have all focused on the theory that there is some sudden change in the autonomic nervous system that occurs at the physiological "limits" of compensation to orthostatic stress (for discussion, see [6,17]). In this context the key question is, What neural signals evoke these sudden changes? As noted in the earlier parts of this chapter, stimulation of carotid and cardiopulmonary afferent receptors by the stress associated with either elevated blood pressure or increased central blood volume are sympatho-inhibitory. One idea is that during orthostatic stress there is somehow paradoxical stimulation of these mechanosensitive afferents so that the brainstem cardiovascular centers receive increased inhibitory afferent information, similar to that which might be seen when the heart is full and blood pressure is high. This concept has been termed the "empty heart syndrome." For example, perhaps when the heart is empty, the reduced central blood volume and rapid heart beat can cause paradoxical activation of the cardiopulmonary afferents that are normally active when the heart is full. Similar ideas have also been advanced about the carotid receptors. If this occurred, the brainstem would be receiving sensory information suggestive of adequate or even high levels of cardiac filling and/or arterial pressure and thus might evoke the sudden sympathetic withdrawal.

Although this concept is certainly attractive, it should be pointed out again that individuals with largely de-innervated hearts as a result of cardiac transplantation experience vasovagal syncope that includes marked peripheral dilation and bradycardia in the atrial remnants [15]. This suggests that vasovagal syncope is possible in the absence of afferent information associated with an empty heart, that the carotid receptors are responsible, or that there is sufficient sympatho-inhibitory information coming from some area of the heart that was not de-innervated during cardiac transplantation.

As noted earlier, there is sometimes a sigh or deep breathing just before a faint. In this context, if ventilation remained normal or was increased while cardiac output was falling during orthostatic stress, it is possible that there might be a relative increase in cerebral vascular resistance as a result of the hypocapnia. Additionally, during the presyncopal period, blood pressure can sometimes become unstable, and recent information suggests that although there are powerful autoregulatory forces that tend to preserve brain blood flow, these mechanisms may not be as dominant as once believed. Additionally, it is possible that there is more sympathetic vasoconstriction in the cerebrovascular circulation than previously thought or that circulating vasoconstricting substances might also alter brain blood flow. Along these lines, it is possible that a collection of physiological events occur during the presyncopal period that cause cerebral blood flow to fall below some critical level, at least in some regions, and the reduced cerebral blood flow is the event which triggers the loss of consciousness, bradycardia, and sympathetic withdrawal. Whatever the mechanism, a fall in cerebral blood flow below some critical level is a key feature of vasovagal syncope. Whether a fall in cerebral blood flow is the primary initiating event in syncope or whether it occurs concurrently or just after the hemodynamic collapse remains unclear [22,23].

VI. Who Faints, How Often, and When?

Almost all humans under selected circumstances can experience vasovagal syncope; however, some individuals are much more susceptible than others. Epidemiological evidence suggests that up to 50% of teenage girls faint occasionally, and the fainting is typically associated with situations linked to viral illnesses, dehydration, fatigue, and

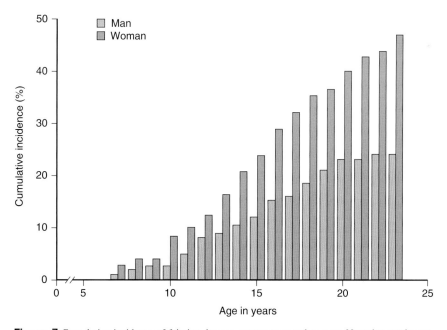

Figure 7 Cumulative incidence of fainting, by age, among men and women. Note that nearly one-quarter of men and almost half of young women report that they have fainted. In almost every case, these faints are common vasovagal syncope associated with the upright posture, venous pooling, and dehydration. Additionally, recent viral illnesses, fatigue, and perhaps hunger can contribute. (Reprinted, with permission, from Coleman, N., Nahm, K., Ganzeboom, K. S., Shen, W. K., Reitsma, J., Linzer, M., Wieling, W., and Kaufmann, H. (2004). Epidemiology of reflex syncope. *Clin. Auton. Res.* **1(Suppl)**, 9–17.)

perhaps hunger ([4] and Fig. 7). Additionally, during experimental combinations of lower-body negative pressure and tilting, vasovagal syncope can be induced in almost all subjects [22]. Although discussion of diagnosis and treatment of orthostatic intolerance is beyond the scope of this chapter, it should be emphasized that most episodes of syncope in otherwise healthy young individuals are vasovagal syncope and that with careful history-taking it is possible to identify patients who are experiencing vasovagal syncope and thus avoid expensive, extensive, and anxiety-provoking medical evaluations [5,7,24].

Explanation of what is likely happening, reassurance, advice that includes staying well-hydrated and increasing dietary sodium, and instruction in physical counter-maneuvers that can be used during presyncope are typically highly effective for many subjects with a tendency toward fainting ([24–26] and Fig. 8).

VII. Summary

In summary, arterial blood pressure is the main short-term regulated variable in the cardiovascular system. Sensory afferents located in the aorta and carotid arteries provide information to the brainstem cardiovascular centers related to arterial blood pressure, and sensory afferents in the heart, great vessels, and thoracic cavity provide these centers with information about "central blood volume." This information is integrated, and under normal circumstances appropriate adjustments in cardiac output (most notably heart rate) and vascular resistance are made so that perfusion pressure remains in an acceptable range during changes in posture and with activities of daily living. In this context, orthostatic stress is one of the most profound but also routine challenges to the blood pressure–regulating system, and although this challenge is met under most circumstances, under some circumstances regulation and compensation fail and syncope occurs. The most common form of syncope is so-called vasovagal or neurocardiogenic syncope, which is marked by profound bradycardia and peripheral sympathetic withdrawal. Although the syndrome has been well described for many years, the precise initiating events are still subject to debate, but inadequate cerebral blood flow either just before or during a syncopal episode would appear to be a "final common pathway" associated with a faint, no matter the initiating series of events. Common vasovagal fainting can frequently be diagnosed with some certainty via a comprehensive history, and the simplest, safest, and generally most effective therapies appear to be education and practice of physiologically based long- and short-term counter-maneuvers. Understanding the physiology of blood pressure regulation and the pathophysiology of fainting are keys to appreciating, diagnosing, and treating common fainting in humans [4–6,8,9,24–26].

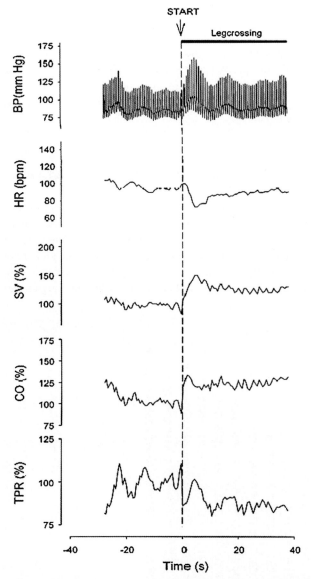

Figure 8 Effects of leg-crossing on hemodynamic responses in a subject with recurrent vasovagal symptoms just before syncope. Leg-crossing increased venous return and decreased venous pooling. This led to an increase in stroke volume (SV) and cardiac output (CO), thereby aborting the faint. TPR, total peripheral resistance; HR, heart rate; BP, blood pressure. (Reprinted, with permission, from van Dijk, N., de Bruin, I. G., Gisolf, J., de Bruin-Bon, H. A., Linzer, M., van Lieshout, J. J., and Wieling, W. (2004). Hemodynamic effects of leg-crossing and skeletal muscle tensing during free standing in patients with vasovagal syncope. *J. Appl. Physiol.* **98**, 584–590.)

References

1. Rowell, L. B. (1986). "Human Circulation: Regulation During Physical Stress." Oxford University Press, New York.
2. Shepherd, J. T., Vanhoutte, P. M., and Joyner, M. J. (1991). The sensory systems involved in cardiovascular regulation. *In:* "Mayo

Clinic Textbook of Cardiology" (E. R. Giuliani, V. Fuster, B. J. Gersh, M. D. McGoon, and D. C. McGoon, eds), 2nd edition, pp. 166–186. Mosby-Year Book, St. Louis.
3. Shepherd, J. T., and Vanhoutte, P. M. (1979). "The Human Cardiovascular System: Facts and Concepts." Raven Press, New York.
4. Coleman, N., Nahm, K., Ganzeboom, K. S., Shen, W. K., Reitsma, J., Linzer, M., Wieling, W., and Kaufmann, H. (2004). Epidemiology of reflex syncope. *Clin. Auton. Res.* **1(Suppl)**, 9–17.
5. Colman, N., Nahm, K., van Dijk, J. G., Reitsma, J. B., Wieling, W., and Kaufmann, H. (2004). Diagnostic value of history taking in reflex syncope. *Clin. Auton. Res.* **14(Suppl 1)**, 37–44.
6. Hainsworth, R. (2003). Syncope: What is the trigger? *Heart* **89**, 23–124.
7. The Task Force on Syncope, European Society of Cardiology. (2004). Guidelines on management (diagnosis and treatment) of syncope: update 2004. Executive summary. *Eur. Heart. J.* **25**, 2054–2072.
8. Thijs, R. D., Benditt, D. G., Mathias, C. J., Schondorf, R., Sutton, R., Wieling. W., and van Dijk, J. G. (2005). Unconscious confusion: a literature search for definitions of syncope and related disorders. *Clin. Auton. Res.* **15**, 35–39.
9. Thijs, R. D., Wieling, W., Kaufmann, H., and van Dijk, G. (2004). Defining and classifying syncope. *Clin. Auton. Res.* **14(Suppl 1)**, 4–8.
10. Johnson, J. M., Rowell, L. B., Niederberger, M., and Eisman, M. M. (1974). Human splanchnic and forearm vasoconstrictor responses to reductions of right atrial and aortic pressures. *Circ. Res.* **34**, 515–524.
11. Taylor, J. A., Halliwill, J. R., Brown, T. E., Hayano, J., and Eckberg, D. L. (1995). "Non-hypotensive" hypovolaemia reduces ascending aortic dimensions in humans. *J. Physiol.* **483**, 289–298.
12. Wallin, B. G., and Sundlöf, G. (1982). Sympathetic outflow to muscles during vasovagal syncope. *J. Auton. Nerv. Syst.* **6**, 287–291.
13. Lewis, T. (1932). Vasovagal syncope and the carotid sinus mechanism. *Br. Med. J.* **1**, 873–876.
14. El-Bedawi, K. M., Wahbha, M. A., and Hainsworth, R. (1994). Cardiac pacing does not improve orthostatic tolerance in patients with vasovagal syncope. *Clin. Auton. Res.* **4**, 233–23.
15. Fitzpatrick, A. P., Banner, N., Cheng, A., Yacoub, M., and Sutton, R. (1993). Vasovagal reactions may occur after orthoptic heart transplantation. *J. Am. Coll. Cardiol.* **21**, 1132–1137.
16. Barcroft, H., and Edholm, O. G. (1945). On the vasodilatation in human skeletal muscles during post-haemorrhagic fainting. *J. Physiol. (Lond).* **104**, 161–175.
17. Dietz, N. M., Joyner, M. J., and Shepherd, J. T. (1997). Vasovagal syncope and skeletal muscle vasodilatation: the continuing conundrum. *PACE* **20**, 775–780.
18. Hunter, J., and Palmer, J. F. (1837). "The Works of John Hunter, F.R.S.," Volume III, p. 91. Longman, Rees, Orme, Brown, Green, and Longman, London.
19. Goldstein, D. S., Holmes, C., Frank, S. M., Naqibuddin, M., Dendi, R., Snader, S., and Calkins, H. (2003). Sympathoadrenal imbalance before neurocardiogenic syncope. *Am. J. Cardiol.* **91**, 53–58.
20. Garovic, V. D., Joyner, M. J., Dietz, N. M., Boerwinkle, E., and Turner, S. T. (2003). β_2-Adrenergic receptor polymorphism and nitric oxide–dependent forearm blood flow responses to isoproterenol in humans. *J. Physiol.* **546**, 583–589.
21. Cui, J., Zhou, X., Chazaro, I., DeStafano, A. L., Manolis, A. J., Baldwin, C. T., and Gavras, H. (2003). Association of polymorphisms in the promoter region of the PNMT gene with essential hypertension in African Americans but not in whites. *Am. J. Hypertens.* **16**, 859–863.
22. Claydon, V. E., and Hainsworth, R. (2003). Cerebral autoregulation during orthostatic stress in healthy controls and in patients with posturally related syncope. *Clin. Auton. Res.* **13**, 321–329.
23. van Lieshout, J. J., Wieling, W., Karemaker, J. M., Secher, N. H. (2003). Syncope, cerebral perfusion, and oxygenation. *J. Appl. Physiol.* **94**, 833–848.
24. Wieling, W., Colman, N., Krediet, C. T., and Freeman, R. (2004). Nonpharmacological treatment of reflex syncope. *Clin. Auton. Res.* **14(Suppl 1)**, 62–70.
25. Krediet, C. T., van Dijk, N., Linzer, M., van Lieshout, J. J., Wieling, W. (2002). Management of vasovagal syncope: controlling or aborting faints by leg crossing and muscle tensing. *Circulation* **106**, 1684–1689.
26. van Dijk, N., de Bruin, I. G., Gisolf, J., de Bruin-Bon, H. A., Linzer, M., van Lieshout, J. J., and Wieling, W. (2004). Hemodynamic effects of leg crossing and skeletal muscle tensing during free standing in patients with vasovagal syncope. *J. Appl. Physiol.* **98**, 584–590.

92

Neoplasm-Induced Pain

Sebastiano Mercadante, MD

Keywords: *cancer pain, pain mechanisms, pharmacological treatment, opioids*

I. Introduction
II. Etiology
III. Pathogenesis
IV. Pain Characteristics
V. Assessment
VI. Pain Syndromes
VII. Pharmacological Treatment
VIII. Interventional Procedures
References

I. Introduction

Pain is a major problem in cancer. The prevalence of chronic pain is about 30%–50% among patients with cancer who are undergoing active treatment for a solid tumor and 70%–90% among those with advanced disease. Pain is consistently one of the most feared consequences of cancer for both patients and families [1]. Although availability of guidelines and accumulating experience have greatly improved the possibility of satisfactory pain control for most patients with advanced cancer, pain remains less than optimally controlled, due to physicians' poor knowledge and negative attitudes about pain management other than patient-related barriers and pain pathophysiology [2].

II. Etiology

Cancer pain is a complex issue. It is initially a signal of ongoing injury associated with the onset or recrudescence of disease or may be caused by some diagnostic procedures. Commonly it subsides after oncological treatments. In different stages of disease, however, the causes cannot adequately be eliminated and symptoms persist or get worse. At this point cancer pain no longer serves a biological purpose in alerting the organism to the presence of harmful stimuli; rather, it assumes the status of a chronic disease and is characterized by alterations in mood and pain behavior. In approximately two-thirds of patients with cancer, pain is directly related to the presence of

Table 1 Pain Syndromes Associated with Oncological Treatments

A. Post-chemotherapy syndromes
- Chemoembolization; intraperitoneal, pleural, intrathecal perfusions; hyperthermia
- Mucositis
- Polyneuropathy
- Bone pain
- Aseptic necrosis of bone
- Steroid pseudorheumatism

B. Post-surgical neuropathic syndromes
- Post-mastectomy
- Post–axillary dissection
- Post-thoracotomy
- Post–radical neck dissection
- Post–limb amputation
- Post–rectal amputation
- Stump pain

C. Post-radiotherapy syndromes
- Enteritis
- Dermatitis
- Mucositis
- Osteonecrosis
- Myelopathy
- Peripheral neuropathy
- Fibrosis of nervous plexuses

primary cancer or metastases. About one-third of patients develop pain syndromes because of oncological treatments, which are increasing considerably. During the course of the disease, acute, subacute, and chronic pain syndromes may develop as a consequence of surgery, radiotherapy, and chemotherapy and often overlap pain due to cancer [3]. Typical pain syndromes associated with oncological treatment are described in Table 1.

Other related causes include osteoporosis, immobility, and infections. The distinction is not always simple and requires careful evaluation involving imaging techniques and expert neurological assessment.

III. Pathogenesis

Injury discharges, ectopic nerve action potentials, and sensitization of nociceptors from peripheral nerves produce tonic input to the spinal cord. A prolonged stimulus determines a repetitive activation of C fibers, resulting in an augmentation of activity in dorsal horn wide-dynamic-range neurons and a strong augmentation in the magnitude of the response evoked by subsequent stimuli. This phenomenon has been related to a centrally mediated temporal summation. Central sensitization and a wind-up phenomenon at the spinal and supraspinal level have been described to explain the pathophysiological background of chronic pain conditions after nerve injury [4].

The release of peptides such as substance P into the spinal cord following afferent stimulation removes the magnesium block of the channel of the N-methyl-D-aspartate (NMDA) receptor and thus allows glutamate to activate NMDA receptors in the range of persistent pain states. Activation of NMDA receptors, predominantly localized on dorsal horn wide-dynamic-range neurons by excitatory amino acids (EEA), has been associated with the phenomenon of wind-up. The activation of such receptors is involved in the development and maintenance of injury-induced central hyperactive states initiating a variety of intracellular processes, principally consisting of an increase in calcium intracellular concentration, activation of protein kinase (PCK), and the calcium-calmodulin-mediated production of nitric oxide (NO), leading to increased neuronal excitability. PCK translocation and NO production may enhance postsynaptic neuronal excitability, leading to the development of hyperactive states, and may activate presynaptic NMDA receptors localized on primary afferent fibers by removing the magnesium blockade of the NMDA receptor, so that even small amounts of excitatory amino acid ligands may allow the opening of a calcium channel, with further activation of a second pool of PCK. Increases in intracellular calcium and activation of PKC also result in c-fos expression in postsynaptic dorsal horn neurons as well as in supraspinal areas. This is considered a third messenger probably involved in encoding a variety of cellular responses responsible for the neural changes associated with hyperalgesia. This is demonstrated by similar time courses of c-fos expression and the development of hyperalgesia. Moreover, the entry of calcium can also activate phospholipases and lead to the spinal production of prostanoids.

This feedback loop may be important for the induction and maintenance of neuron sensitization and long-term potentiation. Increases in membrane-bound PKC occurring in spinal cord have been shown to be positively correlated with the degree of hyperalgesia. Moreover, persistent activation of EEA receptors within the spinal cord may contribute to central hyperexcitability because irreversible morphological changes may occur with loss of function of spinal cord inhibitory interneurons. These excitotoxic processes may result in disinhibition phenomena, reinforcing the central hyperactivity state. Thus, it is possible that receptive field expansions and spontaneous activity generated in the central nervous system following nerve injury are also mediated by alterations in normal inhibitory processes [5].

IV. Pain Characteristics

Pain related to malignant disease is commonly classified as nociceptive (somatic or visceral) and neuropathic in type. Psychological factors may also have an important influence on these mechanisms. The term *idiopathic* is used when pain is perceived to be excessive for the extent of an organic pathology and a psychological pathogenesis is predominating.

A. Nociceptive Pain

Nociceptive pain is usually subdivided into somatic and visceral types, involves direct activation of nociceptors, and is often a complication of tumor infiltration of tissue or injury of tissues as a consequence of oncological treatment. Nociceptive pain is commensurate with tissue damage associated with an identifiable somatic or visceral lesion and is presumed to be related to ongoing activation of primary afferent neurons responsive to noxious stimuli [4]. Nociceptors are widely distributed in skin, muscle, connective tissue, and viscera. No specific histological structure acts as a nociceptive receptor. A-delta and C nociceptors have been clearly identified in fibers innervating somatic structures. In cutaneous tissue, thermal, mechanical, and chemical stimuli inducing tissue damage produce activation of unmyelinated polymodal transducers attached to A-delta and C fibers. Other high-threshold receptors are involved as the intensity of the stimulus increases. Thus, a repeated and intense stimulus induces the release of several inflammatory mediators, which reduces the threshold for activation, increases the response to a given stimulus, or induces the appearance of spontaneous activity. Various chemicals are released into damaged tissue cells. Sustained stimuli or damage to the nerve can alter the profile of several peptides, such as substance P, contained within primary afferents. Substance P is able to induce the production of nitrous oxide, a vasodilator, and the degranulation of mast cells, with further vasodilatation and subsequent extravasation and release of bradykinin. Bradykinin is a powerful algogenic substance that also sensitizes nociceptors by means of prostaglandins E2. Other factors, such as cytokines, are released after tissue damage under the influence of bradykinin. These substances have an important role in inflammatory processes. Although prostaglandins are weak algogens, they have a major role in the sensitization of nociceptors to other substances. The concerted effects of these mediators at the site of tissue damage underlie peripheral hyperalgesia, which accounts for much of the peripheral sensitization of nociceptors.

Deep pain originating from bone and visceral structures is more common than cutaneous pain. Muscle and visceral nociceptors exist in almost all organs and appear to have different properties than do cutaneous nociceptors, including the property of referred pain.

Experiments using distension most likely measure the effects of a mixture of visceral stimuli, including the stimulation of stretch receptors in the intestinal wall and a concomitant excitation of mesenteric mechanoreceptors [6]. There is good correlation between the size of the response and the stimulus intensity. Mesentery traction is another suitable method for investigating acute visceral nociception.

Two distinct classes of nociceptive sensory receptors that innervate internal organs have been proposed: high-threshold receptors, mostly mechanical, activated by stimuli within the noxious range, and low-threshold receptors, activated in the range of stimulation intensity from innocuous to noxious. According to this theory, high-threshold receptors would contribute to the peripheral encoding of noxious events in the viscera. Other prolonged and intense stimuli, such as hypoxia or tissue inflammation, would result in the sensitization of these receptors and bring into play previously silent receptors, normally unresponsive to innocuous stimuli. The activity of afferents and the excitability of dorsal horn neurons are increased with repeated or persistent stimuli because of a process of sensitization of afferents that occurs secondary to local release of chemical substances in tissues that are damaged or ischemic. A critical level of preceding activity in the afferents is required to induce facilitation of dorsal horn neuronal responses via central mechanisms. Moreover, the normal pattern of motility and secretion is altered by the local damage produced by inflammation, producing changes in the environment surrounding the nociceptor endings. This is consistent with the continuous and increasing input in the presence of an abdominal cancer disease.

B. Neuropathic Pain

Neuropathic cancer pain most commonly occurs as a consequence of tumor compression or infiltration of peripheral nerves or the spinal cord. Trauma, chemical, or radiation-induced injury as a result of surgery, chemotherapy, or radiotherapy may also result in this type of pain.

Neuropathic pain may be a complication of injury to the peripheral or central nervous system and is sustained by aberrant somatosensory processing. The pain cannot be explained by continuing tissue injury, but rather is ascribed to foci of disease in the peripheral and central nervous system. It results from changes in the physiological response of neurons in the central or peripheral somatosensory system due to persistent stimulation or to a lesion of the nervous tissue caused by a tumor itself, surgery, or chemotherapy. Neuropathic pain is most strongly suggested when dysesthesia, which is an abnormal and unfamiliar pain sensation, occurs in an area of motor, sensory, or autonomic dysfunction attributable to a neurological lesion. Subjective perception often includes burning or stabbing sensations associated with negative signs, such as hypoesthesia, and positive findings, such as dysesthesia, allodynia, and hyperalgesia. Clinical implications of a diagnosis of neuropathic pain are important, as this kind of pain is usually less responsive to opioids.

However, the diagnosis may be challenging since associated symptoms may be missed. Moreover, the labeling of a pain according to its inferred pathophysiology is a simplification of very complex processes that can involve multiple interacting mechanisms that can evolve and change over time.

Although nociceptive and neuropathic pain depend on separate peripheral mechanisms, they are both significantly influenced by changes in central nervous function. It is now clear that cancer pain is a more complex entity, particularly regarding the response to analgesics, where numerous factors play a role. It remains unclear whether cancer pain is a unique type of pain or merely a subtype of inflammatory or neuropathic pain, because in cancer pain models there are changes in transmitters commonly produced in neuropathic or inflammatory pain states. Thus, there is general acceptance of classification by inferred pathophysiology, but the clinical utility in cancer pain syndromes cannot be precisely determined. Nevertheless, specific pain-related phenomena may yield information that has direct relevance to patient care, as some syndromes are predictably less responsive to treatment and may suggest the need to improve clinical monitoring and provide a more careful evaluation and possible alternative treatments.

Experimental pathophysiological models of neuropathic pain have been important in the development of new methods for its prevention and treatment. In spite of the extensive interest in such underlying neural mechanisms, these phenomena largely remain unexplained.

Such pain states, often observed in cancer patients, are difficult to treat and require higher doses of opioids. Several common factors in neuropathic pain and tolerance have been found, although some central changes at the spinal cord level typically present in neuropathic pain states are not found with chronic exposure to morphine. Activation of spinal cord NMDA receptors, PKC activation, and NO production are biochemical steps equally capable of leading to a state of hyperalgesia or the development of morphine tolerance. Substrate phosphorylation by PKC and NO may result in enduring enhancement of synaptic activity at the spinal cord level and decreasing mu-receptor activity, uncoupling of G-protein with the mu receptor, or facilitating its desensitization. Thus, the reduction of morphine analgesia induced by nerve injury may share a mechanism similar to that of morphine tolerance. Reduction of morphine antinociception occurs after nerve injury–induced hyperalgesia in the absence of daily exposure to morphine. A given dose of morphine produced less antinociception in nerve-injured rats, and the antinociceptive dose-response curve was shifted to the right after the intervention. Thus, nerve injury produces a reduction of morphine nociception prior to exposure to morphine itself. As a consequence, tolerance may also develop before exposition to morphine, and a neuropathic pain state may be equivalent to that of the development of morphine tolerance [7].

However, it is well known that opioids at analgesic doses produce inhibition of or block excitation of dorsal horn cells by depression of the firing. These controversial observations could be resolved by considering the different effects produced by single or limited administration and prolonged or repeated administration of morphine on the intracellular cascade, the relationship between the time of nerve injury and the administration of the drug, the kind and site of the experimental injury, the doses and route of opioid administration, and the different types of scoring systems and evaluations. Globally, neuropathic pain appears more resistant to opioid treatment.

V. Assessment

Pain assessment should be included at each routine visit. A careful chronological review of the medical and oncological history may help in placing the pain complaint in context. If the patient is either unable or unwilling to describe the pain, a family member may need to be questioned. Responses to previous disease-modifying and analgesic treatments should be evaluated. Every kind of pain should be considered, as multiple pain problems are common and should be assessed independently. Consequences of pain, including impairment in daily living activities, psychological and social dysfunction, decreased appetite, sleep disorders, and possible symptoms associated with pain, must be assessed. Psychological status, level of anxiety, depression, and pervasive attitudes can usually be detected through careful questioning. Data collection is followed by a physical examination, particularly from a neurological perspective. Identification of the underlying etiology of the pain problem and the relationship of the pain complaint to the disease are necessary parts of the initial pain assessment. Careful review of previous imaging studies can provide important information about the cause of the pain and the extent of the underlying disease. Additional investigations, including imaging studies, are often required to clarify areas of uncertainty in a way that is proportional to the patient's general status and goals of care [3].

Pain assessment is commonly assessed by simple validated methods, such as numerical scales from 0 to 10, verbal descriptors, or visual analog scales, which can be repeated over time for an appropriate evaluation of the effects of an analgesic treatment.

Patients with cancer pain commonly experience pain at multiple sites. The topographical distribution of a specific pain has implications for diagnosis and treatment and often clarifies the relationship to the underlying organic lesion.

Referred pain is a term applied when the distribution is remote to the lesion, unlike focal pain, which is in the region of the underlying lesion. Somatic and visceral pains are often associated with characteristic pain-referral patterns. Neuropathic pain has typical characteristics, like pain referred in the distribution of a peripheral nerve, or in the dermatome innervated by a damaged root, as in the case of radicular pain.

From a temporal perspective, acute pain is associated with a well-defined onset and an identifiable cause, commonly an anticancer treatment, and subsides after the treatment. Chronic pain is usually insidious and progressive, also characterized by fluctuations in intensity. Transitory exacerbations of severe pain over a baseline of moderate pain have been described as breakthrough or episodic pain. These exacerbations may be precipitated by volitional actions, such as movement, or nonvolitional events. In some cases no identifiable precipitant is identifiable. This kind of pain represents a negative prognostic indicator and poses complex therapeutic problems.

Finally, patients with cancer pain present with a variety of other symptoms, including anorexia, nausea, weakness, dyspnea, and drowsiness. These symptoms can affect the expression of pain and may be aggravated by the pain. On the other hand, these symptoms may occur as a consequence of the use of analgesic drugs. Thus, symptom monitoring is mandatory for the evaluation of a pharmacological pain treatment.

VI. Pain Syndromes

Most chronic cancer-related pains are caused directly by the tumor. Numerous syndromes related to direct tumor involvement have been identified. Bone pain due to bone metastases, compression of neural structures, and visceral organ involvement are the most prevalent conditions.

A. Bone Pain

Skeletal involvement is a frequent and troublesome complication affecting many patients with neoplastic disease. It is the third most common metastatic site after the lung and liver. Bone metastases are more often seen with cancer of the lung and the prostate in males and breast cancer in females. The presence of a bone metastasis frequently gives rise to complications that have an important impact on the patient's quality of life, as it is the major source of pain, causes difficulty in ambulation or immobility, and causes neurological deficits and pathological fractures [8]. Animal models have been developed in to examine the mechanisms involved in the generation and maintenance of bone cancer pain. Osteolytic sarcoma cells were injected and confined by a dental amalgam into the intramedullary space of the femur. As the tumor grows, the number of osteoclasts increases and bone destruction becomes radiologically evident. Correspondingly, ongoing and touch-evoked pain–related behavior develops, as it occurs in humans [9]. These studies have shown that metastatic bone pain could have some correlation, in terms of opioid responsiveness, with neuropathic pain. The injection of cancer cells was coincident with the development of mechanical allodynia and a reduction in the paw withdrawal threshold, and the doses of morphine required to block bone cancer pain–related behaviors were 10 times those required to block peak inflammatory pain behaviors of comparable magnitude. This means that similar pain intensities induced by different stimuli (cancer cells injected in bone or inflammatory stimulus) may have different sensibility to opioids. Thus, from a pathophysiological point of view, bone pain elicited by movement corresponds to a mechanical allodynia (pain induced by a non-noxious stimulus, such as movement), which indicates a state of hypersensitivity, requiring higher opioid doses than those sufficient to control basal pain [10].

The presence of bone metastases has been found to be the most common cause of cancer-related pain. Because of its intermittent nature, bone pain responds poorly to single therapy with opioids and therefore can be difficult to control. Incidental pain, mostly associated with bone metastases, reduces the possibility of pharmacological pain control and is considered a negative prognostic factor.

The pathophysiological mechanism of pain in patients with bone metastases in the absence of a fracture is poorly understood. The presence of pain is not correlated with the type of tumor; the location, number, or size of the metastases; or the gender or age of patients. Although about 80% of patients with breast cancer will develop osteolytic or osteoblastic metastases, about two-thirds of demonstrated sites of bone metastases are painless. The resorption of bone due to increased osteoclastic activation decreases bone density and disrupts skeletal architecture, either at focal sites or throughout the skeleton. Many nerves are found in the periosteum, and others enter bones via the blood vessels. Microfractures occur in bony trabeculae at the site of metastases, resulting in bone distortion. Stretching of the periosteum by tumor expansion, mechanical stress of the weakened bone, nerve entrapment by the tumor, and direct destruction of the bone with consequent collapse are possibly associated mechanisms. The weakening of bone trabeculate and the cytokines, which mediate osteoclastic bone destruction, may activate pain receptors. The release of algesic chemicals within the marrow probably accounts for the observation that pain produced by tumors is often disproportionate to their size or degree of bone involvement. A secondary pain may be caused by reactive muscle spasm Nerve root infiltration and compression of nerves by the collapse of osteolytic vertebrae are other sources of pain.

Osteolytic bone metastases are commonly present with bone pain, pathological fractures, hypercalcemia, or (more rarely) swelling or neurological complaints. The five most frequently involved sites are the vertebrae, pelvis, ribs, femur, and skull. Pain develops gradually during a period of weeks or months, becoming progressively more severe. The pain usually is localized in a particular area and is often felt at night or with weight-bearing. Pain is characteristically described as dull, constant, and gradually progressive in intensity. Pain increases with pressure on the area of involvement. Continuous pain may be moderate on resting but may be exacerbated by different movements or positions, such as standing, walking, or sitting. Breakthrough pain is a very difficult challenge and represents a serious problem to manage, because its temporal characteristics render it not always preventable. It involves intermittent exacerbations of pain that can occur spontaneously or in relation to specific activity (incident pain), especially at the end of the dosing interval of a regularly scheduled analgesic. It can be caused by weight-bearing or by instability due to incipient or actual pathological fractures. Pain is aggravated suddenly as a result of movement. Moreover, mixed syndromes with neuropathic or visceral components due to involvement of the spinal cord and nerves or the presence of intra-abdominal tumors may develop. Pain from bony metastases can produce a variety of symptoms. There may be referred pain, muscle spasm, or paroxysms of stabbing pain, particularly when bony lesions are accompanied by nerve compression.

Specific syndromes associated with neurological involvement have been described, including metastases to the skull and vertebral syndromes [11]. A metastatic spinal cord compression is a relevant complication of bony malignant metastases. The dorsal, lumbar, and cervical spine is affected, in descending order of frequency. In most instances mechanical compression of the spinal cord is caused by the growth of bony metastatic deposits in a vertebral body. The mass can grow posteriorly and extend to the epidural space. The pressure is transmitted to the cord and its vascular supply, resulting in mechanical injury, ischemia, venous stasis, and infarction. Compression leads to progressive neurological deficits below the site of the lesion. The onset of progressive neurological symptoms is often insidious. Vague complaints of back pain, leg weakness, and dysesthesias should be noted because early detection and intervention determine the functional outcome. Radicular pain is unilateral when arising in the cervical or lumbosacral spine, and bilateral when originating in the thoracic spine. Pain is exacerbated by recumbency, neck flexion, straight leg raising, coughing, and local pressure, and it may be relieved by sitting up or lying absolutely still. Weakness, sphincter impairment, and sensory loss are uncommon at presentation, but they develop when the disease progresses in the compressive phase. Compressive lesions in the cervical region cause tetraplegia, whereas thoracic lesions result in paraplegia. Lumbar compression determines the cauda equina syndrome. Once neurological symptoms have become established, they tend to develop rapidly, and the likelihood of a good outcome of treatment is progressively reduced.

Hypercalcemia is a frequent consequence of bone resorption. Symptoms occur with calcium values exceeding 3 mmol/l, and their severity is correlated with higher values. Hypercalcemia is associated with pain, nausea, vomiting, anorexia, constipation, weakness, dehydration and polyuria, mental disturbances, and confusion. Serum levels of alkaline phosphatase and osteocalcin reflect osteoblast activity.

B. Visceral Pain

Visceral cancer pain originates from a primary or metastatic lesion involving the abdominal or pelvic viscera, including luminal organs of the gastrointestinal or genitourinary tracts and the parenchymal organs. Mechanical stimuli, such as torsion or traction of mesentery, distension of hollow organs, stretch of serosal and mucosal surfaces, and compression of some organs, produce pain in humans. These conditions are frequently observed in cancer patients with abdominal diseases and intraperitoneal masses.

Obstruction or inflammation within the biliary tract or pancreatic duct induces pain directly related to an increased intraluminal pressure with consequent inflammation and release of pain-producing substances. Capsular stretching of the liver due to cancer growth produces pain. Distension or traction of the gallbladder leads to deep epigastric pain, inspiratory distress, and vomiting. Renal colic is commonly secondary to ureteral obstruction and subsequent distension of the ureter and renal pelvis. This may be evident in circumstances in which an abdominal-pelvic mass compresses or invades the ureter, as often occurs in gynecological cancers. Reports of pain appear to be directly related to the pressure of the urinary bladder. Bladder distension may also activate mechanisms related to the phenomena of counter-irritation. Etiology of pain from visceral tissues may also be related to ischemia, particularly in metastatic or recently damaged tissues (postsurgical). Ischemia may act as a modulator of mechanoreceptive visceral inputs. The variability in responses to ischemia may be due to the preexisting pathology or cancer-related mechanical distortion of the viscera secondary to local changes.

Visceral pain is difficult to localize and often is referred to other areas of the body [6]. Pain originating from any viscera cannot be easily differentiated from pain originating in another viscus, although descriptions of some visceral pains have been associated with specific etiologies, such as in cases of pancreatic pain or peptic ulcer.

Better localization of the stimulus occurs when the disease extends to a somatically innervated structure such as the parietal peritoneum. Thus, initially, visceral pain is poorly localized and dull because of the wide divergence of visceral afferents in the spinal cord. Poorly localized visceral pain becomes localized as visceral afferent input increases because of spinal facilitation or activation of visceral nociceptors. Somatic structures may also be involved with an indirect mechanism. For example, diaphragmatic irritation due to abdominal distension produced by large subdiaphragmatic masses may induce shoulder pain associated with hiccup.

The models of visceral pain are useful in explaining the outcome of some neurolytic blocks for abdominal cancer pain and the response to analgesic drugs. Failure or partial success of celiac or hypogastric plexus block may be attributed to the fact that the tumor has metastasized beyond the nerves that conduct pain via the celiac plexus and the component nerves that form it. Somatic pain may arise from parietal peritoneum and abdominal wall involvement as well as retroperitoneal nodal involvement or bone metastases. Similarly, a sympathetic block will not be useful for pain from radiculopathy related to retroperitoneal spread, chemotherapy-induced mucositis or liver embolization, and post-radiation pain.

These phenomena occur generally in the late stage of pancreatic disease but may occur earlier with other abdominal malignancies. Clinical problems concerning neurolytic superior hypogastric block are nearly the same as for celiac plexus block. Both control visceral pain. However, the less favorable results obtained with the hypogastric plexus block may be due to the greater tendency of pelvic tumors to infiltrate somatic structures and nerves in comparison with pancreatic tumors, where the celiac plexus block is usually used. Other than with visceral pain, pelvic cancers are more often associated with myofascial involvement, and as a consequence, with a somatic pain mechanism, although pain seems prevalently caused by pressure on the sciatic nerve. Thus, pelvic cancer pain can arise, among other reasons, from visceral and peritoneal involvement as well as from the involvement of muscles in the perineal floor with nerve entrapment and muscle necrosis. Because of the location of the piriformis muscle deep in the pelvic floor, pain may be a complication of a primary tumor or relapse, surgery, or radiotherapy. Buttock or rectal pain with or without posterior thigh pain, aggravated by sitting or activity, may be due to compression of the piriformis muscle. As a consequence, it may be associated with referred/localized pain of somatic origin. Motor reflexes may lead to muscle spasm and an additional component of somatic pain. The clinical picture is often mixed. In pelvic cancer, lumbar pain may also be due to iliopsoas muscle involvement. Nerve trunk pain, often radiating to the lower limbs, may be observed and is due to the involvement of lumbosacral plexus involvement in the presacral area or radiculopathy related to retroperitoneal spread. As a consequence, candidates for hypogastric plexus block, a sympathetic block performed for visceral pelvic pain, should be strictly selected. On the other hand, retroperitoneal invasion may result in a limited spread of neurolytic solutions.

C. Neuropathic Pain Syndromes

The term is applied to a large number of pain syndromes in which the pathophysiology is related to aberrant somatosensory processes that originate with a lesion in the peripheral or central nervous system. Neuropathic pain involving the peripheral nervous system is a common and clinically challenging problem in cancer patients. The syndromes include painful radiculopathy, plexopathy, mononeuropathy, or peripheral neuropathy [3].

Painful radiculopathy is caused by processes that compress, distort, or inflame nerve roots. Epidural and leptomeningeal metastases often produce painful radiculopathy. Postherpetic neuralgia is more frequent in the cancer population and is characterized by a typical radicular pain syndrome in the region of zoster infection.

Tumor invasion or compression can cause painful plexopathy at different levels. At the cervical level, plexus injury is frequently due to tumor or consequent to treatment, such as surgery or radiotherapy. Pain refers to the lateral part of the face or head, para-auricularly, to the anterior neck, or to the shoulder. In some cases pain, aching, or burning, exacerbated by movement, may be associated with Horner's syndrome. The overlap in pain referral areas may relate to the close anatomic relationship of cervical afferents and cranial nerves.

Malignant brachial plexopathy is most commonly observed in patients with lung cancer, lymphoma, or breast cancer. Lower plexus involvement, reflected by the pain and sensory disturbance distribution to the elbow, medial forearm, and external fingers, is frequent, whereas tumor infiltration of the upper plexus is less common and is characterized by pain in the shoulder, lateral arm, and hand. Patients with malignant brachial plexopathy are at high risk for epidural extension of the tumor, as the cancer mass grows medially and invades nerve roots and vertebrae. This process is associated with the evidence of panplexopathy and the development of Horner's syndrome. Radiation-induced brachial plexopathy may be transient or progressive with a delayed onset, months after a course of radiotherapy. Predominant symptoms are sensory changes and weakness, associated with skin lesions and lymphoedema, rather than pain.

Lumbosacral plexopathy is usually produced by several causes, including neoplastic infiltration or compression, radiotherapy or surgical trauma, and cytotoxic damage. Colorectal and cervical cancers and metastases from distant organs involve the plexus by direct extension from intrapelvic masses. Pain localization may vary considerably, in the lower abdomen, inguinal region, buttock, or leg. Sacral plexopathy may occur from direct extension of a sacral lesion or a presacral mass. Symptoms and pain distribution will depend on the involvement of upper, lower, and sacral nerves. A panplexopathy is not rare (about 20% of cases) and is frequently associated with leg edema. Imaging studies generally confirm these clinical features and may allow differential diagnosis with radiation-induced lumbosacral plexopathy, characterized by nonspecific fibrotic infiltration of the tissue. This plexopathy has a typical course from months to years following radiation treatment. The prominent features are slowly progressive bilateral weakness and leg swelling. Lumbosacral plexopathy may also occur following embolization techniques or conditions such as hemorrhage or abscess in the iliopsoas muscle.

Compression or infiltration of nerves from a tumor arising in an adjacent bone usually produces a tumor-related mononeuropathy. Typical examples are cranial neuralgias, sciatalgia, peroneal nerve palsy, and intercostal nerve injury. Other causes include postsurgical syndromes and nerve entrapment due to edema.

Peripheral neuropathies have multiple causes, not always recognizable in the clinical setting. Chemotherapy-induced peripheral neuropathy is the most common form and is typically manifested by distal painful paresthesias and sensory loss during and after treatment with vinca alkaloids, cisplatinum, and paclitaxel. Underlying malignancy may produce undefined neuropathies possibly related to some injury to the dorsal root ganglion or peripheral nerves, probably due to autoimmune-inflammatory mechanisms. These paraneoplastic forms, usually associated with small-cell lung cancer, are sensory neuropathies characterized by paresthesias, sensory loss, and sensory ataxia [12]. The course of the syndrome is typically independent and may develop before tumor diagnosis [12]. A painful sensorimotor peripheral neuropathy is also observed in hematological malignancies and is thought to be due to antibodies reacting with constituents or peripheral nerves.

VII. Pharmacological Treatment

Most pain in cancer responds to pharmacological management with orally administered analgesics. Current treatments are based on the analgesic ladder, which involves a stepwise approach to the use of analgesic drugs and is essentially a framework of principles rather than a rigid protocol [13]. The feasibility and efficacy of the analgesic ladder have been reported in different studies. However, undertreatment of cancer pain persists despite efforts to provide clinicians with information about the use of analgesics [1].

Nonsteroidal anti-inflammatory drugs (NSAIDs) are commonly used as the first step of the analgesic ladder. They have been found to be helpful both as this first step and in combination with opioids, regardless of the pain mechanism involved. However, prolonged use of this class of drugs is of concern, particularly in a population at risk, such as the elderly, receiving multiple drug regimens. The role of so-called weak opioids as the second step of the analgesic ladder has been questioned, and it has been speculated that this step could be bypassed. Strong opioids used at equivalent lower doses can be helpful.

Studies validating the World Health Organization's analgesic ladder had methodological limitations, including the circumstances under which assessments were made, small sample sizes, retrospective analyses, high rates of exclusion and dropout, inadequate follow-up, and lack comparison with levels of analgesia before introduction of the analgesic ladder [14]. Two types of pain should be considered from a temporal perspective. Ongoing pain requires the administration of regular doses of opioids, given at intervals dependent on the pharmacokinetics and duration of effect of each drug.

Despite effective pain control during most hours of the day, pain flares, commonly defined as breakthrough or episodic pain, may occur during opioid titration before optimal analgesia is achieved, or pain flares may develop during the course of the illness, with unpredictable temporal patterns and durations, or as a consequence of incident pain associated with bone metastases. In this case, extra doses of opioids with rapid onset and fast routes of administration are needed (see later discussion).

A. The Use of Opioids

When cancer patients experience severe pain, opioids are the gold standard of therapy. There is a large variety of options for the delivery of opioids in the management of cancer pain. The oral route is the most common, least invasive, and easiest route of opioid administration for most patients with cancer pain. For all patients who can take oral medications, this route should be considered first [15].

The main problem with the oral route is the first-pass biotransformation of opioids in the liver. All opioids given orally are absorbed via the gastric and duodenal mucosa and then transported to the liver via the portal venous system. In the liver, these medications undergo "first-pass metabolism" before entering the systemic circulation. This has a major impact on the systemic plasma concentrations of drugs. Bioavailability is defined as the percentage of administered medication that reaches the systemic circulation. For example, the dose of morphine given orally to a patient with cancer pain must be three times the intravenous or intramuscular dose. Oxycodone, although roughly equipotent to morphine if given parenterally, appears to be approximately twice as potent as morphine when given orally because of less first-pass metabolism.

Morphine, the most commonly used medication in the world to treat cancer pain, has a terminal elimination plasma half-life of about 3 hours. To provide longer-lasting analgesia, several preparations have become available. The bioavailability of these slow-release preparations is the same as that of immediate-release preparations, but the time to peak plasma drug concentrations is longer, and peak plasma concentrations are decreased. These preparations are recommended by the manufacturers to be administered every 12 hours. Clinicians occasionally use an 8-hour schedule, if necessary, to provide adequate analgesia. Other preparations, such as a morphine pellet coated with a polymer, are manufactured to be administered once every 24 hours. If additional analgesia is needed for breakthrough pain, doses of a fast-onset, short-acting opioid preparation should be available to the patient. However, immediate-release oral opioid preparations usually require approximately 30 minutes to become analgesically active when taken on an empty stomach, and faster routes may be required. Oxycodone, methadone, and hydromorphone are possible alternatives to oral morphine.

1. Alternative Routes

a. Parenteral Route Many patients will develop tolerance to most of the undesirable side effects of opioids (such as nausea/vomiting or sedation) over a period of several days. However, certain patients may not be able to tolerate oral medications because of esophageal motility problems or gastrointestinal obstruction (e.g., head and neck or esophageal cancer, bowel obstruction) or may have nausea and vomiting, limiting the utility of the oral route. Finally, some patients are unable to swallow because of the site of their cancer or because they are neurologically impaired. In these cases, an alternative form of analgesia must be used. Alternative routes, including the intravenous and subcutaneous, as well as the transdermal ones, have been advocated for such circumstances [16].

The intravenous route of administration is indicated for those patients whose pain cannot be controlled by a less invasive route or already having a central venous access. The major disadvantage of this route is that it is more complex to manage, especially at home, and requires some expertise. On the other hand, this route is the faster one, allowing for an immediate effect in emergency conditions. Different opioids are available as an intravenous solution in the majority of countries, including morphine, hydromorphone, fentanyl, alfentanil, sufentanil, and methadone. Fentanyl is approximately 80–100 times more potent than morphine, and sufentanil is approximately 1000 times more potent. The main drawback to their use in the practice of oncology is their high cost in comparison with morphine and methadone.

For patients requiring parenteral opioids who do not have in-dwelling intravenous access, the subcutaneous route can be used. This simple method of parenteral administration involves inserting a small plastic cannula on an area of the chest, abdomen, upper arms, or thighs and attaching the tubing to an infusion pump. The limiting factor is the volume of fluid that can be injected per hour; often more concentrated solutions are required. Most drugs used by intravenous route can also be used by subcutaneous infusion, except methadone, which could induce local toxicity. The main advantages of subcutaneous over intravenous patient-controlled analgesia (PCA) are that there is no need for vascular access, changing sites can be easily accomplished, and problems associated with in-dwelling intravenous catheters are avoided. The oral-parenteral ratio for morphine is 2:1 or 3:1. Intravenous or subcutaneous opioid infusions can be given as continuous infusions or by a PCA device, which provides continuous infusion plus on-demand boluses. Confused or uncooperative patients may not be the best candidates for PCA use.

b. Transdermal Route The transdermal route is a very comfortable means of administration. For patients who are unable to take oral medications, the transdermal route is a priority option for maintaining continuous plasma concentrations of opioids. Two drugs are available, fentanyl and buprenorphine, because of their potency and lipophilicity. The delivery system consists of a reservoir of fentanyl and alcohol, or a matrix where fentanyl is dissolved that contains a 3-day supply of drug. The fentanyl patch releases the drug at a constant rate until the reservoir is depleted. Upon initial application of the patch, a subcutaneous "depot" is formed as fentanyl saturates the subcutaneous fat beneath the patch. After approximately 12 hours, steady-state plasma fentanyl concentrations are reached, which are maintained for about 72 hours. Fentanyl patches are currently available in 25, 50, 75, and 100 μg/hour dosages. The bioavailability of transdermal fentanyl is very high (approximately 90%).

Buprenorphine is dissolved in a matrix, releasing a constant amount of drug for 3 days. Available dosages are 35 μg/hour (0.8 mg/day), 52.5 μg/hour (1.2 mg/day), and 70 μg/hour (1.6 mg/day).

Because of the slow depot formation and slow rise in plasma concentrations, these systems are not suitable for rapid titration against pain and are best suited for patients with stable pain for whom the 24-hour opioid requirement has already been determined.

Problems arise from conversion to fentanyl, as no clear protocols have been established. It has been suggested to use a conversion transdermal fentanyl–oral morphine ratio of 1:70–100. For buprenorphine, the buprenorphine–oral morphine conversion ratio ranges from 1:60 to 1:80.

c. Transmucosal/Sublingual Routes Sublingual administration of opioids is particularly beneficial in the patient with cancer who is unable to tolerate oral administration because of nausea/vomiting or dysphagia. It may also be attractive for patients who cannot receive parenteral opioids because of the lack of venous access or the presentation of typical contraindications for subcutaneous drug administration. Because sublingual venous drainage is systemic rather than portal, hepatic first-pass elimination can be avoided. On the other hand, the transmucosal or sublingual route also offers the potential for more rapid absorption and onset of action relative to the oral route. This is particularly useful for treating breakthrough pain, which requires a fast opioid effect. Lipophilic drugs are better absorbed than hydrophilic drugs. For these reasons, fentanyl and buprenorphine primarily use this route. Transmucosal fentanyl is the only medication that has been found to be very useful in the management of breakthrough pain in cancer patients, in various controlled studies.

d. Rectal Route This route may be a simple alternative when the oral route is not possible because of vomiting, obstruction, or altered consciousness. Its principal advantage is that it is independent of gastrointestinal tract motility and rate of gastric emptying. However, the rectal route also has several disadvantages, including a great deal of variation among individuals, the amount of drug drained from the rectum, and absorption changed by the small surface area of the rectum or interrupted by defecation. Finally, the rectal route is uncomfortable for prolonged use and contraindicated if the patient has painful anal conditions such as fissures or inflamed hemorrhoids. The usual recommendation is that initial doses of morphine and most other opioids given rectally be the same as for those given orally.

2. Treatment of Breakthrough Pain

Breakthrough pain has been defined as a transitory increase in pain intensity from a baseline pain of moderate intensity in patients receiving regularly administered analgesic treatment. However, transitory pain may occur in patients with no baseline pain at all. Moreover, pain exacerbations may also occur in patients with severe baseline pain in uncontrolled situations. The intermittent pain may be induced by movement, defined as incident pain, or not related to activity and less predictable [17].

A rescue dose of opioid can provide a means to treat breakthrough pain in patients already stabilized on a baseline opioid regimen. The use of a short-half-life opioid, such as immediate-release morphine, is suggested. The most effective dosage remains unknown, although clinicians suggest a dose roughly equivalent to about 15% of the total daily opioid dose, administered as needed every 2–3 hours. Titration of the rescue dose according to the characteristics of breakthrough pain should be individualized to be most appropriate, because the approach to opioid supplemental dosing has been based solely on anecdotal experience. However, the onset of action of an oral dose may be too slow (longer than 30 minutes), and better results may be obtained with a parenteral rescue dose. Although the intravenous route is the fastest, subcutaneous administration is associated with an acceptable onset of effect and should be considered equivalent in terms of efficacy.

PCA is an interesting modality to deliver drugs as needed. It appears that the demand dose is more important to the success of PCA, and the initial set-up of the patient-controlled analgesia system will influence the therapeutic outcome. There are limitations to the use of such pumps. Some patients are unable to operate a PCA device because of cognitive impairment or because they are overwhelmed by the technical aspects of care, and patients with drug-seeking behavior should be considered inappropriate candidates for PCA. Other drawbacks include the invasiveness of and complications inherent in the long-term use of subcutaneous needles or intravenous lines. The pump may limit patient mobility, and technical expertise is required.

Oral transmucosal dosing is a recent noninvasive approach to the rapid onset of analgesia. Highly lipophilic agents may pass rapidly through the oral mucosa, avoiding the first-pass metabolism and achieving active plasma concentrations within minutes. Fentanyl, incorporated in a hard matrix on a handle, is rapidly absorbed. It has been shown to have an onset of pain-relief action similar to that of intravenous morphine, within 10–15 minutes. When the fentanyl matrix dissolves, approximately 25% of the total fentanyl concentration crosses the buccal mucosa and enters the bloodstream. The remaining amount is swallowed and about one-third of this part is absorbed, thus achieving a total bioavailability of 50%. Different controlled studies have shown the effectiveness of oral transmucosal fentanyl for treating episodes of breakthrough pain. Of importance, the effective dose was not correlated with the basal analgesic regimen, a circumstance stressing the need to individualize the dose [18].

3. Opioid Switching

A substantial number of patients treated with an opioid do not have a successful outcome because of excessive adverse effects, inadequate analgesia, or a combination of both. Individual patients vary greatly in their response to different opioids. Patients who obtain poor analgesic efficacy or tolerability with one opioid will frequently tolerate another opioid, or even the same opioid administered by another route. Sequential opioid trials, so-called opioid switching, may be needed to identify the drug that yields the most favorable balance between analgesia and adverse effects. The biological basis for the individual variability in sensitivity to opioids is multifactorial, and some aspects remain unclear. The frequency of opioid switching is increasing as a consequence of better knowledge and improved assessment and monitoring of opioid-induced symptom intensity in patients who receive higher doses of opioids. Relatively large series involving opioid switching have shown a clear benefit to using different opioid sequences, with an improvement in adverse effects in 70–80% of cases, at lower doses of opioids than expected, in light of the questioned tables of conversion.

Available equianalgesic tables should assist physicians in the calculation of doses when switching opioids. However, these tables derive from the results of earlier studies or relative potency ratios with use of single-dose crossover design. Moreover, contrary to what occurs in cancer management, these studies focused on patients with limited opioid exposure, both in duration and dose. Thus, the applicability of these ratios to the setting of chronic opioid administration has been questioned in the past several years on the basis of a better knowledge of the basic mechanism of pharmacodynamic tolerance.

Proper conversion is dependent upon a variety of factors: drug dosage, cross-tolerance among opioids, and physiologic differences in drug metabolism. The conversion process must take into account such factors as the amount of residual drug in the patient's system and the time to achieve steady-state blood levels with the new drug, as well as individual patient responses during the conversion process. When converting from one chronically (around-the-clock) administered opioid to another, you must first calculate the 24-hour equivalent of the drug from which you are converting. The dosing interval for the new drug should be also taken into account and the 24-hour dose divided as appropriate for the new dosing interval. This dose and dosing interval can serve as the target for the new drug, but not usually as the initial order in the conversion process. Overdosage may occur if an equivalent dose, based on a conversion table, is used while the old opioid is still in the patient's system. Initiation of administration of the new drug, especially at the basal dose, may be delayed if significant residual drug is in the body. In calculating a new basal dose, when controversy exists as to the correct conversion value, it is safest initially to use a conversion value that will result in a new drug dose that is more conservative than the old dose. On the basis of the estimated wash-out period for residual opioids and the individual patient's response in the conversion process, the new basal dose should be increased, if necessary, according to the reported pain scores and the use of breakthrough doses.

All equivalencies listed in Table 2 are approximations for possible starting doses. There is no universal agreement on equivalent doses, and individual dose adjustment is essential. Particular caution should be used in converting from one opioid to another at high doses, particularly when switching to methadone. As opioids may differ significantly in terms of their mechanisms of action (for example, methadone's NMDA antagonism) and the degree of cross-tolerance and metabolism, conversion tables may be inaccurate

Table 2 Initial Conversion Ratios Among Opioids Suggested for Opioid Switching

Drug Route	Dosage (mg)
Morphine, parenteral	20
Morphine, oral	60
Methadone, oral	6–12
Hydromorphone, oral	12
Oxycodone, oral	40
Oxycodone, parenteral	20
Fentanyl, transdermal	0.6
Buprenorphine, transdermal	0.8

for calculating true equivalent doses, which risks overdosage with the new drug (or occasionally underdosage). This has been found to be particularly true in converting from high doses of opioids such as morphine and hydromorphone to methadone; that is, a lower methadone dose than is suggested by most published tables is appropriate. Recent articles have demonstrated the complexities of opioid conversion and argue against simplistic reliance on tables.

Thus, opioid conversion should not be a mere mathematical calculation, but a part of a more comprehensive assessment of the actual opioid therapy, evaluating the underlying clinical situation, pain and adverse effect intensity, comorbidity, and concomitant drugs and excluding any possible pharmacokinetic factor limiting the effectiveness of a certain drug. Given the wide conversion ratios reported in the literature, the process of reaching an optimal dose should be highly individualized. On the basis of data published so far, the initial ratio between morphine and hydromorphone is predictable, although an individual response may change the initial conversion ratio. A morphine-to-fentanyl ratio of 100 should be an acceptable compromise, with a time reduction to achieve stabilization. The conversion ratio between morphine and methadone should be more prudent, although this poses the risk of a delay in achieving the goal [19].

B. Adjuvants

One approach to pain that is poorly responsive to opioids is coadministration of a nonopioid analgesic [20]. Antidepressants may alleviate depression, enhance sleep, and provide decreases in perception of pain. The analgesic efficacy of the tricyclic antidepressants has been established in many painful disorders. The evidence supporting analgesic effects is particularly strong for amitriptyline. The analgesic effect of tricyclics is not directly related to antidepressant activity. Common side effects of tricyclic compounds include antimuscarinic effects, such as dry mouth; impaired visual accommodation; urinary retention and constipation; antihistaminic effects (sedation); and anti-alpha-adrenergic effects (orthostatic hypotension). The analgesic response is usually observed within 5 days. Alternative drugs with lower incidences of side effects should be considered for patients predisposed to the sedative, anticholinergic, or hypotensive effects of amitriptyline. Despite the frequent use of amitriptyline for neuropathic cancer pain, its effectiveness has not been demonstrated appropriately in this context. Small studies have shown that the potential benefits of amitriptyline are associated with a high rate of adverse effects, particularly intense in advanced disease.

It has been suggested that an anomaly in ion channels plays a role in the molecular mechanism of neuropathic pain. Sodium channel–blocking agents, such as systemic local anesthetics (e.g., mexiletine, carbamazepine, phenytoin, and open sodium channel blockers) have been reported to relieve neuropathic pain states. Although the exact mechanism of these drugs is not known, they all inhibit sodium channels of hyperactive and depolarized nerves while not interfering with normal sensory function. Although cancer patients with neuropathic pain have been reported to benefit from lidocaine, one double-blind, crossover, placebo-controlled study failed to demonstrate any benefit from 5 mg/kg of lidocaine administered as an intravenous infusion over 30 minutes for neuropathic pain syndromes related to cancer. Although sodium channel–blocking agents are useful for the management of chronic neuropathic pain, no conclusive clinical study has statistically verified these observations in cancer pain.

Anticonvulsants, such as carbamazepine, phenytoin, valproate, and clonazepam, have been reported to relieve pain in numerous peripheral and central neuropathic pain conditions, although contradictory results have been found. The efficacy of these drugs could be explained by the inhibitory effects exerted on NMDA receptors and other mechanisms, including sodium channel blockade. No measurable differences in the analgesic benefit of anticonvulsants and antidepressants were evidenced in a systematic review of available trials involving neuropathic pain. Gabapentin is promising as an adjuvant to opioid analgesia for neuropathic cancer pain. The addition of gabapentin to the therapeutic regimen for cancer patients decreased the pain score and some typical sensations associated with neuropathic pain.

A number of studies have documented the positive effects of corticosteroids on various cancer-related symptoms, including pain, appetite, energy level, food consumption, general well-being, and depression. Although analgesia in diverse pain syndromes has been reported, most of the evidence of these effects is anecdotal.

The bisphosphonates are pyrophosphate analogues, characterized by a P-C-P bond that render them resistant to hydrolysis by phosphates. Bisphosphonates potentiate the effects of analgesics in metastatic bone pain. Pamidronate, a potent bisphosphate, significantly reduced morbidity caused by bone metastases, including a 30–50% reduction in pain, impending pathological fractures, and the need for radiotherapy. Best results are obtained with doses of 60 mg or 90 mg. This treatment is generally well tolerated; side effects include transient low-grade fever, nausea, myalgia, and mild infusion site reactions. Intravenous application has been preferred by most investigators, mostly because of gastrointestinal adverse events and the urgency of the situation.

According to the developing theory of neuroplasticity previously described, agents that block the activity of NMDA receptors may provide new tools for the treatment of poorly responsive pain syndromes, particularly neuropathic ones. Ketamine is a noncompetitive NMDA receptor blocker that exerts its primary effect when the NMDA-receptor-controlled ion channel has been opened by a nociceptive barrage. A synergic effect between ketamine and opioids has been observed in cancer pain patients who had lost an analgesic response to high doses of morphine. Single 2.5-mg doses of ketamine given intravenously lessened previously uncontrollable pain in most patients. Doses of 110 mg a day produced excellent analgesia, even though daily morphine doses were halved. Ketamine doses of 42–720 mg daily were effective in patients whose pain had not been relieved by conventional treatment. In some patients, ketamine was administered at very high doses and for prolonged periods. Ketamine should be given at an initial starting dose of 100–150 mg daily, while the dose of opioids should be reduced by 50%, with the dose being titrated against the effect.

VIII. Interventional Procedures

A. Spinal Route

A small number of patients may still not obtain adequate analgesia despite large systemic opioid doses, or they may suffer from uncontrollable side effects such as nausea, vomiting, or oversedation. These patients may be candidates for the administration of a combination of opioids, local anesthetics, and clonidine via the spinal (epidural or intrathecal) route [21]. The goal of spinal opioid therapy is to place a small dose of an opioid and/or local anesthetic close to the spinal opioid receptors located in the dorsal horn of the spinal cord to enhance analgesia and reduce systemic side effects by decreasing the systemic daily opioid dose. Use of this route to deliver opioids requires placing a catheter into the epidural or intrathecal space and using an external or implantable infusion pump to deliver the medications. Deciding between epidural vs intrathecal placement or an external vs implantable pump to deliver the opioid involves consideration of multiple factors, including duration of therapy, type and location of the pain, disease extent and central nervous system involvement, opioid requirement, and individual experience. The daily epidural opioid requirement is approximately 10 times that of intrathecal administration. Intrathecal opioid administration has the advantage of allowing a higher concentration of drug to be localized at the receptor site while minimizing systemic absorption, thus possibly decreasing drug-related side effects. Morphine remains the drug of choice for the spinal route, because of its relatively low lipid solubility. It has a slow onset of action, but a long duration of analgesia when given via intermittent bolus. The starting dose is quite difficult to calculate, and various factors, including the previous opioid dose, the age, and the pain mechanism, should be taken into consideration. Adding a local anesthetic (bupivacaine or ropivacaine) to morphine via the spinal route has been successful in providing good analgesia for patients whose pain was resistant to epidural morphine alone, despite high doses. Further clinical studies and trials will still be needed to judge the safety, efficacy, and extended role of the spinal route in chronic cancer pain and, more important, to define for which patients this technique is best indicated.

Clonidine, an alpha-adrenergic agonist that acts at the dorsal horn of the spinal cord to produce analgesia, has been used in cancer patients in combination with epidural (or intrathecal) morphine infusions. There is some evidence to suggest that neuropathic pain may be somewhat more responsive to the combination of clonidine/morphine than to morphine alone, although orthostatic hypotension is of concern. Procedural and surgical complications, system malfunction, and pharmacological adverse effects are the main categories of complications associated with spinal drug delivery [21].

B. Sympathetic Blocks

Visceral pain is transmitted to the brain via sympathetic fibers that run through the visceral plexus more or less near the abdominal organs or viscera. Analgesia to the abdominal organs is possible because the afferent fibers innervating these structures travel in the sympathetic nerves. Chemical sympathectomy for visceral pain is analogous to peripheral nerve neurolysis for somatic pain. Sympathetic blocks, either of the celiac plexus or of the superior hypogastric plexus, are useful if pain pathways are predominantly or solely visceral. Post-residual pain after either procedure is mainly due to the presence of somatic mechanisms [22].

The benefits of neurolytic block of the celiac plexus in the treatment of abdominal visceral pain is well established. Celiac plexus block is very helpful in most patients suffering from upper abdominal pain secondary to carcinoma of the pancreas, liver, or stomach. Its role, duration, and efficacy have been debated. In controlled studies, less opioid consumption and fewer adverse effects were found in comparison with traditional analgesic regimens. Therapeutic failure may reflect involvement of somatic or nervous structures in the retroperitoneal space. Thus, it is necessary to define the tumor extension before performing the block. The superior hypogastric plexus is situated in proximity to the bifurcation of the common iliac vessels at the level of the fifth lumbar vertebral body and the first sacral vertebral body. It is in continuity with the celiac plexus and the lumbar sympathetic chains and innervates pelvic viscera via the hypogastric nerves. Similarly, the sympathetic component in the perineal region derives from the ganglion impar, located at the level of the sacrococcygeal junction.

The efficacy of a percutaneous neurolytic superior hypogastric plexus block for pelvic pain associated with cancer has been demonstrated. Visceral pelvic pain emanating from the descending colon and rectum, vaginal fundus and bladder, prostate, uterus, and ovary may be effectively relieved by superior hypogastric plexus block, particularly when the etiologic process is a gynecologic cancer. However, visceral pain alone is present at a relatively early stage of pelvic cancer. A more complex syndrome, including the infiltration of tissues that have somatic afferent fiber and the presence of somatic and neuropathic pain mechanisms, is more likely to appear during the advanced stage of the disease because of the greater tendency of pelvic tumors to infiltrate the nearest structures. A careful selection of patients is required to recognize the type of nociceptive afferent pathways.

C. Other Interventional Procedures

A percutaneous cordotomy is the interruption of the ascending spinothalamic tract, usually at the cervical level. A percutaneous cervical cordotomy by radiofrequency has been utilized in patients with unilateral bone pain below the C5 dermatome. Cordotomy may be indicated in a selected group of patients with refractory breakthrough pain due to bone metastases—for example, in neck femur fractures—but it is not indicated in the presence of neuropathic lesions. The risk is high for serious complications, including mirror pain, general fatigue or hemiparesis, and respiratory failure, with deterioration of performance status. Pituitary ablation has a potential role in patients with widely disseminated pain of bony metastatic origin and in patients with a hormonally responsive primary tumor. Although the success rate has been quoted as 74%–94%, long-term follow-up to death has not been carried out successfully. A high mortality rate has been reported with transient morbidity regarding rhinorrhea, meningitis, visual disturbances, diabetes insipidus, headache, and hypothalamic disturbances [8].

In conclusion, for the patient with cancer pain, the oral route of opioid delivery should be the first choice. If the oral route cannot be used because of gastrointestinal obstruction and/or severe nausea/vomiting, the rectal is equivalent, although unsuitable for prolonged use. Another noninvasive alternative to the oral route is the transdermal route.

For those patients for whom oral or transdermal opioids are not appropriate, intravenous or subcutaneous administration is effective, the latter being relatively easier. For treatment of breakthrough pain, short-onset opioids given orally or a transmucosal preparation of fentanyl is indicated, the latter being significantly more rapid and effective. Switching opioids or the route of administration, or both, may improve the opioid response in patients with an unfavorable balance between analgesia and adverse effects.

The spinal route can be attempted when the oral and other, parenteral routes have been unsuccessful. This route may be most successful when opioids and local anaesthetics and/or clonidine is used in combination. Whatever route is used, administration of opioids to manage cancer pain requires knowledge of potency relative to morphine and bioavailability of the route chosen. Dose-equivalent tables are only close approximations, and substantial interpatient variability is often observed. Therefore, patients should be closely followed and doses titrated to minimize side effects whenever the opioid, route, or dosage is changed. Neurolytic procedures may be indicated for very selected patients.

References

1. Cleeland, C. S., Gonin, R., Hatfield, A. K., Edmonson, J. H., Blum, R. H., Stewart, J. A., and Pandya, K. J. (1994). Pain and its treatment in outpatients with metastatic cancer. *N. Engl. J. Med.* **330**, 592–596.
2. Bruera, E., and Kim, H. N. (2003). Cancer pain. *JAMA* **290**, 2476–2479.
3. Portenoy, R. K., and Conn, M. (2003). Cancer pain syndromes. *In:* "Cancer Pain, Assessment and Management" (E. Bruera and R. K. Portenoy, eds), pp. 38–50. Cambridge University Press, London.
4. Besson, J. M. (1999). The neurobiology of pain. *Lancet* **353**, 1610–1615.
5. Coderre, T. J., Katz, J., Vaccarino, A. L., and Melzack, R. (1993). Contribution of central neuroplasticity to pathological pain: review of clinical and experimental evidence. *Pain* **52**, 259–285.
6. Cervero, F., and Laird, J. M. A. (1999). Visceral pain. *Lancet* **353**, 2145–2148.
7. Mao, J., Price, D., and Mayer, D. J. (1995). Experimental mononeuropathy reduces the antinociceptive effects of morphine: implications for common intracellular mechanisms involved in morphine tolerance and neuropathic pain. *Pain* **61**, 353–364.
8. Mercadante, S. (1997). Malignant bone pain: physiopathology, assessment and treatment. *Pain* **69**, 1–18.
9. Luger, N. M., Sabino, M. A., Schwei, M. J., Mach, D. B., Pomonis, J. D., Keyser, C. P., Rathbun, M., Clohisy, D. R., Honore, P., Yaksh, T. L., and Mantyh, P. W. (2002). Efficacy of systemic morphine suggests a fundamental difference in the mechanisms that generate bone cancer vs. inflammatory pain. *Pain* **99**, 397–406.
10. Medhurst, S. J., Walker, K., and Bowes, M. (2002). A rat model of bone cancer pain. *Pain* **96**, 129–140.
11. Caraceni, A., and Weinstein, S. M. (2001). Classification of cancer pain syndromes. *Oncology* **15**, 1627–1640.
12. Falah, M., Schiff, D., and Burns, T. M. (2005). Neuromuscular complications of cancer diagnosis and treatment. *J. Support. Oncol.* **3**, 271–282.
13. World Health Organization. (1990). "Cancer Pain Relief and Palliative Care." World Health Organization, Geneva.
14. Jadad, A. R., and Browman, G. P. (1995). The WHO analgesic ladder for cancer pain management. *JAMA* **274**, 1870–1873.
15. Hanks, G. W., and Expert Working Group of the Research Network of the EAPC. (2001). Morphine and alternative opioids in cancer pain: the EAPC recommendations. *Br. J. Cancer* **84**, 587–593.
16. Cherny, N. J., Chang, V., Frager, G., Ingham, J. M., Tiseo, P. J., Popp, B., Portenoy, R. K., and Foley, K. M. (1995). Opioid pharmacotherapy in the management of cancer pain. *Cancer* **76**, 1288–1293.
17. Portenoy, R. K., and Hagen, N. A. (1990). Breakthrough pain: definition, prevalence, and characteristics. *Pain* **41**, 273–281.
18. Mercadante, S., and Steering Committee of the European Association for Palliative Care (EAPC) Research Network. (2002). Episodic (breakthrough) pain: consensus conference of an expert working group of the European Association for Palliative Care. *Cancer* **94**, 832–839.
19. Mercadante, S. (1999). Opioid rotation in cancer pain: rationale and clinical aspects. *Cancer* **86**, 1856–1866.
20. Portenoy, R. K., and Rowe, G. (2003). Adjuvant analgesic drugs. *In:* "Cancer Pain: Assessment and Management" (E. Bruera and R. K. Portenoy, eds.), pp. 188–198. Cambridge University Press, Cambridge.
21. Mercadante, S. (1999). Problems of long-term spinal opioid treatment in advanced cancer patients. *Pain* **79**, 1–13.
22. Mercadante, S., and Nicosia, F. (1998). Celiac plexus block: a reappraisal. *Reg. Anesth. Pain Med.* **23**, 37–48.

93

Pain Associated with the Autonomic Nervous System

Wilfrid Jänig

Keywords: *autonomic nervous system, body protection, cytokines, hyperalgesia, inflammation, neuropathic pain, pain, sympathetic-afferent coupling, sympathetic nervous system*

I. Introduction
II. Pain Generated by the Sympathetic Nervous System: Evidence from Experimental Investigations of Human Patients
III. Excitation of Afferent Neurons Dependent on Activity in Sympathetic Neurons
IV. Sensitization of Nociceptors Mediated by Sympathetic Terminals Independent of Excitation and Release of Noradrenaline
V. Sensitization of Nociceptors, Nerve Growth Factor, and Cytokines Mediated by Sympathetic Terminals
VI. Mechanical Hyperalgesic Behavior in the Rat Controlled by the Sympathoadrenal (SA) System
VII. Synopsis
References

I. Introduction

Any tissue-damaging event, with and without lesion of nerves, that leads to pain, hyperalgesia and inflammation affects the autonomic nervous system. The reactions of this system are protective for the body and must be considered to prevent further damage in the acute and subacute stage following the tissue damage and to further healing from this damage in the chronic stage. The autonomic reflexes and reactions occurring during both time domains following tissue damage are organized at the level of the spinal cord, brainstem, and hypothalamus and are under the control of the cerebral hemispheres. The neural mechanisms underlying this control are only partially known. In this context it is important to emphasize that the sympathetic and parasympathetic nervous system have a distinct organization in the periphery, spinal cord, brainstem, and hypothalamus. This is reflected in their neuroanatomy, neurochemistry, and physiology [1,2].

1. Each group of autonomic target cells is innervated and regulated by only one sympathetic or parasympathetic pathway, and a few by one of each.
2. Each peripheral autonomic pathway is functionally distinct and transmits the centrally generated signals faithfully in the autonomic ganglia and via neuroeffector junctions to the autonomic target cells.
3. Each peripheral autonomic pathway is connected to distinct reflex circuits in the spinal cord, brainstem, and hypothalamus that are responsible for the typical reflex discharge patterns present in the (preganglionic and postganglionic) neurons of the peripheral autonomic pathways.

Spinal and vagal visceral afferent neurons, although indispensable for the regulation of activity in the sympathetic and parasympathetic pathways, strictly speaking do not belong to the autonomic nervous system.

In this article I will not discuss the role of the autonomic nervous system in the generation of protective reactions during pain and hyperalgesia. I will focus on the question of whether and in which way the autonomic nervous system is involved in the *generation of pain*. I will restrict this question further to the sympathetic nervous system because there is no evidence that the parasympathetic system is involved in the generation of pain. I will distinguish three aspects (Table 1, position 1):

1. mechanisms underlying coupling (cross-talk) from sympathetic (noradrenergic) neurons to afferent neurons in the generation of pain, involving excitation of the sympathetic postganglionic neurons and release of noradrenaline;
2. mechanisms underlying coupling (cross-talk) from sympathetic (noradrenergic) neurons to afferent neurons in the generation of pain, probably independent of excitation but involving inflammatory mediators such as prostanoids, nerve growth factor, and cytokines; and
3. the role of the sympathoadrenal system in mechanical hyperalgesia.

The argumentation that the sympathetic nervous system may be involved in the generation of pain is based on clinical observations of patients with pain and hyperalgesia (in particular, patients with complex regional pain syndrome [CRPS] [7]), experimental investigations of these patients, and animal experiments using *in vivo* and *in vitro* models. The role of the sympathetic nervous system and its transmitters in acute inflammation (e.g., bradykinin-induced synovial plasma extravasation) or chronic inflammation (such as rheumatoid arthritis) and its interaction with

Table 1 Sympathetic Nervous System in the Generation of Pain

1. Role of the sympathetic nervous system in the generation of pain [3–6,8,9]: Sympathetic-afferent coupling in the periphery
   - Coupling after nerve lesion (noradrenaline, α-adrenoceptors)
   - Coupling via the micromilieu of the nociceptor and the vascular bed
   - Sensitization of nociceptors mediated by sympathetic terminals independent of excitation and release of noradrenaline
   - Sensitization of nociceptors initiated by cytokines or nerve growth factor and mediated by sympathetic terminals
   - Sympathoadrenal system and nociceptor sensitization

2. Sympathetic nervous system and central mechanisms [6–8,10,11]: Control of inflammation and hyperalgesia by sympathetic and neuroendocrine mechanisms
   - Sympathetic nervous system and CRPS
   - Sympathetic nervous system and immune system
   - Sympathetic nervous system and rheumatic diseases

peptidergic afferent innervation, as well as the hypothalamo-pituitary-adrenal system, will not be discussed (see Table 1, position 2). We are just at the beginning of the experimental investigation of these control mechanisms [4,11].

II. Pain Generated by the Sympathetic Nervous System: Evidence from Experimental Investigations of Human Patients

Pain being dependent on activity in sympathetic neurons is called sympathetically maintained pain (SMP). SMP is a symptom and includes generically spontaneous pain and pain evoked by mechanical or thermal stimuli. It may be present in CRPS type I and type II and in other neuropathic pain syndromes [12]. The idea about the involvement of the (efferent) sympathetic nervous system in pain is based on various clinical observations that have been documented in the literature for decades. Representative of these multiple observations on patients with SMP are the results of quantitative experimental investigations [7,10,13]. These experiments demonstrate that

1. sympathetic postganglionic neurons can be involved in the generation of pain;
2. blockade of the sympathetic activity can relieve the pain;
3. noradrenaline injected intracutaneously is able to rekindle the pain; and
4. α-adrenoceptor blockers or guanethidine (which depletes noradrenaline from its stores) may relieve the pain.

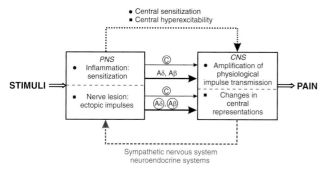

Figure 1 Concept of generation of peripheral and central hyperexcitability during inflammatory pain and neuropathic pain. The upper interrupted arrow indicates that the central changes are generated (and possibly maintained) by persistent activation of nociceptors with C-fibers (e.g., during chronic inflammation), called here "central sensitization," and after trauma with nerve lesion by ectopic activity and other changes in lesioned afferent neurons, called here "central hyperexcitability." The lower interrupted red arrow indicates the efferent feedback via the sympathetic nervous system and neuroendocrine systems (e.g., the sympathoadrenal system). Primary afferent nociceptive neurons (in particular those with C-fibers) are sensitized during inflammation. After nerve lesion *all* lesioned primary afferent neurons (unmyelinated as well as myelinated ones) undergo biochemical, physiological, and morphological changes, which become irreversible with time. These peripheral changes entail changes of the central representation (of the somatosensory system), which become irreversible if no regeneration of primary afferent neurons to their target tissue occurs. The central changes, induced by persistent activity in afferent nociceptive neurons or after nerve lesions, are also reflected in the efferent feedback systems that may establish positive feedback to the primary afferent neurons. PNS, peripheral nervous system; CNS, central nervous system.

These data can be interpreted as follows. Nociceptors are excited and possibly sensitized by noradrenaline released by the sympathetic fibers. Either the nociceptors have expressed adrenoceptors and/or the excitatory effect is generated indirectly, e.g., via changes in blood flow. Sympathetically maintained activity in nociceptive neurons may generate a state of central sensitization/hyperexcitability leading to spontaneous pain and secondary evoked pain (mechanical and possibly cold allodynia [6]; Fig. 1).

III. Excitation of Afferent Neurons Dependent on Activity in Sympathetic Neurons

Sympathetic postganglionic neurons do not influence sensory neurons projecting to skin and deep somatic tissues in mammals under physiological conditions. The effects that have been measured under experimental conditions on receptors with myelinated and unmyelinated axons were weak and can in part be explained by changes of the effector organs (erector pili muscles, blood vessels) induced by the activation of sympathetic neurons [3,5].

A. Direct Coupling Following Nerve Lesion

Following partial or complete nerve lesion, afferent and sympathetic postganglionic fibers start to sprout at the lesion site, in the nerve proximal to the lesion site and in the dorsal root ganglion. This sprouting is maintained over months or longer and leads to changes of the relation between sympathetic postganglionic nerve fibers and primary afferent neurons. Now the sympathetic fibers may influence the afferent neurons by release of noradrenaline resulting in their excitation. The coupling from sympathetic nerve fibers to afferent neurons may occur at the lesion site, distal to the lesion site (when the fibers regenerate), and proximal to the lesion site in the dorsal root ganglion, as well as in the lesioned nerve (Fig. 2A). Coupling at or distal to the lesion site has been shown to occur (see later discussion). Coupling proximal to the lesion site in the nerve has not been investigated so far. Coupling in the dorsal root ganglion does occur, mainly or exclusively to cell bodies of muscle afferent neurons with small-diameter myelinated afferents; however, it is questionable whether this coupling is pathophysiologically important [3,5,6].

Following nerve cut or ligation, coupling may occur between sympathetic fibers and afferent terminals in a neuroma. Some myelinated as well as unmyelinated nerve fibers in the neuroma can be excited following electrical stimulation of the sympathetic supply or systemic injection of noradrenaline or adrenaline. This coupling is chemical and occurs via noradrenaline acting on α-adrenoceptors. Whether other mediators are involved is unknown. The coupling has been observed in young neuromas but less so in old ones weeks and months after nerve lesion. This is compatible with clinical observations showing that neuroma pain is usually not dependent on sympathetic activity. It is also compatible with histological observations showing that catecholamine-containing axon profiles are rare within a neuroma many weeks after cutting and ligating the nerve.

The situation is different when afferent and sympathetic fibers are allowed to regenerate to the target tissue. More than a year after cross-union of nerves (proximal stump of the sural or superficial peroneal nerve to the distal stump of the tibial nerve, in the cat) and after reinnervation of appropriate and inappropriate target tissues, unmyelinated afferent fibers may be vigorously excited by electrical low-frequency stimulation of the sympathetic supply. This excitation is adrenoceptor-mediated (Fig. 3) and can be mimicked by adrenaline or noradrenaline and blocked by α-adrenoceptor blockers.

The cellular mechanisms underlying the increased sensitivity to noradrenaline are unknown. Novel expression or upregulation of adrenoceptors in the afferent neurons occurs or normally present adrenoreceptors that are not functional become uncovered and effective as a response to damage. The subtype of α-adrenoreceptor being involved in the sympathetic–afferent coupling in the different rat models is

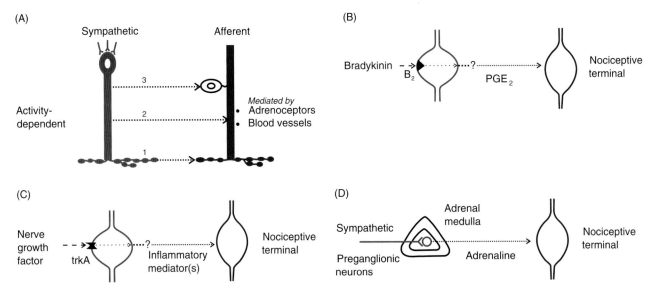

Figure 2 Ways hypothesized to couple sympathetic and primary afferent neurons following peripheral nerve lesion (**A**) or during inflammation (**B–D**). (**A**) These types of coupling depend on the activity in the sympathetic neurons and on the expression of functional adrenoceptors by the afferent neurons or mediation indirectly via the blood vessels (blood flow). It can occur in the periphery (1), in the dorsal root ganglion (3), or possibly also in the lesioned nerve (2). (**B**) The inflammatory mediator bradykinin acts at B_2 receptors in the membrane of the sympathetic varicosities or in cells upstream of these varicosities, inducing release of prostaglandin E_2 (PGE_2) and sensitization of nociceptors. This way of coupling is probably *not dependent* on activity in the sympathetic neurons. (**C**) Nerve growth factor released during an experimental inflammation reacts with its high-affinity receptor trkA in the membrane of the sympathetic varicosities, inducing release of inflammatory mediators and sensitization of nociceptors. This effect is probably *not* dependent on activity in the sympathetic neurons. (**D**) Activation of the adrenal medulla by sympathetic preganglionic neurons leads to release of adrenaline, which generates sensitization of nociceptors. The question mark in **B** and **C** indicates that PGE_2 or other inflammatory mediators may be released by cells other than in the sympathetic varicosities. Modified from Jänig, W., and Häbler, H. J. (2000). Sympathetic nervous system: contribution to chronic pain. *Prog. Brain. Res.* **129**, 451–468.

predominantly α_2 but also α_1. Knowledge about the subtypes of adrenoceptor following nerve trauma may lead to the design of more specific treatment modalities for neuropathic pain conditions involving sympathetic efferent activity [5].

B. Indirect Sympathetic-Afferent Coupling Following Tissue Trauma via Blood Vessels

Neural control of blood vessels can change dramatically after trauma with nerve lesions, but possibly also after trauma without lesions of nerves [14].

1. Cutaneous blood vessels that are reinnervated after a nerve lesion might exhibit stronger-than-normal vasoconstriction to impulses in sympathetic neurons. *In vitro* investigation of the rat tail artery has shown that the functional recovery of neurovascular transmission remains permanently disturbed after a nerve lesion [15].
2. The sympathetically reinnervated cutaneous blood vessels, now hyperactive, may show stronger-than-normal vasoconstrictions to systemic catecholamines.
3. The blood vessels may exhibit a decreased vasodilation to activation of reinnervated unmyelinated peptidergic afferents via the so-called axon reflex. The reinnervated blood vessels may therefore be under stronger-than-normal vasoconstrictor influence, which can no longer be counteracted by an afferent-mediated vasodilation.

The altered neural and nonneural control of blood vessels will contribute to abnormal regulation of microcirculation following trauma due to nerve lesions. These changes may be a permissive factor in the generation of afferent nociceptive impulse activity, in the sensitization of nociceptors, and therefore in the generation of pain in skin, deep somatic tissues, and viscera [3].

IV. Sensitization of Nociceptors Mediated by Sympathetic Terminals Independent of Excitation and Release of Noradrenaline

In the rat, the paw-withdrawal threshold (PWT) to mechanical stimulation of the dorsum of the hind paw decreases dose-dependently after intradermal injection of the inflammatory mediator bradykinin (BK, an octapeptide cleaved from plasma α_2-globulins, by kallikreins circulating in the plasma; Fig. 2B and Fig. 4). Following a single injection of BK, the mechanical hyperalgesic behavior lasts

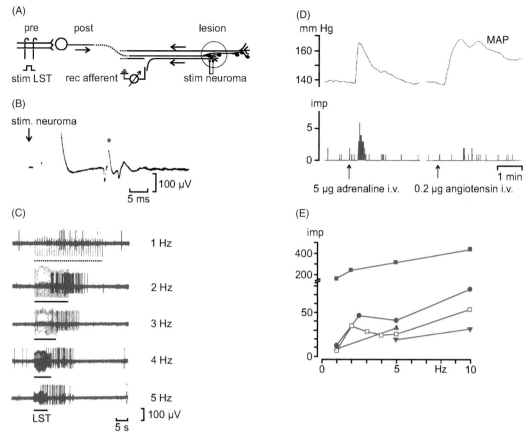

Figure 3 Sympathetic-afferent coupling after nerve lesion with subsequent fiber regeneration. Excitation of unmyelinated afferent units by electrical stimulation of sympathetic fibers following nerve injury. Unmyelinated primary afferents were recorded in cats 11–20 months following a nerve lesion. The central cut stump of a cutaneous nerve innervating hairy skin (sural or superficial peroneal nerve) had been adapted to the distal stump of a transected mixed nerve (tibial nerve). There was a neuroma-in-continuity at the site of the lesion, and cutaneous nerve fibers had regenerated into skin and deep somatic tissue supplied by the mixed nerve. (**A**) Experimental set-up (pre, preganglionic; LST, lumbar sympathetic trunk). (**B**) The afferent fibers were identified as unmyelinated by electrical stimulation of the neuroma with single impulses. The signal indicated by dot was same as in C; the afferent fiber conducted at 1.3 m/s. (**C**) Record from a single unmyelinated afferent unit. Supramaximal stimulation of the LST with trains of 30 pulses at 1–5 Hz (trains and stimulation artifacts indicated by bars). Note that the afferent unit had low ongoing activity (impulses before the trains at 1 and 4 Hz). (**D**) Adrenaline (5 μg injected intravenously) activated the fiber. Angiotensin (0.2 μg injected intravenously) generated a large increase in blood pressure (MAP, mean arterial blood pressure) but did not activate the afferent fiber. (**E**) Stimulus-response relation of a single afferent C-fiber (open squares) and 4 strands with 2–4 or more units in response to stimulation of trains of 30 pulses at frequencies of 0.5 to 10 Hz (abscissa scale) delivered to the LST. Ordinate scale, impulses/stimulus train. Modified from Häbler, H., Jänig, W., and Koltzenberg, H. (1987). Activation of unmyelinated afferents in chronically lesioned nerves by adrenaline and excitation of sympathetic efferents in the cat. *Neurosci. Lett.* **82**, 35–40.

for more than 1 hour. It is mediated by the B_2 BK-receptor and blocked by the cyclooxygenase inhibitor indomethacin and therefore mediated by a prostaglandin (probably PGE_2) that sensitizes nociceptors for mechanical stimulation (see Fig. 4).

The decrease in PWT provided by BK is significantly attenuated after surgical sympathectomy. This shows that the sympathetic innervation of the skin is involved in the sensitization of nociceptors for mechanical stimulation. Decentralization of the lumbar paravertebral sympathetic ganglia (denervating the postganglionic neurons by cutting the preganglionic sympathetic axons) does not abolish BK-induced mechanical hyperalgesic behavior (see Fig. 4), indicating that the sensitizing effect of BK is not dependent on activity in the sympathetic neurons innervating skin and therefore is not dependent on release of noradrenaline [5,16]. BK stimulates the release of prostaglandin in association with the sympathetic terminals. The underlying cellular mechanisms are unknown.

Mechanical hyperalgesic behavior generated by intracutaneous injection of BK is an interesting phenomenon. However, the conventional explanation that this mechanical

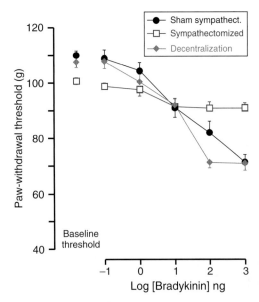

Figure 4 Mechanical hyperalgesic behavior and sympathetic innervation. Bradykinin-induced hyperalgesia in sympathectomized (red squares; n = 12 hind paws) and sham sympathectomized (circles; n = 6 hind paws) rats and in rats with decentralized lumbar sympathetic chains (preganglionic axons in lumbar sympathetic chain interrupted 8 days before, blue diamonds; n = 10 paws). Sham sympathectomy and sympathetic decentralization were both significantly different from sympathectomy groups ($P < 0.01$). After Khasar, S. G., Miao, F. J.-P., Jänig, W., and Levine, J. D. (1998a). Modulation of bradykinin-induced mechanical hyperalgesia in the rat by activity in abdominal vagal afferents. *Eur. J. Neurosci.* **10**, 435–444.

hyperalgesia is due to prostanoids sensitizing nociceptors because indomethacin prevents it appears to be too simple. The reasons are

1. the novel finding that sympathetic fibers mediate this hyperalgesia independent of neural activity and release of noradrenaline;

2. indomethacin does not block this behavior under certain conditions (e.g., when the adrenal medullae are activated after vagotomy [16]; see later); and

3. sensitization of nociceptors to mechanical stimulation by BK is weak or absent, as shown in neurophysiological experiments [5]. Thus, this phenomenon must be reinvestigated with a rigorous experimental approach.

V. Sensitization of Nociceptors, Nerve Growth Factor, and Cytokines Mediated by Sympathetic Terminals

A. Nerve Growth Factor

Systemic injection of nerve growth factor (NGF) is followed by a transient thermal and mechanical hyperalgesia in rats and humans. During experimental inflammation (evoked by Freund's adjuvant in the rat hind paw), NGF increases in the inflamed tissue, paralleled by the development of thermal and mechanical hyperalgesia. Both are prevented by anti-NGF antibodies. The mechanisms responsible are sensitization of nociceptors via high-affinity NGF receptors (trkA receptors) and an induction of increased synthesis of peptides (calcitonin-gene-related peptide [CGRP] and substance P) in the afferent cell bodies by NGF, taken up by the afferent terminals and transported to the cell bodies. The NGF-induced sensitization of nociceptors also seems to be mediated indirectly by the sympathetic postganglionic terminals. Heat and mechanical hyperalgesic behavior generated by local injection of NGF into the skin is prevented or significantly reduced after chemical or surgical sympathectomy [17,18]. Thus, NGF released during inflammation by inflammatory cells may act on the sympathetic terminals via high-affinity trkA receptors, inducing the release of inflammatory mediators and subsequently sensitization of nociceptors for mechanical and heat stimuli (Fig. 2C). It is unclear whether this sensitization of nociceptors mediated by terminal sympathetic nerve fibers is dependent on activity in the sympathetic neurons or release of noradrenaline and adrenoceptors expressed in the nociceptive afferent neurons or is independent of activity and release of noradrenaline [5,8].

B. Proinflammatory Cytokines

Based on behavioral experiments with rats (to study mechanical and heat hyperalgesic behavior), it has been shown that tissue injury, injection of the bacterial cell wall endotoxin lipopolysaccharide, or injection of carrageenan (a plant polysaccharide) stimulates tissue inflammation and leads to sensitization of nociceptors. Systematic pharmacological interventions using blockers or inhibitors of the various mediators demonstrate that the proinflammatory cytokines, tumor necrosis factor α (TNFα), interleukin (IL)-1, IL-6, and IL-8 may be involved in this process of sensitization and therefore in the generation of hyperalgesia. Pathogenic stimuli lead to activation of resident cells and the release of the inflammatory mediator BK and other mediators. The inflammatory mediators and the pathogenic stimuli themselves activate macrophages, monocytes, and other immune-related cells. These cells release TNFα, which generates sensitization of nociceptors by two possible pathways (Fig. 5).

1. It induces production of IL-6 and IL-1β by immune cells, whereby IL-6 enhances the production of IL-1β. These interleukins stimulate cyclooxygenase 2 (COX 2) and the production of prostaglandins (PGE_2, PGI_2) which in turn react with the nociceptive terminal via E-type prostaglandin receptors.

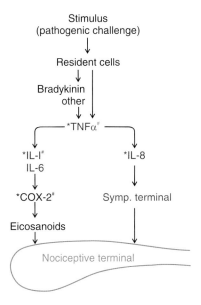

Figure 5 Role of cytokines in sensitization of nociceptors during inflammation and the underlying putative mechanisms leading to hyperalgesia. Pathogenic stimuli activate resident cells and lead to release of inflammatory mediators (such as bradykinin). Pro-inflammatory cytokines are synthesized and released by macrophages and other immune or immune-related cells. Nociceptors are postulated to be sensitized by two pathways involving the cytokines: (1) Tumor necrosis factor α (TNFα) induces synthesis and release of interleukin 1 (IL-1) and IL-6, which, in turn, induce the release of eicosanoids (prostaglandin E_2 and I_2 [PGE_2, PGI_2]) by activating the cyclooxygenase-2 (COX-2); (2) TNFα induces synthesis and release of IL-8. IL-8 activates sympathetic terminals that sensitize nociceptors via β_2-adrenoceptors. Glucocorticoids inhibit the synthesis of the cytokines and the activation of COX-2 (indicated by asterisks). Anti-inflammatory cytokines (such as IL-4 and IL-10) that are also synthesized and released by immune cells inhibit the synthesis and release of pro-inflammatory cytokines (indicated by #). This scheme is fully dependent on behavioral experiments and pharmacological interventions. The different steps will need to be verified in neurophysiological experiments. After [19].

2. It induces the release of IL-8 from endothelial cells and macrophages. IL-8 reacts with the sympathetic terminals that are supposed to mediate sensitization of nociceptive afferent terminals by release of noradrenaline to act via β_2-adrenoceptors. These two peripheral pathways by which nociceptive afferents can be sensitized involving cytokines are under inhibitory control of circulating glucocorticoids (indicated by asterisks in Fig. 5) and of other, anti-inflammatory interleukins (e.g., IL-4 and IL-10, indicated by # in Fig. 5) [8,19].

The mechanisms involving the interaction of NGF, proinflammatory cytokines, and noradrenergic sympathetic postganglionic fibers in the sensitization of nociceptive afferents have been deduced on the basis of behavioral and pharmacological experiments. Such interaction has not been studied directly by recording the activity from nociceptors by electrophysiological techniques.

VI. Mechanical Hyperalgesic Behavior in the Rat Controlled by the Sympathoadrenal (SA) System

The PWT to mechanical stimulation of the dorsal skin of the rat hind paw with a linearly increasing mechanical stimulus decreases dose-dependently after intradermal injection of BK (Fig. 6A; BK-induced mechanical hyperalgesic behavior). The decrease of PWT is generated by sensitization of cutaneous nociceptors by BK. The BK-induced mechanical hyperalgesic behavior is significantly enhanced in rats with subdiaphragmatic vagotomy, performed ≥ 7 days before the measurements (Fig. 6A) and the baseline threshold is decreased. The largest part of this enhancement is generated by activation of the adrenal medulla (AM) after vagotomy, and a small part is generated by removal of central inhibition acting in the dorsal horn (Fig. 6C). After denervation of the AM, subdiaphragmatic vagotomy has only a small effect on BK-induced decrease of PWT (Fig. 5B). This experiment argues that vagotomy leads to release of adrenaline (probably by activation of sympathetic premotor neurons), which in turn sensitizes nociceptors and enhances the sensitization of nociceptors by BK for mechanical stimulation. Two groups of experiments support this conclusion [20,21].

1. In vagotomized rats, the decreased PWT to intracutaneously injected BK and the decreased baseline PWT are reversed after denervation of the AM (section of the preganglionic axons innervating the AM). The development of enhanced mechanical hyperalgesic behavior and its reversal have a slow time course of 1 to 2 weeks (Fig. 6C).

2. Chronic subcutaneous administration of adrenaline generates the same effect as vagotomy: the BK-induced PWT to mechanical stimulation significantly decreases, with a slow time course of 1 to 2 weeks. Chronic infusion of a β_2-adrenoceptor blocker significantly attenuates the decrease of PWT following vagotomy. Plasma levels of adrenaline following vagotomy significantly increase 3, 7, and 14 days after subdiaphragmatic vagotomy.

These results suggest that vagotomy leads to an increased release of adrenaline from the AM, with an increased adrenaline level in the plasma, possibly by removal of ongoing central inhibition of sympathetic premotor neurons and subsequent activation of preganglionic neurons innervating the AM. Interruption of the sympathetic preganglionic axons innervating the AM stops the release of adrenaline and therefore prevents or reverses the decrease

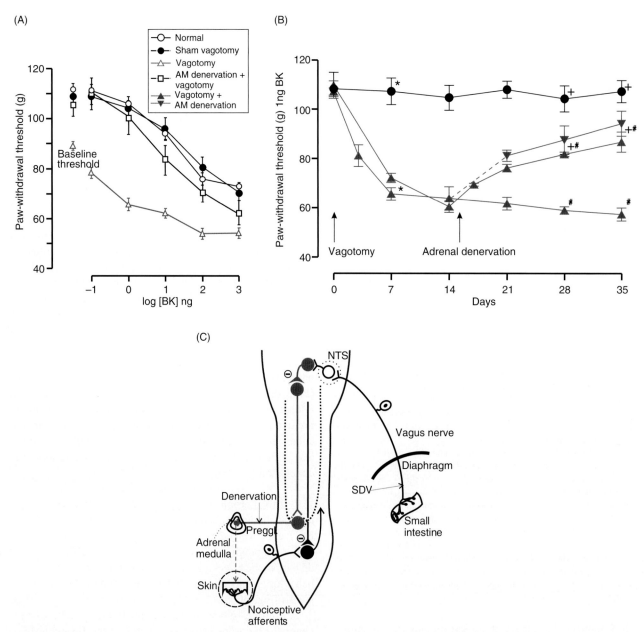

Figure 6 Long-term enhancement of bradykinin (BK)-induced mechanical hyperalgesia after vagotomy and its disappearance after denervation of the adrenal medullae (AM). **(A)** Decrease of paw withdrawal threshold (PWT) to mechanical stimulation of the dorsum of the rat hind paw induced by BK (BK-induced behavioral mechanical hyperalgesia), in naïve control (O, n = 26), vagotomized (Δ, n = 16; 7 days after subdiaphragmatic vagotomy), sham-vagotomized (●, n=18), and vagotomized rats with denervated AM (□, n = 6). Cutaneous nociceptors on the dorsum of the paw are stimulated by a linearly increasing mechanical force. PWT is defined as the mean (± S.E.M.) minimum force (in grams, ordinate scale) at which the rat withdraws its paw. Abscissa scale: log dose of BK (in ng) injected in a volume of 2.5 μL of saline into the dermis of the skin. **(B)** Total change of PWT in response to intradermal injection of 1 ng BK in rats before and 7 to 35 days after vagotomy (Δ, n = 6), before and 7 to 35 days after sham-vagotomy (●, n=8), and in rats whose AMs are denervated 14 days after vagotomy and have measurements taken up to 35 days after initial surgery (▲ rats tested after vagotomy and denervation of the AM [n = 6]; ▼ rats tested only after additional denervation of the AM [n = 4]). *$P < 0.01$, ▲/▲ vs ● on day 7; +$P > 0.05$ ▲/▼ vs ● on days 28 and 35; # $P < 0.01$, ▲/▼ vs Δ on days 28 and 35. **(C)** Schematic diagram showing the proposed neural circuits in spinal cord and brainstem that modulate nociceptor sensitivity via the sympathoadrenal system (AM). Sensitivity of cutaneous nociceptors for mechanical stimulation is modulated by adrenaline from the AM. Activation of the AM increases the sensitivity of the nociceptors. Activity in preganglionic neurons innervating the AM depends on activity in vagal afferents from the small intestine, which has an inhibitory influence on the central pathways to these preganglionic (pregg1) neurons. Thus, interruption of the vagal afferents leads to activation of the AM. It is hypothesized that these neuronal (reflex) circuits in the brainstem are under the control of the upper brainstem, hypothalamus, and telencephalon. Dotted thin lines: Axons of sympathetic premotor neurons in the brainstem, which project through the dorsolateral funiculi of the spinal cord to the preganglionic neurons of the AM. For details see text (+, excitation; −, inhibition). After Khasar, S. G., Miao, F. J.-P., Jänig, W., and Levine, J. D. (1998b). Vagotomy-induced enhancement of mechanical hyperalgesia in the rat is sympathoadrenal-mediated. *J. Neurosci.* **18**, 3043–3049.

of baseline mechanical PWT and the enhancement of BK-induced decrease of PWT to mechanical stimulation. This novel finding implies that the sensitivity of nociceptors to mechanical stimulation is potentially under control of the SA system and that nociceptor sensitivity can be regulated from remote body domains by way of the brain and this neuroendocrine pathway.

The novel mechanism has several implications [9].

1. The vagal afferents that are involved in modulation of hyperalgesic behavior project through the celiac branches of the abdominal vagus nerves [16] and supply the small and large intestines and may monitor toxic and other events at the inner defense line of the body (the "gut associated lymphoid tissue"?) [2,4].

2. Adrenaline obviously must act over a long period of time to induce changes in the micromilieu of the nociceptor population, which in turn leads to their sensitization [19]. It probably does not act directly on the nociceptors but on other cells (e.g., macrophages, mast cells, keratinocytes), which then release substances that generate the sensitization.

3. The change of sensitivity of nociceptors generated by adrenaline is a novel mechanism of nociceptor sensitization involving the SA system, which differs from mechanisms that lead to activation and/or sensitization of nociceptors by sympathetic-afferent coupling under pathophysiological conditions. This mechanism may operate in ill-defined pain syndromes such as chronic fatigue syndrome, fibromyalgia, functional dyspepsia, and irritable bowel syndrome [4,9].

4. This mechanism is under the control of the brain. However, the central pathways leading to activation of preganglionic sympathetic neurons that innervate the AM and their inhibitory control by vagal afferent activity need to be determined.

sympathetic innervation. Several pathophysiological mechanisms may be involved in this process:

- abnormal coupling of noradrenergic postganglionic fibers to primary afferent neurons;
- disturbance of the micromilieu of afferent receptors by changes of neurovascular transmission and by development of hyperreactivity of blood vessels; and
- disturbance of the micromilieu of nociceptors by interference of noradrenergic fibers with nonneural inflammatory and immune-competent cells.

B. Novel ways of involvement of sympathetic postganglionic neurons in the sensitization of nociceptors and subsequent generation of pain and hyperalgesia have been studied only in animal behavioral models.

- Terminals of sympathetic neurons may mediate sensitization of nociceptors involving nerve growth factor, cytokines or bradykinin. This function of the sympathetic fiber may be independent of its excitability and release of noradrenaline.
- Adrenaline released by the adrenal medulla sensitizes nociceptors for mechanical stimulation. This sensitizing effect takes one to two weeks to develop.

C. The peripheral changes related to the communication from the sympathoneural and sympathoadrenal system to the somatovisceral afferent neurons must be seen in the context of control of both systems by the brain.

D. The complexity of the somatosensory and autonomic abnormalities observed in patients with chronic pain that may actively involve, in some way or another, the sympathetic nervous system indicates that several pathobiological processes operate in parallel on the sensory site as well as the efferent site, and that the actual clinical phenomenology may be dependent on the predominance of one type of pathological mechanism.

VII. Synopsis

Peripheral injury (trauma with and without obvious nerve lesions; chronic inflammation) leads to nociceptor sensitization or ectopically generated afferent impulse traffic to the CNS. The peripheral changes entail changes of central neurons (globally described here as "central sensitization" or "central hyperexcitability"), resulting in distorted sensations and distorted autonomic, somatomotor, and endocrine reactions.

A. The sympathetic outflow to the affected peripheral part of the body may be actively involved in the generation of pain and associated changes by way of a positive feedback loop. Stimuli that normally are nonpainful may now elicit painful reactions that are dependent on an intact

References

1. Jänig, W., and McLachlan, E. M. (2002). Neurobiology of the autonomic nervous system. *In:* "Autonomic Failure" (C. J. Mathias and R. Bannister, eds.), 4th edition, pp. 3–15. Oxford University Press, Oxford.
2. Jänig, W. (2006). "The Integrative Action of the Autonomic Nervous System. Neurobiology of Homeostasis." Cambridge University Press, Cambridge, New York.
3. Jänig, W. (2002). Pain and the sympathetic nervous system: pathophysiological mechanisms. *In:* "Autonomic Failure" (C. J. Mathias and R. Bannister, eds.), 4th edition, pp. 99–108. Oxford University Press, Oxford.
4. Jänig, W. (2005). Vagal afferents and visceral pain. *In:* "Advances in Vagal Afferent Neurobiology" (B. Undem and D. Weinreich, eds.), pp. 461–489. CRC Press, Boca Raton, Florida.
5. Jänig, W., and Häbler, H. J. (2000). Sympathetic nervous system: contribution to chronic pain. *Prog. Brain. Res.* **129**, 451–468.

6. Jänig, W., and Baron, R. (2001). The role of the sympathetic nervous system in neuropathic pain: clinical observations and animal models. *In:* "Neuropathic Pain: Pathophysiological and Treatment" (P. T. Hansson, H. L. Fields, R. G. Hill, and P. Marchettini, eds.), pp. 125–149. IASP Press, Seattle.
7. Jänig, W., and Baron, R. (2003). Complex regional pain syndrome: mystery explained? *Lancet Neurol.* **2**, 687–697.
8. Jänig, W., and Levine, J. D. (2005). Autonomic-endocrine-immune responses in acute and chronic pain. *In:* "Wall and Melzack's Textbook of Pain" (S. B. McMahon and M. Koltzenburg, eds.), 5th edition, pp. 205–218. Elsevier Churchill Livingstone, Amsterdam and Edinburgh.
9. Jänig, W., Khasar, S. G., Levine, J. D., and Miao, F. J.-P. (2000). The role of vagal visceral afferents in the control of nociception. *Prog. Brain Res.* **122**, 273–287.
10. R. N. Harden, R. Baron, and W. Jänig, eds. "Complex Regional Pain Syndrome" (2001). IASP Press, Seattle.
11. Straub, R. H., and Härle, P. (2005). Sympathetic transmitters in joint inflammation. *Rheum. Dis. Clin. North Am.* **31**, 43–59.
12. Stanton-Hicks, M., Jänig, W., Hassenbusch, S., Haddox, J. D., Boas, R., and Wilson, P. (1995). Reflex sympathetic dystrophy: changing concepts and taxonomy. *Pain* **63**, 127–133.
13. Baron, R., Schattschneider, J., Binder, A., Siebrecht, D., and Wasner, G. (2002). Relation between sympathetic vasoconstrictor activity and pain and hyperalgesia in complex regional pain syndromes: a case-control study. *Lancet* **359**, 1655–1660.
14. Koltzenburg, M., Häbler, H.-J., and Jänig, W. (1995). Functional reinnervation of the vasculature of the adult cat paw by axons originally innervating vessels in hairy skin. *Neuroscience.* **67**, 245–252.
15. Jobling, P., McLachlan, E. M., Jänig, W., and Anderson, C. R. (1992). Electrophysiological responses in the rat tail artery during reinnervation following lesions of the sympathetic supply. *J. Physiol. Lond.* **454**, 107–128.
16. Khasar, S. G., Miao, F. J.-P., Jänig, W., and Levine, J. D. (1998a). Modulation of bradykinin-induced mechanical hyperalgesia in the rat by activity in abdominal vagal afferents. *Eur. J. Neurosci.* **10**, 435–44.
17. McMahon, S. B. (1996). NGF as a mediator of inflammatory pain. *Phil. Trans. R. Soc. Lond. B. Biol. Sci.* **351**, 431–440.
18. Woolf, C. J. (1996). Phenotypic modification of primary sensory neurons: the role of nerve growth factor in the production of persistent pain. *Phil. Trans. R. Soc. Lond. B.* **351**, 441–448.
19. Poole, S., and Woolf, C. J. (1999). Cytokine-nerve growth factor interactions in inflammatory hyperalgesia. *In:* "Cytokines and Pain" (L. R. Watkins and S. F. Maier, eds.), pp. 89–132. Birkhäuser Verlag, Basel, Boston.
20. Khasar, S. G., Miao, F. J.-P., Jänig, W., and Levine, J. D. (1998b). Vagotomy-induced enhancement of mechanical hyperalgesia in the rat is sympathoadrenal-mediated. *J. Neurosci.* **18**, 3043–3049.
21. Khasar, S. G., Green, P. G., Miao, F. J.-P., and Levine, J. D. (2003). Vagal modulation of nociception is mediated by adrenomedullary epinephrine in the rat. *Eur. J. Neurosci.* **17**, 909–915.

94

Postherpetic Neuralgia

Misha-Miroslav Backonja, MD

Keywords: *varicella zoster virus, neuropathic pain, peripheral sensory nerves, spinal cord, degeneration, peripheral sensitization, central sensitization*

I. Brief History and Nomenclature
II. Etiology and Molecular Mechanisms
III. Pathology and Natural History
IV. Pathophysiology, Biochemistry, and Pharmacology
V. Explanation of Symptoms in Relation to Pathophysiology
   References

I. Brief History and Nomenclature

Postherpetic neuralgia (PHN) is a chronic painful sequela of peripheral sensory nerve injury following acute shingles, which in turn is the result of the reactivation of herpes zoster (HZ) in the dorsal root ganglia (DRG) of the affected nerves. In addition to the pain of PHN, which is one of the most easily recognized types of neuropathic pain, patients with this chronic painful disorder also experience other sequelae, such as a variable degree of sensory loss and paresthesia in the distribution of the affected nerve roots, all of which reflect the underlying pathology and pathophysiological mechanisms.

The story of herpes zoster, shingles, and PHN is the most illustrative example of the link between the etiology of underlying disease (in this case viral infection), neurological localization (peripheral nerve root sensory dermatomal innervation), and the neuropathic pain mechanisms (including those from ongoing spontaneous pain to hyperalgesia). The following brief historical review depicts how this story developed. The first person to describe a link between herpes zoster and peripheral nerves was Richard Bright, who in 1831 correctly associated the characteristic distribution of the zoster rash with segmental nerve involvement [1]. Romberg was first to note the relationship between the rash and neuralgia, in 1855 [2], and in 1862 Von Barensprung demonstrated at autopsy the damage within the DRG and sensory afferent nerves [3]. In 1900, Head and Cambell published a series of 20 detailed postmortem exams of patients recently afflicted with HZ [4]. Lewis and Marvin proposed in 1927 that "the virus causing the mischief in the root ganglion spreads along the sensory tract to the skin, there setting up a distinct inflammatory change" [5]. In the 1950s and 1960s, Hope-Simpson suggested zoster represented the reactivation of latent varicella-zoster virus (VZV) [6], and he was the first to provide evidence demonstrating that the incidence of PHN was increased among the elderly [7]. Weller and Coons, in 1954, showed that varicella and zoster were both caused by VZV [8]. Straus and co-workers fully proved Hope-Simpson's theory to be correct by showing in 1984 that virus isolates recovered during varicella and zoster outbreaks are identical [9]. Observations by Noordenbos, who reported predominant loss of thicker myelinated fibers [10], were the basis for the idea that interaction between different types of fibers might influence pain and the concept that "fast blocks slow," which stimulated development of the gate control theory of pain [11]. The well-documented etiology and anatomy provided an opportunity for pain researchers to investigate poorly understood mechanisms of neuropathic pain by carefully examining PHN phenomenology. Psychophysical, pharmacological, and histopathological studies performed by a few research groups were the basis for modern concepts of neuropathic pain and its complexity, leading to still-evolving theories that are attempting to further define the role of many factors. These factors include deafferentation, irritable nociceptors, the influence of injured peripheral nerve fibers on uninjured ones, ipsilateral and contralateral dorsal horn pathological changes, loss of inhibitory interneurons, and central sensitization [12].

It is a conceptual and practical challenge to define PHN in relation to acute HZ eruption, or shingles (Fig. 1), because of the wide variability in the HZ presentation and the course of shingles pain and its transition into PHN as the sequela (Fig. 2). The pain of acute HZ in most cases persists as a continuum into PHN. When viewed in this context, there is no need to identify the point at which PHN begins. In this case, PHN would be defined as pain that persists 4 months following the onset of rash. On the other hand, according to traditional views, PHN is defined as pain that persists beyond 3 months following healing of the HZ rash. In either case, PHN is anchored to acute HZ; in the former case it is anchored to the onset, and in the latter, to the healing of the HZ rash, so definitive positive identification of the rash is paramount in both cases.

II. Etiology and Molecular Mechanisms

VZV is a human neurotropic alpha-herpesvirus that causes the primary infection of chicken pox (varicella) and then becomes dormant in the DRG of the entire neuraxis. The virus tends to reactivate predominantly in the thoracic and ophthalmic DRG, in the form of shingles (see Fig. 1). VZV is one of eight herpesviruses that can cause human diseases, and the closest form among those is herpes simplex virus type 1 (HSV-1). VZV is the smallest of the alpha-herpesviruses, with a genome consisting of 125 kilobase pairs (kb). The VZV genome exhibits relative genetic stability. A number of VZV genes can be depleted

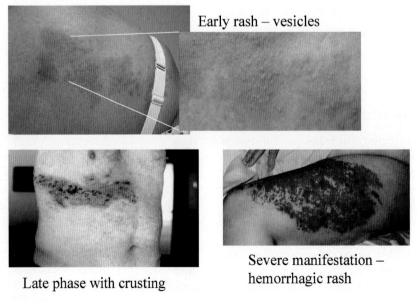

Figure 1 Acute Shingles. Top photograph is the distribution of acute early phase of zoster rash, with close-up demonstrating vesicular nature of the rash. Bottom left photograph is late phases of the zoster rash with crust formations. Bottom right photograph is an example of severe rash with evidence of hemorrhage. (Source: M. Backonja, University of Wisconsin.)

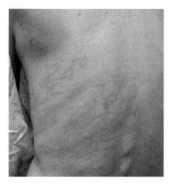
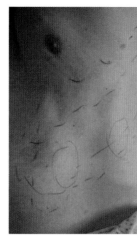

PHN – discoloration at the site of scarring

PHN Examination:

Green – sensory deficit

Red – light brush allodynia

Black – ongoing spontaneous pain

Figure 2 Postherpetic neuralgia (PHN). This is an example of discoloration in the dermatome previously affected by the zoster rash and now the location of persistent pain. Sensory examination revealed multiple abnormalities.

without altering the kinetics of VZV replication in cell culture and without changing its cytopathic effects. VZV is most infectious when the virion envelope is preserved. The replication is temperature-dependent, with optimal growth at 32–33°C. The functions of some VZV proteins have been defined, but the contributions of others are deduced from the sequence homologies with herpes simplex virus (HSV) proteins of known function. Two genes that encode glycoproteins, ORF5 (open reading frames) and ORF68, are required for replication. The major nucleocapsid protein of VZV is encoded by ORF40 and ORF33.5, and it is a putative assembly protein. The glycoproteins of the VZV envelope are presumed to mediate cellular attachment and penetration, and they also participate in fusion of the infected cell membranes and in cell-to-cell spread of the virus. ORF9A and the associated protein are not essential but are important for syncytia formation [13].

VZV can infect activated peripheral blood mononuclear cells (PBMCs) and primary cells derived from nervous system tissue, including supporting nonneural cells such as Schwann cells and astrocytes, which are more susceptible to cytopathic effects than neurons. Primary infection must begin with viral inoculation of mucous membranes, most probably of the respiratory tract, and from there it is transferred into the regional lymph nodes. Entry is followed by primary viremia, with transport of the virus to the liver and other cells of the reticuloendothelial system. Secondary viremia starts after 10–21 days of incubation and is associated with dissemination of the virus to the cutaneous epithelial cells; this is when the varicella rash appears. By this time, in addition to skin cells, other types of cells such as T cells, neurons, and satellite cells of the DRG are infected by the virus. Latent infection of the DRG cells is the result of primary VZV infection: virus is transported from the mucocutaneous sites of replication by the neuronal cell axons to the DRG cells, in addition to transport of virus by the PBMCs, independent of cutaneous replication [13]. It is known that the ganglionic VZV burden during latency is low. Two of the key questions that have been addressed are the cellular site of latent VZV and the identity of the viral genes that are transcribed during latency. There is now a consensus that latent VZV resides predominantly in ganglionic neurons with less frequent infection of non-neuronal satellite cells. There is considerable evidence to show that at least five viral genes are transcribed during latency. Viral protein expression has been demonstrated during VZV latency [14]. Varicellazoster virus (VZV) open reading frame 63 (ORF63) protein is expressed during latency in human sensory ganglia. The ability of ORF63 to downregulate ORF62 transcription may play an important role in virus replication and latency [15].

The relationship between the infecting strain of wild-type VZV and its virulence on the manifestation of the disease in the human host is not established. Modern theories propose that host factors are more-critical determinants of the outcome of primary infection than is the infecting strain and that HZ results when the host responses are perturbed, rather than as a result of differences in the propensity of VZV strains to reactivate from latency. Future immunization with the attenuated Oka vaccine might enable investigators to establish the contributions of viral as well as host factors to VZV-related diseases, including PHN [16].

III. Pathology and Natural History

As discussed earlier, PHN has to be viewed together with acute HZ as the inciting event. Consequently, the pathology and pathophysiology of HZ have a direct impact on the clinical manifestations and natural course of PHN.

Many risk factors have been identified with the increased incidence of acute HZ. These include increasing age, acquired immunodeficiency syndrome, cancer, organ transplantation, systemic lupus erythematosus, and immunosuppressive therapy. Other factors have been analyzed but were found not to be risk factors for HZ, including race, gender, physical trauma, and psychosocial stressors. In contrast to the many risk factors for HZ, it has been established that HZ itself is not a risk factor for or indicator of those or any other disorders.

Many risk factors have been identified with the increased incidence of PHN, including older age, female gender, greater acute HZ pain severity, greater rash severity, greater neurological abnormalities, the presence of prodromal symptoms, fever during acute HZ, psychosocial stressors, and a more pronounced immune response. One factor that is possibly associated with an increased risk for PHN is ophthalmic (i.e., trigeminal) distribution of the rash [17].

The histopathologic findings of VZV infections, whether chicken pox or HZ, are virtually identical. The vesicles involve the corium and dermis of the skin. With the progression of viral replication, the epithelial cells undergo degenerative changes characterized by ballooning, with the subsequent appearance of multinucleated giant cells and prominent eosinophilic intranuclear inclusions. Infrequently, necrosis and hemorrhage may appear in the upper portion of the dermis. As the vesicle evolves, the collected fluid becomes cloudy as a consequence of the appearance of polymorphonuclear leukocytes, degenerated cells, and fibrin. Either the vesicles rupture and release infectious fluid or the fluid gradually becomes reabsorbed. VZV characteristically becomes latent after chicken pox [18]. The virus establishes latency within the DRG. Reactivation of the latent virus leads to HZ rash, manifesting as a sporadic disease. Histopathologic examination of the nerve root and DRG after infection with VZV demonstrates characteristics indicative of infection, namely, satellitosis, lymphocytic infiltration in the nerve root, and degeneration of the ganglia cells. Intranuclear inclusions are found within the ganglia cells [19–22]. The molecular mechanism by which VZV establishes latency still remains unknown.

Neurological complications of HZ reactivation can present in many forms of peripheral nervous system (PNS) or central nervous system (CNS) symptoms and signs, all of which are significantly less frequent than the most common chronic complication of PHN pain. These other complications are most commonly the result of VZV spread into the cerebrospinal fluid spaces and invasion of PNS and CNS elements. Motor neuron involvement could be subclinical but is detectable by positive electromyography and rarely results in complete paralysis. Cranial neuropathies tend to be multiple. Best known is facial palsy with associated auricular rash, which can lead to poor prognosis for recovery unless treated early with antiviral agents. The resultant central neurological complications range from aseptic meningitis and myelitis to encephalitis, cerebral angiitis, and delayed hemiplegia. Neuropathological analysis of encephalitis has revealed a tendency for the virus to localize to the gray-white matter junction, with viral invasion of small intraparenchymal arterioles, which leads to ischemia, necrosis, and demyelination. In cases of myelitis, direct viral invasion is regarded as the major cause of neurological manifestations, although vasculitis has been proposed, and the dorsal horn is most affected at the level of the rash. Many CNS complications can occur in the absence of rash, and the method for confirmation is VZV polymerase chain reaction [23].

Neuropathological studies of PHN have revealed loss of peripheral nerve fibers in skin biopsy and staining with PGP9.5 (protein gene product 9.5), as well as post-mortem peripheral nerve analysis, which also showed Wallerian degeneration, direct inflammatory changes, and fibrosis. On the basis of a limited number of autopsies, it was demonstrated that patients with PHN had cell loss in DRG and atrophy of roots and the dorsal horn. Specifically, in a small series of autopsies on five patients (three with severe PHN and two without persistent pain), there was evidence of dorsal horn atrophy (Fig. 3) and of cell, axon, and myelin loss with fibrosis in the DRG, which was found only in patients with persistent PHN pain. Severe loss of myelin and axons in the nerve and sensory root did not differentiate between cases with and without pain because those findings were common to both. Cell loss in the DRG was a prominent feature in spinal ganglia but not as prominent in trigeminal ganglia in cases of facial PHN. Bilateral pathologic changes were demonstrated at the spinal level in the form of inflammatory infiltrates and as the loss of fibers in nerve roots on autopsies and in peripheral nerves on skin biopsies, in spite of the fact that all of these patients had unilateral rash and PHN pain [24,25].

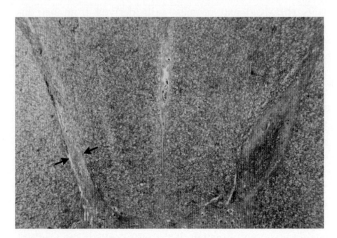

Figure 3 Histological analysis revealed loss of fibers of the dorsal horn of the spinal cord. This image is used with the kind permission of Dr. C. N. P. Watson.

IV. Pathophysiology, Biochemistry, and Pharmacology

Because PHN is an exclusively human condition, information about the pathophysiology of persistent pain and hyperalgesia due to nerve injury by reactivation of VZV comes solely from human clinical studies, although input from animal models has been crucial in elucidating the underlying pain mechanisms in a more general sense. Human studies are based on psychophysics and histology, from skin biopsies and autopsies; animal studies, which primarily use a traumatic model rather than viral infection, employ a full range of approaches, from behavior to electrophysiology and biochemistry. Similarities between the human psychophysics and animal behavior of hyperalgesia have been the basis for the translation from one to the other.

A few psychophysical studies of patients with PHN pain have contributed to improved insight about the significance of the symptoms and signs in PHN. In addition, pharmacological probes such as application of local anesthetics and systemic medications complemented the efforts to elucidate mechanisms. The resulting information was the basis for the development of theories about neuropathic pain, with PHN as a model. The complex and dynamic nature of acute HZ pain and its transition to PHN pain has demonstrated that most patients have a multitude of symptoms and signs, which tend to evolve over time but stabilize, with the predominance of a specific set of symptoms and signs for each particular patient. Symptoms range from ongoing spontaneous burning, aching, and itching to sharp and stabbing pains to the most common type, hypersensitivity and exacerbation of pain from a light touch and brushing of the PHN-affected skin area. Sensory signs range from variable degrees of deficits, predominantly to warm stimuli, to variable types and degrees of allodynia and hyperalgesia to mechanical and thermal stimuli; the most frequent is dynamic mechanical allodynia (pain from nonpainful stimulation, produced by moving a soft object such as clothing or a soft hairbrush across the area of stimulation) (see Fig. 2) [25,26]. Observations that (1) a subgroup of patients had marked dynamic mechanical allodynia and no loss of warm sensation, which is conducted by small fiber afferents, and responded to topical administration of local anesthetics and (2) another subgroup of patients had ongoing pain without allodynia, with no effect of local anesthetic, led to the theory of a spectrum of pain with which PHN pain can present. At one end of that spectrum were patients who had preserved small-diameter afferents that were irritable (so-called irritable nociceptors), and on the other end of the spectrum were patients with loss of small-fiber function due to deafferentation. This theory provides a very elegant explanation about possible underlying mechanisms and places the accent on a peripheral mechanism because most of the observations from the available human studies allowed this much to be concluded. According to this theory, proposed mechanisms responsible for the manifestations are: a) sensitization and lowering of the response thresholds of small-fiber afferents as well as ectopic discharges underlying manifestation of irritable nociceptors; and b) persistent central sensitization due to barrage of nociceptive impluses during pain of acute zoster plus loss of input from large-fiber diameter fibers and loss of inhibitory interneurons as manifestation of deafferentation associated pain. There are still many questions that need to be addressed in future studies, such as (1) the interaction of allodynia and deafferentation, because the majority of patients have a combination of both; (2) the role of ongoing inflammation; and (3) the role of central processes, including central sensitization, central facilitation, and loss of inhibition, which take place not only at the spinal level but also at the supraspinal level.

Regarding the influence of factors at the central spinal synaptic level, there was no evidence that concentrations of the synaptic receptors that participate in pain transmission and modulation, such as substance P, serotonin, noradrenalin, and opioids (Fig. 4), were decreased when compared with the PHN-unaffected side, although postmortem studies of patients revealed atrophy of the affected dorsal horns. Preliminary biochemical analysis of the primary afferents of PHN-affected skin from a single patient revealed increased innervation of the epidermis, with small-diameter fibers expressing VR1 receptors, which would account for the increase in pain from the application of capsaicin [12,28]. Further research at this level is certainly needed.

PHN has been used as a prototypical example of neuropathic pain (NP) which is defined as pain initiated or caused by a primary lesion or dysfunction in the nervous system

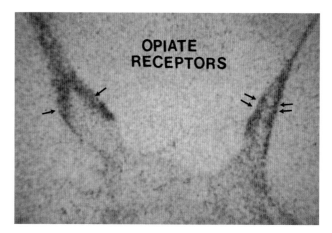

Figure 4 Histochemical staining for opioid receptors with [H3]etorphine revealed symmetrical distribution, although there is evidence of atrophy, as also seen in Figure 3. The finding of no difference in opiate receptors with evident loss of neurons would suggest that remaining neurons are likely to respond to opioid therapy, as it has been demonstrated in clinical trails. This image is used with the kind permission of Dr. C. N. P. Watson.

[29], with proposed peripheral and central mechanisms, as discussed earlier in this chapter. The central mechanisms of neuropathic pain (NP) have been elicited primarily on the basis of animal studies demonstrating, among other findings, the expansion of receptive fields, increased response to stimuli, and exaggeration of response to repeated stimuli (so-called wind-up). A few human studies confirm existence of these mechanisms in patients with PHN pain [30].

Only one human study investigated the possible effect of immune mechanisms in patients with PHN pain, and it demonstrated an increase in a pro-inflammatory chemokine, IL8, in acute HZ and PHN. Furthermore, increased IL-8 during acute HZ was the strongest predictor of the development of PHN pain [21].

Pharmacological studies, especially with topical application of local anesthetics and capsaicin (as described earlier), have certainly demonstrated the role of distal terminals of small-fiber afferents in transduction and transmission of PHN pain. In addition, a few systemically acting pharmacological agents have been studied in randomized trials and clearly were associated with positive outcomes; these include categories such as tricyclic antidepressants, which have multiple modes of action (e.g., descending inhibition through noradrenergic and serotonergic effects and probably systemic local anesthetic effects); gabapentin and pregabalin, which modulate N-type calcium channel activity of hyperexcitable nociceptive neurons; and opioids, which exert central and possibly peripheral opioid analgesia [32]. All of these findings support the concept that multiple pathophysiological and biochemical mechanisms contribute to the generation and maintenance of NP—in this case, PHN pain.

V. Explanation of Symptoms in Relation to Pathophysiology

It is important to acknowledge the possibility of circular logic when explaining PHN symptoms because a significant amount of information about the pathophysiology of PHN was learned on the basis of psychophysical studies of patients with PHN by eliciting NP signs and measuring NP symptoms. However, in spite of this shortcoming, it is possible to draw a few relatively straightforward conclusions. Sensory abnormalities, primarily deficits and spontaneous pain, always follow distribution of specific nerve roots and associated peripheral sensory nerves that are damaged during reactivation of VZV in the corresponding DRG. The presence of mechanical dynamic allodynia is indicative of preserved but sensitized primary afferents. The extension of allodynia beyond the primary distribution borders of the rash is due to central sensitization mechanisms. However, the complexity and dynamic nature of PHN will continue to be a challenge and an inspiration for NP research.

In summary, the lessons from the neurobiology of PHN pain represent the best example of how clinical pain research cannot only contribute to but also lead the field of pain research in general. Establishment of efficacy of the vaccine for development of PHN is another important clinical development which will probably have significant impact for treatment of this painful disorder [16].

References

1. Bright, R. (1831). "Reports of Medical Cases. London, Vol. 2, Part 1." p. 383.
2. Romberg, M. H. (1955). "A Manual of the Nervous Diseases of Man." The Sydenham Society, London.
3. Von Barensprung, F. G. F. (1862). Beitrage zur Kenntnis des Zoster. *Ann. Char.-Krankenh. Berlin* **10**, 96–104.
4. Head, H., and Campbell, A. W. (1900). The pathology of herpes zoster and its bearing on sensory location. *Brain* **22**, 353–523. (An abridged 12 page summary: *Med. Virol.* 1997; **7**, 131–143.)
5. Lewis, T., and Marvin, H. M. (1927). Observations relating to vasodilatation arising from antidromic impulses, to herpes zoster and trophic effects. *Heart* **14**, 27.
6. Hope-Simpson, R. E. (1965). The nature of herpes zoster: a long-term study and a new hypothesis. *Proc. R. Soc. Med.* **58**, 9–20.
7. Hope-Simpson, R. E. (1967). Herpes zoster in elderly. *Geriatrics* **22**, 151–159.
8. Weller, T. H., and Coons, A. H. (1954) Fluorescent antibody studies with agents of varicella and herpes zoster propagated in vitro. *Proc. Soc. Exp. Biol. Med.* **86(4)**, 789–794.
9. Straus, S. E., Reinhold, W., Smith, H. A., Ruyechan, W. T., Henderson, D. K., Blaese, R. M., and Hay, J. (1984). Endonuclease analysis of viral DNA from varicella and subsequent zoster infection in the same patient. *N. Engl. J. Med.* **311**, 1362–1364.
10. Noordenbos, W. (1959). "Pain: Problems Pertaining to the Transmission of Nerve Impulses Which Give Rise to Pain." Elsevier, Amsterdam.
11. Melzack, R., and Wall, P. D. (1965). Pain mechanisms: a new theory. *Science* **150**, 971–979.
12. Fields, H. L., Rowbotham, M., and Baron, R. (1998). Postherpetic neuralgia: irritable nociceptors and deafferentation. *Neurobiol. Dis.* **5**, 209–227.
13. Arvin, A. (2001). The varicella-zoster virus. In: "Herpes Zoster and Postherpetic Neuralgia" (C. P. N. Watson and A. A. Gershon, eds), 2nd revised and enlarged edition, pp. 25–38, Elsevier, Amsterdam.
14. Kennedy, P. G. E. (2002). Varicella-zoster virus latency in human ganglia. *Reviews in Medical Virology* **12(5)**, 327–334.
15. Hoover, S. E., Cohrs, R. J., Rangel, Z. G., Gilden, D. H., Munson, P., and Cohen, J. I. (2006). Downregulation of varicella-zoster virus (VZV) immediate-early ORF62 transcription by VZV ORF63 correlates with virus replication in vitro and with latency. *J. Virol.* **80(7)**, 3459–3468.
16. Oxman, M. N., Levin, M. J., Johnson, G. R., Schmader, K. E., Straus, S. E., Gelb, L. D., Arbeit, R. D., Simberkoff, M. S., Gershon, A. A., Davis, L. E., Weinberg, A., Boardman, K. D., Williams, H. M., Zhang, J. H., Peduzzi, P. N., Beisel, C. E., Morrison, V. A., Guatelli, J. C., Brooks, P. A., Kauffman, C. A., Pachucki, C. T., Neuzil, K. M., Betts, R. F., Wright, P. F., Griffin, M. R., Brunell, P., Soto, N. E., Marques, A. R., Keay, S. K., Goodman, R. P., Cotton, D. J., Gnann Jr., J. W., Loutit, J., Holodniy, M., Keitel, W. A., Crawford, G. E., Yeh, S. S., Lobo, Z., Toney, J. F., Greenberg, R. N., Keller, P. M., Harbecke, R., Hayward, A. R., Irwin, M. R., Kyriakides, T. C., Chan, C. Y., Chan, I. S., Wang, W. W.,

Annunziato, P. W., Silber, J. L., and Shingles Prevention Study Group (2005). A vaccine to prevent herpes zoster and postherpetic neuralgia in older adults. *N. Engl. J. Med.* **352**, 2271–2284.
17. Dworkin, R. H., and Schmader, K. E. (2001). The epidemiology and natural history of herpes zoster and postherpetic neuralgia. *In:* "Herpes Zoster and Postherpetic Neuralgia" (C. P. N. Watson and A. A. Gershon, eds), 2nd revised and enlarged edition, pp. 39–64, Elsevier, Amsterdam.
18. Lycka, B. A. S., Williamson, D., and Sibbald, R. G. (2001). Dermatological aspects of herpes zoster. *In:* "Herpes Zoster and Postherpetic Neuralgia" 2nd Revised and Enlarged Edition (Watson, C. P. N., and Gershon, A. A., eds.) pp. 97–106. Elsevier, Amsterdam.
19. Watson, C. P., Deck, J. H., Morshead, C., Van der Kooy, D., and Evans, R. J. (1991). Post-herpetic neuralgia: further post-mortem studies of cases with and without pain. Pain **44(2)**, 105–117.
20. Watson, C. P., Evans, R. J., Watt, V. R., and Birkett, N. (1988). Post-herpetic neuralgia: 208 cases. *Pain* **35(3)**, 289–297.
21. Watson, C. P., Morshead, C., Van der Kooy, D., Deck, J., and Evans, R. J. Post-herpetic neuralgia: post-mortem analysis of a case. *Pain* **34(2)**, 129–138.
22. Kleinschmidt-DeMasters, B. K., and Gilden, D. H. (2001). Varicella-Zoster virus infections of the nervous system: clinical and pathologic correlates. *Arch Pathol Lab Med.* **125(6)**, 770–780.
23. Gilden, D. H., Kleinschmidt-DeMasters, B. K., LaGuardia, J. J., Mahalingam, R., and Cohrs, R. J. (2000). Neurologic complications of the reactivation of varicella-zoster virus. *N. Engl. J. Med.* **342**, 635–645.
24. Watson, C. P. N., Oaklander, A. L., and Deck, J. (2001). The neuropathology of herpes zoster with particular reference to postherpetic neuralgia and its pathogenesis. *In:* "Herpes Zoster and Postherpetic Neuralgia" (C. P. N. Watson and A. A. Gershon, eds), 2nd revised and enlarged edition, pp. 167–182, Elsevier, Amsterdam.
25. Oaklander, A. L. (2001). The density of remaining nerve endings in human skin with and without postherpetic neuralgia after shingles. *Pain.* **92**, 139–145.
26. Bowsher, D. (1995). Pathophysiology of postherpetic neuralgia: towards a rational treatment. *Neurology.* **45**, 53–56.
27. Haanpaa, M., Laippala, P., and Nurmikko, T. (1999). Pain and somatosensory dysfunction in acute herpes zoster [see comment]. *Clin. J. Pain.* **15**, 78–84.
28. Petersen, K. L., Fields, H. L., Brennum, J., Sandroni, P., and Rowbotham, M. C. (2000). Capsaicin evoked pain and allodynia in post-herpetic neuralgia. *Pain.* **88**, 125–133.
29. International Association for the Study of Pain. IASP Pain Terminology, http://www.iasp-pain.org/terms-p.html#Neuropathic%20pain.
30. Rowbotham, M. C., Baron, R., Petersen, K. L., and Fields, H. L. (2001). Spectrum of pain mechanisms contributing to PHN. *In:* "Herpes Zoster and Postherpetic Neuralgia" (C. P. N. Watson and A. A. Gershon, eds), 2nd revised and enlarged edition, pp. 181–195, Elsevier, Amsterdam.
31. Kotani, N., Kudo, R., Sakurai, Y., Sawamura, D., Sessler, D. I., Okada, H., Nakayama, H., Yamagata, T., Yasujima, M., and Matsuki, A. (2004). Cerebrospinal fluid interleukin 8 concentrations and the subsequent development of postherpetic neuralgia. *Am. J. Med.* **116**, 318–324.
32. Dworkin, R. H., Backonja, M., Rowbotham, M.C., Allen, R. R., Argoff, C. R., Bennett, G. J., Bushnell, M. C., Farrar, J. T., Galer, B. S., Haythornthwaite, J. A., Hewitt, D. J., Loeser, J. D., Max, M. B., Saltarelli, M., Schmader, K. E., Stein, C., Thompson, D., Turk, D. C., Wallace, M. S., Watkins, L. R., and Weinstein, S. M. (2003). Advances in neuropathic pain: diagnosis, mechanisms, and treatment recommendations. *Arch. Neurol.* **60**, 1524–1534.

95

Central Post-Stroke Pain

David Bowsher, MD, ScD, PhD, FRCPEd, FRCPath

Keywords: *central post-stroke pain (CPSP), features, incidence and prevalence, possible mechanisms, treatment*

I. Introduction
II. Incidence and Prevalence
III. Pain Onset
IV. Pain Characteristics
V. Somatosensory Deficit
VI. Autonomic Changes
VII. Natural Course of the Disease
VIII. Treatment
IX. Possible Mechanisms of CPSP
    References

I. Introduction

In a seminal publication of 1906, Dejerine and Roussy [1] described three cases of what they (in the article title) and most subsequent authors for many years called thalamic syndrome. The patients all had spontaneous pain following a stroke, involving the diencephalon and adjacent regions. It was elaborated by Roussy in his thesis (1906) as being characterized by "spontaneous pain, hemianesthesia, little or no hemiplegia, hemiataxia and choreo-athetoid movements." Head and Holmes [2] attributed the pain to lesions of the lateral nucleus of the thalamus (Figs. 1–3) and denied the involvement of the cerebral cortex in pain sensation.[1] However, such involvement was hinted at by Piéron [3].

In 1927, Foix and colleagues [4] reported a case due to pure cortical damage, a phenomenon elaborated on by Retif et al. [5], Schmahmann and Leifer [6], and Greenspan

---

[1]This was responsible for students being taught (by teachers who must have believed it) that "pain comes to consciousness in the thalamus" until at least 15 years after WW2. It was also ultimately responsible for the nonsense about "protopathic" and "epicritic" sensations, so ably and wittily refuted by Walshe [60].

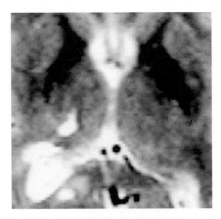

Figure 1 Posteroventrolateral thalamus (VPL) infarct. Aching pain. Raised thresholds for touch (von Frey), sharpness, warmth, and cold; no change in hot pain threshold; no allodynia.

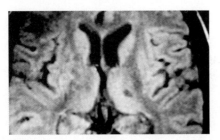

Figure 2 Posteroventrolateral thalamus (VPL) infarct. Aching pain. Raised perception thresholds for touch (von Frey) and skinfold pinch; slight increases in thermal thresholds (probably not significant); tactile allodynia.

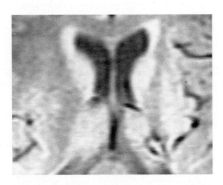

Figure 3 Posteroventrolateral thalamus (VPL) lacune. Burning pain. Raised perception thresholds for touch (von Frey), warmth, cold, and hot pain; no increase in mechanical pain threshold; no allodynia.

cordotomy (Lapresle and Guiot [10], 1953; White and Sweet [11], 1969).

A similar condition following brainstem stroke with crossed symptoms involving the medulla oblongata (Figs. 5 and 6) had earlier been described by Wallenberg (1895) [12]. Symptoms from infarcts in the upper brainstem

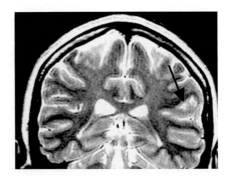

Figure 4 Lesion in upper bank of sylvian fissure (SII), pain in upper limb; slight tactile (von Frey) and cold deficits; severe deficits for warmth, hot pain, and mechanical pain; mild tactile allodynia.

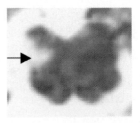

Figure 5 Right medullary infarct; deficits for all modalities. Right forehead, left limbs, aching pain, allodynia.

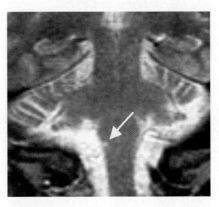

Figure 6 Very small rostral medullary lesion; left face and right body half affected, burning pain, no allodynia. No von Frey or skinfold pinch pain deficit; marked deficits for sharpness, warmth, and hot pain.

and Winfield [7] (Fig. 4). Central pain had been reported to occur in the syringomyelia of the spinal cord [8]; Garcin [9] stated that such pain was clinically indistinguishable from that due to thalamic lesions. Central pain was reported to occur in patients who had undergone anterolateral spinal

(Figs. 7 and 8) are usually not crossed (Garcin [9], 1937; Riddoch [13], 1938), and uncrossed symptomatology can rarely occur with lesions in the medulla oblongata [14]. Central pain occurs in many cases of multiple sclerosis and sometimes in Parkinson's disease. It is also frequent in spinal cord injury (SCI) [15,16]. Davison and Schick [17] drew attention to the fact that the pain and sensory loss in central pain were similar to those found in causalgia affecting peripheral nerves, as described by Weir Mitchell in 1872 [18].

A consensus (more or less) has now emerged, expressed by Boivie and colleagues [19], who renamed the syndrome central post-stroke pain (CPSP). Essentially, CPSP can be defined as pain within (but smaller than) an area of somatosensory impairment [20]. If the syndrome is to be called post-*stroke* pain, there must obviously have been a stroke, thus ruling out many other causes of central pain; but many of the other features originally described, such as choreo-athetosis (rare) and motor symptoms (rarely severe), are not necessary for diagnosis.

In fact, central pain of CPSP type can be caused by a lesion anywhere along the spinothalamocortical ("pain") pathway [21]. When it is due to a thalamic lesion, the infarct occurs within the territory of the inferolateral thalamic artery [22].

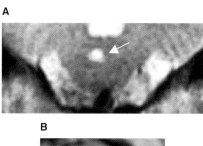

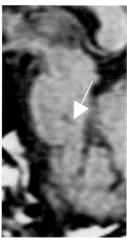

Figure 7 Right pontine infarct; right pain. No allodynia. Deficits for all modalities except von Frey; sharpness not tested.

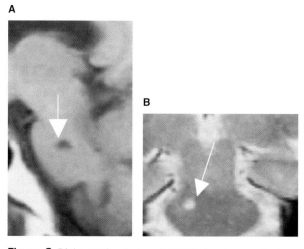

Figure 8 Right pontine lesion, parasagittal and coronal views. Aching pain in lower half of body; tactile and auditory startle allodynia; deficit for all modalities except sharpness.

II. Incidence and Prevalence

About a half-million strokes occur every year in the United States. One-third of the victims die within 3 months. A prospective study in Denmark [23] suggested that about 8% of stroke survivors develop CPSP. A retrospective study of more than a thousand elderly people in England suggested a similar (if slightly higher) incidence [24]. Thus, at a very conservative estimate, some quarter-million people in the United States develop CPSP every year. No studies have been carried out to establish the prevalence of CPSP. However, because motor deficits in CPSP patients are frequently minimal, most clinicians dealing with CPSP seem to be of the view that patients with the condition survive for at least 5 years; thus, an estimate of prevalence in the United States would be 1.25 million people. Regrettably, many of these patients remain "hidden" (see later discussion).

Dejerine and Roussy called the condition "thalamic syndrome" because all three of their patients had lesions in the posteroventrolateral thalamus (VPL), as described by Head and Holmes [2]. However, it turns out that only a minority of patients with VPL lesions have CPSP; Bogousslavsky et al. [22] have shown, using magnetic resonance imaging (MRI), that only 25% of patients with VPL infarcts develop CPSP—a point to which we shall return later.

III. Pain Onset

Pain begins immediately after stroke in just under one-half of CPSP patients (45%) and within 3 months in another third. This leaves more than one-fifth (22%) whose pain does not appear for a longer time, in some cases up to

2 years after stroke. This is just one of the reasons why CPSP so frequently remains undiagnosed and untreated.

IV. Pain Characteristics

The epoch-making paper by Dejerine and Roussy, published 100 years ago, forms the basis beyond which very few medical teachers have progressed. Thus we have nearly all been taught that the pain of CPSP (and other neuropathic pains) is "burning," with the implied corollary that if it isn't burning, it isn't CPSP. Burning pain, when it occurs, is certainly a very striking feature because it is usually an "ice-burn"[1]—burning and freezing at the same time, "like nothing else I ever felt before, doctor." This descriptor becomes memorable to physicians who have encountered it, apparently displacing other, commoner descriptions from the medical mind. Among 96 of my cases, only 54% of the patients complained of burning and/or freezing pain, whereas 30% complained of aching or throbbing pain (such as may occur in tissue damage, infection, etc.), and 16% of other types, such as stabbing pain; some CPSP pains were described as "paroxysmal" by Head and Holmes [2]. Pain is described as burning much more frequently by men than by women, at a ratio of 2.7:1 (χ^2 $P = 0.009$). Perhaps surprisingly, burning pain is not perceived as being more intense than nonburning pain.

Pain intensity may fluctuate, not only over time but from site to site within the body. If the face is involved in the painful area, there is a 5:1 chance that the lesion is infratentorial (χ^2 $P = 0.01$), and if the extremities are involved, there is a 5:1 chance that it is supratentorial (χ^2 $P = 0.04$); if the worst pain is in an extremity, there is a 16:1 chance that the lesion is supratentorial (χ^2 $P = 0.003$).

V. Somatosensory Deficit

It cannot be too strongly emphasized that somatosensory abnormalities are a *sine qua non* of neuropathic pain, especially CPSP; this was established by Head and Holmes [2], who pointed out that the pain and sensory loss were due to lesions in the lateral thalamus, whereas other (e.g., motor) symptoms are due to damage in adjacent structures.

Many of these properties can be tested at the bedside, in the patient's home, without having recourse to fancy instrumentation that will actually measure somatosensory thresholds. But whether bedside or instrumental methods are used, it is always essential to compare the affected area with its unaffected contralateral mirror-image area, for CPSP is always unilateral. Tactile sensation can be evaluated with a wisp of cotton wool or, preferably, a camel-hair paintbrush. Sharpness sensation is measured by testing whether the patient can distinguish between the head and the point of a pin lightly applied to the skin—*not* by driving in and perhaps twisting the point of a pin, which is a noxious stimulus. Mechanical pain is tested by pinching a fold of the patient's skin between the examiner's thumb and forefinger. If the unaffected side is tested first, it will be easy to get a subjective estimate of mechanical pain threshold. Warmth is tested with the patient's eyes shut, by the ability to distinguish (in terms of temperature) between the examiner's warm finger and a cold metal object such as a spoon or tuning fork. Thermal deficits are within the physiological range, and it is thus totally inappropriate to use a test tube filled with boiling water or ice cubes to perform these tests. Last, painful heat sensation can be tested with use of an object at no more than about 50° C, to avoid tissue irritation or damage. Tactile sensation is frequently unaffected by lesions *outside* the VPL (especially in the lower brainstem), as is the mechanical pain threshold in lesions *confined to* the VPL, but there is *always* some somatosensory deficit in CPSP.

In trying to dissect the subtleties of central lesions in different locations and the differences between lesions accompanied by central pain and those in which no pain is found, quantitative instrumental measurement is exceedingly useful (see, e.g., Gracely et al. [25]). When, for example, a group of patients with CPSP affecting the VPL or brainstem was compared with a control group of stroke patients with sensory deficit but no pain (lesion site not specified), it was found that deficits for sharpness and cold but not for touch (von Frey) or warmth were significantly greater for patients with CPSP in the VPL than for controls, whereas patients with brainstem CPSP had significantly greater deficits than controls with regard to sharpness, warmth, cold, and hot pain [26]. More-recent analyses have shown large differences in the intensity of some modalities between CPSP due to lesions in the brainstem, VPL, medial thalamus, and SII-insular cortex [27]. For example, in VPL lesions the affected-unaffected difference in skinfold pinch threshold is 0.03 ± 0.7 kg ($P = 0.4$; NS), whereas that for heat pain is $2.3 \pm 3.1°$ C ($P = 0.0008$). However, for patients with medial thalamic lesions, the affected-unaffected threshold difference for skinfold pinch pain is 1.8 ± 0.9 kg ($P = 0.005$), but for heat pain it is $0.15 \pm 1.8°$ C ($P = 0.4$; NS). Subjects who have undergone anterolateral spinal cordotomy for relief of pain have an enormous rise in the mechanical (as well as heat) pain threshold (e.g., White and Sweet [28,29]), whereas patients with lesions confined to the VPL do not have any significant change in mechanical pain threshold [30]. CPSP-producing lesions at many sites along the supraspinal part of the "pain pathway" show many associations and dissociations of somatosensory modalities [27]. Not all modalities are necessarily affected

---

[1] cf. scalding, sunburn.

in CPSP. Sharpness and innocuous temperature thresholds are the most consistently diminished, although the latter seem to be little affected in medial thalamic lesions.

Allodynia, defined as "pain due to a stimulus which does not normally provoke pain," is a pathognomonic feature of neuropathic pain when it occurs—which it does not do in every case. It is most frequently provoked by light brushing or stroking (dynamic mechanical allodynia) or light pressure on the skin (static mechanical allodynia); allodynia due to contact with cold objects (cold allodynia) and to stimulation of stretch receptors in muscle or tendon (movement allodynia [20]) is also found. Patients may have more than one form of allodynia (see, e.g., Fig. 8). However, allodynia occurs in only about two-thirds of CPSP patients (71% in our own series of 122 patients) (cf. Figs. 1 and 2). Pure mechanical allodynia was found in 28%, and that in association with other forms in 13%; movement allodynia and cold allodynia each occurred in 8% of patients in our series.

We found [26] that CPSP patients with tactile allodynia had a higher threshold for warmth perception ($P = 0.001$) than did CPSP patients without tactile allodynia. There is thus a subtle interplay between allodynia and somatosensory deficits, the nature of which we are just beginning to understand. Although the generator point for allodynia is usually found within the area of pain and sensory deficit, it may occasionally be remote from it, as described by White and Sweet [11] and confirmed by my own experience. Although there may be attenuation or exacerbation of allodynic pain on repetitive stimulation in neuropathic pain, apparently there is never any noteworthy refractory period [31], as in trigeminal neuralgia [32].

VI. Autonomic Changes

Autonomic disturbance is fairly common in CPSP. Thus, in many patients cutaneous thermography shows some vasoconstriction in the affected area. When the stroke affects the lower limb, a number of patients say that "the limb feels cold"; this also occurs in stroke patients without CPSP. The vasoconstriction (but not the pain) can be alleviated with a calcium channel blocker such as nifedipine [33]. Hyperhidrosis has also been observed in affected areas of CPSP patients.

VII. Natural Course of the Disease

The course of the disease is highly variable, and, unfortunately, totally unpredictable. Some pains remain unchanged for years; some get worse; some get better. Sometimes a further stroke relieves CPSP. This latter observation led some neurosurgeons to attempt to relieve the condition by stereotactically enlarging thalamic lesions, but the practice was soon abandoned for obvious reasons. Occasional cases have been seen, when followed over several years, in which the sensory deficit diminishes but the pain does not, or vice versa. Usually, however, pain and sensory deficit march *pari passu*. In those patients for whom treatment is ineffective, even when continued over a long period (sometimes years), the course of the disease appears to be unaltered.

VIII. Treatment

Central pain was at first treated with conventional analgesics, frequently to no avail—although in many cases, doses were increased until the wretched victims became agonized junkies. In the 1960s, it was serendipitously found that the pain of trigeminal neuralgia yielded, at least initially, to carbamazepine [34]. This started a line of thinking that *all* neuropathic pains (the distinction of which from nocigenic pains had begun in the 1970s [35]) were epileptogenic—a belief that persists, with some justification (see later discussion). Anticonvulsants were indiscriminately prescribed for CPSP and all other neuropathic pains.

Also in the 1960s, Woodforde and colleagues [36] started treating post-herpetic neuralgia (PHN) with amitriptyline, on the grounds that the patients were or might be depressed. Treatment with amitriptyline soon became the standard for PHN [37], although it had been shown that actions on depression and neuropathic pain were independent of one another; Leijon and Boivie [38] showed that amitriptyline alone was very much more effective against CPSP than carbamazepine on its own. Sindrup and Jensen [39] reviewed the effectiveness of drugs then in use for neuropathic pain; for CPSP, they reported that the number needed to treat (NNT) to obtain one "cure" in CPSP was 1.7 (1.1–3.0) for adrenergically active tricyclic antidepressants (TCAs) such as amitriptyline or nortriptyline, but 3.4 (1.7–105.0) for carbamazepine. The older TCAs, such as amitriptyline and nortriptyline, have two major disadvantages:

1. they induce drowsiness and even cognitive impairment; and
2. they frequently cause weight gain.

Investigators have reported some success in the treatment of CPSP with venlafaxine, a more recently introduced tricyclic antidepressant with fewer side effects.

A later anticonvulsant, sodium valproate or valproic acid, was used successfully in combination with amitriptyline for the neuropathic pain of post-herpetic neuralgia (PHN) [40] and also with some success for CPSP. Even more recent anticonvulsants are being found to be much more successful than carbamazepine. Among these are lamotrigine [41] and gabapentin and its successor, pregabalin.

For several decades, neuropathic pains, including CPSP, were considered to be opioid-resistant, but these drugs have now come back into favor. A consensus recommendation has been published by Kalso et al. [42], and 15 randomized placebo-controlled trials were reviewed by Kalso et al. [43]. There has been a tendency to lump all neuropathic pains together without distinction in these and many other reviews. A number of opioids have been recommended by different clinical research groups. The mild opioid tramadol is favorably viewed as a result of a Cochrane review [44]; morphine was used by Attal et al. [45], levorphanol by Rowbotham et al. [46], oxycodone by Watson et al. [47], and methadone by Morley et al. [48].

A combination of drugs, selected according to individual response, is recommended by Hansson [49], and a combination of morphine and gabapentin by Gilron et al. [50]. A fairly typical combination is gabapentin (up to 1200 mg three times per day) plus levorphanol (up to a maximum of 240 mg daily, or until side effects appear) [51].

IX. Possible Mechanisms of CPSP

An explanation must be found for a condition that differentially and variably affects somatosensory submodalities and autonomic function; which follows insult to the central (or peripheral) nervous system, but not ineluctably (indeed, in only a minority of cases); and in which the sensory change may or may not be accompanied by neuropathic pain (although, as discussed previously, the presence or absence of pain does appear to depend on the intensity of innocuous thermal loss, particularly for warmth).

The cardinal fact about CPSP is that only a minority of individuals who suffer apparently appropriate insult to the central (or peripheral) nervous system actually develop neuropathic pain. This is equally true, it would seem, of nerve ligation experiments in animals, designed to produce neuropathic pain of peripheral origin (but central mechanism); by no means do all nerve-ligated animals show signs or symptoms of neuropathic pain. It was suggested earlier [38,52] that the best-fit theoretical model for central pain would appear to be one in which a widely distributed transmitter/ligand in the central nervous system and/or its specific receptors may become depleted; possible changes in receptor function were also mentioned by Siddall and Loeser [16], specifically in relation to spinal cord injury. It has been known for a long time that some transmitters, such as serotonin, have both a global function, disturbance of which may be reflected in some psychiatric disorders, and a specific one, disturbance of which is seen in particular "focal" conditions such as migraine. Enhanced pain sensitivity following injury is regulated by NK1 serotonin receptors [53], whereas spinal 5HT3 receptors mediate descending excitatory controls on spinal neurons activated in some neuropathic pain states [54].

Although we still have little idea what the transmitter or receptor concerned with central pain may be, recent developments in the field lend some support to this type of argument. Willoch et al. [55], in a positron emission tomography study of five CPSP patients, have shown that opioid binding is depleted in a number of sites associated with CPSP pathology. However, this cannot be the whole story because therapeutic restitution of opioid levels alone does not alleviate the pain in most cases. It has been shown in the NMDA system that presynaptic transmitters and postsynaptic receptors may be present in varying quantities (concentrations, densities) in the central nervous system [56]. Another relevant observation is that ubiquitin C-terminal hydrolase is up-regulated in rats with sciatic nerve constriction injuries [57]. Wang et al. [58] have described as many as 148 genes that are up- or down-regulated in the dorsal root ganglia of neuropathic rats.

It may therefore be suggested that changes in transmitter concentration and/or receptor density, as either up- or down-regulation, may occur following nervous system injury. Following appropriate insult, transmitters and/or receptors may undergo sudden and massive depletion, leading to immediate onset of central neuropathic pain, or one or both may deplete slowly, giving rise to later onset of pain. They may recover their original levels/concentrations/density, so that the pain "spontaneously" disappears; fluctuant recovery may account for fluctuating degrees of central pain.

However, if this hypothesis is even partly valid, there are a number of unexplained phenomena, among which are the following.

1. Why, in what is apparently the same condition, does one form of therapy succeed in some cases but fail in others, in which another form of therapy (a different drug) is effective? (One example is TCAs, which act on serotonergic and adrenergic systems, versus gabapentin, which acts on subunits of voltage-dependent Ca^{++} channels [59]).

2. It is fairly widely reported that pain in patients with CPSP may be alleviated by a second stroke. Such events are hardly going to increase levels of transmitter X or densities of receptor Y!

References

1. Dejerine, J., and Roussy, J. (1906). Le syndrome thalamique. *Rev. Neurol. (Paris).* **14**, 521–532.
2. Head, H., and Holmes, G. (1911). Sensory disturbances from cerebral lesions. *Brain.* **34**, 102–254.
3. Piéron, H. (1923). "Le Cerveau et la Pensée." Félix Alcan, Paris.

4. Foix, Ch., Chavany, J.-A., and Lévy, M. (1927). Syndrome pseudo-thalamique d'origine pariétale. Lésion de l'artère du sillon interpariétal (Pa P1P2 antérieures, petit territoire insulo-capsulaire). *Rev. Neurol. (Paris)* **35**, 68–76.
5. Retif, J., Brihaye, J., and Vanderhaegen, J. J. (1967). Syndrome douloureux "thalamique" et lésion pariétale: a propos de trois observations de tumeur à localisation pariétale, s'étant accompagnées de douleurs spontanées de l'hemicorps contralatéral. *Neurochirurg.* **13**, 375–384.
6. Schmahmann, J. D., and Leifer, D. (1992). Parietal pseudothalamic pain syndrome: clinical features and autonomic correlates. *Arch. Neurol.* **49**, 1032–1037.
7. Greenspan, J. D., and Winfield, J. A. (1992). Reversible pain and tactile deficits associated with a cerebral tumor compressing the posterior insula and parietal operculum. *Pain* **50**, 29–39.
8. Spiller, W. G. (1923). Central pain in syringomyelia and dysesthesia and overreaction to sensory stimuli in lesions below the optic thalamus. *Arch. Neurol. Psychiat.* **10**, 491–499.
9. Garcin, R. (1937). La douleur dans les affections organiques du système nerveux central. *Rev. Neurol.* **68**, 105–153.
10. Lapresle, J., and Guiot, G. (1953). Etude des résultats éloignés, et en particulier des sequelles neurologiques à type de "douleur centrale," dans 8 cas de cordotomie antéro-latérale pour coxarthrose. *Sem. Hôp. Paris* **29**, 2189–2198.
11. White, J. C., and Sweet, W. H. (1969). "Pain and the Neurosurgeon." C. C Thomas, Springfield, Illinois.
12. Wallenberg, A. (1895). Acute Bulbäraffection (Embolie der Arteria cerebelli posterior inferior sinistra?) *Arch. Psychiat. Nervenkr. Berlin* **27**, 504–540.
13. Riddoch, G. (1938). The clinical features of central pain. *Lancet* **1**, 1093–1098; 1150–1156; 1205–1209.
14. Chia, L. G., and Shen, W. C. (1993). Wallenberg's lateral medullary syndrome with loss of pain and temperature sensation on the contralateral face: clinical, MRI and electrophysiological studies. *J. Neurol.* **240**, 462–467.
15. Beric, A., Dimitrijevic, M. R., and Lindblom, U. (1988). Central dysesthesia syndrome in spinal cord injury patients. *Pain* **34**, 109–116.
16. Siddall, P. J., and Loeser, J. D. (2001). Pain following spinal cord injury. *Spinal Cord* **39**, 63–73.
17. Davison, C., and Schick, W. (1935). Spontaneous pain and other sensory disturbances. *Arch. Neurol. Psychiat. (Chicago)* **34**, 1204–1237.
18. Mitchell, S. W. (1872). "Injuries of Nerves and Their Consequences." Lippincott, Philadelphia, 377 pp.
19. Boivie, J., Leijon, G., and Johansson, I. (1989). Central post-stroke pain: a study of the mechanisms through analyses of the sensory abnormalities. *Pain* **37**, 173–185.
20. Bowsher, D. (1996a). Central pain: clinical and physiological characteristics. *J. Neurol. Neurosurg. Psychiat.* **61**, 62–69.
21. Cassinari, V., and Pagni, C. A. (1969). "Central Pain: A Neurosurgical Survey." Harvard University Press, Cambridge, Massachusetts.
22. Bogousslavsky, J., Regli, F., and Uske, A. (1988). Thalamic infarcts: clinical syndromes, etiology, and prognosis. *Neurology* **38**, 837–848.
23. Andersen, G., Vestergaard, K., Ingeman-Nielsen, M., and Jensen, T. S. (1995). Incidence of central post-stroke pain. *Pain* **61**, 187–194.
24. Bowsher, D. (2001). Stroke and central post-stroke pain in an elderly population. *J. Pain* **2**, 258–261.
25. Gracely, R. H., Eliav, E., and Hansson, P. (2003). Quantitative sensory testing: clinical considerations and new methods. *Prog. Pain Res. Manag.* **24**, 589–600.
26. Bowsher, D., Leijon, G., and Thuomas, K.-A. (1998). Central post-stroke pain: correlation of magnetic resonance imaging with clinical pain characteristics and sensory abnormalities. *Neurology* **51**, 1352–1358.
27. Bowsher, D. (2005a). Representation of somatosensory modalities in pathways ascending from the spinal anterolateral funiculus to the thalamus demonstrated by lesions in man. *Eur. Neurol.* **54**, 14–22.
28. White, J. C., and Sweet, W. H. (1955). "Pain: Its Mechanisms and Neurosurgical Control". C. C. Thomas, Springfield, Illinois.
29. Lahuerta, J., Bowsher, D., Campbell, J. A., and Lipton, S. (1990). Clinical and instrumental evaluation of sensory function before and after percutaneous anterolateral cordotomy at cervical level in man. *Pain* **42**, 23–30.
30. Bowsher, D. (1996b). Functional differences in spinal and supraspinal portions of the spinothalamic pathway in man. *J. Physiol.* **495**, 18P.
31. Bowsher, D. (2005b). Dynamic mechanical allodynia in neuropathic pain. *Pain* **116**, 164–165.
32. Kugelberg, E., and Lindblom, U. (1959). The mechanism of the pain in trigeminal neuralgia. *J. Neurol. Neurosurg. Psychiat.* **22**, 36–43.
33. Bowsher, D. (1995a). The management of central post-stroke pain. *Postgrad. Med. J.* **71**, 598–604.
34. Blom, S. (1962). Trigeminal neuralgia: its treatment with a new anticonvulsant drug (G32883). *Lancet* **i**, 839–840.
35. Bowsher, D. (1987). Neurogenic pain. *J. Intract. Pain Soc.* **5**, 23–27.
36. Woodforde, J. M., Dwyer, B., McEwen, B. W., De Wilde, F. W., Bleasel, K., Connelley, T. J., and Ho, C. Y. (1965). The treatment of postherpetic neuralgia. *Med. J. Aust.* **2**, 869–872.
37. Watson, C. P., Evans, R. J., Reed, K., Merskey, H., Goldsmith, L., and Warsh, J. (1982). Amitriptyline versus placebo in postherpetic neuralgia. *Neurology* **32**, 671–673.
38. Leijon, G., and Boivie, J. (1989). Central post-stroke pain: a controlled trial of amitriptyline and carbamazepine. *Pain* **36**, 27–36.
39. Sindrup, S. H., and Jensen, T. S. (1999). Efficacy of pharmacological treatments of neuropathic pain: an update and effect related to mechanism of drug action. *Pain* **83**, 389–400.
40. Raftery, H. (1975). The management of postherpetic pain using sodium valproate and amitriptyline. *Irish Med. J.* **72**, 399–401.
41. Vestergaard, K., Andersen, G., Gottrup, H., Kristensen, B. T., and Jensen, T. S. (2001). Lamotrigine for central poststroke pain: a randomised controlled trial. *Neurology* **56**, 184–190.
42. Kalso, E., Allan, L., Dellemijn, P. L., Faura, C. C., Ilias, W. K., Jensen, T. S., Perrot, S., Plaghki, L. H., and Zenz, M. (2003). Recommendations for using opioids in chronic non-cancer pain. *Eur. J. Pain.* **7**, 381–386.
43. Kalso, E., Edwards, J. E., Moore, R. A., and McQuay, H. J. (2004). Opioids in chronic non-cancer pain: systematic review of efficacy and safety. *Pain* **112**, 372–380.
44. Duhmke, R. M., Cornblath, D. D., and Hollingshead, J. R. (2004). Tramadol for neuropathic pain. *Cochrane Database Syst. Rev.* **2**, CD003726.
45. Attal, N., Guirimand, F., Brasseur, L., Gaude, V., Chauvin, M., and Bouhassira, D. (2002). Effects of IV morphine in central pain: a randomized placebo-controlled study. *Neurology* **58**, 554–563.
46. Rowbotham, M. C., Twilling, L., Davies, P. S., Reisner, L., Taylor, K., and Mohr, D. (2003). Oral opioid therapy for chronic peripheral and central neuropathic pain. *N. Engl. J. Med.* **348**, 1223–1232.
47. Watson, C. P., Watt-Watson, J. H., and Chipman, M. L. (2004). Chronic noncancer pain and the long term utility of opioids. *Pain Res. Manag.* **9**, 19–24.
48. Morley, J. S., Bridson, J., Nash, T. P., Miles, J. B., White, S., and Makin, M. K. (2003). Low-dose methadone has an analgesic effect in neuropathic pain: a double-blind randomized controlled crossover trial. *Pailliat. Med.* **17**, 576–587.
49. Hansson, P. (2004). Post-stroke pain case study: clinical characteristics, therapeutic options and long-term follow-up. *Eur. J. Neurol. Suppl.* **1**, 22–30.
50. Gilron, I., Bailey, J. M., Tu, D., Holden, R. R., Weaver, D. F., Houlden, R. L. (2005). Morphine, gabapentin, or their combination for neuropathic pain. *N. Engl. J. Med.* **352**, 1324–1334.
51. Sang, C. (2003). Personal communication.
52. Bowsher, D. (1995b). Central pain. *Pain Rev.* **2**, 175–186.
53. Suzuki, R., Morcuende, S., Webber, M., Hunt, S. P., and Dickenson, A. H. (2002). Superficial NK1-expressing neurons control spinal

excitability through activation of descending pathways. *Nat. Neurosci.* **5**, 1319–1326.

54. McCleane, G. J., Suzuki, R., and Dickenson, A. H. (2003). Does a single intravenous injection of 5HT3 receptor antagonist ondansetron have an analgesic effect in neuropathic pain? A double-blinded, placebo-controlled cross-over study. *Anesth. Analg.* 97, 1474–1478.

55. Willoch, F., Schindler, F., Wester, H. J., Empl, M., Straube, A., Schwaiger, M., Conrad, B., and Tolle, T. R. (2004). Central post-stroke pain and reduced opioid receptor binding within pain processing circuitries: a [11C] diprenorphine PET study. *Pain* **108**, 213–220.

56. Yu, X.-M., and Salter, M. W. (1999). Src, a molecular switch governing gain control of synaptic transmission mediated by N-methyl-D-aspartate receptors. *Proc. Natl. Acad. Sci. USA.* **96**, 7697–7704.

57. Moss, A., Blackburn-Munro, G., Garry, E. M., Blakemore, J. A., Dickinson, T., Rosie, R., Mitchell, R., and Fleetwood-Walker, S. M. (2002). A role of the ubiquitin-proteasome system in neuropathic pain. *J. Neurosci.* **22**, 1363–1372.

58. Wang, H., Sun, H., Della Penna, K., Benz, R. J., Xu, J., Gerhold, D. L., Holder, D. J., and Koblan, K. S. (2002). Chronic neuropathic pain is accompanied by global changes in gene expression and shares pathobiology with neurodegenerative diseases. *Neuroscience* **114**, 529–546.

59. Taylor, C. P., Gee, N. S., Su, T.-Z., Koesis, J. D., Welty, D. F., Brown, J. P., Dooley, D. J., Boden, P., and Singh, L. (1998). A summary of mechanistic hypotheses of gabapentin pharmacology. *Epilepsy Res.* **29**, 233–249.

60. Walshe, F. M. R. (1942). The anatomy and physiology of cutaneous sensibility. *Brain* **65**, 48–112.

Index

Note: Page numbers followed by "f" denote figures; "t," tables.

1-bromopropane, 751t
2-bromobutyric acid, 751t
2-bromopropionic acid, 751t
2-butenenitrile, 751t
2-chloropropionic acid, 751t
2-hexanol, 751t
2-Mercaptopropionic acid, 751t
2-Methoxyethanol, 751t
2,4-dichlorophenoxyacetic acid, 751t
2,4-dinitrophenol, 754
2,5-hexanedione, 751t, 754
3-nitropropionic acid, 754
3,3-iminodipropionitrile, 751t
5-HT, 973t, 976
5-HTP-induced myoclonus, 312

A-process, 772
A11 system, 727–728
A-a gradient, 682
AAN, 165, 981–982, 994–995
AB variant, 2, 8
ABC-like DLBCL, 398, 399f
Aberrant learning model, 774
Absence seizures. *See* Idiopathic generalized epilepsy (IGE)
Acanthamoebae, 460
Acclimatization, 687
Acetogenins, 106
Acetylcholine (ACh), 935–936, 973t, 975
Acetylcholine esterase (AChE), 935
Acetylcholine receptor (AChR), 935–940
ACh, 935–936, 973t, 975
AChE, 935
AChE deficiency, 943, 944t
AChR, 935–940
AChR deficiency syndrome, 943–944
AChR fast channel syndrome, 944t

AChR slow channel syndrome, 943, 944t
Acid lipase, 6
Acid maltase deficiency (AMD), 950
Acquired inflammatory demyelinating and axonal neuropathies
  acute inflammatory neuropathies, 860–864
  AIDP, 860–864
  AMAN, 860–864
  AMSAN, 860–864
  chronic inflammatory neuropathies, 864
  CIDP, 864–866
  DADS, 865
  FS, 860–864
  MADSAM, 865
  MMN, 865
  neuropathies associated with monoclonal gammopathies, 866–867
  treatment, 867–868
Acquired neuromyotonia, 942–943
Acrylamide, 756
Acrylonitrile, 751t
Actinomyces, 447t
Activated B cell-like (ABC) DLBCL, 398, 399f
Activated protein C (APC), 277–278
Acute hypoxia, 686
Acute inflammatory demyelinating polyradiculoneuropathy (AIDP), 206t, 208t, 860–864
Acute inflammatory neuropathies, 860–864
Acute intermittent porphyria (AIP), 655, 696
Acute motor and sensory neuropathy (AMSAN), 206t, 208t, 860–864
Acute motor axonal neuropathy (AMAN), 206t, 208t, 860–864
Acute myocardial infarction (AMI), 237
Acute panautonomic neuropathy, 981
Acute pandysautonomia, 206t, 208t, 981

Acute porphyrias, 655–656
Acyclovir, 665, 666
Acyl-CoA oxidase deficiency, 21, 24
AD, 69–81. *See also* Alzheimer's disease (AD)
AD HSP, 542–543
ADAMTS13, 656–657
ADC, 798
ADC image, 798
ADCA, 97t
Addison's disease, 676
Adenosine, 973t, 976
Adenosine triphosphate (ATP), 973t, 977
Adenovirus, 189
ADHD, 631–639. *See also* Attention-deficit hyperactivity disorder (ADH)
Adjuvants, 1017–1018
ADNFLE, 330
ADPFEAF, 361
Adrenal insufficiency, 676
Adrenoleukodystrophy (ALD), 19–32, 43–45. *See also* Peroxisomal disorders
Adrenomyeloneuropathy (AMN), 19–32, 44. *See also* Peroxisomal disorders
Adult Refsum's disease (ARD), 21
Advanced glycation end-products (AGEs), 673, 855
AF, 236–237
AGAT deficiency, 33–41. *See also* Creatine deficiency syndrome (CDS)
AGE, 673, 855
Agrypnia excitata, 712–713
AIDP, 206t, 208t, 860–864
AIDS, 486, 490f, 493–495
AINMT, 331
AIP, 655, 696
Akt, 439
ALA, 696
ALA dehydratase deficiency, 655
Albumin, 277
Alcoholics, 699
Alcoholism. *See* Drug addiction
ALD, 19–32, 43–45. *See also* Peroxisomal disorders
Aldolase deficiency, 951
Aldose reductase inhibitors (ARIs), 673
Aldose reductase pathway, 672–673
ALDP, 44
Alexander's disease, 46–47
Aliphatic chemicals, 751t
Allodynia, 1043
Allyl chloride, 751t
Allyl nitrile, 751t
Alpha interferon signature, 173
α-Antitrypsin deficiency, 257–258
α-Dystroglycan, 930
α-Interferon, 876

α-Mannosidosis, 3t, 13, 15f
α-SNAP, 642
α-Synuclein, 55–56, 115–118
ALPL gene, 388
ALS, 521–535. *See also* ALS-like syndromes
   ALS2, 522–523
   ALS4, 523
   anti-retrovirals/protease inhibitors, 530
   antiinflammatory agents, 530
   antioxidants, 529
   apoptotic factors, 525–526
   bulbar, 523–524
   defects in axonal transport, 527
   epidemiology, 522
   etiology/pathogenesis, 522–523
   glutamate transporters, 527–528
   growth factors, 528
   history/nomenclature, 521–522
   immunomodulatory therapies, 530
   inherited, 522–523
   intermediate filaments, 526–527
   mitochondrial dysfunction, 528–829
   molecular mechanisms, 524–529
   neuroinflammation, 529
   neurotrophins, 530
   pathophysiology, 523
   pharmacology, 529–530
   SOD1 mutations, 522, 524–525
   spinal, 523
   structural detail, 524
   symptoms, 523–524
   treatment, 529–531
ALS-like syndromes, 513–519
   ALS-like syndromes and lymphoma, paraproteinemia, and other lymphoproliferative disorders, 515–519
   anti-Hu antibodies, 514–518
   anti-Purkinje cell antibodies, 515, 516, 518, 567
   history/nomenclature, 513–514
   other antineuronal antibodies, 515, 516, 518
   UMN syndrome and breast cancer, 515, 517, 518, 519
ALS-like syndromes and lymphoma, paraproteinemia, and other lymphoproliferative disorders, 515–519
ALS2, 522–523
ALS4, 523
Alsin mutations, 522–523
Aluminum, 761–762
Alveolar-arterial oxygen difference, 682
Alzheimer, Alois, 69, 112
Alzheimer's disease (AD), 69–81, 112–115
   biochemistry/molecular mechanisms, 76
   cholinergic system, 76
   clinicopathological correlates, 80
   dopaminergic system, 77
   epidemiology, 70

etiology/risk factors, 70
fMRI, 845
GABAergic system, 78
glutamatergic system, 77
history/nomenclature, 69–70
management of, 81
MCI, 75–78
mild AD, 78
MMSE, 78, 79, 79f
moderate AD, 78–79
natural history, 80
neuropathology, 72–74
neuropeptides, 78
noradrenergic system, 77
pathogenesis, 70–72
pathophysiology, 75–76
PET scan, 827
pharmacotherapy, 81
serotonergic system, 77
severe AD, 80
theoretical development curves, 78f
AMAN, 206t, 208t, 860–864
Amantadine, 666
Amaurotic family idiocy, 1
AMD, 950
Amebic brain abscess, 462
Amebic granulomatous encephalitis, 460, 461f
AMI, 237
Amino acid neurotransmitters, 788
Aminolevulinic acid (ALA) dehydratase deficiency porphyria, 655
Amitriptyline, 1017, 1018, 1043
Ammon's horn sclerosis, 350f, 351f, 352
AMN, 19–32, 44. *See also* Peroxisomal disorders
Amnestic deficits, 844–845
Amoebiasis, 460–462
AMPA receptors, 11, 528
AMSAN, 206t, 208t, 860–864
Amygdala, 353, 970
Amyloid autonomic neuropathy, 983
Amyloid hypothesis, 117
Amyotrophic lateral sclerosis, 521–535. *See also* ALS
ANA test, 172
Anaplastic astrocytomas, 435, 436f
Anaplastic meningioma, 383f, 384f
Anaplastic mixed oligoastrocytomas, 436
Anaplastic oligodendrogliomas, 435–436, 436f
Andersen-Tawil syndrome (ATS), 327
Andersen's disease, 950
Andersen's syndrome (AS), 145–147
Androgen receptor, 553–554, 559
Androgens, 560
Anemic hypoxia, 682t, 683
Aneurysm CTA, 805t, 810–813

Aneurysm pathogenesis, 266–267
Aneurysm rupture, 267
Angelman syndrome, 542
Angiogenesis
  CNS metastases, 375
  glioma, 440, 441f
  meningioma, 391
Angiogenic switch, 375
Angiomatous meningioma, 383f, 384f
Angiostrongyliasis, 468t
Angiostrongylus cantonensis, 468t
Angiotensin II, 973t, 976
Anhidrosis and hyperhidrosis (CIA), 993–994
ANS and pain, 1021–1030. *See also* Autonomic nervous system and pain
Antecollis, 90
Anterior cingulate, 970
Anti-CD52 antibody, 202
Anti-Hu antibodies, 514–518
Anti-MAG neuropathy, 867
Anti-NR2 antibodies, 178–179
Anti-phospholipid syndrome (APLS), 177
Anti-Purkinje cell antibodies, 515, 516, 518, 567
Anti-retrovirals, 530
Anti-ribosomal P protein antibodies, 177–178
Anticipation, 931
Anticonvulsants, 1018
Antiepileptic drugs, 666, 876
Antimicrobial agents, 873
Antineoplastic drugs, 876–882
Antioxidants, 277, 529
Antiphospholipid-antibody syndrome, 908
Antitubulin drugs, 879–881
APC, 277–278
Aphasia, 841
APLS, 177
Apolipoprotein A1 FAP, 983
Apoptosis, 274–275, 674, 853–854
APP, 70–71, 74
Apparent diffusion coefficient (ADC), 798
Apraxia, 842
A-process, 772
AR HSP, 541–542
AR pathway, 672–673
ARAS, 738–739
Arcuate nucleus (ARN), 972
ARD, 21
Arginine:glycine amidinotransferase (AGAT) deficiency, 33–41. *See also* Creatine deficiency syndrome (CDS)
ARIs, 673
ARN, 972
Aromatic hydrocarbons, 751t
Arsenic, 762
Arterial blood pressure, 997–1005. *See also* Blood pressure

Arthrogryposis multiplex congenita, 940
ARX, 578
Aryl hydrocarbon receptor nuclear translocator, 684
AS, 145–147
Aβ, 70–71, 74
ASA, 238
Ascending reticular activating system (ARAS), 738–739
Aspartylglucosaminuria, 3t
Asperger, Hans, 582
Asperger's disorder, 583t. *See also* Autism
Aspergillus spp., 447t
Aspergillus vasculitis, 260f
Asterixis, 311, 695
Astrocytes, 376, 764
Astrom, K. E., 185
Asymptomatic brain infarcts, 227–228
Ataxia, 97t, 143–144
Ataxia with vitamin E deficiency (AVED), 699
Ataxic disorders, 97t
Ataxin-1[82Q]-A776, 152, 153
Atlastin, 542
ATP, 973t, 977
ATP1A2 gene, 341
Atrial fibrillation (AF), 236–237
Atrial myxoma, 238
Atrial septal aneurysm (ASA), 238
ATRT, 427
ATS, 327
Attention-deficit hyperactivity disorder (ADH), 631–639
   biochemistry, 636
   diagnostic criteria, 633t
   etiology, 632–634
   history/nomenclature, 631–632
   medication, 638t
   molecular mechanisms, 636
   natural history, 636–639
   neurological findings, 634–636
   neurotransmitters, 637t
   pathogenesis/pathophysiology, 634
   pharmacology, 636, 637t, 638t
   subtypes, 632, 633t
Atypical absence seizures, 291–292
Atypical meningioma, 383f, 384f
Atypical teratoid/rhabdoid tumor (ATRT), 427
Autism, 581–591
   biochemistry, 588–589
   brain structure, 587–588
   DSM-IV, 583–584t
   etiology, 582
   excitation:inhibition ratio, 588–589
   historical review/nomenclature, 581–582
   idiopathic, 584
   immune system abnormalities, 589
   molecular mechanisms, 588–589
   natural history, 590
   neuropeptide abnormalities, 589
   pathogenesis, 582–585
   pathophysiology, 585–587
   pharmacology, 588
   regressive/nonregressive, 584
   symptoms, 589–590
Autistic disorder, 583t
Autoimmune autonomic neuropathy (AAN), 165, 981–982, 994–995
Autoimmune hypothesis, 203f
Autoimmune inflammatory myopathies, 957–968
   autoantibodies, 960
   chemokines, 964
   clinicohistological features, 958–960
   cytokine signaling, 964
   epidemiology, 958
   immunogenetics, 958
   immunoopathogenesis, 960–967
   inflammatory process, 961–964
   MHC-I antigens, 964–967
   signaling pathways, 961–964
   T cell transmigration, 964
   treatment, 967–968
   viruses, 964
Autoimmune leukocytoclastic vasculitis, 261
Autoimmune neuromyotonia (AINMT), 331
Autonomic nervous system and pain, 1021–1030
   direct coupling following nerve lesion, 1023–1024
   excitation of efferent neurons, 1023–1024
   experiments on humans, 1022–1023
   mechanical hyperalgesic behavior (SA system), 1027–1029
   NGF, 1026
   overview, 1022, 1022t, 1029
   proinflammatory cytokines, 1026–1027
   sensitization of nociceptors, 1024–1027
   sympathetic-afferent coupling, 1024
Autonomic nervous system failure, 857t
Autonomic neuropathies, 979–986
   AAN, 981–982
   amyloid autonomic neuropathy, 983
   biochemistry, 983
   classification, 980t
   DAN, 982
   etiology, 980
   history/nomenclature, 979–980
   inherited autonomic neuropathy, 982–983
   molecular mechanisms, 983
   natural history, 984
   pathogenesis/pathophysiology, 981
   pharmacology, 983
   structural basis, 983
   symptoms, 984

Autonomic neuropathy, gastrointestinal dysmotility, 164t
Autophagocytic vacuole, 11
Autophagosome vacuole, 11
Autosomal-dominant ataxias, 149
Autosomal-dominant cerebellar ataxia (ADCA), 97t
Autosomal dominant HSP, 542–543
Autosomal-dominant nocturnal frontal lobe epilepsy (ADNFLE), 330
Autosomal dominant partial epilepsy with auditory features (ADPFEAF), 361
Autosomal recessive HSP, 541–542
AVED, 699
Axonal spheroid formation, 15f, 16
Azathioprine, 179

B-process, 772
Babinski's sign, 450
Bacillus anthracis, 447t
Back-averaging technique, 306
Bacterial infections, 445–451. *See also* CNS infections
Bacteroides, 447t
Bailey, Percival, 425
Balint's syndrome, 842
Baltic myoclonus, 97t
Baroreceptor reflex (BR), 975
Baroregulation, 971
Barth syndrome, 955
Basal forebrain, 970
Basal ganglia, 57
Batten, Fredrick, 1
Batten disease, 4t, 6
BBB, 273–274, 375, 487
BDNF, 856
Becker's disease, 145–147, 323–324
Becker's muscular dystrophy (BMD), 926–927
Bed nucleus of stria terminalis, 970
Behçet's syndrome, 262, 664, 908
Benign familial neonatal convulsions (BFNC), 289–296, 330. *See also* Idiopathic generalized epilepsy (IGE)
Benign familial neonatal-infantile seizures (BFNIS), 137
Benzodiazepines, 141
Beriberi, 698–699
Beta-carotene, 698
β-interferon, 876
β-oxidation defects, 952
β-CIT, 65
β-galactosidase, 2
β-guanidinopropionate, 39
β-hexosaminidase, 2
β-Mannosidosis, 3t
BFNC, 289–296, 330. *See also* Idiopathic generalized epilepsy (IGE)
BFNIS, 137

Bilharz, Theodor, 470
Bilharzia, 470
Birth asphyxia, 604
Bismuth intoxication, 762
Bisphosphonates, 1018
BK virus (BKV), 186–189
Blaw, Michael, 21
Blindness, 698
Bloch, Felix, 794
Blood-brain barrier (BBB), 273–274, 375, 487
Blood diseases. *See* Hematological disease
Blood oxygenation level-dependent (BOLD) effect, 840
Blood pressure, 997–1005
    history/nomenclature, 997–998
    orthostatic stress, 1000
    physiological problem of upright posture, 999–1000
    regulation of, 998–999
    syncope, 1000–1004
Blood-tumor barrier (BTB), 375
BMD, 926–927
BMT, 17
Body temperature, 987–996. *See also* Thermoregulation
BOLD effect, 840
BOLD-fMRI, 248–249
Bone marrow transplantation (BMT), 17
Bone pain, 1011–1012
Borderline lepromatous leprosy, 915t
Borderline leprosy, 915
Borderline tuberculoid leprosy, 915t
Borrelia burgdorferi, 447t
Bortezomib, 882
Botulism, 922t
Bouillaud, Jean-Baptiste, 841
Bovine spongiform encephalopathy (BSE), 118, 119, 476
B-process, 772
BR, 975
Bradley, Charles, 632
Brain, Lord, 217
Brain-derived neurotrophic factor (BDNF), 856
Brain iron sufficiency, 725–726
Brain metastases. *See* CNS metastases
Brain turgor, 643
Brainstem, 972
Brainstem encephalitis, 164t
Branching enzyme deficiency, 950
Breaking of the tectum, 616
Breakthrough pain, 1016
Breast cancer, 373, 515, 517, 518, 519
Bright, Richard, 1031
Broca, Paul Pierre, 841
Broca's aphasia, 841
Broca's area, 846
Brodmann's areas 44 and 45, 846
2-bromobutyric acid, 751t

1-bromopropane, 751t
2-bromopropionic acid, 751t
BRT, 17
Brucellosis, 922t
Brudzinski's sign, 450
BSCL2, 543
BSE, 118, 119, 476
BTB, 375
Buerger's disease, 262
Bulbar ALS, 523–524
Buprenorphine, 1015
Burton, Eric, 443
2-butenenitrile, 751t
Byproduct replacement therapy (BRT), 17

C. jejuni, 207, 862, 863
^{11}C-DTBZ, 827
c-Jun-NH2-terminal kinases (JNKs), 854
C-peptide deficiency, 674
CABG, 229
CACNA1A gene, 340
CACNA1S, 326
CADASIL, 47, 225–226, 257, 343
CAE, 289–296. *See also* Idiopathic generalized epilepsy (IGE)
Café-au-lait spots, 414, 415f
CAG triplet diseases, 553–561
cAIS, 555
Cajal bodies, 504
Calcium channels, 138, 139b
Calcium metabolism disorders, 678–679
California encephalitis virus, 487t
Calpain-3, 929
Campath-1H, 202
Camptocormia, 90
Campylobacter jejuni, 207, 862, 863
CAN, 670, 969–977. *See also* Central autonomic network (CAN)
Canavan's disease, 46
Cancer metastasis. *See* CNS metastases
Cancer pain. *See* Neoplasm-induced pain
Candida albicans, 448t
CANOMAD, 867
Carbamazepine, 876, 1018, 1043
Carbidopa, 63
Carbon monoxide (CO), 973t, 977
Carboplatin peripheral neurotoxicity, 877–878
Cardioembolism, 230, 235–240
  AF, 236–237
  AMI, 237
  ASA, 238
  cardiomyopathy, 237
  etiology, 235–236
  infective endocarditis, 239

left atrial myxoma, 238
MAC, 238–239
mitral stenosis, 239
molecular mechanisms, 239–240
NBTE, 239
nomenclature, 235
papillary fibroelastoma, 239
PFO, 237–238
sources, 236t
valvular disease, 238–239
Cardiolipin, 955
Cardiomyopathy, 237
Cardiovascular autonomic neuropathy (CAN), 670
Cardiovascular drugs, 874
Carnitine deficiency, 951
Carnitine palmitoyltransferase (CPT) deficiency, 952
Carotenoids, 698
Carotid artery stenosis, 229
Carotid stenosis grading system, 812t
Carpal tunnel syndrome, 677
Castleman's disease, 212
Catalase, 19
Cataplexy, 716
Caudal intralaminar thalamo-striatal pathways, 623f
Caudal ventrolateral medulla (CVLM), 975
Caveolin-3, 928
CBF, 272, 683
CCD, 328
CCTG, 126, 127
CCUG, 128
CD, 692–693
CD4+ T cells, 198–203
CD8+ T cells, 198–203
CDCA, 46
CDD, 584t. *See also* Autism
CDDE, 662
CDS, 33–41. *See also* Creatine deficiency syndrome (CDS)
Celecoxib, 530
Celiac plexus block, 1019
Celiac sprue, 690
Central autonomic network (CAN), 969–977
  baroregulation, 971
  basic information, 969
  energy balance, 972
  functional anatomy, 970–975
  integration centers, 969
  neurotransmitters, 973t, 975–977
  stress response modulation, 971
  thermoregulation, 971–972
Central core disease (CCD), 328
Central hyperexcitability, 1029
Central nervous system (CNS), 198
Central nervous system hypoxia, 681–688. *See also* CNS hypoxia

Central nervous system infections, 445–451. *See also* CNS infections
Central nervous system metastases, 371–379. *See also* CNS metastases
Central nervous system viral infections, 485–499. *See also* CNS viral infections
Central pontine myelinosis (CPM), 664
Central post-stroke pain (CPSP), 1039–1046
　autonomic changes, 1043
　cardinal fact, 1044
　defined, 1041
　historical overview, 1039–1041
　incidence/prevalence, 1041
　natural course of disease, 1043
　pain characteristics, 1042
　pain onset, 1041–1042
　possible mechanisms of CPSP, 1044
　somatosensory deficit, 1042–1043
　treatment, 1043–1044
Central sensitization, 1029
Cephalosporins, 666
Cerebellum, 972
Cerebral blood flow (CBF), 272, 683
Cerebral cortex, 970
Cerebral energy metabolism, 682
Cerebral hypoxia, 681–688. *See also* CNS hypoxia
Cerebral ischemia, 271–279
　albumin, 277
　antioxidants, 277
　APC, 277–278
　BBB, 273–274
　cellular/biochemical responses, 272–273
　EPO, 277
　gene expression, 275–276
　global, 278
　hyperglycemia, 276
　hypothermia, 278
　inflammatory response, 276
　ischemic care, 272
　molecular events, 275
　neurovascular unit, 273, 277
　nitrone spin trap agents, 277
　postischemic necrosis/apoptosis, 274–275
　prenumbra, 272
　stem cell therapies, 278
　survival signaling pathways, 278
　t-PA, 276–277
　treatment, 276–278
Cerebral localization of function vs. integration of function, 846
Cerebral malaria, 455–457
Cerebral oxygen, 682
Cerebral palsy (CP), 575–580
　biochemistry, 579
　defined, 576t
　etiology, 576–578
　historical overview/nomenclature, 575–576
　molecular mechanisms, 579
　natural history, 579
　pathogenesis, 578
　pathophysiology, 578
　pharmacology, 579
　risk factors, 577t
　structural detail, 578–579
　symptoms, 579
Cerebral paragonimiasis, 471
Cerebral salt wasting syndrome, 269
Cerebral small vessel disease, 230
Cerebral vasculopathies, 255–263. *See also* Nonatherosclerotic cerebral vasculopathies
Cerebral vasospasm pathophysiology, 268–269
Cerebral venous thrombosis, 815–816
Cerebroretinal vasculopathy, 257
Cerebrospinal fluid (CSF), 641–642
Cerebrotendinous xanthomatosis (CTX), 45–46
CEROVIVE, 277
Cervicocephalic arterial dissection, 256
Cestodes, 464–466, 467t
CGA, 903–904
Chagas disease, 922t
Channelopathies of nervous system, 319–332. *See also* Ion channel disorders
　ADNFLE, 330
　AINMT, 331
　ATS, 327
　BFNC, 330
　channelopathies, 319–323
　EA-1, 328–329
　EA-2, 328–329
　familial hyperekplexia, 328
　FHM, 330
　GEFS+, 330–331
　HyperPP, 325
　HypoPP, 325–327
　inherited epilepsy syndromes, 330–331
　MH, 327–328
　myotonia congenita, 323–324
　non-dystrophic myotonias, 323–324
　PAM, 325
　paroxysmal and progressive ataxias, 328–330
　PMC, 325
　SCA-6, 329–330
Charcot, Jean-Martin, 349
Charcot-Marie-Tooth disease (CMT), 132, 885–899. *See also* Inherited peripheral neuropathies
Charcot-Marie-Tooth disease type X-linked dominant (CMTX), 890–891
Charlevoix-Saguenay syndrome, 539t

ChAT, 62–63, 76, 78
Chemical shift imaging (CSI), 782
Chemotherapy, PCNSL, 406–408
Chenodeoxycholic acid (CDCA), 46
Chiari malformation, 614, 615–616, 617
Childhood absence epilepsy (CAE), 289–296. *See also* Idiopathic generalized epilepsy (IGE)
Childhood ataxia with CNS hypomyelination, 47
Childhood disintegrative disorder (CDD), 584t. *See also* Autism
Chloramphenicol, 694, 873
Chlordecone, 751t
Chloride channels, 140, 140f, 321
2-chloropropionic acid, 751t
Chloroquine, 875
CHN, 890
Cholesterol ester storage disease, 4t, 6
Choline acetylase deficiency, 944t
Cholinergic hypothesis, 76
Cholines, 787
Cholines signal, 787
Chordoid meningioma, 383f, 384f
Chorea, 164t
Chorioretinitis, 463, 464
Chromosome 1, 388
Chromosome 9, 389
Chromosome 10, 389
Chromosome 14, 389
Chromosome 17, 389–390
Chromosome 18, 387–388
Chromosome 22, 386–387
Chronic carotid artery steno-occlusive disease, 808–810
Chronic dialysis-dependent encephalopathy (CDDE), 662
Chronic hydrocephalus, 644
Chronic hypoxia, 686
Chronic idiopathic anhidrosis (CIA), 993–994
Chronic inflammatory demyelinating polyneuropathy (CIDP), 205, 209–211, 652, 864–866
Chronic inflammatory neuropathies, 864
Chronic kidney disease (CKD), 659, 660t. *See also* Renal disease
Chronic obstructive lung disease (COPD), 686–687
Chronic progressive external ophthalmoplegia (CPEO), 158
Chronic wasting disease (CWD), 118, 119, 475–476
Churg-Strauss syndrome (CSS), 260, 905
Chvostek's sign, 679
CIA, 993–994
CIDP, 205, 209–211, 652, 864–866
CIN, 803
Cisplatin, 762, 877
CJD, 118, 119, 476–481. *See also* Prion diseases

CKD, 659, 660t. *See also* Renal disease
Classic Ammon's horn sclerosis, 350f, 351f, 352
CLC-1 channel, 321, 321f
Clear cell meningioma, 383f, 384f
CLN subsets, 315, 316t
CLN1 disease, 4t, 6
CLN2 disease, 4t, 6
CLN3 disease, 4t, 6
CLN4 disease, 4t
CLN5 disease, 4t
CLN6 disease, 4t
CLN7 disease, 4t
CLN8 disease, 4t
Clonazepam, 141, 1018
CMDs, 927–931, 929–931
CMT neuropathies, 885–899. *See also* Inherited peripheral neuropathies
CMT of the Russe type, 894
CMT1 and subtypes, 886–890
CMT1B, 889–890
CMT1C, 890
CMT1D, 890
CMT1F, 890
CMT2 and subtypes, 892–893
CMT2A, 892
CMT2B, 892
CMT2C, 892
CMT2D, 892
CMT2E, 892
CMT4 and subtypes, 893–894
CMT4A, 893
CMT4Ba, 893–894
CMT4C, 894
CMT4C1, 894
CMT4C2, 894
CMT4C3, 894
CMT4F, 894
CMTX, 890–891
CNS, 198
CNS hypoxia, 681–688
  acute hypoxia, 686
  anemic hypoxia, 682t, 683
  cerebral blood flow, 683
  cerebral energy metabolism, 682
  cerebral oxygen, 682
  chronic hypoxia, 686
  COPD, 686–687
  high-altitude sickness, 687
  histotoxic hypoxia, 682t, 683
  hypoperfusion hypoxia, 682t, 683
  hypoxemic hypoxic, 682–683, 682t
  hypoxic-ischemic injury, 684–685
  oxygen-sensing mechanisms, 683–684
  ventilatory control, 683

CNS infections, 445–451
  anatomy, 448f
  etiology, 446
  history/nomenclature, 445–446
  natural history, 450–451
  organisms/syndromes, 447–478t
  pathogenesis, 446
  pathophysiology, 449
  pharmacology, 449
  signs/symptoms, 449–450
  structural detail, 446
  viral infections, 485–499. See also CNS viral infections
CNS metastases, 371–379
  BBB, 375
  chemokines/cytokines, 374–375
  clinical symptoms, 377–378
  etiology/pathogenesis, 373–375
  Her-2, 377
  history/nomenclature, 371–373
  local brain invasion, 374
  mean survival, 378
  metastasis suppressors, 376
  metastatic cascade, 372f, 373
  natural history, 373
  neoangiogenesis, 375
  pathophysiology, 375–378
  peritumoral brain events, 373–374
CNS vasculitides, 258
CNS vasculitis, 258
CNS viral infections, 485–499
  diagnostic considerations, 491
  future perspectives, 495
  HAD, 493–495
  HSE, 491–493
  neuropathogenesis, 486–491
CO, 973t, 977
Coasting, 877
Cobalamin, 649
Cobalamin deficiency, 649–651
Coccidiodes immitis, 448t
Coenzyme Q10 (CoQ10), 955
Cogan's syndrome, 262
Cognitive impairment. See Vascular cognitive impairment (VCI)
Colchicine, 875
Colchicine myoneuropathy, 875
Collision task, 843f
Communicating hydrocephalus, 642
Compatible osmolyte hypothesis, 672
Complete androgen insensitivity syndrome (cAIS), 555
Complex motor tics, 620
Complicated HSP, 539, 539t
Computed tomography angiography (CTA). See Neurovascular computed tomography angiography

Computed tomography perfusion (CTP), 803. See also Neurovascular computed tomography angiography
Computed tomography venography (CTV), 805t, 815–816
COMT, 847
Congenital hydrocephalus, 641–647, 643
  biochemistry, 645
  CSF circulation, 643
  etiology, 642
  history/nomenclature, 641–642
  molecular mechanisms, 645
  natural history, 645–646
  pathogenesis, 642–644
  pathophysiology, 644–645
  pharmacology, 645
  structural details, 645
Congenital hypomyelinating neuropathy (CHN), 890
Congenital muscular dystrophies (CMDs), 929–931
Congenital myasthenic syndromes, 943–944
Contrast-induced nephropathy (CIN), 803
Convulsive syndromes, 290
COPD, 686–687
Copper, 697
Coprolalia, 620
Coproporphyria, 655
CoQ10, 955
Cordotomy, 1019
Core-signaling pathway, 251
Coronary artery bypass grafting (CABG), 229
Cortical Lewy body, 55
Cortical negative myoclonus, 311
Cortical positive myoclonus (CPM), 306–308
Cortico-striatal pathways, 621
Cortico-striatal-thalamocortical (CSTC) circuits, 621–624
Corticosteroids, 177
COX-2 up-regulation, 674
COX2 inhibitors, 530
Coxsackie virus, 487t
CP, 575–580. See also Cerebral palsy (CP)
CPEO, 158–159
CPM, 306–308, 664
CPSP, 1039–1046. See also Central post-stroke pain (CPSP)
CPT deficiency, 952
CPT I deficiency, 952
CPT II deficiency, 952
Cramp-fasciculation syndrome, 331
Cranial nerve palsies, 705–706
Cranial parasympathetic afferent system, 973
CRASH syndrome, 539t, 540
Creatine deficiency syndrome (CDS), 33–41
  accumulation of substances other than creatine, 37–40
  creatine metabolism, 34–35
  disjunction of creatine synthesis, 36
  etiology/biochemical pathogenesis, 34–36

Creatine deficiency syndrome (CDS) (*continued*)
   genetic basis, 35
   history/nomenclature, 33–34
   intracellular creatine depletion, 36–37
   natural history, 40
   pathophysiology, 36–40
   prospects, 40
CREST syndrome, 907
Creutzfeldt-Jakob disease (CJD), 118, 119, 476–481. *See also* Prion diseases
Crohn's disease (CD), 692–693
Cross-correction, 17
Crowe, Frank, 414
Cryoglobulinemic neuropathy, 214
Cryoglobulinemic vasculitis, 906
Cryptococcus neoformans, 447t
CSD, 343
CSI, 782
CSS, 260, 905
CSTC circuits, 621–624
CTA source images, 807
CTG, 126, 127
CTL, 165–167
CTV, 805t, 815–816
CTX, 45–46
Cu/Zn superoxide dismutase (SOD1) mutations, 524–525
CUG, 127, 128
CUGBP, 127, 128
Cushing, Harvey, 425
Cushing's disease, 675–676
CVLM, 975
CWD, 118, 119, 475–476
Cyclophosphamide, 179
Cysticercosis, 464
Cystinosis, 4t, 8
Cytomegalovirus, 488t, 923t
Cytoskeletal microtubule-associated tau protein, 56
Cytotoxic drugs, 177
Cytotoxic (CD8+) T-lymphocyte (CTL), 165–167

DA, 63, 628, 637t, 731, 772–773, 774
DA abnormalities, 727–728
da Vinci, Leonardo, 115
DADS, 212, 865
Damadian, Raymond, 794
DAN, 670, 671, 982, 857
Dandy-Walker syndrome (DWS), 645
Danon disease, 3t, 6, 9, 11
Dapsone, 873
DAT imaging, 822–823
D-BP deficiency, 21, 24
DCCT, 670–671
DCI, 267–268, 269
DCs, 198

DD, 663
DDS, 662–663
de Duve, Christian, 2
Deferoxamine, 284
Degos-Köhlmeier disease, 261
Dejerine-Sottas disease (DSD), 890
Delayed cerebral ischemia (DCI), 267–268, 269
Dendritic cells (DCs), 198
Dendritic spine abnormalities, 564f
Denny-Brown, Derek, 52
Dentatorubral pallidoluysian atrophy (DRPLA), 97, 315–316
Dermatomyositis (DM), 164t, 957–968. *See also* Autoimmune inflammatory myopathies
Desai, Hemali, 816
dHMN, 894–896
dHMN I, 894–895
dHMN II, 895
dHMN III, 895
dHMN IV, 895
dHMN of the Jerash type, 895–896
dHMN pyramidal/amyotrophic lateral sclerosis-4, 895
dHMN V, 895
dHMN VI, 895
dHMN VII, 895
DI-CMT, 891–892
Diabetes and endocrine disorders
   adrenal insufficiency, 676
   calcium metabolism disorders, 678–679
   diabetes mellitus, 669
   DNs, 669–675. *See also* Diabetic neuropathies (DNs)
   GH excess, 678
   glucocorticoid excess states, 675–676
   hyperparathyroidism, 678–679
   hyperthyroidism, 676–677
   hypoparathyroidism, 679
   hypothyroidism, 677–678
   osteomalacia, 678
   pseudohypoparathyroidism, 679
   thyroid disorders, 676–678
Diabetes Control and Complications Trial (DCCT), 670–671
Diabetes mellitus, 669, 849–858. *See also* Impaired glucose regulation (IGR)
Diabetic autonomic neuropathy (DAN), 670, 671, 982, 857
Diabetic neuropathies (DNs), 669–675
   AGE, 673
   apoptosis, 674
   AR pathway, 672–673
   C-peptide deficiency, 674
   COX-2 activation, 674
   endothelial dysfunction, 673
   etiology/pathogenesis, 670
   growth factors, 674–675

history/nomenclature, 670
hyperglycemia, 670–671
immune mechanisms, 674
nitrosative stress, 671–672
oxidative stress, 671
pathological structural changes, 675
pathophysiology/molecular mechanisms, 671–673
PKC activation, 673
Diabetic symmetrical sensory polyneuropathy, 856
Diagnostic imaging. *See* Neuroimaging
Dialysis dementia (DD), 663
Dialysis-dependent encephalopathy, 662
Dialysis disequilibrium syndrome (DDS), 662–663
Dialysis-related complications, 662–664
Diaschisis, 244
Dichloroacetic acid, 751t
Dichloroacetylene, 751t
2,4-dichlorophenoxyacetic acid, 751t
Didanosine, 873
Diffuse astrocytomas, 435, 436f
Diffuse large B-cell lymphoma (DLBCL), 397–398, 399f
Diffusion tensor imaging (DTI)
    neonatal brain injuries, 606
    WMLs, 229
Diffusion tensor imaging data, 799f
Diffusion-weighted imaging (DWI), 798
Digestive diseases. *See* Gastrointestinal and hepatic diseases
Dihydropteridine reductase (DHPR), 663
Diketones, 749
DiMauro, Salvatore, 947
2,4-dinitrophenol, 754
Diphtheria, 917–918
Diphyllobothrium latum, 467t
Direct coupling following nerve lesion, 1023–1024
Disimmune sensory ganglionopathies, 213
Dissecting aneurysms, 266
Distal acquired demyelinating symmetric (DADS), 212, 865
Distal symmetrical peripheral neuropathy (DPN), 670, 671
Disulfide, 751t
Disulfiram, 751t, 874–875
DLB, 115
DLBCL, 397–398, 399f
DM (dermatomyositis), 164t, 957–968. *See also* Autoimmune inflammatory myopathies
DM (myotonic dystrophy), 126–129, 931–932
    etiology, 126
    history/nomenclature, 126
    molecular pathogenesis, 126–128
    natural history, 129
    pathophysiology, 128
    symptoms, 129
DM-CIDP, 214–215

DM1, 126–129, 931
DM2, 126–129, 932
DMD, 926–927
DOA, 159–160
Dominant intermediate Charcot-Marie-Tooth disease (DI-CMT), 891–892
Dominant optic atrophy (DOA), 159–160
Donepezil, 80
Dopamine (DA), 63, 628, 637t, 731, 772–773, 774
Dopamine cell implantation, 826
Dopamine transporter (DAT), 822–823
Dopaminergic abnormalities, 627–629, 727–728
Dopaminergic tonic-phasic model hypothesis, 628–629
Dorsal nucleus of vagus, 974
Doxorubicin, 882
Down syndrome (DS), 565–566
DPN, 670, 671
Dronabinol, 697
DRPLA, 97, 315–316
Drug addiction, 771–779
    aberrant learning model, 774
    complexity, 775–776
    decision-making/impulse inhibition, 775
    dopamine, 772–773, 774
    drug dependence, 772
    excitatory transmission, 775
    hedonic model, 772
    incentive salience model, 773–774
    initiation of, 774–775
    relapse, 777–778
    stress, 777
Drug dependence, 772
Drug induced neuropathies, 871–883. *See also* Toxic and drug-induced neuropathies
Drug toxicities, 665–666
Dry beriberi, 699
DS, 565–566
DSD, 890
DTI. *See* Diffusion tensor imaging (DTI)
Duchenne's muscular dystrophy (DMD), 926–927
DWI, 798
DWS, 645
Dysarthria, 90
Dysautonomia, 89–90
Dysferlin, 928
Dyslexia, 593–598
    behavioral models, 594–595
    definition/history, 593–594
    fMRI, 595–596
    MRI, 595
    natural history, 594
    neurobiology, 595–596
    PET scan, 595–596
    reading behavior, 596–597

Dyslipidoses, 2
Dystrophia myotonica, 126–129. *See also* DM

E. histolytica, 462
E-cadherin-catenin complex, 374
E/L system, 11
EA-1, 143, 144, 328–329
EA-2, 143, 144, 328–329
EAAN, 994–995
EAAT2, 527–528
EAE, 198–203
EAN, 206
Early onset AD (EOAD), 70, 72, 74
Early onset cerebellar ataxia (EOCA), 97t
Early-onset RLS, 732
Eaton, Lee McKendre, 941
EBV, 396, 488t
Echinococcosis, 467t
Echinococcus granulosum, 467t
Echovirus, 487t
ECM, 374
Ectopic dendritogenesis, 15
EDIC study, 670
EDMD, 929
EET, 17
EF, 237
EGFR, 441–442
Ejection fraction (EF), 237
Electrocorticography, 845
Electrolyte imbalances, 664–665
Elephant Man, 413
Encephalopathy score, 605t
Endfolium, 352
Endfolium sclerosis, 351f, 352
Endocrine disorders, 669–680. *See also* Diabetes and endocrine disorders
Endothelial dysfunction, 673
Energy balance, 972
ENM, 311
Entamoeba histolytica, 462
Enzyme enhancement therapy (EET), 17
Enzyme replacement therapy (ERT), 17
EOAD, 70, 72, 74
EOCA, 97t
Eosinophilic granulomatosis, 260
Epidural infections, 446
Epilepsy, 136–143. *See also* Idiopathic generalized epilepsy (IGE); Temporal lobe epilepsy
    etiology, 136–137
    history/nomenclature, 136
    ion channelopathies, 137t
    MRS, 790
    natural history, 142–143
    pathogenesis/pathophysiology, 137–138

    pharmacology, 140–141
    structural detail, 138–140
    symptoms, 141–142
Epileptic negative myoclonus (ENM), 311
Epinephrine, 178
Episodic ataxia type 1 (EA-1), 328–329
Episodic ataxia type 2 (EA-2), 328–329
Episodic hypothermia, 993
Epitope spreading, 199
EPM1, 313–314
EPO, 277, 665
EPO toxicity, 665
Epothilones, 882
Epstein-Barr virus (EBV), 396, 488t
Equilibrium potential, 321
Equine encephalitis viruses, 487t
Ergot alkaloids, 258
ERKs, 854
ERM proteins, 387
Ernst, Richard, 794
ERT, 17
Erythromelalgia, 657
Erythropoietin (EPO), 277, 665
Escherichia coli, 447t
"Essay on the Shaking Palsy, An" (Parkinson), 51
Essential cryoglobulinemic vasculitis, 906
Essential thrombocythemia (ET), 657–658
Ethambutol, 873
Ethanol, 666, 750–751, 751t
Ethidium bromide, 753
Ethionamide, 873
Ethosuximide, 141
Ethyl chloride, 751t
Ethylene oxide, 751t
Ewing, James, 372
Excitation:inhibition ratio, 588–589
Excitotoxicity, 57, 284, 603
Exon 7 inclusion, 509
Experimental autoimmune autonomic neuropathy (EAAN), 994–995
Experimental autoimmune encephalomyelitis (EAE), 198–203
Experimental autoimmune neuritis (EAN), 206
External hydrocephalus, 642
External ophthalmoplegia (PEO), 158
Extracellular matrix (ECM), 374
Extracellular signal-related kinases (ERKs), 854
Extrahippocampal pathology, 353, 354t

Fabry disease, 3t, 4–5, 257
Facioscapulohumeral muscular dystrophy (FSHD), 932–933
Fainting, 1000–1004
FALDH, 46

Familial amyloid polyneuropathy (FAP), 983
Familial CJD (fCJD), 118, 477t, 478
Familial dysautonomia, 896
Familial EOAD (FEOAD), 71, 74, 76
Familial hemiplegic migraine (FHM), 330, 340–343
Familial hyperekplexia, 328
Familial inflammatory IBM, 959
Familial OPCA (FOPCA), 95–104
Familial paroxysmal dystonia, 301
FAP, 983
Farber disease, 3t, 5
Fast inactivation, 145
Fatal familial insomnia (FFI), 118, 119, 477t, 478, 972
fCJD, 118, 477t, 478
FCMD, 930
FDG-PET, 790, 822–827
FDOPA-PET, 822, 825, 826
FDOPA uptake, 822–826
Fe, 694–695
Fentanyl, 1015
Fenton reaction, 284
FEOAD, 71, 74, 76
Ferritin, 695, 728
Fetal ventriculomegaly, 645
FFI, 118, 119, 477t, 478, 972
FHM, 330, 340–343
Fialuridine, 873
Fibromuscular dysplasia (FMD), 255–256, 810
Fibrous meningioma, 383f, 384f
Fingerprint bodies, 13
Finnish snowballs, 13
First-pass cine slab technique, 807
Fisher, C. Miller, 808
Fisher syndrome (FS), 860–864
5-HT, 973t, 976
5-HTP-induced myoclonus, 312
FK506, 875–876
FKRP gene mutations, 931
Fludarabine, 652
Flutamide, 560
FMD, 255–256, 810
fMRI, 839–848
 AD, 845
 BOLD effect, 840
 cerebral localization of function vs. integration of function, 846
 dyslexia, 595–596
 language and aphasia, 841–842
 memory and amnestic deficits, 844–845
 methodology, 840–841
 neuronavigation, 845–846
 phenotyping and genotyping, 847
 praxis and apraxia, 842

 recovery of function, 247–249, 846–847
 spatial cognition and neglect, 842–844
 stroke recovery, 247–249, 846–847
 Tourette's syndrome, 625, 626
Focal cerebral ischemia, 271–279. See also Cerebral ischemia
Foerster, Otfrid, 845
FOPCA, 95–104
4,1B protein, 387–388
14q, 389
Fragile X-associated tumor/ataxia syndrome, 129–132. See also FXTAS
Fragile X syndrome (FXS), 568–569
FRAP, 440
Frataxin, 160, 695
FRDA, 97t, 160, 695
Free radicals, 285
Friedrich's ataxia (FRDA), 97t, 160, 695
Frizzled, 429
Froment's maneuver, 60
Frontotemporal lobar dementia (FTLD), 524
FS, 860–864
FSHD, 932–933
FTLD, 524
Fucosidosis, 3t
Fukuyama congenital muscular dystrophy (FCMD), 930
Functional MRI. See fMRI
Fungal infections, 445–451. See also CNS infections
FXS, 568–569
FXTAS, 129–132
 epidemiology, 129–130
 history/nomenclature, 129
 molecular pathogenesis, 130–131
 natural history, 132
 neuropathology, 130
 pathophysiology, 131
 symptoms, 131–132

GA, 202
GABA, 312, 788
GABHS, 629
Gadolinium computed tomography angiography, 803f
GAERS, 291–294
Gajdusek, Carleton, 473
Galactocerebrosidase, 45
Galactosialidosis, 3t, 8
Galactosylceramide, 45
Galactosylceramide lipidosis, 5
Galactosylsphingosine, 45
Galantamine, 80
Gall, Franz Joseph, 841, 846
γ-Guanidinobutyrate, 38
γ-Aminobutyric acid, 284

GAMT deficiency, 33–41. *See also* Creatine deficiency syndrome (CDS)
Ganglioside antigens, 211
Gangliosides, 862
Garin-Bujadoux-Bannwarth syndrome, 914
Garrod, Archibald, 2
Gastrointestinal and hepatic diseases, 689–701
   AIP, 696
   alimentary tract diseases, 690–695
   Campylobacter-associated GBS, 694
   celiac disease, 690–692
   cerebellar ataxia, 691
   epilepsy, 691–692
   HE, 695–696
   hepatic diseases, 695–698
   IBD, 692–693
   iron overload, 694–695
   pruritus of cholestasis, 696–697
   vitamin deficiencies, 698–700
   Whipple's disease, 693–694
   Wilson's disease, 697–698
   zinc deficiency, 700
Gaucher, Phillipe, 1
Gaucher disease, 3t, 4, 8
Gaze palsy, 108–109
GBS, 206–209, 694, 860–864
GC-like DLBCL, 398, 399f
GCIs, 84, 85
GEFS+, 137–141, 330–331, 362
Gelsolin amyloidosis, 983
Gene therapy, 17
Generalized epilepsy with febrile seizures plus (GEFS+), 137–141, 330–331, 362
Generalized glycogen storage diseases, 950–951
Genetic absence epilepsy rat from Strasbourg (GAERS), 291–294
Genetic mental retardation, 563–573
   dendritic/synaptic abnormalities, 564–565
   Down syndrome (DS), 565–566
   fragile X syndrome (FXS), 568–569
   Rett syndrome (RTT), 569–571
   Rubinstein-Taybi syndrome (RTS), 572
   Williams syndrome (WS), 572
   X-linked MR (XLMR), 567–571
Genetically epilepsy-prone rat type 3 (GEPR-3), 312
Genotyping, 847
GEPR-3, 312
Gerstmann-Sträussler-Scheinker (GSS) syndrome, 118, 119, 478
GH excess, 678
Ghrelin, 973t, 977
Giant cell arteritis (GCA), 259t, 261, 903–904
Giant cell glioblastoma, 436
Gibbs, Frederick and Erna, 351

Gilles de la Tourette, Georges, 620
Ginger-Jake syndrome, 747
Glatiramer acetate (GA), 202
Glial cytoplasmic inclusions (GCIs), 84, 85
Glioblastoma, 436, 436f
Glioma, 433–444
   Akt, 439
   angiogenesis, 440, 441f
   cell of origin, 438–439
   cellular migration/invasion, 440
   combination therapy, 442
   epidemiology, 434
   etiology, 434–435
   future therapeutic interventions, 442–443
   genetic alterations, 436–438
   high-grade, 435–436
   locations/symptoms, 434
   low-grade, 435
   mTOR, 440
   mTOR inhibitors, 442
   nomenclature, 434
   P13-K, 439
   pharmacology, 441–442
   PTEN, 439–440
   RAS/MAPK pathway, 442
   RAS/RAF/MEK/ERK signaling, 440
   receptor tyrosine kinase inhibitors, 441–442
   signaling pathways, 439–440
   targeted molecular therapy, 441
Global cerebral ischemia, 278
Globoid cell, 5
Globoid cell leukodystrophy, 3t, 5, 45
Glucocorticoid, 967
Glucocorticoid excess states, 675–676
Glutamate, 284, 578, 788
Glutamate transporters, 527–528
Glutamine, 788
Gluten, 690
Gluten ataxia, 691
Gluten-sensitive enteropathy, 690
Glycemic control, 850
Glycine, 313
Glycogen metabolism, defects of, 949–951
Glycogen storage diseases, 6
Glycogenoses, 3t
Glycoproteinoses, 3t, 6
Glycosaminoglycans, 2, 5
Glycosphingolipids (GSLs), 5
Glycosylation, disorders, 930–931
GM_1 gangliosidoses, 2, 3t
G_{M2} activator deficiency, 3t
GM_2 gangliosidosis, 4t, 8
Gnathostoma spinigerum, 468t
Gnathostomiasis, 468t

Gold salts, 762, 875
Golgi/TGN, 9, 12, 14
GOM, 47
Gonadotrophin agonists, 560
Gonzalez, R. Gilberto, 816
Gorlin's syndrome, 428
Gould, Stephen, 20
Gowers, William R., 349
GRACILE, 954
Granular osmiophilic material (GOM), 47
Granular spheroids, 16
Granule cell dispersion, 354, 356f
Granulomatous amebic encephalitis, 460, 461f
Growth hormone (GH) excess, 678
GSLs, 5
GSS, 477t, 478
GSS syndrome, 118, 119
Guanidinoacetate, 36–39
Guanidinoacetate methyltransferase (GAMT) deficiency, 33–41. *See also* Creatine deficiency syndrome (CDS)
Guanidinosuccinate, 38, 39
Guillain-Barré syndrome (GBS), 206–209, 694, 860–864

H-ferritin, 728
h-IBM, 959
H1/H1 tau genotype, 106
^{1}H-MRS, 175–176, 784–790
Haber, Fritz, 757
Haber's law, 757
HAD, 493–495
Hadlow, William, 473
Hallervorden-Spatz syndrome, 115
Halogenated aromatics, 751t
Hamilton, Alice, 746
Hansen's disease, 446
Harding's classification of hereditary ataxias, 97t
Harris, Gordon, 816
Hashimoto encephalopathy (HE), 217–221
　clinical course, 220
　clinical features, 218
　etiology, 218–220
　history/nomenclature, 217–218
　pathogenesis, 220
HAT, 457–460
Hb SS, 654
HBO, 278
HBPN, 897
HCAA, 226–227
HCV-associated neuropathies, 920–921
HD-MTX, 406
HDAC inhibitors, 507–509
HE, 217–221, 695–696. *See also* Hashimoto encephalopathy (HE)
HE1, 5, 8

Heat shock protein (HSP), 57, 558
Hedonic model, 772
Helminths, 464
　cestodes, 464–466, 467t
　nematodes, 466, 468–469t
　trematodes, 470–471
Hematogenous metastasis, 373
Hematological disease, 649–658
　acute porphyrias, 655–656
　ET, 657–658
　hemoglobinopathies, 653–655
　MGUS, 652
　multiple myeloma, 651
　myeloproliferative syndromes, 657–658
　paraproteinemias, 651–653
　POEMS syndrome, 653
　primary amyloidosis, 653
　PV, 657–658
　sickle cell disease, 653–655
　thalassemia, 655
　TTP, 656–657
　vitamin B_{12} deficiency, 649–651
　WM, 651–652
Hemodynamic response function (HRF), 840, 841
Hemoglobin, 284
Hemoglobinopathies, 653–655
Hemophilus influenzae type B, 447t
Hemorrhagic stroke
　clot-induced toxicity, 283
　complement, 284
　erythrocyte lysis, 284
　etiology, 282
　fibrosis, 285
　free radicals, 285
　γ-aminobutyric acid, 284
　glutamate, 284
　hemoglobin, 284
　inflammation, 284, 285
　iron, 284
　ischemia, 283
　pathophysiology, 282–283
　physical disruption/mass effect, 283, 285
　protective pathways, 284
　SAH. *See* Subarachnoid hemorrhage (SAH)
　therapeutic interventions, 286
　thrombin, 283–284
Henoch-Schönlein purpura (HSP), 261, 905
Hepatic diseases, 689–701. *See also* Gastrointestinal and hepatic diseases
Hepatic encephalopathy (HE), 695–696
Hepatic porphyrias, 655–656
Hepatitis C virus-related peripheral neuropathies, 920–921
Hepatolenticular degeneration, 697
Her-2, 377

Hereditary brachial plexus neuropathy (HBPN), 897
Hereditary cerebral amyloid angiopathy (HCAA), 226–227
Hereditary coproporphyria, 655
Hereditary endotheliopathy with retinopathy nephropathy and stroke, 257
Hereditary gelsolin amyloidosis, 983
Hereditary inclusion body myopathy (h-IBM), 959
Hereditary motor and sensory neuropathy-Russe (HMSNR), 894
Hereditary motor neuropathies (HMNs), 894–896
Hereditary neuropathy with liability to pressure palsies (HNPP), 891
Hereditary peripheral neuropathies, 885–899. See also Inherited peripheral neuropathies
Hereditary sensory and autonomic neuropathies, 896–897
Hereditary sensory autonomic neuropathy (HSAN), 896–897, 982–983
Hereditary spastic paraplegia (HSP), 97t, 537–544. See also HSP and PLS
Hereditary vascular retinopathy, 257
Herpes simplex encephalitis (HSE), 491–493, 834
Herpes zoster (HZ), 923t, 1031–1037. See also Postherpetic neuralgia (PHN)
Hers, H. G., 2, 6
Heterocycles, 746
Hexachlorophene, 751t
2,5-hexanedione, 751t, 754
2-hexanol, 751t
HIF-1, 684
HIF-1 system, 684–685
High-altitude sickness, 687
High-grade glioma, 435–436
Hilar cell loss, 355–357
Hinshelwood, James, 593
Hippocampal sclerosis, 352–353
Histone deacetylase (HDAC) inhibitors, 507–509
Histotoxic hypoxia, 682t, 683
HIV/AIDS, 486, 490f, 493–495
HIV-associated dementia (HAD), 493–495
HIV-associated distal symmetrical polyneuropathy (HIV-DSP), 918–920
HIV-associated symptomatic distal symmetrical polyneuropathy (HIV-SDSP), 919–920
HIV-DSP, 918–920
HIV encephalopathy, 834
HIV-SDSP, 919–920
HLA-DQB1*0602, 717, 720
HMG-CoA reductase inhibitors, 874
HMNs, 894–896
HMSN of the Lom type, 894
HMSNR, 894
HNPP, 891
Hodgkin-Katz-Goldman equation, 321

Hoffman, Heinrich, 631
Hoffman, Johann, 501
Homocysteine, 662
Homocystinuria, 257
Horton, B. T., 903
HRF, 840, 841
HSAN, 896–897, 982–983
HSAN1, 896, 982
HSAN2, 896
HSAN3, 896, 982–983
HSAN4, 896–897
HSAN5, 896–897
HSE, 491–493
HSP, 57, 97t, 558, 905
HSP and PLS, 537–544
  autosomal dominant HSP, 542–543
  autosomal recessive HSP, 541–542
  clinical features of HSP, 538–539
  common molecular mechanisms, 543–544
  complicated HSP, 539, 539t
  definitions, 537–538
  diagnostic criteria of PLS, 543t
  epidemiology, 538
  genetic classification of HSP, 541t
  historical overview, 538
  neuropathology of HSP, 539–540
  nosology of PLS, 543
  x-linked HSP, 540
Hsp60, 542–543
HTLV, 488t, 491
HTLV-1, 923t
Human African trypanosomiasis (HAT), 457–460
Human T-cell leukemia virus, 488t
Hunter's disease, 3t
Huntington's disease, 827
Hurler/Scheie disease, 3t
Hurler's disease, 3t
Hurler's syndrome, 642
Hydralazine, 874
Hydrocephalus, 267, 614, 616, 641–647. See also Congenital hydrocephalus
Hydrocephalus ex vacuo, 645
Hydroxyurea, 658
Hyperacute MRI, 798
Hyperbaric oxygen (HBO), 278
Hypercalcemia, 1012
Hyperekplexia, 309
Hyperglycemia, 276, 670–671
Hyperhidrosis, 993–994
Hyperkalemia, 665
Hyperkalemic periodic paralysis (HyperPP), 145–147, 325
Hypernatremia, 665
Hyperparathyroidism, 678–679
HyperPP, 145–147, 325

Hypersensitivity vasculitis, 259t, 261
Hypersomnia, 715–722. *See also* Narcolepsy and hypersomnia
Hyperthyroidism, 676–677
Hypocretin, 718, 719, 973t, 976
Hypokalemia, 665
Hypokalemic periodic paralysis (HypoPP), 145–147, 325–327
Hyponatremia, 269, 664
Hypoparathyroidism, 679
Hypoperfusion hypoxia, 682t, 683
HypoPP, 145–147, 325–327
Hypothalamus, 970
Hypothermia, 278, 992–993
Hypothyroidism, 677–678
Hypoxemic hypoxic, 682–683, 682t
Hypoxia, 682–683. *See also* CNS hypoxia
Hypoxia-inducible factor 1 (HIF-1), 684
Hypoxic-ischemic encephalopathy, 604
Hypoxic-ischemic injury, 684–685
HZ. *See* Postherpetic neuralgia (PHN)

I-cell disease, 3t, 5, 6
Iatrogenic CJD (iCJD), 118, 477t, 478
ICEGTC, 137
ICH, 267, 281–287. *See also* Hemorrhagic stroke
iCJD, 118, 477t, 478
Idiopathic autism, 584
Idiopathic autonomic neuropathy, 981
Idiopathic generalized epilepsy (IGE), 289–296. *See also* Epilepsy
  animal models, 290
  clinical overview, 289–296
  concluding remarks, 296
  genetic models, 291
  genetics, 294–296
  pharmacological models, 291–292
  thalamocortical synchronized activity, 292–294
Idiopathic hypersomnia, 721–722
Idiopathic late-onset cerebellar ataxia (ILOCA), 95–104
Idiopathic RBD, 711–713
IFG, 596
IFN-α, 173
IFN-β, 202
IGE, 289–296. *See also* Idiopathic generalized epilepsy (IGE)
IGF-1, 674
IGF-I, 856
IgG and IgA paraprotein-related chronic inflammatory neuropathies, 867
IgM MGUS, 652
IgM monoclonal gammopathies, 866–867
IgM-PN, 212–213
IGR, 849–858. *See also* Impaired glucose regulation (IGR)

IGT, 849
IL-6, 1026
IL-8, 1027
ILOCA, 95–104
Imaging techniques. *See* Neuroimaging
3,3-iminodipropionitrile, 751t
Immune-mediated neuropathies, 205–215
  acute neuropathies, 206–209
  chronic neuropathies, 209–213
  CIDP, 209–211
  classification, 206t
  cryoglobulinemic neuropathy, 214
  disimmune sensory ganglionopathies, 213
  DM-CIDP, 214–215
  EAN, 206
  GBS, 206–209
  hereditary neuropathies, 214
  IgM-PN, 212–213
  MGUS, 212
  MMN, 211–212
  vasculitic neuropathies, 213–214
Immunoglobulin G and A monoclonal gammopathies, 867
Immunoglobulin M monoclonal gammopathy, 866–867
Impaired glucose regulation (IGR), 849–858
  apoptosis, 853–854
  diabetic autonomic neuropathy, 857
  diabetic symmetrical sensory polyneuropathy, 856
  diet and exercise, 851
  ERKs, 854
  glycemic control, 850
  growth factors, 855–856
  JNKs, 854
  MARKs, 854
  mG/uRs, 855
  natural history, 857
  nitric oxide/nitrosative injury, 851–852
  oxidative injury, 852–853, 855
  PI3K, 854
  p53, 854
  PKC, 854
  redox potential, 851
  small-fiber neuropathy, 856
  spectrum of risk factors, 850f
  UCPs, 852–853
Impaired glucose tolerance (IGT), 849
Implicit memory, 251
Imprinting memory, 670
Incentive salience model, 773–774
Inclusion body myositis (IBM), 957–968. *See also* Autoimmune inflammatory myopathies
Incomplete lupus, 172
Infantile AMD, 950
Infantile brainstem syndrome, 617
Infantile free sialic acid storage (ISSD), 4t, 8

Infantile NCL, 316t
Infantile Refsum's disease (IRD), 21
Infections, 445–472. *See also* CNS infections; Neuropathies associated with infections; Parasitic infections
Infectious disease hypothesis, 203f
Infectious inflammatory vasculopathies, 258t, 259
Infective endocarditis, 239
Inferior frontal gyrus (IFG), 596
Inflammatory bowel disease (IBD), 692–693
Inflammatory neuropathies, 859–869. *See also* Acquired inflammatory demyelinating and axonal neuropathies
Inherited ALS, 522–523
Inherited amyloid neuropathy, 983
Inherited autonomic neuropathy, 982–983
Inherited CJD, 477t, 478
Inherited epilepsy syndromes, 330–331
Inherited myoclonus dystonia, 313
Inherited peripheral neuropathies, 885–899
   CHN, 890
   classification, 886, 887t
   CMT1 and subtypes, 886–890
   CMT2 and subtypes, 892–893
   CMT4 and subtypes, 893–894
   CMTX, 890–891
   Dejerine-Sottas disease (DSD), 890
   dHMNs, 894–896
   DI-CMT, 891–892
   HBPN, 897
   HMNs, 894–896
   HMSNR, 894
   HNPP, 891
   HSANs, 896–897
   rare forms, 897
Initial precipitating events, 357
Insomnia, 735–744
   defined, 736–737
   ICSD-II general criteria, 736f
   integrative perspective, 741–742
   neurocognitive model, 737–738
   neuroendocrine measures, 740–741
   neuroimaging measures, 741
   neurophysiologic measures, 740
   plasma measures, 740–741
   psychobiological inhibition model, 738
   urinary measures, 740
Insula, 970
Insulin-like growth factor (IGF)-1, 674
Interactive robotic devices, 247f
Interferon, 658, 876
Interferon-β (IFN-β), 202
Intergenomic communication, defects of, 955
Internal carotid and vertebral artery dissections, 810
Internal hydrocephalus, 642

Intracerebral acidosis, 663
Intracerebral hemorrhage (ICH), 267, 281–287. *See also* Hemorrhagic stroke
Intracortical inhibition, 624
Intractable childhood epilepsy with generalized tonic-clonic seizures (ICEGTC), 137
Intraventricular hemorrhage (IVH), 267, 281–287. *See also* Hemorrhagic stroke
Inward rectification, 322–323
Ion channel, 135–136, 319–323
Ion channel disorders, 135–148. *See also* Channelopathies of nervous system
   ataxia, 97t, 143–144
   epilepsy, 136–143. *See also* Epilepsy
   neuromuscular disorders, 144–147
   nondystrophic myotonia, 144–147
   periodic paralysis, 144–147
IRA domains, 420
IRD, 21
Iron (Fe), 284, 694–695
Iron deficiency, 724–725
Iron deficiency anemia, 655
Iron metabolism abnormalities, 728–729
Isaac's syndrome, 331, 942
Ischemic care, 272
Ischemic stroke, 235
Isolated angiitis of CNS, 906
Isoniazid, 873
ISSD, 4t, 8
IVH, 267

Jackson, John Hughlings, 349
Jake paralysis, 747
JC virus (JCV), 185–195, 488t
JCV, 185–195, 488t
Jennings, G. H., 903
JME, 290. *See also* Idiopathic generalized epilepsy (IGE)
JNKs, 854
Johnston, Michael, 563
Juvenile absence epilepsy (JAE), 289–296. *See also* Idiopathic generalized epilepsy (IGE)
Juvenile myoclonic epilepsy (JME), 290. *See also* Idiopathic generalized epilepsy (IGE)
Juvenile NCL, 315, 316t

Kaminski, Alexander, 858
Kanner, Leo, 581
Katayama fever, 470
Kawasaki disease, 904–905
Kawasaki's syndrome, 262
Kayser-Fleischer ring, 697
KCNQ mutations, 361
Kearns-Sayre syndrome (KSS), 158, 953, 954

Kennedy's disease, 553–561
  clinical features, 554
  female carriers, 558
  GM animal models, 559
  ligand binding, 558
  molecular pathogenesis, 555–558
  muscle, 555
  neurons, 554–555
  pathological features, 554–555
  pharmacology, 558–559
  polyQ expansion, 555–558
  post mortem findings, 555f
  symptoms, 559
  therapy, 559–560
Kernig's sign, 450
Ketamine, 1018
Kidney disease. *See* Renal disease
KIF5A, 542–543
Kjellin syndrome, 539t
Kleine-Levin syndrome, 716, 716t
Knudson, Alfred, 420
Knudson two-hit hypothesis, 420, 420f
Korsakoff's psychosis, 699
Krabbe disease, 3t, 5, 43, 45
KSS, 158–159, 953, 954
Kufs disease, 4t
Kugelberg-Welander disease, 502
Kuru, 118

L-carnitine, 951
L-dopa, 65, 666
L-ferritin, 728
L-S endocarditis, 239
L1, 540
L1CAM, 540, 642
Lactate, 787–788
Lafora's disease (LD), 314–315
Laforin, 315
Lambert, Edward H., 941
Lambert-Eaton myasthenic syndrome (LEMS), 941–942
Lambert-Eaton syndrome, 165, 331
Lamin A/C, 132
Lamivudine, 873
Language and aphasia, 841–842
Large vessel disease, 229–230
Large-vessel vasculitis, 903–904
Late infantile NCL, 316t
Late-onset AD (LOAD), 70, 72, 74
Late-onset RLS, 732
Lauterbur, Paul, 794
LB, 54–55, 115
LC, 972
LCHAD deficiency, 952

LCMD2C-F: "sarcoglycanopathies", 927–928
LD, 314–315
Lead, 760
Lead poisoning, 666
Learning disabilities, 418
Leber's hereditary optic neuropathy (LHON), 158
Leflunomide, 875
Left atrial myxoma, 238
Leigh disease, 954
Leigh's syndrome, 699
LEMS, 941–942
Lennox, William G., 351
Lepromatous leprosy (LL), 915–917
Leprosy, 445, 915–917
Leptin, 973t, 976
Leukotrophies, 43–49
  ALD, 43–45
  Alexander's disease, 46–47
  CADASIL, 47
  Canavan's disease, 46
  CTX, 45–46
  globoid cell leukodystrophy, 45
  lipid metabolism disorders, 43–46
  MLD, 45
  myelin protein disorders, 46
  organic acid disorders, 46
  PMD, 46
  Sjögren-Larsson syndrome, 46
  VWM, 47
Levinthal paradox, 111–112
Levodopa, 63, 90, 103
Lewis, Sir Thomas, 1000
Lewis Sumner syndrome, 206t, 865
Lewy, Frits Heinrich, 52
Lewy body (LB), 54–55, 115
LGDM1C, 928
LGMD1A, 928
LGMD1B, 929
LGMD2A, 929
LGMD2B, 928
LGMD2G, 929
LGMD2H, 929
LGMD2I, 931
LGMD2J, 929
LGMDs, 927–929
Lhermitte's sign, 877
LHON, 158
Li-Fraumeni syndrome, 428
Libman-Sacks (L-S) endocarditis, 239
Lieberman, David N., 563
Ligand binding, 558
Ligand-gated ion channels, 135–136

Limb-girdle muscular dystrophies (LGMDs), 927–929
Limbic encephalitis, 165
LIMK1, 572
Lipid metabolism, defects of, 951–953
Lipid metabolism disorders, 43–46
Lipid milieu, defects of, 955
Lipid rafts, 119
Lipomyelomeningocele, 615f
Lipopolysaccharide (LPS), 178
Lisch nodules, 415, 415f
Listeria monocytogenes, 447t
Lithium, 665, 760–761
Little, Sir William, 575
Liver-X-receptor (LXR), 14
LL, 915–917
LMN, 513–519
LOAD, 70, 72, 74
Local brain invasion, 374
Localizationism, 846
Localized vasculitis, 906–907
Locus coeruleus (LC), 972
Long-QT syndrome type 7 (LQT7), 327
Long-standing ventriculomegaly (LOVA), 644
Lorenzo's oil, 45
Lou Gehrig's disease, 521–535. *See also* ALS
LOVA, 644
Low-grade glioma, 435
Lower motor neuron (LMN), 513–519
LPS, 178
LQT7, 327
Luft, Rolf, 947
Lumbosacral plexopathy, 1013
Lupus. *See* Systemic lupus erythematosus (SLE)
Lupus nephritis, 664
LXR, 14
Lyme disease, 913–915
Lymphomatoid granulomatosis, 260
Lymphoplasmacytic-rich meningioma, 383f, 384f
Lyndol, 747
Lysosomal disorders, 1–18
    classification, 2–9
    clinicopathological features, 2–6
    defective proteins, 6–9
    neuron metabolism/function, 9–12
    pathogenic cascades, 12–16
    therapy, 16–17
Lysosomes, 9

M-protein associated neuropathy, 651–653
MAC, 238–239
Mad cow disease, 118, 119, 476
MAD deficiency, 952–953
Mad Hatter syndrome, 760

MADSAM, 206t, 865
MAG, 206, 212–213
Magnetic resonance angiography (MRA), 798f
Magnetic resonance imaging. *See* MRI
Magnetic resonance spectroscopic imaging (MRSI), 782
Magnetic resonance spectroscopy. *See* MRS
Magnetization transfer, 799
Malaria, 455–457
Malignant atrophic papulosis, 261
Malignant brachial plexopathy, 1013
Malignant hyperthermia (MH), 327–328, 992–993
Malignant peripheral nerve sheath tumor (MPNST), 416, 418, 422
Mancall, E., 185
Manganese, 761
Mannitol, 754
Mansfield, Peter, 794
MAP, 998
MAPKs, 854
MAPT, 106
Marantic endocarditis, 239
Marie, Pierre, 149
Marinesco-Sjögren syndrome, 97t
MARKs, 854
Maroteau-Lamy disease, 3t
MASA syndrome, 539t
Maspardin, 541–542
Mast syndrome, 542
Matrix metalloproteinases (MMPs), 107, 374–375
MBNL, 127, 128
MC, 145–147, 323–324
MCAD, 952
McArdle, Brian, 947
McArdle's disease, 948, 951
MCBs, 13
MCI, 75, 78
MCMD, 493
MD, 313, 925–934. *See also* Muscular dystrophies
MDC1A, 929–930
MDC1C, 931
MDC1D, 931
Mean arterial pressure (MAP), 998
Mean transit time (MTT), 798f
Measles virus, 488t
MEB disease, 930
MeCP2, 571
Medial prefrontal cortex (MPC), 970
Medial preoptic (MPO) region, 988–990
Medication administration, 1015–1016
Medication overdose, 664–665
Medium-chain acyl-CoA dehydrogenase (MCAD), 952
Medium-sized-vessel vasculitis, 904–905
Medullary raphe nuclei (MRN), 975

Medulloblastoma, 425–432
    ATRT, contrasted, 427
    developmental signaling pathways, 428–429
    histopathology, 426–427
    historical overview, 426
    inherited cancer syndromes, 428
    molecular genetics, 427–428
    molecular profiling, 429–431
    neuroembryogenesis of cerebellum, 426
    prognosis, 429–431
    treatment, 431
Meganeurites, 15
MELAS, 159, 257, 954
Memantine, 80
Membrane potential, 321
Membranous cytoplasmic bodies (MCBs), 13
Memory and amnestic deficits, 844–845
Meningioma, 381–393
    angiogenesis, 391
    cell lines/animal models, 392
    chromosome 1, 388
    chromosome 9, 389
    chromosome 10, 389
    chromosome 14, 389
    chromosome 17, 389–390
    chromosome 18, 387–388
    chromosome 22, 386–387
    clinical aspects, 385–386
    clinical presentation, 385
    edema, 391
    future directions, 392–393
    genetics, 386–390
    growth factors, 391–392
    historical overview, 381–382
    hormones, 392
    immunohistochemistry, 384
    incidence/etiology, 382
    multiple, 390–391
    neuroradiology, 385
    neurotransmitters, 392
    NF2-associated, 391
    pathophysiology, 391–392
    pediatric, 391
    prognosis, 385–386
    RIM, 390
    telomerase, 390
    treatment, 386
    WHO classification, 382–384, 382t
Meningothelial meningioma, 383f, 384f
Mental retardation (MR), 563–573. *See also* Genetic mental retardation
Meperidine, 666
MEPPs, 936, 937
2-Mercaptopropionic acid, 751t

Mercury, 760
Merlin, 386–388
Merosin-deficient CMD, 929–930
MERRF, 159, 315, 954
Merrick, Joseph, 413
Mesial temporal sclerosis (MTS), 351f, 352–353
Metabolic memory, 670
Metabolic myopathies, 947–956
    etiology, 948
    glycogen metabolism, defects of, 949–951
    history/nomenclature, 947–948
    lipid metabolism, defects of, 951–953
    lipid milieu, defects of, 955
    mitochondrial respiratory chain defects, 953–955
    natural history, 955
    pathogenesis, 948
    pathophysiology, 948–949
    symptoms, 955
Metabolic salvage, 12
Metabotropic glutamate receptors (mGluRs), 855
Metachromatic leukodystrophy (MLD), 3t, 5, 45
Metal induced neurotoxicity, 759–769
    age-related variables, 768
    aluminum, 761–762
    arsenic, 762
    biochemical and molecular mechanisms, 764–766
    bismuth intoxication, 762
    cisplatin, 762
    energy deficits, 765–766
    gold salts, 762
    lead, 760
    lithium, 760–761
    manganese, 761
    mercury, 760
    metal ions and trace elements, 764–765
    myelin, 764
    nature of response, 767
    neurons, 763
    neurotransmitter/second messenger systems, 765
    oxidative stress, 766
    pathophysiology, 762–763
    progress of insult and recovery, 767–768
    susceptibility factors, 766–767
    targets, 763–764
    thallium, 761
    tin, 762, 763t
Metallic tin, 762, 763t
Metaplastic meningioma, 383f, 384f
Metastasis suppressors, 376
Metastatic brain tumors, 371–379. *See also* CNS metastases
Metastatic cascade, 372f, 373
Methanol, 751t
Methemoglobin, 268

Methionine synthase (MS), 397
Methotrexate, 406
2-Methoxyethanol, 751t
Methyl bromide, 751t
Methyl chloride, 751t
Methyl iodide, 751t
Methyl *n*-butyl ketone, 751t
Methylmercury, 751t, 760, 768
Metoclopramide, 666
Metronidazole, 873
Mexiletine, 1018
MFS, 206t, 208t
MG, 935–946. *See also* Myasthenia gravis and myasthenia syndromes
mG/uRs, 855
MGUS, 212, 652, 866, 867
MH, 327–328, 992–993
MHC-I, 962–966
Microbodies, 19
Microcystic meningioma, 383f, 384f
Microscopic polyangiitis (MPA), 905
Microtubule-associated protein tau (MAPT), 106
Migraine, 333–348
    association studies, 344
    aura, 335
    CADASIL, 343
    central connections, 336–337
    central modulation of trigeminal pain, 338
    cerebral ionopathy, as, 342
    classification, 334
    clinical phases/pathophysiology, 334
    features, 334t
    FHM, 340–343
    functional consequences (FHM gene mutations), 342–343
    genetic epidemiology, 339
    headache phase, 335
    higher-order processing, 337–338
    linkage studies, 343–344
    neuropeptide studies, 336
    PET scans, 338f
    plasma protein extravasation, 335–336
    premonitory phase, 335
    sensitization, 336
    SHM, 341–342
    stroke, 257
    symptoms, 338–339
    thalamus, 337–338
    trigeminocervical complex, 337
    triggers, 334t, 339, 343
    vascular retinopathy, 343
Migraine aura, 335
Migraine with aura, 334
Migraine without aura, 334

Mild AD, 78
Mild cognitive impairment (MCI), 75, 78
Miller Fisher syndrome (MFS), 206t, 208t
Mini-Mental State Examination (MMSE), 78, 79, 79f
Miniature end plate potentials (MEPPs), 936, 937
Minor cognitive and motor deficit (MCMD), 493
Mirror neuron system, 842
Misonidazole, 882
Mitochondrial β-oxidation matrix system, 952
Mitochondrial deoxyribonucleic acid mutations, 953–954
Mitochondrial encephalomyopathy, lactic acidosis, and stroke-like episodes (MELAS), 159, 954
Mitochondrial genetic diseases, 157–161
    CPEO, 158–159
    DOA, 159–160
    FRDA, 160
    KSS, 158–159
    LHON, 158
    MELAS, 159
    MERRF, 159
    NARP, 158
    PEO, 158
Mitochondrial neurogastrointestinal encephalomyopathy (MNGIE), 955
Mitochondrial respiratory chain defects, 953–955
Mitogen-activated protein kinases (MAPKs), 854
Mitral annulus calcification (MAC), 238–239
Mitral stenosis, 239
Mixed connective tissue disease, 907
Mixed cryoglobulinemia, 261
ML II, 3t, 5–7
ML III, 3t, 5–7
MLD, 3t, 5, 45
MLs, 3t, 5–8
MMN, 211–212, 865
MMPs, 107
MMSE, 78, 79, 79f
MND, 513–519
MNGIE, 955
Moderate AD, 78–79
Mohamed, N., 687
Molecular mimicry, 199, 207, 862
Monoclonal gammopathies, 866–867
Monoclonal gammopathy of undetermined significance (MGUS), 652, 866, 867
Monoclonal IgM, 651–653
Mononeuritis multiplex, 164t
Morgan, W. Pringle, 593
Morphine, 1015
Morquio disease, 3t
Morvan's syndrome, 165, 331
Mossy fiber sprouting, 353, 355f
Motor neuron disease (MND), 513–519

Motor recovery program, 241–253. *See also* Stroke recovery
Motor tics, 620
Moyamoya, 256–257
MPA, 905
MPC, 970
MPNST, 416, 418, 422
MPO region, 988–990
MPP+, 751
MPS diseases, 3t, 5
MPTP, 751–752, 755–757
MR, 563–573. *See also* Genetic mental retardation
MRA, 798f
MRI, 793–800
  basic clinical imaging, 797
  basic principles, 795–796
  DWI, 798
  dyslexia, 595
  exogenous contrast agents, 799
  hardware/scanner, 794–795
  historical overview, 794
  image analysis, 796–797
  imaging nuclei, 800
  magnetization transfer, 799
  meningioma, 385
  MS, 193f
  neonatal brain injuries, 606
  NF1, 418
  noninflammatory arteriopathies, 257
  PCNSL, 403–404
  PML, 193f
  PWI, 798
  SLE, 175
  WMLs, 228–229
MRN, 975
MRS, 781–792
  amino acid neurotransmitters, 788
  cholines, 787
  epilepsy, 790
  history/nomenclature, 781–782
  lactate, 787–788
  multiple sclerosis (MS), 790
  NAA, 786–787
  neonatal brain injuries, 606–607
  neoplasia, 789–790
  SLE, 175–176
  stroke, 790
  technological concepts, 782–786
MRSI, 782
MRTIs, 873–874
MS
  methionine synthase, 397
  multiple sclerosis. *See* Multiple sclerosis (MS)
MSA, 83–93. *See also* Multiple system atrophy (MSA)

MSA-C, 85, 89, 90
MSA-P, 88, 89
MSD, 4t, 5, 8
mtDNA mutations, 157–161, 953–954. *See also* Mitochondrial genetic diseases
mTOR, 440
mTOR inhibitors, 442
MTS, 351f, 352–353
MTT, 798f
MTX, 406
Mucolipidoses (MLs), 3t, 5–8
Mucopolysaccharidoses, 3t, 5
Mucor, 448t
Multidetector row computed tomography, 801, 802. *See also* Neurovascular computed tomography angiography
Multifocal acquired demyelinating sensory and motor neuropathy (MADSAM), 865
Multifocal motor neuropathy (MMN), 211–212
Multiple acyl-CoA dehydrogenase (MAD) deficiency, 952–953
Multiple meningioma, 390–391
Multiple myeloma, 651
Multiple sclerosis (MS), 197–204
  animal models of CNS inflammation, 198–200
  autoimmune hypothesis, 203f
  CNS, 198
  EAE, 198–203
  humoral immunity, 201
  IBD, 693
  immune-mediated neurodegeneration, 201
  immunopathogenic concept, 202–204
  immunotherapy, 202
  infectious disease hypothesis, 203f
  MRI, 193f
  MRS, 790
  neurodegeneration hypothesis, 203f
  PML, compared, 193f
  T cells, 200–201
Multiple sulfatase deficiency (MSD), 4t, 5, 8
Multiple system atrophy (MSA), 83–93. *See also* OPCA
  additional sites of pathology, 87
  animal models, 88
  antecollis, 90
  autonomic failure, 86–87
  biochemical findings, 87–88
  camptocormia, 90
  cellular inclusions, 85
  cerebellar disorder, 90
  clinical diagnosis, 91
  clinical picture, 88–91
  dysarthria, 90
  dysautonomia, 89–90
  epidemiology, 84

Multiple system atrophy (MSA) (*continued*)
   gross neuropathology, 85
   history/nomenclature, 83–84
   molecular biology, 88
   MSA-C, 85, 89, 90
   MSA-P, 88, 89
   natural history, 91
   neuropharmacological findings, 88
   nighttime respiratory stridor, 90
   olivopontocerebellar atrophy, 86
   onset, 91
   Parkinsonism, 89
   pathogenesis, 84–85
   Pisa syndrome, 90
   progression, 91
   pyramidal signs, 90
   REM sleep behavior disorder, 90
   striatonigral degeneration, 85–86
   tremulous myoclonic jerks, 90
   UMSARS, 84
Multiple-system atrophy (MSA), 995–996
Multivesicular bodies, 11
Multivoxel MRS, 782
Mumps virus, 487t
Murphy, Erin K., 816
Muscle-eye-brain (MEB) disease, 930
Muscle phosphofructokinase deficiency, 951
Muscle-specific kinase antibody-associated myasthenia gravis, 940–941
Muscular dystrophies, 925–934
   BMD, 926–927
   CMDs, 927–931
   DMD, 926–927
   FSHD, 932–933
   glycosylation, disorders, 930–931
   LGMDs, 927–929
   MDC1A, 929–930
   myotonic dystrophy, 931–932
   overview, 926t
   Ullrich disease, 930
MuSK-MG, 940–941
Mutable motor map, 242–245
Myasthenia gravis, 164t
Myasthenia gravis and myasthenia syndromes, 935–946
   acquired neuromyotonia, 942–943
   congenital myasthenic syndromes, 943–944
   LEMS, 941–942
   MG, 937–940
   MuSK-MG, 940–941
   neuromuscular junction, 935–937
   NMT, 942
   phenotypic variability, 944–945
Myasthenia syndromes, 935–946. *See also* Myasthenia gravis and myasthenia syndromes

Mycobacterium avium complex, 447t
Mycobacterium leprae, 447t
Mycobacterium tuberculosis, 447t
Mycophenolic mofetil, 179
Mycotic aneurysms, 266
Myelin, 764
Myelin-associated glycoprotein (MAG), 206, 212–213
Myelin damage, 753
Myelin protein disorders, 46
Myelin proteins, 211f
Myelomeningocele (MMC), 611–618. *See also* Spina bifida
Myeloproliferative syndromes, 657–658
Myoclonic epilepsy with ragged red fibers (MERRF), 159, 315, 954
Myoclonus, 305–317
   asterixis, 311
   classification, 306
   CPM, 306–308
   definition, 305–306
   DRPLA, 315–316
   ENM, 311
   EPM1, 313–314
   5-HTP, 312
   GABA, 312
   genetics, 313–316
   GEPR-3, 312
   glycine, 313
   history, 305
   hyperekplexia, 309
   LD, 314–315
   MD, 313
   MERRF, 315
   natural history, 316–317
   NCL, 315, 316t
   negative, 306, 310–311
   pharmacology, 311–313
   PME, 313–315
   positive, 305–306, 306–310
   RRM, 308–309
   serotonin, 311–312
   spinal, 309–310
   symptoms, 316
Myoclonus-dystonia (MD), 313
Myoclonus epilepsy ragged-red fibers (MERRF), 159, 315, 954
Myophosphorylase deficiency, 951
Myotilin, 928
Myotonia congenita (MC), 145–147, 323–324
Myotonia fluctuans, 325
Myotonia permanens, 325
Myotonic dystrophy, 126–129, 931–932. *See also* DM (myotonic dystrophy)

N-acetylaspartate (NAA), 607, 786–787
n-hexane, 751t
n-hexane-induced axonal degeneration, 755
NAA, 46, 607, 786–787
NAA:Cho ratio, 176
NAA:Cr ratio, 176
NAC, 115
Naegleria fowleri, 461
NAIM, 218
NALD, 21, 23, 23f
Narcolepsy and hypersomnia, 715–722
   ICSD classification, 716t
   idiopathic hypersomnia, 721–722
   narcolepsy-cataplexy, 716–720
   narcolepsy without cataplexy, 716t, 720–721
Narcolepsy-cataplexy, 716–720
NARP, 158
Natalizumab, 693
Natural history, 857
*N*B-DNJ, 17
NBCCS, 428
NBIA, 115
NBTE, 239
NCL, 315, 316t
NCL disorders, 4t, 6
NCLs, 4t, 6
nDNA mutations, 954
NE, 637t, 973t, 975
Neck CTA, 805t, 808–810
Necrotizing autoimmune vasculitides, 259
Neel, James, 414
Negative myoclonus, 306, 310–311
Negative selection, 173
Neisseria meningitidis, 447t
Nematodes, 466, 468–469t
Neoangiogenesis, 375
Neonatal adrenoleukodystrophy (NALD), 21, 23, 23f
Neonatal brain injuries, 599–609
   etiology and pathogenesis, 600
   excitotoxicity, 603
   genetic effects, 603
   inflammation, 603
   neonatal encephalopathy, 604–605
   neonatal seizures, 605–606
   neonatal stroke, 606
   neuroimaging, 606–607
   nomenclature, 599–600
   oxidative stress, 602
   pathophysiology, 601–604
   patterns of brain injury, 601
   selective cellular vulnerability, 601–602
   subtle neonatal syndromes, 606
   therapy, 603–604
Neonatal encephalopathy, 604–605

Neonatal myasthenia gravis, 940
Neonatal seizures, 605–606
Neonatal stroke, 606
Neoplasia, 789–790
Neoplasm-induced pain, 1007–1020
   adjuvants, 1017–1018
   bone pain, 1011–1012
   breakthrough pain, 1016
   etiology, 1007–1008
   interventional routes, 1018–1020
   neuropathic pain, 1009–1010
   neuropathic pain syndromes, 1013–1014
   nociceptive pain, 1009
   opioids, 1014–1017
   pain assessment, 1010–1011
   pathogenesis, 1008
   pharmacological treatment, 1014–1018
   routes of medication administration, 1015–1016
   spinal opioid therapy, 1018–1019
   sympathetic blocks, 1019
   visceral pain, 1012–1013
Nerve growth factor (NGF), 674, 856, 1026
Neural tube defect (NTD), 611–618. *See also* Spina bifida
Neuroaxonal dystrophy, 15–16, 15f
Neurocysticercosis, 464–466
Neurodegeneration hypothesis, 203f
Neuroferritinopathy, 695
Neurofibromas, 415–416
Neurofibromatosis 1, 413–423. *See also* NF1
Neurofibromin, 419f, 421
Neurofilament (NF) proteins, 526
Neurogenic bladder and bowel, 614–615
Neurogenic orthostatic hypotension, 998
Neuroimaging
   CTA. *See* Neurovascular computed tomography angiography
   DTI. *See* Diffusion tensor imaging (DTI)
   fMRI. *See* fMRI
   migraine, 338f
   MRI. *See* MRI
   MRS. *See* MRS
   neonatal brain injuries, 606–607
   PCNSL, 403–404
   PET scans. *See* PET scans
   SLE, 175–176
   SPECT. *See* SPECT
   stroke recovery, 247–249
Neuroinvasion, 480, 487
Neurokinin A, 973t, 976
Neurokinin B, 973t, 976
Neuroleptic malignant syndrome (NMS), 993
Neuromelanin, 55
Neuromuscular junction, 935–937
Neuromyotonia (NMT), 164t, 165, 942

Neuronal ceroid lipofuscinoses (NCLs), 4t, 6, 315, 316t
Neuronal NO, 851
Neuronavigation, 845–846
Neuropathic pain, 1009–1010
Neuropathic pain syndromes, 1013–1014
Neuropathies associated with infections, 913–924
   diphtheria, 917–918
   HCV-associated neuropathies, 920–921
   HIV-DSP, 918–920
   leprosy, 915–917
   Lyme disease, 913–915
   miscellaneous neuropathies, 922t, 923t
   WNV-associated flaccid paralysis, 921–922
Neuropathies associated with monoclonal gammopathies, 866–867
Neuropathy, ataxia, and retinitis pigmentosa (NARP), 158
Neuropathy, multifocal motor neuropathy (MMN), 865
Neuropeptide Y (NPY), 973t, 976
Neuropsychiatric SLE (NPSLE), 174–177
Neurosarcoidosis, 703–707
   cause, 704
   CNS, 706
   cranial nerve palsies, 705–706
   incidence, 703–704
   pathology, 704–705
   PNS, 706
   sarcoid myopathy, 706–707
   sites of involvement, 705
   skull and vertebrae, 705
   spinal cord, 706
Neurosusceptibility, 490
Neurotransmitters, 973t, 975–977
Neurotrophic growth factors, 674–675
Neurotrophins, 530
Neurotrophins (NTs), 375
Neurotropic viruses, 487
Neurotropism, 489
Neurovascular computed tomography angiography, 801–819
   cerebral venous thrombosis, 815–816
   chronic carotid artery steno-occlusive disease, 808–810
   CIN, 803
   contrast agent safety, 802–803
   CTV, 805t, 815–816
   fibromuscular dysplasia, 810
   future directions, 816
   historical overview, 802
   internal carotid and vertebral artery dissections, 810
   neck, 805t, 808–810
   pitfalls, 803f, 808f
   radiation/image quality, 802
   strategies for optimizing contrast enhancement, 803–804, 804t
   stroke, 804–808
   subarachnoid hemorrhage, 805t, 810–813
   vasculitis, 815
   vasospasm, 805t, 813–814
Neurovirulence, 489
Neurovirulent viruses, 487
Nevoid basal cell carcinoma syndrome (NBCCS), 428
NF proteins, 526
NF1, 413–423
   biochemistry/molecular mechanisms, 420–421
   brain abnormalities, 417f
   children, 417
   etiology, 414
   evolution of tumors, 421f
   historical overview, 413–414
   MPNST, 416, 418, 422
   natural history, 418–419
   neurofibromas, 415–416
   NIH consensus diagnostic criteria, 415t
   nomenclature, 414
   OPG, 416–417
   pathogenesis, 414–418
   pigmentary features, 414–415, 415f
   plexiform neurofibroma, 416
   structural detail, 419–420
   symptoms, 421–422
   two-hit hypothesis, 420, 420f
   UBDs, 418
NF1 gene, 419, 420
NF2-associated meningioma, 391
NF2 gene, 386, 390–391
NGF, 674, 856, 1026
Niacin, 699
Niacin deficiency, 699
Nicotinic acetylcholine receptor, 140, 141f
Niemann, Albert, 1
Niemann-Pick disorders, 3t, 5, 8, 14, 16, 17
Night blindness, 698
Nighttime respiratory stridor, 90
9p, 389
NIPA1, 542
Nipah virus, 487t
Nitric oxide (NO), 851–852, 973t, 976
Nitriles, 751t
Nitrofurantoin, 873
Nitrone spin trap agents, 277
3-nitropropionic acid, 754
Nitrosative stress, 671–672
NKH, 642
NMDA receptors, 11, 178
NMR imaging, 794. *See also* MRI
NMS, 993
NMT, 164t, 165, 942
NO, 851–852, 973t, 976

Nocardia, 447t
Nociceptive pain, 1009
Non-cystic PVL, 577
Non-GC-like DLBCL, 398, 399f
Nonatherosclerotic cerebral vasculopathies, 255–263
　autoimmune leukocytoclastic vasculitis, 261
　Behçet's disease, 262
　Buerger's disease, 262
　CADASIL, 257
　cervicocephalic arterial dissection, 256
　Churg-Strauss syndrome, 260
　CNS vasculitis, 258
　Cogan's syndrome, 262
　Crohn's disease, 261
　Fabry's disease, 257
　FMD, 255–256
　giant cell arteritis, 259t, 261
　Henoch-Schonlein purpura, 261
　homocystinuria, 257
　hypersensitivity vasculitis, 259t, 261
　infectious inflammatory vasculopathies, 258t, 259
　inflammatory arteriopathies, 258–262
　Kawasaki's syndrome, 262
　lymphomatoid granulomatosis, 260
　malignant atrophic papulosis, 261
　metabolic disorders, 257–258
　migraine, 257
　mixed cryoglobulinemia, 261
　moyamoya, 256–257
　necrotizing autoimmune vasculitides, 259
　noninfectious inflammatory vasculopathies, 259–262
　noninflammatory arteriopathies, 255–258
　P. elasticum, 257
　PAN, 260
　primary isolated CNS angitis, 262
　relapsing polychondritis, 262
　rheumatoid arthritis, 261
　sarcoidosis, 261
　scleroderma, 260
　Sjögren's syndrome, 260–261
　SLE, 260. See also Systemic lupus
　　erythematosus (SLE)
　Sneddon syndrome, 258
　Susac syndrome, 261
　Takayasu's arteritis, 261
　WG, 259–260
Nonbacterial thrombotic endocarditis (NBTE), 239
Nonchoroidal CSF, 643
Noncommunicating hydrocephalus, 642
Nondeclarative memory, 251
Nondystrophic myotonia, 144–147
Noninfectious inflammatory vasculopathies, 259–262
Nonketotic hyperglycinemia (NKH), 642
Nonnutritional reperfusion, 832

Nonsteroidal anti-inflammatory drugs (NSAIDs), 1014
Nonsyndromic XLMR, 567
Nonsystemic vasculitis neuropathy, 906–907
Nontropical sprue, 690
Norepinephrine (NE), 637t, 973t, 975
Normal-pressure hydrocephalus, 644
Northern epilepsy (CLN8), 4t
Nortriptyline, 1043
NPC disease, 8, 16, 17
NPC1, 5, 8
NPC2, 5, 8
NPSLE, 174–177
NPY, 973t, 976
NSAIDs, 1014
NTD, 611–618. See also Spina bifida
NTS, 972–973
NTs, 375
Nuclear deoxyribonucleic acid mutations, 954
Nuclear magnetic resonance imaging, 794. See also MRI
Nucleoside reverse transcriptase inhibitors (NRTIs), 873–874
Nucleus ambiguous, 974
Nucleus tractus solitarius (NTS), 972–973

Occipitotemporal (OT) region, 596
Ocular lymphoma, 404, 405f
Ocular motor abnormalities, 108–109
Ocular myasthenia gravis, 939–940
Oculofacial-skeletal myorhythmias (OFSM), 694
Oculomasticatory myorhythmia (OMM), 694
OFSM, 694
Oligoastrocytoma, 435
Oligodendroglioma, 435, 436f, 788f
Olivopontocerebellar atrophy, 95–104. See also OPCA
Olszewski, Jerry, 105
OMM, 694
Oncological treatment, pain. See Neoplasm-induced pain
1-bromopropane, 751t
1p, 388
OPA1, 159
OPCA, 95–104
　algorithm, 103f
　diagnosis, 102–103
　etiology/pathogenesis, 97–102
　Harding's classification, 97t
　history/nomenclature, 95–97
　MSA, 86
　natural history, 103
　treatment, 103–104
OPG, 416–417
OPIDN, 755
Opioids, 1014–1017
Opsoclonus-myoclonus, 164t
Optic pathway glioma (OPG), 416–417

Orexins, 973t, 976
Organic acid disorders, 46
Organic chemicals, 745–758
   affected nervous system functions, 750t
   biochemistry/molecular mechanisms, 754–756
   criteria for investigation, 748t
   epidemiology, 746–747
   Haber's law, 757
   history/nomenclature, 745–746
   natural history, 757
   neuropathy, 752
   neurotoxicity, management of, 758
   pathogenesis, 749–752
   pathophysiology, 753–754
   risk factors, 748–749
   symptoms, 756–757
Organic tin compounds, 762, 763t
Organometals, 751t
Organophosphate-induced delayed neuropathy (OPIDN), 755
Organophosphorus compounds, 751t
Organosulfur compounds, 751t
Ornithine, 39
Orthostatic stress, 1000
Orton, Samuel, 594
Osmotic hypothesis, 672
Osteomalacia, 678
Osteosclerotic myeloma, 212
OT region, 596
Other antineuronal antibodies, 515, 516, 518
Oxaliplatin neurotoxicity, 878
Oxidative injury, 852–853, 855
Oxidative stress, 602, 671, 766
Oxycodone, 1014
Oxygen-sensing mechanisms, 683–684
Oxyhemoglobin, 268
Oxysterols, 14
Oxytocin, 973t, 976

P. elasticum, 257
P. falciparum, 455–457
P13K, 854
p53, 428, 854
PA, 699
Paclitaxel, 879–880
$PaCO_2$, 684
Padgett, B. L., 185
PAG, 990
Paget, Stephen, 372
Pain, 1007–1046. *See also* individual subentries
   ANS, 1021–1030
   central post-stroke, 1039–1046
   neoplasm-induced, 1007–1020
   postherpetic neuralgia, 1031–1037

Pain assessment, 1010–1011
Painful radiculopathy, 1013
Palatal myoclonus, 308
PAM, 145–147, 325
Pamidronate, 1018
PAN, 213, 260, 904
Panarteritis nodosa (PAN), 213
PANDAS, 629
PaO_2, 684
Papillary fibroelastoma, 239
Papillary meningioma, 383f, 384f
Parabrachial nuclei (PBN), 974–975
Paradoxical embolism, 238
Paragonimus spp, 471
Paralytic rabies, 923t
Paramyotonia congenita (PMC), 145–147, 325
Paraneoplastic neurological disorder (PND), 163–169
   etiology/pathogenesis, 164–165
   history/nomenclature, 164
   neuronal autoantibodies, 166t
   pathophysiology, 165–167
   symptoms/natural history, 168
Paraneoplastic vasculitis, 908
Paraplegin, 541
Paraproteinemia, 212, 651–653
Parasitic infections, 453–472
   amoebiasis, 460–462
   cerebral malaria, 455–457
   HAT, 457–460
   neurocysticercosis, 464–466
   other cestodes, 467t
   other nematodes, 466, 468–469t
   other trematodes, 471
   overview, 453t
   schistosomiasis, 470–471
   strongyloidiasis, 466
   toxoplasmosis, 462–464
Parasomnia overlap disorder, 712
Parathyroid hormone (PTH), 660, 678
Paraventricular nucleus (PVN), 970–972
Parenchymal brain metastasis, 377
Parkinson, James, 51, 115
Parkinson's disease (PD), 51–68, 115–118
   α-synuclein, 55–56
   bradykinesia, 61
   cholinergic system, 61–62
   dopaminergic imaging, 822–824
   dopaminergic system, 62
   epidemiology, 52–54
   general pathology, 54–55
   histaminergic system, 63–64
   historical overview, 51–52
   idiopathic PD/atypical Parkinsonism, contrasted, 824
   natural history, 64

neurofilaments, 56
neuromelanin, 55
pathogenesis, 55–57
pathophysiology, 57–61
PET scans, 821–828
rest tremor, 60
rigidity, 60–61
serotonergic system, 63
spatial covariance patterns, 824–825
surgical history, 52
ubiquitin-proteasome pathway, 56–57
Parkinson's disease-related metabolic pattern (PDRP), 824–826
Paroxysmal and progressive ataxias, 328–330
Paroxysmal dyskinesia, 297–304
  history/nomenclature, 297–298
  PED, 298t, 299t, 302
  PHD, 298t, 299t, 302–303
  PKD, 298–300
  PNKD, 298t, 299t, 301–302
Paroxysmal dystonic choreoathetosis, 301
Paroxysmal exercise-induced dystonia (PED), 298t, 299t, 302
Paroxysmal hypnogenic dyskinesia (PHD), 298t, 299t, 302–303
Paroxysmal kinesigenic choreoathetosis, 298
Paroxysmal kinesigenic dyskinesia (PKD), 298–300
Paroxysmal nonkinesigenic dyskinesia (PNKD), 298t, 299t, 301–302
Patau syndrome, 564f, 564t
Patent foramen ovale (PFO), 237–238
Pathoclisis, 357
Pathological aging, 75
PBC, 696
PBD, 20–21
PBG, 696
PBGD, 696
PBK deficiency, 950–951
PBN, 974–975
PCNSL, 395–412
  biomarkers of prognosis, 401–402
  chemotherapy, 406–408
  clinical presentation, 403
  CSF, 404
  diagnostic evaluation, 403–405
  DLBCL, 397–398, 399f
  epidemiology, 396
  genetic alterations, 400–401
  genetic risk factors, 397
  immunocompromised patient, 405–406
  immunohistochemistry, 400t
  immunological risk factor, 396–397
  immunophenotype, 398–399
  incidence rates, 396f
  IPCG guidelines for baseline evaluation, 403t
  neuroimaging, 403–404
  neurotoxicity, 408, 409f
  ocular examination, 404, 405f
  pathology/biology, 397–402
  prognostic scoring, 404–405
  radiation, 406
  salvage therapy, 407–408
  surgery, 406
  treatment, 405–408
PD, 51–68. See also Parkinson's disease (PD)
PDD, 583–584t. See also Autism
PDD-NOS, 583t
PDE_4 inhibitor, 251
PDGFR, 442
PDRP, 824–826
Pediatric meningioma, 391
Pedunculopontine nucleus (PPN), 60, 61
Pelizaeus-Merzbacher disease (PMD), 46, 540
Pellagra, 699
Penfield, Wilder, 845
Penicillamine, 698
Penicillin, 666
Penumbra, 272
PEO, 158
Percutaneous cordotomy, 1019
Perfusion-weighted imaging (PWI), 798
Periaqueductal gray (PAG), 990
Periaqueductal gray matter (PGM), 972
Periictal hemiplegia, 833f
Periodic leg movement (PLM), 724
Periodic leg movements in sleep (PLMS), 723–724, 731–732
Periodic leg movements of wake (PLMW), 724
Periodic paralysis, 144–147
Peripheral neurofibromatosis, 414
Peritumoral brain events, 373–374
Periventricular leukomalacia (PVL), 576, 577, 579
Pernicious anemia (PA), 699
Peroxisomal biogenesis disorders (PBDs), 20–21
Peroxisomal disorders, 19–32
  clinicopathological correlations, 31
  etiology, 21
  groups, 20–21
  history/nomenclature, 19–20
  human pathology/pathogenesis, 23–30
  molecular pathogenesis/pathophysiology, 21–23
  mouse models and pathogenesis, 30–31
Peroxisome, 19–20
Peroxynitrite, 525

Pervasive developmental disorder (PDD), 583–584t.
    See also Autism
PET scans
    Alzheimer's disease, 827
    dyslexia, 595–596
    Huntington's disease, 827
    insomnia, 741
    migraine, 338f
    Parkinson's disease, 821–828
PEX genes, 20–22
PfEMP-1, 455
PFO, 237–238
PGM, 972
Phasic dopamine, 628
PHD, 298t, 299t, 302–303
Phenobarbital, 666, 876
Phenothiazines, 666
Phenotyping, 847
Phenylpropanolamine, 258
Phenytoin, 1018
Phenytoin neuropathy, 876
PHN, 1031–1037. *See also* Postherpetic
    neuralgia (PHN)
Phonic tics, 620
Phosphoglycerate mutase deficiency, 951
Phosphorylase *b* kinase (PBK) deficiency, 950–951
Phosphorylase-limit dextrin, 950
PHVD, 283–286
Phytanic acid storage disease, 21
PI3-K, 439, 442
Pick, Ludwig, 1
PICSS, 238
Pilocytic astrocytomas, 435, 436f
Pisa syndrome, 90
Pituitary ablation, 1019
PKC, 673, 854
PKD, 298–300
Plasmodium falciparum, 455
Platinum-based drugs, 877–879
Plessner, W., 746
Plexiform neurofibroma, 416
PLM, 724
PLMS, 723–724, 731–732
PLMW, 724
PLP, 540
PLP gene, 46
PM, 957–968. *See also* Autoimmune inflammatory
    myopathies
PMC, 145–147, 325
PMD, 46, 540
PME, 313–315
PML, 185–195
    etiology, 186–188
    history/nomenclature, 185–186
    lymphocyte control on latency, 189–190
    molecular mechanisms, 192
    MRI, 193f
    MS, compared, 193f
    natural history, 193–194
    pathogenesis, 188–189
    pathophysiology, 190, 191t
    symptoms, 192–193
    timeline, 186f
PND, 163–169. *See also* Paraneoplastic neurological
    disorder (PND)
PNETs, 425–432. *See also* Medulloblastoma
PNKD, 298t, 299t, 301–302
PNS Lyme disease, 914–915
POEMS syndrome, 206t, 653, 867
Polio picornaviridae, 923t
Polio virus, 487t
Poliomyelitis, 545–551
    background, 545–546
    bulbar dysfunction, 550
    epidemiology, 546–547
    fatigue, 548–549
    pain, 549
    pathophysiology, 547–548
    Salk vaccine, 546
    treatment, 548–550
    weakness, 549–550
Polyarteritis nodosa (PAN), 260, 904
Polycythemia vera (PV), 657–658
Polyglutamine expansion diseases, 315
Polyminimyoclonus, 306
Polymyositis (PM), 957–968. *See also* Autoimmune
    inflammatory myopathies
PolyQ expansion, 555–558
Polyradiculoneuropathy, 164t
Pompe disease, 6, 950
Porphobilinogen (PBG), 696
Porphobilinogen deaminase (PBGD), 696
Porphyrias, 655–656
Positive myoclonus, 305–306, 306–310
Positron emission tomography. *See* PET scans
Post-polio syndrome, 545–551
Post-stroke dementia, 227
Post-treatment reactive encephalopathy (PTRE), 459
Posterior reversible leukoencephalopathy syndrome
    (PRES), 665
Posteroventrolateral thalamus (VPL) infarct, 1040f
Posthemorrhagic ventricular dilatation (PHVD),
    283–286
Postherpetic neuralgia (PHN), 1031–1037
    biochemistry, 1035
    etiology/molecular mechanisms, 1032–1033
    history/nomenclature, 1031–1032
    pain, 1035–1036

pathology/natural history, 1033–1034
pathophysiology, 1035–1036
pharmacology, 1036
symptoms, 1036
Posthypoxic audiogenic myoclonus, 312
Postinfectious meningoencephalomyelitis, 486
Postural instability/falls, 108
Potassium aggravated myotonia (PAM), 145–147, 325
Potassium channels, 138, 320–321
Potassium levels, 665
PPCA, 8
PPN, 60, 61
Prader-Willi syndrome, 542
Praxis and apraxia, 842
Preclinical AD, 75
Preferred direction, 247
PRES, 665
Primary amebic meningoencephalitis, 461–462
Primary amyloidosis, 653
Primary biliary cirrhosis (PBC), 696
Primary CNS lymphoma, 395–412. *See also* PCNSL
Primary isolated CNS angitis, 262
Primary lateral sclerosis (PLS), 537–544. *See also* HSP and PLS
Primary sclerosing cholangitis (PSC), 696
Primary vasculitides, 903–907
Primidone, 666
Primitive neuroectodermal tumors (PNETs), 426. *See also* Medulloblastoma
Prion diseases, 118–119, 473–483
  animal models, 479–480
  BSE, 476
  CJD, 476–481
  CWD, 475–476
  fCJD, 477t, 478
  FFI, 477t, 478
  GSS, 477t, 478
  historical overview, 473–475
  iCJD, 477t, 478
  peripheral prison pathogenesis, 480–481
  properties of prion, 479
  protein-only hypothesis, 475
  sCJD, 477t, 478
  scrapie, 475
  therapy, 481–482
  timeline, 474t
  vCJD, 477t, 478
Prion protein (PrP), 118, 119
Progesterone, 988
Progressive multifocal leukoencephalopathy, 185–195. *See also* PML
Progressive myoclonic epilepsy (PME), 313–315
Progressive myoclonus epilepsy with polyglucosan bodies, 314

Progressive supranuclear palsy (PSP), 105–110
  behavioral/cognitive frontal features, 109
  biochemistry, 108
  environmental aspects, 106
  etiology, 106
  extrapyramidal signs, 109
  genetic aspects, 106
  history/nomenclature, 105–106
  inflammation, 107
  natural history, 109
  ocular motor abnormalities, 108–109
  oxidative injury, 107
  pathogenesis, 107
  pharmacology, 108
  postural instability/falls, 108
  speech, swallowing, 109
  structural detail, 107
Proinflammatory cytokines, 1026–1027
Propriospinal myoclonus, 309–310
Protease inhibitors, 530
Proteasome inhibitors, 882
Protective protein/cathepsin A (PPCA), 8
Protein aggregation disorders, 111–123
  AD, 112–115
  Levinthal paradox, 111–112
  PD, 115–118
  prion diseases, 118–119
  protein folding/misfolding, 111–112
  structural determinants, 119–120
  synucleinopathies, 115–118
  therapeutic intervention, 120
Protein kinase C (PKC), 855
Protein-only hypothesis, 118, 475
Protozoans, 465–474. *See also* Parasitic infections
PrP, 118–119
PrP^C, 479
PRP^{Sc}, 118, 119
PrP^{Sc}, 479
Prusiner, Stanley B., 474, 475
PS-341, 882
Psammomatous meningioma, 383f, 384f
PSC, 696
Pseudo-Hurler polydystrophy, 3t, 5, 6
Pseudohypoparathyroidism, 679
Pseudomonas aeruginosa, 450
Pseudoxanthoma elasticum, 257
PSP, 105–110. *See also* Progressive supranuclear palsy (PSP)
PSP-parkinsonism, 106
Psychomotor epilepsy, 351
Psychosine, 45
PTCH, 428
PTEN, 439–440
PTH, 660, 678

PTK787, 375
PTRE, 459
Purcell, Edward, 794
Purkinje cells, 143, 151
PV, 657–658
PVL, 576, 577, 579
PVN, 970–972
PWI, 798
Pyridinethione salts, 751t
Pyridoxine, 699, 873
Pyrogens, 988, 991

Rabies virus, 487t
RAC-PET, 823–824
Radiation-induced carotid stenosis, 258
Radiation-induced meningioma (RIM), 390
Radicular pain, 1012
Radio frequency (RF) field, 782
RAFT, 440
Raft, 12
Raft log jams, 13
Ramsay Hunt syndrome, 97t, 100
Raphe nucleus, 63
RAPT, 440
RAS-GAP, 420–421
RAS/MAPK pathway, 442
RBD, 709–714. *See also* REM sleep behavior disorder (RBD)
Reading behavior, 596–597
Recovery from stroke, 241–253. *See also* Stroke recovery
Rectification function, 323
Recycling endosomes, 9
Redox hypothesis, 672
Refractory period, 322
Refsum's disease, 21
Relapsing polychondritis, 262
RELN, 354
REM sleep behavior disorder (RBD), 90, 709–714
  agrypnia excitata, 712–713
  biochemistry, 713
  diagnostic criteria, 710t
  etiology, 710–712
  history/nomenclature, 709–710
  molecular mechanisms, 713
  natural history, 714
  neurological disorders, 711t
  parasomnia overlap disorder, 712
  pathogenesis, 712–713
  pathophysiology, 713
  pharmacology, 713
  status dissociatus, 712
  symptoms, 713–714

Renal disease, 659–667
  Behçet's syndrome, 664
  DD, 663
  DDS, 662–663
  dialysis-dependent encephalopathy, 662
  dialysis-related complications, 662–664
  disorders of potassium metabolism, 665
  drug toxicities, 665–666
  electrolyte imbalances, 664–665
  hypernatremia, 665
  hyponatremia, 664
  lupus nephritis, 664
  TTP, 664
  uremic encephalopathy, 660–661
  uremic polyneuropathy, 661–662
Renal Fanconi syndrome, 951
Rest tremor, 60
Resting membrane potential, 321
Restless legs syndrome (RLS), 723–733
  A11 system, 727–728
  brain iron sufficiency, 725–726
  circadian pattern, 730–731
  daytime arousal, 731
  dopamine sensory and motor regulation, 731
  dopaminergic abnormalities, 727–728
  early-onset/late-onset, 732
  environment factors, 724–725
  etiology, 724–727
  genetic factors, 726–727
  iron deficiency, 724–725
  iron metabolism abnormalities, 728–729
  natural history, 724
  PLMS, 723–724, 731–732
  symptoms, 729–731
Reticular formation (RF), 974
Reticular reflex myoclonus (RRM), 308–309
Retinoic acids, 698, 755
Retinol, 698
Retrosomes, 11
Rett syndrome (RTT), 569–571, 583t
Reverse osmotic shift, 663
RF, 974
RF field, 782
Rhabdoid meningioma, 383f, 384f
Rheumatoid arthritis, 261, 907
Rhizomelic chondrodysplasia punctata (RCDP), 21
Rhizopus, 448t
Richardson, E. P., 185
Richardson, J. Clifford, 105
Richardson's syndrome, 105–110
Rickettsii, 447t
Rifaximin, 696
Right pontine infarct, 1041f
Riley-Day syndrome, 896

Riluzole, 509, 529
RIM, 390
Rippling muscle disease, 331
Rituximab, 179, 652
Rivastigmine, 80
RLS, 723–733. *See also* Restless legs syndrome (RLS)
RNA-based disorders, 125–133
    DM, 126–129
    FXTAS, 129–132
    other neuromuscular disorders, 132–133
    overview, 125–126
Rojas, Lissette, 66
Rolipram, 251
Romberg's sign, 699
Romero, Javier, 816
Rostral ventrolateral medulla (RVLM), 975
Roth, Bedrich, 721
Roussy-Levy syndrome, 886
Routes of medication administration, 1015–1016
RRM, 308–309
RTS, 572
RTT, 569–571, 583t
Rubella virus, 487t
Rubinstein-Taybi syndrome (RTS), 572
Rutter, Michael, 582
RVLM, 975

S. mansoni, 470
SA system, 1027–1029
Saccular aneurysms, 266
SAH, 265–270. *See also* Subarachnoid hemorrhage (SAH)
Salla disease, 4t, 8
Sandhoff disease, 4, 14
SANDO, 955
Sanfilippo disease, 3t
SAP, 8
SAP deficiencies, 3t
Sarcoglycans, 927
Sarcoid myopathy, 706–707
Sarcoidosis, 261, 703–707. *See also* Neurosarcoidosis
SCA subtypes
    ataxia, 143–144
    OPCA, 95–104
    SCA1, 149–155
SCA1, 149–155. *See also* Spinocerebellar ataxia type 1 (SCA1)
SCA-6, 329–330
Scheie's disease, 3t
Schindler-Kanzaki disease, 3t
Schistosoma intercalutum, 470
Schistosoma japonicum, 470
Schistosoma mekongi, 470
Schistosomiasis, 470–471

Schull, William, 414
Schwann cells, 416
Schwannomin, 386
sCJD, 118, 477t, 478
SCLC, 164, 165
Scleroatonic muscular dystrophy, 930
Scleroderma, 260, 907
SCNA1 gene, 341
Scoliosis, 417
Scrapie, 118, 119, 475
SDS. *See* Multiple system atrophy (MSA)
Second wind phenomenon, 951
Secondary vasculitides, 907–908
Secretory meningioma, 383f, 384f
Seed and soil hypothesis, 372
Seipin, 543
Seizure
    absence. *See* Idiopathic generalized epilepsy (IGE)
    BFNIS, 137
    dialysis patients, 664–665
    drug toxicity, 665–666
    epilepsy. *See* Epilepsy
    GEFS+, 137–141, 330–331, 362
    IBD, 693
    ICEGTC, 137
    neonatal, 605–606
    SPECT, 835–836
    spike-wave (absence), 289–296
Semiautomated carotid stenosis evaluation software, 811f
Senataxin mutations, 523
Sensory ganglionopathies, 213
Serotonin, 311–312, 637t, 973t, 976
17p, 389–390
Severe AD, 79
Severe myoclonic epilepsy of infancy (SMEI), 137
Shapiro's syndrome, 993
Sheikh, Shams, 816
SHH signaling pathway, 426, 426f, 428–429
Shingles, 1031–1037. *See also* Postherpetic neuralgia (PHN)
SHM, 341–342
sHSPs, 895
Shy-Drager syndrome (SDS). *See* Multiple system atrophy (MSA)
SI, 782
Sialidosis, 3t, 5, 6
Sialin, 8
Sickle cell disease, 653–655
Signaling endosomes, 11
Silent brain infarcts, 227–228
Silver syndrome, 539t, 543
Simple aromatic hydrocarbons, 751t
Simple motor tics, 620
Simple phonic tics, 620

Single-photon emission computed tomography. *See* SPECT
Sjögren-Larsson syndrome, 46, 539t
Sjögren's syndrome, 213, 260–261, 907
Skinfold freckling, 415
SLC6A8, 33–41. *See also* Creatine deficiency syndrome (CDS)
SLE. *See* Systemic lupus erythematosus (SLE)
Sleep disorders, 709–744. *See also* individual subentries
   idiopathic hypersomnia, 721–722
   insomnia, 735–744
   narcolepsy, 715–722
   periodic limb movements in sleep (PLMS), 731–732
   REM sleep behavior disorder (RBD), 709–714
   restless legs syndrome (RLS), 723–733
Sleeping sickness, 457–460
SLI, 594
Slit ventricle syndrome, 644
Slow inactivation, 145
Sly disease, 3t
SMA. *See* Spinal muscular atrophy (SMA)
Small cell lung carcinoma (SCLC), 164, 165
Small-fiber neuropathy, 856
Small vessel disease, 230–231
Small-vessel vasculitis, 905–906
SMEI, 137
SMN gene, 502–503
SMN protein, 504–505, 509
SMP, 1022
SND. *See* Multiple system atrophy (MSA)
Sneddon syndrome, 258
snRNAs, 506f
SNS and pain, 1021–1030. *See also* Autonomic nervous system and pain
SOD1, 524
SOD1 mutations, 522, 524–525
Sodium channel-blocking agents, 1018
Sodium channels, 138, 138b, 145, 321
Solid epithelial cancer, 377
Somatic pain, 1013
SOPCA, 95–104
Sorting endosomes, 9
Spartin, 541–542
Spastin, 542
Spatial cognition and neglect, 842–844
Species barrier, 479
Specific language impairment (SLI), 594
SPECT, 829–837
   AD, 836
   basal ganglia disorders, 836–837
   death, 837
   head trauma, 834–835
   herpes simplex encephalitis, 834
   historical overview, 829
   HIV encephalopathy, 834
   insomnia, 741
   methodology, 829–831
   neoplasms, 834
   SAH, 831–832
   seizure disorders, 835–836
   stroke, 831–834
Spectroscopic imaging (SI), 782
SPG1, 540
SPG2, 540
SPG3, 542
SPG4, 542
SPG6, 542
SPG7, 541
SPG10, 542–543
SPG13, 542–543
SPG17, 543
SPG20, 541–542
SPG21, 541–542
Spheroids, 16
Sphingolipid activator protein (SAP), 8
Sphingolipid activator protein (SAP) deficiencies, 3t
Spike and wave discharge (SWD), 291–294
Spike-wave (absence) seizures, 289–296
Spin, 795
Spina bifida
   biochemistry, 616–617
   Chiari malformation, 614, 615–616, 617
   etiology, 612–613
   functional levels (MMC), 616t
   historical overview/nomenclature, 611–612
   hydrocephalus, 614, 616
   molecular mechanisms, 616–617
   natural history MMC), 617–618
   neurogenic bladder and bowel, 614–615
   orthopedic deformities, 615
   pathogenesis, 613
   pathophysiology, 613–615
   pharmacology, 616
   structural details, 615–616
   syrinx formation, 614
   tethered spinal cord (TSC), 614
Spinal ALS, 523
Spinal CMT, 894–896
Spinal dysraphism, 612, 612t
Spinal epidural abscesses, 446
Spinal muscular atrophy (SMA), 501–511, 895
   animal models, 505–507
   clinical features, 501–502
   disease course, 502
   Drosophila, 505
   exon 7 inclusion, 509
   gene therapy/cell replacement, 509–510
   genetic basis, 502–503
   HDAC inhibitors, 507–509

mouse, 506–507
neuroprotection, 509
pathological features, 502
remaining questions, 510
SMN gene expression, 503–504
SMN protein, 504–505, 509
stabilizing SMN protein, 509
therapeutic strategies, 507–510
zebrafish, 506
Spinal myoclonus, 309–310
Spinal opioid therapy, 1018–1019
Spinal segmental myoclonus, 309
Spinobulbar muscular atrophy (SMBA), 553–561. *See also* Kennedy's disease
Spinocerebellar ataxia type 1 (SCA1), 149–155
concluding remarks, 154–155
linking pathology to pathophysiology, 154
molecular mechanisms, 151–153
pathogenesis, 150–153
recovery, 153–154
regional involvement, 150
SCA1 gene, 150
Spinocerebellar ataxia type 6 (SCA-6), 329–330
Spirometra species, 467t
Spliceosomal small nuclear RNAs (snRNAs), 506f
Spontaneous periodic hypothermia, 993
Sporadic CJD (sCJD), 118, 477t, 478
Sporadic hemiplegic migraine (SHM), 341–342
Sporadic multiple system atrophy, 996f
Sporadic OPCA (SOPCA), 95–104
SREAT, 217
SRO syndrome, 105–110
SRT, 17
Staphylococcus aureus, 449
Staphylococcus epidermidis, 447t, 449
Startle syndromes (hyperekplexia), 309
Statins, 874
Status dissociatus, 712
Stavudine, 873
STAZN, 277
Steele, John C., 105
Steele-Richardson-Olszewski (SRO) syndrome, 105–110
Stem cell therapies, 278
Steroid myopathy, 675–676
Stiff-person and stiff-limb syndromes, 164t
STN, 58–59
Storage diseases, 2. *See also* Lysosomal disorders
"Story of Fidgety Phillip, The" (Hoffman), 631
"Story of Johnny Head-in-the-Air, The" (Hoffman), 631–632
Strephosymbolia, 594
Streptococcus agalactiae, 447t
Streptococcus milleri group, 447t
Streptococcus pneumoniae, 447t

Stress response modulation, 971
Striato-thalamic pathways, 621–622
Striatonigral degeneration (SND). *See* Multiple system atrophy (MSA)
Striatum, 57, 63, 621
Stroke, 227–229
cardioembolism, 235–240
cerebral ischemia. *See* Cerebral ischemia
CTA, 804–808
dialysis patients, 664
hemorrhagic, 281–287
MRS, 790
neonatal, 606
pain, 1039–1046. *See also* Central post-stroke pain (CPSP)
risk of dementia, 226t
silent brain infarcts, 227–228
SPECT, 831–834
TIA, 236
WMLs, 228–229
Stroke recovery, 241–253
experimental precedents, 245–247
fMRI, 846–847
historical overview, 242
mutable motor map, 242–245
neuroimaging, 247–249
pharmacological interventions, 251
pharmacological targets/CNS synapse, 250f
TMS, 249–251
unmasking, 245
Strongyloides stercoralis, 466
Strongyloidiasis, 466
Stroop effect, 231
Strümpell-Lorrain syndrome, 537
Sturge-Weber syndrome (SWS), 691–692
Subacute motor neuropathy, 515
Subarachnoid hemorrhage (SAH), 265–270
aneurysm pathogenesis, 266–267
aneurysm rupture, 267
cerebral salt wasting syndrome, 269
cerebral vasospasm pathophysiology, 268–269
CTA, 810–813
DCI, 267–268, 269
epidemiology, 266
etiology, 265–266
genetics, 266
hydrocephalus, 267
ICH, 267
IVH, 267
SPECT, 831–832
Subdural empyema, 446
Substance P, 973t, 976
Substrate reduction therapy (SRT), 17
Subthalamic nucleus (STN), 58–59

Subtle brain injury syndrome, 606
Sufentanil, 1015
Superior hypogastric plexus block, 1019
Superoxide dismutase mutations, 522, 524–525
Supranuclear gaze palsy, 108
Suramin, 881
Survival motor neuron gene (SMN), 502–503
Survival signaling pathways, 278
Susac syndrome, 261
SWD, 291–294
Sweeney, Sherry, 495
SWS, 691–692
Sympathetic-afferent coupling, 1024
Sympathetic afferent system, 973
Sympathetic blocks, 1019
Sympathetic nervous system and pain, 1021–1030.
    See also Autonomic nervous system and pain
Sympathetically maintained pain (SMP), 1022
Sympathoadrenal (SA) system, 1027–1029
Syncope, 1000–1004
Syndromic XLMR, 567
Synucleinopathies, 115–118, 711
Syrinx formation, 614
Systemic lupus erythematosus (SLE), 171–184, 260, 907
    ACR classification criteria, 172t
    anti-NR2 antibodies, 178–179
    anti-ribosomal P protein antibodies, 177–178
    APLS, 177
    epidemiology, etiology, pathogenesis, 172–174
    neuroimaging, 175–176
    NPSLE, 174–175, 176–177
    therapy, 179
Systemic sclerosis, 260
Systemic vasculitis, 903–906

T. b. gambiense, 458–460
T. b. rhodesiense, 458–460
T. canis, 469t
T. cati, 469t
T. gondii, 462–464
T. spiralis, 469t
t-PA, 276–277
Tachykinins, 973t, 976
Tacrolimus, 875–876
Taenia multiceps, 467t
Taenia solium, 464
Taeniasis, 464
Takayasu's arteritis, 261, 815f, 904
Tarui's disease, 951
Taxanes, 879–880
Tay-Sachs disease, 1, 3t
TCAs, 1017, 1043
Telethonin, 929
Telomerase, 390

Temozolomide, 442
Temporal lobe epilepsy, 349–369. *See also* Epilepsy
    Ammon's horn sclerosis, 350f, 351f, 352
    endfolium sclerosis, 351f, 352
    extrahippocampal pathology, 353, 354t
    future directions, 364
    genetic disorder, as, 361–362
    granule cell dispersion, 354, 356f
    hilar cell loss, 355–357
    hippocampal sclerosis, 352–353
    historical background, 349–351
    mossy fiber sprouting, 353, 355f
    MTS, 351f, 352–353
    neurodevelopmental disorder, as, 362–364
    pathology, 351–357
    pathophysiological mechanisms, 357
    three-stage hypothesis, 357–361
10q, 389
Tethered spinal cord (TSC), 614
Tetraethyl lead, 751t
Tetrahydroisoquinolones (TIQs), 106
Tetramethylthiuram, 751t
Tetrathiomolybdate, 698
Thalamic syndrome, 1039, 1041
Thalamo-cortical pathway, 623
Thalamo-striatal pathway, 623
Thalamocortical synchronized activity, 292–294
Thalamus, 972
Thalassemia, 655
Thalidomide, 881–882
Thallium, 761
Thermoregulation, 971–972, 987–996
    AAN, 994–995
    aging, 991–992
    anhidrosis and hyperhidrosis (CIA), 993–994
    degenerative disorders, 995–996
    episodic hypothermia, 993
    fever, 991, 992f
    how it works, 988–991
    MH, 992–993
    MSA, 995–996
Thiamine, 698
Thiamine deficiency, 699
Thiophene, 751t
Thomsen's disease, 145–147, 323–324
Three-stage hypothesis, 357–361
3-nitropropionic acid, 754
3,3-iminodipropionitrile, 751t
Thrombin, 283–284
Thrombin preconditioning, 285
Thrombotic thrombocytopenic purpura (TTP), 656–657, 664
Thyroid disorders, 676–678
Thyrotoxic periodic paralysis (TPP), 677

Index

Tick paralysis, 922t
Tics, 620
TIMPs, 107
Tin, 762, 763t
TIQs, 106
Tissue inhibitors of metalloproteinases (TIMPs), 107
Tissue plasminogen activator (t-PA), 276–277
Titin, 928–929
TMS, 249–251, 624
TNFα, 1026
Tocopherol, 699
TOCP, 747, 757
Toluene, 751t
Tonabersat, 335
Tonic dopamine, 628
Topiramate, 335
Total Ammon's horn sclerosis, 350
Total cholines signal, 787
Tourette syndrome (TS), 619–630
   autoimmunity, 629
   CSTC circuits, 621–624
   diagnostic criteria, 620t
   dopaminergic abnormalities, 627–629
   frontal lobe dysfunction, 624–625
   GABHS, 629
   historical overview/nomenclature, 620
   location of primary dysfunction, 624–627
   natural history, 620
   neurochemical basis, 627–629
   PANDAS, 629
   striatal dysfunction, 626
   thalamic dysfunction, 626–627
Toxic and drug-induced neuropathies, 871–873
   antiepileptic drugs, 876
   antimicrobial agents, 873
   antineoplastic drugs, 876–882
   antitubulin drugs, 879–881
   cardiovascular drugs, 874
   chloroquine, 875
   colchicine, 875
   disulfiram, 874–875
   doxorubicin, 882
   epothilones, 882
   gold salts, 875
   interferons, 876
   leflunomide, 875
   misonidazole, 882
   MRTIs, 873–874
   platinum-based drugs, 877–879
   proteasome inhibitors, 882
   statins, 874
   suramin, 881
   tacrolimus, 875–876
   taxanes, 879–880
   thalidomide, 881–882
   vinca alkaloids, 880–881
Toxicants, 745
Toxins, 745
Toxocara canis, 469t
Toxocara cati, 469t
Toxocariasis, 469t
Toxoplasma gondii, 462–464
Toxoplasmic lymphadenitis, 463, 464
Toxoplasmosis, 462–464
TP53, 389
TPP, 677
Trace PBG kit, 656
Trans-crotononitrile, 751t
Transcranial magnetic stimulation (TMS), 249–251, 624
Transient ischemic attack (TIA), 236
Transitional meningioma, 383f, 384f
Transmissible spongiform encephalopathies (TSEs), 118, 119, 473. *See also* Prion diseases
Trastuzumab, 377
Trematodes, 470–471
Tremulous myoclonic jerks, 90
Treponema pallidum, 447t
Trialkyltins, 762, 763t
Trichinella spiralis, 469t
Trichinosis, 469t
Trichloroethylene, 751t
Tricyclic antidepressants (TCAs), 1017, 1043
Trientine, 698
Triethyl tin, 751t, 763t
Trigeminocervical complex, 337
TRIM32, 929
Trimethyl tin, 751t, 763t
Tropical sprue (TS), 698
Trousseau's sign, 679
Troyer syndrome, 541–542
Trypanosoma brucei gambiense, 458–460
Trypanosoma brucei rhodesiense, 458–460
TS, 619–630, 698. *See also* Tourette syndrome (TS)
TSC, 614
TSEs, 118, 119, 473. *See also* Prion diseases
TT, 915–917
TTP, 656–657, 664
TTR-related FAP, 983
Tuberculoid leprosy (TT), 915–917
Tumor necrosis factor α (TNFα), 1026
Turcot's syndrome, 428
Two-hit hypothesis, 420, 420f
2-bromobutyric acid, 751t
2-bromopropionic acid, 751t
2-butenenitrile, 751t
2-chloropropionic acid, 751t
2-hexanol, 751t
2-Mercaptopropionic acid, 751t

2-Methoxyethanol, 751t
2,4-dichlorophenoxyacetic acid, 751t
2,4-dinitrophenol, 754
2,5-hexanedione, 751t, 754
22q, 387
Tylman, Vislava T., 65

UBDs, 418
Ubiquitin, 57
Ubiquitin-proteasome pathway, 882
UC, 692
UCPs, 852–853
Ulcerative colitis (UC), 692
Ullrich disease, 930
UMN, 513–519
UMN syndrome and breast cancer, 515, 517, 518, 519
UMSARS, 84
Uncoupling proteins (UCPs), 852–853
Unidentified bright spots (UBOs), 418
Unified MSA Rating Scale (UMSARS), 84
Unverricht-Lundborg disease (EPM1), 313–314
uPA, 375
Upper motor neuron (UMN), 513–519
Upper motor neuron syndrome and breast cancer, 515, 517–519
Uremia, 661t
Uremic encephalopathy, 660–661
Uremic polyneuropathy, 661–662
Urokinase-type plasminogen activator (uPA), 375

VaD, 224, 225t
Valproate, 1018
Valvular disease, 238–239
Vanishing white-matter disease, 47
Variant CJD (vCJD), 118, 477t, 478
Varicella vasculopathy, 259
Varicella-zoster virus (VZV), 259, 488t, 1031–1037. See also Postherpetic neuralgia (PHN)
Variegate porphyria, 655
Vascular CIND, 224
Vascular cognitive impairment (VCI), 223–233
  CADASIL, 225–226
  cardioembolism, 230. See also Cardioembolism
  cardiovascular causes, 229
  epidemiology, 224
  etiology, 225
  HCAA, 226–227
  large vessel disease, 229–230
  NINDS-AIREN criteria, 225t
  nomenclature, 224
  pathophysiology, 229–230
  risk of, after stroke, 226t
  silent brain infarcts, 227–228
  small vessel disease, 230–231
  stroke, 227–229
  vascular disease/neurodegeneration, 230
  WMLs, 228–229
Vascular dementia (VaD), 224, 225t
Vascular endothelial cell growth factor (VEGF), 528, 675
Vascular endothelial growth factor-A (VEGF-A), 391
Vascular permeability factor, 391
Vasculitic neuropathies, 213–214
Vasculitis, 815, 901–911
  CGA, 903–904
  classification, 903t
  CNS manifestations, 908–909
  CSS, 905
  essential cryoglobulinemic vasculitis, 906
  history/nomenclature, 902
  HSP, 905
  isolated angiitis of CNS, 906
  Kawasaki disease, 904–905
  large-vessel, 903–904
  localized, 906–907
  medium-sized-vessel, 904–905
  MPA, 905
  nonsystemic vasculitis neuropathy, 906–907
  PAN, 904
  pathogenesis, 902–903
  PNS manifestations, 909–910
  primary vasculitides, 903–907
  secondary vasculitides, 907–908
  small-vessel, 905–906
  systemic, 903–906
  Takayasu's arteritis, 904
  WG, 905
Vasopressin, 973t, 976
Vasospasm CTA, 805t, 813–814
Vasovagal syncope, 1000–1004
VCI, 223–233. See also Vascular cognitive impairment (VCI)
vCJD, 118, 477t, 478
VEGF, 528, 675
VEGF-A, 391
Venous sinus thrombosis, 815f
Ventrolateral medulla, 975
Ventrolateral preoptic area (VLPO), 739
Vertical gaze palsy, 108
Very long chain acyl-CoA dehydrogenase (VLCAD) deficiency, 952
Very long chain fatty acid (VLCFA), 21–31
Very long chain saturated fatty acid (VLCSFA), 44
Vesicular monoamine transporter (VMAT), 823
VGCC, 138, 139b, 935, 941–942
VGKC, 138, 942
VGKC-associated syndromes, 331
Vicariation, 245
Vinblastine, 880

Vinca alkaloids, 880–881
Vincristine, 754, 880
Vinyl chloride, 751t
Viral infections, 485–499. *See also* CNS viral infections
Virchow-Robin space, 449
Visceral pain, 1012–1013
Visuospatial judgments, 844f
Vitamin A deficiency, 698
Vitamin B_1 deficiency, 698–699
Vitamin B_3, 699
Vitamin B_3 deficiency, 699
Vitamin B_6, 699
Vitamin B_6 deficiency, 699
Vitamin B_{12} deficiency, 649–651, 699
Vitamin E, 529, 698, 699
Vitamin E deficiency, 699–700
VLCFA, 21–31
VLCSFA, 44
VLPO, 739
VMAT, 823
Voltage-gated calcium channels (VGCCs), 138, 139b, 935, 941–942
Voltage-gated chloride channels, 140, 140f
Voltage-gated ion channels, 135
Voltage-gated potassium channel monomer, 139f
Voltage-gated potassium channels (VGKC), 138, 942
Voltage-gated sodium channels, 138, 138b
von Recklinghausen, Friedrich Daniel, 413
von Recklinghausen disease, 414
von Willebrand's factor (vWF), 656, 664
VP1, 186, 187
VPL lesions, 1040f, 1041
vWF, 656, 664
VWM, 47
VZV, 259, 488t, 1031–1037. *See also* Postherpetic neuralgia (PHN)

Wada's test, 845
WAG/Rij, 291–294
Waldenström's disease, 213, 214
Waldenström's macroglobulinopathy (WM), 651–652
Walker, D. L., 185
Walker, Mary, 937
Walker-Warburg syndrome, 930
Wegener's granulomatosis (WG), 259–260, 894–896, 905
Wehrli, Felix, 800

Werdnig, Guido, 501
Werdnig-Hoffman disease, 502
Wernicke, Karl, 841
Wernicke disease, 699
Wernicke-Korsakoff syndrome, 699
West Nile virus, 487t, 490
West Nile virus-associated flaccid paralysis, 921–922
Wet beriberi, 699
WG, 259–260, 905
Whipple, George Hoyt, 693
White-matter lesions (WMLs), 228–229
Wickner, Reed, 119
Williams syndrome (WS), 572
Willis, Thomas, 724, 937
Wilson, S. A. Kinnier, 52
Windmill artifact, 802f
Wingless (WNT) signaling pathway, 429, 430f
WM, 651–652
WMLs, 228–229
WNT signaling pathway, 429, 430f
WNV, 487t, 490
WNV-associated flaccid paralysis, 921–922
Wöhler, Friedrich, 745
Wolman disease, 4t, 6
WS, 572

X-EDMD, 929
X-linked adreno-leukodystrophy (XALD), 21
X-linked Emery-Dreifuss muscular dystrophy (X-EDMD), 929
X-linked HSP, 540
X-linked MR (XLMR), 567–571
XALD, 21
XLMR, 567–571

Yeast prions, 119

Zalcitabine, 873
Zebra bodies, 13
Zellweger syndrome (ZS), 19–32. *See also* Peroxisomal disorders
Zellweger spectrum, 21
Zinc, 698, 700
ZS, 19–32. *See also* Peroxisomal disorders
ZuRhein, G. M., 185